HOLT GEOMETRY

Teacher's Edition

Eugene D. Nichols
Mervine L. Edwards
E. Henry Garland
Sylvia A. Hoffman
Albert Mamary
William F. Palmer

HOLT, RINEHART AND WINSTON, PUBLISHERS
New York • Toronto • Mexico City • London • Sydney • Tokyo

Staff Credits

Editorial Development	Everett T. Draper, Earl D. Paisley, Antoine Y. Maksoud
Product Manager	Daniel M. Loch
Field Advisory Board	Douglas A. Nash, Sam Sherwood, Robert Wolff, Wendell Anthony, Gary Crump, Jack M. Custer, Roy Eliason, Kenneth C. Scupp, Dennis Spurgeon, Jeffra Ann Nicholson
Marketing Research	Erica S. Felman, Linda A. DeLora
Editorial Processing	Margaret M. Byrne, Holly L. Massey
Art and Design	Carol Steinberg, Mary Ciuffitelli
Production	Joan McNeil, Heidi Henney
Photo Resources	Linda Sykes, Rita Longabucco

Printed in the United States of America

ISBN: 0-03-002187-1

5678901234 032 98765432

AS YOU EXAMINE HOLT GEOMETRY

Please Notice

HOLT GEOMETRY provides a complete, balanced course suitable for students of all abilities. All of the basic concepts usually studied in a high school geometry course are found in HOLT GEOMETRY.

Format

HOLT GEOMETRY's proven four-part lesson format facilitates student learning:

- Lesson objectives appear on the pupil page
- Worked-out examples demonstrate solution process
- Hints and suggestions on the pupil page make learning difficult concepts easier
- Graded exercises provide for continuous skill development

Contents

1. The course begins with an introduction to the basic concepts of geometry (see pp. 1—7). As the course progresses, students study a complete and balanced course in geometry, including transformational, coordinate, and space geometry (see pp. 59, 94, 297, 459, 491, 510, 524).

2. Worked-out examples focus on teaching the solution process without complicated explanations (see pp. 250, 251, 315, 316).

3. Suggestions, hints, and comments encourage students to attempt more difficult skills and concepts (see pp. 174—175).

4. Calculator and computer activities familiarize students with accepted methods of logical thought processes (see pp. 20, 192, 381, 565).

5. Proofs, one of the most challenging tasks students undertake in geometry, are introduced in a gradual, step-by-step way that assures students' success (see pp. 33—43).

6. Students review algebraic concepts periodically (see pp. 253, 317).

7. Oral exercises appear before written exercise sets so that the teacher can test understanding and readiness quickly (see pp. 126, 172).

8. Exercises are plentiful and are graded according to three levels of difficulty. Exercises within each level are also graded from easiest to most challenging. Examples presented in the lesson match every type of written exercise in levels A and B (see pp. 172—173, 370—372).

9. Reading in Geometry helps students analyze and better understand the idiosyncrasies of the language of geometry (see pp. 105, 142).

10. Each chapter ends with a Chapter Review and Chapter Test. The Review is diagnostic and keys items to lesson sections in which skills and concepts necessary to solve the item are taught.

11. Special topics pages vary the pace and enrich the course. They include practical applications of geometry in the real world, as well as an introduction to careers that use mathematics (see pp. 22, 68, 202, 207, 238).

12. Worked-out solutions to every problem in the text are available in the Solutions Manual.

13. A teacher's edition with a commentary for each lesson, a complete set of answers, and daily assignments with three levels of difficulty designed to reinforce the current lesson while maintaining the previous two lessons makes teaching HOLT GEOMETRY easy.

Supplementary Materials on Duplicating Masters

- **HOLT GEOMETRY Skillmasters** provide two parallel sections of additional exercises for every two or three lessons and cumulative reviews.
- **HOLT GEOMETRY Testmasters** provide an alternate form of the Chapter Tests and Cumulative and Final Tests.
- **HOLT GEOMETRY Teacher's Resource Package** contains Enrichment masters, Testmasters, and Skillmasters in blackline form.

BENEFITS

HOLT GEOMETRY is designed to help students obtain the skills necessary for success in subsequent mathematics courses. The book gives the student a strong foundation in geometry and logic.

FOUR-PART LESSON FORMAT	A consistent four-part lesson format enhances the gradual development of skills and concepts.
PROBLEM-SOLVING STRAND	A clear, logical problem-solving strategy, special problem-solving application lessons, and non-routine problems provide a meaningful approach to problem solving.
ENRICHMENT	Calculator Activities, Algebra Reviews, and a special Logic Section provide a wide selection of experiences for students, depending on each student's readiness and learning aptitude.
COMPUTER STRAND	Computer Application pages at the end of chapters provide hands-on exercises in BASIC. References are made to further computer applications of geometric concepts.
TESTING AND REVIEW	Chapter Reviews, Chapter Tests, and Cumulative Reviews geared to the material previously taught keep students and teachers aware of progress.
TEACHER'S ANNOTATED EDITION	Teacher's Edition includes full-size annotated pupil's pages, chapter objectives, Cumulative Reviews, Activities, additional chalkboard work, and enrichment, as well as detailed teaching suggestions.
MOTIVATIONAL FEATURES	College Prep Tests acquaint students with College Board test-taking strategies. In addition to an expanded use of calculators and computers, historical puzzles and information regarding related careers also furnish motivation. Special photographic sections pique students' interest and accentuate the relationship of geometry to our technological world.
TEACHER'S RESOURCE PACKAGE	Contains Blackline Masters for Skillmasters, Testmasters, Basic Competency Tests, Enrichment, Non-Routine Problems, and Lesson Aids.
SUPPLEMENTARY MATERIALS	Skillmasters (duplicating masters) for review and Extra Practice. Each Skillmaster, with two parallel forms (A and B) can be used as a pretest or a posttest, as a diagnostic tool, as extra practice for skill mastery, or as a regular assignment.
	Testmasters (duplicating masters) for review, Extra Practice, diagnosing areas of difficulty, and supplementary testing, as well as a complete testing program containing chapter tests, cumulative tests, and final tests.

FOUR-PART LESSON FORMAT

HOLT GEOMETRY uses a carefully structured four-part lesson format to involve students actively in the learning process.

Objectives tell the students what they will learn in each lesson.

Hints, Comments, and Suggestions guide the student through the solution process.

Examples develop the mathematical concepts being taught in a series of carefully structured steps.

AREAS OF PARALLELOGRAMS 12.3

Objective **To apply the formula for the area of a parallelogram**

The Area Addition Postulate and other area postulates can be used to develop a formula for the area of a parallelogram.

Theorem 12.5 **Area $\square = bh$** The area of any parallelogram is the product of the length of a base and the length of a corresponding altitude.

Given: $\square ABCD$, length of base $\overline{AB} = b$, and length of altitude $\overline{CE} = h$

Prove: Area $\square ABCD = bh$

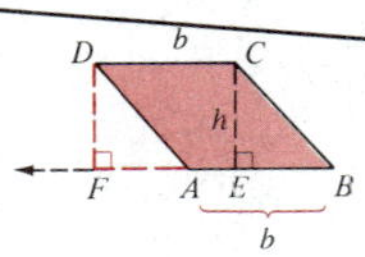

ANALYSIS
Show $\triangle FAD \cong \triangle EBC$. Then area $\square ABCD$ = area rect. $FECD$.

Proof

Statements	Reasons
1. $\square ABCD$ with altitude $\overline{CE}$, $AB = b$, $CE = h$	1. Given
2. Draw $\overline{DF} \perp \overrightarrow{BA}$.	2. There is exactly one $\perp$ from a pt. to a line.
3. $\overline{DC} \parallel \overline{BA}$	3. Opp. sides of a $\square$ are $\parallel$.
4. $\overline{DF} \cong \overline{CE}$	4. $\parallel$ lines are equidistant at all points.
5. $\overline{AD} \cong \overline{BC}$	5. Opp. sides of a $\square$ are $\cong$.
6. $\triangle FAD \cong \triangle EBC$	6. Def. of rt. $\triangle$ and HL for $\cong$ rt. $\triangle$'s
7. Area $\triangle FAD$ = area $\triangle EBC$	7. $\cong$ figures have the same area.
8. Area $\triangle FAD$ + area $AECD$ = area $\triangle EBC$ + area $AECD$	8. Add. prop. of equality
9. Area rect. $FECD$ = area $\square ABCD$	9. Area Add. Post. and substitution property
10. Area rect. $FECD = bh$	10. Area $\square = bh$
11. $\therefore$ Area $\square ABCD = bh$	11. Substitution property

Example 1 Find the area of $\square ABCD$.

$b = 9.2$ and $h = 4.5$
$A = bh = 9.2\,(4.5) = 41.4\text{ cm}^2$

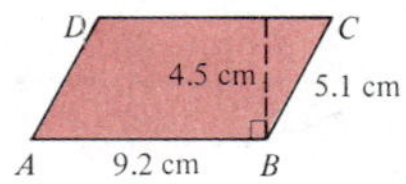

Example 2 The area of a parallelogram is 36 m². The length of a base is 8 m. Find the length of a corresponding altitude.

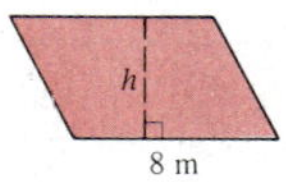

$A = bh$
$36 = 8h$ ◄ *Substitute 36 for A and 8 for b.*
$4.5 = h$
Thus, the length of a corresponding altitude is 4.5 m.

Areas of Parallelograms **395**

Graded Exercises (levels A, B, and C) promote continuous development and challenge for students of all abilities.

Oral Exercises

In each case, tell whether enough information is given to find the area of the parallelogram.

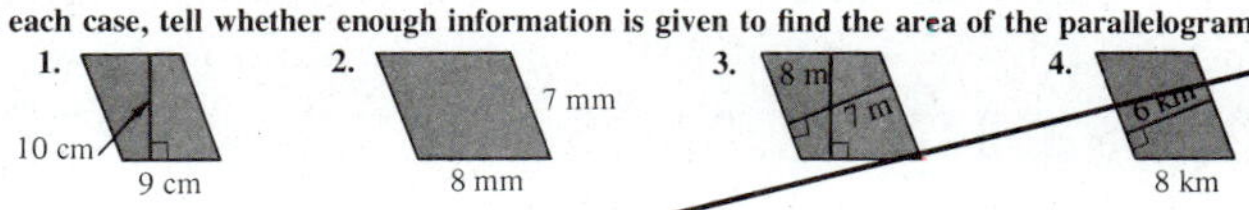

Written Exercises

Find the area of each parallelogram.

A

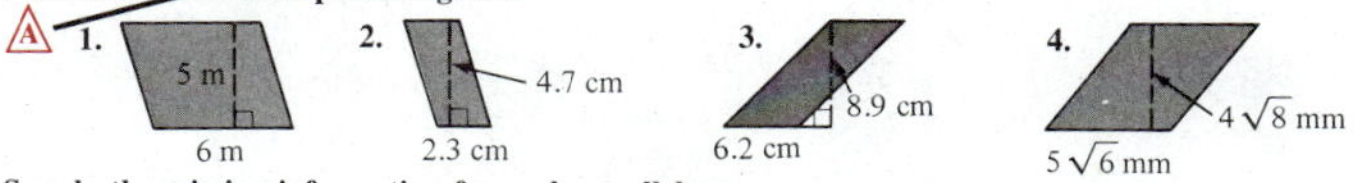

Supply the missing information for each parallelogram.

	Length of base (b)	Length of altitude (h)	Area (A)
5.	9 cm	?	81 cm²
6.	?	0.9 mm	0.72 mm²
7.	0.05 m	?	0.1695 m²
8.	?	$\sqrt{7}$ km	7 km²

In $\square ACEG$, $\overline{FB} \perp \overline{AC}$ and $\overline{GD} \perp \overline{CE}$. Find each indicated length.

B **9.** Given: $AC = 18$, $BF = 6$, and $CE = 12$. Find DG.

10. Given: $AC = 24$, $BF = 8$, and $DG = 12$. Find CE.

11. Given: $AC = 35$, $CE = 24$, and $DG = 21$. Find BF.

12. Given: $CE = 12$, $DG = \sqrt{6}$, and $BF = \sqrt{8}$. Find AC.

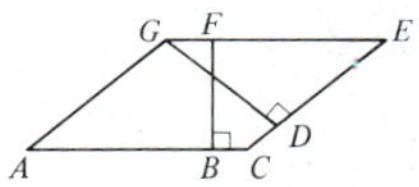

Find the area of $\square ABCD$ in each case. (Ex. 13–16)

13. Given: $AB = 10$, $AD = 6$, and $m\angle A = 30$

14. Given: $DC = 9$, $BC = 8$, and $m\angle C = 60$

15. Given: $AB = 12$, $BC = 10$, and $m\angle A = 45$

16. Given: $DC = 16$, $AD = 12$, and $m\angle D = 135$

C **17.** Prove that the area of a parallelogram can be found by the formula Area $= ab \sin A$.

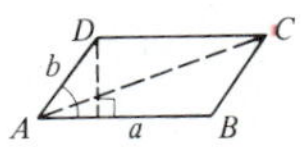

18. Prove that the area of $\triangle ABC$ can be found by the formula Area $= \frac{1}{2}ab \sin A$, where $\angle A$ is acute.

19. Find the area of $\square ABCD$ if $a = 16$, $b = 5$, and $m\angle A = 42$.

20. Find the area of $\triangle ABC$ if $a = 5$, $b = 6$, and $m\angle A = 56$.

A variety of features guide students through activities designed to stimulate thinking.

Reading in Geometry helps students develop the skills needed to read, interpret, and use the language of geometry effectively.

Chapter Openers provide opportunities to practice problem formulation, estimation, and application. Related careers and historical puzzles also furnish motivation.

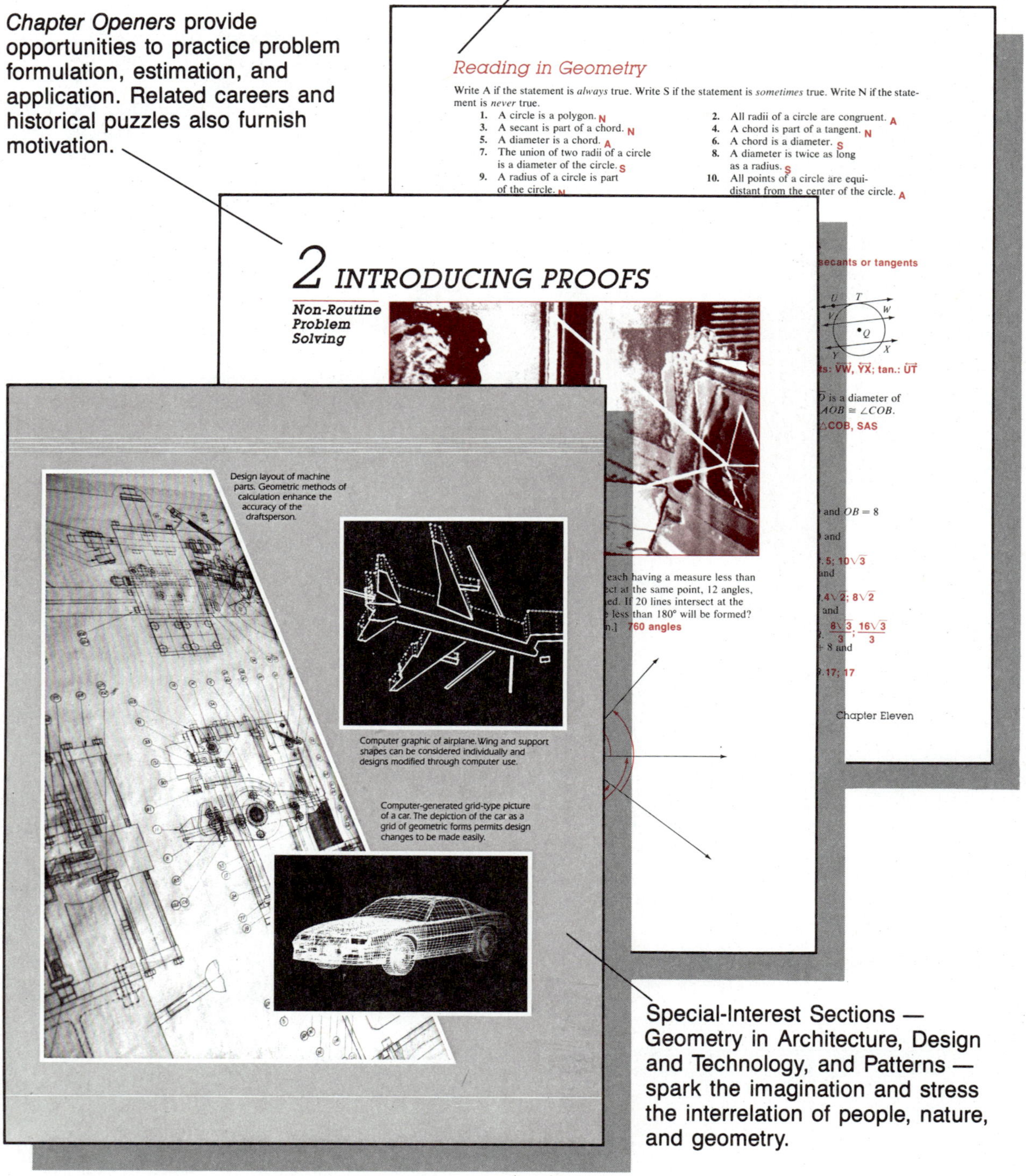

Special-Interest Sections — Geometry in Architecture, Design and Technology, and Patterns — spark the imagination and stress the interrelation of people, nature, and geometry.

ENRICHMENT MATERIALS

HOLT GEOMETRY provides many resources for enriching the curriculum, depending on the students' readiness and learning aptitude.

A special section in Logic, along with Logic Reviews, develops reasoning and challenges the student.

Algebra Review provides algebraic application with continuous cumulative review of previously learned concepts.

Calculator exercises expand students' practical proficiency in computation and estimation.

LOGIC SECTION

Logic Section

FLOW PROOFS

Objective **To write flow proofs**

m ∠A + m ∠B
m ∠D + m ∠C
∠B

560

Which of the following can be lengths of sides of a right triangle? (Ex. 16–21)

16. $3x, 4x, 5x$ **yes** 17. $n, n, 2n$ **no** 18. $5y^2, 12y^2, 13y^2$ **yes**
19. $n, m, n+m$ **no** 20. $s, \sqrt{t^2-s^2}, t$ **yes** 21. $\frac{1}{a}, \frac{1}{b}, \frac{1}{\sqrt{a^2+b^2}}$ **no**

22. Given: $\overline{CD} \perp \overline{AB}$ and CD is the geometric mean of AD and DB. Prove: $\triangle ABC$ is a rt. $\triangle$.

23. Given: $\overline{CD} \perp \overline{AB}$ and AC is the geometric mean of AD and AB. Prove: $\triangle ABC$ is a rt. $\triangle$.

24. Given: $a^2 + b^2 \neq c^2$. ($\overline{AB}$ is the longest side.) Prove: $\triangle ABC$ is not a right $\triangle$.

25. Given: $\triangle ABC$ is not a right

26. Prove Theorem 10.3.

CUMULATIVE REV

1. The measure of a base ... celes triangle is 70°. Fi... each of the other two a... **70°, 40°**

ALGEBRA REVIEW

Objective **To perform**

To add or su
To multiply
To divide ra

Examples Simplify.

$5\sqrt{3} + 4\sqrt{}$
$5\sqrt{3} + 4\sqrt{}$
$5\sqrt{3} + 4 \cdot$
$5\sqrt{3} + 8\sqrt{}$

Simplify.

1. $8\sqrt{2} + 7\sqrt{2}$ **$15\sqrt{2}$**
4. $\sqrt{72} - \sqrt{50}$ **$\sqrt{2}$**
7. $\sqrt{5} + \sqrt{45} + \sqrt{80}$ **$8\sqrt{}$**
10. $\sqrt{5} \cdot \sqrt{5}$ **5**
13. $\frac{12\sqrt{20}}{3\sqrt{5}}$ **8**

Converse of the Pythagorean Th

NON-ROUTINE PROBLEMS

The triple of numbers a, b, c is called a Pythagorean Triple if $a^2 + b^2 = c^2$ and $a, b,$ and c are positive integers. For example, 3, 4, 5 is a Pythagorean Triple since $3^2 + 4^2 = 5^2$. Each set of such triples can be the lengths of the sides of a right triangle. **1. $c = b + 1$ 2. $b = \frac{n^2-1}{2}$; $c = \frac{n^2+1}{2}$**

Pythagorean Triple			Check	
a	b	c	$a^2 + b^2$	c^2
3	4	5	9 + 16	25
5	12	13	25 + 144	169
7	24	25	49 + 576	625

1. Examine the three sets of triples in the table. What is true of the first number, a in each triple? How do b and c compare? **It is an odd integer greater than one.**
2. If a is any positive odd integer n, derive a formula for b and c in terms of n.
3. Let $n = 9, 11, 13 \ldots$ Do your formulas for b and c give Pythagorean Triples? **yes**

CALCULATOR ACTIVITIES

A calculator can help you to apply the Pythagorean Theorem:
$a^2 + b^2 = c^2$ or $c^2 = a^2 + b^2$.

Since a, b, and c represent positive numbers, properties of algebra lead to:
(1) $c^2 = a^2 + b^2$ and $c = \sqrt{a^2 + b^2}$
(2) $a^2 = c^2 - b^2$ and $a = \sqrt{c^2 - b^2}$
(3) $b^2 = c^2 - a^2$ and $b = \sqrt{c^2 - a^2}$

For example, if $a = 18$ and $c = 36$, you can find b by using $b = \sqrt{c^2 - a^2}$.

Find c^2. ► Press 36 ⊗ 36 ⊜ Display 1296 ◄ *Write down and clear.*
Find a^2. ► Press 18 ⊗ 18 ⊜ Display 324 ◄ *Write down and clear.*
Find $c^2 - a^2$. ► Press 1296 ⊖ 324 ⊜ Display 972
$b = \sqrt{c^2 - a^2}$ ► **Thus,** $b = \sqrt{972}$, or 31.2.

For each right triangle, find the indicated length to the nearest tenth.
1. $a = ?$ $b = 40$, $c = 41$ **9.0** 2. $a = 15$, $b = 36$, $c = ?$ **39.0** 3. $a = 60$, $b = ?$, $c = 61$ **11.0**
4. $a = ?$ $b = 75$, $c = 85$ **40.0** 5. $a = 17$, $b = ?$, $c = 34$ **29.4** 6. $a = 16$, $b = 32$, $c = ?$ **35.8**
7. $a = 24$, $b = 72$, $c = ?$ **75.9** 8. $a = ?$, $b = 18$, $c = 90$ **88.2** 9. $a = 32$, $b = ?$, $c = 96$ **90.5**

314 Chapter Ten

COMPUTER STRAND

Computer lessons throughout the book include computer applications of geometry content. Hands-on programming exercises in BASIC encourage the student with computer expertise to extend the geometric concepts further.

COMPUTER ACTIVITIES

Concentric Circles

Objective: **To find the length of a chord of one circle when that chord is tangent to a smaller concentric circle, using a computer program that inputs the lengths of the radii of the circles**

Illustrated at the right are two concentric circles. $\overline{AB}$ is a chord of the larger circle and is tangent to the smaller circle. Radius $\overline{OC}$ is drawn to the point of tangency. $\overline{OC}$ is perpendicular to the chord and bisects it. This creates 2 congruent right triangles, which makes it possible to determine the length of the chord when the lengths of the two radii are known.

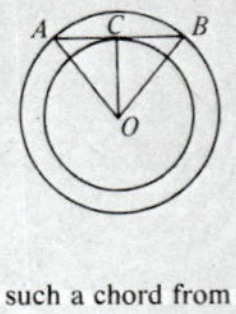

The following computer program calculates the length of such a chord from input of the two radii.

```
...TRIC CIRCLES, THIS"
...D THE LENGTH OF A CHORD"
...RCLE WHICH IS TANGENT"
...US OF THE LARGER";
...US OF THE SMALLER";
...R2 ^ 2)
...HE CHORD IS ";C"."
```

...n beginning on page 565 for more information.

...roblem graded quizzes.

...tes the length of the smaller radius ...nd the length of the chord are

...gth of the larger radius when the

381

TRIGONOMETRY AND THE COMPUTER

Objectives: **To make angle measurement conversions from degrees to radians and vice versa**
To solve trigonometric problems using computer programs

Most versions of BASIC contain three trig functions – sine, cosine, and tangent, denoted as SIN, COS, and TAN. An example of the format used in a program is SIN(X). To use BASIC trig functions, however, you must become familiar with another way to measure angles.

The X in SIN(X) is not in terms of degrees, but in terms of another unit for measuring angles called *radians*. Without going into an extensive explanation of radian measure, this lesson will introduce the relationship between degree measure and radian measure, so that you can readily convert one to the other.

Recall that if a circle is divided into 360 equal parts, then one degree consists of exactly one of those parts. Radian measure is based on the idea of a circle with radius 1. Such a circle has a circumference of 2π (or approximately 6.2832) units. If the circle is divided into 2π (that is, 6.2832) equal parts, then one radian consists of exactly one of those parts. The complete circle is 2π radians. So, an angle measuring 2π radians is the same as an angle measuring 360°. Similarly, half a circle is π (or 3.1416) radians, which is equivalent to 180°. Also, a quarter of a circle is $\pi/2$ (or 1.5708) radians, and $\pi/2$ radians equals 90°. The illustration below shows a few values of radians and their approximate locations along with several corresponding degree measurements.

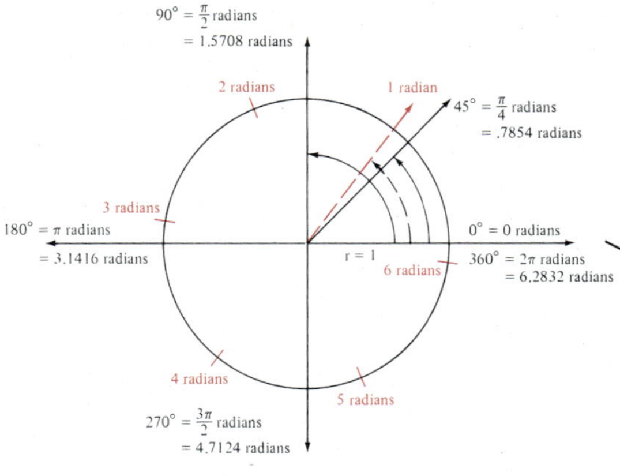

576 Computer Section

The back of the book contains a 13-page Computer Section designed to give students experiences in programming that relate to exploring ideas in geometry. Looping commands, as well as special computer functions, are employed.

PROBLEM-SOLVING STRAND

Problem-Solving Applications in real-world careers and in mathematics usage offer abundant motivation to practice problem-solving strategies.

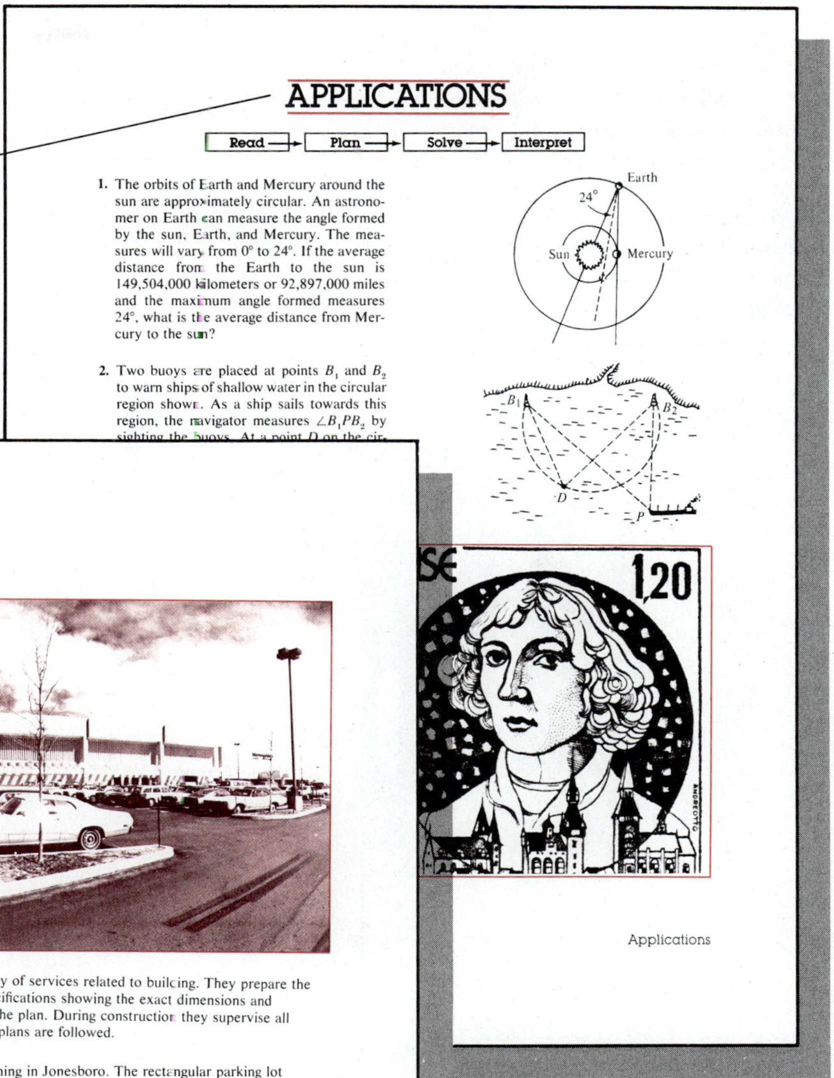

APPLICATIONS

Read → Plan → Solve → Interpret

1. The orbits of Earth and Mercury around the sun are approximately circular. An astronomer on Earth can measure the angle formed by the sun, Earth, and Mercury. The measures will vary from 0° to 24°. If the average distance from the Earth to the sun is 149,504,000 kilometers or 92,897,000 miles and the maximum angle formed measures 24°, what is the average distance from Mercury to the sun?

2. Two buoys are placed at points B_1 and B_2 to warn ships of shallow water in the circular region shown. As a ship sails towards this region, the navigator measures $\angle B_1PB_2$ by sighting the buoys. At a point D on the cir-

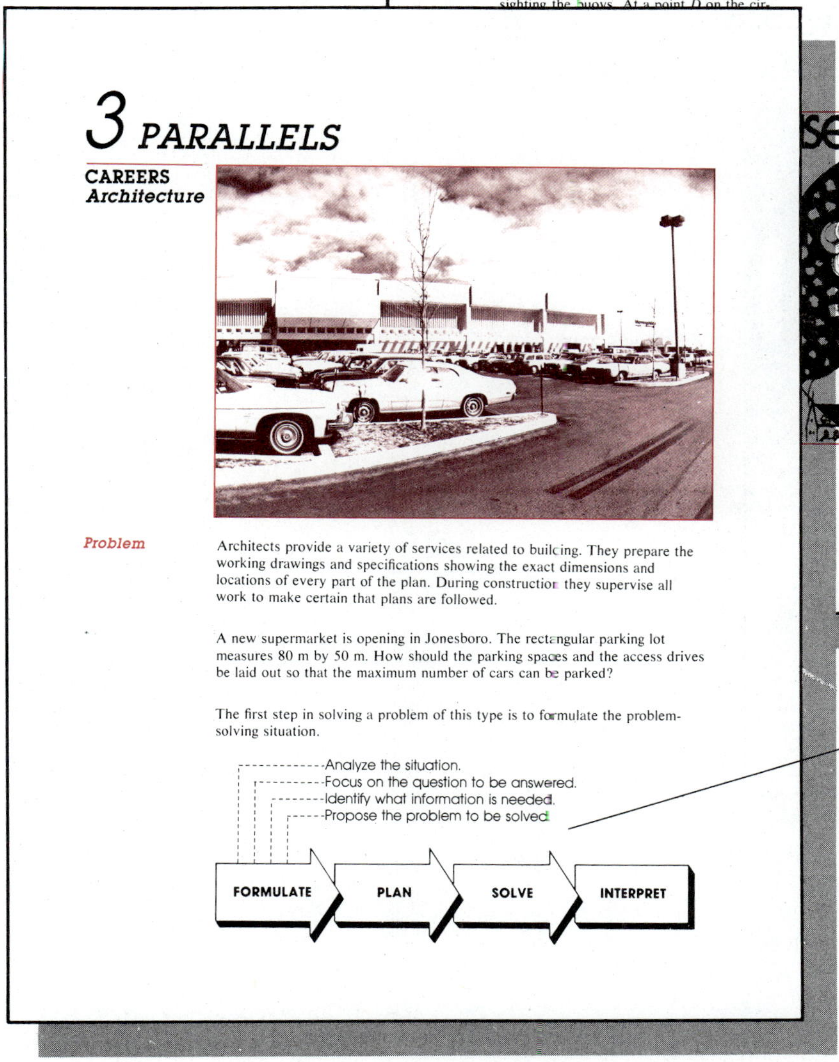

3 PARALLELS

CAREERS
Architecture

Problem Architects provide a variety of services related to building. They prepare the working drawings and specifications showing the exact dimensions and locations of every part of the plan. During construction they supervise all work to make certain that plans are followed.

A new supermarket is opening in Jonesboro. The rectangular parking lot measures 80 m by 50 m. How should the parking spaces and the access drives be laid out so that the maximum number of cars can be parked?

The first step in solving a problem of this type is to formulate the problem-solving situation.

A four-part problem-solving strategy, continually highlighted, encourages students to approach problem solving in a logical manner.

REINFORCING SKILLS

Summary provides a quick review of the key concepts and procedures taught in the lesson.

Non-Routine Problems provide students with the opportunity to solve unusual problems using non-standard aproaches or strategies.

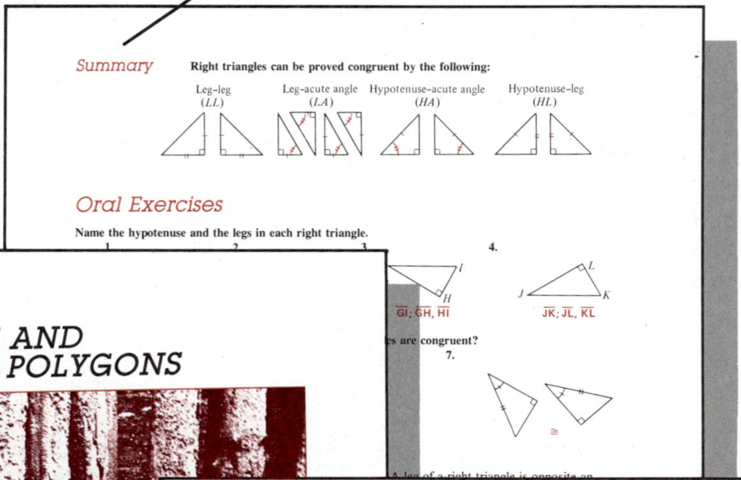

Summary — Right triangles can be proved congruent by the following:

Leg-leg (*LL*) Leg-acute angle (*LA*) Hypotenuse-acute angle (*HA*) Hypotenuse-leg (*HL*)

Oral Exercises

Name the hypotenuse and the legs in each right triangle.

GI; GH, HI

JK; JL, KL

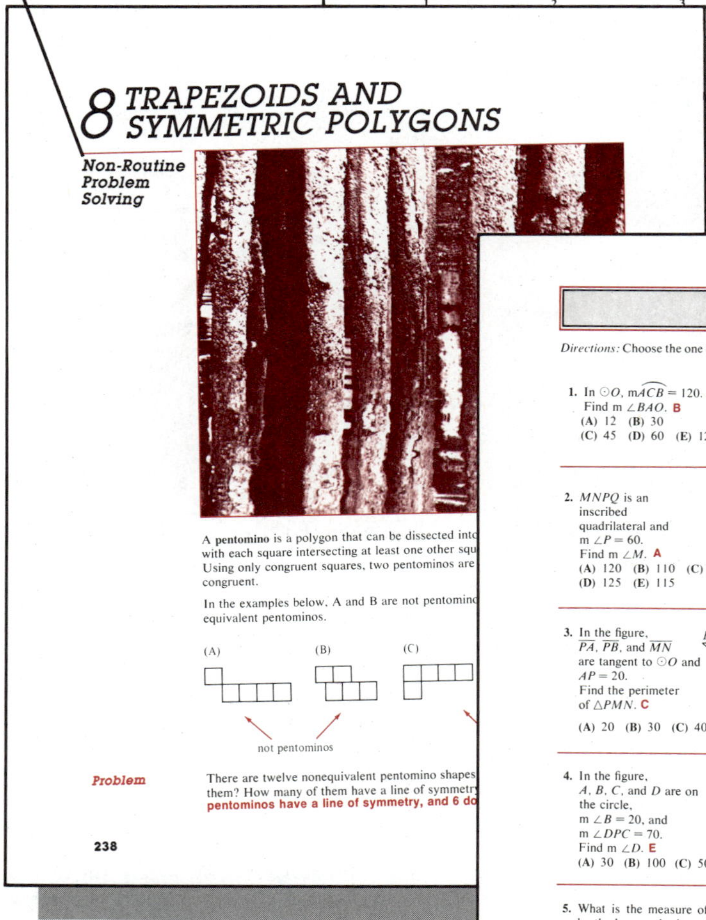

8 TRAPEZOIDS AND SYMMETRIC POLYGONS

Non-Routine Problem Solving

A **pentomino** is a polygon that can be dissected into ... with each square intersecting at least one other squ... Using only congruent squares, two pentominos are ... congruent.

In the examples below, A and B are not pentomino... equivalent pentominos.

(A) (B) (C)

not pentominos

Problem — There are twelve nonequivalent pentomino shapes... them? How many of them have a line of symmetry... **pentominos have a line of symmetry, and 6 do...**

238

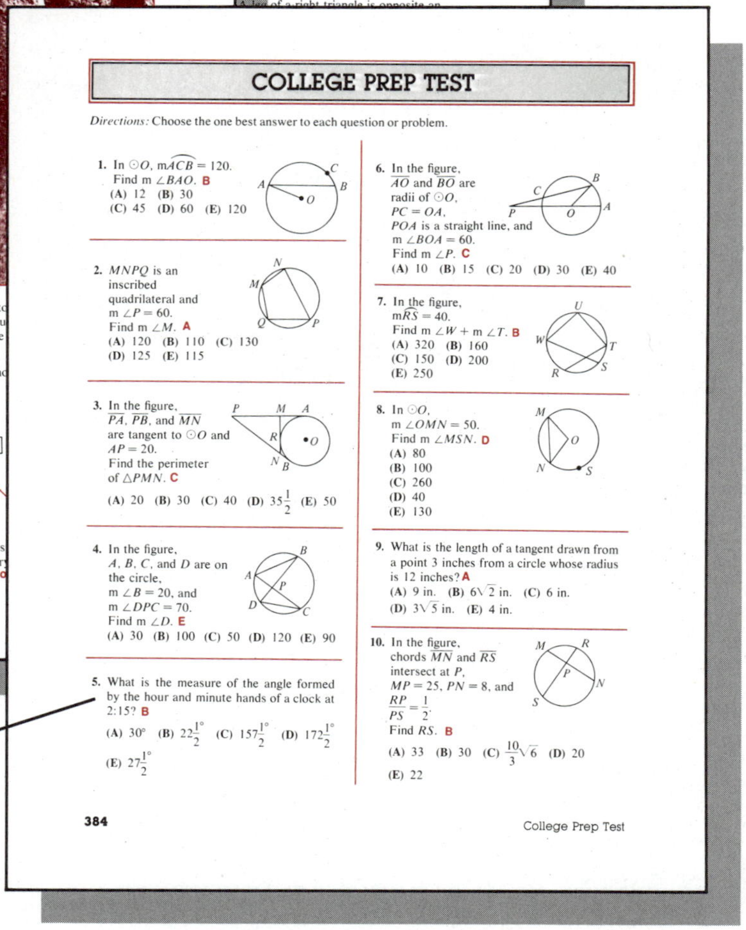

COLLEGE PREP TEST

Directions: Choose the one best answer to each question or problem.

1. In ⊙O, m$\widehat{ACB} = 120$. Find m $\angle BAO$. **B**
(A) 12 (B) 30 (C) 45 (D) 60 (E) 120

2. $MNPQ$ is an inscribed quadrilateral and m $\angle P = 60$. Find m $\angle M$. **A**
(A) 120 (B) 110 (C) 130 (D) 125 (E) 115

3. In the figure, $\overline{PA}$, $\overline{PB}$, and $\overline{MN}$ are tangent to ⊙O and $AP = 20$. Find the perimeter of $\triangle PMN$. **C**
(A) 20 (B) 30 (C) 40 (D) $35\frac{1}{2}$ (E) 50

4. In the figure, A, B, C, and D are on the circle, m $\angle B = 20$, and m $\angle DPC = 70$. Find m $\angle D$. **E**
(A) 30 (B) 100 (C) 50 (D) 120 (E) 90

5. What is the measure of the angle formed by the hour and minute hands of a clock at 2:15? **B**
(A) 30° (B) $22\frac{1}{2}°$ (C) $157\frac{1}{2}°$ (D) $172\frac{1}{2}°$ (E) $27\frac{1}{2}°$

6. In the figure, $\overline{AO}$ and $\overline{BO}$ are radii of ⊙O, $PC = OA$, POA is a straight line, and m $\angle BOA = 60$. Find m $\angle P$. **C**
(A) 10 (B) 15 (C) 20 (D) 30 (E) 40

7. In the figure, m$\widehat{RS} = 40$. Find m $\angle W$ + m $\angle T$. **B**
(A) 320 (B) 160 (C) 150 (D) 200 (E) 250

8. In ⊙O, m $\angle OMN = 50$. Find m $\angle MSN$. **D**
(A) 80 (B) 100 (C) 260 (D) 40 (E) 130

9. What is the length of a tangent drawn from a point 3 inches from a circle whose radius is 12 inches? **A**
(A) 9 in. (B) $6\sqrt{2}$ in. (C) 6 in. (D) $3\sqrt{5}$ in. (E) 4 in.

10. In the figure, chords $\overline{MN}$ and $\overline{RS}$ intersect at P, $MP = 25$, $PN = 8$, and $\frac{RP}{PS} = \frac{1}{2}$. Find RS. **B**
(A) 33 (B) 30 (C) $\frac{10}{3}\sqrt{6}$ (D) 20 (E) 22

384 College Prep Test

College Prep Tests give students practice and experience in strategies they will use in taking multiple-choice tests that they will encounter in the future.

TESTING AND REVIEW

Tests and Reviews assist the teacher in measuring the students' understanding of the major skills and concepts taught.

Cumulative Reviews test the students' understanding of geometric concepts and corresponding problem-solving skills that have been previously taught.

Chapter Tests are designed to measure the students' mastery of the chapter objectives. The difficulty level is indicated.

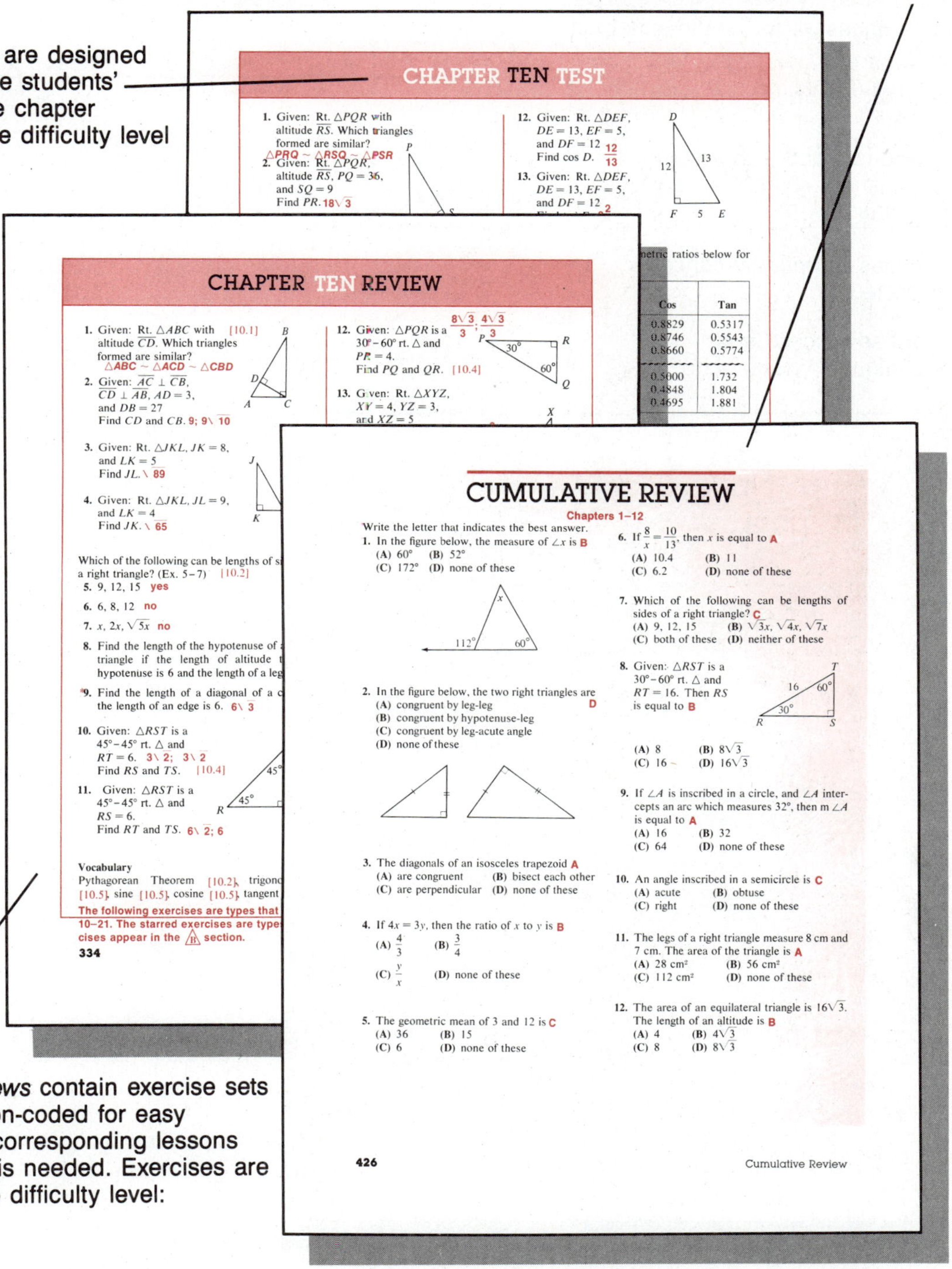

CHAPTER TEN TEST

1. Given: Rt. $\triangle PQR$ with altitude $\overline{RS}$. Which triangles formed are similar? $\triangle PRQ \sim \triangle RSQ \sim \triangle PSR$
2. Given: Rt. $\triangle PQR$, altitude $\overline{RS}$, $PQ = 36$, and $SQ = 9$. Find PR. $18\sqrt{3}$
12. Given: Rt. $\triangle DEF$, $DE = 13$, $EF = 5$, and $DF = 12$. Find cos D. $\frac{12}{13}$
13. Given: Rt. $\triangle DEF$, $DE = 13$, $EF = 5$, and $DF = 12$

Cos	Tan
0.8829	0.5317
0.8746	0.5543
0.8660	0.5774
0.5000	1.732
0.4848	1.804
0.4695	1.881

CHAPTER TEN REVIEW

1. Given: Rt. $\triangle ABC$ with altitude $\overline{CD}$. Which triangles formed are similar? [10.1] $\triangle ABC \sim \triangle ACD \sim \triangle CBD$
2. Given: $\overline{AC} \perp \overline{CB}$, $\overline{CD} \perp \overline{AB}$, $AD = 3$, and $DB = 27$. Find CD and CB. 9; $9\sqrt{10}$
3. Given: Rt. $\triangle JKL$, $JK = 8$, and $LK = 5$. Find JL. $\sqrt{89}$
4. Given: Rt. $\triangle JKL$, $JL = 9$, and $LK = 4$. Find JK. $\sqrt{65}$

Which of the following can be lengths of sides of a right triangle? (Ex. 5–7) [10.2]

5. 9, 12, 15 yes
6. 6, 8, 12 no
7. x, $2x$, $\sqrt{5x}$ no
8. Find the length of the hypotenuse of a triangle if the length of altitude to the hypotenuse is 6 and the length of a leg
9. Find the length of a diagonal of a cube if the length of an edge is 6. $6\sqrt{3}$
10. Given: $\triangle RST$ is a 45°–45° rt. $\triangle$ and $RT = 6$. Find RS and TS. [10.4] $3\sqrt{2}$; $3\sqrt{2}$
11. Given: $\triangle RST$ is a 45°–45° rt. $\triangle$ and $RS = 6$. Find RT and TS. $6\sqrt{2}$; 6
12. Given: $\triangle PQR$ is a 30°–60° rt. $\triangle$ and $PR = 4$. Find PQ and QR. [10.4] $\frac{8\sqrt{3}}{3}$; $\frac{4\sqrt{3}}{3}$
13. Given: Rt. $\triangle XYZ$, $XY = 4$, $YZ = 3$, and $XZ = 5$

Vocabulary
Pythagorean Theorem [10.2], trigono... [10.5], sine [10.5], cosine [10.5], tangent

The following exercises are types that ... 10–21. The starred exercises are type... cises appear in the B section.

334

CUMULATIVE REVIEW

Chapters 1–12

Write the letter that indicates the best answer.

1. In the figure below, the measure of $\angle x$ is B
(A) 60° (B) 52°
(C) 172° (D) none of these
2. In the figure below, the two right triangles are D
(A) congruent by leg-leg
(B) congruent by hypotenuse-leg
(C) congruent by leg-acute angle
(D) none of these
3. The diagonals of an isosceles trapezoid A
(A) are congruent (B) bisect each other
(C) are perpendicular (D) none of these
4. If $4x = 3y$, then the ratio of x to y is B
(A) $\frac{4}{3}$ (B) $\frac{3}{4}$
(C) $\frac{y}{x}$ (D) none of these
5. The geometric mean of 3 and 12 is C
(A) 36 (B) 15
(C) 6 (D) none of these
6. If $\frac{8}{x} = \frac{10}{13}$, then x is equal to A
(A) 10.4 (B) 11
(C) 6.2 (D) none of these
7. Which of the following can be lengths of sides of a right triangle? C
(A) 9, 12, 15 (B) $\sqrt{3x}$, $\sqrt{4x}$, $\sqrt{7x}$
(C) both of these (D) neither of these
8. Given: $\triangle RST$ is a 30°–60° rt. $\triangle$ and $RT = 16$. Then RS is equal to B
(A) 8 (B) $8\sqrt{3}$
(C) 16 (D) $16\sqrt{3}$
9. If $\angle A$ is inscribed in a circle, and $\angle A$ intercepts an arc which measures 32°, then m $\angle A$ is equal to A
(A) 16 (B) 32
(C) 64 (D) none of these
10. An angle inscribed in a semicircle is C
(A) acute (B) obtuse
(C) right (D) none of these
11. The legs of a right triangle measure 8 cm and 7 cm. The area of the triangle is A
(A) 28 cm² (B) 56 cm²
(C) 112 cm² (D) none of these
12. The area of an equilateral triangle is $16\sqrt{3}$. The length of an altitude is B
(A) 4 (B) $4\sqrt{3}$
(C) 8 (D) $8\sqrt{3}$

426 Cumulative Review

Chapter Reviews contain exercise sets that are lesson-coded for easy reference to corresponding lessons when review is needed. Exercises are also coded to difficulty level: A, B, or C.

TEACHER'S EDITION

The Teacher's Edition of *HOLT GEOMETRY* provides an annotated full-size reproduction of each pupil page and includes detailed chapter and lesson commentaries.

References all special feature pages.

Highlights vocabulary.

Reinforces the lesson skills.

Provides practice in problem formulation, estimation, and application.

Focuses on the key concepts to be taught.

Clarifies process and techniques for students.

Special Features

Non-Routine Problem Solving (p. 238)

Solution: There are 6 non-equivalent pentominos which have a line of symmetry and 6 which do not. Those with symmetry:

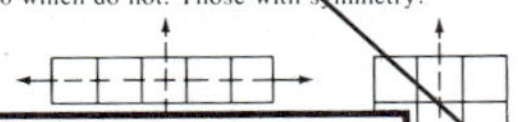

8.1 Trapezoids (pages 239–241)

Vocabulary

trapezoid

Suggestions

Review the idea of a counterexample. Remind your students that one example does not prove a statement is always true, but one counterexample is sufficient to disprove a statement. Exercises 2 and 3 anticipate the theorems about isosceles trapezoids in section 8.3. Make sure your students realize that the base angles can be proved congruent only if given that the nonparallel sides of the trapezoid are congruent. For Exercise 17, the fact that the segments joining the midpoints of opposite sides of a trapezoid bisect each other may be proved by first recalling that the segments joining the midpoints of consecutive sides of a quadrilateral form a parallelogram. Then use the property that the diagonals of a parallelogram bisect each other. This proof holds for all quadrilaterals, not just for trapezoids.

For Additional Chalkboard Work

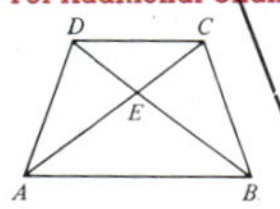

Given: Trapezoid $ABCD$

Which angles are known to be congruent?

$\angle ABD \cong \angle BDC$, $\angle BAC \cong \angle ACD$. Vertical angles formed by the diagonals are also $\cong$.

Cumulative Review

1. Which pairs of angles of parallel lines cut by a transversal are congruent?
2. In using the side-angle-side congruence pattern for triangles, what is meant by the included angle?

1. Alternate interior, corresponding, and alternate exterior angles 2. It is the angle whose vertex is the intersection of the two sides.

7.7 Proving Rhombi, Rectangles, and Squares (pages 230–232)

Background

Students must realize that a quadrilateral may be proved a rhombus, rectangle, or square by using its definition.

Suggestions

Ask your students to sketch a quadrilateral with perpendicular diagonals that is not a rhombus (Figure 1). Have them sketch a quadrilateral with a right angle that is not a rectangle (Figure 2) and a quadrilateral with congruent diagonals that is not a rectangle (Figure 3).

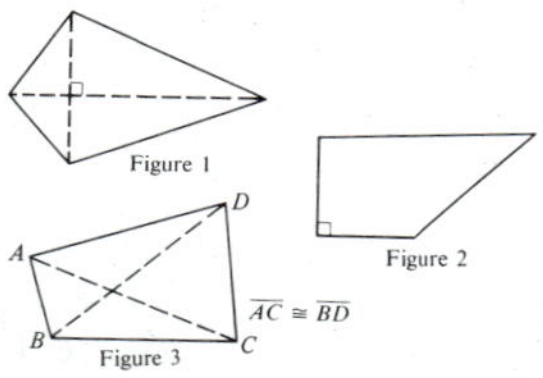

For Additional Chalkboard Work

A contractor is laying out the foundations of a rectangular building. How can the contractor be sure that the lines really do form a rectangle?

Measure so that opposite sides are the same length and diagonals are the same length.

Cumulative Review

1. Is a square a rhombus? a rectangle?
2. If in two congruent triangles, $\angle A \cong \angle T$, $\angle B \cong \angle U$, and $\angle C \cong \angle V$, which sides of the triangles are congruent?

1. Both 2. $\overline{AB} \cong \overline{TU}$, $\overline{BC} \cong \overline{UV}$, $\overline{AC} \cong \overline{TV}$

Computer Activity (p. 236)

Writing a flowchart from a program is not the ordinary sequence of events, but the practice in writing the chart is worthwhile.

8 Chapter and Lesson Commentaries

Chapter 8 Trapezoids and Symmetric Polygons

Chapter Objectives

To identify trapezoids and isosceles trapezoids
To investigate and prove properties of trapezoids and isosceles trapezoids
To prove and apply theorems about the median and an altitude of a trapezoid
To identify congruent quadrilaterals
To prove theorems about congruent quadrilaterals
To construct quadrilaterals with given properties
To identify symmetric polygons
To determine and apply congruence properties of symmetric polygons
To identify reflections
To prove and apply theorems relating reflections and congruence

Chapter Overview

This chapter defines trapezoids and isosceles trapezoids and investigates and proves properties about them. Theorems about the median and altitude of a trapezoid are also considered. Congruent quadrilaterals and their properties are discussed. Construction with given properties is continued in this chapter with quadrilaterals. Identifying symmetric polygons, applying congruence properties, identifying reflections and relating their properties are also introduced in this chapter.

Gives ideas for presentation and individualization.

Indicates mathematical content.

Keeps previously learned skills current.

SUPPLEMENTARY MATERIALS

HOLT GEOMETRY 's supplementary materials include Testmasters, Skillmasters, and a Teacher's Resource Package with Blackline Masters so that the teacher can instruct according to each student's needs and skill levels.

Skillmasters	contain two parallel sections for every two lessons, which can be used as pre- or posttests for diagnosis, extra practice, quizzes, and assignments. Cumulative Review for every few lessons.
Testmasters	supply Chapter Tests, Cumulative Tests, and Final Tests.
Teacher's Resource Package	contains Skillmasters and Testmasters in blackline form, Enrichment masters, Competency Tests, and Lesson Aids. The Enrichment Pages provide additional material that can be used to challenge all students: transformational geometry, applications, history, and non-routine problems. The Competency Tests contain the type of material found on standard competency tests used in many states. The Lesson Aids are a set of extra materials and projects that can be used in teaching the lesson.
Solutions Manual	provides worked-out solutions to every problem and exercise in the text.

ABOUT HOLT GEOMETRY

Success Pattern HOLT GEOMETRY guides students through the process of developing their mathematical abilities by helping them experience a pattern of successful learning that builds self-confidence. First, students examine worked-out examples to see how conclusions can be drawn from assumptions. Then they provide reasons for selected steps in proofs. Next, they write out simple proofs based on familiar concepts. As students become more familiar with proofs, they learn to prove increasingly difficult theorems. This pattern of instruction is characteristic of the approach used throughout HOLT GEOMETRY.

Organization of the Course HOLT GEOMETRY contains all of the major geometry concepts necessary to meet the instructional demands in your school, as well as evaluation materials, which include tests for each unit with items keyed to lesson objectives. HOLT GEOMETRY also provides supplementary materials in logic, computers, calculators, practice for college admission tests, and algebra reviews. Together these elements provide an enriched basal course that can be easily adapted to the needs of your students.

Instructional Advantages The instructional advantages of Holt's approach are clear and well documented. First, frequent evaluation — Chapter Reviews, tests at the end of chapters, and Cumulative Reviews every four chapters — provides continuous feedback for both you and the student. Second, a varied instructional technique provides students with the opportunity to participate actively in learning and to find an instructional method that best suits their needs. Students who learn quickly and can work independently need the challenge provided by the supplements and extensions to the basic course. Others benefit from selected problems that challenge them at their own levels of competence. HOLT GEOMETRY addresses the needs of both types of students.

Students with Handicapping Conditions Instructional materials are also available for helping students who need to overcome learning problems that may be due to physical disabilities (orthopedic or visual impairments), communicative disabilities (hearing loss, dyslexia), mild mental retardation, or disabling emotional disturbance. Teachers should be on the alert to such possibilities so that students with undiagnosed disabilities can be treated professionally.

Whenever possible, students with handicapping conditions should participate either partially or totally in regular school programs and classes, as part of the requirement of Federal Law PL94-142, which states that handicapped children must be educated in the least restrictive environment possible.

By using the chapter and lesson objectives in the book, the teacher can select appropriate lesson goals for each student and communicate them easily to parents or special-education teachers who may be assisting the handicapped. The Skillmasters and Testmasters in the Teacher's Resource Package are valuable tools for assessing each individual student's level of comprehension and skill in a given lesson or topic.

Success of HOLT GEOMETRY HOLT GEOMETRY has been used successfully with a variety of students. It has been found to be stimulating, highly teachable, and motivational, with a good balance of skill, theory, and application. This new edition of HOLT GEOMETRY, with its expanded and new features, is based on suggestions from users of the previous editions.

Reference Chart for Geometry Skillmasters

LESSON PAGES	SKILLMASTER NUMBER	LESSON PAGES	SKILLMASTER NUMBER
1, 5	1	278, 282, 285	25
8, 12	2	290, 294	26
23, 27	3	306, 310	27
29, 33, 36	4	315, 318, 321	28
40, 44	5	324, 326, 329	29
49, 55, 59	6	338, 342	30
69, 72	7	345, 351, 355	31
75, 80	8	359, 364	32
85, 87	9	368, 373, 377	33
91, 93	10	387, 391, 395	34
103, 107	11	398, 402	35
110, 114	12	406, 409, 412	36
118, 123	13	415, 418	37
136, 139	14	429, 432, 435	38
153, 157	15	438, 442, 446	39
162, 166, 169	16	449, 453, 456	40
183, 188	17	467, 470, 475	41
193, 196, 199	18	483, 487	42
208, 211	19	499, 503, 507	43
215, 219, 222	20	511, 516	44
226, 230	21	525, 529, 532	45
239, 242	22	535, 539	46
246, 250, 254	23	544	47
270, 275	24	549	48

Acknowledgments

Holt is grateful to thousands of teachers and to hundreds of thousands of students who have used our geometry texts in the past and who have contributed their comments and constructive criticism to help make the teachable textbooks of the past even more teachable and enjoyable.

We are also grateful to teachers, supervisors, and administrators across the country who have participated in systematic surveys so that Holt can give America's high school students the kind of materials that best facilitate their learning.

1 Chapter and Lesson Commentaries

Chapter 1 Figures and Measures

Chapter Objectives

To identify and name geometric figures such as points, lines, planes, collinear points, and coplanar points
To find the distance between two points on a number line
To identify and construct congruent segments
To construct the midpoint of a segment
To classify angles by their measures
To construct congruent angles and the bisector of an angle

Chapter Overview

Basic sets of points, such as lines, planes, and space, are introduced. Some of these, such as point, line, and plane, are identified as undefined terms; others are definitions. Sets of collinear and coplanar points are identified. Coordinates of points on a line are used to find the distance between two points. This is related to the length of the segment determined by these points. Agreements are made for the proper naming of angles. Angles are classified by their measures. *Betweenness* and *interior* are used in defining midpoints of segments and bisectors of angles. Congruence of segments and angles is defined in terms of equal measures. Basic straightedge and compass constructions are presented. No formal proofs are developed at this time.

Special Features

Non-Routine Problem Solving p. x
Reading in Geometry pp. 3, 15
Algebra Review p. 4
Non-Routine Problems p. 11
Applications p. 17
Chapter Review p. 18
Chapter Test p. 19
Computer Activities p. 20
College Prep. Test p. 21

Non-Routine Problem Solving (p. x)

The "Non-Routine Problem Solving" sections appear on chapter opener pages throughout the text. Encourage the use of various non-routine techniques in solving the problems.

Reading in Geometry (pp. 3, 15)

The "Reading in Geometry" sections that are found periodically throughout this text are meant to be used as a followup activity. You may find it feasible to give your students class time to read the particular section. Then use the "Reading in Geometry" as an individual activity in class. A class discussion would evolve after this. In some cases, you may decide to use the reading activity as a source of a teacher-centered activity.

1.1 Points, Lines, and Planes (pages 1–4)

Vocabulary
point, line, plane, geometric figure, collinear, coplanar, absolute value

Background
The important distinction between undefined terms and definitions is stressed as basic geometric figures are introduced. Proper identification of geometric figures and their properties should be emphasized.

Suggestions
Have students keep a notebook. Pages should be reserved for vocabulary lists and lists for postulates and proved theorems. Have students attempt to define *point, line,* and *plane*. A good definition should include all that belongs but exclude all that does not belong. In the process, students will eventually return to original words. At this point, introduce the concept of undefined terms. Try to identify physical models of geometric figures in the classroom. Students should begin to make drawings of geometric figures. Proper naming of figures is necessary. Do not depend solely on pictures for the identification of collinear and coplanar points. Use models made from sticks and strings. Commercial toys such as Tinkertoys and D-Stix are excellent. A supply of cardboard sheets, dowel rods or other sticks, and string is helpful. In any discussion of number lines and distance, information on absolute value is necessary (Lesson 1.2). The Algebra Review provides such review.

For Additional Chalkboard Work
Write several words such as the following on the chalkboard: *number, one, love, hate.* Ask students to try to define these words so that if they did not already know what they meant, the meanings would be clear. Point out how difficult this is and that in many cases, words have to be accepted as undefined.
Answers will vary.

1.2 Distance and Length (pages 5–7)

Vocabulary
coordinate, number line, distance, length, segment, ray, between, endpoint

Background
The distance between two points is the difference of their coordinates. To avoid negative distances, the absolute value of this difference is used.

Betweenness emphasizes collinearity. Betweenness and straightness are different aspects of collinearity and should be so related.

Suggestions
Stress the naming of segments and rays by their endpoints and the proper use of the symbols for each figure.

For Additional Chalkboard Work
Place two marks on the chalkboard. Place a standard yardstick on these marks and read the coordinates. Compute the distance between the points. Have the stick moved so that different coordinates represent the points. Does the distance change? Repeat this using a meter stick. Are the measures the same? Why not?
No; no; different units result in different measures of the same distance.

Enrichment
You may wish to introduce the half-line and the difference between it and a ray. The half-line can be thought of as a ray without the endpoint.

Cumulative Review
1. What are collinear points?
2. Can you identify three points in your classroom that are noncoplanar?
3. Choose any two corners of your classroom as two points. Where can a third point be located so as to be collinear with the two points?

1. Points on the same line
2. Answers will vary.
3. On the line containing the two corners

1.3 Congruent Segments (pages 8–11)

Vocabulary

congruent segments, midpoint, bisector, equidistant

Background

Congruence of segments is defined as equality of length. The congruence symbol, ≅, expresses a relationship between geometric figures, not numbers. The equals symbol, =, expresses a relationship between numbers. The midpoint of a segment is introduced using an intuitive approach.

Constructions for a segment congruent to a given segment and for the midpoint of a segment are given.

Proofs for constructions cannot be given at this time.

Suggestions

You may wish to introduce congruent segments by comparing sticks of the same length. Stress that sets of points are congruent but numbers are equal. AB is the correct way to represent the measure of $\overline{AB}$.

For Additional Chalkboard Work

Write a formula for finding the coordinate of the midpoint of any segment that has one endpoint with a coordinate twice that of the other endpoint.

For p and $2p$, then $m = \frac{3p}{2}$.

Enrichment

For more advanced students, you may wish to introduce the midpoint formula, $m = \frac{a+b}{2}$.

A coordinate treasure hunt can be used as a fun activity. Give a set of directions for the computation of each step, such as "Start at 3, go 5 to the right, go back half the distance."

Cumulative Review

1. What is the absolute value of −5?
2. Find the length of a segment whose endpoints have the coordinates 4 and 12.
3. What is the difference between a segment and a ray?

1. 5 2. 8 3. A segment has two endpoints; a ray has one endpoint.

1.4 Angles (pages 12–16)

Vocabulary

angle, congruent angles, acute angle, right angle, obtuse angle, straight angle, interior, exterior, angle bisector

Background

We limit the measures of angles to $0 < m \leq 180$. Reflex angles that have measures greater than 180° are not considered. Straight angles are included, even though there may be valid arguments against their inclusion, such as the lack of unique interiors and the possibility of many different vertices. A zero measure angle, however, is not included. Acute, right, and obtuse angles are defined.

Suggestions

Have students estimate, measure, and classify angles in the classroom. Stress accuracy, especially placement of the protractor. *Note:* $m\angle AOB = 65$, but $\angle AOB$ is a 65° angle. The degree (°) is not used when the notation $m\angle AOB$ is used. Review the techniques for using a straightedge and compass.

For Additional Chalkboard Work

Find the measure of the angle formed by the bisector of a 108° angle, a 52° angle, and a 75° angle. 54°, 26°, 37.5°

Cumulative Review

1. Find the coordinate of the midpoint of a segment whose endpoints are −4 and 10.
2. Give an example of noncoplanar points.
3. What are congruent segments?

1. 3 2. Answers will vary.

3. Segments with equal lengths

Computer Activity (p. 20)

If this is your students' first exposure to computers, line 30 of the first program is simple but needs clarification. They need to understand why the quotes are closed, reopened, closed again, etc. The shift from printing words to printing the value of variables and back again to words makes this necessary.

2 Chapter and Lesson Commentaries

Chapter 2 Introducing Proofs

Chapter Objectives

To identify angles as adjacent or vertical
To identify supplementary and complementary angles and find their measures
To write the conjunction and disjunction of two statements
To write the converse, inverse, and contrapositive of a statement
To draw conclusions about angles from given information
To write reasons for statements
To write simple geometric proofs using properties of equations
To state the difference between postulates and theorems and to use postulates in proving theorems
To identify perpendicular lines, rays, segments, and planes and to prove basic properties about them
To perform basic compass and straightedge constructions dealing with concepts of this unit
To set up a proof from a verbal statement
To prove simple theorems
To identify dihedral angles

Chapter Overview

This chapter provides a gradual introduction to proofs. The first section of the chapter deals with pairs of angles. Angles are classified by position (adjacent and vertical) and by measure relation to others (complementary and supplementary). Various types of statements (conjunction, disjunction, converse, inverse, contrapositive) are presented. Practice in drawing conclusions is used as an introduction to proof. Basic introductory postulates are used to prove some simple theorems.

Special Features

Non-Routine Problem Solving p. 22
Cumulative Review pp. 35, 58, 60
Algebra Review pp. 43, 48, 64
Non-Routine Problems p. 48
Applications p. 54
Chapter Two Review p. 65
Chapter Two Test p. 66
College Prep Test p. 67

Non-Routine Problem Solving (p. 22)

Solution: To solve the problem, the students may discover a pattern:

Lines		*Angles*
2	(×2)	4
3	(×4)	12
4	(×6)	24
5	(×8)	40
⋮	⋮	⋮
20	(×38)	760

2.1 Pairs of Angles (pages 23–26)

Vocabulary

adjacent, complement, complementary, linear pair, supplement, supplementary

Background

Pairs of angles can be classified by their measures (complementary or supplementary) or by their relative positions (adjacent or vertical). Complementary and supplementary angles do not have to touch or be adjacent.

Suggestions

Continue to stress that angles are sets of points. Stress that both adjacent and vertical angles are coplanar. However, complementary and supplementary angles do not have to be coplanar. Although linear pairs are adjacent, not all complementary and supplementary angles are adjacent. As a mnemonic device, suggest that supplementary goes with straight and complementary goes with right angles.

For Additional Chalkboard Work

Set up the equation to solve the following: The average of the measures of the complement of an angle and the supplement is the measure of the angle itself. What is the measure of the angle? $\frac{(90 - m) + (180 - m)}{2} = m$; $67\frac{1}{2}^\circ$

Enrichment

Continue the discussion of the language of geometry versus everyday language. For example, what do we mean when we say that two people *complement* each other? Is this different from saying that I *compliment* you? If a person *supplements* his or her income, what does that mean?

Cumulative Review

1. Can a line ever be contained in the interior of a figure?
2. Give an example of a plane figure that has an interior.
3. Find the length of a segment whose endpoints have the coordinates 7 and −9.

1. No 2. Angle 3. 16

2.2 Writing Statements (pages 27–28)

Vocabulary

conjunction, disjunction, negation, statement

Background

The process of proving theorems often involves translating mathematical statements into symbolic form. It is often necessary to distinguish between the original statement and one of the possible derived statements, such as the negation, or to analyze a statement that is a combination of two or more simpler statements.

Suggestions

Stress the different ways in which the negation of a statement can be worded. Practice writing negations in different ways and interpreting them. The word *or* can cause problems. Stress the two distinct meanings of *or* and the fact that the student must make a choice as to the intended meaning. Examine "or" statements found in newspaper or magazine articles to try to determine which meaning is intended. Truth tables provide a convenient reference for establishing whether various types of statements are true or false. The basic truth tables given are actually definitions of the meaning of the connective used. They cannot be proved.

For Additional Chalkboard Work

Complete the truth tables for each of the following:

$p \vee \sim p$

$p \wedge \sim p$

p	$\sim p$	$p \vee \sim p$	$p \wedge \sim p$
T	F	T	F
F	T	T	F

Cumulative Review

1. What is the range of measures of acute angles?
2. Given collinear points X, Y, and Z, give three names for the line containing these points.
3. If the coordinates of the endpoints of two segments are 6 and 2, and 7 and 3, are the segments congruent?

1. $0 < m < 90$ 2. $\overleftrightarrow{XY}$, $\overleftrightarrow{XZ}$, $\overleftrightarrow{YZ}$
3. Yes

2.3 Conditional Statements (pages 29–32)

Vocabulary

biconditional, conditional statement, hypothesis, conclusion, contrapositive, converse, inverse.

Background

Many theorem statements must be translated into "If . . ., then . . ." form before a proof can be organized. Students must learn to think in terms of what is the hypothesis of a statement and what is the conclusion. They must also realize that a statement and its converse are not logically equivalent, although they appear to be similar. It often is helpful to think in terms of the contrapositive of a statement, rather than in terms of the original statement. Truth tables are a graphic way of showing the equivalence or lack of equivalence of such statements.

Suggestions

Begin classroom discussion with such questions as "Do the following sentences mean the same thing?" and "How do we know?" Use as examples some everyday statements such as "If I study hard, then I will learn mathematics." Does this mean that I will not learn mathematics if I do not study hard?

Although truth tables are not necessary for the development of the concepts introduced, many students enjoy them. Work through several with the class, discussing the truth values and trying to agree on them. Have students make up sentences that follow the truth table examples.

For Additional Chalkboard Work

Let $p =$ "John loves Mary" and $q =$ "Mary loves John." Write the sentences that correspond to the following symbolic statements: $p \rightarrow q$, $p \wedge \sim q$, $p \leftrightarrow q$, $\sim p \vee \sim q$.

Cumulative Review

1. Write the negation of "I like mathematics."
2. What is the range of measures of obtuse angles?
3. Find the measure of a supplement of a 65° angle.

1. I do not like mathematics.
2. $90 < m < 180$ 3. 115°

2.4 Drawing Conclusions (pages 33–35)

Vocabulary

transitive, substitution

Background

Prior to the introduction of formal proofs, students must develop skill in drawing valid conclusions from given information. Patterns of formal proof basically involve two types of steps: substitution of equivalent data or the manipulation of a given equation. Many substitution patterns involve the transitive property. For equations, this states that if $a = b$ and $b = c$, then $a = c$. The symmetric property, if $a = b$, then $b = a$, is not formally used in this book. Combining symmetry with transitivity, however, is a generalized substitution property which is used in this and later chapters.

Suggestions

Begin class discussion with questions such as "If you have an argument with a friend, how do you convince the person to come to the same conclusion that you have?" and "Can you think of situations in which you are willing to accept a substitution?" Give other everyday examples dealing with drawing conclusions or the transitive and nontransitive properties.

For Additional Chalkboard Work

For which statement is the logical conclusion that the final angle is acute? For which is it obtuse? For which is it uncertain?

1. The bisector of an acute angle forms two ____ angles.
2. The bisector of a right angle forms two ____ angles.
3. The bisector of an ____ angle forms two acute angles.

statement 3 could be either acute or obtuse; statements 1 and 2 acute

Cumulative Review

1. Find the measure of a complement of a 40° angle.
2. What is the measure of a straight angle?
3. What are congruent angles?

1. 50° 2. 180° 3. Angles with the same measure

2.5 Writing Proofs (pages 36–39)

Background

Formal proofs using the traditional format are introduced. A proof in mathematics is essentially a sequence of steps starting with what is known or assumed. The equivalent statements substituted are generally supported by undefined terms, definitions, postulates, or theorems.

Suggestions

Begin classroom discussion with such questions as "What do you mean when you say to someone that you are going to 'prove something to them?' "; "How can you show someone that what you say is true?"; and "What kinds of reasons can you give in such an argument?"

Solving an equation is actually a proof.

"Linkage arrows" are used in this book to show consecutive steps. Students should eventually know that what is given is usually included in the first steps of a proof and that the last step should always be what is to be proved. Stress that as you prove theorems, you begin to develop a core of experiences that tells you that certain steps are more likely to lead to the desired results than others.

For Additional Chalkboard Work

Give reasons for each step.

$3x + 5 = -2$

$3x \quad = -7$ (?)

$x \quad = -\frac{7}{3}$ (?)

Subtraction prop; Div. prop.

Cumulative Review

1. True or false: It is possible for two rays to form an angle?
2. Find the measure of a supplement of a 60° angle.
3. What is the bisector of an angle?

1. True 2. 120° 3. A ray in the interior of the angle that forms two congruent angles.

2.6 Postulates and Theorems (pages 40–43)

Vocabulary

postulate, theorem

Background

The development of proof continues in this lesson using simple informal arguments. The first postulates and theorems are presented. It is important for students to realize that not everything can be proved. The assumptions are those understandings that mathematicians have agreed over the years to be basic ideas. This lesson also continues a development of the various relationships of points, lines, and planes.

Suggestions

Most students need physical models to understand the relationships among points, lines, and planes. For example, if you rest a piece of board on one marble, there is a number of positions that the board can take. If you rest it on three marbles of any size placed at the vertices of a triangle, however, the board can assume only one position. This illustrates that any three noncollinear points are contained in exactly one plane. Look for other classroom examples.

Cumulative Review

1. Give the converse: If two angles are congruent, then their complements are congruent.
2. Give an example of the subtraction property of equations.
3. If $\angle X \cong \angle Y$ and $\angle X$ and $\angle Y$ are supplementary, what conclusion can you draw?

1. If the complements of two angles are congruent, then the angles are congruent. 2. Answers may vary. 3. $m\angle X = m\angle Y = 90$

2.7 Angle and Segment Addition (pages 44–48)

Vocabulary
reflexive

Background
Proofs involving the geometric properties of betweenness as well as algebraic properties are presented in this lesson. Although, technically, you add numbers instead of segments or angles, it is useful to define angle and segment addition and betweenness.

The postulates in this section give a theoretical background for practical measurement.

Suggestions
Discuss using the ruler properly to guarantee accurate measurements. Give examples of proofs with steps left out for students to complete.

For Additional Chalkboard Work
True or false?

1. An angle whose measure is 35° is formed by the bisector of a 70° angle.
2. An angle whose measure is 89° is formed by the bisector of a 178° angle.
3. An angle with a measure of 43° cannot be formed by the bisector of any angle.
4. An angle with a measure of 91° cannot be formed by the bisector of any angle.

1. T 2. T 3. F 4. T

Enrichment
Discuss measurement situations in which ordinary addition does not work. For example, if I walk 4 miles due north and then 3 miles due east, how far am I from where I started?

Cumulative Review
1. Give an example using the transitive property.
2. True or false? An angle (other than a straight angle) is contained in a single plane.
3. Find the absolute value: $|9-17|$.

1. Answers may vary. 2. True 3. 8

2.8 Perpendiculars (pages 49–53)

Vocabulary
perpendicular, perpendicular bisector

Background
Perpendicular lines must intersect and are coplanar. Perpendicular segments and rays, being subsets of lines, may or may not intersect, but they will be coplanar. Postulate 11 introduces the important concept that the sum of the measures of the angles with the same vertex on one side of a line and with no interior points in common is 180° and that the sum of all such angles around a point is 360°.

Suggestions
If students are keeping notebooks, have them arrange sections for postulates, definitions, theorems proved, and basic constructions. Have students find examples of perpendiculars in the classroom and other locations. You may wish to show students how to form right angles by folding paper, as is described in this lesson. Remind students of the difference between drawing a right angle and constructing it. Several constructions are introduced involving perpendiculars. Repeated use of the construction for an angle bisector gives angles with such measures as 45°, 135°, and $22\frac{1}{2}^\circ$.

For Additional Chalkboard Work
Draw several angles on the chalkboard, including some that are right angles. In which of these angles are there perpendicular rays?
Answers will vary.

Cumulative Review
1. Mark collinear points A, B, and C on your paper in that order. If $AB = 7$ cm and $BC = 4$ cm, find AC.
2. Find the coordinate of the midpoint of the segment whose endpoints have coordinates -2 and 7.
3. What are the properties of adjacent angles?

1. 11 cm 2. 2.5 3. Coplanar, have common side and common vertex, no interior points in common.

2.9 Proving Theorems About Angles (pages 55–58)

Vocabulary

corollary

Suggestions

Begin discussion by asking, "What are some of the different ways to determine if two angles are congruent?" Then ask, "How do you know this?" Exercises on drawing conclusions are open-ended. Let students take these as far as they wish, as long as their conclusions are correct ones.

For Additional Chalkboard Work

Give a reason for each step in the proof.

Given: $\angle ABF \cong \angle DCE$

Prove: $\angle CBF \cong \angle BCE$

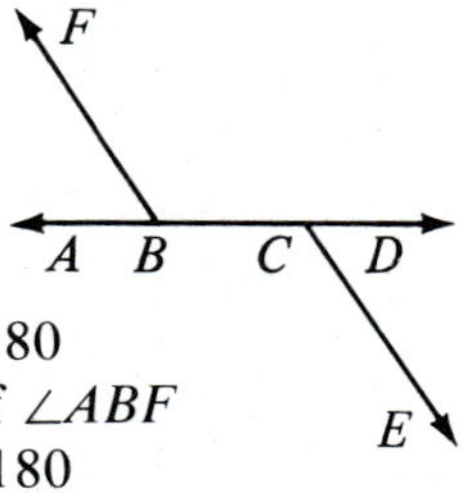

Proof

1. $\angle ABF \cong \angle DCE$
2. $m\angle ABF + m\angle CBF = 180$
3. $\angle CBF$ is a supplement of $\angle ABF$
4. $m\angle DCE + m\angle BCE = 180$
5. $\angle BCE$ is a supplement of $\angle DCE$
6. $\angle CBF \cong \angle BCE$

Reasons

(1) Given (2) The sum of the measures of angles with the same vertex on one side of a line is 180°. (3) Definition of supplementary angles (4) Reason 2 (5) Reason 3 (6) Supplements of congruent angles are congruent.

Cumulative Review

1. If two sides of two adjacent angles are collinear, what is the sum of the measures of the two angles?
2. Give an example of the addition property of equations.
3. An angle measures 113°. Is the angle an acute angle, a right angle, an obtuse angle, or a straight angle?

1. 180° 2. Answers may vary.
3. Obtuse

2.10 Setting Up Proofs (pages 59, 60)

Background

The process of translating a written sentence into a usable symbolic form suitable for a proof is introduced in this lesson. This involves drawing a picture to fit the information given. The picture is not a proof however.

Suggestions

Ask students if they can think of situations in which one must take a written instruction manual and translate it into specific steps. In this discussion, begin an example of the translation process studied in this lesson. Statements of theorems usually follow one of two formats. Some are written in an "If, then" format. Generally, the "If, then" pattern should have the "If" part of the sentence translated as the Given and the "then" part as the Proof.

2.11 Dihedral Angles (Optional) (pp. 61–64)

Vocabulary

dihedral angle, plane angle

Background

This lesson is optional; it is not necessary for developments in later lessons. Many of the properties of spatial figures are analogous to properties of plane figures.

Suggestions

If students are keeping notebooks, have them list in parallel columns the spatial figures studied and their planar counterparts.

Cumulative Review

1. Draw an example of two angles which share a common side but which are not adjacent.
2. How is the interior of an angle defined?

1. Answers may vary. 2. It is the part of the plane that can contain no lines.

3 Chapter and Lesson Commentaries

Chapter 3 Parallels

Chapter Objectives

To identify parallel lines, rays, segments, and planes
To identify angles formed by lines and their transversals
To form converses of statements
To prove and apply theorems about angles formed by parallel lines and a transversal
To prove that certain lines are parallel
To apply the parallel postulate
To prove and apply the theorem for the sum of the angle measures of a triangle
To apply the method of indirect proof
To construct parallel lines
To prove statements about parallel lines and planes

Chapter Overview

In this chapter, parallel lines and some of their basic properties are introduced. The definitions of parallel and skew lines are presented along with practice in recognizing them. Transversals and special angles related to transversals are identified. These special angles include alternate interior, corresponding, and alternate exterior angles. The special properties of these angles, when formed by parallel lines, are developed. The formation of converses is introduced, leading to the properties dealing with

Chapter Overview cont'd.

the determination of whether or not two lines are parallel. The parallel postulate is presented, together with some of its consequences. Theorems are proved using properties of parallel lines, including the important theorem that the sum of the measures of the angles of a triangle is 180°. Some elementary examples of indirect proofs are introduced. Practice is provided in applying properties of parallel lines.

Special Features

Careers: Architecture p. 68
Reading in Geometry pp. 70, 78
Cumulative Review p. 71
Algebra Review p. 74
Non-Routine Problems pp. 84, 86, 92, 94
Logic Review p. 92
Applications p. 98
Chapter Review p. 99
Chapter Test p. 100
College Prep Test 101

Careers: Architecture (p. 68)

Geometry involves problem solving, as do all other branches of mathematics. Most problems in the real world do not have nice neat solutions but involve finding the best possible solution. There may be many possible solutions.

In analyzing the parking lot problem, a number of factors should be considered.

3.1 Parallel and Skew Lines (pp. 69–71)

Vocabulary
parallel lines, skew lines

Background
The main concern of this lesson is not the logical development of properties of parallelism, but the recognition of parallel lines and skew lines. Sometimes parallel lines are defined as lines that are equidistant from each other. Later you can prove that this is a consequence of the definition given in this book.

Suggestions
Ask students to describe and give some examples of parallel and skew lines. Prepare several drawings or overhead transparencies of parallel and nonparallel lines. Stress that parallel lines must be coplanar.

For Additional Chalkboard Work
Tell why the following lines or segments are not parallel.

1. 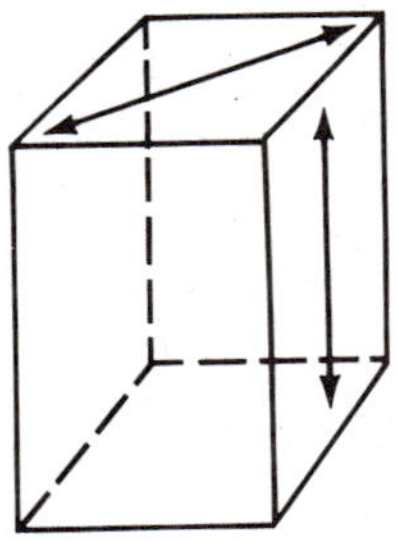

2. 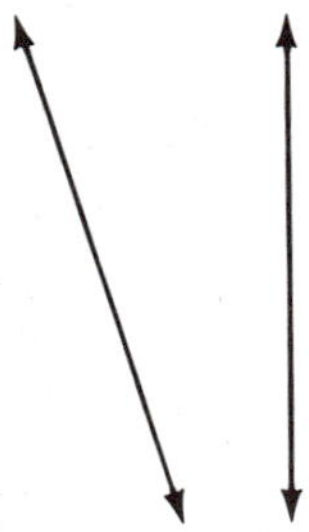

1. The lines are not coplanar. 2. The lines intersect.

Enrichment
Discuss with students some of the visual problems of perspective. For example, if you look at a straight railroad track, it appears that the rails approach each other.

Cumulative Review
1. Draw a figure with an interior.
2. Draw a segment 12 cm long and construct its midpoint.
3. Draw an obtuse angle.

1,2,3. Answers will vary.

3.2 Transversals and Special Angles (pages 72–74)

Vocabulary
transversal, alternate interior angles, alternate exterior angles, corresponding angles

Background
Angles are formed by intersecting lines or rays. In many situations, pairs of angles are formed by a transversal intersecting two or more lines. In this lesson, the primary emphasis is on identification of transversals and the related angles: alternate interior, corresponding, and alternate exterior.

Suggestions
An easy way to show many possible intersections is to use small straight wires, such as florist wires, on the overhead projector.

After defining alternate interior, corresponding, and alternate exterior angles, have students practice the proper identification of them.

For Additional Chalkboard Work
Identify each pair of angles as alternate interior, corresponding, or alternate exterior.

1. $\angle 1$ and $\angle 9$ 2. $\angle 4$ and $\angle 5$
3. $\angle 3$ and $\angle 8$ 4. $\angle 5$ and $\angle 9$

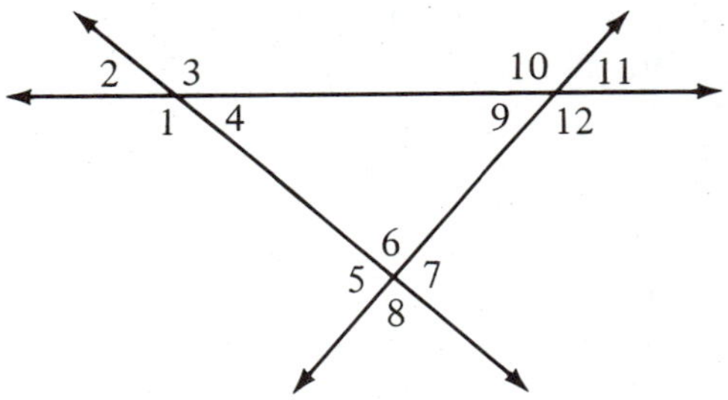

1. Corresponding 2. Alt. interior
3. Alt. exterior 4. Corresponding

Cumulative Review
1. On a number line, plot points A and B so that $AB = 5$ units.
2. On a number line, plot points A, B, C, and D so that $AB = CD$.
3. Find the coordinate of the midpoint of the segment whose endpoints have the coordinates -6 and 14.

1,2. Answers will vary. 3. 4

3.3 Transversals and Parallel Lines (pp. 75–79)

Background
When a transversal intersects parallel lines, certain pairs of angles can be shown to be congruent or supplementary. In this lesson, the use of abbreviations in proofs is begun. Students should understand these abbreviations and their uses.

Suggestions
The ability to recognize congruent or supplementary pairs of angles is crucial in this lesson. Begin the lesson with several measurement activities involving angles formed by parallel lines and transversals.

If you have a set of drafting parallel rulers or a flexible parallelogram model, illustrate how the alternate interior angles stay congruent as the angles are changed.

Continue exercises in which students give reasons for steps or fill in missing steps in proofs.

For Additional Chalkboard Work
Why can you state with certainty that the lines l and m are not parallel?

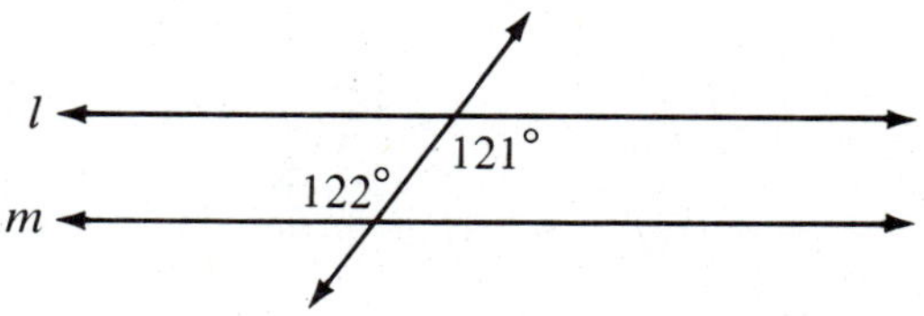

If they were, the alternate interior angles would be congruent.

Enrichment
Have students make models out of meter sticks or other sticks and bolts to illustrate the properties of parallel lines and congruent angles.

Cumulative Review
1. Draw an angle of 52° with a protractor.
2. What is the measure of a supplement of an 83° angle?
3. Draw an angle of 134° using a protractor. Construct another angle congruent to it.

1. Answers will vary. 2. 97°
3. Answers will vary.

3.4 Proving Lines Parallel (pages 80–84)

Vocabulary
converse

Background
The truth of a statement is often assumed to imply the truth of the converse of the statement. This does not always follow, however. The converse of a theorem must be proved independently of the original theorem. Although converses are treated in this lesson, other logical variations of statements such as the inverse and contrapositive are not discussed.

Postulate 14, which is the converse of Postulate 13, is assumed as a postulate rather than proved.

Suggestions
Begin the lesson with exercises of identification of sets of parallel lines based on the properties introduced in the lesson. Have students make up some examples in which the lines are parallel and some in which the lines are not parallel. Provide exercises in which reasons are needed. Students should begin to find several different ways of proving theorems.

For Additional Chalkboard Work
In which of the following applications do you think that either alternate interior angles, corresponding angles, or alternate exterior angles are used to obtain parallel lines?
1. yard lines on a football field
2. floor joists in a building
3. the edges of a page of a book

1. Yes 2. Yes 3. Yes. All use congruent right angles.

Cumulative Review
1. Draw two adjacent angles.
2. Find the measure of the complement of an angle with a measure of 47°.
3. Draw two acute angles $\angle ABC$ and $\angle DEF$. Construct an angle whose measure is $m\angle ABC + m\angle DEF$.

1. Answers will vary. 2. 43°
3. Answers will vary.

3.5 Using the Parallel Postulate (pages 85, 86)

Background

Historically, the parallel postulate is one of the most famous of all mathematical assumptions. Postulate 15 can be proved using postulate 13 and some properties of inequalities. It is included as a postulate rather than a theorem, however.

Suggestions

Discuss real models of parallel lines and problems of visual perspective. Examine such problems as "If two people in different parts of the world started flying in a straight path toward the north pole and then kept on flying straight, they would again meet at the south pole." Point out that this seems to contradict the parallel postulate. The reason is that the earth is not a plane.

For Additional Chalkboard Work

Complete each statement with =, ≠, or cannot tell.

1. If $\overleftrightarrow{EH} \parallel \overleftrightarrow{FJ}$ and m $\angle ABE = 78$, then m $\angle ACF$ ____ 78.
2. If $\overleftrightarrow{EH} \parallel \overleftrightarrow{FJ}$, and m $\angle ABE = 78$, then m $\angle ACG$ ____ 78. 1. = 2. ≠

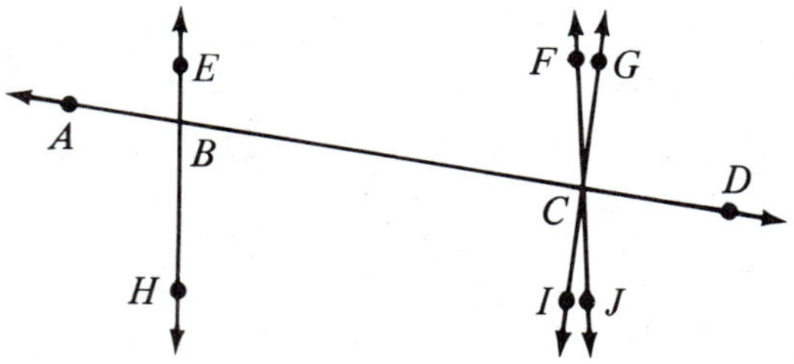

Enrichment

You may wish to have students read some of the history of the parallel postulate and of non-Euclidean geometry.

Cumulative Review

1. Find an angle whose measure is 22° less than that of its complement.
2. Show three points that are collinear.
3. Show three points that are not collinear.

1. 34° 2,3. Answers may vary.

3.6 Measuring Angles in a Triangle (pages 87–90)

Vocabulary

auxiliary line

Background

One of the most useful geometric facts is that the sum of the measures of the angles of a triangle is 180°. The proof of this depends upon an auxiliary line.

Suggestions

Have students draw several triangles of different sizes and shapes. Using a protractor, have them measure the angles and find the total of the measures for each triangle. Ask them to draw a conclusion.

For Additional Chalkboard Work

Determine whether or not contradictory information is given. If so, what information is contradictory?

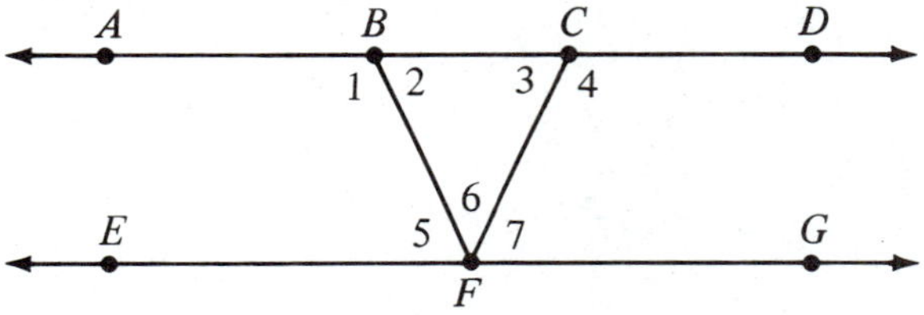

Given: $\overleftrightarrow{AD} \parallel \overleftrightarrow{EG}$, m $\angle 5 = 62$, m $\angle 6 = 59$, m $\angle 3 = 61$ If $\overleftrightarrow{AD} \parallel \overleftrightarrow{EG}$, then m $\angle 5$ cannot equal 62 or one of the other angle measures of the triangle is incorrect.

Enrichment

Have students investigate navigational techniques used by pilots on aircraft to move from one point to another on the earth's surface.

Cumulative Review

1. If $\overrightarrow{EG}$ is in the interior of $\angle DEF$, m $\angle DEG = 30$, and m $\angle DEF = 65$, find m $\angle GEF$.
2. Give a geometric example of the substitution property.
3. Construct a line perpendicular to $\overleftrightarrow{AB}$ through a point C on $\overleftrightarrow{AB}$.

1. 35 2. Answers may vary. 3. See construction page 51.

3.7 Indirect Proof (pages 91, 92)

Vocabulary

direct proof, indirect proof

Background

An informal format for indirect proofs is used in this lesson. The indirect method, as applied here, hinges on listing only the two possibilities that can occur, then eliminating one of them by showing that it leads to a contradiction.

Suggestions

Have students list real-world situations in which there are only two possible alternatives. For example, they may list life or death. Discuss the logic of the fact that if one of a pair is known to be, then the other cannot be, and vice versa. Note that Th. 3.7 can be proved using a direct proof as well as the given indirect proof.

For Additional Chalkboard Work

Give an indirect proof:
Given: $\overleftrightarrow{AD} \parallel \overleftrightarrow{EG}$, m $\angle 6 = 59$, m $\angle 3 = 61$
Prove: m $\angle 5 \neq 62$

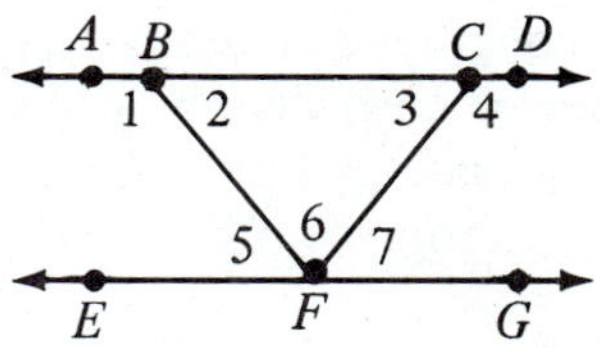

Assume m $\angle 5 = 62$. Then m $\angle 5 =$ m $\angle 2 = 62$ (Post. 13); m $\angle 2 +$ m $\angle 3 +$ m $\angle 6 = 62 + 61 + 59 = 182$; $182 \neq 180$; (Sum of the angles in a triangle = 180) $\therefore$ m $\angle 5 \neq 62$.

Cumulative Review

1. Draw two lines and a transversal on the board. Have students identify the alternate interior angles and the corresponding angles,
2. Draw a line and a point not on the line. Construct a perpendicular through the point to the given line.
3. Define skew lines

1,2. Answers will vary. 3. Noncoplanar lines

3.8 Constructing Parallels (pages 93, 94)

Background

Postulate 14 and Theorem 3.3 lead to ways of constructing parallel lines. In both cases, a transversal must first be drawn. Congruent alternate interior or corresponding angles are then constructed to obtain the desired parallel lines.

Suggestions

Before demonstrating the methods of constructing parallel lines, review the various theorems that guarantee parallel lines. From these develop alternative ways of constructing a line parallel to a given line through a given point. Stress that a transversal must be drawn for all of these alternatives. Note, however, that the exact location of the transversal is not critical. Remind students of the difference between drawing angles and constructing them. You may need to review the method for constructing an angle congruent to a given angle.

For Additional Chalkboard Work

Construct a line parallel to a given line through a point not on the given line using alternate interior angles.

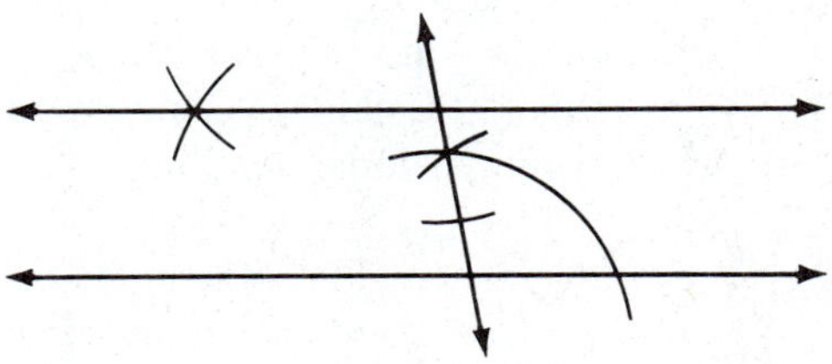

Cumulative Review

1. Draw an acute angle. Construct an angle congruent to it.
2. What are three ways to prove that two lines are parallel?
3. What is a postulate?

1. Answers will vary. 2. Congruent alternate interior angles, congruent corresponding angles, supplementary interior angles on one side of a transversal, 3. A statement accepted without proof.

3.9 Parallel Planes (Optional) (pages 95–97)

Vocabulary
parallel planes

Background
The concepts of parallelism developed for lines have corresponding concepts involving planes in space. Parallel lines were defined as coplanar lines that do not intersect.

Suggestions
Three-dimensional models are helpful in showing the relationships developed in this lesson. Have students practice drawing planes and parallel planes. Remind them that parallelograms are usually used to represent planes.

For Additional Chalkboard Work
How can you make sure that the ceiling of a room under construction is parallel to the floor?

If corner posts are perpendicular to floor, they should be perpendicular to the ceiling.

Enrichment
Geometric models with more than three dimensions have numerous applications in engineering and physics. Suggest that students investigate the language and applications of higher dimensional geometry.

Cumulative Review
1. Write the conjunction and disjunction of the two statements:
   Collinear points lie on the same line.
   Coplanar points lie in the same plane.
2. Write a negation of the following statement:
   Space contains infinitely many points.

   1. Collinear points lie on the same line and coplanar points lie in the same plane. Collinear points lie on the same line or coplanar points lie in the same plane.
   2. Space does not contain infinitely many points.

4 Chapter and Lesson Commentaries

Chapter 4 Proving Triangles Congruent

Chapter Objectives
To classify triangles by the lengths of their sides and by the measures of their angles

To find the measures of exterior and remote interior angles of triangles

To identify congruent triangles

To find unknown measures in congruent triangles

To construct congruent triangles by each of the proved congruence patterns

To prove triangles congruent by each of the proved congruence patterns

To recognize patterns for proving triangles congruent

To identify congruent triangles and prove them congruent

Chapter Overview
In this chapter, proofs involving triangle congruence are introduced. After classifying triangles by the measures of their sides and angles, basic relationships between the angle measures of a triangle and the measures of the exterior angles are developed. Triangle congruence is defined. Experiments are used to introduce each of the standard congruence patterns. Constructions of congruent triangles are used as an exploratory tool. The side-side-

Chapter Overview cont'd.

side pattern is presented first as a postulate followed by the side-angle-side and the angle-side-angle patterns. Other possible congruence patterns are discussed, with counterexamples presented for side-side-angle and angle-angle-angle patterns.

Special Features

Non-Routine Problem Solving p. 102
Reading in Geometry p. 105
Non-Routine Problems pp. 106, 127
Logic Review p. 109
Algebra Review p. 113
Cumulative Review p. 122
Applications p. 128
Computer Activities p. 129
Chapter Review p. 130
Chapter Test p. 131
College Prep. Test p. 132
Cumulative Review
(Chapters 1–4) p. 133

Non-Routine Problem Solving (p. 102)

Solution: There is more than one way to solve this problem. The solution given is based upon (1) the sum of the measures of the angles of a triangle is 180°; and (2) angles that form a linear pair are supplementary. Measure of $\angle ADE$ is 100° (180° − 80°); measure of $\angle AED$ (and of $\angle A$) is 40°; measure of $\angle DEC$ is 140°, so measure of $\angle BEC$ is 40°; thus, measure of $\angle ACD$ is 95° (180° − 85°, using $\triangle EBC$).

4.1 Classifying Triangles (pages 103–106)

Vocabulary

triangle, vertex, side, equilateral, isosceles, scalene, acute, right, obtuse, equiangular

Background

A triangle is composed of three segments. A cardboard cutout represents a triangular region, not a triangle, whereas a wire or stick model properly represents the three segments or sides of the triangle. Triangles can be classified by the measures of their sides or their angles. Triangles can also be classified by combinations of sides and angles.

Suggestions

Practice visual identification of triangles. Then use rulers and protractors to make more accurate judgments. Finally, introduce the analysis using only given information. Stress that the given information can be in the form of "tic marks" on the picture or as a verbal statement of given properties. Have students draw triangles with certain given combinations of properties. Include some that cannot be drawn, such as an obtuse equilateral triangle.

For Additional Chalkboard Work

Would the triangles involved in these applications most likely be acute, right, obtuse, scalene, isosceles, or equilateral?

1. The framing of the roof of a house.
2. The braces on a basketball backboard.
3. The edges of the top of the Washington Monument.

1. Isosceles 2. Right 3. Isosceles

Cumulative Review

1. What is the basis for the construction of a line parallel to a given line through a point not on the given line?
2. True or False? Any two lines intersected by a transversal are parallel.
3. (Draw two intersecting lines on the chalkboard. Label the angles.) Which angles are vertical angles? Why are they congruent?

1. Post. 14 or Th. 3.3 2. False
3. Answers will depend upon the labels used. Supplements of the same angle are congruent.

4.2 Exterior Angles and Interior Angles of Triangles (pages 107–109)

Vocabulary
exterior angle, remote interior angle

Background
Theorems about the measures of angles of triangles continue to be an important concern of geometry. In this lesson, exterior angles are considered and their measures are related to the measures of the angles of the triangle, sometimes called the interior angles. The development of these measurement properties depends primarily on supplementary angles.

Suggestions
Begin by reviewing the definition of supplementary angles. Have students list all of the different ways to prove that angles are supplementary and ways to use supplementary angles. This list should include the fact that supplements of congruent angles or of the same angle are congruent. Cut a triangle out of paper, leaving the rest of the paper in one piece. On that remaining piece, draw the exterior angles at one vertex. Tear off the other two corners of the triangle. They should fit exactly into one of the exterior angles.

For Additional Chalkboard Work
In $\triangle ABC$, m $\angle A =$ m $\angle B$ and m $\angle C = 40$. $\angle 4$ is exterior angle at A. Find m $\angle 4$.
110°

Cumulative Review
1. How many acute angles does an acute triangle have?
2. How many obtuse angles does an obtuse triangle have?
3. If the measure of an angle is 100° and the measure of its adjacent angle is 100°, what is the measure of the angle formed by the noncommon rays?

1. 3; 2. 1; 3. 160°

4.3 Congruent Triangles (pages 110–113)

Vocabulary
congruent triangles

Background
Congruent figures have the same size and shape. Size depends upon lengths, whereas shape depends upon angle measures. A more formal way to describe a congruence is as a distance preserving transformation. If *all* corresponding distances are preserved, then so are all corresponding angle measures. Two triangles are congruent if all pairs of corresponding sides and corresponding angles are congruent. The main emphasis is on finding unknown measures of angles or sides, using congruent triangles.

Suggestions
Some students may find it helpful to have physical examples of congruent triangles. Make cardboard or paper cutouts of pairs of congruent triangles with different vertex labels. Mix these up and have students match them. Have students discover that a pair of isosceles triangles can be fitted together in more than one way. Stress the importance of correct congruence statements.

For Additional Chalkboard Work
Using the given information, complete the correspondence statement.
Given: $AC = 5$, $AB = 12$, $BC = 13$, $QP = 12$, $RQ = 5$, $PR = 13$, m $\angle A = 90$, m $\angle B = 67$, m $\angle R = 23$, m $\angle Q = 90$.
$\triangle ABC \cong \triangle QPR$

Cumulative Review
1. How do the measures of an exterior angle of a triangle and a remote interior angle compare?
2. Draw two parallel lines on the chalkboard. Draw a third line intersecting one of these two lines. Can the third line drawn be parallel to the second line?
3. Give an example of the addition property for equations.

1. The measure of the exterior angle is greater. 2. No 3. Answers will vary.

4.4 SSS Congruence Proofs (pages 114–117)

Background

Although two triangles were defined to be congruent if all pairs of corresponding sides and angles are congruent, it is not necessary to show all of these side and angle correspondences to prove that the triangles are congruent. If the side-angle-side pattern or its equivalent is assumed as a postulate, it is possible to prove all of the other standard congruence patterns as theorems.

Suggestions

Models made from sticks of different lengths are a most effective demonstration of the side-side-side congruence pattern. Show that if you take two sets of three sticks of corresponding lengths, then regardless of how the sticks are put together, the triangles will be congruent.

For Additional Chalkboard Work

Give the reason for each statement in the following proof.

Given: $\overline{AE} \cong \overline{BC}$, and $\overline{BE} \cong \overline{AC}$

Prove: $\triangle ABE \cong \triangle BAC$

Statement	Reason
1. $\overline{AE} \cong \overline{BC}$	1. ? Given
2. $\overline{BE} \cong \overline{AC}$	2. ? Given
3. $\overline{AB} \cong \overline{BA}$	3. ? Reflex prop
4. $\triangle ABE \cong \triangle BAC$	4. ? SSS for $\cong$ $\triangle$'s.

Enrichment

Set up a series of experiments on rigidity of shape using the commercially available D-stix. Have students build models out of these sticks, trying to obtain a completely rigid structure. Analyze what is necessary to obtain rigidity.

Cumulative Review

1. What are congruent figures?
2. What is a scalene triangle?
3. Can a supplement and a complement of the same angle ever be congruent?

1. They have the same size and shape.
2. A triangle with no two sides congruent.
3. No

4.5 SAS and ASA Congruence Proofs (pages 118–122)

Vocabulary

included angle, included side

Background

Side-angle-side and the angle-side-angle patterns are assumed as postulates. Either pattern could be used to prove the other. This is not done because of the difficulty of the proofs. Constructions are used to introduce both the SAS and the ASA postulates.

Suggestions

Models made from dowel rods for segments and bent wires for angles can be used to introduce the uniqueness of both the SAS and the ASA congruency patterns.

If students are keeping notebooks, this may be a good time to have them organize their lists of different ways to prove that pairs of segments and pairs of angles are congruent.

For Additional Chalkboard Work

Complete the following statements:

1. To prove that $\triangle ABC \cong \triangle DEF$ by SAS, you can show that $\overline{BC} \cong \overline{EF}$, $\angle C \cong \angle F$, and ____ $\cong$ ____.
2. To prove that $\triangle ABC \cong \triangle DEF$ by ASA, you can show that $\angle A \cong \angle D$, $\overline{AC} \cong \overline{DF}$, and ____ $\cong$ ____.

1. $\overline{CA} \cong \overline{FD}$ 2. $\angle C \cong \angle F$

Cumulative Review

1. Given three segments, how many shapes and sizes of triangles can have the three segments as sides?
2. If the measure of one angle of a triangle is 50° and the measure of a second angle is 60°, what is the measure of the third angle?
3. Two parallel lines are cut by a transversal so that a pair of interior angles on the same side of the transversal has degree measures $2x + 40$ and $x + 20$ respectively. Find the measure of the smaller angle.

1. At most, one triangle. 2. 70° 3. 60°

4.6 Other Congruence Proofs (pages 123–127)

Suggestions

Have students list all of the different combinations of three angles or sides that they can think of. Have them determine which patterns involve included sides or angles.

Have students draw a triangle and then try to construct a congruent triangle with three angles congruent to the three angles of the given triangle. This should demonstrate that AAA is not a usable congruence pattern. Students often have difficulty accepting the AAS pattern as a valid one. Perhaps the best justification involves constructing the third angle and then using the ASA pattern.

For Additional Chalkboard Work

Complete the following statements:

1. To prove that $\triangle ABC \cong \triangle DEF$ by AAS, you can show that $\angle A \cong \angle D$, $\overline{AB} \cong \overline{DE}$, and ____ $\cong$ ____.
2. If you were trying to prove that $\triangle ABC \cong \triangle DEF$, an SSA pattern of $\overline{AB} \cong \overline{DE}$, $\overline{BC} \cong \overline{EF}$, and ____ $\cong$ ____ would not necessarily prove the congruence.

1. $\angle C \cong \angle F$ 2. $\angle A \cong \angle D$ or $\angle C \cong \angle F$

Cumulative Review

1. List two different ways to prove triangles congruent.
2. True or false? Any two segments that do not intersect are parallel.
3. What are adjacent angles?

1. SSS, SAS, ASA, or AAS 2. False
3. Two coplanar angles having a common side and vertex, but no common interior points.

Computer Activity (p. 129)

You will find flowcharts written in which the diamond shape does not always have the "yes" branch go off to the side and the "no" branch go down. This convention is used here because it simplifies the writing of the program from the flowchart.

5 Chapter and Lesson Commentaries

Chapter 5 Applying Congruent Triangles

Chapter Objectives

To identify corresponding parts of congruent triangles
To use corresponding parts of congruent triangles in proofs
To prove and apply theorems about isosceles triangles
To identify overlapping triangles
To write proofs involving overlapping triangles
To apply right triangle congruence theorems
To prove and apply theorems about bisectors
To identify altitudes and medians of triangles
To prove and apply theorems about altitudes and medians of triangles
To construct altitudes and medians of triangles

Chapter Overview

In this chapter the basic patterns of proving triangles congruent are applied to more complex situations. Corresponding parts of congruent triangles are identified and later used in proving other triangles congruent. Isosceles triangles are reviewed and the parts identified, with basic properties proved. An optional sequence of transformation activities is included, involving experiments with reflections, slides, and rotations. Activities involving the identification of overlapping triangles and proofs using them are begun. Altitudes, angle bisectors, perpendicular bisectors, and medians are introduced, with theorems about them proved and constructions involving them developed.

Special Features

Euclid p. 135
Reading in Geometry p. 142
Algebra Review pp. 143, 156, 176
Transformational Geometry pp. 144–152
Cumulative Review pp. 161, 168, 176
Non-Routine Problems pp. 161, 165
Applications p. 177
Chapter Review p. 178
Chapter Test p. 179
Computer Activities p. 180
College Prep. Test p. 181

Euclid (p. 135)

Euclid is often called the "father of geometry." It would be worthwhile for students to look into the life and works of this extraordinary mathematician. Much of the structure of **Holt Geometry** is based on Euclid's work.

5.1 Corresponding Parts (pages 136–138)

Background

Congruent triangles were defined to be triangles with all corresponding sides and corresponding angles congruent. The corresponding sides and corresponding angles can be thought of as corresponding parts. To simplify the writing of reasons in proofs, the abbreviation CPCTC will be used for "Corresponding parts of congruent triangles are congruent."

Suggestions

A necessary skill for proving complicated theorems is the ability to recognize corresponding parts of congruent figures. Adequate practice should be provided so that students become proficient in such identification. If additional practice is needed, simple card games can be made.

For Additional Chalkboard Work

If $\triangle SGK \cong \triangle LHJ$, name the congruent corresponding parts.

$\overline{SG} \cong \overline{LH}$, $\overline{GK} \cong \overline{HJ}$, $\overline{KS} \cong \overline{JL}$, $\angle S \cong \angle L$, $\angle G \cong \angle H$, $\angle K \cong \angle J$

Enrichment

Have students find and describe real-world examples of the use of congruent corresponding parts. For example, replacement of automobile parts is based upon the principle of congruent parts and interchangeability.

Cumulative Review

1. Give two patterns that cannot be used to prove two triangles congruent.
2. Give two ways to prove two lines parallel.
3. Give two ways to prove two angles congruent.

1. SSA, AAA 2. Show that a pair of alternate interior angles, corresponding angles, or alternate exterior angles are congruent. 3. CPCTC, vertical angles, supplements or complements of congruent angles, alternate interior or corresponding angles formed by a transversal intersecting two parallel lines

5.2 Isosceles Triangles (pages 139–143)

Vocabulary

base, vertex, base angle, perimeter

Background

An isosceles triangle was defined to be a triangle with at least two congruent sides. By this definition an equilateral triangle is also isosceles, and, therefore, all theorems proved about isosceles triangles apply to equilateral triangles. In this lesson theorems about isosceles triangles are proved using congruent triangles formed by auxiliary lines. Several of the theorems used are converses of other theorems already proved. It is important to remember that the converse of a true theorem is not necessarily true. The converse itself must be proved.

Suggestions

Students should be able to recognize triangles that are isosceles regardless of the position in which they are drawn. A cutout of an isosceles triangle can be folded in half to show that the corresponding sides coincide.

For Additional Chalkboard Work

Alternate proof to Theorem 5.1.
Given: $\overline{AC} \cong \overline{BC}$
Prove: $\angle A \cong \angle B$

PROOF

Statements	**Reasons**
1. $\overline{AC} \cong \overline{BC}$	1. Given
2. $\overline{BC} \cong \overline{AC}$	2. Symmetric prop.
3. $\angle C \cong \angle C$	3. Reflexive property
4. $\triangle ACB \cong \triangle BCA$	4. SAS
5. $\angle A \cong \angle B$	5. CPCTC

Cumulative Review

1. If $\triangle XYZ \cong \triangle ABC$, which pairs of sides are congruent?
2. In order to prove that $ABC \cong \triangle A'B'C'$ by the side-side-side pattern, what must be shown?
3. Draw two parallel lines and a transversal. Label the angles. Identify all pairs of corresponding angles.

1. $\overline{XY}$ and $\overline{AB}$, $\overline{XZ}$ and $\overline{AC}$, $\overline{YZ}$ and $\overline{BC}$
2. Show that $\overline{AB} \cong \overline{A'B'}$, $\overline{AC} \cong \overline{A'C'}$, and $\overline{BC} \cong \overline{B'C'}$. 3. Answers will depend upon drawing.

Reflections (pages 144–146)

This lesson on reflections is the first of several optional lessons on geometric transformations. The reflection is the same distance from the mirror as the original figure. Mirror lines or lines of reflection can be found using special mirrors, by paper folding, or by formal straightedge and compass constructions.

Slides (pages 147, 148)

Another basic transformation is the slide or translation. Intuitively, a slide is a motion in which all points move along parallel paths. It takes two instructions to describe a slide: how far and in which direction. These instructions can be provided in terms of specifications of coordinates of points.

Have students list different real-world situations in which a slide motion exists. For example, the movement of an automobile down a straight highway is essentially a slide motion, as is a walk along a straight path.

The overhead projector can be used effectively to demonstrate slides, that is, using sheets of clear plastic instead of tracing paper. Show students that parallel slide arrows of the same length that point in the same direction will produce the same final results. If a figure and its slide image are shown, one possible slide arrow can be drawn by connecting a pair of corresponding points. All other parallel arrows of the same length pointing in the same direction will produce the same results.

Rotations (pages 149–151)

A third basic transformation is the rotation or turn. Perhaps the most familiar example is the movement of the hands of a clock.

It takes two instructions to describe a rotation: the location of the turn center and the amount of the turn. This latter can be expressed as an angle or arc measure.

Paper folding and formal straightedge and compass constructions can be used to find centers of rotation. The point of intersection of the perpendicular bisectors of the segments joining two pairs of corresponding points of the original figure and its image will be the rotation center.

Have students list different real-world situations in which a turn motion exists. The movement of any type of wheel is an example, as is the rotation of the earth around its axis.

Transformations and Congruent Triangles (page 152)

The three basic transformations, reflections, slides, and rotations can be combined into composite transformations. It can be proved that if two figures are congruent, then one of them can be shown to be the image of the other by a series of at most two of these basic transformations. If two transformations are required, then one of them will be a reflection. In many cases, a single transformation, either a slide or a rotation, will be sufficient. At this introductory level, much of the procedure for finding the appropriate combination is basically a trial-and-error process. Sophisticated procedures do exist, but they are not appropriate at this time.

Have students make paper cutouts of the figures involved. Moving these cutouts may help them to visualize the solution.

5.3 Overlapping Triangles (pages 153–156)

Vocabulary

overlapping triangles

Background

Many triangles share common parts. These triangles are called overlapping triangles. Problems involving overlapping triangles often involve separating the figure into its simpler parts.

Suggestions

Many students have visual perception weaknesses that become especially obvious when working with these exercises. It is often helpful to redraw the figures in different colors showing the overlapping triangles as separate figures.

For Additional Chalkboard Work

Name all of the different triangles.

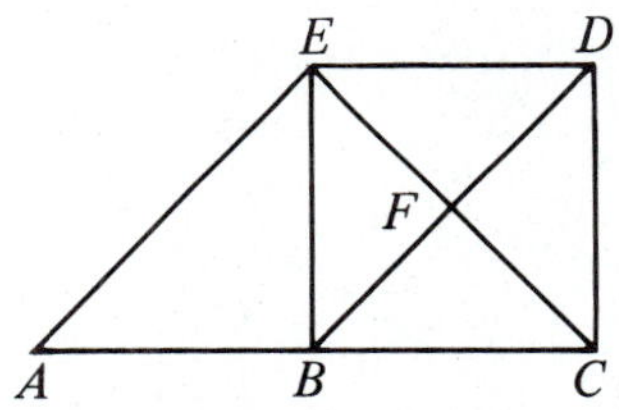

△ABE, △BCF, △BCD, △BCE, △CDE, △CDF, △DFE, △DBE, △BEF

Enrichment

Part C exercises in this lesson are examples of a more sophisticated type of problem involving the description of the intersection of various triangular regions.

Cumulative Review

1. The vertex angle of an isosceles triangle measures 40°. Find the measure of a base angle.
2. If the measures of two angles of a triangle are 50° and 60°, find the measure of the opposite exterior angle.
3. If two coplanar lines are perpendicular to the same line, what is their relationship to each other?

1. 70° 2. 110° 3. They are parallel.

5.4 Congruent Right Triangles (pages 157–161)

Vocabulary
hypotenuse, leg

Background
All congruence patterns that are true for triangles in general are true for right triangles. However, there is one pattern that is true for right triangles that is not true for other triangles: the side-side-angle pattern (HL).

Suggestions
Make certain that students can identify the parts of right triangles, especially the hypotenuse. Have students make a chart showing the original triangle congruence pattern name together with the special name for right triangles.

For Additional Chalkboard Work
Using each set of given information, is rt. $\triangle XYZ \cong$ rt. $\triangle WUV$? Give a reason for each answer.

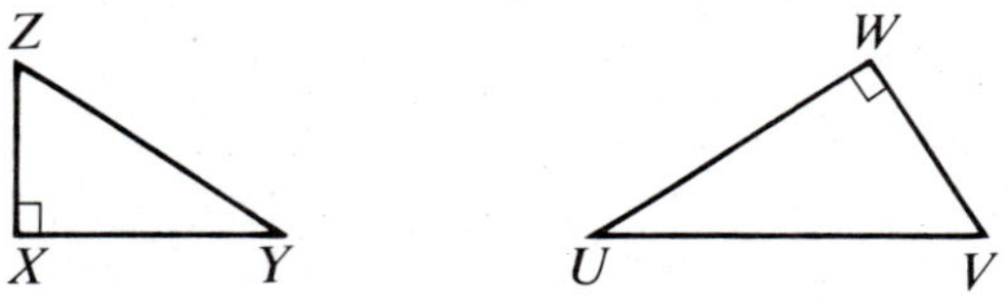

1. $\angle Z \cong \angle V$, and $\overline{ZY} \cong \overline{VU}$ 1. $\cong$, HA
2. $\overline{XZ} \cong \overline{WV}$, and $\overline{YZ} \cong \overline{UV}$ 2. $\cong$, HL

Cumulative Review
1. Draw a set of overlapping triangles. Label them. Name the overlapping triangles in the figure.
2. Draw congruent triangles ABC and $A'B'C'$. Which sides and which angles must be shown to be congruent to prove that $ABC \cong A'B'C'$ by the side-angle-side pattern?
3. Do perpendicular rays have to intersect?

1. Answer depends upon drawing.
2. $\overline{AB} \cong \overline{A'B'}$, $\angle B \cong \angle B'$, $\overline{BC} \cong \overline{B'C'}$: Answers may vary. 3. No

5.5 Using Corresponding Parts of Congruent Triangles (pages 162–165)

Suggestions
In this lesson, the abbreviation *CPCTC* is used in the "Reasons" column to indicate that corresponding parts of congruent triangles are used. It is helpful to label steps in proof with (S) or (A). Point out that arrows on the left of the proof statements are a guide to steps that follow.

For Additional Chalkboard Work
Complete the missing steps in the proof:

Given: $\triangle ABD \cong \triangle BAC$

Prove: $\angle DAE \cong \angle CBE$

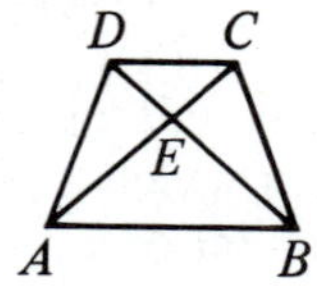

PROOF

Statements	Reasons
1. ?	1. Given
2. $\overline{AD} \cong \overline{BC}$	2. ?
3. $\angle ADB \cong \angle ?$	3. CPCTC
4. $\angle ? \cong \angle BEC$	4. ?
5. $\triangle AED \cong \triangle ?$	5. ?
6. $\angle DAE \cong \angle CBE$	6. ?

1. $\triangle ABD \cong \triangle BAC$ 2. CPCTC 3. *BCA*
4. *AED*, vert. $\angle$'s are $\cong$ 5. *BEC*, AAS
6. CPCTC

Cumulative Review
1. What congruence pattern that is true for right triangles is not true for triangles in general?
2. What is the definition of congruent triangles?
3. Give the converse of the statement "If two parallel lines are intersected by a transversal, then alternate interior angles are congruent."

1. SSA 2. Triangles in which the corresponding sides and angles are congruent. 3. If two lines are intersected by a transversal and the alternate interior angles are congruent, then the lines are parallel.

5.6 Bisectors (pages 166–168)

Vocabulary
perpendicular bisector, angle bisector

Background
In this lesson, two special bisectors, the perpendicular bisector and the angle bisector, are examined.

Suggestions
Paper folding can be used to illustrate some bisectors. Draw a segment on a piece of paper. Fold the paper so that the two ends of the segment coincide. Crease the paper. If done carefully, the crease will be the perpendicular bisector of the segment. A similar method can be used to find an angle bisector.

For Additional Chalkboard Work
Many of the bisector theorems for a plane have spatial counterparts. Describe each of these.

1. In space, where are all points that are equidistant from the endpoints of a segment?
2. What would be the bisector of a dihedral angle?

1. A plane 2. A half-plane

Enrichment
Three-dimensional problems can be introduced as an extension of the basic work of this lesson. For example, the perpendicular bisector of a segment in space is a plane.

Cumulative Review
1. Is a side-side-angle congruence pattern true for right triangles?
2. Is an angle-angle-angle congruence pattern true for right triangles?
3. Draw a line. Construct a line parallel to it.

1. Yes 2. No 3. Answers will vary.

5.7 Altitudes and Medians of Triangles (pages 169–173)

Vocabulary
altitude, median

Background
The word *altitude* sometimes refers to a segment and sometimes to the length of the segment. In this book it refers to the segment.

Suggestions
Have students list different ways in which the word *altitude* is used in everyday conversation. A similar activity for the word *median* may also be helpful.

Errors Students Make
It may confuse some students that an altitude of a triangle can sometimes be outside the triangle. If an obtuse triangle of cardboard is placed on its side, it is clear what to measure to find the height of the triangle.

For Additional Chalkboard Work
Determine if the following statement is a true statement. If so, prove it. The length of the median to the hypotenuse of an isosceles right triangle is half the length of the hypotenuse.

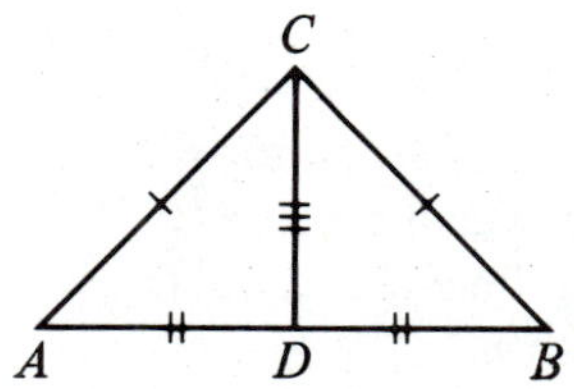

Show $\triangle ADC \cong \triangle BDC$ (SSS) and $\overline{CD} \perp \overline{AB}$. Then show $\angle A \cong \angle ACD$ and $\overline{AD} \cong \overline{CD}$.

Cumulative Review
1. If $\overrightarrow{OP}$ bisects $\angle AOB$ and the measure of $\angle AOB$ is 80°, what is the measure of $\angle AOP$?
2. If $\triangle DEF \cong \triangle XYZ$, which angles are congruent?
3. What is a geometric figure?

1. 40° 2. $\angle D \cong \angle X$, $\angle E \cong \angle Y$, $\angle F \cong \angle Z$
3. A set of points.

5.8 Constructions of Altitudes and Bisectors (pages 174–176)

Suggestions

You may need to review basic constructions with some students before attempting the ones in this lesson. The constructions for bisecting an angle and for constructing perpendiculars are needed.

For Additional Chalkboard Work

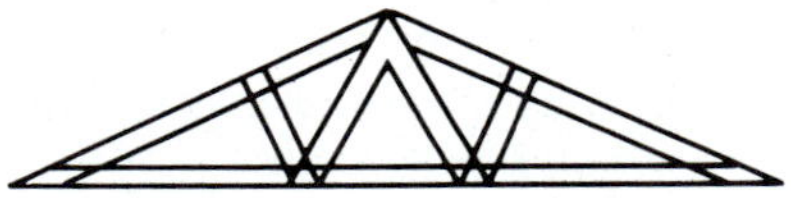

A carpenter needs to find the midpoint of the bottom of the truss above. Assuming that it is in the shape of an isosceles triangle, how can the carpenter use a carpenter's square to find this midpoint?
The altitude is also the median of the triangle.

Enrichment

Although proofs of constructions are dealt with in a later chapter, you may wish to have students justify the constructions used in this chapter. A justification can be an informal discussion of the logic behind the construction.

Cumulative Review

1. Is a side of a triangle ever an altitude?
2. Is a side of a triangle ever a median?
3. Draw two intersecting lines l and m. Draw a third line t parallel to m.
   Is line t always, sometimes, or never parallel to l? 1. Yes 2. No 3. Never

Computer Activity (p. 180)

This program should be studied in conjunction with page 570 of the Computer Section where character strings are described in more detail. Students should understand that using these strings as a response to a question requires that the question be answered exactly as the test for that answer built into the program. Even a slight misspelling will cause the computer to reject the answer.

6 Chapter and Lesson Commentaries

Chapter 6 Exploring Polygons

Objectives

To identify and name polygons and their parts
To identify convex and concave polygons
To prove and apply the theorem about the sum of the angle measures of a convex quadrilateral
To develop and apply a formula for the sum of the angle measures of any convex polygon
To identify the exterior angles of a polygon
To develop and apply a formula for the sum of the measures of the exterior angles of a polygon
To identify regular polygons
To determine angle measures and the number of sides of regular polygons
To construct regular polygons with 3, 4, and 6 sides
To construct angles with certain measures

Chapter Overview

The chapter opens with naming the more common convex polygons. Concave polygons are mentioned to distinguish them from convex polygons. A definition of a diagonal of a polygon is inserted for clarification of proofs to follow later. The sum of the angles of a convex quadrilateral is introduced before a formula for the sum of the angles for any convex polygon is developed.

Chapter Overview cont'd

The theorem on the sum of the measures of the exterior angles of any convex polygon is easily proved. Emphasis on identifying regular polygons and determining angle measures and number of sides from specifically developed formulae is necessary for a thorough understanding of regular polygons.

Special Features

Careers: Surveyors p. 182
Reading in Geometry p. 186
Cumulative Review pp. 187, 201
Using the Calculator pp. 192, 198
Algebra Review p. 195
Applications p. 202
Computer Activities p. 203
Chapter Review p. 204
Chapter Test p. 205
College Prep. Test p. 206

Careers: Surveyors (p. 182)

If simple surveying equipment is available, you may wish to organize some outdoor surveying activities. Many students can gain significant insight into geometry from such applications.

Perhaps a surveyor could speak to the class about the various aspects of the job. A surveyor must make precise measurements to determine contours of the earth's surface and to measure distances between points. Surveyors must be able to make mathematical calculations with great speed and accuracy.

A transit is used to make angle measurements. Once the surveyor has obtained the angle and distance measurements that are needed, this collected data is transferred to carefully constructed diagrams. These diagrams are then used to make decisions about land areas.

Solution to Problem

$\angle Y$ and $\angle A$ are both right angles and thus are congruent. $\angle ACB$ and $\angle XCY$ are vertical angles and thus are congruent. $YC = CA$ by construction. $\triangle YCA \cong \triangle ACB$ by ASA. Thus, $\overline{XY} \cong \overline{AB}$. The distance from X to Y can be found by measuring the distance from A to B.

6.1 Convex and Concave Polygons (pages 183–187)

Vocabulary

polygon, diagonal of a polygon, convex, concave, quadrilateral, hexagon, pentagon, octagon, decagon, regular polygon

Suggestions

Ask your students which of the following are convex sets: a line; a plane; a set consisting of only two points; all points except one in the interior of a triangle; and a half-plane.

For Additional Chalkboard Work

Prove: The intersection of two convex regions is a convex region.

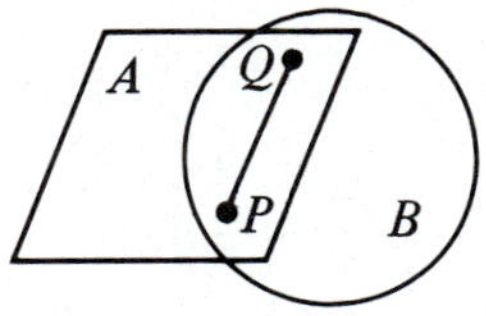

Outline of proof: 1. P is in $A \cap B$ and Q is in $A \cap B$. 2. P is in both A and B; Q is in both A and B. 3. $\overline{PQ}$ is in A; $\overline{PQ}$ is in B. 4. $\overline{PQ}$ is in $A \cap B$. 5. $\therefore A \cap B$ is a convex region.

Enrichment

Ask your students to diagram two concave regions whose union is convex.

Cumulative Review

1. Construct a 45° angle.
2. (Draw quadrilateral $ABCD$ with the diagonals $\overline{AC}$ and $\overline{BD}$.) If $\overline{AC} \cong \overline{BD}$ and $\overline{AD} \cong \overline{BC}$, what are the overlapping triangles that you would prove congruent in order to show that $\angle DAB \cong \angle CBA$?
3. What theorem about parallel lines is the proof of the theorem about the sum of the measures of the angles of a triangle based upon?

1. See p. 51. 2. $\triangle DAB$ and $\triangle CBA$
3. Alternate interior angles of parallel lines are congruent.

6.2 Angles of Convex Polygons (pages 188–192)

Background

Only convex polygons are considered in developing the method for finding the sum of the angle measures of a polygon. This restriction is necessary because we have limited the measure of an angle to being less than (or equal to), 180°. In concave polygon *ABCD*, angle *BCD* is not an interior angle of the polygon.

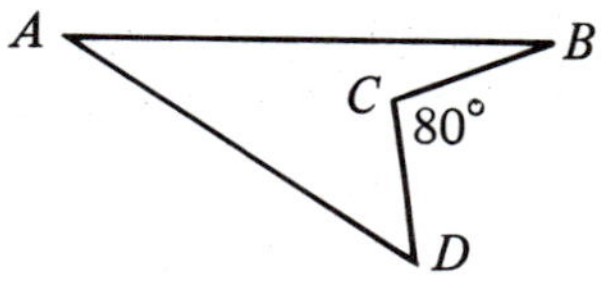

Suggestions

Have each student draw a convex pentagon, measure the interior angles, and find the sum. See if any students can find the sum for other convex polygons without measuring. Develop the formula $(n - 2) \cdot 180°$, where n = the number of sides.

Enrichment

Ask students what the sum would be for a convex polygon with 17 sides; 102 sides; n sides.
2,700°, 18,000°; (n − 2) 180°

Cumulative Review

1. Can a polygon ever be concave?
2. If the coordinate of P is −5 and the coordinate of Q is 7, what is the coordinate of the midpoint of $\overline{PQ}$?
3. Are all nonintersecting lines skew lines?

1. Yes 2. 1 3. No

6.3 Exterior Angles of Convex Polygons (pages 193–195)

Vocabulary

exterior angle of a convex polygon

Suggestions

Help your students discover the theorem about exterior angles of polygons by sketching a number of different kinds of polygons and measuring the exterior angles at each vertex.

For Additional Chalkboard Work

Proof of Theorem 6.4

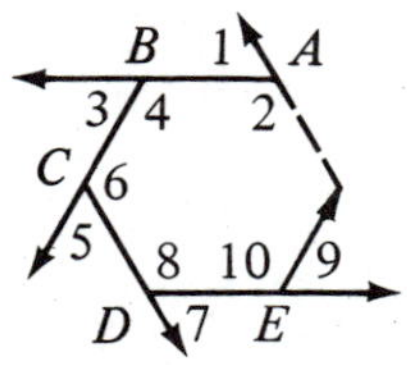

Given: Polygon *ABCDE* . . . with n sides.
Prove: $m\angle 1 + m\angle 3 + m\angle 5 + m\angle 7 + \ldots + m\angle(2n - 1) = 360$

PROOF

Statements	**Reasons**
1. $m\angle 1 + m\angle 2 = 180$ $m\angle 3 + m\angle 4 = 180$ $m\angle(2n - 1) + m\angle 2n = 180$	1. Sum of measures of ∠'s on one side of line = 180°.
2. $[m\angle 1 + m\angle 2] + \ldots + m[\angle(2n - 1) + m\angle 2n] = n(180)$ or $180n$	2. Addition property
3. $m\angle 2 + m\angle 4 + m\angle 6 + \ldots + m\angle 2n = (n - 2)180$	3. Sum of meas. of ∠'s of polygon = $(n - 2)180°$.
4. $m\angle 1 + m\angle 3 + m\angle 5 + m\angle 7 + \ldots + m(\angle 2n - 1) = 180n - (n - 2)180 = 2(180) = 360$	4. Subtraction property

Cumulative Review

1. What is the sum of the measures of the interior angles of a square?
2. Draw an example of a union of four segments that is not a polygon.
3. If $AB = 5$ and $BC = 6$, is AC always, sometimes, or never 11?

1. 360° 2. Answers may vary.
3. Sometimes

6.4 Regular Polygons (pages 196–198)

Vocabulary

regular polygon

Background

Regular polygons are always convex, and all sides and all interior angles are congruent. The sum of the exterior angle measures for *any* polygon, regular or irregular, is 360°. The general method for finding the measure of one exterior angle of a regular polygon is 360° divided by the number of sides.

Suggestions

Review with your students the method for finding the sum of the angle measures for any polygon. Your students may recall proving that each angle of an equilateral triangle has a measure of 60°. Ask them to find the measure of one angle of a regular hexagon and of a regular decagon. Establish the general method as the sum of the angle measures of the polygon divided by the number of angles (or sides). They should then be able to write the formula for a regular polygon of n sides.

For Additional Chalkboard Work

Name each regular polygon whose angle measures are factors of 360°. Draw a regular polygon of each type. Then show that several of them can be placed around a point.

Enrichment

Interested students may want to do a research project on *figurate numbers,* originated by the Pythagoreans. These numbers are illustrated by dots in certain geometric configurations.

Cumulative Review

1. What is the sum of the measures of the exterior angles of a square?
2. The hypotenuse-angle congruence correspondence for right triangles is actually which general pattern?
3. What is the measure of the complement of a 100° angle?

1. 360° 2. SAA 3. It does not exist.

6.5 Constructing Regular Polygons (pages 199–201)

Suggestions

Discuss with your students measures of angles that can easily be constructed. Help them make a list and construct many of the angles, using only integral measures.

Errors Students Make

In constructing a regular hexagon, point out that the radius of the circle has the same length as chord $\overline{AB}$, not arc $\widehat{AB}$. When working with circumference in Chapter 12, students may think the circumference should be six times the radius.

Enrichment

Construct an equilateral triangle, given an altitude that is congruent to segment $\overline{PQ}$. Construct $\angle VPQ = 30°$. Construct line l perpendicular to $\overline{QP}$ at Q, intersecting $\overrightarrow{PV}$ at R. Construct $\overline{QS}$ congruent to $\overline{QR}$, intersecting l so that Q is between R and S. $\triangle RPS$ is equilateral.

Cumulative Review

1. Are all equilateral triangles also equiangular?
2. Are all equiangular polygons also equilateral?
3. How many obtuse angles does an obtuse triangle have?

1. Yes 2. No 3. 1

Computer Activity (p. 203)

The program on this page is a general format that can be used for any combination of numbers and character strings. For example, replacing the items in the data line with 1, I, 5, V, 10, X, etc., could make this a test on knowledge of Roman numerals. Other variations are possible.

7 Chapter and Lesson Commentaries

Chapter 7 Quadrilaterals

Objectives

To identify diagonals, consecutive and opposite sides, and angles of quadrilaterals
To investigate properties of quadrilaterals
To prove and apply theorems about parallelograms
To prove that some quadrilaterals are parallelograms
To prove and apply the theorem about the segment joining the midpoints of two sides of a triangle
To use properties of parallelograms to prove theorems about parallel lines
To divide a segment into a given number of congruent segments
To identify rhombi, rectangles, and squares
To prove and apply properties of rhombi, rectangles, and squares
To prove that a quadrilateral is a rhombus, a rectangle, or a square

Chapter Overview

There are many properties to investigate in quadrilaterals and parallelograms. Yet quick identification of diagonals, consecutive and opposite sides, and angles is emphasized in detail for later use in this chapter and in subsequent chapters. Drawings are utilized in preliminary phases to direct and stimulate thinking on proofs of theorems. Theorems on

Chapter Overview cont'd.

the properties of parallelograms are treated, as well as theorems on quadrilaterals that are parallelograms. The theorem dealing with the segment joining the midpoints of two sides of a triangle is emphasized. Rhombus and rectangle are defined from the point of view of parallelograms.

Special Features

Amalie Noether p. 207
Non-Routine Problems p. 210
Cumulative Review pp. 214, 229
Reading in Geometry pp. 217, 228
Applications p. 233
Chapter Review p. 234
Chapter Test p. 235
Computer Activities p. 236
College Prep. Test p. 237

Amalie Noether (p. 207)

Some students may wish to prepare a report on Amalie Noether and other women mathematicians. You may refer them to *Women in Mathematics* by Lynn M. Osen, 1974, MIT Press.

7.1 Finding Properties of Quadrilaterals (pages 208–210)

Vocabulary

quadrilateral, opposite sides, consecutive sides, opposite angles, consecutive angles, diagonal

Background

Identifying diagonals, opposite and consecutive sides, and angles of quadrilaterals is important. Drawings are useful but do not comprise a proof. Students may not interpret "a quadrilateral with congruent opposite sides" to mean *both pairs* of opposite sides are congruent.

Suggestions

You may want to teach this section as an exploratory lesson in which your students make a number of drawings in order to discover basic properties of quadrilaterals. Encourage them to make reasonably large and accurate drawings but do not require that the drawings be constructed.

For Additional Chalkboard Work

A carpenter is building a section of wall. The edges of this section form a quadrilateral. Without some kind of diagonal bracing, the wall will not keep its shape. How many diagonal braces must be used to make the wall rigid?
One

Enrichment

Have students try all conclusions, proofs of theorems, and exercises on concave quadrilaterals as well as convex quadrilaterals.

Cumulative Review

1. Construct an equilateral triangle.
2. What is the measure of each interior angle of a square?
3. What is the measure of each exterior angle of a square?

1. On a reference line, mark off $\overline{AB}$; measure $\overline{AB}$ with compass. At A draw an arc; keeping distance the same, draw an arc from B. Connect $\overline{AB}$ to point of intersection.
2. 90°   3. 90°

7.2 Finding Properties of Parallelograms (pages 211–214)

Vocabulary

parallelogram

Background

If students are given that a figure is a parallelogram, then the only truth the definition assures is that both pairs of opposite sides are parallel.

Suggestions

You may want to investigate the properties of parallelograms informally before proving they hold. Have each of your students draw a parallelogram and list as many statements as he or she can that appear to be true about the sides, the angles, and the diagonals. Emphasize the importance of the properties of parallelograms. They are used throughout the rest of the text.

For Additional Chalkboard Work

Given: $\square XYZW$ with diagonals $\overline{XZ}$ and $\overline{WY}$. List all pairs of congruent segments.
$\overline{WZ} \cong \overline{XY}$, $\overline{WX} \cong \overline{ZY}$, $\overline{WO} \cong \overline{OY}$, $\overline{XO} \cong \overline{OZ}$

Enrichment

A *skew quadrilateral* is a quadrilateral with noncoplanar vertices. The diagonals of a skew quadrilateral do not intersect.

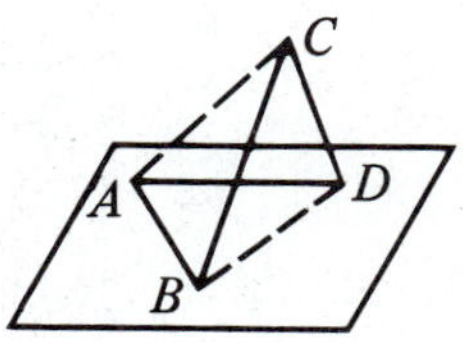

ABCD is a skew quadrilateral

Diagonal $\overline{AC}$ does not intersect $\overline{DB}$

Cumulative Review

1. If quadrilateral *ABCD* is congruent to quadrilateral *WXYZ*, which sides are congruent?
2. Give an example of a curve that separates a plane into two parts but that does not have an interior.

1. $\overline{AB} \cong \overline{WX}$, $\overline{BC} \cong \overline{XY}$, $\overline{DC} \cong \overline{ZY}$, $\overline{AD} \cong \overline{WZ}$
2. Answers may vary.

7.3 Proving Parallelograms (pages 215–218)

Background

The converse of a true statement is not necessarily true. Converses that are to be theorems in our geometry must be proved before they are accepted. [Theorem 7.6] The diagonals of a parallelogram bisect each other, is proved under the assumption that the diagonals intersect. In section 7.1, attention was called to a concave quadrilateral whose diagonals did not intersect.

Suggestions

Write the properties that have been proved about parallelograms on the chalkboard in "if-then" form; for example, "If a quadrilateral is a parallelogram, then both pairs of opposite sides are congruent." Ask your students to write the converse of these theorems. Then examine each of the converses with your class and decide whether or not it can be proved.

Be sure your students understand that the *same* pair of sides must be both parallel and congruent in Theorem 7.5.

For Additional Chalkboard Work

Outline of proof of Theorem 7.4
Given: $\angle A \cong \angle C$ and $\angle B \cong \angle D$.
Prove: $ABCD$ is a parallelogram.

Outline of proof:

1. $m\angle A + m\angle B + m\angle C + m\angle D = 360$
2. $m\angle A = m\angle C$ and $m\angle B = m\angle D$
3. $m\angle A + m\angle B + m\angle A + m\angle B = 360$
4. $2m\angle A + 2m\angle B = 360$
5. $m\angle A + m\angle B = 180$
6. $\overline{BC} \parallel \overline{AD}$
7. Similarly, $\overline{AD} \parallel \overline{BC}$
8. $\therefore ABCD$ is a parallelogram.

Cumulative Review

1. Which quadrilateral has congruent opposite angles?
2. True or false? The diagonals of any quadrilateral bisect each other.
3. What is a transversal?

1. Parallelogram 2. False 3. A line intersecting two other lines.

7.4 A Special Segment (pages 219–221)

Background

When proving Theorem 7.7, call attention to the use of both the properties of parallelograms and the use of one of the methods for proving a quadrilateral is a parallelogram.

Suggestions

Have students construct the midpoints of two sides of a triangle, and connect these midpoints with a straight line. See if they can draw any conclusions. Students usually agree that the third side is parallel to the connecting segment, but they do not realize that the latter is one-half the length of the third side.

For Additional Chalkboard Work

Prove or disprove: If a segment joining points on two sides of a triangle has a length of one-half that of the third side, then it is parallel to the third side.

False; many counterexamples can be drawn.

Enrichment

A model for illustrating Theorem 7.7 can be made with slats of wood, bolts, wing nuts, and protractors. Cut grooves in the slats representing $\overrightarrow{DE}$ and $\overrightarrow{AB}$. Attach bolts with wing nuts at points E and B. When the wing nuts are loosened, B can slide toward A or away from A, but length DE will always be one-half the length AB.

Cumulative Review

1. What are two different ways by which a quadrilateral can be proved to be a parallelogram?
2. According to the definition of congruent triangles, how many pairs of sides and angles must be shown to be congruent to know that the triangles are congruent?
3. According to our theorems on congruent triangles, how many pairs of sides and angles must be shown to be congruent to know that the triangles are congruent?

1. Answers may vary. 2. 6 3. 3

7.5 Parallel Lines and Congruent Segments (pages 222–225)

Background

One pair of parallel lines is constructed carefully, according to the diagram below. The segment they intercept can be used to mark off the other parts. That is, if $\overline{ED} \parallel \overline{BC}$, then $EB = \frac{1}{5}AB$ and $\overline{EB}$ can be used to find the other points that divide $\overline{AB}$ into five congruent parts.

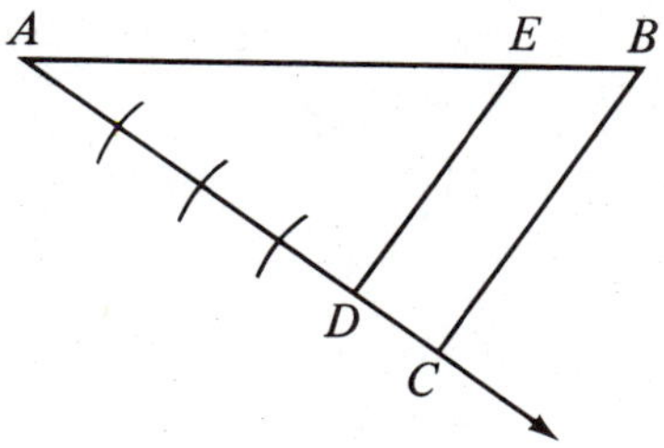

Suggestions

Help students see that if they consider the segment as part of a transversal, then strategically placed parallel lines could divide the segment into congruent parts.

Enrichment

Have students make a "tomahawk." They should construct $RS = ST = TU$, a semicircle with diameter $\overline{SU}$, and $\overline{SV} \perp \overline{RU}$. Place the tomahawk on $\angle ABC$ so that R is on $\overrightarrow{BA}$, $\overline{SV}$ contains B, and the semicircle is tangent to $\overrightarrow{BC}$. Then $\overrightarrow{BS}$ and $\overrightarrow{BT}$ will trisect $\angle ABC$.

Cumulative Review

1. What is known about the segment that joins the midpoints of two sides of any triangle?
2. What is the difference between a regular polygon and an equilateral polygon?
3. Can two quadrilaterals be proved congruent by an angle-angle-angle-angle correspondence?

1. It is parallel to the third side and $\frac{1}{2}$ its length. 2. The regular polygon is both equilateral and equiangular. 3. No

7.6 Rhombi, Rectangles, and Squares (pages 226–229)

Vocabulary

rhombus, rectangle, square

Background

A rhombus and a rectangle are frequently defined from the point of view of parallelograms. A square is often defined in terms of a rectangle.

Suggestions

Discuss the relationships among parallelograms, rectangles, rhombi, and squares in terms of sets.

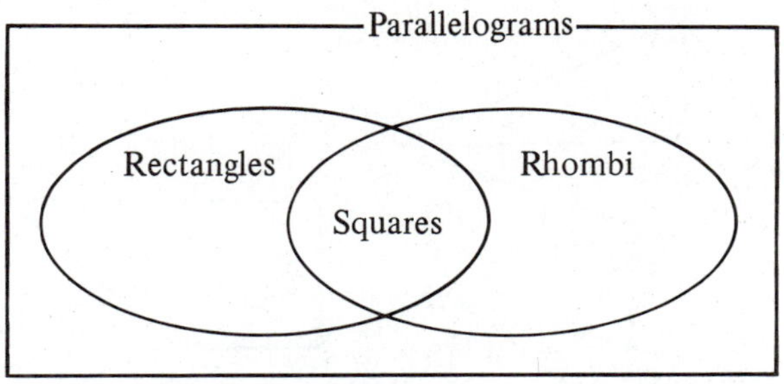

Because each is a parallelogram, all of the properties of parallelograms hold for each of these figures.

For Additional Chalkboard Work

A kite is a quadrilateral with two pairs of adjacent congruent sides. Prove that the diagonals of a kite are perpendicular.

Same proof as example 5

Enrichment

Example 5 gives a simple but elegant method of proving the diagonals of a rhombus are perpendicular by using the perpendicular bisector theorem. Encourage students to use this technique.

Cumulative Review

1. What is true about the legs of an isosceles trapezoid?
2. Find the distance between two points with coordinates −9 and −16.
3. If a transversal intersects nonparallel lines, are alternate interior angles formed?

1. They are congruent. 2. 7 3. Yes

7.7 Proving Rhombi, Rectangles, and Squares (pages 230–232)

Background

Students must realize that a quadrilateral may be proved a rhombus, rectangle, or square by using its definition.

Suggestions

Ask your students to sketch a quadrilateral with perpendicular diagonals that is not a rhombus (Figure 1). Have them sketch a quadrilateral with a right angle that is not a rectangle (Figure 2) and a quadrilateral with congruent diagonals that is not a rectangle (Figure 3).

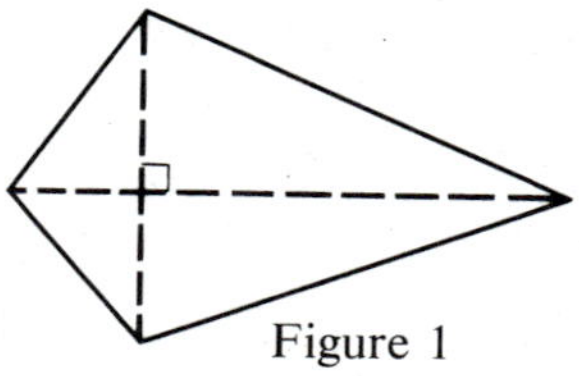

Figure 1

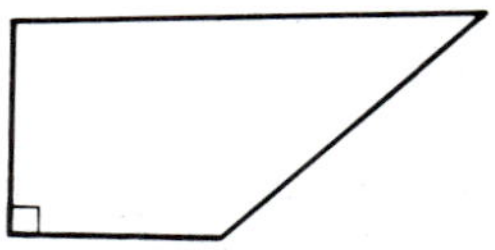

Figure 2

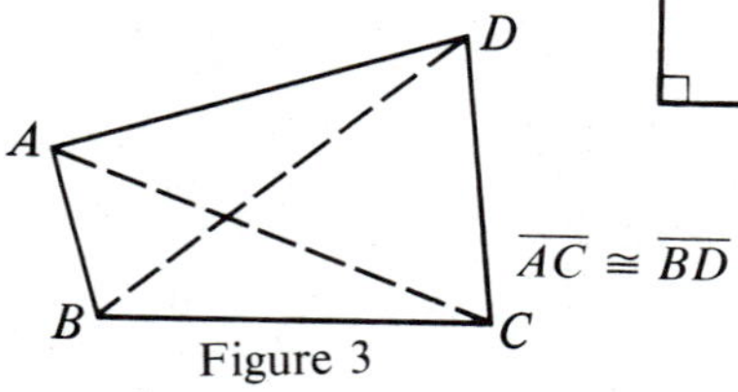

Figure 3

For Additional Chalkboard Work

A contractor is laying out the foundations of a rectangular building. How can the contractor be sure that the lines really do form a rectangle?

Measure so that opposite sides are the same length and diagonals are the same length.

Cumulative Review

1. Is a square a rhombus? a rectangle?
2. If in two congruent triangles, $\angle A \cong \angle T$, $\angle B \cong \angle U$, and $\angle C \cong \angle V$, which sides of the triangles are congruent?

1. Both 2. $\overline{AB} \cong \overline{TU}$, $\overline{BC} \cong \overline{UV}$, $\overline{AC} \cong \overline{TV}$

Computer Activity (p. 236)

Writing a flowchart from a program is not the ordinary sequence of events, but the practice in writing the chart is worthwhile.

8 Chapter and Lesson Commentaries

Chapter 8 Trapezoids and Symmetric Polygons

Chapter Objectives

To identify trapezoids and isosceles trapezoids
To investigate and prove properties of trapezoids and isosceles trapezoids
To prove and apply theorems about the median and an altitude of a trapezoid
To identify congruent quadrilaterals
To prove theorems about congruent quadrilaterals
To construct quadrilaterals with given properties
To identify symmetric polygons
To determine and apply congruence properties of symmetric polygons
To identify reflections
To prove and apply theorems relating reflections and congruence

Chapter Overview

This chapter defines trapezoids and isosceles trapezoids and investigates and proves properties about them. Theorems about the median and altitude of a trapezoid are also considered. Congruent quadrilaterals and their properties are discussed. Construction with given properties is continued in this chapter with quadrilaterals. Identifying symmetric polygons, applying congruence properties, identifying reflections and relating their properties are also introduced in this chapter.

Special Features

Non-Routine Problem Solving p. 238
Cumulative Review pp. 241, 249
Algebra Review pp. 241, 253
Logic Review p. 245
Non-Routine Problems p. 249
Applications p. 262
Computer Activities p. 263
Chapter Review p. 264
Chapter Test p. 265
College Prep. Test p. 266
Cumulative Review (Chapters 1 – 8) p. 267

Non-Routine Problem Solving (p. 238)

Solution: There are 6 non-equivalent pentominos which have a line of symmetry and 6 which do not. Those with symmetry:

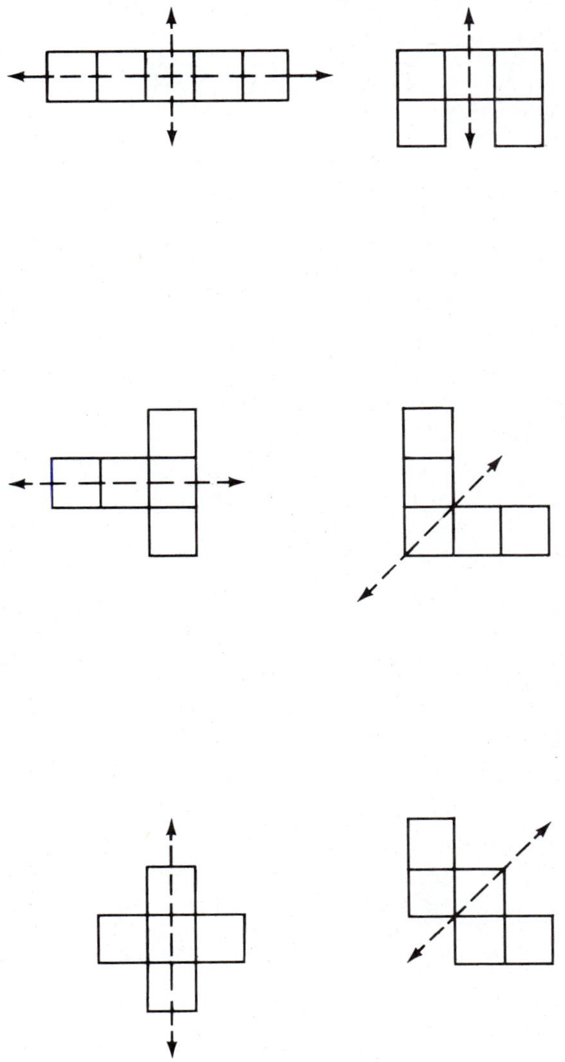

8.1 Trapezoids (pages 239–241)

Vocabulary

trapezoid

Suggestions

Review the idea of a counterexample. Remind your students that one example does not prove a statement is always true, but one counterexample is sufficient to disprove a statement. Exercises 2 and 3 anticipate the theorems about isosceles trapezoids in section 8.3. Make sure your students realize that the base angles can be proved congruent only if given that the nonparallel sides of the trapezoid are congruent. For Exercise 17, the fact that the segments joining the midpoints of opposite sides of a trapezoid bisect each other may be proved by first recalling that the segments joining the midpoints of consecutive sides of a quadrilateral form a parallelogram. Then use the property that the diagonals of a parallelogram bisect each other. This proof holds for all quadrilaterals, not just for trapezoids.

For Additional Chalkboard Work

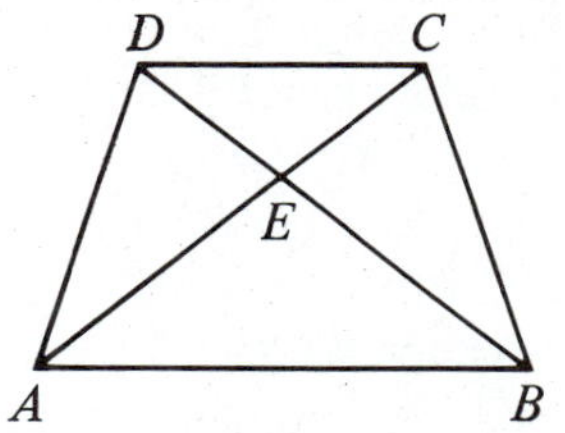

Given: Trapezoid $ABCD$
Which angles are known to be congruent?
$\angle ABD \cong \angle BDC$, $\angle BAC \cong \angle ACD$. Vertical angles formed by the diagonals are also $\cong$.

Cumulative Review

1. Which pairs of angles of parallel lines cut by a transversal are congruent?
2. In using the side-angle-side congruence pattern for triangles, what is meant by the included angle?

1. Alternate interior, corresponding, and alternate exterior angles 2. It is the angle whose vertex is the intersection of the two sides.

8.2 Altitudes and Medians of Trapezoids (pages 242–245)

Vocabulary

altitude, median

Suggestions

Compare and contrast the triangle and the trapezoid. A trapezoid may have many altitudes. They are all congruent. The three altitudes of a triangle, however, may have different lengths. An altitude of a trapezoid, like an altitude of a triangle, need not lie completely in the interior of the trapezoid.

For Additional Chalkboard Work

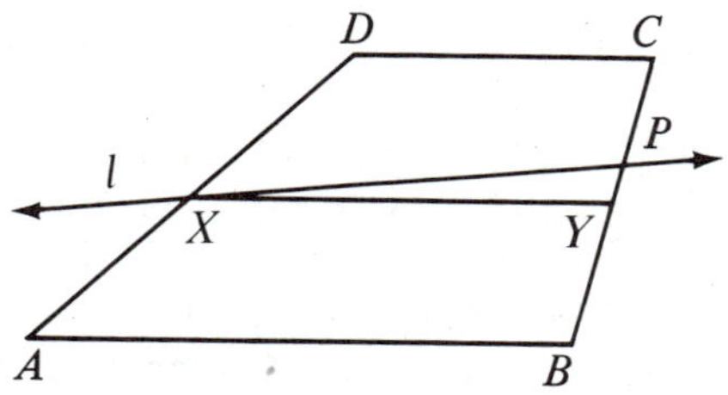

Indirect proof of part of Theorem 8.1: Suppose $\overline{XY}$ is not parallel to $\overline{AB}$. Then, by the Parallel Postulate, there exists some line l containing X and parallel to $\overline{AB}$. Let P be the point of intersection of l with $\overline{CB}$. Then P is the midpoint of $\overline{CB}$, because $\overline{PC} \cong \overline{PB}$ by the theorem that states that if three parallel lines cut off congruent segments on one transversal, then they cut off congruent segments on every transversal. This contradicts the fact that a segment has only one midpoint. Therefore, the assumption that $\overline{XY}$ is not parallel to $\overline{AB}$ is incorrect, so $\overline{XY}$ is parallel to $\overline{AB}$.

Cumulative Review

1. What is an altitude of a triangle?
2. What is the difference between a parallelogram and a trapezoid?
3. Are the opposite angles of a parallelogram supplementary?

1. A segment from a vertex perpendicular to the opposite side. 2. A parallelogram must have both pairs of opposite sides parallel. 3. Not necessarily

8.3 Isosceles Trapezoids (pages 246–249)

Vocabulary

isosceles trapezoid

Suggestions

Remind the students of the definition of an isosceles triangle and ask them to formulate a definition for an isosceles trapezoid. Be sure they understand that the congruent sides are the nonparallel sides. Theorems 8.2 and 8.3 are converses of each other, and Theorems 8.4 and 8.5 are also converses.

A counterexample for Exercise 14 can easily be drawn by making angles Q and S right angles. Then angles P and R must also be supplementary, because the sum of the measures of the angles of any quadrilateral is 360°.

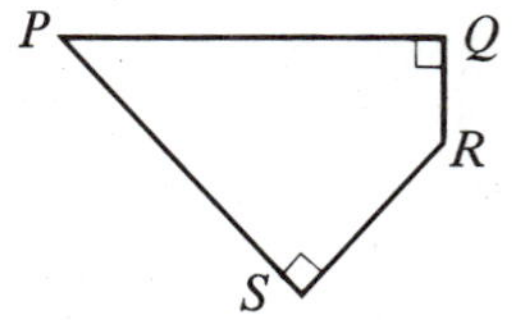

For Additional Chalkboard Work

Prove: The segments whose endpoints are the midpoints of consecutive sides of an isosceles trapezoid form a rhombus.

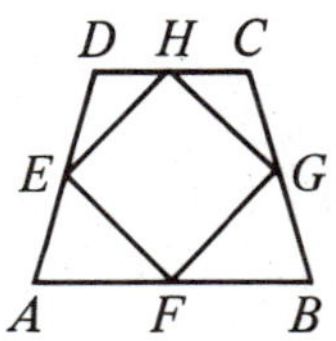

Outline of proof: $EFGH$ is a ▱. Show $\triangle AFE \cong \triangle BFG$. $\overline{EF} \cong \overline{GF}$. ∴ $EFGH$ is a rhombus.

Cumulative Review

1. What is a symmetric polygon?
2. What are the legs of a trapezoid?
3. Can two triangles be proved congruent by a side-side-angle correspondence?

1. A polygon that is its own reflection.
2. The nonparallel sides. 3. No

8.4 Congruent Quadrilaterals (pages 250–253)

Vocabulary

congruent quadrilaterals

Suggestions

Make sure the students realize that the definition of congruent quadrilaterals means that each pair of corresponding sides is congruent and each pair of corresponding angles is congruent. When writing the congruence statements for congruent quadrilaterals, have the students match up the letters in the order of the congruence. The statement Quad $PQRS \cong$ Quad $JKLM$ means $\angle P \cong \angle J$, $\overline{PQ} \cong \overline{JK}$, and so on.

Quad $ABCD$ should be labeled like this.

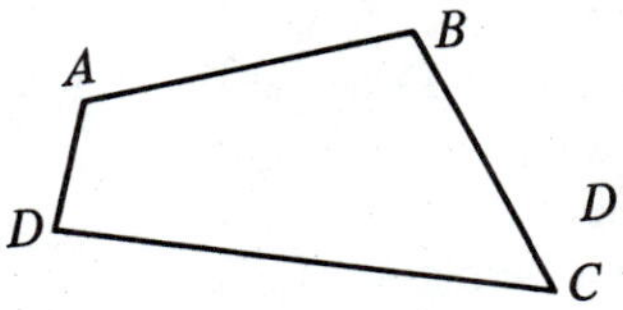

not like this:

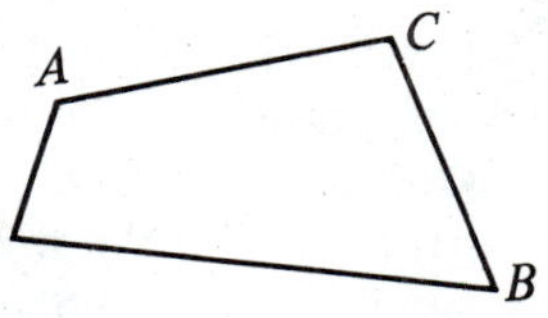

The diagonals should be $\overline{AC}$ and $\overline{BD}$, not $\overline{AB}$ and $\overline{CD}$.

In the proof of Theorem 8.6, students frequently want to add triangles ABD and BDC to obtain Quad $ABCD$.

Cumulative Review

1. Draw and label a quadrilateral. Which sides are the consecutive sides?
2. True or false? Every angle of a triangle has two corresponding exterior angles.
3. What theorem is the basic reason in the proof that an equilateral triangle is also equiangular?

1. Answer depends on the drawing.
2. True   3. Base angles of an isosceles triangle are congruent.

8.5 Constructing Quadrilaterals (pages 254–255)

Background

The compass and the straightedge are known as Euclidean tools. The first three propositions of Book I of Euclid's Elements are actually construction problems. The Euclidean compass collapsed if either leg was lifted from the paper. Constructions required some ingenuity with a Euclidean compass.

Suggestions

Many students enjoy constructions. Review the properties of quadrilaterals as well as the basic constructions previously covered. Proofs of the constructions are not called for in this lesson, but the students should be able to justify their constructions.

Review the methods used to prove a quadrilateral is a rectangle, rhombus, square, parallelogram, or isosceles trapezoid. These methods can be used in the construction of each figure. For example, proving the diagonals of a quadrilateral are perpendicular bisectors of each other proves the quadrilateral is a rhombus. Constructing diagonals that are perpendicular bisectors of each other assures that one has constructed a rhombus. Sometimes more than one construction will produce the same results, but one construction is often easier.

For Additional Chalkboard Work

Which quadrilateral congruence pattern(s) could be used to construct a parallelogram congruent to a given parallelogram?

Any of them

Cumulative Review

1. Tell one way to prove that a quadrilateral is a rhombus.
2. What is the distance between two parallel lines?
3. What is the measure of a straight angle?

1. Show that the diagonals are perpendicular bisectors of each other.   2. The distance from any point on one line to the other line.   3. 180°

8.6 Symmetric Polygons (pages 256–258)

Vocabulary
symmetric polygon, line of symmetry

Background
Biologists use the term *bilateral symmetry* for the symmetry around a line in such animals as mammals. Starfish have radial symmetry. These lines describe only external features, not internal organs.

In algebra, students will learn that the graph of a relation and its inverse are symmetric around the line $y = x$. Folding the graph of a relation along the $y = x$ line gives a quick way to graph its inverse.

Suggestions
Symmetry lines provide a basis for alternate definitions of some polygons. An isosceles triangle can be defined as "a triangle with at least one line of symmetry." An isosceles trapezoid is "a trapezoid that is symmetric about a line through the midpoints of its bases." Have the students find other figures that could be defined by lines of symmetry.

To test a polygon to see if it has a line of symmetry, suggest the students sketch in the line and fold the diagram on that line. If the two parts of the figure fit, then it is a line of symmetry.

For Additional Chalkboard Work
Which figures studied so far are symmetric figures?

segment, line, isosceles triangle, equilateral triangle, rhombus, rectangle, square, isosceles trapezoid, angle

Cumulative Review
1. What is a median of a triangle?
2. What is a median of a trapezoid?
3. Are all quadrilaterals with diagonals that bisect each other parallelograms?

1. A segment from a vertex to the midpoint of the opposite side. 2. A segment joining the midpoints of the legs. 3. Yes

8.7 Congruence and Reflections (pages 259–261)

Vocabulary
reflection

Suggestions
Introduce reflections by having the students experiment with mirrors or with folding and tracing. Students to discover: The line of symmetry is the perpendicular bisector of the segment joining pairs of corresponding points; the reflections of collinear points seem to be collinear; the distance between two points seems to be equal to the distance between their images; if one point is between two others, then its image is between the images of the others; and an angle seems to have the same measure as its image.

For Additional Chalkboard Work
An experiment with mirrors: Draw a segment on a piece of paper. Can you arrange two mirrors so that the reflection forms a triangle? a square? a pentagon? a hexagon? How must the mirrors be placed for each of these?

Measure of angle between mirrors must equal 360° divided by the number of sides wanted.

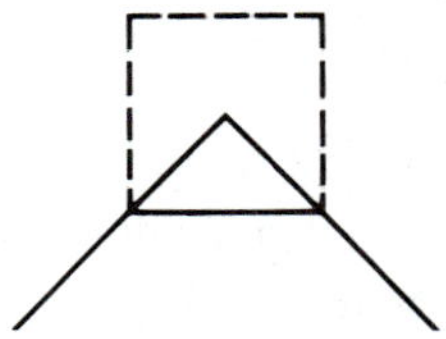

Cumulative Review
1. Construct a square with sides congruent to a given segment.
2. Find the measure of the vertex angle of an isosceles triangle if each base angle measures 20°.

1. See p. 200. 2. 140°

Computer Activity (p. 263)
In lines 20 and 30, the bases a and b are tested and switched if necessary so that b − a, the length of the base of the side triangles, will be positive. In actual fact, that is not necessary. See if your students can explain why (i.e., since the quantity is squared, it does not matter if it is positive or negative).

9 Chapter and Lesson Commentaries

Chapter 9 Similarity

Chapter Objectives

To identify, write, and solve proportions
To find the geometric mean of two numbers
To find the measures of sides and angles of similar polygons
To recognize and apply the angle-angle similarity pattern
To write and solve proportions when a line is parallel to a side of a triangle
To recognize and apply the SAS and SSS similarity patterns
To prove and apply theorems about proportional segments in triangles
To prove products of lengths equal using proportions
To prove and apply theorems about proportions and similarity in space

Chapter Overview

This chapter reviews algebraic properties of ratio and proportion before solving proportions to find the lengths of sides of similar polygons and to prove products of lengths equal. The AA, SAS, and SSS patterns of similarity are introduced. These patterns are applied to numerical and algebraic examples, as well as to proofs. Proportions formed by a line parallel to one side of a triangle, by medians and altitudes in similar triangles, and by an angle bisector in a triangle are also considered. Similarity in space is included as an optional lesson.

Special Features

Non-Routine Problem Solving (p. 269)

Shadows provide an interesting example of similarity when the object and the shadow are parallel. The shape of the shadow will be the same as the shape of the object, but the size of the shadow will be proportionately larger. This can be demonstrated using a projector and a screen.

9.1 Using Proportions (pages 270–274)

Vocabulary

ratio, proportion, means, extremes, geometric mean

Background

It is important to stress the distinction between ratio and proportion. A ratio is a number that can be expressed as $\frac{a}{b}$ or $a{:}b$. A proportion is an equation stating that two ratios are equal. Proportions may be written in the form $\frac{a}{b}=\frac{c}{d}$ or $a{:}b = c{:}d$.

Suggestions

Give other uses of ratios that include examples to help students relate ratios to fractions, such as: Mary is $\frac{3}{4}$ as old as Jane. Have students find ratios in real-life situations.

For Additional Chalkboard Work

Have students identify each of the following as a ratio, a proportion, or neither.

1. $\frac{x}{8}$ Ratio
2. $\frac{2}{3}=\frac{4}{8}$ Proportion
3. $\frac{x}{3}+1=\frac{5}{8}+6$ Neither

Enrichment

Ratios can be used to help find better buys in the grocery store. For example, if w pounds of sugar sell for x cents, whereas y pounds sell for z cents, the ratios $\frac{w}{x}$ and $\frac{y}{z}$ can be used. If $wz > xy$, then w pounds for x cents is the better buy. Have students use this ratio method for comparing buys of products in a grocery store.

Cumulative Review

A rectangle measures 3 cm by 5 cm. Find

1. The ratio of the length to the width.
2. The ratio of the width to the length.
3. The ratio of the perimeter to the length.

1. $\frac{5}{3}$ 2. $\frac{3}{5}$ 3. $\frac{16}{5}$

9.2 Similar Polygons (pages 275–277)

Vocabulary

similar polygons

Suggestions

Discuss the differences and common traits between pairs of congruent polygons and pairs of similar polygons, as illustrated at the top of page 275. Stress that squares and rectangles are not similar. Although the corresponding angles are congruent, pairs of corresponding sides are not proportional. Note that congruent polygons are also similar, but similar polygons are not necessarily congruent.

For Example 3, $\frac{18}{12}$ is reduced to $\frac{3}{2}$.

Encourage the student to reduce each ratio to simplest terms to make the calculation easier. ($\frac{3}{2}$ is sometimes referred to as the ratio of similitude of the two polygons.)

For Additional Chalkboard Work

Show that for squares and rhombuses, pairs of corresponding sides are proportional but the angles are not congruent.

Enrichment

Some students may be able to use set terminology to express the relationship between congruent polygons and similar polygons. In set terminology, the set of pairs of congruent polygons is a subset of the set of pairs of similar polygons.

Cumulative Review

Have the students reduce each ratio to simplest form.

1. $\frac{24}{21}$
2. $\frac{144}{60}$
3. $\frac{56}{7}$

1. $\frac{8}{7}$ 2. $\frac{12}{5}$ 3. 8

9.3 Angle-Angle Similarity (pages 278–281)

Background

In this textbook we choose to assume angle-angle for similar triangles as a postulate. It is possible, however, to prove Postulate 22 provided that Theorem 9.3 on page 285 is given first as a postulate. Because each one of these two statements depends upon the other, one must come first as a postulate and the other may then be proved as a theorem.

Suggestions

Begin by reviewing the definition of similar polygons. Encourage students to suggest possible minimal conditions for proving two triangles similar. Have a student do the construction on page 278 at the chalkboard, or have all the students do the construction at their seats. Point out that when two pairs of corresponding angles of two triangles are congruent, the third pair of angles is also congruent by Theorem 3.6. Have students measure the pairs of corresponding sides to show that they are proportional.

In Examples 3 and 4, students must go one step further. In Example 3 they find unknown lengths, and in Example 4 they apply the definition of similar polygons to prove that two pairs of corresponding sides are proportional.

For Additional Chalkboard Work

Using any three of the dots below as the three vertices, how many different sizes of right triangles can be drawn? 3

Cumulative Review

Find the geometric mean of each pair of numbers.

1. 4 and 9
2. 6 and 7
3. 12 and 15

1. 6 2. $\sqrt{42}$ 3. $\sqrt{180}$ or $6\sqrt{5}$

9.4 Proportionality in Triangles (pages 282–284)

Background

A preliminary theorem to Theorem 9.1 is as follows: If a line is parallel to one side of a triangle and intersects the other two sides, it forms a triangle similar to the original one. This theorem is actually proven in the first three steps in the proof of Theorem 9.1.

Suggestions

Be sure to go over the three possible proportions that can be stated as a result of Theorem 9.1. Have students state these proportions as related to the figure below.

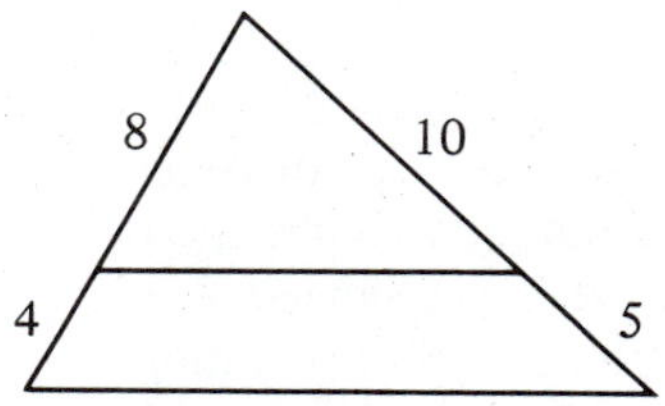

$\frac{4}{8} = \frac{5}{10}, \frac{8}{12} = \frac{10}{15}, \frac{4}{12} = \frac{5}{15}$

Some students may find these additional proportions.

$\frac{8}{4} = \frac{10}{5}, \frac{12}{8} = \frac{15}{10}, \frac{12}{4} = \frac{15}{5}$

For Additional Chalkboard Work

Given: $\overline{DE} \parallel \overline{AB}$, $AD = 2$, $DC = 5$, and $DE = 8$

Find AB.

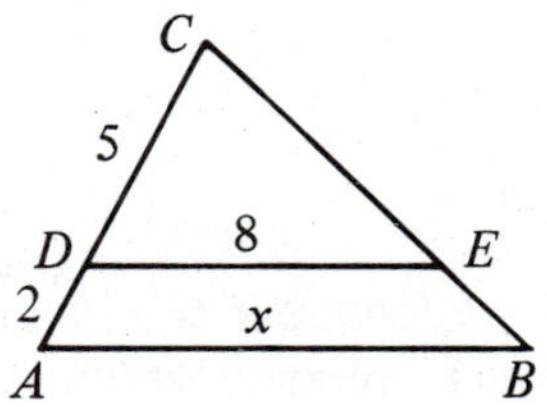

$\frac{DC}{AC} = \frac{DE}{AB}$

$\frac{5}{7} = \frac{8}{x}$

$5x = 56$

$x = 11\frac{1}{5}$

Cumulative Review

$\triangle ABC \sim \triangle DEF$: m $\angle A = 20$, m $\angle B = 80$.

Find the measure of each angle.

1. $\angle D$
2. $\angle E$
3. $\angle F$

What kind of triangles are $\triangle ABC$ and $\triangle DEF$?

1. 20° 2. 80° 3. 80°; isosceles

9.5 SAS and SSS Similarity (pages 285–289)

Suggestions

Have a student do the constructions on pages 285 and 286 at the chalkboard, or have all the students do them at their seats. Students might measure the corresponding parts of the triangles to verify that the triangles are indeed similar.

Carefully review the summary at the bottom of page 286 to be certain that all three similarity patterns are thoroughly understood.

For Example 4, students must first prove that the two triangles are similar. Then they prove that two pairs of corresponding sides are proportional.

Enrichment

The pantograph is used to enlarge or reduce in exact proportions. To reduce a figure the pen and the tracing point are interchanged. Have students construct a pantograph by bolting four pieces of wood or metal together as shown. Then have them use it to enlarge or reduce several drawings.

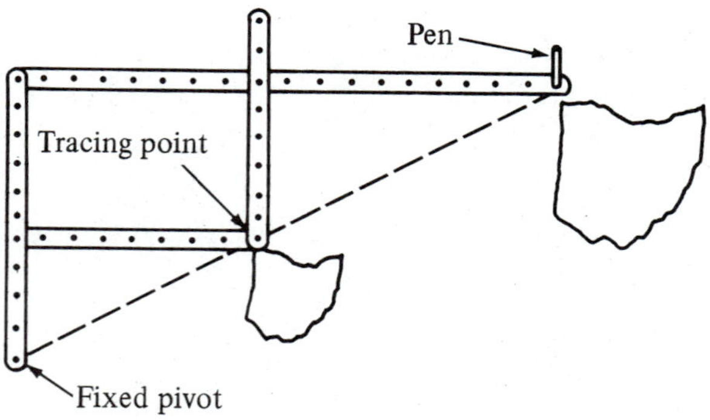

Cumulative Review

Which of the following statements are true?

1. The altitude of a triangle is perpendicular to the base of the triangle.
2. The angle bisector bisects the opposite side of the triangle.
3. The median of a triangle bisects the opposite side.

1. True 2. False 3. True

9.6 Other Proportional Segments (pages 290–293)

Suggestions

While discussing Theorems 9.5 and 9.6, have students prove the following:
Corresponding angle bisectors of similar triangles are proportional to corresponding sides.

For Additional Chalkboard Work

The construction on page 292 may also be done as follows:
Given: Segment $\overline{AB}$

Construct: P on $\overline{AB}$ so that

$$\frac{AP}{PB} = \frac{2}{3}$$

1. Draw a ray through A. Mark off 5 segments of equal length on the ray.

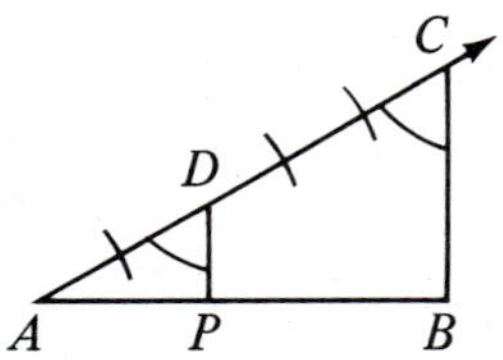

2. Name the last endpoint of the last segment as point C. Draw $\overline{BC}$.
3. Name the second endpoint of the second segment point D. Through D construct a line parallel to $\overline{BC}$ intersecting $\overline{AB}$ at P.

   Thus, $\frac{AP}{PB} = \frac{2}{3}$ by Theorem 9.1.

Cumulative Review

Write two different proportions for each equation.

1. $3 \cdot 6 = 2 \cdot 9$
2. $2 \cdot 3 = x \cdot (90 - x)$
3. $7x = 5y$

1. $\frac{3}{2} = \frac{9}{6}; \frac{6}{9} = \frac{2}{3}$ 2. $\frac{2}{x} = \frac{90-x}{3}; \frac{3}{90-x} = \frac{x}{2}$ 3. $\frac{7}{5} = \frac{y}{x}; \frac{x}{5} = \frac{y}{7}$

9.7 Products of Lengths (pages 294–296)

Vocabulary

cross-product

Background

Note that in Example 5, students are actually proving the following statement: In a right triangle, the length of the altitude to the hypotenuse is the geometric mean of the lengths of the segments of the hypotenuse. This statement will be treated as a corollary on page 306.

Suggestions

As students work with Example 1, you might want to point out that two other proportions can also be written.

$$\frac{EG}{PQ} = \frac{QR}{FE} \quad \text{or} \quad \frac{PQ}{FE} = \frac{EG}{QR}$$

Point out that Example 4 is a numerical application of Example 3.

For Additional Chalkboard Work

Have students examine Example 5 and relate it to the statement given in the Background.

Enrichment

Early Greek geometers were interested in certain ratios and proportions that they believed had mystical meaning. Look up and report on the golden proportionality.

Cumulative Review

In a scale drawing of a room, $\frac{1}{8}$ inch equals 1 foot. What are the dimensions of:

1. a rectangular room that is $2\frac{7}{8}$ in. by $3\frac{1}{4}$ in. on the scale drawing?
2. a rug that is $1\frac{1}{8}$ in. by $1\frac{1}{2}$ in. on the scale drawing?
3. a door that is $\frac{5}{16}$ in. by $\frac{13}{16}$ in. on the scale drawing?

1. 23 ft by 26 ft 2. 9 ft by 12 ft

3. $2\frac{1}{2}$ ft by $6\frac{1}{2}$ ft

9.8 Similarity in Space (optional) (pages 297–299)

Vocabulary

parallel planes, tetrahedron, base, faces

Background

It is interesting to note that if a prism is cut by a plane parallel to its base, then the figure formed is *congruent* to the base. The same is true for a cylinder.

If a tetrahedron is cut by a plane parallel to its base, then the triangle formed by the intersection is *similar* to the base. The same is true for a cone.

Suggestions

Before students work with Example 1, have them examine this problem.
Given: l, m, n parallel lines
Find x.

$$\frac{5}{9} = \frac{a}{b} \text{ and } \frac{a}{b} = \frac{x}{12}$$
$$\frac{5}{9} = \frac{x}{12}$$
$$9x = 60$$
$$x = 6\frac{2}{3}$$

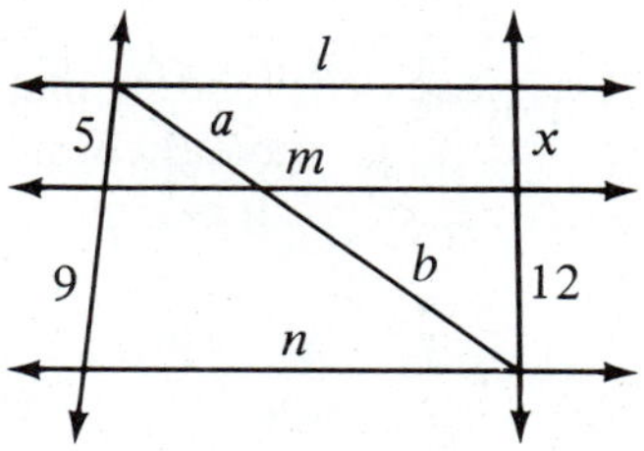

For Additional Chalkboard Work

The following general statement can be proved: Parallel lines intersect proportional segments on any two transversals.

Cumulative Review

Have students consider the problem of making a scale drawing of several of the rooms in their homes. Have them choose an appropriate scale, such as $\frac{1}{8}$ inch = 1 foot, or 1 centimeter = 1 meter.
Answers will vary.

Computer Activity (p. 301)

This program rounds answers to the nearest one hundredth. A commonly used rule in rounding is that a 5 is rounded to an even number (e.g., 0.035 rounds up to 0.04, but 0.045 rounds back to 0.04). Challenge one of your students to write a program using this rule for rounding.

10 Chapter and Lesson Commentaries

Chapter 10 Applying Similar Triangles

Chapter Objectives

To find the lengths of sides and altitudes of right triangles
To apply the Pythagorean Theorem
To find the lengths of segments in a 45°–45° (isosceles) right triangle and a 30°–60° right triangle
To compute the sine, cosine, and tangent of an acute angle of a right triangle
To use a table to find sin *A*, cos *A*, and tan *A*, given m $\angle A$
To use a table to find m $\angle A$, given sin *A*, cos *A*, or tan *A*
To find measures of sides and angles in right triangles by using sine, cosine, and tangent ratios
To solve problems by trigonometric ratios

Chapter Overview

Applying the properties of similar triangles to the case of a right triangle and its altitude to the hypotenuse leads to the proof of the Pythagorean Theorem and its converse. Applications of the Pythagorean Theorem involve not only right triangles, but also rectangles, squares, rhombi, and equilateral triangles. Pythagorean Triples are considered in Non-Routine Problems, and the use of a calculator in applying the Pythagorean Theorem is introduced in the Calculator Activities. The special cases of the 45°–45° (isosceles) right triangle and the 30°–60° right triangle are discussed. Trigonometric ratios are defined and used to find the missing measures of sides and angles of right triangles. Trigonometry is applied to real-life situations.

Special Features

Non-Routine Problem Solving p. 305
Reading in Geometry pp. 308, 322
Non-Routine Problems p. 314
Calculator Activites pp. 314, 328
Cumulative Review p. 317
Algebra Review p. 317
Logic Review p. 323
Applications p. 332
Computer Activities p. 333
Chapter Review p. 334
Chapter Test p. 335
College Prep. Test p. 336

Non-Routine Problem Solving (p. 305)

The Greeks used the golden ratio in the design of their temples. A golden rectangle is said to have the proportions that are most pleasing to the eye.

An interesting point is that many of the parts of the human body are divided into the ratio of the golden rectangle. For example, the distance from the floor to your waist is about .618 of your height. The distance from your fingertips to your elbow is about .618 of the length of your arm. The various joints in the fingers and hand are also in that same ratio.

The numbers 1, 2, 3, 5, 8, 13, . . . form a Fibonacci sequence. Interested students might want to explore the work of Fibonacci.

10.1 Similarity in Right Triangles (pages 306–309)

Suggestions

Adjacent segments have not been introduced previously. Have students identify adjacent segments in the illustration on p. 306. Students often have trouble visualizing the similarity match-ups in the three triangles in Theorem 10.1. The triangles shown on p. 306 can be rearranged in the following way:

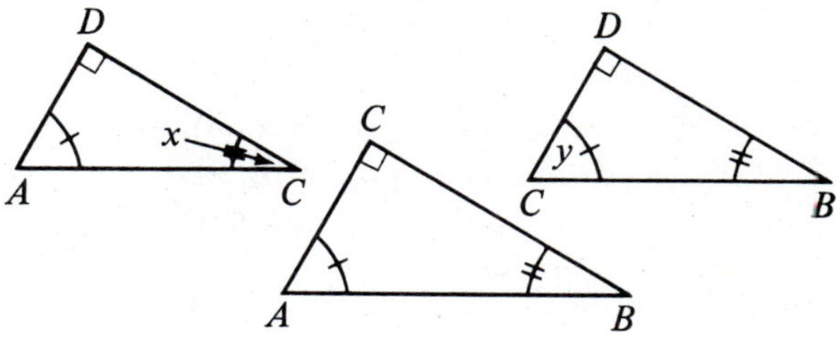

For Additional Chalkboard Work

Given: $\overline{AC} \perp \overline{CB}$,
$\overline{CD} \perp \overline{AB}$
$AD = 6$, and
$CD = 12$
Find CB. $12\sqrt{5}$

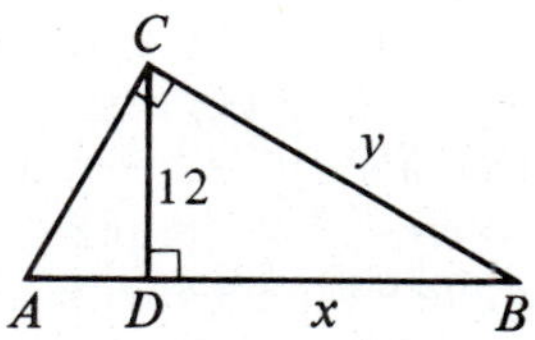

Have students use Corollary 1 to find x equal to 24 and then Corollary 2 to find y.

Enrichment

Have students report on the historical background for the proof that $\sqrt{2}$ is irrational, and the scandal it caused for the Pythagoreans.

Cumulative Review

1. What must be true if two triangles are similar?
2. If the measure of an angle is 38°, find the measure of a complement.
3. What congruence pattern is true for right triangles that is not true for triangles in general?

1. Corresponding angles are congruent and corresponding sides are in proportion.
2. 52° 3. SSA

10.2 The Pythagorean Theorem (pages 310–314)

Background

One of the proofs of the Pythagorean Theorem involves areas. The sum of the areas of the squares on the legs is shown to equal the area of the square on the hypotenuse. (Euclid used this diagram in his proof.)

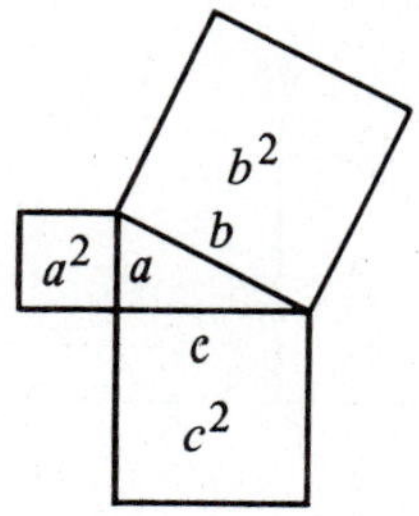

Suggestions

Students will use the Pythagorean Theorem frequently throughout the remainder of the course. Make sure they understand it thoroughly. Students often make the error of setting the unknown in the equation to be the hypotenuse. Be sure your examples include cases of finding the length of a leg. Exercises 21, 22, 27, and 28 anticipate the theorems on special right triangles of section 10.4.

For Additional Chalkboard Work

A bug crawled 3 in. north, then 6 in. east. How far was it from its starting point?
$3\sqrt{5}$, or 6.71 in.

Enrichment

There are over 100 different proofs and demonstrations of the Pythagorean Theorem. Have students report on the proof drawn by the Hindu mathematician Bhāskara.

Cumulative Review

1. What is true about the triangles formed by the altitude to the hypotenuse of a right triangle?
2. What is the hypotenuse of a right triangle?
3. Find the geometric mean of 4 and 9.

1. They are similar. 2. The side opposite the right angle. 3. 6

10.3 Converse of the Pythagorean Theorem (pages 315–317)

Suggestions

Proving the Pythagorean Theorem does not prove its converse. The converse must be proved as a separate theorem. The proof may be assigned as Exercise 26. It can be proved in this way:
Given $\triangle ABC$ with $a^2 + b^2 = c^2$.

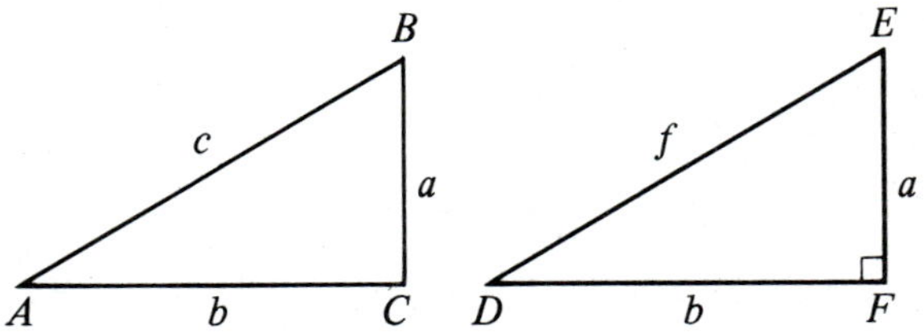

Construct a $\triangle DEF$ with legs b and a, right angle at F. In $\triangle DEF$, $a^2 + b^2 = f^2$ by the Pythagorean Theorem. Because it was given that $a^2 + b^2 = c^2$, then $f^2 = c^2$ and $f = c$ (because both must be positive). $\triangle ABC \cong \triangle DEF$ by SSS, therefore $\angle C \cong \angle F$, and $\angle C$ must be a right angle.

For Additional Chalkboard Work

Have students determine whether these are right triangles. Remind them that $(\sqrt{x})^2 = x$ if $x \geq 0$.

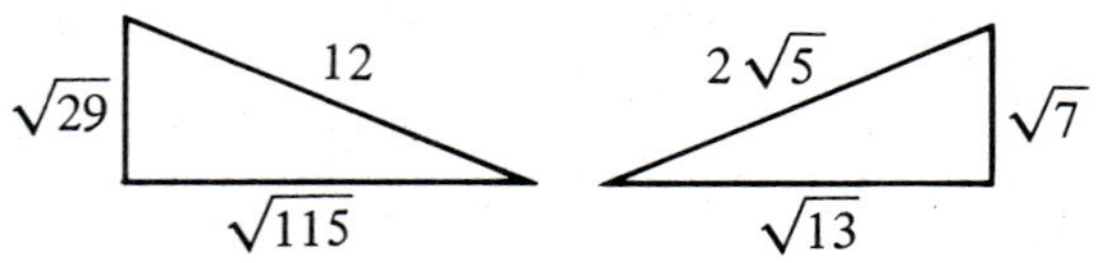

$(\sqrt{29})^2 + (\sqrt{115})^2 = 12^2$ $\quad$ $(\sqrt{13})^2 + (\sqrt{7})^2 = (2\sqrt{5})^2$
$29 + 115 = 144$ $\quad$ $13 + 7 = 20$
yes $\quad$ yes

Cumulative Review

1. What kind of triangle does the Pythagorean Theorem deal with?
2. $\triangle ABC \sim \triangle DEF$. If $AB = 6$, $DE = 18$, and $AC = 3$, find DF.
3. In $\triangle PQR$, m $\angle P = 50$, m $\angle Q = 30$. Find m $\angle R$.

1. Right triangle 2. 9 3. 100

10.4 Special Right Triangles (pages 318–320)

Vocabulary

isosceles right triangle

Background

Theorems 10.4 and 10.5 give shortcuts for finding the measures of the sides of special triangles. These relationships are important for those who plan to take a course in trigonometry.

Suggestions

Draw an isosceles right triangle on the chalkboard or overhead projector. Start with simple measures for the legs and have the students find the length of the hypotenuse using the Pythagorean Theorem. Keep a chart of the results and encourage students to look for a pattern. Prove Theorem 10.4 with the students or assign the proof as Exercise 21.

Some students find it difficult to find the length of a leg when the length of the hypotenuse is given as in Example 1. It helps to write out the phrase, hypotenuse = leg $\cdot \sqrt{2}$. Give examples with hypotenuses of $7\sqrt{2}$, $15\sqrt{2}$, 9, and 17. Ask students to find the length of the legs.

For Additional Chalkboard Work

Construct an isosceles right triangle with each leg measuring 1 unit. Continue constructing right triangles with a leg measuring 1 unit and the other leg the hypotenuse of the previous triangle. Have the students find the lengths of each resulting hypotenuse. Ask for the pattern. hypotenuse increases by 1

Cumulative Review

1. Is a triangle whose sides measure 2, 5, and 6 a right triangle?
2. How does the side-side-side theorem for similarity of triangles differ from the corresponding theorem for congruence?
3. What is a convex polygonal region?

1. No 2. For similarity the sides must be proportional, not congruent. 3. The segment joining any two points of the region lies in the region.

10.5 Trigonometric Ratios (pages 321–323)

Vocabulary
trigonometry, sine, cosine, tangent

Suggestions
Demonstrate the fact that the value of the ratios is dependent on the measure of the angle, not on the length of the sides, as is shown below.

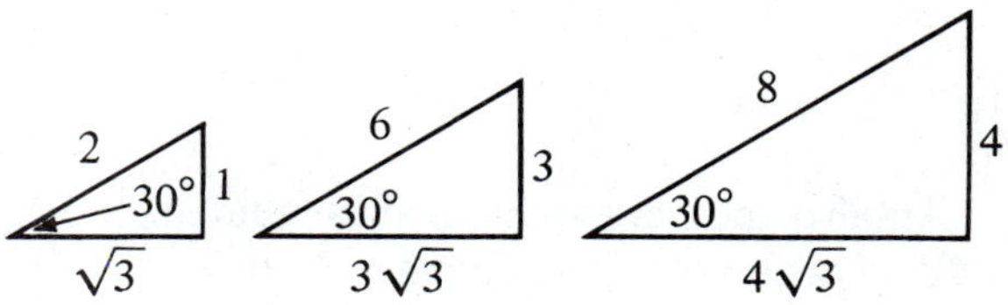

$\sin 30° = \frac{1}{2}$ $\quad \sin 30° = \frac{3}{6} = \frac{1}{2}$ $\quad \sin 30° = \frac{4}{8} = \frac{1}{2}$

Students should memorize the ratios so they will be able to use them in the applications presented later in the chapter.

For Additional Chalkboard Work
Use trigonometric ratios to show that

$$\frac{x}{y} = \frac{a}{b} \text{ and } \frac{x}{z} = \frac{b}{a}$$

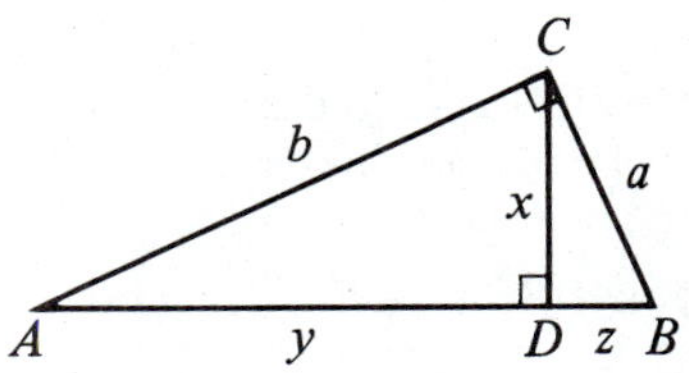

Label congruent angles and use the tangent ratio to prove the proportions.

Cumulative Review
1. Find the length of a leg of an isosceles right triangle if the length of its hypotenuse is 10.
2. True or false? The bisector of an angle of an obtuse triangle divides the opposite side into segments proportional to the other two sides.
3. Why is a convex polygon not a convex set?

1. $5\sqrt{2}$ 2. True 3. The segment joining two points of the polygon contains points not on the polygon.

10.6 Trigonometric Tables (pages 324–325)

Background
The trigonometric tables used in this textbook give the function values for whole degree measures rounded to four decimal places.

Suggestions
Give your students adequate practice in reading the trig tables before working problems applying the ratios. Some minicalculators have access keys for the trigonometric functions. Do not allow your students to use their minicalculators for this purpose. Students are responsible for reading and using tables.

For Additional Chalkboard Work
For right triangle ABC, cotangent A is defined as follows:

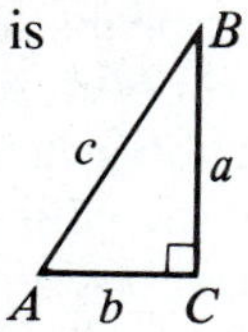

$$\cot A = \frac{\text{length of adj leg}}{\text{length of opp leg}}$$

Show that $\cot A = \frac{1}{\tan A}$.

Enrichment
Some of the exercises in sections B and C anticipate basic ideas in trigonometry. Have students prove which of the following are true for all values of $\angle A$ and $\angle B$, the acute angles of any right triangle.

1. $(\sin A)^2 + (\sin B)^2 = 1$
2. If $m \angle a > m \angle B$, then $\cos A > \cos B$
3. $\tan A = \frac{\sin A}{\cos A}$
4. $\tan B = \frac{1}{\tan A}$

Cumulative Review
1. What is a ratio?
2. True or false? The diagonal of a parallelogram bisects an angle of the parallelogram.
3. What is the tangent of an angle?

1. A quotient of two numbers with denominator $\neq 0$ 2. False 3. The ratio of the lengths of the opposite side and adjacent side in a right triangle containing that angle

10.7 Measures in Right Triangles (pages 326–328)

Background

This is a simple introduction to trigonometry. Significant digits and precision of answers are not covered. In a course in trigonometry, the student will learn that if the measures of sides are accurate to two digits, the measures of angles are accurate to the nearest degree; sides to three digits, then angles to the nearest 10 minutes; sides to four digits, then angles to the nearest minute.

Suggestions

Make sure the students are proficient in setting up trigonometric ratios and in using the tables before they begin to work with computational problems. Review algebraic strategies with proportions if needed. Many of the problems involve time-consuming computation. If you allow your students to use minicalculators, have them show the equations they set up and the values they get from the table. Review rounding of decimal fractions. The exercises require the measures of sides rounded to tenths and of the angles to the nearest degree.

Most of the problems can be solved with a choice of functions or angles. Encourage students to use the function that they find easiest to use. Some students may discover that the co-named functions (sine and cosine) of complementary angles are equal. This is a very useful relationship in trigonometry.

For Additional Chalkboard Work

If $a = 4$ and $b = 6$, find m $\angle B$ to the nearest degree.

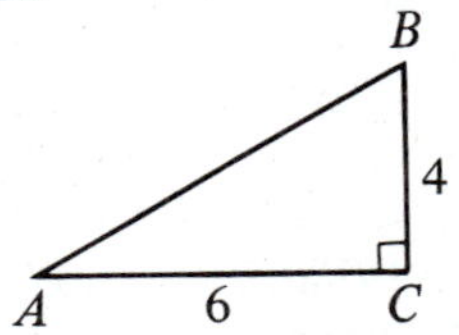

$\tan B = \frac{b}{a} = \frac{6}{4} = 1.500;$

$\tan 56° \doteq 1.483$; Thus, m $\angle B \doteq 56$.

Cumulative Review

1. What is the measure of each interior angle of a regular pentagon?
2. Find sin 50°.
3. Find the length of an altitude of an equilateral triangle with sides 100 cm long.

1. 108°   2. .766   3. $50\sqrt{3}$ cm

10.8 Applications of Trigonometry (pages 329–331)

Vocabulary

angle of elevation, angle of depression

For Additional Chalkboard Work

A 15-m ladder must not form an angle greater than 68° with the ground to be considered safe. What is the maximum height such a ladder will safely reach on a wall?

$\sin 68° = \frac{x}{15}$ $0.9272 \doteq \frac{x}{15}$

$0.9272 \cdot 15 \doteq x$ $13.9 \doteq x$

14 m

An airplane pilot sights a ship at an angle of depression of 27°. The plane is at an altitude of 900 m. How far is the plane from the ship?

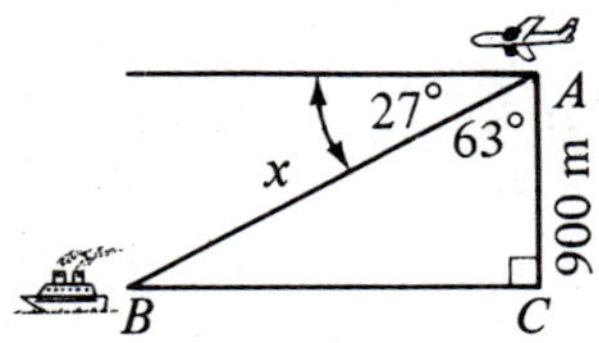

$\cos 63° = \frac{900}{x}$

$0.4540 \cdot x \doteq 900$

$x \doteq \frac{900}{0.4540} \doteq 1982$

$\doteq 2{,}000$ m

Cumulative Review

1. The tangent ratio involves which two sides of a right triangle?
2. Can a pentagon and a hexagon ever be similar?
3. What does the parallel postulate say?

1. Side opposite and side adjacent to the given angle   2. No   3. Through a point not on a line there is only one line parallel to it.

Computer Activity (p. 333)

The accuracy of the answers to these problems can be increased by incorporating into this program the method of rounding used on page 301. Try it rounded to tenths or to hundredths.

11 Chapter and Lesson Commentaries

Chapter 11 Circles and Lines

Chapter Objectives
To identify segments and lines related to circles
To prove and apply theorems about radii, chords, and tangents to circles
To find degree measures of minor arcs, major arcs, and semicircles
To identify concentric circles
To prove and apply relationships among arcs, chords, and central angles
To apply measurement properties for inscribed angles
To apply measurement properties for angles formed by secants and/or tangents
To find ratios and products of lengths of segments related to a circle
To perform constructions related to circles

Chapter Overview
Basic lines and segments of a circle—a radius, chord, diameter, secant, and tangent—are introduced, and the relationships among these various segments and lines are explored. Properties of tangents are considered. Special emphasis is given to the relationship between a tangent and the radius drawn to the point of tangency. This relationship is used to justify the construction of a tangent to a circle from a point in the exterior of the circle.

Theorems related to congruent circles, their arcs, chords, and central angles are proved.

Chapter Overview cont'd.
Methods for finding the measure of arcs and angles formed by chords, secants, and tangents are developed. Pairs of similar triangles are used to show the relationship between the product of lengths of segments. Inscribed and circumscribed polygons are introduced, and the construction of an inscribed regular hexagon is shown. The circle is used to construct proportional line segments.

Special Features
Benjamin Banneker p. 337
Reading in Geometry pp. 340, 343, 348, 361
Cumulative Review pp. 341, 350, 354, 363, 376
Non-Routine Problems p. 367
Calculator Activities p. 372
Applications p. 380
Computer Activities p. 381
Chapter Review p. 382
Chapter Test p. 383
College Prep. Test p. 384

Benjamin Banneker (p. 337)
You may want to point out to your students that the term *ephemeris* means that Banneker's almanac contained astronomical information such as positions of heavenly bodies on particular dates. Encourage interested students to do additional research on the life of Benjamin Banneker.

11.1 Circles and Lines (pages 338–341)

Vocabulary

circle, center, radius, chord, diameter, secant, tangent

Suggestions

Stress the importance of the basic definitions related to circles. Theorems and problems throughout the chapter will be dependent on them. Have students select and label a point *c* in the plane of a sheet of paper. Have them draw in all the points in the plane 4 centimeters from *c*. Define a circle. Discuss the basic definitions.

For Additional Chalkboard Work

1. A chord 16 mm long is 15 mm from the center of a circle. Find the length of the radius. 17

Enrichment

$\widehat{RST}$ is called a circular arch. $\overline{RT}$ is the span and $\overline{SQ}$ is the altitude of the arch. Construct a scale drawing of a circular arch with a span of 18 ft and an altitude of 6 ft. Use the scale $\frac{1}{8}$ in. = 1 ft.

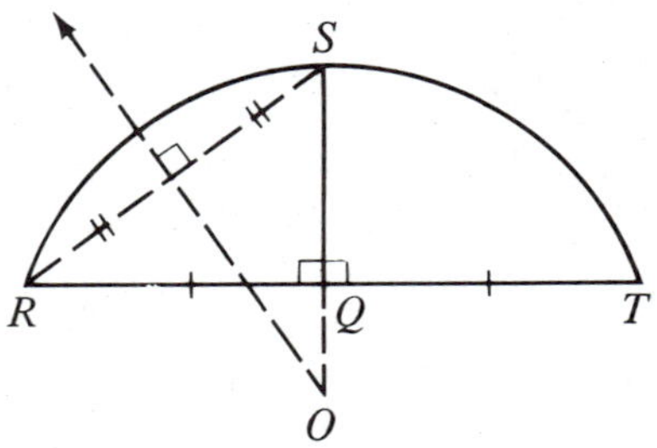

Cumulative Review

1. In right triangle ABC with rt. $\angle C$, if $\sin A = 0.298$ and $AB = 10$, find BC.
2. What is the relationship of the length of the altitude to the hypotenuse of a right triangle and the segments of the hypotenuse formed by the altitude?
3. If the diagonals of a quadrilateral bisect each other, what special quadrilateral is it?

1. 2.98 2. The length of the altitude is the geometric mean. 3. Parallelogram

11.2 Congruent Circles (pages 342–344)

Vocabulary

congruent circles, inscribed polygon, circumscribed circle

Suggestions

Introduce Theorem 11.3. Ask your students if the theorem would still hold if the two congruent chords had been in the same circle. Make sure they understand that the theorem includes both cases. After this theorem has been proved, ask your students to form the converse, and to suggest ways to prove it, or assign the proof as Exercise 24.

For Additional Chalkboard Work

Have the students do the following examples to be sure they can apply the theorems; O is the center in each example.

1. $JL = 17$, $OK = 10$, $ON = 10$. Find MG
   17

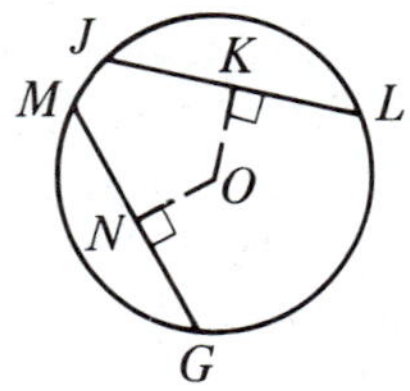

2. $\triangle ABC$ is isosceles. $\overline{AB} \cong \overline{BC}$, $PO = 5$. Find RO. 5

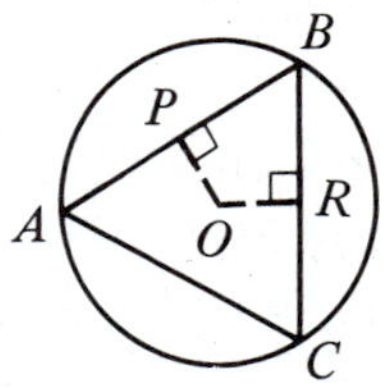

Cumulative Review

1. What is a circle?
2. Do perpendicular rays have to intersect each other?
3. Find the measure of each exterior angle of a regular decagon.

1. The set of points equidistant from a given point in a plane 2. No 3. 36°

11.3 Properties of Tangents (pages 345–350)

Vocabulary

point of tangency, tangent segment, circumscribed polygon, inscribed circle, common tangent, tangent circles

Suggestions

The proof of Theorem 11.5 is indirect. Review the steps in an indirect proof.

For Additional Chalkboard Work

Given: $\overline{PQ}$ is tangent to $\odot O$, $PQ = 12$, and $PR = 8$
Find: OQ

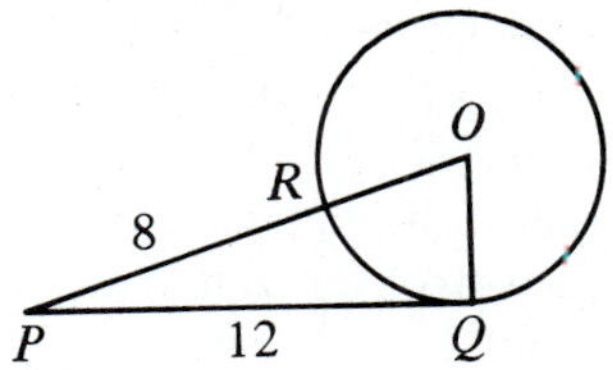

Let $OQ = x$, then $OR = x$ and $OP = x + 8$. Use the Pythagorean theorem to find $OQ = 5$.

Enrichment

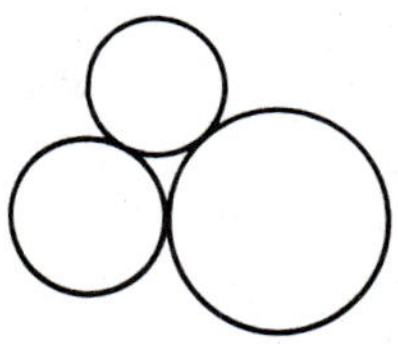

This shows three circles, each tangent to the other two. Ask students to show other arrangements of three circles, each tangent to the other two. (See below.)

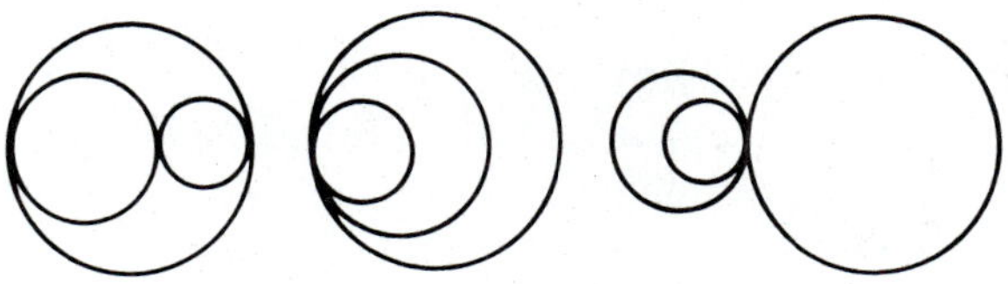

Cumulative Review

1. What is a secant of a circle?
2. If an 8 in. chord is 3 in. from the center of a circle, find the length of a radius.
3. If two triangles are congruent, are they similar?

1. A line intersecting the circle in two points. 2. 5 in. 3. Yes

11.4 Measures of Arcs (pages 351–354)

Vocabulary

arc, central angle, minor arc, major arc, semicircle, concentric circles

Background

Ancient Babylonians were the first to divide the circumference of a circle into 360 equal parts. The Babylonian measure of long distances was roughly 7 of our miles, or 11 kilometers. A traveler was able to walk this *Babylonian mile* in one-twelfth of a day. For convenience, the Babylonian mile had been divided into 30 equal parts. So there were 12×30 or 360 equal parts in one day.

Suggestions

Give students practice in identifying minor arcs, major arcs, and semicircles. Then define the measure of a minor arc to be the measure of its central angle and develop the methods for finding the degree measures for major arcs and semicircles.

The Arc Addition Postulate is comparable to the Angle Addition Postulate. Make sure your students understand that it is necessary for point P to be on $\widehat{AB}$.

For Additional Chalkboard Work

Given $\odot O$ is tangent to $\odot P$ at T, and $\overline{RS}$ is tangent to $\odot O$ at R and to $\odot P$ at S. $OR = 9$ and $PS = 16$. Find RS. 24

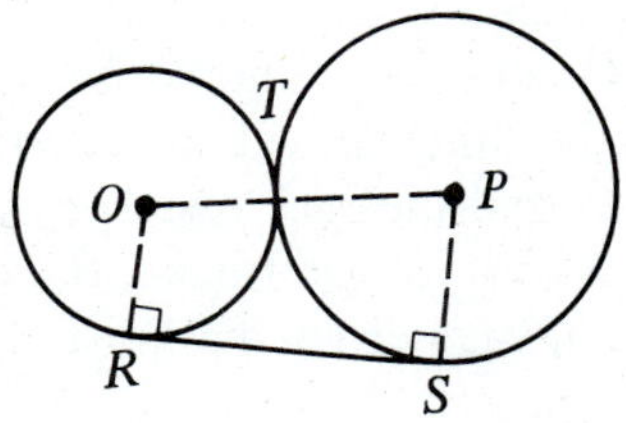

Cumulative Review

1. What is a radius of a circle?
2. What are congruent circles?
3. How many lines of symmetry does a square have?

1. A segment from the center to any point on the circle. 2. Circles with congruent radii. 3. 4

11.5 Arcs, Chords, and Central Angles (pages 355–358)

Suggestions

Point out that when it is stated that a chord determines an arc, the reference is to the minor arc. This terminology is used for chords, arcs, and central angles.

Demonstrate that it is not sufficient for congruent arcs to have the same degree measure. They must also be in the same or congruent circles.

Begin with diagrams of two noncongruent circles. Measure a 50° central angle in each, using colored chalk to highlight the intercepted arcs. Ask your students if the two arcs have the same measure.

For Additional Chalkboard Work

In the figure below, both $\angle x$ and $\angle y$ determine chord $\overline{AB}$. Explain why $\angle x \not\cong \angle y$.

The circles are not congruent.

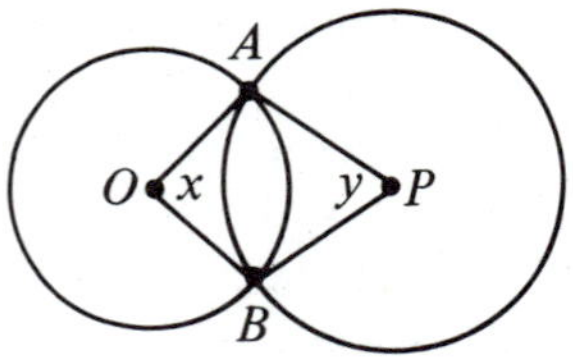

Enrichment

The cycloid is an interesting curve to study. It is a curve traced by a point on the circumference of a circle as the circle rolls along a straight line. Let some students build two tracks on which a marble or ball bearing could roll down. One of these should be a 90° arc of a circle, the other a cycloid of the same length. If two marbles are released simultaneously at the top of the tracks, the one on the cycloid will always cross the finish line first.

Cumulative Review

1. What is the difference between a chord and a secant of a circle?
2. In how many points does a tangent to a circle intersect the circle?
3. What are the possible intersections of two coplanar segments?

1. A chord is a segment joining two points of the circle. A secant is a line intersecting the circle in two points. 2. 1 3. No intersection, a point, or a segment

11.6 Inscribed Angles (pages 359–363)

Vocabulary

inscribed angle

Suggestions

Use a guided discovery lesson to help students find the relationship between an inscribed angle and its intercepted arc. Make sure the students know the difference between a central angle and an inscribed angle. Prepare a worksheet with five circles, each with an inscribed angle. Have the centers of the circles labeled. Let two of the angles intercept minor arcs, two intercept major arcs, and one intercept a semicircle. This will insure having acute, obtuse, and right angles. Give the students time to measure the angles and arcs and record the results on a chart provided on the worksheet. Some students may need to be reminded that in order to measure the arcs, they must first draw in the central angles. When measuring major arcs, they must remember to subtract the measure of the minor arc from 360°.

For Additional Chalkboard Work

In the figure below, explain why $m\,\angle x = \frac{1}{4}\,m\,\angle z$.

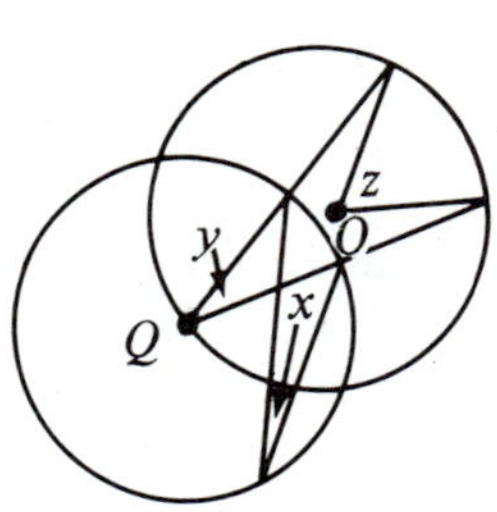

$m\,\angle y = \frac{1}{2}\,m\,\angle z$

$m\,\angle x = \frac{1}{2}\,m\,\angle y$

$m\,\angle x = \frac{1}{4}\,m\,\angle z$

Cumulative Review

1. How is the measure of a central angle of a circle determined?
2. Tell one way to prove that chords of a circle are congruent.
3. Find the measure of an altitude of an equilateral triangle whose sides measure 10 cm.

1. By the measure of its intercepted arc
2. Show they are the same distance from the center. 3. $5\sqrt{3}$ cm

11.7 Angles Formed by Secants (pages 364–367)

Suggestions

Review the definition of a secant. Ask each student to draw a circle with secants $\overleftrightarrow{RS}$ and $\overleftrightarrow{TV}$. Ask how the angle measures could be found if the arc measures are known. Ignore parallel secants, because no angles are formed. When two secants intersect on the circle, inscribed angles are formed and their measure is one-half the measure of the intercepted arc. This is Case 1 in the text. Discuss the case where two secants intersect in the interior of the circle. Select illustrations where the point of intersection is not the center of the circle (central angles).

Enrichment

In the figure below, the congruent circles are tangent to each other as shown. If a radius of each circle measures 1 unit, how many units long is $\overline{AB}$? $2\sqrt{3}$ units

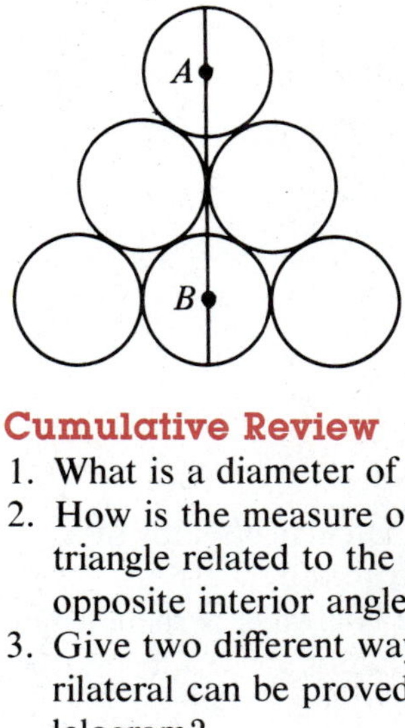

Cumulative Review

1. What is a diameter of a circle?
2. How is the measure of an exterior angle of a triangle related to the measure of the opposite interior angles of the triangle?
3. Give two different ways by which a quadrilateral can be proved to be a parallelogram?

1. A chord passing through the center
2. It is their sum. 3. Show both pairs of opposite sides congruent. Show that the diagonals bisect each other.

11.8 Tangents, Secants, and Angles (pages 368–372)

Background

The proofs of the theorems in this section require rather careful algebraic manipulation. Go over each of the steps of Theorem 11.14 as outlined in the text. Recall that a radius is perpendicular to a tangent at the point of tangency. Review the methods for finding the measures of central angles and inscribed angles if needed. Provide the proofs for Theorems 11.15 and 11.16 or assign them as Exercises 24 and 25.

Suggestions

Review the different kinds of angles related to circles and the methods for finding their measures. Bulletin board material or posters illustrating the different kinds of angles and their measures would be useful.

Ask the students to show the different ways a tangent and a secant or two tangents intersect. Encourage your students to make sketches.

For Additional Chalkboard Work

In the figure below, find m $\angle y$. 10

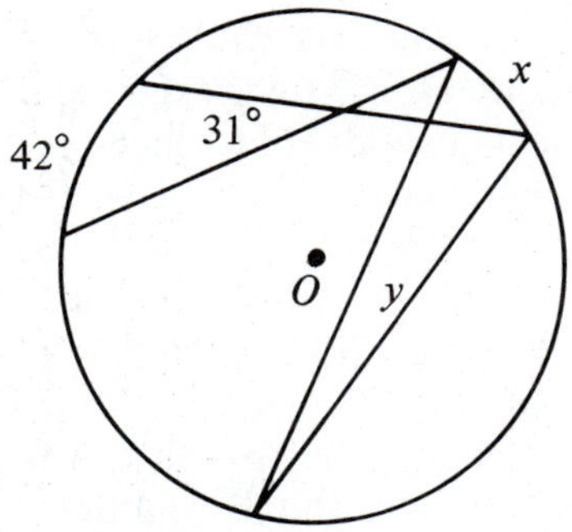

Cumulative Review

1. What is an inscribed angle?
2. If an exterior angle of a triangle measures 30°, what is the measure of the adjacent interior angle?
3. Tell two ways to prove lines are parallel.

1. An angle whose vertex lies on the circle and whose sides are chords of the circle.
2. 150° 3. Show a pair of alternate interior angles congruent. Show a pair of corresponding angles congruent.

11.9 Products of Lengths (pages 373–376)

Background

In this section, the ratios of the corresponding sides of similar triangles are used to find the lengths of segments when secants or tangents intersect. Students must be able to prove triangles similar and to set up the ratios of the sides of similar triangles. They must be proficient in the algebra used to solve proportions and be able to work with square roots in some exercises.

Suggestions

Begin by asking students to show the different ways two secants, two tangents, or a secant and a tangent can intersect. Make sketches of the possibilities on the chalkboard or make worksheets.

As you consider each case, introduce the terminology for identifying the segments you will be using.

Errors Students Make

Students frequently have difficulty determining which angles are congruent and which triangles are similar in Theorems 11.18 and 11.19. The use of colored chalk at the chalkboard or of overlays on the overhead projector can help. Provide practice examples using each theorem. Make sure the students can identify which segments are involved in each case.

Cumulative Review

1. How is the length of a radius of a circle related to the length of a diameter?
2. The median of a trapezoid has length 7 cm. If one base measures 6 cm, find the length of the other base.
3. The length of the altitude to the hypotenuse of a right triangle is the geometric mean between what other lengths?

1. $r = \frac{1}{2}d$ 2. 8 cm 3. The segments of the hypotenuse.

11.10 Constructions (pages 377–379)

Background

Karl Friedrich Gauss is often ranked as one of the three greatest mathematicians of all time. After discovering the construction of a regular 17-sided polygon at age 19, he decided to devote his life to the study of mathematics.

Suggestions

Begin by having students construct a semicircle with a diameter $\overline{AB}$ and to select any point P on the semicircle. Lines $\overleftrightarrow{PA}$ and $\overleftrightarrow{PB}$ will be perpendicular. Ask why. (An angle inscribed in a semicircle is a right angle.)

Enrichment

Students may prove that if $ABCDE$ is a regular pentagon, AFG is a regular triangle, and both are inscribed in circle $\odot$, then chord $\overline{FC}$ will be the side of a regular 15-gon inscribed in the circle. Students can use a similar construction for a regular 20-gon or 12-gon.

For Additional Chalkboard Work

Given: Tangent $\overleftrightarrow{TM}$ and secant $\overleftrightarrow{MP}$, $MR = 6$, $MU = 15$, and $MS = 4$. Find MP. $22\frac{1}{2}$

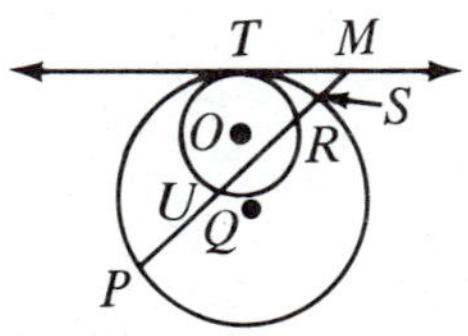

Cumulative Review

1. Construct the perpendicular bisector of a segment. See construction, p. 51.
2. A tangent to a circle is always perpendicular to which segment? Radius drawn to pt. of tangency
3. What are congruent circles? Circles with congruent radii

Computer Activity (p. 381)

This is the first program in which we have used anything as a variable other than a single letter. R1 and R2 are used to emphasize that both values are radii. These are, strictly speaking, not subscripted variables. In BASIC language, subscripted variables are written R(1), R(2), etc.

12 Chapter and Lesson Commentaries

Chapter 12 Areas of Geometric Figures

Chapter Objectives

To apply the area formulas for the following: rectangle, triangle, rhombus, parallelogram, trapezoid, regular polygon, circle

To find the perimeter of a polygon

To find ratios of areas and ratios of perimeters of similar polygons

To find lengths of segments related to a regular polygon

To apply the formula for the circumference of a circle and for the length of an arc

To find the area of a sector and a segment of a circle

To find the area of the ring bounded by two concentric circles

Chapter Overview

Basic area properties are presented together with the formula for the area of a rectangle. On the basis of these assumptions the area formulas are proved for right triangles, general triangles, equilateral triangles, rhombi, parallelograms, and trapezoids. Many applications are given for each of these formulas. In the calculator exercises, the formula for the area of a trapezoid is broken down for use with a calculator. Perimeter is introduced and ratios of areas and ratios of perimeters of similar polygons are applied. The relationship between the length of an apothem and the length of a radius in several regular polygons is discussed.

Chapter Overview cont'd.

The formula for the area of a regular polygon is developed. A regular polygon with an increasing number of sides is used to develop the concept of circumference and area of a circle.

Arc length is introduced as a part of the circumference of a circle and the area of a sector is introduced as a part of the area of a circle. The area of a segment of a circle and the area of a ring bounded by two concentric circles are also explored.

Special Features

Non-Routine Problem Solving p. 385
Non-Routine Problems pp. 386, 405
Algebra Review pp. 390, 417
Reading in Geometry pp. 396, 411
Calculator Activities pp. 401, 414
Cumulative Review pp. 401, 420
Applications p. 421
Chapter Review p. 422
Chapter Test p. 423
Computer Activities p. 424
College Prep Test p. 425
Cumulative Review (Chapters 1 – 12) p. 426

Non-Routine Problem Solving (p. 385)

The Chapter Opener deals with one of the most famous proofs of the Pythagorean Theorem, one introduced by the Hindu mathematician Bhāskara. Another proof by President Garfield is presented on page 386.

12.1 Areas of Rectangles and Squares (pages 387–390)

Vocabulary
polygonal region, circular region, area, square unit

Background
In this textbook we have chosen to take the area of a rectangle as a postulate. The area of a square then becomes an easy corollary, the special case in which both the base and height of the rectangle are of length *s*.

Suggestions
Polygons are the union of segments, and as such have no area.

Polygonal region is the union of a polygon and its interior. Polygonal regions have area. Because there is rarely any misunderstanding, the text uses the simple expression "area of a polygon."

When working with the area of a rectangle, point out that any side may be used as the base and that more than one altitude may be drawn to a base. The term *a corresponding altitude* refers to an altitude drawn to that base. The notation cm^2 is preferred to sq. cm.

The Area Addition Postulate provides a method for finding the area of irregularly shaped figures, which can be subdivided into regions whose areas can be determined. Remind the students the regions must be nonoverlapping.

Enrichment
Have students plan to cover the floor of the classroom with square tiles. (They must measure the floor, find its area, select the tile, and compute the costs of covering the floor.)

Cumulative Review
1. What are congruent quadrilaterals?
2. What is the interior of a polygon?
3. Show a union of segments that is not a polygon.

1. Corresponding sides are congruent and corresponding angles are congruent.
2. The set of points enclosed by the polygon
3. Answers may vary.

12.2 Areas of Triangles and Rhombi (pages 391–394)

Suggestions
Postulate 26 can be written in if-then form: "If two figures are congruent, then they have the same area." Ask the students to write the converse, "If two figures have the same area, then they are congruent."

Errors Students Make
Students frequently have trouble with altitudes that lie in the exterior of triangles. Example 1 covers all three cases, an altitude that coincides with a side of the triangle, one that is in the interior of the triangle, and one that lies in the exterior of the triangle. Discuss these three possibilities with your students. Stress that $\triangle HIJ$ is an obtuse triangle and that its altitude lies outside the triangle.

Make sure the students understand that any side of a triangle can be the base, as long as they use the altitude to that side. Students sometimes think that the base must be drawn horizontally or that the base must be the longest side.

For Additional Chalkboard Work
A leg of a right triangle measures 8 cm and the hypotenuse measures 17 cm. Find the area of the triangle. 60 cm^2

Cumulative Review
1. How many congruent angles does an isosceles triangle have?
2. How is a rectangle related to a parallelogram?
3. How are the triangles formed by the diagonal of a rectangle related?

1. 2 2. It is a parallelogram with a right angle. 3. They are congruent.

12.3 Areas of Parallelograms (pages 395–397)

Suggestions

A model constructed of wooden strips hinged with wing nuts is useful in helping to dispel the common notion that the area of a parallelogram is the product of two sides. Hold the model in various positions and ask the students if the resulting parallelograms have the same area.

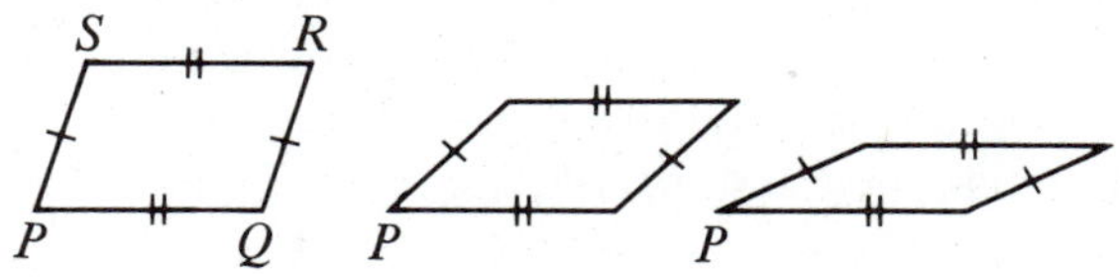

After proving the formula for the area of the parallelogram, use some examples that give the student a choice of bases and corresponding altitudes. Compute the areas of these two examples, first with one base, then with the other, and show that the resulting areas are the same.

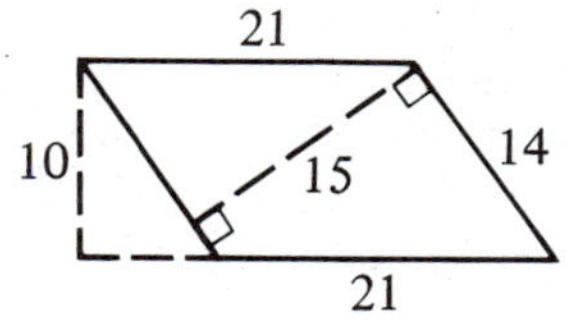

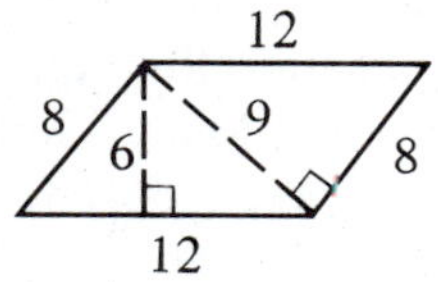

For Additional Chalkboard Work

1. *ABCD* is a rhombus with diagonals of lengths 10 cm and 24 cm. Find the length of $\overline{BC}$ and $\overline{BF}$.

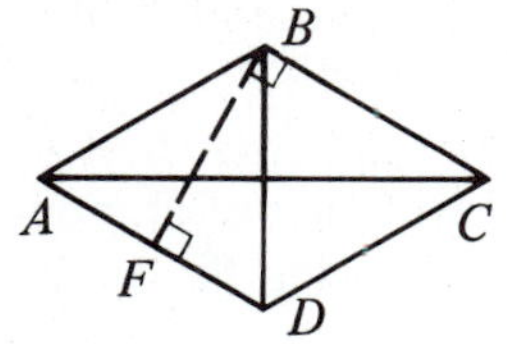

$BC = 13$, $BF = \frac{120}{13} = 9\frac{3}{13}$

2. Given: ▱*ABCD* with $m\angle A = 45$, $DC = 16$, and $AD = 10$. Find the area of *ABCD*. $80\sqrt{2}$

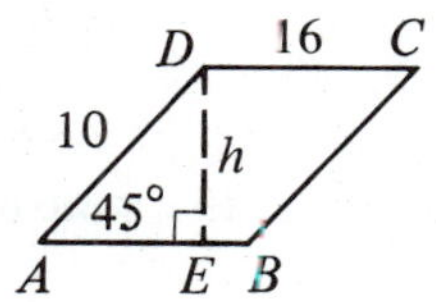

Cumulative Review

1. Describe two ways to prove that two triangles are congruent.
2. Find the length of a leg of a right triangle whose hypotenuse measures 17 and whose other leg measures 8.
3. How is the area of a rectangle found?

1. SSS, SAS, ASA, AAS 2. 15 3. Multiply length by width.

12.4 Areas of Trapezoids (pages 398–401)

Suggestions

Before developing the general area formula for trapezoids, introduce a specific case. Ask the students to find the area of a trapezoid *PQRS* with specific dimensions. Suggest they look for a way to subdivide it into regions whose areas can be found. If someone suggests introducing diagonal $\overline{PR}$, do so, and have the students find the areas of $\triangle SPR$ and $\triangle PQR$.

After finding the area for a specific trapezoid, substitute variables and develop the formula for finding the area of trapezoids in general.

For Additional Chalkboard Work

A lawn sprinkler covers an area that approximates the shaded area below. How much area does it cover? 72 ft^2

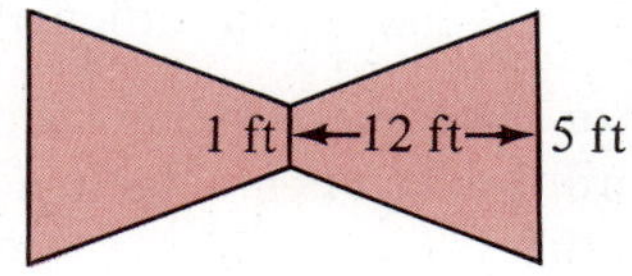

Enrichment

In the next section the students will find the ratio of areas of similar triangles. Have the students find the ratio of the areas of the triangles in the diagram below. Emphasize the fact that if two triangles have congruent bases and congruent altitudes, their areas are equal but the triangles need not be congruent. Encourage the students to write the ratios in fraction form.

Cumulative Review

1. True or false? Congruent triangles are similar.
2. What is a symmetric figure?
3. What is the formula for the area of a triangle?

1. True 2. A figure that has a line of symmetry 3. $A = \frac{1}{2}bh$

12.5 Similar Polygons: Areas and Perimeters (pages 402–405)

Vocabulary
perimeter

Suggestions
Begin this lesson with the following construction.

1. Given $\triangle ABC$, on $\overrightarrow{AB}$ construct D so that $\overline{BD} \cong \overline{AB}$. On $\overrightarrow{AC}$ construct E so that $\overline{CE} \cong \overline{AC}$. Draw $\overline{DE}$. $\triangle ABC$ and $\triangle ADE$ are similar. Ask the students for the ratio of the sides $\left(\frac{1}{2}\right)$, and for the ratio of the areas.
2. Have them construct M, the midpoint of $\overline{DE}$. Then draw $\overline{BM}$ and $\overline{CM}$.

The four triangles can be proved congruent and their areas equal. Students can easily see that the ratio of the areas is $\frac{1}{4}$. Extend the sides of the original construction, creating similar triangles with a $\frac{1}{3}$ and $\frac{1}{4}$ ratio of lengths of sides. Ask for the corresponding ratio of the areas. Record the results in a table and ask for the pattern.

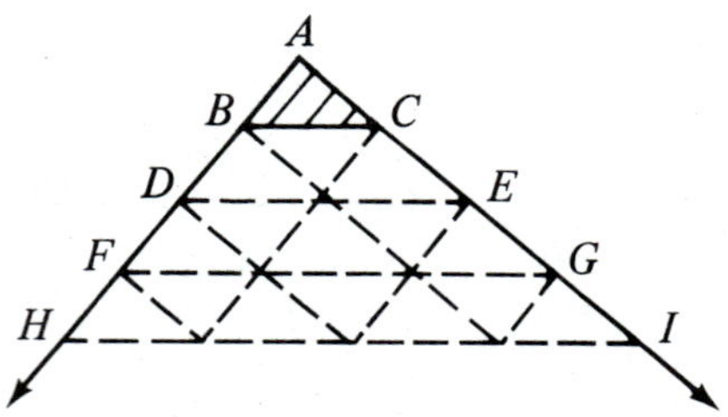

For Additional Chalkboard Work
Have students construct a scale drawing using the scale 1 cm = 2 m and then compute the actual area and the area of the scale drawing to show that Theorem 12.7 applies.

Cumulative Review
1. What are overlapping triangles?
2. Find the sum of the measures of the exterior angles of a hexagon.

1. Part of the interior of one triangle is contained within the interior of the other.
2. 360°

12.6 Areas of Regular Polygons (pages 406–408)

Vocabulary
center of a regular polygon, apothem of a regular polygon, radius of a regular polygon

Suggestions
Review the definition of a regular polygon, calling attention to the criteria of being both equilateral and equiangular. Diagrams of regular polygons for the overhead projector will be helpful. Use overlays to show the circumscribed and inscribed circles. Then introduce the new vocabulary words.

If needed, review the 45°–45° and the 30°–60° right triangle relationships. Now have the students find the perimeters for some polygons. Work out the formula for the perimeter of a regular polygon, $P = n \cdot s$, which will be needed in the proof for the area of a regular polygon. Then ask the students to find the area of a regular octagon with specific dimensions. If needed, suggest that they partition the octagon into regions whose areas can be found. Some students may point out that they need only to find the area of one of the triangles and multiply it by the number of sides. Encourage them to try this strategy on another regular polygon. Develop the formula for finding the area of any regular polygon.

For Additional Chalkboard Work
Each side of an equilateral triangle is 4 cm long. Find the length of an apothem (a) and the length of a radius (r).

$a = \frac{1}{3}h = \frac{2}{3}\sqrt{3}$; $r = \frac{2}{3}h = \frac{4}{3}\sqrt{3}$

Cumulative Review
1. Find the measure of each exterior angle of a regular octagon.
2. If the diagonals of a parallelogram are perpendicular, what special quadrilateral is it?
3. The construction of a tangent to a circle from a point outside the circle is based on which property of tangents?

1. 45° 2. Rhombus 3. A line perpendicular to a radius at its endpoint on the circle is tangent to the circle.

12.7 Circumferences of Circles (pages 409–411)

Vocabulary

circumference, π

Background

The classical method for computing π involved calculating the perimeters of regular inscribed and circumscribed polygons. As the number of sides increased, closer approximations for π were obtained. Archimedes computed the value of π to two decimal places. Vieta (French) used polygons of 393,216 sides in 1579 to find π to 9 decimal places, and in 1610, Van Ceulen (German) found π correct to 35 places using polygons having 2^{62} sides.

The symbol π was first used to denote the ratio of the circumference to the diameter by the English writer, William Jones, in 1706.

Suggestions

Emphasize the definition of π as the ratio of the circumference of *any* circle to the diameter of the same circle. Call attention to the fact that π is an irrational number and that 3.14 is only an approximation.

Enrichment

Six congruent circular discs with radii measuring 1 cm are placed as shown. A string is wound around them and pulled as tight as possible. How long is the string?

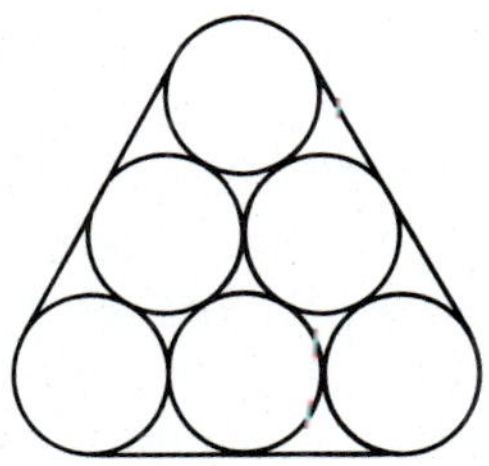

$2\pi + 12$ cm

Cumulative Review

1. What is the apothem of a regular polygon?
2. If two parallelograms have equal areas, are they necessarily congruent?
3. What is the cosine of a 30° angle?

1. The length of a radius of the inscribed circle. 2. No. 3. $\frac{\sqrt{3}}{2}$

12.8 Areas of Circles (pages 412–414)

Background

Introduce one of the paradoxes of Zeno (Greek, ca. 450 B.C.) in a discussion of limits. Zeno contended that one would never be able to walk out the door. For in order to travel the distance, one must first walk one-half the distance, then one must walk one-half the remaining distance, and so on.

Use the formula for the area of a regular polygon to derive the formula for the area of a circle. Suggest that *n,* the number of sides of the regular polygon, be increased without limit. Ask the students to determine what happens to the other variables in the formula ($a \rightarrow r$ and $p \rightarrow c$). Introduce the formula for the area of a circle.

Discuss the strategy for finding the area of a region between circles and/or polygons.

For Additional Chalkboard Work

The circumference of a circle with a 2-cm radius is equal to the perimeter of a square. How long is each side of the square? Find the area of each figure. Which has the greater area? Give answers to the nearest hundredth.

3.14 cm; 12.56 cm^2 and 9.86 cm^2; the circle

Enrichment

One of the three famous problems posed by the Greeks, "the quadrature of the circle," was the problem of constructing (with compass and straightedge) a square having the same area as a given circle. Although the construction was proved impossible in the nineteenth century, pose this problem to the students and encourage them to analyze the requirements (construction of a segment whose length is $\sqrt{\pi}$).

Cumulative Review

1. How is π defined?
2. In a circle, chords $\overline{AB}$ and $\overline{CD}$ intersect at E. If $AE = 3$, $BE = 4$, and $CE = 6$, find DE.
3. What is a central angle of a circle?

1. The ratio of the circumference of a circle to the length of a diameter. 2. 2 3. An angle with vertex at the center.

12.9 Arc Lengths (pages 415–417)

Suggestions

Review the concept of arc measure, reminding the students that arcs are measured in degrees. This lesson deals with arc length, or the number of linear units in the curve. Students frequently confuse the degree measure of arcs with the length of arcs.

Draw a circle on the chalkboard and assign its radius a length of 8 cm. Ask your students to find its circumference. (16π cm) Shade in a semicircle and ask your class to find the length of the semicircle. (8π cm) Draw in a 90° central angle and shade its intercepted arc. Ask your students to find the length of the arc. (4π cm)

Use another simple example, a circle with a radius of 12 m and an arc measure of 60°. Organize the results in a chart such as this:

arc measure	*fractional part*	*circumference*	*arc length*
180°	$\frac{1}{2}$	24π	12π
90°	$\frac{1}{4}$	24π	6π
60°	$\frac{1}{6}$	24π	4π

Establish the pattern for multiplying the fractional part of the circumference and the circumference to find the arc length. Provide problems requiring more complicated fractions.

For Additional Chalkboard Work

A dog is tied at the square corner of a fenced-in yard. The rope is 6 ft long. Find the perimeter of the plot of land in which the dog can move. 21.42 ft

Cumulative Review

1. What is the circumference of a circle with an 8-in. radius?
2. Find the area of a circle with an 8-in. diameter.
3. If you were to forget the formula for the area of a trapezoid, how might you figure out the area?

1. 16π 2. 16π 3. Use a diagonal to divide the trapezoid into two triangles.

12.10 Areas of Sectors and Segments (pages 418–420)

Vocabulary

sector of a circle, segment of a circle

Suggestions

In defining sector of a circle, draw examples bounded by major arcs as well as minor arcs. Use a lesson plan similar to the one in section 12.9 to derive a method for finding the areas of sectors.

Enrichment

For any right triangle, the area of the square on the hypotenuse equals the sum of the areas of the squares on the legs. Ask students to prove an analogous theorem for circles: For any right triangle, the area of the semicircle on the hypotenuse equals the sum of the areas of the semicircles on the legs.

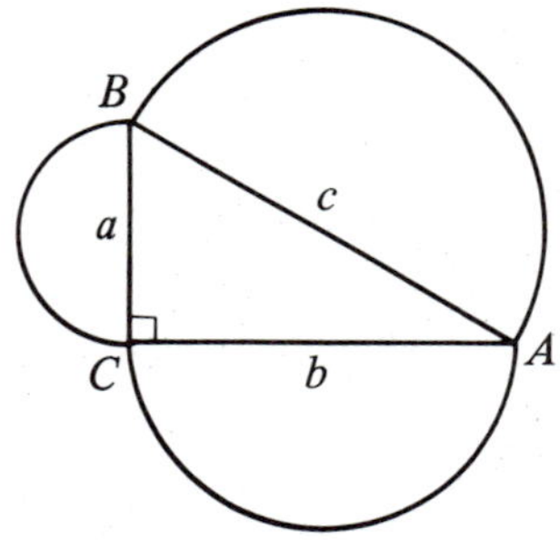

Cumulative Review

1. If a circle has a radius of length 10, what is the area of the region bounded by a semicircle and a diameter?
2. True or false? All arcs with the same degree measure are congruent.
3. If two triangles are similar, what is known about the lengths of their sides?

1. 50π 2. False 3. They are in proportion.

Computer Activity (p. 424)

This program gives students the opportunity to input their own answers and then check. The ON . . . GOTO command used in this program can be studied or reviewed on page 584.

13 Chapter and Lesson Commentaries

Chapter 13 Coordinate Geometry

Chapter Objectives

To graph ordered pairs of numbers in a coordinate plane

To determine if a line is vertical or horizontal, given the coordinates of two of its points

To apply Midpoint and Distance formulas

To find the slope of a line and determine the position of the line from its slope

To draw the graph of a linear equation of the form $ax + by = c$

To draw a line and write its equation in standard form, given the coordinates of one point and the slope of the line

To draw a line and write its equation in standard form, given the coordinates of two points on the line

To draw a line and write its equation by using its slope and y-intercept

To determine if lines are parallel or perpendicular from their slopes

To use coordinate geometry to write proofs

To apply equations of circles and to write an equation

To find the slope of a line parallel or perpendicular to a given line

Chapter Overview

The coordinate plane and the terminology related to the coordinate plane are presented in this chapter. The Distance formula is developed as an application of the Pythagorean theorem. The Midpoint formula is developed and applied.

Special Features

Problem Solving Formulation (p. 428)

In many practical applications involving area, the regions involved unfortunately do not fit the standard patterns for which area formulas have been developed. Land being surveyed is seldom in the form of rectangular regions or even trapezoidal regions.

The areas involved in the aerial photograph problem can be approximated several different ways. The trapezoidal rule can be used. Most forest rangers would use a standard transparent overlay with squares drawn to a predetermined size. They would then simply count the number of squares on the the grid that fall within the boundaries of the plot of land. Many of these grids are spaced so that each square represents one acre. Pick's formula could also be used. Using a geoboard, any figure that can be made by stretching a rubber band on the nails has an area $A = \frac{1}{2}N - 1 + N_i$, where N is the number of nails that the rubber band actually touches, and N_i is the number of nails on the board in the interior of the figure. If a dot grid overlay were used on the aerial photograph, its boundaries could be approximated by such a polygon.

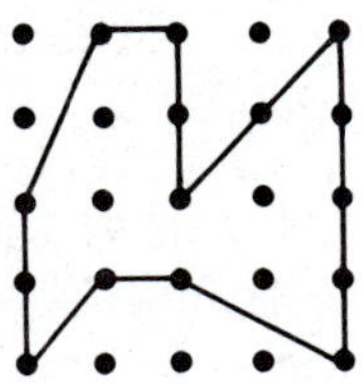

$N = 15$

$N_i = 4$

$A = \frac{1}{2}(15) - 1 + 4$

$A = 10\frac{1}{2}$

13.1 The Coordinate Plane (pages 429–431)

Vocabulary

coordinate plane, origin

Background

The French mathematician and philosopher, René Descartes (1596–1650), is one of the founders of modern analytic geometry. Due to poor health in his early life, Descartes spent long hours in bed. As he lay in bed one morning watching a fly on the ceiling, he worked out a scheme for locating the fly's exact position with an ordered pair of numbers. He realized he could use an equation to describe the path of the fly. Descartes, together with Pierre Fermat, was the first to establish a correspondence between plane curves and algebraic equations, and thus to reduce geometry to algebra and analysis.

Suggestions

Review the basic vocabulary for graphing and make sure the students are able to plot points and identify quadrants. Examples 2 and 3 emphasize the common properties of points on the same horizontal or vertical line. These are important and should be stressed.

For Additional Chalkboard Work

A (−2, 5) and B (4, 5) are the endpoints of a side of an equilateral triangle. Give the coordinates of two possible points for the third vertex. $(1, 5 + 3\sqrt{3})$; $(1, 5 - 3\sqrt{3})$

Cumulative Review

1. True or false? Coplanar points are points on the same line.
2. What is an angle?
3. If the diagonals of a rhombus measure 10 in. and 24 in. respectively, find the length of a side.

1. False 2. The union of two rays with a common endpoint. 3. 13 in.

13.2 The Distance Formula (pages 432–434)

Suggestions

A grid that can be used at the chalkboard is a necessity for this chapter. A grid can be made on a transparency for the overhead projector and projected on the chalkboard.

In Chapter 1 the distance between two points A and B with coordinates a and b was found to be $|b - a|$ or $|a - b|$. Use this to establish the formulas for finding the lengths of horizontal and vertical segments. Review the relationship between distance and length, that is, the *distance* between two points is the *length* of the corresponding segment.

Graph two points such as $P(2, 1)$ and $Q(6, 4)$ and ask your class to find the distance PQ. Lead the students to use the Pythagorean theorem.

Introduce the standard variables (x_1y_1) and (x_2y_2) to represent the coordinates for P and Q. Develop the general formula for finding the distance between $P(x_1y_1)$ and $Q(x_2y_2)$. Review subtraction of integers and simplification of radicals if necessary.

Enrichment

The roads in Smithville are laid out in a rectangular grid and are numbered consecutively. Avenues run north and south and streets run east and west. How many blocks must you walk to get from the corner of 8th Avenue and 53rd Street to the corner of 3rd Avenue and 41st Street? How many blocks apart are the two corners if you could walk through the blocks? 17 blocks; 13 blocks

Cumulative Review

1. Draw a segment. Construct its midpoint.
2. Find the coordinate of the midpoint of the segment whose endpoints have coordinates −3 and 9.
3. What is a trapezoid?

1. See p. 10. 2. 3 3. A quadrilateral with exactly two parallel sides.

13.3 The Midpoint Theorem (pages 435–437)

Background

Pierre Fermat, a contemporary of Descartes, is considered to be a cofounder of analytic geometry. Fermat worked for the parliament at Toulouse, but spent his leisure time pursuing mathematics as a hobby. He published very little, but carried on extensive correspondence with the mathematicians of his day. Our knowledge of his mathematical discoveries comes from this correspondence.

Descartes usually began with a locus and then found its equation, whereas Fermat began with an equation and studied its locus. Both approaches are aspects of analytic geometry.

Suggestions

To develop the midpoint formula, start with graphs of horizontal and vertical segments. Let students find the midpoints first by counting. Then encourage them to look for a method they could use given only the coordinates of the endpoints. After they have developed a method for horizontal and vertical segments, graph some oblique segments and have the class test their methods. Include some examples with negative coordinates.

If students need help, ask them how to find the number halfway between two numbers. Draw the analogy of the coordinates of the midpoint as being the numbers halfway between the *x*-coordinates and *y*-coordinates of the endpoints.

Enrichment

If a hen and a half lays an egg and a half in a day and a half, how many eggs can 6 hens lay in 6 days? 24

Cumulative Review

1. True or false? If m $\angle A = 35$ and m $\angle B = 35$, then $\angle A \cong \angle B$.
2. Use a protractor to draw an angle whose measure is 147°.
3. The altitude to the hypotenuse of a right triangle is drawn. What is the relationship between the triangles formed?

1. True 2. See p. 12. 3. They are similar.

13.4 Slope of a Line (pages 438–441)

Vocabulary

slope of a line

Suggestions

In developing the concept of slope, the expression slope $= \frac{\text{rise}}{\text{run}}$ can be used. This reminds the user that the *y*-distance is the numerator and the *x*-distance is the denominator. In working with slopes, *the most common student error* is interchanging the *x*- and *y*-distances in the slope formula. Draw a large set of axes and graph lines containing the origin with slopes of 1, 2, 3, 5, and 10, and $\frac{1}{2}, \frac{1}{3}, \frac{1}{5}$, and $\frac{1}{10}$. Students will see that the slope is the steepness of the line; the greater the slope, the steeper the line.

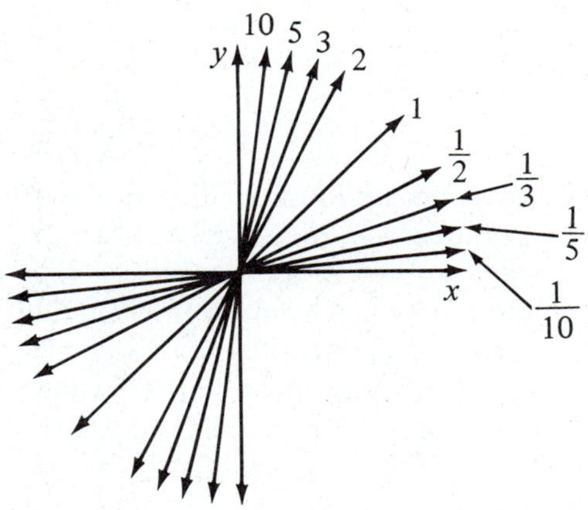

For Additional Chalkboard Work

Present the problem of constructing a ramp to cover a set of stairs in the school building. (To find the slope of the ramp form the ratio of the number of inches of vertical rise to the number of inches of horizontal run for one of the stairs).

Cumulative Review

1. Write the inverse: If an angle is obtuse, then its supplement is acute.
2. $\overrightarrow{BD}$ bisects $\angle ABC$. m $\angle ABD = 7x + 5$ and m $\angle DBC = 6x + 11$. Find m $\angle ABD$.
3. Draw right triangle ABC with m $\angle C = 90$, $AB = 6$, and m $\angle A = 45$. Find AC.

1. If an angle is not obtuse, then its supplement is not acute.
2. 47 3. $3\sqrt{2}$

13.5 Equation of a Line (pages 442–445)

Vocabulary

linear equation

Suggestions

Stress the basic relationship between a graph and its equation. The graph is the set of points associated with the ordered pairs that satisfy the equation. The graph of a linear equation is a line.

Errors Students Make

Introduce graphing a linear equation, given the slope and the coordinates of a point on the line.

Students tend to become confused about the direction in which to move. Stress that the numerator of a slope ratio tells the vertical direction in which to move, whereas the denominator tells the horizontal direction. (See Example 2.) For a line with slope of $-\frac{a}{b}$, remind students that $-\frac{a}{b}$ is equivalent to either $\frac{-a}{b}$ or $\frac{a}{-b}$, but not $\frac{-a}{-b}$. Tell students that on a graph we count up for a positive rise and down for a negative rise, to the right for a positive run and to the left for a negative run.

Develop the point-slope form of the equation.

Give the class the coordinates of two points and ask them to write the equation that would contain the two points. Ask students to write the equation given the slope and one point.

For Additional Chalkboard Work

Show that the slope of a line that makes an angle of 45° with the y-axis is either 1 or -1.

Cumulative Review

1. Draw a pair of adjacent complementary angles. If one of the angles measures 35°, what is the measure of the other angle?
2. True or false? An angle is obtuse if its measure is less than 180°.
3. True or false? Two adjacent angles cannot be vertical angles.

1. 55° 2. False 3. True

13.6 Slope–Intercept Form of an Equation (pages 446–448)

Vocabulary

y-intercept, slope-intercept form of a linear equation

Suggestions

Take an equation in standard form such as $2x - 5y = 15$, and graph it by determining two (or more) ordered pairs that satisfy the equation. Introduce the term y-intercept. Ask the students to solve the equation for y $\left(y = \frac{2}{5}x - 3\right)$. See if they can spot a relationship between the equation written in this form and the graph (the y-intercept is -3 and the slope $\frac{2}{5}$). Try another example such as $3x + 4y = 8$, where the slope is negative and the y-intercept is positive. Work out the general case using the procedure presented in the textbook. Point out to the students that it is easier to draw the graph of a line if the slope and a point on the line can be read directly from the equation than when the equation is in standard form.

For Additional Chalkboard Work

Explain why the line with the equation $y = 3x - 4$ and the line with the equation $y = \frac{1}{2}x - 4$ will intersect at the point with coordinates $(0, -4)$.

Both lines have the same y-intercept, −4. Thus both pass through (0 − 4).

Enrichment

Have students prove that the equation $\frac{x}{a} + \frac{y}{b} = 1$ is a linear equation of a line with y-intercept b and x-intercept a, $a \neq 0$, $b \neq 0$.

Cumulative Review

1. Find the measure of an angle if its measure is 15° less than that of its complement.
2. Find the measure of an angle if its measure is five times that of its supplement.
3. In a circle, an inscribed angle measures 42°. What is the measure of its intercepted arc?

1. $37\frac{1}{2}°$ 2. 150° 3. 84°

13.7 Slopes of Parallel and Perpendicular Lines (pages 449–452)

Suggestions

Introduce the theorems on parallel and perpendicular lines by having students graph pairs of parallel and perpendicular lines and investigate the relationships between their slopes. Review reciprocals if necessary.

Have students select a point P at random and graph two lines containing point P with slopes of $\frac{-2}{5}$ and $\frac{5}{2}$. Ask them what appears to be true about the lines. Have them measure the angle formed with a protractor. Then have them graph lines containing any point Q with slopes of -4 and $\frac{1}{4}$. Use as many examples as you think are needed to show that the slopes of two lines are negative reciprocals if the lines are perpendicular.

Vertical lines are excluded in the statement of Theorem 13.6, since the slopes of vertical lines are undefined. Be sure students understand that any two vertical lines are parallel. Also, you will need to state that any vertical line is perpendicular to any horizontal line. Because the slope of a horizontal line is 0, which has no reciprocal, and a vertical line has no slope, they are not included in Theorem 13.9.

For Additional Chalkboard Work

Which of the following lines, if any, are parallel and which are perpendicular?

1. line a: $y = -2x + 5$
2. line b: $x - 2y = 6$
3. line c: $6x - 3y = 12$
4. line d: $8x - 4y + 1 = 0$

($a \perp b$, $c \parallel d$)

Cumulative Review

1. Draw a segment. Construct its perpendicular bisector.
2. Draw an angle. Construct its bisector.
3. Find the area of a trapezoid with bases 7 in. and 13 in. long and with an altitude 5 in. long.

1. See p. 51. 2. See p. 15. 3. 50 in.^{2}

13.8 Using Coordinates in Proofs (pages 453–455)

Suggestions

In this section students learn to use the methods of coordinate geometry in proofs. Such proofs are frequently shorter or simpler than traditional proofs. An appropriate placement of a polygon on the axes and the appropriate selection of the coordinates of the vertices of the polygon can simplify a coordinate geometry proof immensely. Spend time helping students to place a number of different polygons in the coordinate plane and to identify the coordinates of the vertices. Explain the necessity for using variables instead of numbers for the coordinates of the vertices. The diagram represents a general polygon, not a specific one. Help them set up a square, a rectangle, a general triangle, a right triangle, an isosceles triangle, an isosceles right triangle, and a parallelogram. The coordinates of the fourth vertex of the parallelogram will need to be explained.

Problems often require the student to find the midpoint of a segment. Suggest that in such problems, students set the coordinates of their points as $(2a, 2b)$ and $(2c, 2d)$, rather than (a, b) and (c, d). This avoids working with fractions.

Review simplification of square roots. A common error is to simplify $\sqrt{a^2 + b^2}$ as $a + b$. Have them show that $(a + b)^2$ is not equivalent to $a^2 + b^2$ and, therefore, $\sqrt{a^2 + b^2}$ is not equivalent to $a + b$.

For Additional Chalkboard Work

Show that an equation of the line containing all points that are equidistant from $P(a, b)$ and $Q(c, d)$ is $y - \frac{b + d}{2} = \frac{a - c}{b - d}\left(x - \frac{a + c}{2}\right)$.

Cumulative Review

$\angle AOB$ and $\angle COD$ are vertical angles.

1. If m $\angle AOB = 10x$ and m $\angle COD = 8x + 26$, find m $\angle AOB$.
2. If m $\angle AOB = 2x + 23$ and m $\angle BOC = 3x + 7$, find m $\angle BOC$.
3. If m $\angle COD = 6x + 33$ and m $\angle DOA = 9x + 12$, find the measure of $\angle AOB$.

1. 130 2. 97 3. 145°

13.9 Equation of a Circle (pages 456–458)

Suggestions

Select a point Q in the plane of the chalkboard. Ask your students to visualize the set of all points in the plane 3 units from the given point Q. They should be able to identify this set as a circle. Pick two points on the circle and ask the class what the points have in common. (They are each 3 units from Q.)

Superimpose a coordinate system on your point and circle. For example, let Q be (5, 4) and add the axes appropriately. Select a point $P(x, y)$ to represent any point on the circle and ask your class what would hold for any point P on the circle. (Distance from Q to P is 3.) Translate this into the equation: $PQ = 3$. Then use the distance formula to produce $PQ = \sqrt{(x-5)^2 + (y-4)^2} = 3$. Square both sides to get $(x-5)^2 + (y-4)^2 = 9$, the equation of the circle with the coordinates of the center (5, 4) and a radius of length 3.

Use a similar approach to derive the general equation for a circle with (h, k) the coordinates of the center and radius of length r.

Give students adequate practice both writing the equations, given the center and radius, and determining the center and radius, given the factored form of the equation.

Algebra students usually learn to complete the square when solving quadratic equations. Problems such as Example 4 and exercises 41–44 require this procedure.

For Additional Chalkboard Work

Give the coordinates of the point at which the line whose equation is $y = -5$ is tangent to the circle whose equation is $x^2 + y^2 = 25$.
(0, −5)

Cumulative Review

True or false?

1. Two lines that are not parallel are skew.
2. If two lines are intersected by a transversal, then the alternate interior angles are congruent.
3. In a plane, two lines perpendicular to the same line are parallel to each other.

1. False   2. False   3. True

13.10 Coordinates in Space (Optional) (pages 459–460)

Suggestions

Remind the students that in graphing on a number line only one coordinate is needed to describe the location of a point. When this was extended to graphing on a plane, two coordinates were necessary. Help the class to realize that a third coordinate is needed to locate a point in space.

Here is one way to give a depth perspective to three-dimensional graphs: To graph the point $P(2, 4, -3)$, first mark segments on the x- and y-axis of lengths 2 and 4, as in Figure 1. Then complete the parallelogram that has those segments as sides. From each vertex of the parallelogram, draw a vertical segment of −3 (shown 3 units). Complete the rectangular solid.

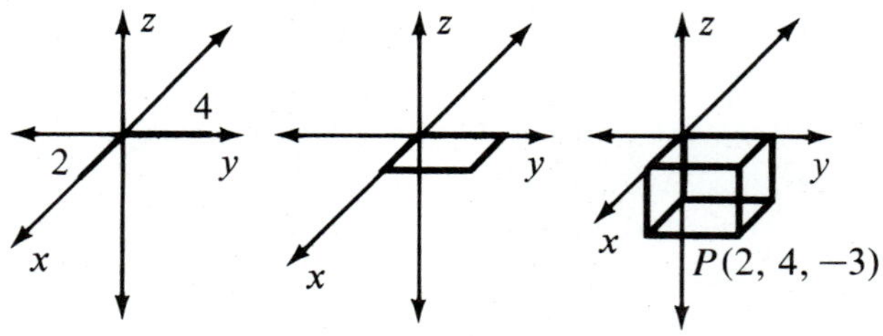

figure 1   figure 2   figure 3

For Additional Chalkboard Work

Graph each ordered triple on a three-dimensional coordinate system.

1. (4, 3, 1)
2. (−2, −2, −2)
3. (1, −1, 1)
4. (−3, 5, −4)

Cumulative Review

1. Find PQ. $P(5, 3)$ $Q(-2, 0)$
2. Find the coordinates of the midpoint of $\overline{PQ}$.
3. Find the slope of $\overline{PQ}$.

1. $\sqrt{58}$ 2. $\left(\frac{3}{2}, \frac{3}{2}\right)$ 3. $\frac{3}{7}$

Computer Activity (p. 464)

You may wish to have some students combine the given program and the program in Exercise 1, using an ON . . . GOTO statement to branch to the appropriate form of the program.

14 Chapter and Lesson Commentaries

Chapter 14 Geometric Loci

Chapter Objectives

To sketch and describe geometric figures that satisfy given conditions
To state theorems about loci and identify what is to be proved
To sketch and describe loci determined by more than one specific condition
To sketch and describe loci determined by two or more general conditions
To prove and apply concurrence theorems about perpendicular bisectors, angle bisectors, altitudes, and medians of a triangle
To sketch and describe loci in space

Chapter Overview

In this chapter the concept of locus of points is introduced. The locus of a point (1) at a given distance from a given point, (2) at a given distance from a given line, and (3) equidistant from two given parallel lines is described and introduced as postulates. The locus of a point (1) equidistant from two given points, and (2) equidistant from the sides of an angle is described and introduced as theorems. These postulates and theorems are then applied to loci problems. Loci determined by two compound loci or more specific conditions are discussed, and an optional lesson on loci determined by two or more general conditions is also given. The conic sections and their locus definitions are treated in Non-Routine Problems.

Chapter Overview cont'd.

Two lessons dealing with the concurrency theorems for the perpendicular bisectors, the angle bisectors, the altitudes, and the medians of a triangle are given, and students apply these theorems to constructions.

An optional lesson on loci in space is also included.

Special Features

Careers: Computer Graphics Artists p. 466
Cumulative Review pp. 469, 478
Reading in Geometry p. 489
Non-Routine Problems pp. 474, 482
Algebra Review pp. 473, 474, 482
Applications p. 493
Chapter Review p. 494
Chapter Test p. 495
Computer Activities p. 496
College Prep. Test p. 497

Careers: Computer Graphics Artists (p. 466)

Computer graphics can automate the drawing of schematics and eliminate much of the guesswork from design. This gives the designer more time for creativity. These displays show how the computer can automatically analyze lines, planes, and surfaces of a building and then project these images from any angle inside or out. Encourage interested students to do research on computer applications in design-related fields.

14.1 Locus of Points (pages 467–469)

Vocabulary

locus

Suggestions

Introduce locus by giving the conditions for a circle. Have students sketch all the points in the plane of their paper that are 5 cm from a given point *P*. Ask them to check for two things: (1) Do all the points sketched satisfy the condition? (2) Are these all of the points that satisfy the condition? Introduce the word *locus,* emphasizing that a locus meets both requirements. Have the students write a description of the curve that satisfies the condition.

In Exercises 9–12, students will be working with inequalities. The students probably graph inequalities such as $y > 2x - 3$ in algebra. This provides experience with graphing the locus of inequalities.

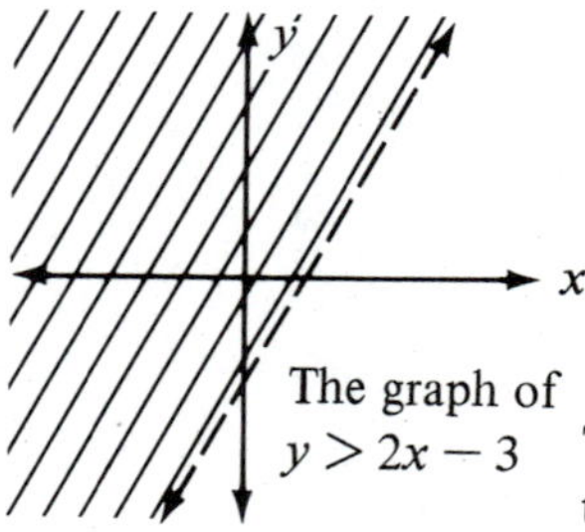

The graph of $y > 2x - 3$

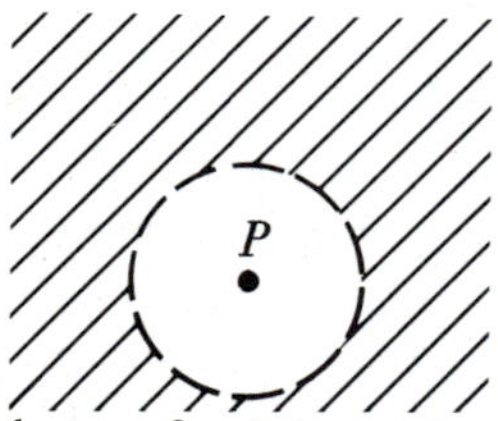

The locus of points greater than 5 cm from *P*

For Additional Chalkboard Work

Sketch and describe:

1. The locus of points 3 cm from a square whose sides are 5 cm each.
2. The locus of points 2 cm from the sides of a 45° angle.

Check students' drawings.

Cumulative Review

1. Draw a triangle that is both obtuse and isosceles.
2. In $\triangle ABC$, m $\angle A = 36$ and m $\angle B = 58$. Find the measure of exterior $\angle ACD$.
3. $\triangle EFG$ is isosceles with $EG = FG$. m $\angle E = 41$. Find m $\angle G$.

1. Answers may vary. 2. 94° 3. 98

14.2 Locus Theorems (pages 470–474)

Suggestions

Frequently students do not understand why it is necessary to prove both parts of Theorem 14.1, the perpendicular bisector theorem. Proving a locus theorem involves establishing the equality of two sets of points. Set $A =$ Set B if and only if Set A is a subset of Set B and Set B is a subset of Set A. For example,
Set $E =$ set of equilateral triangles
Set $F =$ set of equiangular triangles
$E \subseteq F$ because every triangle that is a member of Set E is also in Set F, and $F \subseteq E$ because every triangle that is a member of Set F is also in E. Therefore, Set $E =$ Set F.

In Theorem 14.1 the equality of the set of points equidistant from the endpoints of a segment and the set of points on the perpendicular bisector of the segment is to be proved.

In Theorem 14.2 the equality of the set of points equidistant from the sides of an angle and set of points on the angle bisector is proved.

In each case, start with any point in the first set and prove it must be a member of the second set. Then select a random point in the second set and prove it must be a member of the first set. This establishes the equality of the two sets.

For Additional Chalkboard Work

Write an equation for the locus of points 6 units from point $P(3, -1)$ in a coordinate plane.
$(x - 3)^2 + (y + 1)^2 = 36$

Cumulative Review

1. $\triangle ABC \cong \triangle DEF$. Which pairs of sides and which pairs of angles are congruent?
2. Draw a segment. Construct an equilateral triangle with sides of the same length as the segment.
3. Find the area of a regular hexagon if its side is 6 cm long.

1. $\overline{AB} \cong \overline{DE}$, $\overline{AC} \cong \overline{DF}$, $\overline{BC} \cong \overline{EF}$, $\angle A \cong \angle D$, $\angle B \cong \angle E$, $\angle C \cong \angle F$ 2. See p. 199.
3. $54\sqrt{3}$ cm^2

14.3 Specific Conditions for Loci (pages 475–478)

Suggestions

Be sure to emphasize that the distance between two sets of points is the length of the shortest path between the sets of points. This path may or may not be a path perpendicular to one of the sets.

When working with intersections of two distinct loci, have the students sketch each set independently, perhaps in different colors, before trying to determine the intersection set.

Be sure to use at least one example in which the intersection set is empty, such as Example 6 in the textbook.

Here is an example in which the intersection set is a line: $p \parallel q \parallel r$ with the indicated distances. Sketch and describe the locus of points equidistant from p and q and 5 cm from r (line t).

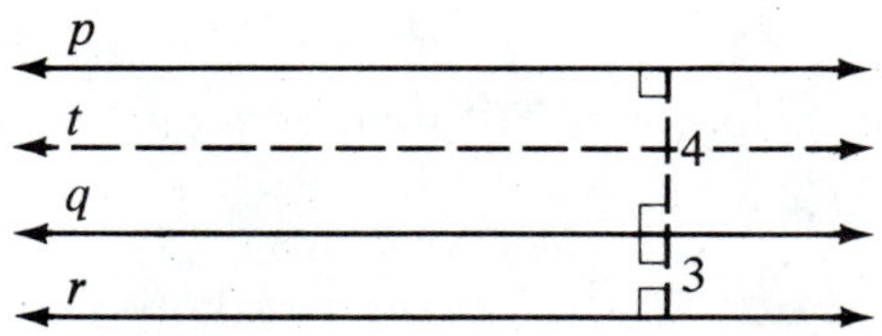

For Additional Chalkboard Work

In the yard pictured below, a dog is tied at point D and a cat is tied at point C. If the dog's rope is 40 ft long and the cat's rope is 30 ft long, use a compass to sketch the area where both the dog and the cat can play.

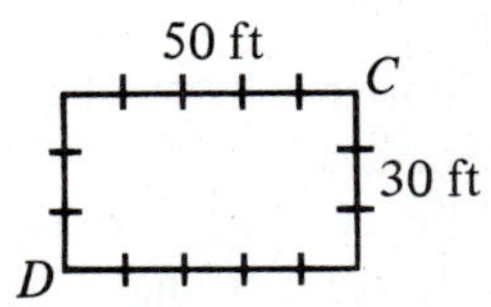

Cumulative Review

1. True or false? Two triangles are congruent if two sides and an angle of one are congruent respectively to two sides and an angle of the other.
2. In ABC, which is the included angle for $\overline{BC}$ and $\overline{CA}$?
3. True or false? An equilateral triangle is also isosceles.

1. False 2. $\angle C$ 3. True

14.4 General Conditions for Loci (pages 479–482)

Suggestions

Students who have no problems working with specific distances frequently run into difficulty when variables are used in representing distances. It often helps to have students substitute specific numbers for the distance variables. Encourage them to vary the numbers they choose to see what happens as the values of the variable increase or decrease.

Use Examples 1, 2, and 3 of section 14.3 to help the students understand Example 1 of this section. Go over the examples using specific numbers for the radius and distances, looking at the three different resulting loci. Then suggest to the students that the length of the radius be called r, to represent any number. Ask the class under what conditions the loci would be a circle concentric to the given circle. Help the students understand that regardless of the length r of the radius, there would be three possibilities of comparison with the distance d, $d > r$, $d = r$, or $d < r$.

Example 4 of section 14.3 could be used to lead into Example 2 of this section. In a similar manner, refer to Example 5 of section 14.3 to help the students understand Example 3.

Enrichment

Bushes are to be planted on both sides of a circular running path with radius 500 ft. Each bush is to be 6 ft from the edge of the path. They are to be planted about 8 ft apart. How many should be ordered? 785 bushes

Cumulative Review

True or False?

1. Two right triangles are congruent if a leg and acute angle of one are congruent to the corresponding leg and acute angle of the other.
2. A leg of a right triangle is opposite an acute angle.
3. A leg of a right triangle may be longer than the hypotenuse.

1. True 2. True 3. False

14.5 Concurrent Bisectors in a Triangle (pages 483–486)

Vocabulary

concurrent lines, circumcircle, circumcenter, incircle, incenter

Suggestions

Have the students draw a triangle ABC and construct the perpendicular bisectors of the sides. Have them put their compass on P, the point of concurrency, open the compass the distance PA, and draw a circle. Ask why the perpendicular bisectors are concurrent and why point P is the center of a circle containing A, B, and C. Help the students use the Perpendicular Bisector theorem to prove the construction. Compare the location of the points of concurrency in acute, right, and obtuse triangles.

Example 1 asks the student to circumscribe a circle about a triangle. In the method suggested, only two of the perpendicular bisectors are constructed. Ask why only two are needed.

The concurrency of the angle bisectors of a triangle can also be handled through constructions. Students sometimes fail to see the necessity for constructing the perpendicular segment from the incenter to one of the sides. This construction gives a radius of the incircle.

Make sure students realize that the circumcenter and incenter are not necessarily the same point.

For Additional Chalkboard Work

Have students investigate under what conditions the incenter and circumcenter are the same point (a triangle in which the perpendicular bisectors of the sides and angle bisectors are the same lines).

Cumulative Review

1. How many vertices does a hexagon have?
2. Find the measure of each angle of quadrilateral $ABCD$ if m $\angle A = 3x$, m $\angle B = 2x + 5$, m $\angle C = 4x - 2$, and m $\angle D = x + 7$.
3. What is an apothem of a polygon?

1. 6 2. m $\angle A = 105$, m $\angle B = 75$, m $\angle C = 138$, m $\angle D = 42$. 3. The length of a radius of the inscribed circle.

14.6 Concurrent Altitudes and Medians (pages 487–490)

Vocabulary

orthocenter, centroid

Suggestions

Have the students construct the medians of an acute triangle, a right triangle, and an obtuse triangle. (See rt. $\triangle ABC$ below.) Have them open their compasses the length OE. Ask them to compare this length with the length BO. They should be able to mark off two segments congruent to $\overline{OE}$ on segment $\overline{BO}$. Let them compare the other medians in a similar way.

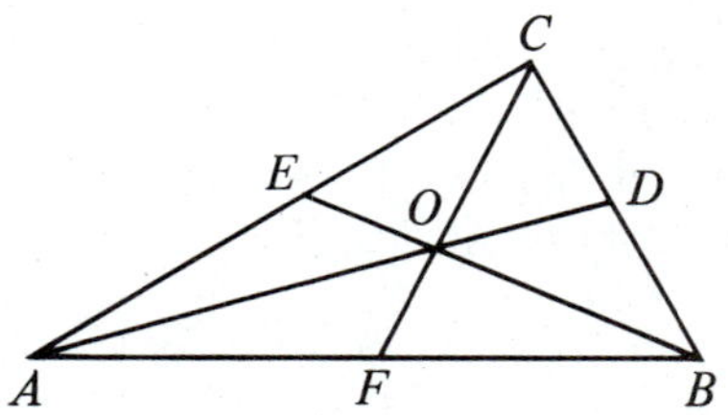

Complete the proof of Theorem 14.6 on page 490 or assign it as Exercise 19.

For Additional Chalkboard Work

Consider $\triangle ABC$ with median $\overline{CD}$, centroid P, and altitude $\overline{CE}$.

1. If $CE = 12$ and $DE = 5$, find CP. $8\frac{2}{3}$
2. If $PD = 10$ and $CE = 24$, find DE. 18
3. If $DP = 6\frac{2}{3}$ and $DE = 12$, find CE. 16

Enrichment

Interested students might want to explore the fact that the centroid is the center of gravity of a triangle. They will need to show that the three medians divide a triangle into six triangles of equal area.

Cumulative Review

1. What is the measure of each interior angle of a regular octagon?
2. What is the sum of the measures of the exterior angles, one at each vertex, of a convex polygon?

1. 135° 2. 360°

14.7 Loci in Space (pages 491–492)

Suggestions

The examples used in the textbook are the three-dimensional counterparts of Postulate 27, Postulate 29, and Theorem 14.1. Review the two-dimensional concepts and extend them to three-dimensions.

Students frequently think a sphere includes the points in its interior. Remind them that just as a circle does not include the points in its interior, neither does a sphere.

To illustrate the examples, select points in the classroom that satisfy the given conditions. To illustrate Example 3, have each of two students hold up a pencil. Let the ends of the pencils be the two points, A and B. Make sure all the points they choose are not collinear. Ask the class what figure they seem to determine. (A plane that is the perpendicular bisector of $\overline{AB}$.) Hold a large piece of plywood equidistant from A and B to represent part of the plane that is the perpendicular bisector of $\overline{AB}$. Then select random points on the plane to see that they are equidistant from A and B.

For Additional Chalkboard Work

Points A and B are 8 cm apart. Sketch and describe the locus of points in space equidistant from points A and B and 6 cm from point B. (A circle that is the intersection of the plane, which is the $\perp$ bisector of $\overline{AB}$, and a sphere with center B and a 6-cm radius.)

Computer Activity (p. 496)

Lines 100 to 150 are written as they are so that the exact coordinates of the centroid will be given. If the values are not integral, printing them as M/3 and N/3 would give them as a rounded decimal.

15 Chapter and Lesson Commentaries

Chapter 15 Inequalities

Chapter Objectives

To apply the properties of inequality to measures of segments and angles
To prove statements involving inequalities
To apply the Triangle Inequality theorem
To compare the measures of sides and the measures of angles in a scalene triangle
To apply the Hinge theorem and its converse to pairs of triangles
To apply inequality relations to circles

Chapter Overview

The basic algebraic properties of inequality are reviewed and applied to the lengths of segments and measures of angles. The Triangle Inequality theorem is introduced and applied. The inequality relationships between the measures of the angles in a scalene triangle are considered. The Hinge theorem and its converse are applied to pairs of triangles. Inequality relations related to a circle are discussed.

Special Features

Non-Routine Problem Solving p. 498
Algebra Review pp. 502, 515
Cumulative Review pp. 506, 515
Non-Routine Problems p. 506
Transformational Geometry p. 510
Reading in Geometry p. 513
Applications p. 519
Chapter Review p. 520
Chapter Test p. 521
Computer Activities p. 522
College Prep. Test p. 523

Non-Routine Problem Solving (p. 498)

Solutions:
Problem 1 – When the diagonals are perpendicular, the figure is a rhombus and the perimeter is 40. As the measure of the angle between the diagonals decreases, the sides approach 0 and 16. Therefore, the minimum perimeter is 32. The range for the perimeter is 32 units to 40 units.

Problem 2 – The sum of the measures of $\overset{\frown}{AE}$ and $\overset{\frown}{BC} = 110°$, and therefore the measure of $\overset{\frown}{CD} = 40°$. Since the measure of $\angle P = \frac{1}{2}(\text{m } \overset{\frown}{AE} - \text{m } \overset{\frown}{CD})$, and since the measure of $\overset{\frown}{CD}$ is constant, the measure of $\angle P$ is greatest when $\overset{\frown}{AE}$ is its maximum. $\overset{\frown}{AE}$ is its maximum when measure of $\overset{\frown}{BC}$ is 0°, at which time the measure of $\overset{\frown}{AE} = 110°$. Thus, the maximum value for the measure of $\angle P$ is 35°.

15.1 Properties of Inequality (pages 499–502)

Suggestions

Review the properties of inequalities. The subtraction of unequals from equals may be new to the students. Give the students adequate explanation and practice with real numbers and algebraic inequalities before applying the properties to the measures of segments and angles.

The restriction that $c > 0$ is included in multiplication and division properties of inequalities. Because the numbers used in geometry problems are the measures of segments and angles, they are all positive. With some students, review the multiplication and division with negatives so the students will not think $a > b \rightarrow ac > bc$ is true for all real values of a, b, and c.

Students tend to confuse the Subtraction Property of Inequality given on page 499 with the Equals-Unequals Property on page 500. Point out the differences between these properties as you introduce them. Give a numerical example of the subtraction of unequals from equals: $a = 3 + 7$, $b = 10$, $c = 7$, and $d = 4$. Then, if $3 + 7 = 10$ and $7 > 4$, $3 + 7 - 7 < 10 - 4$, or $3 < 6$.

To introduce the Trichotomy Property, have each of two students think of, but not divulge, a number. Ask the class how the two numbers compare. Lead them to analyze the possibilities that one of these must hold: $a > b$ or $a = b$ or $a < b$.

For Additional Chalkboard Work

Reviewing properties of inequalities. Ask the students to solve the following:

1. $x + 5 > 2$ $x > -3$
2. $x - 7 < 3$ $x < 10$
3. $3x < 12$ $x < 4$
4. $4x > -20$ $x > -5$
5. $5x - 9 < -24$ $x < -3$
6. $2x + 3 < 11$ $x < 4$

Cumulative Review

1. List five ways to prove that a quadrilateral is a parallelogram.
2. In $\triangle ABC$, M is the midpoint of $\overline{AC}$ and N is the midpoint of $\overline{BC}$. $AB = 18$, $AC = 21$, and $BC = 24$. Find MN.

1. See p. 216. 2. 9

15.2 The Triangle Inequality (pages 503–506)

Suggestions

The inequality property of "Addition of unequals to unequals" will probably be new to your students. Tell your class that the "same order" means that the inequalities you are adding must both be "greater than" or both be "less than." Use numerical examples to illustrate. For example, if $2 < 5$ and $6 < 8$, then $2 + 6 < 5 + 8$, or $8 < 13$.

Enrichment

Prove: $AB + BC + CD > AD$

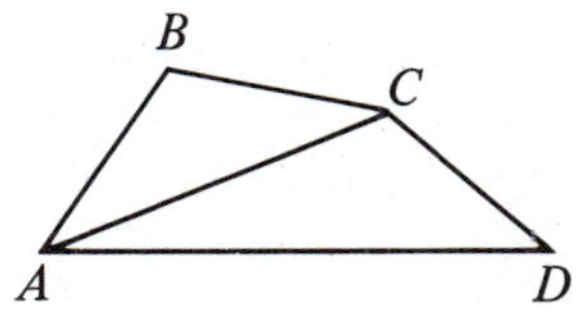

$AB + BC > AC$ by the Triangle Inequality theorem
$AC + CD > AD$ by the Triangle Inequality theorem
$AB + BC + AC + CD > AD + AC$ Addition of unequals to unequals
$AB + BC + CD > AD$ Addition of equals $(-AC)$ to unequals

For Additional Chalkboard Work

An isosceles triangle must have a perimeter of 20 cm. Each side must be at least 3 cm long. The length of each side must be a whole number of centimeters. List the lengths of the sides of all triangles that meet these requirements.
4 cm, 8 cm, 8 cm; 6 cm, 7 cm, 7 cm; 8 cm, 6 cm, 6 cm

Cumulative Review

1. In trapezoid $ABCD$, $AB \parallel CD$, $AB = 28$, and the median of the trapezoid measures 20. Find CD.
2. True or false? All rhombi are squares.
3. True or false? All squares are rhombi.

1. 12 2. False 3. True

15.3 Inequalities in a Scalene Triangle (pages 507–509)

Suggestions

Use the Isosceles Triangle theorem to lead into a discussion of inequalities in scalene triangles. For isosceles triangles, the relationship between the sides and angles of the triangle is important. The theorems of inequalities of scalene triangles are also concerned with the relationship between the sides and the opposite angles. Have your students sketch several triangles with sides of unequal lengths and have them measure the opposite angles with a protractor. Compare the lengths of the sides. Compare the measures of the angles. Introduce Theorem 15.2 and Theorem 15.3.

The proof of Theorem 15.3 is an excellent example of an indirect proof. The cases are clearly drawn and the contradictions are easily obtained.

Given: $\triangle ABC$ with
$m\angle A > m\angle B$
Prove: $BC > AC$

Using the Trichotomy Property, there are three possibilities for BC and AC: $BC < AC$, $BC = AC$, or $BC > AC$. If it is true that $BC < AC$, then $m\angle A < m\angle B$ by Theorem 15.2. This contradicts the given, so it is impossible for $BC < AC$. If $BC = AC$, then $m\angle A = m\angle B$ by the Isosceles Triangle theorem. This also contradicts the given, so $BC = AC$ is false. The only remaining possibility is that $BC > AC$. Theorem 15.3 can be used to prove that the hypotenuse is the longest side of a right triangle. You may want to ask your students to prove this.

Cumulative Review

1. Give the contrapositive: If two lines are parallel, then they do not intersect.
2. In $\triangle ABC$, $\overrightarrow{CD}$ bisects $\angle C$ and intersects $\overline{AB}$ at D. $AB = 24$, $BC = 18$, and $CA = 30$. Find AD and DB.

1. If two lines intersect, then they are not parallel. 2. 15; 9

15.4 Inequalities for Two Triangles (pages 511–515)

Suggestions

The Hinge theorem may be graphically illustrated with a model similar to the one described in the textbook. Protractors can be mounted at the vertices of the angles included between the slats.

Be sure your students understand that the Hinge theorem and its converse involve a comparison between two triangles. Students often confuse these theorems with the theorems in section 15.3 that considered the unequal sides and angles in a single triangle.

Although the idea in the Hinge theorem seems perfectly reasonable, the proof is rather intricate. The analysis suggests that $\triangle ACG$ be constructed congruent to $\triangle DFE$. This can be done by introducing ray $\overrightarrow{CG}$ so that $\angle ACG \cong \angle DFE$. Then select point G so that $CG = FE$. Now the triangles are congruent by SAS. Let $\overrightarrow{CH}$ bisect $\angle GCB$. Then $\triangle GCH \cong \triangle BCH$ by SAS and $HB = HG$. Now apply the Triangle Inequality theorem to $\triangle AGH$, $AH + HG > AG$. Because $HG = HB$ and $AG = DE$, we can substitute and get $AH + HB > DE$, or $AB > DE$. The diagram in the text assumes that G and C lie in opposite half-planes determined by $\overleftrightarrow{AB}$. Actually, C and G could be in the same half-plane or G could be a point of $\overleftrightarrow{AB}$. A complete proof would need to consider both of these cases.

For Additional Chalkboard Work

If $AD = DB = CD = 8$ in. and m $\angle ADC = 60$, which is longer, $\overline{AC}$ or $\overline{CB}$? How much longer? $\overline{CB}$; about 5.9 inches longer

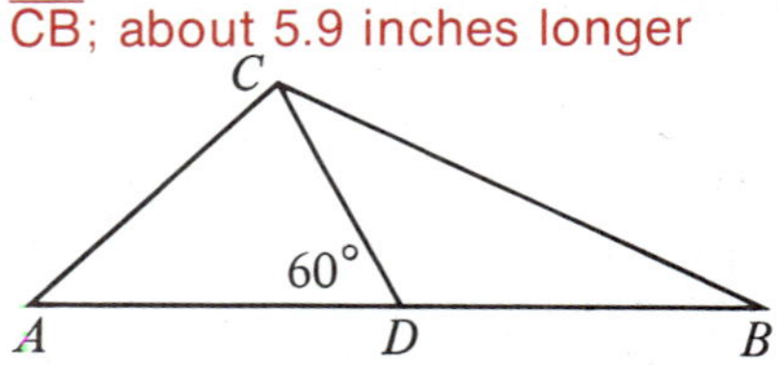

Cumulative Review

1. In $\triangle ABC$, $\angle C$ is a right angle. $AB = 25$, and $CB = 24$. Find AC.
2. The diagonals of a rhombus are 8 cm and 12 cm long. Find the measure of a side.
3. The sides of a triangle measure 9 in., 40 in., and 41 in. Is the triangle a right triangle?

1. 7 2. $2\sqrt{13}$cm 3. Yes

15.5 Inequalities in Circles (pages 516–518)

Suggestions

Begin by sketching a circle on the chalkboard and then asking a student to draw in the longest chord (a diameter). Ask another student to add a shorter chord, and ask a third student to add an even shorter cord. Have the class compare the distances of the chords from the center of the circle. Review the process determining the distance from a point to a segment or line, if necessary. Students should realize that if a first chord is longer than a second, the first is nearer the center of the circle; and conversely, if one chord is nearer the center than a second chord, it is longer than the second.

Use a number of numerical examples before assigning exercises involving proofs.

For Additional Chalkboard Work

The blades on a pair of scissors measure 6 in. If the blades are opened to form an angle measuring 90°, how far apart are the two blades at their tips? $6\sqrt{2}$, or about 8.5 in.

Enrichment

Encourage the students to investigate and extend the concepts that were developed for circles to spheres.

Cumulative Review

1. Each side of an equilateral triangle measures 8 ft. Find the length of an altitude.
2. Each side of a square is 9 m long. Find the length of a diagonal.
3. In a 30°–60° right triangle, the hypotenuse measures 10. Find the measures of the two legs.

1. $4\sqrt{3}$ 2. $9\sqrt{2}$ 3. 5, $5\sqrt{3}$

Computer Activity (p. 522)

This program uses true subscripted variables, not A1, A2, etc., as shown on page 381. If these subscripted variables were not needed, it would not be possible to use the loops in lines 20 and 30. The variables A1, A2 would not be recognized in a loop.

16 Chapter and Lesson Commentaries

Chapter 16 Figures in Space

Chapter Objectives

To identify and count the parts of a prism
To sketch various types of prisms
To find the lateral area and the total area of a right prism, a cylinder, a regular pyramid, and a cone
To find the volume of a prism, a cylinder, a pyramid, and a cone
To find the area and the volume of a sphere

Chapter Overview

Concepts of geometric solids are presented in this chapter. Students learn to identify and count the parts of a prism and to sketch various types of prisms. Formulas for the lateral area and the total area of a right prism are developed. After the concept of volume is explored, a formula for the volume of any prism in developed. The basic similarities between prisms and cylinders are used to develop formulas for the lateral area, total area, and volume of a cylinder. Pyramids are introduced, with special attention given to regular pyramids. The formulas for the lateral area and the total area of a regular pyramid, as well as the formula for the volume of any pyramid, are developed. The similarities between regular pyramids and cones are then used to develop comparable formulas for a cone. The formulas for the area and the volume of a sphere are also developed.

Special Features

Non-Routine Problem Solving (p. 524)

The five polyhedrons illustrated on the opening page of this chapter are the only *regular* polyhedrons. They are sometimes referred to as the Platonic solids because they were described by Plato. The tetrahedron, cube, and octahedron occur in nature as the crystals of sodium sulphantimoniate, table salt, and chrome alum, respectively. Skeletons of the microscopic sea animal *Radialaria* have been found in the shape of the dodecahedron and icosahedron.

The formula relating the faces, edges, and vertices of any convex polyhedron is $v - e + f = 2$. This formula, independently discovered by Descartes in 1635 and Euler in 1752, is often called the Euler-Descartes formula.

16.1 Prisms (pages 525–528)

Vocabulary

solids, polyhedrons, prism, regular polyhedrons, base, face of a prism, edge of a prism, lateral face, lateral edge, right prism, oblique prism, vertex

Background

The traditional definition of a solid is, "a geometric solid is a closed surface completely inclosing a portion of space" (Welchons Krickenberger, *Solid Geometry,* 1950).

Suggestions

Durable models of many geometric solids can be easily constructed. The models will be extremely useful in identifying the parts of the polyhedrons discussed throughout this chapter.

Much of the terminology such as vertices, bases, edges, and lateral faces will be used throughout this chapter. It is important that students learn the vocabulary in this first section.

Illustrate, with models or diagrams, the difference between a right prism and oblique prism. The complete identification of a prism should indicate whether it is right or oblique and give the type of base.

For Additional Chalkboard Work

What type of prism has twice as many edges as faces?

quadrangular

Cumulative Review

1. In $\odot O$, $\overline{OE} \perp$ chord $\overline{AB}$. $OE = 15$ and $OA = 25$. Find AB.
2. In $\odot O$, what is true of all chords that are the same distance from point O?
3. $\overline{PT}$ is tangent to $\odot O$ at point T. If $PO = 34$ and $TP = 30$, find OT.

1. 40 2. They are all the same length.
3. 16

16.2 Surface Areas of Prisms (pages 529–531)

Vocabulary

lateral area, total area, altitude, rectangular solid, cube

Suggestions

Throughout this chapter students will be working with volume, surface area, and perimeter. They need to understand the differences among these concepts. Remind them that perimeter means "distance around," area is concerned with "surface," and volume with "capacity." Perimeter and circumference are measured in linear units, area in square units, and volume in cubic units.

You may want to have a model of a right prism that can be unfolded to show the large rectangle formed in developing the "Second Method." Although this "Second Method" involves less mathematical computation, there are students who are more comfortable finding the areas of all of the lateral faces and taking their sum. If this helps them understand the concept of lateral area, do not insist that they use the short cut. Be sure your students understand the difference between lateral area and total area.

For Additional Chalkboard Work

Have students consider the problem of painting the classroom walls and ceiling. Have them estimate the cost by measuring the room, computing the surface area, and finding the number of gallons of paint needed for the job.

Cumulative Review

1. If a central angle and an inscribed angle intercept the same arc of a circle, what is the ratio of the measure of the central angle to the measure of the inscribed angle?
2. In $\odot O$, chord $\overline{AB}$ measures 16 in. and m $\angle AOB = 60$. Find OA.
3. Square $ABCD$ is inscribed in circle O. Find m$\widehat{AB}$.

1. 2:1 2. 16 in. 3. 90

16.3 Volumes of Prisms (pages 532–534)

Vocabulary
volume, cubic unit

Suggestions
Discuss volume as a measure of capacity, or the number of cubic units contained in a solid. Emphasize that this will hold for any prism, right or oblique, regardless of the polygonal region forming the base. Be sure your students understand that the altitude of the oblique prism is not the lateral edge, but that it is a perpendicular segment between the planes containing the two bases.

Discuss metric units. One liter is equivalent to 1,000 cm^3; thus 1 milliliter = 1 cm^3. A cube with a 10-cm edge has a volume of 1,000 cm^3, or 1 liter.

For Additional Chalkboard Work
The number of cubic centimeters in the volume of a cube is equal to the number of square centimeters in its area. How long is each edge of the cube? 6 cm

Enrichment
Ask the students to solve the following dilemma. The people of Delos were stricken by a plague. Upon consulting the oracle at Delphi, they were told that the god Apollo was angry and would be appeased only if they doubled the size of his altar, a cube of solid gold. The townspeople erected a new altar with each edge twice as long as the original edge. When the plague worsened, they realized that the volume of the new cube was 8 times the original. How long must each edge be so that the volume is exactly twice the original?
$\sqrt[3]{2}$ times the edge of the original.

Cumulative Review
1. Two secants meet at point P outside circle O. They intercept arcs of 116° and 94°. Find the measure of the angle formed by the two secants.
2. Two chords meet at point Q inside circle O. They intercept arcs of 85° and 51°. Find the measure of an angle formed by the two chords.

1. 11° 2. 68°

16.4 Cylinders (pages 535–538)

Vocabulary
cylinder, right circular cylinder, oblique circular cylinder, axis, cylinder of revolution

Suggestions
Cylinders can be oblique as well as right, and can have bases other than circles. Only right circular cylinders are treated in this textbook.

To develop the method for finding the lateral surface area of the cylinder, show the students a model of a cylinder and ask them how they can cut a sheet of paper the correct size to "wallpaper" the lateral surface.

The idea of total area is a simple one, but the formula $T = 2\pi r(h + r)$ will appear rather formidable to some students. They may find the problems easier if they simply understand the concept that total area is the sum of the lateral area and the areas of the two bases.

The volume of the cylinder may be developed by thinking of the cylinder as a "circular prism." Start by thinking of a cylinder with an altitude of one unit. Then add additional layers as the altitude increases.

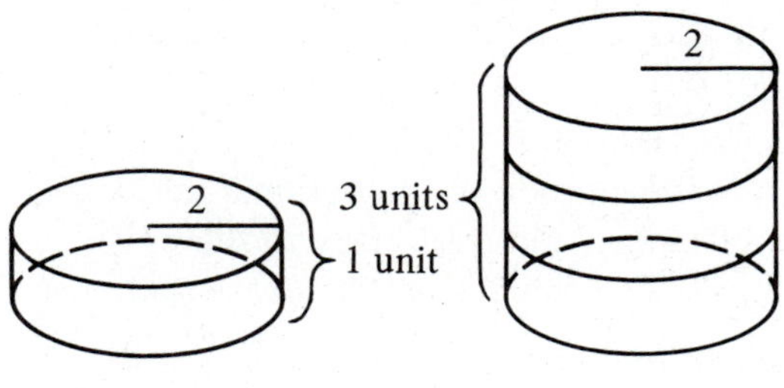

Enrichment
The liquid from a rectangular cardboard container 3 in. by 3 in. by 8 in. is to be transferred to a cylindrical container 4 in. in diameter. How high will the liquid rise in the container? Use 3.14 for π. 5.73 in.

Cumulative Review
1. Find the area of a triangle whose base measures 11 in. and whose corresponding altitude measures 8 in.
2. Find the area of an equilateral triangle whose side measures 6 m.

1. 44 $in.^2$ 2. $9\sqrt{3}$ m^2

16.5 Pyramids (pages 539–543)

Vocabulary

pyramid, regular pyramid, slant height

Suggestions

Compare pyramids with prisms. Much of the terminology is the same. Point out some differences. The formula for the lateral area pertains only to regular pyramids. The lateral area of nonregular pyramids can be found by finding the areas of each face individually and taking the sum of the areas. The formulas for the total area and the volume pertain to all pyramids.

For Additional Chalkboard Work

A regular tetrahedron is a triangular pyramid with all edges congruent. If e represents the length of an edge of a regular tetrahedron, find the length of an altitude in terms of e.

$\frac{e\sqrt{6}}{3}$

Enrichment

A horizontal cross-section is the intersection of a polyhedron with a plane parallel to the plane of the base. Have the students prove the following:

1. A cross-section is similar to the base.
2. If h is the altitude of the pyramid and k is the distance from the vertex to the cross-section, then the area of the cross-section is equal to $\frac{k^2}{h^2}$ times the area of the base.
3. The volume of the small pyramid is $\frac{k^3}{h^3}$ times the volume of the large pyramid.

Cumulative Review

1. The bases of a trapezoid measure 10 ft and 15 ft, and the altitude measures 8 ft. Find the area.
2. The sides of two squares are in a ratio of 1:3. What is the ratio of the area of the smaller square to the area of the larger square?
3. Find the area of a circle whose radius measures 6 cm.

1. 100 ft^2 2. 1:9   3. 36π cm^2

16.6 Cones (pages 544–548)

Vocabulary

cone, right circular cone, cone of revolution

For Additional Chalkboard Work

A paper cup is shaped like a cone. The diameter measures 8 cm and the altitude measures 12 cm. If water is poured into the cup so that 2 cm remain unfilled at the top, how many cm^3 of water are in the cup? Use 3.14 for π.

116.30 cm^3

The formula for the surface area pertains only to the right cone. The formula for the volume of the cone applies to the right cone as well as the oblique cone.

Enrichment

Give the following problem:
A circle of radius 15 cm has a 288° sector. This sector is to be folded to form a cone. (Radius $\overline{OA}$ is moved to radius $\overline{OB}$.) Find the volume of the cone.

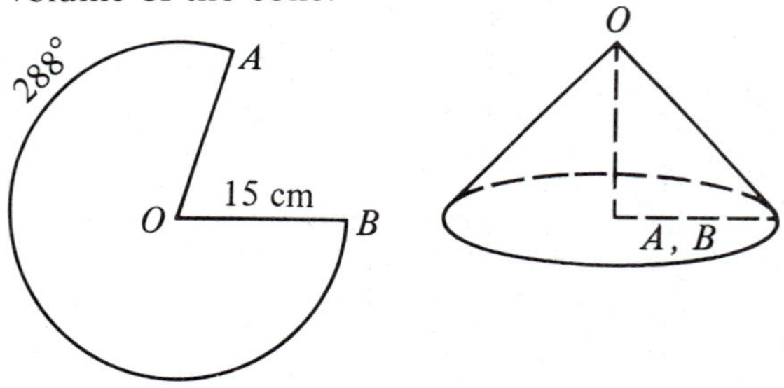

Solution: The slant height of the cone is equal to the radius of the sector or 15. The arc length of the sector is the circumference of the base of the cone.

Circumference = arc length = $\frac{288}{360} \cdot 30\pi = 24\pi$. Therefore, the radius of the base of the cone is 12. Using the Pythagorean theorem, the altitude of the cone is 9.

Volume $= \frac{1}{3} \cdot \pi \cdot 144 \cdot 9 = 432\pi$

Cumulative Review

1. The circumference of a circle is 26π. Find the area.
2. The radius of a circle measures 12 ft. Find the length of an arc which is intercepted by a central angle which measures 60°.

1. 169π units^2 2. 4π ft

16.7 Spheres (pages 549–552)

Vocabulary

sphere, great circle, quadrants

Suggestions

Treat the theorems in this section informally. Their proofs are too complicated for students in an elementary geometry class.

Use models of a sphere, a cylinder, and a cone with the same radius and an altitude that is twice the radius to illustrate volume of cylinder – volume of cone = volume of sphere, or $2\pi r^3 - \frac{2}{3}\pi r^3 = \frac{4}{3}\pi r^3$.

For Additional Chalkboard Work

Have students compute the surface area and volume of a baseball, a basketball, and so on.

Enrichment

Introduce spherical geometry. If great circles are considered to be "lines" on the sphere, then any two lines intersect in two points and parallel lines do not exist. Have students determine the shortest distance between any two points on the sphere, the sum of the angles of a triangle drawn on the surface, and whether triangles may have more than one right angle. Have students prove their findings and develop other theorems in spherical geometry.

Cumulative Review

1. Find the length of a radius of a circle whose area is 75π m^2.
2. An equilateral triangle is inscribed in a circle whose radius measures 14 ft. Find the length of an apothem of the triangle.
3. Find the area of a regular hexagon whose side measures 8 yd.

1. $5\sqrt{3}$ m 2. 7 ft 3. $96\sqrt{3}$ yd^2

Computer Activity (p. 556)

The program on this page is probably one of the simplest, even though it is one of the longer, programs in this book. It is 90% PRINT statements. Values are input and a result printed. The exercises assigned to the students should therefore be very easy to write.

Logic Section (pages 560–564)

The idea of flow proofs may present an interesting alternative to the rigid two-column proof format. Determining the validity of an argument requires considerable expertise with truth tables.

Computer Section (pages 565–577)

To get the best coordination between the geometry portion and the computer portion of this text, the following pairing is suggested.

Computer	Geometry
pp. 565–566	Chap. 1
p. 567	Chap. 2
p. 568	Chap. 3
p. 569	Chap. 4
pp. 570–571	Chap. 5
p. 572	Chap. 6
pp. 573–575	Chap. 9
pp. 576–577	Chap. 10

As your students do the first few pages of the computer section, you should point out that while BASIC language has a much more extensive vocabulary, there is great power in just the two commands LET and PRINT. You may want to get into a discussion of how much can be done with those two alone.

With regard to the LET command, X = 4 without the LET will do the same thing. The students will find that out either by accident or by reading other books. Have them use the LET command as long as you can since it emphasizes that X = 4 in the computer is not a statement of equality but an assignment of value.

The greater part of computer knowledge comes from use and not from reading. Encourage your students to get on-line and use the computers, and learn by doing. In the long run, it will provide the best teaching.

Assignments for Reinforcement and Maintenance

Following is a suggested schedule of assignments for 170 school days. It is designed to provide a program of reinforcement and maintenance of the major skills and concepts of Geometry and to allow flexibility of testing days.

Each day's assignments provide a double dimension.

In a spiral view, there is provision for:

1. REINFORCEMENT of the current lesson (major part),
2. MAINTENANCE of the previous lesson,
3. MAINTENANCE of the lesson taught two days ago.

In a lateral view, there is provision for:

A BASIC COURSE,
B AVERAGE COURSE,
C ENRICHED COURSE.

The reinforcement and maintenance program is provided as a guide for you to adopt and use in the way that you discern to be best for your class.

Included in the text are optional materials that are ancillary to the program. You may use these materials in a variety of ways:

1. Additional examples for teaching the lesson,
2. Practice exercises for seat work or chalkboard work,
3. Items for quizzes,
4. Practice exercises for extra help outside of class time,
5. Extra testing,
6. Enrichment exercises.

OPTIONAL MATERIALS

Practice	Testing	Enrichment
Cumulative Reviews in PE*	Cumulative Review Tests (Ch. 1–4, 1–8, 1–12, and 1–16) in PE	Calculator Activities in PE
Chalkboard Exercises in TE**	College Prep Tests in PE	Non-Routine Problems in PE
Algebra Reviews in PE	Cumulative Reviews in TE	Computer Activities in PE
Logic Reviews in PE		Chapter Opener Pages in PE
		Logic Section in PE
		Enrichment in TE

*PE: Pupil's Edition **TE: Teacher's Edition

Assignment Chart

Day	Page	Level A	Page	Level B	Page	Level C
1	3–4	Reading Ex 1–6 Orals All Ex 1–15 Algebra Review	3–4	Orals All Ex Odd 1–15 Algebra Review	3–4	Ex Odd 1–21, 22, 23
2	3–4 7	Ex 16, 17 Ex Odd 1–21	3–4 7	Ex 16, 17 Ex Odd 1–29	3–4 7	Ex 14, 16 Ex Odd 1–31
3	3–4 7 10–11	Ex 18, 19 Ex 14, 16 Ex 1–18	3–4 7 10–11	Ex 18, 19 Ex 22, 24 Ex Odd 1–31	3–4 7 10–11	Ex 18, 20 Ex 24, 26 Ex Odd 1–35 Non-Routine
4	7 10–11 15–16	Ex 18, 20 Ex 19, 21 Reading Ex 1–6 Orals All Ex 1–12	7 10–11 15–16	Ex 26, 28 Ex 24, 26 Orals All Ex Odd 1–19	7 10–11 15–16	Ex 28, 30 Ex 28, 30 Ex Odd 1–21
5	10–11 15–16 17	Ex 20, 22 Ex 13, 15 Applications	10–11 15–16 17	Ex 28, 30 Ex 14, 16 Applications	10–11 15–16 17 18	Ex 32, 34 Ex 16, 18 Applications Chapter Review
6	18	Chapter Review	18	Chapter Review	19	Chapter Test
7	19	Chapter Test	19	Chapter Test	25–26	Ex Odd 1–39
8	25–26	Orals All Ex Odd 1–27	25–26	Orals All Ex Odd 1–31	25–26 31–32	Ex 30, 32 Ex Odd 1–37
9	25–26 31–32	Ex 22, 24 Ex Odd 1–19	25–26 31–32	Ex 24, 26 Ex Odd 1–27	25–26 31–32 35	Ex 34, 36 Ex 30, 34 Ex Odd 1–17
10	25–26 31–32 35	Ex 26, 28 Ex 21, 23 Reading All Ex 1–8	25–26 31–32 35	Ex 28, 30 Ex 22, 24 Reading All Ex Odd 1–13	31–32 35 38–39	Ex 32, 36 Ex 10, 12 Ex Odd 1–15
11	31–32 35 38–39	Ex 25, 27 Ex 9, 11 Orals All Ex Odd 1–7	31–32 35 38–39	Ex 26–28 Ex 10, 12 Orals All Ex Odd 1–13	35 38–39 43	Ex 14, 16 Ex 10, 12 Ex Odd 1–23

Day	Page	Level A	Page	Level B	Page	Level C
12	35	Ex 10, 12	35	Ex 14, 16	38–39	Ex 14, 16
	38–39	Ex 2, 4	38–39	Ex 8, 10	43	Ex 16, 18
	43	Ex 1–13	43	Ex Odd 1–19	47–48	Ex Odd 1–27
		Algebra Review		Algebra Review		Non-Routine
13	38–39	Ex 6, 8	38–39	Ex 12, 14	43	Ex 20, 22
	43	Ex 14,16	43	Ex 20, 22	47–48	Ex 20, 22
	47–48	Ex Odd 1–19	47–48	Ex Odd 1–23	53	Ex Odd 1–11
		Algebra Review		Algebra Review	54	Applications
14	43	Ex 15, 17	43	Ex 21, 23	47–48	Ex 24, 26
	47–48	Ex 12, 14	47–48	Ex 16, 18	53	Ex 6, 8
	53	Orals All	53	Orals All	58	Ex Odd 1–11
		Ex 1–7		Ex 1–8		
15	47–48	Ex 16, 18	47–48	Ex 20, 22	53	Ex 10, 12
	53	Ex 8, 9	53	Ex 9, 10	58	Ex 4, 6
	54	Applications	54	Applications	60	Ex Odd 1–9
16	53	Ex 10, 11	53	Ex 11, 12	58	Ex 8, 10
	57–58	Orals All	57–58	Orals All	60	Ex 4, 6
		Ex Odd 1–7		Ex Odd 1–9	63–64	Ex Odd 1–23
17	57–58	Ex 4, 6	57–58	Ex 4, 6	60	Ex 8, 10
	60	Ex Odd 1–5	60	Ex Odd 1–9	63–64	Ex 16, 18
					65	Chapter Review
18	57–58	Ex 8, 9	57–58	Ex 8, 10	63–64	Ex 20, 22
	60	Ex 2, 4	60	Ex 4, 6	66	Chapter Test
	63–64	Ex 1–8	63–64	Ex Odd 1–13		
		Algebra Review		Algebra Review		
19	60	Ex 6, 7	60	Ex 8, 10	71	Ex Odd 1–17
	63–64	Ex 9, 11	63–64	Ex 8, 10		
	65	Chapter Review	65	Chapter Review		
20	63–64	Ex 13, 14	63–64	Ex 12, 14	71	Ex 10, 12
	66	Chapter Test	66	Chapter Test	73–74	Ex Odd 1–27
21	71	Orals All	71	Orals All	71	Ex 14, 16
		Ex 1–12		Ex Odd 1–15	73–74	Ex 22, 24
					78–79	Ex Odd 1–21
22	71	Ex 13, 15	71	Ex 10, 12	73–74	Ex 26, 28
	73–74	Ex Odd 1–17	73–74	Ex Odd 1–23	78–79	Ex 14, 16
					83–84	Ex Odd 1–21
						Non-Routine
23	71	Ex 14, 16	71	Ex 14, 16	78–79	Ex 18, 20
	73–74	Ex 16, 18	73–74	Ex 22, 24	83–84	Ex 14, 16
	78–79	Reading Ex 1–4	78–79	Reading Ex 1–4	86	Ex Odd 1–15
		Ex Odd 1–9		Ex Odd 1–17		

Day	Page	Level A	Page	Level B	Page	Level C
24	73–74 78–79 82–84	Ex 20, 22 Ex 4, 6 Orals All Ex 1–8	73–74 78–79 82–84	Ex 25, 27 Ex 10, 12 Orals All Ex Odd 1–17	83–84 86 89–90	Ex 18, 20 Ex 10, 12 Ex Odd 1–19
25	78–79 83–84 85–86	Ex 8, 10 Ex 9, 11 Orals All Ex 1–8	78–79 83–84 85–86	Ex 14, 16 Ex 10, 12 Orals All Ex Odd 1–13	86 89–90 92	Ex 14, 16 Ex 14, 16 Ex Odd 1–15 Non-Routine
26	83–84 86 89–90	Ex 13, 15 Ex 9, 11 Orals All Ex Odd 1–9	83–84 86 89–90	Ex 14, 16 Ex 8, 10 Orals All Ex Odd 1–17	89–90 92 94	Ex 18, 20 Ex 8, 10 Ex 1, 3 Non-Routine
27	86 89–90 92	Ex 10, 12 Ex 2, 4 Ex Odd 1–11	86 89–90 92	Ex 12, 14 Ex 12–14 Ex Odd 1–13	92 94 97 98	Ex 12, 14 Ex 2, 4 Ex Odd 1–19 Applications All
28	89–90 92 94	Ex 6, 8 Ex 6, 8 Ex 1, 2	89–90 92 94	Ex 16, 18 Ex 8, 10 Ex 1, 2	94 97 99	Ex 5, 6 Ex 12, 14 Chapter Review
29	92 94 97	Ex 10, 12 Ex 3, 4 Ex 1–8	92 94 97	Ex 12, 14 Ex 3, 4 Ex Odd 1–13	100	Chapter Test
30	94 97 99	Ex 5, 6 Ex 9, 11 Chapter Review	94 97 99	Ex 5, 6 Ex 12, 14 Chapter Review	105–106	Ex Odd 1–23 Non-Routine
31	100	Chapter Test	100	Chapter Test	105–106 109	Ex 18–20 Ex Even 2–8
32	105–106	Reading Ex 1–4 Orals All Ex Odd 1–15	105–106	Orals All Ex Odd 1–19	105–106 109 112–113	Ex 22, 24 Ex 1, 3 Ex Odd 1–17
33	105–106 109	Ex 10, 12 Ex 1, 2	105–106 109	Ex 14, 16 Ex 1–3	109 112–113 116–117	Ex 5, 7 Ex 10, 12 Ex Odd 1–23
34	105–106 109 112–113	Ex 14, 16 Ex 3, 4 Ex Odd 1–11	105–106 109 112–113	Ex 18, 20 Ex 4, 5 Ex Odd 1–15	112–113 116–117 121–122	Ex 14, 16 Ex 16, 18 Ex Odd 1–23

Day	Page	Level A	Page	Level B	Page	Level C
35	109	Ex 5, 6	109	Ex 6, 7	116–117	Ex 20, 22
	112–113	Ex 6, 8	112–113	Ex 8, 10	121–122	Ex 16, 18
	115–117	Orals All	115–117	Orals All	126–127	Ex Odd 1–21
		Ex Odd 1–11		Ex Odd 1–17		Non-Routine
36	112–113	Ex 10, 12	112–113	Ex 12, 14	121–122	Ex 20, 22
	116–117	Ex 6, 8	116–117	Ex 10, 12	126–127	Ex 14, 16
	121–122	Orals All	121–122	Orals All	128	Applications
		Ex Odd 1–15		Ex Odd 1–21	130	Chapter Review
37	116–117	Ex 10, 12	116–117	Ex 14, 16	126–127	Ex 18, 20
	121–122	Ex 10, 12	121–122	Ex 14, 16	131	Chapter Test
	126–127	Orals All	126–127	Orals All		
		Ex Odd 1–15		Ex Odd 1–19		
38	121–122	Ex 14, 16	121–122	Ex 18, 20	138	Ex Odd 1–15
	126–127	Ex 10, 12	126–127	Ex 12, 14		
	128	Applications	128	Applications		
	130	Chapter Review	130	Chapter Review		
39	126–127	Ex 14, 16	126–127	Ex 16, 18	138	Ex 8, 10
	131	Chapter Test	131	Chapter Test	142–143	Ex Odd 1–25
40	137–138	Orals All	137–138	Orals All	138	Ex 12, 14
		Ex 1–8		Ex Odd 1–11	142–143	Ex 18, 20
					144–146	Ex Odd 1–11
					147–148	Ex Odd 1–7
41	138	Ex 9, 10	138	Ex 6, 8	142–143	Ex 22, 24
	142–143	Reading Ex 1–6	142–143	Orals All	144–146	Ex 2, 4
		Orals All		Ex Odd 1–23	147–148	Ex 2, 4
		Ex Odd 1–13		Algebra Review	149–151	Ex Odd 1–11
		Algebra Review				
42	138	Ex 11, 12	138	Ex 10, 12	144–146	Ex 6, 8
	142–143	Ex 8, 10	142–143	Ex 16, 18	147–148	Ex 6, 8
	144–146	Ex Odd 1–11	144–146	Ex Odd 1–11	149–151	Ex 2, 4
			147–148	Ex Odd 1–7	152	Ex All
					155–156	Ex Odd 1–27
43	142–143	Ex 12, 14	142–143	Ex 20, 22	149–151	Ex 6, 8
	144–146	Ex 2, 4	144–146	Ex 2, 4	155–156	Ex 20, 22
	147–148	Ex Odd 1–7	147–148	Ex 2, 4	160–161	Ex Odd 1–19
			149–151	Ex Odd 1–11		Non-Routine
44	144–146	Ex 6, 8	144–146	Ex 6, 8	155–156	Ex 24, 26
	147–148	Ex 2, 4	147–148	Ex 6, 8	160–161	Ex 14, 16
	149–151	Ex Odd 1–11	149–151	Ex 2, 4	163–165	Ex Odd 15–29
			152	Ex All		Non-Routine
			155–156	Ex Odd 1–11		

Day	Page	Level A	Page	Level B	Page	Level C
45	147–148	Ex 6, 8	149–151	Ex 6, 8	160–161	Ex 18, 20
	149–151	Ex 2, 4	155–156	Ex 6, 8	163–165	Ex 16, 18
	154–155	Orals All	160–161	Ex Odd 1–13	168	Ex Odd 1–19
		Ex Odd 1–9				
46	149–151	Ex 6, 8	155–156	Ex 10, 12	163–165	Ex 20, 22
	154–155	Ex 2, 4	160–161	Ex 8, 10	168	Ex 12, 14
	160–161	Orals All	163–164	Ex Odd 15–25	172–173	Ex Odd 11–23
		Ex Odd 1–7				
47	154–155	Ex 6, 8	160–161	Ex 12, 14	168	Ex 16, 18
	160–161	Ex 2, 4	163–164	Ex 16, 18	172–173	Ex 25, 27
	163–164	Ex Odd 1–13	168	Ex Odd 1–13	175–176	Ex Odd 1–17
48	160–161	Ex 6, 8	163–164	Ex 20, 22	172–173	Ex 29, 31
	163–164	Ex 6, 8	168	Ex 6, 8	175–176	Ex 12, 14
	168	Odd 1–7	172–173	Orals All	177	Applications
				Ex Odd 1–15	178	Chapter Review
49	163–164	Ex 10, 12	168	Ex 10, 12	179	Chapter Test
	168	Ex 2, 4	172–173	Ex 10, 12		
	172–173	Orals All	175–176	Ex Odd 1–13		
		Ex Odd 1–9		Algebra Review		
50	168	Ex 6, 8	172–173	Ex 17, 19	187	Ex Odd 1–25
	172–173	Ex 2, 4	175–176	Ex 8, 10		
	175–176	Ex 1–6	177	Applications		
		Algebra Review				
51	172–173	Ex 6, 8	178	Chapter Review	187	Ex 18, 20
	175–176	Ex 7, 8			190–191	Ex Odd 1–39
	177	Applications				
52	178	Chapter Review	179	Chapter Test	187	Ex 22, 24
					190–191	Ex 32, 34
					195	Ex Odd 1–9
						Algebra Review
53	179	Chapter Text	186	Orals All	190–191	Ex 36, 38
			187	Ex Odd 1–21	195	Ex 4, 6
					197–198	Ex Odd 1–43
54	186–187	Orals All	187	Ex 14, 16	195	Ex 8, 10
		Ex Odd 1–21	190–191	Orals All	197–198	Ex 32, 38
				Ex Odd 1–33	201	Ex Odd 1–25
55	187	Ex 14, 16	187	Ex 18, 20	197–198	Ex 40, 42
	190–191	Orals All	190–191	Ex 28, 30	201	Ex 22, 24
		Ex Odd 1–29	194–195	Orals All	202	Applications
				Ex Odd 1–9	204	Chapter Review
				Algebra Review		

Day	Page	Level A	Page	Level B	Page	Level C
56	187	Ex 18, 20	190–191	Ex 32, 34	205	Chapter Test
	190–191	Ex 24, 26	195	Ex 4, 6		
	194–195	Orals All	197–198	Ex Odd 1–39		
		Ex 1, 3, 5				
		Algebra Review				
57	190–191	Ex 28, 30	195	Ex 8, 10	210	Ex Odd 1–15
	195	Ex 2, 4	197–198	Ex 30, 32		Non-Routine
	197–198	Ex Odd 1–27	201	Ex Odd 1–17		
58	195	Ex 8, 10	197–198	Ex 38, 40	210	Ex 10, 12
	197–198	Ex 22, 24	201	Ex 14, 16	213–214	Ex Odd 1–19
	201	Ex Odd 1–15	202	Applications		
			204	Chapter Review		
59	197–198	Ex 26, 28	205	Chapter Test	210	Ex 14, 16
	201	Ex 12, 14			213–214	Ex 14, 16
	202	Applications			218	Ex Odd 1–27
	204	Chapter Review				
60	205	Chapter Test	210	Orals All	213–214	Ex 18, 20
				Ex Odd 1–13	218	Ex 22, 24
					220–221	Ex Odd 11–17
61	210	Orals All	210	Ex 8, 10	218	Ex 26, 28
		Ex Odd 1–9	213–214	Orals All	220–221	Ex 12, 14
				Ex Odd 1–19	225	Ex Odd 1–29
62	210	Ex 4, 6	210	Ex 12, 14	220–221	Ex 16, 18
	213–214	Orals All	213–214	Ex 12, 14	225	Ex 24, 26
		Ex Odd 1–13	217–218	Reading Ex 1–7	229	Ex Odd 1–25
				Ex Odd 1–25		
63	210	Ex 8, 10	213–214	Ex 16, 18	225	Ex 28, 30
	213–214	Ex 8, 10	218	Ex 20, 22	229	Ex 20, 22
	217–218	Reading Ex 1–7	220–221	Orals All	232	Ex Odd 1–23
		Ex Odd 1–15		Ex Odd 1–13		
64	213–214	Ex 12, 14	218	Ex 24, 26	229	Ex 24, 26
	218	Ex 10, 12	220–221	Ex 8, 10	232	Ex 18, 20
	220–221	Orals All	224–225	Orals All	233	Applications
		Ex Odd 1–9		Ex Odd 1–25	234	Chapter Review
65	218	Ex 14, 16	220–221	Ex 12, 14	235	Chapter Test
	220–221	Ex 4, 6	225	Ex 20, 22		
	224–225	Orals All	228–229	Orals All		
		Ex Odd 1–19		Ex Odd 1–19		

Day	Page	Level A	Page	Level B	Page	Level C
66	220–221	Ex 8, 10	225	Ex 24, 26	240–241	Ex Odd 1–19
	225	Ex 14, 16	228–229	Ex 14, 16		
	228–229	Orals All	232	Ex Odd 1–15		
		Ex Odd 1–15				
67	225	Ex 18, 20	228–229	Ex 18, 20	240–241	Ex 12, 14
	228–229	Ex 14, 16	232	Ex 14, 16	244–245	Ex Odd 1–27
	232	Orals All	233	Applications		
		Ex Odd 1–13	234	Chapter Review		
68	228–229	Ex 17, 18	235	Chapter Test	240–241	Ex 16, 18
	232	Ex 12, 14			244–245	Ex 20, 22
	233	Applications			248–249	Ex Odd 1–17
	234	Chapter Review				Non-Routine
69	235	Chapter Test	240–241	Orals All	244–245	Ex 24, 26
				Ex Odd 1–15	248–249	Ex 12, 14
				Algebra Review	252–253	Ex Odd 9–19
70	240–241	Orals All	240–241	Ex 8, 10	248–249	Ex 16, 18
		Ex Odd 1–11	244–245	Ex Odd 1–19	252–253	Ex 12, 14
		Algebra Review			258	Ex Odd 1–21
71	240–241	Ex 4, 6	240–241	Ex 12, 14	252–253	Ex 16, 18
	244–245	Ex Odd 1–15	244–245	Ex 12, 14	258	Ex 14, 16
			248–249	Orals All	261	Ex Odd 1–19
				Ex Odd 1–15		
72	240–241	Ex 8, 10	244–245	Ex 16, 18	258	Ex 18, 20
	244–245	Ex 8, 10	248–249	Ex 10, 12	261	Ex 16, 18
	248–249	Orals All	252–253	Orals All	262	Applications
		Ex Odd 1–7		Ex Odd 1–15	264	Chapter Review
				Algebra Review		
73	244–245	Ex 12, 14	248–249	Ex 14, 16	265	Chapter Test
	248–249	Ex 2, 4	252–253	Ex 10, 12		
	252–253	Orals All	257–258	Orals All		
		Ex Odd 1–7		Ex Odd 1–15		
		Algebra Review				
74	248–249	Ex 6, 8	252–253	Ex 14, 16	273	Ex Odd 1–43
	252–253	Ex 2, 4	257–258	Ex 8, 10		
	257–258	Orals All	261	Ex Odd 1–15		
		Ex Odd 1–11				
75	252–253	Ex 6, 8	257–258	Ex 12, 14	273	Ex Odd 45–61
	257–258	Ex 4, 6	261	Ex 10, 12		
	261	Ex Odd 1–9	262	Applications		
			264	Chapter Review		

Day	Page	Level A	Page	Level B	Page	Level C
76	257–258	Ex 8, 10	265	Chapter Test	273	Ex 42, 44
	261	Ex 6, 8			276–277	Ex Odd 1–21
	262	Applications				
	264	Chapter Review				
77	265	Chapter Test	272–273	Orals All	273	Ex 60, 62
				Ex Odd 1–49	276–277	Ex 14, 16
					280–281	Ex Odd 1–27
						Non-Routine
78	272–273	Orals All	273	Ex 44, 46	276–277	Ex 18, 20
		Ex Odd 1–31	276–277	Ex Odd 1–17	280–281	Ex 22, 24
					284	Ex Odd 1–21
79	273	Ex 26, 28	273	Ex 48, 50	280–281	Ex 26, 28
	276–277	Ex Odd 1–9	276–277	Ex 12, 14	284	Ex 16, 18
			279–281	Reading Ex 1–8	287–289	Ex Odd 1–37
				Ex Odd 1–23		
80	273	Ex 30, 32	276–277	Ex 16, 18	284	Ex 20, 22
	276–277	Ex 4, 6	279–281	Ex 16, 18	288–289	Ex 34, 36
	279–281	Reading Ex 1–8	283–284	Orals All	293	Ex Odd 1–19
		Ex Odd 1–11		Ex Odd 1–19		
81	276–277	Ex 8, 10	279–281	Ex 20, 22	288–289	Ex 38, 39
	279–281	Ex 4, 6	283–284	Ex 12, 14	293	Ex 14, 16
	283–284	Orals All	287–289	Reading Ex 1–8	295–296	Ex Odd 1–19
		Ex Odd 1–15		Ex Odd 1–21		
82	279–281	Ex 8, 10	284	Ex 16, 18	293	Ex 18, 20
	284	Ex 10, 12	288–289	Ex Odd 23–27	295–296	Ex 12, 14
	287–289	Reading Ex 1–8			299	Ex Odd 1–11
		Ex Odd 1–21				
83	284	Ex 14, 16	288–289	Ex 18, 20	295–296	Ex 16, 18
	287–289	Ex Odd 23–27	292–293	Orals All	299	Ex 8, 10
				Ex Odd 1–15	300	Applications
					302	Chapter Review
84	289	Ex 14, 16	288–289	Ex 24, 26	303	Chapter Test
	292–293	Orals All	293	Ex 8, 10		
		Ex Odd 1–7	295–296	Ex Odd 1–17		
				Algebra Review		
85	289	Ex 18, 22	293	Ex 12, 14	308–309	Reading Ex 1–13
	293	Ex 2, 4	295–296	Ex 12, 14		Ex Odd 1–19
	295–296	Ex Odd 1–11	299	Ex Odd 1–7		
		Algebra Review				

Day	Page	Level A	Page	Level B	Page	Level C
86	293 295–296 299	Ex 6, 8 Ex 4, 6 Ex 1–4	295–296 299 300 302	Ex 16, 18 Ex 4, 6 Applications Chapter Review	308–309	Ex Odd 21–31
87	295–296 300 302	Ex 8, 10 Applications Chapter Review	303	Chapter Test	308–309 312–314	Ex 22, 24 Ex Odd 1–25 Non-Routine
88	303	Chapter Test	308–309	Reading Ex 1–13 Ex Odd 1–25	308–309 312–314	Ex 26, 28 Ex Odd 27–37
89	308–309	Reading Ex 1–13 Ex Odd 1–15	308–309 312–313	Ex 14, 16 Ex Odd 1–27	312–314 316–317	Ex 24, 26 Ex Odd 1–25 Algebra Review
90	308–309 312–313	Ex 10, 12 Ex Odd 1–19	308–309 312–313 316–317	Ex 24, 26 Ex 22, 24 Ex Odd 1–21 Algebra Review	312–314 316–317 320	Ex 28, 30 Ex 18, 20 Ex Odd 1–25
91	308–309 312–313 316–317	Ex 14, 16 Ex 14, 16 Ex Odd 1–13 Algebra Review	312–313 316–317 320	Ex 26, 28 Ex 14, 16 Ex Odd 1–19	316–317 320 322–323	Ex 22, 24 Ex 20, 22 Ex Odd 1–23
92	312–313 316–317 320	Ex 18, 20 Ex 6, 8 Ex Odd 1–15	316–317 320 322–323	Ex 18, 20 Ex 14, 16 Ex Odd 1–17	320 322–323 325	Ex 24, 26 Ex 18, 20 Ex Odd 1–29
93	316–317 320 322–323	Ex 10, 12 Ex 10, 12 Ex Odd 1–11	320 322–323 325	Ex 18, 20 Ex 12, 14 Orals All Ex Odd 1–25	322–323 325 327	Ex 22, 24 Ex 24, 26 Ex Odd 1–11
94	320 322–323 325	Ex 14, 16 Ex 6, 8 Ex Odd 1–11	322–323 325 327	Ex 16, 18 Ex 20, 22 Ex Odd 1–11	325 327–328	Ex 28, 30 Ex Odd 13–33
95	322–323 325 327	Ex 10, 12 Ex 6, 8 Ex Odd 1–9	325 327–328	Ex 24, 26 Ex Odd 13–29	328 330–331	Ex 28, 30 Ex Odd 1–15
96	325 327	Ex 10, 12 Ex Odd 11–23	328 330–331	Ex 24, 26 Ex 1–8	328 330–331	Ex 34, 36 Ex Odd 17–27
97	327 330–331	Ex 10, 12 Ex Odd 1–7	328 330–331	Ex 28, 30 Ex Odd 9–21	330–331 332	Ex 20, 22 Applications

Day	Page	Level A	Page	Level B	Page	Level C
98	327	Ex 14, 16	330–331	Ex 10, 12	330–331	Ex 24, 26
	330–331	Ex 9–12	332	Applications	334	Chapter Review
99	330–331	Ex 2, 4	330–331	Ex 14, 16	335	Chapter Test
	332	Applications	334	Chapter Review		
100	330–331	Ex 6, 8	335	Chapter Test	340–341	Ex Odd 1–21
	334	Chapter Review				
101	335	Chapter Test	340–341	Reading Ex 1–10	340–341	Ex 23, 25
				Ex Odd 1–15	343–344	Ex Odd 1–15
102	340–341	Reading Ex 1–10	340–341	Ex 17, 19	340–341	Ex 27, 29
		Ex Odd 1–17	343–344	Reading Ex 1–6	343–344	Ex Odd 17–27
				Ex Odd 1–15		
103	340–341	Ex 10, 12	340–341	Ex 21, 23	343–344	Ex 20, 22
	343–344	Reading Ex 1–6	343–344	Ex Odd 17–23	349–350	Ex Odd 1–9
		Ex 1–8				
104	340–341	Ex 14, 16	343–344	Ex 18, 20	343–344	Ex 24, 26
	343–344	Ex Odd 9–15	348–350	Reading Ex 1–4	349–350	Ex Odd 11–19
				Orals All		
				Ex Odd 1–9		
105	343–344	Ex 12, 14	343–344	Ex 22, 24	349–350	Ex 12, 14
	348–349	Reading Ex 1–4	348–350	Ex Odd 11, 13	353–354	Ex Odd 1–25
		Orals All				
		Ex 1–4				
106	343–344	Ex 16, 18	348–350	Ex 6, 8	349–350	Ex 16, 18
	348–349	Ex 5–7	353–354	Ex Odd 1–17	353–354	Ex Odd 27–35
107	348–349	Ex 8	348–350	Ex 10, 12	353–354	Ex 28, 30
	353–354	Ex Odd 1–11	353–354	Ex 19–27	357–358	Ex Odd 1–25
108	348–349	Ex 9	353–354	Ex 28, 29	353–354	Ex 32, 34
	353–354	Ex Odd 13–25	357–358	Ex Odd 1–19	357–358	Ex 18, 20
					362–363	Ex Odd 1–21
109	353–354	Ex 18, 20	357–358	Ex 14, 16	357–358	Ex 22, 24
	357–358	Ex Odd 1–13	361–363	Reading Ex 1–6	362–363	Ex Odd 23–39
				Orals All		
				Ex Odd 1–21		
101	353–354	Ex 22, 24	357–358	Ex 18, 20	362–363	Ex 26, 28
	357–358	Ex 8, 10	362–363	Ex Odd 23–33	366–367	Ex Odd 1–27
	361–362	Reading Ex 1–6				Non-Routine
		Orals All				
		Ex Odd 1–11				

Day	Page	Level A	Page	Level B	Page	Level C
111	357–358	Ex 12, 14	363	Ex 28, 30	366–367	Ex 24, 26
	362	Ex 13–21	366–367	Ex Odd 1–23	370–371	Ex Odd 1–23
112	362	Ex 14, 16	363	Ex 32, 34	366–367	Ex 20, 22
	366	Ex Odd 1–17	367	Ex 16, 18	370–372	Ex Odd 25–35
			370–371	Ex Odd 1–19		
113	362	Ex 18, 20	367	Ex 20, 22	370–372	Ex 28, 30
	366	Ex 10, 12	370–371	Ex Odd 21–29	375–376	Ex Odd 1–23
	370–371	Ex Odd 1–11				
114	366	Ex 14, 16	370–371	Ex 22, 24	370–372	Ex 32, 34
	371	Ex Odd 13–23	375–376	Ex Odd 1–15	376	Ex 16, 18
					379	Ex Odd 1–23
115	371	Ex 16, 18	370–371	Ex 26, 28	376	Ex 20, 22
	375	Ex Odd 1–9	375–376	Ex 10, 12	379	Ex 18, 20
			379	Ex Odd 1–17	380	Applications
					382	Chapter Review
116	375	Ex 2, 4	376	Ex 14, 16	383	Chapter Test
	379	Ex Odd 1–9	379	Ex 10, 12		
			380	Applications		
			382	Chapter Review		
117	375	Ex 6, 8	383	Chapter Test	389–390	Ex Odd 1–35
	379	Ex 8, 10				Algebra Review
	380	Applications				
118	382	Chapter Review	389–390	Orals All	389–390	Ex 28, 30
				Ex Odd 1–17	393–394	Ex Odd 1–35
				Algebra Review		
119	383	Chapter Test	389–390	Ex Odd 19–29	389–390	Ex 32, 34
					394	Ex 26, 28
					397	Ex Odd 1–19
120	389–390	Orals All	389–390	Ex 22, 24	394	Ex 30, 32
		Ex Odd 1–17	393–394	Ex Odd 1–25	397	Ex 14, 16
		Algebra Review			400–401	Ex Odd 1–25
121	389	Ex 12, 14	389–390	Ex 26, 28	397	Ex 18, 20
	393–394	Ex Odd 1–15	394	Ex 18, 20	400–401	Ex 18, 20
			396–397	Reading Ex 1–8	404–405	Ex Odd 1–25
				Orals All		Non-Routine
				Ex Odd 1–17		

Day	Page	Level A	Page	Level B	Page	Level C
122	389	Ex 16, 18	394	Ex 22, 24	400–401	Ex 22, 24
	393–394	Ex 8, 10	397	Ex 12, 14	404–405	Ex 20, 22
	396–397	Reading Ex 1–8	400	Orals All	407–408	Ex Odd 1–31
		Orals All		Ex Odd 1–17		
		Ex Odd 1–7				
123	394	Ex 12, 14	397	Ex 16, 18	404–405	Ex 24, 26
	397	Ex Odd 9–15	400	Ex 10, 12	407–408	Ex 26, 28
			404	Ex Odd 1–19	411	Ex Odd 1–19
124	397	Ex 10, 12	400	Ex 14, 16	407–408	Ex 30, 32
	400–401	Orals All	404	Ex 14, 16	411	Ex 14, 16
		Ex Odd 1–9	407–408	Ex Odd 1–21	413–414	Ex Odd 1–29
125	397	Ex 14, 16	404	Ex 18, 20	411	Ex 18, 20
	400	Ex 4, 6	408	Ex 14, 16	413–414	Ex Odd 31–41
	404	Ex Odd 1–11	411	Reading Ex 1–4		
				Ex Odd 1–15		
126	400	Ex 8, 10	408	Ex 18, 20	413–414	Ex 34, 36
	404	Ex 6, 8	411	Ex 10, 12	416–417	Ex Odd 1–21
	407	Ex Odd 1–9	413–414	Ex Odd 1–27		Algebra Review
127	404	Ex 10, 12	411	Ex 14, 16	413–414	Ex 38, 40
	407	Ex 4, 6	413–414	Ex Odd 29–37	416–417	Ex 16, 18
	411	Reading Ex 1–4			419–420	Ex Odd 1–29
		Ex Odd 1–15				
128	407	Ex 8, 10	413–414	Ex 30, 32	416–417	Ex 20, 22
	411	Ex 2, 4	416–417	Ex Odd 1–17	419–420	Ex 24, 26
	413–414	Ex Odd 1–15		Algebra Review	421	Applications
					422	Chapter Review
129	411	Ex 6, 8	413–414	Ex 34, 36	423	Chapter Test
	413	Ex Odd 17–27	416–417	Ex 12, 14		
			419–420	Ex Odd 1–23		
130	413	Ex 22, 24	416–417	Ex 16, 18	431	Ex Odd 1–33
	416–417	Ex Odd 1–9	419–420	Ex 16, 18		
		Algebra Review	421	Applications		
			422	Chapter Review		
131	413	Ex 26, 28	423	Chapter Test	431	Ex 28, 30
	417	Ex 2, 4			434	Ex Odd 1–29
	419–420	Ex Odd 1–17				
132	417	Ex 6, 8	431	Orals All	431	Ex 32, 34
	420	Ex 14, 16		Ex Odd 1–29	434	Ex 22, 24
	421	Applications			437	Ex Odd 1–31

Day	Page	Level A	Page	Level B	Page	Level C
133	422	Chapter Review	431	Ex 20, 22	434	Ex 26, 28
			434	Ex Odd 1–23	437	Ex 24, 26
					440–441	Ex Odd 1–31
						Algebra Review
134	423	Chapter Test	431	Ex 24, 26	437	Ex 28, 30
			434	Ex 18, 20	440–441	Ex 24, 26
			437	Reading Ex 1–4	445	Ex Odd 1–25
				Ex Odd 1–27		
135	431	Orals All	434	Ex 22, 24	440–441	Ex 28, 30
		Ex Odd 1–19	437	Ex 20, 22	445	Ex Odd 27–45
			440–441	Orals All		
				Ex Odd 1–27		
				Algebra Review		
136	431	Ex 14, 16	437	Ex 24, 26	445	Ex 40, 42
	434	Ex Odd 1–15	440	Ex 20, 22	448	Ex Odd 1–41
			444–445	Orals All		
				Ex Odd 1–27		
137	431	Ex 18, 20	440	Ex 24, 26	445	Ex 44, 46
	434	Ex 8, 10	445	Ex Odd 29–39	448	Ex 36, 38
	437	Reading Ex 1–4			451–452	Ex Odd 1–27
		Ex Odd 1–19				Non-Routine
138	434	Ex 12, 14	445	Ex 34, 36	448	Ex 40, 42
	437	Ex 14, 16	448	Orals All	451–452	Ex 22, 24
	440–441	Orals All		Ex Odd 1–39	454–455	Ex Odd 1–15
		Ex Odd 1–17				Algebra Review
		Algebra Review				
139	437	Ex 18, 20	445	Ex 38, 40	452	Ex 26, 28
	440	Ex 10, 12	448	Ex 34, 36	455	Ex 10, 12
	444–445	Orals All	451–452	Reading Ex 1–8	458	Ex Odd 1–27
		Ex Odd 1–17		Orals All		
				Ex Odd 1–21		
140	440	Ex 14, 16	448	Ex 38, 40	455	Ex 14, 16
	445	Ex Odd 19–27	451–452	Ex 16, 18	458	Ex Odd 29–47
			454–455	Ex Odd 1–11		
				Algebra Review		
141	445	Ex 20, 22	452	Ex 20, 22	458	Ex 42, 44
	448	Orals All	454	Ex 6, 8	460	Ex Odd 1–29
		Ex Odd 1–27	457–458	Orals All		
				Ex Odd 1–27		

Day	Page	Level A	Page	Level B	Page	Level C
142	445	Ex 24, 26	455	Ex 10, 12	460	Ex 26, 28
	448	Ex 22, 24	458	Ex Odd 29–39	461	Applications
	451	Reading Ex 1–8			462	Chapter Review
		Orals All				
		Ex Odd 1–11				
143	448	Ex 26, 28	458	Ex 34, 36	463	Chapter Test
	451	Ex 6, 8	460	Ex Odd 1–25		
	454–455	Ex Odd 1–7				
		Algebra Review				
144	451	Ex 10, 12	458	Ex 38, 40	469	Ex Odd 1–15
	454	Ex 2, 4	460	Ex 18, 20		
	457–458	Orals All	461	Applications		
		Ex Odd 1–15	462	Chapter Review		
145	454	Ex 6, 8	463	Chapter Test	469	Ex 10, 12
	458	Ex Odd 17–27			473–474	Ex Odd 1–15
						Non-Routine
						Algebra Reviews
146	458	Ex 22, 24	469	Ex Odd 1–11	469	Ex 14, 16
	460	Ex Odd 1–13			473–474	Ex 8, 10
					477–478	Ex Odd 1–23
147	458	Ex 26, 28	469	Ex 6, 8	473–474	Ex 12, 14
	460	Ex 10, 12	473	Ex Odd 1–11	477–478	Ex 18, 20
	461	Applications		Algebra Reviews	481–482	Ex Odd 1–17
	462	Chapter Review				Non-Routine
						Algebra Review
148	463	Chapter Test	469	Ex 10, 12	477–478	Ex 22, 24
			473	Ex 6, 8	481–482	Ex 12, 14
			477–478	Ex Odd 1–19	486	Ex Odd 1–21
149	469	Ex Odd 1–7	473	Ex 10, 12	481–482	Ex 16, 18
			478	Ex 14, 16	486	Ex 14, 16
			481–482	Orals All	489–490	Ex Odd 1–27
				Ex Odd 1–13		
				Algebra Review		
150	469	Ex 2, 4	478	Ex 18, 20	486	Ex 18, 20
	473–474	Ex Odd 1–7	481	Ex 8, 10	489–490	Ex 20, 22
		Algebra Reviews	486	Ex Odd 1–17	492	Ex Odd 1–15
151	469	Ex 6, 8	481	Ex 12, 14	489–490	Ex 24, 26
	473	Ex 2, 4	486	Ex 12, 14	492	Ex 12, 14
	477–478	Ex Odd 1–13	489–490	Reading Ex 1–10	493	Applications
				Ex Odd 1–21	494	Chapter Review

Day	Page	Level A	Page	Level B	Page	Level C
152	473	Ex 6, 8	486	Ex 16, 18	495	Chapter Test
	478	Ex 8, 10	489–490	Ex 14, 16		
	481–482	Orals All	492	Ex Odd 1–13		
		Ex Odd 1–5				
		Algebra Review				
153	478	Ex 12, 14	489–490	Ex 18, 20	501–502	Ex Odd 1–15
	481	Ex 2, 4	492	Ex 8, 10		Algebra Review
	486	Ex Odd 1–7	493	Applications		
			494	Chapter Review		
154	481	Ex 6, 8	495	Chapter Test	501–502	Ex 8, 10
	486	Ex 2, 4			505–506	Ex Odd 1–15
	489	Reading Ex 1–10				Non-Routine
		Ex Odd 1–17				
155	486	Ex 6, 8	501–502	Ex Odd 1–11	501–502	Ex 12, 14
	489	Ex 12, 14		Algebra Review	505–506	Ex 8, 10
	492	Ex Odd 1–9			509	Ex Odd 1–11
156	489	Ex 16, 18	501	Ex 6, 8	505–506	Ex 12, 14
	492	Ex 4, 6	505–506	Orals All	509	Ex 6, 8
	493	Applications		Ex Odd 1–11	514–515	Ex Odd 1–11
	494	Chapter Review				Algebra Review
157	495	Chapter Test	501–502	Ex 10, 12	509	Ex 10, 12
			505	Ex 6, 8	514–515	Ex 6, 8
			508–509	Orals All	518	Ex Odd 1–13
				Ex Odd 1–9		
158	501	Ex Odd 1–9	505–506	Ex 10, 12	514–515	Ex 10, 12
	502	Algebra Review	509	Ex 2, 4	518	Ex 12, 14
			513–514	Reading Ex 1–6	519	Applications
				Orals All	520	Chapter Review
				Ex Odd 1–9		
159	501	Ex 2, 4	509	Ex 6, 8	521	Chapter Test
	505–506	Orals All	514	Ex 2, 4		
		Ex Odd 1–11	515	Algebra Review		
			517–518	Orals All		
				Ex Odd 1–11		
160	501	Ex 6, 8	514	Ex 6, 8	527–528	Ex Odd 1–21
	505–506	Ex 4, 6	518	Ex 8, 10		Non-Routine
	508–509	Orals All	519	Applications		
		Ex 1, 3, 5	520	Chapter Review		

Day	Page	Level A	Page	Level B	Page	Level C
161	505–506	Ex 8, 10	521	Chapter Test	527–528	Ex 14, 16
	509	Ex 2, 4			531	Ex Odd 1–17
	513–515	Reading Ex 1–6				
		Orals All				
		Ex Odd 1–5				
		Algebra Review				
162	509	Ex 6, 8	527	Orals All	527–528	Ex 18, 20
	514	Ex 2, 4		Ex Odd 1–17	531	Ex 12, 14
	517–518	Orals All			534	Ex Odd 1–23
		Ex 1, 3, 5				
163	514	Ex 6, 8	527	Ex 10, 12	531	Ex 12, 18
	518	Ex 2, 4	531	Ex Odd 1–13	534	Ex 18, 20
	519	Applications			537–538	Ex Odd 1–19
	520	Chapter Review				
164	521	Chapter Test	527	Ex 14, 16	534	Ex 22, 24
			531	Ex 8, 10	537–538	Ex Even 2–14
			533–534	Orals All		
				Ex Odd 1–19		
165	527	Orals All	531	Ex 12, 14	537–538	Ex 16, 18
		Ex Odd 1–11	534	Ex 12, 14	542–543	Ex Odd 1–19
			537–538	Reading Ex 1–11		Algebra Review
				Ex Odd 1–15		
166	527	Ex 4, 6	534	Ex 16, 18	542–543	Ex 14, 16
	531	Ex Odd 1–9	538	Ex 10, 12	546–547	Ex Odd 1–21
			542–543	Ex Odd 1–15		
				Algebra Review		
167	527	Ex 8, 10	538	Ex 14, 16	542–543	Ex 18, 20
	531	Ex 4, 6	542	Ex 10, 12	546–547	Ex 16, 18
	533–534	Orals All	546–547	Reading Ex 1–4	551–552	Ex Odd 1–19
		Ex Odd 1–13		Ex Odd 1–17		Non-Routine
168	531	Ex 8, 10	542	Ex 14, 16	546–547	Ex 20, 22
	534	Ex 6, 8	547	Ex 12, 14	551–552	Ex Even 2–16
	537–538	Reading Ex 1–11	551	Ex Odd 1–15		
		Ex Odd 1–9				
169	534	Ex 10, 12	547	Ex 16, 18	551–552	Ex 18, 20
	537–538	Ex 4, 6	551	Ex 14, 16	553	Applications
	542–543	Ex Odd 1–9	553	Applications	554	Chapter Review
		Algebra Review	554	Chapter Review		
170	538	Ex 8, 10	555	Chapter Test	555	Chapter Test
	542	Ex 8, 10				
	546–547	Reading Ex 1–4				
		Ex Odd 1–11				

Answers to Applications Pages

Page 17 **1.** 58 mi **2.** 135 mi **3.** 57 **4.** 168 **5.** 115 **6.** 160° **7.** 10 **8.** 34, 28

Page 54 **1.** Check to see that the angles with the floor on two adjacent sides are right angles. **2.** 3 pipes and 2 pipes; Post. 2 and Th. 2.2

Page 98 **1.** The two angle measures would have to total 180°. (Th. 3.4) **2.** Assume that the bottom edge of the board is the transversal. Corresponding angles must be cut to be congruent. **3.** A vertical line makes a right angle with a plane tangent to the surface of the earth at the point of intersection. Vertical lines are not parallel. They intersect at the center of the earth.

Page 128 **1.** Assume that the post forms a right angle with the level ground. By turning the rods and sighting on land, we get congruent right triangles (ASA). Distance on land equals distance from ship to shore. **2.** The top design is the correct one. There must be tension on the ropes to keep the triangle rigid.

Page 177 **1.** By Th. 4.1, angle at pole 1 has a measure of $d°$. The triangle is isosceles, so side opposite pole 1 will have the same length as distance between poles 1 and 2. **2.** Place homemade level on surface; let bob hang. Plumb will always form $\perp$ to horizontal because of gravity.

Page 202 **1.** Sum of $\angle$ measures about a common vertex is 360°; symmetric properties of squares **2.** No; the measure of interior $\angle$ must be divisor of 360°. Only regular triangles, quadrilaterals, and hexagons can be used.

Page 233 **1.** Mark off equal distances along the sidelines. Parallelograms are formed (opp. sides parallel and congruent). **2.** A parallelogram is formed by the parts of the frame. The light will be parallel to the opposite side.

Page 262 **1.** There are a number of ways that the tables can be arranged. They cannot be arranged to form a regular octagon if the edges touch. The angles would need to be changed from 120° to 135° to form the octagon. **2.** No; the top edge is 8 ft, the bottom edge is 10 ft, and the height is $\sqrt{3}$ ft.

Page 300 **1.** Count the number of times the pencil can be moved up the building and multiply by the height of the helper. The triangles formed are approximately similar. **2.** $\frac{5}{8} = \frac{h}{48}$; 30 ft tall; the distances from the mirror to the observer and from the mirror to the tree, and the height above ground at eye level must be known. The triangles formed are similar.

Page 332 **1.** The shortest path is from the corner to the midpoint of the opposite sides; 64 m. **2.** This spacing of knots forms a rt $\triangle$ with sides of lengths 3-4-5.

Page 380 **1.** $\angle SME$ is a rt $\angle$ (Th. 11.5). $\sin 24° = \frac{SM}{92{,}897{,}000}$; $SM = 37{,}781{,}000$ mi or 60,803,000 km. **2.** Pt. outside $\odot$ creates $\angle$ formed by 2 secants, m = one-half diff. of int. arcs; pt. D on $\odot$ gives inscrib. $\angle$, m = one-half int. arc; pt. inside $\odot$ gives $\angle$ formed by two chords, m = one-half sum of int. arcs; $\overset{\frown}{B_1B_2}$ is constant; hence m $\angle B_1PB_2$ increases as ship moves into shallow water.

Page 421 **1.** $A = \frac{1}{2}(17) \cdot [(0 + 17) + (17 + 21) + (21 + 22) + (22 + 19) + (19 + 0)]$; $A = 1{,}343$ m^2 **2.** $A = \frac{1}{2}(10) \cdot [(0 + 32) + (32 + 26) + (26 + 28) + (28 + 34) + (34 + 30) + (30 + 28)]$; $A = 1{,}640$ m^2 or 0.164 hectare

Page 461 **1.** 6 blocks **2.** $2\sqrt{5}$ blocks **3.** 60 in. **4.** $\frac{2}{16}$ or 0.125

Page 493 **1.** Check students' sketches. **2.** Check students' drawings.

Page 519 **1.** The 45 ft remaining (100 − 55) is not sufficient to complete the triangle (Th. 15.1). **2.** The cable lengths will be unequal. $a^2 + b_1^2 = c_1^2$, $a^2 + b_2^2 = c_2^2$; if $b_1 < b_2$, then $c_1 < c_2$.

Page 553 **1.** $V = 50 \cdot 1.5 \cdot 1 = 75$ cu. ft; $75 \div 27 = 2.8$ cu. yd. **2.** Volume cone = $\frac{1}{3}\pi(3)^2(12)$; 36π cm^3; volume cylinder = $\pi(3.5)^2(4)$; 49π cm^3; cylinder holds 13π or 40.8 cm^3 more **3.** $V_L = \frac{1}{3}\pi(1.5)^2(5)$; 3.75π in.^{3}; $V_R = \frac{1}{3}\pi(2.5)^2(3)$; 6.25 in.^{3}; $V_R - V_L = 2.5\pi$ in.^{3}; use the cup on the left

Annotated Answers

(Answers too long to be included on the student pages.)

Page 16

9.–12. Draw angles as on p 12; then follow constr on pp 14–15.

Page 20

1.

```
10 PRINT "MIDPOINT":: INPUT M
20 PRINT "LENGTH OF SEGMENT":: INPUT L
30 PRINT "THE TWO ENDPOINTS ARE ":M + L / 2;" A
   ND ":M - L / 2
```

Page 28

1. It is snowing and I am going home. It is snowing or I am going home. **2.** Triangles have three sides and squares have four sides. Triangles have three sides or squares have four sides. **3.** Parallel lines never intersect and right angles are not acute. Parallel lines never intersect or right angles are not acute. **4.** The measure of an acute angle is less than 90° and the measure of a straight angle is 180°. The measure of an acute angle is less than 90° or the measure of a straight angle is 180°. **5.** A triangle does not have three sides. **6.** Two lines cannot intersect in one point. **7.** A triangle can have four sides. **8.** Parallel lines intersect.

Pages 31–32

1. If $5 = 4 + 1$, then $6 = 5 + 1$. **2.** If $8 = 4 \cdot 2$, then $9 = 3 \cdot 3$. **3.** If two segments are congruent, then they have the same length. **4.** If a ray in the interior of an angle divides the angle into two congruent angles, then it bisects the angle. **5.** If lines are parallel, then they do not intersect. **6.** If a figure is a square, then it has four right angles. **7.** If lines are parallel, then they do not intersect. **8.** If angles are complementary, then they are not obtuse. **9.** If two segments are equal in length, then they are congruent. **10.** If two angles are congruent, then they are equal in degree measure. **11.** If an angle is a right angle, then it has a measure of 90°. **12.** If two angles are right angles, then they are supplementary. **13.** If an angle is acute, then its measure is less than 90°. **14.** If a figure is a triangle, then it has three sides. **15.** If two angles are congruent, then they are acute; F, F **16.** If two angles are acute, then they are complementary; T, F **17.** If tomorrow is Tuesday, then today is Monday; T, T **18.** If next month is March, then this month is January; F, F **19.** An angle measures 180° if and only if it is a straight angle. **20.** Two points are collinear if and only if they lie on the same line.

21. Converse: If $4 + 3 = 7$, then $7 = 4 + 3$. True; Inverse: If $7 \neq 4 + 3$, then $4 + 3 \neq 7$. True; Contrapositive: If $4 + 3 \neq 7$, then $7 \neq 4 + 3$. True
22. Converse: If $72 \div 9 = 8$, then $9 \cdot 8 = 72$. True; Inverse: If $9 \cdot 8 \neq 72$, then $72 \div 9 \neq 8$. True; Contrapositive: If $72 \div 9 \neq 8$, then $9 \cdot 8 \neq 72$. True
23. Converse: If points lie in the same plane, then they are coplanar. True; Inverse: If points are not coplanar, then they do not lie in the same plane. True; Contrapositive: If points do not lie in the same plane, then they are not coplanar. True
24. Converse: If two angles are straight angles, then they are congruent. True; Inverse: If two angles are not congruent, then they are not straight angles. True; Contrapositive: If two angles are not straight angles, then they are not congruent. False
25. Converse: If it is a good swimmer, then it is a fish. False; Inverse: If it is not a fish, then it is not a good swimmer. False; Contrapositive: If it is not a good swimmer, then it is not a fish. True
26. Converse: If it is formed by a set of points, then it is a geometric figure. True; Inverse: If it is not a geometric figure, then it is not formed by a set of points. True; Contrapositive: If it is not formed by a set of points, then it is not a geometric figure. True
27. Converse: If angles are congruent, then they are obtuse. False; Inverse: If angles are not obtuse,

then they are not congruent. False; Contrapositive: If angles are not congruent, then they are not obtuse. False **28.** Converse: If it is a whole number, then it is an integer. True; Inverse: If it is not an integer, then it is not a whole number. True; Contrapositive: If it is not a whole number, then it is not an integer. False

29.

p	q	$p \to q$	$q \to p$	$(p \to q) \vee (q \to p)$
T	T	T	T	T
T	F	F	T	T
F	T	T	F	T
F	F	T	T	T

30.

p	q	$p \to q$	$q \to p$	$(p \to q) \wedge (q \to p)$
T	T	T	T	T
T	F	F	T	F
F	T	T	F	F
F	F	T	T	T

31.

p	q	$\sim p$	$p \to q$	$\sim p \wedge (p \to q)$
T	T	F	T	F
T	F	F	F	F
F	T	T	T	T
F	F	T	T	T

32.

p	q	$\sim p$	$p \vee q$	$\sim p \to (p \vee q)$
T	T	F	T	T
T	F	F	T	T
F	T	T	T	T
F	F	T	F	F

Note abbreviated form used for ex. 33–37.

33. $[(p \to q) \wedge p] \to q$

T T T T T
T F F F T
F T T F T
F T F F T

34. $[(p \vee q) \wedge \sim p] \to q$

T T T F F T
T T F F F T
F T T T T T
F F F F T T

35. $\sim (p \wedge q) \to (\sim p \vee \sim q)$

F T T T T F F F
T T F F T F T T
T F F T T T T F
T F F F T T T T

36. $\sim (p \vee q) \to (\sim p \wedge \sim q)$

F T T T T F F F
F T T F T F F T
F F T T T T F F
T F F F T T T T

37. $p \leftrightarrow q$ means $(p \to q) \wedge (q \to p)$

T T T T T
T F F F T
F T T F F
F T F T T

Page 39

5. m $\angle A$ = m $\angle B$ (Given); m $\angle B$ = m $\angle C$ (Given); m $\angle A$ = m $\angle C$ (Trans prop); $\angle A \cong \angle C$ (Def of $\cong \angle$s); **6.** m $\angle D = 79$ (Given); m $\angle E = 79$ (Given); m $\angle D$ = m $\angle E$ (Trans prop); $\angle D \cong \angle E$ (Def of $\cong \angle$s); **7.** m $\angle WOX$ + m $\angle XOY$ = m $\angle ZOY$ + m $\angle XOY$ (Given); m $\angle WOX$ = m $\angle ZOY$ (Subtr prop); $\angle WOX \cong \angle ZOY$ (Def of $\cong \angle$s) **8.** $\angle WOX \cong \angle ZOY$ (Given); m $\angle WOX$ = m $\angle ZOY$ (Def of $\cong \angle$s); m $\angle WOX$ + m $\angle XOY$ = m $\angle ZOY$ + m $\angle XOY$ (Add prop)
9. m $\angle A$ + m $\angle B$ = 103 (Given); m $\angle B$ + m$\angle C$ = 103 (Given); m $\angle A$ + m $\angle B$ = m$\angle B$ + m$\angle C$ (Trans prop); m $\angle A$ = *m* $\angle C$ (Subtr prop)

10. m $\angle D$ − m $\angle E$ = 13 (Given); m $\angle F$ − m $\angle E$ = 13 (Given); m $\angle D$ − m $\angle E$ = m $\angle F$ − m $\angle E$ (Trans prop); m $\angle D$ = m $\angle F$ (Add prop) **11.** m $\angle G$ = 35 (Given); m $\angle H$ = 2(m $\angle G$) (Given); m $\angle I$ = 70 (Given); m $\angle H$ = 2(35) = 70 (Subst prop); m $\angle H$ = m $\angle I$ (Trans prop); $\angle H \cong \angle I$ (Def of $\cong \angle$s) **12.** $JK + KL = 18$ (Given); $KL + LM = 18$ (Given); $JK + KL = KL + LM$ (Trans prop); $JK = LM$ (Subtr prop) **13.** $NO - RS = 25$ (Given); $PQ - RS = 25$ (Given); $NO - RS = PQ - RS$ (Trans prop); $NO = PQ$ (Add prop)
14. $AB = 7$ (Given); $CD = 2(AB)$ (Given); $EF = 14$ (Given); $CD = 2(7) = 14$ (Subst prop); $CD = EF$ (Trans prop); $\overline{CD} \cong \overline{EF}$ (Def of $\cong$ seg)
15. $\angle WYZ \cong \angle QST$ (Given); m $\angle WYZ$ = *m* $\angle QST$ (Def of $\cong \angle$s); m $\angle XYZ = \frac{1}{2}$(m$\angle WYZ$) (Given); m$\angle XYZ = \frac{1}{2}$(m $\angle QST$) (Subst prop); m $\angle RST = \frac{1}{2}$(m $\angle QST$) (Given); m $\angle XYZ$ = m $\angle RST$ (Trans prop); $\angle XYZ \cong \angle RST$ (Def of $\cong \angle$s) **16.** $\overline{LM} \cong \overline{PQ}$ (Given); $LM = PQ$ (Def of $\cong$ seg); $JK = \frac{1}{2}(LM)$ (Given); $NO = \frac{1}{2}(PQ)$ (Given); $NO = \frac{1}{2}(LM)$ (Subst prop); $JK = NO$ (Trans prop); $\overline{JK} \cong \overline{NO}$ (Def of $\cong$ seg)

Page 43

22. Assume line l intersects pl p in 2 pt. Then l is contained in p (Post 3). This contradicts the given.
23. Assume line x lies in a plane z (Post 3); Assume line y intersects line x in one pt. The pt of intersection lies in plane z and line y contains an infinite no. of collinear pt (Post 5); 3 noncollinear pt all lie on plane z (Post 2).

Pages 47–48

18. $BD = BD$ (reflex prop); $AB + BD = BD + DE$ (add prop); $AD = AB + BD$ and $BE = BD + DE$ (def of between); $AD = BE$ (subst prop)
19. $BC = BC$ (reflex prop); $AC - BC = BD - BC$ (subtr prop); $AC = AB + BC$ and $BD = BC + CD$ (def of between); $AB = AC - BC$ and $CD = BD - BC$ (subtr prop); $AB = CD$ (subst prop)
20. m $\angle 1$ = m $\angle 3$ (def of $\cong \angle$s); m $\angle 2$ = m $\angle 2$ (reflex prop); m $\angle 1$ + m $\angle 2$ = m $\angle 2$ + m $\angle 3$ (add prop); m $\angle AOC$ = m $\angle 1$ + m $\angle 2$ and m $\angle DOB$ = m $\angle 2$ + m $\angle 3$ (Post 10); m $\angle AOC$ = m $\angle DOB$ (subst prop); $\angle AOC \cong \angle DOB$ (def of $\cong \angle$s)
21. m $\angle AOC$ = m $\angle DOB$ (def of $\cong \angle$s); m $\angle AOC$ = m $\angle 1$ + m $\angle 2$ and m $\angle DOB$ = m $\angle 2$ + m $\angle 3$ (Post 10); m $\angle 1$ + m $\angle 2$ = m $\angle 2$ + m $\angle 3$ (subst prop); m $\angle 2$ = m $\angle 2$ (reflex prop); m $\angle 1$ = m $\angle 3$ (subtr prop); $\angle 1 \cong \angle 3$ (def of $\cong \angle$s)
22. $EH = FI$ and $HG = IG$ (def of $\cong$ seg); $EH + HG = FI + IG$ (add prop); $EG = EH + HG$ and $FG = FI + IG$ (def of betw); $EG = FG$ (subst prop); $\overline{EG} \cong \overline{FG}$ (def of $\cong$ seg) **23.** $EG = FG$ and $EH = FI$ (def of $\cong$ seg); $EG = EH + HG$ and $FG = FI + IG$ (def of betw); $EH + HG = FI + IG$ (subst prop); $HG = IG$ (subtr prop); $\overline{HG} \cong \overline{IG}$ (def of $\cong$ seg) **24.** $JL = NL$ (def of $\cong$ seg); $\frac{1}{2}JL = \frac{1}{2}NL$ (div prop); $JK = NM$ (subst prop); $\overline{JK} \cong \overline{NM}$ (def of $\cong$ seg) **25.** $JK = NM$ (def of $\cong$ seg); $\frac{1}{2}JL = \frac{1}{2}NL$ (subst prop); $JL = NL$ (mult prop); $\overline{JL} \cong \overline{NL}$ (def of $\cong$ seg) **26.** $\overline{OP} \cong \overline{RS}$ (def of $\cong$ seg); $OP = 2OQ$ and $RS = 2TS$ (def of seg bis); $2OQ = 2TS$ (subst prop); $OQ = TS$ (div prop); $\overline{OQ} \cong \overline{TS}$ (def of $\cong$ seg) **27.** $OQ = QP$, $OQ = RT$ and $OP = KS$ (def of $\cong$ seg); $OP = OQ + QP$ and $RS + TS$ (def of betw); $OQ + QP = RT + TS$ (subst prop); $QP = TS$ (subtr prop); $RT = TS$ (subst prop); T bisects $\overline{RS}$ (def of bis)

Page 53

6. m $\angle AOB$ + m $\angle BOC = 90$ (def of compl $\angle$s); m $\angle AOC$ = m $\angle AOB$ + m $\angle BOC$ (Post 10); m $\angle AOC = 90$ (subst prop); $\angle AOC$ is a rt $\angle$ (def of rt $\angle$); $\overrightarrow{OC} \perp \overleftrightarrow{AD}$ (def of $\perp$) **7.** m $\angle XYT$ + m $\angle WZV = 180$ (def of suppl $\angle$s); m $\angle XYT$ = m $\angle WZV$ (def of $\cong$ $\angle$s); m $\angle WZV$ + m $\angle WZV$ = 180, or 2m $\angle WZV = 180$ (subst prop); m $\angle WZV$ = 90 (div prop); $\overrightarrow{ZV} \perp \overleftrightarrow{XW}$ (def of $\perp$)
11. $\angle AEC \cong \angle AED$ (Th 2.6, def of $\cong$ $\angle$s)
12. $\angle AED$ and $\angle DEB$ are rt $\angle$s); m $\angle AED = 90$, m $\angle DEB = 90$ (def of rt $\angle$s); m $\angle AEC$ + m $\angle DEB = 180$ (add prop)

Page 58

10. $\angle 1 \cong \angle 3$ (Th 2.10 vert $\angle$s are $\cong$); $\angle 1 \cong \angle 5$ (subst prop) **11.** m $\angle 1$ + m $\angle 2 = 180$ (Post 11); $\angle 2$ is suppl to $\angle 1$ (def of suppl $\angle$s); $\angle 5$ is suppl to $\angle 2$ (subst prop)

Page 60

1. Given: $\angle DOC \cong \angle COB$; $\angle AOB \cong \angle COB$; prove: $\angle DOC \cong \angle AOB$ **2.** Given: $\overline{WX} \cong \overline{XY}$; $\overline{YZ} \cong \overline{XY}$; prove: $\overline{WX} \cong \overline{YZ}$ **3.** Given: $\overleftrightarrow{AB}$ and $\overleftrightarrow{CD}$ intersect at E forming vert $\angle$s, $\angle 1$ and $\angle 2$, $\angle 3$ is suppl to $\angle 1$, $\angle 4$ is suppl to $\angle 2$; Prove: $\angle 3 \cong \angle 4$
4. Given: $\overleftrightarrow{AB}$ and $\overleftrightarrow{CD}$ intersect at E forming vert $\angle$s, $\angle 1$ and $\angle 2$, $\angle 3$ is compl to $\angle 1$, $\angle 4$ is compl to $\angle 2$; Prove: $\angle 3 \cong \angle 4$. **5.** Given: $\angle 3$ is adj and suppl to $\angle 1$, $\angle 2$ is adj and suppl to $\angle 1$; Prove: $\angle 3$ and $\angle 2$ are vert $\angle$s. **6.** Given: $\overleftrightarrow{GI}$ and $\overleftrightarrow{JH}$ intersect at O forming vert $\angle$s, $\angle 1$ and $\angle 2$; Prove: $\angle 1$ and $\angle 2$ are adj and suppl to $\angle 3$. **7.** Given: $\angle 2$ is compl to $\angle 1$, $\angle 5$ is compl to $\angle 4$, $\angle 2 \cong \angle 5$, $(\angle 3 + \angle 2)$ is suppl to $\angle 1$, $(\angle 6 + \angle 5)$ is suppl to $\angle 4$; Prove: $\angle 3 + \angle 2 \cong \angle 6 + \angle 5$. $\angle 1 \cong \angle 4$ (Th 2.9); $(\angle 3 + \angle 2) \cong (\angle 6 + \angle 5)$ (Th 2.8) **8.** Given: $(\angle 3 + \angle 2)$ is suppl to $\angle 1$, $(\angle 6 + \angle 5)$ is suppl to $\angle 4$, $(\angle 3 + \angle 2) \cong (\angle 6 + \angle 5)$, $\angle 2$ is compl to $\angle 1$, $\angle 5$ is compl to $\angle 4$; Prove: $\angle 2 \cong \angle 5$. $\angle 1 \cong \angle 4$ (Th 2.8); $\angle 2 \cong 5$ (Th 2.9) **9.** Given: $\angle 1$ and $\angle 2$ are compl, $\angle 1 = \angle 2$; Prove: m $\angle 1 = 45$; m $\angle 2 = 45$. m $\angle 1$ + m $\angle 2 = 90$ (def of comp $\angle$s); m $\angle 1$ = m $\angle 2$ (def of $\cong$ $\angle$s); 2 m $\angle 1 = 90$ and 2 m $\angle 2 = 90$ (subst prop); m $\angle 1 = 45$ and m $\angle 2 = 45$ (div prop) **10.** Given: $\overleftrightarrow{AC} \perp \overleftrightarrow{BD}$ forming adj $\angle$s 1, 2, 3, 4; Prove: $\angle 1 \cong \angle 2 \cong \angle 3 \cong \angle 4$. $\angle 1$, $\angle 2$, $\angle 3$, $\angle 4$ are rt $\angle$s (def of $\perp$); m $\angle 1 = 90$, m $\angle 2 = 90$, m $\angle 3 = 90$, m $\angle 4 = 90$ (def of rt $\angle$s); $\angle 1 \cong \angle 2 \cong \angle 3 \cong \angle 4$ (subst prop and def of $\cong$ $\angle$s)

Page 65

20. Given: $\overrightarrow{AD} \perp \overrightarrow{BD}$; Prove: $\angle 1$ and $\angle 2$ are compl.
21. Given: $\overleftrightarrow{AB}$ and $\overleftrightarrow{CD}$ intersect at E; Prove: $\overleftrightarrow{AB}$ and $\overleftrightarrow{CD}$ are coplanar. **22.** $BC = BC$ (reflex prop); $AB + BC = CD + BC$ (add prop) **23.** m $\angle 1$ = m $\angle 2$, and m $\angle 2$ = m $\angle 3$ (def of $\cong$ $\angle$s); m $\angle 1$ = m $\angle 3$ (trans prop); $\angle 1 \cong \angle 3$ (def of $\cong$ $\angle$s)
24. Given: $\angle 1$ is supp to $\angle 3$, $\angle 2$ is supp to $\angle 3$; Prove: $\angle 1 \cong \angle 2$; m $\angle 1$ + m $\angle 3 = 180$ and m $\angle 2$ + m $\angle 3 = 180$ (def of supp $\angle$s); m $\angle 1$ + m $\angle 3$ = m $\angle 2$ + m $\angle 3$ (subst prop); m $\angle 3$ = m $\angle 3$ (reflex prop); m $\angle 1$ = m $\angle 2$ (subtr prop); $\angle 1 \cong \angle 2$ (def of $\cong$ $\angle$s)

Page 66

20. $\angle POS \cong \angle QOS$ (Th 2.6); m $\angle POS$ = m $\angle QOS$ (def of $\cong$ $\angle$s); m $\angle POS$ = m $\angle POR$ + m $\angle ROS$ and m $\angle QOS$ = m $\angle SOT$ + m $\angle TOQ$ (add prop); m $\angle POR$ = m $\angle SOT$ (def of $\cong$ $\angle$s); m $\angle POR$ + m $\angle ROS$ = m $\angle SOT$ + m $\angle TOQ$ (subtr prop); m $\angle ROS$ = m $\angle TOQ$ (subtr prop); $\angle ROS \cong \angle TOQ$ (def of $\cong$ $\angle$s) **21.** Given: $\angle 1$ is compl to $\angle 2$, $\angle 3$ is compl to $\angle 2$; Prove: $\angle 1 \cong \angle 3$. m $\angle 1$ + m $\angle 2 = 90$ and m $\angle 3$ + m $\angle 2 = 90$ (def of compl $\angle$s); m $\angle 1$ + m $\angle 2$ = m $\angle 3$ + m $\angle 2$ (subst prop); m $\angle 2$ = m $\angle 2$ (reflex prop); m $\angle 1$ = m $\angle 3$ (subtr prop); $\angle 1 \cong \angle 3$ (def of $\cong$ $\angle$s)

Page 71

1. 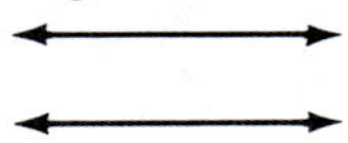

2. 

3. 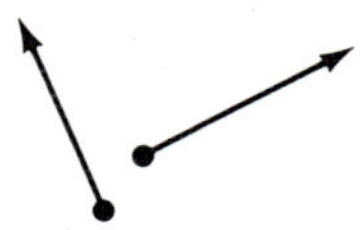

4. 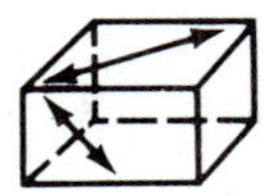

Page 79

11. m $\angle 1$ + m $\angle 2 = 180$ and m $\angle 3$ + m $\angle 4 = 180$ (Th 3.2) m $\angle 1$ + m $\angle 2$ + m $\angle 3$ + m $\angle 4 = 360$ (add prop) **12.** Follow steps of proof of Th 3.5, p. 86
13. m $\angle 1$ + m $\angle 2 = 180$ (Post 11); $\angle 1$ and $\angle 2$ are suppl (def of suppl $\angle$s); $\angle 2 \cong \angle 3$ (Post 13); $\angle 1$ and $\angle 3$ are suppl (subst prop) **14.** Given: $l \parallel m$; Prove: $\angle 3 \cong \angle 4$. $\angle 1 \cong \angle 2$ (Post 13); $\angle 1 \cong \angle 3$ and $\angle 2 \cong \angle 4$ (Th 2.10); $\angle 3 \cong \angle 4$ (subst prop)
15. Given: $\overleftrightarrow{AB} \parallel \overleftrightarrow{CD}$, $\overleftrightarrow{EF} \perp \overleftrightarrow{AB}$; Prove: $\overleftrightarrow{EF} \perp \overleftrightarrow{CD}$. $\angle 1$ is a rt $\angle$ (Th 2.6); $\angle 1 \cong \angle 2$ (Post 13); $\angle 2$ is a rt $\angle$ (subst prop); $\overleftrightarrow{EF} \perp \overleftrightarrow{CD}$ (def of $\perp$ lines)

Pages 83–84

7. $\angle 1 \cong \angle 2$ (Th 2.10); $\angle 2 \cong \angle 4$ (trans prop); $\angle 2$ suppl to $\angle 3$ (Th 3.2); $\angle 4$ suppl to $\angle 3$ (subst prop); t $\parallel$ s (Th 3.4) **8.** $\angle ADF \cong \angle C$ (Th 3.1); $\angle F$ is suppl to $\angle ADF$ (subst prop); $\overrightarrow{CA} \parallel \overrightarrow{FE}$ (Th 3.4) **15.** $\angle ADC$ suppl to $\angle A$ (Th 3.2); $\angle ABC$ suppl to $\angle A$ (subst prop); $\overline{BC} \parallel \overline{AD}$ (Th 3.4) **16.** $\angle 2 \cong \angle 4$ (trans prop); $\angle 4 \cong \angle 5$

(def of $\angle$ bis); $\angle 2 \cong \angle 5$ (trans prop); $\overleftrightarrow{EG} \parallel \overleftrightarrow{HK}$ (Post 14) **17.** m $\angle 2$ + m $\angle 3 = 180$ (Post 11); $\angle 2$ and $\angle 3$ are suppl (def of supp $\angle$s); $\angle 1 \cong \angle 3$ (corr suppl of the same $\angle$ are $\cong$); $\overleftrightarrow{AB} \parallel \overleftrightarrow{CD}$ (Th 3.3)
18. Given: $\overleftrightarrow{AB} \parallel \overleftrightarrow{CD}$, $\overrightarrow{EH}$ bisects $\angle BEF$, $\overrightarrow{FG}$ bisects $\angle CFE$; Prove: $\overrightarrow{EH} \parallel \overrightarrow{FG}$. $\angle BEF \cong \angle CFE$ (Post. 13); m $\angle BEF$ = m $\angle CFE$ (def $\cong$ $\angle$s); $\frac{1}{2}$m $\angle BEF = \frac{1}{2}$m $\angle CFE$ (div prop); m $\angle 1 = \frac{1}{2}$m $\angle BEF$ and m $\angle 2 = \frac{1}{2}$m $\angle CFE$ (def of bis); m $\angle 1$ = m $\angle 2$ (subst prop); $\angle 1 \cong \angle 2$ (def of $\cong$ $\angle$s); $\overrightarrow{EH} \parallel \overrightarrow{FG}$ (Post 14) **19.** Given: $AB \parallel CD$, EG bisects $\angle IEB$, FH bisects $\angle EFD$; Prove: $\overrightarrow{EG} \parallel \overrightarrow{FH}$. $\angle IEB \cong \angle EFD$ (Th 3.2); m $\angle IEB$ = m $\angle EFD$ (def of $\cong$ $\angle$s); $\frac{1}{2}$m $\angle IEB = \frac{1}{2}$m $\angle EFD$ (div prop); m $\angle 1 = \frac{1}{2}$m $\angle IEB$ and m $\angle 2 = \frac{1}{2}$m $\angle EFD$ (def of $\angle$ bis); m $\angle 1$ = m $\angle 2$ (subst prop); $\angle 1 \cong \angle 2$ (def of $\cong$ $\angle$s); $\overrightarrow{EG} \parallel \overrightarrow{FH}$ (Th 3.3) **20.** Given: $\overrightarrow{EG} \parallel \overrightarrow{FH}$, $\overrightarrow{EG}$ bisects $\angle IEB$, $\overrightarrow{FH}$ bisects $\angle EFD$; Prove: $\overleftrightarrow{AB} \parallel \overleftrightarrow{CD}$. $\angle 1 \cong \angle 2$ (Th 3.2); m $\angle 1$ = m $\angle 2$ (def of $\cong$ $\angle$s); m $\angle 1 = \frac{1}{2}$m $\angle IEB$ and m $\angle 2 = \frac{1}{2}$m $\angle EFD$ (def of $\angle$ bis); $\frac{1}{2}$m $\angle IEB = \frac{1}{2}$m $\angle EFD$ (subst prop); m $\angle IEB$ = m $\angle EFD$ (mult prop); $\angle IEB \cong \angle EFD$ (def of $\cong$ $\angle$s); $\overleftrightarrow{AB} \parallel \overleftrightarrow{CD}$ (Th 3.3)
21. Given: $EH \parallel FG$, IH bisects $\angle BEF$, $\overrightarrow{EF}$ bisects $\angle BEF$; Prove: $AB \parallel CD$. $\angle 1 \cong \angle 2$ (Post 13); m $\angle 1$ = m $\angle 2$ (def of $\cong$ $\angle$s); m $\angle 1 = \frac{1}{2}$m $\angle BEF$ and m $\angle 2 = \frac{1}{2}\angle CFE$ (def of bis); $\frac{1}{2}$m $\angle BEF = \frac{1}{2}$m $\angle CFE$ (subst prop); m $\angle BEF$ = m $\angle CFE$ (mult prop); $\angle BEF \cong \angle CFE$ (def of $\cong$ $\angle$s); $\overleftrightarrow{AB} \parallel \overleftrightarrow{CD}$ (Post 14)

Page 90

7. $\angle C$ is a rt $\angle$ (def of $\perp$ lines); m $\angle C = 90$ (def of a rt $\angle$); m $\angle A$ + m $\angle B$ + m $\angle C = 180$ (Th 3.5); m $\angle A$ + m $\angle B = 90$ (subst prop) **8.** m $\angle x$ + m $\angle y$ + m $\angle r = 180$ (Th 3.5); m $\angle r$ + m $\angle s = 180$ (Post 11); m $\angle x$ + m $\angle y$ + m $\angle r$ = m $\angle r$ + m $\angle s$ (trans prop); m $\angle s$ = m $\angle x$ + m $\angle y$ (subtr prop)
9. $\angle DBA \cong \angle CBE$ (Th 2.10); $\angle A \cong \angle E$ (Th 3.6)
10. $\angle CDA \cong \angle CDB$ (Th 2.6); $\angle ACD \cong \angle BCD$ (Th 3.6); $\overrightarrow{CD}$ bisects $\angle ACB$ (def of bis)
11. m $\angle C$ + m $\angle CDE$ + m $\angle CED = 180$ and m $\angle C$ + m $\angle A$ + m $\angle B = 180$ (Th 3.5); m $\angle C$ + m $\angle CDE$ + m $\angle CED$ = m $\angle C$ + m $\angle A$ + m $\angle B$ (trans prop); m $\angle CDE$ + m $\angle CED$ = m $\angle A$ + m $\angle B$ (subtr prop); 2 m $\angle CDE$ = 2 m $\angle A$ (subst prop); m $\angle CDE$ = m $\angle A$ (div prop); $\angle CDE \cong \angle A$ (def of $\cong$ $\angle$s); $\overline{DE} \parallel \overline{AB}$ (Th 3.3) **12.** $\angle FDB \cong \angle EDA$ (Th 2.6); $\angle AED \cong \angle F$ (Th 3.6); $\angle CEF \cong \angle AED$ (Th 2.10); $\angle F \cong \angle CEF$ (trans prop)

Page 92

1. Suppose $\angle 1 \cong \angle 2$. Then $l \parallel m$ (Post 14). This contradicts given $\therefore \angle 1 \ncong \angle 2$. **2.** Suppose $l \parallel m$. Then $\angle 1 \cong \angle 2$ (Post 13). This contradicts given $\therefore l \nparallel m$. **3.** Suppose $t \perp s$. Then $\angle 3 \cong \angle 4$ (Th 2.6). This contradicts given $\therefore t \not\perp s$. **4.** Suppose $\angle 3 \cong \angle 4$. Then $t \perp s$ (Th 2.7). This contradicts given $\therefore \angle 3 \ncong \angle 4$. **5.** Suppose $r \parallel s$. Then $\angle 2 \cong \angle 3$ (Post 13). $\angle 2$ is a rt $\angle$ (Th 2.6). $\angle 3$ is a rt $\angle$ (subst prop). $s \perp t$ (def of $\perp$ lines). This contradicts given $\therefore r \nparallel s$. **6.** Suppose $s \perp t$, then $r \parallel s$. (Cor 3.3) **7.** Suppose $l \parallel m$. Then $\angle 1$ and $\angle 2$ are suppl (Th 3.2). m $\angle 1$ + m $\angle 2 = 180$ (def of suppl $\angle$s). This contradicts the given $\therefore l \nparallel m$.
8. Suppose m $\angle 1$ + m $\angle 2 = 180$. Then $\angle 1$ and $\angle 2$ are suppl (def of suppl $\angle$s). $l \parallel m$ (Th 3.4). This contradicts the given $\therefore$ m $\angle 1$ + m $\angle 2 \neq 180$.
9. Given: $\triangle ABC$; Prove: $\angle A$ and $\angle B$ cannot both be obtuse $\angle$s. Suppose $\angle A$ and $\angle B$ are both obtuse $\angle$s. Then m $\angle A$ + m $\angle B > 180$ and m $\angle A$ + m $\angle B$ + m $\angle C > 180$. This contradicts Th 3.5 $\therefore \angle A$ and $\angle B$ cannot be obtuse. **10.** Given: $\triangle ABC$; Prove: $\angle A$ cannot be a rt $\angle$ when $\angle B$ is an obtuse $\angle$. Suppose $\angle A$ is a rt $\angle$ and $\angle B$ is an obtuse $\angle$. Then m $\angle A$ + m $\angle B > 180$ and m $\angle A$ + m $\angle B$ + m $\angle C > 180$. This contradicts Th 3.5 $\therefore \angle A$ cannot be a rt $\angle$ when $\angle B$ is an obtuse $\angle$. **11.** Given: $\triangle ABC$; Prove: $\overline{AC} \nparallel \overline{BC}$. Suppose $\overline{AC} \parallel \overline{BC}$. Then $\angle A$ and $\angle B$ are suppl (Th 3.2). m $\angle A$ + m $\angle B$ = 180 (def of suppl $\angle$s). m $\angle A$ + m $\angle B$ + m $\angle C >$ 180. This contradicts Th 3.5 $\therefore \overline{AC} \parallel \overline{BC}$.
12. Given: $\overleftrightarrow{AB}$ and $\overleftrightarrow{CD}$ are skew; Prove: $\overleftrightarrow{AB}$ and $\overleftrightarrow{CD}$ do not intersect. Suppose $\overleftrightarrow{AB}$ and $\overleftrightarrow{CD}$ do intersect. Then they would be coplanar. This contradicts def of skew lines $\therefore \overleftrightarrow{AB}$ and $\overleftrightarrow{BC}$ do not intersect. **13.** Suppose $r \parallel t$. Then $s \parallel t$ (Th 3.7). This contradicts given $\therefore r \nparallel t$. **14.** Suppose $r \parallel l$. Then $s \parallel l$ (Th 3.7). This contradicts given since $s \perp l \therefore r \nparallel l$. **15.** Suppose $\angle C \ncong \angle F$. Then m $\angle A$ + m $\angle B$ + m $\angle C \neq$ m $\angle D$ + m $\angle E$ + m $\angle F$ (add prop of ineq). This contradicts m $\angle A$ + m $\angle B$ + m $\angle C = 180$ and m $\angle D$ + m $\angle E$ + m $\angle F =$ 180. (Th. 3.5: The sum of the angle measures of a triangle is 180°.) Therefore $\angle C \cong \angle F$.

Page 97

9–12. Answers may vary. **13.** p and q have no pt in common (def of $\parallel$ planes); $l \parallel p$ (l and p have no pt in common. Def of line $\parallel$ to plane) **14.** Suppose $l \nparallel t$. Then they would intersect at pt x. Line l and plane q would have a pt in common and $l \parallel q$, which contradicts given $\therefore l \parallel t$. **15.** Given: l intersects plane p at pt x, m is on p; Prove: $l \nparallel m$. Suppose $l \parallel m$. Then l and m are coplanar. But only pt x is

contained in p ∴ $l \nparallel m$. **16.** Given: $p \parallel q$, x is on p; Prove: x is not on q. Suppose x is on q. Then $p \nparallel q$. This contradicts given ∴ x is not on q.
17. Given: $p \parallel q$, t intersects p; Prove: t intersects q. Suppose t does not intersect q. Then $t \parallel q$. Then $t \parallel p$ which contradicts given ∴ t intersects q.
18. Given: l is in p, m is in q, l and m intersect at pt x; Prove: p and q intersect. Suppose p and q do not intersect. Then they have no pts in common. Since l is in p and m is in q, then l and m have no pts in common. This contradicts given ∴ p and q intersect.

Page 100

6. Follow constr p 93. **14.** See p 87. **15.** See p 92, ex 9. **16.** $\angle 1 \cong \angle 4$ (Post 13); $\angle 2 \cong \angle 3$ (subst prop); $\overline{AD} \parallel \overline{BC}$ (Post 14)

Pages 105–106

1. Must have an obtuse ∠ and no side ≅.
2. Must be a 45°–45°–90° △ **3.** Must have 3 acute ∠s and 2 sides ≅ **4.** Must have a right ∠ and no sides ≅ **5.** Must have 3 acute ∠s and no sides ≅ **6.** Must be a 60°–60° = 60° △

21.

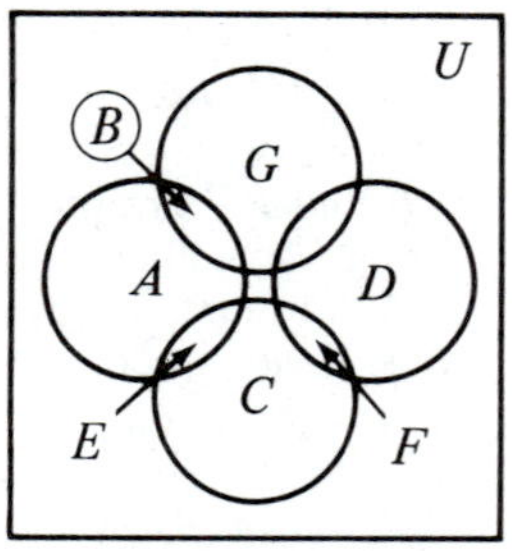

G represents all acute △s.

22.

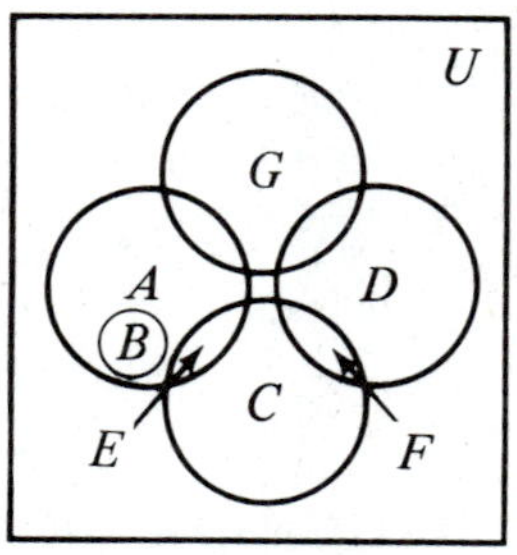

G represents all obtuse △s.

23. Given: △ABC is equiangular; Prove: △ABC is not obtuse. Assume △ABC is obtuse. (This is one of two possibilities): ∠A is obtuse. (Def of obtuse △): ∠A, ∠B, ∠C, are obtuse. (Def of equiangular △): ∠s. m ∠A + m ∠B + m ∠C > 180 (This contradicts Th 3.5); △ABC is not obtuse (Only other possibility)
24. Given: △ABC is scalene; Prove: △ABC is not isosceles. Assume △ABC is isos (This is one of two possibilities); $\overline{AB} \cong \overline{BC}$, $\overline{AB} \cong \overline{AC}$, or $\overline{AC} \cong \overline{BC}$ (Def of isos △. This contradicts given); ∴ △ABC is not isos. (Only other possibility)

Page 109

5. ∠2 ≅ ∠3 (Th 3.6); ∠1 ≅ ∠4 (Th 2.8) **6.** ∠2 ≅ ∠3 (Th 2.8); ∠A ≅ ∠D (Th 3.6) **7.** Given: △ABC, exterior ∠2; Prove: m ∠2 = m ∠A + m ∠B. ∠A + ∠B + ∠1 = 180 (Th 3.5); m ∠1 + m ∠2 = 180 (Post 11); m ∠A + m ∠B + m ∠1 = m ∠1 + m ∠2 (trans prop); m ∠2 = m ∠A + m ∠ B (subst prop) **8.** Given: $\overline{CD} \perp \overleftrightarrow{AB}$; Prove: $\overline{CD}$ is the only seg ⊥ to $\overleftrightarrow{AB}$ through C. Assume $\overline{CE} \perp \overleftrightarrow{AB}$ (This is one of two possibilities): m ∠2 = 90 (Def of ⊥ lines); m ∠1 = m ∠2 (subst prop). This contradicts Cor 4.1): ∴ $\overline{CD}$ is the only seg ⊥ to $\overleftrightarrow{AB}$ through C (only other possibility)

Page 117

9–10. Follow constr p. 115 **11–12.** Follow constr p. 115. Use seg ≅ sides of given △s.
13. Follow constr p. 115. Use 3 seg ≅ $\overline{MN}$.
14. Follow constr p. 115. Use 1 seg ≅ $\overline{AB}$, 2 seg ≅ $\overline{CD}$ **15.** $\overline{AB} \cong \overline{BC} \cong \overline{CA}$ and $\overline{DE} \cong \overline{EF} \cong \overline{FD}$ (def of equilateral △): $\overline{AB} \cong \overline{BC} \cong \overline{CA} \cong \overline{DE} \cong \overline{EF} \cong \overline{FD}$ (subst prop): △ABC ≅ △DEF (SSS for ≅ △s) **16.** $\overline{GI} \cong \overline{GH} \cong \overline{IH}$ (def of equilateral △): $\overline{JI} \cong \overline{JH} \cong \overline{IH}$ (subst prop): △JHI is equilateral (def of equilateral △) **17.** $\overline{LM} \cong \overline{LM}$ (reflex prop); $KL + LM = LM + NM$, or $\overline{KM} \cong \overline{LN}$ (add prop): △KMQ ≅ △NLO (SSS for ≅ △s)
18. $\overline{BC} = \overline{BC}$ (reflex prop); $AB + BC = BC + CD$, or $\overline{AC} \cong \overline{BD}$ (add prop); △ACE ≅ △DBF (SSS for ≅ △s) **19.** $\overline{AB} \cong \overline{BC}$ (def of bis); $\overline{DB} \cong \overline{DB}$ (reflex prop); △ADB ≅ △CBD (SSS for ≅ △s)
20. $\overline{XZ} \cong \overline{YZ}$ (def of equilateral △); $\overline{ZW} \cong \overline{ZW}$ (reflex prop): $\overline{XW} \cong \overline{YW}$ (def of midpt): △XWZ ≅ △YWZ (SSS for ≇ △s) **21.** $\overline{XZ} \cong \overline{XY} \cong \overline{YZ}$ and $\overline{XZ} \cong \overline{XW} \cong \overline{ZW}$ and $\overline{YZ} \cong \overline{ZW} \cong \overline{YW}$ and $\overline{XW} \cong \overline{YW} \cong \overline{XY}$ (def of equilateral △); $\overline{XZ} \cong \overline{XY} \cong \overline{YZ} \cong \overline{ZW} \cong \overline{XW} \cong \overline{YW}$ (subst prop): $\overline{XZ} \cong \overline{XZ}$, $\overline{XY} \cong \overline{XY}$, $\overline{YZ} = \overline{YZ}$, $\overline{ZW} \cong \overline{ZW}$, $\overline{XW} \cong \overline{XW}$, $\overline{YW} \cong \overline{YW}$ (reflex prop): △XYZ ≅ △XYW ≅ △XWZ ≅ △YZW (SSS for ≅ △s) **22.** $\overline{XZ} \cong \overline{XY} \cong \overline{ZY}$ (def of equilateral △); $\overline{XW} \cong \overline{XW}$, $\overline{YW} \cong \overline{YW}$, $\overline{ZW} \cong \overline{ZW}$ (reflex prop); △XWY ≅ △YWZ ≅ △ZWX (SSS for ≅ △s)

Page 122

14. ∠AED ≅ ∠CEB (Th 2.10); △ADE ≅ △CBE (SAS) **15.** $\overline{FJ} \cong \overline{JG}$ (def of bis); ∠FJH ≅ ∠IJG (Th 2.10); △FJH ≅ △GJI (ASA for ≅ △s)
16. $\overline{NL} \cong \overline{NL}$ (reflex prop); △KNL ≅ △MNL (SAS for ≅ △s) **17.** ∠ACD ≅ ∠BCD (def of bis); $\overline{CD} \cong \overline{CD}$ (reflex prop); △ADC ≅ △BDC (SAS for ≅ △s) **18.** $\overline{GF} \cong \overline{FE}$ (def of bis); ∠GFH ≅ ∠EFH (Th 2.6); △GFH ≅ △EFH (SAS for ≅ △s)
19. ∠LKI ≅ ∠JIK and ∠LIK ≅ ∠JKI (Post 13); $\overline{IK} \cong \overline{IK}$ (reflex prop); △LKI ≅ △JIK (ASA for ≅ △s) **20.** Follow constr p. 118. Constr side ≅ $\overline{OP}$, angle ≅ ∠Q, then constr side ≅ $\overline{MN}$ on ray of ∠Q. Draw 3rd side of △ **21.** Follow ex 3, p. 119. **22.** ∠SVU ≅ ∠TVU (Th 2.6); $\overline{UV} \cong \overline{UV}$ (reflex prop.); △SVU ≅ △TVU (ASA

for $\cong \triangle$s) **23.** $m\angle UST = m\angle UTS = m\angle SUT = m\angle SUV = m\angle UVS = m\angle VSU = m\angle VUT = m\angle UTV = m\angle TVU = m\angle SVT = m\angle VTS = m\angle TSV = 60$ (def of equiangular $\triangle$); $\angle UST \cong \angle UTS \cong \angle SUT \cong \angle SUV \cong \angle UVS \cong \angle VSU \cong \angle VUT \cong \angle UTV \cong \angle TVU \cong \angle SVT \cong \angle VTS \cong \angle TSV$ (def of $\cong \angle$s and subst prop); $\overline{SU} \cong \overline{SU}$; $\overline{UT} \cong \overline{UT}$; $\overline{ST} \cong \overline{ST}$; $\overline{UV} \cong \overline{UV}$; $\overline{SV} \cong \overline{SV}$; $\overline{TV} \cong \overline{TV}$ (reflex prop); $\triangle STU \cong \triangle SVU \cong \triangle TVU \cong \triangle STV$ (ASA for $\cong \triangle$s)

Pages 126–127

9. $\angle MNJ \cong \angle KNL$ (Th 2.10); $\triangle JNM \cong \triangle LNK$ (ASA for $\cong \triangle$s) **10.** $\angle MNJ \cong \angle KNL$ (Th. 2.10); $\triangle JNM \cong \triangle KNL$ (AAS for $\cong \triangle$s) **11.** $\angle ACB \cong \angle DCE$ (Th 2.10); $\overline{EC} \cong \overline{CB}$ (def of mdpt); $\triangle ABC \cong \triangle DEC$ (ASA for $\cong \triangle s$) **12.** $\angle FIG \cong \angle HIG$ (def of bis); $\overline{IG} \cong \overline{IG}$ (reflex prop); $\triangle FGI \cong \triangle HGI$ (AAS for $\cong \triangle$s) **13.** $\overline{KM} \cong \overline{KM}$ (reflex prop); $\triangle MLK \cong \triangle KJM$ (SSS for $\cong \triangle$s) **14.** $\angle ROQ \cong \angle NOP$ (Th 2.10); $\triangle ROQ \cong \triangle NOP$ (ASA for $\cong \triangle$s) **15.** $\overline{UV} \cong \overline{VS}$ (def of bis); $\angle UVT \cong \angle SVT$ (Th 2.6 and Th 2.5); $\overline{VT} \cong \overline{VT}$ (reflex prop); $\triangle UVT \cong \triangle SVT$ (SAS for $\cong \triangle$s) **16.** $\angle YZX \cong \angle WXZ$ (Post 13); $\overline{ZX} = \overline{ZX}$ (reflex prop); $\triangle ZWX \cong \triangle XYZ$ (AAS for $\cong \triangle$s) **17.** $\angle DEC \cong \angle ECB$ (Post 13); $\angle ECB \cong \angle FBA$ (Th 3.1); $\angle DEC \cong \angle FBA$ (subst prop); $\triangle ABF \cong DEC$ (AAS for $\cong \triangle$s) **18.** $m\angle ABF + m\angle FBC = 180$, $m\angle DEC + m\angle CEF = 180$ (Post 11); $\angle ABF$ supp to $\angle FBC$, $\angle DEC$ supp to $\angle CEF$ (def of suppl $\angle$s); $\angle ABF \cong \angle DEC$ (Th 2.8); $\triangle ABF \cong \triangle DEC$ (SAS for $\cong \triangle$s) **19.** $\angle C \cong \angle C'$ (Th 3.6); $\triangle ABC \cong \triangle A'B'C'$ (ASA for $\cong \triangle$s) **20.** $\angle IHG \cong \angle IHJ$ (Th 2.6 and Th 2.5); $\overline{IH} \cong \overline{IH}$ (reflex prop); $\triangle GHI \cong \triangle JHI$ (AAS for $\cong \triangle$s) **21.** $\overline{GH} \cong \overline{JH}$ (trans prop); $\overline{IH} \cong \overline{IH}$ (reflex prop), $\angle IHG \cong \angle IHJ$ (Th 2.6 and Th 2.5); $\triangle GHI \cong \triangle JHI$ (SAS for $\cong \triangle$s)

Page 131

7–8. Answers may vary. **9.** Draw an obtuse $\triangle$. Follow constr p. 119. **16.** $\overline{VY} \cong \overline{YZ}$, $\overline{XY} \cong \overline{YW}$ (def of mdpt); $\angle VYW \cong \angle ZYX$ (Th 2.10) $\triangle VYW \cong \triangle ZYX$ (SAS) **17.** $\angle UTR \cong \angle SRT$, $\angle URT \cong \angle STR$ (Alt. int $\angle$s are =); $\overline{RT} \cong \overline{RT}$ (reflex prop); $\triangle RTU \cong \triangle TRS$ (ASA)

Page 134

32. $l \parallel m$ (Post 14); $\angle 3 \cong \angle 4$ (Post 13) **33.** $\overline{DC} \cong \overline{DC}$ (reflex prop); $AD + DC = DC + CF$ (add prop); $\triangle ACB \cong \triangle FDE$ (SAS for $\cong \triangle$s) **34.** $\overline{AB} \cong \overline{AB}$ (reflex prop); $\angle CBA \cong \angle DBA$ (Th 2.6); $\triangle ABC \cong \triangle ABC$ (AAS for $\cong \triangle$s)

Page 138

7. $\overline{AC} \cong \overline{AC}$ (reflex prop); $\triangle ADC \cong \triangle CBA$ (SSS for $\cong \triangle$s); $\angle D \cong \angle B$ (CPCTC) **8.** $\angle 4 \cong \angle 2$ (Post 13); $\overline{AC} \cong \overline{AC}$ (reflex prop); $\triangle ADC \cong \triangle CBA$ (ASA for $\cong \triangle$s); $\overline{AD} \cong \overline{CB}$ (CPCTC) **11.** $\overline{XZ} \cong \overline{XZ}$ (reflex prop); $\triangle WXZ \cong \triangle YZX$ (SAS for $\cong \triangle$s); $\angle 1 \cong \angle 3$ (CPCTC); $\overline{XW} \parallel \overline{YZ}$ (Post 14) **12.** $\overline{XZ} \cong \overline{XZ}$ (reflex prop); $\triangle WXZ \cong \triangle YZX$ (ASA for $\cong \triangle$s); $\angle W$ is a rt $\angle$ (def of $\perp$ lines); $\angle Y \cong \angle W$ (CPCTC); $\angle Y$ is a rt $\angle$ (subst prop); $ZY \perp XY$ (def of $\perp$ lines) **13.** $\overline{CD} \cong \overline{CD}$ (reflex prop); $\angle CDA \cong \angle CDB$ (Th 2.6); $\triangle CDA \cong \triangle CDB$ (SAS for $\cong \triangle$s); $\overline{AC} \cong \overline{BC}$ (CPCTC); $\triangle ABC$ is isos (def of isos $\triangle$) **14.** $\angle GHE \cong \angle GHF$ (Th 2.6); $\overline{GH} \cong \overline{GH}$ (reflex prop); $\triangle GHE \cong \triangle GHF$ (AAS for $\cong \triangle$s); $\overline{EH} \cong \overline{FH}$ (CPCTC); $\triangle EHF$ is isos (def of isos $\triangle$) **15.** $\overline{IK} \cong \overline{IK}$ (reflex prop); $\triangle KLI \cong KJI$ (SSS for $\cong \triangle$s); $\angle LKI$ is a rt $\angle$ (def of $\perp$ lines); $\angle JKI \cong \angle LKI$ (CPCTC); $\angle JKI$ is a rt $\angle$ (subst prop); $\overline{JK} \perp \overline{IK}$ (def of $\perp$ lines)

Page 143

13. $\angle B \cong \angle 1$ (Th 5.1); $\angle 1 \cong \angle 2$ (Th 2.10); $\angle B \cong \angle 2$ (subst prop) **14.** $\angle 2 \cong \angle 1$ (Th 2.10); $\angle A \cong \angle 1$ (subst prop); $\overline{AB} \cong \overline{BC}$ (Th 5.2) **15.** $\angle D \cong \angle E$ (Th 5.1); $\angle 2 \cong \angle D$ (Post 13); $\angle 1 \cong \angle E$ (Th 3.1); $\angle 1 \cong \angle 2$ (subst prop) **16.** $\angle 1 \cong \angle I$ (Th 3.1); $\angle 2 \cong \angle J$ (Post 13); $\angle I \cong \angle J$ (subst prop); $\overline{IK} \cong \overline{JK}$ (Th 5.2); $\triangle IJK$ is isos (def of isos $\triangle$) **17.** $m\angle 1 + m\angle ROP = 180$; $m\angle 2 + m\angle RPO = 180$ (Post 11); $\angle 1$ and $\angle ROP$ are supp, $\angle 2$ and $\angle RPO$ are supp (def of supp); $\angle ROP \cong \angle RPO$ (Th 2.8); $\overline{RO} \cong \overline{RP}$ (Th 5.2); $\triangle OPR$ is isos (def of isos $\triangle$) **18.** Draw $\overline{CD}$ bisecting $\angle C$ (every $\angle$ has an $\angle$ bis); $\angle 1 \cong \angle 2$ (def of bis); $\overline{CD} \cong \overline{CD}$ (reflex prop); $\triangle ACD \cong \triangle BCD$ (AAS for $\cong \triangle$s); $\overline{AC} \cong \overline{BC}$ (CPCTC) **19.** Given: $\triangle ABC$ is equiangular; Prove: $\triangle ABC$ is equilateral. $\angle A \cong \angle B \cong \angle C$ (def of equiangular $\triangle$); $\overline{AC} \cong \overline{CB}$; $\overline{AC} \cong \overline{AB}$; $\overline{AB} \cong \overline{CB}$ (Th 5.2); $\overline{AC} \cong \overline{AB} \cong \overline{CB}$ (subst prop); $\triangle ABC$ is equil (def of equilateral $\triangle$) **20.** $\angle S \cong \angle T$ (Th 5.1); $\angle S \cong \angle 1$; $\angle T \cong \angle 4$ (Th 3.1); $\angle 1 \cong \angle 4$ (subst prop); $WV \cong \overline{UV}$ (Th. 5.2) **21.** $\angle 1 \cong \angle 4$ (Th 5.1); $\angle 1 \cong \angle S$; $\angle 4 \cong \angle T$ (Th 3.1); $\angle S \cong \angle T$ (subst prop); $\overline{SV} \cong \overline{TV}$ (Th 5.2); $\overline{SV} - \overline{VW} = \overline{TV} - \overline{VU}$, or $\overline{SW} \cong \overline{TU}$ (subt prop) **24.** $\overline{PX} \cong \overline{PX}$ (reflex prop); $\overline{XY} \cong \overline{XZ}$ (Th 5.2); $\triangle XYP \cong \triangle XZP$ (SAS for $\cong \triangle$s); $\overline{PY} \cong \overline{PZ}$ (CPCTC); $\triangle PYZ$ is isos (def. of isos $\triangle$) **25.** $\overline{PY} \cong \overline{PZ}$ (Th 5.2); $\overline{PX} \cong \overline{PX}$ (reflex prop); $\triangle PXZ \cong \triangle PXY$ (SAS for $\cong \triangle$s); $\overline{XZ} \cong \overline{XY}$ (CPCTC); $\triangle XYZ$ is isos (def of isos $\triangle$)

Page 155

4. $\angle WTU \cong \angle WUT$ (Th 5.1); $\triangle SUW \cong \triangle VTW$ (SAS for $\cong \triangle$s) **5.** $\overline{SW} \cong \overline{VW}$ (Th 5.2); $\triangle SUW \cong \triangle VTW$ (SAS for $\cong \triangle$s) **6.** $\angle Z \cong \angle Z$ (reflex prop); $\triangle ZXB \cong \triangle ZYA$ (ASA for $\cong \triangle$s); $\overline{XB} \cong \overline{YA}$ (CPCTC) **7.** $\overline{XY} \cong \overline{XY}$ (reflex prop); $\triangle AYX \cong \triangle BXY$ (SAS for $\cong \triangle$s); $\angle AYX \cong \angle BXY$ (CPCTC) **8.** $\angle GCD$ is a rt $\angle$, $\angle EDC$ is a rt $\angle$ (def of $\perp$ lines); $\angle GCD \cong \angle EDC$ (rt $\angle$s are $\cong$); $\overline{CD} \cong \overline{CD}$ (reflex prop); $\triangle GCD \cong \triangle EDC$ (SAS for $\cong \triangle$s); $\overline{DG} \cong \overline{CE}$ (CPCTC)
9. $\angle EDC \cong \angle GCD$ (Th 3.6); $\angle GCD$ is a rt $\angle$ (def of $\perp$ lines); $\angle EDC$ is a rt $\angle$ (subst prop); $\overline{ED} \perp \overline{CD}$ (def of $\perp$ lines) **10.** $\overline{IJ} \cong \overline{IJ}$ (reflex prop); $\overline{HJ} \cong \overline{IK}$ (add prop); $\angle H \cong \angle LIK$ (Th 3.1); $\triangle NHJ \cong \triangle LIK$ (SAS for $\cong \triangle$s); $\angle NJH \cong \angle K$ (CPCTC); $\overline{JN} \parallel \overline{KL}$ (Th 3.3) **11.** $\angle 5 + \angle 3 = 180$; $\angle 6 + \angle 4 = 180$ (Post 11); $\angle 5$ and $\angle 3$ is supp, $\angle 6$ and $\angle 4$ is supp (def of supp); $\angle 5 \cong \angle 6$ (Th 2.8); $\overline{SR} = \overline{SR}$ (reflex prop); $\overline{OR} \cong \overline{QS}$ (add prop); $\triangle OPR \cong \triangle QPS$ (ASA for $\cong \triangle$s) **12.** $\angle T \cong \angle ZUW$; $\angle W \cong \angle YVT$ (Th 3.1); $\triangle TVY \cong \triangle WUZ$ (ASA for $\cong \triangle$s)

Pages 160–161

1. $\angle y$, $\angle s$ are rt $\angle s$ (def of $\perp$ lines); $\triangle XYZ$, $\triangle RST$ are rt $\triangle$s (def of rt $\triangle$s); $\triangle XYZ \cong \triangle RST$ (LL for $\cong$ rt $\triangle$s) **2.** $\angle y$, $\angle$s are rt $\angle$s (def of $\perp$ lines); $\triangle XYZ$, $\triangle RST$ are rt $\triangle$s (def of rt $\triangle$s); $\triangle XYZ \cong \triangle RST$ (HA for $\cong$ rt $\triangle$s) **3.** $\angle y$, $\angle s$ are rt $\angle$s 1. (def of $\perp$ lines); $\triangle XYZ$, $\triangle RST$ are rt $\triangle$s 2. (def of rt $\triangle$s); $\triangle XYZ \cong \triangle RST$ 3. (LA for $\cong$ rt $\triangle s$) **4.** $\angle y$, $\angle s$ are rt $\angle$s (def of $\perp$ lines); $\triangle XYZ$, $\triangle RST$ are rt $\triangle$s (def of rt $\triangle$s); $\triangle XYZ \cong \triangle RST$ (HL for $\cong$ rt $\triangle$s) **5.** $\angle ADB$, $\angle BEA$ are rt $\angle$s (def of $\perp$ lines); $\overline{AB} \cong \overline{AB}$ (reflex prop); $\triangle ABE$, $\triangle BAD$ are rt $\triangle$s (def of rt $\triangle$s); $\triangle ABE \cong \triangle BAD$ (HL for $\cong$ rt $\triangle$s)
6. $\angle ADB$, $\angle BEA$ are rt $\angle$s 1. (def of $\perp$ lines); ADB, $\triangle BEA$ are rt $\triangle$s 2. (def of rt $\triangle$s); $\overline{AB} \cong \overline{AB}$ 3. (reflex prop); $\triangle ADB \cong \triangle BEA$ (HA for $\cong$ rt $\triangle$s) **7.** $\angle ADB$, $\angle BEA$ are rt $\angle$s (def of $\perp$ lines); $\triangle AFD$, $\triangle BFE$ are rt $\triangle$s (def of rt $\triangle$s); $\angle DFA \cong \angle EFB$ (Th 2.10); $\triangle AFD \cong \triangle BFE$ (HA for $\cong$ rt $\triangle$s) **8.** $\angle ADB$, $\angle BEA$ are rt $\angle$s (def of $\perp$ lines); $\triangle AFD$, $\triangle BFE$ are rt $\triangle$s (def of rt $\triangle$s); $\angle DFA \cong \angle EFB$ (Th 2.10); $\triangle AFD \cong \triangle BFE$ (LA for $\cong$ rt $\triangle$s)
9. $\angle IJG$, $\angle IJH$ are rt $\angle$s (def of $\perp$ lines); $\triangle IJG$, $\triangle IJH$ are rt $\triangle$s (def rt $\triangle$s); $\overline{GJ} \cong \overline{JH}$ (def of bis); $\overline{IJ} \cong \overline{IJ}$ (reflex prop); $\triangle IJG \cong \triangle IJH$ ($\angle L$ for $\cong$ rt $\triangle$s); $\overline{IG} \cong \overline{IH}$ (CPCTC); $\triangle GHI$ is isos (def of isos $\triangle$) **10.** $\angle IJG$, $\angle IJH$ are rt $\angle$s (def of $\perp$ lines); $\triangle IJG$, $\triangle IJH$ are rt $\triangle$s (def of rt $\triangle$s); $\angle GIJ \cong \angle HIJ$ (def of bis); $\overline{IJ} \cong \overline{IJ}$ (reflex prop); $\triangle IJG \cong \triangle IJH$ (HA for $\cong$ rt $\triangle$s); $\overline{IG} \cong \overline{IH}$ (CPCTC); $\triangle GHI$ is isos (def of isos $\triangle$) **11.** $\angle MOK$, $\angle LNK$ are rt $\angle$s (def. of $\perp$ lines); $\triangle MOK$, $\triangle LNK$ are rt $\triangle$s (def of rt $\triangle$s; $\angle K \cong \angle K$ (reflex prop); $\triangle MOK \cong \triangle LNK$ (HA for $\cong$ rt $\triangle$s); $\overline{MO} \cong \overline{LN}$ (CPCTC) **12.** $\angle MOK$, $\angle LNK$ are rt $\angle$s (def of $\perp$ lines); $\triangle MOK$, $\triangle LNK$ are rt $\triangle$s (def of rt $\triangle$s); $\angle K \cong \angle K$ (reflex prop); $\triangle MOK \cong \triangle LNK$ (LA for $\cong$ rt $\triangle$s); $\overline{MO} \cong \overline{LK}$ (CPCTC) **13.** $\angle RSP$, $\angle VWT$ are rt $\angle$s (def of $\perp$ lines); $\triangle RSP$, $\triangle VWT$ are rt $\triangle$s (def of rt $\triangle$s); $\angle P \cong \angle T$; $\overline{PR} \cong \overline{TV}$ (CPCTC); $\triangle RSP \cong \triangle VWT$ (HA for $\cong$ rt $\triangle$s); $\overline{RS} \cong \overline{VW}$ (CPCTC) **14.** $\angle RSP$, $\angle RSQ$, $\angle VWT$, $\angle VWU$ are rt $\angle$s (def of $\perp$ lines); $\triangle RSP$, $\triangle RSQ$, $\triangle VWT$, $\triangle VWU$ are rt $\triangle$s (def of rt $\triangle$s); $\angle P \cong \angle T$; $\angle Q \cong U$; $\overline{PR} \cong \overline{TV}$; $\overline{QR} \cong \overline{UV}$ (CPCTC) $\triangle RSP \cong \triangle VWT$; $\triangle RSQ \cong \triangle VWU$ (HA for $\cong$ rt $\triangle$s); $\angle PRS \cong \angle QRS$ (def of bis); $\angle TVW \cong \angle PRS$; $\angle UVW \cong \angle QRS$ (CPCTC); $\angle TVW \cong \angle UVW$ (subst prop); $\overline{VW}$ bis $\angle UVT$ (def of bis)
15. $\angle ZAX$, $\angle ZAY$ are rt $\angle$s (def of $\perp$ lines); $\triangle ZAX$, $\triangle ZAY$ are rt $\triangle$s (def of rt $\triangle$s); $\overline{ZA} \cong \overline{ZA}$ (reflex prop); $\triangle XAZ \cong \triangle YAZ$ (HL for $\cong$ rt $\triangle$s)
16. $\angle ZXY \cong \angle ZYX$ (Th 5.1) **17.** Statements 1 – 4 of Proof 15 (reasons 1 – 4 of Proof 15); statement 1 of Proof 16 (reason 1 of Proof 16); $\angle ZXA \cong \angle ZYA$ (CPCTC); $\angle AXY \cong \angle AYX$ (subt prop) **18.** $\angle ZAX$, $\angle ZAY$ are rt $\angle$s (def of $\perp$ lines); $\triangle ZAX$, $\triangle ZAY$ are rt $\triangle$s (def of rt $\triangle$s); $\overline{ZA} \cong \overline{ZA}$ (reflex prop); $\triangle XAZ \cong \triangle YAZ$ (LL for $\cong$ rt $\triangle$s) **19.** Statements 1 – 4 of Proof 18 (reasons 1 – 4 of Proof 18); $\overline{ZX} \cong \overline{ZY}$ (CPCTC); $\angle ZXY \cong \angle ZYX$ (Th 5.1) **20.** Statements 1 – 3 of Proof 19 (reasons 1 – 3 of Proof 19); $\angle ZXA \cong \angle ZYA$ (CPCTC); $\angle AXY \cong \angle AYX$ (subt prop)

Pages 163–165

4. $\angle DEA \cong \angle CEB$ (Th 2.10); $\triangle DEA \cong \triangle CEB$ (AAS for $\cong \triangle$s); $\overline{DE} \cong \overline{CE}$; $\overline{EA} \cong \overline{EB}$ (CPCTC); $\overline{DB} \cong \overline{CA}$ (add prop); $\overline{AB} = \overline{AB}$ (reflex prop); $\triangle ABD \cong \triangle BAC$ (SSS for $\cong \triangle$s) **5.** $\angle FIG$, $\angle GJF$ are rt $\angle$s (def of $\perp$ lines); $\triangle FIG$; $\triangle GJF$, $\triangle FKI$, $\triangle GKJ$ are rt $\triangle$s (def rt $\triangle$s); $\overline{FG} \cong \overline{FG}$ (reflex prop); $\triangle FIG \cong \triangle GJF$ (HL for $\cong$ rt $\triangle$s); $\overline{FI} \cong \overline{GJ}$ (CPCTC); $\angle IKF \cong \angle JKG$ (Th 2.10); $\triangle FKI \cong \triangle GKJ$ (LA for $\cong$ rt $\triangle$s) **8.** $\angle LGH$, $\angle KIJ$ are rt $\angle$s (def of $\perp$ lines); $\triangle LGH$, $\triangle KIH$ are rt $\triangle$s (def of rt $\triangle$s); $\overline{LH} \cong \overline{KI}$; $\angle LHG \cong \angle KJI$ (CPCTC); $\triangle LGH \cong \triangle KIH$ (HA for $\cong$ rt $\triangle$s)
9. $\angle LGF$, $\angle KIH$ are rt $\angle$s (def of $\perp$ lines); $\triangle FGL$, $\triangle HIK$ are rt $\triangle$s (def of rt $\triangle$s); $\overline{LF} \cong \overline{KH}$; $\angle LFG \cong \angle KHI$ (CPCTC); $\triangle FGL \cong \triangle HIK$ (HA for $\cong$ rt $\triangle$s) **10.** $\overline{FL} \cong \overline{HK}$; $\angle F \cong \angle KHJ$ (CPCTC); $\triangle FHL \cong \triangle HJK$ (AAS for $\cong \triangle$s)
11. $\overline{FG} \cong \overline{JI}$ (CPCTC); $FG = JI$, $GH = HI$ (def $\cong$ seg); $FG + GH = JI + HI$ (add prop); $FG + GH = FH$; $JI + HI = JH$ (betweeness); $FH = JH$

(subst); $\overline{FH} \cong \overline{JH}$ (def of ≅ seg); $\angle J \cong \angle F$, $\overline{FL} \cong \overline{JK}$ (CPCTC); $\triangle FHL \cong \triangle JHK$ (SAS) **12.** $\angle QNM$, $\angle QNO$, $\angle RNQ$, $\angle PNQ$ are rt ∠s (def of ⊥ lines); $\triangle QNM$, $\triangle QNO$, $\triangle RNQ$, $\triangle PNQ$ are rt △s (def of rt △s); $\overline{QN} \cong \overline{QN}$ (reflex prop); $\triangle QNM \cong \triangle QNO$ (LL for ≅ rt △s); $\angle RQN \cong \angle PQN$ (CPCTC); $\triangle RNQ \cong \triangle PNQ$ (HA for ≅ rt △s) **13.** Statements 1–3 Proof 12 (reasons 1–3 of Proof 12); $\triangle QNR \cong \triangle QNP$ (HL for ≅ rt △s); $\angle RQN \cong \angle PQN$ (CPCTC); $\triangle QNM \cong \triangle QNO$ (LA for ≅ rt △s); $\overline{NM} \cong \overline{NO}$ (CPCTC) **14.** $\angle DEA$, $\angle CFB$ are rt ∠s (def of ⊥ lines); $\triangle DEA$, $\triangle CFB$ are rt △s (def of rt △s); $\triangle DEA \cong \triangle CFB$ (HL for ≅ rt △s); $\overline{AB} = \overline{AB}$ (reflex prop); $\angle DAB \cong \angle CBA$ (CPCTC); $\triangle ADB \cong \triangle BCA$ (SAS for ≅ △s); $\overline{AC} \cong \overline{DB}$ (CPCTC) **15.** $\angle DEB$, $\angle CFA$ are rt ∠s (def of ⊥ lines); $\triangle DEB$, $\triangle CFA$ are rt △s (def of rt △s); $\triangle DEB \cong \triangle CFA$ (HL for ≅ rt △s); $\overline{AB} \cong \overline{AB}$ (reflex prop); $\angle CAB \cong \angle DBA$ (CPCTC); $\triangle CAB \cong \triangle DBA$ (SAS for ≅ △s); $\overline{AD} \cong \overline{BC}$ (CPCTC)
19. $\angle DAB$, $\angle CBA$ are rt ∠s (def of ⊥ lines); $\triangle DAB$, $\triangle CBA$ are rt △s (def of rt △s); $\overline{AB} \cong \overline{AB}$ (reflex prop); $\triangle DAB \cong \triangle CBA$ (LL for ≅ rt △s); $\overline{AC} \cong \overline{BD}$ (CPCTC) **20.** (Statements 1–4 Proof 19) (reasons 1–4 of Proof 19); ($\angle CAB \cong \angle DBA$) (CPCTC); ($\overline{AE} \cong \overline{BE}$) (Th 5.2); ($\triangle AEB$ is isos) (def of isos △) **21.** Statements 1–5 Proof 19 (reasons 1–5 of Proof 19); statements 2–3 Proof 20 (reasons 2–3 of Proof 20); $\overline{ED} \cong \overline{EC}$ (subt prop) **22.** Statements 1–3 Proof 21 (reasons 1–3) of Proof 21); $\angle EDC \cong \angle ECB$ (Th 5.1); $\angle AEB \cong \angle DEC$ (Th 2.10); m $\angle AEB$ + m $\angle CAB$ + m $\angle DBA$ = 180, m $\angle DEC$ + m $\angle EDC$ + m $\angle ECD$ = 180 (Th 3.5); m $\angle AEB$ + m $\angle CAB$ + m $\angle DBA$ = m $\angle DEC$ + m $\angle EDC$ + m $\angle ECD$ (subst prop); m $\angle CAB$ + m $\angle DBA$ = m $\angle EDC$ + m $\angle ECD$ (subt prop); 2 m $\angle CAB$ = 2 m $\angle ECD$ (subst prop); m $\angle CAB$ = m $\angle ECD$ (div prop); $\angle CAB \cong \angle ECD$ (def of ≅ ∠s); $\overline{AB} \parallel \overline{DC}$ (Post 14) **23.** Statements 1–10 of Proof 22 (reasons 1–10 of Proof 22); $\angle CDA$ and $\angle DAB$ are supp; $\angle DCB$ and $\angle CBA$ are supp (Th 3.2); $\angle CDA$, $\angle DCB$ are rt ∠s (suppl of a rt ∠ is a rt ∠); $\overline{AD} \perp \overline{DC}$; $\overline{BC} \perp \overline{DC}$ (def of ⊥ lines) **24.** Statements 1–4 Proof 23 (reasons 1–4 of Proof 23); $\overline{DA} \cong \overline{DA}$ (reflex prop) $\triangle DAB \cong ADC$ (HL for ≅ rt △s) $\overline{DC} \cong \overline{AB}$ (CPCTC) $\angle CDB \cong \angle DBA$; $\angle DCA \cong \angle CAB$ (Post 13) $\triangle AEB \cong \triangle CED$ (ASA for ≅ △s) **25.** Statements of Proof 24 (reasons of Proof 24) $\overline{AE} \cong \overline{CE}$; $DE \cong \overline{BE}$ (CPCTC) $\overline{AE} \cong \overline{CE} \cong \overline{DE} \cong \overline{BE}$ (subst prop) **26.** $\overline{FG} \cong \overline{GH}$ (def of midpt); $\angle JFH \cong \angle JHF$ (Th 5.1); $\triangle KFG \cong \triangle IHG$ (ASA for ≅ △s); $\overline{FK} \cong \overline{HI}$ (CPCTC); $\overline{FH} \cong \overline{FH}$ (reflex prop); $\triangle FKH \cong \triangle HIF$ (SAS for ≅ △s); $\overline{FI} \cong \overline{HK}$ (CPCTC) **27.** $\angle QRP \cong \angle MRN$ (Th 2.10); $\triangle QRP \cong \triangle MRN$ (SAS for ≅ △s); $\angle PMO \cong \angle NQO$ (CPCTC); $\angle O \cong \angle O$ (reflex prop); $\overline{QN} \cong \overline{MP}$ (add prop); $\triangle PMO \cong \triangle NQO$ (AAS for ≅ △s); $\overline{OQ} \cong \overline{OM}$ (CPCTC)
28. $\overline{VS} \cong \overline{VS}$ (reflex prop); $\triangle TSV \cong \triangle USV$ (SSS for ≅ △s); $\angle TSW \cong \angle USW$ (CPCTC); $\overline{WS} \cong \overline{WS}$ (reflex prop); $\triangle TSW \cong \triangle USW$ (SAS for ≅ △s); $\overline{WU} \cong \overline{WT}$ (CPCTC); $\triangle UWT$ is isos (def of isos △) **29.** $\overline{XA} \cong \overline{XA}$ (reflex prop); $\triangle XAZ \cong \triangle XAY$ (SSS for ≅ △s); $\angle ZXB \cong \angle YXB$ (CPCTC); $\overline{XB} \cong \overline{XB}$ (reflex prop); $\triangle ZXB \cong \triangle YXB$ (SAS for ≅ △s); $\overline{ZB} \cong \overline{YB}$ (CPCTC)
30. $\overline{XA} \cong \overline{XA}$ (reflex prop); $\triangle XAZ \cong \triangle XAY$ (AAS for ≅ △s); $\overline{XZ} \cong \overline{XY}$ (CPCTC); $\overline{XB} \cong \overline{XB}$ (reflex prop); $\triangle ZXB \cong \triangle YXB$ (SAS for ≅ △s); $\angle XZB \cong \angle XYB$ (CPCTC)

Page 168

7. Given: Isos $\triangle ABC$, $\overline{AC} \cong \overline{BC}$, $\overline{CD}$ bis $\angle ACB$ Prove: $\overline{CD}$ bis $\overline{AB}$ $\angle ACD \cong \angle DCB$ (def of bis); $\overline{CD} \cong \overline{CD}$ (reflex prop); $\triangle ACD \cong \triangle BCD$ (SAS); $\overline{AD} \cong \overline{DB}$ (CPCTC); $\overline{CD}$ bis $\overline{AB}$ (def of bis)
8. Given: isos $\triangle ABC$, $\overline{AC} \cong \overline{BC}$, $\overline{CD}$ bis $\overline{AB}$; Prove: $\overline{CD}$ bis $\angle ACB$, $\overline{AO} \cong \overline{DB}$ (def of bis); $\overline{CD} \cong \overline{CD}$ (reflex prop); $\triangle ADC \cong \triangle BDC$ (SSS) $\angle ACD \cong \angle DCB$ (CPCTC); $\overline{CD}$ bis $\angle ACB$ (def of bis) **9.** Given: isos $\triangle ABC$, $\overline{AC} \cong \overline{BC}$, $\overline{CD}$ bis $\angle ACB$; Prove: $\overline{CD}$ is the ⊥ bis of $\overline{AB}$. $\angle ACD \cong \angle DCB$ (def of bis); $\overline{CD} \cong \overline{CD}$ (reflex prop); $\triangle ADC \cong \triangle BDC$ (SAS); $\overline{AD} \cong \overline{DB}$ (CPCTC); $\overline{CD}$ is ⊥ bis of $\overline{AB}$ (Th 5.8) **10.** Given: $\overline{CD} \perp \overline{AB}$, CD bis $\angle ACB$; Prove: $\triangle ABC$ is isos. $\angle ACD \cong \angle DCB$ (def of bis); $\overline{CD} \cong \overline{CD}$ (reflex prop); $\angle CDA \cong \angle CDB$ (Th 2.6); $\triangle ACD \cong \triangle BDC$ (ASA); $\overline{AC} = \overline{BC}$ (CPCTC); $\triangle ABC$ is isos (def of isos △)
11. Given: $\angle ACB$, $\overrightarrow{CP}$ bis $\angle ACB$, $\overline{PA} \perp \overrightarrow{CA}$, $\overline{DB} \perp \overrightarrow{CB}$; Prove: $\overline{PA} \cong \overline{PB}$. $\angle ACP \cong \angle BCP$ (def of bis); $\overline{CP} \cong \overline{CP}$ (reflex prop); $\angle CPA \cong CBP$ (Th 2.6); $\triangle ACD \cong \triangle BDC$ (AAS); $\overline{PA} \cong \overline{PB}$ (CPCTC) **12.** Given: Isos $\triangle ABC$, $\overline{AB} \cong \overline{BC}$, M mdpt $\overline{BC}$; $\overline{MD} \perp \overline{AC}$; N mdpt. $\overline{BC}$; $\overline{NE} \perp \overline{AC}$; Prove: $\overline{MD} \cong \overline{NE}$. $\angle BAC \cong \angle BCA$ (Th 5.1); $\angle MDA$, m $\angle NEC$ are rt ∠s (def of ⊥ lines); $\triangle MDA$, $\triangle NEC$ are rt △s (def of rt △s); $\overline{AB} = 2\overline{MA}$; $\overline{BC} = 2\overline{NC}$ (def of bis); $2\overline{MA} \cong 2\overline{NC}$ (subst prop); $\overline{MA} \cong \overline{NC}$ (div prop); $\triangle MDA \cong \triangle NEC$ (HA); $\overline{MD} \cong \overline{NE}$ (CPCTC)
13. 1. Given **2.** Given **3.** Reflex prop **4.** SSS for ≅ △s **5.** CPCTC **6.** reflex prop **7.** SAS **8.** CPCTC **9.** CPCTC **10.** Post 11 **11.** Th 2.7 **12.** Def of ⊥ bis **14.** Given: Isos $\triangle ABC$, $\overline{AB} \cong \overline{BC}$, $\overline{AE}$ bis $\angle BAC$, $\overline{CD}$ bis $\angle BCA$; Prove: $\overline{DC} \cong \overline{AE}$. $\overline{AC} \cong \overline{AC}$ (reflex prop); $\angle BAC \cong \angle BCA$ (Th 5.1); 2 $\angle EAC = \angle BAC$, 2 $\angle DCA = \angle BCA$ (def of bis) 2 $\angle EAC \cong$ 2 $\angle DCA$

(subst prop); $\angle EAC \cong \angle DCA$ (div prop); $\triangle EAC \cong \triangle DCA$ (ASA); $\overline{DC} \cong \overline{AE}$ (CPCTC) **15.** $\overline{DC} \cong \overline{DA}$, $\overline{BA} \cong \overline{BC}$ (Th 5.7); $\overline{DB} \cong \overline{DB}$ (reflex prop); $\triangle DCB \cong \triangle DAB$ (SSS); $\angle BAD \cong \angle BCD$ (CPCTC) **16.** $\angle 1 \cong \angle 3$ (subst prop); $\overline{FI} \cong \overline{FG}$; $\overline{HI} \cong \overline{HG}$ (Th 5.2); $\overline{FH}$ is $\perp$ bis of $\overline{GI}$ (Th 5.8) **17.** $\angle 2 \cong \angle 4$ (subt prop); $\overline{FI} \cong \overline{FG}$; $\overline{HI} \cong \overline{HG}$ (Th 5.2); $\overline{FH}$ is $\perp$ bis of $\overline{GI}$ (Th 5.8); $\overline{IJ} \cong \overline{JG}$ (def of bis); $\triangle HIJ \cong \triangle HGJ$ (SAS); $\angle IHJ \cong \angle GHJ$ (CPCTC); $\overline{FH}$ bis $\angle IHG$ (def of bis) **18.** $\overline{MN} \cong \overline{MK}$, $\overline{NO} \cong \overline{KO}$ (Th 5.7); $\overline{ML} = \overline{ML}$ (reflex prop); $\angle MLN$, $\angle MLK$ are rt $\angle$s (def of $\perp$ lines); $\triangle MLN$, $\triangle MLK$ are rt $\triangle$s (def of rt $\triangle$s); $\triangle MLN \cong \triangle MLK$ (HL); $\overline{LN} \cong \overline{LK}$ (CPCTC); $\overline{LO}$ is $\perp$ bis of $\overline{NK}$ (Th 5.8)
19. $\overline{MO} \cong \overline{MO}$ (reflex prop); $\angle NMO \cong \angle KMO$ (def of bis); $\triangle NMO \cong \triangle KMO$ (SAS); $\overline{NO} \cong \overline{KO}$ (CPCTC); $\overline{ML} \cong \overline{ML}$ (reflex prop); $\angle MLN$, $\angle MLK$ are rt $\angle$s (def of $\perp$ lines); $\triangle MLN$, $\triangle MLK$ are rt $\triangle$s (def of rt $\triangle$s); $\triangle MLN \cong \triangle MLK$ (HL); $\overline{KL} \cong \overline{NL}$ (CPCTC); $\angle LNO \cong \angle LKO$ (Th 5.1), $\triangle LNO \cong \triangle LKO$ (SAS); $\angle NLO \cong \angle KLO$ (CPCTC); $\overline{LO}$ bis $\angle NLK$ (def of bis)

Pages 172–173

1. $\overline{CD} \perp \overline{AB}$ (def of alt); $\angle CDA$, $\angle CDB$ are rt $\angle$s (def of $\perp$ lines); $\triangle ADC$, $\triangle DBC$ are rt $\triangle$s (def of rt $\triangle$s); $\overline{CD} \cong \overline{CD}$ (reflex prop); $\triangle ADC \cong \triangle BDC$ (LL) **2.** D is mdpt of $\overline{AB}$ (def of median); $\overline{AD} \cong \overline{DB}$ (def of mdpt); $\overline{CD} \cong \overline{CD}$ (reflex prop); $\triangle ADC \cong BDC$ (SSS) **6.** $\overline{ZW} \perp \overline{XY}$ (def of alt); $\angle ZWX$, $\angle ZWY$ are rt $\angle$s (def of $\perp$ lines); $\triangle ZWX$, $\triangle ZWY$ are rt $\triangle$s (def of rt $\triangle$s); $\overline{ZW} \cong \overline{ZW}$ (reflex prop); $\overline{XW} \cong \overline{WY}$ (def of bis); $\triangle XWZ \cong \triangle YWZ$ (LL); $\overline{XZ} \cong \overline{YZ}$ (CPCTC); $\triangle XYZ$ is isos (def of isos $\triangle$) **7.** $\overline{ZW} \perp \overline{XY}$ (def of alt); $\angle ZWX$, $\angle ZWY$ are rt $\angle$s (def of $\perp$ lines); $\triangle ZWX$, $\triangle ZWY$ are rt $\triangle$s (def of rt $\triangle$s); $\overline{ZW} = \overline{ZW}$ (reflex prop); $\angle XZW \cong YZW$ (def of bis); $\triangle ZWX \cong \triangle ZWY$ (LA); $\overline{XZ} \cong \overline{YZ}$ (CPCTC); $\triangle XYZ$ is isos (def of isos $\triangle$ reason) **8.** $\overline{CD} \perp \overline{AB}$ (def of alt); $\angle CDA \cong \angle CDB$ (Th 2.6); m $\angle 1$ + m $\angle 2$ + m $\angle CDA$ = 180, m $\angle 3$ + m $\angle 4$ + m $\angle CDB$ = 180 (Th 3.5); m $\angle 1$ + m $\angle 2$ + m $\angle CDA$ = m $\angle 3$ + m $\angle 4$ + m $\angle CDB$ (subst prop); m $\angle 1$ + m $\angle 2$ = m $\angle 3$ + m $\angle 4$ (subt prop)
9. $\angle ACB$ is a rt $\angle$ (def of $\perp$ lines); $\angle 4 = \angle 4$ (reflex prop); $\angle CDB \cong \angle ACB$ (Th 3.6); $\angle CDB$ is a rt $\angle$ (subst prop); $\overline{CD} \perp \overline{AB}$ (def of $\perp$ lines); $\overline{CD}$ is an alt of $\triangle ABC$ (def of alt) **10.** $\overline{IG} \cong \overline{NL}$ $\angle G \cong \angle L$ (CPCTC); $2\overline{HG} \cong \overline{IG}$, $2\overline{ML} \cong \overline{NL}$ (def of bis); $2\overline{HG} \cong 2\overline{ML}$ (subst prop); $\overline{HG} \cong \overline{ML}$ (div prop); $\angle FHG$, $\angle KML$ are rt $\angle$s (def of $\perp$ lines); $\triangle FHG$, $\triangle KML$ are rt $\triangle$s (def of rt $\triangle$s); $\triangle FHG \cong \triangle KML$ (LA); $\overline{FH} \cong \overline{KM}$ (CPCTC)
11. $\angle O \cong \angle T$; $\overline{OP} \cong \overline{ST}$; $\angle OPQ \cong \angle TSV$ (CPCTC); 2 $\angle OPR \cong \angle OPQ$; 2 $\angle TSU \cong \angle TSV$ (def of bis); 2 $\angle OPR \cong$ 2 $\angle TSU$ (subst prop); $\angle OPR \cong \angle TSU$ (div prop); $\triangle OPR \cong \triangle TSU$ (ASA); $\overline{PR} \cong \overline{SU}$ (CPCTC) **12.** $\angle Y \cong \angle Y$ (reflex prop); $\angle YZB$, $\angle YCA$ are rt $\angle$s (def of $\perp$ lines); $\triangle YZB$, $\triangle YCA$ are rt $\triangle$s (def of rt $\triangle$s) $\triangle YZB \cong \triangle YCA$ (LA); $\overline{YZ} \cong \overline{YC}$ (CPCTC); $\overline{YW} \cong 2\,\overline{YZ}$, $\overline{YX} = 2\,\overline{YC}$ (def of bis); $2\,\overline{YZ} \cong 2\,\overline{YC}$ (mult prop); $\overline{YW} \cong \overline{YX}$ (subst prop)
13. 1. Given **2.** CPCTC **3.** Def of $\cong$ seg **4.** Given **5.** Def of median **6.** Subst prop **7.** Def of $\cong$ seg **8.** CPCTC **9.** CPCTC **10.** *SAS* **11.** CPCTC **14.** Given: $\triangle ABC \cong \triangle DEF$, $\overline{CI}$ alt to $\overline{AB}$. $\overline{FH}$ alt to $\overline{ED}$; Prove: $\overline{CI} \cong \overline{FH}$. $\overline{CI} \perp \overline{AB}$, $\overline{FH} \perp \overline{ED}$ (def of alt); $\angle CIB$, $\angle FHE$ art rt $\angle$s (def of $\perp$ lines); $\triangle CIB$, $\triangle FHE$ are rt $\triangle$s (def of rt $\triangle$s); $\overline{CB} \cong \overline{FE}$, $\angle B \cong \angle E$ (CPCTC); $\triangle CIB \cong \triangle FHE$ (HA); $\overline{CI} \cong \overline{FH}$ (CPCTC)
15. Given: $\triangle ABC$ is equil; $\overline{AE}$ alt to $\overline{CB}$, $\overline{CD}$ alt to $\overline{AB}$; Prove: $\overline{AE} \cong \overline{CD}$. $\overline{AE} \perp \overline{CB}$, $\overline{CD} \perp \overline{AB}$ (def of alt); $\angle AEB$, $\angle CDB$ are rt $\angle$s (def of $\perp$ lines); $\triangle AEB$, $\triangle CDB$ are rt $\triangle$s (def of rt $\triangle$s); $\angle B \cong \angle B$ (reflex prop); $\overline{BC} \cong \overline{AB}$ (def of equil $\triangle$); $\triangle AEB \cong \triangle CDB$ (HA); $\overline{AE} \cong \overline{CD}$ (CPCTC) **16.** Given: $\triangle ABC$ is equil, $\overline{AE}$ is median to $\overline{CB}$, $\overline{BD}$ is median to $\overline{AC}$; Prove: $\overline{AE} \cong \overline{DB}$. D is mdpt of $\overline{AC}$, E is mdpt of $\overline{CB}$ (def of median); $2\overline{AD} \cong \overline{AC}$, $2\,\overline{EB} \cong \overline{CB}$ (def of mdpt); $\overline{AC} \cong \overline{CB}$ (def of equil $\triangle$); $2\,\overline{AD} \cong 2\,\overline{EB}$ (subst prop); $\overline{AD} \cong \overline{EB}$ (div prop); $\overline{AB} \cong \overline{AB}$ (reflex prop); $\angle CAB \cong \angle CBA$ (Th 5.1); $\triangle ADB \cong \triangle BEA$ (SAS); $\overline{AE} \cong \overline{CD}$ (CPCTC)
17. Given: Isos $\triangle ABC$, $\overline{AC} \cong \overline{BC}$, $\overline{CD}$ alt to $\overline{AB}$; Prove: $\overline{CD}$ is median to $\overline{AB}$. $\angle A \cong \angle B$ (Th 5.1); $\overline{CD} \perp \overline{AB}$ (def of alt); $\angle CDA \cong \angle CDB$ (Th 2.6); $\overline{CD} \cong \overline{CD}$ (reflex prop); $\triangle ACD \cong \triangle BCD$ (AAS); $\overline{AD} \cong \overline{DB}$ (CPCTC); D is the mdpt of $\overline{AB}$ (def of mdpt), $\overline{AD} \cong \overline{DB}$ (def of median) **18.** Given: $\triangle ABC$ with alt $\overline{AE}$ to $\overline{CB}$, alt $\overline{DB}$ to $\overline{AC}$, $\overline{AE} \cong \overline{DB}$; Prove: $\triangle ABC$ is isos. $\overline{AE} \perp \overline{CB}$, $\overline{DB} \perp \overline{AC}$ (def of alt); $\angle CDB$, $\angle CEA$ are rt $\angle$s (def of $\perp$ lines); $\triangle CDB$, $\triangle CEA$ are rt $\triangle$s (def of rt $\triangle$s); $\angle C \cong \angle C$ (reflex prop); $\triangle CDB \cong \triangle CEA$ (LA); $\overline{AC} \cong \overline{BC}$ (CPCTC); $\triangle ABC$ is isos (def of isos $\triangle$)
19. Given: Isos $\triangle ABC$, $\overline{AC} \cong \overline{BC}$, $\overline{CD}$ median to $\overline{AB}$; Prove: $\overline{CD}$ bis $\angle ACB$. D is mdpt of $\overline{AB}$ (def of median); $\overline{AD} \cong \overline{DB}$ (def of mdpt); $\overline{CD} \cong \overline{CD}$ (reflex prop); $\triangle ADC \cong \triangle BDC$ (SSS); $\angle ACD \cong \angle DCB$ (CPCTC); $\overline{CD}$ bis $\angle ACB$ (def of bis);
20. Given: Isos $\triangle ABC$, $\overline{AC} \cong \overline{BC}$, $\overline{CD}$ alt to $\overline{AB}$; Prove: $\overline{CD}$ bis $\angle ACB$. $\overline{CD} \perp \overline{AB}$ (def of alt); $\angle CDA \cong \angle CDB$ (Th 2.6); $\angle A \cong \angle B$ (Th 5.1); $\triangle CDA \cong \triangle CDB$ (AAS); $\angle ACD \cong \angle DCB$ (CPCTC); $\overline{CD}$ bis $\angle ACB$ (def of bis)
21. Given: $\triangle ABC$, $\overline{CD}$ median to $\overline{AB}$, $\overline{CD} \perp \overline{AB}$; Prove: $\triangle ABC$ is isos. D is mdpt of $\overline{AB}$ (def of median); $\overline{AD} \cong \overline{DB}$ (def of mdpt); $\angle ADC \cong$

$\angle BDC$ (Th 2.6); $\overline{CD} \cong \overline{CD}$ (reflex prop); $\triangle ADC \cong \triangle BDC$ (SAS); $\overline{AC} \cong \overline{BC}$ (CPCTC); $\triangle ABC$ is isos (def of isos $\triangle$) **22.** Given: $\triangle ABC$, $\overline{CD}$ alt to $\overline{AB}$, $\overline{CD}$ bis $\angle ACB$; Prove: $\triangle ABC$ is isos. $\angle ACD \cong \angle BCD$ (def of bis); $\overline{CD} \cong \overline{CD}$ (reflex prop); $\overline{CD} \perp \overline{AB}$ (def of alt); $\angle CDA \cong \angle CDB$ (Th 2.6); $\triangle CDA \cong \triangle CDB$ (ASA); $\overline{AC} \cong \overline{BC}$ (CPCTC); $\triangle ABC$ is isos (def of isos $\triangle$) **23.** Given: $\triangle ABC$, $\overline{CD}$ median to $\overline{AB}$, $\overline{CD}$ bis $\overline{CD}$ bis $\angle ACB$; Prove: $\triangle ABC$ is Isos. Draw $\overline{DM} \perp \overline{AC}$, $\overline{DN} \perp \overline{BC}$ (Post 12) $\angle AMD, \angle CMD, \angle CND, \angle BND$ are rt $\angle$s. (def of $\perp$ lines); $\triangle CMD, \triangle CND, \triangle AMD, \triangle BND$ are rt $\triangle$s (def of rt $\triangle$s); $\angle MCD \cong \angle NCD$ (def of bis); $\overline{CD} \cong \overline{CD}$ (reflex prop); $\triangle CDM \cong \triangle CDN$ (HA); $\overline{MD} \cong \overline{ND}$ (CPCTC); D is mdpt of $\overline{AB}$ (def of med); $\overline{AD} \cong \overline{DB}$ (def of midpt); $\triangle AMD \cong \triangle BND$ (HL); $\angle A \cong \angle B$ (CPCTC); $\overline{AC} \cong \overline{BC}$ (Th 5.2); $\triangle ABC$ is isos (def of isos $\triangle$) **24.** Given: $\triangle ABC$, $\overline{BM}$ median to $\overline{AC}$, $\overline{AN}$ median to $\overline{BC}$, $\overline{BM}$ bis $\angle CBA$, $\overline{AN}$ bis $\angle CAB$; Prove: $\triangle ABC$ is equil. $\triangle ABC$ is isos with $\overline{AC} \cong \overline{AB}$ (proved in ex 23, AN the median); $\triangle ABC$ is isos with $\overline{AB} \cong \overline{BC}$ (proved in ex 23, BM the median); $\overline{AC} \cong \overline{AB} \cong \overline{BC}$ (subst prop); $\triangle ABC$ is equil (def of equil $\triangle$) **25.** $\angle CDE \cong \angle ADE$ (def of bis); $\overline{ED} \cong \overline{ED}$ (reflex prop); $\triangle CDE \cong \triangle ADE$ (SAS); $\overline{CE} \cong \overline{EA}$ (CPCTC); $\angle DBC, \angle DBA$ are rt $\angle$s (def of $\perp$ lines); $\triangle DBC, \triangle DBA$ are rt $\triangle$s (def of rt $\triangle$s); $\overline{DB} \cong \overline{DB}$ (reflex prop); $\triangle DBC \cong \triangle DBA$ (HL); $\overline{BC} \cong \overline{BA}$ (CPCTC); $\overline{BE} \cong \overline{BE}$ (reflex prop); $\triangle BCE \cong \triangle BAE$ (SSS); $\angle CBE \cong \angle ABE$ (CPCTC); $\overline{BE}$ bis $\angle CBA$ (def of bis) **26.** E is midpt of $\overline{CA}$ (def of median); $\overline{CE} \cong \overline{EA}$ (def of midpt.); $\overline{DE} \perp \overline{CA}$ (def of alt); $\angle DEC \cong \triangle DEA$ (Th 2.6); $\overline{DE} \cong \overline{DE}$ (reflex prop); $\triangle DEC \cong \triangle DEA$ (SAS); $\angle DBC, \angle DBA$ are rt $\angle$s (def of $\perp$ lines); $\triangle DBC, \triangle DBA$ are rt $\triangle$s (def of rt $\triangle$s); $\overline{DB} \cong \overline{DB}$ (reflex prop); $\overline{DC} \cong \overline{DA}$ (CPCTC); $\triangle DBC \cong \triangle DBA$ (HL); $CB \cong BA$ (CPCTC); $EC = EA$, $BC = BA$ (def $\cong$ seg); $\overline{BE} \perp$ bis $\overline{CA}$ (Th 5.8); $\overline{BE}$ alt, med of $\triangle ABC$ (def) **27.** $\overline{JF} \cong \overline{HF}$, $\angle JFG \cong \angle HFG$ (CPCTC); $\overline{FI} \cong \overline{FI}$ (reflex) prop); $\triangle JFI \cong \triangle HFI$ (SAS); $\angle JIF \cong \angle HIF$ (CPCTC); $\overline{JI} \perp \overline{FG}$ (def of alt); $\angle JIF$ is a rt$\angle$ (def of $\perp$ lines); $\angle HIF$ is a rt $\angle$ (subst prop); $\overline{HI} \perp \overline{FG}$ (def of $\perp$ lines); $\overline{HI}$ is alt of $\triangle FHG$ (def of alt) **28.** Statements 1 – 3 of ex 27 (reasons 1 – 3 of ex 27); $\angle FJI \cong \angle FHI$ (CPCTC); $\angle FJG \cong \angle FHG$ (CPCTC); $2\angle FHI \cong \angle FHG$ (def of bis); $2\angle FHI \cong 2\angle FJI$ (mult prop); $2\angle FJI \cong \angle FJG$ (subst prop); $\overline{JI}$ bis $\angle FJG$ (def of bis) **29.** $\angle NML, \angle OML$ are rt $\angle$s (def of $\perp$ lines); $\triangle NML, \triangle OML$ are rt $\triangle$s (def of rt $\triangle$s); $\overline{LM} \cong \overline{LM}$ (reflex prop); $\overline{NM} \cong \overline{MO}$ (def of bis); $\triangle OLM \cong \triangle NLM$ (LL) **30.** $\overline{LN} \cong \overline{LO}$, $\overline{KO} \cong \overline{KN}$ (Th 5.7); $\overline{KL} \cong \overline{KL}$ (reflex prop); $\triangle KLN \cong \triangle KLO$ (SSS) **31.** $\overline{NM} \cong \overline{NM}$ (reflex prop); $\angle KMN, \angle LMN$ are rt $\angle$s (def $\perp$ lines); $\triangle KMN, \triangle LMN$ are rt $\triangle s$ (def of rt $\triangle$s); $\triangle KMN \cong \triangle LMN$ (HL); $\overline{LM} \cong \overline{KM}$ (CPCTC) **32.** $\overline{NO} \perp \overline{KM}$, $\overline{NO} \perp \overline{LM}$ (def line $\perp$ to plane); $\overline{MO} \cong \overline{MO}$ (reflex prop); $\angle KMO, \angle LMO$ are rt $\angle$s (def of $\perp$ lines); $\triangle KMO, \triangle LMO$ are rt $\triangle$s (def of rt $\triangle$s); $\triangle KMO \cong LMO$ (LL); $\overline{LO} \cong \overline{KO}$ (CPCTC)

Page 178

11. $\angle 1 \cong \angle 2$ (def of bis); $\overline{AC} \cong \overline{AC}$ (reflex prop); $\triangle ADC \cong \triangle ABC$ (SAS); $\overline{CD} \cong \overline{CB}$ (CPCTC) **12.** m $\angle 1 =$ m $\angle 2$; m $\angle DCA =$ m $\angle BCA$ (def of bis); $\overline{AC} \cong \overline{AC}$ (reflex prop); $\triangle ADC \cong \triangle ABC$ (ASA); m $\angle 4 =$ m $\angle 3$ (CPCTC); m $\angle 1 +$ m $\angle 4 =$ m $\angle 2 +$ m $\angle 3$ (add prop) **13.** $\overline{GE} \cong \overline{GF}$, $\overline{GH} \cong \overline{GI}$ (CPCTC); $\overline{HE} \cong \overline{IF}$ (subtr prop) **14.** $\overline{HF} \perp \overline{GE}$, $\overline{EI} \perp \overline{GF}$ (def of alt); $\angle EHF, \angle FIE$ are rt $\angle$s (def of $\perp$ lines); $\triangle EHF, \triangle FIE$ are rt $\triangle$s (def of rt $\triangle$s); $\overline{EF} \cong \overline{EF}$ (reflex prop); $\triangle EHF \cong \triangle FIE$ (HA); $\angle HEF \cong \angle IFE$ (CPCTC); $\overline{GE} \cong \overline{GF}$ (Th 5.2); $\triangle GEF$ is isos (def of isos $\triangle$)

Page 179

12. $\angle ACD \cong \angle DCB$ (def of bis); $\overline{CD} \cong \overline{CD}$ (reflex prop); $\angle ADC, \angle CDB$ are rt $\angle$s (def of $\perp$ lines); $\triangle ADC, \triangle CDB$ are rt $\triangle$s (def of rt $\triangle$s); $\triangle ADC \cong \triangle CDB$ (LA); $\overline{AC} \cong \overline{CB}$ (CPCTC) $\triangle ABC$ is isos

Page 191

35. Given: convex pentagon $ABCDE$; Prove: m $\angle A +$ m $\angle B +$ m $\angle C +$ m $\angle D +$ m $\angle E = 540$. Draw $\overline{AC}, \overline{AD}$ (Post 1); m $\angle BAE =$ m $\angle 1 +$ m $\angle 2 +$ m $\angle 3$, m $\angle BCD =$ m $\angle 4 +$ m $\angle 5$, m $\angle EDC =$ m $\angle 6 +$ m $\angle 7$ (Post 10); m $\angle 1 +$ m $\angle B +$ m $\angle 4 = 180$, m $\angle 2 +$ m $\angle 5 +$ m $\angle 6 = 180$, m $\angle 3 +$ m $\angle E +$ m $\angle 7 = 180$ (Th 3.5); m $\angle A +$ m $\angle B +$ m $\angle C +$ m $\angle D +$ m $\angle E = 540$ (subst and add prop) **36.** Given: convex hexagon $ABCDEF$; Prove: m $\angle A +$ m $\angle B +$ m $\angle C +$ m $\angle D +$ m $\angle E +$ m $\angle F = 1080$. Draw $\overline{AC}, \overline{AD}, \overline{AE}$ (Post 1); m $\angle BAF =$ m $\angle 1 +$ m $\angle 2 +$ m $\angle 3 +$ m $\angle 4$, m $\angle BCD =$ m $\angle 9 +$ m $\angle 10$; m $\angle CDE =$ m $\angle 7 +$ m $\angle 8$, m $\angle DEF =$ m $\angle 5 +$ m $\angle 6$ (Post 10); m $\angle 1 +$ m $\angle F +$ m $\angle 5 = 180$, m $\angle 2 +$ m $\angle 6 +$ m $\angle 7 = 180$, m $\angle 3 +$ m $\angle 8 +$ m $\angle 9 = 180$, m $\angle 4 +$ m $\angle B +$ m $\angle 10 = 180$ (Th 3.5); m $\angle A +$ m $\angle B +$ m $\angle C +$ m $\angle D +$ m $\angle E +$ m $\angle F = 1080$ (subst and add prop) **37.** Given: convex quad $ABCD$; Prove: m $\angle A +$ m $\angle B +$ m $\angle C +$ m $\angle D = 360$. m $\angle BAD =$ m $\angle 1 +$ m $\angle 8$, m $\angle ABC =$ m $\angle 2 +$ m $\angle 3$, m $\angle BCD =$ m $\angle 4 +$ m $\angle 5$, m $\angle CDA =$ m $\angle 6 +$ m $\angle 7$ (Post 10) m $\angle 1 +$ m $\angle 10 +$ m $\angle 2 = 180$, m $\angle 3 +$ m $\angle 4 +$

m $\angle 11 = 180$, m $\angle 6$ + m $\angle 5$ + m $\angle 12 = 180$, m $\angle 7$ + m $\angle 9$ + m $\angle 8 = 180$ (Th 3.5); m $\angle 1$ + m $\angle 2$ + m $\angle 3$ + m $\angle 4$ + m $\angle 5$ + m $\angle 6$ + m $\angle 7$ + m $\angle 8$ + m $\angle 9$ + m $\angle 10$ + m $\angle 11$ + m $\angle 12 = 720$ (add prop); m $\angle 9$ + m $\angle 10$ + m $\angle 11$ + m $\angle 12 =$ 360 (Post 11); m $\angle 1$ + m $\angle 2$ + m $\angle 3$ + m $\angle 4$ + m $\angle 5$ + m $\angle 6$ + m $\angle 7$ + m $\angle 8 = 360$ (subt prop); m $\angle A$ + m $\angle B$ + m $\angle C$ + m $\angle D = 360$ (subst prop) **38.** Given: convex pentagon *ABCDE;* Prove: m $\angle A$ + m $\angle B$ + m $\angle C$ + m $\angle D$ + m $\angle E = 540$. m $\angle EAB$ = m $\angle 10$ + m $\angle 1$, m $\angle BCD$ = m $\angle 4$ + m $\angle 5$, m $\angle ABC$ = m $\angle 2$ + m $\angle 3$, m $\angle CDE$ = m $\angle 6$ + m $\angle 7$, m $\angle DEA$ = m $\angle 8$ + m $\angle 9$ (Post 10); m $\angle 1$ + m $\angle 2$ + m $\angle 12 = 180$, m $\angle 3$ + m $\angle 4$ + m $\angle 13 = 180$, m $\angle 5$ + m $\angle 6$ + m $\angle 14 = 180$, m $\angle 7$ + m $\angle 8$ + m $\angle 15 = 180$, m $\angle 9$ + m $\angle 10$ + m $\angle 11 = 180$ (Th 3.5); m $\angle 1$ + m $\angle 2$ + m $\angle 3$ + m $\angle 4$ + m $\angle 5$ + m $\angle 6$ + m $\angle 7$ + m $\angle 8$ + m $\angle 9$ + m $\angle 10$ + m $\angle 11$ + m $\angle 12$ + m $\angle 13$ + m $\angle 14$ + m $\angle 15 = 900$ (add prop); m $\angle 11$ + m $\angle 12$ + m $\angle 13$ + m $\angle 14$ + m $\angle 15 = 360$ (Post 11); m $\angle 1$ + m $\angle 2$ + m $\angle 3$ + m $\angle 4$ + m $\angle 5$ + m $\angle 6$ + m $\angle 7$ + m $\angle 8$ + m $\angle 9$ + m $\angle 10 = 540$ (subst prop); m $\angle A$ + m $\angle B$ + m $\angle C$ + m $\angle D$ + m $\angle E = 540$ (subst prop) **39.** Given: convex hexagon ABCDEF; Prove: m $\angle A$ + m $\angle B$ + m $\angle C$ + m $\angle D$ + m $\angle E$ + m $\angle F = 720$. m $\angle EFA$ = m $\angle 10$ + m $\angle 11$, m $\angle FAB$ = m $\angle 12$ + m $\angle 1$, m $\angle ABC$ = m $\angle 2$ + m $\angle 3$, m $\angle BCD$ = m $\angle 4$ + m $\angle 5$, m $\angle CDE$ = m $\angle 6$ + m $\angle 7$ (Post 10); m $\angle 1$ + m $\angle 2$ + m $\angle 13 = 180$, m $\angle 3$ + m $\angle 4$ + m $\angle 14 = 180$, m $\angle 5$ + m $\angle 6$ + m $\angle 15 = 180$, m $\angle 7$ + m $\angle 8$ + m $\angle 16 = 180$, m $\angle 9$ + m $\angle 10$ + m $\angle 17 = 180$, m $\angle 11$ + m $\angle 12$ + m $\angle 18 = 180$ (Th 3.5); m $\angle 1$ + m $\angle 2$ + m $\angle 3$ + m $\angle 4$ + m $\angle 5$ + m $\angle 6$ + m $\angle 7$ + m $\angle 8$ + m $\angle 9$ + m $\angle 10$ + m $\angle 15$ + m $\angle 16$ + m $\angle 17$ + m $\angle 18 = 1080$ (add prop); m $\angle 13$ + m $\angle 14$ + m $\angle 15$ + m $\angle 16$ + m $\angle 17$ + m $\angle 18 = 360$ (Post 11); m $\angle 1$ + m $\angle 2$ + m $\angle 3$ + m $\angle 4$ + m $\angle 5$ + m $\angle 6$ + m $\angle 7$ + m $\angle 8$ + m $\angle 9$ + m $\angle 10$ + m $\angle 11$ + m $\angle 12 = 720$ (subtr prop); m $\angle A$ + m $\angle B$ + m $\angle C$ + m $\angle D$ + m $\angle E$ + m $\angle F = 720$ (subst prop)

Page 192

40–45. No

Page 195

6. Given: $\triangle ABC$, ext $\angle$s 1, 2, 3; Prove: m $\angle 1$ + m $\angle 2$ + m $\angle 3 = 360$. **1.** m $\angle 1$ + m $\angle 4 = 180$, m $\angle 2$ + m $\angle 5 = 180$, m $\angle 3$ + m $\angle 6 =$ 180 **1.** (Post 11); **2.** m $\angle 1$ + m $\angle 2$ + m $\angle 3$ + m $\angle 4$ + m $\angle 5$ + m $\angle 6 = 540$ **2.** (add prop); **3.** m $\angle 4$ + m $\angle 5$ + m $\angle 6 = 180$ **3.** (Th 3.5); **4.** m $\angle 1$ + m $\angle 2$ + m $\angle 3 = 360$ **4.** (subtr prop) **7.** Given: convex pentagon ABCDE, ext $\angle$s 1, 2, 3, 4, 5; Prove: m $\angle 1$ + m $\angle 2$ + m $\angle 3$ + m $\angle 4$ + m $\angle 5 = 360$. m $\angle 1$ + m $\angle 6 = 180$, m $\angle 2$ + m $\angle 7 = 180$, m $\angle 3$ + m $\angle 8 =$ 180, m $\angle 4$ + m $\angle 9 = 180$, m $\angle 5$ + m $\angle 10 = 180$ (Post 11); m $\angle 1$ + m $\angle 2$ + m $\angle 3$ + m $\angle 4$ + m $\angle 5$ + m $\angle 6$ + m $\angle 7$ + m $\angle 8$ + m $\angle 9$ + m $\angle 10 = 900$ (add prop); m $\angle 6$ + m $\angle 7$ + m $\angle 8$ + m $\angle 9$ + m $\angle 10 = 540$ (Th 6.2); m $\angle 1$ + m $\angle 2$ + m $\angle 3$ + m $\angle 4$ + m $\angle 5 = 360$ (subtr prop)

Page 201

4. Follow steps to constr an equil $\triangle$ on p. 199. **5.** Constr a $\perp$ to a given *l;* 90° $\angle$s are formed. **6.** Constr an equil $\triangle$; extend one side; the exterior $\angle$ formed is 120°. **7.** Follow steps of ex 1 p. 199 **8.** Constr a $\perp$ to a given *l* bis one of the rt $\angle$s; one of its parts is 45° **9.** Constr 30° $\angle$; $\angle$ forming linear pr with this 30° $\angle$ is 150°. **10.** Constr 30° $\angle$, bisect it; one of its parts is 15°. **11.** Constr 15° $\angle$; at vertex constr $\perp$ to one ray of the 15° $\angle$; compl of 15° $\angle$ is the 75° $\angle$. **12.** Constr 15° $\angle$; at vertex constr $\perp$ to one of rays; 90° $\angle$ + 15° $\angle$ gives 105° $\angle$. **13.** Constr 75° $\angle$; bis it; one half is $37\frac{1}{2}$° $\angle$. **14.** Constr 15° $\angle$; the $\angle$ forming a linear pr with the 15° $\angle$ is 165° $\angle$. **15.** Constr 45° $\angle$; bis it forming two $22\frac{1}{2}$° $\angle$s; the $\angle$ forming a linear pr with either $22\frac{1}{2}$° $\angle$ is $157\frac{1}{2}$ $\angle$. **16.** Constr a seg twice as long as the given seg; using this seg, constr an equil $\triangle$. **17.** Bis the given seg; using this seg follow constr p 200. **18.** Constr 135° $\angle$; mark off length $\cong$ to given seg on each side of the $\angle$; at endpts of these seg constr 135° $\angle$s; continue this until reg octagon is complete. **19.** Follow steps of ex 18, except constr 150° $\angle$s. **20.** Constr seg $\cong$ to given seg; constr $\perp$ bis; constr a $\odot$ whose center is the mdpt of the seg and whose radius is $\frac{1}{2}$ the length of the seg; the seg and $\perp$ bis. divide $\odot$ into 4 $\cong$ arcs; connect these arcs forming required square. **21.** Follow steps of ex 18 except constr $157\frac{1}{2}$° $\angle$s. **22.** Constr an 11-cm seg. Follow steps of constr p 200. **23.** Follow steps of constr p 199 using 14-cm seg. **24.** Constr a reg pentagon as shown on p. 201. Constr $\perp$ bis of each side. Constr mdpt. of each seg from center of $\odot$ to mdpt of a side. Connect each vertex of pentagon to mdpt of $\perp$ bis to form five pointed star. **25.** Constr a reg pentagon as shown on p. 201. Constr $\perp$ bis of each side, these $\perp$ bis also bis 5 $\cong$ arcs of $\odot$. Connect 10 $\cong$ arcs to form reg decagon.

Page 204

26. Follow steps constr p 200. **27.** Follow steps constr. p 199; draw only $\angle$, not $\triangle$. **28.** Constr $\perp$ to *l;* bis one of the rt $\angle$s formed. **29.** Constr 60° $\angle$, bis it, then bis of the 30° $\angle$s formed.

Page 205

25. Given: $\triangle ABC$, ext $\angle$s 4, 5, 6; Prove: $m\angle 4 + m\angle 5 + m\angle 6 = 360°$. $m\angle 1 + m\angle 2 + m\angle 3 = 180°$ (Th 3.5); $m\angle 2 + m\angle 4 = 180$, $m\angle 3 + m\angle 5 = 180$, $m\angle 1 + m\angle 6 = 180$ (def of suppl $\angle$s); $m\angle 1 + m\angle 2 + m\angle 3 + m\angle 4 + m\angle 5 + m\angle 6 = 540$ (add prop); $180 + m\angle 4 + m\angle 5 + m\angle 6 = 540$ (subst prop); $m\angle 4 + m\angle 5 + m\angle 6 = 360$ (subtr prop)

Page 214

15. Given: ▱$ABCD$; Prove: $\angle A$ and $\angle B$ suppl, $\angle B$ and $\angle C$ suppl, $\angle C$ and $\angle D$ suppl, $\angle D$ and $\angle A$ suppl. $\overline{AB} \parallel \overline{DC}$, $\overline{AD} \parallel \overline{BC}$ (def of ▱); $\angle A$ and $\angle B$ suppl, $\angle B$ and $\angle C$ suppl, $\angle C$ and $\angle D$ suppl, $\angle D$ and $\angle A$ suppl (Th 3.2) **16.** 1. Def of ▱. 2. Post 13 3. Cor 7.1 4. ASA 5. CPCTC **17.** Cor. 7.1; Given: ▱$ABCD$, diag $\overline{BD}$; Prove: $\overline{AD} \cong \overline{BC}$, $\overline{DC} \cong \overline{AB}$. $\triangle ABD \cong \triangle CDB$ (Th 7.1); $\overline{AD} \cong \overline{BC}$, $\overline{DC} \cong \overline{AB}$ (CPCTC) Cor 7.2; Given: ▱$ABCD$; diag $\overline{BD}$ and $\overline{AC}$; Prove: $\angle A \cong \angle C$, $\angle D \cong \angle B$. $\triangle ABD \cong \triangle CDB$; $\triangle DAC \cong \triangle BCA$ (Th 7.1); $\angle A \cong \angle C$, $\angle D \cong \angle B$ (CPCTC) **18.** $\overline{DG} \cong \overline{GB}$ (Th 7.2); $\overline{DC} \parallel \overline{AB}$ (def of ▱); $\angle EDG \cong \angle FBG$ (Post 13); $\angle DGE \cong \angle BGF$ (Th 2.10); $\triangle DGE \cong \triangle BGF$ (ASA); $\overline{EG} \cong \overline{GF}$ (CPCTC); G is the mdpt of $\overline{EF}$ (def of mdpt) **19.** $\overline{WZ} \parallel \overline{XY}$ (def of ▱); $\angle VXU \cong \angle VZT$, $\angle ZTV \cong \angle VUX$ (Post 13); $\overline{WZ} \cong \overline{XY}$ (Cor 7.1); $WZ = 2\ TZ$, $XY = 2\ XU$ (def of mdpt); $2\ \overline{TZ} \cong 2\ \overline{XU}$ (subst prop); $\overline{TZ} \cong \overline{XU}$ (div prop); $\triangle TVZ \cong \triangle UVX$ (ASA); $\overline{TV} \cong \overline{VU}$, $\overline{ZV} \cong \overline{VX}$ (CPCTC); V is the mdpt of $\overline{TU}$ and $\overline{XZ}$ (def of mdpt) **20.** $\overline{LK} \cong \overline{IJ}$, $\overline{LI} \cong \overline{KJ}$ (Th 3.8); $\angle IJK$ is a ▱ (def of ▱); $\overline{LK} \cong \overline{IJ}$ (Cor 7.1); $\angle ILK \cong \angle IJK$ (Cor 7.1)

Page 218

17. Given: $\angle A \cong \angle C$, $\angle B \cong \angle D$; Prove: $ABCD$ is ▱. $m\angle A + m\angle B + m\angle C + m\angle D = 360$ (Th 6.1); $2\ m\angle A + 2\ m\angle B = 360$, $2\ m\angle C + 2\ m\angle B = 360$ (subst prop); $m\angle A + m\angle B = 180$, $m\angle C + m\angle D = 180$ (div prop); $\angle A$ and $\angle B$ are suppl, $\angle C$ and $\angle B$ are suppl (def of suppl $\angle$s); $\overline{AD} \parallel \overline{BC}$, $\overline{AB} \parallel \overline{DC}$ (Th 3.4) **18.** Given: $\overline{AC}$ and $\overline{DB}$ bis each other. Prove: $ABCD$ is a ▱. $\overline{DO} \cong \overline{OB}$, $\overline{CO} \cong \overline{OA}$ (def of bis); $\angle 3 \cong \angle 4$ (Th 2.10); $\triangle AOB \cong \triangle COD$ (SAS). $\overline{DC} \cong \overline{AB}$, $\angle 1 \cong \angle 2$ (CPCTC); $\overline{DC} \parallel \overline{AB}$ (Post 14); $ABCD$ is a ▱ (Th 7.5) **19.** $\overline{XU} \cong \overline{VZ}$ (CPCTC); $UXVZ$ is a ▱ (Th 7.4) **20.** $\overline{WZ} \parallel \overline{XY}$ (def of ▱); $\overline{WZ} \cong \overline{XY}$ (Cor 7.1); $\overline{UZ} \cong \overline{XV}$ (subtr prop); $UXVZ$ is ▱ (Th 7.5) **21.** $\overline{AD} \cong \overline{AB} \cong \overline{DB}$, $\overline{DC} \cong \overline{BC} \cong \overline{DB}$ (def of equil $\triangle$); $\overline{AD} \cong \overline{BC}$, $\overline{AB} \cong \overline{DC}$ (subst prop); $ABCD$ is a ▱ (Th 7.3)
22. $\triangle ABD$, $\triangle CDB$ are equil (Cor. 5.2);
23. Given: Quad. $ABCD$; $\angle A$ and $\angle B$ suppl, $\angle B$ and $\angle C$ suppl, $\angle C$ and $\angle D$ suppl, $\angle D$ and $\angle A$ suppl; Prove: $ABCD$ is a ▱. $\overline{AD} \parallel \overline{BC}$, $\overline{AB} \parallel \overline{DC}$ (Th 3.4); $ABCD$ is a ▱ (def of ▱) **24–25.** Answers may vary. **26.** Given: $\overline{AC}$ bis $\angle DAB$ and $\angle DCB$, $\overline{DB}$ bis $\angle ADC$ and $\angle ABC$; Prove: $ABCD$ is a ▱. $\angle 1 \cong \angle 2$, $\angle 3 \cong \angle 4$, $\angle 5 \cong \angle 6$, $\angle 7 \cong \angle 8$ (def of bis); $\overline{AC} \cong \overline{AC}$, $\overline{DB} \cong \overline{DB}$ (reflex prop); $\triangle ADB \cong \triangle CDB$; $\triangle ABC \cong \triangle ADC$ (ASA); $\overline{AD} \cong \overline{AB}$, $\overline{DC} \cong \overline{BC}$, $\overline{AD} \cong \overline{DC}$, $\overline{AB} \cong \overline{BC}$ (CPCTC); $\overline{AB} \cong \overline{DC}$, $\overline{AD} \cong \overline{BC}$ (subst prop); $ABCD$ is a ▱ (Th 7.4)
27. Given: $\overline{AE}$ bis $\angle A$, $\overline{CF}$ bis $\angle C$, $AFCE$ is a ▱; Prove: $ABCD$ is a ▱. $\angle 1 \cong \angle 3$ (Th 7.1); $\overline{AE} \cong \overline{FC}$ (Th 7.1); $\angle 1 \cong \angle 2$, $\angle 3 \cong \angle 4$ (def of bis); $\angle 2 \cong \angle 4$ (subst prop); $\overline{EC} \parallel \overline{AF}$ (def of ▱); $\angle 5 \cong \angle 1$, $\angle 6 \cong \angle 3$ (Post 13); $\angle 5 \cong \angle 6$ (subst prop); $\triangle ADE \cong \triangle CBF$ (ASA); $\overline{AD} \cong \overline{BC}$, $\overline{DE} \cong \overline{FB}$ (CPCTC); $\overline{EC} \cong \overline{AF}$ (Th 7.1); $\overline{DC} \cong \overline{AB}$ (add prop); $ABCD$ is a ▱ (Th 7.3) **28.** Answer may vary.

Page 221

9. $\overline{ST} \parallel \overline{RQ}$, $ST = \frac{1}{2}RQ$ (Th 7.7); $RU = \frac{1}{2}RQ$ (def of mdpt); $\overline{ST} \cong \overline{RU}$ (subst); $RSTU$ is a ▱ (Th 7.5)
10. $\overline{ST} \parallel \overline{RQ}$; $ST = \frac{1}{2}RQ$ (Th 7.7); $UQ = \frac{1}{2}RQ$; $\overline{PT} \cong \overline{TQ}$ (def of mdpt); $\overline{UQ} \cong \overline{ST}$ (subst prop); $\angle PTS \cong \angle TQU$ (Post 13); $\triangle SPT \cong \triangle UTO$ (SAS)
11. $\overline{DE} \cong \overline{EA}$; $\overline{DF} \cong \overline{FB}$ (def of mdpt); $\angle DFE \cong \angle CFB$ (Th 2.10); $\triangle DEF \cong \triangle BCF$ (SAS); $\overline{BC} \cong \overline{DE}$, $\angle FBC \cong \angle EDF$ (CPCTC); $\overline{EA} \cong \overline{BC}$ (subst prop); $\overline{AD} \parallel \overline{BC}$ (Post 14); $EABC$ is a ▱ (Th 7.5)
12. $\overline{EF} \parallel \overline{AB}$ (Th 7.7); $ABCE$ is ▱ (Th 7.5); $\overline{AE} \parallel \overline{BC}$ (def of ▱); $\overline{AE} \cong \overline{BC}$ (Cor 7.1); $\overline{AE} \cong \overline{DE}$ (def of mdpt); $\overline{DE} \cong \overline{BC}$ (subst prop); $DEBC$ is a ▱ (Th 7.5) **13.** $\overline{KM} \parallel \overline{GH}$; $KL = \frac{1}{2}GH$; $\overline{KM} \parallel \overline{JI}$; $LM = \frac{1}{2}JI$ (Th 7.7); $\overline{JI} \parallel \overline{GH}$ (Th 3.7); $\overline{KL} \cong \overline{LM}$ (def of mdpt); $\frac{1}{2}\ \overline{GH} \cong \frac{1}{2}\ \overline{JI}$ (subst prop); $\overline{GH} \cong \overline{JI}$ (mult prop); $GHIJ$ is ▱ (Th 7.5) **14.** $\overline{JG} \parallel \overline{IH}$ (def of ▱); $\overline{KL} \parallel \overline{GH}$ (Th 7.7); $GHMK$ is ▱ (def of ▱) **15.** Given: Quad $WXYZ$, A, B, C, D are the mdpts of $\overline{WX}$, $\overline{WZ}$, $\overline{YZ}$, $\overline{XY}$, respectively; Prove: $\overline{AC}$ and $\overline{BD}$ bis each other. Draw $\overline{AB}$, $\overline{BC}$, $\overline{CD}$, $\overline{DA}$, $\overline{XZ}$ (Post 1); $\overline{AB} \parallel \overline{XZ}$; $AB = \frac{1}{2}XZ$; $\overline{DC} \parallel \overline{XZ}$; $DC = \frac{1}{2}XZ$ (Th 7.7); $\overline{AB} \parallel \overline{DC}$ (Th 3.6); $\overline{AB} \cong \overline{DC}$ (subst prop); $ABCD$ is a ▱ (Th 7.5); $\overline{AC}$ and $\overline{BD}$ bis each other (Th 7.2) **16.** Given: $ABCD$ is a quad, $\overline{AE}$, $\overline{BE}$, $\overline{CF}$, $\overline{DF}$ are bis of $\angle$s A, B, C, D, respectively; Prove: $\angle 11$ and $\angle 12$ are suppl, $\angle 13$ and $\angle 14$ are suppl. $\angle 1 \cong \angle 2$, $\angle 3 \cong \angle 4$, $\angle 5 \cong \angle 6$, $\angle 7 \cong \angle 8$ (def of bis); $m\angle 1 + m\angle 2 + m\angle 3 + m\angle 4 + m\angle 5 + m\angle 6 + m\angle 7 + m\angle 8 = 360$ (Th 6.1); $2\ m\angle 2 + 2\ m\angle 3 + 2\ m\angle 6 + 2\ m\angle 7 = 360$ (subst prop); $m\angle 2 + m\angle 3 + m\angle 6 + m\angle 7 = 180$ (div prop); $m\angle 2 + m\angle 3 + m\angle 9 = 180$, $m\angle 6 + m\angle 7 + m\angle 10 = 180$ (Th 3.5); $m\angle 2 + m\angle 3 + m\angle 6 + m\angle 7 + m\angle 9 + m\angle 10 = 360$ (add prop); $m\angle 9 + m\angle 10 = 180$ (subt prop);

$\angle 11 \cong \angle 9$, $\angle 12 \cong \angle 10$ (Th 2.10); $m\angle 11 + m\angle 12 = 180$ (subst prop); $\angle 11$ and $\angle 12$ are supp (def of suppl $\angle$s); $m\angle 11 + m\angle 12 + m\angle 13 + m\angle 14 = 360$ (Th 6.1); $m\angle 13 + m\angle 14 = 180$ (subtr prop); $\angle 13$ and $\angle 14$ are suppl (def of suppl $\angle$s) **17.** $\overline{RS} = \frac{1}{2}\overline{NO}$, $\overline{TS} = \frac{1}{2}\overline{QP}$ (Th 7.7); $\overline{RS} + \overline{TS} = \frac{1}{2}(\overline{QP} + \overline{NO})$ (add prop); $\overline{RT} = \frac{1}{2}(\overline{QP} + \overline{NO})$ (subst prop) **18.** $ACFD$ is a ▱, $ABED$ is ▱, $BEFC$ is a ▱ (Th 7.5); $\overline{DE} \cong \overline{AB}$, $\overline{EF} \cong \overline{BC}$, $\overline{DF} \cong \overline{AC}$ (Cor 7.1); $\triangle ABC \cong \triangle DEF$ (SSS); $\angle EDF \cong \angle BAC$ (CPCTC)

Page 225

15. $\overleftrightarrow{AD} \parallel \overleftrightarrow{BE} \parallel \overleftrightarrow{CF}$ (Th 3.3); $\overline{DE} \cong \overline{EF}$ (Th 7.8)
16. $ABED$ is a ▱ (Th 7.5); $\overleftrightarrow{AD} \parallel \overleftrightarrow{BE}$ (def of ▱); $\overleftrightarrow{AD} \parallel \overleftrightarrow{BE} \parallel \overleftrightarrow{CF}$ (Th 3.7); $\overline{DE} \cong \overline{EF}$ (Th 7.8);
17. $\overleftrightarrow{AD} \parallel \overleftrightarrow{BE}$ (Th 3.3); $\overleftrightarrow{AD} \parallel \overleftrightarrow{BE} \parallel \overleftrightarrow{CF}$ (Th 3.7); $\overline{AB} \cong \overline{BC}$ (Th 7.8) **18. 1.** $\overline{DE} \cong \overline{EF}$ **1.** (Th 7.8); **2.** $\overline{AB} \cong \overline{EF}$ **2.** (subst prop)
19. Follow steps constr p. 223. **20.** Follow steps constr p. 223, making 4 $\cong$ seg **21.** Follow steps constr p. 223, making 5 $\cong$ seg **22.** Follow steps constr p. 223; constr a seg $\cong$ 2 of the 3 seg
23. Follow steps ex 21; constr seg $\cong$ 4 of the 5 seg
24. Follow steps constr p. 223; constr $\cong$ in length to the sum of the lengths of the seg and one of its parts. **25.** Given: $l \parallel m$, $\overline{PX} \perp m$; $\overline{QY} \perp m$; Prove: $\overline{PX} \cong \overline{QY}$. $\overline{PX} \parallel \overline{QY}$ (Cor 3.3); $PXYQ$ is a ▱ (def of ▱); $\overline{PX} \cong \overline{QY}$ (Cor 7.1) **26.** Given: $l \parallel m \parallel n \parallel o$, transversals $\overleftrightarrow{AD}$ and $\overleftrightarrow{RU}$, $\overline{AB} \cong \overline{BC} \cong \overline{CD}$; Prove: $\overline{RS} \cong \overline{ST} \cong \overline{TU}$. $\overline{UT} \cong \overline{TS}$, $\overline{TS} \cong \overline{SR}$ (Th 7.8); $\overline{UT} \cong \overline{TS} \cong \overline{SR}$ (trans prop)
27. counterexample **30.** counterexample

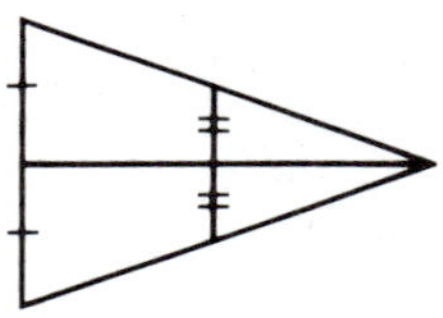

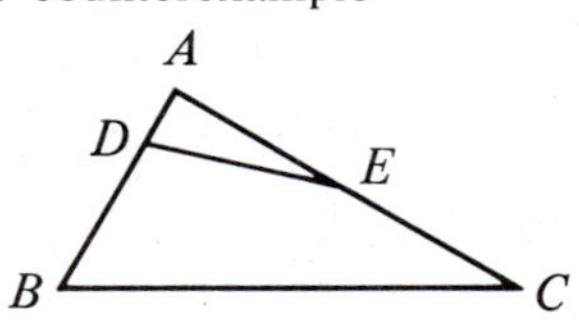

28. Given: $\overline{ED}$ bis $\overline{AB}$, $\overline{ED} \parallel \overline{BC}$; Prove: $\overline{ED}$ bis $\overline{AC}$. Draw $\overleftrightarrow{FAG} \parallel \overline{BC}$ (Post 15); $\overleftrightarrow{FAG} \parallel \overline{DE} \parallel \overline{BC}$ (Th 3.7); $\overline{AE} \cong \overline{EC}$ (Th 7.8) **29.** Given: $\overline{PX} \perp m$, $\overline{QY} \perp m$, $\overline{PX} \cong \overline{QY}$; Prove: $l \parallel m$. $\overline{PX} \parallel \overline{QY}$ (Cor 3.3); $PXYQ$ is a ▱ (Th 7.5); $l \parallel m$ (def of ▱)

Page 229

21. Given: rect $ABCD$, diag $\overline{AC}$ and $\overline{DB}$; Prove: $\overline{AC} \cong \overline{DB}$. $\overline{AD} \cong \overline{CB}$ (Cor 7.1); $\overline{AB} \cong \overline{AB}$ (reflex prop); $\angle DAB$, $\angle CBA$ are rt $\angle$s (def of rect); $\angle DAB \cong \angle CBA$ (Th 2.5); $\triangle DAB \cong \triangle CBA$ (SAS); $\overline{AC} \cong \overline{DB}$ (CPCTC)

22. and 24. counterexample

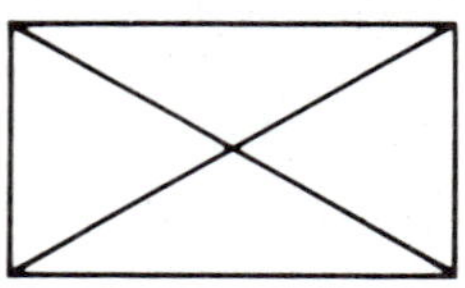

23. Given: Rh $ABCD$, diag $\overline{AC}$ and $\overline{DB}$ intersect at E; Prove: $\overline{AC}$ bis $\angle DAB$, $\angle DCB$, and $\overline{DB}$ bis $\angle ADC$, $\angle ABC$. $\overline{AD} \cong \overline{DC} \cong \overline{CB} \cong \overline{BA}$ (def of rhombus); $\overline{DB} \cong \overline{DB}$, $\overline{AC} \cong \overline{AC}$ (reflex prop); $\triangle ADC \cong \triangle ABC$, $\triangle DAB \cong \triangle DCB$ (SSS); $\angle 1 \cong \angle 2$, $\angle 3 \cong \angle 4$, $\angle 5 \cong \angle 6$, $\angle 7 \cong \angle 8$ (CPCTC); $\overline{AC}$ bis $\angle DAB$, $\angle DCB$ (def of bis); $\overline{DB}$ bis $\angle ADC$, $\angle ABC$ **25.** Given: square $ABCD$, diag $\overline{AC}$ and $\overline{DB}$; Prove: $\overline{AC} \cong \overline{DB}$. Same proof as ex 21
26. Given: square $ABCD$, diag $\overline{AC}$ and $\overline{DB}$ intersect at E; Prove: $\overline{AC}$ and $\overline{DB}$ bis each other. $ABCD$ is ▱ (def of sq); $\overline{AC}$ and $\overline{DB}$ bis each other (Th 7.2)

Page 232

15. $EFGH$ is a ▱ (Th 7.5); $EFGH$ is a rect (If a ▱ has $\cong$ diag, then it is a rect) **16.** $EFGH$ is a ▱ (Th 7.3); $EFGH$ is a rhombus (def of rhombus)
17. Given: ▱$ABCD$ with $\overline{DA} \cong \overline{AB}$; Prove: $ABCD$ is a rhombus. $\overline{DA} \cong \overline{CB}$, $\overline{AB} \cong \overline{DC}$ (Cor 7.1); $\overline{AB} \cong \overline{CB} \cong \overline{DC} \cong \overline{DA}$ (subst prop); $ABCD$ is a rhombus (def of rhombus) **18.** Given: rhombus $ABCD$ with diag $\overline{AC}$ and diag $\overline{DB}$, $\overline{AC} \cong \overline{DB}$; Prove: $ABCD$ is a square. $\overline{AD} \cong \overline{CB}$ (Cor 7.1); $\overline{AB} \cong \overline{AB}$ (reflex prop); $\triangle DAB \cong \triangle CBA$ (SSS for $\cong$ $\triangle$s); $\angle DAB \cong \angle CBA$ (CPCTC); $\overline{AD} \parallel \overline{CB}$ (def of ▱); $\angle DAB$ and $\angle CBA$ are supp (Th 3.2); $m\angle DAB + m\angle CBA = 180$ (def of supp $\angle$s); $2\,m\angle DAB = 180$ (subst prop); $m\angle DAB = 90$ (div prop); $\angle DAB$ is a rt $\angle$ (def of rt $\angle$); $ABCD$ is a square (a rhombus with 1 rt $\angle$ is a square)
19. Given: rhombus $ABCD$, E, F, G, H are mdpts of AB, BC, CD, DA, respectively; Prove: $EFGH$ is a rect. Draw $\overline{AC}$, $\overline{DB}$ (Post 1); $\overline{HG} \parallel \overline{AC}$; $\overline{HG} = \frac{1}{2}\overline{AC}$; (Th 7.7); $\overline{EF} \parallel \overline{AC}$; $\overline{EF} = \frac{1}{2}\overline{AC}$; $\overline{HG} \parallel \overline{EF}$ (Th 3.7); $\overline{HG} \cong \overline{EF}$ (subst prop); $EFGH$ is a ▱ (Th 7.5); $\overline{DB} \perp \overline{AC}$ (diag of a rhom. are $\perp$); $\overline{HE} \parallel \overline{DB}$ (Th 7.7); $\overline{AC} \perp \overline{EH}$, $\overline{EF} \perp \overline{EH}$ (Ex 15, p 78); $\angle DEF$ is a rt $\angle$ (def $\perp$ lines); $EFGH$ is a rect (a ▱ with 1 rt $\angle$ is a rect) **20.** Given: Rect $ABCD$; $\overline{AC} \perp \overline{DB}$; Prove: $ABCD$ is a square. $\angle DAB$ is a rt $\angle$ (def of a rect); $ABCD$ is a ▱ (def of a rect); $\overline{DE} \cong \overline{EB}$ (diag of a ▱ bis each other); $\triangle ADE \cong \triangle ABE$ (LL); $\overline{AD} \cong \overline{AB}$ (CPCTC); $ABCD$ is a square (def of a square) **21.** T. See ex 16, p. 232. **22.** Given: quad $ABCD$, $\angle A \cong \angle B \cong \angle C \cong \angle D$; Prove: $ABCD$ is a rect. $m\angle A + m\angle B + m\angle C + m\angle D = 360$ (Th 6.1); $4\,m\angle A = 360$ (subst prop); $m\angle A = 90$ (div prop); $m\angle A = m\angle B = m\angle C = m\angle D = 90$ (subst prop); $\angle A$, $\angle B$, $\angle C$, $\angle D$ are rt $\angle$s (def of rt $\angle$s); $ABCD$ is a ▱ (Th 7.3); $ABCD$ is a rect (def of rect); **23.** counterexample: parallelogram **24.** Given: quad $ABCD$, $\overline{DB} \perp \overline{AC}$, $\overline{DB}$ and $\overline{AC}$ bis each other; Prove: $ABCD$ is a rhombus. $\angle 1 \cong \angle 2 \cong \angle 3 \cong \angle 4$ (Th 2.6); $\overline{DE} \cong \overline{EB}$, $\overline{AE} \cong \overline{EC}$ (def of bis); $\triangle ADE \cong \triangle CDE \cong \triangle ABE \cong \triangle CBE$ (SAS); $\overline{AD} \cong \overline{AB} \cong \overline{BC} \cong \overline{CD}$ (CPCTC); $ABCD$ is a ▱ (Th 7.3); $ABCD$ is a rhombus (def of rhombus)

Page 234

24. $\overline{BF} \parallel \overline{EC}$, $\overline{BD} \parallel \overline{AE}$ (Th 7.7); $BDEF$ is a ▱ (def of ▱); $2\,\overline{FE} = \overline{EA}$, $2\,\overline{ED} = \overline{EC}$ (def of mdpt); $2\,\overline{FE} \cong 2\,\overline{ED}$ (subst prop); $\overline{FE} \cong \overline{ED}$ (div prop); $\overline{FE} \cong \overline{BD}$, $\overline{ED} \cong \overline{FB}$ (Cor 7.1); $\overline{FE} \cong \overline{ED} \cong \overline{FB} \cong \overline{BD}$ (subst); $BDEF$ is a rhombus (def of rhombus); **25.** Given: $ABCD$ is a ▱; Prove: $\angle A$ and $\angle B$ suppl, $\angle B$ and $\angle C$ suppl, $\angle C$ and $\angle D$ suppl, $\angle D$ and $\angle A$ suppl. $\overline{AD} \parallel \overline{CB}$, $\overline{AB} \parallel \overline{CD}$ (def of ▱); $\angle A$ and $\angle B$ supp, $\angle B$ and $\angle C$ supp, $\angle C$ and $\angle D$ supp, $\angle D$ and $\angle A$ supp (Th 3.2) **26.** (See 214, ex 15.)

Page 235

18. Follow steps constr p. 223. **19.** Follow steps ex 21 p. 225; constr seg $\cong$ 3 of the 5 seg

Page 235

24. Given: $ABCD$ is a square; Prove: $\overline{AC} \cong \overline{DB}$. $\overline{AD} \cong \overline{CB}$ (def of a square); $\overline{AB} \cong \overline{AB}$ (reflex prop); $\angle DAB \cong \angle CBA$ (def of a rect); $\triangle ADB \cong \triangle ACB$ (SAS); $\overline{AC} \cong \overline{DB}$ (CPCTC) **25.** $\overline{WX} \parallel \overline{ZY}$, $\overline{WZ} \parallel \overline{YX}$ (def of $\parallel$); $\overline{WX} \cong \overline{ZY}$, $\overline{WZ} \cong \overline{YX}$ (Cor 7.1); $\angle ZYW \cong \angle YWX$, $\angle ZWY \cong \angle XYW$ (Post 13); $\triangle ZYV \cong \triangle XWU$, $\triangle ZWU \cong \triangle XYV$ (SAS); $\overline{ZV} \cong \overline{UX}$, $\overline{ZU} \cong \overline{XV}$ (CPCTC); $UXVZ$ is a ▱ (Th 7.3)

Page 240

1. $\overline{DE} \parallel \overline{AB}$ (Th 7.7); $ABED$ is a trap (def of trap); **2.** $\overline{IH} \parallel \overline{FG}$ (def of trap); $\overline{IJ} \cong \overline{HK}$ (Th 7.9); $\angle IJF$, $\angle HKG$ are rt $\angle$s (def of $\perp$ lines); $\triangle IJF$, $\triangle HKG$ are rt $\triangle$s (def of rt $\triangle$s); $\triangle IJF \cong \triangle HKG$ (HL); $\angle F \cong \angle G$ (CPCTC) **3.** $\overline{LM} \parallel \overline{ON}$ (def of trap); $\overline{LP} \cong \overline{MQ}$ (Th 7.9); $\angle OPL$, $\angle NQM$ are rt $\angle$s (def of $\perp$ lines); $\triangle OPL$, $\triangle NQM$ are rt $\triangle$s (def of rt $\triangle$s); $\triangle OPL \cong \triangle NQM$ (LA); $\overline{LO} \cong \overline{MN}$ (CPCTC)

Page 245

18. Given: trap $ABCD$ with med $\overline{EH}$, diag $\overline{AC}$, $\overline{DB}$; Prove: EH bis $\overline{AC}$ and $\overline{DB}$. $\overline{EH} \parallel \overline{DC} \parallel \overline{AB}$ (Th 8.1); $\overline{DE} \cong \overline{EA}$ (def of med of trap); $\overline{AF} \cong \overline{FC}$; $\overline{DG} \cong \overline{GB}$ (Th 7.8); EH bis $\overline{AC}$ and $\overline{DB}$ (def of bis) **19.** Given: trap $ABCD$ with med $\overline{EF}$, alt $\overline{HG}$; Prove $\overline{EF}$ bis $\overline{HG}$. $\overline{EF} \parallel \overline{DC} \parallel \overline{AB}$ (Th 8.1); $\overline{DE} \cong \overline{EA}$ (def of med of trap); $\overline{HI} \cong \overline{IG}$ (Th 7.8) **20. 1.** Given **2.** Post 1 **3.** Post 15 **4.** Def of trap **5.** Def of ▱ **6.** Post 13 **7.** Def med of trap **8.** AAS **9.** CPCTC **10.** Def of med of trap **11.** Cor 7.1 **12.** Subst prop **13.** Th 7.5 **14.** Def of ▱ **15.** Follow steps 9–14 **16.** Def of ▱ **17.** Th 3.7 **18.** Th 7.1, Cor 1 **19.** Def of between **20.** Subst **21.** CPCTC **22.** Subst **23.** Th 7.1, Cor 1 **24.** Add prop; subst **25.** Distr prop; def of between; subst **26.** Div prop **21.** False; $AVWX$ can be a rectangle **22.** $\overline{XW} \parallel \overline{UV} \parallel \overline{YZ}$ (Th 8.1); $\overline{WZ} \cong \overline{ZV}$ (def of med of trap); $\overline{UB} \cong \overline{BW}$ (Th 7.8); $XB \perp$ bis $\overline{UW}$ (def of $\perp$ bis); $\overline{XU} \cong \overline{XW}$ (Th 5.7); $\angle XUW \cong \angle XWU$ (Th 5.1); $\angle XWU \cong \angle WBZ$ (Post 13); $\angle WBZ \cong \angle YBU$ (Th 2.10); $\angle XUW \cong \angle YBU$ (subst prop); $\overline{YU} \cong \overline{YB}$ (Th 5.2); $\triangle UBY$ is isos (def of isos $\triangle$) **23–24.** False; answers may vary.

Page 249

7. $BCDE$ is a trap (def of trap); $\angle C \cong \angle B$ (Th 5.1); $BCDE$ is an isos trap (Th 8.3) **8.** $\angle C \cong \angle B$ (Th 8.2); $\overline{AB} \cong \overline{AC}$ (Th 5.2); $\triangle ABC$ is isos (def of isos $\triangle$) **9.** $\overline{DA} \cong \overline{CB}$ (def of isos trap); $\overline{AB} \cong \overline{AB}$ (reflex prop); $\angle DAB \cong \angle CBA$ (Th 8.2); $\triangle ABD \cong \triangle BAC$ (SAS) **10.** $\angle ADB \cong \angle BCA$ (CPCTC); $\angle DEA \cong \angle CEB$ (Th 2.10); $\triangle AED \cong \triangle BEC$ (AAS) **15.** Given: trap $ABCD$ with $\overline{DC} \parallel \overline{AB}$, $\angle A \cong \angle B$, $\angle C \cong \angle D$; Prove: trap $ABCD$ is isos. Draw $\overline{DF} \perp \overline{AB}$, $\overline{CE} \perp \overline{AB}$ (Post 12); $\overline{DF} \cong \overline{CE}$ (Th 7.9); $\angle DFA$, $\angle CEB$ are rt $\angle$s (def of $\perp$ lines); $\triangle DFA$, $\triangle CEB$ are rt. $\triangle$'s. (def of rt $\triangle$s), $\triangle DFA \cong \triangle CEB$ (LA); $\overline{DA} \cong \overline{CB}$ (CPCTC); trap $ABCD$ is isos (def of isos trap) **16.** Given: isos trap $ABCD$ with $\overline{DC} \parallel \overline{AB}$; Prove: $\overline{DB} \cong \overline{AC}$. $\angle DAB \cong \angle CBA$ (Th 8.2); $\overline{AB} \cong \overline{AB}$ (reflex prop); $\triangle DAB \cong \triangle CBA$ (SAS); $\overline{DB} \cong \overline{CA}$ (CPCTC) **17.** $\angle ABF$, $\angle CBF$ are rt $\angle$s (def of $\perp$ lines); $\triangle ABF$, $\triangle CBF$ are rt $\triangle$s (def of rt $\angle$s); $\overline{FB} \cong \overline{FB}$ (reflex prop); $\triangle ABF \cong \triangle CBF$ (LL); $\overline{AF} \cong \overline{FC}$ (CPCTC); $2\,\overline{AM} = \overline{AF}$, $2\,\overline{NC} = \overline{FC}$ (def of mdpt); $2\,\overline{AM} \cong 2\,\overline{NC}$ (subst prop); $\overline{AM} \cong \overline{NC}$ (div prop); $\overline{MN} \parallel \overline{AC}$ (Th 7.7); $ACNM$ is an isos trap (def of isos trap) **18.** $\overline{AF} \cong \overline{FC}$ (CPCTC); $2\,\overline{AM} = \overline{AF}$, $2\,\overline{NC} = \overline{FC}$ (def of mdpt); $2\,\overline{AM} \cong 2\,\overline{NC}$ (subst prop); $\overline{AM} \cong \overline{NC}$ (div prop); $\overline{MN} \parallel \overline{AC}$ (Th 7.7); $ACNM$ is an isos trap (def of isos trap)

Pages 252–253

6. $\overline{AE} \cong \overline{EB}$; $\angle FEA$, $\angle FEB$ are rt $\angle$s (def of $\perp$ bis); $\angle FEA \cong \angle FEB$ (all rt $\angle$s are $\cong$); $\overline{FE} \cong \overline{FE}$ (reflex prop); quad $AEFD \cong$ quad $BEFC$ (SASAS) **7.** $\angle DFE \cong \angle CFE$ (def of bis); $\overline{FE} \cong \overline{FE}$ (reflex prop); $\angle FEA \cong \angle FEB$ (Th 2.6); $\overline{AE} \cong \overline{EB}$ (def of bis); quad $AEFD \cong$ quad $BEFC$ (SASAS) **8.** $\overline{AB} \cong \overline{BE} \cong \overline{EF}$ and $\overline{BC} \cong \overline{BE} \cong \overline{DE}$ (def of Rhom); $\overline{AB} \cong \overline{BC}$ and $\overline{EF} \cong \overline{DE}$ (subst prop); $\overline{BE} \cong \overline{BE}$ (reflex prop); $\angle 1 \cong \angle 2$ (given); $m\angle 1 + m\angle 3 = m\angle 2 + m\angle 4$ (suppl. $\angle$'s); $\angle 3 \cong \angle 4$ (subtr.); Rhom $ABEF \cong$ Rhom $CBED$ (SASAS) **9.** $\overline{AB} \cong \overline{BE} \cong \overline{EF}$ and $\overline{BC} \cong \overline{BE} \cong \overline{DE}$ (def of Rhom); $\overline{AB} \cong \overline{BC}$ and $\overline{EF} \cong \overline{DE}$ (subst prop); $\overline{BE} \cong \overline{BE}$ (reflex prop); $\angle 1$ and $\angle 4$ are suppl (given); $\angle 4$ and $\angle 2$ are suppl (def of ▱); $\angle 1 \cong \angle 2$ (subst prop); Rhom $ABEF \cong$ Rhom $CBED$ (SASAS) **10.** $\overline{AB} \cong \overline{AB}$ (reflex prop); quad $EABC \cong$ quad $FBAD$ (SASAS) **11.** $\overline{DE} \cong \overline{FC}$, $\overline{EG} = \overline{GF}$, $\overline{DG} \cong \overline{GC}$, $\angle D \cong C$, $\angle DEG \cong \angle CFG$ (CPCTC);

$\overline{AD} \cong \overline{BC}$, $\overline{DF} \cong \overline{CE}$ m $\angle AEG$ + m $\angle DEG = 180$, m $\angle BFD$ + m $\angle CFG = 180$ (add prop, Post 11); $\angle AEG$ and $\angle DEG$ are supp, $\angle BFD$ and $\angle CFG$ are supp (def of suppl $\angle$s); $\angle AEG \cong \angle BFD$ (Th 2.8); quad $EABC \cong$ quad $FBAD$ (SASAS) **12.** $\overline{LI} \cong \overline{LI}$ (reflex prop); $\overline{ML} \parallel \overline{HI}$, $\overline{LK} \parallel \overline{IJ}$ (def of trap); $\angle MLI$ and $\angle 1$ are suppl, $\angle KLI$ and $\angle 2$ are suppl (Th 3.2); $\angle MLI \cong \angle KLI$ (Th 2.8); trap $HILM \cong$ trap $JILK$ (ASASA) **13.** $\overline{HM} \cong \overline{JK}$, $\angle M \cong \angle K$, $\overline{ML} \cong \overline{KL}$ (CPCTC); $\angle MLI \cong \angle M$, $\angle KLI \cong \angle K$ (Th 8.2); $\angle MLI \cong \angle KLI$ (subst prop); $\overline{LI} \cong \overline{LI}$ (reflex prop); trap $HILM \cong$ trap $JILK$ (SASAS) **14.** $\overline{FG} \cong \overline{GC}$, $\overline{EG} \cong \overline{GB}$, $\overline{EF} \cong \overline{CB}$, $\angle EFG \cong \angle BCG$, $\angle EGF \cong \angle BGC$ (CPCTC); $\overline{FA} \cong \overline{DC}$ (subtr prop); $\angle EFG + \angle AFG = 180$, $\angle BCG + \angle DCG = 180$ (Post 11); $\angle EFG$, $\angle AFG$ are suppl, $\angle BCG$, $\angle DCG$ are suppl (def of suppl $\angle$s); $\angle FGB \cong \angle CGE$ (Th 2.10); $\angle AFG \cong \angle DCG$, $\angle FGB \cong CGE$ (Th 2.8); quad $ABGF \cong$ quad $DEGC$ (SASAS) **15.** $\angle A \cong \angle D$, $\angle ABE \cong \angle DEB$, $\overline{AB} \cong \overline{ED}$ (CPCTC); $\angle EGC \cong \angle FGB$ (Th 2.10); quad $ABGF \cong$ quad $DEGC$ (ASASA) **16.** $\triangle DAB \cong \triangle HEF$ (SAS for $\cong \triangle$s) **17.** $\angle 3 \cong \angle 7$ (subtr prop); $\angle ADC \cong \angle EHG$ (add prop) **18.** Given: quad $ABCD$, quad $EFGH$, $\angle A \cong \angle E$, $\angle B \cong \angle F$ $\angle C \cong \angle G$, $\overline{AB} \cong \overline{EF}$; $\overline{BC} \cong \overline{FG}$ Prove: quad $ABCD \cong$ quad $EFGH$. Draw $\overline{AC}$, $\overline{EG}$ (Post 1); $\triangle ABC \cong EFG$ (SAS); $\angle 2 \cong \angle 6$, $\angle 4 \cong \angle 8$, $\overline{AC} \cong \overline{EG}$ (CPCTC); $\angle 1 \cong \angle 5$, $\angle 3 \cong \angle 7$ (subtr prop); $\triangle DAC \cong \triangle HEG$ (ASA); $\overline{DC} \cong \overline{HG}$ (CPCTC); quad $ABCD \cong$ quad $EFGH$ (SASAS)

Page 255

1. Draw $\overline{AB}$ 7 cm. At each endpt constr a $\perp$. Make $\overline{CA} \cong \overline{DB} \cong \overline{AB}$. Draw $\overline{CD}$ to complete sq. **2.** Draw seg $\overline{AB}$ and $\overline{DA}$ different lengths. At a pt A and pt B constr $\perp$s. Make these lengths $\cong \overline{DA}$. Connect their endpts to complete rect. **3.** Follow steps ex 1 p. 254. **6.** Follow steps ex 2 p. 254. **7.** Let $\overline{XY} = 10$ cm bis $\overline{XY}$. Call half $\overline{XY}$, $\overline{AB}$, a side of rect. At A and B constr $\perp$s. Using A as a center and $\overline{XY}$ as radius, constr an arc intersecting $\perp$ at B. Similarly using B as a center and $\overline{XY}$ as radius, constr an arc intersecting $\perp$ at A. Connect these 2 pts of intersection for required rect. **8.** Follow steps ex 1 p. 254. **9.** Using $\overline{AB}$ as base, constr an arc at A with radius s and an arc at B with radius s. Using A again constr an arc at A with radius d and at B with radius d. Pts of intersection of arcs connect to form trap. **10.** Using l as length of sides and a 60° $\angle$, follow steps ex 1 p. 254. **11.** Constr $AB = b'$, $AD = e$, $m \angle DAB = 45$; constr $\overline{DC} \parallel \overline{AB}$, $DC = b$; connect C to B. **12.** See ex 9. **13.** Constr $\overline{A'B'} \cong \overline{AB}$. At A' constr $\angle A' \cong \angle A$ and at B' constr $\angle B' \cong \angle B$. Constr $\overline{B'C'} \cong \overline{BC}$. At C' constr $\angle C' \cong \angle C$. Extend rays from pt C' and A'. They meet at D'. **14.** Constr $\overline{A'B'} \cong \overline{AB}$. At B' constr $\angle B' \cong \angle B$. Constr $\overline{B'C'} \cong \overline{BC}$. At C' constr $\angle C' \cong \angle C$. Constr $C'D' \cong CD$. Connect $A'D'$.

Page 258

18. Given: hex $ABCDEF$ with $\overleftrightarrow{FC}$ line of symmetry; Prove: $\angle A \cong \angle E$, $\angle B \cong \angle D$. Draw $\overline{AE}$, $\overline{BD}$ intersecting $\overleftrightarrow{FC}$ at G, H (Post 1); $\overleftrightarrow{FC}$ $\perp$ bis $\overline{AE}$, $\overleftrightarrow{FC}$ $\perp$ bis $\overline{BD}$ (Post 21); $\overline{AF} \cong \overline{FE}$, $\overline{DC} \cong \overline{BC}$ (Th 8.8); $\angle FAG \cong \angle FEG$, $\angle CDH \cong \angle CBH$ (Th 5.1); $\overline{EG} \cong \overline{GA}$, $\overline{DH} \cong \overline{HB}$ (def of bis); $\angle EGH \cong \angle AGH$, $\angle DHG \cong \angle BHG$ (Th 2.6); $\overline{GH} \cong \overline{GH}$ (reflex prop); quad $EGHD \cong$ quad $AGHB$ (SASAS); $\angle GAB \cong \angle GED$, $\angle HBA \cong \angle HDE$ (def of $\cong$ quad); $\angle A \cong \angle E$, $\angle B \cong \angle D$ (add prop) **19.** Given: oct $ABCDEFGH$ with $\overleftrightarrow{AE}$ line of symmetry; Prove: $\angle B \cong \angle H$, $\angle C \cong G$, $\angle F \cong \angle D$. Draw $\overline{BH}$, $\overline{GC}$, $\overline{DF}$ inter $\overleftrightarrow{AE}$ at M, I, J (Post 1); $\overline{AH} \cong \overline{AB}$, $\overline{ED} \cong EF$ (Th 8.8); $\overleftrightarrow{AE}$ bis $\overline{BH}$, $\overleftrightarrow{AE}$ bis $\overline{CG}$, $\overleftrightarrow{AE}$ bis $\overline{DF}$ (Post 21); $\overline{BM} \cong \overline{MH}$, $\overline{CI} \cong \overline{IG}$, $\overline{DJ} \cong \overline{JF}$ (def of bis); $\overline{AM} \cong \overline{AM}$, $\overline{JE} \cong \overline{JE}$ (reflex prop); $\triangle ABM \cong \triangle AHM$, $\triangle EDJ \cong \triangle EFJ$ (SSS); $\angle BAM \cong \angle HAM$, $\angle DEJ \cong \angle FEJ$ (CPCTC); $\overline{AI} \cong \overline{AI}$, $\overline{EI} \cong \overline{EI}$ (reflex prop); $\angle AIC \cong \angle AIG$, $\angle EIC \cong \angle EIG$ (Th 2.6); quad $ABCI \cong$ quad $AHGI$, quad $EDCI \cong$ quad $EFGI$ (SASAS); $\angle B \cong \angle H$, $\angle BCI \cong \angle HGI$, $\angle D \cong \angle F$, $\angle DCI \cong \angle FGI$ (def of $\cong$ quad); $\angle C \cong \angle G$ (add prop)
20. 1. Post 1 **2.** Post 21 **3.** Th 5.7 **4.** Def of bis **5.** Th 2.6 **6.** Reflex prop **7.** Th 2.6 **8.** Def of bis **9.** SASAS **10.** Def of $\cong$ quad **21.** Given: sept. $ABCDEFG$ with $\overleftrightarrow{AH}$ line of symmetry. Prove: $\overline{AG} \cong \overline{AB}$, $\overline{GF} \cong \overline{BC}$, $\overline{FE} \cong \overline{DC}$. Draw $\overline{GB}$, $\overline{FC}$ inters $\overleftrightarrow{AH}$ at I, J (Post 1); $\overleftrightarrow{AH}$ bis $\overline{GB}$, $\overleftrightarrow{AH}$ bis $\overline{FC}$ $\overleftrightarrow{AH}$ bis $\overline{ED}$ (Post 21); $AG \cong \overline{AB}$ (Th 5.7); $\overline{GI} \cong \overline{IB}$, $\overline{FJ} \cong \overline{JC}$, $\overline{EH} \cong HD$ (def of bis); $\overline{IJ} \cong \overline{IJ}$, $\overline{JH} \cong \overline{JH}$ (reflex prop); $\angle GIJ \cong \angle BIJ$, $\angle IJF \cong \angle IJC$, $\angle FJH \cong \angle CJH$, $\angle JHE \cong \angle JHD$ (Th 2.6); quad $GIJF \cong$ quad $BIJC$, quad $FJHE \cong$ quad $CJHD$ (SASAS); $\overline{GF} \cong \overline{BC}$, $\overline{FE} \cong \overline{DC}$ (def of $\cong$ quad)

Page 261

17. Given: quad $ABCD$ and $A'B'C'D'$ are reflections, l is the line of symmetry; Prove: quad $ABCD \cong$ quad $A'B'C'D'$. Draw $\overline{AC}$, $\overline{A'C'}$, $\overline{DB}$, $\overline{D'B'}$ (Post 1); l is $\perp$ bis of $\overline{AA'}$, $\overline{BB'}$, $\overline{CC'}$, $\overline{DD'}$ (def of reflections); $\triangle ADC$ and $\triangle A'D'C'$ are refl, $\triangle DCB$ and $\triangle D'C'B'$ are refl (def of reflections); $\triangle ADC \cong \triangle A'D'C'$, $\triangle DCB \cong \triangle D'C'B'$ (Th 8.10); $\overline{AD} \cong \overline{A'D'}$, $\angle ADC \cong \angle A'D'C'$, $\overline{DC} \cong \overline{D'C'}$, $\angle DCB \cong$

$\angle D'C'B'$, $\overline{BC} \cong B'C'$ (CPCTC); quad $ABCD \cong$ quad $A'B'C'D'$ (SASAS)

Page 264

16. On line l lay off $\overline{AB} \cong a$. At A constr an $\cong$ given $\angle$. At B constr an $\angle \cong \angle A$. At vertex A and vertex B lay off seg $\cong b$. Join the ends of these seg to complete required trap **17.** On line l lay off $\overline{AB} \cong a$. At A constr a $\perp$. Lay off $\overline{AC} \cong b$. At C constr $\overleftrightarrow{CD} \parallel \overleftrightarrow{AB}$. Using A and B as centers and c as a radius constr arcs intersecting $\overleftrightarrow{CD}$. Join these pts of intersection to complete the required trap. **18.** Join 2 corresponding vertices with a seg and constr $\perp$ bis of this seg. **21.** $\overline{FC} \cong \overline{CB}$, $\overline{DC} \cong \overline{CG}$ (def of bis); $\angle FCD \cong \angle GCB$ (Th 2.10); $\angle F = \angle B$, $\angle D \cong \angle G$ (Post 13); $CGAB \cong CDEF$ (ASASA) **22.** $\overline{HI} \cong \overline{HI}$ (reflex prop); $\angle KHI \cong \angle JIH$ (Th 8.2); $\triangle HIK \cong \triangle IHJ$ (SAS)

Page 265

14. On line l lay off $\overline{AB} \cong a$. At A constr $\perp$ and lay off $\overline{AC} \cong c$. At B constr $\perp$ and lay off $\overline{BD} \cong c$. Join $\overleftrightarrow{CD}$. Bisect $\overline{CD}$ and call mdpt E. Bis b. At E on $\overline{CD}$, lay off $\overline{EF} \cong \frac{1}{2} b$ and $\overline{EG} = \frac{1}{2} b$. Join $\overline{AG}$ and $\overline{BF}$ to complete required trap. **15.** There is no common $\perp$ bis. **18.** See p. 249 ex 16. **19.** $\angle FEB \cong \angle DEB$ (def of $\angle$ bis); $\overline{FB} \cong \overline{FB}$ (reflex prop); $\overline{AB} \cong \overline{BC}$ (def of bis); $\angle EBA \cong \angle EBC$ (Th 2.6); quad $ABEF \cong$ quad $CBED$ (SASAS)

Page 277

21. Given: sq $ABCD$ and sq $XYZW$; Prove: sq $ABCD \sim$ sq $XYZW$. $\overline{AB} \cong \overline{BC} \cong \overline{CD} \cong \overline{DE}$, $\overline{XY} \cong \overline{YZ} \cong \overline{ZW} \cong \overline{WX}$ (def of sq); $AB = BC = CD = DE$, $XY = YZ = ZW = WX$ (def of $\cong$); $\frac{AB}{XY} = \frac{BC}{YZ} = \frac{CD}{ZW} = \frac{DE}{WX}$ (div prop and subst); $\angle$s A, B, C, D, X, Y, Z, W are rt $\angle$s (prop of sq); $\angle A \cong \angle X$, $\angle B \cong \angle Y$, $\angle C \cong \angle Z$, $\angle D \cong \angle W$ (all rt $\angle$s are $\cong$); $ABCD \sim XYZW$ (def of $\sim$ poly)

Pages 280–281

7. $\angle D \cong \angle D$ (reflex prop); $\triangle ABD \sim \triangle CED$ (Post 22); **8.** $\angle JHI \cong \angle GHF$ (Th 2.10); $\triangle HFG \sim \triangle HIJ$ (Post 22) **9.** $\angle O \cong \angle M$ (Post 13); $\angle K \cong \angle N$ (Post 13); $\triangle OKL \sim \triangle MNL$ (Post 22) **14.** $\angle EDC \cong \angle EFB$ ($\perp$s form $\cong$ rt $\angle$S); $\angle DEC \cong \angle FEB$ (Th 2.10); $\triangle FBE \sim \triangle DCE$ (Post 22) **15.** $\angle GJI \cong \angle KHI$ ($\perp$s form $\cong$ rt $\angle$s); $\angle I \cong \angle I$ (reflex prop); $\triangle HKI \sim \triangle JGI$ (Post 22) **16.** $\angle OPQ \cong \angle PNM$ (Post 13); $\angle POQ \cong \angle NMQ$ (Post 13); $\triangle MNQ \sim \triangle OPQ$ (Post 22) **17.** $\angle V \cong \angle V$ (reflex prop); $\overline{WT} \parallel \overline{RS}$ (def of $\square$); $\angle VTW \cong \angle RSV$ (Th 3.1); $\triangle RSV \sim \triangle WTV$ (Post 22) **18.** $\angle BED \cong \angle BAC$ ($\perp$s form $\cong$ rt $\angle$s); $\angle B \cong \angle B$ (reflex prop); $\triangle BED \sim \triangle BAC$ (Post 22); $\frac{CB}{DB} = \frac{AB}{EB}$ (corr sides of $\sim$ $\triangle$s are proportional) **19.** $\angle LMN \cong \angle JKN$, $\angle MLN \cong \angle KJN$ (Post 13); $\triangle LMN \sim \triangle JKN$ (Post 22); $\frac{MN}{NK} = \frac{LN}{NJ}$ (corr sides of $\sim$ $\triangle$s are proportional) **24.** $\angle TQC$, $\angle TMC$ are rt $\angle$s (def of $\perp$ lines); m $\angle TQC = 90$, m $\angle TMC = 90$ (def of rt $\angle$s); m $\angle TQC$ + m $\angle QCM$ + m $\angle TMC$ + m $\angle T = 360$ (Th 6.1); 90 + m $\angle QCM$ + 90 + m $\angle T$ = 360 (subst prop); m $\angle QCM$ + m $\angle T = 180$ (subtr prop); $\angle QCM$ and $\angle T$ are suppl (def of suppl); $\angle JCP$ and $\angle JCQ$ are suppl ($\angle PCQ$ is a st. $\angle$); $\angle JCP \cong \angle T$ (Cor 2.7); (In same way as above prove $\angle JPC = \angle N$) $\triangle JCP \sim \triangle XTN$ (Post 22) **25.** $XYZW$ is a $\square$ (Th 7.3); $\overline{WX} \parallel \overline{ZY}$ (def of $\square$); $\angle XWV \cong \angle ZTW$ (Post 13); $\angle WXV \cong \angle TZW$ (Th 7.1, Coro 2); $\triangle WTZ \sim \triangle VWX$ (Post 22) **26.** $\angle Q \cong \angle Q$ (reflex prop); $\angle 1 \cong \angle 2$ (given); $\triangle TQS \sim \triangle RQP$ (Post 22); $\frac{SQ}{TS} = \frac{PQ}{PR}$ (corr sides of $\sim$ $\triangle$s are proportional) **27.** $\angle DEA$ is a rt $\angle$ (def of $\perp$ lines); $\angle DEA \cong \angle BCA$ (all rt $\angle$s are $\cong$); $\angle A \cong \angle A$ (reflex prop); $\triangle ABC \sim \triangle ADE$ (Post 22); $\frac{AD}{AB} = \frac{DE}{BC}$ (corr sides of $\sim$ $\triangle$s are proportional); $\overline{BC} \perp \overline{AC}$ (def $\perp$ lines); $\overline{BC} \parallel \overline{DE}$ (Cor 3.3); $\angle BCD \cong \angle CDE$ (Post 13); $\angle BCD = \angle DCE$ (def $\angle$ bis); $\angle CDE = \angle DCE$ (subst prop); $\overline{DE} \cong \overline{EC}$ (Th 5.2); $\frac{AD}{AB} = \frac{EC}{BC}$ (subst prop) **28.** $\angle DEA$, $\angle BCA$ are rt $\angle$s (def $\perp$ lines); $\angle DEA \cong \angle BCA$ (all rt $\angle$s are $\cong$); $\angle A \cong \angle A$ (reflex prop); $\triangle ABC \sim \triangle ADE$ (Post 22); $\frac{AC}{AE} = \frac{BC}{DE}$ (corr sides of $\sim$ $\triangle$s are proportional); $AC \cdot DE = AE \cdot BC$ (property of proportions); $\overline{ED} \cong \overline{EC}$ (Th 5.2); $ED = EC$ (def of $\cong$); $ED \cdot BC = EC \cdot BC$ (Mult prop); $AC \cdot DE + ED \cdot BC = AE \cdot BC + EC \cdot BC$ (add prop); $AC \cdot DE + ED \cdot BC = BC\,(AE + EC)$ (dist prop); $AC = AE + EC$ (def of between); $AC \cdot DE + ED \cdot BC = BC \cdot AC$ (subst prop); $\frac{1}{BC} + \frac{1}{AC} = \frac{1}{DE}$ (Mult prop)

Page 284

17. $\overline{AC} \parallel \overline{ED}$ (Th 3.3); $\frac{BD}{DC} = \frac{BE}{EA}$ (Th 9.1)

18. $\overline{DE} \parallel \overline{CA}$ (Th 3.3); $\frac{AE}{EB} = \frac{CD}{DB}$ (Th 9.1)

19. $\frac{AE}{EB} = \frac{DG}{GB}$ (Th 9.1); $\overline{DA} \parallel \overline{CB}$ (def of ▱); $\overline{EF} \parallel \overline{CB}$ (Th 3.7); $\frac{DG}{GB} = \frac{DF}{FC}$ (Th 9.1); $\frac{AE}{EB} = \frac{DF}{FG}$ (subst prop) **20.** $\overline{FE} \parallel \overline{AD}$ (Th 3.3); $\frac{AB}{AE} = \frac{DB}{DG}$ (Th 9.1); $\overline{AD} \parallel \overline{CB}$ (def of ▱); $\overline{FE} \parallel \overline{CB}$ (Th 3.7); $\frac{DC}{DF} = \frac{DB}{DG}$ (Th 9.1); $\frac{DC}{DF} = \frac{AB}{AE}$ (subst prop)
21. There are 5 possibilities, all very similar. Given: $\overleftrightarrow{AA'} \parallel \overleftrightarrow{BB'} \parallel \overleftrightarrow{CC'}$; Prove: $\frac{AB}{A'B'} = \frac{BC}{B'C'}$. $\frac{AB}{AP} = \frac{A'B'}{A'P'}$ and $\frac{AC}{AP} = \frac{A'C'}{A'P'}$ (Th 9.1); $AB + BC = AC$ and $A'B' + B'C' = A'C'$ (def of betw); $\frac{AB + BC}{AD} = \frac{A'B' + B'C'}{A'P'}$ (subst); $\frac{AB}{AP} + \frac{BC}{AP} = \frac{A'B'}{A'P'} + \frac{B'C'}{A'P'}$ (subst); $\frac{BC}{AP} = \frac{B'C'}{A'P'}$ (substr); $\frac{AB}{A'B'} = \frac{AP}{A'P'}$ and $\frac{BC}{B'C'} = \frac{AP}{A'P'}$ (prop of propor) $\frac{AB}{A'B'} = \frac{BC}{B'C'}$ (trans prop) **22.** $\frac{GD}{DA} = \frac{GE}{EB}$ (Th 9.1); $\frac{GF}{FC} = \frac{GE}{EB}$ (Th 9.1); $\frac{GD}{DA} = \frac{GF}{FC}$ (trans prop); $\frac{GD}{GD + DA} = \frac{GF}{GF + FC}$ (property of proportion); $GD + DA = GA$, $GF + FC = GC$ (def of between); $\frac{GD}{GA} = \frac{GF}{GC}$ (subst prop); $\triangle DFG \sim \triangle ACG$ (Th 9.3)

Page 289

22. $\angle Z \cong \angle Z$ (reflex prop); $\angle ZTS \cong \angle ZXY$ (def of $\sim$ $\triangle$s); $\triangle TZS \sim \triangle XZY$ (Post 22)
23. $\overline{TS} \parallel \overline{RY}$ (def of ▱); $\angle Y \cong \angle XRT$ (Th 3.1); $\angle ZTS \cong \angle X$, $\angle Y \cong \angle ZST$ (Th 3.1); $\angle XRT \cong \angle ZST$ (trans); $\triangle XRT \cong \triangle ZST$ (Post 22)
24. $\angle QPL \cong \angle QRN$ ($\perp$s form $\cong$ rt $\angle$s); $\angle NQR \cong \angle LQP$ (Th 2.10); $\triangle LPO \sim \triangle NRQ$ (Post 22)
25. $\angle NPM \cong \angle NRQ$ ($\perp$s form $\cong$ rt $\angle$s); $\angle RNQ \cong \angle RNQ$ (reflex prop); $\triangle NQR \sim \triangle NMP$ (Post 22)
26. $\angle M \cong \angle M$ (reflex prop); $\angle LRM \cong \angle NPM$ ($\perp$s form $\cong$ rt $\angle$s); $\triangle LMR \sim \triangle NMP$ (Post 22)
27. $\angle LRN \cong \angle LRM$ ($\perp$s form $\cong$ rt $\angle$s); $\angle LNR \cong \angle LMR$ (given); $\triangle LNR \sim \triangle LMR$ (Post 22)
28. $\overline{AD} \parallel \overline{BC}$, $\overline{DC} \parallel \overline{AB}$ (def of ▱); $\angle AFB \cong \angle CBF$, $\angle CEB \cong \angle FBA$ (Post 13); $\triangle BCE \sim \triangle FAB$ (Post 22) **29.** $\angle R \cong \angle WOV$, $\angle S \cong \angle WVU$ (Th 3.1); $\triangle UVW \sim \triangle RST$ (Post 22)
30. $ED \parallel CA$ (Th 7.9); $\angle BED \cong \angle BCA$, $\angle BDE \cong \angle BAC$ (Th 3.1); $\triangle BDE \sim \triangle BAC$ (Post 22)
31. $\overline{TU} \cong \overline{TV}$, $\overline{TR} \cong TS$ (CPCTC); $\frac{TU}{TR} = \frac{TV}{TS}$ (div prop) **32. 1.** $\angle Z \cong \angle Z$ **1.** (reflex prop); **2.** $\frac{XZ}{MZ} = \frac{YZ}{NZ}$ **2.** (given); **3.** $\triangle ZNM \sim \triangle ZYX$ **3.** (Th 9.3); **4.** $\frac{XZ}{MZ} = \frac{XY}{MN}$ **4.** (corr sides of $\sim$ $\triangle$s are proportional) **33. 1.** $\angle CTQ \cong \angle RJV$ **1.** ($\perp$s form $\cong$ rt $\angle$s); **2.** $\angle CQT \cong \angle CVT$ **2.** (Th 5.1); **3.** $\triangle CTQ \sim \triangle RJV$ **3.** (Post 22); **4.** $\frac{CQ}{RV} = \frac{QT}{JV}$ **4.** (corr sides of $\sim$ $\triangle$s are proportional) **34.** Given: $\triangle ABC$, $\angle C$ a rt $\angle$, $\overline{AC} \cong \overline{BC}$, $\triangle DEF$, $\angle F$ a rt $\angle$, $\overline{DF} \cong \overline{FE}$; Prove: $\triangle ABC \sim \triangle DEF$. **1.** $\angle C \cong \angle F$ **1.** (all rt $\angle$s are $\cong$); **2.** $\frac{AC}{DF} = \frac{BC}{EF}$ **2.** (div prop); **3.** $\triangle ABC \sim \triangle DEF$ **3.** (Th 9.3) **35.** Given: $\overline{AB} \cong \overline{BC}$, $\overline{DE} \cong \overline{EF}$, $\angle B \cong \angle E$; Prove: $\triangle ABC \sim \triangle DEF$. Use div prop and Th 9.3. Given: $\overline{AB} \cong \overline{BC}$, $\overline{DE} \cong \overline{EF}$, $\angle A \cong \angle D$; Prove: $\triangle ABC \sim \triangle DEF$. Use Th 5.1, subst prop, and Post 22 **37.** Given: $\triangle$s ABC, DEF, $\frac{AB}{DE} = \frac{BC}{EF} = \frac{AC}{DF}$ Prove: $\triangle ABC \sim \triangle DEF$, Construct $\overline{DB'} \cong \overline{AB}$, then draw $\overline{B'C'} \parallel \overline{EF}$ (Post 15); $\angle B' \cong \angle E$, $\angle C' \cong \angle F$ (Th 3.1); $\triangle DB'C' \sim \triangle DEF$ (Post 22); $\frac{DB'}{DE} = \frac{DC'}{DF} = \frac{B'C'}{EF}$ (corr sides of $\sim$ $\triangle$s are proportional); $\frac{BC}{EF} = \frac{B'C'}{EF}$; $\frac{DC'}{DF} = \frac{AC}{DF}$ (subst, trans prop); $\overline{AC} \cong \overline{DC'}$, $\overline{BC} \cong \overline{B'C'}$ (mult prop); $\triangle ABC \cong \triangle DB'C'$ (Post 16); $\angle B \cong \angle B'$, $\angle C \cong \angle C'$ (CPCTC); $\triangle ABC \sim \triangle DEF$ (Post 22)

Page 293

11. $\frac{AD}{AB} = \frac{DE}{EB}, \frac{DC}{CB} = \frac{DE}{EB}$ (Th 9.7); $\frac{AD}{AB} = \frac{DC}{CB}$ (subst prop); **12.** $\frac{CD}{GH} = \frac{AB}{EF}$ (Th 9.5); $\frac{\frac{1}{2}AB}{\frac{1}{2}EF} = \frac{AB}{EF}$ (mult prop); $AD = \frac{1}{2}AB$, $EH = \frac{1}{2}EF$ (def of med); $\frac{AB}{EF} = \frac{AD}{EH}$ (subst prop); $\frac{CD}{GH} = \frac{AD}{EH}$ (subst prop);
13. $\angle S \cong \angle W$ (corr $\angle$s of $\sim$ $\triangle$s are $\cong$); $\angle TUS \cong \angle XYW$ ($\perp$s form $\cong$ rt $\angle$s); $\triangle TUS \sim \triangle XYW$ (Post 22); $\frac{TU}{XY} = \frac{US}{YW}$ (corr sides of $\sim$ $\triangle$s are proportional) **16.** $\angle B \cong \angle F$, $\frac{CB}{GF} = \frac{BA}{FE}$ (def $\sim$ $\triangle$s); $BD = \frac{1}{2}BA$, $HF = \frac{1}{2}FE$ (def of med); $\frac{\frac{1}{2}BA}{\frac{1}{2}FE} = \frac{BA}{FE}$ (mult prop); $\frac{CB}{GF} = \frac{BD}{HF}$ (subst prop); $\triangle CBD \sim \triangle GFH$ (Th 9.3); $\frac{CD}{GH} = \frac{CB}{GF}$ (corr sides of

$\sim \triangle$s are proportional) **17.** Given: $\triangle ABC \sim \triangle EFG$, with corr alts $\overline{AD}$ and $\overline{EH}$; Prove: $\frac{AC}{EG} = \frac{AD}{EH}$. $\angle C \cong \angle G$ (def of $\sim \triangle$s); $\angle ADC \cong \angle EHG$ ($\perp$s form $\cong$ rt $\angle$s); $\triangle ADC \sim \triangle EHG$ (Post 22); $\frac{AC}{EG} = \frac{AD}{EH}$ (corr sides of $\sim \triangle$s are proportional) **18.** Draw $\overleftrightarrow{BF} \parallel \overline{DC}$ (Post 15); Draw $\overrightarrow{AC}$ intersecting $\overleftrightarrow{BF}$ at E (Post 1); $\frac{AD}{DB} = \frac{AC}{CE}$ (Th 9.3); $\angle 2 \cong \angle 3$ (Post 13); $\angle 1 \cong \angle 4$ (Th 3.1); $\angle 1 \cong \angle 2$ (def of $\angle$ bis), $\angle 3 \cong \angle 4$ (trans prop); $\overline{BC} \cong \overline{CE}$ (Th 5.2); $\frac{AD}{DB} = \frac{AC}{BC}$ (subst prop)
19. Given: $\triangle ABC \sim \triangle DEF$, $\overline{BG}$, $\overline{EJ}$ alts, $\overline{BH}$, $\overline{EK}$ meds; Prove: $\frac{BH}{EK} = \frac{BG}{EJ}$. $\frac{BH}{EK} = \frac{AB}{ED}$ (Th 9.5); $\frac{BG}{EJ} = \frac{AB}{ED}$ (Th 9.6); $\frac{BH}{EK} = \frac{BG}{EJ}$ (subst prop);
20. Given: Trap $ABCD$; $\overline{AB} \parallel \overline{DC}$; pt. E, the intersection of $\overline{AC}$ and $\overline{DB}$; Prove: $\frac{DE}{EB} = \frac{CE}{AE}$. $\angle 1 \cong \angle 2$, $\angle 3 \cong \angle 4$ (Post 13); $\triangle DCE \sim \triangle BAE$ (Post 22); $\frac{DE}{EB} = \frac{CE}{AE}$ (corr sides of $\sim \triangle$s are proportional)

Pages 295–296

6. $\angle C \cong \angle C$ (reflex prop); $\angle A \cong \angle BDC$ ($\perp$s form $\cong$ rt $\angle$s); $\triangle ACE \sim \triangle DCB$ (Post 22); $\frac{BC}{EC} = \frac{BD}{AE}$ (corr sides of $\sim \triangle$s are proportional); $BC \cdot AE = BD \cdot EC$ (prop of proportionality) **7.** $\angle F \cong \angle I$, $\angle G \cong \angle J$ (Post 13); $\triangle FHG \sim \triangle IHJ$ (Post 22); $\frac{GH}{FH} = \frac{JH}{IH}$ (corr sides of $\sim \triangle$s are proportional); $GH \cdot IH = FH \cdot JH$ (prop of proportionality)
8. $\angle L \cong \angle O$ ($\perp$s form $\cong$ rt $\angle$s); $\angle KML \cong \angle NMO$ (Th 2.8); $\triangle KLM \sim \triangle NOM$ (Post 22); $\frac{LM}{OM} = \frac{KM}{NM}$ (corr sides of $\sim \triangle$s are proportional); $LM \cdot NM = OM \cdot KM$ (prop of proportionality); **9.** $\angle Y \cong \angle X$ ($\perp$s form $\cong$ rt $\angle$s); $\angle YVU \cong \angle XVW$ (Th 2.10); $\triangle YVU \sim \triangle XVW$ (Post 22); $\frac{VY}{VX} = \frac{VU}{VW}$ (corr sides of $\sim \triangle$s are proportional); $VY \cdot VW = VX \cdot VU$ (prop of proportionality) **10.** $\angle RST \cong \angle PQT$, $\angle SRT \cong \angle QPT$ (Post 13); $\triangle RST \sim \triangle PQT$ (Post 22); $\frac{ST}{TQ} = \frac{SR}{PQ}$ (corr sides of $\sim \triangle$s are proportional); $ST \cdot PQ = TQ \cdot SR$ (prop of proportionality) **11.** $\overline{CA}$ bis $\angle ZCB$ (def $\angle$ bis); $\frac{ZA}{AB} = \frac{ZC}{BC}$ (Th 9.7); $ZA \cdot BC = AB \cdot ZC$ (prop of proportionality) **15.** $\angle PRT \cong \angle QRT$ ($\perp$s form $\cong$ rt $\angle$s); $\angle 2 \cong \angle 3$ (subtr prop); $\overline{TR}$ bis $\angle SRU$ (def of $\angle$ bis); $\frac{ST}{TU} = \frac{SR}{RU}$ (Th 9.7); $ST \cdot RU = TU \cdot SR$ (prop of proportionality) **16.** $\angle ZXF \cong \angle PRF$, $\angle XZF \cong \angle RPF$ (Post 13); $\triangle XFZ \sim \triangle RFP$ (Post 22); $\frac{ZF}{FP} = \frac{FX}{FR}$ (corr sides of $\sim \triangle$s are proportional); $ZF \cdot FR = FP \cdot FX$ (prop of proportionality) **17.** $\angle JKY \cong \angle b$ (Post 13); $\angle JKY \cong \angle a$ (trans prop); $\overline{JY} \cong \overline{JK}$ (Th 5.2); $\frac{XK}{KZ} = \frac{XJ}{JY}$ (Th 9.3); $\frac{XK}{KZ} = \frac{XJ}{JK}$ (subst prop); $XK \cdot JK = KZ \cdot XJ$ (prop of proportionality) **18.** $\overline{CD} \parallel \overline{BA}$ (def of ▱); $\angle 2 \cong \angle BEF$ (Post 13); $\angle BEF \cong \angle 1$ (subst prop); $\overline{BC} \cong \overline{BE}$ (Th 5.2); $\overline{BF} \cong \overline{BF}$ (reflex prop); $BF \cdot BC = BF \cdot BE$ (mult prop)
19. $\frac{ED}{DA} = \frac{EF}{FB}$ (Th 9.3); $\overline{FB} \cong \overline{CB}$ (Th 5.2); $\frac{ED}{DA} = \frac{EF}{CB}$ (subst prop); $ED \cdot CB = DA \cdot EF$ (prop of proportionality) **20.** $\overline{MN} \parallel \overline{JK}$ (Th 3.3); $\frac{LM}{MJ} = \frac{LN}{NK}$ (Th 9.3); $\overline{MN} \cong \overline{NK}$ (Th 5.2); $\frac{LM}{MJ} = \frac{LN}{MN}$ (subst prop); $LM \cdot MN = LN \cdot MJ$ (prop of proportionality)

Page 299

7. $\frac{BX}{BA} = \frac{BZ}{ZF}$, $\frac{BY}{BC} = \frac{BZ}{ZF}$ (corr sides of $\sim \triangle$s are proportional); $\frac{BX}{BA} = \frac{BY}{BC}$ (subst prop); $\angle XBY \cong \angle XBY$ (reflex prop); $\triangle XBY \sim \triangle ABC$ (Th 9.3)
8. $\frac{XB}{YB} = \frac{BZ}{BZ}$, $\frac{AB}{CB} = \frac{BF}{BF}$ (corr sides of $\sim \triangle$s are proportional); $\frac{XB}{YB} = 1$, $\frac{AB}{CB} = 1$ (div prop); $\frac{XB}{YB} = \frac{AB}{CB}$ (subst prop); $\angle XBY \cong \angle XBY$ (reflex prop); $\triangle XBY \sim \triangle ABC$ (Th 9.3) **9.** $\angle TAS \cong \angle ZAY$ (Th 2.10); $\triangle ATS \sim \triangle AZY$ (Th 9.3) **10.** $\angle SAR \cong \angle YAX$, $\angle TAR \cong \angle ZAX$ (Th 2.10); $\triangle RAS \sim \triangle XAY$ (Th 9.3); $\triangle TAR \sim \triangle ZAX$ (Th 9.3); $\frac{RS}{YX} = \frac{AR}{AX}$, $\frac{RT}{XZ} = \frac{AR}{AX}$ (corr sides of $\sim \triangle$s are proportional); $\frac{RS}{YX} = \frac{RT}{XZ}$ (subst prop); $\angle SRA \cong \angle YXA$, $\angle TRA \cong \angle ZXA$ (def of $\sim \triangle$s); $\angle TRS \cong \angle ZXY$ (add prop); $\triangle TRS \sim \triangle ZXY$ (Th 9.3) **11.** $\overline{EF} \parallel \overline{AB}$ (Th 3.8); $\angle DEF \cong \angle DAB$, $\angle DFE \cong \angle DBA$ (Th 3.1); $\triangle EDF \sim \triangle ADB$ (Post 22). Similarly, show $\triangle DFG \sim \triangle DBC$ and $\triangle DEG \sim \triangle DAC$. Then $\frac{EF}{AB} = \frac{DF}{DB}$ and $\frac{DF}{DB} = \frac{FG}{BC}$; $\frac{FG}{BC} = \frac{DG}{DC}$ and $\frac{DG}{DC} = \frac{EG}{AC}$

(corr sides of ~ △'s are proportional); $\frac{EF}{AB} = \frac{FG}{BC} = \frac{EG}{AC}$ (trans prop); $\triangle EFG \sim \triangle ABC$ (Th 9.4). **12.** $\frac{ED}{DC} = \frac{FG}{GC}, \frac{AB}{BC} = \frac{FG}{GC}$ (Th 9.3); $\frac{ED}{DC} = \frac{AB}{BC}$ (subst prop)

Page 302

13. $\angle C \cong \angle C$ (refl prop); $\angle PEC$, $\angle CDB$ are rt $\angle$s (def $\perp$ *ls*); $\angle PEC \cong \angle CDB$ (all rt $\angle$s $\cong$); $\triangle CPE \sim \triangle CBD$ (Post 22: AA) **14.** $\angle DCP \cong \angle PCR$ (ex. 23, p 229); $\frac{DP}{PR} = \frac{DC}{RC}$; $DP \cdot RC = PR \cdot DC$

15. $DF = \frac{1}{2}CB$, $DE = \frac{1}{2}AB$ (Th 7.8); $\frac{DF}{CB} = \frac{1}{2}$, $\frac{DE}{AB} = \frac{1}{2}$ (div prop); $\frac{DF}{CB} = \frac{DE}{AB}$ (subst prop); $DE \cdot CB = DF \cdot AB$ (prop of proportion)

Page 303

10. $\angle DEB \cong \angle ABE$; $\angle EDA \cong \angle BAD$ (Post 13); $\triangle ABG \sim \triangle DEG$ (Post 22) **11.** $\angle ABE \cong \angle CEB$; $\angle BAC \cong \angle ECA$ (Post 13); $\triangle CFE \sim \triangle AFB$ (Post 22); $\frac{AB}{CE} = \frac{AF}{CF}$ (def of ~ poly); $\overline{AB} \cong \overline{AE}$ (def of a sq); $\frac{AE}{CE} = \frac{AF}{CF}$ (subst); $AE \cdot CF = CE \cdot AF$ (prop of propor) **12.** Given: $\triangle ABC$, D is mdpt of $\overline{AC}$, E is mdpt of $\overline{BC}$, F is mdpt of $\overline{AB}$; Prove: $\triangle ABC \sim \triangle DEF$. $DF = \frac{1}{2}BC$; $EF = \frac{1}{2}AC$; $DE = \frac{1}{2}AB$ (Th 7.7); $\frac{DF}{BC} = \frac{EF}{AC} = \frac{DE}{AB} = \frac{1}{2}$ (trans prop); $\triangle ABC \sim \triangle DEF$ (Th 9.4)

Page 309

17. 1. $\triangle BEC \sim \triangle CEA$, $\triangle CDE \sim \triangle CEA$ **1.** (Th 10.1); **2.** $\triangle CDE \sim \triangle BEC$ **2.** (Trans prop); **18. 1.** $\triangle BEC \sim \triangle CEA$, $\triangle EDA \sim \triangle CEA$ **1.** (Th 10.1); **2.** $\frac{BE}{ED} = \frac{CE}{AD}$ **2.** (corr sides of ~ △s are proportional); **3.** $BE \cdot AD = ED \cdot CE$ **3.** (prop of proportionality) **19.** Given: $\triangle ABC$, $\angle C$ is a rt $\angle$, Alt CD; Prove: $\frac{AD}{CD} = \frac{CD}{BD}$. $\triangle ACD \sim \triangle CBD$ (Th 10.1); $\frac{AD}{CD} = \frac{CD}{BD}$ (corr sides of ~ △s are proportional) **20.** $\triangle ACB \sim \triangle CBD$ (Th 10.1); $\frac{AB}{CB} = \frac{CB}{DB}$ (corr sides of ~ △s are proportional) **27.** Given: rt $\triangle ABC$, $\overline{CD}$ alt to hyp; Prove: $\frac{AC}{AD} \neq \frac{CB}{DB}$. $\frac{AB}{AC} = \frac{AC}{AD}, \frac{AB}{CB} = \frac{CB}{DB}$ (Th 10.1, cor 2); suppose $\frac{AC}{AD} = \frac{CB}{DB}, \frac{AB}{AC} = \frac{AB}{CB}$; $AB \cdot CB = AC \cdot AB$ (prop of proportionality); $CB = AC$ (div prop); $CB \neq AC$ (given); $\frac{AC}{AD} \neq \frac{CB}{DB}$ (by contradiction) **28.** Given: rt $\triangle XYZ$, $\overline{YW}$ alt to hyp; Prove: $YW \cdot XZ = XY \cdot YZ$. $\triangle XYZ \sim \triangle XWY$ (Th 10.1); $\frac{XY}{XZ} = \frac{YW}{YZ}$ (corr sides of ~ △s are proportional); $YW \cdot XZ = XY \cdot YZ$ (prop of proportionality) **29.** $\frac{r}{h} = \frac{h}{s}$ (Th 10.1); $\frac{u}{h} = \frac{h}{u+v}$ (Th 10.1; $h^2 = r \cdot s$, $h^2 = u(u+v)$ (prop of propor); $r \cdot s = u\,(u+v)$ (trans prop) **30.** $\frac{v}{w} = \frac{w}{u}, \frac{r}{h} = \frac{h}{s}$ (Th 10.1); $h^2 = rs$, $w^2 = v \cdot u$ (prop of propor); $h^2 + w^2 = r \cdot s + u \cdot v$ (add prop) **31.** $\frac{r}{h} = \frac{h}{s}$ Th 10.1; $h^2 = r \cdot s$ (prop of propor); $h^2 + s^2 = r \cdot s + s^2$ (add prop); $h^2 + s^2 = s(r+s)$ (dist prop)

32. $\frac{u}{h} = \frac{h}{u+v}$ (Th 10.1); $h^2 = u(u+v)$ (prop of propor)

Page 313

29. 1. $(AD)^2 = (AE)^2 + (DE)^2$; $(BC)^2 = (EB)^2 + (EC)^2$ **1.** (Th 10.2); **2.** $(AD)^2 + (BC)^2 = (AE)^2 + (DE)^2 + (EB)^2 + (EC)^2$ **2.** (add prop); **3.** $(AE)^2 + (EB)^2 = (AB)^2$, $(DE)^2 + (EC)^2 = (DC)^2$ **3.** (Th 10.2); **4.** $(AD)^2 + (BC)^2 = (AB)^2 + (DC)^2$ **4.** (subst prop); **30. 1.** $(HI)^2 = (HG)^2 + (GI)^2$, $(EF)^2 = (EG)^2 + (GF)^2$ **1.** (Th 10.2); **2.** $(HI)^2 + (EF)^2 = (HG)^2 + (GI)^2 + (EG)^2 + (GF)^2$ **2.** (add prop); **3.** $(HG)^2 + (GF)^2 = (HF)^2$, $(GI)^2 + (EG)^2 = (EI)^2$ **3.** (Th 10.2); **4.** $(HI)^2 + (EF)^2 = (HF)^2 + (EI)^2$ **4.** (subst prop) **31.** Given: $\overline{RT} \perp \overline{TU}$, $\overline{RT} \perp \overline{RS}$; Prove: $(RU)^2 - (TU)^2 = (TS)^2 - (RS)^2$; $(RU)^2 = (RT)^2 + (TU)^2$, $(TS)^2 = (SR)^2 + (RT)^2$ (Th 10.2); $(RU)^2 - (TU)^2 = (RT)^2$, $(TS)^2 - (RS)^2 = (RT)^2$ (subst)

32. $\frac{c}{a} = \frac{a}{n}, \frac{c}{b} = \frac{b}{m}$ (cor 2); $a^2 = cn$, $b^2 = cm$ (prop of propor); $a^2 + b^2 = cn + cm$ (add prop); $a^2 + b^2 = c(n+m)$ (dist prop); $m + n = c$ (seg add); $a^2 + b^2 = c^2$ (subst prop)

Pages 316–317

14. $a^2 = c^2 - b^2$ (mult prop); $a^2 + b^2 = c^2$ (add prop); $\triangle ABC$ is a rt △ (Th 10.3) **15.** $\frac{a^2b^2}{a^2} + \frac{a^2b^2}{b^2} = c^2$

(mult prop); $b^2 + a^2 = c^2$ (alg); $\triangle ABC$ is a rt $\triangle$ (Th 10.3) **22.** $(AC)^2 = (AD)^2 + (DC)^2$, $(CB)^2 = (DB)^2 + (DC)^2$ (Th 10.2); $(AC)^2 + (CB)^2 = (AD)^2 + 2(DC)^2 + (DB)^2$ (add prop); $(AD + DB)^2 = (AD)^2 + 2(AD)(DB) + (DB)^2$ (poly mult); $\frac{AD}{DC} = \frac{DC}{DB}$ (def of geom mean); $(AD)(DB) = (DC)^2$ (prop of propor); $(AC)^2 + (CB)^2 = (AD)^2 + 2(AD)(DB) + (DB)^2$ (subst prop); $(AC)^2 + (CB)^2 = (AD + DB)^2$ (subst prop); $AD + DB = AB$ (def of between); $(AC)^2 + (CB)^2 = (AB)^2$ (subst prop); $\triangle ABC$ is a rt $\triangle$ (Th 10.3)
23. $\frac{AD}{AC} = \frac{AC}{AB}$ (def of geo mean); $(AC)^2 = (AD)(AB)$ (prop of propor); $(AD)^2 + (CD)^2 = (AC)^2$ (Th 10.2); $(AD)^2 + (CD)^2 = (AD)(AB)$ (subst prop); $(CD)^2 = (AD)(AB) - (AD)^2$ (subtr prop); $(CD)^2 = (AD)(AB - AD)$ (dist prop); $AD + DB = AB$ (def of between); $DB = AB - BD$ (subtr); $(CD)^2 = (AD)(DB)$ (subst.); $\frac{AD}{CD} = \frac{CD}{DB}$ (div prop); $\triangle ABC$ is a rt $\triangle$ (ex 22) **24.** Either $\triangle ABC$ is a rt $\triangle$ or not; if $\triangle ABC$ is a rt $\triangle$, $a^2 + b^2 = c^2$; this contradicts given, so $\triangle ABC$ is not a rt $\triangle$ **25.** $a^2 + b^2 = c^2$ or $a^2 + b^2 \neq c^2$; if $a^2 + b^2 = c^2$ $\triangle ABC$ is a rt $\triangle$; this contradicts given so $a^2 + b^2 \neq c^2$. **26.** Construct $\triangle DEF$ with rt $\angle$ at F; $DF = b$, $EF = a$; $\triangle DEF$ is a rt $\triangle$ (def); in $\triangle DEF$, $a^2 + b^2 = f^2$ (Th 10.2); In $\triangle ABC$, $a^2 + b^2 = c^2$ (given); $c^2 = f^2$ (trans); $c = f$ (alg); $\triangle ABC \cong \triangle DEF$ (SSS); $\angle C \cong \angle F$ (CPCTC); m $\angle F = 90$ (constr); m $\angle C = 90$ (trans); $\angle C$ is a rt $\angle$ (def); $\triangle ABC$ is a rt $\triangle$ (def)

Page 320

21. Given: rt $\triangle ABC$, m $\angle B = 45°$, m $\angle C = 45°$, $AB = l$; Prove: $BC = l\sqrt{2}$. $\overline{AB} \cong \overline{AC}$ (Th 5.2); $AB = l$ (given); $AC = l$ (subst prop); $(BC)^2 = l^2 + l^2$ (Th 10.2); $(BC)^2 = 2l^2$ (alg); $BC = l\sqrt{2}$ (alg)
22. Given: rt $\triangle ABC$, m $\angle A = 60$, m $\angle B = 30$; Prove: $AB = 2AC$ and $BC = AC\sqrt{3}$. Constr $CD \cong AC$ on $\overrightarrow{AC}$, draw BD (constrs); $\triangle DCB \cong \triangle ACB$ (Post 17); $\angle D \cong \angle A$, $\angle DBC \cong \angle CBA$ (CPCTC); m $\angle DBA = 60$ (add prop); $\triangle ADC$ is equiangular (def of equiang $\triangle$), $\triangle ADB$ is equilat (Th 5.2); $AD = 2AC$ (seg add); $AB = 2AC$ (subst prop); $(BC)^2 + (AC)^2 = (2AC)^2$ (Th 10.2); $BC = AC\sqrt{3}$ (alg) **23.** $d^2 = s^2 + s^2$ (Th 10.2); $d^2 = 2s^2$ (alg); $d = s\sqrt{2}$ (alg); **24.** Given: rt $\triangle ABC$, $AB = l = BC$, $AC = h$; Prove: $l = \frac{h}{2}\sqrt{2}$. $l^2 + l^2 = h^2$ (Th 10.2); $2l^2 = h^2$ (alg); $l^2 = \frac{h^2}{2}$ (div prop); $l = \frac{h\sqrt{2}}{2}$ (alg)

25. m $\angle FGH = 30$ (diag of $\square$ bis $\angle$s); $\angle FHG$ is a rt $\angle$ (diag of rhombus $\perp$); m $\angle GFD = 60$ (acute $\angle$s of rt $\triangle$s are compl); If $FH = s$, $GF = 2s$, $GH = s\sqrt{3}$ (Th 10.5); $DG = DE = EF = FG = 2s$ (rhombus is equilat); $GE = 25\sqrt{3}$, $DF = 2s$ (diag of $\square$ bis each other); $\frac{GD + DE + EF + FG}{(GE) \cdot (DF)} = \frac{8s}{(2s\sqrt{3})(2s)} = \frac{2}{s\sqrt{3}} \neq \frac{1}{2\sqrt{3}}$ (subst and alg) **26.** $GE = 2s\sqrt{3}$, $DF = 2s$, $GD = 2s$, $FE = 2s$ (ex 25); $(GE)(DF) = 4s^2\sqrt{3}$ (mult prop); $(GD)(FE)\sqrt{3} = 4s^2\sqrt{3}$ (mult prop); $(GE)(DF) = (GD)(FE)\sqrt{3}$ (subst prop)

Page 323

19. $\sin B = \frac{b}{c}$; $\cos A = \frac{b}{c}$; $\sin B = \cos A$

20. $\cos B = \frac{a}{c}$; $\sin A = \frac{a}{c}$; $\cos B = \sin A$

21. $\tan B = \frac{b}{a}$; $\tan A = \frac{a}{b}$; $\frac{1}{\tan A} = \frac{b}{a}$; $\tan b = \frac{1}{\tan A}$

22. $\sin B = \frac{b}{c}$; $\cos B = \frac{a}{c}$; $a^2 + b^2 = c^2$; $b^2 = c^2 - a^2$; $\frac{b^2}{c^2} = 1 - \frac{a^2}{c^2}$; $(\sin B)^2 = 1 - (\cos B)^2$ **23.** $\tan A = \frac{a}{b}$; $\tan (90 - A) = \tan B = \frac{b}{a}$; $\tan A = \frac{1}{\tan B}$; $\tan A = \frac{1}{\tan (90 - A)}$ **24.** $\sin A = \frac{a}{c}$; $\cos (90 - A) = \cos B = \frac{a}{c}$; $\sin A = \cos B$; $\sin A = \cos (90 - A)$

Page 325

27. $\sin A = \frac{a}{c}$; $c > a$; $\frac{a}{c} < 1$ **28.** $\cos A = \frac{b}{c}$; for fixed c, as m $\angle A$ increases b decreases; so $\frac{b}{c}$ decreases. **29.** $\frac{\sin A}{\cos A} = \frac{\frac{a}{c}}{\frac{b}{c}} = \frac{a}{b}$; $\tan A = \frac{a}{b}$; $\tan A = \frac{\sin A}{\cos A}$ **30.** $\tan A = \frac{a}{b}$; for fixed b, as m $\angle A$ increases a increases, so $\frac{a}{b}$ gets infinitely large.

Page 341

17. $\overline{OA} \cong \overline{OB}$ (Th 11.1); $\angle OAB \cong \angle OBA$ (Th 5.1)
18. $\overline{OA} \cong \overline{OB}$ (Th 11.1); $\overline{OX} \cong \overline{OX}$ (reflex prop);

$\triangle AOX \cong \triangle BOX$ (Post 17); $\angle AXO \cong \angle BXO$ (CPCTC); $\overline{OC} \perp \overline{AB}$ (Th 2.5); $\overline{AX} \cong \overline{XB}$ (CPCTC); $\overline{OC}$ $\perp$ bis of $\overline{AB}$ (def $\perp$ bis) **19.** Given: $\odot O$, radii $\overline{OA}, \overline{OB}$; Prove: $\overline{OA} \cong \overline{OB}$. $OA = OB$ (def of $\odot$); $\overline{OA} \cong \overline{OB}$ (def $\cong$ seg) **20.** Given: $\odot O$, rad $\overline{OD}$ bisects chord $\overline{AB}$ at C; Prove: $\overline{OD} \perp \overline{AB}$. Draw rad $\overline{OA}, \overline{OB}$; $\overline{OA} \cong \overline{OB}$ (Th 11.1); $\overline{AC} \cong \overline{CB}$ (def bisector); $\overline{OD} \perp \overline{AB}$ (Th 5.8) **21.** $\angle DOA \cong \angle BOC$ (Th 2.10); $\overline{OD} \cong \overline{OB} \cong \overline{OA} \cong \overline{OC}$ (Th 11.1); $\triangle DOA \cong \triangle BOC$ (Post 17); $\overline{AD} \cong \overline{BC}$ (CPCTC) **22.** $\overline{OD} \cong \overline{OB} \cong \overline{OA} \cong \overline{OC}$ (Th 11.1); $ABCD$ is a ▱ (Th 7.6); $\overline{DB} \cong \overline{AC}$ (add prop); $ABCD$ is a rect (If diags of ▱ $\cong$, fig is a rect) **23.** Draw $\overline{OR}, \overline{OS}$; $\overline{OR} \cong \overline{OS}$ (Th 11.1); $\overline{OM} \cong \overline{OM}$ (reflex prop); $\triangle OMR \cong \triangle OMS$ (Post 16); $\angle OMR \cong \angle OMS$ (CPCTC); $\overline{XY} \perp \overline{RS}$ (Th 2.5); $\angle TNO \cong \angle OMS$ (Post 13); $\angle OMS$ a rt $\angle$ ($\perp$s form rt $\angle$s), $\angle TNO$ is a rt $\angle$ (subst); $\overline{OY} \perp \overline{TU}$ (def $\perp$); $\overline{OY}$ bis $\overline{TU}$ (Th 11.2); N is the mdpt of $\overline{TU}$ (def of mdpt) **24.** $\overline{CE} \cong \overline{CE}$ (reflex); $\angle CEB \cong \angle CEA$ ($\perp$s form $\cong$ rt $\angle$s); $\overline{AE} \cong \overline{EB}$ (Th 11.2); $\triangle CEA \cong \triangle CEB$ (Post 17); $\overline{AC} \cong \overline{BC}$ (CPCTC) **25.** Draw rad $\overline{OC}, \overline{OF}, \overline{OD}, \overline{OG}$; $\overline{OH} \cong \overline{OH}$ (reflex); $\overline{OF} \cong \overline{OG}$ (Th 11.1); $\overline{FH} \cong \overline{HG}$ (def bis); $\triangle OFH \cong \triangle OGH$ (Post 16); $\angle OHF \cong \angle OHG$ (CPCTC); $\overline{AB} \perp \overline{FG}$ (Th 2.5); $\overline{OE} \cong \overline{OE}$ (reflex); $\overline{OC} \cong \overline{OD}$ (Th 11.1); $\overline{CE} \cong \overline{ED}$ (def bis); $\triangle OCE \cong \triangle ODE$ (Post 16); $\angle CEO \cong \angle DEO$ (CPCTC); $\overline{AB} \perp \overline{CD}$ (Th 2.5); $\angle OHF \cong \angle DEO$ ($\perp$s form $\cong$ rt $\angle$s); $\overline{CD} \parallel \overline{FG}$ (Post 14) **27.** Given: $\odot O$, chord $\overline{AB}$, $\overleftrightarrow{DE}$ $\perp$ bis of $\overline{AB}$ at D; Prove: O lies on $\overleftrightarrow{DE}$. Draw $\overline{OA}, \overline{OB}$; $\overline{OA} \cong \overline{OB}$ (Th 11.1); $\overline{AD} \cong \overline{DB}$ (def of bis); $\overline{OD}$ is $\perp$ bis of $\overline{AB}$ (Th 5.8); O lies on $\overline{DE}$ (Post 12) **28.** Given: $\odot O$, chords $\overline{AB}, \overline{BC}$, rad $\overline{OB}$, $\angle 1 \cong \angle 2$; Prove: $\overline{AB} \cong \overline{BC}$. Draw $\overline{OA}, \overline{OC}$; $\overline{OA} \cong \overline{OB} \cong \overline{OC}$ (Th 11.1); $\angle 3 \cong \angle 1$, $\angle 4 \cong \angle 2$ (Th 5.1); $\angle 4 \cong \angle 3$ (trans prop); $\overline{OB} \cong \overline{OB}$ (reflex); $\triangle OAB \cong \triangle OCB$ (Th 4.2); $\overline{AB} \cong \overline{BC}$ (CPCTC) **29.** Given: $\odot$O, $\overline{AD} \cong \overline{AC}$, $\overline{BC} \cong \overline{BD}$; Prove: $\overline{AB}$ is a diam of $\odot O$. $\overline{AB}$ $\perp$ bis of $\overline{CD}$ (Th 5.8); O is on $\overline{AB}$ (ex 27); $\overline{AB}$ is a diam of $\odot O$ (def of diam)

Cumulative Review

1. $\angle SUT$, $\angle SVR$ are rt $\angle$s ($\perp$s form rt $\angle$s); $\angle SUT \cong \angle SVR$ (all rt $\angle$s $\cong$); $\angle S \cong \angle S$ (reflex); $\triangle RSV \sim \triangle TSV$ (Post 22) **2.** $\angle ZWY \cong \angle XYW$, $\angle WZX \cong \angle ZXY$ (Post 13); $\triangle ZWP \sim \triangle XYP$ (Post 22)

Page 344

20. $\overline{OX} \cong \overline{OZ}$, $\overline{OY} \cong \overline{OW}$ (Th 11.1); $\overline{XY} \cong \overline{ZW}$ (Th 11.4), $\triangle XUY \cong \triangle ZUW$ (Post 16)
21. $\overline{OX} \cong \overline{OY} \cong \overline{OW} \cong \overline{OZ}$ (Th 11.1); $\triangle XOY \cong \triangle ZOW$ (Post 17) $\overline{XY} \cong \overline{ZW}$ (CPCTC) $\overline{OT} \cong \overline{OU}$ (Th 11.3) **22.** $\angle XTO \cong \angle ZUO$ ($\perp$s form $\cong$ rt $\angle$s); $\overline{OX} \cong \overline{OZ}$ (Th 11.1); $\triangle XOT \cong \triangle ZOU$ (Th 4.2); $\overline{OT} \cong \overline{OU}$ (CPCTC); $\overline{XY} \cong \overline{WZ}$ (Th 11.4)
23. $\angle XTO \cong \angle ZUO$ ($\perp$s form $\cong$ rt $\angle$s); $\overline{OX} \cong \overline{OZ}$ (Th 11.1); $\triangle XOT \cong \triangle ZOU$ (Th 4.2); $\overline{OT} \cong \overline{OU}$ (CPCTC); $\overline{XY} \cong \overline{WZ}$ (Th 11.4) **24.** Given: $\odot O \cong \odot P$, $\overline{OC} \cong \overline{PF}$, $\overline{OC} \perp \overline{AB}$, $\overline{PF} \perp \overline{DE}$; Prove: $\overline{AB} \cong \overline{DE}$. Draw $\overline{OA}, \overline{OB}, \overline{PD}, \overline{PE}$; $\overline{OA} \cong \overline{PD}$ (def $\cong$ ⊙s); $\angle ACO$, $\angle DFP$ rt $\angle$s ($\perp$s form rt $\angle$s); rt $\triangle ACO \cong$ rt $\triangle DFP$ (Th 5.3); $\overline{AC} \cong \overline{DF}$ (CPCTC); $\overline{AB}$ bis by $\overleftrightarrow{OC}$, $\overline{DE}$ bis by $\overleftrightarrow{PF}$ (Th 11.2); $AC = \frac{1}{2} AB$, $DF = \frac{1}{2} DE$ (def of bis), $\overline{AB} \cong \overline{DE}$ (mult and trans prop) **25.** Given: $\odot O$, $\odot P$, ch. $\overline{AB} \cong$ ch $\overline{DE}$, $\overline{OC} \perp \overline{AB}$, $\overline{PF} \perp \overline{DE}$, $\overline{OC} \cong \overline{PF}$; Prove: $\odot O \cong \odot P$. Draw $\overline{AO}, \overline{DP}$; $\angle ACO \cong \angle DFP$ ($\perp$s form $\cong$ rt $\angle$s); $\overline{OC}$ bis $\overline{AB}$ and $\overline{PF}$ bis $\overline{DE}$ (Th 11.2); $AC = \frac{1}{2} AB$, $DF = \frac{1}{2} DE$ (def bis); $AC = DF$ (trans prop); $\triangle ACO \cong \triangle DFP$ (Post 17); $\overline{OA} \cong \overline{DP}$ (CPCTC) $\odot O \cong \odot P$ (def $\cong$ ⊙s)
26. Given: $\odot O$, rad $\overline{OE}$, $\overline{CE} \perp$ rad $\overline{OA}$, $\overline{DE} \perp$ rad $\overline{OB}$, $\overline{CE} \cong \overline{DE}$; Prove: $\overline{OE}$ bisects $\angle AOB$. $\overline{OE} \cong \overline{OE}$ (reflex prop); $\angle OCE$, $\angle ODE$ are rt $\angle$s ($\perp$s form rt $\angle$s); $\triangle OCE \cong \triangle ODE$ (Th 5.6); $\angle AOE \cong \angle BOE$ (CPCTC); $\overline{OE}$ bis $\angle AOB$ (def) **27.** Given: $\triangle ABC$ inscr in $\odot O$, $\overline{OD} \perp \overline{AB}$, $\overline{OE} \perp \overline{BC}$, $\overline{OF} \perp \overline{AC}$, $\overline{OD} \cong \overline{OE} \cong \overline{OF}$; Prove: $\triangle ABC$ is equil. $\overline{AB} \cong \overline{BC} \cong \overline{AC}$ (Th 11.4); $\triangle ABC$ is equil (def)
28. Given: $\odot O$, diam $\overline{AB}$, ch $\overline{BC}$, m $\angle ABC = 30$, $\overline{OD} \perp$ ch $\overline{BC}$; Prove: $OD = \frac{1}{4} AB$. $BO = \frac{1}{2} AB$ (rad of $\odot = \frac{1}{2}$ diam); $(OD = \frac{1}{2} OB)$ (prop 30–60–90△); $OD = \frac{1}{2}(\frac{1}{2} B)$, or $\frac{1}{4} AB$ (subst and mult)

Page 350

10. $\overline{TB} \cong \overline{TC}$ (Th 11.7); Draw $\overline{OB}, \overline{OC}$; $\overline{OB} \cong \overline{OC}$ (Th 11.1); $\overline{TB} \perp \overline{OB}$, $\overline{TC} \perp \overline{OC}$ (Th 11.5); $\angle OBT \cong \angle OCT$ ($\perp$s form $\cong$ rt $\angle$s); $\triangle OBT \cong \triangle OCT$ (Post 17); $\angle BTO \cong \angle CTO$ (CPCTC); $\overrightarrow{TO}$ bis $\angle BTC$ (def $\angle$ bis) **11.** Draw $\overline{OX}$; $\overline{OX} \perp \overline{AB}$ (Th 11.5); $\overline{OX}$ bis $\overline{AB}$ (Th 11.2); $\overline{AX} \cong \overline{BX}$ (def of seg bis) **12.** $\angle CTP$ suppl to $\angle DPT$ (Th 3.2); m $\angle CTP$ + m $\angle DPT = 180$ (def suppl $\angle$s); $\overline{TO}$ bis $\angle CTP$, $\overline{PO}$ bis $\angle DPT$ (ex 10); m $\angle PTO = \frac{1}{2}$m $\angle CTP$, m $\angle TPO = \frac{1}{2}$ m $\angle DPT$ (def $\angle$ bis); m $\angle PTO$ + m $\angle TPO = 90$ (subst, div); $\angle POT$ is a rt $\angle$ (Th 3.5); $\overline{TO} \perp \overline{PO}$ (def of $\perp$ lines)
13. $\overline{AP} \cong \overline{BP}$ (Th 11.7); $\angle PAB \cong \angle PBA$ (Th 5.1); $\overline{OA} \cong \overline{OB}$ (Th 11.1); $\angle OAB \cong \angle OBA$ (Th 5.1); $\overline{OA} \perp \overline{AP}$, $\overline{OB} \perp \overline{BP}$ (Th 11.5); $\angle OAP$, $\angle OBP$ are rt $\angle$s ($\perp$s form rt $\angle$s), m $\angle OAB$ + m $\angle PAB = 90$, m $\angle OBA$ + m $\angle PBA = 90$ ($\angle$add'n, subst); m $\angle OAB$ + m $\angle OBA$ + m $\angle PAB$ + m $\angle PBA = 180$ (add'n); m $\angle P$ + m $\angle PAB$ + m $\angle PBA = 180$ (Th 3.5); m $\angle OAB$ + m $\angle OBA$ + m $\angle PAB$ + m $\angle PBA$ = m $\angle P$ + m $\angle PAB$ + m $\angle PBA$ (trans); m $\angle P$ = m $\angle OAB$ + m $\angle OBA$ (subtr); m $\angle P$ = 2 m $\angle OAB$ (subst) **14.** $\angle OSU$ compl of $\angle OUS$ (ac $\angle$s of rt $\triangle$ are compl); m $\angle OSU$ + m $\angle OUS$ =

90 (def compl); $\angle RSO \cong \angle OSU$, $\angle TUO \cong \angle OUS$ (ex 10); m $\angle RSO$ + m $\angle TUO = 90$ (subst); m $\angle RSO$ + m $\angle OSU$ + m $\angle OUS$ + m $\angle TUO$ = 180 (add'n); m $\angle RSO$ + m $\angle OSU$ = m $\angle RSU$, m $\angle OUS$ + m $\angle TUO$ = m $\angle TUS$ ($\angle$ add'n); m $\angle RSU$ + m $\angle TUS$ = 180 (subst); $\angle RSU$ suppl of $\angle TUS$ (def suppl); $\overleftrightarrow{RS} \parallel \overleftrightarrow{TU}$ (Th 3.4) **15.** $\overline{OA} \cong \overline{OD}$ (Th 11.1); $\angle OAC \cong \angle ODC$ (Th 5.1); $\overline{OD} \perp \overline{DE}$ (Th 11.5); $\angle ODE$ a rt $\angle$ ($\perp$s form rt $\angle$s); $\angle ODC$ compl of $\angle EDC$ (def of compl $\angle$s; $\angle OAC$ compl of $\angle ACO$ (acute $\angle$s of rt $\triangle$ are compl); $\angle ACO \cong \angle EDC$ (Th 2.9); $\angle ACO \cong \angle ECD$ (Th 2.10); $\angle EDC \cong \angle ECD$ (subs) **16.** Let E be any pt on l other than D; draw OE; $OE > OD$ (hypot is longest side of rt $\triangle$); E is in exterior of $\odot O$ (def of ext of $\odot$); l is tangent to $\odot O$ at D (def of tan)

Page 354

28. Draw $\overline{OA}$; $\overline{OA} \perp \overline{AP}$ (Th 11.5); $\angle OAP$ is a rt $\angle$ ($\perp$s form rt $\angle$s); $\angle AOP$ is compl of $\angle P$ (ac $\angle$s of rt $\triangle$ are compl); m $\angle AOP$ + m $\angle P = 90$ (def compl $\angle$s); m $\angle AOP = 90 -$ m $\angle P$ (subt) m $\angle AOP$ = m$\widehat{AB}$ (cent $\angle$ has same deg meas as its int arc); m$\widehat{AB} = 90 - x$ (subst) **29.** Draw $\overline{OC}$, $\overline{OD}$; $\overline{OC} \perp \overline{CP}$, $\overline{OD} \perp \overline{DP}$ (Th 11.5); $\angle$s OCP, ODP are rt $\angle$s ($\perp$s form rt $\angle$s); m $\angle COD + x + 90 + 90 = 360$ (Th 6.1); m $\angle COD = 180 - x$ (subt); m $\angle COD$ = m$\widehat{CD}$; m$\widehat{CD} = 180 - x$ (subst) **30.** $\angle S$ compl of $\angle T$ (ac $\angle$s of rt $\triangle$ are compl), m $\angle S$ + m $\angle T = 90$ (def compl $\angle$s) m$\widehat{AB} = 180 -$ m $\angle S$, m$\widehat{BC} = 180 -$ m $\angle T$ (ex 29); m$\widehat{ABC}$ = m$\widehat{AB}$ + m$\widehat{BC}$ (Post 23); m$\widehat{ABC} = 180 -$ m $\angle S$ + $180 -$ m $\angle T$, or $360 -$ (m $\angle S$ + m $\angle T$) (subst); m$\widehat{ABC} = 360 - 90$, or 270 (subst)

31. $\tan 66° = \frac{BC}{8}$; 18.0 **32.** $\tan \angle AOC = \frac{14}{9}$; m $\angle AOC = 57°$; m$\widehat{AB} = 114$ **33.** m$\widehat{TSQ} = 180$ (meas semi $\odot = 180$); m$\widehat{TSQ}$ = m$\widehat{TS}$ + m$\widehat{SQ}$ (Post 23); m$\widehat{SQ} = 180 -$ m $\angle P$ (ex 29); m$\widehat{TS}$ + $180 -$ m $\angle P = 180$ (subst); m$\widehat{TS}$ = m $\angle P$ (add'n) **34.** Draw $\overline{OA}$, $\overline{QB}$; $\overline{OA} \perp \overline{AB}$, $\overline{QB} \perp \overline{AB}$ (Th 11.5); $\angle s$ OAB, ABQ rt $\angle$s ($\perp$s form rt $\angle$s); m $\angle OAB$ + m $\angle ABQ$ + m $\angle BQO$ + m $\angle QOA = 360$ (Th 6.1); 180 + m $\angle BQO$ + m $\angle QOA = 360$ (subst); m $\angle BQO$ = m$\widehat{BD}$, m $\angle OAB$ = m$\widehat{AC}$ (meas cent $\angle$ = meas int arc); 180 + m$\widehat{AC}$ + m$\widehat{BD} = 360$ (subst); m$\widehat{AC}$ + m$\widehat{BD} = 180$ (subtr) **35.** m$\widehat{RS} = 180 -$ m $\angle RPS$, m$\widehat{ST} = 180 -$ m $\angle TPS$ (ex 29); m $\angle RPT$ = m $\angle RPS$ + m $\angle TPS$ (Post 9); m$\widehat{RS}$ + m$\widehat{ST} = 360 -$ m $\angle RPT$ (add'n, subst)

Page 358

11. $\overline{RS} \cong \overline{RT}$ (Th 11.8); $\triangle RST$ is isos (defn)
12. $\overline{RS} \cong \overline{RT}$ (Th 5.2); $\widehat{RS} \cong \widehat{RT}$ (Th 11.9)
13. $\overline{XZ} \cong \overline{YZ}$ (Th 5.2); $\widehat{XZ} \cong \widehat{YZ}$ (Th 11.9)
14. $\overline{OX} \cong \overline{OY}$ (Th 11.1); $\overline{OZ} \cong \overline{OZ}$ (reflex); $\overline{XZ} \cong \overline{YZ}$ (Th 11.8); $\triangle XOZ \cong \triangle YOZ$ (SSS)
15. $\overline{KL} \cong \overline{LM}$ (Th 11.8); $\angle KOL \cong \angle MOL$ (Th 11.8); m $\angle KOL$ + m $\angle NOK = 180$, m $\angle NOM$ + m $\angle MOL = 180$ (Post 11); $\angle KOL$ suppl to $\angle NOK$, $\angle MOL$ suppl to $\angle NOM$ (def); $\angle NOK \cong \angle NOM$ (Th 2.6); $\overline{NK} \cong \overline{NM}$ (Th 11.10); $\triangle NKL \cong \triangle NML$ (Post 16); $\angle KNL \cong \angle MNL$ (CPCTC)
16. $\overline{NO} \cong \overline{KO} \cong \overline{MO}$ (Th 11.1); $\angle NKO \cong \angle KNL$, $\angle NMO \cong \angle MNL$ (Th 5.1); $\angle NKO \cong \angle NMO$ (trans); $\overline{NO} \cong \overline{NO}$ (reflex); $\triangle NOK \cong \triangle NOM$ (Th 4.2); $\overline{KN} \cong \overline{MN}$ (CPCTC); $\overline{NL} \cong \overline{NL}$ (reflex); $\triangle NKL \cong \triangle NML$ (Post 17); $\overline{KL} \cong \overline{LM}$ (CPCTC); $\widehat{KL} \cong \widehat{ML}$ (Th 11.9) **17.** Given: $\odot O \cong \odot P$, ch $\overline{AB} \cong$ ch $\overline{CD}$; Prove: $\angle O \cong \angle P$, $\widehat{AB} \cong \widehat{CD}$, $\overline{AO} \cong \overline{CP} \cong \overline{OB} \cong \overline{PD}$ (def $\cong$ ⓢ); $\triangle OAB \cong \triangle CPD$ (Post 16); $\angle O \cong \angle P$ (CPCTC); m$\widehat{AB}$ = m $\angle O$, m$\widehat{CD}$ = m $\angle P$ (meas of cent $\angle$ = meas int arc); m$\widehat{AB}$ = m$\widehat{CD}$ (subst); $\widehat{AB} \cong \widehat{CD}$ (def $\cong$ $\widehat{s}$) **18.** Given: $\odot O \cong \odot P$, $\angle AOB \cong \angle CPD$; Prove: $\widehat{AB} \cong \widehat{CD}$ and $\overline{AB} \cong \overline{CD}$. $\overline{AO} \cong \overline{CP} \cong \overline{OB} \cong \overline{DP}$ (def $\cong$ ⓢ); $\triangle AOB \cong \triangle CPO$ (Post 17); $\overline{AB} \cong \overline{CD}$ (CPCTC); $\widehat{AB} \cong \widehat{CD}$ (Th 11.9) **19.** Draw $\overline{OS}$; $\angle SUO \cong \angle SVO$ ($\perp$s form $\cong$ rt $\angle$s); $\angle ROS \cong \angle TOS$ (Th 11.8); $\overline{OS} \cong \overline{OS}$ (reflex); $\triangle OSU \cong \triangle OSV$ (Th 4.2); $\overline{SU} \cong \overline{SV}$ (CPCTC) **20.** Draw $\overline{OS}$. $\angle SUO \cong \angle SVO$ ($\perp$s form $\cong$ rt $\angle$s); $\overline{OS} \cong \overline{OS}$ (reflex); $\triangle OSU \cong \triangle OSV$ (Th 5.3); $\angle ROS \cong \angle TOS$ (CPCTC); $\widehat{RS} \cong \widehat{TS}$ (Th 11.10) **21.** $x\sqrt{3}$
22. $\sin 40 = \frac{x}{5}$; $x = 3.2$; $AC = 6.4$ **23.** Draw $\overline{AD}$, $\overline{BC}$; m$\widehat{AD}$ = m$\widehat{AC}$ + m$\widehat{CD}$, m$\widehat{CB}$ = m$\widehat{CD}$ + m$\widehat{DB}$ (Post 23); m$\widehat{AC}$ + m$\widehat{CD}$ = m$\widehat{DB}$ + m$\widehat{CD}$ (add); m$\widehat{AD}$ = m$\widehat{CB}$ (subst); $\overline{AD} \cong \overline{CB}$, $\overline{AC} \cong \overline{BD}$ (Th 11.8); $\overline{CD} \cong \overline{CD}$ (refl); $\triangle ACD \cong \triangle BDC$ (Post 16); $\angle ACD \cong \angle BDC$ (CPCTC) **24.** Draw $\overline{OC}$, $\overline{OD}$, $\overline{OC} \cong \overline{OD}$ (Th 11.1); $\angle DCO \cong \angle CDO$ (Th 5.1); $\angle DCO \cong \angle COB$, $\angle CDO \cong \angle DOA$ (Post 13); m $\angle AOC$ = m $\angle AOD$ + m $\angle DOC$, m $\angle DOB$ = m $\angle BOC$ + m $\angle DOC$ (Post 10); m $\angle AOC$ = m $\angle DOB$ (add'n, subst); $\overline{AC} \cong \overline{DB}$ (Th 11.10) **25.** Draw $\overline{DO}$; $\angle EOD \cong \angle D$ (Post 13); $\angle AOE \cong \angle B$ (Th 3.1); $\overline{DO} \cong \overline{BO}$ (Th 11.1); $\angle D \cong \angle B$ (Th 5.1); $\angle AOE \cong \angle EOD$ (subst); $\widehat{AE} \cong \widehat{ED}$ (Th 11.10)

Page 361

Reading (Answers may vary.)

Page 363

25. $\overline{OQ} \cong \overline{OR}$ (Th 11.1); $\angle PRQ \cong \angle SQR$ (Th 5.1)
26. m $\angle ABD = \frac{1}{2}$ m$\widehat{AD}$ (Th 11.11); m $\angle DOC$ = m$\widehat{DC}$ (def of $\frown$ meas); $\frac{1}{2}$ m$\widehat{AD} = \frac{1}{2}$ m$\widehat{DC}$ (subst); m$\widehat{AD}$ = m$\widehat{DC}$ (mult); $\widehat{AD} \cong \widehat{DC}$ (def $\cong$ $\widehat{s}$)
27. $\widehat{RU} \cong \widehat{ST}$ (Th 11.11, cor); $\widehat{UT} \cong \widehat{UT}$ (reflex); m$\widehat{RU}$ + m$\widehat{UT}$ = m$\widehat{RT}$, m$\widehat{ST}$ + m$\widehat{TU}$ = m$\widehat{SU}$

(Post 23); $m\widehat{RU} + m\widehat{UT} = m\widehat{ST} + m\widehat{TU}$ (addn); $m\widehat{RT} = m\widehat{SU}$ (subst); $\widehat{RT} \cong \widehat{SU}$ (def $\cong$ s); $\overline{RT} \cong \overline{SU}$ (Th 11.8) **28.** $\angle ACB$ is a rt $\angle$ (Th 11.11, cor); $\triangle ADC \sim \triangle ACB$ (Th 9.1) **29.** Follow ex 28; then $\frac{AD}{AC} = \frac{CD}{CB}$ (def $\sim$ poly); $AD \cdot CB = AC \cdot CD$ (prop of propor) **30.** Draw diam $\overline{PC}$; $m \angle APC = \frac{1}{2} m\widehat{AC}$, $m \angle CPB = \frac{1}{2} m\widehat{CB}$ (Case I); $m \angle APC + m \angle CPB = \frac{1}{2} m\widehat{AC} + \frac{1}{2} m\widehat{CB}$ (addn); $m \angle APC + m \angle CPB = m \angle APB$ (Post 10); $m \angle APB = \frac{1}{2} m(\widehat{AC} + \widehat{CB})$ (subst); $m\widehat{AC} + m\widehat{CB} = m\widehat{AB}$ (Post 23); $m \angle APB = \frac{1}{2} m\widehat{AB}$ (subst) **31.** Draw diam $\overline{PC}$; $m \angle CPB = \frac{1}{2} m\widehat{CB}$, $m \angle APC = \frac{1}{2} m\widehat{CA}$ (Case I); $m \angle CPB - m \angle APC = \frac{1}{2} m\widehat{CB} - \frac{1}{2} m\widehat{CA}$ (subtr); $m \angle CPB - m \angle APC = m \angle APB$ (Post 10 and subtr.); $m \angle APB = \frac{1}{2}(m\widehat{CB} - m\widehat{CA})$ (subst); $m\widehat{CB} - m\widehat{CA} = m\widehat{AB}$ (Post 23); $m \angle APB = \frac{1}{2} m\widehat{AB}$ (subst) **32.** $m \angle CAD = \frac{1}{2} m\widehat{CD}$, $m \angle CBD = \frac{1}{2} m\widehat{CD}$ (Th 11.11); $m \angle CAD = m \angle CBD$ (subst); $\angle CAD \cong \angle CBD$ (def $\cong$ $\angle$s) **33.** $m \angle ABC = \frac{1}{2} m\widehat{AC}$ (Th 11.11); $m\widehat{AC} = 180$ (semi $\odot$ measures 180); $m \angle ABC = \frac{1}{2}(180)$, or 90 (subst); $\angle ABC$ is a rt $\angle$ (an $\angle$ of 90° is rt $\angle$) **34.** See ex 32 (proof of Cor 1) **35.** Given: $\odot O$, $\angle ABC$ inscr in maj $\widehat{ABC}$; Prove $m \angle ABC < 90$. $m\widehat{ABC} > 180$ (def maj $\frown$); $m\widehat{ABC} + m\widehat{AC} = 360$ ($\odot$ contains 360°); $m\widehat{AC} < 180$ (subtr); $m \angle ABC = \frac{1}{2} m\widehat{AC}$ (Th 11.11); $m \angle \widehat{ABC} < 90$ (subst) **36.** Given: $\odot O$, $\angle ABC$ inscr in minor $\widehat{ABC}$; Disprove $m \angle ABC < 90$. $m\widehat{ABC} < 180$ (def minor $\frown$); $m\widehat{ABC} + m\widehat{AC} = 360$ ($\odot$ cont 360°); $m\widehat{AC} > 180$ (subt); $m \angle ABC = \frac{1}{2} m\widehat{AC}$ (Th 11.11); $m \angle ABC > 90$ (subst)
37. Given: $\odot O \cong \odot P$, $\triangle ABC$ inscr in $\odot O$, $\triangle DEF$ inscr in $\odot P$, $\angle A \cong \angle D$, $\angle B \cong \angle E$; Prove: $\triangle ABC \cong \triangle DEF$. $m \angle B = \frac{1}{2} m\widehat{AC}$, $m \angle E = \frac{1}{2} m\widehat{DF}$ (Th 11.11); $\frac{1}{2} m\widehat{AC} = \frac{1}{2} m\widehat{DF}$ (subst); $m\widehat{AC} = m\widehat{DF}$ (mult); $\widehat{AC} \cong \widehat{DF}$ (def $\cong$ $\frown$s); $\overline{AC} \cong \overline{DF}$ (Th 11.8); $\triangle ABC \cong \triangle DEF$ (Th 4.2) **38.** Given: $\odot O$, $\widehat{AC} \cong \widehat{BD}$; Prove: $AB \parallel CD$. $m \angle 1 = \frac{1}{2} m\widehat{BD}$; $m \angle 2 = \frac{1}{2} m\widehat{AC}$ (Th 11.11); $\frac{1}{2} m\widehat{BD} = \frac{1}{2} m\widehat{AC}$ (div prop); $m \angle 1 \cong m \angle 2$ (trans prop); $AB \parallel CD$(Post 13) **39.** false, rectangle **40.** False. Given $\odot O$ and chord $\overline{AB}$. Consider $\triangle$'s ABC, with vertex C in any position on $\odot$ other than pts A and B. $\triangle$'s meet given conditions, but are not $\cong$.

Page 367

21. $m \angle STQ = 90$ ($\perp$s form rt $\angle$s); $m \angle STQ = \frac{1}{2}(m\widehat{PR} + m\widehat{QS})$ (Th 11.12); $90 = \frac{1}{2}(m\widehat{PR} + m\widehat{QS})$ (subst); $m\widehat{PR} + m\widehat{QS} = 180$ (mult) **22.** Draw $\overline{UR}$ and $\overline{TS}$ (two pts determine exactly one line); $\widehat{RT} \cong \widehat{URS}$ (Th 11.9); $\angle RUT \cong \angle UTS$, $\angle URT \cong \angle UST$ (Th 11.11, Cor); $\overline{UT} \cong \overline{UT}$ (reflex); $\triangle URT \cong \triangle TSU$ (Th 4.2); $\overline{RU} \cong \overline{ST}$ (CPCTC) **23.** $\angle N \cong \angle Q$ (Th 11.11, Cor); $\angle L \cong \angle L$ (reflex); $\triangle LQM \sim \triangle LNP$ (Post 21) **24.** $m \angle y = \frac{1}{2} m\widehat{ABC} - \frac{1}{2} m\widehat{DE}$ (Th 11.13); $m \angle x = \frac{1}{2} m\widehat{AB} + \frac{1}{2} m\widehat{DE}$ (Th 11.12); $m \angle x + m \angle y = \frac{1}{2} m\widehat{ABC} + \frac{1}{2} m\widehat{AB}$ (add'n); $m\widehat{ABC} = m\widehat{AB} + m\widehat{BC}$ (Post 23); $m \angle x + m \angle y = \frac{1}{2} m\widehat{AB} + \frac{1}{2} m\widehat{BC} + \frac{1}{2} m\widehat{AB}$ (subst); $m \angle x + m \angle y = \frac{3}{2} m\widehat{AB}$ (subst, addn) **27.** Given: $\odot O$ and $\odot P$ tang at T, O lies on $\odot P$, chord $\overline{AT}$ int $\odot P$ at B; Prove: $\overline{TB} \cong \overline{AB}$. $\overline{OT}$ is diam of $\odot P$ ($\overline{OT}$ and $\overline{PT}$ are $\perp$ to same line; Th 11.5) Draw $\overline{OB}$; $\angle OBT$ is a rt $\angle$ (Th 11.11, cor); $\overline{OB} \perp \overline{AT}$ (def of $\perp$ lines) OB bis $\overline{AT}$ (Th 11.2); $\overline{TB} \cong \overline{AB}$ (def)

Page 371

24. Draw $\overline{AB}$; $m \angle z = m \angle x + m \angle y$ (Th 4.1); $m \angle x = m \angle z - m \angle y$ (subtr); $m \angle z = \frac{1}{2} m\widehat{AB}$ (Th 11.14); $m \angle y = \frac{1}{2} m\widehat{AC}$ (Th 11.11); $m \angle x = \frac{1}{2} m\widehat{AB} - \frac{1}{2} m\widehat{AC}$ (subst); $m \angle x = \frac{1}{2}(m\widehat{AB} - m\widehat{AC})$ (distrib prop) **25.** Draw $\overline{AB}$; $m \angle z = m \angle x + m \angle y$ (Th 4.2); $m \angle x = m \angle z - m \angle y$ (subtr); $m \angle z = \frac{1}{2} m\widehat{ACB}$, $m \angle y = \frac{1}{2} m\widehat{AB}$ (Th 11.14); $m \angle x = \frac{1}{2} m\widehat{ACB} - \frac{1}{2} m\widehat{AB}$ (subst); $m \angle x = \frac{1}{2}(m\widehat{ACB} - m\widehat{AB})$ (distrib prop) **26.** $\widehat{US} \cong \widehat{VS}$ (Th 11.9); $m \angle RSU = \frac{1}{2} m\widehat{US}$ (Th 11.14); $m \angle SUV = \frac{1}{2} m\widehat{VS}$ (Th 11.1); $\frac{1}{2} m\widehat{US} = \frac{1}{2} m\widehat{VS}$ (div prop); $m \angle RSU = m \angle SUV$ (subst); $\angle RSU \cong \angle SUV$ (def $\cong$ $\angle$s); $\overline{UV} \parallel \overline{RT}$ (Post 14)
27. $\angle UVS \cong \angle VST$ (Post 13); $m \angle UVS = m \angle VST$ (def $\cong$ $\angle$s); $m \angle UVS = \frac{1}{2} m\widehat{US}$ (Th 11.11); $m \angle VST = \frac{1}{2} m\widehat{VS}$ (Th 11.14); $\frac{1}{2} m\widehat{US} = \frac{1}{2} m\widehat{VS}$ (subst); $m\widehat{US} = m\widehat{VS}$ (mult); $\widehat{US} \cong \widehat{VS}$ (def $\cong$ $\widehat{S}$ in same $\odot$); $\overline{US} \cong \overline{VS}$ (Th 11.8) **28.** Given: $\overleftrightarrow{PA}$, $\overleftrightarrow{TB}$ tan to $\odot U$ at A and B, $\overleftrightarrow{PA} \parallel \overleftrightarrow{TB}$; Prove: $\widehat{ACB} \cong \widehat{ADB}$. Draw sec $\overleftrightarrow{AB}$; $\angle 1 \cong \angle 2$ (Post 13); $m \angle 1 = \frac{1}{2} m\widehat{ACB}$, $m \angle 2 = \frac{1}{2} m\widehat{ADB}$ (Th 11.14); $\frac{1}{2} m\widehat{ACB} = \frac{1}{2} m\widehat{ADB}$ (subst); $m\widehat{ACB} = m\widehat{ADB}$ (mult); $\widehat{ACB} \cong \widehat{ADB}$ (def $\cong$ $\frown$s)
29. Given: $\odot O$, $\odot Q$ internally tan at P, $\overleftrightarrow{PE}$ the common tan, $\overline{PA}$ int $\odot Q$ at B; Prove: $m\widehat{PA} = m\widehat{PB}$. $m \angle APE \cong m \angle BPE$ (reflex); $m \angle APE = \frac{1}{2} m\widehat{PA}$, $m \angle BPE = \frac{1}{2} \widehat{PB}$ (Th 11.14); $\frac{1}{2} m\widehat{PA} = \frac{1}{2} m\widehat{PB}$ (subst); $m\widehat{PA} = m\widehat{PB}$ (mult) **30.** Given: $\overleftrightarrow{PA}$ tan to $\odot O$ at P, $\overrightarrow{PC}$ bisector of $\angle BPA$; Prove: $\widehat{BC} \cong \widehat{CP}$. $\angle 1 \cong \angle 2$ (def bis); $m \angle 1 = \frac{1}{2} m\widehat{BC}$ (Th 11.11); $m \angle 2 = \frac{1}{2} m\widehat{CP}$ (Th 11.14); $\frac{1}{2} m\widehat{BC} = \frac{1}{2} m\widehat{CP}$ (subst); $m\widehat{BC} = m\widehat{CP}$ (mult); $\widehat{BC} \cong \widehat{CP}$ (def $\cong$ $\frown$s) **31.** $\angle LMS \cong \angle RMN$ (Th 2.10); $m \angle LMS = \frac{1}{2} m\widehat{LM}$, $m \angle RMN = \frac{1}{2} m\widehat{NM}$ (Th 11.14); $\frac{1}{2} m\widehat{LM} = \frac{1}{2} m\widehat{NM}$ (subst); $m\widehat{LM} = m\widehat{NM}$ (mult) **32.** Follow Ex 31; then $m\widehat{LMK} = 180$, $m\widehat{QMN} = 180$ ($180 =$ meas of semi$\odot$); $m\widehat{LMK} = m\widehat{QMN}$ (subst); $m\widehat{LMK} = m\widehat{LM} + m\widehat{MK}$; $m\widehat{QMN} = m\widehat{QM} + m\widehat{MN}$; $m\widehat{KM} = m\widehat{QM}$ (subt); $m \angle L = \frac{1}{2} m\widehat{KM}$; $m \angle N = \frac{1}{2} m\widehat{QM}$ (Th 11.11);

m $\angle L =$ m $\angle N$ (subst); $\angle L \cong \angle N$ (def'n $\cong \angle$'s; $\overline{KL} \parallel \overline{QN}$ (Post 14) **33.** Given: $ABCD$ inscr in $\odot O$, $\angle EAD$ is an ext $\angle$ of $ABCD$; Prove: $\angle EAD \cong \angle C$. $\angle BCD$ suppl to $\angle BAD$ (Th 11.11, cor 4); m $\angle EAD +$ m $\angle BAD = 180$ (Post 12); $\angle EAD$ suppl to $\angle BAD$ (defn suppl $\angle$s); $\angle EAD \cong \angle C$ (Th 2.8) **34.** Given: $\odot O$ and $\odot P$ int at R and S, $\overline{RS}$ the common chord, $\overline{SU}$, $\overline{ST}$ diams; Prove: $\angle R$ is a st $\angle$. $\angle$s 1 and 2 are rt $\angle$s ($\angle$ inscr in semi$\odot$ is a rt $\angle$); m $\angle 1 = 90$, m $\angle 2 = 90$ (def of rt $\angle$); m $\angle 1 +$ m $\angle 2 = 180$ (add'n); m $\angle URT =$ m $\angle 1 +$ m $\angle 2$ (Post 10); m $\angle URT = 180$ (subst); $\angle URT$ is a st $\angle$ (def of st $\angle$) **35.** Given: $\odot O$, $\odot Q$, ext tang at P, $\overleftrightarrow{PT}$ tan to $\odot O$ and $\odot Q$ at P, $\overleftrightarrow{AB}$ a common ext tang to $\odot O$ and $\odot Q$, chs $\overline{AP}$, $\overline{PB}$; Prove: $\angle APB$ is a rt $\angle$. $\overline{TA} \cong \overline{TP}$, $\overline{TB} \cong \overline{TP}$ (Th 11.7); $\overline{TA} \cong \overline{TP} \cong \overline{TB}$ (trans); T is center of a $\odot$ with rad. $\overline{AT}$, $\overline{PT}$, $\overline{BT}$ (def of $\odot$); $\angle APB$ is a rt $\angle$ ($\angle$ inscr in semi$\odot$ is a rt $\angle$) **36.** Given: $\odot O$, $\odot Q$ tan internally at P, $\overleftrightarrow{PT}$ the common ext tan, $\overline{PA}$, $\overline{PB}$ chords of $\odot Q$, intersecting $\odot O$ at C and D; Prove: $\overline{AB} \parallel \overline{CD}$. m $\angle 1 = \frac{1}{2}$m$\widehat{DP}$, m $\angle 1 = \frac{1}{2}$m$\widehat{BP}$ (Th 11.14); $\frac{1}{2}$m$\widehat{DP} = \frac{1}{2}m\widehat{BP}$ (subst); m $\angle 2 = \frac{1}{2}$m$\widehat{DP}$, m $\angle 3 = \frac{1}{2}$m$\widehat{BP}$ (Th 11.11); m $\angle 2 =$ m $\angle 3$ (subst); $\angle 2 \cong \angle 3$ (def'n); $\overline{AB} \parallel \overline{CB}$ (Th 3.3)

Page 376

16. Draw $\overline{AD}$ and $\overline{BC}$; $\angle P \cong \angle P$ (reflex); $\angle x \cong \angle y$ (Th 11.11, Cor); $\triangle PAD \sim \triangle PCB$ (Post 22); $\frac{PA}{PC} = \frac{PD}{PB}$ (def $\sim$ poly), $PA \cdot PB = PC \cdot PD$ (prop of propor) **17.** Draw $\overline{AB}$, $\overline{AC}$; $\angle P \cong \angle P$ (reflex); m $\angle PAC = \frac{1}{2}$m$\widehat{AC}$ (Th 11.14); m $\angle B = \frac{1}{2}$m$\widehat{AC}$ (Th 11.11); m $\angle PAC =$ m $\angle B$ (subst); $\angle PAC \cong \angle B$ (def $\cong \angle$'s); $\triangle PAB \sim \triangle PCA$ (Post 22); $\frac{PA}{PB} = \frac{PC}{PA}$ (def $\sim$ poly); $(PA)^2 = PB \cdot PC$ (prop of propor)
18. $\angle T \cong \angle T$ (reflex); m $\angle YXT = \frac{1}{2}$m$\widehat{XY}$ (Th 11.14); m $\angle XZT = \frac{1}{2}$m$\widehat{XY}$ (Th 11.11); m $\angle YXT =$ m $\angle XZT$ (subst); $\angle YXT \cong \angle XZT$ (def); $\triangle TXY \sim \triangle TZX$ (Post 22); $\frac{XT}{YT} = \frac{XZ}{XY}$ (def $\sim$ poly); $XT \cdot XY = XZ \cdot YT$ (prop of propor)
19. $\angle T \cong \angle T$ (reflex); $\angle K \cong \angle M$ (Th 11.11, Cor); $\triangle KLT \sim \triangle MJT$ (Post 22); $\frac{KL}{MJ} = \frac{TL}{TJ}$ (def $\sim$ poly); $KL \cdot TJ = TL \cdot MJ$ (prop of propor) **20.** $\overline{GH} \perp \overline{JH}$ (Th 11.5); $\angle GHJ$ a rt $\angle$ ($\perp$s form rt $\angle$s); $\triangle GHJ$ a rt $\triangle$ (defn); $\angle GIH$ is a rt $\angle$ (Th 11.11, Cor); $\overline{HI}$ is alt to hyp $\overline{GJ}$; $\frac{IG}{IH} = \frac{IH}{IJ}$ (Th 10.1, Cor); $(IH)^2 = IJ \cdot IG$ (prop of propor); $IH = \sqrt{IJ \cdot IG}$ (alg)
21. $(TM)^2 = MS \cdot MP$, $(TM)^2 = MR \cdot MQ$ (Th 11.19); $MS \cdot MP = MR \cdot MQ$ (subst)
22. $PA \cdot PB = PD \cdot PE$, $PB \cdot PC = PE \cdot PF$ (Th 11.18); $\frac{PA \cdot PB}{PB \cdot PC} = \frac{PD \cdot PE}{PE \cdot PF}$ (div); $\frac{PA}{PC} = \frac{PD}{PF}$ (alg) **23.** Draw $\overline{AB}$; m $\angle DGE = \frac{1}{2}$(m$\widehat{ED}$ + m$\widehat{BC}$); m $\angle ABD = \frac{1}{2}$m$\widehat{AED}$ (Th 11.11); m$\widehat{AED}$ = m$\widehat{AE}$ + m$\widehat{ED}$ (Post 23); m $\angle ABD = \frac{1}{2}$(m$\widehat{AE}$ + m$\widehat{ED}$) (subst); m $\angle ABD = \frac{1}{2}$(m$\widehat{BC}$ + m$\widehat{ED}$) (subst); m $\angle DGE =$ m $\angle ABD$ (subst); $\angle DGE \cong \angle ABD$ (def'n); $\angle D \cong \angle D$ (reflex); $\triangle DGF \sim \triangle DBA$ (Post 21); $\frac{AD}{BD} = \frac{GD}{FD}$ (def $\sim$ poly); $AD \cdot FD = BD \cdot GD$ (prop of propor)

Page 379

Exercises **1.** Open compass to measure the 2 cm seg; use this rad and construct a $\odot$. **2.** Construct $\perp$ bisector of 5 cm seg; use mdpt as center, half the length of seg as rad; draw $\odot$ **3.** Use rad 3 cm; draw $\odot$; draw a rad of $\odot$, and extend it; constr $\perp$ to rad at its endpt on $\odot$. **4.** Use rad 2 cm; draw $\odot$; then follow steps on page 377
5a. Constr a $\odot$; with same rad, mark off 6 consec $\cong$ arcs; from each pt as ctr, with same rad, draw the 6 arcs in $\odot$. **5b.** Constr a $\odot$ and mark off 6 $\cong$ $\widehat{s}$ as in 5a; draw 3 diams (use alternate pts); bisect a rad; use $\frac{1}{2}$ r as rad and mdpts of radii as ctrs, draw 3 Ⓢ. **6.** Follow steps on p. 378 to inscr hexagon; connect alt vertices of hex to get inscr equil $\triangle$.
7. Constr $\perp$ to seg at given pt; measure given seg with compass; mark off this length on $\perp$; use this pt as ctr, and same rad to draw $\odot$. **8.** Follow steps on p. 378. **9.** Draw diags; use pt of int as ctr, $\frac{1}{2}s$ as rad, and draw $\odot$. **10.** In $\odot O$, draw diam $\overline{AC}$. Constr $\perp$ to $\overline{AC}$ at O, intersecting $\odot$ at pts B and D. Draw chords $\overline{AB}$, $\overline{BC}$, $\overline{CD}$, and $\overline{DA}$. **11.** Follow constr of tans to $\odot$ from ext pt, p 377. Let R be the pt of int of $\overline{OP}$ with $\odot O$. Construct $\overleftrightarrow{RQ} \perp \overline{OP}$, int $\overline{TP}$ at Q and $\overline{SP}$ at N. Construct $\overrightarrow{QV}$, the bis of $\angle PQN$. Call the int of $\overrightarrow{QV}$ with $\overline{OP}$ pt U. With U as ctr, and $\overline{UR}$ as radius, draw required $\odot$.
12. Meas of shorter seg must be $\leq \frac{1}{2}$ meas longer seg, otherwise constr impossible. Constr seg $\overline{AB} \cong$ longer seg. Constr $\perp$ bis of $\overline{AB}$; call int with $\overline{AB}$ pt O. With O as ctr and $\overline{OA}$ as rad, draw semi $\odot$ $\widehat{AB}$. Mark off $\overline{OD} \cong$ shorter seg on $\perp$ bis of $\overline{AB}$ (D in inter of semi $\odot$). Thru D, constr $\overleftrightarrow{ED} \parallel \overline{AB}$, intersecting semi $\odot$ at pts F and G. Connect pts A and B either to F or G. $\triangle AFB$ and $\triangle AGB$ both meet given cond. **13.** Inscr a sq in a $\odot$ (ex 10). Constr $\perp$ bis of each side of sq, giving 4 pts of int with $\odot$. Draw 8 chords connecting 8 adj pts on $\odot$.
14. Follow instr on p. 378. Connect opp vertices. 6 $\cong$ equilat $\triangle$s result. **15.** Inscr reg hex; constr $\perp$ bis of each side; where these int $\odot$, connect pts

to vert of hex. **16.** Follow instr on p. 378; draw $\overline{CF}, \overline{FE}; \frac{CD}{CF} = \frac{CF}{DE}$ (Th 10.1, cor) **17.** Pts P, D, F, noncoll. Through a pt P, draw $\overrightarrow{PD}$ and $\overrightarrow{PF}$; on $\overrightarrow{PD}$, mark $PA = a$, $AB = b$; on $\overrightarrow{PF}$ mark $PC = c$; draw $\overline{AC}$; constr $\overleftrightarrow{BE} \parallel \overline{AC}$, to int $\overrightarrow{PF}$ at E; $CE = x$ **18.** Draw semi ⊙, diam $\overline{AOB}$. At A, constr $\overrightarrow{AX} \perp \overline{AB}$. On $\overrightarrow{AX}$, mark off $AC = AB$. Draw $\overline{OC}$, int semi ⊙ at pt P. Thru P, draw $\overline{PQ} \parallel \overline{AB}$, int semi ⊙ at pt Q. From P and from Q, constr ⊥'s to $\overline{AB}$, inter at pts S and R, respectively. $PQRS$ is required sq. **19.** Follow steps on p. 378; $x = \sqrt{ab}$ **20.** Repeat Ex 19, replace a by $2a$ **21.** Ex 19 with a replaced by $4a$ **22.** Ex 19 with a replaced by $\frac{1}{2}a$ **23.** Ex 19 with b replaced by $\frac{1}{2}a$ **24.** Constr rt △ with legs of length a, b, hypot $= \sqrt{a^2 + b^2}$

Page 382

19. Follow steps p. 377 **20.** Constr seg $= c$; bis seg; constr ⊙ with = diam; from endpt of diam use rad a to find pt on ⊙; connect pt to other end of diam; this is required seg. **21.** $\odot O \cong \odot P$ (def); $\overline{OA} \cong \overline{OB}$, $\overline{PC} \cong \overline{PD}$ (Th 11.1); $\angle A \cong \angle B$, $\angle C \cong \angle D$ (Th 5.1); $\angle B \cong \angle D$ (subst); $\angle O \cong \angle P$ (Th 3.6) $\widehat{AB} \cong \widehat{CD}$ (Th 11.10) **22.** Given: $\odot O$ int $\odot P$ at D, E; $\overleftrightarrow{AB}$ common tan; $\overleftrightarrow{ED}$ int $\overline{AB}$ at C; Prove: $\overline{AC} \cong \overline{BC}$. $(AC)^2 = CE \cdot CD$, $(BC)^2 = CE \cdot CD$ (Th 11.19); $(AC)^2 = (BC)^2$ (subst); $AC = BC$ (alg); $\overline{AC} \cong \overline{BC}$ (def'n)

Page 383

13. Follow instr on p. 378 with $\frac{1}{2}a$ for a; $\frac{1}{2}b$ for b. **14.** $\angle ABC$, $\angle ADC$ are rt ∠s (Th 11.11); $\overline{DC} \cong \overline{AB}$ (Th 11.8); m $\angle DAC =$ m $\angle BCA$ (Th 11.11); $\triangle ABC \cong \triangle CDA$ (Th 4.2); $\angle DCA \cong \angle BAC$ (CPCTC); $\overline{DC} \parallel \overline{AB}$ (Post 14); $ABCD$ is a ▱ (Th 7.5); $ABCD$ is a rect (p 230, ex 2) **15.** $\overline{XZ} \cong \overline{TZ} \cong \overline{YZ}$ (Th 11.7); $XZ = YZ$, $TX = TY$ (Def of ≅ seg); $\overline{TZ}$ is ⊥ bis $\overline{XY}$ (Th 5.8)

Page 390

32. $A = bh$ and $A' = b'h'$ (Post 24); $\frac{A}{A'} = \frac{bh}{b'h'}$ (div prop) **33.** $A = bh$, $A' = b'h'$ (Post 24); $A' = b'h$ (subst); $\frac{A}{A'} = \frac{bh}{b'h}$ (div); $\frac{A}{A'} = \frac{b}{b'}$ (alg) **34,** $A = bh$, $A' = (2b)h$ (Post 24); $A' = 2(bh)$ (alg); $A' = 2A$ (subst) **35.** $A = s^2$, $A' = (2s)^2$ (cor); $A' = 4s^2$ (alg); $A' = 4A$ (subst) the statement is disproved.

Page 394

22. Given rt $\triangle ABC$, alt h, base b; Prove Area $\triangle ABC = \frac{1}{2}bh$. Draw $\overleftrightarrow{BD} \parallel \overleftrightarrow{AC}$, $\overleftrightarrow{AD} \parallel \overleftrightarrow{BC}$ (Post 15); $ABCD$ is a rectangle (def of rect); Area $ABCD = bh$ (Post 24); $\triangle ABC \cong \triangle BAD$ (Th 7.1); Area $\triangle ABC =$ Area $\triangle BAD$ (Post 26); area $ACBD =$ Area $ABC +$ Area BAD (Post 25); Area $ABCD = 2 \cdot$ Area $\triangle ABC$ (subst); Area $\triangle ABC = \frac{1}{2}$ Area $ACBD$ (mult prop); Area $\triangle ABC = \frac{1}{2}bh$ (subst) **23.** $\overline{DE} \perp \overline{AC}$, $DE = \frac{1}{2}DB$ (diags of rhomb ⊥ bis of each other); Area $\triangle ADC = \frac{1}{2}DE \cdot d_2$ (Th 12.2); Area $\triangle ADC = \frac{1}{2}(\frac{1}{2}d_1)\, d_2$ or $\frac{1}{4}d_1d_2$ (subst); $\triangle ADC \cong \triangle CBA$ (Th 7.1); Area $\triangle ADC =$ Area $\triangle CBA$ (Post 26); Area $ABCD =$ Area $ADC +$ Area CBA (Post 27); Area $ABCD = 2 \cdot$ Area ADC (subst); Area $ABCD = 2 \cdot \frac{1}{4}d_1d_2$ (subst); Area $ABCD = \frac{1}{2}d_1d_2$ (alg); **24.** m $\angle A =$ m $\angle B =$ m $\angle C = 60$ (Th 5.1, Cor 2); $\triangle ADC$ is 30-60-90 △ (def); $h = \frac{1}{2}s\sqrt{3}$ (Th 10.5); Area $ABC = \frac{1}{2}sh$ (Th 12.2); Area $ABC = \frac{1}{2}s \cdot \frac{s}{2}\sqrt{3}$ (subst); Area $ABC = \frac{s^2}{4}\sqrt{3}$ (alg) **25.** Given sq $ABCD$, $AC = d$, Prove Area $ABCD = \frac{1}{2}d^2$; $\overline{AC} \cong \overline{DB}$ (diags of sq ≅); $DB = d$ (subst); $ABCD$ is a rhombus (def); Area $ABCD = \frac{1}{2}AC \cdot DB$ (Th 12.3); Area $ABCD = \frac{1}{2}d^2$ (subst) **26.** $s = \frac{1}{2}(5 + 12 + 13)$; $s = 15$; $A = \sqrt{15(15 - 5)(15 - 12)(15 - 13)}$; 30 cm² **27.** $s = \frac{1}{2}(4 + 5 + 6)$; $s = 7\frac{1}{2}$; $A = \sqrt{\frac{15}{2} \cdot \frac{7}{2} \cdot \frac{5}{2} \cdot \frac{3}{2}}$; $A = \frac{15}{4}\sqrt{7}$; $\frac{15}{4}\sqrt{7}$ m² **28.** $s = \frac{1}{2}(4 + 5 + 8)$; $s = 8\frac{1}{2}$; $A = \sqrt{\frac{17}{2} \cdot \frac{9}{2} \cdot \frac{7}{2} \cdot \frac{1}{2}}$; $A = \frac{3}{4}\sqrt{119}$; $\frac{3}{4}\sqrt{119}$ km² **29.** Given $\triangle ACB$, base b, alt h, Prove Area $\triangle ACB = \frac{1}{2}bh$. Area $\triangle ACB +$ Area $\triangle CDA =$ Area $\triangle ADB$ (Post 27); Area $\triangle ADC = \frac{1}{2}xh$, Area $\triangle ADB = \frac{1}{2}h(b + x)$ (Th 12.1); Area $\triangle ACB + \frac{1}{2}xh = \frac{1}{2}(b + x)h$ (subst); Area $\triangle ACB = \frac{1}{2}(b + x)h - \frac{1}{2} \times h$ (subst); Area $\triangle ACB = \frac{1}{2}bh$ (alg) **30.** Given $\triangle ABC \sim \triangle A'B'C'$, Prove $\frac{\text{Area } ABC}{\text{Area } A'B'C'} = \frac{b^2}{b'^2}$; Area $ABC = \frac{1}{2}bh$, Area $A'B'C' = \frac{1}{2}b'h'$ (Th 12.2); $\frac{b}{b'} = \frac{h}{h'}$ (Th 9.6); $\frac{\text{Area } ABC}{\text{Area } A'B'C'} = \frac{\frac{1}{2}bh}{\frac{1}{2}b'h'}$ (div prop); $\frac{\text{Area } ABC}{\text{Area } A'B'C'} = \frac{b}{b'} \cdot \frac{b}{b'}$ or $\frac{b^2}{b'^2}$ (subst) **31.** False. Same as Ex 30, except that the last step becomes $\frac{\text{Area } ABC}{\text{Area } A'B'C'} = \frac{h}{h'} \cdot \frac{h}{h'}$ or $\frac{h^2}{h'^2}$, not $\frac{h}{h'}$ **32.** Counterexample: $b_1 = 8$, $h_1 = 12$, Area $\triangle_1 = \frac{1}{2} \cdot 8 \cdot 12 = 48$; $b_2 = 24$, $h_2 = 4$, Area $\triangle_2 = \frac{1}{2} \cdot 24 \cdot 4 = 48$; $\frac{b_1}{b_2} = \frac{8}{24}$, or $\frac{1}{3}$, $\frac{h_1}{h_2} = \frac{12}{4}$, or 3; $\frac{1}{3} \neq 3$, so $\frac{b_1}{b_2} \neq \frac{h_1}{h_2}$ **33.** Given: $\triangle ABC$, Alt h, base b, $\triangle A'B'C'$, alt h'; base b', $b = b'$. Prove: $\frac{\text{Area } ABC}{\text{Area } A'B'C'} = \frac{h}{h'}$. Area $ABC = \frac{1}{2}bh$, Area $A'B'C' =$

$\frac{1}{2}b'h'$ (Th 12.2); $\frac{\text{Area } ABC'}{\text{Area } A'B'C'} = \frac{\frac{1}{2}bh}{\frac{1}{2}b'h'}$ (div), $\frac{\text{Area } ABC}{\text{Area } A'B'C'} = \frac{h}{h'}$ (alg) **34.** Given $\triangle ABC$, $\overrightarrow{CD}$ bis $\angle C$, $\overline{BA} \perp \overline{AC}$, Compare Areas $\triangle ADC$ and $\triangle BDC$; Draw $\overline{DE} \perp \overline{BC}$; $\overline{DE} \cong \overline{DA}$; $\triangle ADC \cong \triangle EDC$; Area $\triangle BCD >$ Area $\triangle ADC$ **35.** Given $\triangle ABC$, median $\overline{CM}$, Prove Area $\triangle ACM =$ Area $\triangle BCM$; $\overline{AM} \cong \overline{MB}$ (def median); $AM = MB$ (def $\cong$); Area ACM $= \frac{1}{2}AM \cdot h$; area $BCM = \frac{1}{2}MB \cdot h$ (Th 12.2); Area $ACM =$ Area BCM (subst)

Page 401

22. Given trap $ABCD$, alt h, bases a, b; trap $EFGH$, alt h', bases $a'b'$, Area $ABCD =$ Area $EFGH$; Prove $\frac{a+b}{a'+b'} = \frac{h'}{h}$. Area $ABCD = \frac{1}{2}h(a+b)$, Area $EFGH = \frac{1}{2}h'(a'+b')$ (Th 12.6): $\frac{h(a+b)}{2} = \frac{h'(a'+b')}{2}$ (subst); $\frac{a+b}{a'+b'} = \frac{h'}{h}$ (mult prop) **23.** Given: trap $ABCD$, $RSTU$, bases $\overline{AB} \cong \overline{RS}$, $\overline{DC} \cong \overline{UT}$, alts h, h'; Prove: $\frac{\text{Area } ABCD}{\text{Area } RSTU} = \frac{h}{h'}$, $AB = RS = b_1$, $DC = UT = b_2$ (def $\cong$ seg): Area $ABCD = \frac{1}{2}h(b_1 + b_2)$, Area $RSTU = \frac{1}{2}h'(b_1 + b_2)$ (Th 12.6); $\frac{\text{Area } ABCD}{\text{Area } RSTU} = \frac{\frac{1}{2}h(b_1+b_2)}{\frac{1}{2}h'(b_1+b_2)}$ (div); $\frac{\text{Area } ABCD}{\text{Area } RSTU} = \frac{h}{h'}$ (alg)
24. Given trap $ABCD$, with $\overline{AB}$ the longer base; constr rect $EFGH$ with same alt as $ABCD$ and $\overline{EF} \cong \overline{AB}$; on $\overrightarrow{HG}$ mark off $HI = \frac{1}{2}(AB - CD)$; on $\overrightarrow{GH}$ mark off $GJ = \frac{1}{2}(AB - CD)$; $EFJI$ is the reg trap **25.** Constr a seg of length = sum of bases of given trap; bisect this seg to get length of median; constr rect with base $\cong$ to med and alt $\cong$ to alt of given trap.

Page 414

38. $A_1 = \pi r_1^2$, $A_2 = \pi r_2^2$ (Th 12.11); $\frac{A_1}{A_2} = \frac{\pi r_1^2}{\pi r_2^2}$ (div); $\frac{A_1}{A_2} = \frac{r^2}{r_2^2}$ (alg); $\frac{r_1}{r^2} \neq \frac{r_1^2}{r_2^2}$ **39.** $c_1 = 2\pi r_1$, $c_2 = 2\pi r_2$ (Th 12.10, cor); $\frac{c_1}{c_2} = \frac{2\pi r_1}{2\pi r_2}$ (div); $\frac{c_1^2}{c_2^2} = \frac{r_1^2}{r_2^2}$ (alg); $\frac{A_1}{A_2} = \frac{r_1^2}{r_2^2}$ (ex 38); $\frac{c_1^2}{c_2^2} = \frac{A_1}{A_2}$ (subst) **40.** $r_i = a$, $r_i = 2a$ $\frac{Ai}{Ac} = \frac{\pi a^2}{\pi(2a)^2}$; $\frac{1}{4}$ **41.** $r_i = \frac{a}{2}$; $r_c = \frac{a}{2}\sqrt{2}$;

$$\frac{Ai}{Ac} = \frac{\pi\left(\frac{a}{2}\right)^2}{\pi\left(\frac{a}{2}\sqrt{2}\right)^2} = \frac{1}{2}$$

Page 420

28. $A = \frac{60}{360}\pi r^2$; $\triangle = \frac{r^2}{4}\sqrt{3}$; A of seg $= \frac{\pi r^2}{6} - \frac{r^2}{4}\sqrt{3} = r^2\frac{(2\pi - 3\sqrt{3})}{12}$

Page 434

23. $RS = \sqrt{(-4-6)^2 + (3-1)^2}$; $\sqrt{104}$; $ST = \sqrt{(6-2)^2 + (1+3)^2}$; $\sqrt{32}$; $RT = \sqrt{(-4-2)^2 + (3+3)^2}$; $\sqrt{72}$; $(\sqrt{104})^2 = (\sqrt{32})^2 + (\sqrt{72})^2$; $104 = 32 + 72$
24. $MP = \sqrt{(3-7)^2 + (2+1)^2}$, 5; $NQ = \sqrt{(3-7)^2 + (-1-2)^2}$, 5; $MP = NQ$; $\overline{MP} \cong \overline{NQ}$

Page 437

22. $(-1, -1\frac{1}{2})$, $(0, \frac{1}{2})$, $(-3, 3)$; $\frac{1}{2}\sqrt{173}$.

30. $M = \left(\frac{a+c}{2}, \frac{b+d}{2}\right)$;

$TM = \sqrt{\left(\frac{a+c}{2} - e\right)^2 + \left(\frac{b+d}{2} - f\right)^2}$

31. Given: 2 pts $P(x_1, y_1)$ and $Q(x_2\ y_2)$ with M the mdpt of $\overline{PQ}$; Prove: $M = \left(\frac{x_1+x_2}{2}, \frac{y_1+y_2}{2}\right)$; Locate $R(x_2, y_1)$ by drawing $\overleftrightarrow{QR} \parallel \overleftrightarrow{OY}$ and $\overleftrightarrow{PR} \parallel \overleftrightarrow{OX}$ (Post 15); mdpt of $\overline{QR}$ is $T\left(x_2, \frac{y_1+y_2}{2}\right)$, (def of mdpt of vert seg); mdpt of $\overline{PR}$ is $S\left(\frac{x_1+x_2}{2}, y_1\right)$ (def of mdpt of horiz seg); $\overline{MT} \parallel \overline{PR}$ and $\overline{MS} \parallel \overline{QR}$ (Th 7.7); $M(x, y) = M\left(\frac{x_1+x_2}{2}, \frac{y_1+y_2}{2}\right)$ (props of horiz and vert lines)

Page 441

32. up to rt; $y_2 > y_1$, $y_2 - y_1 > 0$; $x_2 > x_1$, $x_2 - x_1 > 0$; $\frac{y_2 - y_1}{x_2 - x_1} > 0$; dn to rt: $y_2 < y_1$, $y_2 - y_1 < 0$; $x_2 > x_1$, $x_2 - x_1 > 0$; $\frac{y_2 - y_1}{x_2 - x_1} < 0$; horiz: $y_2 = y_1$; $x_2 \neq x_1$; $\frac{y_2 - y_1}{x_2 - x_1} = 0$; vert: $x_2 = x_1$, $y_2 \neq y_1$; $\frac{y_2 - y_1}{x_2 - x_1} = \frac{y_2 - y_1}{0}$ undef.

Page 445

38. For $\overline{DE}$: $m = -4$; $y + 1 = -4(x + 5)$; $4x + y = 19$. For $\overline{EF}$: $m = \frac{3}{7}$; $y - 3 = \frac{3}{7}(x - 4)$; $-3x + 7y = 9$. For $\overline{DF}$; $m = -\frac{1}{8}$; $y + 1 = -\frac{1}{8}(x - 5)$; $x + 8y = -3$.
40. $S = (0, 1)$; $R = (-2, \frac{7}{2})$; $T = (-2, -\frac{5}{2})$
For $\overline{JS}$: $m = \frac{1}{4}$; $y = \frac{1}{4}(x + y)$; $-x + 4y = 4$
For $\overline{KT}$: $m = \frac{19}{4}$; $y - 7 = \frac{19}{4}(x - 0)$; $-19x + 4y = 28$
For $\overline{LR}$: $m = -\frac{17}{4}$; $y + 5 = -\frac{17}{4}(x - 0)$; $17x + 4y = -20$

41. $m = \frac{y_2 - y_1}{x_2 - x_1}$; $y - y_1 = \frac{y_2 - y_1}{x_2 - x_1}(x - x_1)$; $\frac{y - y_1}{x - x_1} = \frac{y_2 - y_1}{x_2 - x_1}$ **42.** Given: line l, with slope m, $P(x_1 y_1)$ on l; Prove: $y - y_1 = m(x - x_1)$ is an equa of line l. Let $R(x, y)$ be another pt on l (Post 1); $m = \frac{y - y_1}{x - x_1}$ (def of slope); $y - y_1 = m(x - x_1)$ (mult) **43.** The $\triangle$s formed are 30–60 rt $\triangle$s. For l_1: $m = \frac{0 - a\sqrt{3}}{a - 0}$; $-\sqrt{3}$. For l_2: $m = \frac{a\sqrt{3} - 0}{0 - (-a)}$; $\sqrt{3}$ **44.** The $\triangle$s formed are 30–60 rt $\triangle$s; For l_1: $m = \frac{0 - a}{a\sqrt{3} - 0}$; $\frac{-1}{\sqrt{3}}$; $\frac{-\sqrt{3}}{3}$. For l_2: $m = \frac{a - 0}{0 - (-a\sqrt{3})}$; $\frac{1}{\sqrt{3}}$; $\frac{\sqrt{3}}{3}$

Page 448

41. Let $Q(0, b)$ be the pt where the nonvert line int. y – axis. Then slope $\overleftrightarrow{PQ} = \frac{0 - b}{a - 0}$, or $\frac{-b}{a}$ (def); equa $\overleftrightarrow{PQ}$ is $y = mx + b$ (Th 13.5); $y = \frac{-b}{a}x + b$ (subst); $ay = -bx + ab$ (mult); $x = \frac{-ay}{b} + a$ (alg) $x = \frac{1}{m}y + a$ (subst) **42.** Slope $\overleftrightarrow{PQ} = \frac{-b}{a}$ (def); $y = mx + b$ (Th 13.5); $y = \frac{-b}{a}x + b$ (subst); $ay = -bx + ab$ (mult); $bx + ay = ab$ (add); $\frac{x}{a} + \frac{y}{b} = 1$ (div)

Page 452

26. Given: l_1, with slope m_1; l_2 with slope m_2; $l_1 \parallel l_2$; Prove: $m_1 = m_2$. Draw $\overleftrightarrow{AC}$, $\overleftrightarrow{A'C'} \parallel x$-axis, $\overleftrightarrow{BC} \parallel y$-axis, $\overleftrightarrow{B'C'} \parallel y$-axis (Post 22); $\frac{BC}{AC} = \frac{B'C'}{A'C'}$ (def ~ poly); $\frac{y_2 - y_1}{x_2 - x_1} = \frac{y_4 - y_3}{x_4 - x_3}$ (subst), $m_1 = \frac{y_2 - y_1}{x_2 - x_1}$; $m_2 = \frac{y_4 - y_3}{x_4 - x_3}$ (def of m); $m_1 = m_2$ (subst) **27.** Given: l_1 with slope m_1; l_2 with slope m_2; $m_1 \cdot m_2 = -1$; Prove: $l_1 \perp l_2$. Draw $\overleftrightarrow{AB} \parallel x$-axis (Post 15) $(AC)^2 = (x_3 - x_1)^2 + (y_3 - y_1)^2$, $(BC)^2 = (x_3 - x_2)^2 + (y_3 - y_2)^2$, $(AB)^2 = (x_2 - x_1)^2$ (Th 13.1); $(AC)^2 + (BC)^2 = (x_3 - x_1)^2 + (y_3 - y_1)^2 + (x_3 - x_2)^2 + (y_3 - y_2)^2$ (add); $m_1 = \frac{y_3 - y_1}{x_3 - x_1}$, $m_2 = \frac{y_3 - y_2}{x_3 - y_2}$ (def); $\frac{y_3 - y_1}{x_3 - x_1} \cdot \frac{y_3 - y_2}{x_3 - x_2} = -1$ (subst); $(y_3 - y_1)(y_3 - y_2) = -1(x_3 - x_1)(x_3 - x_2)(y_3 - y_1)^2 - 2(y_3 - y_1)(y_3 - y_2) + (y_3 - y_2)^2 = [(y_3 - y_1) - (y_3 - y_2)]^2 = 0\ (y_3 - y_1)^2 + (y_3 - y_2)^2 = 2(y_3 - y_1)(y_3 - y_2) = -2(x_3 - x_1)(x_3 - x_2)$ (mult); $(AC)^2 + (BC)^2 = (x_3 - x_1)^2 - 2(x_3 - x_1)(x_3 - x_2) + (x_3 - x_2)^2$ (subst); $(AC)^2 + (BC)^2 = [(x_3 - x_1) - (x_3 - x_2)]^2$ (distr); $(AC)^2 + (BC)^2 = (x_2 - x_1)^2$ (alg); $(AC)^2 + (BC)^2 = (AB)^2$ (subst); $m\angle C = 90$ (Th 10.3); $l_1 \perp l_2$ (def of rt $\angle$) **28.** Given: l_1 with slope m_1; l_2 with slope l_2; $l_1 \perp l_2$; Prove: $m_1 \cdot m_2 = -1$. Let $A(x_1, 0)$ and $B(x_2, 0)$ be intersections of l_1, l_2 with x-axis; Draw $\overline{CD} \perp \overline{AB}$ at $D\ (x_3, 0)$ (Post 12); $\frac{AD}{CD} = \frac{CD}{DB}$ (Th 10.1, cor); $\frac{x_3 - x_1}{y_3} = \frac{y_3}{x_2 - x_3}$ (subst); $y_3^2 = (x_3 - x_1)(x_2 - x_3)$ (mult); $m_1 = \frac{y_3}{x_3 - x_1}$, $m_2 = \frac{y_3}{x_3 - x_2}$ (def of m); $m_1 \cdot m_2 = \frac{y_3^2}{(x_3 - x_1)(x_3 - x_2)}$ (mult); $= \frac{(x_3 - x_1)(x_2 - x_3)}{(x_3 - x_1)(x_3 - x_2)}$ (subst); $m_1 \cdot m_2 = -1$ (alg)

Page 455

9. mdpt $\overline{AC} = \left(\frac{a + b}{2}, \frac{c}{2}\right)$; mdpt $\overline{BD} = \left(\frac{a + b}{2}, \frac{c}{2}\right)$; M is mdpt of both diags, so they bisect each other. **10.** $M = \left(\frac{b}{2}, \frac{c}{2}\right)$; $N = \left(\frac{a + b}{2}, \frac{c}{2}\right)$; slope $\overline{MN} = 0$; $\overline{MN} \parallel \overline{AB}$; $MN = \frac{a + b}{2} - \frac{b}{2} = \frac{a}{2}$; $AB = a$; $MN = \frac{1}{2}AB$ **11.** Given: isos trap $ABCD$, $\overline{AB} \parallel \overline{CD}$, $\overline{AD} \cong \overline{BC}$, diags $\overline{AC}$, $\overline{BD}$; Prove: $\overline{AC} \cong \overline{BD}$. $AC = \sqrt{(a - b - 0)^2 + (c - 0)^2}$, $AC = \sqrt{(a - b)^2 + c^2}$; $BD = \sqrt{(a - b)^2 + (0 - c)^2}$, $BD = \sqrt{(a - b)^2 + c^2}$. $AC = BD$; $\overline{AC} \cong \overline{BD}$ **12.** Given: rt $\triangle ABC$, $M\left(\frac{a}{2}, \frac{b}{2}\right)$ midpt of hypot $\overline{BC}$; Prove: $AM = BM = CM$. $AM = \sqrt{\left(\frac{a}{2} - 0\right)^2 + \left(\frac{b}{2} - 0\right)^2}$, $AM = \sqrt{\frac{a^2}{4} + \frac{b^2}{4}}$; $BM = \sqrt{\left(a - \frac{a}{2}\right)^2 + \left(0 - \frac{b}{2}\right)^2}$; $BM = \sqrt{\frac{a^2}{4} + \frac{b^2}{4}}$; $CM = \sqrt{\left(\frac{a}{2} - 0\right)^2 + \left(\frac{b}{2} - b\right)^2}$; $CM = \sqrt{\frac{a^2}{4} + \frac{b^2}{4}}$; $AM = BM = CM$ **13.** Given: $ABCD$, mdpts $Z\left(\frac{a}{2}, 0\right)$, $W\left(\frac{a + d}{2}, \frac{e}{2}\right)$, $X\left(\frac{b + d}{2}, \frac{c + e}{2}\right)$, $Y\left(\frac{b}{2}, \frac{c}{2}\right)$; Prove: $\overline{XZ}$, $\overline{WY}$ bisect each other. Mdpt of $\overline{WY} = \left(\frac{\frac{a + d}{2} + \frac{b}{2}}{2}, \frac{\frac{e}{2} + \frac{c}{2}}{2}\right)$; $\left(\frac{a + b + d}{4}, \frac{c + e}{4}\right)$; mdpt

$\overline{XZ} = \left(\dfrac{\frac{a}{2} + \frac{b+d}{2}}{2}, \dfrac{0 + \frac{c+e}{2}}{2}\right); \left(\dfrac{a+b+d}{4}, \dfrac{c+e}{4}\right)$; since both mdpts have same coords, diags bisect each other. **14.** Given: trap $ABCD$, $\overline{AB} \parallel \overline{CD}$, $M\left(\frac{b}{2}, \frac{c}{2}\right)$ mdpt of $\overline{AD}$, $N\left(\frac{a+d}{2}, \frac{c}{2}\right)$ mdpt of $\overline{BC}$ Prove: $MN = \frac{1}{2}(AB + CD)$. $MN = \frac{a+d}{2} - \frac{b}{2}$; $MN = \frac{a-b+d}{2}$; $AB = a$; $CD = d - b$; $AB + CD = a - b + d$; $MN = \frac{1}{2}(AB + CD)$

15. Given: trap $ABCD$, $\overline{AB} \parallel CD$, $M\left(\frac{d}{2}, \frac{c}{2}\right)$ mdpt diag $\overline{AC}$, $N\left(\frac{a+b}{2}, \frac{c}{2}\right)$ mdpt diag $\overline{BD}$; Prove: $MN = \frac{1}{2}(AB - CD)$. $MN = \frac{a+b}{2} - \frac{d}{2}$, $MN = \frac{a+b-d}{2}$, $AB = a$; $CD = d - b$; $AB - CD = a - (d - b)$, $AB - CD = a + b - d$; $MN = \frac{1}{2}(AB - CD)$ **16.** Given: $\triangle ABC$, $R\left(\frac{a}{2}, 0\right)$ mdpt $\overline{AB}$, $S\left(\frac{a+b}{2}, \frac{c}{2}\right)$ mdpt $\overline{BC}$, $T\left(\frac{b}{2}, \frac{c}{2}\right)$ mdpt $\overline{AC}$; Prove: $\triangle RST \sim \triangle CAB$. $m(\overline{AB}) = 0$ $m(\overline{TS}) = 0$; $\overline{TS} \parallel \overline{AB}$;

$m(\overline{RS}) = \dfrac{\frac{c}{2} - 0}{\frac{a+b}{2} - \frac{a}{2}}$; $m(\overline{RS}) = \frac{c}{b}$; $m(\overline{AC}) = \frac{c-0}{b-0}$;

$m(\overline{AC}) = \frac{c}{b}$; $\overline{RS} \parallel \overline{AC}$; $m(\overline{RT}) = \dfrac{\frac{c}{2} - 0}{\frac{b}{2} - \frac{a}{2}}$;

$m(\overline{RT}) = \frac{c}{b-a}$; $m(\overline{BC}) = \frac{c-0}{b-a}$, $m(\overline{BC}) = \frac{c}{b-a}$; $\overline{RT} \parallel \overline{BC}$; $ARST$, $RBST$, $RSCT$ are ▱s; $\angle C \cong \angle TRS$, $\angle A \cong \angle TSR$, $\angle B \cong \angle RTS$; $\triangle RST \sim \triangle CAB$

Page 460

30. Given: $A(x_1, y_1, z_1)$ and $B(x_2, y_2, z_2)$; $AB = d$; Prove: $d = \sqrt{(x_2 - x_1)^2 + (y_2 - y_1)^2 + (z_2 - z_1)^2}$. $(AC)^2 + (BC)^2 = (AB)^2$ (Th 10.2); $AC = (z_2 - z_1)$ (def); $BC = \sqrt{(x_2 - x_1)^2 + (y_2 - y_1)^2}$ (Th 13.1); $(x_2 - x_1)^2 + (y_2 - y_1)^2 + (z_2 - z_1)^2 = (AB)^2$ (subst); $d = \sqrt{(x_2 - x_1)^2 + (y_2 - y_1)^2 + (z_2 - z_1)^2}$ (alg)

Page 463

12. $M = \left(\frac{-2+6}{2}, \frac{5+5}{2}\right)$; $(2, 5)$; $N = \left(\frac{6+4}{2}, \frac{5+-1}{2}\right)$; $(5, 2)$; $P = \left(\frac{-6+4}{2}, \frac{1+-1}{2}\right)$; $(-1, 0)$; $Q = \left(\frac{-6+(-2)}{2}, \frac{1+5}{2}\right)$; $(-4, 3)$; $MN = \sqrt{(5-2)^2 + (2-5)^2}$; $3\sqrt{2}$; $PQ = \sqrt{[-1-(-4)]^2 + (0-3)^2} = 3\sqrt{2}$; $MN = PQ$

26. Given: ▱ ABCD; Prove: $AB = CD$, $AD = BC$. $AB = a$, $CD = (a + b) - b$ (Th 13.1); $CD = a$ (alg); $AB = CD$ (subst); $AD = \sqrt{b^2 + c^2}$; $BC = \sqrt{(a + b - a)^2 + c^2}$ (Th 13.1); $AD = BC$ (subst)

27. $(x + 4)^2 + (y - 1)^2 = 4$ **28.** $x^2 - 10x + 25 + y^2 + 2y + 1 = 39 + 26$; $(x - 5)^2 + (y + 1)^2 = 65$ $(5, -1)$, $\sqrt{65}$

Page 473

9. The locus of mdpts of all radii of a circle is a concentric circle with radius half that of the given circle. I. The mdpt of a rad of a ⊙ is on the ⊙ concentric to given ⊙ and having a rad $\frac{1}{2}$ the rad of giv ⊙ **II.** Every pt of a ⊙ concentric to a given ⊙ and of rad $= \frac{1}{2}$ rad of the given ⊙ is a mdpt of a rad of the given ⊙. **10.** The locus of points equi dist from 2 concentric ⊙s is a ⊙ concentric to them with rad $= \frac{1}{2}$ the sum of the rad of the 2 given ⊙s. I. If a pt is equidist from 2 concentric ⊙s, it is on the ⊙ concentric to the 2 ⊙s, with rad $= \frac{1}{2}$ the sum of the rad of the 2 ⊙s. II. Any pt on the ⊙ concentric to 2 concentric ⊙s and of rad $= \frac{1}{2}$ the sum of their radii is equidistant from the 2 ⊙s. **11.** The locus of the centers of all ⊙s tan to a given l at a given pt, excluding the given pt, is a $l \perp$ given l at the given pt. **I.** The ctr of a ⊙ tan to a given line at a given pt is on the $\perp$ to the l at the given pt. **II.** Any pt (except the given pt) on the $\perp$ to a l at a given pt is the ctr of a ⊙ tan to the l at the given pt. **12.** The locus of the mdpts of all chords drawn from a given pt of a ⊙ is a ⊙ (minus the given pt) tan internally to the given ⊙ at the given pt, with diam = rad of the given ⊙ **I.** The midpt of a chord drawn from a given pt of a ⊙ is on the ⊙ tan internally to given ⊙ at the given pt, with rad $= \frac{1}{2}$ diam of given ⊙ **II.** Any pt of the ⊙ tan internally to a given ⊙ at a given pt, with rad $= \frac{1}{2}$ diam of given ⊙ is mdpt of a chord of given ⊙ drawn from the given pt. **13.** Given: ⊙O, O is pt on $\overleftrightarrow{QP}$, $\overleftrightarrow{QP} \perp l$ at P; Prove: l tangent to ⊙O at P. Proof: Since $l \perp$ rad $\overline{OP}$ at its endpt on the ⊙, l tan ⊙O (Th 11.6). II. Given: ⊙O, rad $\overline{OP}$, l tan ⊙O at P;

Prove: $\overline{OP} \perp l$. Proof: Since l tan $\odot O$ at P, $\overline{OP} \perp l$ (Th 11.5) **14.** I. Given: $\odot O$, diam $\overline{AB}$, inscr $\angle ACB$; Prove: $\angle ACB$ is a rt $\angle$. Proof: $\widehat{ACB}$ is a semi$\odot$ (def'n); $\angle C$ is a rt $\angle$ (Th 11.11) **II.** Given: $\odot O$, $\triangle ABC$, hypot $\overline{AB}$ diam of $\odot O$; $\angle C$ is a rt $\angle$. Prove C is on $\odot O$. Draw $\overline{OC}$, draw median $\overline{OD}$ (Post 1); $\overline{AO} \cong \overline{OB}$ (rad of $\odot \cong$); $\overline{AD} \cong \overline{CD}$ (def'n of med); $\overline{OD} \parallel \overline{BC}$ (Th 7.7); $\angle ADQ \cong \angle C$ (Th 3.1); $\angle ADO$ a rt $\angle$ (subst); $\overline{DO} \perp$ bis of $\overline{AC}$ (defn); $\overline{AO} \cong \overline{OC}$ (Th 5.7); C is on $\odot O$ (defn of $\odot$)
15. Part I: $\overline{PA} \cong \overline{PB}$ by Th 5.7 Part II: Draw $\overline{PD}$ to mdpt D of $\overline{AB}$; then $\overleftrightarrow{PD}$ is the $\perp$ bis of $\overline{AB}$ by Th 5.8.

Page 490

26. Use method of constr 7.1, p. 222 to obtain segments of length $\frac{2}{3}m_1$ and $\frac{2}{3}m_2$; then use SSS method to construct a $\triangle$ with sides $\frac{2}{3}m_1$, $\frac{2}{3}m_2$, and $\overline{AB}$ (the given side); let the intersection of $\frac{2}{3}m_1$ and $\frac{2}{3}m_2$ be pt O; extend $\overrightarrow{AO}$ and $\overrightarrow{BO}$; on $\overrightarrow{AO}$ constr $\overline{AD} \cong m_1$; on $\overrightarrow{BO}$ constr $\overline{BE} \cong m_2$; draw $\overleftrightarrow{BD}$ and $\overleftrightarrow{AE}$, which meet at C; $\triangle ABC$ is the req $\triangle$
27. Constr a ▱ with sides s_1, s_2 by constructing a $\triangle$ with sides s_1, s_2, and $2m$; the seg $2m$ is diag d_1 of ▱; complete the ▱; draw other diag, d_2; $\triangle$ of sides s_1, s_2 and d_2 is req $\triangle$.

Page 501

1. $RM < PN$ (If $a > b$, $c > 0$, then $ac > bc$
2. $m\angle x > m\angle z$ (If $a > b$, $c = d$, then $a - c > b - d$) **3.** $AC > BD$ (If $a > b$, $c = d$, then $a + c > b + d$) **4.** $XF > YH$ (If $a = b$, $c < d$, then $a - c > b - d$) **5.** $m\angle x = m\angle y$, $m\angle x < m\angle y$, $m\angle x > m\angle y$ (trichotomy) **6.** $AB < CD$ (if $a < b$, $b < c$, then $a < c$) **7.** $m\angle z < m\angle w$ (equal-unequal) **8.** $m\angle ABC > m\angle ADC$ (add'n prop of $\neq$) **9.** $AB < BC$ (add'n prop of $\neq$)
10. $\angle CAB \cong \angle CBA$ (Th 5.1), $m\angle CAB = m\angle CBA$ (def'n $\cong \angle$s); $m\angle CAB = m\angle x + m\angle w$, $m\angle CBA = m\angle y + m\angle z$ (Post 10); $m\angle z > m\angle w$ (subtr prop $\neq$) **11.** $\frac{1}{2}AB > \frac{1}{2}CD$ (mult prop $\neq$); $CF = \frac{1}{2}CD$, $AE = \frac{1}{2}AB$ (Th 11.2); $AE > CF$ (subst); $CF < AE$ (def'n) **12.** $\angle BAC \cong \angle BCA$ (Th 5.1); $m\angle BAC = m\angle BCA$ (def'n $\cong \angle$s); $m\angle BAC + m\angle DAC = m\angle BAD$, $m\angle BCA + m\angle DCA = m\angle BCD$ (Post 10); $m\angle BCD > m\angle BAD$ (add'n prop $\neq$) **13.** $m\angle D + m\angle B = 180$, $m\angle C + m\angle A = 180$ (Th 11.11, cor 4); $m\angle D + m\angle B = m\angle C + m\angle A$ (subst); $m\angle D < m\angle C$ (subtr prop $\neq$) **14.** $m\angle x = m\angle CDB$ (Th 5.1); $m\angle CDB = m\angle y + m\angle z$ (Th 4.1); $m\angle x = m\angle y + m\angle z$ (subst); $m\angle x - m\angle y = m\angle z$ (subtr); $m\angle x - 2\,m\angle y < m\angle z - m\angle y$ (subtr prop $\neq$); $m\angle z - m\angle y < o$ (subtr prop $\neq$); $m\angle x - 2\,m\angle y < 0$ (trans prop); $m\angle x < 2\,m\angle y$ (addn prop $\neq$) **15.** $m\angle DAB = m\angle DBA$ (Th 5.1); $m\angle DAE < m\angle EBD$ (subtr prop $\neq$); $m\angle DAE < m\angle DBF$ (trans prop $\neq$); $m\angle FBC = m\angle FCB$ (Th 5.1); $m\angle DBF < m\angle FCD$ (subtr prop $\neq$); $m\angle DAE < m\angle FCD$ (trans prop $\neq$)

Page 505

5. $AB + BC > AC$, $AD + DC > AC$ (Th 15.1); $AB + BC + AD + DC > 2AC$ (add'n prop of $\neq$); $AB + AD > BD$, $BC + DC > BD$ (Th 15.1); $AB + BC + AD + DC > 2BD$ (add'n prop of $\neq$); $2(AB + BC + AD + DC) > 2(AC) + 2(BD)$ (add'n); $AB + BC + AD + DC > AC + BD$ (div prop of $\neq$) **6.** $AB < AC + BC$ (Th 15.1); $AB - BC < AC$ (subtr prop $\neq$) **7.** $AB + AD > BD$, $BE + EC > BC$ (Th 15.1); $BD = BE + ED$ (Bet'ness) $BD = BE + EC$ (subst); $BD > BC$ (subst); $AB + AD > BC$ (trans prop of $\neq$)
8. $AD + DB = AB$ (Bet'ness) $CD < CA + AD$, $CD < DB + BC$ (Th 15.1); $2CD < CA + AD + DB + BC$ (add'n prop of $\neq$); $2CD < AB + BC + CA$ (subst) **9.** $AP + BP > AB$, $BP + CP > BC$, $AP + CP > CA$ (Th 15.1); $AP + BP + BP + CP + AP + CP > AB + BC + CA$ (add'n prop of $\neq$); $2(AP + BP + CP) > AB + BC + CA$ (subst); $AP + BP + CP > \frac{1}{2}(AB + BC + CA)$ (mult prop of $\neq$) **10.** $AC + CE > AE$, $EB + DE > DB$ (Th 15.1); $AC + CE + EB + DE > AE + DB$ (add'n prop of $\neq$); $CE + EB = CB$, $AD + DE = AE$ (Bet'ness); $AC + CB + DE > AD + DE + DB$ (subst); $AC + CB > AD + DB$ (subtr);
11. $AB + AD > BD$, $BD + CB > DC$ (Th 15.1); $BD > DC - CB$ (subt prop of $\neq$) $AB + AD > DC - CB$ (trans prop of $\neq$) **12.** Draw $\overleftrightarrow{CPB}$; $m\angle APD > m\angle ACP$, $m\angle BPD > m\angle BCP$ (Th 4.1, cor); $m\angle APD + m\angle BPD > m\angle ACP + m\angle BCP$ (add'n prop of $\neq$); $m\angle APD + m\angle BPD = m\angle APB$, $m\angle ACP + m\angle BCP = m\angle C$ (Post 10); $m\angle APB > m\angle C$ (subst) **13.** Given: $\triangle ABC$ with P on $\overline{AB}$; Prove: $CP < \frac{1}{2}(AB + BC + AC)$. $AP + PB = AB$ (Bet'ness), $CP < CA + AP$, $CP < BC + PB$ (Th 15.1); $2(CP) < CA + AP + BC + PB$ (add'n prop of $\neq$); $2(CP) < CA + BC + AB$ (subst) $CP < \frac{1}{2}(AB + BC + AC)$ (div prop of $\neq$)
14. Given: quadr $ABCD$, diags $\overline{AC}$, $\overline{BD}$; Prove: $AB + BC + CD + AD > AC + BD$. $AB + BC > AC$, $BC + CD > DB$, $AD + DC > AC$, $AD + AB > DB$ (Th 15.1); $AB + BC + BC + CD + AD + DC + AD + AB > AC + DB + AC + DB$ (add'n prop of $\neq$); $2(AB + BC + CD + AD) > 2(AC + DB)$ (subst); $AB + BC + CD + AD > AC + DB$ (div)
15. $\angle APD \cong \angle BPF$ (prop of reflected light); $\angle APD \cong \angle EPF$ (Th 2.10); $\overline{PF} \cong \overline{PF}$ (reflex); rt $\triangle PFB \cong$ rt $\triangle PFE$ (Th 5.4); $\overline{BF} \cong \overline{FE}$, $\overline{PE} \cong \overline{PB}$ (CPCTC); $\overline{PF} \perp$ bis of $\overline{BE}$ (Th 5.8) $QB = QE$ (Th 5.7); $AP + PE = AE$ (Bet'ness) $AQ + QE > AE$ (Th 15.1); $AQ + QB > AP + PE$ (subst); (Th 5.7); $AQ + QB > AP + PB$ (subst)

Page 509

1. $m\angle A > m\angle B$ (Th 15.2); $m\angle C > m\angle B$ (trans prop $\neq$) **2.** $\angle A \cong \angle ACB$ (Th 5.1, cor 1); $m\angle ACB + m\angle BCD = m\angle ACD$ (Post 10); $m\angle ACD > m\angle ACB$ def'n of $>$); $m\angle ACD > m\angle A$ (subst); $AD > CD$ (Th 15.3) **3.** $m\angle B > 90$ (defn); $m\angle B + m\angle A + m\angle C = 180$ (Th 3.5); $m\angle A + m\angle C < 90$ (subt prop of $\neq$); $m\angle A < 90$, $m\angle C < 90$ (def'n $<$ and angle add) $m\angle B > m\angle A$, $m\angle B > m\angle C$ (trans prop $>$); $AC > BC$, $AC > AB$ (Th 15.3) **4.** $m\angle A > m\angle EBA$, $m\angle DBC > m\angle D$ (Th 15.2); $m\angle EBA = m\angle DBC$ (Th 2.10); $m\angle A > m\angle D$ (trans prop $\neq$)
5. $m\angle CBA > m\angle CAB$ (Th 15.2); $\frac{1}{2}m\angle CBA > \frac{1}{2}m\angle CAB$ (mult prop $\neq$); $m\angle DBA = \frac{1}{2}m\angle CBA$, $m\angle DAB = \frac{1}{2}m\angle CAB$ (def $\angle$ bis); $m\angle DBA > m\angle DAB$ (subst); $AD > DB$ (Th 15.3) **6.** $CB < CA$ (Th 15.3); $CD < CA$ (trans prop of $\neq$); $m\angle A < m\angle CDA$ (Th 15.2) **7.** $m\angle DBC > m\angle A$ (Th 4.1, cor); $m\angle DBC > m\angle C$ (trans prop of $\neq$); $DC > DB$ (Th 15.3) **8.** $m\angle z > m\angle y$ (Th 4.1, cor); $m\angle x = m\angle y$ (def $\angle$ bis); $m\angle z > m\angle x$ (subst); $AC > DC$ (Th 15.3); $DB > DC$ (trans prop of $\neq$). **9.** $m\angle CAB = m\angle CBA$ (Th 5.1); $m\angle CAB = m\angle w + m\angle x$, $m<CBA = m\angle y + m\angle z$ (Post 10); $m\angle w + m\angle x = m\angle y + m\angle z$ (subst); $m\angle x > m\angle y$ (Th 15.2); $m\angle w < m\angle z$ (subtr prop of $\neq$) **10.** $m\angle DBA < m\angle A$ (Th 15.2); $m\angle DBC = m\angle C$ (Th 5.1); $m\angle ABC = m\angle DBA + m\angle DBC$ (Post 10); $m\angle A + m\angle C > m\angle DBA + m\angle DBC$ (add'n prop of $\neq$); $m\angle A + m\angle C > m\angle ABC$ (subst) **11.** $m\angle BCD = m\angle D$ (Th 5.1); $AB < BC$ (subst); $m\angle ACB < m\angle A$ (Th 15.2); $m\angle ACB + m\angle BCD < m\angle D + m\angle A$ (add'n prop of $\neq$); $m\angle ABC + m\angle BCD = m\angle ACD$ (Post 10); $m\angle ACD < m\angle D + m\angle A$ (subst) **12.** $BC < AC$, $BC = AC$, or $BC > AC$ (trichotomy prop); If $BC = AC$, $m\angle A = m\angle B$ by Th 5.1, but $m\angle A > m\angle B$; if $BC < AC$, then $m\angle A < m\angle B$ by Th 15.2, but $m\angle A > m\angle B$; thus, the only remaining possibility is $BC > AC$.

Page 514

1. $\overline{AB} \cong \overline{CD}$, $\overline{AD} \cong \overline{BC}$ (Th 7.1, cor 1); $BD > AC$ (Th 15.4) **2.** $\overline{OA} \cong \overline{OC}$, $\overline{OB} \cong \overline{OD}$ (Th 11.1); $m\angle x > \angle\ m\angle y$ (Th 15.5) **3.** $\overline{BD} \cong \overline{BD}$ (reflex); $m\angle ADB > m\angle CDB$ (Th 15.5) **4.** D mdpt $\overline{AB}$ (def med); $\overline{AD} \cong \overline{DB}$ (def'n mdpt); $\overline{CD} \cong \overline{CD}$ (reflex); $m\angle y > m\angle x$ (Th 15.5) **5.** $\overline{CD} \cong \overline{CD}$ (reflex); $AD > DB$ (Th 15.4) **6.** $ABCD$ a $\square$ (Th 7.5); $\overline{AC}$, $\overline{BD}$ bis ea other (Th 7.2); $\overline{AE} \cong \overline{EC}$ (def'n bis); $\overline{DE} \cong \overline{DE}$ (reflex); $AD < DC$ (Th 15.4) **7.** $\overline{RU} \cong \overline{SU}$ (defn med); $\overline{TU} \cong \overline{TU}$ (reflex); $RT > ST$ (Th 15.4); $m\angle S > m\angle R$ (Th 15.2) **8.** $m\angle x + m\angle CEA = 180$, $m\angle y + m\angle CEB = 180$ (Post 11); $m\angle x + m\angle CEA = m\angle y + m\angle CEB$ (subst); $m\angle CEA > m\angle CEB$ (subst prop $\neq$); $AC > CB$ (Th 15.4) **9.** $\overline{AB} \cong \overline{AB}$ (reflex); $\overline{AD} \cong \overline{EB}$ (defn $\cong$); $CB > CA$ (Th 15.3) **10.** $\angle x \cong \angle y$ (def'n $\cong$); $\overline{AD} \cong \overline{CD}$ (Th 5.2); $\overline{BD} \cong \overline{BD}$ (reflex); $m\angle BDC > m\angle BDA$ (Th 15.5) **11.** $CD < AC$ (Th 15.3); $\overline{AB} \cong \overline{DB}$ (Th 5.2); $\overline{CB} \cong \overline{CB}$ (reflex); $m\angle DBC < m\angle ABC$ (Th 15.4) **12.** Constr $\triangle ACG \cong \triangle DFE$, and $\overrightarrow{CH}$ bisector $\angle GCB$; $AH + HG > AG$ (Th 15.1); $\overline{GC} \cong \overline{FE}$ (CPCTC); $\overline{CB} \cong \overline{GC}$ (subst); $\angle x \cong \angle y$ (def $\angle$ bis); $\overline{CH} \cong \overline{CH}$ (reflex); $\triangle GCH \cong \triangle BCH$ (SAS); $AH + HB = AB$ (Bet'ness); $\overline{HB} \cong \overline{HG}$ (CPCTC); $HB = HG$ (def'n $\cong$); $AH + HB > AG$ (subst); $AB > AG$ (subst); $\overline{AG} \cong \overline{DE}$ (CPCTC); $AG \cong DE$ (def'n $\cong$); $AB > DE$ (subst)

Page 518

1. $\overline{OC} \cong \overline{OA}$, $\overline{OD} \cong \overline{OB}$ (Th 11.1); $CD < AB$ (Th 15.4); $OF > OE$ (Th 15.6) **2.** $m\widehat{AB} = m\angle x$, $m\widehat{CD} = m\angle y$ (def'n); $m\angle x < m\angle y$ (subst) **3.** $\triangle OFE$ a rt $\triangle$ (def'n); $OE > OF$ (Th 15.3); $AB < CD$ (Th 15.7) **4.** $AE = \frac{1}{2}AB$, $BF = \frac{1}{2}BC$ (Th 11.2); $\frac{1}{2}AB > \frac{1}{2}BC$ (subst); $AB > BC$ (mult); $OE < OF$ (Th 15.6) **5.** $m\angle B = \frac{1}{2}\widehat{AC}$; $m\angle A = \frac{1}{2}\widehat{BC}$ (Th 11.11); $\frac{1}{2}m\widehat{AC} > \frac{1}{2}m\widehat{BC}$ (subst); $m\widehat{AC} > m\widehat{BC}$ (mult) **6.** $\overline{OB} \cong \overline{OB}$ (reflex); $OF > OE$ (Th 15.6); $m\angle x > m\angle y$ (Th 15.5) **7.** $\overline{OA} \cong \overline{OB} \cong \overline{OC}$ (def'n); $m\angle AOB = m\widehat{AB}$, $m\angle BOC = m\widehat{BC}$ (def'n); $m\angle AOB < m\angle BOC$ (subst); $AB < BC$ (Th 15.4); $OD > OE$ (Th 15.6) **8.** $\overline{OA} \cong \overline{OA}$ (reflex); $OF > OE$ (Th 15.4); $AB > CD$ (Th 15.6) **9.** Draw $\overline{OD} \perp \overline{AC}$; $\overline{OD}$ bis $\overline{AC}$ (Th 11.2); $AD = \frac{1}{2}AC$ (def'n); $AO = \frac{1}{2}AB$ (def'n); $\angle ADO$ a rt $\angle$ (defn $\perp$); $AO > AD$ (Th 15.3); $\frac{1}{2}AB > \frac{1}{2}AC$ (subst); $AB > AC$ (mult) **10.** apoth of $\triangle = \frac{1}{2}r$; apoth of $\square = \frac{1}{2}r\sqrt{2}$ (Pyth); $\frac{1}{2}r\sqrt{2} > \frac{1}{2}r$ (mult fact); apoth $\square >$ apoth $\triangle$ (subst) **11.** Given: $\odot O$, P a noncentered interior pt; ch $\overline{AB} \perp \overline{OP}$, ch $\overline{CPD}$; Prove: $AB < CD$. Draw $\overline{OE} \perp \overline{CD}$; then see ex 3, above. **12.** Given: $\odot O$, $\overline{OE} \perp \overline{AB}$, $\overline{OF} \perp \overline{CD}$, $OE < OF$; Prove: $AB > CD$. $AB < CD$, $AB = CD$, or $AB > CD$ (trich prop). If $AB < CD$, $OE > OF$ (Th 15.6); this contradicts $OE < OF$; If $AB = CD$, $OE = OF$ (Th 11.3); this also contradicts $OE < OF$; $AB > CD$ (trich prop) **13.** Given: $\odot O \cong \odot O'$, $AB > CD$, $\overline{OE} \perp \overline{AB}$, $\overline{O'F} \perp \overline{CD}$; Prove: $OE < O'F$. In $\odot O$, draw $\overline{AG} \cong \overline{CD}$, $\overline{OH} \perp \overline{AG}$; $AB > AG$ (subst); $OE < OH$ (Th 15.6); $OH = O'F$ (Th 11.3); $OE < O'F$ (subst) **14.** Given: $\odot O \cong \odot O'$, $\overline{OE} \perp \overline{AB}$, $\overline{O'F} \perp \overline{CD}$, $OE < O'F$; Prove: $AB > CD$. $AB < CD$, $AB = CD$, or $AB > CD$ (trich prop); if $AB < CD$, $OE > OF$ (ex 13); this contradicts $OE < O'F$ if $AB = CD$, $OE = OF$ (Th 11.4); this contradicts $OE < O'F$; $AB > CD$ (trich prop)

Page 520

12. $AB + AD > BD$, $BC + CD > BD$, $AB + BC > AC$, $AD + DC > AC$ (Th 15.1); $2(AB + BC + CD +$

$AD) > 2(AC + BD)$ (add'n prop ≠); $AB + BC + CD + AD > AC + BD$ (div prop ≠) **13.** $AD < AE + ED$, $EB < EC + CB$ (Th 15.1); $AD - AE < ED$ (subtr prop of ≠); $EB = ED + DB$ (Bet'ness) $ED + DB < CB + EC$ (subst); $ED < CB + EC - DB$ (subt prop); $AD - AE < CB + EC - DB$ (trans prop ≠); $AD + DB < CB + EC + AE$ (addn prop ≠); $EC + AE = AC$ (Bet'ness); $AD + DB < CB + AC$ (subst) **14.** $AE = \frac{1}{2}AB$, $CF = \frac{1}{2}CD$ (Th 11.2); $\frac{1}{2}AB < \frac{1}{2}CD$ (subst); $AB < CD$ (mult prop ≠); $\overline{OE} > \overline{OF}$ (Th 15.6) **15.** $AB = DB$ (Th 5.2); $\overline{BC} = \overline{BC}$ (reflex); $AC > CD$ (Th 15.3); m $\angle ABC >$ m $\angle DBC$ (Th 15.5)

Page 521

11. $PQ = PQ$ (reflex); $PS = QR$ (Th 7.1, cor 1); $SQ < PR$ (Th 15.4) **12.** m $\angle A >$ m $\angle B$ (Th 15.2); m $\angle EAB = \frac{1}{2}$m $\angle CAB$, m $\angle EBA = \frac{1}{2}$m $\angle CBA$ (def ∠ bis); $\frac{1}{2}$m $\angle CAB > \frac{1}{2}$m $\angle CBA$ (mult prop ≠); m $\angle EAB >$ m $\angle EBA$ (subst); $AE < EB$ (Th 15.3) **13.** m $\angle P >$ m $\angle SQP$ (Th 15.2); m $\angle R =$ m $\angle SQR$ (Th 5.1); m $\angle P +$ m $\angle R >$ m $\angle SQP +$ m $\angle SQR$ (add'n prop ≠); m $\angle SQP +$ m $\angle SQR =$ m $\angle PQR$ (Post 10); m $\angle P +$ m $\angle R >$ m $\angle PQR$ (subst)

Page 522

1.

```
5  PRINT "THIS PROGRAM GIVES CORRECT ANSWERS FOR "
6  PRINT "SCALENE TRIANGLES ONLY.": PRINT
10  PRINT "WHAT IS THE NAME OF YOUR TRIANGLE":: INPUT N$
20  FOR I = 1 TO 3:A$(I) =  MID$ (N$,I,1): NEXT I
25 B$(1) = A$(2) + A$(3):B$(2) = A$(1) + A$(3)
26 B$(3) = A$(1) + A$(2)
30  FOR I = 1 TO 3: PRINT "WHAT IS THE MEASURE OF SIDE ";B$(I);
40  INPUT A(I): NEXT I
50  IF A(1) > A(2) THEN 110
60  IF A(1) > A(3) THEN 100
70  IF A(2) > A(3) THEN 90
80 B$(1) = A$(3):B$(2) = A$(2):B$(3) = A$(1)
81  GOTO 160
90 B$(1) = A$(2):B$(2) = A$(3):B$(3) = A$(1)
91  GOTO 160
100 B$(1) = A$(2):B$(2) = A$(1):B$(3) = A$(3)
101  GOTO 160
110  IF A(2) > A(3) THEN 150
120  IF A(1) > A(3) THEN 140
130 B$(1) = A$(3):B$(2) = A$(1):B$(3) = A$(2)
131  GOTO 160
140 B$(1) = A$(1):B$(2) = A$(3):B$(3) = A$(2)
141  GOTO 160
150 B$(1) = A$(1):B$(2) = A$(2):B$(3) = A$(3)
151 B$(3) = A$(1) + A$(2)
160  PRINT "THE BIGGEST ANGLE IS ";B$(1)"."
170  PRINT "THE MIDDLE ANGLE IS ";B$(2)"."
180  PRINT "THE SMALLEST ANGLE IS ";B$(3)"."
190  PRINT : GOTO 10
```

Page 531

18. Given: rt prism $ABC\ A'B'C' \cdots$, with $p =$ perimeter of base, $h =$ length of alt; Prove: Lat. Area $= h \cdot p$. Area of lat faces is $a \cdot h$, $b \cdot h$, $c \cdot h$, etc (def'n area of rect); Lat. Area $= ah + bh + ch + \cdots$ (def'n lat. Area); Lat. Area $= h(a + b + c + \cdots)$ (distrib prop); $a + b + c + \cdots = p$ (def'n of perimeter); Lat Area $= h \cdot p$ (subst)

Page 534

24. $T = 6B$; $B = \frac{T}{6}$; $h = \sqrt{\frac{T}{6}}$; $V = Bh$; $V = \frac{T}{6}\sqrt{\frac{T}{6}}$; $V = \frac{T}{6}\sqrt{\frac{T}{6} \cdot \frac{6}{6}}$; $V = \frac{T}{36}\sqrt{6T}$

Page 556

1.

```
10  PRINT "WHAT IS THE RADIUS OF THE CYLINDER";: INPUT R
20  PRINT "WHAT IS THE HEIGHT OF THE CYLINDER";: INPUT H
30  PRINT : PRINT "THE VOLUME IS ";3.1416 * R ^ 2 * H"."
40  PRINT "THE LATERAL AREA IS ";2 * 3.1416 * R * H"."
50  PRINT "THE TOTAL AREA IS ";6.2832 * R * H + 6.2832 * R ^ 2"."
60  PRINT : GOTO 10
```

2.

```
10  PRINT "WHAT IS THE RADIUS OF THE SPHERE";: INPUT R
20  PRINT "THE SURFACE AREA IS ";4 * 3.14168R ^ 2"."
30  PRINT "THE VOLUME IS ";4 * 3.1416 * R ^ 3 / 3"."
40  PRINT : GOTO 10
```

Computer Activities Answers

Page 180

1.
```
10  PRINT "GIVE ME AN ANGLE OF AN ISOSCELES TRIA
     NGLE."
15  PRINT "DEGREES ";: INPUT D: PRINT "MINUTES "
     ;: INPUT M
20  PRINT "IS THIS ANGLE BASE OR VERTEX?"
30  INPUT A$
40  IF A$ = "VERTEX" THEN 70
50  PRINT "THE THREE ANGLES ARE"
51  PRINT D;" DEGREES ";M;" MINUTES"
52  PRINT D;" DEGREES ";M;" MINUTES"
53  IF 2 * M < 60 THEN 65
54  PRINT 178 - 2 * D;" DEGREES ";120 - 2 * M;"
     MINUTES"
60  PRINT : GOTO 10
65  PRINT 179 - 2 * D;" DEGREES ";60 - 2 * M;" M
     INUTES"
66  PRINT : GOTO 10
70  PRINT " THE THREE ANGLES ARE"
71  PRINT D;" DEGREES ";M;" MINUTES
72 A = 0:T = (10800 - D * 60 - M) / 2
73  FOR I = 1 TO 180
74  IF T < 60 THEN 80
75 A = A + 1:T = T - 60
76  NEXT I
80  PRINT A;" DEGREES ";T;" MINUTES"
90  PRINT A;" DEGREES ";T;" MINUTES"
95  PRINT : GOTO 10
97  PRINT : GOTO 10
```

2.
```
10  PRINT "WHAT IS THE PERIMETER?"
15  PRINT "FEET";: INPUT F: PRINT "INCHES";: INPUT
     I
20  PRINT "WHAT IS THE KNOWN SIDE?"
25  PRINT "FEET ";: INPUT G: PRINT "INCHES";: INPUT
     J
30  PRINT "IS THAT SIDE A LEG OR THE BASE?"
35 A = 0
40  INPUT A$
50 H = F * 12 + I:L = G * 12 + J
60  IF A$ = "LEG" THEN 75
70 D = (H - L) / 2: GOTO 80
75 D = H - 2 * L
80  FOR N = 1 TO 100
90  IF D < 12 THEN 115
100 A = A + 1:D = D - 12
110  NEXT N
115  IF A$ = "LEG" THEN 150
120  PRINT "THE OTHER 2 SIDES ARE EACH"
125  PRINT A;" FEET ";D;" INCHES."
150  PRINT "THE OTHER 2 SIDES ARE ";A;" FEET ";D
     ;" INCHES"
160  PRINT  TAB( 18);" AND ";G;" FEET ";J;" INCH
     ES.": GOTO 10
```

Page 203

1.
```
110  PRINT "THIS PROGRAM TESTS YOUR KNOWLEDGE OF
      THE NAMES OF POLYGONS."
130  FOR I = 1 TO 10: READ A,A$
150  PRINT "HOW MANY SIDES DOES A ";A$;" HAVE";:
      INPUT N
155  PRINT
160  IF A = N THEN 190
170  PRINT "SORRY, THE ANSWER IS "A$"."
180  GOTO 200
190  PRINT "VERY GOOD!"
195 R = R + 1
200  PRINT : NEXT I
220  DATA  3,TRIANGLE,4,QUADRILATERAL,5,PENTAGON
     ,6,HEXAGON,7,HEPTAGON,8
230  DATA  OCTAGON,9,NONAGON,10,DECAGON,11,UNDEC
     AGON,12,DODECAGON
240  PRINT "YOU GOT ";R;" RIGHT OUT OF 10."
```

2.
```
110  PRINT "THIS PROGRAM TESTS YOUR KNOWLEDGE OF
      THE NAMES AND ANGLES ";
120  PRINT "OF POLYGONS.": PRINT
130  FOR I = 1 TO 10: READ A,A$
140 A = (A - 2) * 180
150  PRINT "WHAT IS THE SUM OF THE INTERIOR ANGL
     ES OF A ";A$"?"
155  INPUT N
160  IF A = N THEN 190
170  PRINT "SORRY, THE ANSWER IS ";A"."
180  GOTO 200
190  PRINT "VERY GOOD!"
195 R = R + 1
200  PRINT : NEXT I
220  DATA  3,TRIANGLE,4,QUADRILATERAL,5,PENTAGON
     ,6,HEXAGON,7,HEPTAGON,8
230  DATA  OCTAGON,9,NONAGON,10,DECAGON,11,UNDEC
     AGON,12,DODECAGON
240  PRINT "YOU GOT ";R;" RIGHT OUT OF 10."
```

3.
```
110  PRINT "THIS PROGRAM TESTS YOUR KNOWLEDGE OF
      THE NAMES AND ANGLES ";
```

```
120  PRINT "OF POLYGONS.": PRINT
130  FOR I = 1 TO 10: READ A,A$
140 A = (A - 2) * 180 / A
150  PRINT "IF EACH INTERIOR ANGLE OF A POLYGON
     MEASURES ";A
151  PRINT "WHAT IS THE NAME OF THE POLYGON";
155  INPUT N$
160  IF A$ = N$ THEN 190
165  IF A$ = "QUADRILATERAL" THEN 185
170  PRINT "SORRY, THE ANSWER IS ";A$"."
180  GOTO 200
185  IF N$ = "RECTANGLE" THEN 190
186  GOTO 170
190  PRINT "VERY GOOD!"
195 R = R + 1
200  PRINT : NEXT I
220  DATA  3,TRIANGLE,6,HEXAGON,8,OCTAGON,12,DOD
     ECAGON,4,QUADRILTERAL
230  DATA  7,HEPTAGON,11,UNDECAGON,5,PENTAGON,10
     ,DECAGON,9,NONAGON
240  PRINT "YOU GOT ";R;" RIGHT OUT OF 10."
```

Page 236

3.
```
110  PRINT "PROGRAM DETERMINES TYPE OF PARALLELO
     GRAM."
120  PRINT "ENTER MEASURE OF ANGLE W."
130  INPUT W
140  PRINT "ENTER MEASURES OF TWO ADJACENT SIDES
     ."
150  INPUT A,B
160  IF W <  > 90 THEN 110
170  PRINT "PARALLELOGRAM IS A RECTANGLE."
180  IF A <  > B THEN 150
190  PRINT "PARALLELOGRAM IS A SQUARE."
200  GOTO 150
210  IF A <  > B THEN 140
```

4.
```
110  PRINT "PROGRAM TO PROVE TWO QUADRILATERALS
     CONGRUENT."
120  PRINT "INPUT SIDES OF EACH IN ORDER."
130  INPUT A,B,C,D,W,X,Y,Z
140  PRINT "INPUT ANGLES OF EACH IN ORDER."
150  INPUT A1,B1,C1,D1,W1,X1,Y1,Z1
160 K1 = W
170  IF A <  > W THEN 210
180  IF B <  > X THEN 210
190  IF C <  > Y THEN 210
200  GOTO 240
210 K2 = W:W = X: IF W = K1 THEN 290
220 X = 4:Y = 2:Z = K2:K3 = X1:X1 = Y1:Y1 = Z1:Z
     1 = K3
230  GOTO 270
240  IF B1 <  > X1 THEN 290
250  IF C1 <  > Y1 THEN 290
260  IF D1 <  > Z1 THEN 290
270  PRINT "THE QUADRILATERALS ARE CONGRUENT."
280  GOTO 110
290  PRINT "THE QUADRILATERALS ARE NOT CONGRUENT
     ."
300  GOTO 110
```

Page 263

1.
```
10  PRINT "WHAT IS ONE OF THE BASES";: INPUT B
20  PRINT "WHAT IS THE MEDIAN";: INPUT M
30 B2 = 2 * M - B
40  PRINT "THE OTHER BASE IS ";B2"."
50  PRINT : GOTO 10
```

2.
```
5  FOR N = 1 TO 10
10 B =  INT ( RND (1) * 11) + 10
20 M =  INT ( RND (1) * 11) +  INT (B / 2) + 1
25  IF M = B THEN 20
30  PRINT "IF ONE BASE OF A TRAPEZOID IS ";B
40  PRINT "AND IF THE MEDIAN IS ";M
50  PRINT "WHAT IS THE OTHER BASE";: INPUT A
60 B2 = 2 * M - B
70  IF A = B2 THEN 100
80  PRINT "INCORRECT. THE ANSWER IS ";B2"."
90  GOTO 120
100  PRINT "CORRECT.":R = R + 1
120  PRINT : NEXT N
130  PRINT : PRINT "YOUR GRADE IS ";R * 10
```

3.
```
10  PRINT "WHAT ARE THE TWO BASES";: INPUT A,B
20  PRINT "WHAT IS THE AREA";: INPUT R
30  PRINT "THE ALTITUDE OF THIS TRAPEZOID IS ";2
     * R / (A + B)"."
40  PRINT : GOTO 10
```

Page 301

1.
```
110  PRINT "SOLVE THESE PROBLEMS GIVING ANSWERS"
120  PRINT "ROUNDED TO THE NEAREST HUNDREDTH"
130  PRINT "IF NECESSARY.": PRINT
140 Q =  INT ( RND (1) * 3) + 1
150  ON Q GOTO 160,240,300
160 M =  INT ( RND (1) * 11) + 60:S =  INT ( RND
     (1) * 61) + 40:T =  INT ( RND (1) * 121) +
     80
```

```
170  PRINT "HOW TALL IS A TREE WHICH CASTS A "
180  PRINT "SHADOW ";T;" INCHES LONG, IF A MAN "
     ;M
190  PRINT "INCHES TALL CASTS A ";S;" INCH SHADO
     W";: INPUT A
200 Z =  INT (M * T) / S * 100 + .5 / 100: IF  ABS
     (A - Z) < .01 THEN 230
210  PRINT "NO. THE ANSWER IS ": INT (M * T / S *
     100 + .5) / 100".": PRINT : GOTO 140
230  PRINT "YES.": PRINT : GOTO 140
240 M =  INT ( RND (1) * 11) + 20
250 S =  INT ( RND (1) * 41) + 20:T =  INT ( RND
     (1) * 671) + 30
260  PRINT "A ";M;" FOOT FLAGPOLE CASTS A SHADOW
     "
270  PRINT "WHICH IS ";S;" FEET.  HOW HIGH IS A
     "
280  PRINT "BUILDING WHOSE SHADOW IS ";T;" FEET"
290  INPUT A: GOTO 200
300 M =  INT ( RND (1) * 10) + 6
310 S =  INT ( RND (1) * 18) + 3:T =  INT ( RND
     (1) * 111) + 60
320  PRINT "A ";S;" FOOT SHADOW IS CAST BY A ";M
     ;" FOOT TALL TREE."
330  PRINT "HOW TALL IS A BUILDING WHOSE SHADOW
     IS ";T;" FEET";
340  INPUT A: GOTO 200
```

2.

```
10  PRINT "SOLVE THESE PROBLEMS GIVING ANSWERS"
20  PRINT "ROUNDED TO THE NEAREST HUNDREDTH"
30  PRINT "IF NECESSARY.": PRINT
35  FOR I = 1 TO 12
40 Q =  INT ( RND (1) * 3) + 1
50  ON Q GOTO 60,140,200
60 M =  INT ( RND (1) * 11) + 60:S =  INT ( RND
     (1) * 61) + 40:T =  INT ( RND (1) * 121) +
     80
70  PRINT "HOW TALL IS A TREE WHICH CASTS A"
80  PRINT "SHADOW ";T;" INCHES LONG, IF A MAN ";
     M
90  PRINT "INCHES TALL CASTS A ";S;" INCH SHADOW
     ";: INPUT A
100 Z =  INT (M * T / S * 100 + .5) / 100: IF  ABS
     (A - Z) < .01 THEN 130
110  PRINT "NO. THE ANSWER IS ": INT (M * T / S *
     100 + .5) / 100".": GOTO 250
130  PRINT "YES.":R = R + 1: GOTO 250
140 M =  INT ( RND (1) * 11) + 20
150 S =  INT ( RND (1) * 41) + 20:T =  INT ( RND
     (1) * 671) + 30
160  PRINT "A ";M;" FOOT FLAGPOLE CASTS A SHADOW"
170  PRINT "WHICH IS ";S;" FEET.  HOW HIGH IS A"
180  PRINT "BUILDING WHOSE SHADOW IS ";T;" FEET"
190  INPUT A: GOTO 100
200 M =  INT ( RND (1) * 10) + 6
210 S =  INT ( RND (1) * 18) + 3:T =  INT ( RND
     (1) * 111) + 60
220  PRINT "A ";S;" FOOT SHADOW IS CAST BY A(N)
     ";M;" FOOT TALL TREE."
230  PRINT "HOW TALL IS A BUILDING WHOSE SHADOW
     IS ";T;" FEET";
240  INPUT A: GOTO 100
250  PRINT : NEXT I
260  PRINT : PRINT "YOUR SCORE IS ": INT (R * 10
     0 / 12 + .5)"."
```

Page 333

```
10  FOR I = 1 TO 5
20 Q =  INT ( RND (1) * 2) + 1
30  ON Q GOTO 40,110
40 A =  INT ( RND (1) * 36) + 5
50 H =  INT ( RND (1) * 31) + 20
60 R = A * 6.2832 / 360
70 D =  INT (D *  TAN (R) + .5)
80  PRINT "A BOY IS STANDING ";H;" FEET FROM A T
     REE"
90  PRINT "THE ANGLE OF ELEVATION TO THE TOP OF
     THE TREE IS ";A
100  PRINT "HOW HIGH IS THE TREE";: INPUT P
105  IF D = P THEN 200
106  GOTO 190
110 A =  INT ( RND (1) * 21) + 5
120 H =  INT ( RND (1) * 7) + 6
130 R = A * 6.2832 / 360
140 D =  INT (H /  TAN (R) + .5)
150  PRINT "A LIFEGUARD IS SITTING ATOP A ";H;"
     FOOT PLATFORM"
160  PRINT "HE SPOTS A SWIMMER IN THE WATER WITH
      AN ANGLE OF DEPRESSION OF ";A
170  PRINT "HOW FAR OUT IN THE WATER IS THE SWIM
     MER";: INPUT P
180  IF P = D THEN 200
190  PRINT "WRONG. THE ANSWER WAS ";D: GOTO 240
200 C = C + 1: PRINT "CORRECT"
240  PRINT : NEXT I
250  PRINT : PRINT "YOUR SCORE IS ";C * 20
```

Page 381

1.
```
10  PRINT "GIVEN TWO CONCENTRIC CIRCLES, THIS"
20  PRINT "IS A TEST TO FIND THE LENGTH OF A CHO
     RD"
30  PRINT "OF THE LARGER CIRCLE WHICH IS TANGENT
     "
40  PRINT "TO THE SMALLER."
45  FOR I = 1 TO 8
50 R2 =  INT ( RND (1) * 8) + 1
55 R1 =  INT ( RND (1) * 10) + R2 + 1
60  PRINT "THE TWO RADII ARE ";R1;" AND ";R2"."
65  PRINT "WHAT IS THE LENGTH OF THE CHORD";: INPUT
     D
90 C =  INT (2 *  SQR (R1 ^ 2 - R2 ^ 2) * 100 +
     .5) / 100
100  IF  ABS (C - D) < .005 THEN 120
110  PRINT "INCORRECT.  THE CORRECT ANSWER IS ";
     C".": GOTO 130
120 R = R + 1: PRINT "CORRECT"
130  PRINT : NEXT I
```

2.
```
10  PRINT "GIVEN TWO CONCENTRIC CIRCLES, THIS"
20  PRINT "PROGRAM WILL FIND THE RADIUS OF THE"
25  PRINT "SMALLER CIRCLE GIVEN THE RADIUS OF TH
     E"
30  PRINT "LARGER AND THE LENGTH OF THE CHORD OF
     "
40  PRINT "THE LARGER WHICH IS TANGENT TO THE "
50  PRINT "SMALLER."
55  PRINT "WHAT IS THE RADIUS OF THE LARGER";
60  INPUT R1
70  PRINT "WHAT IS THE LENGTH OF THE CHORD";
80  INPUT D
90 R2 =  SQR (R1 ^ 2 - (D / 2) ^ 2)
100  PRINT "THE SMALLER RADIUS IS ";R2"."
110  PRINT : GOTO 55
```

3.
```
110  PRINT "GIVEN TWO CONCENTRIC CIRCLES, THIS"
120  PRINT "PROGRAM WILL FIND THE RADIUS OF THE"
125  PRINT "LARGER CIRCLE GIVEN THE RADIUS OF TH
     E "
130  PRINT "SMALLER AND THE LENGTH OF THE CHORD
     OF"
140  PRINT "THE LARGER WHICH IS TANGENT TO THE "
150  PRINT "SMALLER."
155  PRINT "WHAT IS THE RADIUS OF THE SMALLER ";
160  INPUT R1
170  PRINT "WHAT IS THE LENGTH OF THE CHORD";
180  INPUT D
190 R2 =  SQR (R1 ^ 2 + (D / 2) ^ 2)
200  PRINT "THE LARGER RADIUS IS ";R2"."
210  PRINT : GOTO 155
```

Page 424

```
10 A =  INT ( RND (1) * 12) + 3
20 P = 3 * A +  INT ( RND (1) * 6)
30 R = .5 * A * P
40 Q =  INT ( RND (1) * 3) + 1
50  ON Q GOTO 60,60,160
60  PRINT "THE APOTHEM OF A REGULAR POLYGON IS "
     ;A"."
70  ON Q GOTO 80,120
80  PRINT "THE PERIMETER IS ";P"."
90  PRINT "WHAT IS THE AREA";: INPUT Z
100  IF Z = R THEN 210
110  PRINT "INCORRECT. THE ANSWER IS ";R".": PRINT
     : GOTO 10
120  PRINT "THE AREA IS ";R"."
130  PRINT "WHAT IS THE PERIMETER ";: INPUT Z
140  IF Z = P THEN 210
150  PRINT "INCORRECT. THE ANSWER IS ";P".": PRINT
     : GOTO 10
160  PRINT "THE PERIMETER OF A REGULAR POLYGON I
     S ";P"."
170  PRINT "THE AREA IS ";R"."
180  PRINT "WHAT IS THE APOTHEM";: INPUT Z
190  IF A = Z THEN 210
200  PRINT "INCORRECT. THE ANSWER IS ";A".": PRINT
     : GOTO 10
210  PRINT "CORRECT.": PRINT : GOTO 10
```

Page 464

1.
```
10  PRINT "WHAT ARE THE COORDINATES OF THE CENTE
     R"
20  PRINT "OF THE CIRCLE";: INPUT H,K
30  PRINT "WHAT IS THE RADIUS";: INPUT R
40  PRINT : PRINT "THE EQUATION OF THE CIRCLE IS
     "
50  PRINT "X^2 + Y^2 + "; - 2 * H;"X + "; - 2 *
     K;"Y = ";
60  PRINT R ^ 2 - H ^ 2 - K ^ 2
70  PRINT : GOTO 10
```

2.
```
10  PRINT "INPUT THE COORDINATES OF THE ENDPOINT
     S"
20  PRINT "OF A DIAMETER OF A CIRCLE AND I WILL"
30  PRINT "GIVE YOU THE EQUATION.": PRINT
40  PRINT "ONE ENDPOINT";: INPUT A,B
50  PRINT "OTHER ENDPOINT";: INPUT C,D
```

```
60 H = (A + C) / 2:K = (B + D) / 2
70 R =  SQR ((D - B) * (D - B) + (C - A) * (C -
     A)) / 2
80  PRINT : PRINT "THE EQUATION OF THE CIRCLE IS
     "
90  PRINT "X^2 + Y^2 + "; - 2 * H;"X + "; - 2 *
     K;"Y = ";
100  PRINT R * R - H * H - K * K
110  PRINT : GOTO 10
```

3.
```
10  PRINT "FOR THE LENGTH OF A SEGMENT IN A THRE
     E-"
20  PRINT "DIMENSIONAL SYSTEM, INPUT THE"
30  PRINT "COORDINATES OF ONE POINT.";: INPUT A,
     B,C
40  PRINT "THE OTHER POINT";: INPUT D,E,F
50  PRINT "THE LENGTH OF THE SEGMENT IS ";
60  PRINT  INT ( SQR ((A - D) ^ 2 + (B - E) ^ 2 +
     (C - F) ^ 2) * 100 + .5) / 100
70  PRINT : GOTO 10
```

Page 496

2. a. The midpoint opposite 4, 7 is 5, $\frac{1}{2}$. The equation of the median through the two points is $y - 4 = \frac{6\frac{1}{2}}{-1}(x - 4)$. This simplifies to $13x + 2y = 66$. The midpoint opposite 2, 0 is 6, 4. The equation of the median through those two points is $y - 4 = \frac{4}{4}(x - 6)$, which simplifies to $x - y = 2$. The solution of the two is $\frac{14}{3}, \frac{8}{3}$.
The midpoint opposite 5, 7 is the point 5, 1, which makes the equation $x = 5$. The midpoint opposite 2, 1 is $6\frac{1}{2}$, 4. The equation of the line through the two points is $y - 1 = \frac{3}{4\frac{1}{2}}(x - 2)$, which simplifies to $2x - 3y = 1$. Substitute $x = 5$ and you get the ordered pair (5, 3).

Computer Section Answers

Page 566

1.
```
10  PRINT "3*8+4*6-5 = ":3 * 8 + 4 * 6 - 5
20  PRINT "2(8-6)(2+7) = ":2 * (8 - 6) * (2 + 7)
30  PRINT "(3^2-4*6)/(8+2) = ":(3 ^ 2 - 4 * 6) /
    (8 + 2)
```

2.
```
10  PRINT "|5-3| = ": ABS (5 - 3)
20  PRINT "|5-13| = ": ABS (5 - 13)
30  PRINT "|0-(-4)| = ": ABS (0 - ( - 4))
40  PRINT "|-9-(-7)| = ": ABS ( - 9 - ( - 7))
50  PRINT "|-16-9| = ": ABS ( - 16 - 9)
60  PRINT "|-5-(-5)| = ": ABS ( - 5 - ( - 5))
```

Page 567

1.
```
10  PRINT "GIVE ME AN ACUTE ANGLE IN DEGREES, MI
    NUTES, AND SECONDS."
20  PRINT "NUMBER OF DEGREES":: INPUT D
30  PRINT "NUMBER OF MINUTES":: INPUT M
35  PRINT "NUMBER OF SECONDS":: INPUT S
40  PRINT "THE COMPLEMENT IS ";89 - D;" DEGREES
    ";59 - M;" MINUTES ";60 - S;" SECONDS."
50  PRINT "THE SUPPLEMENT IS ";179 - D;" DEGREES
    ";59 - M;" MINUTES ";60 - S;" SECONDS."
60  PRINT : GOTO 10
```

Page 568

1.
```
10  READ A,B,C
20  IF A =  - 99 THEN  STOP
30  PRINT A;" , ";B;" , AND ";C;
40  IF A + B + C = 180 THEN 60
50  PRINT " DO NOT FORM A TRIANGLE": GOTO 10
60  PRINT " CAN BE A TRIANGLE": GOTO 10
70  DATA 30,60,90,45,45,90,10,10,70,20,20,40,1,1
    ,178,-99,0,0
```

Page 569

1.
```
110  PRINT "WHAT ARE THE THREE ANGLES":: INPUT A
     ,B,C
120  IF A < B THEN 170
130  IF A < C THEN 210
140  IF A > 90 THEN 270
150  IF A = 90 THEN 250
160  GOTO 230
170  IF B < C THEN 210
180  IF B > 90 THEN 270
190  IF B = 90 THEN 250
200  GOTO 230
210  IF C > 90 THEN 270
220  IF C = 90 THEN 250
230  PRINT "THE TRIANGLE IS ACUTE."
240  GOTO 110
250  GOTO 230
260  GOTO 110
270  PRINT "THE TRIANGLE IS OBTUSE."
280  GOTO 110
```

Page 571

1.
```
10  FOR I = 1 TO 7
20  READ Q$,A$
30  PRINT "IS THE NAME ";Q$;" FOR TRIANGLES BASE
    D ON SIDES OR ANGLES?"
40  INPUT R$
50  IF R$ = A$ THEN 70
60  PRINT "NOT CORRECT.": GOTO 90
70 R = R + 1: PRINT "CORRECT."
90  PRINT : NEXT I
100  PRINT "YOUR SCORE IS ": INT (R * 100 / 7 +
     .5)"."
200  DATA SCALENE,SIDES,RIGHT,ANGLES,ACUTE,ANGLE
     S,ISOSCELES,SIDES,OBTUSE,ANGLES
210  DATA EQUIANGULAR,ANGLES,EQUILATERAL,SIDES
```

2.
```
10  FOR I = 1 TO 10
20  READ A,B,C,A$
30  PRINT "WHAT KIND OF TRIANGLE HAS SIDES ";A;"
    ,";B;", AND ";C;
40  INPUT R$: IF R$ = A$ THEN 80
50  IF A$ = "RIGHT" THEN 70
60  PRINT "NOT CORRECT": GOTO 90
70  IF R$ <  > "SCALENE" THEN 50
80 R = R + 1: PRINT "CORRECT"
90  PRINT : NEXT I
100  PRINT "YOUR SCORE IS ";R * 10
200  DATA 5,6,7,SCALENE,6,6,8,ISOSCELES,6,8,10,R
     IGHT,6,6,6,EQUILATERAL
210  DATA 5,12,13,RIGHT,6,8,9,SCALENE,2,3,3,ISOS
     CELES,2.5,2.5,2.5,EQUILATERAL
220  DATA 3,4,5,RIGHT,12,16,20,RIGHT
```

Page 572

1.
```
10  PRINT "WHAT IS THE FIRST NUMBER":: INPUT A
20  PRINT "LAST NUMBER ":: INPUT B
30  FOR I = A TO B
40  PRINT "EACH EXTERIOR ANGLE OF A REGULAR POLY
    GON OF ";I;" SIDES IS ";
50  PRINT 360 / I"."
60  NEXT I
```

2.
```
10  PRINT "SIDES INTERIOR    EXTERIOR    SUM OF
     INTERIOR"
20  FOR I = 3 TO 15
30  PRINT  TAB( 2);I; TAB( 7);(I - 2) * 180 / I;
40  PRINT  TAB( 20);360 / I; TAB( 32);(I - 2) *
     180
50  NEXT I
```

3.
```
10  PRINT "ANGLE   COMPLEMENT   SUPPLEMENT"
20  FOR I = 10 TO 170 STEP 10
30  PRINT  TAB( 2);I;
40  IF I > 89 THEN 60
50  PRINT  TAB( 12);90 - I;
60  PRINT  TAB( 26);180 - I
70  NEXT I
```

Page 575

1.
```
10 G =  INT ( RND (1) * 16) + 20
11 H =  INT ( RND (1) * 16) + 20
12 I =  INT ( RND (1) * 16) + 20
20 A = G:B = H:C = I: GOSUB 200
30 G = A:H = B:I = C
35  PRINT "MY TRIANGLE HAS SIDES ";G;", ";H;". A
     ND ";I"."
40  PRINT "WHAT ARE THE SIDES OF ANY TRIANGLE SI
     MILAR TO THAT ONE";: INPUT D,E,F
50 A = D:B = E:C = F: GOSUB 200
60 D = A:E = B:C = F
65  REM  ABS IS USED IN TESTS BELOW SINCE THE DI
     VISION MIGHT ROUND OFF DIFFERENTLY.
70  IF  ABS (G / D - H / E) > .0001 THEN 110
80  IF  ABS (G / D - I / F) > .0001 THEN 110
90  IF  ABS (H / E - I / F) > .0001 THEN 110
100  PRINT "THESE TRIANGLES ARE SIMILAR."
105  PRINT : GOTO 10
110  PRINT "THESE TRIANGLES ARE NOT SIMILAR.": PRINT
     : GOTO 10
200  IF A < B THEN 260: REM   SUBROUTINE STARTS
     HERE
210  IF A < C THEN 240
220  IF B < C THEN 250
230 Q = A:A = C:C = Q: GOTO 310
240 Q = A:A = B:B = Q: GOTO 310
250 Q = A:A = B:B = C:C = Q: GOTO 310
260  IF B < C THEN 290
270  IF A < C THEN 300
280 Q = A:A = C:C = B:B = Q: GOTO 310
290  GOTO 310
300 Q = C:C = B:B = Q: GOTO 310
310  RETURN : REM        SENDS PROGRAM BACK
```

2.
```
10  FOR I = 1 TO 10
20 Q =  INT ( RND (1) * 2) + 1
30  ON Q GOTO 40,120
40 X =  INT ( RND (1) * 10) + 1
50 Y =  INT ( RND (1) * 15) + 5
60  IF Y = X THEN 50
70  PRINT "ONE ENDPOINT OF A SEGMENT IS ";X", "
80  PRINT "AND THE MIDPOINT IS ";Y;". WHAT IS TH
     E OTHER ENDPOINT";: INPUT A
100  IF A = 2 * Y - X THEN 220
110  GOTO 210
120 L =  INT ( RND (1) * 12) + 4
130 M =  INT ( RND (1) * 21) - 10
140  PRINT "IF THE MIDPOINT OF A SEGMENT IS ";M;
     ", AND THE LENGTH IS ";L;","
160  PRINT "WHAT IS THE LEFT ENDPOINT";: INPUT A
170  IF A = M - L / 2 THEN 190
180  GOTO 210
190  PRINT "WHAT IS THE RIGHT ENDPOINT ";: INPUT
     A
200  IF A = M + L / 2 THEN 220
210  PRINT "INCORRECT.": GOTO 230
220 R = R + 1: PRINT "CORRECT."
230  PRINT : NEXT I
240  PRINT : PRINT "YOUR SCORE IS ";R * 10"."
```

3.
```
5  FOR I = 1 TO 10
10 A =  INT ( RND (1) * 10) + 1
20 B =  INT ( RND (1) * 10) + 1
30 C =  INT ( RND (1) * 10) + 1
40  PRINT "CAN A TRIANGLE HAVE SIDES ";A;",";B;"
     ,";C;
50  IF C =  > A + B THEN 90
60  IF B =  > A + C THEN 90
70  IF A =  > B + C THEN 90
80 R$ = "YES": GOTO 100
90 R$ = "NO"
100  INPUT A$: IF A$ = R$ THEN 120
110  PRINT "INCORRECT.": GOTO 210
120  PRINT "CORRECT.": IF R$ = "NO" THEN 210
130  PRINT "IS IT A RIGHT TRIANGLE";: INPUT A$
140  IF A ^ 2 + B ^ 2 = C ^ 2 THEN 180
150  IF A ^ 2 + C ^ 2 = B ^ 2 THEN 180
160  IF B ^ 2 + C ^ 2 = A ^ 2 THEN 180
170 R$ = "NO": GOTO 185
180 R$ = "YES"
185  IF A$ = R$ THEN 200
190  PRINT "INCORRECT.": GOTO 210
200 R = R + 1: PRINT "CORRECT."
210  PRINT : NEXT I
220  PRINT : PRINT "YOUR SCORE IS ";R * 10"."
```

4.
```
10  PRINT "NUMERATOR";: INPUT N
20  PRINT "DENOMINATOR";: INPUT D
25  PRINT N;"/";D;" = ";
30  FOR I = 2 TO N
35  IF N < I THEN 120
40  IF N / I =  INT (N / I) THEN 60
50  GOTO 100
60  IF D / I =  INT (D / I) THEN 80
70  GOTO 100
80 N = N / I:D = D / I
90  GOTO 40
100  NEXT I
120  PRINT N;"/";D
130  PRINT : GOTO 10
```

5.
```
10  PRINT "HOW MANY INCHES";: INPUT N
20  IF N < 12 THEN 70
25  IF N / 12 =  INT (N / 12) THEN 90
30 F =  INT (N / 12)
40 I = N - F * 12
50  PRINT N;" INCHES = ";F;" FEET, ";I;" INCHES"
60  PRINT : GOTO 10
70  PRINT N;" INCHES CANNOT BE SIMPLIFIED."
80  PRINT : GOTO 10
90  PRINT N;" INCHES = ";N / 12;" FEET"
95  PRINT : GOTO 10
```

6.
```
10  PRINT "WHAT IS THE PERIMETER?"
15  PRINT "FEET";: INPUT F: PRINT "INCHES";: INPUT
     I
20  PRINT "WHAT IS THE KNOWN SIDE?"
25  PRINT "FEET";: INPUT G: PRINT "INCHES";: INPUT
     J
30  PRINT "IS THAT SIDE A LEG OR THE BASE?"
35 A = 0
40  INPUT A$
50 H = F * 12 + I:L = G * 12 + J
60  IF A$ = "LEG" THEN 75
70 D = (H - L) / 2: GOTO 80
75 D = H - 2 * L
80 A =  INT (D / 12)
90 D = D - A * 12
115  IF A$ = "LEG" THEN 150
120  PRINT "THE OTHER 2 SIDES ARE EACH"
125  PRINT A;" FEET ";D;" INCHES.": GOTO 10
150  PRINT "THE OTHER 2 SIDES ARE ";A;" FEET ";D
     ;" INCHES"
160  PRINT  TAB( 18);" AND ";G;" FEET ";J;" INCH
     ES.": GOTO 10
```

Page 577

1.
```
10  FOR I = 32 TO 41
20 R = I * 6.2832 / 360
30  PRINT I: TAB( 10): INT ( SIN (R) * 1000 + .5
     ) / 1000;
35  PRINT  TAB( 20); INT ( COS (R) * 1000 + .5) /
     1000;
40  PRINT  TAB( 30); INT ( TAN (R) * 1000 + .5) /
     1000
50  NEXT I
```

2.
```
10  PRINT "WHAT IS ONE OF THE ACUTE ANGLES";:
     INPUT A
20  PRINT "WHAT IS ONE LEG";: INPUT L
30 R = A * 6.2832 / 360
40  PRINT "IS THAT SIDE OPPOSITE THE ANGLE";:
     INPUT R$
50  IF R$ = "YES" THEN 70
60 S =  TAN (R) * L: GOTO 80
70 S = L /  TAN (R)
80  PRINT "THE OTHER LEG IS "; INT (S * 100 + .5
     ) / 100"."
90  PRINT "THE HYPOTENUSE IS "; INT ( SQR (L ^ 2
      + S ^ 2) * 100 + .5) / 100"."
95  PRINT : GOTO 10
```

HOLT GEOMETRY

The three photographic special interest sections in this book are designed to spark the imagination and emphasize the interdependence of progress and mathematics.

Architecture shows a range of instruments, drawings, and constructions that feature geometric shapes in buildings and structures from early civilizations through modern skyscrapers.

Geometry in Design and Technology illustrates applications of construction and technological advance in design. Such applications as the hydrofoil, helicopter, various propellers, and jets are pictured along with the effective uses of lasers and computer graphics in engineering schemes.

Geometry in Patterns highlights the beauty of mathematics evident in the natural and human-made world. Geometric design is illustrated in art, nature, construction, and architecture, by aerial photography and computer graphics.

Any one of the photographs may serve as a starting point for further investigation about the subject and the mathematics that supports it. You may find it useful to ask for volunteers, either alone or in small groups, to pursue topics illustrated, formulate problems, and look for solutions. Papers can be written and displays made and presented to the class or put on bulletin boards.

About the Authors

Eugene D. Nichols
Robert O. Lawton Distinguished Professor
of Mathematics Education
Florida State University
Tallahassee, Florida

Mervine L. Edwards
Chairman of Mathematics Department
Shore Regional High School
West Long Branch, New Jersey

E. Henry Garland
Head of the Mathematics Department
Developmental Research School
DRS Professor
Florida State University
Tallahassee, Florida

Sylvia A. Hoffman
Resource Consultant in Mathematics
Illinois State Board of Education
State of Illinois

Albert Mamary
Superintendent of Schools for Instruction
Johnson City Central School District
Johnson City, New York

William F. Palmer
Professor of Education and Director
Center for Mathematics and Science Education
Catawba College
Salisbury, North Carolina

Computer Consultant

Michael Murphy
Mathematics Department Chairman
Bonny Eagle High School
West Buxton, Maine

HOLT GEOMETRY

Eugene D. Nichols
Mervine L. Edwards
E. Henry Garland
Sylvia A. Hoffman
Albert Mamary
William F. Palmer

HOLT, RINEHART AND WINSTON, PUBLISHERS
New York • Toronto • Mexico City • London • Sydney • Tokyo

Acknowledgments for Photographs

Cover: TransAmerica Building, Mike Mazzaschi/Stock, Boston.

Page x, David Plowden/Photo Researchers; *17*, Owen Franken/Stock, Boston; *22*, Sonia Sheridan/Photo Researchers; *54*, Jean-Claude Le Jeune/Stock, Boston; *68*, Ellis Herwig/Stock, Boston; *98*, Karen Bussolini; *102*, Elizabeth Crews; *128*, Phil Degginger/Bruce Coleman; *135*, Granger; *177*, Ellis Herwig/Stock, Boston; *202*, Bonnie Rauch/Image Bank; *207*, Bryn Mawr College Archives; *233*, Lawrence Fried/Image Bank; *238*, Moos-Hake/Greenberg/Peter Arnold; *262*, Frank Siteman/Stock, Boston; *269*, Bruno Zehnder/Peter Arnold; *300*, James R. Holland/Stock, Boston; *305*, Art Resource; *332*, E.R. Degginger/Bruce Coleman; *337*, Maryland Historical Society, Baltimore; *380*, Granger; *385*, Robert Phillips/Image Bank; *421*, Dan Esgro/Image Bank; *428*, Keith D. Henley/Photo Researchers; *461*, Wendell Metzen/Bruce Coleman; *493*, Tim Bieber/Image Bank; *498*, Anthony Edgeworth/Image Bank; *519*, Claudia Parks/Stock Market; *553*, O. Cameron Davidson/Bruce Coleman.

Insert 1:
Page 1, Nicholas DeSciose/Stock Broker; *2*, *t*, Granger; *b*, Runk/Schoenberger/Grant Heilman; *3*, *tl*, Granger; *tr*, Art Resource; *b*, Harvey Olsen; *4*, *tl*, Charles Gupton/Southern Light; *tr*, C. Bonington/Woodfin Camp & Assoc.; *b*, Granger; *5*, *t*, Alinari/Art Resource; *b*, Dan McCoy/Rainbow; *6*, *t*, Elizabeth Crews; *b*, Coco McCoy/Rainbow; *7*, *l*, Robert Frerck/Woodfin Camp & Assoc.; *tr*, Marty Katz; *br*, Dick Durrance II/Woodfin Camp & Assoc.; *8*, *tl*, Mike Yamashita/Woodfin Camp & Assoc.; *r*, Ian Griffiths/Robert Harding Picture Library, London; *bl*, Scala/Art Resource.

Insert 2:
Page 1, Derek Berwin/Image Bank; *2*, *t*, Dr. Gary Settles/Photo Researchers; *b*, Jean Pragen/Image Bank; *3*, *tl*, Marty Katz; *tr*, Granger; *b*, Stock Shop; *4*, *l*, Bohdan Hrynewych/Southern Light; *tr*, Dan McCoy/Rainbow; *br*, General Motors Research Laboratories; *5*, *tl*, Ray Ellis/Photo Researchers; *tr*, Granger; *b*, Harvey Olsen; *6*, *l*, Wolf Von Dem Bussche/Image Bank; *tr*, Sue Thorpe/*Cruise Magazine; br*, P. Paz/Image Bank; *7*, *t*, Al Satterwhite/Image Bank; *br*, Don Landwehrle/Image Bank; *bl*, Charles Gupton/Southern Light; *8*, *tl*, Mitchell Funk/Image Bank; *tr*, John Zoiner/Peter Arnold; *b*, Grant Heilman.

Insert 3:
Page 1, Adolahe/Southern Light; *2*, *tl*, Art Resource; *r*, Manfred Kage/Peter Arnold; *bl*, Hans Pfletschinger/Peter Arnold; *3*, *l*, Grant Heilman; *tr*, Paul Brierly/Robert Harding Picture Library; *br*, Grant Heilman; *4*, *tl*, Peter Eising/Image Bank; *r*, Craig Aurness/West Light; *bl*, Scala/Art Resource; *5*, *t*, Geoffrey Gove; *b*, Dr. E.P. Miles/Miles Color Art; *6*, *tl*, detail of sculpture by Alexander Calder, Zorino/Art Resource; *tr*, Adam Woolfitt/Woodfin Camp & Assoc.; *b*, Jeffery L. Cook/Stock Broker; *7*, *l*, W.H. Hodge/Peter Arnold; *tr*, Robert P. Carr/Bruce Coleman; *br*, W.H. Hodge/Peter Arnold; *8*, *t*, Charles Weckler/Image Bank; *br*, Tim Bieber/Image Bank; *bl*, Spence McConnell/Bruce Coleman.

Printed in the United States of America

ISBN: 0-03-002184-7

5678901234 032 98765432

Contents

1 Figures and Measures x

2 Introducing Proofs 22

3 Parallels 68

4 Proving Triangles Congruent 102

5 Applying Congruent Triangles 135

6 Exploring Polygons 182

7 Quadrilaterals 207

8 Trapezoids and Symmetric Polygons 238

Symbol List

1 FIGURES AND MEASURES

Non-Routine Problem Solving

Problem 1. Ten posts are arranged in a straight line with 10 meters between consecutive posts. Mr. Jones must nail a sign on each of the posts. He wants to do as much walking as possible while doing this. As long as he walks only on the straight line of the posts, starting at any post he chooses, what would be the sequence of posts that would give the longest possible walk? **Start at post number 5. The sequence of posts would be 5-10-1-9-2-8-3-7-4-6. He has to walk 490 meters.**

Problem 2. There are 5 lines in a plane. None of the lines are parallel, and no three of the lines are concurrent. (*Concurrent* means that the lines pass through the same point.) These 5 lines determine 15 regions, some of which have a closed boundary and some of which have an open boundary. How many regions have a closed boundary? How many have an open boundary? **5 regions with closed boundary; 10 regions with open boundary**

POINTS, LINES, AND PLANES 1.1

Objective **To identify and name geometric figures such as points, lines, planes, collinear points, and coplanar points**

Geometry is the study of sets of points. When beginning any field of study, you must pay special attention to the meanings of the words you use. When defining words, you must express them in terms of simpler words whose meanings you already know. This means that you must begin with a few simple words that remain undefined but whose meanings are nevertheless understood. Some **undefined terms** of geometry are **point, line,** and **plane.** These terms serve as a basis for defining other geometric terms.

Even though point, line, and plane are not formally defined, the chart below will give you information about identifying and naming them.

Undefined term	Picture	How to name	Other information
point	dot •	capital letter • P Read: point *P*	has no length, width, or thickness; occupies no space
line	straight mark with two arrowheads 	two capital letters with a double arrowhead above them, or a lowercase letter  $\overleftrightarrow{AB}$, $\overleftrightarrow{BA}$, or l Read: line *AB*, line *BA*, or line *l*	has length, but no width or thickness; is straight; is an infinite set of points that extends indefinitely in two directions
plane	slanted four-sided figure 	lowercase letter  Read: plane *m*	has length and width, but no thickness; is a flat surface; is an infinite set of points that extends indefinitely in all directions

Example 1

Give seven possible names for the line shown below.

The line can be named $\overleftrightarrow{RS}$, $\overleftrightarrow{SR}$, $\overleftrightarrow{RT}$, $\overleftrightarrow{TR}$, $\overleftrightarrow{ST}$, $\overleftrightarrow{TS}$, or n.

Once you have accepted a few basic undefined terms, you can use them to define other terms. A good definition tells what is being defined. It uses only accepted undefined terms or previously defined terms. The definition should be worded so as to include all elements that belong to the set being defined and to exclude all elements that do not belong.

Definitions:

Space — *Space* is the set of all points.

Geometric figure — A *geometric figure* is a set of points.

Collinear points — *Collinear points* are points that lie on the same line. Points that do not lie on the same line are ***noncollinear.***

Coplanar points — *Coplanar points* are points that lie in the same plane. Points that do not lie in the same plane are ***noncoplanar.***

Example 2

In the figure, find 3 collinear points, 3 coplanar points, 3 noncollinear points, and 4 noncoplanar points.

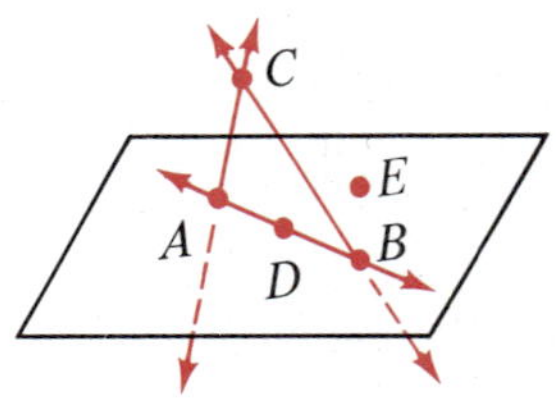

A, D, and B are collinear.
A, B, and E are coplanar.
A, B, and C are noncollinear.
A, B, C, and E are noncoplanar. (Other answers are possible.)

Observe the different ways in which points, lines, and planes can be positioned. Notice that point P *lies on* line l, but point Q does not. Also, point P *is contained in* line l, or line l *contains* point P.

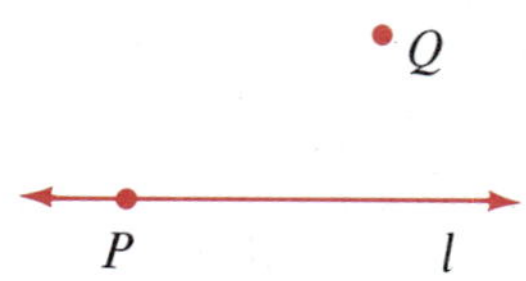

Notice that lines m and n *intersect* at point A, but lines r and s do not intersect. The *intersection* of lines m and n is point A.

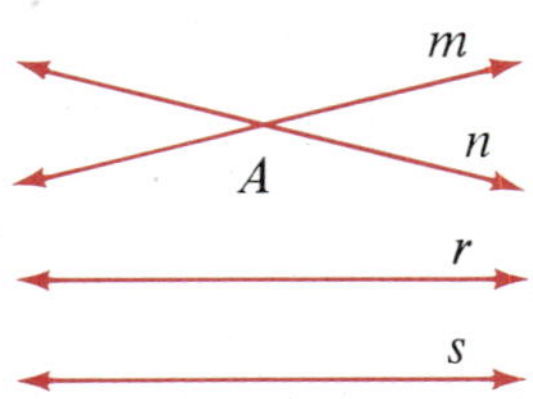

Line *t lies in* plane *p*, or line *t is contained in* plane *p*. Line *u intersects* plane *p* in point *C*. Line *v* does not intersect plane *p*.

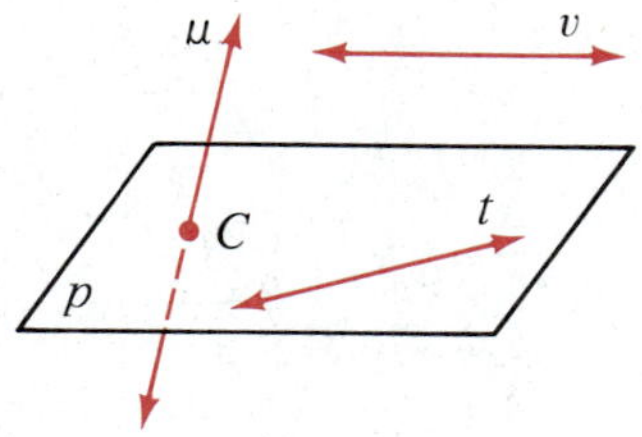

Reading in Geometry

Match each term on the right with its description on the left.

1. Points that lie on the same line **e**
2. The set of all points **f**
3. Occupies no space **d**
4. A flat surface **c**
5. Points that lie in the same plane **b**
6. Has length, but no width or thickness **a**

a. line
b. coplanar points
c. plane
d. point
e. collinear points
f. space

Oral Exercises

Give all possible names for each figure shown below.

1.

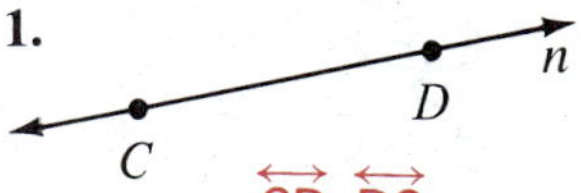

$\overleftrightarrow{CD}$, $\overleftrightarrow{DC}$, n

2.

point S

3.

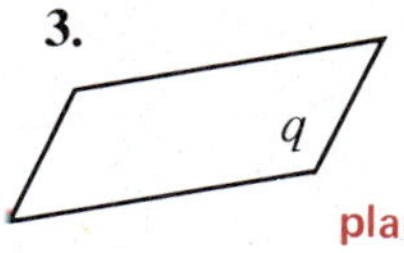

plane q

4.

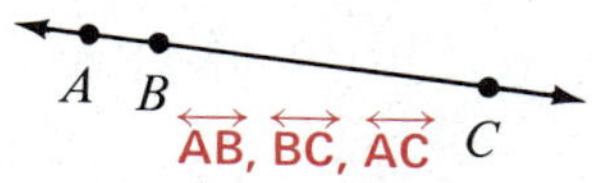

$\overleftrightarrow{AB}$, $\overleftrightarrow{BC}$, $\overleftrightarrow{AC}$
$\overleftrightarrow{CA}$, $\overleftrightarrow{CB}$, $\overleftrightarrow{BA}$

What geometric figure is suggested by each of the following?

5. A laser beam **line**
6. A tabletop **plane**
7. A wall of a room **plane**
8. The tip of a sharp pencil **point**

Written Exercises

Identify the following points as *collinear, coplanar but noncollinear,* or *noncoplanar*.

A

1. *R, U, W* **collinear**
2. *S, T, U, R* **noncoplanar**
3. S, U, V **coplanar but noncollinear**
4. *V, U, S, W* **noncoplanar**
5. *T, U, V* **collinear**
6. *S, T, U* **coplanar but noncollinear**

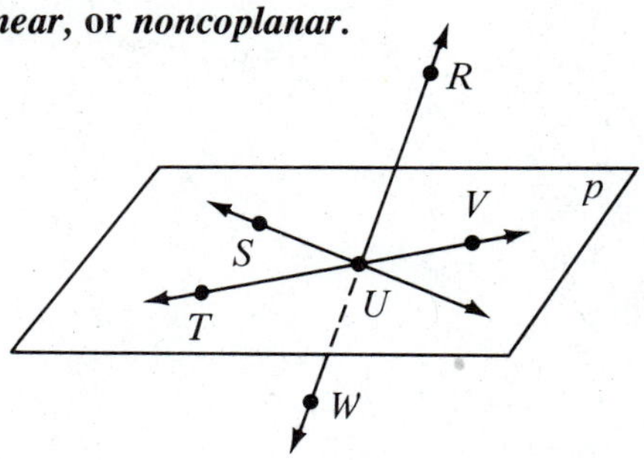

True or false?

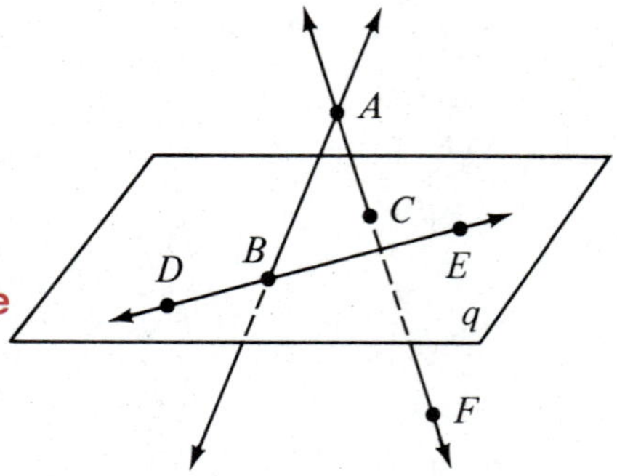

7. $\overleftrightarrow{DE}$ is contained in plane q. **true**
8. $\overleftrightarrow{AC}$ and $\overleftrightarrow{FA}$ are the same line. **true**
9. $\overleftrightarrow{AB}$ and $\overleftrightarrow{DE}$ intersect in point C. **false**
10. B, C, and E are noncoplanar. **false**
11. Point B is the intersection of $\overleftrightarrow{AB}$ and plane q. **true**
12. The intersection of $\overleftrightarrow{AF}$ and plane q is point C. **true**
13. $\overleftrightarrow{CF}$ passes through point E. **false**
14. Point A and plane q do not intersect. **true**
15. $\overleftrightarrow{DE}$ contains point B. **true**

Draw and label each figure described below. Use a straightedge.

B 16. $\overleftrightarrow{MN}$ lying in plane r and point P in plane r but not on $\overleftrightarrow{MN}$

17. $\overleftrightarrow{TU}$ intersecting plane q in point V

18. $\overleftrightarrow{AB}$ and $\overleftrightarrow{CD}$ lying in plane q such that $\overleftrightarrow{AB}$ and $\overleftrightarrow{CD}$ intersect at point E

19. $\overleftrightarrow{EF}$, $\overleftrightarrow{EG}$, and $\overleftrightarrow{EH}$ intersecting plane p in points F, G, and H, respectively

20. $\overleftrightarrow{MN}$ not intersecting plane r with $\overleftrightarrow{MQ}$ intersecting plane r in point Q

21. noncollinear points A, B, and C lying in plane p with $\overleftrightarrow{DC}$ intersecting plane p

Give a brief explanation for each of the following.

C 22. Why is it necessary to begin the study of geometry with undefined terms?
See development on p. 1.

23. Why is it impossible for four noncoplanar points to be collinear?
If 3 pts are collinear, then all 4 will be coplanar. At most only 2 pts can be collinear since the 4 pts given are noncoplanar.

ALGEBRA REVIEW

Objective

To simplify expressions involving absolute value

$|n|$ means the *absolute value* of n.

$|n| = n$, if n is a positive number, or 0.
$|n| = -n$, if n is a negative number.

Thus, $|5| = 5$, $|0| = 0$, and $|-4| = -(-4)$, or 4.
Notice that the absolute value of a number cannot be negative.

Example

Simplify $|-5 - (-3)|$.

$|-5 - (-3)| = |-5 + 3| = |-2| = -(-2)$, or 2

Simplify.

1. $|5 - 3|$ **2**
2. $|9 - 4|$ **5**
3. $|5 - 13|$ **8**
4. $|12 - 18|$ **6**
5. $|0 - (-4)|$ **4**
6. $|13 - (-9)|$ **22**
7. $|-9 - (-7)|$ **2**
8. $|-4 - (-12)|$ **8**
9. $|-16 - 9|$ **25**
10. $|-5 - 5|$ **10**
11. $|-5 - (-5)|$ **0**
12. $|-8 + 13|$ **5**

DISTANCE AND LENGTH 1.2

Objectives

To find the coordinate of a point on a number line
To find the distance between two points on a number line
To identify and name segments and rays
To measure a segment to the nearest centimeter or millimeter

Sometimes it is useful to assign numbers to the various points of a line. When this is done, the line is called a **number line.** The number assigned to each point is called the **coordinate** of the point.

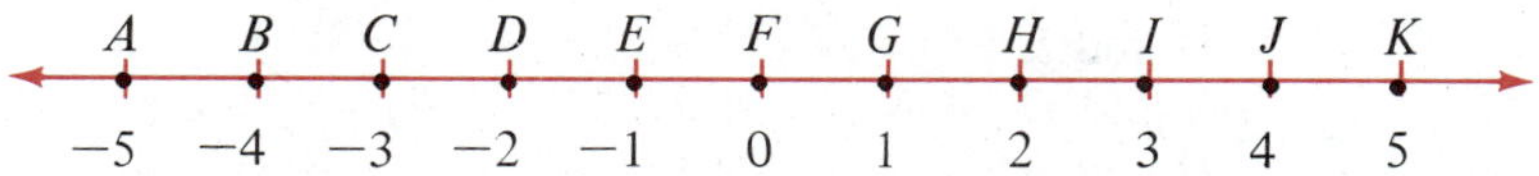

On the number line above, point D corresponds to -2, so -2 is the coordinate of D. On a given number line, no two points have the same coordinate. There is a **one-to-one correspondence** between the points and the coordinates. The symbol $\leftrightarrow$ is used to show the correspondence between a point and its coordinate. $A \leftrightarrow -5$ means "point A corresponds to the number -5."

Example 1

For each point, A, B, C, D, find its corresponding coordinate.

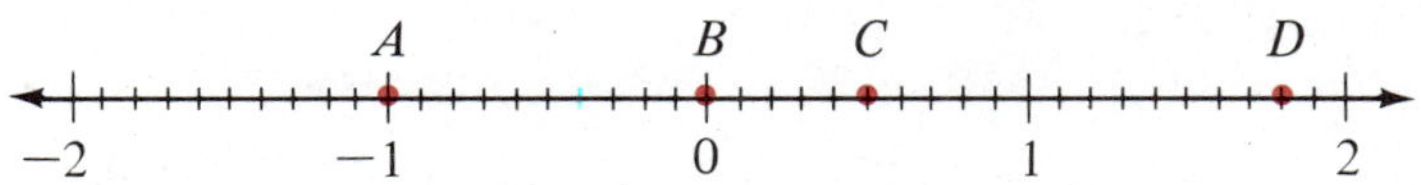

$A \leftrightarrow -1$ $B \leftrightarrow 0$ $C \leftrightarrow 0.5$ $D \leftrightarrow 1.8$

Different one-to-one correspondences between points and numbers can be set up if different number lines are used. On the first number line below, point W corresponds to -5. On the second, W corresponds to -3.

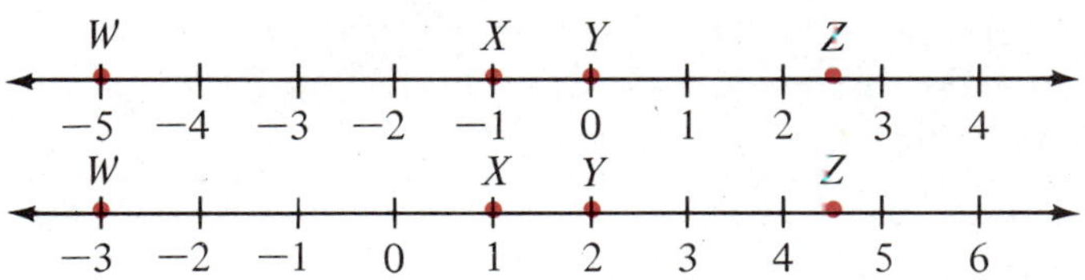

$W \leftrightarrow -5$ $X \leftrightarrow -1$ $Y \leftrightarrow 0$ $Z \leftrightarrow 2.5$

$W \leftrightarrow -3$ $X \leftrightarrow 1$ $Y \leftrightarrow 2$ $Z \leftrightarrow 4.5$

A ruler is like a number line since a number (coordinate) is assigned to every point of the ruler. You can use a ruler to measure the distance between two points. In the diagrams at the right, the distance between points P and Q is 3 cm, no matter which way the ruler is placed. Note that $3 - 0 = 3$ and $7 - 4 = 3$. The distance is the same as the difference between the coordinates.

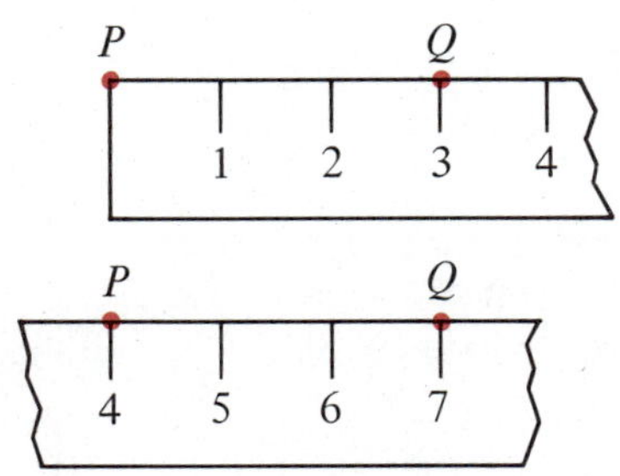

Definition: Distance The *distance between two points*, A and B, with coordinates a and b, is $|b - a|$ or $|a - b|$. The symbol AB means "the distance between points A and B."

Example 2

Find the distance between each pair of points listed below.

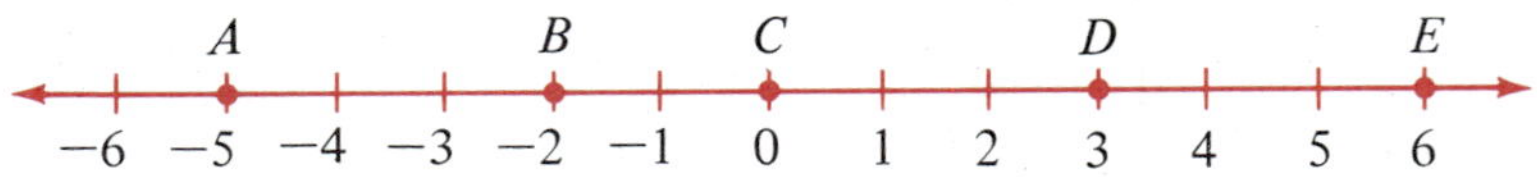

Points	**Distance**
D and E	$DE = \|6 - 3\| = \|3\| = 3$
B and D	$BD = \|3 - (-2)\| = \|5\| = 5$
D and A	$DA = \|-5 - 3\| = \|-8\| = 8$
B and C	$BC = \|0 - (-2)\| = \|2\| = 2$
B and A	$BA = \|-5 - (-2)\| = \|-3\| = 3$

In the figure at the right, point C is between points A and B. Neither D nor E is between A and B. The term *between* can be defined in relation to distances.

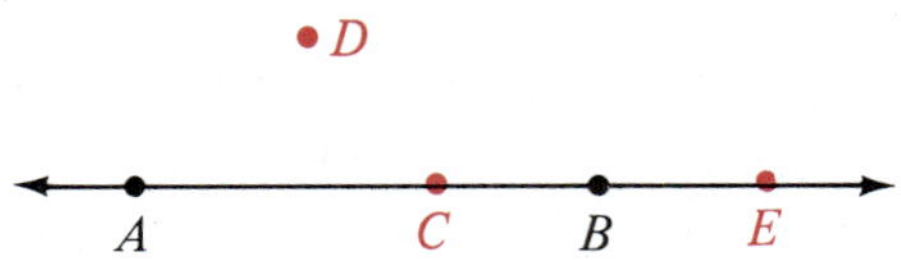

Definition: Between Point C is *between* points A and B if C *is on* $\overleftrightarrow{AB}$ and $AC + CB = AB$.

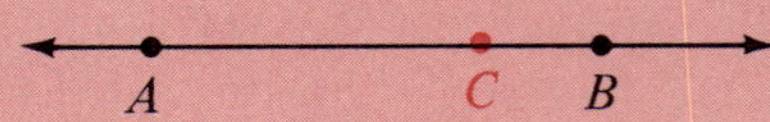

Two other important geometric figures can now be defined.

Definitions: Segment

Segment AB is the set of points consisting of A, B, and all points between A and B. A and B are called the endpoints of the segment. Either $\overline{AB}$ or $\overline{BA}$ can be used to name segment AB.

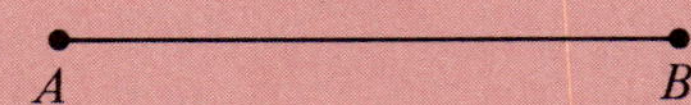

Ray

Ray AB is the set of points consisting of $\overline{AB}$ and all points P such that B is between A and P. A is called the endpoint of the ray. The symbol $\overrightarrow{AB}$ is used to name ray AB.

Example 3

Give all other names for each figure: $\overline{PQ}$, $\overline{SQ}$, $\overrightarrow{QS}$, $\overrightarrow{SR}$.

Other names: $\overline{PQ} = \overline{QP}$ $\quad\overline{SQ} = \overline{QS}$ $\quad\overrightarrow{QS} = \overrightarrow{QR}$ $\quad\overrightarrow{SR} = \overrightarrow{SQ} = \overrightarrow{SP}$

Definition: Length of segment The *length* (or *measure*) of $\overline{AB}$ is the distance between points A and B. The symbol AB means "the length of $\overline{AB}$."

Example 4

Find the length of $\overline{XY}$ to the nearest centimeter (cm) and to the nearest millimeter (mm).

$\overline{XY}$ is about 4 cm long.
$\overline{XY}$ is about 42 mm long.

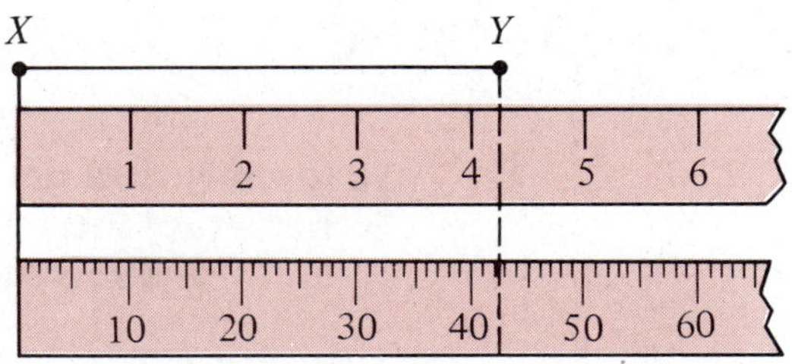

Written Exercises

Give the coordinate of each point.

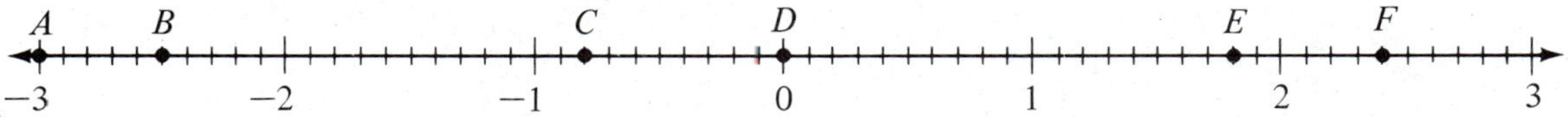

A 1. *A* **−3** 2. *B* **−2.5** 3. *C* **−0.8** 4. *D* **0** 5. *E* **1.8** 6. *F* **2.4**

Find each indicated distance. Use the number line.

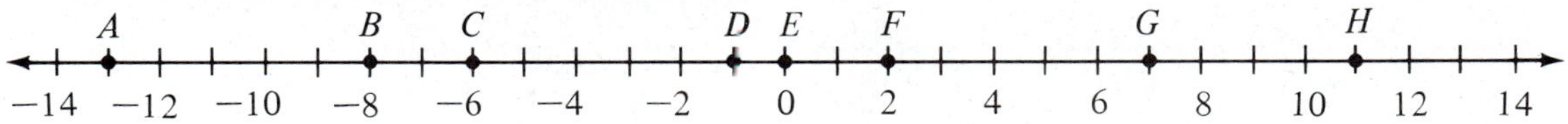

7. *EG* **7** 8. *BE* **8** 9. *FG* **5** 10. *AC* **7** 11. *BG* **15** 12. *HA* **24**

Give all other names for each figure named below. Use the number line.

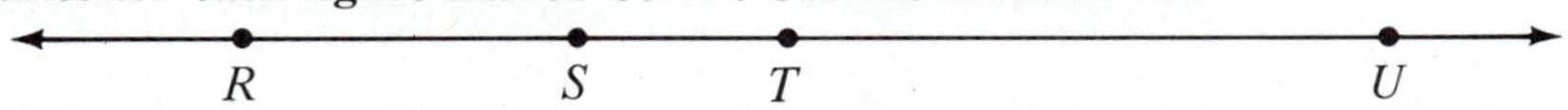

13. $\overline{TS}$ $\overline{ST}$ 14. $\overrightarrow{SU}$ $\overrightarrow{ST}$ 15. $\overline{SU}$ $\overline{US}$ 16. $\overrightarrow{TR}$ $\overrightarrow{TS}$ 17. $\overrightarrow{RU}$ $\overrightarrow{RS}$; $\overrightarrow{RT}$ 18. $\overleftrightarrow{RT}$
$\overleftrightarrow{RS}$; $\overleftrightarrow{RU}$; $\overleftrightarrow{ST}$; $\overleftrightarrow{SU}$; $\overleftrightarrow{TU}$; $\overleftrightarrow{UT}$; $\overleftrightarrow{US}$; $\overleftrightarrow{TS}$; $\overleftrightarrow{UR}$; $\overleftrightarrow{SR}$; $\overleftrightarrow{TR}$

Measure each segment to the nearest centimeter and to the nearest millimeter.

19. 

3 cm; 28 mm

20. 

3 cm; 34 mm

21. 

1 cm; 12 mm

Complete each statement. Use the number line.

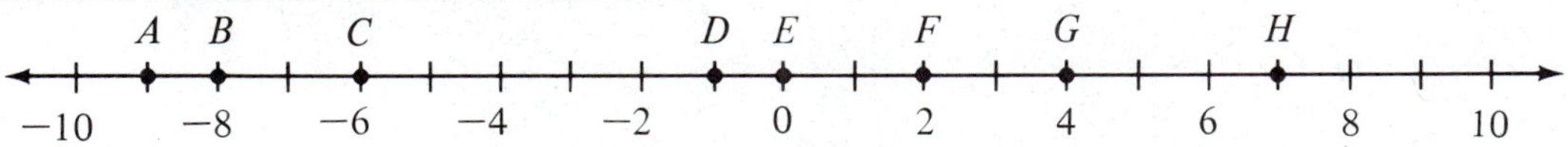

B 22. $AC = D?$ **F** 23. $HG = A?$ **C** 24. $CD = H?$ **F** 25. $AD = E?$ **B**

The coordinate of *A* is given. Find all possible coordinates of *B* so that $\overline{AB}$ has the indicated length.

26. $A \leftrightarrow 6$; $AB = 8$ **14; −2** 27. $A \leftrightarrow -3$; $AB = 4$ **1, −7** 28. $A \leftrightarrow 1.2$; $AB = 7$ **8.2, −5.8** 29. $A \leftrightarrow -3.4$; $AB = 7.8$ **4.4, −11.2**

Explain each of the following. Use the definition of *between*.

C 30. Why is point *C* not between points *A* and *B*? **not on $\overleftrightarrow{AB}$**

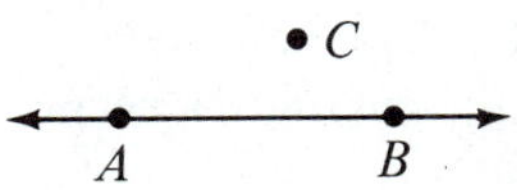

31. Why is point *R* not between points *P* and *Q*? **$PR + RQ \neq PQ$**

CONGRUENT SEGMENTS 1.3

Objectives

To identify and construct congruent segments
To construct the midpoint of a segment
To find the coordinate of the midpoint of a segment

Two segments can have the same length. In the figure below, $\overline{AB}$ and $\overline{CD}$ are each 2 cm long. $\overline{AB}$ and $\overline{CD}$ are examples of **congruent** segments.

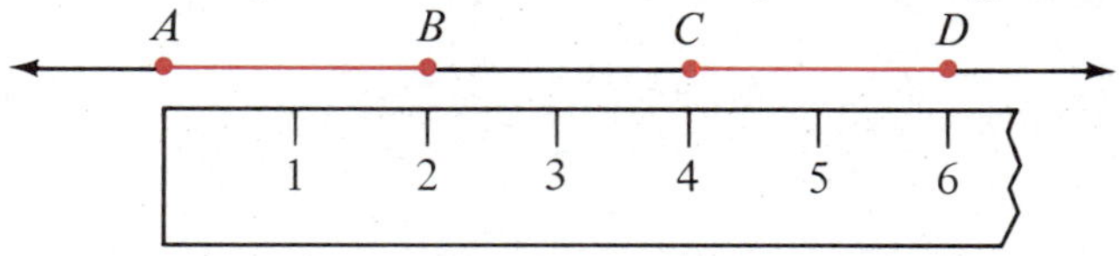

Definition: Congruent segments

Two *segments* are *congruent* if they have the same length. $\overline{AB} \cong \overline{CD}$ means "$\overline{AB}$ is congruent to $\overline{CD}$." Similar markings are used to indicate congruent segments.

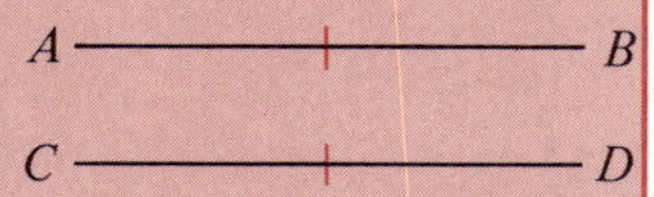

Example 1 Show that each pair of segments is congruent. Use coordinates.

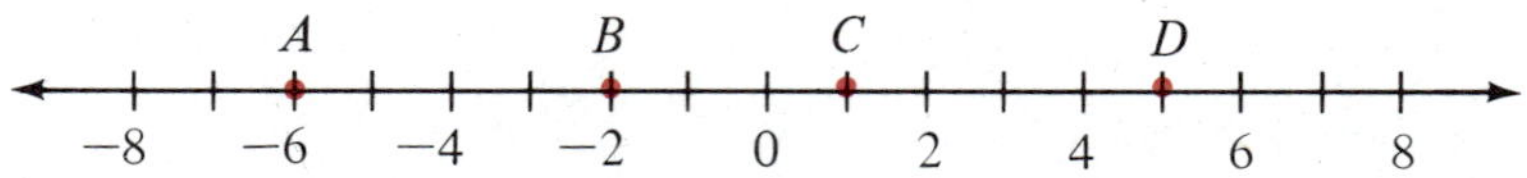

$\overline{AB}$ and $\overline{CD}$

$AB = |-2 - (-6)| = 4$
$CD = |5 - 1| = 4$
The lengths are the same, so $\overline{AB} \cong \overline{CD}$.

$\overline{AC}$ and $\overline{BD}$

$AC = |1 - (-6)| = 7$
$BD = |5 - (-2)| = 7$
The lengths are the same, so $\overline{AC} \cong \overline{BD}$.

Definition: Midpoint

M is the *midpoint* of $\overline{AB}$ if M lies on $\overline{AB}$ and $\overline{AM} \cong \overline{MB}$. The midpoint *bisects* the segment.

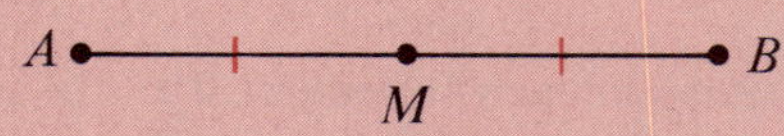

Any geometric figure that contains the midpoint of a segment is called a **bisector** of the segment. Thus, a segment may be bisected by a point, a line, a ray, another segment, or any geometric figure. In the diagram, line l bisects $\overline{AB}$.

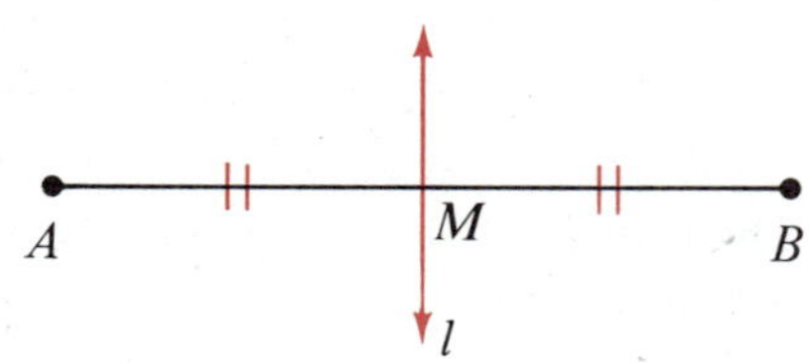

A compass and a straightedge are the basic tools for constructing geometric figures. A straightedge is used for drawing lines, but it need not have the coordinate markings of a ruler. You can use a compass and a straightedge to construct a segment congruent to a given segment.

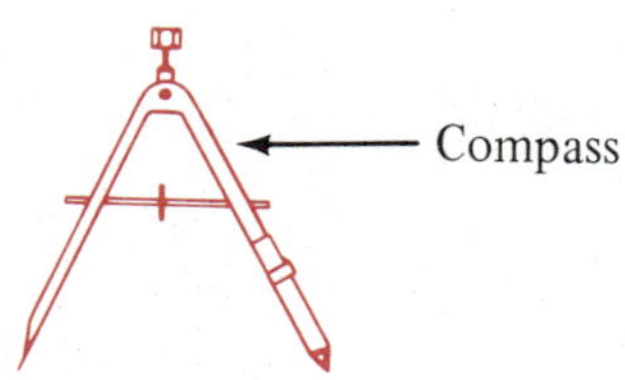

Construction A segment congruent to a given segment

Given: $\overline{AB}$
Construct: $\overline{CD} \cong \overline{AB}$ on line l

Adjust the compass to correspond to A and B.

Place compass point on any point C of line l. Mark D with compass pencil.

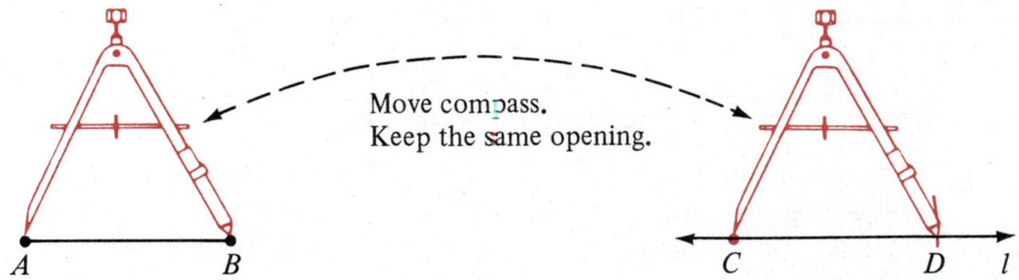

Conclusion: $\overline{CD} \cong \overline{AB}$

Similar procedures can be used to carry out more complicated constructions.

Example 2

Construct a segment with a length equal to the sum of the lengths of two given segments, $\overline{AB}$ and $\overline{CD}$.

Construct $\overline{PO} \cong \overline{AB}$, as shown at the right. Then construct $\overline{OQ} \cong \overline{CD}$.

$PO = AB$ and $OQ = CD$

Thus, $PQ = AB + CD$.

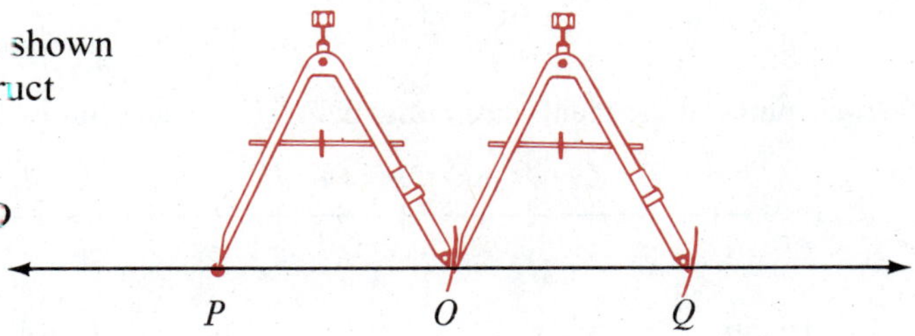

Equidistant means "the same distance from."

Example 3

Construct a point P equidistant from two given points.

Let X and Y be two given points. Open your compass wider than one-half of XY. With the compass point on X, draw an arc. With the same compass opening and with the compass point on Y, draw a second arc that intersects the first arc at P.

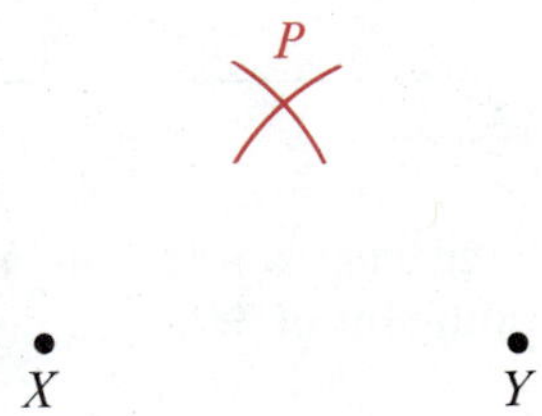

Conclusion: $XP = YP$

Construction The midpoint of a given segment

Given: $\overline{AB}$

Construct: M, the midpoint of $\overline{AB}$

Construct P so that $AP = BP$.

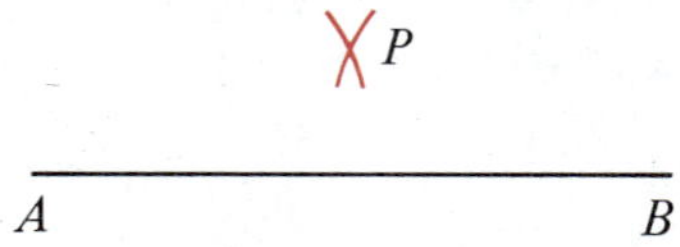

Construct Q different from P so that $AQ = BQ$.

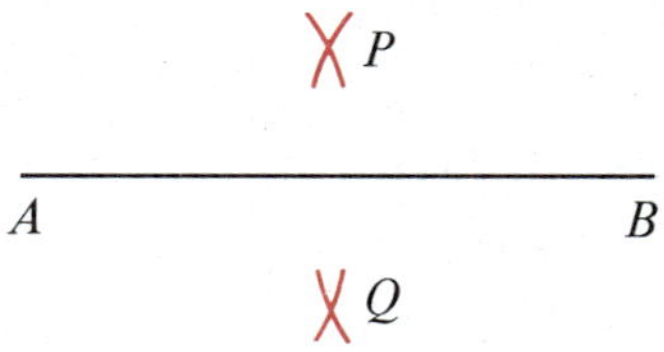

Draw $\overleftrightarrow{PQ}$ intersecting $\overline{AB}$ at M.

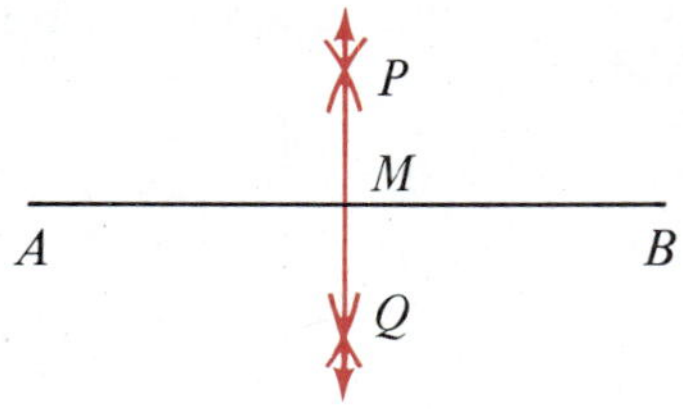

Conclusion: M is the midpoint of $\overline{AB}$.

Written Exercises

Which pairs of segments are congruent? Use a ruler to decide.

A **1.**

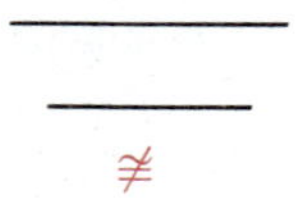

2.

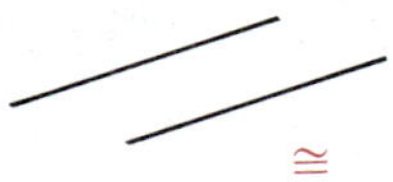

3.

4.

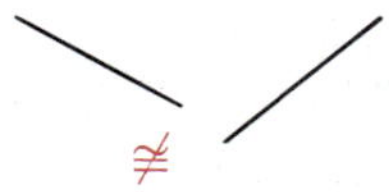

Which pairs of segments are congruent? Use a compass to decide.

5.

6.

7.

8.

Which pairs of segments are congruent? Use coordinates to decide.

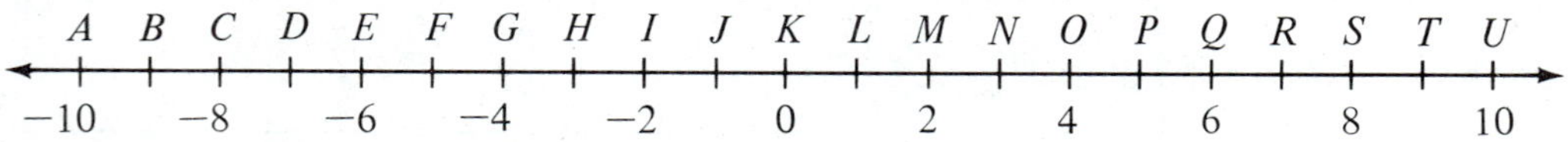

9. $\overline{AE}$ and $\overline{GK}$ ≅

10. $\overline{GM}$ and $\overline{MQ}$ ≇

≅ **11.** $\overline{BH}$ and $\overline{KQ}$

12. $\overline{GN}$ and $\overline{LR}$ ≇

13. $\overline{BP}$ and $\overline{UG}$ ≅

≅ **14.** $\overline{QD}$ and $\overline{TG}$

Construct a segment $\overline{AB}$ congruent to each given segment. Then construct a point P equidistant from the endpoints of $\overline{AB}$.

15.

16.

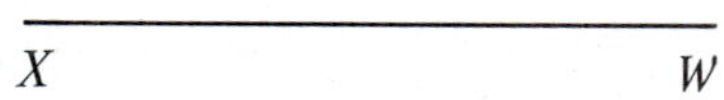

15.–16. Follow the constr., then Ex. 3, p. 9.

Construct a segment $\overline{MN}$ congruent to each given segment. Then construct the midpoint of $\overline{MN}$.

17.

18.

17.–18. Follow the constr. on p. 9, then constr. of midpoint, p. 10.

Find the coordinate of the midpoint of the segment whose endpoints have the given coordinates.

Example

-3 and 8

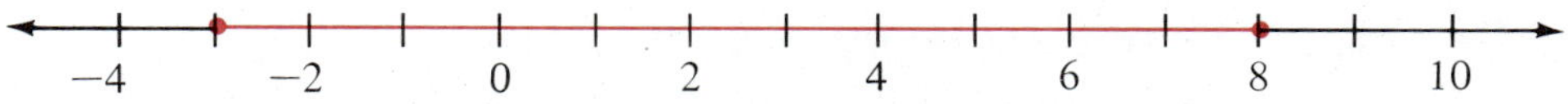

Step 1 Find the distance between the points with coordinates -3 and 8: $|-3-8| = 11$

Step 2 Find one-half of the distance: $\frac{1}{2}(11) = 5\frac{1}{2}$

Step 3 Add it to the smaller coordinate: $-3 + 5\frac{1}{2} = 2\frac{1}{2}$

Thus, $2\frac{1}{2}$ is the coordinate of the midpoint.

B

19. 5 and 17 **11**

20. 5 and -5 **0**

21. 0 and 12 **6**

22. -9 and -4 **-6.5**

23. -5.6 and -8.4 **-7**

24. -4.6 and 3.5 **-0.55**

25. $2\frac{1}{2}$ and $3\frac{1}{4}$ **$2\frac{7}{8}$**

26. $-5\frac{1}{3}$ and $6\frac{2}{3}$ **$\frac{2}{3}$**

27. $-3\frac{1}{8}$ and $-7\frac{3}{4}$ **$-5\frac{7}{16}$**

Construct each of the following.

28. A segment whose length is $MN + OP$

$\overline{M \quad N}$ $\qquad$ $\overline{O \quad P}$

Follow Ex. 2, p. 9

29. A segment whose length is $AB - CD$

$\overline{A \quad B}$ $\qquad$ $\overline{C \quad D}$

Constr. $\overline{MX} \cong \overline{AB}$. Constr. $\overline{MA} \cong \overline{CD}$ with A betw. M and X.

30. A segment twice as long as $\overline{XY}$.

$\overline{X \quad Y}$

Follow Ex. 2 p. 9.

31. A segment whose length is one fourth the length of $\overline{PQ}$

$\overline{P \quad Q}$

Constr. seg. $\cong$ PQ. Then constr. midpt., p. 10, and repeat.

Write a formula for finding each of the following.

C

32. The coordinate (m) of the midpoint of a segment whose endpoints have the coordinates p and q **$m = \frac{1}{2}(p + q)$**

33. The coordinate (q) of the second endpoint of a segment whose first endpoint has the coordinate p and whose midpoint has the coordinate m **$q = 2m - p$**

34. The coordinate (r) of the point $\frac{1}{4}$ of the way from the point whose coordinate is p to the point whose coordinate is q if $p < q$ **$r = p + \frac{1}{4}(q - p)$, or $r = \frac{3}{4}p + \frac{1}{4}q$**

35. The coordinate (r) of the point $\frac{3}{4}$ of the way from the point whose coordinate is p to the point whose coordinate is q if $p < q$ **$r = p + \frac{3}{4}(q - p)$, or $r = \frac{1}{4}p + \frac{3}{4}q$**

NON-ROUTINE PROBLEMS

Examine the figure at the right.

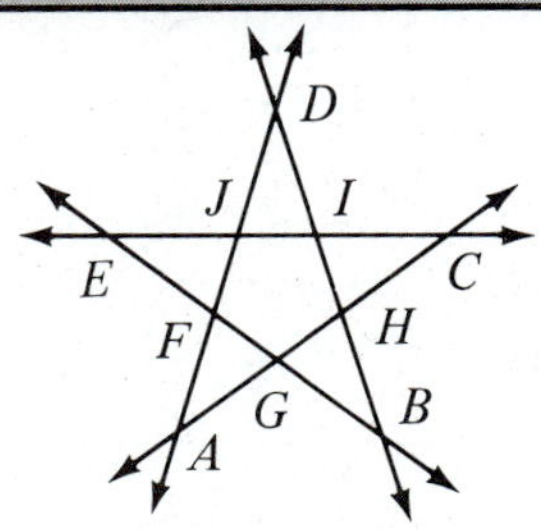

1. How many different lines are determined? **5**
2. How many different segments are determined? **30**
3. How many different rays are determined? **40**

ANGLES

1.4

Objectives

To classify angles by their measures
To construct congruent angles and the bisector of an angle
To apply the definition of the bisector of an angle

Two rays are joined to form an *angle*.

Definition: Angle	An *angle* is the union of two rays with a common endpoint. The rays are the *sides* of the angle. The common endpoint is the *vertex* of the angle.

The symbol ∠ means "angle." Usually three capital letters are used to name an angle. The middle letter names the vertex. The other two letters name points on the two sides of the angle. An angle may also be named by the vertex alone or by a lowercase letter written between the rays.

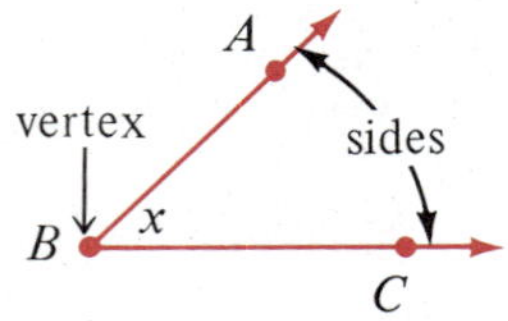

$\angle ABC$, $\angle CBA$, $\angle B$, or $\angle x$

A **protractor** is used to measure an angle. The protractor assigns a number from 0 through 180 to the angle. The number is called the **measure** of the angle. Every angle has a degree measure, m, such that $0 < m \leq 180$.

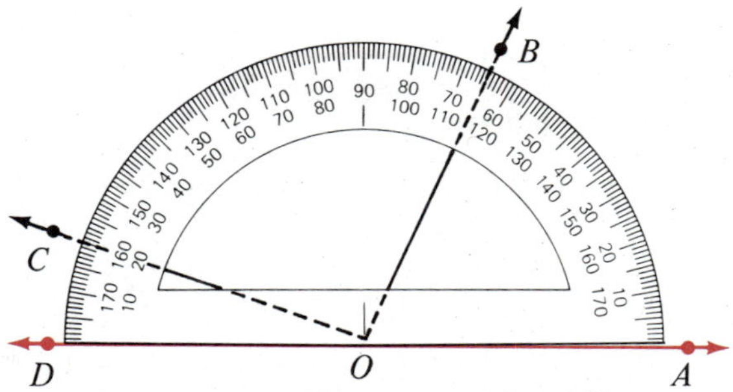

$m\angle AOB = 65$ $\quad$ $m\angle AOC = 160$

A protractor can also be used to draw an angle with a given measure. To draw $\angle XYZ$ so that $m\angle XYZ = 130$, follow these steps:

Step 1 Draw a ray, $\overrightarrow{YX}$.
Step 2 Place the protractor center on Y with $\overrightarrow{YX}$ pointing to 0°.
Step 3 Mark Z at 130°.
Step 4 Draw $\overrightarrow{YZ}$.

$m\angle XYZ = 130$

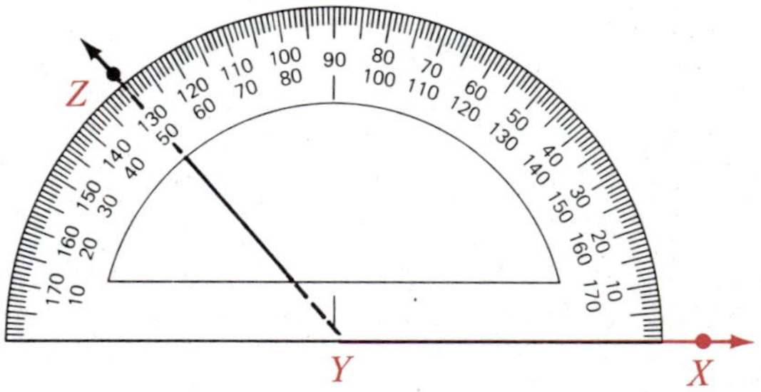

The degree measure m of an angle must be greater than 0 and less than or equal to 180. You can classify angles into four categories by their measures.

Definitions: Acute angle, right angle, obtuse angle, straight angle

An *acute angle* is an angle with degree measure m such that $0 < m < 90$.
A *right angle* is an angle with degree measure $m = 90$.
An *obtuse angle* is an angle with degree measure m such that $90 < m < 180$.
A *straight angle* is an angle with degree measure $m = 180$.

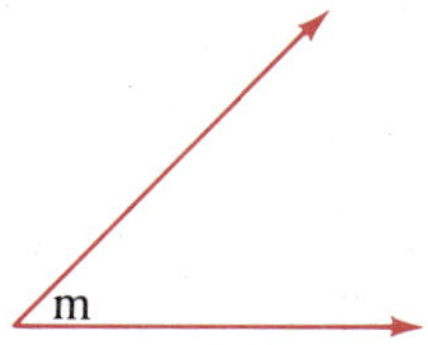

$0 < m < 90$
Acute angle

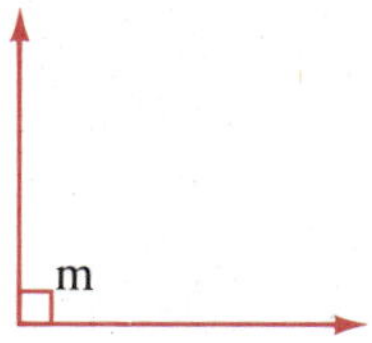

$m = 90$
Right angle

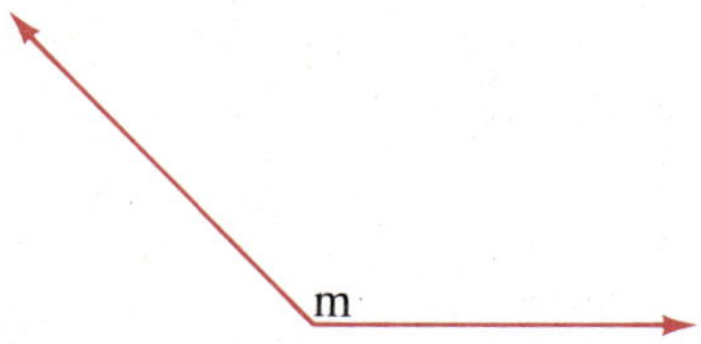

$90 < m < 180$
Obtuse angle

$m = 180$
Straight angle

Notice how the symbol ┐ is used to indicate a right angle. This symbol will be used throughout this book. The picture of the straight angle is also the picture of a line. You may find it useful sometimes to think of a line as a type of angle.

Example 1

Find the measure of $\angle COD$, $\angle BOC$, $\angle AOC$, and $\angle AOD$. Classify each angle as acute, right, obtuse, or straight.

$m\angle COD = 70$	acute
$m\angle BOC = 120 - 70 = 50$	acute
$m\angle AOC = 110$	obtuse
$m\angle AOD = 180$	straight

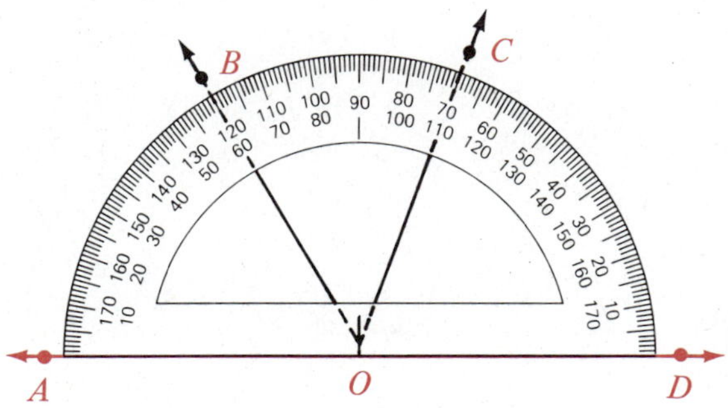

Definition: Congruent angles

Two angles are *congruent* if they have the same measure. $\angle ABC \cong \angle DEF$ means "$\angle ABC$ is congruent to $\angle DEF$." Similar markings are used to indicate congruent angles.

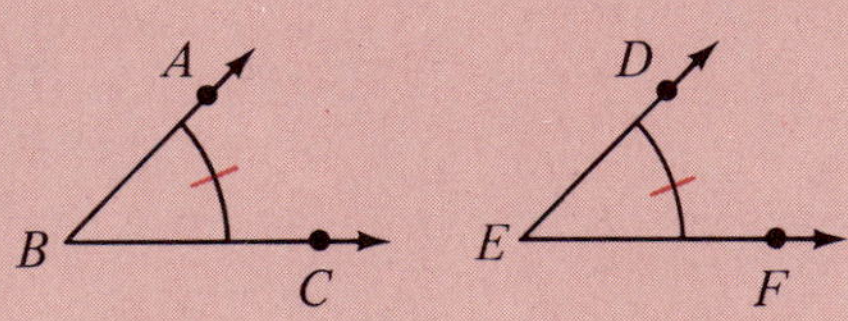

You can construct an angle congruent to a given angle using only a compass and a straightedge.

Construction

An angle congruent to a given angle

Given: $\angle ABC$
Construct: $\angle DEF \cong \angle ABC$

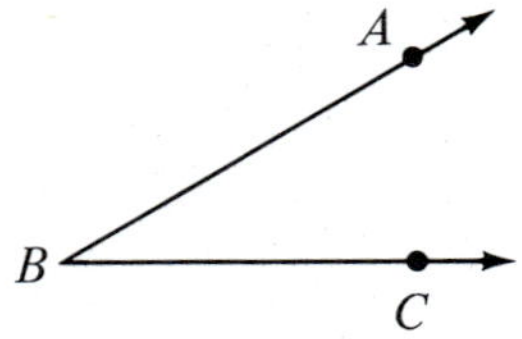

Construct an arc with center B. Use any compass opening.

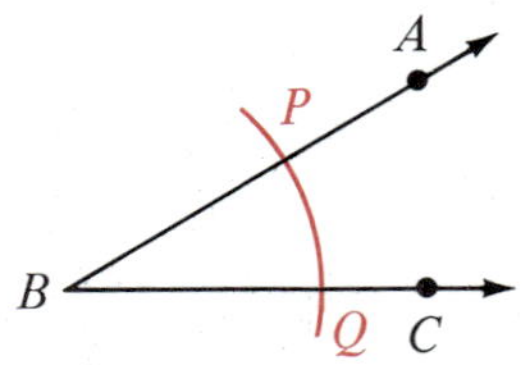

On a line, pick a point E. Using the same compass opening, construct an arc with center E.

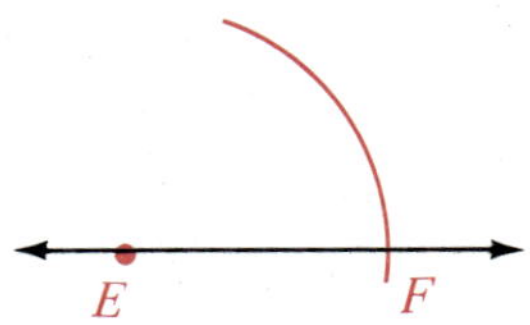

Open the compass to the length PQ. Construct an arc with center F. Draw $\overrightarrow{ED}$.

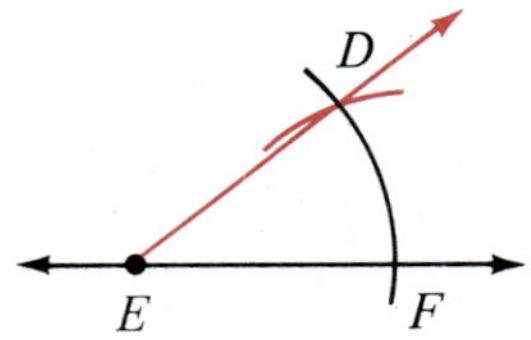

Conclusion: $\angle DEF \cong \angle ABC$

An angle in a plane (other than a straight angle) separates the plane into three parts: the **angle** itself, the **interior** of the angle, and the **exterior** of the angle. The interior of the angle is the part that contains no lines. The exterior is the other part.

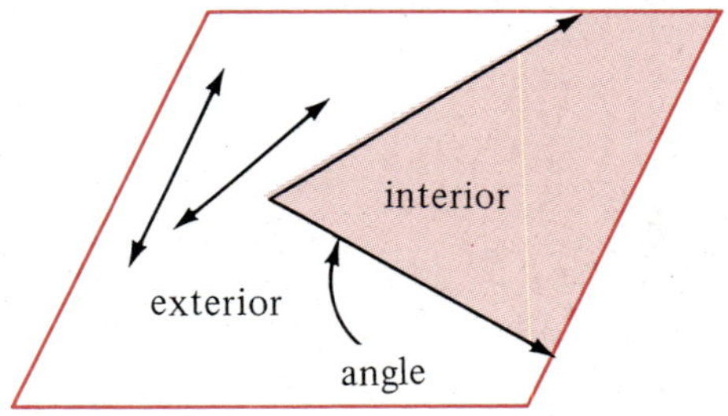

Definition: Bisector of angle

$\overrightarrow{OP}$ *bisects* $\angle AOB$ if P is in the interior of $\angle AOB$ and $\angle AOP \cong \angle POB$. $\overrightarrow{OP}$ is the *bisector* of $\angle AOB$.

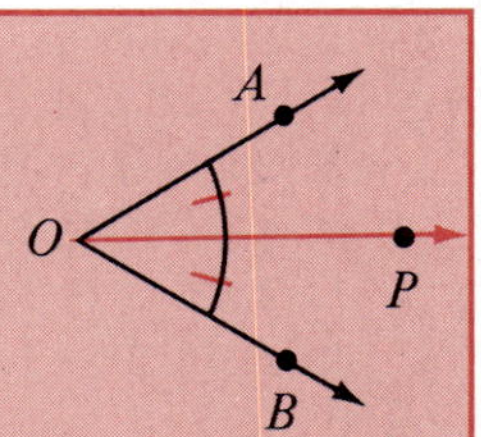

Example 2

Given: $\overrightarrow{BD}$ bisects $\angle ABC$; m $\angle ABD = 2x + 5$, and m $\angle DBC = 3x - 12$.
Find m $\angle ABD$.

$\angle ABD \cong \angle DBC$, so m $\angle ABD =$ m $\angle DBC$

$$2x + 5 = 3x - 12$$
$$x = 17$$

Thus, m $\angle ABD = 2x + 5 = 2 \cdot 17 + 5$, or 39.

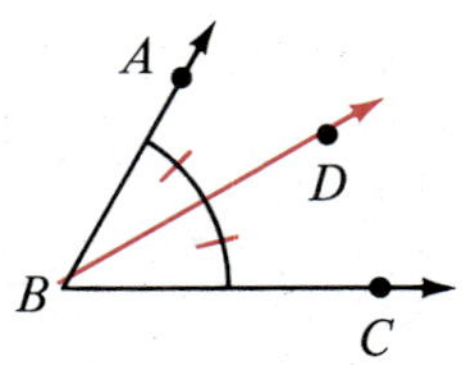

Construction — The bisector of a given angle

Given: $\angle XOY$
Construct: $\overrightarrow{OP}$, the bisector of $\angle XOY$

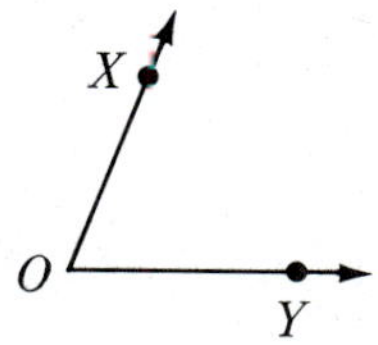

Draw an arc with center O.

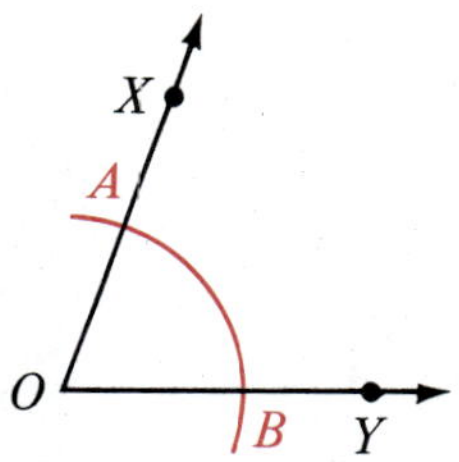

Open the compass to the length AB.
Draw an arc with center A.
Draw an arc with center B.

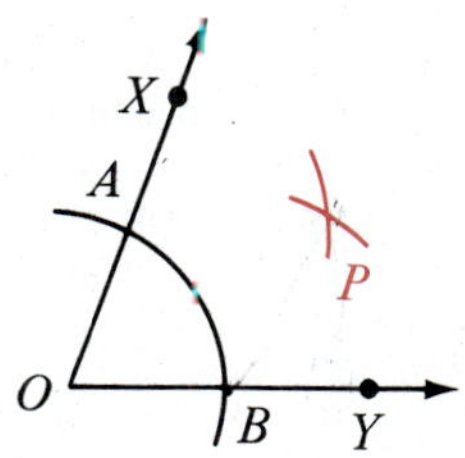

Draw $\overrightarrow{OP}$.

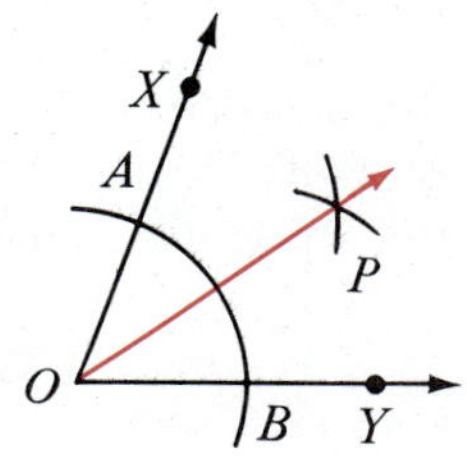

Conclusion: $\overrightarrow{OP}$ is the bisector of $\angle XOY$.

Reading in Geometry

Translate each sentence into symbols.

1. The measure of angle RST is equal to 72 degrees. **$m\angle RST = 72$**
2. The measure of angle C is equal to the measure of angle D. **$m\angle C = m\angle D$**
3. Angle ABC is congruent to angle XYZ.
4. Point W corresponds to negative 4. **$W \leftrightarrow -4$**
5. Segment EF is congruent to segment GH.
   3. $\angle ABC \cong \angle XYZ$ **$\overline{EF} \cong \overline{GH}$**
6. The length of segment PQ is equal to the length of segment RS. **$PQ = RS$**

Oral Exercises

1. Name the vertex and the sides of $\angle PQR$. **Q; $\overrightarrow{QP}$, $\overrightarrow{QR}$**

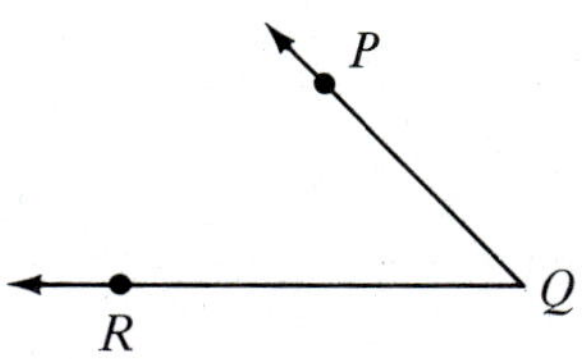

2. Give ten names for the angle.

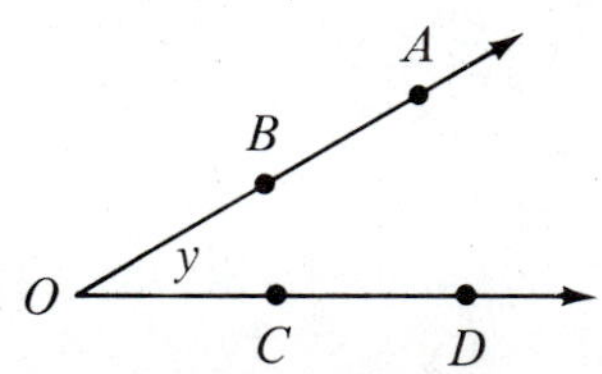

$\angle AOD$; $\angle BOD$ $\angle AOC$; $\angle BOC$; $\angle y$; $\angle O$; $\angle DOA$; $\angle DOB$; $\angle COB$; $\angle COA$

Written Exercises

Find the measure of each angle. Tell whether each angle is acute, right, obtuse, or straight.

A

1. $\angle AOB$ 30°; acute
2. $\angle EOG$ 40°; acute
3. $\angle BOD$ 76°; acute
4. $\angle COG$ 115°; obtuse
5. $\angle EOB$ 110°; obtuse
6. $\angle AOF$ 155°; obtuse
7. $\angle AOG$ 180°; straight
8. $\angle DOB$ 76°; acute

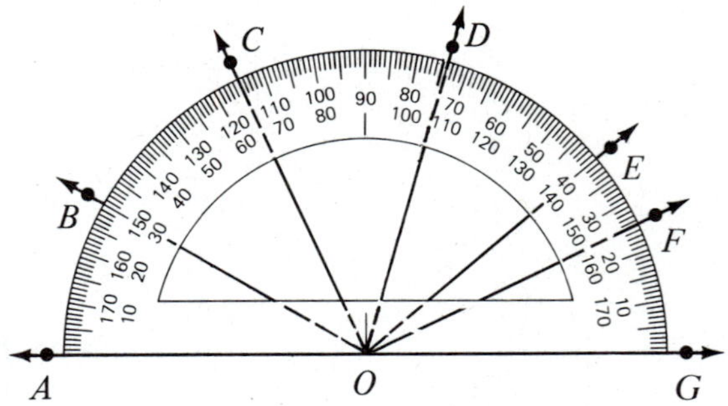

Use a protractor to draw an angle with each given measure. Use a compass and straightedge to construct an angle congruent to the given angle. Then use a compass and straightedge to bisect each angle. (Ex. 9–12)

9. m $\angle ABC = 38$ 10. m $\angle DEF = 76$ 11. m $\angle PQR = 90$ 12. m $\angle JKL = 172$

B

13. Draw an angle with a measure of 100°. Construct an angle with $\frac{1}{4}$ of this measure. Follow Constr., p. 15; then bis. one of the parts to get $\frac{1}{4}$ m∠.
14. Draw an angle with a measure of 136°. Construct an angle with $\frac{1}{8}$ of this measure. Repeat Ex. 13; then bisect one of the parts to get $\frac{1}{8}$ m∠.
15. Draw an angle with a measure of 70°. Construct an angle with $\frac{3}{4}$ of this measure. Repeat Ex. 13. The required ∠ is $\frac{1}{2}$ m∠ + $\frac{1}{4}$ m∠.
16. Draw an angle with a measure of 136°. Construct an angle with $\frac{5}{8}$ of this measure. Repeat Ex. 14. The required ∠ is $\frac{1}{2}$ m∠ + $\frac{1}{8}$ m∠.

Draw and label each figure. Then answer each question.

17. Draw $\angle ABC \cong \angle CBD$ so that C is in the interior of $\angle ABD$. Is $\overrightarrow{BC}$ the bisector of $\angle ABD$? Why or why not? Yes; def. of ∠ bisector
18. Draw $\angle ABC \cong \angle CBD$ so that C is in the exterior of $\angle ABD$. Is $\overrightarrow{BC}$ the bisector of $\angle ABD$? Why or why not? No; C is not in the interior of ∠ABD.

Find the measure of each indicated angle. (Ex. 19–20)

19. Given: $\overrightarrow{OB}$ bisects $\angle AOC$, m $\angle AOB = 5x + 2$, and m $\angle BOC = 6x - 3$. Find m $\angle AOB$. 27

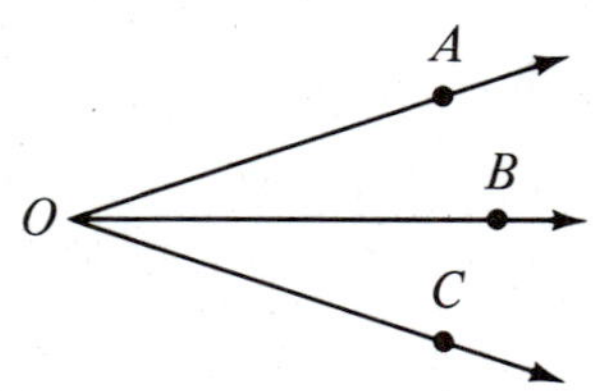

20. Given: $\overrightarrow{OB}$ bisects $\angle AOC$, m $\angle AOB = 8x + 3$, and m $\angle BOC = 9x - 4$. Find m $\angle BOC$. 59

C

21. Can the interior and exterior of an angle with a measure of 180° be defined? Explain. No. Both parts contain lines.
22. Can three coplanar angles have a common side? Three noncoplanar angles? Illustrate, if possible. yes; yes

APPLICATIONS

Read → Plan → Solve → Interpret

Mileposts on interstate highways indicate the number of miles you are from some point of reference, usually the state line.

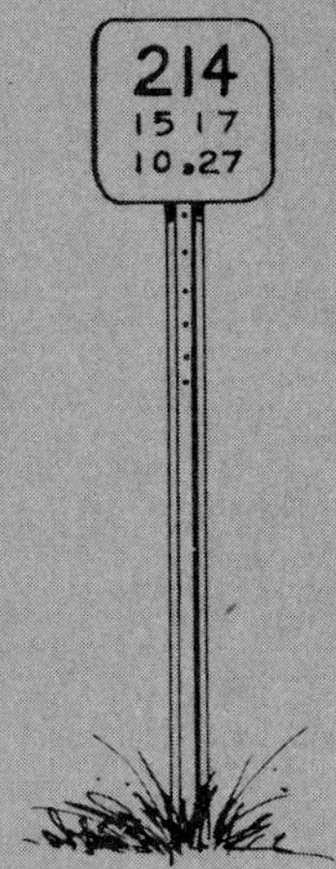

1. How far is it from Milepost 38 to Milepost 96?
2. How far is it from Milepost 214 to Milepost 79?
3. If you drive 57 miles from Milepost 0, at what milepost will you be?
4. If you drive 78 miles toward the state line from Milepost 246, at what milepost will you be?
5. Jonesboro is halfway between Milepost 37 and Milepost 193. At what milepost will you be when you get to Jonesboro?

Airport runways are numbered to indicate angle measures with North. Clockwise from North, the degree measure is 10 times the runway number. For example, Runway 3 is 30° from North.

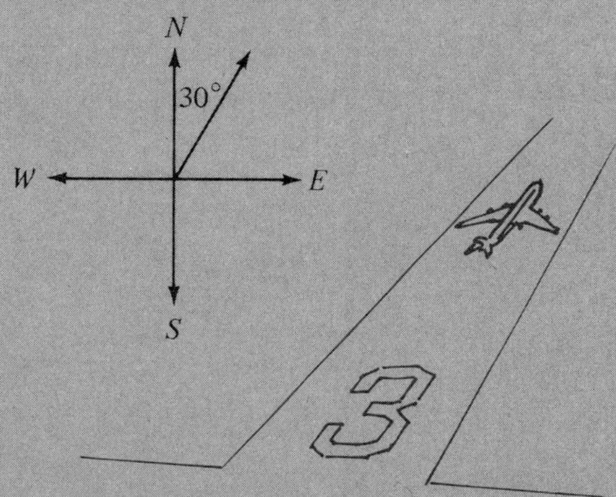

6. What is the measure of the angle that Runway 16 makes with North?
7. What is the runway number if the angle measure from North is 100°?
8. If each runway in Exercises 6 and 7 is approached from the opposite direction, what would each runway number be?

CHAPTER ONE REVIEW

A's are 1–11, 14, 15, 17–22; B's are 12, 13, 23; C's are starred.

Vocabulary

point [1.1], line [1.1], plane [1.1], space [1.1], geometric figure [1.1], collinear [1.1], coplanar [1.1], coordinate [1.2], distance [1.2], segment [1.2], ray [1.2], length [1.2], congruent [1.3], midpoint [1.3], angle [1.4], acute [1.4], right [1.4], obtuse [1.4], straight [1.4]

Give all possible names for each figure. [1.1]

1. $\overrightarrow{AB}$; $\overrightarrow{AC}$

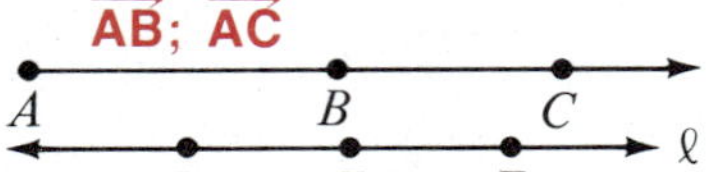

2. $\overleftrightarrow{DE}$; $\overleftrightarrow{DF}$; $\overleftrightarrow{EF}$; $\overleftrightarrow{FE}$; $\overleftrightarrow{FD}$; $\overleftrightarrow{ED}$; ℓ

Identify each set of points as collinear, coplanar but noncollinear, or noncoplanar. [1.1]

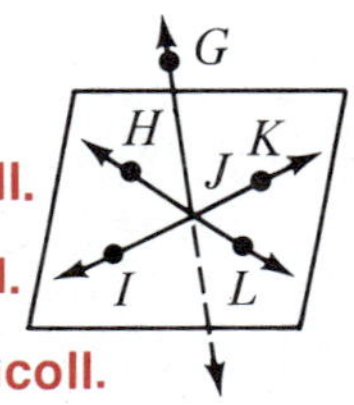

3. *H, L* coll.

4. *G, K* coll.

5. *H, I, L* copl. but noncoll.

6. *H, I, G* cop. but noncoll.

7. *H, I, K, L* copl. but noncoll.

8. *H, J, K, G* noncopl.

Use the number line. (Ex. 9–15) [1.2, 1.3]

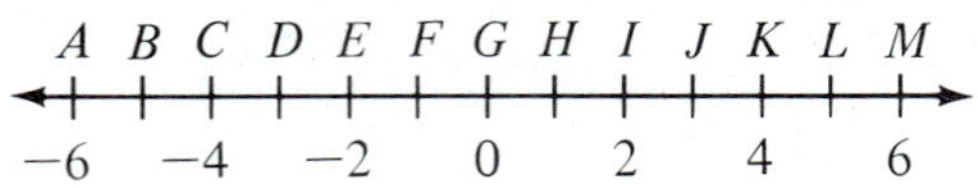

9. Find *FJ*. 4

10. Find *GK*. 4

11. Find *BH*. 6

12. Find the coordinate of the midpoint of $\overline{BL}$. 0

13. Find the coordinate of the midpoint of $\overline{AG}$. −3

14. Is $\overline{AE}$ congruent to $\overline{EJ}$? no

15. Is $\overline{CG}$ congruent to $\overline{EI}$? yes

* **16.** Find the coordinate (r) of the point $\frac{2}{3}$ of the distance from the point with coordinate x to the point with coordinate y if $x < y$. [1.3]

$r = x + \frac{2}{3}(y - x)$ or $r = \frac{1}{3}x + \frac{2}{3}y$

Use a protractor to find the measure of each angle. Tell whether each angle is acute, right, obtuse, or straight. (Ex. 17–18) [1.4]

17. obtuse

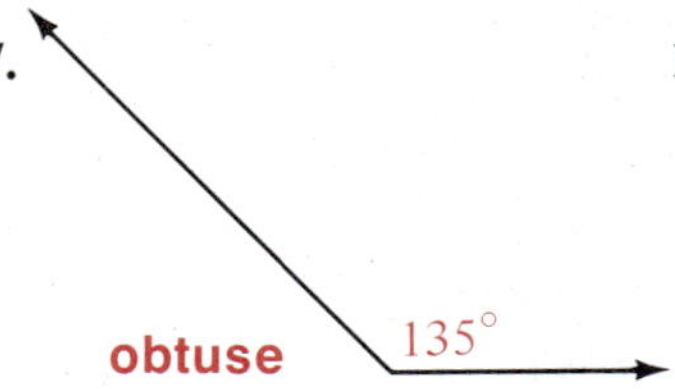

18. acute

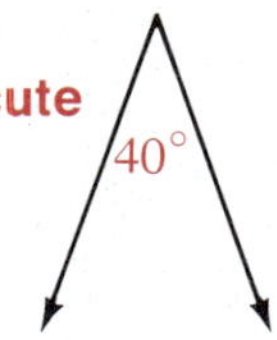

19. Construct a segment congruent to $\overline{AB}$. [1.3]
See constr., p. 9.

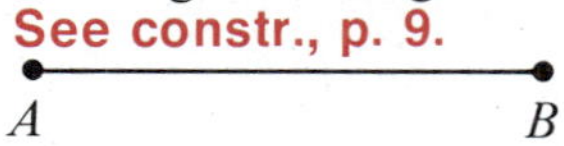

20. Construct the midpoint of $\overline{CD}$. [1.3]
See constr., p. 10.

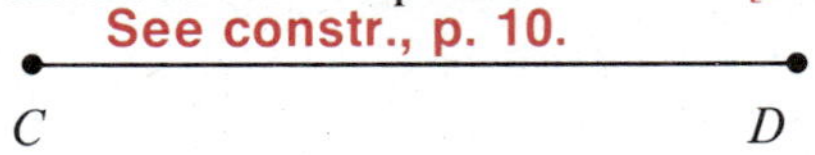

21. Construct an angle congruent to $\angle EFG$. [1.4]
See constr., p. 14.

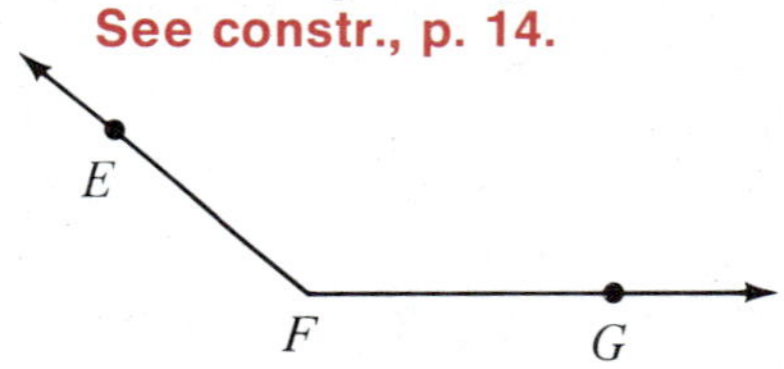

22. Construct the bisector of $\angle HIJ$. [1.4]
See constr., p. 15.

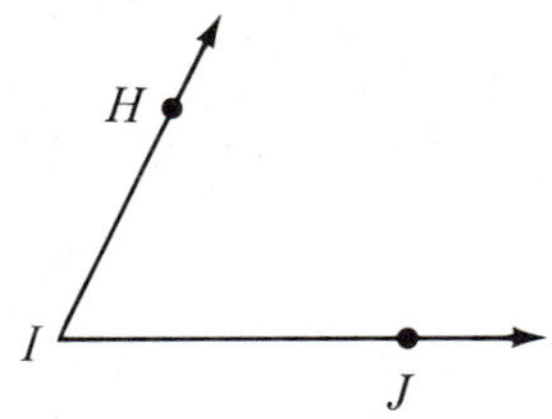

23. Given: $\overrightarrow{KL}$ bisects $\angle MKN$, m $\angle MKL = 5x + 3$, and m $\angle LKN = 6x - 4$. Find m $\angle MKL$. [1.4] 38

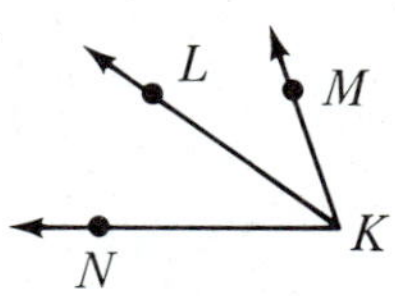

CHAPTER ONE TEST

A's are 1–11, 14–21; B's are 12, 13, 22; C's are starred.

Give all possible names for each figure.

1. 

$\overleftrightarrow{XY}$; $\overleftrightarrow{XZ}$; $\overleftrightarrow{YZ}$; $\overleftrightarrow{ZY}$; $\overleftrightarrow{YX}$; $\overleftrightarrow{ZX}$

2. 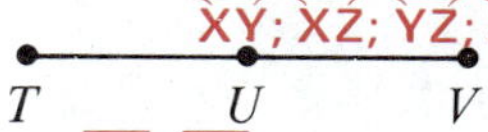

$\overline{TV}$, $\overline{VT}$

Identify each set of points as collinear, coplanar but noncollinear, or noncoplanar.

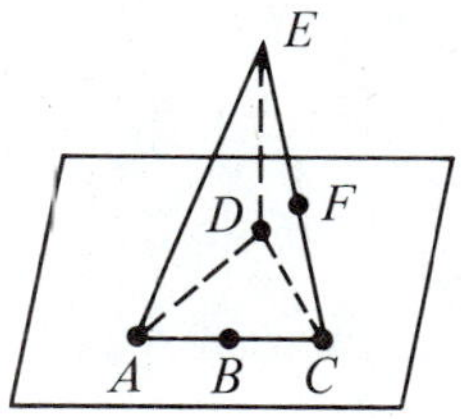

3. A, D coll.
4. B, F coll.
5. A, B, C coll.
6. A, F, C copl. but noncoll.
7. A, B, C, D copl. but noncoll.
8. A, C, D, E noncopl.

Use the number line for Items 9-15.

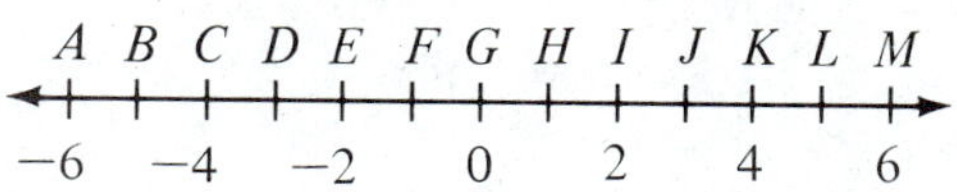

9. Find GM. 6
10. Find IL. 3
11. Find BJ. 8
12. Find the coordinate of the midpoint of $\overline{GM}$. 3
13. Find the coordinate of the midpoint of $\overline{BJ}$. −1
14. Is $\overline{BF}$ congruent to $\overline{HM}$? no
15. Is $\overline{BJ}$ congruent to $\overline{DM}$? no

Use a protractor to find the measure of each angle. Tell whether each angle is acute, right, obtuse, or straight. (Items 16-17)

16. 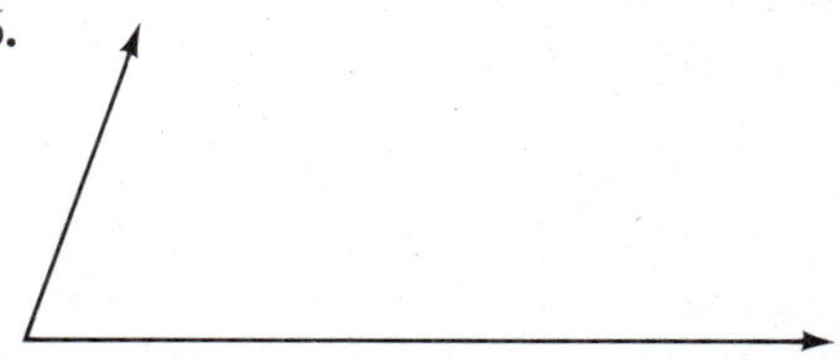

70°; acute

17. 

42°; acute

18. Construct a segment congruent to $\overline{XY}$.

See Constr., p. 9.

19. Construct an angle congruent to $\angle UVW$.
See Constr., p. 14.

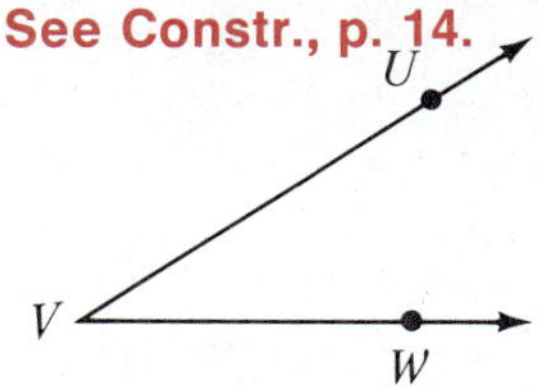

See Constr., p. 10.

20. Construct the midpoint of $\overline{ST}$.

21. Construct the bisector of $\angle PQR$.
See Constr., p. 15.

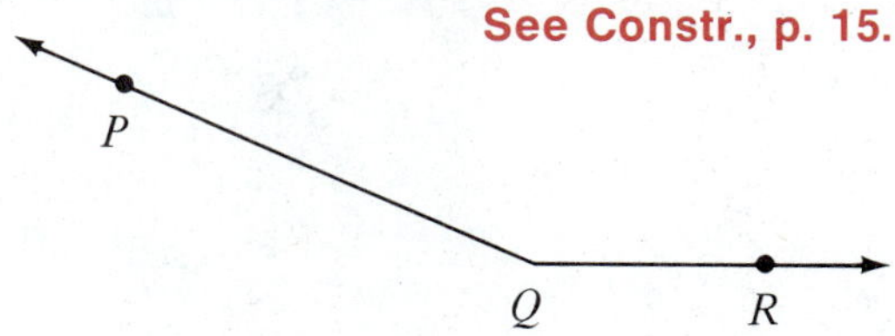

22. Given: $\overrightarrow{IJ}$ *bisects* $\angle HIK$, m $\angle HIJ = 5x - 3$, and m $\angle JIK = 3x + 15$.
Find m $\angle JIK$. 42

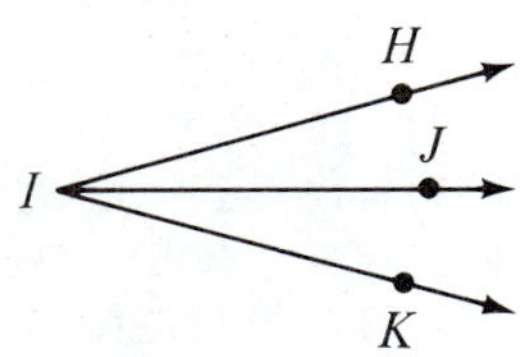

*23. Find the coordinate (r) of the point $\frac{1}{3}$ of the way from the point whose coordinate is p to the point whose coordinate is q if $p < q$.

$r = p + \frac{1}{3}(q - p)$, or $r = \frac{2}{3}p + \frac{1}{3}q$

COMPUTER ACTIVITIES

Midpoint of Segment

Objectives: **To find the coordinate of the midpoint of a segment**
To find the coordinate of the second endpoint of a segment, using a computer program, given the coordinates of one endpoint and the midpoint

In Exercise 32 on page 11, you wrote a formula for finding the coordinate of the midpoint of a segment, given the coordinates of its endpoints. Below is a simple program that uses that formula. (See page 565 for an explanation of the LET and PRINT commands used in this program.) Lines 10 and 20 assign the coordinates of the segment's endpoints to the variables X and Y. To find the coordinate of the midpoint of other segments, simply type in new lines 10 and 20 assigning the new endpoint coordinates to X and Y.

```
10  LET X = 6
20  LET Y = 17
30  PRINT "THE MIDPOINT OF THE SEGMENT WITH ENDPOINTS
     "X" AND "Y" IS ";
40  PRINT (X + Y) / 2"."
```

The program above can be changed so that it will find the coordinate of the second endpoint of a segment, given the coordinates of the segment's first endpoint and its midpoint. To do this, the following program first calculates the *direct distance* from the endpoint to the midpoint. The program does this by subtracting the coordinate of the endpoint from the coordinate of the midpoint. The program then adds this direct distance to the coordinate of the midpoint to find the coordinate of the other endpoint. Note that a direct distance may be a negative number. Consequently, this program does not use the absolute value function, ABS, as the program on page 566 does.

Enter and RUN the program below. Assign different coordinates for segments' endpoints and midpoints by altering lines 10 and 20.

```
10  LET X = 2
20  LET Y = 5
30  LET D = Y - X
40  PRINT "IF ONE ENDPOINT OF A SEGMENT IS "X","
50  PRINT "AND THE MIDPOINT IS "Y","
60  PRINT "THEN THE OTHER ENDPOINT IS "Y + D"."
```

EXERCISES: See the Computer Section beginning on page 565 for more information.

Write a program that finds the coordinates of the two endpoints of a segment, given the length of the segment and the coordinate of the midpoint.

COLLEGE PREP TEST

Directions: Choose the one best answer to each question or problem.

1. Which of the following is true of a ray but not true of a segment? **E**

(A) Can be named by two points
(B) Extends indefinitely in two directions
(C) Has a definite length
(D) Has two endpoints
(E) Has only one endpoint

2. Which of the following is another name for $\overrightarrow{BE}$? **B**

(A) $\overrightarrow{EB}$
(B) $\overrightarrow{BC}$
(C) $\overrightarrow{AB}$
(D) $\overrightarrow{BD}$
(E) $\overrightarrow{BA}$

3. In the figure shown, which of the following is not true? **D**

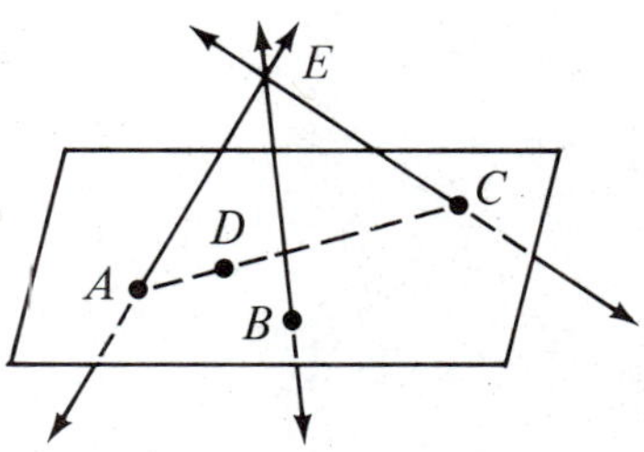

(A) A, B, and C are coplanar.
(B) A, D, and C are coplanar.
(C) A and B are collinear.
(D) A, B, C, and E are coplanar.
(E) A, E, and C are coplanar.

4. 53 m is equivalent to **E**

(A) .53 cm **(B)** .53 mm **(C)** .53 km
(D) 53 mm **(E)** None of these

5. Which of the following are true? Use the number line. **D**

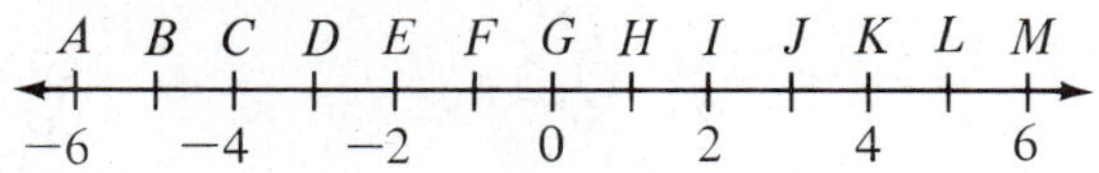

$\overline{BH}$ is congruent to
I. $\overline{DJ}$
II. $\overline{GM}$
III. $\overline{EL}$

(A) I only **(B)** II only **(C)** III only
(D) I and II only **(E)** I, II, and III

6. Which of the following are not true? Use the number line for exercise 5. **D**
I. $CI = -2$
II. $CI = 6$
III. The midpoint of $\overline{CI}$ is 0.

(A) I only **(B)** II only **(C)** III only
(D) I and III only **(E)** II and III only

7. Which of the following figures has an interior? **D**

(A) An angle **(B)** A circle
(C) A closed figure **(D)** All of these
(E) None of these

8.

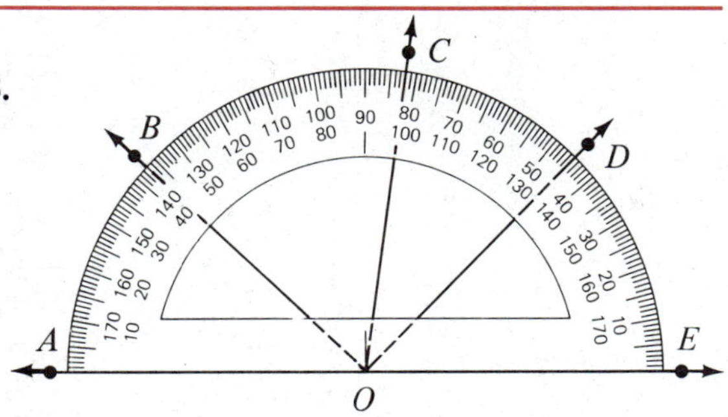

Which of the following is true? **E**

(A) $\angle AOB \cong \angle BOC$ **(B)** $\angle BOC \cong \angle COD$
(C) $\angle COD \cong \angle DOE$ **(D)** $\angle AOB \cong \angle COD$
(E) None of these

2 INTRODUCING PROOFS

Non-Routine Problem Solving

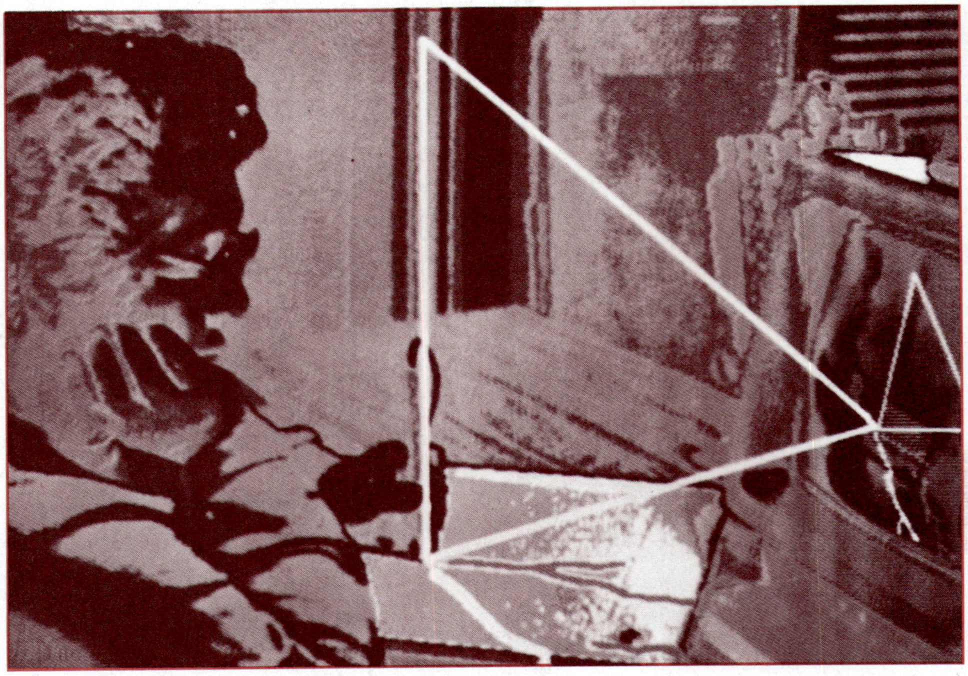

Problem

When two lines intersect, 4 angles are formed, each having a measure less than 180°. As shown below, when three lines intersect at the same point, 12 angles, each having a measure less than 180°, are formed. If 20 lines intersect at the same point, how many angles having a measure less than 180° will be formed? [*Hint:* Try several examples. Look for a pattern.] **760 angles**

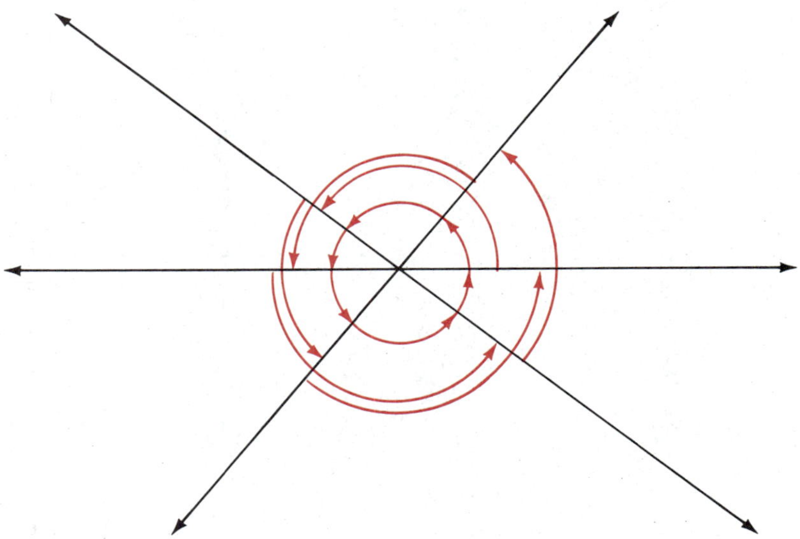

PAIRS OF ANGLES 2.1

Objectives

To identify angles as adjacent or vertical
To identify supplementary and complementary angles and find their measures

Sometimes pairs of angles are classified by their positions with respect to each other. **Adjacent angles** and **vertical angles** are such angles.

Definition: Adjacent angles	*Adjacent angles* are two coplanar angles that have a common vertex and a common side but no common interior points.

Adjacent angles must have a common side and also a common vertex. Below, $\angle XOY$ and $\angle YOZ$ share a common side $\overrightarrow{OY}$ and are adjacent. Although $\overrightarrow{IS}$ is shared by $\angle RAS$ and $\angle SIT$, $\overrightarrow{AS}$ is not shared. The two angles do not have a common vertex. They are not adjacent angles.

Adjacent angles must have no interior points in common. P is in the interior of both $\angle QIS$ and $\angle QIR$. So, $\angle QIS$ and $\angle QIR$ are not adjacent.

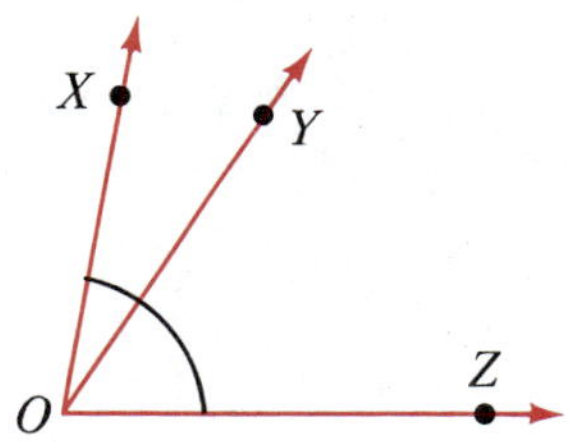

$\angle XOY$ and $\angle YOZ$ are adjacent.

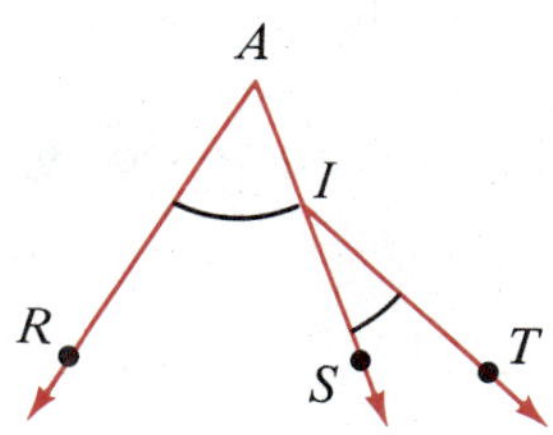

$\angle RAS$ and $\angle SIT$ are not adjacent.

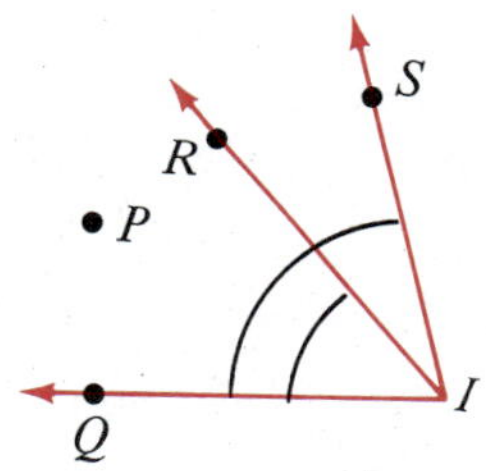

$\angle QIS$ and $\angle QIR$ are not adjacent

Example 1

Is each of the given pairs of angles adjacent? If not, why not?

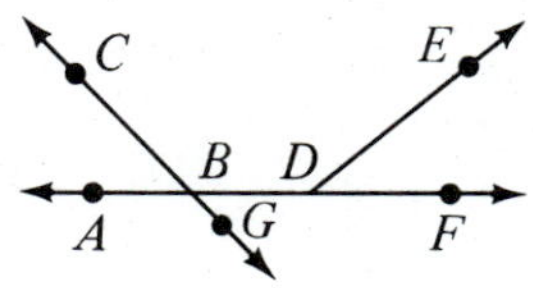

	Answers
$\angle ADE$ and $\angle EDF$	Adjacent
$\angle CBA$ and $\angle FBG$	Not adjacent; no common side

Definition: Vertical angles	*Vertical angles* are nonadjacent angles formed by two intersecting lines.

$\angle XOW$ and $\angle YOZ$ are vertical angles. So are $\angle WOZ$ and $\angle XOY$. $\angle AOB$ and $\angle COD$ are not vertical angles, since they are not formed by intersecting lines. Neither are $\angle BOC$ and $\angle DOA$.

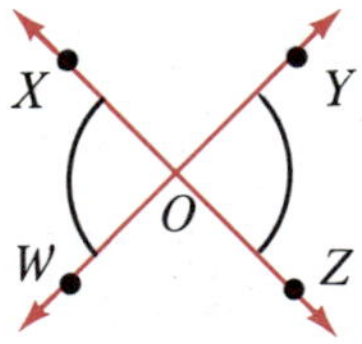

Vertical angles

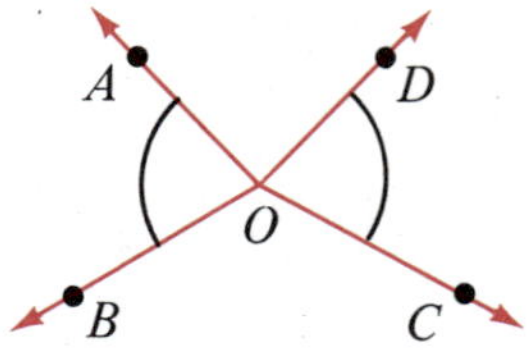

Not vertical angles

Example 2

Name the second angle in each pair of vertical angles.

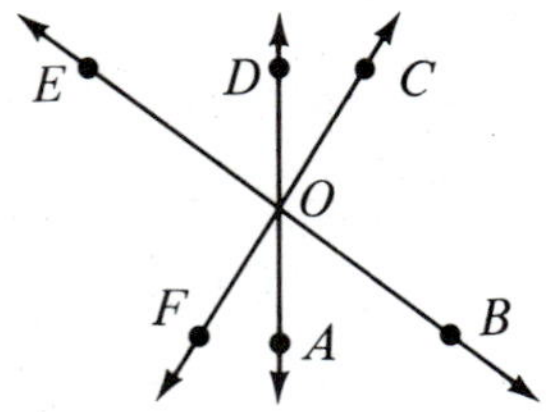

	Answers
$\angle BOC$	$\angle EOF$
$\angle AOE$	$\angle DOB$
$\angle COA$	$\angle FOD$

You learned that a straight angle has a measure of 180° and a right angle has a measure of 90°. Some pairs of angles, called **supplementary** or **complementary**, are related to straight angles or right angles, respectively.

Definitions: Supplementary angles Linear pair

Two angles are *supplementary* if the sum of their measures is 180°. Each angle is called a *supplement* of the other. If the angles are adjacent and supplementary, then they are called a *linear pair*.

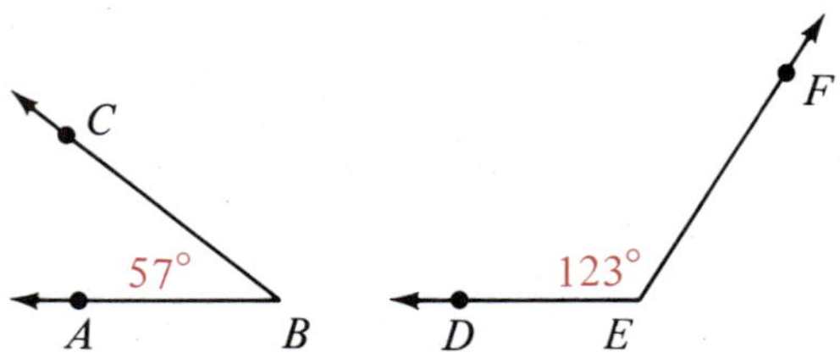

$m\angle ABC + m\angle DEF = 180$.
$\angle ABC$ is a supplement of $\angle DEF$.

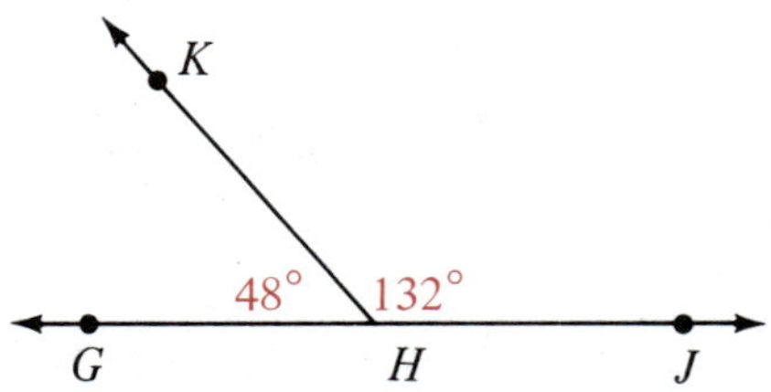

$m\angle GHK + m\angle KHJ = 180$.
$\angle GHK$ and $\angle KHJ$ are a linear pair.

Complementary angles are related to right angles.

Definition: Complementary angles

Two angles are *complementary* if the sum of their measures is 90°. Each angle is called a *complement* of the other.

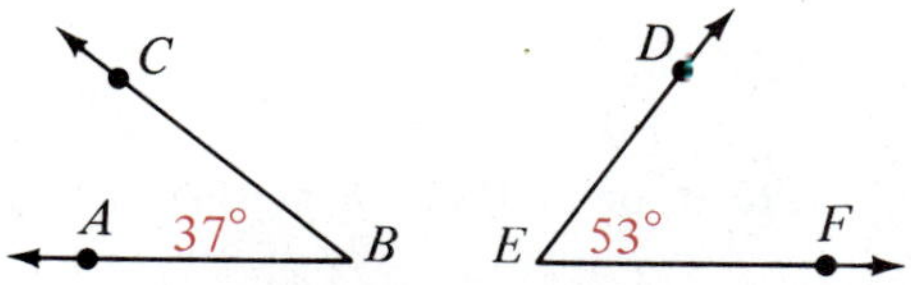

m∠ABC + m∠DEF = 90
∠ABC is a complement of ∠DEF.

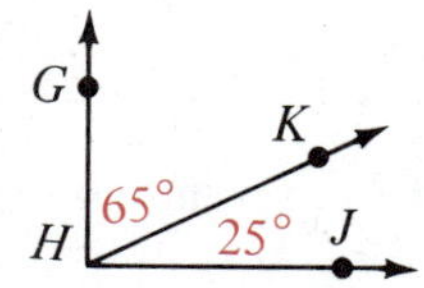

m∠GHK + m∠KHJ = 90
∠GHK is a complement of ∠KHJ.

Neither supplementary nor complementary angles have to be adjacent. If supplementary angles are adjacent, then the sides of the two angles that are not the common side form a straight angle. If complementary angles are adjacent, then the sides of the two angles that are not the common side form a right angle.

Example 3 Which are measures of supplementary angles? Of complementary angles?

	Answers
30° and 160°	Neither
86° and 94°	Supplementary
83° and 7°	Complementary

Example 4 Find the measure of an angle if its measure is 50° more than the measure of its supplement.

Let m = measure of the angle

The sum of the measures is 180°. ► $180 - m$ = measure of its supplement

$$m = (180 - m) + 50$$
$$2m = 230$$
$m = 115$ **Thus,** the angle measure is 115°.

Example 5 Find the measure of an angle if its measure is 40° less than four times the measure of its complement.

Let m = measure of the angle

$90 - m$ = measure of its complement

$$m = 4(90 - m) - 40$$
$$m = 360 - 4m - 40$$
$$5m = 320$$
$m = 64$ **Thus,** the angle measure is 64°.

Oral Exercises

Which are measures of complementary angles? supplementary angles? neither?

1. 60° and 30° comp. **2.** 130° and 50° supp. **3.** 114° and 66° supp. **4.** 92° and 2°
5. 53° and 47° neither **6.** 87° and 87° neither **7.** 65° and 25° comp. **8.** 69° and 21°
9. 57° and 122° neither **10.** 97° and 93° neither **11.** 45° and 45° comp. **12.** 26° and 154°

4. neither 8. comp. 12. supp.

Written Exercises

Is each of the given pairs of angles adjacent, vertical, or neither? If neither, why?

A 1. $\angle GIF$ and $\angle HIJ$ **vert.**
2. $\angle LPK$ and $\angle NPK$ **adj.**
3. $\angle QSR$ and $\angle RST$ **adj.**
4. $\angle VUW$ and $\angle WUX$ **adj.**
5. $\angle FIJ$ and $\angle FIG$ **adj.**
6. $\angle MON$ and $\angle KPN$ **neither; no common vertex**
7. $\angle QST$ and $\angle RST$ **adj.**
8. $\angle WUX$ and $\angle XUY$ **neither; not in same plane**

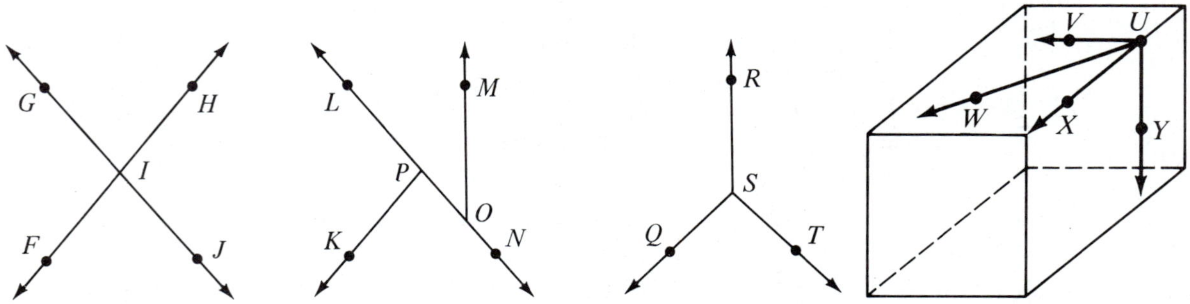

Find the measure of a complement and a supplement, if possible, of each angle whose degree measure is given.

9. 30° **60°, 150°**
10. 75° **15°, 105°**
11. 5° **85°, 175°**
12. 98° **not poss., 82°**
13. 128° **not poss., 52°**
14. 180° **not poss., not poss.**
15. 140° **not poss., 40°**
16. x **$90 - x$, $180 - x$**
17. 22.5° **67.5°, 157.5°**
18. 85.2° **4.8°, 94.8°**
19. $m - 5$ **$95 - m$, $185 - m$**
20. $10 - x$ **$80 + x$, $170 + x$**

Find the measure of the angle if its measure is as given.

21. 50° more than that of its complement **70°**
22. 60° more than that of its supplement **120°**
23. 20° less than that of its supplement **80°**
24. 78° less than that of its complement **6°**
25. twice that of its complement **60°**
26. three times that of its supplement **135°**
27. equal to that of its complement **45°**
28. equal to that of its supplement **90°**

True or false? For each false statement, draw a picture showing why it is false.

B 29. A right angle and an acute angle cannot be adjacent. **F**
30. An obtuse angle and a right angle cannot be vertical angles. **T**
31. Two lines that form vertical angles also form adjacent angles. **T**
32. Two right angles cannot be vertical. **F**

Find the measure of the angle if the measure of its supplement or complement is as given.

33. The measure of its complement is one fourth that of its supplement. **60°**
34. The measure of its supplement is 3 times that of its complement. **45°**
35. Its measure is 24° less than twice that of its complement. **52°**
36. Its measure is 81° more than twice that of its supplement. **147°**
37. The measure of its supplement is 52° more than twice that of its complement. **52°**
38. The measure of its supplement is 10° more than 3 times that of its complement. **50°**
39. Five times the measure of its complement is 24° less than twice the measure of its supplement. **38°**
40. Four times the measure of its complement is 12° more than twice the difference between the measures of its supplement and its complement. **42°**

WRITING STATEMENTS 2.2

Objectives

To write the negation of a statement
To write the conjunction of two statements
To write the disjunction of two statements

In mathematics, a **statement** is a sentence that is either true or false, but not both. A sentence such as $3 = 2 + 1$ is a true statement. $2 + 1 = 4$ is a false statement.

Definition: Negation The *negation* of a statement is the statement formed by denying the original statement.

Statement	**Negation**
$\angle A$ is a right angle.	$\angle A$ is *not* a right angle.

Sometimes it is helpful to use a variable to represent a statement or part of a statement. Often the letter p is used to represent a statement and $\sim p$ (read: not p) is used to represent its negation. The **truth table** below shows the relationship between a statement and its negation. T and F designate the truth values of the statements. When a statement is true, its negation is false. When a statement is false, its negation is true.

p	$\sim p$
T	F
F	T

Example 1

Write a negation of the following statement.

Statement: An isosceles trapezoid does not contain a right angle.
Negation: An isosceles trapezoid contains a right angle.

Two statements may be connected by the word *and*.
A triangle has three sides and a rectangle has four sides.
A square is a rectangle and a rectangle is a square.

Definition: Conjunction If p and q are statements, the statement "p and q" (written $p \wedge q$) is called the *conjunction* of p and q.

Not all conjunctions are true. The truth table for a conjunction shows this. A conjunction is true only if each part of the conjunction is true.

p	q	$p \wedge q$
T	T	T
T	F	F
F	T	F
F	F	F

Example 2 Write a conjunction and tell whether it is true or false. Right angles have a measure of 90°. Supplementary angles are congruent.

Conjunction: Right angles have a measure of 90° and supplementary angles are congruent. False. Supplementary angles may not be congruent.

Two statements may also be connected by the word *or*. In everyday language, you use the word *or* in two different ways. Sometimes you mean that exactly one of the two statements is true. You probably mean this when you say, "I want a hamburger or I want a hot dog." This use of *or* is sometimes called the *exclusive or*. Sometimes you mean that at least one of the statements is true, and possibly both are. If you say, "The figure is a rectangle or it is a square," you agree that both could be true. This use of *or* is often called the *inclusive or*. In mathematics, *or* usually has the inclusive meaning. This means that the combined statement is true except when the two original statements are both false.

Definition: Disjunction If p and q are statements, the statement "p or q" (written $p \vee q$) is called the *disjunction* of p and q.

The truth table for a disjunction is given at the right.

p	q	$p \vee q$
T	T	T
T	F	T
F	T	T
F	F	F

Example 3 Write a disjunction and tell whether it is true or false. Right angles have a measure of 90°. Supplementary angles are congruent.

Disjunction: Right angles have a measure of 90° or supplementary angles are congruent. True. The first part of the disjunction is true.

Written Exercises **Check students' answers for Ex. 1–8.**

A **Write the conjunction and disjunction of each pair of statements.**

1. It is snowing.
   I am going home.
2. Triangles have three sides.
   Squares have four sides.
3. Parallel lines never intersect.
   Right angles are not acute.
4. The measure of an acute angle is less than 90°.
   The measure of a straight angle is 180°.

Write the negation of each statement.

5. A triangle has three sides.
6. Two lines can intersect at one point.
7. A triangle cannot have four sides.
8. Parallel lines do not intersect.

B **Tell which of the following is a negation of the statement "A square has five sides."**

9. A square has four sides. **no**
10. A square has six sides. **no**
11. A square does not have five sides. **yes**
12. A pentagon has five sides. **no**
13. A square has no sides. **no**
14. No square has five sides. **yes**

C **Complete the truth table for each statement.** **Check students' answers.**

15. $p \vee \sim q$ **TTFT**
16. $\sim p \wedge q$ **FFTF**
17. $\sim p \wedge \sim q$ **FFFT**
18. $\sim p \vee \sim q$ **FTTT**

CONDITIONAL STATEMENTS 2.3

Objectives **To write statements in "If . . ., then . . ." form**
To write the converse, inverse, and contrapositive of a statement
To write a biconditional of two given statements

Many statements in geometry are **conditional statements**.

Definitions:
Conditional statement
Hypothesis
Conclusion

A *conditional statement* is a statement that can be written in "If . . ., then . . ." form. The "If" part of the statement is the *hypothesis*. The "then" part of the statement is the *conclusion.*

"If two angles are right angles, then they are congruent" is a conditional statement. "Two angles are right angles" is the hypothesis. "They are congruent" is the conclusion. The hypothesis tells what you are to assume. The conclusion tells what follows from this assumption.

Conditional statements may be written in different equivalent forms. For example:

If two angles are right angles, then they are congruent.
Two right angles are congruent.
Two angles are congruent if they are right angles.
Two angles are right angles only if they are congruent.

Example 1 Write the following statement in "If . . ., then . . ." form. Underline the hypothesis once and the conclusion twice.

All congruent segments have the same length.
If <u>segments are congruent</u>, then <u><u>they have the same length</u></u>.

The letter p is often used to represent the hypothesis of a conditional statement, and the letter q is used to represent the conclusion. The statement "If p, then q" is written $p \rightarrow q$ (read: p implies q). $p \rightarrow q$ is false only when p is true and q is false. The truth table below shows this.

p	q	$p \rightarrow q$
T	T	T
T	F	F
F	T	T
F	F	T

Other statements can be formed from a conditional statement. You should be able to write the **converse** of a conditional statement.

Definition: Converse The *converse* of a conditional statement is the statement formed by interchanging the hypothesis and the conclusion.

Conditional statement:	If two angles are right angles, then the angles are congruent. $p \to q$
Converse:	If two angles are congruent, then the angles are right angles. $q \to p$

Notice that although the original statement is true, the converse is false. Congruent angles are not necessarily right angles. The following truth table shows that a statement and its converse do not have the same truth values. Therefore, they are not equivalent statements.

p	q	$p \to q$	$q \to p$
T	T	T	T
T	F	F	T
F	T	T	F
F	F	T	T

A conditional statement and its converse can be combined to form a **biconditional** statement. Biconditional statements have the form "p if and only if q." This is written $p \leftrightarrow q$.

Example 2 Write the converse of the given statement. Then write a biconditional.

Statement:	If an angle is a right angle, then its measure is 90°.
Converse:	If the measure of an angle is 90°, then it is a right angle.
Biconditional:	An angle is a right angle if and only if its measure is 90°.

The **inverse** and the **contrapositive** are two other statements that can be formed from a given conditional statement.

Definitions: Inverse Contrapositive The *inverse* of a conditional statement is the statement formed by negating the hypothesis and the conclusion.
The *contrapositive* of a conditional statement is the statement formed by interchanging the hypothesis and the conclusion and negating each of them.

Statement:	If two angles are straight angles, then they are congruent.
Converse:	If two angles are congruent, then they are straight angles.
Inverse:	If two angles are not straight angles, then they are not congruent.
Contrapositive:	If two angles are not congruent, then they are not straight angles.

You can write each of these statements symbolically as follows:

Statement	Converse	Inverse	Contrapositive
$p \rightarrow q$	$q \rightarrow p$	$\sim p \rightarrow \sim q$	$\sim q \rightarrow \sim p$

The following truth table shows the truth values of the inverse and the contrapositive of a statement.

p	q	$p \rightarrow q$	$\sim p$	$\sim q$	$\sim p \rightarrow \sim q$	$\sim q \rightarrow \sim p$
T	T	T	F	F	T	T
T	F	F	F	T	T	F
F	T	T	T	F	F	T
F	F	T	T	T	T	T

Notice that the contrapositive of a true statement is always true. The contrapositive of a false statement is always false. This means that a statement and its contrapositive are **logically equivalent**. The converse and the inverse are also logically equivalent. If the converse is true, then the inverse is true. If the converse is false, then the inverse is false.

Example 3 Write the given statement in "If . . ., then . . ." form. Write the converse, the inverse, and the contrapositive. Tell whether each is true or false.

Statement:	An acute angle is not an obtuse angle.	
Rewritten statement:	If an angle is acute, then it is not obtuse.	True
Converse:	If an angle is not obtuse, then it is acute.	False
Inverse:	If an angle is not acute, then it is obtuse.	False
Contrapositive:	If an angle is obtuse, then it is not acute.	True

There are several ways in which an inverse, converse, or contrapositive can be written. In Example 3, the converse could be written as "An angle that is not obtuse is acute." The inverse could be written as "A nonacute angle is obtuse." Later in your study of geometry, you will find it helpful to rewrite statements in equivalent forms.

Written Exercises

Underline each hypothesis once and each conclusion twice.

A **1.** If $5 = 4 + 1$, then $6 = 5 + 1$.

2. If $8 = 4 \cdot 2$, then $9 = 3 \cdot 3$.

3. If two segments are congruent, then they have the same length.

4. If a ray in the interior of an angle divides the angle into two congruent angles, then it bisects the angle.

5. If lines are parallel, then they do not intersect.

6. If a figure is a square, then it has four right angles.

Rewrite each statement in "If . . ., then . . ." form. Check students' answers for Ex. 7–28.

7. Parallel lines do not intersect.
8. Complementary angles are not obtuse.
9. Two segments equal in length are congruent.
10. Two congruent angles are equal in degree measure.
11. All right angles have a measure of 90°.
12. Two right angles are supplementary.
13. An acute angle has a measure less than 90°.
14. A triangle has three sides.

Write the converse of each statement. Label each statement and its converse as true or false.

15. If two angles are acute, then they are congruent.
16. If two angles are complementary, then they are acute.
17. If today is Monday, then tomorrow is Tuesday.
18. If this month is January, then next month is March.

Write the converse of each statement. Then write a biconditional.

19. If an angle measures 180°, then it is a straight angle.
20. If two points are collinear, then they lie on the same line.

Write the converse, inverse, and contrapositive of each statement. Label each of these as true or false.

B

21. If $7 = 4 + 3$, then $4 + 3 = 7$.
22. If $9 \cdot 8 = 72$, then $72 \div 9 = 8$.
23. If points are coplanar, then they lie in the same plane.
24. If two angles are congruent, then they are straight angles.
25. All fish are good swimmers.
26. All geometric figures are formed by sets of points.
27. All obtuse angles are congruent.
28. All integers are whole numbers.

Complete each of the truth tables.

C

29.

p	q	$p \to q$	$q \to p$	$(p \to q) \vee (q \to p)$
T	T	T	T	? T
T	F	F	T	? T
F	T	? T	? F	? T
F	F	? T	? T	? T

30.

p	q	$p \to q$	$q \to p$	$(p \to q) \wedge (q \to p)$
T	T	T	T	? T
T	F	F	T	? F
F	T	? T	? F	? F
F	F	? T	? T	? T

31.

p	q	$\sim p$	$p \to q$	$\sim p \wedge (p \to q)$
T	T	F	T	F
T	F	? F	? F	? F
F	T	? T	? T	? T
F	F	T	? T	? T

32.

p	q	$\sim p$	$p \vee q$	$\sim p \to (p \vee q)$
T	T	? F	T	T
T	F	? F	? T	? T
F	T	? T	? T	? T
F	F	? T	F	? F

Statements that are always true are called *tautologies*. Using a truth table, show that each of the given statements is a tautology. (Ex. 33–36) Check students' answers for Ex. 33–37.

33. $[(p \to q) \wedge p] \to q$
34. $[(p \vee q) \wedge \sim p] \to q$
35. $\sim(p \wedge q) \to (\sim p \vee \sim q)$
36. $\sim(p \vee q) \to (\sim p \wedge \sim q)$
37. Write the truth table for $p \leftrightarrow q$.

DRAWING CONCLUSIONS 2.4

Objectives **To draw conclusions about angles using given information**
To use the transitive and substitution properties of congruence

If you know certain things about angles, you can often discover other things about them. For example, suppose that m $\angle ABC = 105$ and m $\angle DEF = 105$. Since the angle measures are equal, you can conclude that the two angles are congruent.

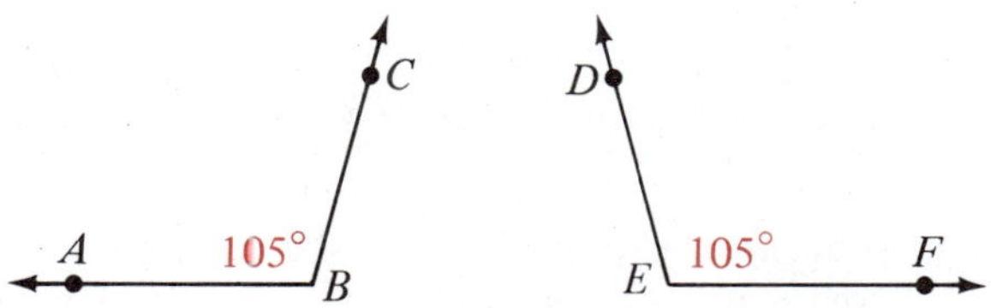

Since m $\angle ABC = 105$ and m $\angle DEF = 105$, $\angle ABC \cong \angle DEF$.

Example 1 $\overrightarrow{HJ}$ bisects $\angle GHI$. What conclusions can you draw about the angles formed?

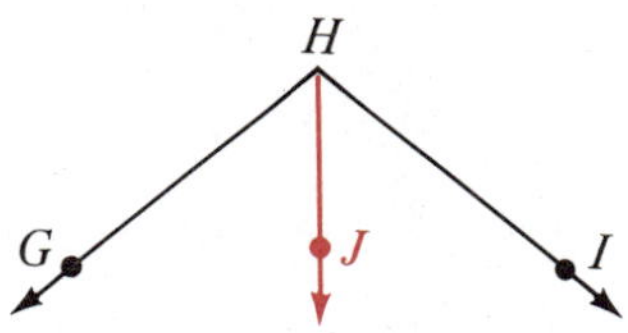

A bisector forms two congruent angles.

Thus, $\angle GHJ \cong \angle JHI$ and m $\angle GHJ =$ m $\angle JHI$.

Conclusions in geometry may be based on definitions previously accepted as being true. In Example 1, the definition of *angle bisector* was used. Conclusions may also be based on other accepted agreements. In algebra, you learned about the *transitive property of equality*. This states that if $x = y$ and $y = z$, then $x = z$, for all numbers x, y, and z. Since the measures of angles and segments are numbers, the transitive property is true for them.

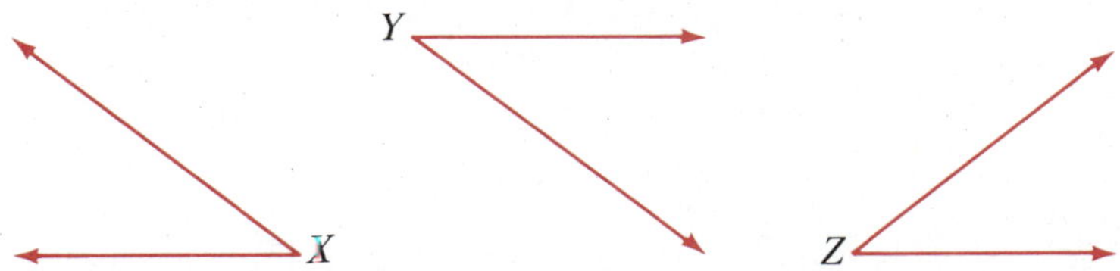

Suppose that $\angle X \cong \angle Y$ and $\angle Y \cong \angle Z$. This means that m $\angle X =$ m $\angle Y$ and m $\angle Y =$ m $\angle Z$. By the transitive property, m $\angle X =$ m $\angle Z$. Since the measures are equal, $\angle X \cong \angle Z$. A similar argument can be given to show that if $\overline{AB} \cong \overline{CD}$ and $\overline{CD} \cong \overline{EF}$, then $\overline{AB} \cong \overline{EF}$. This leads to the transitive property of congruence, which is true for all angles or segments.

Transitive property of congruence	If $\angle X \cong \angle Y$ and $\angle Y \cong \angle Z$, then $\angle X \cong \angle Z$. If $\overline{AB} \cong \overline{CD}$ and $\overline{CD} \cong \overline{EF}$, then $\overline{AB} \cong \overline{EF}$.

Example 2

If $\angle AOB \cong \angle BOC$ and $\angle BOC \cong \angle COD$, which other angles are congruent?

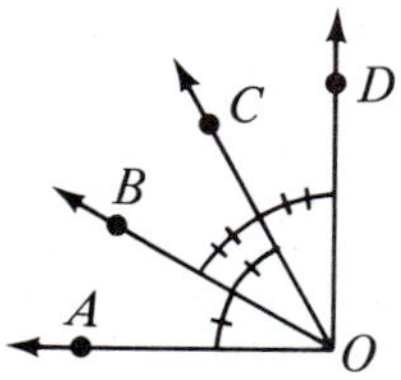

Use the transitive property. ► $\angle AOB \cong \angle COD$

The *substitution property of equality* states that if $a = b$, then a can be substituted for b in any equation. For example, if $y = 9$, you can substitute in the equation $3x + 4 = y$ to get $3x + 4 = 9$. Since any congruence statement such as $\angle A \cong \angle B$ or $\overline{AB} \cong \overline{CD}$ can be changed into a statement of equality ($m\angle A = m\angle B$ or $AB = CD$), there is a similar substitution property of congruence.

Substitution property of congruence	If $\angle A \cong \angle B$, then $\angle A$ can be substituted for $\angle B$ in any congruence statement. If $\overline{AB} \cong \overline{CD}$, then $\overline{AB}$ can be substituted for $\overline{CD}$ in any congruence statement.

Example 3

Given: $\angle T \cong \angle V$ and $m\angle T = 53$. What conclusion can be drawn? Give a reason for your conclusion.

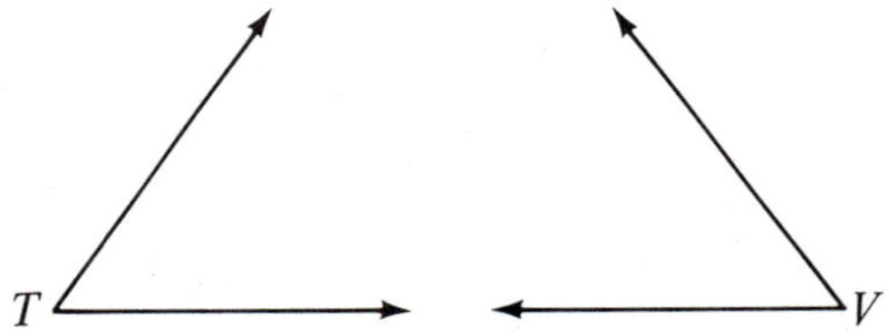

$\angle T \cong \angle V$ and $m\angle T = 53$

$m\angle T = m\angle V$ ◄ *Definition of congruent angles*

Conclusion: $53 = m\angle V$, or $m\angle V = 53$ ◄ *Substitution property*

Example 4

Based on the given information, is the stated conclusion correct? Why or why not?

Given: $\angle x$ and $\angle y$ are complementary.
Conclusion: $m\angle x = 45$.

No. The conclusion assumes that $\angle x \cong \angle y$.

Written Exercises

In each of the following, which other angle measures are equal?

(A) 1. m $\angle AOB$ = m $\angle AOD$, m $\angle AOD$ = m $\angle DOC$, m $\angle DOC$ = m $\angle BOC$

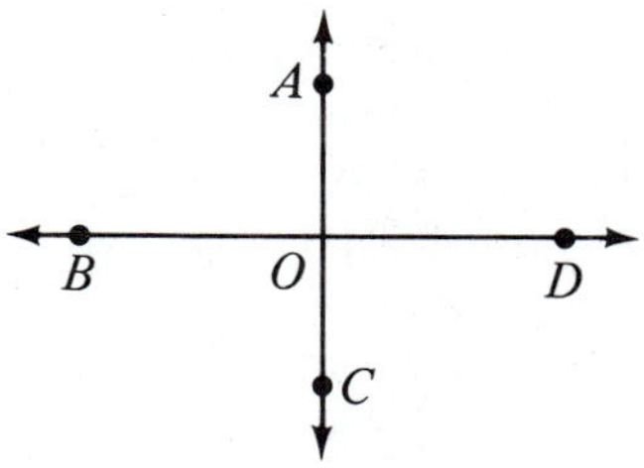

m ∠AOB = m ∠DOC, m ∠AOB = m ∠BOC, m ∠AOD = m ∠BOC

2. m $\angle EOF$ = m $\angle HOI$, m $\angle FOG$ = m $\angle GOH$, m $\angle FOG$ = m $\angle HOI$.

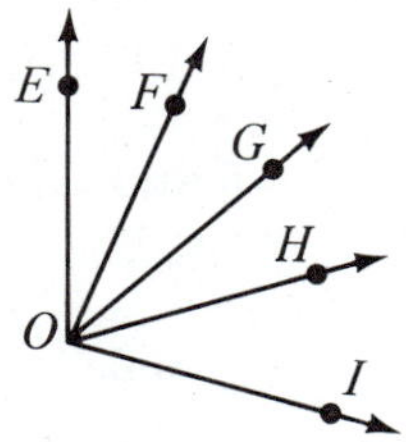

m ∠EOF = m ∠GOH, m ∠EOF = m ∠FOG, m ∠HOI = m ∠GOH

What conclusion can you draw based on each set of given information? Give a reason for your conclusion.

3. $\overrightarrow{KL}$ bisects $\angle JKM$.

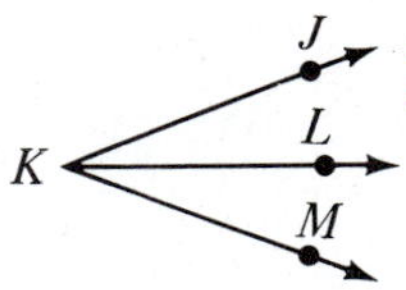

∠JKL ≅ ∠LKM; def. of ∠ bis.

4. $\angle NOP \cong \angle POQ$, $\angle POQ \cong \angle QOR$. **∠NOP ≅ ∠QOR; trans prop.**

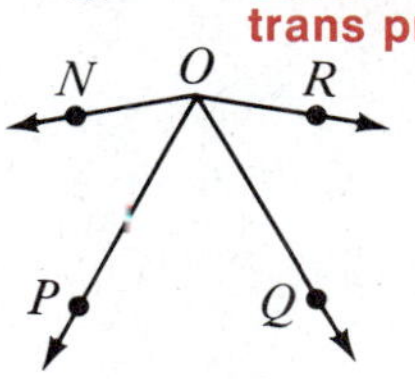

5. U is in the interior of $\angle STV$, $\angle STU \cong \angle UTV$.

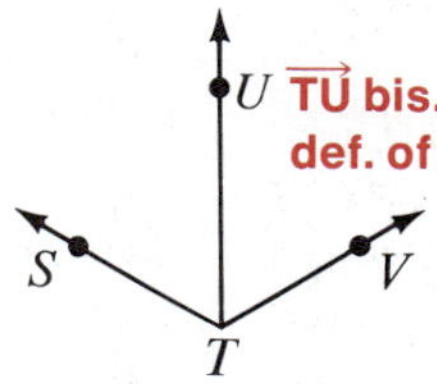

$\overrightarrow{TU}$ bis. ∠STV; def. of ∠bis.

6. B is the midpoint of $\overline{AC}$.

$\overline{AB} \cong \overline{BC}$; def. of midpt.

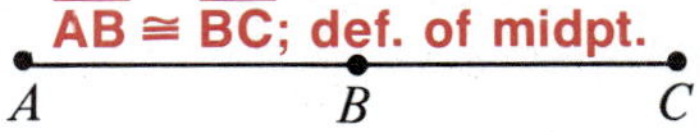

7. E is between D and F, $\overline{DE} \cong \overline{EF}$.

E is midpt. of $\overline{DF}$; def. of midpt.

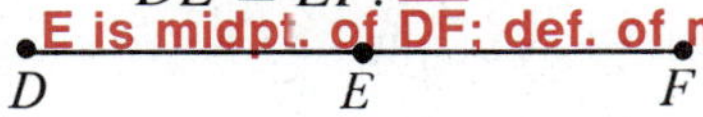

8. $\overline{GH} \cong \overline{HI}$, $\overline{HI} \cong \overline{IJ}$.

$\overline{GH} \cong \overline{IJ}$; trans. prop.

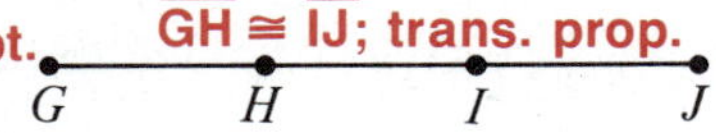

Draw a diagram for each of the following. What conclusion can you draw based on the given information. Give a reason for your conclusion.

12. m ∠A = m ∠B = 45; def. of comp. ∠'s.
13. m ∠C = m ∠D = 90; def. of supp. ∠'s.

(B) 9. m $\angle a$ = 48, $\angle a$ and $\angle b$ are complementary.

10. m $\angle c$ = 83, $\angle c$ and $\angle d$ are adjacent angles.

11. m $\angle e$ = 104, $\angle e$ and $\angle f$ are supplementary.

12. $\angle A \cong \angle B$, $\angle A$ and $\angle B$ are complementary.

13. $\angle C \cong \angle D$, $\angle C$ and $\angle D$ are supplementary.

14. m $\angle E$ = 38, $\angle E \cong \angle F$.

11. m ∠f = 76; def. of supp. ∠'s.
9. m ∠b = 42; def. of comp. ∠'s **10. no conclusion** **14. m ∠F = 38; def. of ≅ ∠'s.**

15. m $\angle XYZ$ = 80, m $\angle ZYW$ = 80. Conclusion: $\overrightarrow{YZ}$ bisects $\angle XYW$. **See below.**

16. m $\angle M$ = 71, $\angle M \cong \angle N$, $\angle N$ and $\angle O$ are supplementary. Conclusion: m $\angle O$ = 109.

17. m $\angle S$ = 45, $\angle S$ and $\angle T$ are complementary. Conclusion: $\angle S \cong \angle T$.

16. Yes; def. of ≅ ∠'s and supp ∠'s, subst. and subtr. prop.
17. Yes; def. of comp. ∠'s and ≅ ∠'s, subst. and subtr. prop.

CUMULATIVE REVIEW

1. Find the measure of a supplement of a 70° angle. **110°**
2. What is the measure of a right angle? **90°**
3. What are congruent segments? **seg. that have the same length**

15. No; pt. Z not in interior of ∠XYW.

WRITING PROOFS 2.5

Objectives

To write reasons for statements
To write simple geometric proofs using properties of equations

In geometry you should be able to give a reason for any statement that you make or for any conclusion that you draw. There are many types of reasons that you may use. Sometimes the reason for a statement may be a geometric property or definition that you have learned. Other reasons may involve algebraic properties or methods, or arithmetic facts. Later you will learn about other types of reasons.

In algebra you learned that if you add the same number to both sides of an equation, or subtract the same number from both sides, the resulting equation is still true. Similarly, if you multiply both sides by the same number or divide by the same number (except 0), the equality is unchanged. Since you will use these equation properties throughout your work in geometry, they are stated below.

Equation Properties

Addition: If $a = b$, then $a + c = b + c$ for all numbers a, b, and c.

Subtraction: If $a = b$, then $a - c = b - c$ for all numbers a, b, and c.

Multiplication: If $a = b$, then $a \cdot c = b \cdot c$ for all numbers a, b, and c.

Division: If $a = b$, then $\frac{a}{c} = \frac{b}{c}$ for all numbers a, b, and c ($c \neq 0$).

Example 1

What reason can you give for the given conclusion?
If m $\angle AOB =$ m $\angle COD$,
then m $\angle AOB +$ m $\angle BOC =$ m $\angle COD +$ m $\angle BOC$.

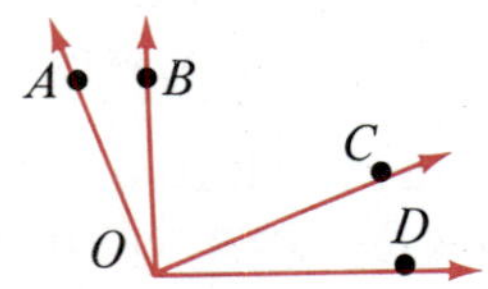

Addition property of equality.
(m $\angle BOC$ is added to both sides of the equation.)

Example 2

What reason can you give for the given conclusion?
If 2m $\angle PML = 80$, then m $\angle PML = 40$.

Division property of equality.
(Both sides of the equation were divided by 2.)

The reasons for any statements or conclusions should be written and labeled clearly. The symbol $\therefore$ is used to mean *therefore* and usually indicates the conclusion. Example 3 shows you how geometric statements or conclusions and their corresponding reasons should be written.

Example 3

What conclusion can you draw based on the given information? Give a reason for each statement.

Given: $\overrightarrow{OP}$ bisects $\angle AOB$.

Conclusion: $\angle AOP \cong \angle POB$

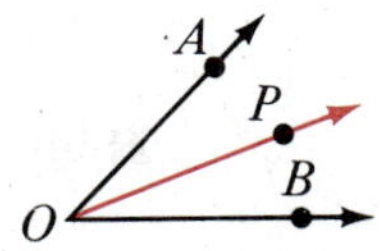

	Statements	Reasons
	1. $\overrightarrow{OP}$ bisects $\angle AOB$	1. Given
∴ means therefore. ►	2. $\therefore \angle AOP \cong \angle POB$	2. Definition of $\angle$ bisector

A **proof** is a formal argument that shows how a conclusion follows logically from other statements. A standard style or format is usually used in writing a geometric proof. You should write what is known or given, followed by steps which lead to the desired conclusion. You will usually use a two-column pattern. In this book, colored arrows will be used to show which step or steps in a proof lead to succeeding steps.

Example 4

Write a complete proof of the following.

Given: $m\angle X = 38$, $m\angle Y = 38$
Prove: $\angle X \cong \angle Y$

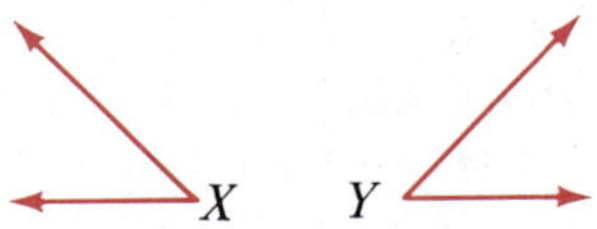

Proof

The colored arrows show which statements follow from other statements.

Statements	Reasons
1. $m\angle X = 38$	1. Given
2. $m\angle Y = 38$	2. Given
3. $m\angle X = m\angle Y$	3. Substitution property
4. $\therefore \angle X \cong \angle Y$	4. Definition of $\cong$ angles

Example 5

Give a reason for each step in the following proof.

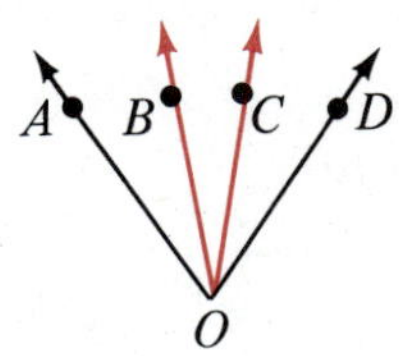

Given: $m\angle AOB + m\angle BOC = m\angle COD + m\angle BOC$
Prove: $\angle AOB \cong \angle COD$

Proof

	Statements	Reasons
	1. $m\angle AOB + m\angle BOC = m\angle COD + m\angle BOC$	1. Given
Subtract m∠BOC. ►	2. $m\angle AOB = m\angle COD$	2. Subtraction property of equality
	3. $\therefore \angle AOB \cong \angle COD$	3. Definition of $\cong$ angles

Example 6 Draw a diagram based on the given information. Write the missing statements in the proof.

Given: $\overline{AB} \cong \overline{BC}$, $\overline{BC} \cong \overline{CD}$, and $AB = 21$
Prove: $CD = 21$

Proof

Statements	Reasons
1. $\overline{AB} \cong \overline{BC}$	1. Given
2. $\overline{BC} \cong \overline{CD}$	2. Given
3. $\overline{AB} \cong \overline{CD}$	3. Transitive property
4. $AB = CD$	4. Definition of $\cong$ segments
5. $AB = 21$	5. Given
6. $\therefore CD = 21$	6. Substitution property

Oral Exercises

If each statement in the first column is given information, what is the reason for each corresponding conclusion?

	Given	Conclusion
1.	$XY = ZW$ and $ZW = AB$	$XY = AB$ **trans. prop.**
2.	$AB = CD$ and $AB = 8$	$CD = 8$ **subst. prop.**
3.	$\angle XYZ \cong \angle ABC$ and $\angle ABC \cong \angle RPG$	$\angle XYZ \cong \angle RPG$ **trans. prop.**
4.	$m\angle ROA = m\angle POW$ and $m\angle POW = 78$	$m\angle ROA = 78$ **trans. prop.**
5.	$RT + LM = 19$ and $LM = 7$	$RT = 12$ **subst. prop. and subtr. prop.**
6.	$m\angle SLC + m\angle RTU = 127$ and $m\angle SLC = 18$	$m\angle RTU = 109$ **subst prop. and subtr. prop.**
7.	$m\angle RTU = 18$ and $m\angle HJK = 37$	$m\angle RTU + m\angle HJK = 55$ **add prop.**
8.	$m\angle NMJ = 83$	$2(m\angle NMJ) = 166$ **mult. prop.**
9.	$AZ = 26$	$\frac{AZ}{2} = 13$ **div. prop.**
10.	$\overrightarrow{LM}$ bisects $\angle ALZ$.	$\angle ALM \cong \angle MLZ$ **def. of $\angle$bis.**

Written Exercises

Write the missing statements in each proof.

A **1.** Given: $AB = 9$, $CD = 9$
Prove: $\overline{AB} \cong \overline{CD}$

Proof

Statements	Reasons
1. $AB = 9$	1. Given
2. $CD = 9$	2. Given
3. $AB = ?$ **CD**	3. Substitution property
4. $\therefore ?$ **$\overline{AB} \cong \overline{CD}$**	4. Definition of $\cong$ segments

2. Given: $\overline{AB} \cong \overline{CD}$, $CD = 12$
Prove: $AB = 12$

Proof

Statements	Reasons
1. $\overline{AB} \cong \overline{CD}$	1. Given
2. ? **$AB = CD$**	2. Definition of $\cong$ segments
3. $CD = 12$	3. Given
4. $\therefore ?$ **$AB = 12$**	4. Substitution property

Give a reason for each statement in each proof.

3. Given: $AB + BC = 12$ and $\overline{AB} \cong \overline{CD}$
Prove: $CD + BC = 12$

Proof

Statements	Reasons
1. $\overline{AB} \cong \overline{CD}$	1. ? **Given**
2. $AB = CD$	2. ? **Def. of ≅ seg.**
3. $AB + BC = 12$	3. ? **Given**
4. $\therefore CD + BC = 12$	4. ? **Subst. prop.**

4. Given: $AB = 8$, $\overline{AB} \cong \overline{BC}$ and $\overline{BC} \cong \overline{CD}$
Prove: $CD = 8$

Proof

Statements	Reasons
1. $\overline{AB} \cong \overline{BC}$	1. ? **Given**
2. $\overline{BC} \cong \overline{CD}$	2. ? **Given**
3. $\overline{AB} \cong \overline{CD}$	3. ? **Trans. prop.**
4. $AB = CD$	4. ? **Def. of ≅ seg.**
5. $AB = 8$	5. ? **Given**
6. $\therefore CD = 8$	6. ? **Subst. prop.**

Write a proof for each, using the pattern shown in this lesson.

5. Given: $m\angle A = m\angle B$ and $m\angle B = m\angle C$
Prove: $\angle A \cong \angle C$

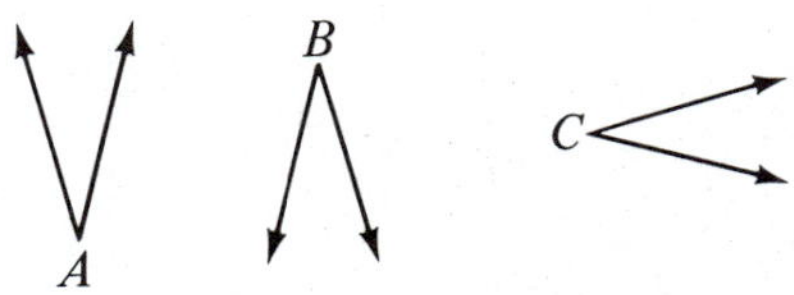

6. Given: $m\angle D = 79$ and $m\angle E = 79$
Prove: $\angle D \cong \angle E$

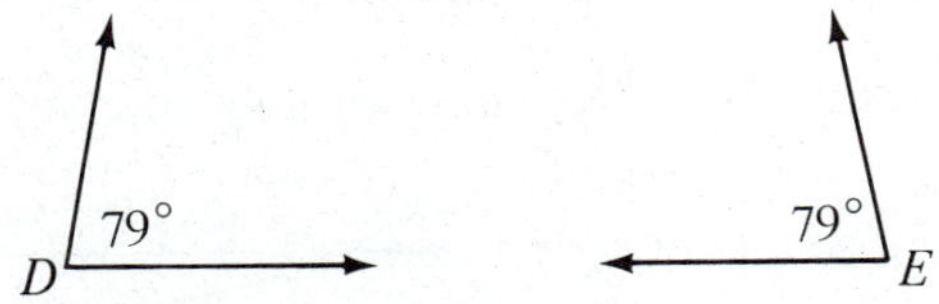

7. Given: $m\angle WOX + m\angle XOY = m\angle ZOY + m\angle XOY$
Prove: $\angle WOX \cong \angle ZOY$

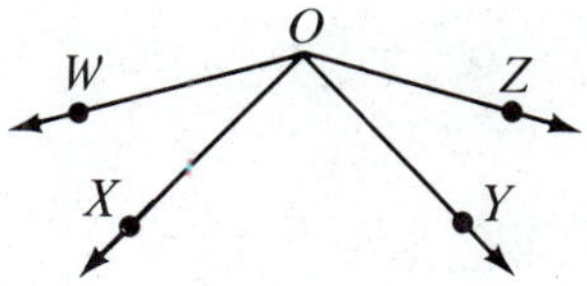

8. Given: $\angle WOX \cong \angle ZOY$
Prove: $m\angle WOX + m\angle XOY = m\angle ZOY + m\angle XOY$

Draw a diagram based on the given information. Then write a proof for each.

B

9. Given: $m\angle A + m\angle B = 103$, and $m\angle B + m\angle C = 103$
Prove: $m\angle A = m\angle C$

10. Given: $m\angle D - m\angle E = 13$, and $m\angle F - m\angle E = 13$
Prove: $m\angle D = m\angle F$

11. Given: $m\angle G = 35$, $m\angle H = 2(m\angle G)$, and $m\angle I = 70$
Prove: $\angle H \cong \angle I$

12. Given: $JK + KL = 18$ and $KL + LM = 18$
Prove: $JK = LM$

13. Given: $NO - RS = 25$ and $PQ - RS = 25$
Prove: $NO = PQ$

14. Given: $AB = 7$, $CD = 2(AB)$, and $EF = 14$
Prove: $\overline{CD} \cong \overline{EF}$

C

15. Given: $m\angle XYZ = \frac{1}{2}(m\angle WYZ)$, $m\angle RST = \frac{1}{2}(m\angle QST)$, and $\angle WYZ \cong \angle QST$
Prove: $\angle XYZ \cong \angle RST$

16. Given: $JK = \frac{1}{2}(LM)$, $NO = \frac{1}{2}(PQ)$, and $\overline{LM} \cong \overline{PQ}$
Prove: $\overline{JK} \cong \overline{NO}$

POSTULATES AND THEOREMS 2.6

Objective **To use postulates and theorems involving points, lines, and planes**

In geometry, it is impossible to prove every statement. You must accept some basic statements as reasonable without a formal argument to show that they are true. Statements that are accepted without proof are called **postulates**.

Postulate 1 For any two points, there is exactly one line containing them.

It is accepted without proof that two given points are contained in exactly one line. From this, you can give an argument to prove another statement.

Example 1 Can two lines intersect in two points?

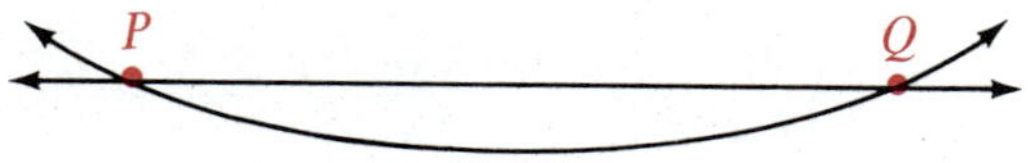

No. If they did, then there would be two different lines containing the two points. This contradicts postulate 1.

Such statements that are proved are called **theorems.** Most of the proofs of theorems in this book will use the formal pattern consisting of statements and reasons as presented in the last lesson. However, some proofs will be given using an informal, indirect way. The argument given in Example 1 is an indirect proof of your first theorem.

Theorem 2.1 Two lines intersect in at most one point.

The next postulate involves points and a plane.

Postulate 2 Three noncollinear points are contained in exactly one plane.

Many planes contain two given points, but only one plane contains three noncollinear points.

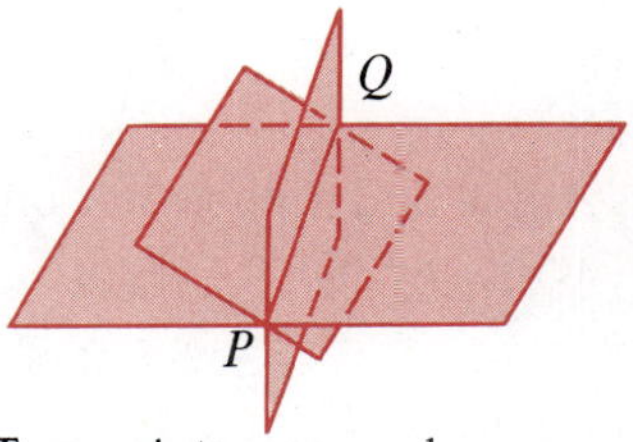

Two points—many planes

Three noncollinear points—one plane

Three other postulates involving points, lines, and planes are generally accepted.

Postulate 3 If two points of a line are in a given plane, then the line itself is in the plane.

Postulate 4 If two planes intersect, then they intersect in exactly one line.

Postulate 5 Space is determined by at least four points not all in the same plane.

Example 2 What are the possible intersections of a line and a plane? Draw a picture of each.

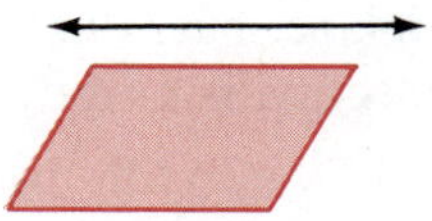
No points

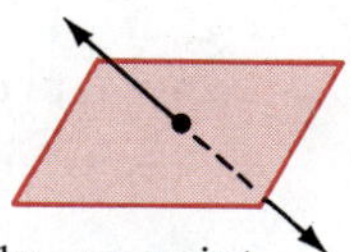
Exactly one point

(Postulate 3 suggests these.)

The line itself

Example 3 What are the possible intersections of two planes? Draw a picture of each.

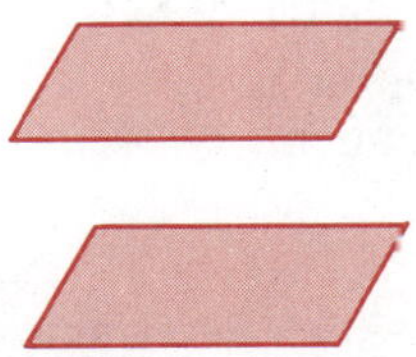
Planes do not intersect.

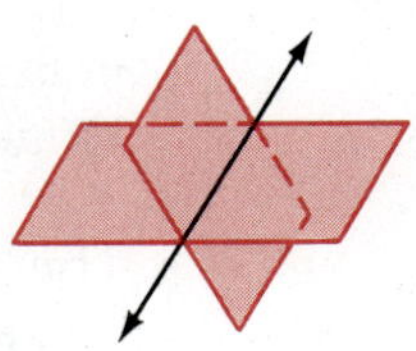
Planes intersect in a line.

(Postulate 4 suggests these.)

You can use these postulates to prove other theorems about points, lines, and planes.

Theorem 2.2 A line and a point not on the line are contained in exactly one plane.

Given: Line l and point P not on l
Prove: Exactly one plane contains l and P.

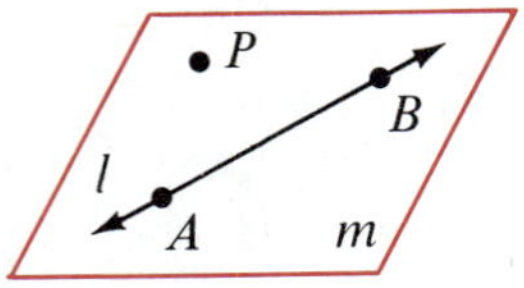

Proof

Statements	Reasons
1. Line l and point P not on l	1. Given
2. l contains at least 2 points. Name them A and B.	2. Postulate 5
3. P, A, and B are noncollinear	3. Definition of noncollinear points
4. P, A, and B are in exactly one plane. Name it m.	4. Postulate 2
5. l is in m.	5. Postulate 3
6. ∴ Exactly one plane contains l and P.	6. Steps 4 and 5

The proofs of two other theorems are similar to the proofs of Theorems 2.1 and 2.2, respectively. You should complete these proofs as Exercises 22 and 23.

Theorem 2.3 If a line intersects a plane, but is not contained in the plane, then the intersection is exactly one point. (See Exercise 22.)

Theorem 2.4 Two intersecting lines are contained in exactly one plane. (See Exercise 23.)

Example 4 True or false? An angle (other than a straight angle) is contained in exactly one plane.
Draw a picture and give an argument.

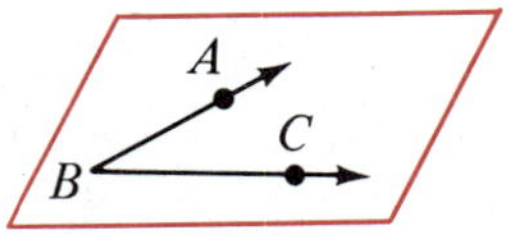

True.
The angle consists of two rays which determine two intersecting lines.
By Theorem 2.4, the lines are contained in exactly one plane.

Written Exercises

True or false?

A 1. Two points are in exactly one line. T

2. Three noncollinear points are in exactly one plane. T

3. Two points are in exactly one plane. F

4. Four points are in exactly one plane. F

5. Two planes can intersect in exactly one line. T

6. Two planes can intersect in exactly one point. F

7. Two planes must intersect in exactly one line. F

8. A line and a plane must intersect in exactly one point. F

9. Two lines must intersect in exactly one point. F

10. Two lines can intersect in exactly one point. T

11. Postulates are statements to be proved. F

12. Since postulates are not proved, they may not be used as reasons in proofs. F

13. Theorems are statements to be proved. T

Answer each question. Draw a picture to illustrate. (Ex. 14–21)

14. Can a curve intersect a line in exactly two points? Yes

15. Can a curve intersect a plane in exactly two points? Yes

16. Can more than one plane contain three given points? Yes

17. Can three points ever be noncoplanar? No

18. How many different lines are determined by four noncollinear points? 6

19. How many different planes are determined by four noncoplanar points? 4

20. Is it possible for exactly two lines to be determined by three points? No

21. Is it possible to have an angle that is not in a single plane? No

B 22. Prove Theorem 2.3. Use postulates 3 and 5.

23. Prove Theorem 2.4. Use postulates 2, 3, and 5.

ALGEBRA REVIEW

Objective

To evaluate algebraic expressions

To evaluate an algebraic expression:
(1) Substitute the given value for each variable.
(2) Simplify any powers.
(3) Do multiplications and divisions.
(4) Do additions and subtractions.

Example

Evaluate $x^3 + 3x^2 + 4$ if $x = -3$.

$(-3)^3 + 3(-3)^2 + 4$

$-27 + 3(9) + 4$

$-27 + 27 + 4$, or 4

1. 7 2. 52 3. 4 4. 5 5. −12 6. 25

Evaluate.

1. $x^2 - 2x + 7$ if $x = 2$
2. $3b^2 - 6b + 7$ if $b = -3$
3. $2t^3 + t^2 - 3t + 4$ if $t = 1$
4. $(a + 2)^2 + 4$ if $a = -3$
5. $n^3 + (n - 3)(n + 4)$ if $n = 0$
6. $2(y - 1)^2 + y + 3$ if $y = 4$

ANGLE AND SEGMENT ADDITION 2.7

Objectives **To use the angle addition postulate**
To use segment addition

When you measure segments or angles, you are assuming that there are segments congruent to those marked on your ruler and that there are angles congruent to those indicated on your protractor. Two postulates guarantee that angles and segments of any given measure do exist.

Postulate 6 On every line, there is a segment with a given point as an endpoint congruent to any given segment.

Postulate 7 For every ray, there is an angle with the given ray as a side congruent to any given angle.

When you bisect a segment or an angle, you are assuming that there actually is a midpoint of the segment or a bisector of the angle.

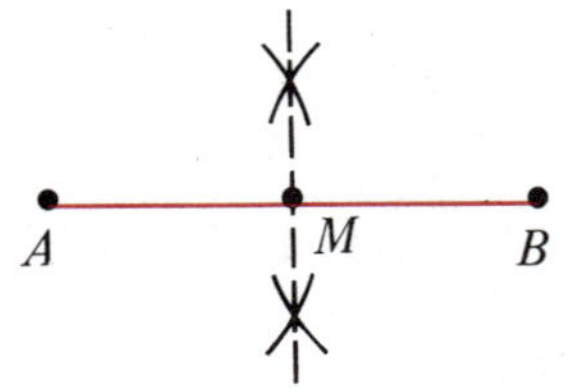

M is the midpoint of $\overline{AB}$.

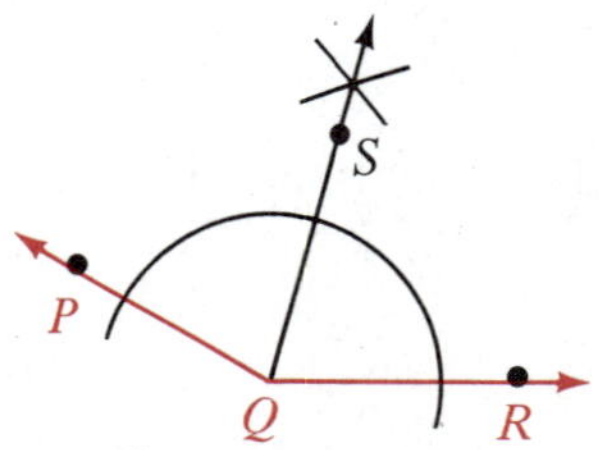

$\overrightarrow{QS}$ bisects $\angle PQR$.

Postulate 8 Every segment has exactly one midpoint.

Postulate 9 Every angle, except a straight angle, has exactly one bisector.

You may need to find the measure of an angle from what is known about the measures of other angles. You may know that m $\angle XYZ = 40$ and m $\angle ZYW = 60$. Using a protractor, m $\angle XYW$ is found to be 100. In this case, the sum of the measures of the two smaller angles is the measure of the larger angle. This leads to the statement of the next postulate.

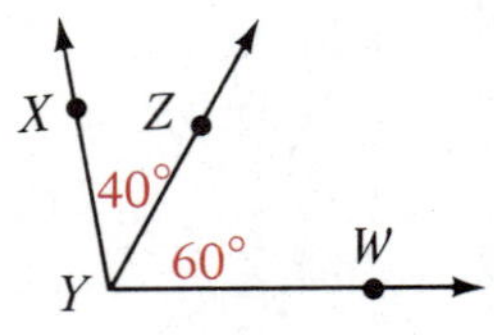

Postulate 10 Angle Addition Postulate

If D is in the interior of $\angle ABC$,
then $m\angle ABC = m\angle ABD + m\angle DBC$.

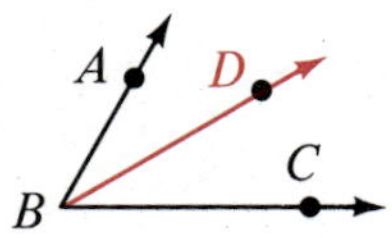

In the figure below, $m\angle ABD + m\angle DBC$ is not equal to $m\angle ABC$. D is not in the interior of $\angle ABC$. In this case, the Angle Addition Postulate does not apply.

$m\angle ABD = 160$, $m\angle DBC = 155$.
But, $m\angle ABC = 45$, not 315.

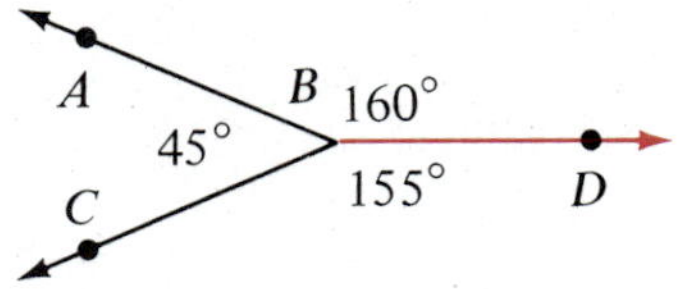

Example 1

Given: $m\angle DOF = 124$, $m\angle DOE = 36$.
Find $m\angle EOF$

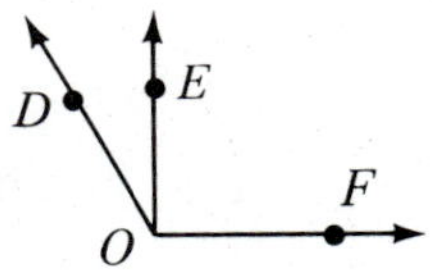

$m\angle DOE + m\angle EOF = m\angle DOF$

$m\angle EOF = m\angle DOF - m\angle DOE$

Thus, $m\angle EOF = 124 - 36$, or 88.

A geometric figure can be rather complicated. Several angles in the figure may have the same vertex. In such cases, it often is helpful to number the angles, as shown below. Thus, $\angle AOB$ can also be called $\angle 1$. Labeling angles in this way helps avoid confusion and saves time in writing proofs.

$\angle 1 = \angle AOB$
$\angle 2 = \angle BOC$
$\angle 3 = \angle COD$

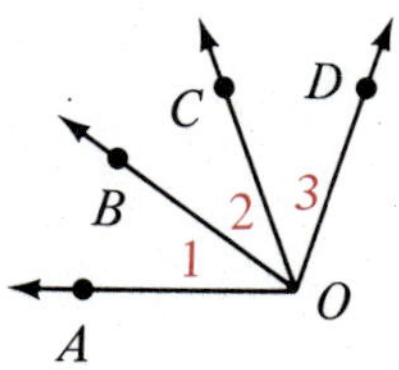

The *reflexive property of equality* states that a number is equal to itself: $a = a$, for any number a. Since congruence statements are actually statements about measures, which are numbers, there is a similar reflexive property for congruence statements.

Reflexive property of congruence	$\angle A \cong \angle A$ and $\overline{AB} \cong \overline{AB}$, for any angle A and any segment $\overline{AB}$.

You will use the reflexive property in many proofs.

Example 2

Give a reason for each step in the following proof.

Given: m $\angle 1 = $ m $\angle 3$
Prove: $\angle WOY \cong \angle ZOX$

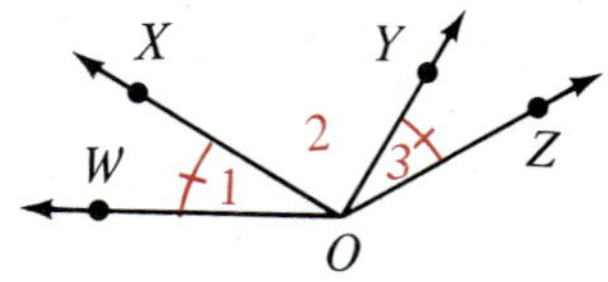

Proof

Add m $\angle 2$ to both sides of equation. ► (step 3)

Statements	Reasons
1. m $\angle 1 = $ m $\angle 3$	1. Given
2. m $\angle 2 = $ m $\angle 2$	2. Reflexive property
3. m $\angle 1 + $ m $\angle 2 = $ m $\angle 3 + $ m $\angle 2$	3. Addition property of equality
4. m $\angle WOY = $ m $\angle 1 + $ m $\angle 2$, m $\angle ZOX = $ m $\angle 3 + $ m $\angle 2$	4. Angle addition postulate
5. m $\angle WOY = $ m $\angle ZOX$	5. Substitution property
6. $\therefore \angle WOY \cong \angle ZOX$	6. Definition of $\cong$ angles

You learned to add the measures of angles, using the angle addition postulate. The definition of *between* accepted in Chapter 1 allows the same type of addition with segments. If point P is between points A and B, then $AP + PB = AB$. You can assume betweenness from the picture used with the proof.

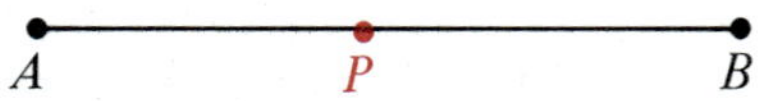

$AP + PB = AB$

If P is not between A and B, then $AP + PB \neq AB$. Below, if $AP = 15$, and $PB = 6$, then $AB = 9$, not 21.

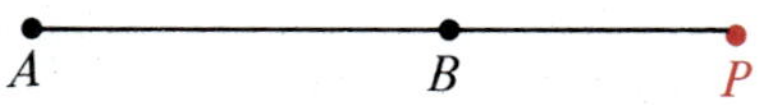

Example 3

Given: $AB = CD$
Prove: $AC = BD$

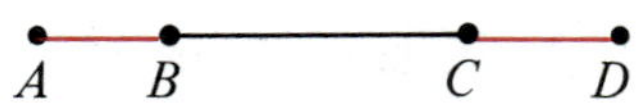

Proof

Statements	Reasons
1. $AB = CD$; $BC = BC$	1. Given; reflexive property
2. $AB + BC = CD + BC$	2. Addition property of equality
3. $AC = AB + BC$	3. Definition of between
4. $BD = BC + CD$	4. Definition of between
5. $\therefore AC = BD$	5. Substitution property

Written Exercises

Give another name for each angle.

A 1. $\angle 1$ $\angle AOB$

2. $\angle BOC$ $\angle 2$

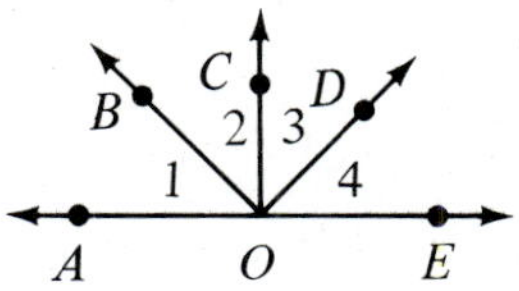

3. $\angle 4$ $\angle DOE$

4. $\angle COD$ $\angle 3$

Find each indicated length.

5. $XY = 7$ cm, $YZ = 4$ cm, $XZ = ?$ 11 cm

6. $XZ = 9$ cm, $XY = 6$ cm, $YZ = ?$ 3 cm

7. $XY = 5$ m, $YZ = 3$ m, $XZ = ?$ 8 m

8. $XZ = 15$ m, $YZ = 6$ m, $XY = ?$ 9 m

***W* is in the interior of $\angle XYZ$. Find each indicated angle measure. (Ex. 9–11)**

9. Given: m $\angle XYW = 40$ and m $\angle WYZ = 60$ Find m $\angle XYZ$. 100

10. Given: m $\angle XYZ = 113$ and m $\angle XYW = 42$ Find m $\angle WYZ$. 71

11. Given: m $\angle XYZ = 94$ and m $\angle WYZ = 32$ Find m $\angle XYW$. 62

12. Given: m $\angle 1 = 55$, m $\angle 2 = 20$, and m $1 =$ m$\angle 3$ Find m $\angle GOI$. 75

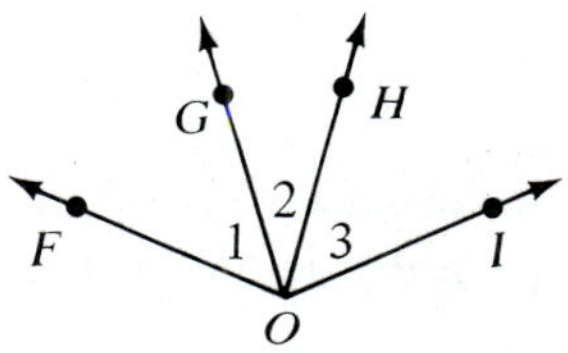

13. Given: m $\angle 1 =$ m $\angle 3$ and m $\angle GOI = 78$ Find m $\angle FOH$. 78

14. Given: $JK = 4$ cm, $KL = 7$ cm, and $JK = LM$ Find KM. 11 cm

15. Given: $JK = LM$ and $JL = 20$ mm Find KM. 20 mm

Write the missing statements or reasons in each proof. (Ex. 16–17)

16. Given: m $\angle 1 =$ m $\angle 3$
Prove: m $\angle NRP =$ m $\angle ORQ$

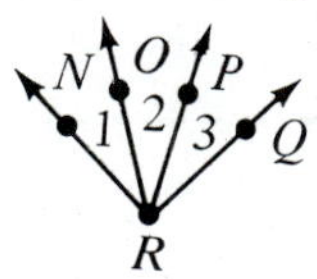

17. Given: $SU = TV$
Prove: $ST = UV$

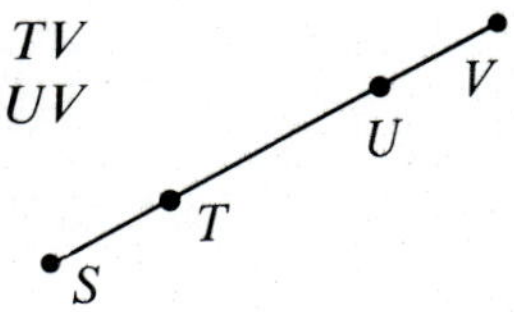

Proof

Statements	Reasons
1. ? m $\angle 1 =$ m $\angle 3$	1. Given
2. m $\angle 1 +$ m $\angle 2 =$ m $\angle 3 +$ m $\angle 2$	2. ? Add. prop.
3. m $\angle NRP =$ m $\angle 1 +$ m $\angle 2$	3. ? $\angle$add. post.
4. ?	4. Angle addition postulate
5. $\therefore$?	5. Substitution property

4. m $\angle ORQ =$ m $\angle 3 +$ m $\angle 2$ 5. m $\angle NRP =$ m $\angle ORQ$.

Proof

Statements	Reasons
1. $SU = TV$	1. ? Given
2. $ST + TU = SU$	2. ? Def. of between
3. ? TU + UV = TV	3. Definition of between
4. $ST + TU = TU + UV$	4. ? Subst. prop.
5. $\therefore$? ST = UV	5. Subtraction property of equality

18. Given: $AB = DE$
Prove: $AD = BE$

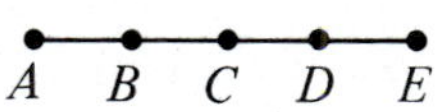

19. Given: $AC = BD$
Prove: $AB = CD$

B **20.** Given: $\angle 1 \cong \angle 3$
Prove: $\angle AOC \cong \angle DOB$

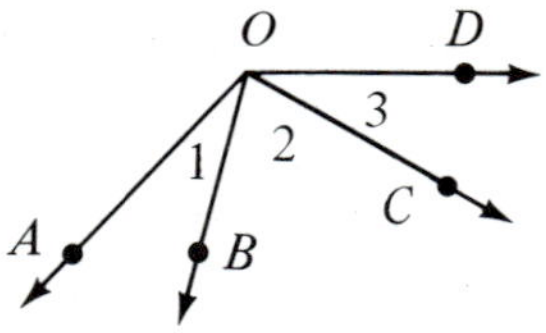

21. Given: $\angle AOC \cong \angle DOB$
Prove: $\angle 1 \cong \angle 3$

22. Given: $\overline{EH} \cong \overline{FI}$ and $\overline{HG} \cong \overline{IG}$
Prove: $\overline{EG} \cong \overline{FG}$

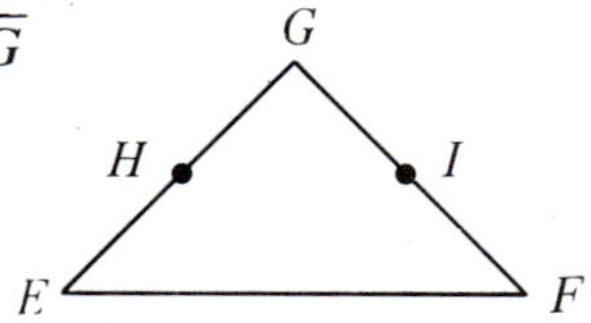

23. Given: $\overline{EG} \cong \overline{FG}$ and $\overline{EH} \cong \overline{FI}$
Prove: $\overline{HG} \cong \overline{IG}$

24. Given: $\overline{JL} \cong \overline{NL}$,
$JK = \frac{1}{2}JL$, and $NM = \frac{1}{2}NL$
Prove: $\overline{JK} \cong \overline{NM}$

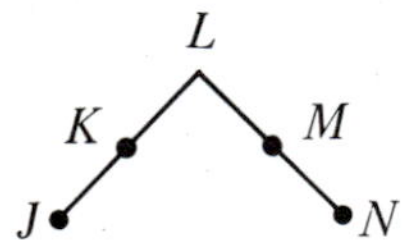

25. Given: $\overline{JK} \cong \overline{NM}$,
$JK = \frac{1}{2}JL$, and $NM = \frac{1}{2}NL$
Prove: $\overline{JL} \cong \overline{NL}$

C **26.** Given: Q bisects $\overline{OP}$, T bisects $\overline{RS}$, and $\overline{OP} \cong \overline{RS}$
Prove: $\overline{OQ} \cong \overline{TS}$

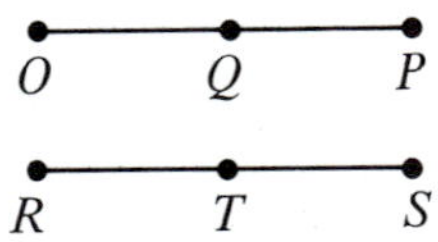

27. Given: $\overline{OQ} \cong \overline{QP}$, $\overline{OQ} \cong \overline{RT}$, and $\overline{OP} \cong \overline{RS}$
Prove: T bisects $\overline{RS}$

NON-ROUTINE PROBLEMS

Mona has to drive from Metropolis to Century Village. According to the highway map, there are several different routes. How many different routes are there? What is the shortest route?

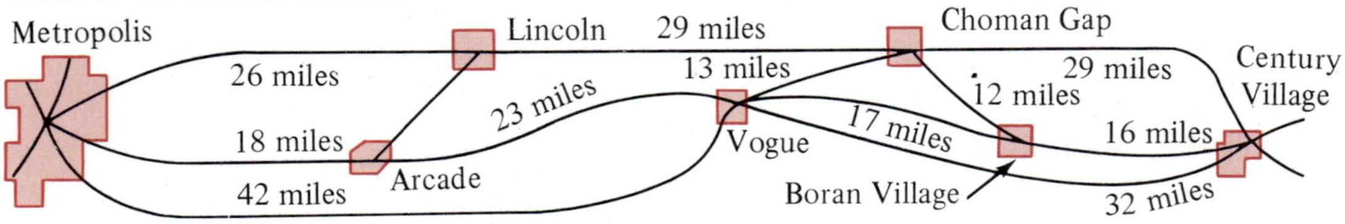

20 routes: Metropolis to Arcade to Vogue to Century Village

ALGEBRA REVIEW

Objective

To multiply polynomials

1. $3x + 12$ **2.** $12a^3 + 18a^2$ **3.** $8x^5 - 10x^2$
4. $-15t + 10s$ **5.** $-24x^5 + 16x^4$ **6.** $2z^2 - 5z - 3$
7. $6x^2 + x - 15$ **8.** $4r^2 + 12r + 9$ **9.** $p^2 - q^2$
10. $9c^2 - 16$ **11.** $16c^4 - 9b^2$

To multiply polynomials, use the distributive property: $a(b + c) = ab + ac$.

12. $25x^2 + 30xy^2 + 9y^4$

Examples

Multiply $3x(x^2 + 5)$.
$3x \cdot x^2 + 3x \cdot 5$
$3x^3 + 15x$

Multiply $(3y + 4)(y - 3)$.
$(3y + 4) \cdot y + (3y + 4) \cdot (-3)$
$3y^2 + 4y - 9y - 12$
$3y^2 - 5y - 12$

Multiply. **(See above.)**

1. $3(x + 4)$	**2.** $6a(2a^2 + 3a)$	**3.** $2x^2(4x^3 - 5)$	**4.** $(3t - 2s)(-5)$
5. $(6x^2 - 4x)(-4x^3)$	**6.** $(2z + 1)(z - 3)$	**7.** $(3x + 5)(2x - 3)$	**8.** $(2r + 3)(2r + 3)$
9. $(p - q)(p + q)$	**10.** $(3c - 4)(3c + 4)$	**11.** $(4c^2 - 3b)(4c^2 + 3b)$	**12.** $(5x + 3y^2)(5x + 3y^2)$

PERPENDICULARS 2.8

Objectives

To identify perpendicular lines, rays, and segments
To prove statements about perpendiculars
To construct perpendicular lines

You learned earlier that a right angle has a measure of 90°.

Definition: Perpendicular Two lines are *perpendicular* if they intersect to form a right angle.

$\overleftrightarrow{AB} \perp \overleftrightarrow{XY}$ means that $\overleftrightarrow{AB}$ is perpendicular to $\overleftrightarrow{XY}$. The symbol $\not\perp$ means *is not perpendicular to*. In the figure, the symbol ⌉ is used to show that the lines are perpendicular. This is the same symbol used to indicate a right angle.

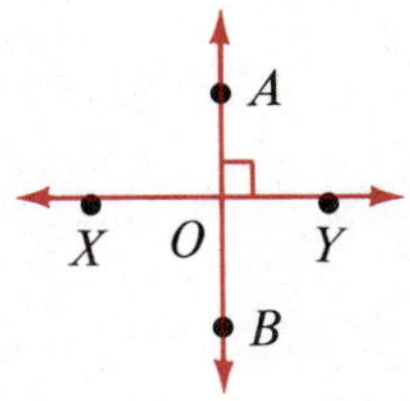

Since the measure of a straight angle is 180°, it seems reasonable to assume that the sum of the measures of certain angles on one side of a line is 180°.

Postulate 11

The sum of the measures of the angles with the same vertex on one side of a line and with no interior points in common is 180°. The sum of the measures of all angles around a common vertex and with no interior points in common is 360°.

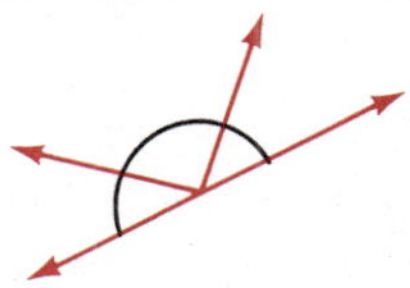

You should try to write proofs of the following theorems.

Theorem 2.5 All right angles are congruent.

Theorem 2.6 Two perpendicular lines form four congruent right angles.

Segments and *rays* are *perpendicular* if the lines that contain them are perpendicular.

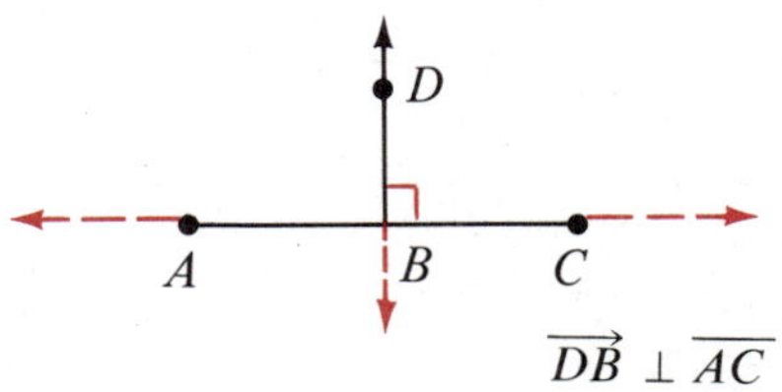

$\overrightarrow{DB} \perp \overline{AC}$

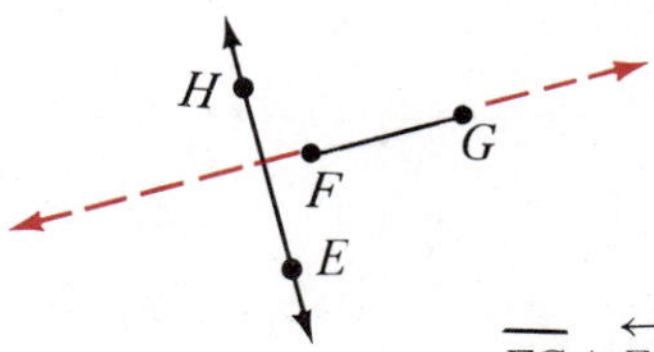

$\overline{FG} \perp \overleftrightarrow{EH}$

Example 1

Which of the following appear to be perpendicular?

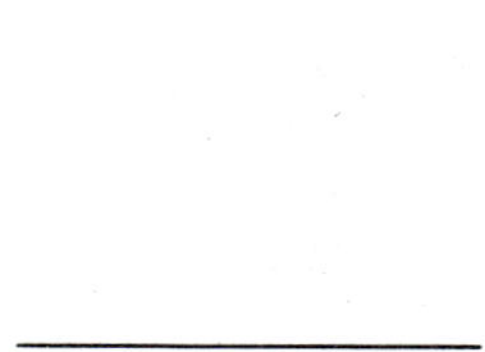

perpendicular

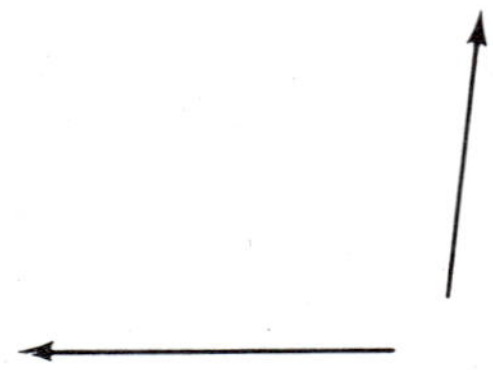

not perpendicular

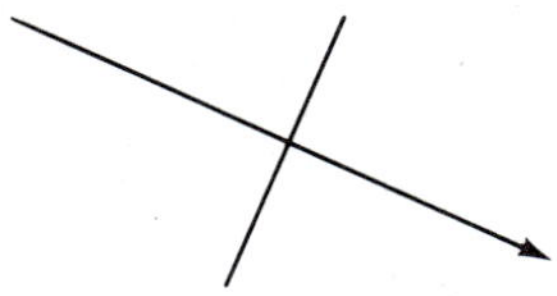

perpendicular

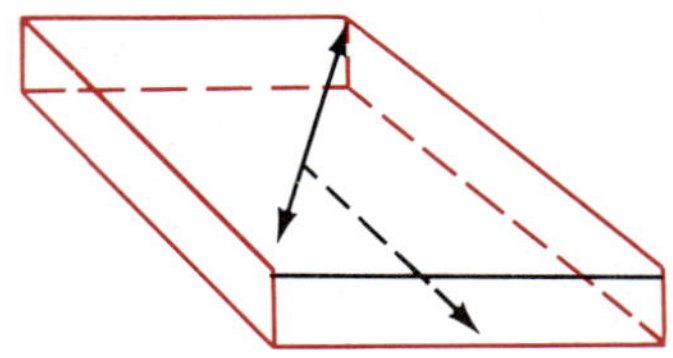

not perpendicular

A line can be constructed perpendicular to a given line.

Construction

A line perpendicular to a given line through a point not on the line

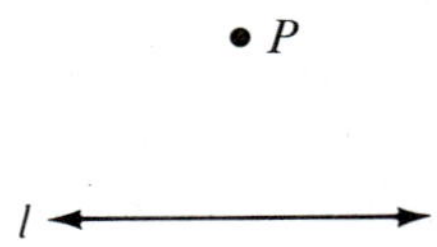

Given: Line l and P not on l
Construct: $\overleftrightarrow{PQ} \perp l$

Draw an arc with center P intersecting l at A and B.

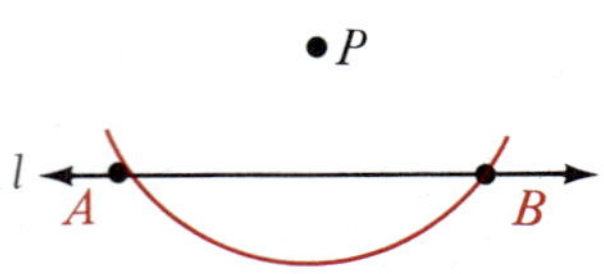

Using A and B as centers, and keeping the same opening, draw two arcs intersecting at Q.

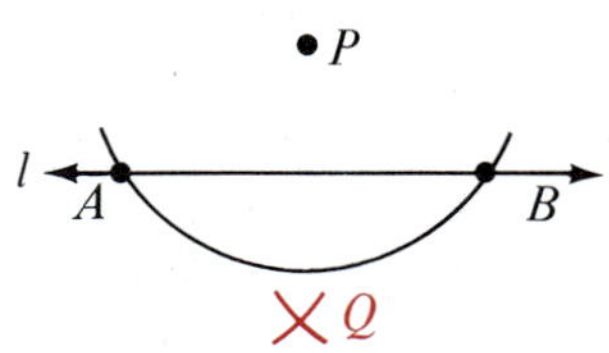

Draw $\overleftrightarrow{PQ}$.

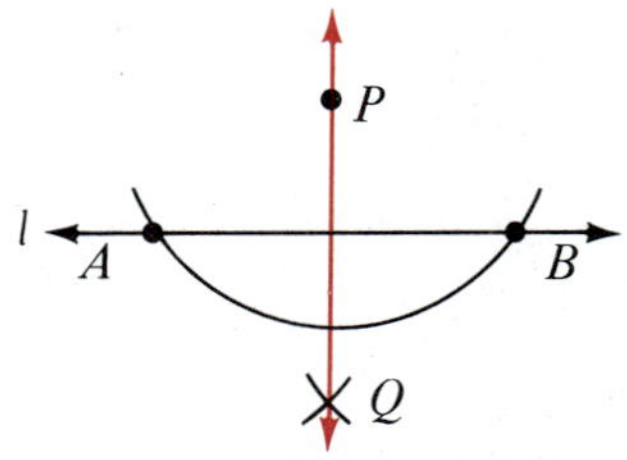

Conclusion: $\overleftrightarrow{PQ} \perp l$

The construction above involves a perpendicular from a point not on the line. You can also construct a perpendicular through a point on a given line.

Construction

A line perpendicular to a given line through a point on the line

Given: Line l and point P on l
Construct: $\overleftrightarrow{QP} \perp l$

Draw an arc with center P intersecting l at A and B.

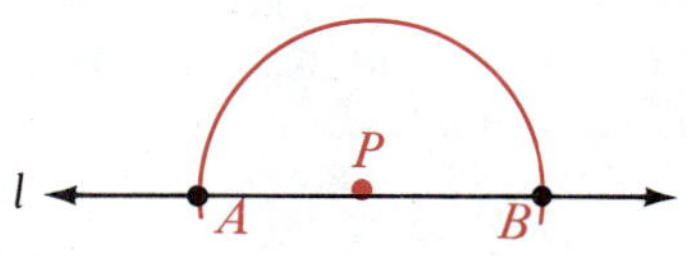

Using A and B as centers, and keeping the same opening, draw two arcs intersecting at Q.

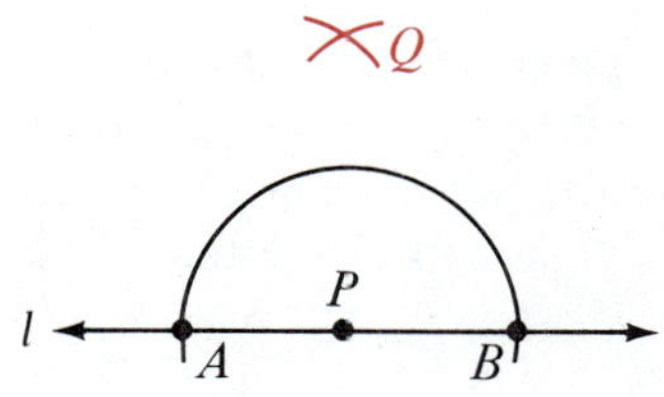

Draw $\overleftrightarrow{QP}$.

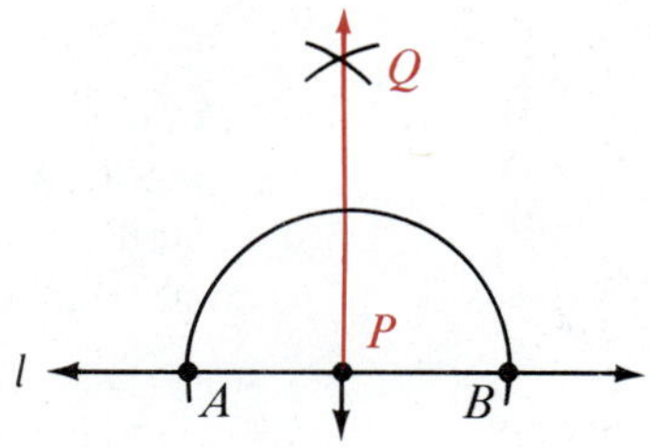

Conclusion: $\overleftrightarrow{QP} \perp l$

It seems reasonable to conclude that there is exactly one line through any given point which is perpendicular to a given line.

Postulate 12

For a given point and a line in a plane, there is exactly one line through the point which is perpendicular to the given line.

In the construction above and the one on page 50, $\overleftrightarrow{QP}$ is the **perpendicular bisector** of $\overline{AB}$. This means that it both bisects the segment $\overline{AB}$ and is perpendicular to it. You should modify these constructions to complete the following.

Construction

The perpendicular bisector of a segment
(See Exercise 5.)

Example 2

Construct a 45° angle.

Construct $\overleftrightarrow{AO} \perp \overleftrightarrow{OB}$
Bisect $\angle AOB$. ▶

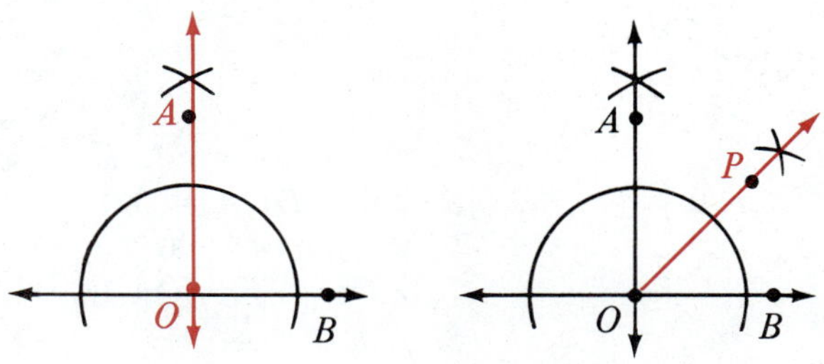

m $\angle POB = 45$

Example 3

In the given figure, which pairs of rays are perpendicular? Which pairs form a straight angle?

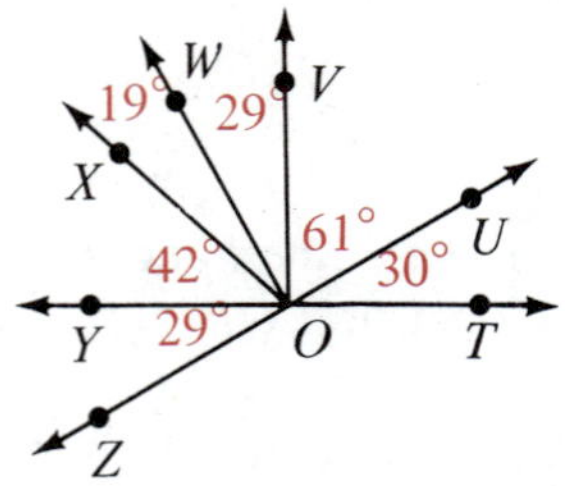

m $\angle ZOW = 90$ ▶ $\overrightarrow{OZ} \perp \overrightarrow{OW}$
m $\angle YOV = 90$ ▶ $\overrightarrow{OY} \perp \overrightarrow{OV}$
m $\angle WOU = 90$ ▶ $\overrightarrow{OW} \perp \overrightarrow{OU}$
m $\angle ZOU = 180$ ▶ $\overrightarrow{OZ}$ and $\overrightarrow{OU}$ form a straight angle.

You can fold a piece of paper to form perpendicular edges. Fold any piece of paper and crease it to form a straight edge. Fold this edge at some point back on itself and crease the paper again. You have formed two congruent angles which are a linear pair. You can prove that these must be right angles.

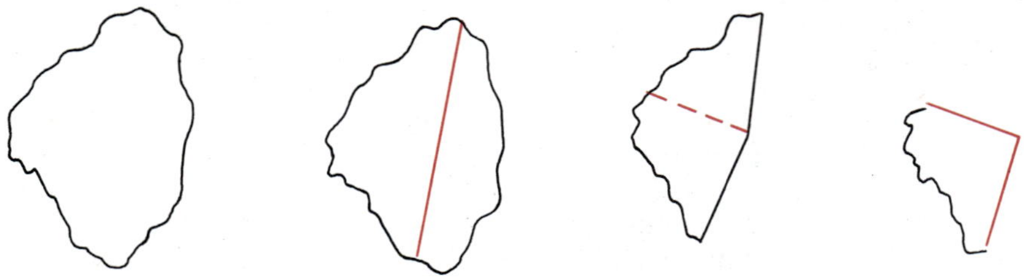

Theorem 2.7

If two lines form congruent adjacent angles, then the lines are perpendicular.

Given: $\angle DOA \cong \angle DOC$
Prove: $\overleftrightarrow{DB} \perp \overleftrightarrow{AC}$

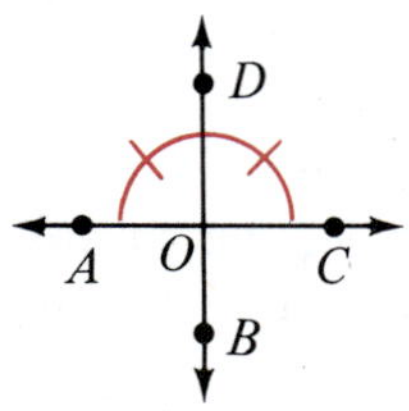

ANALYSIS

Lines are $\perp$ if a right angle is formed.

Proof

Statements	Reasons
1. $\angle DOA \cong \angle DOC$	1. Given
2. m $\angle DOA =$ m $\angle DOC$	2. Definition of congruent angles
3. m $\angle DOA +$ m $\angle DOC = 180$	3. Sum of measures of angles with the same vertex on one side of a line and with no interior points in common is 180°.
4. m $\angle DOA +$ m $\angle DOA = 180$	4. Substitution property
5. 2 m $\angle DOA = 180$	5. Addition fact; distributive
6. m $\angle DOA = 90$	6. Division property of equality
7. $\angle DOA$ is a right angle.	7. Definition of right angle
8. $\therefore \overleftrightarrow{DB} \perp \overleftrightarrow{AC}$	8. Definition of perpendicular

Oral Exercises

Which of the following appear to be perpendicular?

1. $\perp$
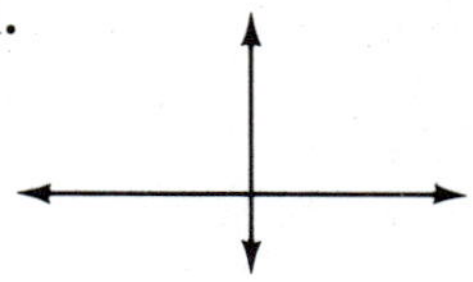

2. $\not\perp$

3. $\perp$
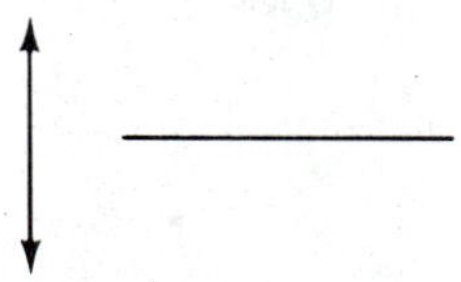

4. $\not\perp$
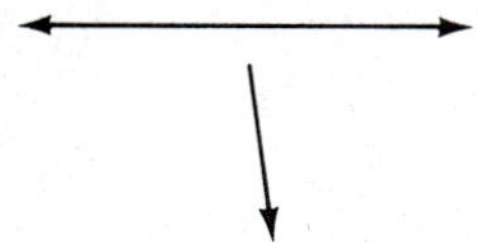

5. $\perp$
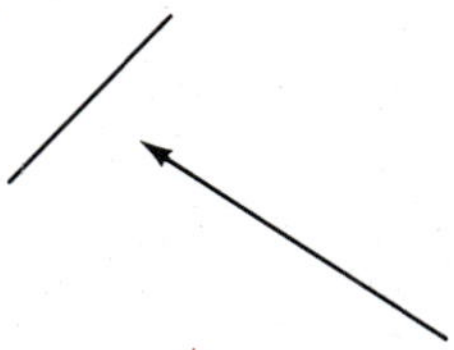

6. $\perp$
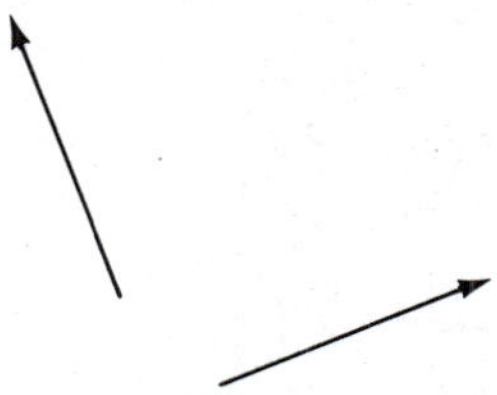

7. $\perp$
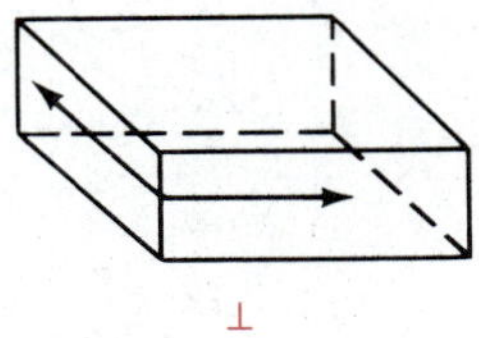

8. $\not\perp$
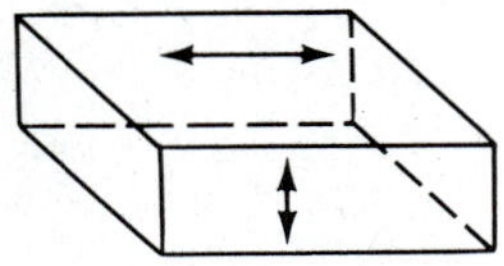

Written Exercises

Which pairs of rays are perpendicular? Which pairs form a straight angle? (Ex. 1–2)

A 1.
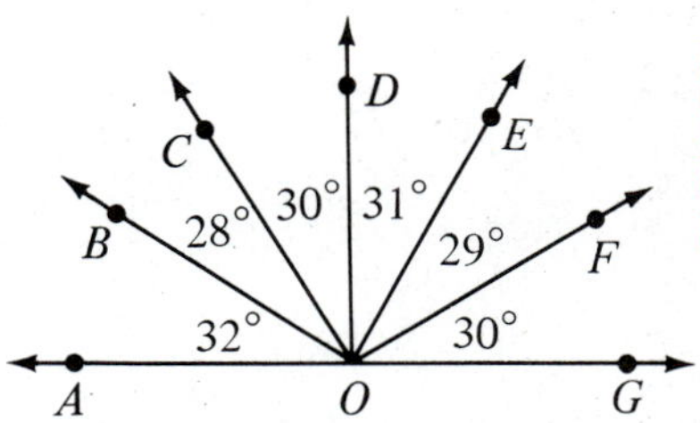

$\overrightarrow{OA} \perp \overrightarrow{OD}$
$\overrightarrow{OD} \perp \overrightarrow{OG}$
$\overrightarrow{OC} \perp \overrightarrow{OF}$
$\overrightarrow{OA}$ and $\overrightarrow{OG}$ form a st. $\angle$

2.
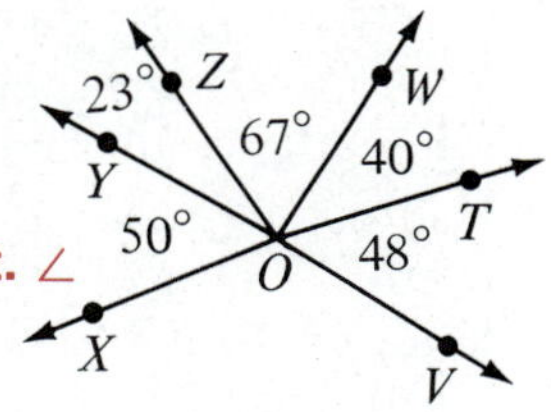

$\overrightarrow{OY} \perp \overrightarrow{OW}$
$\overrightarrow{OX}$ and $\overrightarrow{OT}$ form a st. $\angle$

3. Draw a line and a point not on the line. Construct a perpendicular to the line through the point. Follow constr., p. 50.

4. Draw a line. Mark a point on the line. Construct a perpendicular to the line through the point. Follow constr., top of p. 51.

5. Draw a segment. Construct its perpendicular bisector. Constr. midpt. of seg. then follow constr., top of p. 51.

6. Given: $\angle AOB$ and $\angle BOC$ are complementary. See TE Answer Section.
Prove: $\overrightarrow{OC} \perp \overleftrightarrow{AD}$

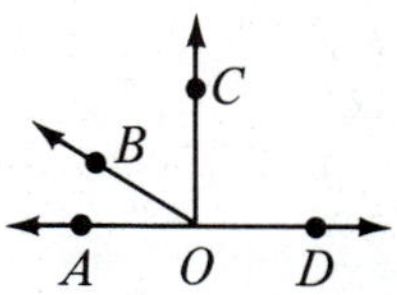

7. Given: $\angle XYT \cong \angle WZV$ and $\angle XYT$ and $\angle WZV$ are supplementary.
Prove: $\overrightarrow{ZV} \perp \overleftrightarrow{XW}$ See TE Answer Section.

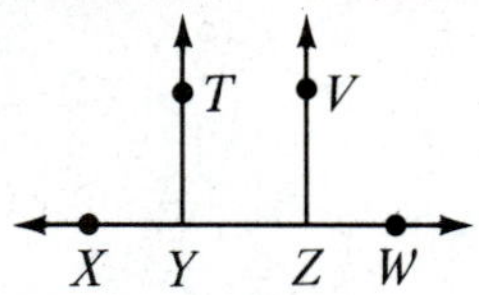

Construct an angle for each of the given measures.

B 8. 45°   9. 135°   10. $22\frac{1}{2}°$

8. Follow Ex. 2, p. 51.

10. Follow Ex. 2, p. 51. Then bis. $\angle POB$.

Draw a picture and write a proof for each of the following.

11. Given: $\overleftrightarrow{AB} \perp \overleftrightarrow{CD}$ intersecting at E
Prove: $\angle AEC \cong \angle AED$

12. Given: $\overleftrightarrow{AB} \perp \overleftrightarrow{CD}$ intersecting at E
Prove: m $\angle AED$ + m $\angle DEB = 180$

9. Follow Ex. 2., p. 51. Meas. of required $\angle$ is (90° + 45°).

APPLICATIONS

Read → Plan → Solve → Interpret

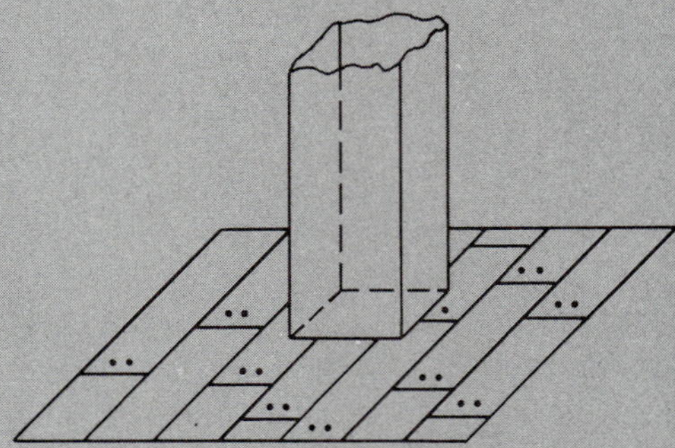

1. A carpenter must place a post so that it is perpendicular to the floor. If the only tool available is a framing square, how can the job be done accurately? How many measurements with the square are necessary if the floor is perfectly flat?

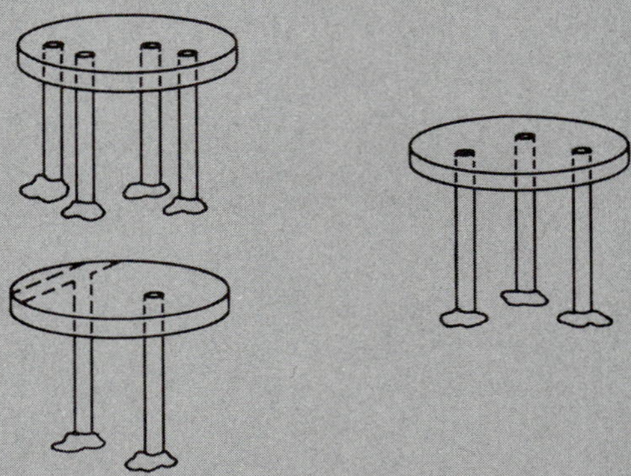

2. A base has to be made out of the pipes for an outdoor round tabletop. The top does not have to be perfectly level, but the table should not wobble. The legs can be set in concrete. The top will be a heavy piece of slate resting on the pipes. Which combination of legs shown will guarantee that the table will sit rigidly and not wobble? Why?

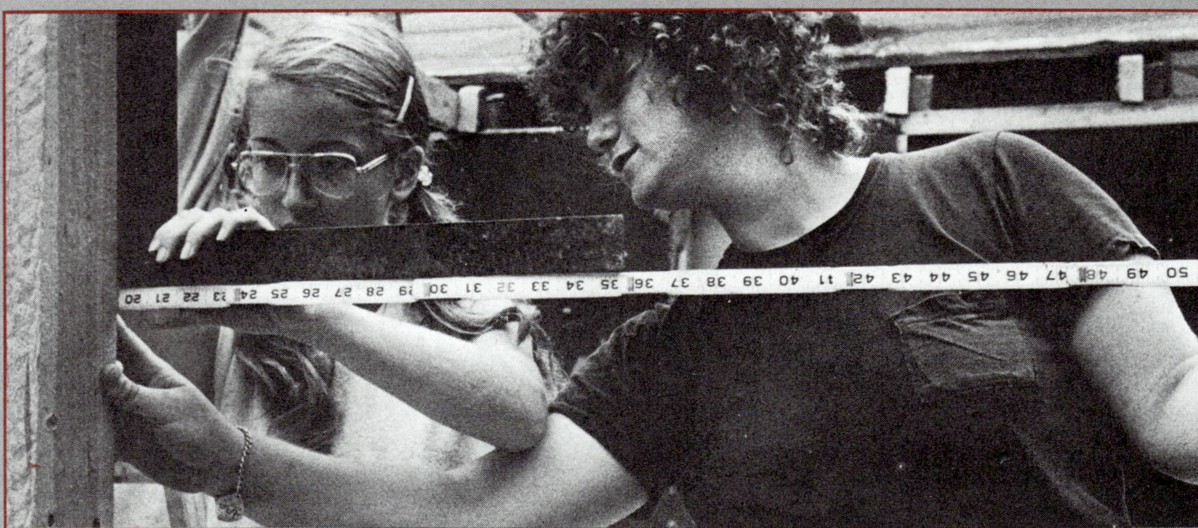

ARCHITECTURE

An architect "shapes" space by organizing materials into geometric forms that are logical in relation to one another as well as aesthetically pleasing.

An architect preparing detailed plans of a building.
These geometric diagrams serve as instructions to the contractors.

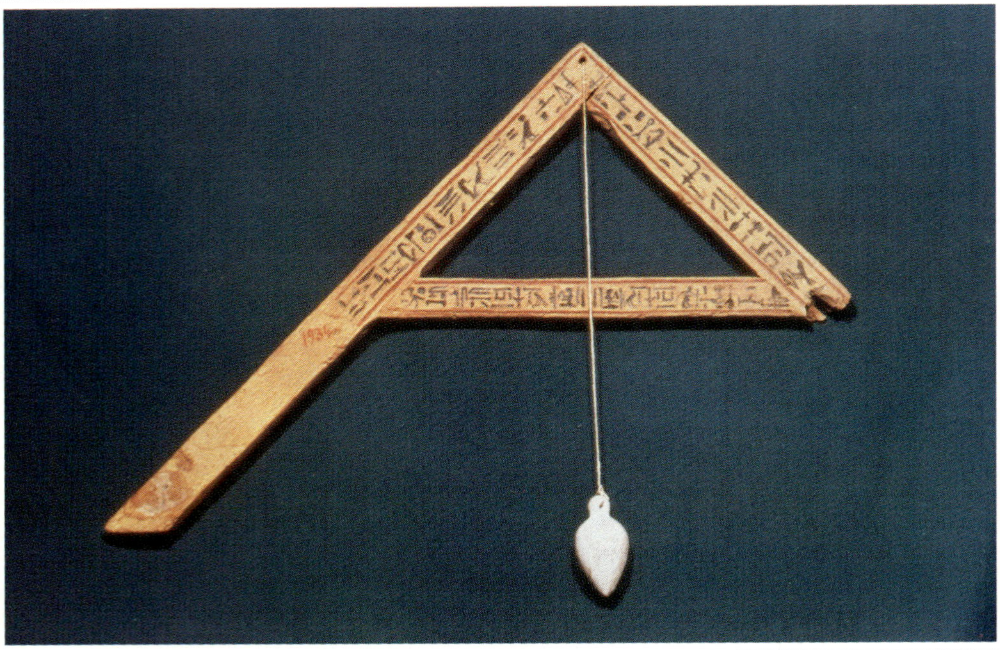

Plumb line found in the tomb of a 21st dynasty Egyptian. The configuration of the plumb line and triangle allowed the builder to read angles accurately.

A ruler with both English and metric calibrations and a slide rule converter are both used to make architectural drawings. Besides these tools, architects often use triangular scales for creating scale drawings.

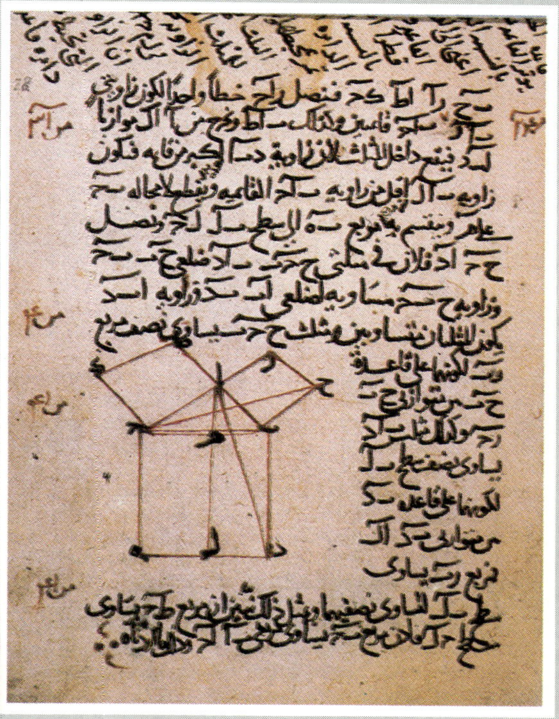

Arabic rendition of Euclid's proof of the Pythagorean theorem. The text of this proof dealing with right triangles has been preserved in Arabic and Greek manuscripts dating from the 9th century.

Leonardo da Vinci drawing of a church floor plan. The use of regular polygons was evident in da Vinci's architectural drawings, which also included scientific notes on stresses, arches, walls, and columns.

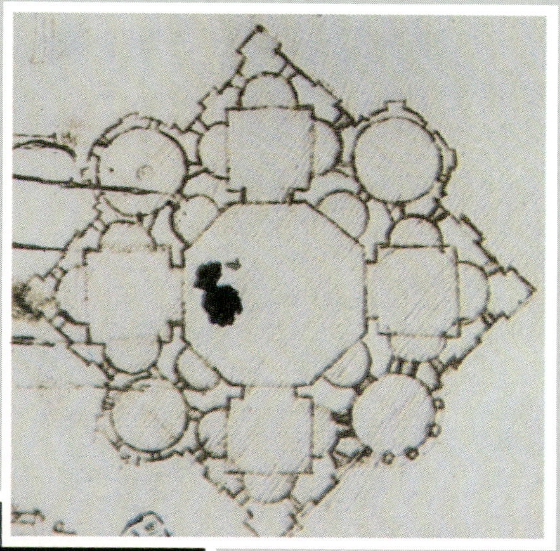

Computer graphic drawing of four buildings. Computer-aided design facilitates the architect's ability to experiment with various shapes within the drawings.

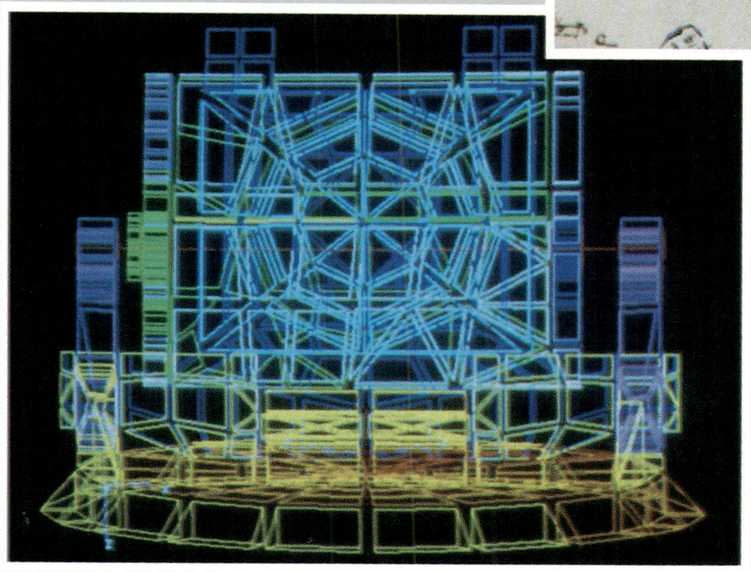

Carpenter constructing a house. The angles and dimensions for building a house are specified in the contract that the builder uses.

Eskimo igloo at the Arctic Circle. Eskimos stack blocks cut from snow in a continuous, circular row that winds upward in smaller and smaller circles to form a dome-shaped house.

15th century depiction of the building of a fortified town. Unlike that of a castle, the space within the walls of a fortified town was occupied by blocks of houses, which had to be planned so as not to impede the movement of troops.

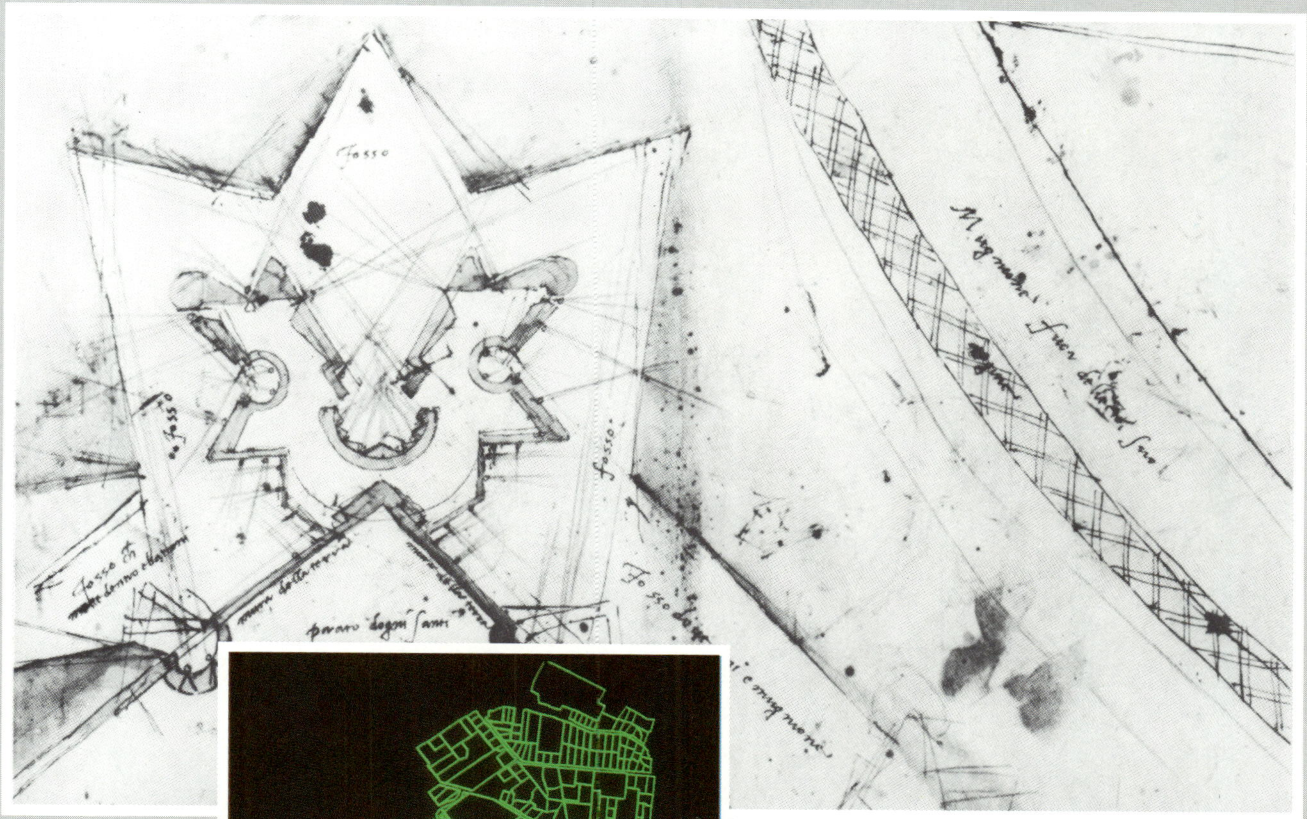

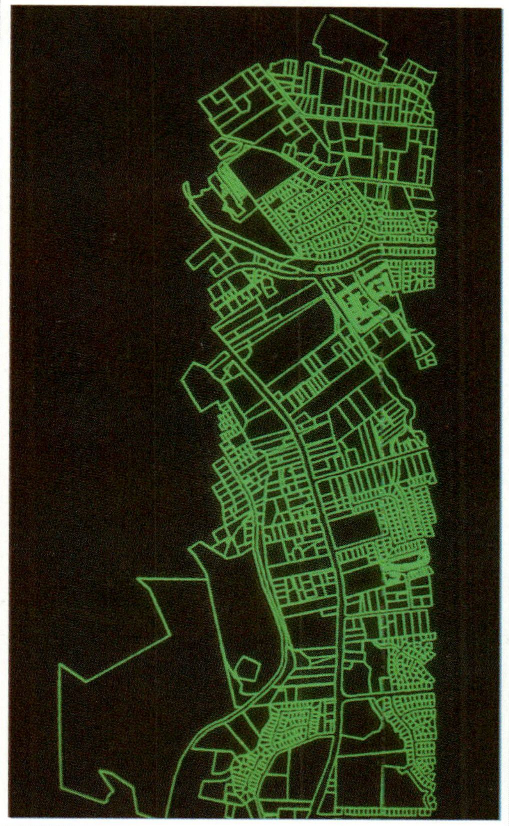

Michelangelo plan for the fortification of the walls of Florence. Michelangelo designed the shape of these walls to maximize the action of missiles and trajectories of cannons fired from them.

Town planning by computer. Computers allow urban planners to have accurate, detailed maps reflecting any measurable data. Densities of housing, crime, traffic, etc., can be displayed on a terminal instantly.

The Pueblo of Taos in northern New Mexico. Adobe structures such as this have been inhabited by the Pueblo Indians since the 1500s. Access to the cube-like rooms is through trapdoors in the roofs, reached by ladders.

Row houses in Baltimore. Attached rectangular structures are common urban solutions to the shortage of relatively low-cost land.

Modern building with triangular shape. Sometimes the shape of a building is dictated by the site upon which it is being built, other times by the vision of the architect.

The facades of old and new buildings in the city of Chicago. In 1884, William Le Baron Jenney built the first skyscraper, a 10-story rectangular structure, the Home Insurance Building, in Chicago.

Pyramids at Giza. The Giza pyramids were erected by the kings of the 4th Egyptian dynasty—Khufu, Khafre, and Menkaure—who brought pyramid building to its apogee.

Hyperbolic curve of a roller coaster at Knott's Berry Farm in California. Most students of roller coasters believe that the origin of the rides can be traced to ice slides in 15th century Russia.

"Spiral of Progress" monument in Yaounde, Cameroon. The spiral has long been a favorite architectural element and was used by Frank Lloyd Wright in the Guggenheim Museum in New York.

Vista Del Minarete in Samarra, Iraq. The ziggurat is a temple of Sumerian origin and is characterized by the broad ascent winding round the structure.

PROVING THEOREMS ABOUT ANGLES 2.9

Objective **To prove and use theorems about angles**

Suppose that $\angle X$ and $\angle Y$ are supplements of two angles whose measures are 73° each. $\angle X$ and $\angle Y$ must each have a measure of 107°. $\angle X$ and $\angle Y$ are congruent. This result is generalized in the following theorem.

Theorem 2.8 Supplements of congruent angles are congruent.

Given: $\angle A$ is a supplement of $\angle B$,
$\angle D$ is a supplement of $\angle C$,
$\angle B \cong \angle C$
Prove: $\angle A \cong \angle D$

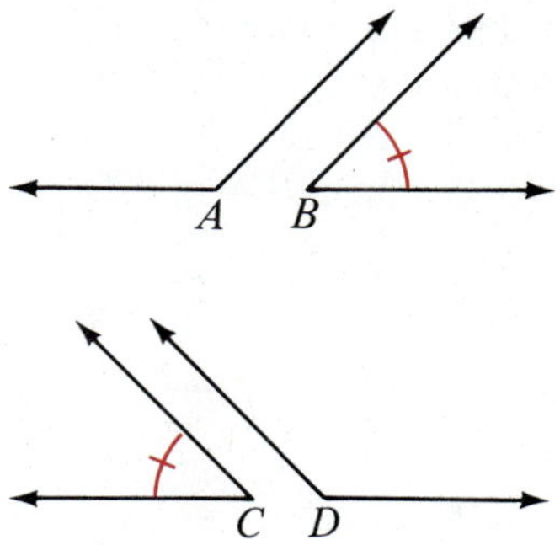

Proof

Statements	Reasons
1. $\angle A$ is a supplement of $\angle B$.	1. Given
2. $\angle D$ is a supplement of $\angle C$.	2. Given
3. $m\angle A + m\angle B = 180$, $m\angle D + m\angle C = 180$	3. Definition of supplementary angles
4. $m\angle A + m\angle B = m\angle D + m\angle C$	4. Substitution property
5. $\angle B \cong \angle C$	5. Given
6. $m\angle B = m\angle C$	6. Definition of congruent angles
7. $m\angle A = m\angle D$	7. Substitution and subtraction properties
8. $\therefore \angle A \cong \angle D$	8. Definition of congruent angles

A **corollary** of a theorem is a theorem whose proof requires only a few simple statements in addition to the proof of the original theorem. The statement below is a corollary of Theorem 2.8.

Corollary Supplements of the same angle are congruent.

Two angles are complements of each other if their measures total 90°. An argument similar to that of Theorem 2.8 can be given to prove that complements of congruent angles are also congruent.

Theorem 2.9 Complements of congruent angles are congruent.

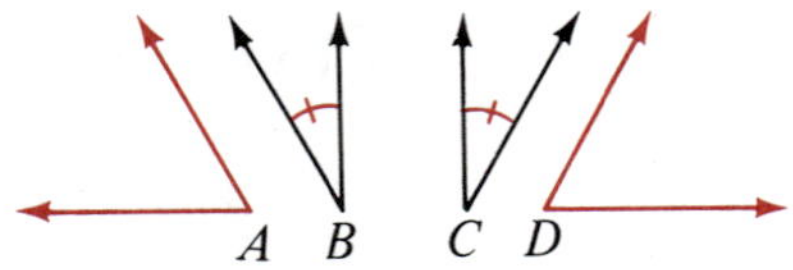

Given: $\angle A$ is a complement of $\angle B$,
$\angle D$ is a complement of $\angle C$, and
$\angle B \cong \angle C$
Prove: $\angle A \cong \angle D$

Proof

Statements	Reasons
1. $\angle A$ is a complement of $\angle B$.	1. Given
2. $\angle D$ is a complement of $\angle C$.	2. Given
3. $m\angle A + m\angle B = 90$, $m\angle D + m\angle C = 90$	3. Definition of complementary angles
4. $m\angle A + m\angle B = m\angle D + m\angle C$	4. Substitution property
5. $\angle B \cong \angle C$	5. Given
6. $m\angle B = m\angle C$	6. Definition of congruent angles
7. $m\angle A = m\angle D$	7. Substitution and subtraction properties
8. $\therefore \angle A \cong \angle D$	8. Definition of congruent angles

Corollary Complements of the same angle are congruent.

Example 1

Given: $m\angle BOC = 40$, $m\angle COD = 40$,
$\overrightarrow{OC} \perp \overleftrightarrow{AE}$
What conclusion can you draw about $\angle AOB$ and $\angle DOE$?

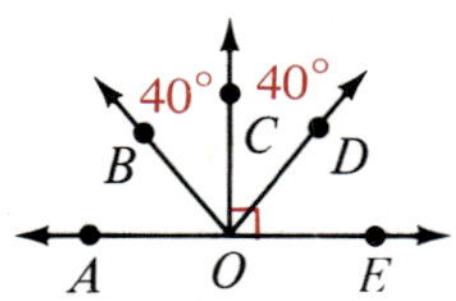

$\angle AOB \cong \angle DOE$ ◄ *Complements of congruent angles are congruent.*

Example 2

Give a reason for each step in the following proof.
Given: $m\angle HOI = 40$
Prove: $\angle GOH \cong \angle FOI$

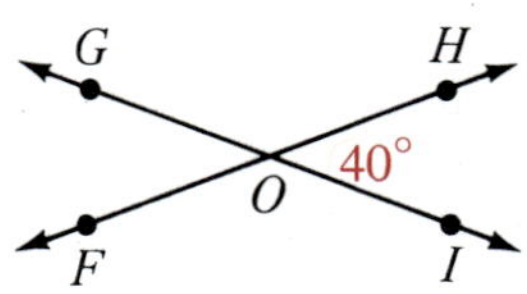

Proof

Statements	Reasons
1. $m\angle HOI = 40$	1. Given
2. $m\angle HOI + m\angle GOH = 180$ $m\angle HOI + m\angle FOI = 180$	2. The sum of the measures of angles with the same vertex on one side of a line is 180°.
3. $m\angle GOH = 140$ $m\angle FOI = 140$	3. Substitution and subtraction properties of equality
4. $\therefore \angle GOH \cong \angle FOI$	4. Definition of congruent angles

Example 2 suggests the following theorem about vertical angles.

Theorem 2.10 Vertical angles are congruent.

Given: $\angle 1$ and $\angle 3$ are vertical angles.
Prove: $\angle 1 \cong \angle 3$

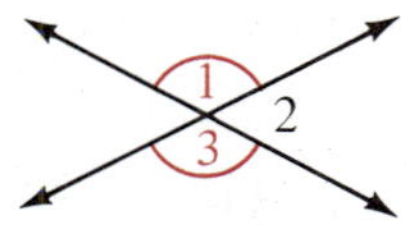

ANALYSIS
$\angle 1$ and $\angle 3$ are both supplements of $\angle 2$.

Proof

Statements	Reasons
1. $\angle 1$ and $\angle 3$ are vertical angles.	1. Given
2. $m\angle 1 + m\angle 2 = 180$ $m\angle 3 + m\angle 2 = 180$	2. Sum of measures of angles with the same vertex on one side of a line is 180°.
3. $\angle 1$ is a supplement of $\angle 2$ $\angle 3$ is a supplement of $\angle 2$	3. Definition of supplementary $\angle$'s
4. $\therefore \angle 1 \cong \angle 3$	4. Supplements of the same angle are congruent.

Example 3 Given: $m\angle RHM = 3x$,
$m\angle AHQ = 4x - 23$
Find $m\angle RHM$.

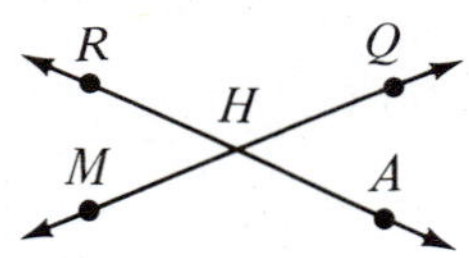

$m\angle AHQ = m\angle RHM$ ◄ *Vertical angles are congruent.*
$4x - 23 = 3x$
$x = 23$
$3x = 69$ **Thus,** $m\angle RHM = 69$.

Oral Exercises

Complete each statement. Give a reason for your answer.

1. $\angle DOA \cong$?
$\angle$COB; vert. $\angle$'s are ≅.

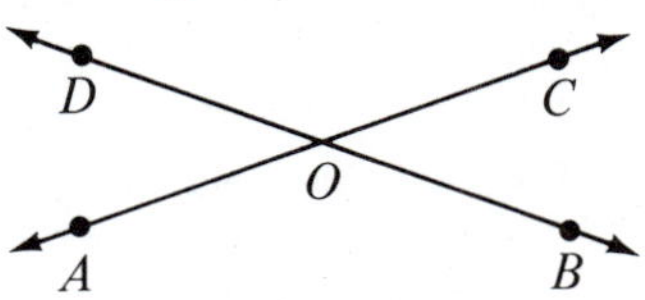

2. $\angle HOI \cong$?
$\angle$GOF; comp. of ≅ $\angle$'s are ≅.

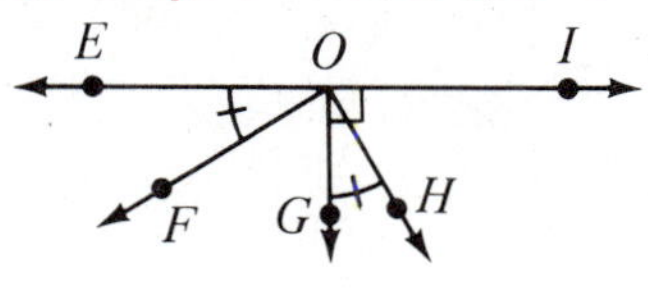

3. $\angle JKN \cong$?
$\angle$JKL; supp. of ≅ $\angle$'s are ≅.

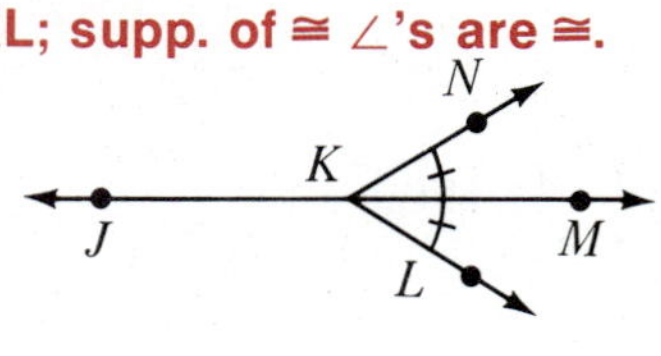

4. $m\angle POQ =$?
25; def. comp. $\angle$'s.

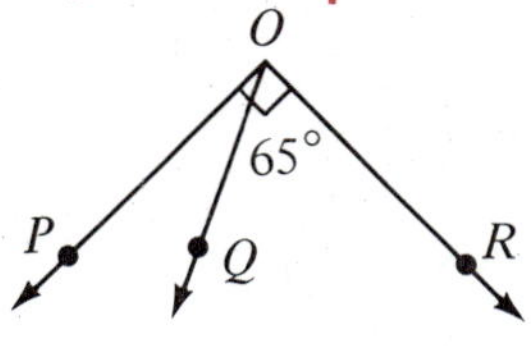

5. $m\angle STV =$? **116**
def. supp. $\angle$'s.

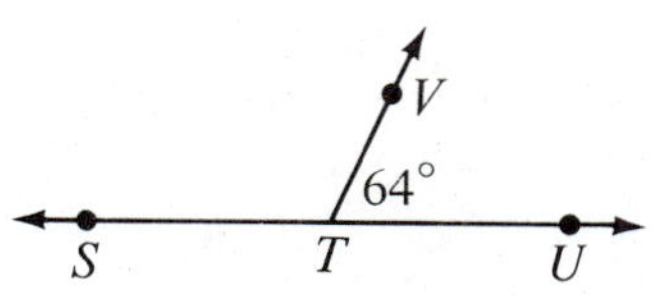

6. $m\angle ZOW =$?
44; vert. $\angle$'s are ≅.

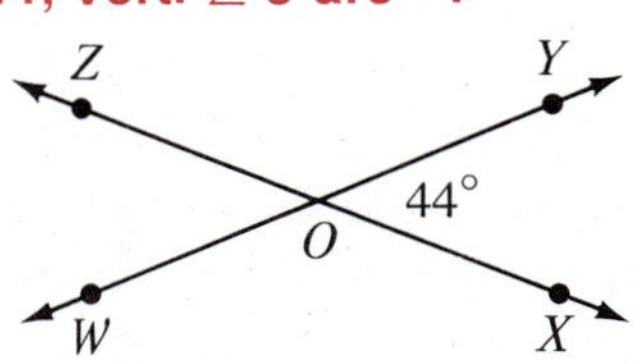

Written Exercises

A **1.** Given: m $\angle SOP = 5x$ and m $\angle ROQ = 4x + 10$
Find m $\angle SOP$. **50**

2. Given: m $\angle SOP = 7x + 9$, m $\angle ROQ = 8x$
Find m $\angle ROQ$. **72**

3. Given: m $\angle SOR = 7y -$ 12 and m $\angle POQ =$ $5y + 12$
Find m $\angle SOR$. **72**

4. Given: m $\angle SOR = 8y + 8$ and m $\angle POQ = 10y - 12$
Find m $\angle POQ$. **88**

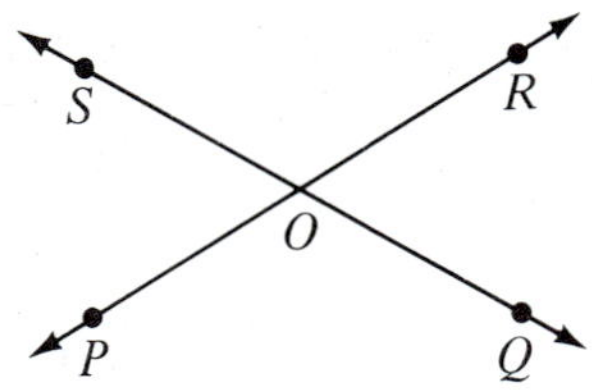

What conclusions can you draw about the angles in each figure? (Ex. 5–7)

5. Given: $\angle 1 \cong \angle 4$ **$\angle 2 \cong \angle 3$**

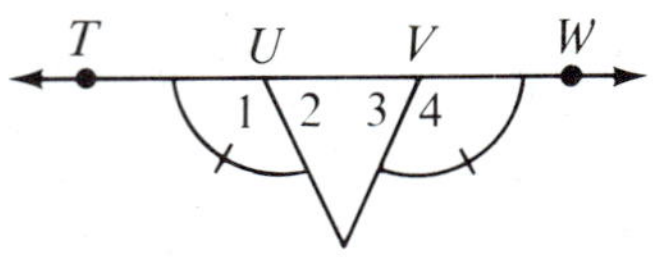

6. Given: $\angle 1 \cong \angle 3$
$\angle JKN \cong \angle MKL$

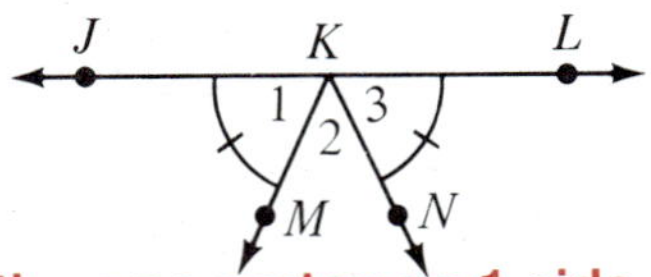

7. Given: $\angle 1 \cong \angle 4$ **$\angle 2 \cong \angle 3$**

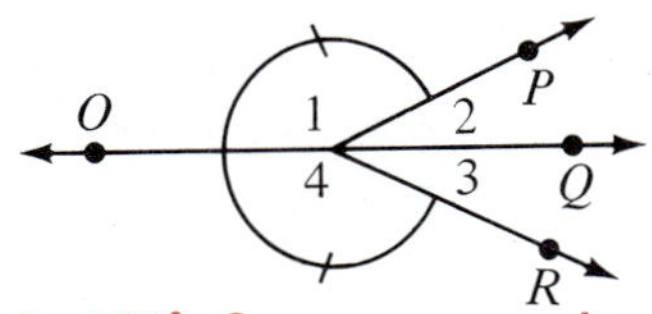

8. Given; Sum of means of $\angle$'s with same vertex on 1 side of line is 180°; Same as previous reason; Def of supp $\angle$'s; Same as previous reason; Supp of $\cong$ $\angle$'s are $\cong$.

Give a reason for each step in each proof. (Ex. 8–9)

8. Given: $\angle TOP \cong \angle TOQ$
Prove: $\angle POS \cong \angle QOS$

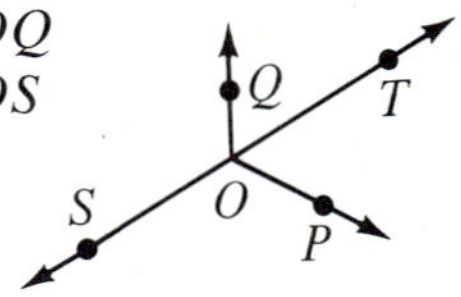

9. Given: $\angle ABD \cong \angle CBF$, m $\angle ABD$ + m $\angle DBE = 90$, and m $\angle CBF$ + m $\angle FBE = 90$
Prove: $\angle DBE \cong \angle FBE$

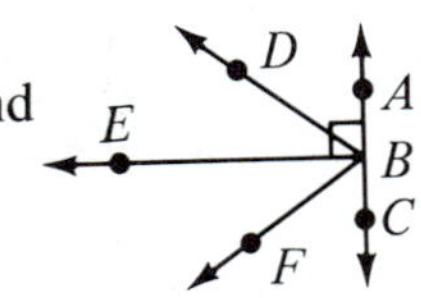

9. Given; Given; Given; Def of comp $\angle$'s; Same as 4; Comp of $\cong$ $\angle$'s are $\cong$.

Proof

Statements	Reasons
1. $\angle TOP \cong \angle TOQ$	1. ?
2. m $\angle QOS$ + m $\angle TOQ = 180$	2. ?
3. m $\angle POS$ + m $\angle TOP = 180$	3. ?
4. $\angle QOS$ is a supplement of $\angle TOQ$.	4. ?
5. $\angle POS$ is a supplement of $\angle TOP$.	5. ?
6. $\therefore$ $\angle POS \cong \angle QOS$	6. ?

Proof

Statements	Reasons
1. $\angle ABD \cong \angle CBF$	1. ?
2. m $\angle ABD$ + m $\angle DBE = 90$	2. ?
3. m $\angle CBF$ + m $\angle FBE = 90$	3. ?
4. $\angle ABD$ is a complement of $\angle DBE$.	4. ?
5. $\angle CBF$ is a complement of $\angle FBE$.	5. ?
6. $\therefore$ $\angle DBE \cong \angle FBE$	6. ?

B **10.** Given: $\angle 3 \cong \angle 5$
Prove: $\angle 1 \cong \angle 5$

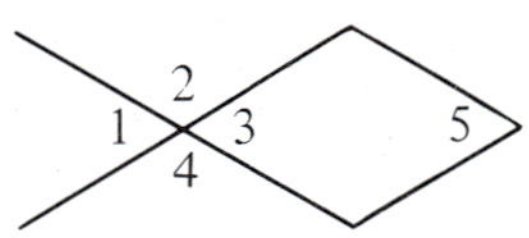

11. Given: $\angle 1 \cong \angle 5$
Prove: $\angle 2$ is a supplement of $\angle 5$.

CUMULATIVE REVIEW

1. Give an example of the division property of equality. **Answers will vary.**
2. An angle has a measure of 89°. Is the angle an acute angle, a right angle, an obtuse angle, or a straight angle? **acute**
3. If the noncommon sides of two adjacent angles are perpendicular, what is the sum of the measures of the two angles? **90°**

SETTING UP PROOFS 2.10

Objectives

To set up a proof from a verbal statement
To prove theorems

Theorems are statements that can be proved. Before beginning to write a proof for a given theorem, you should draw a diagram, label it, and then restate the theorem in terms of the diagram. The restatement should indicate what is *given* and what is to be *proved.* It should use the specific segments, rays, angles, and other geometric parts from your diagram.

The statement of a theorem is a conditional statement. If it is written in "If . . ., then . . ." form, it is obvious which part is the hypothesis and which part is the conclusion. The "If" part, or hypothesis, of the sentence is the *Given;* the "then" part, or conclusion, is the *Prove.*

Example 1

Draw a labeled diagram. Restate the theorem in terms of the diagram. State what is given and what is to be proved.

If two congruent angles are supplementary, then they are right angles.

Given: $\angle ABC \cong \angle FED$, $\angle ABC$ and $\angle FED$ are supplementary.
Prove: $\angle ABC$ and $\angle FED$ are right angles.

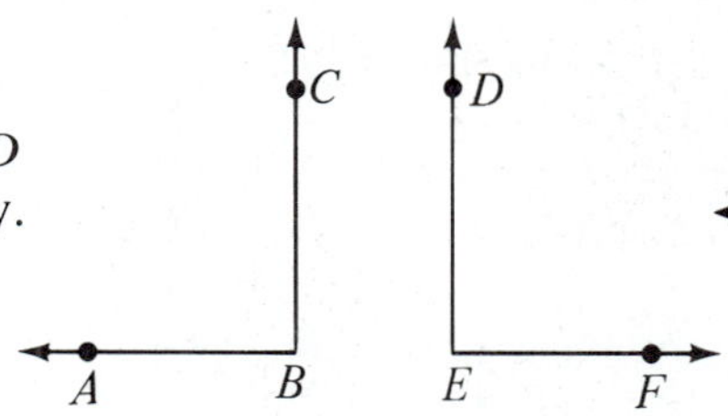

The diagram includes two angles which are congruent and supplementary, and are right angles.

Sometimes the statement of a theorem is written in a general way as a simple declarative sentence. In that case, the subject of the sentence becomes the hypothesis or *Given,* and the remainder of the sentence becomes the conclusion or *Prove.* In some cases, you may find it helpful to rewrite the sentence in "If . . ., then . . ." form.

Example 2

Draw a labeled diagram. Restate the theorem in terms of the diagram. State what is given and what is to be proved.

The bisectors of two adjacent supplementary angles are perpendicular.

Given: $\angle AOC$ and $\angle COB$ are supplementary, $\overrightarrow{OD}$ bisects $\angle AOC$, $\overrightarrow{OE}$ bisects $\angle COB$.
Prove: $\overrightarrow{OD} \perp \overrightarrow{OE}$

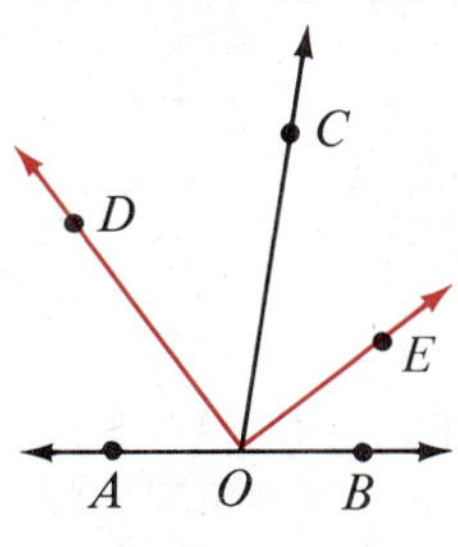

The diagram includes adjacent angles which are supplementary and bisectors which are $\perp$.

You may find it helpful to write an *analysis* of the proof before you try to prove a theorem. The analysis is a statement of key steps necessary to prove your final outcome. You may wish to list several possible ways of showing this final result. Essentially, in an *analysis* you work backwards from what is to be proved to try to see what will work.

Example 3 Write an analysis for the given theorem.

If two lines are perpendicular, then they form congruent angles.

Analysis: Show that two angles have equal measures.
Measures are equal if both are right angles.

An analysis is provided in many of the proofs in this book.

Written Exercises

Drawings and answers will vary.

Draw a labeled diagram for each of the following. Restate the theorem in terms of the diagram. State what is given and what is to be proved.

A 1. If two angles are each congruent to a third angle, then they are congruent to each other.

2. If two segments are each congruent to a third segment, then they are congruent to each other.

3. Supplements of vertical angles are congruent.

4. Complements of vertical angles are congruent.

5. If two angles are adjacent to and supplements of the same angle, then the two angles are vertical angles.

6. If two angles are vertical angles, then they are adjacent to and supplements of the same angle.

Draw a labeled diagram for each of the following. Restate the theorem in terms of the diagram. State what is given and what is to be proved.
Write an analysis and prove each theorem.

B 7. If the complements of two given angles are congruent, then the supplements of the given angles are also congruent.

8. If the supplements of two given angles are congruent, then the complements of the given angles are also congruent.

9. If two angles are complementary and congruent, then each angle has a measure of 45°.

10. If two lines are perpendicular, then they form congruent adjacent angles.

CUMULATIVE REVIEW

1. Simplify $|3 - 7|$. **4**
2. Is it possible for two planes to intersect in exactly one line? **Yes**
3. True or false: An angle is a coplanar figure. **True**

DIHEDRAL ANGLES (optional) 2.11

Objectives **To identify and measure dihedral angles**
To recognize perpendicular lines and planes in space

An angle was defined as the union of two rays with a common endpoint. There are angles in space which share some of the properties of angles in a single plane. These are called **dihedral angles**. Two lines intersect to form four angles; two planes intersect to form four dihedral angles.

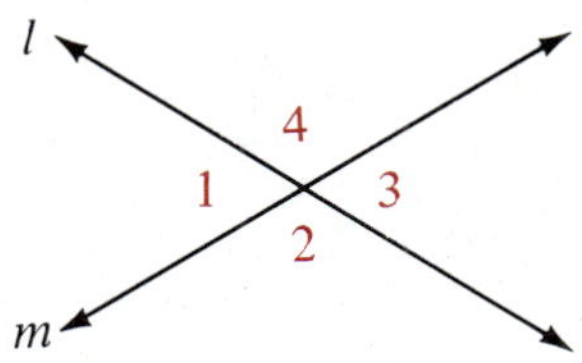

Two lines—four angles

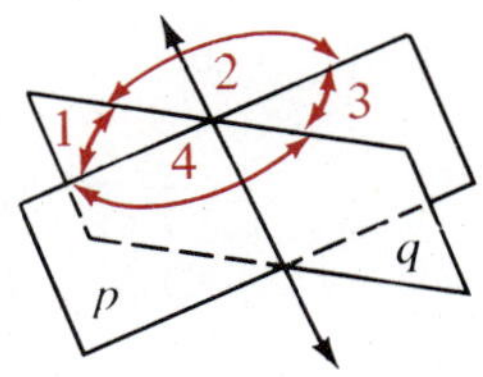

Two planes—four dihedral angles

Each dihedral angle has two faces and an edge. Each face is a half-plane. Each edge is a line.

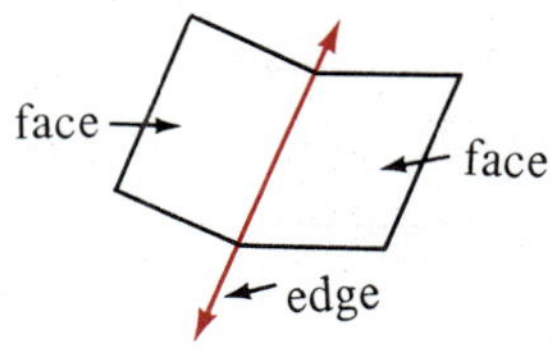

Dihedral angles are named by a point in each face and by the edge.

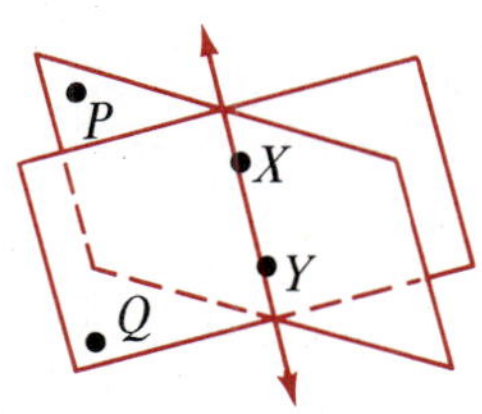

$\angle P - \overleftrightarrow{XY} - Q$

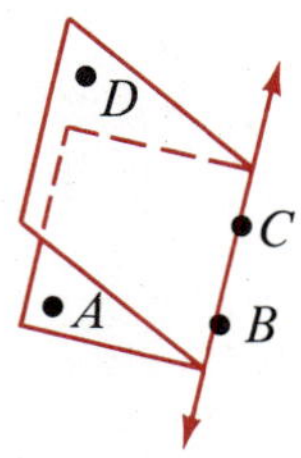

$\angle A - \overleftrightarrow{BC} - D$

It is possible to have *adjacent dihedral angles*. These share a common face and a common edge.

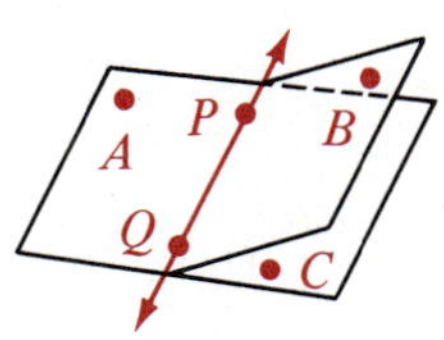

$\angle A - \overleftrightarrow{PQ} - B$ and $\angle B - \overleftrightarrow{PQ} - C$ are adjacent dihedral angles.

Many common objects are models of dihedral angles. Two pages of a book and the inside part of a cake with a slice cut out are two examples.

A dihedral angle is measured by its **plane angles**. A *plane angle* is the union of the two rays, one in each face, which are perpendicular to the edge at the same point.

$\overrightarrow{VU} \perp \overleftrightarrow{VY}$ at V.
$\overrightarrow{YX} \perp \overleftrightarrow{VY}$ at Y.

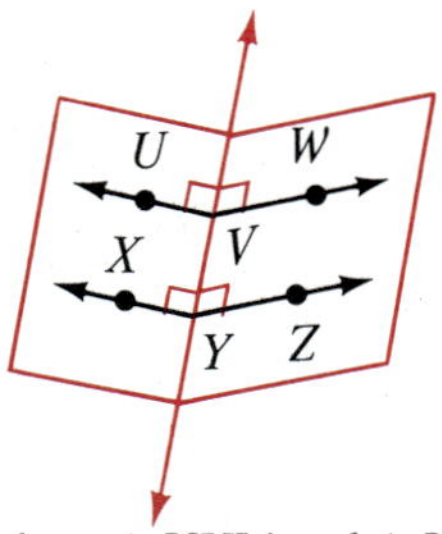

$\overrightarrow{VW} \perp \overleftrightarrow{VY}$ at V.
$\overrightarrow{YZ} \perp \overleftrightarrow{VY}$ at Y.

Plane angles: $\angle UVW$ and $\angle XYZ$

The following theorem about dihedral angles can be proved. The proof involves some ideas you have not yet studied, so it will not be given here.

Theorem 2.11 Any two plane angles of a dihedral angle are congruent.

The measure of a dihedral angle is the measure of any one of its plane angles.

Example 1 Given: The measure of plane angle ABC is 130.
Find the measure of plane angle DEF.

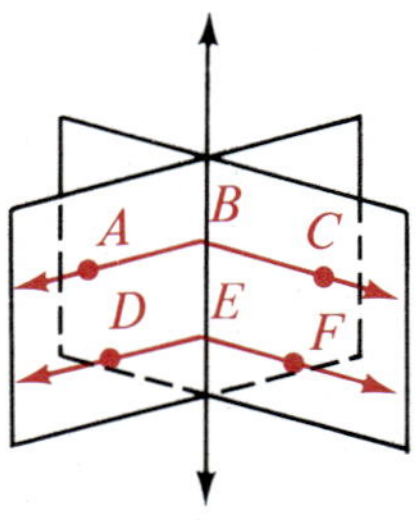

m $\angle DEF = 130$ ◄ *Any two plane angles of a dihedral angle are ≅.*

Example 2 If $\angle ABC$ is a plane angle of $\angle C - \overleftrightarrow{BE} - D$ and m $\angle ABC = 85$, find the measure of $\angle C - \overleftrightarrow{BE} - D$.

m $\angle C - \overleftrightarrow{BE} - D = 85$ ◄ *The measure of a dihedral angle is the measure of one of its plane angles.*

Definition: Perpendicular planes	Two *planes* are *perpendicular* if they intersect to form a right dihedral angle.

If plane p is perpendicular to plane q, then any one of the dihedral angles formed is a right dihedral angle. This also means that any of the plane angles formed are right angles.

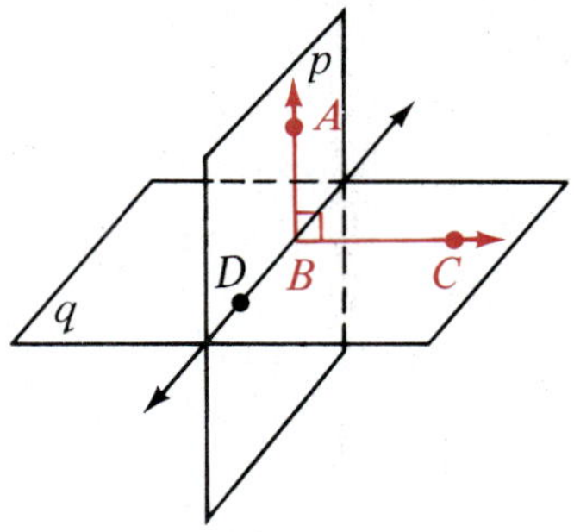

If $p \perp q$, then m $\angle A - \overleftrightarrow{BD} - C = 90$ and m $\angle ABC = 90$.

Definition: Line perpendicular to a plane	A line is *perpendicular to a plane* if it is perpendicular to all lines in the plane which intersect it.

Later, you will be able to prove that if l is perpendicular to any two lines in plane p, it is perpendicular to all the lines in p that intersect it.

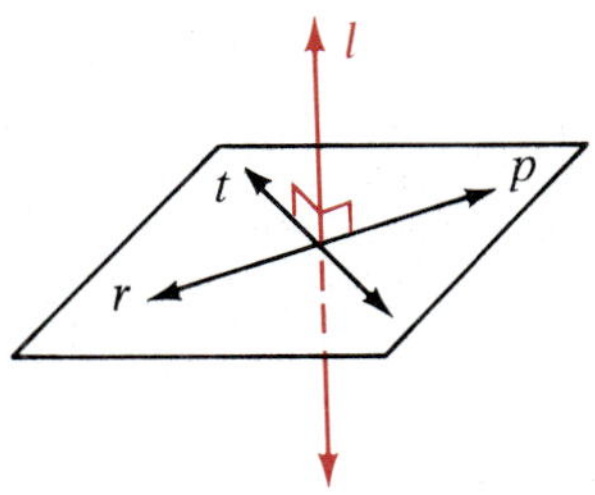

Written Exercises

If $\angle UYV$ is a plane angle of $\angle C - \overleftrightarrow{XZ} - B$, find each of the following:

A

1. Given: m $\angle UYV = 82$
   Find m $\angle C - \overleftrightarrow{XZ} - B$. **82**
2. Given: m $\angle TYU = 127$
   Find m $\angle A - \overleftrightarrow{XZ} - C$. **127**
3. Given: m $\angle C - \overleftrightarrow{XZ} - B = 81$
   Find m $\angle UYV$. **81**
4. Given: m $\angle A - \overleftrightarrow{XZ} - C = 149$
   Find m $\angle TYU$. **149**
5. Given: m $\angle UYV = 69$
   Find m $\angle C - \overleftrightarrow{XZ} - A$. **111**
6. Given: m $\angle TYU = 132$
   Find m $\angle C - \overleftrightarrow{XZ} - B$. **48**
7. Given: m $\angle A - \overleftrightarrow{XZ} - C = 127$
   Find m $\angle UYV$. **53**
8. Given: m $\angle C - \overleftrightarrow{XZ} - B = 59$
   Find m $\angle TYU$. **121**

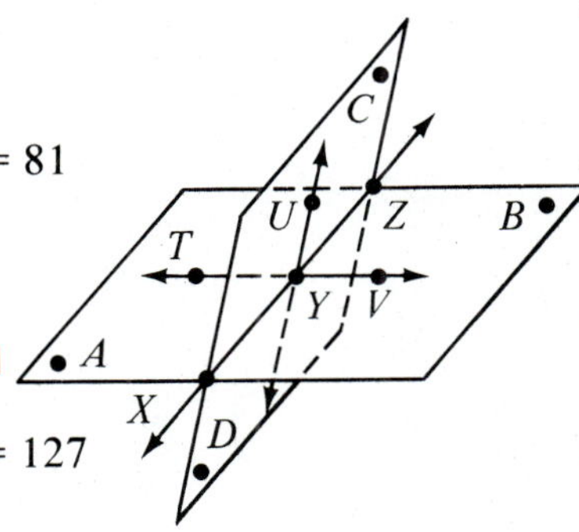

True or false?

Ⓑ **9.** Any angle whose sides are contained in the faces of a dihedral angle is a plane angle of the dihedral angle. **F**

10. Any angle whose vertex is on the edge of a dihedral angle is a plane angle of the dihedral angle. **F**

11. The measures of all plane angles of a dihedral angle are equal. **T**

12. Adjacent dihedral angles can have adjacent plane angles. **T**

13. All angles with sides in the faces of a dihedral angle and vertices on its edge are congruent. **F**

14. If a plane is perpendicular to a second plane, then it is perpendicular to every line in the second plane. **F**

If possible give the measure of each angle. If not enough information is given, state what additional information is needed. (Ex. 15–22)

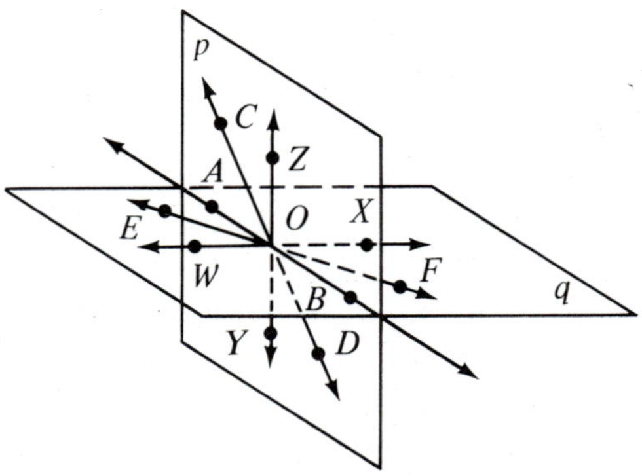

p: contains A, B, C, D, O, Y, Z
q: contains A, B, E, F, O, X, W

15. Given: m $\angle WOZ = 90$
Find m $\angle W - \overleftrightarrow{AB} - Z$.
Too little info.

16. Given: plane $p \perp$ plane q
Find m $\angle ZOF$.
Too little info.

17. Given: plane $\angle ZOX$, and m $\angle ZOX = 90$
Find m $\angle Z - \overleftrightarrow{AB} - X$.
90

18. Given: plane $\angle WOZ$, and m $\angle EOZ = 90$
Find m $\angle Z - \overleftrightarrow{AB} - W$. **90**

19. Given: m $\angle EOW = 30$
Find m $\angle WOF$. **150**

20. Given: plane $p \perp$ plane q, and plane $\angle ZOW$
Find m $\angle ZOW$. **90**

21. Given: plane $\angle WOZ$, and $\overleftrightarrow{EF} \perp \overleftrightarrow{ZY}$
Find m $\angle W - \overleftrightarrow{AB} - Z$. **90**

22. Given: plane $\angle WOZ$, and m $\angle W - \overleftrightarrow{AB} - Z = 88$
Find m $\angle ZOX$. **92**

Ⓒ **23.** Prove: If a line is perpendicular to a plane, then every plane that contains the line is also perpendicular to the given plane.

ALGEBRA REVIEW

Objective **To add and subtract polynomials**

To simplify polynomials, combine like terms.

Example

Simplify. $(5x^4 + 3x^2 + 5) - (2x^3 + 2x^2 - x + 3)$

$5x^4 + 3x^2 + 5 - 2x^3 - 2x^2 + x - 3 = 5x^4 - 2x^3 + x^2 + x + 2$

1. $8x^2 - x + 3$ 2. $6x^3 + 8x^2 - 4x + 7$ 3. $4x^4 + 5x^3 + 2x^2$ 4. $2x^4 + 2x^3 + x^2 + x$
5. $3 + 7x + x^2 + x^3$ 6. $4 - 2x - 2x^2$

Simplify.

1. $(5x^2 + 2x - 1) + (3x^2 - 3x + 4)$

2. $(2x^3 + 5x^2 + 7) + (4x^3 + 3x^2 - 4x)$

3. $(7x^4 + 3x^3 + 4x^2) - (3x^4 - 2x^3 + 2x^2)$

4. $(4x^4 + x^3 + x) - (2x^4 - x^3 - x^2)$

5. $(1 + 3x + x^2) + (2 + 4x + x^3)$

6. $(7 - 2x^2 + x^3) - (3 + 2x + x^3)$

CHAPTER TWO REVIEW

A exercises are 1–8, 10–18, 20–24;

1. Rewrite in "If . . ., then . . ." form: Vertical angles are congruent. [2.2]

2. Write the inverse, converse, and contrapositive of "If two angles are not vertical angles, then they are not congruent." [2.2]

Which are measures of complementary angles? Supplementary angles? Neither? [2.1]

3. 58°; 42° **N** 4. 132°; 48° **S** 5. 72°; 18° **C**

6. 89°; 1° **C** 7. 100°; 90° **N**

Find the degree measure of each angle. [2.1]

8. Its measure is 30° less than that of its complement. **30°**

9. Its measure is 100° less than three times the measure of its supplement. **110°**

True or false?

10. Vertical angles are congruent. [2.9] **T**

11. Exactly one plane contains two given points. [2.6] **F**

12. All adjacent angles are supplementary. [2.1] **F**

13. All right angles are congruent. [2.8] **T**

14. Three points may be in more than one plane. [2.6] **T**

Find the indicated measure of each angle. (Ex. 15–16)

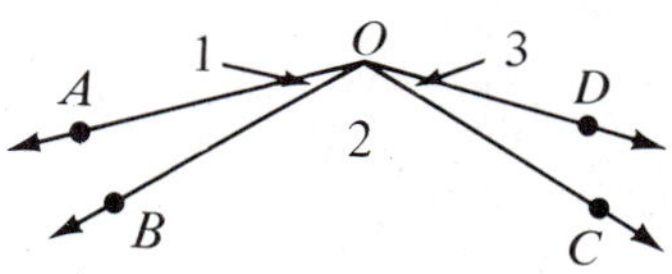

15. Given: $\angle 1 \cong \angle 3$, m $\angle 1 = 23$, and m $\angle 2 = 131$.
Find m $\angle BOD$. [2.4, 2.7] **154**

16. Given: $\angle 1 \cong \angle 3$, m $\angle AOC = 137$, and m $\angle 2 = 98$. Find m $\angle 3$. [2.4, 2.7] **39**

Vocabulary

adjacent angles [2.1], complement [2.1]
perpendicular [2.8], supplement [2.1]
vertical angles [2.1]

B exercises are 9, 19.

17. Give the reason for each statement in the proof. [2.7, 2.9]

Given: $\angle WXU \cong \angle ZYV$
Prove: $\angle YXU \cong \angle XYV$

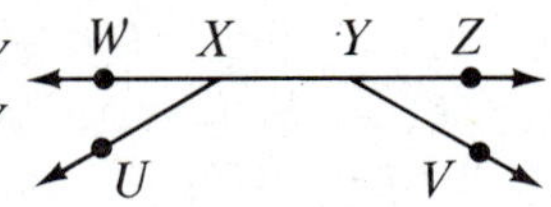

Proof

Statements	Reasons
1. m $\angle WXU$ + m $\angle YXU = 180$	1. ? **Post. 11**
2. m $\angle ZYV$ + m $\angle XYV = 180$	2. ? **same as 1**
3. $\angle WXU$ is a supp. of $\angle YXU$	3. ?
4. $\angle ZYV$ is a supp. of $\angle XYV$	4. ? **same as 3**
5. $\angle WXU \cong \angle ZYV$	5. ? **Given**
6. $\therefore \angle YXU \cong \angle XYV$	6. ? **Supp. of $\cong \angle$'s are $\cong$**

3. Def. of supp. $\angle$'s

Construct each of the following (Ex. 18–19). [2.8]

18. A line perpendicular to a given line through a point on the line. **See constr. p. 51.**

19. An angle with a measure of 135°. **Construct $\perp$, bisect the 90° angle.**

Draw a labeled diagram. State what is given and what is to be proved in terms of the diagram. (Ex. 20–21) **Answers will vary.**

20. If the noncommon sides of two adjacent angles are perpendicular, then the angles are complements. [2.3, 2.10]

21. Two intersecting lines are coplanar. [2.6, 2.10]

22. Given: $AB = CD$ [2.7]
Prove: $AB + BC = CD + BC$

23. Given: $\angle 1 \cong \angle 2$ and $\angle 2 \cong \angle 3$ [2.9]
Prove: $\angle 1 \cong \angle 3$

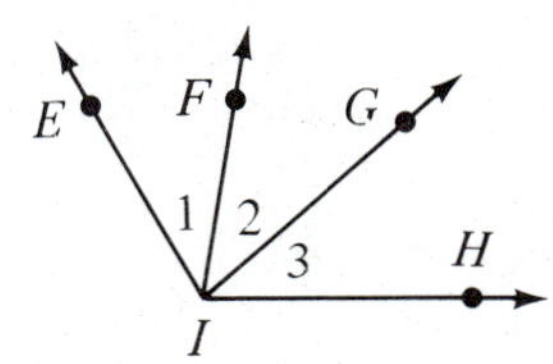

24. Prove: Supplements of the same angle are congruent.

CHAPTER TWO TEST

A exercises are 1–5, 8–21;

1. Rewrite in "If . . ., then . . ." form: Congruent angles have equal measures.

2. Write the inverse, converse, and contrapositive of "If two angles are supplements, then they are not complements."

Which are measures of complementary angles? Supplementary angles? Neither?

3. 27°; 73° N 4. 124°; 66° N 5. 50°; 40° C

Find the measure of each angle.

6. Its measure is 50° less than its supplement. 65°

7. Its measure is 30° more than twice that of its complement. 70°

True or false?

8. Two lines can intersect in exactly two points. F

9. Two planes can intersect in exactly one point. F

10. Two angles that are supplements of each other must be a linear pair. F

11. Vertical angles cannot be adjacent angles. T

12. Perpendicular lines must form four congruent angles. T

13. Complementary angles are also adjacent. F

Find the indicated measure of each angle. (Ex. 14–15)

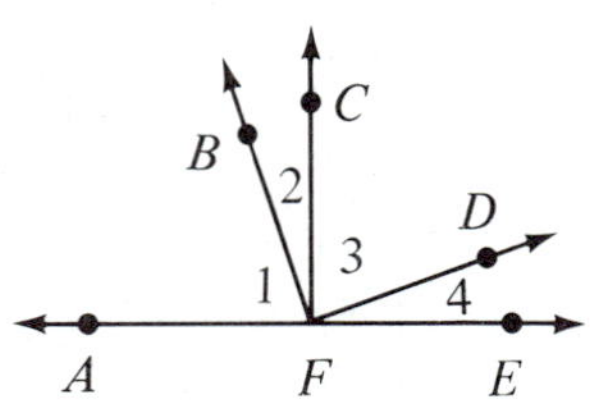

14. Given: $\overrightarrow{FC} \perp \overleftrightarrow{AE}$, m $\angle 1 = 70$, and $\angle 2 \cong \angle 4$ Find m $\angle 3$. 70

15. Given: m $\angle AFC = 90$, m $\angle BFD = 90$, and m $\angle 2 = 20$. Find m $\angle AFD$. 160

B exercises are 6, 7.

16. Give the reason for each statement in the proof.
Given: $\angle XOY \cong \angle YOZ$, $\angle YOZ \cong \angle ZOW$, and m $\angle XOY = 32$
Prove: m $\angle ZOW = 32$

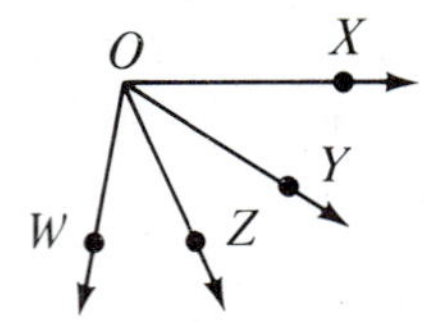

Proof

Statements		Reasons
1. $\angle XOY \cong \angle YOZ$	Given	1. ?
2. $\angle YOZ \cong \angle ZOW$	Given	2. ?
3. $\angle XOY \cong \angle ZOW$	Trans. prop.	3. ?
4. m $\angle XOY =$ m $\angle ZOW$	Def of ≅ ∠'s	4. ?
5. m $\angle XOY = 32$	Given	5. ?
6. ∴ m $\angle ZOW = 32$	Subst. prop.	

Construct each of the following. (Ex. 17–18)

17. Draw a line and a point P not on the line. Construct the perpendicular from the point to the line. See constr., p. 50.

18. Construct an angle with a measure of 45°. See Ex. 2, p. 51.

19. Draw a labeled diagram for the given statement. State what is given and what is to be proved in terms of the diagram.
If two lines form congruent adjacent angles, then they are perpendicular.
Answers will vary.

20. Given: $\overrightarrow{OS} \perp \overleftrightarrow{PQ}$ and $\angle POR \cong \angle SOT$
Prove: $\angle ROS \cong \angle TOQ$

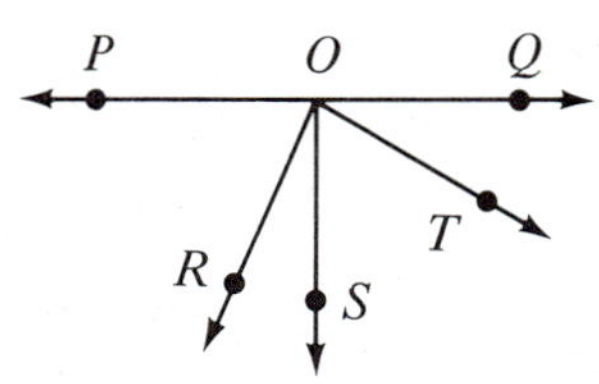

21. Prove. Complements of the same angle are congruent.

COLLEGE PREP TEST

Directions: Choose the one best answer to each question or problem.

1. An angle with a measure of 72.9° is
(A) acute **(B)** right **(C)** obtuse
(D) straight
(E) neither of these A

2. An obtuse angle could have a measure of
I. 89.99°
II. 90.001°
III. 135°

(A) I only **(B)** II only **(C)** III only
(D) I and II only
(E) II and III only E

3. If $\angle A$ and $\angle B$ are supplementary, which of the following could be true?
I. $m\angle A = 71.4$, $m\angle B = 18.6$
II. $m\angle A = 82.4$, $m\angle B = 87.6$
III. $m\angle A = 56.2$, $m\angle B = 123.8$

(A) I only **(B)** II only
(C) III only **(D)** I and II only
(E) II and III only C

4. If $m\angle A = 45$ and $m\angle B = 45$, then
(A) $\angle A$ and $\angle B$ are complementary.
(B) $\angle A$ and $\angle B$ are congruent.
(C) $\angle A$ and $\angle B$ are supplementary.
(D) $\angle A$ and $\angle B$ are congruent and complementary.
(E) $\angle A$ and $\angle B$ are congruent and supplementary. D

5. If the measure of an angle is 30° less than 4 times that of its supplement, then its measure is
(A) 66° **(B)** 75° **(C)** 138° **(D)** 165° C
(E) 250°

6. If P is between A and B, and $AP = 7$ and $PB = 12$, then
(A) $AP - PB = 5$ **(B)** $AB = 5$
(C) $AB = 17$ **(D)** $AB < 17$
(E) $AB > 17$ E

7. Two given angles cannot be both
(A) congruent and complementary.
(B) congruent and supplementary.
(C) vertical and congruent.
(D) adjacent and congruent.
(E) vertical and adjacent. E

8. The intersection of two planes could be
I. the empty set.
II. a point.
III. a line.

(A) I only **(B)** II only
(C) III only **(D)** I and II only
(E) I and III only E

9. Which of the following are always contained in exactly one plane?
I. Any three points
II. Two intersecting lines
III. A line and a point not on the line

(A) I only **(B)** II only
(C) III only **(D)** II and III only D
(E) I, II, and III.

10. In the given proof, what is the missing step?
Given: $\angle 1$ and $\angle 3$ are supplementary.
Prove: $\angle 1 \cong \angle 4$

Proof:
1. $\angle 1$ and $\angle 3$ are supplementary.
2. ?
3. $\therefore \angle 1 \cong \angle 4$

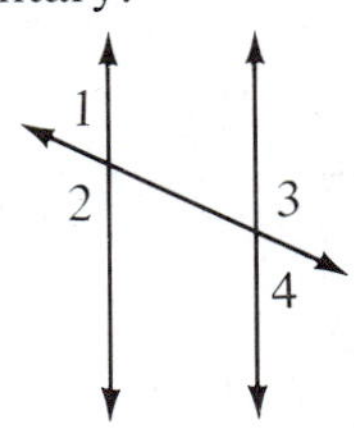

(A) $m\angle 1 + m\angle 3 = 180$
(B) $m\angle 1 + m\angle 3 = 90$
(C) $\angle 1$ and $\angle 2$ are supplementary.
(D) $\angle 3$ and $\angle 4$ are supplementary.
(E) $m\angle 1 = m\angle 4$ D

3 PARALLELS

CAREERS

Architecture

Problem

Architects provide a variety of services related to building. They prepare the working drawings and specifications showing the exact dimensions and locations of every part of the plan. During construction they supervise all work to make certain that plans are followed.

A new supermarket is opening in Jonesboro. The rectangular parking lot measures 80 m by 50 m. How should the parking spaces and the access drives be laid out so that the maximum number of cars can be parked?

The first step in solving a problem of this type is to formulate the problem-solving situation.

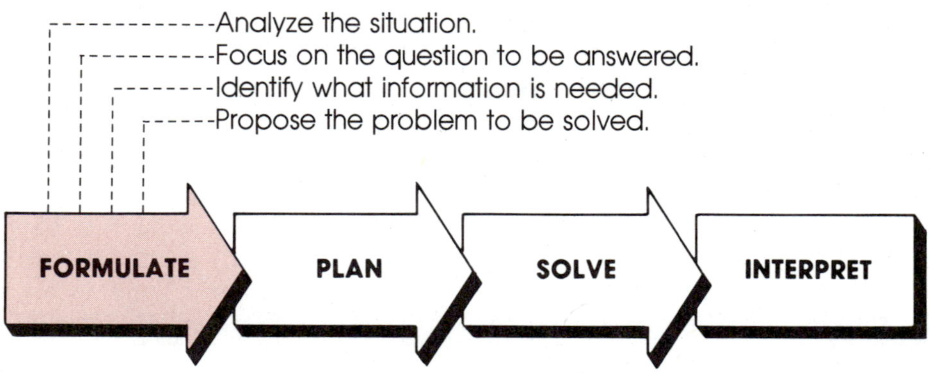

PARALLEL AND SKEW LINES 3.1

Objectives **To identify parallel lines, rays, and segments**
To identify skew lines

Some lines intersect; others do not. Lines that do not intersect may be in the same plane or in different planes.

Definition: Parallel lines	*Parallel lines* are coplanar lines that do not intersect.

$\overleftrightarrow{AB} \parallel \overleftrightarrow{CD}$ means $\overleftrightarrow{AB}$ *is parallel to* $\overleftrightarrow{CD}$. $\overleftrightarrow{EF} \nparallel \overleftrightarrow{GH}$ means $\overleftrightarrow{EF}$ *is not parallel to* $\overleftrightarrow{GH}$.

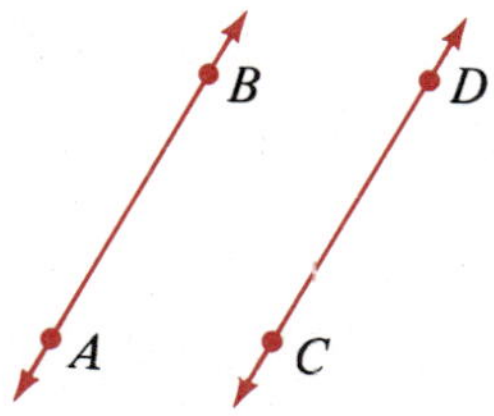

$\overleftrightarrow{AB} \parallel \overleftrightarrow{CD}$

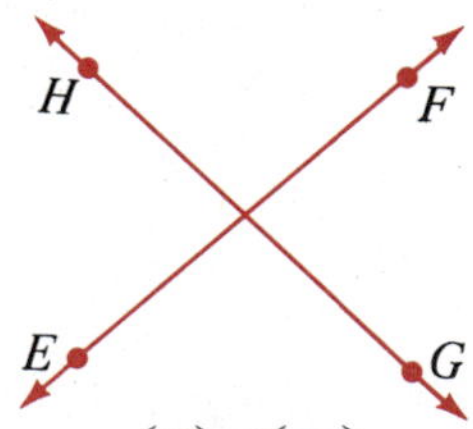

$\overleftrightarrow{EF} \nparallel \overleftrightarrow{GH}$

Lines that do not lie in the same plane have a special name.

Definition: Skew lines	*Skew lines* are noncoplanar lines.

Since skew lines are not in the same plane, they cannot intersect. For the same reason, skew lines cannot be parallel. Parallel lines must be *coplanar* lines.

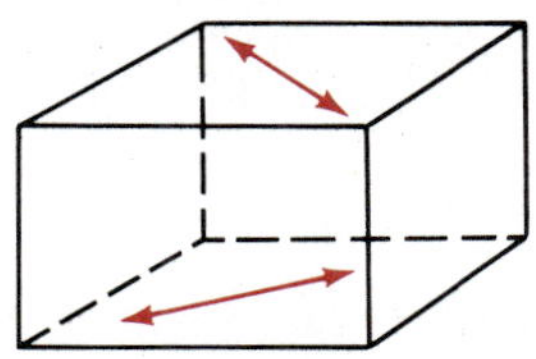

skew lines

Example 1 Which pairs of lines appear to be parallel? Which appear to be skew?

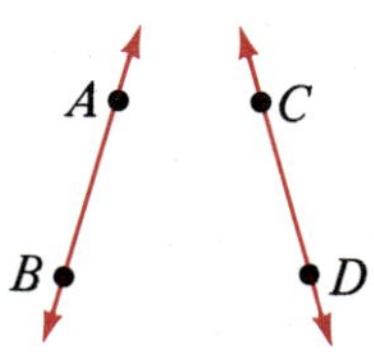

$\overleftrightarrow{AB} \nparallel \overleftrightarrow{CD}$

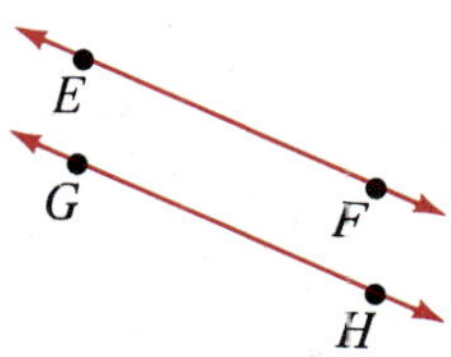

$\overleftrightarrow{EF} \parallel \overleftrightarrow{GH}$

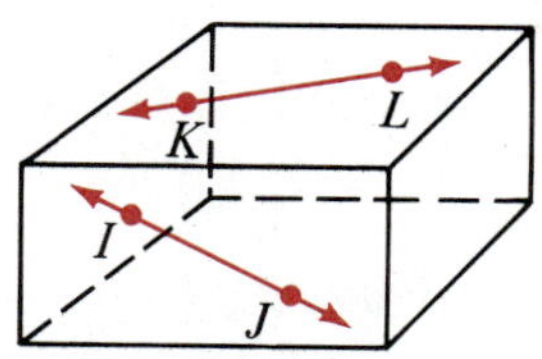

$\overleftrightarrow{IJ}$ and $\overleftrightarrow{KL}$ are skew.

Segments or rays can also be parallel. However, segments or rays that do not intersect are not necessarily parallel. This depends on the lines that contain them.

Definition: Parallel segments and rays

Segments or *rays* are *parallel* if the lines that contain them are parallel.

Even though $\overline{XY}$ and $\overrightarrow{VW}$ do not intersect, they are not parallel since $\overleftrightarrow{XY}$ and $\overleftrightarrow{VW}$ are not parallel. If $\overleftrightarrow{RS} \parallel \overleftrightarrow{TU}$, then $\overline{RS} \parallel \overrightarrow{TU}$.

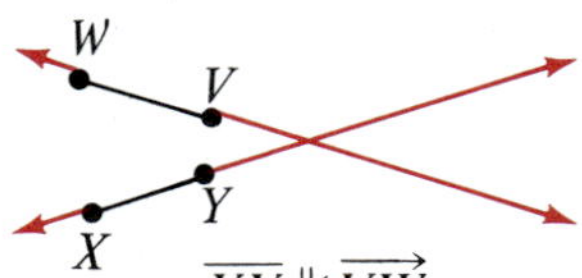

$\overline{XY} \nparallel \overrightarrow{VW}$

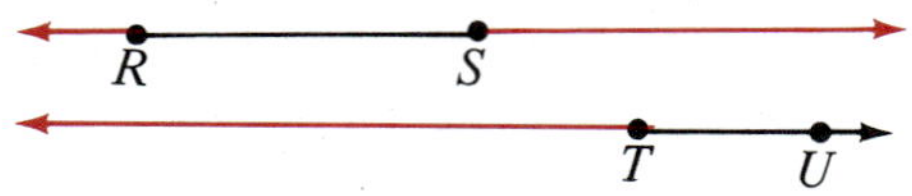

$\overline{RS} \parallel \overrightarrow{TU}$

Example 2 Which pairs of segments or rays appear to be parallel?

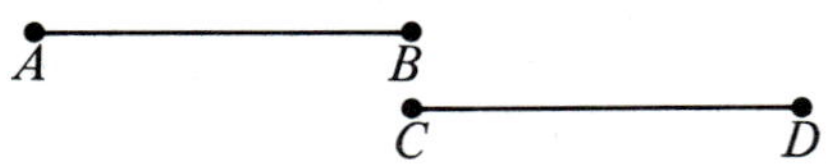

$\overline{AB} \parallel \overline{CD}$

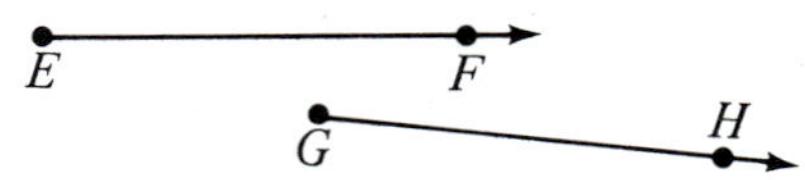

$\overrightarrow{EF} \nparallel \overrightarrow{GH}$

Example 3 True or false? Draw a picture to defend your answer.
Two coplanar rays that do not intersect are parallel.

False $\overrightarrow{AB}$ does not intersect $\overrightarrow{CD}$,
but $\overleftrightarrow{AB}$ does intersect $\overleftrightarrow{CD}$.
$\overrightarrow{AB} \nparallel \overrightarrow{CD}$

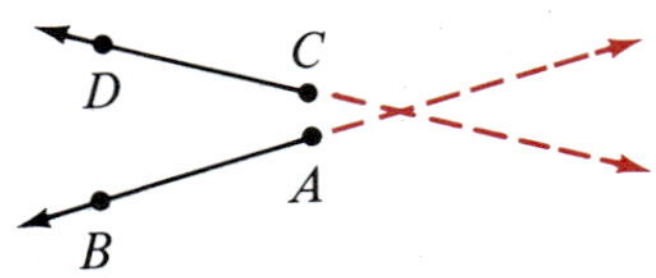

Reading in Geometry

Find a word in this lesson that means each of the following.

1. Lines that are not coplanar. **skew**
2. Lines that are coplanar but not parallel. **intersecting**
3. Lines that are both coplanar and nonintersecting. **parallel**

Explain what each of the following symbols means.

4. $\parallel$ **is parallel to**
5. $\nparallel$ **is not parallel to**

Find each of the following words in a standard dictionary. Which of the definitions listed for each word applies to this lesson?

6. vertical **perpendicular to plane of the horizon**
7. horizontal **parallel to plane of the horizon**

Oral Exercises

Which pairs of lines, rays, or segments appear to be parallel? Which appear to be skew? Which appear to be neither parallel nor skew?

1.

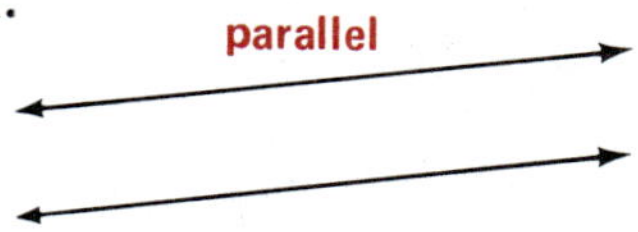

2.

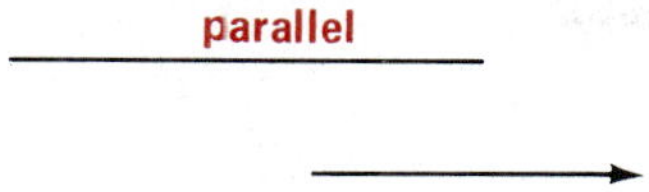

3.

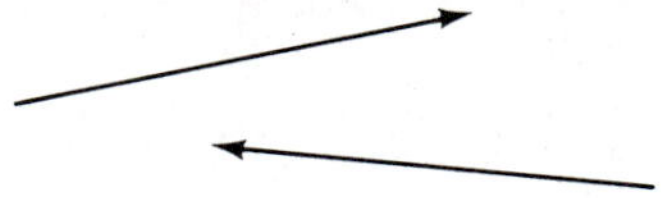

4.

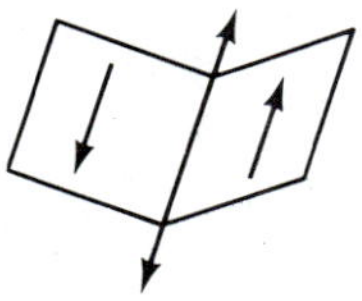

5.

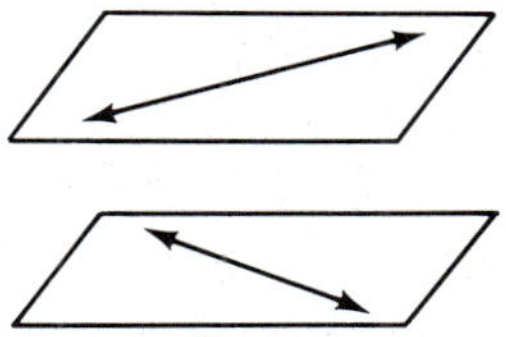

6.

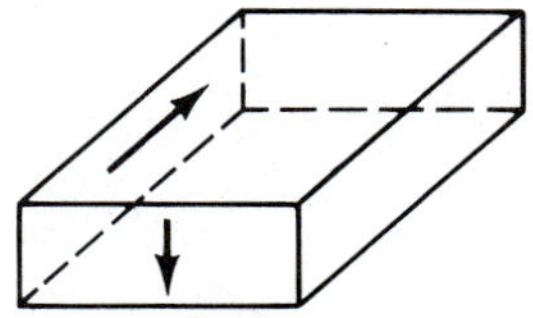

Written Exercises

Draw a picture of each of the following. **Check students' drawings.**

A

1. Two lines that do not intersect.
2. Two parallel rays.
3. Two nonintersecting rays that are not parallel.
4. Two nonintersecting lines that are not parallel.

True or false? Draw a picture to defend your answer.

5. Two parallel lines are coplanar. **T**
6. Two coplanar lines are parallel. **F**
7. Skew lines do not intersect. **T**
8. Nonintersecting segments are parallel. **F**
9. Two rays contained in parallel lines are parallel. **T**
10. Two segments contained in parallel rays are parallel. **T**
11. Two segments that are not skew and that do not intersect are parallel. **F**
12. Two lines that are not skew and that do not intersect are parallel. **T**

Always true, sometimes true, or never true? (Ex. 13–16)

B

13. A segment that does not intersect a line is parallel to the line. **ST**
14. Two lines parallel to a third line are parallel to each other. **AT**
15. Coplanar lines cannot be skew lines. **AT**
16. A line parallel to one of two skew lines is parallel to the other. **NT**

C

17. Define vertical and horizontal lines. Can two vertical lines be parallel? Can two horizontal lines be parallel? **See answers p. 70. Ex 6, 7. No. Yes. See also p. 98, Ex 3. (If vertical lines are defined as *lines with undefined slope,* then vertical lines are parallel. See p. 449.)**

CUMULATIVE REVIEW

1. Draw a figure with an interior and an exterior and whose sides are rays. **an angle**
2. Draw a segment 20 cm long. Construct its midpoint. **See constr. p. 10.**
3. Is it possible for the union of two rays to be a line? Why or why not? **yes; if the rays form a straight angle**

TRANSVERSALS AND SPECIAL ANGLES 3.2

Objectives

To identify transversals
To identify angles formed by lines and their transversals

Line t intersects line l at P and line m at Q. Line t is called a **transversal** of l and m.

Definition: Transversal A *transversal* is a line that intersects two or more coplanar lines in two or more distinct points.

Example 1 Is line t a transversal? Give a reason for each answer.

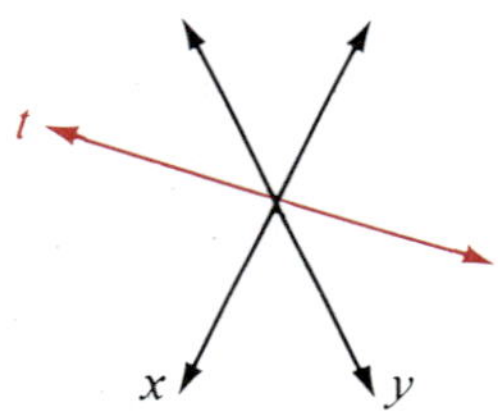

No, line t intersects x and y in one point.

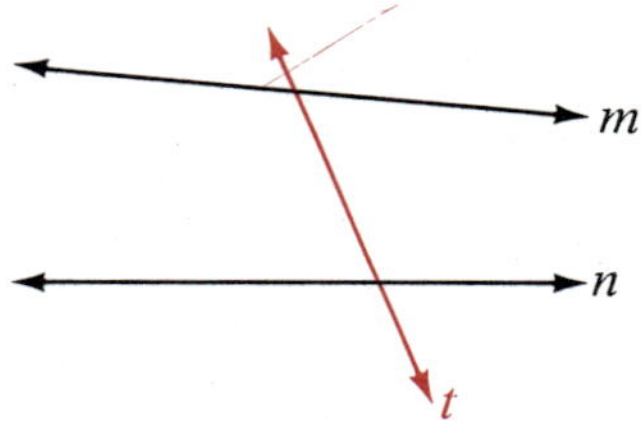

Yes, line t intersects m and n in different points.

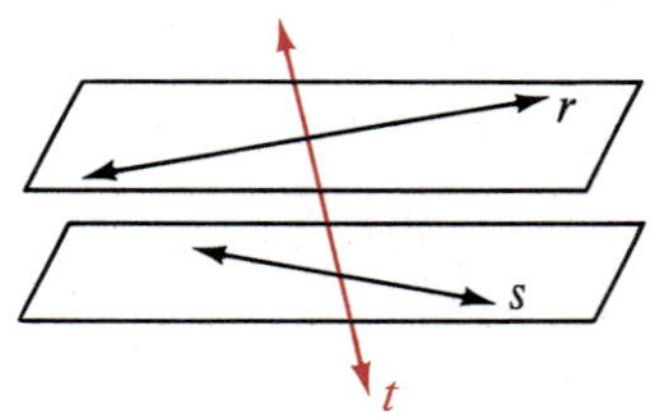

No, lines r and s are not coplanar lines.

Example 2 Identify each transversal in the figure.

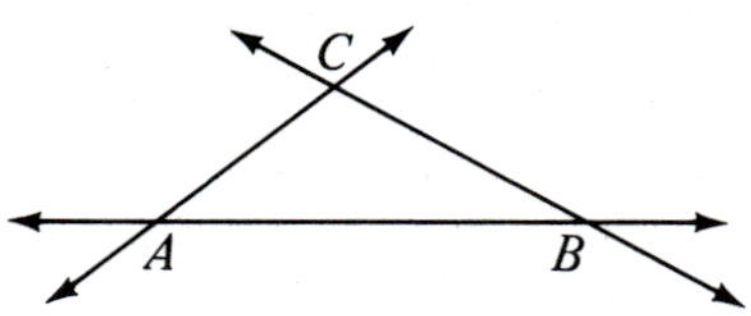

$\overleftrightarrow{AB}$ is a transversal of $\overleftrightarrow{AC}$ and $\overleftrightarrow{BC}$.
$\overleftrightarrow{BC}$ is a transversal of $\overleftrightarrow{AB}$ and $\overleftrightarrow{AC}$.
$\overleftrightarrow{CA}$ is a transversal of $\overleftrightarrow{BC}$ and $\overleftrightarrow{AB}$.

When a transversal intersects two lines, some special pairs of angles are formed.

Special Pairs of Angles
Formed by Two Lines and a Transversal

Alternate Interior Angles	**Corresponding Angles**	**Alternate Exterior Angles**
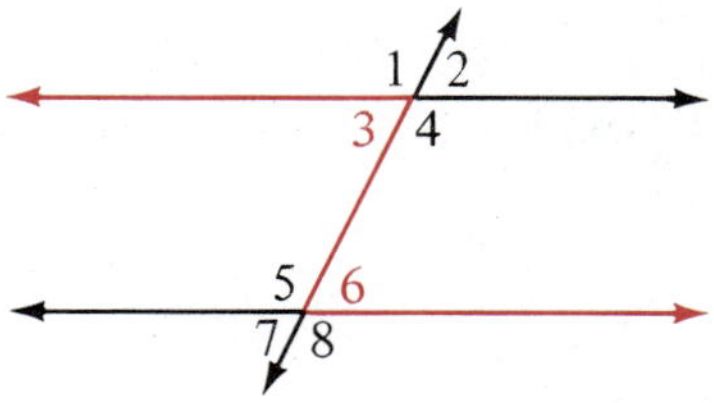	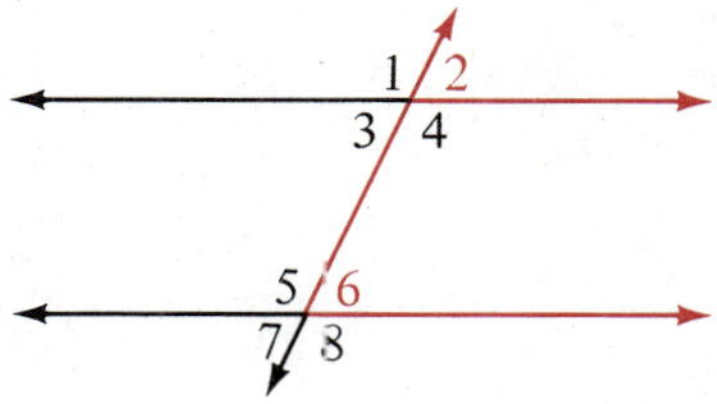	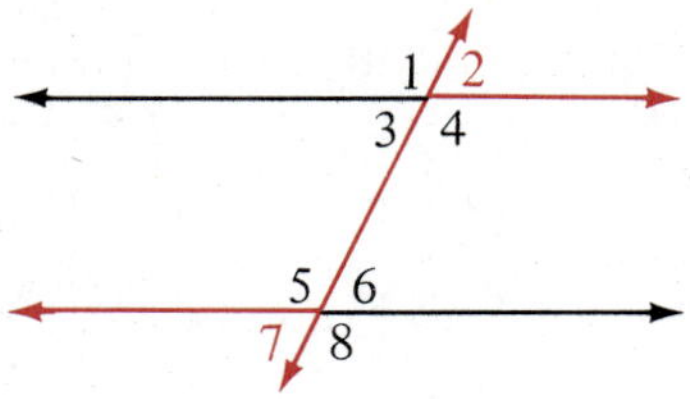
∠3 and ∠6 are alternate interior angles.	∠2 and ∠6 are corresponding angles.	∠2 and ∠7 are alternate exterior angles.
Other pair: ∠4 and ∠5	Other pairs: ∠1 and ∠5, ∠4 and ∠8, ∠3 and ∠7	Other pair: ∠1 and ∠8

Example 3 Identify each given pair of angles as alternate interior, corresponding, alternate exterior, or none of these.

Angles	**Answers**
∠1 and ∠8	alternate exterior
∠3 and ∠7	corresponding
∠4 and ∠5	alternate interior
∠2 and ∠8	none of these
∠3 and ∠5	none of these

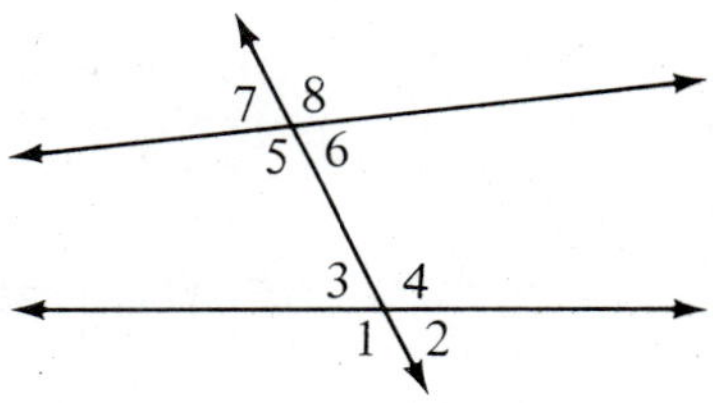

Written Exercises

Identify each transversal in each figure.

A **1.**

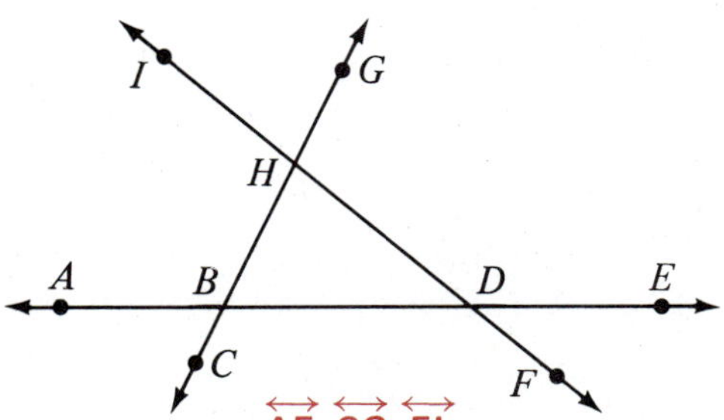

$\overleftrightarrow{AE}$; $\overleftrightarrow{CG}$; $\overleftrightarrow{FI}$

2.

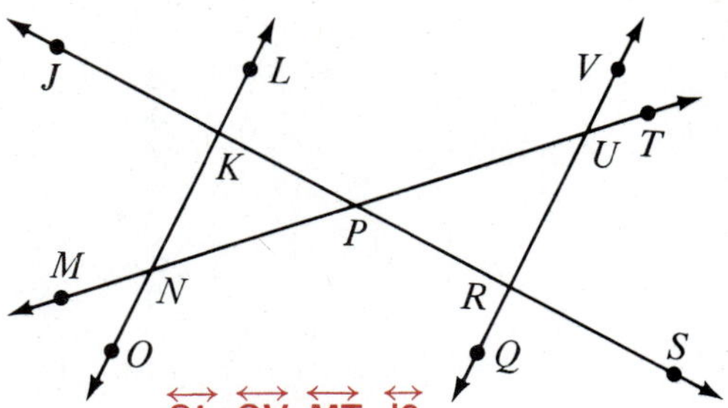

$\overleftrightarrow{OL}$; $\overleftrightarrow{QV}$; $\overleftrightarrow{MT}$; $\overleftrightarrow{JS}$

Identify each given pair of angles as alternate interior, corresponding, alternate exterior, or none of these.

8. alt. int.; 9. corr.; 10. alt ext.; 11. corr.; 12. none of these;

3. ∠2 and ∠4
4. ∠1 and ∠8
5. ∠4 and ∠5
6. ∠1 and ∠5
7. ∠3 and ∠5

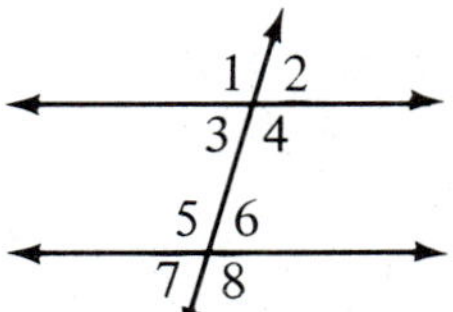

8. ∠2 and ∠7
9. ∠4 and ∠8
10. ∠1 and ∠8
11. ∠1 and ∠5
12. ∠4 and ∠7

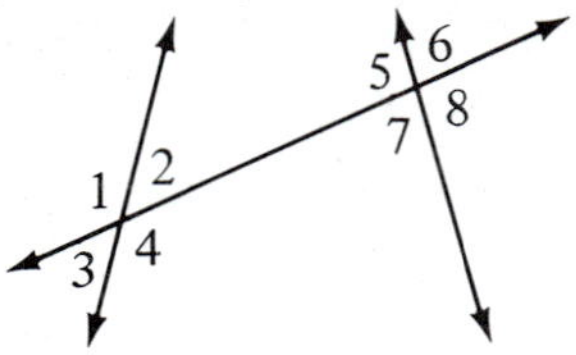

3. none of these; 4. alt. ext.; 5. alt. int.; 6. corr.; 7. none of these;
13. corr.; 14. alt. ext.; 15. alt. int.; 16. alt. int.; 17. alt. ext.; 18. none of these; 19. alt. int.; 20. corr.;

13. ∠1 and ∠10
14. ∠6 and ∠12
15. ∠8 and ∠10
16. ∠4 and ∠11
17. ∠2 and ∠8
18. ∠5 and ∠7

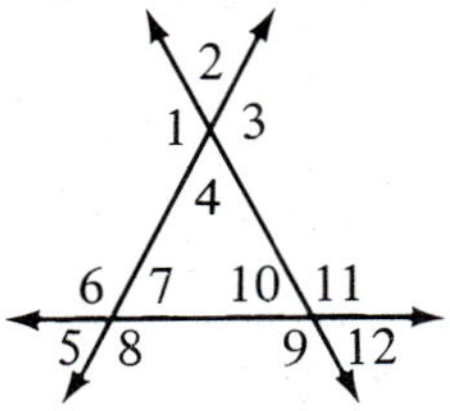

19. ∠1 and ∠6
20. ∠5 and ∠13
21. ∠7 and ∠16
22. ∠8 and ∠14
23. ∠2 and ∠12
24. ∠6 and ∠8

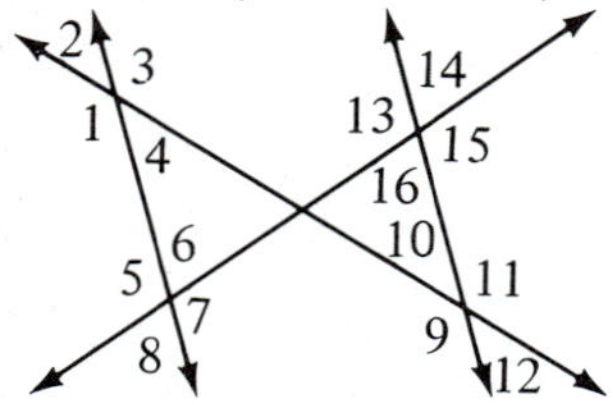

21. none of these; 22. alt. ext.; 23. alt. ext.; 24. none of these

Which of the capital letters listed below contain each of the following?

B 25. A transversal A, E, F, H, M, N, W, Z

26. Corresponding angles A, E, F

27. Parallel segments E, F, H, M, N, W, Z

28. Alternate interior angles A, H, M, N, W, Z

A E F H K L M N T V W X Y Z

ALGEBRA REVIEW

Objective **To solve linear equations containing parentheses**

To solve a linear equation containing parentheses:
(1) Remove parentheses, (2) combine like terms, and (3) get the variable alone.

Example Solve $5(x-3) = 2-(2-2x)$.

$$5x - 15 = 2 - 2 + 2x \quad \blacktriangleleft \; -(2-2x) = -1\,(2-2x) = -2+2x$$
$$5x - 15 = 2x$$
$$3x = 15$$
$$x = 5$$ **Thus,** the solution is 5.

Solve each equation.

1. $3(x-1) = x+7$ 5
2. $4(x+3) = 2x-8$ −10
3. $3(x-1) = 2(x+4)$ 11
4. $2(2x+1) = 3(x-2)$ −8
5. $5x-(2x+1) = x+3$ 2
6. $5(2x-1)-(x+2) = 2$ 1
7. $3(2x+3)-2(x+4) = 6$ $\frac{5}{4}$
8. $4(3-2x)+2(x+1) = 5$ $\frac{3}{2}$
9. $3(2-3x)-(4-x) = 6$ $-\frac{1}{2}$
10. $\frac{1}{2}(4x+6)-\frac{2}{3}(6x+9) = 4$ $-\frac{7}{2}$
11. $\frac{2}{3}(9-6x)-2 = \frac{3}{4}(8-4x)$ −2
12. $5x-\frac{1}{2}(14+2x) = \frac{3}{2}(6x+6)-x$ −4

TRANSVERSALS AND PARALLEL LINES 3.3

Objectives

To apply the alternate interior angle postulate
To prove and apply theorems about angles formed by parallel lines and a transversal.

When a transversal intersects two lines, two pairs of alternate interior angles are formed. At the right, line l is parallel to line m. If you measure the alternate interior angles, $\angle w$ and $\angle z$, you will find that m $\angle w = 57$ and m $\angle z = 57$. Thus, $\angle w$ and $\angle z$ are congruent. Similarly, the alternate interior angles, $\angle x$ and $\angle y$, are congruent.

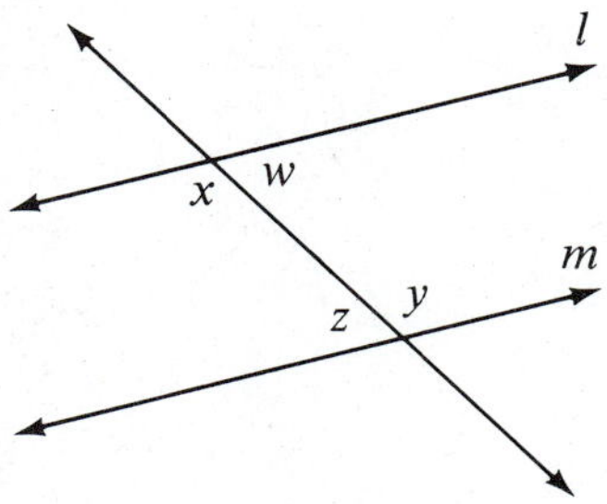

At the right, line l is not parallel to line m. If you measure the alternate interior angles, $\angle w$ and $\angle y$, you will find that m $\angle w = 77$ and m $\angle y = 62$. These angles are not congruent. Similarly, $\angle x$ and $\angle z$ are not congruent, since their measures are 103° and 118°, respectively.

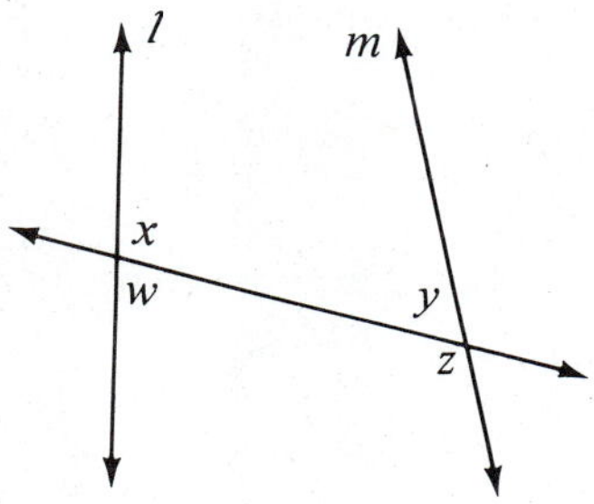

From your measurements, it appears that the pairs of alternate interior angles are congruent only when the two lines intersected by the transversal are parallel.

Postulate 13 Alternate Interior Angle Postulate

If two parallel lines are intersected by a transversal, then the alternate interior angles are congruent.

You can use Postulate 13 to find the measures of certain angles.

Example 1 Given: $l \parallel m$, $r \parallel s$, and m $\angle 1 = 115$.
Find m $\angle 2$ and m $\angle 3$.

$\angle 1 \cong \angle 2$ ◄ *∠1 and ∠2 are alternate interior angles.*
m $\angle 2 =$ m $\angle 1 = 115$

$\angle 1 \cong \angle 3$ ◄ *∠1 and ∠3 are alternate interior angles.*
m $\angle 3 =$ m $\angle 1 = 115$

Thus, m $\angle 2 = 115$ and m $\angle 3 = 115$.

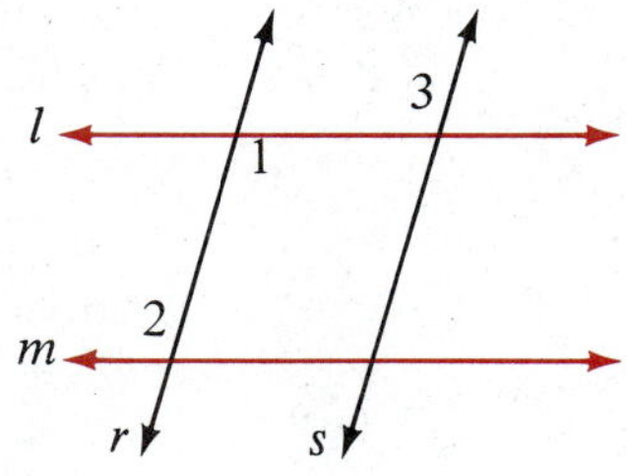

You can use Postulate 13 to show that certain other pairs of angles formed by parallel lines and a transversal are also congruent.

At the right, $l \parallel m$ and m $\angle 1 = 140$. Since $\angle 1$ and $\angle 2$ are alternate interior angles, m $\angle 2 = 140$. Since $\angle 2$ and $\angle 3$ are vertical angles, m $\angle 3 = 140$. Thus, the corresponding angles, $\angle 1$ and $\angle 3$, are congruent. This result is summarized in Theorem 3.1.

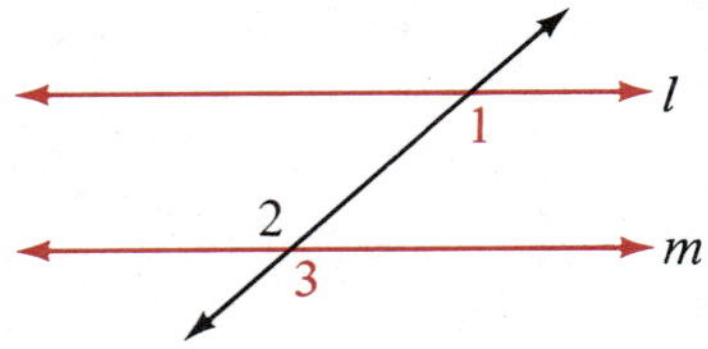

Theorem 3.1

If two parallel lines are intersected by a transversal, then the corresponding angles are congruent.

Given: $\overleftrightarrow{AB} \parallel \overleftrightarrow{CD}$
Prove: $\angle 1 \cong \angle 3$

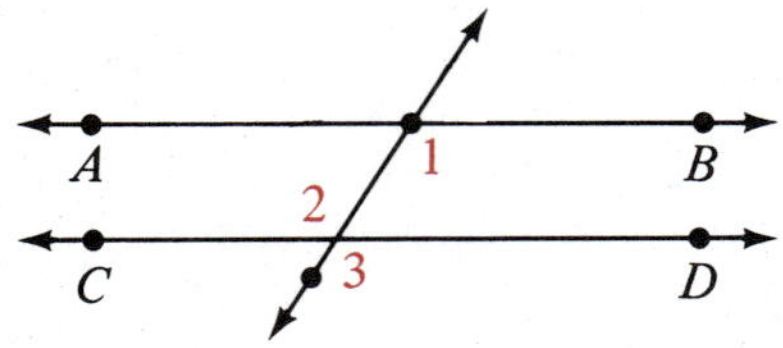

Proof

Statements	Reasons
1. $\overleftrightarrow{AB} \parallel \overleftrightarrow{CD}$	1. Given
2. $\angle 1 \cong \angle 2$	2. Alt. int. $\angle$'s of $\parallel$ lines are $\cong$.
3. $\angle 2 \cong \angle 3$	3. Vert. $\angle$'s are $\cong$.
4. $\therefore \angle 1 \cong \angle 3$	4. Transitive property

In the proof of Theorem 3.1, a previously proved theorem was used as a reason for Step 3. Once a theorem has been proved, it can be used in the proof of another theorem. Abbreviations were also used in writing reasons.

Example 2

Given: $l \parallel m$,
m $\angle 1 = 3x + 15$, and
m $\angle 2 = 2x + 35$.
Find m $\angle 1$.

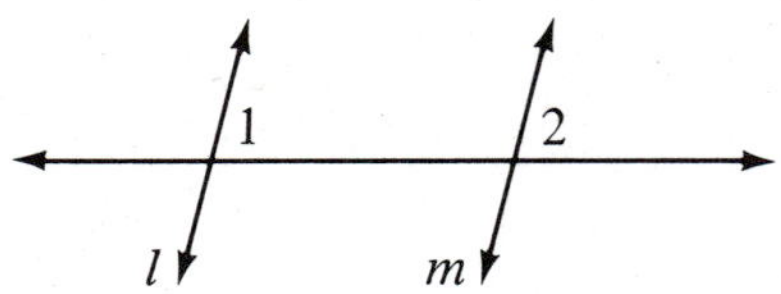

m $\angle 1$ = m $\angle 2$ ◀ *Corr. $\angle$'s of $\parallel$ lines are $\cong$.*
$3x + 15 = 2x + 35$
$x = 20$

Thus, m $\angle 1 = 3 \cdot 20 + 15$, or 75.

Postulate 13 will be used to prove the next theorem, which states a relationship between the interior angles on the same side of a transversal that intersects two parallel lines.

Theorem 3.2

If two parallel lines are intersected by a transversal, then the interior angles on the same side of the transversal are supplementary.

Given: $l \parallel m$
Prove: $\angle 1$ and $\angle 3$ are supplementary.

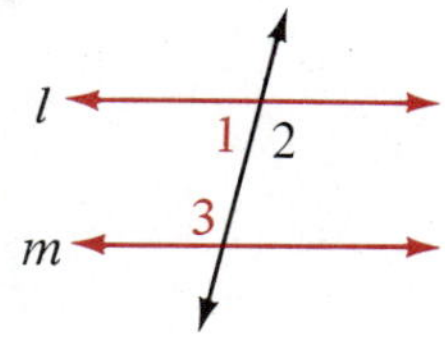

Proof: (See Exercise 13.)

Postulate 13, Theorem 3.1, and Theorem 3.2 can be used to find unknown angle measures and as reasons in other proofs.

Example 3

Given: $\overleftrightarrow{PQ} \parallel \overleftrightarrow{RS}$,
$m\angle y = 4x - 20$, and
$m\angle z = 3x + 25$.
Find $m\angle y$.

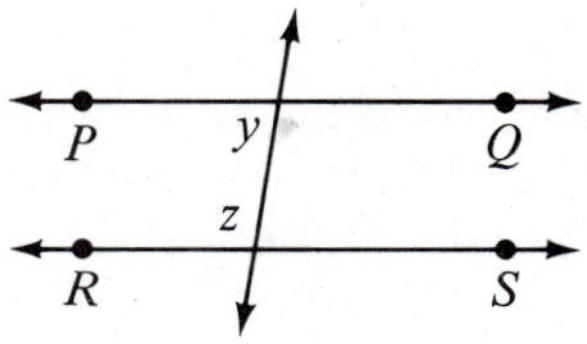

$$\begin{aligned} m\angle y + m\angle z &= 180 \\ (4x - 20) + (3x + 25) &= 180 \\ 7x + 5 &= 180 \\ 7x &= 175 \\ x &= 25 \end{aligned}$$

◀ *Use Theorem 3.2 and def. of suppl. ∠'s.*

Thus, $m\angle y = 4 \cdot 25 - 20$, or 80.

Example 4

Given: $\overleftrightarrow{AD} \parallel \overleftrightarrow{EG}$ and $\overleftrightarrow{BF} \parallel \overleftrightarrow{CH}$
Prove: $\angle 1 \cong \angle 4$

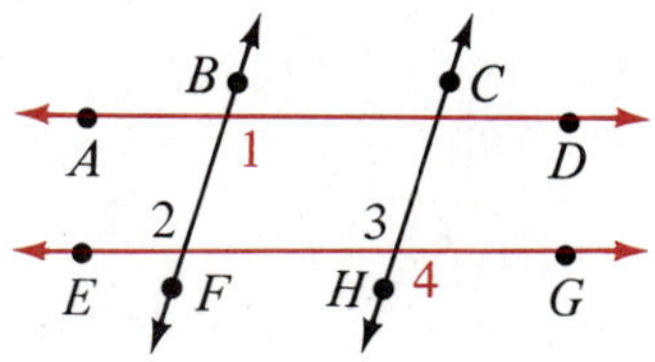

Proof

Statements	Reasons
1. $\overleftrightarrow{AD} \parallel \overleftrightarrow{EG}$ and $\overleftrightarrow{BF} \parallel \overleftrightarrow{CH}$	1. Given
2. $\angle 1 \cong \angle 2$	2. Alt. int. ∠'s of ∥ lines are ≅.
3. $\angle 2 \cong \angle 3$	3. Corr. ∠'s of ∥ lines are ≅.
4. $\angle 3 \cong \angle 4$	4. Vertical ∠'s are ≅.
5. $\therefore \angle 1 \cong \angle 4$	5. Transitive property

You may have to determine if enough information is given to enable you to find the measures of unknown angles. Before a theorem can be used to find such angle measures, you must be sure that all of the conditions of the theorem are satisfied.

Example 5

Based only on the information given, find each indicated angle measure. If not enough information is given, state this fact.

Given: $\overleftrightarrow{AB} \parallel \overleftrightarrow{CD}$ and m $\angle AFG = 102$.
Find m $\angle FGD$ and m $\angle FGC$.

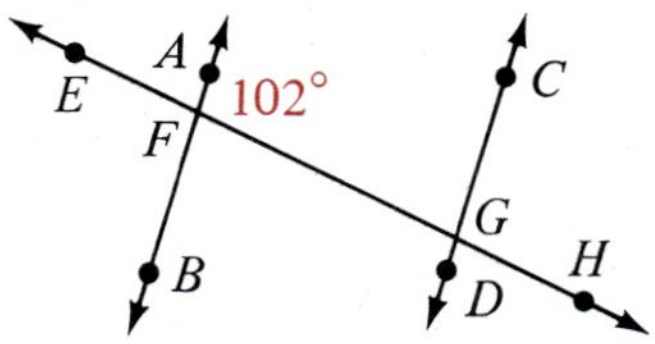

m $\angle FGD = 102$
m $\angle FGC = 78$

Given: m $\angle MPQ = 81$.
Find m $\angle PQY$, m $\angle XQR$, and m $\angle OPN$.

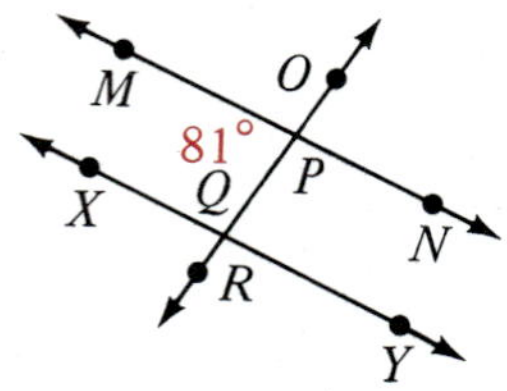

The lines are not known to be $\parallel$.

► m $\angle PQY$: not enough information
► m $\angle XQR$: not enough information
m $\angle OPN = 81$

Reading in Geometry

Abbreviations are often used in writing reasons for steps in proofs. Write out in full the statement that each abbreviation represents. Then indicate whether each statement is true or false.

1. Vert. $\angle$'s are $\cong$. **T**
2. Alt. int. $\angle$'s of $\parallel$ lines are $\cong$. **T**
3. Corr. $\angle$'s of $\parallel$ lines are $\cong$. **T**
4. Int. $\angle$'s on same side of transv. are supp. **F**

Written Exercises

Find each indicated angle measure.

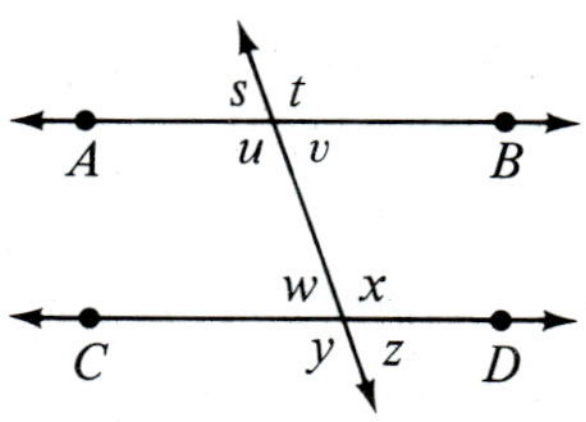

A 1. Given: $\overleftrightarrow{AB} \parallel \overleftrightarrow{CD}$ and m $\angle v = 63$ **63, 63**
Find m $\angle w$ and m $\angle z$.

2. Given: $\overleftrightarrow{AB} \parallel \overleftrightarrow{CD}$ and m $\angle x = 128$ **128, 52**
Find m $\angle t$ and m $\angle v$.

3. Given: $\overleftrightarrow{AB} \parallel \overleftrightarrow{CD}$ and m $\angle s = 71$ **71, 71**
Find m $\angle w$ and m $\angle z$.

4. Given: $\overleftrightarrow{AB} \parallel \overleftrightarrow{CD}$ and m $\angle u = 113$ **67, 67**
Find m $\angle w$ and m $\angle z$.

5. Given: $\overleftrightarrow{AB} \parallel \overleftrightarrow{CD}$,
m $\angle v = 2x + 10$, and
m $\angle w = 3x - 20$
Find m $\angle v$ and m $\angle w$. **70, 70**

6. Given: $\overleftrightarrow{AB} \parallel \overleftrightarrow{CD}$,
m $\angle v = n$, and
m $\angle x = 5n$
Find m $\angle v$ and m $\angle x$. **30, 150**

Based only on the information given, find each indicated angle measure. If not enough information is given, state this fact.

7. Given: $\overleftrightarrow{AC} \parallel \overleftrightarrow{DG}$ and m $\angle BEF +$ m $\angle BFE = 158$.
Find m $\angle EBF$, m $\angle ABE$ and m $\angle CBF$.

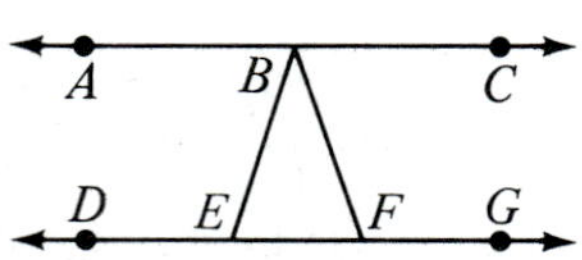

22, not enough information, not enough information

8. Given: $\overleftrightarrow{AD} \parallel \overleftrightarrow{GI}$ and m $\angle EBC = 67$.
Find m $\angle EHI$, m $\angle FCD$ and m $\angle GHJ$.

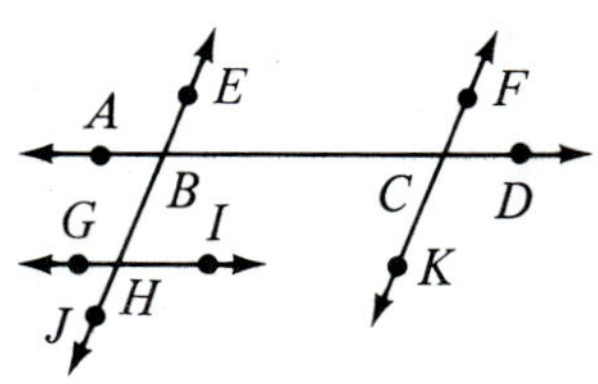

67, not enough information, 67

Complete the steps of each given proof. (Ex. 9–10)

9. Given: $\overleftrightarrow{AD} \parallel \overleftrightarrow{EK}$ and $\overleftrightarrow{CG} \parallel \overleftrightarrow{JH}$

Prove: $\angle ABC \cong \angle EIJ$

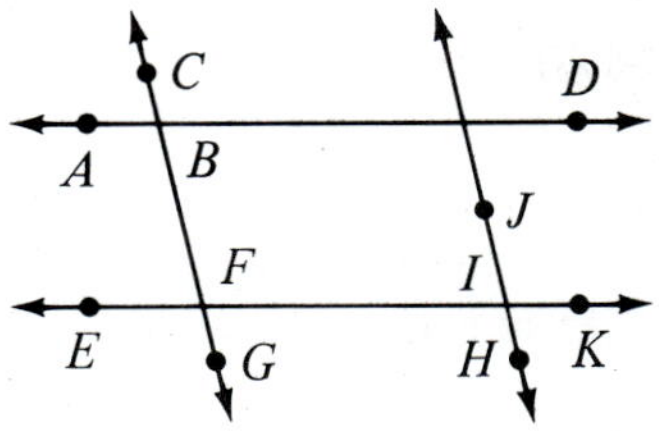

Proof

Statements	Reasons
1. $\overleftrightarrow{AD} \parallel \overleftrightarrow{EK}$ and $\overleftrightarrow{CG} \parallel \overleftrightarrow{JH}$	1. ? **Given**
2. $\angle ABC \cong \angle EFC$	2. ? **Corr. ∠'s of ∥ l's are ≅.**
3. $\angle EFC \cong$? **∠EIJ**	3. Corr. ∠'s of ∥ lines are ≅.
4. ∴ ? **∠ABC ≅ ∠EIJ**	4. ? **Trans. Prop.**

10. Given: $\overleftrightarrow{AC} \parallel \overleftrightarrow{DF}$ and $\overleftrightarrow{DF} \parallel \overleftrightarrow{GJ}$

Prove: $m\angle BEH = m\angle CBE + m\angle EHJ$

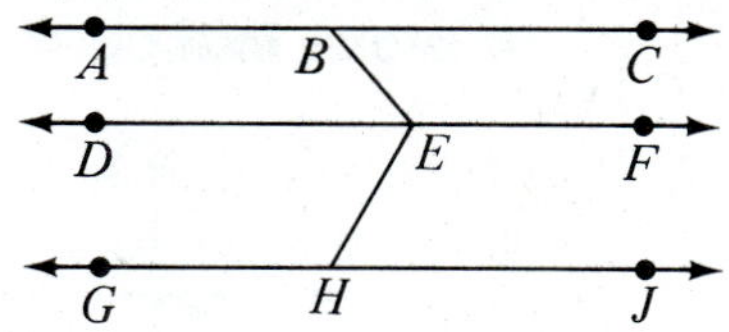

Proof

Statements	Reasons
1. ? **$\overleftrightarrow{AC} \parallel \overleftrightarrow{DF}$, $\overleftrightarrow{DF} \parallel \overleftrightarrow{GJ}$**	1. Given
2. $m\angle BEH = m\angle BED +$? **m ∠DEH**	2. ? **∠Add. prop.**
3. $\angle CBE \cong \angle BED$	3. ? **Alt. int. ∠'s of ∥ l's ≅.**
4. $\angle EHJ \cong$? **∠DEH**	4. Alt. int. ∠'s of ∥ lines are ≅.
5. ∴ ? **m ∠BEH = m ∠CBE + m ∠EHJ**	5. Substitution property

B **11.** Given: $\overrightarrow{AB} \parallel \overrightarrow{CD} \parallel \overrightarrow{EF}$

Prove: $m\angle 1 + m\angle 2 + m\angle 3 + m\angle 4 = 360$

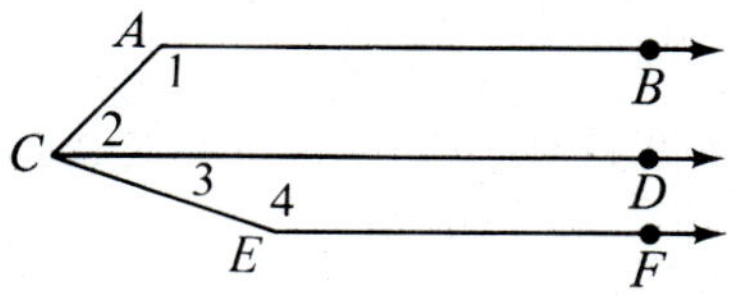

12. Given: $\overleftrightarrow{GI} \parallel \overleftrightarrow{JM}$

Prove: $m\angle 2 + m\angle 4 + m\angle 5 = 180$

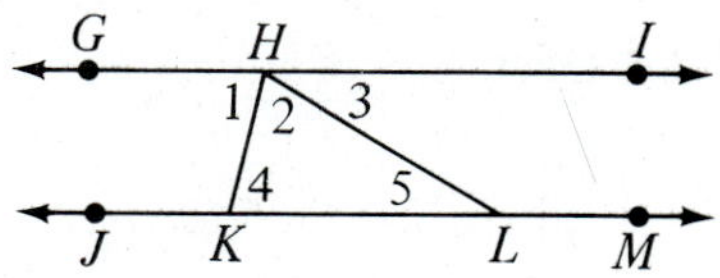

13. Complete the proof of Theorem 3.2.

14. Prove: If two parallel lines are intersected by a transversal, then the alternate exterior angles are congruent.

15. Prove: If a transversal is perpendicular to one of two parallel lines, then it is perpendicular to the other.

Find the measure of each angle using only the information given.

16. Given: $\overleftrightarrow{AB} \parallel \overleftrightarrow{CD}$, $m\angle m = 63$, and $\angle n$ and $\angle q$ are supplementary.

Meas. of l, m, p, q, t, u, x, y = 63.
Meas. of k, n, o, r, s, v, w, z = 117.

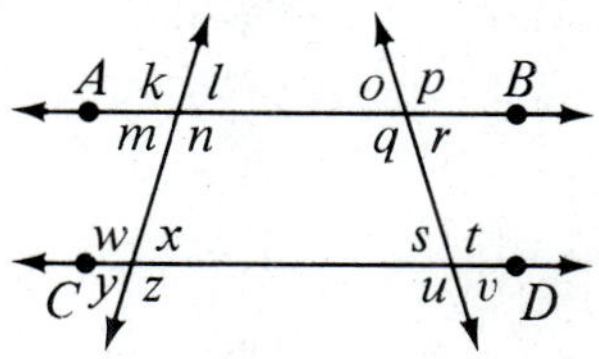

17. Given: $\overleftrightarrow{AB} \parallel \overleftrightarrow{CD}$, $m\angle n = 2a$, $m\angle x = 3a - 10$, $m\angle p = 2b + 15$, and $m\angle t = 3b - 7$.

Meas. of n, k, w, z = 76.
Meas. of l, m, x, y = 104.
Meas. of p, t, q, u = 59.
Meas. of r, o, s, v = 121

Contradictory information is given for each figure below. Determine what is contradictory.

C **18.** Given: $\overleftrightarrow{AF} \parallel \overleftrightarrow{GM}$, $\angle ABC \cong \angle DEF$, $\angle JBF$ is a supplement of $\angle AEK$, and $m\angle MLK = 88$. **m ∠MLK = 88**

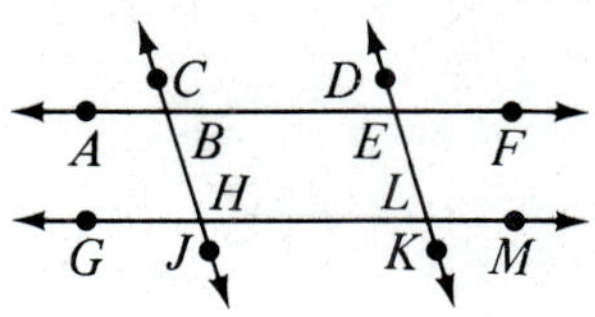

19. Given: $\overleftrightarrow{AF} \parallel \overleftrightarrow{GM}$, $\overleftrightarrow{CJ} \parallel \overleftrightarrow{DK}$, $m\angle ABC = 48$, $m\angle DLM = 132$, and $\angle AED$ is a complement of $\angle JHM$. **∠AED is a comp. of ∠JHM.**

20. Given: $\overrightarrow{BA} \parallel \overrightarrow{CE}$, $\angle 1$ is a complement of $\angle 2$, $m\angle 3 = 48$, and $m\angle 5 = 97$. **m ∠5 = 97**

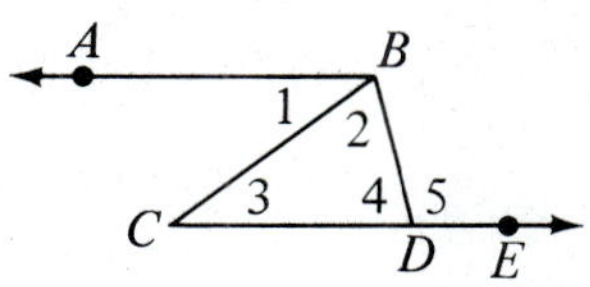

21. Given: $\overrightarrow{BA} \parallel \overrightarrow{CE}$, $\angle 1$ is a complement of $\angle 3$, $m\angle 2 = 42$, and $m\angle 3 = 42$. **m ∠3 = 42**

PROVING LINES PARALLEL 3.4

Objectives

To form converses of statements
To prove that certain lines are parallel

Earlier you learned that the converse of a conditional statement is obtained by interchanging the "if" clause with the "then" clause. A conditional statement and its converse have different meanings. "If it is raining, then the streets are wet" does not mean the same thing as "If the streets are wet, then it is raining." The truth of one of these statements does not guarantee the truth of the other.

Example 1

Write the converse of "If two lines are parallel, then the alternate interior angles are congruent."

Converse: If the alternate interior angles are congruent, then the two lines are parallel. ◄ *Interchange the "if" and the "then" clauses.*

Example 2

Write the converse of "If an animal is a dog, then it has four legs." Is the converse true or false?

Converse: If an animal has four legs, then it is a dog. False.

Example 2 shows that the converse of a true statement is not necessarily true. You must deal with the converse as a completely independent statement. Postulate 14 which follows is the converse of Postulate 13 and is an example of a true statement whose converse is true.

Postulate 14 If two lines are intersected by a transversal so that the alternate interior angles are congruent, then the lines are parallel.

Postulate 14 gives you one way to determine if lines are parallel.

Example 3

Tell if each pair of lines is parallel. Give a reason for your answer.

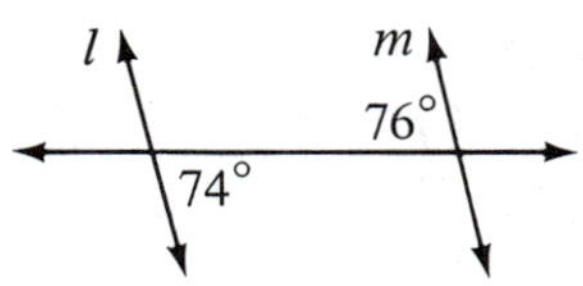

$l \nparallel m$. Alternate interior ∠'s are not ≅.

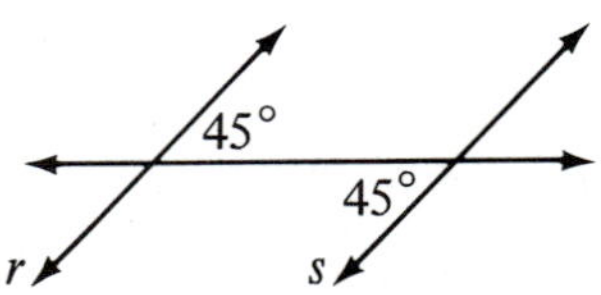

$r \parallel s$. Alternate interior ∠'s are ≅.

You proved that if two parallel lines are intersected by a transversal, then the corresponding angles are congruent. You can use Postulate 14 to prove the converse of this statement.

Theorem 3.3 If two lines are intersected by a transversal so that the corresponding angles are congruent, then the lines are parallel.

Given: $\angle 1 \cong \angle 3$
Prove: $\overleftrightarrow{AB} \parallel \overleftrightarrow{CD}$

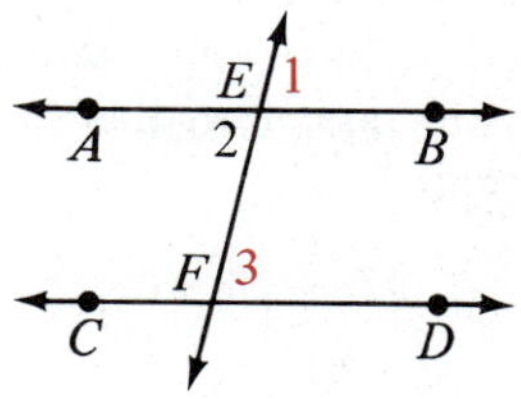

ANALYSIS
Lines are $\parallel$ if alt. int. $\angle$'s are $\cong$.

Proof

Statements	Reasons
1. $\angle 2 \cong \angle 1$	1. Vert. $\angle$'s are $\cong$.
2. $\angle 1 \cong \angle 3$	2. Given
3. $\angle 2 \cong \angle 3$	3. Transitive property
4. $\therefore \overleftrightarrow{AB} \parallel \overleftrightarrow{CD}$	4. If alt. int. $\angle$'s are $\cong$, then lines are $\parallel$.

In Theorem 3.3, if $\overleftrightarrow{AB}$ and $\overleftrightarrow{CD}$ are both perpendicular to $\overleftrightarrow{EF}$, then the corresponding angles formed are congruent right angles and so the two lines are parallel. This special case of Theorem 3.3 is stated in the following corollary.

Corollary In a plane, if two lines are perpendicular to the same line, then they are parallel.

The next theorem is the converse of Theorem 3.2. You can use Theorem 3.3 in its proof.

Theorem 3.4 If two lines are intersected by a transversal so that the interior angles on the same side of the transversal are supplementary, then the lines are parallel. (See Exercise 17.)

Given: $\angle 1$ and $\angle 2$ are supplementary.
Prove: $\overleftrightarrow{AB} \parallel \overleftrightarrow{CD}$

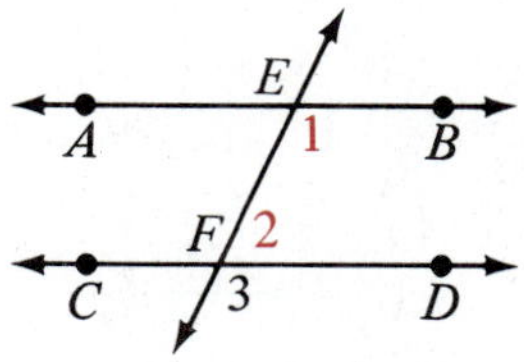

ANALYSIS
Lines are $\parallel$ if corr. $\angle$'s are $\cong$.

Example 4

If m $\angle z = 75$ and m $\angle y = 105$, are lines l and m parallel? Give a reason for your answer.

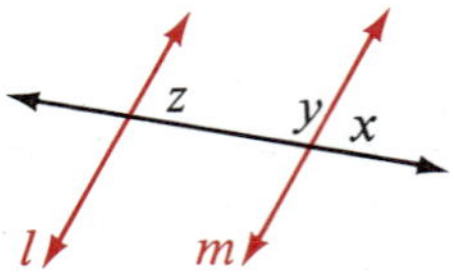

m $\angle z$ + m $\angle y = 75 + 115$ or 180, and
$\angle z$ and $\angle y$ are supplementary.
$\therefore l \parallel m$ by Theorem 3.4.

Certain other relationships between special pairs of angles that are formed by intersecting lines can guarantee that the lines are parallel. Example 5 shows that if exterior angles on the same side of a transversal are supplementary, the lines are parallel.

Example 5

Given: $\angle 1$ and $\angle 4$ are supplementary.
Prove: $l \parallel m$

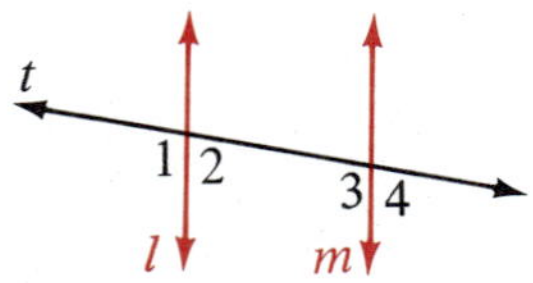

Proof

Statements	Reasons
1. m $\angle 3$ + m $\angle 4 = 180$	1. Sum of meas. of $\angle$'s with same vertex on one side of a line is 180°.
2. $\angle 3$ and $\angle 4$ are supplementary.	2. Def. of suppl. $\angle$'s
3. $\angle 1$ and $\angle 4$ are supplementary.	3. Given
4. $\angle 3 \cong \angle 1$	4. Suppl. of same $\angle$ are $\cong$.
5. $\therefore l \parallel m$	5. If corr. $\angle$'s are $\cong$, then lines are $\parallel$.

Oral Exercises

State the converse of each statement.

1. If Pat and Bobbie practice, then they will become good tennis players.
2. If guitars play, then people dance.
3. If mice appear, then elephants run.

1. If Pat and Bobbie become good tennis players, then they will practice.
2. If people dance, then guitars play.
3. If elephants run, then mice appear.

If possible, tell if each pair of lines is parallel.

4.

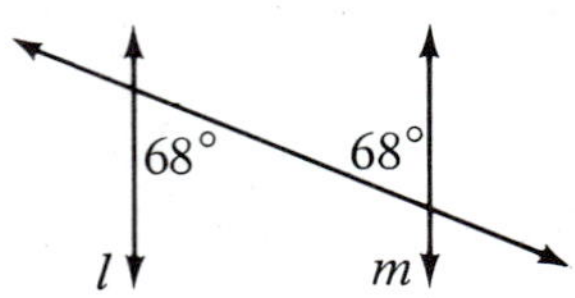

l ∥ m

5.

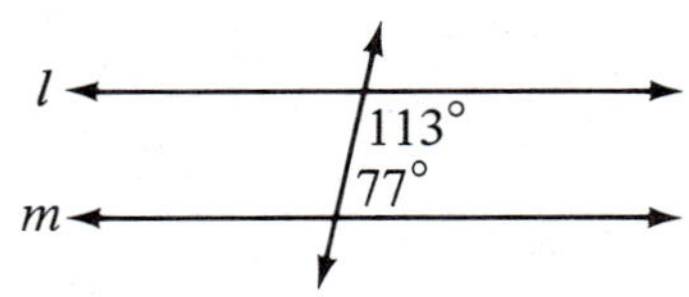

l ∦ m

6.

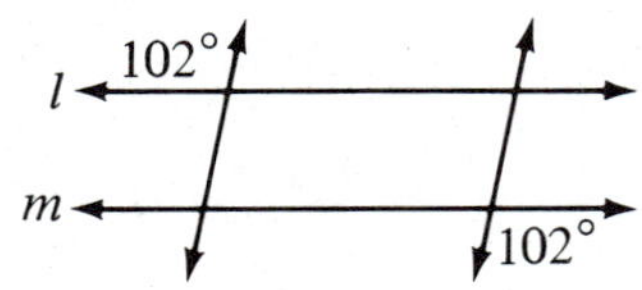

cannot tell

7.

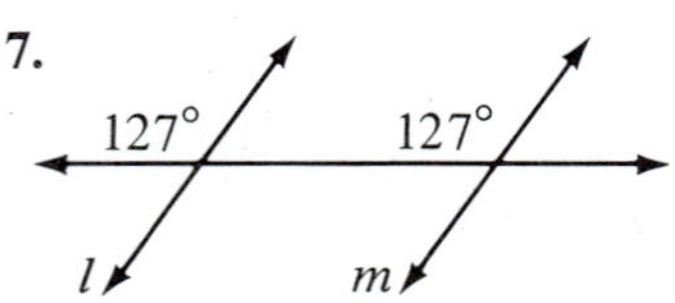

l ∥ m

8.

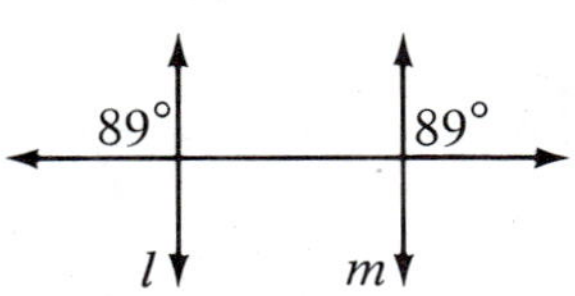

l ∦ m

9.

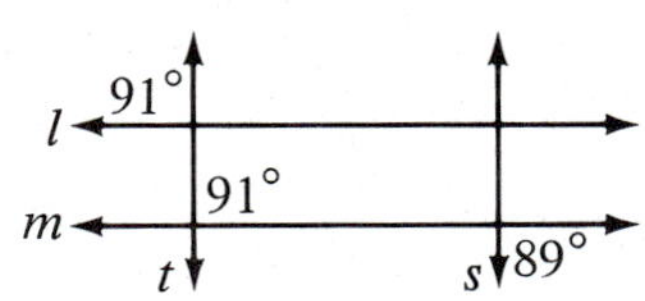

l ∦ m, t ∥ s

Written Exercises

Write the converse of each statement. If possible, determine if the converse is true or false.

A

1. If two angles are complementary, then the sum of their measures is 90°. **true**
2. If two lines are perpendicular, then they form a right angle. **true**
3. If two coplanar lines are perpendicular to the same line, then they are parallel. **false**
4. If two lines are intersected by a transversal, then the lines are parallel. **false**

For each of the following, tell which lines or rays are parallel and which angles are congruent. Give a reason for your answer. (Ex. 5–6)

5. Given: $\overleftrightarrow{BC} \perp \overleftrightarrow{AH}$ and $\overleftrightarrow{EG} \perp \overleftrightarrow{AH}$

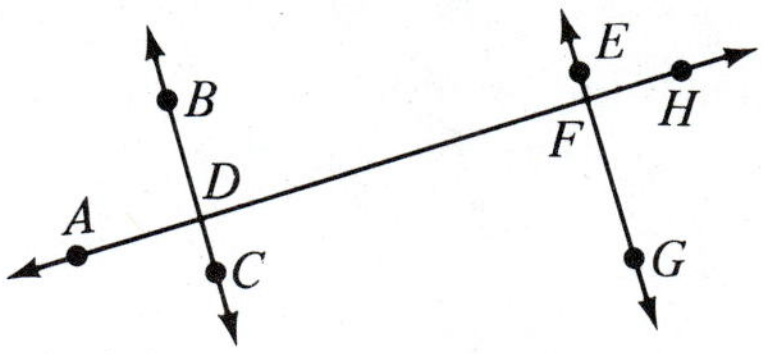

5. $\overleftrightarrow{BC} \parallel \overleftrightarrow{EG}$**; all ∠'s are ≅ (rt. ∠'s).**

6. Given: $\overrightarrow{BG}$ bisects $\angle ABE$, $\overrightarrow{EH}$ bisects $\angle BEF$, and $\overleftrightarrow{AC} \parallel \overleftrightarrow{DF}$.

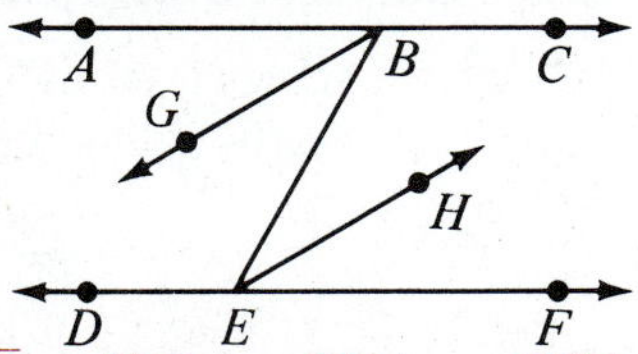

6. $\overrightarrow{BG} \parallel \overrightarrow{EH}$**;** $\angle ABG \cong \angle GBE \cong \angle BEH \cong \angle HEF$**,** $\angle ABE \cong \angle FEB$**,** $\angle EBC \cong \angle DEB$

7. Given: $l \parallel m$, and $\angle 1 \cong \angle 4$
   Prove: $t \parallel s$

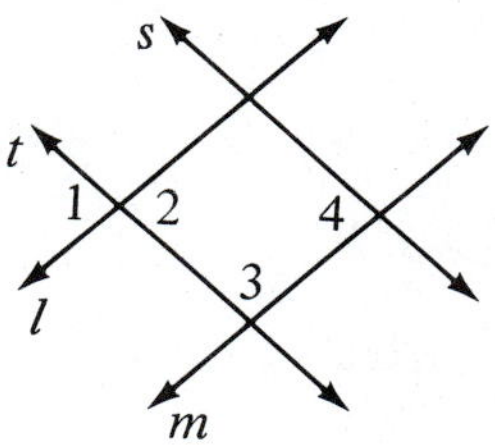

8. Given: $\overrightarrow{FD} \parallel \overrightarrow{CG}$, and $\angle F$ and $\angle C$ are supplements.
   Prove: $\overrightarrow{CA} \parallel \overrightarrow{FE}$

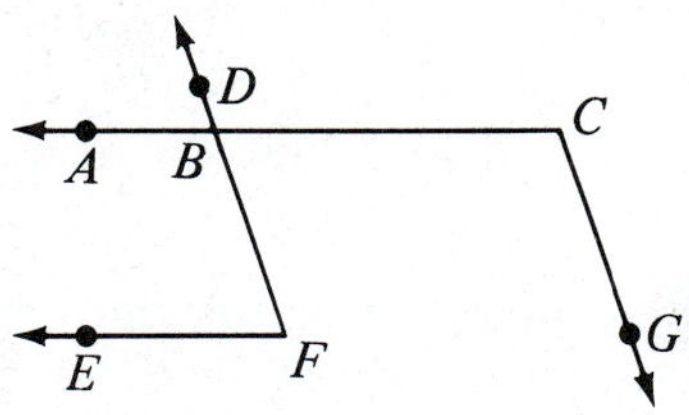

Based on the given information for each of the following, which lines *cannot* be parallel? (Ex. 9–14)

B

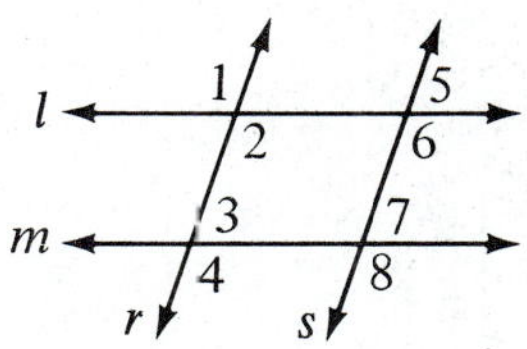

9. m ∠2 = 127, m ∠3 = 63
   l ∦ m
10. m ∠3 = 69, m ∠7 = 71
   r ∦ s
11. m ∠3 = 76, m ∠8 = 114
   r ∦ s
12. m ∠1 = 103, m ∠4 = 105
   l ∦ m
13. m ∠1 = 111, m ∠5 = 59
   r ∦ s
14. m ∠1 = 123, m ∠8 = 132
   l ∥ m or r ∥ s, but not both.

15. Given: $\overline{AB} \parallel \overline{DC}$, and $\angle ADC \cong \angle ABC$.
   Prove: $\overline{BC} \parallel \overline{AD}$

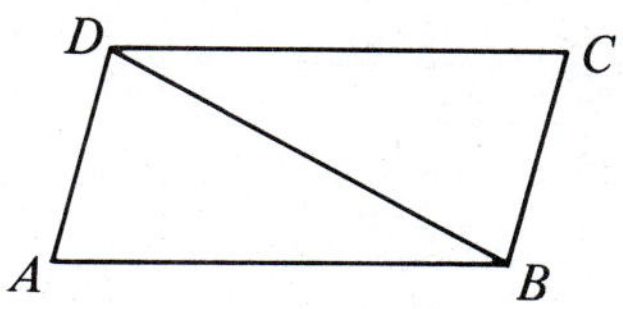

16. Given: $\angle 3 \cong \angle 4$, $\angle 2 \cong \angle 3$, and $\overrightarrow{FJ}$ bisects $\angle IFG$
   Prove: $\overleftrightarrow{EG} \parallel \overleftrightarrow{HK}$

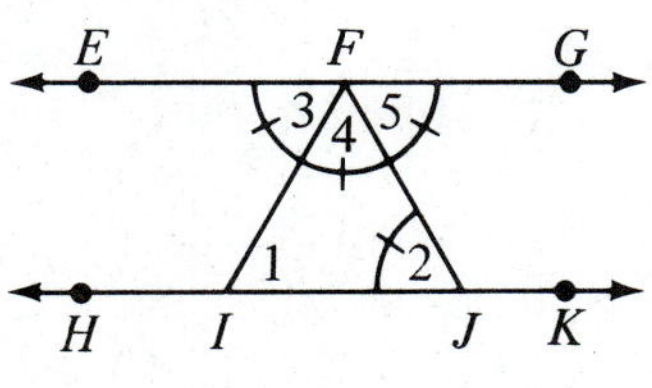

17. Prove Theorem 3.4.

Prove each of the following.

18. If two parallel lines are intersected by a transversal, then the bisectors of a pair of alternate interior angles are parallel.

19. If two parallel lines are intersected by a transversal, then the bisectors of a pair of corresponding angles are parallel.

20. If two lines are intersected by a transversal so that the bisectors of a pair of corresponding angles are parallel, then the lines are parallel.

21. If two lines are intersected by a transversal so that the bisectors of a pair of alternate interior angles are parallel, then the lines are parallel.

NON-ROUTINE PROBLEMS

Below is a scale drawing of a 3 ft-by-4 ft pool table. A ball is shot from corner A along a 45° path, as shown. The ball will eventually drop in the pocket at B.

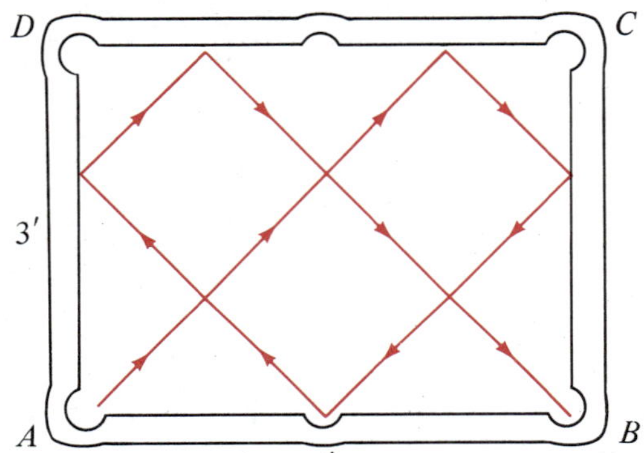

1. Suppose the ball is shot from corner B along a 45° path. Into which pocket will it eventually drop? **A**

Below are scale drawings of pool tables of two different sizes. Assume that a ball is shot from corner A along a 45° path. Into which pocket will it eventually drop? (You may use squared paper to help you decide.)

2. **B**

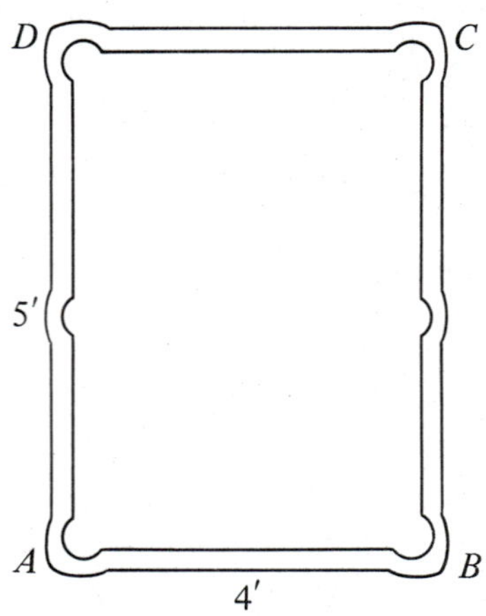

3. **D**

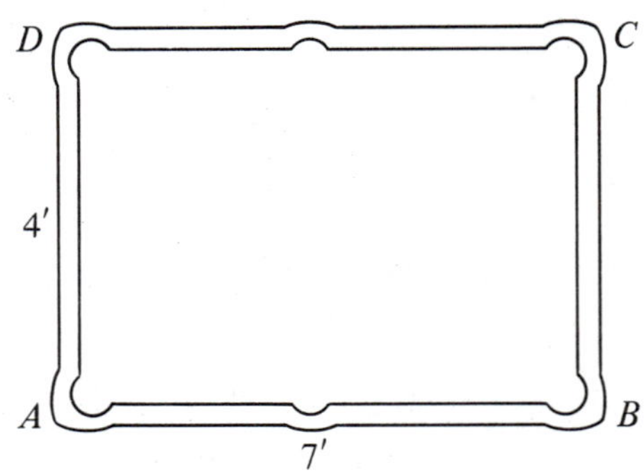

USING THE PARALLEL POSTULATE 3.5

Objective

To apply the parallel postulate

Is it possible for both $\overleftrightarrow{AP}$ and $\overleftrightarrow{BP}$ to be parallel to $\overleftrightarrow{CD}$?

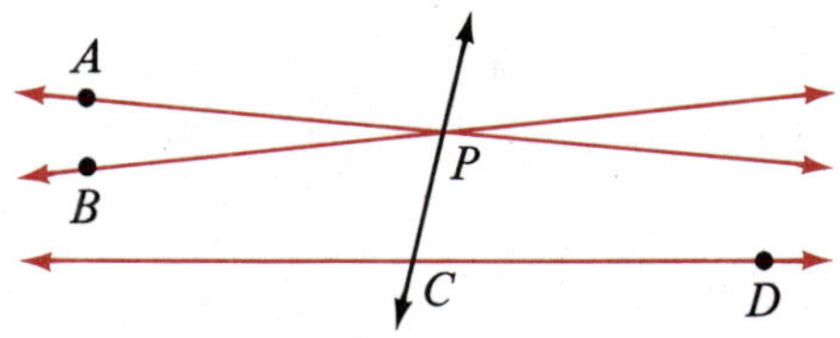

If $\overleftrightarrow{AP} \parallel \overleftrightarrow{CD}$, then $\angle APC \cong \angle PCD$, since parallel lines determine congruent alternate interior angles. If $\overleftrightarrow{BP} \parallel \overleftrightarrow{CD}$, then $\angle BPC \cong \angle PCD$ for the same reason. This means that $\angle APC \cong \angle BPC$. But since B is in the interior of $\angle APC$, $\angle APC$ cannot be congruent to $\angle BPC$. The assumption that $\overleftrightarrow{AP} \parallel \overleftrightarrow{CD}$ and $\overleftrightarrow{BP} \parallel \overleftrightarrow{CD}$ results in a contradiction. This leads to the following postulate.

Postulate 15 The Parallel Postulate

Through a point not on a line, there is exactly one line parallel to the given line.

Using the parallel postulate, you can determine if certain angle measures are possible or if lines can be parallel.

Example 1

Complete the statement with =, ≠, or "cannot tell."

If $l \parallel n$, and m $\angle y = 130$, then m $\angle z$ ______ 130.

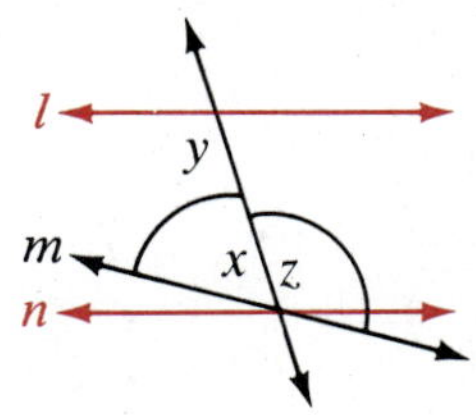

m $\angle z \neq 130$ ◄ $l \nparallel m$

Example 2

Complete the statement with $\parallel$, $\nparallel$, or "cannot tell."

If $\angle 1 \ncong \angle 3$, then v ______ r.

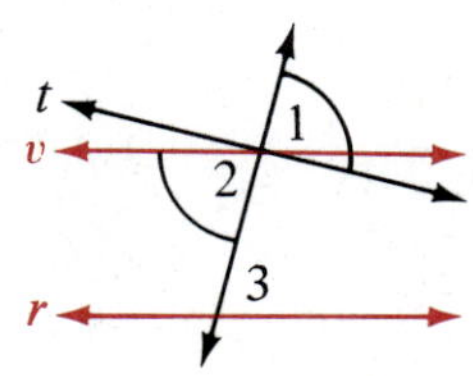

$\angle 2$ could be $\cong$ to $\angle 3$. ► cannot tell

Oral Exercises

Complete each statement with $\parallel$, $\nparallel$, or "cannot tell."

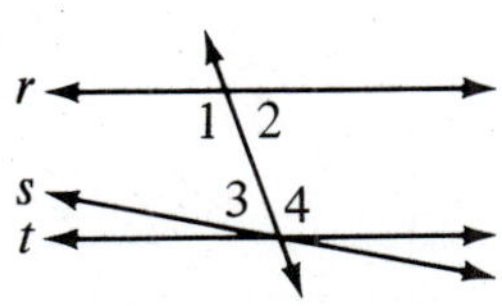

1. If $\angle 2 \cong \angle 4$, then r __ct.__ t.
2. If $\angle 1 \cong \angle 4$, then r __$\parallel$__ t.
3. If $r \parallel s$, then r __$\nparallel$__ t.
4. If $r \nparallel t$, then r __ct.__ s.
5. If $\angle 2 \cong \angle 3$, then r __$\parallel$__ s.
6. If $\angle 2 \ncong \angle 3$, then r __$\nparallel$__ s.

Written Exercises

Complete each statement with =, ≠, or "cannot tell."

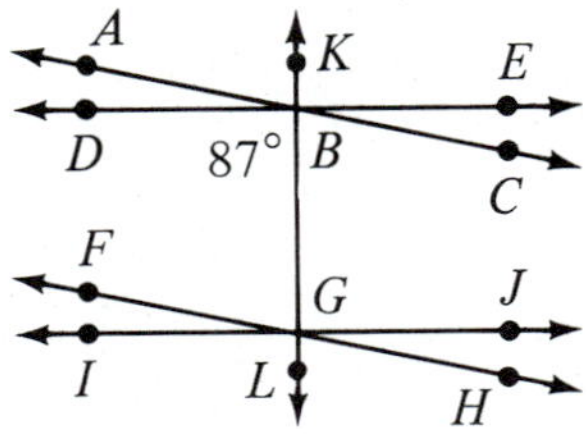

A **1.** If $\overleftrightarrow{AC} \parallel \overleftrightarrow{FH}$, and m $\angle DBL =$ 87, then m $\angle KGJ$ __ct__ 87.

2. If $\overleftrightarrow{DE} \parallel \overleftrightarrow{IJ}$, and m $\angle DBL =$ 87, then m $\angle KGJ$ __=__ 87.

3. If $\overleftrightarrow{AC} \parallel \overleftrightarrow{FH}$, $\overleftrightarrow{DE} \parallel \overleftrightarrow{IJ}$, and m $\angle DBL = 87$, then m $\angle FGL$ __≠__ 87.

4. If $\overleftrightarrow{AC} \parallel \overleftrightarrow{FH}$, $\overleftrightarrow{DE} \parallel \overleftrightarrow{IJ}$, and m $\angle DBL = 87$, then m $\angle FGK$ __≠__ 93.

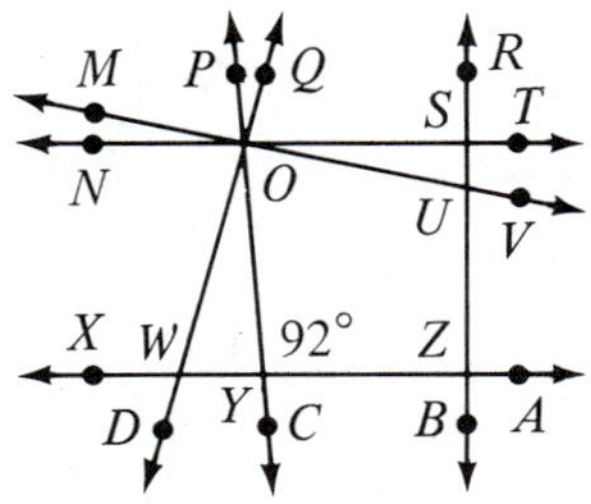

5. If $\overleftrightarrow{NT} \parallel \overleftrightarrow{WA}$, and m $\angle PYA =$ 92, then m $\angle SZA$ __ct__ 92.

6. If $\overleftrightarrow{PC} \parallel \overleftrightarrow{SB}$, and m $\angle PYA =$ 92, then m $\angle SZW$ __=__ 88.

7. If $\overleftrightarrow{NT} \parallel \overleftrightarrow{WA}$, $\overleftrightarrow{PC} \parallel \overleftrightarrow{SB}$, and m $\angle PYA = 92$, then m $\angle VUZ$ __≠__ 88.

8. If $\overleftrightarrow{MV} \parallel \overleftrightarrow{WA}$, $\overleftrightarrow{PC} \parallel \overleftrightarrow{SB}$, and m $\angle PYA = 92$, then m $\angle VUZ$ __=__ 88.

Complete each statement with ∥, ∦, or "cannot tell." (Ex. 9–14)

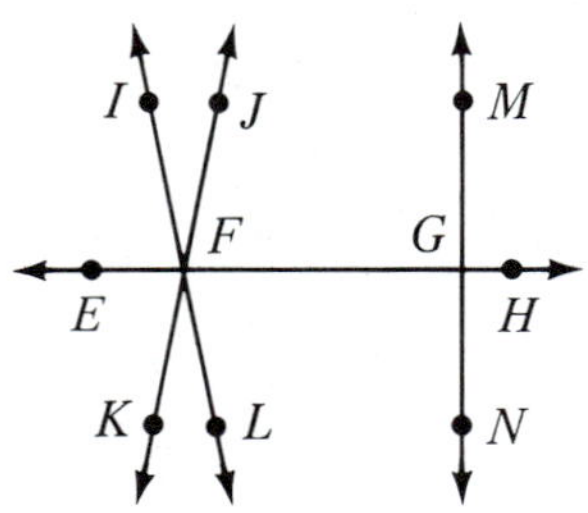

9. If $\overleftrightarrow{IL} \nparallel \overleftrightarrow{MN}$, then $\overleftrightarrow{JK}$ __ct__ $\overleftrightarrow{MN}$

10. If $\overleftrightarrow{MN} \parallel \overleftrightarrow{JK}$, then $\overleftrightarrow{IL}$ __∦__ $\overleftrightarrow{MN}$.

11. If $\angle IFH \cong \angle MGH$, then $\overleftrightarrow{IL}$ __∥__ $\overleftrightarrow{MN}$.

12. If $\angle JFH$ and $\angle EGN$ are supplements, then $\overleftrightarrow{JK}$ __ct__ $\overleftrightarrow{MN}$.

13. If $\angle JFE \cong \angle HGN$, then $\overleftrightarrow{JK}$ __∥__ $\overleftrightarrow{MN}$.

14. If $\overleftrightarrow{JK} \perp \overleftrightarrow{EH}$ and $\overleftrightarrow{MN} \perp \overleftrightarrow{EH}$, then $\overleftrightarrow{JK}$ __∥__ $\overleftrightarrow{MN}$.

C **15.** Prove: In a plane, if one of two parallel lines is intersected by a third line, then the second is also intersected by the third line.

16. Prove Postulate 13 using Postulate 15.

NON-ROUTINE PROBLEMS

Some figures can be divided into four parts, all having the same size and shape. One of these is shown. Can you divide the other figures?

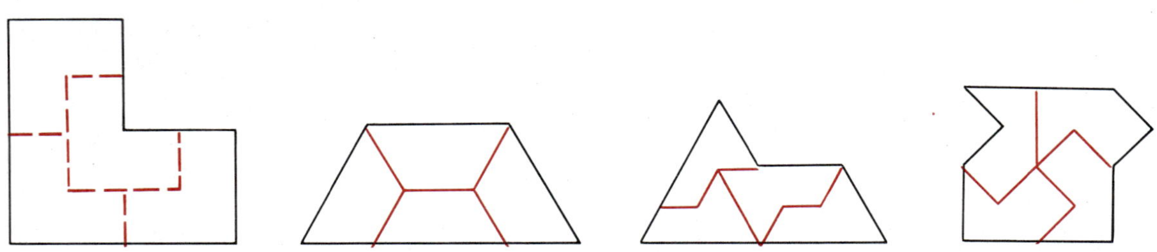

MEASURING ANGLES IN A TRIANGLE 3.6

Objective

To prove and apply the theorem about the sum of the angle measures of a triangle

You can use the properties of angles formed by parallel lines and a transversal to prove other important theorems.

Suppose that $\overleftrightarrow{CD}$ is parallel to the base $\overline{AB}$ of $\triangle ABC$. If m $\angle A = 61$ and m $\angle B = 46$, then the measure of the third angle of the triangle can be found.

$$m\angle 1 + m\angle 2 + m\angle 3 = 180$$
$$\downarrow \qquad\qquad \downarrow$$
$$m\angle A + m\angle 2 + m\angle B = 180$$
$$61 + m\angle 2 + 46 = 180$$

Thus, $m\angle 2 = 73$.

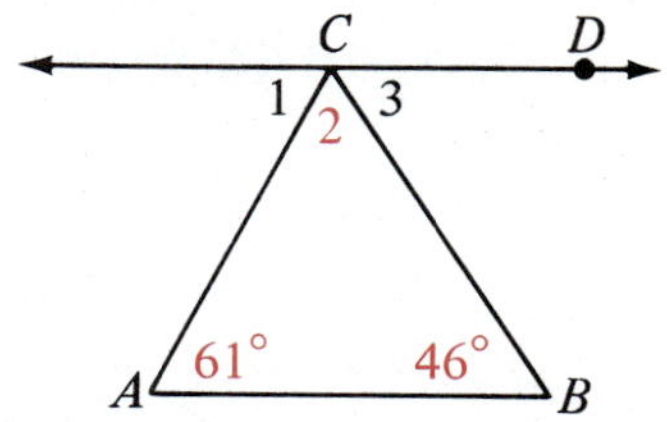

This suggests a proof of one of the most widely used theorems in geometry. The proof of this theorem requires a line which is not part of the original triangle. This line $\overleftrightarrow{DC}$ is called an **auxiliary line** and is indicated with a dashed line in the figure of the proof.

Theorem 3.5

The sum of the measures of the angles of a triangle is 180°.

Given: $\triangle ABC$
Prove: $m\angle x + m\angle y + m\angle z = 180$

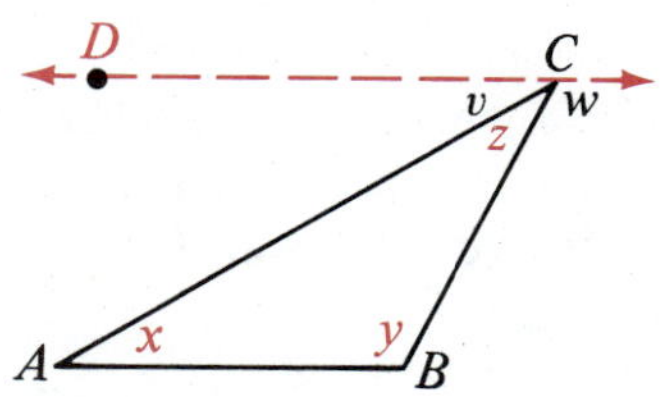

Proof

Statements	Reasons
1. $\triangle ABC$	1. Given
2. Draw $\overleftrightarrow{DC} \parallel \overline{AB}$.	2. Parallel postulate
3. $m\angle x = m\angle v$	3. Alt. int. $\angle$'s of $\parallel$ lines are $\cong$.
4. $m\angle y = m\angle w$	4. Alt. int. $\angle$'s of $\parallel$ lines are $\cong$.
5. $m\angle v + m\angle z + m\angle w = 180$	5. Sum of meas. of $\angle$'s with the same vertex on one side of a line = 180°.
6. $\therefore m\angle x + m\angle y + m\angle z = 180$	6. Substitution

---

You can use Theorem 3.5 to find unknown angle measures in a triangle.

Example 1

Find the measure of the third angle of the triangle.

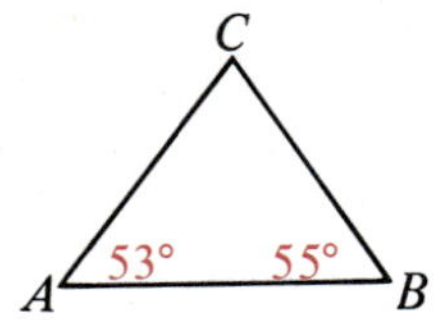

$$53 + 55 + m\angle C = 180$$
$$m\angle C = 180 - 108$$

Thus, $m\angle C = 72$.

Example 2

Given: $m\angle A = x + 2$, $m\angle B = 3x - 10$, and $m\angle C = 4x - 4$.
Find $m\angle B$ and $m\angle C$.

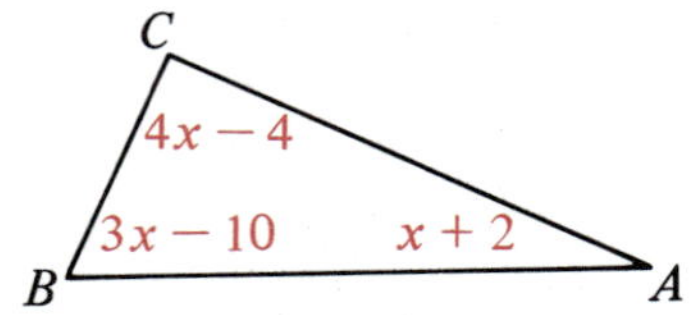

$(x + 2) + (3x - 10) + (4x - 4) = 180$ ◄ *The sum of the measures of all 3 angles is 180°.*

$$8x - 12 = 180$$
$$8x = 192$$
$$x = 24$$

Thus, $m\angle B = 3 \cdot 24 - 10$, or 62, and $m\angle C = 4 \cdot 24 - 4$, or 92.

You can prove the following theorem about the measures of corresponding angles in two triangles, using Theorem 3.5 in the proof.

Theorem 3.6

If two angles of one triangle are congruent to two angles of a second triangle, then the third angles are congruent.

Given: $\triangle ABC$, $\triangle DEF$,
$\angle A \cong \angle D$, and $\angle B \cong \angle E$.
Prove: $\angle C \cong \angle F$

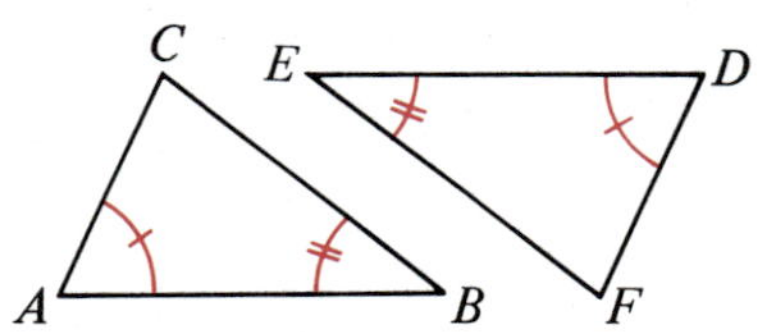

Proof

Statements	Reasons
1. $\triangle ABC$, $\triangle DEF$, $\angle A \cong \angle D$, $\angle B \cong \angle E$	1. Given
2. $m\angle A = m\angle D$, $m\angle B = m\angle E$	2. Def. of $\cong$ $\angle$'s
3. $m\angle A + m\angle B + m\angle C = 180$, $m\angle D + m\angle E + m\angle F = 180$	3. Sum of meas. of $\angle$'s of a $\triangle = 180°$.
4. $m\angle A + m\angle B + m\angle C = m\angle D + m\angle E + m\angle F$	4. Substitution property
5. $m\angle A + m\angle B = m\angle D + m\angle E$	5. Addition property of equality
6. $m\angle C = m\angle F$	6. Subtraction property of equality
7. $\therefore \angle C \cong \angle F$	7. Def. of $\cong$ $\angle$'s

Example 3 Given: $\angle A \cong \angle D$, $\angle G \cong \angle E$,
m $\angle GCA = 2x - 30$,
and m $\angle EBD = x + 5$.
Find m $\angle GCA$.

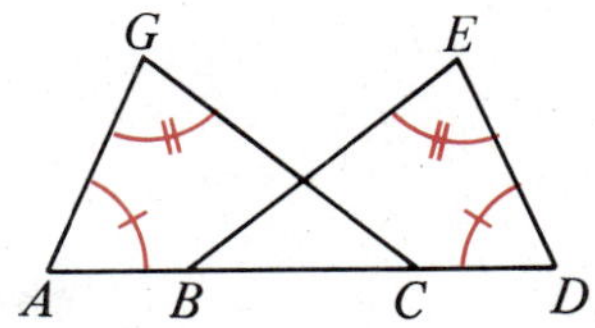

Use Th. 3.6. ► $\angle GCA \cong \angle EBD$

$$2x - 30 = x + 5$$
$$x = 35$$

Thus, m $\angle GCA = 2 \cdot 35 - 30$, or 40.

Oral Exercises

Find the measure of the third angle of $\triangle ABC$.

1. m $\angle A = 60$, m $\angle B = 60$, m $\angle C = ?$ **60**
2. m $\angle A = 45$, m $\angle B = 45$, m $\angle C = ?$ **90**
3. m $\angle A = 30$, m $\angle B = 60$, m $\angle C = ?$ **90**
4. m $\angle A = 100$, m $\angle B = 40$, m $\angle C = ?$ **40**

Find the measure of the third angle of each triangle.

5. **60°**

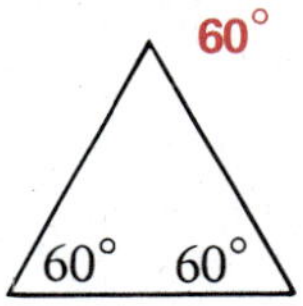

6. **45°**

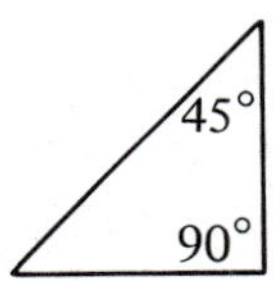

7. **55°**

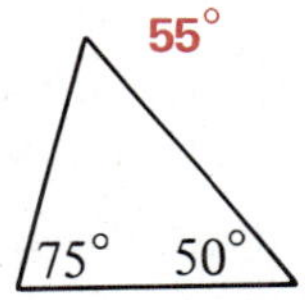

8. **120°**

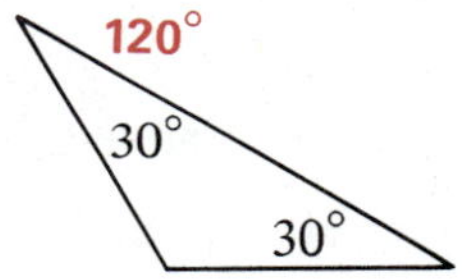

9. **70°**

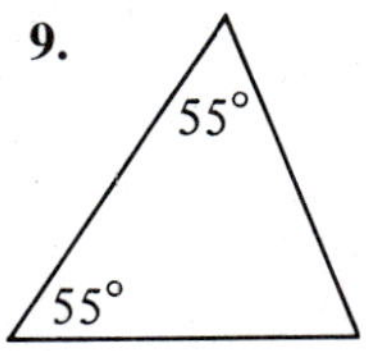

10. **30°**

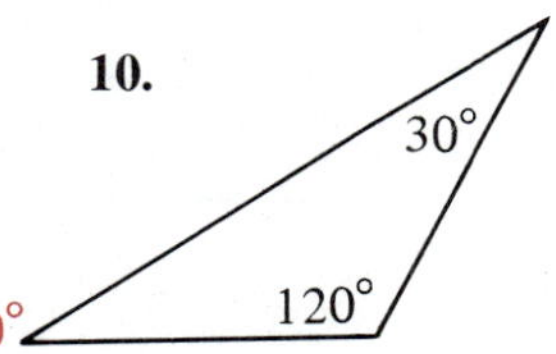

11. **20°**

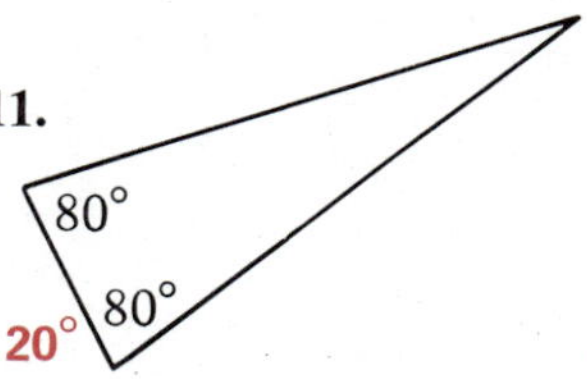

12. **30°**

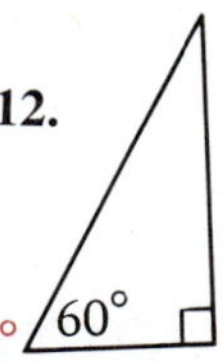

Written Exercises

Find the measure of each angle in $\triangle ABC$. (Ex. 1–4)

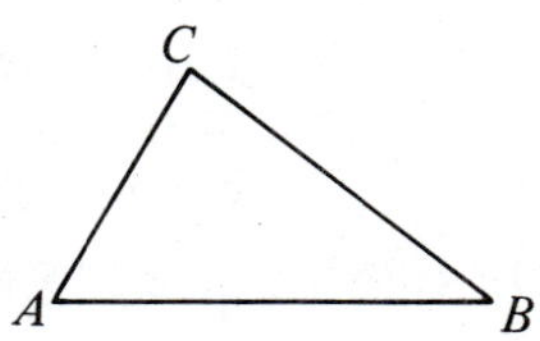

(A) **1.** m $\angle A = x$
m $\angle B = x$
m $\angle C = 2x$
45°, 45°, 90°

2. m $\angle A = 2x + 15$
m $\angle B = 3x - 5$
m $\angle C = 4x + 35$
45°, 40°, 95°

3. m $\angle A = x$
m $\angle B = x + 10$
m $\angle C = x + 20$
50°, 60°, 70°

4. m $\angle A = 2x + 28$
m $\angle B = x + 13$
m $\angle C = 4x - 15$
72°, 35°, 73°

5. Given: $\angle A \cong \angle D$, $\angle B \cong \angle E$,
m $\angle C = x$, and m $\angle F = 2x - 30$
Find m $\angle C$. **30**

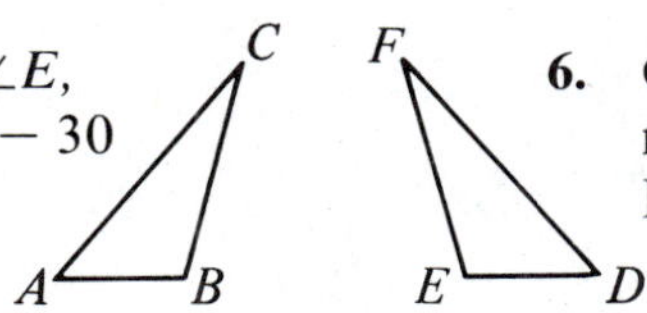

6. Given: $\angle A \cong \angle D$, $\angle C \cong \angle F$,
m $\angle B = 4y + 20$, and m $\angle E = y + 80$
Find m $\angle E$. **100**

7. Given: $\overline{BC} \perp \overline{CA}$
Prove: $m\angle A + m\angle B = 90$

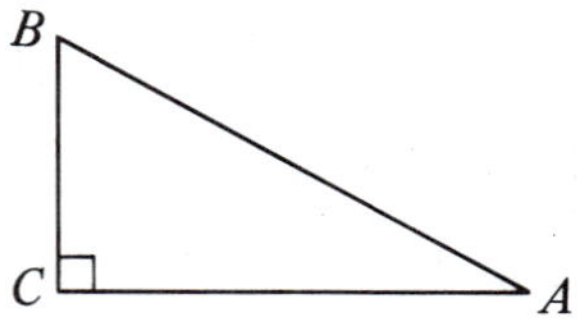

8. Given: A, C, and E are collinear.
Prove: $m\angle s = m\angle x + m\angle y$

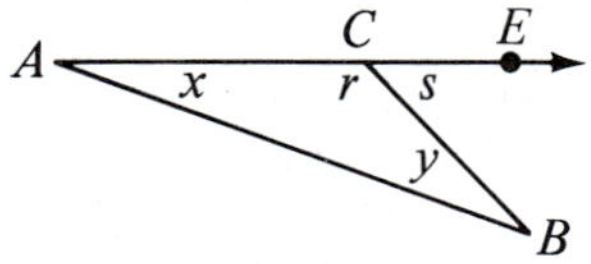

9. Given: $\angle D \cong \angle C$
Prove: $\angle A \cong \angle E$

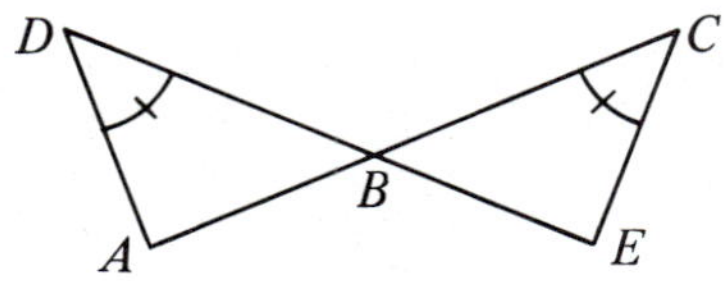

△B 10. Given: $\angle A \cong \angle B$ and $\overrightarrow{CD} \perp \overline{AB}$
Prove: $\overrightarrow{CD}$ bisects $\angle ACB$.

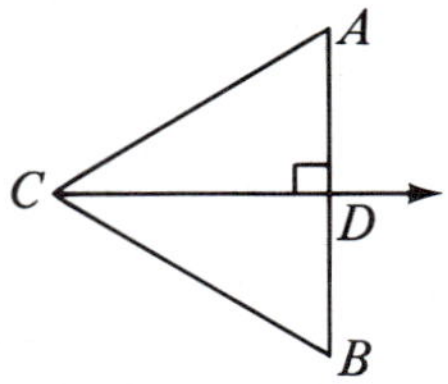

11. Given: $\angle CDE \cong \angle CED$ and $\angle A \cong \angle B$
Prove: $\overline{DE} \parallel \overline{AB}$

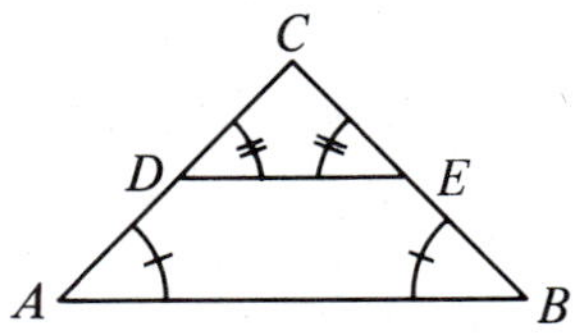

12. Given: $\overline{FD} \perp \overline{AB}$ and $\angle A \cong \angle B$
Prove: $\angle F \cong \angle CEF$

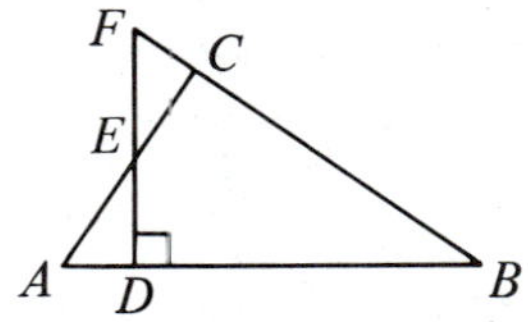

Use the information given to find each indicated angle measure. If not enough information is given, write "too little information."

Given: $m\angle BEG = 61$ and $m\angle DFB = 52$

13. Find $m\angle EBF$. 67
14. Find $m\angle BFG$. 128
15. Find $m\angle ABE$. too little information

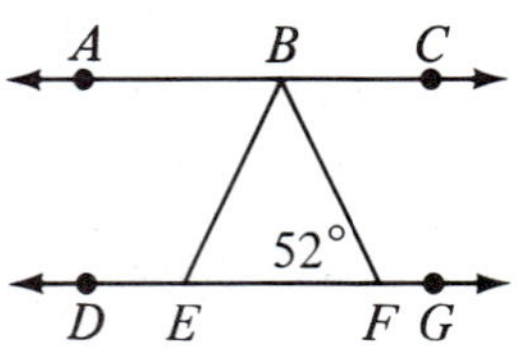

Given: $\overleftrightarrow{AB} \parallel \overleftrightarrow{DE}$, $\overline{BD} \parallel \overline{EC}$, $\overline{EB} \perp \overline{AB}$, and $m\angle BDE = 76$

16. Find $m\angle DBE$. 14
17. Find $m\angle ACE$. 76
18. Find $m\angle ABC$. 180

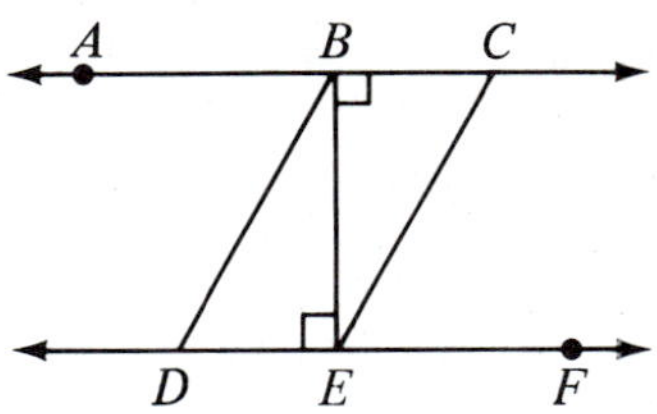

For each figure, determine whether or not contradictory information is given. If so, what information is contradictory?

△C 19. Given: $\overleftrightarrow{AC} \nparallel \overleftrightarrow{FJ}$, $m\angle ABD = 27$, $m\angle HBK = 105$, and $m\angle EGJ = 48$
$\angle ABH \cong \angle EGJ$ contradicts $\overleftrightarrow{AC} \nparallel \overleftrightarrow{FJ}$.

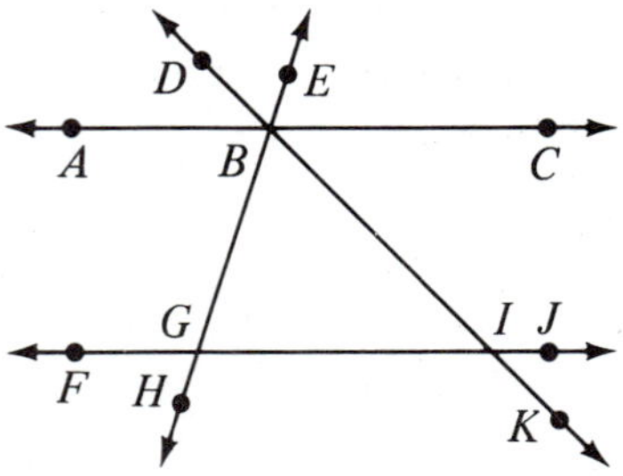

20. Given: $\overleftrightarrow{AE} \parallel \overleftrightarrow{PM}$, $\overleftrightarrow{CL} \perp \overleftrightarrow{AE}$, $\angle CKM$ is a supplement of $\angle CBE$, $m\angle PKC = 92$, and $m\angle QGC + m\angle ADQ = 180$
$m\angle PKC = 92$ contradicts $\overleftrightarrow{AE} \parallel \overleftrightarrow{PM}$ and $\overleftrightarrow{CL} \perp \overleftrightarrow{AE}$.

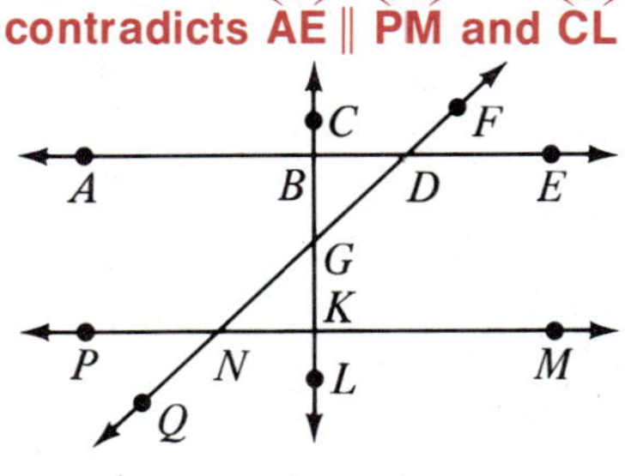

INDIRECT PROOF 3.7

Objective **To apply the method of indirect proof**

The proofs of almost all of the theorems presented so far have been what are called *direct* proofs. Sometimes, however, a theorem can more easily be proved by an **indirect proof**. In an indirect proof, two possibilities exist. To prove that possibility A is true, you must show that possibility B cannot be true. This leaves possibility A as the only true one.

The following proof of Theorem 3.7 uses an indirect proof. Notice that the proof is written in paragraph form, rather than in the usual two-column format. This often is a more natural approach.

Theorem 3.7 In a plane, two lines parallel to the same line are parallel to each other.

Given: $\overleftrightarrow{AB} \parallel \overleftrightarrow{EF}$ and $\overleftrightarrow{CD} \parallel \overleftrightarrow{EF}$; $\overleftrightarrow{AB}$, $\overleftrightarrow{CD}$, and $\overleftrightarrow{EF}$ are coplanar.
Prove: $\overleftrightarrow{AB} \parallel \overleftrightarrow{CD}$

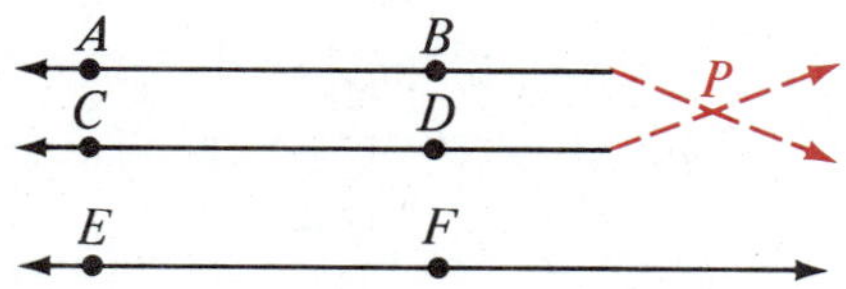

Proof

Suppose that $\overleftrightarrow{AB}$ and $\overleftrightarrow{CD}$ are not parallel. Then they intersect at a point P. But this means that there are two lines through P parallel to $\overleftrightarrow{EF}$. This contradicts postulate 15. Therefore, $\overleftrightarrow{AB}$ and $\overleftrightarrow{CD}$ must be parallel.

Example

Give an indirect proof.
A triangle cannot have two right angles.

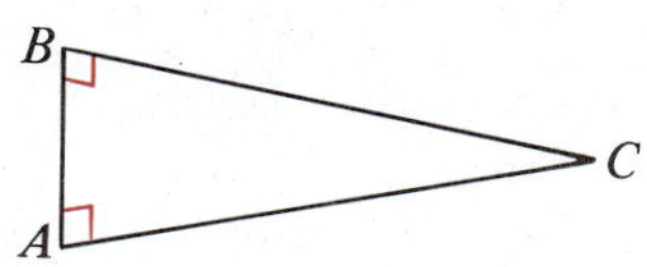

Given: $\triangle ABC$
Prove: $\angle A$ and $\angle B$ cannot both be right angles.

Suppose that $\angle A$ and $\angle B$ are both right angles. Then $m\angle A + m\angle B = 180$ and $m\angle A + m\angle B + m\angle C > 180$. This contradicts Theorem 3.5 which states that the sum of the angle measures of a triangle is 180°.
Therefore, $\angle A$ and $\angle B$ cannot both be right angles.

Written Exercises

Give an indirect proof of each of the following. (Ex. 1–14)

A

1. Given: $l \nparallel m$
   Prove: $\angle 1 \ncong \angle 2$

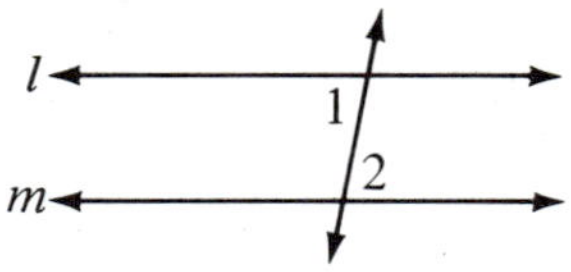

2. Given: $\angle 1 \ncong \angle 2$
   Prove: $l \nparallel m$

3. Given: $\angle 3 \ncong \angle 4$
   Prove: $t \not\perp s$

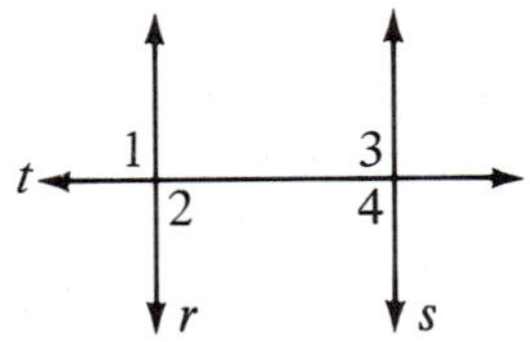

4. Given: $t \not\perp s$
   Prove: $\angle 3 \ncong \angle 4$

5. Given: $r \perp t$, $s \not\perp t$
   Prove: $r \nparallel s$

6. Given: $r \perp t$, $r \nparallel s$
   Prove: $s \not\perp t$

7. Given: $m\angle 1 + m\angle 2 =$ 176
   Prove: $l \nparallel m$

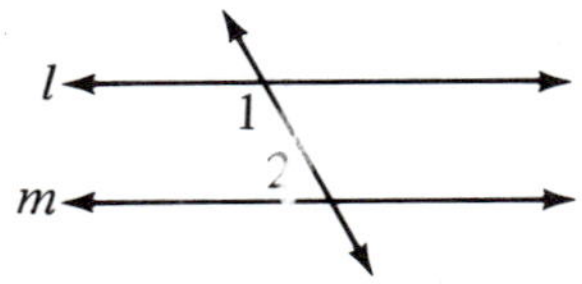

8. Given: $l \nparallel m$
   Prove: $m\angle 1 + m\angle 2 \neq$ 180

9. A triangle cannot have more than one obtuse angle.
10. A triangle cannot have both a right angle and an obtuse angle.
11. A triangle cannot have two parallel sides.
12. Skew lines do not intersect.

B

13. Given: $r \parallel s$, $s \nparallel t$
    Prove: $r \nparallel t$

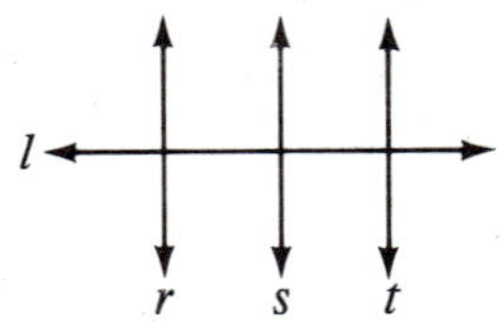

14. Given: $r \parallel s$, $s \perp l$, $t \perp l$
    Prove: $r \nparallel l$

C

15. Give an indirect proof of Theorem 3.6.

LOGIC REVIEW

Rewrite in If . . ., then . . . form.

1. Parallel lines do not intersect.
   If lines are parallel, then they do not intersect.

2. All angles have two rays.
   If it is an angle, then it has two rays.

NON-ROUTINE PROBLEMS

An architect is designing a floor plan as shown with 13 wall sections. By rearranging the rooms to form a square, only twelve wall sections are needed. Design some floor plans including one with five rooms, one with six rooms, one with seven rooms, and one with eight rooms. In each case, each room must share at least one wall section with another room. For each plan, what is the least number and the greatest number of wall sections needed?

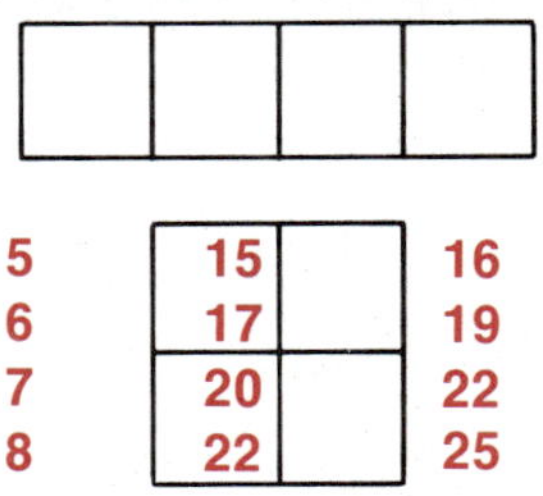

5 15 16
6 17 19
7 20 22
8 22 25

CONSTRUCTING PARALLELS 3.8

Objective **To construct a line parallel to a given line**

You learned earlier how to construct an angle congruent to a given angle. A similar construction can now be used to construct a line parallel to a given line, using congruent corresponding angles.

Construction A line parallel to a given line through a point not on the given line.

Given: line l and point P
Construct: $\overleftrightarrow{PT} \parallel l$

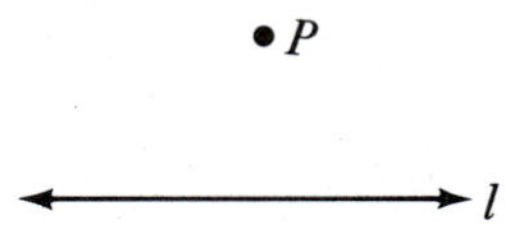

Draw $\overleftrightarrow{PQ}$ through any point Q on line l.

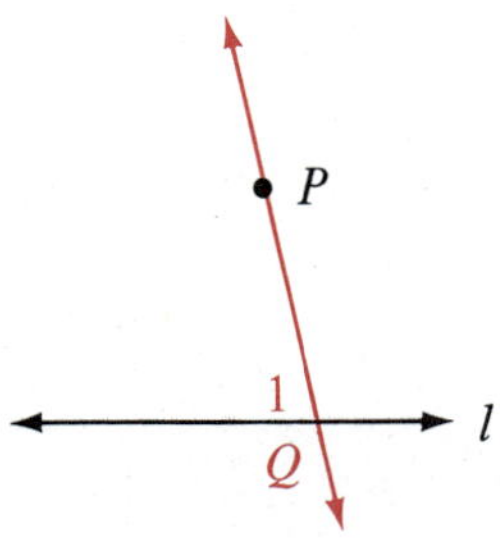

Construct a corresponding angle at P congruent to $\angle 1$.

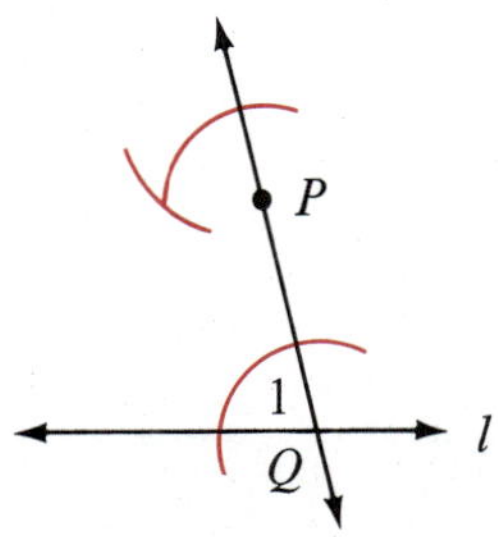

Draw $\overleftrightarrow{PT}$.

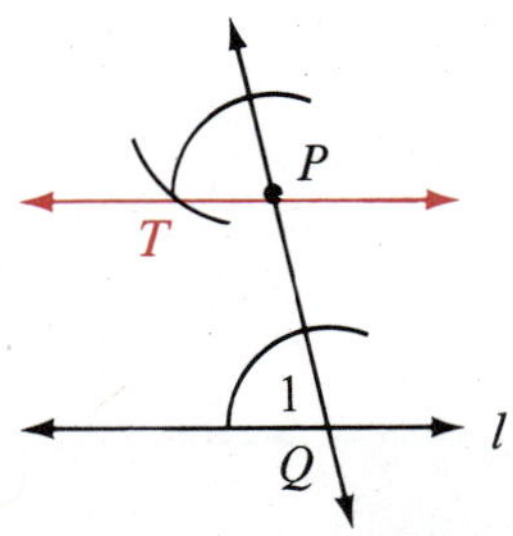

Conclusion: $\overleftrightarrow{PT} \parallel l$

Since both congruent alternate interior angles and congruent corresponding angles assure that two lines will be parallel, you may use either of these to construct parallel lines.

Example 1 Through A, construct a line parallel to $\overline{BC}$ by constructing. congruent alternate interior angles.

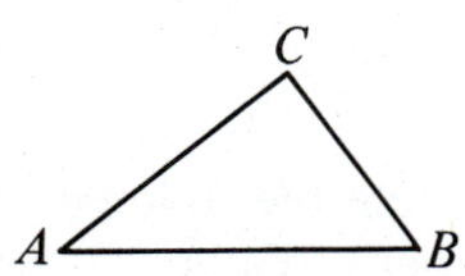

Use $\overline{AC}$ as a transversal. ►

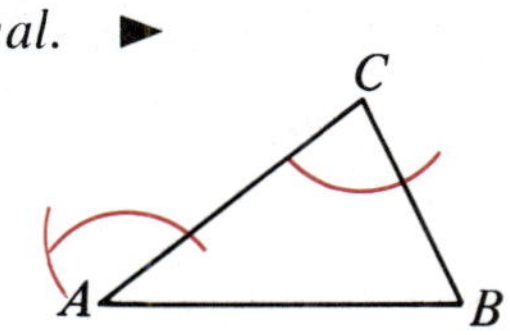

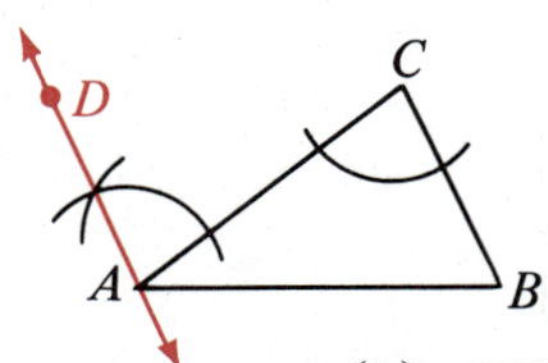

Conclusion: $\overleftrightarrow{AD} \parallel \overline{BC}$

Written Exercises

A **1.** Construct a line through A parallel to l. **See Constr., p. 93.**

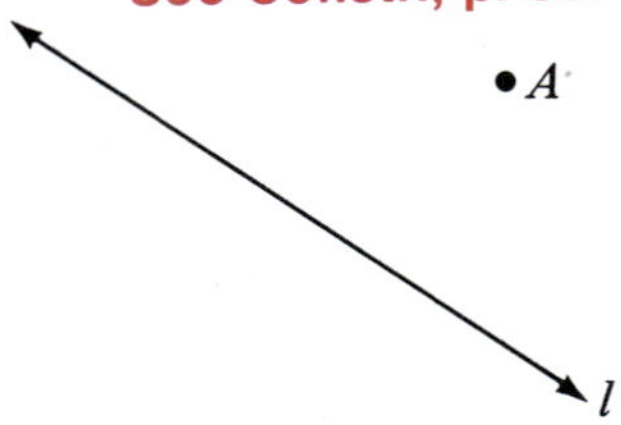

2. Construct a line through C parallel to $\overline{BD}$. **See Ex. 1, p. 93.**

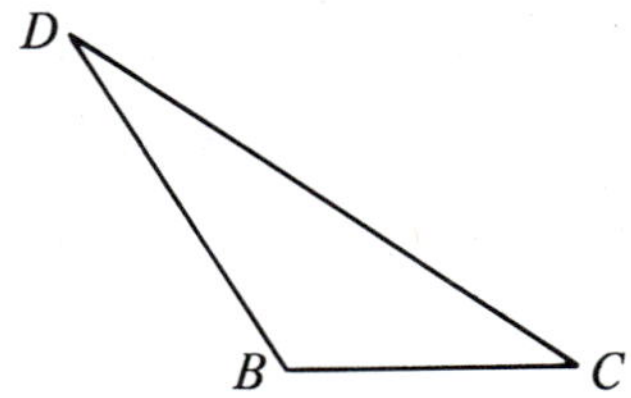

3. Construct a line through E parallel to $\overleftrightarrow{FG}$. **See Ex. 1, p. 93.**

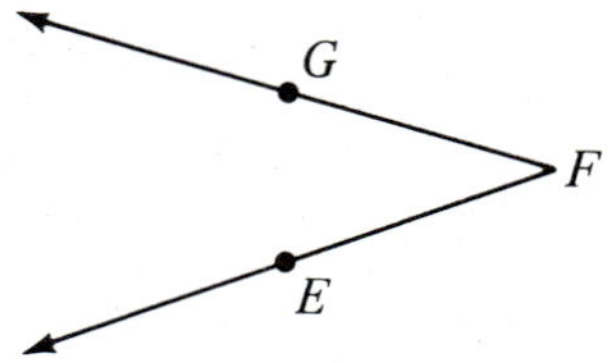

B **4.** Construct $\overleftrightarrow{JK} \parallel \overline{HI}$ and $\overleftrightarrow{HL} \parallel \overline{JI}$. **See Ex. 1, p. 93.**

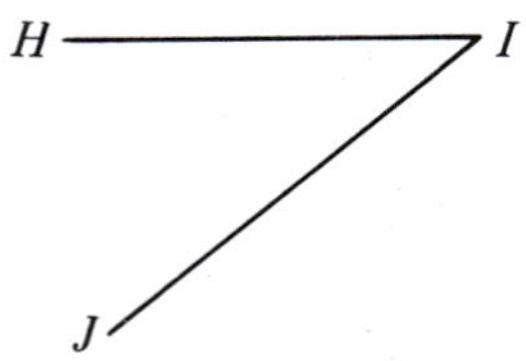

5. Construct M, the midpoint of $\overline{NO}$. Through M, construct $\overleftrightarrow{MQ} \parallel \overline{NP}$ and $\overleftrightarrow{MR} \parallel \overline{OP}$.

See Constr., p. 10 and Constr., p. 93.

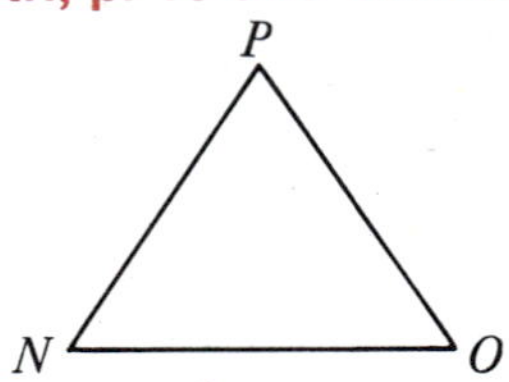

6. Construct $\triangle VWX$ such that $\overline{VW} \parallel \overline{ST}$, $\overline{WX} \parallel \overline{TU}$ and $\overline{XV} \parallel \overline{US}$.

See Constr., p. 93.

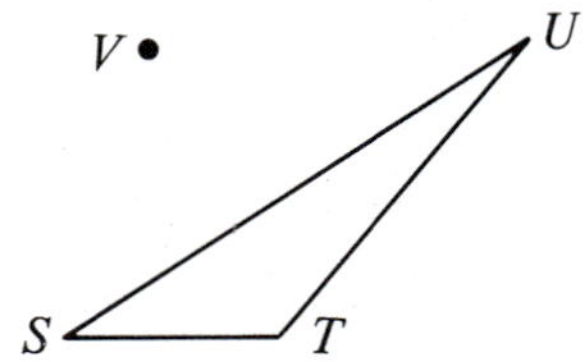

NON-ROUTINE PROBLEMS

A pie can be cut into two pieces with one straight cut. Two straight cuts will form a minimum of three pieces and a maximum of four pieces. Three straight cuts will form a minimum of four pieces and a maximum of seven pieces. How many pieces will there be with four cuts? with five cuts? with n cuts?

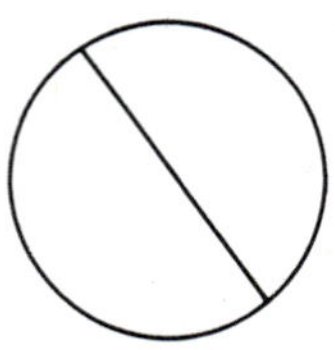

one cut

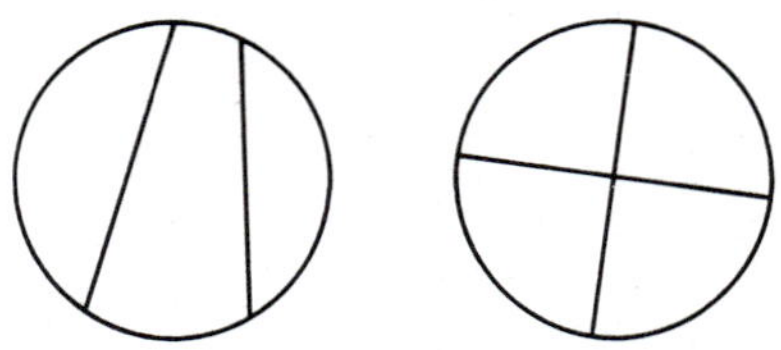

two cuts

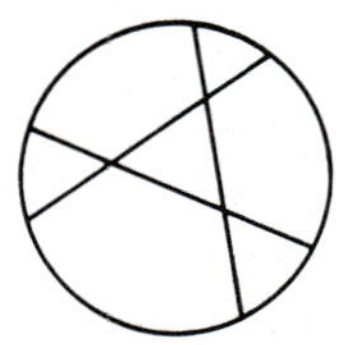

three cuts

Complete the following:

Number of Cuts	Smallest Number of Pieces	Largest Number of Pieces
1	2	2
2	3	4
3	**4**	7
4	**5**	**11**
5	**6**	**16**
.		
n	**n + 1**	$\frac{n(n+1)}{2} + 1$

PARALLEL PLANES (optional) 3.9

Objectives **To identify true statements about parallel lines and planes**
To prove statements about parallel lines and planes

You learned that two given lines can be parallel, intersecting, or skew. Two planes, on the other hand, must be either intersecting or parallel.

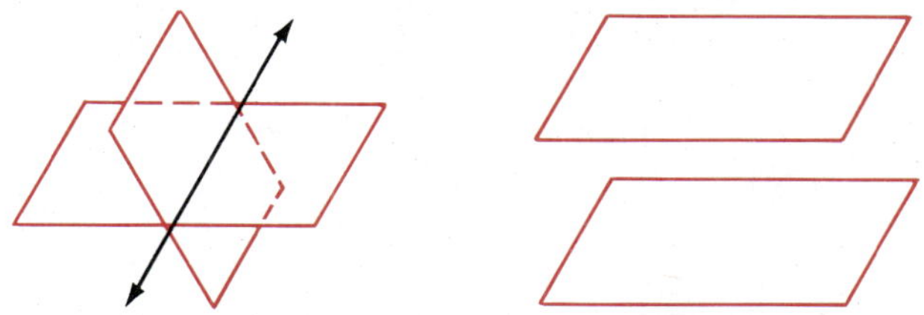

intersecting planes parallel planes

From Postulate 4, you learned that if two planes intersect, then their intersection must be exactly one line. If they do not intersect, then the planes are parallel.

Definition: Parallel planes *Parallel planes* are planes which do not intersect.

There are three possible intersections of a line and a plane.

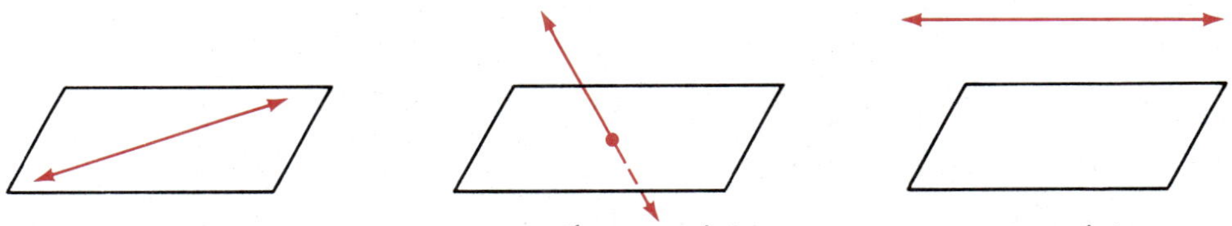

the line itself exactly one point no points

If a line and a plane have no points in common, they do not intersect. In this case, the line and the plane are said to be *parallel.*

Definition: Parallel line and plane A line and a plane are *parallel* if they do not intersect.

Segments or rays can also be parallel to a plane if the line that contains them is parallel to the plane.

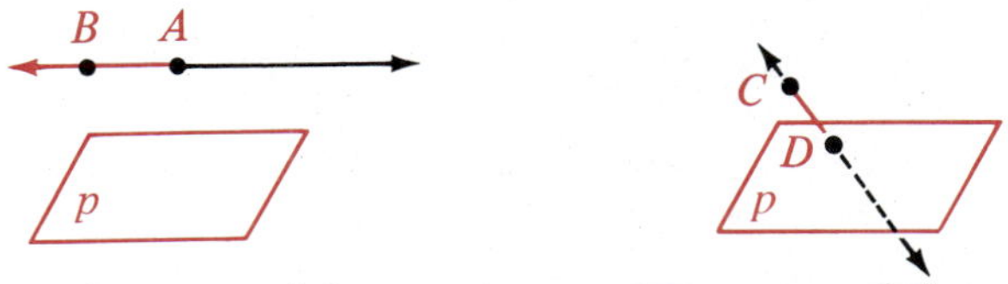

$\overrightarrow{AB} \parallel p$ since $\overleftrightarrow{AB} \parallel p$ $\overline{CD} \nparallel p$ since $\overleftrightarrow{CD} \nparallel p$

Suppose that two parallel planes are intersected by a third plane. The following theorem shows that the intersection consists of two parallel lines. An indirect proof is given.

Theorem 3.8

If two parallel planes are intersected by a third plane, then the lines of intersection are parallel.

Given: plane $p \parallel$ plane q, p intersects r in l, and q intersects r in m.
Prove: $l \parallel m$

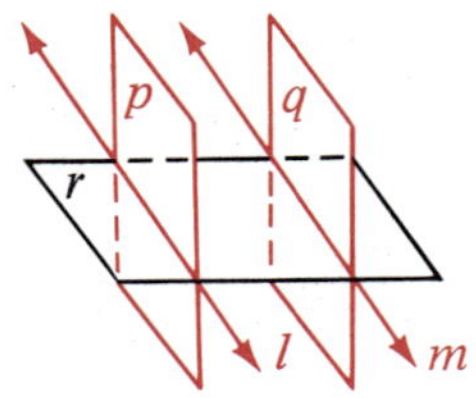

ANALYSIS
Use an indirect proof. Show that l and m are coplanar and cannot intersect.

Proof

I. l and m are coplanar since both are in plane r.

II. Either l and m intersect at some point, call it A, or they do not intersect.
Suppose that they do intersect.
Since A is on l, then A is in plane p.
Since A is on m, then A is in plane q.
If A is in both planes p and q, then p and q are not parallel.
But p and q were given to be parallel.
Thus, l and m do not intersect.

$\therefore$ $l \parallel m$, by the definition of parallel lines.

You may be able to show that some statements about parallels are not true by finding a **counterexample.** A counterexample is an example that contradicts a predicted result.

Example 1

True or false? Draw a picture to support your answer.
If two lines are parallel, then every plane containing one of the lines is parallel to every plane containing the other line.

False: $l \parallel m$, but $p \nparallel q$.

The diagram is a counterexample. ►

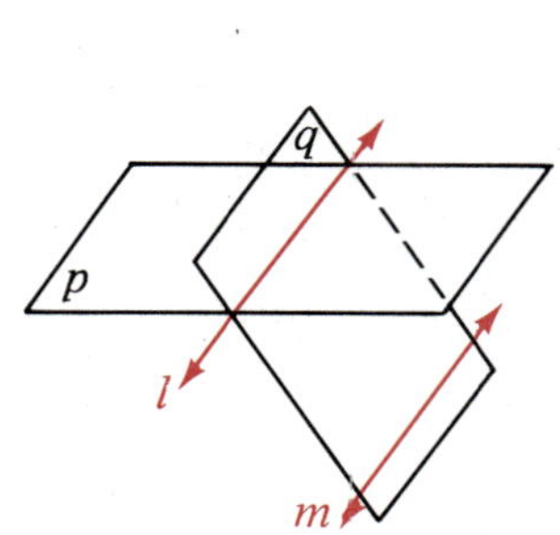

Written Exercises

True or false? Draw a picture to support your answer.

A 1. Two planes parallel to the same line are parallel to each other. **F**

2. Two lines parallel to the same plane are parallel to each other. **F**

3. Two planes perpendicular to the same line are parallel to each other. **T**

4. Two lines perpendicular to the same plane are parallel to each other. **T**

5. Two planes parallel to the same plane are parallel to each other. **T**

6. Two planes perpendicular to the same plane are parallel to each other. **F**

7. A plane parallel to one of two intersecting planes intersects the other plane. **T**

8. Two lines in space parallel to a third line are parallel to each other. **T**

Find a counterexample of each of the following statements. Draw a picture to illustrate. (Ex. 9–12)

B 9. A plane and a line parallel to the same line are parallel to each other.

10. A line parallel to a plane is parallel to a line contained in the plane.

11. A line in one of two parallel planes is parallel to a line in the second plane.

12. A line parallel to a second line is parallel to a plane containing the second line.

13. Given: plane $p \parallel$ plane q, and line l on q
Prove: $l \parallel p$

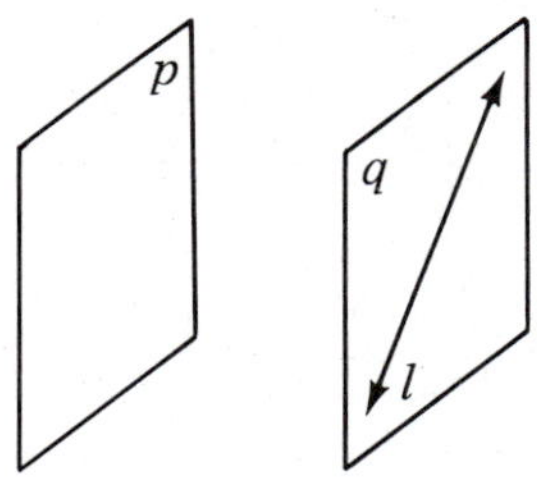

14. Given: line $l \parallel$ plane q, l on plane p, and p intersects q at line t.
Prove: $l \parallel t$

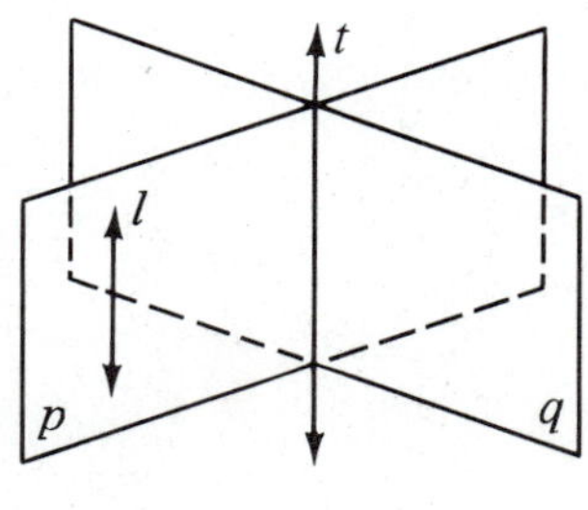

Give an indirect proof of each of the following. (Ex. 15–18)

C 15. If a line intersects a plane in exactly one point, then it is not parallel to any line contained in the plane.

16. There is exactly one plane that contains a given point parallel to a second plane not containing the point.

17. If a plane intersects one of two given parallel planes, then it intersects the second of the two planes.

18. If two lines in two different planes intersect, then the planes that contain the lines intersect.

19. Skew lines were defined. Is there a need to define skew planes? Why or why not?
No. Planes are parallel or they intersect.

APPLICATIONS

Read → Plan → Solve → Interpret

1. Two special rulers with protractors on them can be connected to use in drawing parallel lines. How would the different angle measures have to be adjusted to make sure that the rulers are parallel?

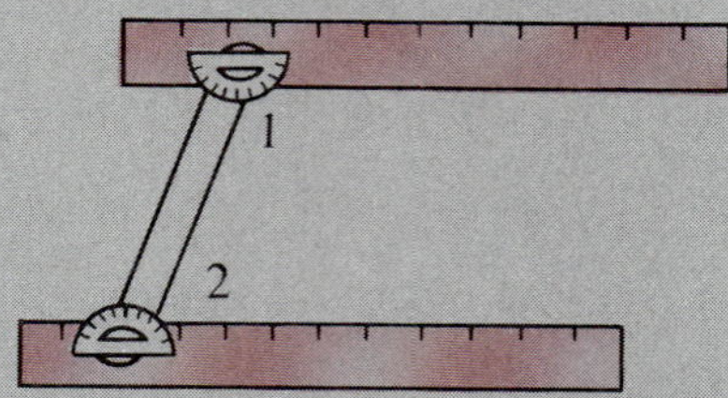

2. A carpenter must cut a board to support the treads in a stairway. What must be measured to guarantee that the treads are all parallel to each other?

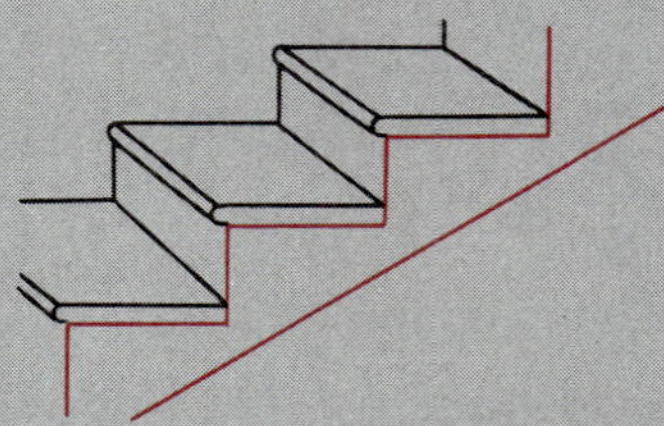

3. A plumb bob hangs so that it lies along a line through the center of the earth. Such lines are called *vertical*. What angle does a vertical line make with the earth's surface? Are all vertical lines parallel? Explain.

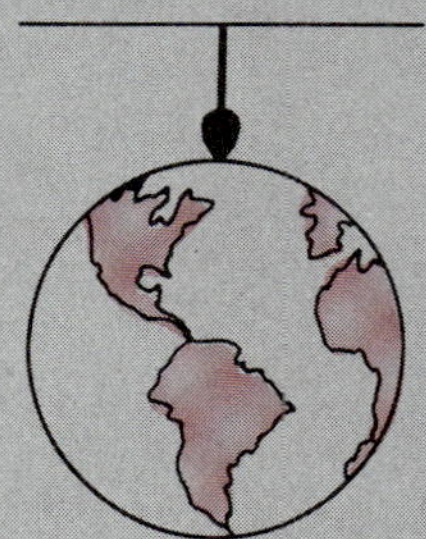

CHAPTER THREE REVIEW

A exercises are 1–4, 7, 8, 11–15 B exercises are 5, 6, 9, 10. C exercises are starred.

1. In the figure below, $l \parallel m$ and $m\angle 3 = 60$. [3.3]
   Find $m\angle 9$ and $m\angle 8$. **60, 120**
2. In the figure below, $l \parallel m$, $m\angle 2 = 145$, and $\angle 4 \cong \angle 6$. $72\frac{1}{2}$
   Find $m\angle 5$.

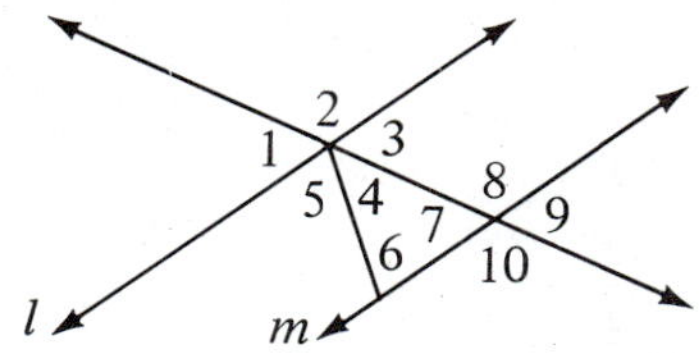

3. In the figure above, $l \parallel m$, $m\angle 1 = 5x - 5$, and $m\angle 7 = 4x + 4$.
   Find $m\angle 1$. **40**
4. In the figure above, $l \parallel m$, $m\angle 3 = 6x + 3$, and $m\angle 10 = 10x - 15$. Find $m\angle 8$. **105**

Based only on the given information, which lines in the figure are parallel? [3.4]

5. $\angle KIJ$ and $\angle JLK$ are supplementary. **none**
6. $m\angle JIK = 120$, $m\angle IKL = 60$, $m\angle IJG = 120$. $\overleftrightarrow{AB} \parallel \overleftrightarrow{CD}$ and $\overleftrightarrow{EF} \parallel \overleftrightarrow{GH}$

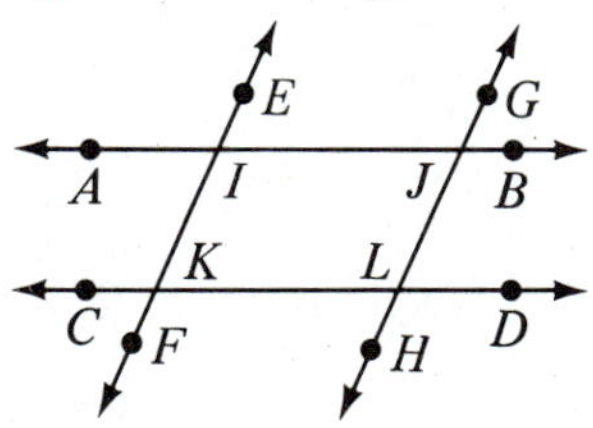

Write the converse of each statement. (Ex. 7–8)

7. If lines are parallel, then they do not intersect. [3.4] **If two lines do not intersect then they are parallel.**
8. If two lines are non-coplanar, then they are skew. **If two lines are skew, then they are non-coplanar.**
9. Draw a line and a point not on the line. Construct a line through the point parallel to the given line using congruent corresponding angles. [3.8]
10. Repeat the construction in Exercise 9, using congruent alternate interior angles.

Vocabulary

alternate exterior angle [3.2], alternate exterior angle [3.2], auxiliary line [3.6], converse [3.4], corresponding angles [3.2], indirect proof [3.7], parallel lines [3.1], skew lines [3.1], transversal [3.2]

True or false?

11. If two lines intersect so that vertical angles are congruent, then the lines are parallel. [3.1] **F**
12. If two parallel lines are intersected by a transversal, then interior angles on the same side of the transversal are complementary. [3.3] **F**

Complete each statement with $\parallel$, $\nparallel$, or "cannot tell." (Ex. 13–14) [3.5]

13. If $\overleftrightarrow{AD} \parallel \overleftrightarrow{EG}$, then $\overleftrightarrow{AD}$ ___$\nparallel$___ $\overleftrightarrow{HI}$
14. If $\overleftrightarrow{AD} \perp \overleftrightarrow{CJ}$ and $\overleftrightarrow{EG} \perp \overleftrightarrow{CJ}$, then $\overleftrightarrow{AD}$ ___$\parallel$___ $\overleftrightarrow{EG}$.

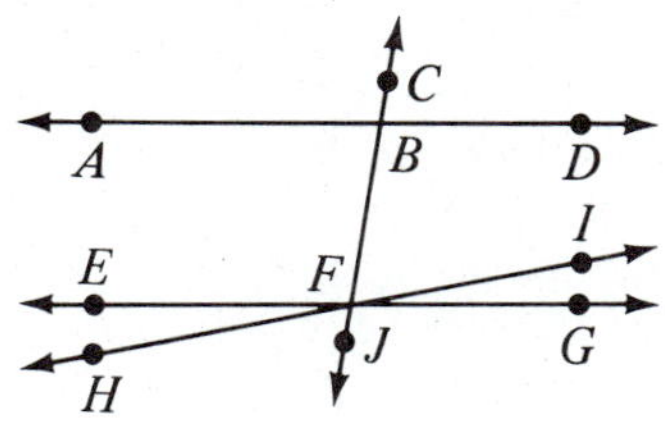

15. Give an indirect proof: A triangle cannot have both a right angle and an obtuse angle. [3.7]

*16. Given: $\angle 2 \cong \angle 5$ and $\angle 1 \cong \angle 4$
   Prove: $\overrightarrow{PL} \parallel \overrightarrow{ON}$

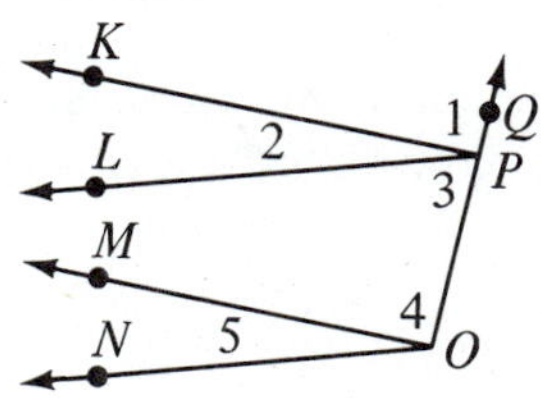

CHAPTER THREE TEST

A exercises are 1–4, 7–15. B exercises are 5, 6. C exercises are starred.

1. In the figure below, $\overleftrightarrow{AC} \parallel \overleftrightarrow{FI}$, m $\angle FGE = 105$, and m $\angle HBC = 68$.
   Find m $\angle JBK$ and m $\angle GHB$.
   m ∠JBK = 37, m ∠GHB = 68

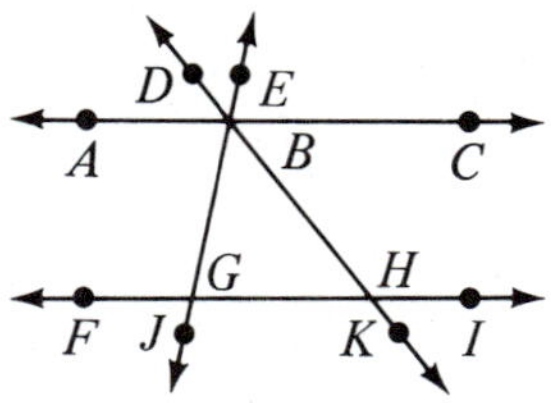

2. In the figure above, $\overleftrightarrow{AC} \parallel \overleftrightarrow{FI}$, $\overleftrightarrow{BK}$ bisects $\angle CBJ$, and m $\angle BGI = 80$.
   Find m $\angle CBK$ and m $\angle JBK$. **50, 50**

3. In the figure below, $\overleftrightarrow{AB} \parallel \overleftrightarrow{CD}$, m $\angle v = 105$, and m $\angle z = 20$.
   Find m $\angle y$. **55**

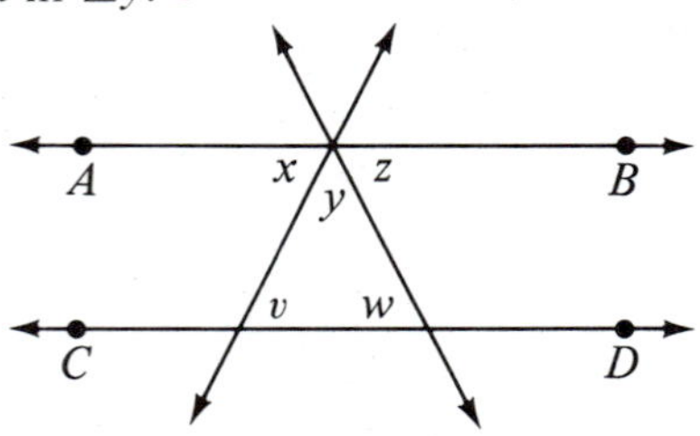

4. In the figure above, $\overleftrightarrow{AB} \parallel \overleftrightarrow{CD}$, m $\angle v = 9t - 1$, m $\angle y = 1 + 6t$, and m $\angle w = 9 + 4t$.
   Find m $\angle z$. **45**

5. Write the converse of "If lines are coplanar, then they are parallel."
   If lines are parallel, they are coplanar.

6. Draw a line and a point not on the line. Construct a line through the point parallel to the given line.

True or false? (Ex. 7–13)

7. Any two lines that are not parallel are skew lines. **F**

8. Any two coplanar segments that do not intersect are parallel. **F**

9. If two lines are intersected by a transveral so that alternate interior angles are congruent, then the lines are parallel. **T**

10. The sums of the measures of the angles of two triangles are equal. **T**

11. Any two noncoplanar lines are skew lines. **T**

12. If two lines are not parallel to a third line, then they are not parallel to each other. **F**

13. Through point P not on line l, there are many lines parallel to l. **F**

14. Prove: The sum of the measures of the angles of a triangle is 180°.

15. Give an indirect proof: A triangle cannot have two obtuse angles.

*16.

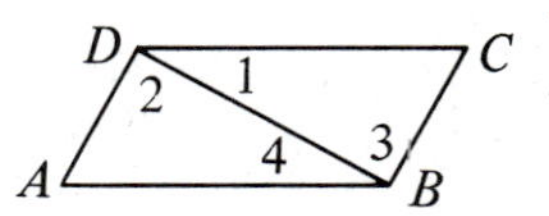

Given: $\overline{AB} \parallel \overline{CD}$ and $\angle ADC \cong \angle ABC$
Prove: $\overline{AD} \parallel \overline{BC}$

COLLEGE PREP TEST

Directions: For each question, choose the best answer. Indicate your answer by writing the letter of this choice.

1. Parallel lines are
I. coplanar
II. noncoplanar
III. nonintersecting

(A) I only
(B) II only
(C) III only
(D) I and III only
(E) II and III only **D**

2.

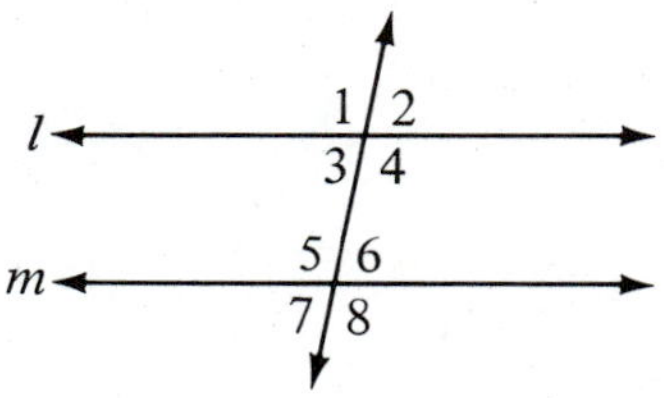

If $l \parallel m$, then
I. $\angle 3 \cong \angle 6$
II. $\angle 1 \cong \angle 5$
III. $\angle 4$ and $\angle 6$ are supplementary.

(A) I only
(B) II only
(C) III only
(D) I and II only
(E) I, II, and III **E**

3.

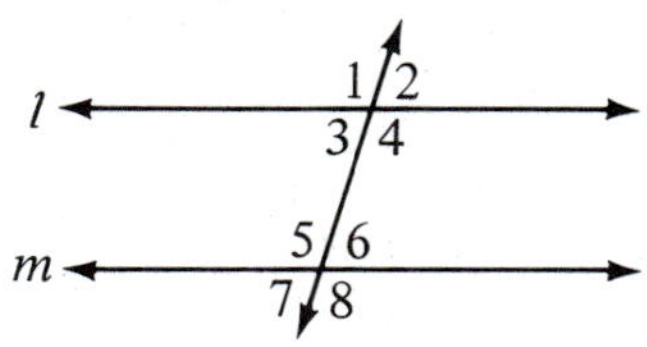

$l \parallel m$ if
(A) $\angle 3 \cong \angle 5$
(B) $\angle 2$ and $\angle 6$ are supplementary.
(C) $\angle 3$ and $\angle 5$ are complementary.
(D) $\angle 4 \cong \angle 6$
(E) none of these **E**

4.

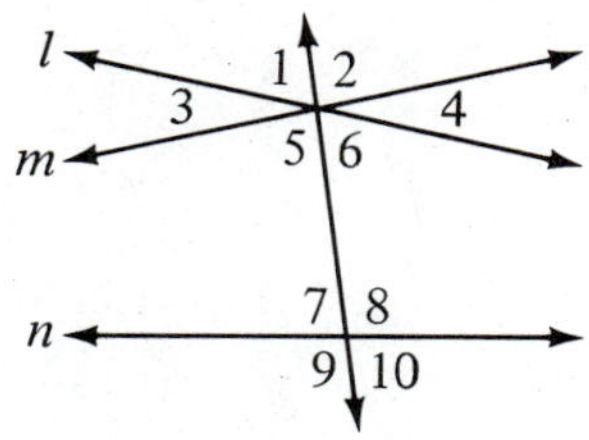

If $l \parallel n$, then
I. $m \nparallel n$
II. $\angle 5$ and $\angle 7$ are supplementary.
III. $m\angle 3 + m\angle 5 = m\angle 8$

(A) I only
(B) II only
(C) III only
(D) I and II only
(E) I and III only **E**

5.

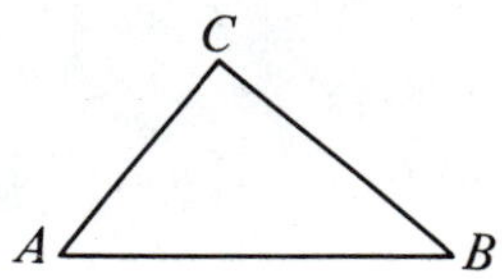

If $m\angle A = 5x - 5$, $m\angle B = 6x - 2$, and $m\angle C = 7x + 7$, then
(A) $m\angle A = 10$
(B) $m\angle A = 20$
(C) $m\angle A = 58$
(D) $m\angle A = 77$
(E) none of these **E**

6. The figure shows the construction of a line through Y parallel to $\overline{XZ}$.

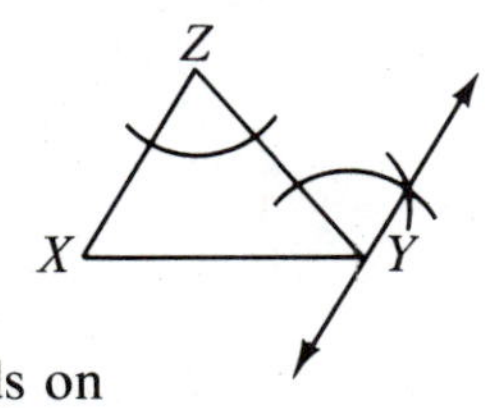

This construction depends on
(A) congruent corresponding angles
(B) congruent vertical angles
(C) congruent alternate interior angles
(D) supplementary angles
(E) none of these **C**

4 PROVING TRIANGLES CONGRUENT

Non-Routine Problem Solving

Problem

In the following figure, the measure of $\angle B$ is 45°, the measure of $\angle BDC$ is 30°, the measure of $\angle ADG$ is 50°, and $\angle A$ is congruent to $\angle AED$. Find the measure of $\angle ACB$. **95°**

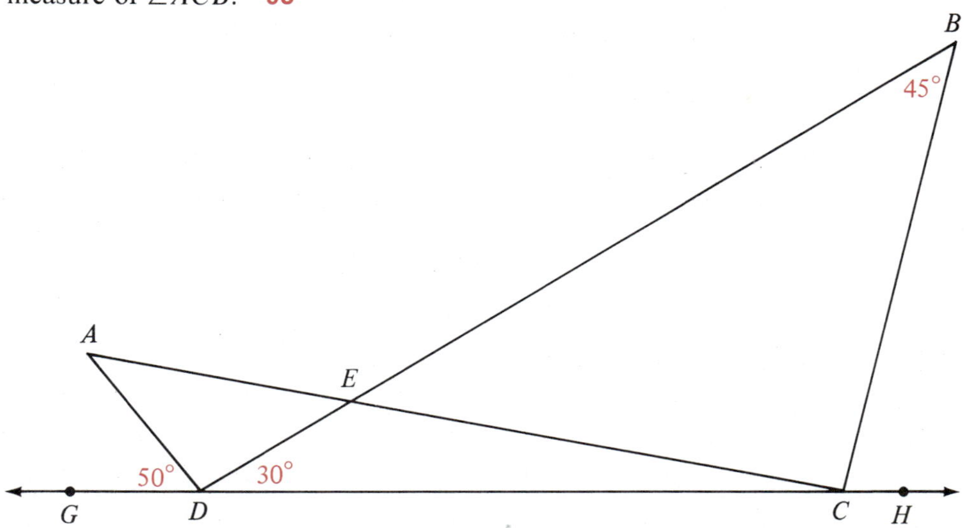

CLASSIFYING TRIANGLES 4.1

Objectives **To classify triangles by the lengths of their sides**
To classify triangles by the measures of their angles

A triangle is one of the basic figures studied in geometry.

Definition: Triangle A *triangle* is a figure consisting of three segments that have three noncollinear points as their endpoints.

The segments that form the triangle are its **sides** and the endpoints of the segments are its **vertices.**

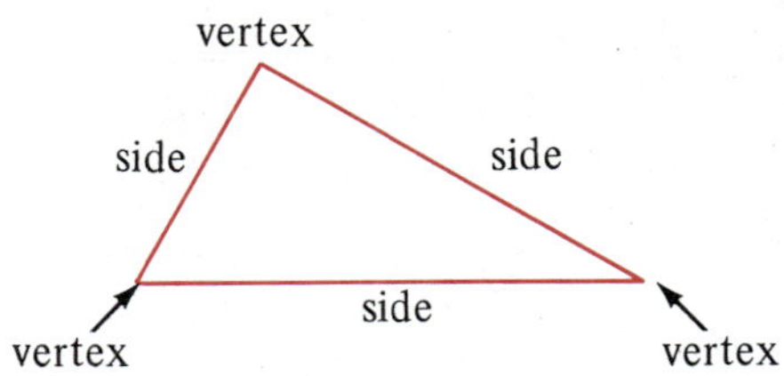

Triangles can be classified in different ways. One method of classification involves the lengths of the sides.

Definitions: Equilateral, isosceles, scalene triangles

An *equilateral triangle* is a triangle with three congruent sides.
An *isosceles triangle* is a triangle with at least two congruent sides.
A *scalene triangle* is a triangle with no congruent sides.

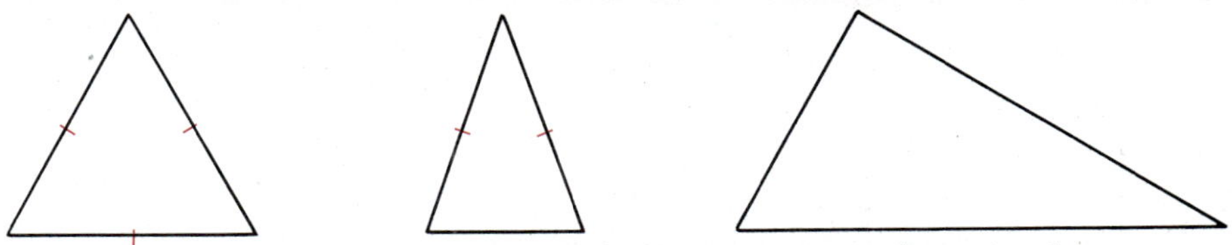

equilateral triangle isosceles triangle scalene triangle

Triangles can also be classified by the measures of their angles.

Definitions: Acute, right, obtuse, equiangular triangles

An *acute triangle* is a triangle with three acute angles.
A *right triangle* is a triangle with one right angle.
An *obtuse triangle* is a triangle with one obtuse angle.
An *equiangular triangle* is a triangle with three congruent angles.

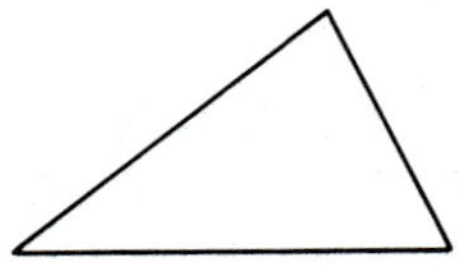

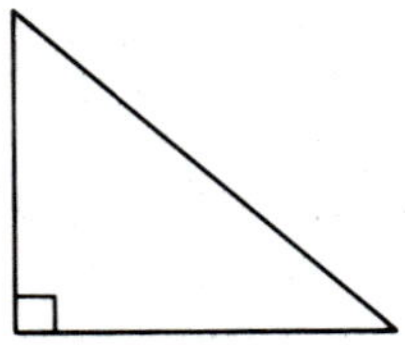

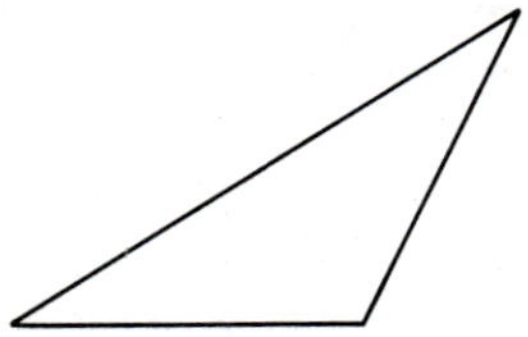

An equiangular triangle has three congruent angles. Later, you will prove that an equiangular triangle is also equilateral.

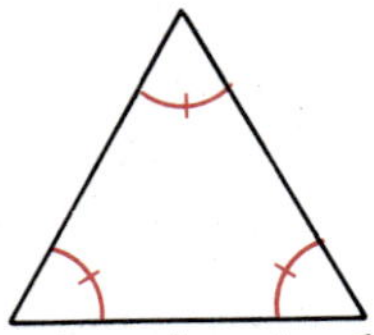

Example 1 Classify each triangle by its sides and by its angles.

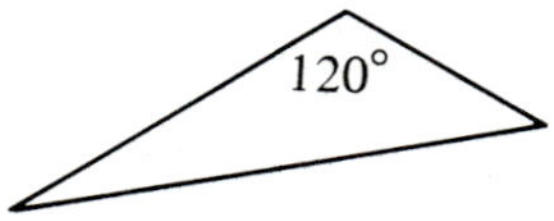

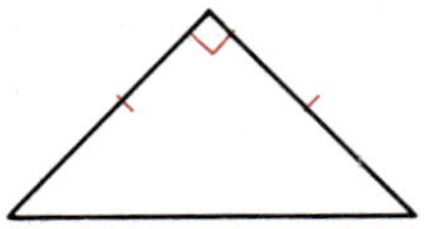

A given triangle can be classified by both the lengths of its sides and the measures of its angles. However, some combinations of sides and angles are possible, while others are not.

Example 2 Draw a triangle that is both isosceles and obtuse.

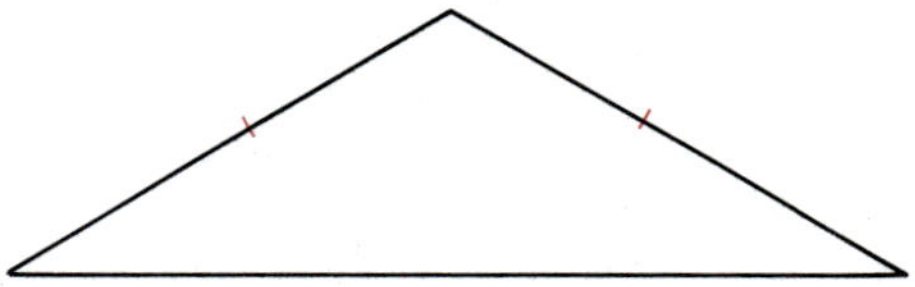

Example 3 Indicate whether the statement is always true, sometimes true, or never true.

An acute triangle is isosceles.

Sometimes true.

Example 4 True or false? Make a sketch to support your answer.

An equiangular triangle is also acute.

True.

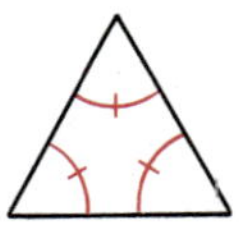

Reading in Geometry

A word may mean one thing in geometry and something different in everyday language. Often, its meaning must be determined by analyzing the entire sentence in which it is used.

In each set of sentences below, a nonsense word is used instead of a mathematical word. The same word is used in each sentence in a different way. Find the best word to substitute for the nonsense word.

1. She had a case of *oble* appendicitis. **acute**
His sense of hearing was *oble*. **acute**
The measure of the angle was 96°, so he knew that it was not an *oble* angle. **acute**

2. She asked, "What's your *blark?*" **angle**
The measure of the *blark* is 36°. **angle**
You should consider this problem from every *blark*. **angle**

3. It was obviously a *bligle* angle. **right**
The directions said to make a *bligle* turn at Seventh Avenue. **right**
She knew that her answer was *bligle*. **right**

4. His friends thought that he was rather *isnenck*. **obtuse**
An *isnenck* triangle cannot be equilateral. **obtuse**
It was an *isnenck* sound. **obtuse**

Oral Exercises

Classify each triangle by its sides.

1.
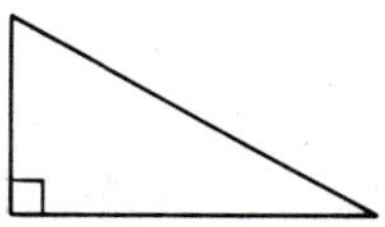
scalene

2.
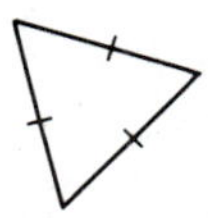
equilateral

3.
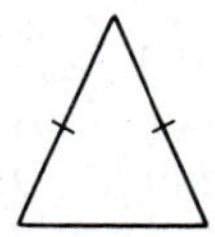
isosceles

4.
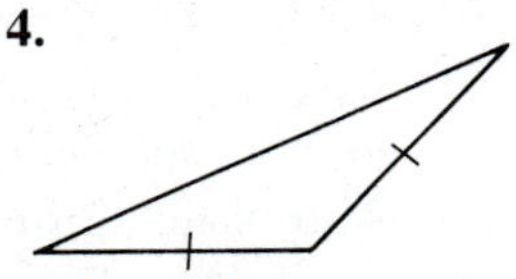
isosceles

Classify each triangle by its angles.

5.
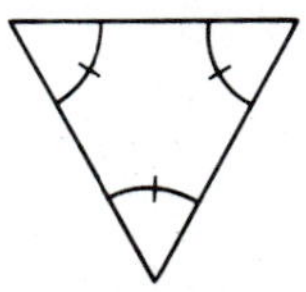
equiangular

6.
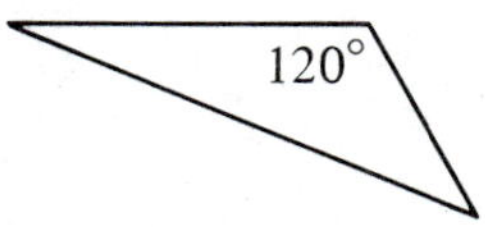

obtuse

7.
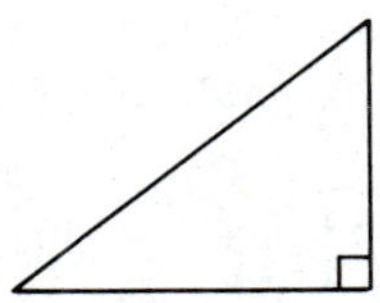
right

8.
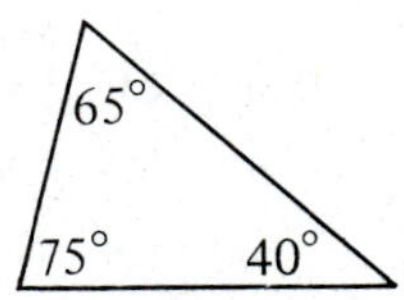

acute

Written Exercises

Draw a triangle with the indicated properties.

A **1.** Both obtuse and scalene
2. Both right and isosceles
3. Both acute and isosceles
4. Both right and scalene
5. Both acute and scalene
6. Both acute and equilateral

Indicate whether each statement is always true, sometimes true, or never true.

7. An equilateral triangle is isosceles. **A**
8. A scalene triangle is isosceles. **N**
9. A right triangle is equiangular. **N**
10. An equiangular triangle is equilateral. **A**
11. An equilateral triangle is obtuse. **N**
12. An equilateral triangle is equiangular. **A**
13. A right triangle is scalene. **S**
14. A right triangle is obtuse. **N**
15. An isosceles triangle is obtuse. **S**
16. A right triangle is isosceles. **S**

Use the diagram to find each indicated set of triangles. (Ex. 17–22).

Suppose that U represents all triangles,
A represents all isosceles triangles,
B represents all equilateral triangles,
C represents all right triangles, and
D represents all scalene triangles.

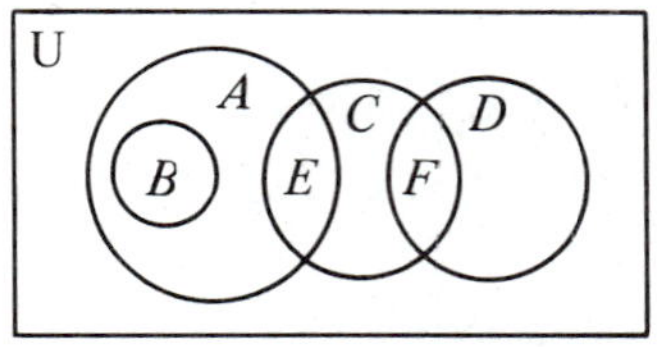

B 17. If E includes triangles that are in both A and C, which triangles are these? **isos rt △'s**

18. If F includes triangles that are in both C and D, which triangles are these? **scal. rt. △'s**

19. To which set, or sets, do acute triangles belong? **U; A; B; D**

20. To which set, or sets, do obtuse triangles belong? **U; A; D**

C 21. Redraw the diagram to show the set of all acute triangles.

22. Redraw the diagram to show the set of all obtuse triangles.

Prove each statement by using an indirect proof.

23. An equiangular triangle is not obtuse.
24. A scalene triangle is not isosceles.

NON-ROUTINE PROBLEMS

The picture below can be drawn without lifting your pencil, retracing any part of the picture, or crossing any part of the picture already drawn. One possible way of doing this is shown at the right of the given picture.

Can each of the following pictures be drawn using the same rules?

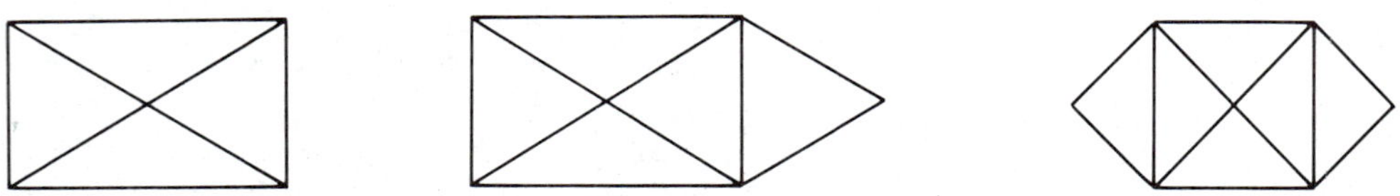

no yes yes

Make several similar diagrams of your own. Can each of these be drawn under these same conditions? Is there a pattern that will enable you to look at any such picture and immediately tell whether or not it is possible to draw it under these conditions? If so, what is the pattern?

If an odd number of segments join at more than two vertices, the figure cannot be drawn without lifting the pencil.

EXTERIOR ANGLES AND INTERIOR ANGLES OF TRIANGLES 4.2

Objective **To find the measures of exterior and remote interior angles of triangles**

The sides of an angle are rays. However, it often is convenient to think of two sides of a triangle as forming an angle. You will use this interpretation in many theorems about triangles. According to this interpretation, a triangle has three angles. A triangle also has **exterior angles** associated with it.

Definitions: Exterior angle remote interior angles

An *exterior angle* of a triangle is an angle that is adjacent and supplementary to one of the angles of the triangle.

The other two angles of the triangle are called the *remote interior angles* of the exterior angle.

$\angle YZW$ is an exterior angle of $\triangle XYZ$. $\angle X$ and $\angle Y$ are the remote interior angles of $\angle YZW$.

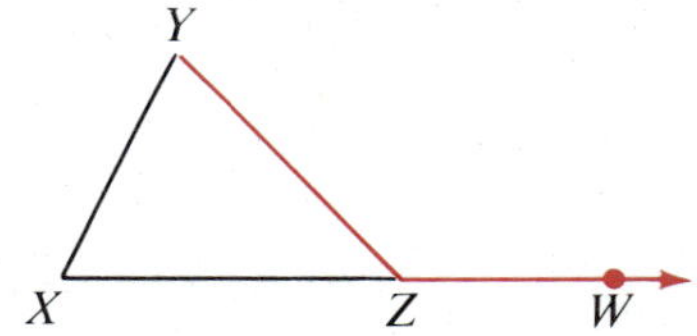

Each triangle has six exterior angles, two at each vertex, as shown in the diagram below. The two exterior angles at each vertex are vertical angles and so they are congruent.

Exterior Angle	Adjacent Angle	Remote Interior Angles
$\angle 1$	$\angle x$	$\angle y$ and $\angle z$
$\angle 2$	$\angle x$	$\angle y$ and $\angle z$
$\angle 3$	$\angle y$	$\angle z$ and $\angle x$
$\angle 4$	$\angle y$	$\angle z$ and $\angle x$
$\angle 5$	$\angle z$	$\angle x$ and $\angle y$
$\angle 6$	$\angle z$	$\angle x$ and $\angle y$

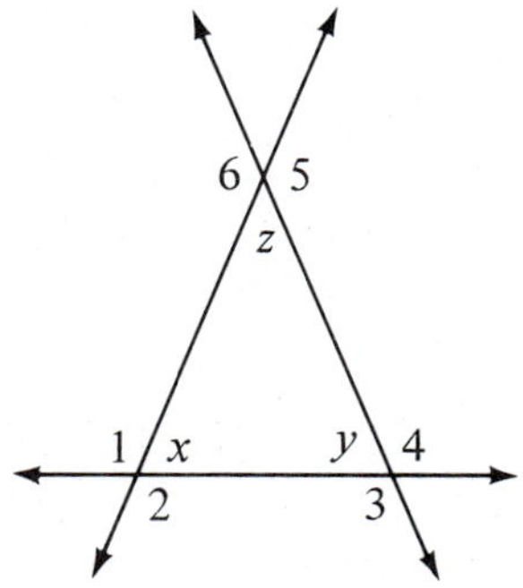

In the picture at the right, you know that m $\angle 1$ + m $\angle B$ + m $\angle A$ = 180.

m $\angle 1 = 180 - (124 + 20)$, or 36.

m $\angle 2 = 180 - 36$, or 144.

Thus, m $\angle 2$ = m $\angle A$ + m $\angle B$.

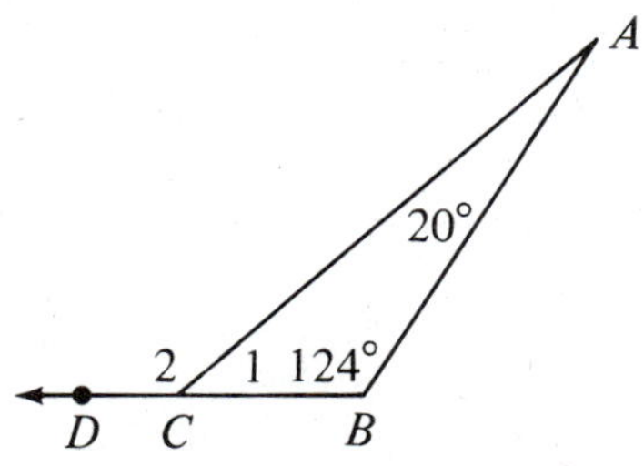

This leads to the following theorem.

Theorem 4.1 The measure of an exterior angle of a triangle is equal to the sum of the measures of its two remote interior angles. (See Exercise 7.)

You can use this theorem to find the measures of interior and exterior angles of triangles.

Example 1 What is the measure of the exterior angle, $\angle x$?

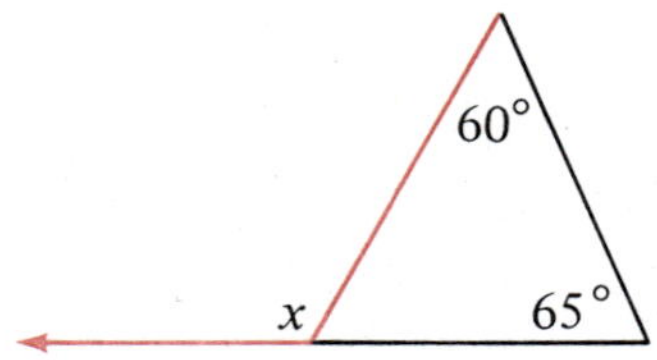

$m\,\angle x = 60 + 65$, or 125.

Example 2 Given: $m\,\angle A = x + 8$,
$m\,\angle B = 2x + 3$,
$m\,\angle BCD = 5x - 11$
Find $m\,\angle BCD$.

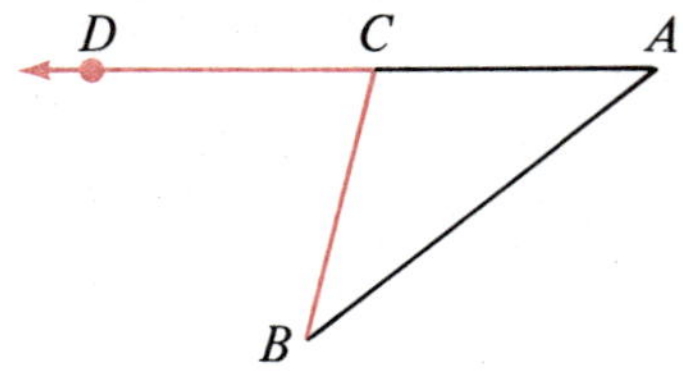

$m\,\angle BCD = m\,\angle A + m\,\angle B$ ◄ *$\angle BCD$ is an exterior $\angle$.*

$$
\begin{aligned}
5x - 11 &= (x + 8) + (2x + 3) \\
5x - 11 &= 3x + 11 \\
2x &= 22 \\
x &= 11
\end{aligned}
$$

Thus, $m\,\angle BCD = 5(11) - 11$, or 44.

An important corollary to Theorem 4.1 is stated below.

Corollary The measure of an exterior angle of a triangle is greater than the measure of either of its remote interior angles.

Example 3 Complete the sentence with $<, =,$ or $>$.

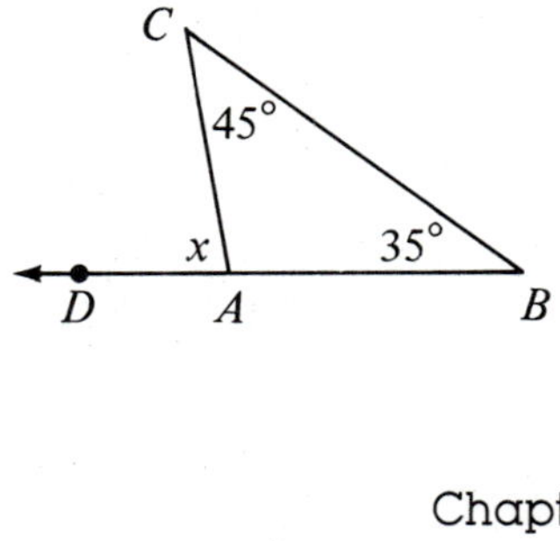

$m\,\angle x$ __________ 45

$m\,\angle x$ ____ $>$ ____ 45

Oral Exercises

What is the measure of each indicated exterior angle?

1.

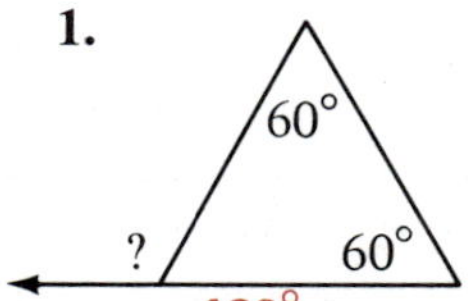

120°

2.

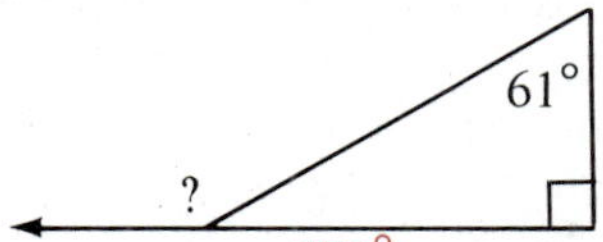

151°

3.

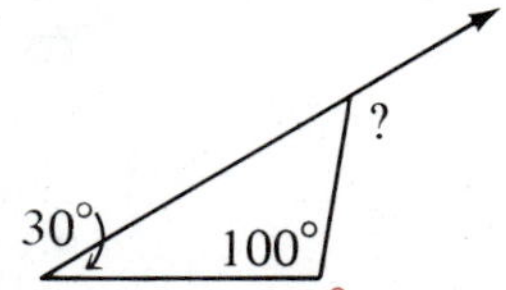

130°

Complete each sentence with <, =, or >.

4. m $\angle x$ ___>___ 60

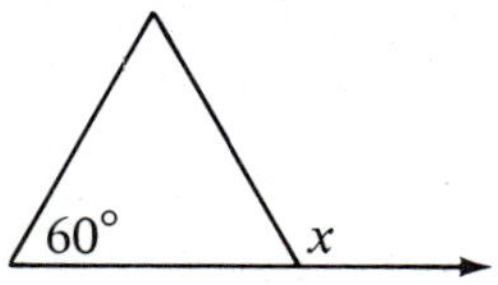

5. m $\angle y$ ___<___ 135

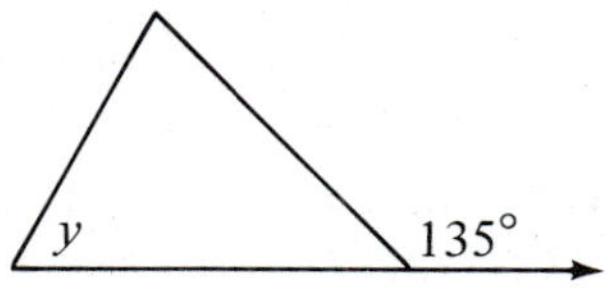

6. m $\angle z$ + m $\angle w$ ___=___ 45

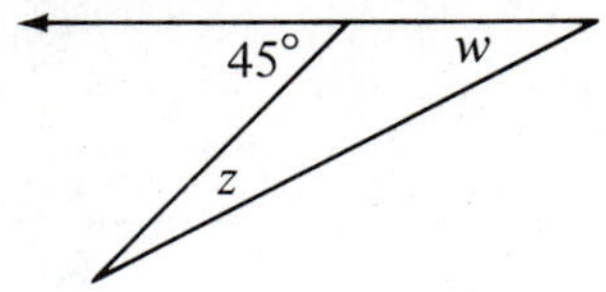

Written Exercises

Find each indicated angle measure.

A **1.** Given: m $\angle 1 = 9x - 15$, m $\angle 3 = 4x$, and m $\angle 4 = 3x + 15$ Find m $\angle 1$. **120**

2. Given: m $\angle 1 = 8x - 6$, m $\angle 3 = 3x + 4$, and m $\angle 4 = 4x + 2$ Find m $\angle 3$. **40**

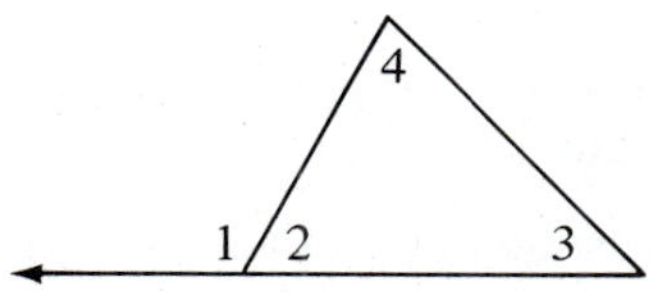

3. Given: m $\angle 1 = 10x$, m $\angle 3 = 4x + 4$, and m $\angle 4 = 5x + 7$ Find m $\angle 4$. **62**

4. Given: m $\angle 1 = 9x + 1$, m $\angle 3 = 4x - 2$, and m $\angle 4 = 6x - 7$ Find m $\angle 1$. **91**

Prove each of the following.

B **5.** Given: $\angle A \cong \angle D$ and $\angle G \cong \angle E$
Prove: $\angle 1 \cong \angle 4$

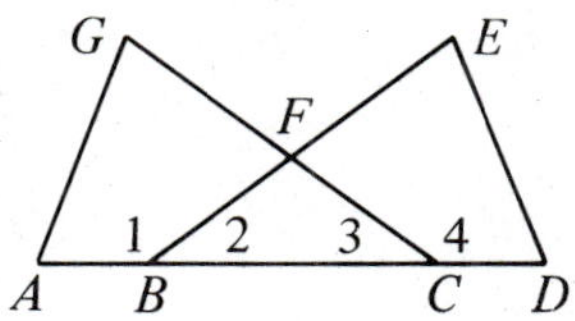

6. Given: $\angle 1 \cong \angle 4$ and $\angle G \cong \angle E$
Prove: $\angle A \cong \angle D$

7. Prove Theorem 4.1.

C **8.** Postulate 12 states that there is exactly one line through a given point perpendicular to a given line. Using the corollary to Theorem 4.1, *but not Postulate 12,* write an indirect proof for: There is at most one line perpendicular to a given line from a given point.

LOGIC REVIEW

Write the converse of each statement. Label the statement and the converse as true or false. **Check students' answers.**

1. If two segments are congruent, then they have the same measure.
statement: true; converse: true

2. If a triangle is isosceles, then it has three congruent sides.
statement: false; converse: true

CONGRUENT TRIANGLES 4.3

Objectives **To identify congruent triangles**
To find unknown measures in congruent triangles

Congruent segments have the same length, and congruent angles have the same measure. Geometric figures in general are congruent if they have the same size and shape. In triangles, the lengths of the sides determine the size, and the measures of the angles determine the shape. This leads to the following definition.

Definition: Congruent triangles Two triangles are *congruent* if corresponding angles are congruent and corresponding sides are congruent.

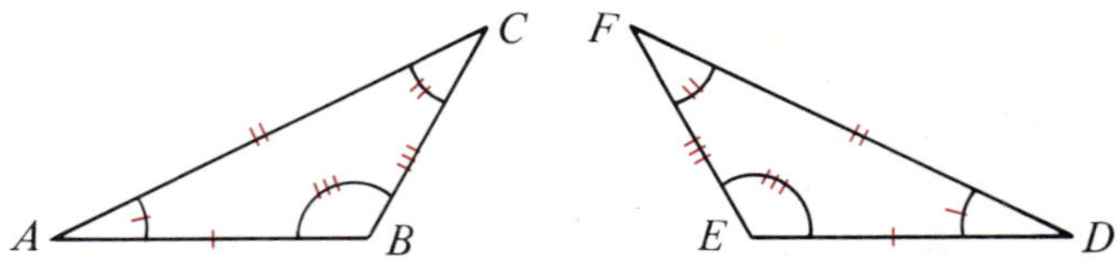

In the diagram above, the markings indicate that the corresponding angles and the corresponding sides are congruent.

Corresponding $\angle$'s are $\cong$.	Corresponding sides are $\cong$.
$\angle A \cong \angle D$	$\overline{AB} \cong \overline{DE}$
$\angle B \cong \angle E$	$\overline{BC} \cong \overline{EF}$
$\angle C \cong \angle F$	$\overline{CA} \cong \overline{FD}$

To indicate that triangle ABC is congruent to triangle DEF, you will write

$$\triangle ABC \cong \triangle DEF.$$

Example 1 Using the markings shown, complete the congruence statement.

$\triangle TJR \cong$ ________

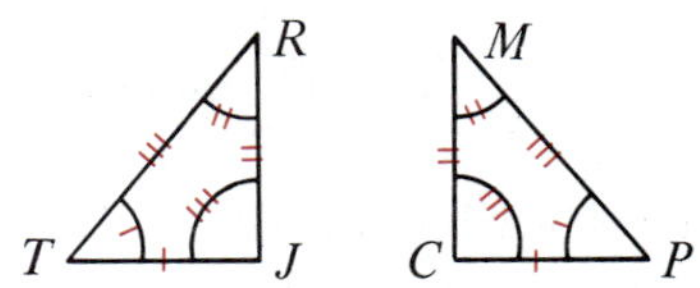

Using the diagram, $\angle T \cong \angle P$ $\quad \overline{TJ} \cong \overline{PC}$
$\angle J \cong \angle C$ and $\overline{JR} \cong \overline{CM}$
$\angle R \cong \angle M$ $\quad \overline{RT} \cong \overline{MP}$

Thus, $\triangle TJR \cong$ $\underline{\triangle PCM}$.

It is possible to tell from a triangle congruence statement which parts correspond. The order of the letters indicates the corresponding vertices.

$$\triangle ABC \cong \triangle DEF$$

In the statement above, $A \leftrightarrow D$, $B \leftrightarrow E$, and $C \leftrightarrow F$.

Example 2 If $\triangle BDF \cong \triangle XVT$, which angles and which sides are corresponding?

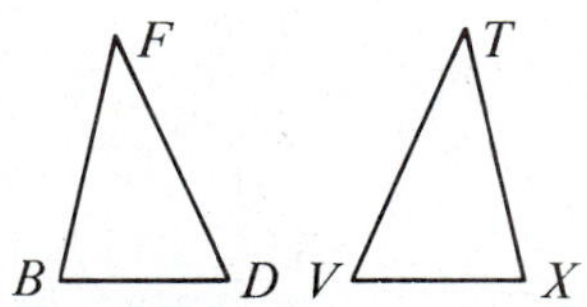

Corresponding angles: $\angle B$ and $\angle X$, $\angle D$ and $\angle V$, $\angle F$ and $\angle T$

Corresponding sides: $\overline{BD}$ and $\overline{XV}$, $\overline{DF}$ and $\overline{VT}$, $\overline{FB}$ and $\overline{TX}$

A special lettering scheme is sometimes used to name the corresponding parts of congruent triangles.

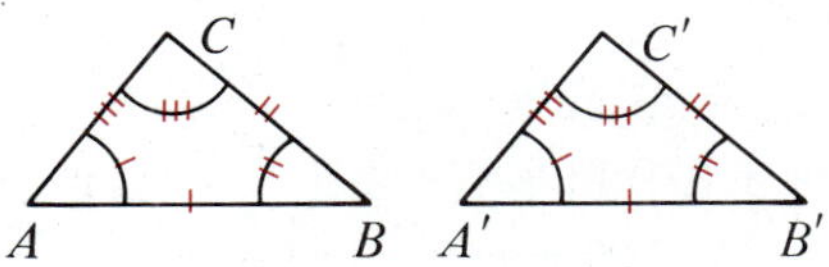

$\triangle ABC \cong \triangle A'B'C'$

This is read "Triangle ABC is congruent to triangle A prime B prime C prime." When this lettering scheme is used, it generally means that A corresponds to A', B to B', and C to C'.

Triangles do not have to be turned the same way to be congruent.

Example 3 Using the markings shown, complete the statement $\triangle XYZ \cong$ ________.

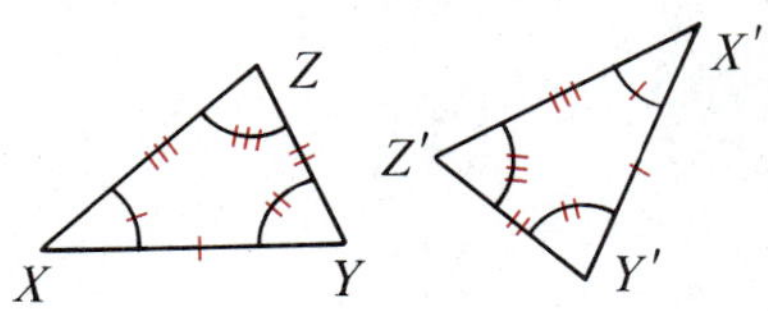

◄ $X \leftrightarrow X'$, $Y \leftrightarrow Y'$, $Z \leftrightarrow Z'$

$\triangle XYZ \cong \underline{\triangle X'Y'Z'}$.

Diagrams showing triangle congruences are not always provided. You may have to use given information to find out if two triangles are congruent.

Example 4 Complete the triangle congruence statement using the given information. Draw and label a diagram to show the correspondences.

Given: $\angle S \cong \angle M$, $\angle B \cong \angle X$, $\angle R \cong \angle T$
$\overline{SB} \cong \overline{MX}$, $\overline{BR} \cong \overline{XT}$, and $\overline{RS} \cong \overline{TM}$.
$\triangle SBR \cong$ ________

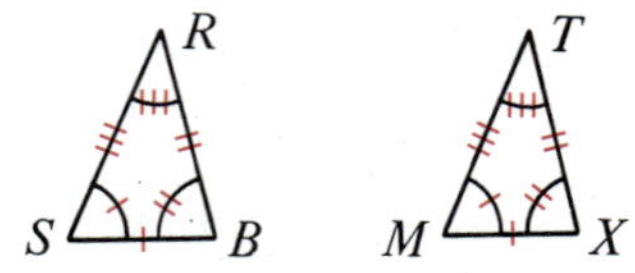

Since $S \leftrightarrow M$, $B \leftrightarrow X$, and $R \leftrightarrow T$,
$\triangle SBR \cong \underline{\triangle MXT}$.

Oral Exercises

1. Which triangle appears to be congruent to $\triangle ABC$? △GHI

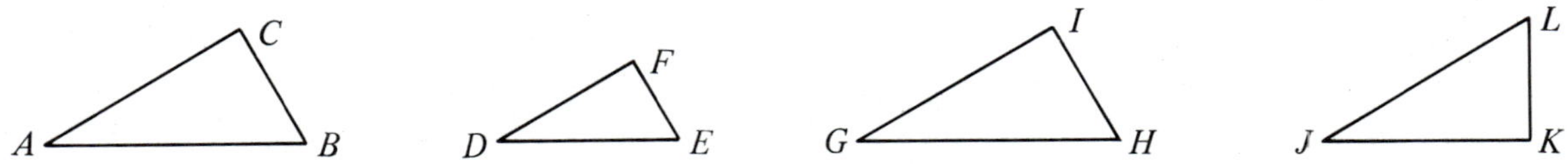

2. Which triangle appears to be congruent to $\triangle GHI$? △JKL

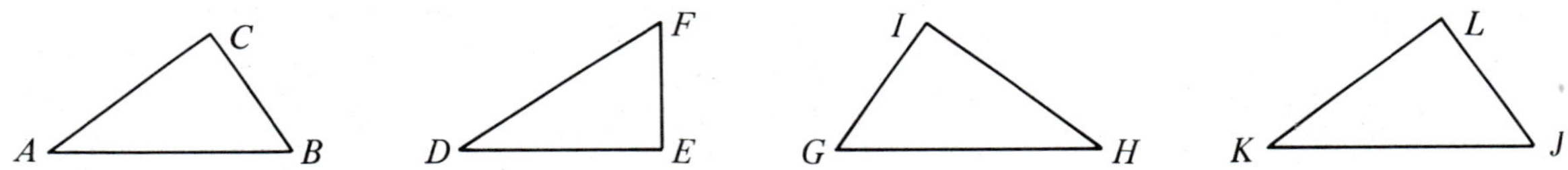

Written Exercises

Using the markings shown, complete each congruence statement.

A **1.**

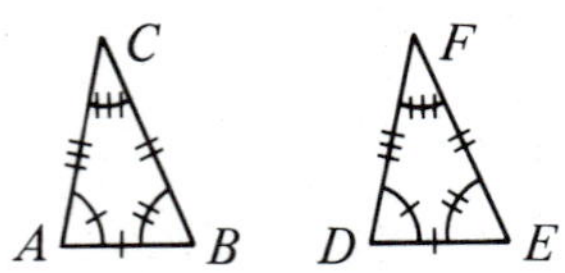

$\triangle ABC \cong$ △ DEF

2.

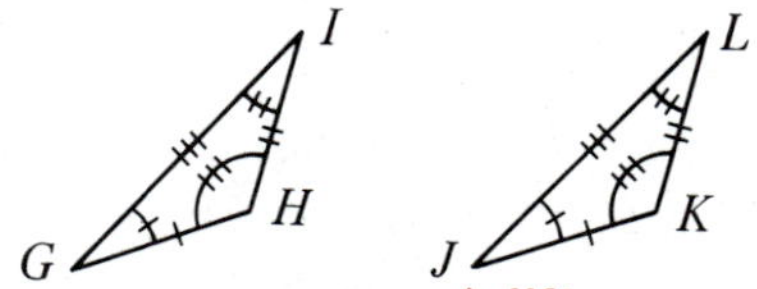

$\triangle GHI \cong$ △ JKL

3.

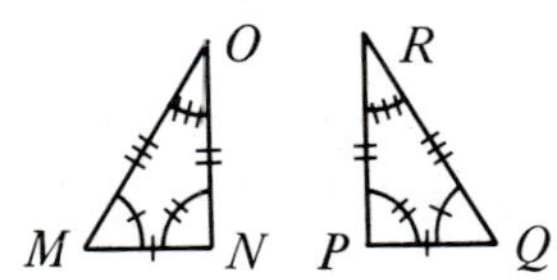

$\triangle MNO \cong$ △ QPR

4.

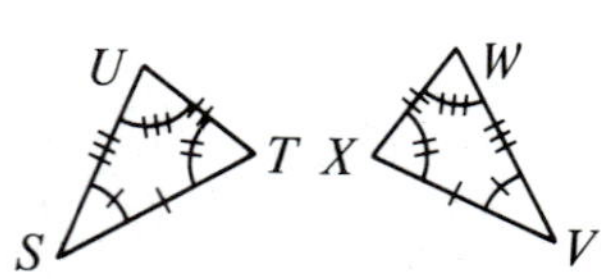

$\triangle STU \cong$ △ VXW

5.

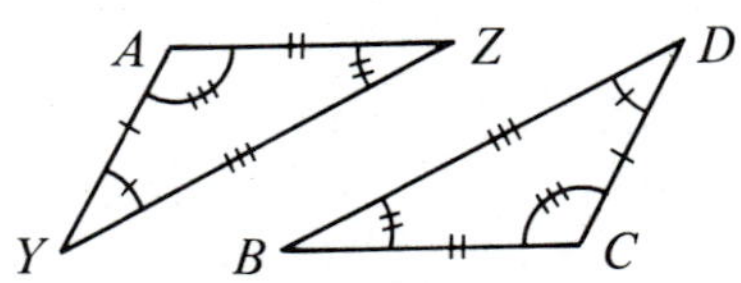

$\triangle YZA \cong$ △ DBC

6.

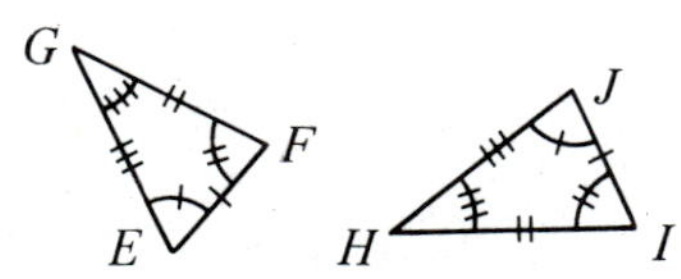

$\triangle EFG \cong$ △ JIH

Complete each triangle congruence statement.
Draw and label a diagram to show the correspondences.

7. $\angle X \cong \angle X'$
$\angle Y \cong \angle Y'$
$\angle Z \cong \angle Z'$
$\overline{XY} \cong \overline{X'Y'}$
$\overline{YZ} \cong \overline{Y'Z'}$
$\overline{ZX} \cong \overline{Z'X'}$
$\triangle XYZ \cong$ __$\triangle X'Y'Z'$__

8. $\angle T \cong \angle P$
$\angle N \cong \angle V$
$\angle L \cong \angle C$
$\overline{TN} \cong \overline{PV}$
$\overline{NL} \cong \overline{VC}$
$\overline{LT} \cong \overline{CP}$
$\triangle TNL \cong$ __$\triangle PVC$__

9. $\angle R \cong \angle C$
$\angle A \cong \angle H$
$\angle Q \cong \angle W$
$\overline{RA} \cong \overline{CH}$
$\overline{AQ} \cong \overline{HW}$
$\overline{QR} \cong \overline{WC}$
__$\triangle ARQ$__ $\cong \triangle HCW$

Draw a labeled diagram to illustrate each of the following. (Ex. 10–12)

10. $\triangle ABC \cong \triangle XYZ$ **11.** $\triangle EFG \cong \triangle E'F'G'$ **12.** $\triangle TJH \cong \triangle RXA$

Write two different congruence statements for each pair of isosceles or equilateral triangles shown.

B **13.**

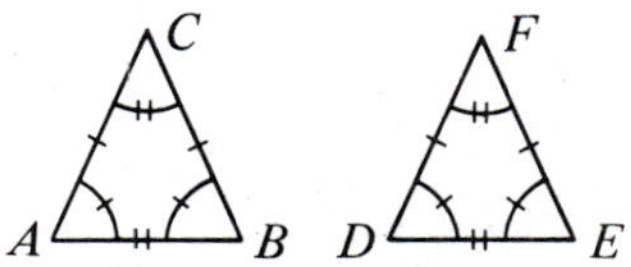

Answers may vary.

14.

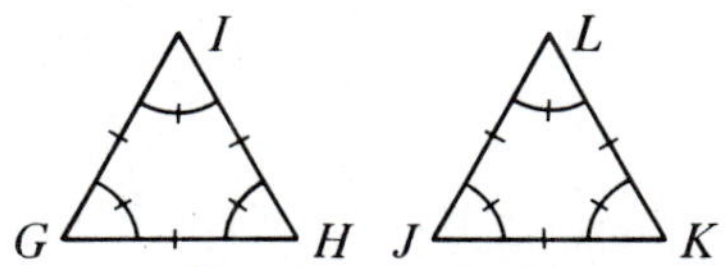

Answers may vary.

15.

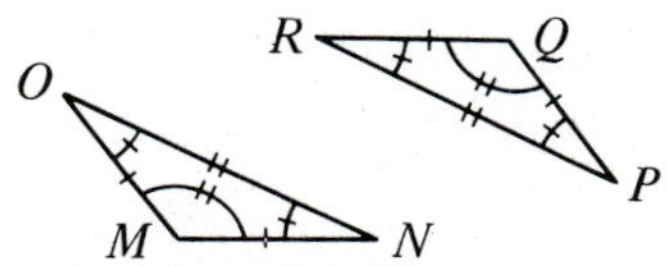

Answers may vary.

16. Given: $\triangle LVC \cong \triangle MPI$, $LV = 3x + 2$, and $MP = 5x - 8$
Find LV. 17

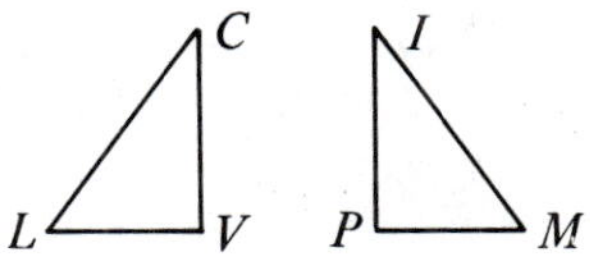

17. Given: $\triangle LVC \cong \triangle MPI$, $m\,\angle C = 3x - 23$, and $m\,\angle I = 2x - 5$
Find $m\,\angle I$. 31

ALGEBRA REVIEW

Objective

To factor a common monomial from a polynomial

To factor out the greatest common monomial:
(1) Factor out the greatest common whole number factor other than 1.
(2) Factor out the greatest common variable factor, if any.

Example

Factor out the *GCF* from $10x^3 - 15x$.

$$10x^3 - 15x = 2 \cdot 5 \cdot x^3 - 3 \cdot 5 \cdot x$$

Greatest common number factor: 5 Greatest common variable factor: x

The *GCF* is $5x$.

Thus, $10x^3 - 15x = 5x(2x^2 - 3)$ ◄ *Distributive property*

Factor out the *GCF*.

1. $6x^3 - 9x$
$3x(2x^2 - 3)$

2. $5y^4 - 15y^3 + 25y^2$
$5y^2(y^2 - 3y + 5)$

3. $8z^3 + 12z + 20$
$4(2z^3 + 3z + 5)$

4. $27a^5 + 36a^3 - 18a^2 + 45a$
$9a(3a^4 + 4a^2 - 2a + 5)$

SSS CONGRUENCE PROOFS 4.4

Objectives **To prove triangles congruent by the side-side-side correspondence**
To construct triangles using the side-side-side pattern

You know that two triangles are congruent if the corresponding angles and the corresponding sides are congruent. However, it is not necessary to show all six of these congruences to prove that two given triangles are congruent, as you will discover in this lesson and the following lessons.

Take three sticks of different lengths. Connect the ends to form a triangle. Rearrange the three sticks to form as many triangles as possible. Each triangle that you make has the same size and shape as the first one.

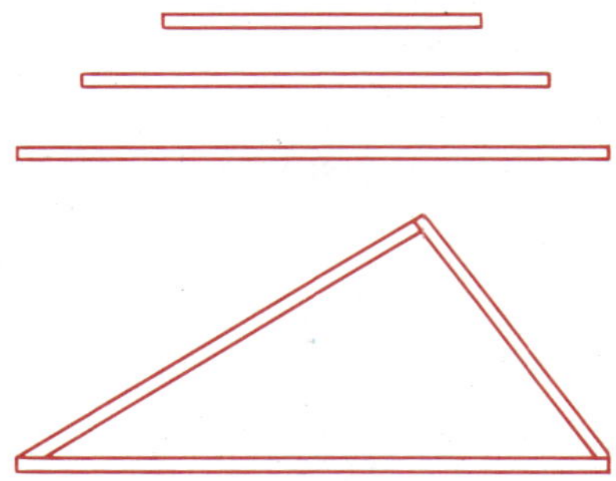

This experiment leads to the following important postulate.

Postulate 16 SSS for $\cong \triangle$'s Two triangles are congruent if the three sides of one are congruent to the corresponding sides of the other.

Postulate 16 is often called the "side-side-side" pattern.

Example 1 From the markings, which triangles are congruent?

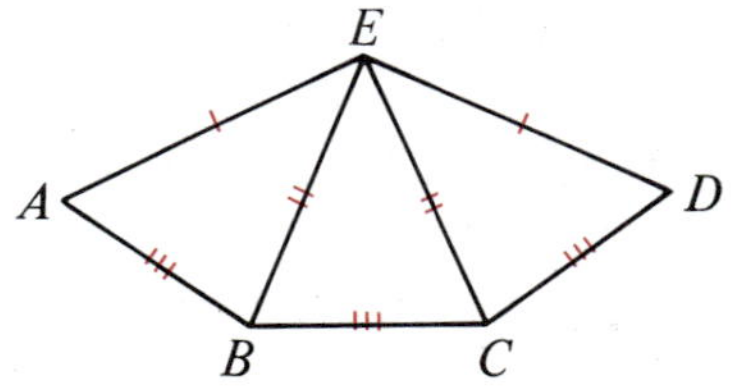

Apply Postulate 16. ► $\overline{AB} \cong \overline{DC}$, $\overline{BE} \cong \overline{CE}$, and $\overline{EA} \cong \overline{ED}$
Thus, $\triangle ABE \cong \triangle DCE$.

Example 2 Write a proper congruence statement for the triangles.
Given: $\overline{XY} \cong \overline{KL}$, $\overline{YZ} \cong \overline{LJ}$, and $\overline{ZX} \cong \overline{JK}$

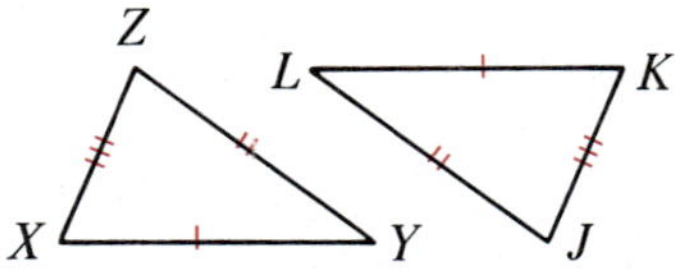

Apply Postulate 16. ► $\triangle XYZ \cong \triangle KLJ$

You can use Postulate 16 in proofs involving congruent triangles.

Example 3 Given: P bisects $\overline{AC}$,
P bisects $\overline{DB}$, and
$\overline{AD} \cong \overline{CB}$.
Prove: $\triangle ADP \cong \triangle CBP$

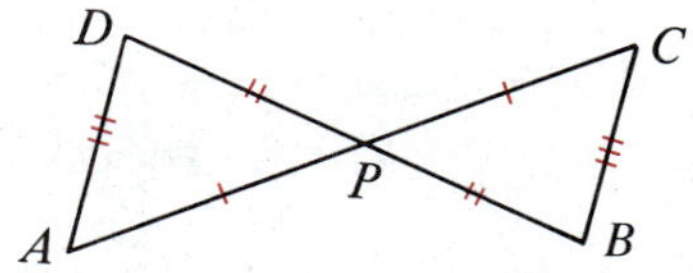

Proof

Statements	Reasons
1. P bisects $\overline{AC}$ and P bisects $\overline{DB}$.	1. Given
2. $\overline{AP} \cong \overline{CP}$	2. Definition of bisector
3. $\overline{DP} \cong \overline{BP}$	3. Definition of bisector
4. $\overline{AD} \cong \overline{CB}$	4. Given
5. $\therefore \triangle ADP \cong \triangle CBP$	5. SSS for $\cong$ $\triangle$'s

Using Postulate 16, you can construct a triangle congruent to a given triangle. (See Exercises 9 and 10.) The construction below shows you how to construct a triangle with sides congruent to three given segments.

Construction A triangle, given segments congruent to the sides.

Given: $\overline{AB}$, $\overline{CD}$, and $\overline{EF}$
Construct: $\triangle A'B'C'$ with sides congruent to $\overline{AB}$, $\overline{CD}$, and $\overline{EF}$

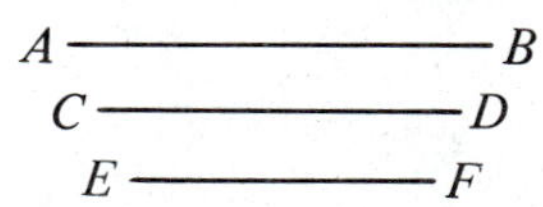

Strategy: Copy the three segments, joining each pair at a common endpoint.

Construct $\overline{A'B'} \cong \overline{AB}$.

Using center A' and radius $\cong \overline{CD}$, draw an arc.
Using center B' and radius $\cong \overline{EF}$, draw another arc.

Draw $\overline{A'C'}$ and $\overline{B'C'}$.

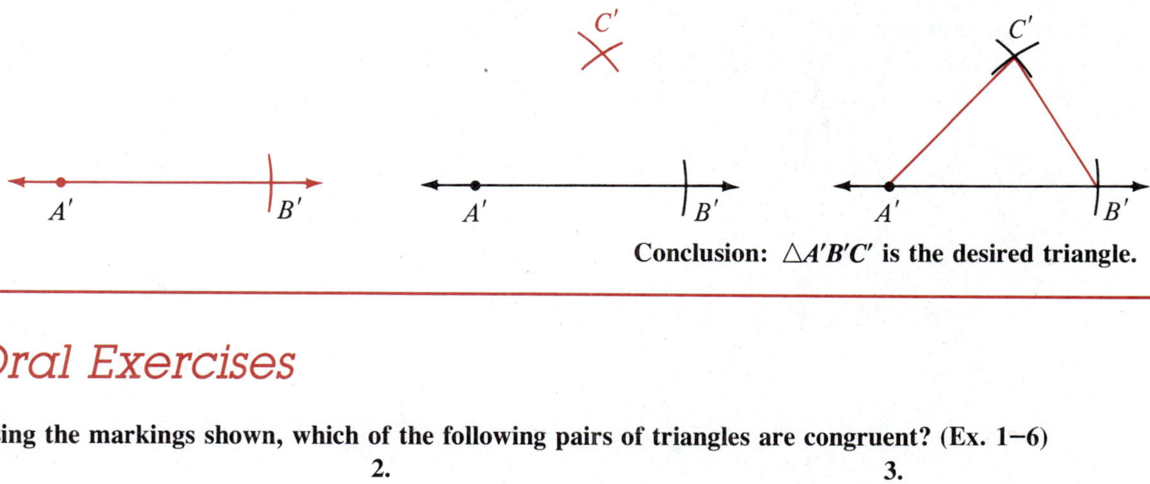

Conclusion: $\triangle A'B'C'$ is the desired triangle.

Oral Exercises

Using the markings shown, which of the following pairs of triangles are congruent? (Ex. 1–6)

1.

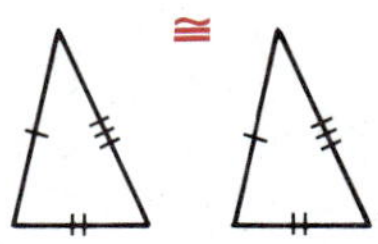

2.

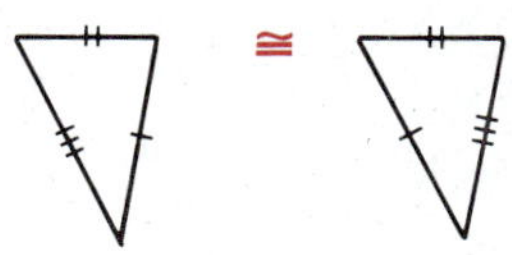

3.

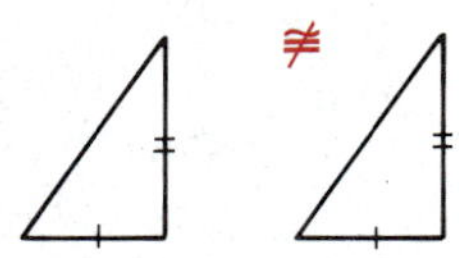

4. 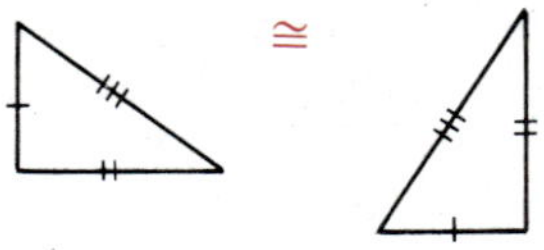 ≅ **5.** 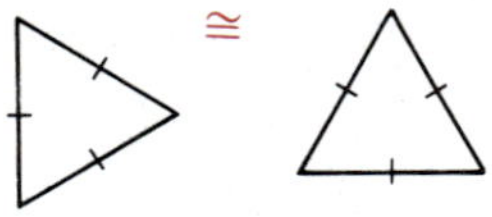 ≅ **6.** 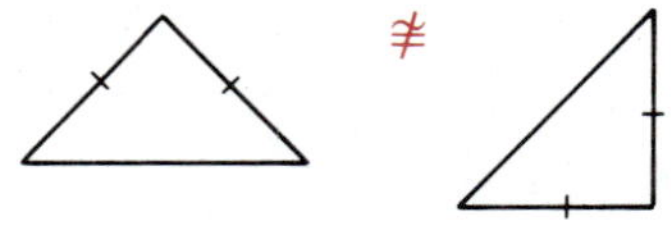 ≇

Written Exercises

Write a proper congruence statement for the triangles.

A 1. Given: $\overline{AB} \cong \overline{DE}$, $\overline{BC} \cong \overline{EF}$, and $\overline{CA} \cong \overline{FD}$

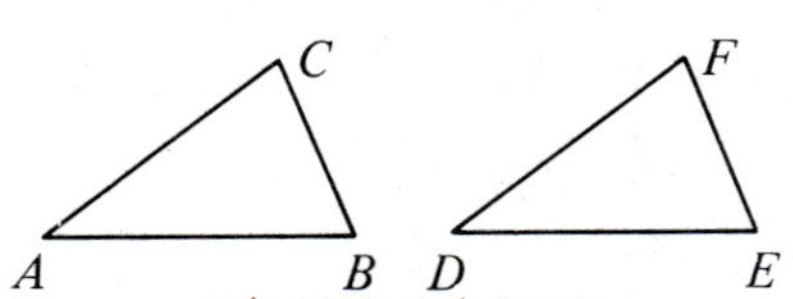

△ ABC ≅ △ DEF

2. Given: $\overline{JK} \cong \overline{RS}$, $\overline{KL} \cong \overline{ST}$, and $\overline{LJ} \cong \overline{TR}$

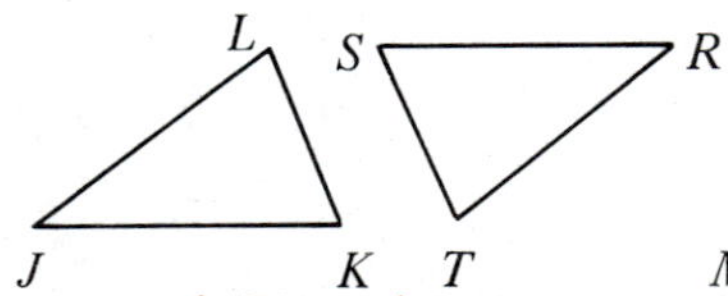

△ JKL ≅ △ RST

3. Given: $\overline{MN} \cong \overline{HF}$, $\overline{NO} \cong \overline{FG}$, and $\overline{OM} \cong \overline{GH}$

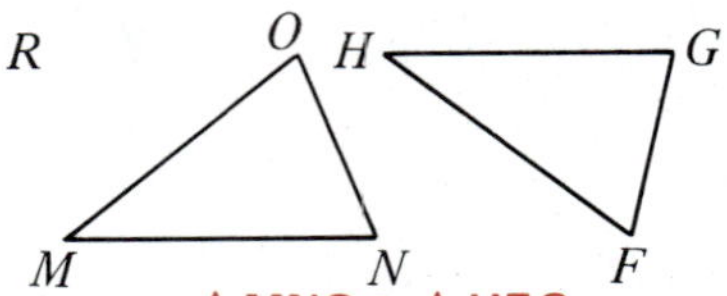

△ MNO ≅ △ HFG

Using the given information, which of the following pairs of triangles are congruent?

4. Given: $\overline{AB} \cong \overline{DE}$, $\overline{BC} \cong \overline{EF}$, and $\overline{CA} \cong \overline{FD}$

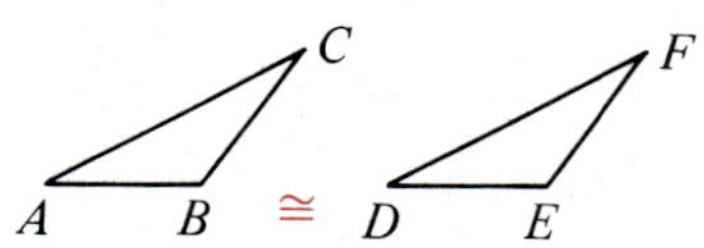

5. Given: $\overline{GI} \cong \overline{JL}$, and $\triangle GHI$ and $\triangle JKL$ are isosceles.

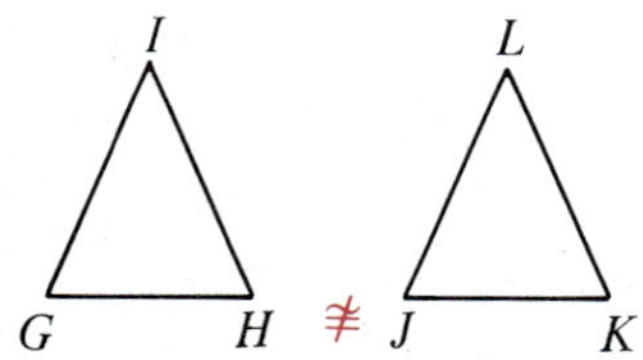

6. Given: $\overline{MN} \cong \overline{QP}$, and O bisects $\overline{NQ}$ and $\overline{MP}$.

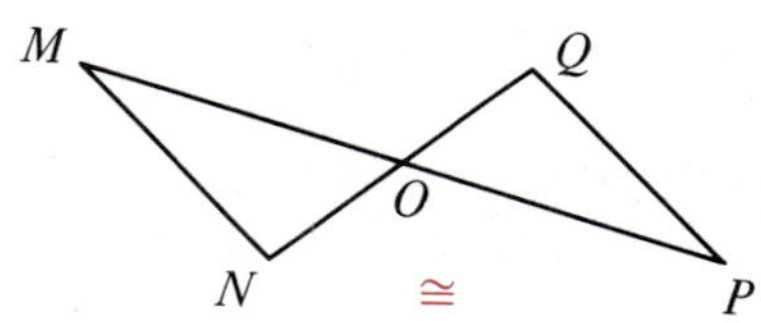

Complete the steps in each proof.

7. Given: $\overline{MP} \cong \overline{ON}$, and $\overline{OP} \cong \overline{MN}$
Prove: $\triangle MNO \cong \triangle OPM$

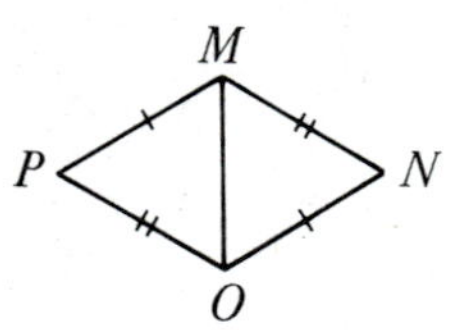

Proof

Statements	Reasons
1. $\overline{MP} \cong \overline{ON}$	1. ? **Given**
2. $\overline{OP} \cong \overline{MN}$	2. ? **Given**
3. ? **MO ≅ MO**	3. Reflexive prop.
4. ∴ $\triangle MNO \cong$? **△OPM**	4. SSS for ≅ △'s

8. Given: $\overline{TR} \cong \overline{SQ}$, U bisects $\overline{TS}$, and U bisects $\overline{RQ}$.
Prove: $\triangle UTR \cong \triangle USQ$

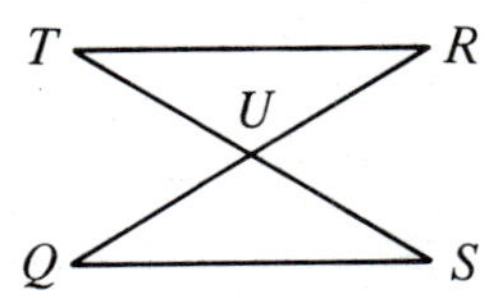

Proof

Statements	Reasons
1. $\overline{TR} \cong$? **SQ**	1. Given
2. ? **U bis. TS, RQ**	2. Given
3. $\overline{TU} \cong$? **SU**	3. Def. of bisector
4. $\overline{RU} \cong$? **QU**	4. Def. of bisector
5. ∴ $\triangle UTR \cong$ **△USQ**	5. ? **SSS for ≅ △'s**

Construct a triangle with sides congruent to the given segments. (Ex. 9–10)

9. A———B C———D E———F

10. G———H I———J K———L

11. Construct a triangle congruent to $\triangle ABC$.

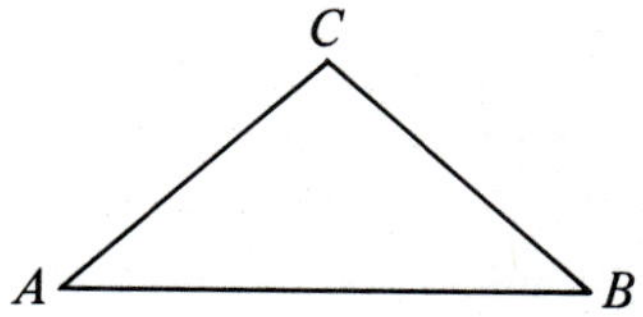

12. Construct a triangle congruent to $\triangle XYZ$.

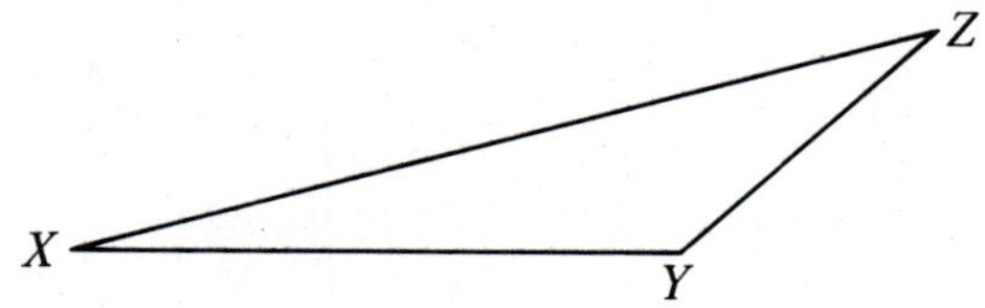

B **13.** Construct an equilateral triangle with sides congruent to $\overline{MN}$.

M———N

14. Construct an isosceles $\triangle PQR$ with $\overline{PQ} \cong \overline{AB}$ and $\overline{QR} \cong \overline{RP} \cong \overline{CD}$.

A———B C———D

15. Given: $\triangle ABC$ and $\triangle DEF$ are equilateral, and $\overline{AB} \cong \overline{DE}$.
Prove: $\triangle ABC \cong \triangle DEF$

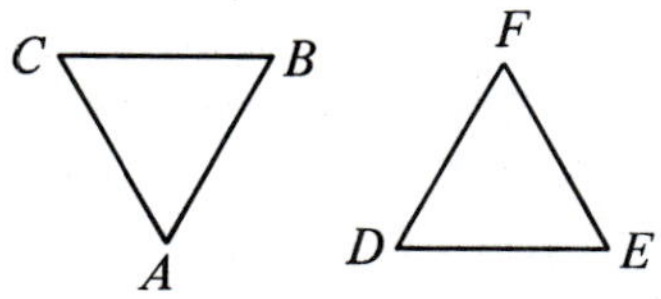

16. Given: $\triangle GHI$ is equilateral, $\overline{GI} \cong \overline{JI}$, and $\overline{GH} \cong \overline{JH}$.
Prove: $\triangle JHI$ is equilateral.

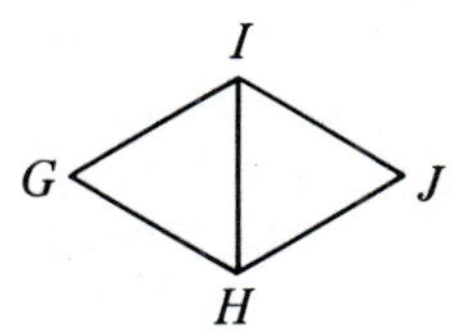

17. Given: $\overline{KL} \cong \overline{NM}$, $\overline{KQ} \cong \overline{NO}$, and $\overline{QM} \cong \overline{OL}$.
Prove: $\triangle KMQ \cong \triangle NLO$

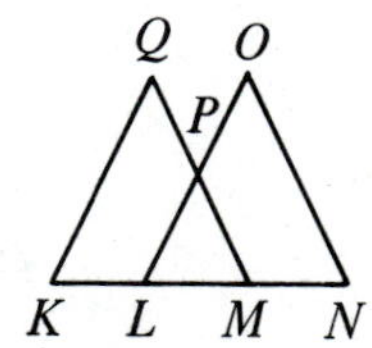

C **18.** Given: $\overline{AB} \cong \overline{DC}$, $\overline{AE} \cong \overline{DF}$, and $\overline{CE} \cong \overline{BF}$
Prove: $\triangle ACE \cong \triangle DBF$

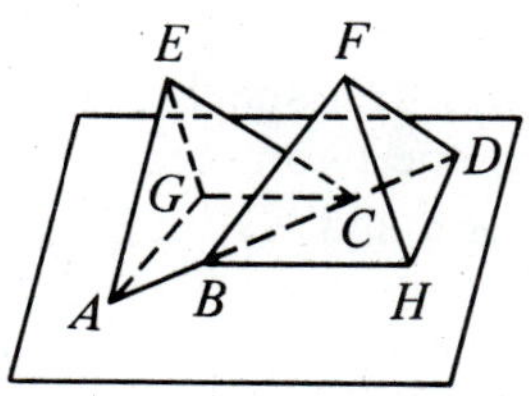

19. Given: $\overline{AD} \cong \overline{CD}$ and $\overline{DB}$ bisects $\overline{AC}$.
Prove: $\triangle ABD \cong \triangle CBD$

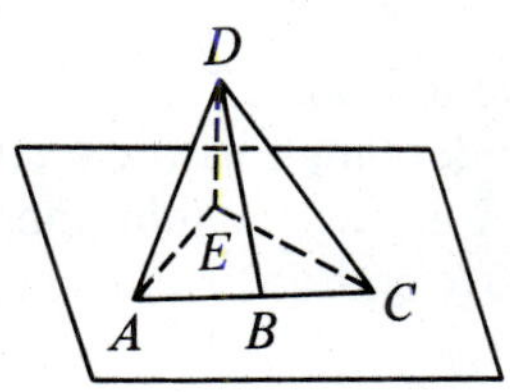

20. Given: Equilateral $\triangle XYZ$ with W the midpoint of $\overline{XY}$
Prove: $\triangle XWZ \cong \triangle YWZ$

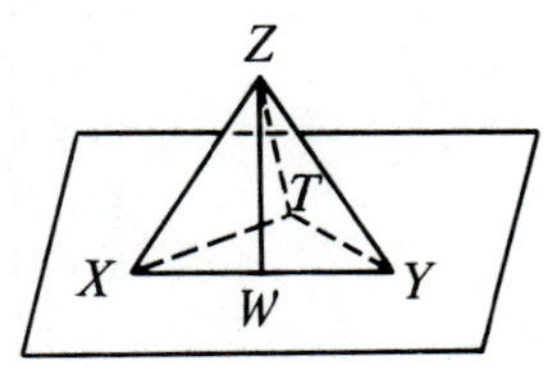

21. Given: $\triangle XYZ$, $\triangle XYW$, $\triangle XWZ$, and $\triangle YZW$ are all equilateral.
Prove: $\triangle XYZ \cong \triangle XYW \cong \triangle XWZ \cong \triangle YZW$

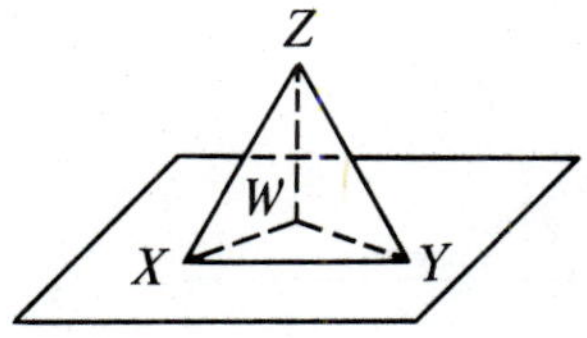

22. Given: Equilateral $\triangle XYZ$ and $\overline{XW} \cong \overline{YW} \cong \overline{ZW}$
Prove: $\triangle XWY \cong \triangle YWZ \cong \triangle ZWX$

23. Can a triangle be constructed with sides congruent to any three given segments? Why or why not? Give examples to defend your answer.
No. Sum of lengths of any two seg. must be greater than the length of the 3rd seg.

SAS AND ASA CONGRUENCE PROOFS 4.5

Objectives **To construct congruent triangles using the side-angle-side and angle-side-angle patterns**
To prove triangles congruent by the side-angle-side and the angle-side-angle correspondences

Two triangles are congruent only if their corresponding angles and sides are congruent. You know, however, that it is possible to prove two triangles congruent by showing that the three corresponding pairs of sides are congruent. There are other ways of proving that two triangles are congruent without showing that all the corresponding parts are congruent.

$\triangle DEF$ below was constructed so that $\overline{DE} \cong \overline{AB}$, $\angle E \cong \angle B$, and $\overline{EF} \cong \overline{BC}$.

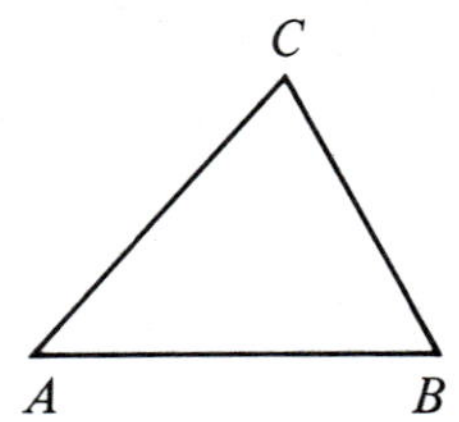

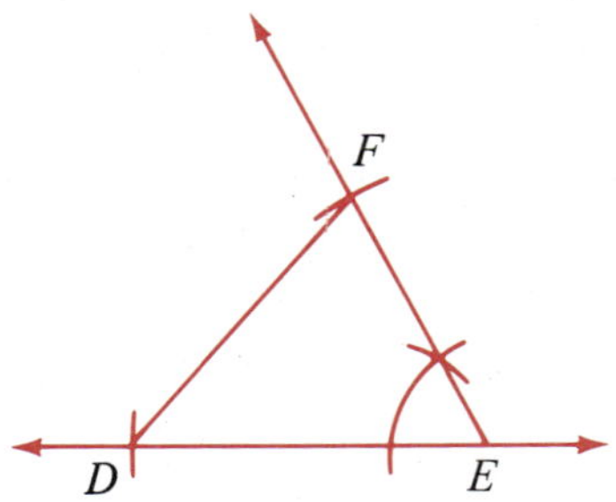

If you were to trace $\triangle DEF$ and fit it on top of $\triangle ABC$, you would see that they both have the same size and shape. This leads to the conclusion that $\triangle ABC \cong \triangle DEF$, which suggests the following postulate.

Postulate 17 SAS for ≅ △'s Two triangles are congruent if two sides and the included angle of one are congruent to the corresponding two sides and included angle of the other.

The term "included angle" in Postulate 17 refers to a specific angle of the triangle. The included angle is the angle formed by the two given sides. In the diagram below, $\angle X$ is the included angle of sides $\overline{XZ}$ and $\overline{XY}$. $\angle Q$ is the included angle of sides $\overline{QR}$ and $\overline{QP}$.

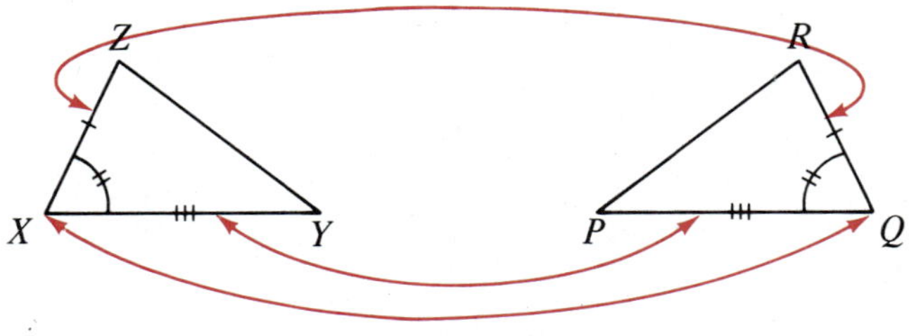

$\triangle ZXY \cong \triangle RQP$

In order to use the side-angle-side congruence postulate, you must be able to identify the appropriate sides and angles.

Example 1 To prove $\triangle ABC \cong \triangle DEF$, which side or angle of $\triangle DEF$ corresponds to the given side or angle of $\triangle ABC$?

Given	Answers
$\overline{CA}$	$\overline{FD}$
$\angle A$	$\angle D$
$\overline{AB}$	$\overline{DE}$

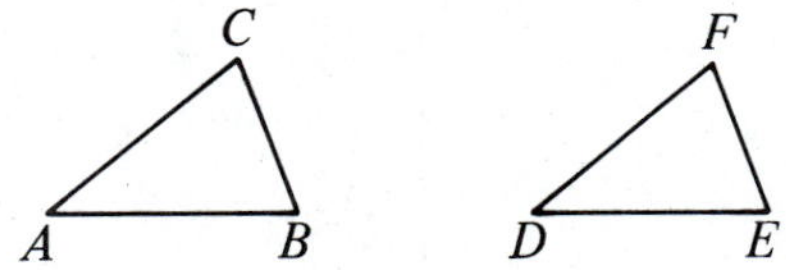

Example 2 Based on the markings shown, are the triangles congruent? If so, write a congruence statement.

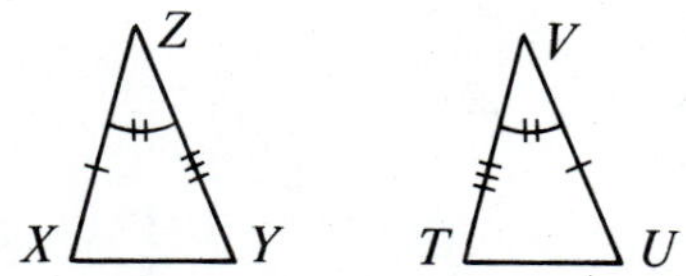

Yes, $\triangle XYZ \cong \triangle UTV$. ◄ *Apply Postulate 17.*

A triangle can be constructed congruent to another triangle by another method. This construction involves an angle-side-angle pattern.

Example 3 Given $\triangle ABC$. Construct $\triangle A'B'C'$ so that $\angle A' \cong \angle A$, $\overline{A'B'} \cong \overline{AB}$, and $\angle B' \cong \angle B$.

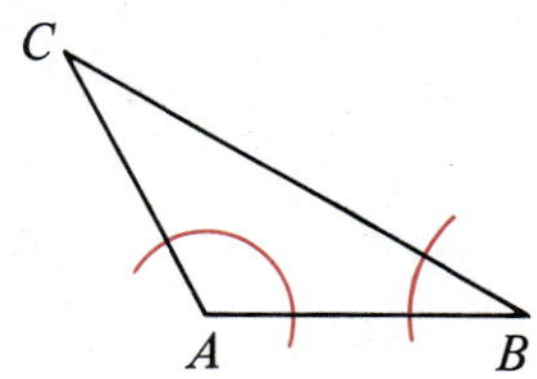

Construct $\overline{A'B'} \cong \overline{AB}$. Construct $\angle PA'B' \cong \angle CAB$. Construct $\angle A'B'C' \cong \angle ABC$.

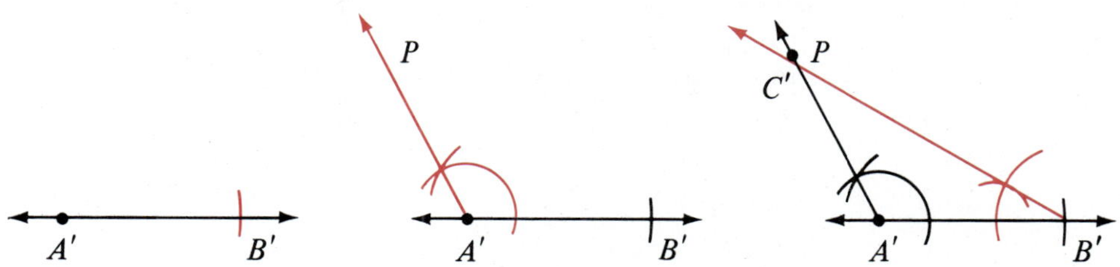

Conclusion: $\triangle A'B'C' \cong \triangle ABC$

In Example 3, if you measure all the angles and sides of $\triangle ABC$ and $\triangle A'B'C'$, you will find that the corresponding angles and sides are congruent.

Postulate 18 ASA for $\cong \triangle$'s Two triangles are congruent if two angles and the included side of one are congruent to the corresponding two angles and included side of the other.

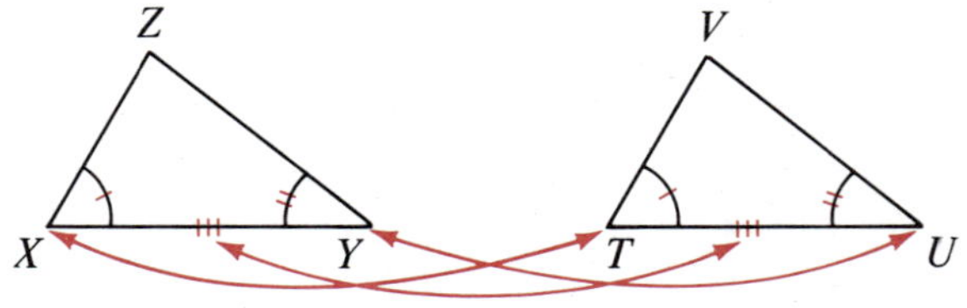

$\triangle XYZ \cong \triangle TUV$

The "included side" mentioned in Postulate 18 is the side whose endpoints are the vertices of the two given angles. In $\triangle XYZ$ above, $\overline{XY}$ is the included side of $\angle X$ and $\angle Y$. $\overline{TU}$ is the included side of $\angle T$ and $\angle U$.

To prove triangles congruent, you must first decide which congruence pattern must be used. Once that is determined, you can then complete the proof.

Example 4 Using the given information, mark the congruent parts on the diagram. State which congruence pattern uses the given information.

Given: $\overline{AC} \cong \overline{EC}$ and
$\overline{BC} \cong \overline{DC}$
Vertical angles are $\cong$. ►

There is a SAS pattern.

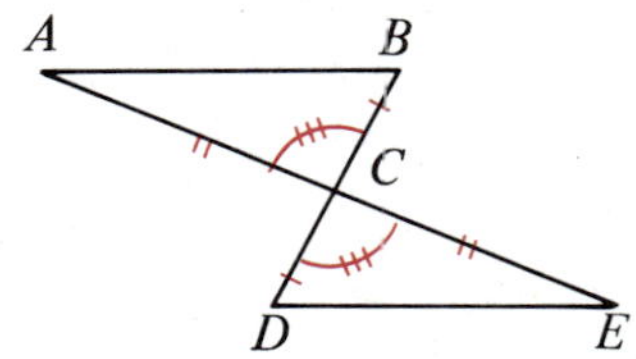

Example 5 Given: $\overline{HI}$ bisects $\angle FHG$, and
$\overline{HI} \perp \overline{FG}$.
Prove: $\triangle FIH \cong \triangle GIH$

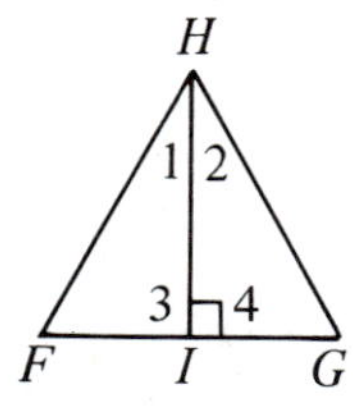

Proof

Statements		Reasons
1. $\overline{HI}$ bisects $\angle FHG$.		1. Given
2. $\angle 1 \cong \angle 2$	(A)	2. Def. of $\angle$ bisector
3. $\overline{HI} \cong \overline{HI}$	(S)	3. Reflexive property
4. $\overline{HI} \perp \overline{FG}$		4. Given
5. $\angle 3 \cong \angle 4$	(A)	5. $\perp$ forms $\cong$ rt. $\angle$'s.
6. $\therefore \triangle FIH \cong \triangle GIH$		6. ASA for $\cong \triangle$'s

Oral Exercises

For each pair of sides of each triangle, identify the included angle.

1\. 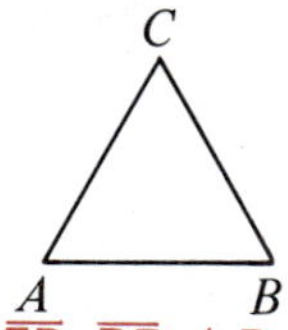

$\overline{CA}, \overline{AB}: \angle A$; $\overline{AB}, \overline{BC}: \angle B$; $\overline{BC}, \overline{AC}: \angle C$

2\. 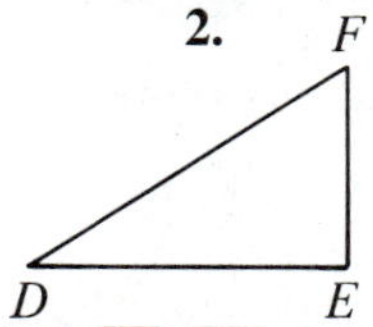

$\overline{FD}, \overline{DE}: \angle D$; $\overline{DF}, \overline{FE}: \angle F$; $\overline{FE}, \overline{DE}: \angle E$

3\. 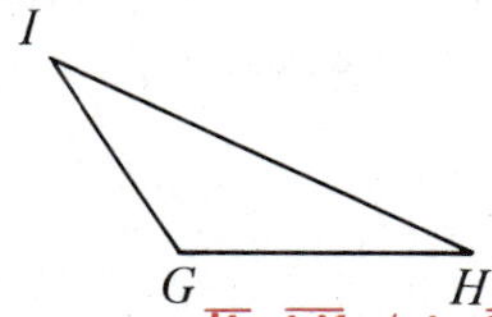

$\overline{HG}, \overline{GI}: \angle G$; $\overline{HI}, \overline{IG}: \angle I$; $\overline{GH}, \overline{IH}: \angle H$

4\. 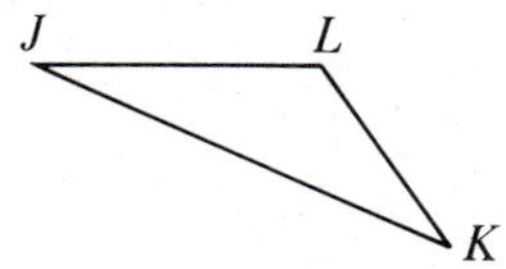

$\overline{JL}, \overline{LK}: \angle L$; $\overline{LK}, \overline{KJ}: \angle K$; $\overline{KJ}, \overline{JL}: \angle J$

For each pair of angles of each triangle, identify the included side.

5\. 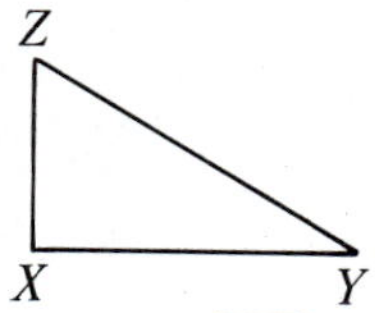

$\angle X, \angle Z: \overline{XZ}$; $\angle Z, \angle Y: \overline{ZY}$; $\angle X, \angle Y: \overline{XY}$

6\. 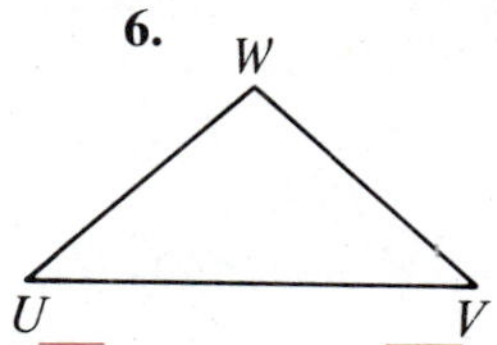

$\angle U, \angle W: \overline{UW}$; $\angle U, \angle V: \overline{UV}$; $\angle V, \angle W: \overline{VW}$

7\. 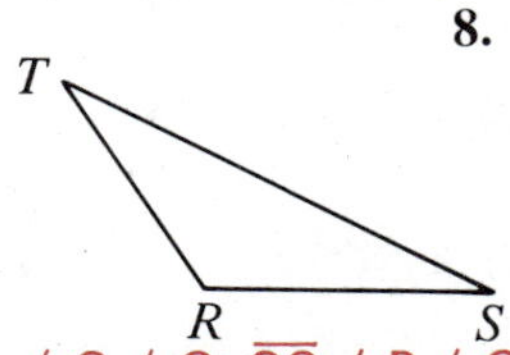

7\. $\angle T, \angle R: \overline{TR}$; $\angle S, \angle T: \overline{ST}$; $\angle S, \angle R: \overline{SR}$

8\. 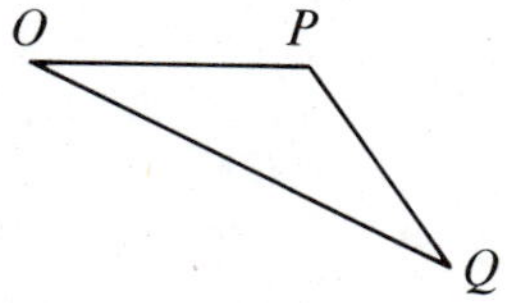

$\angle O, \angle Q: \overline{OQ}$; $\angle P, \angle Q: \overline{PQ}$; $\angle O, \angle P: \overline{OP}$

Written Exercises

A 1. Draw an acute triangle. Construct a triangle congruent to it by SAS; by ASA.

2\. Draw an obtuse triangle. Construct a triangle congruent to it by SAS; by ASA.

1.–2. Follow constr., p. 118, for SAS and Ex. 3, p. 119, for ASA.

Complete each statement.

3\. To prove that $\triangle GOP \cong \triangle ERA$ by SAS, you must show that $\overline{PO} \cong$?, $\angle O \cong$?, and $\overline{OG} \cong$?.
$\overline{AR}$, $\angle R$; $\overline{RE}$

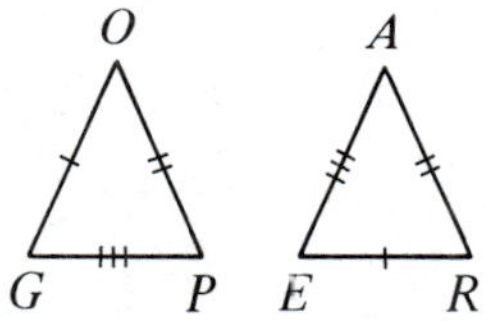

4\. To prove that $\triangle OPG \cong \triangle AER$ by ASA, you must show that $\angle G \cong$?, $\overline{GO} \cong$?, and $\angle O \cong$?.
$\angle R$; $\overline{RA}$; $\angle A$

Based on the markings shown, are the triangles congruent? If so, write a congruence statement.

5\. 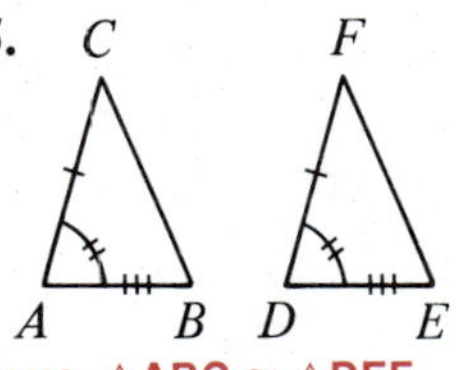

yes, $\triangle ABC \cong \triangle DEF$

6\. 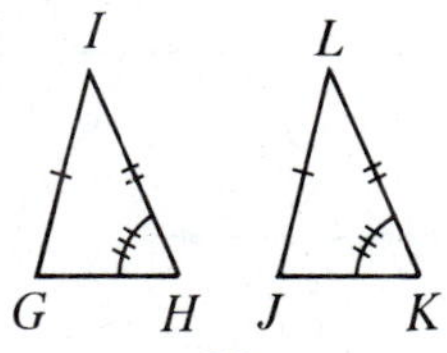

no

7\. 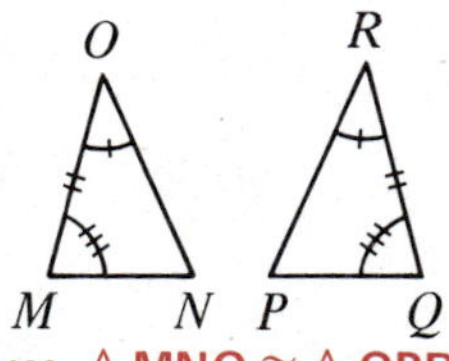

yes, $\triangle MNO \cong \triangle QPR$

Using the given information, mark the congruent parts on the diagram. State which congruence pattern uses the given information.

8\. Given: $\overline{ZY} \perp \overline{XY}$, $\overline{WV} \perp \overline{UV}$, $\angle X \cong \angle U$, and $\overline{XY} \cong \overline{UV}$

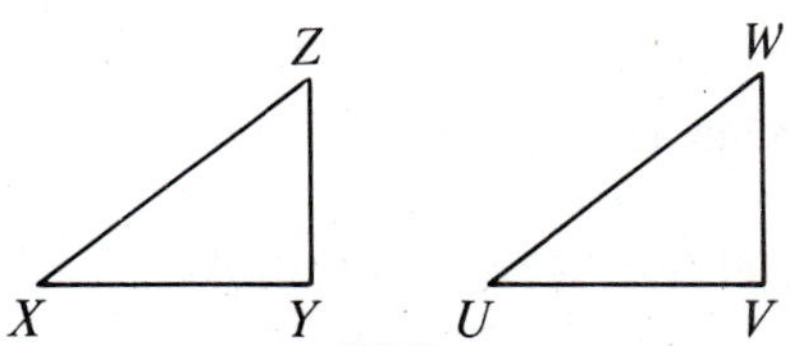

ASA

9\. Given: $\angle C \cong \angle F$, $\overline{AC} \cong \overline{EF}$, and $\overline{BC} \cong \overline{DF}$

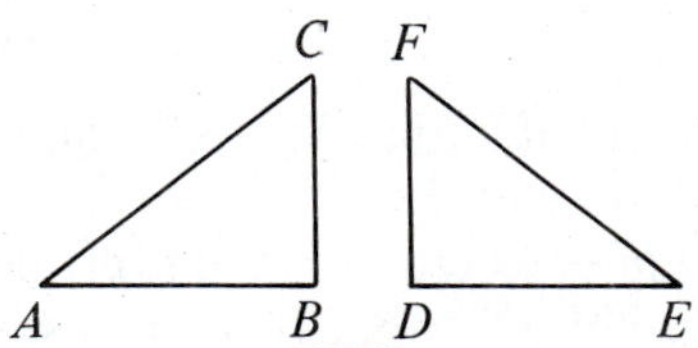

SAS

10\. Given: P bisects $\overline{TR}$ and P bisects $\overline{QS}$.

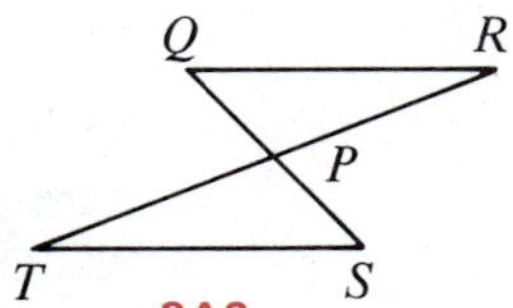

SAS

Based on the given information, are the triangles congruent? Why or why not? (Ex. 11–13)

11. Given: $\overline{AB} \cong \overline{ED}$, $\overline{AC} \cong \overline{FE}$, $\overline{CA} \perp \overline{AB}$, and $\overline{FE} \perp \overline{ED}$

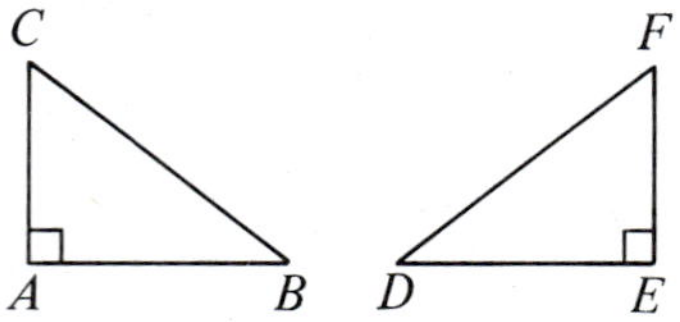

yes, SAS

12. Given: $\overline{GI} \cong \overline{JL}$, $\angle G \cong \angle J$, and $\angle H \cong \angle K$

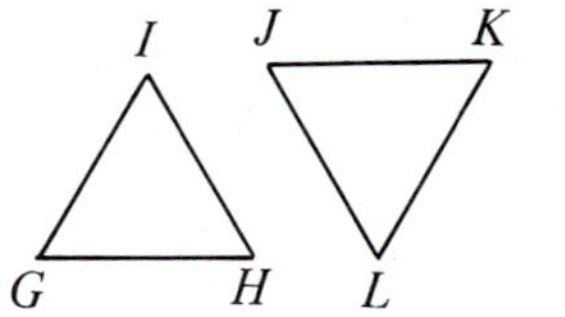

yes, by Th 3.6 and ASA

13. Given: $\triangle MNO$ and $\triangle QPR$ are right $\triangle$'s, m $\angle M = 31$, m $\angle Q = 31$, and $\overline{MO} \cong \overline{QR}$

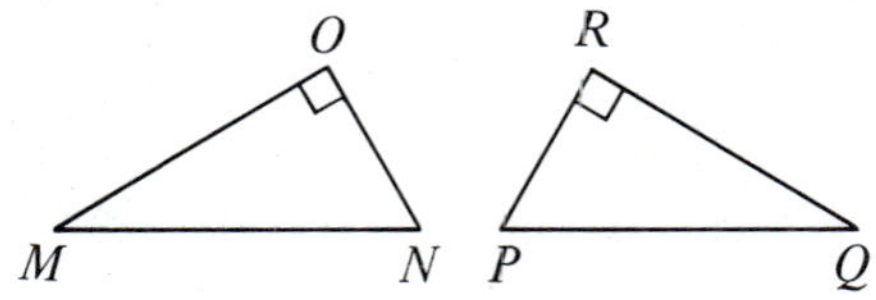

yes, ASA

14. Given: $\overline{AE} \cong \overline{CE}$ and $\overline{DE} \cong \overline{BE}$
Prove: $\triangle ADE \cong \triangle CBE$

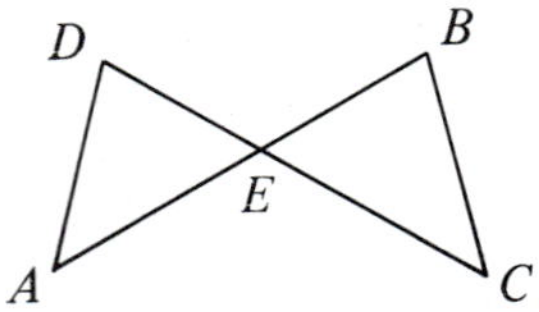

15. Given: J bisects $\overline{FG}$, and $\angle F \cong \angle G$.
Prove: $\triangle FJH \cong \triangle GJI$

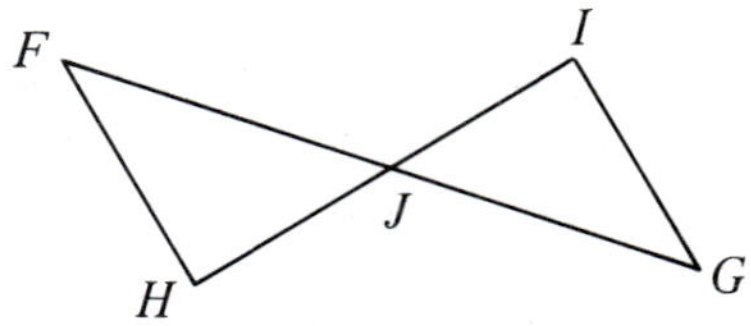

16. Given: $\angle KNL \cong \angle MNL$, and $\overline{KN} \cong \overline{MN}$
Prove: $\triangle KNL \cong \triangle MNL$

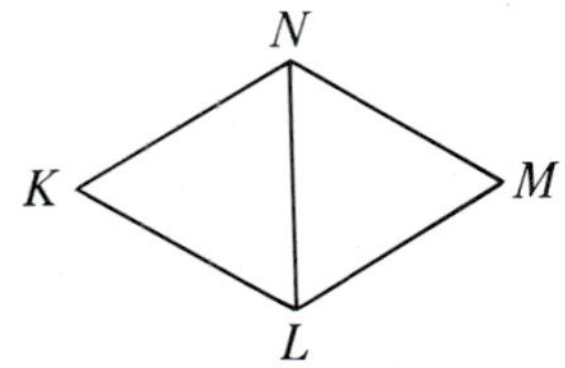

B **17.** Given: $\overline{AC} \cong \overline{BC}$ and $\overline{CD}$ bisects $\angle ACB$.
Prove: $\triangle ADC \cong \triangle BDC$

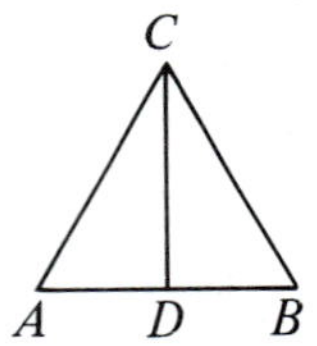

18. Given: $\overline{HF} \perp \overline{GE}$, and $\overline{HF}$ bisects $\overline{GE}$.
Prove: $\triangle GFH \cong \triangle EFH$

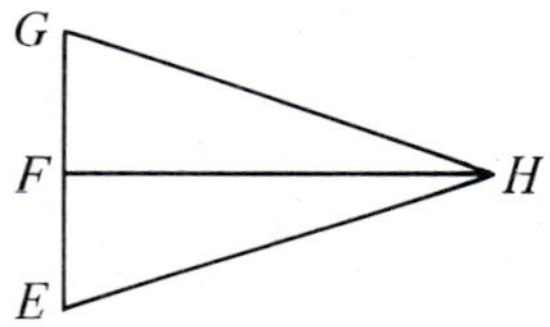

19. Given: $\overline{LK} \parallel \overline{IJ}$, and $\overline{LI} \parallel \overline{KJ}$
Prove: $\triangle LKI \cong \triangle JIK$

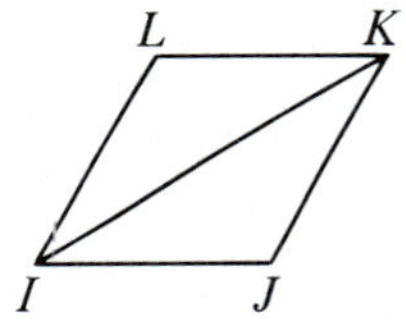

20. Construct a triangle with sides congruent to $\overline{MN}$ and $\overline{OP}$ and the included angle congruent to $\angle Q$.

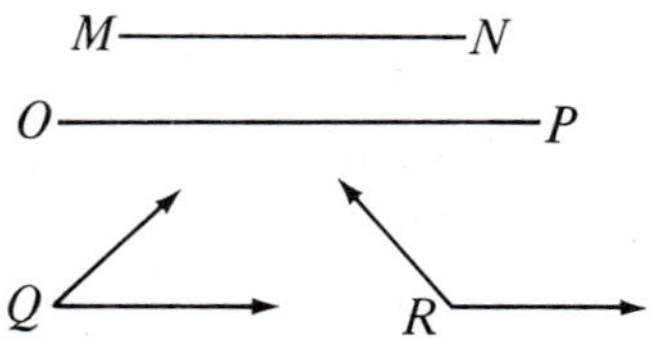

21. Construct a triangle with angles congruent to $\angle Q$ and $\angle R$, and the included side congruent to $\overline{MN}$.

C **22.** Given: $\overline{UV} \perp$ all lines in plane p through V, $\overline{SV}$ and $\overline{TV}$ are in p, and $\angle SUV \cong \angle TUV$.
Prove: $\triangle SVU \cong \triangle TVU$

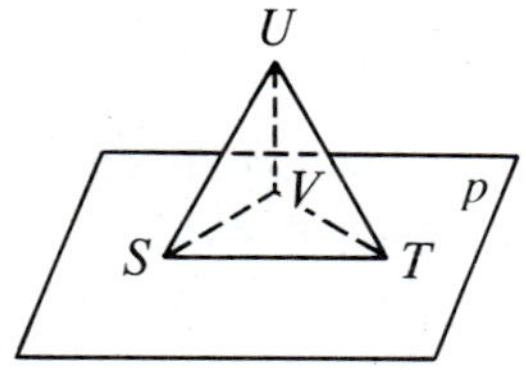

23. Given: $\triangle STU$, $\triangle SVU$, $\triangle TVU$, and $\triangle STV$ are all equiangular.
Prove: $\triangle STU \cong \triangle SVU \cong \triangle TVU \cong \triangle STV$.

CUMULATIVE REVIEW

1. If the measure of one angle of a triangle is 87° and the measure of a second angle is 49°, what is the measure of the third angle? 44°
2. If two parallel lines are intersected by a transversal so that one angle of a pair of corresponding angles has a measure of 67°, what is the measure of the other angle in the pair? 67°

OTHER CONGRUENCE PROOFS 4.6

Objectives **To recognize patterns for proving triangles congruent**
To identify congruent triangles and prove them congruent

So far you have used three different patterns to prove triangles congruent: side-side-side, side-angle-side, and angle-side-angle. There are other possibilities that should be considered. One such possibility is an angle-angle-angle pattern.

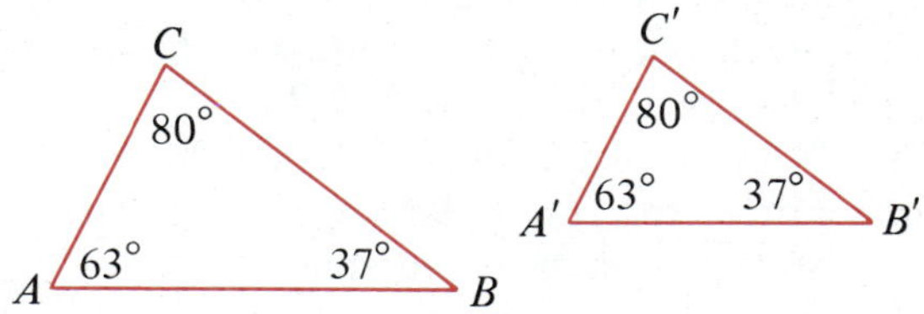

In the two triangles shown, $\angle A \cong \angle A'$, $\angle B \cong \angle B'$, and $\angle C \cong \angle C'$. But $\triangle ABC$ is not congruent to $\triangle A'B'C'$. Although the triangles are the same shape, they obviously are not the same size. The angle-angle-angle correspondence does not guarantee triangle congruence.

Another possibility is a side-side-angle pattern. You can use three sticks to find out if it works. In the diagram below, stick s and stick t form two sides of fixed length and stick t and stick r form a fixed angle. As shown, stick s can be moved so that it meets stick r in two different positions, thus forming two distinct triangles. Even though the lengths of s and t and the measure of the angle are the same in both triangles, the triangles are not congruent. Thus, the side-side-angle correspondence does not guarantee triangle congruence.

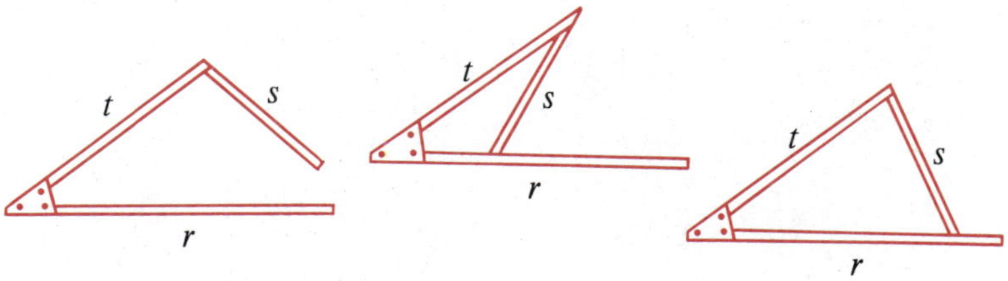

Example 1 Using the markings shown, are the triangles congruent? Why or why not?

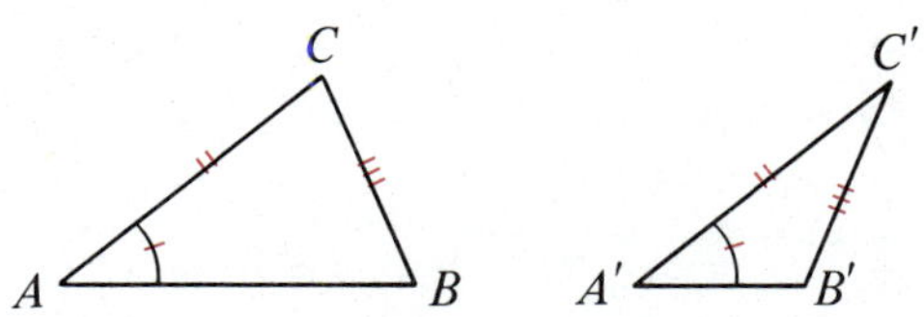

$\triangle ABC \not\cong \triangle A'B'C'$. Side-side-angle is not a triangle congruence pattern.

There is a third possible method for showing that two triangles are congruent: an angle-angle-side pattern. In the two triangles below, $\angle Z \cong \angle B$, $\angle X \cong \angle C$, and $\overline{XY} \cong \overline{CA}$. $\triangle ZXY$ appears to be congruent to $\triangle BCA$.

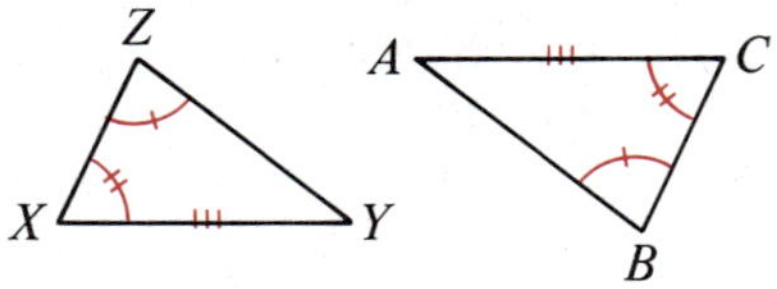

The information given does not fit any congruence pattern already used. The given congruent sides are not the included sides of the given congruent angles. However, since $\angle Z \cong \angle B$ and $\angle X \cong \angle C$, then $\angle Y \cong \angle A$. (See Theorem 3.6.) Now you have an angle-side-angle pattern, which assures you that $\triangle ZXY$ is congruent to $\triangle BCA$. So the AAS pattern does assure triangle congruence. This is stated in the next theorem.

Theorem 4.2 (AAS) Two triangles are congruent if two angles and the sides opposite one of them are congruent to the corresponding two angles and side of the other. (See Exercise 19.)

Given: $\angle A \cong \angle A'$, $\angle B \cong \angle B'$,
and $\overline{BC} \cong \overline{B'C'}$
Prove: $\triangle ABC \cong \triangle A'B'C'$

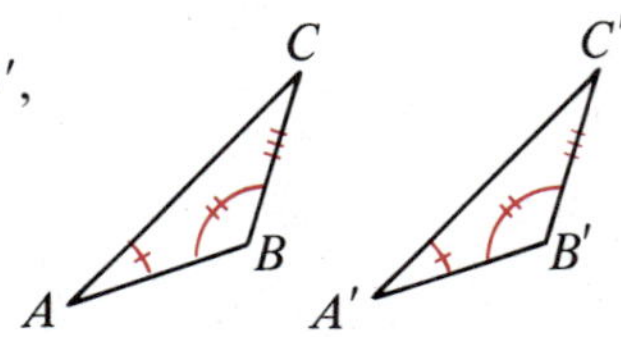

ANALYSIS
Use ASA.

Example 2 Using the markings shown, are the triangles congruent? Why or why not?

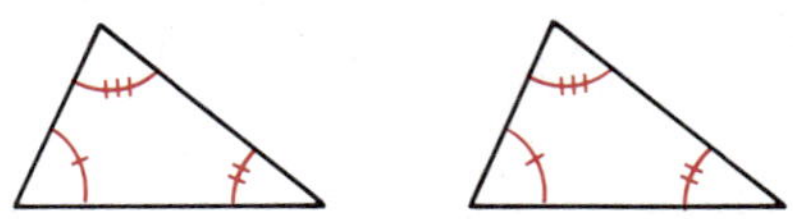

No, AAA is not a congruence pattern.

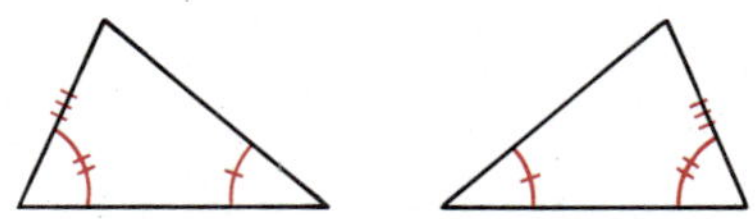

Yes, AAS is a congruence pattern.

Example 3 Based on the given information, is there an ASA, SAS, SSS, AAA, SSA, or AAS correspondence? Are the triangles necessarily congruent?

Given: $\angle Z \cong \angle C$, $\overline{ZY} \cong \overline{CB}$, and
$\angle Y \cong \angle B$

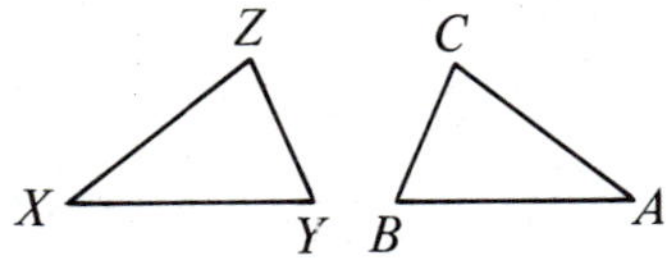

ASA. The triangles are congruent.

Summary

You *can* prove triangles congruent by:

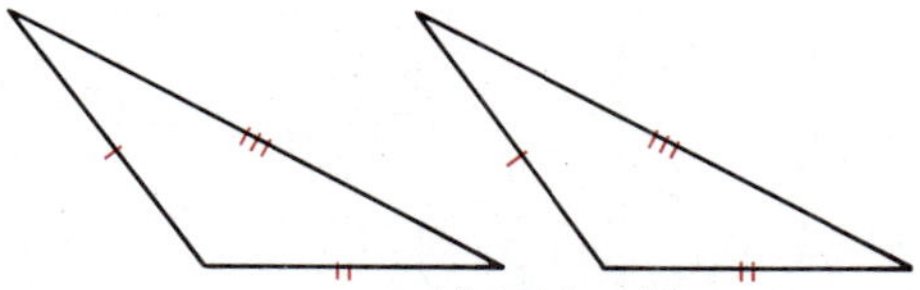

side–side–side

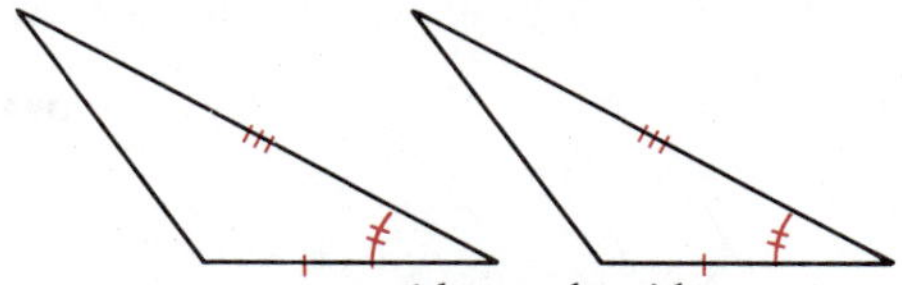

side–angle–side

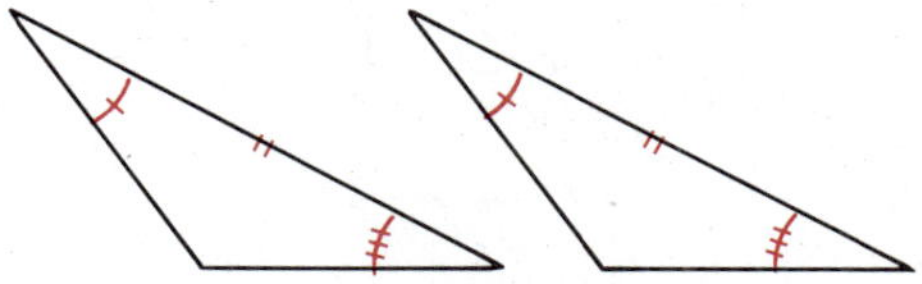

angle–side–angle

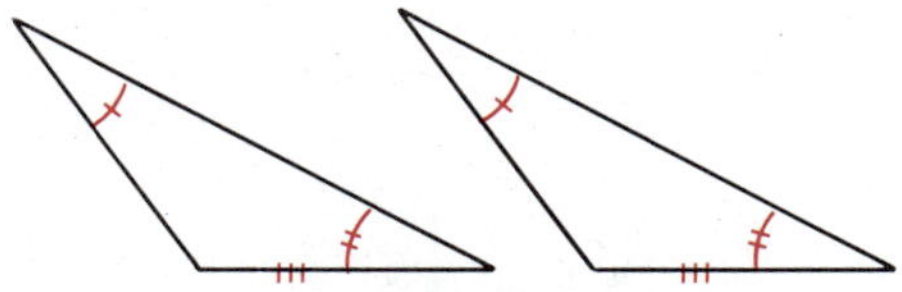

angle–angle–side

You *cannot* prove triangles congruent by:

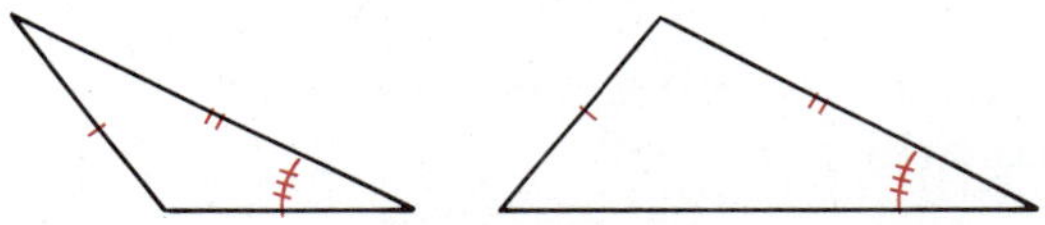

side–side–angle

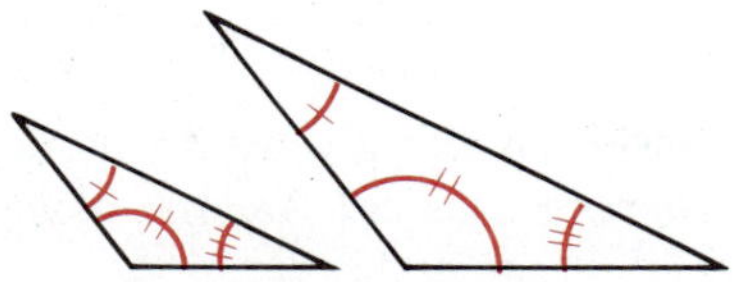

angle–angle–angle

When you are trying to prove that two triangles are congruent, you must decide which congruence pattern fits the information given. To get the necessary congruent sides or angles, you may have to recall previously learned theorems, postulates, definitions, and so on.

Example 4

Given: $\overline{AB} \cong \overline{ED}$ and $\overline{DE} \parallel \overline{AB}$
Prove: $\triangle ABC \cong \triangle EDC$

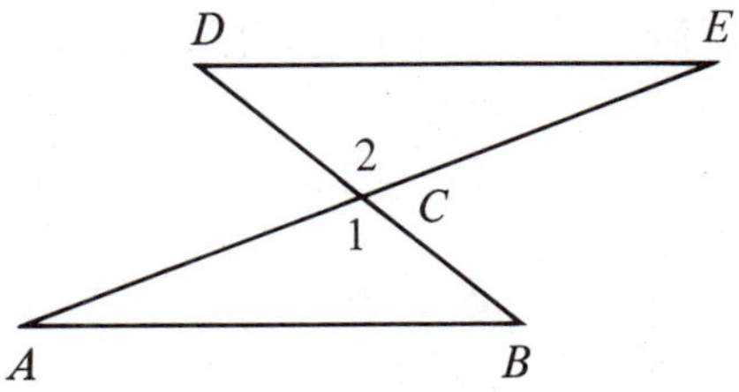

Proof

Statements		Reasons
1. $\overline{DE} \parallel \overline{AB}$		1. Given
2. $\angle A \cong \angle E$	(A)	2. Alt. int. ∠'s of ∥ lines are ≅.
3. $\angle 1 \cong \angle 2$	(A)	3. Vertical ∠'s are ≅.
4. $\overline{AB} \cong \overline{ED}$	(S)	4. Given
5. $\therefore \triangle ABC \cong \triangle EDC$		5. AAS

Oral Exercises

Using the markings shown are the triangles congruent? Why or why not?

1.

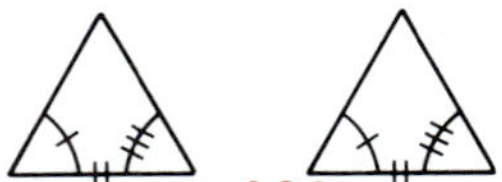

≅; ASA

2.

≅; AAS

3.

not necessarily ≅

4.

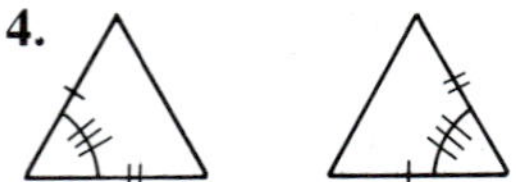

≅; SAS

5.

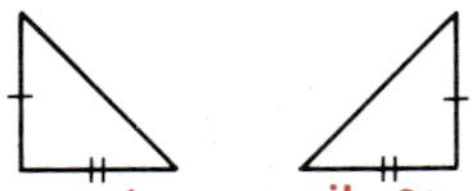

not necessarily ≅

6.

≅; SSS

7.

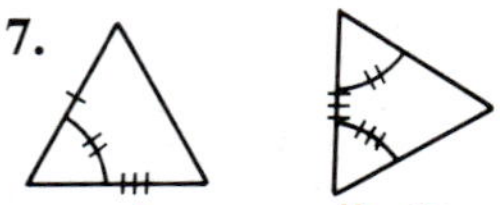

not necessarily ≅

8.

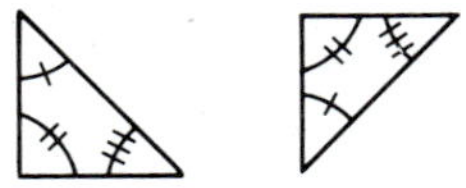

not necessarily ≅

9.

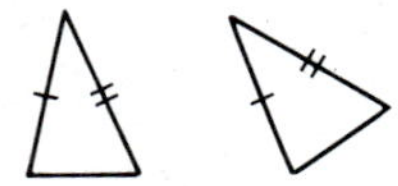

not necessarily ≅

Written Exercises

Based on the given information, is there an ASA, SAS, SSS, SSA, AAA, or AAS correspondence? Are the triangles necessarily congruent?

A

1. Given: $\overline{AB} \cong \overline{DE}$, $\overline{BC} \cong \overline{EF}$, and $\angle B \cong \angle E$ **SAS, yes**
2. Given: $\angle C \cong \angle F$, $\angle A \cong \angle D$, and $\overline{BC} \cong \overline{EF}$ **AAS, yes**
3. Given: $\overline{AC} \cong \overline{DF}$, $\overline{AB} \cong \overline{DE}$, and $\angle C \cong \angle F$ **SSA, no**
4. Given: $\angle A \cong \angle D$, $\angle B \cong \angle E$, and $\angle C \cong \angle F$ **AAA, no**
5. Given: $\angle A \cong \angle D$, $\overline{AC} \cong \overline{DF}$, and $\angle C \cong \angle F$ **ASA, yes**
6. Given: $\overline{AC} \cong \overline{DF}$, $\overline{AB} \cong \overline{DE}$, and $\overline{BC} \cong \overline{EF}$ **SSS, yes**

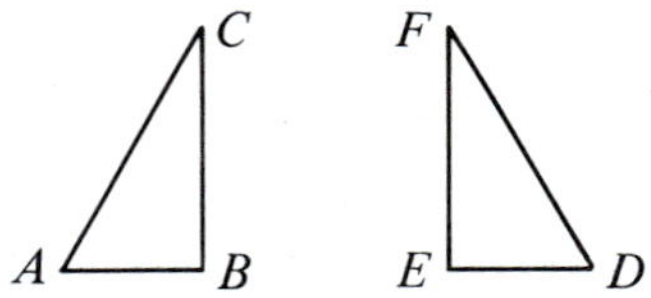

Complete the steps in each proof. (Ex. 7–8)

7. Given: $\angle B \cong \angle D$ and $\overline{DC} \parallel \overline{AB}$
Prove: $\triangle ABC \cong \triangle CDA$.

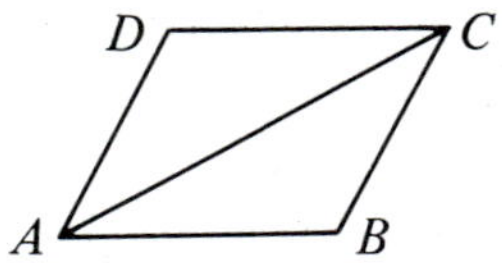

Proof

Statements		Reasons
1. $\angle B \cong \angle D$	(A)	1. ? **Given**
2. ? **DC ∥ AB**		2. Given
3. $\angle BAC \cong$? **3. ∠ACD**	(A)	3. Alt int. ∠'s of ∥ lines are ≅.
4. $\overline{AC} \cong \overline{CA}$	(S)	4. ? **Reflexive prop.**
5. ∴ ? **△ABC ≅ △CDA**		5. ? **AAS**

8. Given: F is the midpoint of $\overline{EH}$ and $\overline{GI}$.
Prove: $\triangle EFI \cong \triangle HFG$.

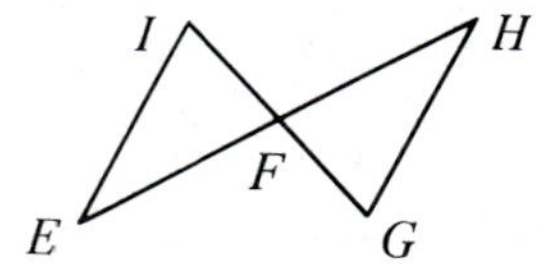

Proof

Statements		Reasons
1. ? **1. F is midpt of $\overline{EH}$ and $\overline{GI}$.**		1. Given
2. $\overline{EF} \cong \overline{HF}$	(S)	2. ? **Def. of midpt.**
3. $\angle EFI \cong$? **3. ∠HFG**	(A)	3. Vert. ∠'s are ≅.
4. $\overline{IF} \cong$? **FG**	(S)	4. ? **Def. of midpt.**
5. ∴ ? **△EFI ≅ △HFG**		5. ? **SAS**

9. Given: $\overline{MN} \cong \overline{KN}$ and $\angle M \cong \angle K$
Prove: $\triangle JNM \cong \triangle LNK$

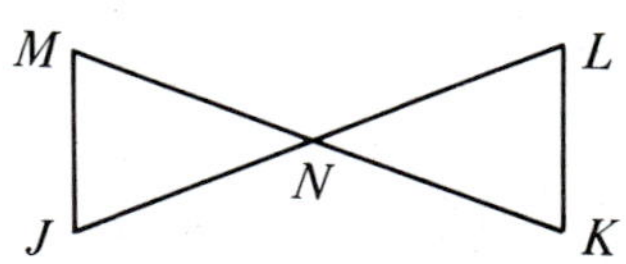

10. Given: $\angle J \cong \angle K$ and $\overline{NM} \cong \overline{NL}$
Prove: $\triangle JNM \cong \triangle KNL$

11. Given: C is the midpoint of $\overline{EB}$ and $\angle B \cong \angle E$.
Prove: $\triangle ABC \cong \triangle DEC$

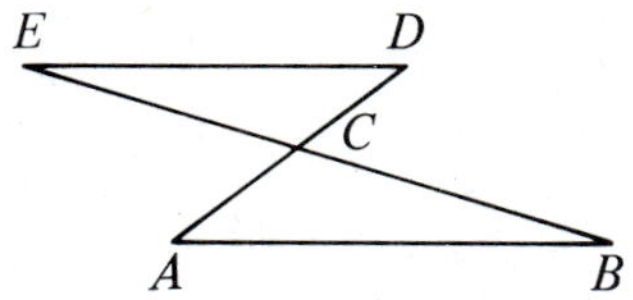

12. Given: $\overline{IG}$ bisects $\angle FIH$ and $\angle F \cong \angle H$.
Prove: $\triangle FGI \cong \triangle HGI$

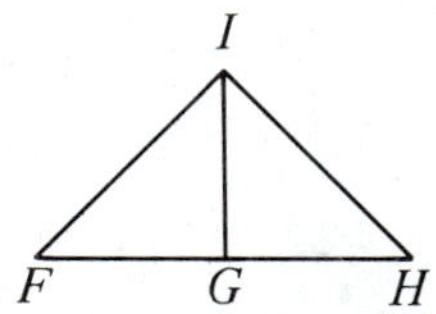

13. Given: $\overline{ML} \cong \overline{KJ}$ and $\overline{KL} \cong \overline{MJ}$
Prove: $\triangle MLK \cong \triangle KJM$

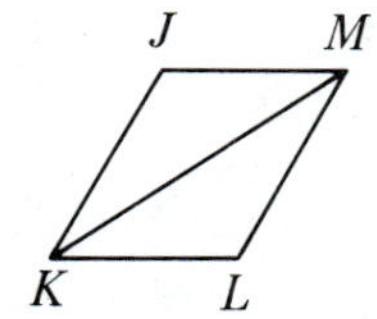

14. Given: $\overline{RO} \cong \overline{NO}$ and $\angle R \cong \angle N$
Prove: $\triangle ROQ \cong \triangle NOP$

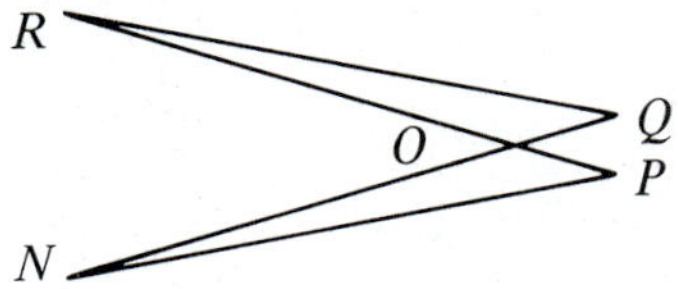

15. Given: $\overline{TV} \perp \overline{US}$ and $\overline{TV}$ bisects $\overline{US}$.
Prove: $\triangle UVT \cong \triangle SVT$

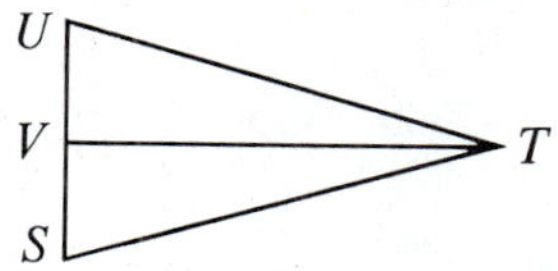

16. Given: $\overline{ZY} \parallel \overline{WX}$ and $\angle W \cong \angle Y$
Prove: $\triangle ZWX \cong \triangle XYZ$

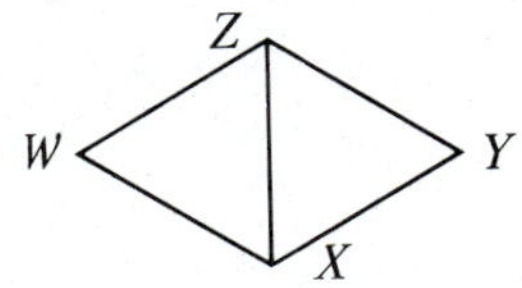

B **17.** Given: $\overline{FD} \parallel \overline{AC}$, $\overline{BF} \parallel \overline{CE}$, $\angle A \cong \angle D$, and $\overline{BF} \cong \overline{EC}$
Prove: $\triangle ABF \cong \triangle DEC$

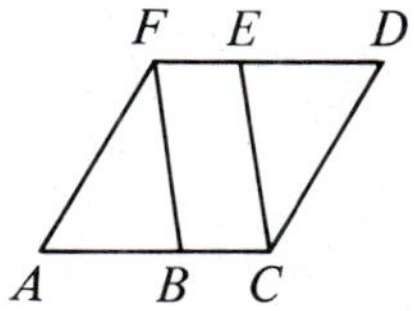

18. Given: $\overline{BF} \cong \overline{EC}$, $\overline{AB} \cong \overline{DE}$, and $\angle FBC \cong \angle CEF$
Prove: $\triangle ABF \cong \triangle DEC$

19. Prove Theorem 4.2.

C **20.** Given: $\overline{IH} \perp$ plane p, $\overline{GH}$ and $\overline{JH}$ are in p, and $\angle G \cong \angle J$.
Prove: $\triangle GHI \cong \triangle JHI$

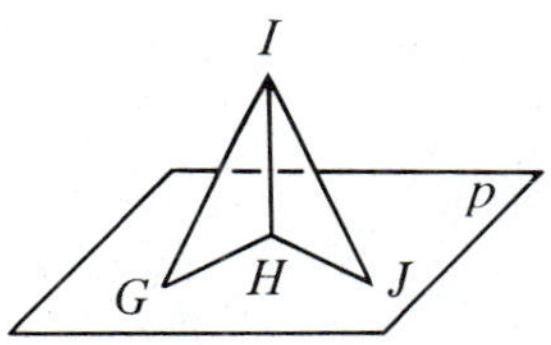

21. Given: $\overline{IH} \perp$ plane p, $\overline{GH}$ and $\overline{JH}$ are in p, $\overline{GH} \cong \overline{IH}$, and $\overline{JH} \cong \overline{IH}$.
Prove: $\triangle GHI \cong \triangle JHI$

NON-ROUTINE PROBLEMS

Several different squares can be seen in this figure. There are nine different ones of size A, four of size B, and one of size C.

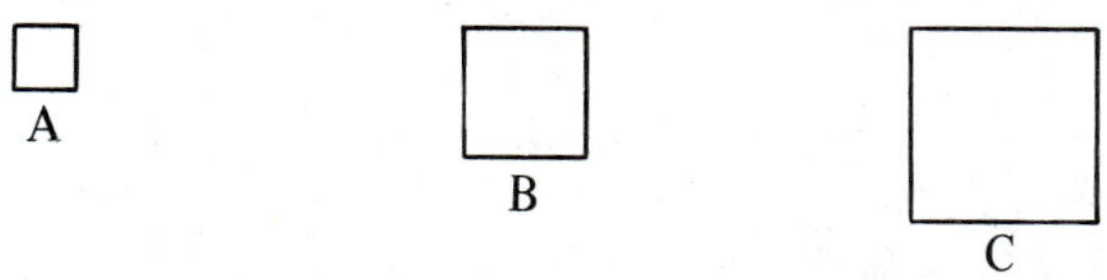

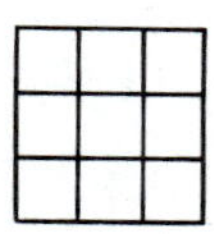

How many different squares are there in each of the following figures? Is there a formula in terms of *n* and *m*, the numbers of units in the sides?

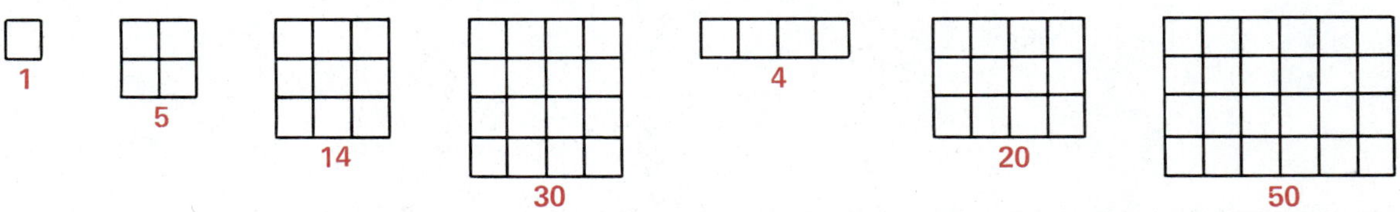

If m < n, $N = \dfrac{-m^3 + 3nm^2 + 3nm + m}{6}$ If m = n, $N = \dfrac{n(n+1)(2n+1)}{6}$

APPLICATIONS

Read → Plan → Solve → Interpret

1. An ancient device was invented for measuring the distance from the edge of the ocean to a ship at sea. The hinged rods can be adjusted to sight the ship. The device can also be turned around on the post to sight a point on land. How could this device be used to find the distance from ship to shore? Why does this work?

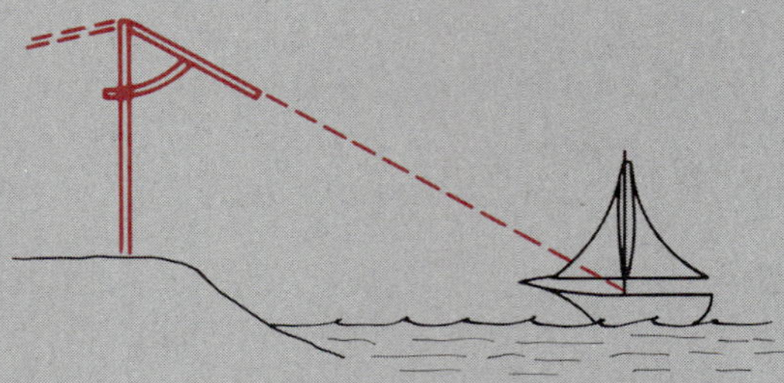

2. A forest ranger needs to build a small bridge over a creek. She does not have any pieces of wood long enough to reach all the way across. She decides to hinge the sides in the middle, with rope stretched to keep the bridge from sagging. Which of the two designs is the correct one to use? Why?

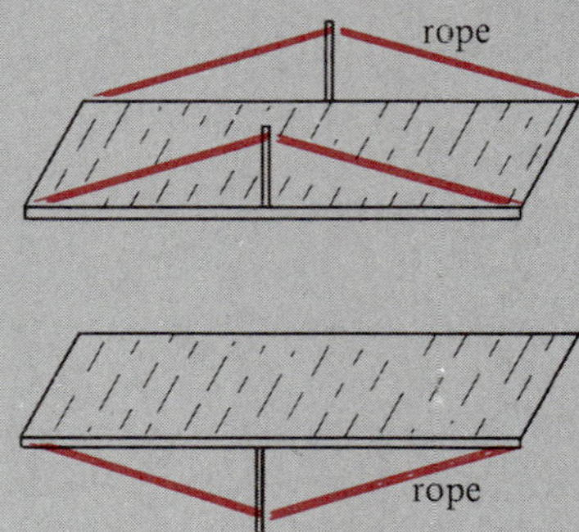

COMPUTER ACTIVITIES

Triangles and Flowcharts

Objective: **To use a flowchart to classify several triangles by the lengths of their sides**

When you are working with a computer program that contains many IF . . . THEN statements (see page 569), a flowchart of the program can help clarify the operation of the program's various branches. Although flowcharts employ many conventions, the main concern now is with only one of them—decision boxes. Decision boxes are the diamond shapes representing yes-no decisions in flowcharts. A yes decision is represented by an arrow branching left or right from a decision box, a no decision by an arrow branching down. Enter the sides of each triangle below into the flowchart and follow each set of inputs through it to see that the flowchart really does classify triangles correctly.

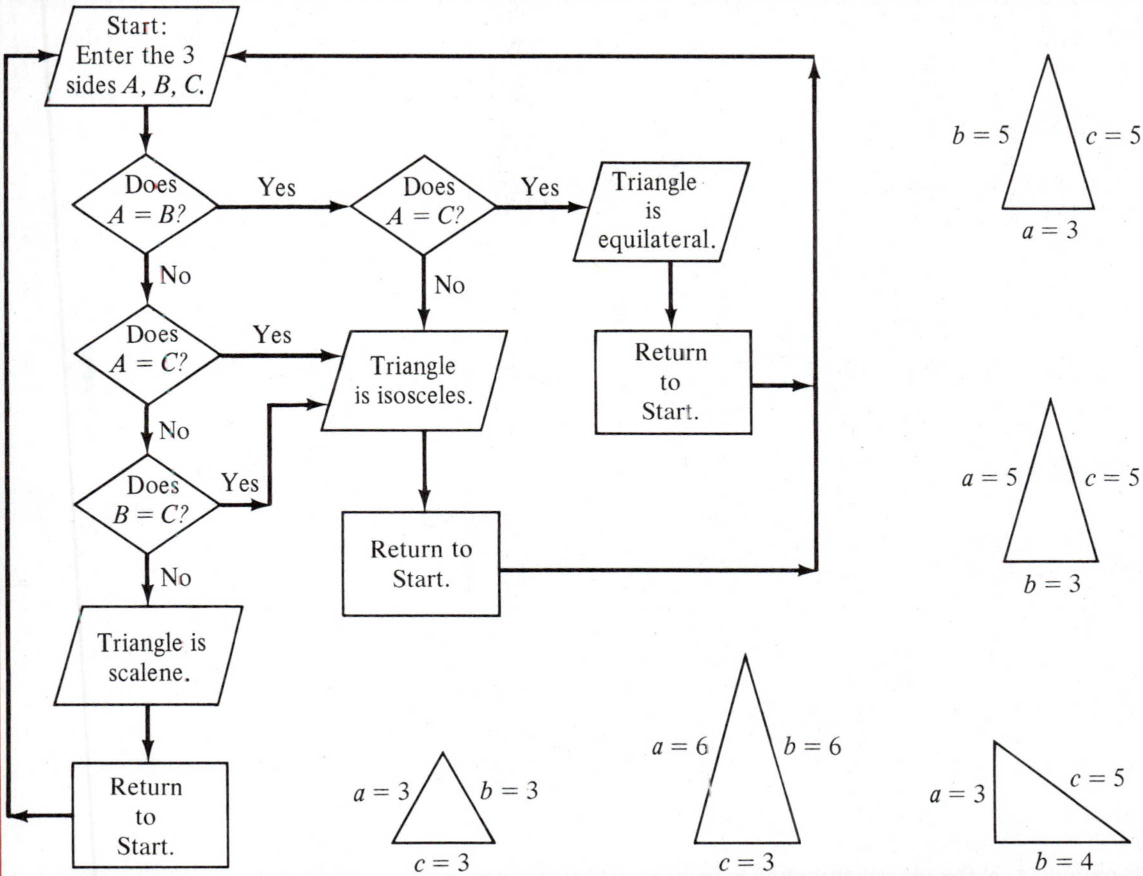

EXERCISES: See the Computer Section beginning on page 565 for more information.

Write a flowchart that classifies several triangles by the measures of their angles. **Check students' charts.**

CHAPTER FOUR REVIEW

(A) exercises are 1–10, 15–18. (B) exercises are 11, 12, 13, 14. (C) exercises are starred.

Classify each triangle by its sides and by its angles. [4.1]

1.

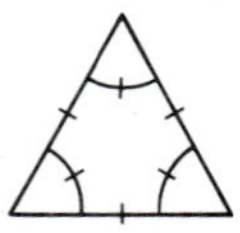

equilat; equiang or acute

2.

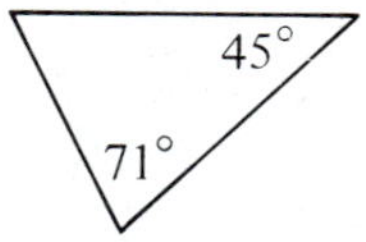

scalene; acute

Find m ∠x in each of the following. [4.2]

3.

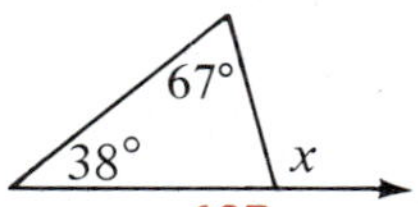

105

4.

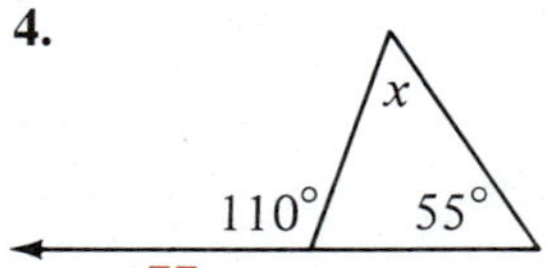

55

Draw each of the given triangles [4.1]

5. An obtuse scalene triangle

6. An acute isosceles triangle

Find the indicated angle measures [4.2]

7. Given: m ∠1 = $7x - 5$,
m ∠2 = $2x$, and
m ∠3 = $4x + 10$
Find m ∠1. 100

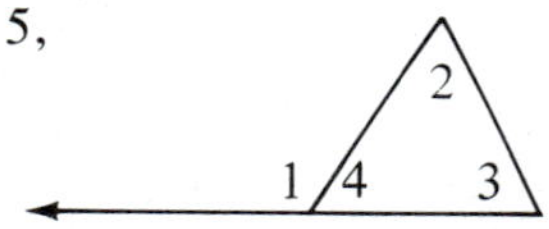

8. Given: m ∠1 = $3x + 4$,
m ∠2 = $4x - 4$, and
m ∠3 = $8x - 8$
Find m ∠1. 28

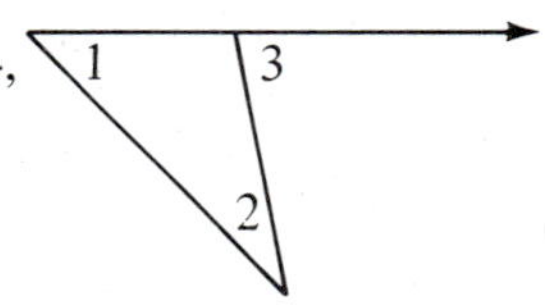

Using the markings shown, are the triangles congruent? Why or why not? [4.3]

9. yes, ASA

10. yes, AAS

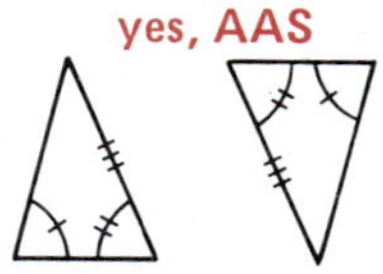

Vocabulary

acute triangle [4.1], congruent triangles [4.3], equiangular [4.1], equilateral [4.1], exterior angle [4.2], included angle [4.5], included side [4.5], isosceles triangle [4.1], obtuse triangle [4.1], remote exterior angle [4.2], right triangle [4.1], scalene triangle [4.1]

Construct each of the following.

11. Draw an obtuse triangle. Construct a triangle congruent to it using a SSS pattern. [4.4]

12. Draw an acute scalene triangle. Construct a triangle congruent to it using an ASA pattern. [4.5]

Complete each statement. [4.5]

13. To prove $\triangle XYZ \cong \triangle RST$, you must show that $\overline{XY} \cong$?, $\angle Y \cong$?, $\overline{YZ} \cong$?. $\overline{RS}$; $\angle S$; $\overline{ST}$

14. Given: $\triangle DEF \cong \triangle GHI$, m $\angle D = 7x + 3$, and m $\angle G = 8x - 4$
m $\angle D$ = ? and m $\angle G$ = ?. 52; 52

Based on the given information, are the triangles congruent? Why or why not? [4.6]

15. Given: $\overline{AC} \cong \overline{EF}$, $\overline{CB} \cong \overline{FD}$, and $\overline{BA} \cong \overline{DE}$

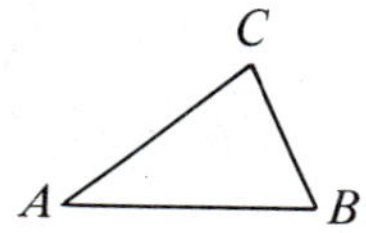

yes, SSS

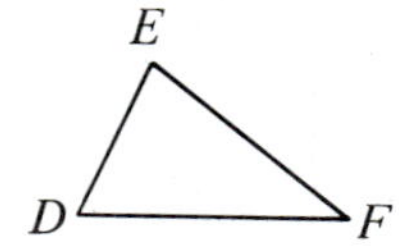

16. Given: $\overline{GH} \cong \overline{JI}$, $\angle G \cong \angle J$, and $\overline{GL} \cong \overline{JK}$

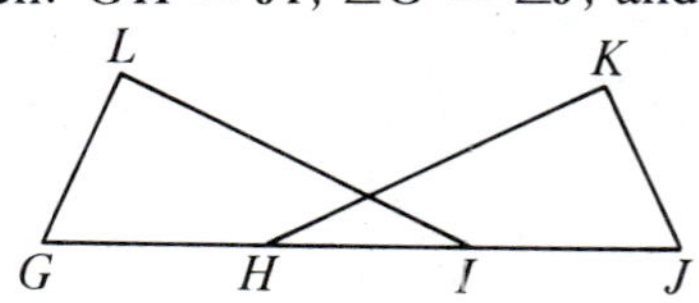

yes, SAS

17. Given: O is the midpoint of $\overline{PN}$, and $\overline{MN} \cong \overline{QP}$.

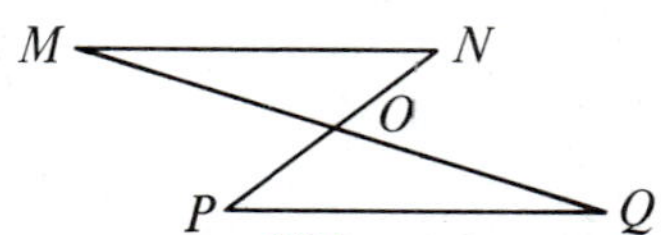

no, SSA not ≅ pattern

Prove each of the following. [4.6]

18. Given: T is the midpoint of $\overline{RU}$ and $\overline{SV}$.
Prove: $\triangle STR \cong \triangle VTU$

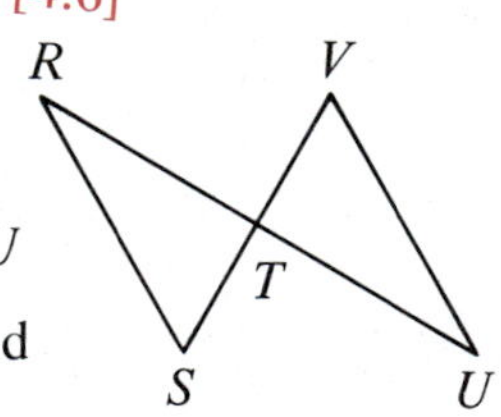

***19.** Given: $\angle R \cong \angle U$ and $\overline{ST} \cong \overline{VT}$
Prove: $\triangle SRT \cong \triangle VUT$

CHAPTER FOUR TEST

△A exercises are 1–6, 10–16. △B exercises are 7–9. △C exercises are starred.

Classify each triangle by its sides and by its angles.

1.

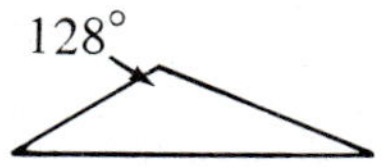

scalene; obtuse

2.

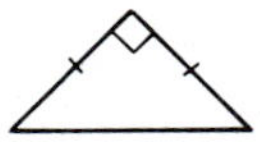

isos; rt

Find m $\angle x$.

3.

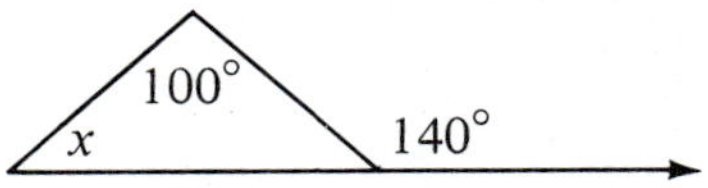

40

4.

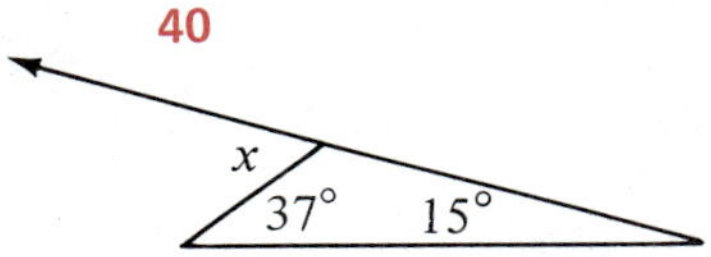

52

5. Given: m $\angle ACD = 5y + 10$, m $\angle t = 40$, and m $\angle z = y + 30$
Find m $\angle x$. 95

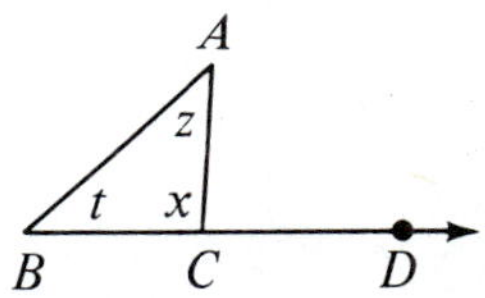

6. Given: $\angle t \cong \angle z$ and m $\angle v = 40$
Find m $\angle x$. 110

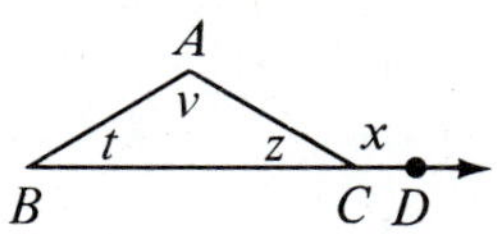

Draw each of the following.

7. A triangle which is both obtuse and isosceles.

8. A triangle which is both right and scalene.

Construct the following:

9. Draw an obtuse triangle. Construct a triangle congruent to this triangle. Use a SAS pattern.

Based on the markings shown, are the triangles congruent? Why or why not?

10.

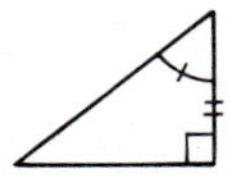

yes, ASA

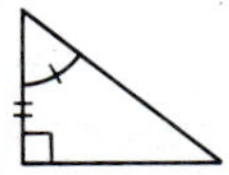

11.

no, SSA not ≅ pattern

12.

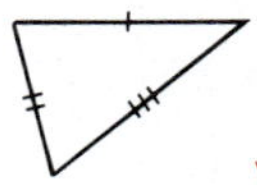

yes, SSS

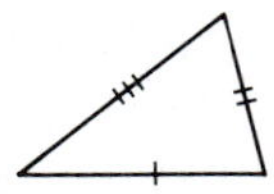

Answer each of the following.

In the figures above, $\angle A \cong \angle M$ and $\overline{AC} \cong \overline{MR}$. Which other parts must be congruent to prove

13. $\triangle ABC \cong \triangle MTR$ by ASA? $\angle C \cong \angle R$

14. $\triangle ABC \cong \triangle MTR$ by SAS? $\overline{AB} \cong \overline{MT}$

15. $\triangle ABC \cong \triangle MTR$ by AAS? $\angle B \cong \angle T$

Prove each of the following.

16. Given: Y is the midpoint of $\overline{VZ}$ and $\overline{XW}$.
Prove: $\triangle VYW \cong \triangle ZYX$

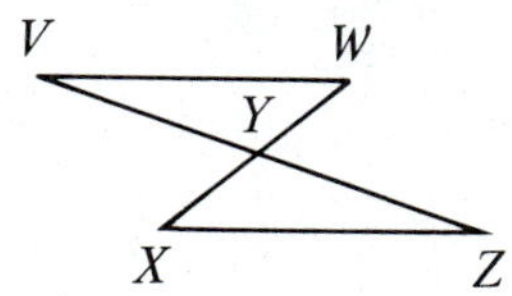

*__17.__ Given: $\overline{UT} \parallel \overline{RS}$ and $\overline{RU} \parallel \overline{ST}$
Prove: $\triangle RTU \cong \triangle TRS$

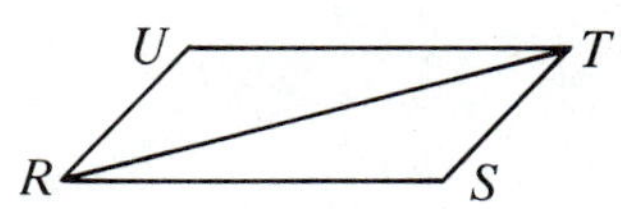

COLLEGE PREP TEST

Directions: For each question, choose the best answer. Indicate your answer by writing the letter of this choice.

1. A triangle cannot be

I. isosceles and right
II. equiangular and scalene
III. equilateral and right E

(A) I only
(B) II only
(C) III only
(D) I and II only
(E) II and III only

---

2. If the measure of one angle of an isosceles triangle is 74°, which of these is possible?

I. Another angle has a measure of 74°.
II. Another angle has a measure of 32°.
III. Another angle has a measure of 53°. E

(A) I only
(B) II only
(C) III only
(D) I and II only
(E) I, II, or III

---

3. m $\angle C = 64$ and m $\angle D = 53$

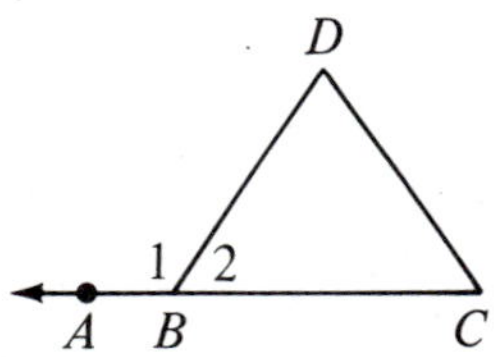

Which of these is *not* possible? A

(A) m $\angle 2 = 73$
(B) m $\angle 1 = 117$
(C) m $\angle 1 > 64$
(D) m $\angle 2 > 53$
(E) None of these

---

4. If $\angle X \cong \angle P$, $\angle Y \cong \angle T$, $\angle Z \cong \angle L$, $\overline{XY} \cong \overline{PT}$, $\overline{YZ} \cong \overline{TL}$, and $\overline{ZX} \cong \overline{LP}$, then B

(A) $\triangle XYZ \cong \triangle TLP$
(B) $\triangle XYZ \cong \triangle PTL$
(C) $\triangle XYZ \cong \triangle LPT$
(D) $\triangle XYZ \cong \triangle LTP$
(E) $\triangle XYZ \cong \triangle PLT$

5.

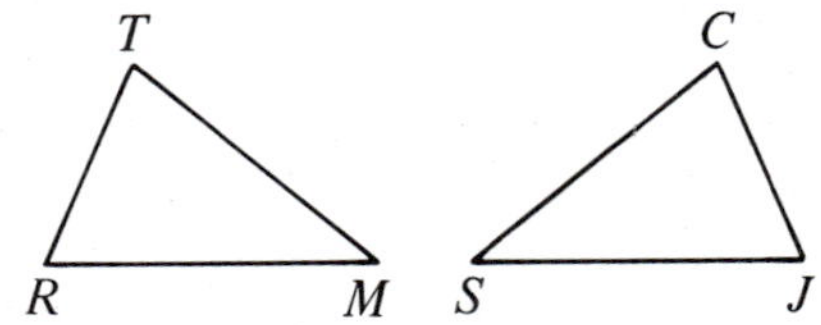

Given: $\overline{RM} \cong \overline{JS}$, $\overline{MT} \cong \overline{SC}$, $\overline{TR} \cong \overline{CJ}$
Conclusion: B

(A) $\triangle RMT \cong \triangle SJC$ by SSS
(B) $\triangle RMT \cong \triangle JSC$ by SSS
(C) $\triangle RMT \cong \triangle CSJ$ by SSS
(D) $\triangle RMT \cong \triangle CJS$ by SSS
(E) The triangles are not necessarily congruent.

6.

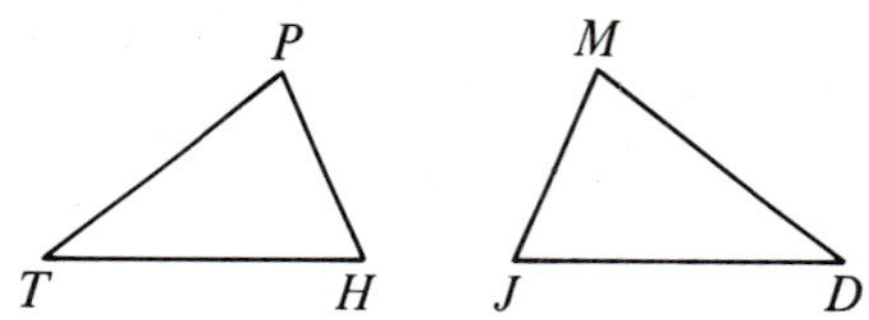

Given: $\angle T \cong \angle D$, $\overline{TP} \cong \overline{DM}$
To show that $\triangle THP \cong \triangle DJM$, you must show that

I. $\angle P \cong \angle M$
II. $\overline{PH} \cong \overline{MJ}$
III. $\overline{TH} \cong \overline{DJ}$ E

(A) I only
(B) II only
(C) III only
(D) I or II only
(E) I or III only

---

7. Given: $\angle A \cong \angle E$, $\angle B \cong \angle D$, $\overline{BC} \cong \overline{DF}$
Conclusion: D

(A) $\triangle ABC \cong \triangle DEF$ by AAS
(B) $\triangle ABC \cong \triangle DEF$ by ASA
(C) $\triangle ABC \cong \triangle DEF$ by SAS
(D) $\triangle ABC \cong \triangle EDF$ by AAS
(E) $\triangle ABC \cong \triangle EFD$ by ASA

CUMULATIVE REVIEW

Chapters 1–4

Write the letter that indicates the best answer.
For Items 1–5, refer to the figure below.

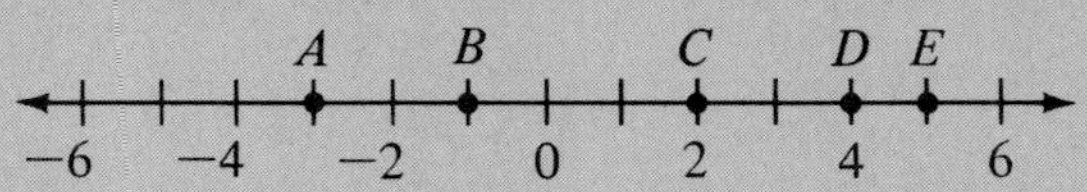

1. The coordinate of point A is B
   (A) −4 (B) −3 (C) 3 (D) −2

2. The distance BC is B
   (A) 2 (B) 3
   (C) −3 (D) none of these

3. All coordinates of F so that $BF = 4$ are B
   (A) −4 and 4 (B) −5 and 3
   (C) −4 and 2 (D) none of these

4. Which pair of segments is congruent? A
   (A) $\overline{BC}$ and $\overline{CE}$ (B) $\overline{AC}$ and $\overline{CE}$
   (C) $\overline{BC}$ and $\overline{CD}$ (D) none of these

5. The coordinate of the midpoint of $\overline{AE}$ is C
   (A) C (B) 2
   (C) 1 (D) none of these

6. If m $\angle ABC = 80$, then $\angle ABC$ is A
   (A) acute (B) right
   (C) obtuse (D) straight

7. Which are measures of complementary angles? D
   (A) 62° and 118° (B) 40° and 60°
   (C) 26° and 65° (D) none of these

8. The measure of an angle is 20° less than four times the measure of its complement. What is the measure of the angle? C
   (A) 14° (B) 22°
   (C) 68° (D) none of these

9. Any two coplanar lines which do not intersect are B
   (A) perpendicular (B) parallel
   (C) skew (D) none of these

10. If two parallel lines are intersected by a transversal, then interior angles on the same side of the transversal are A
    (A) supplementary (B) congruent
    (C) complementary (D) none of these

11. Line l intersects both line m and line n. If $m \parallel n$, $l \perp m$, then l and n are C
    (A) parallel (B) skew
    (C) perpendicular (D) none of these

12. It is impossible for a triangle which is scalene to also be C
    (A) acute (B) obtuse
    (C) isosceles (D) right

13. In the figure below, the measure of $\angle x$ is D
    (A) 85° (B) 23°
    (C) 95° (D) none of these

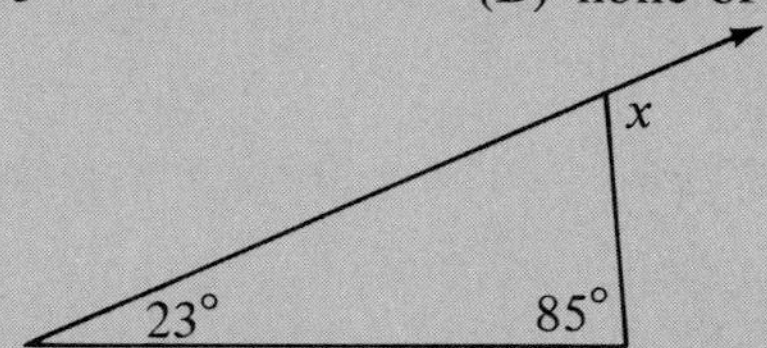

A exercises are 1–4, 6–10, 12–15, 19–26, 28–30, 32. B exercises are 3, 5, 11, 16–18, 27, 31, 33. C exercises are starred.

Find the measure of each angle. Use a protractor.

14.

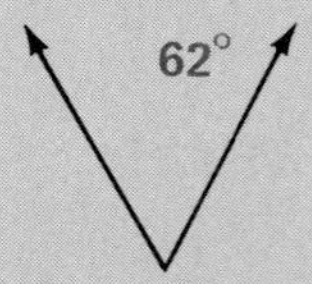

15.

Draw and label the following:

16. $\overleftrightarrow{AB}$ intersecting plane p in point C

The coordinate of point X is given. Find all possible coordinates of Y so that $\overline{XY}$ has the indicated length.

17. $X \leftrightarrow 4$; $XY = 7$ 11, −3

18. $X \leftrightarrow -1.6$; $XY = 3.7$ 2.1, −5.3

Construct each of the following:

19. Construct $\overline{CD}$ congruent to $\overline{AB}$. Then construct the midpoint of $\overline{CD}$.

Follow midpt constr., p. 10.

20. Use a protractor to draw an angle measuring 52°. Then use a compass and straightedge to bisect the angle
Follow angle bisector constr., p. 15.

True or false?

21. Any three points are contained in exactly one plane. F

22. Complementary angles are congruent. F

23. If an angle measures 70°, then its supplement measures 110°. T

24. If two lines form congruent adjacent angles, then they are perpendicular. T

25. A theorem is a statement which cannot be proved. F

26. Vertical angles are supplementary. F

27. The supplement of an acute angle measures 90° more than the complement of the angle. T

Use the figure below to find the following:

28. If $r \parallel s$ and m $\angle 2 = 58$, find m $\angle 1$ and m $\angle 4$. m $\angle 1 = 122$; m $\angle 4 = 58$

29. If m $\angle 3 = 64$ and m $\angle 8 = 49$, find m $\angle 1$ and m $\angle 10$. m $\angle 1 = 113$; m $\angle 10 = 49$

30. If $r \parallel s$ and m $\angle 6 = 132$, find m $\angle 7$ and m $\angle 8$. m $\angle 7 = 48$; m $\angle 8 = 48$

31. If $r \parallel s$, m $\angle 3 =$ m $\angle 8$, and m $\angle 2 = 40$, find m $\angle 7$. m $\angle 7 = 70$

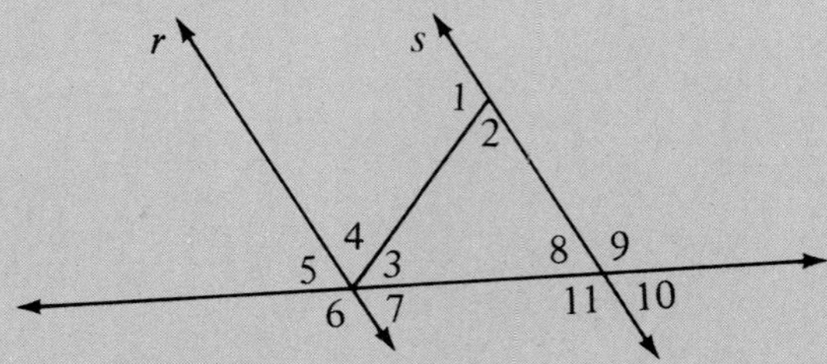

Prove each of the following:

32. Given: $\angle 1 \cong \angle 2$
Prove: $\angle 3 \cong \angle 4$

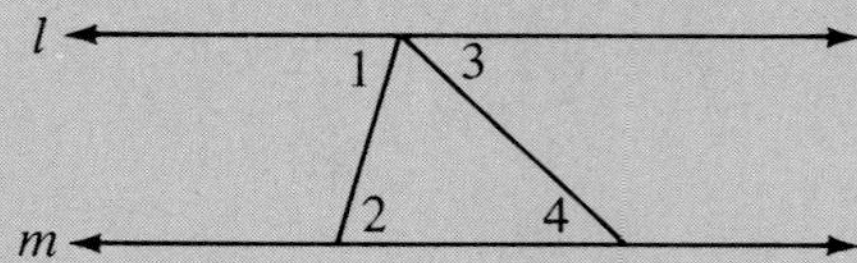

33. Given: $\angle A \cong \angle F$, $\overline{AD} \cong \overline{CF}$, and $\overline{BA} \cong \overline{EF}$
Prove: $\triangle ACB \cong \triangle FDE$

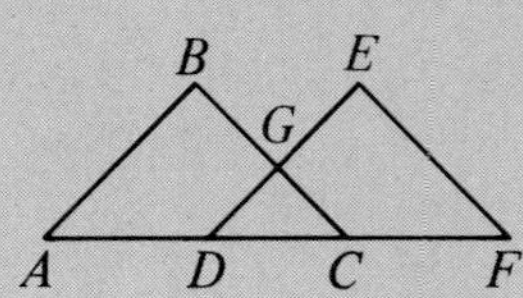

***34.** Given: $\overline{AB} \perp$ plane p, $\overline{CB}$, $\overline{DB}$, and $\overline{CD}$ are in plane p, and $\angle ACB \cong \angle ADB$
Prove: $\triangle ABC \cong \triangle ABD$

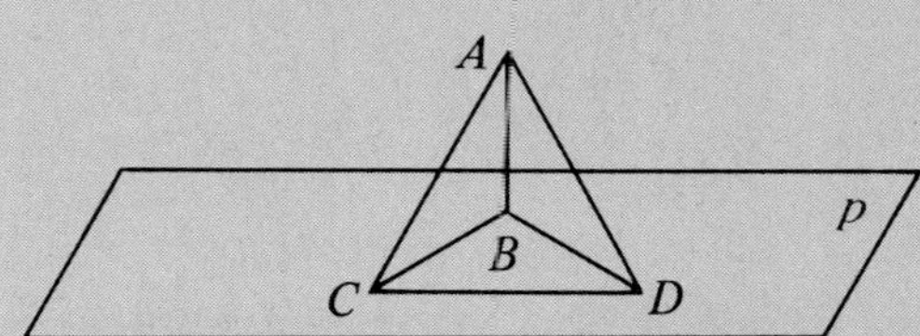

5 APPLYING CONGRUENT TRIANGLES

HISTORY

Euclid

Euclid was one of the most famous mathematicians of all times. Much of the foundation of geometry is based on the work of Euclid.

Living in Greece about 300 B.C., Euclid was the first mathematician known to have proved gometric statements in a logical pattern, starting from a few assumptions or postulates. Euclid was a great organizer. He collected, and arranged logically, essentially all of the important mathematical knowledge of his time. His most famous work is his *Elements*. Many geometry textbooks are based on the *Elements*.

Project

Use reference books and encyclopedias to discover more of Euclid's work in geometry. Contrast Euclidean geometry with non-Euclidean geometries.

CORRESPONDING PARTS 5.1

Objectives **To identify corresponding parts of congruent triangles**
To use corresponding parts in proofs

If two triangles are congruent, then their corresponding sides are congruent and their corresponding angles are congruent. This statement can be simplified into the following property.

Property of corresponding parts (CPCTC)	Corresponding parts of congruent triangles are congruent.

This property is actually a restatement of the definition of *congruent triangles.* You will use it as a reason in proofs so often that it is convenient to use the abbreviation, CPCTC.

When writing proofs that involve congruent triangles, you may need to identify the corresponding congruent parts.

Example 1

Given: $\triangle ABC \cong \triangle ZXY$. Draw a diagram to show this. Name the corresponding congruent parts.

$\angle A \cong \angle Z$ $\quad \overline{AB} \cong \overline{ZX}$
$\angle B \cong \angle X$ $\quad \overline{BC} \cong \overline{XY}$
$\angle C \cong \angle Y$ $\quad \overline{CA} \cong \overline{YZ}$

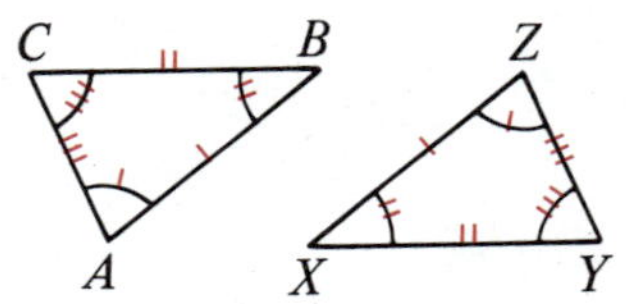

If you want to show that two segments or two angles are congruent, you may be able to do so by showing that they are corresponding parts of congruent triangles.

Example 2

Given: $\angle 2 \cong \angle 4$ and $\angle 3 \cong \angle 1$
Prove: $\overline{BC} \cong \overline{DA}$

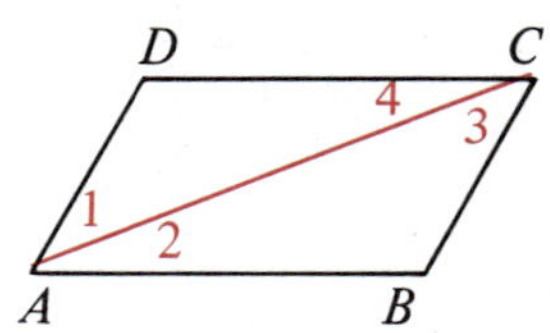

Proof

Statements		Reasons
1. $\angle 2 \cong \angle 4$	(A)	1. Given
2. $\overline{AC} \cong \overline{CA}$	(S)	2. Reflexive property
3. $\angle 3 \cong \angle 1$	(A)	3. Given
4. $\triangle ABC \cong \triangle CDA$		4. ASA
5. $\therefore \overline{BC} \cong \overline{DA}$		5. CPCTC

The property of corresponding parts can often be used to help find the measure of angles or sides of triangles or to show that lines or segments are perpendicular or parallel.

Example 3

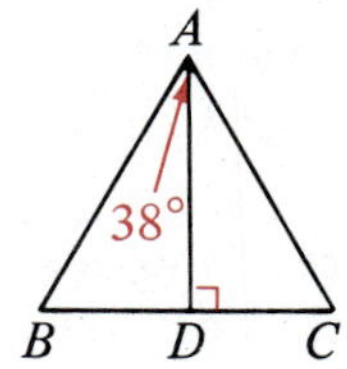

Given: $\triangle BDA \cong \triangle CDA$, m $\angle BAD =$ 38, and $\overline{AD} \perp \overline{BC}$

Find m $\angle C$.

CPCTC ► $\angle C \cong \angle B$

m $\angle B = 180 - (90 + 38)$, or 52

Thus, m $\angle C =$ 52.

Example 4

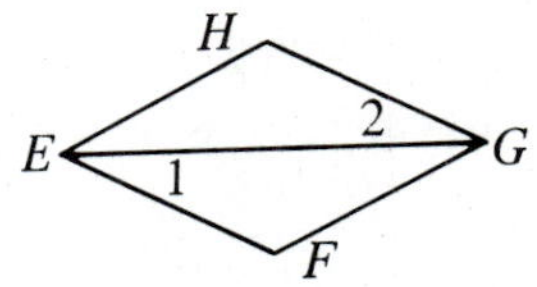

Given: $\overline{EF} \cong \overline{GH}$ and $\overline{FG} \cong \overline{HE}$

Prove: $\overline{EF} \parallel \overline{GH}$

Proof

Statements		Reasons
1. $\overline{EF} \cong \overline{GH}$	(S)	1. Given
2. $\overline{FG} \cong \overline{HE}$	(S)	2. Given
3. $\overline{GE} \cong \overline{EG}$	(S)	3. Reflexive property
4. $\triangle EFG \cong \triangle GHE$		4. SSS
5. $\angle 1 \cong \angle 2$		5. CPCTC
6. $\therefore \overline{EF} \parallel \overline{GH}$		6. Lines are $\parallel$ if alt. int. $\angle$'s are $\cong$.

Oral Exercises

Name the corresponding congruent parts of each pair of congruent triangles.

1.

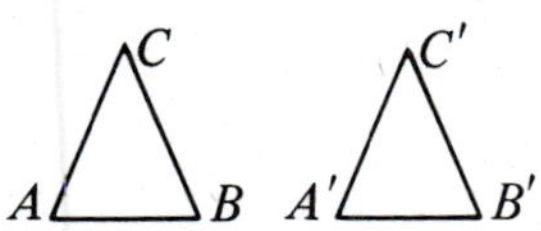

$\angle A \cong \angle A'$, $\angle B \cong \angle B'$, $\angle C \cong \angle C'$

$\triangle ABC \cong \triangle A'B'C'$

$\overline{AB} \cong \overline{A'B'}$, $\overline{BC} \cong \overline{B'C'}$, $\overline{CA} \cong \overline{C'A'}$

2.

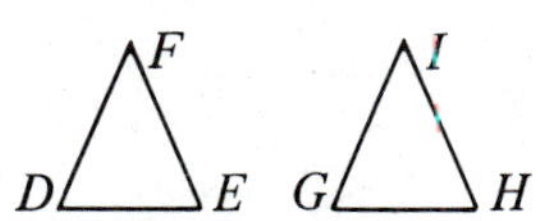

$\angle D \cong \angle G$, $\angle E \cong \angle H$, $\angle F \cong \angle I$

$\triangle DEF \cong \triangle GHI$

$\overline{DE} \cong \overline{GH}$, $\overline{EF} \cong \overline{HI}$, $\overline{FD} \cong \overline{IG}$

3.

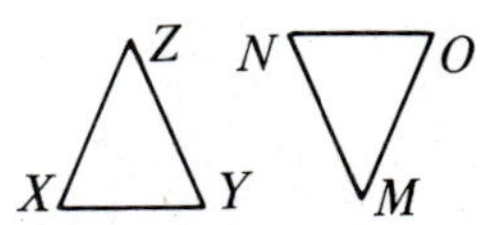

$\angle X \cong \angle O$, $\angle Y \cong \angle N$, $\angle Z \cong \angle M$

$\triangle XYZ \cong \triangle ONM$

$\overline{XY} \cong \overline{ON}$, $\overline{YZ} \cong \overline{NM}$, $\overline{ZX} \cong \overline{MO}$

Complete each statement.

4.

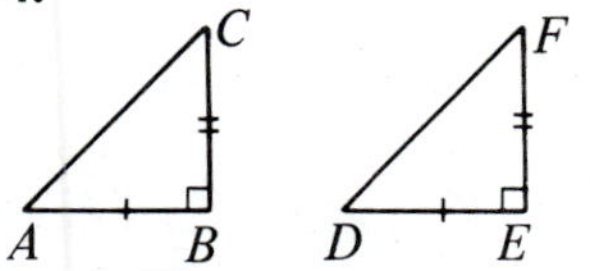

$\overline{AC} \cong$? $\overline{DF}$

5.

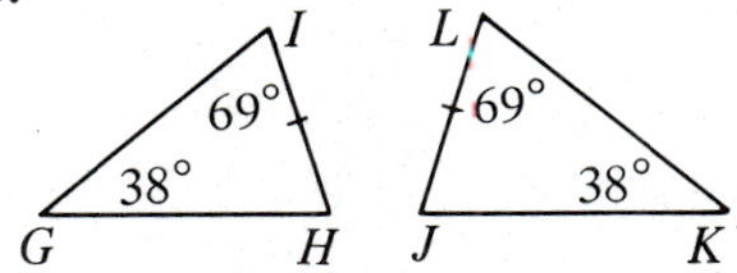

$\overline{GI} \cong$? $\overline{KL}$

6.

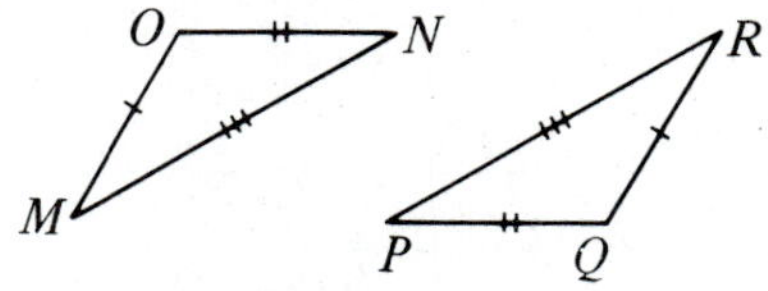

$\angle M \cong$? $\angle R$

Written Exercises

$\angle A \cong \angle D$, $\angle B \cong \angle E$, $\angle C \cong \angle F$ $\quad$ $\angle G \cong \angle X$, $\angle H \cong \angle Y$, $\angle I \cong \angle Z$ $\quad$ $\angle P \cong \angle Q$, $\angle T \cong \angle U$, $\angle L \cong \angle M$

Draw a diagram to show that each given pair of triangles is congruent. Name the corresponding congruent parts.

A

1. $\triangle ABC \cong \triangle DEF$
   $\overline{AB} \cong \overline{DE}$, $\overline{BC} \cong \overline{EF}$, $\overline{CA} \cong \overline{FD}$
2. $\triangle GHI \cong \triangle XYZ$
   $\overline{GH} \cong \overline{XY}$, $\overline{HI} \cong \overline{YZ}$, $\overline{IG} \cong \overline{ZX}$
3. $\triangle PTL \cong \triangle QUM$
   $\overline{PT} \cong \overline{QU}$, $\overline{TL} \cong \overline{UM}$, $\overline{LP} \cong \overline{MQ}$

Find the measure of the corresponding side or angle. (Ex. 4–6)

4. Given: $\triangle ABC \cong \triangle DEF$ and m $\angle A = 72$

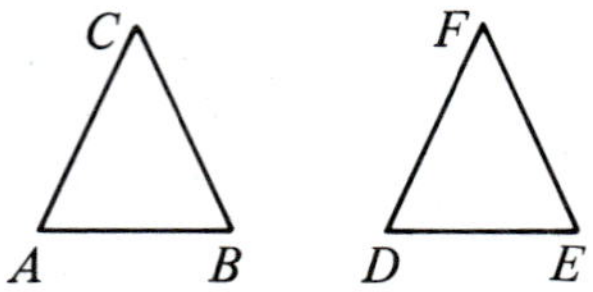

m $\angle D = 72$

5. Given: $\triangle LMN \cong \triangle OPQ$ and $LN = 9$ cm

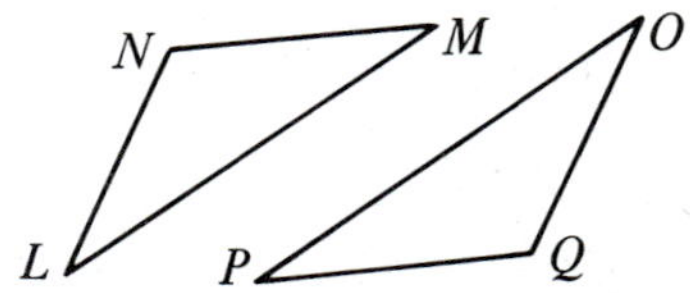

$OQ = 9$ cm

6. Given: $\triangle XYZ \cong \triangle HIG$ and $YZ = 42$ mm

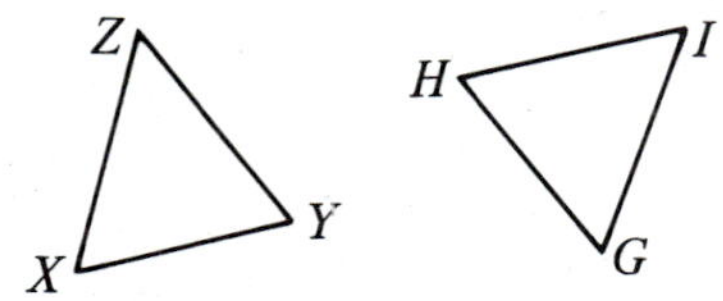

$IG = 42$ mm

7. Given: $\overline{AD} \cong \overline{CB}$ and $\overline{CD} \cong \overline{AB}$
   Prove: $\angle D \cong \angle B$

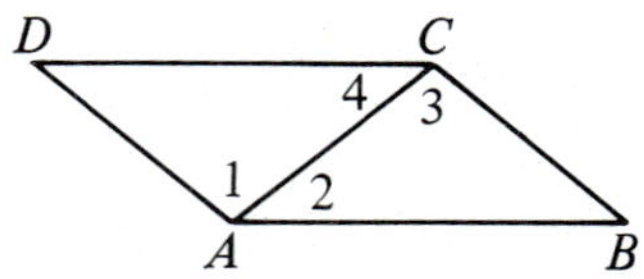

8. Given: $\overline{DC} \parallel \overline{AB}$ and $\angle 1 \cong \angle 3$
   Prove: $\overline{AD} \cong \overline{CB}$

B

9. Given: $\overline{AD} \cong \overline{CD}$, m $\angle ABD = 56$, and $\overline{DB}$ bisects $\angle ADC$.
   Find m $\angle CBD$. 56

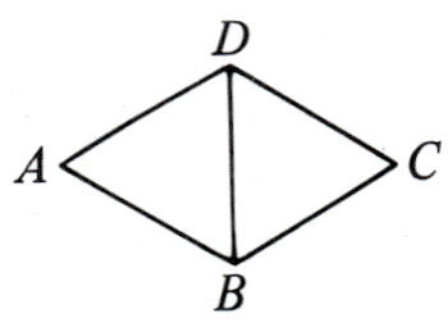

10. Given: $\overline{AD} \cong \overline{CB}$, $\overline{AD} \parallel \overline{BC}$, m $\angle ADB = 61$, and m $\angle ABD = 56$.
    Find m $\angle C$. 63

11. Given: $\overline{WZ} \cong \overline{YX}$ and $\angle 2 \cong \angle 4$
    Prove: $\overline{XW} \parallel \overline{YZ}$

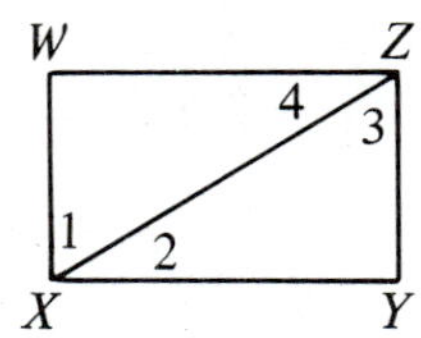

12. Given: $\overline{XW} \perp \overline{WZ}$, $\angle 1 \cong \angle 3$, and $\angle 2 \cong \angle 4$
    Prove: $\overline{ZY} \perp \overline{XY}$

C

13. Given: $\overline{CD} \perp \overline{AD}$, $\overline{CD} \perp \overline{BD}$, and $\overline{AD} \cong \overline{BD}$
    Prove: $\triangle ABC$ is isosceles.

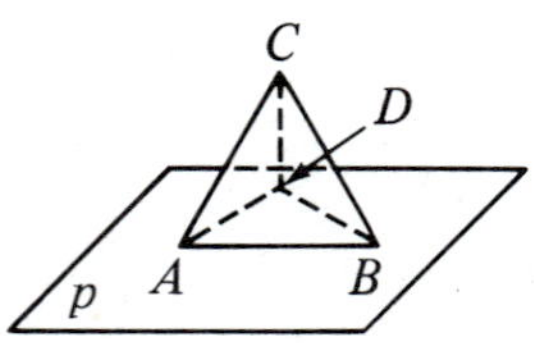

14. Given: $\overline{GH} \perp \overline{EH}$, $\overline{GH} \perp \overline{FH}$, and $\angle GEH \cong \angle GFH$
    Prove: $\triangle EHF$ is isosceles.

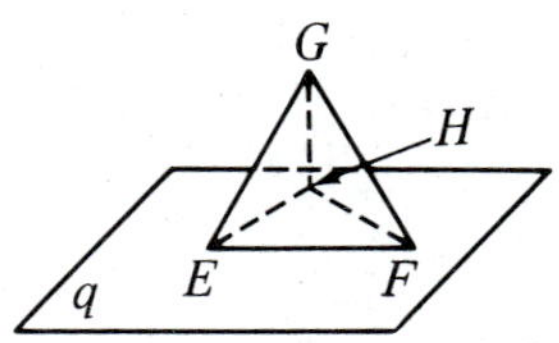

15. Given: $\overline{LK} \perp \overline{IK}$, $\overline{LK} \cong \overline{JK}$, and $\overline{LI} \cong \overline{JI}$
    Prove: $\overline{JK} \perp \overline{IK}$

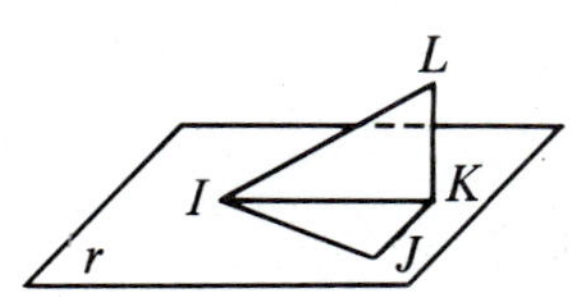

ISOSCELES TRIANGLES 5.2

Objectives

To prove and apply theorems about isosceles triangles
To find perimeters of isosceles triangles

An isosceles triangle has at least two congruent sides, each of which is called a *leg* of the triangle. The angle formed by the legs is called the *vertex angle*. The side opposite the vertex angle is called the *base* of the triangle and the angles opposite the legs are called *base angles*.

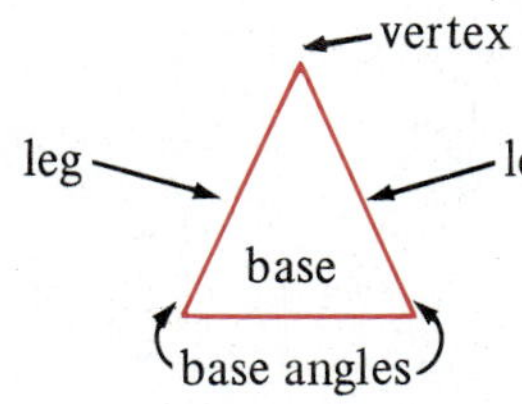

Example 1

Name the vertex angle, legs, base, and base angles of isosceles $\triangle ABC$.

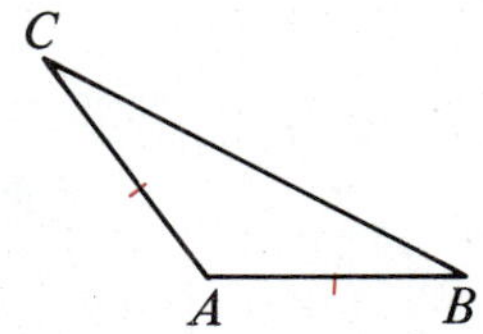

vertex angle: $\angle A$ legs: $\overline{AB}$, $\overline{AC}$
base: $\overline{BC}$ base angles: $\angle B$, $\angle C$

Theorem 5.1 below is proved by using corresponding parts of congruent triangles. Notice that it is necessary to draw an auxiliary line to form the two triangles. You will need to use auxiliary lines in many proofs involving isosceles triangles.

Theorem 5.1 If two sides of a triangle are congruent, then the angles opposite those sides are congruent.

Given: $\triangle ABC$ and $\overline{AC} \cong \overline{BC}$
Prove: $\angle A \cong \angle B$

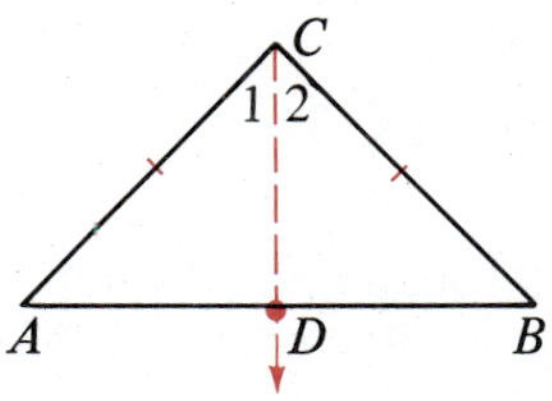

ANALYSIS
Draw an auxiliary line to form 2 $\triangle$'s that can be proved $\cong$.

Proof

Statements		Reasons
1. $\overline{AC} \cong \overline{BC}$	(S)	1. Given
2. Draw $\overrightarrow{CD}$, the bisector of $\angle ACB$ intersecting $\overline{AB}$ at D.		2. Every $\angle$, except a st. $\angle$, has exactly one bisector.
3. $\angle 1 \cong \angle 2$	(A)	3. Def. of $\angle$ bisector
4. $\overline{CD} \cong \overline{CD}$	(S)	4. Reflexive property
5. $\triangle ACD \cong \triangle BCD$		5. SAS
6. $\therefore \angle A \cong \angle B$		6. *CPCTC*

This theorem is sometimes stated "Base angles of an isosceles triangle are congruent."

Theorem 5.1 can be used to find certain angle measures.

Example 2

Given: $\overline{AB} \cong \overline{BC}$,
m $\angle B = x$, and
m $\angle A = 2x - 20$.
Find m $\angle B$.

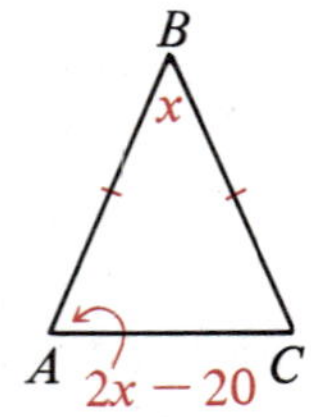

Since $\overline{AB} \cong \overline{BC}$, then $\angle A \cong \angle C$ ◄ *Use Th. 5.1*

$$\begin{aligned} m\angle C &= 2x - 20 \\ m\angle A + m\angle B + m\angle C &= 180 \\ (2x - 20) + x + (2x - 20) &= 180 \\ 5x - 40 &= 180 \\ 5x &= 220 \\ x &= 44 \end{aligned}$$

Thus, m $\angle B = 44$.

Since an equilateral triangle is also isosceles, Theorem 5.1 applies to equilateral triangles. All three sides of an equilateral triangle are congruent. Using Theorem 5.1, it can be shown that all three angles are also congruent (Corollary 1). Since the sum of the measures of the angles of any triangle is 180°, it follows that the measure of each angle of an equilateral or equiangular triangle is 60° (Corollary 2).

Corollary 1 An equilateral triangle is also equiangular.

Corollary 2 The measure of each angle of an equilateral, or equiangular, triangle is 60°.

$\triangle ABC$ is equilateral and equiangular.
m $\angle A$ = m $\angle B$ = m $\angle C$ = 60

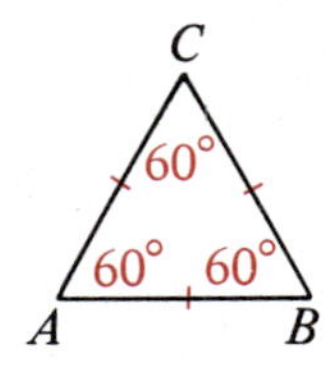

Theorem 5.2 If two angles of a triangle are congruent, then the sides opposite those angles are congruent. (See Exercise 18.)

Given: $\triangle ABC$
and $\angle A \cong \angle B$
Prove: $\overline{AC} \cong \overline{BC}$

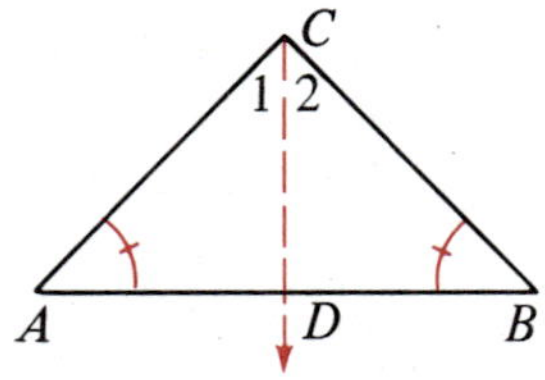

ANALYSIS

Draw an auxiliary line to form 2△'s that can be proved ≅.

You can prove the following corollary of Theorem 5.2.

Corollary An equiangular triangle is also equilateral.

The properties of isosceles triangles are frequently used as reasons in proofs.

Example 3

Given: $\triangle ABC$ and $\angle 1 \cong \angle 3$
Prove: $\overline{AC} \cong \overline{BC}$

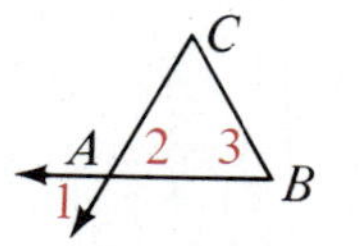

Proof

Statements	Reasons
1. $\angle 1 \cong \angle 3$	1. Given
2. $\angle 1 \cong \angle 2$	2. Vertical $\angle$'s are $\cong$.
3. $\angle 2 \cong \angle 3$	3. Substitution property
4. $\therefore \overline{AC} \cong \overline{BC}$	4. If two $\angle$'s of $\triangle$ are $\cong$, sides opp. the $\angle$'s are $\cong$.

The **perimeter** of a triangle is the sum of the lengths of the sides. You can use the properties of isosceles triangles to help find lengths of sides and perimeters.

Example 4

Find the perimeter of $\triangle ABC$.

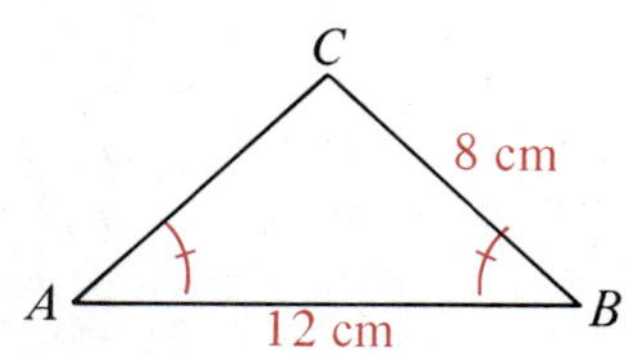

$\angle A \cong \angle B$ ► $AC = BC = 8$ cm
Perimeter $= 8 + 8 + 12 = 28$ cm

Example 5

Given: $\angle M \cong \angle N$,
$MN = 13$ cm, and
the perimeter of $\triangle MNO$ is 43 cm.
Find OM.

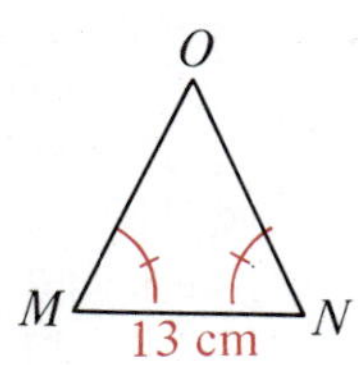

Perimeter ► $MN + NO + OM = 43$
$13 + NO + OM = 43$
$NO + OM = 30$
$2OM = 30$ ◄ *$NO = OM$ by Th. 5.2.*
Thus, $OM = 15$.

Example 6

Given: $\triangle XYZ$ is equiangular and $YZ = 10$ m.
Find the perimeter.

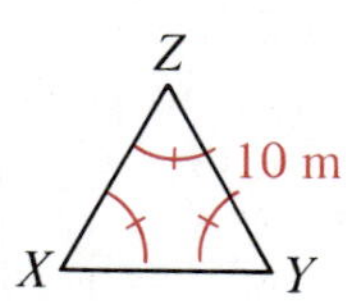

Use Cor. to Th. 5.2. ► Perimeter $= XY + YZ + ZX$
Perimeter $= 10 + 10 + 10$, or 30 m

Reading in Geometry

True or False? Give the page and paragraph in the lesson that support your choice.

1. The base of an isosceles triangle is one of the two congruent sides. F
2. The vertex angle of an isosceles triangle is one of the two congruent angles. F
3. An equilateral triangle is also equiangular. T
4. Base angles of an isosceles triangle are congruent. T
5. An equilateral triangle is also isosceles. T
6. An isosceles triangle can have two 60° angles. T

Oral Exercises

Name the vertex angle, legs, base, and base angles of each triangle.

1.

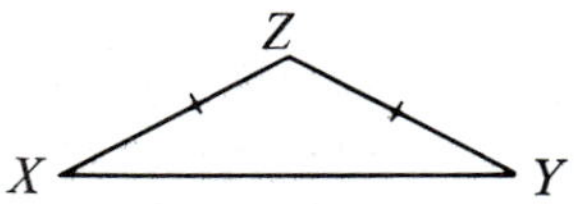

$\angle Z$; $\overline{XZ}$, $\overline{YZ}$; $\overline{XY}$; $\angle X$, $\angle Y$

2.

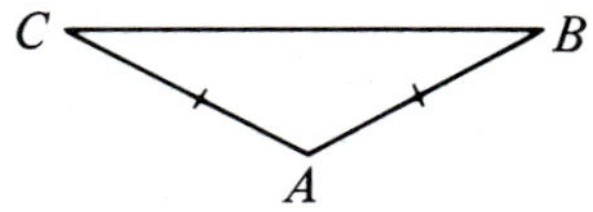

$\angle A$; $\overline{AC}$, $\overline{AB}$; $\overline{BC}$; $\angle B$, $\angle C$

3.

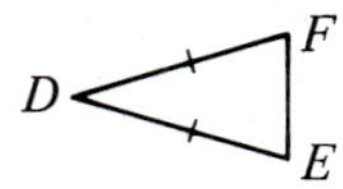

$\angle D$; $\overline{DE}$, $\overline{DF}$; $\overline{EF}$; $\angle E$, $\angle F$

Written Exercises

4. $m\angle Z = 40$, $m\angle X = m\angle Y = 70$ 5. $m\angle X = m\angle Y = m\angle Z = 60$

Find the measure of each angle.

A **1.**

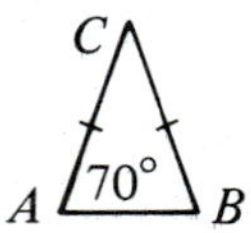

$m\angle B = 70$, $m\angle C = 40$

2.

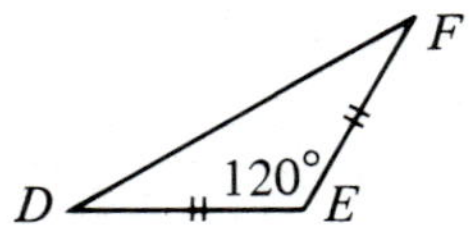

$m\angle D = 30$, $m\angle F = 30$

3.

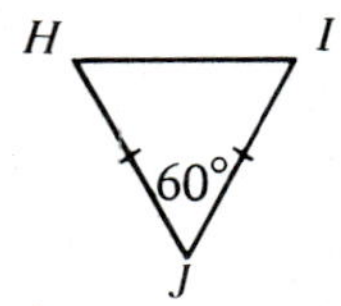

$m\angle H = 60$, $m\angle I = 60$

4. Given: $\overline{XZ} \cong \overline{YZ}$, $m\angle X = 3a + 4$, and $m\angle Z = 2a - 4$

5. Given: $\overline{XZ} \cong \overline{YZ}$, $m\angle Y = 3c + 3$, and $m\angle Z = 4c - 16$

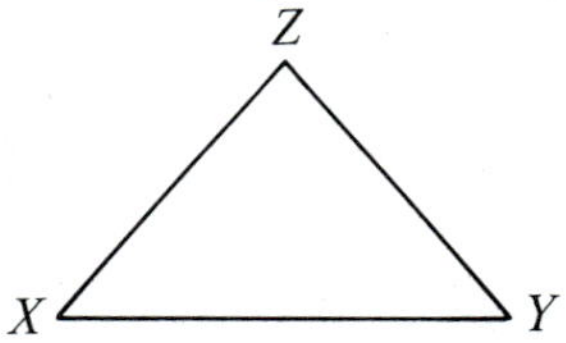

6. Given: $\overline{XZ} \cong \overline{YZ}$, $m\angle X = 4t + 4$, and $m\angle Y = 5t - 9$

7. Given: $\overline{XZ} \cong \overline{YZ}$, $m\angle X = 10e + 12$, and $m\angle Y = 12e - 2$

6. $m\angle X = m\angle Y = 56$, $m\angle Z = 68$ 7. $m\angle X = m\angle Y = 82$, $m\angle Z = 16$

8. Given: $\overline{AB} \cong \overline{BC}$ and $m\angle B = 110$
Find $m\angle A$ and $m\angle 2$.

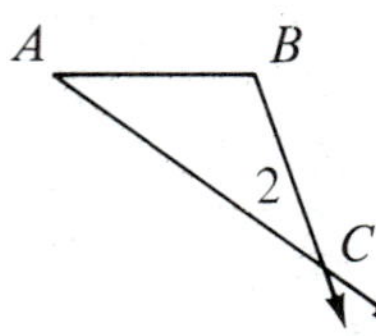

$m\angle A = 35$, $m\angle 2 = 35$

9. Given: $\overline{FE} \perp \overline{DE}$ and $\overline{FE} \cong \overline{DE}$
Find $m\angle D$ and $m\angle F$.

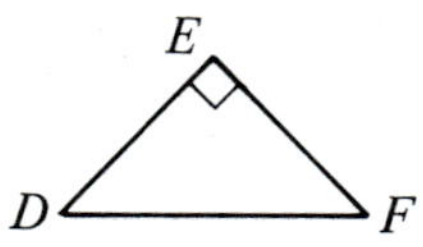

$m\angle D = m\angle F = 45$

10. Given: $\overline{GE} \cong \overline{HI}$ and $m\angle I = 52$
Find $m\angle 2$, $m\angle 3$, and $m\angle 4$.

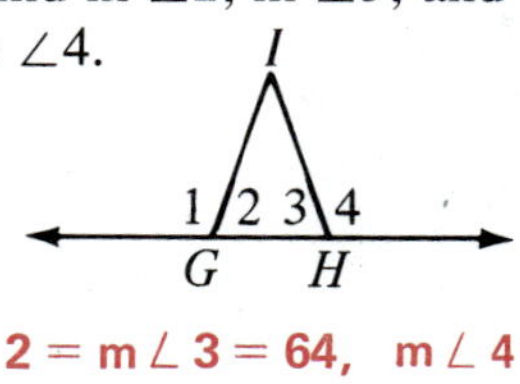

$m\angle 2 = m\angle 3 = 64$, $m\angle 4 = 116$

Find the perimeter of each triangle. (Ex. 11–12)

11. Given: $\overline{AB} \cong \overline{CB}$, $AB = 8$ cm, and $AC = 4$ cm 20 cm

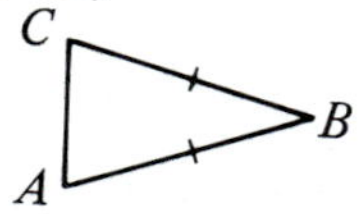

12. Given: $\angle A \cong \angle C$, $BC = 7$ cm, and $AC = 3$ cm 17 cm

13. Given: $\overline{AB} \cong \overline{AC}$
Prove: $\angle B \cong \angle 2$

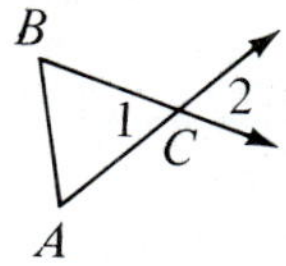

14. Given: $\angle A \cong \angle 2$
Prove: $\overline{AB} \cong \overline{BC}$

B

15. Given: $\overrightarrow{FH} \parallel \overline{DE}$ and $\overline{DF} \cong \overline{EF}$
Prove: $\angle 1 \cong \angle 2$

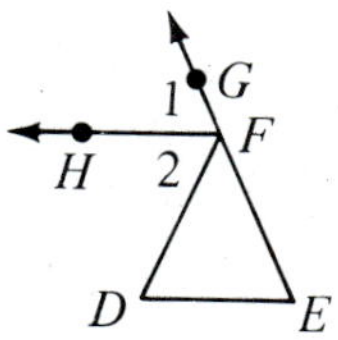

16. Given: $\overrightarrow{KM} \parallel \overline{IJ}$ and $\angle 1 \cong \angle 2$
Prove: $\triangle IJK$ is isosceles.

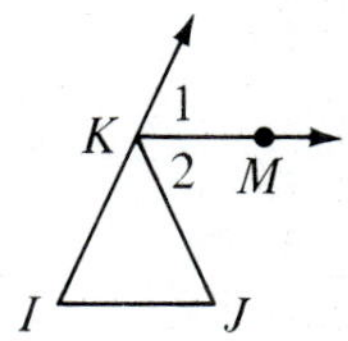

17. Given: $\angle 1 \cong \angle 2$
Prove: $\triangle OPR$ is isosceles.

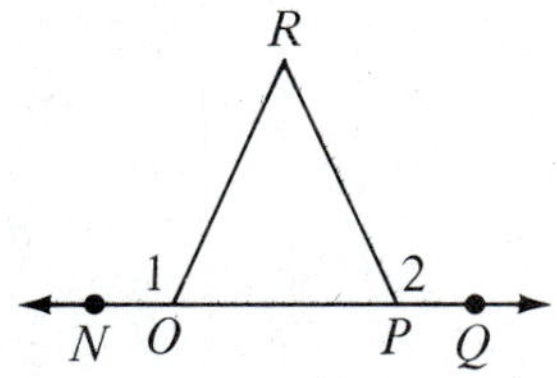

18. Prove Theorem 5.2.

19. Prove Corollary of Theorem 5.2.

20. Given: $\overline{WU} \parallel \overline{ST}$ and $\overline{SV} \cong \overline{TV}$
Prove: $\overline{WV} \cong \overline{UV}$

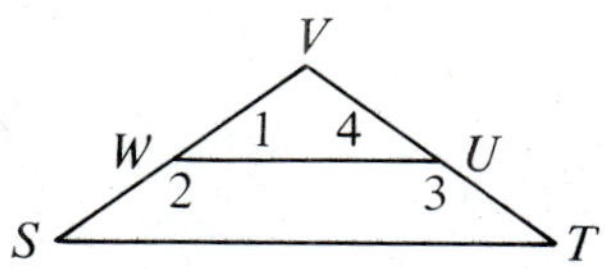

21. Given: $\overline{WU} \parallel \overline{ST}$ and $\overline{VW} \cong \overline{VU}$
Prove: $\overline{SW} \cong \overline{TU}$

22. Given: $\triangle ABC$, m $\angle A = 2x + 10$, m $\angle B = 3x - 15$, m $\angle C = 4(x - 10)$, and $\angle A \cong \angle B$.
Find the perimeter of $\triangle ABC$, if possible.
not possible

23. Given: $\triangle ABC$, $AC = 4x$, $BC = 6x - 6$, $AB = 5x - 8$, and $\angle A \cong \angle B$.
Find the perimeter of $\triangle ABC$, if possible.
31

C

24. Given: $\angle PXY \cong \angle PXZ$, and $\angle XYZ \cong \angle XZY$
Prove: $\triangle PYZ$ is isosceles.

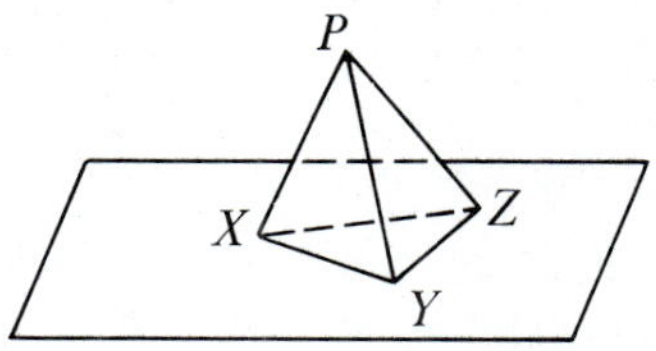

25. Given: $\angle PYZ \cong \angle PZY$, and $\angle XPY \cong \angle XPZ$
Prove: $\triangle XYZ$ is isosceles.

ALGEBRA REVIEW

Objective

To simplify radicals

To simplify a radical, factor to get a perfect square, then use

$$\sqrt{ab} = \sqrt{a} \cdot \sqrt{b}.$$

Example

Simplify $\sqrt{75}$.

25 is a perfect square. ► $\sqrt{25 \cdot 3} = \sqrt{25} \cdot \sqrt{3} = 5\sqrt{3}$

Simplify. (All variables represent positive numbers.)

1. $\sqrt{16}$ **4**
2. $\sqrt{32}$ **$4\sqrt{2}$**
3. $\sqrt{45}$ **$3\sqrt{5}$**
4. $\sqrt{147}$ **$7\sqrt{3}$**
5. $\sqrt{242}$ **$11\sqrt{2}$**
6. $2\sqrt{25}$ **10**
7. $3\sqrt{20}$ **$6\sqrt{5}$**
8. $-4\sqrt{108}$ **$-24\sqrt{3}$**
9. $\sqrt{8x^2}$ **$2x\sqrt{2}$**
10. $\sqrt{144x^3}$ **$12x\sqrt{x}$**

REFLECTIONS

A geometric transformation is a means of relating the points of a figure with the points of a second figure, called its image, so that each point of the figure corresponds to exactly one point of the image and each point of the image corresponds to exactly one point of the figure. One such transformation is the **reflection.**

EXPERIMENT 1

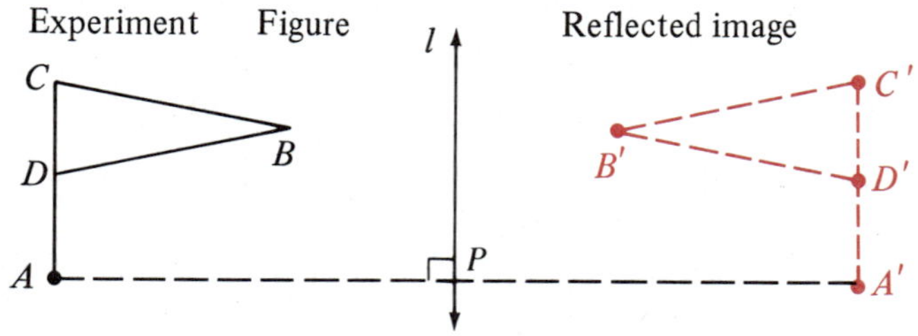

1. Draw $\overline{AP} \perp l$. Extend $\overline{AP}$ to A' so that $\overline{PA'} \cong \overline{AP}$.
2. Repeat this for points B, C, and D.
3. Connect A' and D', D' and B', B' and C', and C' and D'.
4. This image is the reflection of the original figure about line l.
5. Compare the reflected image with the original figure. Are they congruent?

Project Copy each triangle and the reflection line l. Draw the reflected image using the method in the experiment above. Is each reflected image congruent to the original triangle? **yes**

1.

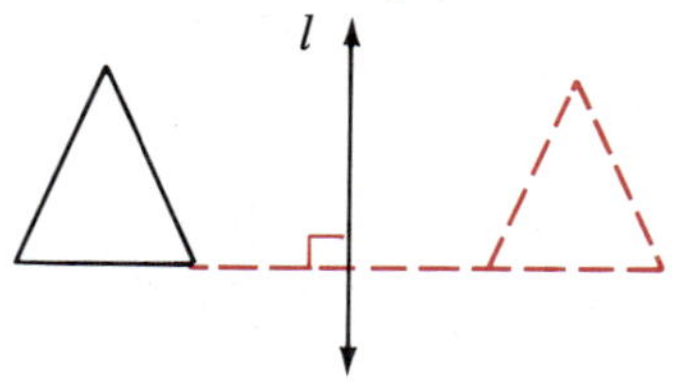

2.

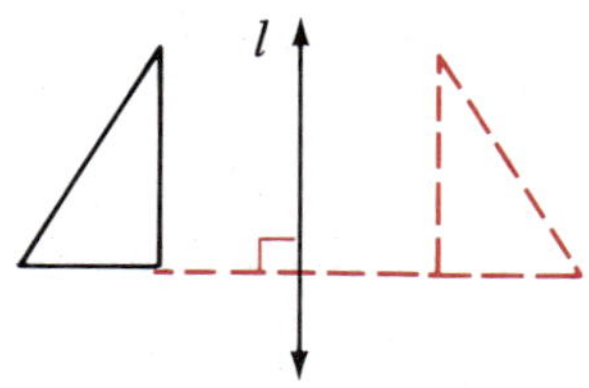

3.

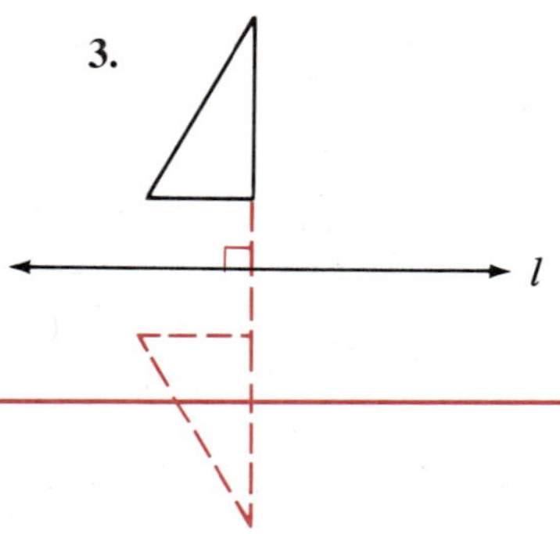

4.

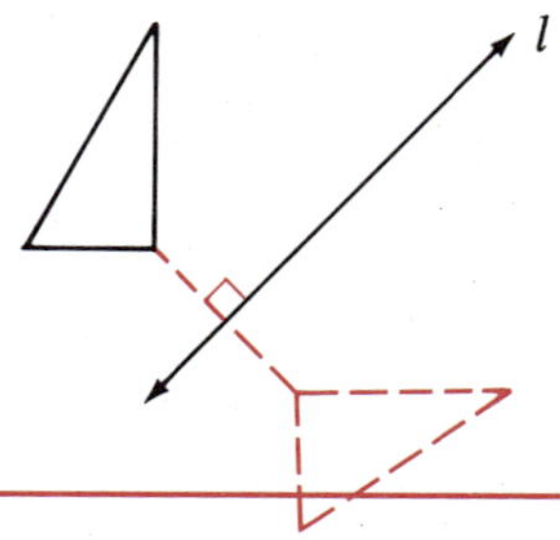

The reflection line is often called the **line of symmetry.** You can find the line of symmetry using geometric constructions.

EXPERIMENT 2

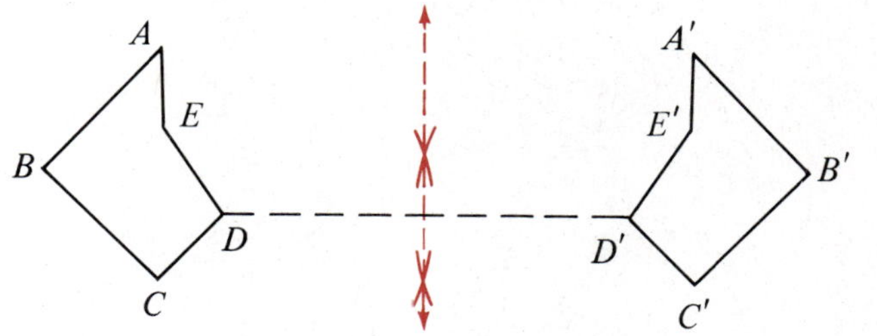

1. $A'B'C'D'E'$ is the reflection of $ABCDE$.
2. Using compass and straightedge, construct the perpendicular bisector of $\overline{DD'}$.
3. Construct the perpendicular bisector of $\overline{CC'}$. What do you notice about this line?
4. The perpendicular bisector of $\overline{DD'}$ is the line of symmetry of the figure and its reflection.

Project

Copy each triangle and its reflection. Construct the line of symmetry.

5.

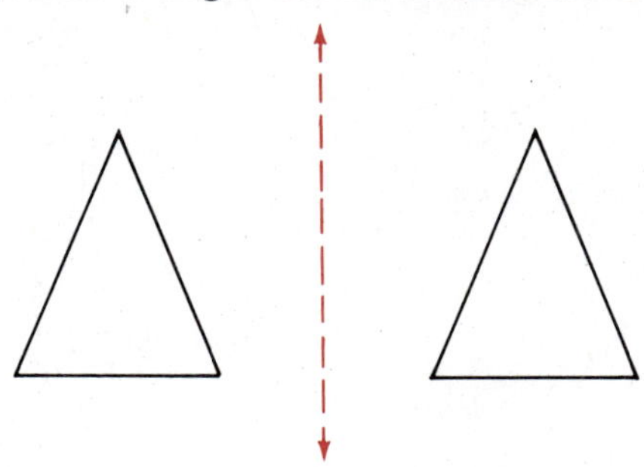

6.

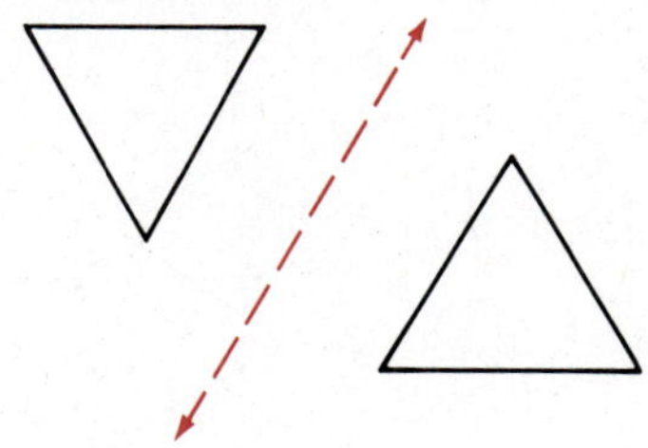

7.

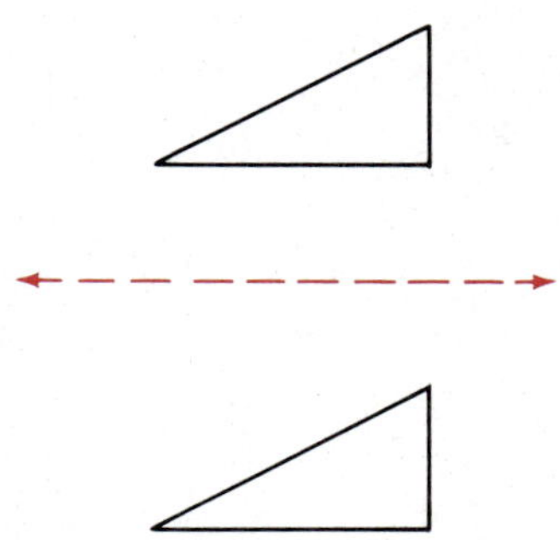

8.

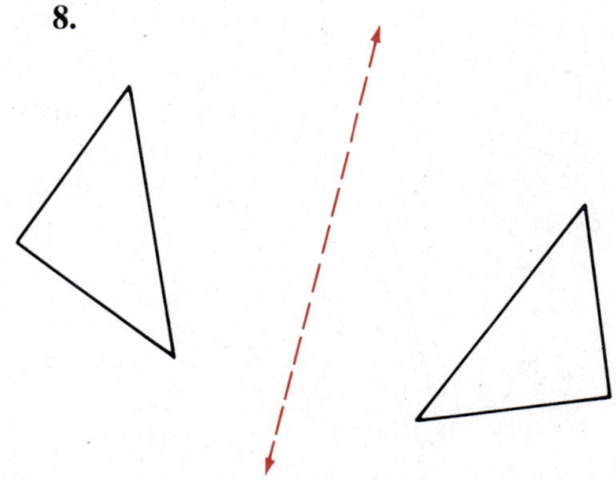

Paper folding can be used to find the line of symmetry of a figure and its reflection, or to determine if two figures are reflections. If they are not reflections, you cannot find a line of symmetry.

EXPERIMENT 3

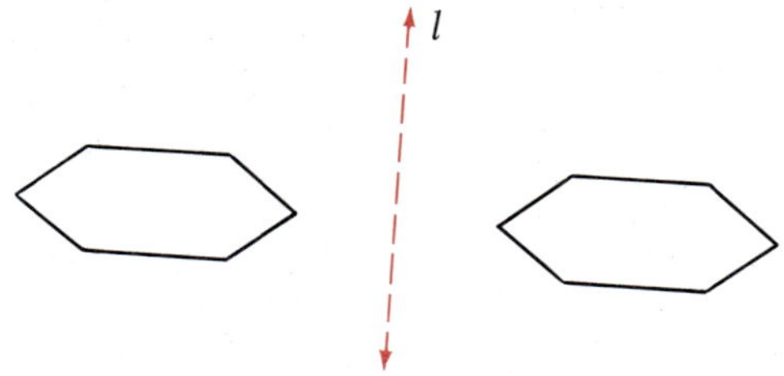

1. Fold your paper so that the two figures coincide.
2. With the two figures together, crease the paper.
3. This crease is the line of symmetry of the two figures.

Project

Copy each pair of triangles on the same sheet of paper. By folding the paper, find out which triangles are reflections of each other.

9.

reflections

10.

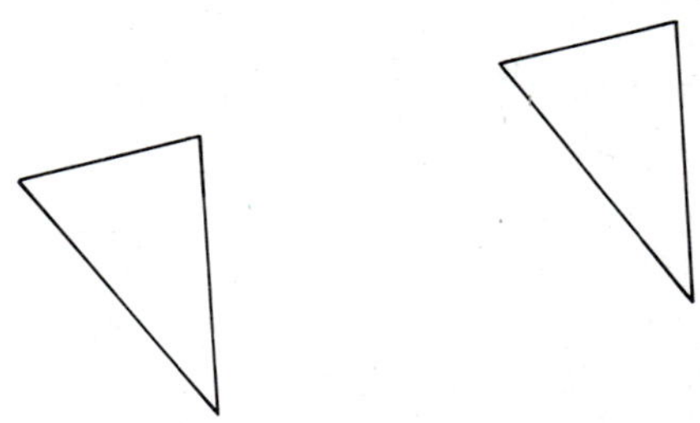

not reflections

11.

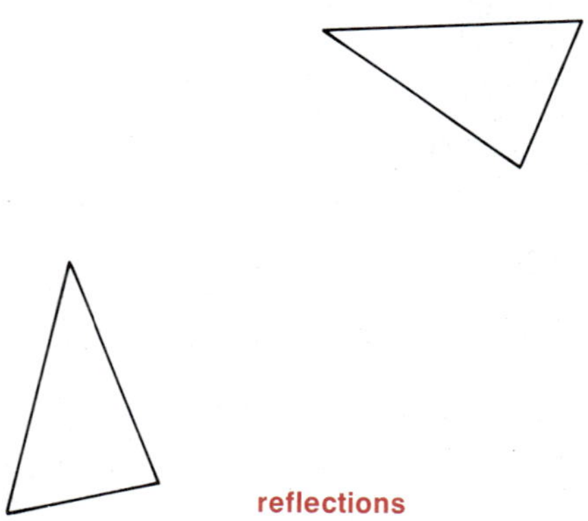

reflections

12.

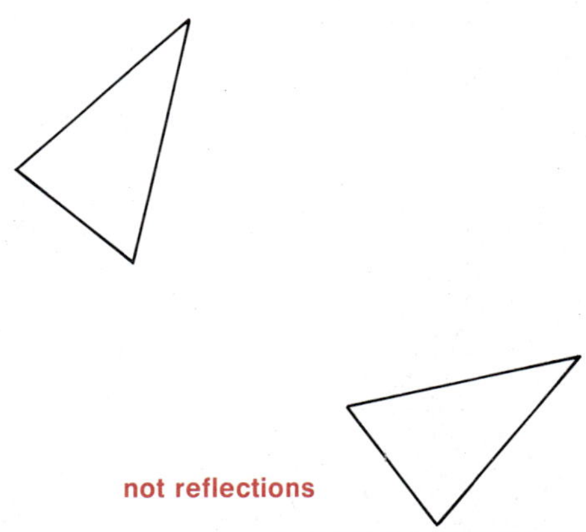

not reflections

SLIDES

Another type of transformation is the **slide.**

EXPERIMENT 1

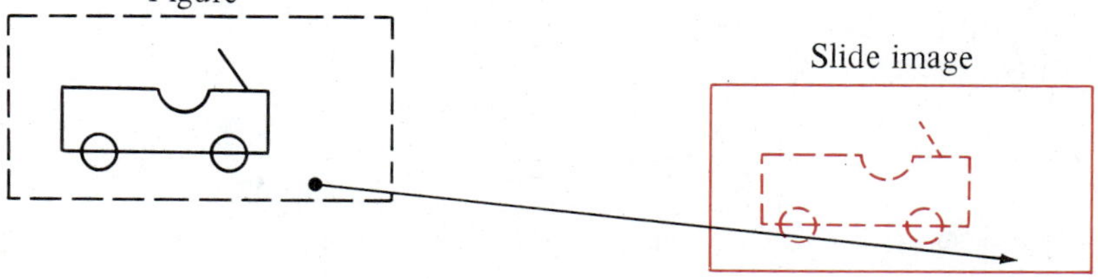

1. Trace the figure on tracing paper. Draw a dot on the tracing at the end of the slide arrow.
2. Slide the dot on the tracing along the slide arrow to the arrow head. Do not turn the paper.
3. Place carbon paper under the tracing. Trace this copy onto this sheet.
4. This figure is the slide image of the original figure with respect to the given slide arrow.

Project Copy each triangle and its slide arrow. Draw the slide image with respect to the given slide arrow. **Check students' drawings.**

1.

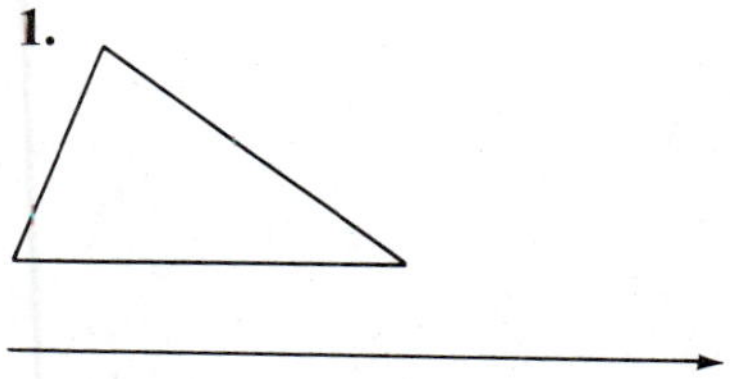

2.

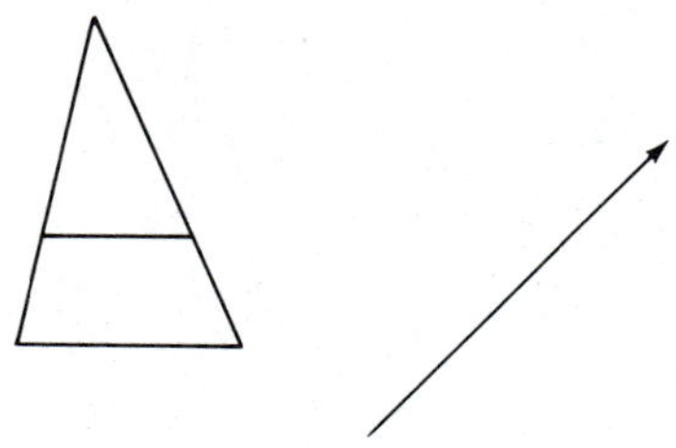

3.

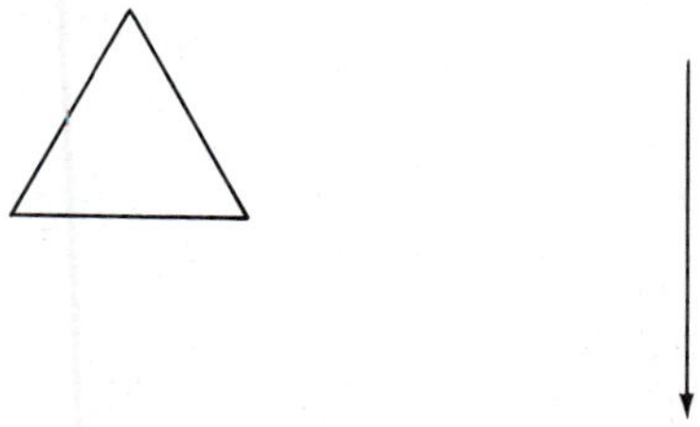

4.

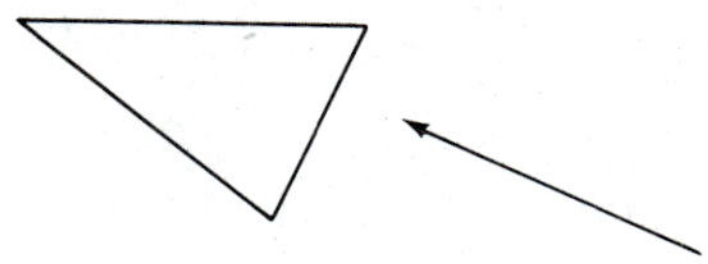

You can follow a slide with a second slide. The two slides can then be combined into a single slide.

EXPERIMENT 2

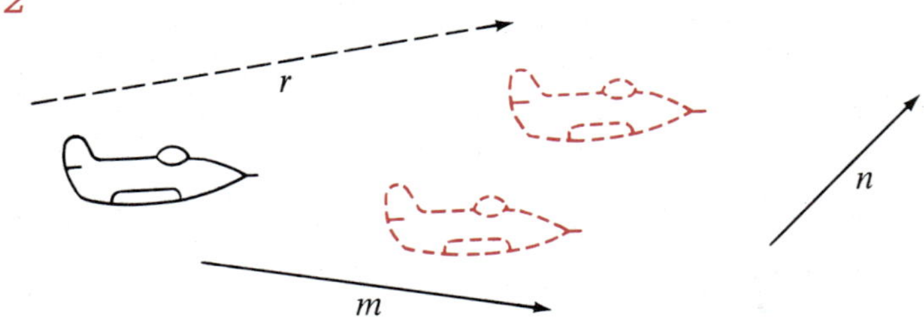

1. Draw the slide image of the original figure with respect to slide arrow *m*.
2. Draw the slide image of this image with respect to slide arrow *n*.
3. Show that this last slide image is also the slide image of the original figure with respect to slide arrow *r*.
4. Slide arrow *r* is the *resultant* of *m* and *n*.

Project

Copy each figure and the two slide arrows shown. For each figure, draw the slide image with respect to slide arrow *m*. Then draw the slide image of this image with respect to slide arrow *n*. Find a single slide arrow *r* which is the resultant of *m* and *n*. This resultant arrow should give the same result in one slide that slide *m* followed by slide *n* produced. **Check students' drawings.**

5.

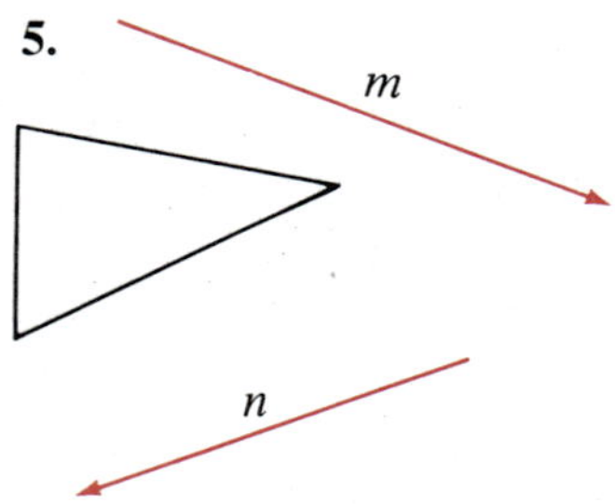

6.

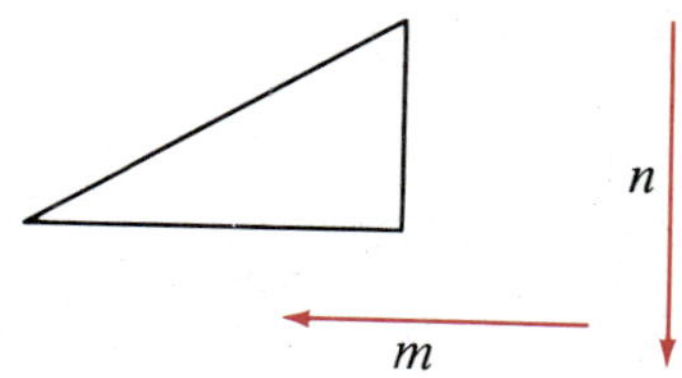

7.

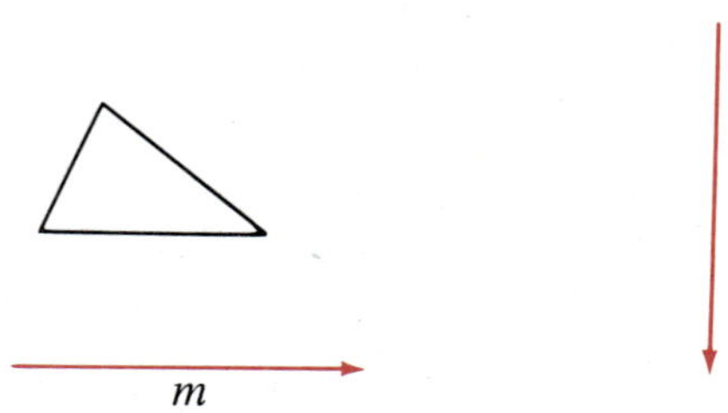

8.

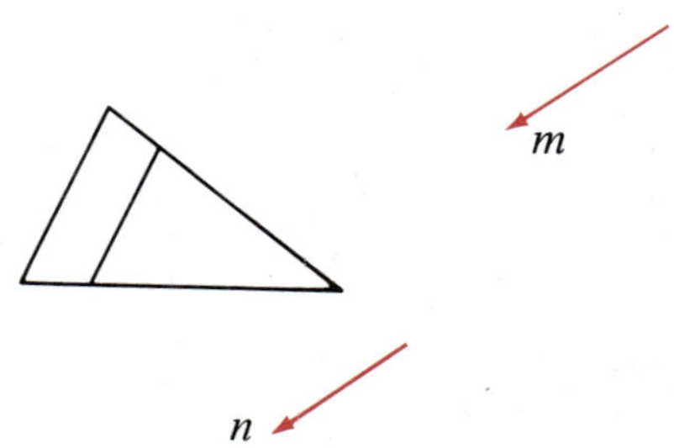

ROTATIONS

A third type of transformation is the **rotation.**

EXPERIMENT 1

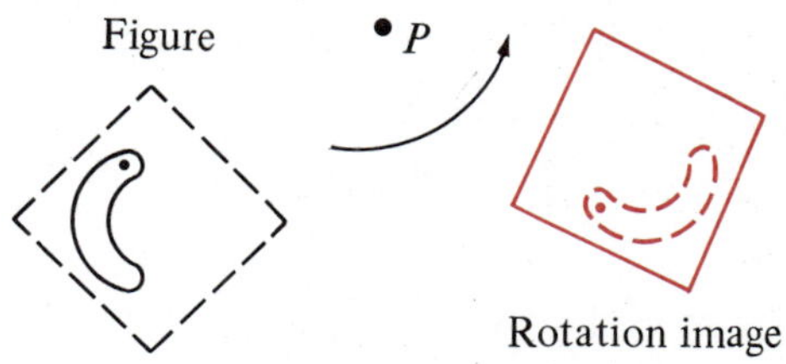

1. Trace the figure on tracing paper. Draw a dot on the copy at the beginning of the rotation arrow.
2. Place a pencil, pen, or other pointed object on the copy at point P. Turn the copy around point P until the dot is at the end of the arrow.
3. Place carbon paper under the copy and trace the copy on this sheet.
4. This figure is the rotation image of the figure with respect to P and the given rotation arrow.

Project Copy each triangle and its rotation arrow. Draw the rotation image about P with respect to the given rotation arrow. **Check students' drawings.**

1.

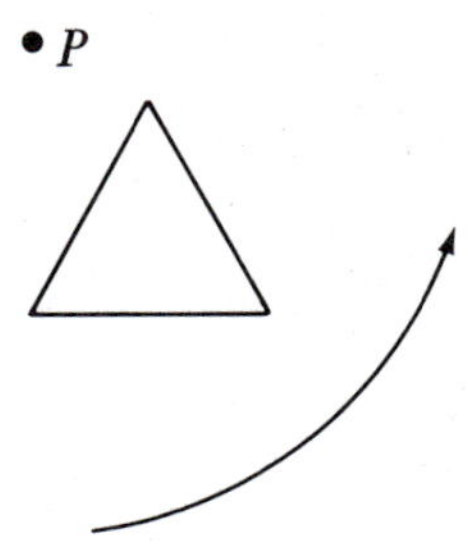

2.

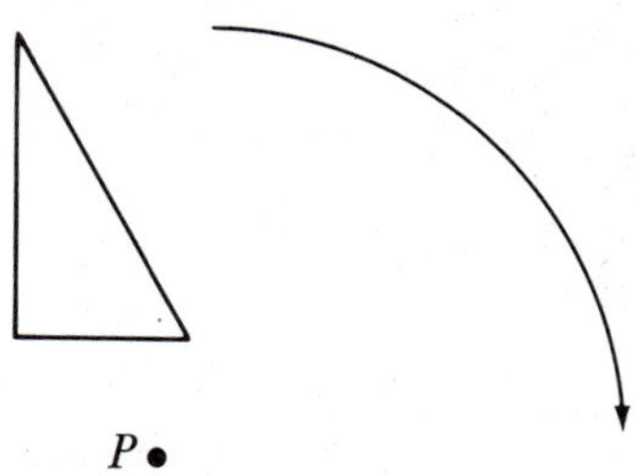

3.

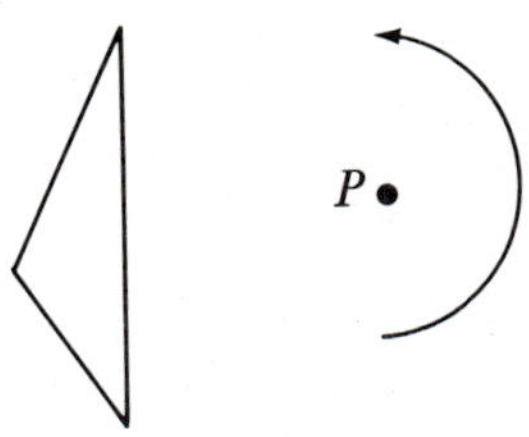

4.

If a figure and its rotation image are given, you can find the rotation center using geometric constructions.

EXPERIMENT 2

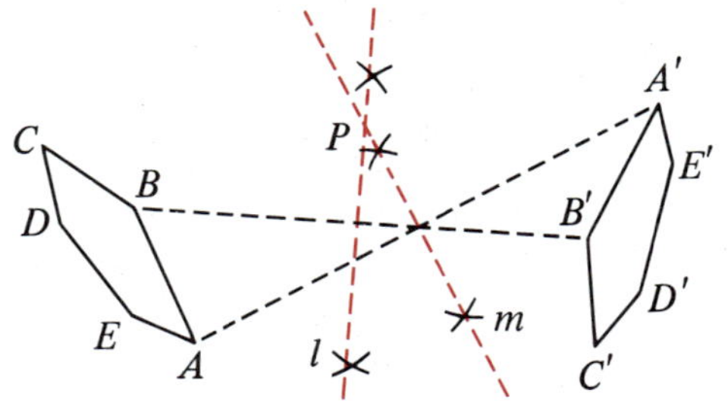

1. Draw $\overline{BB'}$. Construct l, the perpendicular bisector of $\overline{BB'}$.
2. Draw $\overline{AA'}$. Construct m, the perpendicular bisector of $\overline{AA'}$.
3. Find P, the point of intersection of l and m.
4. P is the rotation center for the rotation of ABCDE into $A'B'C'D'E'$.

Project Copy each pair of triangles. Find the rotation center using geometric constructions.
Check students' constructions.

5.

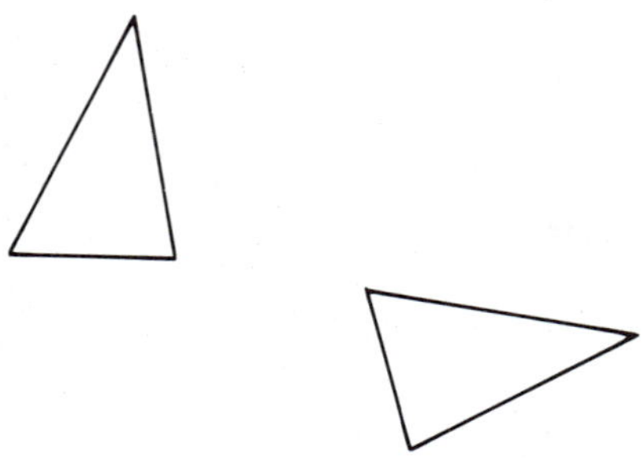

6.

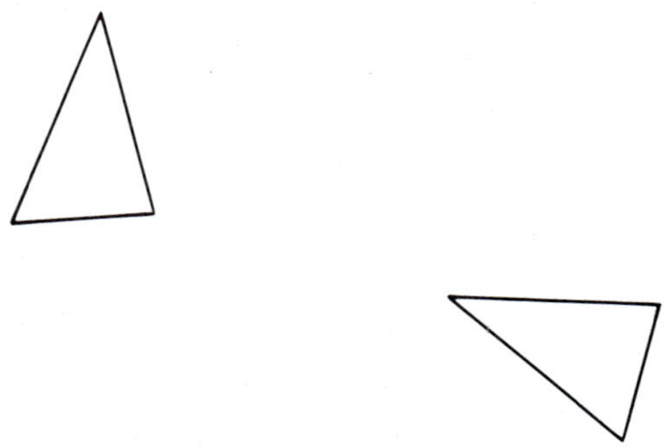

7.

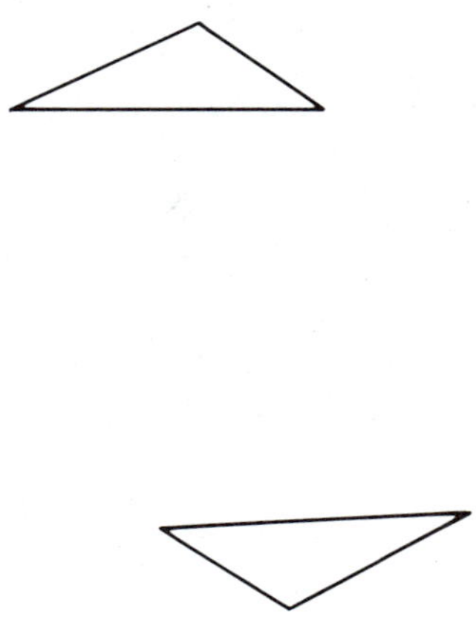

8.

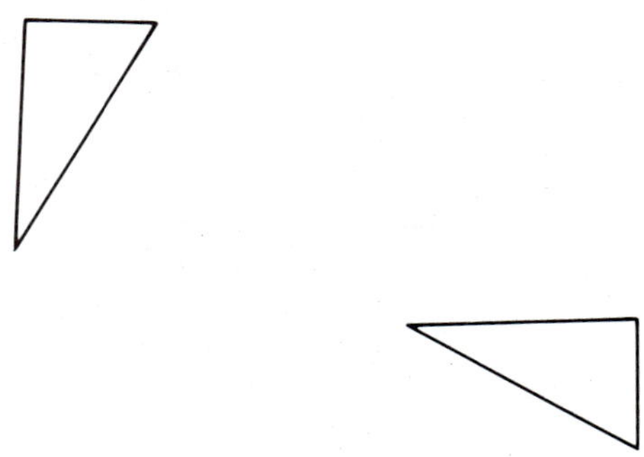

Rotation centers can also be found by paper folding. Some pairs of figures may have more than one possible rotation center.

EXPERIMENT 3

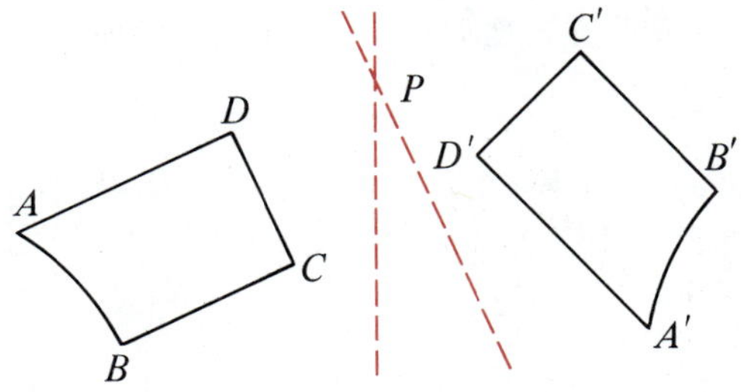

1. Fold your paper so that A and A' coincide. Crease the paper.
2. Refold your paper so that C and C' coincide. Crease the paper with these points together.
3. The point where these two creases intersect is the rotation center for the rotation of $ABCD$ into $A'B'C'D'$.

Project

Copy each pair of triangles. Find the rotation center by using paper folding. Check your accuracy by rotating one figure into the other. Which figures have more than one rotation center? **Check students' answers.**

9.

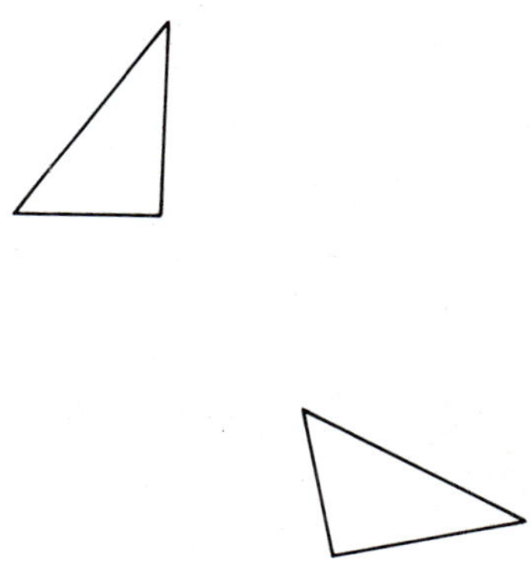

10.

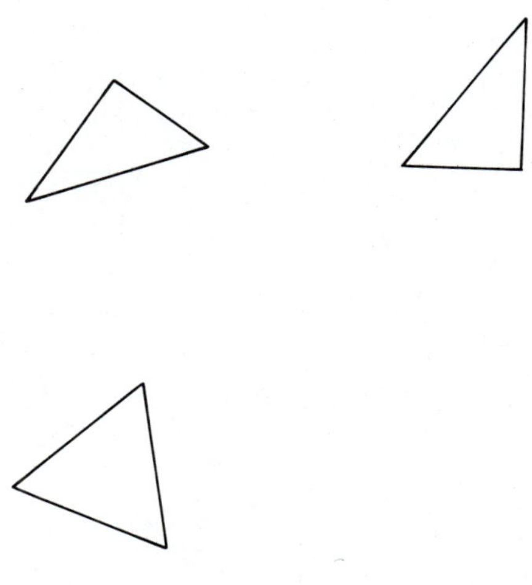

11.

12.

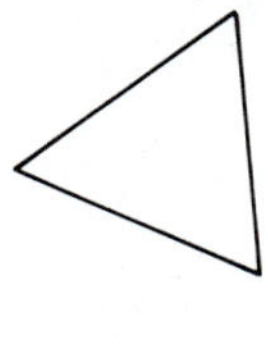

TRANSFORMATIONS: CONGRUENT TRIANGLES

If two triangles are congruent, one of them can be shown to be the image of the other by a reflection, a slide, a rotation, or a combination of two of these, as shown below.

$\triangle DEF$ is the image of $\triangle ABC$ by a slide followed by a reflection.

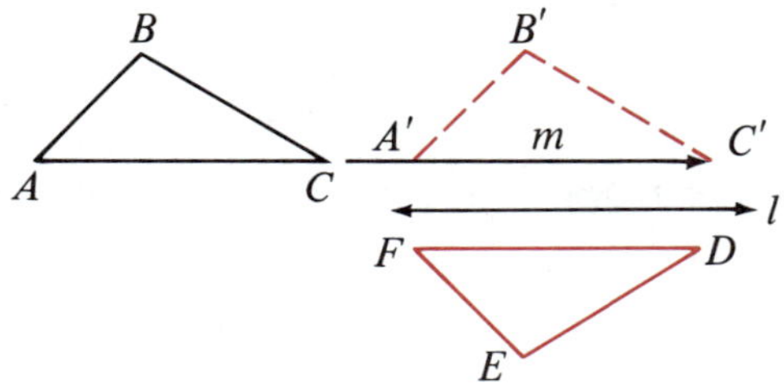

$\triangle UVW$ is the image of $\triangle RST$ by a rotation followed by a slide.

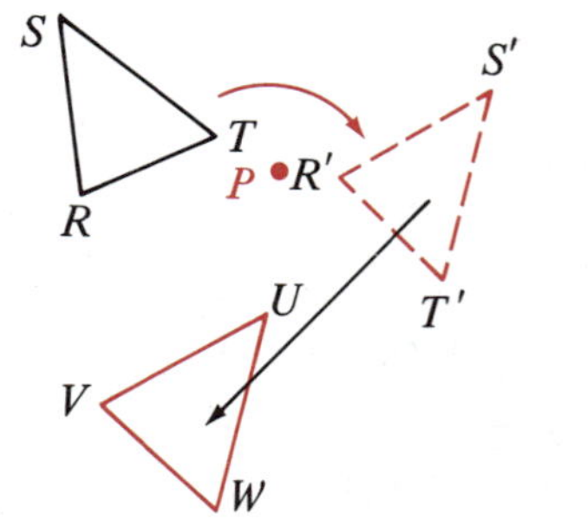

Project Copy each pair of congruent triangles. Find the transformation or pair of transformations that relates the two.

1.

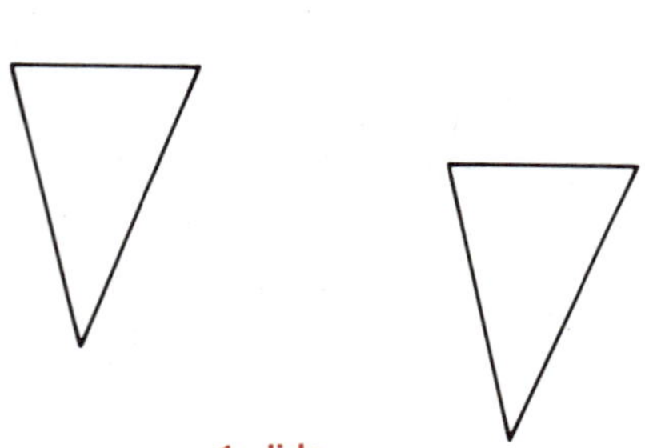

1 slide

2.

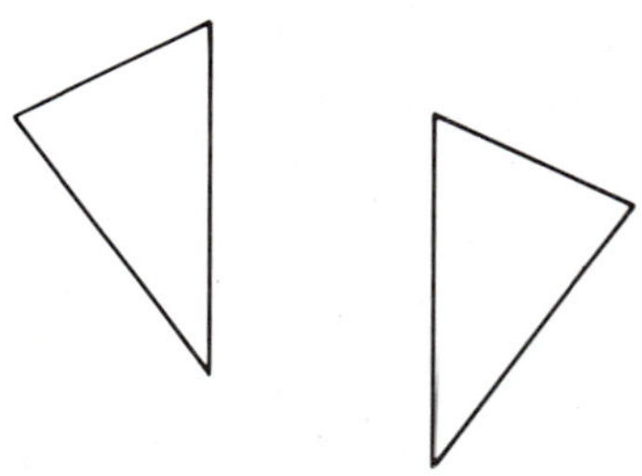

reflection and slide

3.

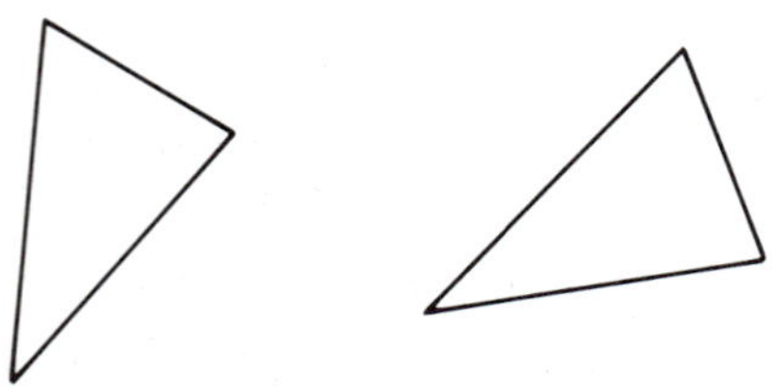

rotation

4.

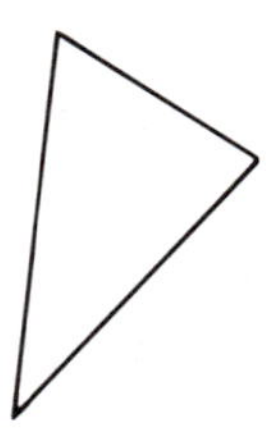

rotation

OVERLAPPING TRIANGLES 5.3

Objectives **To identify and name overlapping triangles**
To write proofs involving overlapping triangles

Sometimes a geometric figure contains two or more triangles that overlap each other. In such cases you will need to "untangle" the figure to identify the triangles involved.

Example 1 Name each triangle in the figure.

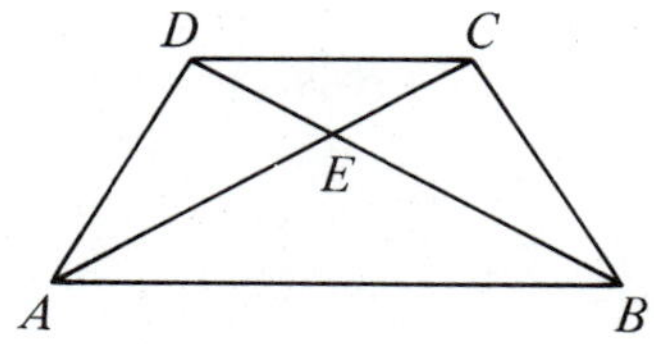

$\triangle ABC$, $\triangle ABD$, $\triangle ABE$, $\triangle ACD$, $\triangle ADE$, $\triangle BCD$, $\triangle BCE$, $\triangle CDE$

Example 2 $\triangle FGI$ and $\triangle FGJ$ are overlapping triangles. Draw two pictures, showing each of these triangles in a different color.

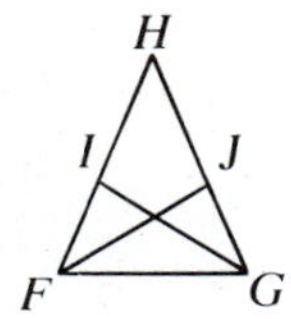

$\triangle FGI$ and $\triangle FGJ$ share side $\overline{FG}$. ►

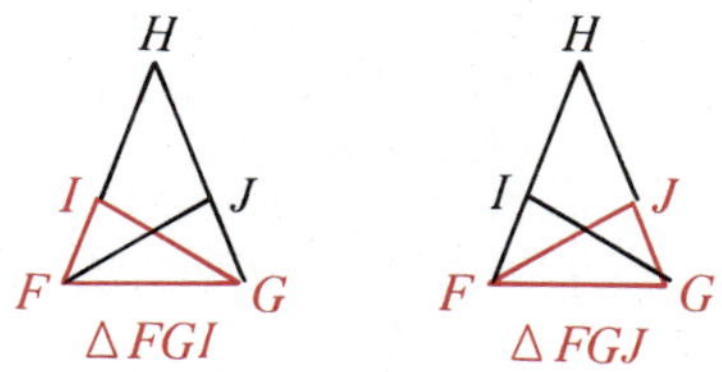

Below is a proof involving overlapping triangles. In proofs of this type, you may find it helpful to redraw the original diagram using two colors, or, more simply, redraw each triangle as a separate figure.

Example 3 Given: $\overline{AC} \cong \overline{AD}$ and $\angle C \cong \angle D$.
Prove: $\overline{AE} \cong \overline{AB}$

Proof

Statements		Reasons
1. $\angle C \cong \angle D$	(A)	1. Given
2. $\overline{AC} \cong \overline{AD}$	(S)	2. Given
3. $\angle A \cong \angle A$	(A)	3. Reflexive property
4. $\triangle ACE \cong \triangle ADB$		4. ASA
5. $\therefore \overline{AE} \cong \overline{AB}$		5. CPCTC

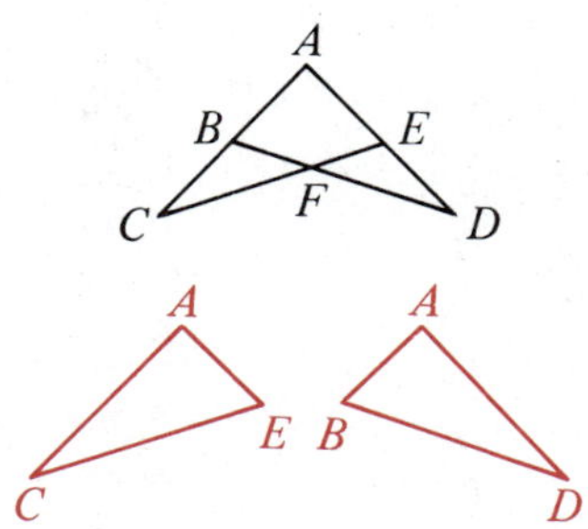

Proofs involving overlapping triangles require a careful analysis of the figure involved.

Example 4

Given: $\overline{YT} \parallel \overline{XU}$, $\overline{YT} \perp \overline{SV}$, $\overline{ST} \cong \overline{VU}$, and $\overline{RS} \cong \overline{RV}$

Prove: $\overline{XU} \cong \overline{YT}$

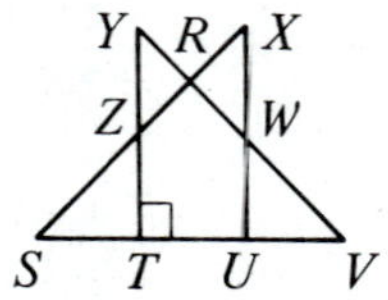

ANALYSIS

Show $\triangle XSU \cong \triangle YVT$.

Redraw each of the △'s separately. ►

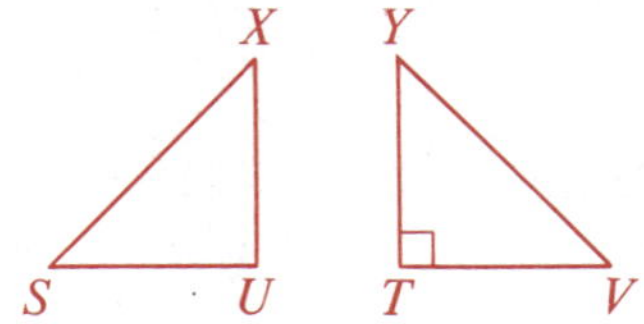

Proof

Statements		Reasons
1. $\overline{YT} \perp \overline{SV}$		1. Given
2. $\angle YTV$ is a rt. $\angle$.		2. Def. of $\perp$
3. $\overline{YT} \parallel \overline{XU}$		3. Given
4. $\angle YTV$ and $\angle XUS$ are supplementary.		4. Int. $\angle$'s of $\parallel$ lines on same side of trans. are suppl.
5. $\angle XUS$ is a rt. $\angle$.		5. Subtraction (m $\angle XUS = 180 - 90$)
6. $\angle XUS \cong \angle YTV$	(A)	6. Rt. $\angle$'s are $\cong$.
7. $\overline{ST} \cong \overline{VU}$		7. Given
8. $ST = VU$		8. Def. of $\cong$ segments
9. $ST + TU = VU + TU$		9. Add. prop. of equality
10. $SU = VT$		10. Def. of between, substitution prop.
11. $\overline{SU} \cong \overline{VT}$	(S)	11. Def. of $\cong$ segments
12. $\overline{RS} \cong \overline{RV}$		12. Given
13. $\angle S \cong \angle V$	(A)	13. $\angle$'s opp. $\cong$ sides of $\triangle$ are $\cong$.
14. $\triangle XSU \cong \triangle YVT$		14. ASA
15. $\therefore \overline{XU} \cong \overline{YT}$		15. CPCTC

Oral Exercises

Name all the different triangles in each figure.

1.

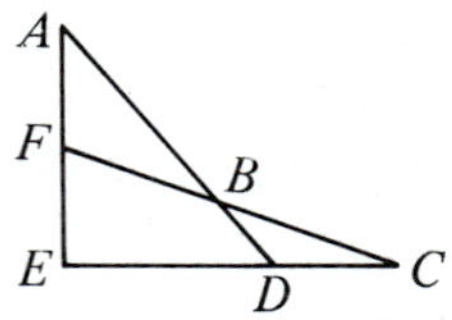

△ ADE, △ EFC, △ BDC, △ ABF

2.

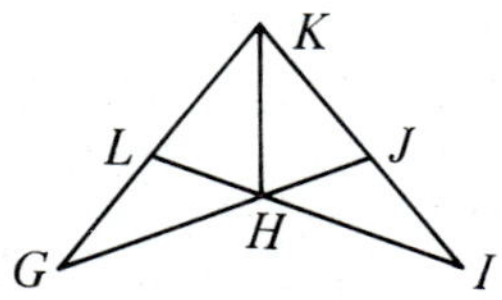

△ GLH, △ LHK, △ GKH, △ HKJ, △ HJI, △ KHI, △ LKI, △ GKJ

3.

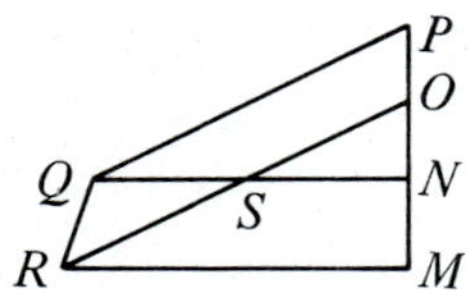

△ QRS, △ NOS, △ NPQ, △ MOR

Written Exercises

Redraw each picture using different colors to show each of the indicated overlapping triangles. (Ex. 1–3)

A

1. $\triangle ABD$ and $\triangle ABC$

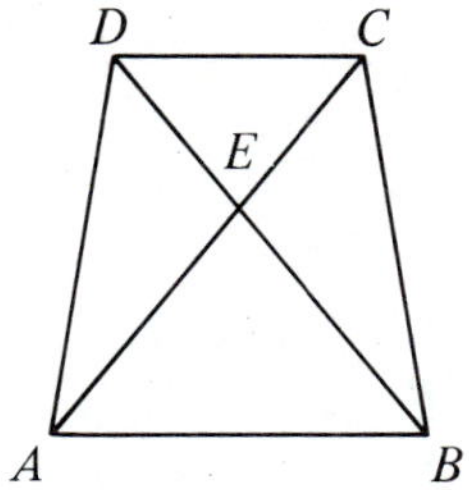

2. $\triangle FHL$ and $\triangle GIJ$

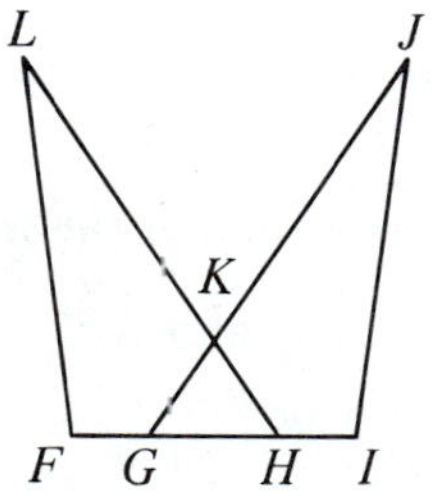

3. $\triangle MOP$ and $\triangle NOQ$

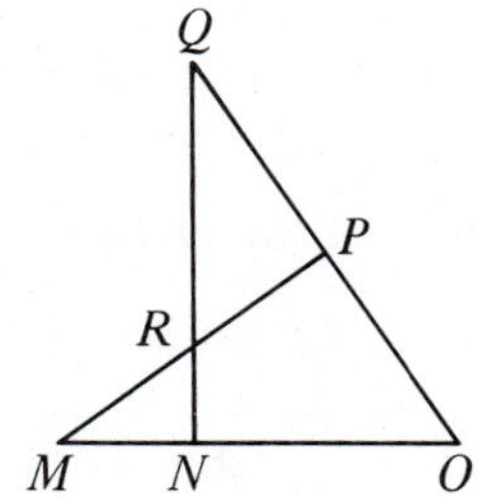

4. Given: $\overline{WT} \cong \overline{WU}$ and $\overline{SU} \cong \overline{VT}$
Prove: $\triangle SUW \cong \triangle VTW$

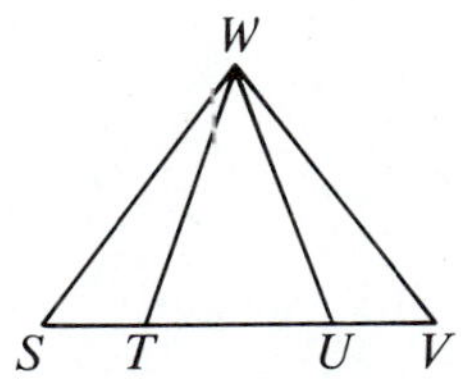

5. Given: $\angle S \cong \angle V$ and $\overline{SU} \cong \overline{VT}$
Prove: $\triangle SUW \cong \triangle VTW$

6. Given: $\overline{XZ} \cong \overline{YZ}$ and $\angle ZXB \cong \angle ZYA$
Prove: $\overline{XB} \cong \overline{YA}$

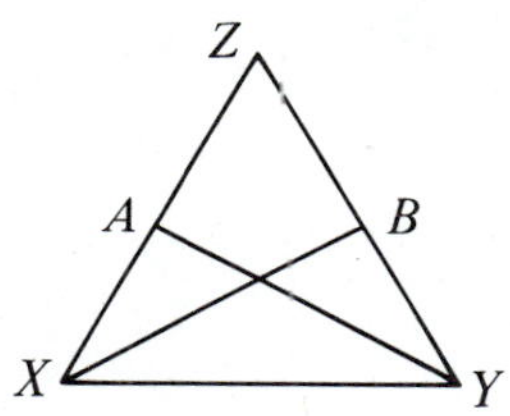

7. Given: $\angle ZXY \cong \angle ZYX$ and $\overline{AX} \cong \overline{BY}$
Prove: $\angle AYX \cong \angle BXY$

8. Given: $\overline{GC} \perp \overline{CD}$, $\overline{ED} \perp \overline{CD}$, and $\overline{CG} \cong \overline{DE}$
Prove: $\overline{DG} \cong \overline{CE}$

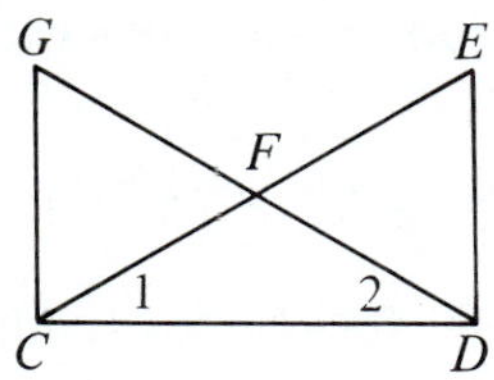

9. Given: $\overline{GC} \perp \overline{CD}$, $\angle 1 \cong \angle 2$, and $\angle G \cong \angle E$
Prove: $\overline{ED} \perp \overline{CD}$

B

10. Given: $\overline{HI} \cong \overline{KJ}$, $\overline{HN} \cong \overline{IL}$, and $\overline{HN} \parallel \overline{IL}$
Prove: $\overline{JN} \parallel \overline{KL}$

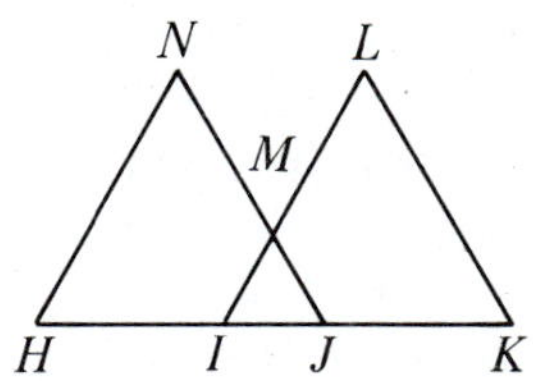

11. Given: $\angle 1 \cong \angle 2$, $\angle 3 \cong \angle 4$, and $\overline{OS} \cong \overline{QR}$
Prove: $\triangle OPR \cong \triangle QPS$

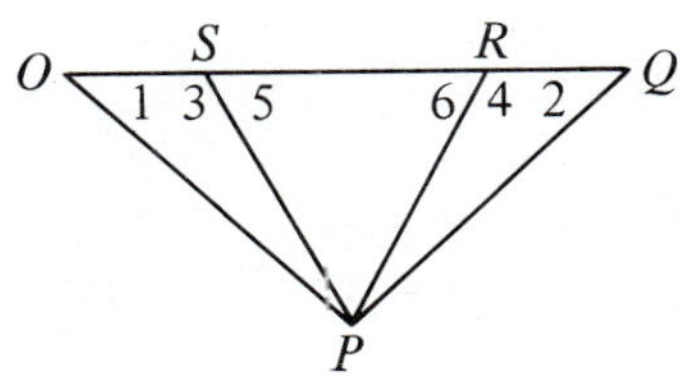

12. Given: $\overline{TX} \parallel \overline{UZ}$, $\overline{WX} \parallel \overline{VY}$, and $\overline{TV} \cong \overline{WU}$
Prove: $\triangle TVY \cong \triangle WUZ$

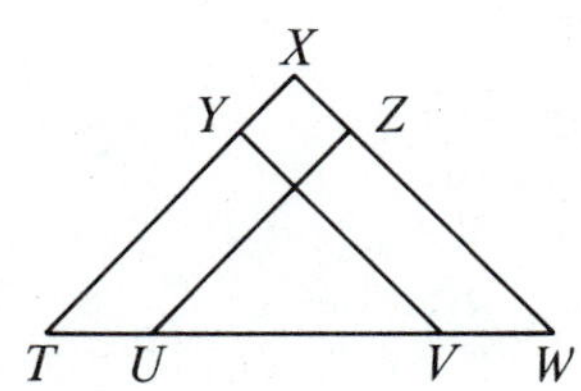

$R_{\triangle ABC}$ means the **region of $\triangle ABC$.** $R_{\triangle ABC}$ includes both $\triangle ABC$ and its interior.

Example

What is the intersection of $R_{\triangle ABC}$ and $R_{\triangle ABE}$?

$R_{\triangle ABF}$ is shaded both colors. ► $R_{\triangle ABF}$

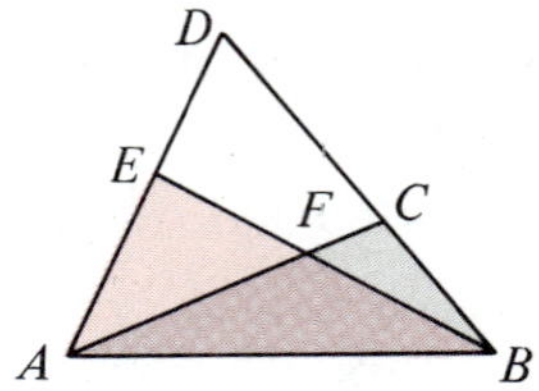

Using each of the given figures, find the intersection of the indicated regions.

C

13. $R_{\triangle FBC}$ and $R_{\triangle AEC}$ **R△GEC**
14. $R_{\triangle AGF}$ and $R_{\triangle FGB}$ **$\overline{FG}$**
15. $R_{\triangle AGC}$ and $R_{\triangle DGC}$ **R△DGC**
16. $R_{\triangle ABD}$ and $R_{\triangle ABE}$ **R△ABG**
17. $R_{\triangle AFC}$ and $R_{\triangle ABG}$ **R△AGF**
18. $R_{\triangle AQE}$ and $R_{\triangle CQE}$ **$\overline{QE}$**
19. $R_{\triangle FOD}$ and $R_{\triangle COA}$ **O**
20. $R_{\triangle FAC}$ and $R_{\triangle FBC}$ **R△FIC**
21. $R_{\triangle FCE}$ and $R_{\triangle DOC}$ **R△SOC**
22. $R_{\triangle ACE}$ and $R_{\triangle BCD}$ **R△JCK**
23. $R_{\triangle ACG}$ and $R_{\triangle BCD}$ **R△BCD**
24. $R_{\triangle ABF}$ and $R_{\triangle ACG}$ **R△ABE**
25. $R_{\triangle AEF}$ and $R_{\triangle ABE}$ **$\overline{AE}$**
26. $R_{\triangle ABE}$ and $R_{\triangle BCD}$ **B**
27. $R_{\triangle GCH}$ and $R_{\triangle BCH}$ **R△HDC**

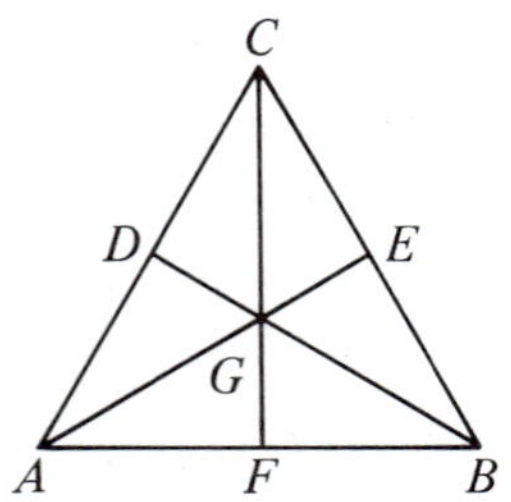

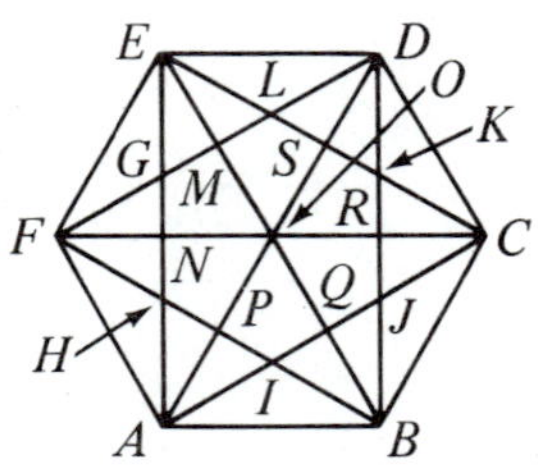

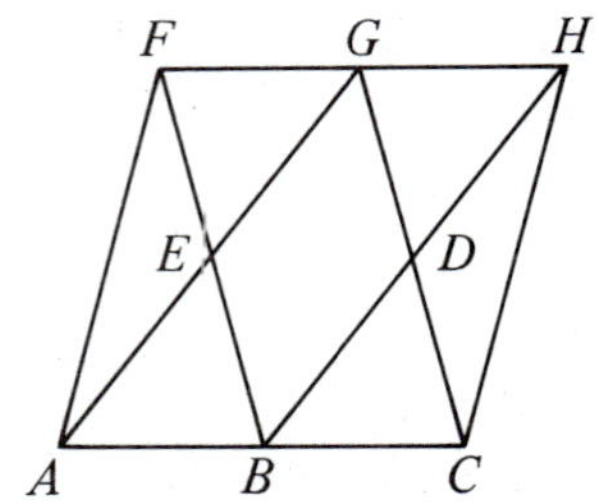

ALGEBRA REVIEW

Objective

To factor a trinomial and the difference of two squares

To factor a trinomial like $x^2 - 7x + 12$:
1. Try factors of x^2 and 12.
2. Determine signs for correct middle terms.

Factor $x^2 - 7x + 12$

$$\begin{array}{ll} 1x \quad \square 3 & (1x)(\boxminus 3) = -3x \\ 1x \quad \square 4 & (\boxminus 4)(1x) = -4x \\ & (-3x) + (-4x) = -7x \end{array}$$

Thus, $x^2 - 7x + 12 = (x - 3)(x - 4)$

Factor $4x^2 - 25$

↑

Difference of squares

$(2x)^2 - (5)^2$

Thus, $4x^2 - 25 = (2x + 5)(2x - 5)$

Factor.

1. $x^2 - 6x + 8$ **(x − 4)(x − 2)**
2. $x^2 + 2x - 15$ **(x + 5)(x − 3)**
3. $x^2 + 9x + 20$ **(x + 4)(x + 5)**
4. $x^2 - 1$ **(x − 1)(x + 1)**
5. $x^2 - 81$ **(x − 9)(x + 9)**
6. $9x^2 - 36$ **6. (3x − 6)(3x + 6), or 9(x − 2)(x + 2)**
7. $6x^2 - x - 2$ **(2x + 1)(3x − 2)**
8. $2x^2 + 7x - 15$ **(x + 5)(2x − 3)**
9. $6x^2 - 13x + 6$ **(3x − 2)(2x − 3)**

CONGRUENT RIGHT TRIANGLES 5.4

Objectives

To identify the parts of a right triangle
To apply right triangle congruence theorems

Since you often will use right triangles or parts of right triangles in your work in geometry, it will be helpful to have names for these different parts.

Definition: Hypotenuse and leg

The *hypotenuse* of a right triangle is the side opposite the right angle. The *legs* are the other two sides.

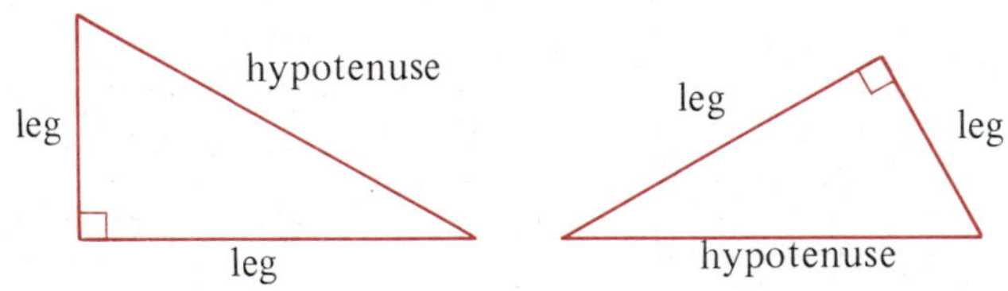

You can prove right triangles congruent by any of the methods used for other triangles. There are, however, some special methods that are used only with right triangles. Three of these special methods are actually restatements of other general congruence patterns. The first of these is proved below. It is usually called the *Leg-Leg,* or *LL,* pattern.

Theorem 5.3 Leg-Leg (LL)

Two right triangles are congruent if the two legs of one are congruent to the corresponding legs of the other.

Given: Rt. △'s ABC and $A'B'C'$;
rt. ∠'s B and B';
$\overline{AB} \cong \overline{A'B'}$ and $\overline{BC} \cong \overline{B'C'}$
Prove: $\triangle ABC \cong \triangle A'B'C'$

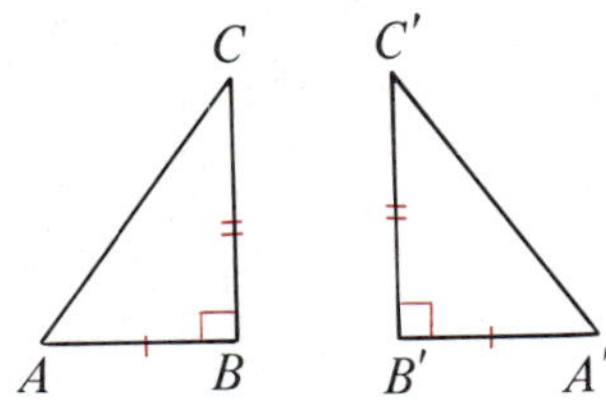

Proof

Statements		Reasons
1. Rt. △'s ABC and $A'B'C'$		1. Given
2. $\overline{AB} \cong \overline{A'B'}$	(S)	2. Given
3. $\angle B$ and $\angle B'$ are rt. ∠'s.		3. Given
4. $\angle B \cong \angle B'$	(A)	4. All rt. ∠'s are ≅.
5. $\overline{BC} \cong \overline{B'C'}$	(S)	5. Given
6. $\therefore \triangle ABC \cong \triangle A'B'C'$		6. SAS

Two other patterns for right triangles are special cases of other general congruence patterns. You should complete proofs of these.

Theorem 5.4 Leg-Acute Angle (LA)

Two right triangles are congruent if a leg and an acute angle of one are congruent to the corresponding leg and acute angle of the other.

Case I

Given: Rt. △'s ABC and $A'B'C'$;
rt. ∠'s B and B';
$\overline{AB} \cong \overline{A'B'}$ and $\angle A \cong \angle A'$
Prove: $\triangle ABC \cong \triangle A'B'C'$

ANALYSIS
Use ASA.

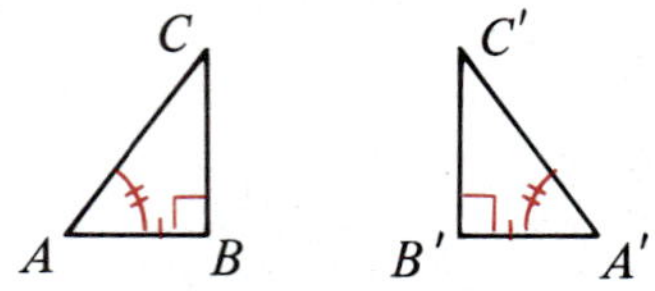

Case II

Given: Rt. △'s ABC and $A'B'C'$;
rt. ∠'s B and B';
$\overline{AB} \cong \overline{A'B'}$ and $\angle C \cong \angle C'$
Prove: $\triangle ABC \cong A'B'C'$

ANALYSIS
Use AAS.

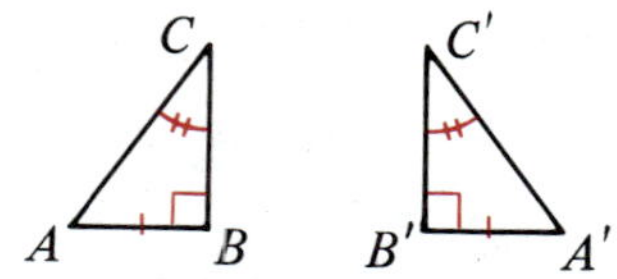

Theorem 5.5 Hypotenuse-Acute Angle (HA)

Two right triangles are congruent if the hypotenuse and an acute angle of one are congruent to the hypotenuse and the corresponding acute angle of the other.

The abbreviations LA and HA will be written when these theorems are used as reasons in proofs.

Example 1

Given: $\overline{AC} \perp \overline{BD}$ and $\overline{AB} \cong \overline{AD}$
Prove: $\triangle BCA \cong \triangle DCA$

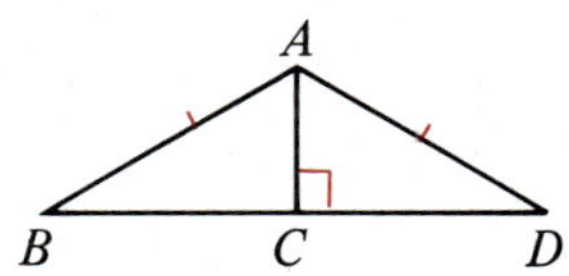

Proof

Statements		Reasons
1. $\overline{AC} \perp \overline{BD}$		1. Given
2. $\angle BCA$ and $\angle DCA$ are rt. ∠'s.		2. ⊥ forms rt. ∠'s.
3. $\triangle BCA$ and $\triangle DCA$ are rt. △'s.		3. Def. of rt. △
4. $\overline{AB} \cong \overline{AD}$	(H)	4. Given
5. $\angle B \cong \angle D$	(A)	5. ∠'s opp. ≅ sides of △ are ≅.
6. ∴ $\triangle BCA \cong \triangle DCA$		6. HA

You found earlier that side-side-angle (SSA) is not a method for proving triangles congruent. However, this pattern can be used to prove right triangles congruent. This is demonstrated in Theorem 5.6. The proof given is an outline of the complete proof. You should complete the details.

Theorem 5.6 Hypotenuse-Leg (HL)

Two right triangles are congruent if the hypotenuse and a leg of one are congruent to the hypotenuse and corresponding leg of the other.

Given: Rt. △'s ABC and $A'B'C'$;
Rt. ∠'s B and B';
$\overline{CA} \cong \overline{C'A'}$, and $\overline{CB} \cong \overline{C'B'}$

Prove: $\triangle ABC \cong \triangle A'B'C'$

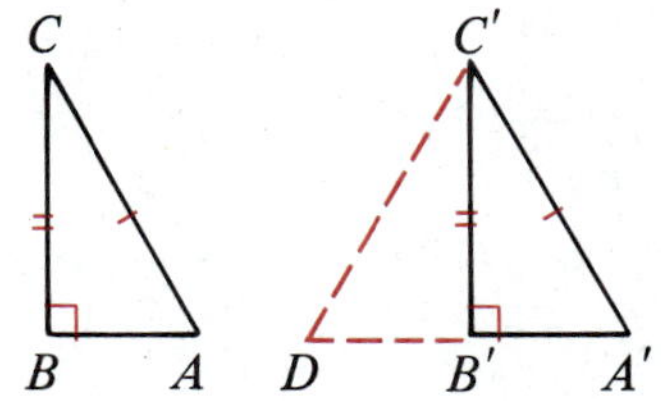

ANALYSIS
Construct $\triangle DC'B'$ with $\overline{B'D} \cong \overline{BA}$. Show $\triangle ABC \cong \triangle DB'C'$, using corresponding parts.

Outline of Proof

1. Extend $\overline{A'B'}$ to D so that $\overline{B'D} \cong \overline{BA}$.
2. Draw $\overline{DC'}$.
3. $m\angle C'B'D + m\angle C'B'A' = 180$
4. $m\angle C'B'D = 90$
5. $\angle CBA$ is a rt. ∠.
6. $\angle CBA \cong \angle C'B'D$
7. $\overline{CB} \cong \overline{C'B'}$
8. $\triangle ABC \cong \triangle DB'C'$
9. $\overline{CA} \cong \overline{C'D}$
10. $\overline{CA} \cong \overline{C'A'}$
11. $\overline{C'D} \cong \overline{C'A'}$
12. $\angle C'A'B' \cong \angle C'DB'$
13. $\triangle DB'C' \cong \triangle A'B'C'$
14. $\triangle ABC \cong \triangle A'B'C'$

You can use the hypotenuse-leg theorem (HL) as a reason in other proofs.

Example 2

Given: $\overline{RS} \perp \overline{PQ}$ and $\overline{PR} \cong \overline{QR}$
Prove: $\overline{RS}$ bisects $\overline{PQ}$.

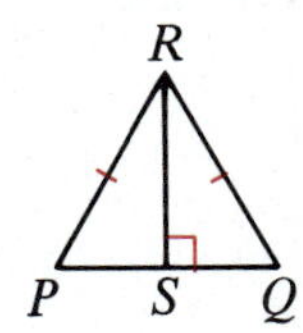

Proof

Statements		Reasons
1. $\overline{RS} \perp \overline{PQ}$		1. Given
2. $\angle PSR$ and $\angle QSR$ are rt. ∠'s.		2. ⊥ forms rt. ∠'s.
3. $\triangle PSR$ and $\triangle QSR$ are rt. △'s.		3. Def. of rt. △
4. $\overline{PR} \cong \overline{QR}$	(H)	4. Given
5. $\overline{RS} \cong \overline{RS}$	(L)	5. Reflexive property
6. $\triangle PSR \cong \triangle QSR$		6. HL
7. $\overline{PS} \cong \overline{QS}$		7. CPCTC
8. ∴ $\overline{RS}$ bisects $\overline{PQ}$.		8. Def. of bisector

Summary

Right triangles can be proved congruent by the following:

Leg-leg (LL)

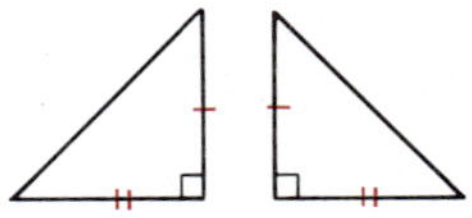

Leg-acute angle (LA)

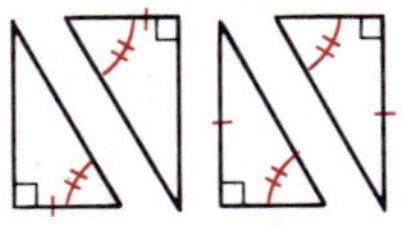

Hypotenuse-acute angle (HA)

Hypotenuse-leg (HL)

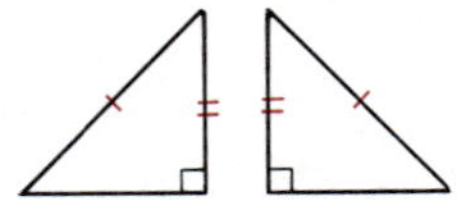

Oral Exercises

Name the hypotenuse and the legs in each right triangle.

1.

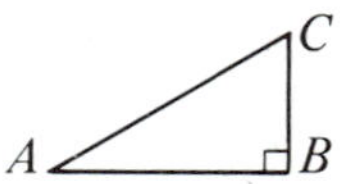

$\overline{AC}$; $\overline{AB}$, $\overline{BC}$

2.

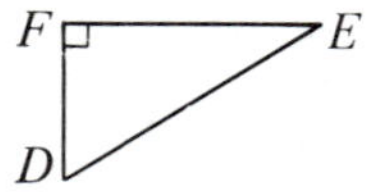

$\overline{DE}$; $\overline{DF}$, $\overline{FE}$

3.

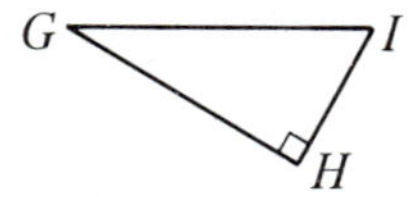

$\overline{GI}$; $\overline{GH}$, $\overline{HI}$

4.

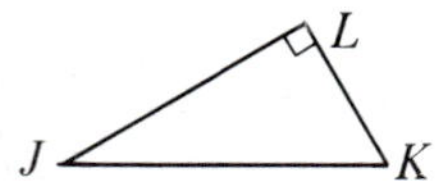

$\overline{JK}$; $\overline{JL}$, $\overline{KL}$

Based on the markings, which of these pairs of right triangles are congruent?

5.

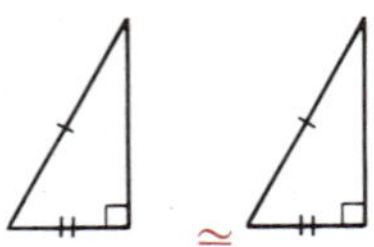

$\cong$

6.

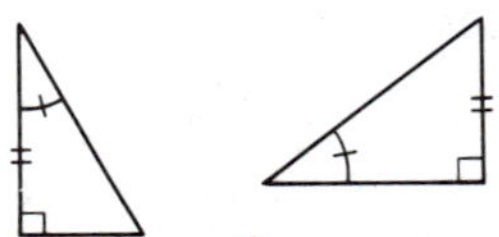

$\not\cong$

7.

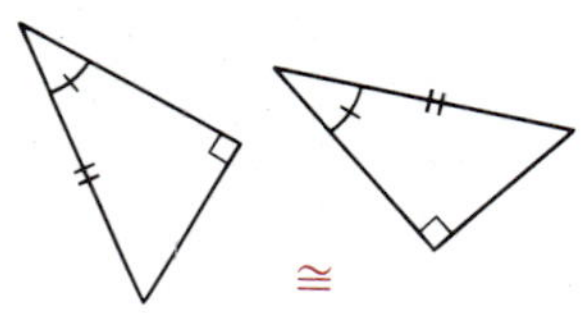

$\cong$

True or false?

8. The hypotenuse of a right triangle is adjacent to the right angle. **F**

9. A leg of a right triangle is opposite an acute angle. **T**

Using each set of given information, is rt. $\triangle ABC \cong$ rt. $\triangle EDF$? Give a reason for your answer.

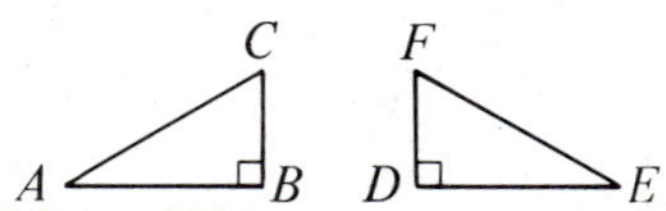

10. $\overline{AB} \cong \overline{ED}$, and $\overline{AC} \cong \overline{EF}$

11. $\angle A \cong \angle E$, and $\overline{AB} \cong \overline{ED}$

12. $\angle A \cong \angle E$, and $\angle C \cong \angle F$

13. $\overline{AB} \cong \overline{ED}$, and $\overline{BC} \cong \overline{DF}$

14. $\overline{AC} \cong \overline{EF}$, and $\angle C \cong \angle F$

15. $\overline{AC} \cong \overline{EF}$, and $\overline{CB} \cong \overline{FD}$

10. $\cong$, HL **12.** $\not\cong$, no $\cong$ sides **14.** $\cong$, HA **11.** $\cong$, LA **13.** $\cong$, LL **15.** $\cong$, HL

Written Exercises

A

1. Given: $\overline{ZY} \perp \overline{XY}$, $\overline{TS} \perp \overline{RS}$, $\overline{XY} \cong \overline{RS}$, and $\overline{ZY} \cong \overline{TS}$
Prove: $\triangle XYZ \cong \triangle RST$

2. Given: $\overline{ZY} \perp \overline{XY}$, $\overline{TS} \perp \overline{RS}$, $\overline{XZ} \cong \overline{RT}$, and $\angle Z \cong \angle T$
Prove: $\triangle XYZ \cong \triangle RST$

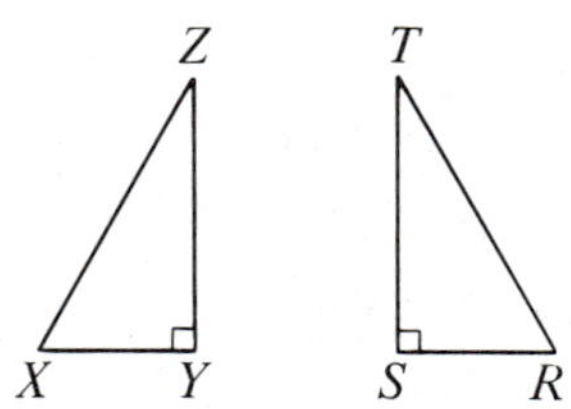

3. Given: $\overline{ZY} \perp \overline{XY}$, $\overline{TS} \perp \overline{RS}$, $\overline{XY} \cong \overline{RS}$, and $\angle Z \cong \angle T$
Prove: $\triangle XYZ \cong \triangle RST$

4. Given: $\overline{ZY} \perp \overline{XY}$, $\overline{TS} \perp \overline{RS}$, $\overline{XZ} \cong \overline{RT}$, and $\overline{XY} \cong \overline{RS}$
Prove: $\triangle XYZ \cong \triangle RST$

5. Given: $\overline{AE} \perp \overline{BC}$, $\overline{BD} \perp \overline{AC}$, and $\overline{AD} \cong \overline{BE}$
Prove: $\triangle ABE \cong \triangle BAD$

7. Given: $\overline{AE} \perp \overline{BC}$, $\overline{BD} \perp \overline{AC}$, and $\overline{AF} \cong \overline{BF}$
Prove: $\triangle AFD \cong \triangle BFE$

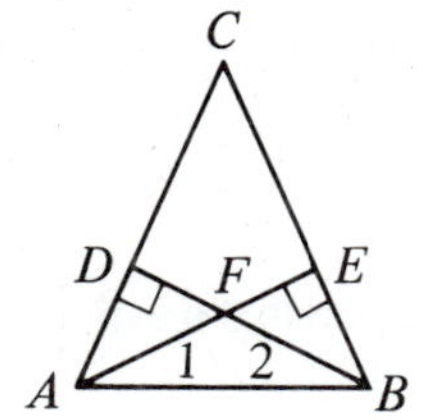

6. Given: $\overline{AE} \perp \overline{BC}$, $\overline{BD} \perp \overline{AC}$, and $\angle 1 \cong \angle 2$
Prove: $\triangle ABE \cong \triangle BAD$

8. Given: $\overline{AE} \perp \overline{BC}$, $\overline{BD} \perp \overline{AC}$, and $\overline{FD} \cong \overline{FE}$
Prove: $\triangle AFD \cong \triangle BFE$

B

9. Given: $\overline{IJ}$ is $\perp$ bisector of $\overline{GH}$
Prove: $\triangle GHI$ is isosceles

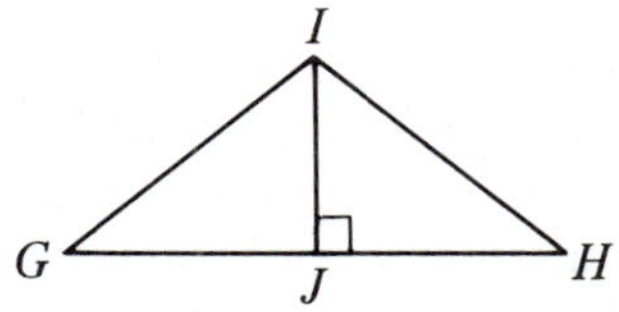

10. Given: $\overline{IJ} \perp$ GH, and $\overline{IJ}$ bisects $\angle GIH$
Prove: $\triangle GHI$ is isosceles

11. Given: $\overline{LN} \perp \overline{MK}$, $\overline{MO} \perp \overline{LK}$, and $\overline{MK} \cong \overline{LK}$
Prove: $\overline{MO} \cong \overline{LN}$

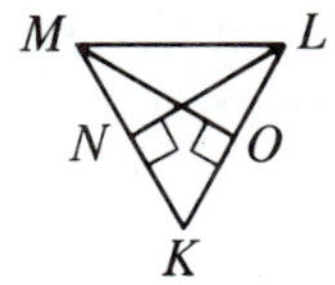

12. Given: $\overline{LN} \perp \overline{MK}$, $\overline{MO} \perp \overline{LK}$, and $\overline{NK} \cong \overline{OK}$
Prove: $\overline{MO} \cong \overline{LN}$

13. Given: $\triangle PQR \cong \triangle UTV$, $\overline{RS} \perp \overline{PQ}$, and $\overline{VW} \perp \overline{UT}$
Prove: $\overline{RS} \cong \overline{VW}$

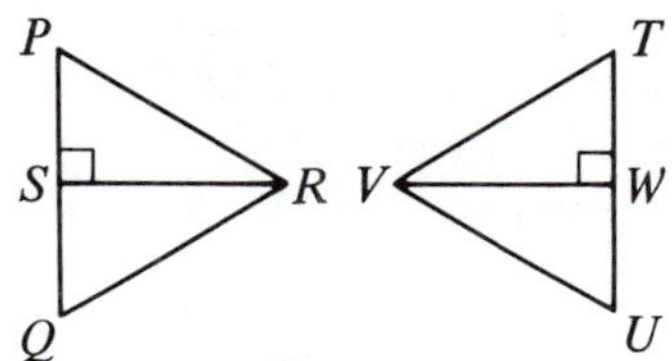

14. Given: $\triangle PQR \cong \triangle UTV$, $\overline{RS} \perp \overline{PQ}$, $\overline{VW} \perp \overline{UT}$, and $\overline{RS}$ bisects $\angle PRQ$
Prove: $\overline{VW}$ bisects $\angle UVT$

C

Given: $\overline{ZA} \perp$ plane P at A, $\overline{AX}$ and $\overline{AY}$ are in P, $\overline{XZ} \cong \overline{YZ}$

15. Prove: $\triangle XAZ \cong \triangle YAZ$
16. Prove: $\angle ZXY \cong \angle ZYX$
17. Prove: $\angle AXY \cong \angle AYX$

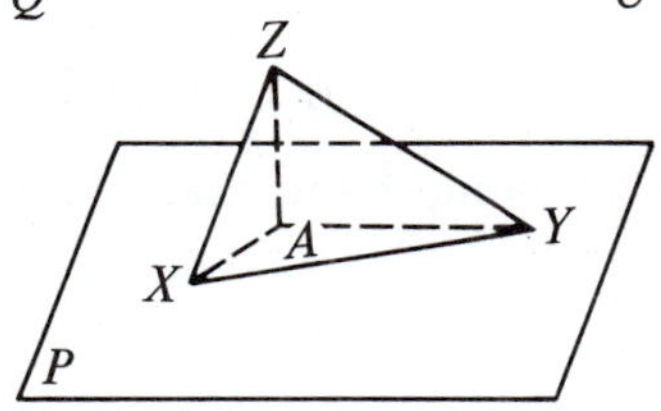

Given: $\overline{ZA} \perp$ plane P at A, $\overline{AX}$ and $\overline{AY}$ are in P, $\overline{XA} \cong \overline{YA}$

18. Prove: $\triangle XAZ \cong \triangle YAZ$
19. Prove: $\angle ZXY \cong \angle ZYX$
20. Prove: $\angle AXY \cong \angle AYX$

CUMULATIVE REVIEW

1. Draw an example of overlapping triangles. Name the triangles. **Answers may vary.**
2. To prove $\triangle XYZ \cong \triangle RJT$ by AAS, which congruent angles and congruent sides can you use?
3. Do perpendicular segments have to intersect? Perpendicular rays? **No; no**

2. $\angle X \cong \angle R$, $\angle Y \cong \angle J$, $\overline{YZ} \cong \overline{JT}$ (Answers may vary.)

NON-ROUTINE PROBLEMS

The floor plans of two houses are shown. Can you walk through the rooms of each house so that you go through each door exactly one time? Draw other floor plans and try them. Is there a way to predict the results without actually trying it?

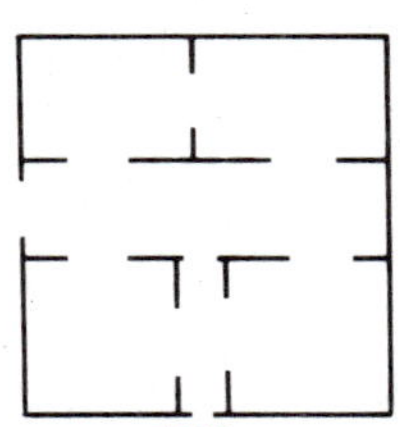

yes

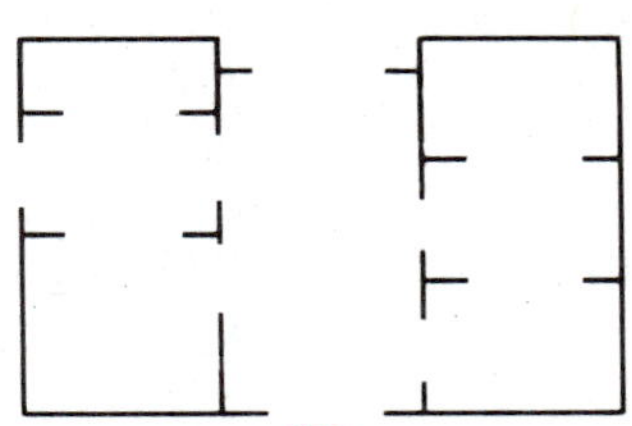

no

USING CORRESPONDING PARTS OF CONGRUENT TRIANGLES 5.5

Objective **To prove triangles congruent using corresponding parts of other congruent triangles**

Sometimes, when trying to prove two triangles congruent, it is impossible to show directly that a given pair of angles or sides is congruent. In such cases, however, you may be able to do the following.

(1) Find another pair of triangles which contain the required angles or sides.
(2) Prove that these triangles are congruent.
(3) Use congruent corresponding parts from these triangles as the needed angles or sides.

Example 1 Given: $\overline{XZ}$ and $\overline{WY}$ bisect each other at V. What combinations of sides and angles can be used to prove $\triangle WYZ \cong \triangle YWX$? Which other triangles contain the needed sides and angles?

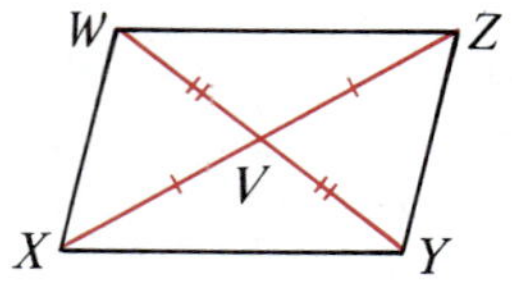

$\overline{WY} \cong \overline{WY}$ (S), $\angle YWZ \cong \angle WYX$ (A), $\overline{WZ} \cong \overline{YX}$ (S)

$\triangle VZW$ and $\triangle VXY$ contain the needed angles and sides.

Example 2 shows you how congruent corresponding parts from congruent triangles can be used to prove another triangle congruency.

Example 2 Complete the missing steps in the following proof.

Given: $\triangle AFD \cong \triangle BFE$,
$\overline{BD} \perp \overline{AC}$ and
$\overline{AE} \perp \overline{BC}$
Prove: $\triangle ABD \cong \triangle BAE$

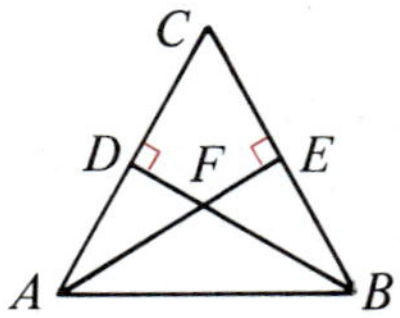

Proof

Statements		Reasons
1. $\triangle AFD \cong \triangle BFE$		1. Given
2. $\overline{AF} \cong \overline{BF}$		2. CPCTC
3. $\angle EAB \cong \angle DBA$	(A)	3. $\angle$'s opp. $\cong$ sides of $\triangle$ are $\cong$.
4. $\overline{BD} \perp \overline{AC}$, $\overline{AE} \perp \overline{BC}$		4. Given
5. $\angle BDA \cong \angle AEB$	(A)	5. $\perp$'s form $\cong$ rt. $\angle$'s.
6. $\overline{AB} \cong \overline{BA}$	(S)	6. Reflexive property
7. $\therefore \triangle ABD \cong \triangle BAE$		7. AAS

Example 3

Given: $\overline{AD} \cong \overline{CB}$, $\overline{DC} \cong \overline{BA}$, and F is the midpoint of $\overline{AC}$.
Prove: $\triangle AGF \cong \triangle CEF$

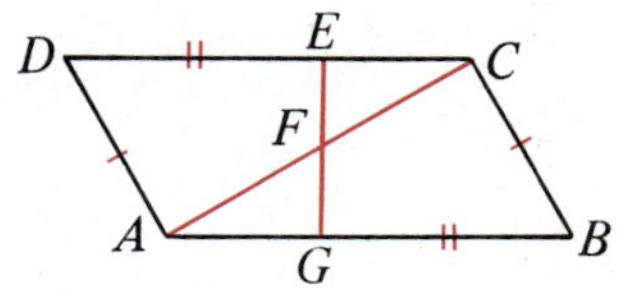

Proof

Statements		Reasons
1. $\overline{AD} \cong \overline{CB}$	(S)	1. Given
2. $\overline{DC} \cong \overline{BA}$	(S)	2. Given
3. $\overline{CA} \cong \overline{AC}$	(S)	3. Reflexive property
4. $\triangle CDA \cong \triangle ABC$		4. SSS
5. $\angle ACD \cong \angle CAB$	(A)	5. CPCTC
6. F is the midpoint of $\overline{AC}$.		6. Given
7. $\overline{AF} \cong \overline{CF}$	(S)	7. Def. of midpoint
8. $\angle AFG \cong \angle CFE$	(A)	8. Vert. $\angle$'s are $\cong$.
9. $\therefore \triangle AGF \cong \triangle CEF$		9. ASA

Written Exercises

For each figure, what combinations of sides and angles can be used to prove the indicated triangle congruency? Which other triangles contain the needed sides and angles? (Ex. 1–3)

A **1.** $\triangle ABC \cong \triangle CDA$

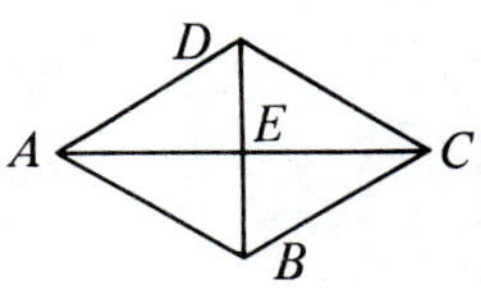

$\overline{AB} \cong \overline{CD}$
$\overline{AC} \cong \overline{CA}$
$\angle BAC \cong \angle DCA$
Use $\triangle$ DCE and $\triangle$ BAE.

2. $\triangle FGJ \cong \triangle IHJ$

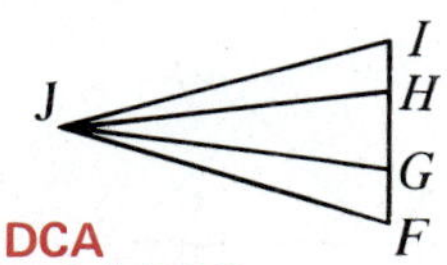

$\overline{FJ} \cong \overline{IJ}$
$\angle JFH \cong \angle JIG$
$\overline{FH} \cong \overline{IG}$
Use $\triangle$ JFH and $\triangle$ JIG.

3. $\triangle KLO \cong \triangle MNO$

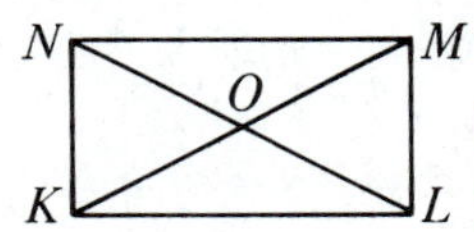

$\overline{KL} \cong \overline{MN}$
$\angle LOK \cong \angle NOM$
$\angle OKL \cong$ OMN
Use $\triangle$ KLM and $\triangle$ MNK.

4. Given: $\overline{AD} \cong \overline{BC}$ and $\angle 1 \cong \angle 2$
Show: $\triangle ABD \cong \triangle BAC$

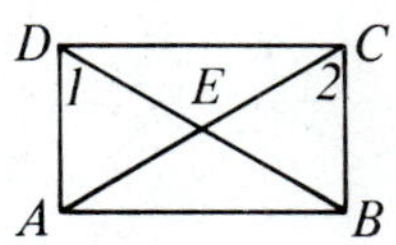

5. Given: $\overline{GI} \perp \overline{FH}$, $\overline{FJ} \perp \overline{GH}$, and $\overline{FJ} \cong \overline{GI}$
Show: $\triangle FKI \cong \triangle GKJ$

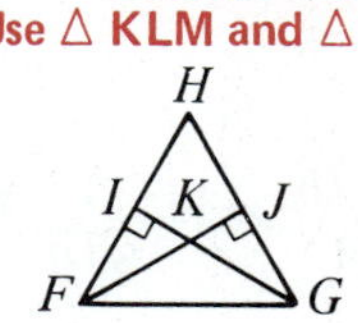

Complete the missing steps in each proof. (Ex. 6–7)

6. Given: $\triangle JFL \cong \triangle KHL$ and $\overline{IH} \parallel \overline{FG}$
Prove: $\triangle FIH \cong \triangle HGF$

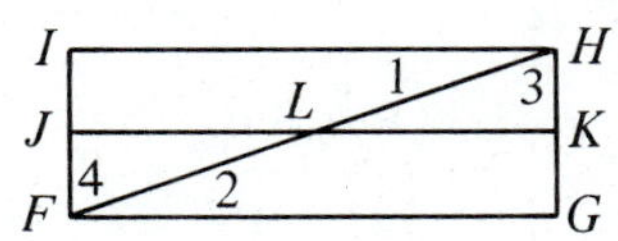

Proof

Statements		Reasons
1. ? $\triangle JFL \cong \triangle KHL$		1. Given
2. $\angle 4 \cong \angle 3$	(A)	2. ? CPCTC
3. ? $\overline{HF} \cong \overline{FH}$	(S)	3. ? Reflexive prop
4. ? $\overline{IH} \parallel \overline{FG}$		4. Given
5. $\angle 1 \cong \angle 2$	(A)	5. ? Alt int $\angle$'s of $\parallel$ lines are $\cong$.
6. $\therefore \triangle FIH \cong \triangle HGF$		6. ? ASA

7. Given: $\overline{AC}$ and $\overline{BD}$ bisect each other at E.
Prove: $\triangle ABC \cong \triangle CDA$

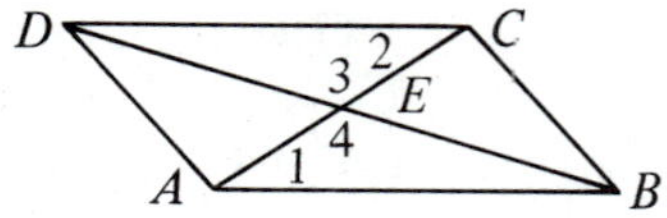

Proof

1. $\overline{AC}$ and $\overline{BD}$ bisect each other at E.

Statements		Reasons
1. ?		1. Given
2. $\overline{AE} \cong \overline{CE}$		2. ? Def. of bisector of seg
3. $\overline{BE} \cong$? DE		3. ? Same as 2
4. $\angle 3 \cong \angle 4$		4. ? Vert $\angle$'s are $\cong$.
5. $\triangle ABE \cong$? $\triangle CDE$		5. SAS
6. $\overline{AB} \cong \overline{CD}$	(S)	6. ? CPCTC
7. ? $\angle 1 \cong \angle 2$	(A)	7. CPCTC
8. $\overline{AC} \cong$? CA	(S)	8. ? Reflexive prop
9. $\therefore \triangle ABC \cong \triangle CDA$		9. ? SAS

8. Given: $\triangle FHL \cong \triangle JHK$, $\overline{LG} \perp \overline{FH}$, and $\overline{KI} \perp \overline{HJ}$
Prove: $\triangle LGH \cong \triangle KIH$

9. Given: $\triangle FHL \cong \triangle HJK$, $\overline{LG} \perp \overline{FH}$, and $\overline{KI} \perp \overline{HJ}$
Prove: $\triangle FGL \cong \triangle HIK$

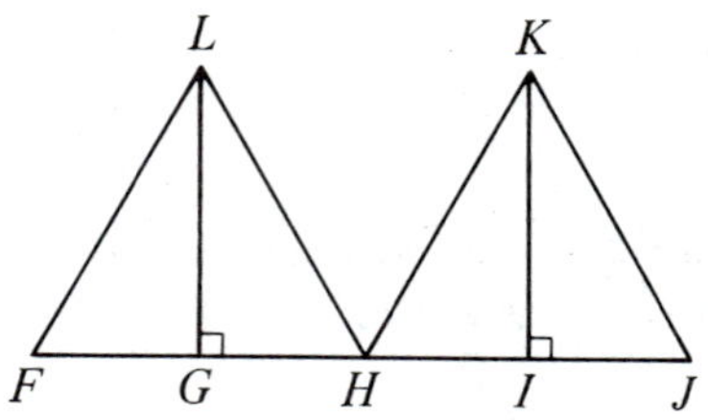

10. Given: $\triangle FGL \cong \triangle HIK$, and $\angle FHL \cong \angle HJK$
Prove: $\triangle FHL \cong \triangle HJK$

11. Given: $\triangle FGL \cong \triangle JIK$, $\overline{LG} \perp \overline{FH}$, $\overline{KI} \perp \overline{HJ}$, $\overline{GH} \cong \overline{HI}$
Prove: $\triangle FHL \cong \triangle JHK$

12. Given: $\overline{QN} \perp \overline{MO}$, $\overline{NR} \perp \overline{MQ}$, $\overline{NP} \perp \overline{OQ}$, and $\overline{NM} \cong \overline{NO}$
Prove: $\overline{NR} \cong \overline{NP}$

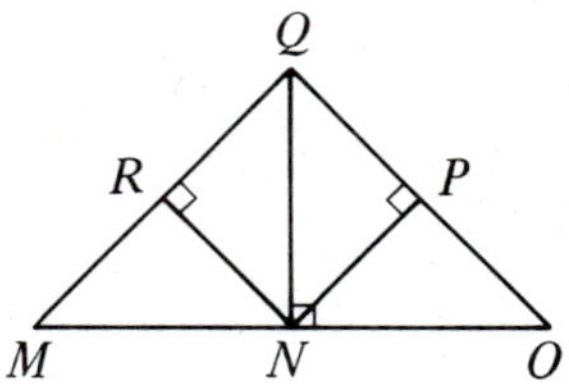

13. Given: $\overline{QN} \perp \overline{MO}$, $\overline{NR} \perp \overline{MQ}$, $\overline{NP} \perp \overline{OQ}$, and $\overline{NR} \cong \overline{NP}$
Prove: $\overline{NM} \cong \overline{NO}$

B 14. Given: $\overline{AD} \cong \overline{BC}$, $\overline{DE} \perp \overline{AB}$, $\overline{CF} \perp \overline{AB}$, and $\overline{DE} \cong \overline{CF}$
Prove: $\overline{AC} \cong \overline{BD}$

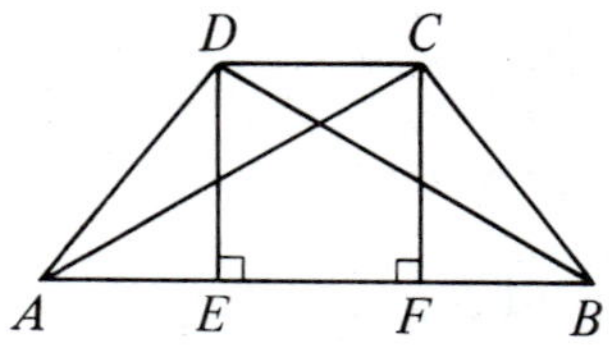

15. Given: $\overline{DE} \perp \overline{AB}$, $\overline{CF} \perp \overline{AB}$, $\overline{DE} \cong \overline{CF}$, and $\overline{AC} \cong \overline{DB}$
Prove: $\overline{AD} \cong \overline{BC}$

16. Given: $\triangle JHG \cong \triangle HJI$, $\overline{JG} \perp \overline{GH}$, and m $\angle JGI = 63$
Find m $\angle HJI$. 27

17. Given: $\triangle PLM \cong \triangle NLQ$, m $\angle L = 88$, and m $\angle P = 26$
Find m $\angle NQL$. 66

18. Given: $\triangle SVW \cong RVU$, $\overline{RW} \perp \overline{ST}$, and m $\angle T = 60$
Find m $\angle RVS$. 120

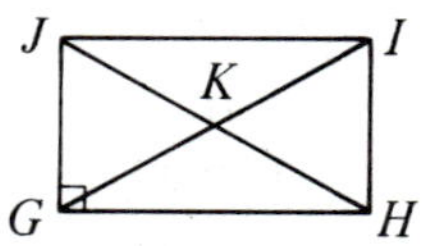

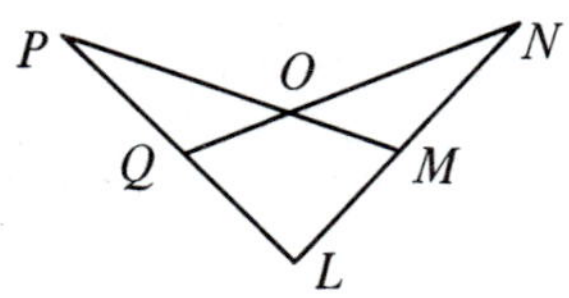

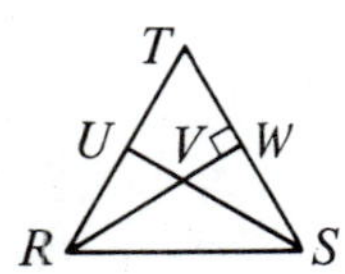

Using the given information, prove each of the following. (Ex. 19–25)

Given: $\overline{AD} \perp \overline{AB}$, $\overline{BC} \perp \overline{AB}$, and $\overline{AD} \cong \overline{BC}$

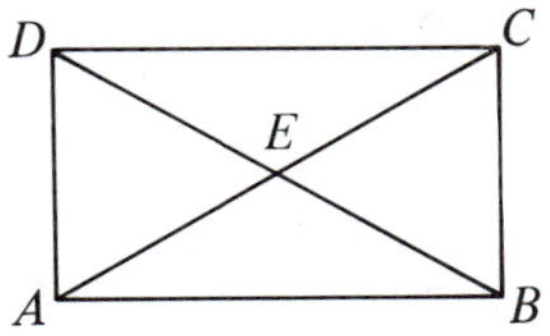

19. Prove: $\overline{AC} \cong \overline{BD}$

20. Prove: $\triangle AEB$ is isosceles.

21. Prove: $\overline{ED} \cong \overline{EC}$

22. Prove: $\overline{AB} \parallel \overline{DC}$

23. Prove: $\overline{AD} \perp \overline{DC}$ and $\overline{BC} \perp \overline{DC}$

24. Prove: $\triangle AEB \cong \triangle CED$

25. Prove: $\overline{AE} \cong \overline{BE} \cong \overline{CE} \cong \overline{DE}$

C **26.** Given: $\overline{FJ} \cong \overline{HJ}$, $\angle 1 \cong \angle 3$, and G is the midpoint of $\overline{FH}$.
Prove: $\overline{FI} \cong \overline{HK}$

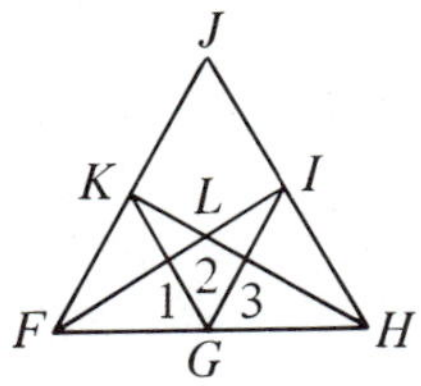

27. Given: $\overline{QR} \cong \overline{MR}$, and $\overline{PR} \cong \overline{NR}$
Prove: $\overline{OQ} \cong \overline{OM}$

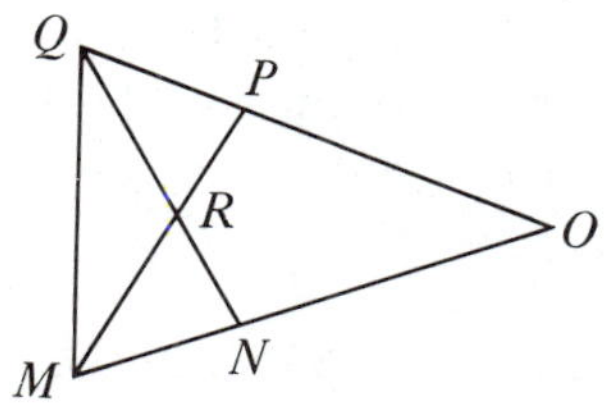

28. Given: $\overline{US} \cong \overline{TS}$, and $\overline{UV} \cong \overline{TV}$
Prove: $\triangle UWT$ is isosceles.

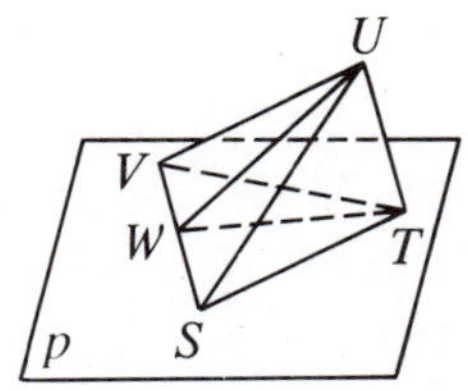

29. Given: $\overline{AZ} \cong \overline{AY}$ and $\overline{XZ} \cong \overline{XY}$
Prove: $\overline{ZB} \cong \overline{YB}$

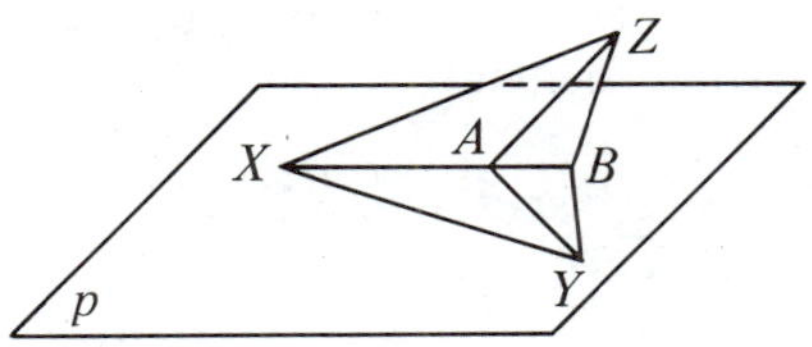

30. Given: $\angle ZXB \cong \angle YXB$ and $\angle XZA \cong \angle XYA$
Prove: $\angle XZB \cong \angle XYB$

NON-ROUTINE PROBLEMS

Many different figures can be made from segments. The intersection of two or more of these segments is a *vertex*. These segments can enclose *regions*. Each figure below has a certain number of segments (S), vertices (V), and regions (R). Is there any relationship between S, V, and R? **V + R = S + 1**

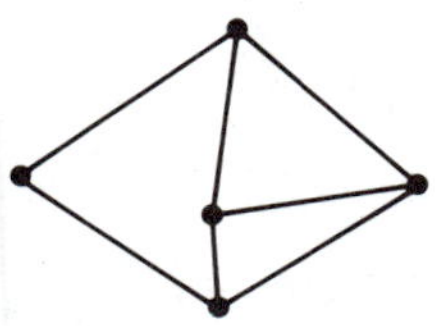

$V = 5, S = 7, R = 3$

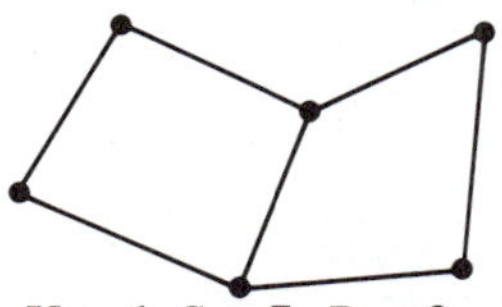

$V = 6, S = 7, R = 2$

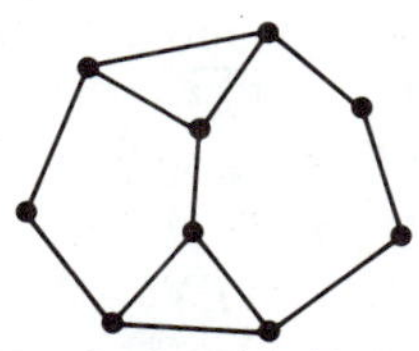

$V = 9, S = 12, R = 4$

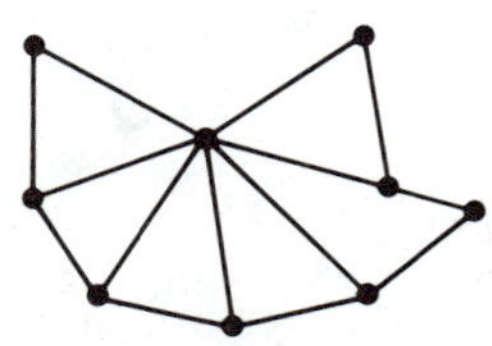

$V = 9, S = 14, R = 6$

How many regions should a figure with 8 vertices and 13 segments have? Can you draw such a figure? **6**

How many segments should be used to connect 10 vertices to form 5 regions? Can you draw such a figure? **14**

How many vertices are needed if 15 segments form 7 regions? Can you draw such a figure? **9**

BISECTORS 5.6

Objective **To prove and apply theorems about bisectors**

You learned earlier that any figure which contains the midpoint of a segment is called a *bisector* of the segment. Since every segment has exactly one midpoint, it has at least one bisector. In the figure at the right, line l is a bisector of $\overline{AB}$.

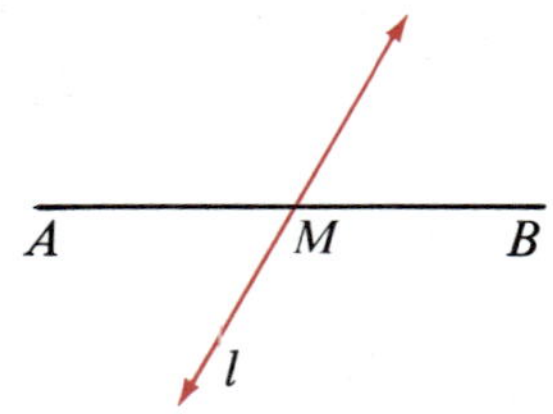

Earlier you learned that one special bisector is the perpendicular bisector. Any segment, ray, or line that is both perpendicular to a segment and also a bisector of the segment is called a **perpendicular bisector** of the segment. You can use congruent triangles to prove the following theorem about perpendicular bisectors.

Theorem 5.7 Perpendicular Bisector Theorem

A point on the perpendicular bisector of a segment is equidistant from the endpoints of the segment.

Given: $\overleftrightarrow{CM}$ is $\perp$ bisector of $\overline{XY}$, and P is on $\overleftrightarrow{CM}$.
Prove: $PX = PY$

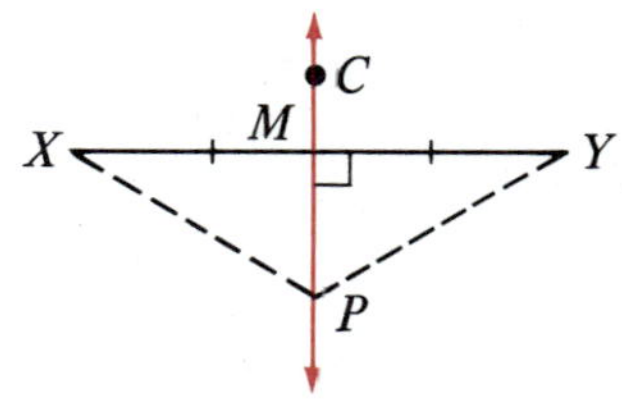

Proof

Statements		Reasons
1. $\overleftrightarrow{CM}$ is $\perp$ bisector of $\overline{XY}$.		1. Given
2. P is on $\overleftrightarrow{CM}$.		2. Given
3. $\overline{XM} \cong \overline{YM}$	(L)	3. Def. of bisector
4. $\triangle XMP$ and $\triangle YMP$ are rt. $\triangle$'s.		4. $\perp$ forms rt. $\angle$'s.
5. $\overline{PM} \cong \overline{PM}$	(L)	5. Reflexive property
6. $\triangle XMP \cong \triangle YMP$		6. LL
7. $\overline{PX} \cong \overline{PY}$		7. CPCTC
8. $\therefore PX = PY$		8. Def. of $\cong$ segments

Example 1

Based on the given information, which segments are congruent?

Given: $\overrightarrow{CP}$ is the $\perp$ bisector of $\overline{AB}$.

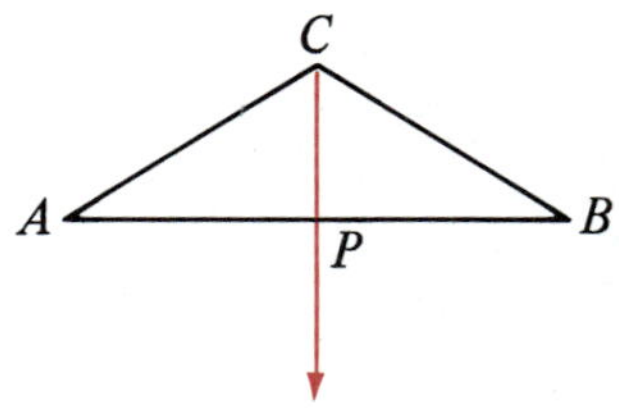

$\overline{AP} \cong \overline{BP}$ and $\overline{AC} \cong \overline{BC}$.

Below is another theorem about perpendicular bisectors. You should complete the details of the proof.

Theorem 5.8 A line containing two points each equidistant from the endpoints of a segment is the perpendicular bisector of the segment. (See Exercise 13.)

Given: $AP = BP$ and $AQ = BQ$
Prove: $\overleftrightarrow{PO}$ is the $\perp$ bisector of $\overline{AB}$.

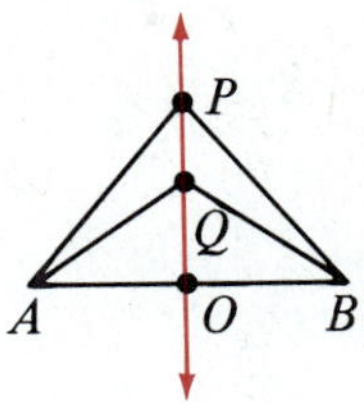

ANALYSIS

Show $\overline{AO} \cong \overline{BO}$ and $\angle AOP \cong \angle BOP$ by corr. parts of $\cong$ $\triangle$'s AOP and BOP.

Outline of Proof

1. $\overline{AP} \cong \overline{BP}$ (S)
2. $\overline{AQ} \cong \overline{BQ}$ (S)
3. $\overline{PQ} \cong \overline{PQ}$ (S)
4. $\triangle APQ \cong \triangle BPQ$
5. $\angle APQ \cong \angle BPQ$ (A)
6. $\overline{PO} \cong \overline{PO}$ (S)
7. $\triangle AOP \cong \triangle BOP$
8. $\overline{AO} \cong \overline{BO}$
9. $\angle AOP \cong \angle BOP$
10. $\angle AOP$ and $\angle BOP$ are supp.
11. $\angle AOP$ and $\angle BOP$ are rt. $\angle$'s.
12. $\therefore \overleftrightarrow{PO}$ is the $\perp$ bisector of $\overline{AB}$.

Every angle has a bisector. In some triangles, an angle bisector is also the perpendicular bisector of the opposite side.

Example 2 Prove: The perpendicular from a vertex of an equilateral triangle to a side of the triangle bisects the side and bisects the angle.

Given: Equil. $\triangle ABC$, $\overline{CD} \perp \overline{AB}$
Prove: $\overline{CD}$ bisects $\overline{AB}$ and $\angle C$.

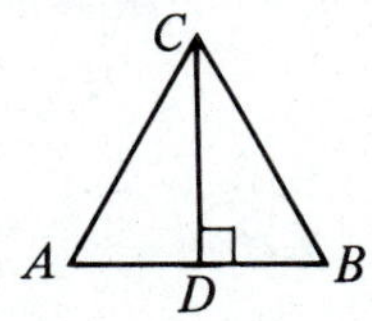

Proof

Statements		Reasons
1. $\triangle ABC$ is equilateral.		1. Given
2. $\overline{CD} \perp \overline{AB}$		2. Given
3. $\triangle ADC$ and $\triangle BDC$ are rt. $\triangle$'s.		3. $\perp$ forms rt. $\angle$'s; Def. of rt. $\triangle$
4. $\overline{AC} \cong \overline{BC}$	(H)	4. Def. of equil. $\triangle$
5. $\overline{CD} \cong \overline{CD}$	(L)	5. Reflexive property
6. $\triangle ADC \cong \triangle BDC$		6. HL
7. $\overline{AD} \cong \overline{BD}$		7. CPCTC
8. $\therefore \overline{CD}$ is the $\perp$ bisector of $\overline{AB}$.		8. Post. 12; Def. of bisector
9. $\angle ACD \cong \angle BCD$		9. CPCTC
10. $\therefore \overline{CD}$ bisects $\angle C$.		10. Def. of angle bisector

Written Exercises

1. $\angle XZW \cong \angle YZW$ 2. $\overline{XW} \cong \overline{YW}$ 4. $\angle ACE \cong \angle ABE$, $\angle DCE \cong \angle DBE$, $\overline{CE} \cong \overline{EB}$

Based on the given information, which segments or angles are congruent?

Ⓐ **1.** Given: $\overline{ZW}$ bisects $\angle XZY$.

2. Given: $\overline{ZW}$ bisects $\overline{XY}$.

3. Given: $\overline{XZ} \cong \overline{YZ}$, and $\overline{XW} \cong \overline{YW}$ 3. all corr parts of $\triangle XZW$ and $\triangle YZW$

4. Given: $\overline{AC} \cong \overline{AB}$ and $\overline{DC} \cong \overline{DB}$

5. Given: $\overline{DE}$ bisects $\angle BDC$. 5. $\angle CDE \cong \angle BDE$

6. Given: $\overleftrightarrow{AE}$ is $\perp$ bisector of $\overline{BC}$. 6. $\overline{AC} \cong \overline{AB}$, $\overline{DC} \cong \overline{DB}$, $\overline{EC} \cong \overline{EB}$, $\angle AEB \cong \angle AEC$, $\angle ACE \cong \angle ABE$

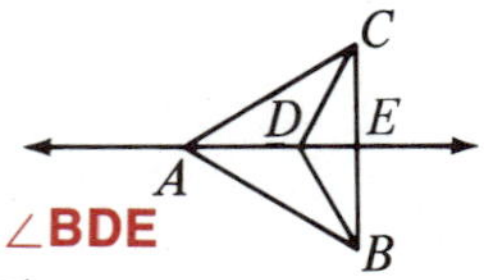

Prove each of the following.

7. The bisector of the vertex angle of an isosceles triangle bisects the base of the triangle.

8. The bisector of the base of an isosceles triangle bisects the vertex angle of the triangle.

Ⓑ **9.** The bisector of the vertex angle of an isosceles triangle is the perpendicular bisector of the base.

10. If the perpendicular from the vertex angle of a triangle to the base bisects the vertex angle, then the triangle is isosceles.

11. All points in the bisector of an angle are equidistant from the sides of the angle. (Note: Each distance is along the perpendicular from the point to the line.)

12. Perpendicular segments from the midpoints of the legs of an isosceles triangle to the base are congruent.

13. Complete the proof of Theorem 5.8.

Ⓒ **14.** Segments of the bisectors of base angles of isosceles triangles cut off by the legs of the triangle are congruent.

15. Given: $\overline{BD}$ is the $\perp$ bisector of $\overline{AC}$. Prove: $\angle BAD \cong \angle BCD$

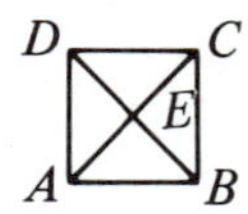

16. Given: $\angle 2 \cong \angle 4$, and $\angle FIH \cong \angle FGH$ Prove: $\overline{FH}$ is a $\perp$ bisector of $\overline{GI}$.

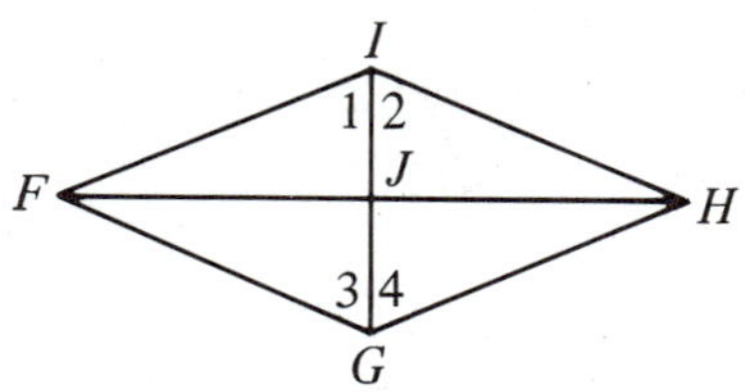

17. Given: $\angle 1 \cong \angle 3$, and $\angle FIH \cong \angle FGH$ Prove: $\overline{FH}$ is the bisector of $\angle IHG$.

18. Given: $\overline{ML} \perp$ plane p, and $\overline{MO}$ is a $\perp$ bisector of $\overline{NK}$. Prove: $\overline{LO}$ is a $\perp$ bisector of $\overline{NK}$.

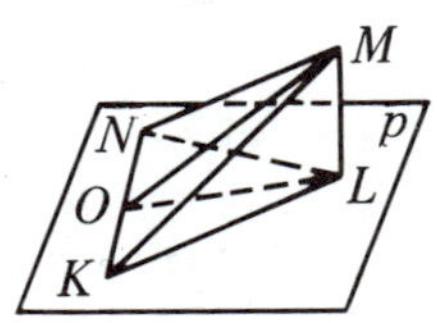

19. Given: $\overline{ML} \perp$ plane p, $\overline{NM} \cong \overline{KM}$, and $\overline{MO}$ bisects $\angle NMK$. Prove: $\overline{LO}$ bisects $\angle NLK$.

CUMULATIVE REVIEW

1. List four congruence patterns for triangles. ASA, SSS, AAS, SAS
2. List four congruence patterns for right triangles. HL, HA, LL, LA
3. Draw an angle. Construct the bisector of the angle. See constr., p. 15.

ALTITUDES AND MEDIANS OF TRIANGLES 5.7

Objectives

To identify altitudes and medians of triangles
To prove and apply theorems about altitudes and medians of triangles

There are two special types of segments related to triangles: altitudes and medians.

Definition: Altitude of a triangle

An *altitude of a triangle* is a segment from a vertex of the triangle, perpendicular to the line containing the opposite side of the triangle.

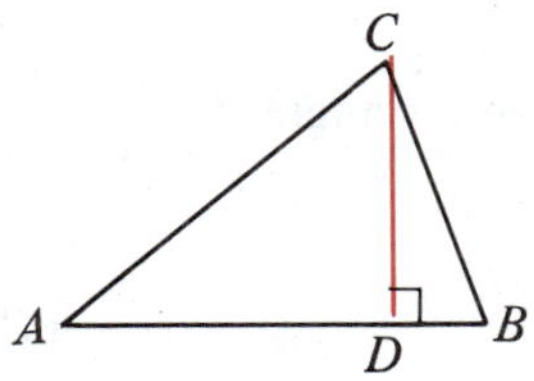

$\overline{CD}$ is an altitude of $\triangle ABC$.

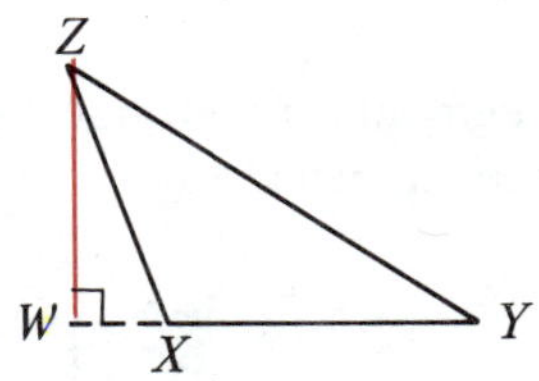

$\overline{ZW}$ is an altitude of $\triangle XYZ$.

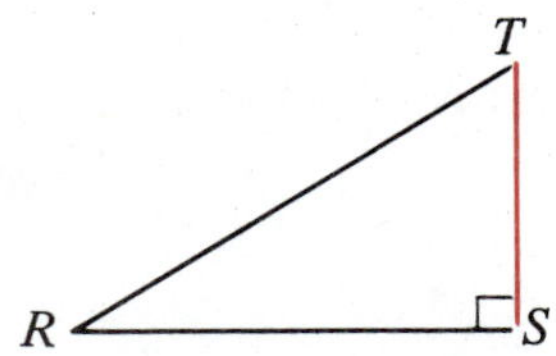

$\overline{TS}$ is an altitude of $\triangle RST$.

As you can see from the figures above, altitudes (except for their endpoints) can be in the interior or exterior of the triangle or on the triangle itself. Although the endpoints of altitude $\overline{CD}$ above are not in the interior of $\triangle ABC$, all other points of the altitude are. For convenience, the statement "$\overline{CD}$ is in the interior of $\triangle ABC$" will be used. You will use a similar agreement about other segments related to triangles.

Every triangle has three altitudes, one from each vertex to the line containing the opposite side. In $\triangle ABC$ below, $\overline{CX}$, $\overline{AY}$, and $\overline{BZ}$ are all altitudes. In this case, all three altitudes are in the interior of the triangle.

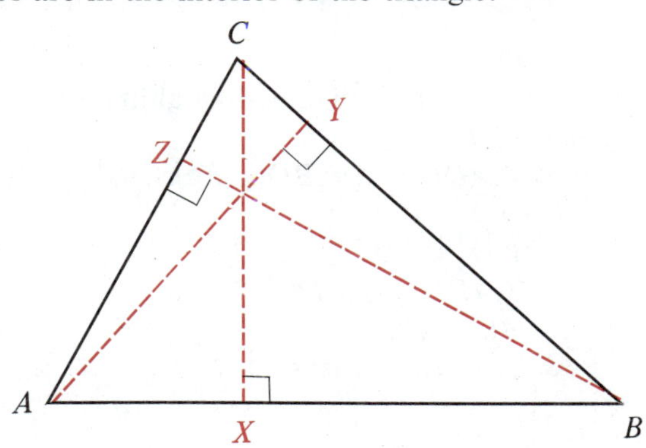

Altitudes $\overline{CX}$, $\overline{AY}$, and $\overline{BZ}$ of $\triangle ABC$

The other special segment related to a triangle is a median.

Definition: Median of a triangle

A *median of a triangle* is a segment from a vertex of the triangle to the midpoint of the opposite side of the triangle.

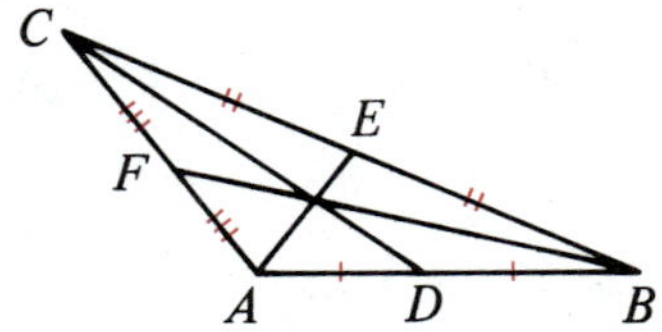

Medians $\overline{AE}$, $\overline{BF}$, and $\overline{CD}$ of $\triangle ABC$

Every triangle has three medians, one from each vertex. As shown in the diagram above, each median bisects the opposite side.

Example 1

Which segments are altitudes of the given triangles?
Which are medians?

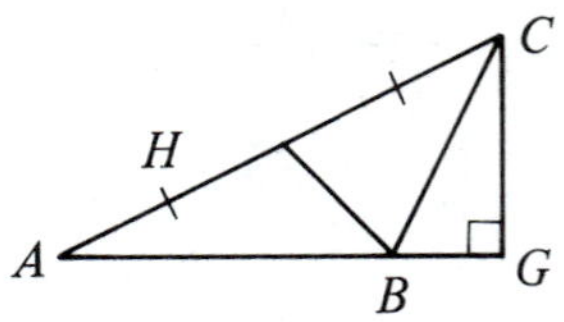

$\overline{AG}$ and $\overline{CG}$ are altitudes.
$\overline{BH}$ is a median.

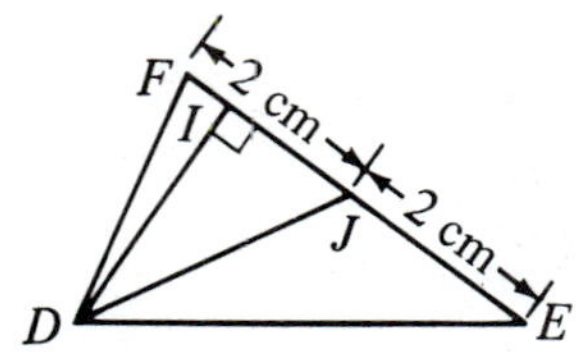

$\overline{DI}$ is an altitude.
$\overline{DJ}$ is a median.

In certain cases, a median is also an altitude.

Example 2

Given: $\overline{CD}$ is a median and an altitude.
Prove: $\triangle ABC$ is isosceles.

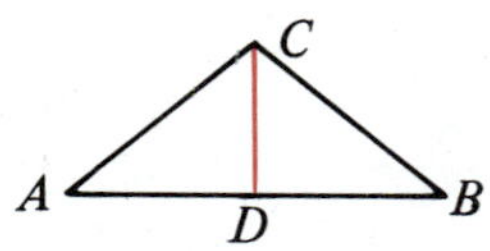

Proof

Statements		Reasons
1. $\overline{CD}$ is a median and an altitude.		1. Given
2. $\overline{CD} \perp \overline{AB}$		2. Def. of altitude
3. $\triangle ADC$ and $\triangle BDC$ are rt. $\triangle$'s.		3. $\perp$ forms rt. $\angle$'s; Def. of rt. $\triangle$'s
4. $\overline{AD} \cong \overline{BD}$	(L)	4. Def. of median
5. $\overline{CD} \cong \overline{CD}$	(L)	5. Reflexive property
6. $\triangle ADC \cong \triangle BDC$		6. LL
7. $\overline{AC} \cong \overline{BC}$		7. CPCTC
8. $\therefore \triangle ABC$ is isosceles.		8. Def. of isosceles $\triangle$

Corresponding medians of congruent triangles are medians that are drawn to corresponding congruent sides of the triangles. You can prove that these corresponding medians are congruent.

Theorem 5.9 Corresponding medians of congruent triangles are congruent. (See Exercise 13.)

Given: $\triangle ABC \cong \triangle A'B'C'$, $\overline{CD}$ is a median of $\triangle ABC$, and $\overline{C'D'}$ is a median of $\triangle A'B'C'$.
Prove: $\overline{CD} \cong \overline{C'D'}$

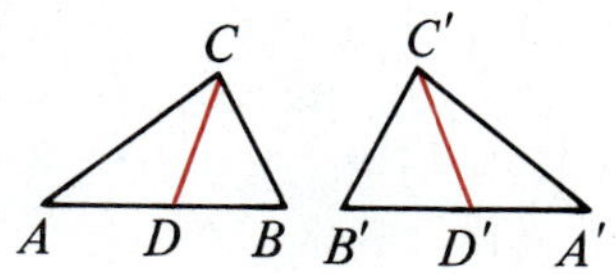

Outline of Proof

1. $\triangle ABC \cong \triangle A'B'C'$
2. $\overline{AB} \cong \overline{A'B'}$
3. $AB = A'B'$
4. $\overline{CD}$ and $\overline{C'D'}$ are medians.
5. $AD = \frac{1}{2}AB$, $A'D' = \frac{1}{2}A'B'$
6. $AD = A'D'$
7. $\overline{AD} \cong \overline{A'D'}$ (S)
8. $\angle A \cong \angle A'$ (A)
9. $\overline{CA} \cong \overline{C'A'}$ (S)
10. $\triangle CAD \cong \triangle C'A'D'$
11. $\therefore \overline{CD} \cong \overline{C'D'}$

Theorem 5.10 Corresponding altitudes of congruent triangles are congruent. (See Exercise 14.)

You can use Theorems 5.9 and 5.10 to find measures of corresponding altitudes or medians.

Example 3 Given: $\triangle EGH \cong \triangle KIJ$, $\overline{HF}$ and $\overline{JL}$ are corresponding altitudes, $HF = 12x - 4$, and $JL = 10x + 10$.
Find HF.

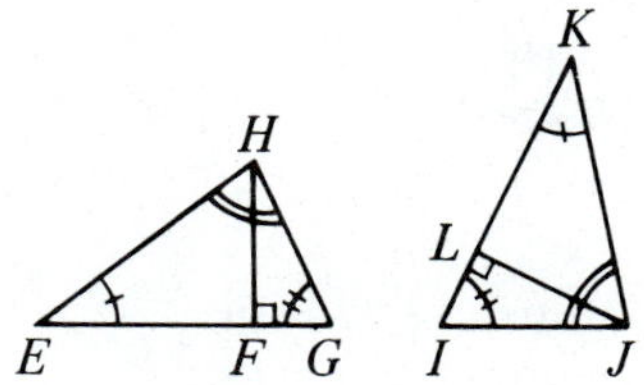

$HF = JL$ ◄ *Use Th. 5.10.*
$12x - 4 = 10x + 10$
$2x = 14$
$x = 7$

Thus, $HF = 12(7) - 4$ or 80.

Oral Exercises

Which segments are altitudes? Which are medians?

1.

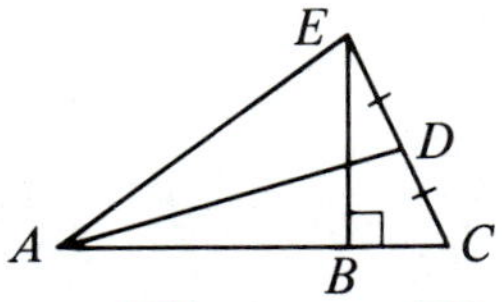

Alt: $\overline{EB}$; median: $\overline{AD}$

2.

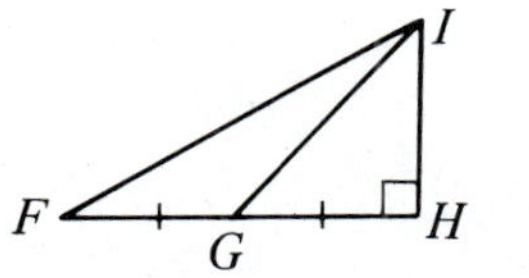

Alt: $\overline{FH}$, $\overline{IH}$; median: $\overline{IG}$

3.

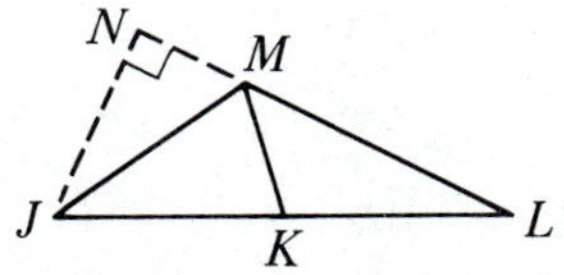

Alt: $\overline{JN}$; median: none

Written Exercises

Prove each of the following.

A **1.** Given: $\overline{CD}$ is an altitude and $\overline{AD} \cong \overline{BD}$.
Prove: $\triangle ADC \cong \triangle BDC$

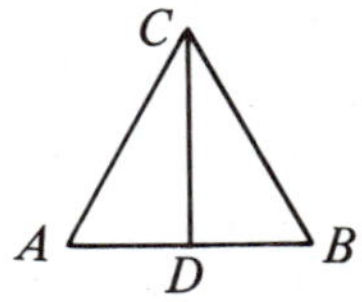

2. Given: $\overline{CD}$ is a median and $\overline{AC} \cong \overline{BC}$.
Prove: $\triangle ADC \cong \triangle BDC$

Find each indicated length. (Ex. 3–5)

3. Given: $\triangle ABC \cong \triangle EFG$, $\overline{CD}$ and $\overline{GH}$ are corresponding medians, $CD = 6s + 2$, and $GH = 8s - 14$. Find CD. **50**

4. Given: $\triangle IJK \cong \triangle NMO$, $\overline{IL}$ and $\overline{NP}$ are corresponding altitudes, $IL = 3(t + 3)$, and $NP = 4(2t + 1)$. Find IL. **12**

5. Given: $\triangle QRS \cong \triangle WVU$, $\overline{ST}$ and $\overline{UX}$ are corresponding medians, $ST = 9a - 2$, and $UX = 4(a + 2)$. Find ST. **16**

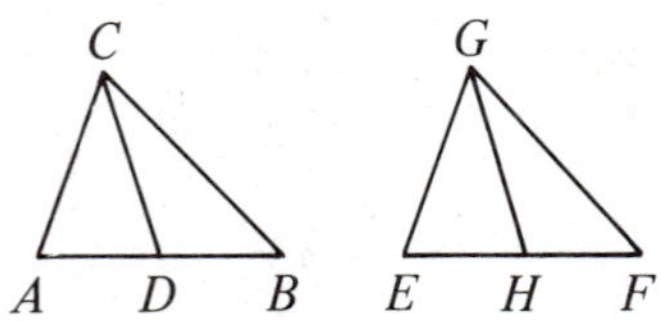

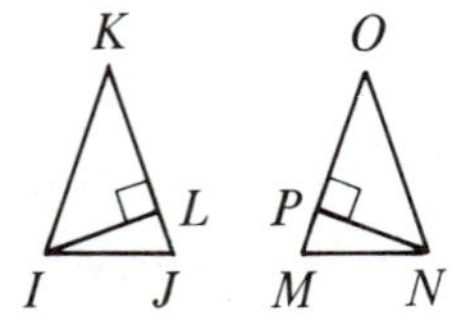

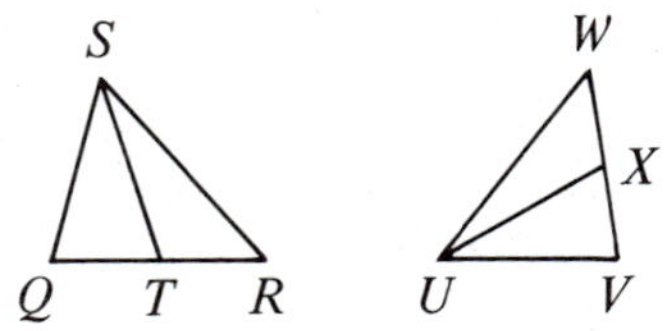

6. Given: $\overline{ZW}$ is an altitude of $\triangle XYZ$ and a bisector of $\overline{XY}$.
Prove: $\triangle XYZ$ is isosceles.

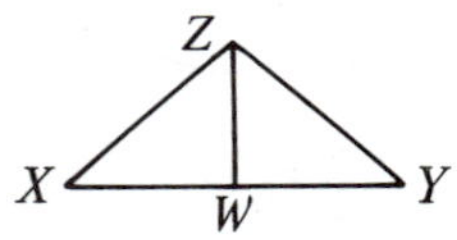

7. Given: $\overline{ZW}$ is an altitude of $\triangle XYZ$ and bisects $\angle XZY$.
Prove: $\triangle XYZ$ is isosceles.

8. Given: $\overline{CD}$ is an altitude of $\triangle ABC$.
Prove: $m\angle 1 + m\angle 2 = m\angle 3 + m\angle 4$

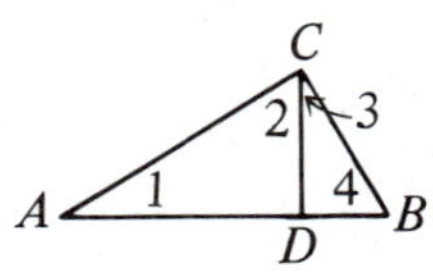

9. Given: $\overline{AC} \perp \overline{CB}$ and $\angle 1 \cong \angle 3$
Prove: $\overline{CD}$ is an altitude of $\triangle ABC$.

B **10.** Given: $\triangle EGI \cong \triangle JLN$, $\overleftrightarrow{FH}$ is the $\perp$ bisector of $\overline{GI}$, and $\overleftrightarrow{KM}$ is the $\perp$ bisector of $\overline{LN}$. Prove: $\overline{FH} \cong \overline{KM}$

11. Given: $\triangle OPQ \cong \triangle TSV$, $\overrightarrow{PR}$ bisects $\angle OPQ$, and $\overrightarrow{SU}$ bisects $\angle VST$.
Prove: $\overline{PR} \cong \overline{SU}$

12. Given: $\overline{BZ} \cong \overline{AC}$, $\overline{BZ}$ is the $\perp$ bisector of $\overline{YW}$, and $\overline{AC}$ is the $\perp$ bisector of $\overline{YX}$.
Prove: $\overline{YW} \cong \overline{YX}$

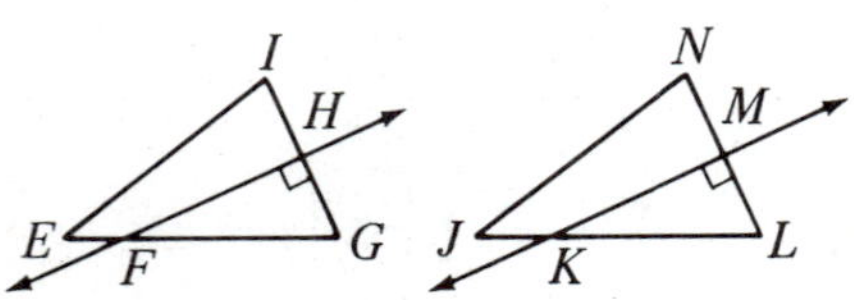

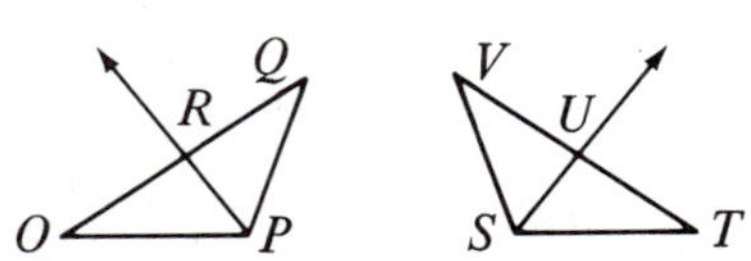

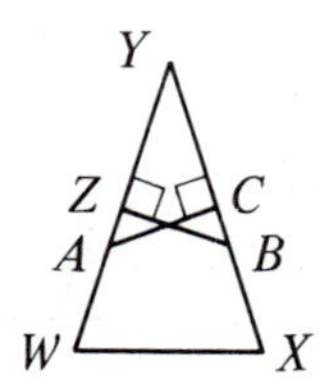

13. Complete the proof of Theorem 5.9.

14. Prove Theorem 5.10.

For each statement, draw a diagram and label it. State what is given and what is to be proved in terms of the diagram. Prove the statement.

15. Any two altitudes of an equilateral triangle are congruent.

16. Any two medians of an equilateral triangle are congruent.

17. The altitude to the base of an isosceles triangle is also a median.

18. If two altitudes of a triangle are congruent, then the triangle is isosceles.

19. The median to the base of an isosceles triangle is the bisector of the vertex angle.

20. The altitude to the base of an isosceles triangle is the bisector of the vertex angle.

Determine if each of the following is a true statement. If so, prove it. If not, show a counterexample. (Ex. 21–24)

21. If a median of a triangle is perpendicular to the side to which it is drawn, then the triangle is isosceles. T

22. If an altitude of a triangle also bisects an angle of the triangle, then the triangle is isosceles. T

23. If a median of a triangle bisects the vertex angle, then the triangle is isosceles. T

24. If two medians of a triangle are also angle bisectors, then the triangle is equilateral. T

25. Given: $\overline{DB} \perp$ plane p, $\overline{DE}$ bisects $\angle CDA$, and $\overline{CD} \cong \overline{AD}$.
Prove: $\overline{BE}$ bisects $\angle CBA$.

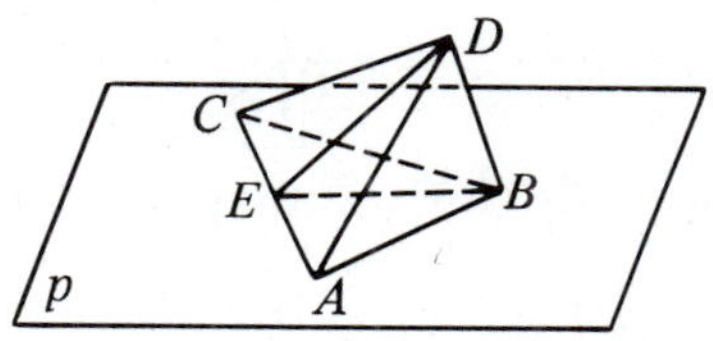

26. Given: $\overline{DB} \perp$ plane p, and $\overline{DE}$ is a median and an altitude of $\triangle CDA$.
Prove: $\overline{BE}$ is a median and an altitude of $\triangle CBA$.

27. Given: $\triangle FJG \cong \triangle FHG$, and $\overline{JI}$ is an altitude of $\triangle FJG$.
Prove: $\overline{HI}$ is an altitude of $\triangle FHG$.

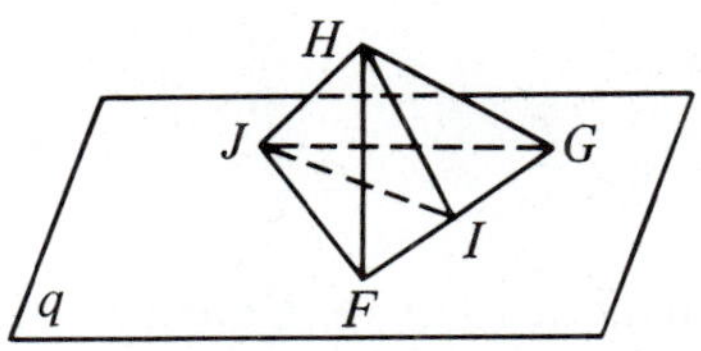

28. Given: $\triangle FJG \cong \triangle FHG$, and $\overline{HI}$ bisects $\angle FHG$.
Prove: $\overline{JI}$ bisects $\angle FJG$.

Given: $\overline{KM}$ is $\perp$ bisector of $\overline{NO}$, and $\overline{LM}$ is $\perp$ bisector of $\overline{NO}$.

29. Prove: $\triangle OLM \cong \triangle NLM$

30. Prove: $\triangle KLN \cong \triangle KLO$

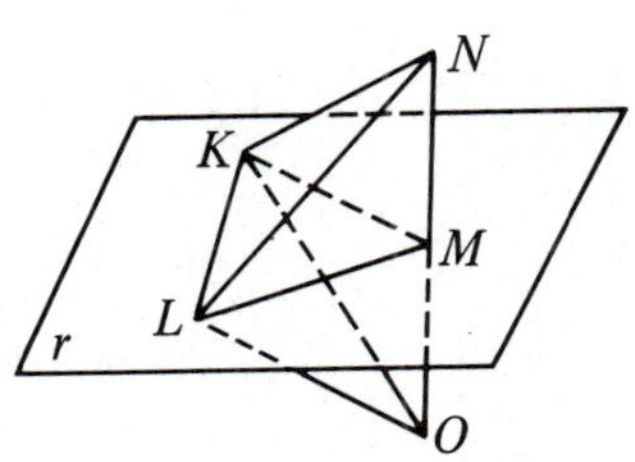

Given: $\triangle KLM$ in plane r, $\overline{NO} \perp r$ at M, $\overline{NM} \cong \overline{OM}$, and $\overline{LN} \cong \overline{KN}$.

31. Prove: $\overline{LM} \cong \overline{KM}$

32. Prove: $\overline{LO} \cong \overline{KO}$

CONSTRUCTIONS OF ALTITUDES AND BISECTORS 5.8

Objectives **To construct altitudes and medians of triangles**
To use constructions to determine if a statement is true or false

You have already learned several basic geometric constructions. These constructions can now be used as parts of other constructions. In Example 1, the midpoint of a segment is constructed to get a median of the triangle. In Example 2, a perpendicular is constructed to obtain an altitude of the triangle, and in Example 3, the procedure for bisecting an angle is applied three times to construct the bisector of each angle of the triangle. You may want to review these three basic constructions before studying the examples.

Example 1 Draw an obtuse $\triangle XYZ$, with obtuse $\angle X$. Construct the median to side $\overline{YZ}$.

Construct the midpoint M of $\overline{YZ}$. Draw $\overline{XM}$. ►

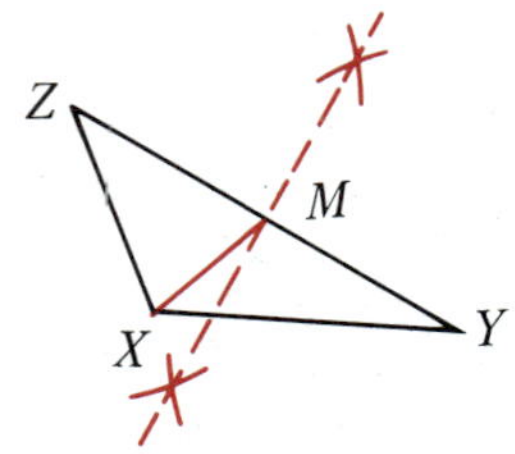

Conclusion: $\overline{XM}$ is the required median.

Example 2 Draw an acute $\triangle ABC$. Construct the altitude to side $\overline{AB}$.

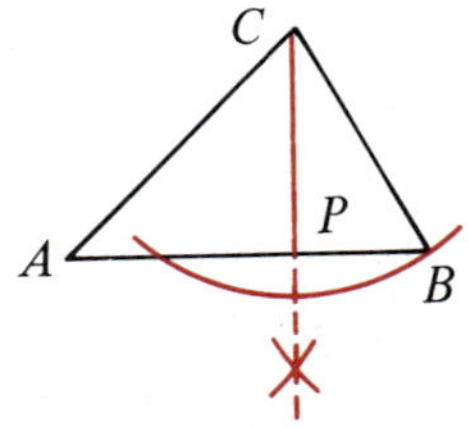

◄ *Construct a $\perp$ to $\overline{AB}$ from point C.*

Conclusion: $\overline{CP}$ is the required altitude.

Example 3 Draw an equilateral triangle. Construct the bisectors of all three angles. Where do they meet?

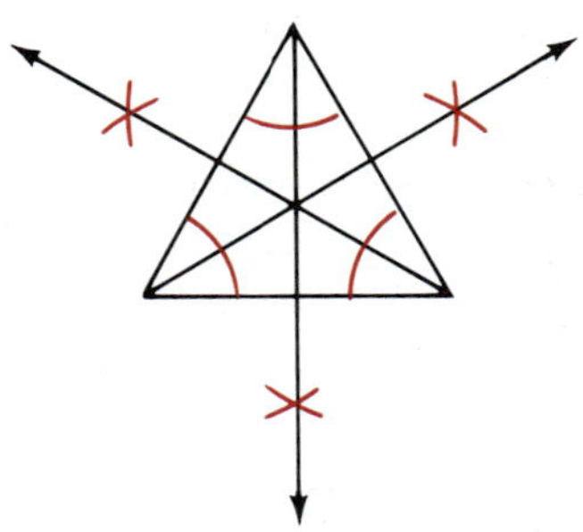

Conclusion: All three meet in a single point in the interior of the triangle.

A construction can sometimes be used to determine if a given statement is true or false. The construction will usually show if a statement is false, but it cannot prove conclusively that it is true.

Example 4 True or false? "The bisector of the vertex angle of an isosceles triangle bisects the base." Check by making the necessary construction.

Construct $\overrightarrow{CD}$ bisecting $\angle C$. ►

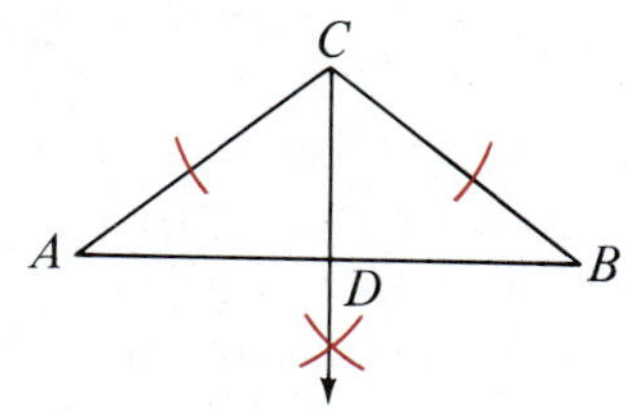

$\overline{AD}$ and $\overline{BD}$ appear to be congruent. **Thus,** the statement seems to be true.

Written Exercises

Copy the given triangle. Construct each of the following.

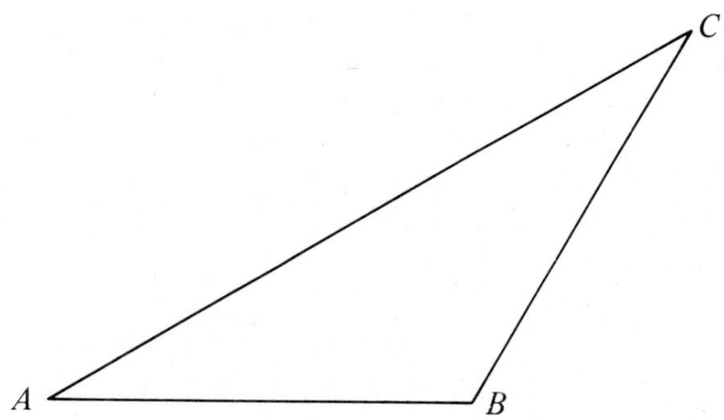

A
1. The bisector of $\angle ABC$.
   Follow constr. in Ex. 3, p. 174.
2. The median to $\overline{AB}$.
   Follow constr. in Ex. 1, p. 174.
3. The perpendicular bisector of $\overline{AB}$.
   Follow constr. in Ex. 5, p. 51.
4. The altitude to $\overline{AC}$.
   Follow constr. in Ex. 2, p. 174.
5. The altitude to $\overline{AB}$.
   Follow constr. in Ex. 2, p. 174.
6. The median to $\overline{AC}$.
   Follow constr. in Ex. 1, p. 174.

True or false? Check by making the necessary constructions. (Ex. 7–14)

B
7. A median of a scalene triangle can never be an altitude. **T**
8. A median of a scalene triangle can never be an angle bisector. **T**
9. A median of a scalene triangle can never be a perpendicular bisector of a side of the triangle. **T**
10. An altitude of a scalene triangle can never be a perpendicular bisector of a side of the triangle. **T**
11. A median of an isosceles triangle can never be an altitude. **F**
12. A median of an isosceles triangle can never be an angle bisector. **F**
13. An altitude of an isosceles triangle can never be a perpendicular bisector of a side of the triangle. **F**
14. An angle bisector of an isosceles triangle can never be a perpendicular bisector of a side of the triangle. **F**

15. Draw three triangles—one acute, one obtuse, and one right. Construct the three altitudes of each triangle. What conclusions can you draw about the intersections of these altitudes?

16. Repeat Exercise 15, constructing the three medians.

17. Repeat Exercise 15, constructing the three angle bisectors.

15. Follow constr in Ex 2, p. 174.
16. Follow constr in Ex 1, p. 174.
17. Follow constr in Ex 3, p. 174

CUMULATIVE REVIEW

1. Can the altitude of a triangle be located in the exterior of the triangle? If so, for what kind of triangle? **Yes, obtuse triangle**
2. Can the median of a triangle be located in the exterior of the triangle? If so, for what kind of triangle? **No**
3. Draw two parallel lines. Draw a third line intersecting one of these lines. Will this third line ever intersect the other line? Explain. **yes; The third line could be a transversal.**

ALGEBRA REVIEW

Objective

To multiply and divide fractions

To multiply fractions:
1. Use $\frac{a}{b} \cdot \frac{c}{d} = \frac{a \cdot c}{b \cdot d}$.
2. Factor both numerator and denominator.
3. Divide out common factors.

To divide fractions:
1. Use the reciprocal of the divisor.
2. Multiply.

Example

Divide $\frac{x^2-4}{x+4} \div \frac{x-2}{x^2+x-12}$.

Use the reciprocal of $\frac{x-2}{x^2+x-12}$ and multiply. $\frac{x^2-4}{x+4} \cdot \frac{x^2+x-12}{x-2}$

Use $\frac{a}{b} \cdot \frac{c}{d} = \frac{a \cdot c}{b \cdot d}$. $\frac{(x^2-4)(x^2+x-12)}{(x+4)(x-2)}$

Factor and divide out common factors. $\frac{\overset{1}{\cancel{(x-2)}}(x+2)\overset{1}{\cancel{(x+4)}}(x-3)}{\underset{1}{\cancel{(x+4)}}\underset{1}{\cancel{(x-2)}}}$

$\frac{(x+2)(x-3)}{1}$, or x^2-x-6

Multiply or divide as indicated.

1. $\frac{x^2-4}{4x+12} \cdot \frac{2x+6}{4x+8}$ **$\frac{x-2}{8}$**
2. $\frac{9x^2-25}{x^2-16} \div \frac{3x-5}{2x-8}$ **$\frac{6x+10}{x+4}$**
3. $\frac{y^2+3y+2}{y^2-3y-10} \div \frac{y^2+8y+7}{y^2-6y+5}$ **$\frac{y-1}{y+7}$**
4. $\frac{9-y^2}{4+2y} \cdot \frac{8+4y}{9-3y}$ **$\frac{6+2y}{3}$**
5. $\frac{z^2-4z+3}{z^2+7z+6} \cdot \frac{2z^2+6z+4}{z^2-z-6}$ **$\frac{2z-2}{z+6}$**
6. $\frac{6z^2+5z-6}{3z^2+13z-10} \div \frac{4z^2-9}{2z^2+7z-15}$ **1**

APPLICATIONS

Read → Plan → Solve → Interpret

1. Electric wires are to be strung between two poles across a gulch. The power company needs to know how much wire is needed, but it is impossible to measure directly across the gulch. How could this distance be found using the given diagram? Why does this work?

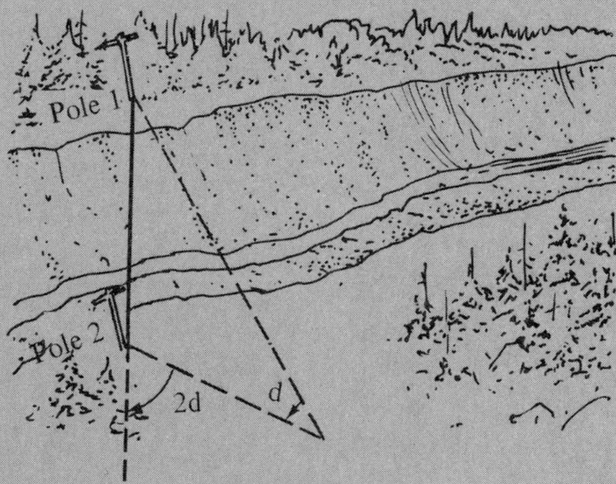

2. A homemade carpenter's level has an isosceles triangular frame, with a plumb bob hanging from the vertex. The plumb bob should always hang vertically. How can this be used to show whether or not a surface is horizontal? Why does this work?

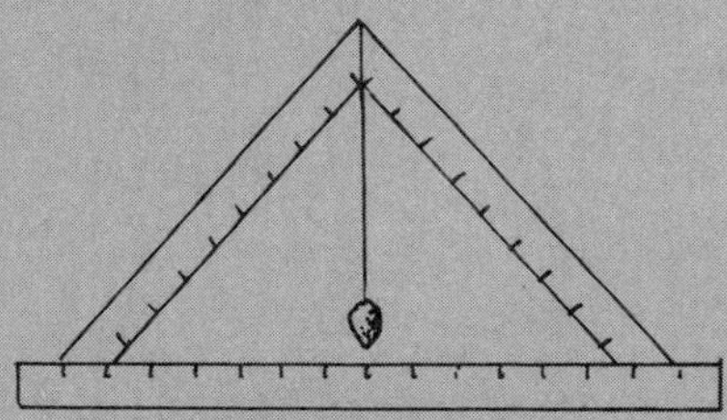

CHAPTER FIVE REVIEW

A exercises are 1–11, 13, 15, 16. B exercise is 12. C exercises are starred.

Vocabulary

altitude [5.7], base angles [5.2], corresponding parts [5.1], hypotenuse [5.4], leg [5.2], median [5.7], perimeter [5.2], perpendicular bisector [5.6], vertex angle [5.2]

If $\triangle ABC \cong \triangle EDF$, name the following. [5.1]

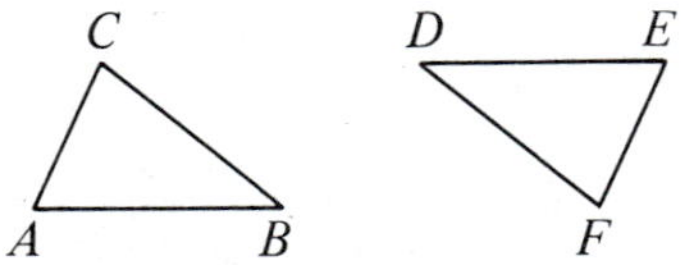

$\overline{AB} \cong \overline{ED}$; $\overline{BC} \cong \overline{DF}$; $\overline{CA} \cong \overline{FE}$

1. The pairs of congruent sides
2. The pairs of congruent angles
   $\angle A \cong \angle E$; $\angle B \cong \angle D$, $\angle C \cong \angle F$

Identify the following. [5.4]

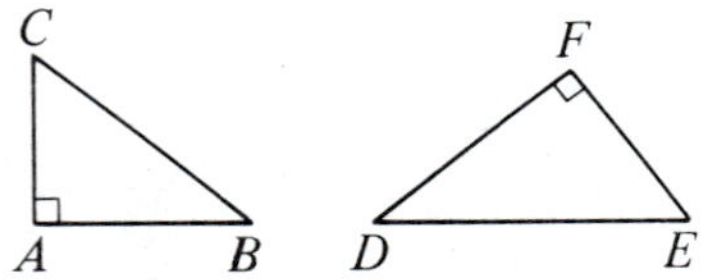

3. The hypotenuse of $\triangle DEF$
   $\overline{DE}$
4. The legs of $\triangle ABC$
   $\overline{AC}$ $\overline{AB}$

Which pairs of right triangles are congruent? What is the reason? [5.4]

5. 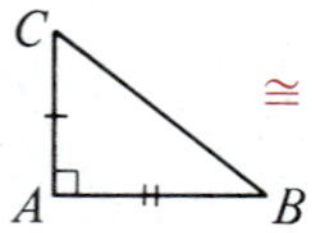 ≅ (LL) 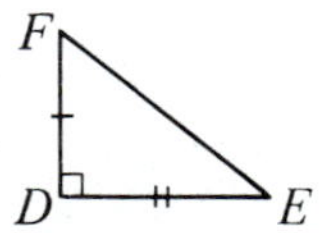

6. 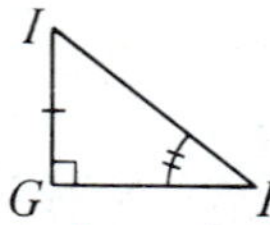 ≇ 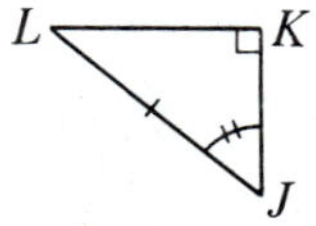

Sides do not correspond.

7. 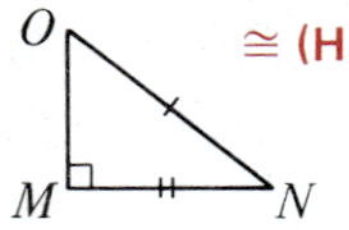 ≅ (HL) 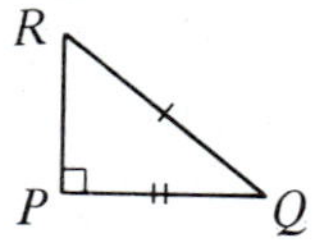

Tell whether $\overline{AX}$ is an altitude, a median, or neither. (Ex. 8–10) [5.7]

median altitude neither

8. 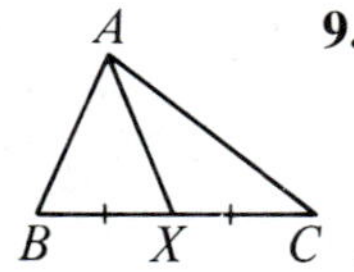

9. 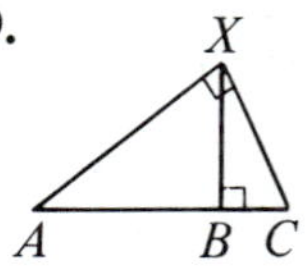

10. 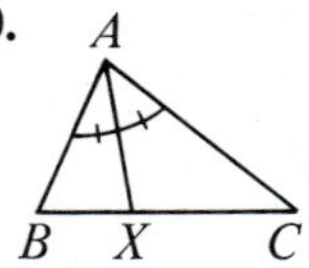

11. Given: $\overline{AD} \cong \overline{AB}$, and $\overline{AC}$ bisects $\angle DAB$
    Prove: $\overline{CD} \cong \overline{CB}$ [5.6]

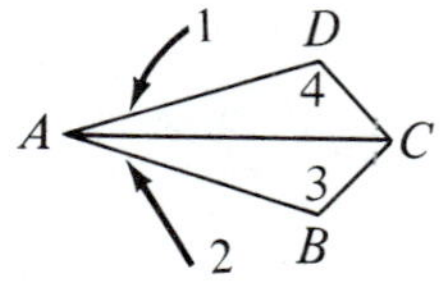

12. Given: $\overline{AC}$ bisects $\angle DCB$. [5.6]
    Prove: $m\angle 1 + m\angle 4 = m\angle 2 + m\angle 3$
13. Given: $\triangle GHF \cong \triangle GIE$. [5.2, 5.5]
    Prove: $\overline{HE} \cong \overline{IF}$

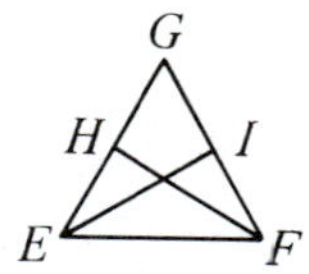

*14. Given: $\overline{FH}$ and $\overline{EI}$ are altitudes of $\triangle GEF$ and $\angle HFE \cong \angle IEF$.
    Prove: $\triangle GEF$ is isosceles. [5.7]

15. Given: Isosceles $\triangle LJK$, $\overline{LJ} \cong \overline{LK}$, $m\angle J = 5x$, and $m\angle K = 6x - 11$.
    Find $m\angle J$. **55** [5.2]

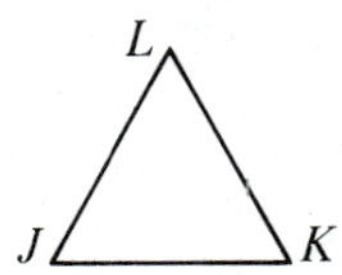

16. Given: $\angle J \cong \angle K$, $JL = 2x + 7$, and $KL = 4x + 1$
    Find KL. **13** [5.2]

*17. Draw an obtuse triangle. Construct the altitude from the vertex of one of the acute angles. **See Ex 2, p. 174.** [5.8]

CHAPTER FIVE TEST

A exercises are 1–11. B exercise is 14. C exercises are starred.

If $\triangle UVW \cong \triangle XYZ$, name the following.

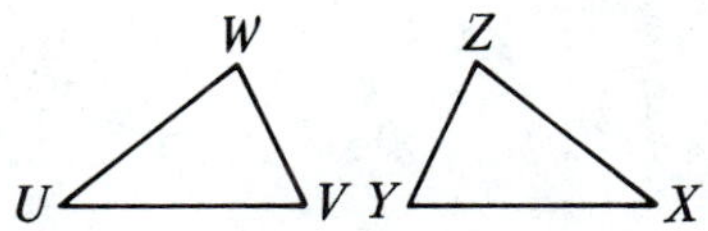

1. The pairs of congruent angles
$\angle U \cong \angle X$; $\angle V \cong \angle Y$; $\angle W \cong \angle Z$

2. The pairs of congruent sides
$\overline{UV} \cong \overline{XY}$; $\overline{VW} \cong \overline{YZ}$; $\overline{WU} \cong \overline{ZX}$

Identify the following.

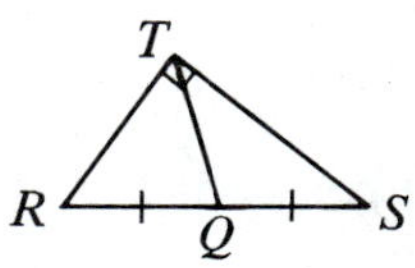

$\overline{RS}$; $\overline{RT}$, $\overline{TS}$

3. The hypotenuse and legs of $\triangle RST$

4. An altitude and a median of $\triangle RST$
$\overline{RT}$, $\overline{ST}$: altitudes; $\overline{TQ}$: median

Which pairs of right triangles are congruent? What is the reason?

5.

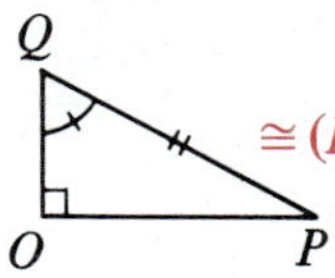

$\cong$ (*HA*)

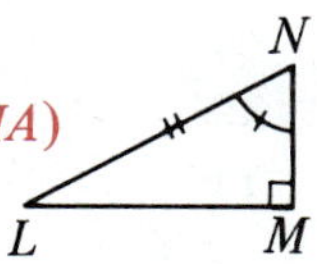

6.

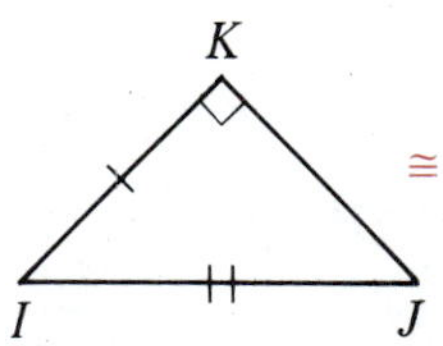

$\cong$ (*HL*)

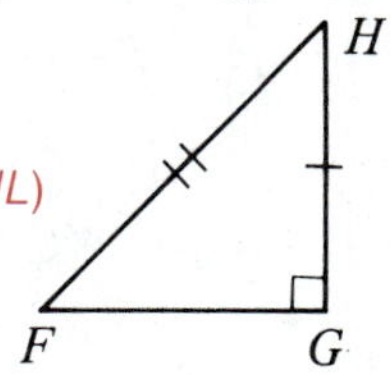

True or false? (Ex. 7–8)

7. Any two altitudes of an isosceles triangle are congruent. F

8. Two right triangles are congruent if a leg and hypotenuse of one are congruent to the corresponding leg and hypotenuse of the other. T

9. Given: Isosceles $\triangle CED$, $\overline{CE} \cong \overline{DE}$, $m\angle C = 8x - 16$, and $m\angle D = 6x + 2$.
Find $m\angle D$. $8x - 16 = 6x + 2$; 56

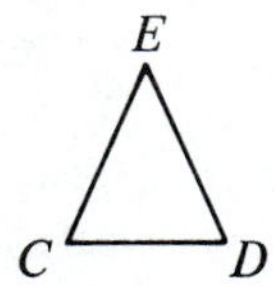

10. Given: $\angle C \cong \angle D$, $CE = 4y + 8$, and $DE = 6y - 2$ $4y + 8 = 6y - 2$; 28
Find DE.

11. Draw an obtuse triangle. Construct the median to the longest side. See Ex. 1, pg. 174

*__12.__ Given: $\overline{CD}$ bisects $\angle ACB$, and $\overline{CD} \perp \overline{AB}$.
Prove: $\triangle ABC$ is isosceles.

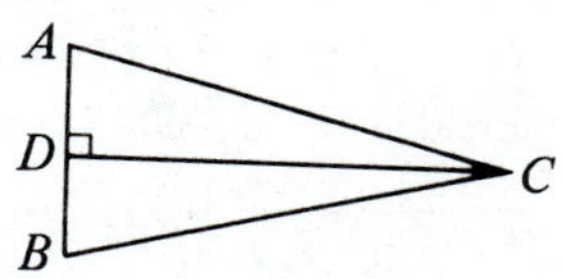

*__13.__ Given: $\triangle EFG \cong \triangle JIK$, and $\overline{GH}$ and $\overline{KL}$ are altitudes.
Prove: $\overline{HF} \cong \overline{LI}$

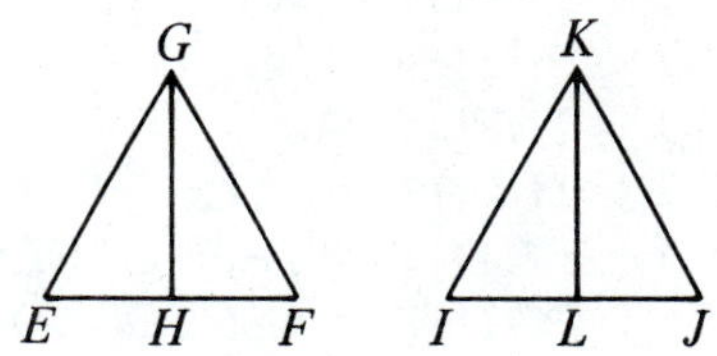

14. Given: $\overline{MP}$ is an altitude of $\triangle MNO$.
Prove: $m\angle 1 + m\angle 4 = m\angle 2 + m\angle 3$

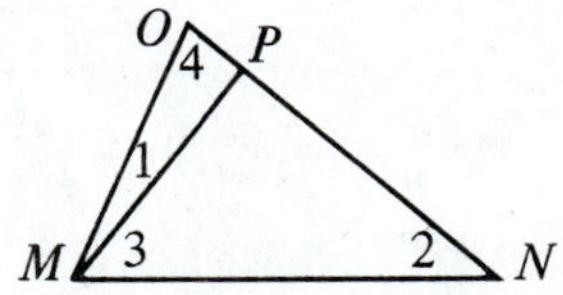

COMPUTER ACTIVITIES

Isosceles Triangles and the Computer

Objective: **To determine the measures of parts of isosceles triangles, using computer programs**

For an isosceles triangle, given the measure of one angle, and given its identity as a base or vertex angle, you can write a program that determines the measures of the other two angles. The following program performs this operation, using character strings (see page 570) in doing so. From your inputs of the measure of an angle and its identification, the computer determines the measures of the other two angles.

```
10  PRINT "ENTER AN ANGLE OF AN ISOSCELES TRIANGLE."
15  INPUT A
20  PRINT "IS THIS ANGLE BASE OR VERTEX?"
30  INPUT A$
40  IF A$ = "VERTEX" THEN 70
50  PRINT "THE THREE ANGLES ARE "A", "A", AND "180 -
     2 * A"."
60  PRINT : GOTO 10
70  PRINT "THE THREE ANGLES ARE "A", "(180 - A) / 2"
     , AND "(180 - A) / 2"."
80  PRINT : GOTO 10
```

From inputs of the length of one side, the perimeter, and the identity of the side as a leg or base, the following program determines the lengths of the remaining two sides of an isosceles triangle. Enter and RUN it.

```
10  INPUT "WHAT IS THE PERIMETER? ";P
20  INPUT "WHAT IS THE KNOWN SIDE? ";S
30  PRINT "IS THAT SIDE A LEG OR THE BASE?"
40  INPUT A$
50  IF A$ = "LEG" THEN 80
60  PRINT "THE OTHER 2 SIDES ARE EACH "(P - S) / 2"."
70  PRINT : GOTO 10
80  PRINT "THE OTHER SIDES ARE "S" AND "P - 2 * S"."
90  PRINT : GOTO 10
```

EXERCISES: See the Computer Section beginning on page 565 for more information.

1. Rewrite the first program above so that it will accept angle input and produce output in terms of degrees and minutes.

2. Rewrite the second program above so that it will accept measurement inputs and produce output in terms of feet and inches.

COLLEGE PREP TEST

For each item, choose the best answer. Indicate your answer by writing the letter of this choice.

1. In an isosceles $\triangle RST$, if m $\angle R = 38$, then which of the following cannot be true? E

(A) m $\angle S = 71$ **(B)** m $\angle S = 38$
(C) m $\angle T = 71$ **(D)** m $\angle T = 104$
(E) None of these

2. Given: $\triangle AFD \cong \triangle BFE$, $\overline{BD} \perp \overline{AC}$, m $\angle C = 50$. What is m $\angle AFB$? E

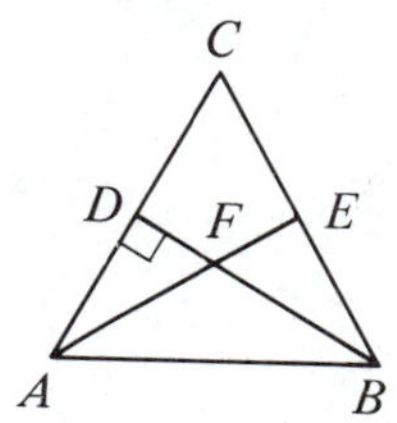

(A) 40 **(B)** 50 **(C)** 90 **(D)** 100 **(E)** 130

3.

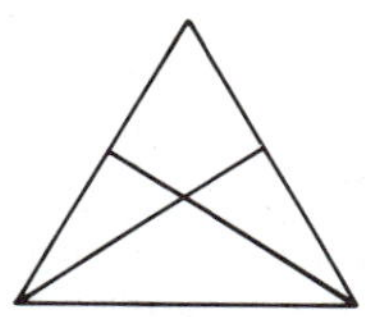

How many triangles are there in the figure above? E

(A) 1 **(B)** 3 **(C)** 4 **(D)** 6 **(E)** 8

4. Right triangles cannot be proved congruent by D

(A) HL **(B)** SSA **(C)** LL
(D) AA **(E)** LA

5. In $\triangle ABC$, $\overline{AB} \cong \overline{CB}$, m $\angle A = 2x - 10$, and m $\angle B = 3x + 25$. Which of the following is true? E

(A) m $\angle A = 25$
(B) m $\angle A = 100$
(C) m $\angle B = 40$
(D) m $\angle C = 25$
(E) m $\angle C = 40$

6.

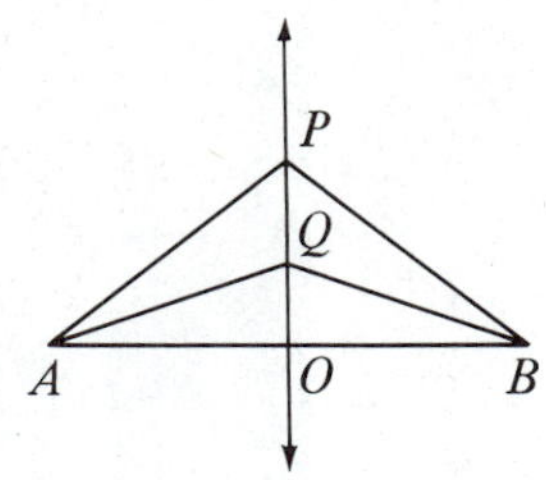

To prove that "two points each equidistant from the endpoints of a segment determine the perpendicular bisector of the segment," it is necessary to show that D

(A) $\triangle AOQ \cong \triangle BOQ$ by SSS
(B) $\triangle AOP \cong \triangle BOP$ by SAS
(C) $\triangle AOP \cong \triangle BOP$ by ASA
(D) $\triangle AQP \cong \triangle BQP$ by SSS
(E) $\triangle AQP \cong \triangle BQP$ by AAS

7. Which of the following cannot be true? D

(A) A median of a triangle is also an altitude.
(B) Corresponding medians of congruent triangles are congruent.
(C) A median of a triangle is also an angle bisector.
(D) A median is outside the triangle.
(E) An altitude is outside the triangle.

8. The construction of the median of a triangle involves the construction of B

(A) an angle bisector.
(B) the midpoint of a segment.
(C) a segment congruent to a segment.
(D) an angle congruent to an angle.
(E) an equilateral triangle.

6 EXPLORING POLYGONS

CAREERS

Surveyors

Mathematics helps surveyors make accurate measurements of the earth. This surveyor is is sighting through an instrument called a transit.

A surveyor drives a stake at A so that $\angle XYA$ is a right angle. The stake at C is halfway between Y and A. The stake at B is on the line of sight from C to X so that $\angle YAB$ is a right angle. How can the distance from Y to X be determined without crossing the river?

Analyze the information.

Decide on the property that can be used to solve. **CPCTC**

Write a mathematical sentence.
$\overline{XY} \cong \overline{AB}$

Solve the sentence by measuring.
Measure the distance from A to B.

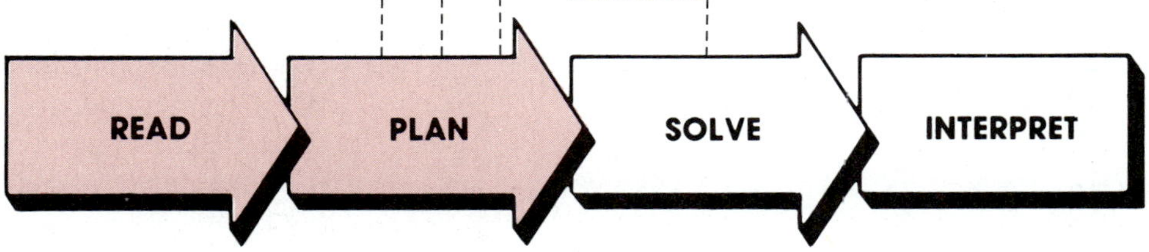

CONVEX AND CONCAVE POLYGONS 6.1

Objectives

To identify and name polygons and their parts
To identify convex and concave polygons

In this lesson you will work with a special type of geometric figure called a **polygon**. You have already studied one type of polygon – the triangle. All properties that are true for polygons in general apply to triangles.

Definition: Polygon

A *polygon* is the union of three or more coplanar segments which intersect only at each endpoint, with each endpoint shared by exactly two segments.

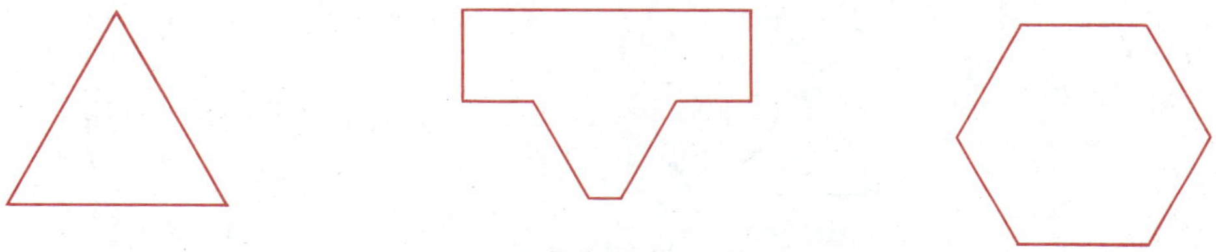

Polygons

Some coplanar figures which are the union of segments are not polygons; the segments may not intersect in the proper way.

Example 1

Which of the following figures are polygons? Indicate why the others are not.

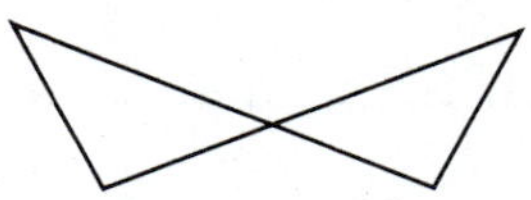

Not a polygon. Two of the segments intersect at a point that is not an endpoint.

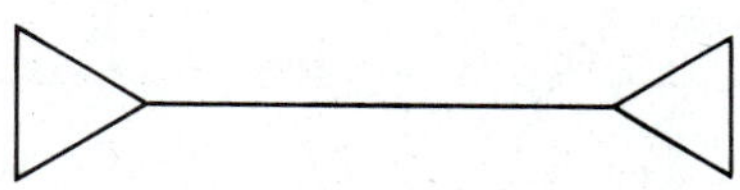

Not a polygon. The endpoints of some of the segments are shared by three segments.

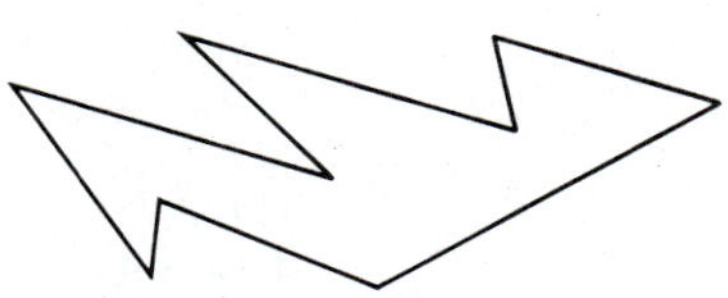

A polygon

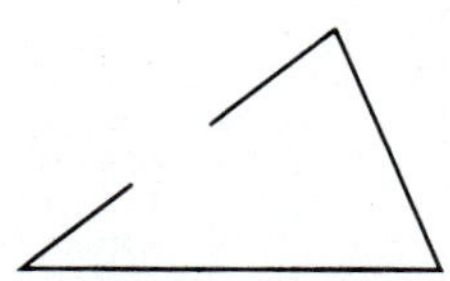

Not a polygon. Two endpoints are not shared by a second segment.

The segments that make up a polygon are called the *sides* of the polygon and the endpoints of the segments are called the *vertices* of the polygon. The number of sides is always equal to the number of vertices. Polygons are classified according to the number of sides they contain. A polygon with five sides is called a 5-gon; a polygon with 17 sides is called a 17-gon. However, certain polygons have special names.

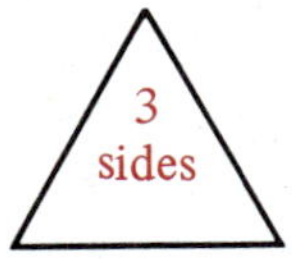

Triangle

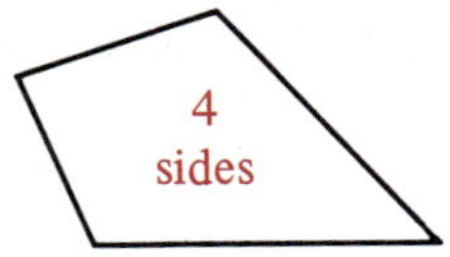

Quadrilateral

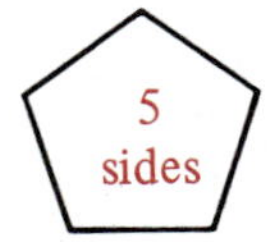

Pentagon

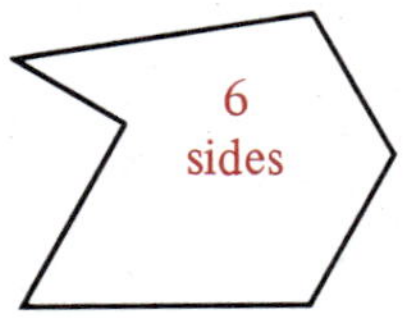

Hexagon

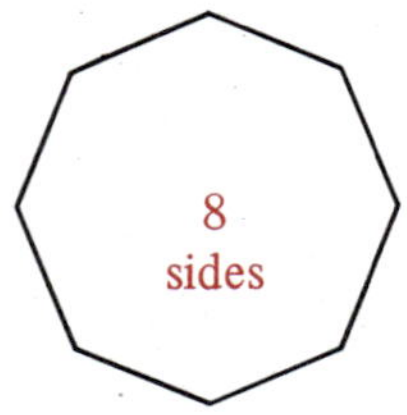

Octagon

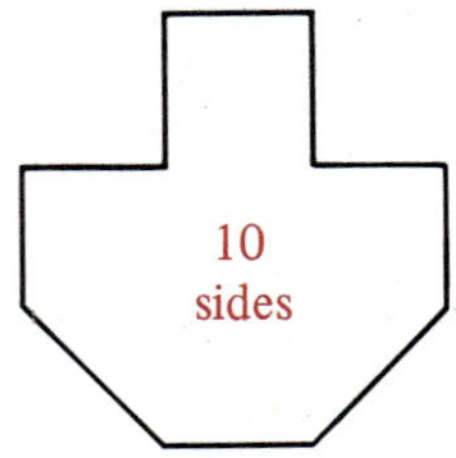

Decagon

A specific polygon is often named by listing its vertices in order. The polygon below is polygon ABCDEF. The segment, $\overline{EB}$, is a **diagonal** of polygon *ABCDEF*.

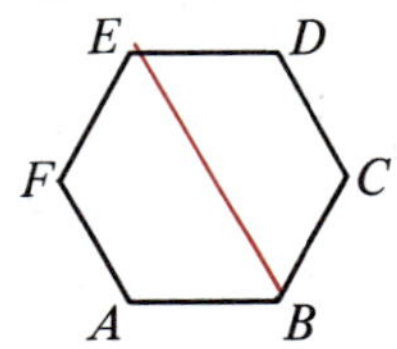

Definition: Diagonal of a polygon

A *diagonal* of a polygon is a segment which connects any two nonconsecutive vertices.

Example 2

For each polygon, list the sides and vertices and classify the polygon. How many diagonals can be drawn for each polygon?

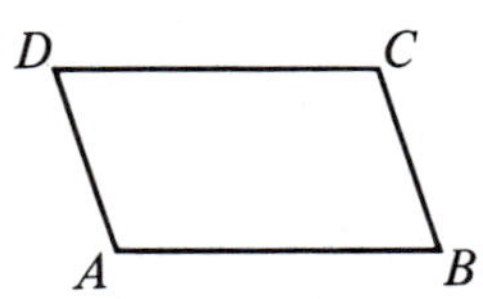

Sides: $\overline{AB}, \overline{BC}, \overline{CD}, \overline{DA}$
Vertices: A, B, C, D
Type: quadrilateral
Diagonals: 2

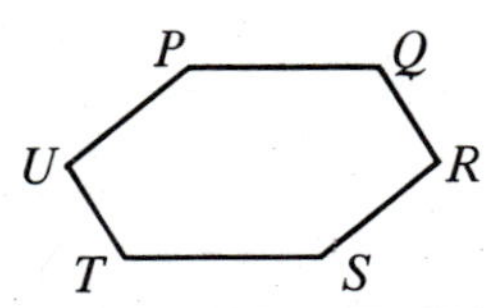

Sides: $\overline{PQ}, \overline{QR}, \overline{RS}, \overline{ST}, \overline{TU}, \overline{UP}$
Vertices: P, Q, R, S, T, U
Type: hexagon
Diagonals: 9

In Chapter 1 you learned that an angle separates the plane that contains it into three sets of points: the angle itself, the interior of the angle, and the exterior of the angle. A polygon separates the plane that contains it in a similar manner. In the diagram below, point P is in the interior of the polygon and point Q is in the exterior. The exterior of a polygon can contain lines but the interior cannot.

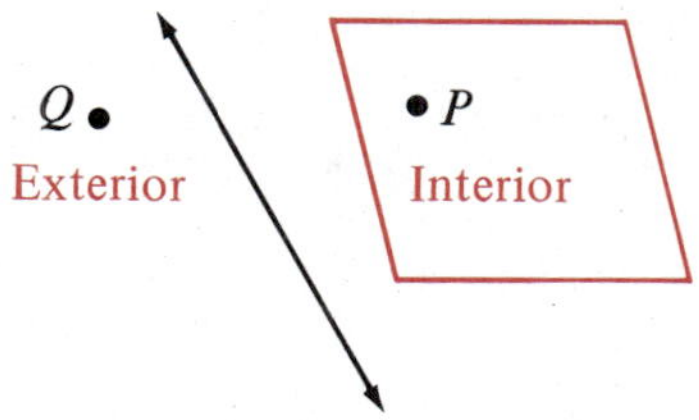

The union of a polygon and its interior is called a polygonal region. Some polygonal regions "bend inward" while others do not. Those that bend inward are called **concave** and those that do not are called **convex**. A more formal definition of convex and concave regions is presented below.

Definition: Convex and concave regions

A polygonal region is *convex* if the segment $\overline{XY}$ joining any two points X and Y of the region is part of the region. If the region is not convex, then it is *concave*.

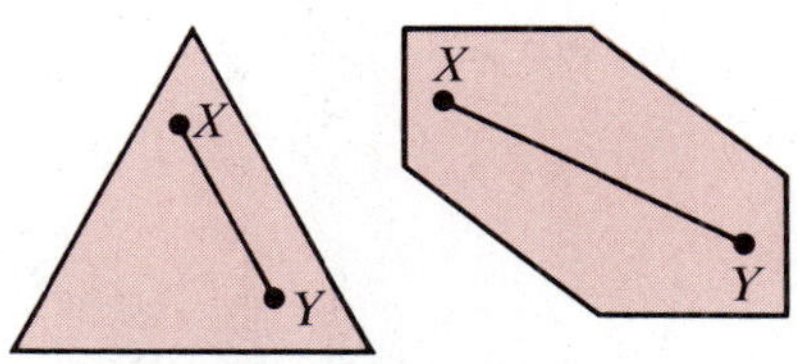

Convex polygonal regions

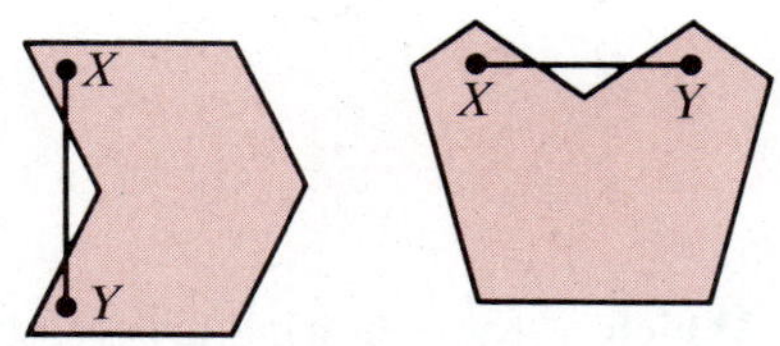

Concave polygonal regions

A polygon that determines a convex region is called a **convex polygon.** A polygon which determines a concave region is called a **concave polygon.** Most of the polygons that you will study will be convex. In this book, when *polygon* is used it will mean *convex polygon.*

Example 3

Which of the following are convex polygonal regions? Show why or why not.

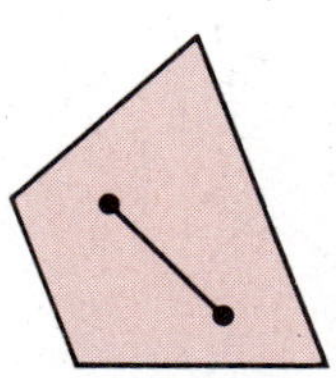

Convex

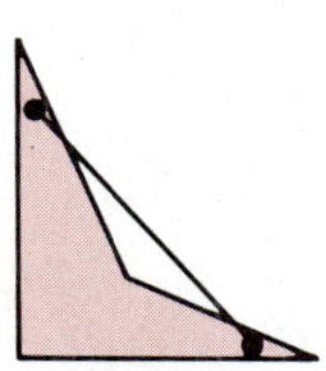

Concave

Which of the following are convex polygons?

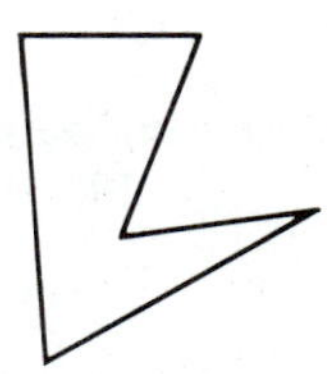

Concave

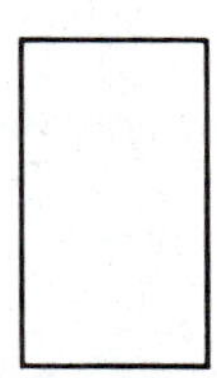

Convex

Example 4 Draw each of the following.

A convex pentagon

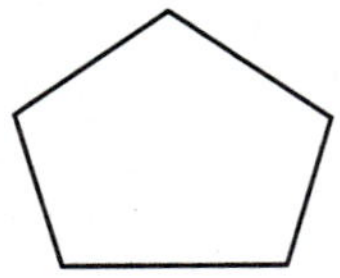

A concave hexagon

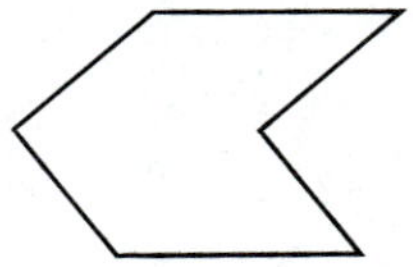

Reading in Geometry

Many words used in geometry are derived from Latin or Greek root words. The word *polygon* has two root words: *poly,* meaning "many," and *gon* meaning "angle."

Look up each of the following words in a dictionary. Write the root words and their meanings.

1. triangle **3, corner**
2. quadrilateral **4, side**
3. pentagon **5, angle**
4. hexagon **6, angle**
5. octagon **8, angle**
6. decagon **10, angle**
7. bisector **2, cut**
8. perpendicular **through, hang**
9. transversal **across, turned**
10. hypotenuse **under, to stretch**
11. congruent **to suit**
12. isosceles **equal, leg**

Oral Exercises

Which of the following are polygons?

1.

no

2.

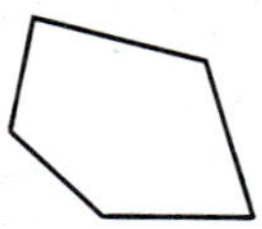

yes

3.

no

4.

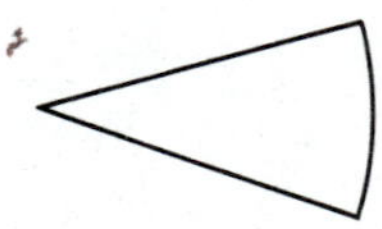

no

Which of the following are convex polygons?

5.

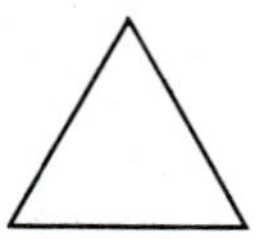

convex

6.

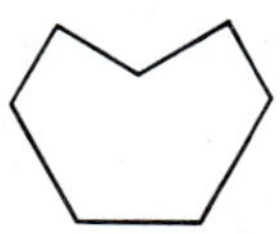

concave

7.

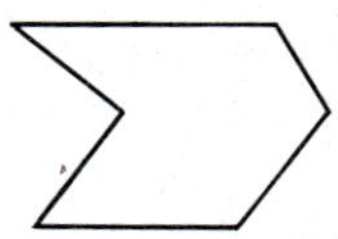

concave

8.

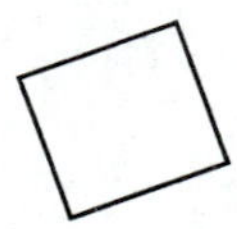

convex

Name each of the following polygons.

9.

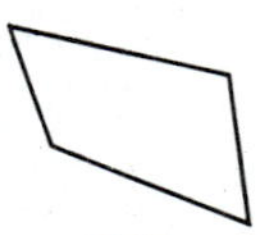

quadrilateral

10.

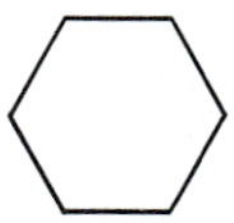

hexagon

11.

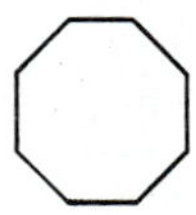

octagon

12.

hexagon

Written Exercises

Which of the following are polygons? Indicate why the others are not.

A **1.**

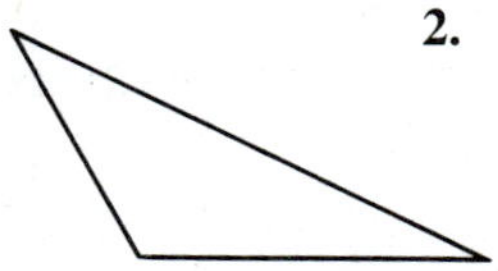

yes

2.

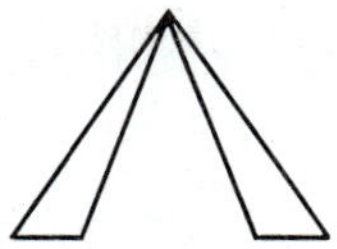

no

3.

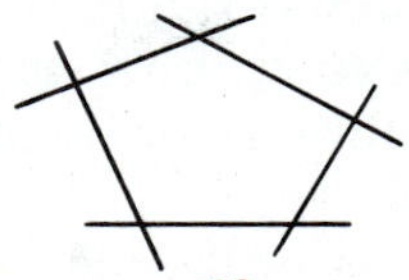

no

4.

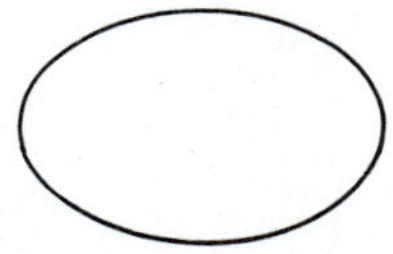

no

5.

no

6.

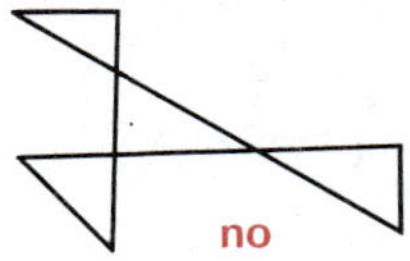

no

7.

no

8.

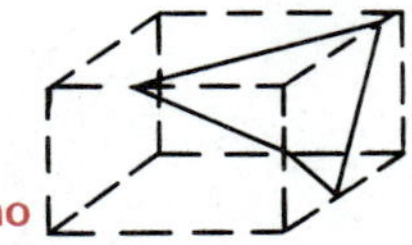

no

For each polygon, list the sides and vertices and classify the polygon. How many diagonals can be drawn for each polygon?

9.

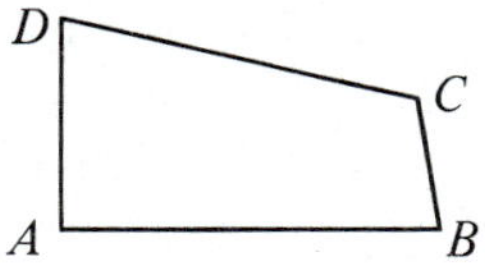

quadrilateral; 2

10.

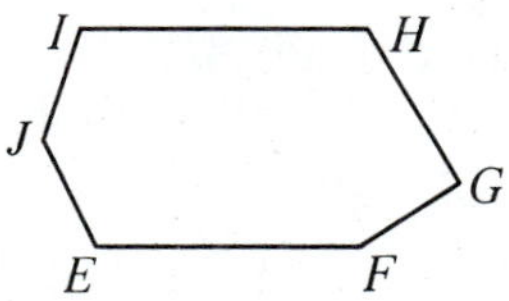

hexagon; 9

11.

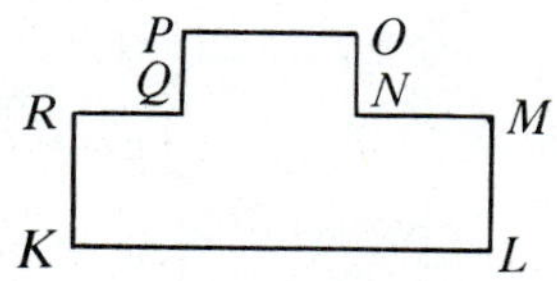

octagon; 20

Which of the following are convex polygonal regions? Show why or why not.

12.

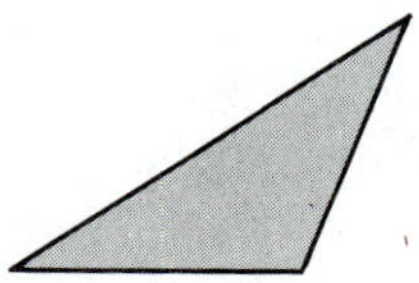

convex

13.

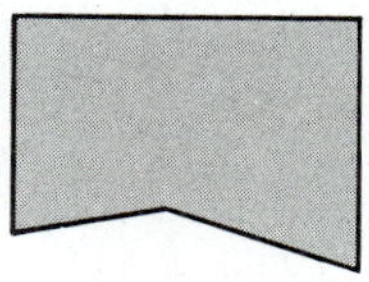

concave

14.

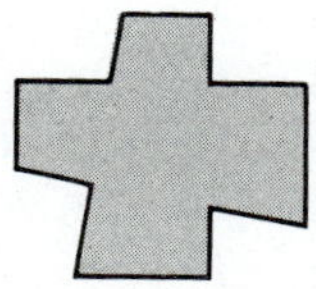

concave

15.

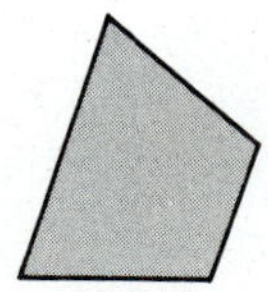

convex

Draw each of the following.

16. A convex quadrilateral
17. A convex hexagon
18. A concave decagon
19. A concave pentagon
20. A concave octagon
21. A concave quadrilateral

How many sides does each type of polygon have? (Use a dictionary, if necessary.) Draw a picture of each type.

B **22.** Heptagon 7
23. Undecagon 11
24. Nonagon 9
25. Dodecagon 12

CUMULATIVE REVIEW

1. Prove that the base angles of an isosceles triangle are congruent. **See Theorem 5.1, p. 139.**
2. Draw three different angles. Estimate their measures, check the measures with a protractor.
3. Construct a 45° angle. **See Ex. 2, p. 51.**

ANGLES OF CONVEX POLYGONS 6.2

Objectives

To prove and apply the theorem about the sum of the angle measures of a convex quadrilateral
To develop and apply a formula for the sum of the angle measures of any convex polygon

The sum of the measures of the angles of a triangle is 180°. If you draw diagonal $\overline{AC}$ in quadrilateral *ABCD,* two triangles are formed. The sum of the angle measures of the quadrilateral should be that of the two triangles, or 360°. This is proved in the following theorem.

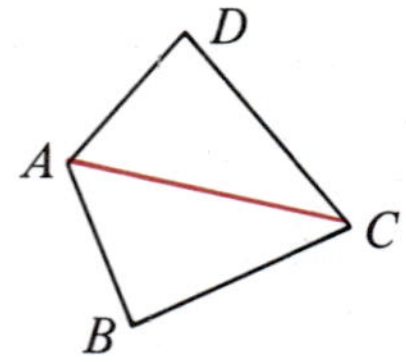

Theorem 6.1 The sum of the measures of the angles of a convex quadrilateral is 360°.

Given: Convex quad. *ABCD*
Prove: $m\angle A + m\angle B + m\angle C + m\angle D = 360$

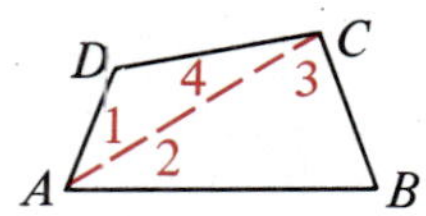

Proof

Statements	Reasons
1. Convex quad. *ABCD*	1. Given
2. Draw $\overline{AC}$.	2. Two points are contained in exactly one line.
3. $m\angle A = m\angle 1 + m\angle 2$	3. Angle addition postulate
4. $m\angle C = m\angle 3 + m\angle 4$	4. Angle addition postulate
5. $m\angle A + m\angle B + m\angle C + m\angle D = m\angle 1 + m\angle 2 + m\angle B + m\angle 3 + m\angle 4 + m\angle D$	5. Substitution property
6. $m\angle 2 + m\angle B + m\angle 3 = 180$	6. Sum of meas. of ∠'s of a △ = 180°.
7. $m\angle 1 + m\angle D + m\angle 4 = 180$	7. Sum of meas. of ∠'s of a △ = 180°.
8. $m\angle 1 + m\angle 2 + m\angle B + m\angle 3 + m\angle 4 + m\angle D = 360$	8. Addition prop. of equality and substitution property
9. $\therefore m\angle A + m\angle B + m\angle C + m\angle D = 360$	9. Substitution property

Example 1

Given: $m\angle X = 142$,
$m\angle Y = 63$, and
$m\angle Z = 134$.
Find $m\angle W$.

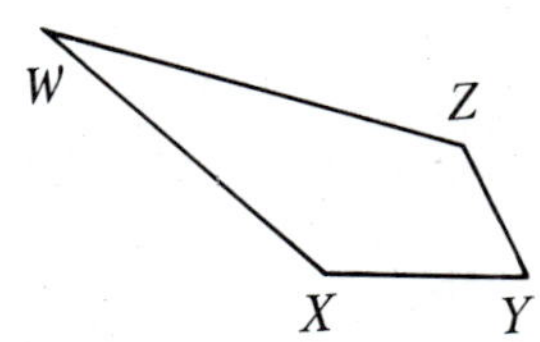

$142 + 63 + 134 + m\angle W = 360$
$m\angle W = 360 - 339$, or 21

Example 2

Given: m $\angle A = 5x + 10$,
m $\angle B = 6x - 2$,
m $\angle C = 4x + 17$, and
m $\angle D = 4x + 12$.
Find m $\angle A$ and m $\angle C$.

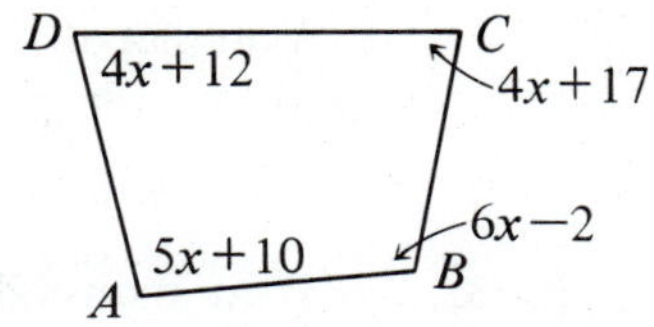

$$(5x + 10) + (6x - 2) + (4x + 17) + (4x + 12) = 360$$
$$19x + 37 = 360$$
$$19x = 323$$
$$x = 17$$

Thus, m $\angle A = 5(17) + 10$, or 95,
and m $\angle C = 4(17) + 17$, or 85.

If all possible diagonals of a given polygon are drawn from a single vertex, triangles are formed. The number of triangles formed is two less than the number of sides. If the polygon has n sides, then there will be $n - 2$ triangles.

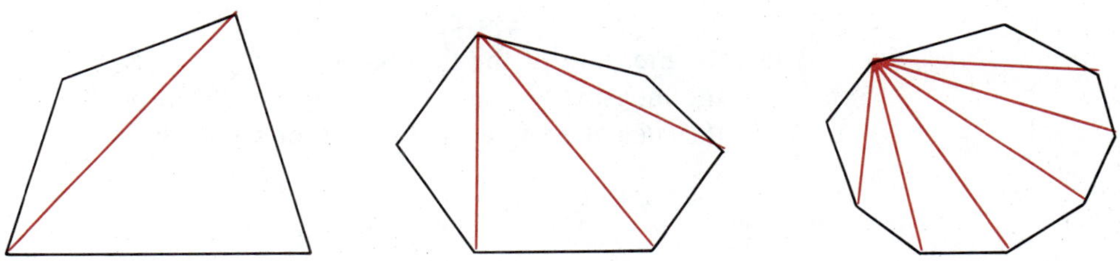

4 sides
2 triangles

6 sides
4 triangles

9 sides
7 triangles

The following theorem is based on this pattern.

Theorem 6.2 The sum of the measures of the angles of a convex polygon with n sides is $(n - 2)\ 180°$.
(See Exercises 35 and 36.)

Theorem 6.2 can be proved for any particular integer n, where $n \geq 3$. You can use it to find the sum of the angle measures of any convex polygon.

Example 3

Find the sum of the measures of the angles of a convex hexagon.

A hexagon has 6 sides. ► $(6 - 2)\ 180 = 720$

Thus, the sum of the angle measures is 720°.

You can also use Theorem 6.2 to determine the number of sides in a polygon and to find the measure of one of the angles.

Example 4 How many sides does a convex polygon have if the sum of the measures of its angles is 1980°?

$$(n-2)\,180 = 1980 \quad \blacktriangleleft \quad n = \textit{number of sides}$$
$$180n - 360 = 1980$$
$$180n = 2340$$
$$n = 13$$

Thus, the polygon has 13 sides.

Example 5 The measure of the second angle of a pentagon is twice the measure of the first angle. The measure of each of the other three angles is three times the measure of the first. What is the measure of each angle?

$$\text{Sum} = (5-2)180, \text{ or } 540$$
$$x + 2x + 3x + 3x + 3x = 540 \quad \blacktriangleleft \quad \text{Let } x = \textit{measure of first angle.}$$
$$12x = 540$$
$$x = 45$$

Thus, the measure of the first angle is 45°,
the measure of the second angle is 2(45), or 90°, and
the measure of each of the other angles is 3(45), or 135°.

Oral Exercises

How many triangles are formed by drawing all of the diagonals from a single vertex of each of the following polygons?

1.

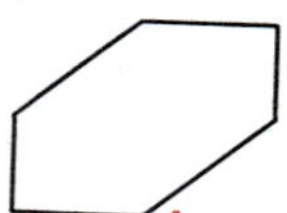

4

2.

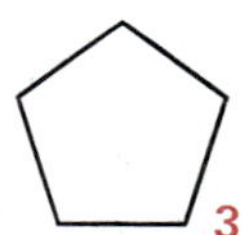

3

3.

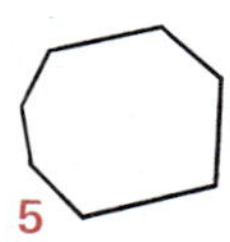

5

4.

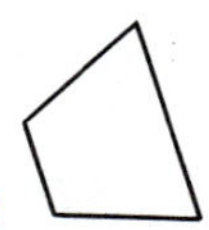

2

How many triangles are formed by drawing all of the diagonals from a single vertex of a polygon with each given number of sides?

5. 7 **5** **6.** 10 **8** **7.** 14 **12** **8.** 20 **18** **9.** 42 **40** **10.** 134 **132**

Written Exercises

For each of the following convex quadrilaterals, the measures of three angles are given. Find the measure of the fourth angle.

A **1.** 42°, 88°, 112° **118°** **2.** 87°, 123°, 131° **19°** **3.** 90°, 90°, 90° **90°**

4. 8°, 52°, 203° **97°** **5.** 59°, 60°, 161°, **80°** **6.** 120°, 60°, 30° **150°**

Find the measure of each angle of convex quadrilateral ABCD.

7. m $\angle A = 8x$, m $\angle B = 7x + 10$, m $\angle C = 11x - 10$, and m $\angle D = 10x$
80°, 80°, 100°, 100°

8. m $\angle A = 9x + 5$, m $\angle B = 11x + 5$, m $\angle C = 7x - 8$, and m $\angle D = 14x - 11$
86°, 104°, 55°, 115°

9. m $\angle A = 8x + 1$, m $\angle B = 9x + 1$, m $\angle C = 15x - 7$, and m $\angle D = 13x + 5$
65°, 73°, 113°, 109°

Find the sum of the measures of the angles of each convex polygon.

10. A pentagon 540°
11. An octagon 1080°
12. A dodecagon 1800°
13. A forty-gon 6840°
14. A 52-gon 9000°
15. A hundred-gon 17640°

Draw a picture of each of the following convex polygons. Using a protractor, measure each angle. Add these to determine the sum of the angle measures of the polygon. Compare this with the sum found by the formula of Theorem 6.2.

16. A pentagon
Sum = 540°
17. A heptagon
Sum = 900°
18. An octagon
Sum = 1,080°

The sum of the measures of the angles of a convex polygon is given. Find the number of sides of the polygon.

19. 540° 5
20. 1080° 8
21. 1800° 12
22. 1620° 11
23. 2340° 15
24. 3600° 22
25. 2880° 18
26. 7560° 44

Which of the following cannot be the sum of the angle measures of a convex polygon? (Ex. 27–30)

27. 1530° cannot
28. 3420° can
29. 6480° can
30. 4500° can

B **31.** If all of the angles of a hexagon are congruent, then what is the measure of each angle? 120°

32. If all of the angles of an octagon are congruent, then what is the measure of each angle? 135°

33. Three angles of a hexagon are congruent. The other angles are also congruent, each with a measure twice that of the first three. What is the measure of each angle? 80°, 80°, 80°, 160°, 160°, 160°

34. The measure of one angle of an octagon is twice that of the first. All of the other angles are congruent, each with a measure three times that of the first. What is the measure of each angle?
$51\frac{3}{7}°$, $102\frac{6}{7}°$, other 6 ∠'s measure $154\frac{2}{7}°$

Prove each of the following.

C **35.** Prove Theorem 6.2 for convex pentagons.

36. Prove Theorem 6.2 for convex hexagons.

Another approach can be used to prove Theorem 6.2. Prove each of the following, based on the diagram shown.

37. Prove Theorem 6.1, using the diagram.

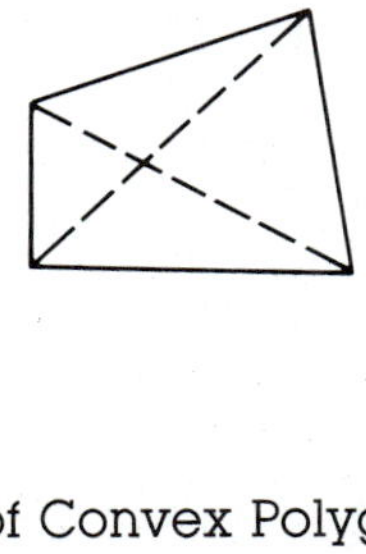

38. Prove Theorem 6.2 for convex pentagons, using the diagram.

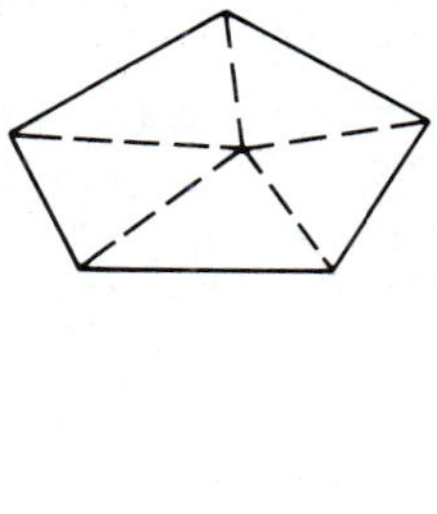

39. Prove Theorem 6.2 for convex hexagons, using the diagram.

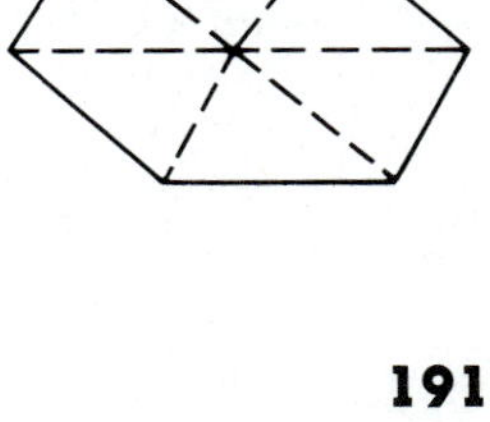

Using a protractor, measure each angle of each concave quadrilateral. Add these to determine the sum of the angle measures. Does the formula of Theorem 6.1 apply? (*Hint:* Every angle has a degree measure m such that $0 < m \leq 180$.)

40.

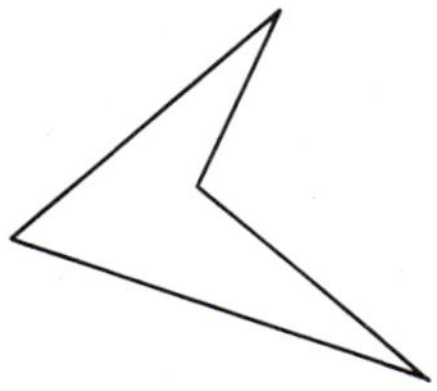

41.

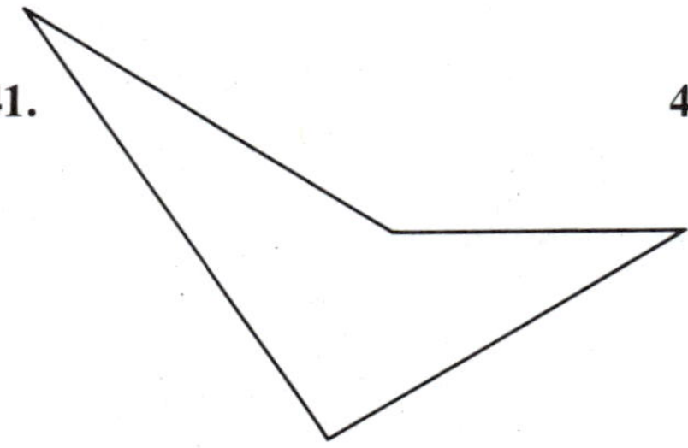

42.

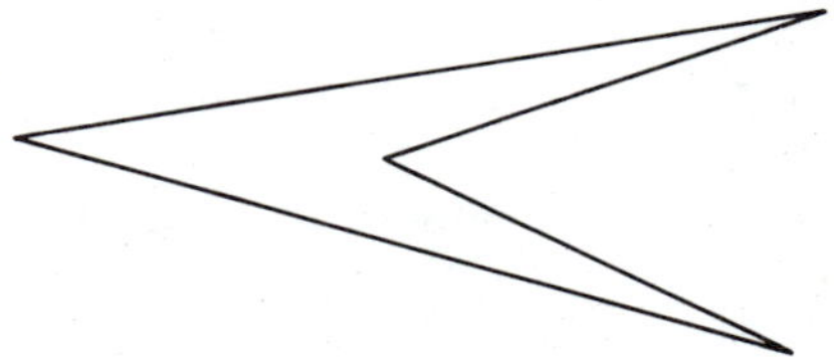

Using a protractor, measure each angle of each concave polygon. Add these to determine the sum of the angle measures. Does the formula of Theorem 6.2 apply? (*Hint:* Every angle has a degree measure m such that $0 < m \leq 180$.)

43.

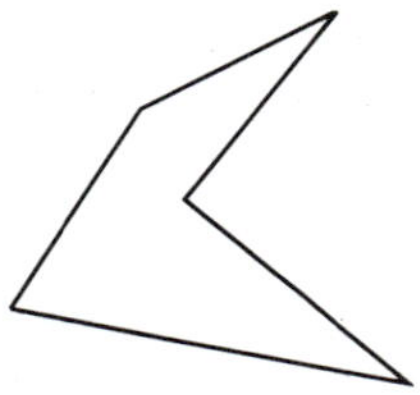

44.

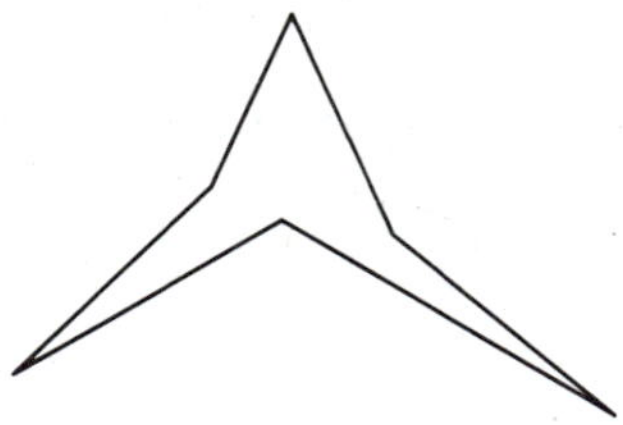

45.

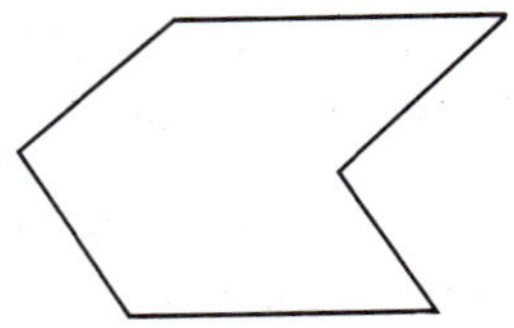

CALCULATOR ACTIVITIES

Example 1 Find the sum of the measures of the angles of a convex 792-gon.

Press 792 [−] 2 [=] [×] 180 [=].
Display is 142200.

The answer is 142200°.

Example 2 Can a convex polygon have an angle measure sum of 29810°?

Press 29810 [÷] 180 [=] [+] 2 [=].
Display is 167.61111.

If the display is a whole number, the answer is "yes."
Since 167.61111 is not a whole number, the answer is "no."

Find the sum of the measures of the angles of each convex polygon.

1. 313-gon **55,980°** **2.** 477-gon **85,500°** **3.** 1029-gon **184,860°** **4.** 1346-gon **241,920°**

Can a convex polygon have an angle measure sum of each of the following?

5. 10380° **no** **6.** 97000° **no** **7.** 18000° **yes** **8.** 18180° **yes**

EXTERIOR ANGLES OF CONVEX POLYGONS

6.3

Objectives

To identify the exterior angles of a convex polygon
To develop and apply a formula for the sum of the measures of the exterior angles of a convex polygon

Every convex polygon has two congruent **exterior angles** at each vertex. Quadrilateral $ABCD$ has eight exterior angles, angles 1 through 8. The two exterior angles at each vertex are congruent, since they are vertical angles.

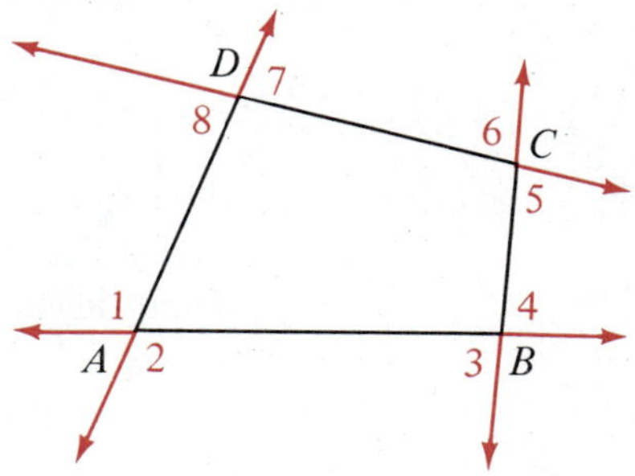

You can easily find the sum of the measures of the exterior angles, one at each vertex, by measuring with a protractor.

Example 1

Measure each exterior angle shown, using a protractor. What is the sum of the measures?

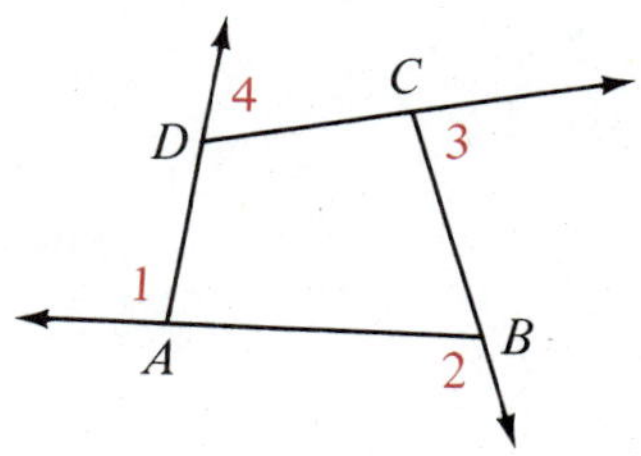

$m\angle 1 = 98$, $m\angle 2 = 113$,
$m\angle 3 = 81$, and $m\angle 4 = 68$.
Thus, $m\angle 1 + m\angle 2 + m\angle 3 + m\angle 4 = 360$.

The result of Example 1 can be shown to be true for any convex quadrilateral. An outline of the proof is given.

Theorem 6.3 The sum of the measures of the exterior angles, one at each vertex, of any convex quadrilateral is 360°.

Given: Convex quad. $ABCD$, and exterior ∠'s 1, 2, 3, and 4.
Prove: $m\angle 1 + m\angle 2 + m\angle 3 + m\angle 4 = 360$

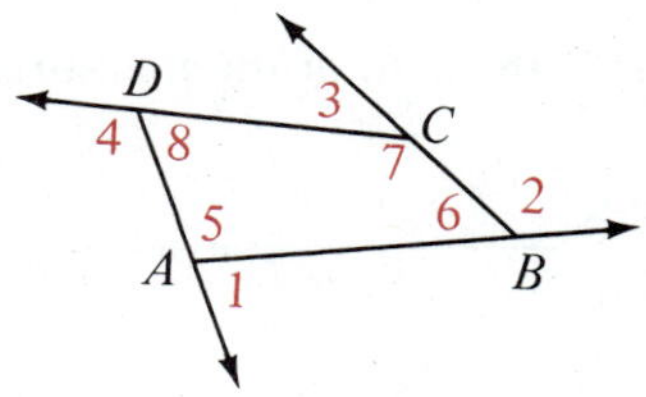

Outline of Proof

1. $m\angle 1 + m\angle 5 = 180$
2. $m\angle 2 + m\angle 6 = 180$
3. $m\angle 3 + m\angle 7 = 180$
4. $m\angle 4 + m\angle 8 = 180$
5. $m\angle 1 + m\angle 2 + m\angle 3 + m\angle 4 + m\angle 5 + m\angle 6 + m\angle 7 + m\angle 8 = 720$
6. $m\angle 5 + m\angle 6 + m\angle 7 + m\angle 8 = 360$
7. $\therefore m\angle 1 + m\angle 2 + m\angle 3 + m\angle 4 = 360$

Theorem 6.3 can be shown to be true for any convex polygon.

Theorem 6.4 The sum of the measures of the exterior angles, one at each vertex, of any convex polygon is 360°. (See Exercises 6 and 7.)

Example 2 What is the sum of the measures of the exterior angles, one at each vertex, of a convex 138-gon?

The sum is 360°. ◄ *Use Theorem 6.4.*

Example 3 The measures of the exterior angles, one at each vertex, of a convex quadrilateral are m $\angle x = a + 2$, m $\angle y = a + 4$, m $\angle z = a + 6$, and m $\angle w = a + 8$. Find the measure of each angle.

$$\begin{aligned} m\angle x + m\angle y + m\angle z + m\angle w &= 360 \\ (a+2)+(a+4)+(a+6)+(a+8) &= 360 \\ 4a + 20 &= 360 \\ 4a &= 340 \\ a &= 85 \end{aligned}$$

Thus, m $\angle x = 85 + 2$ or 87, m $\angle y = 85 + 4$, or 89, m $\angle z = 85 + 6$, or 91, and m $\angle w = 85 + 8$, or 93.

Oral Exercises

Which of the angles shown in each figure are exterior angles?

1. 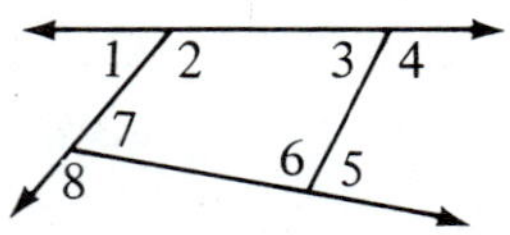

∠1, ∠4, ∠5, ∠8

2. 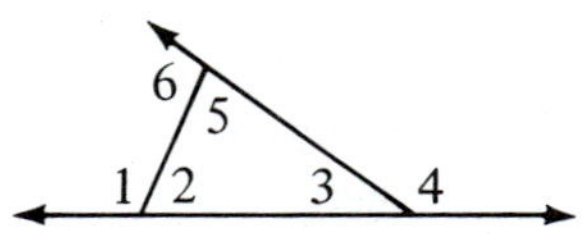

∠1, ∠4, ∠6

3. 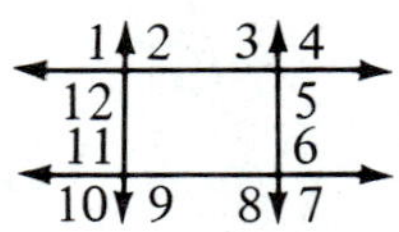

∠2, ∠3, ∠5, ∠6, ∠8, ∠9, ∠11, ∠12

What is the sum of the measures of the exterior angles shown in each figure?

4. 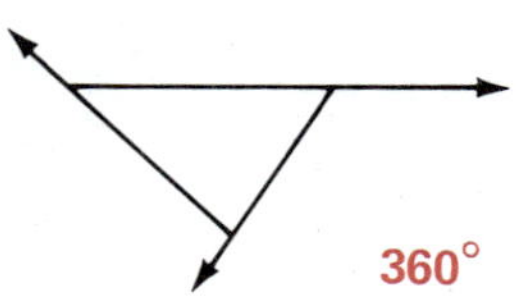

360°

5. 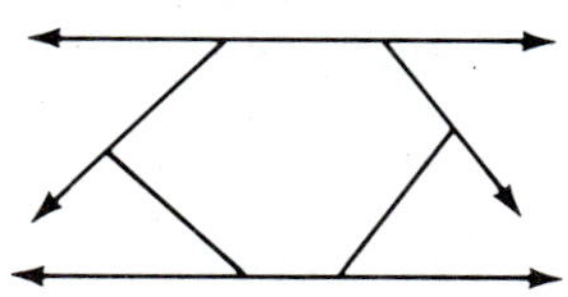

360°

6. 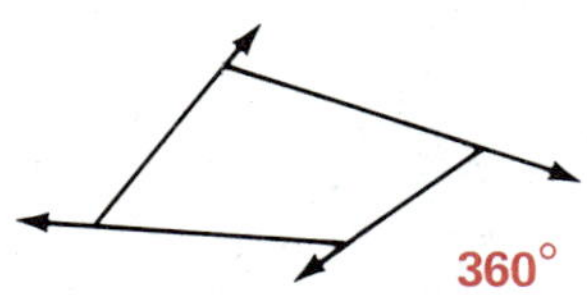

360°

What is the sum of the measures of the exterior angles, one at each vertex, of each of the given convex polygons?

7. An octagon 360°
8. A nonagon 360°
9. A dodecagon 360°
10. A 21-gon 360°
11. A 312-gon 360°
12. A 978-gon 360°

Written Exercises

Copy each polygon. Draw an exterior angle at each vertex. Use a protractor to measure each exterior angle drawn. What is the sum of these measures? Does the result agree with Theorem 6.4? (Ex. 1–3) 360°; yes

A 1.

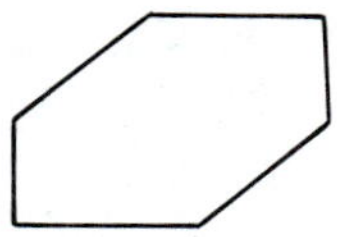

2.

3.

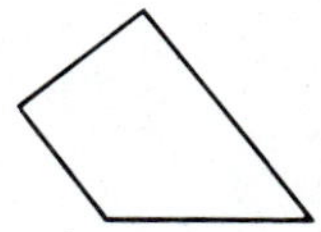

4. The measures of the exterior angles of a convex pentagon, one at each vertex, are $m\angle p = x$, $m\angle q = 2x$, $m\angle r = 3x$, $m\angle s = 4x$, and $m\angle t = 5x$. Find the measure of each exterior angle. 24°, 48°, 72°, 96°, 120°

5. The measures of the exterior angles of a convex hexagon, one at each vertex, are $m\angle 1 = 6x - 1$, $m\angle 2 = 10x + 2$, $m\angle 3 = 8x + 2$, $m\angle 4 = 9x - 3$, $m\angle 5 = 5x + 4$, and $m\angle 6 = 12x + 6$. Find the measure of each exterior angle.
41°, 72°, 58°, 60°, 39°, 90°

B 6. Prove Theorem 6.4 for triangles.

7. Prove Theorem 6.4 for convex pentagons.

For each convex polygon, compute the sum of the measures of all pairs of interior angles of the polygon and their corresponding exterior angles, one at each vertex. Compute the sum of the measures of the interior angles of the polygon. Using these sums, compute the sum of the measures of the exterior angles, one at each vertex.

8. A triangle
540 − 180 = 360; 360°

9. A heptagon
1,260 − 900 = 360; 360°

10. A 30-gon
5,400 − 5,040 = 360; 360°

ALGEBRA REVIEW

Objective

To solve fractional equations of the form $\frac{ax + b}{cx} = d$

To solve a fractional equation of the form $\frac{ax + b}{cx} = b$:

1. Multiply both sides of the equation by cx.
2. Solve for x, using appropriate equation properties.

Example

Solve $\frac{7x + 9}{2x} = 5$

$7x + 9 = 10x$ ◄ *Mult. by 2x.*

$-3x = -9$ ◄ *Add −10x to both sides.*

$x = 3$

Solve.

1. $\frac{5x + 32}{3x} = 7$ 2
2. $\frac{4x - 8}{3x} = 2$ −4
3. $\frac{9x - 7}{5x} = 2$ −7
4. $\frac{8x + 7}{5x} = 3$ 1
5. $\frac{2x - 31}{4x} = 7$ $\frac{-31}{26}$
6. $\frac{2x - 9}{3x} = 8$ $\frac{-9}{22}$

REGULAR POLYGONS 6.4

Objectives

To identify regular polygons
To determine angle measures and the number of sides of regular polygons

A polygon is **equilateral** if all of its sides are congruent. If all of its angles are congruent, then the polygon is **equiangular**.

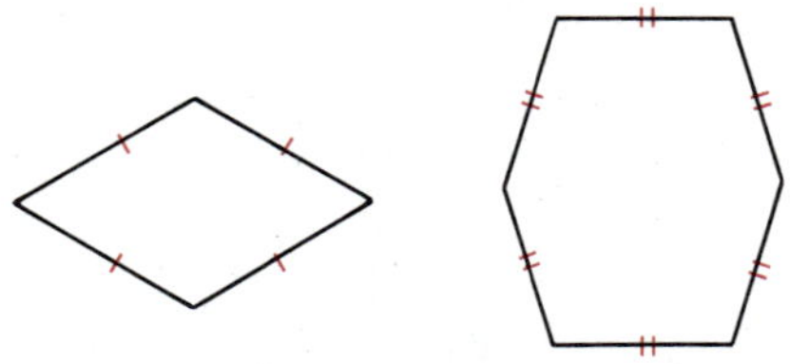

Equilateral polygons

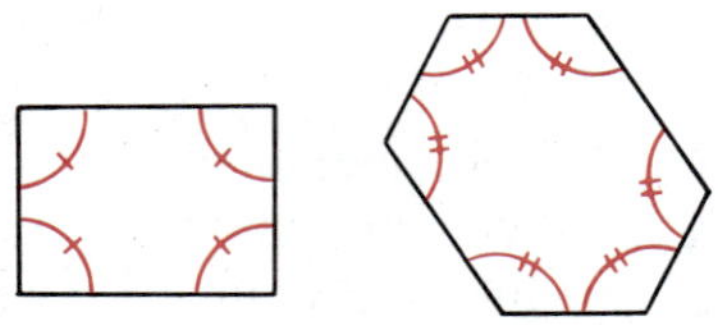

Equiangular polygons

Some polygons are both equilateral and equiangular.

Definition: Regular polygon A *regular polygon* is a convex polygon which is both equilateral and equiangular.

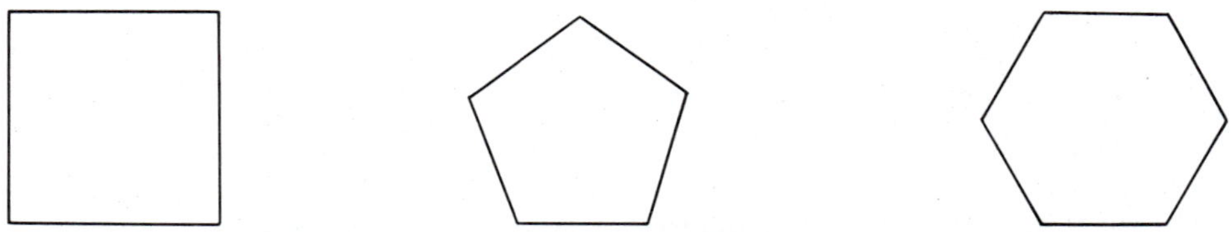

Some regular polygons

You can find the measure of an angle of a regular polygon if you know how many sides it has.

Example 1 Find the measure of each angle of a regular hexagon.

Use Theorem 6.2. ► The sum of the measures is $(6-2)(180)$, or 720.

Thus, the measure of each angle is $\frac{720}{6}$, or 120°.

Example 1 can be generalized in the following corollary to Theorem 6.2.

Corollary The measure of an angle of a regular polygon with n sides is $\frac{(n-2)\,180°}{n}$.

A similar corollary of Theorem 6.4 can be used to find the measure of an exterior angle of a regular polygon.

Corollary The measure of an exterior angle of a regular polygon with n sides is $\frac{360°}{n}$.

Example 2 Find the measure of each exterior angle of a regular octagon.

$$\text{Measure} = \frac{360}{8}, \text{ or } 45°$$

If you know the measure of one angle or one exterior angle of a regular polygon, you can find the number of sides of the polygon.

Example 3 How many sides does a regular polygon have if the measure of each angle is 144°?

$$144 = \frac{(n-2)\,180}{n} \quad \blacktriangleleft \; \textit{The measure of each } \angle \textit{ is } \frac{(n-2)\,180°}{n}.$$
$$144n = (n-2)\,180$$
$$144n = 180n - 360$$
$$-36n = -360$$
$$n = 10$$

A 10-sided polygon is a decagon. ► **Thus,** the polygon has 10 sides.

Example 4 How many sides does a regular polygon have if the measure of each exterior angle is 72°?

$$72 = \frac{360}{n} \quad \blacktriangleleft \; \textit{The measure of each exterior } \angle \textit{ is } \frac{360°}{n}.$$
$$72n = 360$$
$$n = 5$$

Thus, the polygon has 5 sides.

Written Exercises

Draw each indicated type of regular polygon and label the congruent parts.

A **1.** Triangle **2.** Quadrilateral **3.** Hexagon **4.** Octagon

Find the measure of an angle of each of the following regular polygons.

5. Triangle **60°** **6.** Quadrilateral **90°** **7.** Pentagon **108°** **8.** Octagon **135°**
9. Nonagon **140°** **10.** Decagon **144°** **11.** 30-gon **168°** **12.** 100-gon **176.4°**

Find the measure of an exterior angle of each of the following regular polygons.

13. Triangle 120° **14.** Quadrilateral 90° **15.** Pentagon 72° **16.** Octagon 45°
17. Decagon 36° **18.** 30-gon 12° **19.** 50-gon 7.2° **20.** 100-gon 3.6°

Find the number of sides in each regular polygon with angles of the given measure.

21. 60° 3 **22.** 135° 8 **23.** 150° 12 **24.** 171° 40

Find the number of sides in each regular polygon with exterior angles of the given measure.

25. 72° 5 **26.** 36° 10 **27.** 30° 12 **28.** 15° 24

Which of the following are possible measures of the interior angles of a regular polygon? If the measures are possible, how many sides does the polygon have?

B **29.** 90° poss, 4 **30.** 100° not poss **31.** 110° not poss **32.** 125° not poss **33.** 150° poss, 12 **34.** 175° poss, 72

Which of the following are possible measures of exterior angles of a regular polygon? If the measures are possible, how many sides does the polygon have?

35. 180° not poss **36.** 150° not poss **37.** 120° poss, 3 **38.** 100° not poss **39.** 50° not poss **40.** 30° poss, 12

True or false? Draw pictures to illustrate. (Ex. 41–42)

C **41.** An angle of a regular polygon cannot be congruent to an exterior angle. F

42. For no value of n is an equilateral n-gon equiangular. F

43. A regular polygon was defined to be a convex polygon. Are there any concave polygons which have all angles and all sides congruent? Sketch and investigate. Yes

CALCULATOR ACTIVITIES

You can use a calculator to find the measure of each angle of a regular polygon with a large number of sides, such as a 213-gon. The process is shown below.

Press 213 [−] 2 [=] [×] 180 [=] [÷] 213 [=] .
The answer is 178.30986°.

Certain answers may vary, depending upon type of calculator used.

Find the measure of an angle of each regular polygon.

1. 814-gon 179.55774° **2.** 697-gon 179.4835° **3.** 1024-gon 179.64844°
4. 613-gon 179.41272° **5.** 4213-gon 179.91455° **6.** 9027-gon 179.96012°

Find the measure of an exterior angle of each regular polygon.

7. 896-gon 0.40179° **8.** 211-gon 1.70616° **9.** 45-gon 8°
10. 102-gon 3.52941° **11.** 379-gon 0.94987° **12.** 2135-gon 0.16862°

CONSTRUCTING REGULAR POLYGONS 6.5

Objectives **To construct regular polygons with 3, 4, or 6 sides**
To construct angles with certain specified measures

A regular polygon with three sides is an equilateral triangle. You can construct an equilateral triangle as shown below.

Construction An equilateral triangle

Given: $\overline{AB}$

Construct: An equilateral triangle with sides $\cong \overline{AB}$.

Open the compass to length AB. Draw an arc with center A.

Keep the same compass opening. Draw an arc with center B, intersecting the first arc at C.

Draw $\overline{AC}$ and $\overline{BC}$.

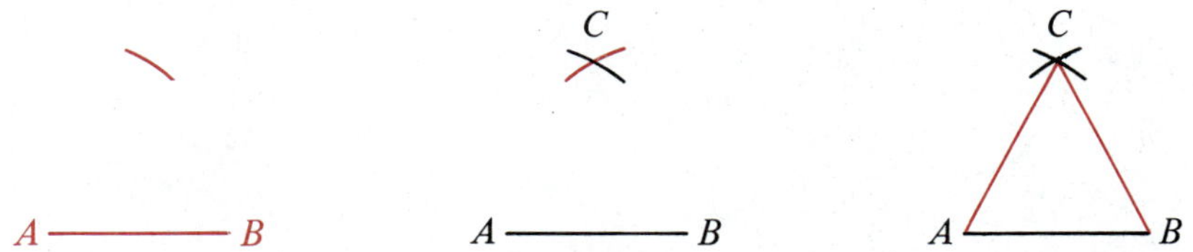

Conclusion: $\triangle ABC$ is equilateral.

Each angle of an equilateral triangle has a measure of 60°. Therefore, you can construct a 30° angle by first constructing an equilateral triangle and then bisecting one of its angles.

Example 1 Construct a 30° angle.

Construct equil. $\triangle ABC$. ►
Bisect $\angle CAB$.

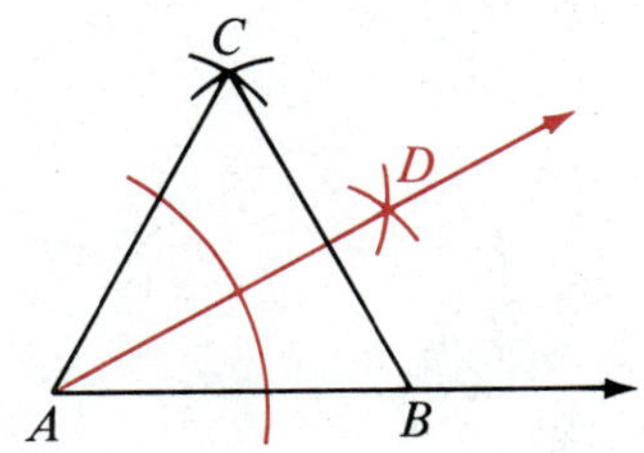

Conclusion: m $\angle BAD = 30$

A regular hexagon can be formed by joining six equilateral triangles as shown. This is the basis for the following construction.

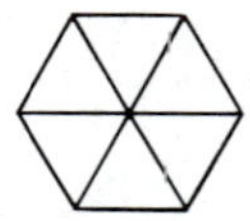

Construction A regular hexagon

Given: $\overline{AB}$

Construct: A regular hexagon with sides $\cong \overline{AB}$

Open the compass to length AB. Draw a circle with center P.

Keep the same compass opening. Starting with any point A on the circle, draw a series of arcs on the circle.

Draw the segments.

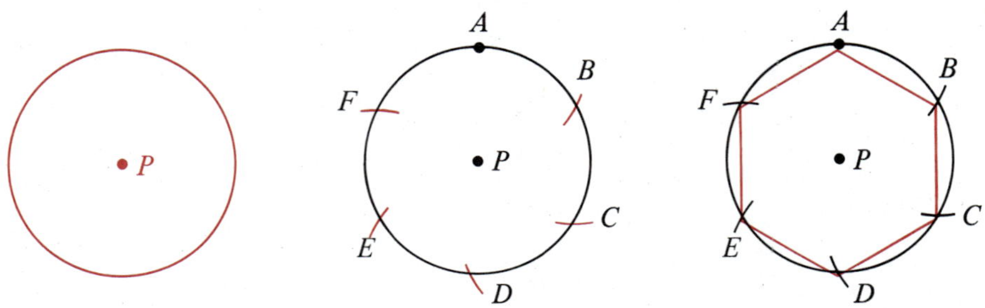

Conclusion: *ABCDEF* is a regular hexagon.

Using procedures similar to the above for constructing a regular hexagon, many other types of regular polygons can be constructed. Also, by bisecting angles of various measures, angles of other measures can be constructed. The example below shows two such constructions.

Example 2

Construct each of the following.

A square with sides $\cong \overline{AB}$

A 150° angle

Construst ⊥'s at A and B.

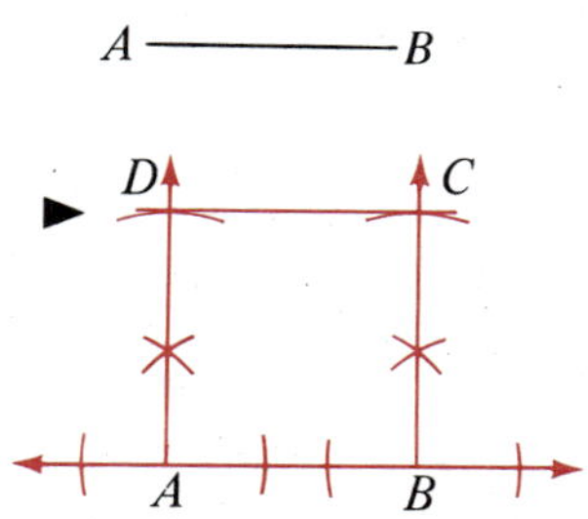

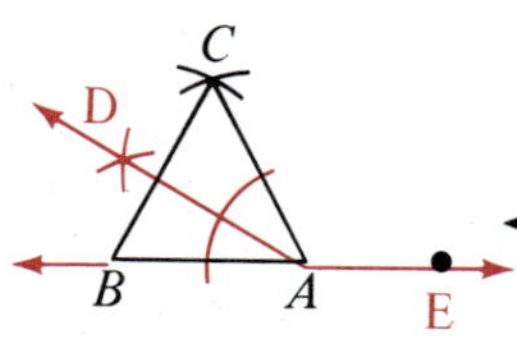

Bisect ∠CAB of equil. △ABC.

Conclusion: *ABCD* is a square.

Conclusion: m ∠*DAE* = 150

Written Exercises

Draw a segment. Construct the following with sides congruent to the segment.

A 1. An equilateral triangle **Follow constr, top of p. 199.**

2. A square **Follow constr in Ex 2, p. 200.**

3. A regular hexagon **Follow constr, top of p. 200.**

Construct an angle with each given measure.

4. 60° 5. 90° 6. 120° 7. 30° 8. 45° 9. 150°

B 10. 15° 11. 75° 12. 105° 13. $37\frac{1}{2}°$ 14. 165° 15. $157\frac{1}{2}°$

Construct each of the following.

16. An equilateral triangle with sides twice as long as a given segment
17. A regular hexagon with sides half as long as a given segment

C 18. A regular octagon with sides congruent to a given segment

19. A regular dodecagon with sides congruent to a given segment
20. A square with a diagonal congruent to a given segment
21. A regular 16-gon with sides congruent to a given segment
22. A regular hexagon with an 11-cm perimeter
23. An equilateral triangle with a 14-cm perimeter

A regular pentagon can also be constructed.

Draw diameter $\overline{AB}$ in circle O. Construct $\overline{CO}$, the $\perp$ bisector of $\overline{AB}$. Construct D, the midpoint of $\overline{OB}$.

With D as center and $\overline{CD}$ as a radius, draw arc $\widehat{CE}$. Draw $\overline{CE}$.

Open compass to length CE. Starting at C, draw a series of arcs on the circle. Draw the segments.

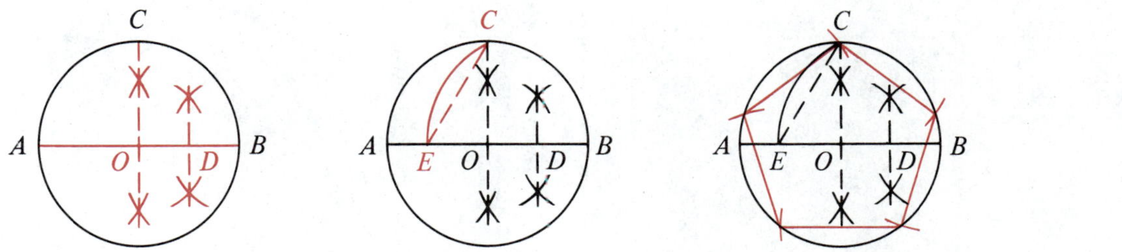

24. Construct a five-pointed star.
25. Construct a regular decagon.

CUMULATIVE REVIEW

1. If the coordinate of p is -12 and the coordinate of M, the midpoint of $\overline{PQ}$, is 3, what is the coordinate of Q? **18**
2. What is the sum of the measures of the angles of a regular heptagon? **900°**
3. What congruence pattern for right triangles is not true for triangles in general? **SSA (HL for rt. △'s)**

APPLICATIONS

Read → Plan → Solve → Interpret

1. Square carpet tiles can be used to cover a floor. The tiles will fit together with no gaps or overlapping. What geometric properties insure that this is true?

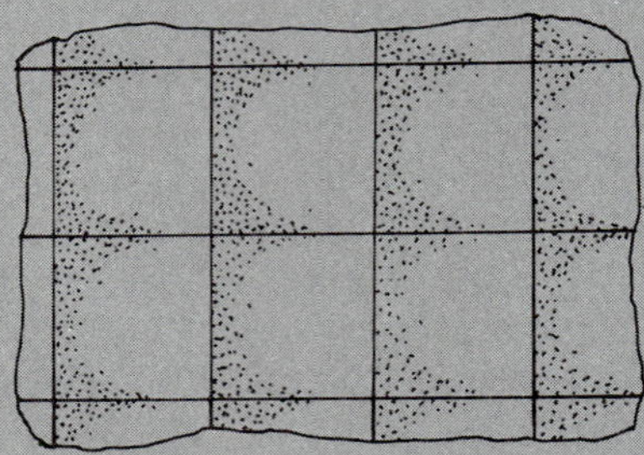

2. Many bathroom floors are made of ceramic tiles in the shape of regular hexagons. Can regular pentagons or regular octagons be used for such a floor? Why? Make paper cutouts to test your answers. What other types of regular polygons can be used for such floor tiles?

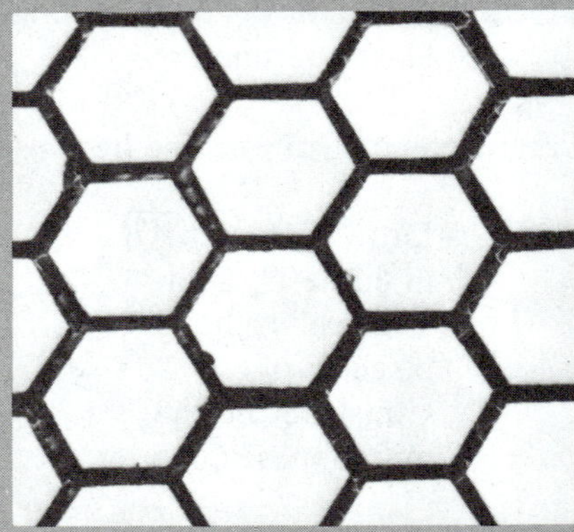

COMPUTER ACTIVITIES

A Computerized Quiz on Polygons

Objective: **To use computer-generated quizzes to identify a polygon from the number of its sides, and vice versa**

In this chapter you have been introduced to the names of many polygons. You can use the following program to test your knowledge of those names. (For another example of a computer-generated quiz, see page 571.) This program uses READ and DATA lines for inputting both questions and answers. This program also uses a FOR . . . NEXT loop so that it will generate exactly ten questions.

```
110  PRINT "THIS PROGRAM TESTS YOUR KNOWLEDGE OF THE
      NAMES OF POLYGONS."
130  FOR I = 1 TO 10: READ A,A$
150  PRINT "NAME THE POLYGON WITH "A" SIDES.": INPUT
     P$
155  PRINT
160  IF P$ = A$ THEN 190
170  PRINT "SORRY, THE ANSWER IS "A$"."
180  GOTO 200
190  PRINT "VERY GOOD!"
195 R = R + 1
200  PRINT : NEXT I
220  DATA   3,TRIANGLE,4,QUADRILATERAL,5,PENTAGON,6,
     HEXAGON,7,HEPTAGON,8
230  DATA   OCTAGON,9,NONAGON,10,DECAGON,11,UNDECAGO
     N,12,DODECAGON
240  PRINT "YOU GOT "R" RIGHT OUT OF 10."
```

EXERCISES: See the Computer Section beginning on page 565 for more information.

1. Rewrite the program above so that the computer gives you the name of the polygon and asks you for the number of sides.
2. Rewrite the program above so that the computer gives you the name of the polygon and asks you for the sum of the measures of the interior angles. Do not change the DATA lines when you rewrite this program.
3. Without changing the DATA lines, alter the program above so that the computer tells you the number of degrees in each interior angle of a regular polygon and then asks you for the name of the polygon.

CHAPTER SIX REVIEW

Which of the figures below are polygons? Why are the others not polygons? [6.1]

1. 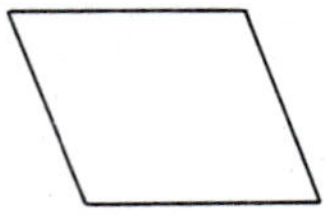
yes

2. 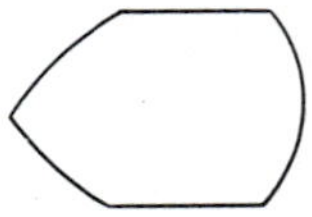
no

3. 
no

4. 
yes

Which polygons are convex? Which are concave? [6.1]

5. 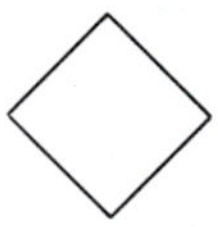
convex

6. 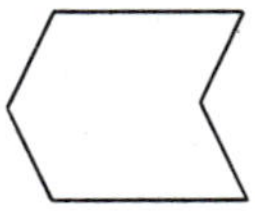
concave

7. 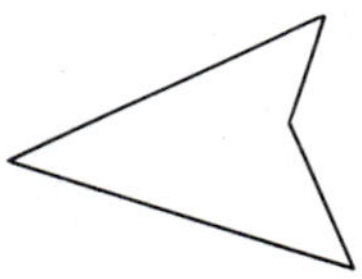
concave

8. 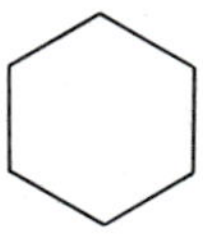
convex

Find each indicated angle measure. [6.2]

9. Find the sum of the measures of the angles of a pentagon. 540°

10. Find the sum of the measures of the angles of a 20-gon. 3240°

11. Find the sum of the measures of the exterior angles, one at each vertex, of a dodecagon. [6.3] 360°

12. Find the sum of the measures of the exterior angles, one at each vertex, of a 29-gon. 360°

13. Find the measure of each angle of a regular pentagon. [6.4] 108°

14. Find the measure of each exterior angle of a regular dodecagon. 30°

***15.** Prove: The sum of the measures of the angles of a pentagon is 540°. [6.2]

True or false?

16. Two sides of a polygon always intersect at a vertex. [6.1] T

17. If each exterior angle of a regular polygon has a measure of 90°, then the polygon is a square. [6.4] T

***18.** All the sides of an equiangular polygon are congruent. [6.4] F

19. A dodecagon has ten sides. [6.1] F

20. The sum of the measures of the angles of a regular polygon is 360°. [6.2] F

21. A polygon must be coplanar. [6.1] T

Find each indicated angle measure.

22. In quadrilateral $ABCD$, $m\angle A = 4x + 9$, $m\angle B = 5x + 2$, $m\angle C = 9x - 1$, and $m\angle D = 7x$. Find the measure of each angle. [6.2]
65°, 72°, 125°, 98°

23. The measures of the exterior angles, one at each vertex, of a pentagon are $m\angle a = 3t + 7$, $m\angle b = 4t + 4$, $m\angle c = 6t - 2$, $m\angle d = 5t + 4$, and $m\angle e = 6t - 13$. Find the measure of each exterior angle. [6.3]
52°, 64°, 88°, 79°, 77°

Find each of the following.

24. Find the number of sides in a regular polygon each of whose angles measures 144°. [6.4]
10

25. Find the number of sides in a regular polygon, each of whose exterior angles measures 14.4°. [6.4] 25

Construct each of the following.

26. Draw a segment 5 cm long. Construct a regular hexagon with sides congruent to the segment. [6.5]

27. Construct a 60° angle.

28. Construct a 45° angle.

29. Construct a 15° angle.

Vocabulary

concave [6.1], convex [6.1], decagon [6.1], diagonal [6.1], exterior angle [6.3], hexagon [6.1], octagon [6.1], pentagon [6.1], polygon [6.1], regular polygon [6.4]

(A) exercises are 1–14, 16, 17, 19, 21, 24–28. (B) exercises are 20, 22, 23, 29.

(C) exercises are starred.

CHAPTER SIX TEST

Which of the figures below are polygons? Why are the others not polygons?

1.

no

2.

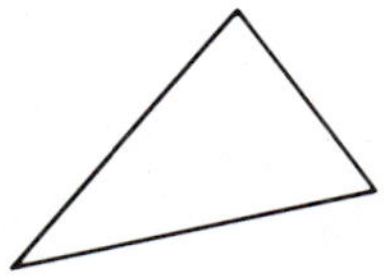

yes

3.

no

4.

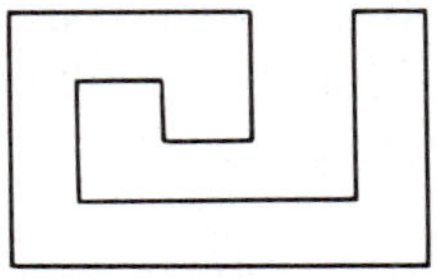

yes

Which polygons are convex? Which are concave?

5.

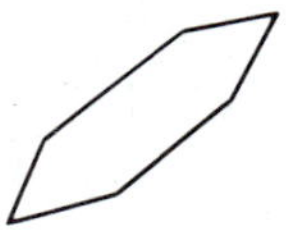

convex

6.

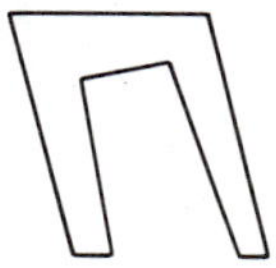

concave

7.

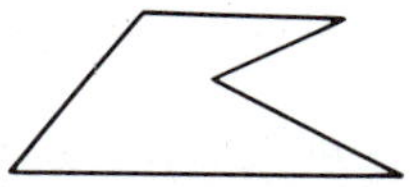

concave

8.

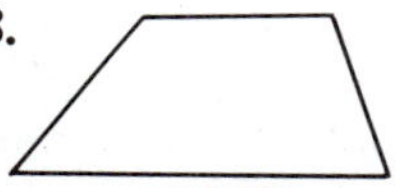

convex

Find each indicated angle measure.

9. Find the sum of the measures of the angles of a 32-gon. **5400°**

10. Find the sum of the measures of the exterior angles, one at each vertex, of a 27-gon. **360°**

11. Find the measure of each angle of a regular hexagon. **120°**

12. Find the measure of each exterior angle of a regular 30-gon. **12°**

13. In quadrilateral $ABCD$, $m\angle A = 7x - 4$, $m\angle B = 8x - 1$, $m\angle C = 8x + 4$, and $m\angle D = 7x + 1$. Find the measure of each angle. **80°, 95°, 100°, 85°**

14. The measures of the exterior angles, one at each vertex, of a triangle are $m\angle 1 = 2y + 10$, $m\angle 2 = 3y - 30$, and $m\angle 3 = y + 80$. Find the measure of each angle. **110°, 120°, 130°**

Find each of the following.

15. Find the number of sides of a regular polygon each of whose angles measures 140°. **9**

16. Find the number of sides of a regular polygon each of whose exterior angles measures 18°. **20**

True or false?

17. A decagon is a polygon with 12 sides. **F**

18. If each angle of a regular polygon has a measure of 60°, then the polygon is a triangle. **T**

* **19.** An equilateral polygon is also equiangular. **F**

20. A regular polygon cannot have an angle whose measure is 150°. **F**

21. A concave polygon cannot be equilateral. **F**

Construct each of the following.

22. A regular hexagon with sides congruent to $\overline{AB}$.

A ———————————— B

23. A 30° angle.

24. A 135° angle.

Prove the following.

* **25.** The sum of the measures of the exterior angles of a triangle is 360°.

A exercises are 1–12, 15–18, 20–23. B exercises are 13, 14, 24. C exercises are starred.

COLLEGE PREP TEST

For each item, choose the best answer. Indicate your answer by writing the letter of this choice.

1. Which of the following is a polygon? **E**

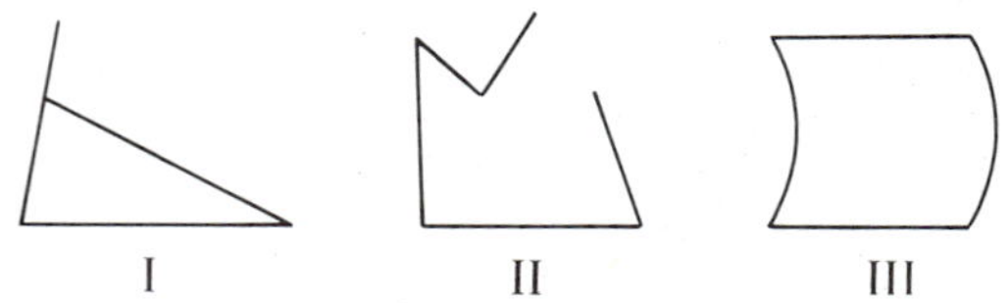

(A) I only
(B) II only
(C) III only
(D) I and III only
(E) None of these

2. Which of the following is a convex polygon? **B**

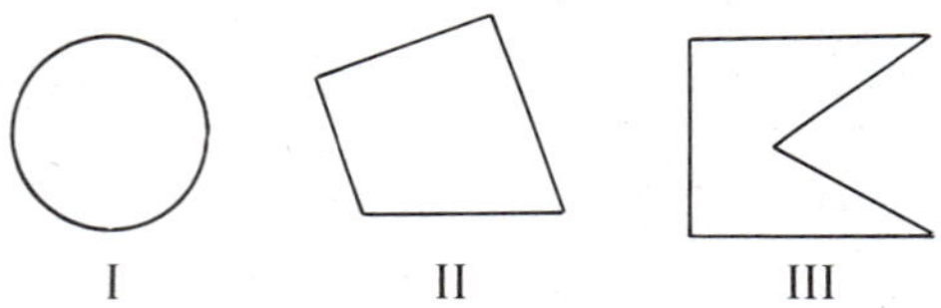

(A) I only
(B) II only
(C) III only
(D) I and II only
(E) II and III only

3. The sum of the measures of the angles of a convex decagon is **B**

(A) 360 **(B)** 1440
(C) 1800 **(D)** 2160
(E) None of these

4. The sum of the measures of the exterior angles of a convex 20-gon is **A**

(A) 360 **(B)** 3240
(C) 3600 **(D)** 3960
(E) None of these

5. Each angle of a regular pentagon has a measure of **C**

(A) 72 **(B)** 90
(C) 108 **(D)** 135
(E) None of these

6. If an angle of a regular polygon has a measure of 156°, then the polygon has **B**

(A) 13 sides **(B)** 15 sides
(C) 16 sides **(D)** 17 sides
(E) None of these

7. The angles of a quadrilateral have measures of $5x$, $3x + 20$, $4x + 5$, and $5x - 5$. The measure of one of the angles is **D**

(A) 20 **(B)** 50
(C) 90 **(D)** 95
(E) 105

8. Which of the following is not necessary for a figure to be a polygon? **B**

I. Contains 3 or more segments
II. Is convex
III. Is coplanar

(A) I only
(B) II only
(C) III only
(D) I and II only
(E) I and III only

9. To construct a 30° angle, you must first construct a **B**

(A) Right triangle
(B) An equilateral triangle
(C) A square
(D) An octagon
(E) A perpendicular bisector

7 QUADRILATERALS

HISTORY

Amalie Noether

Amalie Noether was born in Germany. Much of her work was in the area of abstract algebra.

Amalie Noether was well known as a teacher and mathematician. She began her advanced education studying French and English. Her interests soon turned to mathematics, and she became one of the most famous abstract algebraists in the world.

Because of oppression in her native Germany, Noether came to the United States in 1933. She taught at Bryn Mawr College. She died at the peak of her career on April 14, 1935.

Project

Write a report on Amalie Noether, emphasizing her work in abstract algebra.

FINDING PROPERTIES OF QUADRILATERALS

7.1

Objectives

To identify diagonals and consecutive and opposite sides and angles of quadrilaterals
To investigate properties of quadrilaterals

As you learned in the last chapter, some polygons are given special names according to the number of sides they have. A **quadrilateral** is a polygon with four sides. The *sides* of a quadrilateral are segments, and each endpoint of each segment is a *vertex* of the quadrilateral.

Quadrilaterals have *opposite* and *consecutive sides* and *opposite* and *consecutive angles*. The consecutive sides of a quadrilateral intersect to share a common vertex, while opposite sides do not intersect. The vertices of consecutive angles are endpoints of the same side of the quadrilateral. Vertices of opposite angles are not endpoints of the same side.

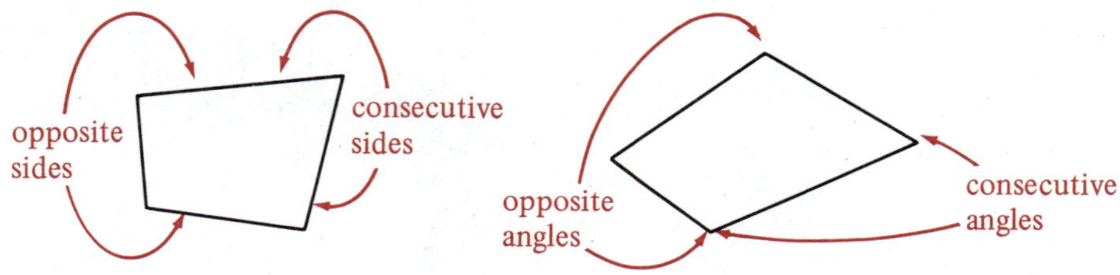

Example 1

List the pairs of consecutive sides and the pairs of consecutive angles of quadrilateral $ABCD$.

Consecutive sides	**Consecutive angles**
$\overline{AB}$ and $\overline{BC}$	$\angle A$ and $\angle B$
$\overline{BC}$ and $\overline{CD}$	$\angle B$ and $\angle C$
$\overline{CD}$ and $\overline{DA}$	$\angle C$ and $\angle D$
$\overline{DA}$ and $\overline{AB}$	$\angle D$ and $\angle A$

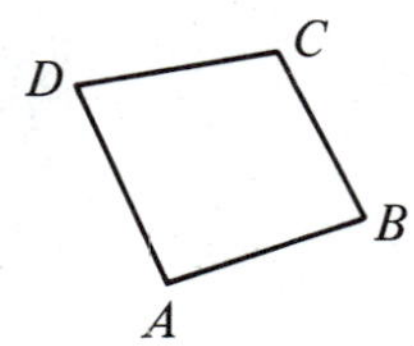

Example 2

Which side is opposite each given side? Which angle is opposite each given angle?

$\overline{XY}$, $\overline{YZ}$, $\angle X$, $\angle W$

$\overline{WZ}$ is opposite $\overline{XY}$.
$\overline{WX}$ is opposite $\overline{YZ}$.
$\angle Z$ is opposite $\angle X$.
$\angle Y$ is opposite $\angle W$.

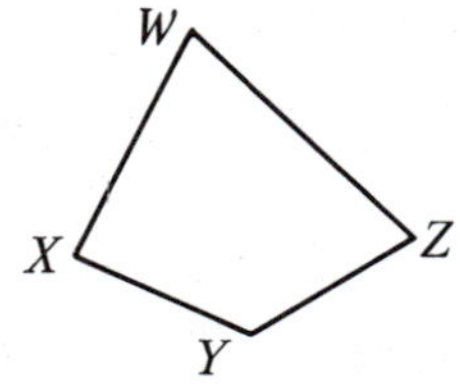

Quadrilaterals, like other polygons, have diagonals. A **diagonal** of a quadrilateral always connects opposite vertices. The diagonals of some quadrilaterals are congruent while others are not. The diagonals, $\overline{AC}$ and $\overline{BD}$, of quadrilateral $ABCD$ below are congruent, but the diagonals, $\overline{EG}$ and $\overline{FH}$, of quadrilateral $EFGH$ are not.

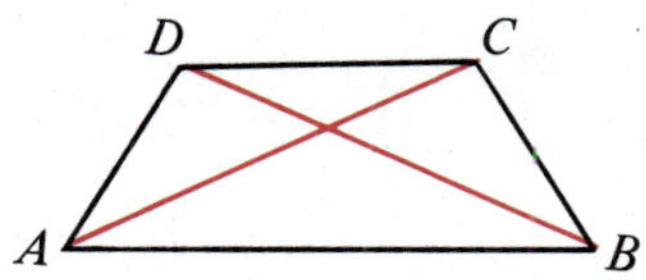

$AC = 3$ cm, $BD = 3$ cm

$\overline{AC} \cong \overline{BD}$

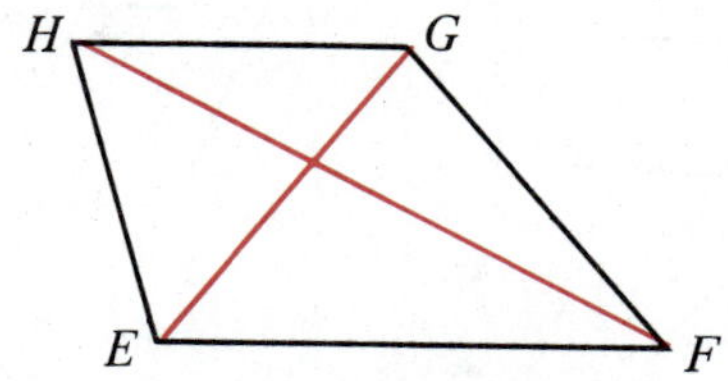

$EG = 25$ mm, $FH = 45$ mm

$\overline{EG} \not\cong \overline{FH}$

You should make drawings to investigate other statements of the properties of quadrilaterals, including statements about sides, angles, and diagonals.

Example 3

Do the diagonals of a quadrilateral bisect each other? Draw pictures to defend your answer.

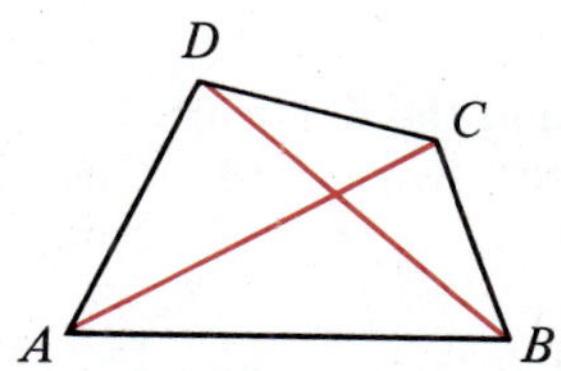

$\overline{AC}$ and $\overline{BD}$ do not bisect each other.

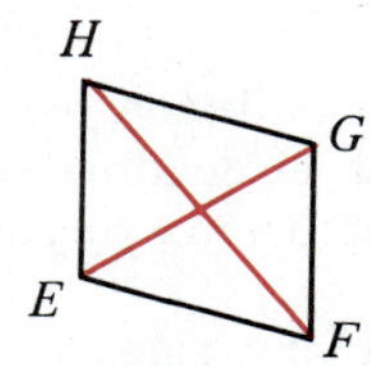

$\overline{EG}$ and $\overline{FH}$ seem to bisect each other.

Thus, the diagonals of a quadrilateral do not necessarily bisect each other.

Example 4

Are opposite angles of a quadrilateral congruent? Draw pictures to defend your answer.

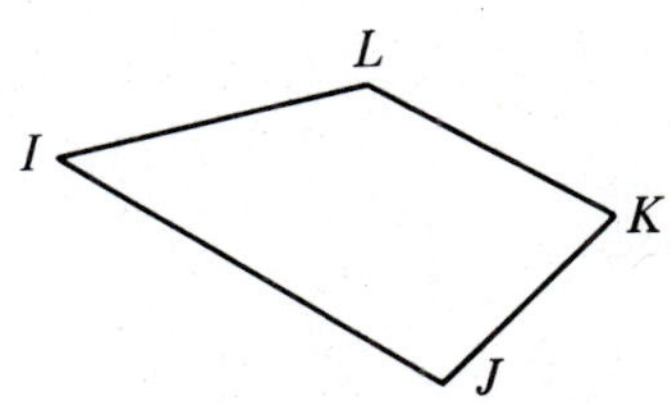

$\angle I \not\cong \angle K$, $\angle J \not\cong \angle L$

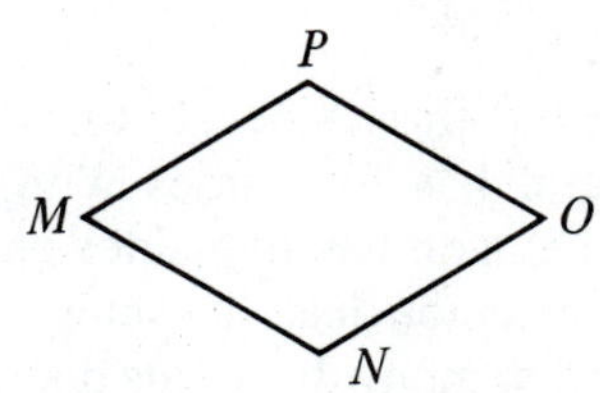

$\angle M$ appears congruent to $\angle O$.
$\angle N$ appears congruent to $\angle P$.

Thus, the opposite angles of a quadrilateral are not necessarily congruent.

Oral Exercises

Name the pairs of consecutive sides and opposite sides of each quadrilateral.

Check students' answers.

1.

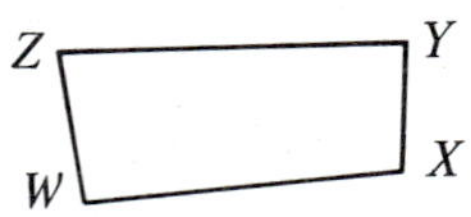

2.

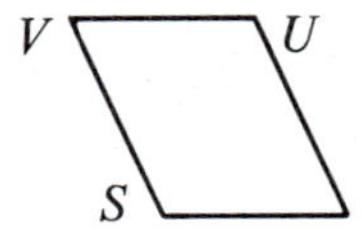

Name the pairs of consecutive angles and opposite angles of each quadrilateral.

3.

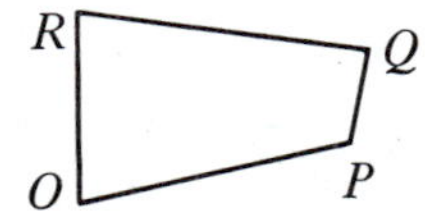

4.

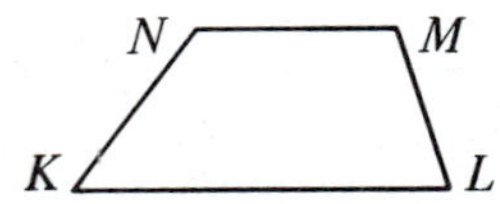

Written Exercises

Name the diagonals of each quadrilateral. Do the diagonals intersect?

A 1.

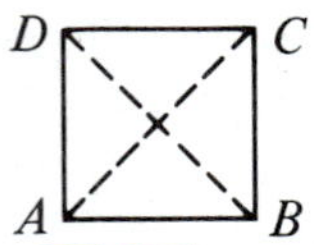

$\overline{AC}$, $\overline{BD}$, yes

2.

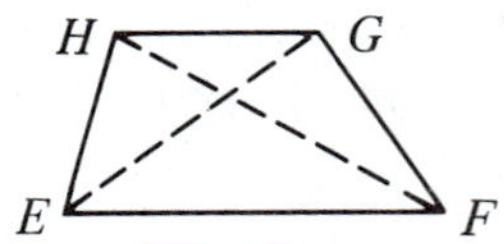

$\overline{EG}$, $\overline{FH}$, yes

3.

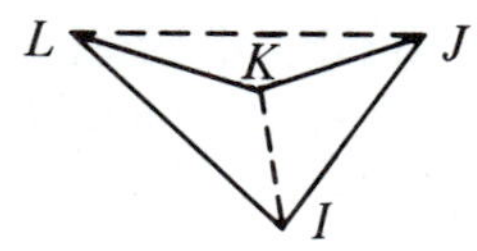

$\overline{IK}$, $\overline{JL}$, no

4.

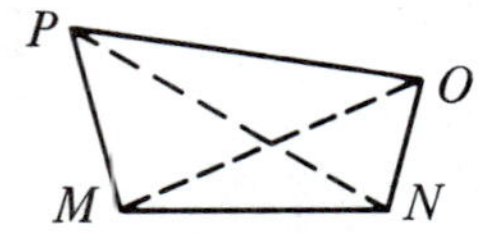

$\overline{MO}$, $\overline{NP}$, yes

For each of the following, draw a quadrilateral to fit the description.

5. Congruent opposite sides
6. Congruent consecutive sides
7. No congruent sides
8. No congruent angles

Draw pictures to defend your answer to each of the following.

9. Can a quadrilateral have exactly two congruent consecutive sides? yes
10. Can a quadrilateral have exactly two congruent consecutive angles? yes

B 11. Are any sides of a quadrilateral necessarily congruent? no
12. Do the diagonals of a quadrilateral bisect its angles? not necessarily
13. Are congruent triangles formed by the diagonals of a quadrilateral? not necessarily
14. Are opposite sides of a quadrilateral parallel? sometimes

C 15. Do the segments that join the midpoints of consecutive sides of a quadrilateral have any special properties? Opp. seg. are ∥ Opp. seg. are ≅.
16. Do the segments that join the midpoints of opposite sides of a quadrilateral have any special properties? They bis. each other.

NON-ROUTINE PROBLEMS

A triangle has three sides, but no diagonals. A quadrilateral has four sides with two diagonals. A pentagon has five sides and five diagonals. A hexagon has six sides and nine diagonals. How many diagonals does a heptagon have? An octagon? How many diagonals does an n-gon have? Can you write a formula for this?

14; 20; $\frac{n(n-1)}{2} - n$; no. diag. $= \frac{n(n-1)}{2} - n$

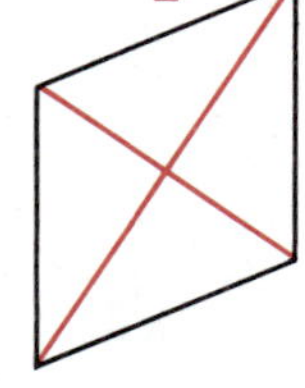

4 sides – 2 diagonals

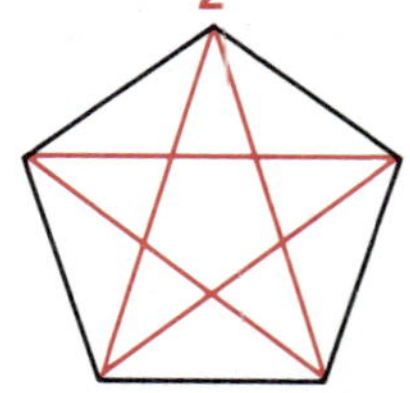

5 sides – 5 diagonals

FINDING PROPERTIES OF PARALLELOGRAMS 7.2

Objective

To prove and apply theorems about parallelograms

Some quadrilaterals have special properties and special names. One of these is the **parallelogram**.

Definition: Parallelogram A *parallelogram* is a quadrilateral in which both pairs of opposite sides are parallel. (Parallelogram $ABCD$ is written $\square ABCD$.)

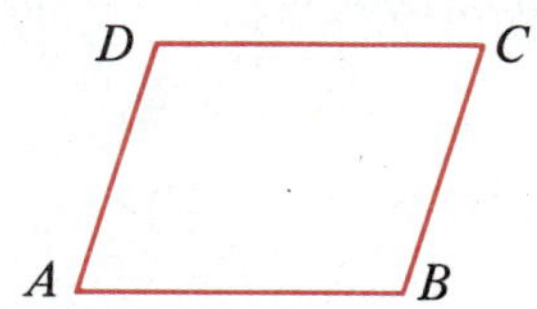

$ABCD$ is a parallelogram.

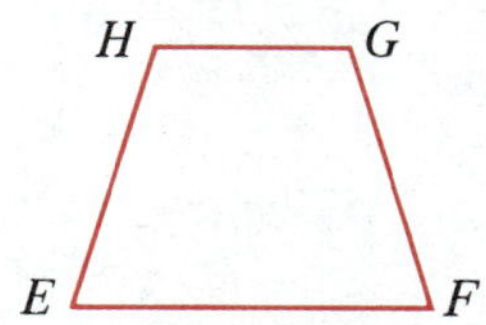

$EFGH$ is not a parallelogram. ($\overline{FG}$ and $\overline{EH}$ are not parallel.)

The following theorem will help you prove other properties of parallelograms.

Theorem 7.1 A diagonal of a parallelogram forms two congruent triangles.

Given: $\square ABCD$ with diagonal $\overline{BD}$

Prove: $\triangle DAB \cong \triangle BCD$

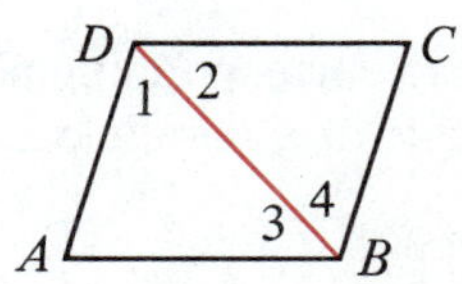

ANALYSIS
Use ASA. Alt. int. $\angle$'s of $\parallel$ lines are $\cong$.

Proof

Statements		Reasons
1. $\square ABCD$, with diagonal $\overline{BD}$		1. Given
2. $\overline{AB} \parallel \overline{DC}$ and $\overline{AD} \parallel \overline{BC}$		2. Def. of $\square$
3. $\angle 1 \cong \angle 4$	(A)	3. Alt. int. $\angle$'s of $\parallel$ lines are $\cong$.
4. $\overline{BD} \cong \overline{DB}$	(S)	4. Reflexive property
5. $\angle 3 \cong \angle 2$	(A)	5. Alt. int. $\angle$'s of $\parallel$ lines are $\cong$.
6. $\therefore \triangle DAB \cong \triangle BCD$		6. ASA

The definition tells you that the opposite sides of a parallelogram are parallel. Since corresponding parts of congruent triangles are congruent, you can easily show that opposite sides and opposite angles of parallelograms are congruent. These properties are stated in Corollaries 1 and 2.

Corollary 1 Opposite sides of a parallelogram are congruent.

Corollary 2 Opposite angles of a parallelogram are congruent.

You can use Theorem 7.1 and its corollaries to find unknown lengths of sides and measures of angles of parallelograms.

Example 1

Given: $\square ABCD$, m $\angle A = 120$
Find the measure of each of the other angles.

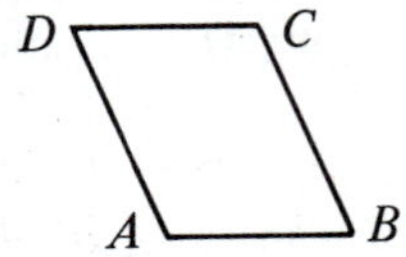

m $\angle C = 120$ ◀ *Use Corollary 2.*
m $\angle B = 60$
m $\angle D = 60$ ◀ *Interior ∠'s of ∥ lines on the same side of transversal are supplementary.*

Example 2

Given: $\square EFGH$ with $EF = 3x - 4$ and
$GH = 2x + 2$
Find EF.

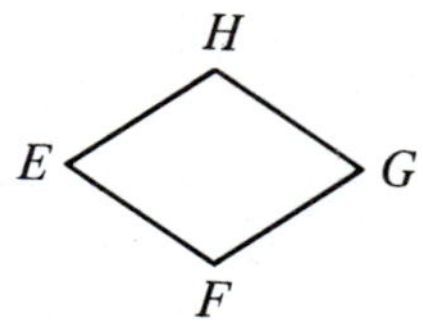

$EF = GH$ ◀ *Use Corollary 1.*
$3x - 4 = 2x + 2$
$x = 6$
Thus, $EF = 3 \cdot 6 - 4$, or 14.

You can use corresponding parts of congruent triangles to prove a theorem about the diagonals of a parallelogram.

Theorem 7.2 The diagonals of a parallelogram bisect each other.

Given: $\square WXYZ$ with diagonals $\overline{XZ}$ and $\overline{WY}$
Prove: $\overline{XO} \cong \overline{ZO}$ and $\overline{YO} \cong \overline{WO}$

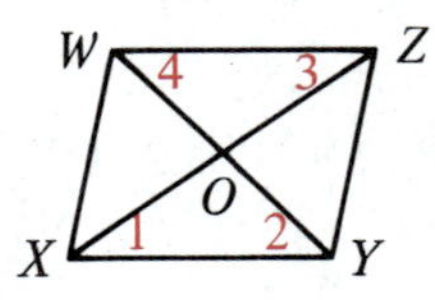

ANALYSIS
Show that $\overline{XO}$ and $\overline{ZO}$, $\overline{WO}$ and $\overline{YO}$ are corresponding sides of $\cong \triangle$'s.

Outline of proof

1. $\overline{WZ} \parallel \overline{XY}$
2. $\angle 1 \cong \angle 3$, $\angle 2 \cong \angle 4$
3. $\overline{XY} \cong \overline{ZW}$
4. $\triangle XYO \cong \triangle ZWO$
5. $\therefore \overline{XO} \cong \overline{ZO}$ and $\overline{YO} \cong \overline{WO}$

(See Exercise 16.)

Example 3

Given: $\square ABCD$ with $AC = 16$ and $BD = 12$
Find AO and BO.

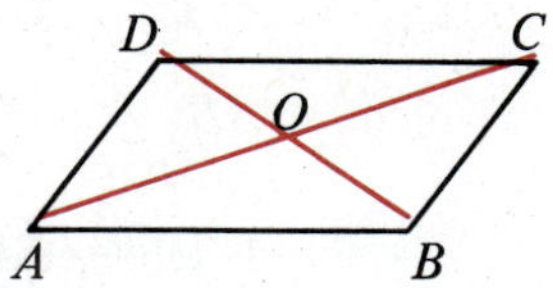

Apply Th. 7.2. ► $AO = \frac{1}{2}(AC)$, or 8

$BO = \frac{1}{2}(BD)$, or 6

Example 4

$EFGH$ is a parallelogram. Which of the following pairs of triangles are congruent? Why?

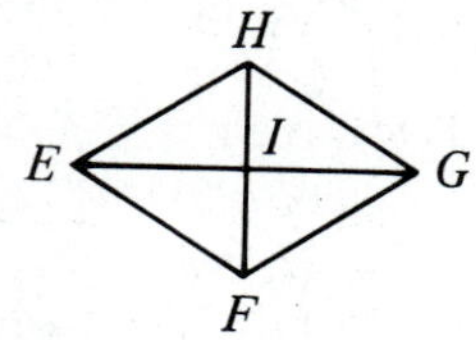

$\triangle EFG$ and $\triangle GHE$
Congruent by Th. 7.1.

$\triangle EIF$ and $\triangle GIF$
Not necessarily congruent

You have proved several important theorems and corollaries about a parallelogram. These, along with the definition, provide you with the basic properties of a parallelogram, as summarized below.

Properties of a Parallelogram

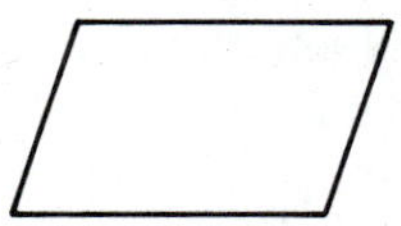

Opposite sides are parallel.

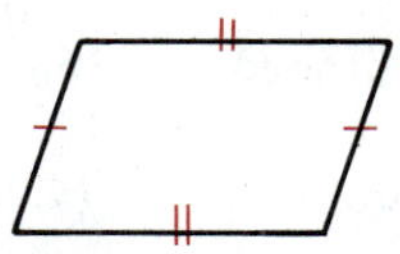

Opposite sides are congruent.

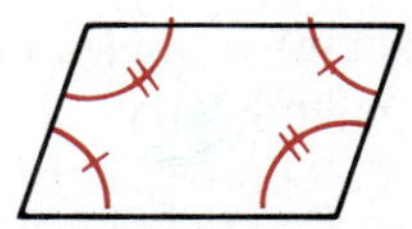

Opposite angles are congruent.

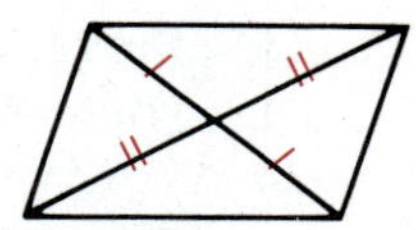

Diagonals bisect each other.

Oral Exercises

Which of the following polygons are parallelograms? (Ex. 1–4)

1.

no

2.

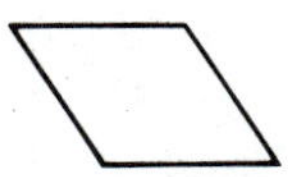

yes

3.

yes

4.

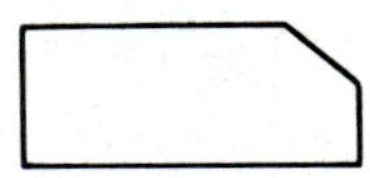

no

5. Which sides of $\square ABCD$ are parallel? **$\overline{AB} \parallel \overline{DC}$, $\overline{AD} \parallel \overline{BC}$**

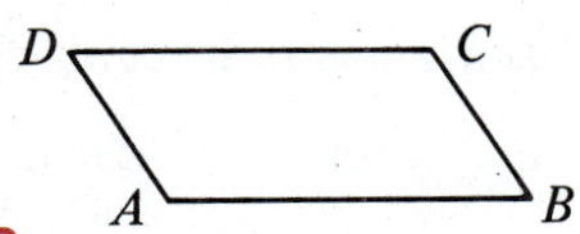

6. Which sides of $\square ABCD$ are congruent? **$\overline{AB} \cong \overline{DC}$, $\overline{AD} \cong \overline{BC}$**

7. Which angles of $\square ABCD$ are congruent? **$\angle A \cong \angle C$, $\angle B \cong \angle D$**

8. Which angles of $\square ABCD$ are supplementary?
$\angle A$, $\angle B$; $\angle B$, $\angle C$; $\angle C$, $\angle D$; $\angle D$, $\angle A$

Written Exercises

A

1. Given: $\square ABCD$ and $m\angle A = 120$ Find the measure of each of the other angles. **60°, 120°, 60°**

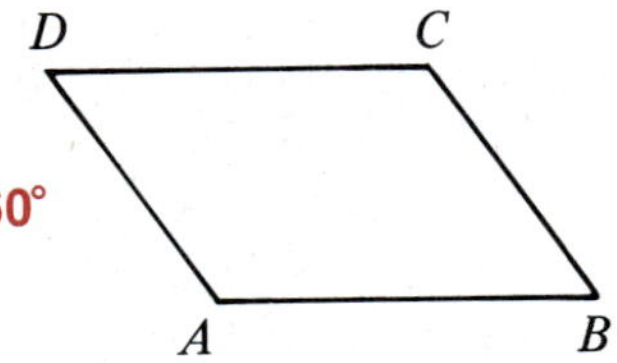

2. Given: $\square ABCD$ and $m\angle B = 53$ Find the measure of each of the other angles. **127°, 53°, 127°**

3. Given: $\square ABCD$ and $m\angle A = 3\ m\angle B$ Find $m\angle A$ and $m\angle B$. **135, 45**

4. Given: $\square ABCD$, $m\angle B = 4x - 5$, and $m\angle D = 3x + 10$ **55, 125** Find $m\angle B$ and $m\angle C$.

5. Given: $\square EFGH$, $EF = 5x - 7$, and $GH = 3x + 1$ Find GH. **13**

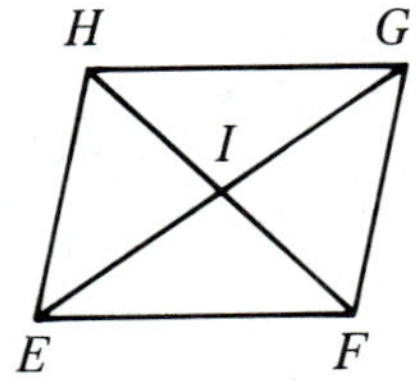

6. Given: $\square EFGH$, $EH = 3x - 5$, and $FG = 2x + 2$ Find EH. **16**

7. Given: $\square EFGH$, $EI = 7x - 4$, and $GI = 5x + 2$ Find EI. **17**

8. Given: $\square EFGH$, $FI = 4x - 6$, and $HI = 3x + 1$ Find HI. **22**

***JKLM* is a parallelogram. Which of the following pairs of triangles are congruent? Why? (Ex. 9–14)**

9. $\triangle MNJ$ and $\triangle KNL$ **≅; SAS**

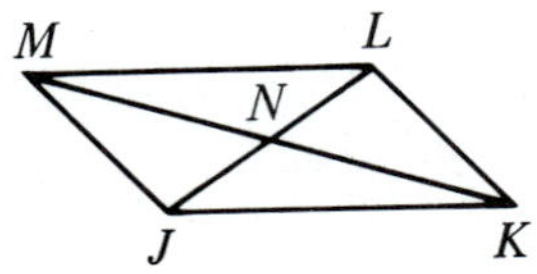

10. $\triangle MJK$ and $\triangle LKJ$ **not necess.**

11. $\triangle MNL$ and $\triangle KNJ$ **≅; SAS**

12. $\triangle MJK$ and $\triangle KLM$ **≅, Th. 7.1**

13. $\triangle MNJ$ and $\triangle KNJ$ **not necess.**

14. $\triangle JKL$ and $\triangle LMJ$ **≅, Th. 7.1**

B

15. Prove: Consecutive angles of a parallelogram are supplementary.

16. Complete the proof of Theorem 7.2.

17. Prove Corollaries 1 and 2.

18. Given: $\square ABCD$; diagonals $\overline{AC}$ and $\overline{BD}$ intersect at G.
Prove: G is the midpoint of $\overline{EF}$.

19. Given: $\square XYZW$; T is the midpoint of $\overline{WZ}$; and U is the midpoint of $\overline{XY}$.
Prove: V is the midpoint of $\overline{TU}$ and $\overline{XZ}$.

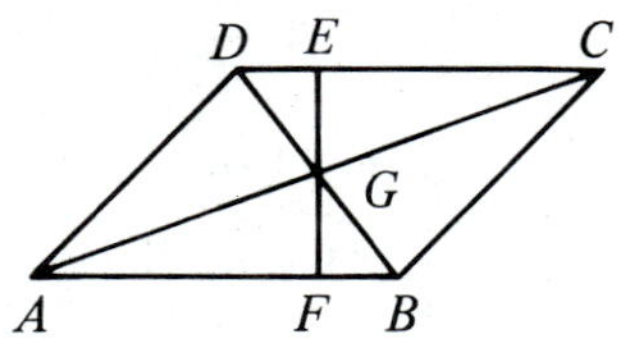

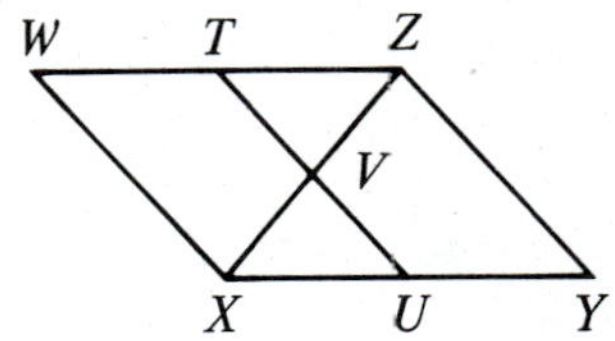

C

20. Given: $\square ABCD \parallel \square EFGH$; $\square ADHE \parallel \square BCGF$; $\square ABFE \parallel \square DCGH$; and I, J, K, and L are coplanar.
Prove: $\overline{LK} \cong \overline{IJ}$, and $\angle ILK \cong \angle IJK$

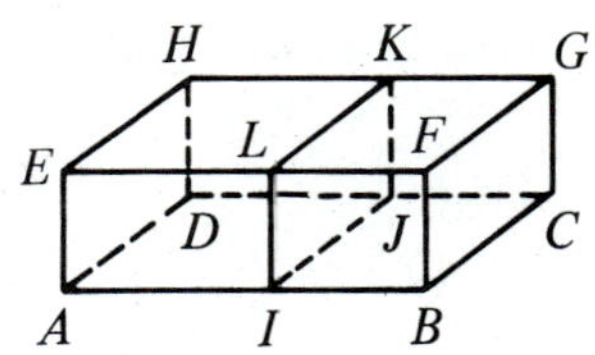

Cumulative Review

1. What is the measure of each exterior angle of a regular hexagon? $\frac{360°}{6}$; **60°**

2. What is the measure of each angle of a regular dodecagon? $\frac{(12-2)\,180°}{12}$; **150°**

3. Is a regular pentagon a convex or a concave polygon? **convex**

GEOMETRY IN DESIGN AND TECHNOLOGY

Applied geometry provides a tool for both construction and technological advance in design. Experimentation with geometric shapes allows designers to maximize the speed and the efficiency of their technological products.

Tower Bridge and Hydrofoil. A winglike device lifts the hull of a hydrofoil above the water to reduce friction and drag, which limit speed.

Shuttle, Schlieren Wind Tunnel Test. A Schlieren test using light refraction allows a designer to study the effectiveness of wing and fuselage shapes.

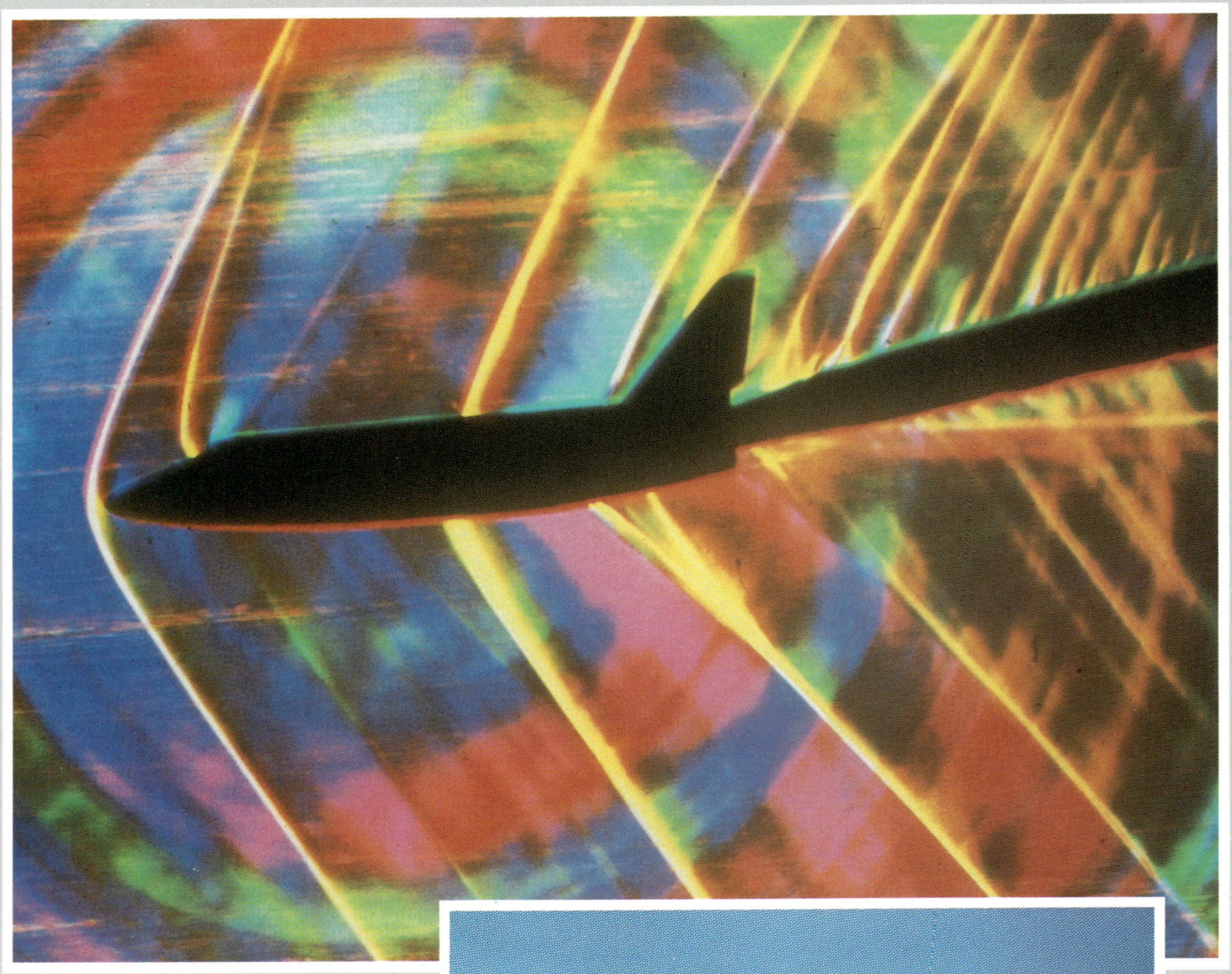

French Concorde. The French Concorde was the first supersonic airliner used for passenger service.

Bobsled. The rounded bottom of the bobsled reduces drag to near zero, which in turn causes the sled to leave the ground at very high speeds.

The Blue Flame. The Blue Flame, whose driver described it as "a long piece of pipe," holds the official land speed record of 622 mph.

Bullet Train, Japan. Japan's Shinkansen (New Railways), which travel in excess of 100 mph, possess a bulletlike styling.

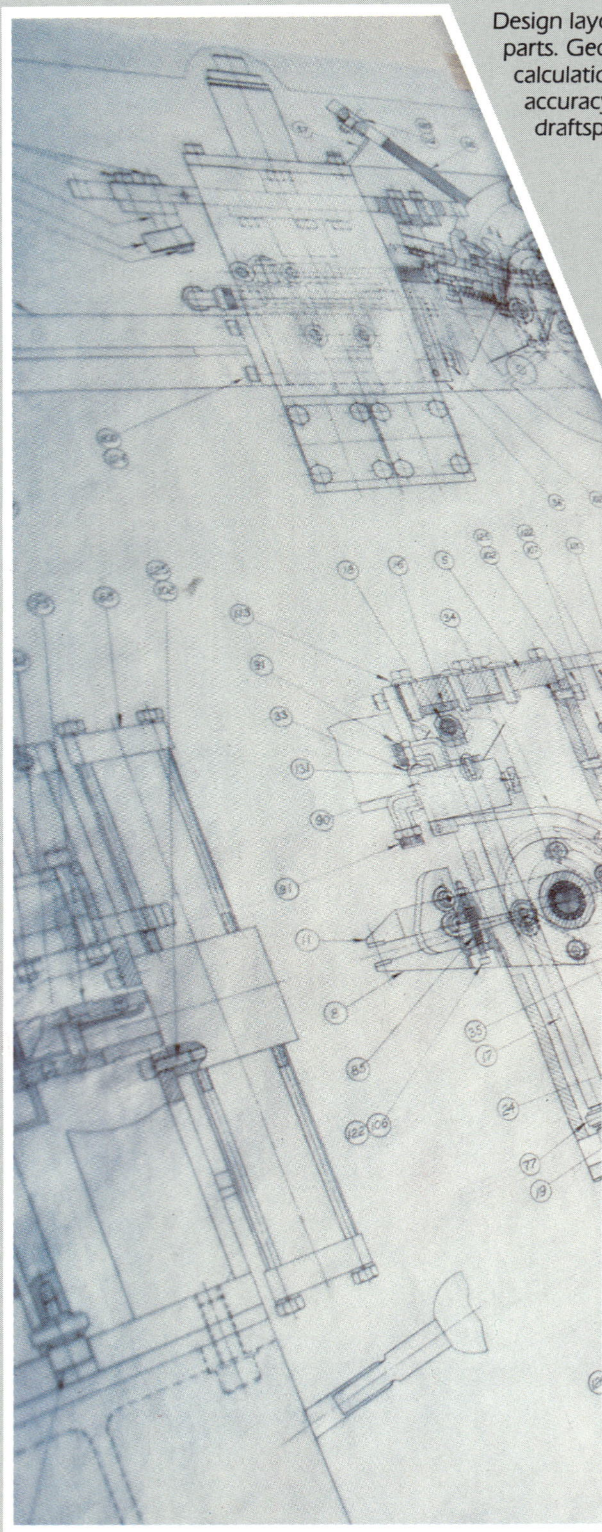

Design layout of machine parts. Geometric methods of calculation enhance the accuracy of the draftsperson.

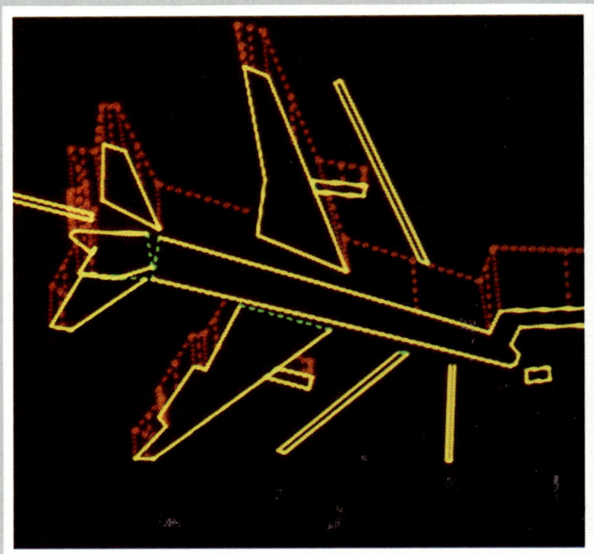

Computer graphic of airplane. Wing and support shapes can be considered individually and designs modified through computer use.

Computer-generated grid-type picture of a car. The depiction of the car as a grid of geometric forms permits design changes to be made easily.

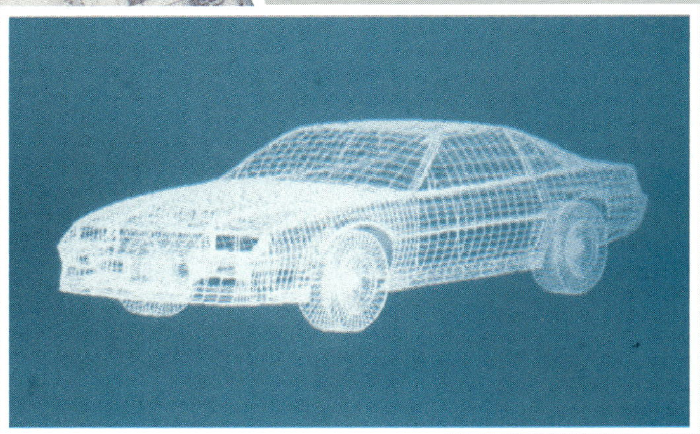

Engine pod in wind tunnel. Smaller versions of airplane prototypes are tested for aerodynamic stability in wind tunnels.

Da Vinci model of a helical screw helicopter. Da Vinci's helicopter used a helical screw in place of rotor blades.

Laser used in car design. The laser shows the flow pattern of light over the exterior of the car to check the effectiveness of its shape.

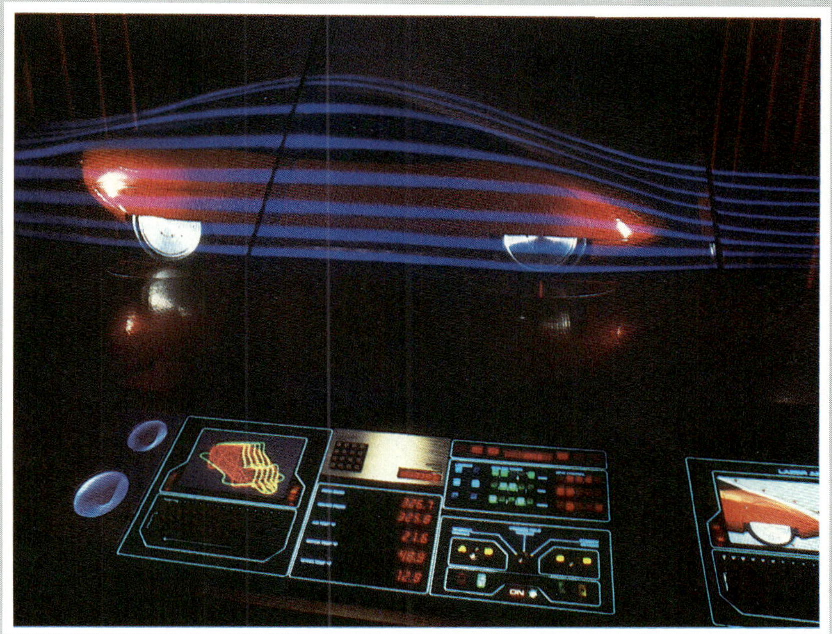

Keel of the Australian *Challenger*. The winner of the 1983 America's Cup had a bulb-nosed keel sporting swept-back delta-type fins.

Oil tanker propeller. Merchant ships usually have four-blade propellers designed to maximize the forward thrust of the ship.

Ship's hull. Hull design concentrates on insuring the stability of the ship.

Gliders in the desert. Gliders, which fly with less vibration than airplanes, are used to study air currents to benefit airplane design.

Airplane propeller. The blades of a propeller, also called an air screw, are shaped like an airplane wing.

Jet trails over jet wing. A wing's shape helps produce lift and varies with the type of plane for which it is designed.

Gas storage tanks. Cylindrical storage tanks withstand the pressures of the stored volatile gases.

Oil retaining tank. Spherical tanks contain the maximum amount of volume with the minimum amount of surface area.

Concrete silos. Early silos were square or octagonal, but round silos resist the pressure of tightly packed feed best.

PROVING PARALLELOGRAMS 7.3

Objective **To prove that some quadrilaterals are parallelograms**

You already know that certain quadrilaterals are also parallelograms. One way to prove that a quadrilateral is a parallelogram is to show that the definition of a parallelogram is satisfied. In other words, you must show that each pair of opposite sides of the quadrilateral is parallel.

For example, if you show that $\overline{AB} \parallel \overline{DC}$ and $\overline{AD} \parallel \overline{BC}$, then you have proved that quadrilateral $ABCD$ is a parallelogram.

Two additional ways of proving that a quadrilateral is a parallelogram follow.

Theorem 7.3 If both pairs of opposite sides of a quadrilateral are congruent, then the quadrilateral is a parallelogram.

Given: $\overline{AB} \cong \overline{CD}$ and $\overline{AD} \cong \overline{CB}$
Prove: $ABCD$ is a parallelogram.

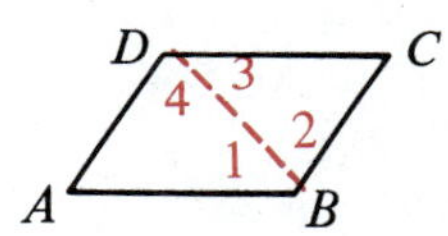

ANALYSIS
Draw $\overline{BD}$. Show opposite sides $\parallel$ by using $\cong$ alt. int. $\angle$'s.

Proof

Statements		**Reasons**
1. Draw $\overline{BD}$.		1. Two points are contained in exactly one line.
2. $\overline{AB} \cong \overline{CD}$	(S)	2. Given
3. $\overline{AD} \cong \overline{CB}$	(S)	3. Given
4. $\overline{BD} \cong \overline{DB}$	(S)	4. Reflexive property
5. $\triangle ABD \cong \triangle CDB$		5. SSS
6. $\angle 1 \cong \angle 3$ and $\angle 2 \cong \angle 4$		6. CPCTC
7. $\overline{AB} \parallel \overline{DC}$ and $\overline{AD} \parallel \overline{BC}$		7. If alt. int. $\angle$'s are $\cong$, then lines are $\parallel$.
8. $\therefore ABCD$ is a ▱.		8. Def. of ▱

Theorem 7.4 If both pairs of opposite angles of a quadrilateral are congruent, then the quadrilateral is a parallelogram.

Given: $\angle A \cong \angle C$ and $\angle B \cong \angle D$
Prove: $ABCD$ is a parallelogram.

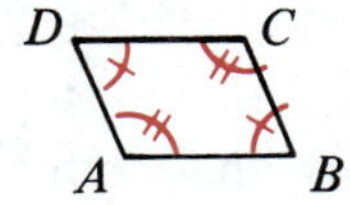

(See Exercise 17.)

ANALYSIS
Show opposite sides $\parallel$ by proving that the adj. $\angle$'s are supplementary.

A fourth way to prove that a quadrilateral is a parallelogram is to show that two sides are both parallel and congruent.

Theorem 7.5

If two sides of a quadrilateral are parallel and congruent, then the quadrilateral is a parallelogram.

Given: $\overline{AB} \parallel \overline{CD}$ and $\overline{AB} \cong \overline{CD}$
Prove: $ABCD$ is a parallelogram.

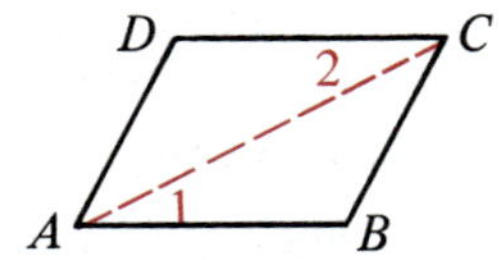

ANALYSIS
Show that $\overline{BC} \cong \overline{DA}$ by proving $\triangle ABC \cong \triangle CDA$. Apply Th. 7.3.

Proof

Statements		Reasons
1. Draw $\overline{AC}$.		1. Two points are contained in exactly one line.
2. $\overline{AB} \cong \overline{CD}$	(S)	2. Given
3. $\overline{AB} \parallel \overline{CD}$		3. Given
4. $\angle 1 \cong \angle 2$	(A)	4. Alt. int. $\angle$'s of $\parallel$ lines are $\cong$.
5. $\overline{AC} \cong \overline{CA}$	(S)	5. Reflexive property
6. $\triangle ABC \cong \triangle CDA$		6. SAS
7. $\overline{BC} \cong \overline{DA}$		7. CPCTC
8. $\therefore ABCD$ is a $\square$.		8. If both pairs of opp. sides are $\cong$, then quad. is a $\square$.

You can also prove that a quadrilateral is a parallelogram by showing that the diagonals bisect each other. You should complete the proof of this theorem as an exercise.

Theorem 7.6

If the diagonals of a quadrilateral bisect each other, then the quadrilateral is a parallelogram.
(See Exercise 18.)

Summarized below are five ways to prove that a quadrilateral is a parallelogram. Each of these is a *sufficient condition* for proving that a given quadrilateral is a parallelogram; a quadrilateral that satisfies any one of these five conditions is a parallelogram.

Summary

To prove that a quadrilateral is a parallelogram:

1. Show that both pairs of opposite sides are parallel.
2. Show that both pairs of opposite sides are congruent.
3. Show that both pairs of opposite angles are congruent.
4. Show that two sides are congruent and parallel.
5. Show that the diagonals bisect each other.

You can now use the methods just summarized to determine if a given quadrilateral is a parallelogram.

Example 1 If a diagonal of a quadrilateral forms two congruent triangles, is the quadrilateral a parallelogram?

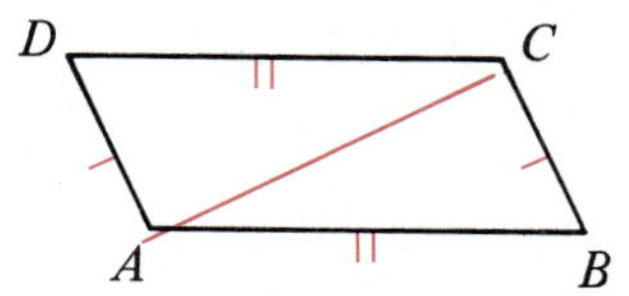

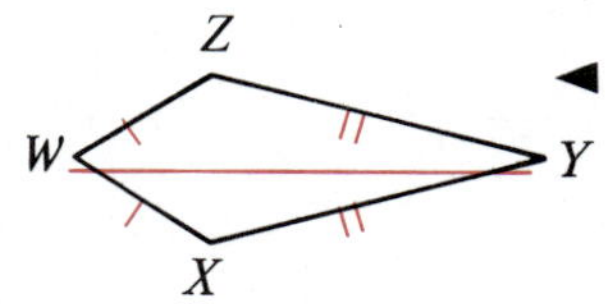

◄ *WXYZ is called a kite.*

$\triangle ABC \cong \triangle CDA$
$ABCD$ is a parallelogram.

$\triangle WXY \cong \triangle WZY$
$WXYZ$ is not a parallelogram.

Thus, the quadrilateral is not necessarily a parallelogram.

Example 2 In quadrilateral $ABCD$, $AO = CO = 8$ and $BO = DO = 7$. Is $ABCD$ a parallelogram? Why?

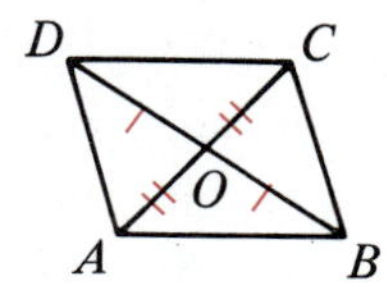

Since $AO = CO$ and $BO = DO$, the diagonals bisect each other.

Thus, $ABCD$ is a parallelogram. ◄ *Use Th. 7.6.*

Reading in Geometry

You learned that the converse of a statement is formed by interchanging the "if" and the "then" clauses. Write the converse of each of the seven statements below. Pick out the statements and converses that were proved as theorems. Give the theorem numbers of each.

1. If the diagonals of a quadrilateral bisect each other, then the quadrilateral is a parallelogram.

If a quad. is a ▱, then the diag. of the quad. bis. each other. S:Th. 7.6 C: Th. 7.2

2. If both pairs of opposite angles of a quadrilateral are congruent, then the quadrilateral is a parallelogram.

If a quad. is a ▱, then both pairs of opp. ∠'s of the quad. are ≅ S:Th. 7.4 C: Th. 7.1 Corr. 2

3. If a diagonal of a quadrilateral forms two congruent triangles, then the quadrilateral is a parallelogram.

If a quad. is a ▱, then a diag. of the quad. forms two ≅ △'s. S: not proved; C: Th. 7.1

4. If both pairs of opposite sides of a quadrilateral are congruent, then the quadrilateral is a parallelogram.

If a quad. is a ▱, then both pairs of opp. S's of the quad. are ≅. S: Th. 7.3 C: Th. 7.1 Corr. 1

5. If the diagonals of a quadrilateral are perpendicular, then the quadrilateral is a parallelogram.

If a quad. is a ▱, then the diag. of the quad. are ⊥. S: not proved; C: not proved

6. If two sides of a quadrilateral are congruent and parallel, then the quadrilateral is a parallelogram.

If a quad. is a ▱, then two S's of the quad. are ∥ and ≅. S: Th. 7.5.

7. If the diagonals of a quadrilateral are congruent, then the quadrilateral is a parallelogram.

If a quad. is a ▱, then the diag. of the quad. are ≅. S: not proved; C: not proved

Written Exercises

Based only on the markings shown, which of these quadrilaterals are parallelograms?

A

1.

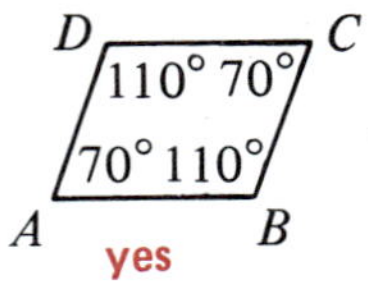

yes

2.

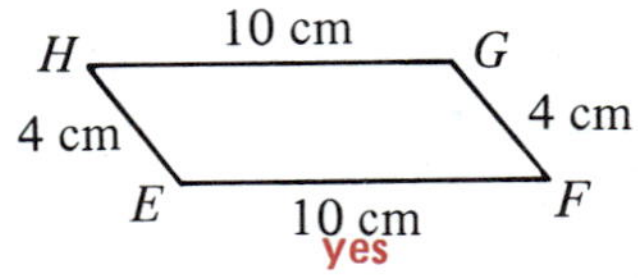

yes

3.

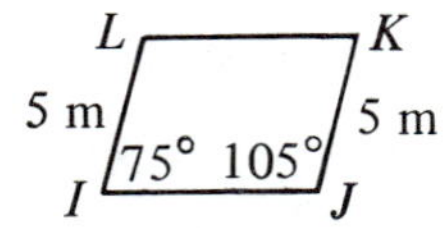

yes

4.

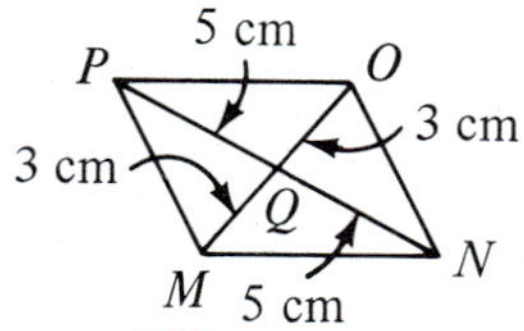

yes

5.

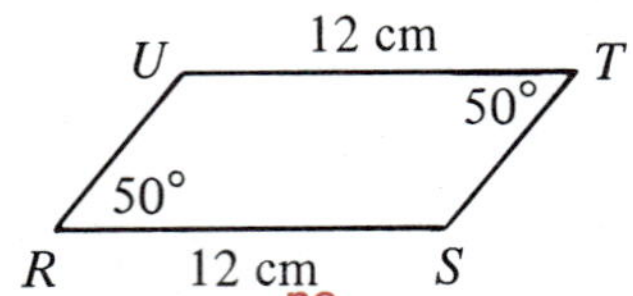

no

6.

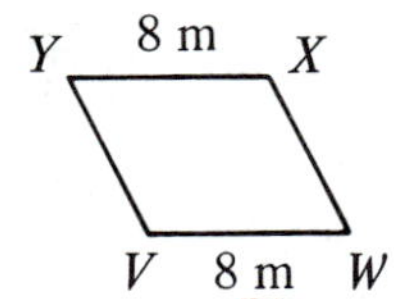

no

Based only on the given information, is $ABCD$ a parallelogram? Give a reason for your answer. (Ex. 7–16)

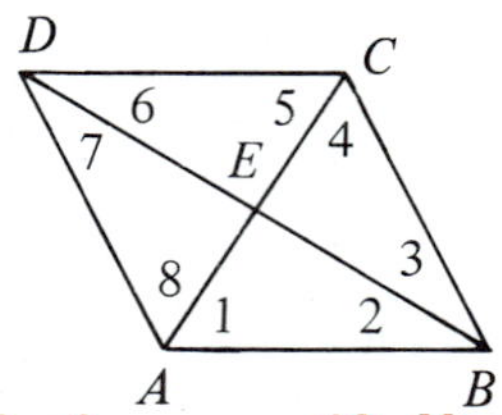

7. $m\angle DAB = 120$ and $m\angle BCD = 120$ No
8. $AB = 12$, $DC = 12$, $\overline{AB} \parallel \overline{DC}$ Yes, Th 7.5
9. $\overline{AB} \cong \overline{DC}$ and $\overline{AB} \parallel \overline{DC}$
10. $\angle 1 \cong \angle 5$ and $\angle 2 \cong \angle 6$ No
11. $\angle 4 \cong \angle 8$ and $\angle 1 \cong \angle 5$
12. $\triangle ABD \cong \triangle ADC$ No
13. $\overline{AB} \parallel \overline{DC}$ and $\overline{AD} \parallel \overline{BC}$
14. $\overline{AB} \cong \overline{DC}$ and $\overline{AD} \cong \overline{BC}$ Yes, Th 7.3
15. $\overline{AE} \cong \overline{EC}$ and $\overline{BE} \cong \overline{ED}$
16. $\triangle AEB \cong \triangle CED$ Yes, Th. 7.6

9. Yes, Th. 7.5 11. Yes, Def. of ▱ 13. Yes, Def. of ▱ 15. Yes, Th. 7.6

B

17. Prove Theorem 7.4.

18. Prove Theorem 7.6.

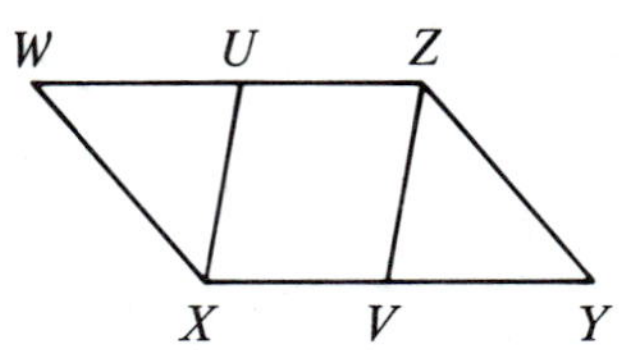

19. Given: $\overline{UZ} \cong \overline{XV}$ and $\triangle XUW \cong \triangle ZVY$
Prove: $UXVZ$ is a ▱.

20. Given: ▱$XYZW$ and $\overline{WU} \cong \overline{YV}$
Prove: $UXVZ$ is a ▱.

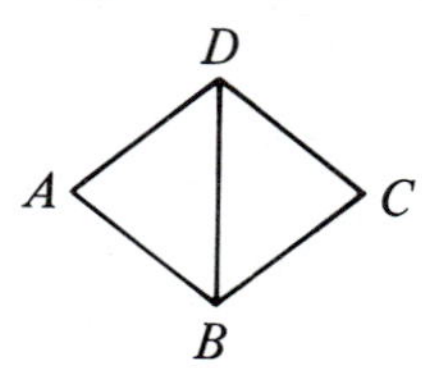

21. Given: $\triangle ABD$ is equilateral and $\triangle CDB$ is equilateral
Prove: $ABCD$ is a ▱.

22. Given: $\triangle ABD$ is equiangular and $\triangle CDB$ is equiangular
Prove: $ABCD$ is a ▱.

Give a proof or counterexample to defend your answer.

23. If consecutive angles of a quadrilateral are supplementary, is the quadrilateral a parallelogram? yes

24. If the diagonals of a quadrilateral are perpendicular, is the quadrilateral a parallelogram? not necessarily

25. If the diagonals of a quadrilateral are congruent, is the quadrilateral a parallelogram? not necessarily

26. If the diagonals of a quadrilateral bisect its angles, is the quadrilateral a parallelogram? yes

C

27. If the bisectors of a pair of opposite angles of a quadrilateral form a parallelogram with two sides of the quadrilateral, is the quadrilateral a parallelogram? yes

28. If the segments joining the midpoints of consecutive sides of a quadrilateral form a parallelogram, is the quadrilateral a parallelogram? not necessarily

A SPECIAL SEGMENT 7.4

Objective

To prove and apply the theorem about the segment joining the midpoints of two sides of a triangle

You can use the properties of parallelograms to prove a theorem about a special segment in a triangle.

Theorem 7.7

The segment joining the midpoints of two sides of a triangle is parallel to the third side, and its length is half the length of the third side.

Given: $\triangle ABC$, D is the midpoint of $\overline{AC}$, and E is the midpoint of $\overline{BC}$.
Prove: $\overline{DE} \parallel \overline{AB}$,
$DE = \frac{1}{2}AB$

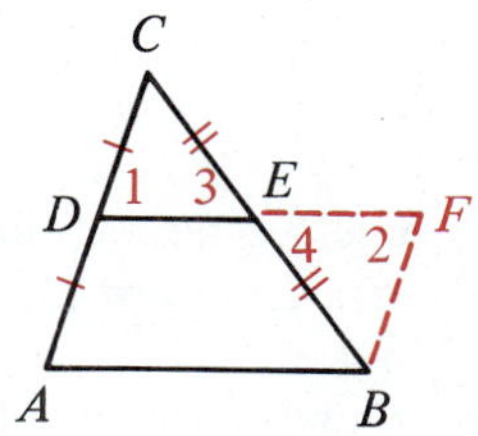

ANALYSIS
Form ▱ and use properties of ▱'s.

Proof

Statements		Reasons
1. $\triangle ABC$		1. Given
2. Draw $\overline{EF} \cong \overline{DE}$ so E is between D and F.	(S)	2. $\cong$ segment postulate
3. Draw $\overline{BF}$.		3. Two pt. are contained in exactly 1 line.
4. $\angle 4 \cong \angle 3$	(A)	4. Vertical $\angle$'s are $\cong$.
5. E is the midpoint of $\overline{BC}$.		5. Given
6. $\overline{BE} \cong \overline{CE}$	(S)	6. Def. of midpoint
7. $\triangle FEB \cong \triangle DEC$		7. SAS
8. $\angle 1 \cong \angle 2$ and $\overline{CD} \cong \overline{BF}$		8. CPCTC
9. D is the midpoint of $\overline{AC}$.		9. Given
10. $\overline{CD} \cong \overline{DA}$		10. Def. of midpoint
11. $\overline{BF} \cong \overline{DA}$		11. Substitution
12. $\overline{BF} \parallel \overline{DA}$		12. Lines are $\parallel$ if alt. int. $\angle$'s are $\cong$.
13. $ABFD$ is a ▱.		13. Quad. is ▱ if 2 sides $\parallel$ and $\cong$.
14. $\overline{DF} \parallel \overline{AB}$ and $\therefore$ $\overline{DE} \parallel \overline{AB}$		14. Def. of ▱
15. $\overline{DF} \cong \overline{AB}$ and $DF = AB$		15. Opp. sides of ▱ are $\cong$; def. of $\cong$ seg.
16. $DF = DE + EF$		16. Def. of *between*
17. $DE = \frac{1}{2}DF$		17. Step 2, step 16, algebraic properties
18. $\therefore$ $DE = \frac{1}{2}AB$		18. Substitution

You can now use Theorem 7.7 to find the lengths of segments associated with triangles. Examples 1 and 2 illustrate two such applications.

Example 1 Given: E is the midpoint of $\overline{AB}$,
D is the midpoint of $\overline{AC}$, and $ED = 6$.
Find BC.

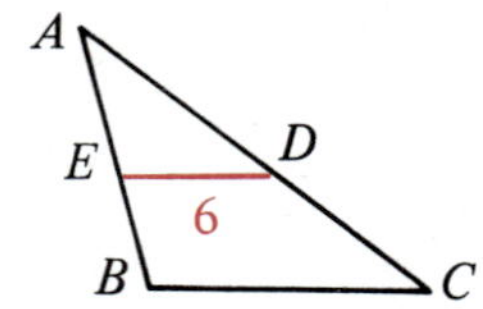

Use Th. 7.7. ► $6 = \frac{1}{2}(BC)$ **Thus,** $BC = 12$.

Example 2 Given: $\overline{AD} \cong \overline{DB}$, $\overline{AE} \cong \overline{EC}$,
$DE = 2x + 3$, and $BC = 5x + 4$.
Find DE and BC.

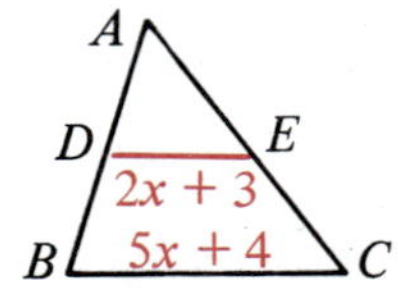

Use Th. 7.7. ►

$$2x + 3 = \frac{1}{2}(5x + 4)$$
$$2(2x + 3) = 5x + 4$$
$$4x + 6 = 5x + 4$$
$$2 = x$$

Thus, $DE = 7$ and $BC = 14$.

Example 3 Prove: The segments joining the midpoints of consecutive sides of a quadrilateral form a parallelogram.

Given: Quad. $ABCD$ with E, F, G, and H the midpoints of $\overline{AB}$, $\overline{BC}$, $\overline{CD}$, and $\overline{DA}$, respectively.
Prove: $EFGH$ is a ▱.

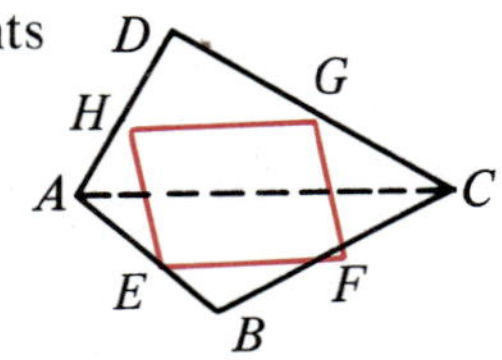

Proof

Statements	Reasons
1. Draw $\overline{AC}$, forming $\triangle ABC$ and $\triangle ADC$.	1. Two pts. are contained in exactly one line; def. of △
2. Quad. $ABCD$ with E, F, G, H the midpoints of $\overline{AB}$, $\overline{BC}$, $\overline{CD}$, and $\overline{DA}$, resp.	2. Given
3. $\overline{HG} \parallel \overline{AC}$ and $\overline{EF} \parallel \overline{AC}$	3. Seg. joining midpts. of 2 sides of △ is ∥ to 3rd side.
4. $\overline{HG} \parallel \overline{EF}$	4. Lines ∥ to same line are ∥.
5. $HG = \frac{1}{2} AC$ and $EF = \frac{1}{2} AC$	5. Seg. joining midpts. of 2 sides of △ has length $\frac{1}{2}$ length of 3rd side.
6. $HG = EF$ and $\overline{HG} \cong \overline{EF}$	6. Substitution and def. of ≅ segs.
7. ∴ $EFGH$ is a ▱.	7. Quad. is ▱ if 2 sides are ∥ and ≅.

Oral Exercises

Find each indicated length if V is the midpoint of $\overline{XZ}$ and W is the midpoint of $\overline{YZ}$.

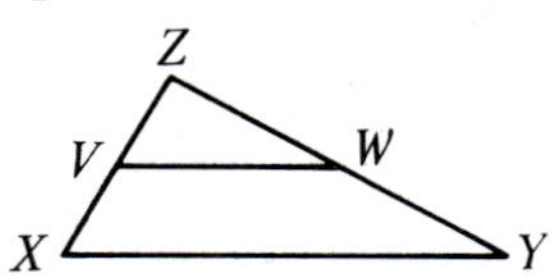

1. If $VW = 8$, then $XY = ?$ **16**
2. If $XY = 18$, then $VW = ?$ **9**
3. If $VW = 3$, then $XY = ?$ **6**
4. If $XY = 4$, then $VW = ?$ **2**
5. If $VW = 2\frac{1}{2}$, then $XY = ?$ **5**
6. If $XY = 9$, then $VW = ?$ **$4\frac{1}{2}$**

Written Exercises

If W, V, and U are the midpoints of the sides of $\triangle XYZ$, then complete each statement. (Ex. 1–6)

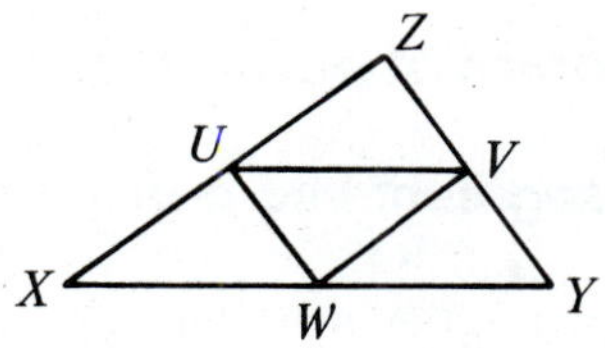

A 1. If $XZ = 10$, then $WV = ?$ 5

2. If $UW = t$, then $YZ = ?$ 2t

3. $\overline{VW} \parallel ?$ $\overline{XZ}$

4. $\overline{XY} \parallel ?$ $\overline{UV}$

5. If $XW = 2$, then $UV = ?$ 2

6. If $ZV = t$, then $UW = ?$ t

7. Given: M and N are the midpoints of $\overline{JL}$ and $\overline{KL}$. $JK = 6x - 4$, and $MN = 2x + 1$
Find MN. 7

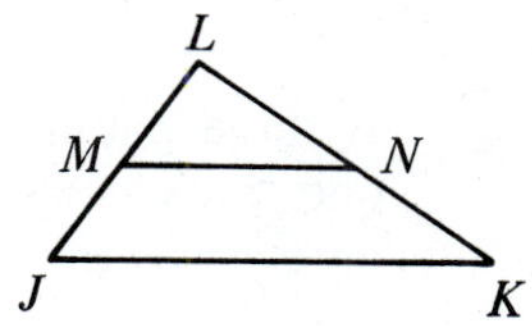

8. Given: M and N are the midpoints of $\overline{JL}$ and $\overline{KL}$, $JK = 3x - 8$, and $MN = x + 4$
Find JK. 40

9. Given: S, T, and U are the midpoints of $\overline{RP}$, $\overline{PQ}$, and $\overline{QR}$, respectively.
Prove: $RSTU$ is a ▱.

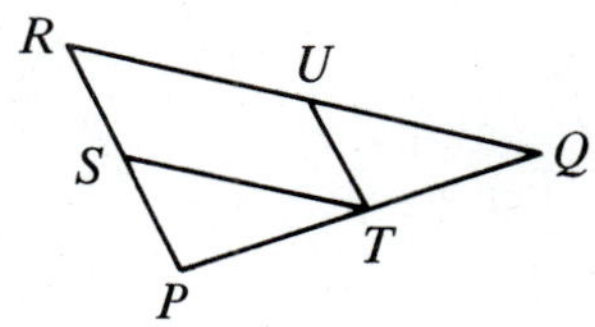

10. Given: S, T, and U are the midpoints of $\overline{RP}$, $\overline{PQ}$, and $\overline{QR}$, respectively.
Prove: $\triangle SPT \cong \triangle UTQ$.

B 11. Given: E and F are the midpoints of $\overline{DA}$ and $\overline{DB}$, respectively. $\overline{FC} \cong \overline{EF}$.
Prove: $EABC$ is a ▱.

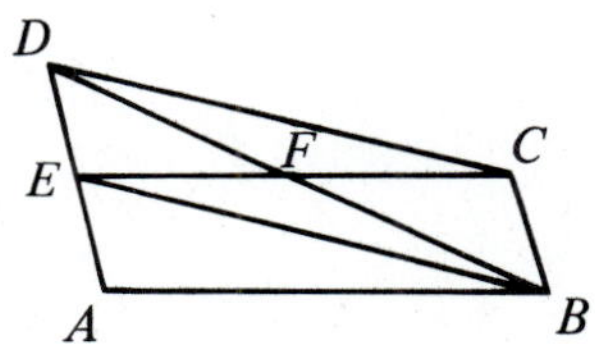

12. Given: E and F are the midpoints of $\overline{DA}$ and $\overline{DB}$, respectively. $\overline{EC} \cong \overline{AB}$.
Prove: $DEBC$ is a ▱.

13. Given: K and M are the midpoints of $\overline{JG}$ and $\overline{IH}$, respectively, and L is the midpoint of $\overline{KM}$ and $\overline{JH}$.
Prove: $GHIJ$ is a ▱.

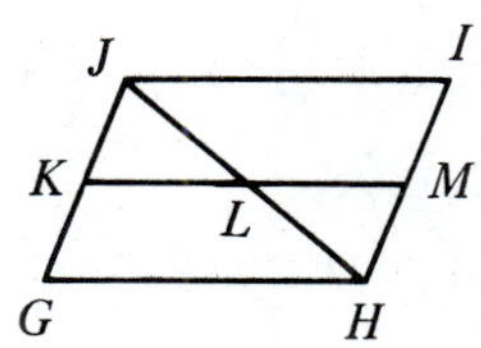

14. Given: ▱$GHIJ$, K and L are the midpoints of $\overline{JG}$ and $\overline{JH}$, respectively.
Prove: $GHMK$ is a ▱.

C 15. Prove: The segments joining the midpoints of the opposite sides of a quadrilateral bisect each other.

16. Prove: The opposite angles of the quadrilateral formed by the bisectors of the angles of a convex quadrilateral are supplementary.

17. Given: $\overline{QP} \parallel \overline{NO}$, R, S, and T are midpoints of $\overline{QN}$, $\overline{QO}$, and $\overline{PO}$, respectively.
Prove: $RT = \frac{1}{2}(QP + NO)$

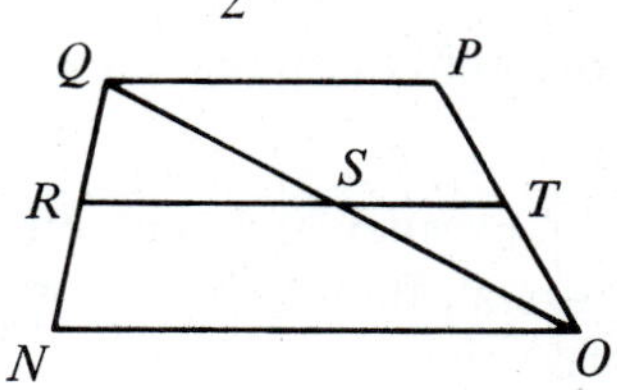

18. Given: $\overline{AD} \parallel \overline{BE} \parallel \overline{CF}$, $\overline{AD} \cong \overline{BE} \cong \overline{CF}$
Prove: $\angle EDF \cong \angle BAC$

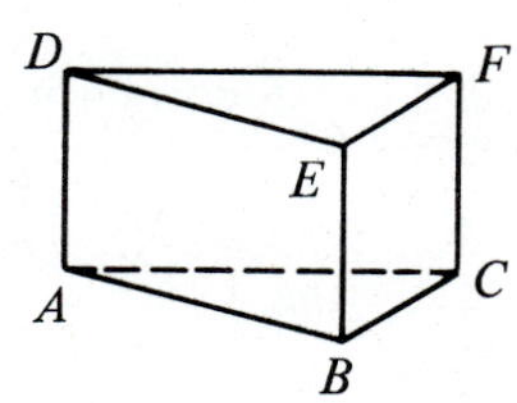

PARALLEL LINES AND CONGRUENT SEGMENTS 7.5

Objectives

To use properties of parallelograms to prove theorems about parallel lines
To divide a segment into a given number of congruent segments

You can use properties of parallelograms to prove theorems about parallel lines.

Theorem 7.8 If three parallel lines cut off congruent segments on one transversal, then they cut off congruent segments on every transversal.

Given: $l \parallel m \parallel n$, transversals $\overleftrightarrow{AC}$ and $\overleftrightarrow{DF}$, and $\overline{CB} \cong \overline{BA}$
Prove: $\overline{EF} \cong \overline{DE}$

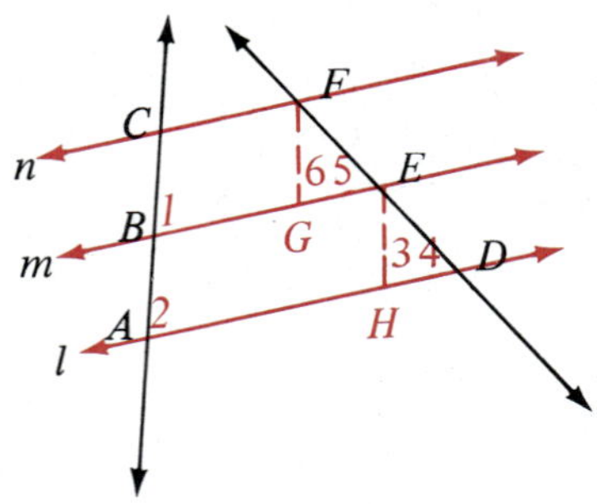

ANALYSIS
Make ▱'s by drawing $\overline{FG} \parallel \overline{CB}$ and $\overline{EH} \parallel \overline{BA}$. Show $\triangle FGE \cong \triangle EHD$.

Proof

Statements		Reasons
1. $l \parallel m \parallel n$, trans. $\overleftrightarrow{AC}$ and $\overleftrightarrow{DF}$		1. Given
2. Draw $\overline{FG} \parallel \overline{CB}$ and $\overline{EH} \parallel \overline{BA}$.		2. Through a given pt., there is exactly one line $\parallel$ to a given line.
3. $AHEB$ and $BGFC$ are ▱'s.		3. Def. of ▱
4. $\overline{FG} \cong \overline{CB}$ and $\overline{EH} \cong \overline{BA}$		4. Opp. sides of a ▱ are ≅.
5. $\overline{CB} \cong \overline{BA}$		5. Given
6. $\overline{EH} \cong \overline{FG}$	(S)	6. Substitution property
7. $\angle 6 \cong \angle 1$, $\angle 3 \cong \angle 2$		7. Corr. ∠'s of $\parallel$ lines are ≅.
8. $\angle 1 \cong \angle 2$		8. Corr. ∠'s of $\parallel$ lines are ≅.
9. $\angle 6 \cong \angle 3$	(A)	9. Substitution property
10. $\angle 5 \cong \angle 4$	(A)	10. Corr. ∠'s of $\parallel$ lines are ≅.
11. $\triangle FGE \cong \triangle EHD$		11. AAS
12. $\therefore \overline{EF} \cong \overline{DE}$		12. CPCTC

A corollary of Theorem 7.8 extends the theorem to more than three parallel lines.

Corollary If any given number of parallel lines cut off congruent segments on one transversal, then they cut off congruent segments on every transversal.

One application of Theorem 7.8 and its corollary is to find unknown lengths of segments cut off by parallel lines.

Example 1 Given: $\overleftrightarrow{AD} \parallel \overleftrightarrow{BE} \parallel \overleftrightarrow{CF}$, $AB = 8$, $BC = 8$, and $DE = 7$.
Find DF.

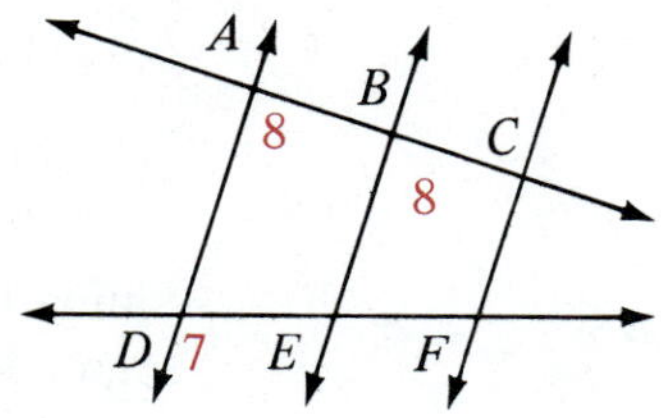

$EF = DE = 7$ ◄ *Use Th. 7.8.*
Thus, $DF = 7 + 7$, or 14.

Example 2 Given: $\overleftrightarrow{GJ} \parallel \overleftrightarrow{HK} \parallel \overleftrightarrow{IL}$, $GH = 15$, $HI = 15$, $JK = 3x - 9$, and $KL = 2x - 2$.
Find JL.

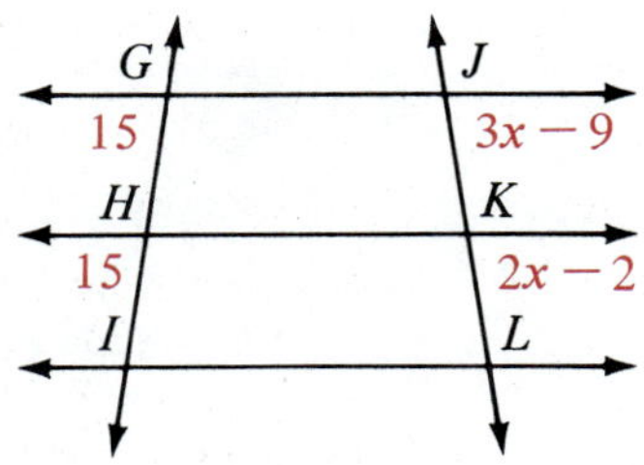

Use Th. 7.8. ► $3x - 9 = 2x - 2$
$x = 7$

$JK = 3 \cdot 7 - 9$, or 12
$KL = 2 \cdot 7 - 2$, or 12
Thus, $JL = 12 + 12$, or 24.

Another application of Theorem 7.8 and its corollary is a construction in which a given segment can be divided into any specified number of congruent segments.

Construction To divide $\overline{AB}$ into three congruent segments.

Given: $\overline{AB}$
Construct: Points P and Q on $\overline{AB}$ so that $\overline{AP} \cong \overline{PQ} \cong \overline{QB}$.

Strategy: Apply corollary to Th. 7.8 by constructing parallel lines.

Draw $\overrightarrow{AX}$.

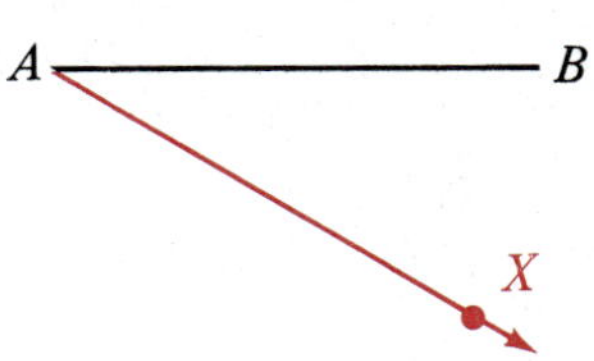

Set compass at any convenient opening. On $\overrightarrow{AX}$, construct $\overline{AC} \cong \overline{CD} \cong \overline{DE}$.

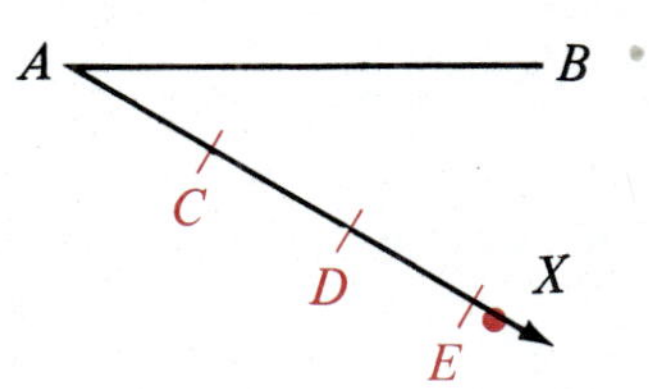

Draw $\overline{BE}$. Construct $\overline{CP} \parallel \overline{BE}$ and $\overline{DQ} \parallel \overline{BE}$.

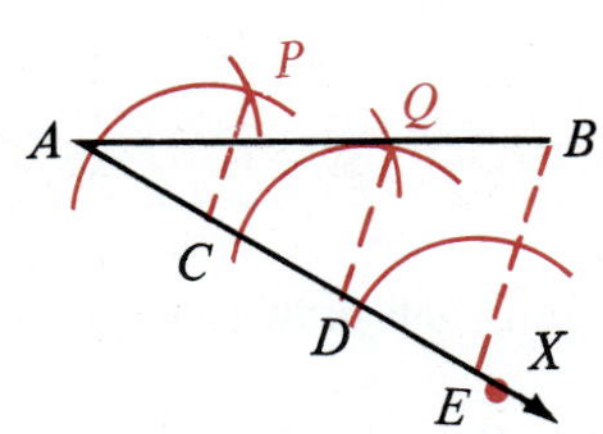

Conclusion: $\overline{AP} \cong \overline{PQ} \cong \overline{QB}$

If you were asked to find the shortest path between points A and B, you probably would select the straight one.

Postulate 19 The shortest path joining two points is a segment.

Similarly, if you were asked to find the shortest segment from point P to line l, you would choose $\overline{PQ}$.

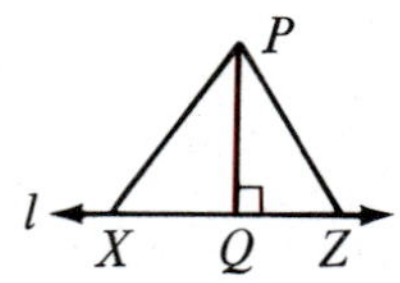

Postulate 20 The shortest segment joining a point and a line is the segment perpendicular to the line.

The length of this shortest segment is called the **distance** from a point to a line. This can be used to define the distance between two parallel lines.

Definition: Distance between parallel lines The *distance between two parallel lines* is the distance from a point on one line to the other line.

You should complete the proof of the following theorem.

Theorem 7.9 Parallel lines are equidistant at all *points*.

Given: $l \parallel m$, $\overline{PX} \perp m$ and $\overline{QY} \perp m$
Prove: $PX = QY$

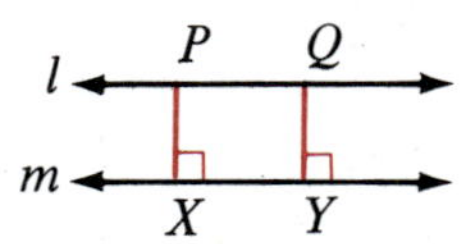

(See Exercise 25.)

ANALYSIS
$\overline{PX}$ and $\overline{QY}$ are opp. sides of a ▱.

Oral Exercises

Find each indicated length if $l \parallel m \parallel n$, $AD = 6$, and $DG = 6$.

1. If $CF = 5$, then $FI = ?$ 5
2. If $BE = 6$, then $EH = ?$ 6
3. If $BE = 4$, then $BH = ?$ 8
4. If $CF = 7$, then $CI = ?$ 14
5. If $FI = 9$, then $CF = ?$ 9
6. If $EH = 3$, then $BH = ?$ 6
7. If $CI = 12$, then $CF = ?$ 6
8. If $BH = 9$, then $BE = ?$ 4.5

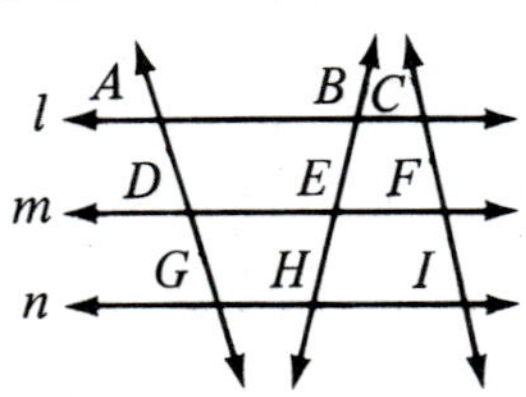

Written Exercises

Find each indicated length if $l \parallel m \parallel n \parallel r \parallel s \parallel t$ and $\overline{LK} \cong \overline{KJ} \cong \overline{JI} \cong \overline{IH} \cong \overline{HG}$. (Ex. 1–10)

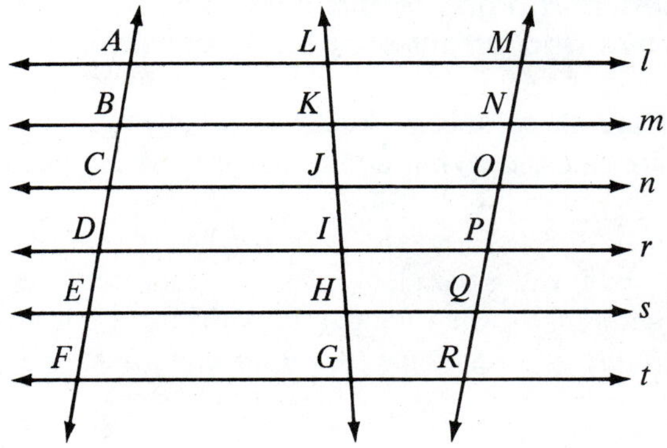

A

1. If $AB = 4$, then $CD = ?$ 4
2. If $OP = 9$, then $QR = ?$ 9
3. If $NO = 6$, then $NP = ?$ 12
4. If $EF = 8$, then $CF = ?$ 24
5. If $CD = 3$, then $BF = ?$ 12
6. If $PQ = 5$, then $MR = ?$ 25
7. If $BF = 36$, then $CD = ?$ 9
8. If $NQ = 15$, then $OQ = ?$ 10
9. If $CE = 5$, then $AD = ?$ $7\frac{1}{2}$
10. If $MP = 8$, then $NR = ?$ $10\frac{2}{3}$

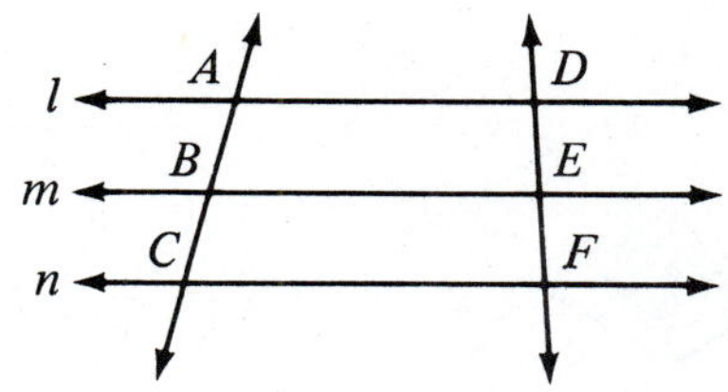

11. Given: $l \parallel m \parallel n$, $\overline{AB} \cong \overline{BC}$, $DE = 4x - 7$, and $EF = 2x + 1$
Find DE. 9
12. Given: $l \parallel m \parallel n$, $\overline{AB} \cong \overline{BC}$, $DE = 2x + 13$, and $EF = 5x + 1$
Find DF. 42
13. Given: $l \parallel m \parallel n$, $\overline{DE} \cong \overline{EF}$, $AC = 7x - 2$, and $BC = 3x + 4$
Find AB. 34
14. Given: $l \parallel m \parallel n$, $\overline{DE} \cong \overline{EF}$, $AB = 2x - 1$, and $AC = 7x - 8$
Find AC. 6

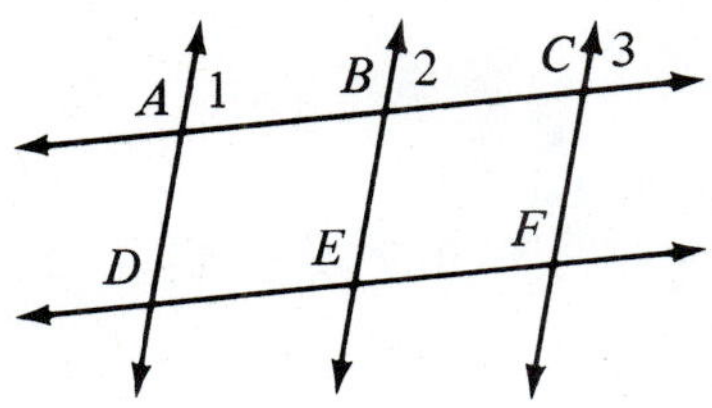

15. Given: $\angle 1 \cong \angle 2 \cong \angle 3$ and $\overline{AB} \cong \overline{BC}$
Prove: $\overline{DE} \cong \overline{EF}$.
16. Given: $\overline{AB} \parallel \overline{DE}$, $\overline{BE} \parallel \overline{CF}$, $\overline{AB} \cong \overline{DE}$, and $\overline{AB} \cong \overline{BC}$
Prove: $\overline{DE} \cong \overline{EF}$.
17. Given: $\angle 1 \cong \angle 2$, $\overleftrightarrow{BE} \parallel \overleftrightarrow{CF}$, and $\overline{DE} \cong \overline{EF}$
Prove: $\overline{AB} \cong \overline{BC}$.
18. Given: $\overleftrightarrow{AD} \parallel \overleftrightarrow{BE} \parallel \overleftrightarrow{CF}$, $\overline{AB} \cong \overline{BC}$, and $\overline{AB} \cong \overline{DE}$
Prove: $\overline{AB} \cong \overline{EF}$.
19. Draw a segment 10 cm long. Divide it into three congruent segments.
20. Draw a segment 10 cm long. Divide it into four congruent segments.
21. Draw a segment 10 cm long. Divide it into five congruent segments.

B

22. Draw a segment 8 cm long. Construct a segment two-thirds as long.
23. Draw a segment 12 cm long. Construct a segment four-fifths as long.
24. Draw a segment 8 cm long. Construct a segment four-thirds as long.
25. Prove Theorem 7.9.
26. Prove Corollary 7.8 for four parallel lines.

True or false? If true, prove the statement. If false, show a counterexample.

C

27. If three lines cut off congruent segments on each of two transversals, then the lines are parallel. F
28. If a segment is parallel to one side of a triangle and bisects a second side, then it bisects the third side. T
29. If two lines are equidistant at all points, then the lines are parallel. T
30. If a segment joining a point on each of two sides of a triangle is half as long as the third side, then it is parallel to the third side. F

RHOMBI, RECTANGLES, AND SQUARES 7.6

Objectives **To identify rhombi, rectangles, and squares**
To prove and apply properties of rhombi, rectangles, and squares

There are three special kinds of parallelograms. While all three have parallel opposite sides, each has additional properties which distinguish it from the others.

Definition: Rhombus, rectangle, square

A *rhombus* is a parallelogram with four congruent sides.
A *rectangle* is a parallelogram with four right angles.
A *square* is a rectangle with four congruent sides.

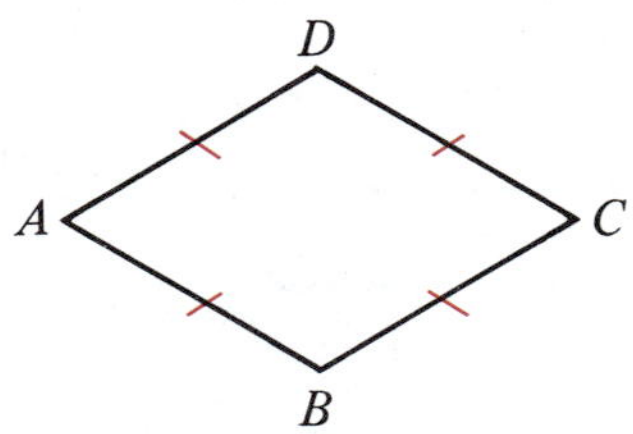

Rhombus $ABCD$

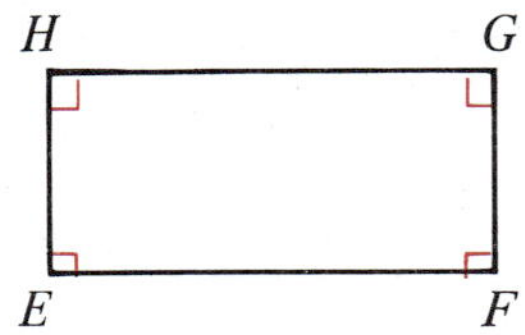

Rectangle $EFGH$

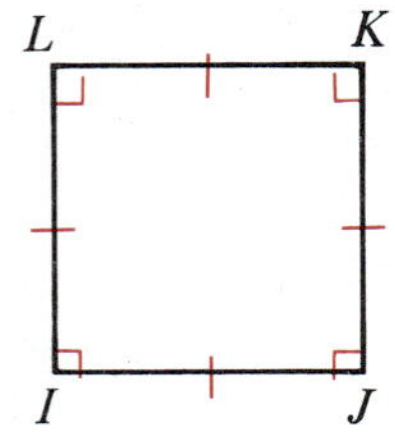

Square $IJKL$

You should be able to use the properties of a rhombus (plural: *rhombi*), a rectangle, and a square to identify each of them.

Example 1 True or false?

Statements	Answers
Some rhombi are squares.	True
All rhombi are parallelograms.	True
A square is both a rhombus and a rectangle.	True
All rhombi are rectangles.	False

Example 2 True or false? If true, prove the statement. If false, give a counterexample.

Consecutive angles of a rhombus are congruent.

False.
$m\angle A = 60$
$m\angle B = 120$
$m\angle C = 60$
$m\angle D = 120$

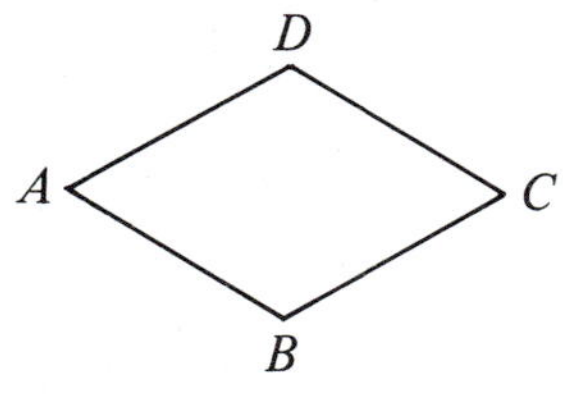

In the examples that follow, you will examine certain properties involving the diagonals of parallelograms, rhombi, rectangles, and squares.

Example 3

Which of the following figures have congruent diagonals: a parallelogram, a rhombus, a rectangle, a square? Draw pictures to show this.

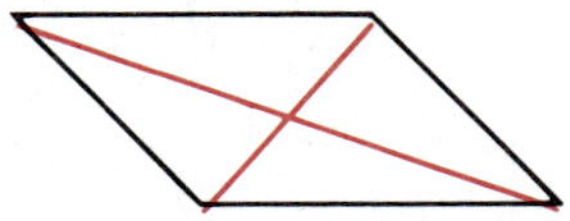
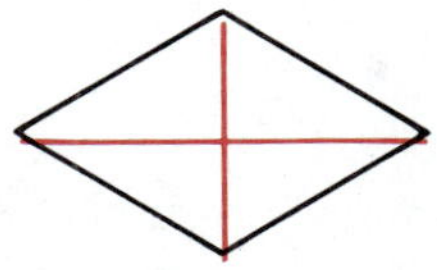
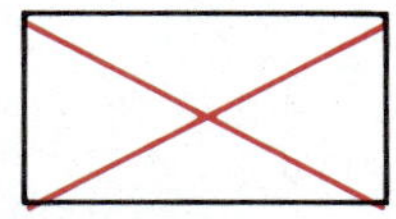

$\ncong$ $\ncong$ $\cong$ $\cong$

Thus, the rectangle and square appear to have congruent diagonals.

Example 4

Given: Rect. $ABCD$ with diagonals $\overline{AC}$ and $\overline{BD}$.
Prove: $\overline{BD} \cong \overline{AC}$

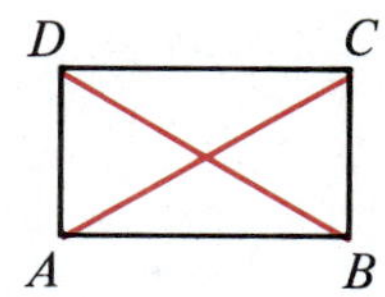

Proof

Statements		Reasons
1. Rect. $ABCD$, diagonals $\overline{AC}$ and $\overline{BD}$		1. Given
2. $\overline{AD} \cong \overline{BC}$	(S)	2. Opp. sides of ▱ are $\cong$.
3. $\angle DAB \cong \angle CBA$	(A)	3. A rect. has four $\cong$ rt. $\angle$'s.
4. $\overline{AB} \cong \overline{BA}$	(S)	4. Reflexive property
5. $\triangle DAB \cong \triangle CBA$		5. SAS
6. $\therefore\ \overline{BD} \cong \overline{AC}$		6. CPCTC

Example 4 shows that the diagonals of a rectangle are congruent. A square is a rectangle by definition, so the diagonals of a square are also congruent.

Example 5

Prove: The diagonals of a rhombus are perpendicular.

Given: Rhombus $XYZW$ with diagonals $\overline{XZ}$ and $\overline{WY}$
Prove: $\overline{WY} \perp \overline{XZ}$

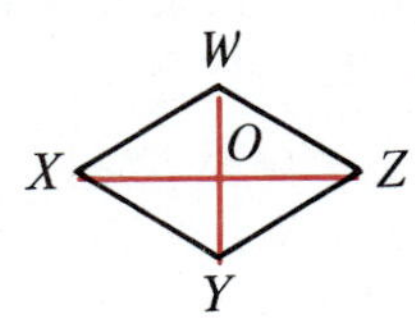

Proof

Statements	Reasons
1. Rhombus $XYZW$, diagonals $\overline{XZ}$ and $\overline{WY}$	1. Given
2. $\overline{XW} \cong \overline{ZW}$ and $\overline{XY} \cong \overline{ZY}$	2. Def. of rhombus
3. $\therefore\ \overline{WY} \perp \overline{XZ}$	3. Line containing 2 pts. each equidistant from endpoints of a seg. is the $\perp$ bisector of the seg.

Summary

Special properties of rhombi, rectangles, and squares.

	Rhombus	Rectangle	Square
All sides $\cong$	X		X
$4 \cong$ rt. $\angle$'s		X	X
$\cong$ diagonals		X	X
$\perp$ diagonals	X		X

From this summary, you should observe that a square has all the special properties of both a rhombus and a rectangle.

Example 6

Given: Rectangle $ABCD$, $AC = 3x + 2$, and $BD = 5x - 8$
Find BD.

$5x - 8 = 3x + 2$
$2x = 10$ ◄ *The diagonals of a rect. are* $\cong$.
$x = 5$

Thus, $BD = 5(5) - 8$, or 17.

Reading in Geometry

Copy each sentence. Underline once the words that are part of the "Given." Underline twice the words that are part of the "Prove."

1. If all four sides of a quadrilateral are congruent, then the quadrilateral is a rhombus.
2. If all four angles of a quadrilateral are congruent, then the quadrilateral is a rectangle.
3. If the diagonals of a parallelogram are perpendicular, then the parallelogram is a rhombus.
4. The opposite sides of a parallelogram are congruent.
5. The adjacent sides of a square are congruent.
6. The diagonals of a rhombus are perpendicular.

Oral Exercises

Tell whether each figure appears to be a rhombus, a rectangle, a square, or none of these. Some figures may correctly be identified in more than one way.

1.

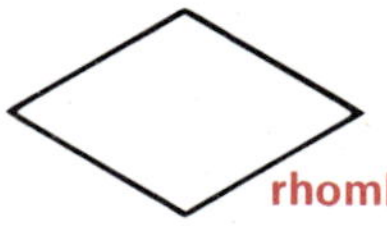

rhombus

2.

rhombus, rectangle, square

3.

rectangle

4.

none of these

Based only on the markings shown, identify each figure as a rhombus, a rectangle, a square, or none of these. Some figures may correctly be identified in more than one way.

5. 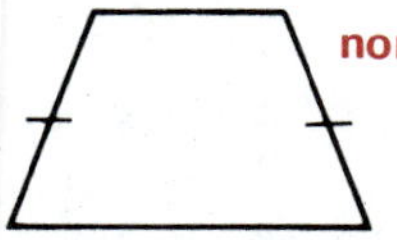 none of these

6.  rectangle

7. 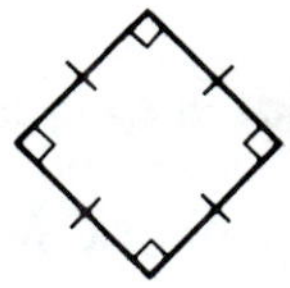 rhombus, rectangle, square

8. 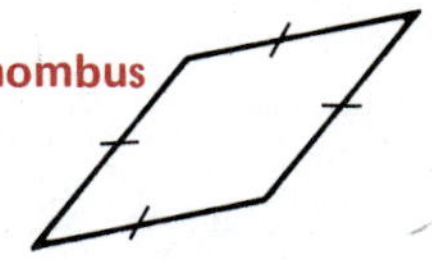 rhombus

Written Exercises

True or false?

A

1. All squares are rectangles. T
2. All squares are parallelograms. T
3. Some rectangles are rhombi. T
4. Some rhombi are rectangles. T
5. All rhombi are squares. F
6. Some parallelograms are rectangles. T

For each given property, determine which of the following figures have that property: a parallelogram, a rhombus, a rectangle, or a square. (Ex. 7–16)

7. Opposite sides are parallel. P, Rh, R, S
8. Opposite sides are congruent. P, Rh, R, S
9. Opposite angles are congruent. P, Rh, R, S
10. Opposite angles are supplements. R, S
11. Diagonals are congruent. R, S
12. Diagonals are perpendicular. Rh, S
13. Diagonals bisect each other. P, Rh, R, S
14. Consecutive sides are congruent. Rh, S
15. Consecutive angles are congruent. R, S
16. Consecutive angles are supplements. P, Rh, R, S

17. Given: Rhombus $ABCD$, and $m\,\angle A = 75$
Find $m\,\angle B$ and $m\,\angle C$. 105 75

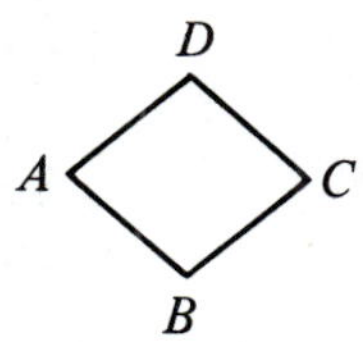

18. Given: Rhombus $ABCD$, and $BC = 12$
Find AB and CD. 12

19. Given: Square $EFGH$, $EG = 8x$, and $FH = 12x - 4$
Find EG. 8

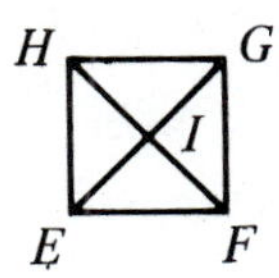

20. Given: Square $EFGH$, $EI = 5x - 1$, and $IG = 4x + 1$
Find EI. 9

True or false? If true, prove the statement. If false, give a counterexample.

B

21. The diagonals of a rectangle are congruent. T
22. The diagonals of a rectangle are perpendicular. F
23. The diagonals of a rhombus bisect the vertex angles. T
24. The diagonals of a rectangle bisect the vertex angles. F
25. The diagonals of a square are congruent. T
26. The diagonals of a square bisect each other. T

CUMULATIVE REVIEW

1. Find the distance between two points with coordinates −8 and 17. 25
2. Can two triangles be proved congruent by a side-side-angle correspondence? no; rt. △'s only
3. What is the median of a triangle? the seg. from a vert. to the midpt. of the opp. s.

PROVING RHOMBI, RECTANGLES, AND SQUARES 7.7

Objective **To prove that a quadrilateral is a rhombus, a rectangle, or a square**

Although a rhombus was defined to be a parallelogram with four congruent sides, you can prove that a figure is a rhombus based on other properties. Similarly, you can prove that a figure is a rectangle without showing that it has four right angles.

Example 1 Prove: A parallelogram with perpendicular diagonals is a rhombus.

Given: $\square ABCD$ and $\overline{AC} \perp \overline{BD}$
Prove: $ABCD$ is a rhombus.

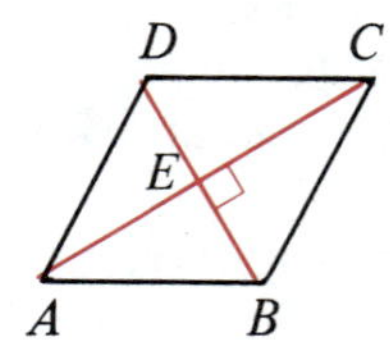

Proof

Statements		Reasons
1. $\overline{DE} \cong \overline{BE}$	(S)	1. Diag. of a $\square$ bisect each other.
2. $\overline{AC} \perp \overline{BD}$		2. Given
3. $\angle DEA \cong \angle BEA$	(A)	3. $\perp$ lines form $\cong$ rt. $\angle$'s.
4. $\overline{AE} \cong \overline{AE}$	(S)	4. Reflexive property
5. $\triangle AED \cong \triangle AEB$		5. SAS
6. $\overline{AD} \cong \overline{AB}$		6. CPCTC
7. $\overline{AD} \cong \overline{CB}$ and $\overline{AB} \cong \overline{CD}$		7. Opp. sides of a $\square$ are $\cong$.
8. $\overline{AB} \cong \overline{CB} \cong \overline{CD} \cong \overline{AD}$		8. Substitution
9. $\therefore ABCD$ is a rhombus.		9. Def. of rhombus

Example 2 Prove: A parallelogram with at least one right angle is a rectangle.

Given: $\square FGHI$ and rt. $\angle F$
Prove: $FGHI$ is a rectangle.

Proof

Statements	Reasons
1. $\square FGHI$, rt. $\angle F$	1. Given
2. $\angle H \cong \angle F$	2. Opp. $\angle$'s of a $\square$ are $\cong$.
3. $\overline{IF} \parallel \overline{HG}$	3. Def. of $\square$
4. $\angle F$ and $\angle G$ are supplementary.	4. Int. $\angle$'s of $\parallel$ lines on same side of trans. are supplementary.
5. m $\angle F = 90$	5. Def. of rt. $\angle$.
6. m $\angle G = 90$	6. Def. of supp. $\angle$'s, subst., subtraction
7. m $\angle H = 90$	7. Steps 2, 5 and def. of $\cong$ $\angle$'s
8. $\angle I \cong \angle G$	8. Opp. $\angle$'s of a $\square$ are $\cong$.
9. m $\angle I = 90$	9. Steps 6, 8 and def. of $\cong$ $\angle$'s
10. $\therefore FGHI$ is a rectangle.	10. Def. of rectangle

Example 3

Prove: A parallelogram with congruent diagonals is a rectangle.

Given: $\square PQRS$ and $\overline{PR} \cong \overline{SQ}$
Prove: $PQRS$ is a rectangle

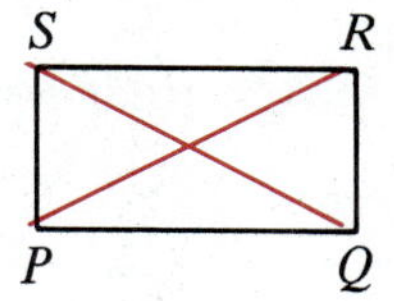

Proof

Statements		Reasons
1. $\square PQRS$ and $\overline{PR} \cong \overline{SQ}$	(S)	1. Given
2. $\overline{RQ} \cong \overline{SP}$	(S)	2. Opp. sides of a $\square$ are $\cong$.
3. $\overline{PQ} \cong \overline{QP}$	(S)	3. Reflexive property
4. $\triangle PQR \cong \triangle QPS$		4. SSS
5. $\angle PQR \cong \angle QPS$		5. CPCTC
6. $\overline{RQ} \parallel \overline{SP}$		6. Def. of $\square$
7. $\angle PQR$ and $\angle QPS$ are supplementary.		7. Int. $\angle$'s of $\parallel$ lines on same side of trans. are supp.
8. $\angle PQR$ is a rt. $\angle$.		8. $\cong$ supp $\angle$'s are rt $\angle$'s.
9. $\therefore$ $PQRS$ is a rectangle.		9. $\square$ with 1 rt. $\angle$ is a rectangle.

Summarized below are methods for proving that a parallelogram is a rhombus, a rectangle, or a square.

Summary

To show $\square ABCD$ is a rectangle, show that it has a rt. $\angle$, or $\cong$ diagonals.

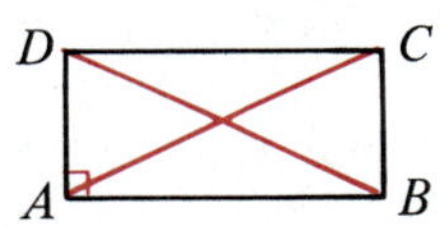

To show $\square ABCD$ is a rhombus, show that 2 adj. sides are $\cong$, or the diag. are $\perp$.

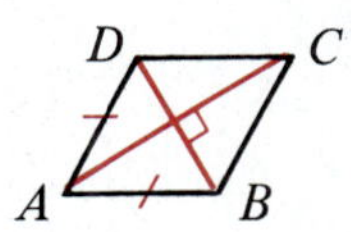

To show $\square ABCD$ is a square, show that it is a rect. with 2 $\cong$ adj. sides, or a rhombus with a rt. $\angle$.

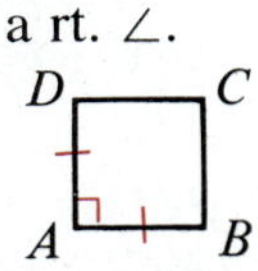

It is also possible to determine whether a quadrilateral not specifically identified as a parallelogram is a rhombus, a rectangle, or a square.

Example 4

True or False? Draw a picture to defend your answer. A quadrilateral with four congruent sides is a rhombus.

True. The quadrilateral is a parallelogram, since both pairs of opposite sides are $\cong$.

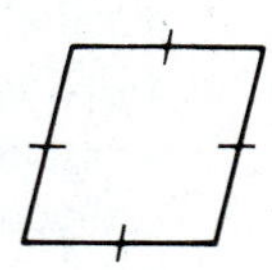

Example 5

For the given property, tell whether quadrilateral $EFGH$ is a rhombus, a rectangle, a square, or cannot tell.

Given: $\overline{EG} \cong \overline{FH}$

Cannot tell.

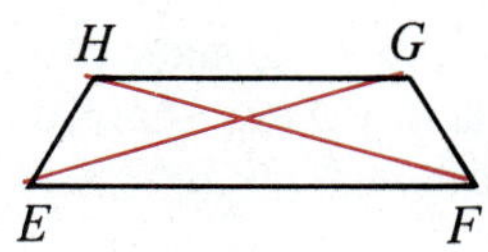

Oral Exercises

Tell whether each figure is a rhombus, a rectangle, a square, or cannot tell.

1.

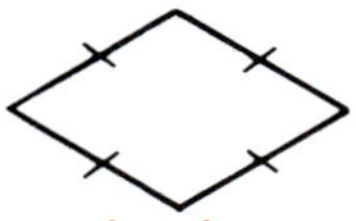

rhombus

2.

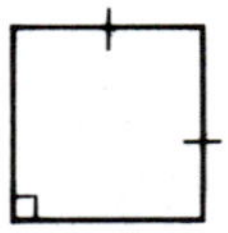

cannot tell

3.

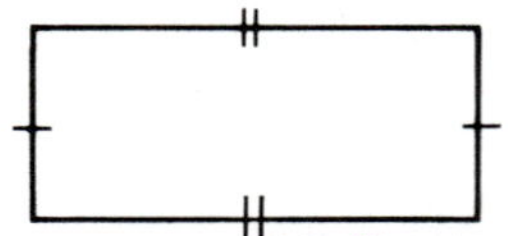

cannot tell

4.

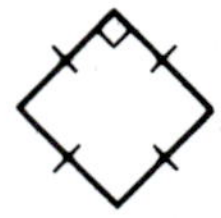

square

5.

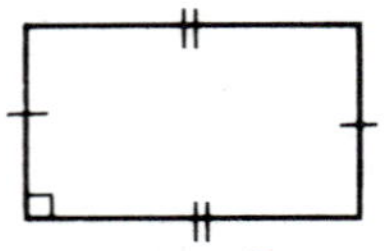

rectangle

6.

cannot tell

7.

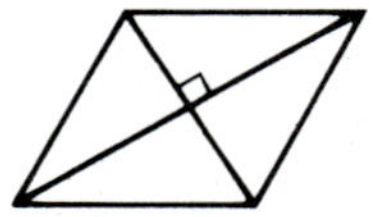

cannot tell

8.

rectangle

Written Exercises

True or false? Draw a picture to defend your answer.

A

1. A parallelogram with two right angles is a rectangle. T
2. A rhombus with at least one right angle is a square. T
3. A quadrilateral with perpendicular diagonals is a rhombus. F
4. A quadrilateral with congruent diagonals is a rectangle. F

For each given set of properties, tell whether quadrilateral *WXYZ* is a rhombus, a rectangle, a square, or cannot tell. (Ex. 5–14) 5. rectangle 7. rectangle 9. cannot tell

5. $\square WXYZ$ and $\overline{WX} \perp \overline{XY}$
6. $\overline{XZ} \cong \overline{WY}$ cannot tell
7. $\square WXYZ$ and $\overline{XZ} \cong \overline{WY}$
8. $\overline{XV} \cong \overline{VZ}$ and $\overline{WV} \cong \overline{VY}$ cannot tell
9. $\overline{WX} \perp \overline{XY}$ and $\overline{XY} \perp \overline{YZ}$
10. $\square WXYZ$ and $\overline{XZ} \perp \overline{WY}$ rhombus
11. $\overline{WV} \cong \overline{XV} \cong \overline{YV} \cong \overline{ZV}$

11. rectangle 13. rhombus

12. $\square WXYZ$, $\overline{XV} \cong \overline{VZ}$, and $\overline{WV} \cong \overline{VY}$ cannot tell
13. $\square WXYZ$ and $\overline{WX} \cong \overline{XY}$
14. $\square WXYZ$, $\overline{WX} \perp \overline{XY}$, and $\overline{XY} \perp \overline{YZ}$ rectangle

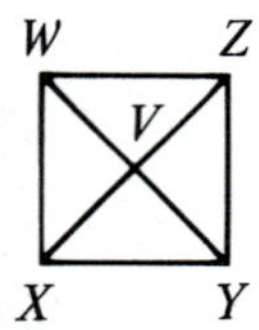

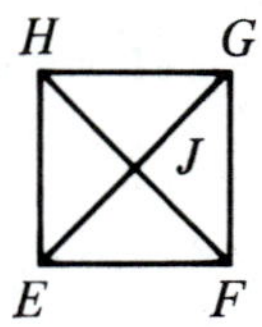

15. Given: $\overline{HG} \cong \overline{EF}$, $\overline{HG} \parallel \overline{EF}$, and $\overline{HF} \cong \overline{EG}$
    Prove: *EFGH* is a rectangle.
16. Given: $\overline{EF} \cong \overline{FG} \cong \overline{GH} \cong \overline{HE}$
    Prove: *EFGH* is a rhombus.

B

17. Prove: A parallelogram with two consec- congruent sides is a rhombus.
18. Prove: If the diagonals of a rhombus are congruent, then the rhombus is a square.
19. Prove: The segments connecting the midpoints of consecutive sides of a rhombus form a rectangle.
20. Prove: If the diagonals of a rectangle are perpendicular, then the rectangle is a square.

True or false? If true, prove the statement. If false, give a counterexample.

21. If all sides of a quadrilateral are congruent, then the quadrilateral is a rhombus. T
22. If all angles of a quadrilateral are congruent, then the quadrilateral is a rectangle. T
23. If the diagonals of a quadrilateral bisect each other, then the quadrilateral is a rectangle. F
24. If the diagonals of a quadrilateral are perpendicular bisectors of each other, then the quadrilateral is a rhombus. T

APPLICATIONS

Read → Plan → Solve → Interpret

1. The yard lines on a football field must be marked off for the first game of the season. A permanent marker is at the end of each goal line. There are no other markers on the field. How can the coach lay off the yard lines so that they are parallel to the goal line and parallel to each other? Why does this work?

2. The lamp shown can be raised or lowered using the center or base hinges. Why will the light always point in the same direction when raised or lowered by the center or base hinges?

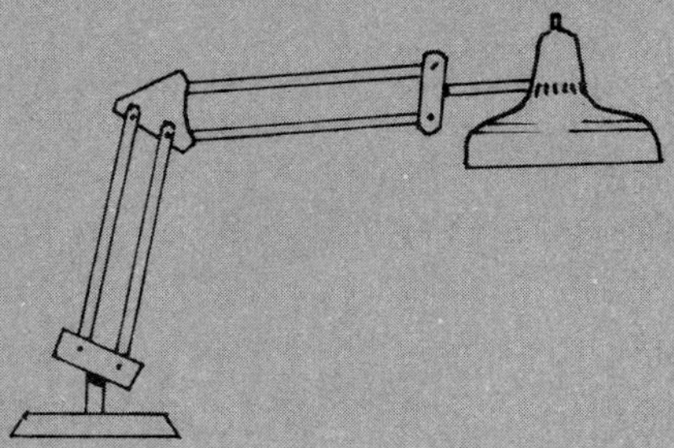

CHAPTER SEVEN REVIEW

Based only on the markings shown, tell whether each figure is a parallelogram, a rhombus, a rectangle, a square, or none of these. [7.6]

1.

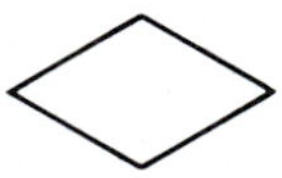

none of these

2.

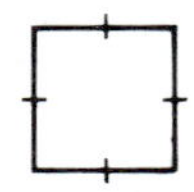

rhombus

3.

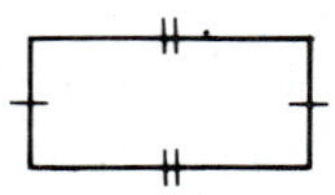

parallelogram

4.

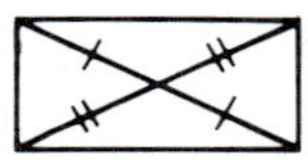

parallelogram

For each given set of properties, tell whether quadrilateral $ABCD$ is a parallelogram, a rhombus, a rectangle, a square, or cannot tell.

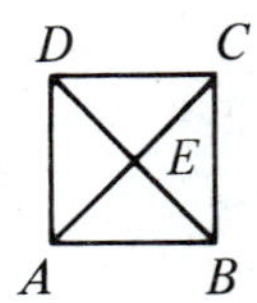

5. $\overline{AC} \cong \overline{BD}$ [7.3, 7.7] cannot tell

6. $\overline{AB} \cong \overline{DC}$, $\overline{AB} \parallel \overline{DC}$ parallelogram

7. $\angle A \cong \angle B \cong \angle C \cong \angle D$ rectangle

8. $\overline{AE} \cong \overline{CE}$, $\overline{BE} \cong \overline{DE}$ parallelogram

9. $\triangle ABC \cong \triangle ADC$ cannot tell

10. $\overline{AB} \cong \overline{BC} \cong \overline{CD} \cong \overline{DA}$, and $\overline{DA} \perp \overline{AB}$ square

True or false?

11. A quadrilateral can have no congruent sides. [7.1] T

12. A parallelogram can have no congruent sides. [7.2] F

13. A quadrilateral can have no congruent angles. [7.1] T

14. A parallelogram can have no congruent angles. [7.2] F

15. A square is a rhombus. [7.6] T

16. A rhombus is a rectangle. F

17. No rectangles are rhombi. F

Vocabulary

consecutive angles [7.1], consecutive sides [7.1], diagonal [7.1], opposite angles [7.1], opposite sides [7.1], parallelogram [7.2], quadrilateral [7.1], rectangle [7.6], rhombus [7.6], square [7.6]

Construct each of the following. (Ex. 18–19)

18. Draw a segment 10 cm long. Divide it into three congruent segments. [7.5]
See constr. p. 223

19. Draw a segment 12 cm long. Construct a segment three-fifths as long.
See constr. p. 223

20. Given: $l \parallel m \parallel n$, $AB = 5$, $AC = 10$, $DE = 4$
Find EF. 4 [7.5]

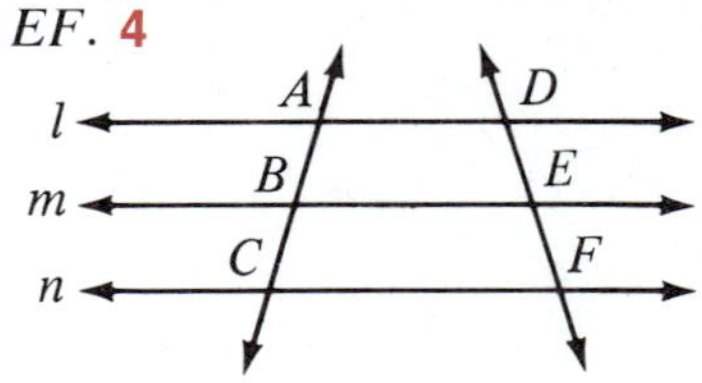

21. Given: $l \parallel m \parallel n$, $AB = 7$, $BC = 7$, $DE = 5x - 3$, $EF = 3x + 1$ [7.5]
Find DE. 7

22. Given: K is the midpoint of $\overline{GJ}$, I is the midpoint of $\overline{JH}$, $KI = 7$ [7.4]
Find GH. 14

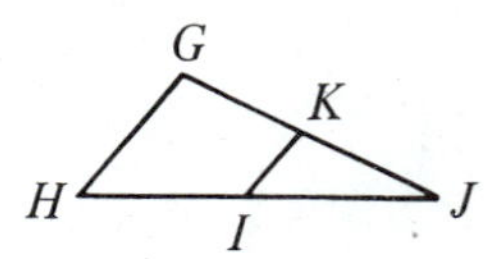

23. Given: $\overline{GK} \cong \overline{KJ}$, $\overline{HI} \cong \overline{IJ}$, $GH = 3x + 3$, $KI = 2x - 2$
Find KI. 12

24. Given: $\overline{EA} \cong \overline{EC}$; B, D, and F are midpoints of $\overline{AC}$, $\overline{CE}$, and $\overline{EA}$. [7.7]
Prove: $BDEF$ is a rhombus.

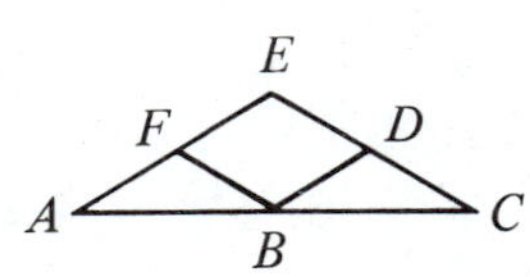

25. Prove: Consecutive angles of a parallelogram are supplementary. [7.2]

***26.** Prove: The diagonals of a square bisect each other. [7.4]

The following exercises are types that appear in the A section of the lesson's exercises: 1–18, 20–24. The starred exercises are types that appear in the C section. The remaining exercises appear in the B section.

CHAPTER SEVEN TEST

Based only on the markings shown, tell whether each figure is a parallelogram, a rhombus, a rectangle, a square, or none of these.

1.

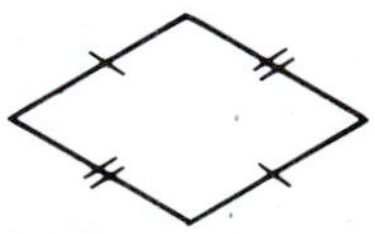

parallelogram

2.

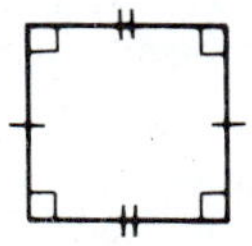

rectangle

3.

parallelogram

4.

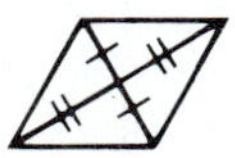

parallelogram

For each given set of properties, tell whether quadrilateral $ABCD$ is a parallelogram, a rhombus, a rectangle, a square, or cannot tell.

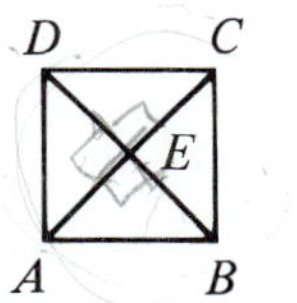

5. $\overline{AB} \parallel \overline{DC}$, $\overline{AD} \parallel \overline{BC}$ **parallelogram**
6. $\overline{AB} \cong \overline{BC}$, $\overline{CD} \cong \overline{DA}$ **cannot tell**
7. $\triangle ABD \cong \triangle CDB$ **parallelogram**
8. $\overline{AB} \cong \overline{DC}$, $\overline{BC} \cong \overline{AD}$, $\overline{AC} \cong \overline{BD}$ **rectangle**
9. $\overline{AE} \cong \overline{EC}$, $\overline{BE} \cong \overline{DE}$, $\overline{AC} \perp \overline{BD}$ at E **rhombus**
10. $\overline{AB} \cong \overline{BC} \cong \overline{CD}$, $\overline{CB} \perp \overline{AB}$ **cannot tell**

True or false?

11. The diagonals of a parallelogram are congruent. **F**
12. The opposite angles of a rhombus are congruent. **T**
13. A square is a rectangle. **T**
14. A rectangle is a parallelogram. **T**
15. A quadrilateral whose diagonals are perpendicular bisectors of each other is a rhombus. **T**
16. A quadrilateral whose diagonals are congruent is a rectangle. **F** T error
17. No rhombi are rectangles. **F**

Construct each of the following. (Ex. 18–19)

18. Copy $\overline{XY}$. Divide $\overline{XY}$ into three congruent segments.

X ———————————————— Y

19. Copy $\overline{PQ}$. Construct a segment three-fifths as long as $\overline{PQ}$.

P ———————————————— Q

20. Given: $t \parallel r \parallel s$, $PQ = 9$, $QR = 9$, $ST = 8$
Find SU. **16**

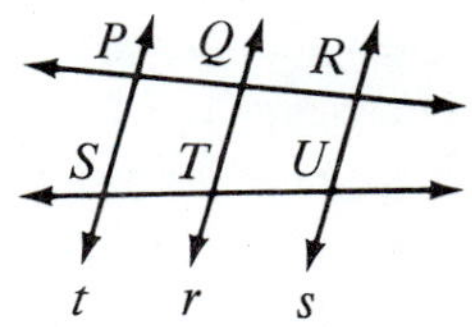

21. Given: $t \parallel r \parallel s$, $PQ = 7$, $PR = 14$, $ST = 4x - 3$, $TU = x + 6$
Find TU. **9**

22. Given: K and I are the midpoints of $\overline{GJ}$ and $\overline{HJ}$, respectively, and $KI = 9$.
Find GH. **18**

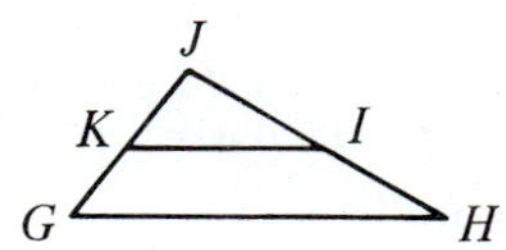

23. Given: $\overline{GK} \cong \overline{KJ}$, $\overline{HI} \cong \overline{IJ}$, $KI = 2x - 1$, $GH = 3x + 3$
Find KI. **9**

*24. Prove: The diagonals of a square are congruent.

25. Given: ▱$WXYZ$, $\overline{WU} \cong \overline{YV}$
Prove: $UXVZ$ is a ▱.

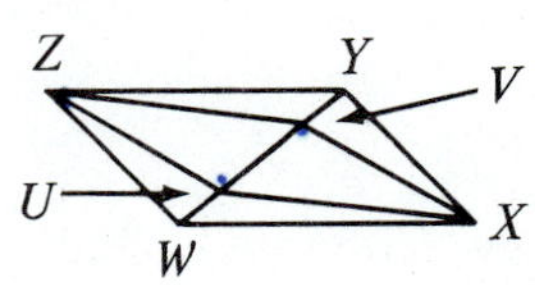

The following exercises are types that appear in the A section of the lesson's exercises: 1–18, 20–23. The starred exercises are types that appear in the C section. The remaining exercises appear in the B section.

COMPUTER ACTIVITIES

Parallelograms

Objective: **To classify parallelograms by using a computer program that inputs the measures of their sides and diagonals**

The program below will ask you to enter the measures of two adjacent sides and the diagonals of a parallelogram. The computer will determine whether the figure is a square, rectangle, rhombus, or none of these.

Line 130 tests the lengths of the sides. If they are not equal, the computer branches to line 160, which tests the lengths of the diagonals. Line 135, executed only when the sides are equal, also tests the lengths of the diagonals.

```
105  PRINT "THIS PROGRAM DETERMINES TYPES OF
     PARALLELOGRAMS."
110  PRINT "ENTER MEASURES OF SIDES A,B."
115  INPUT A,B
120  PRINT "ENTER MEASURES OF DIAGONALS D1, D2."
125  INPUT D1,D2
130  IF A <  > B THEN 160
135  IF D1 <  > D2 THEN 150
140  PRINT "THE PARALLELOGRAM IS A SQUARE."
145  GOTO 180
150  PRINT "THE PARALLELOGRAM IS A RHOMBUS."
155  GOTO 180
160  IF D1 <  > D2 THEN 175
165  PRINT "THE PARALLELOGRAM IS A RECTANGLE."
170  GOTO 180
175  PRINT "THE FIGURE IS NOT A RECTANGLE, SQUARE,
     OR RHOMBUS."
180  END
```

EXERCISES: See the Computer Section beginning on page 565 for more information.

Run the program for the following values. (Ex. 1–2)

1. A = 8, B = 4, D1 = 3, D2 = 5
2. A = 2, B = 2, D1 = 2.8, D2 = 2.8
3. Rewrite the program to identify a particular parallelogram by right angles and by the measures of two adjacent sides.
4. Write a program that determines whether two quadrilaterals are congruent, given three corresponding sides and the corresponding included angle of each.
5. Write a flowchart for the program above.

COLLEGE PREP TEST

For each item, choose the best answer. Indicate your answer by writing the letter of this choice.

1. For quadrilateral $GHDQ$, a pair of consecutive sides is **E**

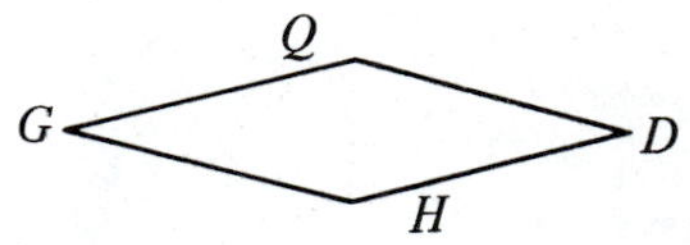

I $\overline{GH}$ and $\overline{HD}$
II $\overline{GH}$ and $\overline{DQ}$
III $\overline{GH}$ and $\overline{QG}$

(A) I only
(B) II only
(C) III only
(D) I and II only
(E) I and III only

2. Which of the following is *not* true about a parallelogram? **B**
(A) Opposite sides are congruent.
(B) Adjacent sides are congruent.
(C) Opposite angles are congruent.
(D) Opposite sides are parallel.
(E) Diagonals bisect each other.

3. A quadrilateral is a parallelogram if **D**
(A) Diagonals are perpendicular.
(B) Diagonals are congruent.
(C) Two adjacent sides are congruent.
(D) Adjacent angles are supplementary.
(E) A diagonal forms two congruent triangles.

4. If $\overline{AD} \cong \overline{DC}$ and $\overline{BE} \cong \overline{EC}$, which of the following is true? **D**

I If $AB = 10$, then $DE = 5$.
II $\overline{DE} \parallel \overline{AB}$
III If $BC = 16$, then $BE = 8$.

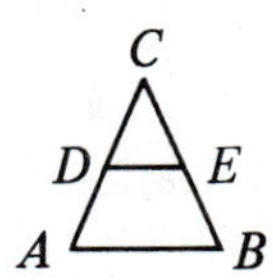

(A) I only
(B) II only
(C) III only
(D) I, II, and III
(E) I and II only

5. If $l \parallel m \parallel n$, $AB = 5$, $AC = 10$, $DE = 3x + 4$, and $DF = 10x - 4$, then **E**

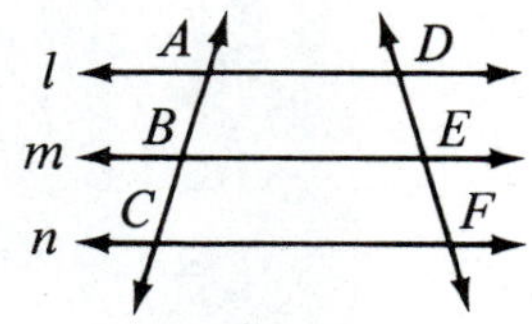

(A) $DE = 5$ and $DF = 10$
(B) $DE = 7\frac{3}{7}$
(C) $DE = 6\frac{2}{7}$
(D) $DE = 3$
(E) None of these

6. Which of the following is *not* true about a rectangle? **D**
(A) The diagonals are congruent.
(B) All angles are congruent.
(C) The diagonals bisect each other.
(D) The diagonals are perpendicular.
(E) Adjacent angles are supplementary.

7. Which of the following is *not* sufficient to prove that a parallelogram is a rectangle? **B**
(A) The diagonals are congruent.
(B) The diagonals are perpendicular.
(C) Two adjacent angles are congruent.
(D) One angle is a right angle.
(E) All angles are congruent.

8. Which of the following is sufficient to prove that a quadrilateral is a rhombus? **D**
(A) The diagonals are perpendicular.
(B) The diagonals are congruent.
(C) The diagonals bisect each other.
(D) Each diagonal is a perpendicular bisector of the other.
(E) Two adjacent sides are congruent.

8 TRAPEZOIDS AND SYMMETRIC POLYGONS

Non-Routine Problem Solving

A **pentomino** is a polygon that can be dissected into five congruent squares, with each square intersecting at least one other square along a complete edge. Using only congruent squares, two pentominos are equivalent if they are congruent.

In the examples below, A and B are not pentominos. C, D, and E are equivalent pentominos.

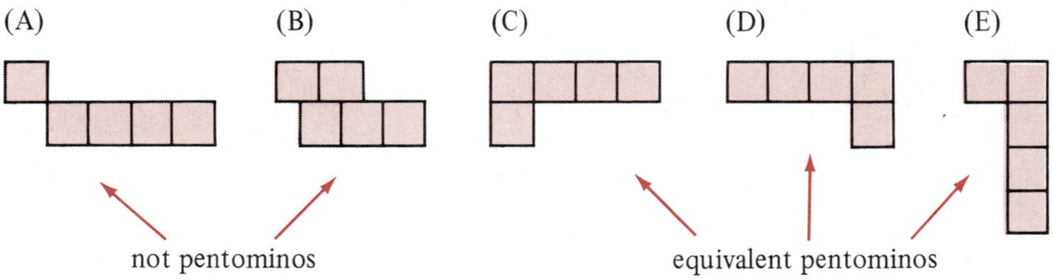

Problem There are twelve nonequivalent pentomino shapes. Can you draw each of them? How many of them have a line of symmetry? **6 nonequivalent pentominos have a line of symmetry, and 6 do not.**

TRAPEZOIDS

8.1

Objectives

To identify trapezoids
To investigate and prove properties of trapezoids

You have learned that a parallelogram is a quadrilateral with both pairs of opposite sides parallel. Another special kind of quadrilateral, a **trapezoid**, is defined below.

Definition: Trapezoid A *trapezoid* is a quadrilateral with exactly one pair of parallel sides. (Trapezoid $ABCD$ is written *trap.* $ABCD$.)

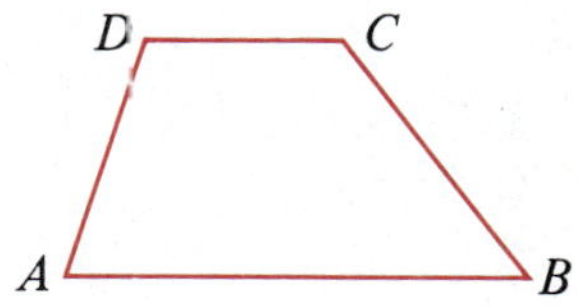

trap. $ABCD$

Example 1 Does the figure appear to be a trapezoid?

You can investigate some of the properties of trapezoids by making careful drawings.

Example 2 Make drawings to determine whether or not the diagonals of a trapezoid are congruent.

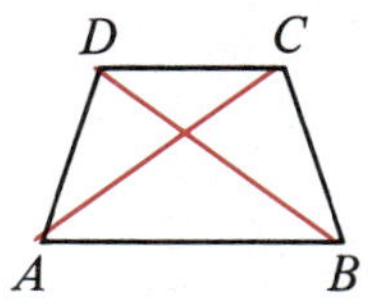

$\overline{AC} \cong \overline{BD}$

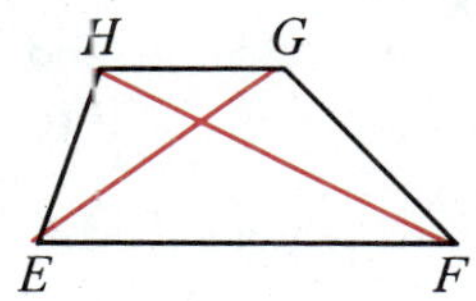

$\overline{EG} \not\cong \overline{FH}$

Thus, the diagonals of a trapezoid are not necessarily congruent.

The definition of trapezoid is used in the proof in Example 3.

Example 3

Given: Trap. $ABCD$, $\overline{AD} \cong \overline{BC}$, $\overline{DE} \perp \overline{AB}$, and $\overline{CF} \perp \overline{AB}$
Prove: $\angle 1 \cong \angle 2$

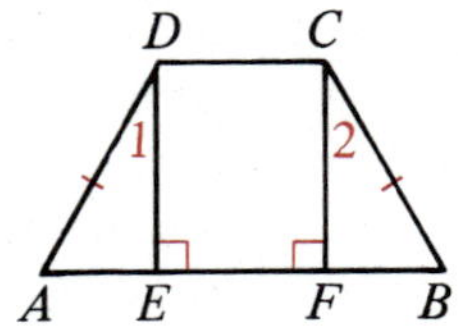

Proof

Statements		Reasons
1. Trap. $ABCD$		1. Given
2. $\overline{AB} \parallel \overline{DC}$		2. Def. of trapezoid
3. $\overline{DE} \perp \overline{AB}$, $\overline{CF} \perp \overline{AB}$		3. Given
4. $\overline{DE} \cong \overline{CF}$	(L)	4. $\parallel$ lines are equidistant.
5. $\overline{AD} \cong \overline{BC}$	(H)	5. Given
6. $\triangle DEA$ and $\triangle CFB$ are rt. $\triangle$'s.		6. $\perp$'s form rt. $\angle$'s; def. of rt. $\triangle$
7. $\triangle DEA \cong \triangle CFB$		7. HL
8. $\therefore \angle 1 \cong \angle 2$		8. CPCTC

Oral Exercises

Which of the following appear to be trapezoids?

1.
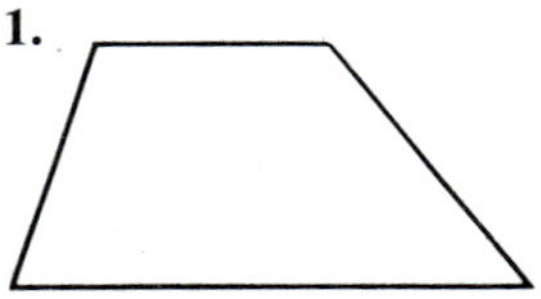
yes

2.
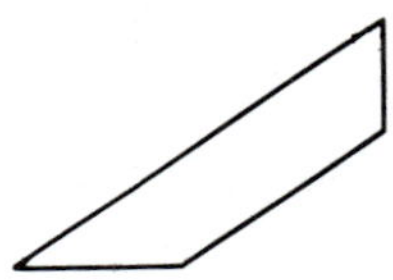
yes

3.
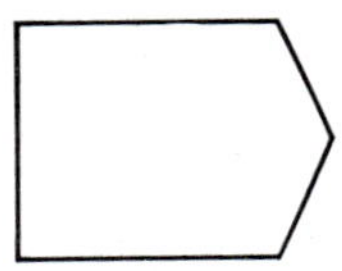
no

4.
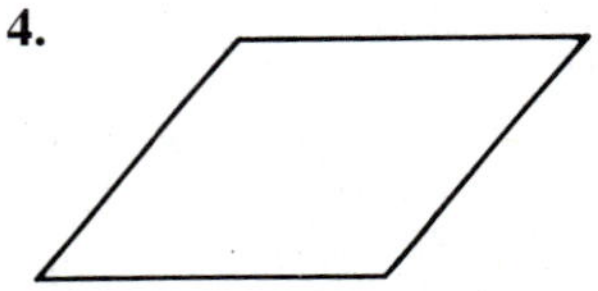
no

5.
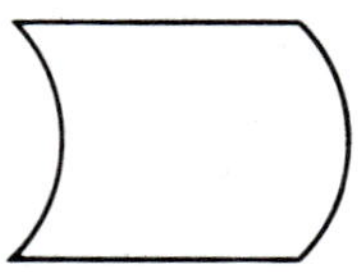
no

6.
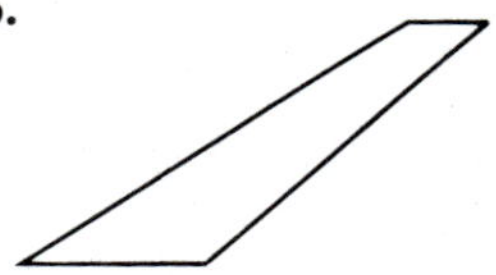
yes

7.
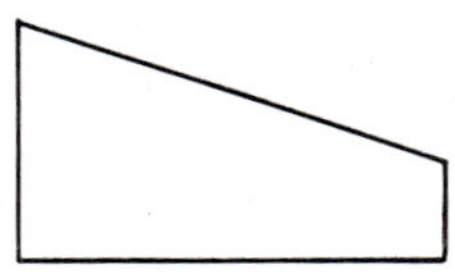
yes

8.
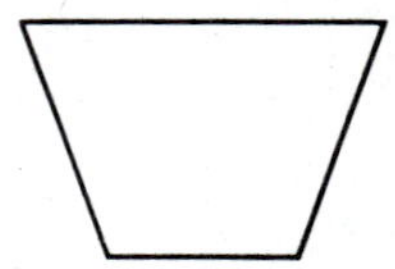
yes

Written Exercises

A

1. Given: D is midpoint of $\overline{AC}$, and E is midpoint of $\overline{BC}$
Prove: $ABED$ is a trapezoid.

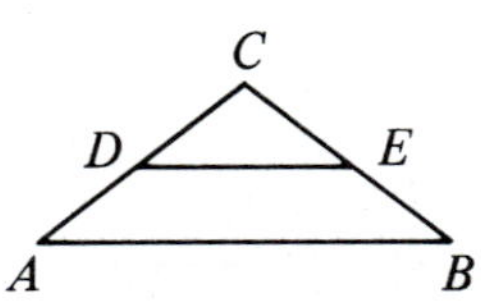

2. Given: Trap. $FGHI$, $\overline{FI} \cong \overline{GH}$, $\overline{IJ} \perp \overline{FG}$, and $\overline{HK} \perp \overline{FG}$
Prove: $\angle F \cong \angle G$

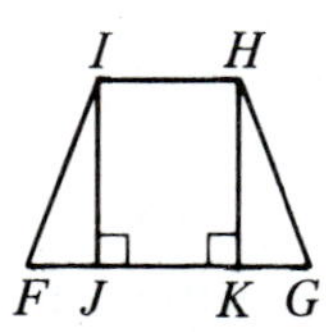

3. Given: Trap. $ONML$, $\angle O \cong \angle N$, $\overline{LP} \perp \overline{ON}$, and $\overline{MQ} \perp \overline{ON}$
Prove: $\overline{LO} \cong \overline{MN}$

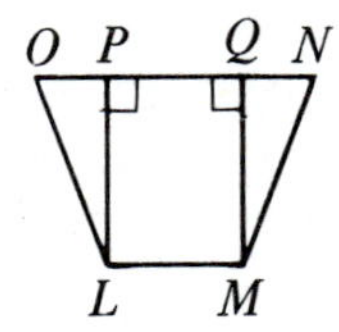

Make drawings to determine each of the following.

4. Can a trapezoid have two congruent sides? yes
5. Can a trapezoid have two congruent angles? yes
6. Can a trapezoid have congruent diagonals? yes
7. Can a trapezoid have diagonals which bisect each other? no
8. Can a trapezoid have two right angles? yes
9. Can a trapezoid have diagonals which are perpendicular? yes
10. Can a trapezoid have three congruent sides? yes
11. Can a trapezoid have three congruent angles? no

True or false? If the statement is true, prove it. If false, show a counterexample.

B 12. Opposite angles of a trapezoid are congruent. F
13. Opposite angles of a trapezoid are supplementary. F
14. Consecutive angles of a trapezoid are supplementary. F
15. Diagonals of a trapezoid bisect the angles of the trapezoid. F
16. The segments joining the midpoints of adjacent sides of a trapezoid form a trapezoid. F
17. The segments joining the midpoints of opposite sides of a trapezoid bisect each other. T
18. If the diagonals of a quadrilateral bisect each other, then the quadrilateral is a trapezoid. F
19. If the diagonals of a quadrilateral are congruent, then the quadrilateral is a trapezoid. F

CUMULATIVE REVIEW

1. a rt. △ with ≅ legs 2. a post. is accepted; a th. is proved.

1. Draw an isosceles triangle which is also a right triangle.
2. What is the difference between a postulate and a theorem?
3. In using the angle-side-angle congruence pattern for triangles, what is meant by the included side? 3. the side whose endpoints are vertices of the two angles

ALGEBRA REVIEW

Objective **To add and subtract algebraic fractions.**

Example Simplify $\frac{4x}{x^2-9} - \frac{5}{x+3}$

$$\frac{4x}{(x-3)(x+3)} + \frac{-1\,(5)(x-3)}{(x-3)(x+3)} \quad \blacktriangleleft \quad \frac{a}{b} - \frac{c}{d} \textit{ means } \frac{a}{b} + \frac{-1\,(c)}{d}$$

$$LCD = (x-3)(x+3) \quad \blacktriangleright \quad \frac{4x - 5\,(x-3)}{(x-3)(x+3)}, \text{ or } \frac{-x+15}{x^2-9}$$

Simplify.

1. $\frac{3}{a-3} + \frac{6a+1}{a^2-9}$
2. $\frac{3x+1}{x^2-4} - \frac{4}{x-2}$
3. $\frac{3}{y} - \frac{2}{y+6}$
4. $4 + \frac{6c-2}{3}$
5. $4 - \frac{3z-5}{5z+3}$
6. $\frac{4f+3}{2f+1} + \frac{2f+4}{2f-1}$
7. $\frac{w+1}{w+2} + 3w$
8. $\frac{3b-2}{2b^2-b} - \frac{2b}{4b^2-1}$

1. $\frac{9a+10}{a^2-9}$ 2. $\frac{-x-7}{x^2-4}$ 3. $\frac{y+18}{y^2+6y}$ 4. $\frac{6c+10}{3}$

5. $\frac{17z+17}{5z+3}$ 6. $\frac{12f^2+12f+1}{4f^2-1}$ 7. $\frac{3w^2+7w+1}{w+2}$ 8. $\frac{4b^2-b-2}{4b^3-b}$

ALTITUDES AND MEDIANS OF TRAPEZOIDS 8.2

Objectives

To identify the bases and legs of a trapezoid
To prove and apply theorems about the median and an altitude of a trapezoid

The sides of a trapezoid are usually given specific names.

> **Definitions: Bases and legs of trapezoid**
>
> The *bases* of a trapezoid are the two parallel sides.
> The *legs* of a trapezoid are the two nonparallel sides.

Example 1 Which sides of each trapezoid are the bases?

$\overline{AB}$ and $\overline{DC}$

$\overline{EF}$ and $\overline{HG}$

$\overline{IL}$ and $\overline{JK}$

A special segment of a trapezoid which is frequently used is the **altitude**.

> **Definition: Altitude of trapezoid**
>
> An *altitude* of a trapezoid is any segment from a point on one base perpendicular to the line containing the other base.

A triangle can have exactly three altitudes. However, a trapezoid can have many. $\overline{DE}$, $\overline{JF}$, $\overline{IG}$, and $\overline{HB}$ are all altitudes of trap. $ABCD$.

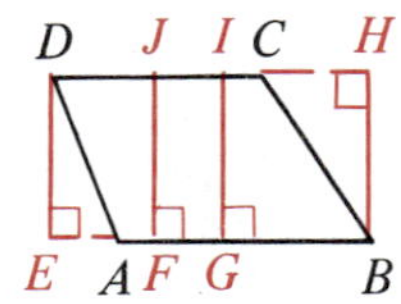

Example 2 Prove: All altitudes of a trapezoid are congruent.

Given: Trap. $ABCD$ with altitudes $\overline{EH}$ and $\overline{FG}$
Prove: $\overline{EH} \cong \overline{FG}$

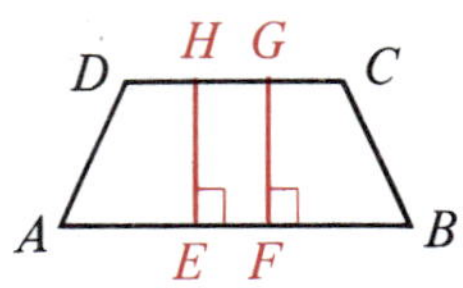

Proof

Statements	**Reasons**
1. Trap. $ABCD$, altitudes $\overline{EH}$ and $\overline{FG}$	1. Given
2. $\overline{DC} \parallel \overline{AB}$	2. Def. of trapezoid
3. $\overline{EH} \perp \overline{AB}$ and $\overline{FG} \perp \overline{AB}$	3. Def. of altitude
4. $\therefore EH = FG$ and $\overline{EH} \cong \overline{FG}$	4. $\parallel$ lines are equidistant.; def. of $\cong$ segs.

Another special segment in a trapezoid is the **median**.

Definition: Median of trapezoid

The *median* of a trapezoid is the segment that joins the midpoints of the legs.

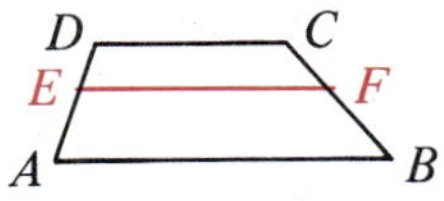

Median $\overline{EF}$

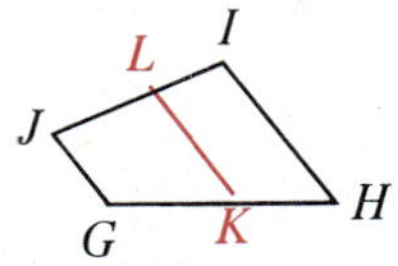

Median $\overline{LK}$

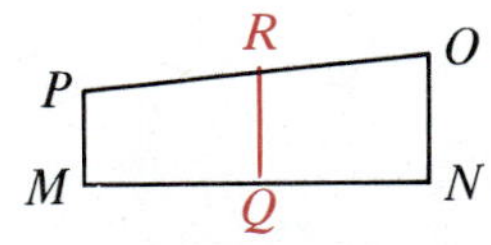

Median $\overline{QR}$

Theorem 8.1

The median of a trapezoid is parallel to its bases, and its length is half the sum of the lengths of the bases.

Given: Trap. $ABCD$, $\overline{AB} \parallel \overline{DC}$, and median $\overline{XY}$

Prove: $\overline{XY} \parallel \overline{DC} \parallel \overline{AB}$ and $XY = \frac{1}{2}(DC + AB)$

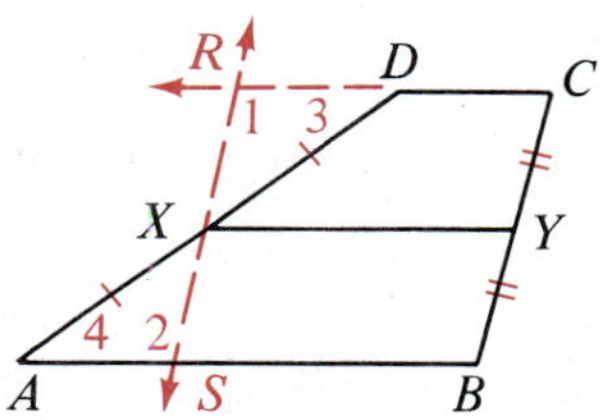

ANALYSIS

Through X, draw $\overline{RS} \parallel \overline{CB}$. Show that $RSBC$, $RXYC$, and $XYBS$ are ▱'s.

Outline of Proof

I. Prove $\overline{XY} \parallel \overline{DC} \parallel \overline{AB}$

1. Trap. $ABCD$, $\overline{AB} \parallel \overline{DC}$, and median $\overline{XY}$
2. Draw $\overrightarrow{CD}$.
3. Draw $\overline{RS} \parallel \overline{CB}$ through X.
4. $\overline{SB} \parallel \overline{RC}$
5. $RSBC$ is a ▱.
6. $\angle 1 \cong \angle 2$ and $\angle 3 \cong \angle 4$
7. $\overline{DX} \cong \overline{XA}$
8. $\triangle RXD \cong \triangle SXA$
9. $RX = SX$ and $RX = \frac{1}{2}(RS)$
10. $CY = YB$ and $CY = \frac{1}{2}(CB)$
11. $RS = CB$
12. $RX = CY$
13. $RXYC$ is a ▱.
14. $\therefore \overline{XY} \parallel \overline{DC}$
15. Similarly, $XYBS$ is a ▱.
16. $\overline{XY} \parallel \overline{AB}$
17. $\therefore \overline{XY} \parallel \overline{DC} \parallel \overline{AB}$.

II. Prove $XY = \frac{1}{2}(DC + AB)$

18. $XY = RC$
19. $RC = DC + RD$
20. $XY = DC + RD$
21. $RD = SA$
22. $XY = DC + SA$
23. $XY = SB$
24. $XY + XY = DC + SA + SB$
25. $2(XY) = DC + AB$
26. $\therefore XY = \frac{1}{2}(DC + AB)$

(See Exercise 20.)

Example 3 Given: $ABCD$ is a trapezoid.
Find the length of median $\overline{EF}$.

$EF = \frac{1}{2}(8 + 12)$, or 10.

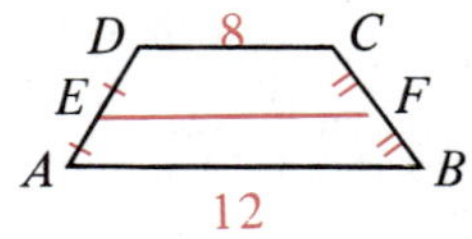

Example 4 Given: Trap. $GHIJ$, median $\overline{KL}$, $GJ = 8$, and $KL = 6$
Find HI.

$6 = \frac{1}{2}(8 + HI)$

$12 = 8 + HI$

Thus, $HI = 4$.

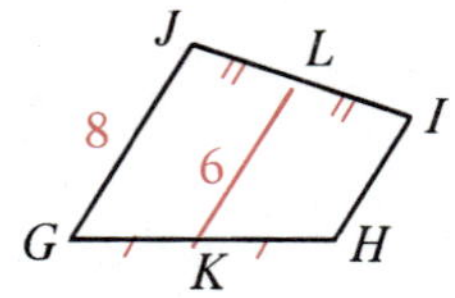

Example 5 Given: Trap. $MNOP$, median $\overline{QR}$, $PO = 2x + 2$, $MN = 7x - 2$, and $QR = 4x + 1$
Find QR.

$4x + 1 = \frac{1}{2}[(2x + 2) + (7x - 2)]$

$8x + 2 = (2x + 2) + (7x - 2)$

$8x + 2 = 9x$

$x = 2$

Thus, $QR = 4(2) + 1$, or 9.

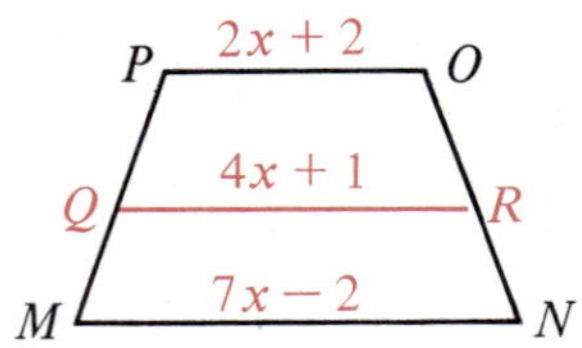

When you are planning a proof involving trapezoids, there frequently are several possible conclusions that can be drawn from the given data. You may need to determine these to help you select the plan for the proof.

Example 6 Write two possible conclusions that can be drawn from the given data.

Given: $\overline{VU} \parallel \overline{ST}$, $\overline{VW} \cong \overline{WS}$, and $\overline{UX} \cong \overline{XT}$

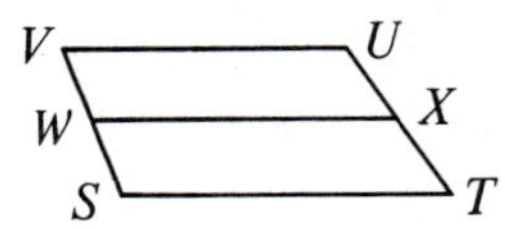

Possible conclusions: $WX = \frac{1}{2}(VU + ST)$ ◄ *Other conclusions are possible.*

$\overline{WX} \parallel \overline{VU}$

Written Exercises

Find the length of the median of each trapezoid.

A **1.**

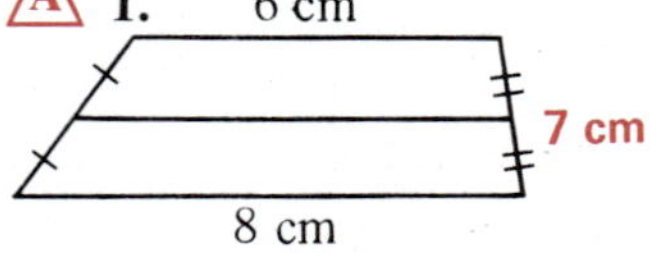

2.

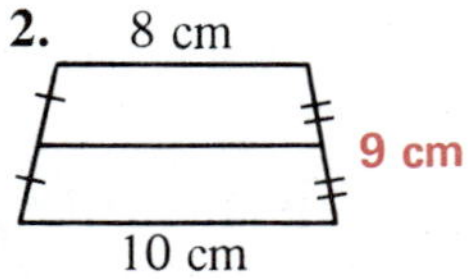

3.

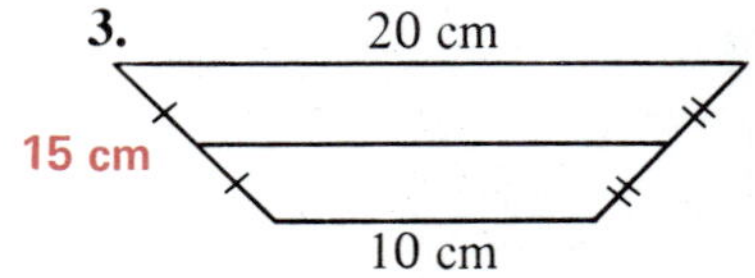

Using each given set of measurements for trap. $ABCD$, find the length of median $\overline{EF}$.

4. $DC = 8$ and $AB = 12$. 10
5. $DC = 9$ and $AB = 11$. 10
6. $DC = 13$ and $AB = 21$. 17
7. $DC = 10$ and $AB = 13$. $11\frac{1}{2}$
8. $DC = 2x - 1$, $AB = 5x + 1$, and $EF = 3x + 2$. 14
9. $DC = x + 3$, $AB = 3x - 5$, and $EF = x + 7$. 15

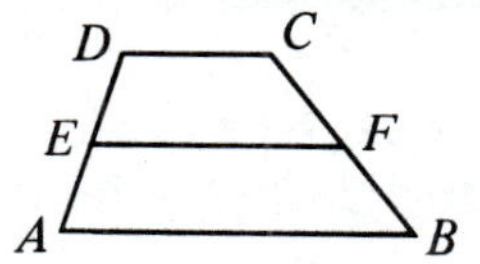

Using each given set of measurements for trap. $GHIJ$ and median $\overline{KL}$, find the indicated base length.

10. $JI = 4$ and $KL = 6$. $GH = ?$ 8
11. $KL = 9$ and $GH = 10$. $JI = ?$ 8
12. $JI = 9$, $KL = 14$. $GH = ?$ 19
13. $KL = 13$ and $GH = 19$. $JI = ?$ 7
14. $JI = x + 5$, $KL = 3x$, and $GH = 4x - 3$. Find GH. 5
15. $JI = x + 5$, $KL = 3x - 7$, and $GH = 3x - 1$. Find JI. 14

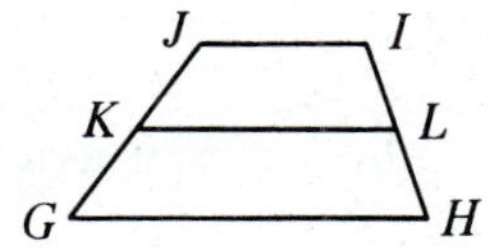

Write two possible conclusions that can be drawn from the given data. (Ex. 16–17)

B 16. Given: Trap. $MNOP$, Q is the midpoint of $\overline{PM}$, and R is the midpoint of $\overline{NO}$.
$QR = \frac{1}{2}(PO + MN)$; others are possible.

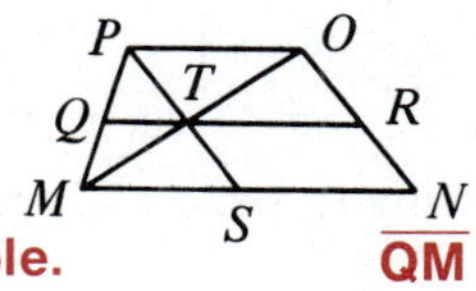

17. Given: Trap $MNOP$ with median $\overline{QR}$ and diagonal $\overline{MO}$.
$\overline{QM} \cong \overline{QP}$, $\overline{QR} \parallel \overline{PO}$; others are possible.

18. Prove: The median of a trapezoid bisects both diagonals of the trapezoid.
19. Prove: The median of a trapezoid bisects each altitude of the trapezoid that intersects it.
20. Complete the proof of Theorem 8.1

Prove each conclusion, or disprove it by giving a counterexample. (Ex. 21–24)

C Given: Trap. $UVWX$ with median $\overline{YZ}$ and diagonal $\overline{UW}$ intersecting both $\overline{YZ}$ and $\overline{XA}$ at B.

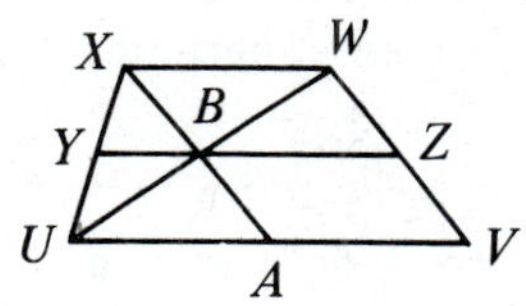

Given: Trap. $UVWX$, median $\overline{YZ}$, diagonal $\overline{UW}$ intersecting both $\overline{YZ}$ and $\overline{XA}$ at B, and $\overline{XB} \perp \overline{UW}$.

21. Conclusion: $AVWX$ is a trapezoid.
22. Conclusion: $\triangle UBY$ is isosceles.
23. Conclusion: $\triangle UAX$ is isosceles.
24. Conclusion: $\triangle BWZ$ is isosceles.
25. Is there any relationship between the lengths of the legs, the bases, and the altitudes of a trapezoid? Give a reason for your answer. no
26. Is there any relationship between the length of the median of a trapezoid and the length of an altitude? Give a reason for your answer. no
27. Does Theorem 8.1 apply to parallelograms? Give a reason for your answer. yes

LOGIC REVIEW

Write the conjunction and the disjunction of each pair of statements. Label each as true or false. Check students' answers.

1. A trapezoid is a quadrilateral. A trapezoid has two congruent sides.
conjunction: false; disjunction: true
2. Opposite sides of a parallelogram are parallel. The bases of a trapezoid are parallel.
conjunction: true; disjunction: true

ISOSCELES TRAPEZOIDS 8.3

Objectives

To identify isosceles trapezoids
To prove and apply theorems about isosceles trapezoids

You have learned earlier that an isosceles triangle has at least two congruent sides. Similarly, an isosceles trapezoid has two congruent legs.

Definition: Isosceles trapezoid An *isosceles trapezoid* is a trapezoid with congruent legs.

You should be able to identify isosceles trapezoids.

Example 1

Which of the following trapezoids appear to be isosceles?

Not isosceles

Isosceles

Not isosceles

Each isosceles trapezoid has two sets of base angles. The definition of isosceles trapezoid tells you that the legs are congruent. The following theorem shows that the base angles are also congruent.

Theorem 8.2 The base angles of an isosceles trapezoid are congruent.

Given: Trap. $ABCD$, $\overline{AB} \parallel \overline{DC}$ and $\overline{AD} \cong \overline{BC}$
Prove: $\angle A \cong \angle B$

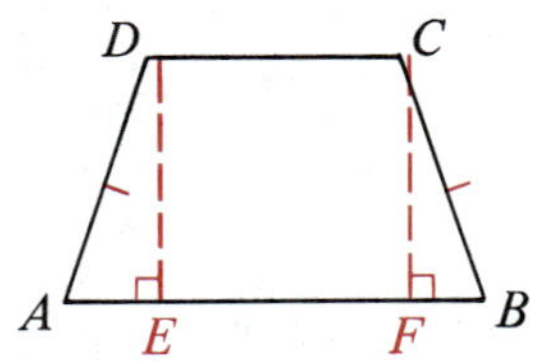

ANALYSIS
Draw $\overline{DE} \perp \overline{AB}$ and $\overline{CF} \perp \overline{AB}$. Show $\angle A$ and $\angle B$ are corr. parts of $\cong$ $\triangle$'s.

Proof

Statements	Reasons
1. Trap. $ABCD$, $\overline{AB} \parallel \overline{DC}$	1. Given
2. Draw $\overline{DE} \perp \overline{AB}$, $\overline{CF} \perp \overline{AB}$.	2. There is exactly one $\perp$ from a pt. to a line.
3. $\angle DEA$ and $\angle CFB$ are rt. $\angle$'s.	3. Def. of $\perp$
4. $\triangle AED$ and $\triangle BFC$ are rt. $\triangle$'s.	4. Def. of rt. $\triangle$
5. $\overline{AD} \cong \overline{BC}$	5. Given
6. $\overline{DE} \cong \overline{CF}$	6. $\parallel$ lines are equidistant.
7. $\triangle AED \cong \triangle BFC$	7. HL
8. $\therefore \angle A \cong \angle B$	8. CPCTC

Example 2

Given: Isosceles trap. $STUV$ and $\overline{VU} \parallel \overline{ST}$
Find m $\angle S$ and m $\angle U$.

$VU \parallel ST$ ► m $\angle S$ + m $\angle V = 180$
m $\angle S = 180 - 108$, or 72.
m $\angle U$ = m $\angle V$ ◄ *Apply. Th. 8.2*
m $\angle U = 108$
Thus, m $\angle S = 72$ and m $\angle U = 108$.

The converse of Theorem 8.2 is also true and is stated below. You should complete the proof of this theorem.

Theorem 8.3 If the base angles of a trapezoid are congruent, then the trapezoid is isosceles.
(See Exercise 15.)

Theorem 8.4 presents another property of isosceles trapezoids.

Theorem 8.4 The diagonals of an isosceles trapezoid are congruent.

Given: Isosceles trap. $WXYZ$, $\overline{WX} \cong \overline{ZY}$, and $\overline{WZ} \parallel \overline{XY}$
Prove: $\overline{XZ} \cong \overline{YW}$

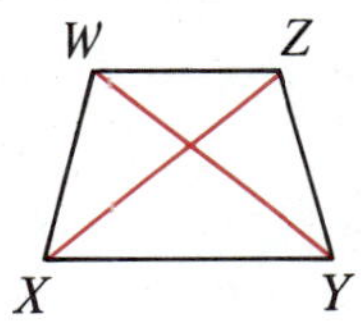

ANALYSIS
Show that $\overline{XZ}$ and $\overline{YW}$ are $\cong$ corr. parts of $\cong$ △'s XYZ and YXW.

(See Exercise 16.)

Although the diagonals of an isosceles trapezoid are congruent, this does not necessarily mean that every quadrilateral with congruent diagonals is an isosceles trapezoid. The figures below show this. However, any trapezoid with congruent diagonals is isosceles and this is proved as the next theorem.

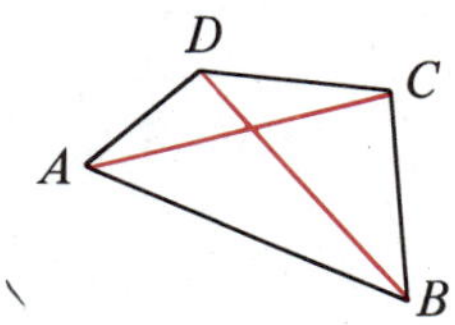

$\overline{AC} \cong \overline{BD}$
$ABCD$ is not a trapezoid.

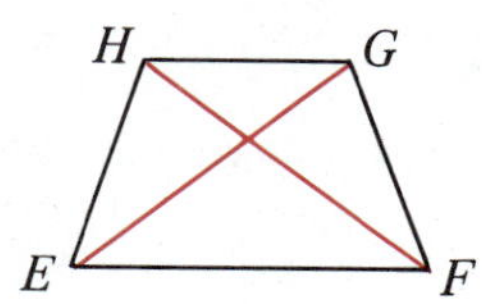

$\overline{EG} \cong \overline{FH}$
$EFGH$ is an isosceles trapezoid.

Theorem 8.5

If the diagonals of a trapezoid are congruent, then the trapezoid is isosceles.

Given: Trap. $ABCD$, $\overline{AB} \parallel DC$, and $\overline{AC} \cong \overline{BD}$

Prove: $\overline{BC} \cong \overline{AD}$

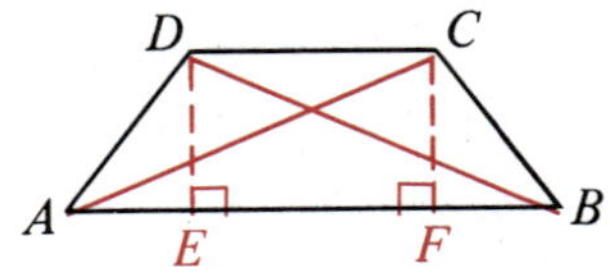

Proof

Statements	Reasons
1. Trap. $ABCD$, $\overline{AB} \parallel \overline{DC}$	1. Given
2. Draw $\overline{CF} \perp \overline{AB}$, $\overline{DE} \perp \overline{AB}$.	2. There is exactly one $\perp$ from a pt. to a line.
3. $\angle AFC$ and $\angle BED$ are rt. $\angle$'s.	3. Def. of $\perp$
4. $\triangle DFC$ and $\triangle BED$ are rt. $\triangle$'s.	4. Def. of rt. $\triangle$
5. $\overline{CF} \cong \overline{DE}$	5. $\parallel$ lines are equidistant.
6. $\overline{AC} \cong \overline{BD}$	6. Given
7. $\triangle AFC \cong \triangle BED$	7. HL
8. $\angle BAC \cong \angle ABD$	8. CPCTC
9. $\overline{AB} \cong \overline{BA}$	9. Reflexive property
10. $\triangle BAC \cong \triangle ABD$	10. SAS
11. $\therefore \overline{BC} \cong \overline{AD}$	11. CPCTC

Summary

To prove that a trapezoid is isosceles:

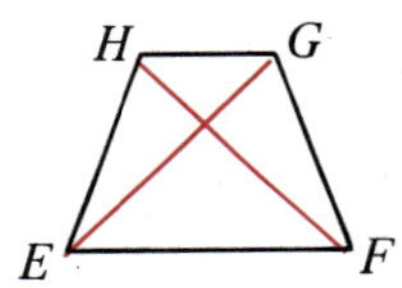

1. Show that the legs are congruent: $\overline{EH} \cong \overline{FG}$.
2. Show that a pair of base angles are congruent: $\angle E \cong \angle F$ or $\angle G \cong \angle H$.
3. Show that the diagonals are congruent: $\overline{EG} \cong \overline{FH}$.

Oral Exercises

Which of the following trapezoids appear to be isosceles?

1.

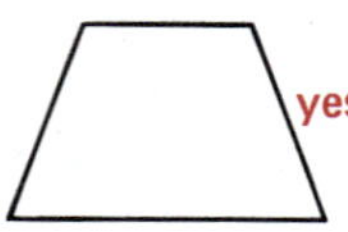

yes

2.

yes

3.

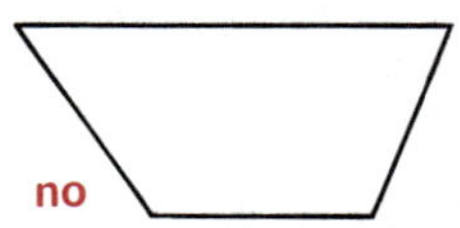

no

4.

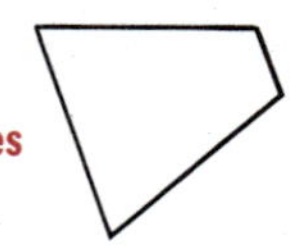

yes

Written Exercises

Find the unknown angle measures in each isosceles trapezoid. (Ex. 1–6)

A **1.**

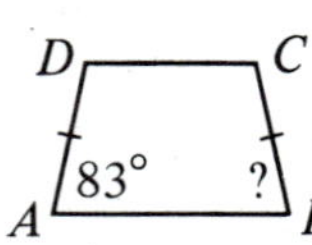

83°

2.

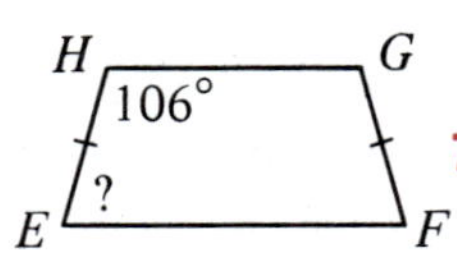

74°

3.

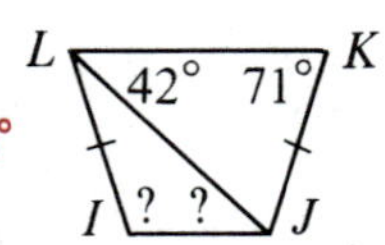

109°; 42°

4. 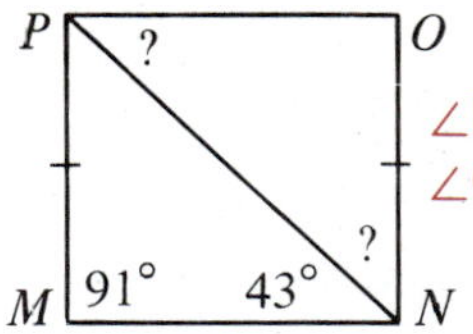

∠PNO: 48°
∠OPN: 43°

5. 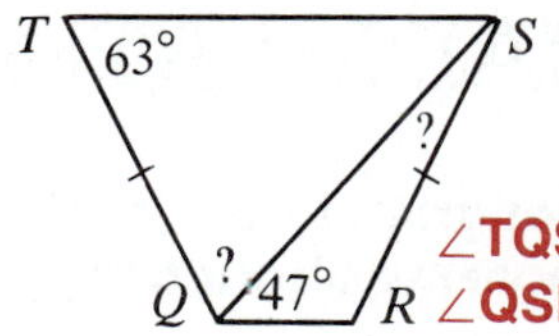

∠TQS: 70°
∠QSR: 16°

6. 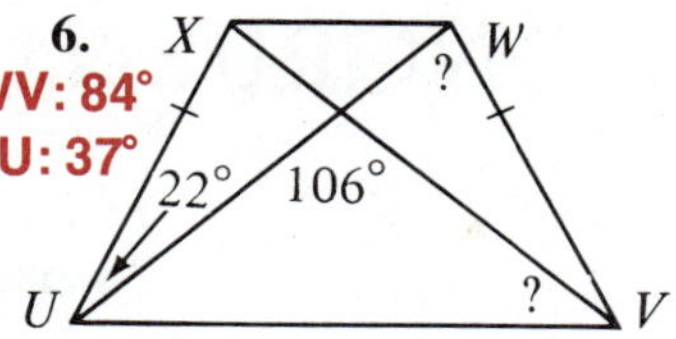

∠UWV: 84°
∠XVU: 37°

7. Given: $\overline{AB} \cong \overline{AC}$ and $\overline{DE} \parallel \overline{BC}$
Prove: $BCDE$ is an isosceles trapezoid.

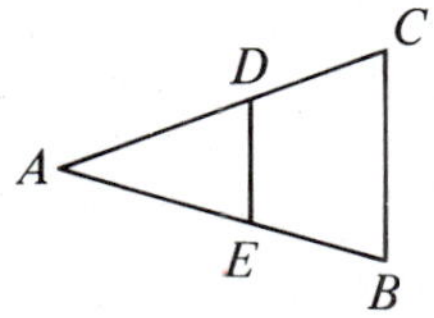

8. Given: $BCDE$ is an isosceles trapezoid with $\overline{EB} \cong \overline{DC}$
Prove: $\triangle ABC$ is isosceles.

B 9. Given: Isos. trap. $ABCD$ and $\overline{DC} \parallel \overline{AB}$
Prove: $\triangle ABD \cong \triangle BAC$

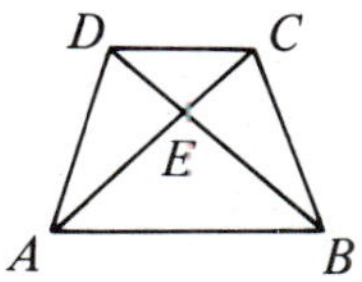

10. Given: Isos. trap. $ABCD$ and $\overline{DC} \parallel \overline{AB}$
Prove: $\triangle AED \cong \triangle BEC$

True or false? If the statement is true, prove it. If false, give a counterexample. (Ex. 11–14)

11. If a trapezoid is isosceles, then opposite angles are supplementary. T
12. If a trapezoid is isosceles, then the diagonals bisect each other. F
13. If opposite angles of a trapezoid are supplementary, then the trapezoid is isosceles. T
14. If the opposite angles of a quadrilateral are supplementary, then the quadrilateral is an isosceles trapezoid. F
15. Prove Theorem 8.3.
16. Prove Theorem 8.4.

C 17. Given: $\overline{FB} \perp \overline{AB}$, $\overline{FB} \perp \overline{BC}$, $\overline{AB} \cong \overline{BC}$, M is midpoint of $\overline{AF}$, and N is midpoint of $\overline{FC}$
Prove: $ACNM$ is an isos. trap.

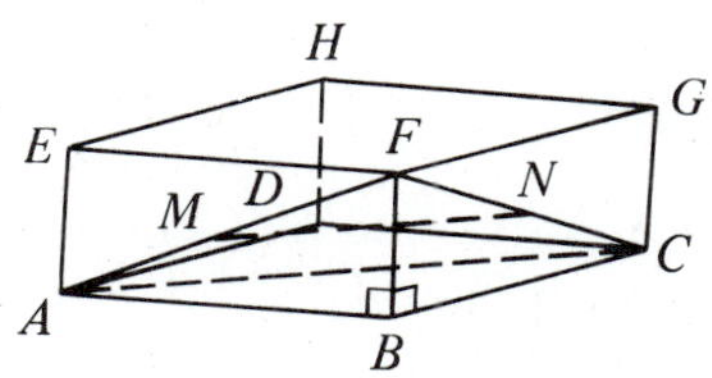

18. Given: $\square ABCD \parallel \square EFGH$, $\overline{FB} \perp \overline{AB}$, $\overline{FB} \perp \overline{BC}$, M is midpoint of $\overline{AF}$, N is midpoint of $\overline{FC}$, and $\triangle AFE \cong \triangle CFG$
Prove: $ACNM$ is an isos. trap.

CUMULATIVE REVIEW

1. Answers may vary.

1. Give an example to show that angle-angle-angle is not a congruence pattern.
2. Why are adjacent angles of a parallelogram supplementary? Theorem 3.2
3. Draw an acute angle. Construct its bisector. See constr., p. 15.

NON-ROUTINE PROBLEMS

A fly is walking from a spot A on the ceiling of a room to a spot B on the floor. What is the shortest path the fly can take? How can you show that it is the shortest possible path?

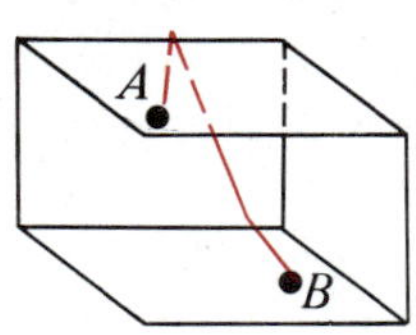

CONGRUENT QUADRILATERALS 8.4

Objectives **To identify congruent quadrilaterals**
To prove theorems about congruent quadrilaterals

Parallelograms, squares, rectangles, rhombi, and trapezoids are specific types of quadrilaterals. In order to prove certain theorems about them, additional properties of quadrilaterals are needed.

Congruent quadrilaterals have the same size and shape. The size of a quadrilateral depends upon the lengths of the sides while the shape depends upon the angle measures.

quad $ABCD \cong$ quad $EFGH$

Definition: Congruent quadrilaterals Two *quadrilaterals* are *congruent* if their corresponding sides are congruent and their corresponding angles are congruent.

A congruence statement specifies which sides and angles correspond.

quad $ABCD \cong$ quad $EFGH$

Example 1 Quad. $JKLM \cong$ quad. $NOPQ$.
Name the pairs of congruent sides.
Name the pairs of congruent angles.

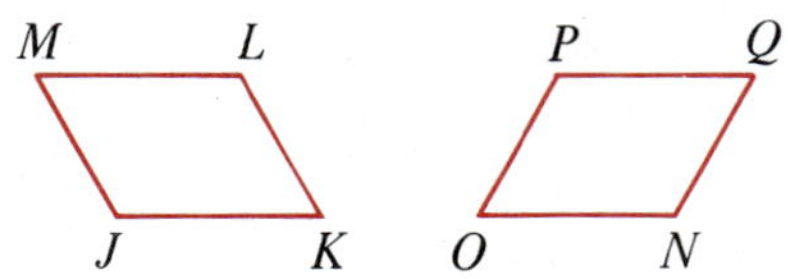

Congruent sides: $\overline{JK} \cong \overline{NO}$, $\overline{KL} \cong \overline{OP}$, $\overline{LM} \cong \overline{PQ}$, $\overline{MJ} \cong \overline{QN}$
Congruent angles: $\angle J \cong \angle N$, $\angle K \cong \angle O$, $\angle L \cong \angle P$, $\angle M \cong \angle Q$

Example 2 According to the markings shown, what is a correct congruence statement for the two trapezoids?

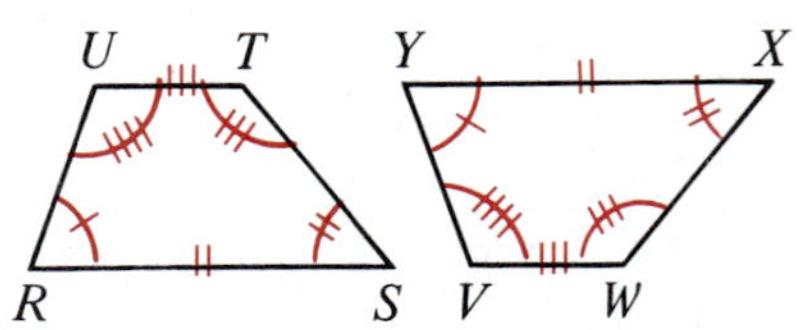

Trap. $RSTU \cong$ trap. $YXWV$

You can prove that two quadrilaterals are congruent without showing that all four pairs of sides and all four pairs of angles are congruent. The following theorem shows one of several possible congruence patterns.

Theorem 8.6 SASAS for ≅ quadrilaterals Two quadrilaterals are congruent if any three sides and the included angles of one are congruent, respectively, to three sides and the included angles of the other.

Given: Quad. $ABCD$ and quad. $EFGH$, $\overline{DA} \cong \overline{HE}$, $\angle A \cong \angle E$, $\overline{AB} \cong \overline{EF}$, $\angle B \cong \angle F$, and $\overline{BC} \cong \overline{FG}$.

Prove: Quad $ABCD \cong$ quad $EFGH$

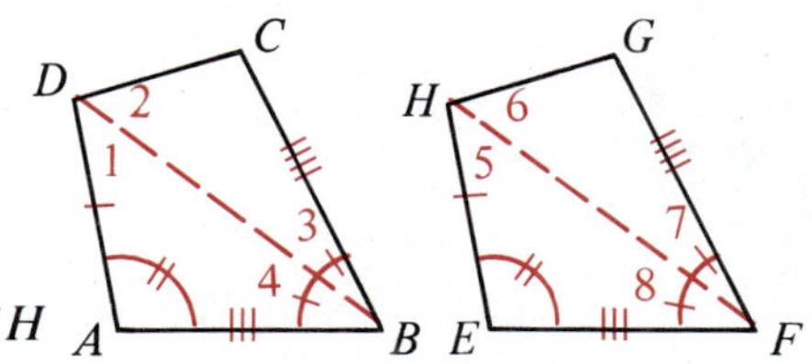

ANALYSIS

Draw diagonals. Show that corr. sides and ∠'s are corr. parts of ≅ △'s.

Outline of proof

1. Draw diagonals $\overline{DB}$ and $\overline{HF}$.
2. $\triangle DAB \cong \triangle HEF$
3. $\angle 1 \cong \angle 5$, $\overline{DB} \cong \overline{HF}$, $\angle 4 \cong \angle 8$
4. $\angle ABC \cong \angle EFG$
5. $\angle 3 \cong \angle 7$
6. $\triangle DBC \cong \triangle HFG$
7. $\overline{DC} \cong \overline{HG}$
8. $\angle C \cong \angle G$, $\angle 2 \cong \angle 6$
9. $\angle ADC \cong \angle EHG$
10. $\therefore$ Quad $ABCD \cong$ quad $EFGH$

In a similar manner, you can prove an angle-side-angle-side-angle congruence pattern.

Theorem 8.7 ASASA for ≅ quadrilaterals Two quadrilaterals are congruent if any three angles and the included sides of one are congruent, respectively, to three angles and the included sides of the other. (See Exercise 18.)

Theorems 8.6 and 8.7 can now be used to prove quadrilaterals congruent.

Example 3

Given: Square $ABEF$ and square $CBED$
Prove: Square $ABEF \cong$ square $CBED$

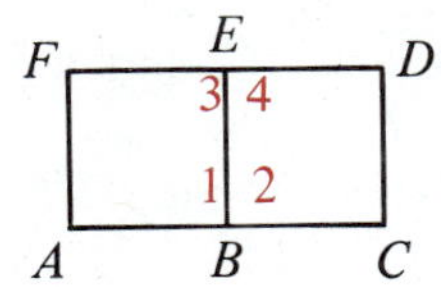

Proof

Statements	Reasons
1. Square $ABEF$ and square $CBED$	1. Given
2. $\overline{AB} \cong \overline{BE} \cong \overline{EF}$ and $\overline{CB} \cong \overline{BE} \cong \overline{ED}$	2. Def. of square
3. $\overline{AB} \cong \overline{CB}$ and $\overline{ED} \cong \overline{EF}$	3. Substitution property
4. $\overline{BE} \cong \overline{BE}$	4. Reflexive property
5. $\angle 1$, $\angle 2$, $\angle 3$, and $\angle 4$ are rt. ∠'s.	5. Def. of square
6. $\angle 1 \cong \angle 2$ and $\angle 3 \cong \angle 4$	6. All rt. ∠'s are ≅.
7. $\therefore$ Square ABEF $\cong$ square $CBED$	7. SASAS for ≅ quads.

Oral Exercises

Which of the following quadrilaterals appear to be congruent?

1.

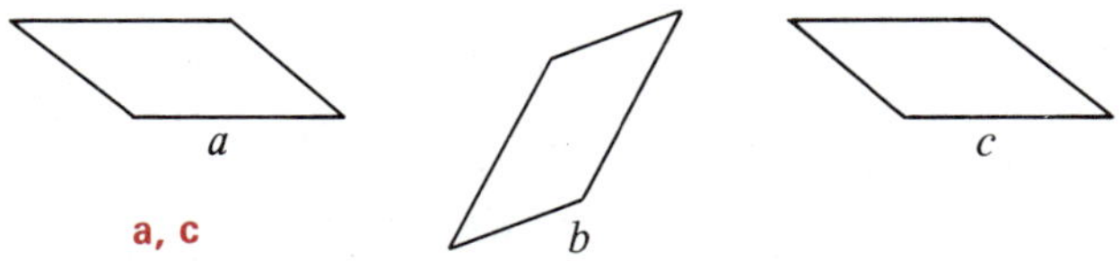

a, c

2.

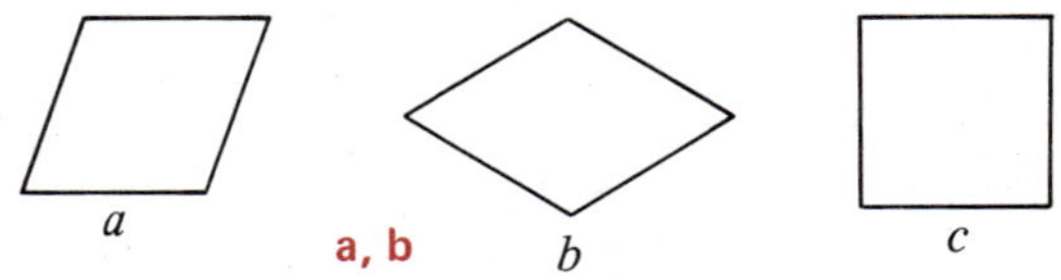

a, b

3.

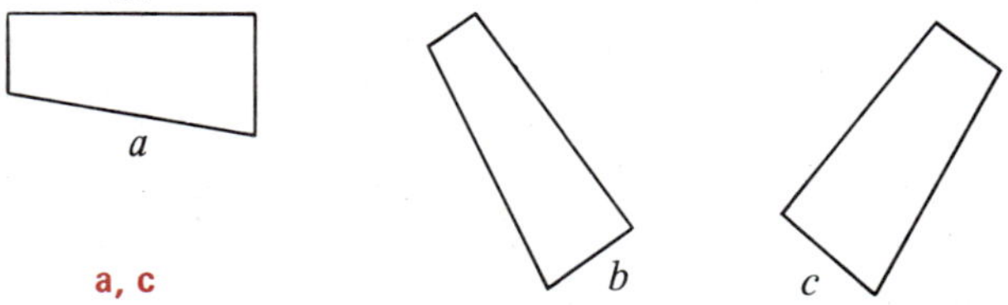

a, c

4.

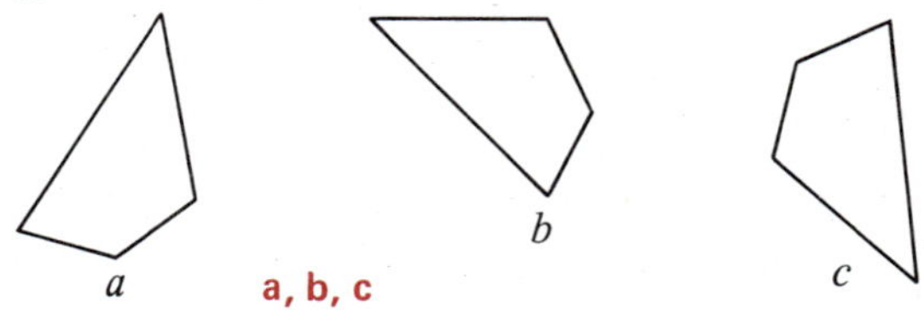

a, b, c

Written Exercises

For each pair of congruent quadrilaterals, name the pairs of congruent sides and the pairs of congruent angles.

1. $\overline{AB}, \overline{JK}; \overline{BC}, \overline{KL}; \overline{CD}, \overline{LM}; \overline{DA}, \overline{MJ}; \angle A, \angle J; \angle B, \angle K; \angle C, \angle L; \angle D, \angle M$

2. $\overline{EF}, \overline{NO}; \overline{FG}, \overline{OP}; \overline{GH}, \overline{PQ}; \overline{HE}, \overline{QN}; \angle E, \angle N; \angle F, \angle O; \angle G, \angle P; \angle H, \angle Q$

1. quad ABCD $\cong$ quad $JKLM$
2. quad $EFGH \cong$ quad $NOPQ$

According to the markings shown, what is a correct congruence statement? (Ex. 3–5)

A

3.

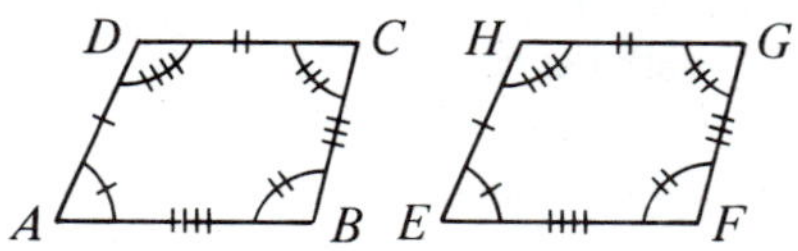

quad. ABCD $\cong$ quad. EFGH

4.

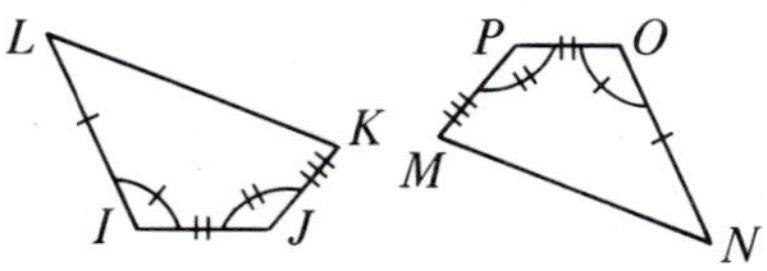

quad. IJKL $\cong$ quad. OPMN

5.

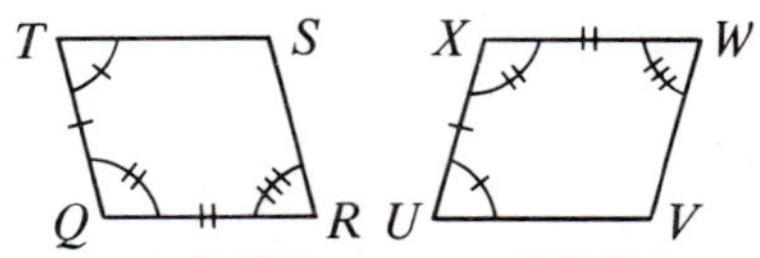

quad. RSTQ $\cong$ quad. WVUX

6. Given: $\overline{FE}$ is the $\perp$ bisector of $\overline{AB}$, $\overline{AD} \cong \overline{BC}$, and $\angle A \cong \angle B$.
Prove: Quad $AEFD \cong$ quad $BEFC$.

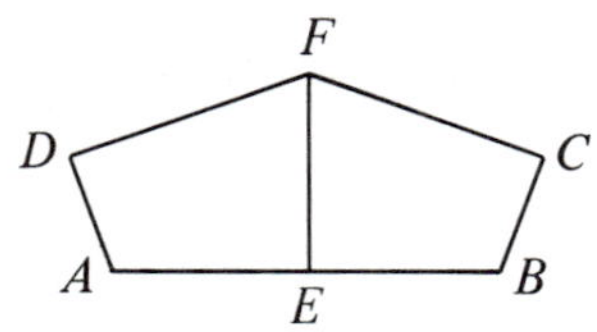

7. Given: $\overline{EF}$ bisects $\angle DFC$, $\overline{DF} \cong \overline{CF}$, and $\overline{FE}$ is the $\perp$ bisector of $\overline{AB}$
Prove: Quad $AEFD \cong$ quad $BEFC$.

B

8. Given: Rhombus $ABEF$ and rhombus $CBED$; $\angle 1 \cong \angle 2$
Prove: Rhombus $ABEF \cong$ rhombus $CBED$

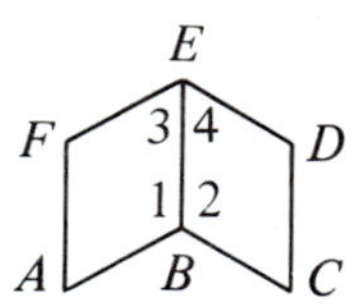

9. Given: Rhombus $ABEF$ and rhombus $CBED$; $\angle 1$ and $\angle 4$ are supplementary.
Prove: Rhombus $ABEF \cong$ rhombus $CBED$

10. Given: $\overline{DA} \cong \overline{CB}$, $\overline{EA} \cong \overline{FB}$, and $\angle A \cong \angle B$
Prove: Quad $EABC \cong$ quad $FBAD$

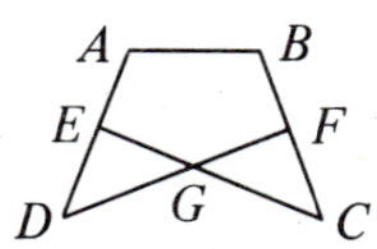

11. Given: $\triangle DEG \cong \triangle CFG$, and $\overline{EA} \cong \overline{FB}$
Prove: Quad $EABC \cong$ quad $FBAD$

12. Given: Trap. *HILM* and trap. *JILK*; $\angle 1 \cong \angle 2$, $\overline{HI} \cong \overline{JI}$, and $\angle MHI \cong \angle KJI$
Prove: Trap. *HILM* $\cong$ trap. *JILK*

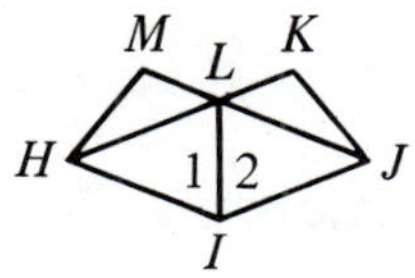

13. Given: Isos. trap. *HILM* and isos. trap. *JILK*; $\triangle HLM \cong \triangle JLK$
Prove: Trap. *HILM* $\cong$ trap. *JILK*

14. Given: $\overline{AE} \cong \overline{BD}$ and $\triangle FEG \cong \triangle CBG$
Prove: Quad *ABGF* $\cong$ quad *DEGC*

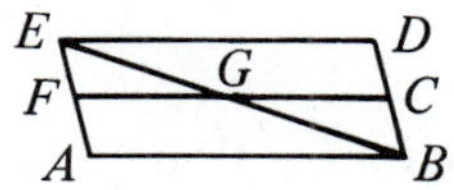

15. Given: $\triangle ABE \cong \triangle DEB$, $\overline{EF} \cong \overline{AF}$, $\overline{DC} \cong \overline{BC}$, and $\overline{EG} \cong \overline{GB}$
Prove: Quad *ABGF* $\cong$ quad *DEGC*

16. Give a complete proof of step 2 in the proof of Theorem 8.6.

17. Give a complete proof of steps 5 and 9 in the proof of Theorem 8.6.

C 18. Prove Theorem 8.7.

19. A SSS congruence correspondence is sufficient to show that two triangles are congruent. Is a similar (SSSS) pattern sufficient to show that two quadrilaterals are congruent? Give an argument in defense of your answer. **no**

20. Which of the following congruence patterns is sufficient to show that two quadrilaterals are congruent: SSASS, AASAA, SSAAS, AASSA, ASSAS? Give an argument in defense of your answer. **all except AASAA and SSASS**

ALGEBRA REVIEW

Objective **To solve fractional equations**

To solve a fractional equation:
(1) Multiply each side by the LCD.
(2) Solve the resulting equation.

Example

Solve $\frac{3x-5}{4} = \frac{2x+3}{3}$.

$$LCD = 12 \quad \blacktriangleright \quad 12\left(\frac{3x-5}{4}\right) = 12\left(\frac{2x+3}{3}\right)$$
$$3(3x-5) = 4(2x+3)$$
$$9x - 15 = 8x + 12$$
$$x = 27$$

Solve.

1. $\frac{4x-2}{3} = \frac{5x+1}{4}$ **11**

2. $\frac{2y-5}{3} = \frac{3y+2}{4}$ **−26**

3. $\frac{z+2}{3} = \frac{4z-3}{6} - \frac{z}{2}$ **−7**

4. $\frac{2w-3}{6} - \frac{w+5}{9} = \frac{w-1}{2}$ **−2**

5. $\frac{4}{t} + \frac{3}{5} = 3$ $\mathbf{\frac{5}{3}}$

6. $\frac{3}{4v} = \frac{5}{6} - \frac{2}{3v}$ $\mathbf{\frac{17}{10}}$

CONSTRUCTING QUADRILATERALS 8.5

Objective **To construct quadrilaterals with given properties**

You can use the unique properties of each type of quadrilateral to construct a given quadrilateral.

Example 1 Construct a parallelogram with adjacent sides congruent to two given segments and the included angle congruent to a given angle. Describe your strategy.

Strategy: Construct congruent opposite sides.

Construct $\angle A \cong$ given $\angle$. Construct $\overline{AB}$ so $AB = a$ and $\overline{AD}$ so $AD = b$.

Construct $\overline{DC}$ and $\overline{BC}$ so $DC = a$ and $BC = b$.

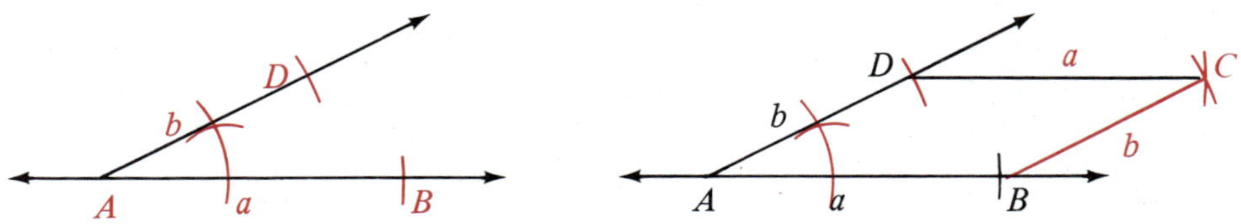

$ABCD$ is the required parallelogram.

Example 2 Construct a rhombus with diagonals congruent to two given segments. Describe your strategy.

Strategy: Construct diagonals that are perpendicular bisectors of each other.

Bisect segment of length b.
Construct $\overline{AB}$ so $AB = a$.
Construct $\perp$ bisector of $\overline{AB}$.

On $\perp$ bisector of $\overline{AB}$, mark off segments of length $\frac{1}{2}b$. Connect the endpoints.

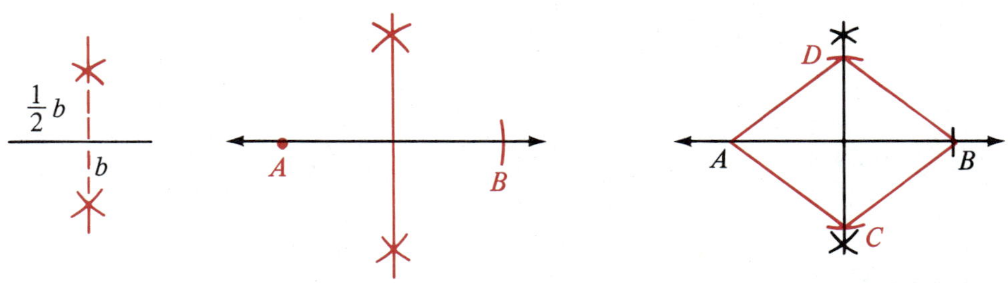

$ACBD$ is the required rhombus.

Written Exercises

Construct each of the following. Describe your strategy. (Ex. 1–14)

A

1. Draw a segment 7 cm long. Construct a square with sides congruent to the segment.

2. Draw two segments with different lengths. Construct a rectangle with sides congruent to these segments.

3. Construct a parallelogram with sides congruent to the given segments and an angle congruent to the given angle.

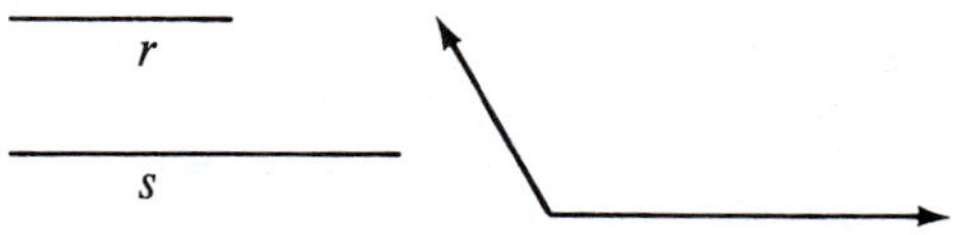

4. Construct a rhombus with sides congruent to the given segment and an angle congruent to the given angle.

5. Draw a segment. Construct a square with diagonals congruent to this segment.

6. Draw two segments. Construct a rhombus with diagonals congruent to these segments.

B

7. Draw a segment 10 cm long. Construct a rectangle with diagonals congruent to this segment and a side with half the length of a diagonal.

8. Draw a segment 12 cm long. Construct a parallelogram with one side congruent to this segment, a side half as long, and a 60° angle.

9. Construct an isosceles trapezoid with three sides of length *s* and diagonal of length *d*.

10. Construct a rhombus with sides and one diagonal congruent to the given segment.

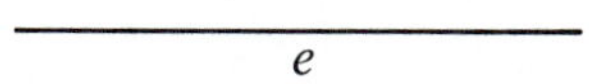

11. Construct a trapezoid with bases and a leg congruent to the given segments and a base angle measuring 45°.

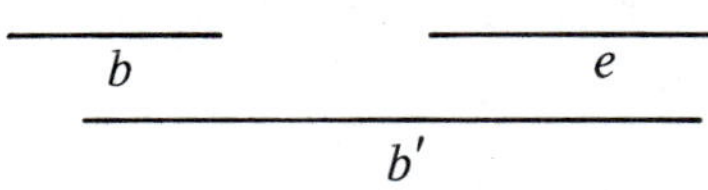

12. Construct an isosceles trapezoid with three sides of length *s* and diagonal of length *d*.

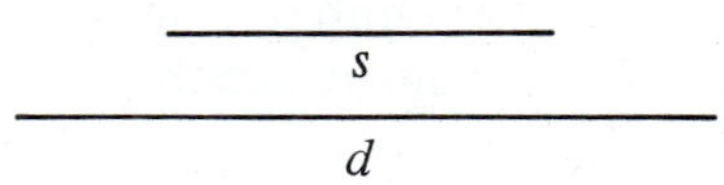

C

13. Construct a quadrilateral congruent to quad. *ABCD* using an angle-side-angle-side-angle pattern.

14. Construct a quadrilateral congruent to quad. *ABCD* using a side-angle-side-angle-side pattern.

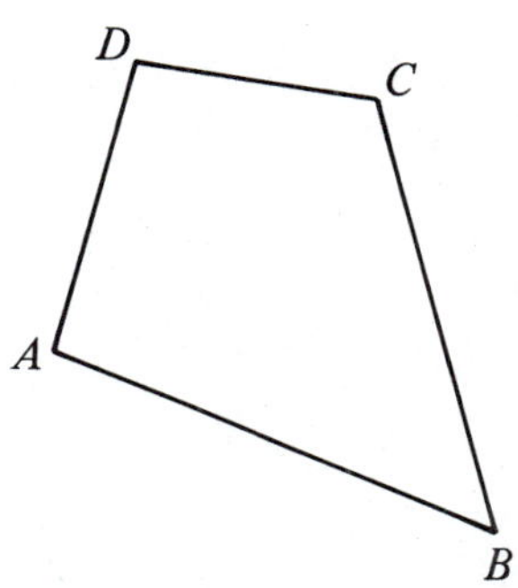

15. Is it possible to construct a quadrilateral congruent to quad. *ABCD* using a side-side-side-side pattern? **no**

16. Is it possible to construct a quadrilateral congruent to quad. *ABCD* using a side-side-side-side-angle pattern? **no**

SYMMETRIC POLYGONS (optional) 8.6

Objectives

To identify symmetric polygons
To determine and apply congruence properties of symmetric polygons

Some polygons are their own reflections. If the polygon were drawn on a piece of paper, you could fold it along a line so that the two halves would coincide. Such a polygon is called a **symmetric polygon**. The line is called a **line of symmetry**. Each of the following figures is a symmetric polygon.

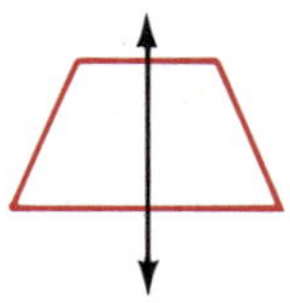
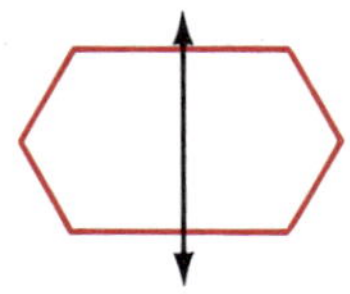
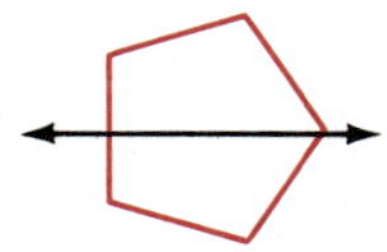

Postulate 21 A line of symmetry of a symmetric polygon is the perpendicular bisector of any segment joining a pair of corresponding points of the polygon.

In quadrilateral $ABCD$, $\overleftrightarrow{AC}$ is the line of symmetry. Points B and D are corresponding points. $\overline{CD}$ and $\overline{CB}$ are corresponding sides, as are $\overline{AD}$ and $\overline{AB}$. According to Postulate 21 the line of symmetry, $\overleftrightarrow{AC}$, bisects $\overline{DB}$ and is perpendicular to it.

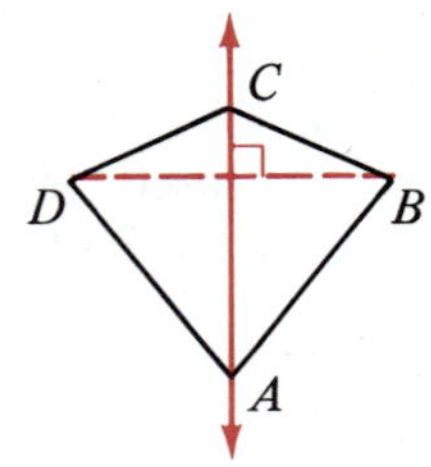

Some polygons have more than one line of symmetry.

Example 1 Draw all lines of symmetry of the symmetric polygon. How many are there?

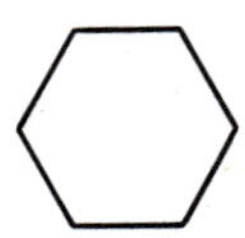
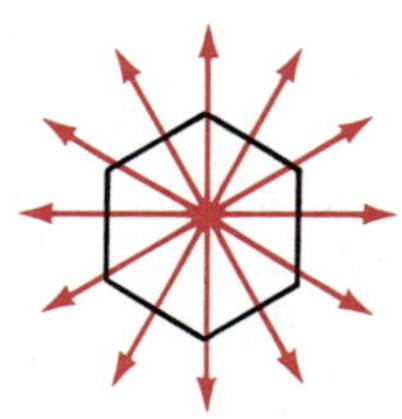

There are six lines of symmetry.

The figure in Example 1 is a regular hexagon. Notice that there are six lines of symmetry for this regular six-sided figure.

The following theorems state two important properties of symmetric polygons. The proof outlined for Theorem 8.8 is for a hexagon.

Theorem 8.8 Corresponding sides of a symmetric polygon are congruent.

Given: Polygon $ABCDEF$, with line of symmetry $\overleftrightarrow{AD}$
Prove: $\overline{AB} \cong \overline{AF}$, $\overline{BC} \cong \overline{FE}$, and $\overline{CD} \cong \overline{ED}$

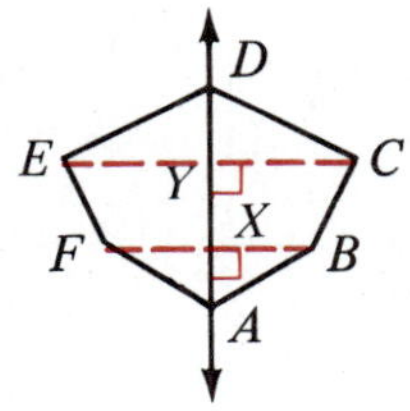

ANALYSIS
Use congruent quadrilaterals.

Outline of proof

1. Draw diagonals $\overline{EC}$ and $\overline{FB}$, intersecting $\overleftrightarrow{AD}$ at Y and X.
2. $\overleftrightarrow{AD}$ is the $\perp$ bisector of $\overline{FB}$ and $\overleftrightarrow{AD}$ is the $\perp$ bisector of $\overline{EC}$.
3. $\therefore \overline{AB} \cong \overline{AF}$ and $\overline{CD} \cong \overline{ED}$
4. $\overline{BX} \cong \overline{FX}$ (S)
5. $\angle BXY \cong \angle FXY$ (A)
6. $\overline{XY} \cong \overline{XY}$ (S)
7. $\angle XYC \cong \angle XYE$ (A)
8. $\overline{CY} \cong \overline{EY}$ (S)
9. Quad. $BXYC \cong$ quad. $FXYE$
10. $\therefore \overline{BC} \cong \overline{FE}$

(See Exercise 20.)

Theorem 8.9 Corresponding angles of a symmetric polygon are congruent. (See Exercises 18 and 19.)

Using these theorems, it is possible to identify which sides and which angles of a symmetric polygon are congruent.

Example 2 If $ABCDE$ is symmetric about $\overleftrightarrow{AX}$, which sides and which angles are congruent?

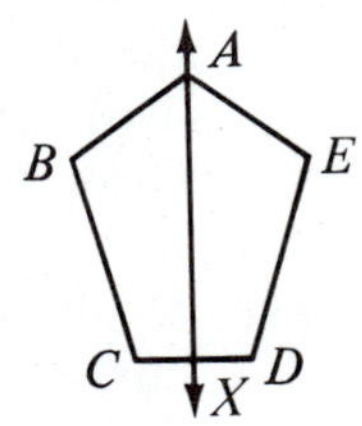

Sides: $\overline{AB} \cong \overline{AE}$, $\overline{BC} \cong \overline{ED}$
Angles: $\angle B \cong \angle E$, $\angle C \cong \angle D$

Oral Exercises

Which of the following polygons appear to be symmetric?

1. 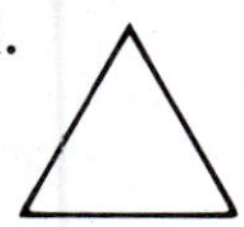 yes

2. 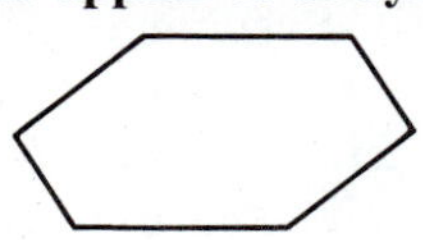 no

3. 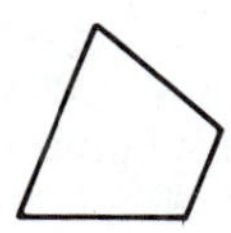 yes

4.  yes

Written Exercises

Copy each figure. Draw all the lines of symmetry of each polygon. How many are there?

A 1.

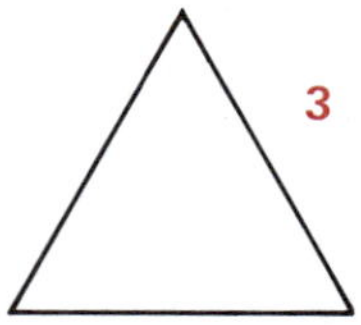

3

2.

4

3.

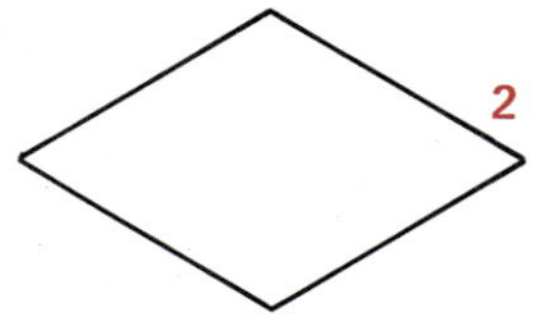

2

4.

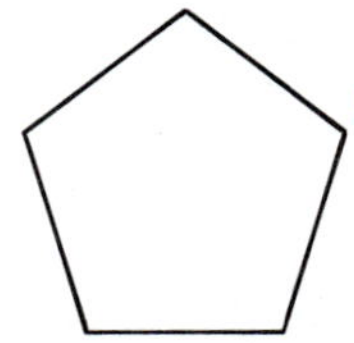

5

Each polygon is symmetric about the line of symmetry shown. Which sides and which angles are congruent?

5.

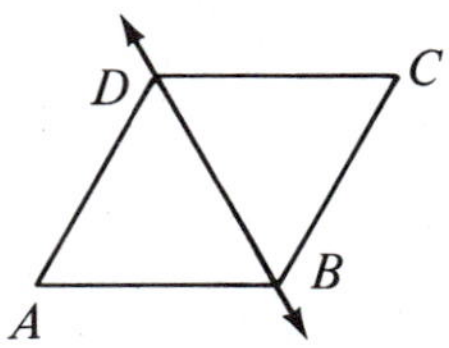

6.

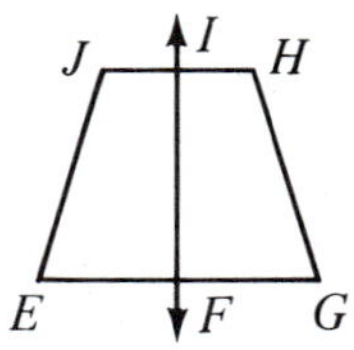

7.

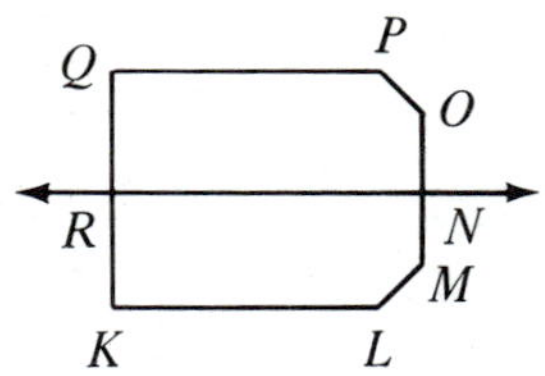

8.

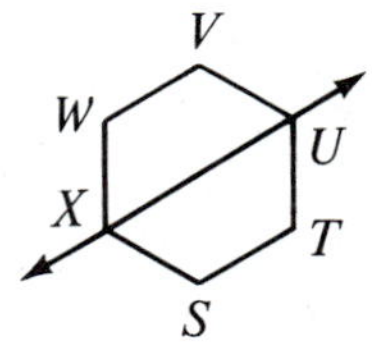

Copy each figure and find the perpendicular bisector of the segment joining each pair of corresponding points. Based on your results, is the figure symmetric? (Ex. 9–11)

B 9.

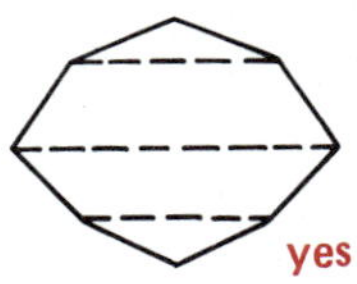

yes

10.

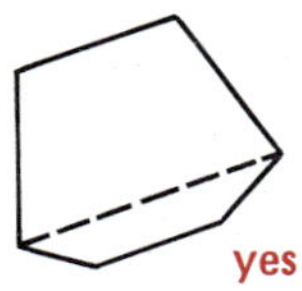

yes

11.

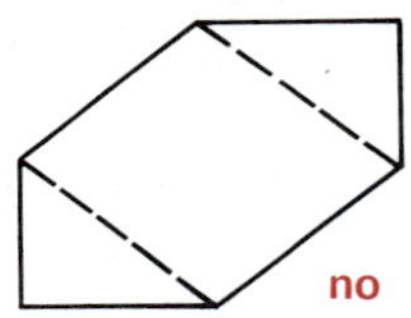

no

12. Is it possible for a quadrilateral to have exactly one line of symmetry? Exactly two? Exactly three? More than three? Draw pictures to defend your answer. **Yes; yes; no; yes**

13. Is it possible for a pentagon to have exactly one line of symmetry? Exactly two? Exactly three? Exactly four? More than four? Draw pictures to defend your answer. **Yes; no; no; no; yes**

14. Is there any relationship between the number of sides of a regular polygon and the number of lines of symmetry? If so, what is it? **Yes, they are the same.**

15. Can a decagon be drawn with exactly one line of symmetry? Exactly two? Draw pictures to defend your answer. **Yes; no**

C 16. If corresponding sides of a polygon are congruent, then is the polygon necessarily symmetric? **no**

17. If corresponding angles of a polygon are congruent, then is the polygon necessarily symmetric? **no**

18. Prove Theorem 8.9 using a hexagon.

19. Prove Theorem 8.9 using an octagon.

20. Complete the proof of Theorem 8.8.

21. Theorem 8.8 was proved using a hexagon. Prove it using a heptagon.

CONGRUENCE AND REFLECTIONS (optional) 8.7

Objectives

To identify reflections
To prove and apply theorems relating reflections and congruence

When you place an object before a mirror, you see the reflected image of that object in the mirror. In geometry, the concept of reflection is similar except that the mirror is replaced by a line of symmetry.

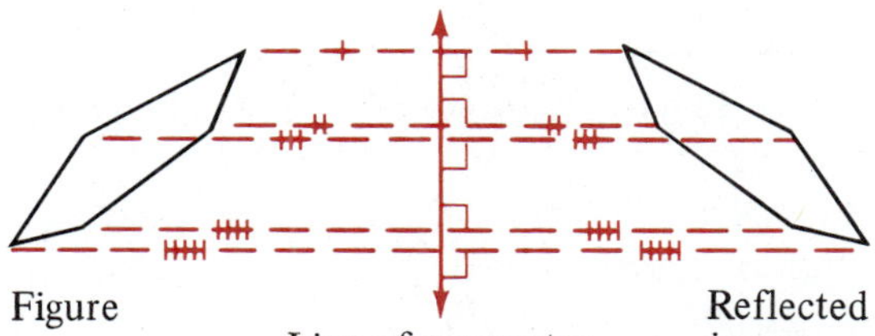

Definition: Reflection Two polygons are *reflections* of each other if all segments joining pairs of corresponding points have the same perpendicular bisector.

You can use this definition to draw reflections and to determine if two polygons are reflections of each other.

Example 1 Draw the reflection of the given figure about the given line of symmetry.

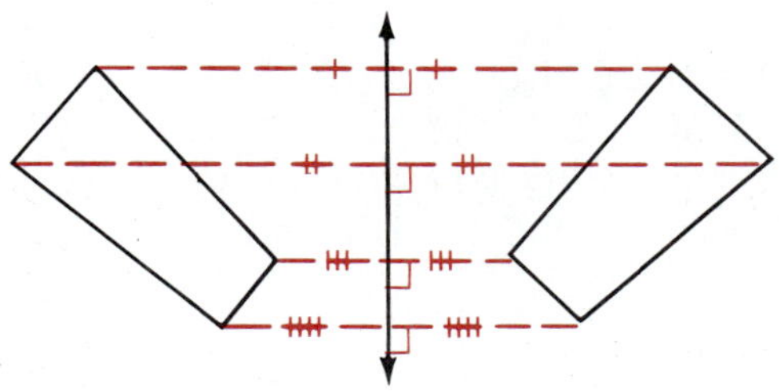

Example 2 Show that the two figures below are *not* reflections of each other.

Use the construction for a ⊥ bisector. ►

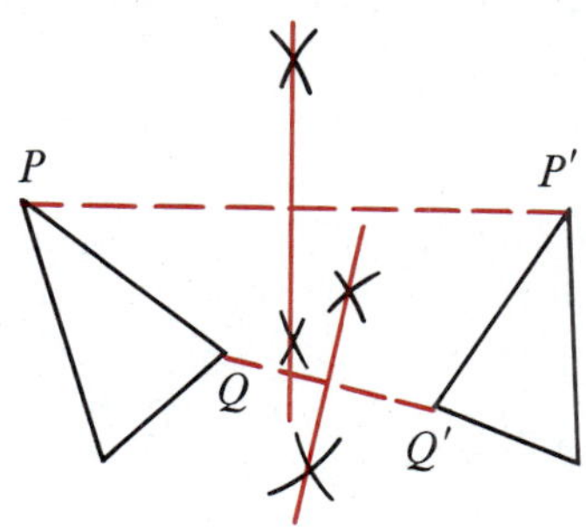

The ⊥ bisectors of $\overline{PP'}$ and $\overline{QQ'}$ are not the same.

If two figures are drawn on a sheet of paper and the paper can be folded so that the figures coincide, then the two figures are reflections of each other. The fold is the line of symmetry. If the paper cannot be folded so that they coincide, then the figures are not reflections of each other.

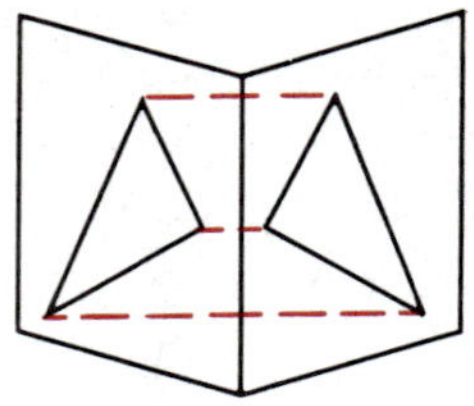

Since reflections drawn on paper can be made to coincide, they must have the same size and shape. This leads to the following theorem and corollary.

Theorem 8.10

If two triangles are reflections, then they are congruent.

Given: $\triangle A'B'C'$ is a reflection of $\triangle ABC$.
Prove: $\triangle ABC \cong \triangle A'B'C'$

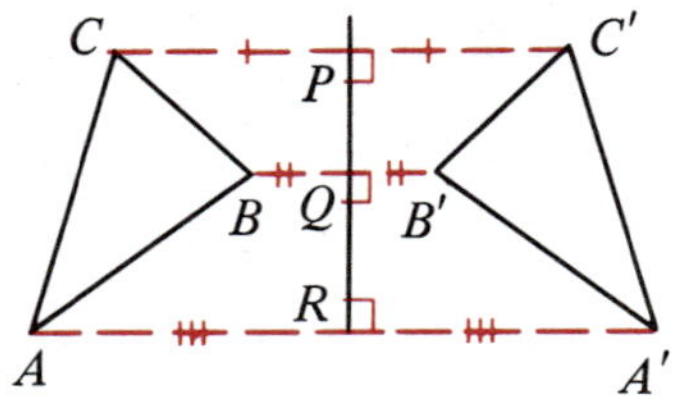

ANALYSIS
Use $\cong$ quadrilaterals to show $\cong$ sides.

Outline of proof

1. $\overline{CP} \cong \overline{C'P}$ (S)
2. $\angle CPQ \cong \angle C'PQ$ (A)
3. $\overline{PQ} \cong \overline{PQ}$ (S)
4. $\angle PQB \cong \angle PQB'$ (A)
5. $\overline{BQ} \cong \overline{B'Q}$ (S)
6. Quad. $CBQP \cong$ quad. $C'B'QP$
7. $\overline{CB} \cong \overline{C'B'}$ and $\angle CBQ \cong \angle C'B'Q$
8. $\overline{BQ} \cong \overline{B'Q}$ (S)
9. $\angle BQR \cong \angle B'QR$ (A)
10. $\overline{QR} \cong \overline{QR}$ (S)
11. $\angle ARQ \cong \angle A'RQ$ (A)
12. $\overline{AR} \cong \overline{A'R}$ (S)
13. Quad. $ARQB \cong$ quad. $A'RQB'$
14. $\overline{AB} \cong \overline{A'B'}$ and $\angle ABQ \cong \angle A'B'Q$
15. $m\angle CBQ = m\angle C'B'Q$; $m\angle QBA = m\angle QB'A'$
16. $m\angle ABC + m\angle CBQ + m\angle QBA = 360$; $m\angle A'B'C' + m\angle C'B'Q + m\angle QB'A' = 360$
17. $m\angle ABC + m\angle CBQ + m\angle QBA = m\angle A'B'C' + m\angle C'B'Q + m\angle QB'A'$
18. $m\angle ABC = m\angle A'B'C'$; $\angle ABC \cong \angle A'B'C'$
19. $\therefore \triangle ABC \cong \triangle A'B'C'$

Corollary

If two quadrilaterals are reflections, then they are congruent. (See Exercise 17.)

Since a figure and its reflections are congruent, you can determine congruent corresponding sides and angles. If $\triangle ABC$ is a reflection of $\triangle EDF$, then $\triangle ABC \cong \triangle EDF$. $\overline{AB} \cong \overline{ED}$, $\overline{BC} \cong \overline{DF}$, and $\overline{CA} \cong \overline{FE}$. $\angle A \cong \angle E$, $\angle B \cong \angle D$, and $\angle C \cong \angle F$.

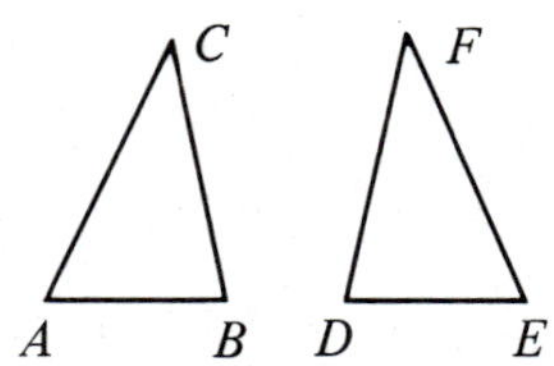

Written Exercises

Copy each picture. Draw the reflection of the given figure about the given line of symmetry.

A **1.**

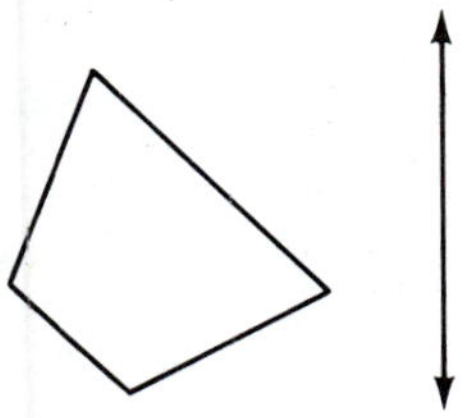

2.

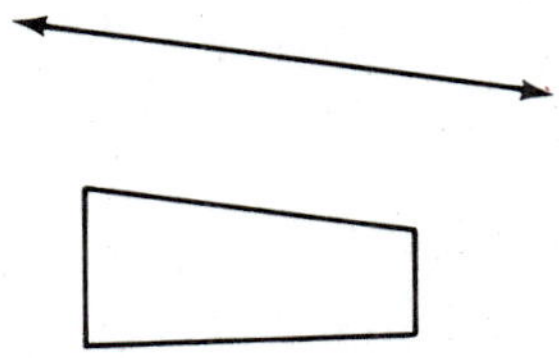

3.

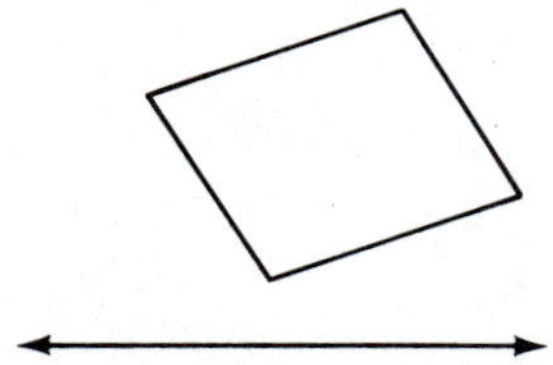

Copy each picture. Determine which are reflections by constructing the line of symmetry. Check your work by folding the paper along the line of symmetry.

4.

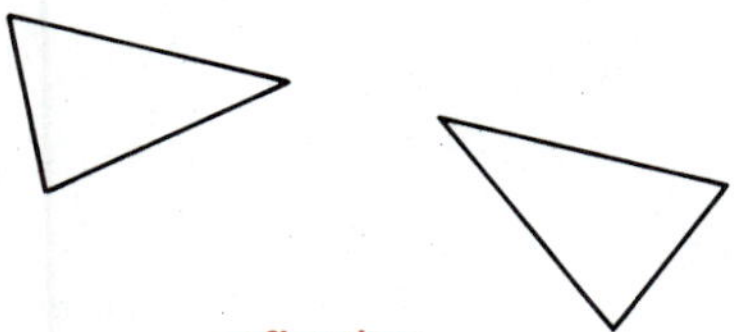

reflection

5.

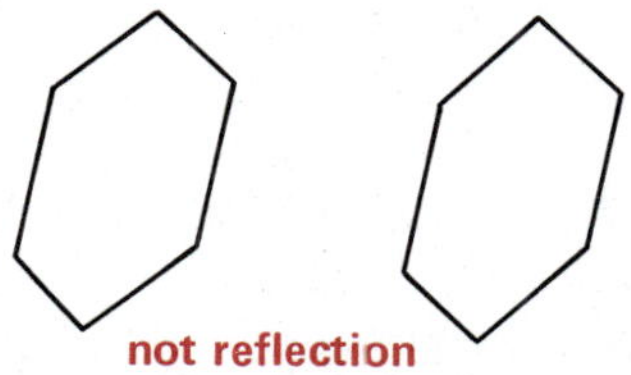

not reflection

6.

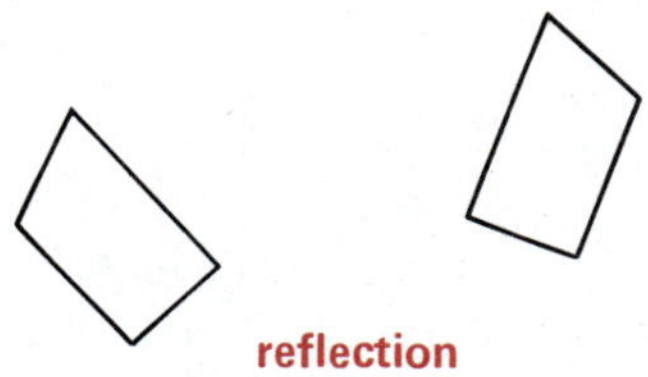

reflection

In each pair of reflections list the corresponding congruent sides and angles. (Ex. 7–9) $\angle C, \angle F; \angle B, \angle D; \angle A, \angle E$ $\angle J, \angle R; \angle I, \angle M; \angle H, \angle N; \angle G, \angle O.$ $\angle S, \angle U; \angle Y, \angle V; \angle W, \angle X$

7.

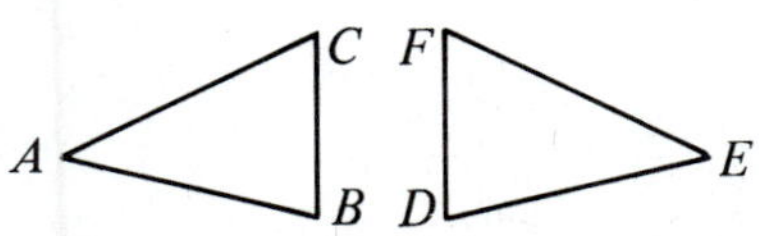

$\overline{AC}, \overline{EF}; \overline{AB}, \overline{ED}; \overline{CB}, \overline{FD}$

8.

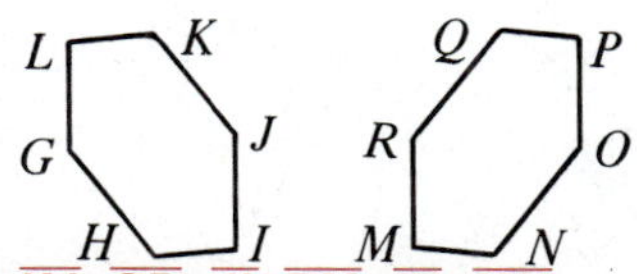

$\overline{LK}, \overline{PQ}; \overline{KJ}, \overline{QR}; \overline{JI}, \overline{RM}; \overline{IH}, \overline{MN}; \overline{HG}, \overline{NO}; \overline{GL}, \overline{OP}; \angle L, \angle P; \angle K, \angle Q$

9.

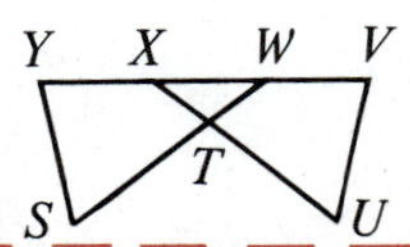

$\overline{YS}, \overline{VU}; \overline{SW}, \overline{UX}; \overline{YW}, \overline{VX}$

B **10.** $\triangle ABC$ is a reflection of $\triangle DEF$. Is $\triangle ABC \cong \triangle DEF$? Why?

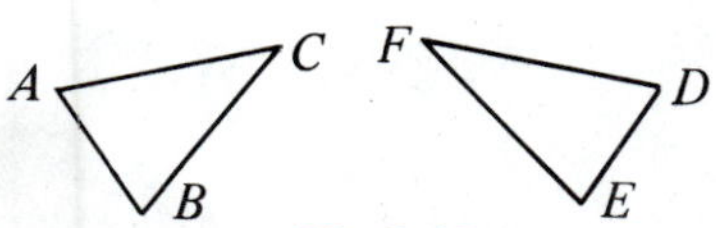

yes, Th. 8.10

11. Trap. *GHIJ* is a reflection of trap. *LKNM*. Describe the line of symmetry.

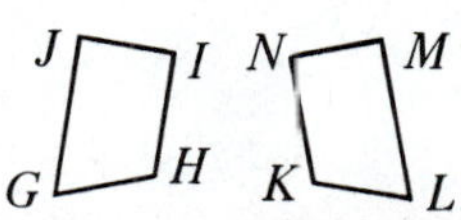

It is common $\perp$ bis. of $\overline{GL}$, $\overline{HK}$, $\overline{IN}$, and $\overline{JM}$.

12. *l* is the perpendicular bisector of $\overline{OS}$, $\overline{PT}$, $\overline{QU}$, and $\overline{RV}$. Is $OPQR \cong STUV$? Why?

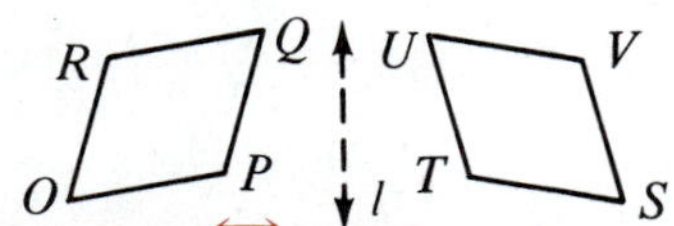

Yes; $\overleftrightarrow{QP}$ is a line of symm; 2 quads. are reflections; Th 8.10, Corr.

True or false?

13. Reflected triangles are congruent. T

14. Reflected polygons are congruent. T

15. Congruent triangles are reflections. F

16. Congruent polygons are reflections. F

C **17.** Prove the corollary to Theorem 8.10.

18. If a triangle is reflected about a line and the image is then reflected about the same line, what is the result?

The second reflected image is the original $\triangle$.

19. Describe a construction for finding the line of symmetry of reflections.

Connect any two corresponding pts, then construct $\perp$ bis. of that segment.

APPLICATIONS

Read → Plan → Solve → Interpret

1. A science teacher has been asked to select new tables that can be arranged in different ways. In what other ways can the tables shown be arranged so that the edges touch? Can eight of these same tables be arranged as an octagon? If not, what differences must be made in the design?

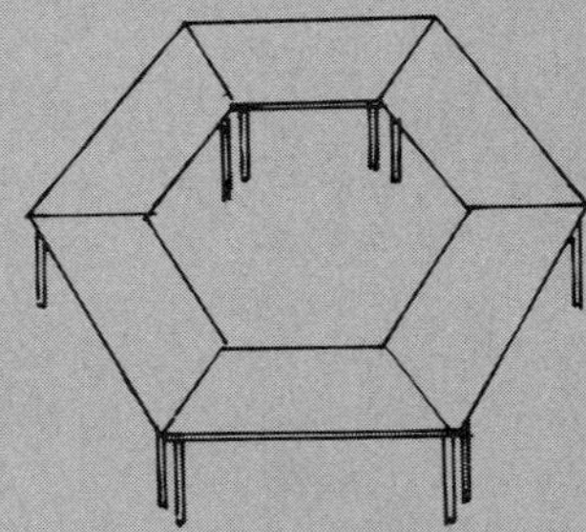

2. Flat trusses are often built in an isosceles trapezoidal pattern. Would such a truss be rigid with no internal braces? If the braces are in the shape of equilateral triangles, with sides 2 ft long, what are the dimensions of the truss?

COMPUTER ACTIVITIES

Isosceles Trapezoids

Objective: **To determine the lengths of the median and altitude as well as the area of an isosceles trapezoid by using a computer program that inputs the lengths of its two bases and one of its nonparallel sides**

At the right is an isosceles trapezoid with bases b_1 and b_2 and with congruent sides of length s. Dropping perpendiculars as shown forms two congruent right triangles with bases $(b_2 - b_1)/2$. From these, you can determine the length of an altitude of the trapezoid. The computer program below uses the BASIC square root function, SQR, to find the length of an altitude and a median as well as the area.

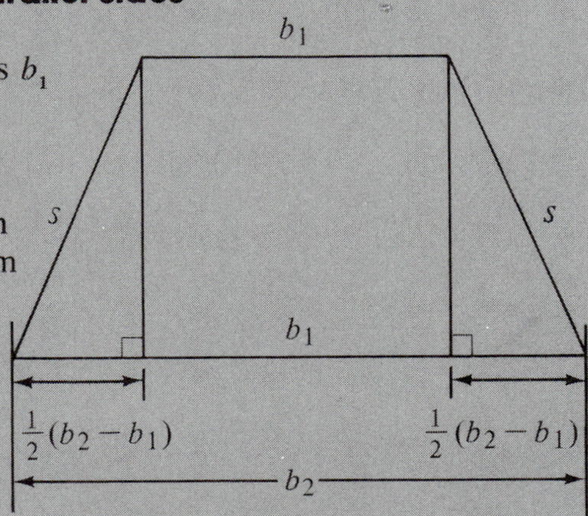

Line 130 tests to see if A < B. If not, the two values are interchanged in line 140. Line 160 calculates the median, line 170 the altitude, and line 180 the area.

```
110  PRINT "WHAT ARE THE 2 BASES OF THE TRAPEZOID?"
120  INPUT A,B
130  IF A < B THEN 150
140 H = A:A = B:B = H
150  PRINT "WHAT IS THE LENGTH OF ONE OF THE CONGRUE
     NT SIDES": INPUT C
160 M = (A + B) / 2
170 H =  SQR (C ^ 2 - ((B - A) / 2) ^ 2)
180 R = M * H
190  PRINT "THE MEDIAN IS ";M"."
200  PRINT "THE ALTITUDE IS ";H"."
210  PRINT "THE AREA IS ";M * H"."
220  PRINT : GOTO 110
```

EXERCISES: See the Computer Section beginning on page 565 for more information.

1. Write a program that calculates the length of the second base of a trapezoid from inputs of the first base and the median.
2. Alter the program you wrote for Exercise 1 so that it gives a quiz on calculating the second base of a trapezoid from randomly generated numbers (see pages 573–574) for the first base and median. Also, make your program print your quiz score.
3. Write a program that calculates the length of an altitude of a trapezoid from inputs of both bases and the area.

CHAPTER EIGHT REVIEW

Identify each of the following. [8.2]

1. The median of trap. $ABCD$ $\overline{EF}$
2. An altitude of trap. $ABCD$ $\overline{GH}$
3. Bases of trap. $ABCD$ $\overline{AB}, \overline{DC}$
4. Legs of trap. $ABCD$ $\overline{AD}, \overline{BC}$

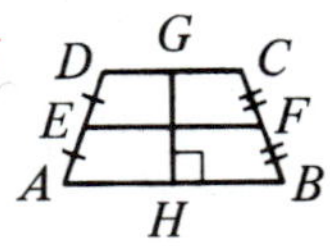

Which of the following appear to be isosceles trapezoids? [8.3]

5. isos.

6. isos.

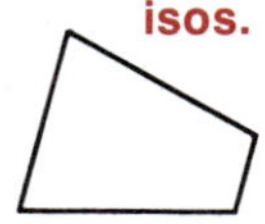

True or false? Draw pictures to defend your answers.

7. Two sides of a trapezoid are congruent. [8.1] F
8. An isosceles trapezoid is a symmetric quadrilateral. [8.3] T
9. Consecutive angles of a trapezoid are supplementary. [8.1] F
10. The median of a trapezoid is parallel to a base. [8.2] T
11. Quadrilaterals can be proved congruent by an ASASA correspondence. [8.4] T
12. All congruent quadrilaterals are reflections of each other. [8.7] F
13. If a quadrilateral has congruent diagonals, then the quadrilateral is a trapezoid. [8.1] F
14. Diagonals of an isosceles trapezoid are congruent. [8.3] T
15. A line of symmetry is the perpendicular bisector of the segment connecting corresponding points of a figure and its reflection. [8.6] T

Vocabulary

altitude [8.2], isosceles trapezoid [8.3], line of symmetry [8.6], median [8.2], symmetric polygon [8.6], trapezoid [8.1]

Construct each of the following. (Ex. 16–17)

16. An isosceles trapezoid with a base of length a, a leg of length b, and a base angle congruent to the given angle. [8.5]

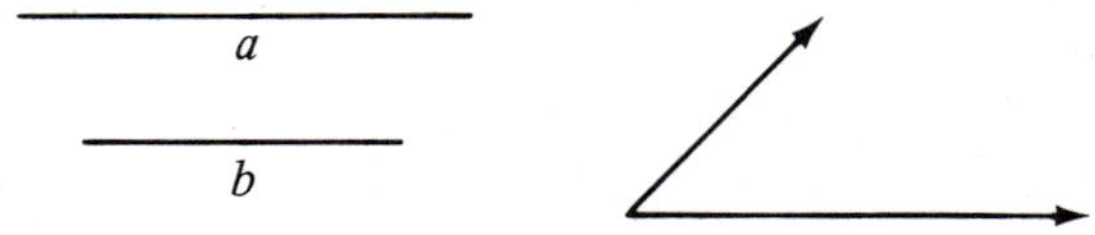

17. An isosceles trapezoid with a base of length a, an altitude of length b, and a leg of length c.

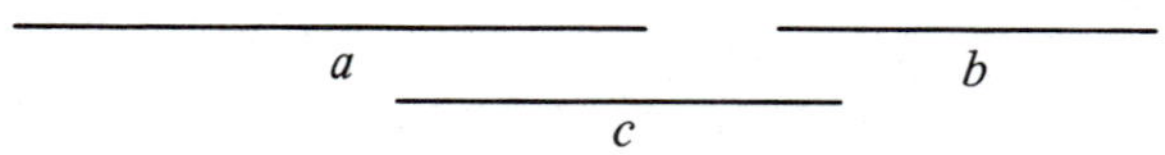

18. Copy the reflections below. Construct their line of symmetry. [8.6]

19. Given: $AB = 13$ and $CD = 9$ [8.2]
    Find EF. 11

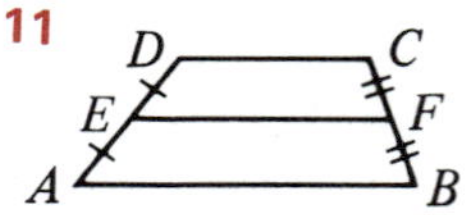

20. Given: $AB = 7x - 5$, $EF = 3x + 1$, and $DC = 2x + 1$ [8.2]
    Find EF. 7

21. Given: $\overline{BF}$ and $\overline{DG}$ bisect each other, $\overline{GA} \parallel \overline{ED}$, and $\overline{AB} \parallel \overline{FE}$.
    Prove: $CGAB \cong CDEF$ [8.4]

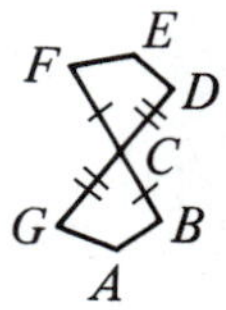

*22. Given: Isos. trap. $HIJK$, $\overline{HK} \cong \overline{IJ}$, and $\overline{KJ} \parallel \overline{HI}$
    Prove: $\triangle HIK \cong \triangle IHJ$ [8.3]

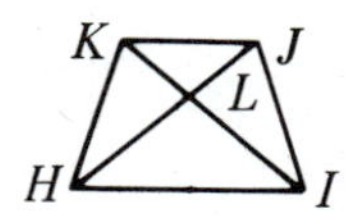

The following exercises are types that appear in the (A) section of the lesson's exercises: 1–7, 18–21. The starred exercises are types that appear in the (C) section. All the remaining exercises appear in the (B) section.

CHAPTER EIGHT TEST

Identify each of the following.

1. A median of trap. $HJLM$ $\overline{DB}$
2. An altitude of trap. $PQRS$ $\overline{PQ}$

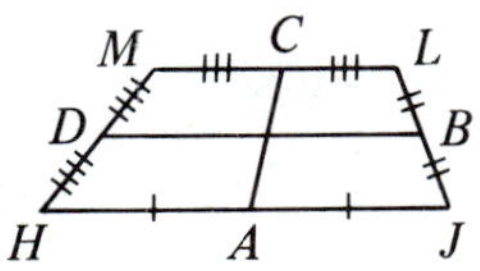

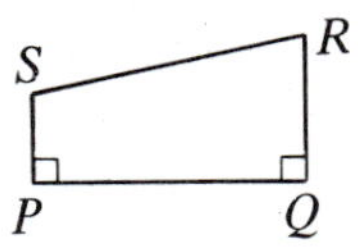

Which of the following appear to be isosceles trapezoids?

3.

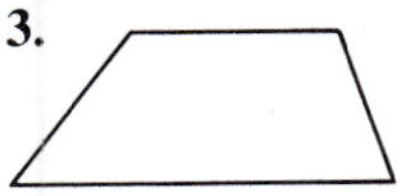

not isos.

4.

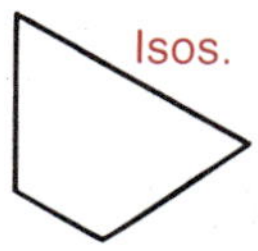

Isos.

True or false? Draw pictures to defend your answers.

5. A trapezoid has one pair of parallel sides. T
6. The diagonals of a trapezoid are congruent. F
7. If a trapezoid has congruent diagonals, then it is isosceles. T
8. Opposite angles of an isosceles trapezoid are supplementary. T
9. Quadrilaterals can be proved congruent by a SSSS correspondence. F
10. A reflection of a figure has the same shape as the figure. T
11. A symmetric quadrilateral may have exactly two lines of symmetry. T
12. The reflection of an isosceles trapezoid is an isosceles trapezoid. T
13. A line of symmetry of a symmetric polygon is the perpendicular bisector of the segment connecting corresponding points of the polygon. T

14. Construct an isosceles trapezoid with one base of length a, a second base of length b, and an altitude of length c.

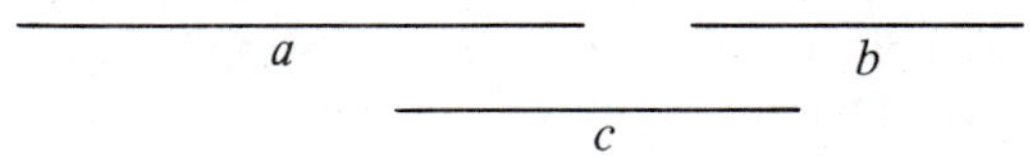

15. Copy the figures below. Show by constructing the perpendicular bisectors of the segments joining two sets of corresponding points that the polygons are not reflections of each other.

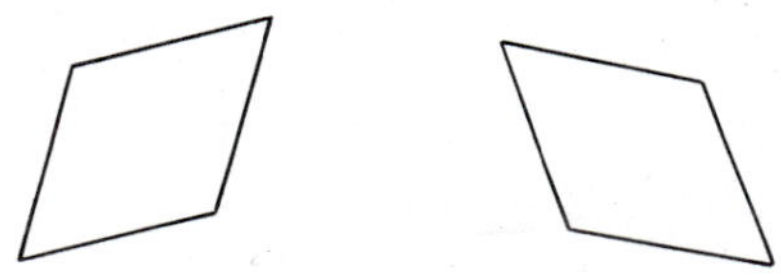

16. Given: $AB = 12$ and $CD = 18$
    Find EF. 15

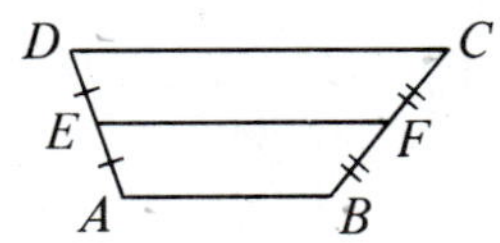

17. Given: $AB = x + 3$, $CD = 4x - 3$, and $EF = 3x - 1$ 5
    Find EF.

18. Prove: The diagonals of an isosceles trapezoid are congruent. **Ex. 16, p. 249 (Th. 8.4)**

*19. Given: $\overline{EB}$ is the $\perp$ bisector of $\overline{AC}$, $\overline{EB}$ bisects $\angle FED$, and $\overline{FE} \cong \overline{DE}$.
    Prove: Quad $ABEF \cong$ quad $CBED$

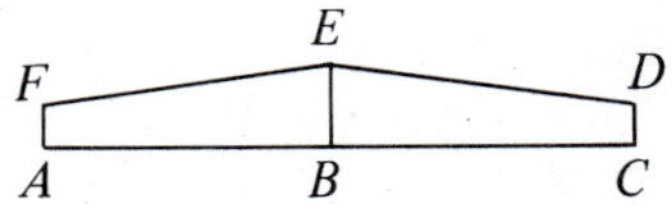

The following exercises are types that appear in the (A) section of the lesson's exercises: 1–7, 15–16. The starred exercises are types that appear in the (C) section. All the remaining exercises appear in the (B) section.

COLLEGE PREP TEST

For each item choose the best answer. Indicate your answer by writing the letter of this choice.

1. Which of these is not a property of all trapezoids? D
- (A) Two sides are parallel.
- (B) Diagonals intersect.
- (C) Two pairs of angles are supplementary.
- (D) Diagonals are congruent.
- (E) None of these

2. Given: Trap. *ABCD*, median $\overline{EF}$, $DC = 20x + 2$, $EF = 26x$, and $AB = 24x + 4$. Which of the statements are true? E

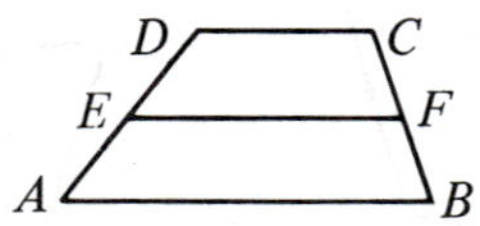

I $AB = 22$

II $EF = 19\frac{1}{2}$

III $DC = 17$

- (A) I only
- (B) II only
- (C) III only
- (D) I and III only
- (E) I, II, and III

3. Which of these is sufficient to prove that a quadrilateral is an isosceles trapezoid? E
- (A) Diagonals are congruent.
- (B) Diagonals are perpendicular.
- (C) Opposite angles are supplementary.
- (D) Two angles are congruent.
- (E) None of these

4. You can prove two quadrilaterals congruent by C
- (A) SSSS
- (B) AAAA
- (C) SASAS
- (D) SAAAA
- (E) SSASS

5. To construct a rhombus given the two diagonals, you must D
- (A) construct a triangle with the diagonals as two of the sides.
- (B) construct a perpendicular to one segment.
- (C) bisect one segment.
- (D) construct a perpendicular bisector of one segment.
- (E) bisect an angle.

6. Which of the following are true? E

I A symmetric triangle can have exactly two lines of symmetry.

II A symmetric quadrilateral can have exactly two lines of symmetry.

III A symmetric quadrilateral can have exactly four lines of symmetry.

- (A) I only
- (B) II only
- (C) III only
- (D) I and II only
- (E) II and III only

7. Which of the following is *not* true? B
- (A) A reflection is congruent to the original figure.
- (B) Corresponding sides of a figure and its reflection are parallel.
- (C) Corresponding sides of a figure and its reflection are congruent.
- (D) The line of symmetry bisects a segment connecting corresponding points of a figure and its reflection.
- (E) The line of symmetry is perpendicular to a segment connecting corresponding points of a figure and its reflection.

CUMULATIVE REVIEW

Chapters 1–8

Write the letter that indicates the best answer.

1. Which are measures of supplementary angles? **C**
   (A) 40° and 50° (B) 32° and 68°
   (C) 95° and 85° (D) none of these

2. If two coplanar lines are perpendicular to the same line, then they are ______ to each other. **B**
   (A) perpendicular (B) parallel
   (C) supplementary (D) none of these

3. It is impossible for a triangle to be **C**
   (A) both scalene and right
   (B) both obtuse and isosceles
   (C) both equilateral and obtuse
   (D) both acute and isosceles

4. Two of the medians of an isosceles triangle are **A**
   (A) congruent (B) altitudes
   (C) perpendicular (D) angle bisectors

5. The legs of an isosceles right triangle are **D**
   (A) perpendicular (B) altitudes
   (C) congruent (D) all of these

6. In $\triangle ABC$, $\overline{AC} \cong \overline{BC}$, $m\angle A = 4x - 19$, and $m\angle B = 2x + 17$. Then $m\angle A$ is equal to **C**

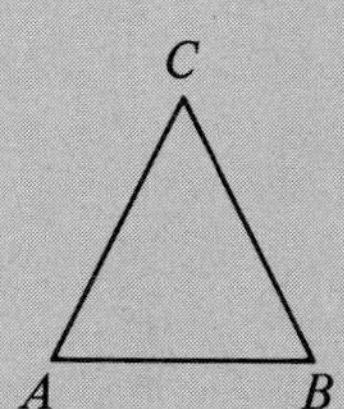

   (A) 18 (B) 36
   (C) 53 (D) 74

7. A polygon with seven sides is a **D**
   (A) hexagon (B) octagon
   (C) nonagon (D) none of these

8. An angle of a regular polygon cannot measure **D**
   (A) 90° (B) 108°
   (C) 120° (D) none of these

9. The sum of the measures of the exterior angles of a hexagon is **A**
   (A) 360° (B) 720°
   (C) 1080° (D) none of these

10. If a parallelogram has a pair of congruent adjacent sides, then it is a **B**
   (A) rectangle (B) rhombus
   (C) square (D) none of these

11. In $\triangle ABC$, D is the midpoint of $\overline{AB}$, E is the midpoint of $\overline{AC}$, and $CB = 12$. Then $ED =$ __**B**__.

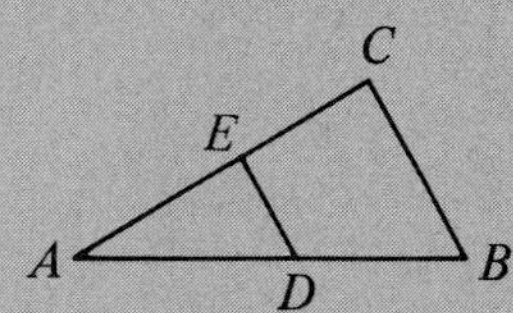

   (A) 3 (B) 6
   (C) 9 (D) none of these

12. The length of the median of a trapezoid is the ______ of the lengths of the bases. **A**
   (A) average (B) sum
   (C) difference (D) none of these

13. In isosceles trapezoid $ABCD$, $\overline{AD} \cong \overline{BC}$. If $\overline{DE} \perp \overline{AB}$ and E is between A and B, then **C**
   (A) $AB = 2(DC)$ (B) $\overline{AD} \cong \overline{DC}$
   (C) $AB = DC + 2(AE)$ (D) none of these

CUMULATIVE REVIEW

Give all other names for each figure named below.

14. $\overline{BC}$ **$\overline{CB}$**

15. $\overrightarrow{AB}$ **$\overrightarrow{AC}$, $\overrightarrow{AD}$**

True or false?

16. Complementary angles are adjacent. **F**

17. Two congruent supplementary angles are right angles. **T**

18. If two lines are parallel to the same line, then they are perpendicular to each other. **F**

19. Adjacent sides of a parallelogram are congruent. **F**

20. An octagon is a polygon with eight sides. **T**

21. Each exterior angle of a regular hexagon measures 120°. **F**

22. The sum of the measures of the interior angles of a pentagon is 540°. **T**

23. If one of the base angles of an isosceles triangle measures 40° then the triangle is obtuse. **T**

24. The bases of an isosceles trapezoid cannot be congruent. **T**

25. If a median of a triangle is also an altitude, then the triangle is isosceles. **T**

26. If the diagonals of a parallelogram are congruent, then the parallelogram is a square. **F**

27. A regular hexagon has exactly three lines of symmetry. **F**

Find each indicated measure.

28. Given: $l \parallel m$, m $\angle 1 = x + 17$, and m $\angle 2 = 2x - 11$. Find m $\angle 1$ and m $\angle 2$. **75, 105**

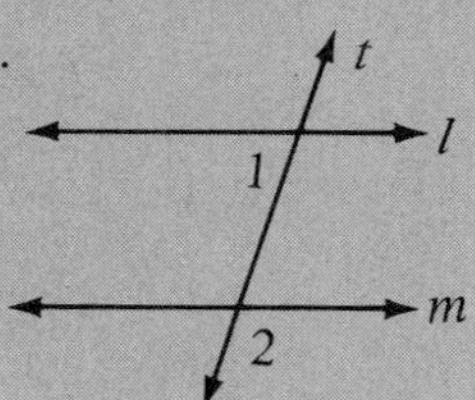

29. Given: $\triangle PQR$ with $\overline{ST} \parallel \overline{PQ}$, $PS = 6$, $SR = 6$, and $ST = 6$. Find PQ. **12**

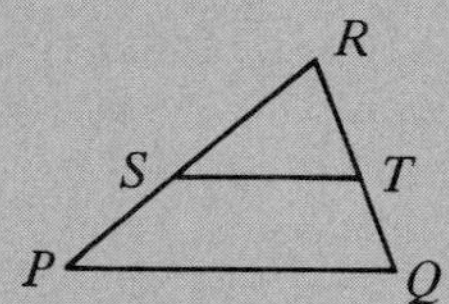

30. Given: Trapezoid $ABCD$ with $\overline{AB} \parallel \overline{CD}$, median $\overline{EF}$, $DC = 9$, and $EF = 12$. Find AB. **15**

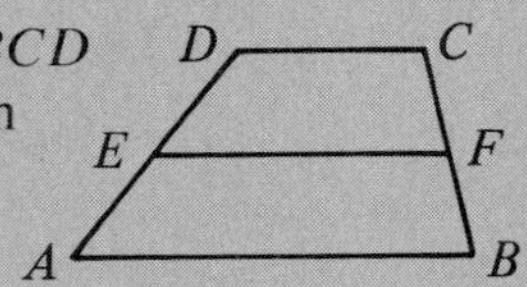

Construct each of the following:

31. A 45° angle.

32. Draw a segment. Divide it into three congruent segments.

33. Draw an acute triangle. Construct an altitude to one of the sides.

Prove each of the following:

34. Given: $\overline{AB} \cong \overline{AC}$ and $\angle B \cong \angle C$
Prove: $\overline{EB} \cong \overline{DC}$

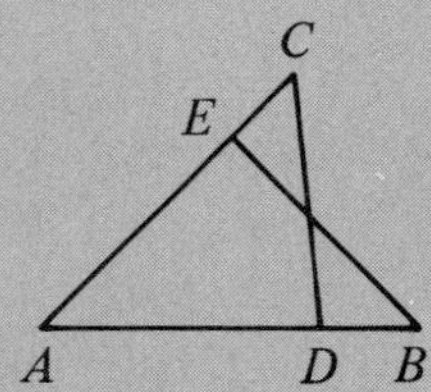

35. Given: $\square RSTU$ with $\angle RVU \cong \angle RSW$
Prove: $VSWU$ is a parallelogram.

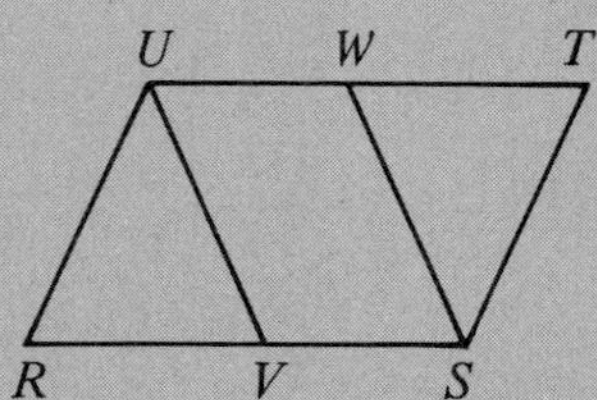

36. Given: Quadrilateral $ABCD$ with E the midpoint of $\overline{AB}$, F the midpoint of $\overline{BC}$, G the midpoint of $\overline{DC}$, and H the midpoint of $\overline{AD}$.
Prove: $\overline{EG}$ and $\overline{HF}$ bisect each other.

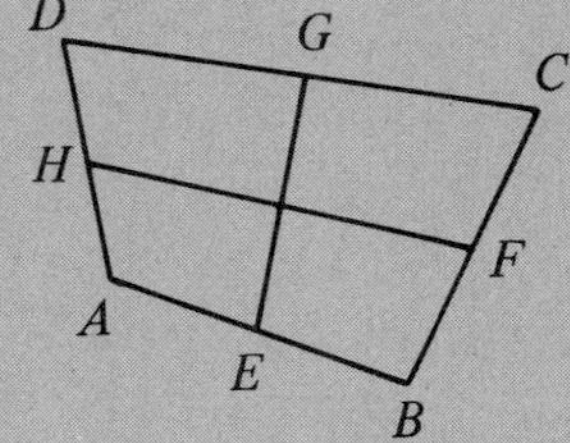

9 SIMILARITY

Non-Routine Problem Solving

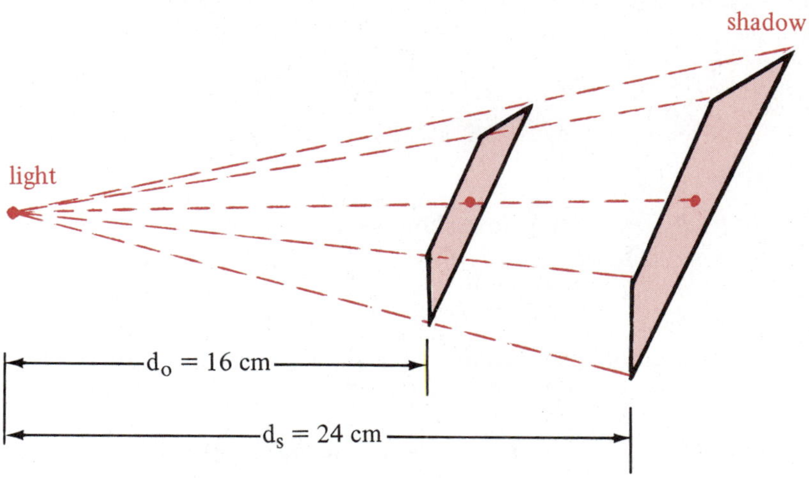

A shadow can involve measurements that form a true proportion. If the shadow is parallel to the object, then it will have the same shape as the object. The lengths of corresponding sides form equal ratios:

$$\frac{\text{HEIGHT OF SHADOW } (h_s)}{\text{HEIGHT OF OBJECT } (h_o)} = \frac{\text{DISTANCE OF SHADOW FROM LIGHT } (d_s)}{\text{DISTANCE OF OBJECT FROM LIGHT } (d_o)}$$

If the measure of h_s is 12 cm, what is the measure of h_o? **8 cm**

USING PROPORTIONS 9.1

Objectives **To identify, write, and solve proportions**
To find the geometric mean of two numbers

The boy at the right is 5 feet tall. The man is 6 feet tall. The boy is $\frac{5}{6}$ as tall as the man. The man is $\frac{6}{5}$ or $1\frac{1}{5}$ times as tall as the boy. The ratio of the boy's height to the man's height is $\frac{5}{6}$ or 5 to 6. The ratio of the man's height to the boy's height is $\frac{6}{5}$ or 6 to 5.

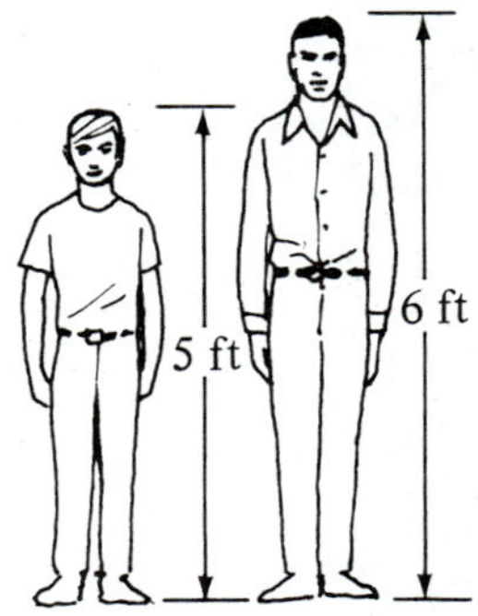

Definition: Ratio The *ratio* of two numbers a and b is the number $\frac{a}{b}$ $(b \neq 0)$.

To compare two measures, you may use a ratio. To solve some types of problems, you may need to use two ratios to form a proportion.

Definitions: Proportion extremes means A *proportion* is an equation of the form $\frac{a}{b} = \frac{c}{d}$ $(b \neq 0,\ d \neq 0)$; a and d are the *extremes* of the proportion; b and c are the *means* of the proportion.

The example below shows you how to write a true proportion and how to find its solution.

Example 1 The numbers 2 and x have the ratio 6 to 9. Write a proportion that states this. Identify the means and the extremes. Solve the proportion for x.

$$\frac{2}{x} = \frac{6}{9}$$ The means are x and 6. The extremes are 2 and 9.

$$(9x)\left(\frac{2}{x}\right) = (9x)\left(\frac{6}{9}\right)$$ ◄ *Multiply by the LCD, 9x.*

$$18 = 6x$$
$$x = 3$$

Thus, the solution is 3.

Proportion property In a true proportion, the product of the extremes equals the product of the means. If $\frac{a}{b}=\frac{c}{d}$ $(b \neq 0, d \neq 0)$, then $ad = bc$.

You will be asked to prove this property of proportions in Exercise 61.

Example 2 Which is a true proportion?

$\frac{3}{7}=\frac{9}{21}$ $\frac{5}{7}=\frac{7}{9}$

$3 \times 21 = 7 \times 9$ ◄ *Use the proportion property.* ► $5 \times 9 \neq 7 \times 7$
$63 = 63$ $45 \neq 49$

Thus, $\frac{3}{7}=\frac{9}{21}$ is a true proportion; $\frac{5}{7}=\frac{7}{9}$ is not a true proportion.

Example 3 Solve the proportion $\frac{2}{3}=\frac{x}{12}$ for x.

The product of the extremes equals the product of the means. ► $24 = 3x$
$x = 8$

Thus, the solution is 8.

The converse of the proportion property is also true: If $ad = bc$, then $\frac{a}{b}=\frac{c}{d}$ $(b \neq 0, d \neq 0)$. You can also show that $\frac{a}{c}=\frac{b}{d}$, $\frac{b}{a}=\frac{d}{c}$, and $\frac{c}{a}=\frac{d}{b}$.

Example 4 If $4x = 5y$, write three different proportions.

Arrange the numbers so that the property of proportions is true. ► $\frac{4}{5}=\frac{y}{x}$, $\frac{5}{4}=\frac{x}{y}$, $\frac{x}{5}=\frac{y}{4}$

In the proportion $\frac{3}{m}=\frac{m}{12}$, both means are represented by the variable m.

You can solve for m.

$$\frac{3}{m}=\frac{m}{12}$$
$$36 = m^2$$
$$m = 6, \text{ or } m = -6$$

The number 6 is called the **geometric mean** of 3 and 12.

Definition: Geometric mean The positive number m is the *geometric mean* of two positive numbers a and b if $\frac{a}{m}=\frac{m}{b}$.

To find the geometric mean of two numbers, you must set up the proportion and then solve for the mean.

Example 5 Find the geometric mean of 2 and 7.

$$\frac{2}{m} = \frac{m}{7}$$
$$14 = m^2$$
$$m = \sqrt{14}, \text{ or } m = -\sqrt{14}$$

Thus, the geometric mean of 2 and 7 is $\sqrt{14}$.

You can write and solve proportions involving geometric properties.

Example 6 The measures of two complementary angles have the ratio 1 to 5. What is the measure of each angle?

Let x = measure of one angle
$90 - x$ = measure of its complement

$$\frac{x}{90 - x} = \frac{1}{5}$$
$$5x = 1(90 - x)$$
$$5x = 90 - x$$
$$6x = 90$$
$$x = 15$$
$$90 - x = 75$$

Thus, the measures are 15° and 75°.

Oral Exercises

Which of the following are ratios?

1. $\frac{3}{5}$ **ratio**
2. $4x$ **not a ratio**
3. $2 + 7$ **not a ratio**
4. $\frac{p}{q}$ **ratio**
5. $x = 7$ **not a ratio**
6. $\frac{y}{5}$ **ratio**
7. mn **not a ratio**
8. $\frac{2c}{d}$ **ratio**

Which of the following are proportions? For each proportion, identify the means and the extremes.

9. $\frac{3}{5}$ **not a prop.**
10. $\frac{x}{6} = \frac{2}{3}$ **6, 2; x, 3**
11. $\frac{2}{9} + \frac{3}{18}$ **not a prop.**
12. $\frac{5x}{7y}$ **not a prop.**
13. $\frac{x}{9} - \frac{3}{12}$ **not a prop.**
14. $\frac{a}{b} = \frac{m}{n}$ **b, m; a, n**
15. $\frac{2}{5} = \frac{x}{10} + \frac{y}{5}$ **not a prop.**
16. $\frac{3}{m} = \frac{m}{12}$ **m, m; 3, 12**

True or false?

17. A proportion is the sum of two ratios. **F**
18. The ratio 5 to 8 is the same as the ratio 8 to 5. **F**

Written Exercises

Which of the following are true proportions?

A 1. $\frac{3}{4}=\frac{9}{12}$ **T** 2. $\frac{5}{6}=\frac{3}{4}$ **F** 3. $\frac{2}{3}=\frac{10}{15}$ **T** 4. $\frac{6}{8}=\frac{39}{52}$ **T**

5. $\frac{1}{2}=\frac{1+3}{2+3}$ **F** 6. $\frac{3}{4}=\frac{4}{3+4}$ **F** 7. $\frac{5}{6}=\frac{4\cdot 5}{4\cdot 6}$ **T** 8. $\frac{3}{4}=\frac{3+3}{4+3}$ **F**

Identify the means and the extremes of each proportion. Solve for x.

9. $\frac{2}{3}=\frac{10}{x}$ **3, 10; 2, x; 15** 10. $\frac{4}{6}=\frac{x}{15}$ **6, x; 4, 15; 10** 11. $\frac{x}{9}=\frac{4}{12}$ **9, 4; x, 12; 3** 12. $\frac{6}{x}=\frac{10}{15}$ **x, 10; 6, 15; 9**

13. $\frac{12}{8}=\frac{9}{x}$ **8, 9; 12, x; 6** 14. $\frac{3}{5}=\frac{7}{x}$ **5, 7; 3, x; $11\frac{2}{3}$** 15. $\frac{4}{9}=\frac{6}{x}$ **9, 6; 4, x; $13\frac{1}{2}$** 16. $\frac{x}{3}=\frac{7}{12}$ **3, 7; x, 12; $1\frac{3}{4}$**

Write three different proportions for each equation. **Answers may vary.**

17. $3\cdot 4=2\cdot 6$ 18. $5\cdot 9=3\cdot 15$ 19. $2\cdot 8=4\cdot 4$ 20. $6\cdot 6=4\cdot 9$

21. $5x=10\cdot 3$ 22. $4n=7m$ 23. $3x=yz$ 24. $ab=cd$

Find the geometric mean of each pair of numbers.

25. 1 and 4 **2** 26. 9 and 16 **12** 27. 5 and 20 **10** 28. 3 and 12 **6**

29. 3 and 7 **$\sqrt{21}$** 30. 5 and 4 **$\sqrt{20}$, or $2\sqrt{5}$** 31. 7 and 11 **$\sqrt{77}$** 32. x and y **$\sqrt{xy}$**

Which are true proportions for all replacements of the variables? (Assume each denominator $\neq 0$.)

B 33. $\frac{am}{bn}=\frac{ar}{bg}$ **F** 34. $\frac{ab}{ad}=\frac{cd}{cb}$ **F** 35. $\frac{ab}{ac}=\frac{bd}{cd}$ **T** 36. $\frac{a+b}{c+b}=\frac{a-b}{c-b}$ **F**

Find the ratio of x to y for each of the following.

37. $7x=9y$ **$\frac{9}{7}$** 38. $13x=12y$ **$\frac{12}{13}$** 39. $5y=7x$ **$\frac{5}{7}$** 40. $4y=6x$ **$\frac{4}{6}$**

41. $14x=21y$ **$\frac{21}{14}$** 42. $35y=25x$ **$\frac{35}{25}$** 43. $ax=by$ **$\frac{b}{a}$** 44. $y=cx$ **$\frac{1}{c}$**

Use proportions to solve each of the following problems.

45. The measures of two complementary angles have the ratio 2 to 3. What is the measure of each angle? **36°, 54°**

46. The measures of two supplementary angles have the ratio 4 to 5. What is the measure of each angle? **80°, 100°**

47. The sum of the lengths of two segments is 242. The ratio of the lengths is 5 to 6. What is the length of each segment? **110; 132**

48. The difference between the lengths of two segments is 6. The ratio of the lengths is 3 to 5. What is the length of each segment? **9; 15**

49. The length and the width of a rectangle have the ratio 3 to 2. The perimeter is 20 cm. Find the length and the width. **6 cm; 4 cm**

50. The length and the width of a rectangle have the ratio 5 to 3. The perimeter is 48 m. Find the length and the width. **15m; 9m**

C 51. The length and the width of a rectangle have the ratio 5 to 6. Their product is 120. Find the length and the width. **10; 12**

52. The length and the width of a rectangle have the ratio 3 to 2. Their product is 54. Find the length and the width. **9; 6**

True or false? If true, give a proof using algebraic properties. If false, give a numerical counterexample. (Assume each denominator $\neq 0$.)

Example If $\frac{a}{b} = \frac{c}{d}$, then $\frac{d}{c} = \frac{b}{a}$.

True:	$\frac{a}{b} = \frac{c}{d}$	Given
	$(bd)\left(\frac{a}{b}\right) = (bd)\left(\frac{c}{d}\right)$	Mult. prop. of equality
	$ad = bc$	
	$\frac{ad}{ac} = \frac{bc}{ac}$	Div. prop. of equality
	$\frac{d}{c} = \frac{b}{a}$	

53. If $\frac{a}{b} = \frac{c}{d}$, then $\frac{a}{c} = \frac{b}{d}$. **T**

54. If $\frac{a}{b} = \frac{c}{d}$, then $\frac{a+n}{b+n} = \frac{c}{d}$. **F**

55. If $\frac{a}{b} = \frac{c}{d}$, then $\frac{a+b}{b} = \frac{c+d}{d}$. **T**

56. If $\frac{a}{b} = \frac{c}{d}$, then $\frac{a-b}{b} = \frac{c-d}{d}$. **T**

57. If $\frac{a}{b} = \frac{c}{d}$, then $\frac{a}{a+b} = \frac{c}{c+d}$. **T**

58. If $\frac{a}{b} = \frac{c}{d}$, then $\frac{a+b}{c+d} = \frac{a+d}{c+b}$. **F**

59. If $\frac{a}{b} = \frac{c}{d}$, then $\frac{an}{bn} = \frac{cm}{dm}$. **T**

60. If $\frac{a}{b} = \frac{c}{d}$, then $\frac{a^n}{b^n} = \frac{c^n}{d^n}$. **T**

61. Prove: If $\frac{a}{b} = \frac{c}{d}$, and $b \neq 0$, $d \neq 0$, then $ad = bc$.

62. Prove: If $ad = bc$, and $b \neq 0$, $d \neq 0$, then $\frac{a}{b} = \frac{c}{d}$.

CALCULATOR ACTIVITIES

A calculator can help you to solve proportions.

For example, in $\frac{24}{42} = \frac{37}{x}$, $x = \frac{42 \cdot 37}{24}$. You can now proceed as shown.

Press 42 ⊗ 37 ⊜
Display 1554
Press ⊕ 24 ⊜
Display 64.75

Thus, $x = 64.8$ to the nearest tenth.

Find x to the nearest tenth.

1. $\frac{46}{59} = \frac{28}{x}$ **35.9**

2. $\frac{37}{89} = \frac{x}{52}$ **21.6**

3. $\frac{x}{123} = \frac{57}{65}$ **107.9**

4. $\frac{59}{x} = \frac{129}{107}$ **48.9**

SIMILAR POLYGONS 9.2

Objectives **To identify similar polygons**
To find the measures of sides and angles of similar polygons

Congruent polygons have the same size and shape; quadrilateral $ABCD$ is congruent to quadrilateral $EFGH$.

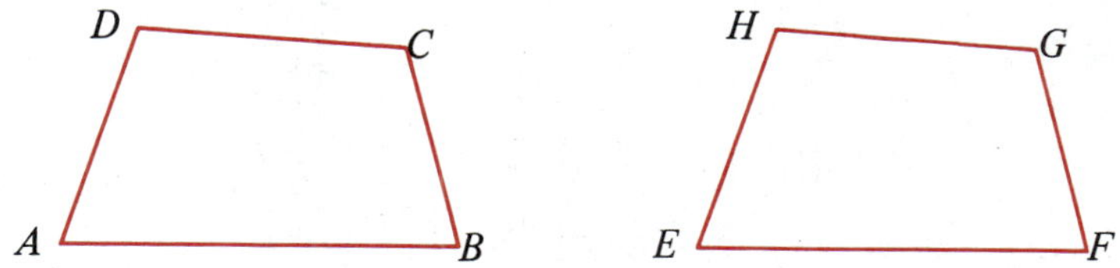

Similar polygons have the same shape but not necessarily the same size.

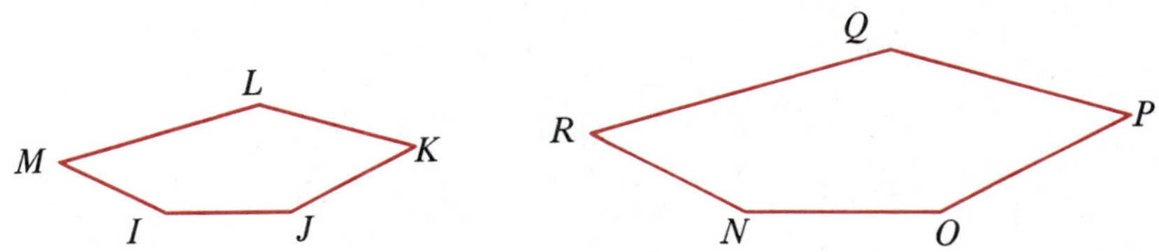

Polygon $IJKLM$ is similar to polygon $NOPQR$.

Definition: Similar polygons Two polygons are *similar* if corresponding angles are congruent and the lengths of corresponding sides are proportional.

If polygon $IJKLM$ is similar to polygon $NOPQR$, then you will write $IJKLM \sim NOPQR$. The symbol $\sim$ means *is similar to.* In this case, $\angle I \cong \angle N$, $\angle J \cong \angle O$, $\angle K \cong \angle P$, $\angle L \cong \angle Q$, and $\angle M \cong \angle R$.
Also $\frac{IJ}{NO} = \frac{JK}{OP} = \frac{KL}{PQ} = \frac{LM}{QR} = \frac{MI}{RN}$. If the lengths of corresponding sides are proportional, you may simply say that the sides are proportional.

Example 1 For similar triangles ABC and DEF, which angles are congruent? Which sides are proportional?

$\angle A \cong \angle D$, $\angle B \cong \angle E$, $\angle C \cong \angle F$ ◄ *Corresponding ∠'s are ≅.*

$\frac{AB}{DE} = \frac{BC}{EF} = \frac{CA}{FD}$. ◄ *Corresponding sides are proportional.*

To determine if two polygons are similar, you must check both the corresponding angles and the corresponding sides.

Example 2 Is square $ABCD$ similar to rectangle $EFGH$? Why or why not?

All corresponding angles are congruent. They are right angles.

All corresponding sides are not proportional. $\frac{4}{6} \neq \frac{4}{3}$.

Thus, square $ABCD \nsim$ rect. $EFGH$. ◄ *$\nsim$ means is not similar to.*

Using the property of proportions, you can find the lengths of sides of similar polygons.

Example 3 Quad. $QRST \sim$ quad. $UVWX$. Find the unknown lengths.

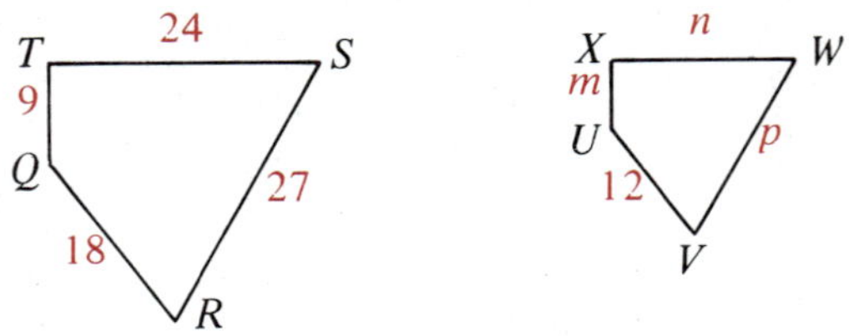

$\frac{QR}{UV} = \frac{18}{12}$, or $\frac{3}{2}$ $\qquad$ $\frac{27}{p} = \frac{3}{2}$ $\qquad$ $\frac{24}{n} = \frac{3}{2}$ $\qquad$ $\frac{9}{m} = \frac{3}{2}$

$3p = 54$ $\qquad$ $3n = 48$ $\qquad$ $3m = 18$

$p = 18$ $\qquad$ $n = 16$ $\qquad$ $m = 6$

Thus, $VW = 18$, $WX = 16$, and $XU = 6$.

Written Exercises

3. yes; all corres. ∠s are ≅ and all corres. sides are prop.

1. no; all corres. sides are not prop. 2. yes; all corres. ∠s are ≅ and all corres. sides are prop.

Are the pairs of polygons similar? Why or why not?

A **1.**

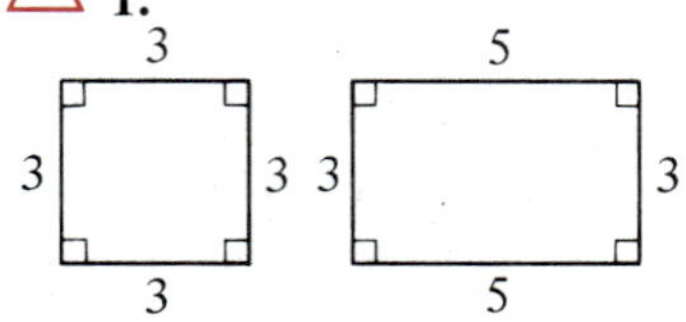

2.

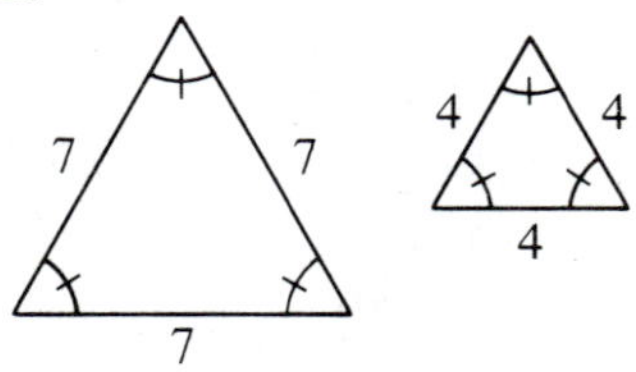

3.

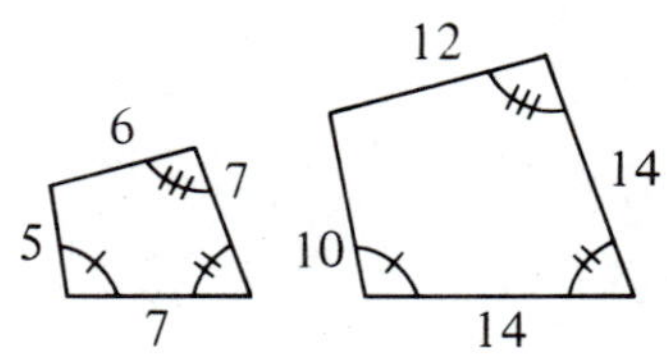

Which angles are congruent? Which sides are proportional? 4. ∠X ≅ ∠M, ∠Y ≅ ∠N, ∠Z ≅ ∠O, $\frac{XY}{MN} = \frac{YZ}{NO} = \frac{XZ}{MO}$

4. $\triangle XYZ \sim \triangle MNO$ $\qquad$ **5.** rect. $ABCD \sim$ rect. $EFGH$ $\qquad$ **6.** $VWXYZ \sim ABCDE$

5. ∠A ≅ ∠E, ∠B ≅ ∠F, ∠C ≅ ∠G, ∠D ≅ ∠H; $\frac{AB}{EF} = \frac{BC}{FG} = \frac{CD}{GH} = \frac{AD}{EH}$

Find the unknown lengths in each of the following.

7. $\triangle ABC \sim \triangle DEF$

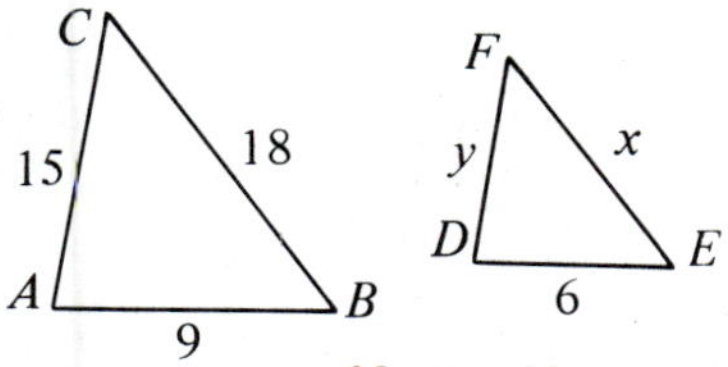

x = 12, y = 10

8. $\triangle GHI \sim \triangle JKL$

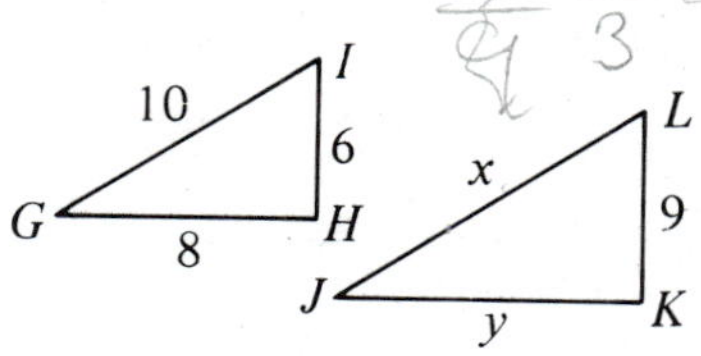

x = 15, y = 12

9. Quad. $MNOP \sim$ quad. $QRST$

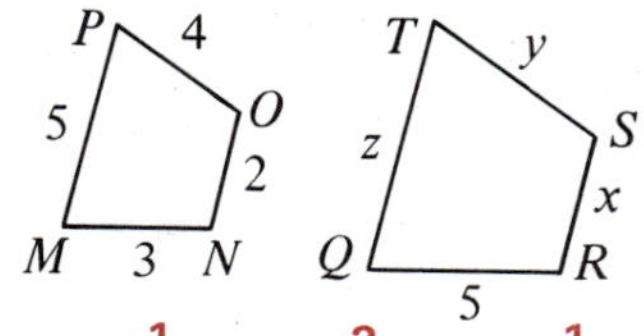

$x = 3\frac{1}{3}$, $y = 6\frac{2}{3}$, $z = 8\frac{1}{3}$

B 10. ▱$RSTU \sim$ ▱$PONM$

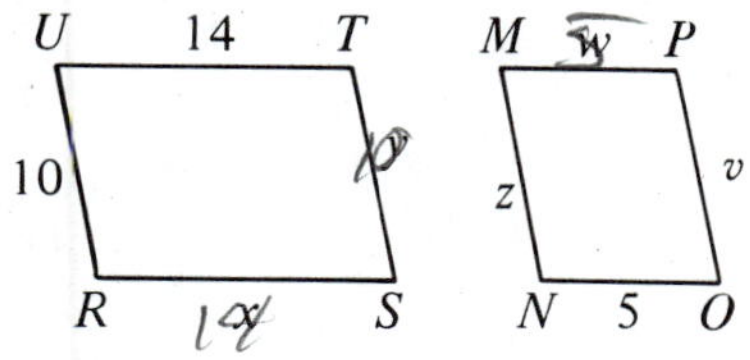

x = 14, y = 10, w = 5, v = 7, z = 7

11. Quad. $ABCD \sim$ quad. $EFGH$

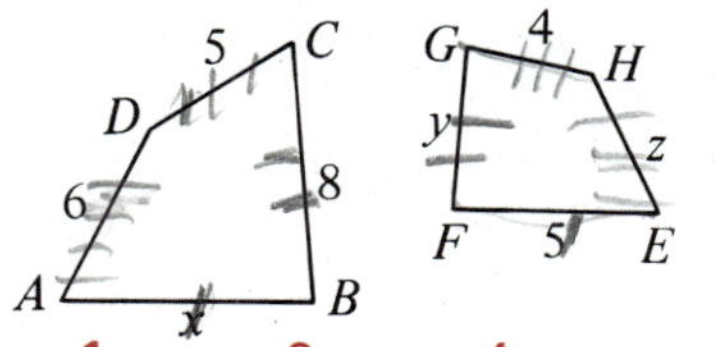

$x = 6\frac{1}{4}$, $y = 6\frac{2}{5}$, $z = 4\frac{4}{5}$

12. $DEFGH \sim IJKLM$

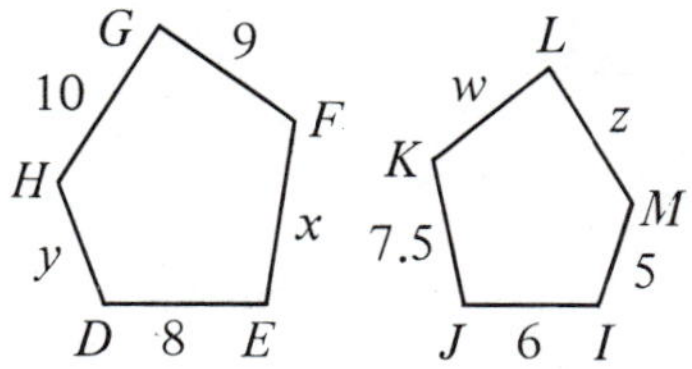

$w = 6\frac{3}{4}$, $x = 10$, $y = 6\frac{2}{3}$, $z = 7\frac{1}{2}$

13. Given: Trap. $ABCD \sim$ trap. $LMNO$, $AB = 4$, $BC = 6$, $CD = 4$, $DA = 2$, and $LM = 2$. Find NO and OL. 2; 1

14. Given: ▱$RSTU \sim$ ▱$HJKL$, $RS = 12$, $UR = 8$, and $JK = 6$. Find HJ and KL. 9; 9

15. Given: Rhombus $ABCD \sim$ rhombus $EFGH$, $AB = 8$, and $FG = 12$. Find BC and GH. 8; 12

16. Given: Quad. $ABCD \sim$ quad. $EFGH$, $m\angle A = 47$, $m\angle B = 39$, and $m\angle C = 112$. Find $m\angle F$ and $m\angle H$. 39; 162

17. Given: Pent. $QRSTU \sim$ pent. $VWXYZ$, $m\angle Q = 68$, $m\angle R = 110$, $m\angle S = 91$, and $m\angle T = 94$. Find $m\angle X$, $m\angle V$, and $m\angle Z$. 91; 68; 177

18. Given: ▱$EFGH \sim$ ▱$IJKL$ and $m\angle E = 70$. Find $m\angle J$ and $m\angle L$. 110; 110

C 19. If two polygons are similar, are they necessarily congruent? Why?
No, although corres. sides are prop., they may not be ≅.

20. If two polygons are congruent, are they necessarily similar? Why?
20. Yes, the corres. ∠'s are ≅ and the corres. sides are prop.

21. Prove: All squares are similar.

6. $\angle V \cong \angle A$, $\angle W \cong \angle B$, $\angle X \cong \angle C$, $\angle Y \cong \angle D$, $\angle Z = \angle E$; $\frac{VW}{AB} = \frac{WX}{BC} = \frac{XY}{CD} = \frac{YZ}{DE} = \frac{VZ}{AE}$

LOGIC REVIEW

Write the negation of each statement. Label the statement and its negation as true or false. Check students' answers.

1. Similar triangles cannot be congruent.
2. Two complementary angles can have the same measure.
3. A pentagon is a quadrilateral.
4. Congruent triangles are similar.

ANGLE-ANGLE SIMILARITY 9.3

Objective **To recognize and apply the angle-angle similarity pattern**

Although all corresponding angles are congruent and all corresponding sides are proportional when two triangles are similar, there are simpler patterns that can be used to prove triangles similar. If you are given $\triangle ABC$, you can construct $\triangle DEF$ so that $\angle D \cong \angle A$ and $\angle E \cong \angle B$.

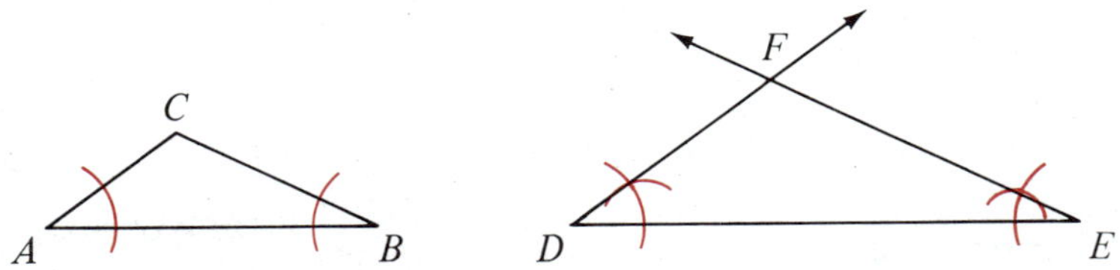

It appears that $\triangle ABC \sim \triangle DEF$. This leads to the following postulate.

Postulate 22 AA for $\sim \triangle$'s If two angles of a triangle are congruent to two angles of a second triangle, then the two triangles are similar.

You can use this postulate in problems that involve similar triangles.

Example 1 Are the triangles similar? Give a reason for your answer.

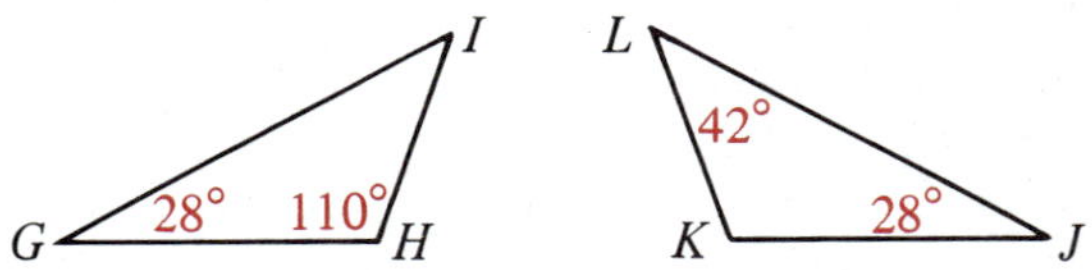

m $\angle I = 180 - (28 + 110) = 180 - 138$, or 42 $\qquad \angle G \cong \angle J$, $\angle I \cong \angle L$

Thus, $\triangle GHI \sim \triangle JKL$ by AA for $\sim \triangle$'s.

Example 2 Given: $\overline{PQ} \parallel \overline{MN}$

Prove: $\triangle MNO \sim \triangle PQO$

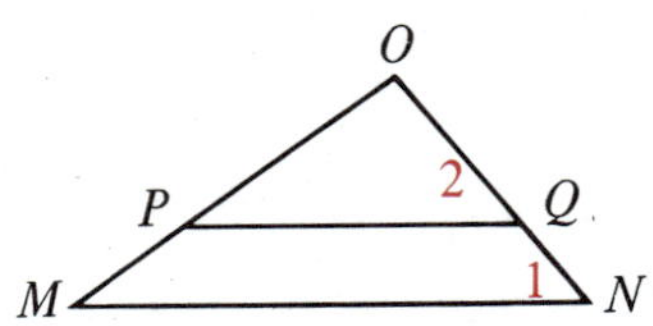

Proof

Statements	**Reasons**
1. $\overline{PQ} \parallel \overline{MN}$	1. Given
2. $\angle 1 \cong \angle 2$	2. Corr. $\angle$'s of $\parallel$ lines are $\cong$.
3. $\angle O \cong \angle O$	3. Reflexive property
4. $\therefore \triangle MNO \sim \triangle PQO$	4. AA for $\sim \triangle$'s

You can find unknown lengths using the proportions that exist between the sides of similar triangles. You can also use similar triangles to prove that lengths of segments are proportional.

Example 3

Given: $\overline{AB} \perp \overline{BC}$ and $\overline{DE} \perp \overline{BC}$.
Find AD.

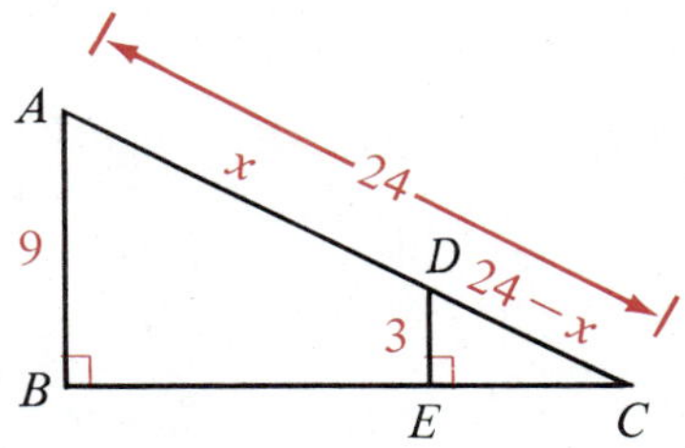

$$\begin{aligned} \angle C &\cong \angle C \\ \angle ABC &\cong \angle DEC && \blacktriangleleft\ \textit{⊥'s form} \cong \textit{rt. ∠'s.} \\ \triangle ABC &\sim \triangle DEC \\ \frac{9}{3} &= \frac{24}{24 - x} && \blacktriangleleft\ \textit{Corresponding sides of} \sim \triangle\textit{'s are proportional.} \\ 9(24 - x) &= 24 \cdot 3 \\ 216 - 9x &= 72 \\ -9x &= -144 \\ x &= 16 \end{aligned}$$

Thus, $AD = 16$.

Example 4

Given: $\overline{DE} \parallel \overline{AB}$

Prove: $\frac{AC}{DC} = \frac{BC}{EC}$

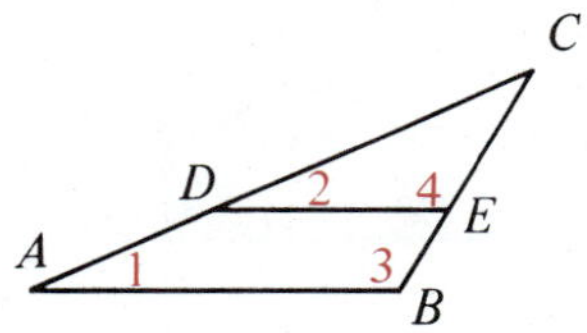

Proof

Statements	Reasons
1. $\overline{DE} \parallel \overline{AB}$	1. Given
2. $\angle 1 \cong \angle 2$, $\angle 3 \cong \angle 4$	2. Corr. ∠'s of ∥ lines are ≅.
3. $\triangle ABC \sim \triangle DEC$	3. AA for ~ △'s
4. $\therefore \frac{AC}{DC} = \frac{BC}{EC}$	4. Corr. sides of ~ △'s are proportional.

Reading in Geometry

Without looking up these words in your book or in a dictionary, indicate which is the correct spelling for each word.

1.	simalar	similur	similar	similiar	similer
2.	proportionul	perportional	preportional	proportional	proportionial
3.	congruent	congruant	congerant	congruunt	congrunt
4.	extreems	extrimes	externs	extrims	extremes
5.	ratuo	ratio	rasho	rashio	rashuo
6.	altutude	altitude	altatude	altitud	altatood
7.	quaderlateral	quadrulateral	quadralateral	quadrilaterial	quadrilateral
8.	paralel	paruhlel	paralell	parallel	parillel

similar; proportional; congruent; extremes; ratio; altitude; quadrilateral; parallel

Written Exercises

Are the triangles similar? Give a reason for your answer. (Ex. 1–6)

A

1.

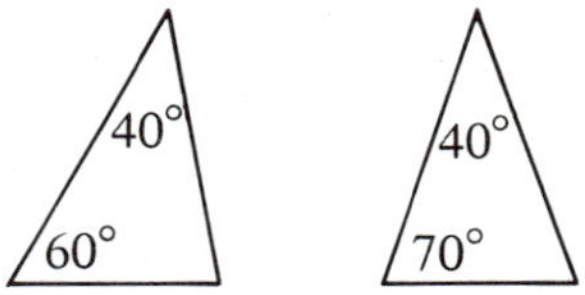

No; only 1 pr. of ∠'s are ≅.

2.

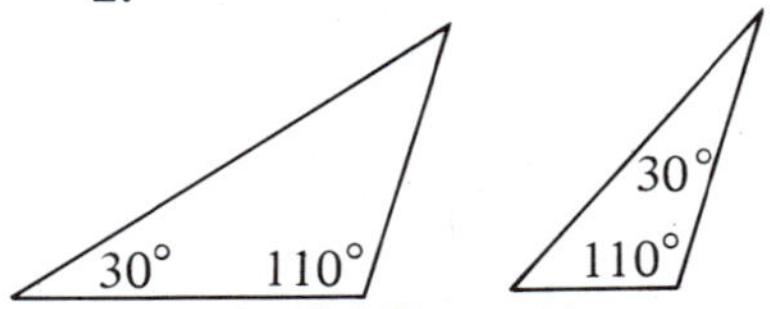

yes; AA for ~ △'s

3.

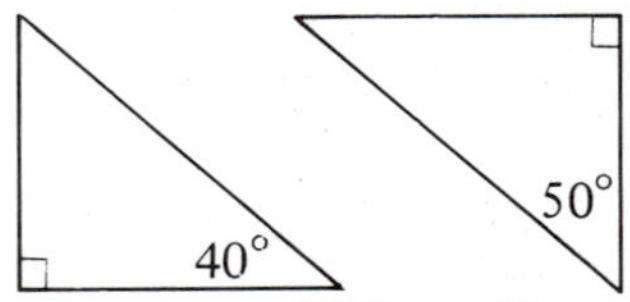

yes; AA for ~ △'s

4.

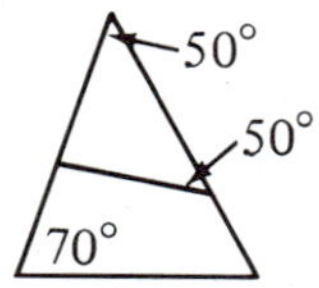

No; only 1 pr. of ∠'s are ≅.

5.

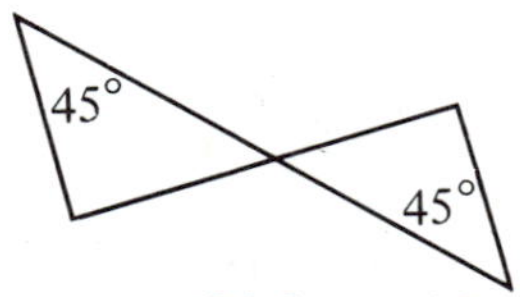

yes; AA for ~ △'s

6.

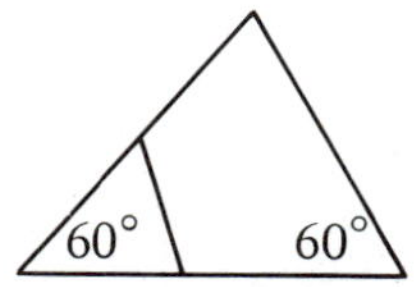

cannot tell

7. Given: $\angle 1 \cong \angle 2$
Prove: $\triangle ABD \sim \triangle CED$

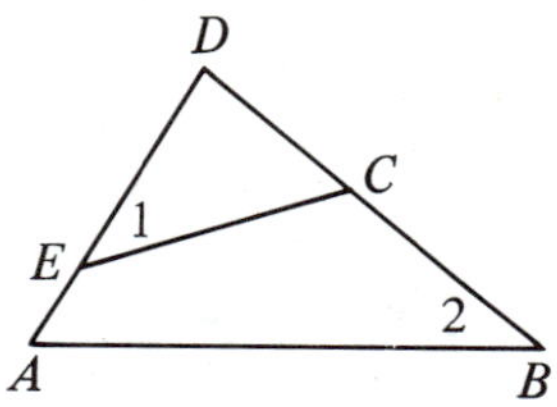

8. Given: $\angle J \cong \angle G$
Prove: $\triangle HFG \sim \triangle HIJ$

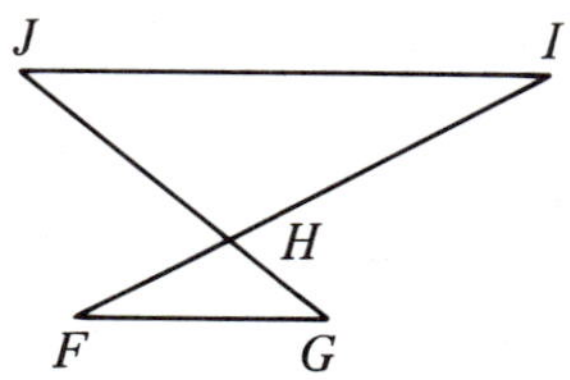

9. Given: $\overline{OK} \parallel \overline{MN}$
Prove: $\triangle OKL \sim \triangle MNL$

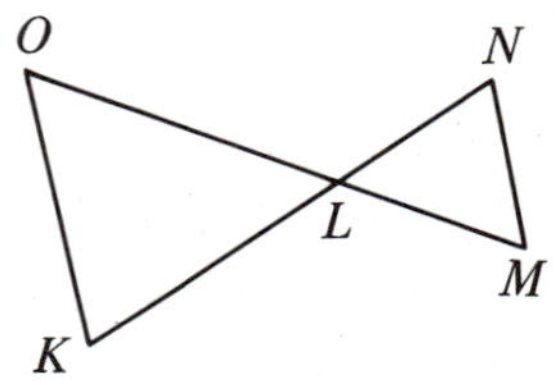

10. Given: $\overline{TQ} \perp \overline{PR}$, $\overline{SR} \perp \overline{PR}$, $TQ = 4$, $SR = 6$, and $PS = 15$
Find PT. 10

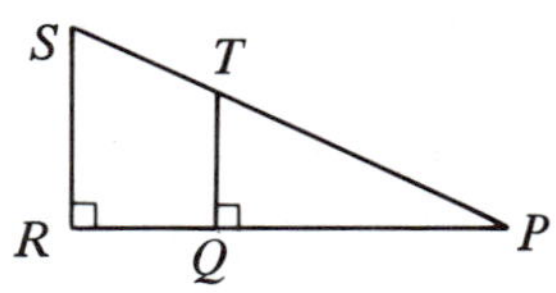

11. Given: $\overline{TQ} \perp \overline{PR}$, $\overline{SR} \perp \overline{PR}$, $PQ = 9$, $QR = 6$, and $TQ = 6$
Find SR. 10

Based on the information given in each case, is $\triangle ABC \sim \triangle DEF$? (Ex. 12–13)

B

12. $m\angle A = 47$, $m\angle B = 85$, $m\angle C = x + 4$, $m\angle E = 2x - 1$, and $m\angle F = x + 5$ no

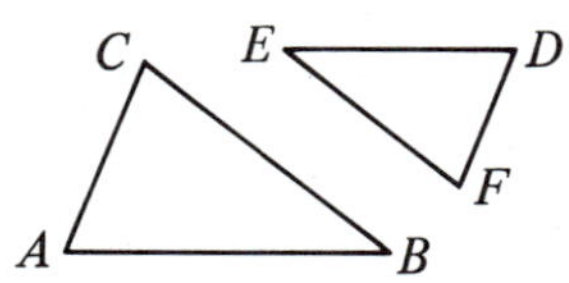

13. $m\angle A = 55$, $m\angle C = 75$, $m\angle D = 2x + 3$, $m\angle F = 3x - 1$, and $m\angle E = x + 22$ no

14. Given: $\overline{CF} \perp \overline{AB}$ and $\overline{BD} \perp \overline{AC}$
Prove: $\triangle FBE \sim \triangle DCE$

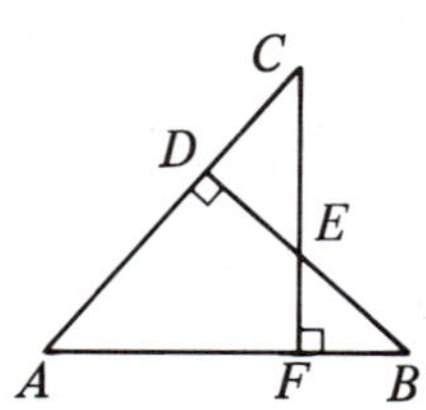

15. Given: $\overline{GJ} \perp \overline{KI}$ and $\overline{KH} \perp \overline{GI}$
Prove: $\triangle HKI \sim \triangle JGI$

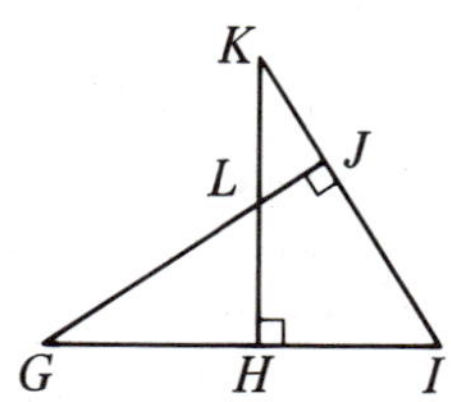

16. Given: $\overline{MN} \parallel \overline{PO}$
Prove: $\triangle MNQ \sim \triangle OPQ$

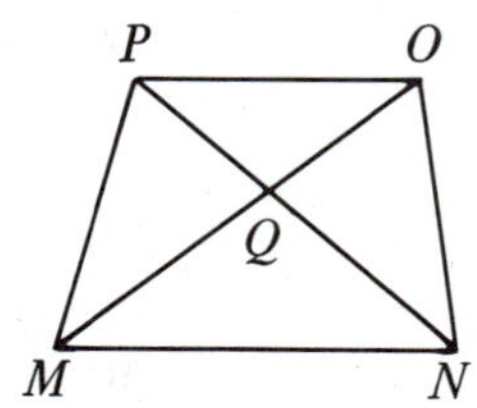

17. Given: ▱$RSTU$
Prove: $\triangle RSV \sim \triangle WTV$

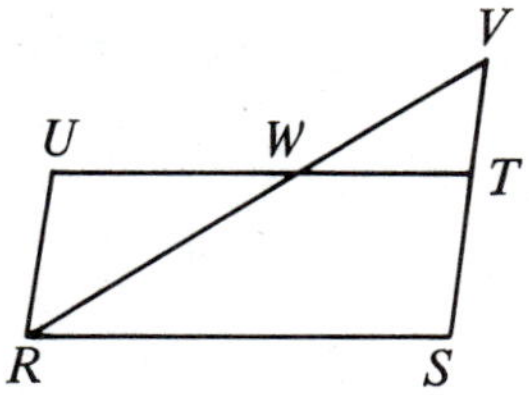

18. Given: $\overline{CA} \perp \overline{AB}$, $\overline{DE} \perp \overline{AB}$
Prove: $\frac{CB}{DB} = \frac{AB}{EB}$

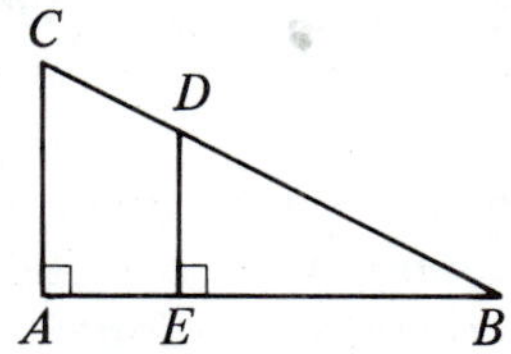

19. Given: $\overline{ML} \parallel \overline{JK}$
Prove: $\frac{MN}{NK} = \frac{LN}{NJ}$

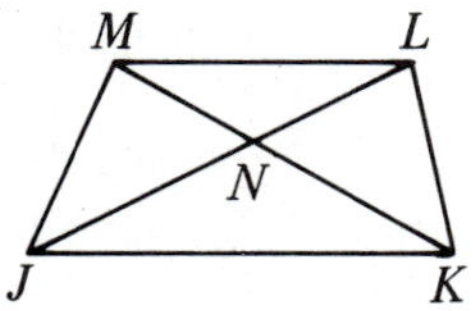

True or false? (Ex. 20–23)

20. All equiangular triangles are similar. **T**

21. All isosceles triangles are similar. **F**

22. An obtuse triangle cannot be similar to an acute triangle. **T**

23. All isosceles triangles with a 40° angle are similar. **F**

(C) **24.** Given: $\overline{CM} \perp \overline{TX}$, $\overline{PQ} \perp \overline{TN}$, $\overline{JL} \perp \overline{XN}$
Prove: $\triangle JCP \sim \triangle XTN$

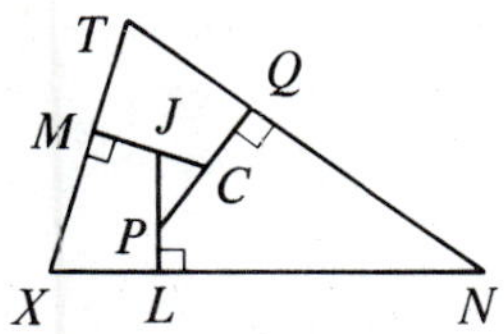

25. Given: $\overline{WZ} \cong \overline{XY}$ and $\overline{WX} \cong \overline{ZY}$
Prove: $\triangle WTZ \sim \triangle VWX$

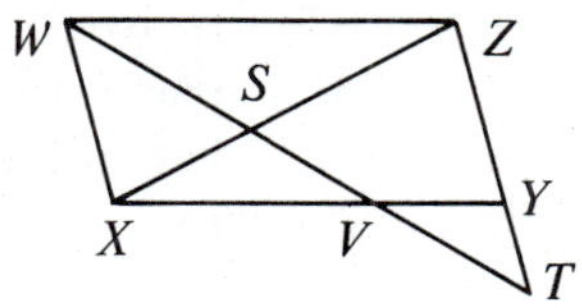

26. Given: $\angle 1 \cong \angle 2$
Prove: $\frac{SQ}{TS} = \frac{PQ}{PR}$

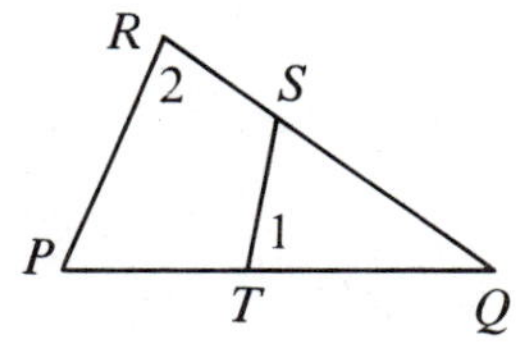

27. Given: $\angle ACB$ is a right $\angle$, $\overline{CD}$ bisects $\angle BCA$, and $\overline{DE} \perp \overline{CA}$.
Prove: $\frac{AD}{AB} = \frac{CE}{BC}$

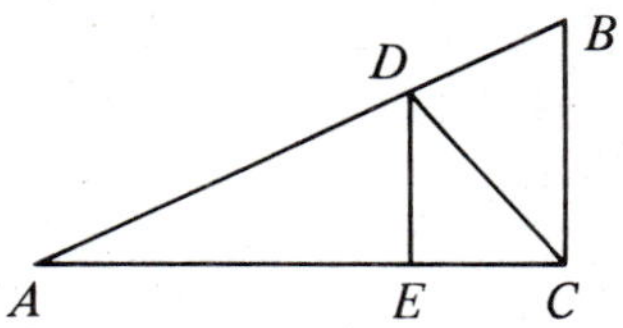

28. Given: $\overline{BC} \perp \overline{AC}$, $\overline{DE} \perp \overline{AC}$, and $\angle EDC \cong \angle DCE$.
Prove: $\frac{1}{BC} + \frac{1}{AC} = \frac{1}{DE}$

CUMULATIVE REVIEW

1. Given: $\overleftrightarrow{AB} \parallel \overleftrightarrow{CD}$, m $\angle x = 70$, m $\angle w = 40$. Find m $\angle z$. **m $\angle z = 70$**

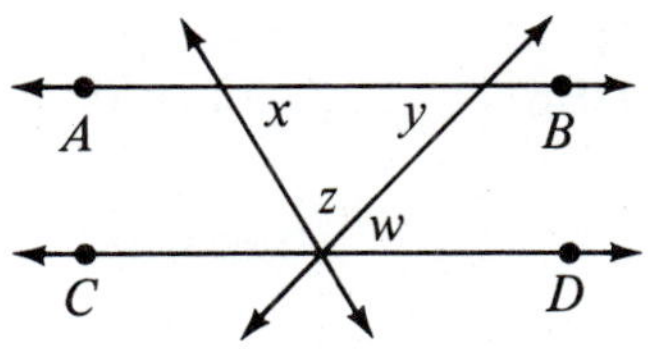

2. Given: $\overleftrightarrow{AB} \parallel \overleftrightarrow{CD}$, m $\angle x = 2r + 23$, m $\angle z = 6r + 14$, and m $\angle y = 12r - 17$. Find m $\angle w$. **m $\angle w = 79$**

NON-ROUTINE PROBLEMS

Copy the three sets of dots on your paper. For each set, how many different sizes of squares can be drawn by connecting four dots? One example is shown for each set of dots. **3; 5; 8**

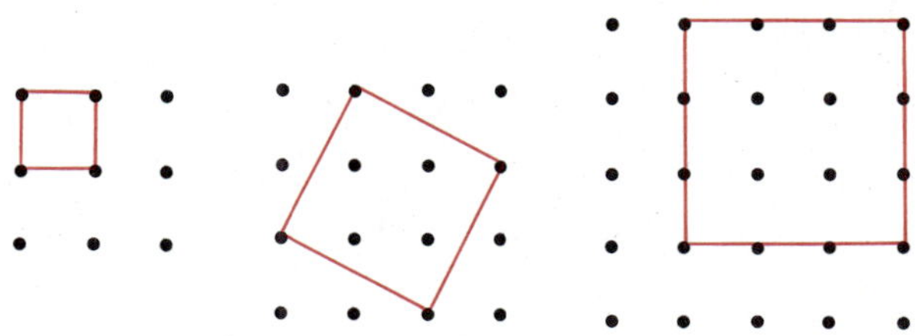

PROPORTIONALITY IN TRIANGLES 9.4

Objective **To write and solve proportions when a line is parallel to a side of a triangle**

Two different segments can be divided so that the lengths of the resulting parts are proportional. Points X on $\overline{AB}$ and Y on $\overline{CD}$ divide the segments proportionally, since $\frac{2}{4}=\frac{3}{6}$. In general, if $\frac{AX}{XB}=\frac{CY}{YD}$, then $\overline{AB}$ and $\overline{CD}$ are divided proportionally.

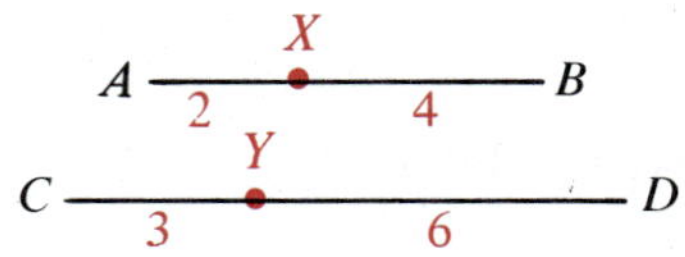

Since the sides of a triangle are segments, you can divide them proportionally. One way is shown in Theorem 9.1.

Theorem 9.1 If a line is parallel to one side of a triangle and intersects the other two sides, then it divides the two sides proportionally.

Given: $\triangle ABC$ with $\overleftrightarrow{DE} \parallel \overline{AB}$

Prove: $\frac{AD}{DC}=\frac{BE}{EC}$

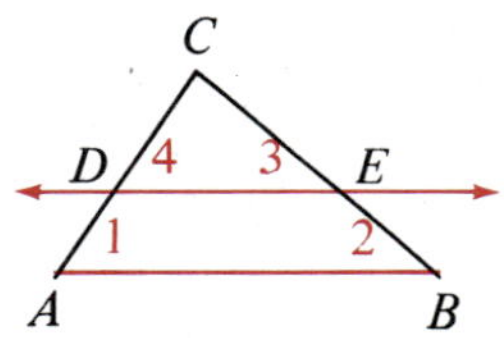

ANALYSIS Show that $\triangle ABC \sim \triangle DEC$. Then use properties of algebra.

Proof

Statements	Reasons
1. $\triangle ABC$ with $\overleftrightarrow{DE} \parallel \overline{AB}$	1. Given
2. $\angle 1 \cong \angle 4$ and $\angle 2 \cong \angle 3$	2. Corr. $\angle$'s of $\parallel$ lines are $\cong$.
3. $\triangle ABC \sim \triangle DEC$	3. AA for $\sim \triangle$'s
4. $\frac{AC}{DC}=\frac{BC}{EC}$	4. Def. of $\sim \triangle$'s
5. $\frac{AC}{DC}-1=\frac{BC}{EC}-1$	5. Subtraction property
6. $\frac{DC}{DC}=1$ and $\frac{EC}{EC}=1$	6. Algebraic fact $\left(\frac{a}{a}=1, a \neq 0\right)$
7. $\frac{AC}{DC}-\frac{DC}{DC}=\frac{BC}{EC}-\frac{EC}{EC}$	7. Substitution property
8. $\frac{AC-DC}{DC}=\frac{BC-EC}{EC}$	8. Algebraic fact $\left(\frac{a}{b}-\frac{c}{b}=\frac{a-c}{b}, b \neq 0\right)$
9. $AC-DC=AD$, $BC-EC=BE$	9. Def. of between; subtraction prop.
10. $\therefore \frac{AD}{DC}=\frac{BE}{EC}$	10. Substitution property

Using the definition of similar triangles or Theorem 9.1, you can show that the following proportions are true: If $\overleftrightarrow{DE} \parallel \overline{AB}$ in $\triangle ABC$, then $\frac{AD}{DC} = \frac{BE}{EC}$, $\frac{DC}{AC} = \frac{EC}{BC}$, and $\frac{AD}{AC} = \frac{BE}{BC}$. You may use these triangle proportionality properties as reasons in proofs or in finding lengths.

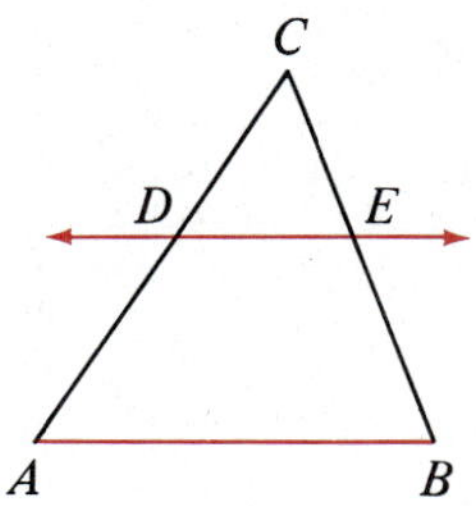

Example 1

Given: $\overline{UV} \parallel \overline{RS}$, $UR = 6$, $TV = 5$, and $VS = 10$
Find TU.

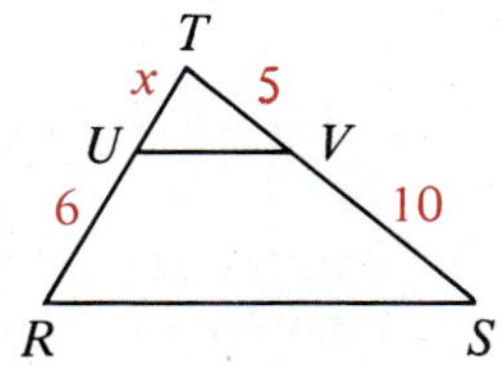

$$\frac{x}{6} = \frac{5}{10} \quad \blacktriangleleft \quad \frac{TU}{UR} = \frac{TV}{VS}$$

$$10x = 30$$
$$x = 3$$

Thus, $TU = 3$.

Example 2

Given: $\overline{JK} \parallel \overline{GH}$, $IG = 12$, $IK = 6$, and $KH = 9$
Find IJ and JG.

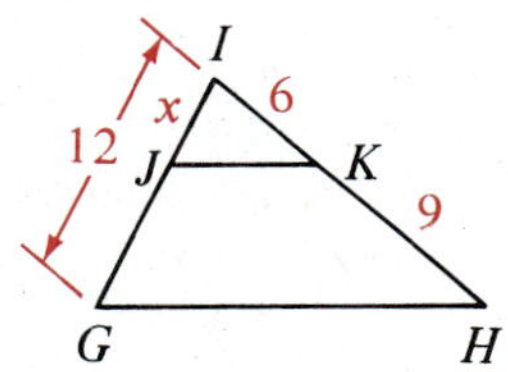

$$\frac{x}{12} = \frac{6}{15} \quad \blacktriangleleft \quad \frac{IJ}{IG} = \frac{IK}{IH}, \; IH = 6 + 9$$

$$15x = 72$$

$$x = 4\frac{12}{15}, \text{ or } 4\frac{4}{5}$$

Thus, $IJ = 4\frac{4}{5}$, and $JG = 12 - 4\frac{4}{5}$, or $7\frac{1}{5}$.

You can prove the converse of Theorem 9.1 using properties of similar triangles.

Theorem 9.2 If a line divides two sides of a triangle proportionally, then it is parallel to the third side of the triangle.

Oral Exercises

In $\triangle PQR$, $\overline{ST} \parallel \overline{PQ}$. Complete each proportion.

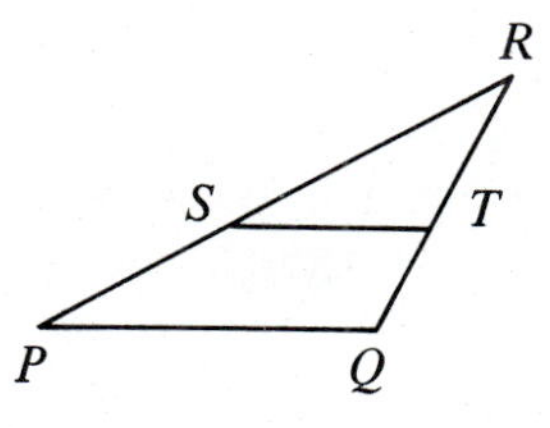

1. $\frac{RS}{SP} = \frac{RT}{?}$ **TQ**

2. $\frac{RP}{SP} = \frac{RQ}{?}$ **TQ**

3. $\frac{TQ}{RQ} = \frac{?}{RP}$ **SP**

4. $\frac{QT}{TR} = \frac{?}{SR}$ **PS**

5. $\frac{RP}{RS} = \frac{RQ}{?}$ **RT**

6. $\frac{RQ}{TQ} = \frac{RP}{?}$ **SP**

Written Exercises

In $\triangle ABC$, $\overline{DE} \parallel \overline{CB}$. Which of the following proportions are always true?

A

1. $\frac{AD}{DC} = \frac{AE}{EB}$ T
2. $\frac{AD}{DC} = \frac{DE}{CB}$ F
3. $\frac{AD}{DE} = \frac{AC}{CB}$ T
4. $\frac{AD}{AC} = \frac{AE}{AB}$ T
5. $\frac{BE}{EA} = \frac{AD}{DC}$ F
6. $\frac{AD}{DC} = \frac{AB}{AE}$ F

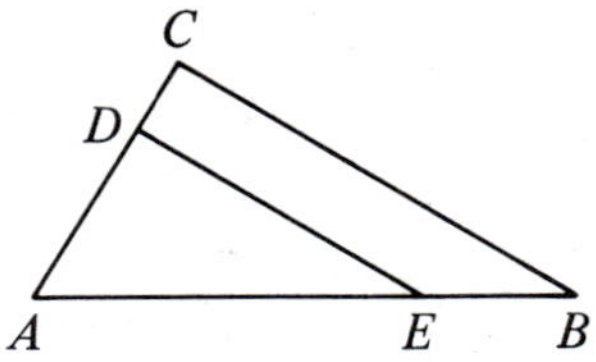

In $\triangle ABE$, $\overleftrightarrow{CD} \parallel \overline{AB}$. Find each indicated length. (Ex. 7–16)

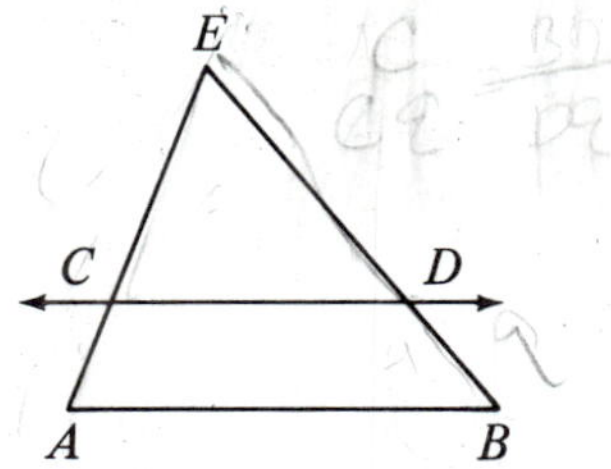

7. $AC = 3$, $CE = 2$, and $BD = 4$ Find DE. $2\frac{2}{3}$
8. $AE = 6$, $CE = 4$, and $BE = 9$ Find BD. 3
9. $AC = 4$, $CE = 7$, and $BD = 5$ Find DE. $8\frac{3}{4}$
10. $EA = 9$, $EC = 5$, and $ED = 12$ Find EB. $21\frac{3}{5}$
11. $EB = 10$, $ED = 4$, and $EA = 14$ Find EC. $5\frac{3}{5}$
12. $AC = 12$, $AE = 20$, and $ED = 10$ Find BD. 15
13. $AC = 9$, $AE = 13$, and $DE = 8$ Find BE. 26
14. $AE = 16$, $EB = 12$, and $DB = 9$ Find EC. 4
15. $EC = 4$, $CA = 8$, and $EB = 18$ Find DB. 12
16. $ED = 5$, $DB = 3$, and $AE = 10$ Find AC and CE. $3\frac{3}{4}$, $6\frac{1}{4}$

B

17. Given: $\angle 1 \cong \angle 2$
    Prove: $\frac{BD}{DC} = \frac{BE}{EA}$

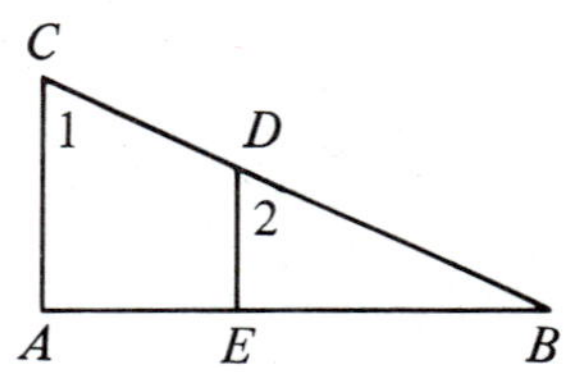

18. Given: $\overline{CA} \perp \overline{AB}$ and $\overline{DE} \perp \overline{AB}$
    Prove: $\frac{AE}{EB} = \frac{CD}{DB}$

19. Given: $\square ABCD$ with $\overline{EF} \parallel \overline{DA}$
    Prove: $\frac{AE}{EB} = \frac{DF}{FC}$

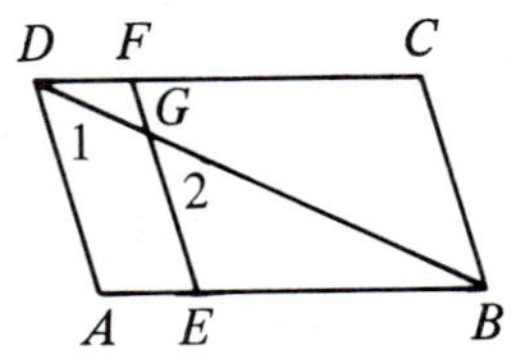

20. Given: $\square ABCD$ with $\angle 1 \cong \angle 2$
    Prove: $\frac{DC}{DF} = \frac{AB}{AE}$

C

21. Prove: If three parallel lines intersect two transversals, then the segments determined are proportional.
22. Given: $\overline{AB} \parallel \overline{DE}$ and $\overline{BC} \parallel \overline{EF}$
    Prove: $\triangle DFG \sim \triangle ACG$

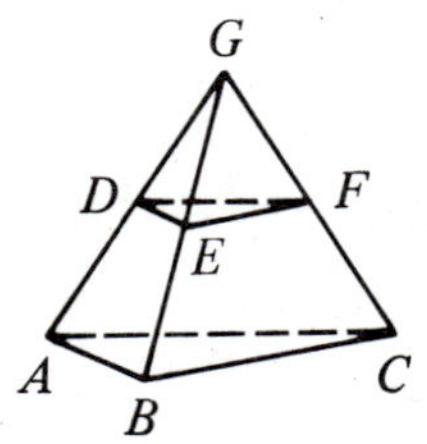

CUMULATIVE REVIEW

1. What is the geometric mean of 8 and 18? 12
2. Draw an obtuse angle. Construct its bisector. Draw an obtuse $\angle$; then see constr., p. 15.
3. What is true about corresponding angles of similar polygons? About corresponding sides? $\cong$; proportional

SAS AND SSS SIMILARITY 9.5

Objective **To recognize and apply the SAS and SSS similarity patterns**

In addition to the angle-angle pattern, you can prove two other methods for showing that triangles are similar. One of these methods involves two pairs of corresponding sides and the angles included by these sides.

$\triangle DEF$ below was constructed so that $\angle D \cong \angle A$ and $\frac{DE}{AB} = \frac{DF}{AC} = \frac{3}{2}$.

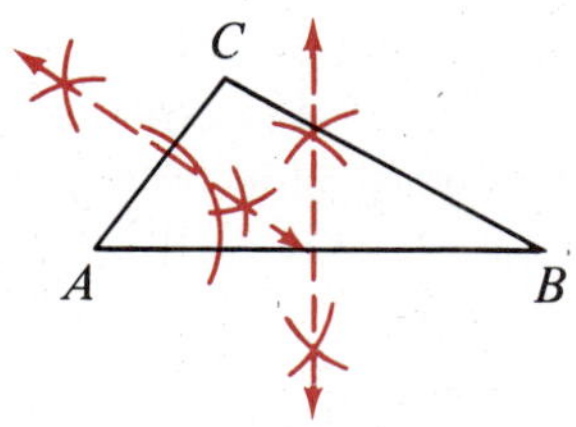

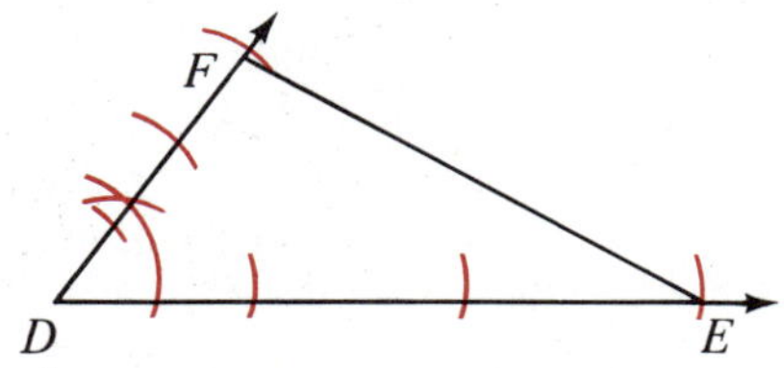

If you measure the other parts of the triangles, you will find that $\angle F \cong \angle C$, $\angle E \cong \angle B$, and $\frac{FE}{CB} = \frac{3}{2}$. Since all pairs of corresponding angles are congruent and all pairs of corresponding sides are proportional, then $\triangle DEF \sim \triangle ABC$. This suggests the following theorem.

Theorem 9.3 (SAS for ~ △'s) If an angle of one triangle is congruent to an angle of another triangle and the corresponding sides that include these angles are proportional, then the triangles are similar.

Given: $\angle Z \cong \angle V$ and $\frac{ZX}{VT} = \frac{YZ}{UV}$

Prove: $\triangle XYZ \sim \triangle TUV$
(See Exercise 36.)

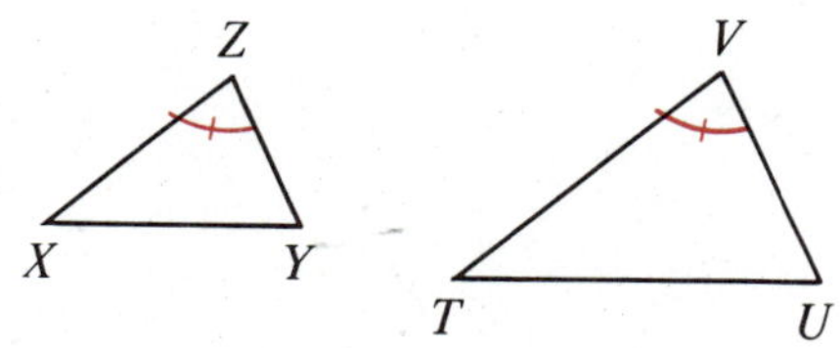

Example 1

Are the two triangles similar?

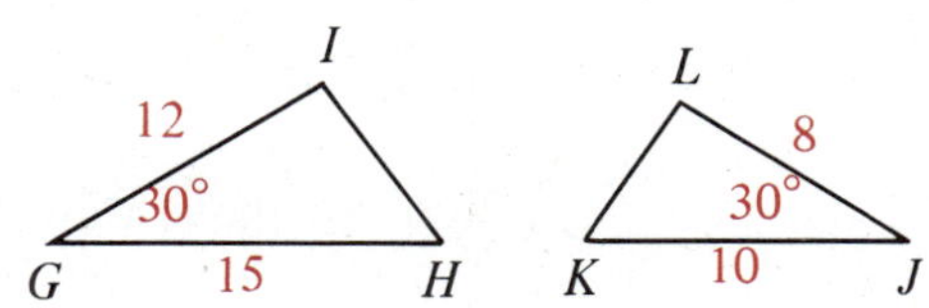

$\frac{GI}{JL} = \frac{12}{8}$ $\frac{GH}{JK} = \frac{15}{10}$

$\frac{12}{8} = \frac{15}{10}$ ◄ *Two pairs of corresponding sides are proportional.*

$m\angle G = 30$ $m\angle J = 30$ ◄ *The included ∠'s are* ≅.

Thus, $\triangle GHI \sim \triangle JKL$ by SAS for ~ △'s.

Another method of proving triangles similar involves the ratios of the lengths of corresponding sides of the triangles. $\triangle RST$ below was constructed so that each side is twice as long as the corresponding side of $\triangle XYZ$.

$$\frac{ST}{YZ} = \frac{RS}{XY} = \frac{RT}{XZ} = \frac{2}{1}$$

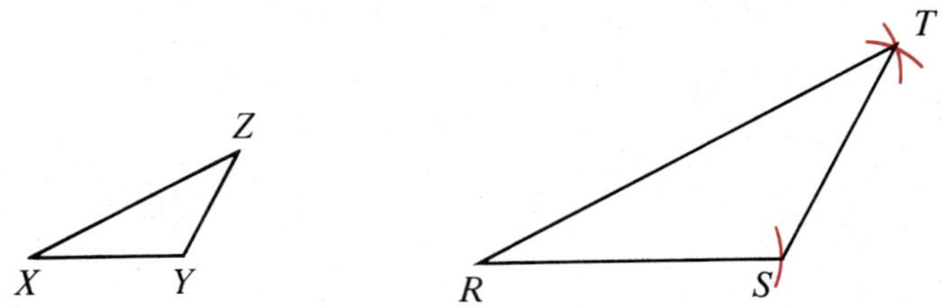

If you measure the angles of the triangles, you will find that the corresponding angles are congruent. Since all pairs of corresponding angles are congruent and all pairs of corresponding sides are proportional, then $\triangle XYZ \sim \triangle RST$. This suggests the following theorem.

Theorem 9.4 (SSS for $\sim$ $\triangle$'s) If the corresponding sides of two triangles are proportional, then the triangles are similar. (See Exercise 37.)

Given: $\frac{AB}{DE} = \frac{BC}{EF} = \frac{CA}{FD}$

Prove: $\triangle ABC \sim \triangle DEF$

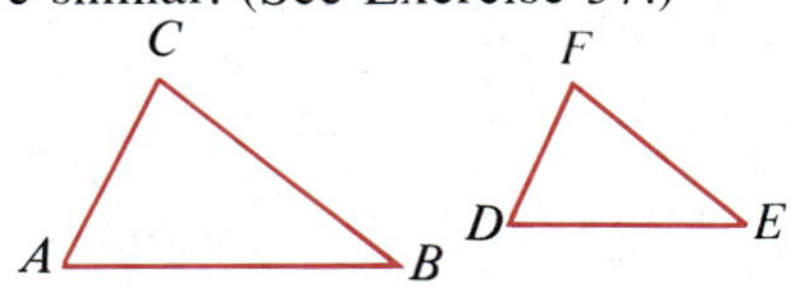

Example 2

Are the two given triangles similar?

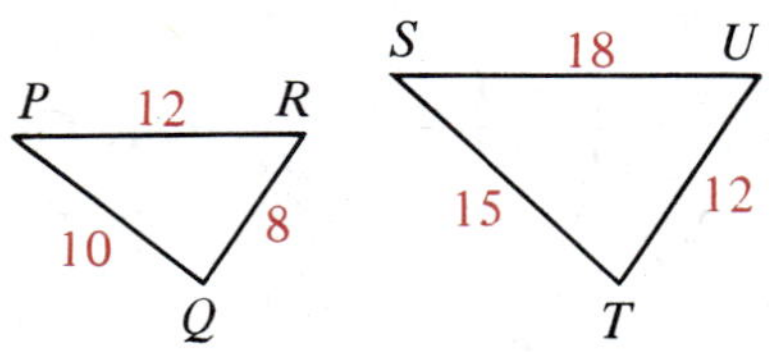

$\frac{8}{12} = \frac{10}{15} = \frac{12}{18}$, or $\frac{QR}{TU} = \frac{PQ}{ST} = \frac{RP}{US}$ ◄ *Corresponding sides are proportional.*

Thus, $\triangle PQR \sim \triangle STU$ by SSS for $\sim$ $\triangle$'s.

You now have three methods for proving triangles similar.

Summary

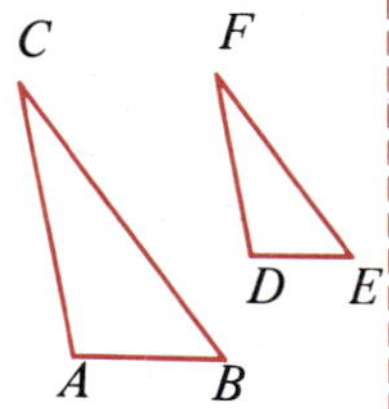

AA for $\sim$ $\triangle$'s	**SAS for $\sim$ $\triangle$'s**	**SSS for $\sim$ $\triangle$'s**
If $\angle A \cong \angle D$ and $\angle B \cong \angle E$, then $\triangle ABC \sim \triangle DEF$.	If $\angle A \cong \angle D$ and $\frac{AC}{DF} = \frac{AB}{DE}$, then $\triangle ABC \sim \triangle DEF$.	If $\frac{AB}{DE} = \frac{BC}{EF} = \frac{CA}{FD}$, then $\triangle ABC \sim \triangle DEF$.

You will use either angle-angle, side-angle-side, or side-side-side in proofs that triangles are similar.

Example 3

Prove: All equilateral triangles are similar.

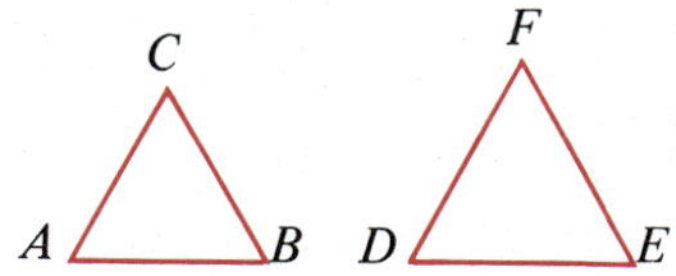

Given: Equilateral $\triangle$'s ABC and DEF
Prove: $\triangle ABC \sim \triangle DEF$

Proof

Statements	Reasons
1. $\triangle ABC$ and $\triangle DEF$ are equilateral.	1. Given
2. $AB = BC = CA$ $DE = EF = FD$	2. Def. of equilat. $\triangle$
3. $\frac{AB}{DE} = \frac{BC}{EF} = \frac{CA}{FD}$	3. Division prop. of equality; subst.
4. $\therefore \triangle ABC \sim \triangle DEF$	4. SSS for $\sim$ $\triangle$'s

Example 4

Given: $\frac{RT}{ST} = \frac{PT}{QT}$

Prove: $\frac{RT}{ST} = \frac{RP}{SQ}$

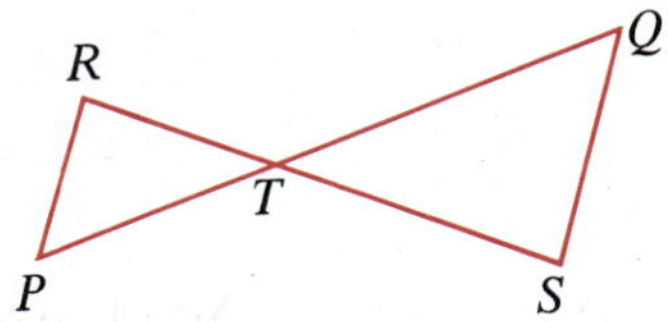

Proof

Statements	Reasons
1. $\frac{RT}{ST} = \frac{PT}{QT}$	1. Given
2. $\angle PTR \cong \angle QTS$	2. Vertical $\angle$'s are $\cong$.
3. $\triangle PTR \sim \triangle QTS$	3. SAS for $\sim$ $\triangle$'s
4. $\therefore \frac{RT}{ST} = \frac{RP}{SQ}$	4. Corr. sides of $\sim$ $\triangle$'s are proportional.

Reading in Geometry

The statement $\triangle ABC \sim \triangle DEF$ implies other statements. If $\triangle ABC \sim \triangle DEF$, which of the following statements must also be true?

1. The ratio of AC to AB is equal to the ratio of DF to DE. T
2. Angle C is congruent to angle E. F
3. Segment CB is congruent to segment FE. F
4. If the measure of angle A is 80°, then the measure of angle D is 80°. T
5. DF is to FE as AC is to CB. T
6. Triangle ABC is congruent to triangle DEF. F
7. If the ratio of AC to DF is 2 to 3, then the ratio of AB to DE is 2 to 3. T
8. The ratio of the measure of angle C to the measure of angle F is 1 to 1. T

Written Exercises

Are the triangles similar? Give a reason for your answer.

A **1.** yes, SSS

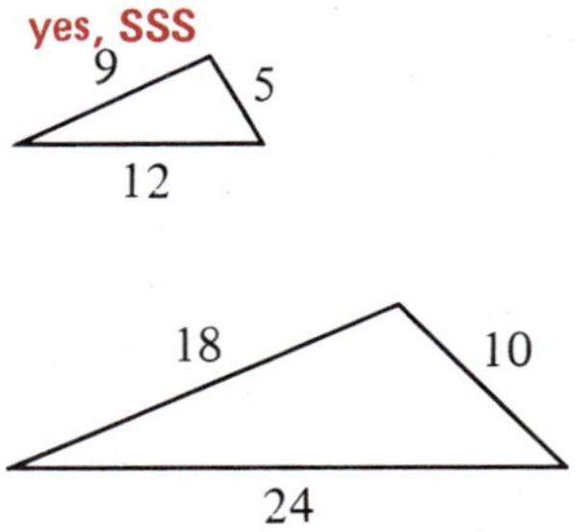

2. yes, SAS

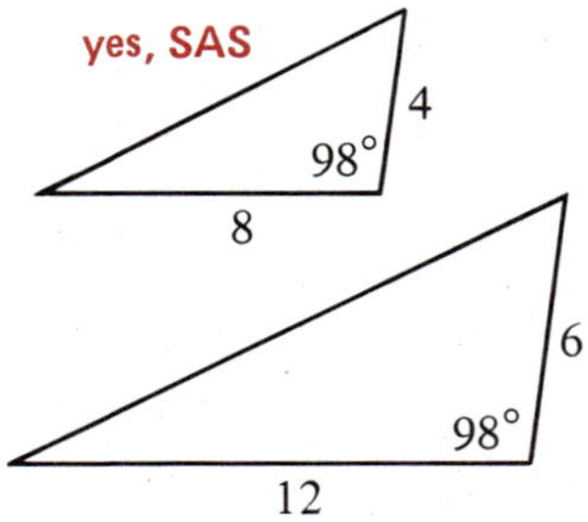

3. yes, AA

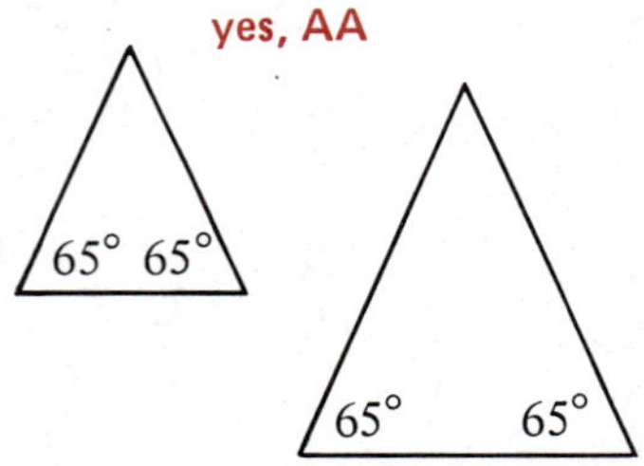

4.

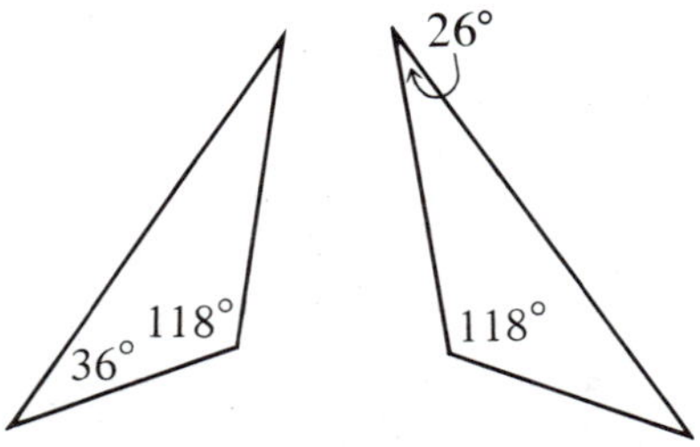

yes, AA

5.

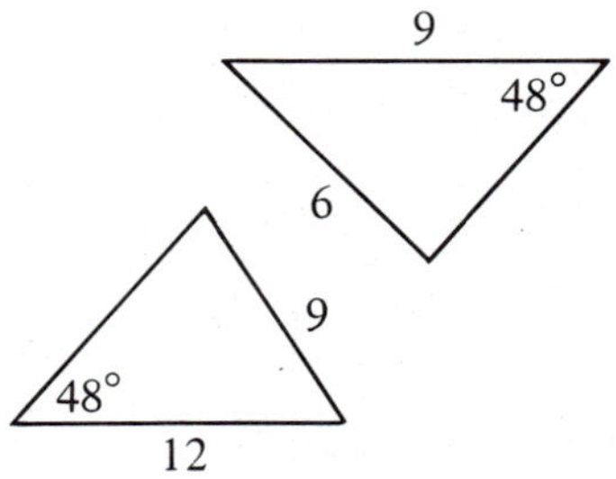

no, ≅ ∠'s are not included ∠'s.

6.

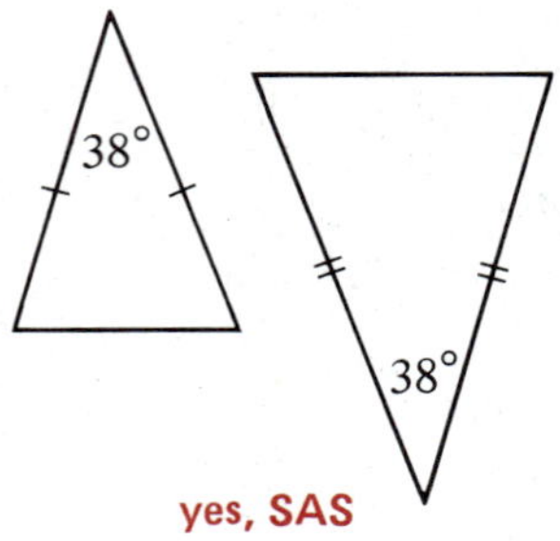

yes, SAS

7. no, one pair of ∠'s are ≅.

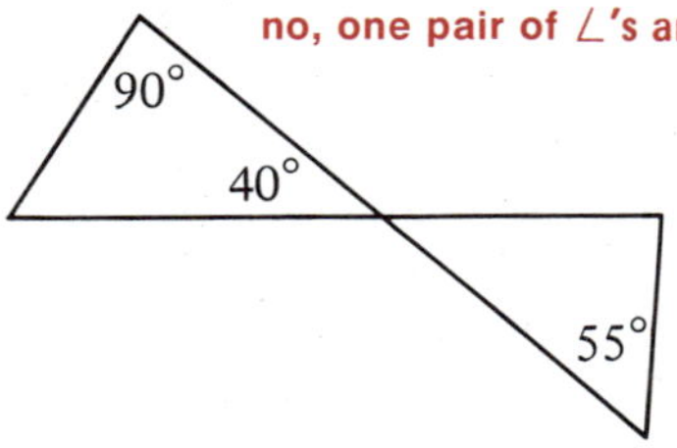

8. no, SSS does not apply.

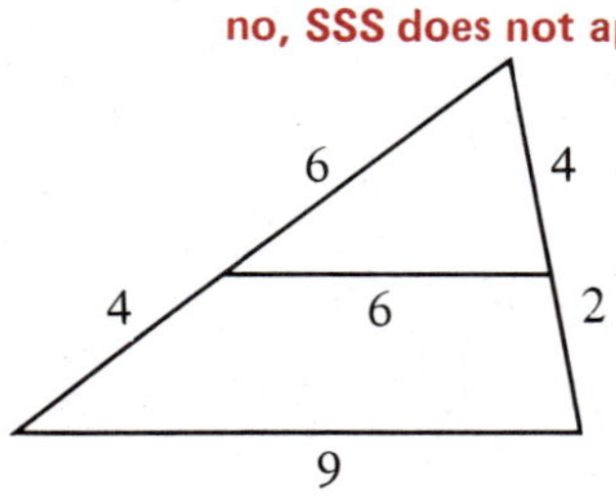

9. yes, SAS

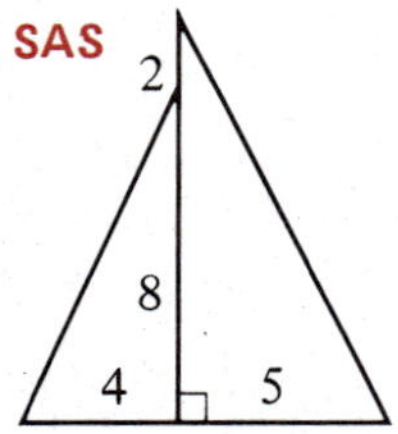

Which of the following are the lengths of sides of similar triangles?

10. 3, 4, 5 and 6, 8, 10 Yes

11. 9, 15, 18 and 6, 10, 12 Yes

12. $\frac{4}{5}, \frac{6}{5}, \frac{8}{5}$ and $\frac{2}{8}, \frac{3}{8}, \frac{4}{8}$ Yes

13. $\frac{3}{4}, 1, \frac{5}{4}$ and $\frac{4}{5}, 1, \frac{6}{5}$ no

14. 3.6, 12, 12.6 and 3, 10, 10.5 Yes

15. 3.5, 4.6, 5.7 and 2.1, 3.3, 4.6 no

In which of the following cases is $\triangle ABC \sim \triangle DEF$?

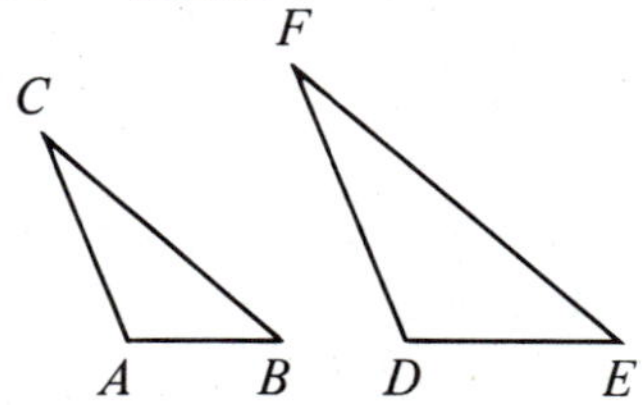

16. $\frac{AB}{DE} = \frac{BC}{EF} = \frac{CA}{FD}$ ~

17. $\frac{AB}{BC} = \frac{DE}{EF}, \angle B \cong \angle E$ ~

18. $\frac{AB}{DE} = \frac{BC}{EF}, \angle C \cong \angle F$ ≁

19. $\angle A \cong \angle D, \angle C \cong \angle E$ ≁

20. $\frac{BC}{EF} = \frac{AC}{DF}, \angle B \cong \angle D$ ≁

21. $\frac{AB}{BC} = \frac{BC}{DE} = \frac{DE}{EF}$ ≁

22. Given: $\triangle TZS \sim \triangle XTR$
Prove: $\triangle TZS \sim \triangle XZY$

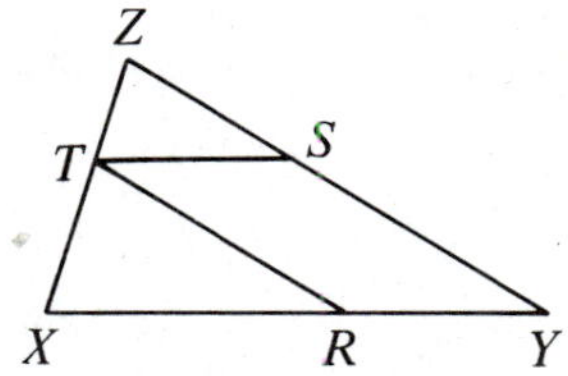

23. Given: $\square TSYR$
Prove: $\triangle XRT \sim \triangle TSZ$

24. Given: $\overline{LR} \perp \overline{NM}$ and $\overline{NP} \perp \overline{LM}$
Prove: $\triangle LPQ \sim \triangle NRQ$

25. Given: $\overline{LR} \perp \overline{NM}$ and $\overline{NP} \perp \overline{LM}$
Prove: $\triangle NQR \sim \triangle NMP$

26. Given: Altitudes $\overline{LR}$ and $\overline{NP}$ of $\triangle LMN$
Prove: $\triangle LMR \sim \triangle NMP$

27. Given: $\overline{LR} \perp \overline{NM}$ and $\angle LNR \cong \angle LMR$
Prove: $\triangle LNR \sim \triangle LMR$

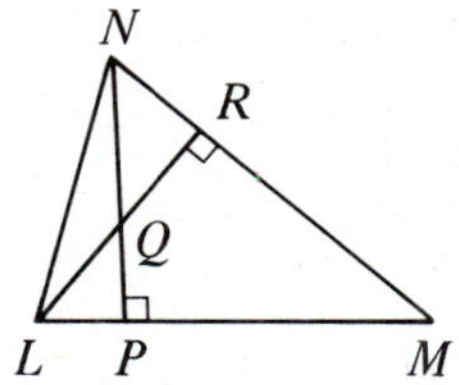

B 28. Given: $\square ABCD$
Prove: $\triangle BCE \sim \triangle FAB$

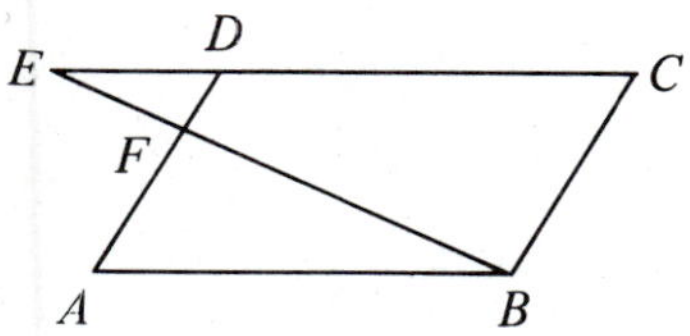

29. Given: $\overline{RT} \parallel \overline{UW}$ and $\overline{TS} \parallel \overline{WV}$
Prove: $\triangle RST \sim \triangle UVW$

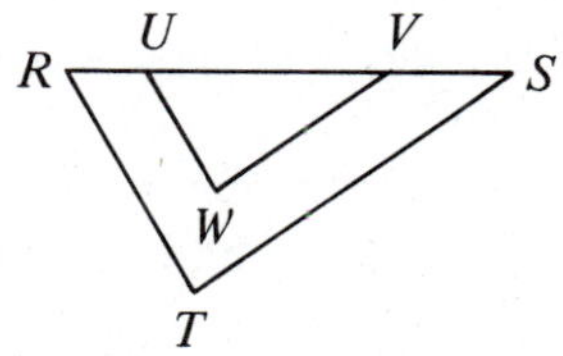

30. Given: $\overline{AD} \cong \overline{DB}$ and $\overline{BE} \cong \overline{EC}$
Prove: $\triangle ABC \sim \triangle DBE$

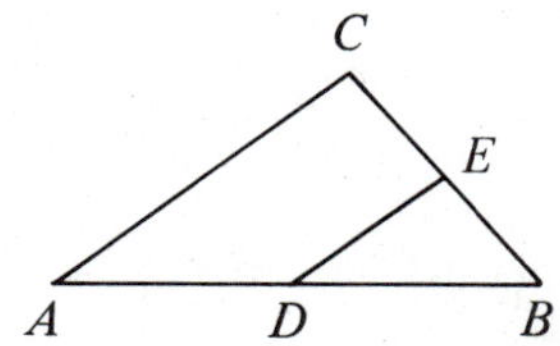

31. Given: $\triangle RVT \cong \triangle SUT$
Prove: $\dfrac{TU}{TR} = \dfrac{TV}{TS}$

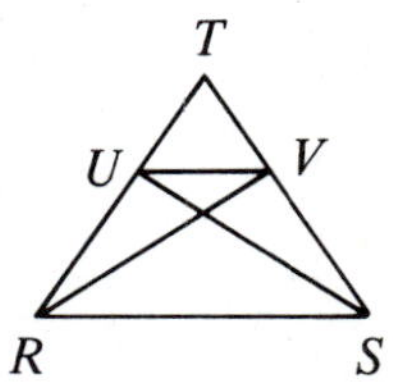

32. Given: $\dfrac{XZ}{MZ} = \dfrac{YZ}{NZ}$
Prove: $\dfrac{XZ}{MZ} = \dfrac{XY}{MN}$

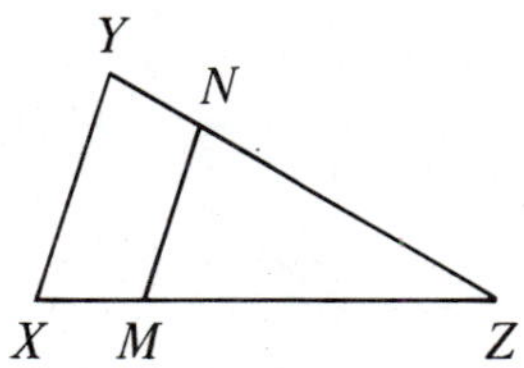

33. Given: $\overline{CQ} \cong \overline{CV}$, $\overline{CT} \perp \overline{RV}$, $\overline{RJ} \perp \overline{CV}$
Prove: $\dfrac{CQ}{RV} = \dfrac{QT}{JV}$

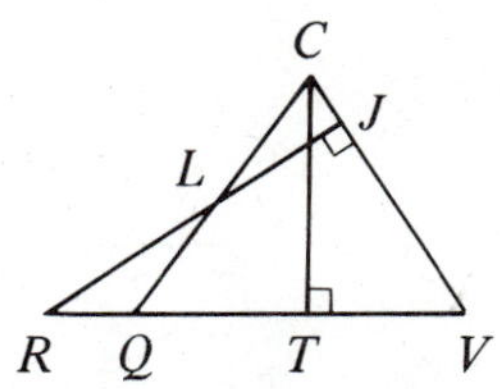

34. Prove: Any two isosceles right triangles are similar.

35. Prove: Two isosceles triangles are similar if an angle of one is congruent to the corresponding angle of the other.

C 36. Prove Theorem 9.3. Use the following pictures and analysis.
Given: $\angle Z \cong \angle V$ and $\dfrac{ZX}{VT} = \dfrac{YZ}{UV}$
Prove: $\triangle XYZ \sim \triangle TUV$

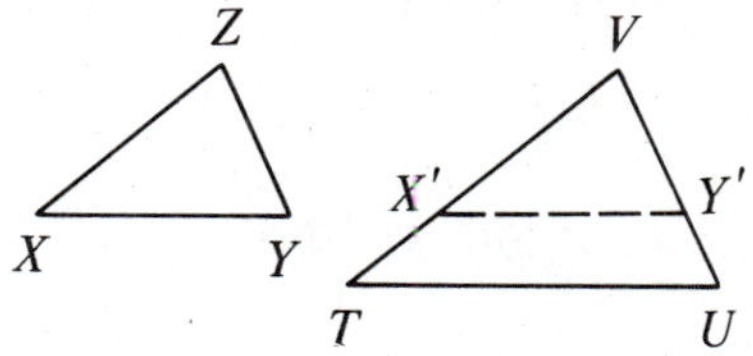

ANALYSIS
Construct $\triangle X'Y'V$ so that $\overline{VX'} \cong \overline{ZX}$ and $\overline{VY'} \cong \overline{ZY}$. Show that $\triangle XYZ \sim \triangle TUV$ by AA.

37. Prove Theorem 9.4.

For each of the following, draw pictures. Which triangles are similar? Which triangles are congruent? State and prove appropriate theorems. **Answers may vary.**

38. Connect the midpoints of the sides of a triangle.

39. Form a triangle from the perpendiculars to the sides of a triangle.

OTHER PROPORTIONAL SEGMENTS 9.6

Objective

To prove and apply theorems about proportional segments in triangles

There are special segments related to a triangle. Three that you have already studied are medians, angle bisectors, and altitudes.

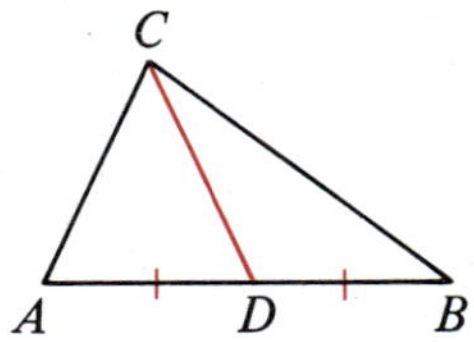

Median $\overline{CD}$
$\overline{AD} \cong \overline{DB}$

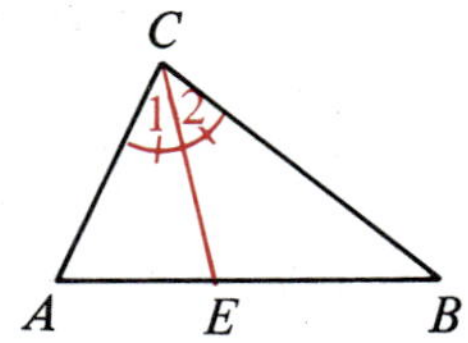

Angle bisector $\overline{CE}$
$\angle 1 \cong \angle 2$

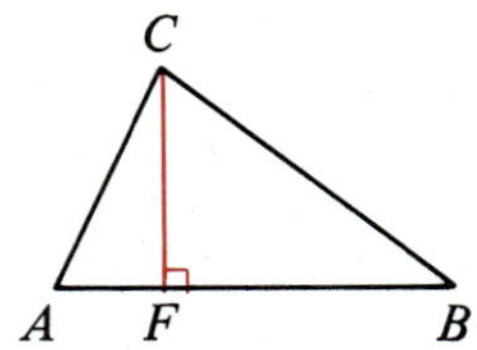

Altitude $\overline{CF}$
$\overline{CF} \perp \overline{AB}$

You can prove that the lengths of these segments within similar triangles are proportional to the lengths of the corresponding sides of the triangles.

Theorem 9.5 Corresponding medians of similar triangles are proportional to corresponding sides. (See Exercise 16.)

Given: $\triangle ABC \sim \triangle EFG$, $\overline{CD}$ is a median to $\overline{AB}$, and $\overline{GH}$ is a median to $\overline{EF}$.

Prove: $\frac{CD}{GH} = \frac{CB}{GF}$

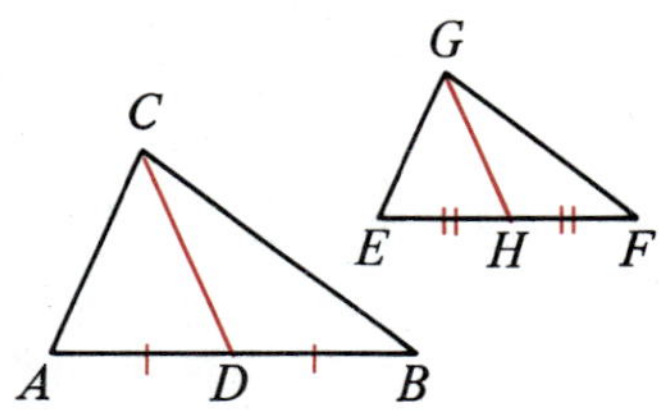

ANALYSIS
Show that $\triangle CDB \sim \triangle GHF$. Corresponding sides of $\sim \triangle$'s are proportional.

Theorem 9.5 can sometimes be used to find unknown lengths of segments.

Example 1

Given: $\triangle STU \sim \triangle WXY$ and medians $\overline{TV}$ and $\overline{XZ}$
Find XZ.

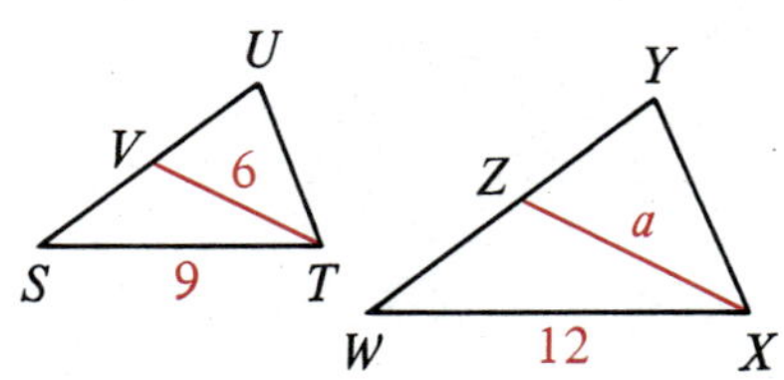

$\frac{TV}{XZ} = \frac{ST}{WX}$ ◄ *Apply Th. 9.5.*

Let $a = XZ$.

$\frac{6}{a} = \frac{9}{12}$

$9a = 72$

$a = 8.$

Thus, $XZ = 8$.

Theorem 9.6

Corresponding altitudes of similar triangles are proportional to corresponding sides. (See Exercise 17.)

Theorem 9.6 can also be used to find unknown lengths of segments.

Example 2

Given: $\triangle ABC \sim \triangle TUV$ and altitudes $\overline{AD}$ and $\overline{TW}$

Find AB.

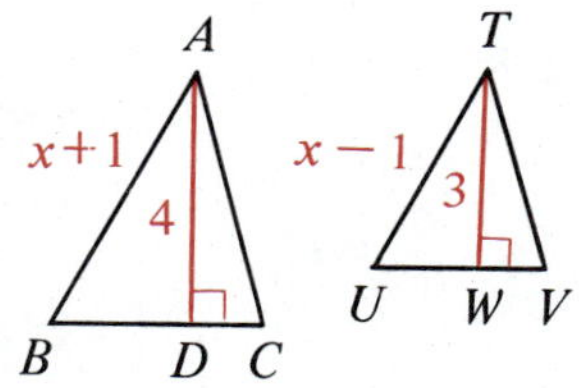

$$\frac{AD}{TW} = \frac{AB}{TU} \quad \blacktriangleleft \textit{ Apply Th. 9.6.}$$

$$\frac{4}{3} = \frac{x+1}{x-1}$$

$$4(x-1) = 3(x+1)$$

$$4x - 4 = 3x + 3$$

$$x = 7$$

Thus, $AB = 7 + 1$, or 8.

Angle bisectors form proportional segments in a triangle. You should complete the proof of the following theorem.

Theorem 9.7

The bisector of an angle of a triangle divides the opposite side into segments proportional to the other sides of the triangle.

Given: $\triangle ABC$ and $\overline{CD}$ bisects $\angle ACB$.

Prove: $\frac{AD}{DB} = \frac{AC}{BC}$

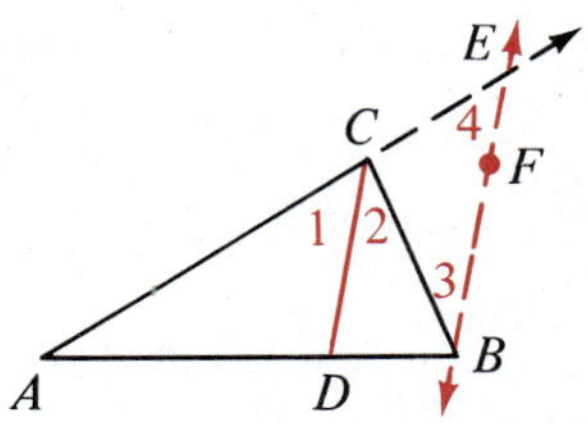

ANALYSIS

Use $\parallel$ auxiliary lines to form proportional segments. Substitute in the resulting proportion.

Outline of proof

1. Draw $\overleftrightarrow{BF} \parallel \overline{DC}$ through B.
2. Draw $\overrightarrow{AC}$ intersecting $\overleftrightarrow{BF}$ at E.
3. $\frac{AD}{DB} = \frac{AC}{CE}$
4. $\angle 2 \cong \angle 3$
5. $\angle 1 \cong \angle 4$
6. $\overline{CD}$ bisects $\angle ACB$.
7. $\angle 1 \cong \angle 2$
8. $\angle 3 \cong \angle 4$
9. $BC = CE$
10. $\therefore \frac{AD}{DB} = \frac{AC}{BC}$

Example 3

Given: $\angle y \cong \angle z$, $RT = 9$, $QT = 12$, and $RQ = 14$
Find RN and NQ.

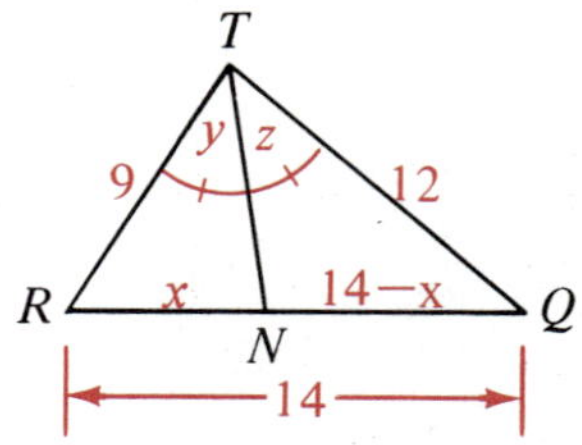

$$\frac{RN}{NQ} = \frac{RT}{TQ} \quad \blacktriangleright \quad \frac{x}{14-x} = \frac{9}{12}$$

$$12x = 9(14-x)$$
$$12x = 126 - 9x$$
$$21x = 126$$
$$x = 6$$

Thus, $RN = 6$ and $NQ = 14 - 6$, or 8.

You can divide a segment into two segments whose lengths have a given ratio by a construction based on Theorem 9.7.

Construction

To divide $\overline{AB}$ into two segments whose lengths have the ratio $\frac{2}{3}$.

Given: Segment $\overline{AB}$
Construct: R on $\overline{AB}$ so that $\frac{AR}{RB} = \frac{2}{3}$.

Strategy: Apply Th. 9.7. Construct $\triangle ABC$ so that $\frac{AC}{BC} = \frac{2}{3}$.

Draw a segment $\overline{PQ}$. Construct segments with lengths 2 PQ and 3 PQ.

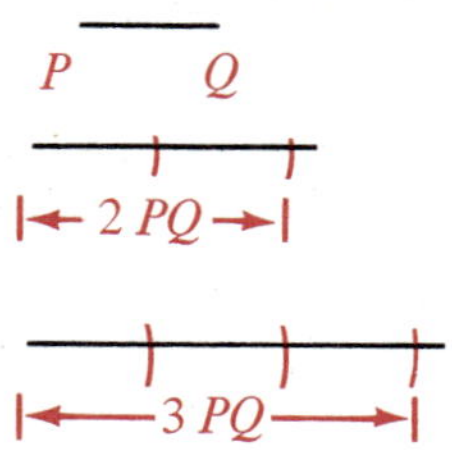

Construct $\triangle ABC$ so that $AC = 2\ PQ$ and $BC = 3\ PQ$.

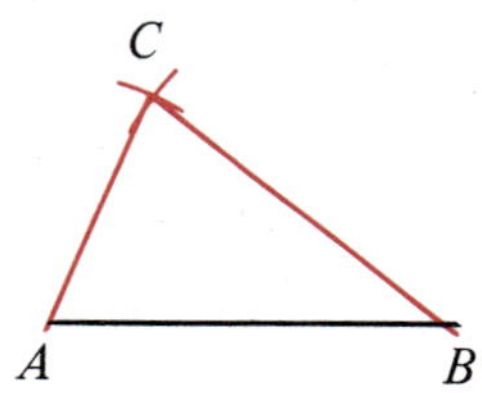

Bisect $\angle ACB$. Draw bisector intersecting $\overline{AB}$ at R.

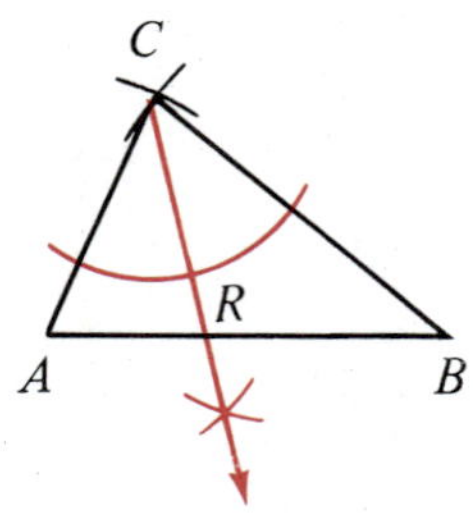

Conclusion: $\frac{AR}{RB} = \frac{AC}{BC} = \frac{2}{3}$

Oral Exercises

In $\triangle MNO$, $\overline{OP}$ bisects $\angle MON$. Complete each proportion.

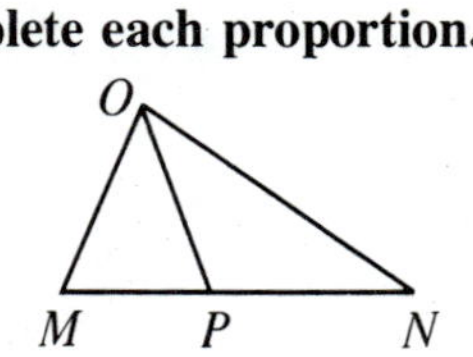

1. If $\frac{MO}{NO} = \frac{5}{6}$, then $\frac{MP}{PN} = ?$ **$\frac{5}{6}$**

2. If $\frac{MP}{PN} = \frac{1}{2}$, then $\frac{MO}{NO} = ?$ **$\frac{1}{2}$**

3. If $\frac{MO}{NO} = \frac{2}{3}$, then $\frac{MP}{PN} = ?$ **$\frac{2}{3}$**

4. If $\frac{MP}{PN} = \frac{3}{4}$, then $\frac{MO}{NO} = ?$ **$\frac{3}{4}$**

Written Exercises

A

1. Given: $\overline{CD}$ bisects $\angle ACB$, $AC = 7$, $AD = 5$, and $DB = 6$. Find CB. $8\frac{2}{5}$

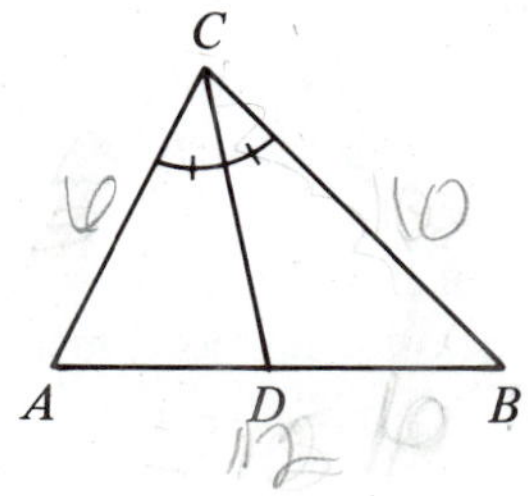

2. Given: $\overline{CD}$ bisects $\angle ACB$, $AC = 5$, $CB = 8$, and $DB = 6$. Find AD. $3\frac{3}{4}$

3. Given: $\overline{CD}$ bisects $\angle ACB$, $AC = 5$, $CB = 10$, and $AB = 12$. Find AD and DB. 4; 8

4. Given: $\overline{CD}$ bisects $\angle ACB$, $AC = 6$, $CB = 10$, and $AB = 12$. Find AD and DB. $4\frac{1}{2}$; $7\frac{1}{2}$

5. Given: $\triangle EGH \sim \triangle IKL$, $\overline{HF}$ and $\overline{LJ}$ are medians, $HE = 9$, $LI = 6$, $HF = x + 2$, and $LJ = 2x - 4$. Find HF. 6

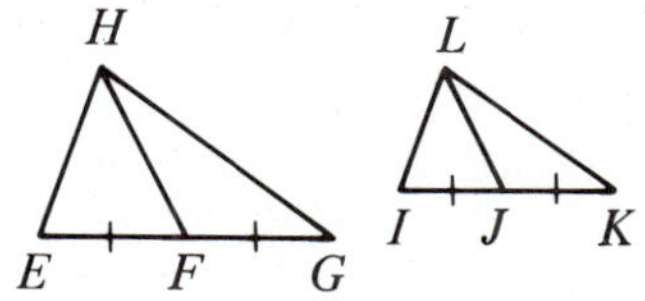

6. Given: $\triangle EGH \sim \triangle IKL$, $\overline{HF}$ and $\overline{LJ}$ are medians, $EF = 9$, $IJ = 6$, $HF = 2x + 1$, and $LJ = x + 3$. Find LJ. 10

7. Given: $\triangle MOP \sim \triangle QST$, $\overline{PN}$ and $\overline{TR}$ are altitudes, $MP = 15$, $PN = 10$, $QT = 2x + 2$, and $TR = x + 3$. Find QT and TR. 12; 8

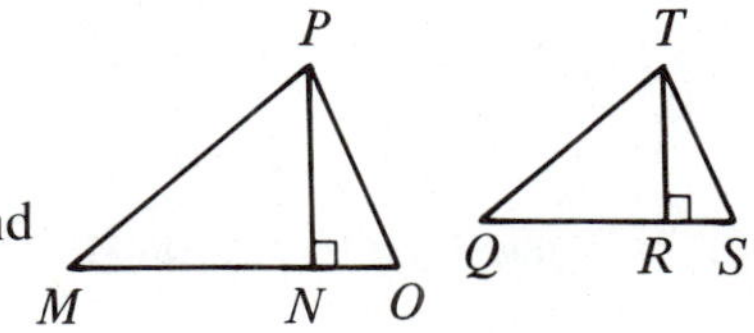

8. Given: $\triangle MOP \sim \triangle QST$, $\overline{PN}$ and $\overline{TR}$ are altitudes, $TR = 3$, $TS = 5$, $PN = 2x + 1$, and $PO = 4x - 5$. Find PN and PO. 21; 35

B

9. Given: $\overline{YZ}$ bisects $\angle WYX$, $WZ = 3$, $ZX = 4$, and the perimeter of $\triangle WXY$ is 21. Find YX. 8

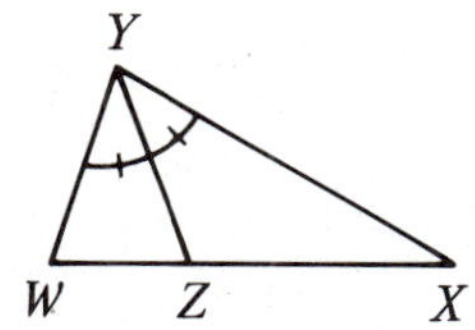

10. Given: $\overline{YZ}$ bisects $\angle WYZ$, $WZ = 4$, $ZX = 5$, and the perimeter of $\triangle WXY$ is 27. Find YW. 8

11. Given: $\triangle ABD$ and $\triangle DBC$, $\overline{AE}$ and $\overline{CE}$ are angle bisectors.

Prove: $\frac{AD}{AB} = \frac{DC}{CB}$

12. Given: $\triangle ABC \sim \triangle EFG$, $\overline{CD}$ and $\overline{GH}$ are medians.

Prove: $\frac{CD}{GH} = \frac{AD}{EH}$

13. Given: $\triangle RST \sim \triangle VWX$, $\overline{TU}$ and $\overline{XY}$ are altitudes.

Prove: $\frac{TU}{XY} = \frac{US}{YW}$

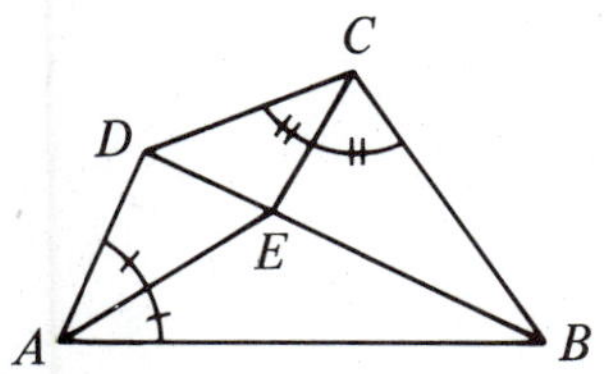

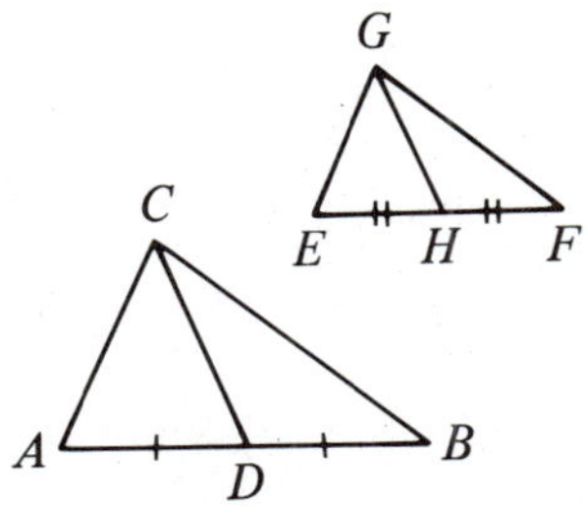

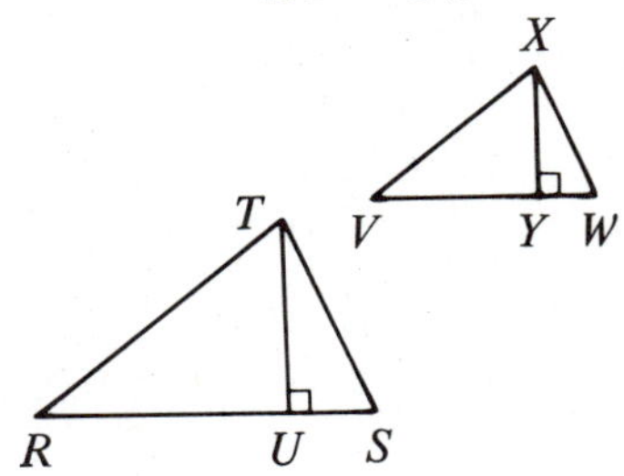

14. Draw $\overline{AB}$. Construct P so that $\frac{AP}{PB} = \frac{3}{4}$. See Constr., p. 292.

15. Draw $\overline{XY}$. Construct Q so that $\frac{XQ}{QY} = \frac{2}{5}$. See Constr., p. 292.

16. Prove Theorem 9.5.

17. Prove Theorem 9.6.

18. Complete the proof of Theorem 9.7.

19. Prove: If two triangles are similar, then corresponding altitudes are proportional to corresponding medians.

20. Prove: The diagonals of a trapezoid divide each other into proportional segments.

PRODUCTS OF LENGTHS 9.7

Objectives

To find products of lengths of sides of triangles
To prove products of lengths equal using proportions

Every proportion has a related pair of cross-products. Given the cross-products, you can write several proportions.

Example 1

Write two proportions that give the product $PQ \cdot QR = FE \cdot EG$.

$\frac{PQ}{EG} = \frac{FE}{QR}$, or $\frac{FE}{PQ} = \frac{QR}{EG}$ ◄ *Make one pair of factors the means; the other pair, the extremes.*

You can use similar triangles to obtain true proportions. These proportions can then be used to find the products of lengths of sides of the triangles.

Example 2

Given: $\triangle ABC \sim \triangle DEF$, $AB = 6$, and $DF = 9$
Find $AC \cdot DE$.

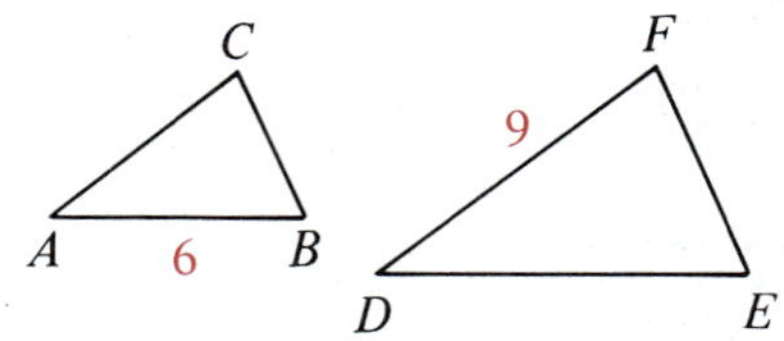

Corr. sides of $\sim \triangle$'s ► $\frac{AC}{DF} = \frac{AB}{DE}$

$AC \cdot DE = DF \cdot AB$ **Thus,** $AC \cdot DE = 9 \cdot 6$, or 54.

Example 3

Given: $\square ABCD$
Prove: $AG \cdot EG = FG \cdot CG$

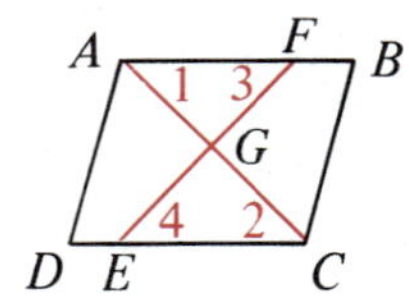

Proof

Statements	**Reasons**
1. $\square ABCD$	1. Given
2. $\overline{AB} \parallel \overline{CD}$	2. Opp. sides of a $\square$ are $\parallel$.
3. $\angle 1 \cong \angle 2$, $\angle 3 \cong \angle 4$	3. Alt. int. $\angle$'s of $\parallel$ lines are $\cong$.
4. $\triangle AGF \sim \triangle CGE$	4. AA for $\sim \triangle$'s
5. $\frac{AG}{CG} = \frac{FG}{EG}$	5. Corr. sides of $\sim \triangle$'s are proportional.
6. $\therefore AG \cdot EG = FG \cdot CG$	6. Property of proportions

Example 4

In Example 3, if $AG = 7$ and $EG = 4$, find $FG \cdot CG$.

$FG \cdot CG = AG \cdot EG$ **Thus,** $FG \cdot CG = 7 \cdot 4$, or 28.

Sometimes you can use similar triangles to find the square of the length of a segment. This is illustrated below.

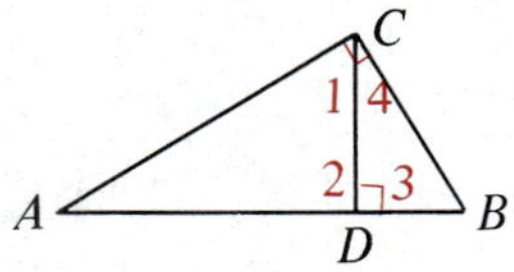

Example 5

Given: $\overline{AC} \perp \overline{BC}$ and altitude $\overline{CD}$
Prove: $AD \cdot BD = (CD)^2$

Proof

Statements	Reasons
1. $\overline{AC} \perp \overline{BC}$ and altitude $\overline{CD}$	1. Given
2. $\angle 2 \cong \angle 3$	2. Def. of altitude; $\perp$ forms $\cong$ rt. $\angle$'s.
3. $\triangle ACB$ $\triangle ADC$, and $\triangle CDB$ are rt. $\triangle$'s.	3. $\perp$'s form rt. $\angle$'s; def. of rt. $\triangle$.
4. $\angle 1$ is a complement of $\angle A$.	4. Acute $\angle$'s of a rt. $\triangle$ are compl.
5. $\angle B$ is a complement of $\angle A$.	5. Same as 4
6. $\angle 1 \cong \angle B$	6. Compl. of same $\angle$ are $\cong$.
7. $\triangle ADC \sim \triangle CDB$	7. AA for $\sim$ $\triangle$'s
8. $\frac{AD}{CD} = \frac{CD}{BD}$	8. Corr. sides of $\sim$ $\triangle$'s are proportional.
9. $\therefore AD \cdot BD = (CD)^2$	9. Property of proportions

Written Exercises

Write two proportions that give each product. (Ex. 1–3) **Answers may vary.**

A

1. $XY \cdot ZW = AB \cdot CD$
$\frac{XY}{AB} = \frac{CD}{ZW}$, $\frac{AB}{XY} = \frac{ZW}{CD}$

2. $RJ \cdot TK = SP \cdot AD$
$\frac{RJ}{SP} = \frac{AD}{TK}$, $\frac{SP}{RJ} = \frac{TK}{AD}$

3. $(RP)^2 = AJ \cdot WF$
$\frac{RP}{AJ} = \frac{WF}{RP}$, $\frac{AJ}{RP} = \frac{RP}{WF}$

4. Given: $\triangle JQT \sim \triangle CPK$, $CP = 8$, and $QT = 12$
Find $JQ \cdot PK$. **96**

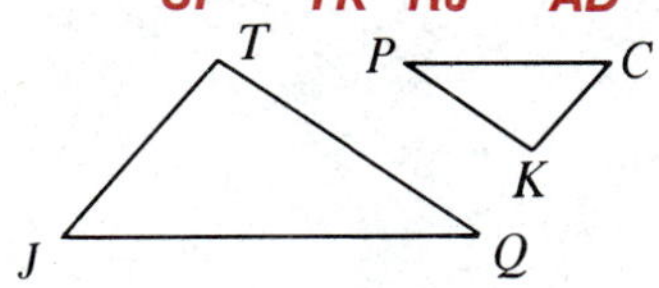

5. Given: $\triangle JQT \sim \triangle CPK$, $JQ = 31$, and $PK = 18$
Find $CP \cdot QT$. **558**

6. Given: $\overline{CA} \perp \overline{AE}$ and $\overline{BD} \perp \overline{EC}$ Prove: $BC \cdot AE = BD \cdot EC$

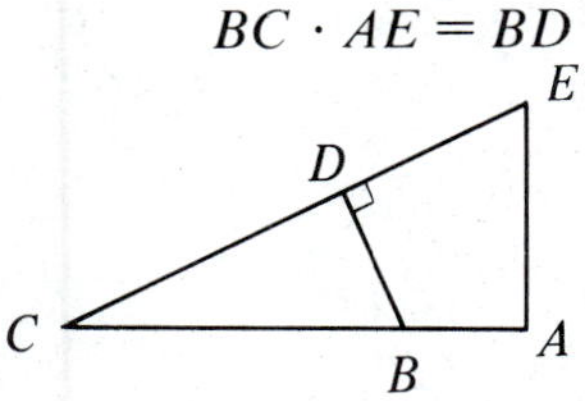

7. Given: $\overline{FG} \parallel \overline{JI}$
Prove: $GH \cdot IH = FH \cdot JH$

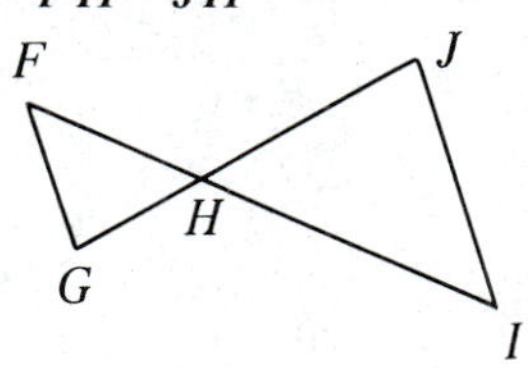

8. Given: $\overline{KL} \perp \overline{LN}$ and $\overline{KO} \perp \overline{ON}$ Prove: $LM \cdot NM = OM \cdot KM$

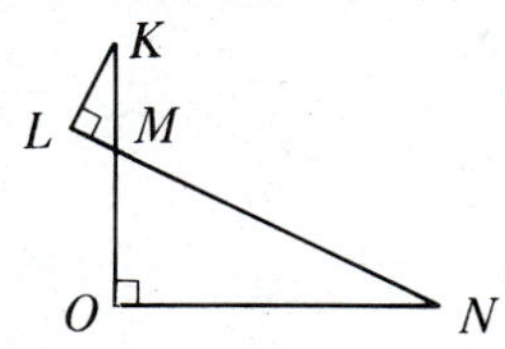

9. Given: $\overline{UY} \perp \overline{YW}$ and $\overline{WX} \perp \overline{UX}$
Prove: $VY \cdot VW = VX \cdot VU$

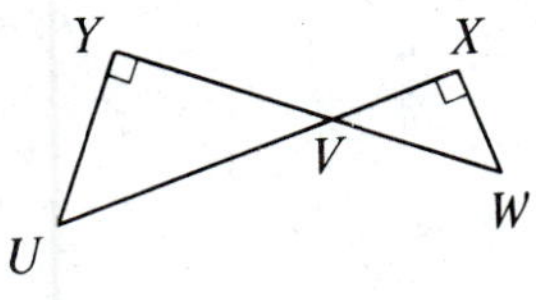

10. Given: $\overline{PQ} \parallel \overline{SR}$
Prove: $ST \cdot PQ = TQ \cdot SR$

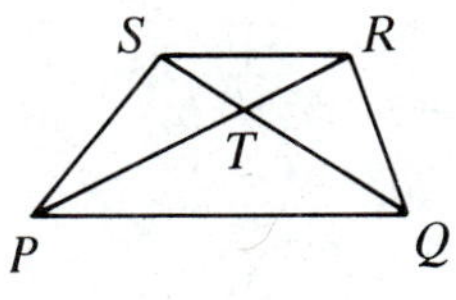

11. Given: $\angle 1 \cong \angle 2$
Prove: $ZA \cdot BC = AB \cdot ZC$

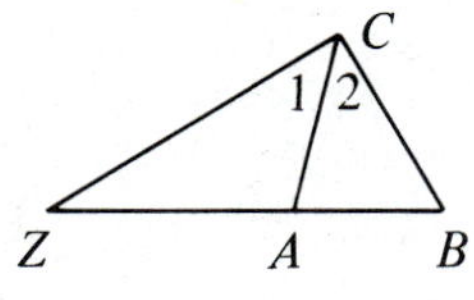

Indicate which information is unnecessary. Then find the product. (Ex. 12–14)

B **12.** Given: $\triangle XYZ \sim \triangle RST$
Find $RT \cdot YZ$.

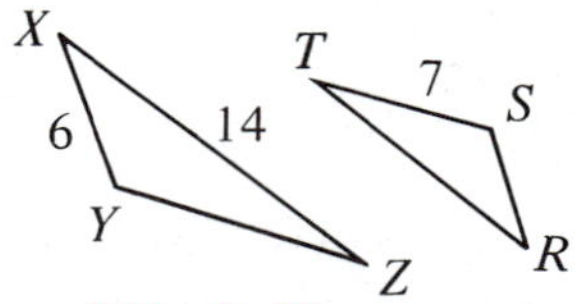

XY = 6; 98

13. Given: $\overline{FD} \perp \overline{CD}$ and $\overline{EB} \perp \overline{FC}$
Find $FB \cdot FC$.

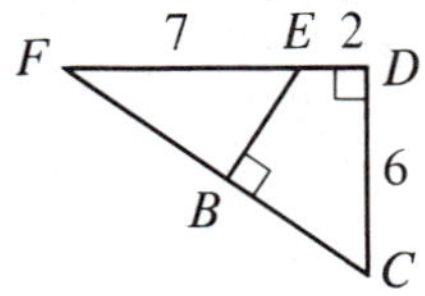

CD = 6; 63

14. Given: $\triangle JNM \sim \triangle MLK$
Find $ML \cdot NM$.

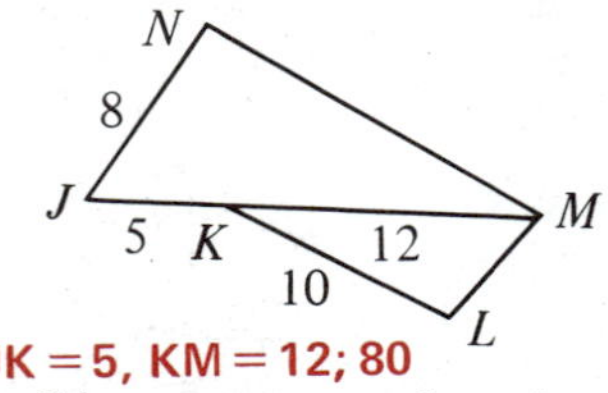

JK = 5, KM = 12; 80

15. Given: $\overline{TR} \perp \overleftrightarrow{PQ}$ and $\angle 1 \cong \angle 4$
Prove: $ST \cdot RU = SR \cdot TU$

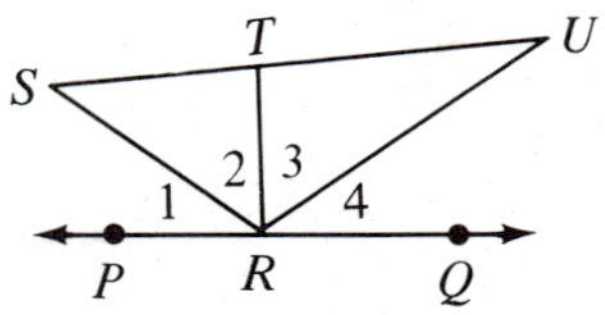

16. Given: $\square RQXZ$
Prove: $ZF \cdot FR = FP \cdot FX$

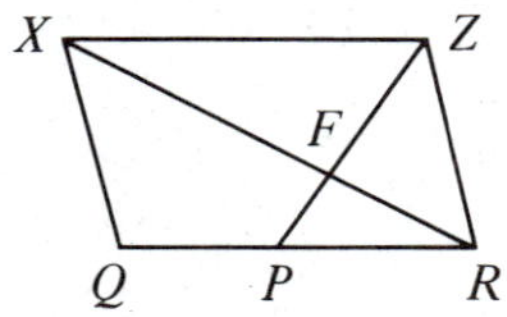

17. Given: $\angle a \cong \angle b$ and $\overline{JK} \parallel \overline{YZ}$
Prove: $XK \cdot JK = KZ \cdot XJ$

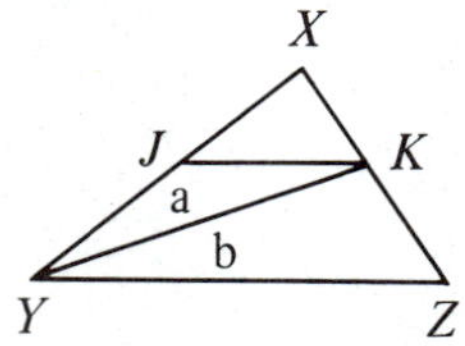

C **18.** Given: $\square ABCD$ and $\angle x \cong \angle y$
Prove: $BF \cdot BE = BF \cdot BC$

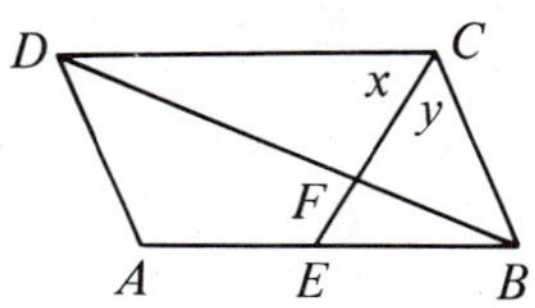

19. Given: $\overline{AB} \parallel \overline{CD}$ and $\angle x \cong \angle y$
Prove: $ED \cdot CB = DA \cdot EF$

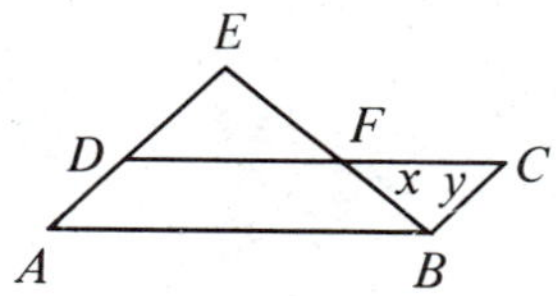

20. Given: $\angle 1 \cong \angle 4$ and $\angle 5 \cong \angle J$
Prove: $LM \cdot NM = LN \cdot MJ$

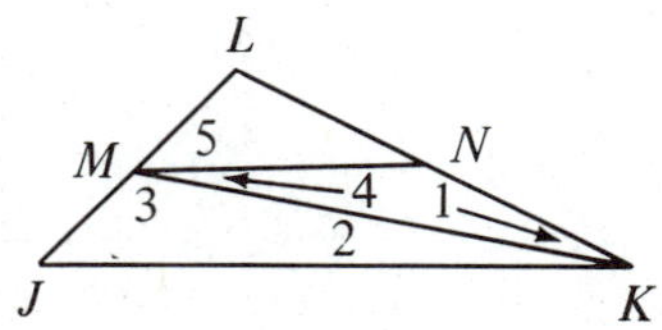

ALGEBRA REVIEW

Objective

To perform operations with radicals

Examples

Simplify.

$5\sqrt{3} + 4\sqrt{12} - 2\sqrt{75}$
$5\sqrt{3} + 4\sqrt{4 \cdot 3} - 2\sqrt{25 \cdot 3}$
$5\sqrt{3} + 8\sqrt{3} - 10\sqrt{3}$
$3\sqrt{3}$

$3\sqrt{2} \cdot \sqrt{2}$
$3\sqrt{2 \cdot 2}$
$3 \cdot 2$
6

$\dfrac{12\sqrt{75}}{4\sqrt{3}}$
$\dfrac{12}{4} \cdot \sqrt{\dfrac{75}{3}}$
$3\sqrt{25} = 15$

Simplify.

1. $\sqrt{2} + \sqrt{50}$ $6\sqrt{2}$
2. $\sqrt{27} + \sqrt{75}$ $8\sqrt{3}$
3. $\sqrt{72} - \sqrt{50}$ $\sqrt{2}$
4. $\sqrt{80} - \sqrt{5}$ $3\sqrt{5}$
5. $\sqrt{12} - \sqrt{48} + \sqrt{3}$ $-\sqrt{3}$
6. $\sqrt{98} - 4\sqrt{8} + 3\sqrt{128}$ 6. $23\sqrt{2}$
7. $\sqrt{5} \cdot \sqrt{5}$ 5
8. $\sqrt{32} \cdot \sqrt{2}$ 8
9. $5\sqrt{8} \cdot 7\sqrt{3}$ $70\sqrt{6}$
10. $\dfrac{12\sqrt{20}}{3\sqrt{5}}$ 8
11. $\dfrac{4\sqrt{48}}{8\sqrt{3}}$ 2
12. $\dfrac{14\sqrt{150}}{7\sqrt{2}}$ $10\sqrt{3}$

SIMILARITY IN SPACE (optional) 9.8

Objective **To prove and apply theorems about proportions and similarity in space**

Remember that two planes are parallel if they do not intersect. Plane $m \parallel$ plane n.

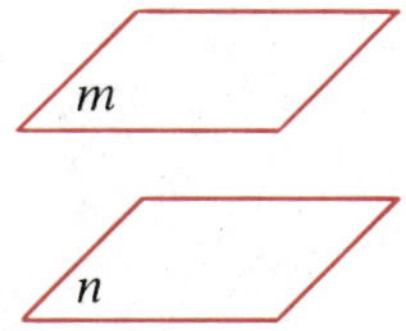

You can often apply properties of similar triangles to figures in space. In many cases, parallel planes are comparable to parallel lines.

Example 1 Prove: If two lines are intersected by three parallel planes, then the corresponding segments that are formed are proportional.

Given: Lines t and q intersected by planes l, m, and n at A, C, F, and B, E, G, respectively; $l \parallel m \parallel n$

Prove: $\frac{AC}{CF} = \frac{BE}{EG}$

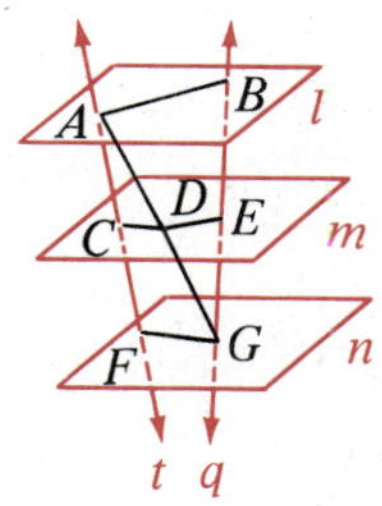

ANALYSIS
A line $\parallel$ to a side of a $\triangle$ divides the other 2 sides proportionally.

Proof

Statements	Reasons
1. $l \parallel m \parallel n$	1. Given
2. Draw $\overline{AG}$, intersecting m at D.	2. Two points are contained in exactly one line.
3. Draw $\overline{AB}$, $\overline{FG}$, $\overline{CD}$, $\overline{DE}$.	3. Same as 2
4. $\overline{AB} \parallel \overline{DE}$	4. $\parallel$ planes intersected by 3rd plane form $\parallel$ lines of intersection.
5. $\frac{AD}{DG} = \frac{BE}{EG}$	5. A line $\parallel$ to a side of a $\triangle$ divides the other 2 sides proportionally.
6. $\overline{CD} \parallel \overline{FG}$	6. Same as 4
7. $\frac{AC}{CF} = \frac{AD}{DG}$	7. Same as 5
8. $\therefore \frac{AC}{CF} = \frac{BE}{EG}$	8. Substitution property

The example above is comparable to Theorem 9.1, which involved a line parallel to a side of a triangle. In each case, proportional segments were determined by the parallel lines or planes.

You can use the results of Example 1 to find unknown lengths.

Example 2 Given: $p \parallel q \parallel r$, $AB = 6$, $BC = 7$, and $DE = 9$
Find EF.

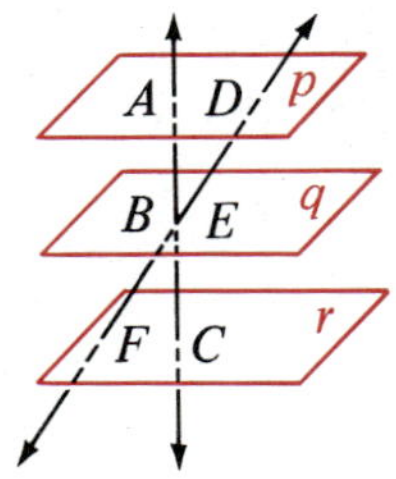

$$\frac{AB}{BC} = \frac{DE}{EF}$$ ◄ *Apply Example 1.*

$$\frac{6}{7} = \frac{9}{EF}$$

$$6 \cdot EF = 63$$

Thus, $EF = 10\frac{1}{2}$.

$ABCD$ is a solid figure called a **tetrahedron.** $\triangle ABC$ is its **base.** $\triangle ABD$, $\triangle BCD$, and $\triangle CAD$ are its **faces.**

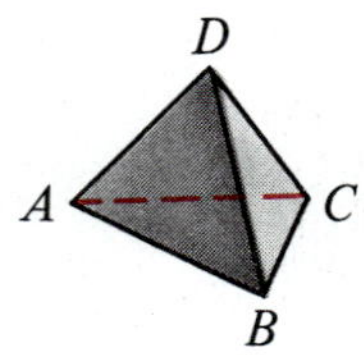

Theorem 9.8 If a tetrahedron is intersected by a plane parallel to its base, then the triangle formed by the intersection is similar to the base.

Given: Tetrahedron $ABCD$ intersected by plane p in $\triangle EFG$, $p \parallel \triangle ABC$
Prove: $\triangle EFG \sim \triangle ABC$

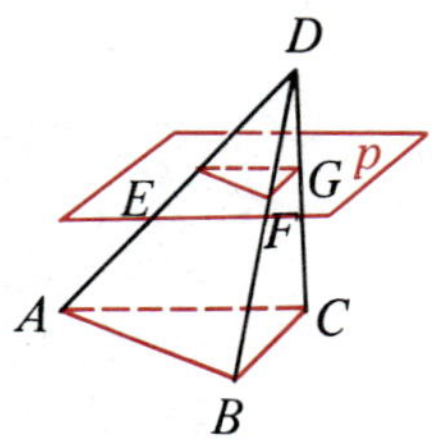

ANALYSIS
Each face of the tetrahedron contains similar triangles. Sides of similar triangles are proportional.

Outline of Proof

$\overline{EF} \parallel \overline{AB}$
$\triangle EFD \sim \triangle ABD$
$$\frac{EF}{AB} = \frac{DF}{DB} = \frac{DE}{DA}$$

$\overline{GE} \parallel \overline{CA}$
$\triangle GED \sim \triangle CAD$
$$\frac{GE}{CA} = \frac{DG}{DC} = \frac{DE}{DA}$$

$$\frac{EF}{AB} = \frac{FG}{BC} = \frac{GE}{CA}$$
$\therefore \triangle EFG \sim \triangle ABC$

Example 3 If $p \parallel \triangle ABC$, find XY.

$$\frac{XY}{15} = \frac{8}{12}$$
$$12 \cdot XY = 120$$

Thus, $XY = 10$.

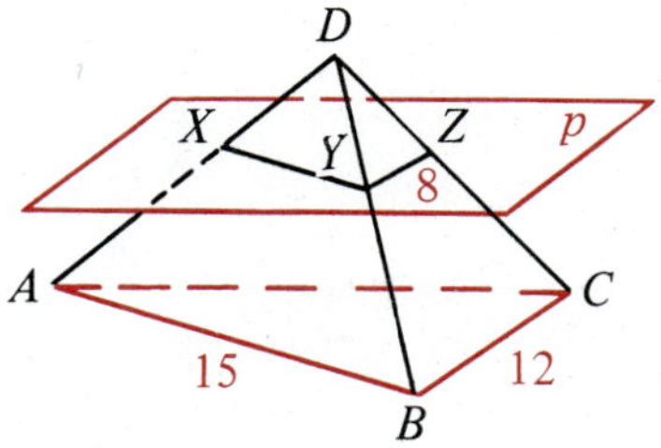

Written Exercises

A

1. Given: $p \parallel q \parallel r$, $AB = 7$, $BC = 9$, and $DE = 8$. Find EF. **$10\frac{2}{7}$**

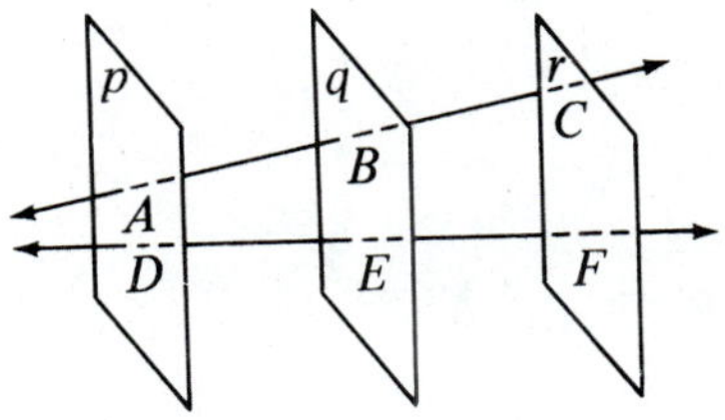

2. Given: $p \parallel q \parallel r$, $AB = 8$, $AC = 18$, and $DE = 10$. Find EF. **$12\frac{1}{2}$**

3. Given: $p \parallel q \parallel r$, $AB = 6$, $BC = 9$, $DE = 2x + 1$, and $EF = 4x - 1$. Find DE and EF. **6; 9**

4. $p \parallel q \parallel r$, $AB = 6$, $AC = 14$, $DE = 3x + 1$, and $EF = 5x - 2$. Find DE and EF. **11; $14\frac{2}{3}$**

B

5. Given: $q \parallel \triangle PCV$, the distance from L to $\triangle PCV = 12$, and $\frac{VP}{NR} = \frac{2}{3}$. Find the distance from L to q. **18**

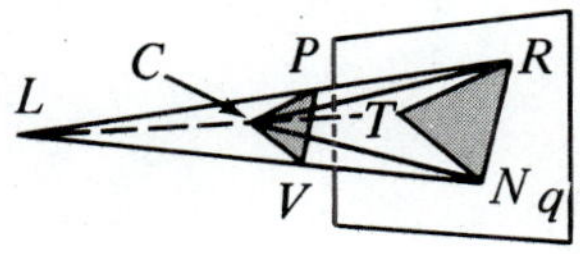

6. Given: $q \parallel \triangle PCV$, the distance from q to $\triangle PCV = 2x + 6$, the distance from L to $\triangle PCV = 4x - 2$, and $\frac{VC}{NT} = \frac{3}{5}$. Find the distance from L to q. **70**

7. Given: $\triangle XBZ \sim \triangle ABF$ and $\triangle YBZ \sim \triangle CBF$
Prove: $\triangle XBY \sim \triangle ABC$

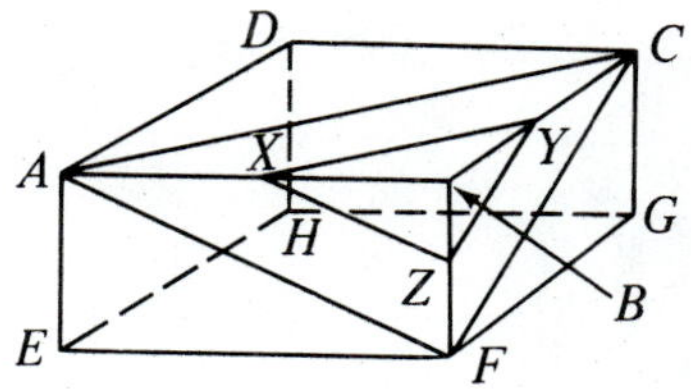

8. Given: $\triangle XBZ \sim \triangle YBZ$ and $\triangle ABF \sim \triangle CBF$
Prove: $\triangle XBY \sim \triangle ABC$

9. Given: $p \parallel q$ and $\frac{AR}{AX} = \frac{AS}{AY} = \frac{AT}{AZ}$
Prove: $\triangle ATS \sim \triangle AZY$

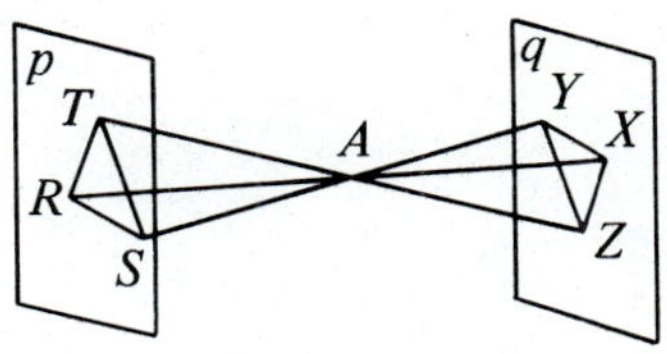

10. Given: $p \parallel q$ and $\frac{AR}{AX} = \frac{AS}{AY} = \frac{AT}{AZ}$
Prove: $\triangle RST \sim \triangle XYZ$

11. Write a complete proof of Theorem 9.8.

12. Given: $\overline{AF} \parallel \overline{BG}$ and $\overline{FE} \parallel \overline{GD}$
Prove: $\frac{ED}{DC} = \frac{AB}{BC}$

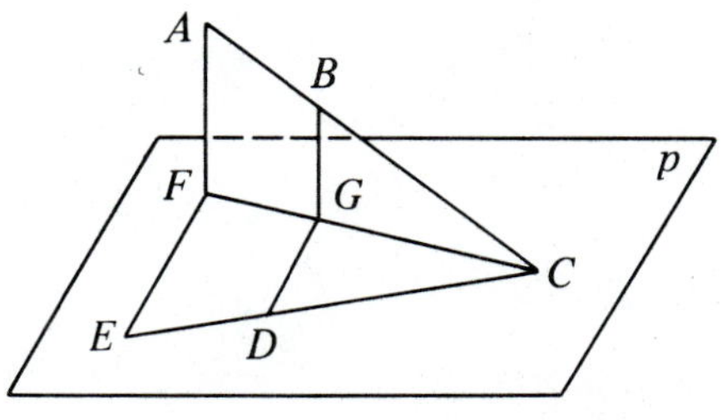

APPLICATIONS

Read → Plan → Solve → Interpret

1. A tax assessor needed a quick estimate of the height of a building. Her helper stands in the doorway of the building. Holding a pencil in her hand with her arm outstretched, the assessor backed off until her helper appeared as tall as the pencil. How will she estimate the height of the building? Why does this work?

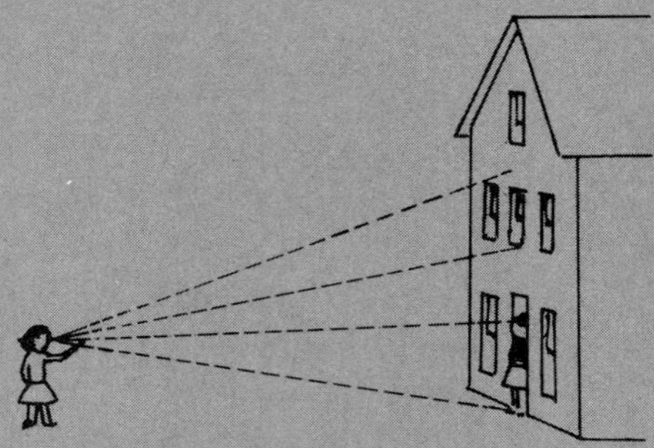

2. The height of a tall object can also be estimated with the help of a mirror placed horizontally on the ground. Suppose that a person's eyes are 5 ft above the ground. Standing 8 ft from the mirror, she can see the top of the tree in the mirror. If the mirror is 48 ft from the foot of the tree, how tall is the tree? Why does this work?

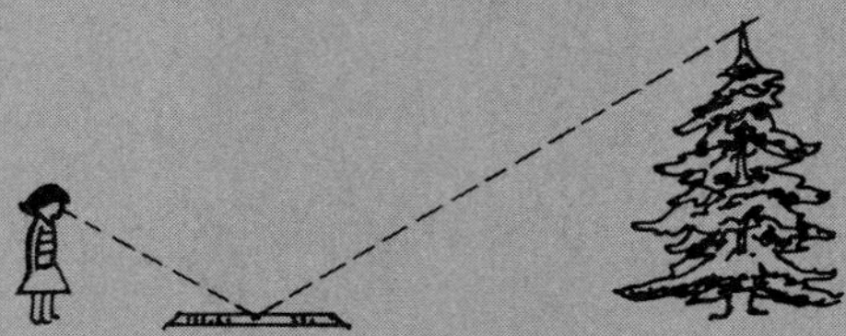

COMPUTER ACTIVITIES

Indirect Measurement

Objective: **To solve computer-generated problems of indirect measurement by means of similar triangles**

If a man is standing next to a tree and both cast shadows from the sun's rays, similar triangles are formed. From the man's height and the lengths of the shadows, you can determine the height of the tree.

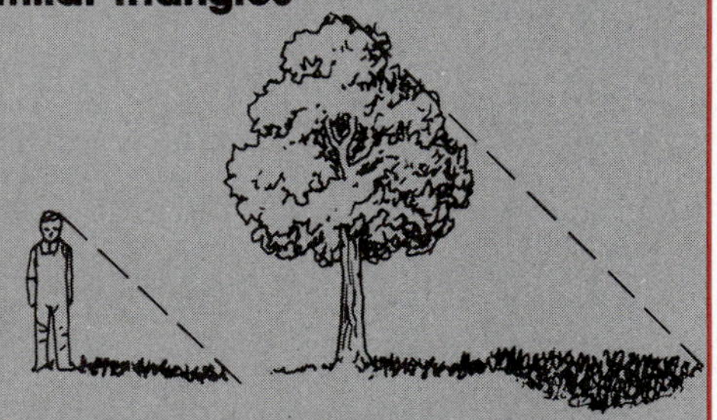

The program below gives two types of verbal problems involving the idea of indirect measurement. Note the ON . . . GOTO command in line 50. When Q = 1, the computer branches to line 60, and when Q = 2, it branches to line 140.

```
10  PRINT "SOLVE THESE PROBLEMS GIVING ANSWERS"
20  PRINT "ROUNDED TO THE NEAREST HUNDREDTH"
30  PRINT "IF NECESSARY.": PRINT
40 Q =  INT ( RND (1) * 2) + 1
50  ON Q GOTO 60,140
60 M =  INT ( RND (1) * 11) + 60:S =  INT ( RND (1) *
    61) + 40:T =  INT ( RND (1) * 121) + 80
70  PRINT "HOW TALL IS A TREE WHICH CASTS A"
80  PRINT "SHADOW ";T;" INCHES LONG, IF A MAN ";M
90  PRINT "INCHES TALL CASTS A ";S;" INCH SHADOW";: INPUT
    A
100  IF A =  INT (M * T / S * 100 + .5) / 100 THEN 130
110  PRINT "NO. THE ANSWER IS "; INT (M * T / S * 10
    0 + .5) / 100: PRINT : GOTO 40
130  PRINT "YES": PRINT : GOTO 40
140 M =  INT ( RND (1) * 11) + 20
150 S =  INT ( RND (1) * 41) + 20:T =  INT ( RND (1)
    * 671) + 30
160  PRINT "A ";M;" FOOT FLAGPOLE CASTS A SHADOW"
170  PRINT "WHICH IS ";S;" FEET. HOW HIGH IS A"
180  PRINT "BUILDING WHOSE SHADOW IS ";T;" FEET"
190  INPUT A: GOTO 100
```

EXERCISES: See the Computer Section beginning on page 565 for more information.

1. Add a third type of problem to the program above. Make any necessary changes in the existing program to allow for your addition. Make sure that the random numbers generated are appropriate for the objects chosen.
2. Alter the program you wrote for Exercise 1 so that it gives a twelve-question graded quiz.

CHAPTER NINE REVIEW

Solve each proportion for x. [9.1]

1. $\frac{8}{12} = \frac{x}{15}$ **10**

2. $\frac{x}{9} = \frac{5}{4}$ **$11\frac{1}{4}$**

Find the ratio of x to y for each of the following.

3. $3x = 8y$ **$\frac{8}{3}$**

4. $ay = bx$ **$\frac{a}{b}$**

Find the geometric mean of each pair of numbers.

5. 9 and 16 **12**

6. 2 and 5 **$\sqrt{10}$**

Use a proportion to solve the following problem.

*__7.__ The length and the width of a rectangle have the ratio 7 to 4. The area of the rectangle is 252. Find the length and the width.
length = 21, width = 12

Are the triangles similar? Give a reason for your answer. (Ex. 8–10) [9.2]

8.

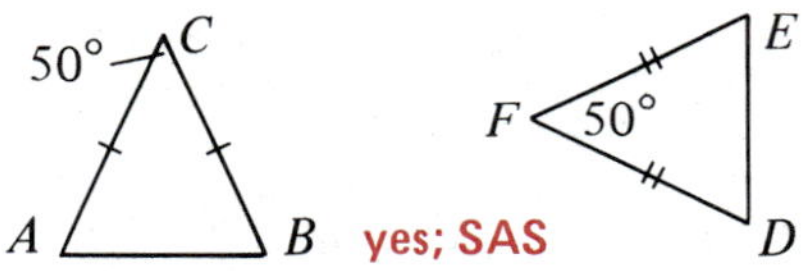

yes; SAS

9.

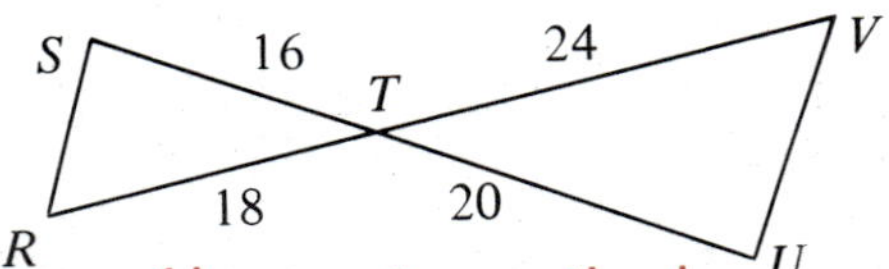

no; sides are not proportional.

10. Given: $\overline{DE} \parallel \overline{AB}$, $CE = 8$, $DE = 12$, and $CB = 15$.
Find AB. [9.4] **$22\frac{1}{2}$**

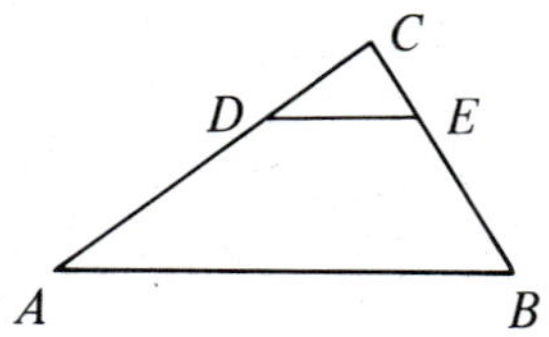

Vocabulary

ratio [9.1], proportion [9.1], geometric mean [9.1], similar polygons [9.2]

11. Given: $\overline{TU}$ bisects $\angle RTS$, $RU = 4$, $US = 6$, and $TR = 12$.
Find TS. [9.6] **18**

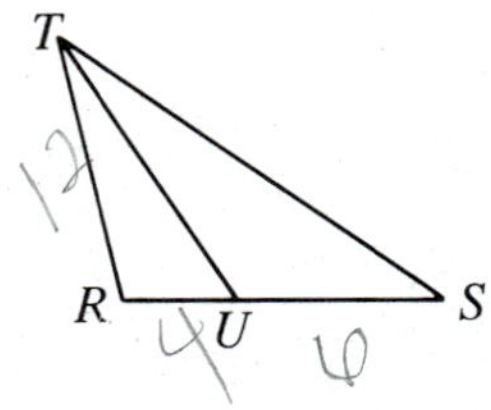

12. Given: $\overline{FH} \perp \overline{DE}$, $\overline{EG} \perp \overline{DG}$, $DH = 8$, $DF = 12$, $DG = 3x - 1$, and $DE = 4x + 2$
Find DG and DE. [9.3]
$DG = 20$; $DE = 30$

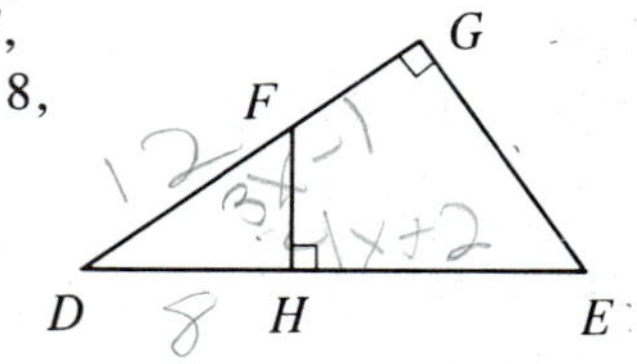

13. Given: $\overline{CD} \perp \overline{AB}$ and $\overline{AE} \perp \overline{CB}$
Prove: $\triangle CPE \sim \triangle CBD$
[9.5]

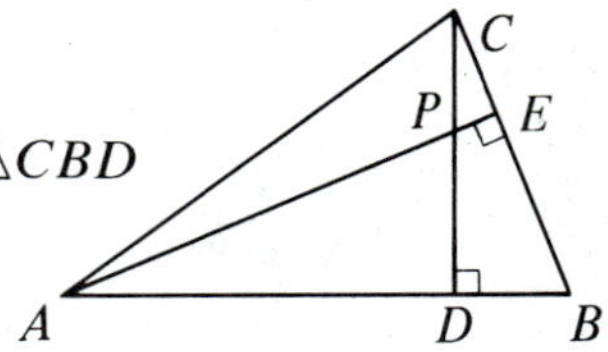

14. Given: Rhombus $ABCD$
Prove: $DP \cdot CR = DC \cdot PR$
[9.5, 9.7]

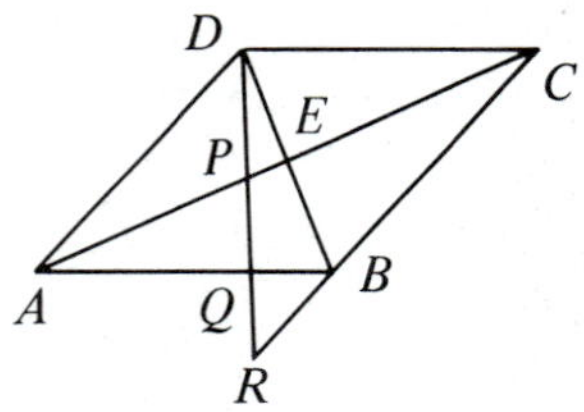

15. Given: D is the midpoint of $\overline{AC}$, E is the midpoint of $\overline{CB}$, and F is the midpoint of $\overline{AB}$.

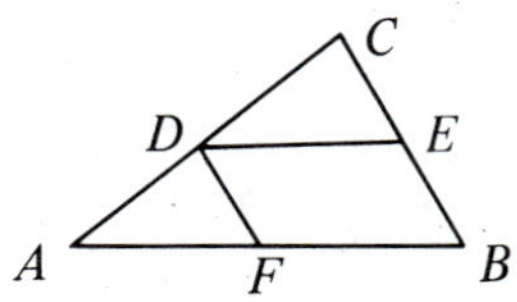

Prove: $DE \cdot CB = DF \cdot AB$ [9.7]

The following exercises are types that appear in the A section of the lessons' exercises: 1–2, 8–13. The starred exercises are types that appear in the C section. All the remaining exercises appear in the B section.

CHAPTER NINE TEST

Solve each proportion for x.

1. $\frac{3}{5} = \frac{7}{x}$ $11\frac{2}{3}$
2. $\frac{12}{x} = \frac{16}{20}$ 15

Find the ratio of x to y.

3. $5x = 4y$ $\frac{4}{5}$

Find the geometric mean of the given pair of numbers.

4. 9 and 36 18

Use a proportion to solve the following problem.

*5. The length and the width of a rectangle have the ratio 4 to 3. Their area is 300. Find the length and the width. length = 20, width = 15

Are the triangles similar? Give a reason for your answer. (Ex. 6–7)

6.

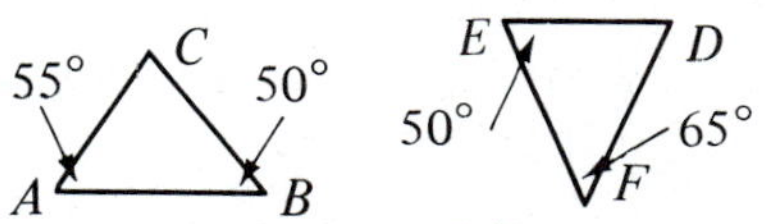

no; only 1 pr. of ∠'s is ≅.

7.

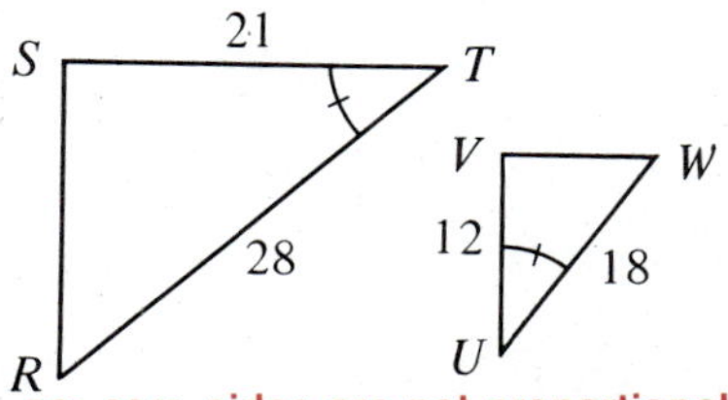

no; corr. sides are not proportional.

8. Given: $\overline{DE} \parallel \overline{AB}$, $AD = 6$, $DC = 10$, and $BC = 12$. Find EC.

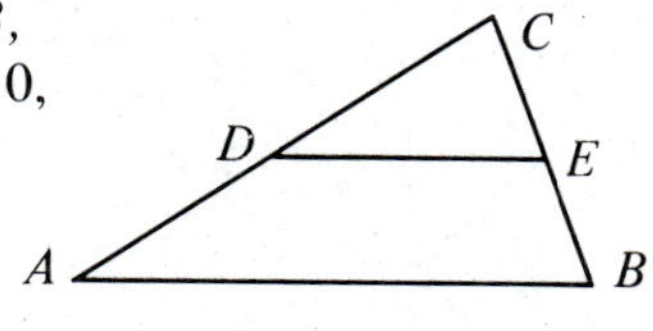

$EC = 7\frac{1}{2}$

9. Given: $\triangle ABC \sim \triangle DEF$, $AB = 15$, $AC = 12$, $DE = 2x - 4$, and $DF = x + 1$. Find DE.

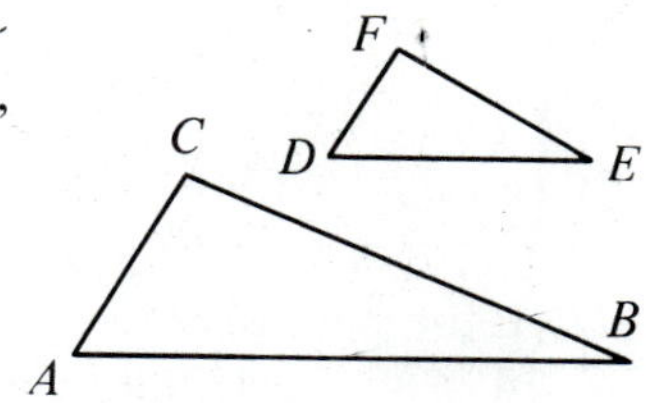

DE = 10

10. Given: $\overline{ED} \parallel \overline{AB}$ Prove: $\triangle ABG \sim \triangle DEG$

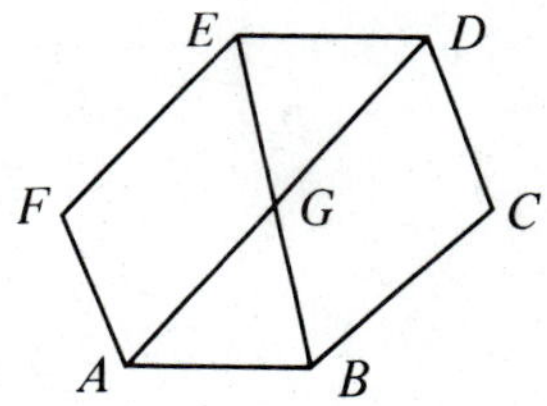

11. Given: $ABDE$ is a square. Prove: $AE \cdot CF = CE \cdot AF$

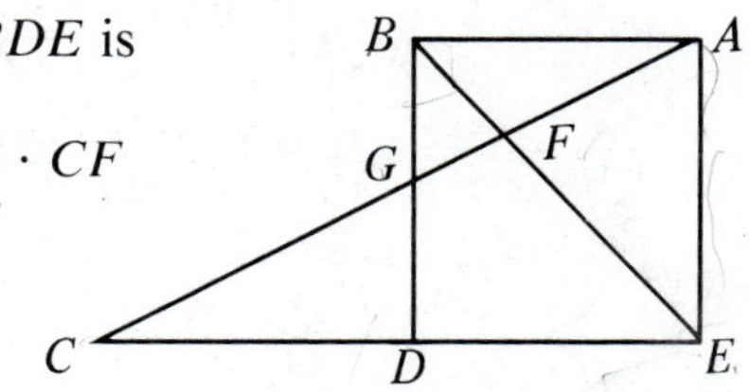

*12. Prove that a triangle whose vertices are the midpoints of the sides of a given triangle is similar to the given triangle.

The following exercises are types that appear in the (A) section of the lessons' exercises: 1–2, 4, 6–10. The starred exercises are types that appear in the (C) section. All the remaining exercises appear in the (B) section.

COLLEGE PREP TEST

Directions: Choose the one best answer to each question or problem.

1. In $\triangle ABC$, $\overline{DE} \parallel \overline{AC}$, $BE = 2$, $CE = 4$, and $DE = 3$.
Find AC.

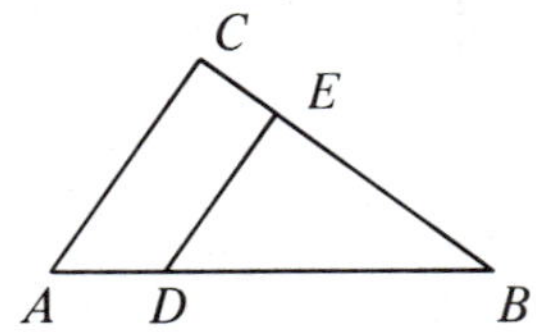

(A) 5.4 **(B)** 6 E **(C)** 7.2 **(D)** 7.6 **(E)** 9

2. In the figure below, $MQ = 6$, $MS = 4$, $SN = 14$, $\overline{PN} \perp \overline{PM}$, and $\overline{QS} \perp \overline{MS}$.
Find QP.

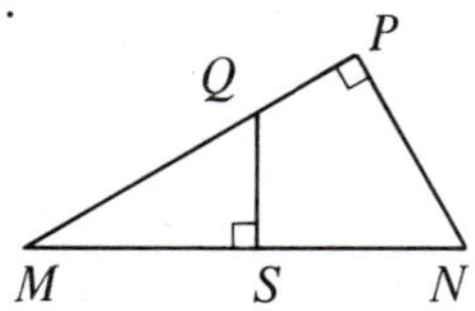

(A) 6 **(B)** 7 A **(C)** 9 **(D)** 10 **(E)** 12

3. If it takes 30 minutes for Jane to type three pages, how many hours will it take her to type 63 pages at the same rate?

(A) 6.3 **(B)** 630 **(C)** 15 **(D)** $10\frac{1}{2}$
(E) none of these D

4. If k pencils cost c cents, how many pencils can be bought for d dollars?

(A) $\frac{kd}{c}$ **(B)** $\frac{kd}{100\,c}$ **(C)** $\frac{100\,kd}{c}$
(D) $\frac{100\,cd}{k}$ C **(E)** $\frac{cd}{100\,k}$

5. A vertical pole 6 feet high casts a shadow 4 feet long. At the same time, a tree casts a shadow 64 feet long. How many feet high is the tree?
(A) 44 **(B)** 96 B **(C)** 72 **(D)** 192 **(E)** 256

6.

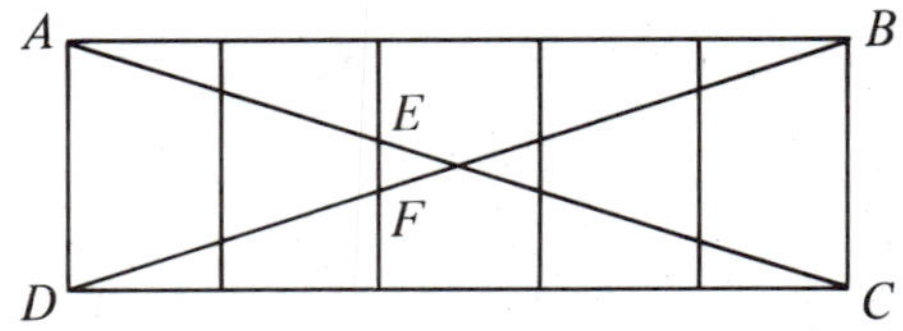

In the figure above, rectangle $ABCD$ is made up of 5 rectangles of equal area. $AD = 45$.
Find EF. A
(A) 9 **(B)** 15 **(C)** 12 **(D)** 10 **(E)** 18

7. In the figure,
$PL = 2KP$,
$OL = 2MO$, and
$OP = 14$.
Find KM.

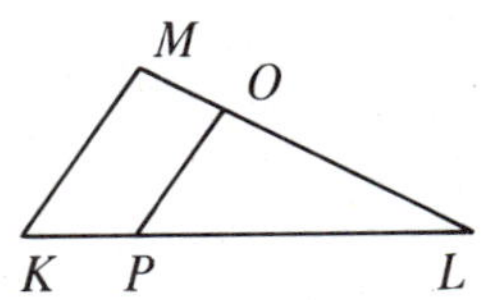

(A) 7 **(B)** 18 **(C)** 21 C **(D)** 24 **(E)** 28

8. On a map drawn to a scale $\frac{1}{2}$ inch = 200 miles, what is the distance between two cities which are $3\frac{1}{2}$ inches apart?
(A) 300 miles **(B)** 350 miles **(C)** 700 miles
(D) 1200 miles E **(E)** 1400 miles

9.

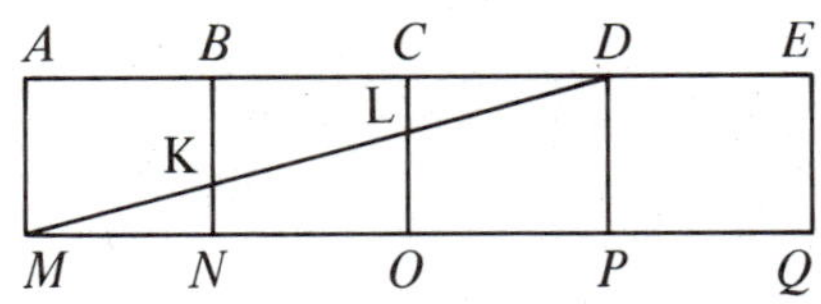

In the figure, $ABNM$, $BCON$, $CDPO$ and $DEQP$ are congruent rectangles. $AM = 21$.
Find KN.
(A) 3 **(B)** 5.25 **(C)** 10.5 D **(D)** 7 **(E)** 14

10 APPLYING SIMILAR TRIANGLES

Non-Routine Problem Solving

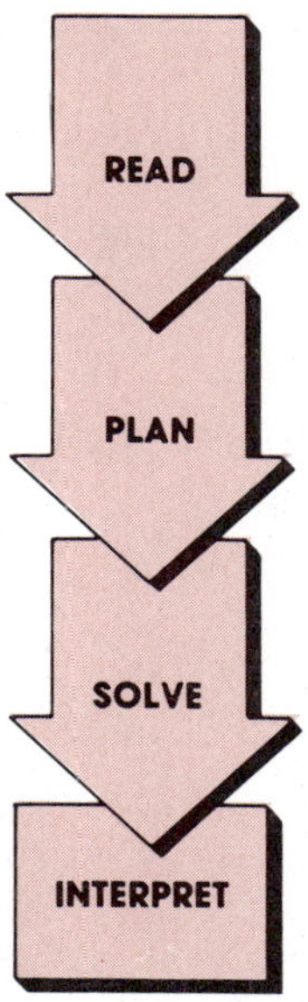

The design of the Parthenon in Athens, Greece, is based on a **golden rectangle**. The proportions of a golden rectangle are said to be the most pleasing. In a golden rectangle, the width divided by the length will always equal about 0.618, the **golden ratio**. This ratio is derived from the proportion below.

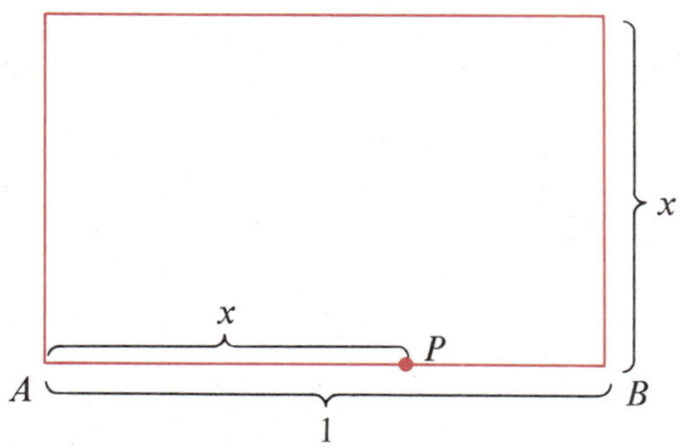

$AB = 1,\ AP = x,\ PB = 1 - x$

$$\frac{PB}{AP} = \frac{AP}{AB}$$

$$(AP)^2 = (PB)(AB)$$

$$x^2 = (1 - x)(1)$$

$$x^2 = 1 - x$$

$$x^2 + x - 1 = 0$$

By the quadratic formula

$$x = \frac{-1 \pm \sqrt{5}}{2}, \text{ or } 0.618$$

SIMILARITY IN RIGHT TRIANGLES 10.1

Objective **To find the lengths of sides and altitudes of right triangles**

When the altitude to the hypotenuse of a right triangle is drawn, three pairs of similar triangles are formed.

Theorem 10.1 In a right triangle, the altitude to the hypotenuse forms two triangles, each similar to the original triangle, and each similar to the other.

Given: Rt. $\triangle ABC$ with rt. $\angle ACB$ and alt. $\overline{CD}$
Prove: $\triangle ADC \sim \triangle ACB$, $\triangle CDB \sim \triangle ACB$, and $\triangle ADC \sim \triangle CDB$

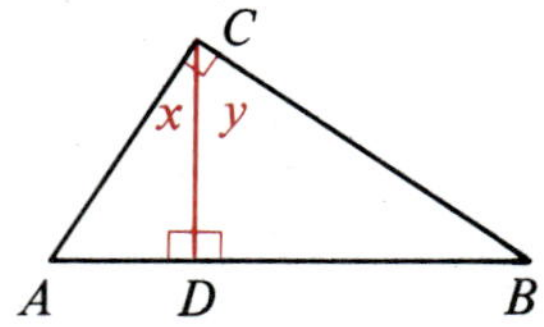

Proof

Statements	Reasons
1. Rt. $\triangle ABC$, rt. $\angle ACB$, alt. $\overline{CD}$	1. Given
2. $\overline{CD} \perp \overline{AB}$	2. Def. of altitude
3. $\angle ADC$ and $\angle CDB$ are rt. $\angle$'s.	3. $\perp$ lines form rt. $\angle$'s.
4. $\triangle ADC$ and $\triangle CDB$ are rt. $\triangle$'s.	4. Def. of rt. $\triangle$'s
5. $\angle ACB \cong \angle ADC \cong \angle CDB$	5. All rt. $\angle$'s are $\cong$.
6. $\angle A \cong \angle A$ and $\angle B \cong \angle B$	6. Reflexive property
7. $\triangle ADC \sim \triangle ACB$	7. AA for $\sim$ $\triangle$'s
8. $\triangle CDB \sim \triangle ACB$	8. AA for $\sim$ $\triangle$'s
9. $\angle A \cong \angle y$ and $\angle B \cong \angle x$	9. Def. of $\sim$ $\triangle$'s
10. $\therefore \triangle ADC \sim \triangle CDB$	10. AA for $\sim$ $\triangle$'s

In the figure for Theorem 10.1, since $\triangle ADC \sim \triangle CDB$, it follows that $\frac{AD}{CD} = \frac{CD}{BD}$.

Thus, CD is the geometric mean of AD and DB. This is stated in Corollary 1.

Corollary 1 In a right triangle, the length of the altitude to the hypotenuse is the geometric mean of the lengths of the segments of the hypotenuse that are formed.

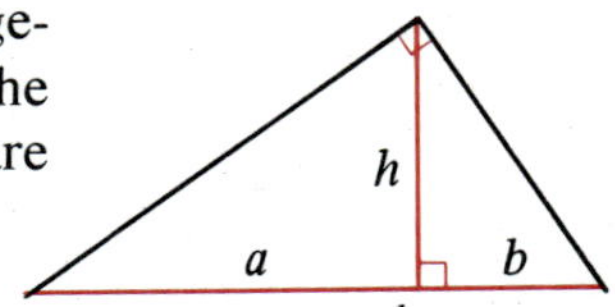

$\frac{a}{h} = \frac{h}{b}$

(See Exercise 19.)

Examples 1 and 2 are applications of Corollary 1.

Example 1

Given: Rt. $\triangle RST$ with altitude $\overline{SU}$, $RU = 10$, and $SU = 8$.
Find TU.

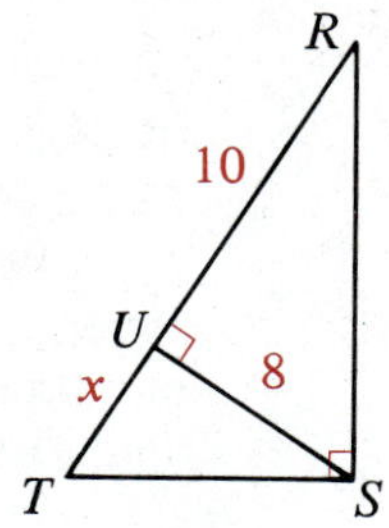

Let $x = TU$.

$\frac{RU}{SU} = \frac{SU}{TU}$ ► $\frac{10}{8} = \frac{8}{x}$ by Corollary 1.

$10x = 64$

$x = 6.4$, or $TU = 6.4$

Example 2

Given: Rt. $\triangle ABC$ with altitude $\overline{CD}$, $AB = 18$, and $DB = 6$.
Find CD.

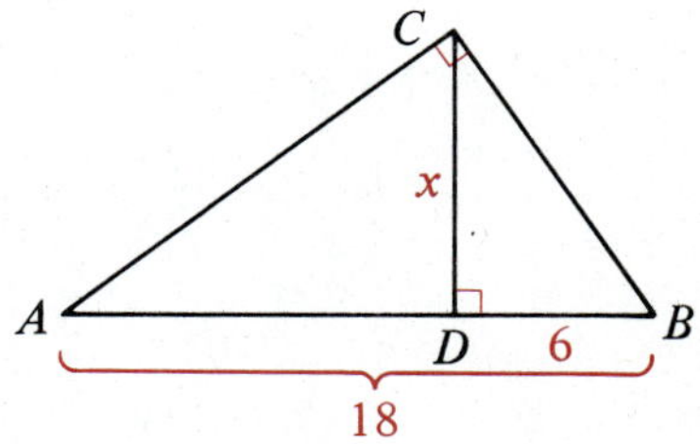

Since $AD = AB - DB$,
$AD = 18 - 6$, or 12.
Let $x = CD$.

$\frac{AD}{CD} = \frac{CD}{DB}$ ► $\frac{12}{x} = \frac{x}{6}$

$x^2 = 72$

$x = \sqrt{72}$, or $6\sqrt{2}$, so $CD = 6\sqrt{2}$.

The altitude to the hypotenuse of right triangle ABC divides the hypotenuse into two segments, $\overline{AD}$ and $\overline{DB}$. $\overline{AD}$ is adjacent to leg $\overline{AC}$.

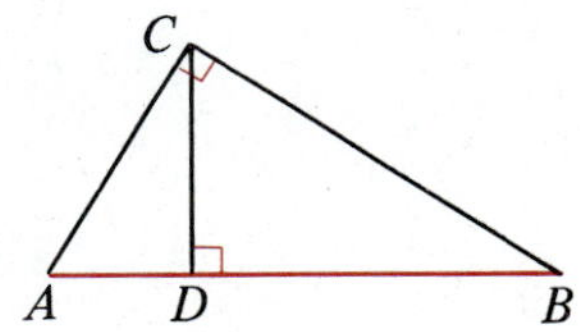

Corollary 2

If the altitude is drawn to the hypotenuse of a right triangle, then the length of either leg is the geometric mean of the length of the hypotenuse and the length of the segment of the hypotenuse adjacent to that leg.

Given: Rt. $\triangle ABC$ with rt. $\angle ACB$ and altitude $\overline{CD}$

Prove: $\frac{AB}{AC} = \frac{AC}{AD}$ and $\frac{AB}{CB} = \frac{CB}{DB}$

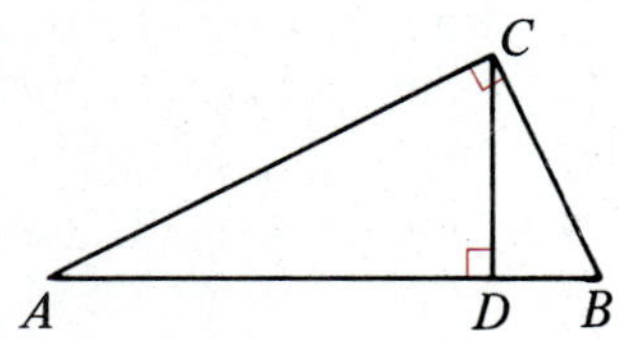

Proof

Statements	**Reasons**
1. Rt. $\triangle ABC$, rt. $\angle ACB$, alt. $\overline{CD}$	1. Given
2. $\triangle ACB \sim \triangle ADC$	2. Alt. to hyp. forms $\sim$ rt. $\triangle$'s.
3. $\therefore \frac{AB}{AC} = \frac{AC}{AD}$	3. Def. of $\sim$ $\triangle$'s

(See Exercise 20 for 2nd part of proof.)

You can use Corollary 2 to find the length of the hypotenuse or the length of either of the legs of a right triangle. You can use Corollary 2 to find the length of the segment of the hypotenuse adjacent to either leg.

Example 3

Given: Rt. $\triangle ABC$ with rt. $\angle ACB$ and altitude $\overline{CD}$, $CA = 7$ and $BA = 9$. Find DA.

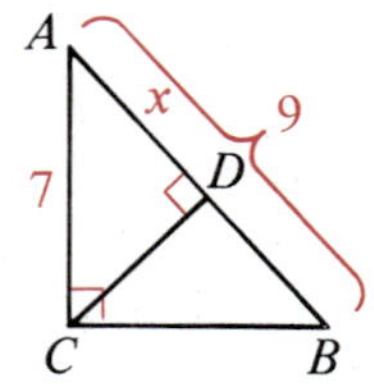

Let $x = DA$.

$\frac{AB}{AC} = \frac{AC}{AD}$ ► $\frac{9}{7} = \frac{7}{x}$ by Corollary 2.

$9x = 49$

$x = \frac{49}{9}$ **Thus,** $DA = \frac{49}{9}$, or $5\frac{4}{9}$.

Example 4

Given: Rt. $\triangle RST$ with rt. $\angle RTS$, $\overline{TU} \perp \overline{RS}$, $RU = 6$, and $US = 3$. Find TS.

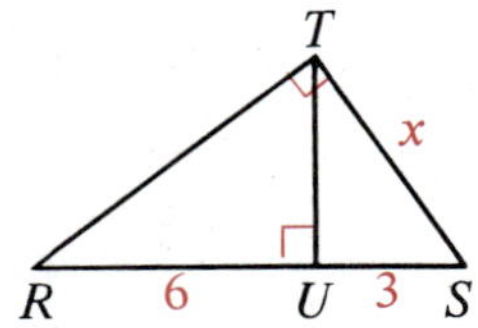

$RS = 6 + 3$, or 9

Let $x = TS$.

$\frac{RS}{TS} = \frac{TS}{US}$ ► $\frac{9}{x} = \frac{x}{3}$

$x^2 = 27$

$x = \sqrt{27}$, or $3\sqrt{3}$

Thus, $TS = 3\sqrt{3}$.

Reading in Geometry

1. $\overline{AC}, \overline{CB}$; 2. $\overline{AB}$; 3. $\overline{CD}$; 4. $\overline{AD}$; 5. $\overline{DB}$; 6. $\overline{AC}$; 7. $\overline{CD}, \overline{DB}$; 8. $\triangle ADC \sim \triangle CDB \sim \triangle ACB$

For Exercises 1–13, $\triangle ABC$ is a rt. $\triangle$ with rt. $\angle ACB$ and $\overline{CD} \perp \overline{AB}$.

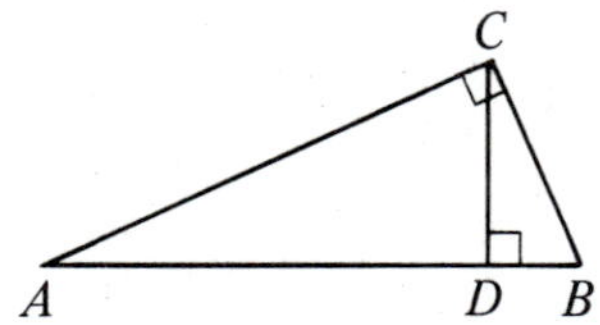

1. Name the legs of rt. $\triangle ABC$.
2. Name the hypotenuse of rt. $\triangle ABC$.
3. Name the altitude to the hypotenuse of rt. $\triangle ABC$.
4. Name the segment of the hypotenuse that is adjacent to leg $\overline{AC}$.
5. Name the segment of the hypotenuse that is adjacent to leg $\overline{CB}$.
6. Name the hypotenuse of rt. $\triangle ADC$.
7. Name the legs of rt. $\triangle CDB$.
8. Name three triangles that are similar to each other.
9. Name an angle that has the same measure as $\angle A$.
10. Name an angle that has the same measure as $\angle B$.
11. CD is the geometric mean of ________ and ________.
12. AC is the geometric mean of ________ and ________.
13. CB is the geometric mean of ________ and ________.

9. $\angle BCD$; 10. $\angle ACD$; 11. AD, DB; 12. AB, AD; 13. AB, DB

Written Exercises

In rt. $\triangle RST$, $\angle RTS$ is a rt. $\angle$ and $\overline{TU}$ is an altitude. Find the indicated lengths.

(A)

1. $RU = 8$, $TU = 4$, $US = ?$ **2**
2. $RS = 32$, $RT = 24$, $RU = ?$ **18**
3. $RU = 27$, $TU = ?$, $US = 3$ **9**
4. $RS = ?$, $US = 4$, $TS = 6$ **9**
5. $RU = ?$, $TU = 2$, $US = 1$ **4**
6. $RS = 16$, $RU = 8$, $RT = ?$ **$8\sqrt{2}$**
7. $RU = 12$, $TU = 6$, $US = ?$ **3**
8. $RS = 25$, $US = 5$, $TS = ?$ **$5\sqrt{5}$**

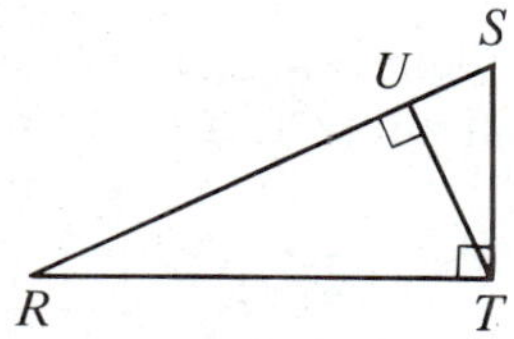

In rt. $\triangle JKL$, $\angle JKL$ is a rt. $\angle$ and $\overline{KM}$ is an altitude. Find the indicated lengths. (Ex. 9–16)

9. $JL = 20$, $ML = 4$, $KM = ?$ **8**
10. $JL = 18$, $KL = 12$, $ML = ?$ **8**
11. $JL = ?$, $JM = 4$, $JK = 6$ **9**
12. $JL = 8$, $JK = ?$, $JM = 2$ **4**
13. $JL = 16$, $ML = 8$, $KL = ?$ **$8\sqrt{2}$**
14. $JL = ?$, $JK = 24$, $JM = 18$ **32**
15. $JL = 25$, $JM = 5$, $KM = ?$ **10**
16. $JL = 20$, $KL = ?$, $ML = 15$ **$10\sqrt{3}$**

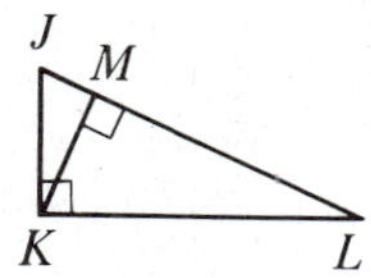

(B)

17. Given: $\overline{BC} \perp \overline{AC}$, $\overline{CE} \perp \overline{AB}$, and $\overline{ED} \perp \overline{AC}$
Prove: $\triangle BEC \sim \triangle CDE$

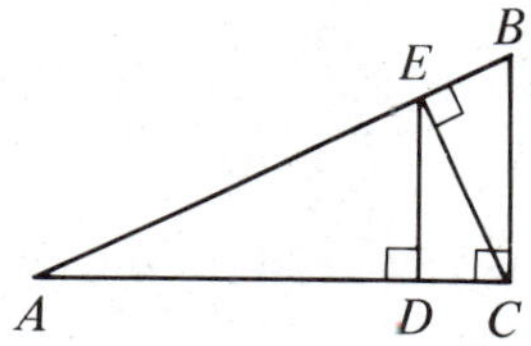

18. Given: $\overline{BC} \perp \overline{AC}$, $\overline{CE} \perp \overline{AB}$, $\overline{ED} \perp \overline{AC}$
Prove: $BE \cdot AD = ED \cdot CE$
(Hint: First prove $\triangle BEC \sim \triangle EDA$.)

19. Prove Corollary 1.

20. Complete the proof of Corollary 2.

$\angle XZY$ is a right angle, $\overline{ZH} \perp \overline{XY}$, and $\overline{HM} \perp \overline{XZ}$. Can the length of each indicated segment be found by methods in this lesson? If so, find the length. (Ex. 21–26)

21. Given: $HY = 4$ and $XH = 16$. Find XZ. **$8\sqrt{5}$**
22. Given: $XM = 16$ and $MZ = 4$. Find MH and ZH. **8; $4\sqrt{5}$**
23. Given: $XH = 9$ and $ZH = 6$. Find HY and ZY. **4; $2\sqrt{13}$**
24. Given: $MZ = 4$ and $MX = 6$. Find ZH and XH. **$2\sqrt{10}$; $2\sqrt{15}$**
25. Given: $XZ = 12$ and $ZH = 7$. Find ZY and HY. **no**
26. Given: $XZ = 9$ and $XY = 12$. Find XH and MH. **$6\frac{3}{4}$; $\frac{27}{16}\sqrt{7}$**

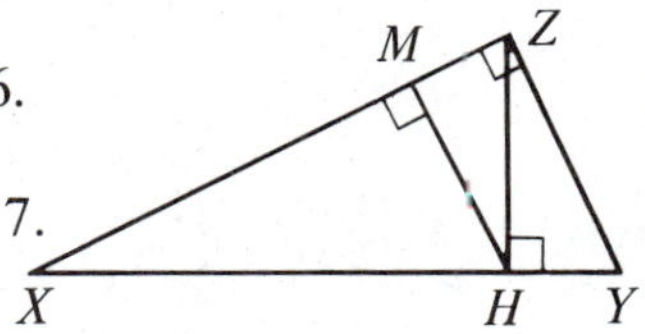

(C)

27. Prove or disprove: In a right triangle, the ratios of the length of each leg to the length of the segment of the hypotenuse determined by the altitude and adjacent to that same leg are equal.

28. Prove or disprove: In a right triangle, the product of the lengths of the hypotenuse and the altitude to the hypotenuse is equal to the product of the lengths of the legs.

$\angle ACB$ is a rt. $\angle$, $\overline{CD} \perp \overline{AB}$, and $\overline{DE} \perp \overline{AC}$. Prove each of the following.

29. $r \cdot s = u(u + v)$
30. $w^2 + h^2 = u \cdot v + r \cdot s$
31. $h^2 + s^2 = s(r + s)$
32. $h^2 = u(v + u)$

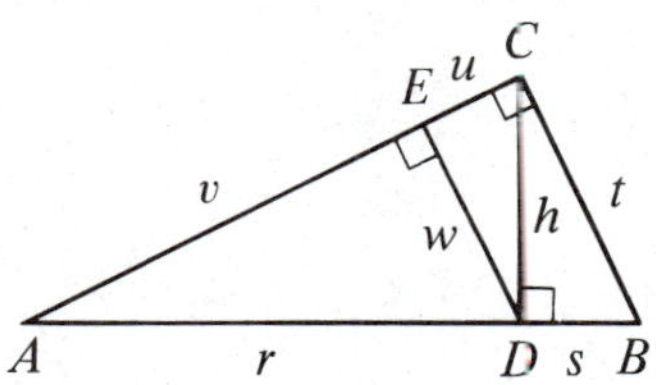

THE PYTHAGOREAN THEOREM 10.2

Objective **To apply the Pythagorean Theorem**

Consider the two right triangles shown below. If the length of each of the three sides is squared, an interesting relationship may be seen for each triangle.

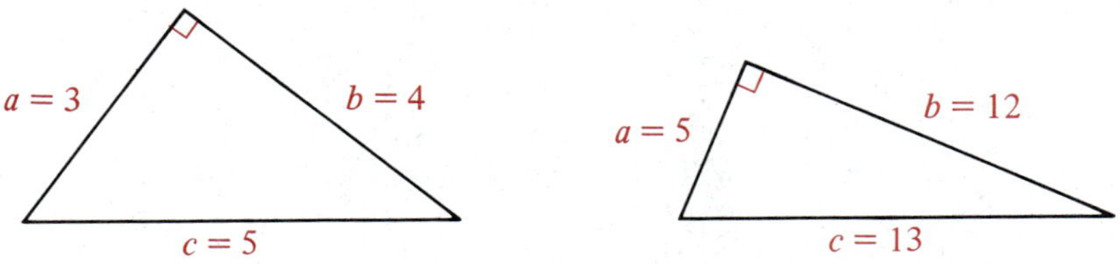

$a = 3$	$b = 4$	$c = 5$	$a = 5$	$b = 12$	$c = 13$
$a^2 = 9$	$b^2 = 16$	$c^2 = 25$	$a^2 = 25$	$b^2 = 144$	$c^2 = 169$

$9 + 16 = 25$ $\qquad$ $25 + 144 = 169$

$a^2 + b^2 = c^2$ ←Relationship→ $a^2 + b^2 = c^2$

It can be proved that the relationship shown above is true for all right triangles. Pythagoras, a famous Greek mathematician and philosopher, is given credit for the first proof of the relationship. Hence, the theorem is called the Pythagorean Theorem. The outline of a more recent proof is given below.

Theorem 10.2 Pythagorean Theorem

In any right triangle, the square of the length of the hypotenuse is equal to the sum of the squares of the lengths of the two legs.

Given: Rt. $\triangle ABC$ with rt. $\angle ACB$

Prove: $a^2 + b^2 = c^2$

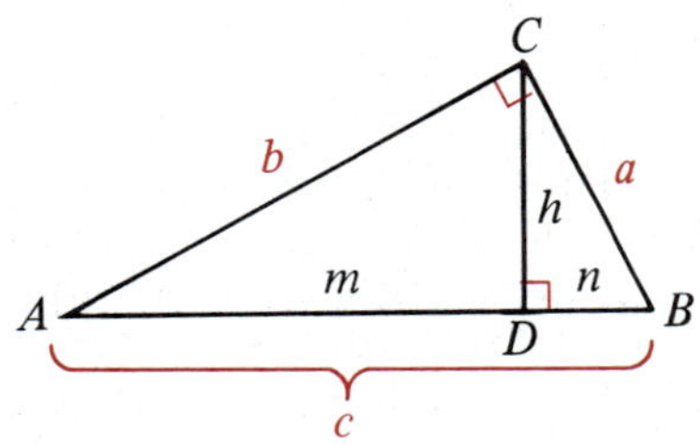

ANALYSIS

Use Theorem 10.1 and its corollaries to write proportions. Then use algebra.

Outline of proof

1. $\frac{c}{a} = \frac{a}{n}$ and $\frac{c}{b} = \frac{b}{m}$
2. $a^2 = cn$ and $b^2 = cm$
3. $a^2 + b^2 = cn + cm$
4. $a^2 + b^2 = c(n + m)$
5. $n + m = c$
6. $\therefore a^2 + b^2 = c \cdot c$, or $a^2 + b^2 = c^2$

The Pythagorean Theorem has many applications. Several are given in the examples that follow.

Example 1

Given: Rt. $\angle C$, $BC = 12$, and $AC = 16$
Find AB.

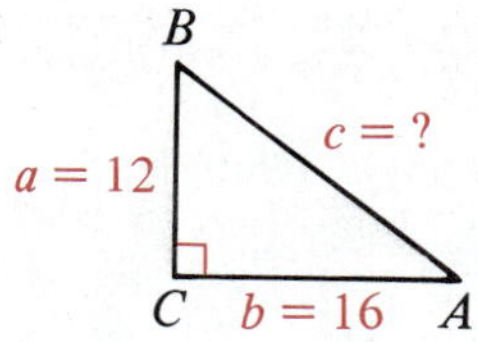

$c^2 = a^2 + b^2$ by the Pythagorean Th.
$c^2 = 12^2 + 16^2$
$c^2 = 144 + 256$
$c^2 = 400$
$c = \sqrt{400}$, or 20
Thus, $AB = 20$.

Example 2

Find the length of a base of a rectangle if an altitude drawn to the base is 10 m long and a diagonal is 26 m long.

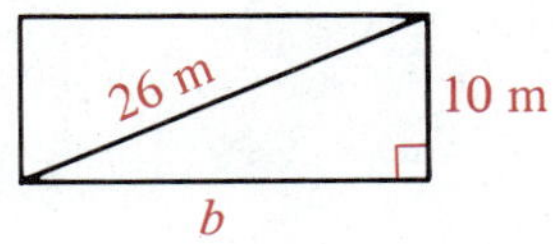

$b^2 + 10^2 = 26^2$
$b^2 + 100 = 676$
$b^2 = 576$
$b = \sqrt{576}$, or 24
Thus, the base is 24 m long.

Example 3

Find the length of the altitude to the hypotenuse of a right triangle if the lengths of the legs are 8 and 15.

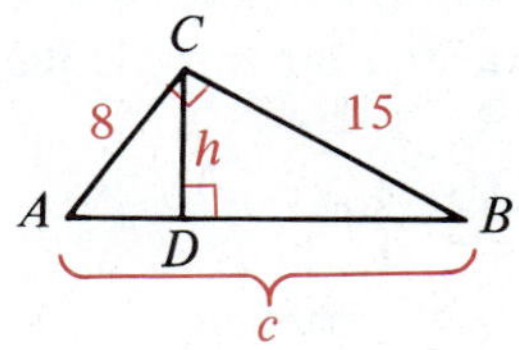

$c^2 = 8^2 + 15^2$
$c^2 = 64 + 225$
$c^2 = 289$
$c = \sqrt{289}$, or 17

Since $\triangle ABC \sim \triangle CBD$, $\frac{AC}{AB} = \frac{CD}{CB}$

$$\frac{8}{17} = \frac{h}{15}$$
$$17h = 120$$
$$h = 7\frac{1}{17}$$

Thus, the length of the altitude to the hypotenuse is $7\frac{1}{17}$.

Since the diagonals of a rhombus are perpendicular bisectors of each other, you can apply the Pythagorean Theorem to the right triangles that are formed by the diagonals.

Example 4

The diagonals of a rhombus are 20 cm and 16 cm long. Find the length of a side.

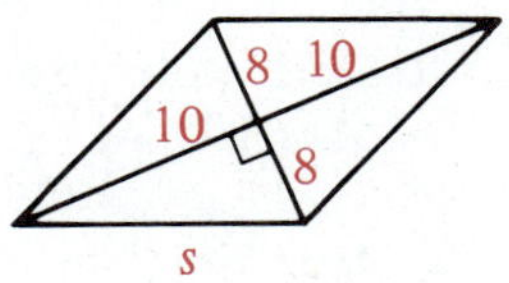

$s^2 = 10^2 + 8^2$
$s^2 = 100 + 64$
$s^2 = 164$
$s = \sqrt{164}$, or $2\sqrt{41}$
Thus, a side is $2\sqrt{41}$ cm long.

Written Exercises

For each right triangle, find the indicated length.

A **1.**

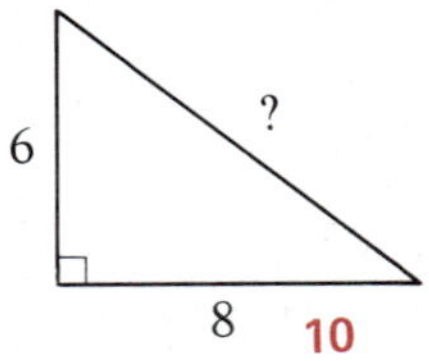

10

2.

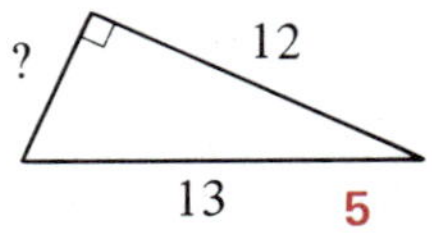

5

3.

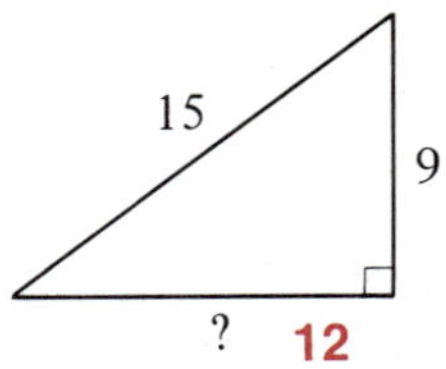

12

4.

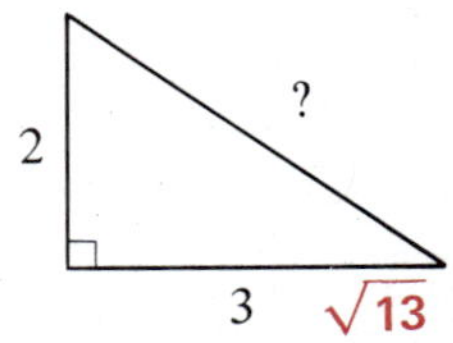

$\sqrt{13}$

5.

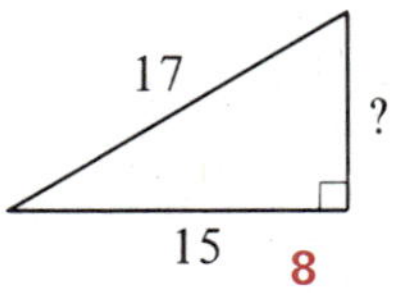

8

6.

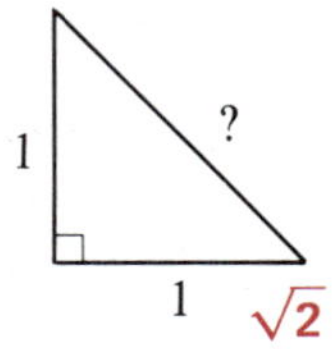

$\sqrt{2}$

7.

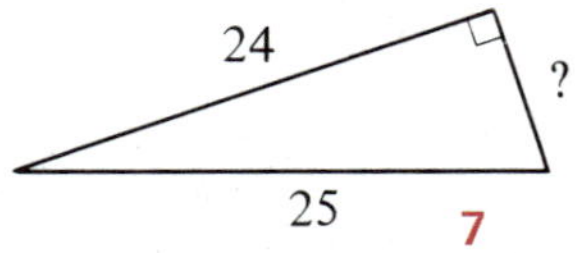

7

8.

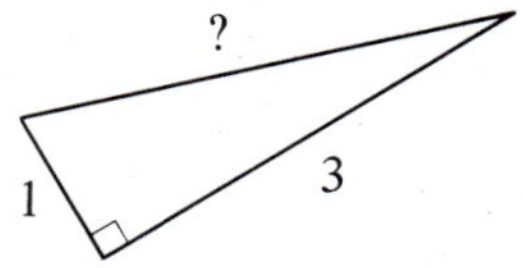

$\sqrt{10}$

In each of the following sets of numbers, the first two are the lengths of legs of a right triangle and the third is the length of a hypotenuse. Find each missing number.

9. 3, 4, ____ 5 **10.** 5, ____, 13 12 **11.** ____, 40, 41 9 **12.** 3, 5, ____ $\sqrt{34}$

13. ____, 24, 25 7 **14.** 5, 5, ____ $5\sqrt{2}$ **15.** $\sqrt{3}$, ____, $\sqrt{5}$ $\sqrt{2}$ **16.** $\sqrt{7}$, ____, 7 $\sqrt{42}$

Solve each problem.

17. Find the length of each side of a rhombus with diagonals 12 mm and 16 mm long. **10 mm**

18. A diagonal of a rhombus is 20 m long. A side is 26 m long. Find the length of the other diagonal. ***48 m***

19. Find the length of a base of a rectangle if an altitude drawn to the base is 15 cm long and a diagonal is 39 cm long. **36 cm**

20. Find the length of a diagonal of a square if a side is 7 cm long. $7\sqrt{2}$ **cm**

B **21.** If the length of a side of a square is s and the length of a diagonal is d, write a formula for d in terms of s. $d = s\sqrt{2}$

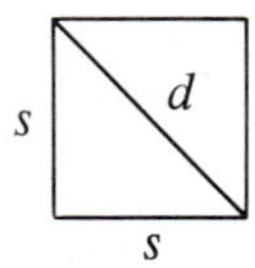

22. If the length of a diagonal of a square is d and the length of a side is s, write a formula for s in terms of d. $s = \frac{\sqrt{2}}{2}d$

23. Find the length of the altitude to the hypotenuse of a right triangle if the lengths of the legs are 10 and 24. $9\frac{3}{13}$

24. Find the length of the altitude to the hypotenuse of a right triangle if the lengths of the legs are 8 and 15. $7\frac{1}{17}$

25. Find the length of the hypotenuse of a right triangle if the length of the altitude to the hypotenuse is 6 and the length of a leg is 10. $12\frac{1}{2}$

26. Find the length of a leg of a right triangle if the length of the other leg is 5 and the length of the altitude to the hypotenuse is 4. $6\frac{2}{3}$

27. Given: Equilateral $\triangle ABC$ with $\overline{CD} \perp \overline{AB}$ and $AC = 10$. Find CD. $5\sqrt{3}$

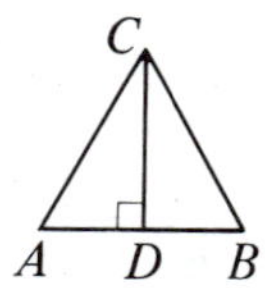

28. Given: Equilateral $\triangle ABC$ with $\overline{CD} \perp \overline{AB}$ and $DB = 4$. Find CD. $4\sqrt{3}$

29. Given: $\overline{DB} \perp \overline{AC}$
Prove: $(AD)^2 + (BC)^2 = (AB)^2 + (DC)^2$

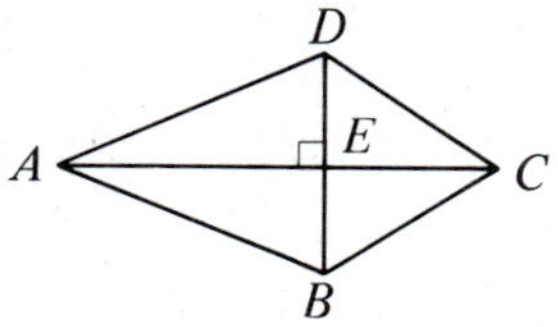

30. Given: $\overline{EG} \perp \overline{GF}$
Prove: $(HI)^2 + (EF)^2 = (HF)^2 + (EI)^2$

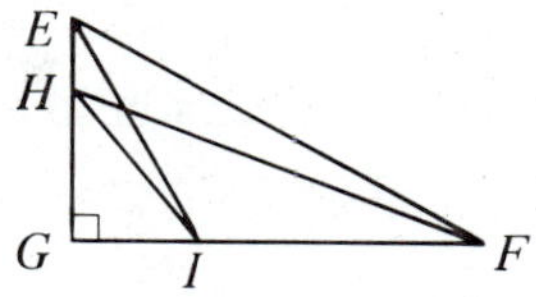

31. Given: $\overline{RT} \perp \overline{TU}$ and $\overline{RT} \perp \overline{RS}$
Prove: $(RU)^2 - (TU)^2 = (TS)^2 - (RS)^2$

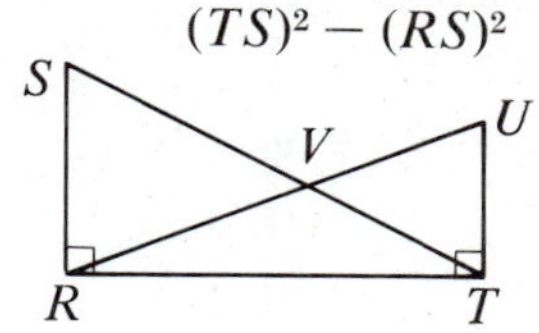

32. Prove Theorem 10.2.

Solve each problem.

Example Each face of a *rectangular solid* is a rectangle. For the one shown, the lengths of the edges are, a, b, and c. Find d, the length of diagonal $\overline{HB}$.

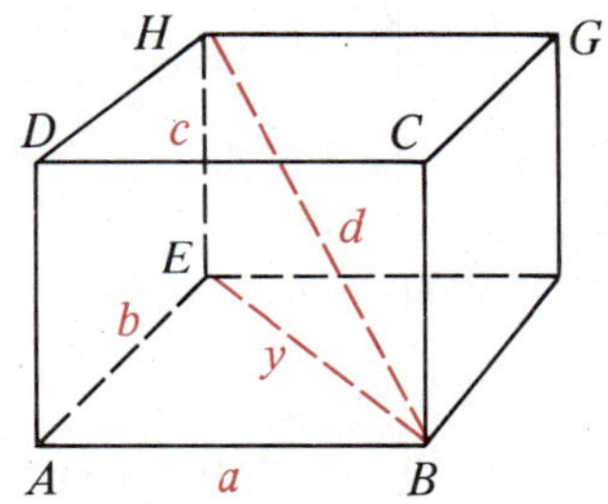

In rt. $\triangle HEB$, $d^2 = c^2 + y^2$.
In rt. $\triangle EAB$, $y^2 = a^2 + b^2$.
So, $d^2 = a^2 + b^2 + c^2$.
Thus, $d = \sqrt{a^2 + b^2 + c^2}$.

33. If the edges of a rectangular solid are 6 cm, 8 cm, and 24 cm long, find the length of a diagonal of the solid. **26 cm**

34. A cube is a rectangular solid with congruent edges. Find the length of a diagonal of a cube if the length of an edge is 7. $7\sqrt{3}$

35. Find a formula for the length of a diagonal of a cube if the length of each edge is s. $d = s\sqrt{3}$

Example In $\triangle RST$, $RS = 6$, $RT = 5$, and $ST = 4$. Find the length of altitude $\overline{TU}$.

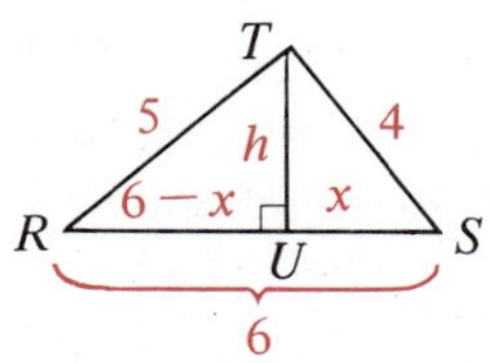

In rt. $\triangle STU$, $x^2 + h^2 = 16$
In rt. $\triangle RUT$, $(6 - x)^2 + h^2 = 25$

$$36 - 12x + x^2 + h^2 = 25$$
$$36 - 12x \quad + 16 = 25$$
$$-12x = -27$$
$$x = \frac{27}{12}, \text{ or } \frac{9}{4}$$
$$h^2 + x^2 = 16$$
$$h^2 = 16 - \left(\frac{9}{4}\right)^2 = 16 - \frac{81}{16} = \frac{175}{16}$$

Thus, $TU = \sqrt{\frac{175}{16}}$, or $\frac{5\sqrt{7}}{4}$.

36. Given: $\overline{CD} \perp \overline{AB}$, $BC = 4$, $CA = 3$, and $AB = 6$. Find CD. $\frac{\sqrt{455}}{12}$

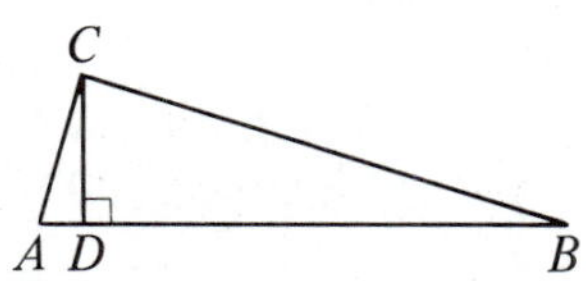

37. Given: $\overline{CD} \perp \overline{AB}$, $BC = 6$, $AC = 4$, and $AB = 9$. Find CD. $\frac{\sqrt{1,463}}{18}$

NON-ROUTINE PROBLEMS

The triple of numbers a, b, c is called a Pythagorean Triple if $a^2 + b^2 = c^2$ and $a, b,$ and c are positive integers. For example, 3, 4, 5 is a Pythagorean Triple since $3^2 + 4^2 = 5^2$. Each set of such triples can be the lengths of the sides of a right triangle. **1. $c = b + 1$** **2. $b = \frac{n^2 - 1}{2}$; $c = \frac{n^2 + 1}{2}$**

Pythagorean Triple			Check	
a	b	c	$a^2 + b^2$	c^2
3	4	5	9 + 16	25
5	12	13	25 + 144	169
7	24	25	49 + 576	625

1. Examine the three sets of triples in the table. What is true of the first number, a in each triple? How do b and c compare? **It is an odd integer greater than one.**
2. If a is any positive odd integer n, derive a formula for b and c in terms of n.
3. Let $n = 9, 11, 13 \ldots$ Do your formulas for b and c give Pythagorean Triples? **yes**

CALCULATOR ACTIVITIES

A calculator can help you to apply the Pythagorean Theorem:

$$a^2 + b^2 = c^2 \quad \text{or} \quad c^2 = a^2 + b^2.$$

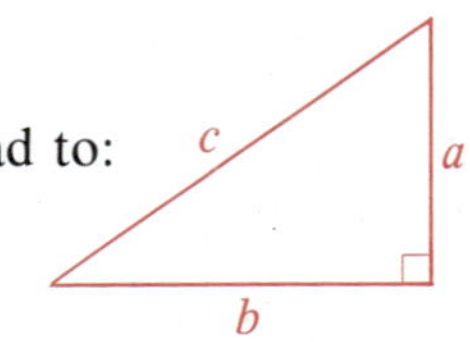

Since a, b, and c represent positive numbers, properties of algebra lead to:

(1) $c^2 = a^2 + b^2$ and $c = \sqrt{a^2 + b^2}$
(2) $a^2 = c^2 - b^2$ and $a = \sqrt{c^2 - b^2}$
(3) $b^2 = c^2 - a^2$ and $b = \sqrt{c^2 - a^2}$

For example, if $a = 18$ and $c = 36$, you can find b by using $b = \sqrt{c^2 - a^2}$.

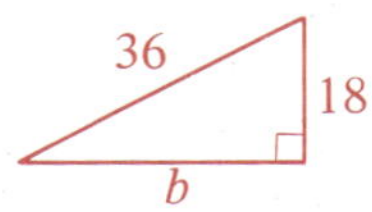

Find c^2.	►	Press 36 ⊗ 36 ⊜		
		Display 1296	◄	*Write down and clear.*
Find a^2.	►	Press 18 ⊗ 18 ⊜		
		Display 324	◄	*Write down and clear.*
Find $c^2 - a^2$.	►	Press 1296 ⊖ 324 ⊜		
		Display 972		
$b = \sqrt{c^2 - a^2}$	►	**Thus,** $b = \sqrt{972}$, or 31.2.		

For each right triangle, find the indicated length to the nearest tenth.

1. $a = ?$ $b = 40$, $c = 41$ **9.0** **2.** $a = 15$, $b = 36$, $c = ?$ **39.0** **3.** $a = 60$, $b = ?$, $c = 61$ **11.0**
4. $a = ?$ $b = 75$, $c = 85$ **40.0** **5.** $a = 17$, $b = ?$, $c = 34$ **29.4** **6.** $a = 16$, $b = 32$, $c = ?$ **35.8**
7. $a = 24$, $b = 72$, $c = ?$ **75.9** **8.** $a = ?$, $b = 18$, $c = 90$ **88.2** **9.** $a = 32$, $b = ?$, $c = 96$ **90.5**

CONVERSE OF THE PYTHAGOREAN THEOREM 10.3

Objective **To apply the converse of the Pythagorean Theorem**

Recall that you can obtain the converse of a statement that is written in the "If . . ., then . . ." form by interchanging the "if" and "then" clauses.

Pythagorean Theorem: If $\triangle ABC$ is a rt. $\triangle$, then $a^2 + b^2 = c^2$.

Converse: If $a^2 + b^2 = c^2$, then $\triangle ABC$ is a rt. $\triangle$.

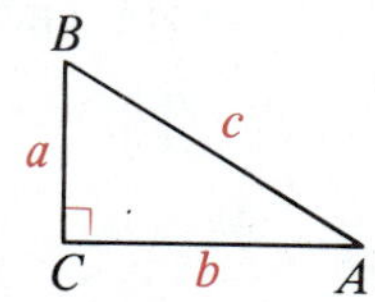

Theorem 10.3 Converse of the Pythagorean Theorem

If the sum of the squares of the lengths of two sides of a triangle is equal to the square of the length of the third side, then the triangle is a right triangle.

Given: $\triangle ABC$ with $a^2 + b^2 = c^2$
Prove: $\triangle ABC$ is a rt. $\triangle$.

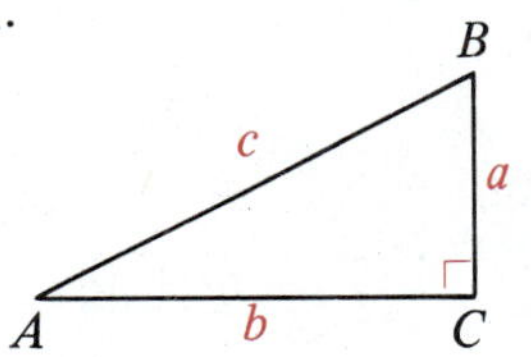

ANALYSIS

Construct a $\triangle DEF$ with rt. $\angle F$, $DF = b$, and $EF = a$. Show that $\triangle ABC \cong \triangle DEF$.

Proof: (See Exercise 26.)

Given the length of each of the three sides of a triangle, you can use the converse of the Pythagorean Theorem to determine whether a triangle is a right triangle. Later you will be able to prove that the hypotenuse of a right triangle is the longest side, so the largest measure is the length of the hypotenuse.

Example 1 Is a triangle whose sides are 11 cm, 60 cm, and 61 cm long a right triangle? If so, which angle is the right angle?

$a^2 + b^2$	c^2
$11^2 + 60^2$	61^2
$121 + 3{,}600$	$3{,}721$
$3{,}721$	

◄ *Use the largest measure, 61, for c. Apply Theorem 10.3.*

Thus, since $a^2 + b^2 = c^2$, the triangle is a right triangle. The right angle lies opposite the side measuring 61 cm.

Recall that if the diagonals of a quadrilateral are perpendicular bisectors of each other, then the quadrilateral is a rhombus. This is applied in the next example.

Example 2

Determine whether quadrilateral $ABCD$ is a rhombus.

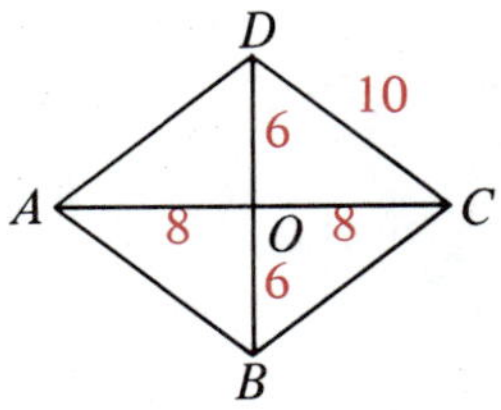

Since $AO = OC = 8$ and $DO = OB = 6$, the diagonals $\overline{AC}$ and $\overline{BD}$ bisect each other.
In $\triangle DOC$, $6^2 + 8^2 = 10^2$, so $\angle DOC$ is a right angle and $\overline{OC} \perp \overline{DO}$. Hence, $\overline{AC} \perp \overline{BD}$.
$\overline{AC}$ and $\overline{BD}$ are perpendicular bisectors of each other.

Thus, $ABCD$ is a rhombus.

Written Exercises

Which of the following are right triangles?

A

1.

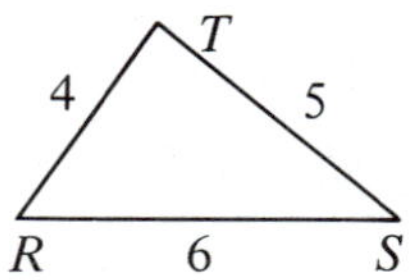

no

2.

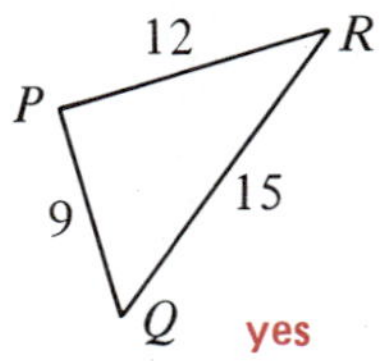

yes

3.

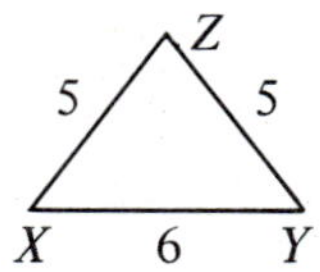

no

4.

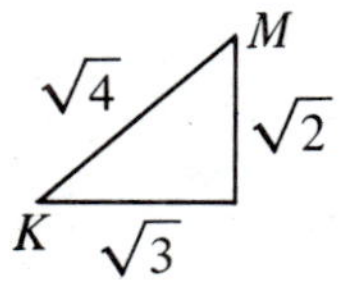

no

Which of the following are lengths of sides of right triangles? For each right triangle, which angle is the right angle?

5. $AB = 3$, $BC = 4$, $CA = 5$ yes; $\angle B$

6. $DC = 0.5$, $CE = 1.2$, $ED = 1.3$ yes; $\angle C$

7. $RT = 4$, $TQ = 7$, $QR = 9$ no

8. $EF = \sqrt{3}$, $FG = \sqrt{4}$, $GE = \sqrt{5}$ no

9. $AB = \sqrt{3}$, $BC = \sqrt{3}$, $CA = \sqrt{6}$ yes; $\angle B$

10. $MN = 3$, $NO = 5$, $OM = \sqrt{34}$ yes; $\angle N$

Determine whether each quadrilateral is a rhombus. (Ex. 11–13)

11.

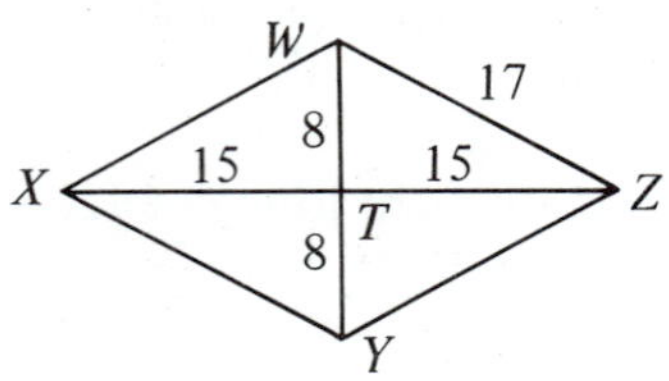

yes

12.

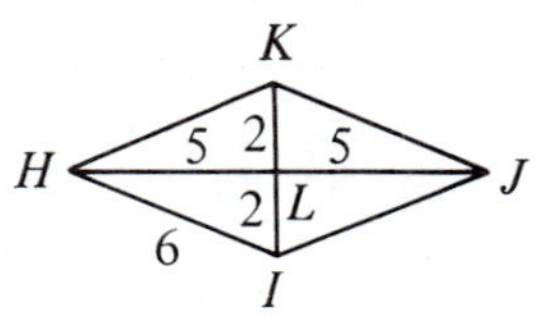

no

13.

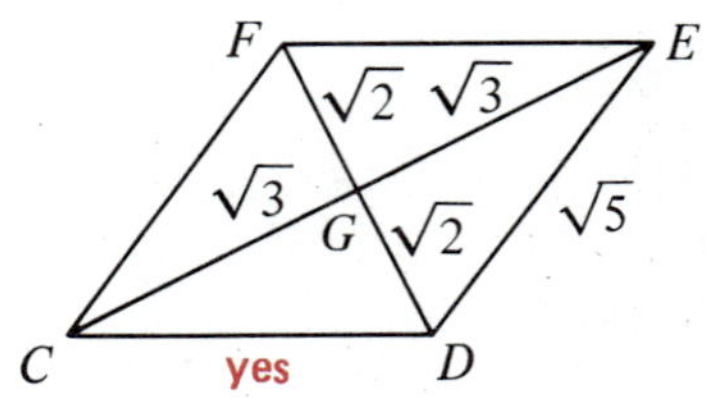

yes

B

14. Given: $a^2 = (c - b)(c + b)$
Prove: $\triangle ABC$ is a rt. $\triangle$.

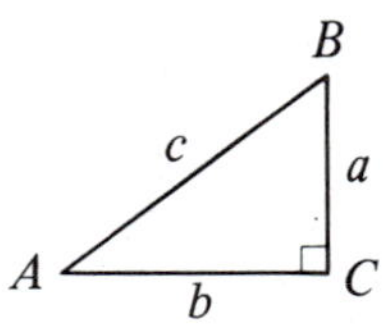

15. Given: $\dfrac{1}{a^2} + \dfrac{1}{b^2} = \dfrac{c^2}{a^2b^2}$
Prove: $\triangle ABC$ is a rt. $\triangle$.

Which of the following can be lengths of sides of a right triangle? (Ex. 16–21)

16. $3x, 4x, 5x$ **yes**

17. $n, n, 2n$ **no**

18. $5y^2, 12y^2, 13y^2$ **yes**

19. $n, m, n + m$ **no**

20. $s, \sqrt{t^2 - s^2}, t$ **yes**

21. $\frac{1}{a}, \frac{1}{b}, \frac{1}{\sqrt{a^2 + b^2}}$ **no**

Ⓒ **22.** Given: $\overline{CD} \perp \overline{AB}$ and CD is the geometric mean of AD and DB.
Prove: $\triangle ABC$ is a rt. $\triangle$.

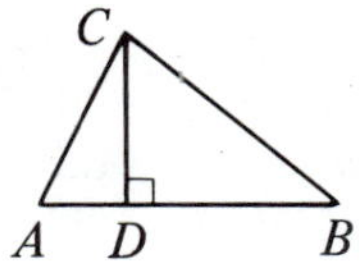

23. Given: $\overline{CD} \perp \overline{AB}$ and AC is the geometric mean of AD and AB.
Prove: $\triangle ABC$ is a rt. $\triangle$.

24. Given: $a^2 + b^2 \neq c^2$. ($\overline{AB}$ is the longest side.)
Prove: $\triangle ABC$ is not a right $\triangle$.

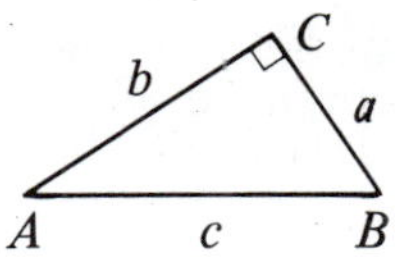

25. Given: $\triangle ABC$ is not a right triangle.
Prove: $a^2 + b^2 \neq c^2$

26. Prove Theorem 10.3.

CUMULATIVE REVIEW

1. The measure of a base angle of an isosceles triangle is 70°. Find the measure of each of the other two angles.
70°, 40°

2. The measure of one of two complementary angles is twice the measure of the other. Find the measure of each angle.
30°, 60°

ALGEBRA REVIEW

Objective **To perform operations with radicals**

To add or subtract radicals: Simplify each radical and combine like terms.
To multiply radicals: Use $\sqrt{a} \cdot \sqrt{b} = \sqrt{ab}$.
To divide radicals: Use $\frac{\sqrt{a}}{\sqrt{b}} = \sqrt{\frac{a}{b}}$.

Examples

Simplify.

$$\begin{aligned} &5\sqrt{3} + 4\sqrt{12} - 2\sqrt{75} \\ &5\sqrt{3} + 4\sqrt{4 \cdot 3} - 2\sqrt{25 \cdot 3} \\ &5\sqrt{3} + 4 \cdot 2\sqrt{3} - 2 \cdot 5\sqrt{3} \\ &5\sqrt{3} + 8\sqrt{3} - 10\sqrt{3} \\ &3\sqrt{3} \end{aligned}$$

$$\begin{aligned} &3\sqrt{2} \cdot \sqrt{2} \\ &3\sqrt{2 \cdot 2} \\ &3 \cdot 2 \\ &6 \end{aligned}$$

$$\begin{aligned} &\frac{12\sqrt{75}}{4\sqrt{3}} \\ &\frac{12}{4} \cdot \sqrt{\frac{75}{3}} \\ &3\sqrt{25} = 15 \end{aligned}$$

Simplify.

1. $8\sqrt{2} + 7\sqrt{2}$ **$15\sqrt{2}$**

2. $\sqrt{2} + \sqrt{50}$ **$6\sqrt{2}$**

3. $\sqrt{27} + \sqrt{75}$ **$8\sqrt{3}$**

4. $\sqrt{72} - \sqrt{50}$ **$\sqrt{2}$**

5. $\sqrt{80} - \sqrt{5}$ **$3\sqrt{5}$**

6. $7\sqrt{2} - \sqrt{2}$ **$6\sqrt{2}$**

7. $\sqrt{5} + \sqrt{45} + \sqrt{80}$ **$8\sqrt{5}$**

8. $\sqrt{12} - \sqrt{48} + \sqrt{3}$ **$-\sqrt{3}$**

9. $\sqrt{98} - 4\sqrt{8} + 3\sqrt{128}$ **$23\sqrt{2}$**

10. $\sqrt{5} \cdot \sqrt{5}$ **5**

11. $\sqrt{32} \cdot \sqrt{2}$ **8**

12. $5\sqrt{8} \cdot 7\sqrt{3}$ **$70\sqrt{6}$**

13. $\frac{12\sqrt{20}}{3\sqrt{5}}$ **8**

14. $\frac{4\sqrt{48}}{8\sqrt{3}}$ **2**

15. $\frac{14\sqrt{150}}{7\sqrt{2}}$ **$10\sqrt{3}$**

SPECIAL RIGHT TRIANGLES 10.4

Objectives

To find the lengths of segments in a 45°–45° (isosceles) right triangle
To find the lengths of segments in a 30°–60° right triangle

Two special right triangles occur frequently in many branches of mathematics. The first one that you will study is the **isosceles right triangle**, often called the **45°–45° right triangle.**

Examine the two isosceles right triangles shown below. In each case, the length of a leg of the triangle is given, so the Pythagorean Theorem can be used to find the length of the hypotenuse.

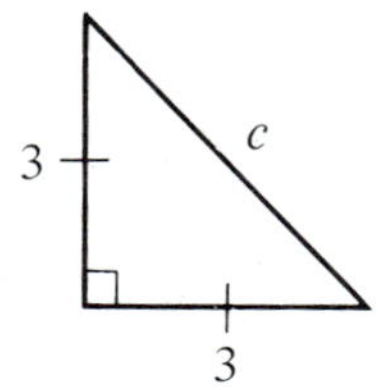

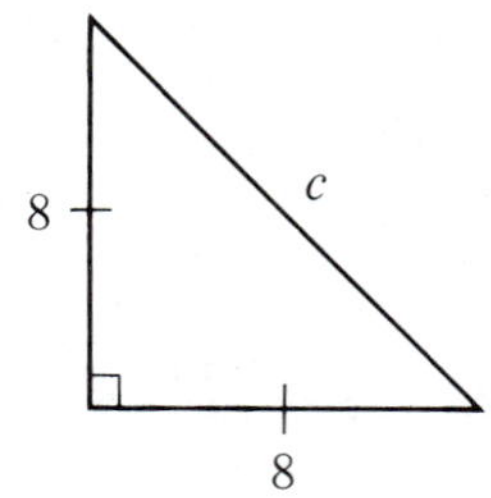

By the Pythagorean Th.,
$$c^2 = 3^2 + 3^2.$$
$$c^2 = 9 + 9$$
$$c^2 = 18$$
$$c = \sqrt{18}, \text{ or } 3\sqrt{2}$$

$$c^2 = 8^2 + 8^2$$
$$c^2 = 64 + 64$$
$$c^2 = 128$$
$$c = \sqrt{128}, \text{ or } 8\sqrt{2}$$

leg	leg	hyp.
3	3	$3\sqrt{2}$

◄ *Note the pattern.* ►

leg	leg	hyp.
8	8	$8\sqrt{2}$

These examples suggest a more general theorem.

Theorem 10.4

In a 45°–45° (isosceles) right triangle, the length of the hypotenuse is equal to the length of a leg times $\sqrt{2}$.

(See Exercise 21.)

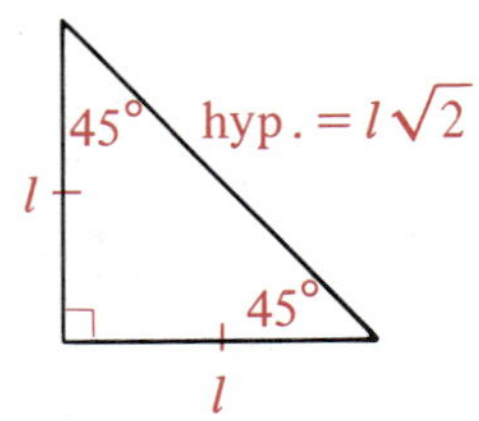

Example 1

$\triangle ABC$ is a 45°–45° right triangle and the length of the hypotenuse is 20. Find the length of a leg.

$$\text{hyp.} = l\sqrt{2}$$
$$20 = l\sqrt{2}$$
$$l = \frac{20}{\sqrt{2}} = \frac{20}{\sqrt{2}} \cdot \frac{\sqrt{2}}{\sqrt{2}} = \frac{20\sqrt{2}}{2} = 10\sqrt{2}$$

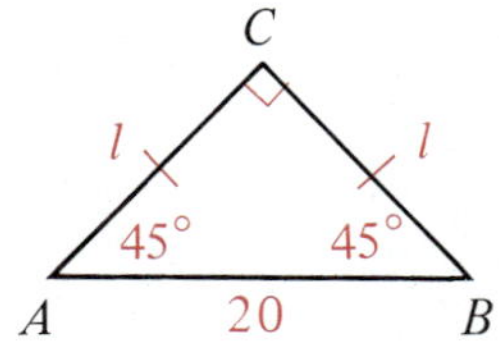

Thus, the length of a leg is $10\sqrt{2}$.

$\triangle ABC$ shown at the right is a 30°–60° right triangle. Suppose that $\overline{AC}$ is extended to D so that $\overline{CA} \cong \overline{CD}$ and $\overline{BD}$ is drawn. You can prove that $\triangle DCB \cong \triangle ACB$. It follows that $\triangle ABD$ is equiangular and therefore equilateral.

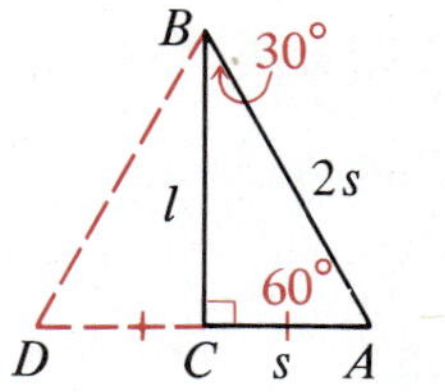

Let $BC = l$ and $CA = s$. Then $AB = AD = BD = 2s$.

$$l^2 + s^2 = (2s)^2$$
$$l^2 = 4s^2 - s^2$$
$$l^2 = 3s^2$$
$$l = \sqrt{3s^2}, \text{ or } s\sqrt{3}$$

This argument suggests the next theorem.

Theorem 10.5

In a 30°–60° right triangle, (1) the hypotenuse is twice as long as the leg opposite the 30° angle, and (2) the leg opposite the 60° angle is $\sqrt{3}$ times as long as the leg opposite the 30° angle.

(See Exercise 22.)

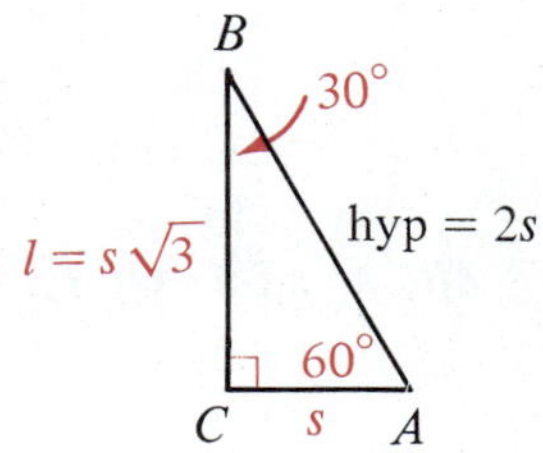

Example 2

The length of the hypotenuse of a 30°–60° right triangle is 10. Find the length of each leg.

Find s.	Find l.
hyp. $= 2s$	$l = s\sqrt{3}$
$10 = 2s$	$l = 5\sqrt{3}$
$s = 5$	

Thus, the lengths of the legs are 5 and $5\sqrt{3}$.

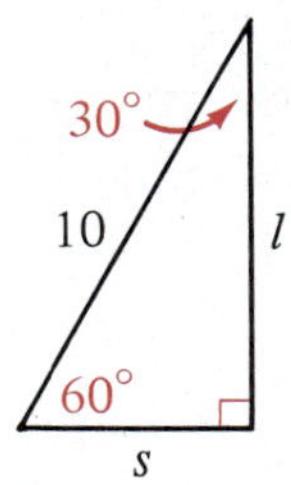

Example 3

The length of the leg opposite the 60° angle of a 30°–60° right triangle is 9. Find the length of the leg opposite the 30° angle and the length of the hypotenuse.

Find s. ► $l = s\sqrt{3}$

$$9 = s\sqrt{3}$$

$$s = \frac{9}{\sqrt{3}} = \frac{9}{\sqrt{3}} \cdot \frac{\sqrt{3}}{\sqrt{3}} = \frac{9\sqrt{3}}{3} = 3\sqrt{3}$$

Find hyp. ► hyp. $= 2s$

hyp. $= 2(3\sqrt{3}) = 6\sqrt{3}$

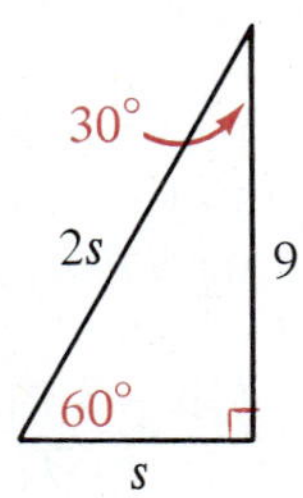

Thus, the length of the leg opposite the 30° angle is $3\sqrt{3}$. The length of the hypotenuse is $6\sqrt{3}$.

Written Exercises

Find the indicated lengths in each right triangle.

A

1.

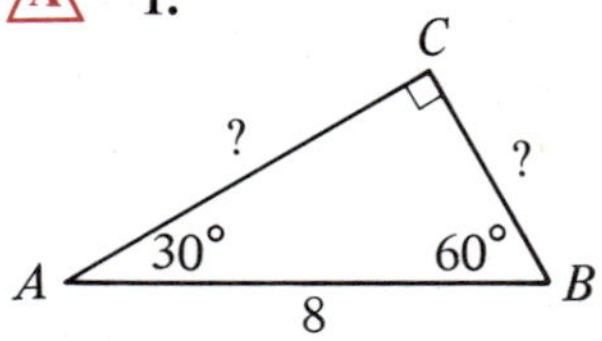

BC = 4; AC = $4\sqrt{3}$

2.

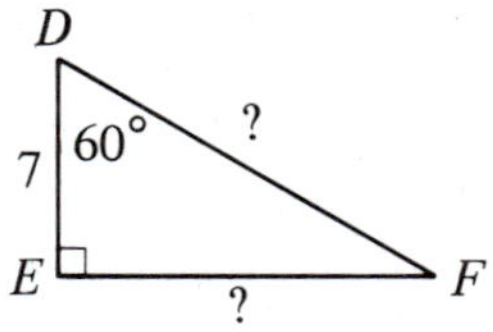

DF = 14; EF = $7\sqrt{3}$

3.

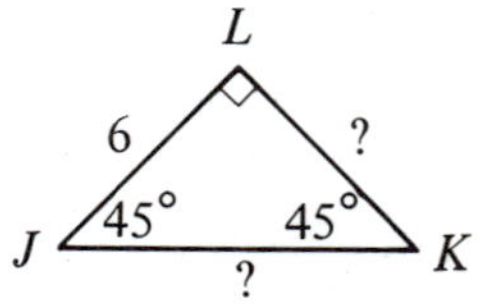

LK = 6; JK = $6\sqrt{2}$

4.

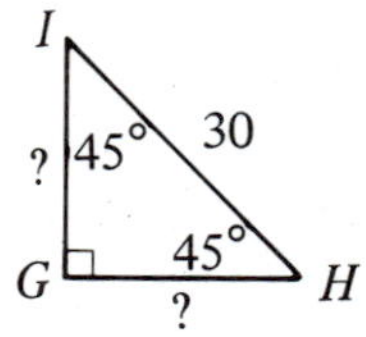

IG = GH = $15\sqrt{2}$

$\triangle XYZ$ is a 45°–45° rt. triangle. Find the indicated lengths.

5. $XZ = YZ = 6\sqrt{2}$; **6.** $XY = XZ = \frac{27\sqrt{2}}{2}$; **7.** $XY = 6$, $YZ = 3\sqrt{2}$ **8.** $XY = 18$, $XZ = 9\sqrt{2}$

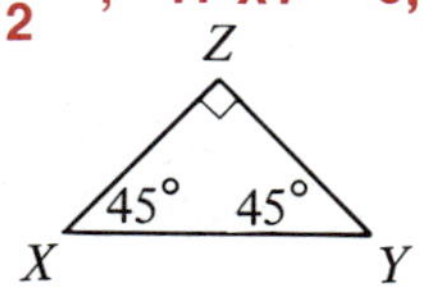

5. $XY = 12$, $XZ = ?$, $YZ = ?$

6. $XY = 27$, $XZ = ?$, $YZ = ?$

7. $XY = ?$, $XZ = 3\sqrt{2}$, $YZ = ?$

8. $XY = ?$, $XZ = ?$, $YZ = 9\sqrt{2}$

9. $XY = ?$, $XZ = ?$, $YZ = 4$

10. $XY = 8\sqrt{2}$, $XZ = ?$, $YZ = ?$

9. $XY = 4\sqrt{2}$, $XZ = 4$

10. $XZ = YZ = 8$

$\triangle ABC$ is a 30°–60° rt. triangle. Find the indicated lengths. (Ex. 11–16)

11. $BC = 18$, $CA = 18\sqrt{3}$; **12.** $BC = 13\frac{1}{2}$, $CA = \frac{27}{2}\sqrt{3}$; **13.** $AB = 18$, $CA = 9\sqrt{3}$

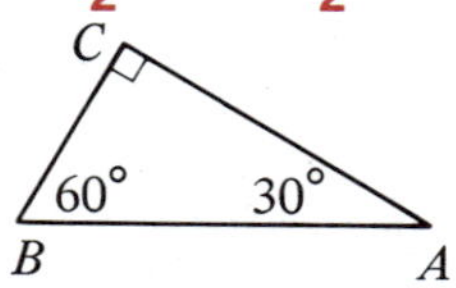

11. $AB = 36$, $BC = ?$, $CA = ?$

12. $AB = 27$, $BC = ?$, $CA = ?$

13. $AB = ?$, $BC = 9$, $CA = ?$

14. $AB = ?$, $BC = 4\sqrt{3}$, $CA = ?$

15. $AB = ?$, $BC = ?$, $CA = 6\sqrt{3}$

16. $AB = ?$, $BC = ?$, $CA = 10$

14. $AB = 8\sqrt{3}$, $CA = 12$;

15. $AB = 12$, $BC = 6$;

16. $AB = \frac{20\sqrt{3}}{3}$, $BC = \frac{10\sqrt{3}}{3}$

B

17. Given: $\triangle DEF$ is equilateral, $\overline{DG}$ is an altitude, and $DE = 18$. Find DG. $9\sqrt{3}$

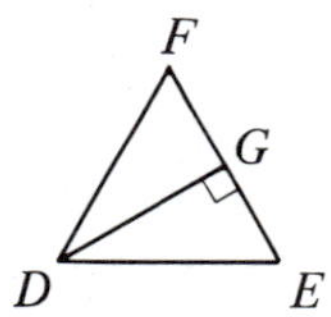

18. Given: $\triangle DEF$ is equilateral, $\overline{DG}$ is an altitude, and $DG = 12$. Find DE. $8\sqrt{3}$

19. Given: Isosceles right $\triangle JKL$ with $\overline{LM}$ the altitude to the hypotenuse and $JL = 16$. Find JK and LM.
$JK = 16\sqrt{2}$; $LM = 8\sqrt{2}$

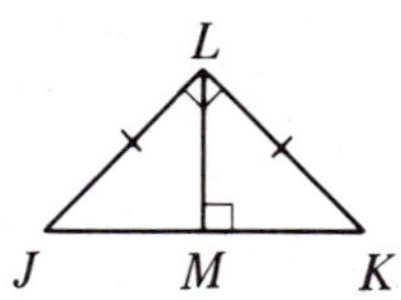

20. Given: Isosceles right $\triangle JKL$ with $\overline{LM}$ the altitude to the hypotenuse and $LM = 7$. Find JK and JL.
$JK = 14$, $JL = 7\sqrt{2}$

C

21. Prove Theorem 10.4.

22. Prove Theorem 10.5.

23. Prove that if d is the length of a diagonal of a square with side s, then $d = s\sqrt{2}$.

24. Prove that in an isosceles right triangle, the length l of a leg is $\frac{\text{hyp.}}{2}\sqrt{2}$.

25. Given: Rhombus $DEFG$ with $m\angle FGD = 60$.
Prove or disprove:
$$\frac{DE + EF + FG + GD}{GE \cdot DF} = \frac{1}{2\sqrt{3}}$$

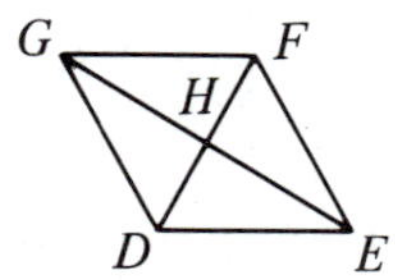

26. Given: Rhombus $DEFG$ with $m\angle FGD = 60$.
Prove or disprove:
$GE \cdot DF = \sqrt{3} \cdot GD \cdot FE$

TRIGONOMETRIC RATIOS 10.5

Objective **To compute the tangent, sine, and cosine of an acute angle**

Right triangle **trigonometry** is a branch of mathematics that deals with the relationship between the sides and the angles in right triangles. The word *trigonometry* is derived from two Greek words meaning "triangle measurement."

Notice that if right triangles have a pair of corresponding acute angles that are congruent, then the triangles are similar. In the case shown below, the right triangles each have a 28° acute angle.

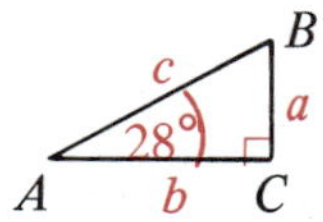

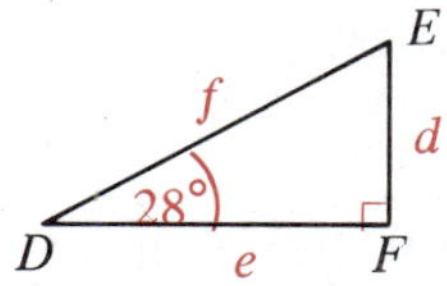

$\triangle ABC \sim \triangle DEF$ by AA for $\sim \triangle$'s.

Thus, $\frac{a}{b} = \frac{d}{e}$, $\frac{a}{c} = \frac{d}{f}$, and $\frac{b}{c} = \frac{e}{f}$.

You can conclude that for all right triangles with a given acute angle A, the ratio $\frac{a}{b}$ is determined. The ratios $\frac{a}{c}$ and $\frac{b}{c}$ are also determined.

Definition: Tangent, sine, cosine

For all right triangles ABC, with acute angle A,

$$\textit{tangent } A = \frac{\text{length of opposite leg}}{\text{length of adjacent leg}} \qquad \tan A = \frac{a}{b}$$

$$\textit{sine } A = \frac{\text{length of opposite leg}}{\text{length of hypotenuse}} \qquad \sin A = \frac{a}{c}$$

$$\textit{cosine } A = \frac{\text{length of adjacent leg}}{\text{length of hypotenuse}} \qquad \cos A = \frac{b}{c}$$

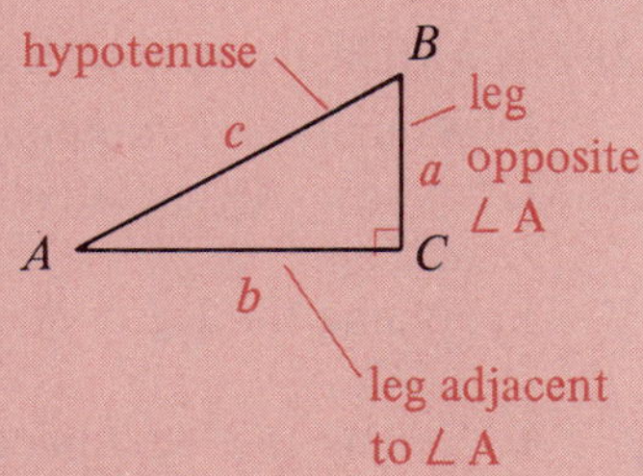

The values of the tangent, sine, and cosine ratios depend on the measure of angle A and not on the size of the right triangle. For this reason, these ratios are often called **functions of $\angle A$.**

Example 1 For rt. $\triangle ABC$, find $\tan A$, $\sin A$, and $\cos A$.

$$\tan A = \frac{\text{opp.}}{\text{adj.}} = \frac{3}{4}, \text{ or } 0.75$$

$$\sin A = \frac{\text{opp.}}{\text{hyp.}} = \frac{3}{5}, \text{ or } 0.6$$

$$\cos A = \frac{\text{adj.}}{\text{hyp.}} = \frac{4}{5}, \text{ or } 0.8$$

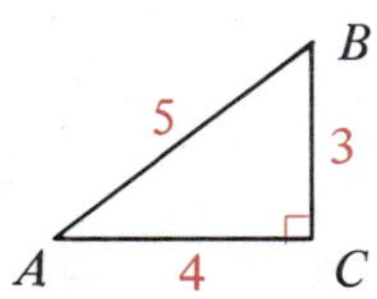

Example 2 For rt. $\triangle ABC$, find tan B, sin B, and cos B correct to three decimal places.

$$\tan B = \frac{\text{opp.}}{\text{adj.}} = \frac{12}{5} = 2.400$$

$\doteq$ *means approx.* =. ► $$\sin B = \frac{\text{opp.}}{\text{hyp.}} = \frac{12}{13} \doteq 0.923$$

$$\cos B = \frac{\text{adj.}}{\text{hyp.}} = \frac{5}{13} \doteq 0.385$$

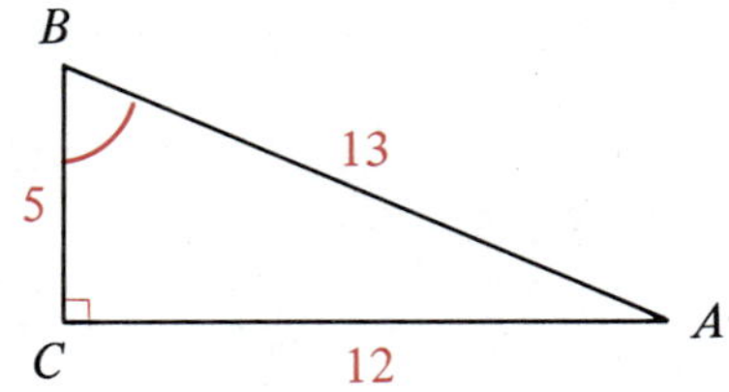

The properties of a 30°–60° right triangle and a 45°–45° right triangle can be used to find certain sine, cosine, and tangent ratios.

Example 3 Use rt. $\triangle ABC$ to find sin 30°, cos 30°, and tan 30°, correct to three decimal places.

$$\sin 30° = \frac{1}{2} = 0.500$$

$$\cos 30° = \frac{\sqrt{3}}{2} \doteq \frac{1.732}{2} \doteq 0.866$$

$$\tan 30° = \frac{1}{\sqrt{3}} = \frac{1}{\sqrt{3}} \cdot \frac{\sqrt{3}}{\sqrt{3}} = \frac{\sqrt{3}}{3} \doteq \frac{1.732}{3} \doteq 0.577$$

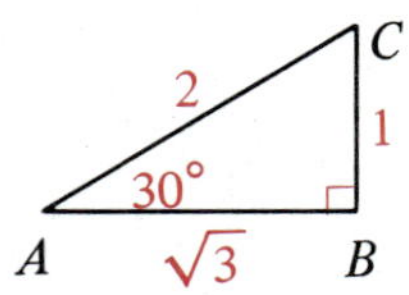

Reading in Geometry

$\triangle JKL$ is a right triangle with right $\angle L$.

1. Name the hypotenuse. $\overline{JK}$
2. Name the leg adjacent to $\angle J$. $\overline{JL}$
3. Name the leg opposite $\angle K$. $\overline{JL}$
4. Name the leg adjacent to $\angle K$. $\overline{KL}$
5. Name the leg opposite $\angle J$. $\overline{KL}$
6. What is sin J? $\frac{j}{l}$
7. What is tan K? $\frac{k}{j}$
8. What is cos J? $\frac{k}{l}$
9. What is sin K? $\frac{k}{l}$
10. What is tan J? $\frac{j}{k}$
11. What is cos K? $\frac{j}{l}$

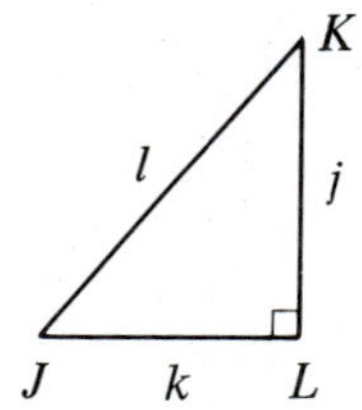

Written Exercises

For each triangle, find tan A, sin A, cos A, tan B, sin B, and cos B, correct to three decimal places. (Ex. 1–6)

A

1. 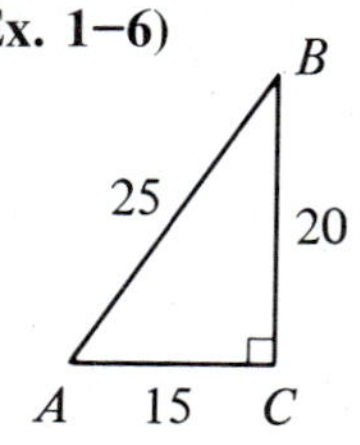

2. 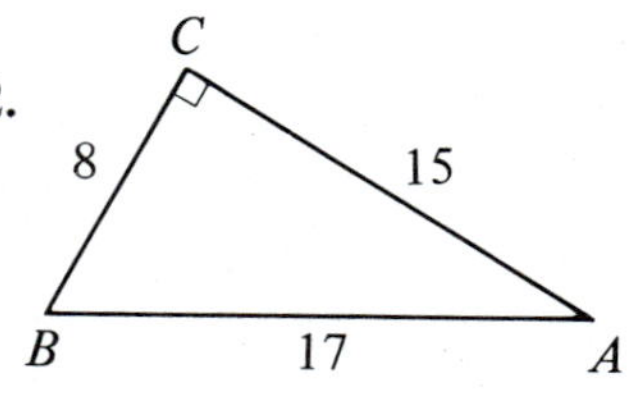

3. 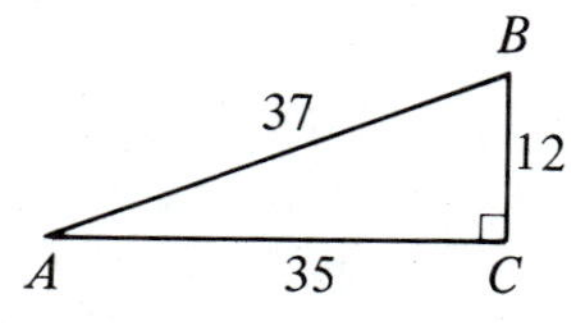

1.333, 0.800, 0.600, 0.750, 0.600, 0.800; 0.533, 0.471, 0.882, 1.875, 0.882, 0.471; 0.343, 0.324, 0.946, 2.917, 0.946, 0.324

4.
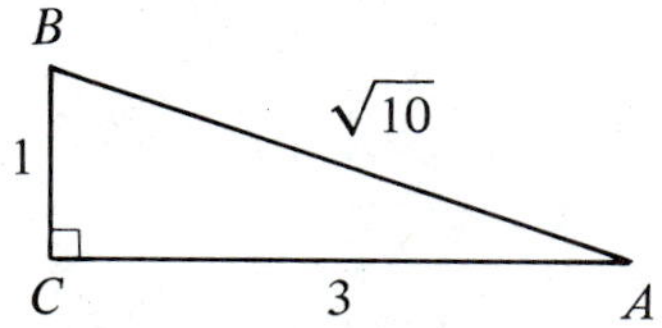

5.
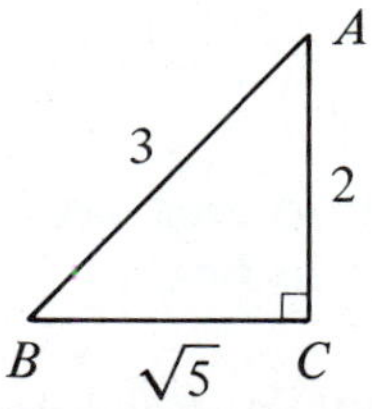

6.
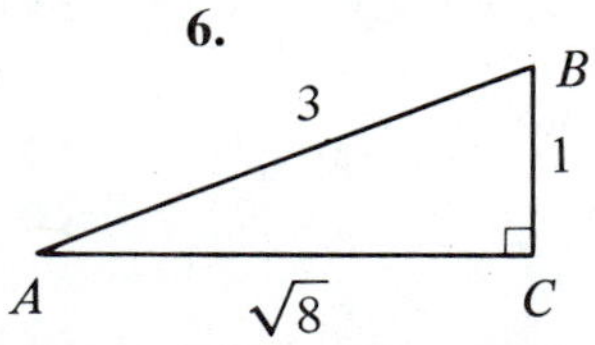

0.333, 0.316, 0.949, 3.000, 0.949, 0.316; 1.118, 0.745, 0.667, 0.894, 0.667, 0.745; 0.354, 0.333, 0.943, 2.828, 0.943, 0.333

Find each value, correct to three decimal places. Use the triangle shown.

7. tan 60° 1.732
8. sin 60° 0.866
9. cos 60° 0.500

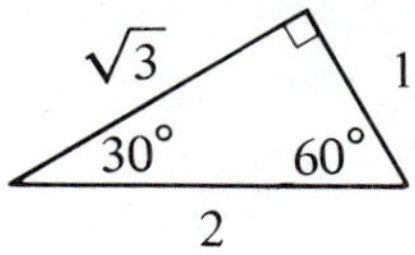

10. tan 45° 1.000
11. sin 45° 0.707
12. cos 45° 0.707

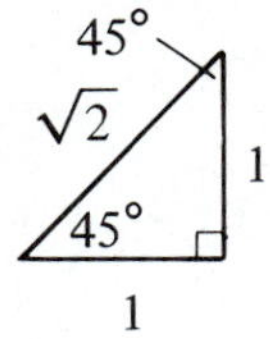

For each triangle, find tan *A*, sin *A*, cos *A*, tan *B*, sin *B*, and cos *B*, correct to three decimal places. [*Hint:* Use the Pythagorean Theorem first.] 0.436, 0.400, 0.917, 2.291, 0.917, 0.400; 1.134, 0.750, 0.662, 0.882, 0.662, 0.750; 0.354, 0.333, 0.943, 2.828, 0.943, 0.333

B

13.
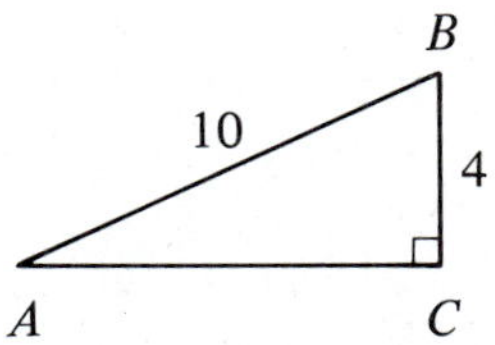

14.
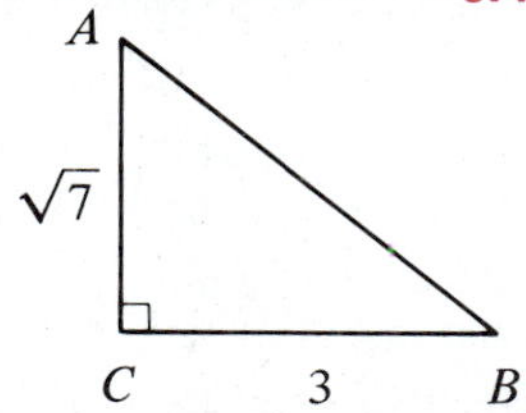

15.
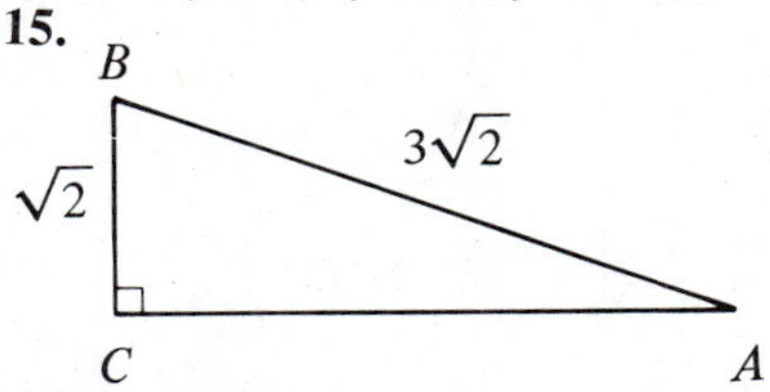

16.
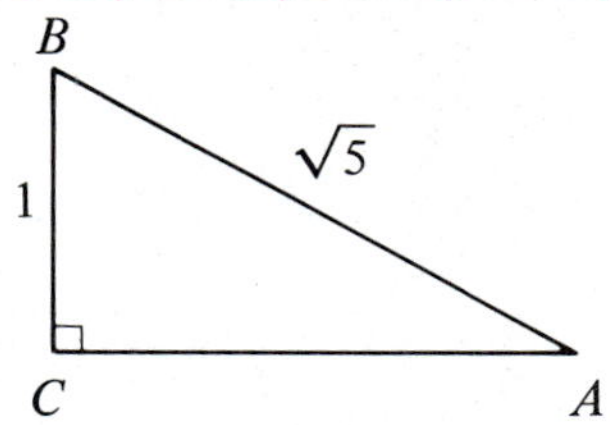

17.
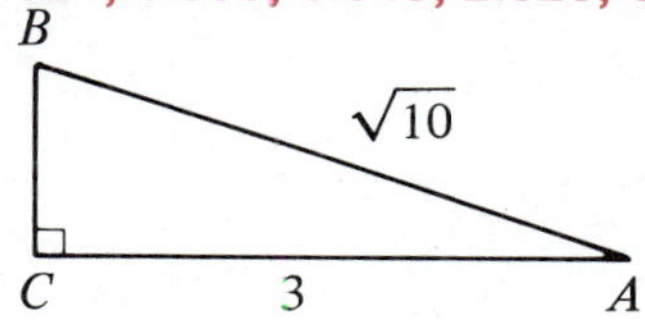

18.
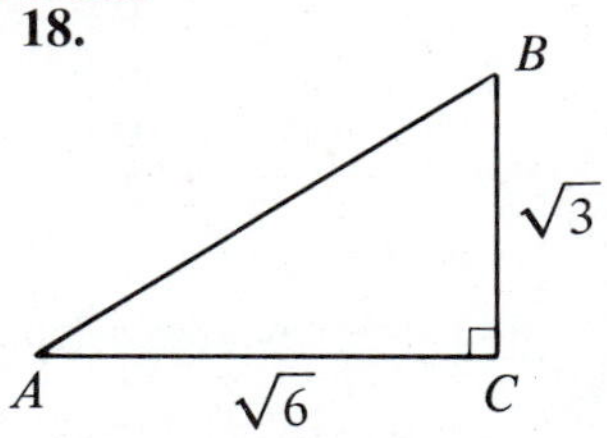

0.500, 0.447, 0.894, 2, 0.894, 0.447; 0.333, 0.316, 0.949, 3, 0.949, 0.316; 0.707, 0.577, 0.816, 1.414, 0.816, 0.577

Show that each statement is true. Use the figure below. (Ex. 19–24)

C

19. $\sin B = \cos A$

21. $\tan B = \dfrac{1}{\tan A}$

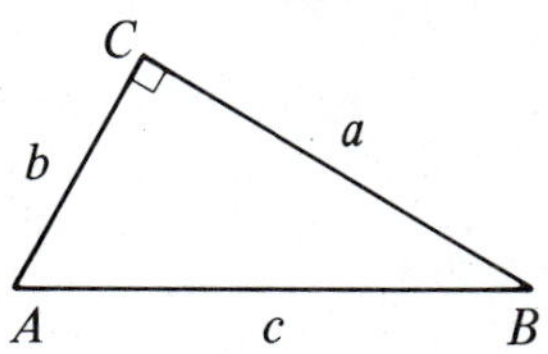

20. $\cos B = \sin A$

22. $(\sin B)^2 = 1 - (\cos B)^2$

[*Hint:* Use the Pythagorean Theorem.]

23. The tangent of an acute angle is equal to the reciprocal of the tangent of its complement.

24. The sine of an acute angle is equal to the cosine of its complement.

LOGIC REVIEW

Write a biconditional for each pair of statements. Label each statement and biconditional as true or false. Check students' answers.

1. If points are collinear, then they lie in the same plane. If points lie in the same plane, then they are collinear.

2. If $\triangle ABC \cong \triangle DEF$, then $\triangle ABC \sim \triangle DEF$. If $\triangle ABC \sim \triangle DEF$, then $\triangle ABC \cong \triangle DEF$.

TRIGONOMETRIC TABLES 10.6

Objectives

To use a table to find sin *A*, cos *A*, and tan *A*, given m ∠*A*
To use a table to find m ∠*A*, given sin *A*, cos *A*, or tan *A*

In the last lesson, you found sine, cosine, and tangent values for 30°, 45°, and 60° angles. You can find approximate values of sine, cosine, and tangent for angles of other degree measures by constructing right triangles, measuring the sides, and computing the ratios. For practical purposes, tables have been compiled to give these values. A portion of such a table is given below.

Angle Meas.	Sin	Cos	Tan	Angle Meas.	Sin	Cos	Tan
32°	0.5299	0.8480	0.6249	**78°**	0.9781	0.2079	4.704
33°	0.5446	0.8387	0.6494	**79°**	0.9816	0.1908	5.145
34°	0.5592	0.8290	0.6745	**80°**	0.9848	0.1736	5.671
35°	0.5736	0.8192	0.7002	**81°**	0.9877	0.1564	6.314
36°	0.5878	0.8090	0.7265	**82°**	0.9903	0.1392	7.115
37°	0.6018	0.7986	0.7536	**83°**	0.9925	0.1219	8.144
38°	0.6157	0.7880	0.7813	**84°**	0.9945	0.1045	9.514
39°	0.6293	0.7771	0.8098	**85°**	0.9962	0.0872	11.43
40°	0.6428	0.7660	0.8391	**86°**	0.9976	0.0698	14.30
41°	0.6561	0.7547	0.8693	**87°**	0.9986	0.0523	19.08

A complete table is found on page 590.

Example 1

Use the table to find sin 35°, cos 35°, and tan 35°.
Look in the appropriate column next to 35°.

Thus, sin 35° = 0.5736, cos 35° = 0.8192, and tan 35° = 0.7002.

Example 2

Find m ∠*A* if cos *A* = 0.1219.
Find 0.1219 in the Cos column.
cos 83° = 0.1219

Thus, m ∠*A* = 83.

Example 3

Find m ∠*B* to the nearest degree, if tan *B* = 4.976.

4.976	5.145
−4.704	−4.976
0.272	0.169

Find the closest value to 4.976 in the Tan column. Tan 78° = 4.704 and tan 79° = 5.145. 4.976 is closer to 5.145 than to 4.704.

Thus, m ∠*B* = 79 to the nearest degree.

Oral Exercises

Find each indicated value. Use the table on page 590.

1. sin 56° 0.8290 **2.** sin 29° 0.4848 **3.** cos 5° 0.9962 **4.** tan 35° 0.7002
5. cos 38° 0.7880 **6.** tan 81° 6.314 **7.** sin 55° 0.8192 **8.** cos 74° 0.2756
9. tan 15° 0.2679 **10.** sin 12° 0.2079 **11.** tan 52° 1.280 **12.** sin 41° 0.6561
13. cos 21° 0.9336 **14.** cos 85° 0.0872 **15.** cos 73° 0.2924 **16.** sin 76° 0.9703

Find m ∠*B*. Use the table on page 590.

17. tan $B = 3.487$ 74° **18.** tan $B = 0.8693$ 41° **19.** cos $B = 0.1736$ 80°
20. cos $B = 0.4695$ 62° **21.** cos $B = 0.7771$ 39° **22.** tan $B = 1.111$ 48°
23. sin $B = 0.9945$ 84° **24.** sin $B = 0.6428$ 40° **25.** sin $B = 0.7986$ 53°
26. cos $B = 0.9659$ 15° **27.** tan $B = 0.4245$ 23° **28.** sin $B = 0.4067$ 24°

Written Exercises

Find m ∠*A* to the nearest degree. Use the table on page 590.

A **1.** tan $A = 0.8721$ 41° **2.** cos $A = 0.9931$ 7° **3.** cos $A = 0.4295$ 65°
4. sin $A = 0.8392$ 57° **5.** tan $A = 1.352$ 54° **6.** cos $A = 0.1196$ 83°
7. cos $A = 0.8516$ 32° **8.** sin $A = 0.9927$ 83° **9.** sin $A = 0.2593$ 15°
10. tan $A = 0.4439$ 24° **11.** sin $A = 0.5036$ 30° **12.** tan $A = 12.57$ 85°

True or false? Use the right triangle at the right and the table on page 590.

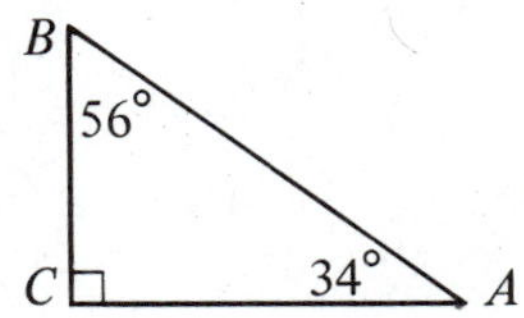

B **13.** $\sin A = \cos B$ T **14.** $\cos A = \cos B$ F
15. $\sin A = \sin B$ F **16.** $\tan B = \dfrac{1}{\tan A}$ T
17. m ∠A + m ∠B = m ∠C T **18.** $\sin A + \sin B = \sin C$ F
19. $\sin A + \cos A = 1$ F **20.** $\tan A \cdot \tan B = 1$ T

△*ABC* is a right triangle with right angle *C*. True or false?

21. m ∠A < m ∠B and m ∠B < m ∠C. F
22. If m ∠A < m ∠B, then $\cos A > \cos B$. T
23. If m ∠A < m ∠B, then $\tan A < \tan B$. T
24. If $\sin A > \sin B$, then m ∠A < m ∠B. F
25. $(\sin A)^2 + (\sin B)^2 = 1$ T
26. $(\sin A)^2 + (\cos A)^2 = 1$ T

C **27.** Explain why sin A cannot be greater than 1.

28. Explain why cos A decreases as m ∠A increases.

29. Show that $\tan A = \dfrac{\sin A}{\cos A}$.
(Hint: Use a, b, and c as the lengths of the sides of rt. △ABC.)

30. Explain why tan A gets infinitely large as m ∠A increases.

MEASURES IN RIGHT TRIANGLES 10.7

Objective **To find measures of sides and angles in right triangles by using sine, cosine, and tangent ratios**

Trigonometric ratios can be used to find the measures of sides and angles in right triangles. In Examples 1 and 2, you will find measures of sides.

Example 1 If m $\angle A = 48$ and $c = 9$, find a and b to the nearest tenth.

$$\sin A = \frac{a}{c} \qquad \cos A = \frac{b}{c}$$

$$\sin 48^\circ = \frac{a}{9} \qquad \cos 48^\circ = \frac{b}{9}$$

$$0.7431 \doteq \frac{a}{9} \qquad 0.6691 \doteq \frac{b}{9}$$

$$0.7431\,(9) \doteq a \qquad 0.6691\,(9) \doteq b$$

$$6.6879 \doteq a \qquad 6.0219 \doteq b$$

$$6.7 \doteq a \qquad 6.0 \doteq b$$

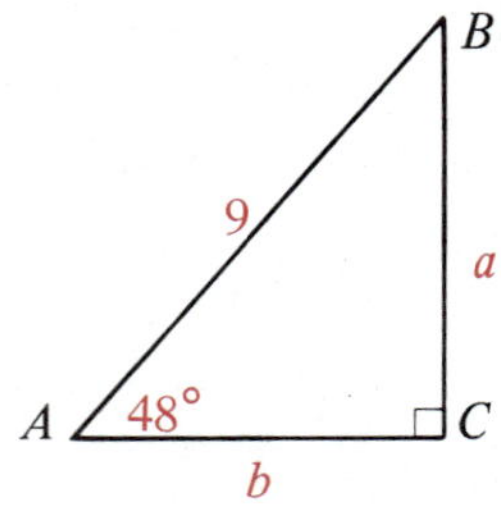

Thus, a is 6.7 and b is 6.0, to the nearest tenth.

Sometimes there is more than one way to solve a problem involving trigonometric ratios. In the next example, you can use either tan A or tan B.

Example 2 If m $\angle A = 23$ and $a = 15$, find b to the nearest tenth.

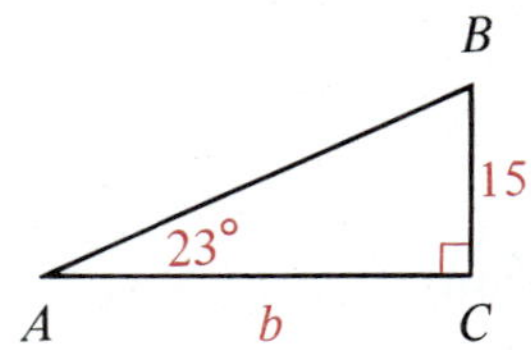

First Way

$$\tan A = \frac{a}{b}$$

$$\tan 23^\circ = \frac{15}{b}$$

$$0.4245 \doteq \frac{15}{b}$$

$$0.4245b \doteq 15$$

$$b \doteq \frac{15}{0.4245}$$

$$b \doteq 35.3$$

Second Way

$$m\,\angle B = 90 - 23 = 67$$

$$\tan B = \frac{b}{15}$$

$$\tan 67^\circ = \frac{b}{15}$$

$$2.356 \doteq \frac{b}{15}$$

$$(2.356)(15) \doteq b$$

$$35.340 \doteq b$$

$$35.3 \doteq b$$

◄ *In the second way, you multiply rather than divide.*

Thus, b is 35.3 to the nearest tenth.

Example 3

In Example 3, you will use trigonometric ratios to find measures of angles.

If $a = 5$ and $c = 15$, find m $\angle B$ and m $\angle A$ to the nearest degree.

$$\cos B = \frac{a}{c}$$

$$\cos B = \frac{5}{15}$$

$$\cos B = \frac{1}{3}$$

$$\cos B \doteq 0.3333$$

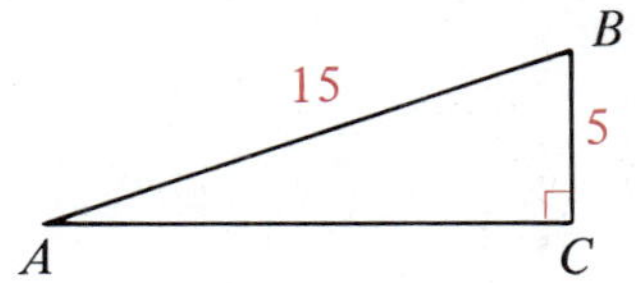

0.3420	0.3333
−0.3333	−0.3256
0.0087	0.0077

$\cos 71° \doteq 0.3256$ $\qquad$ $\cos 70° \doteq 0.3420$

Thus, m $\angle B \doteq 71$ and m $\angle A \doteq 90 - 71$, or 19.

Written Exercises

Find the indicated measure (sides to the nearest tenth and angles to the nearest degree).

A

1.

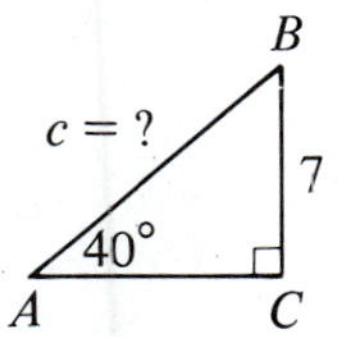

10.9

2.

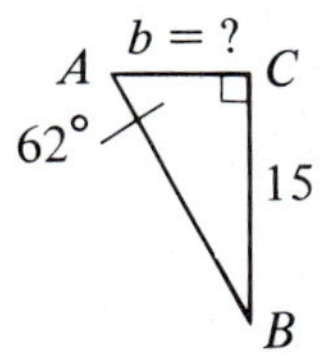

8.0

3.

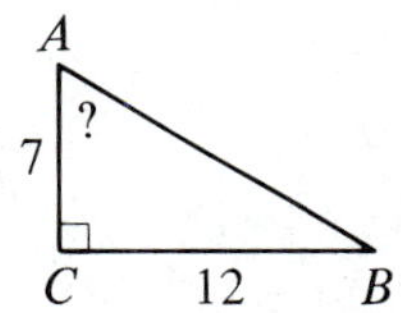

60°

4.

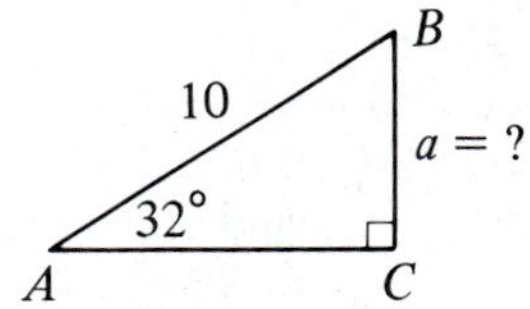

5.3

5.

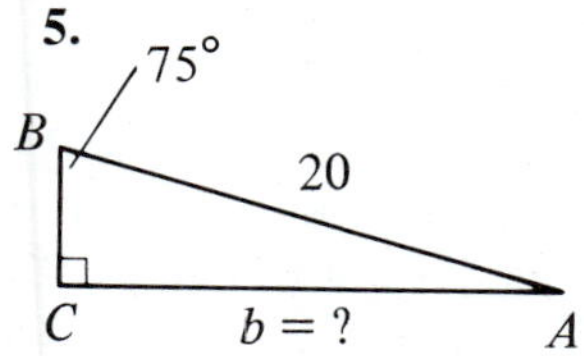

19.3

6.

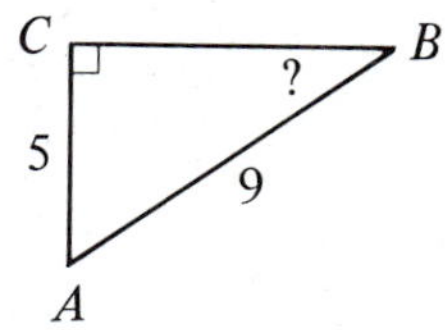

34°

7.

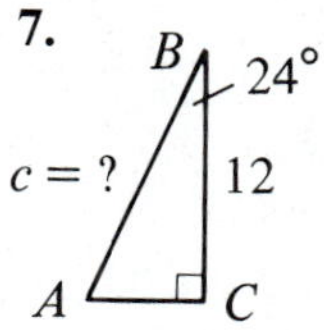

13.1

8.

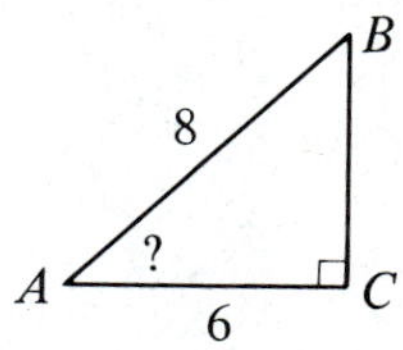

41°

Find each indicated measure (sides to the nearest tenth and angles to the nearest degree). Use the figure at the right.

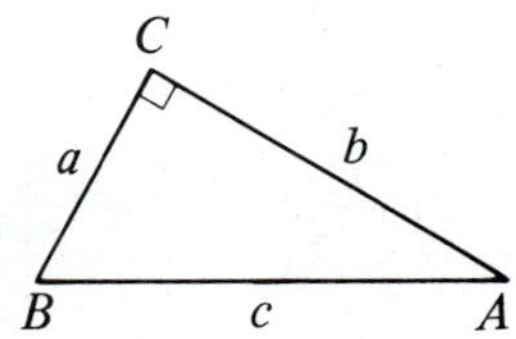

9. If $a = 6$ and m $\angle B = 18$, find b. **1.9**

10. If $a = 12$ and $c = 19$, find m $\angle B$. **51°**

11. If $b = 9$ and $c = 15$, find m $\angle B$. **37°**

12. If $b = 21$ and m $\angle B = 76$, find a. **5.2**

13. If $b = 15$ and $c = 20$, find m $\angle A$. **41°**

14. If $c = 28$ and m $\angle A = 31$, find a. **14.4**

15. If $c = 18$ and m $\angle B = 40$, find a. **13.8**

16. If $a = 9$ and $b = 13$, find m $\angle B$. **55°**

17. If $b = 17$ and m $\angle B = 49$, find c. **22.5**

18. If $c = 26$ and m $\angle B = 62$, find a. **12.2**

19. If $c = 10$ and m $\angle A = 38$, find a. **6.2**

20. If $b = 16$ and m $\angle A = 12$, find c. **16.4**

21. If $a = 3$ and $c = 5$, find m $\angle A$. **37°**

22. If $b = 18$ and m $\angle B = 38$, find a. **23.0**

23. If $a = 14$ and $b = 17$, find m $\angle A$. **39°**

24. If $a = 16$ and m $\angle B = 8$, find c. **16.2**

Find each of the missing measures (sides to the nearest tenth and angles to the nearest degree). (Ex. 25–30)

B 25.

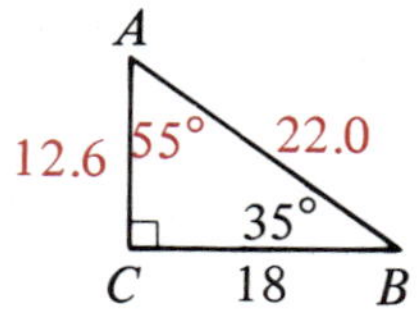

26.

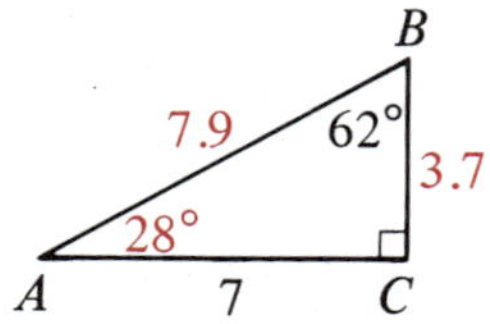

27.

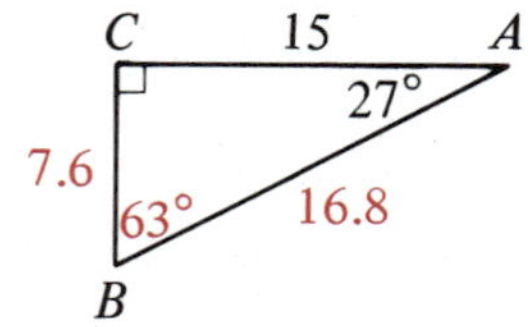

28.

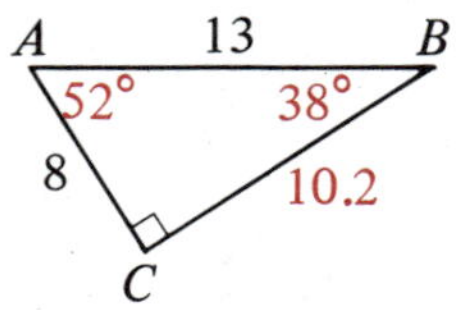

29.

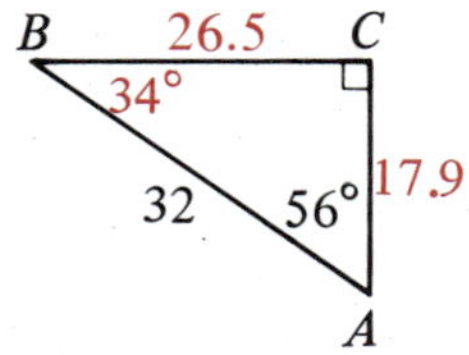

30.

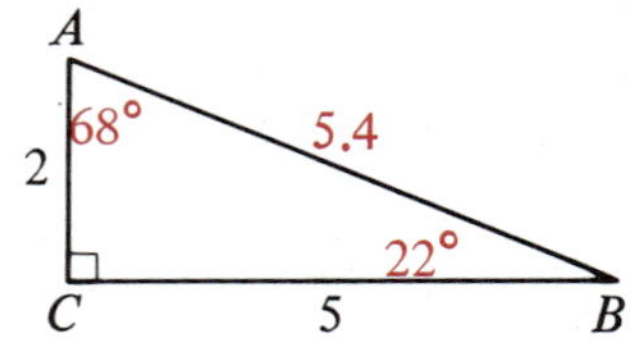

31. The lengths of the legs of a right triangle are in the ratio 1 to 4. Find the measure of each acute angle to the nearest degree. **14°; 76°**

32. One leg of a right triangle is $\frac{2}{3}$ as long as the hypotenuse. Find the measure of each acute angle to the nearest degree. **42°; 48°**

C **33.** Given: $\triangle ABC$ with $\overline{BC} \perp \overline{AC}$, m $\angle A = 27$, and $BC = 10$. Find the perimeter of $\triangle ABC$ to the nearest tenth. **51.6**

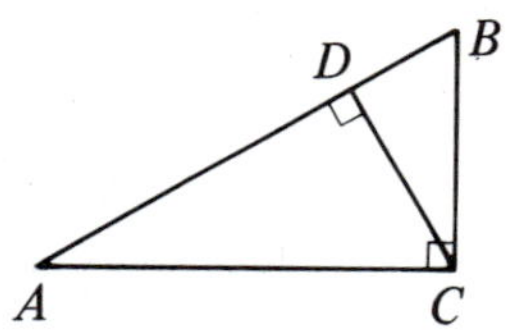

34. Given: $\triangle ABC$ with $\overline{BC} \perp \overline{AC}$, $\overline{CD} \perp \overline{AB}$, $BC = 8$, and m $\angle A = 32$. Find CD to the nearest tenth. **6.8**

35. Given: $\triangle RST$ with $\overline{ST} \perp \overline{TR}$, m $\angle x = 51$, m $\angle R = 28$, and $SR = 20$. Find SU to the nearest tenth. **12.1**

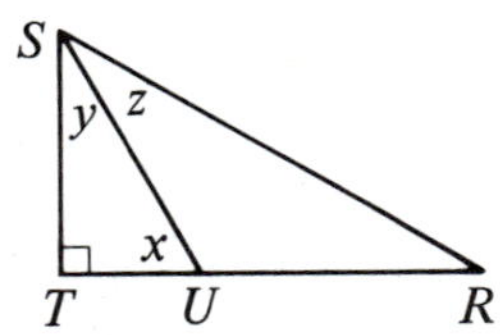

36. Given: $\triangle RST$ with m $\angle T = 90$, m $\angle y = 35$, m $\angle z = 18$, and $TU = 12$. Find SR to the nearest tenth. **28.5**

CALCULATOR ACTIVITIES

$\triangle ABC$ is a right triangle with right $\angle C$ and $\overline{CE} \perp \overline{AB}$. Find each indicated length to the nearest hundredth.

1. Find c. **28.29** **2.** Find b. **21.35** **3.** Find x. **14.01**
4. Find y. **16.11** **5.** Find z. **12.18**

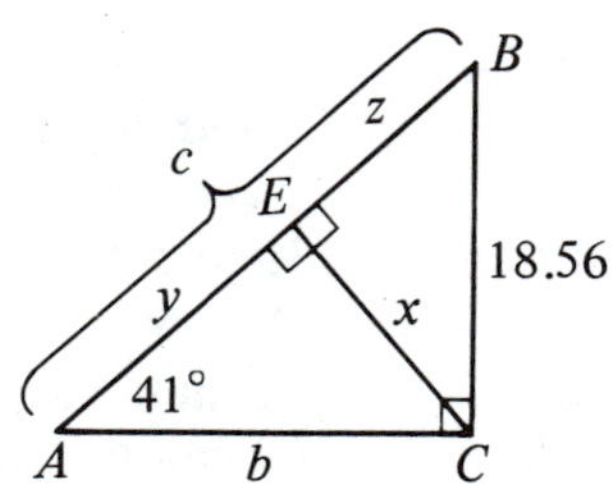

APPLICATIONS OF TRIGONOMETRY 10.8

Objective **To solve problems using trigonometric ratios**

Trigonometric ratios can be used to help solve measurement problems, especially problems involving lengths or distances that cannot be measured directly.

Example 1 A diagonal of a rectangle measures 5 cm. It makes an angle of 71° with a side of the rectangle. Find the dimensions of the rectangle to the nearest tenth of a centimeter.

$$\sin 71° = \frac{b}{5} \qquad \cos 71° = \frac{h}{5}$$

$$0.9455 \doteq \frac{b}{5} \qquad 0.3256 \doteq \frac{h}{5}$$

$$0.9455(5) \doteq b \qquad 0.3256(5) \doteq h$$

$$4.7275 \doteq b \qquad 1.6280 \doteq h$$

$$4.7 \doteq b \qquad 1.6 \doteq h$$

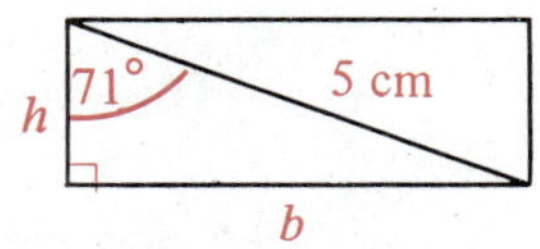

Thus, the length of a base is 4.7 cm and length of an altitude is 1.6 cm.

Surveyors and engineers often measure an *angle of elevation* or an *angle of depression.*

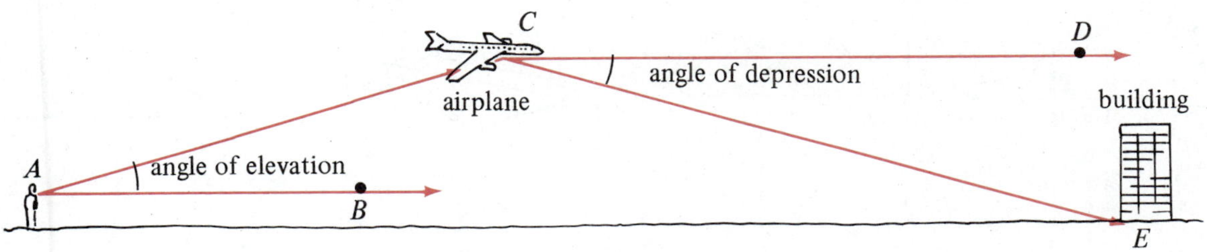

An observer at A, looking up at C, would have to elevate his line of sight from the horizontal $\overrightarrow{AB}$ to the direction $\overrightarrow{AC}$. $\angle CAB$ is called the *angle of elevation of C at A*. If the observer were at C, looking down at E, he would have to depress his line of sight from the horizontal $\overrightarrow{CD}$ to the direction $\overrightarrow{CE}$. $\angle DCE$ is called the *angle of depression of E at C.*

Example 2 When the angle of elevation of the sun measures 42°, a tower casts a 26-meter shadow. How tall is the tower?

$$\tan 42° = \frac{x}{26}$$

$$0.9004 \doteq \frac{x}{26}$$

$$0.9004(26) \doteq x$$

$$23.4104 \doteq x$$

$$23.4 \doteq x$$

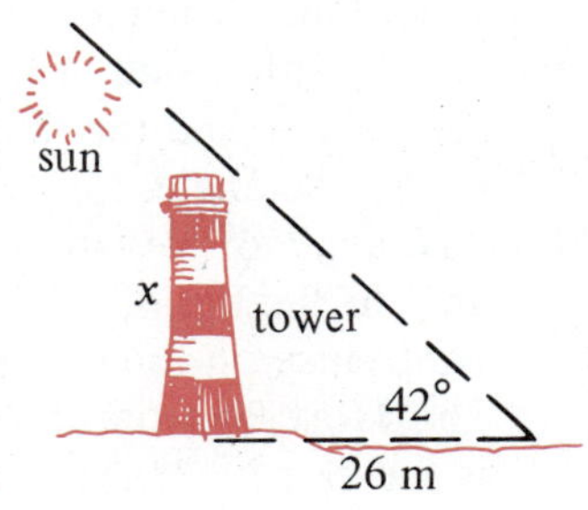

Thus, the tower is about 23.4 m tall.

Example 3 From a treehouse, the angle of depression of the base of a house measures 53°. The tree is 12 m from the base of the house. How far is the treehouse from the base of the house?

$$\cos 53° = \frac{12}{x}$$

$$0.6018 \doteq \frac{12}{x}$$

$$0.6018\,x \doteq 12$$

$$x \doteq \frac{12}{0.6018}$$

$$x \doteq 19.9$$

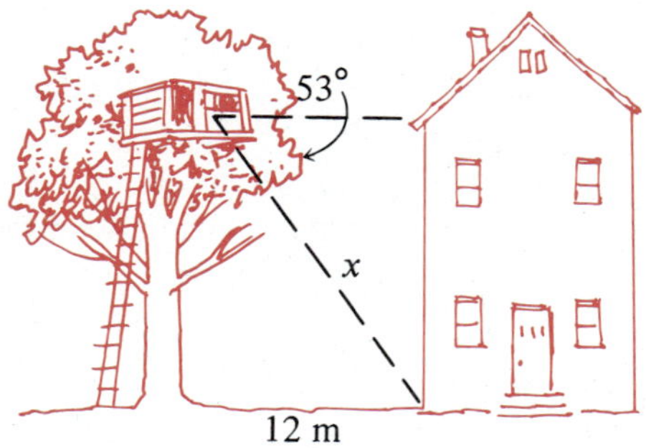

Thus, the treehouse is about 19.9 m from the base of the house.

Written Exercises

Find x to the nearest tenth.

A **1.**

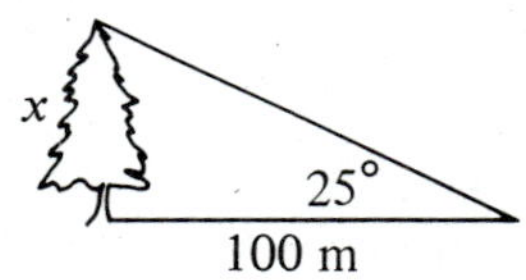

46.6 m

2.

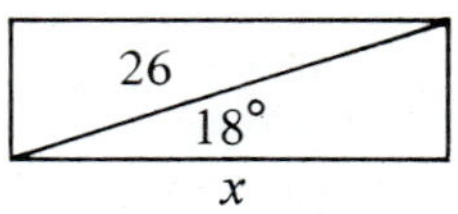

24.7

3.

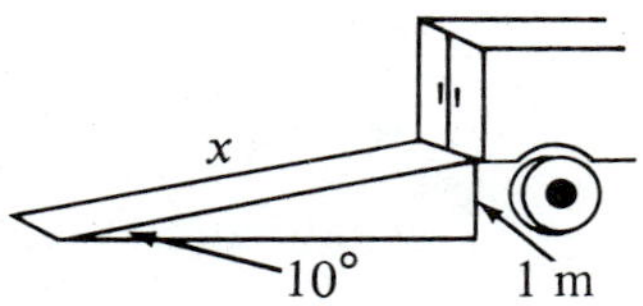

5.8 m

4.

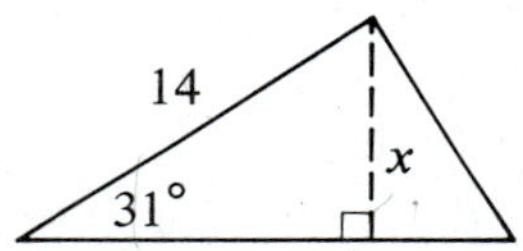

7.2

5.

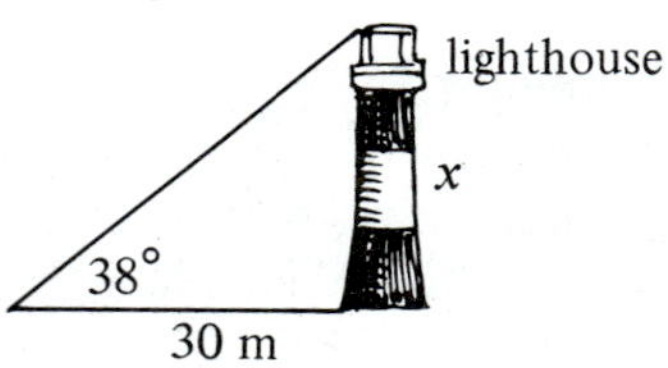

23.4 m

6.

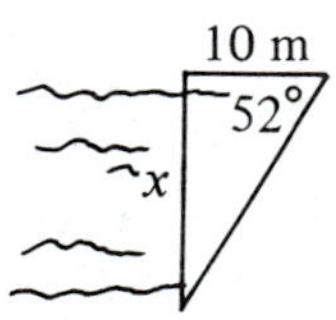

12.8 m

7.

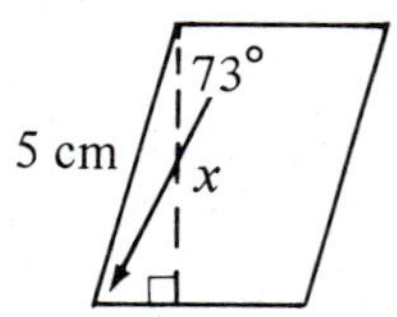

4.8 cm

8.

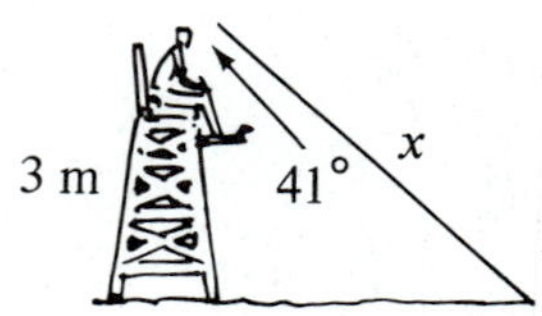

4.0 m

Find lengths to the nearest tenth and angle measures to the nearest degree.

9. The diagonal of a rectangle is 6 cm long. It makes an angle of 55° with a side of the rectangle. Find the dimensions of the rectangle. **3.4 cm, 4.9 cm**

10. The leg opposite the 50° angle in a right triangle measures 8 m. Find the length of the hypotenuse. **10.4 m**

11. The angle of elevation from a ship to the top of a 35-m lighthouse on the coast measures 26°. How far from the coast is the ship? **71.8 m**

12. A ramp is 60 m long. It rises a vertical distance of 8 m. Find the measure of the angle of elevation. **8°**

Find lengths to the nearest tenth and angle measures to the nearest degree.

B 13. A kite is flying at the end of a 150-m string (straight). The string makes an angle of 75° with the ground. How high above the ground is the kite? **144.9 m**

14. A tree casts a 50-m shadow while the angle of elevation of the sun is 48°. How tall is the tree? **55.6 m**

15. Two sides of a triangle measure 8 mm and 11 mm. The included angle measures 34°. Find the measure of the altitude to the 11-mm side. **4.5 mm**

16. A cliff is 90 m above the sea. From the cliff, the angle of depression of a boat measures 46°. How far is the boat from the base of the cliff? **86.9 m**

17. In the trapezoid below, $\overline{AB} \parallel \overline{CD}$ and $\overline{DA} \perp \overline{AB}$. $DC = 12$, $CB = 9$, and $m\angle B = 52$. Find AB. [*Hint:* Draw $\overline{CE} \perp \overline{AB}$.] **17.5**

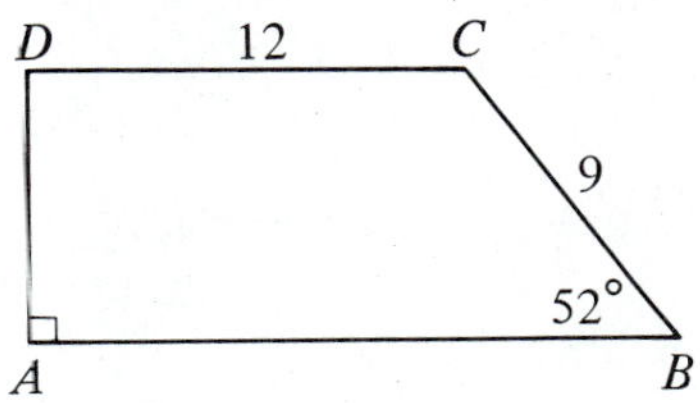

18. The diagonals of the rhombus below measure 8 and 16. Find the measure of each angle of the rhombus.
m ∠A = m ∠C = 54 m ∠B = m ∠D = 126

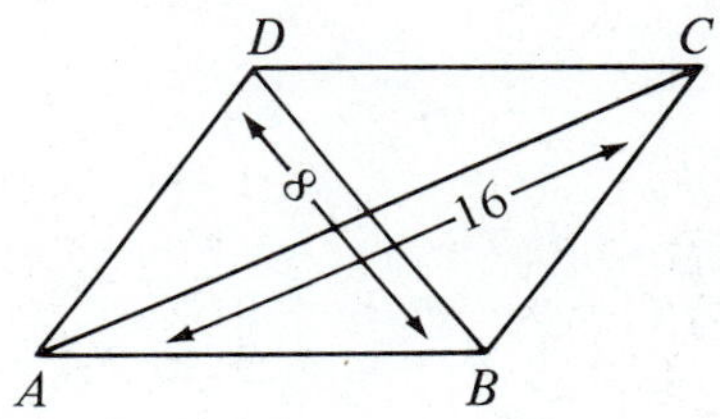

19. The diagonals of a rhombus measure 18 cm and 24 cm. Find the measure of each angle of the rhombus. **74°, 106°, 74°, 106°**

20. Each side of a rhombus is 9 cm long and an angle measures 76°. Find the length of the longer diagonal. **14.2 cm**

21. Two sides of an isosceles triangle each measure 10 cm, and the vertex angle measures 40°. Find the length of the altitude to the base. **9.4 cm**

22. In ▱$ABCD$, $m\angle B = 20$ and $AD = 12$. Find the length of the altitude from C to $\overline{AB}$. **4.1**

C 23. The sides of an isosceles triangle measure 14, 14, and 18. Find the measure of each angle of the triangle. **50°, 50°, 80°**

24. An angle of a rhombus measures 110° and the diagonal opposite that angle measures 30. Find the measure of the other diagonal. **21.0**

25. The sides of a triangle measure 12, 12, and 18. Find the measure of each angle of the triangle. **41°, 41°, 98°**

26. $ABCD$ is an isosceles trapezoid with $\overline{AB} \parallel \overline{CD}$. $AB = 20$, $CD = 14$, and $m\angle A = 72$. Find BC. **9.7**

27. The longer base of a trapezoid measures 21 m, and the other three sides each measure 12 m. Find the measure of each angle of the trapezoid.
112°, 68°, 68°, 112°

28. A diagonal of a rectangle divides one of its angles into two angles whose measures are in the ratio 4 to 5. Express the length of a base of the rectangle as a percentage of the length of a corresponding altitude.
$\frac{w}{l} = 83.91\%$

APPLICATIONS

Read → Plan → Solve → Interpret

1. In a swimming race, each swimmer is to start at one corner of the pool, swim to a point on the opposite side of the pool, and swim back to the other corner on the starting side. Each swimmer wants the shortest possible path. Does it matter where on the opposite side each aims for? If so, how much difference can it make in the distance each must swim?

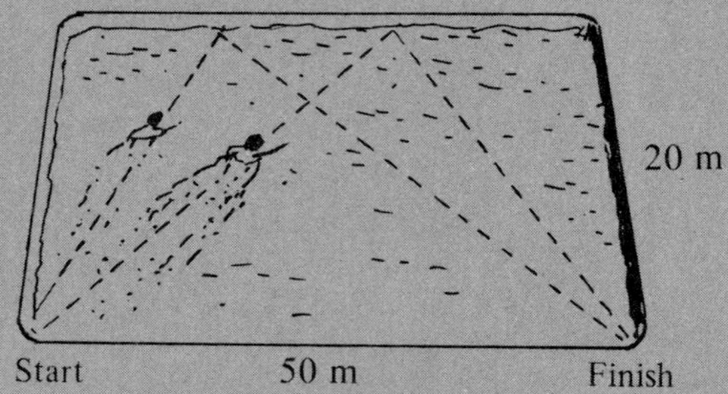

2. Ancient surveyors used a rope with equally spaced knots to lay out right-angled corners on land. By using the third, seventh, and twelfth knots as corners, they formed a right triangle. Why did this work?

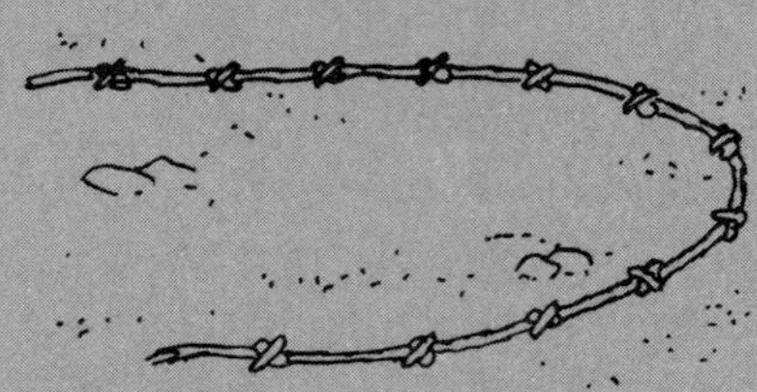

COMPUTER ACTIVITIES

Trigonometry Applications

Objective: **To solve computer-generated problems that involve angles of elevation and depression**

Trigonometric functions can be used to solve a variety of problems that deal with angles of elevation or depression (see pages 329–331). The computer program below tests your ability to solve two different kinds of problems dealing with those topics. Final answers should be rounded to the nearest integer.

```
110  FOR I = 1 TO 5
120  Q =  INT ( RND (1) * 2) + 1
130  ON Q GOTO 140,210
140  A =  INT ( RND (1) * 36) + 5
150  H = 500 * ( INT ( RND (1) * 19) + 2)
160  R = A * 6.2832 / 360
170  D =  INT (H /  SIN (R) + .5)
180  PRINT "A PLANE IS FLYING AT AN ALTITUDE OF "H"
     FEET."
190  PRINT "THE ANGLE OF DEPRESSION TO THE END OF"
195  PRINT "THE RUNWAY IS "A"."
200  INPUT "HOW FAR FROM THE RUNWAY IS THE PLANE IN
     FEET?";P
205  IF D = P THEN 340
206  GOTO 290
210 A =  INT ( RND (1) * 51) + 20
220 H =  INT ( RND (1) * 51) + 10
230 R = A * 6.2832 / 360
240 D =  INT (H /  TAN (R) + .5)
250  PRINT "THE TOP OF A LIGHTHOUSE IS "H" FEET ABOVE
     SEA LEVEL."
260  PRINT "THE ANGLE OF ELEVATION FROM A FISHING BOAT
     IS "A"."
270  INPUT "HOW FAR FROM THE LIGHTHOUSE IS THE BOAT? ";P
280  IF P = D THEN 340
290  PRINT "INCORRECT. THE ANSWER IS "D"."
300  PRINT : NEXT I
340 C = C + 1: PRINT "CORRECT."
350  PRINT : PRINT "YOUR SCORE IS "C * 20"."
```

EXERCISES: See the Computer Section beginning on page 565 for more information.

Write a quiz similar to the one above that replaces the original problems with problems to find the height of a tree and to find the distance out in the water of a swimmer spotted by a lifeguard. (See the illustrations for Exercises 1 and 8 on page 330.)

CHAPTER TEN REVIEW

1. Given: Rt. $\triangle ABC$ with altitude $\overline{CD}$. Which triangles formed are similar? [10.1]
$\triangle ABC \sim \triangle ACD \sim \triangle CBD$

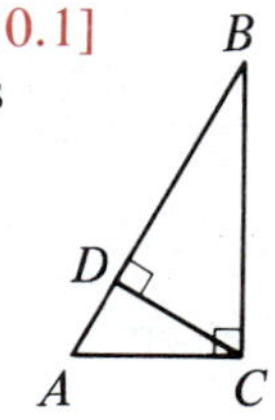

2. Given: $\overline{AC} \perp \overline{CB}$, $\overline{CD} \perp \overline{AB}$, $AD = 3$, and $DB = 27$
Find CD and CB. **9; $9\sqrt{10}$**

3. Given: Rt. $\triangle JKL$, $JK = 8$, and $LK = 5$
Find JL. **$\sqrt{89}$**

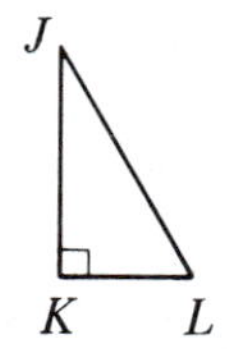

4. Given: Rt. $\triangle JKL$, $JL = 9$, and $LK = 4$
Find JK. **$\sqrt{65}$**

Which of the following can be lengths of sides of a right triangle? (Ex. 5–7) [10.2]

5. 9, 12, 15 **yes**
6. 6, 8, 12 **no**
7. x, $2x$, $\sqrt{5x}$ **no**
8. Find the length of the hypotenuse of a right triangle if the length of altitude to the hypotenuse is 6 and the length of a leg is 10. **$12\frac{1}{2}$**

*9. Find the length of a diagonal of a cube if the length of an edge is 6. **$6\sqrt{3}$**

10. Given: $\triangle RST$ is a 45°–45° rt. $\triangle$ and $RT = 6$. **$3\sqrt{2}$; $3\sqrt{2}$**
Find RS and TS. [10.4]

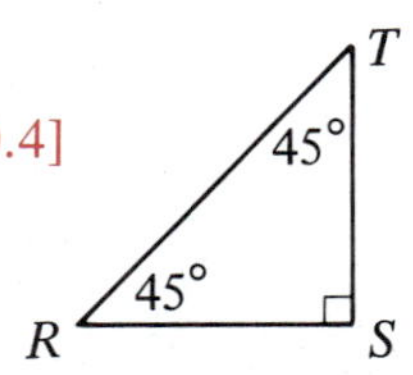

11. Given: $\triangle RST$ is a 45°–45° rt. $\triangle$ and $RS = 6$.
Find RT and TS. **$6\sqrt{2}$; 6**

Vocabulary
Pythagorean Theorem [10.2], trigonometry [10.5], sine [10.5], cosine [10.5], tangent [10.5]

12. Given: $\triangle PQR$ is a 30°–60° rt. $\triangle$ and $PR = 4$. **$\frac{8\sqrt{3}}{3}$; $\frac{4\sqrt{3}}{3}$**
Find PQ and QR. [10.4]

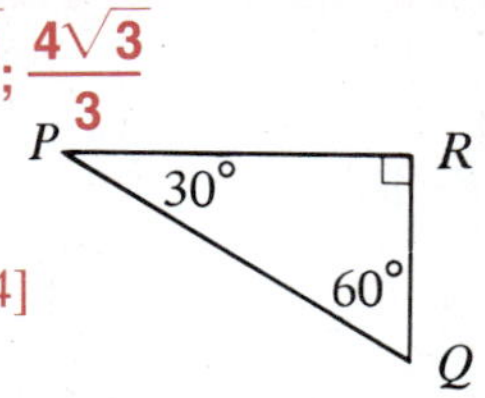

13. Given: Rt. $\triangle XYZ$, $XY = 4$, $YZ = 3$, and $XZ = 5$
Find $\tan X$. [10.5, 10.6] **$\frac{3}{4}$**

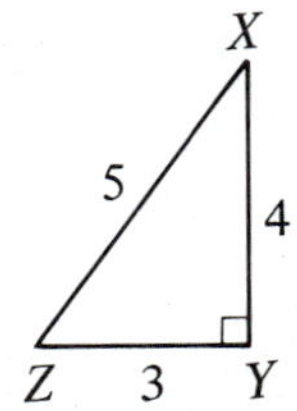

14. Given: Rt. $\triangle XYZ$, $XY = 4$, $YZ = 3$, and $XZ = 5$
Find $\sin Z$. **$\frac{4}{5}$**

Use the table of trigonometric ratios to find each of the following. [10.5, 10.6]

15. $\sin 32°$ **0.5299**
16. $\tan 57°$ **1.5399**

Use the table to find m $\angle A$ to the nearest degree. [10.6]

17. $\tan A = 0.5407$ **28°**
18. $\cos A = 0.6633$ **48°**

Find each indicated measure. [10.7]

19. Given: $\tan K = 0.5543$ and $JK = 10$
Find LJ. **5.543**

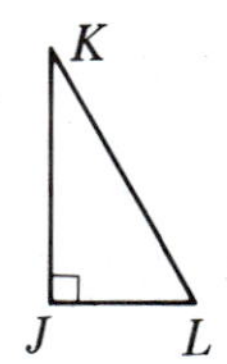

20. Given: $\sin K = 0.6947$ and $JK = 8$
Find LK. **11.12**

Find lengths to the nearest tenth. [10.8]

21. A flagpole casts a 20-meter shadow while the angle of elevation of the sun is 42°. How tall is the flagpole? **18.0 m**
22. Two sides of an isosceles triangle each measure 12 cm, and the vertex angle measures 130°. Find the length of the altitude to the base. **5.1 cm**

*23. $ABCD$ is an isosceles trapezoid with $\overline{AB} \parallel \overline{CD}$. If $AB = 25$, $CD = 17$, and m $\angle A = 52$, find BC. **6.5**

The following exercises are types that appear in the A section of the lessons' exercises: 1–6, 10–21. The starred exercises are types that appear in the C section. All the remaining exercises appear in the B section.

CHAPTER TEN TEST

1. Given: Rt. $\triangle PQR$ with altitude $\overline{RS}$. Which triangles formed are similar?
$\triangle PRQ \sim \triangle RSQ \sim \triangle PSR$

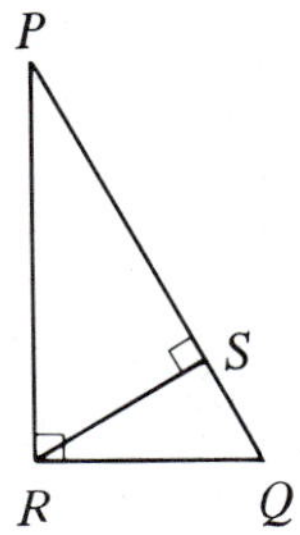

2. Given: Rt. $\triangle PQR$, altitude $\overline{RS}$, $PQ = 36$, and $SQ = 9$
Find PR. $18\sqrt{3}$

3. Given: Rt. $\triangle HJK$, $HK = 4$, and $KJ = 7$
Find HJ. $\sqrt{65}$

4. Given: Rt. $\triangle HJK$, $HJ = 9$, and $KJ = 7$
Find HK. $4\sqrt{2}$

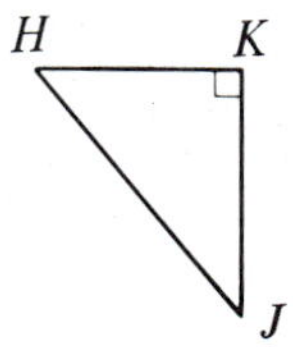

Which of the following can be lengths of sides of a right triangle? (Ex. 5–7)

5. 5, 12, 13 yes
6. 2, 3, 5 no
7. $\sqrt{2}x$, $\sqrt{3}x$, $\sqrt{6}x$ no

8. Find the length of the altitude to the hypotenuse of a right triangle if the lengths of the legs are 10 and 24. 9.2

* 9. If the edges of a rectangular solid are 5 cm, 6 cm, and 6 cm long, find the length of a diagonal of the solid. $\sqrt{97}$ cm

10. Given: $\triangle XYZ$ is a 45°–45° rt. $\triangle$ and $XY = 6$.
Find YZ. $3\sqrt{2}$

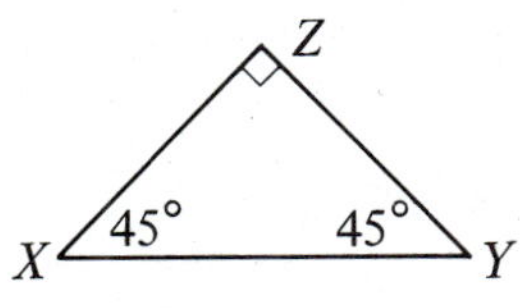

11. Given: $\triangle ABC$ is a 30°–60° rt. $\triangle$ and $AB = 12$.
Find CB and AC. $6\sqrt{3}$; 6

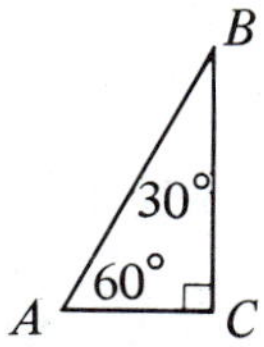

12. Given: Rt. $\triangle DEF$, $DE = 13$, $EF = 5$, and $DF = 12$
Find cos D. $\frac{12}{13}$

13. Given: Rt. $\triangle DEF$, $DE = 13$, $EF = 5$, and $DF = 12$
Find tan E. $2\frac{2}{5}$

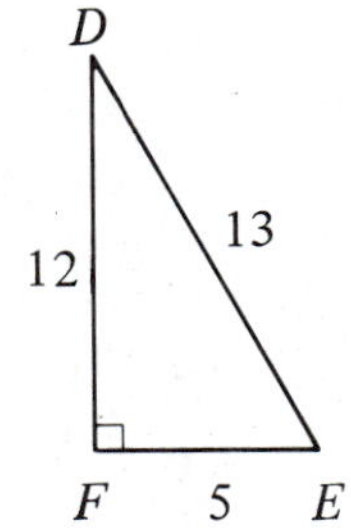

Use the table of trigonometric ratios below for Items 14–20.

Angle Measure	Sin	Cos	Tan
28	0.4695	0.8829	0.5317
29	0.4848	0.8746	0.5543
30	0.5000	0.8660	0.5774
~~	~~~	~~~	~~~
60	0.8660	0.5000	1.732
61	0.8746	0.4848	1.804
62	0.8829	0.4695	1.881

14. Find cos 62°. 0.4695

15. If tan $A = 1.839$, find m $\angle A$ to the nearest degree. 61°

16. Given: Rt. $\triangle RST$, m $\angle R = 29$ and $ST = 10$
Find RT to the nearest tenth. 20.6

17. Given: Rt. $\triangle RST$, m $\angle T = 62$ and $RT = 10$
Find ST to the nearest tenth. 4.7

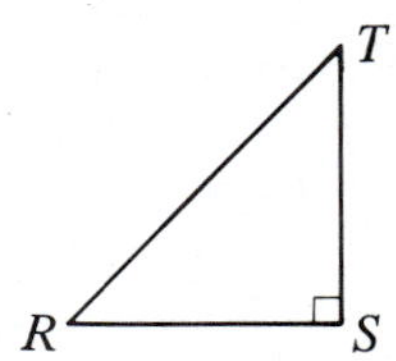

Find lengths to the nearest tenth.

18. In $\square ABCD$, m $\angle B = 29$ and $AD = 15$. Find the length of the altitude from C to $\overline{AB}$. 7.3

* 19. An angle of a rhombus measures 124° and the diagonal opposite that angle measures 12. Find the measure of the other diagonal. 6.4

The following exercises are types that appear in the A section of the lessons' exercises: 1–6, 10–18. The starred exercises are types that appear in the C section. All the remaining exercises appear in the B section.

COLLEGE PREP TEST

Directions: Choose the one best answer to each question or problem.

1. Two highways meet at right angles. A post is 30 ft from one road and 50 ft from the other road. How far is the post from the intersection? D

(A) 40 ft **(B)** 60 ft **(C)** $2\sqrt{85}$ ft
(D) $10\sqrt{34}$ ft **(E)** none of these

2. In the diagram below, $\angle BCA$, $\angle ADC$, and $\angle ADE$ are right angles. B

$AB = 17$
$BC = 8$
$CD = 9$
$DE = 5$

Find EA.

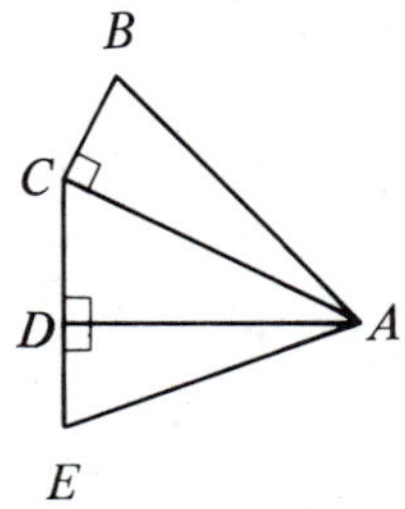

(A) 12 **(B)** 13 **(C)** 15 **(D)** 25 **(E)** 17

3. In the diagram below, $\overline{BC} \perp \overline{AC}$.

$CB = 12$
$AC = 27$
$AD = 1\frac{1}{4}\ CD$

Find the measure of $\angle DBC$. A

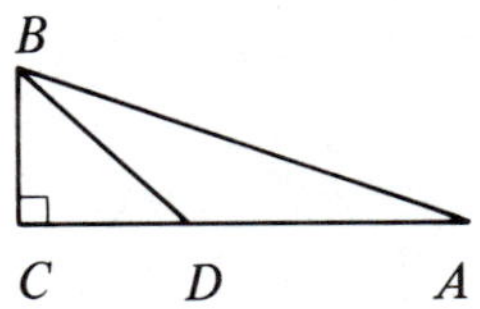

(A) 45 **(B)** 30 **(C)** 60 **(D)** 50
(E) none of these

4. A 25-foot ladder is placed against a vertical wall of a building. The foot of the ladder is 7 ft from the base of the building. If the top of the ladder slips 4 ft, how many feet will the bottom of the ladder slide? D

(A) 9 ft **(B)** 15 ft **(C)** 5 ft **(D)** 8 ft
(E) 4 ft

5. In the figure below, $MP = OP$, $MN = 4$, $NP = 3$, $\overline{ON} \perp \overline{NM}$.
Find MO. C

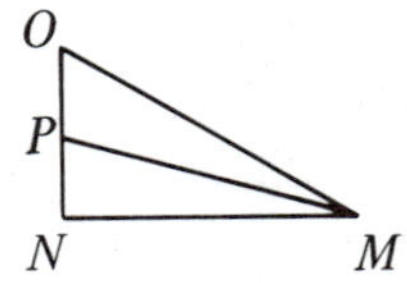

(A) $4\sqrt{3}$ **(B)** 8 **(C)** $4\sqrt{5}$ **(D)** 12
(E) $5\sqrt{2}$

6. $MNPR$ is a rectangle.
$MN = 12$
$MR = 16$
Find NO. E

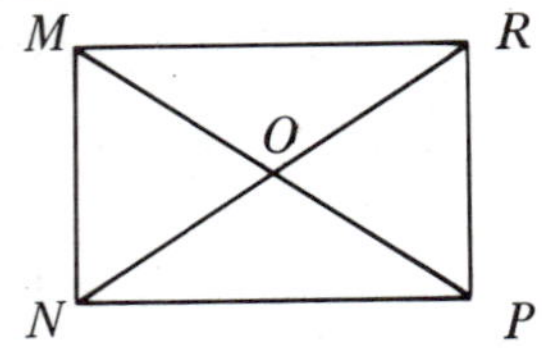

(A) 8 **(B)** 20 **(C)** 14 **(D)** 15 **(E)** 10

7. Given: $\overline{AE} \perp \overline{ED}$,
$\overline{CD} \perp \overline{ED}$,
$\overline{DC} \perp \overline{CB}$,
$ED = 13$, $CD = 3$
$CB = 2$, $AE = 11$
Find AB. A

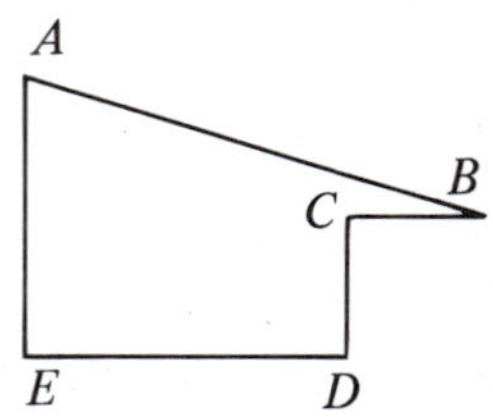

(A) 17 **(B)** 13 **(C)** 14 **(D)** 15 **(E)** 8

8. In right triangle ABC, $\overline{CD}$, the altitude to the hypotenuse $\overline{AB}$, measures 3. If DB exceeds AD by 8, find AB. C

(A) 9 **(B)** $8\frac{1}{2}$ **(C)** 10 **(D)** 12 **(E)** 15

9. Triangle ABC is equilateral.
$\overline{BD} \perp \overline{AC}$
$BD = 6\sqrt{3}$
Find AB. D

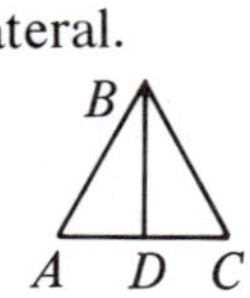

(A) 6 **(B)** $12\sqrt{3}$
(C) $3\sqrt{3}$ **(D)** 12 **(E)** $6\sqrt{3}$

11 CIRCLES AND LINES

HISTORY

Benjamin Banneker

Benjamin Banneker was born in Maryland. He is known for his work in the fields of geology, astronomy, and physics.

Benjamin Banneker was an ambitious scholar who received a formal education at a time when it was extremely difficult for a black person to get an education. He also studied a lot on his own in the fields of astronomy, geology, and physics. In 1789, Banneker was one of a team of surveyors that originally laid out the city of Washington, D.C. **This photo is from the cover of his Almanac, where for some reason *Banneker* is spelled *Bannaker*.**

In 1792, he published an almanac entitled *Benjamin Banneker's Almanac and Ephemeris,* which continued for a period of seven years. His almanacs contained commentaries on school problems as well as factual material.

Project

Write a report on Benjamin Banneker, emphasizing either his work as a surveyor or his publication of the almanac.

CIRCLES AND LINES 11.1

Objectives

To identify segments and lines related to circles
To prove and apply theorems about radii and chords

Segments and lines related to a circle are shown below. The symbol for circle is ⊙.

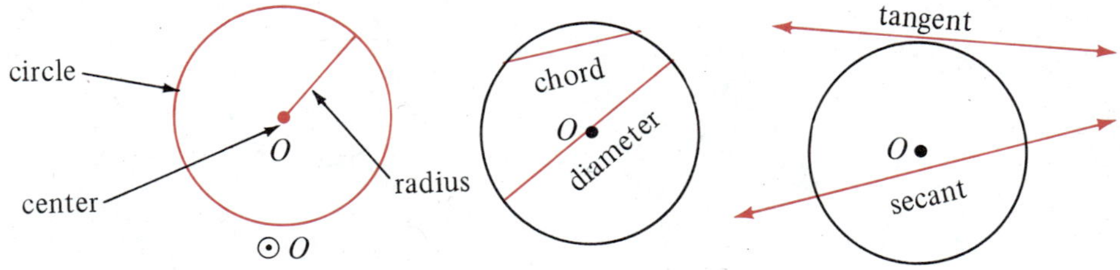

Definitions: Circle, center, radius	A *circle* is the set of all points in a plane that are a given distance from a given point in the plane. The given point is the *center* of the circle. A *radius* (plural: *radii*) of a circle is a segment that joins the center and a point on the circle.
chord	A *chord* is a segment whose endpoints lie on the circle.
diameter	A *diameter* is a chord that contains the center of the circle.
secant	A *secant* is a line that intersects the circle in two points.
tangent	A *tangent* is a line that is coplanar with the circle and intersects the circle in exactly one point.

Example 1

Identify a radius, a diameter, a chord, a secant, and a tangent of ⊙O.

Radius: $\overline{OB}$, $\overline{OC}$, or $\overline{OD}$
Diameter: $\overline{BD}$
Chord: $\overline{EA}$ or $\overline{BD}$
Secant: $\overleftrightarrow{EA}$
Tangent: $\overleftrightarrow{FA}$

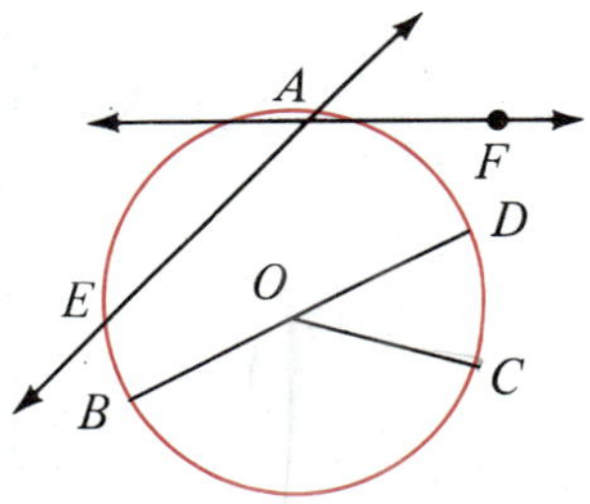

The following theorem follows from the definition of a circle.

Theorem 11.1 All radii of a circle are congruent. (See Exercise 19.)

From Theorem 11.1, you can also conclude that each diameter of a circle is twice as long as a radius of the circle; that is,

$$d = 2r.$$

Thus, if $r = 7$, then $d = 14$.

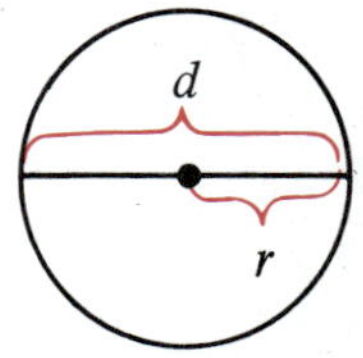

$d = 2r$

The next theorem states an important relationship between radii and chords.

Theorem 11.2

In a circle, a radius perpendicular to a chord bisects the chord.

Given: $\odot O$ with radius $\overline{OA} \perp$ chord $\overline{XY}$ at B
Prove: $\overline{OA}$ bisects $\overline{XY}$.

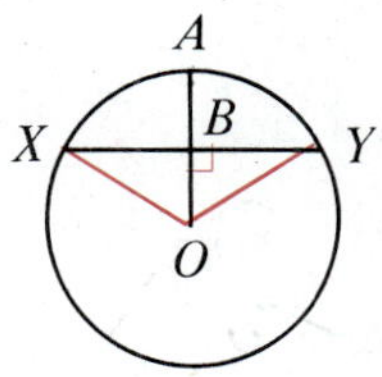

ANALYSIS
Draw radii $\overline{OX}$ and $\overline{OY}$.
Show $\triangle OBX \cong \triangle OBY$.

Proof

Statements		Reasons
1. $\odot O$, with radius $\overline{OA} \perp$ chord $\overline{XY}$ at B		1. Given
2. Draw radii $\overline{OX}$ and $\overline{OY}$.		2. Two points are contained in exactly one line.
3. $\overline{OX} \cong \overline{OY}$	(H)	3. All radii of a $\odot$ are $\cong$.
4. $\overline{OB} \cong \overline{OB}$	(L)	4. Reflexive property
5. $\triangle XOB$ and $\triangle YOB$ are rt. $\triangle$'s.		5. $\perp$'s form rt. $\angle$'s; def. of rt. $\triangle$.
6. $\triangle XOB \cong \triangle YOB$		6. HL for $\cong$ rt. $\triangle$'s
7. $\overline{XB} \cong \overline{YB}$		7. CPCTC
8. $\therefore$ $\overline{OA}$ bisects $\overline{XY}$.		8. Def. of bisector

Example 2

Given: Radius $\overline{OR} \perp$ chord $\overline{PQ}$,
$PS = 3x - 1$, and $SQ = x + 9$.
Find PS and SQ.

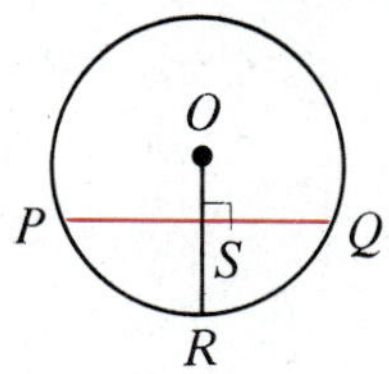

$$PS = SQ \text{ by Th. 11.2}$$
$$3x - 1 = x + 9$$
$$2x = 10$$
$$x = 5$$
$$PS = 3x - 1 = 3 \cdot 5 - 1 = 14; \; SQ = x + 9 = 5 + 9 = 14$$

Thus, $PS = 14$ and $SQ = 14$.

Example 3

In a circle with a radius 10 cm long, a chord is 6 cm from the center. How long is the chord?

$\overline{OD} \perp \overline{AB}$ ◄ *Distance from a point to a line is the length of the $\perp$ segment.*

$$(AO)^2 = (OD)^2 + (AD)^2$$
$$10^2 = 6^2 + (AD)^2$$
$$100 = 36 + (AD)^2$$
$$64 = (AD)^2$$
$$AD = 8$$

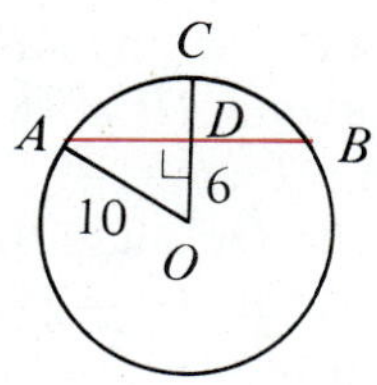

$\overline{OD}$ bisects $\overline{AB}$. ◄ *Th. 11.2*
Thus, $AB = 16$ cm.

Reading in Geometry

Write A if the statement is *always* true. Write S if the statement is *sometimes* true. Write N if the statement is *never* true.

1. A circle is a polygon. N
2. All radii of a circle are congruent. A
3. A secant is part of a chord. N
4. A chord is part of a tangent. N
5. A diameter is a chord. A
6. A chord is a diameter. S
7. The union of two radii of a circle is a diameter of the circle. S
8. A diameter is twice as long as a radius. S
9. A radius of a circle is part of the circle. N
10. All points of a circle are equidistant from the center of the circle. A

Written Exercises

For each $\odot O$, name all radii and diameters. For each $\odot Q$, name all secants, chords, and tangents. 1. radii: $\overline{OA}, \overline{OD}, \overline{OC}, \overline{OB}$; diams: $\overline{AB}, \overline{CD}$ 3. chord: $\overline{AB}$; no secants or tangents

A **1.**

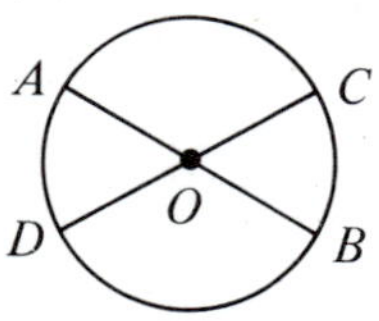

2.

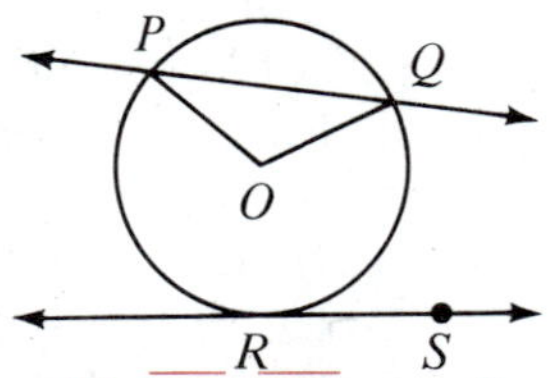

3.

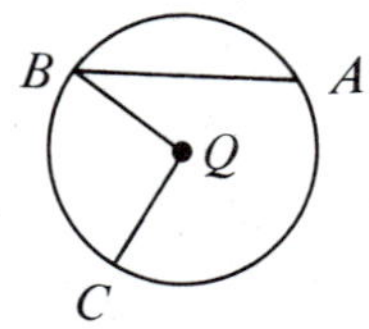

4.

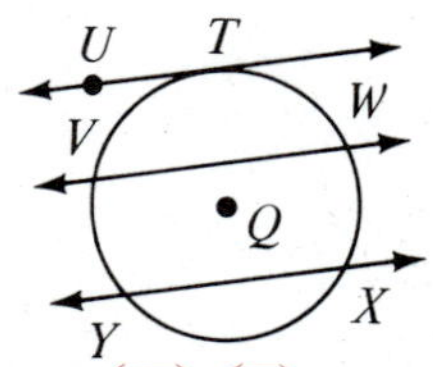

2. radii: $\overline{OP}, \overline{OQ}$; no diam. 4. chords: $\overline{VW}, \overline{YX}$; secants: $\overleftrightarrow{VW}, \overleftrightarrow{YX}$; tan.: $\overleftrightarrow{UT}$

Which triangles are congruent? Why?

5. Given: $\overline{BD}$ is a diameter of $\odot O$ and $\overline{AB} \cong \overline{CB}$.
   $\triangle AOB \cong \triangle COB$, SSS

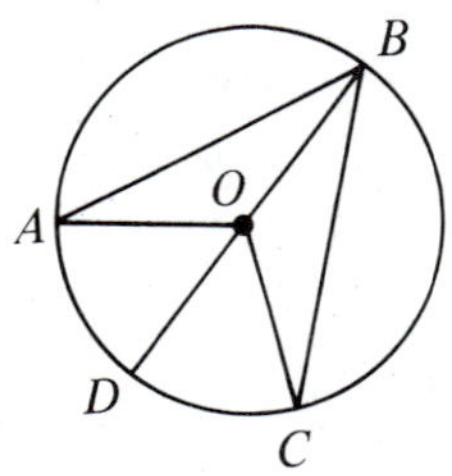

6. Given: $\overline{BD}$ is a diameter of $\odot O$ and $\angle AOB \cong \angle COB$.
   $\triangle AOB \cong \triangle COB$, SAS

In addition to the "given" in Exercises 7–16, $\odot O$ has $\overline{OP} \perp \overline{AB}$.

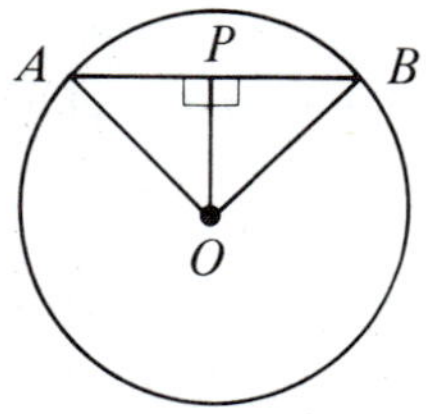

7. Given: $OP = 4$ and $OB = 5$
   Find AB. 6
8. Given: $AB = 10$ and $OB = 8$
   Find OP. $\sqrt{39}$
9. Given: $OP = 6$ and $m\angle ABO = 30$
   Find OB and AB. 12; $12\sqrt{3}$
10. Given: $OB = 10$ and $m\angle POB = 60$
    Find OP and AB. 5; $10\sqrt{3}$
11. Given: $OP = 7$ and $m\angle POB = 45$
    Find OB and AB. $7\sqrt{2}$; 14
12. Given: $OB = 8$ and $m\angle ABO = 45$
    Find OP and AB. $4\sqrt{2}$; $8\sqrt{2}$
13. Given: $AB = 18$ and $m\angle POB = 45$
    Find OP and OB. 9; $9\sqrt{2}$
14. Given: $AB = 16$ and $m\angle ABO = 30$
    Find OP and OB. $\frac{8\sqrt{3}}{3}$; $\frac{16\sqrt{3}}{3}$
15. Given: $AP = 3x + 4$ and $PB = 5x - 2$
    Find AB. 26
16. Given: $AO = x + 8$ and $OB = 2x - 1$
    Find AO and OB. 17; 17

17. Given: Radii $\overline{OA}$ and $\overline{OB}$
Prove: $\angle OAB \cong \angle OBA$

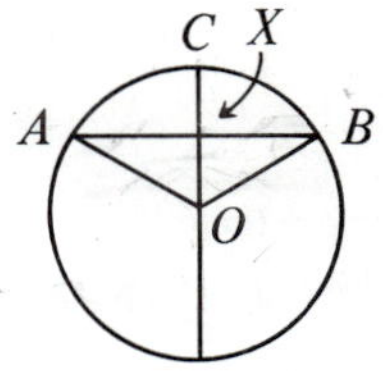

18. Given: $\angle AOC \cong \angle BOC$ and $\overline{OA}$ and $\overline{OB}$ are radii.
Prove: $\overline{OC}$ is the $\perp$ bisector of $\overline{AB}$.

B 19. Prove Theorem 11.1.

20. Prove: In a circle, a radius that bisects a chord which is not a diameter is perpendicular to the chord.

21. Given: Diameters $\overline{AC}$ and $\overline{BD}$ of $\odot O$
Prove: $\overline{AD} \cong \overline{BC}$

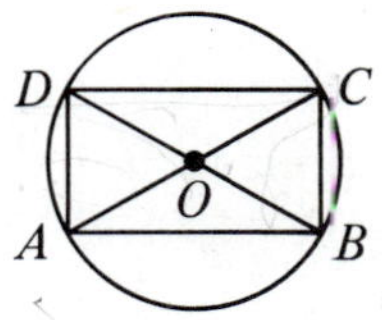

22. Given: Diameters $\overline{AC}$ and $\overline{BD}$ of $\odot O$
Prove: $ABCD$ is a rectangle.

23. Given: Diameter $\overline{XY}$ of $\odot O$, chord $\overline{RS} \parallel$ chord $\overline{TU}$, and $\overline{XY}$ bisects $\overline{RS}$ at M.
Prove: N is the midpoint of $\overline{TU}$.

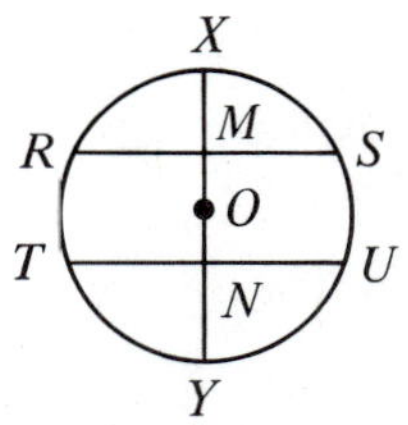

24. Given: $\overline{DC}$ is a diameter of $\odot O$. $\overline{DC} \perp$ chord $\overline{AB}$.
Prove: $\overline{AC} \cong \overline{BC}$

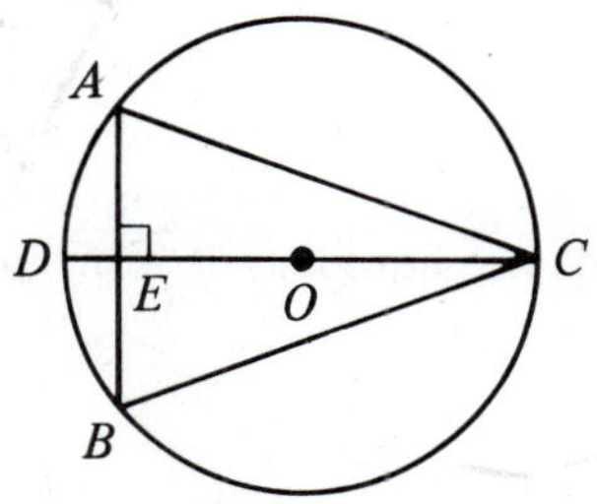

25. Given: Diameter $\overline{AB}$ bisects chords $\overline{CD}$ and $\overline{FG}$.
Prove: $\overline{CD} \parallel \overline{FG}$

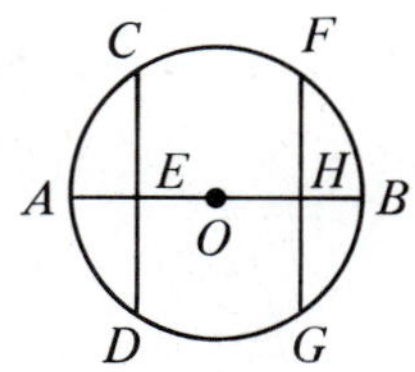

C 26. A chord is as long as its distance from the center of the circle. A second chord of the same circle is twice as long. How far is the second chord from the center?
The second chord is half as far from center as the first chord.

27. Prove: The perpendicular bisector of a chord of a circle in the same plane as the circle goes through the center of the circle.

28. Prove: If two chords of a circle have a common endpoint and form congruent angles with the radius drawn to that endpoint, then the chords are congruent.

29. Prove: Two points of a circle each equidistant from two other points of the circle determine a diameter of the circle.

CUMULATIVE REVIEW

1. Given: $\overline{RV} \perp \overline{ST}$ and $\overline{TU} \perp \overline{SR}$
Prove: $\triangle RSV \sim \triangle TSU$

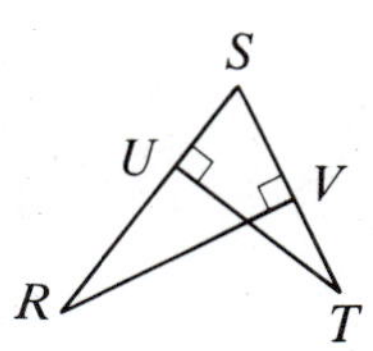

2. Given: $\overline{WZ} \parallel \overline{XY}$
Prove: $\triangle ZWP \sim \triangle XYP$

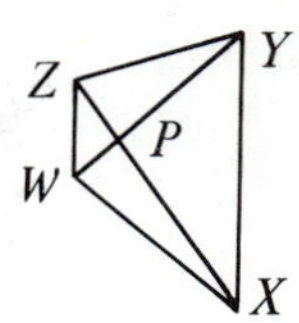

CONGRUENT CIRCLES 11.2

Objectives

To identify congruent circles
To prove and apply theorems about chords and radii in congruent circles

In general, congruent figures are figures that have the same size and shape. Congruent circles are defined more specifically.

Definition: Congruent circles

Congruent circles are circles that have congruent radii. $\odot O \cong \odot P$ is read "circle O is congruent to circle P."

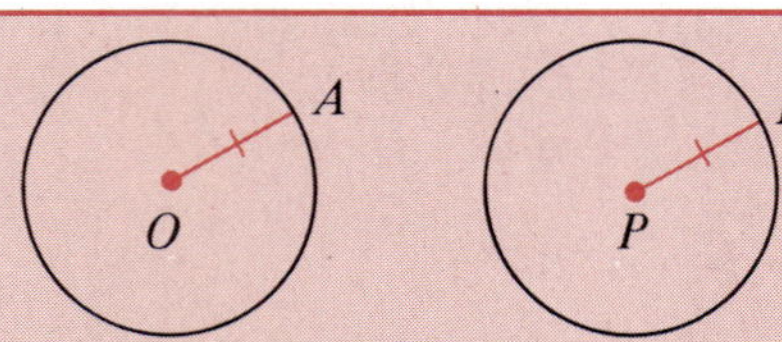

$\odot O \cong \odot P$

In the diagram above, circles O and P are congruent since $\overline{OA} \cong \overline{PB}$. Some theorems in this book apply either to one circle or to congruent circles. Proofs will be presented for congruent circles only.

Theorem 11.3 If chords of a circle or of congruent circles are congruent, then the chords are equidistant from the center(s).

Given: $\odot O \cong \odot P$,
$\overline{OC} \perp \overline{AB}$,
$\overline{PF} \perp \overline{DE}$, and
$\overline{AB} \cong \overline{DE}$
Prove: $OC = PF$

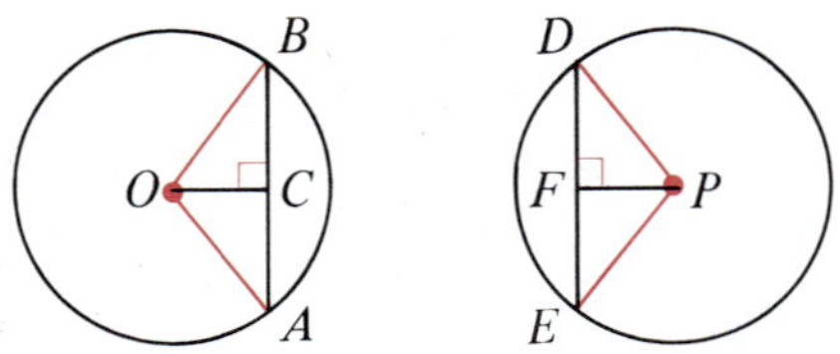

ANALYSIS
Show $\triangle AOB \cong \triangle EPD$. Then $\overline{OC}$ and $\overline{PF}$ are corr. altitudes of $\cong \triangle$'s.

Proof

Statements	Reasons
1. $\odot O \cong \odot P$, $\overline{OC} \perp \overline{AB}$, $\overline{PF} \perp \overline{DE}$, $\overline{AB} \cong \overline{DE}$	1. Given
2. $\overline{OA} \cong \overline{PE}$, $\overline{OB} \cong \overline{PD}$	2. Def. of $\cong$ circles
3. $\triangle AOB \cong \triangle EPD$	3. SSS for $\cong \triangle$'s
4. $\therefore \overline{OC} \cong \overline{PF}$ and $OC = PF$	4. Corr. alt. of $\cong \triangle$'s are $\cong$.

Example 1

Given: $\odot O$ with $JK = 12$, $MS = 6$, $\overline{OS} \perp \overline{MN}$, $\overline{OR} \perp \overline{JK}$, and $OR = 3$.
Find OS.

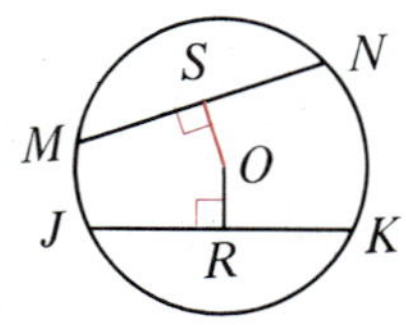

If $MS = 6$, then $MN = 12$ by Th. 11.2.
So $\overline{MN} \cong \overline{JK}$.
$\overline{OR} \cong \overline{OS}$ by Th. 11.3.
Thus, $OS = 3$.

The converse of Theorem 11.3 is stated below as Theorem 11.4.

Theorem 11.4

If chords of a circle or of congruent circles are equidistant from the center(s), then the chords are congruent.

Given: $\odot O \cong \odot P$,
$OC = PF$,
$\overline{OC} \perp \overline{AB}$, and
$\overline{PF} \perp \overline{DE}$
Prove: $\overline{AB} \cong \overline{DE}$

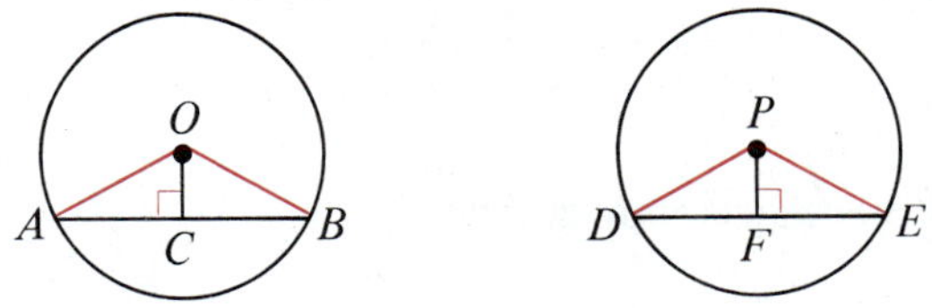

ANALYSIS

Show $\triangle OCA \cong \triangle PFD$. Then $\overline{AC} \cong \overline{DF}$.

Proof: (See Exercise 24.)

Recall that the distance from a point to a line is the length of the perpendicular segment from the point to the line. In Theorems 11.3 and 11.4, the term "equidistant" refers to equal lengths of the segments perpendicular to the chords.

Definitions: Inscribed polygon, circumscribed circle

An *inscribed polygon* is a polygon whose sides are chords of a circle. The polygon is *inscribed* in the circle, and the circle is *circumscribed* about the polygon.

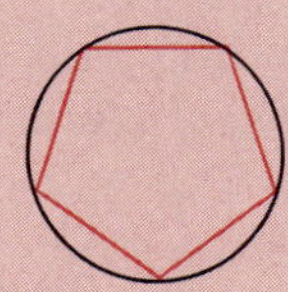

Notice that when a polygon is inscribed in a circle, each vertex of the polygon lies on the circle.

Reading in Geometry

Draw each figure described.

1. An equilateral triangle inscribed in a circle
2. A circle circumscribed about an isosceles trapezoid
3. A scalene triangle inscribed in a circle and circumscribed about another circle
4. A circle with two congruent chords that intersect in a point that is not an endpoint of the chords
5. A circle with two congruent chords that have a common endpoint.
6. Two congruent quadrilaterals inscribed in two congruent circles

Written Exercises

In each case, determine whether $\odot O \cong \odot P$.

A

1. $AO = 5$ cm, and $CP = 5$ cm ≅
2. $AO = 8$ cm, and $DP = 8$ cm ≅
3. $AB = 4$ cm, and $CD = 4$ cm ≅
4. $AB = 10$ cm, and $CP = 5$ cm ≅
5. $AB = 3$ km, and $CD = 3$m ≇
6. $AO = 12$ mm, and $CD = 2.4$ cm ≅
7. $OA = 8$ cm, and $CD = 1.6$ m ≇
8. $OB = 35$ cm, and $CD = 0.7$ m ≅

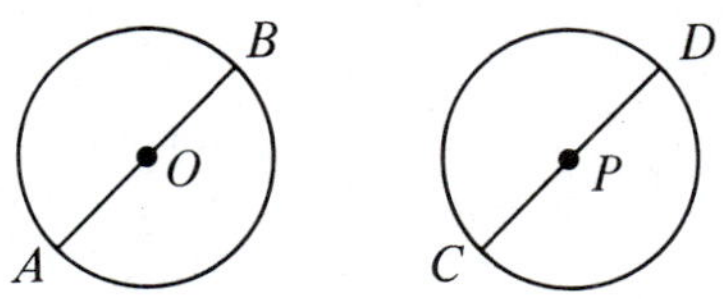

In $\odot O$, $\overline{OD} \perp \overline{AC}$ and $\overline{OE} \perp \overline{BC}$. Find the indicated measures.

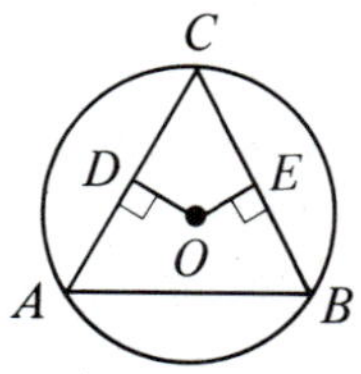

9. Given: $OD = 4$, $AC = 16$. and $OE = 4$ Find BC. **16**

10. Given: $AC = 16$, $BE = 8$, and $OE = 4$ Find OD. **4**

11. Given: $OD = 3$, $OE = 3$, and $m\angle C = 40$ Find $m\angle A$ and $m\angle B$. **70, 70**

12. Given: $m\angle A = 50$, $m\angle B = 50$, and $AC = 30$ Find CE. **15**

$\overline{XY}$ is a diameter of $\odot O$. Find the indicated measures.

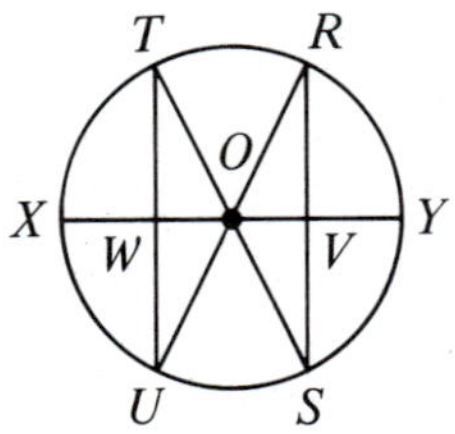

13. Given: $\overline{OY}$ bisects $\angle ROS$, $OR = 12$, $m\angle ROS = 120$, and $\overline{TU} \cong \overline{RS}$ Find WX. **6**

14. Given: $\overline{OY} \perp \overline{RS}$ at V, $\overline{OX} \perp \overline{TU}$ at W, $OV = 12$, $OS = 16$, and $\overline{TU} \cong \overline{RS}$ Find OW. **12**

15. Given: $\overline{OY} \perp \overline{RS}$ at V, $\overline{OX} \perp \overline{TU}$ at W, $\overline{TU} \cong \overline{RS}$, $OW = 5$, and $OT = 9$ Find VY. **4**

16. Given: $\overline{OY} \perp \overline{RS}$ at V, $\overline{OX} \perp \overline{TU}$ at W, $\overline{OW} \cong \overline{OV}$, $m\angle R = 45$, and $OS = 8$ Find TU. **$8\sqrt{2}$**

On the basis of the given information, which segments can you conclude are congruent? (Ex. 17–19)

B 17. Given: $\odot A \cong \odot B$, $\overline{RS} \cong \overline{TU}$, $\overline{AM} \perp \overline{RS}$, and $\overline{BN} \perp \overline{TU}$

18. Given: $\odot O$ with $\overline{OV} \cong \overline{OU}$

19. Given: Radii $\overline{OA} \cong \overline{QC}$, $\overline{OE} \perp \overline{AB}$, and $\overline{QF} \perp \overline{CD}$

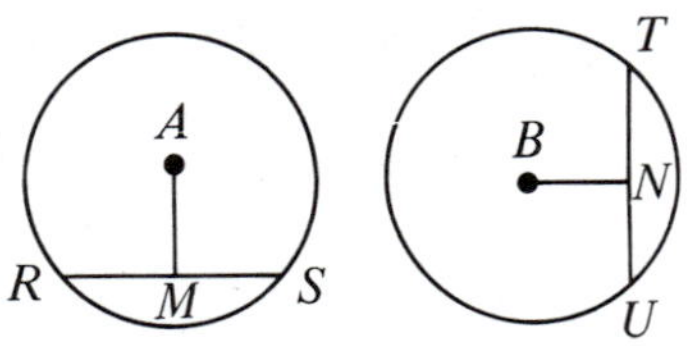

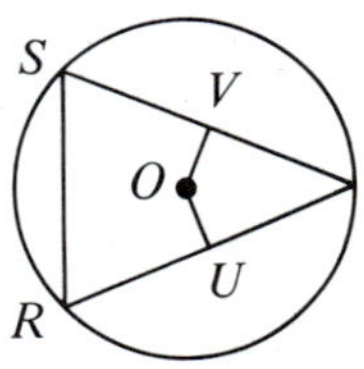

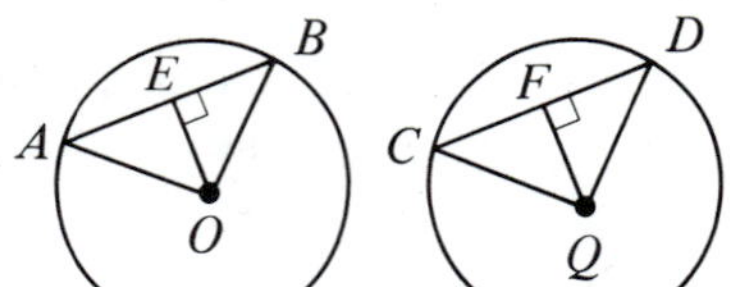

17. $\overline{AM} \cong \overline{BN}$; $\overline{RM} \cong \overline{MS} \cong \overline{UN} \cong \overline{NT}$ 18. none 19. $\overline{OA} \cong \overline{OB} \cong \overline{QC} \cong \overline{QD}$; $\overline{AE} \cong \overline{EB}$; $\overline{CF} \cong \overline{FD}$

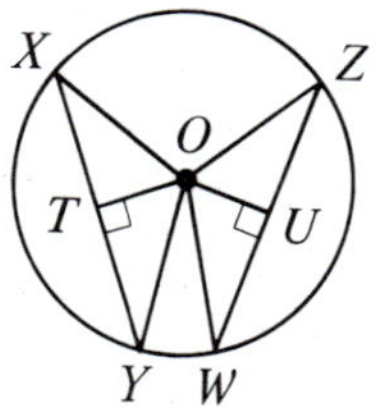

20. Given: $\odot O$ with $\overline{OT} \perp \overline{XY}$, $\overline{OU} \perp \overline{WZ}$, and $\overline{OT} \cong \overline{OU}$ Prove: $\triangle XOY \cong \triangle ZOW$

21. Given: $\odot O$ with $\overline{OT} \perp \overline{XY}$, $\overline{OU} \perp \overline{WZ}$, and $\angle XOY \cong \angle ZOW$ Prove: $\overline{OT} \cong \overline{OU}$

22. Given: $\odot O$ with $\overline{OT} \perp \overline{XY}$, $\overline{OU} \perp \overline{WZ}$, and $\angle X \cong \angle Z$ Prove: $\overline{XY} \cong \overline{WZ}$

23. Given: $\odot O$ with $\overline{OT} \perp \overline{XY}$, $\overline{OU} \perp \overline{WZ}$, and $\angle XOT \cong \angle ZOU$ Prove: $\overline{XY} \cong \overline{WZ}$

24. Prove Theorem 11.4.

Prove each statement.

C 25. If congruent chords are equidistant from the centers of circles, then the circles are congruent.

26. If a point on a circle is equidistant from two radii, then the radius from the point bisects the angle formed by the two given radii.

27. If the center of a circle is equidistant from the sides of an inscribed triangle, then the triangle is equilateral.

28. If a chord of a circle makes an angle of 30° with a diameter, then the distance from the center to the chord is one fourth the length of the diameter.

PROPERTIES OF TANGENTS 11.3

Objective **To prove and apply theorems about tangents to circles**

A circle separates the plane into three sets of points: the circle itself, the interior of the circle, and the exterior of the circle. The exterior of the circle is the set of all points in the plane of the circle whose distances from the center are greater than the length of a radius. Thus, for point Q in the exterior of $\odot O$, $OQ > OR$, where $\overline{OR}$ is a radius.

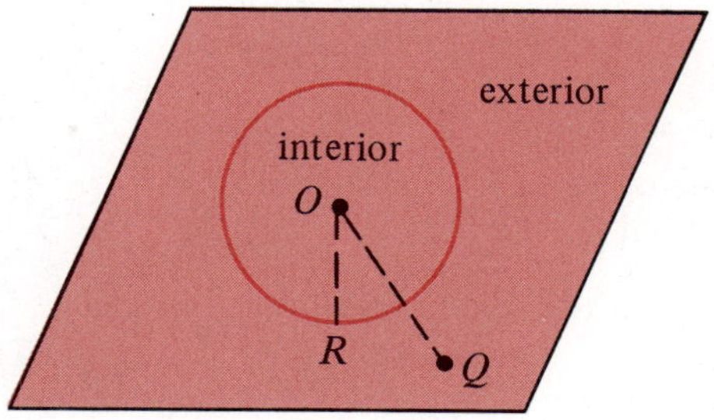

A **tangent** to a circle is a line, coplanar with the circle, that intersects the circle in exactly one point. The point is called the **point of tangency**. In the diagram at the right, $\overleftrightarrow{TB}$ intersects $\odot P$ at point T, so T is the point of tangency.

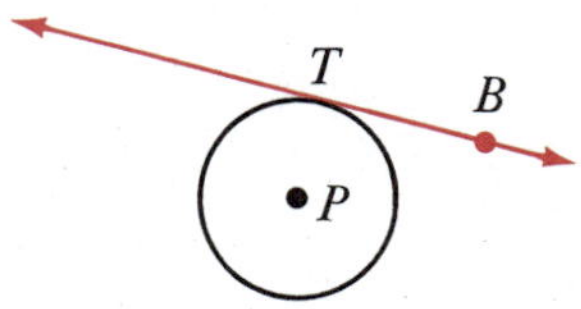

Theorem 11.5 If a line is tangent to a circle, then it is perpendicular to the radius drawn to the point of tangency.

Given: l is tangent to $\odot O$ at A.
Prove: $\overline{OA} \perp l$

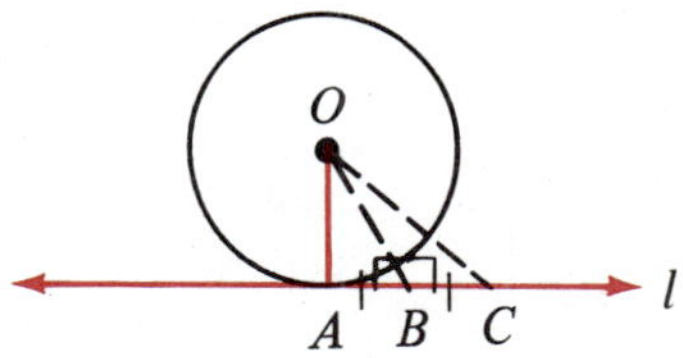

ANALYSIS
Use an indirect proof. Assume $\overline{OA} \not\perp l$. Then some segment such as $\overline{OB} \perp l$. This leads to a contradiction.

Outline of Proof

Assume $\overline{OA} \not\perp l$. Then there is a segment $\overline{OB}$ such that $\overline{OB} \perp l$. On the ray opposite $\overrightarrow{BA}$, choose a point C such that $\overline{BC} \cong \overline{BA}$. $\overline{OB} \cong \overline{OB}$ and rt. $\angle OBA \cong$ rt. $\angle OBC$, so $\triangle OBA \cong \triangle OBC$. Thus, $\overline{OC} \cong \overline{OA}$ and $\overline{OC}$ must also be a radius. Then both points A and C lie on $\odot O$. This contradicts that l is a tangent.

Theorem 11.6 If a line is perpendicular to a radius at its endpoint on the circle, then the line is tangent to the circle.

Given: $\odot O$ with $\overline{OD} \perp l$ at D.
Prove: l is tangent to $\odot O$ at D.

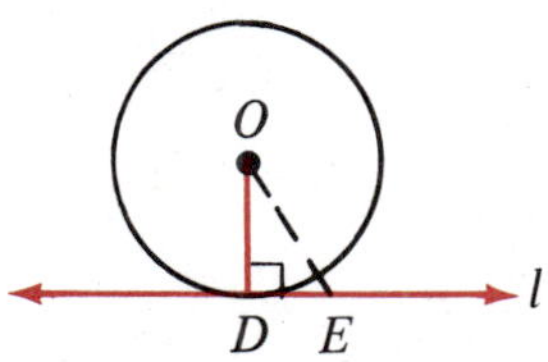

ANALYSIS
Let E be any point on l other than D. Show E is in the exterior of $\odot O$.

Proof: (See Exercise 16.)

A **tangent segment** is a segment that joins the point of tangency and another point on the tangent. In Example 1 below, $\overline{PT}$ is a tangent segment.

Example 1

Given: $\overline{PT}$ tangent to $\odot O$;
$\overline{OT}$ is a radius, $OT = 6$, and $PO = 9$.
Find PT.

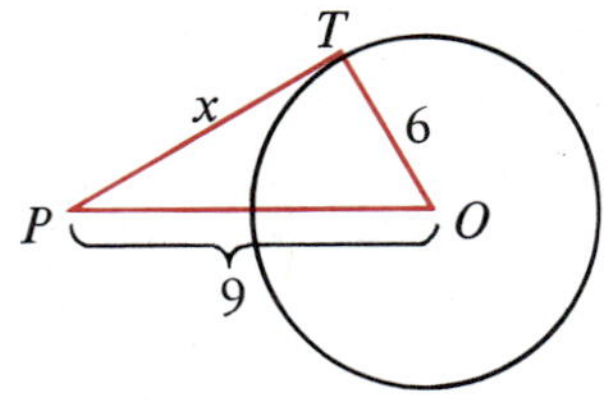

$\triangle OPT$ is a rt. $\triangle$ by Th. 11.5.

$$x^2 + 6^2 = 9^2$$
$$x^2 + 36 = 81$$
$$x^2 = 45$$
$$x = \sqrt{45}, \text{ or } 3\sqrt{5}$$

Thus, PT is $3\sqrt{5}$.

As indicated in Theorem 11.7, two tangent segments may be congruent.

Theorem 11.7

Tangent segments to a circle from the same point are congruent.

Given: $\odot O$ with tangent segments $\overline{BA}$ and $\overline{BC}$
Prove: $\overline{BA} \cong \overline{BC}$

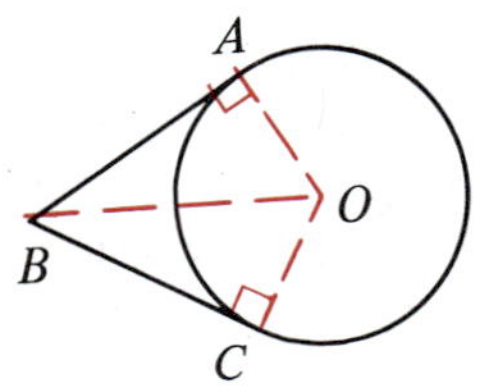

ANALYSIS

Draw radii $\overline{OA}$ and $\overline{OC}$. Draw $\overline{OB}$. Show $\triangle OAB \cong \triangle OCB$.

Proof

Statements	**Reasons**
1. $\odot O$ with tangent segments $\overline{BA}$ and $\overline{BC}$	1. Given
2. Draw radii $\overline{OA}$ and $\overline{OC}$. Draw $\overline{OB}$.	2. Two points are contained in exactly one line.
3. $\overline{OA} \perp \overline{BA}$ and $\overline{OC} \perp \overline{BC}$	3. Radius drawn to tangency pt. is $\perp$.
4. $\triangle AOB$ and $\triangle COB$ are rt. $\triangle$'s.	4. Def. of $\perp$ and def. of rt. $\triangle$
5. $\overline{BO} \cong \overline{BO}$	5. Reflexive property
6. $\overline{OA} \cong \overline{OC}$	6. All radii of a $\odot$ are $\cong$.
7. $\triangle AOB \cong \triangle COB$	7. *HL* for $\cong$ rt. $\triangle$'s
8. $\therefore \overline{BA} \cong \overline{BC}$	8. CPCTC

If a circle is *inscribed* in a polygon, then the sides of the polygon form tangent segments.

Definitions: Circumscribed polygon, inscribed circle

A *circumscribed polygon* is a polygon whose sides are tangent to a circle. The polygon is *circumscribed* about the circle, and the circle is *inscribed* in the polygon.

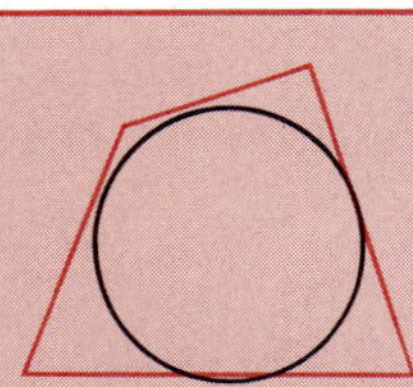

Example 2 Given: $\odot O$ with circumscribed $\triangle ABC$ and chord $\overline{DE}$; m $\angle ADE = 6t - 2$ and m $\angle AED = 4t + 22$.
Find m $\angle A$.

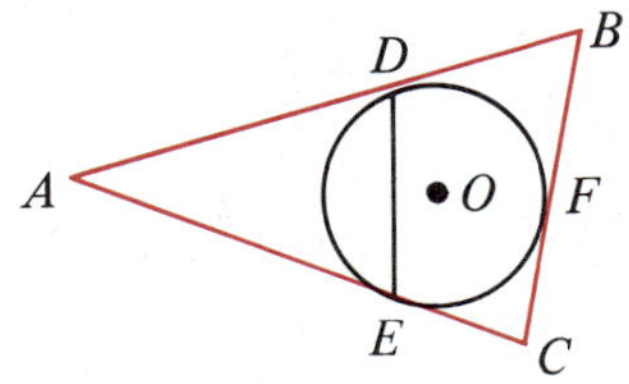

$\overline{AD} \cong \overline{AE}$ by Th. 11.7
So, $\angle ADE \cong \angle AED$. ◄ *∠'s opp. ≅ sides of △* are ≅.
$6t - 2 = 4t + 22$
$t = 12$
m $\angle ADE = 6(12) - 2 = 70$; m $\angle AED = 4(12) + 22 = 70$
Thus, m $\angle A = 180 - (70 + 70)$, or 40.

It is possible for a line to be tangent to two circles.

Definition: Common tangent

A *common tangent* is a line that is tangent to each of two coplanar circles.

A common tangent can be tangent either internally or externally. A **common internal tangent** is a common tangent that intersects the segment that joins the center of the two circles. A **common external tangent** is a common tangent that does not intersect the segment that joins the centers of the two circles. In the diagram below, l is a common internal tangent while m is a common external tangent.

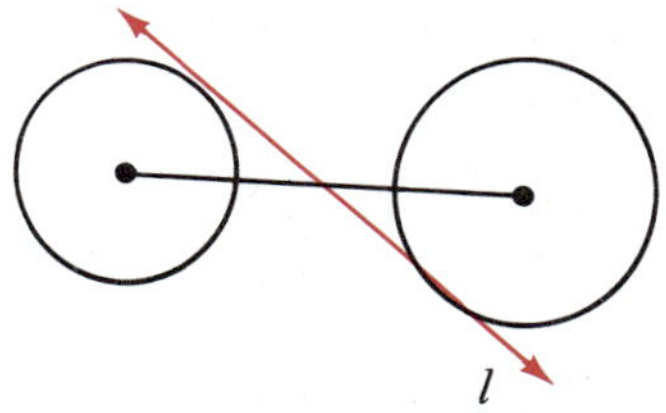

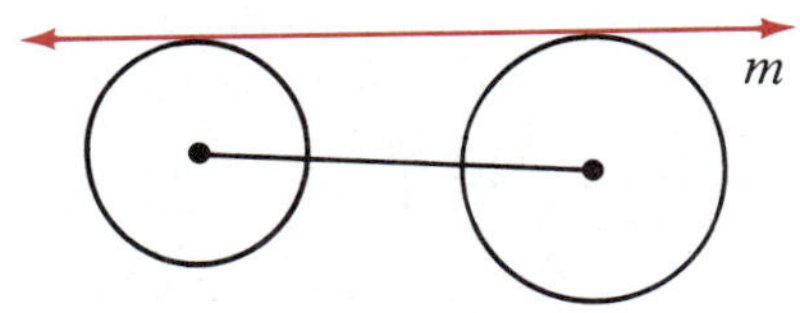

Two circles may also be tangent to each other.

Definition: Tangent circles

Two coplanar *circles* are *tangent* to each other if they are tangent to the same line at the same point.

Tangent circles may be tangent either internally or externally. Tangent circles are **tangent internally** if one lies in the interior of the other (except for the point of tangency). Tangent circles are **tangent externally** if each lies in the exterior of the other (except for the point of tangency). In the diagram below, $\odot P$ and $\odot Q$ are tangent internally while $\odot R$ and $\odot S$ are tangent externally.

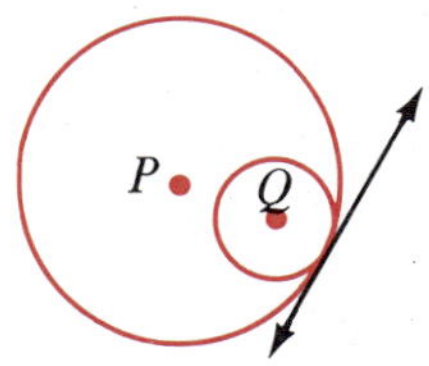

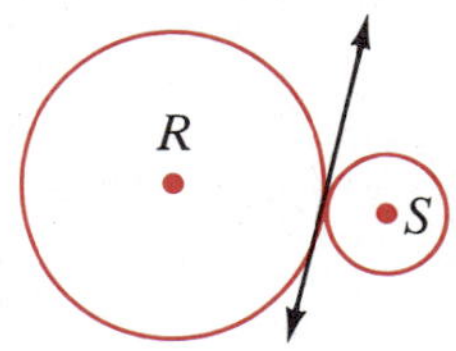

Example 3

Given two circles that are tangent externally, how many common internal tangents and how many common external tangents can be drawn?

Draw two circles that are tangent externally. Then draw all possible common tangents. ►

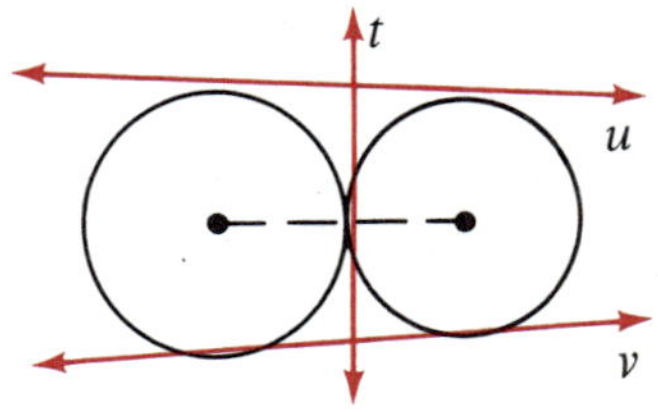

Thus, one common internal tangent, t, and two common external tangents, u and v, can be drawn.

Example 4

Given: $\odot O \cong \odot P$. $\odot O$ and $\odot P$ are tangent to $\overleftrightarrow{ST}$ at T.
Prove: $\overline{OS} \cong \overline{PS}$

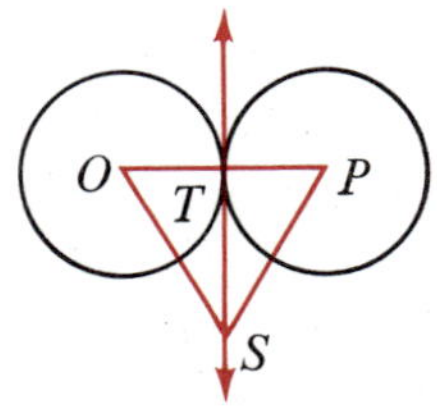

Proof

Statements	**Reasons**
1. $\odot O \cong \odot P$; $\odot O$ and $\odot P$ are tangent to $\overleftrightarrow{ST}$ at T.	1. Given
2. $\overline{OT} \perp \overleftrightarrow{ST}$, $\overline{PT} \perp \overleftrightarrow{ST}$	2. Radius drawn to pt. of tangency is $\perp$.
3. $\angle OTS \cong \angle PTS$	3. $\perp$'s form $\cong$ rt. $\angle$'s.
4. $\overline{OT} \cong \overline{PT}$	4. Radii of $\cong$ $\odot$'s are $\cong$.
5. $\overline{ST} \cong \overline{ST}$	5. Reflexive property
6. $\triangle OTS \cong \triangle PTS$	6. SAS for $\cong$ $\triangle$'s
7. $\therefore \overline{OS} \cong \overline{PS}$	7. CPCTC

Reading in Geometry

Draw each pair of coplanar circles as indicated below. Then tell (a) how many common internal tangents and (b) how many common external tangents can be drawn for each picture.

1. Two circles which are tangent internally **0; 1**
2. Two circles which intersect in two points **0; 2**
3. Two non-intersecting circles, one in the interior of the other **0; 0**
4. Two non-intersecting circles, each in the exterior of the other **2; 2**

Oral Exercises

Tell whether each pair of circles is tangent internally, tangent externally, or neither. Then tell whether the line is a common internal tangent, a common external tangent, or neither.

1.

2.

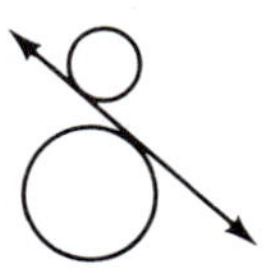

3.

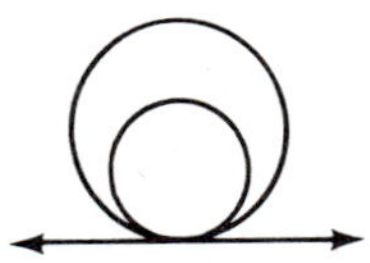

4.

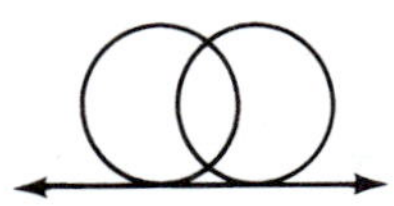

1. Ext.; common int. **2. Neither; common int.** **3. Int.; common ext.** **4. Neither; common ext.**

Tell whether the circle is inscribed in the polygon, the polygon is inscribed in the circle, or neither.

5.

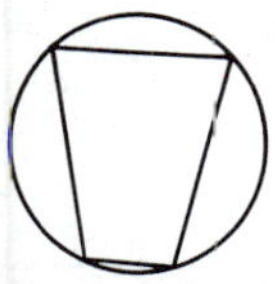

Poly. inscr. in ⊙.

6.

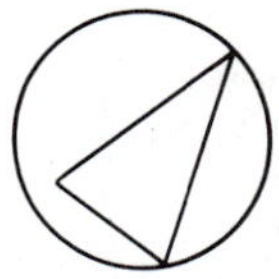

Neither

7.

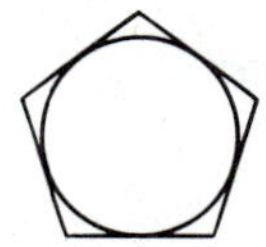

⊙ inscr. in poly.

8.

Neither

Written Exercises

A

1. Given: Tangents $\overrightarrow{PR}$, $\overrightarrow{PS}$, and $\overrightarrow{PT}$, and $PR = 11$
Find PT. **11**

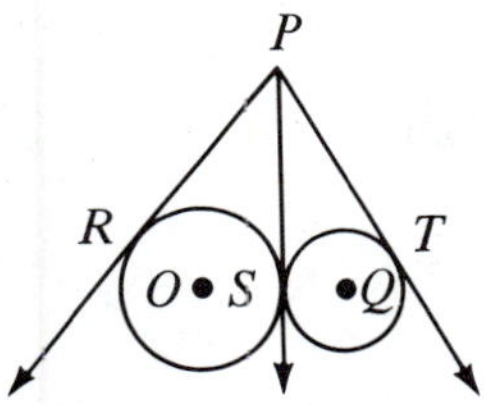

2. Given: Circumscribed $\triangle ABC$, $AD = 4$, $DB = 6$, and $CE = 2$
Find $AB + BC + CA$. **24**

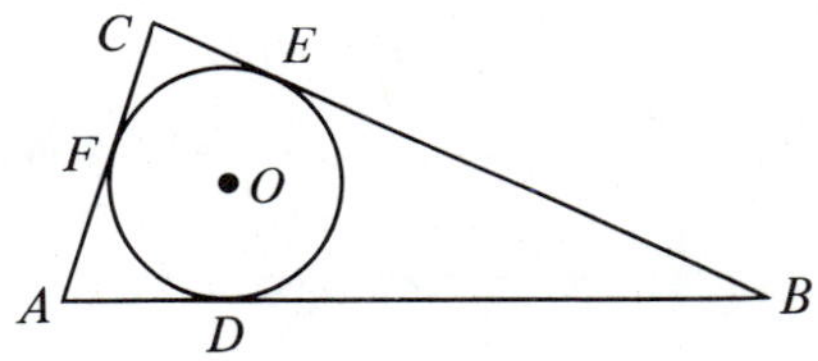

3. Given: $\overrightarrow{PH}$ and $\overrightarrow{PJ}$ tangent to ⊙O and ⊙Q, $IH = 7$, and $PI = 10$
Find PJ. **17**

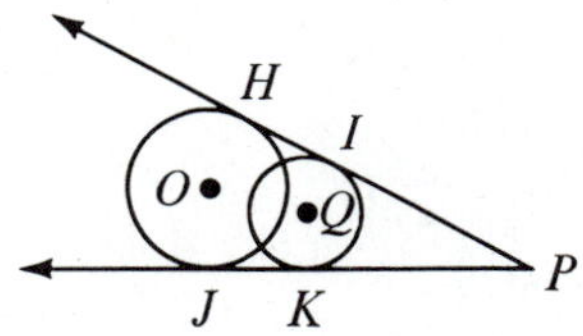

4. Given: Radii $\overline{OA}$ and $\overline{OC}$, tangents $\overleftrightarrow{AP}$ and $\overleftrightarrow{CP}$, and m $\angle AOC = 150$
Find m $\angle APC$. **30**

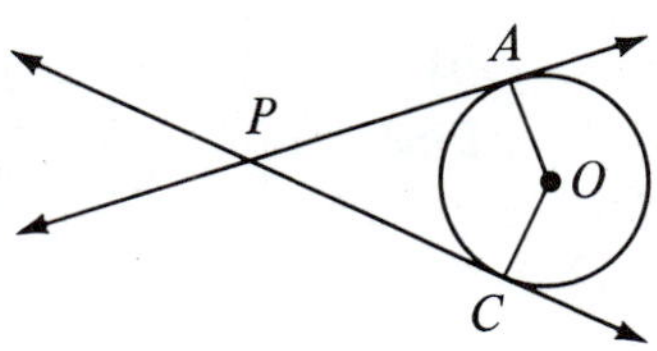

5. Given: Circumscribed polygon $ACEG$, $AB = 5$, $CD = 4$, $DE = 3$, and $FG = 2$
Find $AC + CE + EG + GA$. **28**

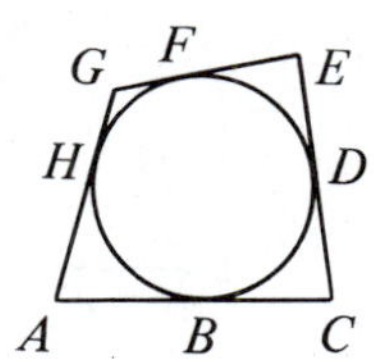

6. Given: Diameter $\overline{XY}$, tangent $\overleftrightarrow{WZ}$, $XY = 10$, and $WZ = 12$
Find YZ. **8**

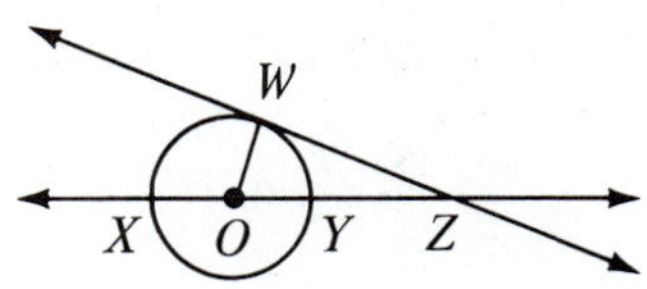

Which segments are congruent? Which angles are congruent? (Ex. 7–9)

7. Given: $\overleftrightarrow{PQ}$ tangent to ⊙O at Q, $\overleftrightarrow{PR}$ tangent to ⊙O and ⊙T at R, $\overleftrightarrow{PS}$ tangent to ⊙T at S

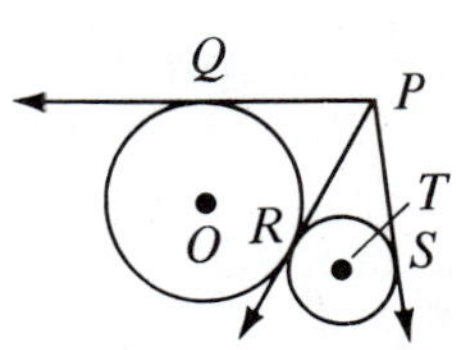

8. Given: $\overleftrightarrow{AB}$ tangent to ⊙O at A and ⊙P at B, $\overleftrightarrow{CD}$ tangent to ⊙O at C and ⊙P at D

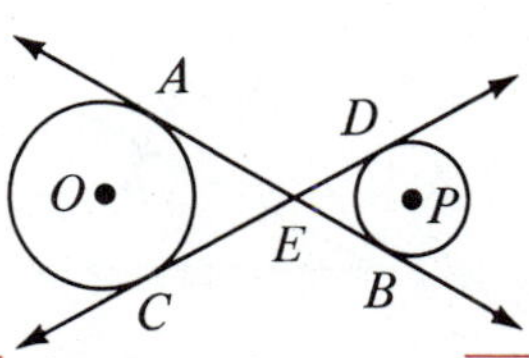

9. Given: $\overline{FI}$ is a diameter of ⊙O, $\overleftrightarrow{EG}$ tangent to ⊙O at F, $\overleftrightarrow{HJ}$ tangent to ⊙O at I

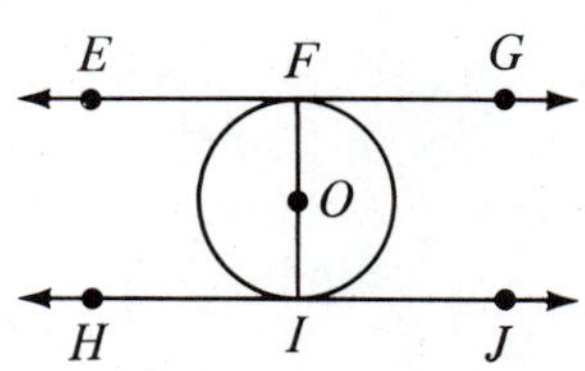

$\overline{QP} \cong \overline{RP} \cong \overline{SP}$; no $\cong \angle$'s

$\overline{EA} \cong \overline{EC}$, $\overline{ED} \cong \overline{EB}$; $\angle AEC \cong \angle DEB$

$\overline{FO} \cong \overline{IO}$; $\angle EFO \cong \angle GFO \cong \angle HIO \cong \angle JIO$

B

10. Given: $\overrightarrow{TB}$ tangent to $\odot O$ at B, and $\overrightarrow{TC}$ tangent to $\odot O$ at C.
Prove: $\overrightarrow{TO}$ bisects $\angle BTC$.

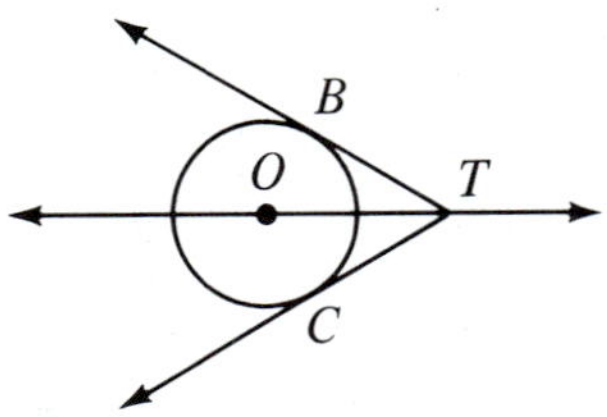

11. Given: Two circles with same center O, and chord $\overline{AB}$ of outer circle tangent to inner circle at X
Prove: $\overline{AX} \cong \overline{BX}$

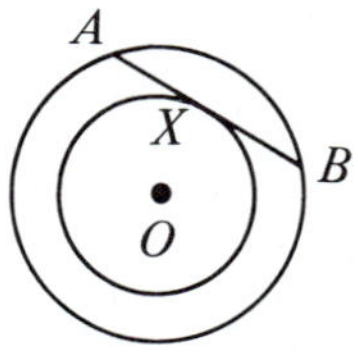

12. Given: $\overleftrightarrow{TC}$, $\overleftrightarrow{PD}$, $\overleftrightarrow{TP}$ tangent to $\odot O$ at C, D, and E respectively; $\overleftrightarrow{TC} \parallel \overleftrightarrow{PD}$.
Prove: $\overline{TO} \perp \overline{PO}$

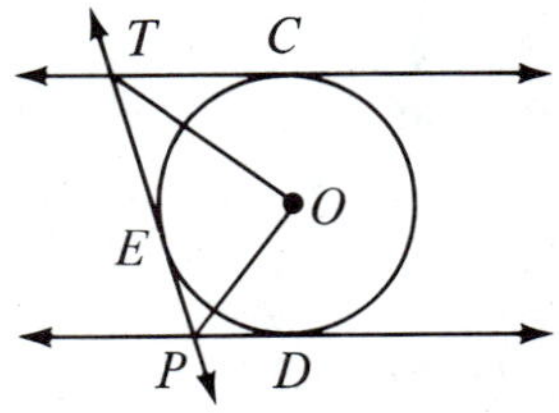

C

13. Given: $\overrightarrow{PA}$ tangent to $\odot O$ at A and $\overrightarrow{PB}$ tangent to $\odot O$ at B
Prove: $m\angle P = 2 \cdot m\angle OAB$

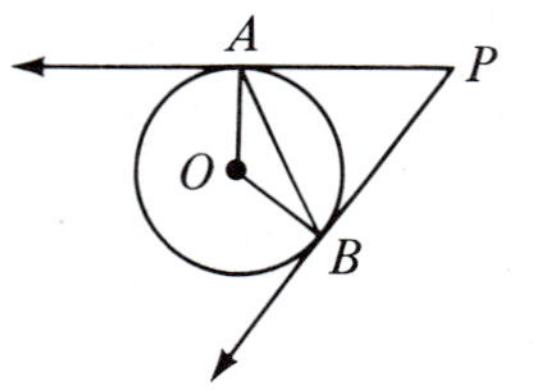

14. Given: $\overleftrightarrow{RS}$ tangent to $\odot O$ at R, $\overleftrightarrow{TU}$ tangent to $\odot O$ at T, $\overleftrightarrow{SU}$ tangent to $\odot O$ at V, and $\overline{OS} \perp \overline{OU}$
Prove: $\overleftrightarrow{RS} \parallel \overleftrightarrow{TU}$

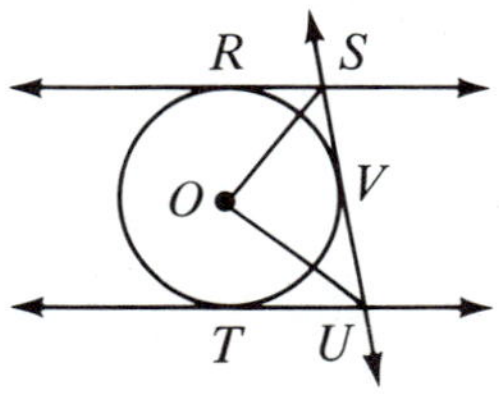

15. Given: Radius $\overline{OA} \perp$ radius $\overline{OB}$ of $\odot O$, chord $\overline{AD}$, $\overleftrightarrow{DE}$ tangent at D, and radius $\overline{OD}$
Prove: $\angle ECD \cong \angle EDC$

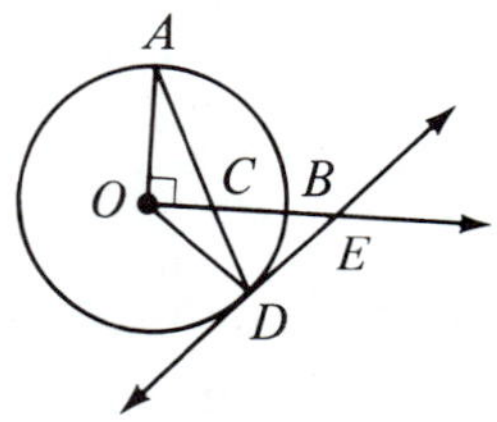

16. Prove Theorem 11.6.

17. Given: Circumscribed $\triangle XYZ$, $XY = 7$, $YZ = 8$, and $ZX = 5$
Find AX, BY, and CZ.

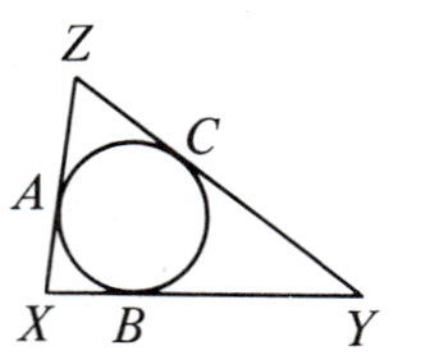

AX = 2; BY = 5; CZ = 3

18. Given: Circumscribed polygon $ABCD$, $AB = 21$, $BC = 24$, and $CD = 30$
Find AD.

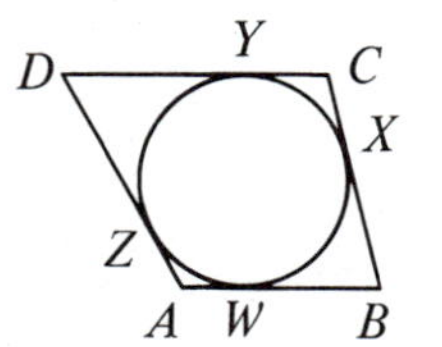

AD = 27

19. Given: $\overleftrightarrow{XY}$ tangent to $\odot O$ and $\odot P$, $OX = 16$, $PY = 6$, and $OP = 26$
Find XY.

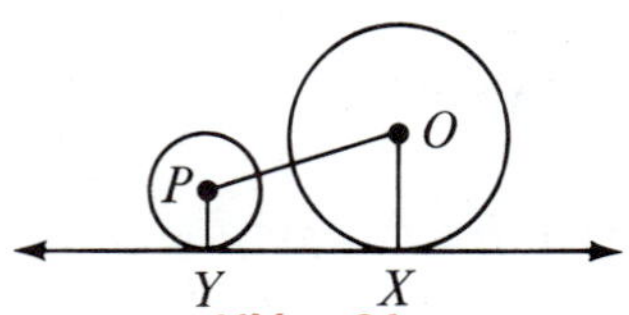

XY = 24

CUMULATIVE REVIEW

1. Given: $m\angle BOD = b$, $m\angle AOC = a$, and $\overrightarrow{OB}$ bisects $\angle AOC$
Find $m\angle AOD$.

$b + \frac{1}{2}a$

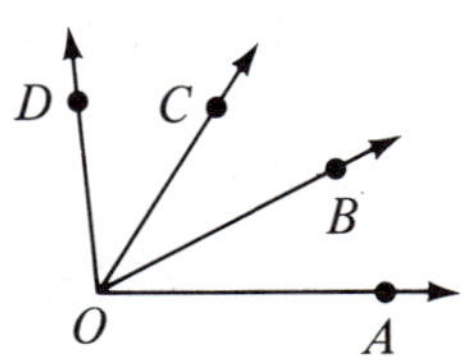

2. Given: $m\angle AOD = d$, $m\angle AOB = b$, and $\overrightarrow{OC}$ bisects $\angle DOB$
Find $m\angle AOC$.

$\frac{b + d}{2}$

MEASURES OF ARCS 11.4

Objectives **To find degree measures of minor arcs, major arcs, and semicircles**
To identify concentric circles

In this lesson, only degree measures of arcs are discussed. In Chapter 12, lengths of arcs will be discussed.

Definitions: Arc central angle

An *arc* is part of a circle. $\widehat{AB}$ is an arc of circle O.
A *central angle* of a circle is an angle whose vertex is the center of the circle. $\angle AOB$ is a central angle of $\odot O$.

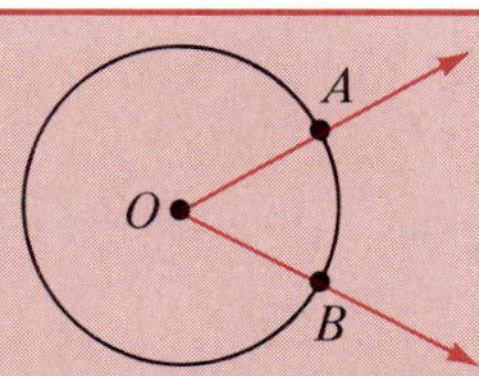

Different types of arcs are shown below. Their definitions follow.

Minor arc

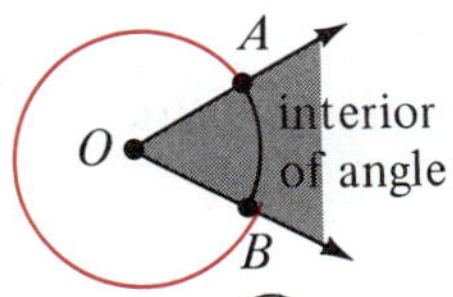

Minor $\widehat{AB}$
Read: Minor arc AB

Major arc

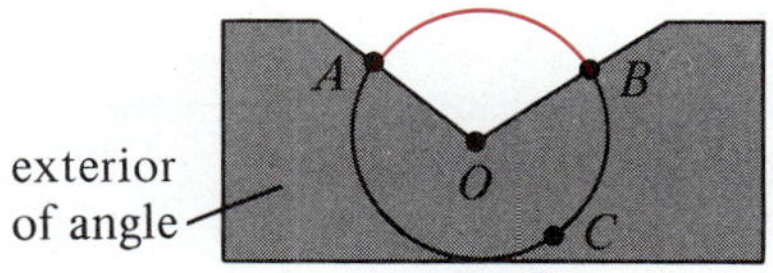

Major $\widehat{ACB}$
Read: Major arc ACB

Semicircle

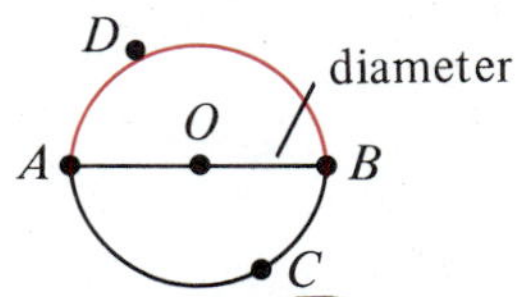

Semicircle $\widehat{ACB}$
Read: Semicircle ACB

Definitions: Minor arc major arc semicircle

Minor $\widehat{AB}$ consists of points A and B and all points of $\odot O$ that are in the interior of central $\angle AOB$.
Major $\widehat{ACB}$ consists of points A and B and all points of $\odot O$ that are in the exterior of central $\angle AOB$.
Semicircle $\widehat{ACB}$ *(or* $\widehat{ADB}$*)* consists of endpoints A and B of diameter $\overline{AB}$ and all points of $\odot O$ that lie on one side of $\overline{AB}$.

The degree measures of arcs are defined in terms of central angles.

Definitions: Degree measure of: minor arc major arc semicircle

The *degree measure of minor* $\widehat{AB}$ is equal to the degree measure of central $\angle AOB$.
The *degree measure of major* $\widehat{ACB}$ is equal to 360 minus the degree measure of central $\angle AOB$.
The *degree measure of semicircle* $\widehat{ACB}$ is equal to 180.

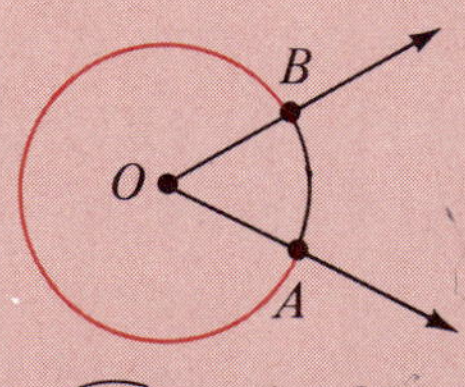

$m\,\widehat{AB} = m\angle AOB$

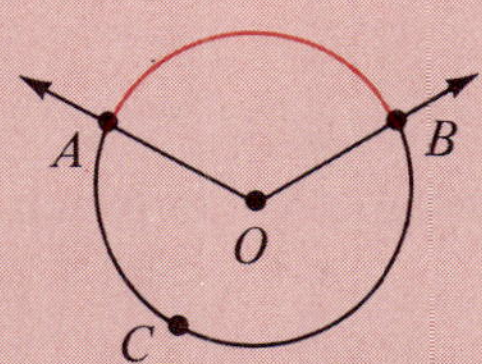

$m\,\widehat{ACB} = 360 - m\angle AOB$

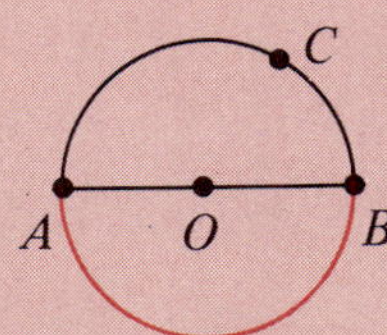

m semicircle $\widehat{ACB} = 180$

Recall that the sum of the measures of the angles on one side of a line is 180°. This, along with the definition of *degree measure* can be used to show that the degree measure of a circle is equal to 360°.

Example 1

Given: m $\angle AOB = 32$
Find m$\widehat{AB}$.

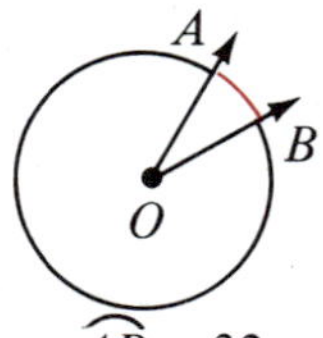

m $\widehat{AB} = 32$

Given: m $\angle QOR = 27$
Find m$\widehat{QR}$.

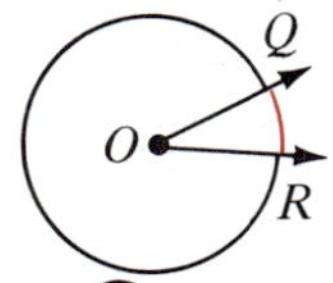

m $\widehat{QR} = 27$

Given: m $\angle BOA = 160$
Find m$\widehat{BCA}$.

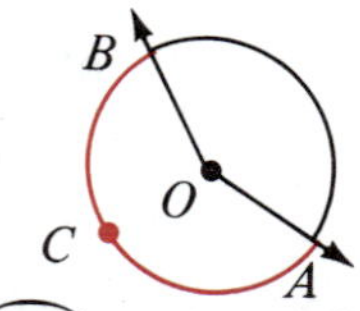

m $\widehat{BCA} = 360 - 160$, or 200

The Arc Addition Postulate stated below applies to major arcs, minor arcs, and semicircles.

Postulate 23 Arc Addition Postulate

If P is a point on $\widehat{AB}$, then m$\widehat{AP}$ + m$\widehat{PB}$ = m$\widehat{APB}$.

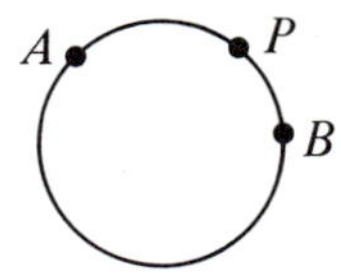

In Example 2, the Arc Addition Postulate is used to add the measures of two arcs that form a circle.

Example 2

Given: m$\widehat{AB} = 62$ and m$\widehat{BC} = 141$
Find m$\widehat{ABC}$ and m$\widehat{ADC}$.

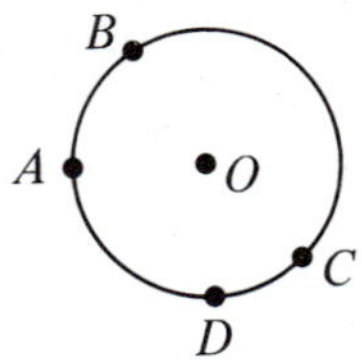

m$\widehat{ABC} = 62 + 141$, or 203 by Post. 23.

m$\widehat{ADC}$ + m$\widehat{ABC}$ = *360* ► m$\widehat{ADC} = 360 -$ m$\widehat{ABC}$
$= 360 - 203$, or 157

Thus, m$\widehat{ABC} = 203$ and m$\widehat{ADC} = 157$.

If you toss a stone into a lake, the waves that are formed represent a model of concentric circles.

Definition: Concentric circles

Concentric circles are coplanar circles with a common center.

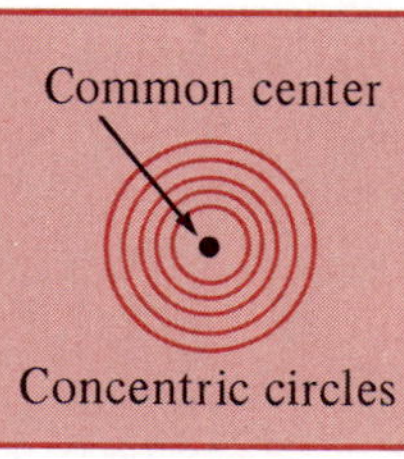

Written Exercises

Are the circles concentric circles?

A **1.** 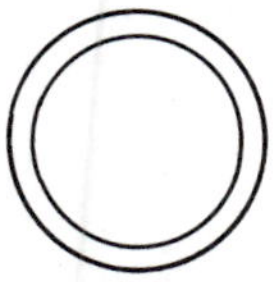
yes

2. 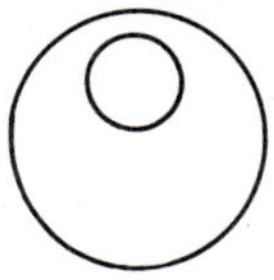
no

3. 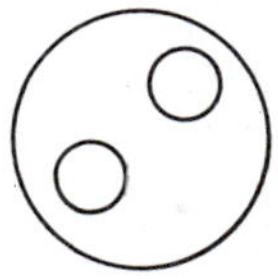
no

4. 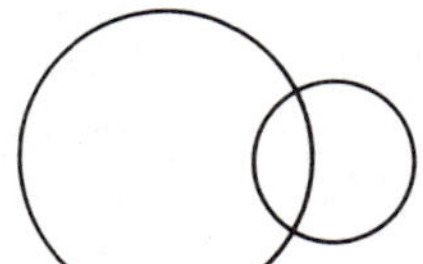
no

Find each indicated arc measure.

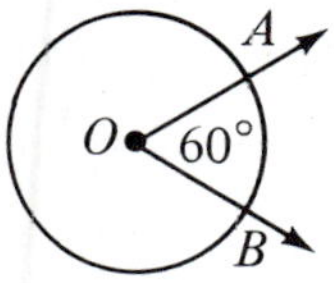

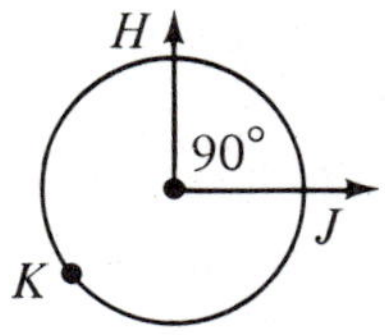

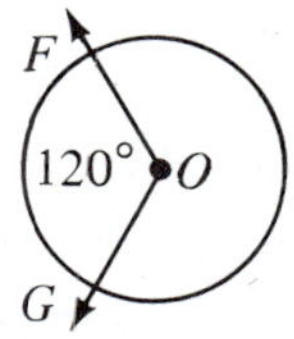

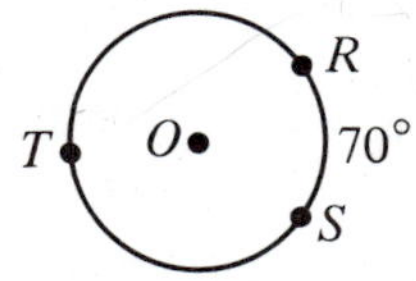

5. $m\overset{\frown}{AB} = ?$ 60 **6.** $m\overset{\frown}{HKJ} = ?$ 270 **7.** $m\overset{\frown}{FG} = ?$ 120 **8.** $m\overset{\frown}{RTS} = ?$ 290

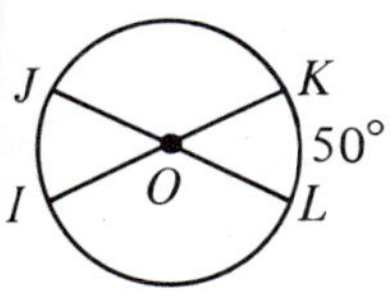

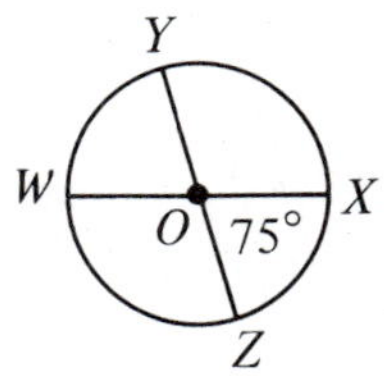

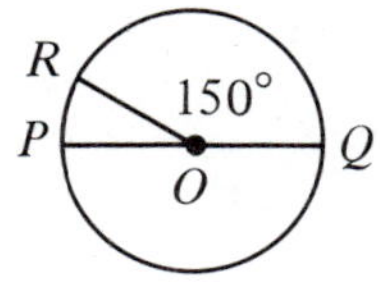

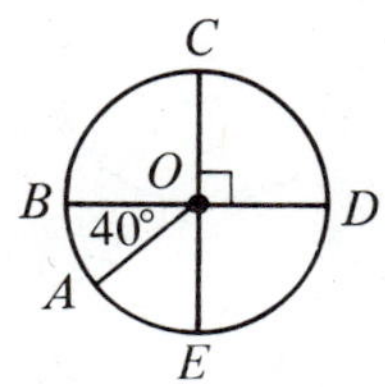

9. $m\overset{\frown}{IJ} = ?$ 50 **10.** $m\overset{\frown}{WZ} = ?$ 105 **11.** $m\overset{\frown}{PR} = ?$ 30 **12.** $m\overset{\frown}{AB} = ?$ 40

For Exercises 13–15, use ⊙*O* with diameter $\overline{BD}$.

$\overset{\frown}{AB}$; $\overset{\frown}{BC}$; $\overset{\frown}{DC}$; $\overset{\frown}{DA}$; $\overset{\frown}{AC}$

13. Identify all minor arcs of ⊙*O*.
14. Identify all major arcs of ⊙*O*.
15. Identify all semicircles of ⊙*O*. $\overset{\frown}{DAB}$; $\overset{\frown}{DCB}$

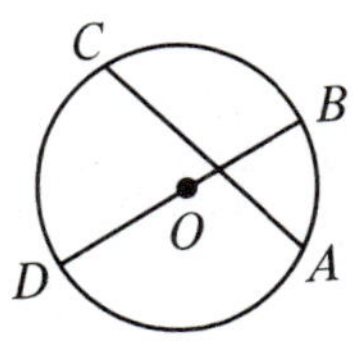

14. $\overset{\frown}{DAC}$; $\overset{\frown}{CAB}$; $\overset{\frown}{CDA}$; $\overset{\frown}{ADB}$; $\overset{\frown}{DBA}$

For Exercises 16–17, use ⊙*Q*.

16. Given: $\overline{UX}$ is a diameter of ⊙*Q*, $m\overset{\frown}{UV} = 62$, and $m\overset{\frown}{WX} = 42$. Find $m\overset{\frown}{VW}$. 76
17. Given: $m\overset{\frown}{TU} = 72$, $m\overset{\frown}{UV} = 48$, $m\overset{\frown}{VW} = 60$, $m\overset{\frown}{WX} = 72$, $m\overset{\frown}{XY} = 56$, and $m\overset{\frown}{YT} = 52$. Which chords (if any) are diameters? $\overline{UX}$; $\overline{TW}$

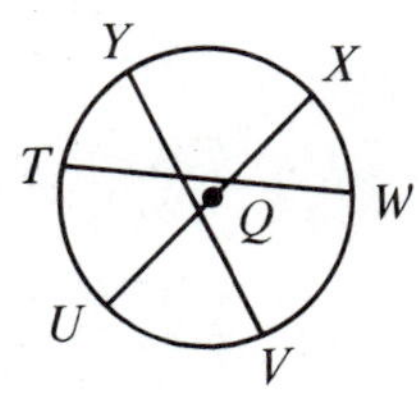

For Exercises 18–21, use ⊙*O* with diameters $\overline{AC}$ and $\overline{BD}$ and $m\overset{\frown}{DC} = 79$.

18. Find $m\angle AOD$. 101
19. Find $m\angle AOB$. 79
20. Find $m\overset{\frown}{ADC}$. 180
21. Find $m\overset{\frown}{BC}$. 101

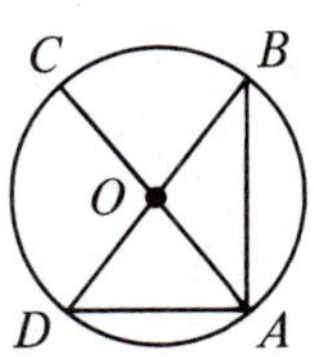

For Exercises 22–25, use ⊙*P* with diameters $\overline{RT}$ and $\overline{SU}$ and $m\angle RPU = 143$.

22. Find $m\angle SPT$. 143
23. Find $m\overset{\frown}{RU}$. 143
24. Find $m\angle RPS$. 37
25. Find $m\angle RSP$. 71.5

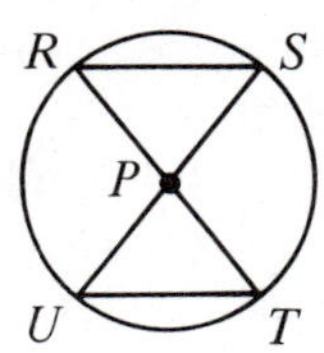

Identify the contradictory information. (Ex. 26–27)

B **26.** $\overline{GJ}$ and $\overline{HK}$ are diameters. $m\overset{\frown}{GL} = 84$, $m\overset{\frown}{LK} = 34$, $m\overset{\frown}{KJ} = 62$, $m\overset{\frown}{JI} = 27$, $m\overset{\frown}{IH} = 79$, and $m\overset{\frown}{HG} = 74$.
HK is not a diameter.

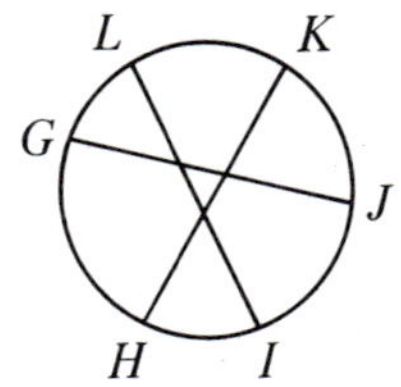

27. $\overline{GJ}$ is a diameter, $m\overset{\frown}{GK} = 110$, $m\overset{\frown}{LK} = 72$, $m\overset{\frown}{LJ} = 142$, $m\overset{\frown}{GI} = 113$, $m\overset{\frown}{HI} = 44$, and $m\overset{\frown}{IJ} = 77$.
Circle cannot have 370°

28. Given: $\odot O$ with tangent $\overline{PA}$ and $m\angle P = x$
Prove: $m\overset{\frown}{AB} = 90 - x$

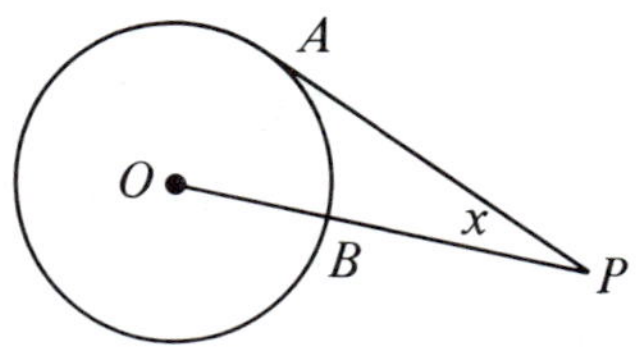

29. Given: $\odot O$ with tangents $\overline{PC}$ and $\overline{PD}$, and $m\angle P = x$
Prove: $m\overset{\frown}{CD} = 180 - x$

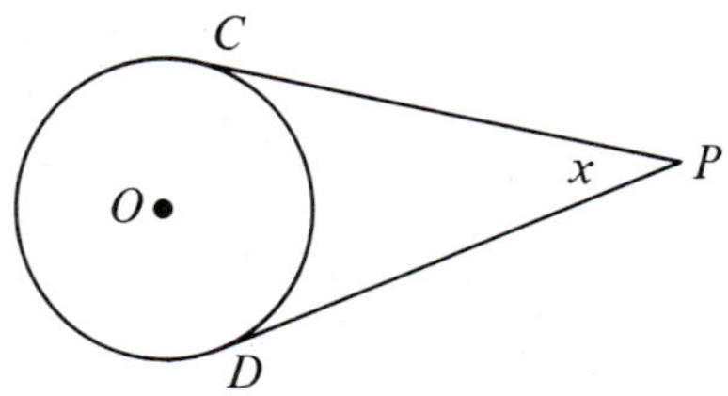

30. Given: $\odot O$ inscribed in $\triangle RST$, A, B, and C points of tangency, and $m\angle R = 90$
Prove: $m\overset{\frown}{ABC} = 270$

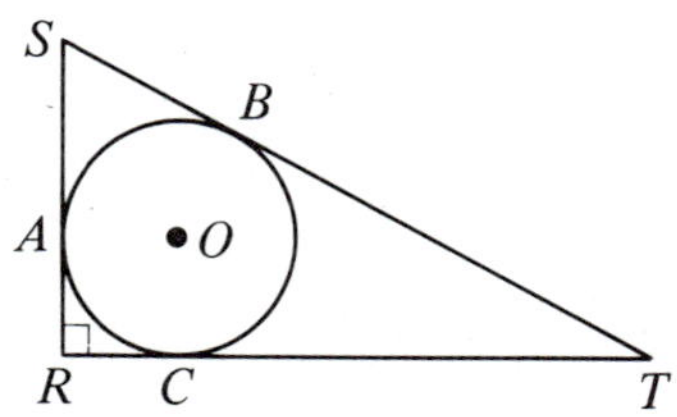

C **31.** Given: $\odot O$ with tangents $\overline{AC}$ and $\overline{BC}$, $m\overset{\frown}{AB} = 132$, and $OA = 8$
Find BC to the nearest tenth. **18.0**

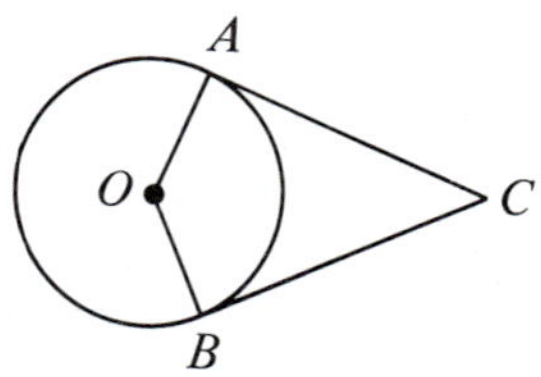

32. Given: $\odot O$ with tangents $\overline{AC}$ and $\overline{BC}$, $OB = 9$, and $AC = 14$
Find $m\overset{\frown}{AB}$ to the nearest degree. **114**

33. Given: $\odot O$ with $\triangle PQR$, and S and Q points of tangency
Prove: $m\angle P = m\overset{\frown}{TS}$

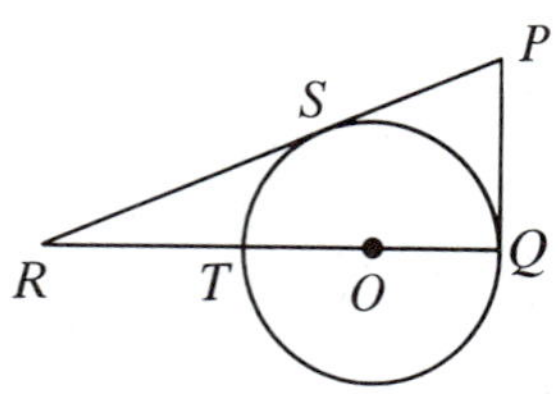

34. Given: $\overleftrightarrow{AB}$ tangent to $\odot O$ at A and to $\odot Q$ at B
Prove: $m\overset{\frown}{AC} + m\overset{\frown}{BD} = 180$

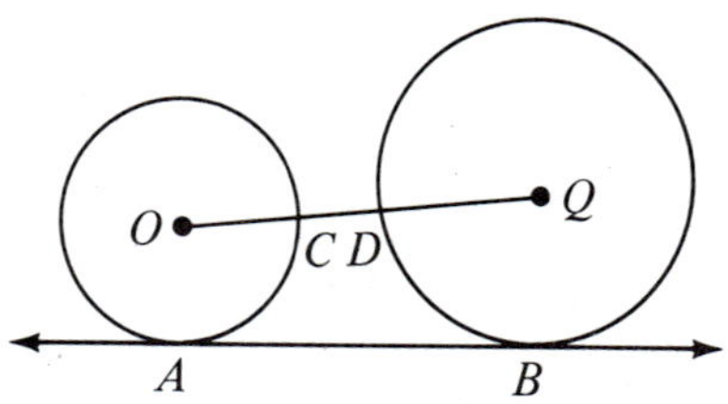

35. Given: $\odot O$ and $\odot Q$ with tangents $\overline{PR}$, $\overline{PS}$, and $\overline{PT}$
Prove: $m\overset{\frown}{RS} + m\overset{\frown}{ST} = 360 - m\angle RPT$

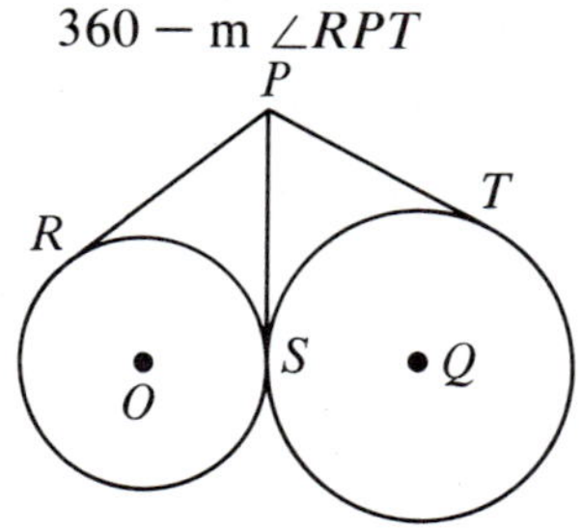

CUMULATIVE REVIEW

1. Find the length of a leg of an isosceles right triangle if the length of the hypotenuse is 10. **$5\sqrt{2}$**

2. Find the length of a diagonal of a square if the length of a side is 7. **$7\sqrt{2}$**

ARCS, CHORDS, AND CENTRAL ANGLES 11.5

Objectives **To identify congruent arcs and congruent chords**
To prove relationships between arcs, chords, and central angles

This lesson deals with relationships between arcs, chords, and central angles of a circle or of congruent circles.

> **Definition: Congruent arcs** In the same circle or in congruent circles, *congruent arcs* are arcs that have the same measure.

Notice that some arcs with the same measure are congruent and some are not.

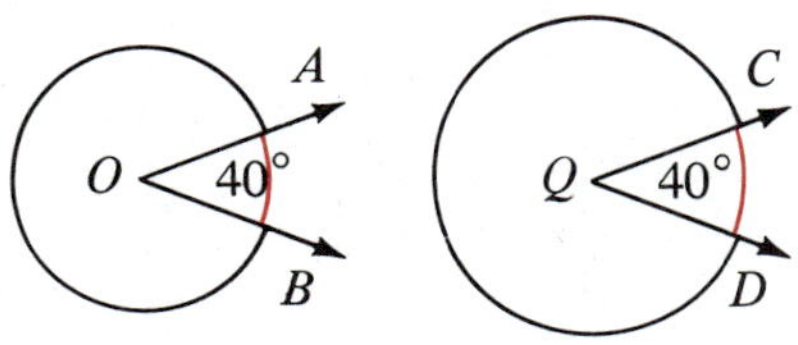

$\odot O \not\cong \odot Q$ and $m\widehat{AB} = m\widehat{CD}$.
But $\widehat{AB} \not\cong \widehat{CD}$.

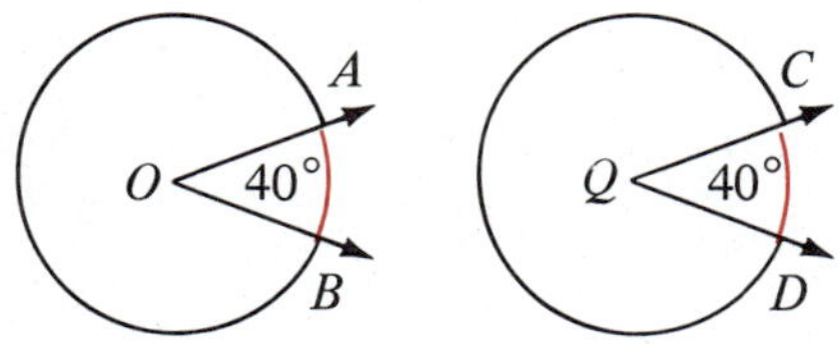

$\odot O \cong \odot Q$ and $m\widehat{AB} = m\widehat{CD}$.
Also, $\widehat{AB} \cong \widehat{CD}$.

Example 1 Based on the given information, which arcs are congruent?

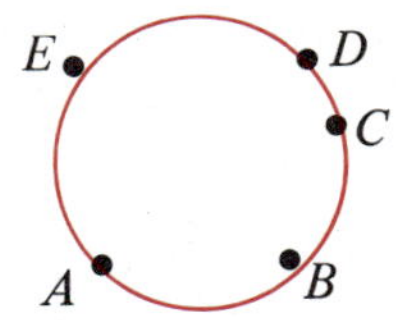

Given: $m\widehat{AB} = 48$, $m\widehat{AC} = 124$,
$m\widehat{ABD} = 208$, and $m\widehat{ABE} = 312$

$m\widehat{EA} = 360 - m\widehat{ABE}$ ► $m\widehat{EA} = 48$ and $m\widehat{AB} = 48$

Thus, since $\widehat{EA}$ and $\widehat{AB}$ are in the same circle, $\widehat{EA} \cong \widehat{AB}$.

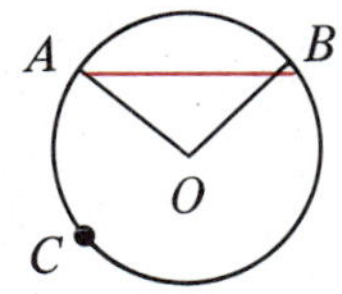

Notice that for chord $\overline{AB}$ in the diagram, there are two arcs with endpoints A and B, minor $\widehat{AB}$ and major $\widehat{ACB}$. When it is stated that a chord determines an arc, the reference is to the *minor* arc. This terminology is used for chords, arcs, and central angles. **Thus,**

(1) $\overline{AB}$ determines $\widehat{AB}$ and $\angle O$.

(2) $\widehat{AB}$ determines $\overline{AB}$ and $\angle O$.

(3) $\angle O$ determines $\overline{AB}$ and $\widehat{AB}$.

Theorem 11.8

In a circle or in congruent circles, congruent arcs determine congruent central angles and congruent chords.

Given: $\odot O \cong \odot Q$ and $\overset{\frown}{AB} \cong \overset{\frown}{CD}$

Prove: $\angle O \cong \angle Q$ and $\overline{AB} \cong \overline{CD}$

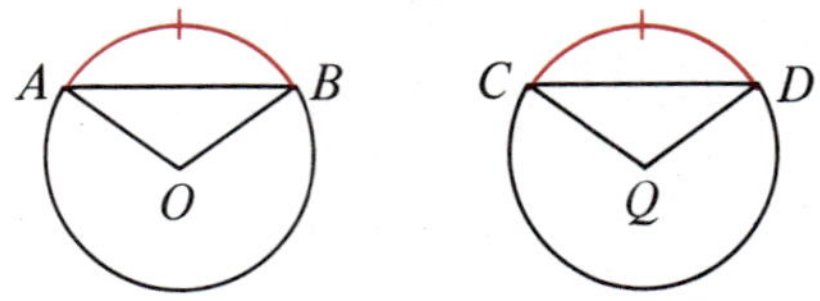

Proof

Statements	**Reasons**
1. $\odot O \cong \odot Q$ and $\overset{\frown}{AB} \cong \overset{\frown}{CD}$	1. Given
2. $m\overset{\frown}{AB} = m\overset{\frown}{CD}$	2. Def. of $\cong$ arcs
3. $m\overset{\frown}{AB} = m\angle AOB$; $m\overset{\frown}{CD} = m\angle CQD$	3. Def. of degree measure of minor arc
4. $m\angle O = m\angle Q$	4. Substitution
5. $\angle O \cong \angle Q$	5. Def. of $\cong$ $\angle$'s
6. $\overline{OA} \cong \overline{QC}$; $\overline{OB} \cong \overline{QD}$	6. Def. of $\cong$ $\odot$'s
7. $\triangle AOB \cong \triangle CQD$	7. SAS
8. $\therefore \overline{AB} \cong \overline{CD}$	8. CPCTC

Example 2

Which chords are congruent?

Use Post. 23. ► $m\overset{\frown}{AD} = 52 + 50 + 36 = 138$
$m\overset{\frown}{FC} = 50 + 36 + 43 = 129$
$m\overset{\frown}{EB} = 36 + 43 + 50 = 129$
$m\overset{\frown}{AB} = 360 - (52 + 50 + 36 + 43 + 50)$
$= 360 - 231 = 129$

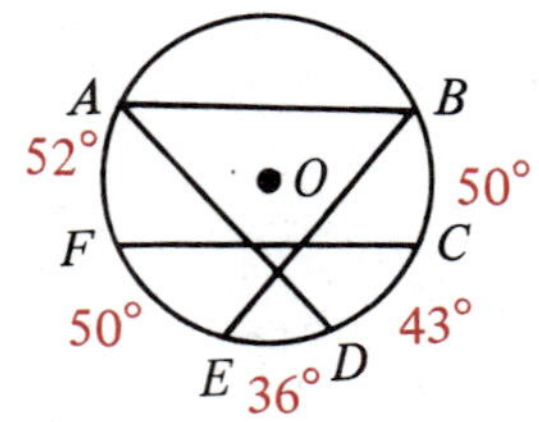

$\overset{\frown}{FC} \cong \overset{\frown}{EB} \cong \overset{\frown}{AB}$ ► **Thus,** $\overline{FC} \cong \overline{EB} \cong \overline{AB}$ by Th. 11.8.

It is possible for two chords to be congruent while the central angles and arcs which they determine are not congruent. For example, $\overline{AB} \cong \overline{CD}$ as shown, but $\angle O \not\cong \angle P$. Also $\overset{\frown}{AB} \not\cong \overset{\frown}{CD}$ since $m\overset{\frown}{AB} = 78$ and $m\overset{\frown}{CD} = 112$. In this case, $\odot O \not\cong \odot P$. This suggests the following theorems.

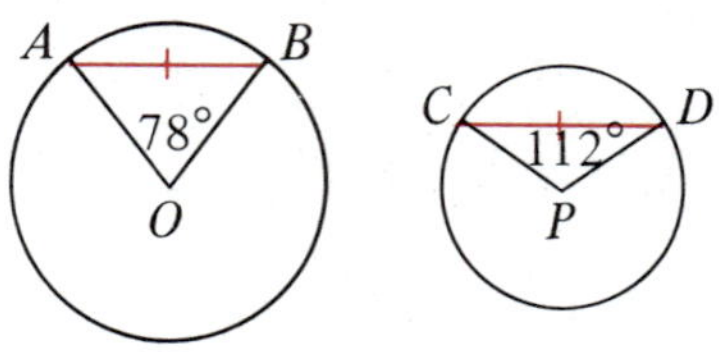

Theorem 11.9

In a circle or in congruent circles, congruent chords determine congruent central angles and congruent arcs. (See Exercise 17.)

Theorem 11.10

In a circle or in congruent circles, congruent central angles determine congruent arcs and congruent chords. (See Exercise 18.)

Example 3

Prove: In a circle, a radius perpendicular to a chord bisects the arc determined by the chord.

Given: $\odot O$ with radius $\overline{OZ} \perp$ chord $\overline{XY}$
Prove: $\overset{\frown}{XZ} \cong \overset{\frown}{ZY}$

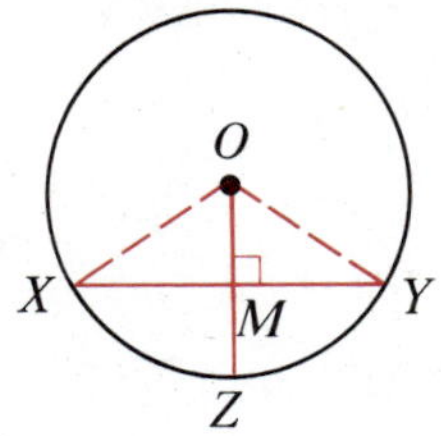

Proof

Statements	Reasons
1. $\odot O$ with radius $\overline{OZ} \perp$ chord $\overline{XY}$	1. Given
2. Draw radii $\overline{OX}$ and $\overline{OY}$.	2. Two pt. are contained in exactly one line.
3. $\overline{XM} \cong \overline{MY}$	3. Radius $\perp$ chord bisects the chord.
4. $\overline{OX} \cong \overline{OY}$	4. Radii of a $\odot$ are $\cong$.
5. $\overline{OM} \cong \overline{OM}$	5. Reflexive property
6. $\triangle XOM \cong \triangle YOM$	6. SSS
7. $\angle XOM \cong \angle YOM$	7. CPCTC
8. $\therefore \overset{\frown}{XZ} \cong \overset{\frown}{ZY}$	8. $\cong$ central $\angle$'s determine $\cong$ arcs.

Written Exercises

In which cases can you conclude that chord $\overline{WX} \cong$ chord $\overline{ST}$?

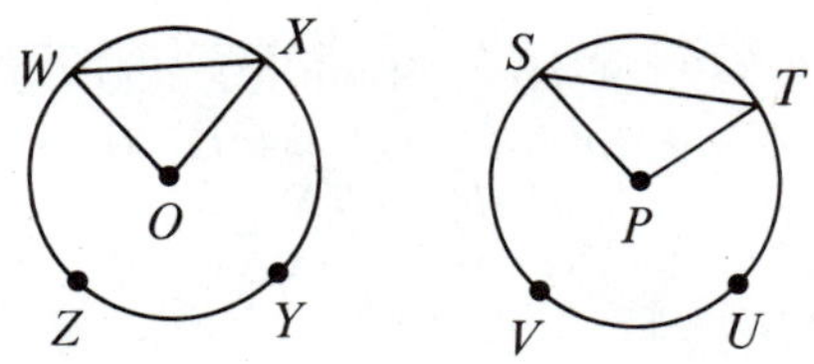

A 1. Given: $\overline{OW} \cong \overline{PS}$ **no**
2. Given: $\overset{\frown}{WX} \cong \overset{\frown}{ST}$ **no**
3. Given: $m\overset{\frown}{WYX} = m\overset{\frown}{SUT}$ **no**

4. In $\odot O$, $m\overset{\frown}{XY} = 85$, $m\overset{\frown}{XYZ} = 190$, and $m\overset{\frown}{XYW} = 300$. In $\odot P$, $m\overset{\frown}{ST} = 95$, $m\overset{\frown}{STU} = 180$, and $m\overset{\frown}{STV} = 240$. Which arcs are congruent if $\odot O \cong \odot P$?
$\overset{\frown}{XY}$, $\overset{\frown}{TU}$; $\overset{\frown}{XWY}$, $\overset{\frown}{TSU}$; $\overset{\frown}{WX}$, $\overset{\frown}{VU}$; $\overset{\frown}{WZX}$, $\overset{\frown}{VSU}$; $\overset{\frown}{WY}$, $\overset{\frown}{VT}$; $\overset{\frown}{WZY}$, $\overset{\frown}{VST}$; $\overset{\frown}{STU}$, $\overset{\frown}{SVU}$

On the basis of the given information, which arcs are congruent?

5. Given: $\overline{CD} \cong \overline{AB}$
$\overset{\frown}{CD}$, $\overset{\frown}{AB}$; $\overset{\frown}{AD}$, $\overset{\frown}{CB}$; $\overset{\frown}{DBC}$, $\overset{\frown}{BDA}$; $\overset{\frown}{ABD}$, $\overset{\frown}{CDB}$

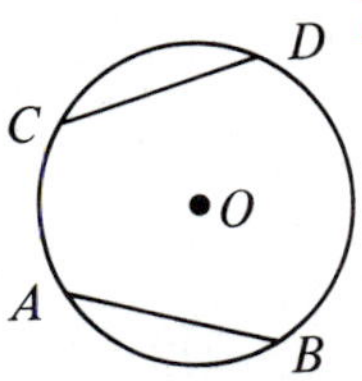

6. Given: Radius $\overline{OE} \cong$ radius $\overline{PH}$ and $\overline{EF} \cong \overline{HI}$
$\overset{\frown}{EF}$, $\overset{\frown}{HI}$, $\overset{\frown}{EGF}$, $\overset{\frown}{HJI}$

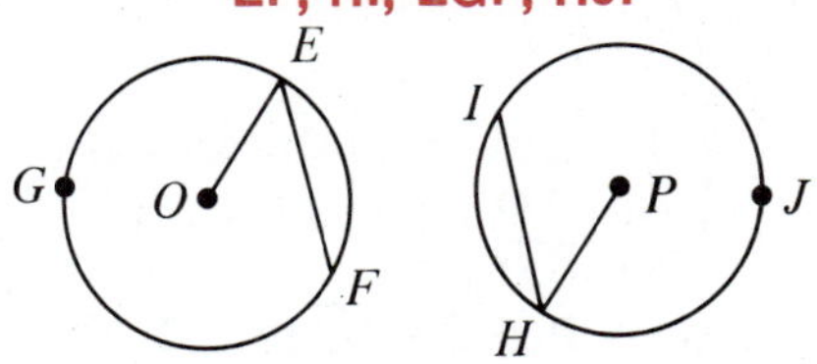

7. Given: $\angle TOU \cong \angle WQV$ **none**

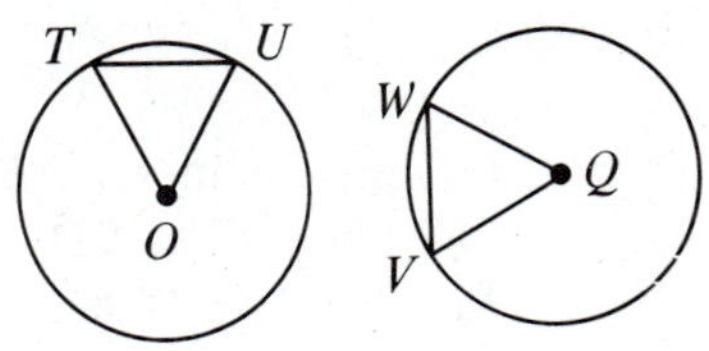

In $\odot O$, $m\overset{\frown}{AB} = 35$, $m\overset{\frown}{AC} = 75$, $m\overset{\frown}{AD} = 110$, $m\overset{\frown}{AE} = 160$, $m\overset{\frown}{ABF} = 180$, $m\overset{\frown}{ADG} = 215$, $m\overset{\frown}{ADH} = 240$, $m\overset{\frown}{ADI} = 280$, and $m\overset{\frown}{ADJ} = 305$.

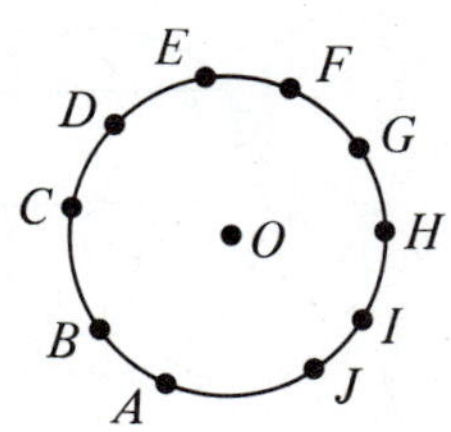

8. Which arcs are congruent to $\overset{\frown}{AB}$? **$\overset{\frown}{AB} \cong \overset{\frown}{CD} \cong \overset{\frown}{FG}$**
9. Which arcs are congruent to $\overset{\frown}{BC}$? **$\overset{\frown}{BC} \cong \overset{\frown}{HI}$**
10. Which arcs are congruent to $\overset{\frown}{DF}$? **none**

11. Given: $\overset{\frown}{RS} \cong \overset{\frown}{RT}$
Prove: $\triangle RST$ is isosceles.

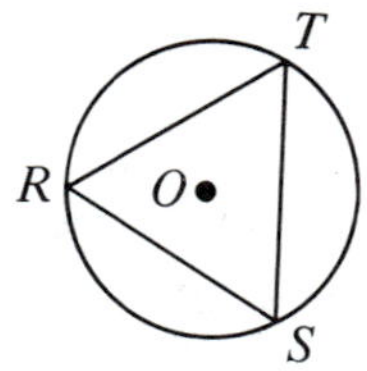

12. Given: $\angle S \cong \angle T$
Prove: $\overset{\frown}{RS} \cong \overset{\frown}{RT}$

13. Given: $\overline{OX}$, $\overline{OY}$, $\overline{OZ}$ are radii of $\odot O$ and $\angle ZXY \cong \angle ZYX$.
Prove: $\overset{\frown}{XZ} \cong \overset{\frown}{YZ}$

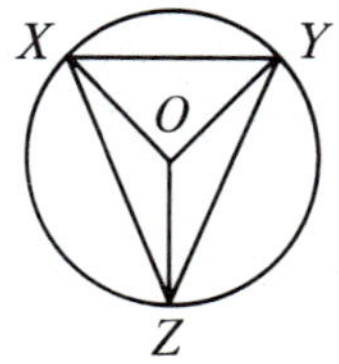

14. Given: $\overline{OX}$, $\overline{OY}$, $\overline{OZ}$ are radii of $\odot O$ and $\overset{\frown}{XZ} \cong \overset{\frown}{YZ}$.
Prove: $\triangle XOZ \cong \triangle YOZ$

B **15.** Given: $\overline{NL}$ is a diameter of $\odot O$ and $\overset{\frown}{KL} \cong \overset{\frown}{ML}$.
Prove: $\angle KNL \cong \angle MNL$

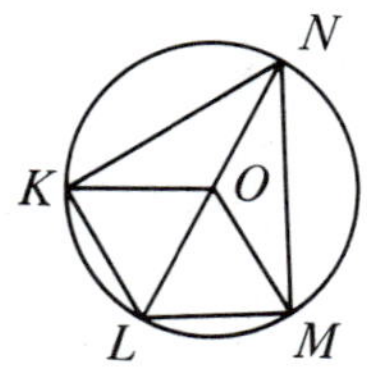

16. Given: $\overline{NL}$ is a diameter of $\odot O$ and $\angle KNL \cong \angle MNL$.
Prove: $\overset{\frown}{KL} \cong \overset{\frown}{ML}$

17. Prove Theorem 11.9.

18. Prove Theorem 11.10.

19. Given: Radii $\overline{OR}$ and $\overline{OT}$ of $\odot O$, $\overline{SU} \perp \overline{OR}$, $\overline{SV} \perp \overline{OT}$, and $\overset{\frown}{RS} \cong \overset{\frown}{TS}$
Prove: $\overline{SU} \cong \overline{SV}$

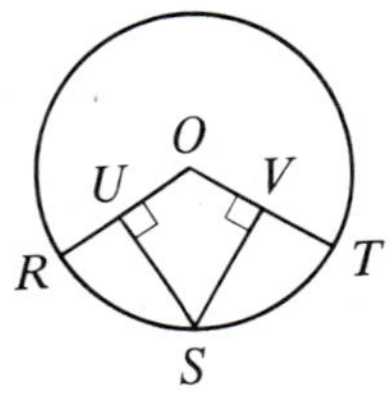

20. Given: Radii $\overline{OR}$ and $\overline{OT}$ of $\odot O$, $\overline{SU} \perp \overline{OR}$, $\overline{SV} \perp \overline{OT}$, and $\overline{SU} \cong \overline{SV}$
Prove: $\overset{\frown}{RS} \cong \overset{\frown}{TS}$.

C **21.** Given: Radii $\overline{OA}$, $\overline{OB}$, and $\overline{OC}$; B bisects $\overset{\frown}{AC}$; $m\overset{\frown}{AC} = 120$; and $AO = x$.
Find AC in terms of x.
$x\sqrt{3}$

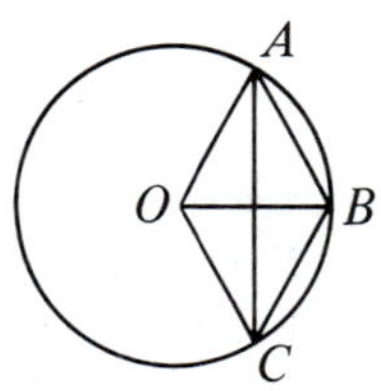

22. Given: Radii $\overline{OA}$, $\overline{OB}$, and $\overline{OC}$; B bisects $\overset{\frown}{AC}$; $m\overset{\frown}{AC} = 80$; and $AO = 5$.
Find AC to the nearest tenth.
6.4

23. Given: Diameter $\overline{AB}$ of $\odot O$ and $\overset{\frown}{AC} \cong \overset{\frown}{BD}$
Prove: $\angle ACD \cong \angle BDC$

24. Given: Diameter $\overline{AB}$ of $\odot O$ with chord $\overline{CD} \parallel \overline{AB}$
Prove: $\overline{AC} \cong \overline{DB}$

25. Given: Diameter $\overline{AB}$ of $\odot O$ with chord $\overline{BD} \parallel \overline{OE}$
Prove: $\overset{\frown}{AE} \cong \overset{\frown}{ED}$

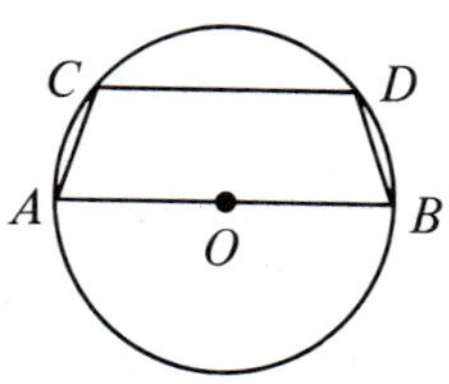

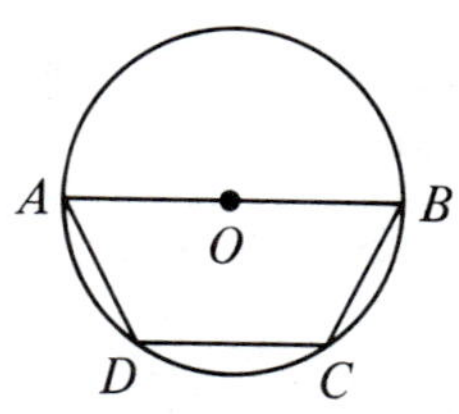

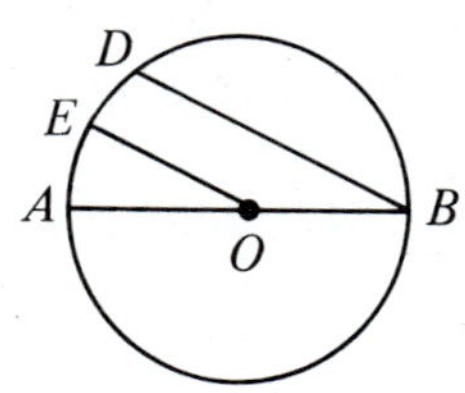

INSCRIBED ANGLES 11.6

Objective **To apply measurement properties of inscribed angles**

$\angle B$ is *inscribed* in minor $\widehat{AC}$ of $\odot O$. $\angle B$ *intercepts* major $\widehat{ADC}$.

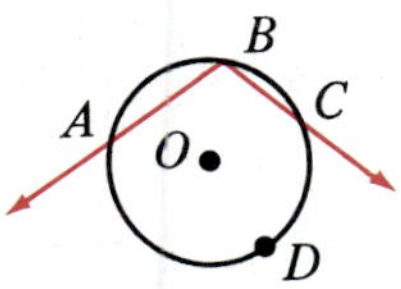

$\angle F$ is *inscribed* in semicircle $\widehat{EFG}$ of $\odot P$. $\angle F$ *intercepts* semicircle $\widehat{EHG}$.

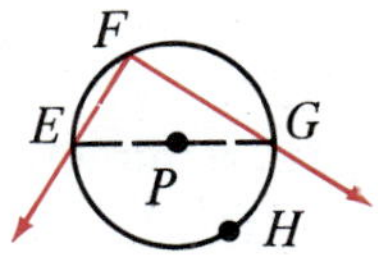

$\angle J$ is *inscribed* in major $\widehat{IJK}$ of $\odot Q$. $\angle J$ *intercepts* minor $\widehat{IK}$.

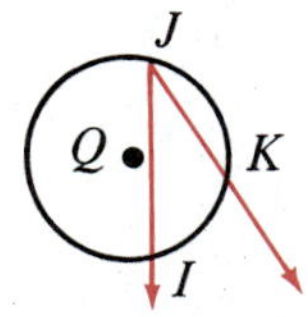

Definition: Inscribed angle An *inscribed angle* is an angle whose vertex lies on a circle and whose sides contain chords of the circle.

There are three possible positions for inscribed angles, as shown in Theorem 11.11 below.

Theorem 11.11 The measure of an inscribed angle is one half the measure of its intercepted arc.

Given: $\odot O$ with inscribed $\angle APB$ Prove: $m\,\angle APB = \frac{1}{2} m\widehat{AB}$

CASE I: O lies on $\angle APB$.

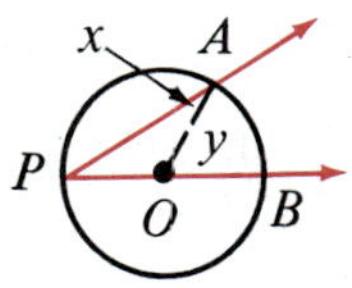

CASE II: O lies in the interior of $\angle APB$. (See Exercise 30.)

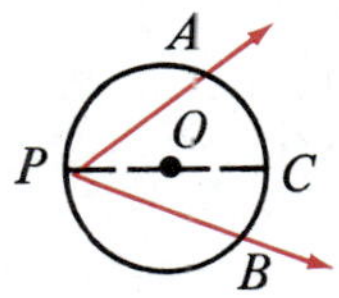

CASE III: O lies in the exterior of $\angle APB$. (See Exercise 31.)

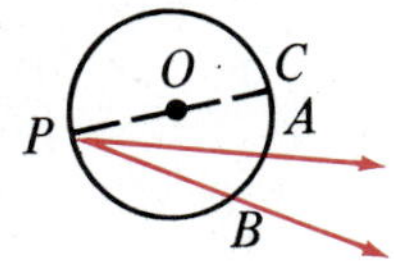

Proof of Case I

Statements	Reasons
1. Draw $\overline{OA}$.	1. Two pt. are contained in exactly one line.
2. $\overline{OA} \cong \overline{OP}$	2. Radii of a $\odot$ are $\cong$.
3. $\angle x \cong \angle P$ and $m\,\angle x = m\,\angle P$	3. $\angle$'s opp. $\cong$ sides of a $\triangle$ are $\cong$.
4. $m\,\angle y = m\,\angle x + m\,\angle P$	4. Meas. of ext. $\angle$ of $\triangle$ = sum of meas. of remote int. $\angle$'s.
5. $m\,\angle y = 2 \cdot m\,\angle P$	5. Substitution property
6. $m\,\angle P = \frac{1}{2} \cdot m\,\angle y$	6. Multiplication property of equality
7. $m\,\angle y = mAB$	7. Meas. of minor arc = meas. of central $\angle$.
8. $\therefore m\,\angle APB = \frac{1}{2} m\widehat{AB}$	8. Substitution property

Example 1

Given: m $\angle ABC = 62$ and m$\widehat{BC} = 110$.
Find m $\angle A$, m $\angle C$, and m$\widehat{AB}$.

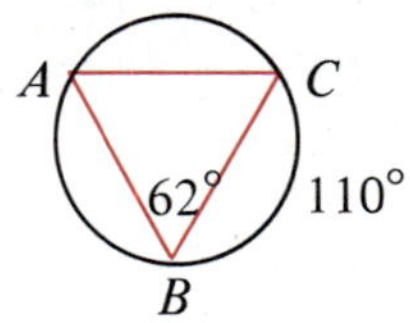

m $\angle A = \frac{1}{2}(110) = 55$ by Th. 11.11.

Sum of meas. of $\angle$'s of a $\triangle$ is 180°. ► m $\angle C = 180 - (62 + 55) = 63$

m$\widehat{AB} = 2(63) = 126$ by Th. 11.11.

The following corollaries are direct results of Theorem 11.11.

Corollary 1

If two inscribed angles intercept the same arc (or congruent arcs), then the angles are congruent. (See Exercise 32.)

Given: $\odot O$ with inscribed $\angle CAD$ and $\angle CBD$
Prove: $\angle CAD \cong \angle CBD$

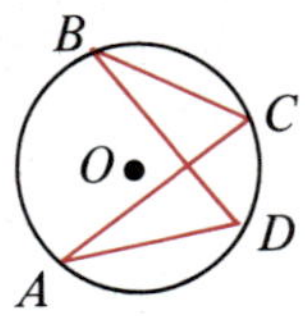

The case of the same arc is shown.

Corollary 2

An angle inscribed in a semicircle is a right angle. (See Exercise 33.)

Given: $\odot O$ with semicircle $\widehat{ABC}$
Prove: $\angle ABC$ is a right angle.

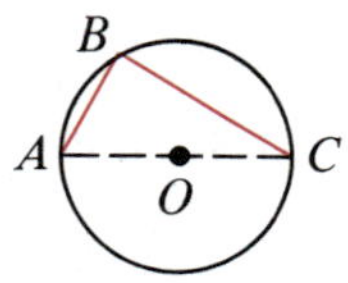

Corollary 3

If two arcs of a circle are included between parallel secants, then the arcs are congruent.

Given: $\odot O$ with secants $\overleftrightarrow{PQ}$ and $\overleftrightarrow{RS}$ and $\overleftrightarrow{PQ} \parallel \overleftrightarrow{RS}$
Prove: $\widehat{PR} \cong \widehat{QS}$

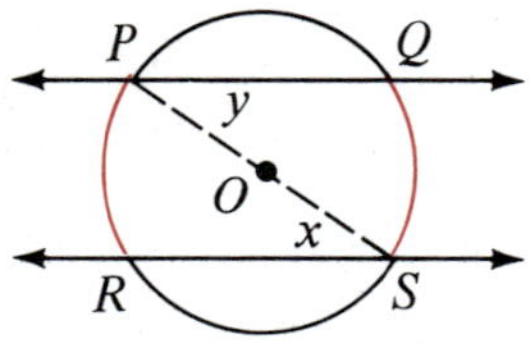

ANALYSIS
Draw chord $\overline{PS}$. Use $\angle x$ and $\angle y$ to show $\widehat{PR} \cong \widehat{QS}$.

Proof

Statements	**Reasons**
1. $\odot O$ with $\overleftrightarrow{PQ} \parallel \overleftrightarrow{RS}$	1. Given
2. Draw chord $\overline{PS}$.	2. Two pts. are contained in exactly one line.
3. m $\angle x$ = m $\angle y$	3. Alt. int. $\angle$'s of $\parallel$ lines are $\cong$.
4. m $\angle x = \frac{1}{2} \cdot$ m$\widehat{PR}$; m $\angle y = \frac{1}{2} \cdot$ m$\widehat{QS}$	4. Th. 11.11
5. $\frac{1}{2} \cdot$ m$\widehat{PR} = \frac{1}{2} \cdot$ m$\widehat{QS}$	5. Substitution property
6. m$\widehat{PR}$ = m$\widehat{QS}$	6. Multiplication property of equality
7. $\therefore \widehat{PR} \cong \widehat{QS}$	7. Def. of $\cong$ arcs

Example 2

Given: $\odot O$ with diameter $\overline{AC}$, $\overline{BC} \parallel \overline{AD}$, and m $\angle COD = 50$.
Find m $\angle ABC$, m$\widehat{AB}$, and m$\widehat{BC}$.

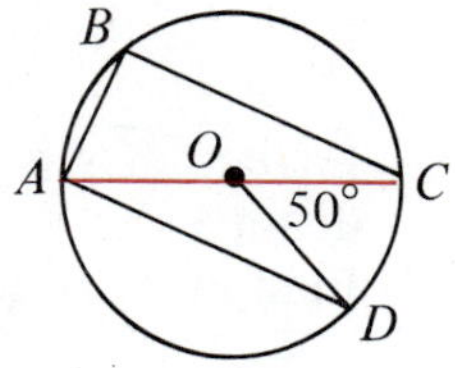

$\widehat{ABC}$ is a semicircle. ► m $\angle ABC = 90$, by Cor. 2
m$\widehat{DC} = 50$, so m$\widehat{AB} = 50$, by Cor. 3.
m$\widehat{BC} = 180 - 50 = 130$

Corollary 4 is also a direct result of Theorem 11.11.

Corollary 4

Opposite angles of an inscribed quadrilateral are supplementary.

Given: Quad. $KLMN$ inscribed in $\odot O$
Prove: $\angle K$ and $\angle M$ are supplementary;
$\angle L$ and $\angle N$ are supplementary.

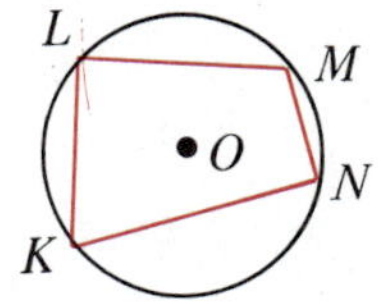

Proof

Statements	Reasons
1. Quad. $KLMN$ inscribed in $\odot O$	1. Given
2. m $\angle K = \frac{1}{2}$ m$\widehat{LMN}$ m $\angle M = \frac{1}{2}$ m$\widehat{LKN}$	2. Meas. of inscribed $\angle = \frac{1}{2}$ meas. of intercepted arc.
3. m $\angle K$ + m $\angle M = \frac{1}{2}$ m$\widehat{LMN} + \frac{1}{2}$ m$\widehat{LKN}$	3. Addition property of equality
4. m $\angle K$ + m $\angle M = \frac{1}{2}$ (m$\widehat{LMN}$ + m$\widehat{LKN}$)	4. Distributive property
5. m$\widehat{LMN}$ + m$\widehat{LKN} = 360$	5. Arc addition postulate
6. m $\angle K$ + m $\angle M = 180$	6. Substitution property
7. $\therefore$ $\angle K$ and $\angle M$ are supplementary.	7. Def. of supplementary $\angle$'s
8. Similarly, $\angle L$ and $\angle N$ are supplementary.	8. Steps 1–6

Reading in Geometry

Draw and label each figure described below.

1. An angle inscribed in a minor arc
2. An angle intercepting a minor arc
3. An angle inscribed in a semicircle
4. Two inscribed angles intercepting the same arc
5. An inscribed angle with the center of the circle on the angle
6. An inscribed angle with the center of the circle in the exterior of the angle

Oral Exercises

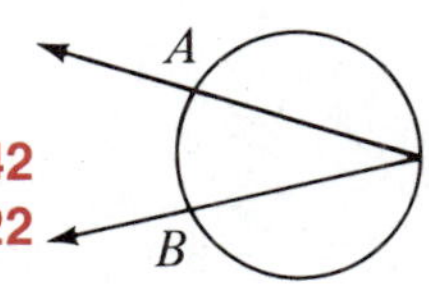

Find m $\angle AXB$ in each case.

1. m$\widehat{AB} = 72$ **36** **2.** m$\widehat{AB} = 84$ **42**
3. m$\widehat{AB} = 110$ **55** **4.** m$\widehat{AB} = 44$ **22**

Find m$\widehat{AB}$ in each case.

50 **5.** m $\angle AXB = 25$ **80** **6.** m $\angle AXB = 40$
126 **7.** m $\angle AXB = 63$ **36** **8.** m $\angle AXB = 18$

Written Exercises

A

1. Given: $\overline{MN}$ is a diameter.
Find m $\angle MPN$.

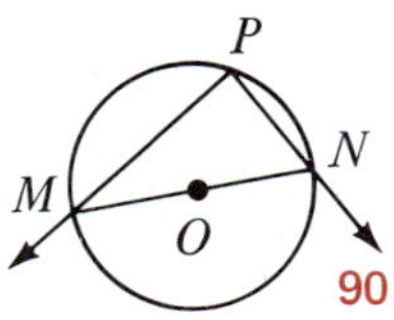

90

2. Given: $\overline{AB} \cong \overline{BC}$ and m$\widehat{AB} = 110$
Find m $\angle ABC$.

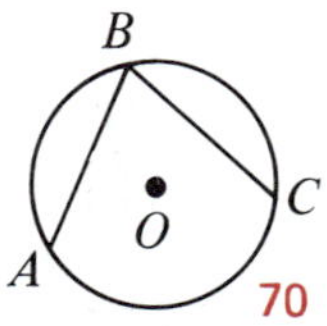

70

3. Given: m$\widehat{HK} =$ 112
Find m $\angle I$ and m $\angle J$.

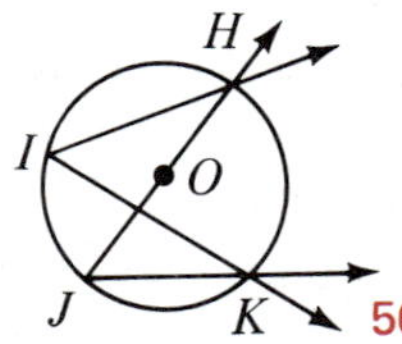

56

4. Given: m$\widehat{SVU} =$ 156
Find m $\angle T$ and m $\angle V$.

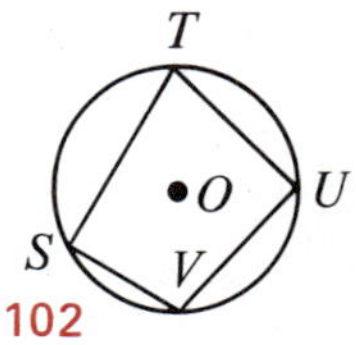

78; 102

5. Given: m $\angle I = 86$ and m $\angle L = 113$
Find m $\angle J$ and m $\angle K$.

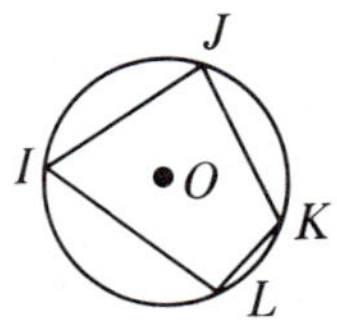

67; 94

6. Given: $\odot O$ with diameter $\overline{RS}$ and m$\widehat{QR} = 84$
Find m $\angle Q$ and m $\angle R$.

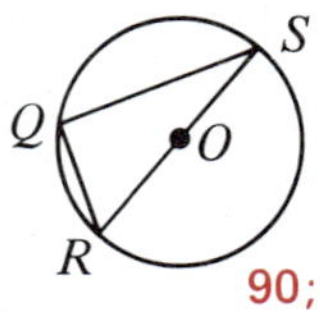

90; 48

7. Given: $\odot O$ with diameter $\overline{AB} \parallel \overline{DC}$ and m$\widehat{AD} = 64$
Find m $\angle ACD$ and m$\widehat{BC}$.

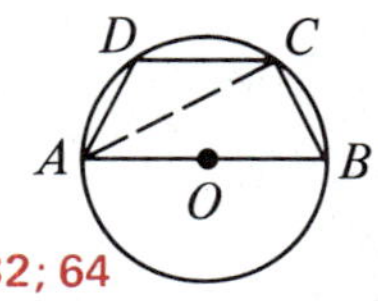

32; 64

In $\odot O$, m$\widehat{RU} = 86$ and m$\widehat{SV} = 62$.

8. Find m $\angle SRV$. 31
9. Find m $\angle RSU$. 43
10. Find m $\angle RSQ$. 137
11. Find m $\angle RQU$. 12

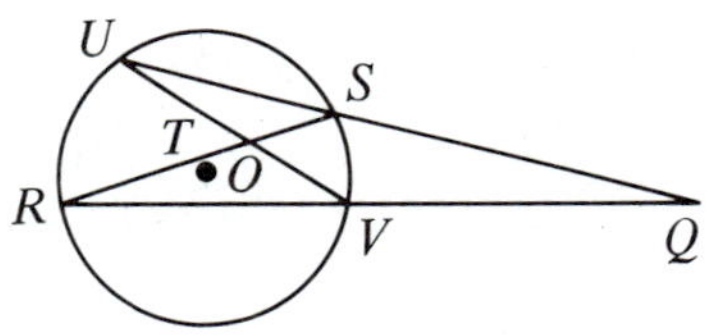

In $\odot O$, m $\angle R = 36$ and m $\angle Q = 42$.

12. Find m $\angle RSU$. 78
13. Find m$\widehat{RU}$. 156
14. Find m$\widehat{SV}$. 72
15. Find m $\angle UTS$. 66

In $\odot O$, $\overleftrightarrow{ST} \parallel$ tangent $\overleftrightarrow{PQ}$, m $\angle TSR = 61$, and $\overline{SR} \cong \overline{TR}$.

16. Find m$\widehat{TR}$. 122
17. Find m$\widehat{SR}$. 122
18. Find m $\angle STR$. 61

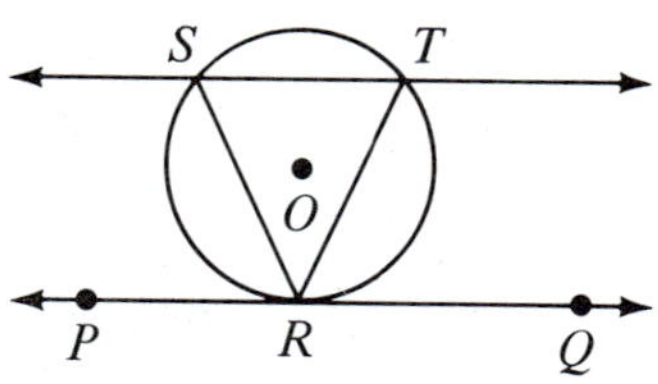

In $\odot O$, $\overleftrightarrow{ST} \parallel$ tangent $\overleftrightarrow{PQ}$, m$\widehat{SR} = 142$, and $\overline{SR} \cong \overline{TR}$.

19. Find m$\widehat{TR}$. 142
20. Find m $\angle STR$. 71
21. Find m $\angle SRT$. 38

B

22. Given: $\odot O$ with $\overline{DC} \parallel \overline{AB}$ and m$\widehat{AD} = 108$
Find m $\angle A$ and m $\angle AOB$.

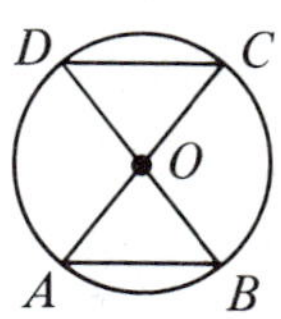

54; 72

23. Given: $\odot O$ with m$\widehat{NM} =$ 65 and $\angle PNO \cong \angle PMO$
Find m $\angle PNO$ and m $\angle ONM$.

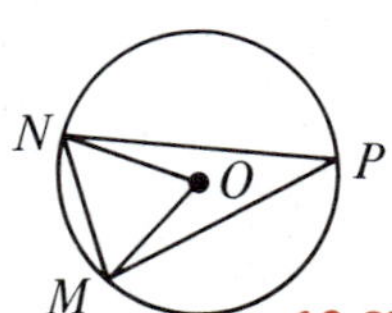

16.25; 57.5

24. Given: $\odot O$ with diameter $\overline{US}$, m$\widehat{RU} = 50$, and m$\widehat{UT} = 30$.
Find m $\angle RUS$ and m $\angle STU$.

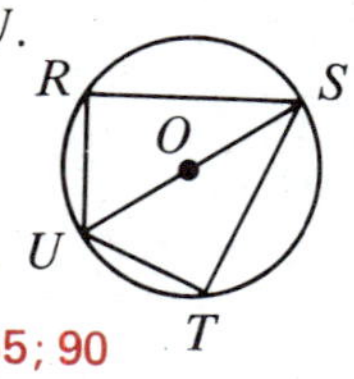

65; 90

25. Given: $\odot O$ with diameters $\overline{SQ}$ and $\overline{PR}$
Prove: $\angle PRQ \cong \angle SQR$

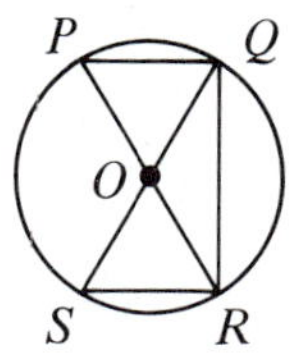

26. Given: $\odot O$ with diameter $\overline{DB}$ and m $\angle ABD = \frac{1}{2} \cdot$ m $\angle DOC$
Prove: $\widehat{AD} \cong \widehat{DC}$

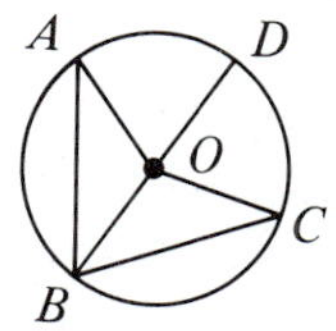

27. Given: $\odot O$ with $\overline{RS} \parallel \overline{UT}$
Prove: $\overline{RT} \cong \overline{SU}$

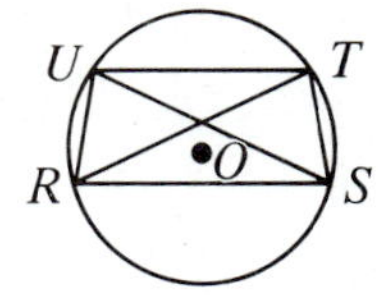

28. Given: $\odot O$ with diameter $\overline{AB}$ and $\overline{CD} \perp \overline{AB}$
Prove: $\triangle ADC \sim \triangle ACB$

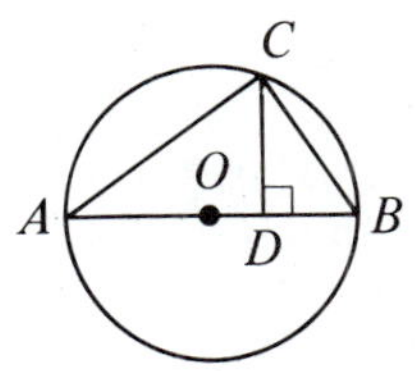

29. Given: $\odot O$ with diameter $\overline{AB}$ and $\overline{CD} \perp \overline{AB}$
Prove: $AD \cdot CB = AC \cdot CD$

30. Prove Case II of Theorem 11.11. [*Hint:* Apply Case I to $\angle APC$ and to $\angle CPB$. Then add the measures.]

31. Prove Case III of Theorem 11.11. [*Hint:* Apply Case I to $\angle CPB$ and to $\angle CPA$. Then subtract the measures.]

32. Prove Corollary 1 to Theorem 11.11.

33. Prove Corollary 2 to Theorem 11.11.

34. Prove that if two angles are inscribed in the same arc, the angles are congruent.

Prove or disprove each of the following.

35. An angle inscribed in a major arc is acute.

36. An angle inscribed in a minor arc is acute.

37. Two triangles inscribed in congruent circles are congruent if two angles of one are congruent to two angles of the other.

38. If two secants intercept congruent arcs on a circle, then the secants are parallel.

39. If two consecutive angles of an inscribed quadrilateral are congruent, then the quadrilateral is an isosceles trapezoid.

40. Two triangles inscribed in congruent circles are congruent if a side and an angle of one are congruent to a side and angle of the other.

CUMULATIVE REVIEW

True or false?

1. The diagonals of a rhombus are congruent. F
2. Each interior angle of a regular hexagon measures 120°. T
3. The diagonals of an isosceles trapezoid are congruent. T
4. The sum of the measures of the exterior angles of an octagon is 360°. T
5. A parallelogram is a rhombus if the diagonals of the parallelogram are perpendicular. T
6. A rhombus with at least one right angle is a square. T

ANGLES FORMED BY SECANTS 11.7

Objective **To apply measurement properties for angles formed by two secants**

Two secants to a circle can intersect in three possible positions:

Case I
On the circle

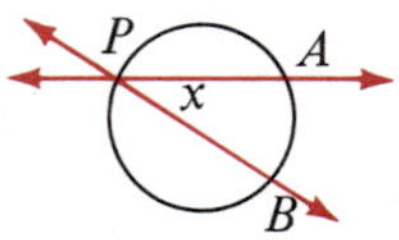

$\widehat{AB}$ is the intercepted arc with respect to $\angle x$.

Case II
In the interior of the circle

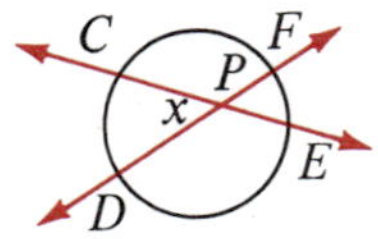

$\widehat{CD}$ and $\widehat{EF}$ are the intercepted arcs with respect to $\angle x$.

Case III
In the exterior of the circle

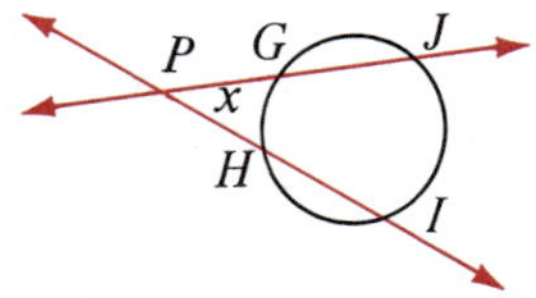

$\widehat{GH}$ and $\widehat{IJ}$ are the intercepted arcs with respect to $\angle x$.

In each case, there is a relationship between the measure of $\angle x$ and the measure(s) of the intercepted arc(s). For Case I, $\angle x$ is an inscribed angle, so its measure is one half the measure of the intercepted arc. Case II is given in Theorem 11.12.

Theorem 11.12

The measure of an angle formed by two secants intersecting in the interior of a circle is one half the sum of the measures of the arcs intercepted by the angle and its vertical angle.

Given: $\odot O$ with secants $\overleftrightarrow{AB}$ and $\overleftrightarrow{CD}$ intersecting at P.

Prove: $m\angle x = \frac{1}{2}(m\widehat{BC} + m\widehat{AD})$

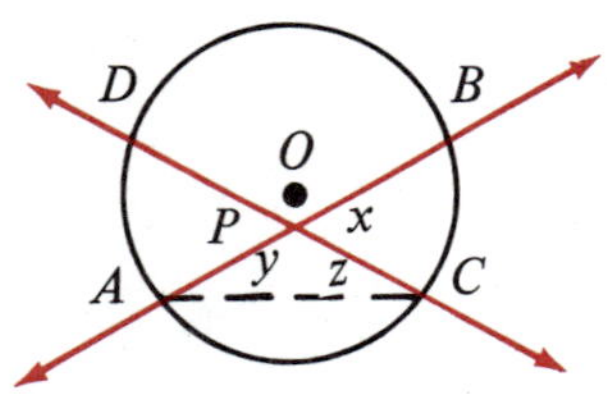

ANALYSIS
Draw $\overline{AC}$. $m\angle x = m\angle y + m\angle z$. Express $m\angle y$ and $m\angle z$ in terms of their intercepted arcs.

Proof

Statements	**Reasons**
1. $\odot O$ with secants $\overleftrightarrow{AB}$ and $\overleftrightarrow{CD}$	1. Given
2. Draw $\overline{AC}$.	2. Two pt. are contained in exactly one line.
3. $m\angle x = m\angle y + m\angle z$	3. Meas. of ext. $\angle$ of $\triangle$ = sum of meas. of remote int. $\angle$'s.
4. $m\angle y = \frac{1}{2}m\widehat{BC}$	4. Meas. of inscribed $\angle = \frac{1}{2}$ meas. of intercepted arc.
5. $m\angle z = \frac{1}{2}m\widehat{AD}$	5. Same as 4
6. $m\angle x = \frac{1}{2}m\widehat{BC} + \frac{1}{2}m\widehat{AD}$	6. Substitution property
7. $\therefore m\angle x = \frac{1}{2}(m\widehat{BC} + m\widehat{AD})$	7. Distributive property

---

In the diagram for Theorem 11.12, notice that the measure of $\angle APD$ is also equal to $\frac{1}{2}(m\widehat{BC} + m\widehat{AD})$.

You can use Theorem 11.12 to find the measures of angles and arcs.

Example 1

Given: $m\widehat{RS} = 58$ and $m\widehat{TU} = 36$
Find $m\ \angle RWS$.

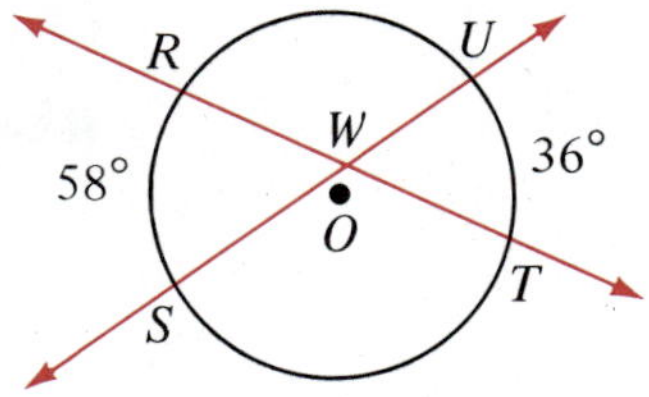

$$m\ \angle RWS = \frac{1}{2}(m\widehat{RS} + m\widehat{TU}) \text{ by Th 11.12}$$

$$m\ \angle RWS = \frac{1}{2}(58 + 36) = \frac{1}{2}(94)$$

Thus, $m\ \angle RWS = 47$.

Example 2

Given: $m\ \angle AEB = 100$ and $m\widehat{CD} = 30$
Find $m\widehat{AB}$.

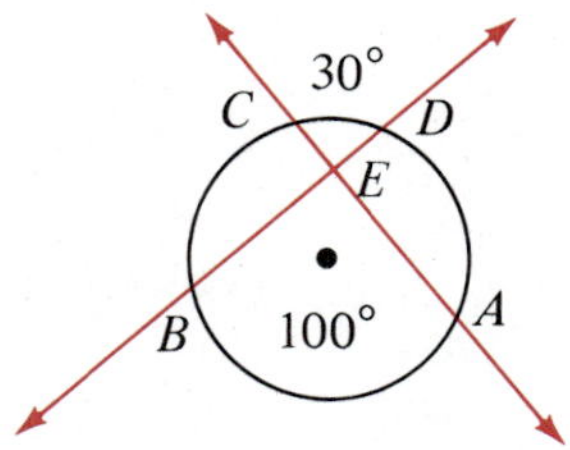

$$m\ \angle AEB = \frac{1}{2}(m\widehat{CD} + m\widehat{AB})$$

$$100 = \frac{1}{2}(30 + m\widehat{AB})$$

$$200 = 30 + m\widehat{AB}$$

Thus, $m\widehat{AB} = 170$.

Recall that in Case III on page 364, the two secants intersect in the exterior of the circle. Theorem 11.13 below deals with Case III.

Theorem 11.13

The measure of an angle formed by two secants intersecting in the exterior of a circle is one half the difference of the measures of the intercepted arcs.

Given: $\odot O$ with secants $\overleftrightarrow{AB}$ and $\overleftrightarrow{CD}$ intersecting at P

Prove: $m\ \angle x = \frac{1}{2}(m\widehat{BD} - m\widehat{AC})$

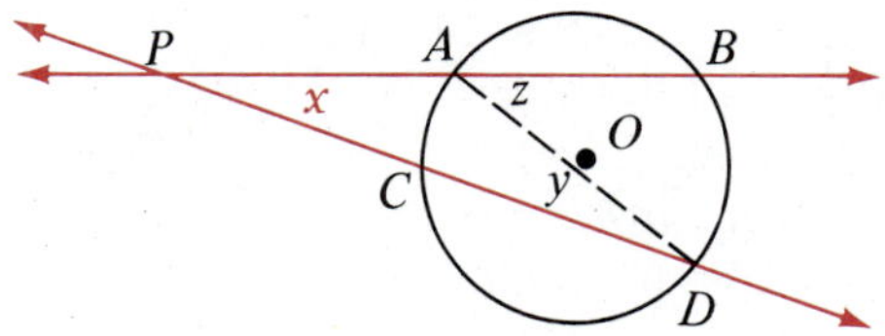

Proof

Statements	**Reasons**
1. $\odot O$ with secants $\overleftrightarrow{AB}$ and $\overleftrightarrow{CD}$	1. Given
2. Draw $\overline{AD}$.	2. Two pts. are contained in exactly one line.
3. $m\ \angle z = m\ \angle x + m\ \angle y$	3. Meas. of ext. ∠ of △ = sum of meas. of remote int. ∠'s.
4. $m\ \angle x = m\ \angle z - m\ \angle y$	4. Subtraction property of equality
5. $m\ \angle z = \frac{1}{2}m\widehat{BD}$; $m\ \angle y = \frac{1}{2}m\widehat{AC}$	5. Meas. of inscribed ∠ = $\frac{1}{2}$ meas. of intercepted arc.
6. $m\ \angle x = \frac{1}{2}m\widehat{BD} - \frac{1}{2}m\widehat{AC}$	6. Substitution property
7. $\therefore m\ \angle x = \frac{1}{2}(m\widehat{BD} - m\widehat{AC})$	7. Distributive property

---

Notice that when two secants intersect in the exterior of a circle, two arcs are formed. As shown in Example 3, if you know the measure of each of these arcs, the measure of the angle can be found.

Example 3 Given: $m\overset{\frown}{AB} = y$, $m\overset{\frown}{BC} = 3y$, $m\overset{\frown}{CD} = y$, and $m\overset{\frown}{DA} = 7y$

Find m $\angle x$.

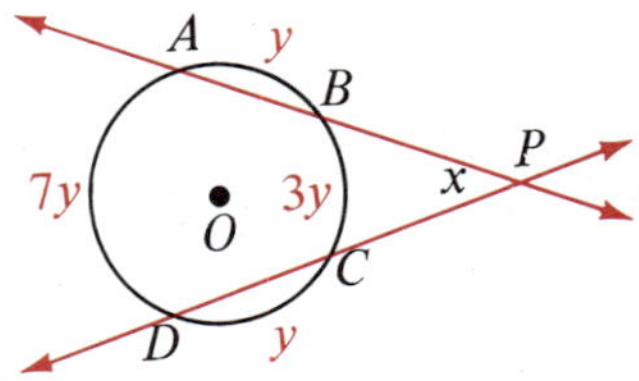

$$y + 3y + y + 7y = 360$$
$$12y = 360$$
$$y = 30$$
$$m\overset{\frown}{DA} = 7(30) = 210; \; m\overset{\frown}{BC} = 3(30) = 90$$
$$m \angle x = \frac{1}{2}(210-90)$$

Thus, m $\angle x = \frac{1}{2} \cdot 120$, or 60.

Written Exercises

A

1. Given: $m\overset{\frown}{RS} = 60$ and $m\overset{\frown}{UT} = 40$ Find m $\angle RPS$. **50**

2. Given: $m\overset{\frown}{RU} = 100$ and $m\overset{\frown}{ST} = 80$ Find m $\angle RPU$. **90**

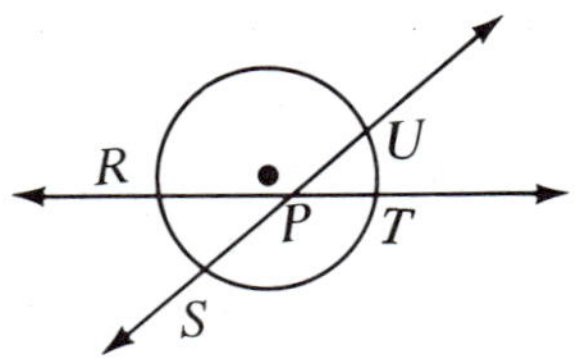

3. Given: $m\overset{\frown}{RS} = 20$ and $m\overset{\frown}{UT} = 100$ Find m $\angle UPT$. **60**

4. Given: $m\overset{\frown}{RU} = 90$ and $m\overset{\frown}{ST} = 110$ Find m $\angle TPS$. **100**

5. Given: $m\overset{\frown}{MJ} = 140$ and $m\overset{\frown}{LK} = 40$ Find m $\angle MPJ$. **50**

6. Given: $m\overset{\frown}{MJ} = 200$ and $m\overset{\frown}{LK} = 100$ Find m $\angle MPJ$. **50**

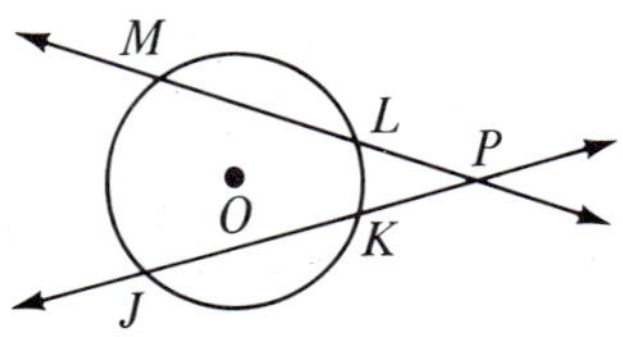

7. Given: $m\overset{\frown}{MJ} = 90$ and $m\overset{\frown}{LK} = 30$ Find m $\angle MPJ$. **30**

8. Given: $m\overset{\frown}{MJ} = 120$ and $m\overset{\frown}{LK} = 60$ Find m $\angle MPJ$. **30**

9. Given: $m\overset{\frown}{AB} = 115$ and m $\angle ATB = 63$ Find $m\overset{\frown}{DC}$. **11**

10. Given: m $\angle APC = 27$ and $m\overset{\frown}{DC} = 27$ Find $m\overset{\frown}{AB}$. **81**

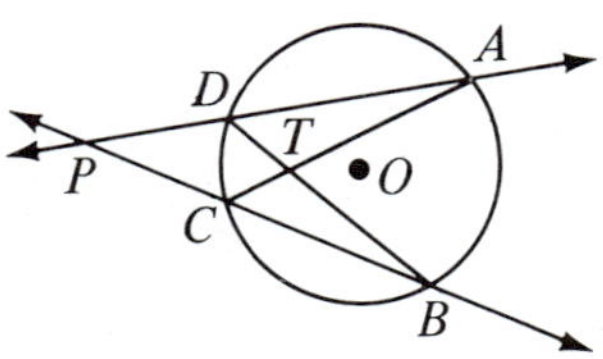

11. Given: $m\overset{\frown}{AB} = 6x$, $mBC = 6x$, $m\overset{\frown}{CD} = x$, and $m\overset{\frown}{DA} = 2x$ Find m $\angle ATB$. **84**

12. Given: $m\overset{\frown}{AB} = 4x$, $m\overset{\frown}{BC} = 3x$, $m\overset{\frown}{CD} = 2x$, and $m\overset{\frown}{DA} = 3x$ Find m $\angle APB$. **30**

13. Given: m $\angle V = 103$, m $\angle VRT = 71$, and $\overline{RS} \parallel \overline{VU}$ Find m $\angle TRS$ and m $\angle U$. **6; 109**

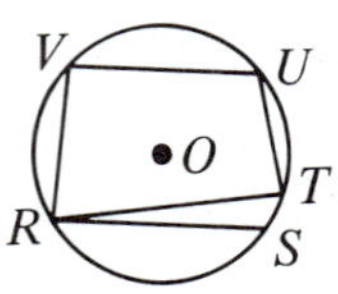

14. Given: Diameter $\overline{EG}$, $m\overset{\frown}{FG} = 41$, and $m\overset{\frown}{EH} = 53$ Find m $\angle E$ and m $\angle GPF$. **20.5; 47**

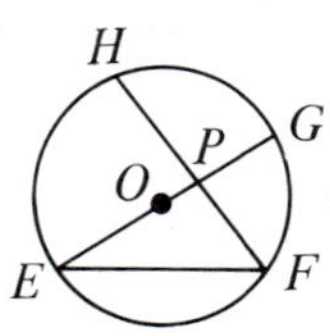

15. Given: $\overleftrightarrow{AB} \parallel \overleftrightarrow{DC}$, $m\overset{\frown}{CB} = 62$, and m $\angle DAB = 104$ Find m $\angle DEA$ and m $\angle ADB$. **62; 45**

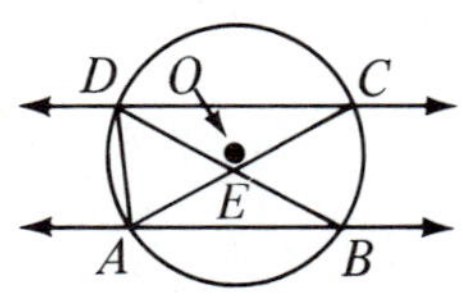

16. Given: m $\angle AOB = 70$ and m $\angle COD = 80$ Find m $\angle CTD$. **75**

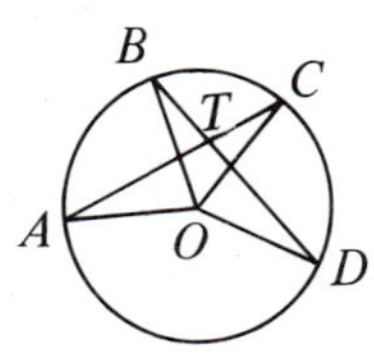

17. True or false? m $\angle CTD$ is equal to the average of $m\overset{\frown}{AB}$ and $m\overset{\frown}{CD}$. **T**

B **18.** Given: $m\overset{\frown}{EH} = 2x + 1$, $m\overset{\frown}{HG} = 3x - 2$, $m\overset{\frown}{GF} = 4x - 5$, and $m\overset{\frown}{FE} = x - 4$
Find $m\angle HIE$. **109**

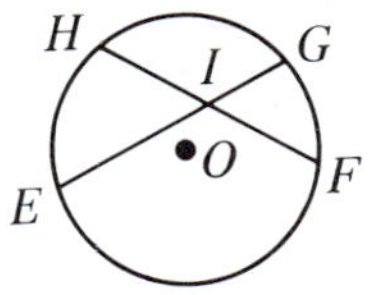

19. Given: $m\overset{\frown}{AB} = 7x + 3$, $m\overset{\frown}{BC} = 2x + 10$, $m\overset{\frown}{CD} = 6x$, and $m\overset{\frown}{DA} = 6x - 10$
Find $m\angle DPA$. **24**

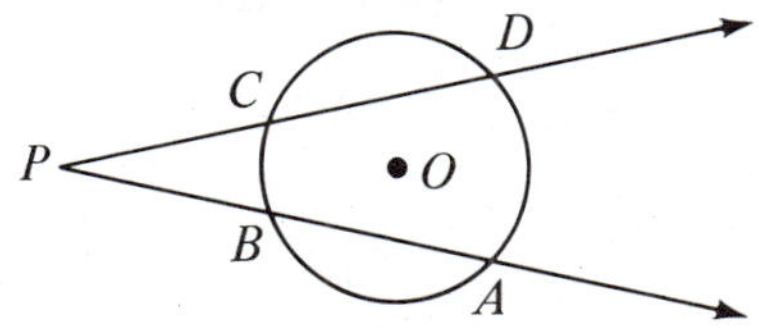

20. Given: $m\angle DEA = 42$ and $m\angle P = 18$
Find $m\overset{\frown}{AD}$ and $m\overset{\frown}{BC}$. **60; 24**

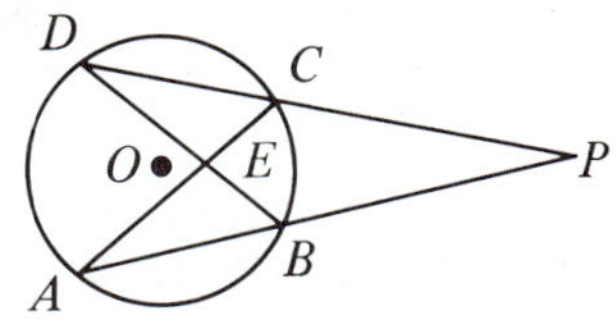

21. Given: $\overline{PQ} \perp \overline{RS}$ at T
Prove: $m\overset{\frown}{PR} + m\overset{\frown}{QS} = 180$

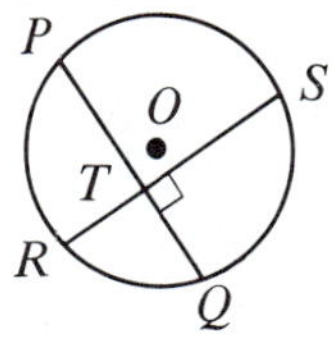

22. Given: $\overline{RT} \cong \overline{US}$
Prove: $\overline{UR} \cong \overline{TS}$

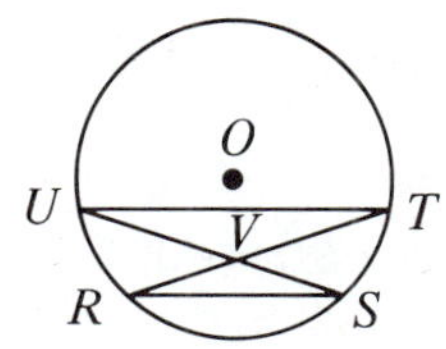

23. Given: Secants $\overleftrightarrow{LN}$ and $\overleftrightarrow{LQ}$
Prove: $\triangle LQM \sim \triangle LNP$

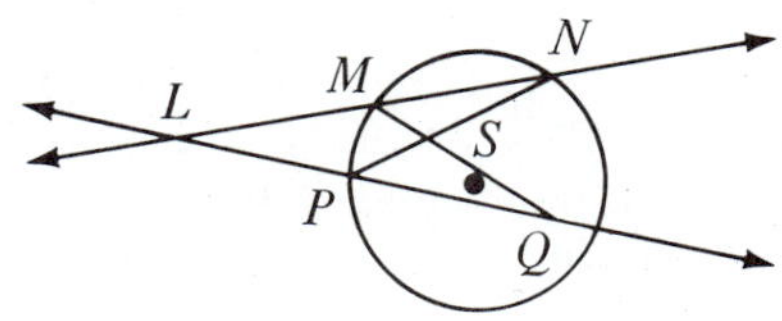

C **24.** Given: $m\overset{\frown}{AB} = m\overset{\frown}{BC}$
Prove: $m\angle x + m\angle y = \frac{3}{2} m\overset{\frown}{AB}$

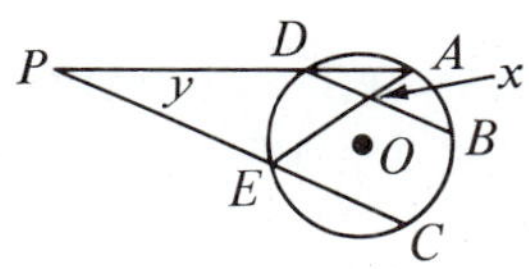

25. Given: $m\overset{\frown}{AB} = m\overset{\frown}{BC}$, $m\angle x = 60$, and $m\angle y = 15$
Find $m\overset{\frown}{AC}$. **100**

26. The vertex of an angle measuring 32° is in the exterior of a circle and its sides contain chords of the circle. If the sum of the measures of the intercepted arcs is 180°, find the measure of each intercepted arc. **122°, 58°**

27. Prove: If two circles are tangent internally and the center of the larger circle lies on the smaller circle, then any chord of the larger circle drawn from the point of tangency is bisected by the smaller circle.

NON-ROUTINE PROBLEMS

Some figures that are constructed with segments appear to be curves. The figures below can be drawn on paper or sewn with yarn or string on heavy paper or cloth. Can you construct each figure?

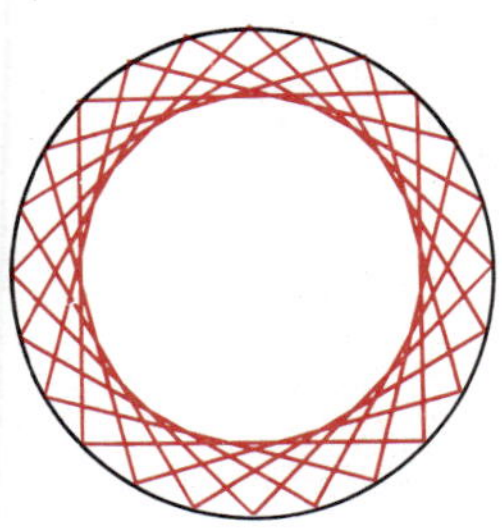

Concentric circles

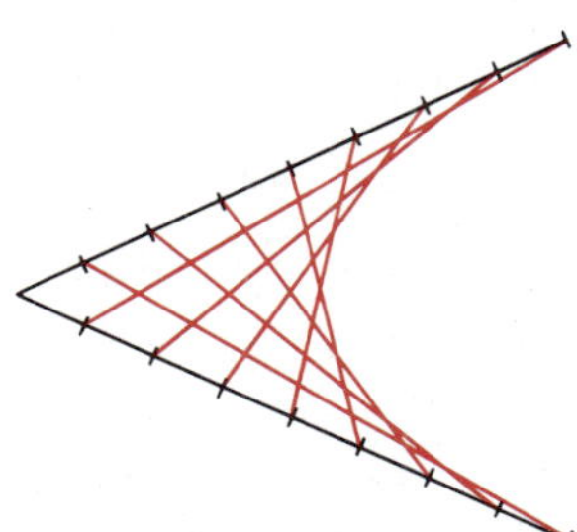

Parabola

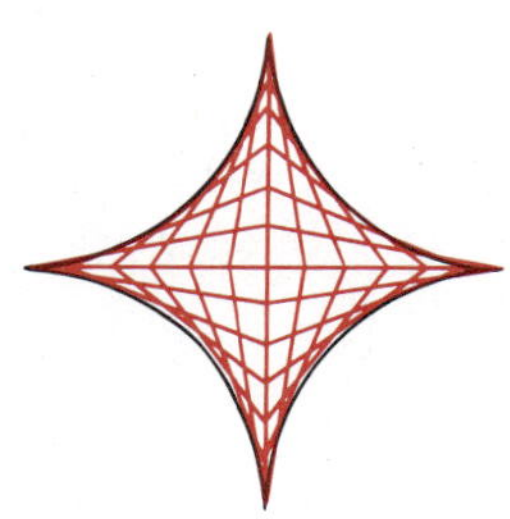

Asteroid

TANGENTS, SECANTS, AND ANGLES 11.8

Objective

To apply measurement properties for angles formed by a secant and a tangent or by two tangents

A secant and a tangent or two tangents can intersect in three possible positions:

Case I
Secant and tangent on the circle

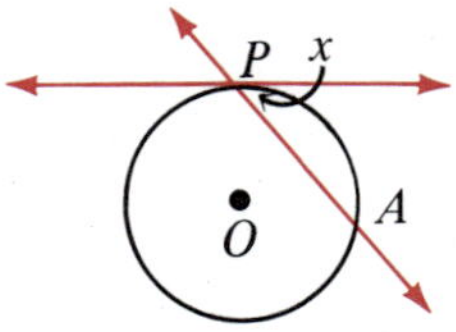

$\widehat{PA}$ is the intercepted arc with respect to $\angle x$.

Case II
Secant and tangent in the exterior of the circle

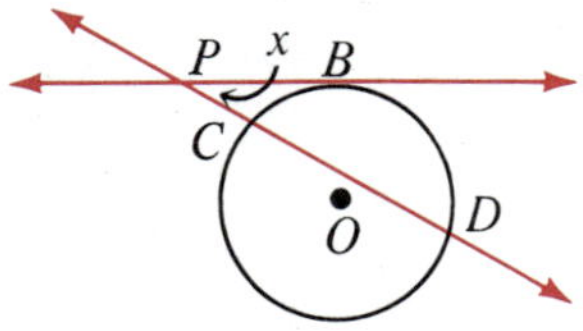

$\widehat{BC}$ and $\widehat{BD}$ are the intercepted arcs with respect to $\angle x$.

Case III
Two tangents in the exterior of the circle

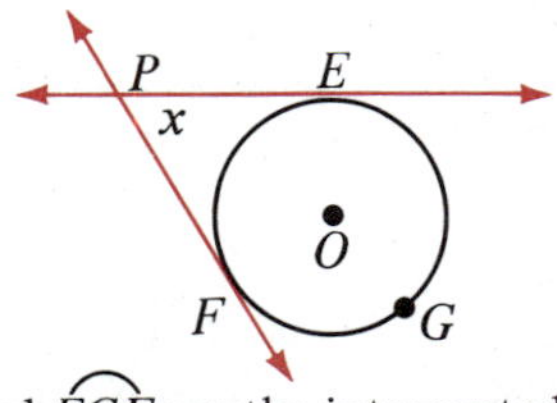

$\widehat{EF}$ and $\widehat{EGF}$ are the intercepted arcs with respect to $\angle x$.

In each case, there is a relationship between the measure of $\angle x$ and the measure(s) of the intercepted arc(s). The relationship for Case I is given in Theorem 11.14.

Theorem 11.14

The measure of an angle formed by a secant and a tangent intersecting at the point of tangency is one half the measure of the intercepted arc.

Given: Secant $\overleftrightarrow{PA}$, tangent $\overleftrightarrow{PB}$, intersecting at point P on $\odot O$

Prove: $m\angle x = \frac{1}{2} m\widehat{PA}$

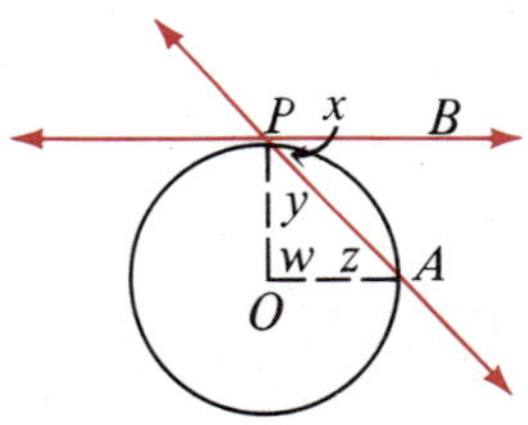

ANALYSIS

Draw radii $\overline{OP}$ and $\overline{OA}$. $m\angle x + m\angle y = 90$. Express $m\angle y$ in terms of $m\angle z$. Then use $m\angle w = m\widehat{PA}$.

Outline of Proof

1. $m\angle x + m\angle y = 90$
2. $m\angle x = 90 - m\angle y$
3. $m\angle y + m\angle w + m\angle z = 180$
4. $m\angle z = m\angle y$
5. $m\angle w = m\widehat{PA}$
6. $m\angle y + m\widehat{PA} + m\angle y = 180$
7. $m\angle y = \frac{1}{2}(180 - m\widehat{PA})$
8. $m\angle y = 90 - \frac{1}{2}m\widehat{PA}$
9. $m\angle x = 90 - \left(90 - \frac{1}{2}m\widehat{PA}\right)$
10. $\therefore m\angle x = \frac{1}{2}m\widehat{PA}$

If a secant and a tangent intersect at the point of tangency, then they intersect on the circle. In Example 1, Theorem 11.14 is applied to find the measure of an angle formed by a secant and a tangent.

Example 1

Given: Tangent $\overleftrightarrow{YZ}$ and $m\widehat{XY} = 124$
Find m $\angle XYZ$.

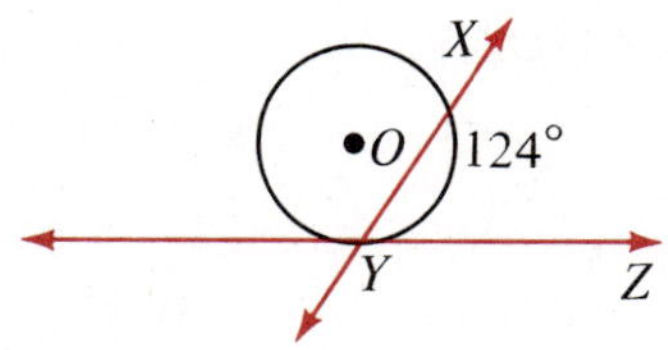

m $\angle XYZ = \frac{1}{2} m\widehat{XY}$, by Th. 11.14.

Thus, m $\angle XYZ = \frac{1}{2} \cdot 124$, or 62.

Theorem 11.15 deals with Case II on page 368. It involves the measure of an angle formed by a secant and a tangent that intersects in the exterior of a circle.

Theorem 11.15 The measure of an angle formed by a secant and a tangent intersecting in the exterior of a circle is one half the difference of the measures of the intercepted arcs.

Given: Tangent $\overleftrightarrow{PA}$ and secant $\overleftrightarrow{PB}$ intersecting at point P in the exterior of $\odot O$

Prove: m $\angle x = \frac{1}{2}(m\widehat{AB} - m\widehat{AC})$

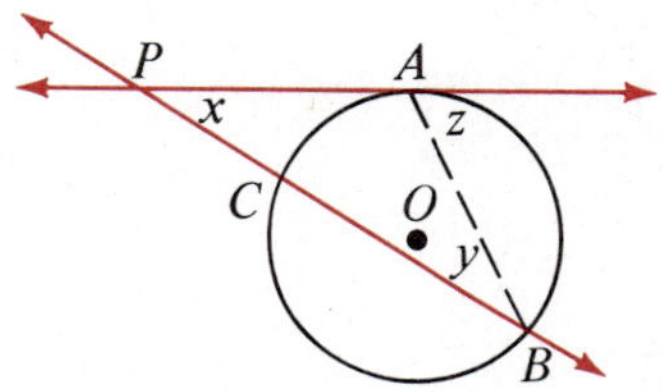

Proof: (See Exercise 24.)

Theorem 11.16 deals with Case III on page 368. It involves the measure of an angle formed by two tangents.

Theorem 11.16 The measure of an angle formed by two tangents to the same circle is one half the difference of the measures of the intercepted arcs.

Given: Tangents $\overleftrightarrow{PA}$ and $\overleftrightarrow{PB}$ to $\odot O$ intersecting at point P

Prove: m $\angle x = \frac{1}{2}(m\widehat{ACB} - m\widehat{AB})$

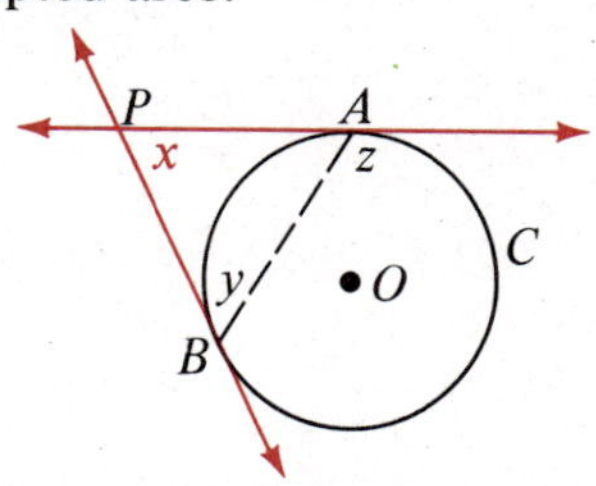

Proof: (See Exercise 25.)

Example 2

Given: Tangent $\overleftrightarrow{PW}$, $m\widehat{TW} = 173$, and $m\widehat{JW} = 69$.
Find m $\angle JPW$.

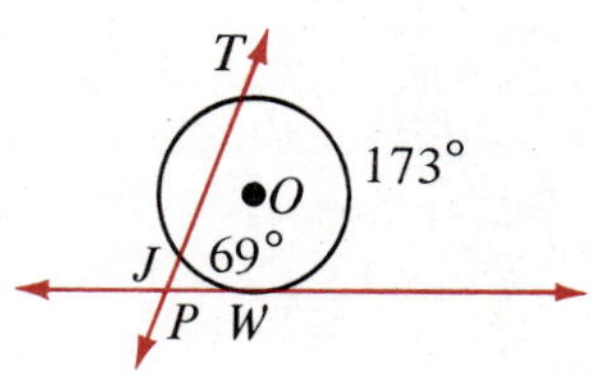

m $\angle JPW = \frac{1}{2}(173 - 69) = \frac{1}{2}(104) = 52$, by Th. 11.15.

Example 3 Given: Tangents $\overrightarrow{PR}$ and $\overrightarrow{PS}$ and m $\angle P = 48$.
Find m$\widehat{RS}$ and m$\widehat{RQS}$.

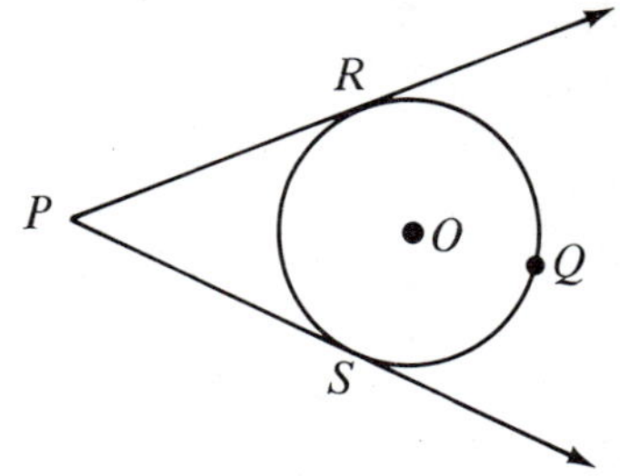

Let $x = m\widehat{RQS}$. Then $360 - x = m\widehat{RS}$.

$$m\ \angle P = \frac{1}{2}(m\widehat{RQS} - m\widehat{RS}) \text{ by Th. 11.16.}$$

$$48 = \frac{1}{2}[x - (360 - x)]$$

$$96 = 2x - 360$$

$$456 = 2x$$

$$x = 228;\ 360 - x = 132$$

Thus, $m\widehat{RS} = 132$ and $m\widehat{RQS} = 228$.

You may use the following summary to help you find the measures of angles formed by intersecting secants or tangents.

Summary **Angles Formed by Secants or Tangents**

If the intersection of two secants is in the interior of the circle, then the measure of the angle is one half the sum of the measures of the intercepted arcs.

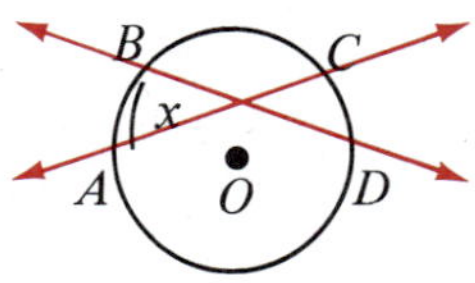

$$m\angle x = \frac{1}{2}(m\widehat{AB} + m\widehat{CD})$$

If the intersection of a secant and a tangent is on the circle, then the measure of the angle is one half the measure of the intercepted arc.

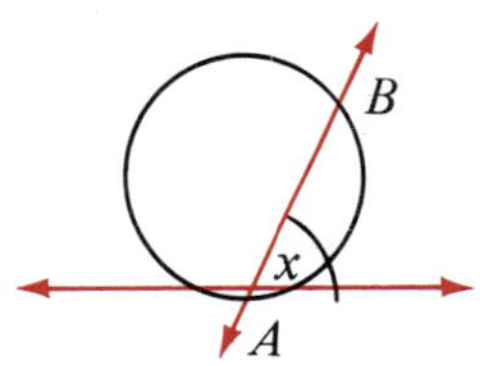

$$m\angle x = \frac{1}{2}(m\widehat{AB})$$

If the intersection of secants or tangents is in the exterior of the circle, then the measure of the angle is one half the difference of the measures of the intercepted arcs.

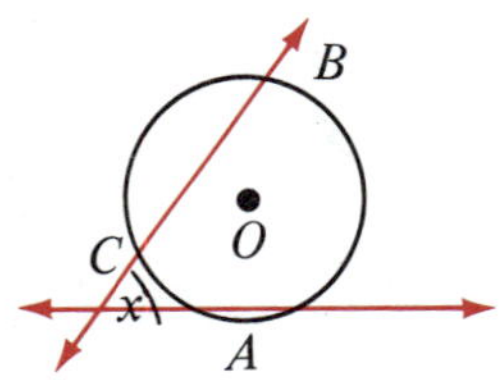

$$m\angle x = \frac{1}{2}(m\widehat{AB} - m\widehat{CA})$$

Written Exercises

For Exercises 1–20, you may assume that lines that appear to be tangents are tangents. Find each indicated measure.

A **1.** Given: $m\widehat{AB} = 104$
Find m $\angle APB$. **76**

2. Given: m $\angle APB = 71$
Find $m\widehat{AB}$. **109**

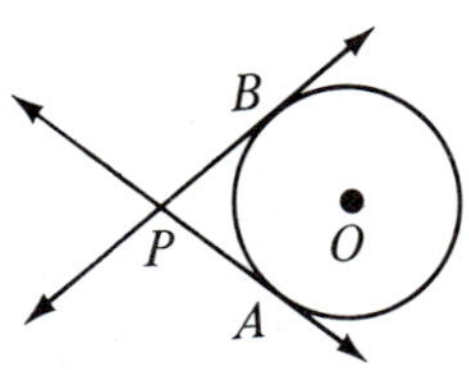

3. Given: $m\widehat{GE} = 42$
and $m\widehat{FE} = 138$
Find m $\angle FDE$. **48**

4. Given: m $\angle FDE = 46$
and $m\widehat{FE} = 121$
Find $m\widehat{GE}$. **29**

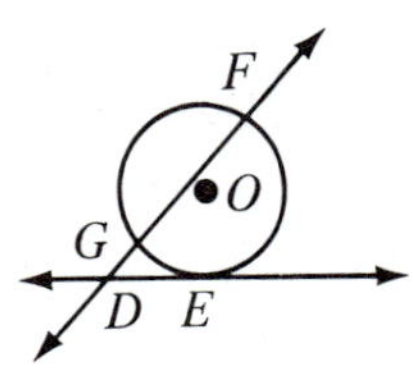

5. Given: $m\widehat{RT} = 142$
Find $m\angle RTS$. **71**

6. Given: $m\angle RTS = 61$
Find $m\widehat{RQT}$. **238**

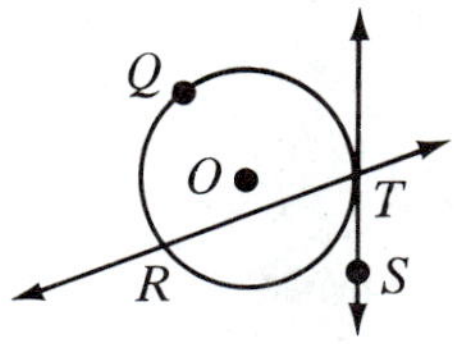

7. Given: $m\widehat{TU} = 40$ and $m\widehat{WV} = 70$
Find $m\angle WPV$. **55**

8. Given: $m\angle TPW = 151$ and $m\widehat{TU} = 30$
Find $m\widehat{WV}$. **28**

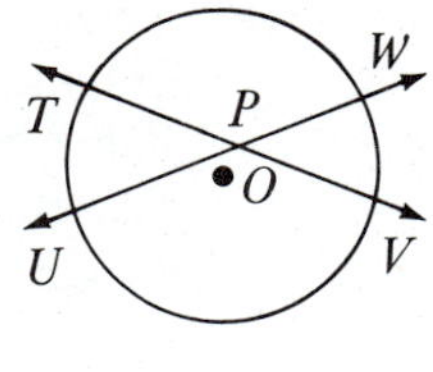

9. Given: $m\widehat{XVZ} = 212$
Find $m\angle XYZ$. **32**

10. Given: $m\widehat{WX} = 60$, $m\widehat{XV} = 130$, and $m\widehat{VZ} = 110$
Find $m\angle VYZ$. **25**

11. Given: $m\widehat{XW} = 47$ and $m\angle XYV = 47$
Find $m\widehat{XV}$. **141**

12. Given: $m\widehat{WZ} = 40$, $m\widehat{WX} = 70$, and $m\widehat{XV} = 140$
Find $m\angle XYZ$. **70**

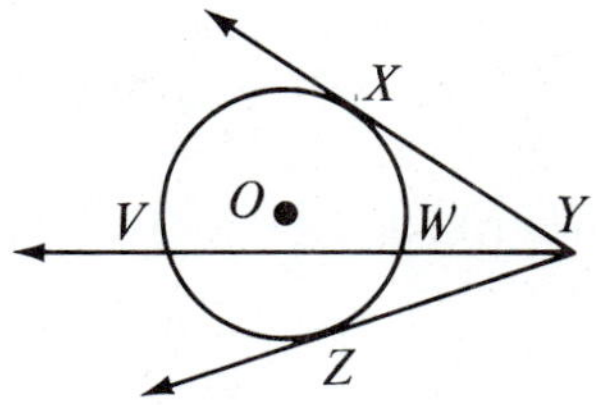

13. Given: $m\widehat{XVZ} = 3 \cdot m\widehat{XZ}$
Find $m\angle XYZ$. **90**

14. Given: $m\angle XYZ = 58$
Find $m\widehat{XZ}$ and $m\widehat{XVZ}$. **122; 238**

15. Given: $m\widehat{NMK} = 162$
Find $m\angle JKN$ **99**

16. Given: $m\widehat{GK} + m\widehat{NM} = 62$
Find $m\angle GHK$. **31**

17. Given: $m\widehat{KGN} = \frac{3}{2} m\widehat{KMN}$
Find $m\angle NKL$. **72**

18. Given: $m\widehat{GN} = 142$ and $m\widehat{KM} = 160$
Find $m\angle NHM$. **29**

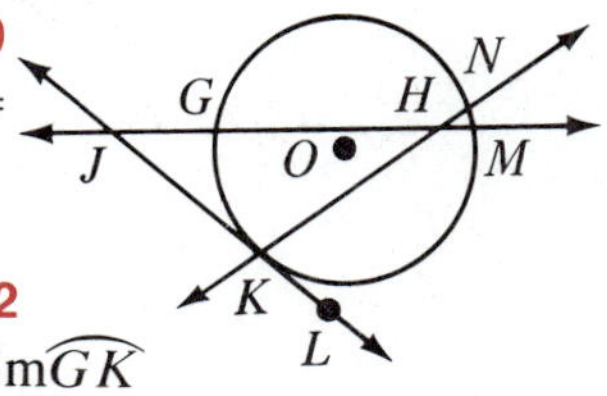

19. Given: $m\widehat{MK} = 3m\widehat{GK}$ and $m\widehat{GNM} = 160$
Find $m\angle MJL$. **50**

20. Given: $m\angle GHK = 35$, $m\widehat{GN} = 154$, and $m\widehat{NM} = 20$
Find $m\angle NKL$. **78**

B 21. Given: Diameter $\overline{AC}$ and $\overline{AD} \cong \overline{AO}$
Find $m\angle CAD$ and $m\angle DBA$. **60; 30**

22. Given: $m\widehat{AD} = 92$, $m\widehat{CD} = 103$, $m\widehat{BC} = 41$, and tangent $\overleftrightarrow{AF}$
Find $m\angle CQD$ and $m\angle ECA$. **113.5; 134**

23. Given: $m\widehat{AB} = 43$, $\overline{AC} \cong \overline{BC}$, and tangent $\overleftrightarrow{DE}$
Find $m\angle CAO$ and $m\angle DAC$. **10.75; 79.25**

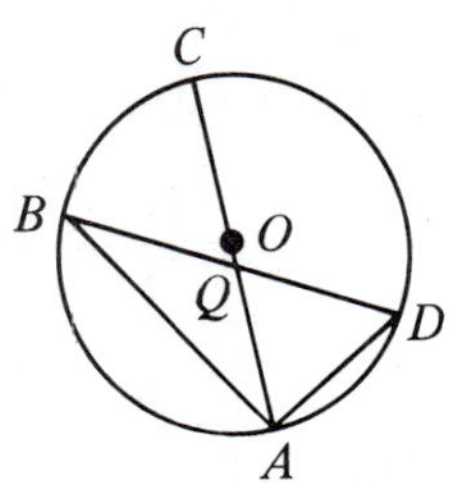

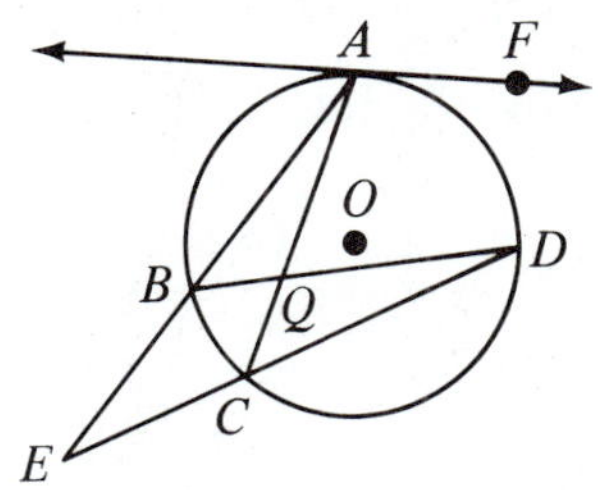

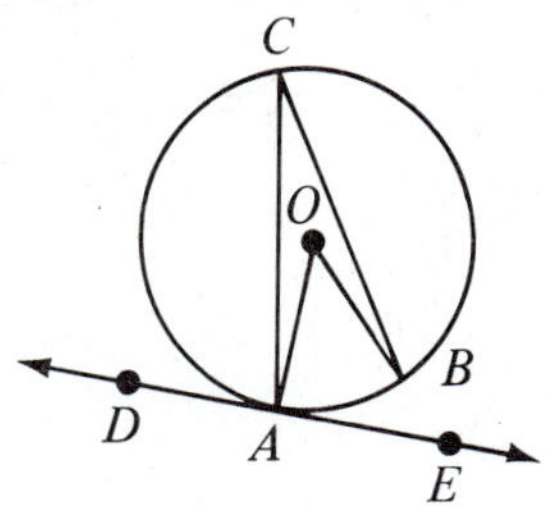

24. Prove Theorem 11.15.

25. Prove Theorem 11.16.

26. Given: $\overline{US} \cong \overline{VS}$ and tangent $\overleftrightarrow{RT}$
Prove: $\overline{UV} \parallel \overleftrightarrow{RT}$

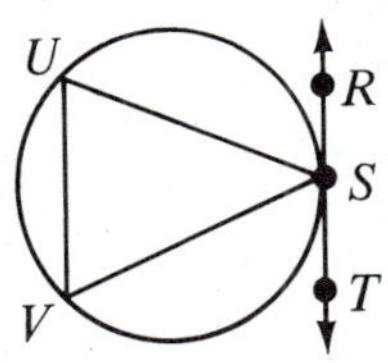

27. Given: Tangent $\overleftrightarrow{RT} \parallel \overline{UV}$
Prove: $\overline{US} \cong \overline{VS}$

28. Prove: If two tangents to a circle are parallel, then the points of tangency divide the circle into two congruent arcs.

29. Prove: If two circles are tangent internally at point P, and chord $\overline{PA}$ of the larger circle intersects the smaller circle at B, then $m\widehat{PA} = m\widehat{PB}$.

30. Prove: The bisector of the angle formed by a tangent and a chord drawn from the point of tangency divides the intercepted arc into two congruent arcs.

31. Given: $\overleftrightarrow{RS}$ tangent to $\odot O$ and $\odot P$ at M, diameter $\overline{KL}$, and diameter $\overline{QN}$
Prove: $m\widehat{LM} = m\widehat{NM}$

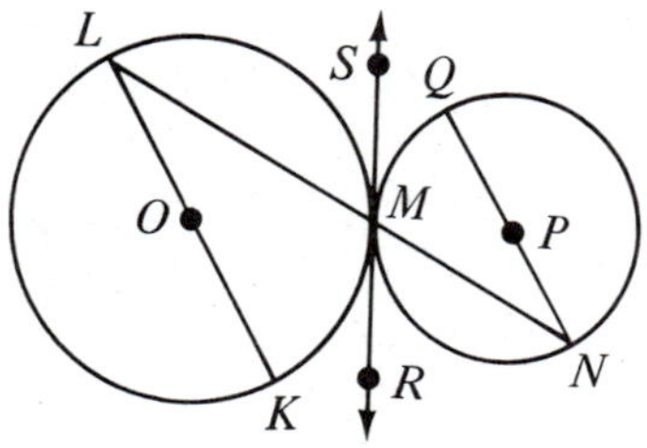

32. Given: $\overleftrightarrow{RS}$ tangent to $\odot O$ and $\odot P$ at M, diameter $\overline{KL}$, and diameter $\overline{QN}$
Prove: $\overline{KL} \parallel \overline{QN}$

33. Prove: If a quadrilateral is inscribed in a circle, then an exterior angle of the quadrilateral is congruent to its opposite interior angle.

34. Prove: If $\overline{RS}$ is the common chord of two intersecting circles and $\overline{ST}$ and $\overline{SU}$ are diameters, then $\angle TRU$ is a straight angle.

35. Prove: If two circles are tangent externally at point P and $\overleftrightarrow{AB}$ is a common external tangent which intersects the two circles at points A and B, respectively, then $\angle APB$ is a right angle.

36. Prove: If two circles are tangent internally at point P and $\overline{PA}$ and $\overline{PB}$ are chords of the larger circle which intersect the smaller circle at points C and D, respectively, then $\overline{AB} \parallel \overline{CD}$.

CALCULATOR ACTIVITIES

A calculator can be helpful to you in finding the measures of angles and arcs formed by intersecting chords, secants, or tangents. See if you can show how each of the formulas below has been translated into the "language" of the calculator.

Part 1 An angle formed by two intersecting chords

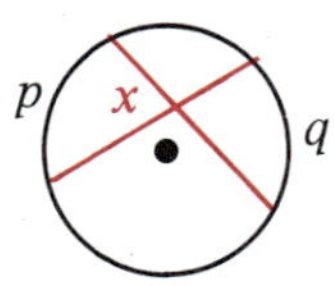

$x = (p + q) \div 2$ or
$p = (2 \times x) - q$ or
$q = (2 \times x) - p$

Part 2 An angle formed by secants or tangents intersecting outside the circle

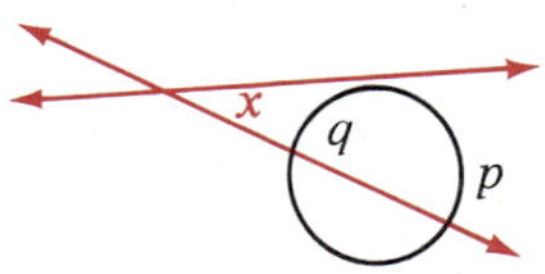

$x = (p - q) \div 2$
$p = (2 \times x) + q$
$q = p - (2 \times x)$ ◄ *(Be careful!)*

Use one of the formulas from Part 1 to find the indicated angle or arc measure.

1. If $p = 68$ and $q = 75$, find x. **71.5**
2. If $x = 37$ and $p = 55$, find q. **19**
3. If $x = 47$ and $q = 82$, find p. **12**

Use one of the formulas from Part 2 to find the indicated angle or arc measure.

4. If $p = 158$ and $q = 65$, find x. **46.5**
5. If $x = 39$ and $q = 73$, find p. **151**
6. If $x = 43$ and $p = 162$, find q. **76**

PRODUCTS OF LENGTHS 11.9

Objective

To find ratios and products of lengths of segments related to a circle

Intersecting chords, secants, and tangents often determine pairs of similar triangles. The resulting ratios can be useful in finding lengths of segments. Theorem 11.17 deals with intersecting chords.

Theorem 11.17

If two chords of a circle intersect, then the product of the lengths of the segments of one chord equals the product of the lengths of the segments of the other.

Given: $\odot O$ with chords $\overline{AB}$ and $\overline{CD}$ intersecting at point P

Prove: $AP \cdot PB = DP \cdot PC$

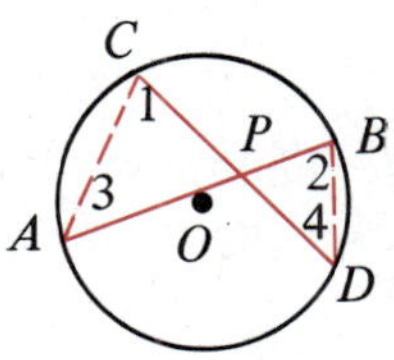

ANALYSIS

Show $\frac{AP}{DP} = \frac{PC}{PB}$ by proving $\triangle APC \sim \triangle DPB$.

Proof

Statements	Reasons
1. $\odot O$ with chords $\overline{AB}$ and $\overline{CD}$	1. Given
2. Draw $\overline{AC}$ and $\overline{BD}$.	2. Two points are contained in exactly one line.
3. $\angle 1 \cong \angle 2$, $\angle 3 \cong \angle 4$	3. Inscribed $\angle$'s that intercept the same arc are $\cong$.
4. $\triangle APC \sim \triangle DPB$	4. AA for $\sim \triangle$'s
5. $\frac{AP}{DP} = \frac{PC}{PB}$	5. Corr. sides of $\sim \triangle$'s are proportional.
6. $\therefore AP \cdot PB = DP \cdot PC$	6. Prod. of extremes = prod. of means.

Example 1

Given: $VS = 4$, $RV = 8$, and $UV = 6$

Find VT.

$UV \cdot VT = RV \cdot VS$, by Th. 11.17

Let $VT = x$. ► $6 \cdot x = 8 \cdot 4$

$6x = 32$

$x = \frac{32}{6} = \frac{16}{3}$

Thus, $VT = \frac{16}{3}$, or $5\frac{1}{3}$.

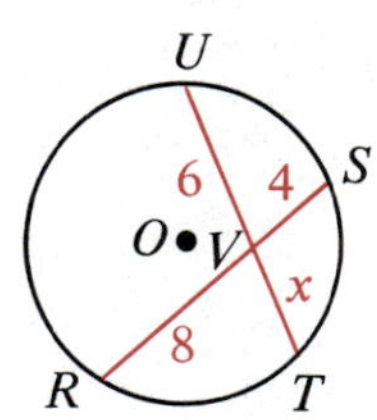

In the figure, $\overline{PA}$ is a **secant segment**, and $\overline{PB}$ is its **external secant segment**. Similarly, $\overline{PC}$ is a secant segment, and $\overline{PD}$ is its external segment. Theorem 11.18 deals with secants that intersect in the exterior of a circle.

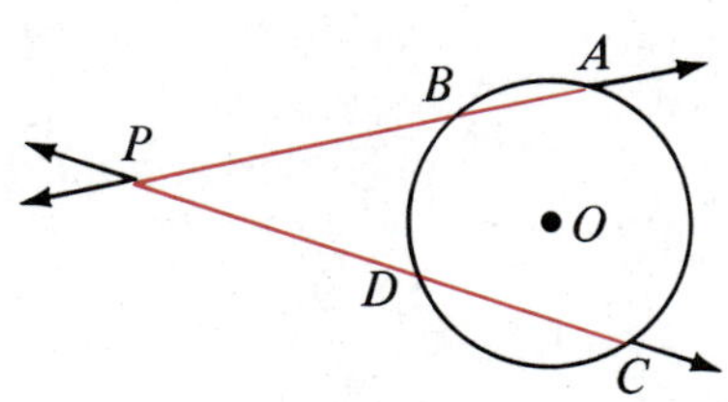

Theorem 11.18

If two secants intersect in the exterior of a circle, then the product of the lengths of one secant segment and its external secant segment equals the product of the lengths of the other secant segment and its external secant segment.

Given: $\odot O$ with secants $\overleftrightarrow{PA}$ and $\overleftrightarrow{PC}$ intersecting at point P

Prove: $PA \cdot PB = PC \cdot PD$

Proof: (See Exercise 16.)

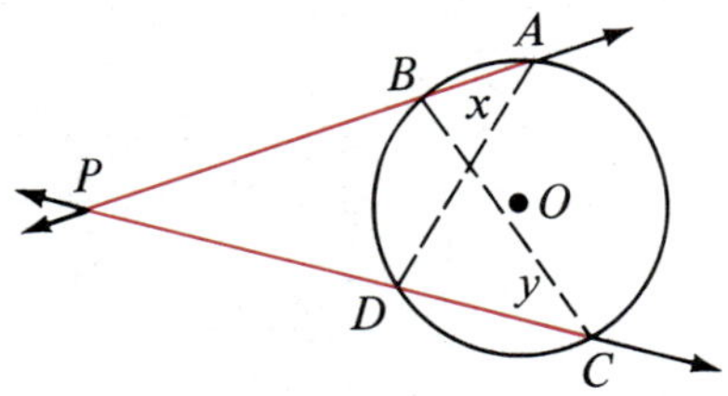

ANALYSIS

Draw $\overline{AD}$ and $\overline{BC}$. Show $\frac{PA}{PC} = \frac{PD}{PB}$ by proving $\triangle PAD \sim \triangle PCB$.

You can use Theorem 11.18 to find the lengths of segments formed by two secants intersecting in the exterior of a circle.

Example 2

Given: $QR = 10$, $PR = 6$, and $PT = 8$
Find TS.

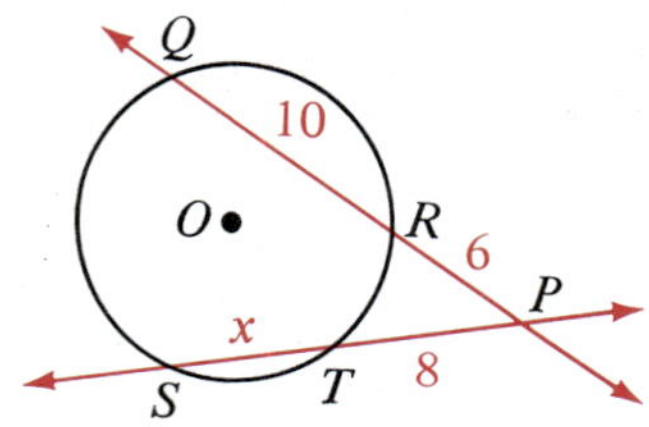

$PS \cdot PT = PQ \cdot PR$, by Th. 11.18.

Let $TS = x$; then $PS = x + 8$. ► $(x + 8)8 = 16 \cdot 6$

$8x + 64 = 96$

$8x = 32$

$x = 4$

Thus, $TS = 4$.

In the figure, $\overleftrightarrow{PA}$ is tangent to $\odot O$, and $\overleftrightarrow{PB}$ is a secant. $\overline{PA}$ is a **tangent segment,** $\overline{PB}$ is a **secant segment,** and $\overline{PC}$ is an **external secant segment.** Theorem 11.19 is a statement about the products of lengths involving a tangent segment, a secant segment, and an external secant segment.

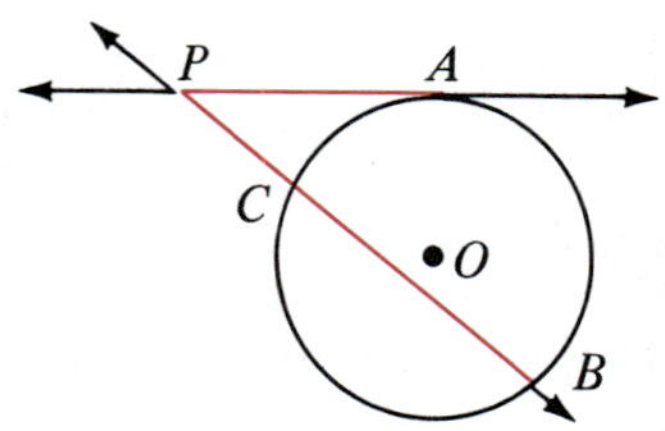

Theorem 11.19

If a tangent and a secant intersect in the exterior of a circle, then the square of the length of the tangent segment equals the product of the lengths of the secant segment and its external secant segment.

Given: $\odot O$ with tangent $\overleftrightarrow{PA}$ and secant $\overleftrightarrow{PB}$ intersecting at point P

Prove: $(PA)^2 = PB \cdot PC$

Proof: (See Exercise 17.)

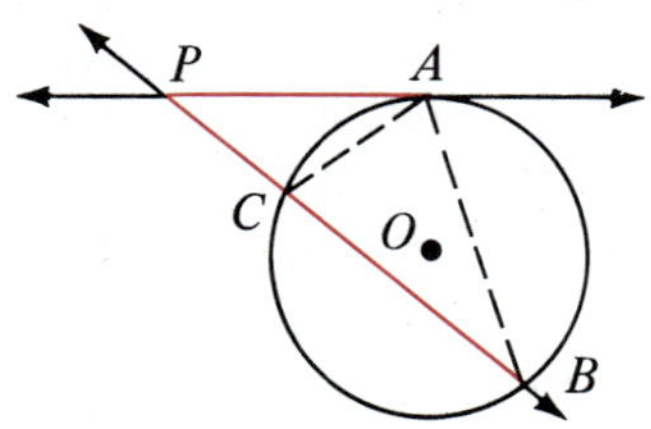

ANALYSIS

Draw $\overline{AB}$ and $\overline{AC}$. Show $\frac{PA}{PB} = \frac{PC}{PA}$ by proving $\triangle PAB \sim \triangle PCA$.

Examples 3 and 4 are applications of Theorem 11.19.

Example 3

Given: Secant $\overleftrightarrow{AB}$, tangent $\overleftrightarrow{AD}$, $AC = 2$, and $CB = 4$

Find AD.

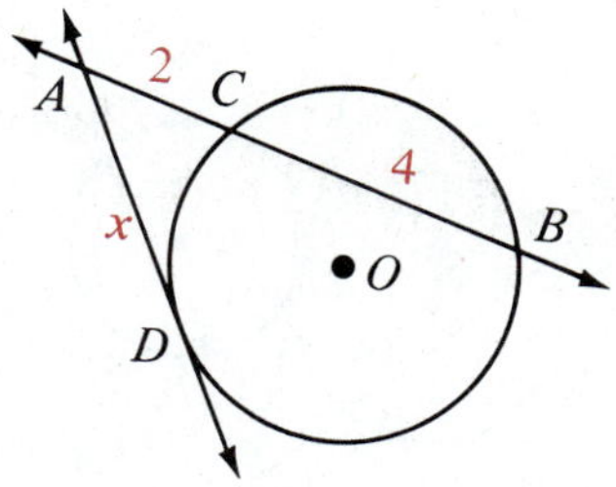

$(AD)^2 = AB \cdot AC$ by Th. 11.19.

Let $AD = x$. ► $x^2 = 6 \cdot 2$

$x^2 = 12$

$x = \sqrt{12}$, or $2\sqrt{3}$

Thus, $AD = 2\sqrt{3}$.

Example 4

A tangent and a secant intersect in the exterior of a circle. The tangent segment is 10 cm long and the secant segment is 25 cm long. Find the length of the external secant segment.

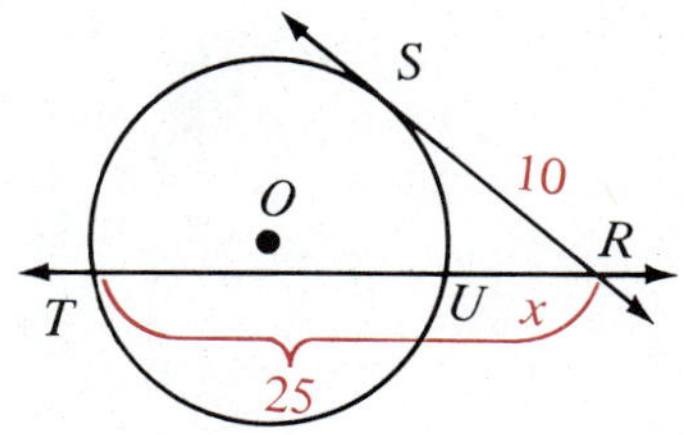

Draw and label a figure.

$(RS)^2 = RT \cdot RU$

Let $RU = x$. ► $10^2 = 25 \cdot x$

$100 = 25x$

$4 = x$

Thus, $RU = 4$ cm.

Written Exercises

For Exercises 1–15, you may assume that lines that appear to be tangents are tangents. Find each indicated measure.

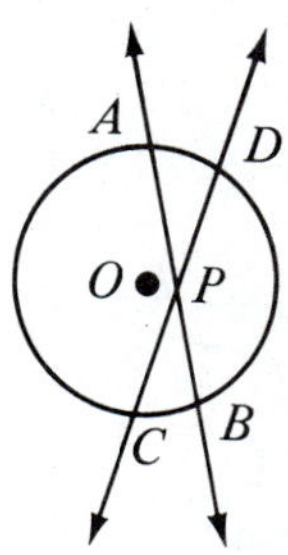

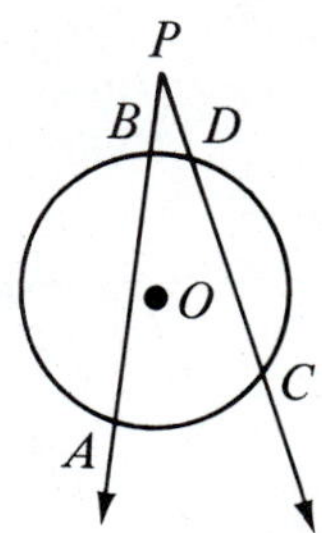

A **1.** Given: $AP = 12$, $PC = 9$, and $PD = 4$ Find PB. **3**

2. Given: $AP = 12$, $PC = 8$, and $PB = 4$ Find PD. **6**

3. Given: $AP = 9$, $AB = 17$, and $CP = 12$ Find CD. **18**

4. Given: $AP = 18$, $BP = 3$, and $CP = 9$ Find CD. **3**

5. Given: $AP = 4x$, $PB = x$, $CP = 6$, and $PD = 6$ Find AP. **12**

6. Given: $AP = 9x$, $BP = 4x$, $CP = 16$, and $DP = 9$ Find BP. **8**

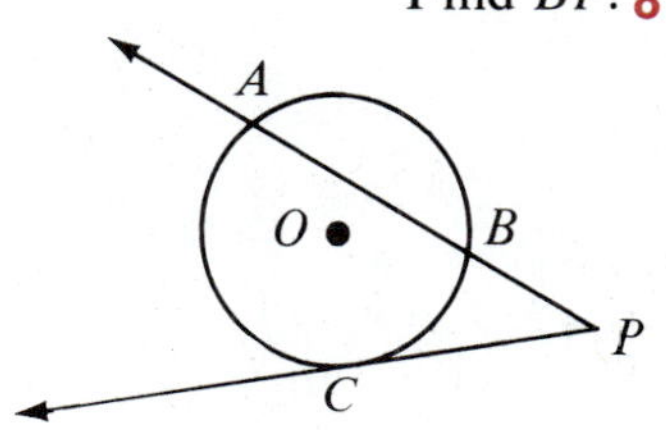

7. Given: $PB = 3$ and $PA = 12$ Find PC. **6**

8. Given: $PA = 16$ and $PC = 8$ Find PB. **4**

9. Given: $PA = 4x$, $PB = x$, and $PC = 6$ Find PA. **12**

10. Given: $PA = 7$ and $PB = 3$ Find PC. $\sqrt{21}$

B

11. Given: $RX = 16$, $RS = 5$, and $RU = 6$. Find RW and SX. $13\frac{1}{3}$; 11

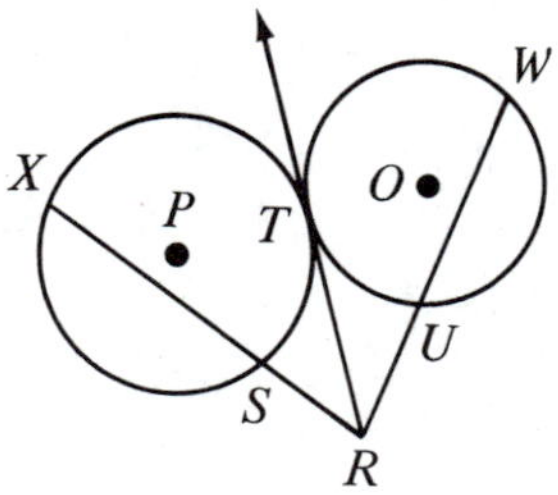

12. Given: $PW + WX = 12$ and $PY = 3$. Find $\frac{PW}{PZ}$. $\frac{1}{4}$

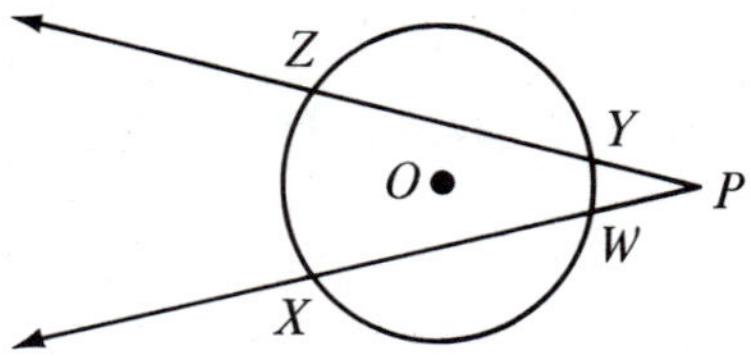

13. Given: $SR = 20$, $WR = 4$, and $XR = 6$. Find YX and SY. $7\frac{1}{3}$; $6\frac{2}{3}$

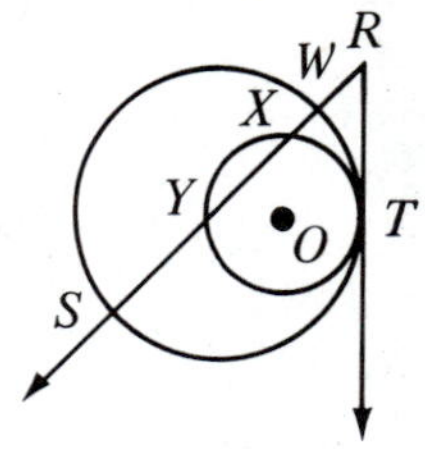

14. Given: $QR = 2x$, $PQ = 6$, $PT = 2x - 1$, and $PS = 8$. Find PR and PT. 28; 21

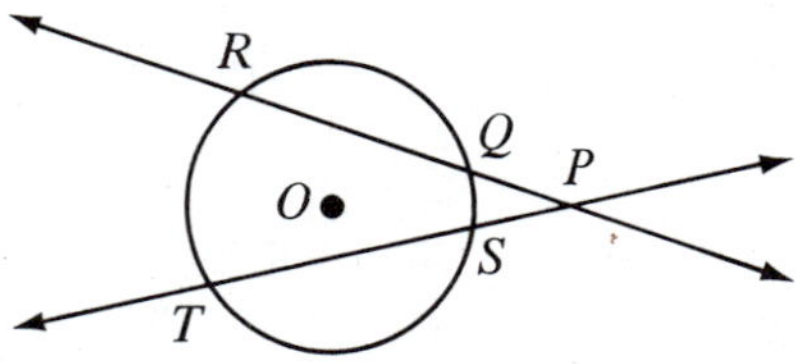

15. Given: $QR = 8$, $ST = 2$, and $PS = 6$. Find PR. 12

16. Prove Theorem 11.18.

17. Prove Theorem 11.19.

C

18. Given: Tangent $\overrightarrow{TX}$ and secant $\overleftrightarrow{TZ}$
Prove: $XT \cdot XY = XZ \cdot YT$

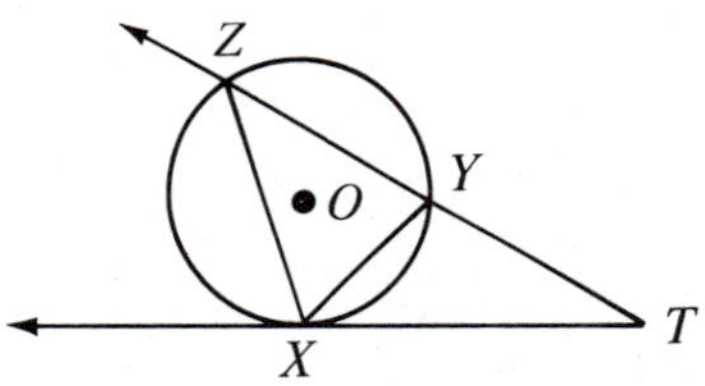

19. Given: Secants $\overrightarrow{TK}$ and $\overrightarrow{TM}$
Prove: $KL \cdot TJ = MJ \cdot TL$

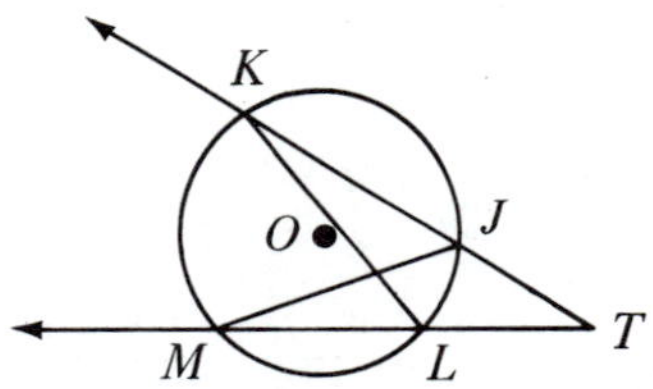

20. Given: Diameter $\overline{GH}$, tangent $\overleftrightarrow{JH}$, and secant $\overleftrightarrow{JG}$
Prove: $IH = \sqrt{IJ \cdot IG}$

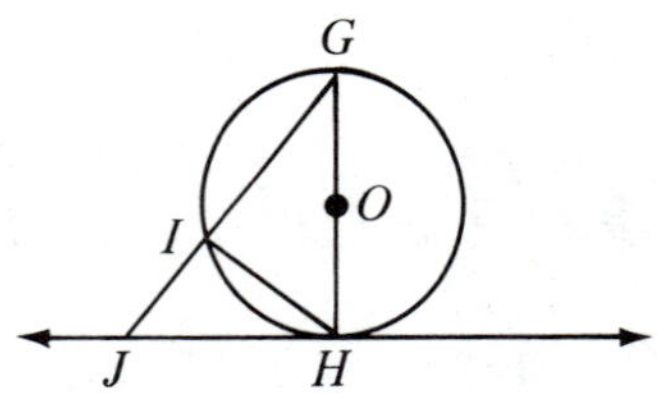

21. Given: Tangent $\overleftrightarrow{TM}$ and secant $\overleftrightarrow{MP}$
Prove: $MS \cdot MP = MR \cdot MQ$

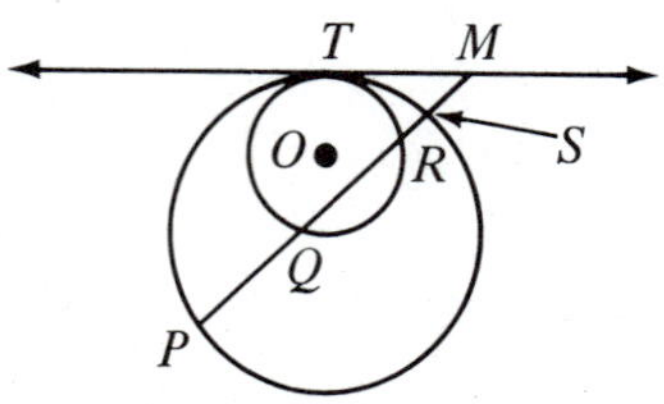

22. Given: Secants $\overleftrightarrow{PC}$ and $\overleftrightarrow{PF}$
Prove: $\frac{PA}{PC} = \frac{PD}{PF}$

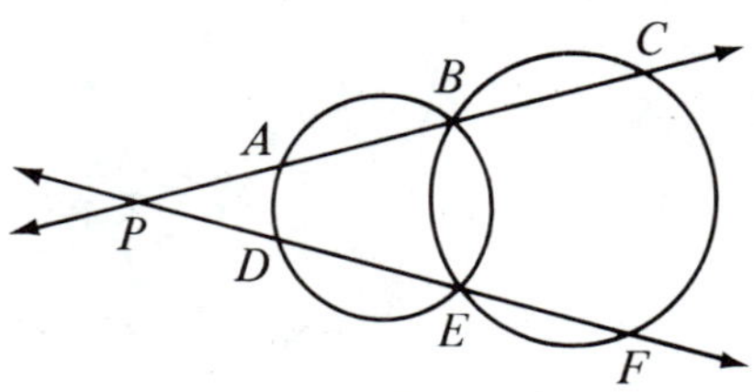

23. Given: $\widehat{ED} \cong \widehat{DC}$ and $\widehat{EA} \cong \widehat{BC}$
Prove: $AD \cdot FD = BD \cdot GD$

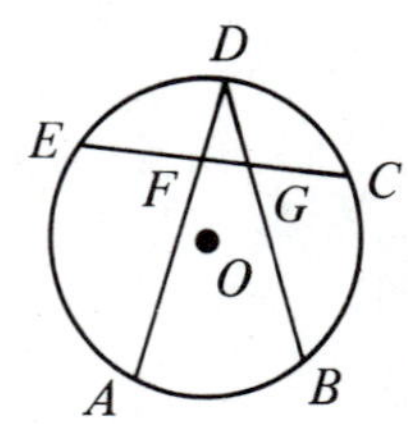

CUMULATIVE REVIEW

1. One base of a trapezoid measures 4 cm and the median measures 7 cm. Find the length of the other base. 10 cm

2. In $\triangle ABC$, $AB = 6$, $BC = 9$, and $CA = 7$; D is the midpoint of $\overline{CA}$ and E is the midpoint of $\overline{BC}$. Find DE. 3

CONSTRUCTIONS 11.10

Objective **To perform constructions related to circles**

Recall that an angle inscribed in a semicircle is a right angle. Also, a line perpendicular to a radius at its endpoint on the circle is tangent to the circle. These two statements are the basis for the construction of tangents to a circle from an exterior point.

Construction Construct the tangents to a circle from a point in the exterior of the circle.

Given: $\odot O$ and point P in the the exterior of $\odot O$
Construct: Two tangents to $\odot O$ from P

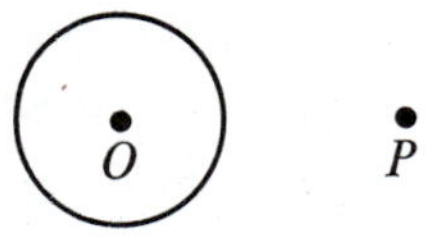

Strategy: Construct a $\odot$ with $\overline{OP}$ as its diameter. Apply Th. 11.7.

Draw $\overline{OP}$. Construct M, the midpoint of $\overline{OP}$.

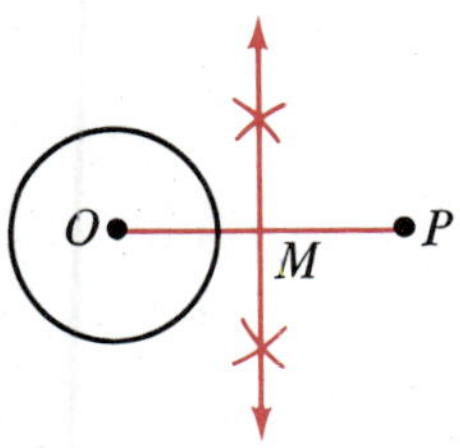

With radius $\overline{MO}$, construct $\odot M$ intersecting $\odot O$ in points T and S.

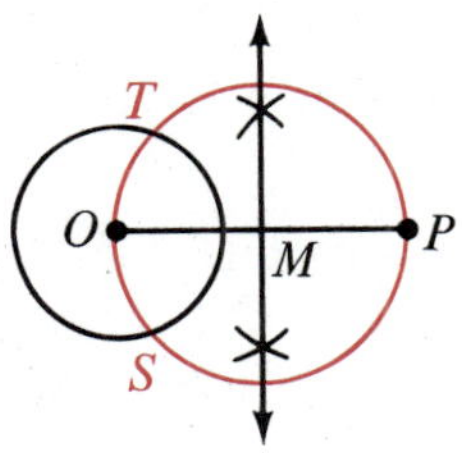

Draw $\overleftrightarrow{PT}$ and $\overleftrightarrow{PS}$.

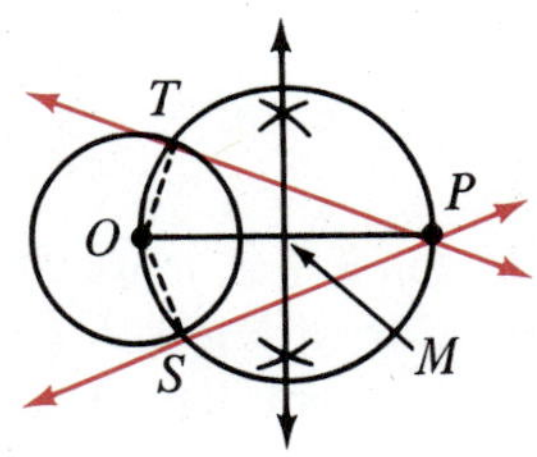

Conclusion: $\overleftrightarrow{PT}$ and $\overleftrightarrow{PS}$ are the desired tangents.

To justify the construction above, draw $\overline{OT}$ and $\overline{OS}$. $\angle OTP$ and $\angle OSP$ are right angles since each is inscribed in a semicircle. So $\overline{OT} \perp \overleftrightarrow{PT}$ and $\overline{OS} \perp \overleftrightarrow{PS}$. **Thus,** $\overleftrightarrow{PT}$ and $\overleftrightarrow{PS}$ are tangents, since each is perpendicular to a radius at its endpoint on the circle.

The next construction provides the basis for many beautiful designs.

Construction

Inscribe a regular hexagon in a circle.

Given: $\odot O$
Construct: Inscribed hexagon $ABCDEF$ in $\odot O$.

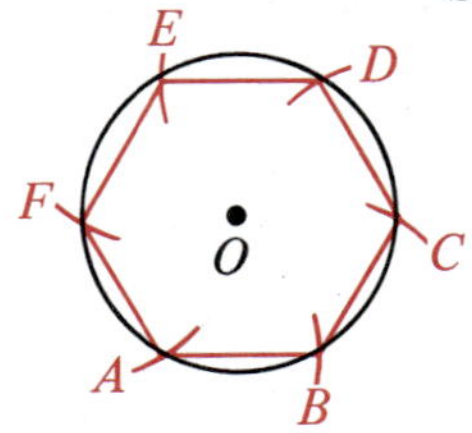

Strategy: Open compass to the length of a radius. Starting at any point A on the circle, mark successive arcs on the circle. Draw chords to form a regular hexagon.

Construction: (See Exercise 14.)

You can use a circle to construct segments whose lengths are in a particular proportion, as shown in the next construction.

Construction

Construct a segment whose length is the geometric mean of the lengths of two given segments.

Given: Segments of lengths a and b
Construct: A segment of length x such that $\frac{a}{x} = \frac{x}{b}$.

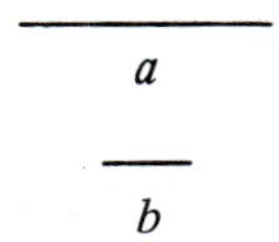

On a line, mark off $\overline{CD}$ with length a and $\overline{DE}$ with length b.

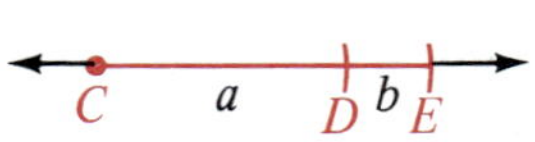

Construct M, the midpoint of $\overline{CE}$. Then construct $\odot M$ with radius $\overline{ME}$.

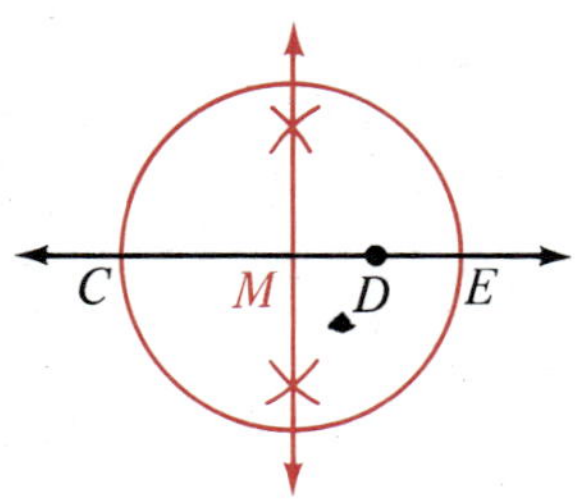

At D, construct a perpendicular to $\overline{CE}$ that intersects $\odot M$ at F. Then $DF = x$.

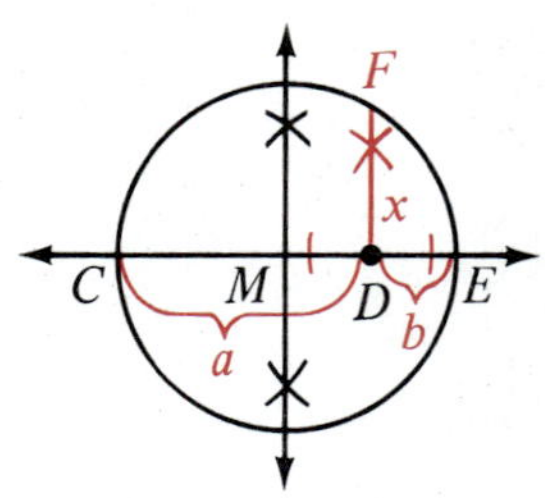

Conclusion: $\overline{DF}$ is the required segment of length x such that $\frac{a}{x} = \frac{x}{b}$.

Written Exercises

A

1. Draw a segment 2 cm long. Construct a circle with a radius congruent to this segment.
2. Draw a segment 5 cm long. Construct a circle with a diameter congruent to this segment.
3. Draw a circle with a diameter approximately 6 cm long. Mark a point on the circle. Construct a tangent to the circle at this point.
4. Draw a circle with a diameter approximately 4 cm long. From a point outside this circle, construct two tangents to the circle.
5. Construct the two designs shown at the bottom of page 377.
6. Inscribe a regular hexagon and an equilateral triangle in the same circle.
7. Draw a segment 4 cm long and a line. Mark a point on the line. At that point, construct a circle tangent to the line with a radius congruent to the given segment.
8. Construct a segment whose length is the geometric mean of lengths p and q.

p q

9. Inscribe a circle in a given square.
10. Inscribe a square in a given circle.

B

11. Draw a circle and a point outside the circle. Construct two tangents to the circle from this point. Construct a second circle tangent both to the given circle and to the two tangents.
12. Draw two segments. Construct a right triangle with the hypotenuse congruent to the longer segment and the altitude to the hypotenuse congruent to the shorter segment. [*Hint:* An angle inscribed in a semicircle is a right angle.]
13. Inscribe a regular octagon in a given circle.
14. Inscribe a regular hexagon in a circle. Then justify your construction. [*Hint:* Draw radii to the vertices. Use equiangular triangles.]
15. Inscribe a regular polygon of 12 sides in a given circle.
16. Construct a segment whose length is the geometric mean of the lengths of two given segments. Then justify your construction.
17. Given segments with lengths a, b, and c, construct a segment with length x such that $\frac{a}{b} = \frac{c}{x}$.

C

18. Inscribe a square in a semicircle.

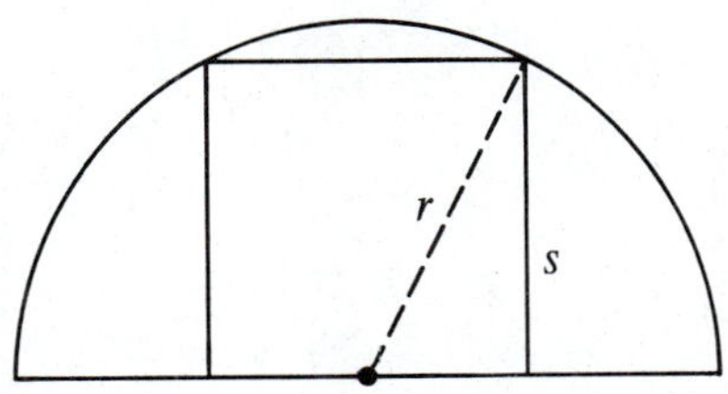

Draw two segments of lengths *a* and *b*. Then construct a segment with each indicated length.

19. $\sqrt{ab}$
20. $\sqrt{2ab}$
21. $2\sqrt{ab}$
22. $\sqrt{\frac{1}{2}ab}$
23. $\sqrt{\frac{a^2}{2}}$
24. $\sqrt{a^2 + b^2}$

APPLICATIONS

Read → Plan → Solve → Interpret

1. The orbits of Earth and Mercury around the sun are approximately circular. An astronomer on Earth can measure the angle formed by the sun, Earth, and Mercury. The measures will vary from 0° to 24°. If the average distance from the Earth to the sun is 149,504,000 kilometers or 92,897,000 miles and the maximum angle formed measures 24°, what is the average distance from Mercury to the sun?

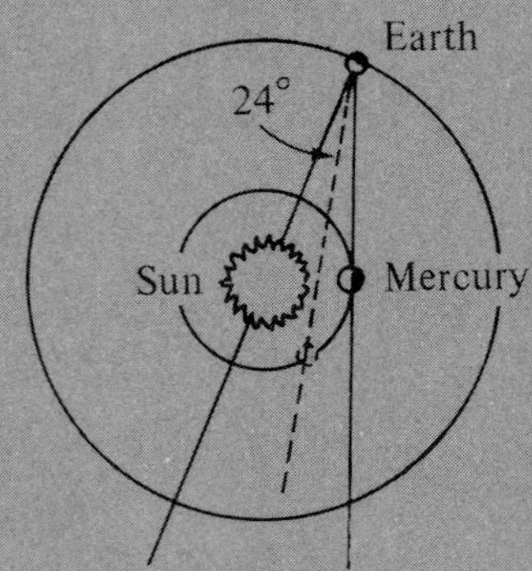

2. Two buoys are placed at points B_1 and B_2 to warn ships of shallow water in the circular region shown. As a ship sails towards this region, the navigator measures $\angle B_1PB_2$ by sighting the buoys. At a point D on the circle, the measure of $\angle B_1DB_2$ is known. The navigator knows that if he is too close to the shallow water, then the size of $\angle B_1PB_2$ will be greater than that of $\angle B_1DB_2$. Why is this so?

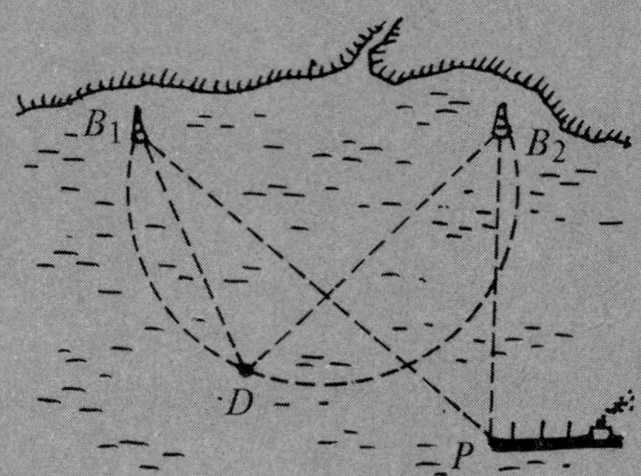

COMPUTER ACTIVITIES

Concentric Circles

Objective: **To find the length of a chord of one circle when that chord is tangent to a smaller concentric circle, using a computer program that inputs the lengths of the radii of the circles**

Illustrated at the right are two concentric circles. $\overline{AB}$ is a chord of the larger circle and is tangent to the smaller circle. Radius $\overline{OC}$ is drawn to the point of tangency. $\overline{OC}$ is perpendicular to the chord and bisects it. This creates 2 congruent right triangles, which makes it possible to determine the length of the chord when the lengths of the two radii are known.

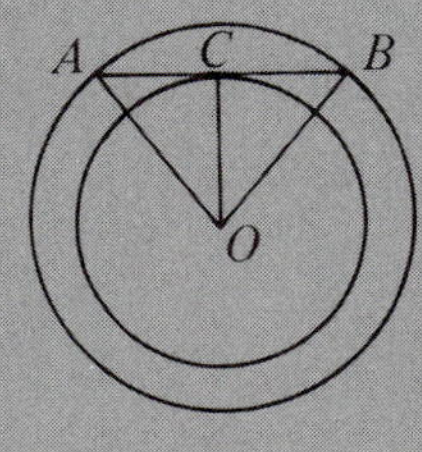

The following computer program calculates the length of such a chord from input of the two radii.

```
110  PRINT "GIVEN TWO CONCENTRIC CIRCLES, THIS"
120  PRINT "PROGRAM WILL FIND THE LENGTH OF A CHORD"
130  PRINT "OF THE LARGER CIRCLE WHICH IS TANGENT"
140  PRINT "TO THE SMALLER."
150  PRINT "WHAT IS THE RADIUS OF THE LARGER";
160  INPUT R1
170  PRINT "WHAT IS THE RADIUS OF THE SMALLER";
180  INPUT R2
190  C = 2 *  SQR (R1 ^ 2 - R2 ^ 2)
200  PRINT "THE LENGTH OF THE CHORD IS ";C"."
210  PRINT : GOTO 150
```

EXERCISES: See the Computer Section beginning on page 565 for more information.

1. Alter the program above to give ten-problem graded quizzes.
2. Rewrite the program so that it calculates the length of the smaller radius when the length of the larger radius and the length of the chord are inputted.
3. Alter the program to calculate the length of the larger radius when the other two values are inputted.

CHAPTER ELEVEN REVIEW

Use $\odot O$ for Exercises 1–6. [11.1]

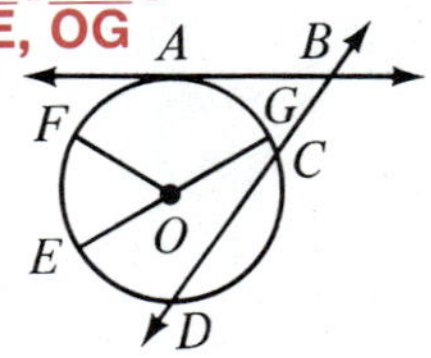

1. Name all radii. **1. $\overline{OF}$, $\overline{OE}$, $\overline{OG}$**
2. Name a secant. **$\overleftrightarrow{DC}$**
3. Name a diameter. **$\overline{EG}$**
4. Name all chords. **$\overline{EG}$, $\overline{DC}$**
5. Name a tangent. **$\overleftrightarrow{AB}$**
6. Name a minor arc. [11.4]
$\widehat{EF}$ is one possible answer.

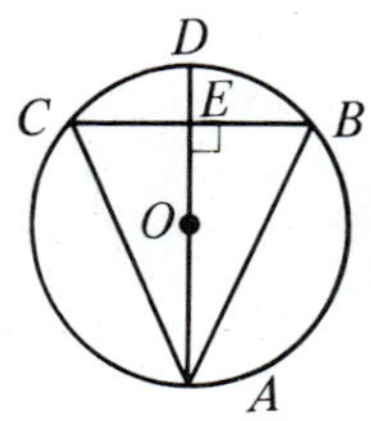

7. Given: $\overline{DA}$ is a diameter of $\odot O$ and chord $\overline{CB} \perp \overline{DA}$
Prove: $\angle C \cong \angle B$ [11.1]

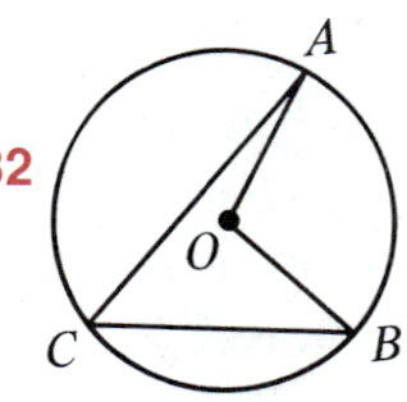

8. Given: $\odot O$ with $m\widehat{AB} = 82$
Find m $\angle AOB$. [11.5] **82**
9. Given: $\odot O$ with m $\angle ACB = 32$
Find $m\widehat{AB}$. [11.6] **64**

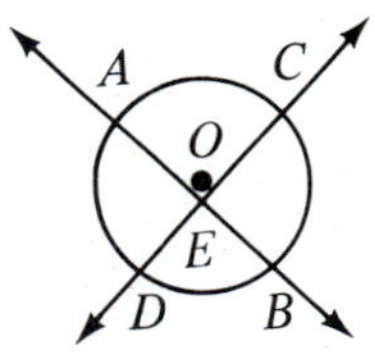

10. Given: $\odot O$ with $m\widehat{AD} = 84$, $m\widehat{BC} = 62$
Find m $\angle AED$. [11.5] **73**
11. Given: $\odot O$ with $m\widehat{AC} = 91$, m $\angle AEC = 97$
Find $m\widehat{BD}$. **103**

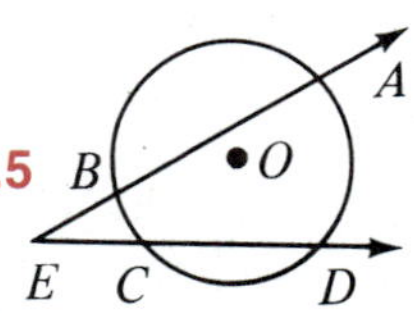

12. Given: $\odot O$ with $m\widehat{AD} = 113$, $m\widehat{BC} = 48$
Find m $\angle AED$. [11.7] **32.5**
13. Given: $\odot O$ with $m\widehat{BC} = 21$, m $\angle AED = 47$
Find $m\widehat{AD}$. **115**

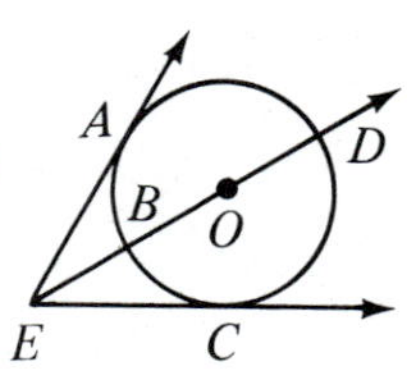

14. Given: $\odot O$ with tangents $\overrightarrow{EA}$ and $\overrightarrow{EC}$ and $m\widehat{AC} = 115$
Find m $\angle AEC$. [11.8] **65**
15. Given: $\odot O$ with tangents $\overrightarrow{EA}$ and $\overrightarrow{EC}$, diameter $\overline{BD}$, and $m\widehat{BC} = 50$
Find m $\angle DEC$. **40**

Vocabulary

circle [11.1], radius [11.1], chord [11.1], diameter [11.1], secant [11.1], tangent [11.1, 11.3], arc [11.4], central angle [11.4], inscribed angle [11.6]

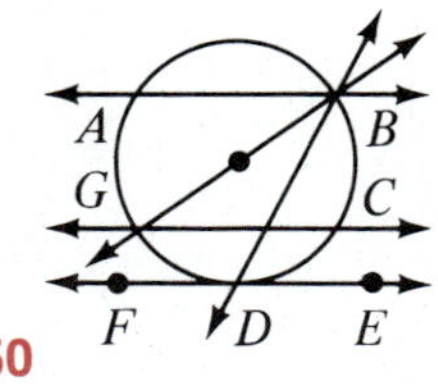

16. Given: $\overleftrightarrow{AB} \parallel \overleftrightarrow{GC} \parallel$ tangent $\overleftrightarrow{FE}$, $\overline{AB} \cong \overline{GC}$, $m\widehat{BD} = 140$, and $m\widehat{AB} = 80$
Find m $\angle FDB$ **110** and m $\angle BGC$. [11.6] **50**

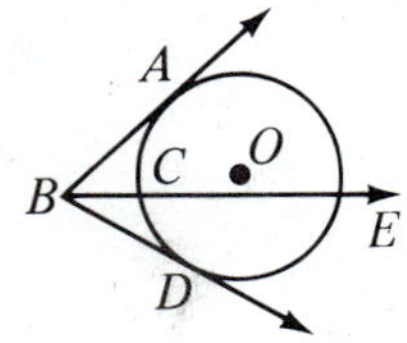

17. Given: $\odot O$ with $AB = 14$
Find BD. [11.3] **14**
18. Given: $\odot O$ with $AB = 12$, $BC = 9$
Find BE. **16**
19. Draw a circle and a point outside the circle. Construct a tangent to the circle through the point. [11.10]

*20. Draw two segments with lengths a and c. Construct a segment with length $\sqrt{c^2 - a^2}$.

21. Given: Radius $\overline{OA} \cong$ radius $\overline{PC}$ and $\angle A \cong \angle C$
Prove: $\widehat{AB} \cong \widehat{CD}$ [11.5]

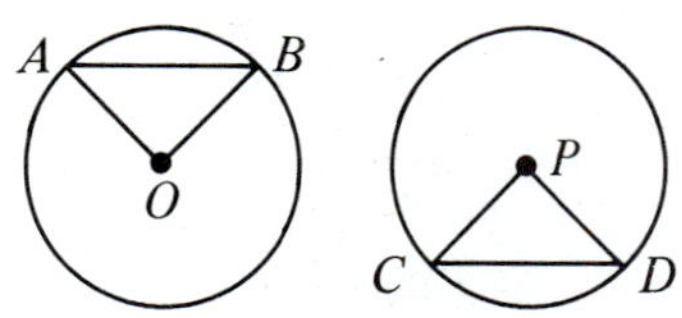

*22. Prove that if two circles intersect in two points, then the line joining those points bisects a common external tangent segment. [11.8]

(A) exercises are 8–18, 21. (B) exercises are 1–7, 19. (C) exercises are starred.

CHAPTER ELEVEN TEST

Use $\odot O$ or $\odot P$ for Items 1–5.

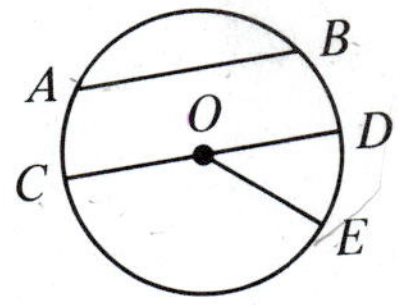

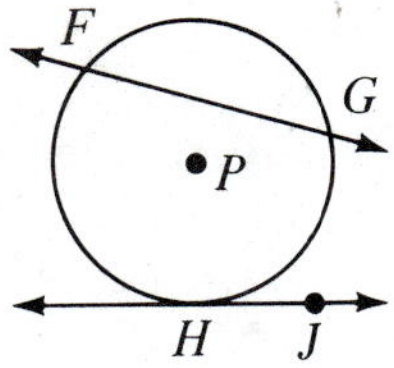

1. Name a diameter of $\odot O$. $\overline{CD}$
2. For which circle is there no secant drawn? O
3. Name a tangent of $\odot P$. $\overleftrightarrow{HJ}$
4. Name a semicircle of $\odot O$. $\overparen{CED}$, $\overparen{CAD}$, $\overparen{CBD}$
5. Name a minor arc of $\odot P$. $\overparen{FG}$, $\overparen{GH}$, $\overparen{FH}$

6. Given: Diameter $\overline{AB}$ bisects chord $\overline{CD}$ and chord $\overline{EF}$.
   Prove: $\overline{CD} \parallel \overline{EF}$

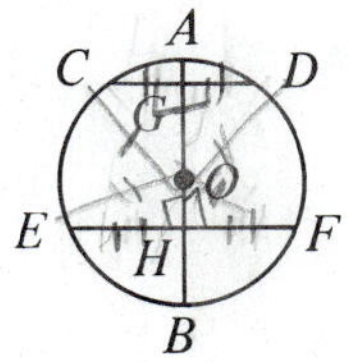

7. Given: $\odot O$ with $m\overparen{SRT} = 200$
   Find m $\angle SOT$. 160

8. Given: $\odot O$ with $m\overparen{RS} = 71$, $m\overparen{UT} = 43$
   Find m $\angle RNS$. 57

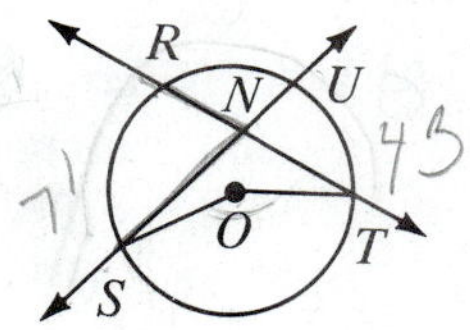

9. Given: $\odot O$ with tangent $\overleftrightarrow{AB} \parallel \overleftrightarrow{CD}$, m $\angle ECD = 80$
   Find $m\overparen{EC}$. 160

10. Given: $\overleftrightarrow{AB}$ tangent to $\odot O$, $\overleftrightarrow{AB} \parallel \overleftrightarrow{CD}$, and $m\overparen{CD} = 140$
    Find m $\angle CEB$. 125

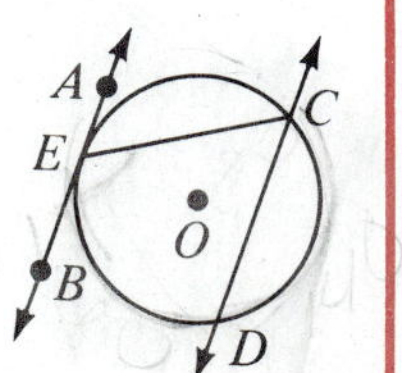

11. Given: $\overline{FH}$ is a diameter of $\odot O$, $m\overparen{FJ} = 100$, and $m\overparen{GF} = 120$.
    Find m $\angle FPJ$. 20

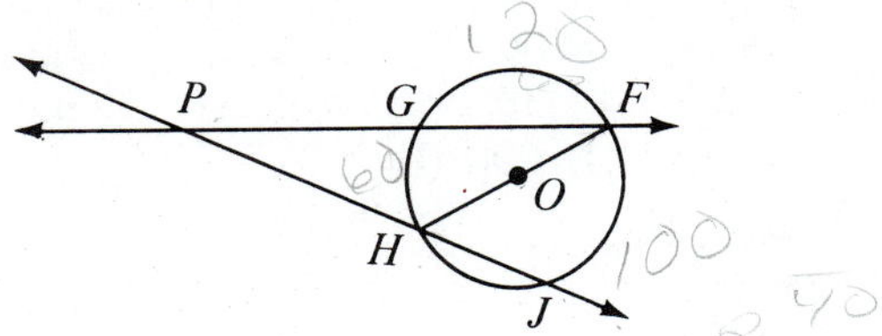

12. Given: $\overrightarrow{WT}$ is tangent to $\odot O$, $WS = 9$, and $WR = 4$.
    Find WT. 6

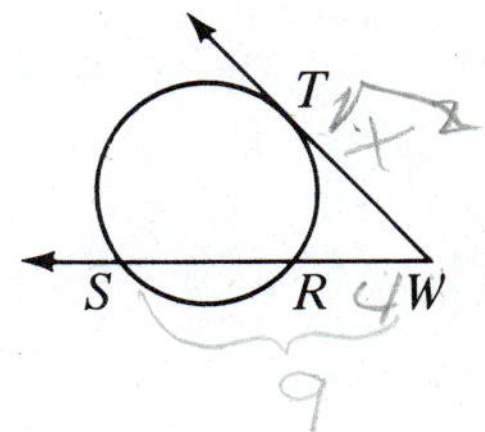

*13. Draw two segments with lengths a and b. Construct a segment with length $\frac{\sqrt{ab}}{2}$.

14. Given: $\odot O$ with diameter $\overline{AC}$ and $\overparen{DC} \cong \overparen{AB}$
    Prove: $ABCD$ is a rectangle.

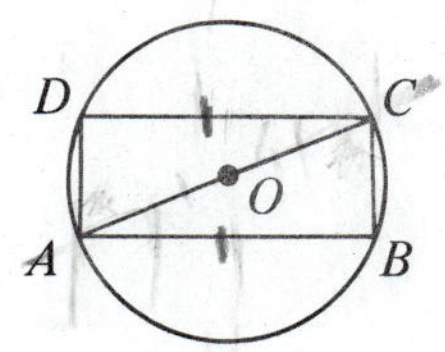

*15. Given: $\overleftrightarrow{XY}$ tangent to $\odot O$ and $\odot P$, $\overleftrightarrow{ZT}$ tangent to $\odot O$ and $\odot P$, and $\overline{XT} \cong \overline{YT}$.
    Prove: $\overleftrightarrow{TZ} \perp \overleftrightarrow{XY}$

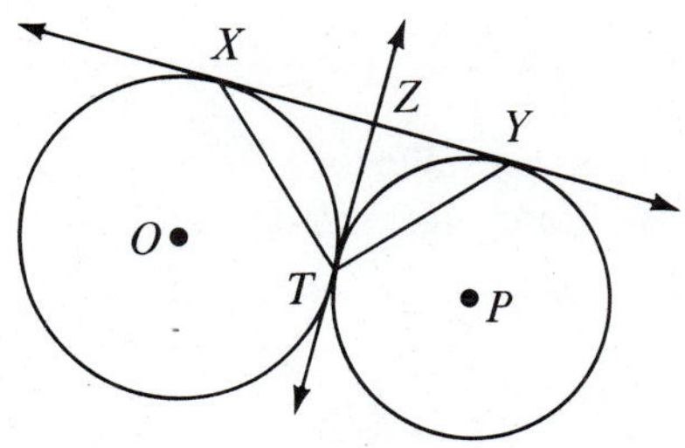

A exercises are 1–5, 7–9, 11–12, 14. B exercises are 6, 10. C exercises are starred.

COLLEGE PREP TEST

Directions: Choose the one best answer to each question or problem.

1. In $\odot O$, $m\widehat{ACB} = 120$.
Find $m\angle BAO$. B

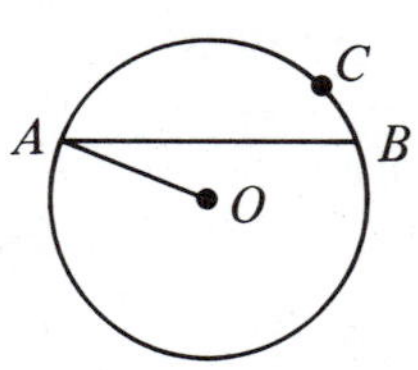

(A) 12 **(B)** 30
(C) 45 **(D)** 60 **(E)** 120

2. $MNPQ$ is an inscribed quadrilateral and $m\angle P = 60$.
Find $m\angle M$. A

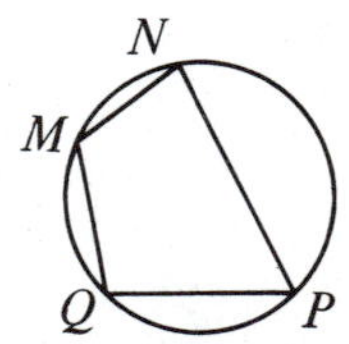

(A) 120 **(B)** 110 **(C)** 130
(D) 125 **(E)** 115

3. In the figure, $\overline{PA}$, $\overline{PB}$, and $\overline{MN}$ are tangent to $\odot O$ and $AP = 20$.
Find the perimeter of $\triangle PMN$. C

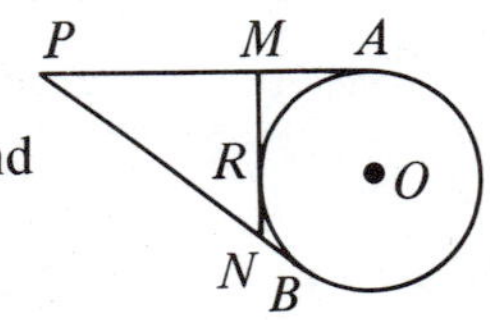

(A) 20 **(B)** 30 **(C)** 40 **(D)** $35\frac{1}{2}$ **(E)** 50

4. In the figure, A, B, C, and D are on the circle, $m\angle B = 20$, and $m\angle DPC = 70$.
Find $m\angle D$. E

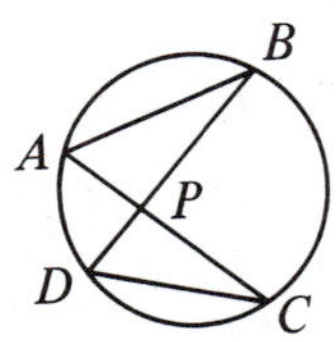

(A) 30 **(B)** 100 **(C)** 50 **(D)** 120 **(E)** 90

5. What is the measure of the angle formed by the hour and minute hands of a clock at 2:15? B

(A) 30° **(B)** $22\frac{1}{2}^\circ$ **(C)** $157\frac{1}{2}^\circ$ **(D)** $172\frac{1}{2}^\circ$

(E) $27\frac{1}{2}^\circ$

6. In the figure, $\overline{AO}$ and $\overline{BO}$ are radii of $\odot O$, $PC = OA$, POA is a straight line, and $m\angle BOA = 60$.
Find $m\angle P$. C

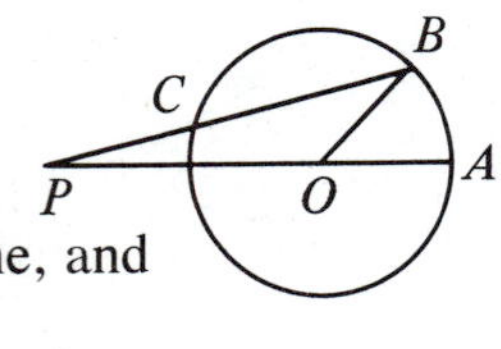

(A) 10 **(B)** 15 **(C)** 20 **(D)** 30 **(E)** 40

7. In the figure, $m\widehat{RS} = 40$.
Find $m\angle W + m\angle T$. B

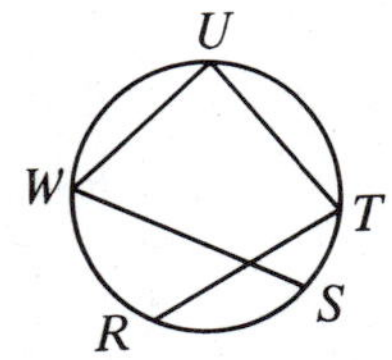

(A) 320 **(B)** 160
(C) 150 **(D)** 200
(E) 250

8. In $\odot O$, $m\angle OMN = 50$.
Find $m\angle MSN$. D

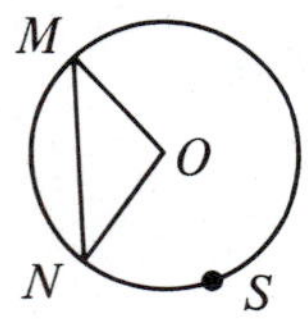

(A) 80
(B) 100
(C) 260
(D) 40
(E) 130

9. What is the length of a tangent drawn from a point 3 inches from a circle whose radius is 12 inches? A

(A) 9 in. **(B)** $6\sqrt{2}$ in. **(C)** 6 in.
(D) $3\sqrt{5}$ in. **(E)** 4 in.

10. In the figure, chords $\overline{MN}$ and $\overline{RS}$ intersect at P, $MP = 25$, $PN = 8$, and $\frac{RP}{PS} = \frac{1}{2}$.
Find RS. B

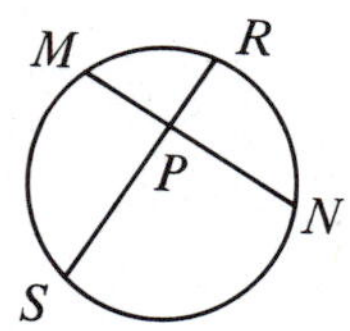

(A) 33 **(B)** 30 **(C)** $\frac{10}{3}\sqrt{6}$ **(D)** 20

(E) 22

12 AREAS OF GEOMETRIC FIGURES

Non-Routine Problem Solving

There are several ways to prove the Pythagorean theorem. Here is one method.

Project: Cut out four identical right triangles.

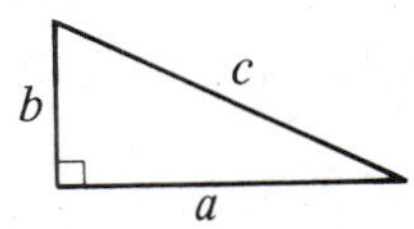

Arrange them as shown at the right. Verify that the figure in the center is a square.

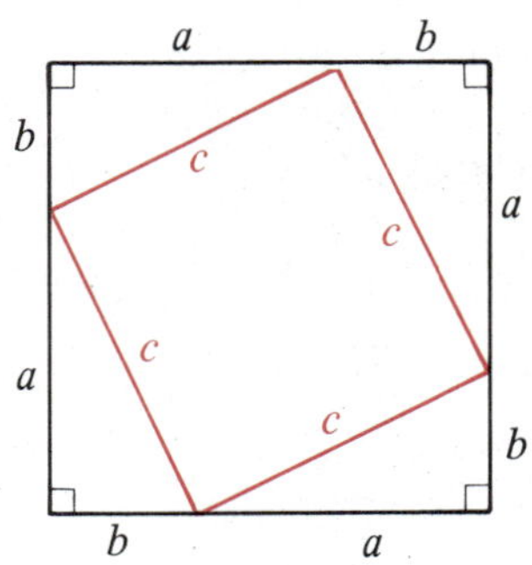

Find the area of the big square in two ways. [(Area of big square) = (length of side) · (length of side); (Area of big square) = (4 times area of triangle) + (area of small square)]. Set the two expressions equal to each other. $(a + b)^2 = 4 \cdot \frac{1}{2}ab + c^2$; $a^2 + 2ab + b^2 = 2ab + c^2$; $a^2 + b^2 = c^2$

NON-ROUTINE PROBLEMS

Prove the Pythagorean Theorem by making the arrangement at the right. [*Hint:* Find the area of the big square in two ways.]

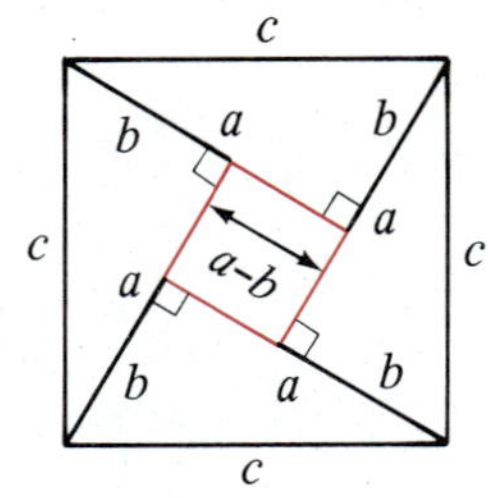

President Garfield discovered another proof of the Pythagorean Theorem by arranging two identical right triangles as shown in the picture.

Follow these steps to prove the Pythagorean Theorem by President Garfield's method.

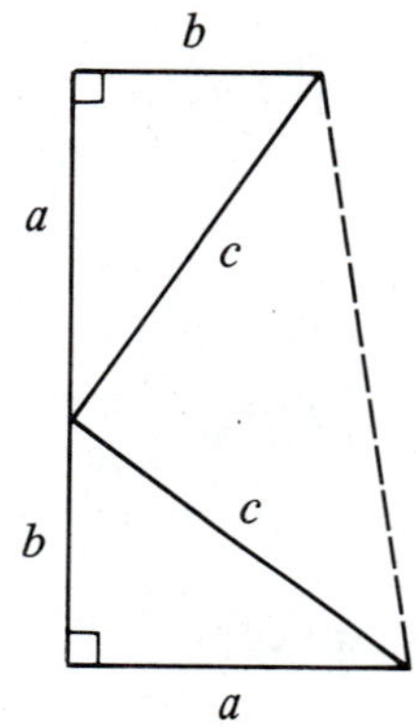

1. Verify that the figure in the center is a right triangle.
2. Verify that the large figure is a trapezoid with a and b the lengths of the bases and $(a + b)$ the length of an altitude.
3. Find the area of the trapezoid in two ways and complete the proof. [*Hint:* See page 398.]

The square at the right is determined by four identical right triangles.

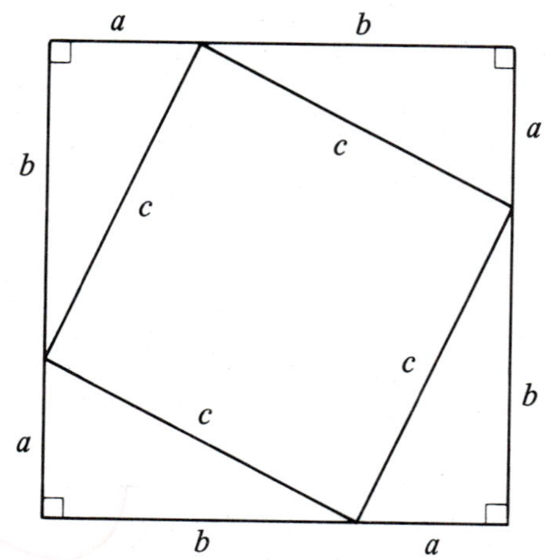

4. Use the figure above to prove the Pythagorean Theorem. [*Hint:* Suppose the four triangles are removed from the big square.]

AREAS OF RECTANGLES AND SQUARES 12.1

Objective **To apply the formula for the area of a rectangle**

A **polygonal region** is the union of a polygon and its interior.
A **circular region** is the union of a circle and its interior.

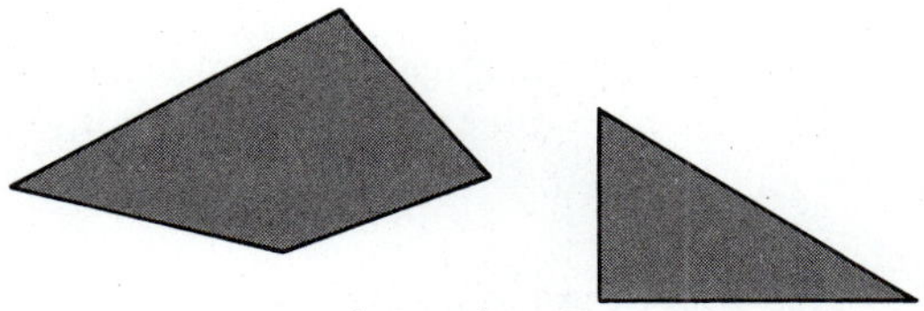

polygonal regions

circular region

One way to measure a polygonal region or a circular region is to find its **area.** To simplify language, the area of a polygonal region or a circular region will be referred to as the area of the polygon or circle.

A **square unit** of area is a region bounded by a square whose sides are each one unit long. For example, a square with sides each 1 centimeter long can be used as a unit of area. It is called a square centimeter (cm^2).

1 square centimeter (cm^2)

To find the *area of a polygon,* you find the number of square units of area which are contained in the region bounded by the polygon.

Example 1 If the small square is the unit of area, find the area of the rectangle shown.

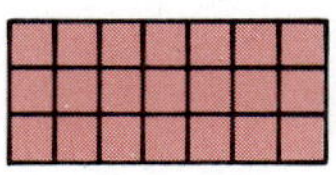

There are 3 rows of 7 unit squares in the region.
The rectangle contains 21 square units.
$(3 \times 7) = 21$

Thus, the area is 21 square units.

Example 2 Find the area of the rectangle in square centimeters.

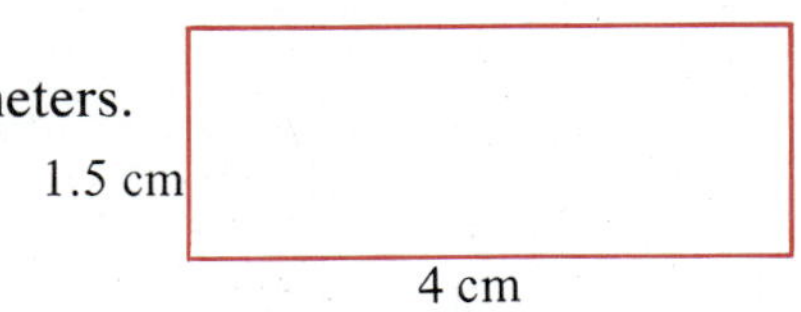

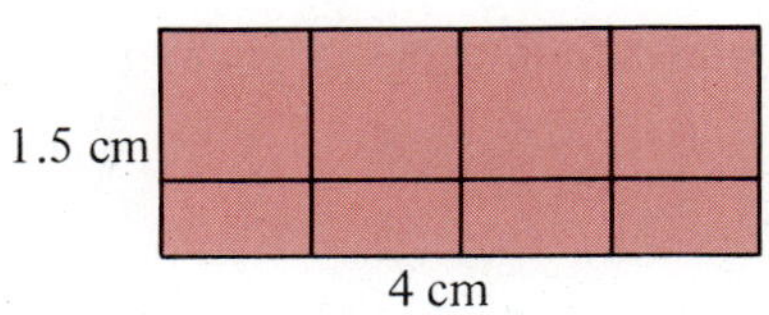

The first row contains 4 squares.
The second row contains 4 half-squares, or 2 squares. The total is 6 squares. $(1.5 \times 4 = 6)$

Thus, the area is 6 cm^2.

Examples 1 and 2 suggest a formula for the area of a rectangle.

Postulate 24 The area of a rectangle is the product of the length of a base and the length of a corresponding altitude.

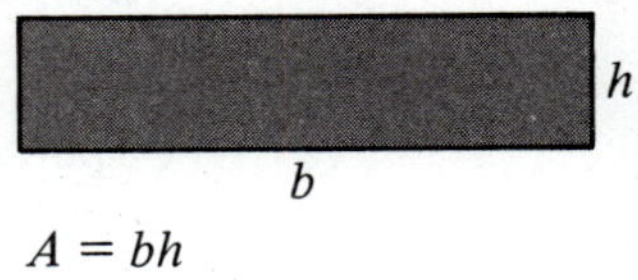

A square is a rectangle, so you can use the formula $A = bh$ to find the area of any square. However, since each side of a square has the same length, there is another formula that can be used.

Corollary The area of a square is the square of the length of a side.

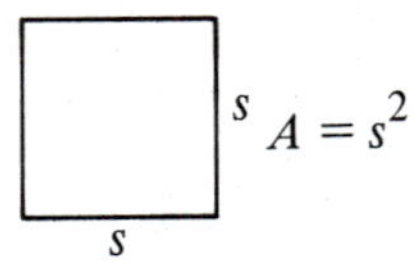

Examples 3 and 4 apply the formula $A = bh$.

Example 3 Find the area of a rectangle with a base of length 15 cm and a diagonal of length 17 cm.

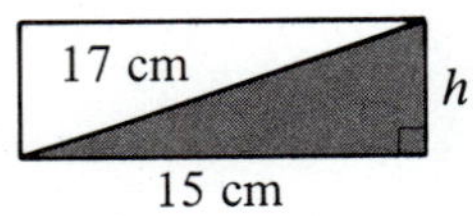

Use the Pythagorean Theorem to find h.

$$h^2 + 15^2 = 17^2$$
$$h^2 + 225 = 289$$
$$h^2 = 64, \text{ or } h = 8$$

$$A = bh = 15 \cdot 8 = 120$$

Thus, the area is 120 cm^2.

Example 4 Find the length of a base of a rectangle if $A = x^2 + 2x - 3$ and $h = x - 1$.

$$A = b\ h$$
$$x^2 + 2x - 3 = (\quad)(x - 1)$$

◀ *Substitute for A and h. Find the missing factor.*

Thus, $b = x + 3$, since $(x + 3)(x - 1) = x^2 + 2x - 3$.

The next postulate enables you to add areas of regions.

Postulate 25 Area Addition Postulate

If a region is the union of two or more nonoverlapping regions, then its area is the sum of the areas of these nonoverlapping regions.

Example 5 Find the area of the polygon. Each angle is a right angle.

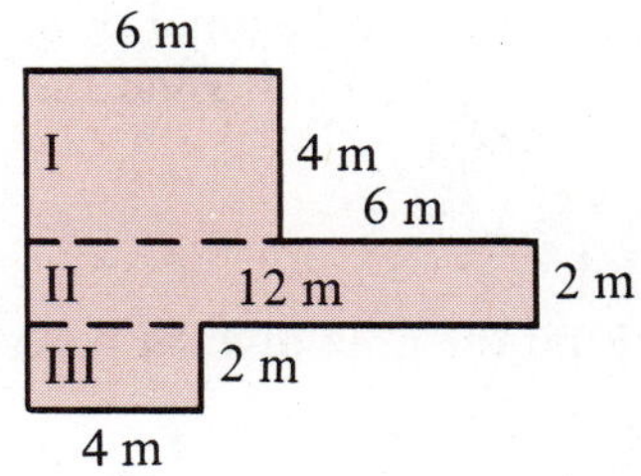

Divide the polygon into rectangles.

Area of rectangle I = 6×4, or 24 m^2
Area of rectangle II = 12×2, or 24 m^2
Area of rectangle III = 4×2, or 8 m^2

Thus, the area of the polygon is $24 + 24 + 8$, or 56 m^2.

Oral Exercises

If the sides of a rectangle are measured in the following units, what is the corresponding unit for the area?

1. millimeters **mm^2**
2. centimeters **cm^2**
3. meters **m^2**
4. kilometers **km^2**

Find the area of each rectangle.

5. $b = 6$ cm, $h = 3$ cm **18 cm^2**
6. $b = 8$ m, $h = 4$ m **32 m^2**
7. $b = 7$ km, $h = 9$ km **63 km^2**
8. $b = 5$ m, $h = 9$ m **45 m^2**
9. $b = 9$ cm, $h = 3$ cm **27 cm^2**
10. $b = 6$ m, $h = 5$ m **30 m^2**

Find the indicated length for each rectangle.

11. $A = 20$ mm^2, $b = 4$ mm, $h = ?$ **5 mm**
12. $A = 63$ km^2, $b = ?$, $h = 7$ km **9 km**
13. $A = 52$ cm^2, $b = ?$, $h = 13$ cm **4 cm**
14. $A = 120$ m^2, $b = 10$ m, $h = ?$ **12 m**

Find the area of each square.

15. $s = 7$ mm **49 mm^2**
16. $s = 9$ cm **81 cm^2**
17. $s = 30$ m **900 m^2**

Written Exercises

1. 22.75 m^2 2. 16.8238 mm^2 3. 24.51 cm^2 4. 7.875 m^2 5. $7\sqrt{5}$ m^2 6. $5\sqrt{7}$ km^2 7. 25 cm^2

Find the area of each rectangle or square.

A

1. $b = 6.5$ m, $h = 3.5$ m
2. $b = 4.27$ mm, $h = 3.94$ mm
3. $b = 5.7$ cm, $h = 4.3$ cm
4. $b = 3.5$ m, $h = 2.25$ m
5. $b = \sqrt{5}$ m, $h = 7$ m
6. $b = 5$ km, $h = \sqrt{7}$ km
7. Square with $s = 5$ cm
8. Square with $s = \sqrt{6}$ cm
9. Square with $s = (x + y)$
10. $b = (x - y)$, $h = (x + y)$
11. $b = (x + y)$, $h = (x + y)$
12. $b = (2x + 3)$, $h = (3x - 5)$

8. 6 cm^2 9. $x^2 + 2xy + y^2$ 10. $x^2 - y^2$ 11. $x^2 + 2xy + y^2$ 12. $6x^2 - x - 15$

Find the length of the base of each rectangle.

13. $A = 78$ cm^2, $h = 1.5$ cm
14. $A = 9.86$ km^2, $h = 3.4$ km
15. $A = 5$ m^2, $h = \sqrt{5}$ m
16. $A = x^2 + 2x + 1$, $h = x + 1$
17. $A = x^2 - 3x - 4$, $h = x - 4$
18. $A = 2x^2 - 5x - 12$, $h = x - 4$

13. 52 cm 14. 2.9 km 15. $\sqrt{5}$ m 16. $x + 1$ 17. $x + 1$ 18. $2x + 3$

Find the area of each rectangle. (Ex. 19–22)

B

19. $b = 6$, length of diagonal = 10 **48**
20. $h = 40$, length of diagonal = 41 **360**
21. $b = 3$, length of diagonal = 7 **$6\sqrt{10}$**
22. $h = \sqrt{2}$, length of diagonal = $\sqrt{3}$ **$\sqrt{2}$**
23. A diagonal of a rectangle makes an angle of 30° with a base. The diagonal measures 12 cm. Find the area of the rectangle. **$36\sqrt{3}$ cm^2**
24. Two consecutive sides of a rectangle are in a ratio of 2 to 3. The area is 216 m^2. Find the lengths of the two sides. **12 m, 18 m**

25. A diagonal of a square measures 16 cm. Find the area. **128 cm²**

26. The dimensions of a rectangle are 4 cm by 5 cm. Find the length of a side of a square with the same area as the rectangle. **$2\sqrt{5}$ cm**

Find the area of each polygon. Each angle is a right angle. (Ex. 27–29)

27.
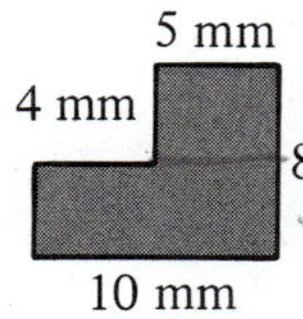

60 mm²

28. **72 cm²**
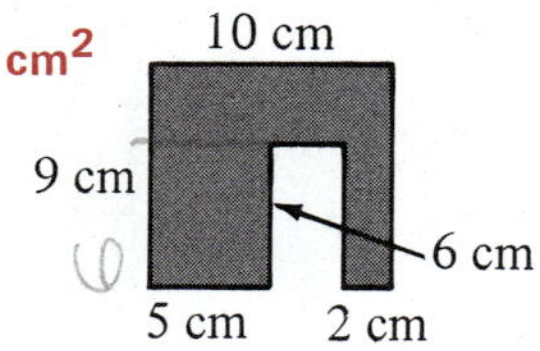

29.
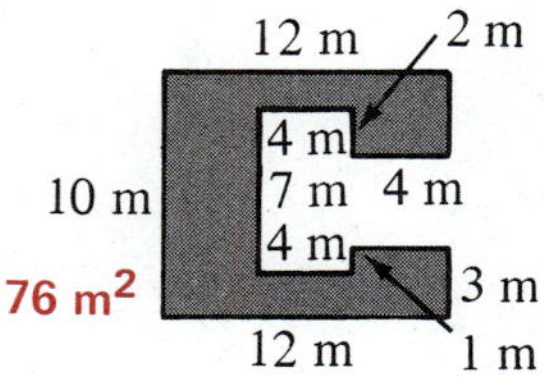

76 m²

30. Find the area of a square inscribed in a circle with a radius 6 cm long. **72 cm²**

31. A circle with a radius 8 cm long is inscribed in a square. Find the area of the square. **256 cm²**

32. Prove or disprove: The ratio of the areas of two rectangles is equal to the ratio of the products of the lengths of their bases and altitudes.

$\frac{A}{A'} = \frac{bh}{b'h'}$

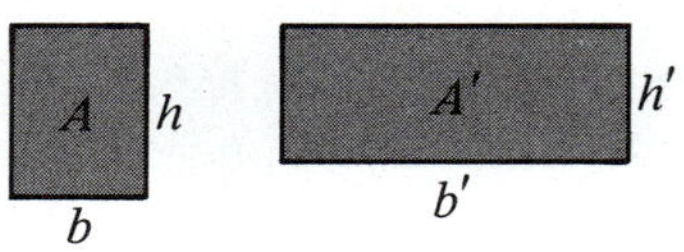

33. Prove or disprove: The ratio of the areas of two rectangles with congruent altitudes is equal to the ratio of the lengths of their bases.

$\frac{A}{A'} = \frac{b}{b'}$

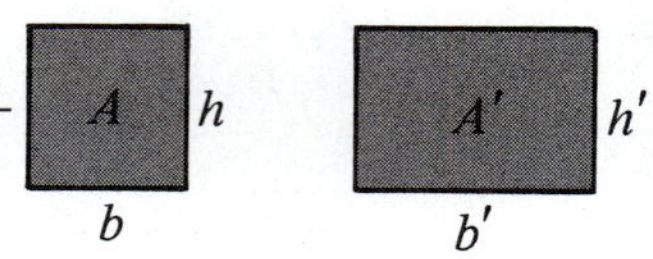

34. Prove or disprove: If the length of a base of a rectangle is doubled and the length of a corresponding altitude remains the same, then the area is doubled.

35. Prove or disprove: If the lengths of the sides of one square are twice the lengths of the sides of another, then the area of the first square is twice the area of the second.

ALGEBRA REVIEW

Objective

To solve an equation for one of its variables

Example

Solve $5xy = 4x + 3z$ for x.

$$5xy = 4x + 3z$$
$$5xy - 4x = 3z \quad \text{◄ } \textit{Add } -4x \textit{ to each side.}$$
$$(5y - 4)x = 3z \quad \text{◄ } \textit{Distributive property}$$
$$x = \frac{3z}{5y - 4} \quad \text{◄ } \textit{Divide each side by } 5y - 4.$$

Solve each equation for the indicated variable.

1. $5x = b$; x **$\frac{b}{5}$**
2. $4 + x = k$; x **$k - 4$**
3. $rx + s = t$; x **$\frac{t - s}{r}$**
4. $9y - 24a = 6a + 4y$; y **$6a$**
5. $5t + 2b = t + 6b$; t **b**
6. $rsx - 2s = 3$; s **$\frac{3}{rx - 2}$**
7. $x + y = 3$; y **$3 - x$**
8. $2x + y = 8$; y **$8 - 2x$**
9. $3x - y = 12$; y **$3x - 12$**
10. $4ab = 3a - 2c$; a **$\frac{-2c}{4b - 3}$**
11. $-6xy + 4z = 3x$; x **$\frac{4z}{6y + 3}$**
12. $5rs - 3t = 7s$; s **$\frac{3t}{5r - 7}$**
13. $-6pq + 2q = 3p$; q **$\frac{3p}{-6p + 2}$**
14. $abc + 2b = 4a$; a **$\frac{-2b}{bc - 4}$**
15. $3x = 2yz - 5xyz$; x **$\frac{2yz}{3 + 5yz}$**

AREAS OF TRIANGLES AND RHOMBI 12.2

Objectives **To apply the formula for the area of a triangle**
To apply the formula for the area of a rhombus

Since congruent figures have the same size and shape, it is reasonable to assume that they have the same area. This is stated in Postulate 26.

Postulate 26 Congruent figures have the same area.

Recall that a diagonal of a rectangle divides the rectangle into two congruent right triangles. This leads to the following theorem.

Theorem 12.1 **Area rt. $\triangle = \frac{1}{2}bh$** The area of a right triangle is one half the product of the lengths of the two legs. (See Exercise 22.)

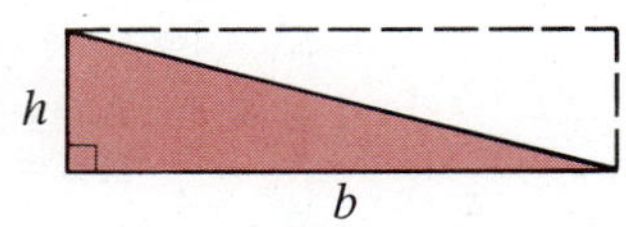

You can use Theorem 12.1 to prove a general theorem about the area of a triangle.

Theorem 12.2 **Area $\triangle = \frac{1}{2}bh$** The area of any triangle is one half the product of the length of a base and the length of a corresponding altitude.

Given: $\triangle XYZ$, length of base $\overline{XY} = b$, and length of altitude $\overline{ZD} = h$

Prove: Area $\triangle XYZ = \frac{1}{2}bh$

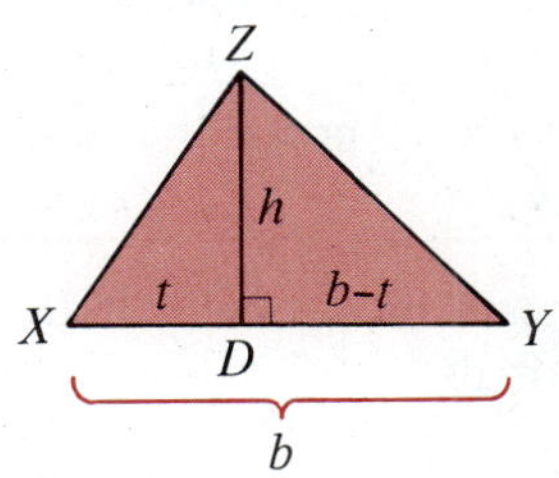

ANALYSIS
Area $\triangle XYZ$ = area rt. $\triangle ZDY$ + area rt. $\triangle DZX$. Use the formula for the area of a rt. $\triangle$ and the area addition postulate.

Proof

Statements	Reasons
1. $\triangle XYZ$ with altitude $\overline{ZD}$, $XY = b$, $ZD = h$	1. Given
2. $\triangle ZDY$ and $\triangle DZX$ are rt. $\triangle$'s.	2. Def. of altitude and def. of rt. $\triangle$
3. Area $\triangle ZDY = \frac{1}{2}(b - t)h$ Area $\triangle DZX = \frac{1}{2}th$	3. Area rt. $\triangle = \frac{1}{2}bh$
4. Area $\triangle XYZ$ = area $\triangle DZX$ + area $\triangle ZDY$	4. Area Addition Postulate
5. Area $\triangle XYZ = \frac{1}{2}th + \frac{1}{2}(b - t)h$	5. Substitution property
6. Area $\triangle XYZ = \frac{1}{2}h\ (t + b - t)$	6. Distributive property
7. $\therefore$ Area $\triangle XYZ = \frac{1}{2}bh$	7. Algebraic properties

Theorem 12.2 can also be proved if the triangle is obtuse. (See Exercise 29.) In Example 1, the area is computed for a right triangle, an acute triangle, and an obtuse triangle.

Example 1

Find the area of each shaded triangular region.

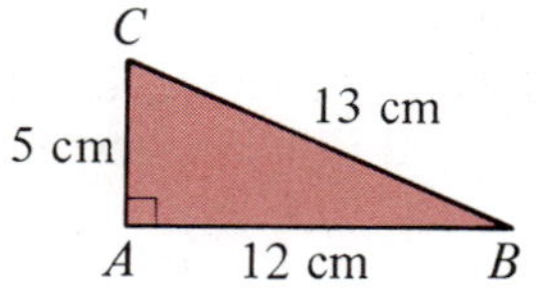

$b = 12, h = 5$

$A = \frac{1}{2}\ bh$

$A = \frac{1}{2}\ (12 \cdot 5) = 30\ \text{cm}^2$

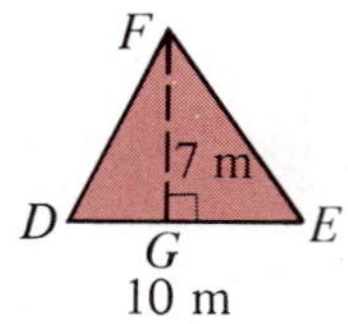

$b = 10, h = 7$

$A = \frac{1}{2}\ bh$

$A = \frac{1}{2}\ (10 \cdot 7) = 35\,m^2$

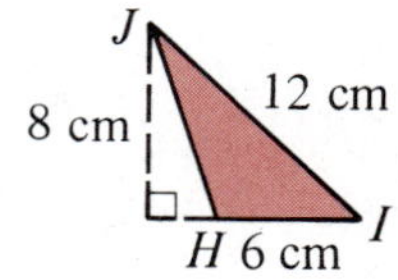

$b = 6, h = 8$

$A = \frac{1}{2}\ bh$

$A = \frac{1}{2}\ (6 \cdot 8) = 24\ \text{cm}^2$

Example 2

Given: $\triangle ABC$ with $\overline{AE} \perp \overline{BC}$, $\overline{CD} \perp \overline{AB}$,
$AB = 8$, $CD = 4$, and $BC = 6$.
Find AE.

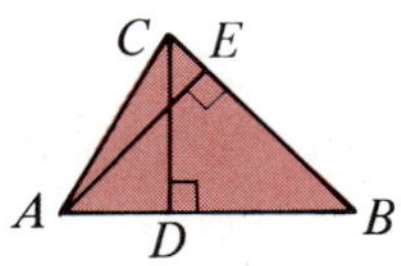

$$\text{Area } \triangle ABC = \frac{1}{2}\ (AB)(CD) \quad \text{or} \quad \text{Area } \triangle ABC = \frac{1}{2}\ (BC)(AE)$$

$$= \frac{1}{2} \cdot 8 \cdot 4 \qquad \rightarrow 16 = \frac{1}{2} \cdot 6\ (AE)$$

$$= 16 \qquad \frac{16}{3} = AE$$

Thus, $AE = \frac{16}{3}$.

Recall that the diagonals of a rhombus are perpendicular bisectors of each other. Also, either diagonal divides a rhombus into two congruent triangles. These reminders should help you to prove the following theorem.

Theorem 12.3

The area of a rhombus is one half the product of the lengths of the diagonals.

Given: Rhombus $ABCD$ with diagonals of lengths d_1 and d_2

Prove: Area rhombus $ABCD = \frac{1}{2}\ d_1\ d_2$

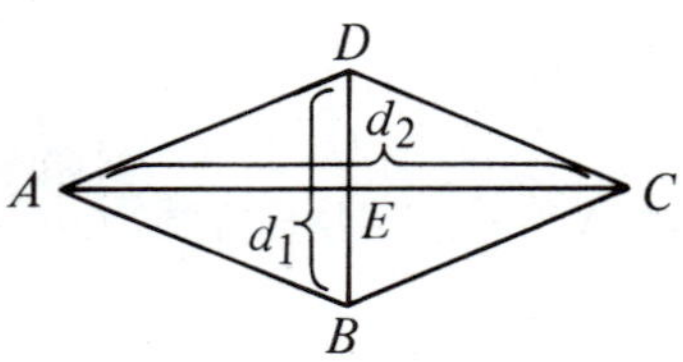

ANALYSIS

Find the area of $\triangle ACD$ in terms of d_2 and d_1. Then double that area to find the area of the rhombus. (See Exercise 23.)

Example 3

One diagonal of a rhombus is 12 cm long. The area is 48 cm². Find the length of the other diagonal.

$$A = \frac{1}{2}\ d_1 \cdot d_2$$

$$48 = \frac{1}{2} \cdot 12 \cdot d_2$$

$$48 = 6 \cdot d_2$$

Thus, $d_2 = 8$ cm.

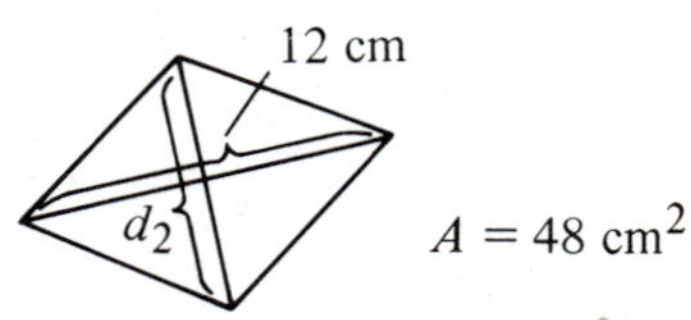

The length of an altitude and the area of an equilateral triangle can be represented in terms of the length of a side, as stated below.

Theorem 12.4

Area equilateral $\triangle = \frac{s^2}{4}\sqrt{3}$ If s is the length of a side of an equilateral triangle, then the length of the altitude is $\frac{s}{2}\sqrt{3}$ and the area is $\frac{s^2}{4}\sqrt{3}$.

Given: Equilateral $\triangle ABC$ with side of length s and altitude $\overline{CD}$ of length h

Prove: $h = \frac{s}{2}\sqrt{3}$ and $A = \frac{s^2}{4}\sqrt{3}$

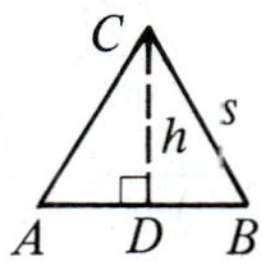

ANALYSIS

$\overline{CD}$ forms a 30°–60° rt. $\triangle$. Find h in terms of s. Then use the formula $A = \frac{1}{2}bh$.

(See Exercise 24.)

Example 4

Find the length of an altitude and the area of an equilateral triangle if each side is 8 cm long.

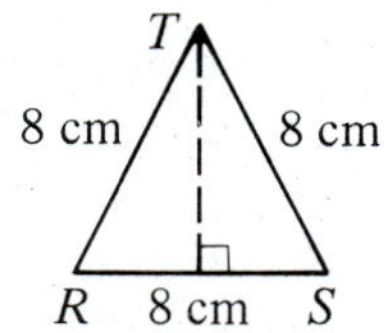

$$h = \frac{s}{2}\sqrt{3} = \frac{8}{2}\sqrt{3} = 4\sqrt{3}$$

$$A = \frac{s^2}{4}\sqrt{3} = \frac{8^2}{4}\sqrt{3} = \frac{64}{4}\sqrt{3} = 16\sqrt{3}$$

Thus, the length of an altitude is $4\sqrt{3}$ cm and the area is $16\sqrt{3}$ cm^2.

Written Exercises

Find the area of each shaded triangular region.

A

1.

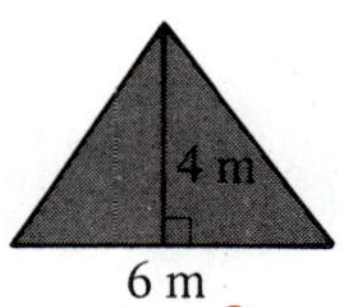

12 m^2

2.

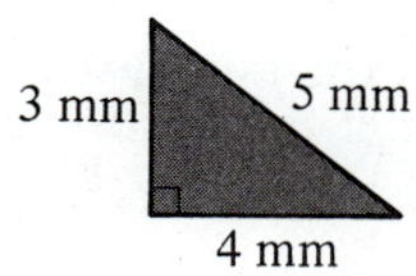

6 mm^2

3.

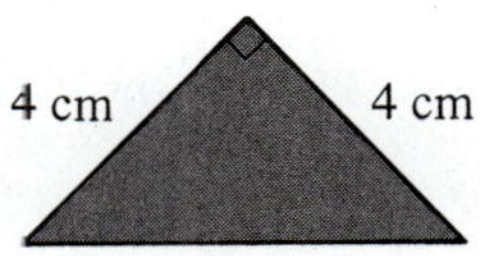

8 cm^2

4.

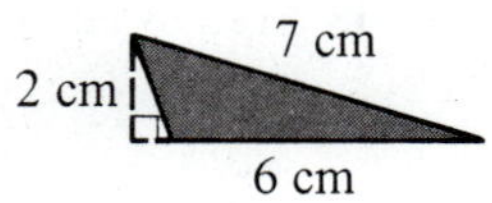

6 cm^2

5.

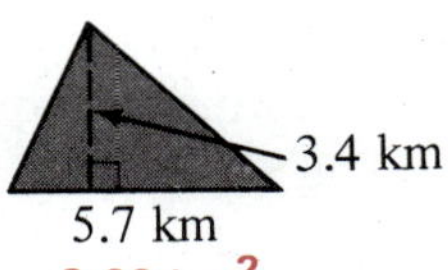

9.69 km^2

6.

$9\frac{3}{16}$ m^2

7.

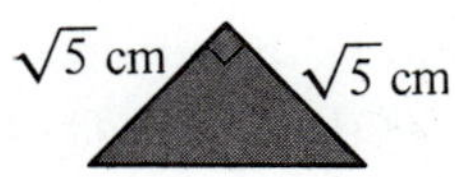

2.5 cm^2

8.

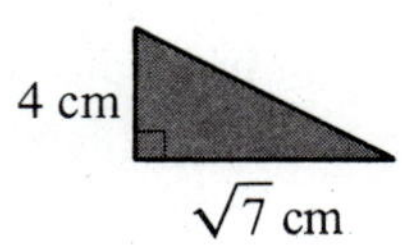

$2\sqrt{7}$ cm^2

Supply the missing information for each triangle.

	Length of base (b)	Length of altitude (h)	Area (A)
9.	9 cm	? 16 cm	72 cm^2
10.	7 m	? 18 m	63 m^2
11.	? 1.6 km	0.7 km	0.56 km^2

12. The diagonals of a rhombus measure 7 m and 10 m. Find the area. **35 m²**

13. One diagonal of a rhombus is 18 cm long. The area is 45 cm². Find the length of the other diagonal. **5 cm**

14. Find the length of an altitude and the area of an equilateral triangle if each side is 10 cm long. **$5\sqrt{3}$ cm; $25\sqrt{3}$ cm²**

15. Find the length of an altitude and the area of an equilateral triangle if each side is 12 cm long. **$6\sqrt{3}$ cm; $36\sqrt{3}$ cm²**

In $\triangle XYZ$, $\overline{XW} \perp \overline{ZY}$ and $\overline{ZT} \perp \overline{XY}$. Find each indicated length.

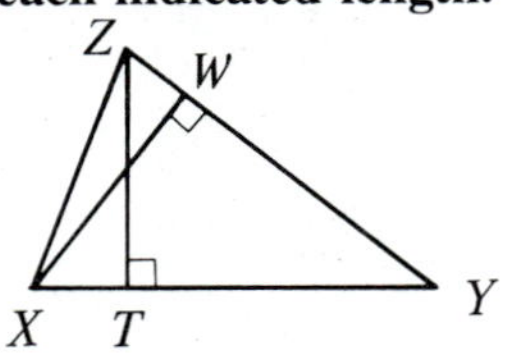

B 16. Given: $XY = 12$, $ZT = 8$, $YZ = 16$. Find XW. **6**

17. Given: $XY = 15$, $ZT = 9$, $XW = 12$. Find YZ. **$11\frac{1}{4}$**

18. Given: $XW = 4$, $YZ = 7$, $ZT = 6$. Find XY. **$4\frac{2}{3}$**

19. Given: $XY = 22$, $YZ = 33$, $XW = 12$. Find ZT. **18**

20. The length of an altitude of an equilateral triangle is $7\sqrt{3}$. Find the length of a side. **14**

21. The area of an equilateral triangle is $9\sqrt{3}$. Find the length of a side. **6**

22. Prove Theorem 12.1.

23. Prove Theorem 12.3.

24. Prove Theorem 12.4.

25. Prove that the area of a square is one half the square of the length of a diagonal.

Hero's Formula can be used to find the area of a triangle if the lengths of the three sides are known.

> **HERO'S FORMULA**
> Area of a triangle $= \sqrt{s(s-a)(s-b)(s-c)}$, where a, b, and c are the lengths of the three sides and $s = \frac{1}{2}(a+b+c)$.

Example Find the area of a triangle with sides 3 m, 4 m, and 5 m long.

$$s = \frac{1}{2}(a+b+c) = \frac{1}{2}(3+4+5) = \frac{1}{2} \cdot 12 = 6$$

$$A = \sqrt{s(s-a)(s-b)(s-c)} = \sqrt{6(6-3)(6-4)(6-5)} = \sqrt{6 \cdot 3 \cdot 2 \cdot 1} = \sqrt{36} = 6$$

Thus, the area is 6 m².

Find the area of each triangle. The lengths of the three sides are given.

C 26. 5 cm, 12 cm, 13 cm **30 cm²**

27. 4 m, 5 m, 6 m **$\frac{15\sqrt{7}}{4}$ m²**

28. 4 km, 5 km, 8 km **$\frac{3}{4}\sqrt{119}$ km²**

29. Prove Theorem 12.2 for an obtuse triangle.

Prove or disprove each of the following.

30. The areas of two similar triangles have the same ratio as the squares of the lengths of any two corresponding sides.

31. The ratio of the areas of two similar triangles is equal to the ratio of the squares of the lengths of the corresponding altitudes.

32. If two triangles have equal areas, then the ratio of the lengths of their bases is equal to the ratio of the lengths of their altitudes.

33. The ratio of the areas of two triangles with congruent bases is equal to the ratio of the lengths of their altitudes.

34. Any angle bisector of a triangle separates the region into two regions with equal areas.

35. Any median of a triangle separates the region into two regions with equal areas.

AREAS OF PARALLELOGRAMS 12.3

Objective

To apply the formula for the area of a parallelogram

The Area Addition Postulate and other area postulates can be used to develop a formula for the area of a parallelogram.

Theorem 12.5 **Area $\square = bh$** The area of any parallelogram is the product of the length of a base and the length of a corresponding altitude.

Given: $\square ABCD$, length of base $\overline{AB} = b$, and length of altitude $\overline{CE} = h$
Prove: Area $\square ABCD = bh$

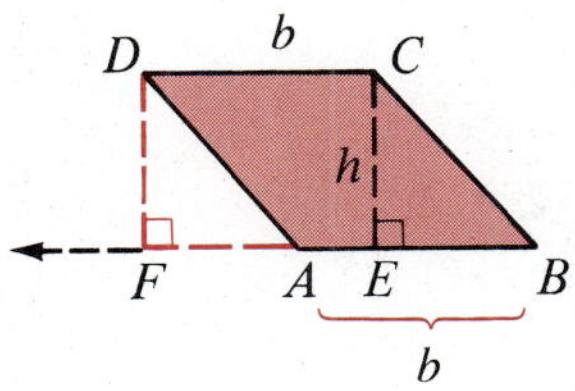

ANALYSIS
Show $\triangle FAD \cong \triangle EBC$. Then area $\square ABCD =$ area rect. $FECD$.

Proof

Statements	Reasons
1. $\square ABCD$ with altitude $\overline{CE}$, $AB = b$, $CE = h$	1. Given
2. Draw $\overline{DF} \perp \overrightarrow{BA}$.	2. There is exactly one $\perp$ from a pt. to a line.
3. $\overline{DC} \parallel \overline{BA}$	3. Opp. sides of a $\square$ are $\parallel$.
4. $\overline{DF} \cong \overline{CE}$	4. $\parallel$ lines are equidistant at all points.
5. $\overline{AD} \cong \overline{BC}$	5. Opp. sides of a $\square$ are $\cong$.
6. $\triangle FAD \cong \triangle EBC$	6. Def. of rt. $\triangle$ and HL for $\cong$ rt. $\triangle$'s
7. Area $\triangle FAD =$ area $\triangle EBC$	7. $\cong$ figures have the same area.
8. Area $\triangle FAD +$ area $AECD =$ area $\triangle EBC +$ area $AECD$	8. Add. prop. of equality
9. Area rect. $FECD =$ area $\square ABCD$	9. Area Add. Post. and substitution property
10. Area rect. $FECD = bh$	10. Area $\square = bh$
11. $\therefore$ Area $\square ABCD = bh$	11. Substitution property

---

Example 1

Find the area of $\square ABCD$.

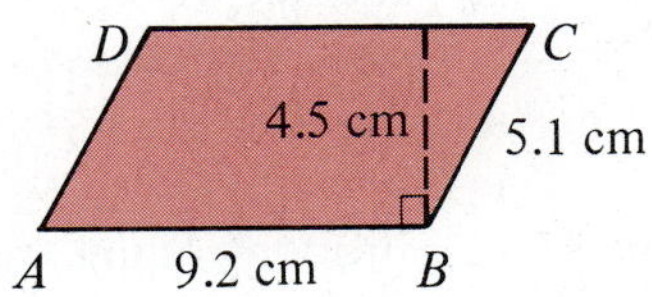

$b = 9.2$ and $h = 4.5$
$A = bh = 9.2\ (4.5) = 41.4\ \text{cm}^2$

Example 2

The area of a parallelogram is 36 m². The length of a base is 8 m. Find the length of a corresponding altitude.

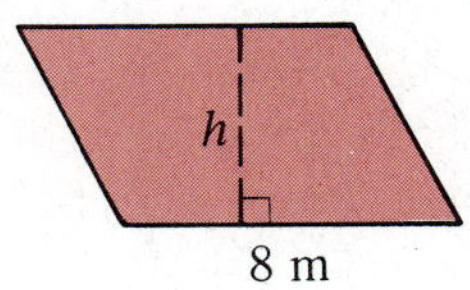

$A = bh$
$36 = 8h$ ◄ *Substitute 36 for A and 8 for b.*
$4.5 = h$
Thus, the length of a corresponding altitude is 4.5 m.

Notice that you can use any of the four sides of a parallelogram as a base. The area can be found by multiplying the length of any base by the length of any altitude drawn to that base.

Example 3

Given: $\square ACEF$ with $\overline{AD} \perp \overline{EC}$, $\overline{EB} \perp \overline{AC}$, $AC = 12$, $BE = 6$, and $CE = 8$
Find AD.

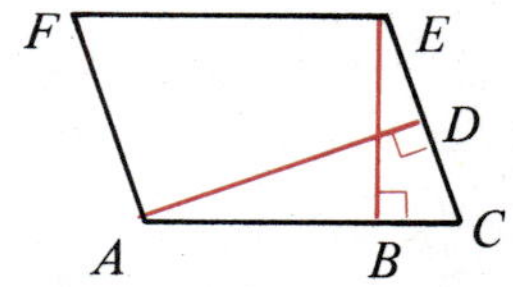

$$\text{Area } \square ACEF = (AC)(BE) = 12 \cdot 6 = 72$$

or

$$\text{Area } \square ACEF = (CE)(AD)$$
$$72 = 8\,(AD)$$
$$9 = AD$$

Thus, $AD = 9$.

In Example 4, an altitude of a parallelogram is drawn so that a 30°–60° right triangle is formed. After the length of the altitude is computed, then the area of the parallelogram is found.

Example 4

The lengths of two consecutive sides of a parallelogram are 12 m and 14 m, respectively. The included angle measures 30°. Find the area of the parallelogram.

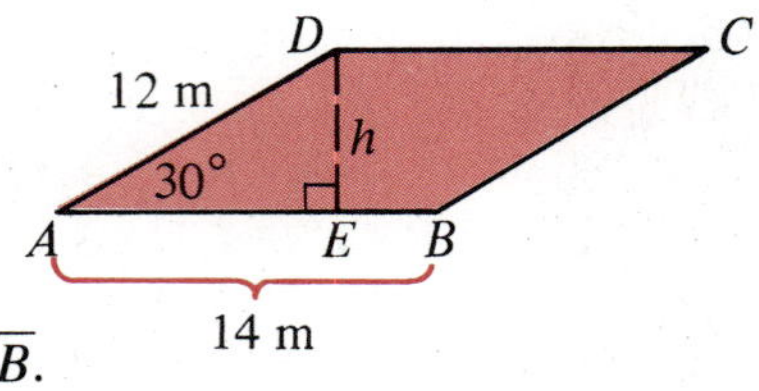

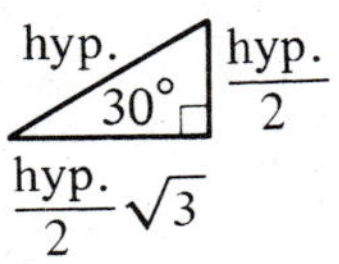

Let $h = DE$, the length of the altitude from D to $\overline{AB}$.
$\triangle AED$ is a 30°–60° rt. $\triangle$, so $h = 6$.
Area $\square ABCD = bh = 14 \cdot 6 = 84$.

Thus, the area is 84 m².

Reading in Geometry

Write *A* if the statement is *always* true, *S* if it is *sometimes* true, and *N* if it is *never* true. Draw a figure, if necessary.

1. Congruent parallelograms have the same area. **A**
2. Parallelograms with the same area are congruent. **S**
3. Parallelograms with equal areas have congruent bases and congruent corresponding altitudes. **S**
4. Parallelograms with congruent bases and congruent corresponding altitudes have the same area. **A**
5. If the measure of an angle of a parallelogram is 30°, then the area is equal to one half the product of the lengths of two consecutive sides. **A**
6. If two consecutive sides of a parallelogram are 5 cm and 8 cm long, then the area is greater than 40 cm². **N**
7. If two sides of a triangle have the same measures as a base and a corresponding altitude of a parallelogram, respectively, then the area of the triangle is equal to one half the area of the parallelogram. **S**
8. If b_1 and b_2 represent the lengths of two consecutive sides of a parallelogram, and h_1 and h_2 represent the lengths of the corresponding altitudes, then $b_1 \cdot h_1 = b_2 \cdot h_2$. **A**

Oral Exercises

In each case, tell whether enough information is given to find the area of the parallelogram.

1. 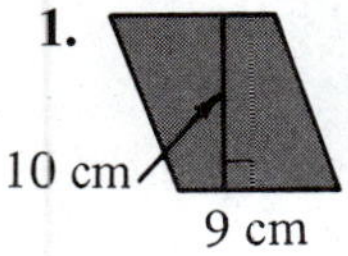

yes

2. 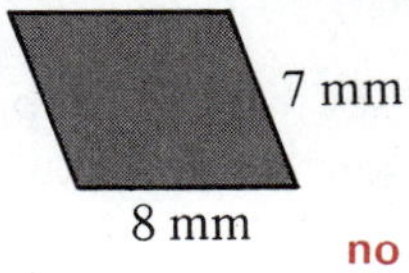

no

3. 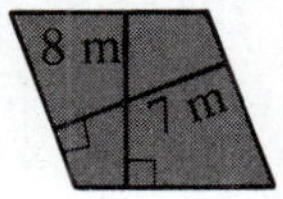

no

4. 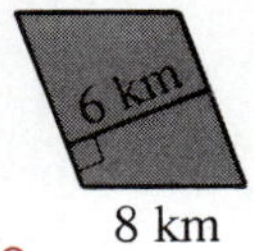

no

Written Exercises

Find the area of each parallelogram.

A **1.** 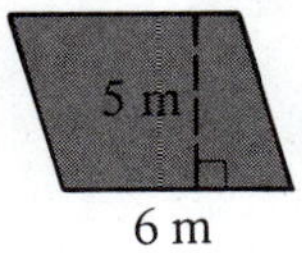

30 m²

2. 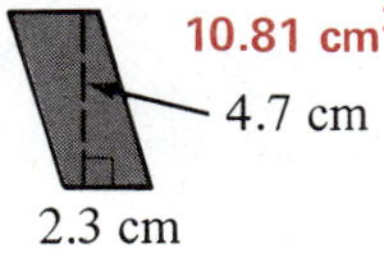

10.81 cm²

3. 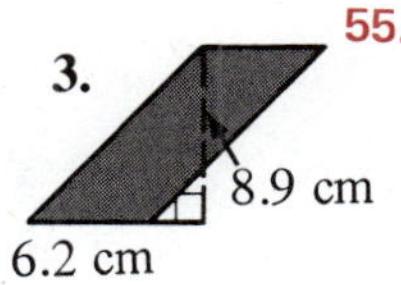

55.18 cm²

4. 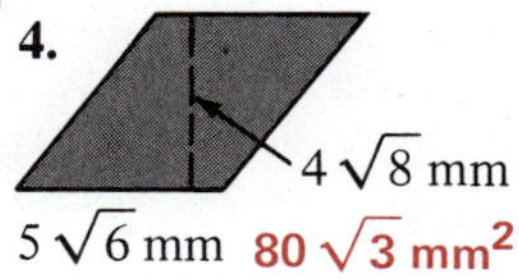

$80\sqrt{3}$ mm²

Supply the missing information for each parallelogram.

	Length of base (b)	Length of altitude (h)	Area (A)
5.	9 cm	? 9 cm	81 cm²
6.	? 0.8 mm	0.9 mm	0.72 mm²
7.	0.05 m	? 3.39 m	0.1695 m²
8.	? $\sqrt{7}$ km	$\sqrt{7}$ km	7 km²

In ▱$ACEG$, $\overline{FB} \perp \overline{AC}$ and $\overline{GD} \perp \overline{CE}$. Find each indicated length.

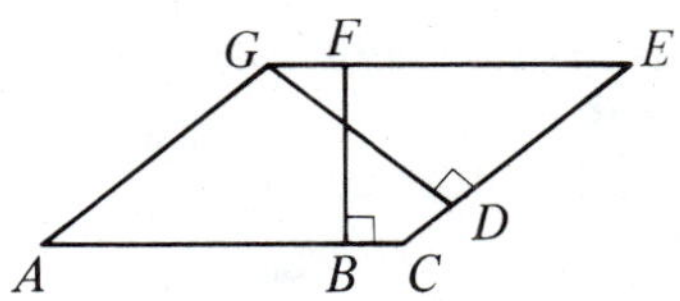

B **9.** Given: $AC = 18$, $BF = 6$, and $CE = 12$. Find DG. 9

10. Given: $AC = 24$, $BF = 8$, and $DG = 12$. Find CE. 16

11. Given: $AC = 35$, $CE = 24$, and $DG = 21$. Find BF. 14.4

12. Given: $CE = 12$, $DG = \sqrt{6}$, and $BF = \sqrt{8}$. Find AC. $6\sqrt{3}$

Find the area of ▱$ABCD$ in each case. (Ex. 13–16)

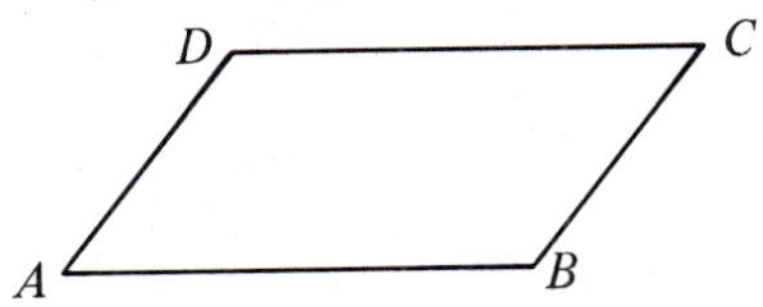

13. Given: $AB = 10$, $AD = 6$, and $m\angle A = 30$ 30

14. Given: $DC = 9$, $BC = 8$, and $m\angle C = 60$ $36\sqrt{3}$

15. Given: $AB = 12$, $BC = 10$, and $m\angle A = 45$ $60\sqrt{2}$

16. Given: $DC = 16$, $AD = 12$, and $m\angle D = 135$ $96\sqrt{2}$

C **17.** Prove that the area of a parallelogram can be found by the formula Area $= ab \sin A$.

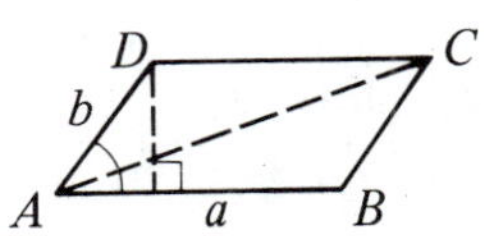

18. Prove that the area of $\triangle ABC$ can be found by the formula Area $= \frac{1}{2}ab \sin A$, where $\angle A$ is acute.

19. Find the area of ▱$ABCD$ if $a = 16$, $b = 5$, and $m\angle A = 42$. 53.528 sq. units

20. Find the area of $\triangle ABC$ if $a = 5$, $b = 6$, and $m\angle A = 56$. 12.4 sq. units

AREAS OF TRAPEZOIDS 12.4

Objective

To apply the formula for the area of a trapezoid

Recall that a trapezoid has two bases which are parallel. An altitude is a perpendicular segment from any point on one base to the line containing the other base.

Consider trapezoid $RSTU$ below. You can find its area by dividing the region into two triangular regions and adding the areas. Notice that $\triangle RSU$ and $\triangle UTS$ have congruent altitudes.

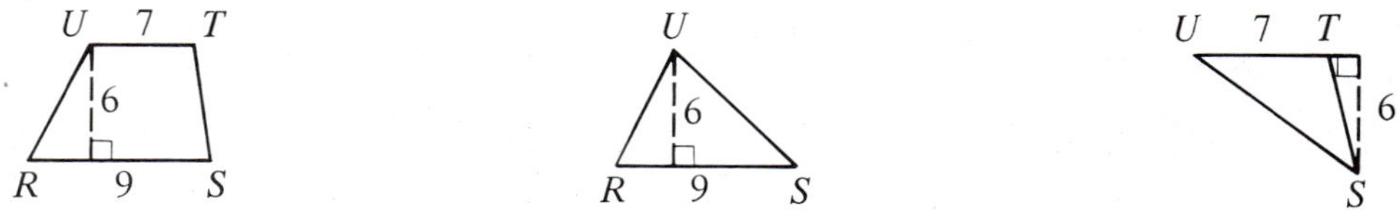

$$\text{Area trap. } RSTU = \text{area } \triangle RSU + \text{area } \triangle UTS$$
$$= \frac{1}{2} \cdot 6 \cdot 9 + \frac{1}{2} \cdot 6 \cdot 7$$
$$= \frac{1}{2} \cdot 6\,(9 + 7) \quad \blacktriangleleft \textit{ Distributive Property}$$

Thus, Area trap. $RSTU = 48$.

The proof of Theorem 12.6 follows the strategy used above.

Theorem 12.6

The area of a trapezoid is one half the product of the length of an altitude and the sum of the lengths of the two bases.

Given: Trap. $ABCD$ with base lengths b_1 and b_2, and altitude length h

Prove: Area trap. $ABCD = \frac{1}{2}h\,(b_1 + b_2)$

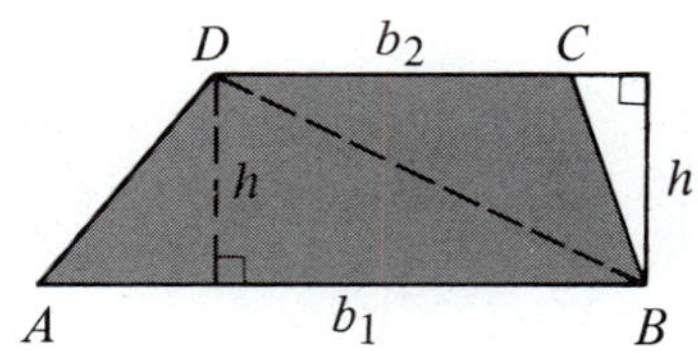

Proof

Statements	Reasons
1. Trap. $ABCD$ with base lengths b_1 and b_2 and altitude length h	1. Given
2. Draw diagonal $\overline{DB}$.	2. Two points are contained in exactly one line.
3. Area $\triangle ABD = \frac{1}{2} b_1 h$	3. Area $\triangle = \frac{1}{2} bh$
4. Area $\triangle DBC = \frac{1}{2} b_2 h$	4. Same as 3
5. Area trap. $ABCD$ = area $\triangle ABD$ + area $\triangle DBC$	5. Area Addition Postulate
6. Area trap. $ABCD = \frac{1}{2} b_1 h + \frac{1}{2} b_2 h$	6. Substitution property
7. $\therefore$ Area trap. $ABCD = \frac{1}{2} h\,(b_1 + b_2)$	7. Distributive property

Example 1 Find the area of trap. $ABCD$.

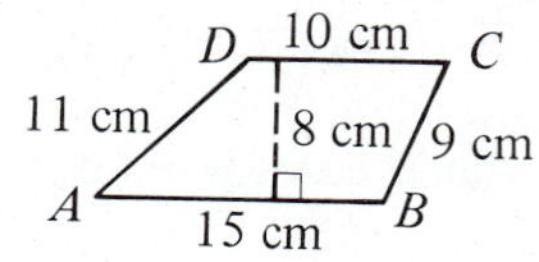

$b_1 = 15,\ b_2 = 10,\ h = 8$

$A = \frac{1}{2}h(b_1 + b_2) = \frac{1}{2} \cdot 8\,(15 + 10) = 4 \cdot 25 = 100$

Thus, the area is 100 cm².

Example 2 The area of a trapezoid is 36 m². The length of one base is 7 m and the length of an altitude is 6 m. Find the length of the other base.

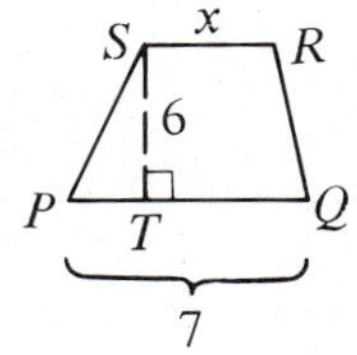

Let $x =$ the length of the other base.

$A = \frac{1}{2}h(b_1 + b_2)$

$36 = \frac{1}{2} \cdot 6\,(7 + x)$

$36 = 3\,(7 + x)$

$12 = 7 + x$

$5 = x$

Thus, the length of the other base is 5 m.

Example 3 The area of a trapezoid is 175 cm². An altitude is 10 cm long. If the length of one base is 8 cm greater than twice the length of the other base, find the length of each base.

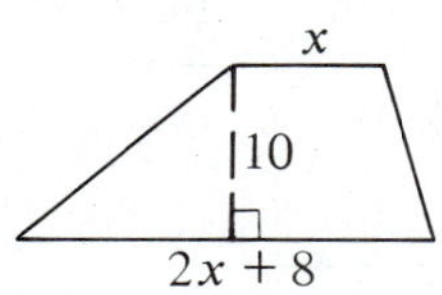

Let $x =$ the length of one base.

$2x + 8 =$ the length of other base.

$$A = \frac{1}{2}h(b_1 + b_2)$$

$$175 = \frac{1}{2} \cdot 10\,[x + (2x + 8)]$$

$$175 = 5\,(3x + 8)$$

$$35 = 3x + 8$$

$$27 = 3x$$

$$9 = x$$

$2x + 8 = 2 \cdot 9 + 8 = 26$

Thus, the lengths of the bases are 9 cm and 26 cm.

Example 4 In trap. $ABCD$, $AB = 11$, $DA = 4$, $DC = 9$, and m $\angle A = 60$. Find the area of trap. $ABCD$.

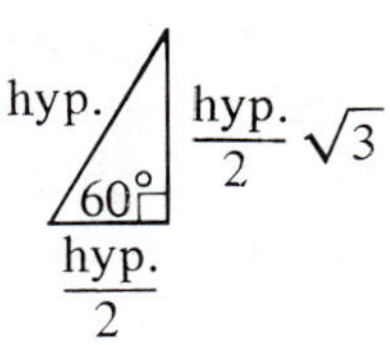

Draw altitude $\overline{DE}$ from D to $\overline{AB}$.

$\triangle ADE$ is a 30°–60° rt. $\triangle$.

So, $DE = \frac{4}{2}\sqrt{3} = 2\sqrt{3}$.

$A = \frac{1}{2}h(b_1 + b_2) = \frac{1}{2} \cdot 2\sqrt{3}(11 + 9)$

$= \sqrt{3} \cdot 20 = 20\sqrt{3}$

Thus, the area is $20\sqrt{3}$.

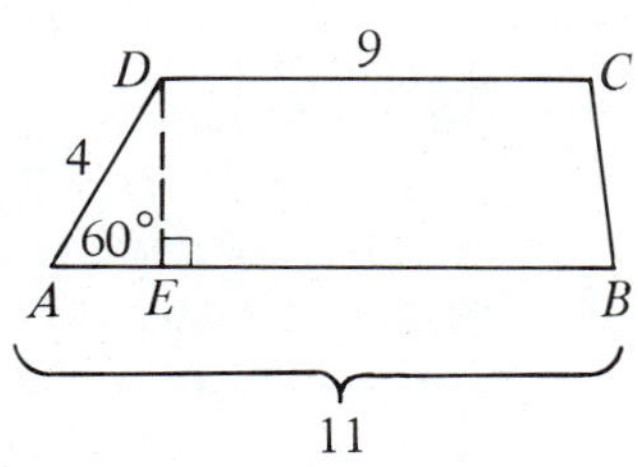

Oral Exercises

In which cases is enough information given to find the area of the trapezoid?

1. no

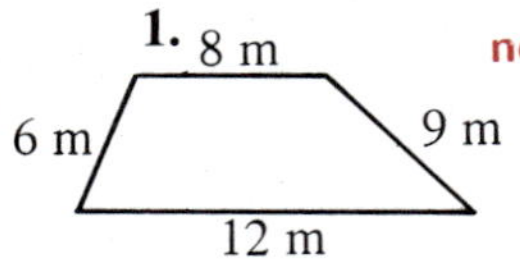

2. yes

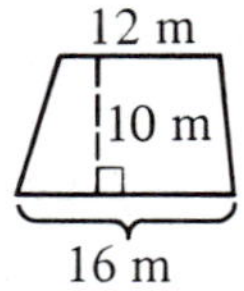

3. no

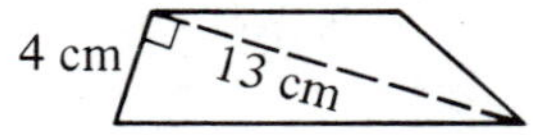

4. yes

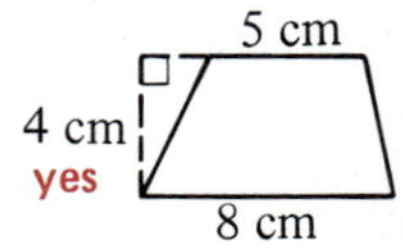

Written Exercises

Find the area of each trapezoid.

A **1.** 28 mm²

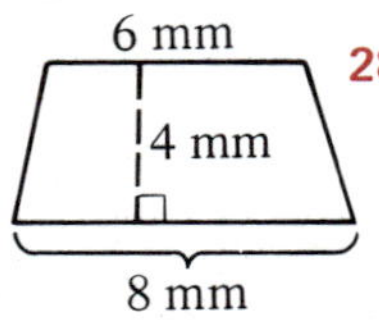

2. 73.5 cm²

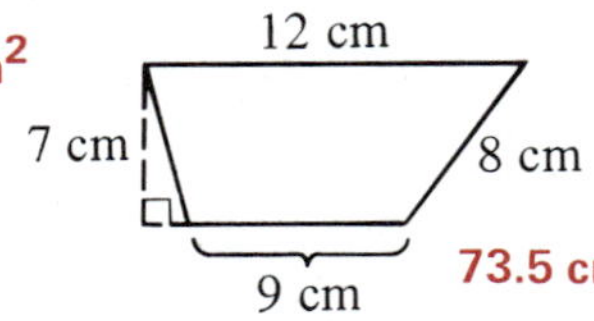

3. 12.6 m²

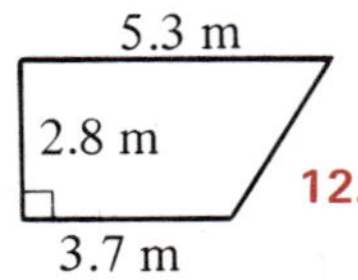

4. 44.5 km²

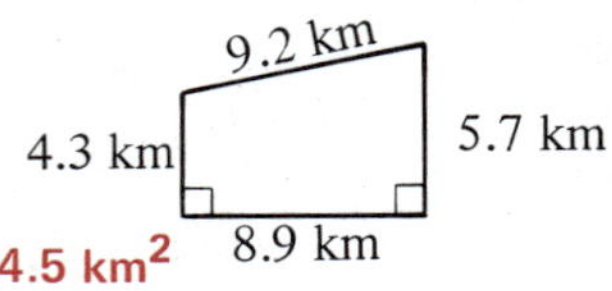

Supply the missing information for each trapezoid. (Ex. 5–13)

	b_1	b_2	h	Area
5.	7 m	9 m	5 m	? 40 m²
6.	12 mm	11 mm	9 mm	? 103.5 mm²
7.	9 cm	15 cm	? 6 cm	72 cm²
8.	12 mm	16 mm	? 4 mm	56 mm²
9.	8 m	? 3 m	10 m	55 m²
10.	4 cm	? 2 cm	7 cm	21 cm²

B **11.**	$x + 2$	$3x - 6$	$2x + 1$	? $4x^2 - 2x - 2$
12.	$3x - 1$	$5x + 7$	? $x - 4$	$4x^2 - 13x - 12$
13.	? $2x - 1$	$2x + 3$	$2x + 1$	$4x^2 + 4x + 1$

14. The area of a trapezoid is 136 cm². An altitude is 8 cm long. If one base is 4 cm longer than the other base, find the length of each base. **19 cm, 15 cm**

15. The area of a trapezoid is 234 m². An altitude is 12 m long. If the length of one base is 9 m less than twice the length of the other base, find the length of each base. **16 m, 23 m**

16. $\overline{AB}$ and $\overline{CD}$ are the bases of trap. $ABCD$. If $AB = 30$, $CD = 20$, $AD = 8$, and $m\angle A = 60$, find the area of the trapezoid. **$100\sqrt{3}$**

17. $\overline{XY}$ and $\overline{ZW}$ are the bases of trap. $XYZW$. If $XY = 15$, $ZW = 3$, $XW = 8$, and $m\angle X = 45$, find the area of the trapezoid. **$36\sqrt{2}$**

C **18.** $\overline{AB}$ and $\overline{CD}$ are the bases of isosceles trapezoid $ABCD$. If $AB = 12$, $CD = 6$, and $AD = 5$, find the area of the trapezoid. **36**

19. $\overline{XY}$ and $\overline{ZW}$ are the bases of isosceles trapezoid $XYZW$. If $XY = 11$, $ZW = 7$, and $m\angle X = 60$, find the area of the trapezoid. **$18\sqrt{3}$**

20. In trapezoid $ABCD$, the lengths of bases $\overline{AB}$ and $\overline{CD}$ are 17 and 13, respectively. The length of $\overline{AD}$ is 4. If m $\angle A = 37$, find the area of trapezoid $ABCD$ to the nearest tenth. **36.1**

21. In trapezoid $XYZW$, the lengths of bases $\overline{XY}$ and $\overline{WZ}$ are 9 and 4, respectively. The length of $\overline{XW}$ is 3. If m $\angle X = 72$, find the area of trapezoid $XYZW$ to the nearest tenth. **18.5**

22. Prove or disprove: If two trapezoids have equal areas, then the sums of the lengths of the bases are inversely proportional to the lengths of the corresponding altitudes.

23. Prove or disprove: The ratio of the areas of two trapezoids with bases of one congruent to bases of the other is equal to the ratio of the lengths of their altitudes.

24. Construct an isosceles trapezoid with the same bases and area as a given nonisosceles trapezoid.

25. Construct a rectangle with the same area as a given trapezoid. [*Hint:* Use the median of the trapezoid.]

Cumulative Review

1. In ▱$ABCD$, if m $\angle A = 120$, find the measure of each angle.
m $\angle$B = 60, m $\angle$C = 120, m $\angle$D = 60

2. If $\triangle ABC \sim \triangle DEF$, write 3 proportions showing the correspondence between the sides. $\frac{AB}{DE} = \frac{BC}{EF}, \frac{AB}{DE} = \frac{CA}{FD}, \frac{BC}{EF} = \frac{CA}{FD}$

CALCULATOR ACTIVITIES

The formula for the area of a trapezoid can be written as follows:

$$\text{Area trap.} = \frac{(b_1 + b_2)\,h}{2}$$

Example

Find the area of trap. $ABCD$ to the nearest tenth.

Formula ► $\text{Area trap.} = \frac{(b_1 + b_2)\,h}{2}$

$$= \frac{(8.2 + 5.7)\,4.8}{2}$$

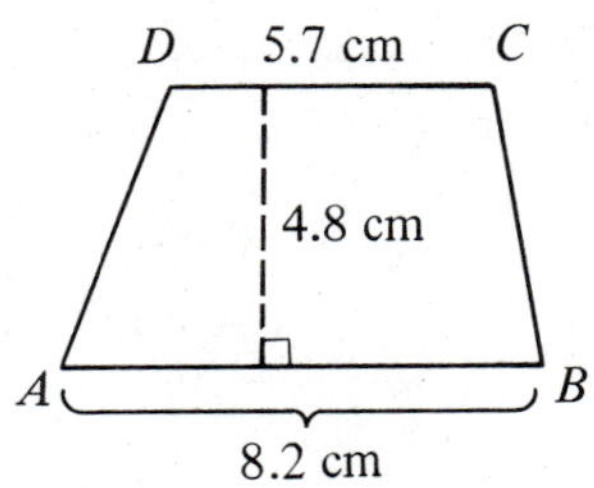

Press ► 8.2 ⊕ 5.7 ⊗ 4.8 ⊖ 2

Display ► 33.36

Thus, the area is 33.4 cm² to the nearest tenth.

Use a calculator to find the area of each trapezoid, to the nearest tenth.

1. 32.4 cm² 2. 117.8 mm² 3. 638.0 m² 4. 26.5 cm² 5. 148.9 m² 6. 161.6 m²

	b_1	b_2	h
1.	6.9 cm	7.2 cm	4.6 cm
2.	10.1 mm	14.7 mm	9.5 mm
3.	23.8 m	17.9 m	30.6 m

	b_1	b_2	h
4.	5.23 cm	6.92 cm	4.37 cm
5.	18.64 m	12.81 m	9.47 m
6.	21.71 m	16.05 m	8.56 m

SIMILAR POLYGONS: AREAS AND PERIMETERS

12.5

Objectives **To find the perimeter of a polygon**
To find ratios of areas of similar polygons
To find ratios of perimeters of similar polygons

Consider the following experiment with similar triangles $\triangle ABC$ and $\triangle DEF$. Find the area of each triangle.

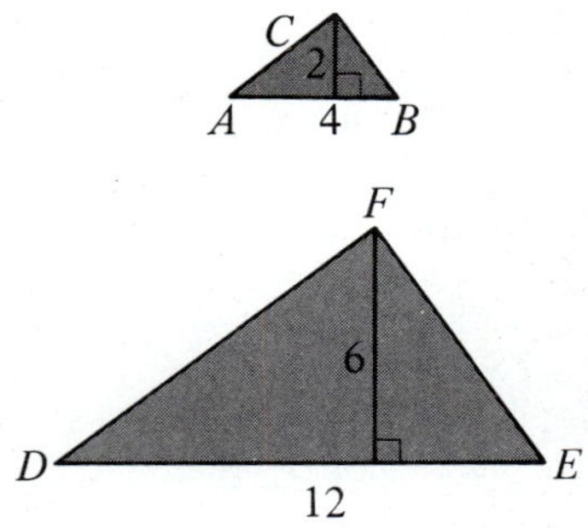

Area $\triangle ABC = \frac{1}{2}bh = \frac{1}{2} \cdot 4 \cdot 2 = 4$

Area $\triangle DEF = \frac{1}{2}bh = \frac{1}{2} \cdot 12 \cdot 6 = 36$

The ratio of the lengths of two corresponding sides is $\frac{4}{12}$, or $\frac{1}{3}$. The ratio of the areas is $\frac{4}{36}$, or $\frac{1}{9}$. This leads to the following theorem.

Theorem 12.7 The ratio of the areas of two similar triangles is the square of the ratio of the lengths of any two corresponding sides. (See Exercise 23.)

You can use Theorem 12.7 to find the ratio of the areas of two similar triangles or to find the length of one of the sides.

Example 1 The ratio of the lengths of two corresponding sides of two similar triangles is $\frac{5}{6}$. Find the ratio of the areas.

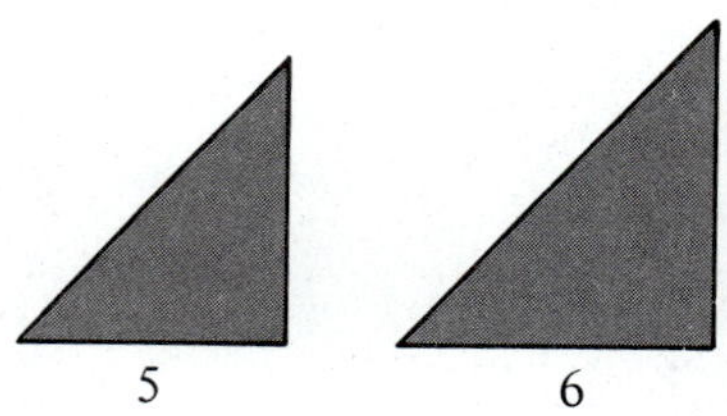

$$\frac{\text{Area of 1st } \triangle}{\text{Area of 2nd } \triangle} = \left(\frac{5}{6}\right)^2 = \frac{25}{36}$$

Example 2 Given: $\triangle ABC \sim \triangle DEF$. The ratio of the areas of the triangles is $\frac{9}{16}$. $BC = 6$. Find EF.

$$\frac{\text{Area } \triangle ABC}{\text{Area } \triangle DEF} = \left(\frac{BC}{EF}\right)^2 \text{ by Th. 12.7}$$

Let $x = EF$. ► $\frac{9}{16} = \left(\frac{6}{x}\right)^2$

$$\frac{9}{16} = \frac{36}{x^2}$$

$$9x^2 = 16 \cdot 36$$

$$x^2 = 64$$

$x = 8$, so $EF = 8$.

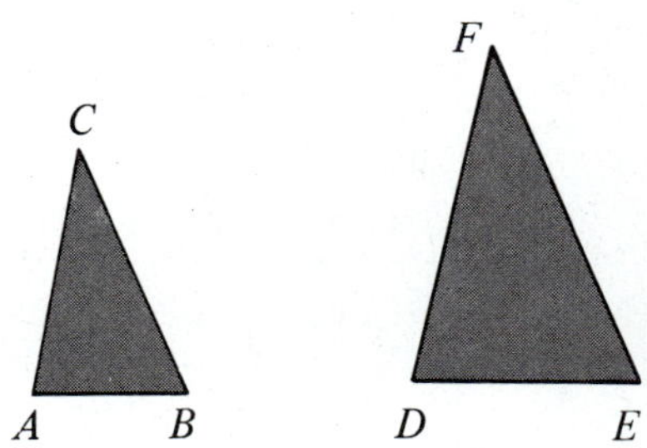

You have learned to find the **perimeter** of a triangle by adding the lengths of the three sides. Now the concept of the perimeter will be extended to apply to any polygon.

Definition: Perimeter The *perimeter* of a polygon is the sum of the lengths of its sides.

The definition of perimeter is applied in Example 3.

Example 3

Find the perimeter of a rhombus with a side 7.2 cm long.

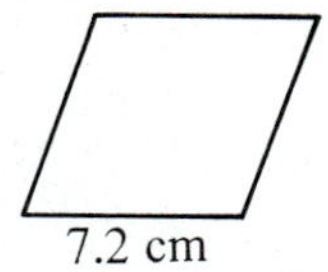

All sides of a rhombus are the same length.
$p = 7.2 + 7.2 + 7.2 + 7.2$, or $p = 4 \cdot 7.2 = 28.8$

Thus, the perimeter is 28.8 cm.

Consider the following experiment with similar quadrilaterals $ABCD$ and $EFGH$ and their perimeters.

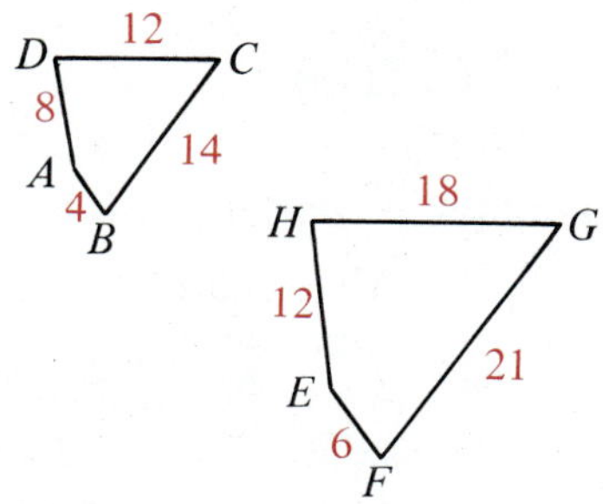

The ratio of the lengths of two corresponding sides is $\frac{8}{12}$, or $\frac{2}{3}$. The ratio of the perimeters is $\frac{8+4+14+12}{12+6+21+18} = \frac{38}{57} = \frac{2}{3}$. This suggests the following theorem.

Theorem 12.8 The ratio of the perimeters of two similar polygons is the same as the ratio of the lengths of any two corresponding sides. (See Exercise 24.)

Theorem 12.8 can sometimes be used to find the perimeter of one of two similar polygons.

Example 4

The ratio of the lengths of corresponding sides of similar triangles is $\frac{2}{5}$. If the perimeter of the larger triangle is 15 cm, find the perimeter of the smaller triangle.

Let p = perimeter of the smaller $\triangle$.

$$\frac{p}{15} = \frac{2}{5} \text{ by Th. 12.8}$$
$$5p = 30$$
$$p = 6$$

Thus, the perimeter of the smaller triangle is 6 cm.

Written Exercises

Find the ratio of the areas and the ratio of the perimeters for each pair of similar triangles.

A 1. Given: $\triangle ABC \sim \triangle DEF$

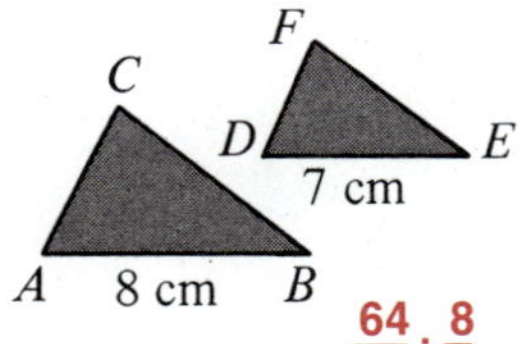

$\frac{64}{49}; \frac{8}{7}$

2. Given: $\triangle GHI \sim \triangle JKL$

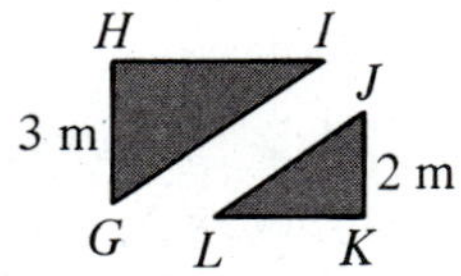

$\frac{9}{4}; \frac{3}{2}$

3. Given: Rt. $\triangle TUV \sim \triangle XYZ$

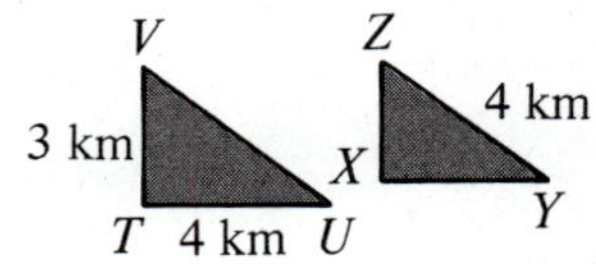

$\frac{25}{16}; \frac{5}{4}$

4. Given: Equilateral $\triangle ABC \sim \triangle DEF$

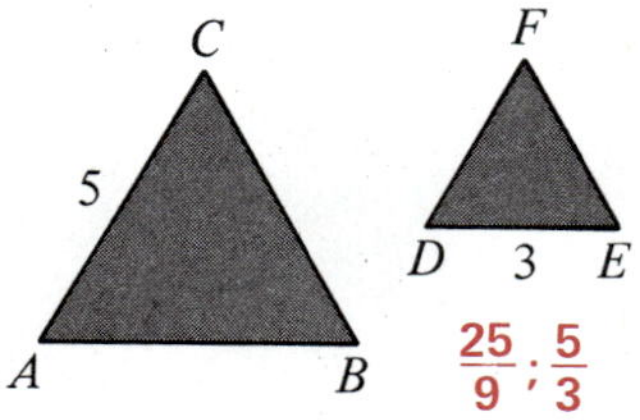

$\frac{25}{9}; \frac{5}{3}$

5. Given: Rt. $\triangle EFG \sim \triangle HIJ$

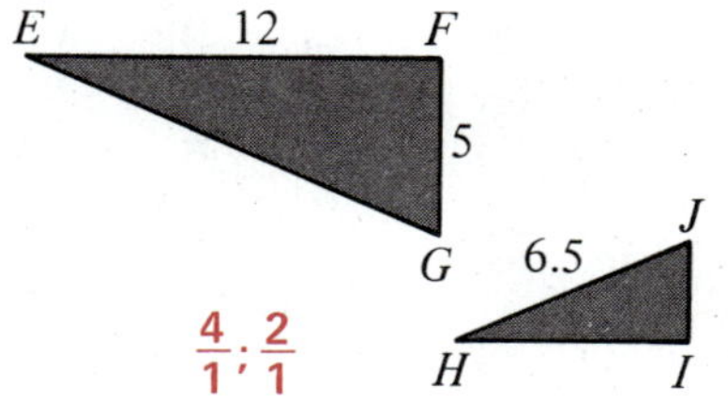

$\frac{4}{1}; \frac{2}{1}$

6. Given: $\triangle PQR \sim \triangle STU$ and $\overline{PQ} \cong \overline{QR}$

$\frac{9}{4}; \frac{3}{2}$

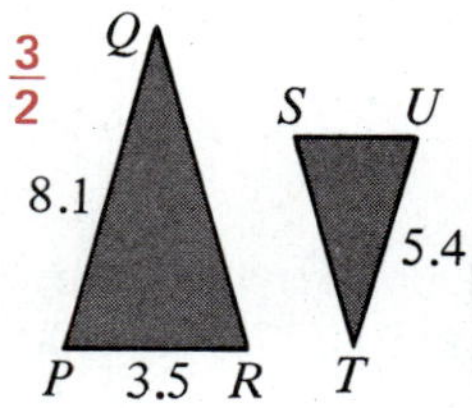

Find the perimeter of each polygon. (Ex. 7–10)

7. An equilateral triangle with a side 7 cm long **21 cm**

8. A square with a side 3 m long **12 m**

9. A parallelogram with consecutive sides 5.2 cm and 3.9 cm long **18.2 cm**

10. A rhombus with a side 6.8 mm long **27.2 mm**

11. The ratio of the lengths of corresponding sides of similar triangles is $\frac{3}{4}$. If the perimeter of the smaller triangle is 51 cm, find the perimeter of the larger triangle. **68 cm**

12. The ratio of the perimeters of two similar polygons is $\frac{5}{8}$. One side of the larger polygon is 28 cm long; find the length of the corresponding side of the smaller polygon. **17.5 cm**

Find the perimeter of each polygon. (Ex. 13–18)

16. $(18 + 6\sqrt{3})$**m**

B 13. A square with a diagonal 8 cm long $16\sqrt{2}$ **cm**

14. A rectangle with a side 8 m long and a diagonal 17 m long **46 m**

15. A right triangle with legs 9 cm and 12 cm long **36 cm**

16. A right triangle with a leg 6 m long and the hypotenuse 12 m long

17. A 30°–60° right triangle with the hypotenuse 14 m long $(21 + 7\sqrt{3})$ **m**

18. An isosceles right triangle with a leg 9 cm long $(18 + 9\sqrt{2})$ **m**

19. Given: $\triangle ABC \sim \triangle DEF$, the ratio of the areas is $\frac{16}{25}$, and $AC = 12$. Find DF. **15**

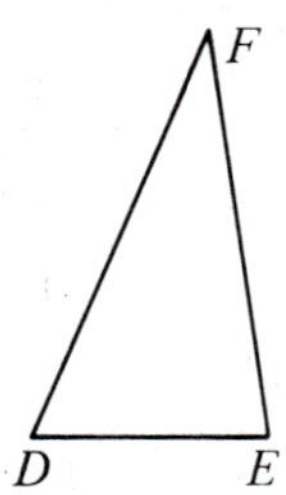

20. Given: $\triangle ABC \sim \triangle DEF$, the ratio of the areas is $\frac{4}{9}$, $AB = 5$, $BC = 9$, and $CA = 10$. Find the perimeter of $\triangle DEF$. **36**

21. The area of one triangle is 9 cm^2. The area of a similar triangle is 4 cm^2. If the perimeter of the first is $5x - 3$ and the perimeter of the second is $2x + 2$, what is the perimeter of each? **12 cm; 8 cm**

22. In two similar triangles, corresponding sides are 4 m and 2 m long. If the area of the smaller is 18 m^2 less than the area of the larger, find the area of the larger. **24 m^2**

23. Prove Theorem 12.7.

24. Prove Theorem 12.8.

25. A side of a triangle measures 1 unit. What is the measure of the corresponding side of a similar triangle with an area three times the area of the original triangle? **$\sqrt{3}$**

26. A side of a triangle measures x units. What is the measure of the corresponding side of a similar triangle with an area y times the area of the original triangle? **$x\sqrt{y}$**

NON-ROUTINE PROBLEMS

Approximating Areas

Two different methods are used to estimate the area of a quadrilateral-shaped land plot. In most cases, the area obtained is only an approximation since the shapes are not exactly rectangles or kites.

Sides Method

To find the approximate area of a quadrilateral that is almost rectangular-shaped, use

$$\textbf{Area} \doteq \left(\frac{AB + CD}{2}\right)\left(\frac{AD + BC}{2}\right)$$

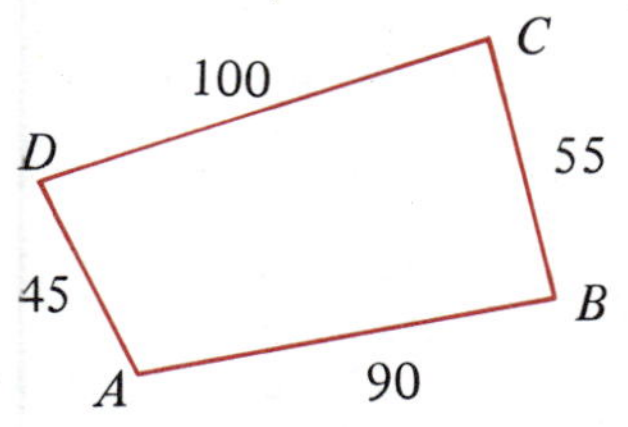

$$\text{Area} \doteq \left(\frac{90 + 100}{2}\right)\left(\frac{45 + 55}{2}\right)$$
$$\doteq (95)(50)$$
$$\doteq 4{,}750 \text{ square units}$$

Diagonal Method

To find the approximate area of a quadrilateral that is almost kite-shaped, use

$$\textbf{Area} \doteq \frac{1}{2}(AC)(DB)$$

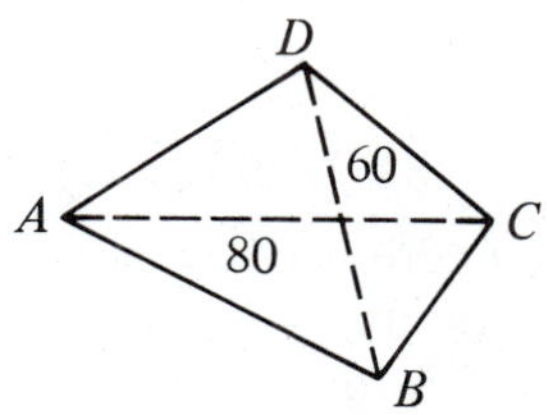

$$\text{Area} \doteq \frac{1}{2}(80)(60)$$
$$\doteq (40)(60)$$
$$\doteq 2{,}400 \text{ square units}$$

For the two figures below, measure carefully in centimeters. Approximate the area of each using both formulas. Which formula gives a better estimate. Why?

1.8 cm^2; 2.2 cm^2. Sides method gives better estimate.

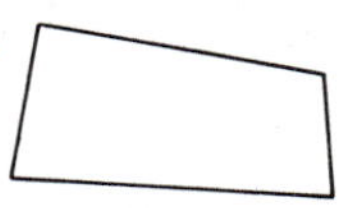

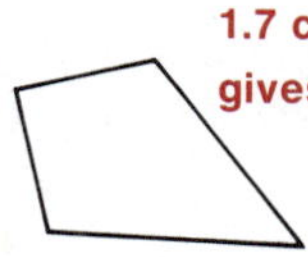

1.7 cm^2; 1.4 cm^2. Diagonal method gives better estimate.

AREAS OF REGULAR POLYGONS 12.6

Objectives **To apply the formula for the area of a regular polygon**
To find lengths of segments related to a regular polygon

Recall that a regular polygon is both equilateral and equiangular. It can be proved that a circle can be inscribed in any regular polygon. Also, a circle can be circumscribed about any regular polygon. The following definitions pertain to a regular polygon.

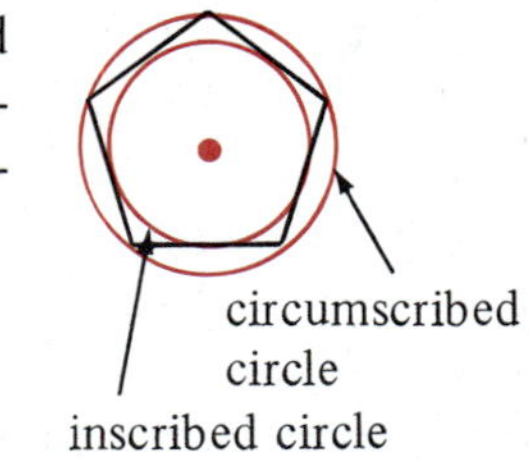

Definitions: Center, apothem, radius of regular polygon

The *center of a regular polygon* is the common center of its inscribed and circumscribed circles.
An *apothem of a regular polygon* is a segment from the center of the polygon perpendicular to a side of the polygon.
A *radius of a regular polygon* is a segment that joins the center of the polygon and a vertex of the polygon.

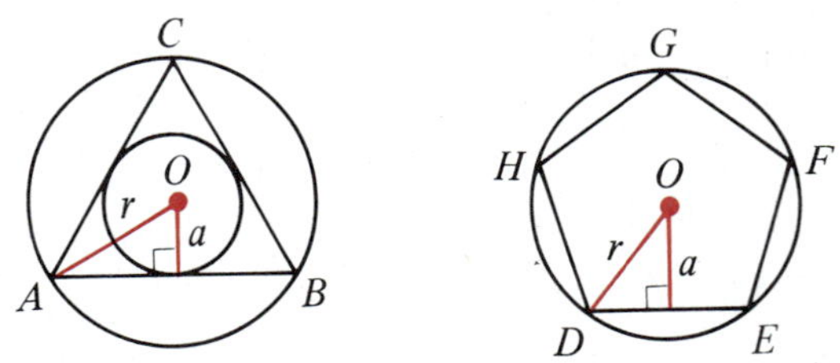

In the figures above, note that an apothem (a) is also a radius of the inscribed circle, and a radius (r) is also a radius of the circumscribed circle.

It can be proved that an apothem of a regular polygon bisects a side of the polygon. (See Exercise 20.) Also, a radius of a regular polygon bisects an angle of the polygon. (See Exercise 21.) These two statements are applied in Example 1.

Example 1 The length of an apothem of a square is 8. Find the length of a radius, the length of a side, and the area of the square.

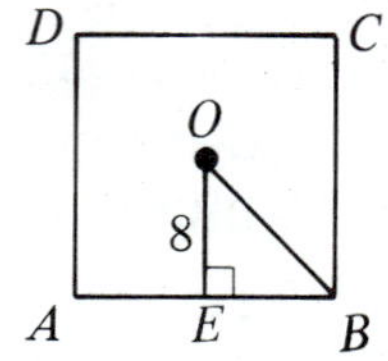

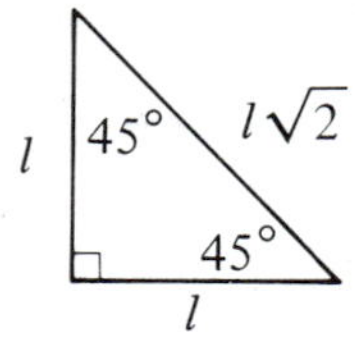

Radius $\overline{OB}$ bisects $\angle ABC$, so m $\angle OBE = 45$.
Then $\triangle OEB$ is a 45°–45° rt. $\triangle$.
So $EB = 8$ and $OB = 8\sqrt{2}$.
Apothem $\overline{OE}$ bisects $\overline{AB}$, so $AB = 16$.
Area of square $ABCD = s^2 = 16^2 = 256$.

Thus, the length of a radius is $8\sqrt{2}$, the length of a side is 16, and the area is 256.

Theorem 12.9

The area of a regular polygon is one half the product of the length of an apothem and the perimeter.

Given: Regular polygon $ABCD \cdots$ with n sides, a is the length of an apothem, and p is the perimeter.

Prove: Area regular polygon $ABCD \cdots = \frac{1}{2}ap$.

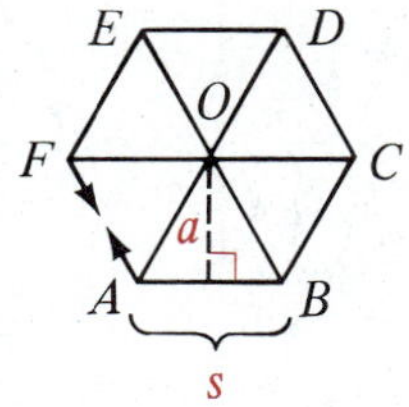

Outline of Proof

For a regular polygon with n sides, its radii divide it into n congruent triangles. If a is the length of an apothem and s is the length of a side, then the area of each triangle is $\frac{1}{2}as$. The area of the polygon is $n \cdot \left(\frac{1}{2}as\right)$, or $\frac{1}{2}a\,(ns)$. But ns is the perimeter of the polygon. **Thus,** the area of the polygon is $\frac{1}{2}ap$.

Example 2

The length of an apothem of a regular hexagon is $5\sqrt{3}$. Find the area.

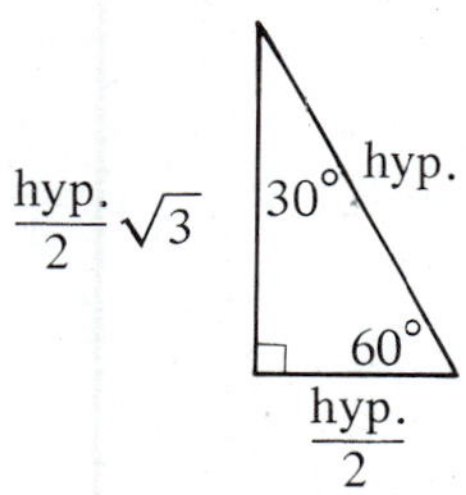

Radius $\overline{OB}$ bisects $\angle ABC$, so m $\angle OBG = 60$.
Then $\triangle OBG$ is a $30°-60°$ rt. $\triangle$ and $GB = 5$.
Apothem $\overline{OG}$ bisects $\overline{AB}$, so $AB = 10$.
Perimeter $= 6 \cdot 10 = 60$

Area $= \frac{1}{2}ap = \frac{1}{2} \cdot 5\sqrt{3} \cdot 60 = 150\sqrt{3}$

Thus, the area is $150\sqrt{3}$.

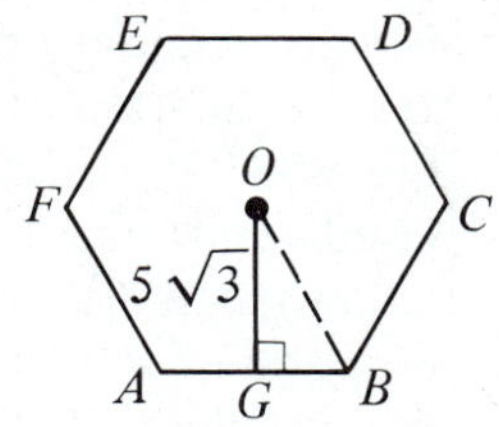

Written Exercises

Find the area of each of the following regular polygons.

	Number of sides	Length of side	Length of apothem	Area
A 1.	5	2	1.4	? 7
2.	5	1.4	1	? 3.5
3.	8	10	12.1	? 484
4.	8	0.5	0.6	? 1.2

Supply the missing information for each regular polygon.

	Number of sides	Length of side	Length of apothem	Length of radius
5.	3	? $14\sqrt{3}$	7	? 14
6.	3	8	? $\frac{4\sqrt{3}}{3}$	? $\frac{8\sqrt{3}}{3}$
7.	4	? 12	6	$6\sqrt{2}$?
8.	4	? 10	? 5	$5\sqrt{2}$
9.	6	10	? $5\sqrt{3}$	? 10
10.	6	? 8	$4\sqrt{3}$	? 8

Find the area of each of the following regular polygons.

	Number of sides	Length	Area	
B 11.	3	apothem: 4 cm	?	$48\sqrt{3}$ cm²
12.	3	radius: 6 m	?	$27\sqrt{3}$ m²
13.	4	apothem: 3 mm	?	36 mm²
14.	4	radius: 8 m	?	128 m²
15.	6	radius: 10 cm	?	$150\sqrt{3}$ cm²
16.	6	apothem: 6 cm	?	$72\sqrt{3}$ cm²

Find the missing measures for each regular polygon.

	Number of sides	Area	Length of apothem	Length of radius	Length of side
17.	3	$12\sqrt{3}$	? 2	? 4	? $4\sqrt{3}$
18.	4	25	? $\frac{5}{2}$	? $\frac{5\sqrt{2}}{2}$	? 5
19.	6	$8\sqrt{3}$	? 2	? $\frac{4\sqrt{3}}{3}$	? $\frac{4\sqrt{3}}{3}$

Prove each of the following.

20. An apothem of a regular polygon bisects a side of the polygon.

21. A radius of a regular polygon bisects an angle of the polygon.

C 22. If an equilateral triangle is inscribed in a circle, and another equilateral triangle is circumscribed about the same circle, what is the ratio of the areas of the two triangles? **1:4**

23. If a square is inscribed in a circle, and another square is circumscribed about the same circle, what is the ratio of the areas of the two squares? **1:2**

Prove or disprove each of the following.

24. The ratio of the areas of two similar regular polygons is equal to the ratio of their perimeters.

25. The ratio of the areas of two similar regular polygons is equal to the square of the ratio of the lengths of their apothems.

For each regular polygon, use sine, cosine, or tangent to find the missing measures. Give each answer to the nearest hundredth.

NOTE: You may wish to have students use a calculator for Ex. 26–32.

	Number of sides	Length of radius	Length of apothem	Area	
26.	5	4	? 3.24	?	38.09
27.	5	? 4.94	4	?	58.12
28.	5	? 1.30	? 1.05	4	
29.	9	6	? 5.64	?	104.16
30.	9	? 10.64	10	?	327.60
31.	9	? 2.03	? 1.91	12	
32.	10	8	? 7.61	?	188.12

CIRCUMFERENCES OF CIRCLES 12.7

Objective **To apply the formula for the circumference of a circle**

Two regular polygons are inscribed in congruent circles, as shown.

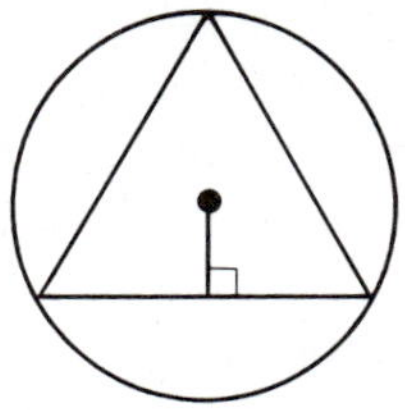

3 sides

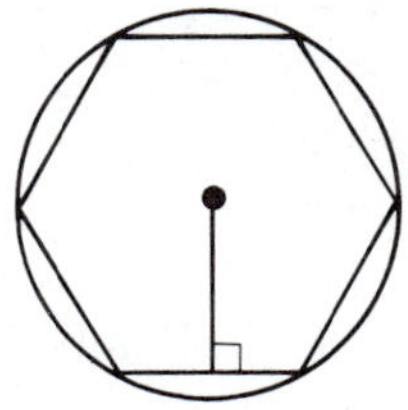

6 sides

If you were to continue the pattern of doubling the number of sides, the results would be described as follows:

1. The length of an apothem approaches the length of a radius of the circle as a limit.
2. The perimeter of the polygons approaches the "distance around" the circle as a limit.
3. The area of the polygons approaches the area of the circle as a limit.

These observations lead to the following definition.

Definition: Circumference The *circumference* of a circle is the limit of the perimeters of inscribed regular polygons of the circle as the number of sides of the polygons increases indefinitely.

Examine the relationship between the perimeter of an inscribed regular hexagon and the length of a diameter of its circumscribed circle. Suppose three regular hexagons are inscribed in circles. You can compute the ratio of the perimeter of each hexagon to the length of a diameter of the circle.

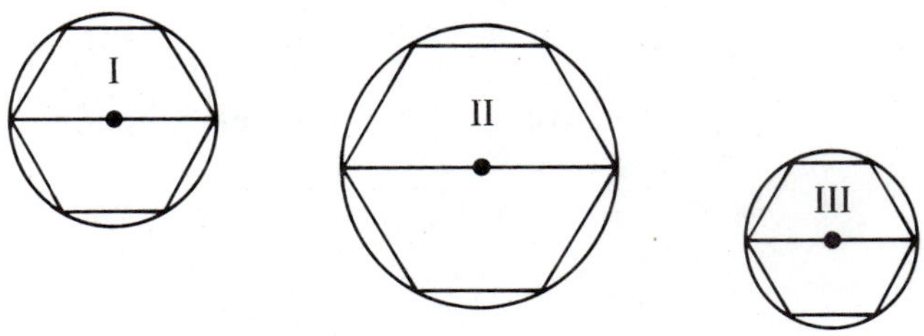

Since a side of a regular hexagon has the same length as a radius of the circle,

$$\frac{\text{Perimeter}}{\text{Length of Diameter}} = \frac{p}{d} = \frac{6r}{2r} = \frac{3}{1}.$$

Thus, in each case, the ratio is a constant, $\frac{3}{1}$.

Now suppose the number of sides of the inscribed regular polygons is increased. Then the perimeters of the polygons approach the circumferences of the circles as a limit. So it is reasonable to assume that the ratio of the circumference of a circle to the length of a diameter is also a constant. This can be proved with the help of a set of postulates and theorems about limits. However, in this book the theorem is presented without proof.

Theorem 12.10 For all circles, the ratio of the circumference to the length of a diameter is the same.

π (pi) is the symbol used for the ratio of the circumference to the length of a diameter of a circle.

$$\frac{c}{d} = \pi$$

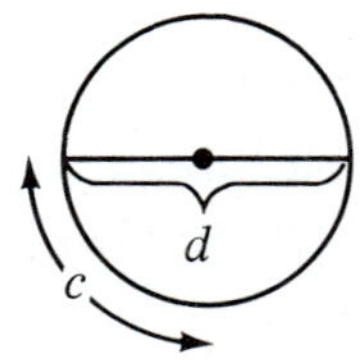

π is an irrational number which can be approximated correct to thousands of decimal places with the help of a computer. $\pi \doteq 3.141592653589793$, correct to fifteen decimal places. Commonly used approximations are 3.14, $\frac{22}{7}$, and 3.1416.

There are two formulas for the circumference of a circle. These formulas are given in the following corollary to Theorem 12.10.

Corollary If c is the circumference of a circle with a diameter of length d and a radius of length r, then $c = \pi d$ or $c = 2\pi r$.

Either $c = \pi d$ or $c = 2\pi r$ can be used in problems involving circumference.

Example 1 If a diameter of a circle measures 8 cm, find the circumference in terms of π.

$c = \pi d = \pi \cdot 8$, or 8π cm

Example 2 If the circumference of a circle is 20π, find the length of a radius.

Substitute 20π for c. ►

$$c = 2\pi r$$
$$20\pi = 2\pi r$$
$$\frac{20\pi}{2\pi} = r$$
$$r = 10$$

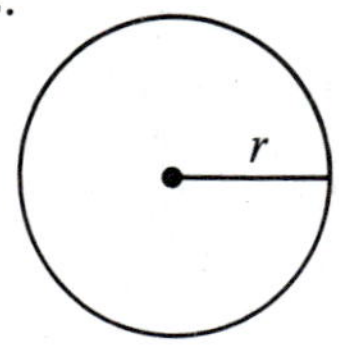

$c = 20\pi$

Reading in Geometry

Translate into symbols. Use *c* for circumference, *d* for length of diameter, *r* for length of radius, and *A* for area.

1. The circumference of a circle is equal to π times the length of a diameter. $c = \pi d$
2. π is equal to the ratio of the circumference of a circle to twice the length of a radius. $\pi = \frac{c}{2r}$
3. One half the circumference of a circle is equal to the length of a radius times π. $\frac{1}{2}c = r\pi$
4. The area of a square inscribed in a circle is twice the square of the length of a radius of the circle. $A = 2r^2$

Written Exercises

Find each missing measurement for each circle. For the last column, use 3.14 for π.

A

	1.	2.	3.	4.	5.	6.	7.	8.
r	2 m	4 cm	6 mm	1 cm	32 mm	7.5 cm	2.6 mm	11.5 cm
d	4 m	8 cm	12 mm	2 cm	64 mm	15 cm	5.2 mm	23 cm
c in terms of π	4π m	8π cm	12π mm	2π cm	64π mm	15π cm	5.2π mm	23π cm
c to two decimal places	12.56 m	25.12 cm	37.68 mm	6.28 cm	200.96 mm	47.1 cm	16.33 mm	72.22 cm

For each given circumference, find the length of a radius and the circumference correct to the nearest hundredth. Use 3.14 for π. (Ex. 9–12)

B

9. $\frac{3}{4}$ m 0.12 m; 0.75 m
10. $\frac{2}{3}$ m 0.11 m; 0.67 m
11. $\sqrt{3}$ mm 0.28 mm; 1.73 mm
12. $\sqrt{7}$ km 0.42 km; 2.65 km
13. Two approximations, $3\frac{1}{7}$ and 3.14 are often used for π. Which is closer to π? $3\frac{1}{7}$
14. Archimedes found π to be between $3\frac{1}{7}$ and $3\frac{10}{71}$. Which is closer to π? $3\frac{10}{71}$
15. If a radius of a circle is increased by *d*, express the increase in the circumference in terms of *d*. $2\pi d$
16. Find the length of the equator if the radius of the earth is 6,400 kilometers. $12{,}800\pi$ km

C

17. Using the Pythagorean theorem, find the length of each side of a square inscribed in a circle with a 1 cm radius. Compare the perimeter of the square to the circumference of the circle. $s = \sqrt{2}$ cm; $c > p$
18. Using the appropriate trigonometric ratio, find the length of each side of an octagon inscribed in a circle with a 1-cm radius. Compare the perimeter of the octagon with the circumference of the circle. $s = .7652$ cm; $c > p$
19. The circumference of a circle is equal to the perimeter of a square with side s. Show that the radius of this circle is $\frac{2s}{\pi}$. $2\pi r = 4s$, $r = 2s/\pi$
20. The circumference of a circle is 20π. Find the perimeter and area of a regular inscribed hexagon. $p = 60$; $A = 150\sqrt{3}$

AREAS OF CIRCLES 12.8

Objective

To apply the formula for the area of a circle

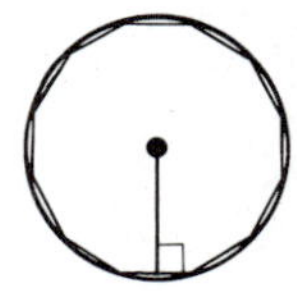

Recall that as you increase the number of sides of a regular polygon inscribed in a circle,

1. The length of an apothem approaches the length of a radius ($a \to r$);
2. The perimeter approaches the circumference ($p \to c$);
3. The area of the polygon approaches the area of the circle (Area polygon $\to$ Area $\odot$).

Consider the formula for the area of a regular polygon, $A = \frac{1}{2}ap$. Suppose you increase the number of sides of the inscribed regular polygon indefinitely. You can derive a formula for the area of a circle as follows:

$$\text{Area regular polygon} = \frac{1}{2}\,a\,p$$

$$\downarrow \qquad \downarrow\downarrow$$

$a \to r$ and $p \to c$. ► $\text{Area } \odot = \frac{1}{2}\,r\,c$ But $c = 2\pi r$.

$$\text{Area } \odot = \frac{1}{2}\,r\,(2\pi r)$$

$$\text{Area } \odot = \pi r^2$$

This leads to the following theorem. It is given without proof.

Theorem 12.11 **Area $\odot = \pi r^2$** The area of a circle with a radius of length r is πr^2.

Since $r = \frac{1}{2}d$, you can also write the area formula in terms of the length of a diameter.

$$A = \pi r^2 = \pi\left(\frac{1}{2}d\right)^2, \text{ or } A = \frac{\pi d^2}{4}$$

Example 1 The length of a radius of a circle is 6 cm. Find the area. (Use 3.14 for π.)

$$A = \pi r^2 = \pi(6)^2 = 36\pi = 36(3.14) = 113.04 \text{ cm}^2$$

Example 2 The area of a circle is 12π m^2. Find the length of a diameter.

$$A = \frac{\pi d^2}{4}$$

Substitute 12π for A. ► $12\pi = \frac{\pi d^2}{4}$

Multiply each side by $\frac{4}{\pi}$. $48 = d^2$

$$d = \sqrt{48} = \sqrt{16 \cdot 3}, \text{ or } 4\sqrt{3} \text{ m}$$

Example 3

Find the area of a circle if the circumference is 5π.

$$c = 2\pi r$$
$$5\pi = 2\pi r$$
$$\frac{5}{2} = r$$
$$A = \pi r^2 = \pi \left(\frac{5}{2}\right)^2 = \frac{25}{4}\pi$$

Example 4

Regular hexagon $ABCDEF$ is inscribed in $\odot O$, $AO = 1$. Find the area of the shaded region in terms of π.

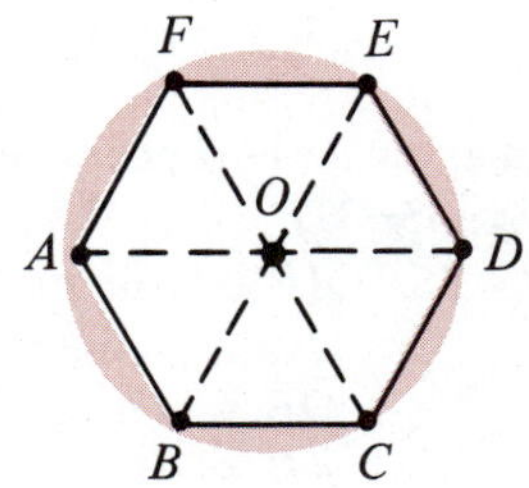

Area shaded region = Area $\odot$ − Area hex.

Area $\odot = \pi r^2 = \pi \cdot 1^2 = \pi$

$\triangle AOB$ is equilateral. ► Area $\triangle AOB = \frac{s^2}{4}\sqrt{3} = \frac{r^2}{4}\sqrt{3}$

Area hex. $= 6 \cdot$ Area $\triangle AOB$

$$= 6 \cdot \frac{r^2}{4}\sqrt{3} = 6 \cdot \frac{(1)^2}{4}\sqrt{3} = \frac{3\sqrt{3}}{2}$$

Thus, the area of the shaded region is $\pi - \frac{3\sqrt{3}}{2}$.

Written Exercises

Find the area of each circle in terms of π.

A 1. $r = 2$ m **4π m²** 2. $r = 1$ mm **π mm²** 3. $d = 4$ km **4π km²** 4. $d = 16$ cm **64π cm²**

Find the length of a radius of each circle whose area is given.

5. 25π m² **5 m** 6. 16π cm² **4 cm** 7. 49π mm² **7 mm** 8. π cm² **1 cm**

Find the length of a diameter of each circle whose area is given.

9. 9π m² **6 m** 10. 25π cm² **10 cm** 11. 144π mm² **24 mm** 12. 36π cm² **12 cm**

Find the area of a circle with each given radius length. (Use 3.14 for π.)

13. $r = 0.6$ m **1.13 m²** 14. $r = 5.6$ mm **98.47 mm²** 15. $r = 0.75$ cm **1.77 cm²** 16. $r = \sqrt{5}$ km **15.70 km²**

Find the area of a circle with each given diameter length. (Use 3.14 for π.)

17. $d = 3$ cm **7.065 cm²** 18. $d = 6.2$ mm **30.1754 mm²** 19. $d = 0.3$ m **0.07065 m²** 20. $d = \sqrt{3}$ m **2.355 m²**

Find the lengths of a radius and a diameter of a circle with each given area. (Use 3.14 for π.)

21. 314 cm² **10 cm; 20 cm** 22. 78.5 m² **5 m; 10 m** 23. 1.1304 mm² **0.6 mm; 1.2 mm** 24. 4.5216 cm² **1.2 cm; 2.4 cm**

Find the area of a circle with each given circumference. (Use 3.14 for π.)

25. $c = 25.12$ mm **50.24 mm²** 26. $c = 6.28$ cm **3.14 cm²** 27. $c = 9$ m **6.449 m²** 28. $c = 100$ cm **796.178 cm²**

B **29.** Find the area of a square with an apothem of length 1 m and a perimeter of 8 m. Compare this to the area of a circle with a 1 m radius. area of square = 4 m² area of circle ≐ 3.14 m²

30. Find the area of a regular octagon with an apothem of length 1 m and a perimeter of 6.6 m. Compare this with the area of a circle with a 1 m radius. area of octagon 3.3 m² area of circle ≐ 3.14 m²

Find the area of the shaded region. Each polygon is regular with sides 2 cm long. (Hint: Find the length of the radius of each circle. Then find the difference between the larger and smaller areas.)

31.

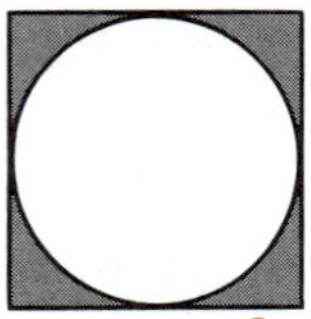

$(4 - \pi)$ cm²

32.

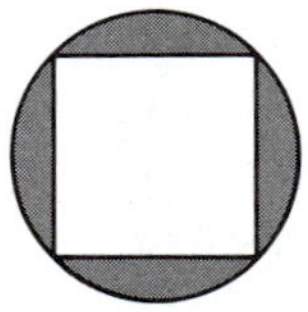

$(2\pi - 4)$ cm²

33.

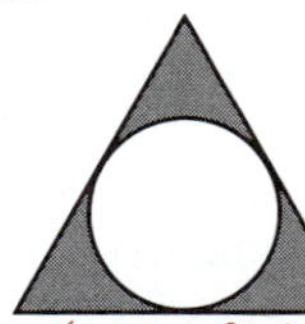

$\left(\sqrt{3} - \frac{1}{3}\pi\right)$cm²

34.

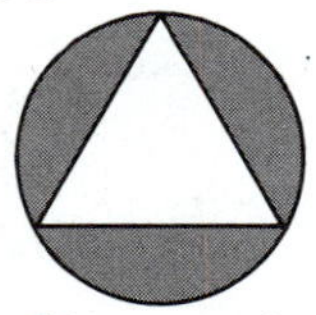

$\left(\frac{4}{3}\pi - \sqrt{3}\right)$cm²

Each circle is tangent to the adjacent circles. Find the area of the shaded region. (Ex. 35–37)

35. $\overline{AB}$ is a diameter of $\odot O$, $\overline{AO}$ is a diameter of $\odot P$, $\overline{OB}$ is a diameter of $\odot Q$. $AB = 10$.

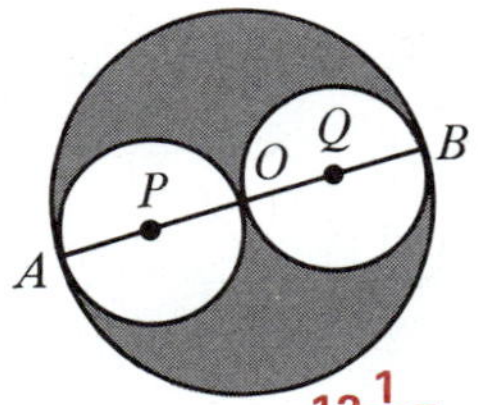

$12\frac{1}{2}\pi$

36. Each circle is tangent to the square, all $\odot$'s are $\cong$, and $AB = 2$.

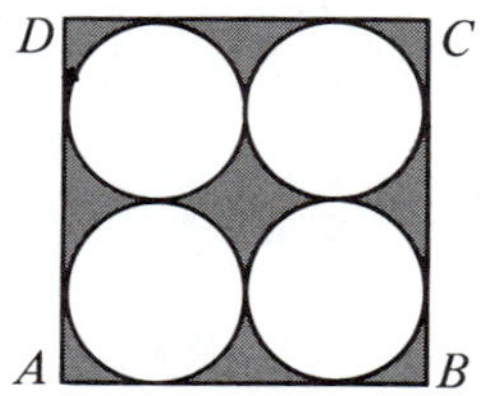

$4 - \pi$

37. All circles except the center circle are tangent to the square, all $\odot$'s are $\cong$, and $AB = 2$.

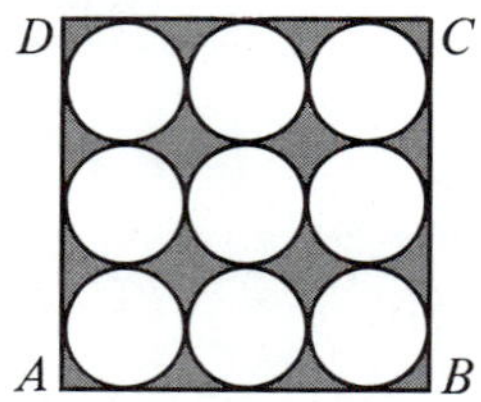

$4 - \pi$

C **38.** Prove or disprove: The ratio of the areas of two circles equals the ratio of the lengths of the radii.

39. Prove or disprove: The ratio of the areas of two circles equals the ratio of the squares of the circumferences.

40. If a circle is inscribed in an equilateral triangle, and another circle is circumscribed about the same triangle, what is the ratio of the areas of the two circles? 1 to 4

41. If a circle is inscribed in a square, and another circle is circumscribed about the same square, what is the ratio of the areas of the two circles? 1 to 2

CALCULATOR ACTIVITIES

Find the area of the shaded region correct to the nearest hundredth. Use 3.1416 for π.

1.

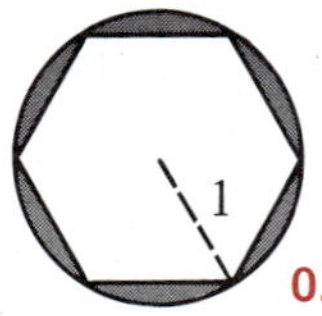

0.54

2.

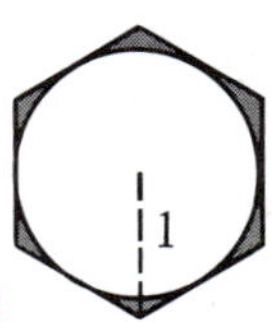

0.24

ARC LENGTHS 12.9

Objective **To apply the formula for the length of an arc**

Recall that two arcs may have the same measure, yet not be congruent. For the two concentric circles shown, $\angle COD$ intercepts $\widehat{AB}$ and $\widehat{CD}$. If $m\angle COD = 35$, then $m\widehat{AB} = 35$ and $m\widehat{CD} = 35$. But $\widehat{CD}$ is not congruent to AB.

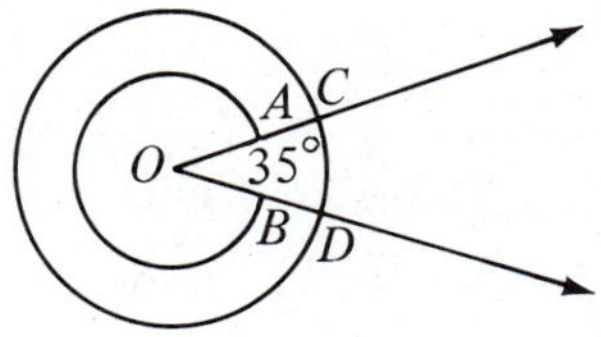

When you find the length of an arc, you are actually finding a part of the circumference of the circle. Consider $\odot Q$ with $m\angle EQF = 60$ and $QE = 6$ cm. $\angle EQF$ intercepts $\frac{60}{360}$, or $\frac{1}{6}$ of the circle, so you can find the length of $\widehat{EF}$ by finding $\frac{1}{6}$ of the circumference.

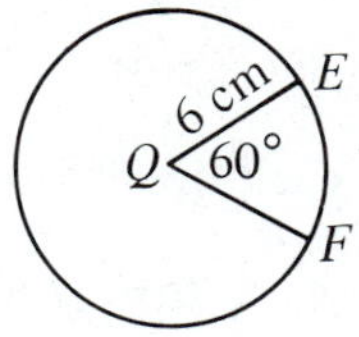

$$c = 2\pi r = 2\pi \cdot 6 = 12\pi \text{ cm}$$

$$\text{length of } \widehat{EF} = \frac{1}{6} \cdot 12\pi = 2\pi \text{ cm}$$

This suggests the following theorem. It is given without proof.

Theorem 12.12 $\mathbf{l = \frac{m}{360} \cdot 2\pi r}$ If the degree measure of an arc is m and the length of a radius of the circle is r, then the length l of the arc is given by the formula

$$l = \frac{m}{360} \cdot 2\pi r.$$

Since $d = 2r$, it is also true that $l = \frac{m}{360} \cdot \pi d$. You can use either of these two formulas to find lengths of arcs, as shown in the following examples.

Example 1 Find the length of $\widehat{RS}$ if $m\widehat{RS} = 90$ and $OS = 2$ cm.

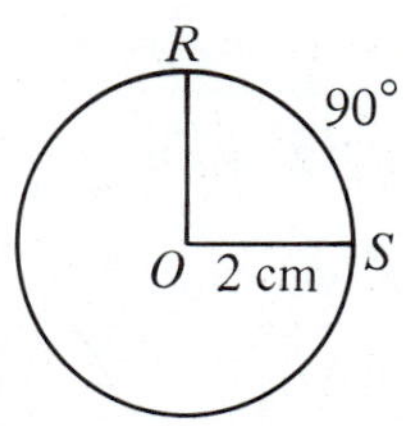

$$l = \frac{m}{360} \cdot 2\pi r = \frac{90}{360} \cdot 2\pi \cdot 2$$

$$= \frac{1}{4} \cdot 4\pi = \pi$$

Thus, the length of $\widehat{RS}$ is π cm.

Example 2 Given: $\odot O$ with diameter $\overline{RP}$, $RP = 16$, and $m\angle PRQ = 60$.
Find the length of $\widehat{PQ}$.

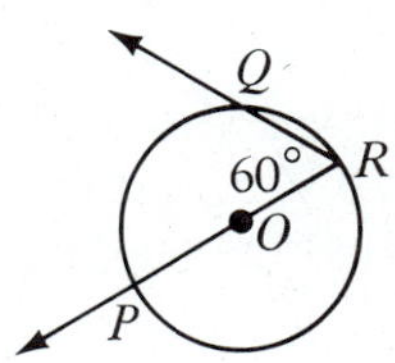

$\angle PRQ$ is inscribed, so $m\widehat{PQ} = 120$. ► $l = \frac{m}{360} \cdot \pi d = \frac{120}{360} \cdot \pi \cdot 16$

$$= \frac{1}{3} \cdot 16\pi = \frac{16}{3}\pi$$

Example 3

Given: $\odot O$ with length of $\widehat{AB} = 0.75\pi$ cm and $OA = 4.5$ cm
Find $m\widehat{AB}$ and m $\angle AOB$.

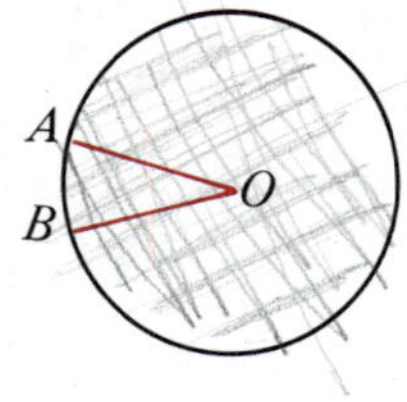

$$l = \frac{m}{360} \cdot 2\pi r \quad \blacktriangleleft \ \textit{m = degree measure of arc.}$$

$$0.75\pi = \frac{m}{360} \cdot 2\pi \cdot 4.5$$

$$0.75\pi = \frac{m}{40}\pi$$

$$0.75\pi\left(\frac{40}{\pi}\right) = m \quad \blacktriangleleft \ \textit{Multiply each side by } \frac{40}{\pi}.$$

$$30 = m$$

Thus, $m\widehat{AB} = 30$ and m $\angle AOB = 30$.

Written Exercises

Find each missing measure. Answers may be left in terms of π.

	Degree Measure of Arc	Length of Radius	Length of Diameter	Length of Arc
A 1.	45	10 mm	? **20 mm**	? **$\frac{5\pi}{2}$ mm**
2.	30	? **3 km**	6 km	? **$\frac{\pi}{2}$ km**
3.	120	2.5 cm	? **5 cm**	? **$\frac{5\pi}{3}$ cm**
4.	270	? **1.6 m**	3.2 m	? **2.4π m**
5.	300	1.8 cm	? **3.6 cm**	? **3π cm**
6.	15	120 m	? **240 m**	? **10π m**
7.	60	12 mm	? **24 mm**	? **4π mm**
8.	75	? **10.8 cm**	21.6 cm	? **4.5π m**

Find each indicated arc length.

9. In a circle with a 6-cm radius, find, correct to the nearest cm, the length of an arc with a measure of 120°. **13 cm**

10. In a circle with a 10-cm diameter, find, correct to the nearest cm, the length of an arc determined by a central angle with a measure of 75°. **7 cm**

Find each missing measure. Answers may be left in terms of π. (Ex. 11–16)

	Degree measure of Arc	Length of Radius	Length of Diameter	Length of Arc
B 11.	180	? **1 km**	? **2 km**	π km
12.	120	? **1.2 cm**	? **2.4 cm**	0.8π cm
13.	? **$22\frac{1}{2}$**	8 m	? **16 m**	π m
14.	? **108**	? **5 cm**	10 cm	3π cm
15.	? **360**	2.5 mm	? **5 mm**	5π mm
16.	? **180**	? **0.4 m**	0.8 m	0.4π m

17. Given: $\odot O$ with $\overline{OB} \perp \overleftrightarrow{AC}$, $OB = 9$, and m $\angle DBC = 60$
Find the length of $\widehat{DEB}$. **12π**

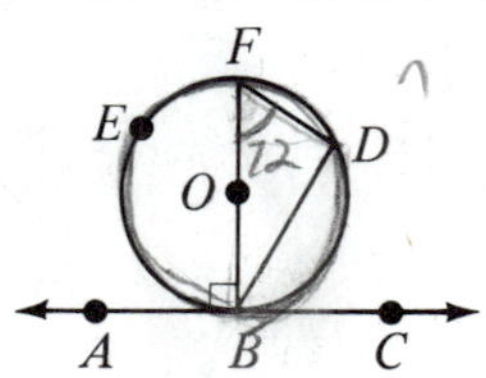

18. Given: $\odot O$ with diameter $\overline{BF}$, $BF = 20$, and m $\angle BFD = 72$
Find the length of $\widehat{BD}$, of $\widehat{FD}$, and $\widehat{BEF}$. **8π, 2π, 10π**

Find each indicated measure.

C **19.** Given: $\overrightarrow{BA}$ and $\overrightarrow{BC}$ are tangent to $\odot O$, m $\angle ABC = 48$, and $OA = 8$. Find the length of $\widehat{AC}$. $\frac{88\pi}{15}$

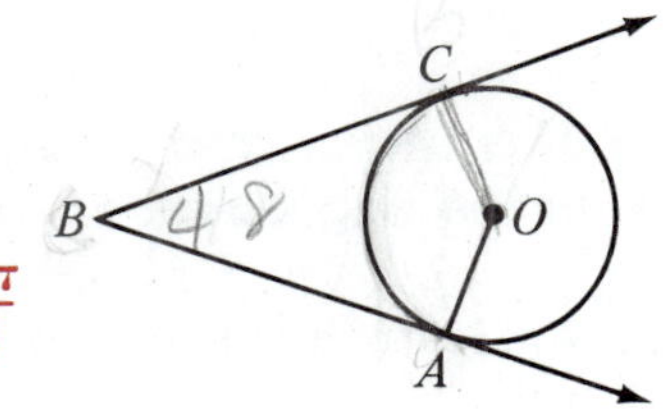

20. Given: $\overrightarrow{BA}$ and $\overrightarrow{BC}$ are tangent to $\odot O$, $OA = 10$, the length of $\widehat{AC} = 12$. Find m $\angle ABC$. (Use 3.14 for π.) 111.21

21. Given: $\odot O$ with m$\widehat{BD} = 30$, m $\angle APC = 50$, and $OB = 10$. Find the length of $\widehat{AC}$. $\frac{35\pi}{9}$

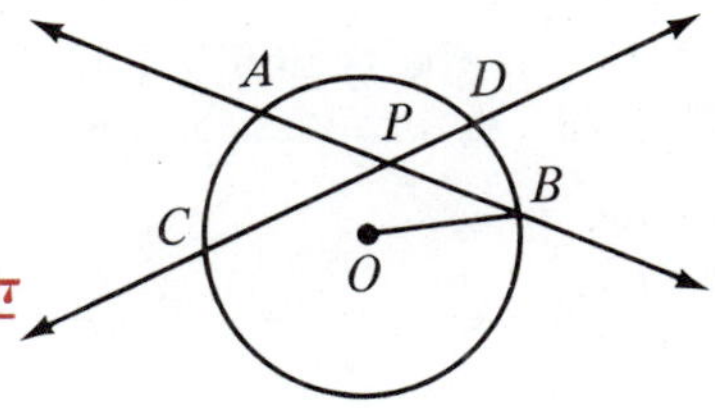

22. Given: $\odot O$ with m$\widehat{BD} = 40$, length of $\widehat{AC} = 12$, and $OB = 10$. Find m $\angle APC$. (Use 3.14 for π.) 54.39

ALGEBRA REVIEW

Objective

To simplify radical expressions by rationalizing the denominator

To simplify expressions like $\frac{3+\sqrt{5}}{\sqrt{5}}$ and $\frac{1+\sqrt{2}}{3-\sqrt{7}}$,

(1) Rationalize the denominator.

(2) Write in simplest form.

Example 1

Simplify $\frac{2+\sqrt{7}}{\sqrt{3}}$.

Make the denominator a perfect square. ►

$$\frac{2+\sqrt{7}}{\sqrt{3}} \cdot \frac{\sqrt{3}}{\sqrt{3}} = \frac{(2+\sqrt{7}) \cdot \sqrt{3}}{\sqrt{3} \cdot \sqrt{3}} = \frac{2\sqrt{3}+\sqrt{21}}{3}$$

Example 2

Simplify $\frac{2}{1+\sqrt{3}}$.

$$\frac{2}{1+\sqrt{3}} \cdot \frac{1-\sqrt{3}}{1-\sqrt{3}} = \frac{2-2\sqrt{3}}{1-3}$$

$$= \frac{2-2\sqrt{3}}{-2}, \text{ or } -1+\sqrt{3}$$

Simplify.

1. $\frac{\sqrt{7}-1}{\sqrt{7}}$ $\frac{7-\sqrt{7}}{7}$
2. $\frac{4\sqrt{3}+2}{\sqrt{3}}$ $\frac{12+2\sqrt{3}}{3}$
3. $\frac{2\sqrt{5}-1}{\sqrt{5}}$ $\frac{10-\sqrt{5}}{5}$
4. $\frac{3\sqrt{6}+2}{\sqrt{2}}$ $3\sqrt{3}+\sqrt{2}$
5. $\frac{3}{1+\sqrt{2}}$ $-3+3\sqrt{2}$
6. $\frac{6}{4-\sqrt{2}}$ $\frac{12+3\sqrt{2}}{7}$
7. $\frac{4}{2+\sqrt{5}}$ $-8+4\sqrt{5}$
8. $\frac{2+\sqrt{3}}{1-\sqrt{6}}$
9. $\frac{1}{2\sqrt{2}+3}$ $3-2\sqrt{2}$
10. $\frac{3}{3\sqrt{2}-3}$ $\sqrt{2}+1$
11. $\frac{3\sqrt{3}}{2\sqrt{6}-1}$ $\frac{18\sqrt{2}+3\sqrt{3}}{23}$
12. $\frac{2\sqrt{3}+1}{2\sqrt{3}-1}$

8. $\frac{2+2\sqrt{6}+\sqrt{3}+3\sqrt{2}}{-5}$ 12. $\frac{13+4\sqrt{3}}{11}$

AREAS OF SECTORS AND SEGMENTS 12.10

Objectives **To find the area of a sector and a segment of a circle**
To find the area of the ring bounded by two concentric circles

In this lesson, you will learn about some other areas in relation to circles.

Definition: Sector of circle A *sector of a circle* is the region bounded by two radii of the circle and the arc determined by the radii.

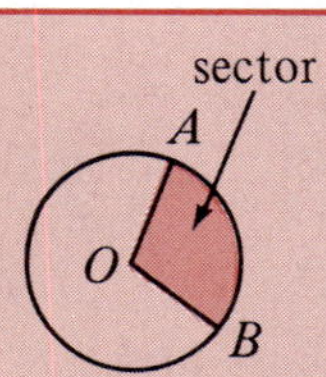

When you find the area of a sector, you are actually finding a part of the area of the circle. Consider $\odot O$ with m $\angle COD =$ 36 and $OC = 10$ cm. $\angle COD$ intercepts $\frac{36}{360}$, or $\frac{1}{10}$ of the circle, so you can find the area of the sector bounded by $\overline{OC}$, $\overline{OD}$, and $\widehat{CD}$ by finding $\frac{1}{10}$ of the area of $\odot O$.

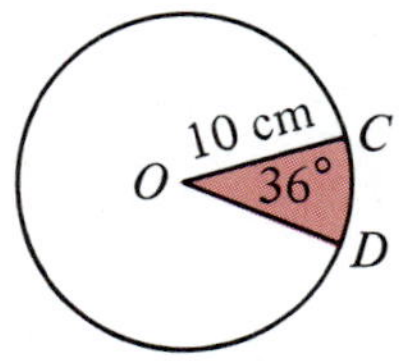

$$\text{Area } \odot O = \pi r^2 = \pi \cdot 10^2 = 100\pi \text{ cm}^2$$
$$\text{Area sector} = \frac{1}{10} \cdot 100\pi = 10\pi \text{ cm}^2$$

This leads to the following theorem, which is given without proof.

Theorem 12.13 **Area sector $= \frac{m}{360} \cdot \pi r^2$** In a circle, the area of a sector bounded by an arc of measure m and radii of length r is given by the formula, Area $= \frac{m}{360} \cdot \pi r^2$.

Example 1 The length of a radius of a circle is 8. The arc of a sector of the circle measures 90°. Find the area of the sector.

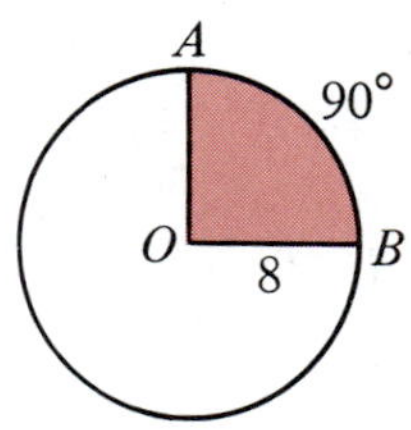

$$\text{Area sector} = \frac{m}{360} \cdot \pi r^2 = \frac{90}{360} \cdot \pi \cdot (8)^2$$
$$= \frac{1}{4} \cdot 64\pi = 16\pi$$

Another region associated with a circle is a segment of the circle.

Definition: Segment of circle

A *segment of a circle* is a region bounded by an arc of the circle and the chord determined by the arc.

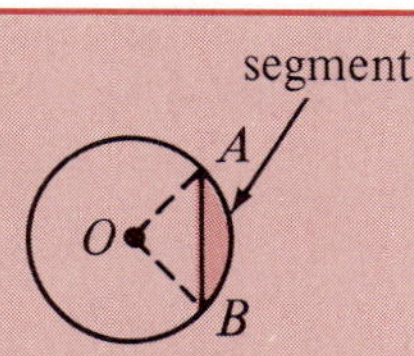

You can find the area of a segment of a circle by subtracting the area of the triangle formed by the radii and the chord from the area of the corresponding sector.

Example 2 The length of a radius of a circle is 10 cm. Find the area of a segment bounded by an arc measuring 60° and the corresponding chord.

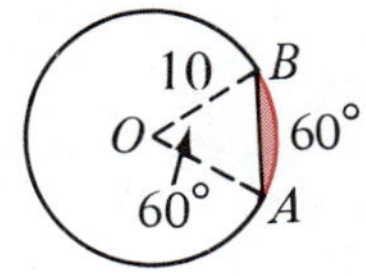

Area segment = Area sector − Area $\triangle AOB$

$$\text{Area sector} = \frac{m}{360}(\pi r^2) = \left(\frac{60}{360}\right)(\pi \cdot 10^2) = \frac{1}{6} \cdot 100\pi = \frac{50}{3}\pi \text{ cm}^2$$

$\triangle AOB$ is equilateral. ► $$\text{Area } \triangle AOB = \frac{s^2}{4}\sqrt{3} = \frac{10^2}{4}\sqrt{3} = \frac{100}{4}\sqrt{3} = 25\sqrt{3} \text{ cm}^2$$

Thus, the area of the segment is $\left(\frac{50}{3}\pi - 25\sqrt{3}\right)$ cm².

Subtraction can also be used to find the area of the region bounded by two concentric circles.

Example 3 Find the area of the ring bounded by two concentric circles with 8-cm and 5-cm radii.

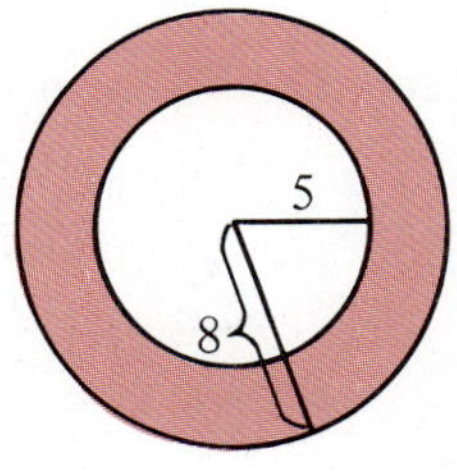

Area outer $\odot = \pi r^2 = \pi \cdot 8^2 = 64\pi$ cm²
Area inner $\odot = \pi r^2 = \pi \cdot 5^2 = 25\pi$ cm²
Thus, the area of the ring is $64\pi - 25\pi$, or 39π cm².

Written Exercises

The length of a radius of a circle is 10. The arc of a sector of the circle has the given measure. What is the area of the sector?

A

1. 36° **10π** **2.** 90° **25π** **3.** 54° **15π** **4.** 72° **20π**
5. 180° **50π** **6.** 60° **$\frac{50\pi}{3}$** **7.** 270° **75π** **8.** 45° **$\frac{25\pi}{2}$**

Find each indicated measure.

9. The length of a radius of a circle is 6 cm. The arc of a sector measures 72°. Find the area of the sector. **7.2π cm²**

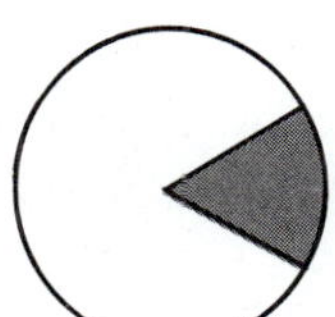

10. The length of a radius of a circle is 5 cm. The arc of a sector measures 60°. Find the area of the sector. **$\frac{25\pi}{6}$ cm²**

11. Find the area of the ring bounded by concentric circles with 10-cm and 8-cm radii. **36π cm²**

12. Find the area of the ring bounded by concentric circles with 8-m and 5-m radii. **39π m²**

Find the missing measures. $\overline{OA}$ and $\overline{OB}$ are radii of $\odot O$. Answers may be left in terms of π. (Ex. 13–17) (See below for Area of Sector and Area of Segment.)

B

	m $\angle O$	Length of Radius r	m$\widehat{AB}$	Area of Sector Shown	Area of Segment Shown
13.	60	5 cm	? 60	?	?
14.	? 90	5 m	90	?	?
15.	120	8 mm	? 120	?	?
16.	? 90	10 cm	? 90	25π cm²	?
17.	? 120	? 6 m	120	12π m²	?

$(12\pi - 9\sqrt{3})$ m²

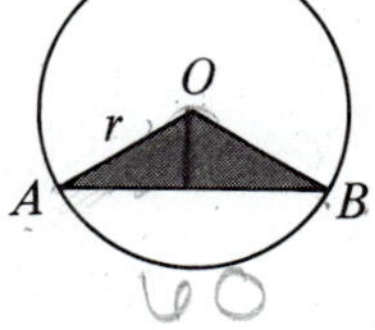

$\frac{25\pi}{6}$ cm² $\frac{25\pi}{4}$ m² $\frac{64\pi}{3}$ mm² $\frac{(50\pi - 75\sqrt{3})}{12}$ cm² $\frac{(25\pi - 50)}{4}$ m² $\frac{(64\pi - 48\sqrt{3})}{3}$ mm² $(25\pi - 50)$ cm²

18. Find the area of a sector of a circle with a 12-cm radius and an arc length of 3π cm. 18π cm²

19. Find the area of a sector of a circle with a 10-cm radius and an arc length of 4π cm. 20π cm²

20. Find the area of a segment of a circle with an 8-cm radius and an arc length of 4π cm. $(16\pi - 32)$cm²

21. Find the area of a segment of a circle with a 9-m radius and an arc length of 3π m. $\frac{54\pi - 81\sqrt{3}}{4}$ m²

22. If the area of a ring is 48π cm² and the outer circle has an 8-cm radius, find the length of a radius of the inner circle. 4 cm

23. If the area of a ring is 48π m² and the inner circle has an 8-m radius, find the length of a radius of the outer circle. $4\sqrt{7}$ m

Find the change in the area of a sector of a circle. (Ex. 24–26)

C

24. The measure of the angle is doubled. Area is doubled.

25. The length of the radius of the circle is doubled. Area is four times as large.

26. The measure of the angle is halved and the length of the radius is doubled. Area is doubled.

27. Find a formula for the area of a ring bounded by two given concentric circles with radii of length R and $r (R > r)$. $A = \pi(R^2 - r^2)$

28. Show that in a circle of radius r the segment cut off by a 60° central angle has area $= \frac{r^2(2\pi - 3\sqrt{3})}{12}$.

29. Find the area of a segment of a circle with a 10-cm radius and an arc measure of 68°. [*Hint:* Use the appropriate trigonometric ratio.] $\left(\frac{170\pi}{9} - 46.36\right)$ cm²

30. Find the area of a segment of a circle with a 12-cm radius and an arc measure of 32°. [*Hint:* Use the appropriate trigonometric ratio.] $\left(\frac{64\pi}{5} - 38.15\right)$ cm²

Cumulative Review

1. Given: $\overline{BC} \perp \overline{AC}$ and $\overline{DE} \perp \overline{AB}$. Find $ED \cdot AB$. 120

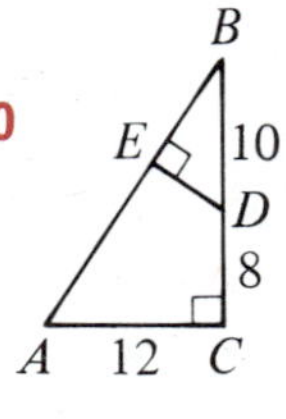

2. Given: $\triangle ABC \sim \triangle BDE$. Find $BE \cdot BC$. 80

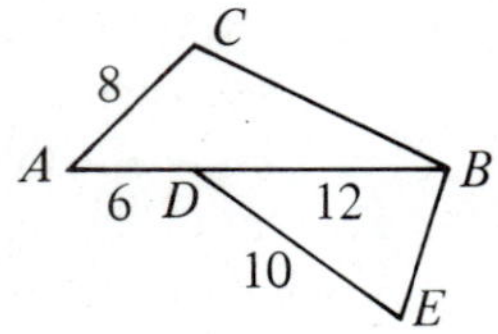

APPLICATIONS

Read → Plan → Solve → Interpret

1. An engineering firm is gathering data for a new water supply for a city. They need to know the water flow in a nearby river. At the site of the water intake, they must find the cross-sectional area of the river. At four points 17 meters apart, the depth of the water is measured. Using this data, approximate the cross-sectional area of the river.

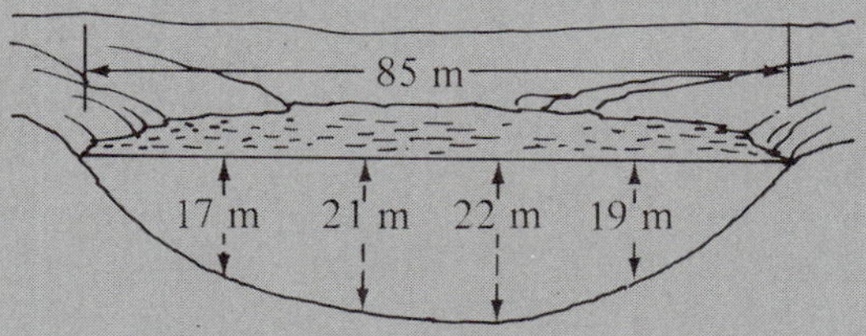

2. The area of an irregularly shaped region can be estimated by dividing the region into narrow strips, all the same width. Each strip is assumed to be a trapezoid. A plot map of some land shows the given dimensions. Approximately how many square meters of land is in the parcel? If a hectare is 10,000 square meters, how many hectares are there?

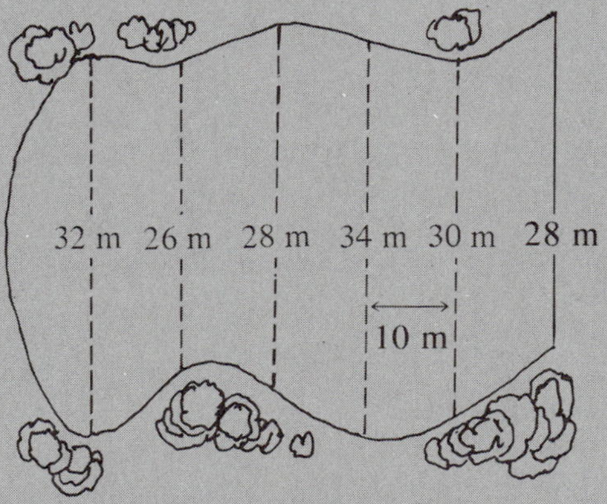

CHAPTER TWELVE REVIEW

Vocabulary

area [12.1]
apothem [12.6]
circumference [12.7]
sector [12.10]
segment [12.10]

Find the area of each figure.

1. Rectangle $ABCD$ with $AB = 15$ m and $BC = 8$ m [12.1] **120 m²**

2. Square $EFGH$ with $EF = 14$ mm [12.1] **196 mm²**

3. Parallelogram $WXYZ$ [12.3] **252 cm²**

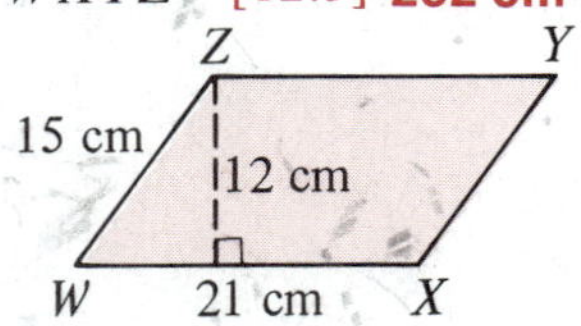

4. Triangle PQR [12.2] **45 m²**

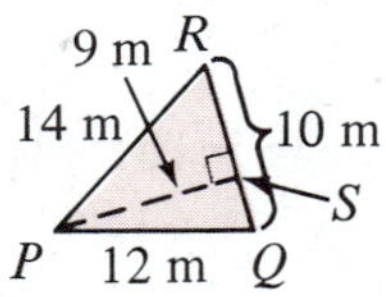

5. Trapezoid $HIJK$ [12.4] **150 cm²**

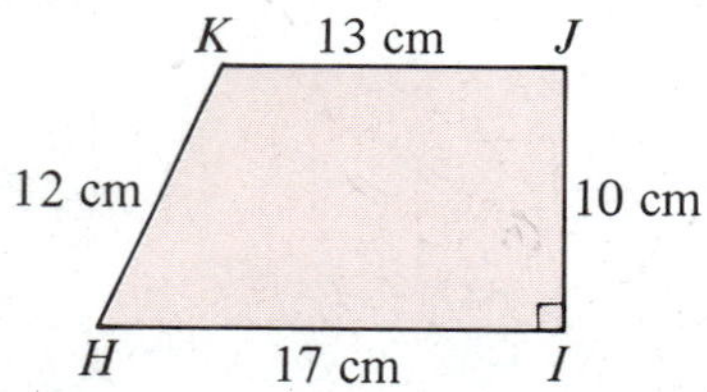

6. Equilateral triangle RST [12.2] **16 $\sqrt{3}$ m²**

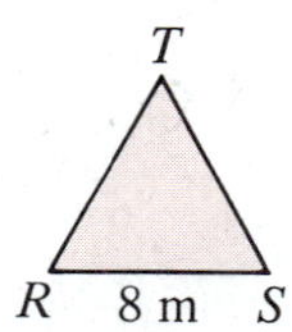

7. Regular pentagon $ABCDE$ [12.6] **28 sq. units**

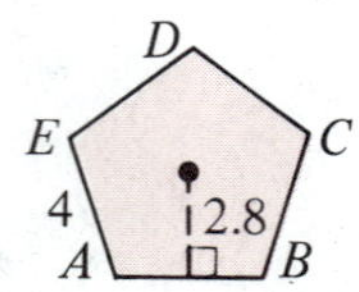

8. Regular hexagon $FGHIJK$ whose perimeter is 18. [12.6] **$\frac{27}{2}\sqrt{3}$ sq. units**

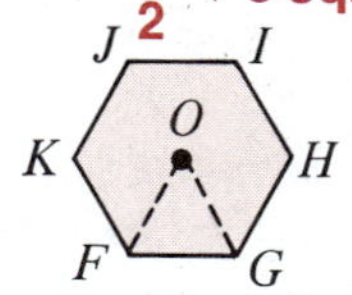

9. Sector AOB [12.10] **$\frac{64\pi}{5}$ cm²**

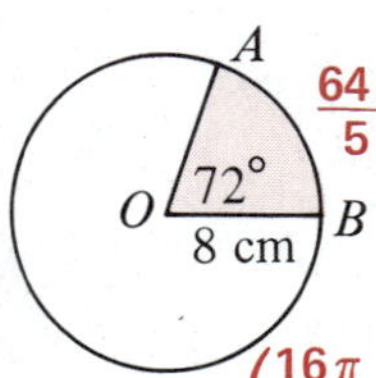

10. Segment CD [12.10] **$\left(\frac{16\pi}{3} - 4\sqrt{3}\right)$ cm²**

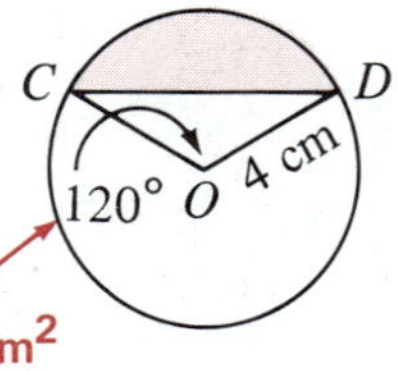

Find each indicated area.

11. Find the area of a rhombus whose diagonals are 8 m and 11 m long. [12.2] **44 m²**

12. Find the area of a circle whose diameter measures 12 cm. [12.8] **36π cm²**

13. Find the area of the ring bounded by circles with 9 cm and 7 cm radii. [12.10] **32π cm²**

*__14.__ $\overline{AB}$ and $\overline{CD}$ are the bases of isosceles trapezoid $ABCD$. If $AB = 9$, $CD = 5$, and $AD = 4$, find the area of the trapezoid. [12.2] **14$\sqrt{3}$ sq. units**

Find each indicated measure.

15. The radius of a circle measures 7 cm. Find the circumference. [12.7] **14π cm**

16. Each side of a rhombus is 23 mm long. Find the perimeter. [12.5] **92 mm**

17. Each side of a regular hexagon measures 6 cm. Find the length of an apothem. [12.6] **3$\sqrt{3}$ cm**

18. The area of an equilateral triangle is $16\sqrt{3}$. Find the length of a side. **8 units**

*__19.__ The circumference of a circle is 18π. Find the perimeter of a regular inscribed hexagon. [12.7] **54 units**

20. In $\odot O$, m $\angle AOB = 50$ and $OA = 8$. Find the length of $\widehat{AB}$. [12.9] **$\frac{20\pi}{9}$**

21. In $\odot P$, the length of $\widehat{CD} = 8\pi$ and $PC = 6$. Find m$\widehat{CD}$. **240**

22. Given: $\triangle RST \sim \triangle XYZ$, $\angle R \cong \angle S$, $RT = 18$, and $ZX = 12$. Find the ratio of the areas. [12.5] **$\frac{9}{4}$**

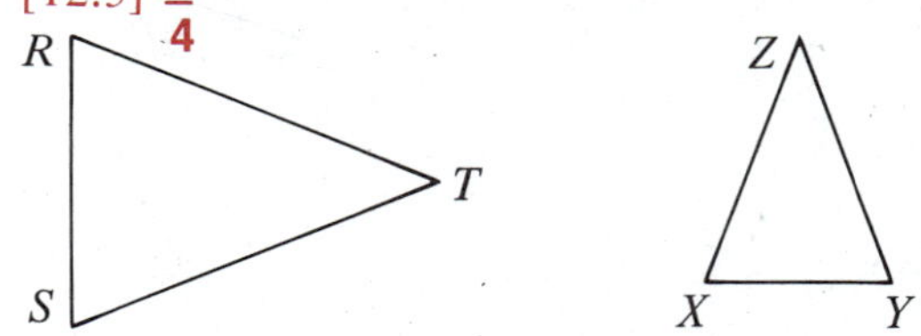

23. Given: In the figure above, $\triangle RST \sim \triangle XYZ$, $\angle R \cong \angle S$, $RT = 21$, $RS = 18$, and $YZ = 14$. Find the ratio of the perimeters. **$\frac{3}{2}$**

(A) exercises are 1–7, 9, 11–13, 16–18, 20–23. (B) exercises are 8, 10, 15. (C) exercises are starred.

CHAPTER TWELVE TEST

Find the area of each shaded figure.

1. Parallelogram $ABCD$

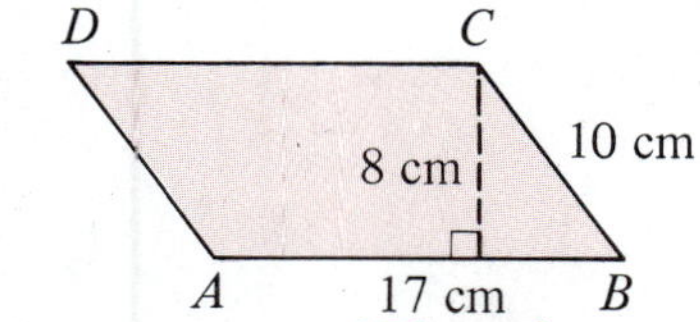

136 cm²

2. Square $RSTU$

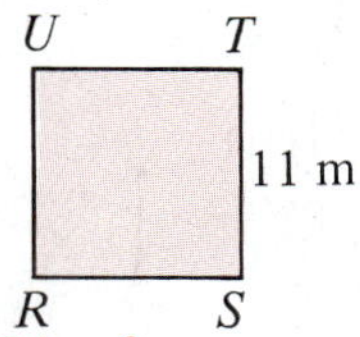

121 m²

3. Triangle XYZ

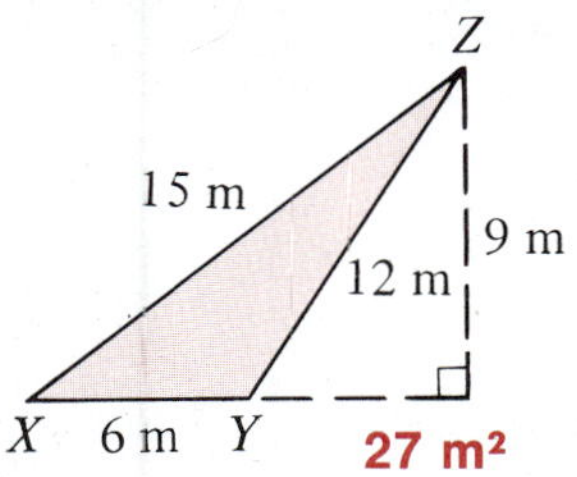

27 m²

4. Rhombus $EFGH$

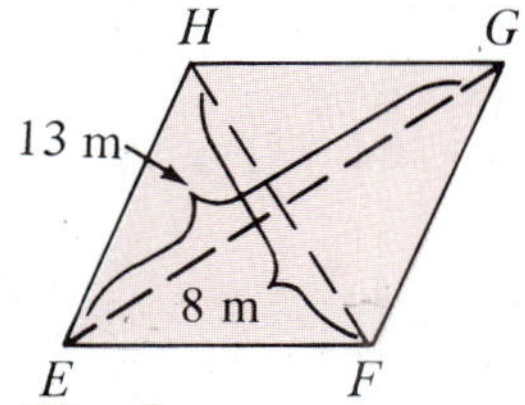

52 m²

5. Equilateral triangle PQR

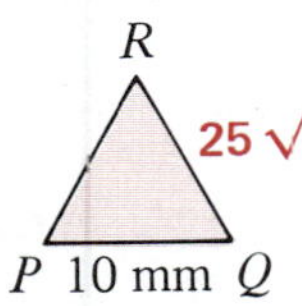

$25\sqrt{3}$ mm²

6. Trapezoid $JKLM$

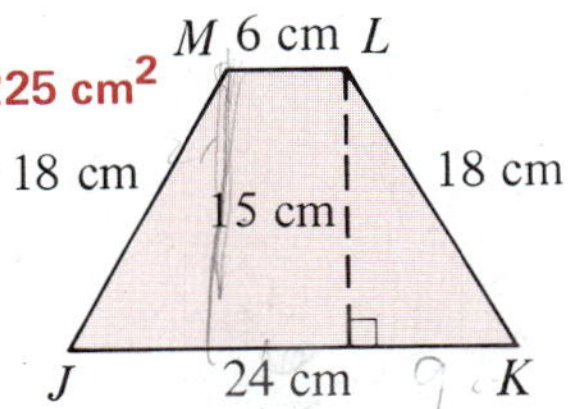

225 cm²

7. Regular octagon $ABCDEFGH$

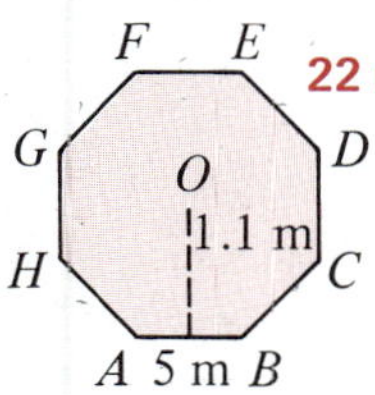

22 m²

8. Regular hexagon $UVWXYZ$

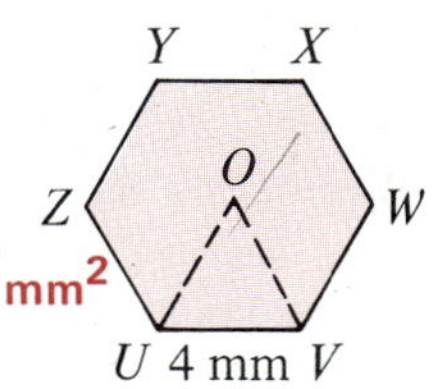

$24\sqrt{3}$ mm²

9. Sector COD

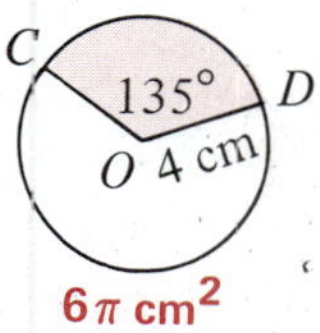

6π cm²

10. Segment AB

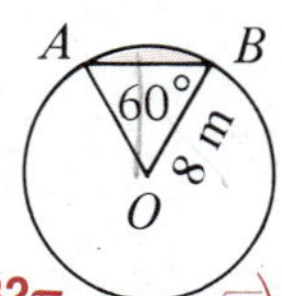

$\left(\frac{32\pi}{3} - 16\sqrt{3}\right)$ m²

Find each indicated area.

11. Find the area of a circle whose radius measures 7 cm. **49π cm²**

12. Find the area of a ring bounded by circles with 10 cm and 5 cm radii. **75π cm²**

***13.** $\overline{AB}$ and $\overline{CD}$ are the bases of isosceles trapezoid $ABCD$. If $AB = 15$, $CD = 9$, and $m\angle A = 45$, find the area of the trapezoid. **36 sq. units**

Find each indicated measure.

14. Each side of a square is 17 cm long. Find the length of an apothem. **8.5 cm**

15. The area of an equilateral triangle is $25\sqrt{3}$. Find the length of an altitude. **$5\sqrt{3}$ units**

***16.** The circumference of a circle is 16π. Find the perimeter of an inscribed square. **$32\sqrt{2}$ units**

17. In $\odot O$, $m\angle COD = 40$ and $CO = 9$. Find the length of $\widehat{CD}$. **2π units**

18. Given: $\triangle ABC \sim \triangle XYZ$, $\overline{CD} \perp \overline{AB}$, $AB = 8$, $CD = 6$, and $WZ = 9$
Find the ratio of the areas of the two triangles. **$\frac{4}{9}$**

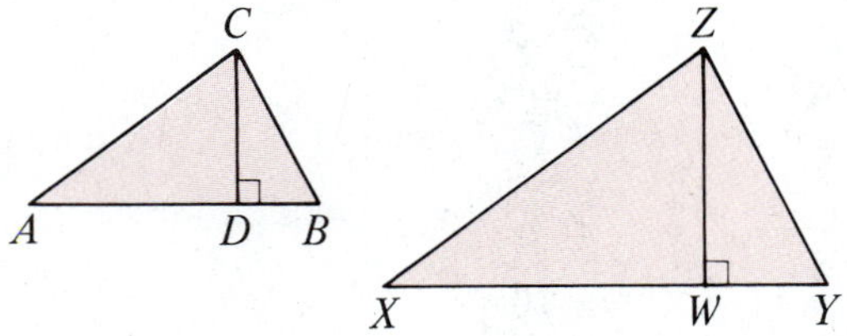

Ⓐ exercises are 1–7, 9, 11, 12, 14, 17, 18. Ⓑ exercises are 8, 10, 15. Ⓒ exercises are starred.

COMPUTER ACTIVITIES

Area of Trapezoids

Objective: **To solve computer-generated problems for any of the variables in the formula for the area of a trapezoid**

The formula for the area of a trapezoid is $A = \frac{1}{2}h(b_1 + b_2)$. As with any formula, this one can be solved for any of its variables.
The following program uses two ON . . . GOTO statements, providing the means by which the computer generates problems for any of the variables in the formula for the area of a trapezoid.

```
110 B1 =  INT ( RND (1) * 110) + 3
120 B2 =  INT ( RND (1) * 110) + B1 + 1
130 H =  INT ( RND (1) * 6) + 4
140 A = (B1 + B2) * H / 2
150  PRINT "IN A TRAPEZOID THE ";
160 Q =  INT ( RND (1) * 2) + 1
170  ON Q GOTO 180,290
180  PRINT "TWO BASES ARE "B1" AND "B2"."
190 Q =  INT ( RND (1) * 2) + 1
200  ON Q GOTO 210,250
210  PRINT "THE ALTITUDE IS "H"."
220  PRINT "WHAT IS THE AREA";: INPUT Z
230  IF A = Z THEN 330
240  PRINT "INCORRECT. THE ANSWER IS "A".": GOTO 340
250  PRINT "THE AREA IS "A"."
260  PRINT "WHAT IS THE ALTITUDE";: INPUT Z
270  IF Z = H THEN 330
280  PRINT "INCORRECT. THE ANSWER IS "H".": GOTO 340
290  PRINT "ALTITUDE IS "H". THE AREA IS "A", AND "
300  PRINT "ONE BASE IS "B1". WHAT IS THE OTHER BASE
     ": INPUT Z
310  IF Z = B2 THEN 330
320  PRINT "INCORRECT. THE ANSWER IS "B2".": GOTO 34
     0
330  PRINT "VERY GOOD.": PRINT : GOTO 340
340  PRINT : GOTO 110
```

EXERCISE: See the Computer Section beginning on page 565 for more information.

The area of a regular polygon equals one-half the product of the perimeter and the length of an apothem ($A = \frac{1}{2}ap$). Write a program capable of solving this formula for any of its three variables.

COLLEGE PREP TEST

Directions: Choose the one best answer to each question or problem.

1. The area of square $GAEF$ is 36 and the area of square $HEBK$ is 81. Find the area of square $ABCD$. **B**

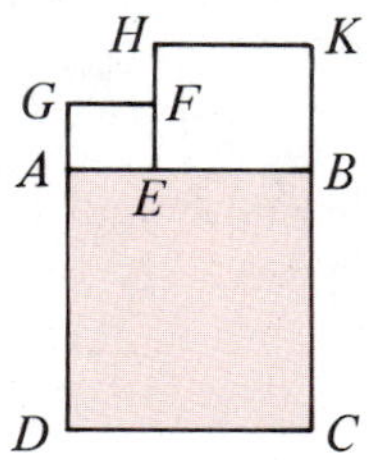

(A) 90
(B) 225
(C) 117
(D) 54
(E) 45

2. In rectangle $MNPR$, $PQ = \frac{1}{3}PR$ and area of $\triangle NPQ = 8$.

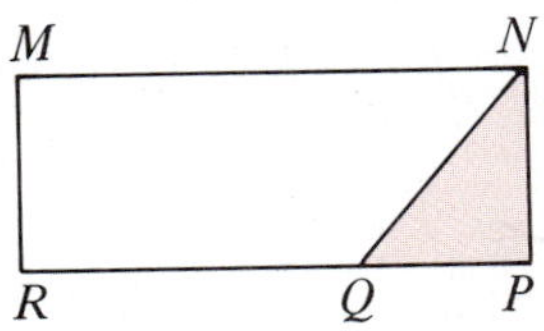

Find the area of rectangle $MNPR$. **C**

(A) 24 **(B)** 32 **(C)** 48 **(D)** 72 **(E)** 96

3. In $\odot O$, $OA = 6$ and $\overline{AO} \perp \overline{OB}$. Find the area of the shaded portion. **D**

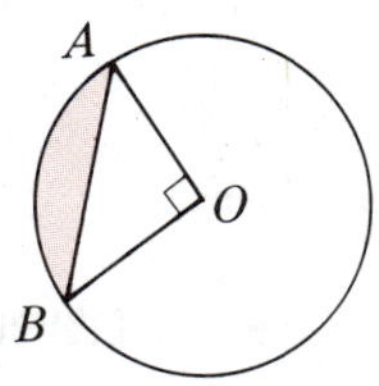

(A) 2π **(B)** $\pi - 2$
(C) $6\pi - 9\sqrt{3}$
(D) $9\pi - 18$ **(E)** $36\pi - 9\sqrt{3}$

4. What is the radius of a circle if it is equal in area to a triangle with an altitude of 4π and a base of 16? **B**

(A) 4 **(B)** $4\sqrt{2}$ **(C)** 8 **(D)** $4\sqrt{2\pi}$ **(E)** 8π

5. Four equal circles with 1 foot diameters touch as shown. What is the area of the shaded portion? **A**

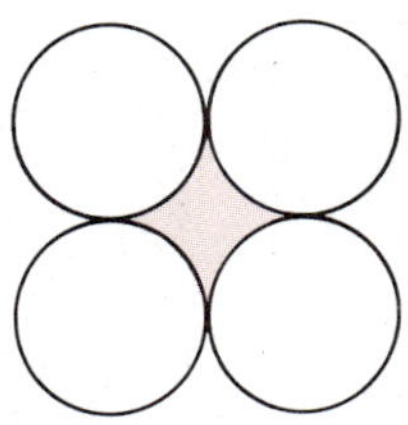

(A) $1 - \frac{\pi}{4}$ **(B)** $1 - \pi$ **(C)** $1 - 4\pi$ **(D)** π **(E)** $\frac{\pi}{4}$

6. A picture in an art museum is 6 feet wide by 8 feet long. If the frame has a width of 6 inches, what is the ratio of the area of the frame to the area of the picture? **A**

(A) $\frac{5}{16}$ **(B)** $\frac{5}{4}$ **(C)** $\frac{4}{5}$ **(D)** $\frac{5}{12}$ **(E)** $\frac{16}{5}$

7. On each side of a regular hexagon, an isosceles right triangle is drawn. If the perimeter of the hexagon is 36, what is the area of the shaded portion? **E**

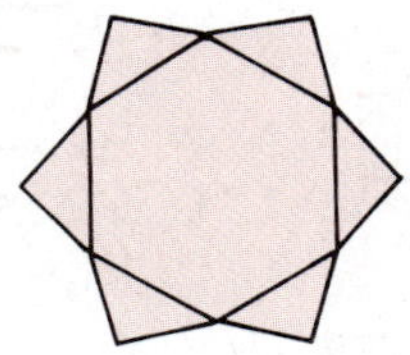

(A) 54 **(B)** $54\sqrt{3}$ **(C)** 108
(D) $108\sqrt{3} + 108$ **(E)** $54 + 54\sqrt{3}$

8. Given: $\overline{AB} \parallel \overline{CD}$, $DC = 8$, $AB = 12$, and $m\angle A = m\angle B = 45$. Find the area of trapezoid $ABCD$. **A**

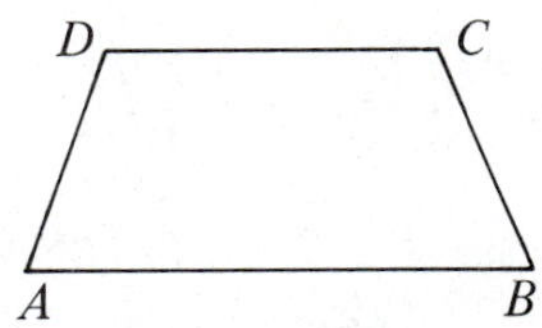

(A) 20 **(B)** 40 **(C)** $20\sqrt{2}$
(D) 32 **(E)** 28

9. A rectangular field is half as wide as it is long and is completely enclosed by x yards of fencing. What is the area of the field? **C**

(A) $\frac{x^2}{2}$ **(B)** $2x^2$ **(C)** $\frac{x^2}{18}$ **(D)** $3\frac{1}{2}$
(E) $\frac{x^2}{72}$

10. What is the area of $PQRS$? **E**

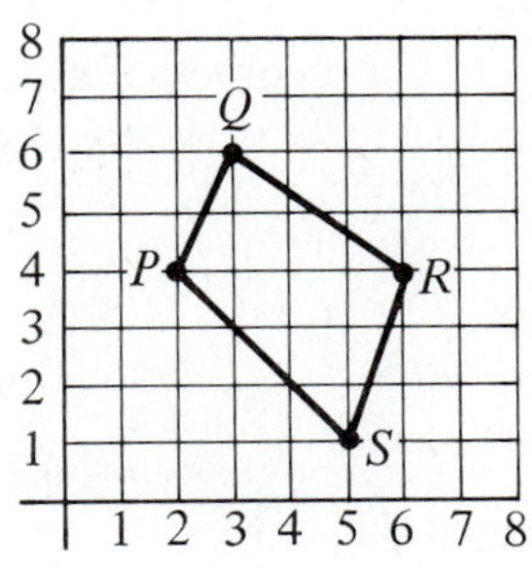

(A) 5 **(B)** 8
(C) 20 **(D)** 16
(E) 10

CUMULATIVE REVIEW

Chapters 1–12

Write the letter that indicates the best answer.

1. In the figure below, the measure of $\angle x$ is **B**
   (A) 60° (B) 52°
   (C) 172° (D) none of these

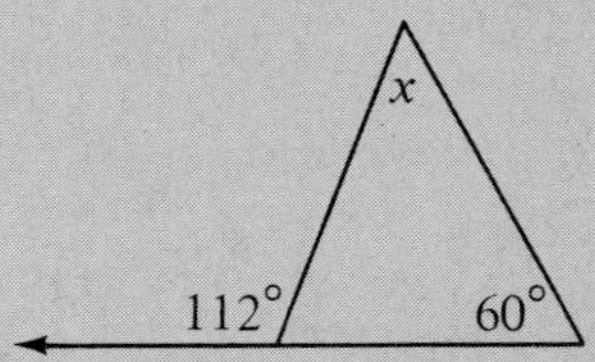

2. In the figure below, the two right triangles are **D**
   (A) congruent by leg-leg
   (B) congruent by hypotenuse-leg
   (C) congruent by leg-acute angle
   (D) none of these

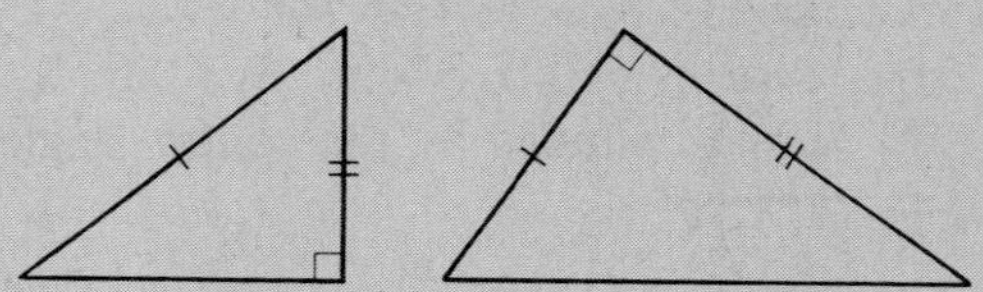

3. The diagonals of an isosceles trapezoid **A**
   (A) are congruent (B) bisect each other
   (C) are perpendicular (D) none of these

4. If $4x = 3y$, then the ratio of x to y is **B**
   (A) $\frac{4}{3}$ (B) $\frac{3}{4}$
   (C) $\frac{y}{x}$ (D) none of these

5. The geometric mean of 3 and 12 is **C**
   (A) 36 (B) 15
   (C) 6 (D) none of these

6. If $\frac{8}{x} = \frac{10}{13}$, then x is equal to **A**
   (A) 10.4 (B) 11
   (C) 6.2 (D) none of these

7. Which of the following can be lengths of sides of a right triangle? **C**
   (A) 9, 12, 15 (B) $\sqrt{3x}$, $\sqrt{4x}$, $\sqrt{7x}$
   (C) both of these (D) neither of these

8. Given: $\triangle RST$ is a 30°–60° rt. $\triangle$ and $RT = 16$. Then RS is equal to **B**

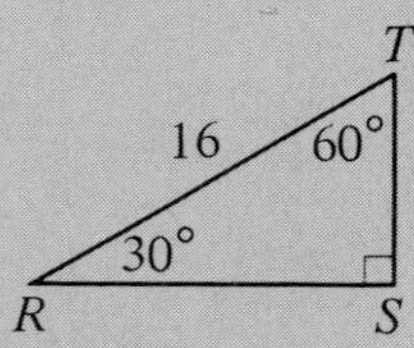

   (A) 8 (B) $8\sqrt{3}$
   (C) 16 (D) $16\sqrt{3}$

9. If $\angle A$ is inscribed in a circle, and $\angle A$ intercepts an arc which measures 32°, then m $\angle A$ is equal to **A**
   (A) 16 (B) 32
   (C) 64 (D) none of these

10. An angle inscribed in a semicircle is **C**
    (A) acute (B) obtuse
    (C) right (D) none of these

11. The legs of a right triangle measure 8 cm and 7 cm. The area of the triangle is **A**
    (A) 28 cm² (B) 56 cm²
    (C) 112 cm² (D) none of these

12. The area of an equilateral triangle is $16\sqrt{3}$. The length of an altitude is **B**
    (A) 4 (B) $4\sqrt{3}$
    (C) 8 (D) $8\sqrt{3}$

True or false.

13. Any two right angles are complementary. F

14. All angles of a rhombus are congruent. F

15. All squares are rectangles. T

16. A pentagon is a polygon with seven sides. F

17. Each interior angle of a regular hexagon measures 120°. T

18. The sum of the measures of the exterior angles of any polygon is 360°. T

19. If an angle bisector of a triangle is also an altitude, then the triangle is equilateral. F

20. An isosceles trapezoid has two pairs of congruent angles. T

21. A square has four lines of symmetry. T

Find each indicated measure.

22. The bases of a trapezoid measure 6 cm and 13 cm. Find the length of the median. 9.5 cm

23. The length and the width of a rectangle have the ratio 5 to 2. The area is 1,440. Find the length and the width. 60, 24

24. One leg of a 45°–45° right triangle measures 7 cm. Find the length of the hypotenuse. $7\sqrt{2}$ cm

*25. If the edges of a rectangular solid are 3 cm, 4 cm, and 8 cm long, find the length of a diagonal of the solid. $\sqrt{89}$ cm

26. Given: ⊙O with $m\widehat{CA} = 50$ and $m\widehat{BD} = 36$ Find m $\angle CEA$. 43

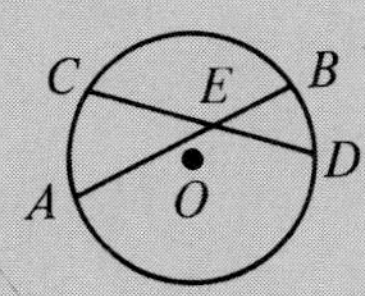

27. Given: ⊙O with tangent $\overrightarrow{PQ}$, $m\widehat{RQ} = 80$, and $m\widehat{RS} = 162$ Find m $\angle P$. 19

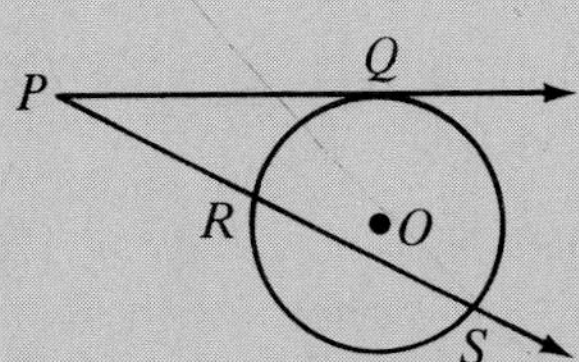

28. In the figure for Exercise 26, $AE = 8$, $EB = 5$, and $CE = 6$. Find ED. $6\frac{2}{3}$

29. In the figure for Exercise 27, $PR = 9$ and $RS = 7$. Find PQ. 12

30. Find the area of a square if each side measures 9 m. 81 m²

31. Two adjacent sides of a parallelogram measure 8 cm and 11 cm. An angle of the parallelogram measures 45°. Find the area of the parallelogram. $44\sqrt{2}$ cm²

32. A side of an equilateral triangle is 5 cm long. Find the area of the triangle. $\frac{25}{4}\sqrt{3}$ cm²

33. The radius of a circle measures 7 m. Find the circumference of the circle. 14π m

34. A central angle of a circle measures 120° and a radius is 6 cm long. Find the area of the sector bounded by the central angle and its intercepted arc. 12π cm²

Find lengths to the nearest tenth. (Ex. 35–37)

35. Find x. 12.4

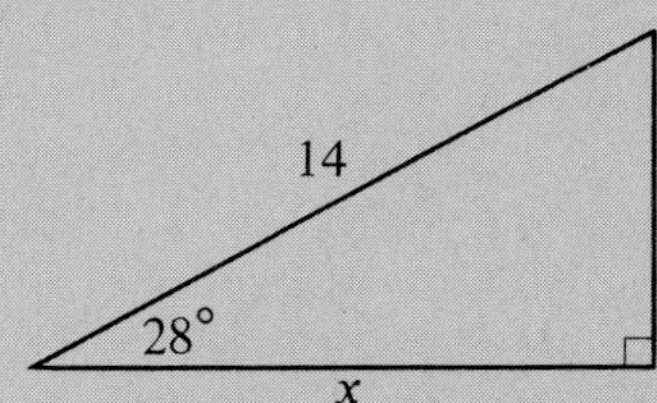

36. In ▱$ABCD$, m $\angle B = 125$ and $BC = 16$. Find the length of the altitude from D to $\overline{AB}$. 13.1

37. An angle of a rhombus measures 154° and the diagonal opposite that angle measures 20. Find the measure of the other diagonal. 4.6

13 COORDINATE GEOMETRY

Problem Solving Formulation

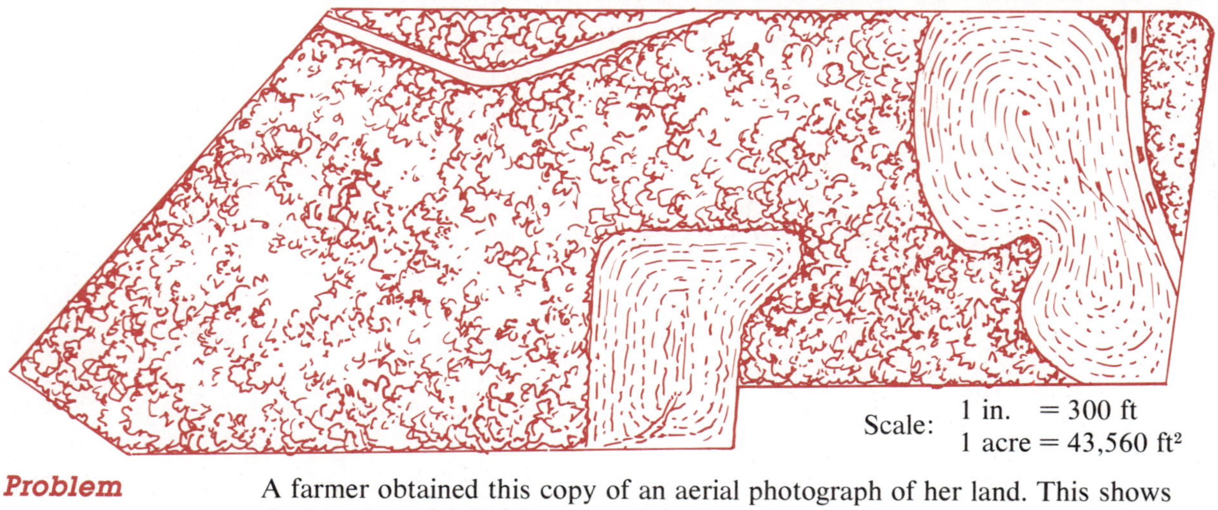

Scale: 1 in. = 300 ft
1 acre = 43,560 ft^2

Problem

A farmer obtained this copy of an aerial photograph of her land. This shows the location of fields, forest, and roads.

An offer of five thousand dollars was made for all usable timber in the forest. It costs one hundred dollars per acre to clear the land in the forest for fields after the trees have been cut. Will the farmer make a net profit or a loss after selling the timber and clearing the land? **profit**

THE COORDINATE PLANE 13.1

Objectives

To give the coordinates of a point in a coordinate plane
To graph ordered pairs of numbers in a coordinate plane
To determine if a line is vertical or horizontal, given the coordinates of two of its points

An ordered pair of numbers such as (2, −8) can be graphed in a **coordinate plane**. In general, an ordered pair is represented by (x, y).

ordered pair

(x, y)

x-coordinate **abscissa** — y-coordinate **ordinate**

A coordinate plane is determined by two perpendicular number lines intersecting at 0. The horizontal line is called the **x-axis**, and the vertical line is called the **y-axis**. The point of intersection is called the **origin**. The axes divide the plane into four **quadrants** labeled with Roman numerals in the diagram.

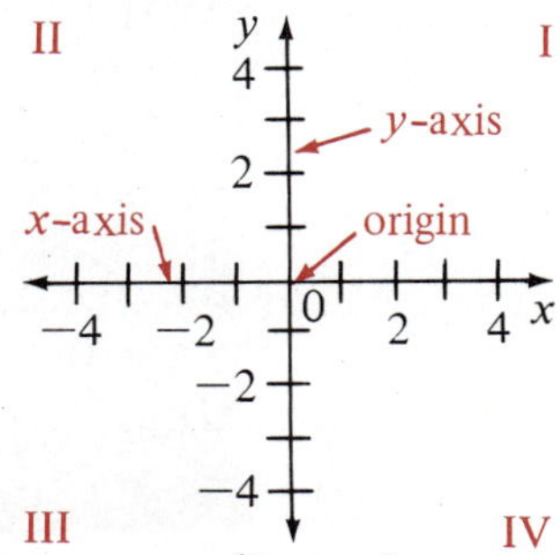

coordinate plane

For every point in a coordinate plane, there is a corresponding ordered pair of numbers, and for every ordered pair, there is a corresponding point. The point is called the **graph** of the ordered pair.

To graph the ordered pair (2, 4) on your paper, start at the origin and move 2 units to the right. Then move 4 units up and mark point A.

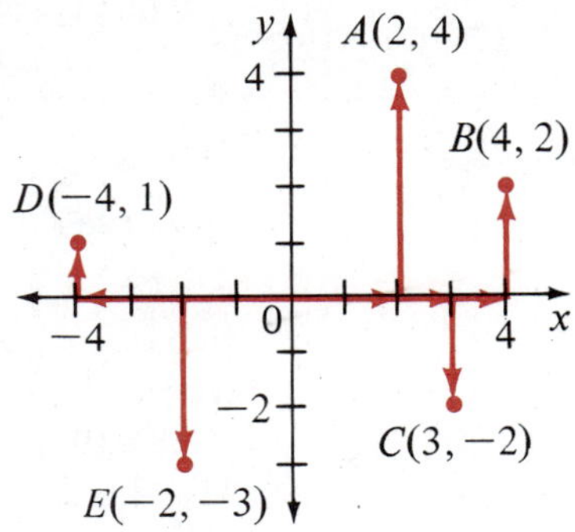

Observe that (2, 4) and (4, 2) are two different ordered pairs. The order is important.

A (2, 4): right 2, up 4

B (4, 2): right 4, up 2

C (3, −2): right 3, down 2

D (−4, 1): left 4, up 1

E (−2, −3): left 2, down 3

Example 1 Give the coordinates of each labeled point. Then tell which quadrant or axis contains each point.

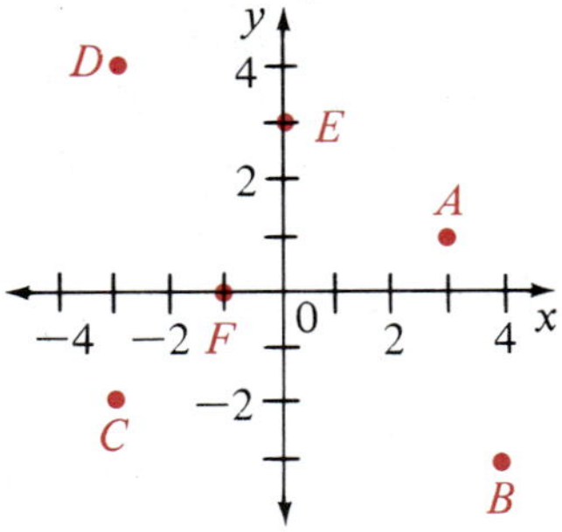

Answers
$A(3, 1)$; Quadrant I
$B(4, -3)$; Quadrant IV
$C(-3, -2)$; Quadrant III
$D(-3, 4)$; Quadrant II
$E(0, 3)$; y-axis
$F(-1, 0)$; x-axis

Examples 2 and 3 below lead to important properties of horizontal and vertical lines.

Example 2 Graph each ordered pair. Then draw a line through the points. What do you observe?

$A(-4, 3)$
$B(-1, 3)$
$C(0, 3)$
$D(2, 3)$

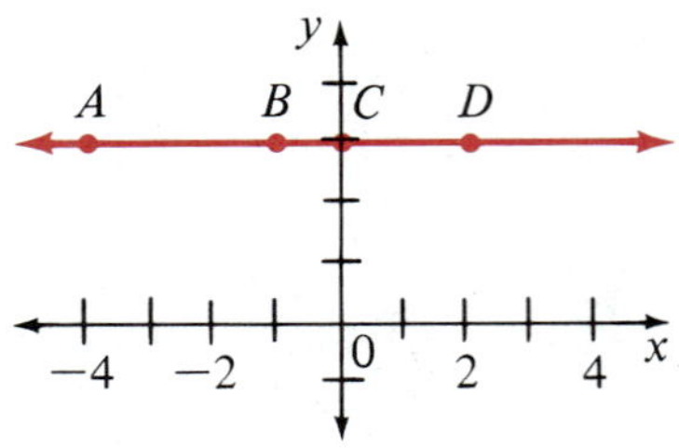

Every point has the same y-coordinate, 3.
The points appear to lie on a horizontal line.

Property of horizontal lines Every point on a horizontal line has the same y-coordinate.

Example 3 Graph each ordered pair. Then draw a line through the points. What do you observe?

$R(-2, 4)$
$S(-2, 0)$
$T(-2, -2)$
$U(-2, -3)$

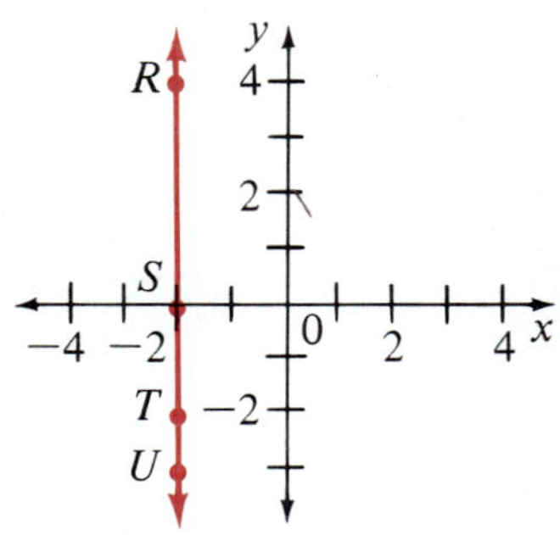

Every point has the same x-coordinate, -2.
The points appear to lie on a vertical line.

Property of vertical lines Every point on a vertical line has the same x-coordinate.

Oral Exercises

Give the coordinates of each labeled point. Then tell which quadrant or axis contains each point.

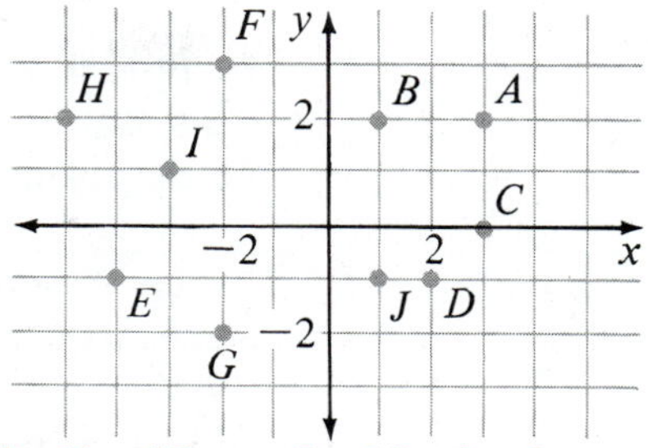

1. A (3, 2); I
2. B (1, 2); I
3. C (3, 0); x-axis
4. D (2, −1); IV
5. E (−4, −1); III
6. F (−2, 3); II
7. G (−2, −2); III
8. H (−5, 2); II
9. I (−3, 1); II
10. J (1, −1); IV

11. rt. 2, up 3 12. left 4, up 2 13. rt. 6, dn. 4 14. left 3, dn. 5, 15. up 8 16. dn. 3

Tell in which directions and how many units to move from the origin in order to graph each ordered pair.

11. $A(2, 3)$ 12. $B(-4, 2)$ 13. $C(6, -4)$ 14. $D(-3, -5)$ 15. $E(0, 8)$ 16. $F(0, -3)$
17. $G(3, 0)$ 18. $H(-5, 0)$ 19. $I(-3, 8)$ 20. $J(0, 0)$ 21. $K(-1, -4)$ 22. $L(6, -2)$

17. rt 3 18. left 5 19. left 3, up 8 20. stay at origin 21. left 1, dn. 4 22. rt. 6, dn. 2

Written Exercises

A

1. On the same coordinate plane, graph and label each ordered pair in Exercises 11–16 above.
2. On the same coordinate plane, graph and label each ordered pair in Exercises 17–22 above.

3. vert. 4. horiz. 5. neither 6. horiz. 7. vert. 8. neither 9. vert. 10. vert.

Is $\overleftrightarrow{PQ}$ a vertical line, a horizontal line, or neither?

3. $P(6, 1)$, $Q(6, -2)$
4. $P(4, 0)$, $Q(1, 0)$
5. $P(2, 6)$, $Q(6, 2)$
6. $P(-3, 4)$, $Q(-2, 4)$
7. $P(-1, 3)$, $Q(-1, -4)$
8. $P(5, 1)$, $Q(1, 2)$
9. $P(-3, 6)$, $Q(-3, -6)$
10. $P(4, -1)$, $Q(4, 1)$
11. $P(-4, 5)$, $Q(4, -5)$
12. $P(1, -1)$, $Q(1, -2)$
13. $P(4, -3)$, $Q(0, -3)$
14. $P(6, -1)$, $Q(1, 6)$

11. neither 12. vert. 13. horiz. 14. neither

For what value of a will $\overleftrightarrow{AB}$ be vertical or horizontal, as indicated?

15. $A(-4, a)$, $B(6, 2)$; horizontal 2
16. $A(a, 6)$, $B(6, -6)$; vertical 6
17. $A(3, 1)$, $B(a, -1)$; vertical 3
18. $A(4, 3)$, $B(a, -1)$; vertical 4
19. $A(7, a)$, $B(-1, -3)$; horizontal −3
20. $A(6, 5)$, $B(-2, a)$; horizontal 5

In which quadrant(s) will a point have the following coordinate(s)?

B

21. Positive y-coordinate I, II
22. Negative x-coordinate II, III
23. Positive x-coordinate and negative y IV
24. Positive y-coordinate and positive x I
25. Negative x-coordinate and positive y II
26. Negative x-coordinate and negative y III

Give the coordinates of D so that $ABCD$ is a rectangle. Graph $ABCD$. (Ex. 27–30)

27. $A(1, 2)$, $B(5, 2)$, $C(5, 5)$, $D(?, ?)$ (1, 5)
28. $A(-1, -6)$, $B(-5, -6)$, $C(-5, -2)$, $D(?, ?)$ (−1, −2)
29. $A(4, 0)$, $B(7, 0)$, $C(7, 3)$, $D(?, ?)$ (4, 3)
30. $A(-3, -2)$, $B(-3, 4)$, $C(4, 4)$, $D(?, ?)$ (4, −2)

C

31. $A(-3, -2)$ and $C(4, 5)$ are endpoints of a diagonal of a square. Give the coordinates of the other two vertices. (−3, 5), (4, −2)
32. $R(1, 3)$, $S(-2, -1)$, and $T(5, -1)$ are vertices of a parallelogram. Give the coordinates of three possible points for the other vertex. (8, 3), (2, −5), or (−6, 3)
33. $E(-1, -2)$ and $F(4, -2)$ are vertices of a square. Give the coordinates of three possible pairs of points for the other two vertices. (4, 3), (−1, 3); (4, −7), (−1, −7); $\left(1\frac{1}{2}, \frac{1}{2}\right)$, $\left(1\frac{1}{2}, -4\frac{1}{2}\right)$
34. $J(-6, 2)$ and $K(3, 2)$ are endpoints of the base of an isosceles triangle. Give the x-coordinate of the third vertex. $-1\frac{1}{2}$

THE DISTANCE FORMULA 13.2

Objectives

To find the distance between pairs of points on a horizontal or a vertical line
To apply the Distance Formula

In this lesson, you will find the distance between pairs of points in a coordinate plane. You will consider points which lie on a horizontal line or a vertical line, and then points which lie on any line. Recall that the distance between two points is the length of the segment that connects them.

Notice that for the points $A(1, 2)$ and $B(6, 2)$, the y-coordinate, 2, is the same. A and B lie on a horizontal line. To find the distance between A and B, you need only to consider the x-coordinates, 1 and 6. By the definition of distance,

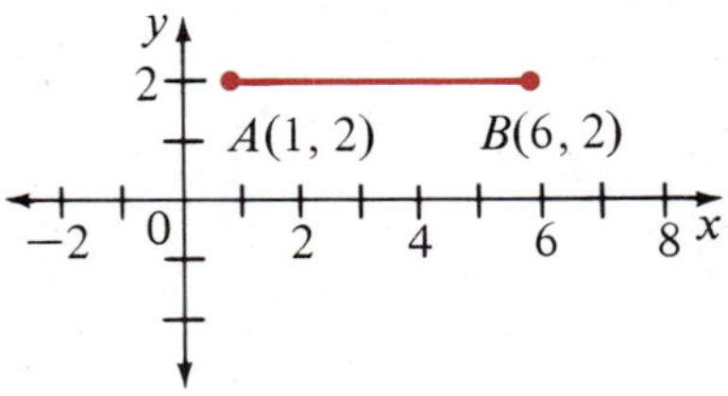

$$AB = |6 - 1| = |5| = 5.$$

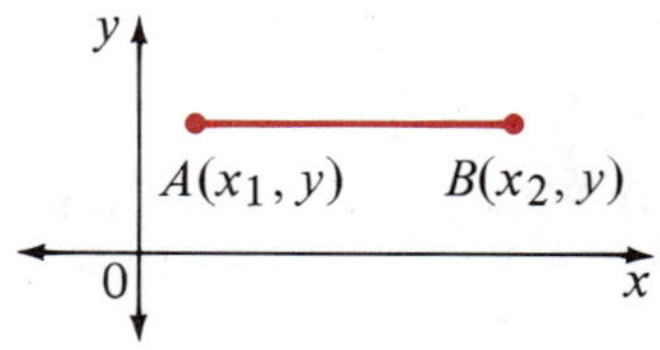

Length of a horizontal segment For a horizontal segment $\overline{AB}$ with $A(x_1, y)$ and $B(x_2, y)$,

$$AB = |x_2 - x_1|.$$

Now examine the vertical segment $\overline{CD}$ with $C(4, 3)$ and $D(4, -1)$. The x-coordinate, 4, is the same. To find the distance between C and D, you need only to consider the y-coordinates, 3 and -1.

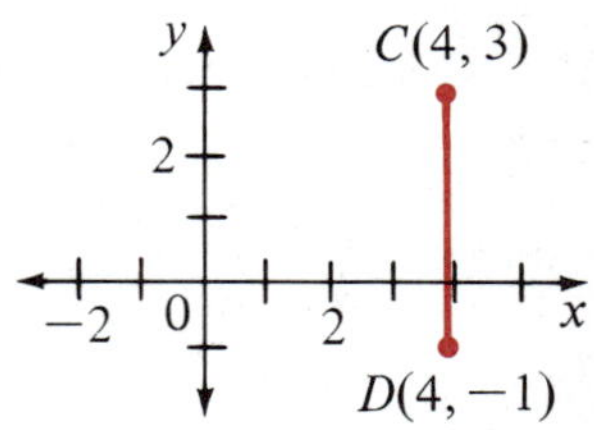

$$CD = |-1 - 3| = |-4| = 4$$

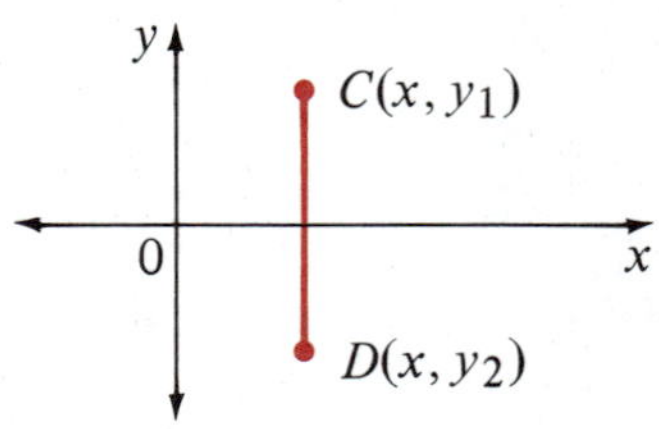

Length of a vertical segment For a vertical segment $\overline{CD}$ with $C(x, y_1)$ and $D(x, y_2)$,

$$CD = |y_2 - y_1|.$$

Example 1

Find RS for $R(6, -2)$ and $S(-1, -2)$. Then find TU for $T(3, 1)$ and $U(3, -5)$.

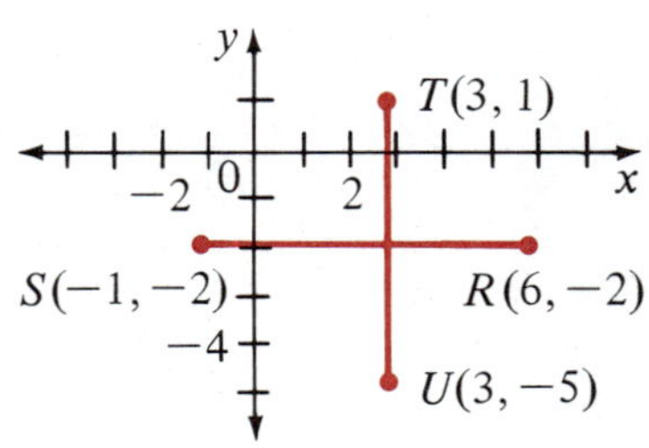

$\overline{RS}$ is horizontal. ► $RS = |x_2 - x_1| = |-1 - 6| = |-7| = 7$

$\overline{TU}$ is vertical. ► $TU = |y_2 - y_1| = |1 - (-5)| = |6| = 6$

Thus, $RS = 7$ and $TU = 6$.

Suppose you want to find the distance between $A(3, 2)$ and $B(7, 5)$ in the coordinate plane at the right. The horizontal segment $\overline{AC}$ intersects the vertical segment $\overline{BC}$ at C. $\triangle ABC$ is a right triangle. You can use the Pythagorean Theorem to find AB. First find AC and BC.

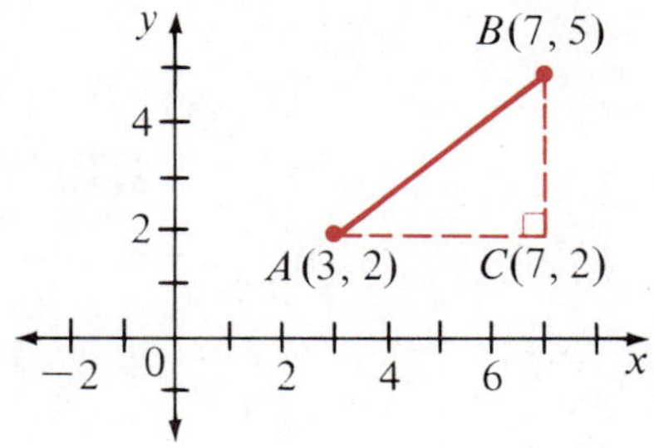

$AC = |7 - 3| = 4 \quad BC = |2 - 5| = 3$

$(AB)^2 = (AC)^2 + (BC)^2$ by the Pythagorean Th.

$$AB = \sqrt{(AC)^2 + (BC)^2} = \sqrt{4^2 + 3^2} = \sqrt{16 + 9} = \sqrt{25} = 5$$

Thus, $AB = 5$. This suggests the following formula for finding the distance between any two points, P and Q, in a coordinate plane.

Theorem 13.1 The Distance Formula

The distance d between $P(x_1, y_1)$ and $Q(x_2, y_2)$ is given by the formula

$$d = \sqrt{(x_2 - x_1)^2 + (y_2 - y_1)^2}.$$

(See Exercise 29.)

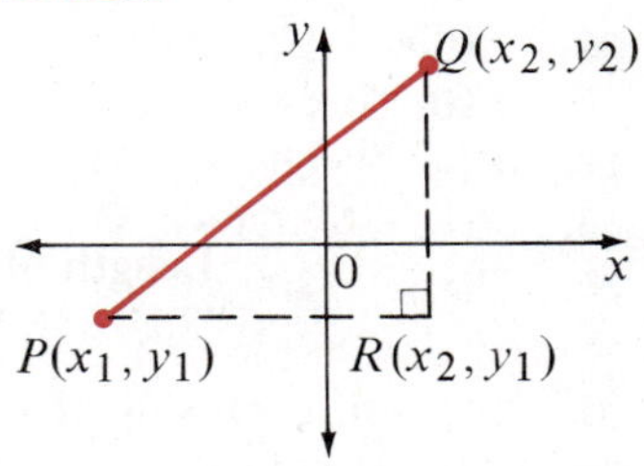

Notice that the formula above can be used to find the distance between any two points in a coordinate plane. However, if the segment connecting the points is horizontal or vertical, you may want to use one of the shorter formulas given on page 432.

Example 2 Use the Distance Formula to find the distance between $P(-2, -5)$ and $Q(2, -3)$.

$$P(-2, -5) \qquad Q(2, -3)$$

(-2 is x_1, -5 is y_1; 2 is x_2, -3 is y_2)

$$\begin{aligned} d &= \sqrt{(x_2 - x_1)^2 + (y_2 - y_1)^2} \\ &= \sqrt{[2 - (-2)]^2 + [-3 - (-5)]^2} \\ &= \sqrt{4^2 + 2^2} \\ &= \sqrt{16 + 4} \\ &= \sqrt{20} \\ &= 2\sqrt{5} \end{aligned}$$

Thus, the distance PQ is $2\sqrt{5}$.

In the next example, you must apply the Distance Formula several times.

Example 3 $\triangle RST$ has vertices $R(-3, 5)$, $S(6, 3)$, and $T(-1, -3)$. Show that $\triangle RST$ is isosceles.

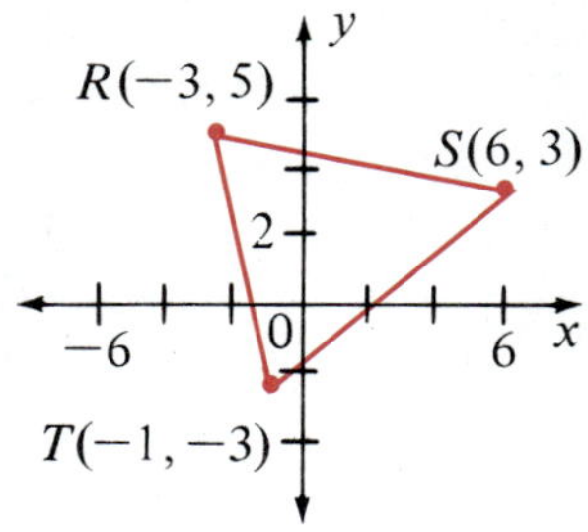

$$\begin{aligned} RS &= \sqrt{[6-(-3)]^2 + (3-5)^2} \\ &= \sqrt{(9)^2 + (-2)^2} \\ &= \sqrt{81 + 4} \\ &= \sqrt{85} \end{aligned}$$

$$\begin{aligned} RT &= \sqrt{[-1-(-3)]^2 + (-3-5)^2} \\ &= \sqrt{(2)^2 + (-8)^2} \\ &= \sqrt{4 + 64} \\ &= \sqrt{68} \end{aligned} \qquad \begin{aligned} ST &= \sqrt{(-1-6)^2 + (-3-3)^2} \\ &= \sqrt{(-7)^2 + (-6)^2} \\ &= \sqrt{49 + 36} \\ &= \sqrt{85} \end{aligned}$$

Thus, since $RS = ST = \sqrt{85}$, $\triangle RST$ is isosceles.

Written Exercises

Find PQ. 1. 2 2. 6 3. 12 4. 2 5. $\sqrt{13}$ 6. $\sqrt{89}$ 7. $2\sqrt{2}$ 8. $\sqrt{85}$ 9. $3\sqrt{10}$ 10. $\sqrt{193}$

A

1. $P(4, 1)$, $Q(2, 1)$
2. $P(2, -1)$, $Q(2, 5)$
3. $P(6, 3)$, $Q(-6, 3)$
4. $P(4, 1)$, $Q(4, -1)$
5. $P(1, 4)$, $Q(4, 2)$
6. $P(-1, 3)$, $Q(7, -2)$
7. $P(6, -3)$, $Q(4, -5)$
8. $P(-1, 7)$, $Q(5, 0)$
9. $P(-3, -2)$, $Q(-6, 7)$
10. $P(4, 7)$, $Q(-3, -5)$
11. $P(-2, 8)$, $Q(6, -4)$
12. $P(0, 0)$, $Q(-8, -1)$
13. $P(6, -2)$, $Q(-7, 1)$
14. $P(1, 9)$, $Q(-1, 7)$
15. $P(2, 5)$, $Q(m, n)$
16. $P(r, s)$, $Q(t, u)$

11. $4\sqrt{13}$ 12. $\sqrt{65}$ 13. $\sqrt{178}$ 14. $2\sqrt{2}$ 15. $\sqrt{(2-m)^2 + (5-n)^2}$ 16. $\sqrt{(r-t)^2 + (s-u)^2}$

Find the length of each side of $\triangle ABC$. Then tell whether $\triangle ABC$ is isosceles or scalene. (Ex. 17–22)

B

17. $A(1, 1)$, $B(3, 6)$, $C(5, 1)$ $AB = BC = \sqrt{29}$, $AC = 4$; isos.
18. $A(-3, 4)$, $B(2, -1)$, $C(4, 5)$ $BC = 2\sqrt{10}$; $AB = AC = 5\sqrt{2}$, isos.
19. $A(3, 4)$, $B(5, -3)$, $C(-2, 2)$ $\sqrt{53}$, $\sqrt{74}$; $\sqrt{29}$; scal.
20. $A(0, 2)$, $B(-3, -2)$, $C(4, -1)$ 20. $AB = AC = 5$; $BC = 5\sqrt{2}$; isos.
21. $A(-6, 2)$, $B(5, -1)$, $C(4, 4)$ $\sqrt{130}$, $\sqrt{26}$, $2\sqrt{26}$; scal.
22. $A(-3, -3)$, $B(1, -6)$, $C(-2, -2)$ $AB = BC = 5$, $AC = \sqrt{2}$; isos.
23. Show that $\triangle RST$ is a right triangle for $R(-5, 4)$, $S(5, 2)$, and $T(1, -2)$. [*Hint:* Use the converse of the Pythag. Th.]
24. Show that the diagonals of $MNPQ$ are congruent for $M(3, 2)$, $N(3, -1)$, $P(7, -1)$, and $Q(7, 2)$.

23. $RS = \sqrt{104}$, $ST = \sqrt{32}$, $RT = \sqrt{72}$; $(\sqrt{104})^2 = (\sqrt{32})^2 + (\sqrt{72})^2$; $104 = 32 + 72$ 24. $MP = NQ = 5$

C

25. $\overline{AB}$ is the hypotenuse of isosceles right triangle ABC for $A(2, 3)$ and $B(7, 3)$. Find all possible coordinates of C. 25. $\left(4\frac{1}{2}, 5\frac{1}{2}\right)$, $\left(4\frac{1}{2}, \frac{1}{2}\right)$
26. Repeat Exercise 25 for $A(-6, -2)$ and $B(-6, 5)$. 26. $\left(-9\frac{1}{2}, 1\frac{1}{2}\right)$, $\left(-2\frac{1}{2}, 1\frac{1}{2}\right)$
27. $Q(2, a)$ is 5 units from $P(-2, -1)$. Find all possible values of a. 2, −4
28. $Q(-1, a)$ is $4\sqrt{5}$ units from $P(3, -2)$. Find all possible values of a. 6, −10
29. Prove Theorem 13.1.

CUMULATIVE REVIEW

1. In $\triangle XYZ$, $XZ = 10$ and $YZ = 15$. A line parallel to $\overline{XY}$ divides $\overline{XZ}$ into segments measuring 2 and 8. Find the measures of the segments into which the line divides $\overline{YZ}$. 3, 12
2. In $\triangle ABC$, $\overline{AC} \perp \overline{BC}$. If $c = 16$ and m $\angle B = 48$, find a to the nearest tenth. (Use a table.) $a = 10.7$

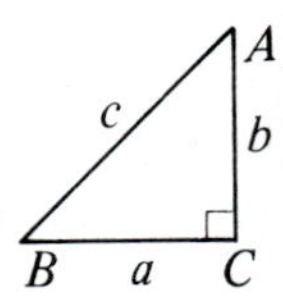

THE MIDPOINT THEOREM 13.3

Objectives

To find the coordinates of the midpoint of a segment in a coordinate plane
To apply the Midpoint Theorem

You can use the coordinates of the endpoints of a segment in a coordinate plane to find the coordinates of the midpoint of the segment. First, examine a horizontal segment and a vertical segment.

Consider points $A(1, 3)$ and $B(9, 3)$. $\overline{AB}$ is horizontal, so the y-coordinate of the midpoint must also be 3. The midpoint lies halfway between points A and B, so the x-coordinate of the midpoint will be the average of the x-coordinates of the endpoints, 1 and 9. **Thus,** the coordinates of the midpoint are $\left(\frac{1+9}{2}, 3\right)$, or $(5, 3)$.

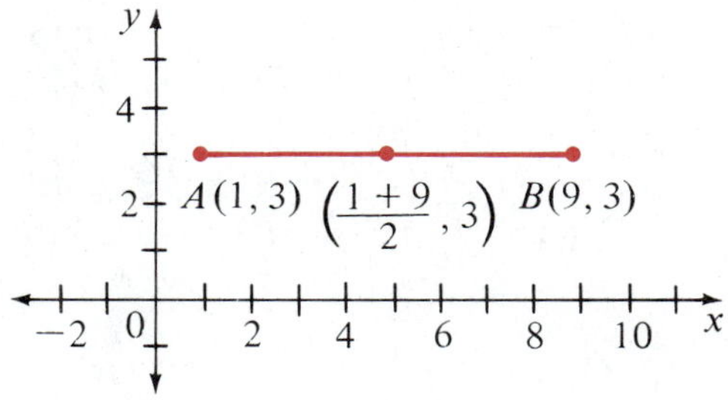

Now consider the vertical segment $\overline{BC}$ with endpoints $B(9, 3)$ and $C(9, 5)$. The x-coordinate of the midpoint must also be 9. The y-coordinate will be the average of the y-coordinates of the endpoints, 5 and 3. **Thus,** the coordinates of the midpoint are $\left(9, \frac{5+3}{2}\right)$, or $(9, 4)$.

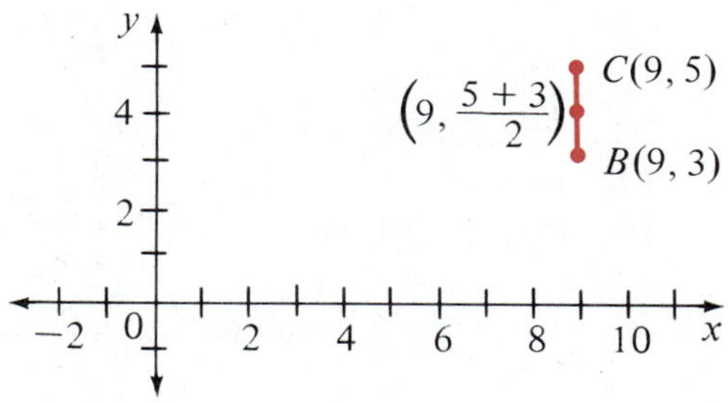

Now examine the segment with endpoints $A(1, 3)$ and $C(9, 5)$. Form a right triangle with $B(9, 3)$. To find the coordinates of M, the midpoint of $\overline{AC}$, use the midpoints of $\overline{AB}$ and $\overline{BC}$ as guides. **Thus,** the coordinates of M are $\left(\frac{1+9}{2}, \frac{5+3}{2}\right)$, or $(5, 4)$. This suggests the following theorem.

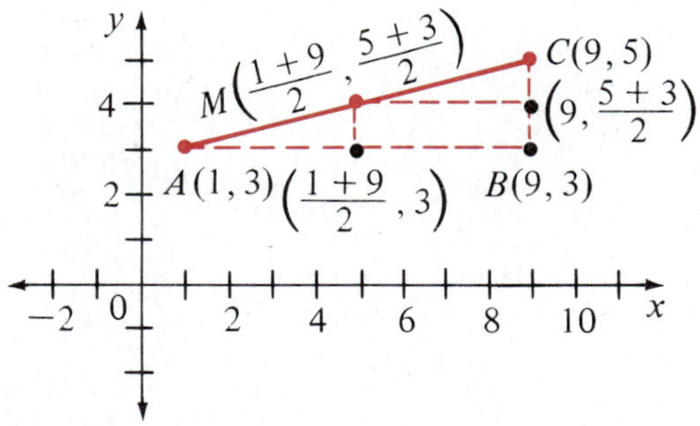

Theorem 13.2 The Midpoint Formula

If $P(x_1, y_1)$ and $Q(x_2, y_2)$ are any two points in a coordinate plane, then the midpoint M of $\overline{PQ}$ has the coordinates $\left(\frac{x_1+x_2}{2}, \frac{y_1+y_2}{2}\right)$.

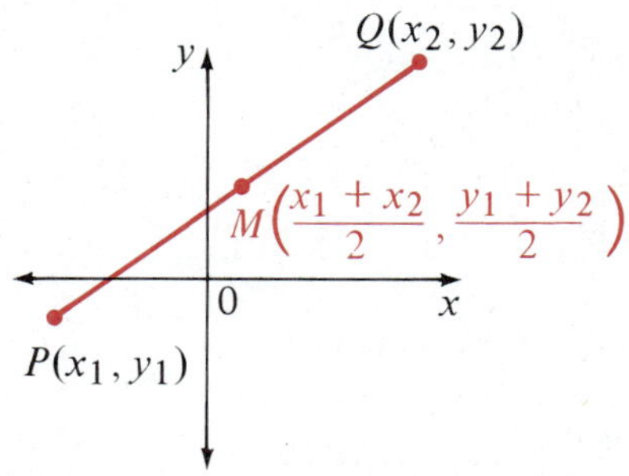

(See Exercise 31.)

Example 1 Find the coordinates of the midpoint of $\overline{AB}$ for $A(5, -4)$ and $B(-3, -6)$.

$$A(5, -4) \qquad B(-3, -6)$$

$$x_1 \quad y_1 \qquad x_2 \quad y_2$$

$$\frac{x_1 + x_2}{2} = \frac{5 + (-3)}{2} = \frac{2}{2} = 1 \qquad \frac{y_1 + y_2}{2} = \frac{-4 + (-6)}{2} = \frac{-10}{2} = -5$$

Thus, the midpoint of $\overline{AB}$ has coordinates $(1, -5)$.

Example 2 M is the midpoint of $\overline{CD}$. Find the coordinates of D for $C(-5, 4)$ and $M(-2, 1)$.

Let C have coordinates (x_1, y_1) and D have coordinates (x_2, y_2).

$$\left(\frac{x_1 + x_2}{2}, \frac{y_1 + y_2}{2}\right) = (-2, 1)$$

$$\frac{-5 + x_2}{2} = -2 \qquad \frac{4 + y_2}{2} = 1$$

$$-5 + x_2 = -4 \qquad 4 + y_2 = 2$$

$$x_2 = 1 \qquad y_2 = -2$$

Thus, D has coordinates $(1, -2)$.

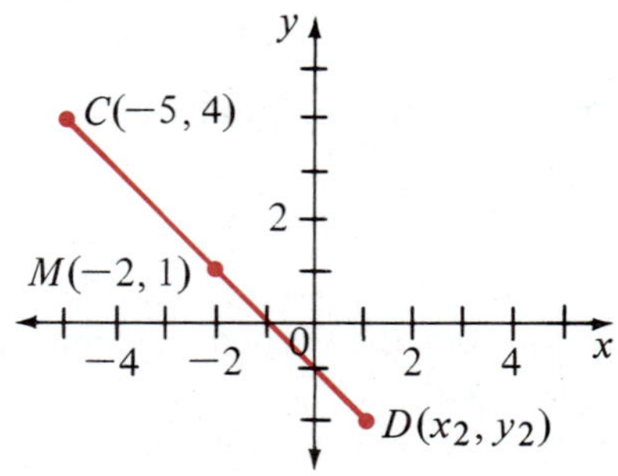

Example 3 requires the use of both the Midpoint Theorem and the Distance Formula.

Example 3 $\triangle ABC$ has vertices $A(-4, -3)$, $B(4, -1)$, and $C(-2, 3)$. Find the length of median $\overline{CM}$.

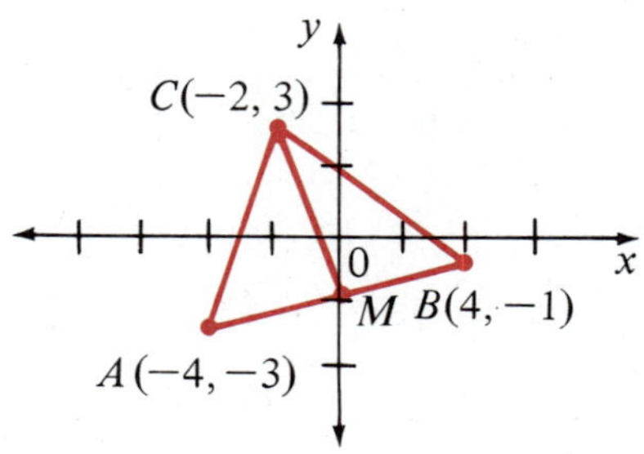

First find the coordinates of M. Use the Midpoint Theorem.

$$\frac{x_1 + x_2}{2} = \frac{-4 + 4}{2} = \frac{0}{2} = 0$$

$$\frac{y_1 + y_2}{2} = \frac{-3 + (-1)}{2} = \frac{-4}{2} = -2$$

M has coordinates $(0, -2)$.

Then find CM for $C(-2, 3)$ and $M(0, -2)$. Use the Distance Formula.

$$d = \sqrt{(x_2 - x_1)^2 + (y_2 - y_1)^2}$$

$$CM = \sqrt{[0 - (-2)]^2 + (-2 - 3)^2}$$

$$= \sqrt{2^2 + (-5)^2}$$

$$= \sqrt{4 + 25}$$

$$= \sqrt{29}$$

Thus, the length of median $\overline{CM}$ is $\sqrt{29}$.

Reading in Geometry

Write *A* if the statement is *always* true. Write *S* if it is *sometimes* true. Write *N* if it is *never* true.

1. The x-coordinate of the midpoint of a segment is the average of the x-coordinates of its endpoints. **A**
2. The y-coordinate of the midpoint of a segment is the average of the y-coordinates of its endpoints. **A**
3. The y-coordinate of the midpoint of a vertical segment is the same as the y-coordinate of an endpoint. **N**
4. The y-coordinate of the midpoint of a horizontal segment is the same as the y-coordinate of an endpoint. **A**

Written Exercises

Find the coordinates of the midpoint of $\overline{AB}$.

1. (2, 3) 2. (1, −2) 3. (1, 6) 4. (−1, 5) 5. $\left(\frac{1}{2}, \frac{1}{2}\right)$ **6.** $\left(4\frac{1}{2}, -\frac{1}{2}\right)$ **7.** $\left(4, 1\frac{1}{2}\right)$ **8. (0, 0)**

A 1. $A(7, 1), B(-3, 5)$ 2. $A(5, 2), B(-3, -6)$ 3. $A(4, 7), B(-2, 5)$ 4. $A(3, 8), B(-5, 2)$
5. $A(5, 0), B(-4, 1)$ 6. $A(5, -8), B(4, 7)$ 7. $A(-1, 5), B(9, -2)$ 8. $A(3, 7), B(-3, -7)$
9. $A(1, 6), B(-1, 3)$ 10. $A(-3, 1), B(8, -5)$ 11. $A(4, -6), B(3, -8)$ 12. $A(-2, 6), B(5, -5)$

M is the midpoint of $\overline{AB}$. Find the coordinates of B. **9.** $\left(0, 4\frac{1}{2}\right)$ **10.** $\left(2\frac{1}{2}, -2\right)$ **11.** $\left(3\frac{1}{2}, -7\right)$ **12.** $\left(1\frac{1}{2}, \frac{1}{2}\right)$

13. $M(-3, 2), A(4, 7)$ 14. $M(6, 1), A(-8, 2)$ 15. $M(-3, -1), A(7, 5)$ 16. $M(-4, 7), A(8, -2)$
17. $M(1, 4), A(-4, 2)$ 18. $M(6, 4), A(-2, -3)$ 19. $M(5, 0), A(-8, 2)$ 20. $M(-6, 1), A(0, -3)$

13. (−10, −3) 14. (20, 0) 15. (−13, −7) 16. (−16, 16) 17. (6, 6) 18. (14, 11) 19. (18, −2) 20. (−12, 5)

Find the coordinates of the midpoints of each side of $\triangle PQR$. Then find the length of the median to $\overline{PQ}$. (Ex. 21–23)

21. (2, −3), (4, 2), (0, 1); 9
22. $\left(-1, -1\frac{1}{2}\right), \left(0, \frac{1}{2}\right), (-3, 3); \frac{1}{2}\sqrt{173}$
23. $\left(1\frac{1}{2}, -1\frac{1}{2}\right), \left(4\frac{1}{2}, 1\right), \left(2, 1\frac{1}{2}\right); \frac{1}{2}\sqrt{170}$

B 21. $P(-2, -4), Q(6, -2), R(2, 6)$ 22. $P(-4, 1), Q(2, -4), R(-2, 5)$ 23. $P(-1, -1), Q(4, -2), R(5, 4)$

24. In $\triangle ABC$, M is the midpoint of $\overline{AB}$. Show that $AM = MB = MC$ for $A(7, 1)$, $B(1, -7)$ and $C(1, 1)$. **AM = MB = MC = 5**
25. In $\triangle PQR$, M is the midpoint of $\overline{PQ}$ and N is the midpoint of $\overline{QR}$. Show that $MN = \frac{1}{2}(PR)$ for $P(-3, -2)$, $Q(2, 2)$ and $R(3, -7)$. **PR =** $\sqrt{61}$**, MN =** $\frac{1}{2}\sqrt{61}$
26. Show that $\overline{AC}$ and $\overline{BD}$ have the same midpoint for $A(-3, -5)$, $B(2, -3)$, $C(3, 5)$, and $D(-2, 3)$. **midpoint = (0, 0)**
27. M is the midpoint of $\overline{AD}$ and N is the midpoint of $\overline{BC}$. Show that $MN = AB = CD$ for $A(-1, -4)$, $B(2, 2)$, $C(-2, 6)$, and $D(-5, 0)$. **MN = AB = CD =** $3\sqrt{5}$

C 28. For $\triangle ABC$ with $A(-5, -4)$, $B(3, -2)$, and $C(-1, 6)$, M is the midpoint of $\overline{AB}$ and N is the midpoint of $\overline{BC}$. Show that $MN = \frac{1}{2}(AC)$. **AC =** $2\sqrt{29}$**, MN =** $\sqrt{29}$
29. For quadrilateral $ABCD$ with $A(-3, -4)$, $B(3, 0)$, $C(5, 6)$ and $D(-7, 2)$, M, N, P, and Q are midpoints of $\overline{AB}$, $\overline{BC}$, $\overline{CD}$, and $\overline{DA}$, respectively. Show that $MN = PQ$ and $PN = QM$. **MN = PQ =** $\sqrt{41}$**; PN = QM =** $\sqrt{26}$
30. For $\triangle RST$ with $R(a, b)$, $S(c, d)$, and $T(e, f)$, M is the midpoint of $\overline{RS}$. Show that $TM = \sqrt{\left(\frac{a+c}{2} - e\right)^2 + \left(\frac{b+d}{2} - f\right)^2}$.
31. Prove Theorem 13.2.

CALCULATOR ACTIVITIES

Find the coordinates of the midpoint of $\overline{CD}$.

(4.95, 6.05) 1. $C(6.1, 4.2), D(3.8, 7.9)$
2. $C(5.4, 7.2), D(-2.3, -3.8)$ **(1.55, 1.7)**
(−1.2, −0.8) 3. $C(-9.6, 3.1), D(7.2, -4.7)$
4. $C(-3.9, -4.1), D(-6.3, -8.9)$ **(−5.1, −6.5)**
(−2.04, .62) 5. $C(12.46, -8.23), D(-16.53, 9.47)$
6. $C(-26.58, -13.53), D(-12.75, -9.23)$ **(−19.67, −11.38)**

SLOPE OF A LINE 13.4

Objectives

To find the slope of a line in a coordinate plane
To determine the position of a line from its slope

Just as the slope of a roof refers to its "steepness," the slope of a line in a coordinate plane refers to its steepness. Consider lines l and m shown at the right. Line l appears to be steeper than line m, or the slope of line l appears to be greater than the slope of line m. The mathematical definition of slope gives you a precise method for assigning a numerical value to the slope of a line.

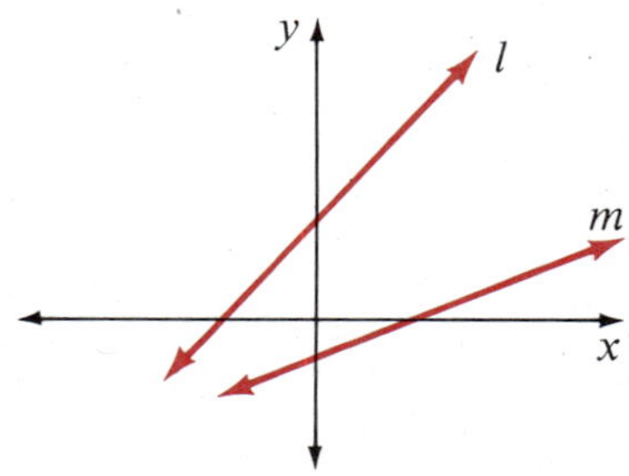

Definition: Slope

The *slope* (m) *of a line* passing through $P(x_1, y_1)$ and $Q(x_2, y_2)$ is defined as follows:

$$m = \frac{y_2 - y_1}{x_2 - x_1} \qquad [x_2 \neq x_1]$$

Slope $(\overleftrightarrow{PQ})$ means the slope of $\overleftrightarrow{PQ}$.

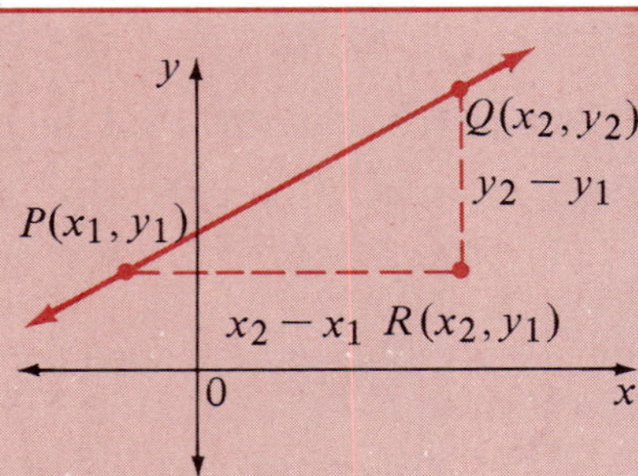

Notice that $y_2 - y_1$ represents the change in the y-coordinates, and $x_2 - x_1$ the change in the x-coordinates. For any two points on a line, the slope of the line is the ratio of the change in the y-coordinates to the change in the x-coordinates.

Example 1

Find slope $(\overleftrightarrow{CD})$ for $C(-3, -1)$ and $D(3, 2)$.

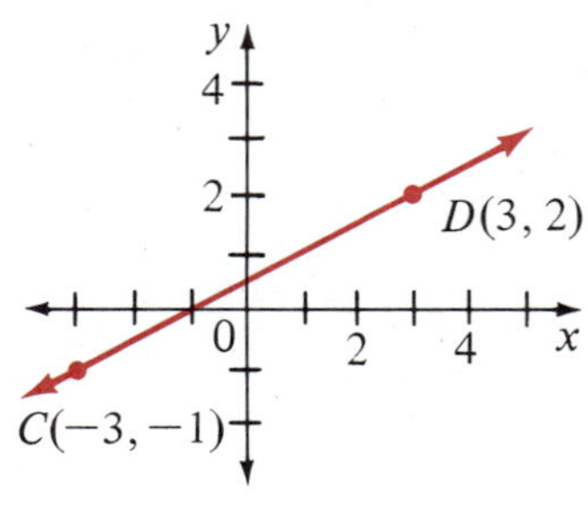

$C(-3, -1)$: $x_1 = -3$, $y_1 = -1$ $\qquad$ $D(3, 2)$: $x_2 = 3$, $y_2 = 2$

$$m = \frac{y_2 - y_1}{x_2 - x_1} = \frac{2 - (-1)}{3 - (-3)} = \frac{3}{6} = \frac{1}{2}$$

Thus, slope $(\overleftrightarrow{CD}) = \frac{1}{2}$.

In Example 1, suppose (x_1, y_1) and (x_2, y_2) were reversed. Then $m = \frac{y_2 - y_1}{x_2 - x_1} = \frac{-1 - 2}{-3 - 3} = \frac{-3}{-6} = \frac{1}{2}$. Notice that the resulting slope is still $\frac{1}{2}$. When computing the slope of a line, any point on the line can be (x_1, y_1) and any other point (x_2, y_2).

In Examples 2 and 3, you will compare the slope of a line with its position in the coordinate plane.

Example 2 Find the slope of $\overleftrightarrow{AB}$ in each case. Then describe the position of $\overleftrightarrow{AB}$ in the coordinate plane.

$A(-2, -4)$ and $B(3, 1)$

$x_1 \quad y_1 \quad x_2 \quad y_2$

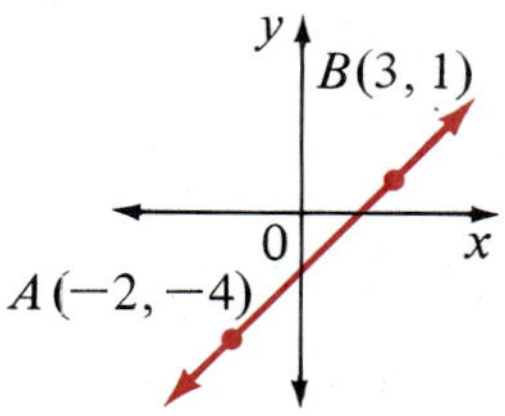

$$m = \frac{y_2 - y_1}{x_2 - x_1} = \frac{1 - (-4)}{3 - (-2)} = \frac{5}{5} = 1$$

Slope ($\overleftrightarrow{AB}$) is positive.
$\overleftrightarrow{AB}$ slants up to the right.

$A(-5, 1)$ and $B(1, -2)$

$x_1 \quad y_1 \quad x_2 \quad y_2$

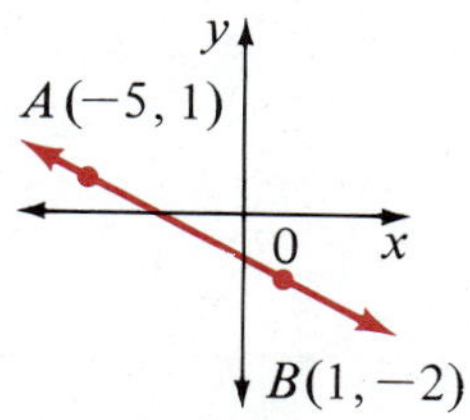

$$m = \frac{y_2 - y_1}{x_2 - x_1} = \frac{-2 - 1}{1 - (-5)} = \frac{-3}{6} = -\frac{1}{2}$$

Slope ($\overleftrightarrow{AB}$) is negative.
$\overleftrightarrow{AB}$ slants down to the right.

Example 3 deals with horizontal and vertical lines.

Example 3 Find the slope of $\overleftrightarrow{AB}$ in each case. Then describe the position of $\overleftrightarrow{AB}$ in the coordinate plane.

$A(-4, 6)$ and $B(6, 6)$

$x_1 \quad y_1 \quad x_2 \quad y_2$

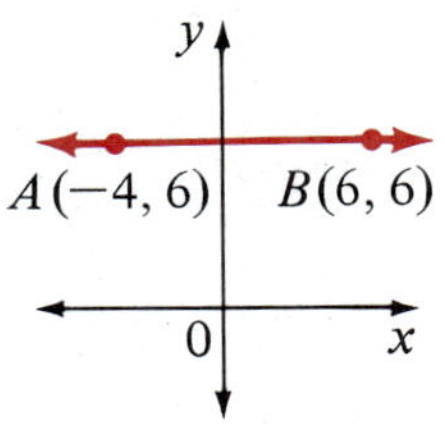

$$m = \frac{y_2 - y_1}{x_2 - x_1} = \frac{6 - 6}{6 - (-4)} = \frac{0}{10} = 0$$

Slope ($\overleftrightarrow{AB}$) is zero.
$\overleftrightarrow{AB}$ is horizontal.

$A(-3, 6)$ and $B(-3, -5)$

$x_1 \quad y_1 \quad x_2 \quad y_2$

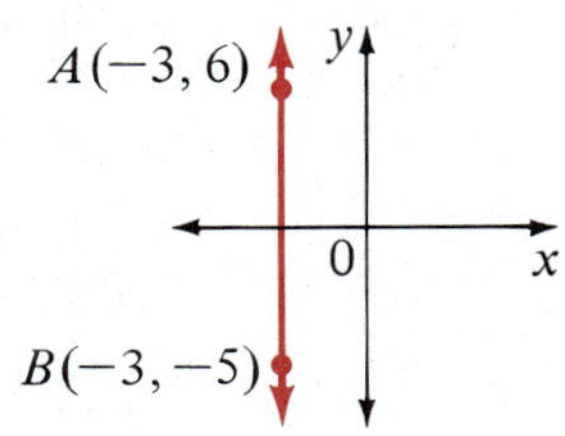

$$m = \frac{y_2 - y_1}{x_2 - x_1} = \frac{-5 - 6}{-3 - (-3)} = \frac{-11}{0}$$

The denominator is 0; division by zero is undefined.
Slope ($\overleftrightarrow{AB}$) is undefined. $\overleftrightarrow{AB}$ is vertical.

Summary

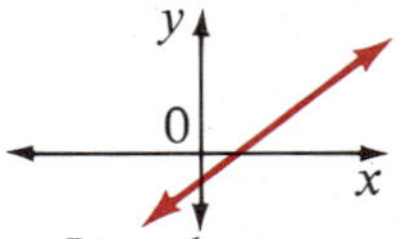

Line slants up to the right: Positive slope

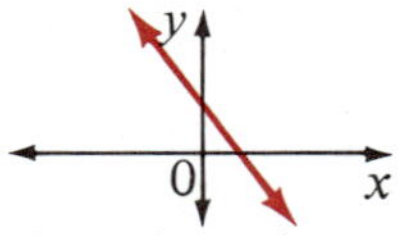

Line slants down to the right: Negative slope

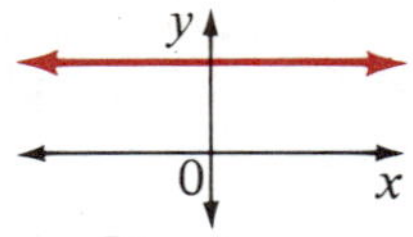

Line is horizontal: Zero slope

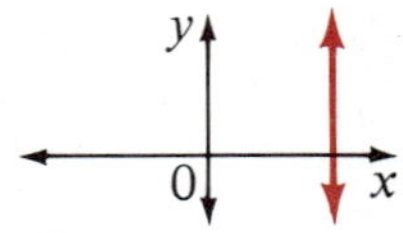

Line is vertical: Undefined slope

Oral Exercises

Classify the slope of each line *l* as positive, negative, zero, or undefined.

1. 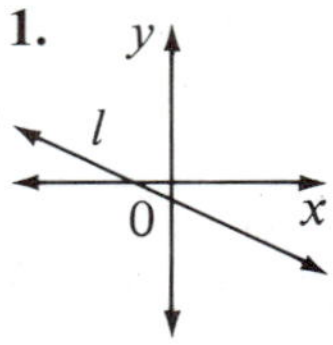

negative

2. 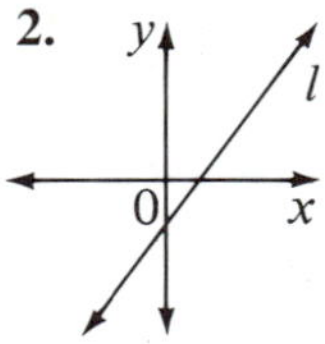

positive

3. 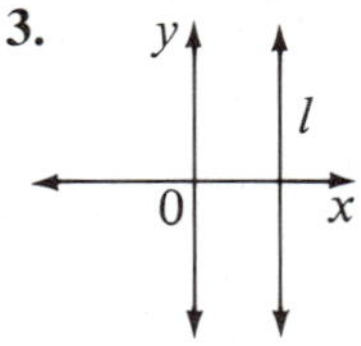

undefined

4. 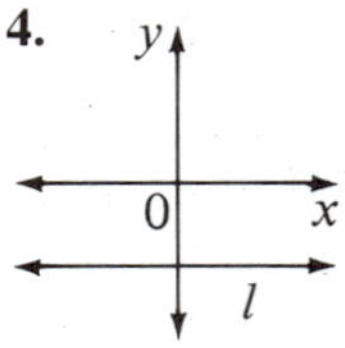

zero

5. 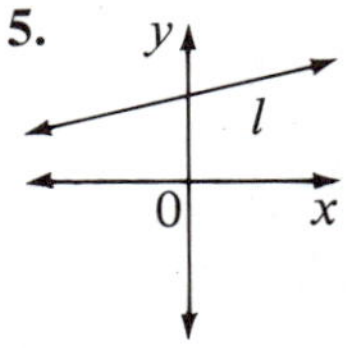

positive

Describe the position of each line whose slope is indicated.

6. 6 up to rt. 7. $\frac{0}{4}$ horiz. 8. -3 dn. to rt. 9. $\frac{-5}{0}$ vert. 10. $\frac{8}{5}$ up to rt. 11. $-\frac{3}{4}$ dn. to rt. 12. $\frac{7}{0}$ vert.

Written Exercises

Find the slope of the line joining each pair of points. Then describe the position of the line in a coordinate plane. 6. $\frac{3}{10}$; up rt. 9. -1; dn. rt. 12. 1; up rt. 15. -1; dn. rt.

A

1. $C(4, -2)$, $D(3, 4)$ -6; dn. rt.
2. $P(-1, 1)$, $Q(7, 4)$ $\frac{3}{8}$; up rt.
3. $A(6, -7)$, $B(-3, -7)$ 0; horiz.
4. $S(1, 4)$, $T(8, 1)$ $-\frac{3}{7}$; dn. rt.
5. $R(-2, -3)$, $S(-2, -6)$ undef.; vert.
6. $E(6, 2)$, $F(-4, -1)$
7. $A(5, -4)$, $B(-5, -4)$ 0; horiz.
8. $G(1, -6)$, $H(5, 3)$ $\frac{9}{4}$; up rt.
9. $K(-1, 3)$, $L(3, -1)$
10. $P(6, -2)$, $Q(-4, 1)$ $-\frac{3}{10}$; dn. rt.
11. $A(1, 7)$, $B(1, -7)$ undef.; vert.
12. $R(-2, 6)$, $S(-6, 2)$
13. $M(5, 2)$, $N(5, -4)$ undef.; vert.
14. $P(4, -1)$, $Q(3, -4)$ 3; up rt.
15. $U(-1, -4)$, $V(-4, -1)$
16. $E(6, -1)$, $F(-2, 8)$ $-\frac{9}{8}$; dn. rt.
17. $C(6, -3)$, $D(-6, 3)$ $\frac{-1}{2}$; dn. rt.
18. $P(-2, 7)$, $Q(-2, 0)$ undef.; vert.

Find *a* so that $\overleftrightarrow{CD}$ will have the given slope.

B

19. $C(4, -2)$, $D(6, a)$; 1 0
20. $C(3, -3)$, $D(-3, a)$; $-\frac{4}{3}$ 5
21. $C(3, a)$, $D(8, 1)$; -2 11
22. $C(-5, a)$, $D(1, -a)$; $-\frac{1}{3}$ 1
23. $C(a, -1)$, $D(-3, 2)$; undefined -3
24. $C(2, a)$, $D(3, -6)$; $\frac{2}{3}$ $-\frac{20}{3}$
25. $C(5, -2)$, $D(-4, a)$; 0 -2
26. $C(2, a)$, $D(-a, 3)$; $\frac{5}{4}$ -22
27. $C(-3, -2)$, $D(1, a)$; 2 6

28. slope ($\overleftrightarrow{AB}$) = slope ($\overleftrightarrow{CD}$) = 1; collinear 29. slope ($\overleftrightarrow{AB}$) = slope ($\overleftrightarrow{CD}$) = $-\frac{3}{4}$; collinear

Graph *A*, *B*, *C*, and *D*. Draw $\overline{AB}$ and $\overline{CD}$. Find slope ($\overleftrightarrow{AB}$) and slope ($\overleftrightarrow{CD}$). What relationship appears to exist between $\overline{AB}$ and $\overline{CD}$ in each case?

C

28. $A(-4, -6)$, $B(-1, -3)$, $C(2, 0)$, $D(6, 4)$
29. $A(-5, 7)$, $B(-1, 4)$, $C(3, 1)$, $D(7, -2)$
30. $A(-3, -3)$, $B(-1, 3)$, $C(1, -5)$, $D(3, 1)$
31. $A(-4, 1)$, $B(2, -3)$, $C(-3, 6)$, $D(3, 2)$

30. slope ($\overleftrightarrow{AB}$) = slope ($\overleftrightarrow{CD}$) = 3; parallel 31. slope ($\overleftrightarrow{AB}$) = slope ($\overleftrightarrow{CD}$) = $-\frac{2}{3}$; parallel

32. Justify each conclusion about the slope in the summary at the top of page 440. [*Hint:* For the line that slants up to the right, $y_2 > y_1$, so $y_2 - y_1 > 0$. Also $x_2 > x_1$, so $x_2 - x_1 > 0$. **Thus,** $\frac{y_2 - y_1}{x_2 - x_1} > 0$.]

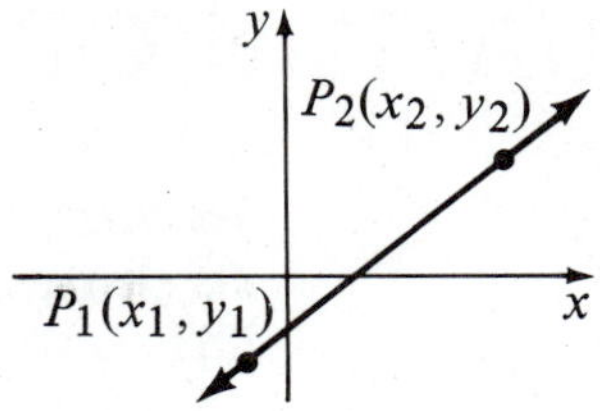

ALGEBRA REVIEW

Objectives

To square a binomial
To complete the square

To square a binomial:
Use $(a + b)^2 = a^2 + 2ab + b^2$.

To complete the square of a binomial such as $x^2 - 8x$:
(1) Take one half of the middle coefficient;
(2) Square it and add the result to the binomial;
(3) The trinomial formed is a perfect square.

Example 1

Simplify $(2x + 3)^2$.

$$(a + b)^2 = a^2 + 2ab + b^2$$

$$(2x + 3)^2 = (2x)^2 + 2(2x)(3) + 3^2$$
$$= 4x^2 + 12x + 9$$

Example 2

Complete the square: $x^2 - 6x + ?$.

Take one half of -6, square it, and add it to $x^2 - 6x$.

$$\left(\frac{-6}{2}\right)^2 = (-3)^2 = 9$$

$x^2 - 6x + 9$ is a perfect square.

Simplify.

1. $(x + 2)^2$ **$x^2 + 4x + 4$**
2. $(x - 3)^2$ **$x^2 - 6x + 9$**
3. $(y + 5)^2$ **$y^2 + 10y + 25$**
4. $(a - 8)^2$ **$a^2 - 16a + 64$**
5. $(2x + 1)^2$ **$4x^2 + 4x + 1$**
6. $(3y - 2)^2$ **$9y^2 - 12y + 4$**
7. $(2c - 5)^2$ **$4c^2 - 20c + 25$**
8. $(4x + 1)^2$ **$16x^2 + 8x + 1$**
9. $(3a + b)^2$ **$9a^2 + 6ab + b^2$**
10. $(x + 3y)^2$ **$x^2 + 6xy + 9y^2$**
11. $(2c - d)^2$ **$4c^2 - 4cd + d^2$**
12. $(5x - 4y)^2$ **$25x^2 - 40xy + 16y^2$**

Complete the square.

13. $x^2 + 2x$ **$x^2 + 2x + 1$**
14. $y^2 - 4y$ **$y^2 - 4y + 4$**
15. $m^2 - 8m$ **$m^2 - 8m + 16$**
16. $a^2 + 12a$ **$a^2 + 12a + 36$**
17. $x^2 + 5x$ **$x^2 + 5x + \frac{25}{4}$**
18. $y^2 - 9y$ **$y^2 - 9y + \frac{81}{4}$**
19. $z^2 - 24z$ **$z^2 - 24z + 144$**
20. $a^2 + 15a$ **$a^2 + 15a + \frac{225}{4}$**
21. $x^2 - 25x$ **$x^2 - 25x + \frac{625}{4}$**
22. $a^2 + 6ab$ **$a^2 + 6ab + 9b^2$**
23. $x^2 - 12xy$ **$x^2 - 12xy + 36y^2$**
24. $c^2 + 7cd$ **$c^2 + 7cd + \frac{49}{4}d^2$**

EQUATION OF A LINE 13.5

Objectives

To draw the graph of a linear equation of the form $ax + by = c$
To draw a line, given the coordinates of one point and the slope of the line, and then write its equation
To write an equation of a line in standard form, given the coordinates of two points on the line

Suppose you could graph all ordered pairs (x, y) that are solutions of the equation $2x + 4y = 10$. A few of the ordered pairs are listed below and graphed on the coordinate plane.

$2x + 4y = 10$

x	y
5	0
0	$\frac{5}{2}$, or $2\frac{1}{2}$
3	1
1	2
−1	3

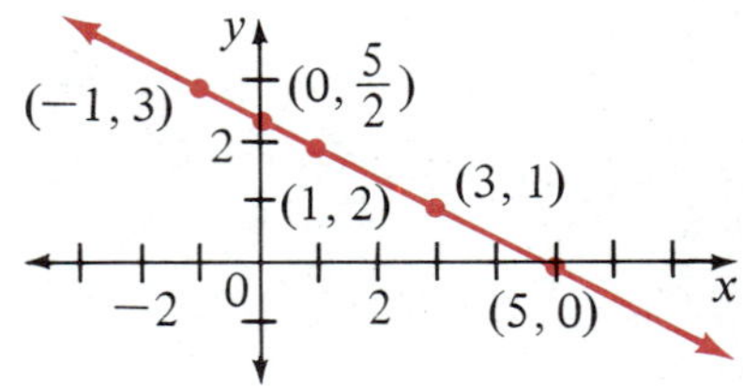

Notice that the points corresponding to the ordered pairs appear to lie on a line. This leads to Theorem 13.3, which is stated without proof.

Theorem 13.3 Standard Form

The graph of an equation that can be written in the form $ax + by = c$ [a and b not both equal to 0] is a line.

An equation whose graph is a line is called a **linear equation**. The form $ax + by = c$ [a, b, and c integers] is called the **standard form** of a linear equation.

Example 1

Write $y + 3 = -\frac{1}{2}(x - 5)$ in standard form.

$y + 3 = -\frac{1}{2}(x - 5)$ ◀ *Multiply each side by −2.*

−x is the same as −1 · x. ▶

$-2y - 6 = x - 5$

$-x - 2y = 1$ is in standard form.

Theorem 13.3 enables you to graph a linear equation such as $2x + 4y = 10$ by graphing only two ordered pairs that satisfy the equation. It is usually wise to graph a third ordered pair as a check. Thus, the first two or three ordered pairs in the chart at the top of the page are sufficient to graph $2x + 4y = 10$.

You can also draw a line if you are given its slope and the coordinates of a point on the line, as shown in Example 2.

Example 2

Draw a line passing through $P(-1, 4)$ and having slope -2.

1. Start at $P(-1, 4)$.
2. $m = \dfrac{-2}{1}$ $\begin{matrix} \leftarrow \text{down } 2 \\ \leftarrow \text{right } 1 \end{matrix}$
   Move down 2 units and to the right 1 unit. Thus, a second point on the line is $Q(0, 2)$.
3. Draw a line through P and Q.

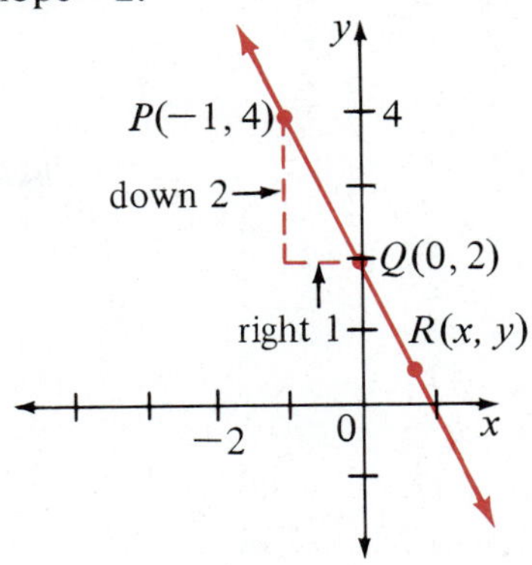

You can write an equation for the line drawn in Example 2. Let $R(x, y)$ be any other point on $\overleftrightarrow{PQ}$. You will use the definition of slope, $m = \dfrac{y_2 - y_1}{x_2 - x_1}$. Consider points P and R and the slope.

$P(-1, 4)$, with $-1 = x_1$ and $4 = y_1$

$R(x, y)$, with $x = x_2$ and $y = y_2$

$m = -2$

$$m = \frac{y_2 - y_1}{x_2 - x_1}$$

$$-2 = \frac{y - 4}{x - (-1)}$$

$$-2 = \frac{y - 4}{x + 1}$$

$$y - 4 = -2(x + 1)$$

The form $y - 4 = -2\,(x + 1)$ is convenient because you can easily determine both the slope of the line and the coordinates of a point on the line simply by examining the equation.

$$y - 4 = -2(x + 1)$$
$$y - 4 = -2\,[x - (-1)]$$

4 ↑ y-coordinate of point; -2 ↑ slope; -1 ↑ x-coordinate of point

Thus, the line has slope -2 and passes through $P(-1, 4)$. This suggests Theorem 13.4.

Theorem 13.4 Point-Slope Form

If a line passes through $P(x_1, y_1)$ and has slope m, then an equation of the line is $y - y_1 = m(x - x_1)$.

(See Exercise 42.)

Example 3

Write an equation of the line with slope -4 and passing through the point $P(3, -5)$.

$P(3, -5)$ $m = -4$ (3 is x_1, -5 is y_1)

► $y - y_1 = m(x - x_1)$ by theorem 13.4.
$y - (-5) = -4(x - 3)$
$y + 5 = -4(x - 3)$

You can also write an equation of a line, given the coordinates of two points on the line, as shown in the next example.

Example 4

Write an equation in standard form of the line through $C(-1, -7)$ and $D(3, 5)$.

$C(-1, -7)$ $D(3, 5)$ (-1 is x_2, -7 is y_2; 3 is x_1, 5 is y_1)

► First find the slope.

$$m = \frac{y_2 - y_1}{x_2 - x_1} = \frac{-7 - 5}{-1 - 3} = \frac{-12}{-4} = 3$$

Then use the Point-Slope Form.

$y - y_1 = m(x - x_1)$
$y - 5 = 3(x - 3)$
$y - 5 = 3x - 9$
$y = 3x + 4$
$-3x + y = 4$

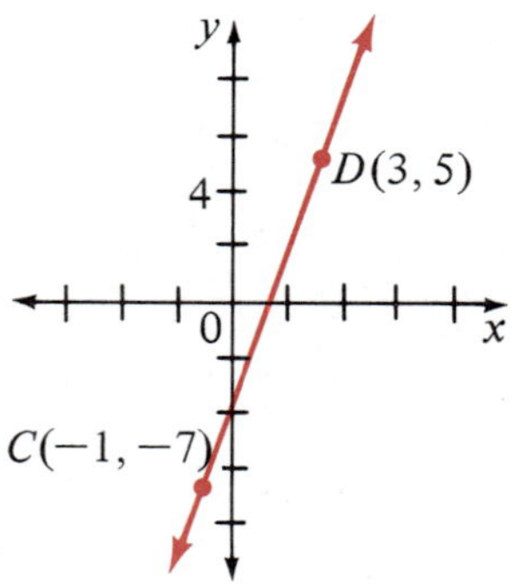

Oral Exercises

Points P and Q lie on the line whose equation is given. Find the missing coordinate for P and for Q.

1. $2x + y = -8$; $P(0, ?)$, $Q(?, 0)$ **(0, −8), (−4, 0)**
2. $x - 3y = 6$; $P(0, ?)$, $Q(?, 0)$ **(0, −2), (6, 0)**
3. $4x + 2y = -4$; $P(0, ?)$, $Q(?, 0)$ **(0, −2), (−1, 0)**
4. $5x - 2y = 10$; $P(0, ?)$, $Q(?, 0)$ **(0, −5), (2, 0)**
5. $y = 3x - 12$; $P(0, ?)$, $Q(?, 0)$ **(0, −12), (4, 0)**
6. $y = -7x$; $P(0, ?)$, $Q(?, 0)$ **(0, 0), (0, 0)**
7. $-3x + 5y = 9$; $P(0, ?)$, $Q(?, 0)$ **$\left(0, \frac{9}{5}\right)$, (−3,0)**
8. $4x - 3y = -8$; $P(0, ?)$, $Q(?, 0)$ **$\left(0, \frac{8}{3}\right)$, (−2,0)**

Give an expression for the slope of $\overleftrightarrow{AB}$.

9. $\frac{y-3}{x-2}$ 10. $\frac{y}{x-1}$ 11. $\frac{y+2}{x-6}$ 12. $\frac{y-4}{x+2}$ 13. $\frac{y+7}{x-5}$ 14. $\frac{y+2}{x-3}$ 15. $\frac{y+4}{x+1}$ 16. $\frac{y}{x}$

9. $A(2, 3)$, $B(x, y)$
10. $A(1, 0)$, $B(x, y)$
11. $A(6, -2)$, $B(x, y)$
12. $A(-2, 4)$, $B(x, y)$
13. $A(5, -7)$, $B(x, y)$
14. $A(3, -2)$, $B(x, y)$
15. $A(-1, -4)$, $B(x, y)$
16. $A(0, 0)$, $B(x, y)$

For each line whose equation is given, find the slope and the coordinates of one point on the line. **Slope is given; coordinates may vary.**

17. $y - 6 = 2(x - 5)$ **2**
18. $y - 3 = -\frac{4}{5}(x - 2)$ **$\frac{-4}{5}$**
19. $y + 4 = \frac{2}{3}(x - 1)$ **$\frac{2}{3}$**
20. $y - \frac{7}{3} = \left(x + \frac{3}{4}\right)$ **1**
21. $y - 7 = -(x + 5)$ **−1**
22. $y = -\frac{4}{3}(x - 2)$ **$-\frac{4}{3}$**
23. $y + 6 = -x + 8$ **−1**
24. $y = x$ **1**
25. $y = -x$ **−1**

Written Exercises

Graph each equation.

A
1. $2x + y = -8$
2. $x - 3y = 6$
3. $4x + 2y = -4$
4. $5x - 2y = 10$
5. $y = 3x - 12$
6. $y = -7x$
7. $-3x + 5y = 9$
8. $4x - 3y = -8$
9. $y = \frac{3}{4}x$

Write each equation in standard form.

10. $6y = -3x - 9$ **10. $3x + 6y = -9$**
11. $y = 4x - 7$ **11. $-4x + y = -7$**
12. $-6y = 2x$ **12. $-2x - 6y = 0$**
13. $y - 8 = 3(x + 2)$ **13. $-3x + y = 14$**
14. $y - 3 = -(x + 7)$ **14. $x + y = -4$**
15. $y + 6 = -\frac{1}{3}(x - 9)$ **15. $-x - 3y = 9$**
16. $y = -\frac{2}{3}x + 8$ **16. $2x + 3y = 24$**
17. $y = -\frac{1}{3}x - \frac{1}{2}$ **17. $2x + 6y = -3$**
18. $y = \frac{2}{5}x - \frac{3}{10}$ **18. $-4x + 10y = -3$**

Draw a line with slope m and passing through point P. Then write an equation for the line.

19. $m = 3$; $P(2, 5)$ **$y - 5 = 3(x - 2)$**
20. $m = -4$; $P(3, -2)$ **$y + 2 = -4(x - 3)$**
21. $m = -6$; $P(-1, -5)$ **$y + 5 = -6(x + 1)$**
22. $m = \frac{1}{2}$; $P(-4, 3)$ **$y - 3 = \frac{1}{2}(x + 4)$**
23. $m = -\frac{4}{3}$; $P(-6, -1)$ **$y + 1 = -\frac{4}{3}(x + 6)$**
24. $m = 0$; $P(2, 4)$ **$y - 4 = 0$**
25. $m = 0$; $P(-1, -6)$ **$y + 6 = 0$**
26. $m = 1$; $P(0, 0)$ **$y = x$**
27. $m = -1$; $P(0, 0)$ **$y = -x$**

Write an equation in standard form of the line through points P and Q.

28. $-x + y = -3$ 29. $x + 2y = -1$ 30. $x + y = 4$ 31. $2x + y = 5$

B
28. $P(6, 3)$, $Q(4, 1)$
29. $P(-5, 2)$, $Q(3, -2)$
30. $P(0, 4)$, $Q(6, -2)$
31. $P(3, -1)$, $Q(5, -5)$
32. $P(4, 7)$, $Q(0, -2)$
33. $P(6, -1)$, $Q(3, -2)$
34. $P(0, 3)$, $Q(-4, 2)$
35. $P(-7, -2)$, $Q(-1, 4)$
36. $P(-3, -1)$, $Q(-8, 2)$

32. $-9x + 4y = -8$ 33. $x - 3y = 9$ 34. $-x + 4y = 12$ 35. $-x + y = 5$ 36. $3x + 5y = -14$

For Exercises 37–40, graph each triangle. Then write the indicated equations in standard form.

37. The vertices of a triangle are $A(-2, -4)$, $B(-3, 1)$, and $C(2, 6)$. Write an equation for the line containing each side.
38. The vertices of a triangle are $D(5, -1)$, $E(4, 3)$, and $F(-3, 0)$. Write an equation for the line containing each side.
39. The vertices of a triangle are $G(0, 0)$, $H(0, 8)$, and $I(6, 4)$. Write an equation for the line containing each median.
40. The vertices of a triangle are $J(-4,0)$, $K(0, 7)$, and $L(0, -5)$. Write an equation for the line containing each median.

C
41. Show that an equation of the line containing points (x_1, y_1) and (x_2, y_2) is $\frac{y - y_1}{x - x_1} = \frac{y_2 - y_1}{x_2 - x_1}$.
42. Prove Theorem 13.4. [*Hint:* Let $R(x, y)$ be another point on the line.]
43. Show that the slope of a line that makes an angle of 30° with the y-axis is either $\sqrt{3}$ or $-\sqrt{3}$.
44. Show that the slope of a line that makes an angle of 30° with the x-axis is either $\frac{\sqrt{3}}{3}$ or $-\frac{\sqrt{3}}{3}$.
45. Write an equation for each line that passes through (4, 3) and makes an angle of 45° with the x-axis. **$-x + y = -1$ or $x + y = 7$**
46. What are the possible slopes of all lines that make an angle of 70° with the x-axis? [*Hint:* Use a trigonometric ratio.] **±2.747**

SLOPE-INTERCEPT FORM OF AN EQUATION 13.6

Objective **To draw a line and write its equation by using its slope and *y*-intercept**

In this lesson you will learn a shortcut for drawing the graph of a linear equation. Consider the equation $2x - 3y = 6$. If you let $x = 0$, then $y = -2$, and $P(0, -2)$ is the point where the graph of $2x - 3y = 6$ intersects the y-axis. The y-coordinate, -2, is called the **y-intercept** of the line.

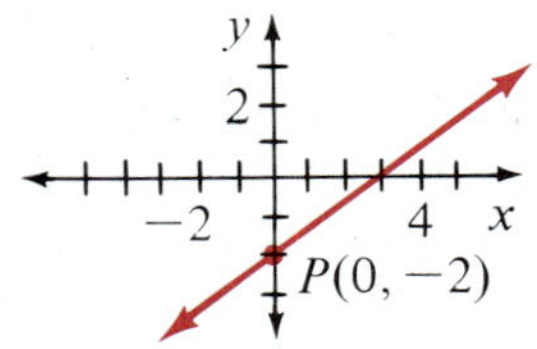

Definition: y-intercept The *y-intercept* of a line is the y-coordinate of the point where the line intersects the y-axis.

Every nonvertical line intersects the y-axis at exactly one point. The x-coordinate of that point is 0 and the y-coordinate is the y-intercept. Generally, b is used to represent the y-intercept of a line. Thus, $(0, b)$ are the coordinates of the point where the line intersects the y-axis. Using the point-slope form of an equation and substituting $(0, b)$ for (x_1, y_1), you can derive another valuable form of the equation of a line.

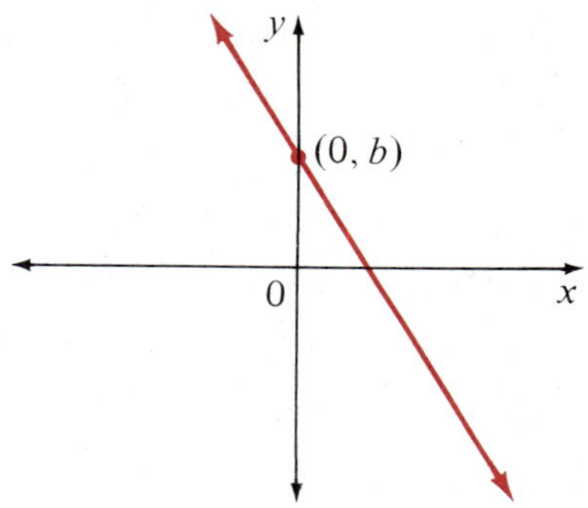

$$y - y_1 = m(x - x_1)$$
$$y - b = m(x - 0)$$
$$y - b = mx$$
$$y = mx + b$$

This argument represents a proof of the next theorem.

Theorem 13.5 Slope-Intercept Form

If a line has slope m and y-intercept b, then an equation of the line is $y = mx + b$.

Example 1 Draw the graph of $y = \frac{2}{3}x - 1$.

$m = \frac{2}{3}$ and $b = -1$.

1. Start at $P(0, -1)$.
2. $m = \frac{2}{3}$ ← up 2, ← right 3

   Move up 2 and to the right 3 to Q.
3. Draw $\overleftrightarrow{PQ}$, the graph of $y = \frac{2}{3}x - 1$.

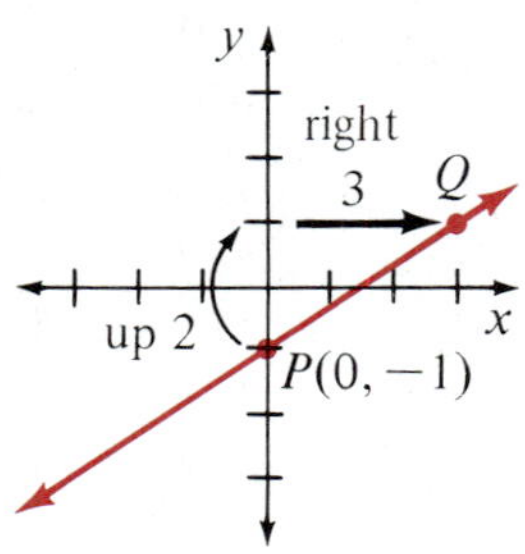

Example 2

The equation of a line is $-2x - 6y = 12$. Find the slope and the y-intercept of the line. Then draw the graph of the equation.

First find the slope and the y-intercept.

$-2x - 6y = 12$

$-6y = 2x + 12$

$y = -\frac{1}{3}x - 2$ ◄ $m = -\frac{1}{3}, b = -2$

Thus, the slope is $-\frac{1}{3}$ and the y-intercept is -2.

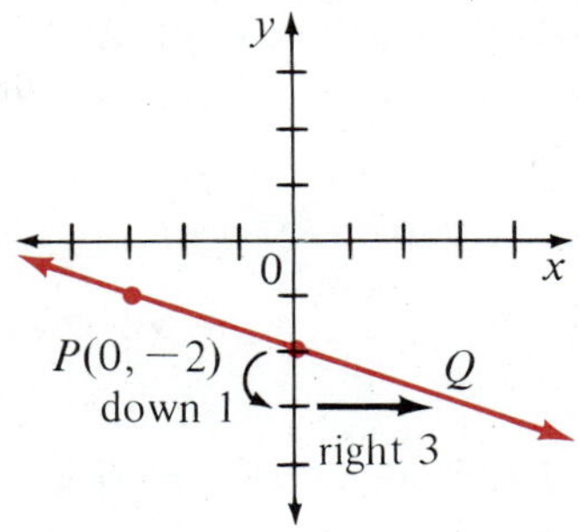

Now draw the graph of the equation.

1. Start at $P(0, -2)$.
2. $m = \frac{-1}{3}$ ← down 1, ← right 3

   Move down 1 and to the right 3 to $Q(3, -3)$.
3. Draw $\overleftrightarrow{PQ}$, the graph of $-2x - 6y = 12$.

In the next two examples, you will consider equations for horizontal and vertical lines.

Example 3

Draw the graph of $y = 3$.

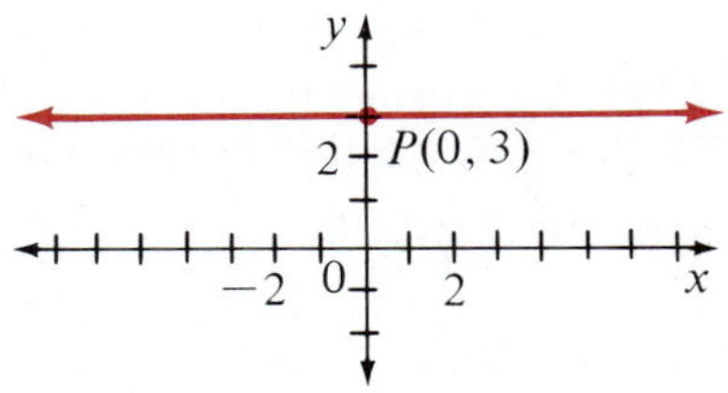

$y = 3$ is the same as $y = 0 \cdot x + 3$.

m = 0 and $b = 3$.

1. Start at $P(0, 3)$.
2. $m = 0$. **Thus,** the line is horizontal. Draw a horizontal line through $P(0, 3)$.

From Example 3, you can conclude that the equation of a horizontal line is $y = b$, where b is the y-intercept. Now consider a vertical line.

Example 4

Draw the graph of $x = -2$.

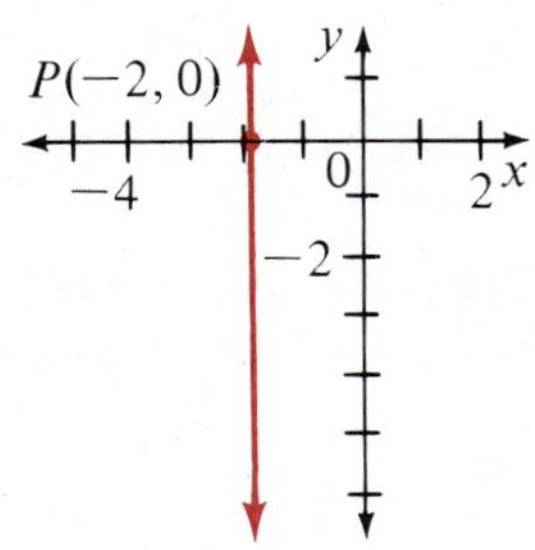

The x-coordinate of each point must be -2, so the line is vertical.

1. Start at $P(-2, 0)$.
2. m is undefined. Draw a vertical line through $P(-2, 0)$.

From Example 4, you can conclude that the equation of a vertical line cannot be in the form $y = mx + b$ since the slope is undefined and there is no y-intercept. In general, the equation of a vertical line is $x = a$, where $(a, 0)$ are the coordinates of the point where the line intersects the x-axis.

Oral Exercises

4. −3; −5 dn. rt. 8. −2/3; 2 dn. rt. 12. 0; −3 horiz. 16. −4/5; 0 dn. rt.

Give the slope and the *y*-intercept of each line whose equation is given. Then tell whether the line slants up to the right, down to the right, is horizontal, or is vertical.

1. $y = 4x - 2$ 4; −2 up rt. **2.** $y = -7x + 3$ −7; 3 dn. rt. **3.** $y = 6x + 4$ 6; 4 up rt. **4.** $y = -3x - 5$

5. $y = \frac{3}{5}x + 4$ 3/5; 4 up rt. **6.** $x = 2$ undef.; no y-int. vert. **7.** $y = 3x$ 3; 0 up rt. **8.** $y = -\frac{2}{3}x + 2$

9. $x = -3$ undef.; no y-int. vert. **10.** $y = -x + 5$ −1; 5 dn. rt. **11.** $y = 0$ 0; 0 horiz. **12.** $y = -3$

13. $y = x - 7$ 1; −7 up rt. **14.** $y = -\frac{7}{2}x + \frac{3}{2}$ −7/2; 3/2 dn. rt. **15.** $y = x$ 1; 0 up rt. **16.** $y = -\frac{4}{5}x$

Tell whether $\overleftrightarrow{PQ}$ is vertical or horizontal in each case. Then give its equation.

17. $P(6, -2), Q(4, -2)$ **18.** $P(3, 8), Q(3, -2)$ **19.** $P(5, 0), Q(4, 0)$ **20.** $P(4, 5), Q(-3, 5)$
21. $P(1, 1), Q(1, 2)$ **22.** $P(0, 1), Q(0, -6)$ **23.** $P(3, 6), Q(-3, 6)$ **24.** $P(4, 7), Q(4, -2)$
25. $P(5, -3), Q(5, 3)$ **26.** $P(3, -1), Q(-1, -1)$ **27.** $P(-6, 2), Q(-6, -2)$ **28.** $P(-4, 4), Q(-4, -4)$

17. H; y = −2 18. V; x = 3 19. H; y = 0 20. H; y = 5 21. V; x = 1 22. V; x = 0
23. H; y = 6 24. V; x = 4 25. V; x = 5 26. H; y = −1 27. V; x = −6 28. V; x = −4

Written Exercises

1. y = 3x + 5 2. y = −3x + 4 3. y = 7x − 2 4. y = −x − 2 5. y = −1/2 6. y = x 7. y = −x + 1/4 8. y = 0

Write an equation of each line with slope *m* and *y*-intercept *b*.

A **1.** $m = 3, b = 5$ **2.** $m = -3, b = 4$ **3.** $m = 7, \ b = -2$ **4.** $m = -1, b = -2$

5. $m = 0, b = -\frac{1}{2}$ **6.** $m = 1, \ b = 0$ **7.** $m = -1, b = \frac{1}{4}$ **8.** $m = 0, \ b = 0$

9. y = 2x + 4; 2; 4 10. y = −1/2x + 3/4; −1/2; 3/4 11. y = −5x − 2; −5; −2 12. y = −3x − 4; −3; −4

Write each equation in the form $y = mx + b$. Identify the slope and the *y*-intercept.

9. $3y = 6x + 12$ **10.** $-4y = 2x - 3$ **11.** $5x + y = -2$ **12.** $-3x - y = 4$

13. $y - 5 = 0$ **14.** $4x - 3y = 0$ **15.** $-2x = 6y + 3$ **16.** $y - 8 = \frac{1}{3}(x + 12)$

13. y = 5; 0; 5 14. y = 4/3x; 4/3; 0 15. y = −1/3x − 1/2; −1/3; −1/2 16. y = 1/3x + 12; 1/3; 12

Draw the graph of each equation.

17. $y = 3x + 1$ **18.** $y = \frac{1}{2}x - 4$ **19.** $y = -\frac{4}{3}x - 2$ **20.** $y = -2x + 3$

21. $y = \frac{1}{3}x$ **22.** $x = -5$ **23.** $y = -3$ **24.** $y = -x - 4$

25. $y = x - 4$ **26.** $y = x$ **27.** $-3x - 2y = 8$ **28.** $4x - 5y = 10$

29. y = −2x + 6; −2, 6 30. y = x − 2; 1, −2 31. y = −x + 3; −1, 3 32. y = 1/3x − 8/3; 1/3, −8/3

Draw $\overleftrightarrow{PQ}$. Then write an equation of $\overleftrightarrow{PQ}$ in the form $y = mx + b$. Identify the slope and the *y*-intercept.

B **29.** $P(3, 0), Q(5, -4)$ **30.** $P(5, 3), Q(3, 1)$ **31.** $P(0, 3), Q(6, -3)$ **32.** $P(5, -1), Q(2, -2)$
33. $P(0, -1), Q(2, -4)$ **34.** $P(-3, 3), Q(3, 3)$ **35.** $P(6, 6), Q(-2, -2)$ **36.** $P(-4, 4), Q(1, -1)$

33. y = −3/2x − 1; −3/2, −1 34. y = 3; 0, 3 35. y = x; 1, 0 36. y = −x; −1, 0

Draw the graph of each pair of equations in the same coordinate plane. Find the coordinates of the point of intersection of the two lines. Check these coordinates in each equation.

37. $y = 6$, $x = -1$ (−1, 6) **38.** $y = 3x - 5$, $y = -\frac{2}{3}x + 6$ (3, 4) **39.** $3x - 2y = 6$, $x - y = 2$ (2, 0) **40.** $y = 2x$, $x + y = 9$ (3, 6)

C **41.** Prove that if $P(a, 0)$ is the point where a nonvertical line intersects the x-axis, then $x = \frac{1}{m} \cdot y + a$.

42. Prove that if a line intersects the x-axis at $P(a, 0)$ and the y-axis at $(0, b)$, then $\frac{x}{a} + \frac{y}{b} = 1$.

SLOPES OF PARALLEL AND PERPENDICULAR LINES 13.7

Objectives

To determine from their slopes if lines are parallel or perpendicular
To find the slope of a line parallel or perpendicular to a given line
To write an equation of a line parallel or perpendicular to a given line

Examine the graphs of the equations $y=\frac{3}{2}x+2$ and $y=\frac{3}{2}x-4$. Notice that both lines have the same slope, $\frac{3}{2}$. It appears that the lines are parallel. This result is generalized in Theorem 13.6.

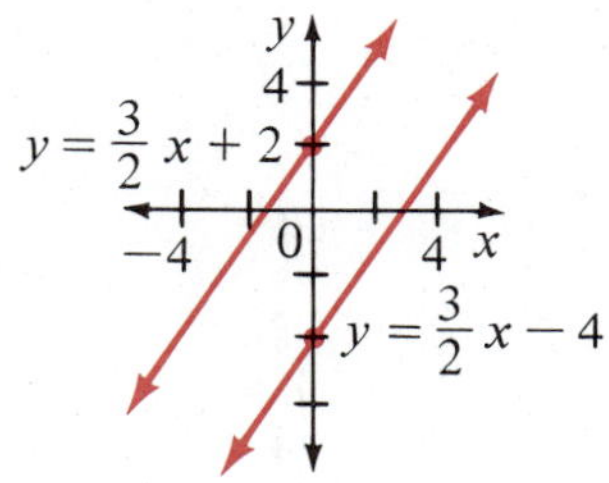

Theorem 13.6

If two nonvertical lines have the same slope, then the lines are parallel.

Lines with slope 0 are horizontal and therefore are parallel. Lines with undefined slope are vertical and are also parallel.

Given: Slope $(\overleftrightarrow{AB})$ = slope $(\overleftrightarrow{DE})$
Prove: $\overleftrightarrow{AB} \parallel \overleftrightarrow{DE}$

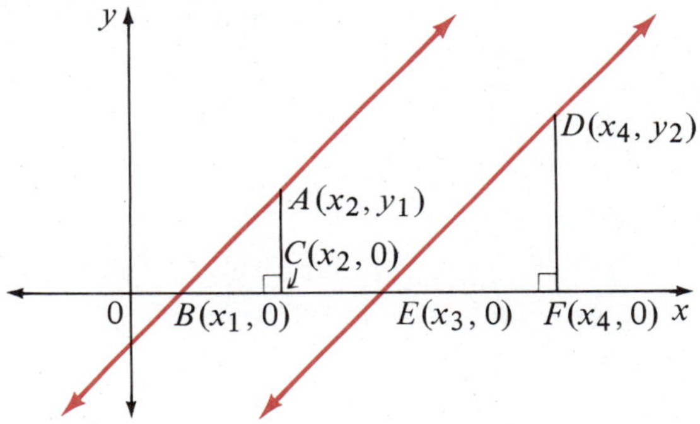

Outline of Proof

1. Draw $\overline{AC}$ and $\overline{DF}$ perpendicular to the x-axis. Let the points have coordinates as shown.
2. $AC = |y_1 - 0| = y_1 \quad (y_1 > 0)$
   $BC = |x_2 - x_1| = x_2 - x_1 \quad (x_2 > x_1)$
   $DF = |y_2 - 0| = y_2 \quad (y_2 > 0)$
   $EF = |x_4 - x_3| = x_4 - x_3 \quad (x_4 > x_3)$
3. Slope $(\overleftrightarrow{AB}) = \frac{y_1 - 0}{x_2 - x_1} = \frac{y_1}{x_2 - x_1}$
   Slope $(\overleftrightarrow{DE}) = \frac{y_2 - 0}{x_4 - x_3} = \frac{y_2}{x_4 - x_3}$
4. slope $(\overleftrightarrow{AB})$ = slope $(\overleftrightarrow{DE})$
5. $\frac{y_1}{x_2 - x_1} = \frac{y_2}{x_4 - x_3}$
6. $\frac{AC}{BC} = \frac{DF}{EF}$
7. $\angle ACB \cong \angle DFE$
8. $\triangle ABC \sim \triangle DEF$
9. $\angle ABC \cong \angle DEF$
10. $\therefore \overleftrightarrow{AB} \parallel \overleftrightarrow{DE}$

The converse of Theorem 13.6 is also true.

Theorem 13.7

If two nonvertical lines are parallel, then the lines have the same slope. (See Exercise 26.)

Example 1 Use slopes to show that $\overleftrightarrow{PQ} \parallel \overleftrightarrow{RS}$ for $P(-3, 2)$, $Q(5, 0)$, $R(-4, -3)$, and $S(0, -4)$.

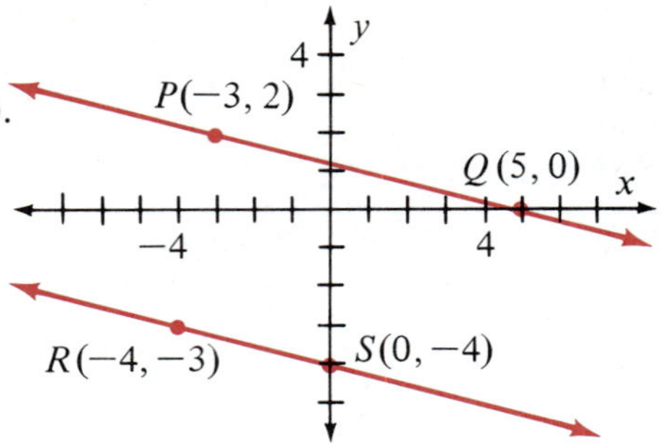

$$\text{Slope } (\overleftrightarrow{PQ}) = \frac{y_2 - y_1}{x_2 - x_1} = \frac{0 - 2}{5 - (-3)} = \frac{-2}{8} = -\frac{1}{4}$$

$$\text{Slope } (\overleftrightarrow{RS}) = \frac{-4 - (-3)}{0 - (-4)} = \frac{-1}{4} = -\frac{1}{4}$$

Slope $(\overleftrightarrow{PQ})$ = slope $(\overleftrightarrow{RS})$

Thus, $\overleftrightarrow{PQ} \parallel \overleftrightarrow{RS}$ by Theorem 13.6.

Examine the graphs of the equations $y = \frac{2}{3}x - 2$ and $y = -\frac{3}{2}x + 1$. Notice that the slopes of the lines are $\frac{2}{3}$ and $-\frac{3}{2}$; $\frac{2}{3}$ and $-\frac{3}{2}$ are negative reciprocals of each other since their product is -1 $\left(\frac{2}{3} \cdot -\frac{3}{2} = -1\right)$. It appears that the lines are perpendicular. This result and converse are stated in Theorems 13.8 and 13.9.

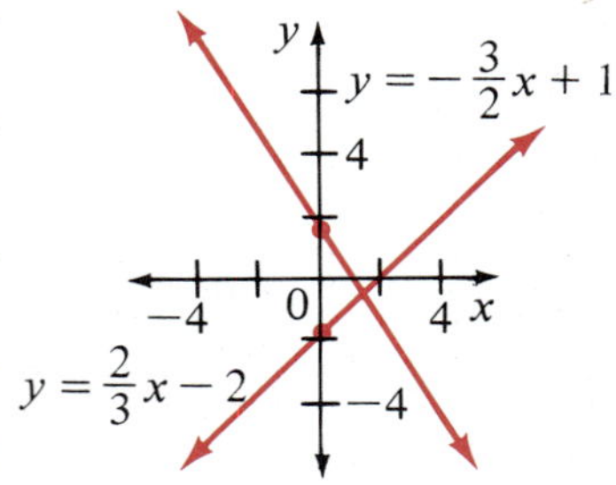

Theorem 13.8 If the slopes of two lines are negative reciprocals, then the lines are perpendicular. (See Exercise 27.)

Theorem 13.9 If two nonvertical lines are perpendicular, then their slopes are negative reciprocals. (See Exercise 28.)

Theorems 13.8 and 13.9 deal with nonvertical lines. A vertical line and a horizontal line in the same coordinate plane are always perpendicular.

Example 2 Show that $ABCD$ is a rectangle for $A(-1, -2)$, $B(3, 2)$, $C(1, 4)$, and $D(-3, 0)$.

$$\text{Slope } (\overline{AD}) = \frac{0 - (-2)}{-3 - (-1)} = \frac{2}{-2} = -1$$

$$\text{Slope } (\overline{BC}) = \frac{4 - 2}{1 - 3} = \frac{2}{-2} = -1$$

$$\text{Slope } (\overline{AB}) = \frac{2 - (-2)}{3 - (-1)} = \frac{4}{4} = 1$$

$$\text{Slope } (\overline{DC}) = \frac{4 - 0}{1 - (-3)} = \frac{4}{4} = 1$$

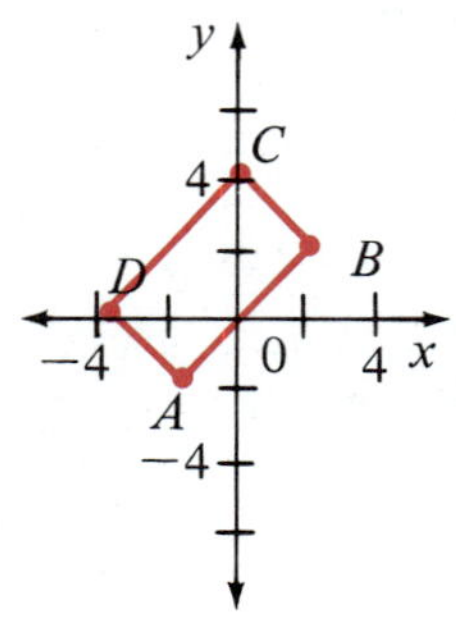

$\overline{AD} \parallel \overline{BC}$ and $\overline{AB} \parallel \overline{DC}$ by Th. 13.6, so $ABCD$ is a parallelogram.
Slopes of consecutive sides of $ABCD$ are negative reciprocals $(-1 \times 1 = -1)$.
$\overline{AD} \perp \overline{DC}$, $\overline{BC} \perp \overline{DC}$, $\overline{AB} \perp \overline{BC}$, and $\overline{AD} \perp \overline{AB}$ by Th. 13.8.
Since four right angles are formed, $ABCD$ is a rectangle.

Reading in Geometry

Copy and complete to make each statement true.

1. The negative reciprocal of -1 is _____. 1
2. The negative reciprocal of $-\frac{1}{5}$ is _____. 5
3. Any two vertical lines are _____ to each other. parallel
4. If the slope of $\overleftrightarrow{AB}$ is 0 and the slope of $\overleftrightarrow{CD}$ is undefined, then $\overleftrightarrow{AB}$ is _____ to $\overleftrightarrow{CD}$. perpendicular
5. Line l has slope 4. Any line parallel to l has slope _____. Any line perpendicular to l has slope _____. $4; -\frac{1}{4}$
6. Line l has slope $-\frac{2}{3}$. Any line parallel to l has slope _____. Any line perpendicular to l has slope _____. $-\frac{2}{3}; \frac{3}{2}$
7. If two lines have the equations $y = 3x - 2$ and $y = -\frac{1}{3}x - 2$, then the lines are _____ and intersect at the point with coordinates _____. perpendicular; (0, −2)
8. If two lines have the equations $y = x$ and $y = -x$, then the lines are _____ and intersect at the point with coordinates _____. perpendicular; (0, 0)

Oral Exercises

Give the negative reciprocal of each slope.

1. $\frac{2}{3}$ $-\frac{3}{2}$ 2. 4 $-\frac{1}{4}$ 3. -7 $\frac{1}{7}$ 4. 1 -1 5. $\frac{5}{2}$ $-\frac{2}{5}$ 6. -1 1 7. $-\frac{3}{4}$ $\frac{4}{3}$ 8. 0 none

Give the slope of a line parallel to the line with the given equation. Then give the slope of a line perpendicular to the line with the given equation.

9. $y = 3x - 1$ $3; -\frac{1}{3}$ 10. $y = -2x + 4$ $-2; \frac{1}{2}$ 11. $y = x + 3$ $1; -1$ 12. $y = -\frac{4}{3}x - \frac{2}{3}$ $-\frac{4}{3}; \frac{3}{4}$

Written Exercises

Give the slope of a line parallel to $\overleftrightarrow{AB}$. Then give the slope of a line perpendicular to $\overleftrightarrow{AB}$.

A

1. $A(0, -5)$, $B(7, 2)$ $1; -1$
2. $A(3, 5)$, $B(-2, 5)$ 0; undef.
3. $A(4, -5)$, $B(6, 0)$ $\frac{5}{2}; -\frac{2}{5}$
4. $A(-3, 2)$, $B(6, -1)$ $-\frac{1}{3}; 3$
5. $A(2, -8)$, $B(2, 9)$ undef.; 0
6. $A(-4, 3)$, $B(7, -1)$ $-\frac{4}{11}; \frac{11}{4}$
7. $A(-3, -1)$, $B(5, -8)$ $-\frac{7}{8}; \frac{8}{7}$
8. $A(1, 6)$, $B(2, 5)$ $-1; 1$

Use slopes to show that $\overleftrightarrow{PR} \perp \overleftrightarrow{SQ}$.

9. $P(2, -1)$, $Q(5, 3)$, $R(1, 6)$, $S(-2, 2)$ $-7 \cdot \frac{1}{7} = -1$
10. $P(-2, -2)$, $Q(6, -1)$, $R(0, 4)$, $S(-6, 3)$ $3\left(-\frac{1}{3}\right) = -1$

Use slopes to show that $\overline{AB} \parallel \overline{CD}$ and $\overline{AD} \parallel \overline{BC}$.

11. $A(-1, 0)$, $B(4, 5)$, $C(3, 9)$, $D(-2, 4)$ slope $(\overline{AB})$ = slope $(\overline{CD})$ = 1; slope $(\overline{AD})$ = slope $(\overline{BC})$ = −4
12. $A(5, -8)$, $B(7, -6)$, $C(6, 1)$, $D(4, -1)$ slope $(\overline{AB})$ = slope $(\overline{CD})$ = 1; slope $(\overline{AD})$ = slope $(\overline{BC})$ = −7

Show that $ABCD$ is a parallelogram.

B

13. $A(-2, -5)$, $B(0, -3)$, $C(-1, 4)$, $D(-3, 2)$ slope $(\overline{AB})$ = slope $(\overline{CD})$ = 1; slope $(\overline{AD})$ = slope $(\overline{BC})$ = −7
14. $A(2, -7)$, $B(7, -2)$, $C(6, 2)$, $D(1, -3)$ slope $(\overline{AB})$ = slope $(\overline{CD})$ = 1; slope $(\overline{AD})$ = slope $(\overline{BC})$ = −4

Show that $EFGH$ is a rectangle.

15. $E(1, 4)$, $F(7, 0)$, $G(9, 3)$, $H(3, 7)$ slope $(\overline{EF})$ = slope $(\overline{GH})$ = $-\frac{2}{3}$; slope $(\overline{EH})$ = slope $(\overline{FG})$ = $\frac{3}{2}$
16. $E(-4, -5)$, $F(1, 0)$, $G(-1, 2)$, $H(-6, -3)$ slope $(\overline{EF})$ = slope $(\overline{GH})$ = 1; slope $(\overline{EH})$ = slope $(\overline{FG})$ = −1

slope $(\overline{MP}) = -7$, slope $(\overline{NQ}) = \frac{1}{7}$, $-7 \cdot \frac{1}{7} = -1$ slope $(\overline{MP}) = \frac{7}{5}$, slope $(\overline{NQ}) = \frac{-5}{7}$; $\frac{7}{5}\left(\frac{-5}{7}\right) = -1$

Show that the diagonals of *MNPQ* are perpendicular.

17. $M(5, -2)$, $N(8, 2)$, $P(4, 5)$, $Q(1, 1)$

18. $M(0, -3)$, $N(6, -2)$, $P(5, 4)$, $Q(-1, 3)$

Write an equation in slope-intercept form of the line passing through the given point and parallel to the line whose equation is given.

19. $(5, -2)$; $y = -6x + 1$ [*Hint:* Use the point-slope form for $m = -6$.] $y = -6x + 28$

20. $(0, 6)$; $2x + 4y = 10$ $y = -\frac{1}{2}x + 6$

Write an equation in slope-intercept form of the line passing through the given point and perpendicular to the line whose equation is given. (Ex. 21–22)

21. $(-3, 1)$; $y = \frac{2}{3}x - 4$ $y = -\frac{3}{2}x - \frac{7}{2}$

22. $(-4, -5)$; $3x + 2y = -7$ $y = \frac{2}{3}x - \frac{7}{3}$

C **23.** Write an equation of the line containing the altitude to $\overline{AB}$ of $\triangle ABC$ with $A(-1, 2)$, $B(-3, -4)$ and $C(-2, 5)$. $x + 3y = 13$

24. Write an equation of the line containing the median to $\overline{BC}$ of $\triangle ABC$ with $A(3, 5)$, $B(-4, 1)$, and $C(-2, -4)$. $-13x + 12y = 21$

25. Write an equation of the perpendicular bisector of $\overline{RS}$ for $R(-2, -3)$ and $S(4, 7)$. $3x + 5y = 13$

26. Prove Theorem 13.7.

27. Prove Theorem 13.8.

28. Prove Theorem 13.9.

NON-ROUTINE PROBLEMS

On one full-page coordinate plane, graph each segment described below. Get the message?

YOU'RE GREAT

1. $y = -2x - 16$ between $(-11, 6)$ and $(-10, 4)$
2. $y = 2x + 24$ between $(-10, 4)$ and $(-9, 6)$
3. $x = -10$ between $(-10, 2)$ and $(-10, 4)$
4. $x = -7$ between $(-7, 6)$ and $(-7, 2)$
5. $y = 6$ between $(-7, 6)$ and $(-5, 6)$
6. $x = -5$ between $(-5, 6)$ and $(-5, 2)$
7. $y = 2$ between $(-7, 2)$ and $(-5, 2)$
8. $x = -3$ between $(-3, 6)$ and $(-3, 2)$
9. $y = 2$ between $(-3, 2)$ and $(-1, 2)$
10. $x = -1$ between $(-1, 6)$ and $(-1, 2)$
11. $x = 1$ between $(1, 7)$ and $(1, 6)$
12. $x = 2$ between $(2, 6)$ and $(2, 2)$
13. $y = 6$ between $(2, 6)$ and $(4, 6)$
14. $x = 4$ between $(4, 6)$ and $(4, 4)$
15. $y = 4$ between $(2, 4)$ and $(4, 4)$
16. $y = -2x + 10$ between $(3, 4)$ and $(4, 2)$
17. $x = 6$ between $(6, 6)$ and $(6, 2)$
18. $y = 6$ between $(6, 6)$ and $(8, 6)$
19. $y = 4$ between $(6, 4)$ and $(7, 4)$
20. $y = 2$ between $(6, 2)$ and $(8, 2)$
21. $y = -2$ between $(-11, -2)$ and $(-9, -2)$
22. $x = -11$ between $(-11, -2)$ and $(-11, -6)$
23. $y = -6$ between $(-11, -6)$ and $(-9, -6)$
24. $x = -9$ between $(-9, -5)$ and $(-9, -6)$
25. $y = -5$ between $(-10, -5)$ and $(-8, -5)$
26. $x = -7$ between $(-7, -2)$ and $(-7, -6)$
27. $y = -2$ between $(-7, -2)$ and $(-5, -2)$
28. $x = -5$ between $(-5, -2)$ and $(-5, -4)$
29. $y = -4$ between $(-7, -4)$ and $(-5, -4)$
30. $y = -2x - 16$ between $(-6, -4)$ and $(-5, -6)$
31. $x = -3$ between $(-3, -2)$ and $(-3, -6)$
32. $y = -2$ between $(-3, -2)$ and $(-1, -2)$
33. $y = -4$ between $(-3, -4)$ and $(-2, -4)$
34. $y = -6$ between $(-3, -6)$ and $(-1, -6)$
35. $x = 1$ between $(1, -2)$ and $(1, -6)$
36. $y = -2$ between $(1, -2)$ and $(3, -2)$
37. $x = 3$ between $(3, -2)$ and $(3, -6)$
38. $y = -4$ between $(1, -4)$ and $(3, -4)$
39. $x = 6$ between $(6, -2)$ and $(6, -6)$
40. $y = -2$ between $(5, -2)$ and $(7, -2)$

USING COORDINATES IN PROOFS 13.8

Objectives **To find an appropriate placement for a polygon in a coordinate plane and assign coordinates to its vertices**
To use coordinate geometry to prove statements

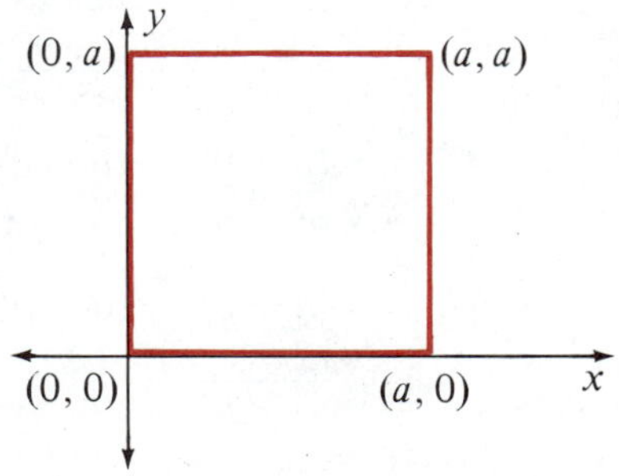

Many statements can be proved quite simply by the use of coordinate geometry. As a first step, the given geometric figure must be placed in a coordinate plane. The placement is crucial. For example, if a proof involves a square, you can rely on the definition of a square in placing it. Since the angles of a square are right angles, you can place the square so that two of its sides lie on the axes. Let a be the length of a side. Since all four sides have the same length, the coordinates of the vertices are as shown at the right.

If the square were in one of the positions shown below, the coordinates of the vertices would become much more complicated.

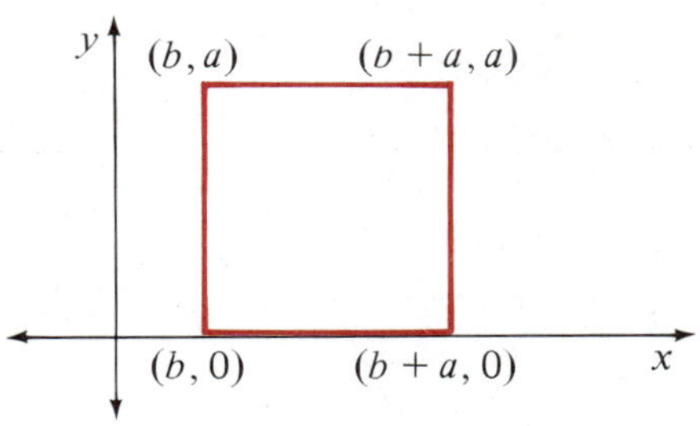

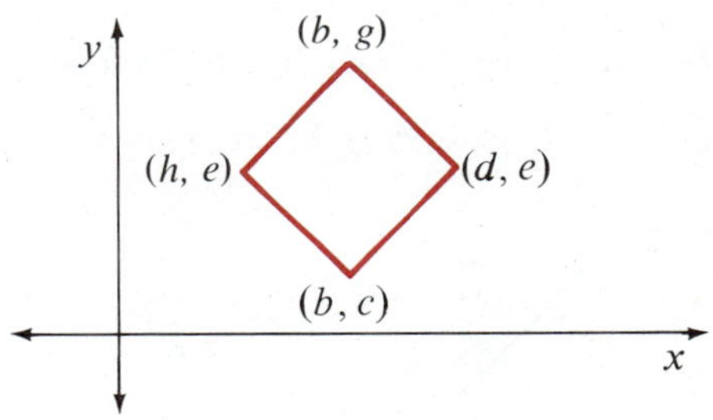

To find an appropriate placement for a polygon in a coordinate plane
(1) Place one vertex at the origin.
(2) Place one side on the x-axis.

In Example 1, notice that the placement of the square in the coordinate plane helps to simplify the proof of the statement.

Example 1 Prove that the diagonals of a square are perpendicular.

Given: Square $ABCD$, diagonals $\overline{AC}$ and $\overline{BD}$
Prove: $\overline{AC} \perp \overline{BD}$

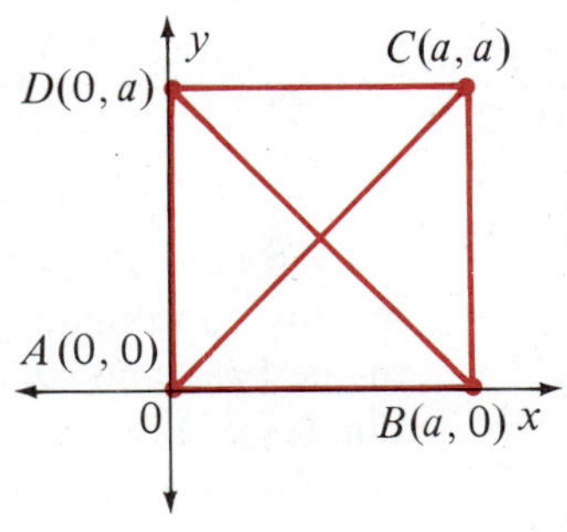

Place square $ABCD$ in a coordinate plane with coordinates as shown.

Slope $(\overleftrightarrow{AC}) = \dfrac{a-0}{a-0} = \dfrac{a}{a} = 1$ ◄

Slope $(\overleftrightarrow{BD}) = \dfrac{0-a}{a-0} = \dfrac{-a}{a} = -1$ ◄ *Def. of slope*

Thus, $\overline{AC} \perp \overline{BD}$ by Theorem 13.8.

Example 2

Prove that the diagonals of a rectangle are congruent.

Given: Rect. $ABCD$ with diagonals $\overline{AC}$ and $\overline{BD}$
Prove: $\overline{AC} \cong \overline{BD}$

Place rect. $ABCD$ in a coordinate plane with coordinates as shown.

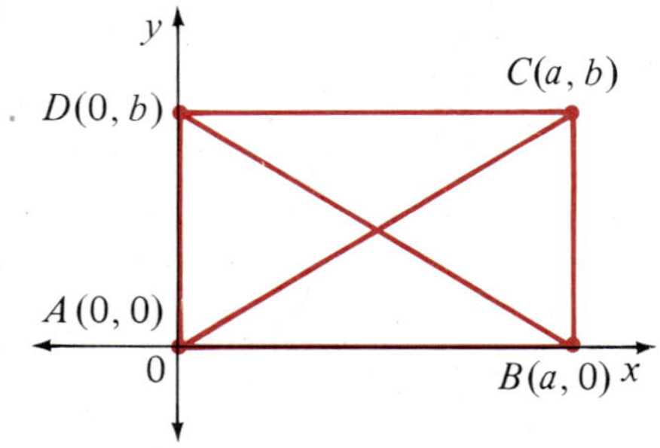

$d = \sqrt{(x_2 - x_1)^2 + (y_2 - y_1)^2}$ ◄ *Dist. formula*

$AC = \sqrt{(a-0)^2 + (b-0)^2}$
$= \sqrt{a^2 + b^2}$

$BD = \sqrt{(a-0)^2 + (0-b)^2}$
$= \sqrt{a^2 + (-b)^2}$
$= \sqrt{a^2 + b^2}$

So $AC = BD$.
Thus, $\overline{AC} \cong \overline{BD}$.

Written Exercises

Give the missing coordinates for each figure. (Ex. 1–6)

A 1. Isosceles right triangle

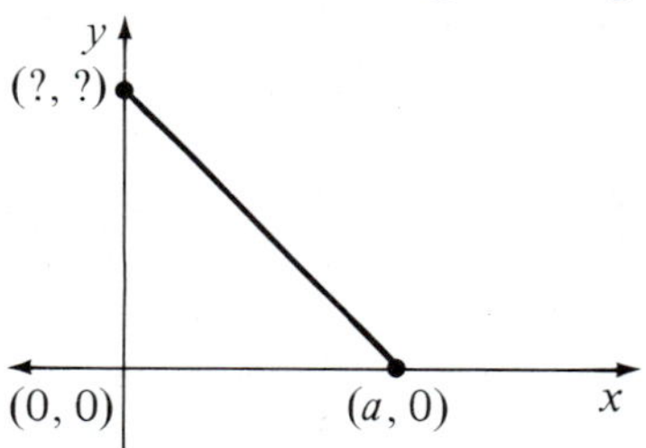

(0, a)

2. Parallelogram

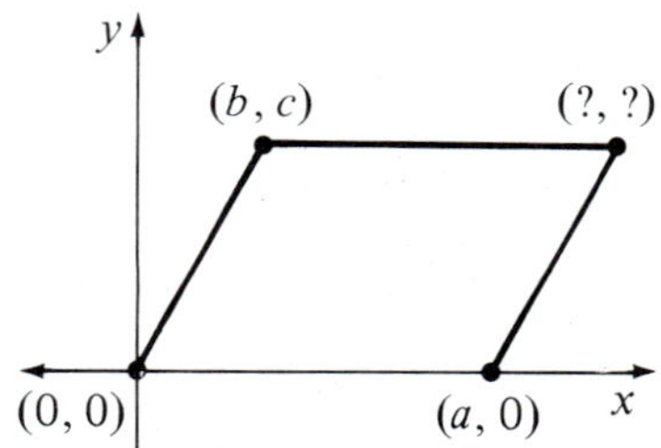

(a + b, c)

3. Isosceles triangle

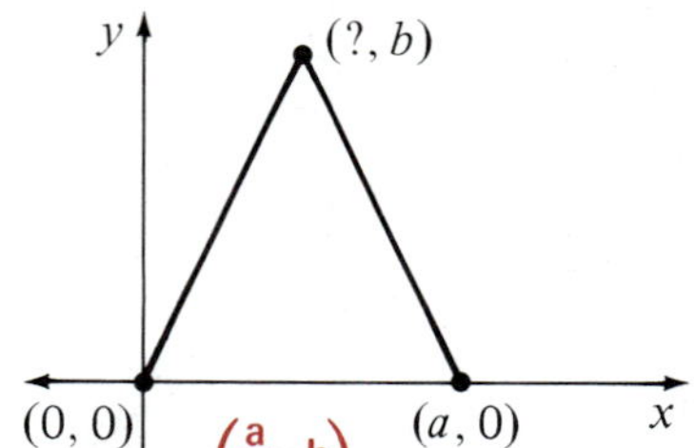

$\left(\frac{a}{2}, b\right)$

4. Isosceles trapezoid

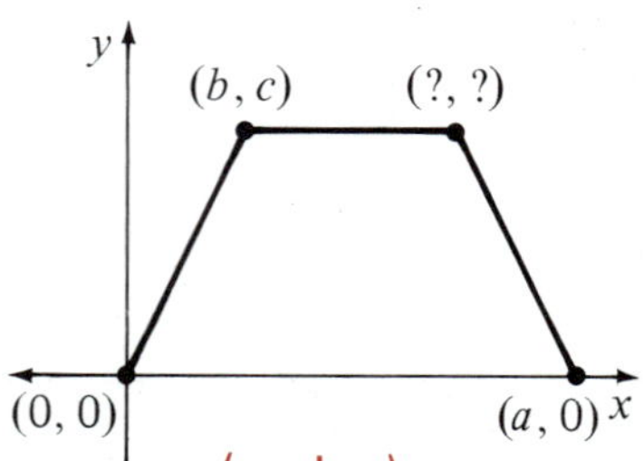

(a − b, c)

5. Midpoints of sides of a triangle

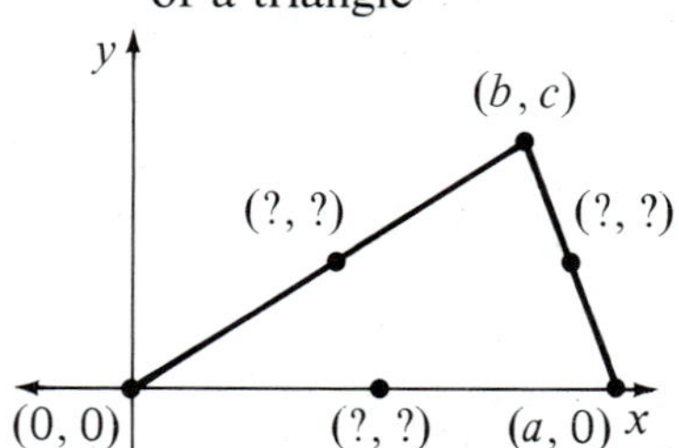

$\left(\frac{a}{2}, 0\right), \left(\frac{a+b}{2}, \frac{c}{2}\right)\left(\frac{b}{2}, \frac{c}{2}\right)$

6. 30°–60° right triangle

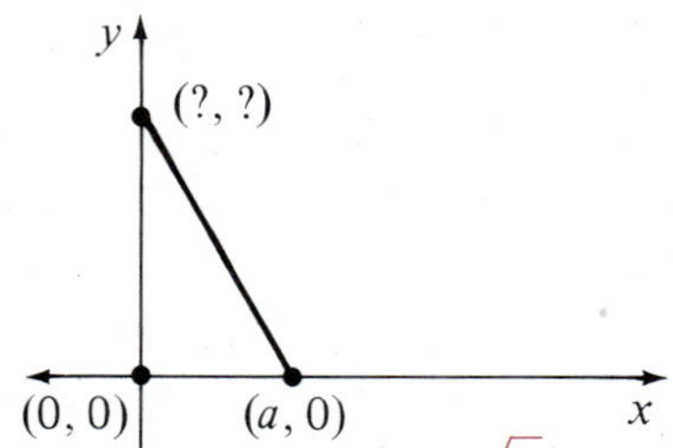

$(0, a\sqrt{3})$

7. For an isosceles triangle, if $(a, 0)$ are the coordinates of an endpoint of the base and $(0, b)$ are the coordinates of the vertex angle, give the coordinates of the other endpoint of the base. **(−a, 0)**

8. If the coordinates of the endpoints of a segment are $(a, 0)$ and $(0, b)$, give the coordinates of the midpoint of the segment. $\left(\frac{a}{2}, \frac{b}{2}\right)$

Use coordinate geometry to prove each statement.

B

9. The diagonals of a parallelogram bisect each other. [*Hint:* Show that the diagonals have the same midpoint.]

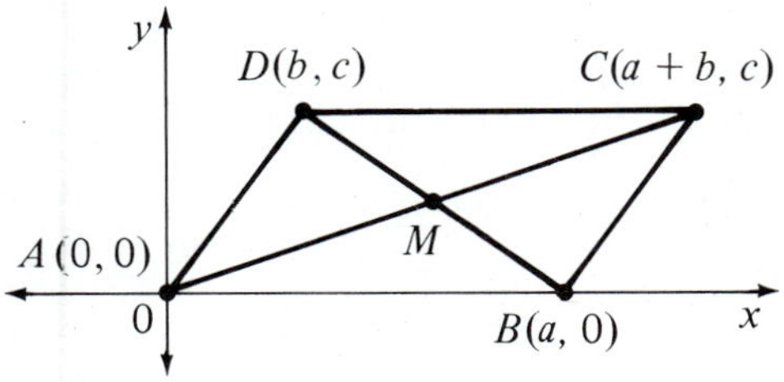

10. The segment joining the midpoint of two sides of a triangle is parallel to the third side and is one half as long.

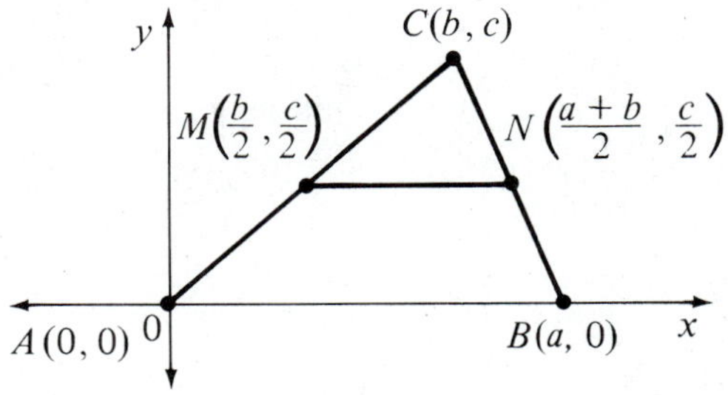

11. The diagonals of an isosceles trapezoid are congruent.

12. The midpoint of the hypotenuse of a right triangle is equidistant from the vertices.

C

13. The segments joining the midpoints of the opposite sides of a quadrilateral bisect each other.

14. The length of the median of a trapezoid is equal to one half the sum of the lengths of the bases.

15. The length of the segment joining the midpoints of the diagonals of a trapezoid is equal to one half the difference of the lengths of the bases.

16. The segments joining the midpoints of the sides of a triangle form another triangle which is similar to the given triangle.

ALGEBRA REVIEW

Objective **To solve equations involving absolute value**

$|a|$ is read *the absolute value of a.*
$|a| = 3$ means $a = 3$ or $a = -3$.

Example

Solve $|2x - 1| = 5$.

$|2x - 1| = 5$ means $\quad 2x - 1 = 5$ or $2x - 1 = -5$

Solve each equation: $\quad 2x = 6 \qquad 2x = -4$

$x = 3 \qquad x = -2$

Check: $|2 \cdot 3 - 1| = |5| = 5 \quad |2(-2) - 1| = |-5| = 5$

Thus, the solutions are 3 and −2.

Solve each equation.

1. $\lvert x - 2\rvert = 7$ **9, −5**	2. $\lvert x + 6\rvert = 10$ **4, −16**	3. $\lvert 7 - x\rvert = 4$ **3, 11**	4. $\lvert 2 + x\rvert = 9$ **7, −11**
5. $\lvert 2x - 6\rvert = 4$ **5, 1**	6. $\lvert 5 - 3y\rvert = 7$ $-\frac{2}{3}$, **4**	7. $\lvert 2z - 14\rvert = 1$ $\frac{15}{2}, \frac{13}{2}$	8. $\lvert 2 - 5y\rvert = 8$ $-\frac{6}{5}$, **2**
9. $\lvert 3k + 2\rvert = 10$ $\frac{8}{3}$, **−4**	10. $\lvert 4 + 2y\rvert = 0$ **−2**	11. $\lvert 5a + 3\rvert = 6$ $\frac{3}{5}, -\frac{9}{5}$	12. $\lvert 6z - 8\rvert = 4$ **2**, $\frac{2}{3}$

EQUATION OF A CIRCLE 13.9

Objective **To apply equations of circles**

Suppose circle Q lies in a coordinate plane with (h, k) the coordinates of its center. Let $P(x, y)$ be any point on the circle and r be the length of a radius. Since every point P is the same distance r from Q, you can use the Distance Formula to find an equation which describes all points $P(x, y)$ on the circle.

$Q(h, k)$ $\quad$ $P(x, y)$ $\quad$ $d = r$

$x_1 \; y_1 \qquad x_2 \; y_2$

$$d = \sqrt{(x_2 - x_1)^2 + (y_2 - y_1)^2}$$

$$r = \sqrt{(x - h)^2 + (y - k)^2}$$

$$r^2 = (x - h)^2 + (y - k)^2$$

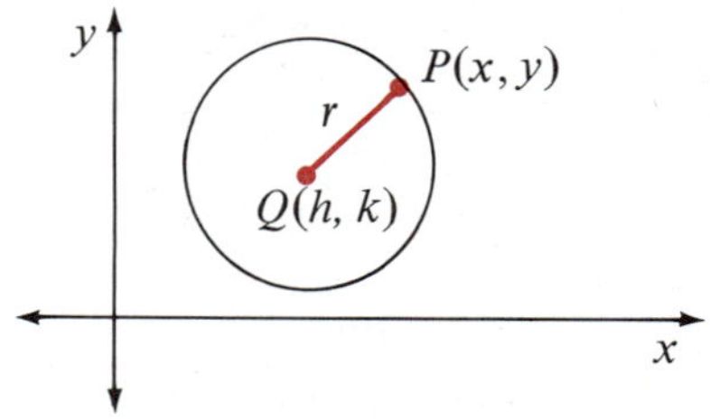

Theorem 13.10 The equation of a circle with (h, k) the coordinates of the center and a radius of length r is

$$(x - h)^2 + (y - k)^2 = r^2.$$

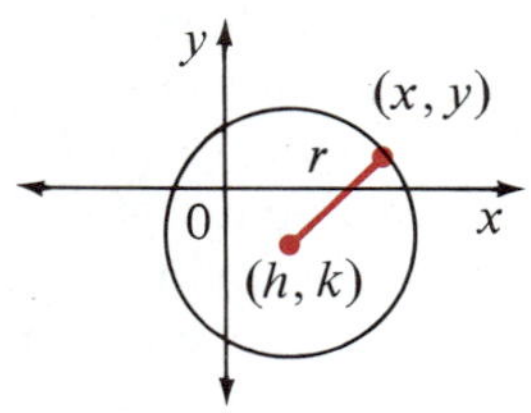

The equation derived above is the general equation of a circle in a coordinate plane. Examples 1 and 2 are applications of Theorem 13.10.

Example 1 Write an equation for a circle with its center at the origin and a radius of length 8.

$(h, k) = (0, 0)$ and $r = 8$.

$(x - h)^2 + (y - k)^2 = r^2$ by Theorem 13.10

$(x - 0)^2 + (y - 0)^2 = 8^2$

$x^2 + y^2 = 64$

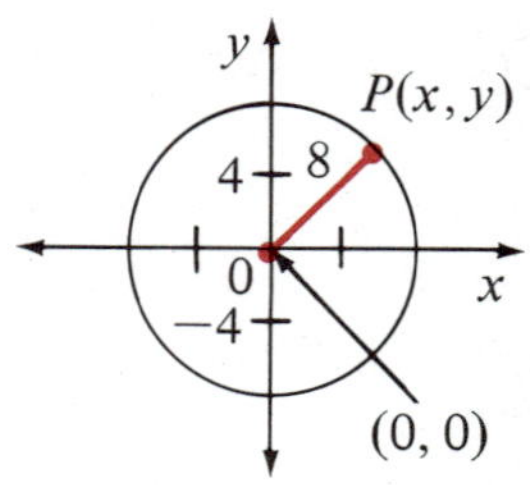

Example 2 Show that $R(-4, 3)$ and $S(1, -2\sqrt{6})$ lie on a circle with equation $x^2 + y^2 = 25$.

$x = -4$ and $y = 3$

$x^2 + y^2$	25
$(-4)^2 + 3^2$ $16 + 9$ 25	25

$x = 1$ and $y = -2\sqrt{6}$

$x^2 + y^2$	25
$1^2 + (-2\sqrt{6})^2$ $1 + 24$ 25	25

Thus, $R(-4, 3)$ and $S(1, -2\sqrt{6})$ lie on the circle.

To simplify an equation of a circle means to express it with only the variable terms on the left side. This is shown in Example 3.

Example 3

Write an equation of a circle with center $Q(-2, 1)$ and radius of length 6. Simplify the equation.

$(h, k) = (-2, 1)$ and $r = 6$.

$$(x - h)^2 + (y - k)^2 = r^2$$
$$[x - (-2)]^2 + (y - 1)^2 = 6^2$$
$$(x + 2)^2 + (y - 1)^2 = 6^2$$
$$x^2 + 4x + 4 + y^2 - 2y + 1 = 36$$
$$x^2 + 4x + y^2 - 2y = 31$$

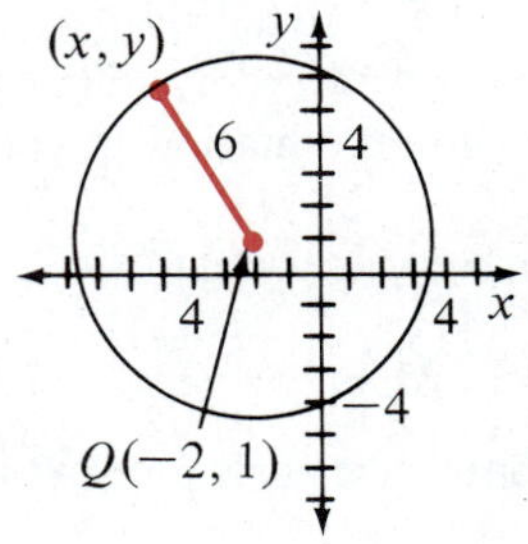

Thus, the equation is $x^2 + 4x + y^2 - 2y = 31$.

The process of completing the square is required in the next example. If necessary, you can refer to the Algebra Review on page 441.

Example 4

Find the coordinates of the center and the length of a radius of a circle whose equation is $x^2 + y^2 + 8x - 14y = 16$.

$$x^2 + y^2 + 8x - 14y = 16$$
$$(x^2 + 8x + \quad) + (y^2 - 14y + \quad) = 16$$
$$(x^2 + 8x + 16) + (y^2 - 14y + 49) = 16 + 16 + 49$$
$$(x + 4)^2 + (y - 7)^2 = 81$$
$$\text{or } [x - (\underset{h}{-4})]^2 + (y - \underset{k}{7})^2 = \underset{r}{9}^2$$

Thus, the coordinates of the center are (−4, 7) and the length of a radius is 9.

Oral Exercises

Identify the coordinates of the center and the length of a radius of each circle whose equation is given.

1. $(x - 1)^2 + (y - 3)^2 = 5^2$ **(1, 3); 5**
2. $(x + 5)^2 + (y - 4)^2 = 2^2$ **(−5, 4); 2**
3. $(x - 2)^2 + (y + 6)^2 = 100$ **(2, −6); 10**
4. $(x + 4)^2 + (y - 6)^2 = 4$ **(−4, 6); 2**
5. $(x + 10)^2 + (y + 5)^2 = 16$ **(−10, −5); 4**
6. $(x - 6)^2 + y^2 = 1$ **(6, 0); 1**
7. $x^2 + (y + 8)^2 = 7$ **(0, −8); $\sqrt{7}$**
8. $x^2 + y^2 = 5$ **(0, 0); $\sqrt{5}$**

Written Exercises

1. $(x - 0)^2 + (y - 0)^2 = 3^2$ 2. $(x - 3)^2 + (y - 1)^2 = 5^2$ 3. $[x - (-1)]^2 + (y - 4)^2 = 1^2$

Write an equation of each circle with center A and radius of length r. Leave the equation in the form $(x - h)^2 + (y - k)^2 = r^2$. 4. $[x - (-2)]^2 + [y - (-6)]^2 = 5^2$ 5. $(x - 6)^2 + [y - (-1)]^2 = 7^2$

A

1. $A(0, 0)$; $r = 3$
2. $A(3, 1)$; $r = 5$
3. $A(-1, 4)$; $r = 1$
4. $A(-2, -6)$; $r = 5$
5. $A(6, -1)$; $r = 7$
6. $A(-3, -8)$; $r = \sqrt{12}$
7. $A(5, 5)$; $r = 8$
8. $A(6, -2)$; $r = \sqrt{10}$

6. $[x - (-3)]^2 + [y - (-8)]^2 = (\sqrt{12})^2$ 7. $(x - 5)^2 + (y - 5)^2 = 8^2$ 8. $(x - 6)^2 + [y - (-2)]^2 = (\sqrt{10})^2$

Write an equation of each circle with center A and radius of length r. Simplify the equation.

9. $x^2 - 6x + y^2 - 2y = 6$ 10. $x^2 - 4x - 10y = -20$ 11. $x^2 + y^2 + 8y = -12$ 12. $x^2 + 2x + y^2 - 10y = 38$

9. $A(3, 1)$; $r = 4$
10. $A(2, 5)$; $r = 3$
11. $A(0, -4)$; $r = 2$
12. $A(-1, 5)$; $r = 8$
13. $A(-8, -1)$; $r = \sqrt{15}$
14. $A(7, 0)$; $r = 8$
15. $A(-2, -9)$; $r = \sqrt{7}$
16. $A(6, -7)$; $r = 20$

13. $x^2 + 16x + y^2 + 2y = -50$ 14. $x^2 - 14x + y^2 = 15$ 15. $x^2 + 4x + y^2 + 18y = -78$ 16. $x^2 - 12x + y^2 + 14y = 315$

Show that each point P lies on a circle with the given equation.

17. $(4)^2 + (0)^2 = 16$ 18. $(5)^2 + (-3)^2 = 34$ 19. $(5 - 1)^2 + (-3)^2 = 25$ 20. $(-1 - 2)^2 + (-1 + 1)^2 = 9$

17. $P(4, 0)$; $x^2 + y^2 = 16$
18. $P(5, -3)$; $x^2 + y^2 = 34$
19. $P(5, -3)$; $(x - 1)^2 + y^2 = 25$
20. $P(-1, -1)$; $(x - 2)^2 + (y + 1)^2 = 9$
21. $P(-\sqrt{3}, -2)$; $x^2 + y^2 + 6y = -5$
22. $P(-3, 4)$; $x^2 + y^2 + 2x - 8y = -13$

21. $(-\sqrt{3})^2 + (-2)^2 + 6(-2) = -5$ 22. $(-3)^2 + (4)^2 + 2(-3) - 8(4) = -13$

Find the coordinates of the center and the length of a radius of the circle with each given equation. Draw the graph of the equation.

23. $x^2 + y^2 = 16$ (0, 0); 4
24. $x^2 + y^2 = 4$ (0, 0); 2
25. $x^2 + y^2 = 10$ (0, 0); $\sqrt{10}$
26. $(x - 3)^2 + y^2 = 9$ (3, 0); 3
27. $(x - 3)^2 + (y + 4)^2 = 16$ (3, −4); 4
28. $(x + 2)^2 + (y - 6)^2 = 64$ (−2, 6); 8

Find the coordinates of the center and the length of a radius of a circle with each given equation.

B

29. $x^2 + y^2 = 18$ (0, 0), $3\sqrt{2}$
30. $x^2 + y^2 + 6x = 7$ (−3, 0); 4
31. $x^2 + y^2 + 2x - 10y = 1$ (−1, 5); $3\sqrt{}$
32. $x^2 + y^2 - 12y = 4$ (0, 6), $2\sqrt{10}$
33. $x^2 + y^2 - 16x - 6y = -72$ (8, 3); 1
34. $x^2 + y^2 + 18x - 20y = -177$ (−9, 10), 2

Write an equation of each circle with center Q and passing through the origin. Draw the graph of the equation.

35. $Q(4, 3)$ $(x - 4)^2 + (y - 3)^2 = 25$
36. $Q(-1, 2)$ $(x + 1)^2 + (y - 2)^2 = 5$
37. $Q(-2, -\sqrt{5})$ $(x + 2)^2 + (y + \sqrt{5})^2 = 9$

Write an equation of each circle with R and S the endpoints of a diameter. Draw the graph of the equation.

38. $R(-4, -1)$, $S(4, 5)$ $x^2 + (y - 2)^2 = 25$
39. $R(-2, -5)$, $S(6, -1)$ $(x - 2)^2 + (y + 3)^2 = 20$
40. $R(1, 2)$, $S(-3, -4)$ $(x + 1)^2 + (y + 1)^2 = 13$

Show that the two circles with the given equations are congruent.

C

41. $x^2 + y^2 + 4x = 32$; $x^2 + y^2 - 2x + 6y = 26$ For each circle, $r = 6$.
42. $x^2 + y^2 - 8y = 1$; $x^2 + y^2 + 10x = -8$ For each circle, $r = \sqrt{17}$.

Show that the two circles with the given equations are concentric.

43. $(x - 3)^2 + (y + 2)^2 = 10$; $x^2 + y^2 - 6x + 4y = 52$ For each circle, coordinates of center are (3, −2).
44. $x^2 + (y - 8)^2 = 7$; $x^2 + y^2 - 16y = 46$ For each circle, coordinates of center are (0, 8).

Find the coordinates of all common points of the two circles with the given equations.

45. $(x - 4)^2 + (y + 5)^2 = 16$; $(x - 4)^2 + (y - 2)^2 = 9$ (4, −1)
46. $x^2 + y^2 = 10$; $(x + 3)^2 + (y + 3)^2 = 4$ (−1, −3), (−3, −1)

The equation of a circle is given. Write an equation of a line which is tangent to the circle at the given point P.

47. $x^2 + y^2 + 4x - 6y = 12$; $P(-6, 6)$ $-4x + 3y = 42$
48. $x^2 + y^2 - 10x + 8y = -28$; $P(7, -8)$ $-x + 2y = -23$

COORDINATES IN SPACE (optional) 13.10

Objectives

To graph ordered triples in a three-dimensional coordinate system
To find the distance between points in a three-dimensional coordinate system

Suppose you take a coordinate plane and add a third axis, called the z-axis, perpendicular to the plane at the origin. You now have a three dimensional coordinate system. On it you can locate points in space by means of **ordered triples** of numbers.

Ordered triple:
$Q(3, -2, 1)$

x-coord.	y-coord.	z-coord.
forward 3	left 2	up 1

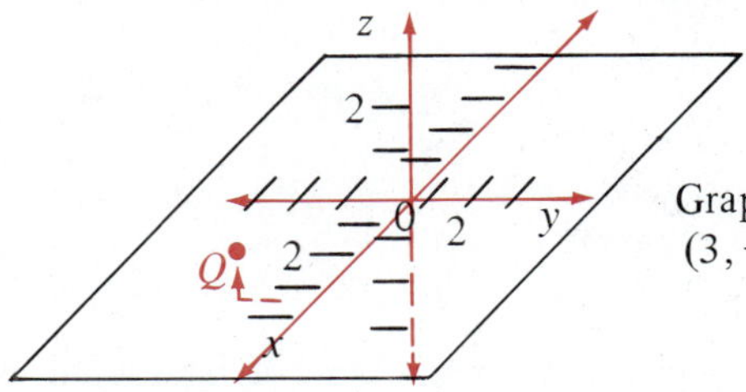

Graph of Q
$(3, -2, 1)$

The chart below can serve as a guide for graphing ordered triples.

GRAPHING ORDERED TRIPLES					
x-coordinate		**y-coordinate**		**z-coordinate**	
Positive	Negative	Positive	Negative	Positive	Negative
Forward	Back	Right	Left	Up	Down

Example 1 Graph $P(3, 2, -1)$.

x-coord.	y-coord.	z-coord.
forward 3	right 2	down 1

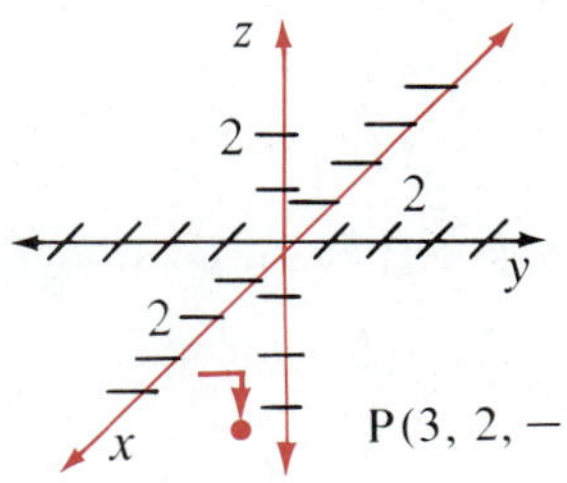

$P(3, 2, -1)$

The next theorem gives a formula for the distance between two points in a three-dimensional coordinate system.

Theorem 13.11 If d is the distance between points $A(x_1, y_1, z_1)$ and $B(x_2, y_2, z_2)$, then
$d = \sqrt{(x_2 - x_1)^2 + (y_2 - y_1)^2 + (z_2 - z_1)^2}$.

(See Exercise 30.)

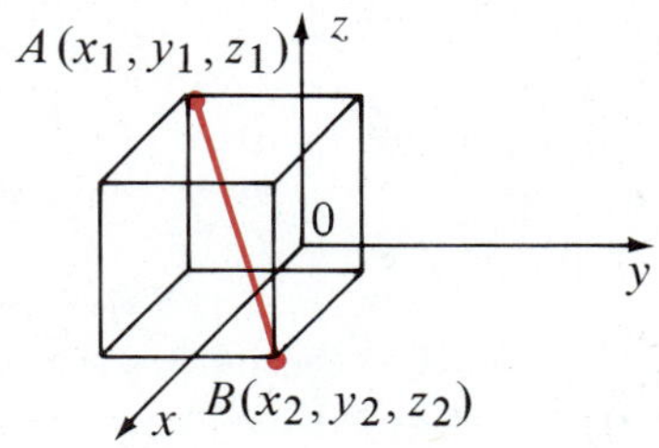

Example 2

Find the distance between $P(-2, -8, 5)$ and $Q(3, -4, 2)$.

$$P(-2, -8, 5) \qquad Q(3, -4, 2)$$

$$\begin{matrix} \uparrow & \uparrow & \uparrow & \qquad & \uparrow & \uparrow & \uparrow \\ x_2 & y_2 & z_2 & & x_1 & y_1 & z_1 \end{matrix}$$

$$d = \sqrt{(x_2 - x_1)^2 + (y_2 - y_1)^2 + (z_2 - z_1)^2}$$

$$PQ = \sqrt{(-2 - 3)^2 + (-8 + 4)^2 + (5 - 2)^2}$$

$$= \sqrt{(-5)^2 + (-4)^2 + 3^2}$$

$$= \sqrt{25 + 16 + 9}$$

$$= \sqrt{50}, \text{ or } 5\sqrt{2}$$

Thus, $PQ = 5\sqrt{2}$.

Written Exercises

Graph each ordered triple on a three-dimensional coordinate system.

A

1. $A(3, 1, 2)$
2. $B(4, -2, 1)$
3. $C(-2, 1, 5)$
4. $D(4, -1, -5)$
5. $E(-3, -2, 1)$
6. $F(-1, 0, 5)$
7. $G(0, -3, 4)$
8. $H(-3, 0, -5)$
9. $J(-1, 0, 0)$
10. $K(0, -4, 0)$

On which axis does each of the following points lie?

11. $P(5, 0, 0)$ **x**
12. $Q(0, -3, 0)$ **y**
13. $R(0, 0, -6)$ **z**

Which pair of axes determines the plane on which each of the following points lies?

B

14. $S(0, 1, -2)$ **y and z**
15. $T(4, -3, 0)$ **x and y**
16. $U(-5, 0, -1)$ **x and z**

Find the distance between each pair of points. (Ex. 17–25)

17. $A(3, 5, 2), B(6, -2, 8)$ $\sqrt{94}$
18. $C(6, 2, 8), D(-3, 4, -2)$ $\sqrt{185}$
19. $E(-7, -3, 0), F(4, -1, 6)$ $\sqrt{161}$
20. $G(6, 2, -9), H(1, -1, -2)$ $\sqrt{83}$
21. $J(-5, 1, 0), K(7, -6, 3)$ $\sqrt{202}$
22. $L(0, 8, 0), M(-5, -2, 1)$ $3\sqrt{14}$
23. $P(4, -6, 5), Q(0, 3, 1)$ $\sqrt{113}$
24. $R(0, -6, 5), S(3, -8, 4)$ $\sqrt{14}$
25. $T(0, -2, -1), U(-3, -7, 0)$ $\sqrt{35}$
26. Derive a formula for the midpoint of $\overline{PQ}$ for $P(x_1, y_1, z_1)$ and $Q(x_2, y_2, z_2)$. **midpoint** $= \left(\frac{x_1 + x_2}{2}, \frac{y_1 + y_2}{2}, \frac{z_1 + z_2}{2}\right)$

Find the coordinates of the midpoint of $\overline{RS}$. (Ex. 27–29)

27. $R(1, -3, 6), S(5, 7, -2)$ **(3, 2, 2)**
28. $R(-3, 6, 0), S(5, -2, 6)$ **(1, 2, 3)**
29. $R(4, 0, -1), S(5, -7, 3)$ $\left(\frac{9}{2}, -\frac{7}{2}, 1\right)$
30. Prove Theorem 13.11.

CUMULATIVE REVIEW

1. In $\triangle ABC$, $AB = 9$, $BC = 8$, and $AC = 13$. Find the perimeter of the triangle formed by joining the midpoints of the three sides of $\triangle ABC$. **15**
2. If each angle of a regular polygon measures 120°, how many sides does the polygon have? **6 sides**

APPLICATIONS

Read → Plan → Solve → Interpret

In some cities, the streets are laid out in a rectangular grid. Those streets running north–south are called avenues; those running east–west are called streets.

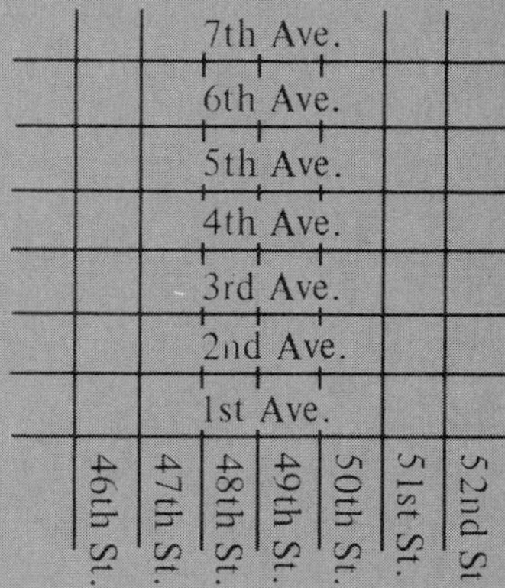

1. If you could not take any shortcuts through the blocks, how many blocks would you walk to go from the corner of 5th Avenue at 47th Street to the corner of 3rd Avenue at 51st Street?
2. If you could walk straight through the blocks, how far would you walk?

A carpenter needs to know the "pitch" or "slope" of a roof. This is usually expressed as the number of inches of vertical rise in 12 in. of horizontal run. A slope of 4 in 12 means a vertical rise of 4 in. for every 12 in. of horizontal run.

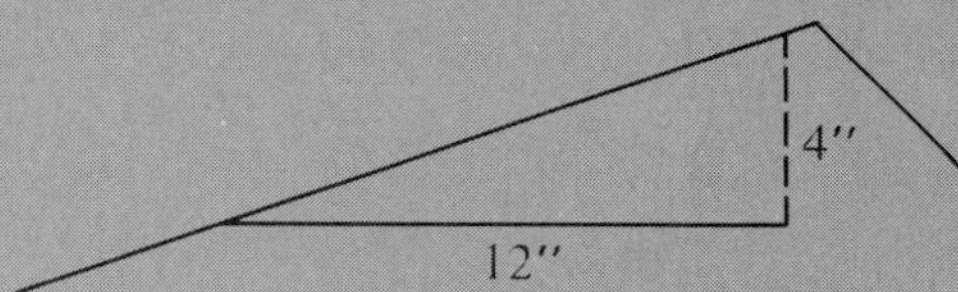

3. How many inches would it rise in a run of 15 ft?
4. If a roof rises 2 ft in a run of 16 ft, what is the slope of the roof?

CHAPTER THIRTEEN REVIEW

On the same coordinate system, graph and label each ordered pair.

1. $P(3, 6)$ 2. $Q(-2, 1)$ [13.1]
3. $R(-4, -3)$ 4. $S(0, -2)$

Is $\overleftrightarrow{AB}$ a vertical line, a horizontal line, or neither?
vert   horiz
5. $A(-3, 1)$, $B(-3, 2)$ 6. $A(1, 5)$, $B(2, 5)$

Find RS. $\sqrt{17}$ $\sqrt{37}$ [13.2]
7. $R(2, 5)$, $S(3, 1)$ 8. $R(4, -6)$, $S(5, 0)$

Find the coordinates of the midpoint of $\overline{AB}$. [13.3] **(2, 4)**   **(−2, 3)**

9. $A(6, 5)$, $B(-2, 3)$ 10. $A(-1, 2)$, $B(-3, 4)$

M is the midpoint of $\overline{AB}$. Find the coordinates of B. (Ex. 11–12) **(−13, 2)**   **(1, −15)**

11. $M(-5, 1)$, $A(3, 0)$ 12. $M(-2, -4)$, $A(-5, 7)$

*13. For $\triangle ABC$ with $A(-3, -2)$, $B(5, 0)$, and $C(1, 8)$, M is the midpoint of $\overline{AB}$ and N is the midpoint of $\overline{BC}$. Show that $MN = \frac{1}{2}(AC)$.

Find the slope of $\overleftrightarrow{RS}$ in each case. Then determine its position in a coordinate plane. (Ex. 14–15) [13.4] **6/5; up to rt**   **−3; down to rt.**

14. $R(4, 3)$, $S(-1, -3)$ 15. $R(-1, 6)$, $S(2, -3)$

16. Find a so that $\overleftrightarrow{CD}$ will have slope 2 for $C(-2, -1)$, $D(2, a)$. **7**

Vocabulary
x-coordinate, y-coordinate [13.1]
x-axis, y-axis [13.1]
origin [13.1]
slope [13.4]
horizontal line, vertical line [13.4]
y-intercept [13.5]

Graph each equation. [13.5]

17. $2x - 3y = 6$ 18. $x + 4y = -2$
19. $-2x = y + 8$ 20. $y = -x + 1$

Draw a line with slope m and passing through point P. Then write an equation for the line. [13.6]

21. $m = 2$; $P(1, 3)$ **21. $y = 2x + 1$**   22. $m = -4$, $P(-2, 1)$ **22. $y = -4x - 7$**

23. $m = \frac{1}{3}$; $P(4, -5)$ **23. $y = \frac{1}{3}x - 6\frac{1}{3}$**   24. $m = -\frac{3}{2}$, $P(-1, 0)$ **24. $y = -\frac{3}{2}x - 1\frac{1}{2}$**

Write an equation in standard form for the line through P and Q.

25. $P(2, 1)$, $Q(-1, -3)$ **$-4x + 3y = -5$**   26. $P(-4, 3)$, $Q(2, 0)$ **$x + 2y = 2$**

Write an equation of each line with slope m and y-intercept b.

27. $m = 3$, $b = -5$ **$y = 3x - 5$**   28. $m = -\frac{1}{2}$, $b = 3$ **$y = -\frac{1}{2}x + 3$**

For each equation, identify the slope m and the y-intercept b. Then draw the graph of the equation. (Ex. 29–30)

29. $y = 6x - 4$ **$m = 6$; $b = -4$**   30. $y = -x$ **$m = -1$; $b = 0$**

31. Give the slope of a line parallel to $\overleftrightarrow{AB}$ for $A(-3, 4)$, $B(2, -1)$. **−1** [13.7]

32. Give the slope of a line perpendicular to $\overleftrightarrow{CD}$ for $C(4, -6)$, $D(-2, 3)$. **$\frac{2}{3}$**

33. Show that the diagonals of $ABCD$ are perpendicular for $A(2, -1)$, $B(8, 0)$, $C(7, 6)$, and $D(1, 5)$. **$\frac{7}{5} \cdot \frac{-5}{7}$ or −1** [13.8]

*34. Use coordinate geometry to prove that the diagonals of a rhombus are perpendicular.

35. Write an equation of the circle with center $A(5, -2)$ and radius of length 3. [13.9] **$(x - 5)^2 + (y + 2)^2 = 9$**

36. Find the coordinates of the center and the length of a radius of a circle with the equation $x^2 + y^2 + 4x - 6y = 30$. **(−2, 3); $\sqrt{43}$**

A exercises are 1–12, 14, 15, 17–24, 27–32, 35.
B exercises are 16, 25, 26, 33, 36.
C exercises are starred.

CHAPTER THIRTEEN TEST

On the same coordinate system, graph and label each ordered pair.

1. $E(2, 3)$
2. $F(-4, -1)$
3. $G(0, -6)$
4. $H(-5, 4)$

Is $\overleftrightarrow{AB}$ a vertical line, a horizontal line, or neither? **neither**

5. $A(7, 0)$, $B(-2, 0)$ **horiz**
6. $A(-2, 3)$, $B(3, -2)$

7. Give the coordinates of D so that $ABCD$ is a rectangle for $A(-2, -4)$, $B(5, -4)$, $C(5, 1)$, and $D(?, ?)$. **(−2, 1)**

Find PQ.

8. $P(-3, 4)$, $Q(2, -7)$ **$\sqrt{146}$**
9. $P(6, 0)$, $Q(-4, 1)$ **$\sqrt{101}$**

10. Find the coordinates of the midpoint of $\overline{RS}$ for $R(3, -6)$, $S(-1, 4)$. **(1, −1)**

11. M is the midpoint of $\overline{AB}$. Find the coordinates of B for $M(4, 2)$ and $A(-1, 5)$. **(9, −1)**

*12. For quadrilateral $ABCD$ with $A(-2, 5)$, $B(6, 5)$, $C(4, -1)$, and $D(-6, 1)$, M, N, P, and Q are midpoints of $\overline{AB}$, $\overline{BC}$, $\overline{CD}$, and $\overline{DA}$, respectively. Show that $MN = PQ$.

Find the slope of $\overleftrightarrow{CD}$ in each case. Then determine its position in a coordinate plane.

13. $C(-2, -4)$, $D(5, 3)$ **1; up to the rt.**
14. $C(0, 6)$, $D(0, -3)$ **vert.**

15. Find b so that $\overleftrightarrow{PQ}$ will have slope -1 for $P(2, 3)$, $Q(-4, b)$. **9**

Graph each equation.

16. $x + 3y = -9$
17. $5x - 2y = 8$

18. Draw a line with slope $\frac{1}{2}$ and passing through point $P(-3, 4)$. Then write an equation for the line. **$y - 4 = \frac{1}{2}(x + 3)$**

19. Write an equation in standard form for the line through $P(-2, -1)$ and $Q(-4, 5)$. **$3x + y = -7$**

Write an equation of each line with slope m and y-intercept b.

20. $m = 4$, $b = -2$ **$y = 4x - 2$**
21. $m = -\frac{2}{3}$, $b = 1$ **$y = -\frac{2}{3}x + 1$**

For each equation, identify the slope m and the y-intercept b. Then draw the graph of the equation.

22. $y = -3x + 2$ **$m = -3$; $b = 2$**
23. $y = x - 3$ **$m = 1$; $b = -3$**

24. Give the slope of a line parallel to $\overleftrightarrow{PQ}$ for $P(2, 1)$, $Q(-3, 4)$. **$-\frac{3}{5}$**

25. Give the slope of a line perpendicular to $\overleftrightarrow{RS}$ for $R(3, 7)$, $S(-4, 2)$. **$-\frac{7}{5}$**

*26. Use coordinate geometry to prove that the opposite sides of a parallelogram have the same length.

27. Write an equation of the circle with center $A(-4, 1)$ and radius of length 2. **$(x + 4)^2 + (y - 1)^2 = 4$**

28. Find the coordinates of the center and the length of a radius of a circle with the equation $x^2 + y^2 - 10x + 2y = 39$. **(5, −1); $\sqrt{65}$**

A exercises are 1–6, 8–11, 13, 14, 16–18, 20–25, 27.
B exercises are 7, 15, 19, 28. **C exercises are starred.**

COMPUTER ACTIVITIES

Equations of a Circle

Objective: **To change one form of the equation of a circle to its alternate form, using a computer program**

There are two common forms of the equation of a circle:

$$(1)\ x^2 + y^2 + ax + bx = c, \text{ and } (2)\ (x - h)^2 + (y - k)^2 = r^2$$

There are various reasons why one or the other of these equations might be desired at a particular time. Therefore, one should be able to use these forms interchangeably. To convert these forms, you simply expand the binomials and combine terms in one case and complete the squares in the other.

The computer program below converts form 1 on the left to form 2 on the right, analyzing resulting values in the process.

```
110  PRINT "IF YOUR EQUATION IS IN THE FORM "
120  PRINT "X^2+Y^2+AX+BY=C, THEN GIVE THE VALUES OF "
130  PRINT "A, B, AND C.": INPUT A,B,C
140  PRINT "THE CENTER OF THE CIRCLE IS (": - A / 2;
     ",": - B / 2;")"
150  PRINT " AND THE RADIUS IS ": SQR (C + A ^ 2 / 4
      + B ^ 2 / 4)"."
155  PRINT
160  PRINT "THE EQUATION COULD BE WRITTEN THUS"
170  PRINT "(X+": - A / 2;")^2 + (Y +"; - B / 2;")^2
      = ";C + A ^ 2 / 4 + B ^ 2 / 4
180  PRINT : GOTO 110
```

EXERCISES: See the Computer Section beginning on page 565 for more information.

1. Write a program that converts form 2 of the equation of a circle to form 1 of the equation.
2. Write a program that produces the equation of a circle from the coordinates of the endpoints of a diameter.
3. Write a program that calculates the length of a segment in a three-dimensional coordinate system.

COLLEGE PREP TEST

Directions: Choose the one best answer to each question or problem.

1. A line segment has one endpoint at (3, −2) and its midpoint at (2, −5). What are the coordinates of the other endpoint? A

(A) (1, −8) **(B)** (5, −7) **(C)** (1, −3)
(D) (1, 8) **(E)** (−1, 8)

2. What is the length of a line segment joining the points whose coordinates are (−2, −7) and (6, 8)? E

(A) 4 **(B)** 5 **(C)** $7\frac{1}{2}$ **(D)** $8\frac{1}{2}$ **(E)** 17

3. A circle whose center is the point (−2, 6) is tangent to the x-axis. The coordinates of the point of tangency are B

(A) (0, 6) **(B)** (−2, 0) **(C)** (0, −2)
(D) (6, 0) **(E)** (−2, −2)

4. Which point lies at the greatest distance from the origin? D

(A) (0, −9) **(B)** (−2, 9) **(C)** (−7, −6)
(D) (8, 5) **(E)** (0, 0)

5. $\triangle AOB$ and $\triangle PCB$ are isosceles right triangles with equal area. What are the coordinates of point P? E

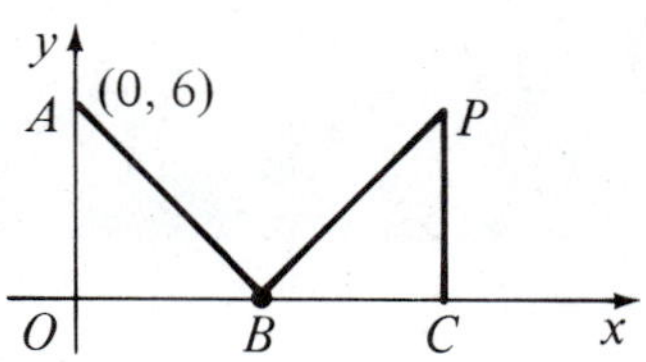

(A) (6, 0) **(B)** (6, 12) **(C)** (12, 0)
(D) (0, 12) **(E)** (12, 6)

6. An equation of the line parallel to the x-axis and passing through point (4, 2) is B

(A) $x = -4$ **(B)** $y = 2$ **(C)** $x = 2$
(D) $x = 4$ **(E)** $y = 0$

7. The endpoints of the diameter of a circle are M and N. The coordinates of M and N are (−1, 3) and (5, 5). What are the coordinates of the center of the circle? A

(A) (2, 4) **(B)** (3, 4) **(C)** (4, 8)
(D) (4, 4) **(E)** (0, 0)

8. The coordinates of A are (4, 0) and of C are (15, 0). Find the area of $\triangle ABC$ if the equation of $\overleftrightarrow{AB}$ is $y = 2x - 8$. B

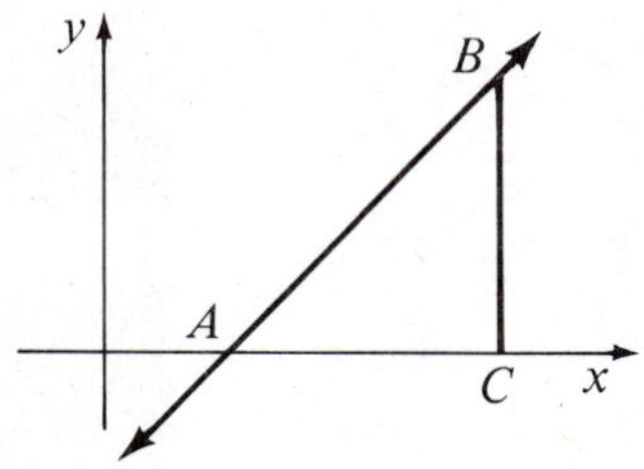

(A) 100 **(B)** 121 **(C)** 144 **(D)** 169
(E) 132

9. What is the equation of the line $\overleftrightarrow{AB}$? A

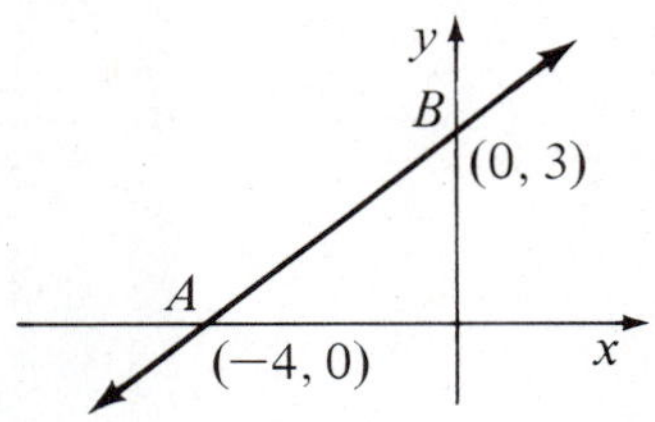

(A) $y = \frac{3}{4}x + 3$

(B) $y = \frac{3}{4}x - 4$

(C) $y = \frac{4}{3}x + 3$

(D) $y = -\frac{4}{3}x + 3$ **(E)** $y = -\frac{3}{4}x + 3$

10. The vertices of rectangle $ABCD$ are the points $A(0, 0)$, $B(8, 0)$, $C(8, k)$, $D(0, 5)$. The value of k is B

(A) 4 **(B)** 5 **(C)** 6 **(D)** 3 **(E)** 2

11. The area of a circle whose center is at (0, 0) is 36π. This circle passes through all of the following points except D

(A) (−6, 0) **(B)** (0, 6) **(C)** (6, 0)
(D) (6, 6) **(E)** (0, −6)

14 GEOMETRIC LOCI

CAREERS

Computer Graphics Artists

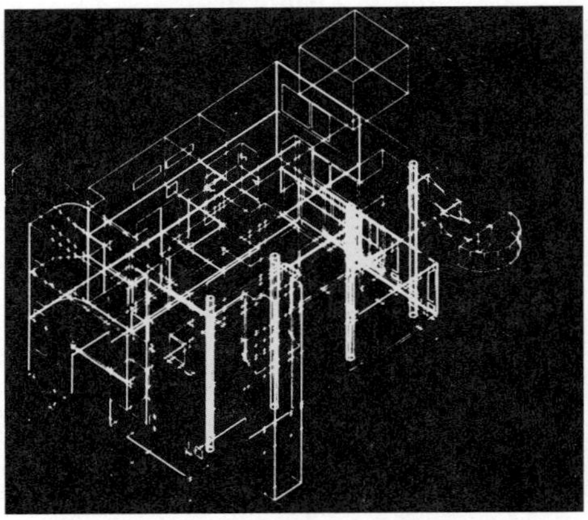

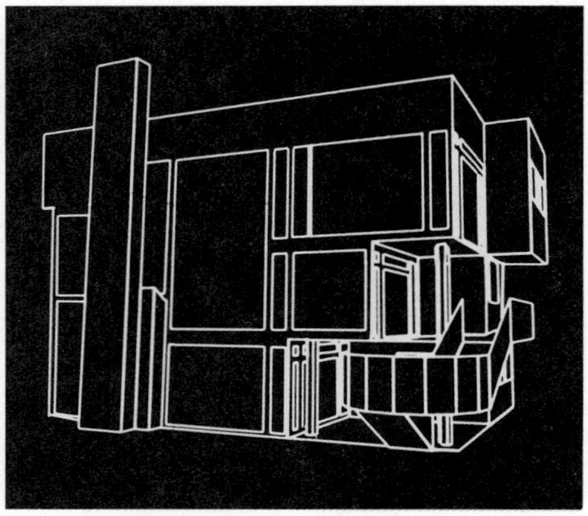

A computer graphics artist must know how to use geometry in design. An artist who specializes in this area must be able to create and manipulate three-dimensional images as a means of developing computer-aided designs. Special skills in drafting, motion transformations, data representation, animation, and programming are needed.

Each of the above displays shows the computer-generated images of a house as rendered by a computer graphics system. The display (top left) shows all the planes of a house. The display (top right) shows some of the exterior planes of the house. The lower display shows the geometric model of the house in its final setting.

LOCUS OF POINTS 14.1

Objective **To sketch and describe geometric figures that satisfy given conditions**

Recall that a geometric figure is a set of points. In this chapter you will be concerned with locating and describing sets of points which satisfy given conditions. You can assume that all points lie in the same plane, unless otherwise stated.

Example 1 Locate all points that are 3 cm from a given point Q. Then describe, as completely as possible, the figure formed.

Start with a given point Q.

Locate several points 3 cm from Q.

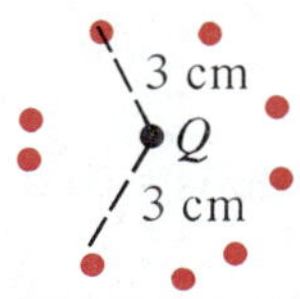

Draw a smooth dashed curve through the points.

The figure formed is a circle with center Q and a 3-cm radius.

In Example 1, the circle formed is called a **locus.**

Definition: Locus A *locus* is a geometric figure containing all points, and only those points, that satisfy given conditions. [plural: loci (lō′sī)]

The word *locus* is a Latin word that means "place" or "location." There are several steps involved in determining a locus.

Summary **To determine a locus**

1. Start with the given figure(s);
2. Locate several points that satisfy the given conditions;
3. Draw a smooth dashed curve through the points;
4. Describe the locus completely by naming the geometric figure formed and indicating its location.

Example 1 leads to the following postulate.

Postulate 27 The locus of points at a given distance d from a given point P is a circle with center P and with d the length of a radius.

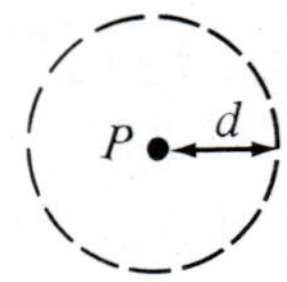

It is possible to locate points that are a given distance from a given line.

Example 2 Sketch and describe the locus of points 2 cm from a given line l.

Start with a given line l.

Locate several points 2 cm from line l.

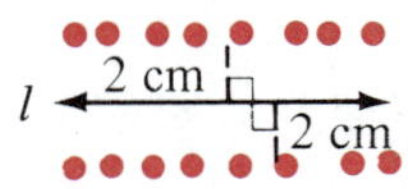

Draw dashed lines through the points.

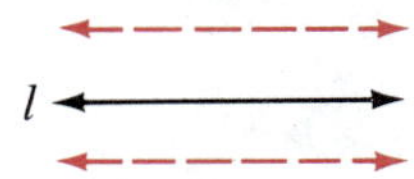

The locus is a pair of lines each parallel to l and 2 cm from it.

The next postulate is based on Example 2.

Postulate 28 The locus of points a given distance d from a given line l is a pair of lines each parallel to l and at the distance d from l.

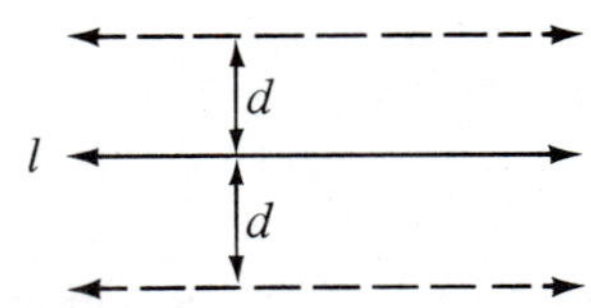

Example 3 presents a different situation from that of Postulate 28. In Example 3, you are given two parallel lines and asked to find the locus of all points that are the same distance from each of the lines.

Example 3 Sketch and describe the locus of points equidistant from two given parallel lines r and s.

Start with two given parallel lines, r and s.

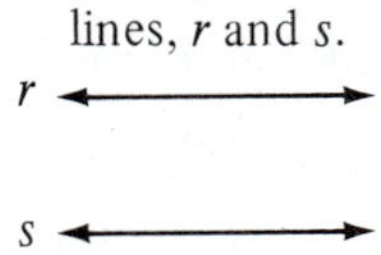

Locate several points the same distance from r as from s.

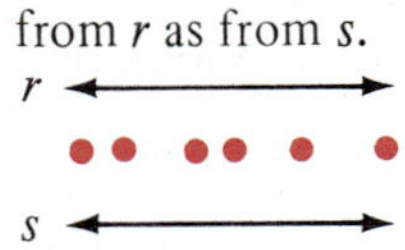

Draw a dashed line through the points.

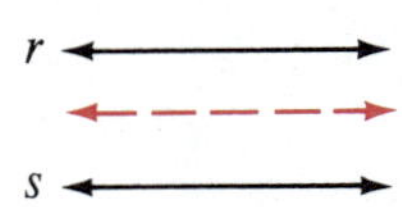

The locus is a line parallel to r and s and midway between them.

Example 3 suggests the next postulate.

Postulate 29

The locus of points equidistant from two given parallel lines is a line parallel to each of the given lines and midway between them.

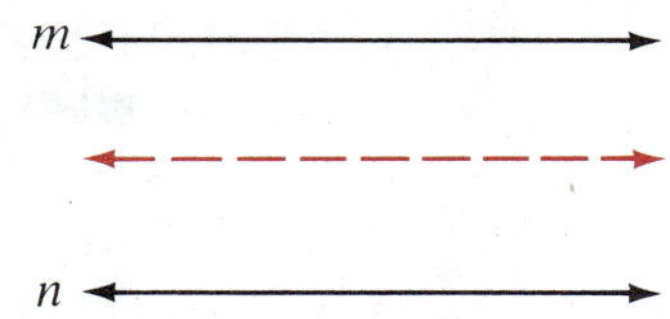

Written Exercises

Sketch and describe each locus.

A 1. The locus of points equidistant from two given parallel lines that are 7 cm apart **See Post. 29.**

2. The locus of points equidistant from points C and D that are 4 cm apart **⊥ bis of $\overline{CD}$**

3. The locus of points 8 mm from a given line **See Post. 28.**

4. The locus of points 5 cm from a given point **See Post. 27.**

5. The locus of points equidistant from two given points R and S **⊥ bis of $\overline{RS}$**

6. The locus of points 6 mm from a given line **See Post. 28.**

7. The locus of points 2.5 cm from a given point G **See Post. 27.**

8. The locus of points equidistant from the sides of a given angle **bis. of given ∠**

9. int. of ⊙ with given pt. as center and 2-cm radius

10. ⊙ with given pt. as center and 30-mm radius and the int. of that ⊙

B 9. The locus of points less than 2 cm from a given point

10. The locus of points less than or equal to 30 mm from a given point

11. The locus of points greater than 4 cm from a given point

12. The locus of points greater than or equal to 2.5 cm from a given point

11. ext. of ⊙ with given pt. as center and 4-cm radius

12. ⊙ with given pt. as center and 2.5-cm radius and the ext. of that ⊙

The *locus of a moving point* is the set of all points traced by the point as it moves according to given conditions. Sketch and describe each locus.

C 13. The locus of the center of a given circle that rolls along a given line and is always tangent to the line

14. The locus of the center of a circle that rolls between two parallel lines and is always tangent to both lines

15. The locus of the center of a given circle that rolls around the outside of a second given circle and is always tangent to it

16. The locus of the center of a given circle that rolls around the inside of a larger given circle and is always tangent to it

13. line ∥ given line such that the dist. of the line = radius of given ⊙

14. a line ∥ the given l's and midway between them

15. ⊙ with same center as stationary ⊙ and radius = $r_1 + r_2$

16. ⊙ with same center as stationary ⊙ and radius = $r_1 - r_2$

CUMULATIVE REVIEW

1. The area of a triangle is 52 cm^2. One side measures 13 cm. Find the length of the corresponding altitude. **8 cm**

2. The bases of a trapezoid measure 5 cm and 8 cm, respectively, and an altitude measures 10 cm. Find the area. **65 cm^2**

LOCUS THEOREMS 14.2

Objectives **To state the postulate or theorem that applies to a locus**
To state theorems about loci and identify what is to be proved

Consider two given points C and D. Observe the result when you sketch the locus of points equidistant from C and D.

Start with two given points C and D.

Locate several points the same distance from C as from D.

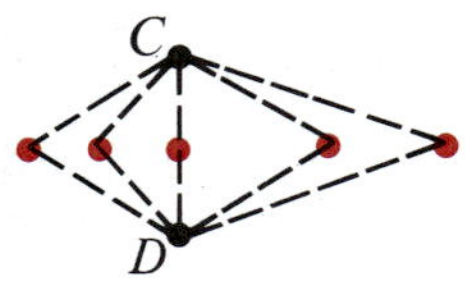

Draw a smooth dashed line through the points.

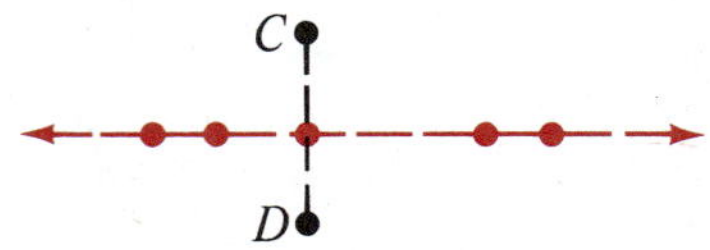

Conclusion: The locus is the perpendicular bisector of $\overline{CD}$.

It is possible to prove the conclusion stated above as a theorem. Recall that a locus of points consists of *all points* and *only those points* that satisfy the given conditions. Thus, there are two parts to a locus theorem that must be proved.

PART I Prove that every point on the locus satisfies the conditions.
PART II Prove that every point that satisfies the conditions lies on the locus.

The conclusion determined above is stated as Theorem 14.1.

Theorem 14.1 The locus of points equidistant from two given points is the perpendicular bisector of the segment joining the two points.

PART I
If a point lies on the perpendicular bisector of a segment, then it is equidistant from the endpoints of the segment.

Given:
P lies on the $\perp$ bisector of $\overline{AB}$.
Prove:
$PA = PB$

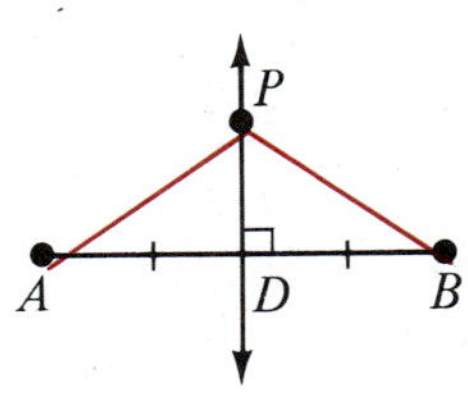

PART II
If a point is equidistant from the endpoints of a segment, then it lies on the perpendicular bisector of the segment.

Given:
$PA = PB$
Prove:
P lies on the $\perp$ bisector of $\overline{AB}$.

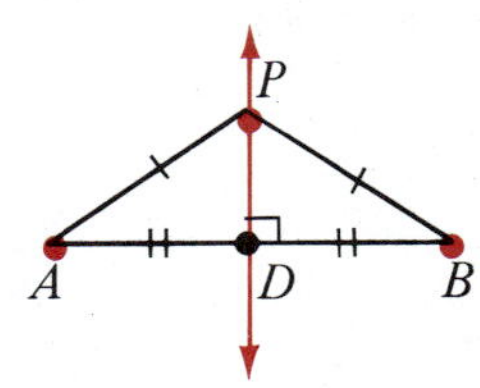

(See Exercise 15.)

Recall that in Exercise 8 on page 469, you found that the locus of points equidistant from the sides of an angle is the bisector of the angle. This conclusion is stated in Theorem 14.2.

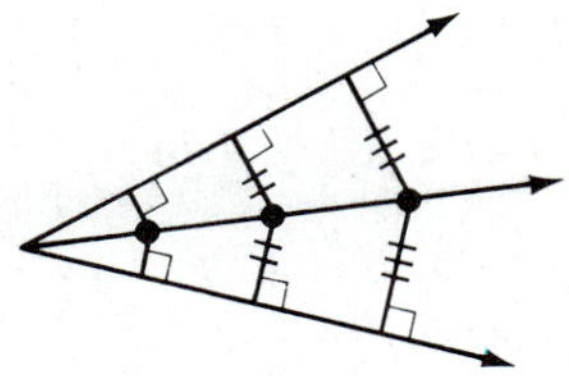

Theorem 14.2 The locus of points equidistant from the sides of an angle is the angle bisector.

PART I If a point is on the bisector of an angle, then it is equidistant from the sides of the angle.

Given: $\overrightarrow{BE}$ bisects $\angle ABC$, P is on $\overrightarrow{BE}$, $\overline{PF} \perp \overrightarrow{BC}$, and $\overline{PG} \perp \overrightarrow{BA}$.
Prove: $PF = PG$

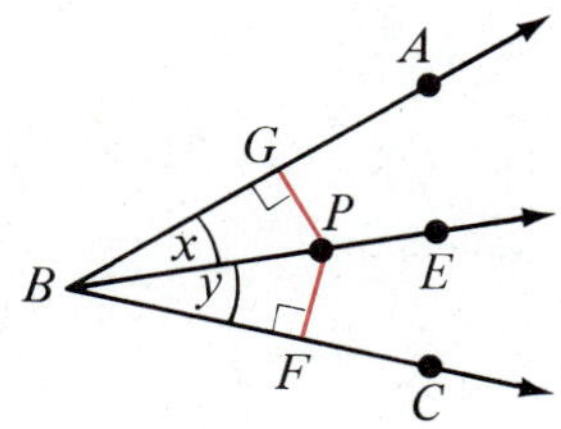

Proof

Statements	Reasons
1. $\overrightarrow{BE}$ bisects $\angle ABC$, $\overline{PF} \perp \overrightarrow{BC}$, $\overline{PG} \perp \overrightarrow{BA}$.	1. Given
2. $\angle PFB \cong \angle PGB$	2. $\perp$'s form $\cong$ rt. $\angle$'s.
3. $\angle x \cong \angle y$	3. Def. of $\angle$ bisector
4. $\overline{BP} \cong \overline{BP}$	4. Reflexive property
5. $\triangle PFB \cong \triangle PGB$	5. AAS
6. $\therefore PF = PG$	6. CPCTC

PART II If a point is equidistant from the sides of an angle, then it lies on the bisector of the angle.

Given: $\angle ABC$, with $\overline{PF} \perp \overrightarrow{BC}$, $\overline{PG} \perp \overrightarrow{BA}$, and $PF = PG$
Prove: $\overrightarrow{BP}$ is the bisector of $\angle ABC$. (P lies on $\overrightarrow{BP}$.)

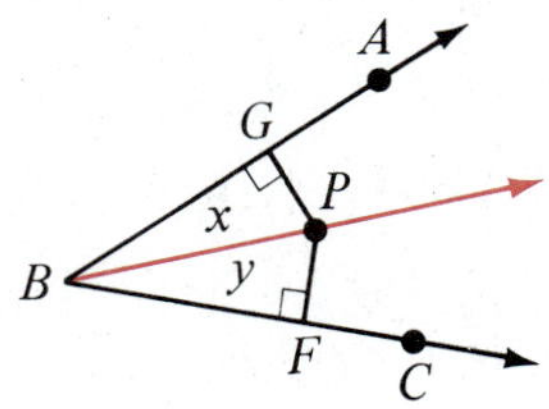

Proof

Statements	Reasons
1. $\angle ABC$, with $\overline{PF} \perp \overrightarrow{BC}$, $\overline{PG} \perp \overrightarrow{BA}$, and $PF = PG$	1. Given
2. $\triangle PFB$ and $\triangle PGB$ are rt. $\triangle$'s.	2. Def. of $\perp$ and def. of rt. $\triangle$
3. $\overline{PF} \cong \overline{PG}$	3. Def. of $\cong$ segments
4. $\overline{BP} \cong \overline{BP}$	4. Reflexive property
5. $\triangle PFB \cong \triangle PGB$	5. HL for $\cong$ rt. $\triangle$'s
6. $\angle x \cong \angle y$	6. CPCTC
7. $\therefore \overrightarrow{BP}$ is the bisector of $\angle ABC$.	7. Def. of $\angle$ bisector

So far, three locus postulates and two locus theorems have been presented. The locus problem given in Example 1 is an application of one of the postulates.

Example 1 Sketch and describe the locus of the midpoints of all radii of a given circle. Then give the postulate that applies.

Start with a given circle O.

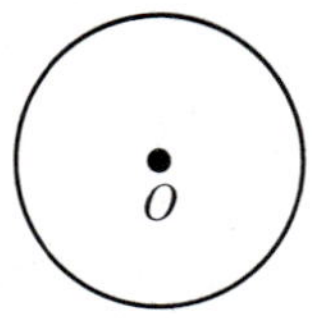

Locate the midpoints of several radii.

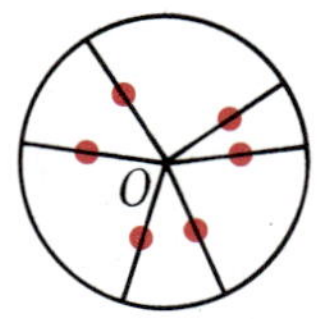

Draw a smooth dashed curve through the midpoints.

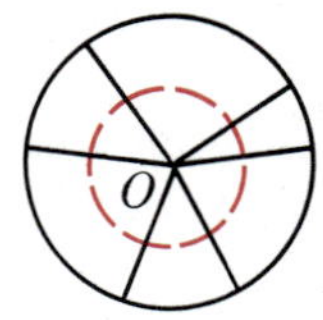

The locus is a circle concentric to the given circle and with a radius one half the length of the radius of the given circle. Since every point on the locus is the same distance from point O, the locus is an application of Postulate 27.

The locus problem given in the next example is an application of one of the locus theorems.

Example 2 Sketch and describe the locus of the centers of all circles that pass through two given points. Then give the theorem that applies.

Start with two given points G and H.

Draw several circles through G and H. Mark their centers.

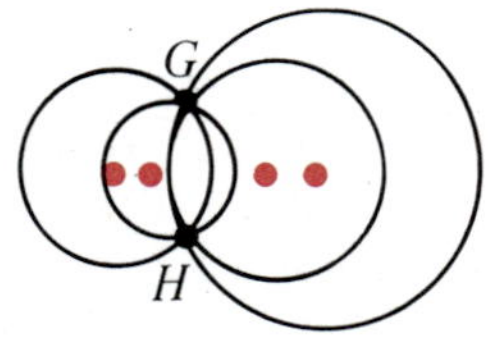

Draw a smooth dashed line through the centers.

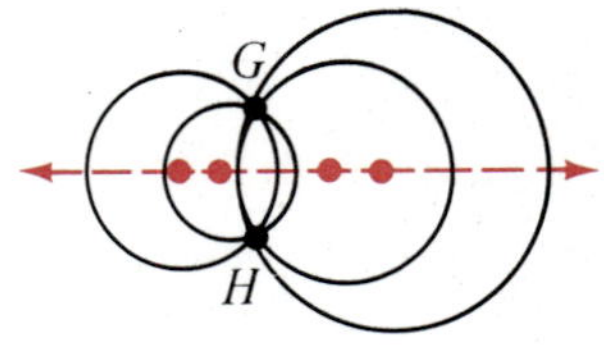

The locus is the perpendicular bisector of $\overline{GH}$. Since every point on the locus is equidistant from G and H, the locus is an application of Theorem 14.1.

Written Exercises

Sketch and describe each locus. Then give the postulate or theorem that applies.

A

1. The locus of the centers of all circles that are tangent to both sides of a given angle **Th. 14.2**
2. The locus of the centers of all circles that are tangent to each of two parallel lines **Post. 29**
3. The locus of the centers of all circles of the same size that are tangent to a given line **Post. 28**
4. The locus of points equidistant from two intersecting lines **Th. 14.2**
5. The locus of the midpoints of all chords parallel to a given chord of a circle **Th. 14.1**
6. The locus of the midpoints of all chords of the same length in a given circle **Post. 27**
7. The locus of points within a circle that are equidistant from the endpoints of a given chord **Th. 14.1**
8. For all triangles with a given base and a given altitude, the locus of the vertices opposite that base **Post. 28**

Sketch and describe each locus. Then write the locus statement as a theorem. Then write Part I and Part II of the theorem.

9. ⊙ concentric to given ⊙ and radius = half the given radius

B

9. The locus of the midpoints of all radii of a given circle
10. The locus of points equidistant from two concentric circles
11. The locus of the centers of all circles tangent to a given line at a given point
12. The locus of the midpoints of all chords drawn from a given point of a circle

11. a line ⊥ given line at the pt. of tangency

12. ⊙ tangent to given ⊙ at the given pt. and radius half the radius of given ⊙

Prove each locus statement as a theorem.

C

13. The locus of the centers of all circles tangent to a given line at a given point is a line perpendicular to the given line at the given point.
14. The locus of the vertex of the right angle of a right triangle with a given hypotenuse is a circle with the hypotenuse as a diameter.
15. Prove Theorem 14.1.

10. ⊙ concentric to given ⊙'s and radius = $\frac{r_1 + r_2}{2}$

ALGEBRA REVIEW

Objective **To solve quadratic equations by factoring**

Example Solve. $6x^2 + x - 2 = 0$

Factor. ► $(3x + 2)(2x - 1) = 0$

Set each factor equal to 0 and solve. ► $3x + 2 = 0$ or $2x - 1 = 0$

$3x = -2 \qquad 2x = 1$

$x = -\frac{2}{3} \qquad x = \frac{1}{2}$

Thus, the solutions are $-\frac{2}{3}$ and $\frac{1}{2}$.

Solve.

1. $x^2 + 3x - 28 = 0$ **−7, 4**
2. $x^2 + 5x + 6 = 0$ **−3, −2**
3. $x^2 - 8x + 15 = 0$ **5, 3**
4. $x^2 - 9x + 14 = 0$ **2, 7**
5. $x^2 + 10x + 25 = 0$ **−5**
6. $x^2 - 49 = 0$ **−7, 7**
7. $2x^2 - 11x + 5 = 0$ **$\frac{1}{2}$, 5**
8. $3x^2 + 8x - 3 = 0$ **−3, $\frac{1}{3}$**
9. $2x^2 + 3x - 9 = 0$ **−3, $\frac{3}{2}$**
10. $x^2 = -5x + 24$ **−8, 3**
11. $24 - 10x = -x^2$ **4, 6**
12. $x^2 + 35 = -12x$ **−7, −5**

NON-ROUTINE PROBLEMS

The figures formed by a plane intersecting the surface of a cone are called **conic sections**.

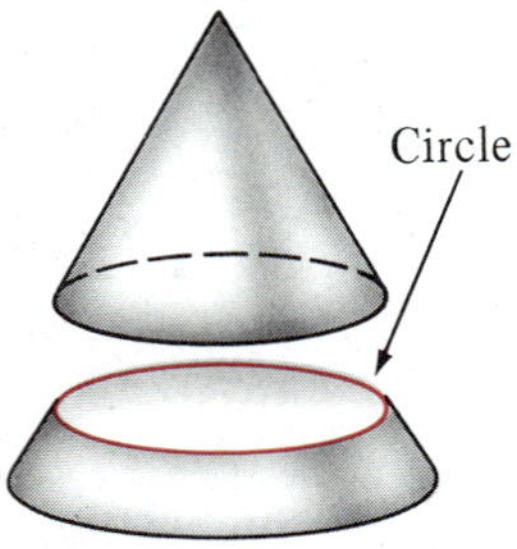

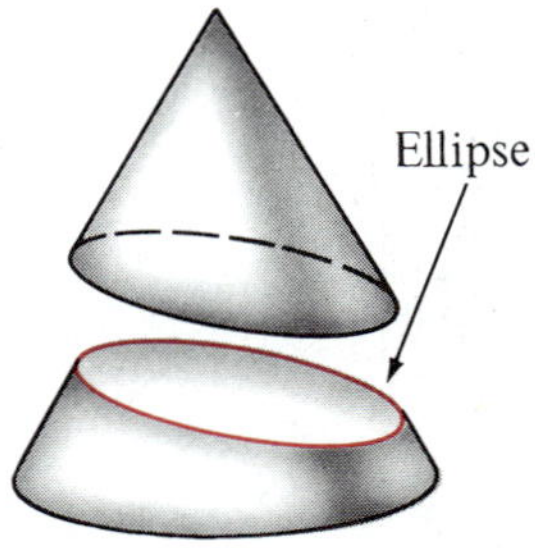

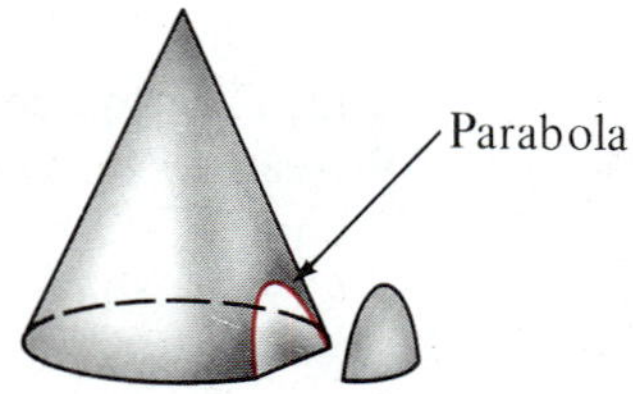

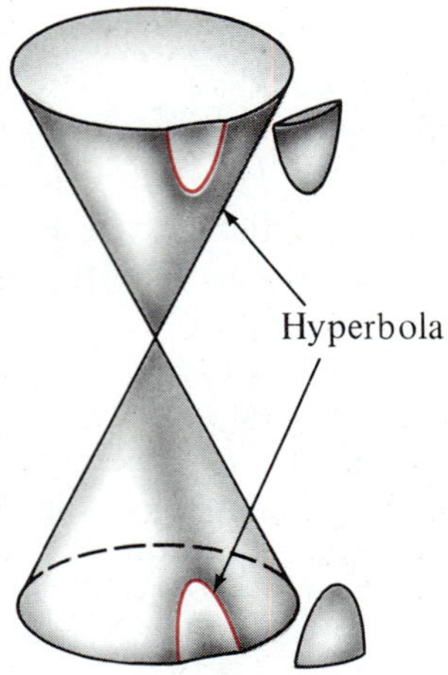

You have already seen that a circle can be interpreted as a locus of points. On page 482, you will see that each of the other three conic sections can also be defined as a locus.

Describe the way in which a cone is cut by a plane to form each of the following conic sections.

1. A circle **pl. $\parallel$ base**
2. An ellipse **pl $\nparallel$ base**
3. A parabola **pl. $\parallel$ generator of cone**
4. A hyperbola **pl. $\parallel$ axis of cone**

ALGEBRA REVIEW

Objective **To solve quadratic equations by completing the square**

Example Solve. $x^2 - 8x - 20 = 0$

Add 20 to each side. ► $x^2 - 8x \quad = 20$

Take one half of −8, square it, and add the result to each side. ► $x^2 - 8x + 16 = 20 + 16$

$(x - 4)^2 = 36$

$x^2 - 8x + 16$ is a perfect square trinomial.

$x - 4 = \sqrt{36}$ or $x - 4 = -\sqrt{36}$

$x - 4 = 6 \qquad x - 4 = -6$

$x = 10 \qquad x = -2$

Thus, the solutions are 10 and −2.

Solve.

1. $x^2 - 8x + 12 = 0$ **2, 6**	**2.** $x^2 + 4x - 5 = 0$ **1, −5**	**3.** $y^2 - 6y + 8 = 0$ **2, 4**
4. $z^2 + 2z - 3 = 0$ **1, −3**	**5.** $x^2 + 2x - 24 = 0$ **4, −6**	**6.** $y^2 + 4y + 3 = 0$ **−3, −1**
7. $x^2 - 6x + 5 = 0$ **1, 5**	**8.** $x^2 - 2x - 35 = 0$ **7, −5**	**9.** $z^2 - 14z + 45 = 0$ **5, 9**
10. $x^2 - 10x = 24$ **12, −2**	**11.** $12y = -y^2 - 32$ **−4, −8**	**12.** $-16z - 36 = -z^2$ **18, −2**

SPECIFIC CONDITIONS FOR LOCI 14.3

Objective **To sketch and describe loci determined by more than one specific condition**

The distance between two objects is generally defined to be the length of the shortest path between the objects. This is also true for the distance from a point to a circle.

Definition: Distance from a point to a circle

The *distance from point P to circle O* is measured along $\overleftrightarrow{PO}$ and is the length of the shortest segment joining P and circle O.

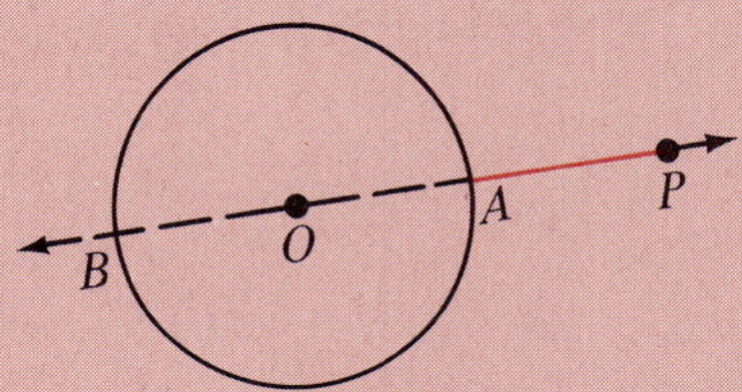

P in exterior of $\odot O$.
PA is the distance.

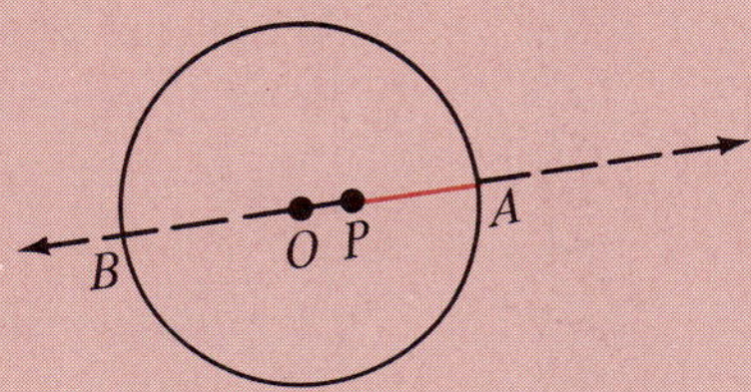

P in interior of $\odot O$.
PA is the distance.

In each figure above, $\overline{PB}$ lies on $\overleftrightarrow{PO}$, but PA is the distance since $PA < PB$.

In each of the next three examples, you will find the locus of points at a given distance from a circle with an 8-cm radius. Notice that changing the given distance changes the resulting locus.

Example 1 Sketch and describe the locus of points 10 cm from a circle with an 8-cm radius.

1. Start with a circle with an 8-cm radius.
2. Locate several points 10 cm from the circle. The points lie in the exterior of the circle.
3. Draw a smooth dashed curve through the points.
4. Describe the locus:
   The locus is a circle concentric to the given circle and with an 18-cm radius.

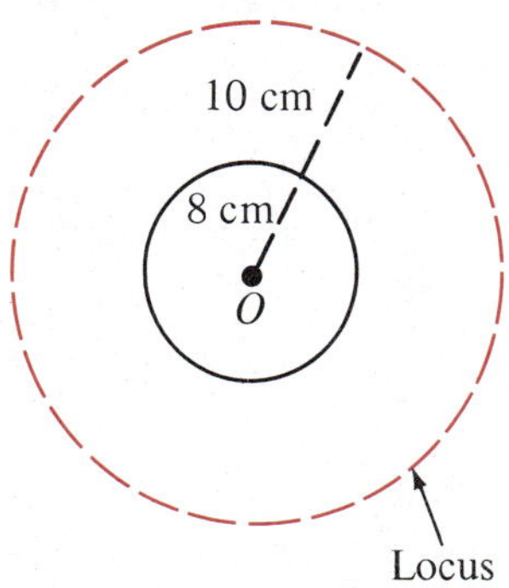

In Example 1, the given distance is greater than the length of a radius. In Examples 2 and 3, notice what happens when the given distance is equal to or less than the length of a radius.

Example 2

Sketch and describe the locus of points 8 cm from a circle with an 8-cm radius.

Follow steps similar to those in Example 1.

Note that the center is also 8 cm from the circle. ►

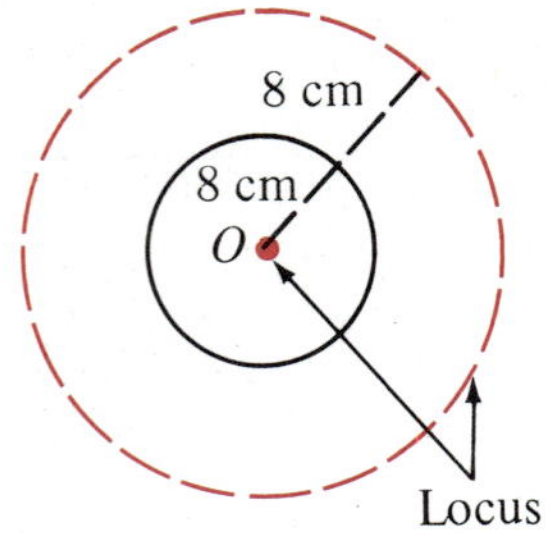

The locus is a circle concentric to the given circle and with a 16-cm radius, plus the center of the given circle.

Example 3

Sketch and describe the locus of points 5 cm from a circle with an 8-cm radius.

The points of the locus form an outer and an inner circle. ►

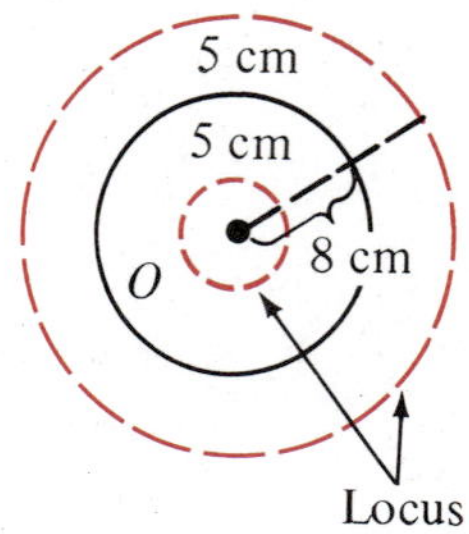

The locus is two circles concentric to the given circle, one with a 13-cm radius and one with a 3-cm radius.

In the next three examples, you will look for the intersection of two distinct loci.

Example 4

Sketch and describe the locus of points 6 cm from line *m* and 8 cm from point *P* on line *m*.

Use Post. 28 for the locus 6 cm from line m. ►
Use Post. 27 for the locus 8 cm from point P. ►
Look for the intersection of the parallel lines and the circle. ►

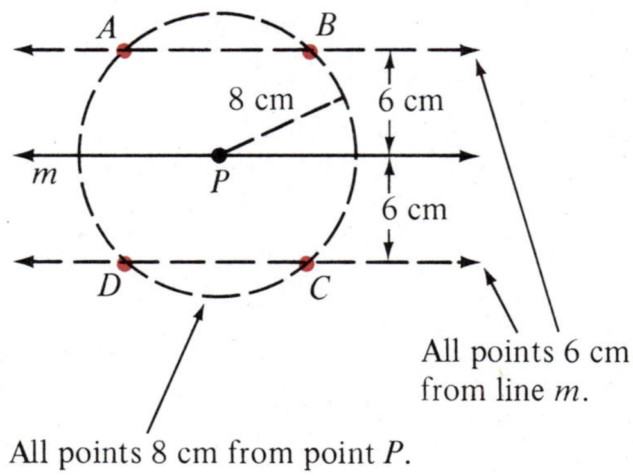

A, *B*, *C*, and *D* are the only points 6 cm from *m* and 8 cm from *P*. **Thus**, the locus is the set of points *A*, *B*, *C*, and *D*.

Example 5

Points A and B are 3 cm apart. Sketch and describe the locus of points 2 cm from A and 2 cm from B.

Use Post. 27 for each locus. ►
Look for the intersection of the two circles. ►

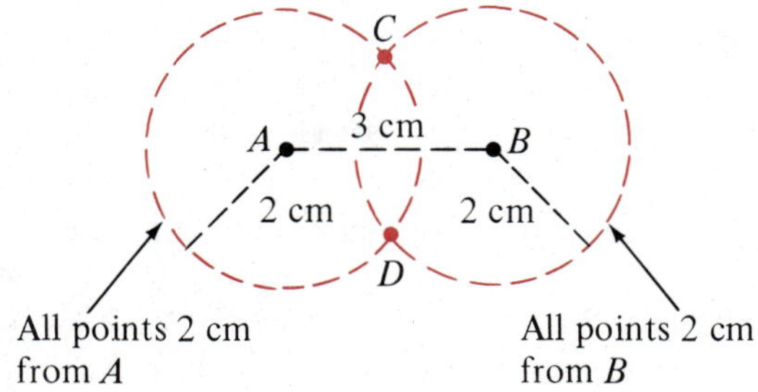

C and D are the only points 2 cm from A and 2 cm from B.
Thus, the locus is the set of points C and D.

Two loci may not intersect at all, as shown in the next example. In such a case, there are no points that meet all of the given conditions.

Example 6

Point Q is 4 cm below line l. Sketch and describe the locus of points 1 cm from Q and 2 cm from l.

Use Post. 27 and Post. 28. Sketch each locus. ►

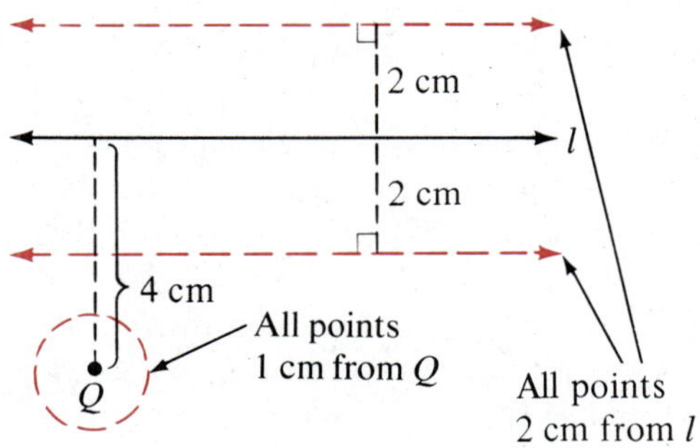

The circle and the parallel lines do not intersect.

Thus, there are no points in the locus.

Written Exercises

1. ⊙concentric to ⊙Q and a 9-cm radius

Sketch and describe each locus.

A

1. The locus of points 6 cm from circle Q
2. The locus of points 3 cm from circle Q
3. The locus of points 1 cm from circle Q

2. ⊙concentric to ⊙Q with radius of 6 cm, and the center of given ⊙

3. two ⊙'s concentric to ⊙Q, one with 2-cm radius, other with 4-cm radius

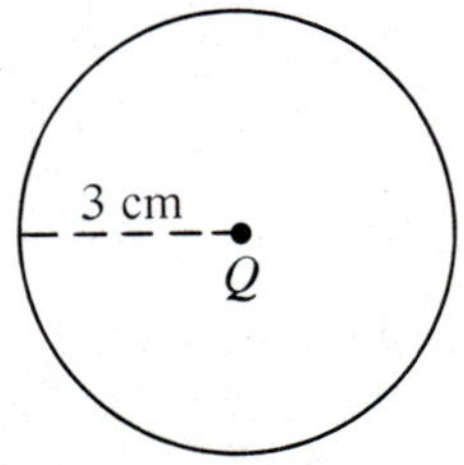

4. The two pt. of inters. of ⊙A, radius 3 cm, and ⊙B, radius 4 cm

4. The locus of points 3 cm from A and 4 cm from B

5. The locus of points 1 cm from A and 5 cm from B

6. The locus of points 1.5 cm from A and 4 cm from B no points

5. The pt. of inters. of ⊙A, radius 1 cm, and ⊙B, radius 5 cm

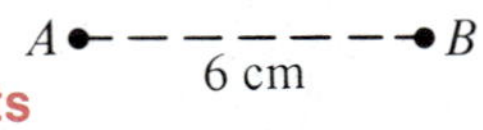

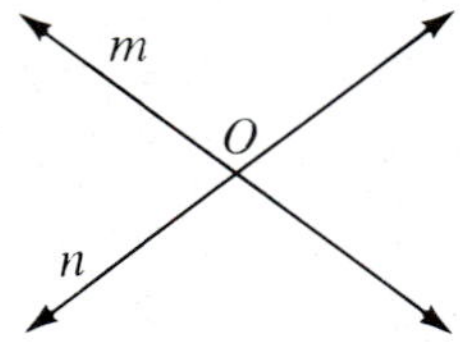

7. The locus of points equidistant from m and n and 2 cm from O

7. 4 pt., one on each ∠ bis. 2 cm from O

8. The 2 pt. of inters. of a line ∥ to both r and s with the ⊙ with P as the center and a 3-cm radius

8. The locus of points equidistant from r and s and 3 cm from P

9. Same as 8 except ⊙ has a 4-cm radius

9. The locus of points equidistant from r and s and 4 cm from P

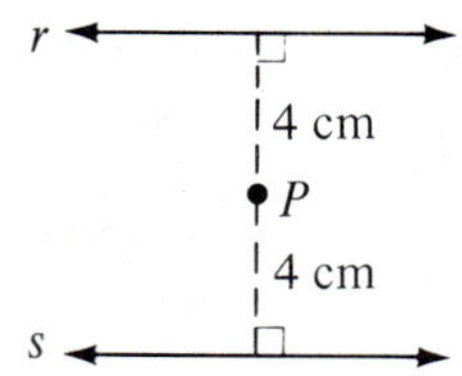

10. The 2 pt. of inters. of the lines 2 cm from l and ∥ l and the ⊙ with center P and 4-cm radius

10. The locus of points 2 cm from l and 4 cm from P

11. The locus of points 2 cm from l and 3 cm from P

11. The pt. of tan. of line 2 cm above l and ⊙ with center P and 3-cm radius

12. The locus of points 2 cm from l and 7 cm from P See below.

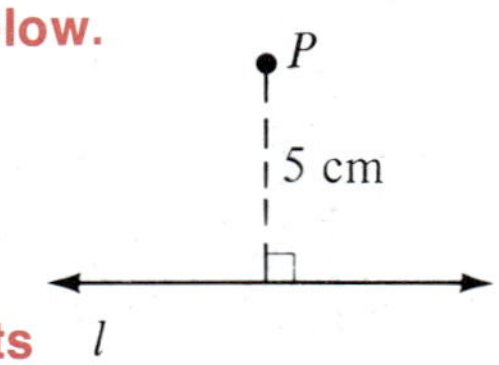

13. The locus of points 2 cm from l and 9 cm from P

The 4 pt. of inters. of lines 2 cm above and below l with ⊙ with center P and 9-cm radius

14. The locus of points 2 cm from l and 2 cm from P no points

B 15. The locus of points 5 cm from a circle with a 3-cm radius See Example 1.

16. The locus of points 6 cm from a circle with a 6-cm radius See Example 2.

17. The locus of points 3 m from a circle with a 7-m radius See Example 3.

18. The locus of points equidistant from two concentric circles See below.

19. The locus of points 3 cm from point A and 2 cm from point B, where A and B are 4 cm apart See Example 5.

20. Repeat Exercise 19, but with A and B 5 cm apart. pt. on $\overline{AB}$ 3 cm from A and 2 cm from B

22. 2 pt.; inters. of ⊙ with center P and 8-cm radius and line ∥ m and n and midway between them

C 21. The locus of points equidistant from intersecting lines r and s, and 3 cm from line l which is 2 cm above r

4 pt.; 2 on ∠ bis. 5 cm above r, 2 on ∠ bis. 1 cm below r

22. The locus of points equidistant from lines m and n and 8 cm from point P, where P is 3 cm above m, $m \parallel n$, and m is 6 cm above n

23. Repeat Exercise 22, but with P 5 cm above m.

pt. of inters. of ⊙ with center P and 8-cm radius and line ∥ m and n and midway between them

24. Repeat Exercise 22, but with P 6 cm above m. no points

18. a concentric ⊙ with radius $\frac{r_1 + r_2}{2}$

12. The 3 pt. of inters. of lines 2 cm above and below l with ⊙ with center P and 7-cm radius

CUMULATIVE REVIEW

1. Each side of an equilateral triangle is 16 cm long. Find the length of an apothem of the triangle. $\frac{8\sqrt{3}}{3}$ cm

2. An apothem of a square is 7 m long. Find the area of the square. 196 m^2

GENERAL CONDITIONS FOR LOCI (optional) 14.4

Objective **To sketch and describe loci determined by two or more general conditions**

In the preceding lesson, you found loci determined by very specific conditions. This lesson deals with more general conditions. In Example 1, you will examine all possible cases of the locus of points at a given distance from a given circle.

Example 1 Sketch and describe the locus of points at a given distance d from a given circle with a radius of length r.

There are three possible cases, depending upon the relationship between d and r.

Case I: $d > r$

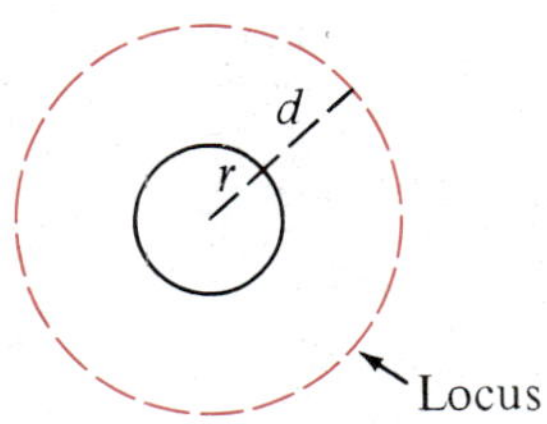

The locus is a circle concentric to the given circle with radius of length $r + d$.

Case II: $d = r$

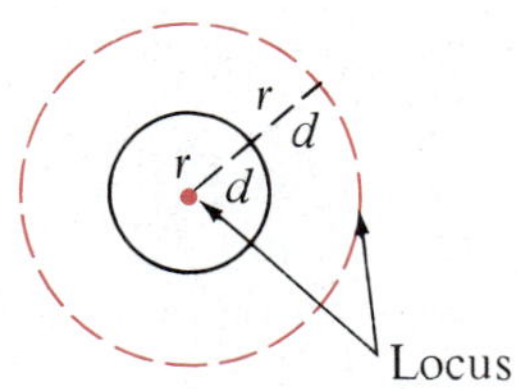

The locus is a circle concentric to the given circle with radius of length $2r$, plus the center of the given circle.

Case III: $d < r$

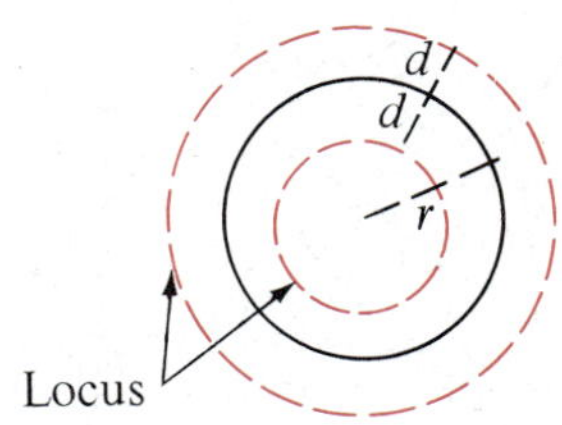

The locus is two circles concentric to the given circle and with radii of lengths $r + d$ and $r - d$, respectively.

Example 2 Sketch and describe the locus of points p cm from line m and q cm from point R on line m.

Case I: $p < q$

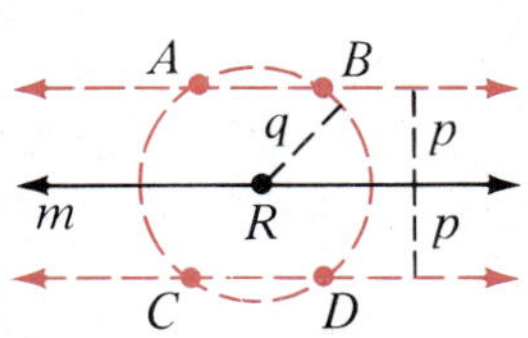

The locus is the set of points A, B, C, and D.

Case II: $p = q$

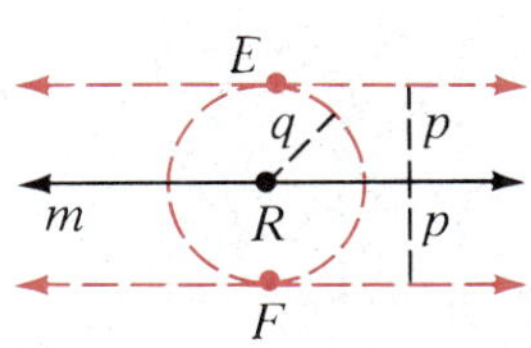

The locus is the set of points E and F.

Case III: $p > q$

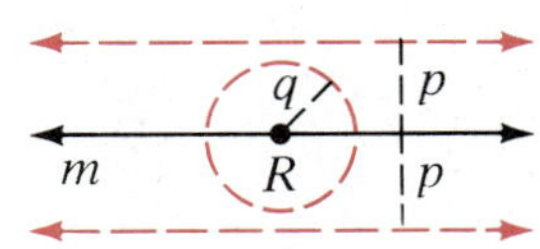

There are no points in the locus.

In Example 2, you found the locus of points at a given distance from a given line and at a given distance from a point on the line. The results varied according to the relationship between the two given distances. The next example also produces three different results depending upon given distances.

Example 3

Points P and Q are d meters apart. Sketch and describe the locus of points r meters from P and s meters from Q.

Case I: $r + s > d$

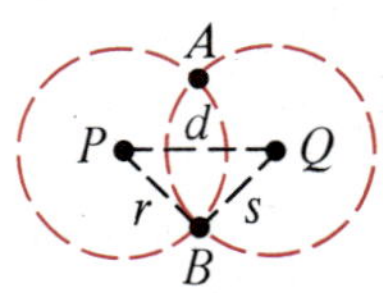

The locus is the set of points A and B.

Case II: $r + s = d$

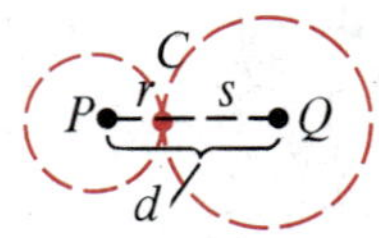

The locus is the point C.

Case III: $r + s < d$

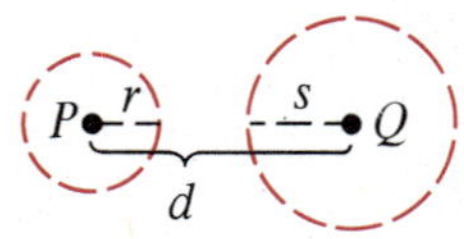

There are no points in the locus.

In Example 3, points P and Q are given to be d meters apart. In the next example, three given points are not restricted in their relative positions. The result is a "free-floating" locus problem.

Example 4

Sketch and describe the locus of points equidistant from two given points and at a given distance from a third given point.

Let A and B be the two given points and P be the third point. Apply Theorem 14.1 and Postulate 27. Then look for all possible intersections of the two loci.

Case I: The circle and the perpendicular bisector intersect in 2 points.

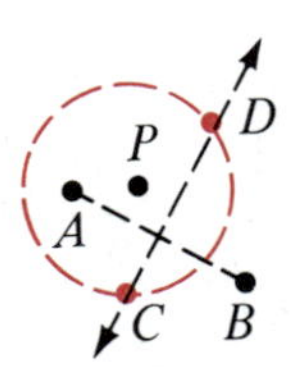

The locus is the set of points C and D.

Case II: The circle and the perpendicular bisector intersect in 1 point.

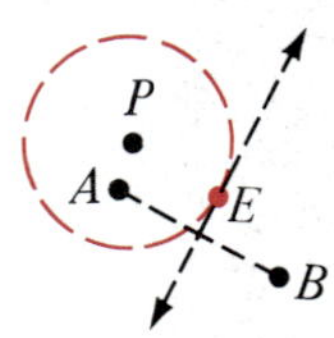

The locus is the point E.

Case III: The circle and the perpendicular bisector do not intersect.

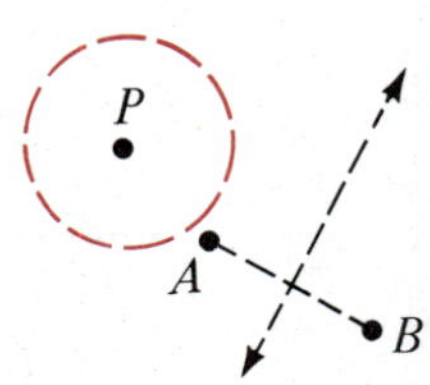

There are no points in the locus.

Oral Exercises

Give all possible intersections of the following figures.

1. Two lines 1 pt., line, no pt.
2. A line and a pair of parallel lines 2 pts, a line, no points
3. A circle and a pair of parallel lines
4. A ray and a circle 1 pt., 2 pt., no pt.
5. Two circles with different radii
6. Two circles with the same radii
7. A line and a pair of concentric circles
8. A line and a pair of intersecting lines

3. 1 pt., 2 pt., 3 pt., 4 pt., no pt.
5. 1 pt., 2 pt., no pt.
7. 1 pt., 2 pt., 3 pt., 4 pt., no pt.
6. 1 pt, 2 pt., a circle, no points
8. 1 pt., 2 pt., a line

Written Exercises

Sketch and describe each locus. Show all possible cases.

A

1. The locus of points equidistant from intersecting lines r and s and y units from line r 4 pt.
2. The locus of points equidistant from the sides of $\angle A$ and d cm from point A pt. on bis. $\angle$A and d cm from A
3. The locus of points equidistant from two parallel lines m and n and d meters from point A on line m, where m and n are f meters apart 2 pt.; 1 pt.; no pt.
4. The locus of points equidistant from intersecting lines p and q and x units from A, the point of intersection of p and q 4 pt.
5. The locus of points r meters from point A and r meters from point B, where A and B are d meters apart See Example 3.
6. The locus of points equidistant from points A and B and h cm from point A, where A and B are d cm apart 2 pt.; 1 pt., no pt.

B

7. The locus of points at a given distance from a given point and equidistant from the sides of a given angle 2 pt., 1 pt., no pt.
8. The locus of points equidistant from two given points and equidistant from two given parallel lines 1 pt.; no pt.; $\perp$ bis.
9. The locus of points equidistant from two given concentric circles and at a given distance from each of two given parallel lines no pt.; 1 pt.; 2 pt.
10. The locus of points at a given distance d_1 from a given point and at a given distance d_2 from a second given point, where $d_1 \neq d_2$ 2 pt.; 1 pt.; no pt.
11. The locus of points equidistant from a pair of intersecting lines and at a given distance from a given point See below.
12. The locus of points equidistant from two given parallel lines and at a given distance from another given line See below.
13. The locus of points equidistant from the sides of a given angle and equidistant from two given parallel lines 1 pt., no pt., ray
14. The locus of points equidistant from a pair of intersecting lines and at a given distance from a given line 2 pt., 3 pt., 4 pt., line and a pt.

C

15. The locus of points equidistant from the sides of a given triangle [*Hint:* Take the sides two at a time.] 1 pt. inters. of $\angle$ bis.
16. The locus of points equidistant from three given points [*Hint:* Take the points two at a time.] See below.
17. The locus of points at the distance of a radius from a given circle and equidistant from two given points 2 pt.; 3 pt.; no pt.; 1 pt.
18. The locus of points at a given distance from a given circle and equidistant from two given lines that intersect at the center of the given circle 4 pt.; 5 pt.; 8 pt.

11. no pt.; 1 pt.; 2 pt.; 3 pt.; 4 pt.
12. 2 pt.; line; no pt.
16. inters. of $\perp$ bis. may be 1 pt. or 0 pt.

NON-ROUTINE PROBLEMS

A circle, an ellipse, a parabola, and a hyperbola are called **conic sections**. (See page 474.) A circle is the locus of points at a given distance from a given point. Each of the other conic sections is defined as a locus below.

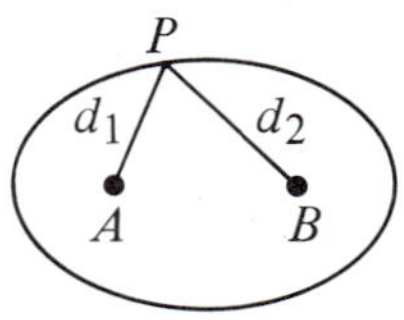

An **ellipse** is the locus of points the sum of whose distances from two given points is a constant. ($d_1 + d_2 =$ constant)

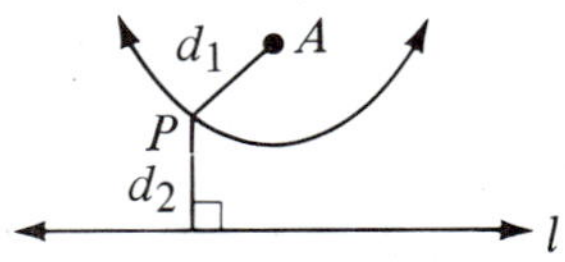

A **parabola** is the locus of points equidistant from a given point and a given line. ($d_1 = d_2$)

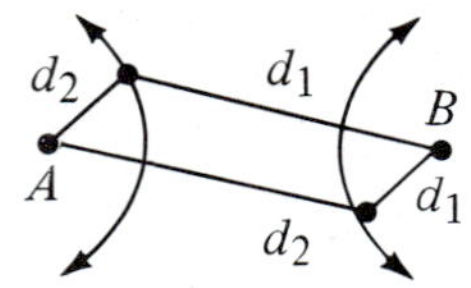

A **hyperbola** is the locus of points the difference of whose distances from two given points is a constant. ($|d_1 - d_2| =$ constant)

A circle can be constructed using a compass. You can also locate any number of points of a parabola or hyperbola by using a compass and straightedge. However, to draw an ellipse you will need a piece of string.

1. Use a compass and straightedge to construct seven points of a parabola and eight points for each branch of a hyperbola.
2. Use a piece of string to draw an ellipse.

ALGEBRA REVIEW

Objective **To solve quadratic equations by the quadratic formula**

Example Solve $2x^2 + 5x - 2 = 0$ by the quadratic formula.

Quadratic formula ► $x = \frac{-b \pm \sqrt{b^2 - 4ac}}{2a}$ $a = 2$ $b = 5$ $c = -2$

$$x = \frac{-5 \pm \sqrt{5^2 - 4(2)(-2)}}{2(2)} = \frac{-5 \pm \sqrt{25 + 16}}{4} = \frac{-5 \pm \sqrt{41}}{4}$$

Solve each equation by the quadratic formula.

1. $x^2 + 2x - 3 = 0$ $-3, 1$
2. $x^2 + x - 6 = 0$ $-3, 2$
3. $x^2 - 12x - 13 = 0$ $13, -1$
4. $2x^2 - 11x + 12 = 0$ $\frac{3}{2}, 4$
5. $3x^2 + 5x - 2 = 0$ $-2, \frac{1}{3}$
6. $5x^2 - 3x - 2 = 0$ $1, -\frac{2}{5}$
7. $x^2 - 6x - 10 = 0$ $3 \pm \sqrt{19}$
8. $2x^2 + 5x - 6 = 0$ $\frac{-5 \pm \sqrt{73}}{4}$
9. $3x^2 - 7x = -3$ $\frac{7 \pm \sqrt{13}}{6}$
10. $-3x^2 = 4x - 2$ $\frac{-2 \mp \sqrt{10}}{3}$
11. $6 - x = 5x^2$ $1, -\frac{6}{5}$
12. $2x - 3x^2 = -5$ $-1, \frac{5}{3}$

CONCURRENT BISECTORS IN A TRIANGLE 14.5

Objective **To prove and apply concurrence theorems about perpendicular bisectors in a triangle**

Three or more lines are **concurrent** if they pass through the same point.

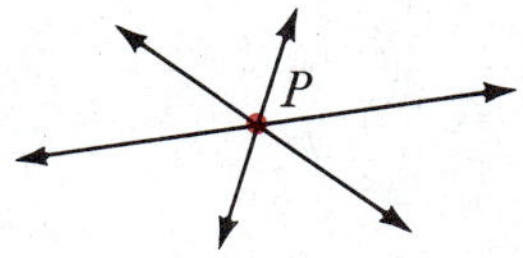

Suppose you were to construct the perpendicular bisectors of all three sides of a triangle. Each perpendicular bisector is the locus of points equidistant from two of the vertices of the triangle. You would find that the three loci appear to be concurrent, as shown in the figures below. This result leads to theorem 14.3.

Acute triangle

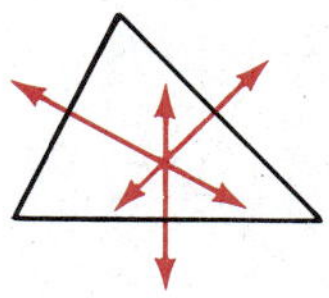

Right triangle

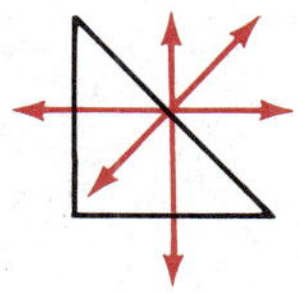

Obtuse triangle

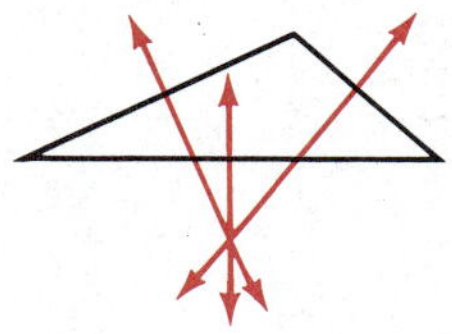

Theorem 14.3 The perpendicular bisectors of the sides of a triangle are concurrent at a point that is equidistant from the vertices of the triangle.

Given: $\triangle ABC$ with l, m, and n the $\perp$ bisectors of $\overline{AB}$, $\overline{AC}$, and $\overline{BC}$, respectively

Prove: l, m, and n are concurrent at a point equidistant from A, B, and C.

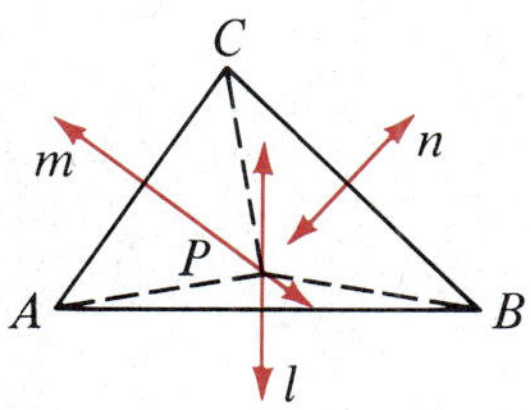

ANALYSIS

Let P be the point where two of the $\perp$ bisectors intersect. Show that the third $\perp$ bisector passes through P.

Proof

Statements	**Reasons**
1. $\triangle ABC$, with l, m, n the $\perp$ bisectors of $\overline{AB}$, $\overline{AC}$, and $\overline{BC}$, respectively	1. Given
2. l and m intersect at some point P.	2. $l \parallel m$ or $l \nparallel m$; l cannot be $\parallel m$ by Th. 3.2.
3. $PB = PA$ and $PA = PC$	3. A point on the $\perp$ bisector of a segment is equidistant from its endpoints.
4. $PB = PC$	4. Transitive property
5. P lies on n.	5. A point equidistant from the endpoints of a segment lies on its $\perp$ bisector.
6. $\therefore$ l, m, and n are concurrent at a point equidistant from A, B, and C.	6. Steps 1–5 and def. of concurrent lines

The construction in Example 1 is an important application of Theorem 14.3.

Example 1

Circumscribe a circle about $\triangle ABC$.

Strategy: Construct the perpendicular bisectors of two of the sides. The point where they intersect is the center of the circumscribed circle.

Start with $\triangle ABC$.

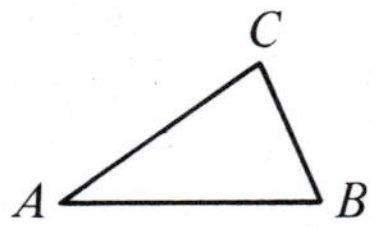

Construct the $\perp$ bisectors of $\overline{AB}$ and of $\overline{AC}$. Let P be the point where they intersect.

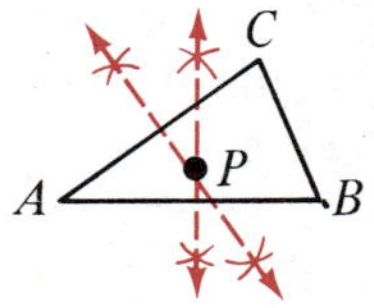

With PA as a radius length, draw circle P.

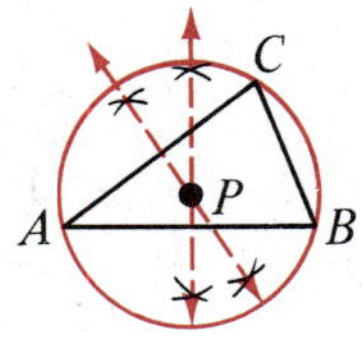

Conclusion: $\odot P$ is the circumscribed circle.

Now suppose you were to construct the bisectors of the three angles of a triangle. As shown below, whether the triangle is acute, right, or obtuse, the angle bisectors appear to be concurrent. This leads to Theorem 14.4.

Acute triangle

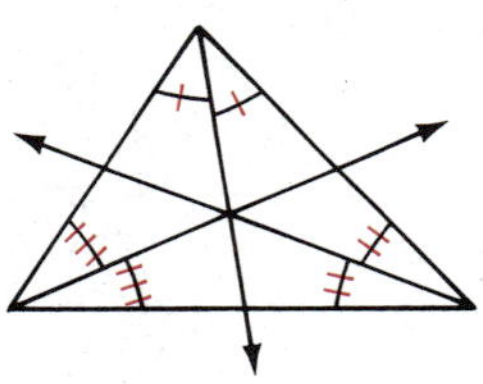

Right triangle

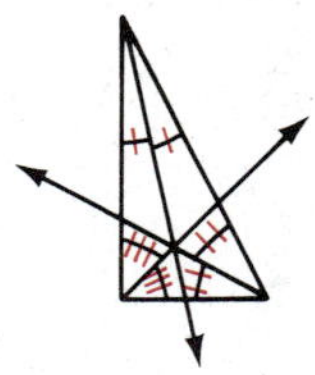

Obtuse triangle

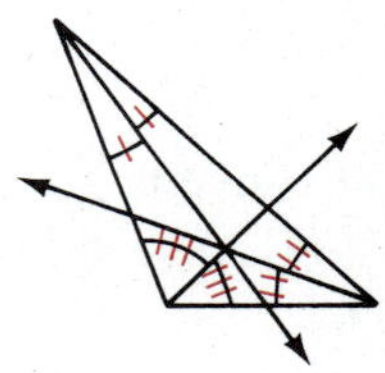

Theorem 14.4

The bisectors of the angles of a triangle are concurrent at a point which is equidistant from the sides of the triangle.

Given: $\triangle ABC$ with rays r, s, and t bisecting $\angle B$, $\angle A$, and $\angle C$, respectively

Prove: r, s, and t are concurrent at a point equidistant from $\overline{AB}$, $\overline{AC}$, and $\overline{BC}$.

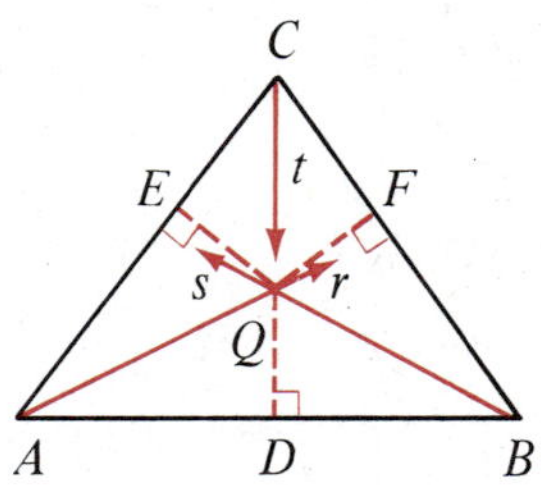

ANALYSIS

Let Q be the point where two of the angle bisectors intersect. Show that the third bisector passes through Q. Use Th. 14.2.

(See Exercise 15.)

Theorem 14.4 is applied in the construction in the next example.

Example 2 Inscribe a circle in $\triangle ABC$.

Strategy: Bisect two angles of the triangle. The point where the bisectors intersect is the center of the inscribed circle.

Start with $\triangle ABC$. Bisect $\angle A$ and $\angle B$. Let Q be the point where the bisectors intersect.

Construct a perpendicular $\overline{QR}$ from Q to $\overline{AB}$.

With QR as a radius length, draw circle Q.

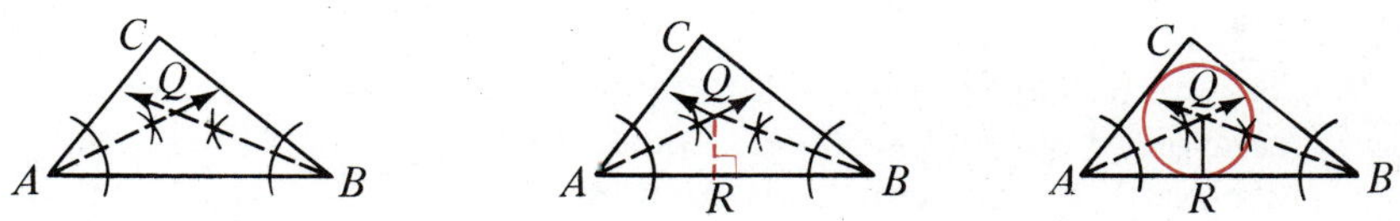

Conclusion: $\odot Q$ is the inscribed circle.

Notice that in Examples 1 and 2, only two bisectors are constructed. In each case, since the bisectors are concurrent, it is not necessary to construct the third bisector.

From Examples 1 and 2, you can conclude that a circle can be circumscribed about any triangle. Also, a circle can be inscribed in any triangle. This leads to the following definitions.

Definitions: Circumcircle and circumcenter

The *circumcircle* of a triangle is the circle that is circumscribed about the triangle. The *circumcenter* of a triangle is the center of the circumcircle.

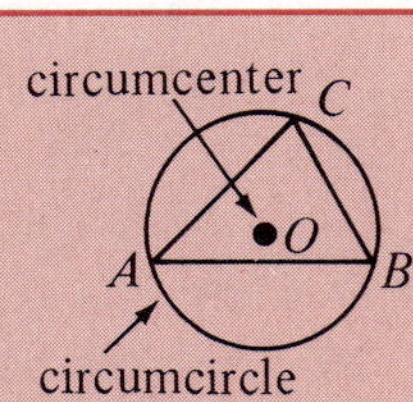

Definitions: Incircle and incenter

The *incircle* of a triangle is the circle that is inscribed in the triangle. The *incenter* of a triangle is the center of the incircle.

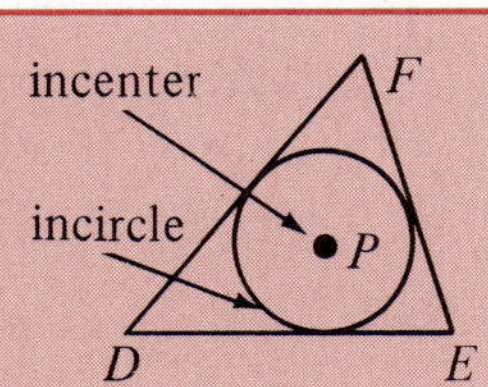

Written Exercises

A

See Constr., pp. 10, 51.

1. Draw a right triangle RST. Construct the perpendicular bisector of each side. Label the point of intersection P.
2. Draw an acute triangle PQR. Construct the bisector of each angle. Label the point of intersection S. See Constr., p. 15.
3. Draw an obtuse triangle EFG. Inscribe a circle in triangle EFG. Label the incenter I. See Example 2.
4. Draw an acute triangle ABC. Circumscribe a circle about triangle ABC. Label the circumcenter C. See Ex 1.
5. Draw a right triangle. Construct a circle that passes through all three vertices. See Ex 1.
6. Draw a triangle PQR. Construct a circle tangent to all three sides. See Ex 2.
7. Construct an equilateral triangle. Construct the incenter and the circumcenter. What appears to be true of the two centers? They coincide.
8. Construct an isosceles triangle. Construct the incenter and the circumcenter. Where do the two centers appear to lie? on $\perp$ bis. of base

B

9. Trace a circle by drawing around a fifty-cent piece or similar object. Locate the center of the circle by construction. [*Hint:* Select three points on the circle.]
   9. pt. of inters. of $\perp$ bis. of 2 chords
10. Consider an acute, a right, and an obtuse triangle. For each triangle, tell whether the circumcenter lies in the interior, on, or in the exterior of the triangle. int.; on; ext.
11. Repeat Exercise 10, but for the incenter instead of the circumcenter. interior in each case
12. Given three noncollinear points, where is the point in their plane equidistant from all three? circumcenter
13. Sketch and describe the locus of points equidistant from the three sides of a triangle. incenter
14. Sketch and describe the locus of points equidistant from three noncollinear points. circumcenter
15. Supply a reason for each step 3–6 in this proof of Theorem 14.4.

Statements	Reasons
1. Rays r, s, and t bisect $\angle B$, $\angle A$, and $\angle C$, respectively.	1. Given
2. Let Q be the point where r and s intersect.	2. $r \parallel s$ or $r \nparallel s$; r cannot be $\parallel$ s by Th. 3.2.
3. Draw $\overline{QD} \perp \overline{AB}$, $\overline{QE} \perp \overline{AC}$, and $\overline{QF} \perp \overline{BC}$.	3. Post. 12
4. $QE = QD$ and $QD = QF$	4. Th. 14.2.
5. $QE = QF$	5. Trans. prop.
6. Q lies on t.	6. Th. 14.2
7. r, s, and t are concurrent at a point equidistant from $\overline{AB}$, $\overline{AC}$, and $\overline{BC}$.	7. Steps 1–6 and def. of concurrent lines

***ABCD* is any convex quadrilateral. (Ex. 16–18)**

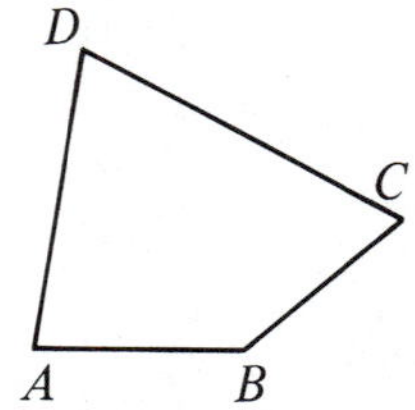

16. Explain how to locate a point equidistant from $\overleftrightarrow{AB}$ and $\overleftrightarrow{AD}$ and also equidistant from D and C. inters. of bis. $\angle$A and $\perp$ bis. $\overline{DC}$
17. Explain how to locate a point equidistant from $\overleftrightarrow{AB}$, $\overleftrightarrow{AD}$, and $\overleftrightarrow{DC}$.
18. Do the points of Exercises 16 and 17 necessarily coincide? no

    17. inters. bis. $\angle$A and bis. $\angle$D

C

19. For an equilateral triangle, what is the ratio of the length of the radius of the incircle to the length of the radius of the circumcircle, $\frac{x}{y}$? $\frac{1}{2}$
20. For an equilateral triangle, what is the ratio of the area of the incircle to the area of the circumcircle? Justify your answer. 1:4
21. A side of an equilateral triangle measures 24 cm. Find the area of the incircle and the area of the circumcircle. 48π cm²; 192π cm²

CONCURRENT ALTITUDES AND MEDIANS 14.6

Objective **To prove and apply concurrence theorems about altitudes and medians of a triangle**

In the preceding lesson, you found that in a triangle the perpendicular bisectors of the sides are concurrent and the angle bisectors are concurrent. In the figure below, observe that the three altitudes of a triangle appear to be concurrent. This leads to Theorem 14.5.

Acute triangle

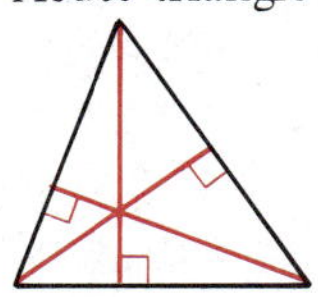

Right triangle

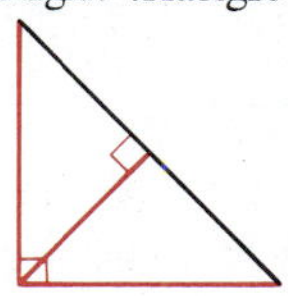

Obtuse triangle

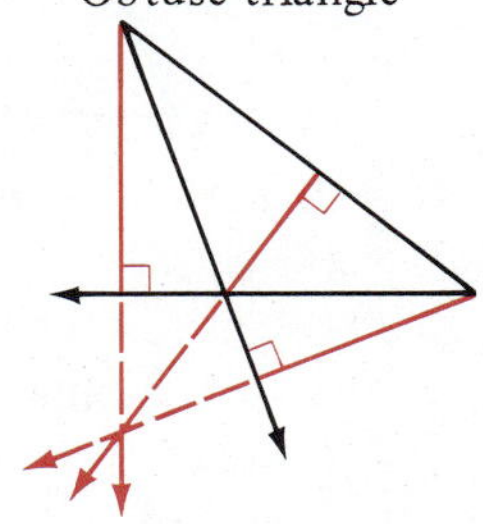

Theorem 14.5 The lines containing the altitudes of a triangle are concurrent.

Given: $\triangle ABC$ with altitudes $\overline{AD}$, $\overline{BE}$, and $\overline{CF}$

Prove: $\overleftrightarrow{AD}$, $\overleftrightarrow{BE}$, and $\overleftrightarrow{CF}$ are concurrent.

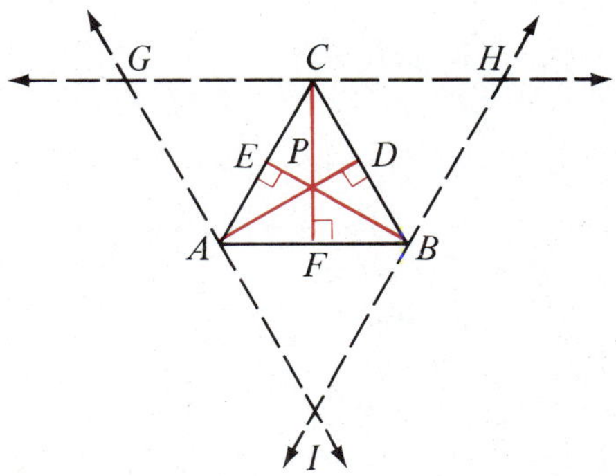

ANALYSIS
Form another $\triangle$ and show that the given altitudes are the $\perp$ bisectors of its sides and are therefore concurrent by Theorem 14.3.

Proof

Statements	**Reasons**
1. $\triangle ABC$ with altitudes $\overline{AD}, \overline{BE}$, and $\overline{CF}$	1. Given
2. Through C draw $\overleftrightarrow{GH} \parallel \overline{AB}$; through B draw $\overleftrightarrow{HI} \parallel \overline{AC}$; through A draw $\overleftrightarrow{GI} \parallel \overline{BC}$.	2. Parallel postulate
3. $ABCG$ and $ABHC$ are $\square$'s.	3. Def. of $\square$
4. $\overline{GC} \cong \overline{AB}$ and $\overline{AB} \cong \overline{CH}$	4. Opp. sides of a $\square$ are $\cong$.
5. $\overline{GC} \cong \overline{CH}$	5. Transitive property
6. $\overline{CF} \perp \overline{AB}$	6. Def. of altitude
7. $\overline{CF} \perp \overline{GH}$	7. If a line is $\perp$ to one of two $\parallel$ lines, it is $\perp$ to the other.
8. $\overleftrightarrow{CF}$ is the $\perp$ bisector of $\overline{GH}$.	8. Def. of $\perp$ bisector
9. Similarly, $\overleftrightarrow{AD}$ is the $\perp$ bisector of $\overline{GI}$; $\overleftrightarrow{BE}$ is the $\perp$ bisector of $\overline{HI}$.	9. Steps 3–8
10. $\overleftrightarrow{AD}$, $\overleftrightarrow{BE}$, and $\overleftrightarrow{CF}$ are concurrent.	10. The $\perp$ bisectors of the sides of a $\triangle$ are concurrent.

Definition: Orthocenter The *orthocenter* of a triangle is the point at which the altitudes of the triangle are concurrent.

From the figures below, it also appears that the three medians of a triangle are concurrent. This is stated in Theorem 14.6.

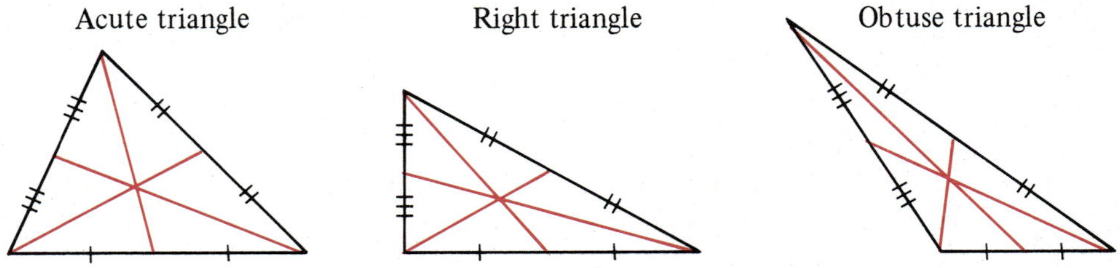

Theorem 14.6

The medians of a triangle are concurrent at a point which is two thirds of the distance from each vertex to the midpoint of the opposite side.

Given: $\triangle ABC$ with medians $\overline{AD}$, $\overline{BE}$, and $\overline{CF}$

Prove: $\overline{AD}$, $\overline{BE}$, and $\overline{CF}$ are concurrent at point O.

$AO = \frac{2}{3}(AD)$;

$BO = \frac{2}{3}(BE)$;

$CO = \frac{2}{3}(CF)$

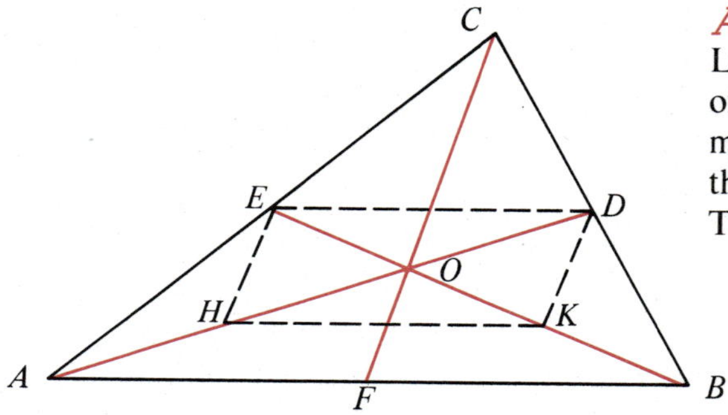

(See Exercise 19.)

ANALYSIS
Let H be the midpoint of $\overline{AO}$ and K be the midpoint of $\overline{BO}$. Show that $HKDE$ is a ▱. Then use Theorem 7.2.

Definition: Centroid The *centroid* of a triangle is the point at which the medians are concurrent.

Example 1

In $\triangle PQR$, medians $\overline{QU}$ and $\overline{PT}$ intersect at S. If $PT = 3x$ and $PS = 8x - 12$, find PT, PS, and ST.

$PS = \frac{2}{3}(PT)$ by Th. 14.6

$8x - 12 = \frac{2}{3} \cdot 3x$

$8x - 12 = 2x$

$6x = 12$

$x = 2$

Thus, $PT = 3 \cdot 2 = 6$

$PS = 8 \cdot 2 - 12 = 4$

$ST = 6 - 4 = 2.$

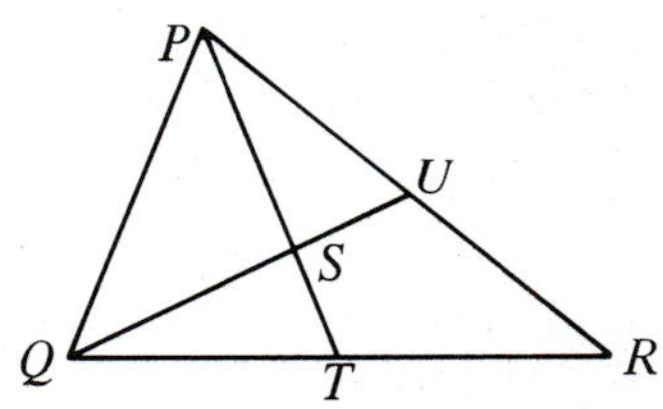

Reading in Geometry

***Ortho* comes from the Greek word *orthos*, meaning "straight" or "right."**

1. Explain why *orthocenter* is an appropriate name for the point at which the altitudes of a triangle are concurrent.

Alt. form rt. ∠'s with s's; seg. go "straight down" to base.

Give the meaning of each word. Use a dictionary, if necessary.

2. orthodontist 3. orthogonal 4. orthorhombic 5. orthognathous

2. person who straightens teeth 3. involving rt. ∠'s or ⊥ lines 4. pertaining to system of crystallization characterized by 3 ≢ axes, inters. at rt. ∠'s 5. straight-jawed

The centroid of a triangle is the center of gravity of the triangle. A triangular shaped board or piece of cardboard will balance at its centroid. The suffix *-oid* comes from the Greek language and means "like" or "in the form of."

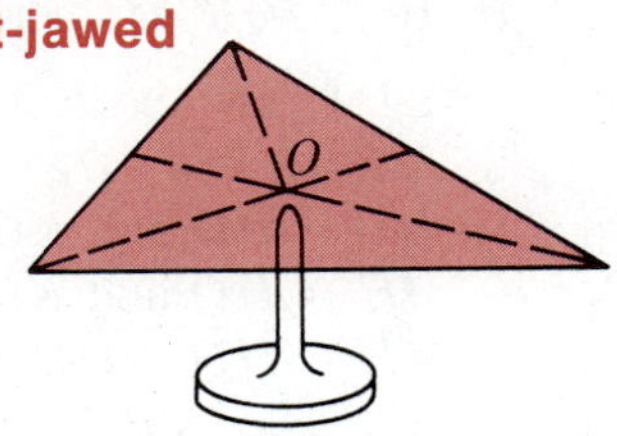

6. Explain why *centroid* is an appropriate name for the point at which the medians of a triangle are concurrent. **It means "centerlike."**

7. a solid of revolution formed by rotating an ellipse about one of its axes

Give the meaning of each word. Use a dictionary, if necessary.

7. spheroid 8. colloid 9. centric 10. centrifugal

8. a substance that when dissolved in a liquid will not diffuse readily 9. pertaining to or situated at the center 10. moving or tending to move away from a center

Written Exercises

A

1. Draw a right triangle. Construct the median to each side. Label the centroid P. **See Example 1, p. 174.**
2. Draw an obtuse triangle. Construct the altitude to each side. Label the orthocenter Q. **See Example 2, p. 174.**
3. Construct an isosceles triangle. Construct the centroid and the orthocenter. Where do the two centers appear to lie? **on ⊥ bis. of base**
4. Construct an equilateral triangle. Construct the centroid. What is true of the centroid, the orthocenter, the incenter, and the circumcenter? **same pt.**
5. Consider an acute, a right, and an obtuse triangle. Tell whether the orthocenter lies in the interior, on, or in the exterior of each triangle. **int.; on; ext.**
6. Repeat Exercise 5, but for the centroid instead of the orthocenter. **interior in each case**

Consider $\triangle ABC$ with medians $\overline{AD}$, $\overline{BE}$, and $\overline{CF}$ concurrent at centroid P.

7. If $AD = 18$, find AP. **12**
8. If $CF = 21$, find PF. **7**
9. If $PD = 5$, find PA. **10**
10. If $PE = 12$, find BE. **36**
11. If $CP = 11$, find CF. **$16\frac{1}{2}$**
12. If $AP = 11$, find PD. **$5\frac{1}{2}$**
13. If $BP = 4x$ and $BE = 9x - 6$, find BP and BE. **BP = 8, BE = 12**
14. If $PF = 3x$ and $PC = 7x - 5$, find PF and PC. **PF = 15, PC = 30**
15. If $PD = 2x + 6$ and $AD = 10x$, find PD and AD. **PD = 15, AD = 45**

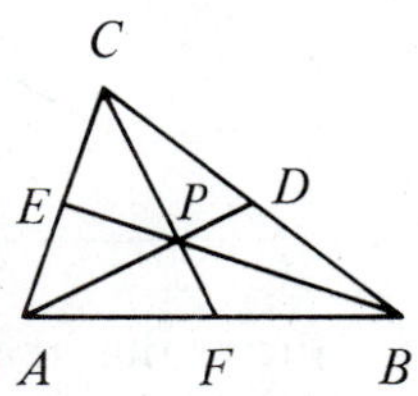

Consider $\triangle RST$ with median $\overline{TM}$, centroid Q, and altitude $\overline{TA}$.

16. If $TA = 15$ and $AM = 8$, find QM. **$5\frac{2}{3}$**
17. If $TQ = 10$ and $TA = 12$, find AM. **9**
18. If $QM = 2$ and $AM = 3$, find TA. **$3\sqrt{3}$**

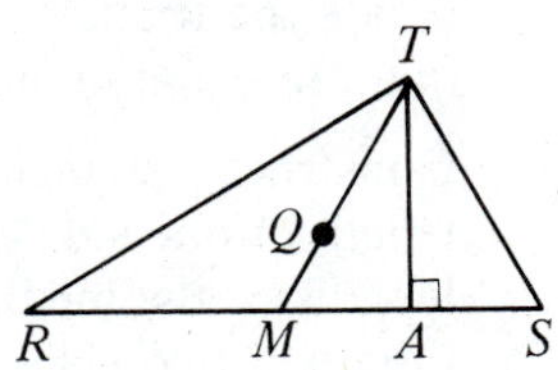

19. Supply the missing reasons in the proof of Theorem 14.6 below.

Statements	Reasons
1. $\triangle ABC$ has medians $\overline{AD}$, $\overline{BE}$, and $\overline{CF}$.	1. **Given**
2. $\overline{AD}$ and $\overline{BE}$ meet in a point O.	2. $\overline{AD} \nparallel \overline{BE}$ by Th. 3.2.
3. Let H be the midpoint of $\overline{AO}$; let K be the midpoint of $\overline{BO}$.	3. **Post. 8**
4. Draw $\overline{HK}$, $\overline{KD}$, $\overline{DE}$, and $\overline{EH}$.	4. **Postulate 1**
5. In $\triangle ABC$, $\overline{ED} \parallel \overline{AB}$ and $ED = \frac{1}{2}(AB)$.	5. **Th. 7.7**
6. In $\triangle AOB$, $\overline{HK} \parallel \overline{AB}$ and $HK = \frac{1}{2}(AB)$.	6. **Same as 5**
7. $\overline{ED} \parallel \overline{HK}$ and $ED = HK$	7. **Th. 3.7 and transitive prop.**
8. $HKDE$ is a $\square$.	8. **Theorem 7.5**
9. $HO = OD$ and $KO = OE$	9. **Diagonals of a $\square$ bisect each other.**
10. $AH = HO$ and $KO = BK$	10. **Def. of midpoint**
11. $AH = HO = OD$ and $BK = KO = OE$	11. **Transitive property**
12. $AO = \frac{2}{3}(AD)$ and $BO = \frac{2}{3}(BE)$	12. Step 11
13. $\overline{AD}$ and $\overline{CF}$ meet in a point P. Let J be the midpoint of $\overline{CP}$. Draw $\square JHFD$. Then $AP = \frac{2}{3}(AD)$ and $CP = \frac{2}{3}(CF)$.	13. Steps 1 – 12
14. P is the only point on $\overline{AD}$ such that $AP = \frac{2}{3}(AD)$. So, $O = P$, and $\overline{AD}$, $\overline{BE}$, and $\overline{CF}$ are concurrent.	14. On a given number line, no two points have the same coordinate; def. of distance.
15. $\therefore CO = \frac{2}{3}(CF)$	15. Substitution

20. An isosceles triangle has sides measuring 18 cm, 15 cm, and 15 cm. How far above the base is the centroid? **4 cm**

21. Each side of an equilateral triangle measures 10 mm. Find the lengths of the radius of the incircle and the circumcircle. $\frac{5\sqrt{3}}{3}$; $\frac{10\sqrt{3}}{3}$

22. Given: $\triangle ABC$ with median $\overline{CD}$
Prove: Area $\triangle ADC$ = area $\triangle BDC$

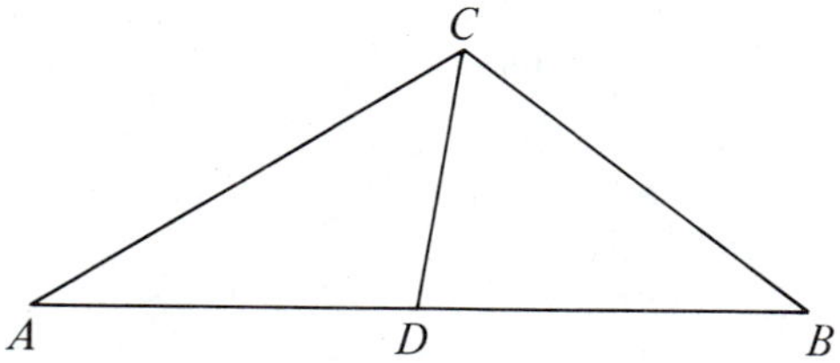

23. Given: $\triangle ABC$ with median $\overline{CD}$, centroid P, and altitudes $\overline{CE}$ and $\overline{PF}$
Prove: $\frac{PF}{CE} = \frac{1}{3}$, or $PF = \frac{1}{3}(CE)$

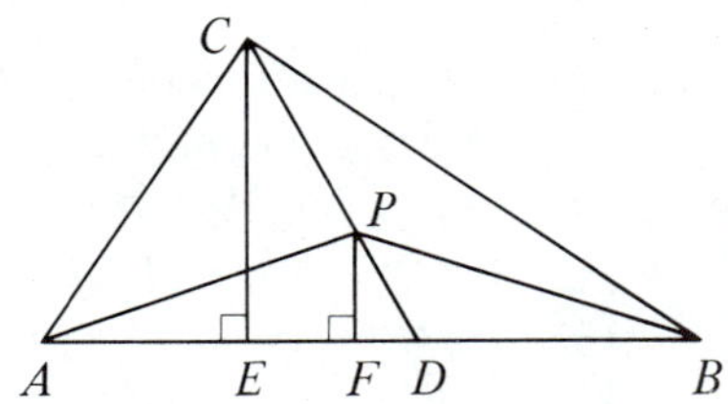

24. Prove that the medians of a triangle divide the triangle into six triangles that are equal in area.

25. Explain why it is logical that the centroid of a triangle is its center of gravity. **All 6 $\triangle$'s (See Ex. 24) have the same area and in a sense the same wt. thus, the wt. is equally balanced around the centroid.**

26. Construct a triangle given the length of one side and the lengths of the medians to the other two sides.

27. Construct a triangle given the lengths of two sides and the length of the median to the third side.

LOCI IN SPACE (optional) 14.7

Objective **To sketch and describe loci in space**

So far the loci that you have studied have involved only coplanar points. In this lesson, you will not be restricted to a single plane. You will be locating points in space and determining loci that are three-dimensional figures.

Recall that the locus of points in a plane at a given distance from a given point is a circle. In Example 1, you will use the same conditions, but you will consider points in space.

Example 1 Sketch and describe the locus of points in space at a given distance d from a given point P.

Start with a given point P.

Locate several points at a distance d from point P.

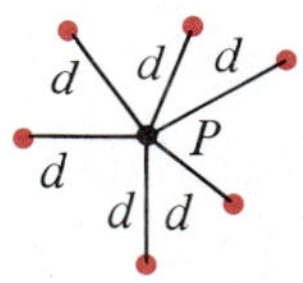

Sketch the figure formed by all such points.

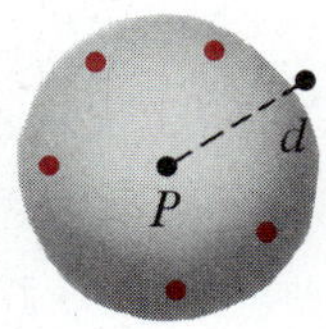

The locus is a sphere with center P and radius of length d.

Example 2 Sketch and describe the locus of points in space equidistant from two parallel planes.

Start with two given parallel planes.

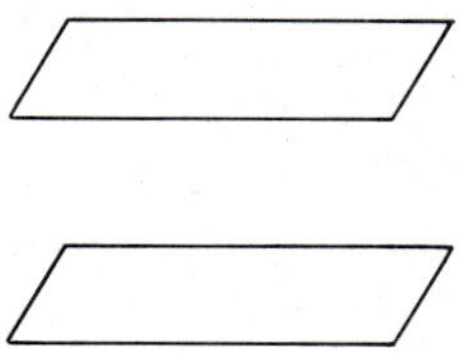

Locate several points the same distance from each plane.

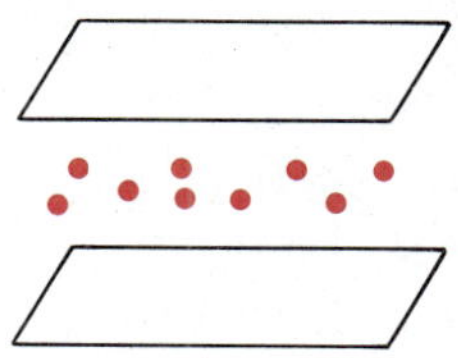

Sketch the figure formed by all such points.

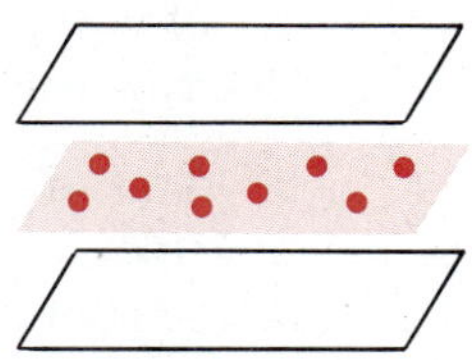

The locus is a plane parallel to each of the given planes and midway between them.

Note that Example 2 is the three-dimensional counterpart of Postulate 29.

Example 3 Sketch and describe the locus of points in space equidistant from two given points.

Start with two given points A and B.

Locate points the same distance from A as from B.

Sketch the figure formed by all such points.

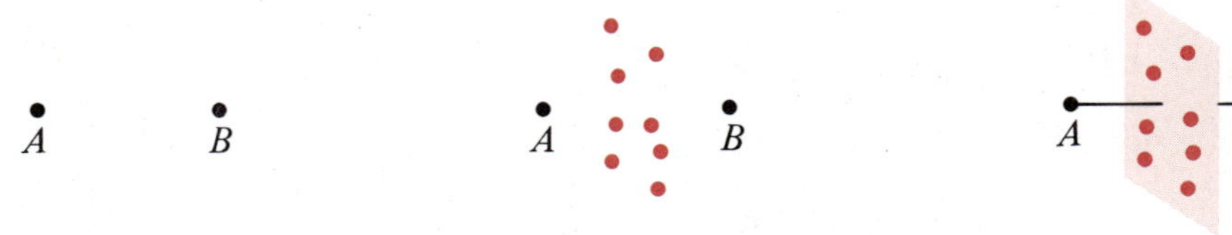

The locus is a plane that is the perpendicular bisector of $\overline{AB}$.

Note that Example 3 is the three-dimensional counterpart of Theorem 14.1.

Written Exercises

7. pl. midway between the l's and ∥ l's 8. an l ⊥ pl. containing the △ at the circumcenter of the △

5. 2 pl. each ∥ given pl. and 3 m from it **6. half pl. that bis. dihedral ∠**

Sketch and describe each locus of points in space.

A

1. The locus of points in space 5 cm from a given point Q **See Example 1.**
2. The locus of points in space equidistant from two given points R and S **See Example 3.**
3. The locus of points equidistant from two given parallel planes r and s which are 2 m apart **See Example 2.**
4. The locus of points in space 4 cm from a given line l **cylinder around l and 4 cm from l on all sides**
5. The locus of points in space 3 m from a given plane
6. The locus of points equidistant from the faces of a dihedral angle
7. The locus of points in space equidistant from two given parallel lines
8. The locus of points in space equidistant from the vertices of a given triangle
9. The locus of points in space equidistant from the sides of a given triangle **a line ⊥ pl. containing the △ at the incenter of the △**
10. The locus of points in space equidistant from all the points of a given circle **line ⊥ pl. containing the ⊙ at center of ⊙**

Sketch and describe each locus of points in space.

11. int. of sphere at the 5-cm radius and P as cente

B

11. The locus of points in space less than 5 cm from a given point P
12. The locus of points in space less than 3 m from a given line l **(See below.)**
13. The locus of the centers of all circles tangent to a given line l at a given point Q on l **pl. ⊥ l at Q, except pt. Q**
14. The locus of the centers of all circles with a given radius r and that pass through a given point S **sphere with center S and radius r**

12. int. of cylinder around l and 3 m from it on all s's

Sketch and describe each locus of points in space.

16. inters. of 2 spheres, one with center A and 3-cm radius other with center B and 4-cm radius

C

15. The locus of points in space 2 cm from a given segment $\overline{AB}$
16. The locus of points in space 3 cm from point A and 4 cm from point B, where A and B are 5 cm apart

15. cylinder plus 2 hemispheres; axis of cyl., $\overline{AB}$; ctrs. of hemis., A and B, r = 2 cm

APPLICATIONS

Read → Plan → Solve → Interpret

1. In wiring an office, an electrician has been asked to keep intercom wires at least 12 in. away from any 120-V electric wires. There are two 120-V wires in the wall where the intercom wires are to be placed. Sketch the wall to show where the intercom wires can be placed.

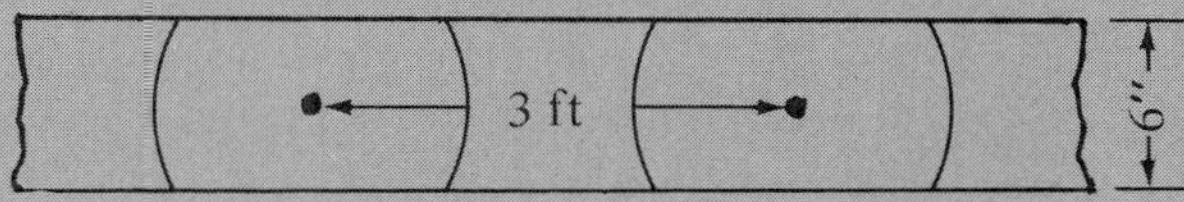

2. A family is planning to build a house. The city building code states that no part of the house can be closer than 30 ft to the street and 10 ft from all property lines. There is a large tree in the back yard that they do not want to cut down. They must build at least 10 ft away from the trunk of the tree. Draw a plot map of the lot to show where the house can be built.

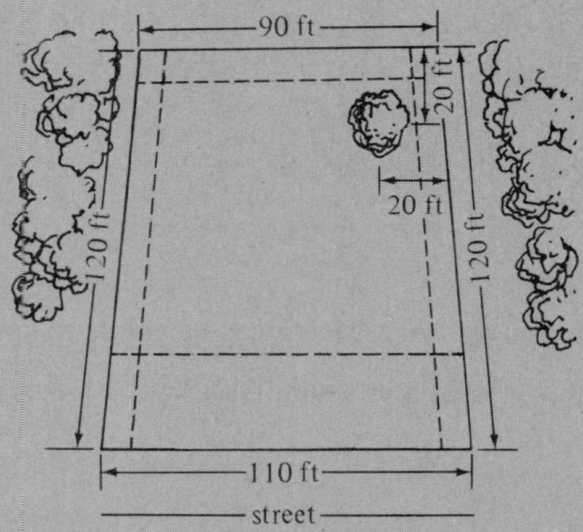

CHAPTER FOURTEEN REVIEW

(A) exercises are 1, 2, 4–7, 11–16. (B) exercises are 3, 8, 17. (C) exercises are starred.

Sketch and describe each locus.

1. The locus of points 1 cm from a given point A [14.1]
⊙ with center A and 1 cm radius

2. The locus of points equidistant from the sides of a 30° angle [14.2] bis. of ∠

3. The locus of points less than 2 cm from a given line [14.3] See Post. 28.

4. The locus of points equidistant from two intersecting lines bis. of angles formed

5. The locus of the centers of all circles tangent to a given line at a given point
⊥ to line at given pt., but excluding the pt.

6. The locus of points 7 cm from P and 2 cm from Q See Ex 5 p. 477.

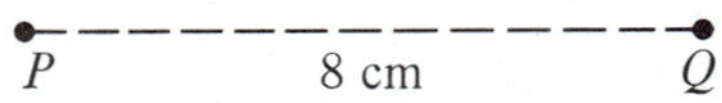

7. The locus of points 4 cm from l and 2 cm from A One pt; pt of tangency of ⊙, ctr. A, rad. = 2 cm and line between l and pt A, ∥ l and 4 cm distant

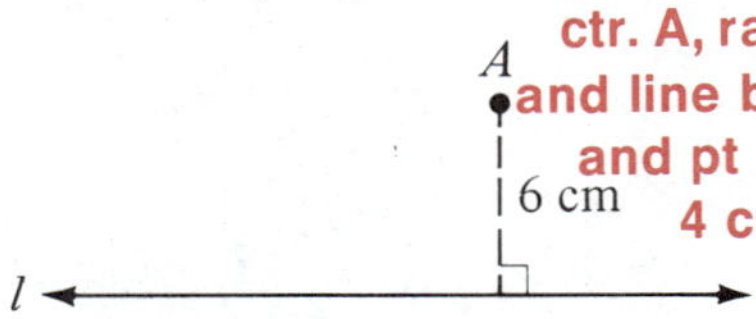

8. The locus of points 3 cm from a circle with a 4-cm radius [14.4]
concentric ⊙'s, r = 1 cm, 7 cm

Sketch and describe each locus. Show all possible cases. (Ex. 9–10)

* 9. The locus of points x cm from point P and y cm from point Q, where P and Q are d cm apart See Ex 3 p. 480.

* 10. The locus of points equidistant from two given concentric circles and at a given distance from a given line
no point, 1 pt, 2 pts, 3 pts, 4 pts.

11. Copy $\triangle ABC$. Locate the incenter of the triangle. [14.5] See Ex 2 p. 485.

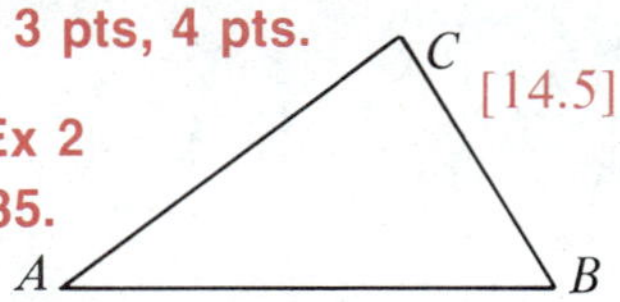

Vocabulary

locus [14.1]
distance from a point to a circle [14.3]
concurrent lines [14.5]
circumcircle [14.5]
circumcenter [14.5]
incircle [14.5]
incenter [14.5]
orthocenter [14.6]
centroid [14.6]

12. Copy $\triangle RST$. Circumscribe a circle about the triangle. [14.5] See Ex 1 p. 484.

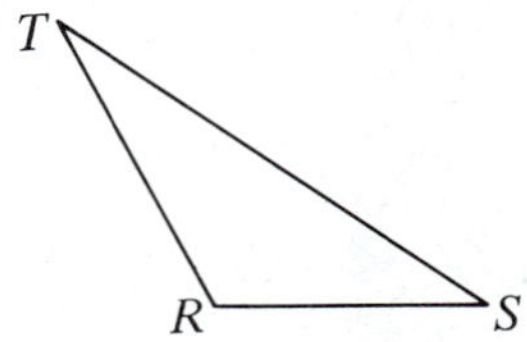

13. Construct a right triangle. Locate the centroid of the triangle. Construct 2 ⊥ lines; form a rt △; construct 2 medians.

In $\triangle PQR$, $\overline{PA}$, $\overline{QB}$, and $\overline{RC}$ are medians. Find each indicated measure. (Ex. 14–16)

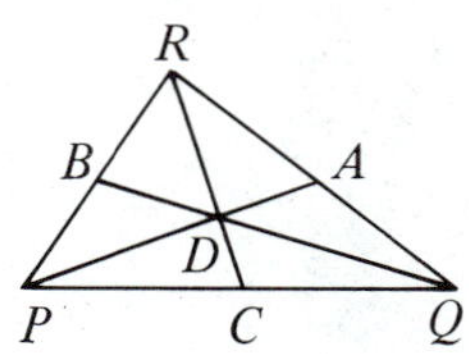

14. If $PA = 18$, find DP. 12

15. If $DB = 8$, find BQ. 24

16. If $DC = 2x$ and $DR = 3x + 5$, find RC. 30

17. An isosceles triangle has sides measuring 10 cm, 10 cm, and 16 cm. How far above the base is the centroid? 2 cm

CHAPTER FOURTEEN TEST

(A) exercises are 1–2, 4–6, 9–12. (B) exercises are 3, 13. (C) exercises are starred.

Sketch and describe each locus.

1. The locus of points equidistant from two given points
⊥ bis. of segment joining them

2. The locus of points 2 cm from a given line
See Post. 28.

3. The locus of points less than 3 cm from a given point See Post. 27.

4. The locus of the centers of all circles tangent to both sides of a given angle
ray which is bis. of ∠ except endpoint

5. The locus of points equidistant from l and m and 2 cm from A
2 pts 2 cm from A equidistant from l and m

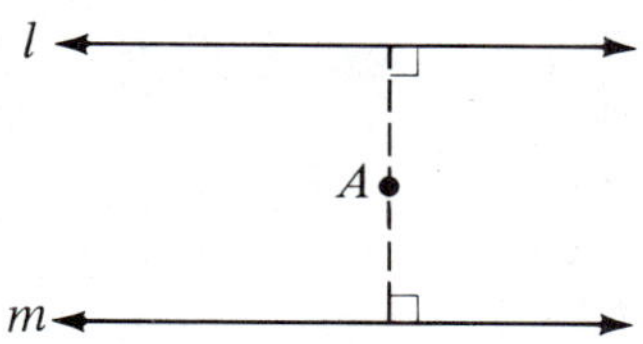

6. The locus of points 4 cm from circle Q
Concentric ⊙ with radius of 8 cm and pt. Q

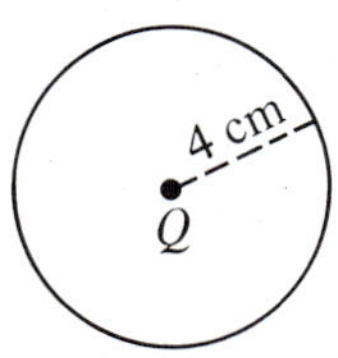

Sketch and describe each locus. Show all possible cases.

* **7.** The locus of points equidistant from intersecting lines t and u and x units from P, the point of intersection of t and u 4 pts

* **8.** The locus of points equidistant from two parallel lines and at a given distance from a given point no pts, 1 pt, 2 pts.

Construct.

9. Copy $\triangle XYZ$. Locate the orthocenter of the triangle Construct 2 alt.; see Ex. 2, p. 174.

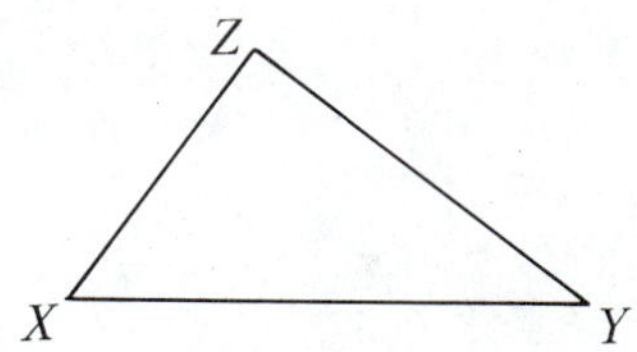

10. Copy $\triangle EFG$. Inscribe a circle in the triangle. See Ex 2 p. 485.

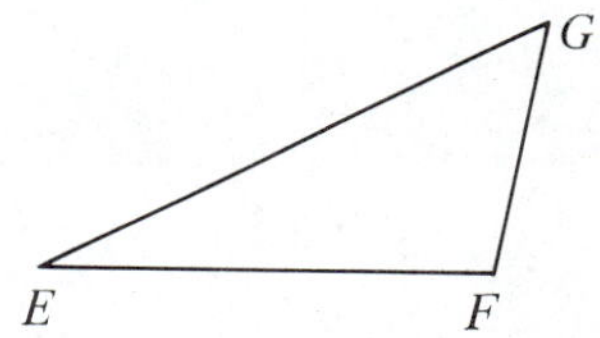

In $\triangle XYZ$, $\overline{XA}$, $\overline{YB}$, and $\overline{ZC}$ are medians. Find each indicated measure.

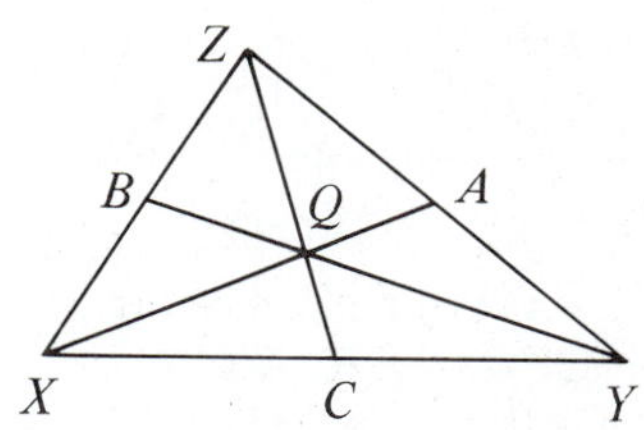

11. If $ZC = 24$, find ZQ. 16

12. If $BQ = x$ and $QY = 4x - 27$, find BY. 40.5

13. Each side of an equilateral triangle measures 12 cm. Find the lengths of a radius of the incircle and a radius of the circumcircle.
$2\sqrt{3}$, $4\sqrt{3}$

COMPUTER ACTIVITIES

Centroid of a Triangle

Objective: **To determine the coordinates of the centroid of a triangle, using a computer program that inputs the coordinates of the vertices**

Given the triangle at the right, you can determine the coordinates of its centroid. First, you find the coordinates of the midpoint of a side, for instance, $\left(\frac{a+c}{2}, \frac{b+d}{2}\right)$. The centroid is 2/3 of the way from (e,f) to that midpoint, or $\left(e+\frac{2}{3}\left[\frac{a+c}{2}-e\right], f+\frac{2}{3}\left[\frac{b+d}{2}-f\right]\right)=\left(\frac{a+c+e}{3}, \frac{b+d+f}{3}\right)$.

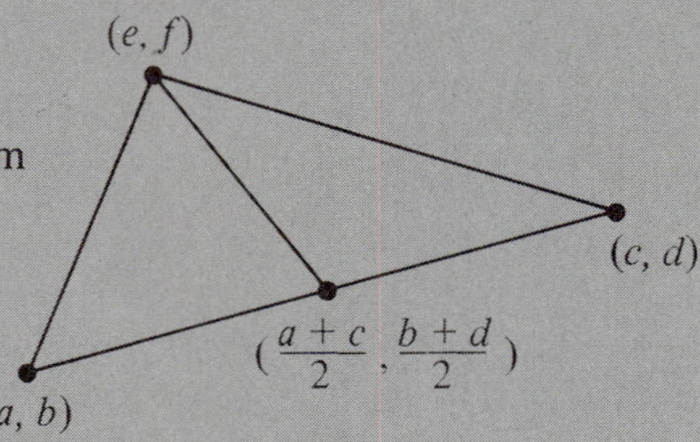

The program below uses this equation to determine the coordinates of the centroid.

```
110  PRINT "GIVEN THE COORDINATES OF THE THREE"
120  PRINT "VERTICES OF A TRIANGLE, THIS PROGRAM"
130  PRINT "WILL DETERMINE THE COORDINATES OF THE"
140  PRINT "CENTROID (INTERSECTION OF THE MEDIANS)."
150  PRINT : PRINT "ONE POINT IS";: INPUT A,B
160  PRINT "SECOND POINT IS";: INPUT C,D
170  PRINT "THIRD POINT IS";: INPUT E,F
180 M = A + C + E:N = B + D + F
190  PRINT : PRINT "THE COORDINATES OF THE CENTROID
     ARE"
200  PRINT "(";: IF M / 3 =  INT (M / 3) THEN 220
210  PRINT M;"/";"3 , ";: GOTO 230
220  PRINT M / 3:" , ";
230  IF N / 3 =  INT (N / 3) THEN 250
240  PRINT N;"/3).": GOTO 260
250  PRINT N / 3:")"
260  GOTO 150
```

EXERCISES: See the Computer Section beginning on page 565 for more information.

1. Run the above program with (2,0), (8,1), (4,7) as the coordinates of one set of vertices and (2,1), (8,1), (5,7) as another.
2. Prove that the answers from the program in Exercise 1 are correct by writing equations for two medians for each set of vertices and solving each pair of equations simultaneously.

COLLEGE PREP TEST

Directions: Choose the one best answer to each question or problem.

1. The locus of points 2 inches from a given line and 3 inches from a point on that line is exactly D

(A) 1 point **(B)** 2 points **(C)** 3 points
(D) 4 points **(E)** 5 points

2. Which of the following is the equation of the locus of points whose ordinates are equal to -5? C

(A) $x = 5$ **(B)** $x = -5$ **(C)** $y = -5$
(D) $y = 5$ **(E)** $y = \pm 5$

3. The locus of points in a plane at a given distance d from a given line in the plane is B

(A) one line **(B)** two lines **(C)** one circle
(D) two circles **(E)** many circles

4. Which of the following is an equation of the locus of points that are at a distance of 5 from the origin? D

(A) $x = 5$ **(B)** $y = 5$ **(C)** $x^2 + y^2 = 5$
(D) $x^2 + y^2 = 25$ **(E)** $x^2 + y^2 = 0$

5. The locus of points in the plane of a given triangle, equidistant from the 3 vertices of the triangle consists of A

(A) one point **(B)** one circle **(C)** one line
(D) two lines **(E)** two circles

6. How many points are there that are equidistant from 2 given points A and B and also 2 inches from the line $\overleftrightarrow{AB}$? C

(A) 0 **(B)** 1 **(C)** 2 **(D)** 3 **(E)** 4

7. The midpoint of the hypotenuse of a right triangle is A
(A) equidistant from all 3 vertices
(B) the intersection of the 3 angle bisectors
(C) the intersection of the 3 medians
(D) the center of the incircle
(E) the incenter

8. Given: $\triangle RST$, with medians $\overline{SL}$, $\overline{RN}$ and $\overline{TM}$ concurrent at P and $SP = 10$. Find PL. B

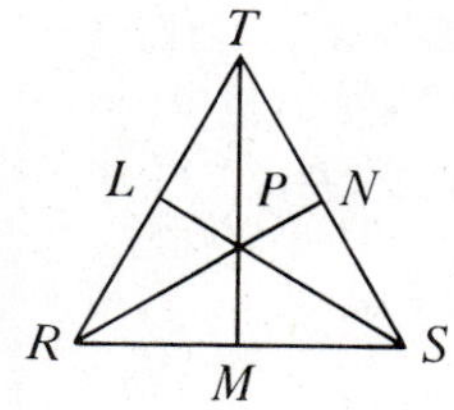

(A) 10 **(B)** 5 **(C)** 15 **(D)** $3\frac{1}{3}$ **(E)** $13\frac{1}{3}$

9. Given points A (0, 2) and B (4, 0). What is the equation of the locus of points equidistant from A and B? D

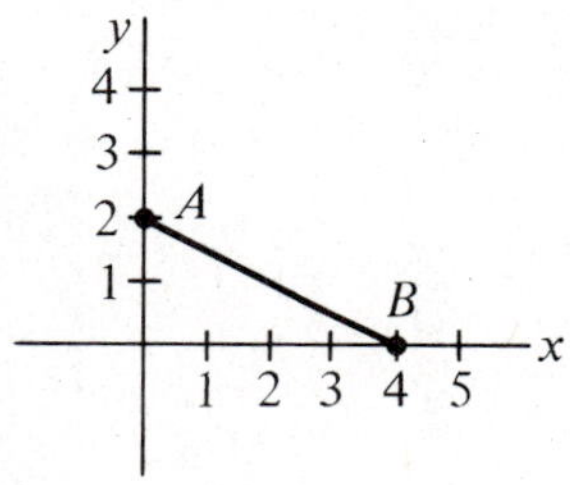

(A) $y = \frac{-1}{2}x + 2$ **(B)** $y = 2x + 2$
(C) $y = -\frac{1}{2}x + 4$ **(D)** $y = 2x - 3$
(E) $y = 2x + 1$

10. Given point P on line l. The number of points that are 3 inches from P and 5 inches from l is A

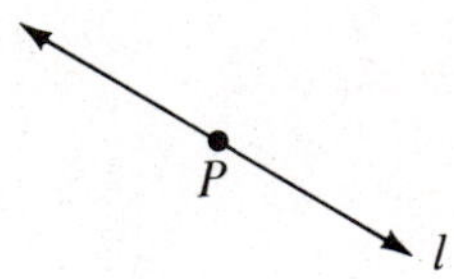

(A) 0 **(B)** 1 **(C)** 2 **(D)** 3 **(D)** 4

11. Two concentric circles have radii of 2 in. and 6 in. Line m is tangent to the smaller circle. How many points are equidistant from the circles and 2 inches from m? D

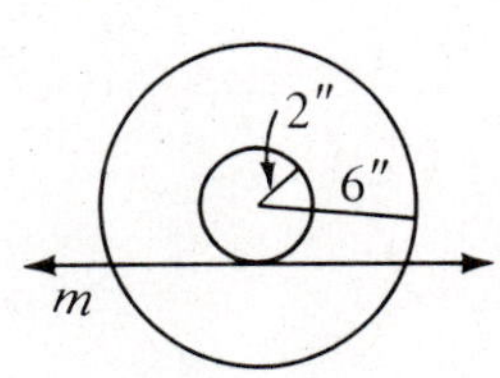

(A) 0 **(B)** 1 **(C)** 2 **(D)** 3 **(E)** 4

15 INEQUALITIES

Non-Routine Problem Solving

Problem 1 A parallelogram has diagonals whose lengths are 12 units and 16 units, respectively. What is the possible range for the perimeter of the parallelogram? **Range: 32 units to 40 units**

Problem 2 In the figure below, the measure of $\widehat{AB} = 110°$, the measure of $\widehat{ED} = 100°$, and the measure of $\angle AFE = 55°$. What is the maximum value for the measure of $\angle P$? **35°**

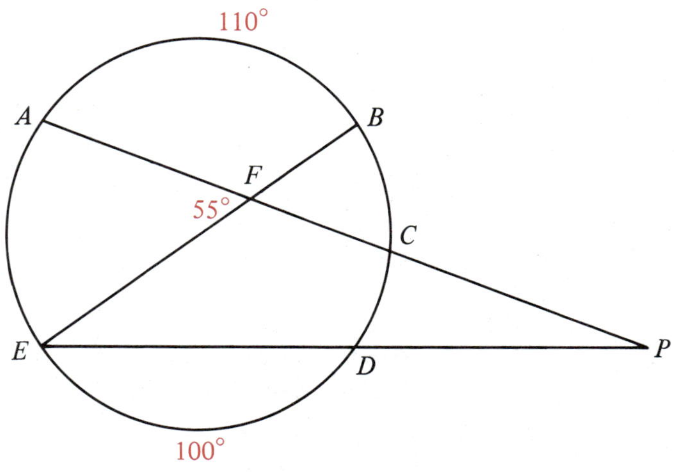

PROPERTIES OF INEQUALITY 15.1

Objectives

To apply the properties of inequality to measures of segments and angles
To prove statements involving inequalities

Recall that the symbol for equality is =. Two symbols for inequality are < (is less than) and > (is greater than).

$\boldsymbol{b > a}$ means that for some positive number c, $b = a + c$. Also, $b > a$ means $a < b$, and similarly, $a < b$ means $b > a$.

In algebra, you used the following properties of inequality. The properties apply for all real numbers, a, b, and c. They are stated for >, but they are also true for <.

Inequality properties:	
Addition	If $a > b$, then $a + c > b + c$.
Subtraction	If $a > b$, then $a - c > b - c$.
Multiplication	If $a > b$ and $c > 0$, then $ac > bc$.
Division	If $a > b$ and $c > 0$, then $\frac{a}{c} > \frac{b}{c}$.

Since measures in geometric figures are numbers, the properties can be applied to measures of segments and angles, as shown below.

Example 1 Given: $AE > CF$, $EB = FD$
Compare AB and CD. Give the property that applies.

$AE > CF$
$EB = FD$
$AE + EB > CF + FD$ by Addition property of inequality
$AB = AE + EB$ and $CD = CF + FD$ by Def. of between
Thus, $AB > CD$ by Substitution property.

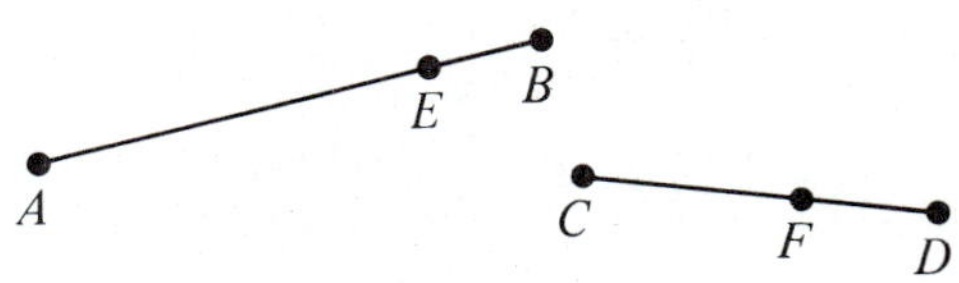

Example 2 Given: m $\angle CBA >$ m $\angle CAB$, m $\angle x =$ m $\angle y$
Compare m $\angle z$ and m $\angle w$. Give the property that applies.

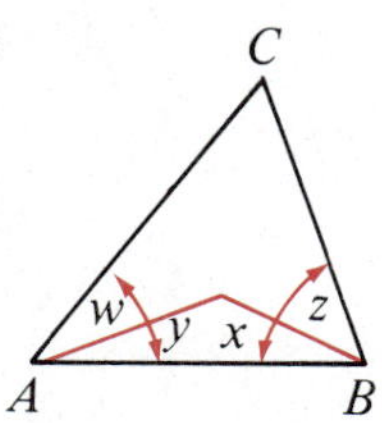

m $\angle CBA >$ m $\angle CAB$
m $\angle x =$ m $\angle y$
m $\angle CBA -$ m $\angle x >$ m $\angle CAB -$ m $\angle y$ by subtraction property of inequality
m $\angle CBA -$ m $\angle x =$ m $\angle z$ and m $\angle CAB -$ m $\angle y =$ m $\angle w$ by {Angle addition postulate and subtraction property of equality.
Thus, m $\angle z >$ m $\angle w$ by Substitution property.

When a segment or angle is bisected, the measure of the segment or angle is multiplied by $\frac{1}{2}$. This is used in the next example.

Example 3 Given: $\odot O$ with $AB < CD$, E bisects $\overline{AB}$, F bisects $\overline{CD}$. Compare CF and AE. Give the property that applies.

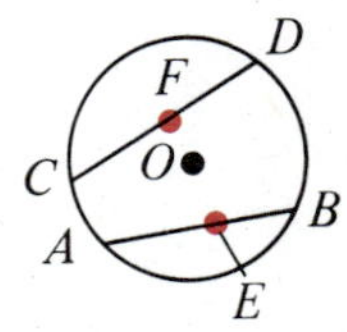

$CD > AB$

$\frac{1}{2}(CD) > \frac{1}{2}(AB)$ by Multiplication property of inequality

Thus, $CF > AE$.

The properties of inequality on page 499 can also be used in proofs, as shown in Example 4.

Example 4 Given: $m\angle x < m\angle z$
Prove: $m\angle ABE < m\angle CBD$

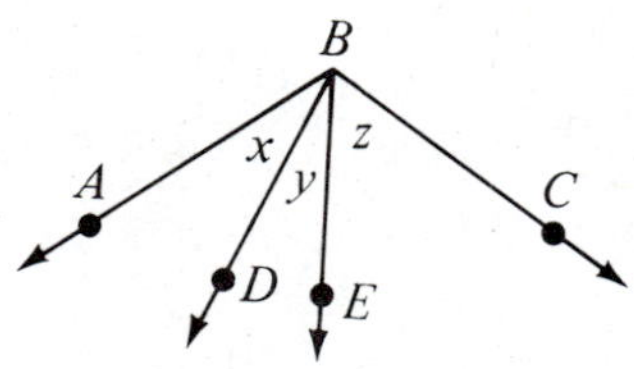

Proof

Statements	Reasons
1. $m\angle x < m\angle z$	1. Given
2. $m\angle z > m\angle x$	2. $a < b$ means $b > a$.
3. $m\angle y = m\angle y$	3. Reflexive property
4. $m\angle z + m\angle y > m\angle x + m\angle y$	4. Add. prop. of inequality
5. $m\angle CBD > m\angle ABE$	5. Angle Add. Post. and Substitution property
6. $m\angle ABE < m\angle CBD$	6. $b > a$ means $a < b$.

Three other properties of inequalities will be useful. These properties also apply to all real numbers a, b, c, and d.

Inequality properties:

Equals-unequals If $a = b$ and $c > d$, then $a - c < b - d$.
If $a = b$ and $c < d$, then $a - c > b - d$.

Transitive If $a < b$ and $b < c$, then $a < c$.
If $a > b$ and $b > c$, then $a > c$.

Trichotomy For any two real numbers a and b, $a < b$ or $a = b$ or $a > b$.

Example 5 Given: $PR = XZ$ and $PQ > XY$
Compare QR and YZ. Give the property that applies.

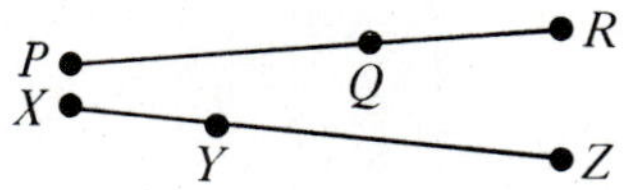

$PR = XZ$
$PQ > XY$
$PR - PQ < XZ - XY$ by subtraction of unequals from equals
$PR - PQ = QR$ and $XZ - XY = YZ$ by def. of between and subtraction property of equality
Thus, $QR < YZ$ by substitution property.

Written Exercises

Compare each of the following as indicated. Then give the property that applies.

A

1. Given: $RS < PQ$
M is the midpoint of $\overline{RS}$.
N is the midpoint of $\overline{PQ}$.
Compare RM and PN.

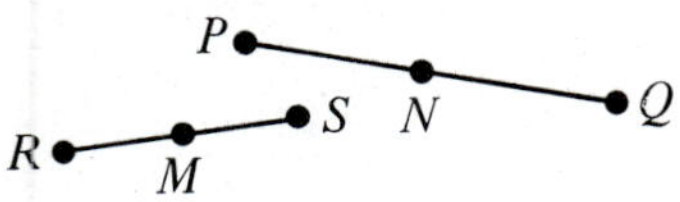

RM < PN, mult. prop. of ineq.

2. Given: $m\angle AEC >$ $m\angle BED$
Compare $m\angle x$ and $m\angle z$.

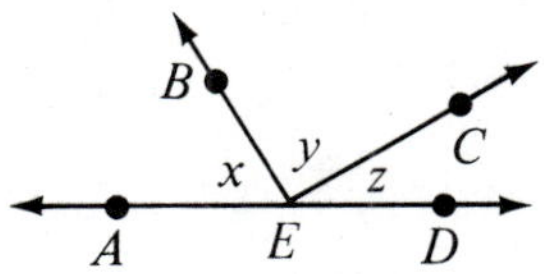

m ∠ x > m ∠ z, sub. prop. of ineq.

3. Given: $AB > CD$
Compare AC and BD.

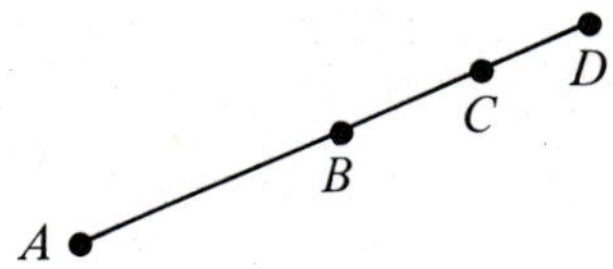

AC > BD, add. prop. of ineq.

4. Given: $EF = GH$ and $EX < GY$
Compare XF and YH.

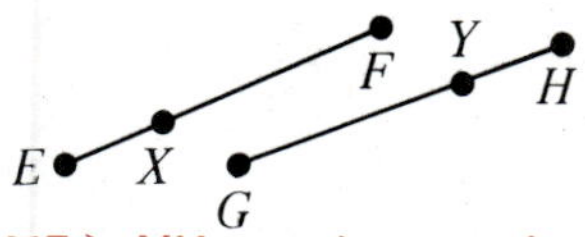

XF > YH, equal-unequal prop. of ineq.

5. Given: $\angle x$ and $\angle y$
Give all possible relationships between $m\angle x$ and $m\angle y$.

m ∠ x < m ∠ y, m ∠ x = m ∠ y, m ∠ x > m ∠ y,
trichotomy prop. of ineq.

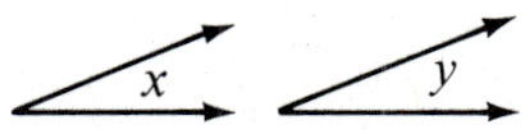

6. Given: $AB < BC$ and $BC < CD$
Compare AB and CD.

AB < CD, trans. prop. of ineq.

7. Given: $m\angle ABC =$ $m\angle BCD$ and $m\angle x > m\angle y$
Compare $m\angle z$ and $m\angle w$.

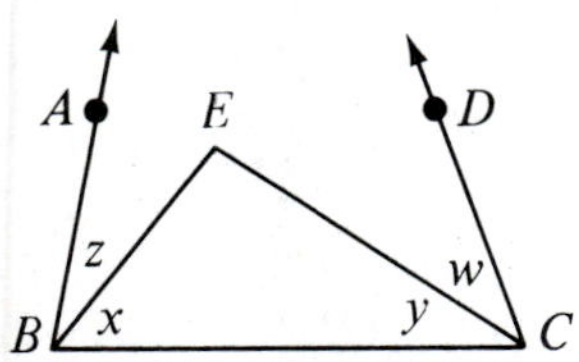

m ∠ z < m ∠ w;
equal-unequal prop. of ineq.

8. Given: $m\angle x = m\angle y$ and $m\angle w < m\angle z$
Compare $m\angle ABC$ and $m\angle ADC$.

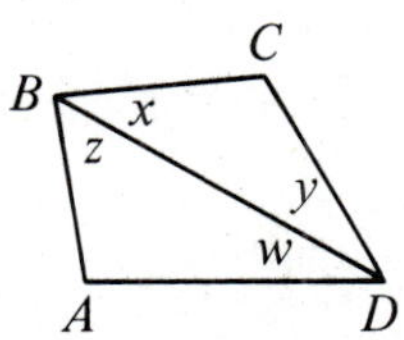

m ∠ ABC > m ∠ ADC;
add. prop. of ineq.

9. Given: $AE < FC$ and $EB = BF$
Compare AB and BC.

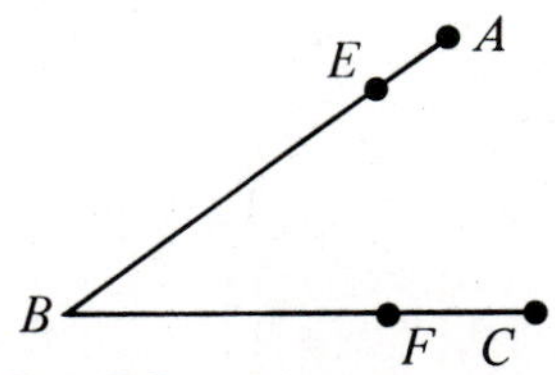

AB < BC; add. prop. of ineq.

B **10.** Given: $\overline{AC} \cong \overline{CB}$ and m $\angle x >$ m $\angle y$
Prove: m $\angle z >$ m $\angle w$

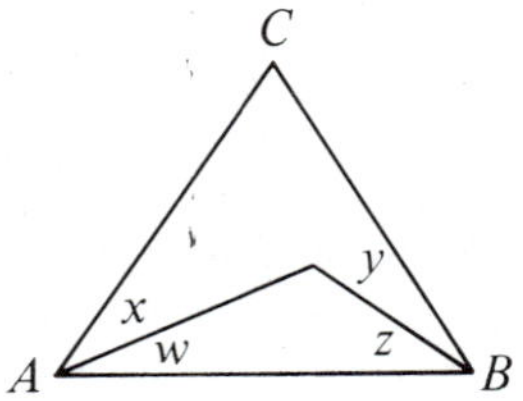

11. Given: $\odot O$ with $AB > CD$, $\overline{OE} \perp \overline{AB}$, and $\overline{OF} \perp \overline{CD}$
Prove: $CF < AE$

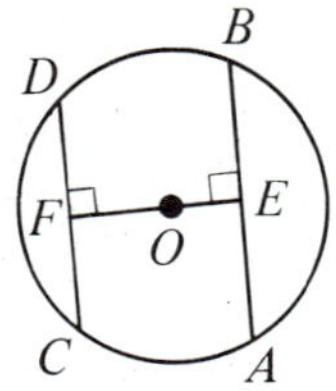

12. Given: $\overline{AB} \cong \overline{BC}$ and m $\angle DAC <$ m $\angle DCA$
Prove: m $\angle BCD >$ m $\angle BAD$

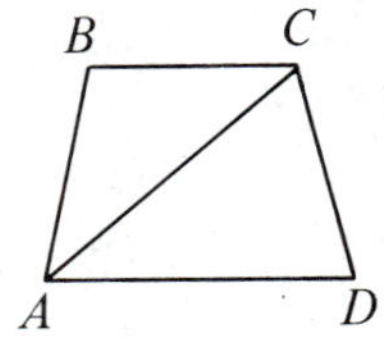

C **13.** Given: $ABCD$ is inscribed in $\odot O$, m $\angle B >$ m $\angle A$
Prove: m $\angle D <$ m $\angle C$

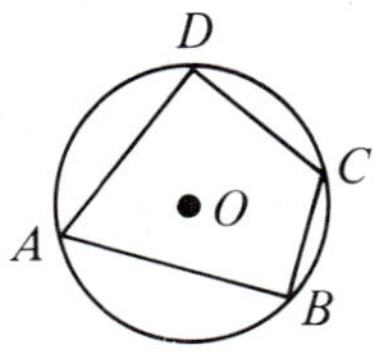

14. Given: $\triangle ABC$, $\overline{DC} \cong \overline{BC}$, m $\angle y >$ m $\angle z$
Prove: m $\angle x < 2 \cdot$ m $\angle y$

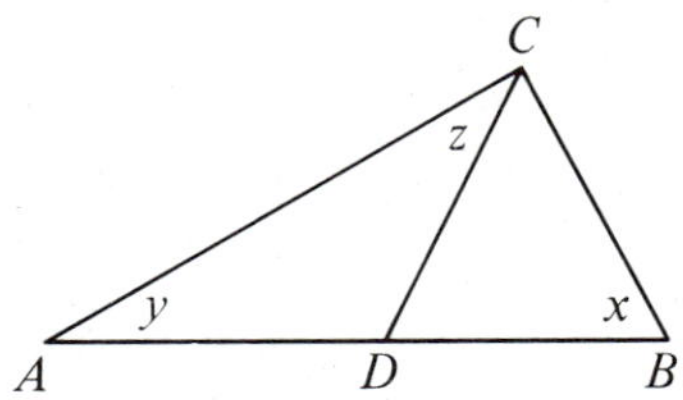

15. Given: $\overline{AD} \cong \overline{DB}$, $\overline{BF} \cong \overline{FC}$, m $\angle EAB >$ m $\angle EBA$, m $\angle DBC <$ m $\angle DCB$, m $\angle EBD <$ m $\angle DBF$ Prove: m $\angle DAE <$ m $\angle FCD$

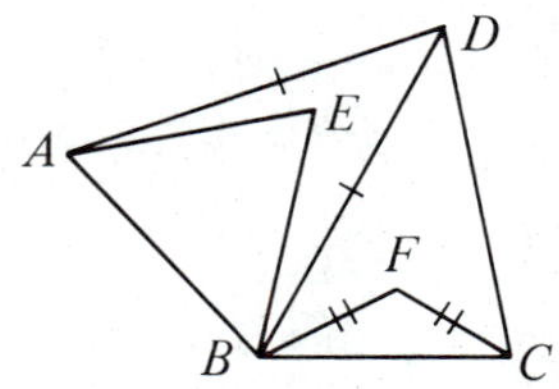

ALGEBRA REVIEW

Objective

To solve inequalities by using the Addition Property and the Division Property

Inequality properties can be used to solve algebraic inequalities. In the example below, the Addition and Division Properties of Inequality are used.

Example

Solve $4x - 5 \leq 19$.

$4x - 5 \leq 19$ ($\leq$ means "is less than or equal to.")

Add 5 to each side. ► $4x \leq 24$

Divide each side by 4. ► $x \leq 6$

Thus, 6 and all real numbers less than 6 are solutions of $4x - 5 \leq 19$.

Solve. 1. $x > -2$ 2. $z < -1$ 3. $x \geq -\frac{1}{3}$ 4. $x < \frac{11}{7}$ 5. $x \leq 6$ 6. $x \geq -6$

1. $5x + 2 > -8$ **2.** $9z - 8 < -17$ **3.** $3x - 7 \geq -8$ **4.** $7x - 18 < -7$

5. $3x - 11 \leq 7$ **6.** $4x + 8 \geq -16$ **7.** $8z + 8 > 8$ **8.** $4x - 4 \leq 12$

9. $2y + 13 > 7$ **10.** $7x - 24 < -3$ **11.** $3z - 18 > -6$ **12.** $5y - 17 < 3$

7. $z > 0$ 8. $x \leq 4$ 9. $y > -3$ 10. $x < 3$ 11. $z > 4$ 12. $y < 4$

GEOMETRY IN PATTERNS

The beauty of mathematics is evident both in the natural and in the human-made world. Mathematical design and geometry are important in many nonscience areas such as art, advertising, and architecture.

Death Valley. The varying elevations of the sand dunes produce a pattern of light and shadow.

Tiled façade and portal, Iran. Persian tile patterns reflect three themes: plant forms, natural forms, and geometric schemes.

Detergent soap bubbles. Soap bubbles always connect at an angle of 120° and form polygons.

Honeycomb. Honeycomb cells are hexagonal compartments. About 25 cells are in a sq. in. of honeycomb

Ammonite fossil. The extinct ammonite was a mollusk with a coiled, chambered shell.

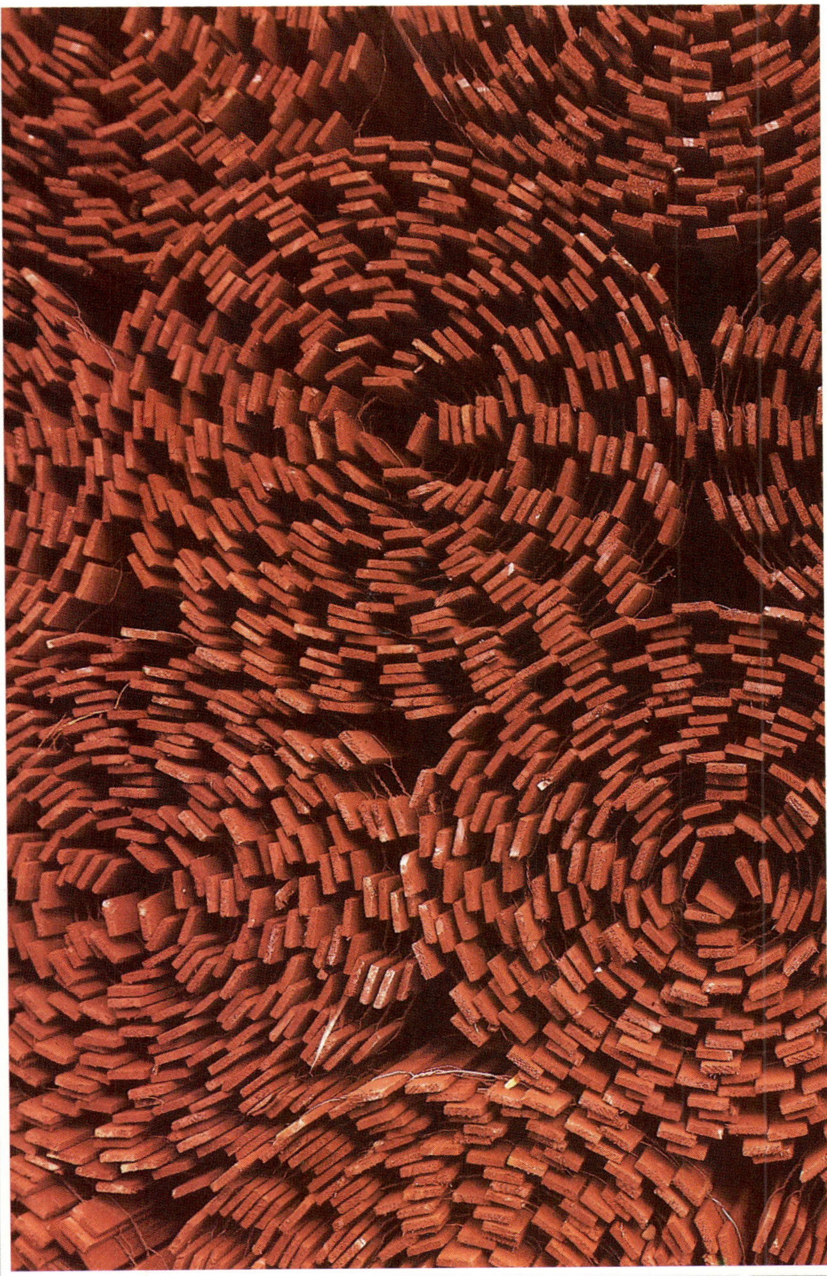

Rolls of snow fence. The spiral-coiled packing of the fencing requires the least space for storage.

Concrete irrigation pipes. The reflection of light within the pipes appears as a spiral design.

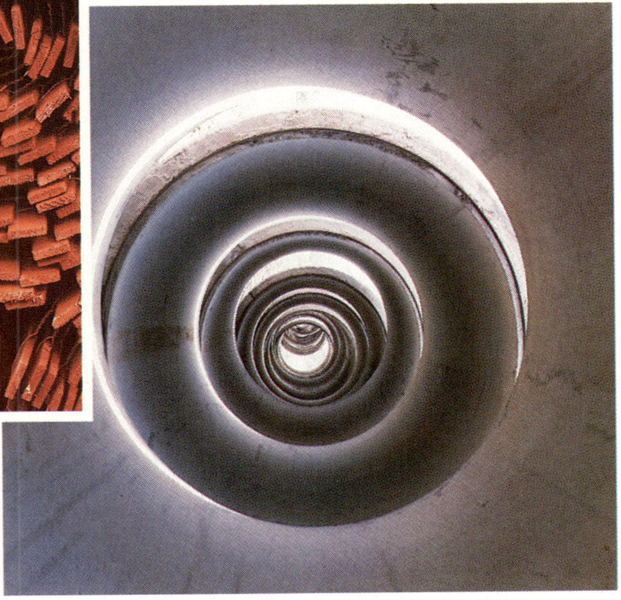

Building detail, New York City. As the perspective of the photograph changes, the shape of the rectangular sections is varied.

Aerial photograph, Lompoc Valley, Calif. Aerial photographs can be plotted as geometric forms in map making and aerial survey.

"Senecio," Paul Klee. This painting has been described as "a marriage of geometry and the color spectrum."

Computer graphics. Data, such as weather information or census population, can be presented as a map of a mathematical function.

Computer design. Mathematical functions can be displayed as computer-generated color block-graphs.

Alhambra. The Alhambra palace in Granada, Spain, is constructed of vaults with interlaced arching of unusual complexity.

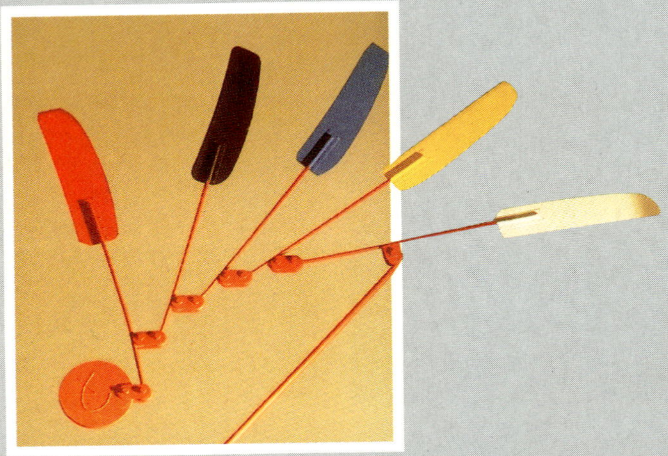

"Indian Feathers" mobile. The shapes and weights of the components of a mobile determine its balance.

Horses. The shape of the horse has evolved from arch-backed to straight-backed over 65 million years.

Prickly pear cactus, evanescent leaves on young pad. Cactus spines and bristles are modified leaves whose patterns reflect sunlight and keep the underlying tissue cooler.

Maple seeds. The thin, oval wings joining two maple seeds resemble an airplane propeller and permit the seeds to be broadcast away from the tree.

Water clover. Like this quatrefoil clover, botanical patterns usually mimic a target to attract bees and other pollinators.

Opera House, Sydney, Australia. The concrete spinnakers and cowled dome of the Sydney Opera House are said to yield perfect acoustics.

Chiang Mai, Thailand. Light emphasizes the patterns of umbrellas in the Chiang Mai open-air market.

Spider web. Orb weavers build the most complicated webs of spokes, which radiate from a spiral held in place by a silk coiling line.

THE TRIANGLE INEQUALITY 15.2

Objective **To apply the Triangle Inequality Theorem**

Suppose you are asked to construct a triangle, given the lengths of the three sides. You will find that certain restrictions exist for the lengths of the three segments. Consider the following three cases.

Case I
Lengths: 8 cm, 5 cm, 2 cm

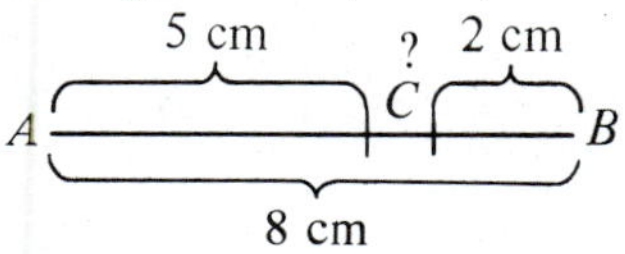

The 5-cm and the 2-cm segments will never meet to form the third vertex C.

$5 + 2 < 8$

Case II
Lengths: 9 cm, 5 cm, 14 cm

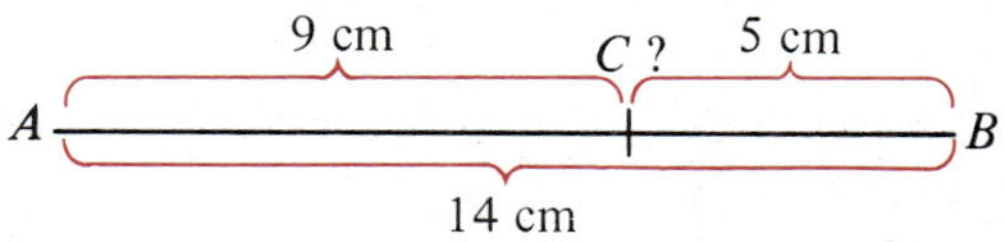

The 9-cm and the 5-cm segments meet on $\overline{AB}$, so the third vertex C lies on $\overline{AB}$.

$9 + 5 = 14$

Case III
Lengths: 2 cm, 3 cm, 4 cm

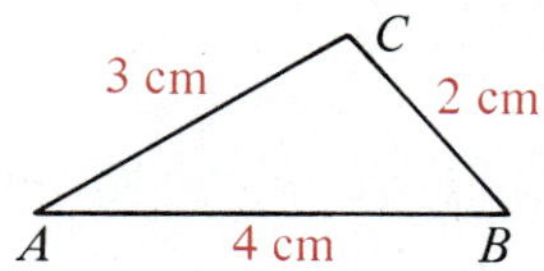

The 2-cm and the 3-cm segments meet to form the third vertex C, which is not on $\overline{AB}$.

$2 + 3 > 4$

These conclusions suggest the following summary and theorem.

Summary

If three segments have lengths such that the sum of any two is greater than the third, then a triangle can be constructed from the segments.

Theorem 15.1 *The Triangle Inequality Theorem*

In a triangle, the sum of the lengths of any two sides is greater than the length of the third side.

Given: $\triangle ABC$
Prove: $AC + CB > AB$
$AB + BC > AC$
$AC + AB > BC$

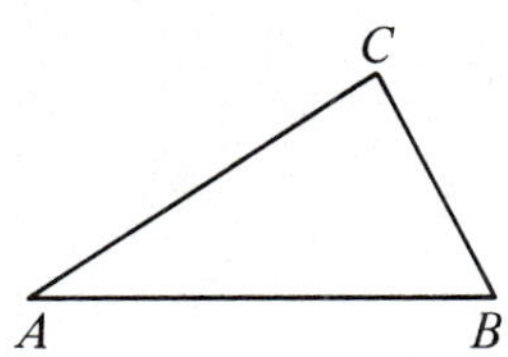

ANALYSIS
Consider the shortest path between each pair of vertices.

Proof

Statements	Reasons
1. $AC + CB > AB$.	1. The shortest path joining two points is a segment.
2. $AB + BC > AC$.	2. Same as 1
3. $AC + AB > BC$.	3. Same as 1

Example 1

Can a triangle be constructed with segments 6 mm, 11 mm, and 8 mm long?

Check the sum of the two smallest lengths: $6 + 8 > 11$

Thus, a triangle can be constructed.

Example 1 applies the summary on page 503. If the sum of the two smallest lengths is greater than the third length, it is not necessary to check the other combinations of lengths.

The next example is an application of the Triangle Inequality Theorem.

Example 2

Given: $\odot O$ with segments $\overline{PO}$ and $\overline{PN}$ and radius $\overline{NO}$
Prove: $PN > PM$

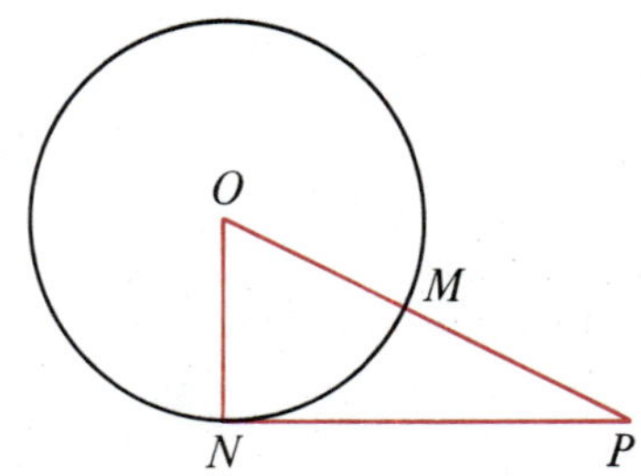

Proof

Statements	Reasons
1. $\odot O$ with radius $\overline{NO}$	1. Given
2. $PN + NO > PO$	2. Triangle Inequality Theorem
3. $PO = PM + MO$	3. Def. of between
4. $PN + NO > PM + MO$	4. Substitution property
5. $NO = MO$	5. Any two radii of a $\odot$ are $\cong$.
6. $PN > PM$	6. Subtraction prop. of inequality

Subtract NO and MO from step 4. ► (step 6)

The following property of inequality is true for all real numbers *a, b, c,* and *d.*

Unequals + unequals, same order

If $a < b$ and $c < d$, then $a + c < b + d$.
If $a > b$ and $c > d$, then $a + c > b + d$.

Example 3

Given: $MN < QR$, $NP < RS$
Compare MP and QS. Give the property that applies.

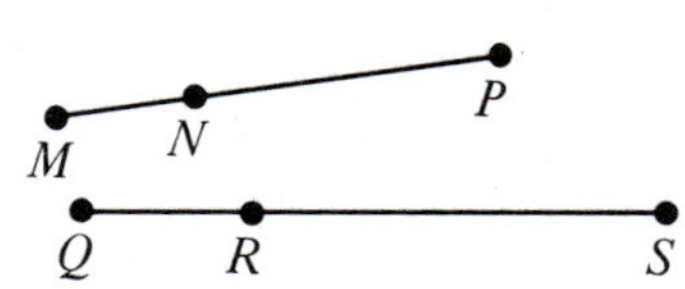

$MN < QR$
$NP < RS$
$MN + NP < QR + RS$ by addition of unequals to unequals, same order
$MN + NP = MP$ and $QR + RS = QS$ by Def. of between

Thus, $MP < QS$ by Substitution property.

Oral Exercises

Tell whether a triangle can be constructed with segments having these lengths.

1. 5, 8, 13 **no**
2. 9, 12, 15 **yes**
3. 9, 10, 20 **no**
4. 4 m, 6 m, 1 m **no**
5. 12 cm, 9 cm, 4 cm **yes**
6. 16 mm, 11 mm, 6 mm **yes**
7. $1\frac{1}{2}, 2\frac{1}{2}, 3\frac{1}{2}$ **yes**
8. 8.2 m, 6.1 m, 2.1 m **no**
9. 6 cm, 9 cm, 1 cm **no**

Written Exercises

Explain why each of the following is true. (Ex. 1–4)

A

1. People often cut across vacant lots rather than go around a corner. **It is shorter dist. to walk by Th. 15.1.**
2. The measure of a diagonal of a rectangle is less than the sum of the measures of two adjacent sides. **Theorem 15.1**
3. If the congruent sides of an isosceles triangle each measure 7 cm, then the base measures less than 14 cm. **Theorem 15.1**
4. If two sides of a triangle have measures 5 and 7, then the measure of the third side is between 2 and 12. **Summary, pg. 504 and Th. 15.1**
5. Given: Quadrilateral $ABCD$
   Prove: $AB + BC + CD + DA > AC + BD$
6. Given: $\triangle ABC$
   Prove: $AB - BC < AC$
7. Given: $ED = EC$
   Prove: $AB + AD > BC$

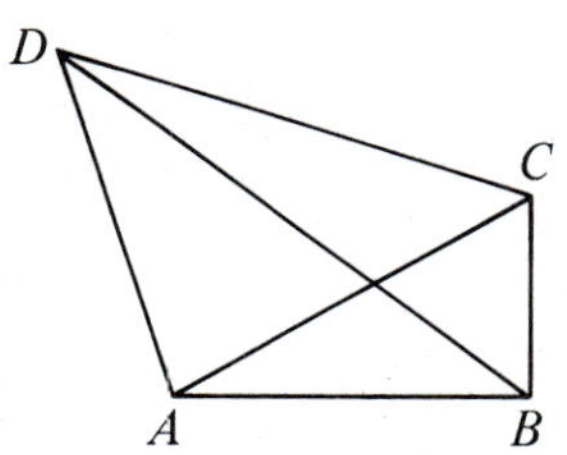

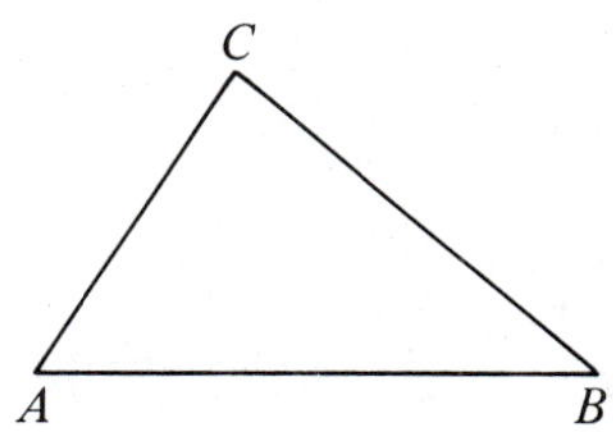

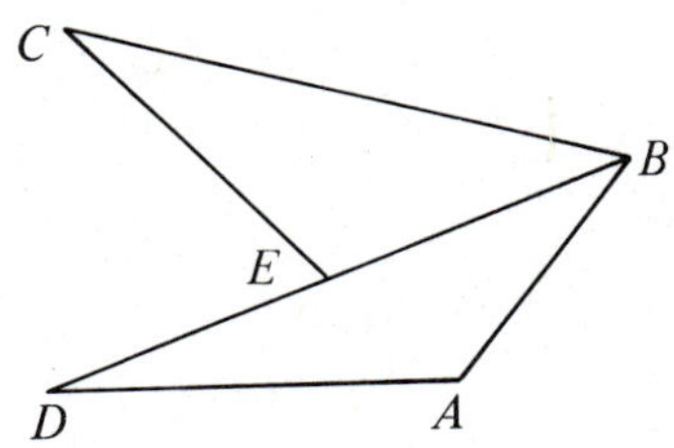

B

8. Given: $\triangle ABC$ with D between A and B
   Prove: $2(CD) < AB + BC + CA$
9. Given: $\triangle ABC$ with P in the interior
   Prove: $AP + BP + CP > \frac{1}{2}(AB + BC + CA)$
10. Given: $\triangle ABC$ with D in the interior
    Prove: $AC + CB > AD + DB$

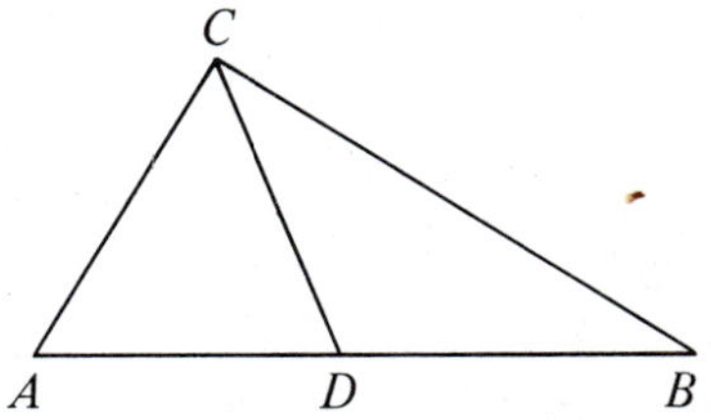

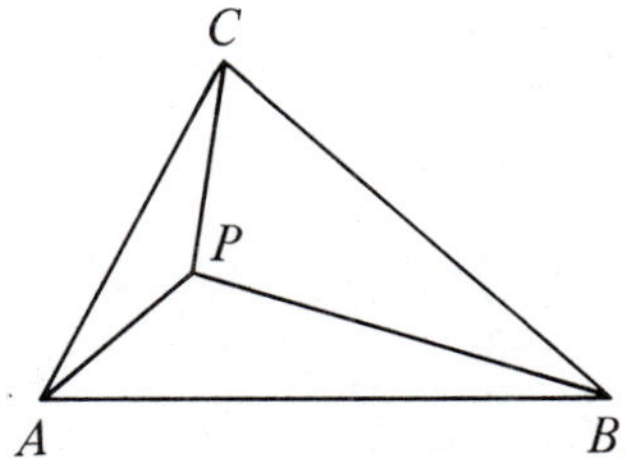

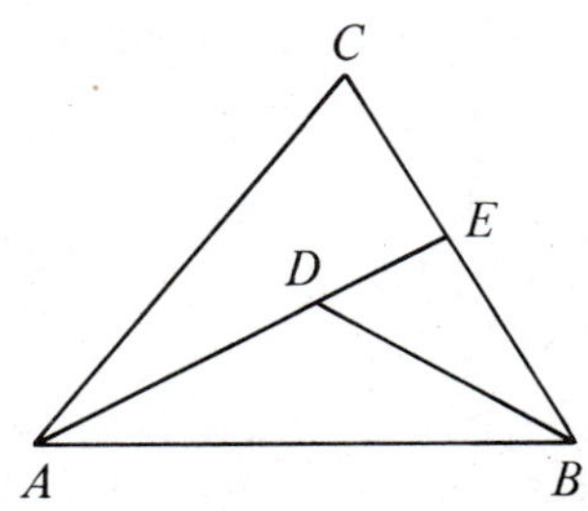

11. Given: Quadrilateral $ABCD$
Prove: $AB + AD > DC - CB$

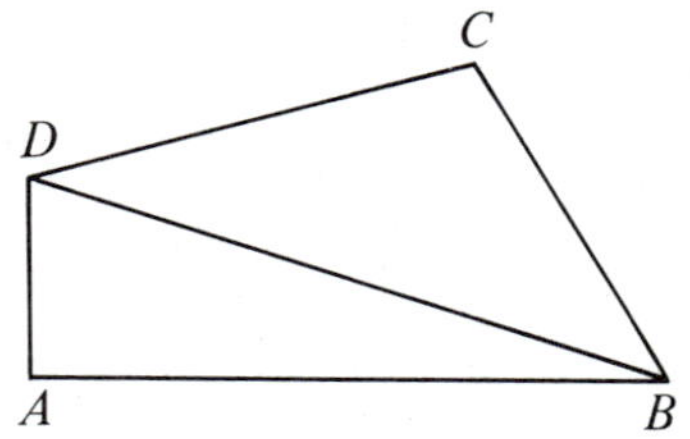

12. Given: $\triangle ABC$ with P in the interior
Prove: $m\angle APB > m\angle C$

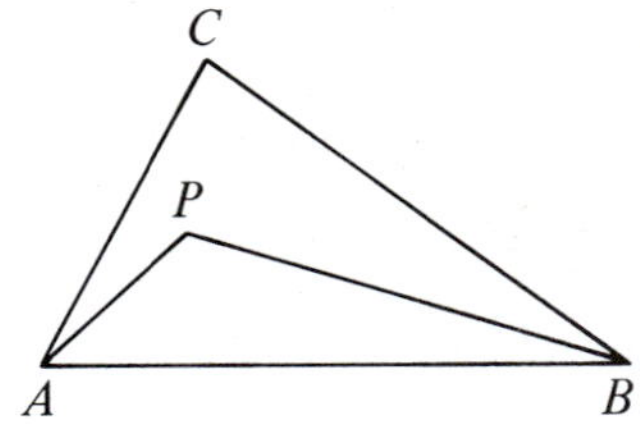

C 13. Prove that if P is a point on $\overline{AB}$ of $\triangle ABC$, then CP is less than one half the perimeter of $\triangle ABC$.

14. Prove that the perimeter of a concave quadrilateral is greater than the sum of the lengths of the two diagonals.

15. Light reflected from a mirror surface $\overline{DF}$ from point A to point B will take the path APB. Show that path APB is shorter than any other path, say path AQB. $\angle APD \cong \angle BPF$.

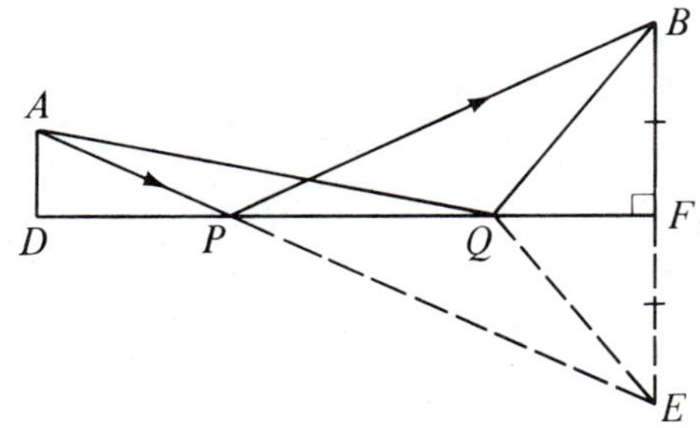

CUMULATIVE REVIEW

1. Given: $\overline{PT}$ is tangent to $\odot O$ at T, $OT = 5$, and $PT = 12$. Find RP. **8**

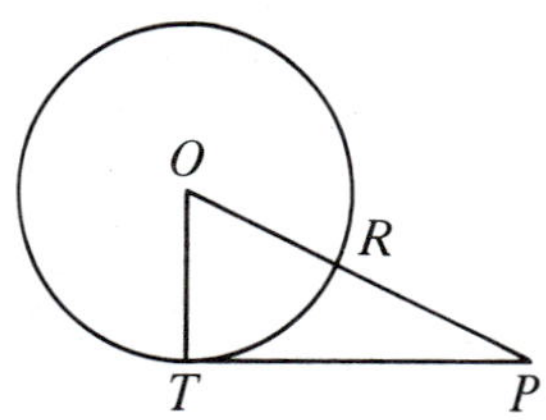

2. Given: $\overline{PT}$ is tangent to $\odot O$ at T, $OT = 8$, and $RP = 9$. Find TP. **15**

NON-ROUTINE PROBLEMS

Prove that the length of a median of a triangle is less than one half the sum of the lengths of the two adjacent sides.
[*Hint:* Extend median $\overline{CD}$ to E so that $CD = DE$.]

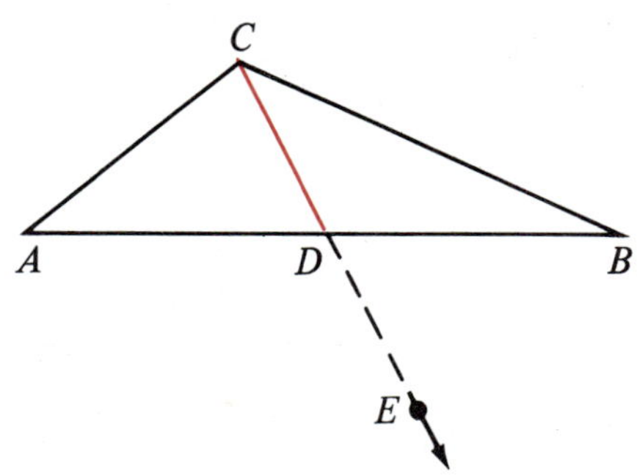

INEQUALITIES IN A SCALENE TRIANGLE 15.3

Objectives

To compare the measures of sides in a scalene triangle
To compare the measures of angles in a scalene triangle

In this lesson, you will discover relationships that exist among the measures of sides and angles of a scalene triangle. For $\triangle ABC$ below, $AC = 3$ cm and $BC = 5$ cm, so $BC > AC$.

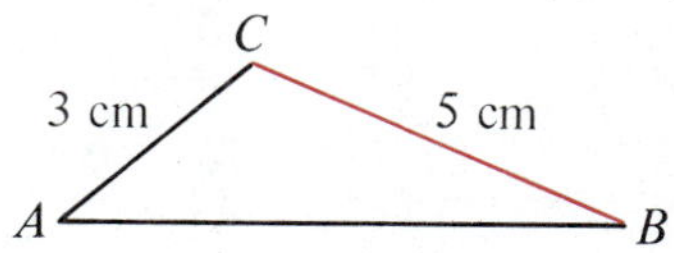

Notice that the measure of the angle opposite $\overline{BC}$ appears to be greater than the measure of the angle opposite $\overline{AC}$, or m $\angle A >$ m $\angle B$. This observation suggests the following theorem.

Theorem 15.2 If one side of a triangle is longer than a second side, then the measure of the angle opposite the first side is greater than the measure of the angle opposite the second side.

Given: $\triangle ABC$ with $BC > AC$
Prove: m $\angle BAC >$ m $\angle B$

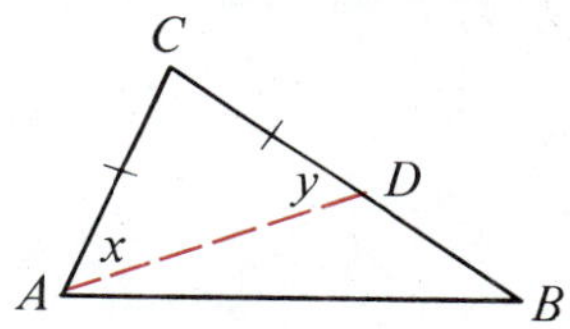

ANALYSIS
Construct isos. $\triangle ACD$. Then, m $\angle x =$ m $\angle y$, m $\angle BAC >$ m $\angle x$, and m $\angle y >$ m $\angle B$.

Proof

Statements	Reasons
1. $\triangle ABC$ with $BC > AC$	1. Given
2. On $\overline{CB}$, let $CD = AC$.	2. Postulate 6
3. Draw $\overline{AD}$.	3. Two points are contained in exactly one line.
4. m $\angle x =$ m $\angle y$	4. Base $\angle$'s of an isos. $\triangle$ are $\cong$.
5. m $\angle BAC =$ m $\angle DAB +$ m $\angle x$	5. Angle addition postulate
6. m $\angle BAC >$ m $\angle x$	6. $b > a$ means for some $c > 0$, $b = a + c$.
7. m $\angle BAC >$ m $\angle y$	7. Substitution property
8. m $\angle y >$ m $\angle B$	8. An exterior $\angle$ of a $\triangle$ has greater measure than either remote interior $\angle$.
9. $\therefore$ m $\angle BAC >$ m $\angle B$	9. Transitive prop. of inequality

---

Now consider $\triangle ABC$ with m $\angle A = 40$ and m $\angle B = 70$, or m $\angle B >$ m $\angle A$. It appears that the measure of the side opposite $\angle B$ is greater than the measure of the side opposite $\angle A$, or $AC > BC$. This observation suggests the next theorem.

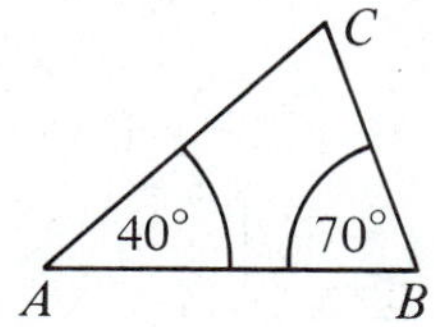

Theorem 15.3

If one angle of a triangle has a greater measure than a second angle, then the side opposite the first angle is longer than the side opposite the second angle.

Given: $\triangle ABC$ with m $\angle A >$ m $\angle B$
Prove: $BC > AC$

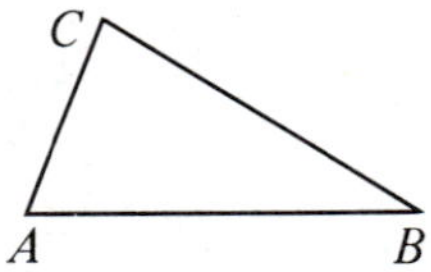

(See Exercise 12.)

ANALYSIS
Use an indirect proof. Either $BC < AC$, $BC = AC$, or $BC > AC$. Show that the first two possibilities lead to contradictions.

Example 1

Given: $\triangle ABC$ with $\overline{CD}$ bisecting $\angle ACB$
Prove: $AC > AD$

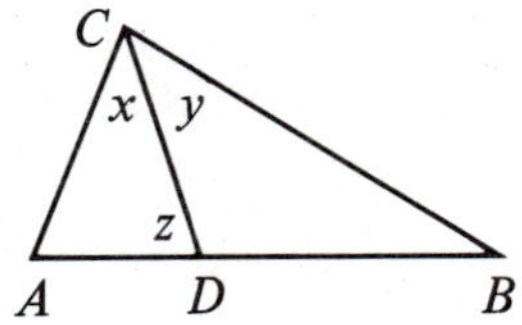

Proof

Statements	Reasons
1. $\overline{CD}$ bisects $\angle ACB$.	1. Given
2. m $\angle x =$ m $\angle y$	2. Def. of $\angle$ bisector and Def. of $\cong$ $\angle$'s
3. m $\angle z >$ m $\angle y$	3. An exterior $\angle$ of a $\triangle$ has greater measure than either remote interior $\angle$.
4. m $\angle z >$ m $\angle x$	4. Substitution property
5. $\therefore AC > AD$	5. Theorem 15.3

$\angle z$ is an ext. $\angle$ of $\triangle DBC$. ► (step 3)

Oral Exercises

Name the angles in order according to their measures, from largest to smallest.

1.

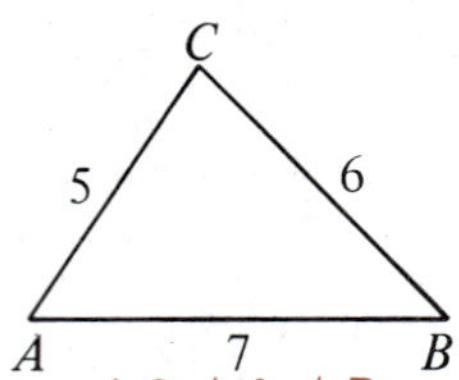

$\angle C, \angle A, \angle B$

2.

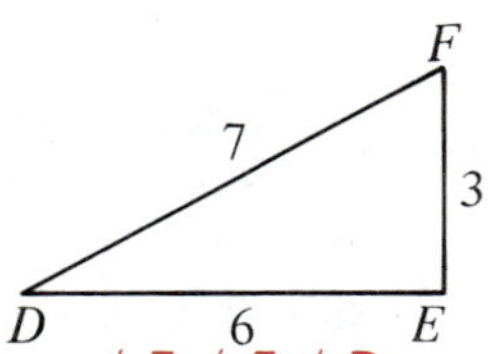

$\angle E, \angle F, \angle D$

3.

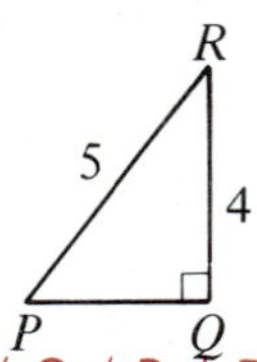

$\angle Q, \angle P, \angle R$

4.

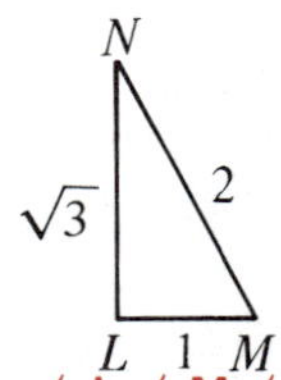

$\angle L, \angle M, \angle N$

Name the sides in order according to their measures, from largest to smallest.

5.

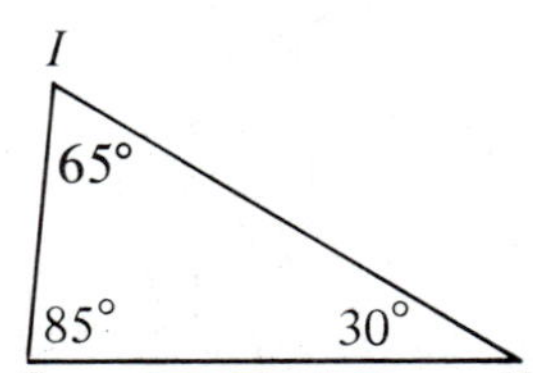

$\overline{IH}, \overline{GH}, \overline{IG}$

6.

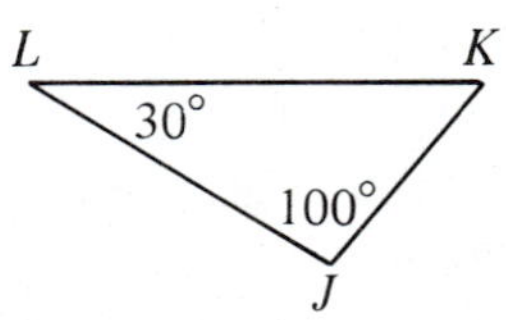

$\overline{LK}, \overline{LJ}, \overline{JK}$

7.

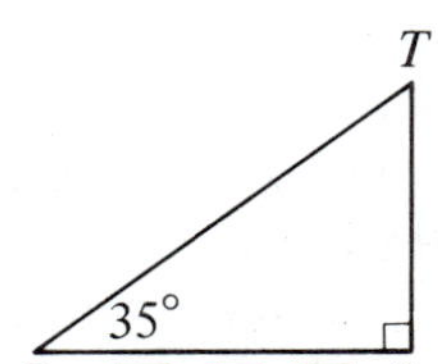

$\overline{RT}, \overline{RS}, \overline{ST}$

8.

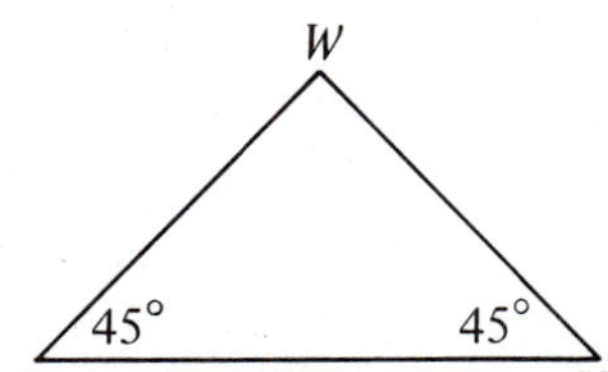

$\overline{UV}$; $\overline{UW} \cong \overline{VW}$

Written Exercises

A

1. Given: $\triangle ABC$ with $BC > AC$ and $m\angle C > m\angle A$
Prove: $m\angle C > m\angle B$

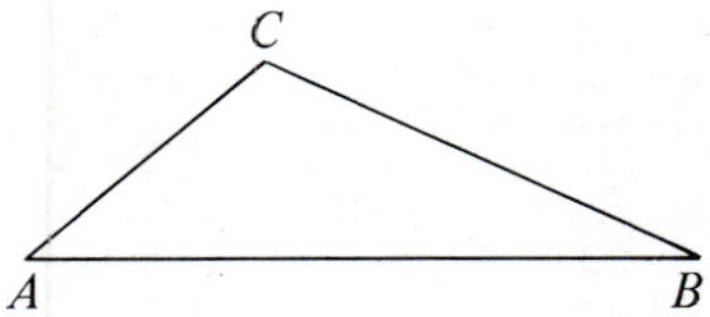

2. Given: Equilateral $\triangle ABC$
Prove: $AD > CD$

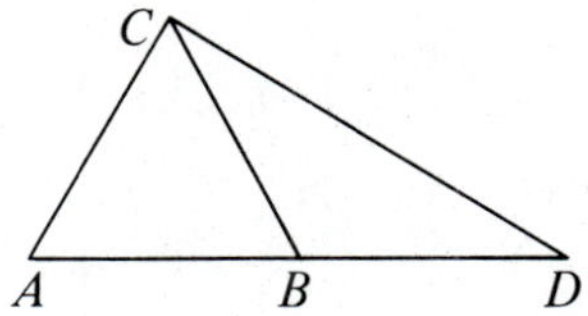

3. Given: $\angle ABC$ is obtuse.
Prove: $AC > AB$ and $AC > BC$

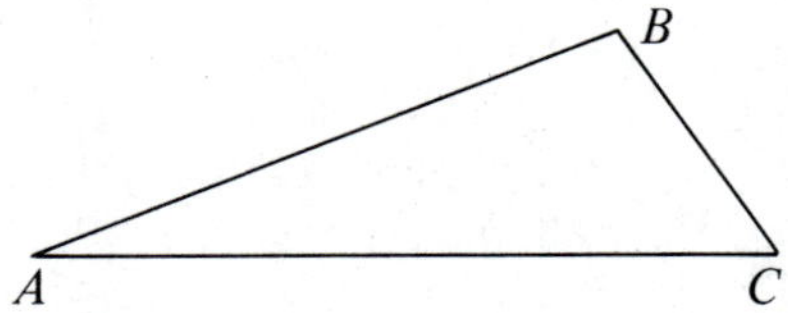

4. Given: $EB > AE$ and $CD > BC$
Prove: $m\angle A > m\angle D$

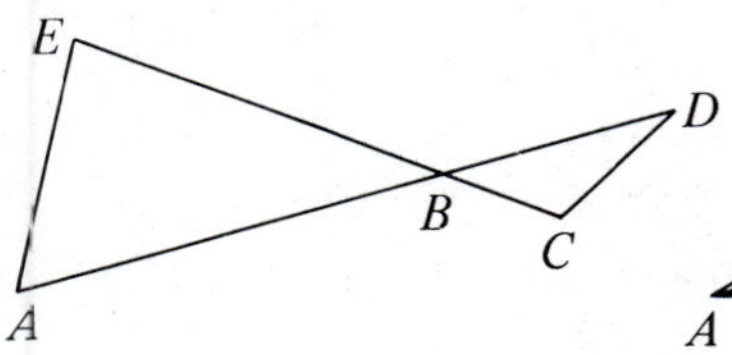

5. Given: $AC > CB$, $\overrightarrow{AD}$ bisects $\angle CAB$, $\overrightarrow{BD}$ bisects $\angle ABC$.
Prove: $AD > DB$

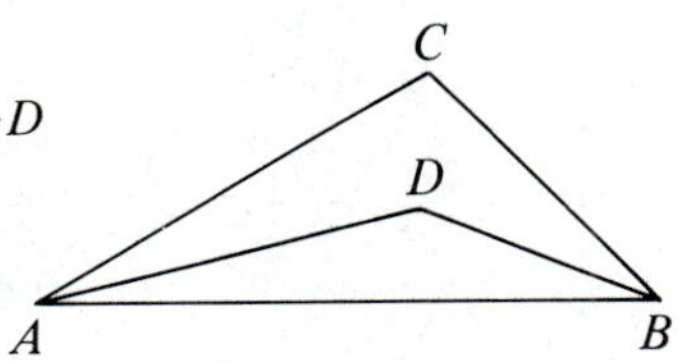

6. Given: $\triangle ABC$ with $m\angle A < m\angle B$ and $CD < CB$
Prove: $m\angle A < m\angle CDA$

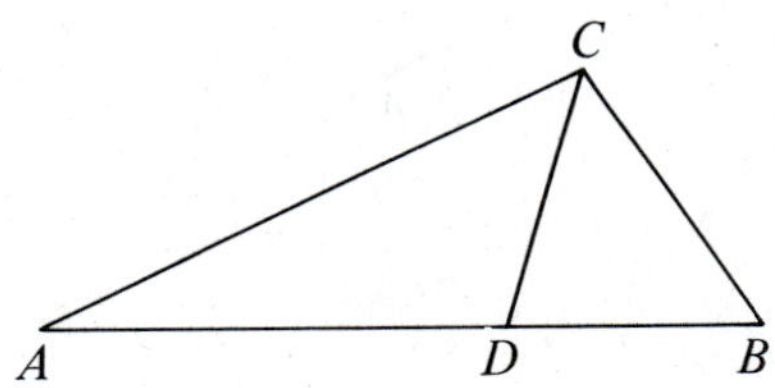

B

7. Given: $m\angle A > m\angle C$
Prove: $DC > DB$

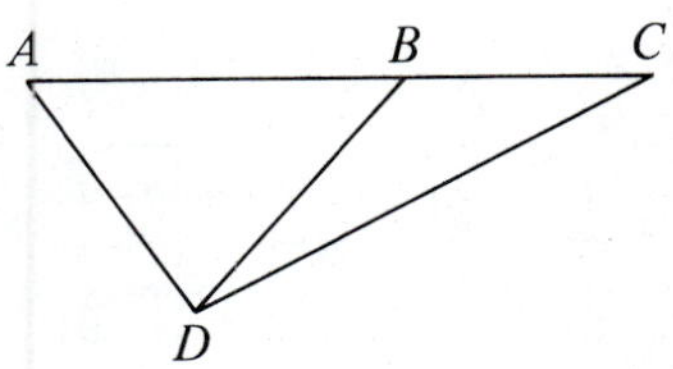

8. Given: $DB > AC$ and $\overrightarrow{AD}$ bisects $\angle CAB$.
Prove: $DB > DC$

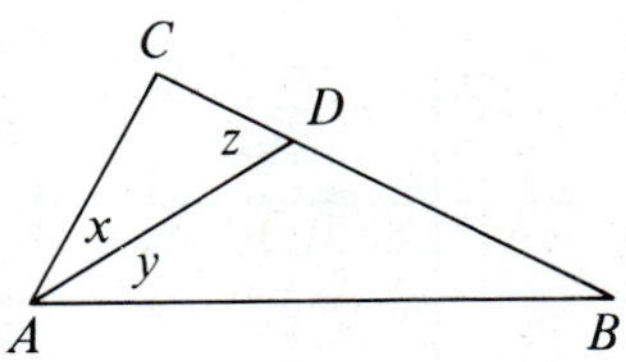

9. Given: $AC = CB$ and $DB > DA$
Prove: $m\angle w < m\angle z$

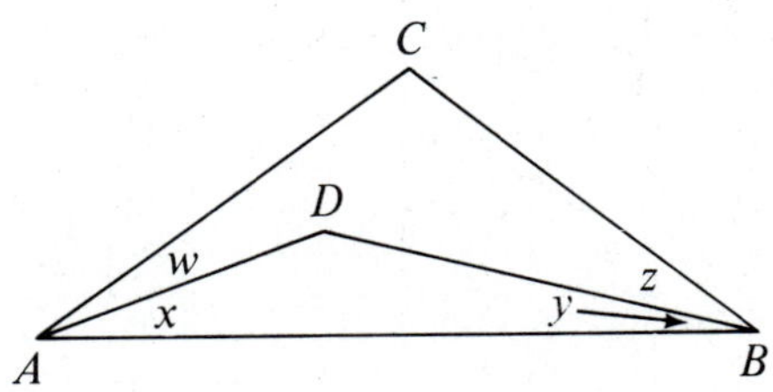

C

10. Given: $AD < DB$, $DB = DC$
Prove: $m\angle A + m\angle C > m\angle ABC$

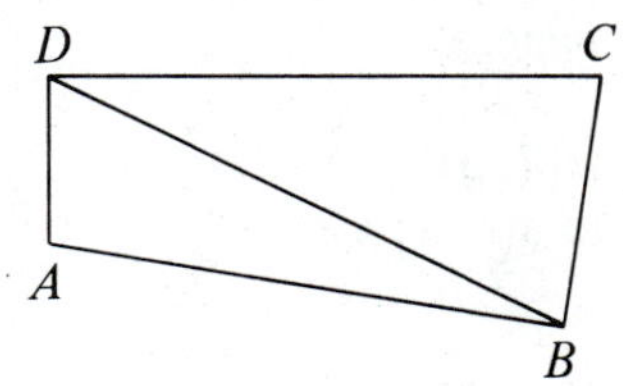

11. Given: $BC = BD$, $BD > AB$
Prove: $m\angle ACD < m\angle D + m\angle A$

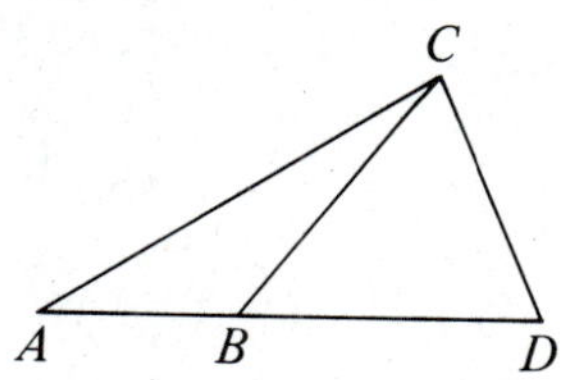

12. Prove Theorem 15.3.

SLIDES AND COORDINATES

EXPERIMENT

On graph paper, copy the axes, $\triangle ABC$, $\triangle A'B'C'$, and the slide arrow, TT'.

1. Are the triangles congruent?
2. Give the coordinates of the vertices of $\triangle ABC$ and $\triangle A'B'C'$. Give the coordinates of the tail and the tip of the slide arrow, TT'.
3. Compare the x-coordinates of A and A', B and B', C and C', T and T'.
4. Similarly, compare the y-coordinates.

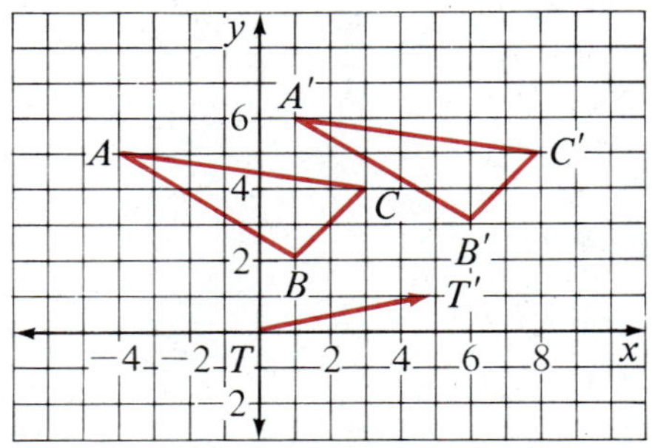

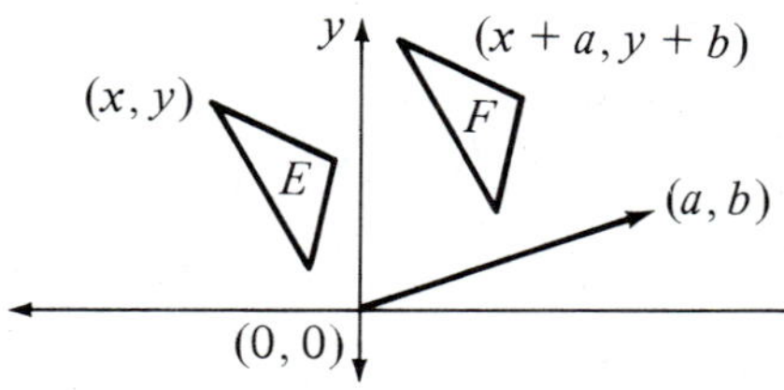

Figure F is a slide image of figure E. The segment from (0, 0) to (a, b) is the slide arrow.

In figure E		In figure F
(x, y)	$\longleftrightarrow$	$(x + a, y + b)$

Project

Copy each figure and the slide arrow on graph paper. Draw the slide image of each.

1.

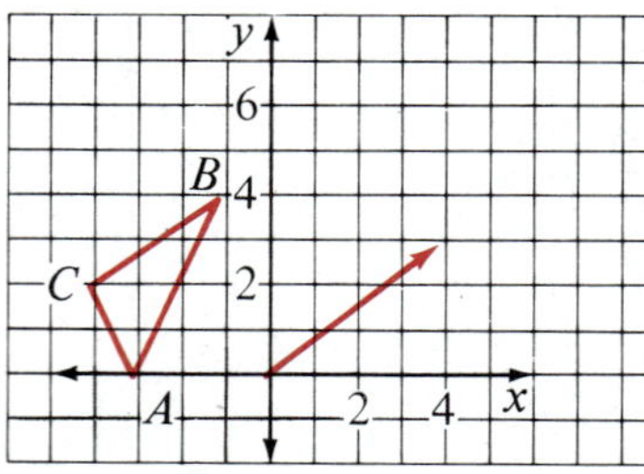

2.

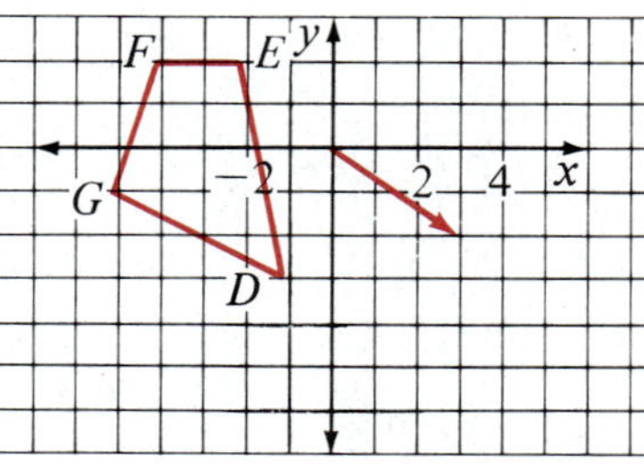

3.

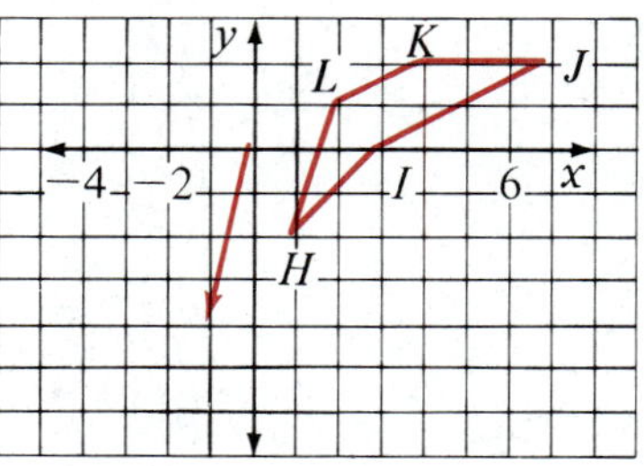

The coordinates of the vertices of a polygon and the tip of a slide arrow are given. Graph and label the polygon, the slide arrow, and the slide image.

4. $A(5, 4)$, $B(3, 1)$, $C(2, 0)$
tip of slide arrow: $(3, -1)$

5. $D(6, 1)$, $E(4, 3)$, $F(2, -3)$, $G(1, -1)$
tip of slide arrow: $(-4, 2)$

INEQUALITIES FOR TWO TRIANGLES 15.4

Objective

To apply the Hinge Theorem and its converse to pairs of triangles

Suppose two sticks are hinged together and connected with an elastic band. The two triangles shown below represent different possibilities for the degree measure of the angle formed by the sticks. Observe the effect on the side opposite that angle as the measure of the angle increases.

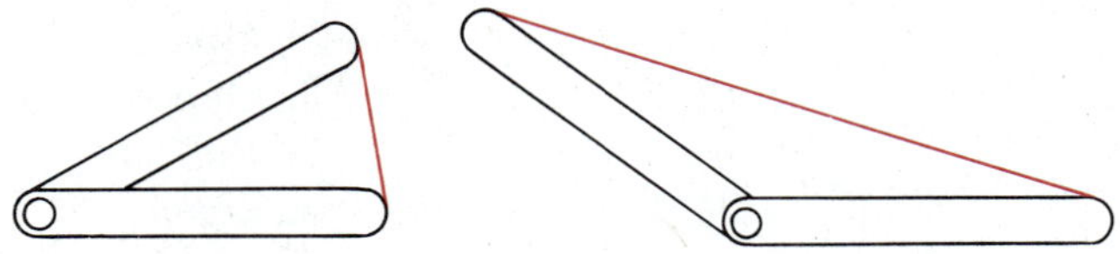

The situation represented above is generalized in Theorem 15.4.

Theorem 15.4 The Hinge Theorem If two sides of one triangle are congruent, respectively, to two sides of another triangle, and the included angle of the first triangle has a greater measure than the included angle of the second triangle, then the third side of the first triangle has a greater measure than the third side of the second triangle.

Given: $\triangle ABC$ and $\triangle DEF$ with $\overline{AC} \cong \overline{DF}$, $\overline{CB} \cong \overline{FE}$, and $m\angle ACB > m\angle F$

Prove: $AB > DE$

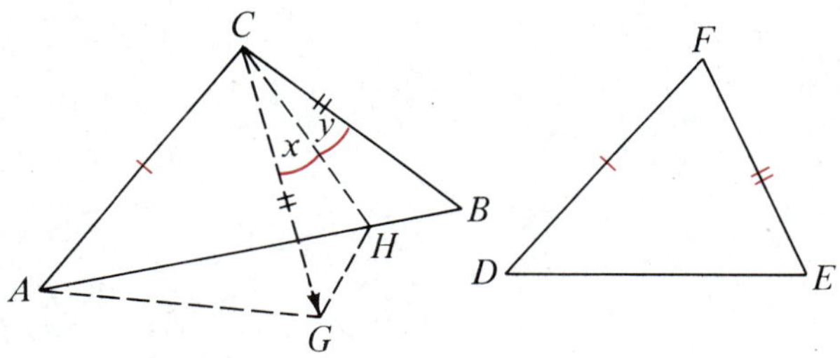

ANALYSIS

Construct $\triangle ACG \cong \triangle DFE$. Let $\overrightarrow{CH}$ bisect $\angle GCB$. $AH + HG > AG$. Prove $\triangle GCH \cong \triangle BCH$ and substitute to show $AB > AG$.

(See Exercise 12.)

You can use the Hinge Theorem to compare lengths of sides in two triangles as shown in Examples 1 and 2.

Example 1 C is the midpoint of $\overline{AD}$, $\overline{AB} \cong \overline{DE}$, $m\angle A = 65$, and $m\angle D = 50$. Compare CE and BC.

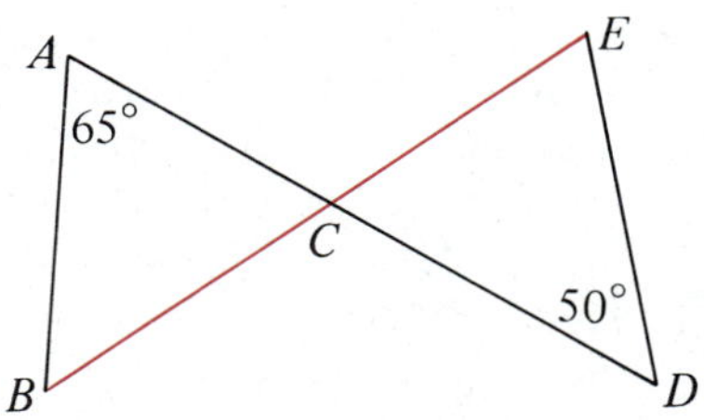

$\overline{AC} \cong \overline{CD}$
$\overline{AB} \cong \overline{DE}$
$m\angle A > m\angle D$
Thus, $BC > CE$, or $CE < BC$ by Theorem 15.4.

Example 2

Given: ▱ $ABCD$ with m $\angle ADC >$ m $\angle DCB$
Prove: $AC > BD$

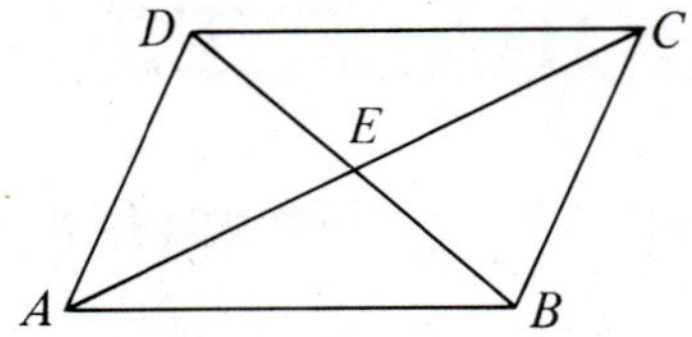

Proof

Statements	Reasons
1. $ABCD$ is a ▱.	1. Given
2. $\overline{AD} \cong \overline{BC}$	2. Opp. sides of a ▱ are $\cong$.
3. $\overline{DC} \cong \overline{DC}$	3. Reflexive property
4. m $\angle ADC >$ m $\angle DCB$	4. Given
5. $\therefore AC > BD$	5. The Hinge Theorem

The next theorem is the converse of Theorem 15.4.

Theorem 15.5 Converse of the Hinge Theorem

If two sides of one triangle are congruent, respectively, to two sides of another triangle, and the third side of the first has a greater measure than the third side of the second, then the angle opposite the third side of the first triangle has a greater measure than the angle opposite the third side of the second triangle.

Given: $\triangle ABC$ and $\triangle DEF$ with $\overline{AC} \cong \overline{DF}$, $\overline{CB} \cong \overline{FE}$, and $AB > DE$
Prove: m $\angle C >$ m $\angle F$

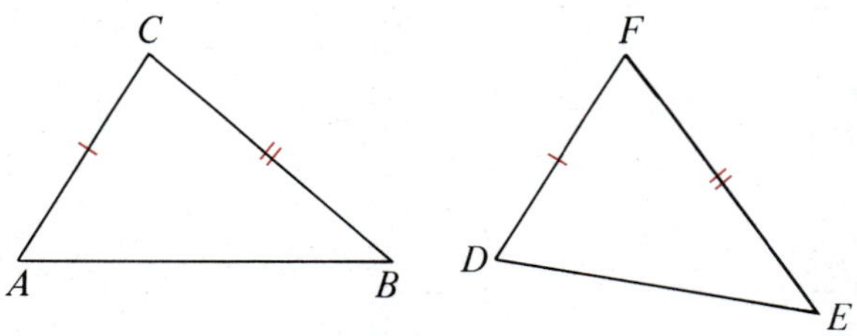

ANALYSIS

Use an indirect proof. Either m $\angle C <$ m $\angle F$, m $\angle C =$ m $\angle F$, or m $\angle C >$ m $\angle F$. Show that the first two cases lead to contradictions.

Proof

Statements	Reasons
1. $\overline{AC} \cong \overline{DF}$, $\overline{CB} \cong \overline{FE}$, $AB > DE$	1. Given
2. m $\angle C <$ m $\angle F$ or m $\angle C =$ m $\angle F$ or m $\angle C >$ m $\angle F$	2. Trichotomy property
3. If m $\angle C <$ m $\angle F$, then $AB < DE$.	3. The Hinge Theorem
4. Thus, m $\angle C$ is *not* $<$ m $\angle F$.	4. Given: $AB > DE$
5. If m $\angle C =$ m $\angle F$, then $\triangle ABC \cong \triangle DEF$.	5. SAS for $\cong \triangle$'s
6. Then $AB = DE$	6. CPCTC
7. Thus, m $\angle C \neq$ m $\angle F$	7. Given: $AB > DE$
8. $\therefore$ m $\angle C >$ m $\angle F$	8. Steps 2, 4, and 7

Example 3

Given: $\triangle ABC$ with M the midpoint of $\overline{AB}$, $\overline{AD} \cong \overline{BE}$, and $DM > EM$
Prove: $BC > AC$

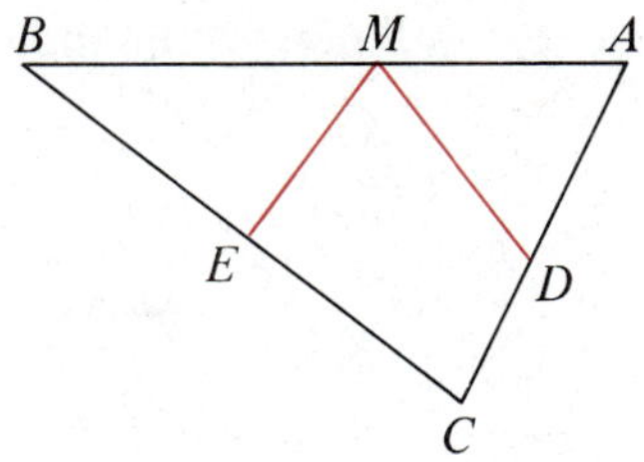

Proof

Statements	**Reasons**
1. M is the midpoint of $\overline{AB}$.	1. Given
2. $\overline{AM} \cong \overline{MB}$	2. Def. of midpoint
3. $\overline{AD} \cong \overline{BE}$	3. Given
4. $DM > EM$	4. Given
5. $m\angle A > m\angle B$	5. Converse of the Hinge Theorem
6. $\therefore BC > AC$	6. If meas. of 1 $\angle$ of a $\triangle$ is $>$ meas. of a 2nd $\angle$, then side opp. 1st $\angle$ is longer than side opp. 2nd $\angle$.

Reading in Geometry

Copy and complete each sentence to make a true statement.

1. In a triangle, the sum of the measures of any two sides is ? than the measure of the third side. **greater**

2. The length of the ? in a right triangle is greater than the length of either ?. **hypotenuse, leg**

3. In $\triangle ABC$, if the measure of $\angle A$ is greater than the measure of $\angle B$, then BC is ? than AC. **greater**

4. The difference between the lengths of two sides of a triangle is ? than the length of the third side. **less**

5. In $\triangle RST$, if RS is less than ST, then the measure of $\angle R$ is ? than the measure of $\angle T$. **greater**

6. In an obtuse triangle, the side opposite the obtuse angle has the ? measure of the three sides. **greatest**

Oral Exercises

Compare the measures of the indicated segments.

1. $\overline{BC}$ and $\overline{CD}$ **BC < CD** **2.** $\overline{AD}$ and $\overline{DC}$ **AD > DC** **3.** $\overline{CD}$ and $\overline{DB}$ **CD < DB** **4.** $\overline{RV}$ and $\overline{UT}$ **RV < UT**

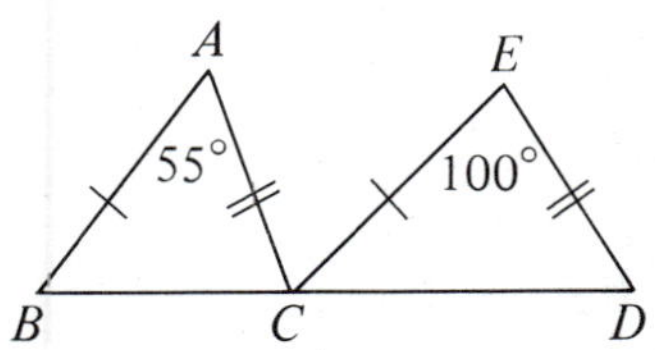

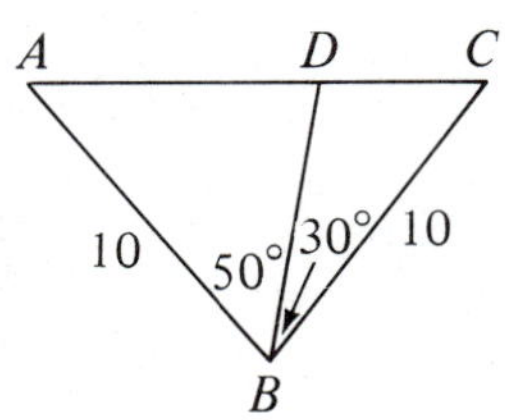

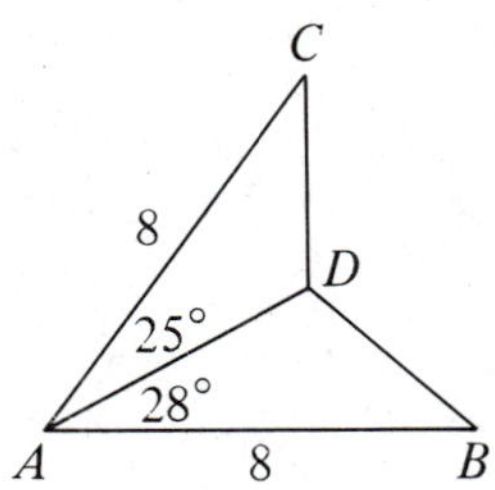

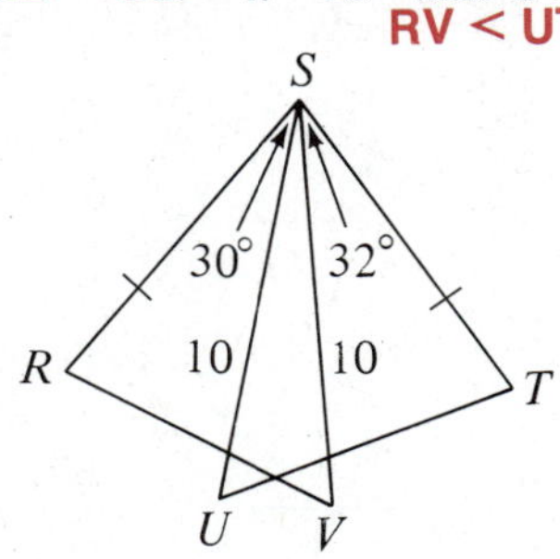

Compare the measures of the indicated angles.

5. $\angle x$ and $\angle y$ — m $\angle x <$ m $\angle y$

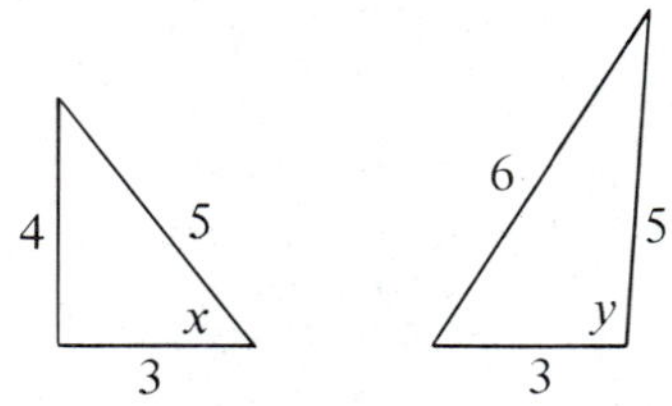

6. $\angle r$ and $\angle s$ — m $\angle r >$ m $\angle s$

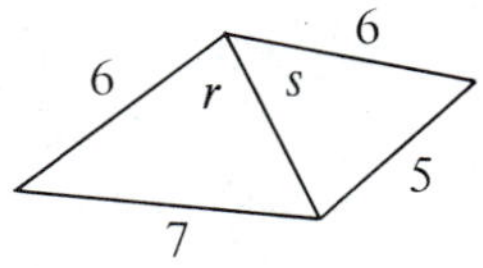

7. $\angle w$ and $\angle z$ — m $\angle w >$ m $\angle z$

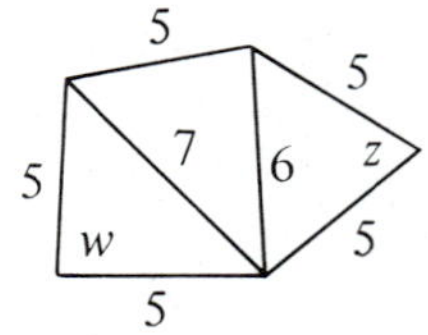

8. $\angle p$ and $\angle q$ — m $\angle p <$ m $\angle q$

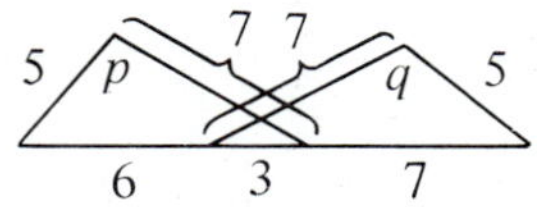

Written Exercises

A

1. Given: ▱ $ABCD$ with m $\angle BAD >$ m $\angle ADC$
   Prove: $BD > AC$

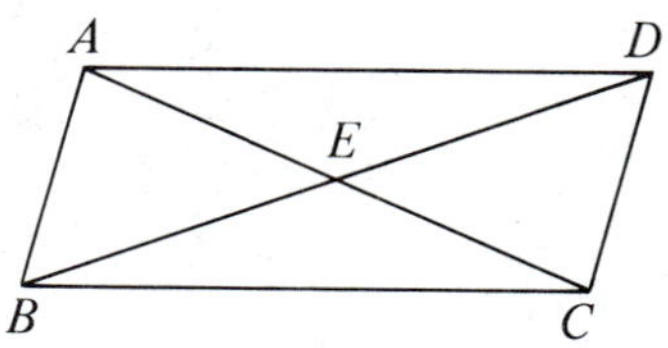

2. Given: $\odot O$ with $AB > CD$
   Prove: m $\angle x >$ m $\angle y$

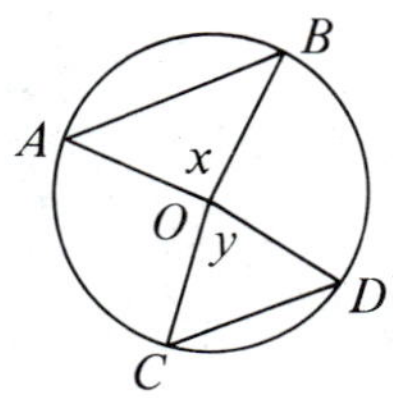

3. Given: $AD = CD$ and $AB > CB$
   Prove: m $\angle ADB >$ m $\angle CDB$

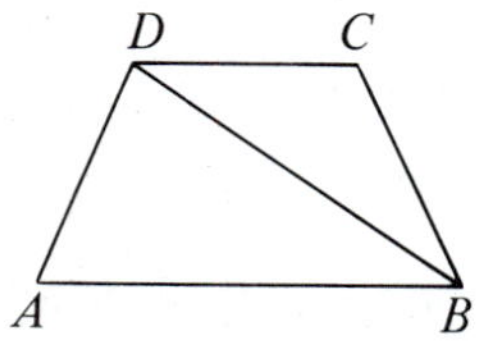

4. Given: $\overline{CD}$ is the median to $\overline{AB}$ and $CB < CA$
   Prove: m $\angle y >$ m $\angle x$

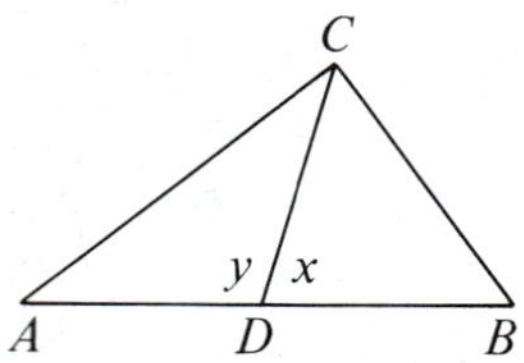

5. Given: $AC = BC$ and m $\angle ACD >$ m $\angle DCB$
   Prove: $AD > DB$

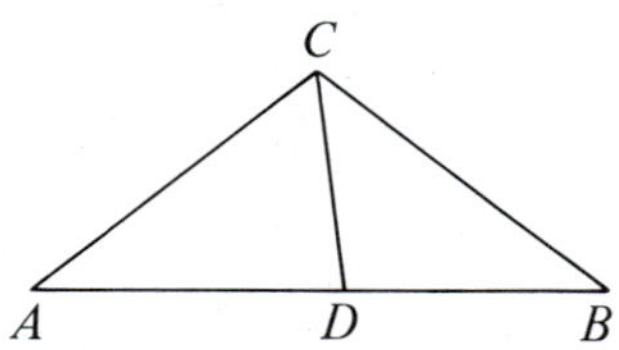

6. Given: $AB = DC$, $\overline{AB} \parallel \overline{DC}$, m $\angle AED <$ m $\angle DEC$
   Prove: $AD < DC$

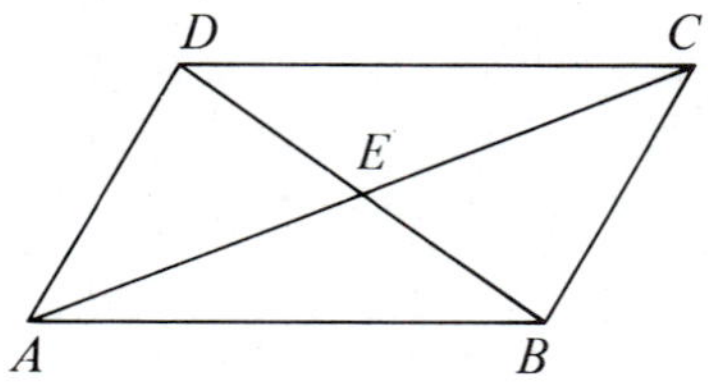

B

7. Given: $\overline{TU}$ is the median to $\overline{RS}$ and m $\angle x >$ m $\angle y$
   Prove: m $\angle S >$ m $\angle R$

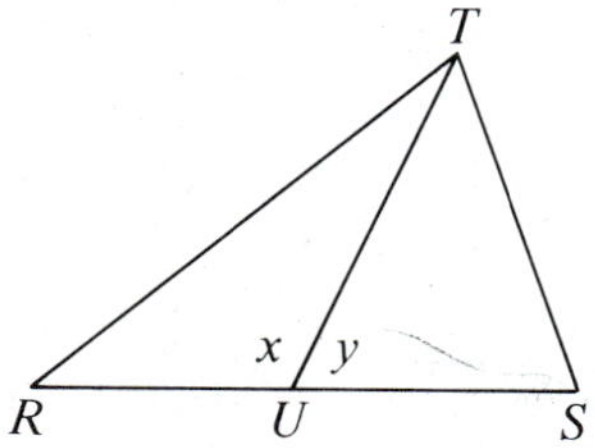

8. Given: $\triangle ABC$ with m $\angle x <$ m $\angle y$, $\overline{AE} \cong \overline{EB}$
   Prove: $AC > CB$

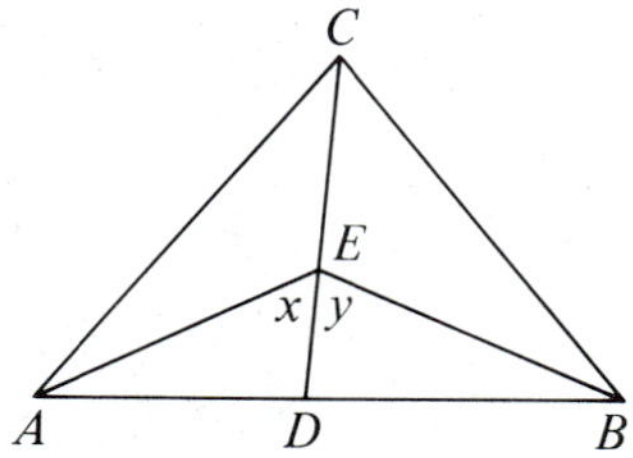

9. Given: $AD = EB$ and $DB > AE$
   Prove: $CB > CA$

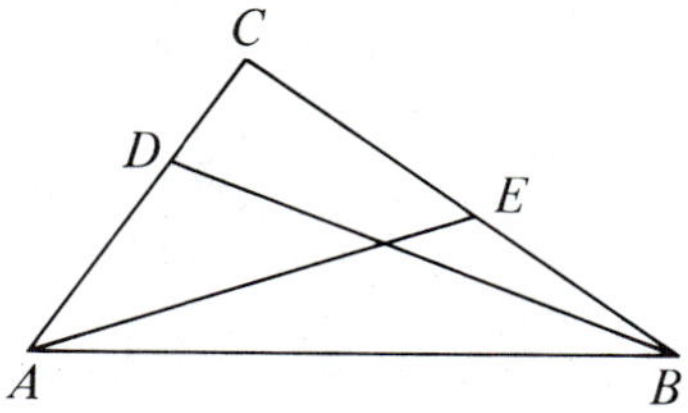

10. Given: m $\angle x =$ m $\angle y$
and $BC > AB$
Prove: m $\angle BDC >$ m $\angle BDA$

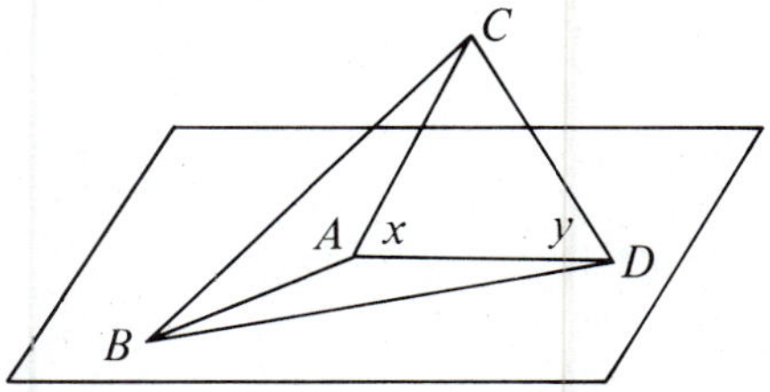

11. Given: m $\angle r <$ m $\angle s$
and m $\angle p =$ m $\angle q$
Prove: m $\angle DBC <$ m $\angle ABC$

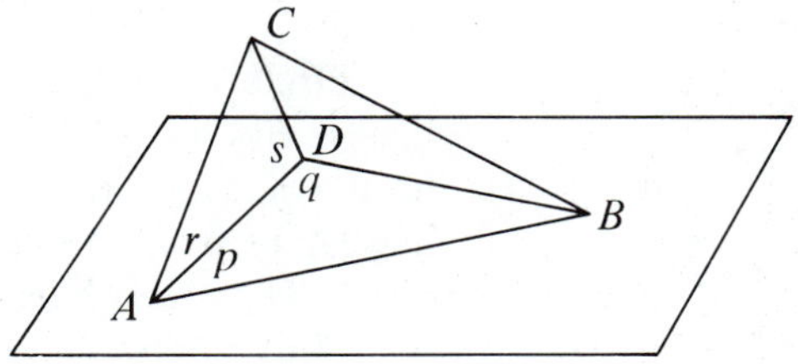

12. Prove Theorem 15.4.

CUMULATIVE REVIEW

1. Find the area of a rectangle with one side 7 cm long and a diagonal $\sqrt{113}$ cm long. **56 cm²**

2. Find the length of a diagonal of a rectangle with one side 6 cm long and an area of 30 cm². **$\sqrt{61}$ cm**

ALGEBRA REVIEW

Objective

To solve inequalities by using inequality properties

When negative numbers occur in algebraic inequalities, the two properties below are sometimes necessary.

Inequality Properties: Multiplication — If $a > b$ and $c < 0$, then $ac < bc$.

Division — If $a > b$ and $c < 0$, then $\frac{a}{c} < \frac{b}{c}$.

Example

Solve $2x + 5 > 8x + 23$.

$2x + 5 > 8x + 23$
$2x > 8x + 18$ ◄ *Add −5 to each side.*
$-6x > 18$ ◄ *Add −8x to each side.*
$x < -3$ ◄ *Divide each side by −6.*

Thus, all real numbers less than −3 are solutions of $2x + 5 > 8x + 23$.

1. $-6z < -96$; $z > 16$ **2. $-x > -5$; $x < 5$** **3. $-8x > 32$; $x < -4$**
4. $-4x < -40$; $x > 10$ **5. $-7x > -63$; $x < 9$** **6. $-7x < 14$; $x > -2$**
7. $-5x \leq -10$; $x \geq 2$ **8. $2x > -14$; $x > -7$**

Solve.

1. $2z + 8 < 8z - 88$ **2.** $x + 15 > 2x + 10$ **3.** $9x - 4 > 17x + 28$
4. $2 + x < 5x - 38$ **5.** $55 + 5x > 12x - 8$ **6.** $-3x - 8 < 6 + 4x$
7. $-7x + 4 \leq -2x - 6$ **8.** $6 > -3x - 8 + x$ **9.** $-x + 8 - 3x < 0$
10. $2x - 3 \geq 5x + 6$ **11.** $-\frac{3}{5}x - 4 \leq 6$ **12.** $6 - \frac{2}{3}x \geq -8$

9. $-4x < -8$; $x > 2$ **10. $-3x \geq 9$; $x \leq -3$** **11. $-\frac{3}{5}x \leq 10$; $x \geq -\frac{50}{3}$** **12. $-\frac{2}{3}x \geq -14$; $x \leq 21$**

INEQUALITIES IN CIRCLES 15.5

Objective

To apply inequality relationships to circles

In a circle there is a relationship between the lengths of chords and their distances from the center. Recall that chords of a given circle that are of the same length are equidistant from the center. For chords $\overline{AB}$ and $\overline{CD}$ at the right, $CD > AB$. Notice that $\overline{CD}$ appears to be nearer to the center of the circle than $\overline{AB}$; that is, $OF < OE$. This observation is generalized in Theorem 15.6.

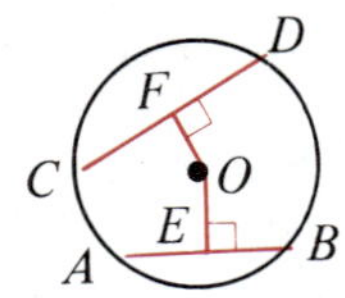

Theorem 15.6

If two chords of a circle are unequal in length, then the longer chord is nearer to the center of the circle.

Given: $\odot O$ with $AB > BC$, $\overline{OF} \perp \overline{BC}$, and $\overline{OE} \perp \overline{AB}$

Prove: $OE < OF$

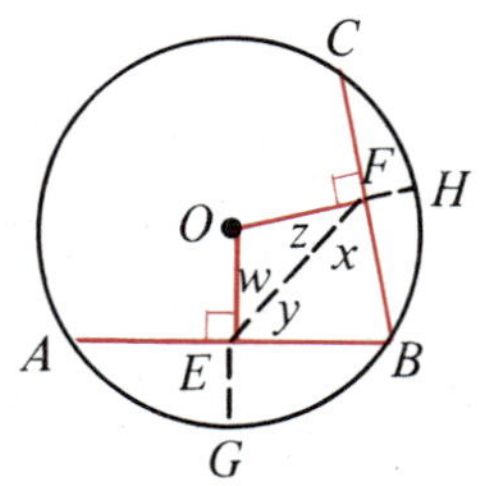

ANALYSIS

Assume that the two chords have a common endpoint. (If not, construct ≅ chords that have a common endpoint.) Use $\triangle OEF$ and $\triangle BEF$.

Proof

Statements	Reasons
1. $\odot O$ with $\overline{OF} \perp \overline{BC}$ and $\overline{OE} \perp \overline{AB}$	1. Given
2. Draw $\overline{EF}$, $\overline{FH}$, and $\overline{EG}$.	2. Two points are contained in exactly one line.
3. $AE = EB$ and $BF = FC$	3. A radius ⊥ to a chord bisects the chord.
4. $AB = AE + EB$ and $BC = BF + FC$	4. Def. of between
5. $AB = 2(EB)$ and $BC = 2(BF)$	5. Substitution and Distributive properties
6. $AB > BC$	6. Given
7. $EB > BF$	7. Substitution and Division prop. of inequality
8. $m\angle x > m\angle y$	8. Theorem 15.2
9. $m\angle OEB = m\angle OFB$	9. ⊥ lines form ≅ rt. ∠'s.
10. $m\angle z + m\angle x = m\angle w + m\angle y$	10. Angle Addition Post.; Substitution
11. $m\angle z < m\angle w$	11. Subtraction of unequals from equals
12. $\therefore OE < OF$	12. Theorem 15.3

Theorem 15.6 is also true for congruent circles. (See Exercise 13.)

Example 1

Compare ON and OM.

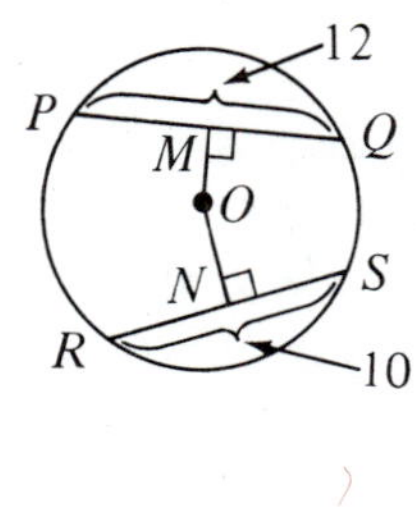

$PQ = 12$ and $RS = 10$, so $PQ > RS$.

Thus $OM < ON$, or $ON > OM$, by Th. 15.6.

The converse of Theorem 15.6 is also true.

Theorem 15.7 If two chords of a circle are unequally distant from the center, then the chord nearer the center is the longer. (See Exercise 12.)

Theorem 15.7 is also true for congruent circles. (See Exercise 14.)

Example 2

Given: $\odot O$ with $\overline{OE} \perp \overline{AB}$, $\overline{OF} \perp \overline{CD}$, and $OF < OE$
Prove: $m\angle COD > m\angle AOB$

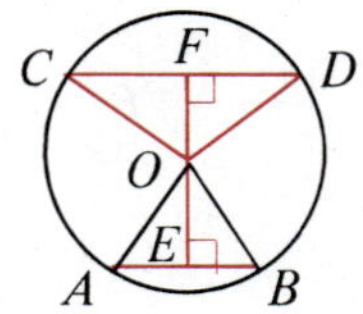

Proof

Statements	Reasons
1. $\overline{OE} \perp \overline{AB}$, $\overline{OF} \perp \overline{CD}$, $OF < OE$	1. Given
2. $CD > AB$	2. Theorem 15.7
3. $\overline{OA} \cong \overline{OC}$ and $\overline{OB} \cong \overline{OD}$	3. Any two radii of a $\odot$ are $\cong$.
4. $m\angle COD > m\angle AOB$	4. Converse of the Hinge Theorem

The proof in Example 3 involves arcs of unequal measure.

Example 3

Given: $\odot O$ with $m\angle x > m\angle y$
Prove: $m\widehat{AB} > m\widehat{CD}$

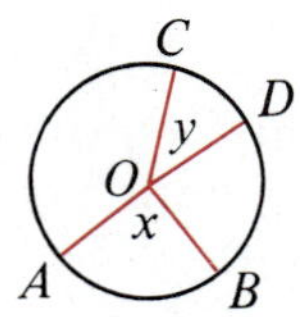

Proof

Statements	Reasons
1. $m\angle x > m\angle y$	1. Given
2. $m\angle x = m\widehat{AB}$; $m\angle y = m\widehat{CD}$	2. Meas. of minor arc = meas. of central $\angle$.
3. $\therefore m\widehat{AB} > m\widehat{CD}$	3. Substitution property

Oral Exercises

Compare the indicated measures.

1. AB and CD, FD and EB
AB > CD; FD < EB

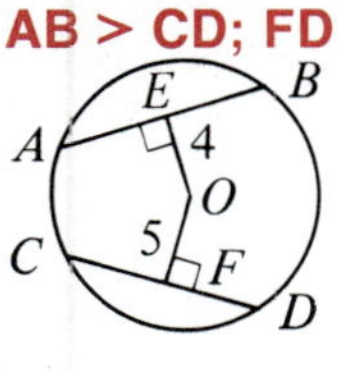

2. AO and CO, $m\widehat{AB}$ and $m\widehat{BC}$
AO = CO; $m\widehat{AB} < m\widehat{BC}$

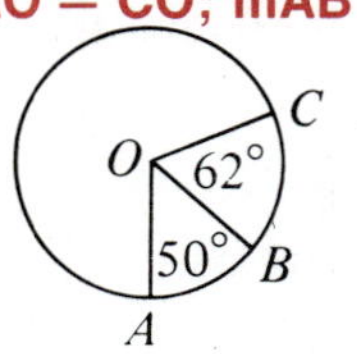

3. OP and OR, PN and NR
OP < OR; PN > NR

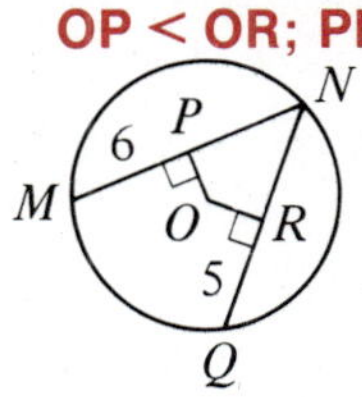

4. RS and TU, WT and VS
RS > TU; WT < VS

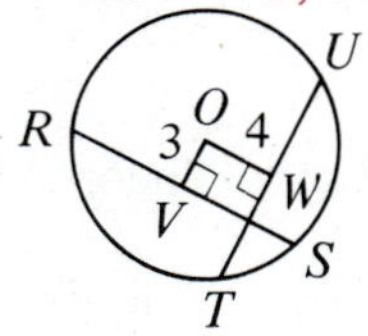

Written Exercises

A **1.** Given: $\odot O$ with $\overline{OE} \perp \overline{AB}$, $\overline{OF} \perp \overline{CD}$, m $\angle COD <$ m $\angle AOB$
Prove: $OF > OE$

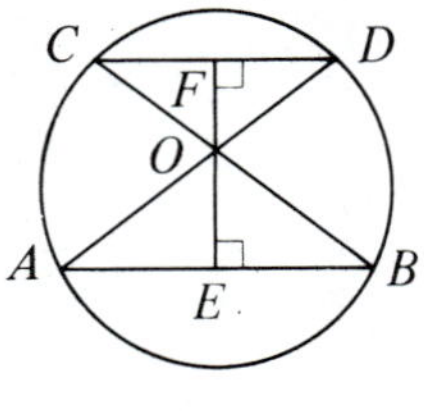

2. Given: $\odot O$ with m$\widehat{AB} <$ mCD
Prove: m $\angle x <$ m $\angle y$

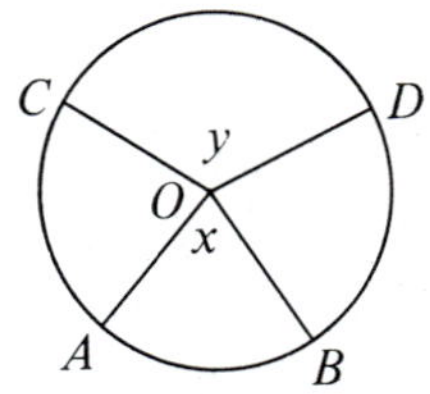

3. Given: $\odot O$ with $\overline{OE} \perp \overline{AB}$ and $\overline{OF} \perp \overline{CD}$
Prove: $AB < CD$

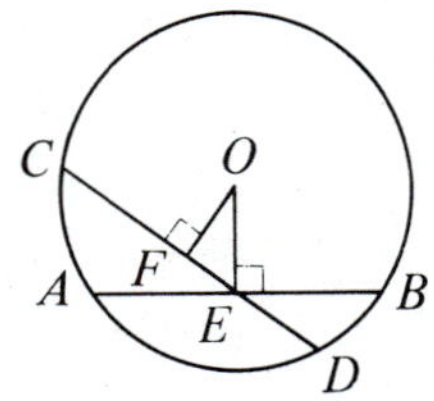

4. Given: $\odot O$ with $\overline{OE} \perp \overline{AB}$, $\overline{OF} \perp \overline{BC}$, $AE > BF$
Prove: $OE < OF$

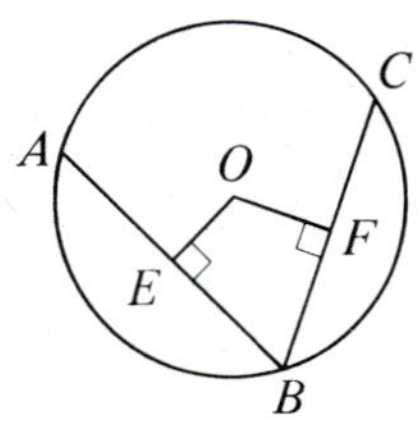

5. Given: $\odot O$ with m $\angle B >$ m $\angle A$
Prove: m$\widehat{AC} >$ m$\widehat{BC}$

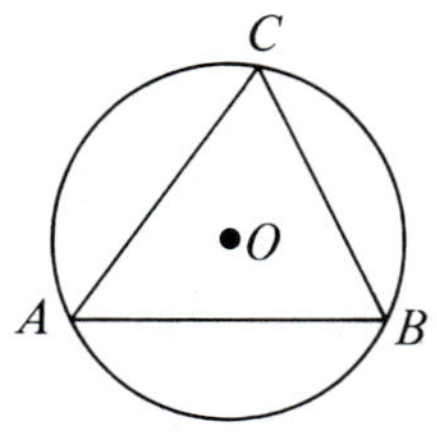

6. Given: $\odot O$ with $AB > CD$, $\overline{OE} \perp \overline{AB}$, $\overline{OF} \perp \overline{CD}$, $\overline{EB} \cong \overline{BF}$
Prove: m $\angle x >$ m $\angle y$

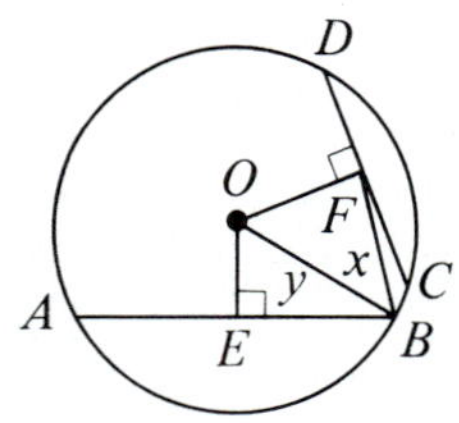

B **7.** Given: $\odot O$ with m$\widehat{AB} <$ m$\widehat{BC}$, $\overline{OD} \perp \overline{AB}$, $\overline{OE} \perp \overline{BC}$
Prove: $OD > OE$

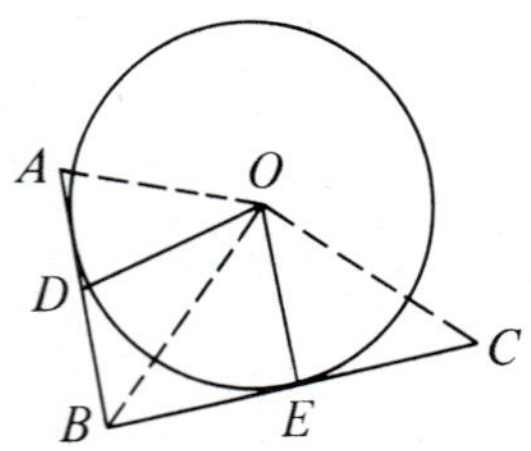

8. Given: $\odot O$ with $\overline{OE} \perp \overline{AB}$, $\overline{OF} \perp \overline{CD}$, m $\angle x >$ m $\angle y$, $\overline{AE} \cong \overline{AF}$
Prove: $AB > CD$

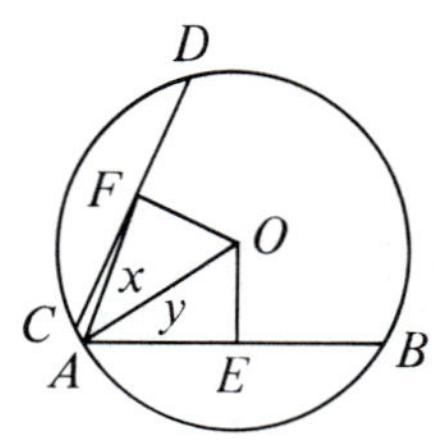

9. Given: $\odot O$ with diameter $\overline{AOB}$ and chord $\overline{AC}$
Prove: $AB > AC$
[*Hint:* Draw $\overline{OD} \perp \overline{AC}$.]

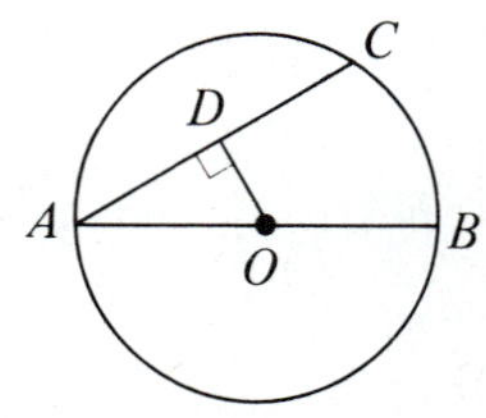

10. Prove that if a square and an equilateral triangle are inscribed in the same circle, the apothem of the square is longer than the apothem of the triangle.

11. Prove that through any point in the interior of a circle, except the center, the chord that is perpendicular to the radius through this point is shorter than any other chord through the point.

C **12.** Prove Theorem 15.7.

13. Prove Theorem 15.6 for congruent circles.

14. Prove Theorem 15.7 for congruent circles.

APPLICATIONS

Read → Plan → Solve → Interpret

1. Two children wish to fence in a corner of the back yard for their dog. They have 100 ft of fence. They first stretched the fence diagonally across the corner, using 55 ft of fence. What can you predict about the rest of their fencing project? Why?

2. Two towns are located on the same highway. The main telephone cable is not parallel to the highway. If the telephone company connects each town to this main cable as shown, will the cable lengths be equal? Why or why not?

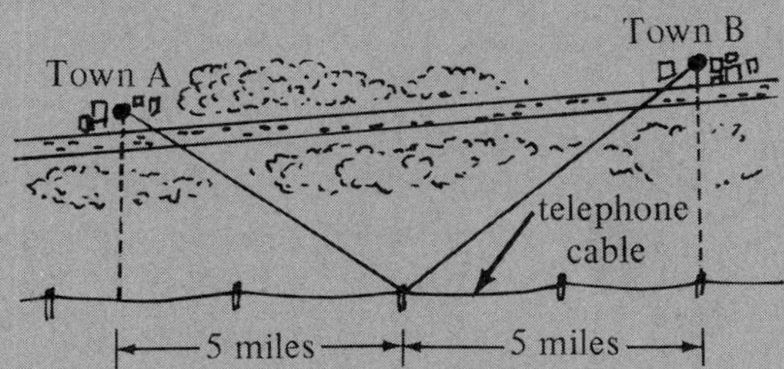

CHAPTER FIFTEEN REVIEW

(A) exercises are 1–12. (B) exercises are 13, 14. (C) exercises are starred.

Compare each of the following as indicated. Then give the property that applies. [15.1]

1. Given: $WY > XZ$
Compare WX and YZ.
WX > YZ, subtr. prop. of ineq.

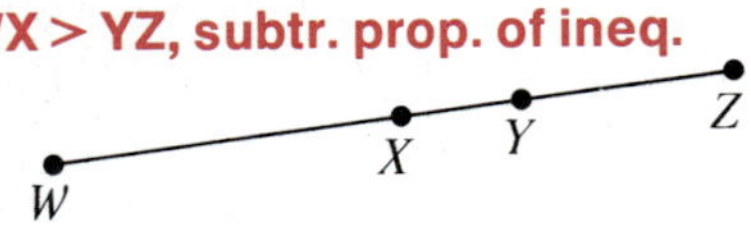

2. Given: $m\angle t < m\angle s$ and $m\angle s < m\angle r$
Compare $m\angle t$ and $m\angle r$.
$m\angle t < m\angle r$, trans. prop. of ineq.

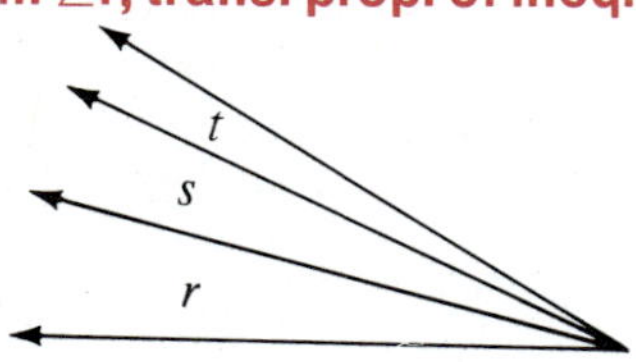

3. Given: $AB > CD$, M is the midpoint of $\overline{AB}$, and N is the midpoint of $\overline{CD}$.
Compare AM and ND.
AM > ND, div. prop. of ineq.

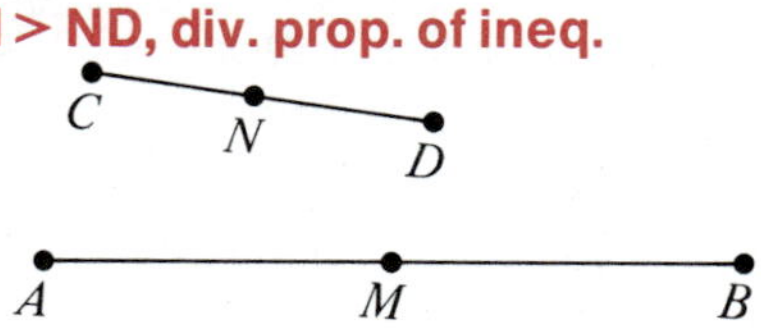

Determine whether a triangle can be constructed with segments having these lengths. [15.2]

4. 7, 9, 15
yes

5. 1 m, 3 m, 5 m
no

Name the angles in order according to their measures, from largest to smallest. [15.3]

6.

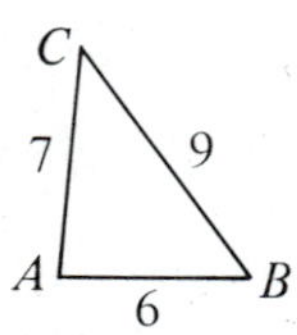

$\angle A, \angle B, \angle C$

7.

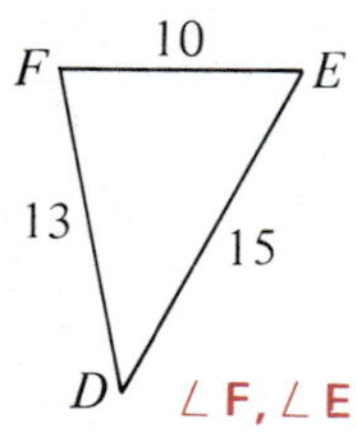

$\angle F, \angle E, \angle D$

Name the sides in order according to their measures, from largest to smallest. [15.3]

8.

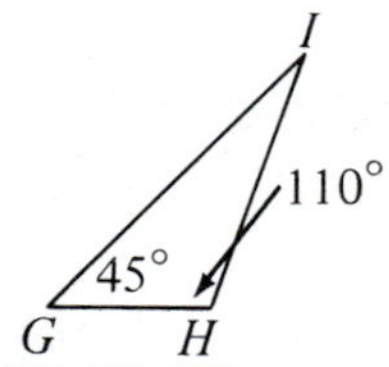

$\overline{GI}, \overline{IH}, \overline{GH}$

9.

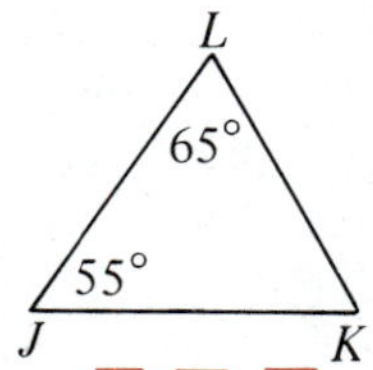

$\overline{JK}, \overline{JL}, \overline{LK}$

Compare the indicated measures.

10. $m\angle x$ and $m\angle y$

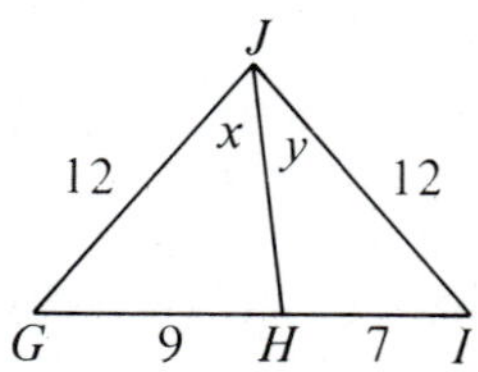

$m\angle x > m\angle y$

11. AB and CD

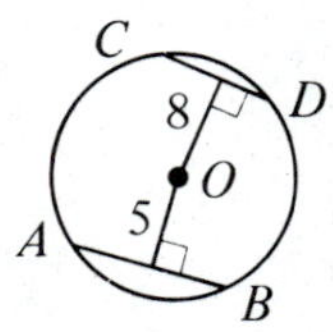

$AB > CD$

Prove each of the following. [15.4]

12. Given: $ABCD$ is a quadrilateral.
Prove: $AC + BD < AB + BC + CD + DA$

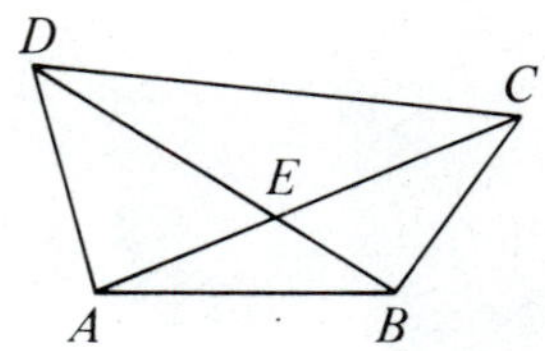

13. Given: $\triangle ABC$ with D in the interior [15.3]
Prove: $AC + BD < AC + CB$

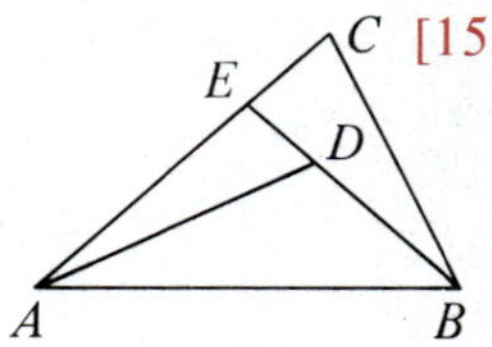

14. Given: $\odot O$ with $\overline{OE} \perp \overline{AB}$, $\overline{OF} \perp \overline{CD}$, and $AE < CF$
Prove: $OE > OF$

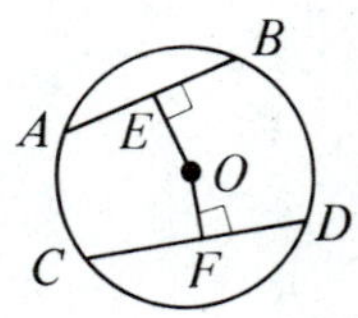

***15.** Given: $m\angle x > m\angle y$ and $m\angle w = m\angle z$ [15.4]
Prove: $m\angle ABC > m\angle DBC$

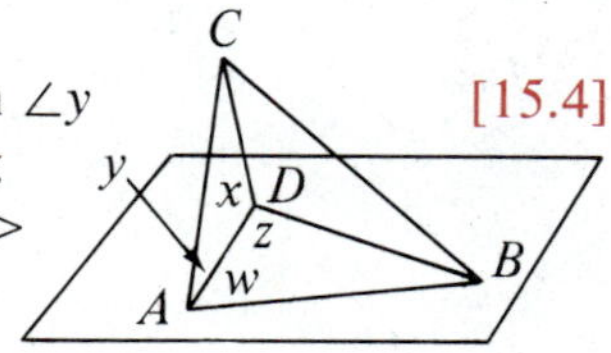

CHAPTER FIFTEEN TEST

A exercises are 1–11. B exercise is 12. C exercise is starred.

Compare each of the following as indicated. Then give the property that applies.

1. Given: m $\angle r >$ m $\angle s$
Compare m $\angle ABD$ and m $\angle CBE$.

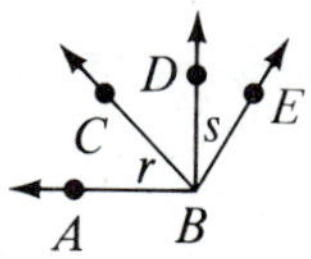

m $\angle$ ABD > m $\angle$ CBE, add. prop. of ineq.

2. Given: $PQ = RS$ and $PT < RU$
Compare TQ and US.

TQ > US, subtr. of unequals from equals

Determine whether a triangle can be constructed with segments having these lengths.

3. 6, 2, 4 **No**

4. 5 m, 6 m, 7 m **Yes**

Name the angles in order according to their measures, from largest to smallest.

5.

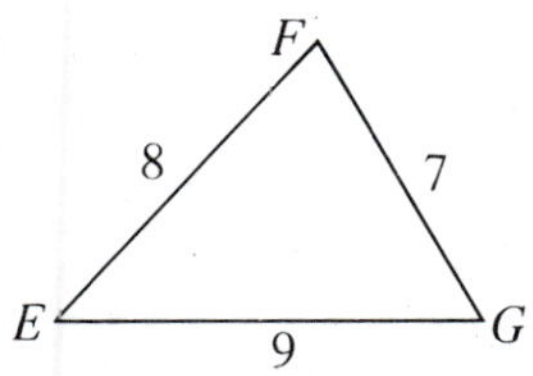

$\angle$ F, $\angle$ G, $\angle$E

6.

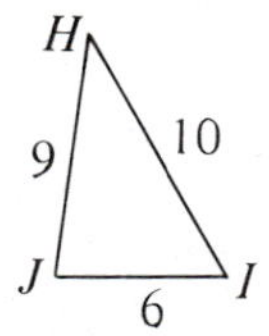

$\angle$ J, $\angle$ I, $\angle$ H

Name the sides in order according to their measures, from largest to smallest.

7.

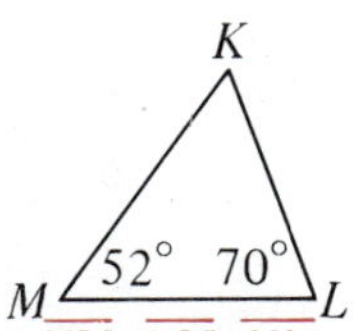

$\overline{KM}$, $\overline{LM}$, $\overline{KL}$

8.

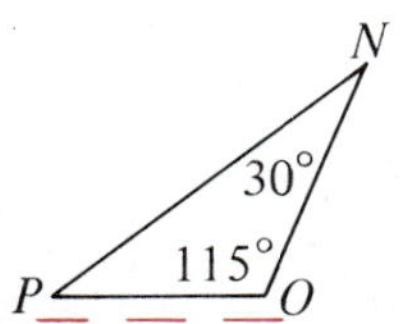

$\overline{PN}$, $\overline{NO}$, $\overline{OP}$

Compare the indicated measures.

9. m $\angle x$ and m $\angle y$

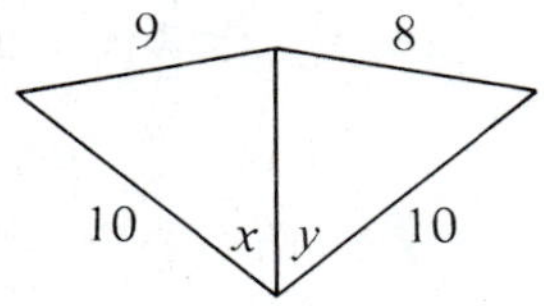

m $\angle$x > m $\angle$ y

10. OF and OE

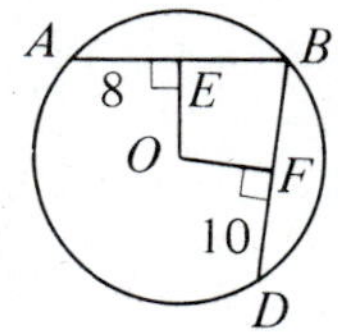

OF $\angle$ OE

Prove each of the following.

11. Given: $\square PQRS$ with m $\angle SPQ <$ m $\angle PQR$
Prove: $SQ < PR$

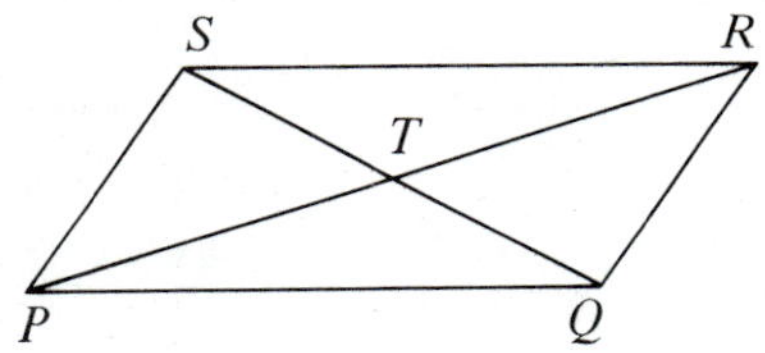

12. Given: $AC < CB$, $\overrightarrow{AE}$ bisects $\angle CAB$, $\overrightarrow{BE}$ bisects $\angle ABC$.
Prove: $AE < EB$

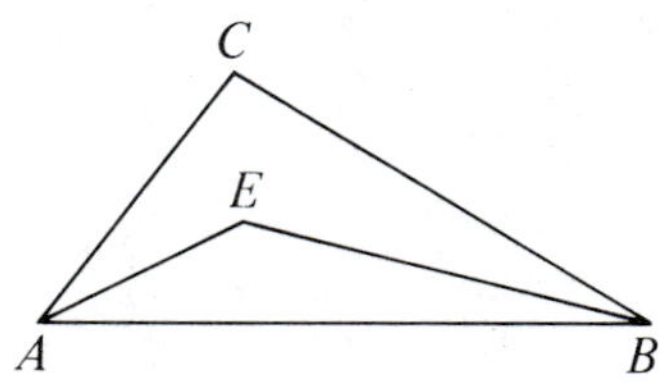

***13.** Given: $SQ > PS$ and $SQ = SR$
Prove: m $\angle P +$ m $\angle R >$ m $\angle PQR$

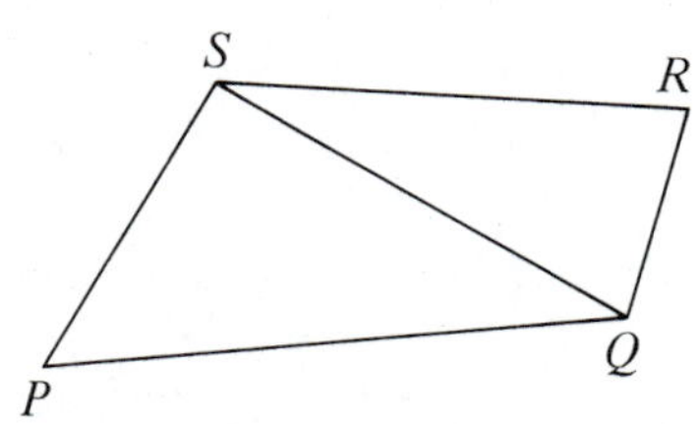

COMPUTER ACTIVITIES

Inequalities in a Triangle

Objective: **To order the sides of a triangle by their lengths, using a computer program that inputs the measures of the angles**
To order the angles of a triangle by their measures, using a computer program that inputs the lengths of the sides

Theorem 15.2 and its converse, Theorem 15.3, permit you to order the angles of a triangle by measure from the lengths of the sides, and vice versa. In the program below, the angles are inputted, and the computer orders the sides by length. Use the values given in Exercises 5–8 on page 508 for test numbers.

```
110  PRINT "THIS PROGRAM GIVES CORRECT ANSWERS FOR"
120  PRINT "SCALENE TRIANGLES ONLY.": PRINT
130  PRINT "WHAT IS THE NAME OF YOUR TRIANGLE";: INPUT
     N$
140  FOR I = 1 TO 3:A$(I) =  MID$ (N$,I,1): NEXT I
150  FOR I = 1 TO 3: PRINT "WHAT IS THE MEASURE OF
     ANGLE ";A$(I);
160  INPUT A(I): NEXT I
170  IF A(1) > A(2) THEN 260
180  IF A(1) > A(3) THEN 240
190  IF A(2) > A(3) THEN 220
200 B$(1) = A$(1) + A$(2):B$(2) = A$(3) + A$(1)
210 B$(3) = A$(3) + A$(2): GOTO 340
220 B$(1) = A$(3) + A$(1):B$(2) = A$(2) + A$(1)
230 B$(3) = A$(2) + A$(3): GOTO 340
240 B$(1) = A$(1) + A$(3):B$(2) = A$(2) + A$(3)
250 B$(3) = A$(2) + A$(1): GOTO 340
260  IF A(2) > A(3) THEN 320
270  IF A(1) > A(3) THEN 300
280 B$(1) = A$(1) + A$(2):B$(2) = A$(3) + B$(2)
290 B$(3) = A$(1) + A$(3): GOTO 340
300 B$(1) = A$(3) + A$(2):B$(2) = A$(1) + A$(2)
310 B$(3) = A$(1) + A$(3): GOTO 340
320 B$(1) = A$(2) + A$(3):B$(2) = A$(1) + A$(3)
330 B$(3) = A$(1) + A$(2)
340  PRINT "THE LONGEST SIDE IS "B$(1)"."
350  PRINT "THE MIDDLE SIDE IS "B$(2)"."
360  PRINT "THE SHORTEST SIDE IS "B$(3)"."
370  PRINT : GOTO 130
```

EXERCISE: See the Computer Section beginning on page 565 for more information.

Write a program that inputs the lengths of the sides of a triangle and outputs an ordering of the angles by measure.

COLLEGE PREP TEST

Directions: In each item you are to compare a quantity in Column 1 with a quantity in Column 2. Write the letter of the the correct answer from these choices:

A—The quantity in Column 1 is greater than the quantity in Column 2.
B—The quantity in Column 2 is greater than the quantity in Column 1.
C—The quantity in Column 1 is equal to the quantity in Column 2.
D—The relationship cannot be determined from the given information.

Notes: Information centered over both columns refers to one or both of the quantities to be compared.
A symbol that appears in both columns has the same meaning in each column and all variables represent real numbers.

Sample Question and Answer

Column 1		Column 2
m $\angle 4$ + m $\angle 1$	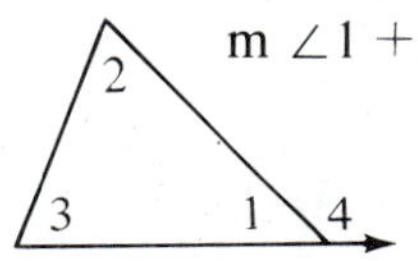	m $\angle 1$ + m $\angle 2$ + m $\angle 3$

Answer: **C** since m $\angle 4$ = m $\angle 2$ + m $\angle 3$ by substitution m $\angle 4$ + m $\angle 1$ = m $\angle 1$ + m $\angle 2$ + m $\angle 3$

Column 1		Column 2
1. $2AC$	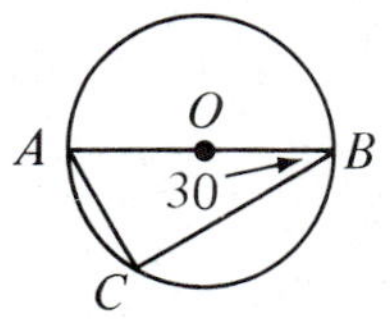	AB **C**
2. m $\angle B$	In $\triangle ABC$, $AC < AB$, $AC > BC$	m $\angle A$ **A**
3. DC	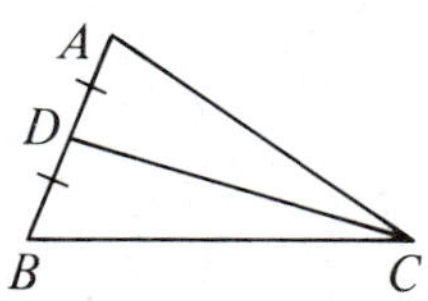	$BC + AD$ **B**
4. $b - a$	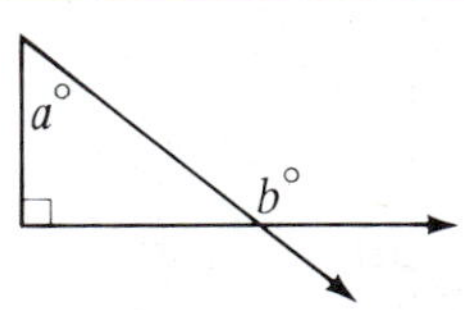	90 **C**
5. BC	In $\triangle ABC$, m $\angle A$ > m $\angle B$, m $\angle C = 60$	AB **A**

Column 1		Column 2
6. y	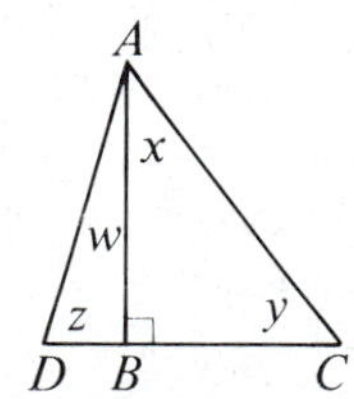	z **D**
7. x		y **C**
8. w		z **D**
9. $w + z$	$\overline{AB} \perp \overline{DC}$, $AB = BC$ 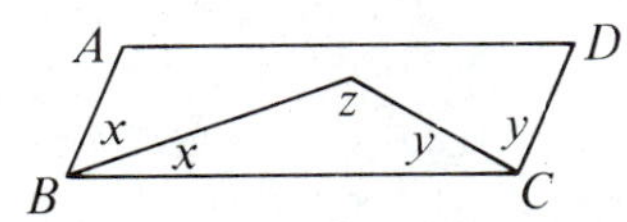	$x + y$ **C**
10. z	$\overline{AB} \parallel \overline{CD}$, $\overline{AD} \parallel \overline{BC}$, $y = 50$	90 **C**
11. $3x$	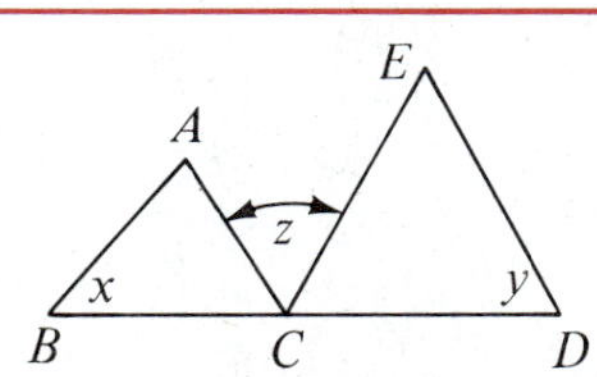	$2z$ **C**
12. y	$AB = AC$, $EC = ED$, $x = 40$, $y = 80$	z **A**
13. BC	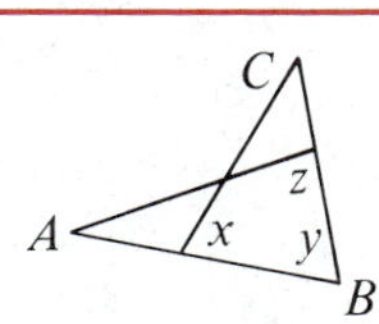	AB **D**
14. m $\angle A$	$x = y = z$	m $\angle C$ **C**

16 FIGURES IN SPACE

Non-Routine Problem Solving

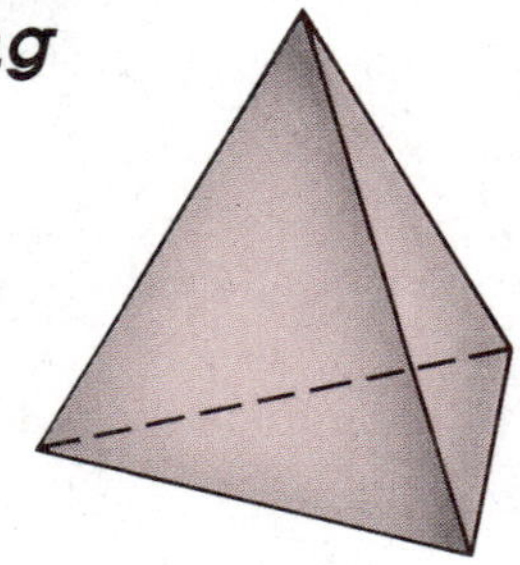

Tetrahedron
4 faces that are equilateral triangular regions

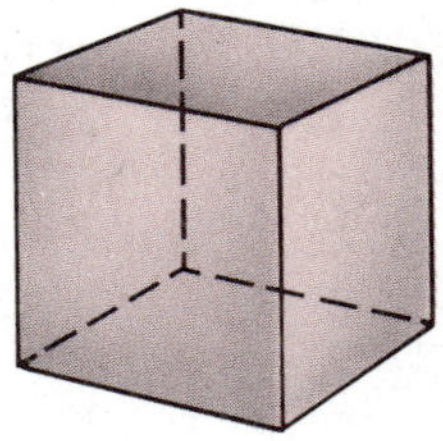

Hexahedron
6 faces that are square regions

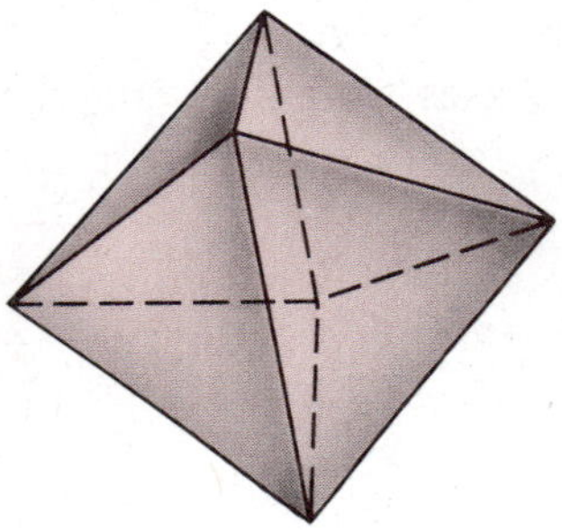

Octahedron
8 faces that are equilateral triangular regions

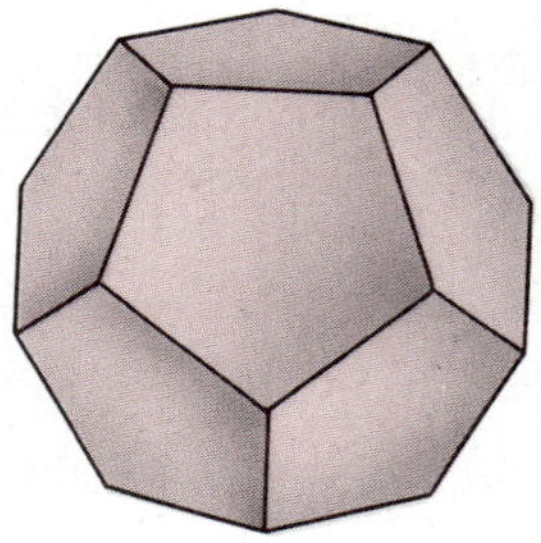

Dodecahedron
12 faces that are regular pentagonal regions

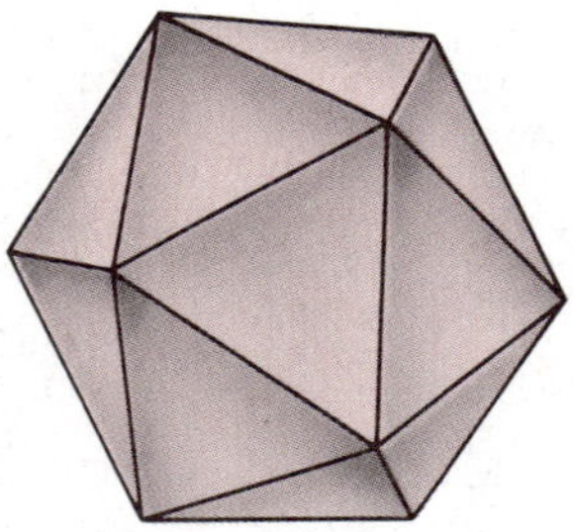

Icosahedron
20 faces that are equilateral triangular regions

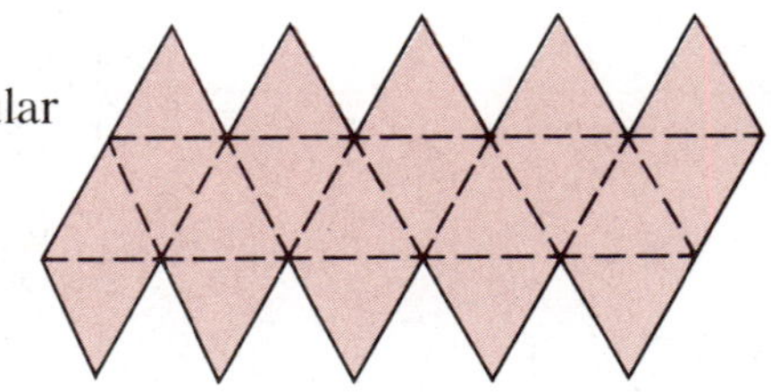

Each of the regular polyhedrons shown above can be constructed from congruent regular polygonal regions. One pattern for construction is shown on the bottom right.

Problem 1 For each polyhedron, count the number of faces (F), edges (E), and vertices (V). Find a formula relating F, E, and V. **$E + 2 = F + V$**

Problem 2 Construct a model of one of these regular polyhedrons. Draw a pattern first.

PRISMS

16.1

Objectives

To identify and count the parts of a prism
To sketch various types of prisms

Recall that all of the points that make up a plane figure lie in the same plane. On the other hand, all of the points that make up a space figure do not lie in the same plane. Space figures are often called **solids.** Several types of solids are shown below.

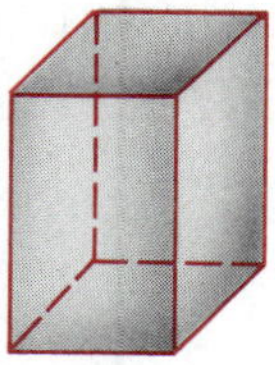
Prism

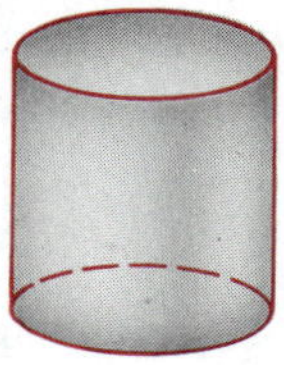
Cylinder

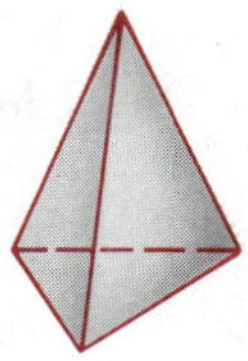
Pyramid

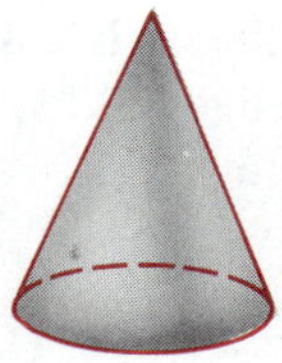
Cone

Sphere

A prism is a special type of solid. A prism has two bases which are congruent polygonal regions lying in parallel planes. A prism is named according to its bases. For example, a triangular prism has two bases which are congruent triangular regions.

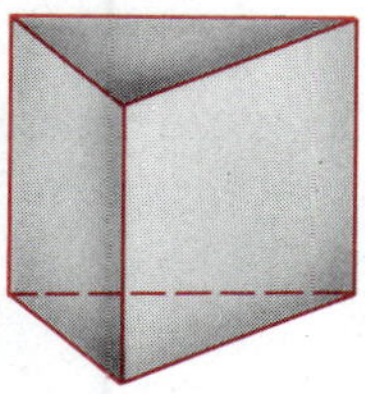
Triangular Prism

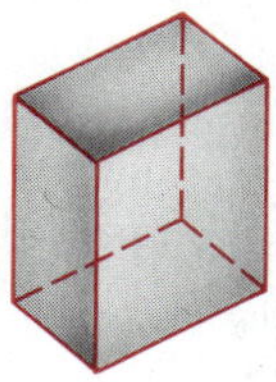
Quadrangular Prism

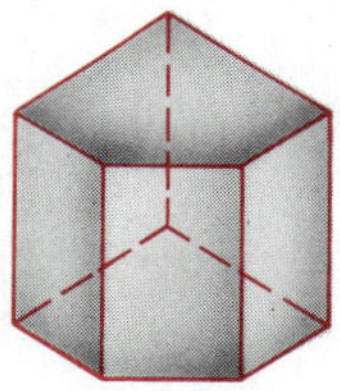
Pentagonal Prism

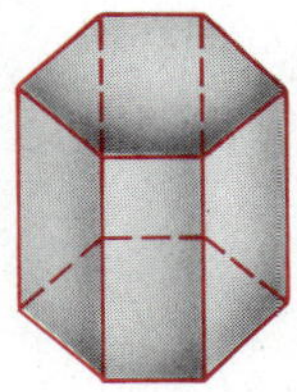
Hexagonal Prism

Example 1

Give the type of base of each prism.

Prism	Type of Base
Quadrangular prism	Quadrilateral region
Pentagonal prism	Pentagonal region
Octagonal prism	Octagonal region

The polygonal regions that make up a prism are called its **faces.** For the sake of simplicity, the faces will be referred to as polygons rather than polygonal regions. The bases of a prism are two of its faces. The faces intersect in segments which are called the **edges** of the prism. The edges intersect in points which are called the **vertices** of the prism. A triangular prism has 5 faces, 9 edges, and 6 vertices.

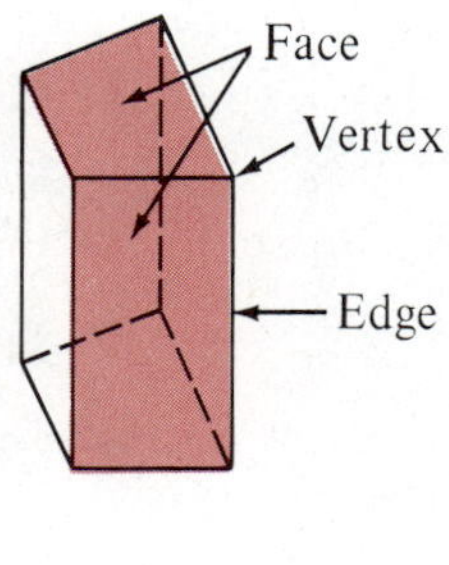

Example 2 Give the number of edges, faces, and vertices of each type of prism.

Prism	Edges	Faces	Vertices
Quadrangular prism	12	6	8
Pentagonal prism	15	7	10
Hexagonal prism	18	8	12

The faces of a prism that are not bases are called the **lateral faces.** The lateral faces of a prism are parallelograms. The edges that are not sides of the bases are called the **lateral edges.** A triangular prism has 3 lateral faces and 3 lateral edges.

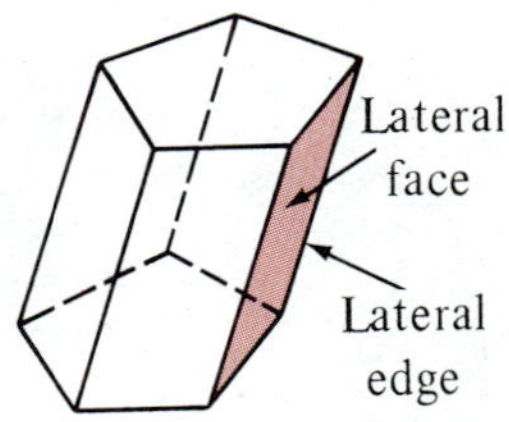

Example 3 Give the number of lateral faces and lateral edges for each type of prism.

Prism	Lateral faces	Lateral edges
Quadrangular prism	4	4
Pentagonal prism	5	5
Hexagonal prism	6	6

In a **right prism**, the lateral faces and lateral edges are perpendicular to the bases. The lateral faces of a right prism are rectangles. In an **oblique prism**, the lateral faces and lateral edges are *not* perpendicular to the bases.

Right Prisms

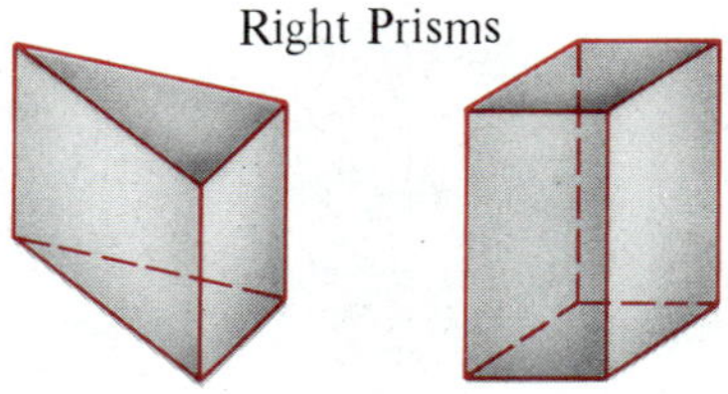

Oblique Prisms

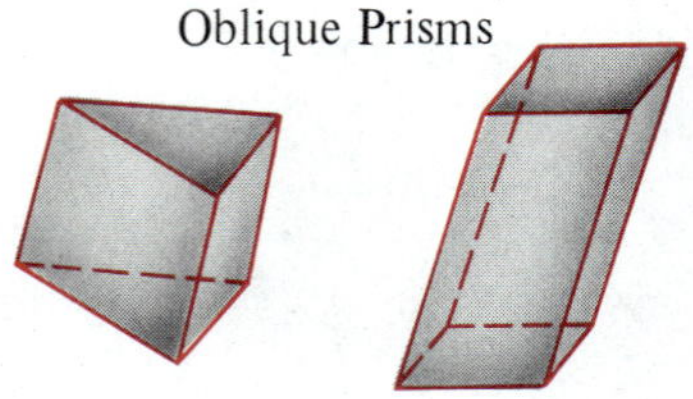

Example 4 Sketch a right pentagonal prism.

Draw 2 congruent pentagons with parallel sides.

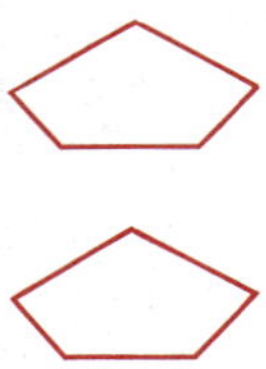

Connect the vertices.

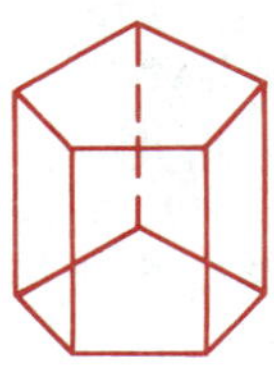

Be sure that all invisible edges are dashed.

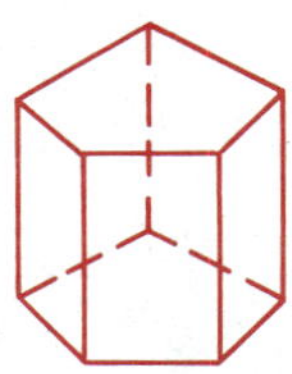

Oral Exercises

All pts. of pl. figure lie in 1 pl; Space figure contains pts. of more than 1 pl.

1. Explain the difference between a plane figure and a space figure.
2. Name five plane figures. Name five space figures.
Answers may vary.

Identify each figure as completely as you can.

3. **sphere**

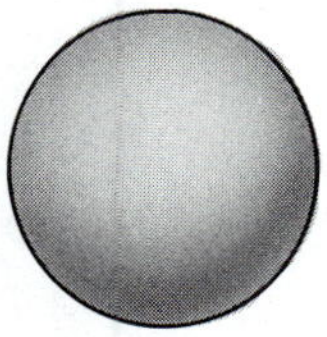

4. **pyramid**

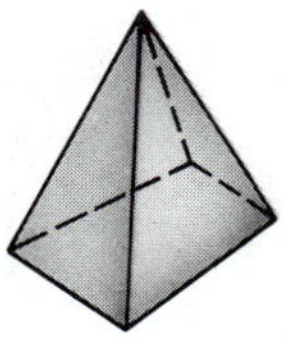

5. **cylinder**

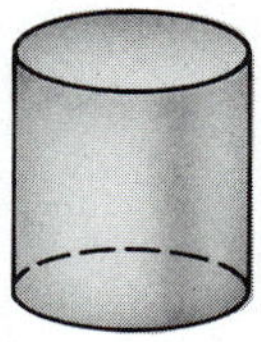

6. **rt. hex. prism**

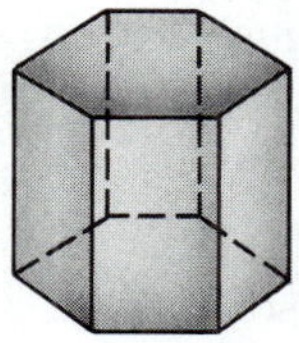

7. **cone**

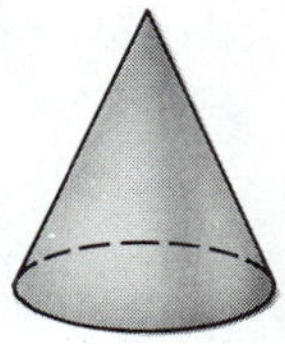

8. **oblique triang. prism**

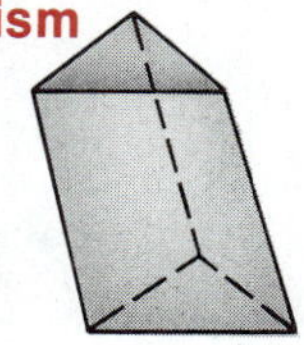

Written Exercises

Copy and complete the chart. Then sketch each prism.

	Type of Prism	No. of Lateral Edges	No. of Edges	No. of Vertices	No. of Bases	No. of Lateral Faces	No. of Faces	Type of Base
A 1.	Quadrangular	**4**	**12**	8	**2**	**4**	**6**	**quadrilateral**
2.	Pentagonal	5	**15**	**10**	**2**	**5**	7	**pentagon**
3.	Hexagonal	**6**	**18**	**12**	2	**6**	**8**	**hexagon**
4.	Octagonal	**8**	24	**16**	**2**	**8**	**10**	octagon
5.	Decagonal	**10**	**30**	**20**	**2**	10	**12**	**decagon**

Sketch each of the following.

6. A right triangular prism whose bases are right triangles
7. A right triangular prism whose bases are equilateral triangles
8. A right quadrangular prism whose bases are rectangles (a rectangular solid)
9. A rectangular solid with all edges congruent (a cube)
10. A right octagonal prism
11. An oblique hexagonal prism

True or false? Refer to the chart above.

B 12. A prism has as many bases as lateral faces. **F**

13. A prism has two more lateral edges than faces. **F**

14. A prism has twice as many vertices as lateral faces. **T**

15. A prism has the same number of lateral faces and lateral edges. **T**

16. In a prism, the ratio of the number of vertices to the number of edges is 2:3. **T**

17. A hexagonal prism has twice as many edges as a triangular prism. **T**

V + F = E + 2

18. Let V be the number of vertices, F be the number of faces, and E be the number of edges of a prism. What relationship exists between V, F, and E?

19. Suppose a prism has n lateral faces. Represent V, F, and E in terms of n and prove what you discovered in Exercise 18.

$V = 2n$; $F = n + 2$; $E = 3n$; $V + F = 2n + n + 2$; $V + F = 3n + 2$; $V + F = E + 2$

20. What is the measure of a dihedral angle formed by two adjacent lateral faces of a right prism whose bases are regular hexagons? **120°**

21. What is the measure of a dihedral angle formed by a base and a lateral face of a right prism whose bases are pentagons? **90°**

CUMULATIVE REVIEW

1. In $\odot O$, m $\angle AOB = 60$ and $AO = 12$. Find the length of $\widehat{AB}$. (Leave answer in terms of π.) **4π**

2. The length of the hypotenuse of an isosceles right triangle is 8. Find the length of a leg. **$4\sqrt{2}$**

NON-ROUTINE PROBLEMS

Inscribed Regular Polyhedrons

A regular tetrahedron is a space figure consisting of 4 faces that are congruent equilateral triangles. (See page 524.) A regular tetrahedron can be inscribed in a larger regular tetrahedron, as shown at the right. Each vertex of the smaller one is the center of a face of the larger one.

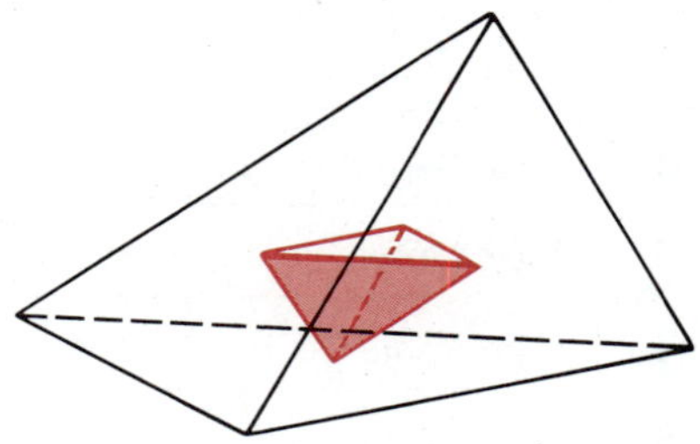

1. Draw a regular octahedron inscribed in a regular hexahedron.
2. Draw a regular hexahedron inscribed in a regular octahedron.
3. Copy and complete the chart. Is there a pattern that tells which polyhedrons can be inscribed in each other? If so, what is the pattern?

Polyhedron	Number of Faces	Number of Vertices
Tetrahedron	4	**4**
Hexahedron	**6**	8
Octahedron	**8**	**6**
Dodecahedron	**12**	20
Icosahedron	20	**12**

If a polyhedron has the same number of vertices as another polyhedron has faces, then one can be inscribed inside the other.

SURFACE AREAS OF PRISMS 16.2

Objective **To find the lateral area and the total area of a *right* prism**

Since the faces of a prism consist of polygons, you can define the surface area of a prism as the sum of the areas of the faces of the prism. The **lateral area** *(L)* of a prism is the sum of the areas of the lateral faces. The **total area** *(T)* of a prism is the sum of the lateral area and the areas of the two bases. An **altitude** of a prism is a segment perpendicular to the plane of each base with endpoints in the planes of the bases. In a right prism, each lateral edge is also an altitude.

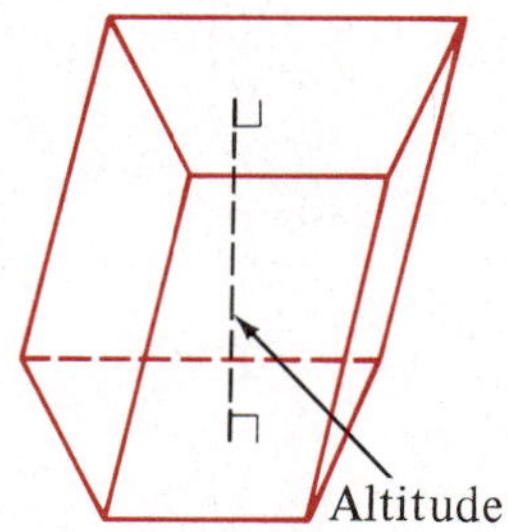

For the right prism below, two methods for finding the lateral area are shown.

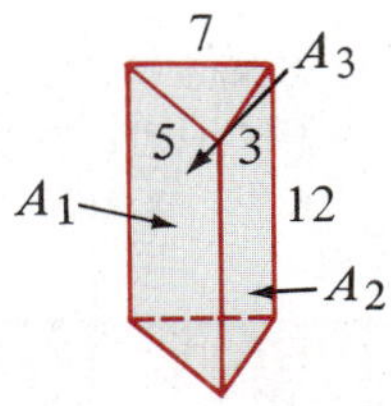

First Method

Each lateral face is a rectangle. Find the three areas, A_1, A_2, and A_3, and add.

Use the formula for the area of a rectangle:

$$\begin{aligned} A &= bh \\ A_1 &= 5 \cdot 12 = 60 \\ A_2 &= 3 \cdot 12 = 36 \\ A_3 &= 7 \cdot 12 = 84 \\ A_1 + A_2 + A_3 &= 180 \\ L &= 180 \end{aligned}$$

Second Method

"Unfold" the lateral faces. Find the area of the large rectangle formed.

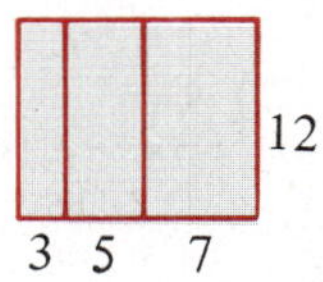

$$\begin{aligned} L &= ph \\ &= \begin{pmatrix}\text{perimeter}\\ \text{of}\\ \text{base}\end{pmatrix} \cdot \begin{pmatrix}\text{length of}\\ \text{altitude}\\ \text{of prism}\end{pmatrix} \\ &= (3+5+7) \cdot 12 \\ &= 15 \cdot 12 \\ L &= 180 \end{aligned}$$

Thus, the lateral area is 180 square units.
The second method above suggests the following theorem.

Theorem 16.1 For any right prism, lateral area = perimeter of base × length of an altitude.

$$L = ph$$

(See Exercise 18.)

Example 1

Find the lateral area of the right prism shown.

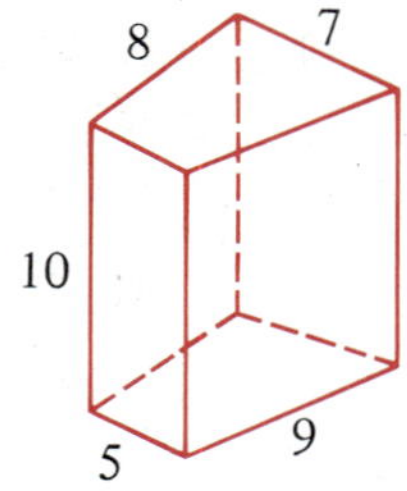

$L = ph \qquad p = 5 + 9 + 7 + 8 = 29$

$L = 29 \cdot 10 \qquad h = 10$

$L = 290$

Thus, the lateral area is 290 square units.

Theorem 16.2 follows directly from the definition of total area.

Theorem 16.2 For any prism, total area = lateral area + 2 times base area. $T = L + 2B$

A **rectangular solid** is a right prism whose bases are rectangles. A **cube** is a rectangular solid all of whose edges are congruent. In Examples 2 and 3, you will use Theorem 16.2 to find the total area of a rectangular solid and a right triangular prism, respectively.

Example 2

Find the total area of the rectangular solid shown.

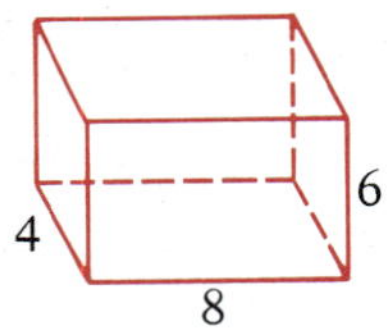

$T = L + 2B$

$L = ph \qquad p = 4 + 8 + 4 + 8 = 24 \qquad B = 4 \cdot 8$

$L = 24 \cdot 6 \qquad h = 6 \qquad B = 32$

$L = 144$

$T = L + 2B = 144 + 2(32) = 144 + 64 = 208$

Thus, the total area is 208 square units.

Example 3

Find the total area of a right triangular prism with an altitude measuring 7 cm and whose bases are equilateral triangles with sides measuring 8 cm.

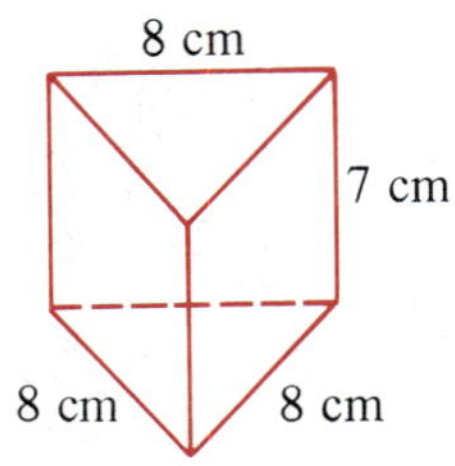

$T = L + 2B$

$L = ph \qquad p = 8 + 8 + 8 = 24 \qquad B = \frac{s^2}{4}\sqrt{3}$

$L = 24 \cdot 7 \qquad h = 7$

$L = 168 \qquad B = \frac{8^2}{4}\sqrt{3}$

$B = 16\sqrt{3}$

$T = L + 2B = 168 + 2(16\sqrt{3}) = 168 + 32\sqrt{3}$

Thus, the total area is $168 + 32\sqrt{3}$ cm^2.

Written Exercises

Find the lateral area of each right prism. (Ex. 1–4)

A

1\.

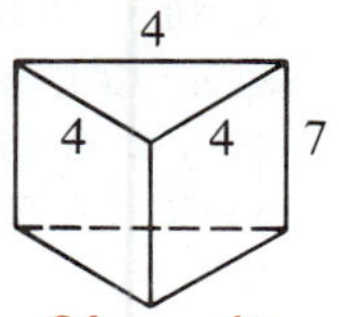

84 sq units

2\.

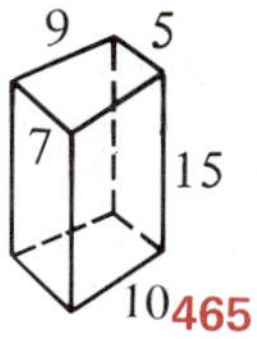

465 sq units

3\.

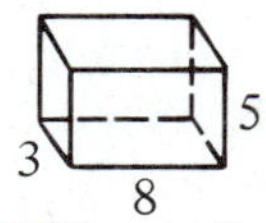

110 sq units

4\.

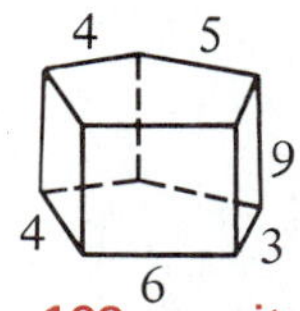

198 sq units

5\. A lateral edge of a right prism measures 6 cm and the perimeter of a base is 25 cm. Find the lateral area. **150 cm²**

6\. Find the total area of a cube with each edge 7 cm long. **294 cm²**

7\. Find the total area of a right prism whose bases are equilateral triangles with sides measuring 8 cm and with an altitude measuring 10 cm. **$240 + 32\sqrt{3}$ cm²**

8\. The bases of a prism are regular pentagons with sides measuring 7 mm. An altitude measures 12 mm. Find the lateral area of the prism. **420 mm²**

9\. Find the total area of a rectangular solid whose dimensions are 6 mm, 8 mm, and 11 mm. **404 mm²**

10\. The walls and ceiling of a warehouse are to be painted. How many square meters must be covered if the warehouse is 120 m by 96 m with a ceiling 3 m high? **12,816 m²**

B

11\. An altitude of a right prism measures 14 m, and each base is a right triangle with legs measuring 6 m and 8 m. Find the total area. **384 m²**

12\. A base of a prism is a rhombus with diagonals measuring 10 cm and 24 cm. An altitude measures 16 cm. Find the lateral area. **832 cm²**

13\. Find the total area of a prism whose bases are regular hexagons with 8-cm sides and whose altitudes measure 16 cm. **$768 + 192\sqrt{3}$ cm²**

14\. The total area of a right prism is 210 square cm. Each base is a square 5 cm on a side. Find the length of an altitude of the prism. **8 cm**

C

15\. In a truncated prism, the bases are not parallel. Find the total area of the truncated right prism shown. **156 sq. units**

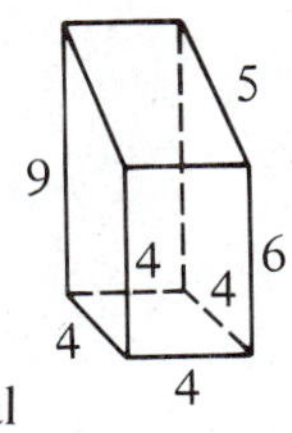

16\. A truncated right prism has a square base with 5-cm sides. Two of the lateral edges measure 8 cm and the other two measure 20 cm. Find the lateral area and the total area. **280 cm²; 370 cm²**

17\. A truncated right prism has an equilateral triangular base with 12-unit sides. Two of the lateral edges measure 14 units and the third measures 9 units. Find the total area. **$36\sqrt{3} + 6\sqrt{133} + 444$ sq units**

18\. Prove Theorem 16.1.

LOGIC REVIEW

Write the converse, inverse, and contrapositive of each given statement. Label each of these as true or false. **Check students' answers.**

1\. If a polygon is a triangle, then it has three congruent sides.

2\. If a prism is triangular, then it has exactly three rectangular lateral faces.

VOLUMES OF PRISMS 16.3

Objective **To find the volume of a prism**

To find any measure, an appropriate unit must be used. To find the surface area of a solid, a square unit is used. To find the **volume** of a solid, a **cubic unit** is used. A cubic unit of measure is a cube whose edges each measure 1 unit of length. For example, a cube with edges 1 centimeter long is called a *cubic centimeter* (cm^3).

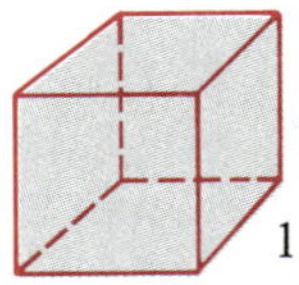

Cubic unit

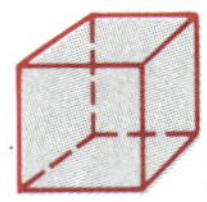

Cubic centimeter
cm^3

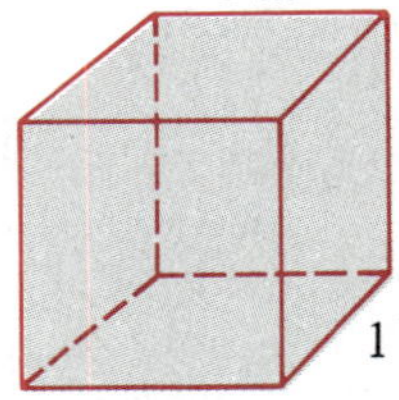

Cubic meter
m^3

Example 1 The *volume* of a solid is the number of cubic units that can be contained in the solid. First, you will consider the volume of a rectangular solid.

Find the volume of the rectangular solid below.

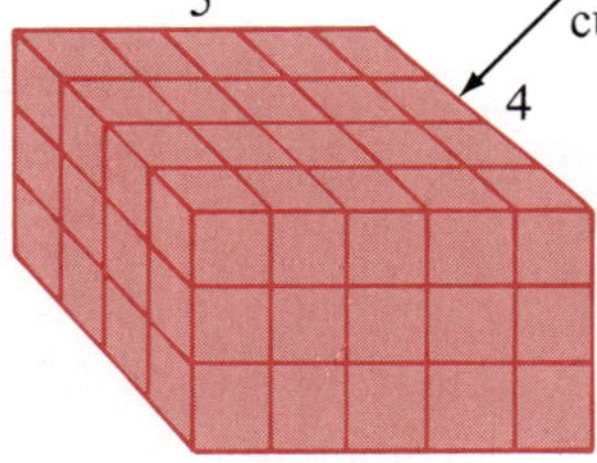

$$\begin{aligned} V &= \begin{pmatrix}\text{no. of unit cubes}\\ \text{per layer}\end{pmatrix} \cdot \begin{pmatrix}\text{no. of}\\ \text{layers}\end{pmatrix} \\ &= \quad \text{Base area} \quad \cdot \quad \text{length of an altitude} \\ & \qquad\quad 20 \qquad\quad \cdot \quad 3 \\ &= \qquad\quad 60 \end{aligned}$$

Thus, the volume is 60 cubic units.

The process used for finding the volume of the prism in Example 1 can be applied to all prisms, both right and oblique. The result is stated in Theorem 16.3, which is given without proof.

Theorem 16.3 For any prism, volume = base area × length of an altitude. $V = Bh$

Example 2 Find the volume of the right prism shown below.

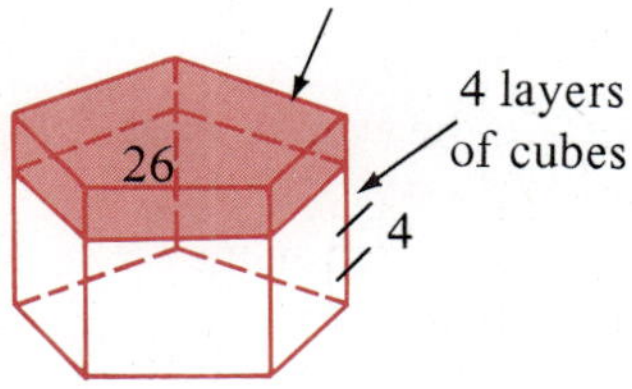

$V = Bh = 26 \cdot 4 = 104$

Thus, the volume is 104 cubic units.

Example 3 The bases of a prism are equilateral triangles with each side measuring 4 cm. An altitude measures 9 cm. Find the volume of the prism.

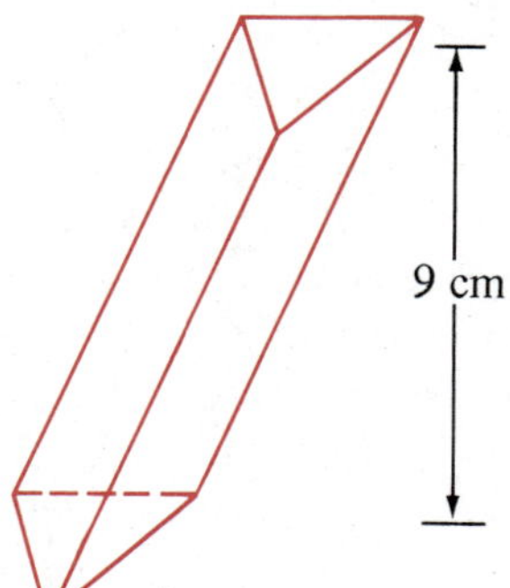

First, find the area of a base.

$$B = \frac{s^2}{4}\sqrt{3}$$
$$= \frac{4^2}{4}\sqrt{3}$$
$$= 4\sqrt{3} \qquad h = 9$$

$V = Bh = (4\sqrt{3})(9) = 36\sqrt{3}$

Thus, the volume is $36\sqrt{3}$ cm³.

Oral Exercises

1. Let l, w, and h be the dimensions of a rectangular solid. Give a formula for the volume. **$V = lwh$**

2. Let s be the length of an edge of a cube. Give a formula for the volume. **$V = s^3$**

What is the volume of each prism? The area of the base *(B)* and the altitude length *(h)* are given.

3. $B = 8$ cm², $h = 3$ cm **24 cm³**

4. $B = 16$ mm², $h = 4$ mm **64 mm³**

5. $B = 10$ cm², $h = 14$ cm **140 cm³**

6. $B = 9$ m², $h = 6$ m **54 m³**

7. $B = 9\sqrt{2}$ cm², $h = 5$ cm **$45\sqrt{2}$ cm³**

8. $B = 5\sqrt{3}$ m², $h = 4$ m **$20\sqrt{3}$ m³**

Written Exercises

Find the volume of each right prism. (Ex. 1–5)

A

1.

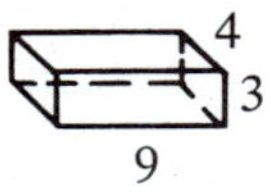

108 cu units

2.

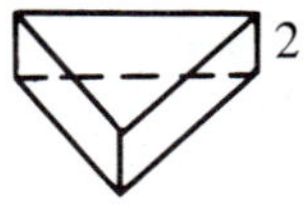

$B = 25$ sq u

50 cu units

3.

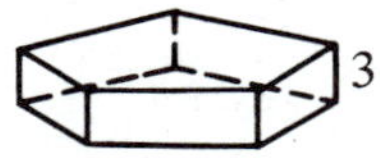

$B = 63$ sq u

189 cu units

4.

$B = 10$ sq u

60 cu units

5.

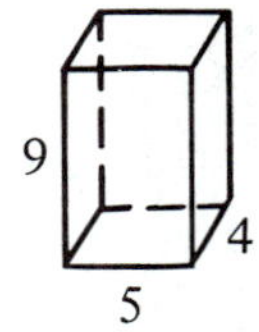

180 cu units

6. Find the volume of a cube whose edges each measure 5 cm. **125 cm³**

7. Find the volume of a rectangular solid whose dimensions are 4, 7, and 11. **308 cu units**

8. A lateral edge of a right prism measures 8 m. The area of the base is 45 m². Find the volume. **360 m³**

9. Find the volume of a prism whose bases are squares, 6 cm on a side, if an altitude measures 11 cm. **396 cm³**

10. The bases of a prism are equilateral triangles with sides measuring 8 cm. An altitude measures 5 cm. Find the volume. **$80\sqrt{3}$ cm³**

11. The bases of a prism are isosceles right triangles with legs measuring 10 m. An altitude measures 16 m. Find the volume. **800 m³**

12. A block is 18 mm wide, 30 mm long, and 9 mm high. Find its volume in cubic centimeters. [1 cm³ = 1,000 mm³] **4.86 cm³**

13. Find the weight of a stone block 1 meter by 1.5 m, by 2 m if the stone weighs 560 kilograms per cubic meter. **1,680 kg**

B

14. The total area of a cube is 294 cm². Find the volume. **343 cm³**

15. Each base of a rectangular solid measures 7 by 12. Find the length of an altitude if the volume is 504 cubic units. **6 units**

16. An altitude of a prism measures 9 cm and the base is a right triangle with one leg measuring 5 cm and the hypotenuse measuring 13 cm. Find the volume. **270 cm³**

17. How many bricks 20 cm by 10 cm by 5 cm are needed to build a wall 6 m long, 2 m high, and 0.5 m thick, if 10 percent of the wall is mortar? **5,400**

18. A cistern in the form of a rectangular solid measures 2 m by 2 m by 3 m. How many liters of water will it hold? [1 m³ = 1,000 L] **12,000 L**

19. Find the volume of a prism whose bases are regular hexagons with 10-cm sides and whose altitude measures 12 cm. **$1{,}800\sqrt{3}$ cm³**

C

20. The volume of a rectangular solid is 432 cubic units. Each side of its square base is $\frac{1}{2}$ the measure of an altitude. Find the area of the base. **36 sq. units**

21. Each lateral edge of a prism measures 14 cm and makes an angle of 45° with the plane of the base. The area of the base is 62 cm². Find the volume. **$434\sqrt{2}$ cm³**

22. A diagonal of a face of a cube measures 8 cm. Find the volume of the cube.

23. The dimensions of a rectangular solid are in a ratio of 2:4:5. A diagonal measures $6\sqrt{5}$ cm. Find the volume.

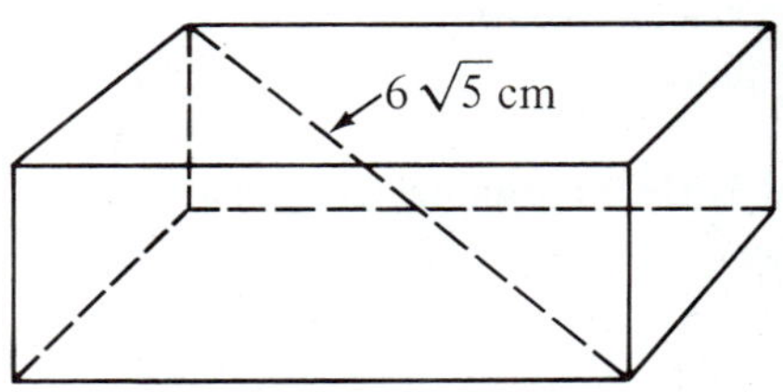

24. If V is the volume of a cube and T is its total area, show that $V = \frac{T}{36}\sqrt{6T}$.

22. $128\sqrt{2}$ cm³ 23. 320 cm³

CYLINDERS

16.4

Objectives

To find the lateral area, total area, and volume of a cylinder

A **cylinder** is a solid that has many properties in common with a prism. A **circular cylinder** has two *bases* which are congruent circular regions lying in parallel planes. The *axis* is the line segment joining the centers of the two circles. An *altitude* is a segment perpendicular to the plane of each base, with endpoints in the planes of the bases.

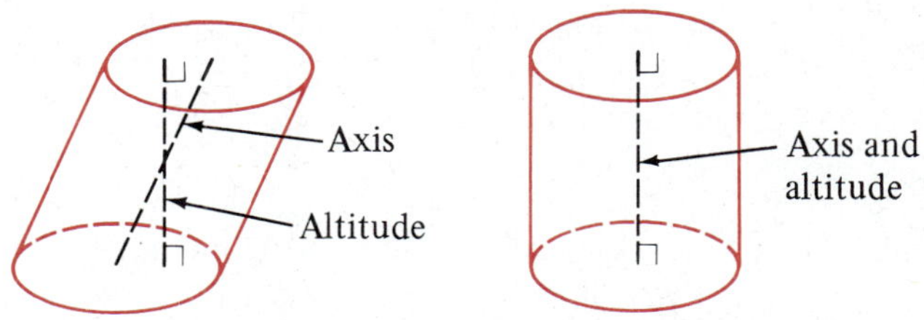

Oblique circular cylinder — Right circular cylinder

The cylinder on the left above is an **oblique circular cylinder**. Notice that in the cylinder on the right, the axis is also an altitude. When this is true, the cylinder is a **right circular cylinder**. From this point on, when the word *cylinder* is used, it will mean "right circular cylinder."

The lateral area (L) of a cylinder can be found in a way similar to the way the lateral area of a prism is found.

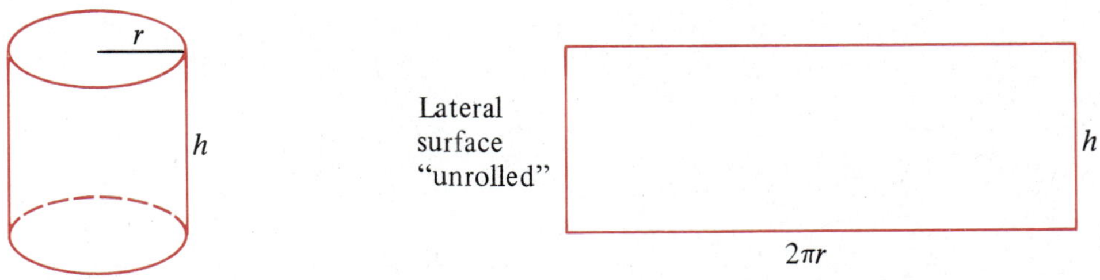

If the lateral surface of a cylinder is "unrolled," it becomes a rectangle. The length of the rectangle is equal to the circumference of a base. The width is equal to the measure of an altitude of the cylinder. Thus, the formula for the lateral area is $L = 2\pi r \cdot h$. Note that this formula is comparable to the formula for the lateral area of a prism, $L = ph$. This argument justifies the following theorem.

Theorem 16.4 For a cylinder, lateral area = circumference of base × length of an altitude.

$$L = 2\pi rh$$

The total area (T) of a cylinder is equal to the lateral area plus the areas of the two bases, as stated in Theorem 16.5.

Theorem 16.5 For a cylinder, total area = lateral area + 2 times base area.

$$T = L + 2B$$
$$T = 2\pi rh + 2\pi r^2$$
$$T = 2\pi r(h + r)$$

The formula $T = 2\pi r(h + r)$ is often the most convenient to use to find total area of a cylinder.

Example 1 The radius of a base of a cylinder is 6 cm long and an altitude is 8 cm long. Find the lateral area and the total area.

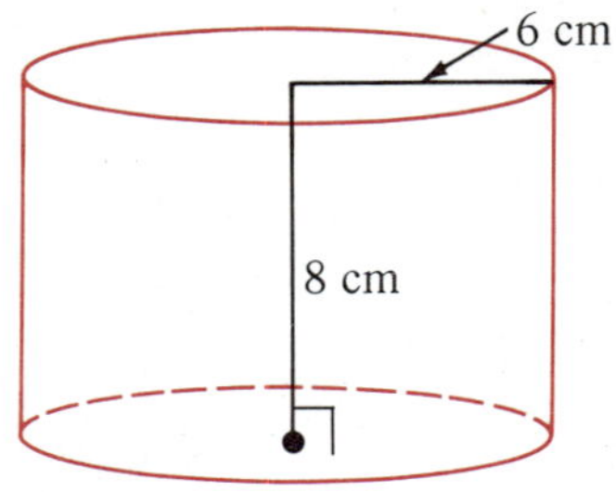

$L = 2\pi rh$
$= 2\pi \cdot 6 \cdot 8$
$= 96\pi$

$T = 2\pi r(h + r)$
$= 2\pi \cdot 6(8 + 6)$
$= 12\pi \cdot 14$
$= 168\pi$

Thus, the lateral area is 96π cm² and the total area is 168π cm².

Recall that the volume of a prism is the product of the area of a base and the length of an altitude. The same formula can be used to find the volume of a cylinder, where the area of a base is πr^2.

Theorem 16.6 For a cylinder, volume = base area × length of an altitude. $V = \pi r^2 h$

Example 2 A water tank is in the shape of a large cylinder. The diameter of the base measures 3 m and an altitude measures 6 m. Find the volume.

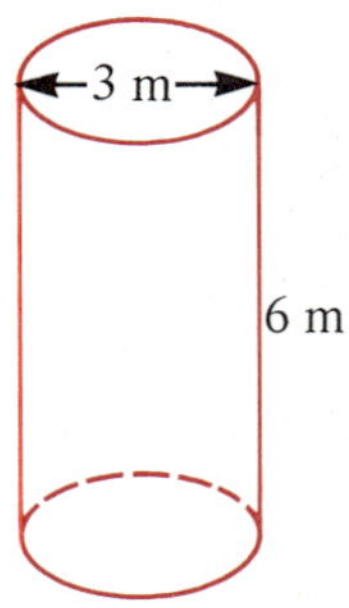

$V = \pi r^2 h \qquad d = 3$, so $r = 1.5$.
$V = \pi(1.5)^2 \cdot 6$
$V = \pi \cdot 2.25 \cdot 6$
$V = 13.5\pi$

Thus, the volume is 13.5π m³.

A **cylinder of revolution** is formed when a rectangle is revolved about one of its sides as an axis.

Example 3 Find the volume of the cylinder of revolution formed when a rectangle 3 cm by 8 cm is revolved about an 8-cm side.

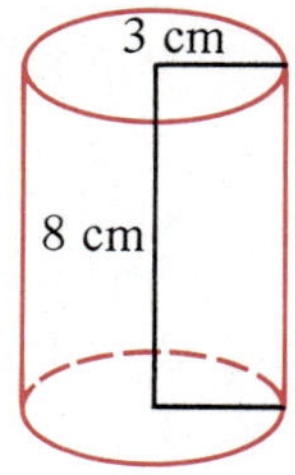

$$\begin{aligned} V &= \pi r^2 h \\ &= \pi \cdot 3^2 \cdot 8 \\ &= \pi \cdot 9 \cdot 8 \\ &= 72\pi \end{aligned}$$

Thus, the volume is 72π cm^3.

Summary **Formulas for a Cylinder**

Lateral area: $L = 2\pi rh$
Total area: $T = 2\pi r(h + r)$
Volume: $V = \pi r^2 h$

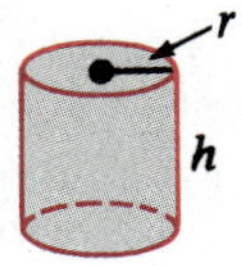

Reading in Geometry

Replace the question mark with the correct expression from the column at the right.

1. If you unrolled a cylinder, it would look like a ?. g
2. The area (A) of a rectangle is ?. j
3. The length of an unrolled cylinder is the same as the ? of a circle. a
4. The circumference (C) of a circle is ?. e
5. The lateral surface (L) of a cylinder is ?. f
6. The top of a cylinder is in the shape of a ?. b
7. The area of a circle (A) is ?. k
8. There are ? circles found on a cylinder. c
9. The total area (T) of a cylinder is ?. d
10. The ? of the bottom circle is πr^2. l
11. The volume (V) of a cylinder is ?. h

a. circumference
b. circle
c. 2
d. $2\pi rh + 2\pi r^2$
e. $2\pi r$
f. $2\pi rh$
g. rectangle
h. $\pi r^2 h$
i. layers
j. length × width
k. πr^2
l. area

Written Exercises

Find the lateral area, total area, and volume of each cylinder. Leave answers in terms of π.

A **1.**

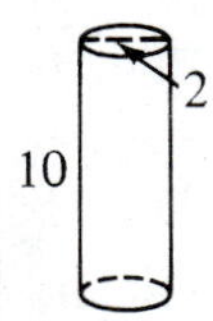

20π; 22π; 10π

2.

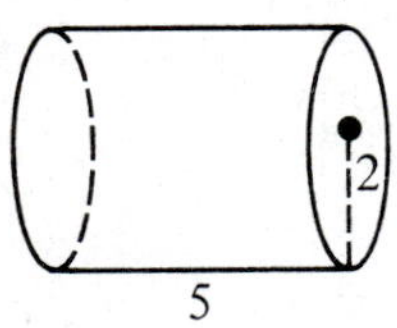

20π; 28π; 20π

3.

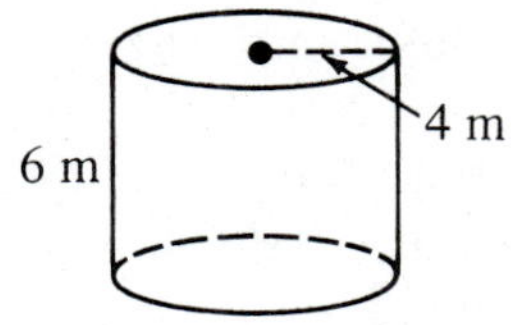

48π m^2; 80π m^2; 96π m^3

4.

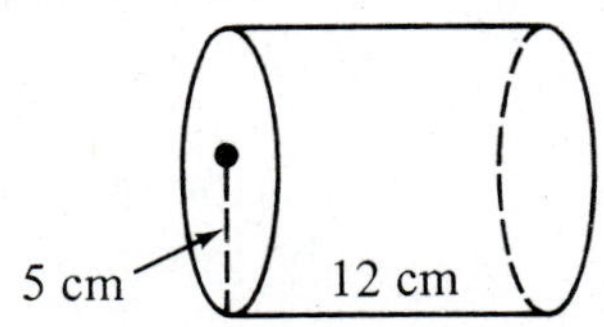

120π cm^2; 170π cm^2; 300π cm^2

Leave answers in terms of π, whenever possible.

5. A cylindrical-oil tank is 20 m high and the radius of its base is 5 m long. Find the volume. **500π m³**

6. Find the total area of a cylinder whose altitude measures 7 m and whose base has a radius 4 m long. **88π m²**

7. The diameter of the base of a cylinder is 16 cm long and its altitude measures 12 cm. Find the lateral area. **192π cm²**

8. A cylindrical pail is 10 cm deep and the radius of its base is 5 cm long. Find the volume. **250π cm³**

9. How many square centimeters of tin are required to make the pail in Exercise 8? **125π cm²**

10. A concrete roller is 3 m long. Its diameter is 2 m long. How much area will it cover in 400 revolutions? **$2{,}400\pi$ m²**

B 11. A diameter of a cylindrical oil tank is 1 m long. The tank is 60 cm long. About how many liters of oil will it hold? [1 L = 1,000 cm³] **150π L**

12. When a piece of iron is submerged in a cylindrical tank, the water level rises 6 cm. The diameter of the tank is 18 cm long. What is the volume of the piece of iron? **486π cm³**

13. Suppose the rectangle in Example 3 is revolved about a 3-cm side. Compare the volume of the resulting cylinder of revolution to the volume of the one in the Example 3. Sketch the figure. **8 : 3**

14. Consider the cylinder of revolution in Example 3 and the one in Exercise 13. Compare their lateral areas. **They are equal.**

15. The measures of the sides of a rectangle are in the ratio 3 : 1. It is possible to form two cylinders of revolution with the rectangle. What is the ratio of the lateral areas of the two cylinders? [*Hint:* Let x and $3x$ represent the sides.] **1 : 1**

16. Consider the two cylinders of revolution in Exercise 15. What is the ratio of the volumes of the two cylinders? **3 : 1, or 1 : 3**

Refer to the figures at the right for Exercises 17–19.

C 17. A cylinder may be inscribed in a cube, as shown. What is the ratio of the volume of the cylinder to the volume of the cube?

18. A cylinder may be circumscribed about a cube as shown. What is the ratio of the volume of the cube to the volume of the cylinder?

19. Find the volume of the cylinder and the cube in Exercise 17 if $r = 3$. Then do the same for Exercise 18.

Cylinder inscribed in cube

Cylinder circumscribed about cube

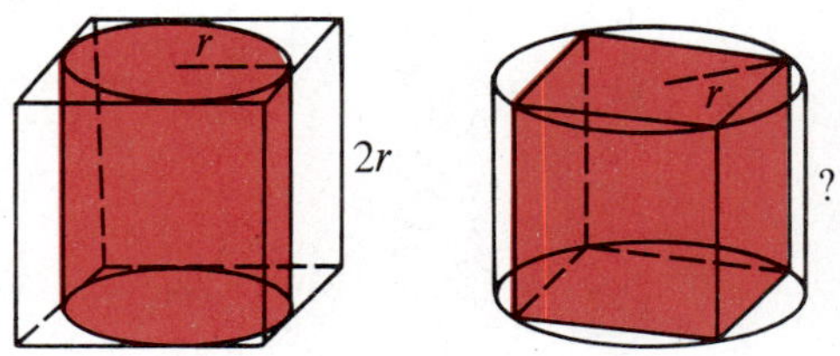

17. π : 4 18. 2 : π 19. cyl in 17, 54π cu units; cube, 216 cu units cyl in 18, $27\sqrt{2}\pi$ cu units; cube, $54\sqrt{2}$ cu units

PYRAMIDS

16.5

Objectives

To find the lateral area and the total area of a regular pyramid
To find the volume of a pyramid

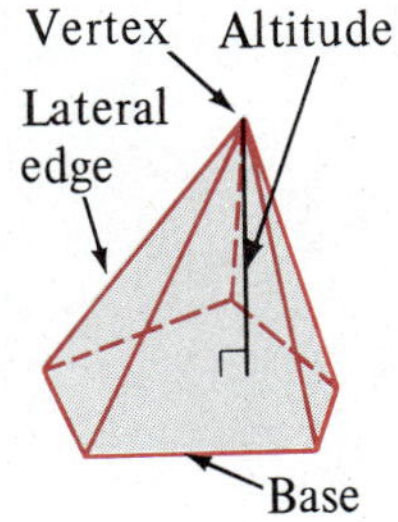

A **pyramid** is a solid whose base is a polygonal region; its *lateral faces* are triangular regions that share a common point called the *vertex*. The lateral faces intersect in segments called the lateral edges of the pyramid. The *altitude* of a pyramid is the segment from the vertex perpendicular to the plane of the base.

For the sake of simplicity, the faces of a pyramid will be referred to as polygons, even though they are polygonal regions.

A **regular pyramid** is a pyramid with the following special properties:
(1) The base is a regular polygon.
(2) The lateral faces are congruent isosceles triangles.
(3) The altitude meets the base at its center.

The altitude of each lateral face of a regular pyramid is called the **slant height** of the pyramid.

Regular Pyramids

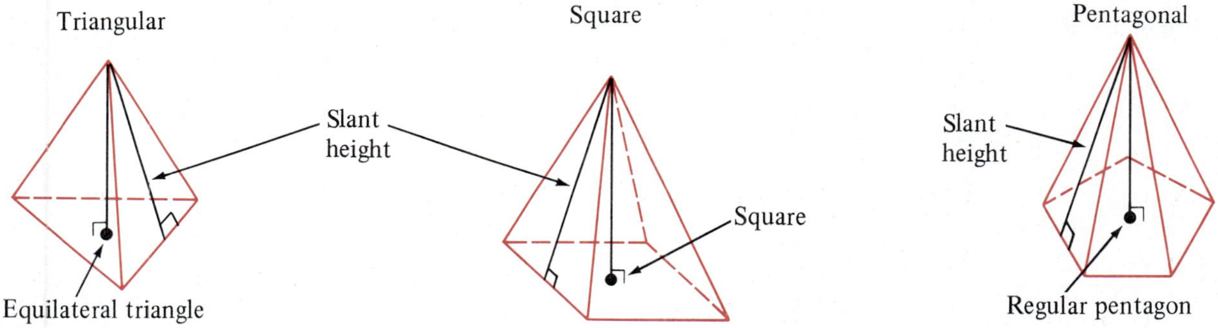

A formula for the lateral area of a regular pyramid can be derived as follows: Suppose the base of a regular pyramid has n sides. Let b be the length of each side of the base and l be the length of the slant height. Then the area of each lateral face is $\frac{1}{2}bl$. Since there are n lateral faces, the lateral area of the pyramid can be found by multiplying the area of a lateral face by n, as shown below.

$$\text{Lateral area} = \begin{pmatrix}\text{area of a}\\ \text{lateral face}\end{pmatrix} \cdot \begin{pmatrix}\text{number of sides}\\ \text{of the base}\end{pmatrix}$$

$$L = \frac{1}{2}bl \qquad \cdot \quad n$$

$$L = \frac{1}{2}l \cdot bn$$

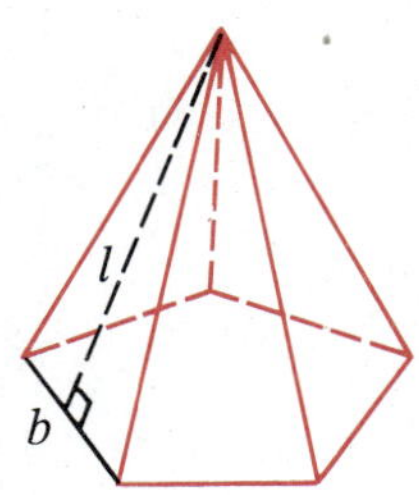

But bn is the perimeter of the base.

So, $$L = \frac{1}{2}lp.$$

Theorem 16.7 For a regular pyramid, lateral area $= \frac{1}{2} \times$ length of slant height $\times$ perimeter of base.

$$L = \frac{1}{2} lp$$

Total area = lateral area + base area.

$$T = L + B$$

The formula used to find the base area (B) depends upon the type of base that the pyramid has. In Example 1, the base is a regular pentagon, and in Example 2, the base is a square.

Example 1 Find the lateral area and the total area of a regular pyramid whose slant height measures 11 and whose base is a regular pentagon with sides measuring 8 and apothem measuring 6.

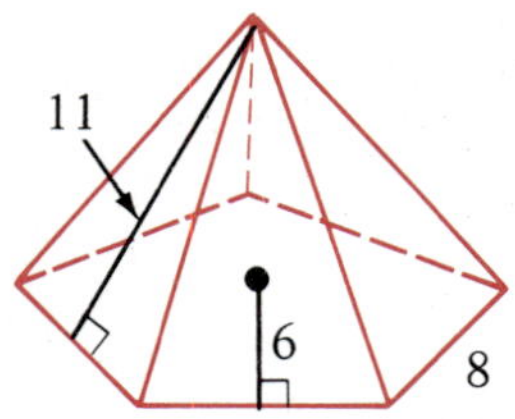

$L = \frac{1}{2} lp$

A regular pentagon has 5 congruent sides, so $p = 5 \times 8$. ▶

$= \frac{1}{2} \cdot 11 \cdot (5 \times 8)$

$= \frac{1}{2} \cdot 11 \cdot 40$

$= 220$

$T = L + B$

$B = \frac{1}{2} ap$ ◀ *Formula for area of a regular polygon*

$= \frac{1}{2} \cdot 6 \cdot 40$

$= 120$

$$T = 220 + 120 = 340$$

Thus, the lateral area is 220 square units and the total area is 340 square units.

Example 2 Each edge of the base of a regular square pyramid measures 16 m and each lateral edge measures 17 m. Find the total area.

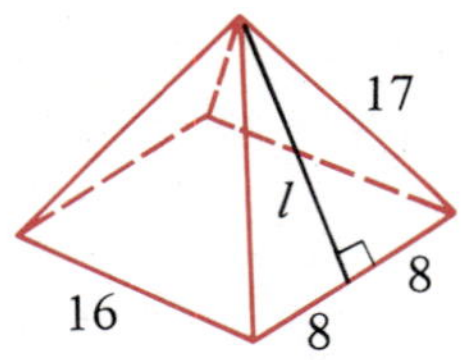

$T = L + B$

Find l, the length of the slant height.

Use the Pythagorean Theorem.

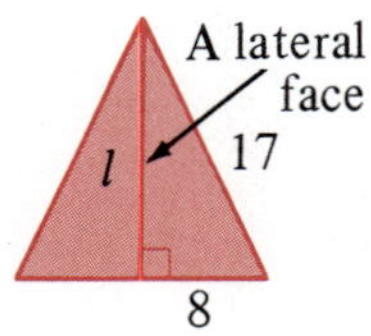

$l^2 + 8^2 = 17^2$

$l^2 + 64 = 289$

$l^2 = 225$

$l = 15$

$L = \frac{1}{2} lp \longrightarrow p = 4 \times 16 = 64$

$L = \frac{1}{2} \cdot 15 \cdot 64$

$L = 480$

$B = s^2$

$= 16^2$ ◀ *Formula for area of a square: $B = s^2$*

$= 256$

$$T = L + B = 480 + 256 = 736$$

Thus, the total area is 736 m^2.

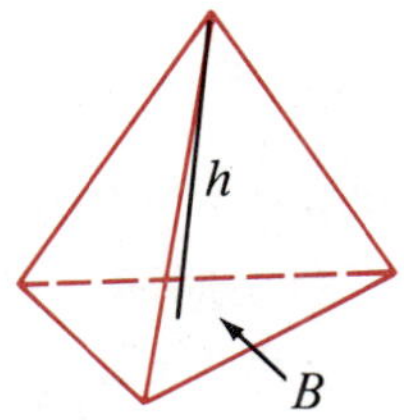

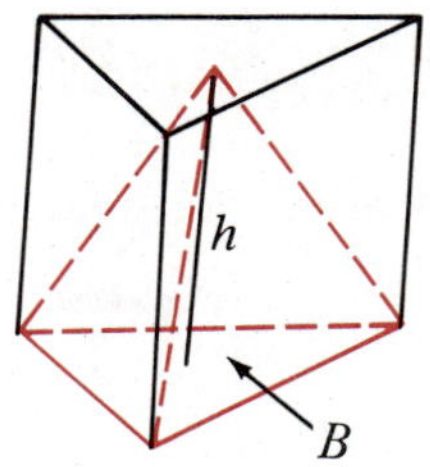

Suppose a pyramid and a prism have the same base and altitude, as shown in the figures above. The volume of the pyramid will be one third the volume of the prism. This leads to the following theorem, which is given without proof.

Theorem 16.8 For any pyramid, volume $= \frac{1}{3} \times$ base area $\times$ length of altitude.

$$V = \frac{1}{3}Bh$$

In Example 3, the formula $V = \frac{1}{3}Bh$ is used to find the volume of a regular triangular pyramid. The base of a regular triangular pyramid is an equilateral triangle.

Example 3 Each edge of the base of a regular triangular pyramid is 6 cm long and the altitude is 5 cm long. Find the volume.

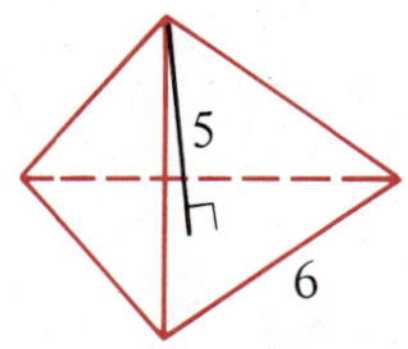

$V = \frac{1}{3}Bh$

$B = \frac{s^2}{4}\sqrt{3}$ ◄ *Formula for area of an equilateral △*

$B = \frac{6^2}{4}\sqrt{3}$

$B = 9\sqrt{3}$

$V = \frac{1}{3}Bh \longrightarrow h = 5$

$V = \frac{1}{3}(9\sqrt{3})\,5$

$V = 15\sqrt{3}$

Thus, the volume is $15\sqrt{3}$ cm³.

Summary

Formulas for a Pyramid

Lateral Area: $L = \frac{1}{2}lp$ {Regular pyramid only

Total Area: $T = L + B$

Volume: $V = \frac{1}{3}Bh$

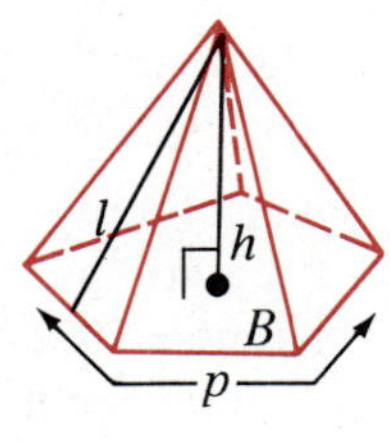

Written Exercises

Find the lateral area, the total area, and the volume of each regular pyramid. (Ex. 1–4)

A **1.**

135; 192; 152

2.

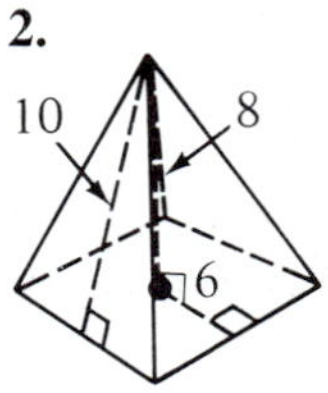

240; 384; 384

3.

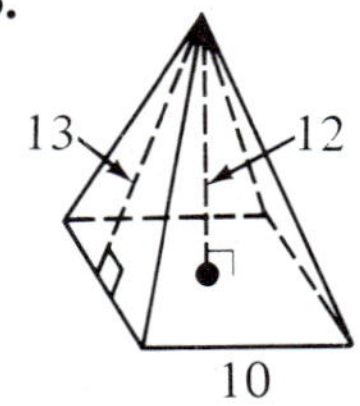

260; 360; 400

4.

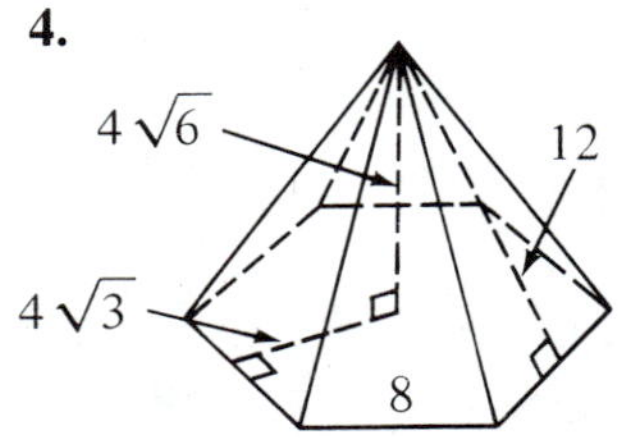

288; $288 + 96\sqrt{3}$; $384\sqrt{2}$

5. Find the lateral area of a regular pyramid whose slant height is 7 cm long and whose base has a perimeter of 20 cm. **70 cm²**

6. Find the volume of a pyramid whose base is a square with sides 6 cm long and whose altitude measures 15 cm. **180 cm³**

7. Each side of the base of a regular triangular pyramid measures 6 m. Each lateral edge measures 5 m. Find the total area of the pyramid. **$36 + 9\sqrt{3}$ m²**

8. The altitude of a pyramid measures 12 m. The base is a right triangle with legs measuring 8 m and 11 m. Find the volume. **176 m³**

9. Each side of the base of a regular square pyramid measures 12 cm and each lateral edge measures 10 cm. Find the lateral area and the total area of the pyramid. **L = 192 cm²; T = 336 cm²**

10. The base of a pyramid is a rhombus with diagonals measuring 7 and 8. The altitude of the pyramid measures 21. Find the volume. **196 cu units**

B **11.** The perimeter of the base of a regular pyramid is 56 m and the lateral area is 224 m². How long is the slant height? **8 m**

12. The lateral area of a regular pyramid is 476 square units and the slant height measures 14 units. Find the perimeter of the base. **68 units**

13. The slant height of a regular square pyramid measures 26 cm and a side of the base measures 20 cm. Find the lateral area, total area, and volume. **1,040 cm²; 1,440 cm²; 3,200 cm³**

14. Find the lateral area and the total area of a regular hexagonal pyramid whose lateral edges measure 25 cm and whose base has sides measuring 14 cm. **1,008 cm²; $(1{,}008 + 294\sqrt{3})$ cm²**

15. The altitude of a regular square pyramid measures 4 cm and a side of the base measures 6 cm. Find the lateral area and the total area. **60 cm²; 96 cm²**

16. The altitude of a regular triangular pyramid measures 8 cm and each side of the base measures 18 cm. Find the lateral area and the total area. **$27\sqrt{91}$ cm²; $(27\sqrt{91} + 81\sqrt{3})$ cm²**

C **17.** Each side of the base of a regular triangular pyramid measures 12 units. The slant height makes an angle of 45° with the altitude. Find the total area and the volume of the pyramid. **$(36\sqrt{6} + 36\sqrt{3})$ sq units; 72 cu units**

18. Derive a formula for the total area of a regular pyramid in terms of its slant height, apothem, and perimeter of the base. **$T = \frac{1}{2}p\,(l + a)$**

19. The base of a regular square pyramid is inscribed in the base of a cylinder. The two figures have altitudes of the same length. Find the ratio of the volume of the pyramid to the volume of the cylinder. [*Hint:* Express s in terms of r.] **$2:3\pi$**

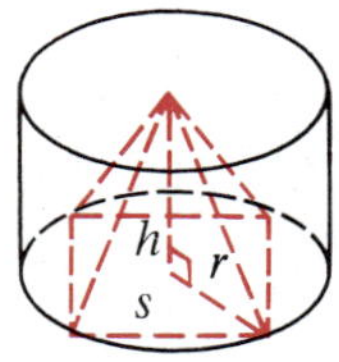

20. Repeat Exercise 19 for a regular triangular pyramid. $\frac{\sqrt{3}}{4} : \pi$

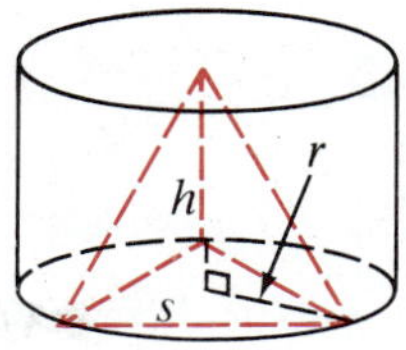

CUMULATIVE REVIEW

Find the area of each polygon.

1. An equilateral triangle whose perimeter is 24 **$16\sqrt{3}$ sq units**
2. An equilateral triangle whose altitude measures $5\sqrt{3}$ **$25\sqrt{3}$ sq units**
3. A square whose apothem measures 7 **196 sq units**
4. A square whose diagonal measures 20 **200 sq units**

ALGEBRA REVIEW

Objective **To solve systems of equations**

To solve a system of equations:
(1) Get one equation with only one variable;
(2) Solve for the variable;
(3) Substitute the value in one of the original equations and solve for the other variable.

Example

Solve this system. $5x - 3y = 2$
$4x + 2y = 6$

Multiply each side by 2. ► $2(5x - 3y) = 2(2)$
Multiply each side by 3. ► $3(4x + 2y) = 3(6)$

$10x - 6y = 4$ ◄ *Now the coefficients of y are opposites.*
$12x + 6y = 18$

Add the equations. ► $22x = 22$
$x = 1$

$5x - 3y = 2$
Substitute 1 for x. ► $5(1) - 3y = 2$
$5 - 3y = 2$
$-3y = -3$
$y = 1$ **Thus,** (1, 1) is the solution.

Solve.

1. $x - y = 7$, $x + y = 9$ **(8, 1)**
2. $-2x + y = 6$, $2x + 3y = 10$ **(−1, 4)**
3. $3x - y = 5$, $2x + y = 5$ **(2, 1)**
4. $3x - 4y = 4$, $2x + 4y = 16$ **(4, 2)**
5. $-x + y = -4$, $3x - 2y = 12$ **(4, 0)**
6. $2x + y = 9$, $3x + 2y = 16$ **(2, 5)**
7. $5x + 2y = 14$, $2x - y = 3$ **$\left(\frac{20}{9}, \frac{13}{9}\right)$**
8. $-12x - y = 14$, $5x + 3y = -11$ **(−1, −2)**

CONES

16.6

Objectives **To find the lateral area, the total area, and the volume of a cone**

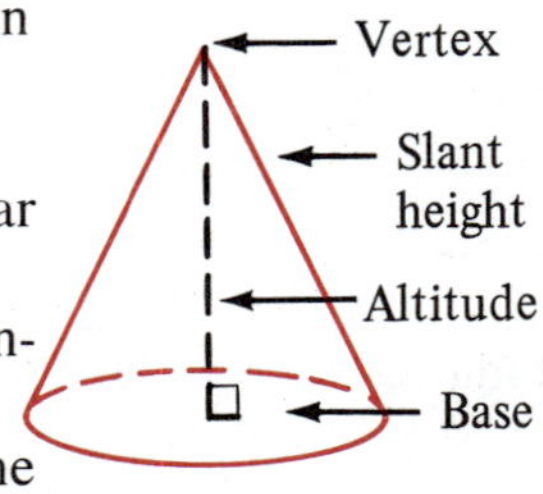

A **right circular cone** is a solid that has many properties in common with regular pyramids.

1. The **base** is a circular region.
2. The **vertex** is a point that lies on the line perpendicular to the plane of the base at its center.
3. The **altitude** is a segment joining the vertex to the center of the base.
4. A **slant height**, or *element,* is a segment joining the vertex to a point on the circle which bounds the base.
5. The **lateral surface** is the union of all the slant heights. From this point on, the word *cone* will mean "right circular cone." Notice that the regular pyramids below resemble the cone.

Regular pyramids | Cone

6 sides | 12 sides | 24 sides | Circle

A cone is comparable to a regular pyramid with an almost circular base. The circumference of the base of a cone corresponds to the perimeter of the base of a regular pyramid. The table below shows how the formulas for a cone are derived from the formulas for a regular pyramid. The formulas are summarized in Theorem 16.9, which is given without proof.

Formulas	Regular Pyramid	Cone
Lateral area	$L = \frac{1}{2}lp$	$L = \frac{1}{2}lc$ $L = \frac{1}{2}l(2\pi r)$ $L = \pi rl$
Total area	$T = L + B$	$T = L + B$ $T = \pi rl + \pi r^2$ $T = \pi r(l + r)$
Volume	$V = \frac{1}{3}Bh$	$V = \frac{1}{3}Bh$ $V = \frac{1}{3}(\pi r^2)h$ $V = \frac{1}{3}\pi r^2 h$

Note:
$c = 2\pi r$
$B = \pi r^2$
h = length of altitude
l = length of slant height

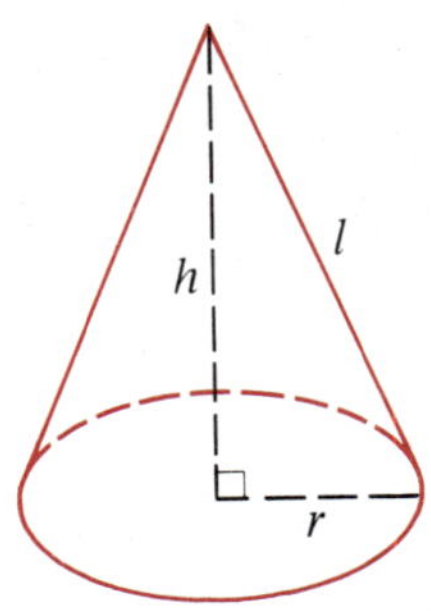

Theorem 16.9 For a cone, lateral area (L) is πrl, total area (T) is $\pi r(l + r)$, volume (V) is $\frac{1}{3}\pi r^2h$.

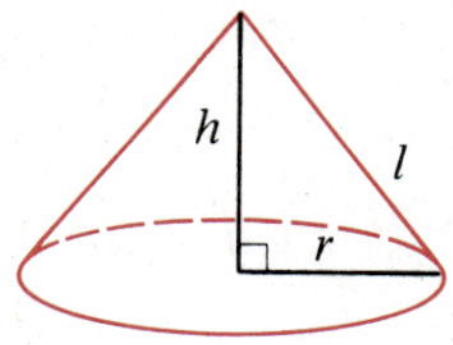

Examples 1 and 2 below are applications of these formulas for a cone.

Example 1 Find the lateral area, the total area, and the volume of the cone shown at the right.

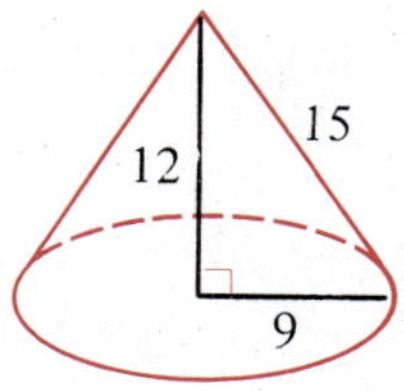

$$L = \pi rl.$$
$$= \pi \cdot 9 \cdot 15$$
$$= 135\pi$$

$$T = \pi r(l + r)$$
$$= \pi \cdot 9(15 + 9)$$
$$= \pi \cdot 9 \cdot 24$$
$$= 216\pi$$

$$V = \frac{1}{3}\pi r^2h$$
$$= \frac{1}{3}\pi \cdot 9^2 \cdot 12$$
$$= \frac{1}{3}\pi \cdot 81 \cdot 12$$
$$= 324\pi$$

Thus, the lateral area is 135π square units, the total area is 216π square units, and the volume is 324π cubic units.

In Example 2, the length of the radius of the base and the length of the slant height of a cone are given. The Pythagorean Theorem must be used to find the length of the altitude.

Example 2 The radius of the base of a cone measures 9 cm and the slant height measures 18 cm. Find the lateral area, total area, and volume.

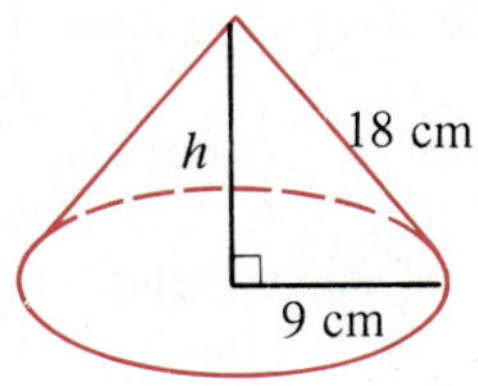

Find h: $h^2 + 9^2 = 18^2$
$$h^2 + 81 = 324$$
$$h^2 = 243$$
$$h = \sqrt{243}, \text{ or } 9\sqrt{3}$$

$$L = \pi rl$$
$$= \pi \cdot 9 \cdot 18$$
$$= 162\pi$$

$$T = \pi r(l + r)$$
$$= \pi \cdot 9(18 + 9)$$
$$= \pi \cdot 9 \cdot 27$$
$$= 243\pi$$

$$V = \frac{1}{3}\pi r^2h$$
$$= \frac{1}{3}\pi \cdot 9^2 \cdot 9\sqrt{3}$$
$$= \frac{1}{3}\pi \cdot 81 \cdot 9\sqrt{3}$$
$$= 243\sqrt{3}\pi$$

Thus, the lateral area is 162π cm^2, the total area is 243π cm^2, and the volume is $243\sqrt{3}\pi$ cm^3.

A **cone of revolution** is formed when a right triangle is revolved about one of its legs as an axis.

Example 3 Find the volume of the cone of revolution formed when a right triangle with legs measuring 4 and 9 is revolved about the 9-unit leg.

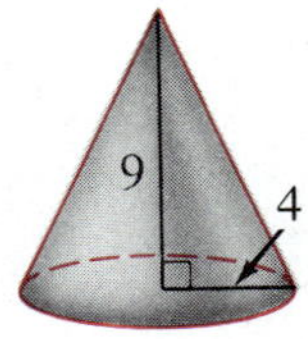

Substitute 4 for r and 9 for h. ► $V = \frac{1}{3}\pi r^2 h$

$= \frac{1}{3}\pi \cdot 4^2 \cdot 9$

$= \frac{1}{3}\pi \cdot 16 \cdot 9$

$= 48\pi$

Thus, the volume is 48π cubic units.

Summary

Formulas for a Cone

Lateral area: $L = \pi r l$

Total area: $T = \pi r(l + r)$

Volume: $V = \frac{1}{3}\pi r^2 h$

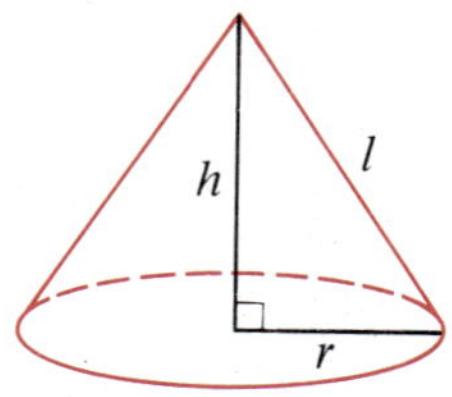

Reading in Geometry

Explain the difference between each pair of terms.

2. lateral area = area of lateral faces, total area = lateral area + area of base

1. A cone and a regular pyramid
2. The lateral area and the total area of a cone

1. Cone's base is a ⊙, pyramid's base is a polygon

3. An element of a cone and the slant height of a cone
4. An element of a cone and the altitude of a cone

3. Both terms are names for a segment joining the vertex to a point on the circle which bounds the base. 4. See Ex 3 for element, altitude is a segment joining the vertex to the center of the base.

Written Exercises

Find the lateral area, total area, and volume of each cone. Leave answers in terms of π.

Ⓐ **1.**

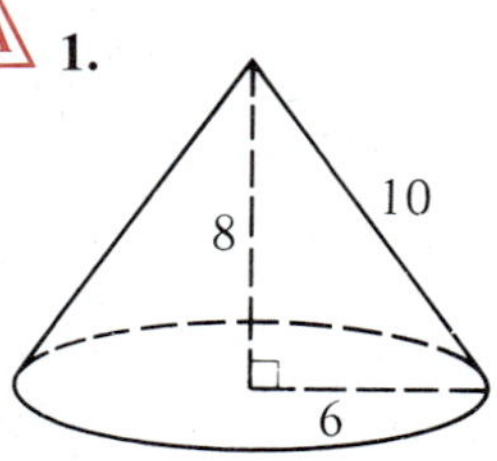

60π; 96π; 96π

2.

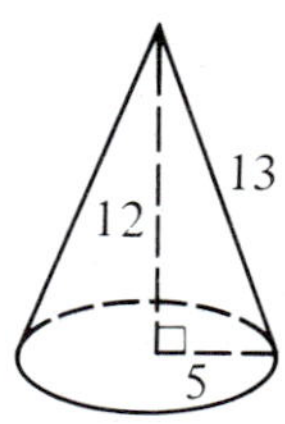

65π; 90π; 100π

3.

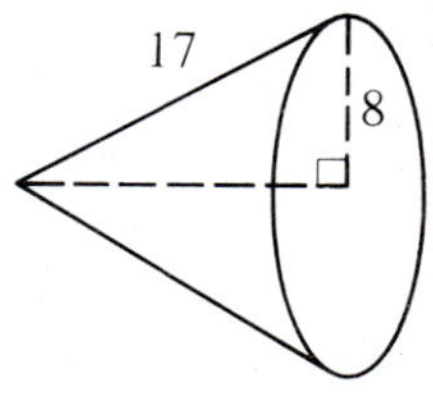

136π; 200π; 320π

4.

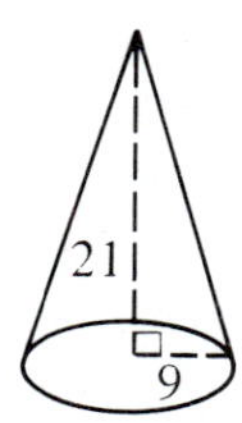

$27\sqrt{58}\pi$; $9(3\sqrt{58} + 9)\pi$; 567π

Leave answers in terms of π, wherever possible.

5. The slant height of a cone measures 18 m and the altitude measures 9 m. Find the length of the radius of the base. **$9\sqrt{3}$ m**

6. The slant height of a cone measures 16 cm and the radius of the base measures 5 cm. Find the lateral area and the total area. **80π cm²; 105π cm²**

7. The altitude of a cone measures 12 cm and the radius of the base measures 6 cm. Find the length of the slant height. **$6\sqrt{5}$ cm**

8. The radius of the base of a cone measures 3 units and the altitude measures 7 units. Find the volume. **21π cu units**

9. The slant height of a cone measures 20 m and the radius of the base measures 12 m. Find the volume. **768π m³**

10. The altitude of a cone measures 24 units and the slant height measures 26 units. Find the lateral area and the total area. **260π sq units; 360π sq units**

11. A cone and a cylinder have congruent bases, and their altitudes have the same measure. Compare their volumes. **vol. cone $= \frac{1}{3}$ vol. cyl.**

12. A cone and a cylinder have congruent bases. The altitude of the cone is three times as long as the altitude of the cylinder. Compare their volumes. **They are equal.**

B 13. The area of the base of a cone is 49π square units and the slant height measures 20 units. Find the length of the altitude. **$3\sqrt{39}$**

14. The volume of a cone is 320π cm³ and the altitude measures 15 cm. Find the length of the radius of the base. **8 cm**

15. Suppose the triangle in Example 3 is revolved about the 4-unit leg. Compare the volume of the resulting cone of revolution with the volume of the cone in Example 3. **9 : 4**

16. Consider the cone of revolution in Example 3 and the cone in Exercise 15. Compare their lateral areas. **4 : 9**

17. The measures of the legs of a right triangle are in a ratio of 4 : 1. It is possible to form two cones of revolution with the right triangle. What is the ratio of the lateral areas of the two cones? [*Hint:* Let x and $4x$ represent the legs.] **4 : 1**

18. Consider the two cones of revolution in Exercise 17. What is the ratio of the volumes of the two cones? **4 : 1**

C 19. Suppose a square with 6-cm sides is revolved about a diagonal as an axis. Find the volume. **$36\pi\sqrt{2}$ cm³**

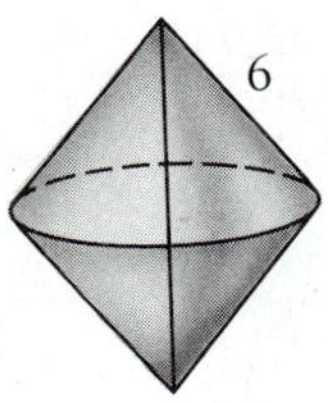

20. Find the surface area of the figure in Exercise 19. **$36\pi\sqrt{2}$ cm²**

21. Suppose a 3-4-5 right triangle is revolved about its hypotenuse as an axis. Find the volume. **$\frac{48}{5}\pi$ cu units**

22. Find the surface area of the figure in Exercise 21. **$\frac{84}{5}\pi$ sq units**

23. In the figure shown, the length of the altitude is twice the length of the radius of the base. Find the ratio of the volume of the cone to the volume of the cylinder. **1:3**

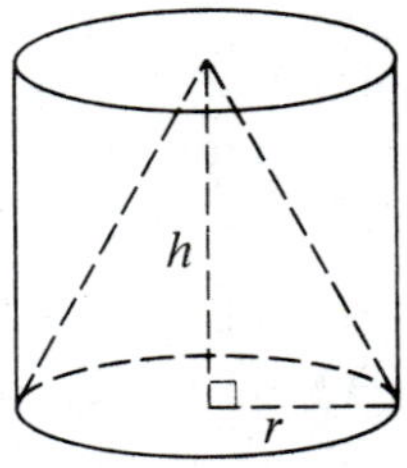

24. For the cone and cylinder in Exercise 23, what is the ratio of their lateral areas? $\frac{\sqrt{5}}{4}$

25. What is the ratio of their total areas? $(\sqrt{5}+1):6$

CALCULATOR ACTIVITIES

A calculator can be useful in obtaining approximations of areas and volumes where π is involved. The formulas for a cone can be rewritten in the "language" of the calculator as follows:

$$L = rl\pi \qquad T = (l + r)r\pi \qquad V = \frac{r \cdot r \cdot h}{3}\pi$$

If your calculator does not have a π button, you may omit the part of each exercise which requires a π button.

Example Find three *different* approximations for the volume of a cone if $r = 7.5$ and $h = 6.3$. Use 3.14, 3.1416, and the π button on your calculator.

$$V = \frac{r \cdot r \cdot h}{3}\pi = \frac{(7.5)(7.5)(6.3)}{3}\pi$$

Press → 7.5 ⊗ 7.5 ⊗ 6.3 ⊕ 3 ⊜
Display → 118.125 ← Write down.
Press → ⊗ 3.14 ⊜
Display → 370.9125 ← Write down and clear.

Press → 118.125 ⊗ 3.1416 ⊜
Display → 371.1015 ← Write down and clear.

Press → 118.125 ⊗ π ⊜
Display → 371.10064 ← Write down and clear.

Thus, approximations for the volume are
370.9125 if $\pi \doteq 3.14$,
371.1015 if $\pi \doteq 3.1416$
371.10064 if the π button is pressed.

◄ *Notice that the π button approximation lies between the other two.*

Find three different approximations for the lateral area, total area, and volume of each cone.

	Find	r	l	h	3.14 for π	3.1416 for π	π button
1.	Lateral Area	3.2	6.5	–	65.312	65.34528	65.345127
2.	Total Area	5.8	4.5	–	187.5836	187.67918	187.67875
3.	Volume	7.6	–	8.3	501.78037	502.03606	502.03489

SPHERES 16.7

Objective **To find the area and the volume of a sphere**

Recall that a circle is the set of all points in a plane at a given distance from a fixed point called the center. By removing the restriction "in a plane" and moving into space, you can define a sphere in a similar manner. A **sphere** is the set of *all* points at a given distance from a given point called the *center*. Most terms defined for circles also apply to spheres. For example, a *radius* of a sphere is a segment that joins the center and any point on the sphere.

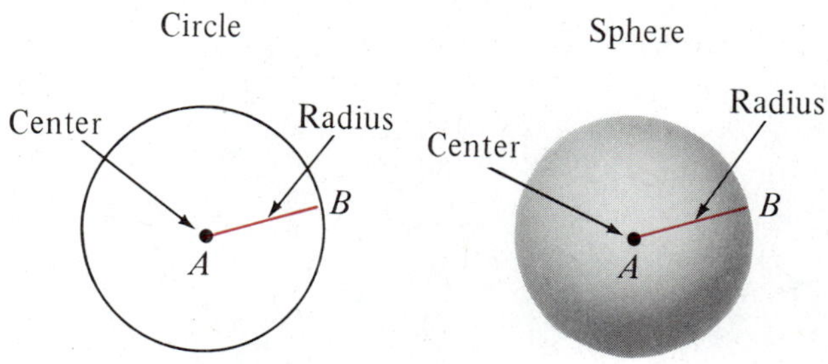

A **great circle of a sphere** is the intersection of the sphere and any plane that contains the center of the sphere. Thus, a radius of a great circle is also a radius of the sphere.

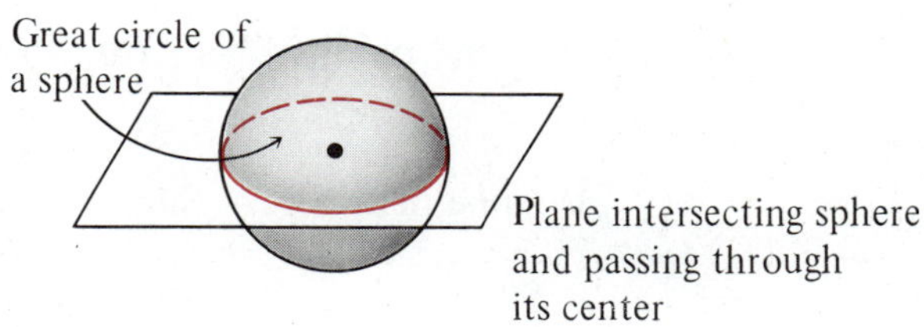

Example 1 Find the area of a great circle of a sphere whose radius measures 7.

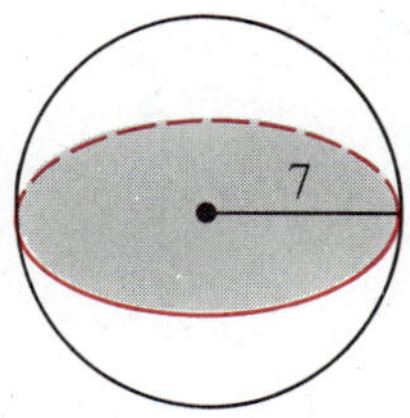

$A = \pi r^2$
$A = \pi \cdot 7^2$
$A = 49\pi$ square units

Two intersecting great circles formed by perpendicular planes divide a sphere into four quadrants. It can be proved in advanced geometry that the surface area of a quadrant is equal to the area of a great circle. Thus, the area of a quadrant is πr^2. This suggests the next theorem.

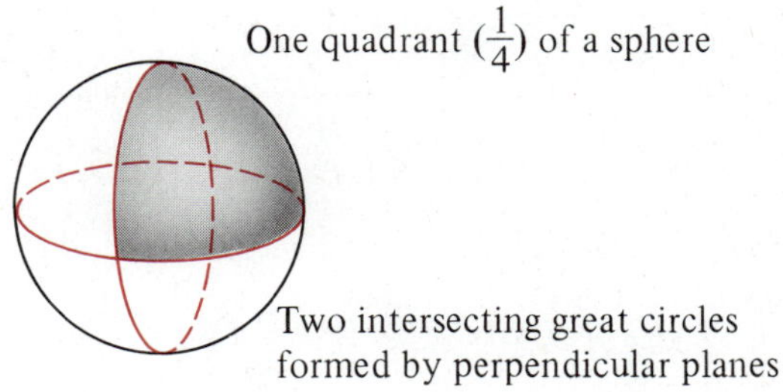

Theorem 16.10 For a sphere, area = 4 × area of great circle.

$$A = 4\pi r^2$$

Example 2 Find the area of a sphere with a 5-cm radius.

$A = 4\pi r^2 = 4\pi \cdot 5^2 = 4\pi \cdot 25 = 100\pi$

Thus, the area is 100π cm^2.

Suppose you cut a solid sphere into pyramidlike figures. Use enough "pyramids" to cover the entire sphere. (Only 3 are shown.) Let $B_1, B_2, B_3, \ldots, B_n$ be the areas of the bases of the pyramids.

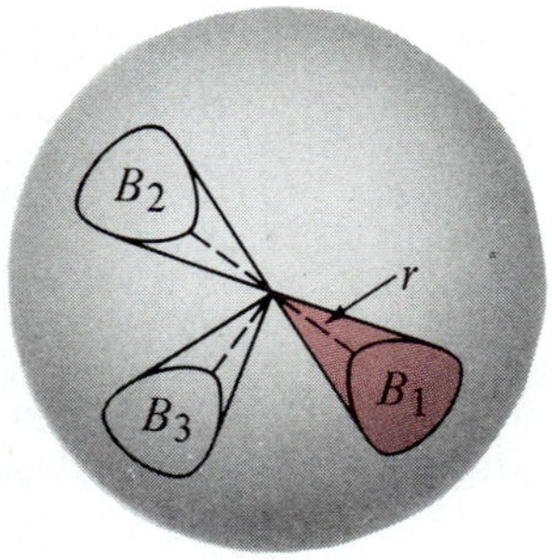

Volume of a pyramid $= \frac{1}{3} Bh$.

Volume of shaded pyramid $= \frac{1}{3} B_1 r$.

The volume of each "pyramid" is $\frac{1}{3} Bh$, where h is the length of the radius of the sphere.

$$\begin{aligned}\text{Volume of sphere} &= \text{Sum of volumes of pyramids}\\ V &= \frac{1}{3}B_1 r + \frac{1}{3}B_2 r + \frac{1}{3}B_3 r + \cdots + \frac{1}{3}B_n r\\ &= \frac{1}{3}r(B_1 + B_2 + B_3 + \cdots + B_n)\\ &= \frac{1}{3}r\ (\text{area of sphere})\\ &= \frac{1}{3}r(4\pi r^2)\\ &= \frac{4}{3}\pi r^3\end{aligned}$$

Thus, you have derived a formula for the volume of a sphere, as stated in Theorem 16.11.

Theorem 16.11 For a sphere, the volume is $\frac{4}{3}\pi r^3$.

$$V = \frac{4}{3}\pi r^3$$

Example 3

Find the volume of a sphere with a 9-cm radius.

$$V = \frac{4}{3}\pi r^3 = \frac{4}{3}\pi \cdot 9^3 = \frac{4}{3}\pi \cdot 729 = 972\pi$$

Thus, the volume is 972π cm³.

Summary

Formulas for a Sphere

Area: $A = 4\pi r^2$

Volume: $V = \frac{4}{3}\pi r^3$

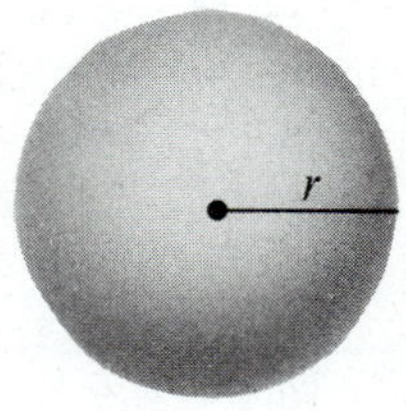

Written Exercises

Find the area and volume of each sphere. Leave answers in terms of π. (Ex. 1–4)

A

1.

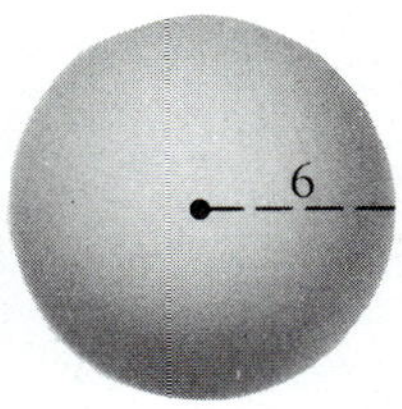

144π sq units; 288π cu units

2.

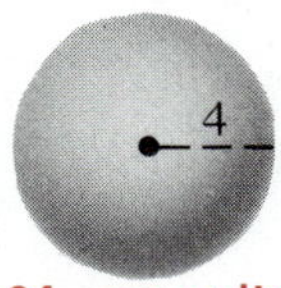

64π sq units; $\frac{256}{3}\pi$ cu units

3.

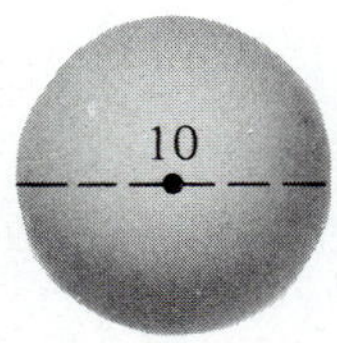

100π sq units; $\frac{500}{3}\pi$ cu units

4.

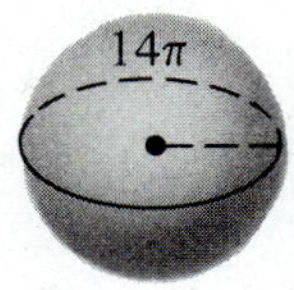

196π sq units; $\frac{1,372}{3}\pi$ cu units

Leave answers in terms of π, whenever possible.

5. A baseball has a radius 7 cm long. Find its surface area. 196π cm²

6. Find the volume of a sphere with a radius 3 cm long. 36π cm³

7. The area of a great circle of a sphere is 54π cm². 216π cm² Find the area of the sphere.

8. How much air can be pumped into a basketball if its maximum diameter measures 20 cm? $\frac{4,000\pi}{3}$ cm³

9. The diameter of a sphere is 8 m long. Find the area and the volume. 64π m²; $\frac{256\pi}{3}$ m³

10. Two spheres have radii of lengths 1 mm and 4 mm. What is the ratio of their surface areas? 1:16

B

11. Find the length of a radius of a sphere whose area is 196π cm². 7 cm

12. A great circle of a sphere has an area of 64π square units. Find the volume. $\frac{2,048}{3}\pi$ cu units

13. Find the length of a radius of a sphere whose volume is 972π cm³. 9 cm

14. If the length of the radius of a sphere is doubled, what happens to the area of the sphere? to the volume?

14. 4 times as large; 8 times as large

15. If the number of square units in the area of a sphere is equal to the number of cubic units in its volume, find the length of a radius of the sphere. 3 units

16. Find the volume of metal in a hollow metal ball if its outside diameter measures 12 cm and its inside diameter measures 10 cm. $\frac{364\pi}{3}$ cm³

17. The radius of a sphere is 13 cm long. How far from the center of the sphere should a plane be passed so as to intersect the sphere in a circle whose area is 144π cm^2? **5 cm**

18. Two spheres resting on a table touch each other. One sphere has a radius that is 5 cm long and the other has a radius that is 8 cm long. How far apart are the centers of the spheres. **$4\sqrt{10}$ cm**

19. If a sphere with radius of length r is inscribed in a cube, what is the ratio of the volume of the sphere to the volume of the cube? [*Hint:* The height of the cube equals the length of a diameter of the sphere.] **π:6**

20. If a sphere with radius of length r is inscribed in a cylinder, what is the ratio of the volume of the sphere to the volume of the cylinder? [*Hint:* The height of the cylinder equals the length of a diameter of the sphere.] **2:3**

NON-ROUTINE PROBLEMS

It can be shown that in the picture below, the volume of the cylinder is equal to the volume of the sphere plus the volume of the cone. In the mobile shown, the cylinder will balance the sphere and the cone combined if all are solid figures constructed from the same material.

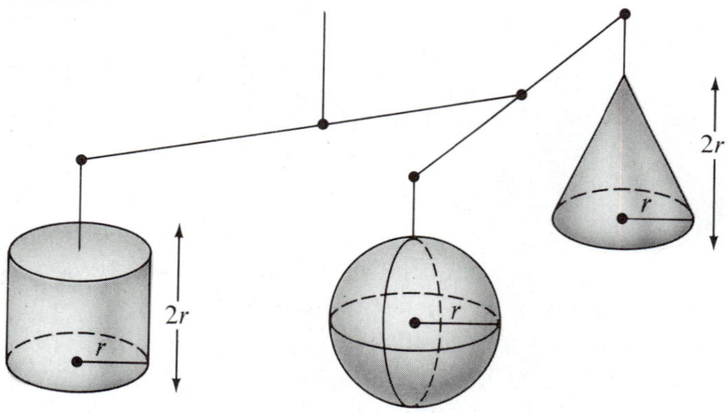

Volume of cylinder = Volume of sphere + Volume of cone

$V = \pi r^2 h \longleftarrow h = 2r$ $\qquad V = \frac{4}{3}\pi r^3$ $\qquad V = \frac{1}{3}\pi r^2 h \longleftarrow h = 2r$

$V = \pi r^2(2r)$ $\qquad V = \frac{1}{3}\pi r^2(2r)$

$V = 2\pi r^3$ $\qquad V = \frac{2}{3}\pi r^3$

$$2\pi r^3 = \frac{4}{3}\pi r^3 + \frac{2}{3}\pi r^3$$

Construct a mobile with figures similar to the ones shown. Why might the balance system need to be adjusted?

APPLICATIONS

Read → Plan → Solve → Interpret

1. Concrete is usually sold by the yard. A yard of concrete is really one cubic yard in volume. How many yards of concrete must be ordered to pour the rectangular section shown?

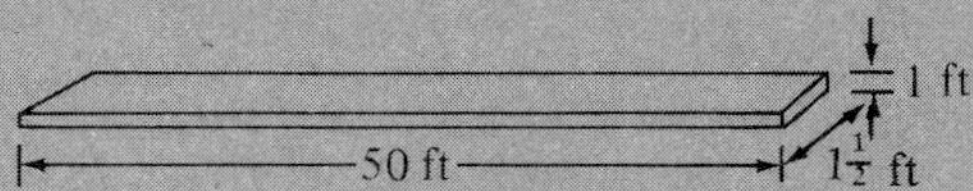

2. "Fancy Pack" treats containing nuts and dried fruit are packed in cones 12 cm tall and 6 cm in diameter or in cylinders 7 cm in diameter and 4 cm deep. Which contains more? How much more?

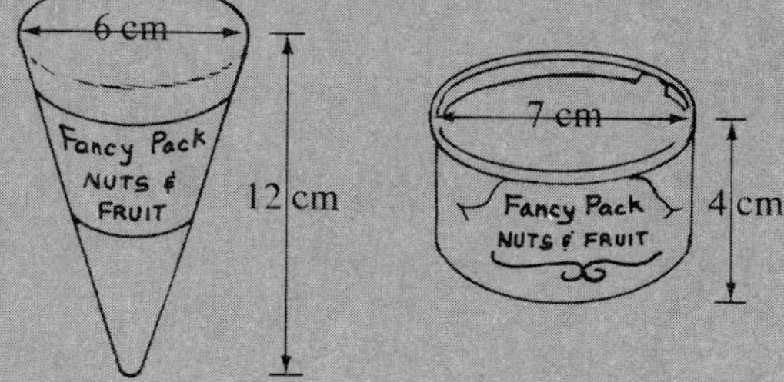

3. A restaurant owner wishes to use cups for her expensive desserts that look large but contain a small amount. Which of the cups pictured should she use for the expensive dessert? Use your calculator to find the difference.

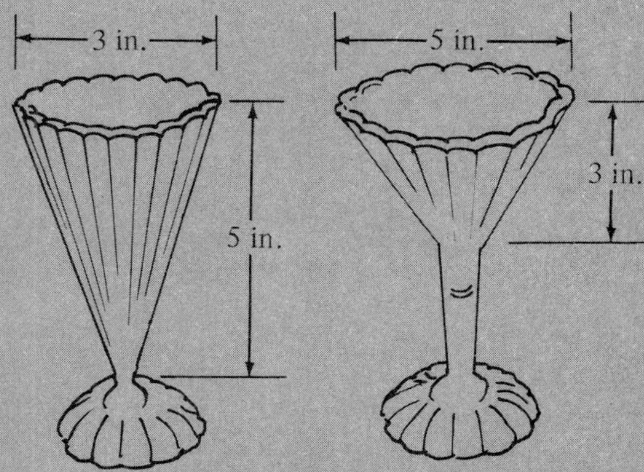

CHAPTER SIXTEEN REVIEW

A exercises are 1–7, 9, 11, 12. B exercises are 8, 10, 13, 14. C exercises are starred.

Sketch each of the following. Then state how many edges, vertices, and faces each has. [16.1]

1. a cube **12, 8, 6**
2. a right hexagonal prism **18, 12, 8**

Find the lateral area, total area, and volume. Leave answers in terms of π. [16.2, 16.3]

3. rectangular solid
L = 380 sq. units
T = 500 sq. units
V = 600 cu. units

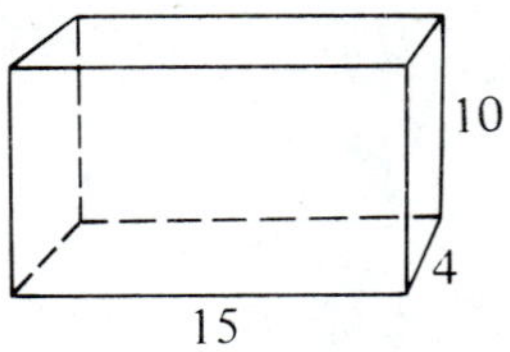

4. right prism
L = 240 sq. units
T = 390 sq. units
V = 450 cu. units

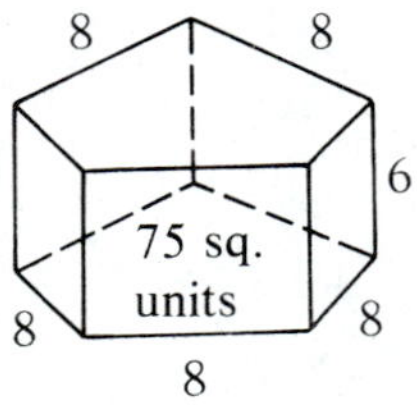

5. right prism
L = 369 m²
T = 468 m²
V = 540 m³

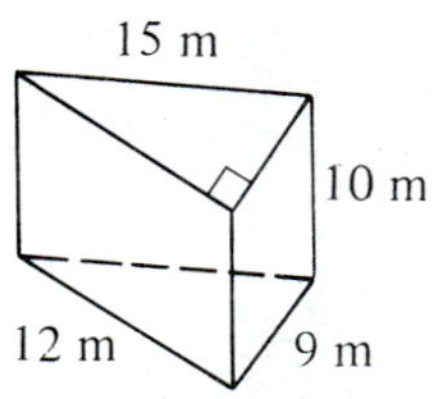

6. cylinder [16.4]

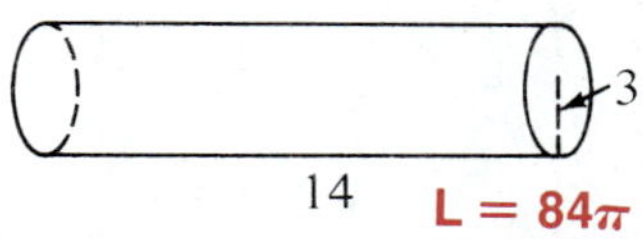

L = 84π sq. u
T = 102π sq u
V = 126π cu u

7. regular pyramid [16.5]
L = 360 cm²
T = 432 cm²
V = 384 cm³

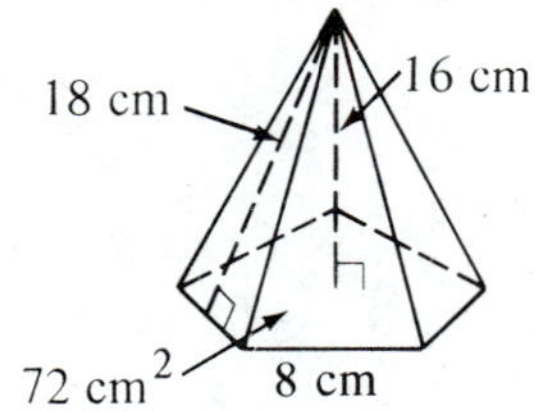

8. regular pyramid
L = 240 sq units
T = 384 sq units
V = 384 cu units

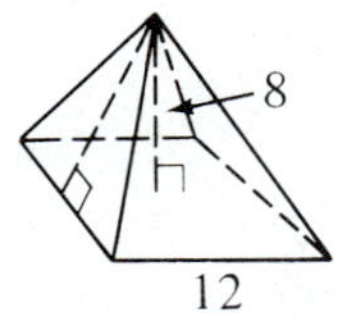

9. cone [16.6]
L = 65π sq. units
T = 90π sq. units
V = 100π cu units

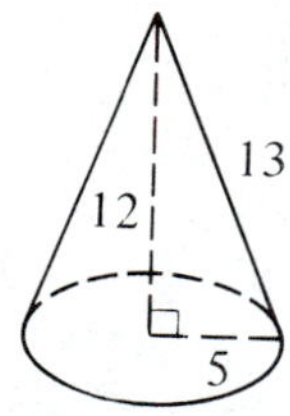

10. cone
L = 24π sq. units
T = 40π sq. units
V = $\frac{32}{3}\sqrt{5}\pi$ cu. units

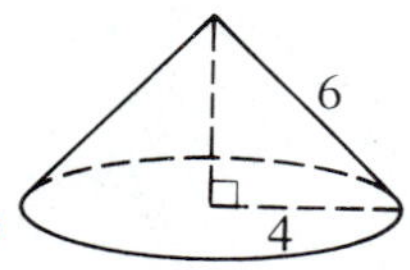

Leave answers in terms of π.

11. The radius of a sphere measures 6 cm. Find the area. [16.7] **144π cm²**
12. The diameter of a sphere measures 4 cm. Find the volume. **$\frac{32}{3}\pi$ cm³**
13. The volume of a rectangular solid is 300 m³. The base is a rectangle 4 m by 15 m. Find the length of an altitude. [16.3] **5 m**
14. A rectangle 7 cm by 9 cm is revolved about a 9-cm side as an axis. Find the total area of the cylinder of revolution formed. [16.4] **224π cm²**

*15. Each side of the base of a regular triangular pyramid is 12 cm long. A slant height makes an angle of 60° with the altitude. Find the total area and volume of the pyramid. [16.5]
T = $72 + 36\sqrt{3}$ cm² **V = $24\sqrt{3}$ cm³**

*16. Suppose a square with 8-cm sides is revolved about a diagonal as an axis. Find the volume of the figure formed. [16.6]
$\frac{256}{3}\sqrt{2}\pi$ cm³

Vocabulary

prism [16.1]
face [16.1]
edge [16.1]
vertex [16.1]
altitude [16.2]
volume [16.3]
cylinder [16.4]
pyramid [16.5]
slant height [16.5]
cone [16.6]
sphere [16.7]

CHAPTER SIXTEEN TEST

A exercises are 1–5, 7, 9, 10. B exercises are 6, 8, 11. C exercises are starred.

Sketch each of the following. Then state how many edges, faces, and vertices each has.

1. a rectangular solid **12, 6, 8**

2. a right pentagonal prism **15, 7, 10**

Find the lateral area, total area, and volume. Leave answers in terms of π.

3. rectangular solid
$L = 450\text{ cm}^2$
$T = 562\text{ cm}^2$
$V = 840\text{ cm}^3$

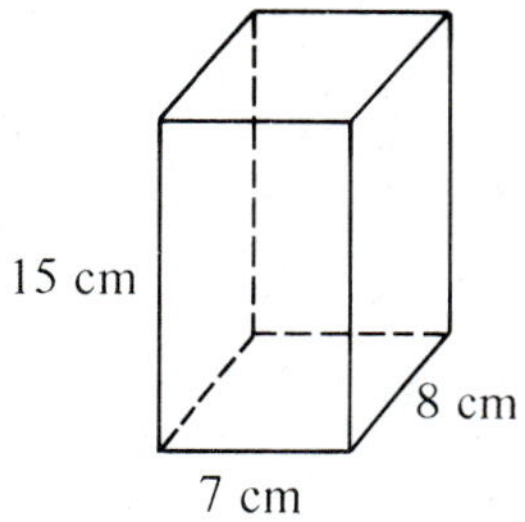

4. right prism
$L = 72$ sq. units
$T = (72 + 8\sqrt{3})$ sq. units
$V = 24\sqrt{3}$ cu. units

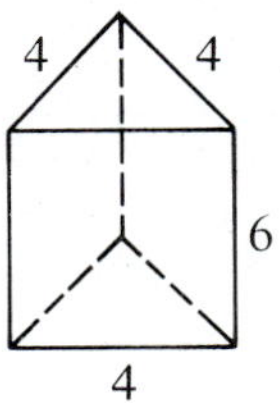

5. cylinder
$L = 120\pi\text{ m}^2$
$T = 170\pi\text{ m}^2$
$V = 300\pi\text{ m}^3$

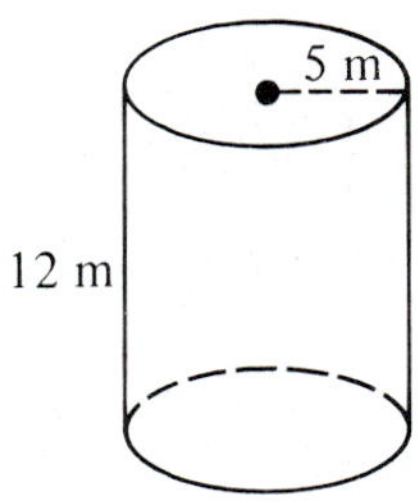

6. regular pyramid
$L = 540$ sq units
$T = 864$ sq units
$V = 1{,}296$ cu units

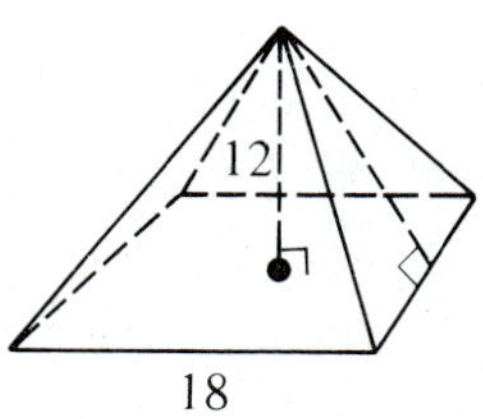

7. cone
$L = 175\pi$ sq units
$T = 224\pi$ sq. units
$V = 392\pi$ cu units

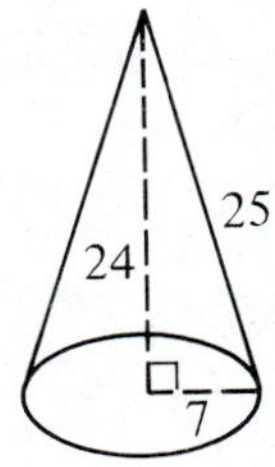

8. cone
$L = 50\pi\text{ cm}^2$
$T = 75\pi\text{ cm}^2$
$V = \frac{125}{3}\sqrt{3}\pi\text{ cm}^3$

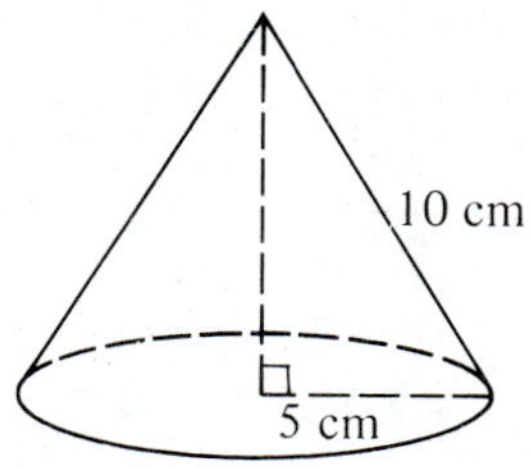

Leave answers in terms of π.

9. The diameter of a sphere measures 4 cm. Find the area. $16\pi\text{ cm}^2$

10. The radius of a sphere measures 3 cm. Find the volume. $36\pi\text{ cm}^3$

11. Find the volume of a prism whose bases are regular hexagons with 4-cm sides and whose altitude measures 10 cm. $240\sqrt{3}\text{ cm}^3$

***12.** Each side of the base of a regular square pyramid is 8 cm long. A slant height makes an angle of 45° with the altitude. Find the volume of the pyramid. $\frac{256}{3}\text{ cm}^3$

***13.** Suppose a square with 12-cm sides is revolved about a diagonal as an axis. Find the area of the figure formed. $144\sqrt{2}\pi\text{ cm}^2$

COMPUTER ACTIVITIES

Geometric Solids

Objective: **To find the surface area and volume of geometric solids, using a computer program**

Chapter 16 has been devoted to a discussion of the surface areas and volumes of various geometric solids. It is possible to use any of the formulas from this chapter in a computer program. The following program makes use of one of the formulas from this chapter. It will determine the lateral area and volume of a cone, square pyramid, or triangular pyramid.

```
110  PRINT "IS THIS A PYRAMID OR A CONE";
120  INPUT P$
130  IF P$ = "PYRAMID" THEN 190
140  PRINT "WHAT IS THE RADIUS";: INPUT R
150  PRINT "WHAT IS THE ALTITUDE";: INPUT A
160  PRINT "THE VOLUME IS ";3.1416 * R ^ 2 * A / 3"."
170  PRINT "THE LATERAL AREA IS ";3.1416 * R *  SQR
     (A ^ 2 + R ^ 2)"."
180  PRINT : GOTO 110
190  PRINT "IS ITS BASE SQUARE OR TRIANGULAR"
200  INPUT P$
210  IF P$ = "TRIANGULAR" THEN 270
220  PRINT "WHAT IS THE SIDE OF THE SQUARE";: INPUT
     S
230  PRINT "WHAT IS THE ALTITUDE OF THE PYRAMID";: INPUT
     A
240  PRINT "THE VOLUME IS ";S ^ 2 * A / 3"."
250  PRINT "THE LATERAL AREA IS "2 * S *  SQR (A ^ 2
      + (S / 2) ^ 2)"."
260  PRINT : GOTO 110
270  PRINT "WHAT IS THE SIDE OF THE TRIANGLE";: INPUT
     S
280  PRINT "WHAT IS THE ALTITUDE OF THE PYRAMID";: INPUT
     A
290  PRINT "THE VOLUME IS ";A * S ^ 2 *  SQR (3) / 12"."
300  PRINT "THE LATERAL AREA IS ";
310  PRINT 3 * S *  SQR (A ^ 2 + S ^ 2 / 12) / 2"."
320  PRINT : GOTO 110
```

EXERCISES: See the Computer Section beginning on page 565 for more information.

1. Write a program that determines the lateral area, total area, and volume of a cylinder.
2. Write a program that determines the surface area and volume of a sphere.

COLLEGE PREP TEST

Directions: **In each item you are to compare a quantity in Column 1 with a quantity in Column 2. Write the letter of the correct answer from these choices:**

A—The quantity in Column 1 is greater than the quantity in Column 2.
B—The quantity in Column 2 is greater than the quantity in Column 1.
C—The quantity in Column 1 is equal to the quantity in Column 2.
D—The relationship cannot be determined from the given information.

Notes: Information centered over both columns refers to one or both of the quantities to be compared.
A symbol that appears in both columns has the same meaning in each column and all variables represent real numbers.

Sample Question and Answer.

Column 1		Column 2
Area $\odot O$	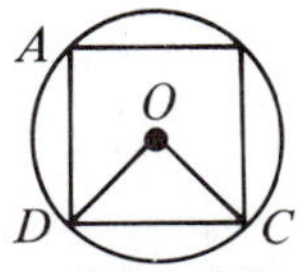	25π

$DO \perp OC$
Area $\triangle DOC = 12.5$

Answer: C. Since $\overline{DO} \perp \overline{OC}$, Area $\triangle DOC = \frac{1}{2} DO \cdot OC = 12.5$ and $DO \cdot OC = 25$.
Area $\odot O = \pi r^2 = \pi(DO \cdot OC) = 25\pi$.

Column 1		Column 2
1. Volume of cylinder A	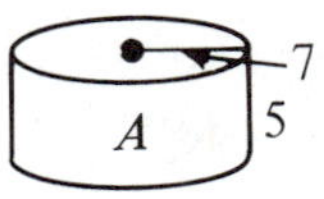	Volume of cylinder B

Answer: A

2. Total Area of A	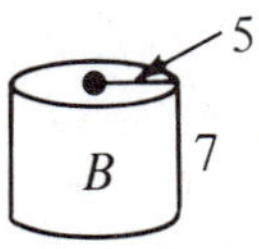	Total Area of B

Answer: A

3. Volume of Cone — $\frac{\text{height of Cone}}{\text{height of Cylinder}} = \frac{3}{1}$, $\frac{\text{base of Cone}}{\text{base of Cylinder}} = 1$ — Volume of Cylinder
Answer: C

4. No. of edges in a triangular prism. — No. of vertices in a hexagonal prism.
Answer: B

Column 1	Column 2
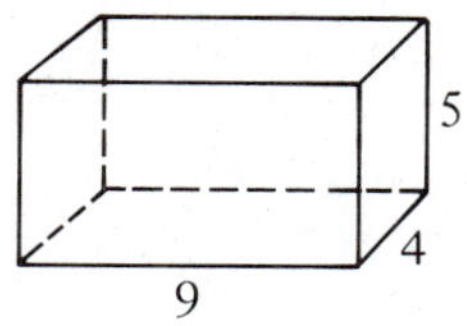	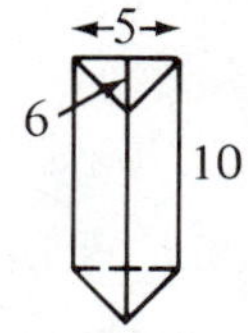

5. Volume of above Prism **Answer: A** — Volume of above Prism

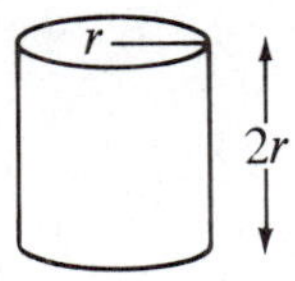

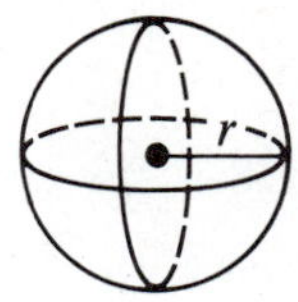

6. Volume of cylinder — Volume of sphere
Answer: A

7. Lateral area of cylinder — Area of sphere
Answer: C

8. Area of $ABCD$ — Area of Circle O

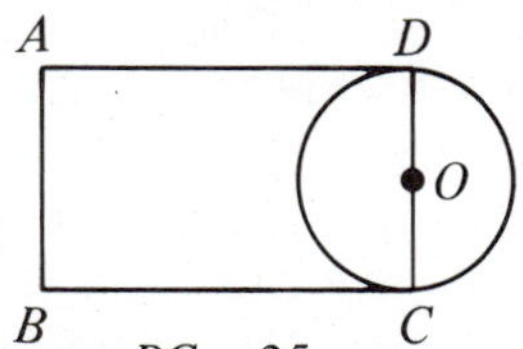

$BC = 25\pi$
$DC = 10$

Answer: A

9. Perimeter of $ABCD$ — 20
Answer: C

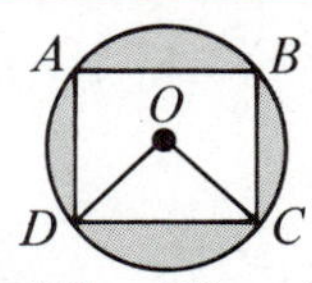

10. Area of shaded portion — $25\pi - 20$

$AB = BC = CD = AD = OD$
Perimeter $\Delta ODC = 15$

Answer: B

CUMULATIVE REVIEW

Chapters 1–16

Write the letter that indicates the best answer.

1. If the measure of an angle is 10° more than the measure of its supplement, then the measure of the angle is **B**

(A) 85° (B) 95°
(C) 170° (D) none of these

2. It is possible for a triangle to be **C**

(A) both scalene and obtuse
(B) both equilateral and acute
(C) both of these
(D) neither of these

3. Each interior angle of a regular polygon with nine sides measures **D**

(A) 120° (B) 144°
(C) $147\frac{3}{11}°$ (D) none of these

4. 8 is the geometric mean of 2 and **B**

(A) 16 (B) 32
(C) 64 (D) none of these

5. If the hypotenuse of a 45°–45° right triangle is 10 m long, then the length of a leg is **B**

(A) 5 m (B) $5\sqrt{2}$ m
(C) 10 m (D) $10\sqrt{2}$ m

6. A central angle and an inscribed angle of a circle intercept the same arc. If the inscribed angle measures 35°, then the central angle measures **C**

(A) $17\frac{1}{2}°$ (B) 35°
(C) 70° (D) none of these

7. For points P(−5, 2) and Q(3, −4), PQ is equal to **B**

(A) $2\sqrt{2}$ (B) 10
(C) 100 (D) none of these

8. An equation in standard form for the line through $A(3, 1)$ and $B(-2, 5)$ is **A**

(A) $4x + 5y = 17$ (B) $-4x + 5y = 7$
(C) $y - 1 = -\frac{4}{5}(x - 3)$ (D) none of these

9. The locus of points equidistant from the sides of an angle is **D**

(A) a point (B) a line
(C) a circle (D) none of these

10. In $\triangle ABC$, $\overline{AC} \cong \overline{BC}$, AC is 10 cm, and AB is 16 cm. The centroid of $\triangle ABC$ is ______ from $\overline{AB}$. **A**

(A) 2 cm (B) 3 cm
(C) 6 cm (D) none of these

11. In $\odot O$, if $OF < OE$, then **D**

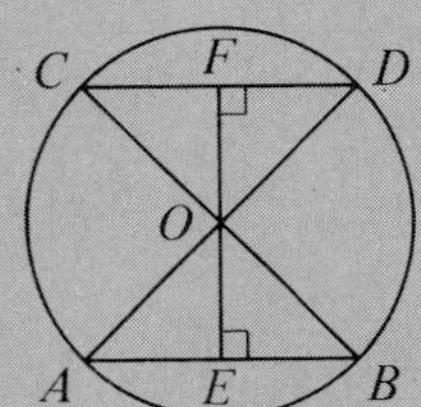

(A) $AB < CD$
(B) $<COD > \angle AOB$
(C) $CF > EB$
(D) all of these

12. If a radius of a sphere measures 6 cm, then the volume is **C**

(A) 72π (B) 144π
(C) 288π (D) none of these

13. The total area of the figure shown is **B**

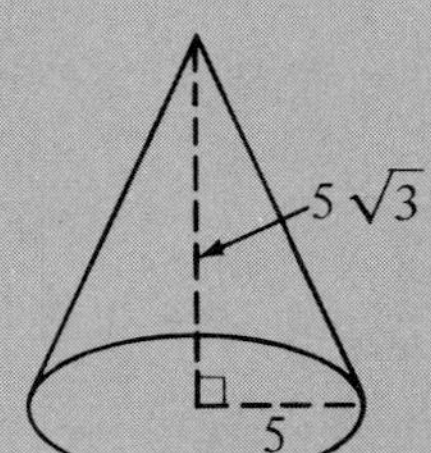

(A) 50π
(B) 75π
(C) $50\sqrt{3}\pi$
(D) $25(\pi + \sqrt{3})$

True or false?

14. The distance between two points on a number line is the absolute value of the difference of the coordinates of the two points. **T**

15. A postulate is a statement to be proved. **F**

16. If two lines are parallel, then they are not skew. **T**

17. A concave polygon cannot be equiangular. **T**

18. The diagonals of a rhombus are congruent. **F**

19. If the measure of a base of a trapezoid is 3 units longer than the measure of the median, then the measure of the other base is 3 units shorter than the measure of the median. **T**

Find each indicated measure.

20. Given: $\overline{ST} \parallel \overline{PQ}$, $PS = 5$, $SR = 4$, and $ST = 6$. Find PQ. **13.5**

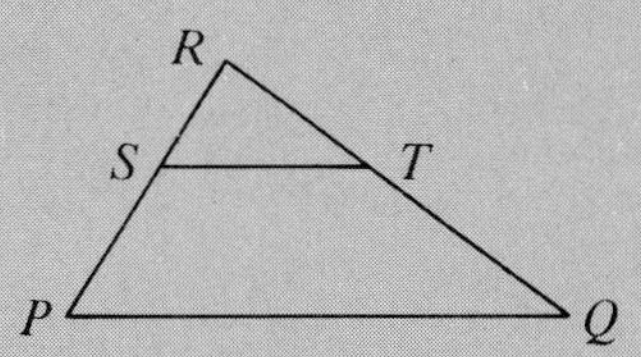

21. The angle of elevation from a spot on the ground to the top of a tree measures 48°. The spot is 5 m from the base of the tree. How tall is the tree, to the nearest tenth of a meter? **5.6 m**

22. Given: $\odot O$ with $m\widehat{QR} = 100$, $m\widehat{QS} = 30$, and $m\widehat{ST} = 136$. Find m $\angle P$. **32**

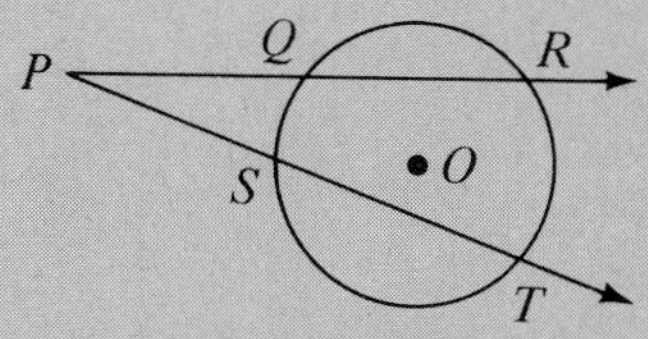

23. Each side of an equilateral triangle measures 9. Find the area. $\mathbf{\frac{81}{4}\sqrt{3}}$

24. Give the slope of the line perpendicular to $\overleftrightarrow{AB}$ for $A(-3, 4)$, $B(-2, -1)$. $\mathbf{\frac{1}{5}}$

25. Find the coordinates of the center and the length of a radius of a circle with the equation $x^2 + y^2 - 6x + 10y = 15$. **(3, −5); 7**

Sketch and describe each locus. Show all possible cases.

26. The locus of points 4 cm from a given line **See Post. 28.**

27. The locus of points less than 2 cm from a given point **Int. of ⊙, $r = 2$ cm, ctr. is given pt.**

28. The locus of points equidistant from two given points and at a given distance from a given point **See Th. 14.1 and Post. 27.**

Prove the following:

29. Given: $\overline{WX} \cong \overline{ZY}$, $\overline{WX} \parallel \overline{ZY}$, and m $\angle ZVW <$ m $\angle ZVY$
Prove: $ZW < ZY$

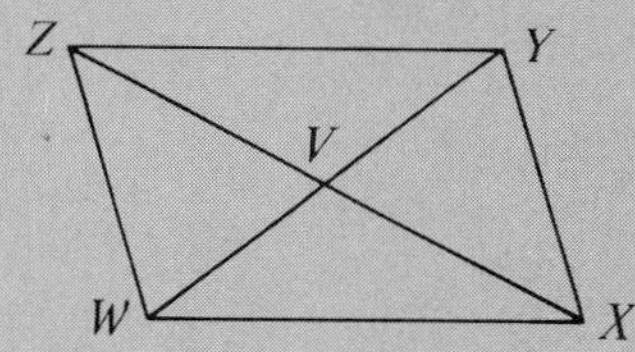

Leave answers in terms of π.

30. Find the lateral area, the total area, and the volume of the cylinder shown.
L = 40 π cm²; T = 72 π cm²; V = 80 π cm³

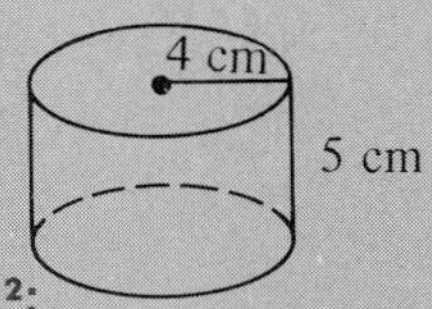

31. Find the lateral area of the regular pyramid shown. **260 sq. units**

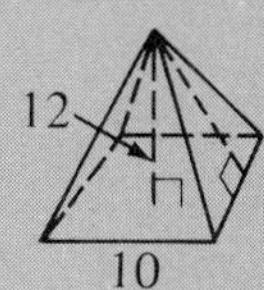

32. Suppose a rhombus with 5-cm sides is revolved about a diagonal which is 6 cm long. Find the volume of the figure formed. **32 π cm³**

Logic Section

FLOW PROOFS

Objective **To write flow proofs**

The proofs that you have written so far have been organized in a two-column format. You can use a flow-diagram format or **flow proof** instead. In Theorem 2.9, you proved that complements of congruent angles are congruent. Compare the flow proof of this theorem with its original proof on page 56.

Given: $\angle A$ is a complement of $\angle B$, $\angle D$ is a complement of $\angle C$, and $\angle B \cong \angle C$.
Prove: $\angle A \cong \angle D$

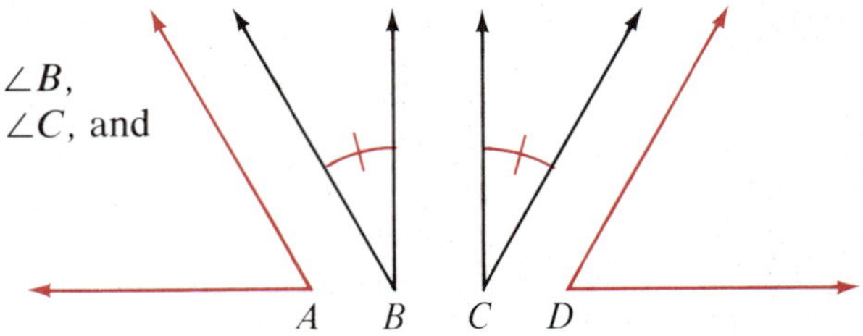

$m\angle A + m\angle B = 90$, $m\angle D + m\angle C = 90$ → $m\angle A + m\angle B = m\angle D + m\angle C$

$\angle B \cong \angle C$ → $m\angle B = m\angle C$

$m\angle A + m\angle B = m\angle D + m\angle C$ and $m\angle B = m\angle C$ → $m\angle A = m\angle D$ → $\angle A \cong \angle D$

In the proof above, the reasons for the steps were not written. If you wish to write the reasons, they can be written below the statements. The flow proof of Theorem 3.1 illustrates this.

Given: $\overleftrightarrow{AB} \parallel \overleftrightarrow{CD}$
Prove: $\angle 1 \cong \angle 3$

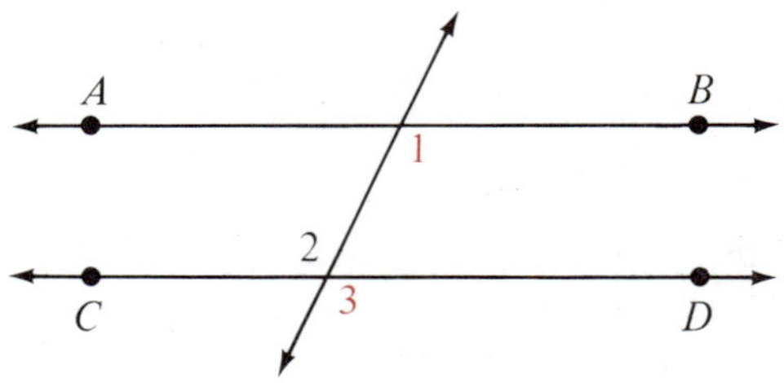

$\overleftrightarrow{AB} \parallel \overleftrightarrow{CD}$ (Given) → $\angle 1 \cong \angle 2$ (Alt. int. $\angle$'s of $\parallel$ lines are $\cong$.)

$\angle 2$ and $\angle 3$ are vert. $\angle$'s. (Intersecting lines form vert. $\angle$'s.) → $\angle 2 \cong \angle 3$ (Vert. $\angle$'s are $\cong$.)

$\angle 1 \cong \angle 2$ and $\angle 2 \cong \angle 3$ → $\angle 1 \cong \angle 3$ (Transitive property)

Flow diagrams are very helpful in organizing complicated proofs. The flow diagram shows clearly how each step relates to all other steps. In organizing such a proof, you should know the reason for each step, but you may not wish to write it.

Example 1

Write a flow proof for Example 4 on page 154.

Given: $\overline{YT} \parallel \overline{XU}$, $\overline{YT} \perp \overline{SV}$, $\overline{ST} \cong \overline{VU}$, and $\overline{RS} \cong \overline{RV}$

Prove: $\overline{XU} \cong \overline{YT}$

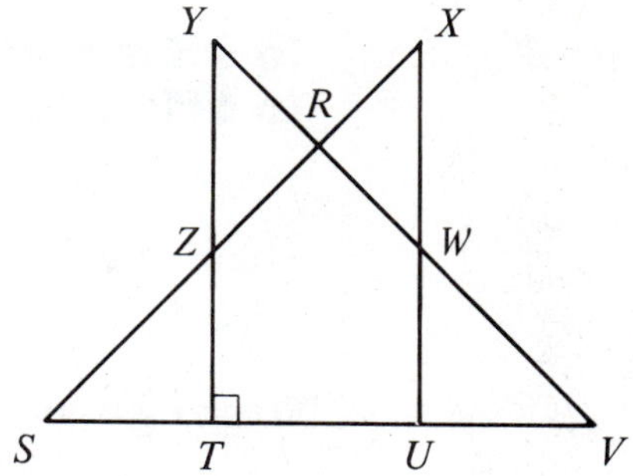

$\angle YTV$ is a rt. $\angle$. → $\angle XUS$ is a rt. $\angle$.

$\angle YTV$ and $\angle XUS$ are suppl. → $\angle XUS$ is a rt. $\angle$.

$\angle YTV$ is a rt. $\angle$. and $\angle XUS$ is a rt. $\angle$. → $\angle XUS \cong \angle YTV$

$ST = VU$ → $ST + TU = VU + TU$ → $SU = VT$ → $\overline{SU} \cong \overline{VT}$

$\overline{RS} \cong \overline{RV}$ → $\angle S \cong \angle V$

$\angle XUS \cong \angle YTV$, $\overline{SU} \cong \overline{VT}$, $\angle S \cong \angle V$ → $\triangle XSU \cong \triangle YVT$ → $\overline{XU} \cong \overline{YT}$

Example 2

Organize the steps of the proof of Theorem 5.7 on page 166 according to the flow diagram outlined below.

XXX, XXX → XXX, XXX

XXX, XXX, XXX → XXX → XXX → XXX

P is on $\overleftrightarrow{CM}$. and $\overleftrightarrow{CM}$ is $\perp$ bis. of $\overline{XY}$. → $\overline{XM} \cong \overline{YM}$, $\triangle XMP$ and $\triangle YMP$ are rt. $\triangle$'s.

$\overline{XM} \cong \overline{YM}$, $\triangle XMP$ and $\triangle YMP$ are rt. $\triangle$'s., $\overline{PM} \cong \overline{PM}$ → $\triangle XMP \cong \triangle YMP$ → $\overline{PX} \cong \overline{PY}$ → $PX = PY$

Written Exercises **Check students' proofs.**

1. Write a flow proof for Theorem 2.8 on page 55.
2. Write a flow proof for Example 2 on page 162.

TRUTH TABLES AND ARGUMENTS

Objectives **To write truth tables for arguments**
To determine whether an argument is valid

Truth tables can be used to show that statements are equivalent.

Example 1 **Prepare a truth table to show that $\sim(p \lor q)$ and $\sim p \land \sim q$ are equivalent statements.**

p	q	$p \lor q$	$\sim(p \lor q)$	$\sim p$	$\sim q$	$\sim p \land \sim q$
T	T	T	F	F	F	F
T	F	T	F	F	T	F
F	T	T	F	T	F	F
F	F	F	T	T	T	T

Since their truth tables are the same, you know that the statement $\sim(p \lor q)$ and $\sim p \land \sim q$ are equivalent.

Truth tables can also be used to show the relationships between more than two statements.

Example 2 **Write the truth table for $(p \to q) \to r$.**

p	q	r	$p \to q$	$(p \to q) \to r$
T	T	T	T	T
T	T	F	T	F
T	F	T	F	T
T	F	F	F	T
F	T	T	T	T
F	T	F	T	F
F	F	T	T	T
F	F	F	T	F

A statement that is always true is called a **tautology**. A truth table shows that the statement $[(p \land q) \land \sim p] \to q$ is a tautology

p	q	$p \land q$	$\sim p$	$(p \land q) \land \sim p$	$[(p \land q) \land \sim p] \to q$
T	T	T	F	F	T
T	F	F	F	F	T
F	T	F	T	F	T
F	F	F	T	F	T

A **valid argument** is an argument that is always true. This means that if you can write the argument as a statement that is a tautology, then it is valid.

Consider the following argument:

Hypothesis: John likes hamburgers, or Mary likes hot dogs.
But Mary does not like hot dogs.
Conclusion: John likes hamburgers.

To analyze the argument, first code the statements.

Let p be "John likes hamburgers."
Let q be "Mary likes hot dogs."

Hypothesis: $p \vee q, \sim q$
Conclusion: $\therefore p$

The argument can then be written as $[(p \vee q) \wedge \sim q] \rightarrow p$. The truth table for this statement shows that it is a tautology.

p	q	$p \vee q$	$\sim q$	$(p \vee q) \wedge \sim q$	$[(p \vee q) \wedge \sim q] \rightarrow p$
T	T	T	F	F	T
T	F	T	T	T	T
F	T	T	F	F	T
F	F	F	T	F	T

Since it is a tautology, the argument is valid.

Example 3 **Show that the following is not a valid argument:**

If a triangle is isosceles, then it is a right triangle.
The triangle is a right triangle.
Therefore, it is isosceles.

Let p be "A triangle is isosceles."
Let q be "The triangle is a right triangle."

Hypothesis: $p \rightarrow q, q$
Conclusion: $\therefore p$
or $[(p \rightarrow q) \wedge q] \rightarrow p$

p	q	$p \rightarrow q$	$(p \rightarrow q) \wedge q$	$[(p \rightarrow q) \wedge q] \rightarrow p$
T	T	T	T	T
T	F	F	F	T
F	T	T	T	F
F	F	T	F	T

Since the statement is not a tautology, the argument is not valid.

Many proofs in geometry depend upon the following type of argument.

If quadrilateral $ABCD$ has opposite parallel sides, then it is a parallelogram.
Quadrilateral $ABCD$ has opposite parallel sides.
Therefore, $ABCD$ is a parallelogram.

This argument is of the form:

$$\left.\begin{array}{ll}\text{Hypothesis:} & p \rightarrow q,\ p \\ \text{Conclusion:} & \therefore q\end{array}\right\} \text{ or } [(p \rightarrow q) \wedge p] \rightarrow q$$

The following truth table shows that this is a valid argument.

p	q	$p \rightarrow q$	$(p \rightarrow q) \wedge p$	$[(p \rightarrow q) \wedge p] \rightarrow q$
T	T	T	T	T
T	F	F	F	T
F	T	T	F	T
F	F	T	F	T

Written Exercises

Use a truth table to determine if the two statements in each pair are equivalent.

1. $\sim(\sim p)$; p **yes**
2. $p \rightarrow q$; $\sim p \rightarrow \sim q$ **no**
3. $\sim(p \rightarrow q)$; $\sim p \rightarrow \sim q$ **no**
4. $\sim(p \rightarrow q)$; $\sim q \rightarrow \sim p$ **no**
5. $\sim(p \wedge q)$; $\sim p \vee \sim q$ **yes**
6. $\sim(p \rightarrow q)$; $p \wedge \sim q$ **no**

Use a truth table to determine if each statement is a tautology.

7. $\sim(p \wedge \sim p)$ **yes**
8. $(p \wedge q) \rightarrow p$ **yes**
9. $p \rightarrow (p \vee q)$ **yes**
10. $[(p \wedge q) \rightarrow r] \rightarrow [p \rightarrow (q \rightarrow r)]$ **yes**
11. $[p \rightarrow (q \rightarrow r)] \rightarrow [(p \wedge q) \rightarrow r]$ **yes**

Use a truth table to determine which arguments are valid.

12. $p \rightarrow q$; $\sim p$; $\therefore \sim q$ **not valid**
13. $p \wedge q$; $\sim p$; $\therefore \sim q$ **valid**
14. $p \rightarrow \sim q$; $p \vee q$; $\therefore \sim q$ **not valid**

Translate each argument into symbolic form. Tell whether it is a valid argument.

15. Hypothesis: If $PQRS$ is a parallelogram, then $PQRS$ is a quadrilateral.
$PQRS$ is a quadrilateral.
Conclusion: $PQRS$ is a parallelogram. **$[(p \rightarrow q) \wedge q] \rightarrow p$; not valid**
16. Hypothesis: $\overleftrightarrow{AB} \parallel \overleftrightarrow{CD}$ and $\overleftrightarrow{CD} \perp \overleftrightarrow{EF}$
Conclusion: $\overleftrightarrow{AB} \perp \overleftrightarrow{EF}$ **$(p \wedge q) \rightarrow r$; not valid**

Computer Section

BEGINNING WITH BASIC

Objective: **To find the length of a segment on a number line by means of a computer program**

In this section, you will learn more about computer programming with specific applications to geometry. Programs in this section will progress from the simple to the more complex as additional programming concepts and BASIC language commands are introduced. The simple programs early in this section may be written more efficiently, and students with previous knowledge of BASIC are encouraged to elaborate on the programs presented.

BASIC is one of many computer languages in existence. It is the one most commonly used in schools. All programs in this section are written in a version of BASIC that is common to most computers. If some word does not "compute" with your computer, then consult the manual for that machine.

A computer program is a numbered series of instructions that directs a computer to execute particular operations. It is intriguing to use the fewest possible statements to make the computer perform meaningful tasks.

The BASIC operation symbols are as follows: + for addition and − for subtraction, * for multiplication and / for division, ∧ for exponentiation (raising to a power), and parentheses as the symbols of grouping. In addition, the computer obeys all the algebraic rules for order of operations. For example, the computer calculates 3 + 2 * 5 as 13, not 25.

You can write a computer program that uses only two different BASIC commands: PRINT and RUN. Note that these, and all other BASIC statements, are capitalized. Therefore, when typing BASIC programs, use the CAPS LOCK key on your keyboard. PRINT can be used to perform computations. For example, PRINT 3 + 2 * 5 will compute and print the result. PRINT can also be used to print phrases and statements, by enclosing them within quotation marks. For example, PRINT "3 + 2 * 5" will print 3 + 2 * 5.

COMPUTER SECTION

Type in the following one-line program. Try to predict exactly what the computer will print before you test the program.

```
10 PRINT "3 + 2 * 5 = " 3 + 2 * 5
```

By typing in the program line above and pressing the RETURN key, you have entered the line into the computer's memory. (NOTE: You must press RETURN after entering a program line.) To get the computer to execute your one-line program, you must type in the word RUN (without a line number) and press RETURN. Type in RUN now and check your prediction.

If your prediction was not correct, you may have typed the program line incorrectly. To check, type LIST and press RETURN. The computer will display a LISTing. To correct any errors, retype the entire line, including the line number. The computer will replace the incorrect line with the new one.

The following program uses two more BASIC statements: LET and ABS. LET is used to assign values to a variable. LET X = 2 means assign the value 2 to the variable X. (NOTE: LET 2 = X does *not* accomplish the same thing.) Variables are designated by single letters. The program finds the length of a segment, which requires the use of absolute value. The BASIC command ABS( ) finds the absolute value of the expression in parentheses. The semicolon at the end of line 30 makes the next PRINT statement's output appear on the same line as line 30's output. To find the lengths of other segments, change lines 10 and 20 to insert different values for the segment's endpoints. RUN this program for a variety of endpoints of segments.

```
10  LET X = 6
20  LET Y = 17
30  PRINT "LENGTH OF THE SEGMENT WITH ENDPOINTS "X"
     AND "Y" IS ";
40  PRINT  ABS (X - Y)"."
```

Exercises:

1. Write a multiline program that evaluates the following algebraic expressions. Make your program output each expression and its value, as in the first program above.

   $3 \cdot 8 + 4 \cdot 6 - 5 \qquad 2 \cdot (8 - 6) \cdot (2 + 7) \qquad (3^2 - 4 \cdot 6)/(8 + 2)$

2. Write a 6-line program that computes the answers to the Odd Exercises in the Algebra Review on page 4.

INPUT AND GOTO

Objective: **To determine the complement and supplement of an angle that has been INPUT into a program**

To assign different values to a variable in previous programs, you had to type in new LET statements. You can avoid this tedious process by using the BASIC command, INPUT. For example, when a program is RUN, INPUT X displays a question mark, signaling the user to enter a value for X.

Both of this lesson's programs contain the unconditional branching command, GOTO. When either program is RUN, GOTO 10 instructs the computer to branch back to line 10 and resume program execution at that point. These GOTO statements allow you to INPUT as many angles as you desire without continually typing in RUN and pressing RETURN. To stop program execution, hold down the CTRL key and press the C key.

The following program asks the user to INPUT an angle, whereupon it outputs the complement and supplement of the angle. The program outputs complete sentences in reply to your input. It is good practice to make all your programs do so. Also note the colon in line 50; it allows you to put two or more statements in one program line.

```
10  PRINT "ENTER AN ACUTE ANGLE."
20  INPUT A
30  PRINT "THE COMPLEMENT OF "A" IS "90 - A"."
40  PRINT "THE SUPPLEMENT OF "A" IS "180 - A"."
50  PRINT : GOTO 10
```

The next program accept inputs and displays answers in terms of degrees and minutes. The INPUT statements in lines 20 and 30 print messages in addition to requesting information.

```
10  PRINT "ENTER AN ACUTE ANGLE IN DEGREES AND
     MINUTES."
20  INPUT "NUMBER OF DEGREES ";D
30  INPUT "NUMBER OF MINUTES ";M
40  PRINT "THE COMPLEMENT IS "89 - D" DEGREES
     "60 - M" MINUTES."
50  PRINT "THE SUPPLEMENT IS "179 - D" DEGREES
     "60 - M" MINUTES."
60  GOTO 10
```

Exercise: Change the second program above so that it will accept input and produce output in terms of degrees, minutes, and seconds. (A degree consists of 60 minutes. A minute consists of 60 seconds.)

READ WITH DATA

Objective: **To determine the measure of the third angle of a triangle, using a program that utilizes READ and DATA statements**

It is convenient to use an INPUT statement in a program when you do not know all the values to be entered into the program. When you do know all the values to be entered into a program, it is often more efficient to use the dual commands, READ and DATA.

To do so, you write READ X in place of INPUT X. READ X instructs the computer to find a program line labeled DATA, and to assign the first value in the DATA line to the variable X. Thereafter, every time READ X is executed, successive values in the DATA line are assigned to X one at a time, until all the values have been used.

READ can assign values to more than one variable at a time, as in the program below. You simply write READ X, Y. This statement instructs the computer to retrieve two values at a time from the DATA line and to assign the first value to X and the second to Y.

Whenever you use a READ command in a program, you must also include a DATA line in the program. The DATA line may appear anywhere in a program. However, it is often convenient to place DATA lines at the end of a program, where they are easy to find and alter, if necessary.

The following program READS the values for two angles of a triangle, and then calculates the third angle. Enter and RUN it.

```
10  READ A,B
20  PRINT "TWO ANGLES OF A TRIANGLE ARE ";A;" AND
     ";B"."
30  PRINT "THE THIRD ANGLE IS ";180 - AB"."
40  PRINT : GOTO 10
50  DATA  10,20,30,40,10,78,25,25
```

When you RUN the program above, an error message appears at the end of your output. This occurs because the computer used up all of the DATA. At this time, you need not be concerned about such error messages. In later lessons, you will learn techniques for avoiding them.

Exercise: Write a program that READs the values for three angles at a time from a DATA line, and determines whether or not the angles can possibly form a triangle.

IF . . . THEN

Objective: **To classify triangles by the lengths of their sides, using a computer program that contains IF . . . THEN statements**

The computer can compare two values, and make a decision based on the result of that comparison. The BASIC statement IF . . . THEN performs this function. The IF portion contains the comparison, and the THEN portion contains the action to be taken when the comparison in the IF statement is true.

The IF comparison is made between two values in accordance with the following BASIC relations: =, <, >, < =, > =, < >. Usually, THEN is followed by a line number. IF X = 5 THEN 100 instructs the computer to branch to line 100 when X = 5. If X = 5, the computer executes the very next program line.

The program below, which classifies triangles by the lengths of their sides, is based on the flowchart found on page 129. If you haven't looked at that page yet, do so now.

Lines 110 through 160 in the program correspond with the left-hand column of six boxes in the flowchart. Lines 170, 180, and 190 correspond with the middle column of three boxes, and lines 200 and 210 correspond with the last two boxes in the flowchart.

Compare the program below with the chart on page 129 and see if you can follow the transition from one to the other.

```
110  INPUT "WHAT ARE THE THREE SIDES?";A,B,C
120  IF A = B THEN 170
130  IF A = C THEN 180
140  IF B = C THEN 180
150  PRINT "THE TRIANGLE IS SCALENE."
160  PRINT : GOTO 110
170  IF A = C THEN 200
180  PRINT "THE TRIANGLE IS ISOSCELES."
190  PRINT : GOTO 110
200  PRINT "THE TRIANGLE IS EQUILATERAL."
210  PRINT : GOTO 110
```

Exercise: Write a program corresponding to the flowchart you wrote for the Exercise on page 129.

CHARACTER STRINGS

Objective: **To determine the congruent parts of two congruent triangles, using a computer program that employs character strings**

Up to this point in the Computer Section, you have used only numerical variables, represented by single letters. It is also possible, however, to store words and symbols in BASIC *string variables*. String variables are denoted by single letters followed by a $. For example, you can assign the word BANANA to string variable X$ by using the following statement: LET X$ = "BANANA". Similarly, LET Y$ = " BREAD" is another such assignment. If you then entered the command PRINT X$ Y$, the computer would print BANANA BREAD.

The program on the next page will ask you to enter the name of a triangle, which will be INPUT into character string T$. Then the program will ask you to enter the name of a congruent triangle, which will be stored in C$. The program will then print out the six pairs of congruent parts of the two triangles. To do this, the computer must pick individual letters out of the character strings T$ and C$. The BASIC function MID$ (called SEG in some versions of BASIC) enables the computer to pick characters out of character strings. Consider the operation of MID$ on string X$ defined above. MID$(X$,3,1) instructs the computer to choose 1 letter, beginning with the 3rd letter of X$. Therefore, MID$(X$,3,1) is N. Similarly, for Y$ defined above, MID$(Y$,2,4) instructs the computer to pick 4 letters, beginning with the 2nd letter of Y$. In this case, MID$(Y$,2,4) is READ.

Character strings can be strung together into longer character strings by using a plus sign. For example, for X$ and Y$ above, the statement LET W$ = X$ + Y$ creates a new string, W$. In this case, the command PRINT W$ would output BANANA BREAD.

The following short program demonstrates the use of character strings and the MID$ function.

```
10   INPUT "WHAT IS YOUR FIRST NAME? ";F$
20   INPUT "WHAT IS YOUR LAST NAME? ";S$
30   LET N$ = F$ + " " + S$
40   PRINT "THANK YOU, "N$"."
50   PRINT " YOUR INITIALS ARE ";
60   PRINT  MID$ (F$,1,1)"." MID$ (S$,1,1)"."
```

Note the space inserted between F$ and S$ in line 30. Without the space, line 40 would print your first and last names with no space in between.

The following program was described on the previous page. Enter and RUN it. The LET statements are in lines 35 and 60, and the IF . . . THEN statements are in lines 50 and 75. They cause the computer to execute line 40 three times and line 70 twice. In computer terminology, the computer *loops* through lines 35 to 50 three times and through lines 60 to 75 twice. You will learn another way to create program loops in the next lesson.

```
10  INPUT "NAME A TRIANGLE BY ITS 3 VERTICES.";T$
20  INPUT "NAME A CONGRUENT TRIANGLE IN CORRECT
     ORDER.";C$
30  PRINT "IF TRIANGLE "T$" IS CONGRUENT TO TRIANGLE
     "C$" THEN:"
35  LET I = I + 1
40  PRINT "ANGLE " MID$ (T$,I,1)" IS CONGRUENT TO
     ANGLE " MID$ (C$,I,1)"."
50  IF I < 3 THEN 35
60  LET J = J + 1
70  PRINT " SIDE " MID$ (T$,J,2)" IS CONGRUENT TO
     SIDE " MID$ (C$,J,2)"."
75  IF J <  = 2 THEN 60
80  PRINT " SIDE " MID$ (T$,1,1) MID$ (T$,3,1)" IS
     CONGRUENT TO SIDE " MID$ (C$,1,1) MID$
     (C$,3,1)"."
90  PRINT : GOTO 10
```

The following program gives a quiz on congruence theorems. Character strings are entered into the program by means of READ and DATA statements. Line 20 READs two strings at a time from the DATA lines. Each pair of strings constitute a question and its answer. Line 40 tests your answers, line 60 tallies correct replies, and line 90 computes and displays your score.

```
10  LET I = I + 1
20  READ Q$,A$
30  PRINT "IS "Q$" A CONGRUENCE PATTERN?": INPUT R$
40  IF R$ = A$ THEN 60
50  PRINT "NO, THAT IS INCORRECT.": GOTO 80
60  LET R = R + 1
70  PRINT "CORRECT."
80  IF I < 13 THEN 10
90  PRINT "THANK YOU. YOUR GRADE IS "; INT (R * 100 /
     13 + .5)"."
95  DATA  SSS,YES,LL,YES,SAS,YES,AA,NO,AAS,YES,SSA,
     NO,LA,YES,HL,YES
97  DATA  ASA,YES,AAA,NO,ASA,YES,HH,NO,HA,YES
```

Exercise: Write a quiz program similar to the one above that tests whether classifications of triangles are based on the lengths of their sides or the measures of their angles.

FOR . . . NEXT LOOPS

Objective: **To determine the measures of angles of regular polygons using a computer program that employs a FOR . . . NEXT loop**

In this lesson, you will learn an easier way to write loops using FOR and NEXT statements. To begin a FOR . . . NEXT loop, you use a FOR statement, such as FOR X = 1 TO 10. To complete this FOR . . . NEXT loop, you use the statement, NEXT X. The same variable must be specified in the FOR statement and in the NEXT statement. This particular loop would be executed 10 times.

The following program will determine the measures of angles of regular polygons for a series of polygons. The range of polygons' numbers of sides is specified by you.

```
10  PRINT "WHAT IS THE FIRST NUMBER";: INPUT A
20  PRINT "LAST NUMBER";: INPUT B
30  FOR I = A TO B
40  PRINT "EACH ANGLE OF A REGULAR POLYGON OF ";I;"
     SIDES IS ";
50  PRINT (I - 2) * 180 / I"."
60  NEXT I
```

The values in a FOR . . . NEXT loop need not always be incremented by ones, as in the program above. By using the STEP option in FOR statements, you can make the computer count by any multiple you desire. You can even make the computer count backwards. For example, the statement FOR X = 5 TO 100 STEP 5 will make the computer count 5,10,15, . . .,95,100.

Enter and RUN the following short program.

```
10  FOR X = 10 TO 0 STEP  - 1
20  PRINT X;" ";: NEXT X
30  PRINT "BLAST-OFF"
```

Exercises:

1. Change the first program above so that it outputs measures of exterior angles of polygons.
2. Write a program that prints a chart for a series of regular polygons, listing the measures of each polygon's interior and exterior angles along with the sum of its interior angles.
3. Write a program that prints a table, listing angles by multiples of 10 degrees, and each angle's complement and supplement (when they exist).

MISCELLANEOUS COMPUTER FUNCTIONS

Objective: **To use ratios of corresponding sides in a computer program to determine the similarity of triangles**

It is often helpful to insert notes in computer programs to explain (to others and yourself) the purpose of various program lines. The BASIC word, REM, standing for REMark, allows you to do that. When the computer sees REM in a program, it ignores statements following it, understanding that they are REMarks and are not part of the working program. You will find some REM statements in the program below.

In writing programs, there are times when you will want execution to branch from the main body of the program to a smaller section of the program in order to perform a particular operation, and to return to the main program where execution left it. In computer terminology, the small program portion is called a *subroutine.* The BASIC command GOSUB, followed by a line number, instructs the computer to branch to a subroutine beginning with the line number indicated.

The program below inputs the lengths of the sides of two triangles. The program uses a subroutine from lines 240 to 350 to order the sides of both triangles from smallest to largest. The GOSUB 240 statement in lines 120 and 150 makes execution branch to the subroutine for both triangles. The RETURN command at the end of the subroutine RETURNs execution to the line following each GOSUB that directed execution to the subroutine. Once both sets of inputs are properly ordered, lines 180 to 200 test the ratios of corresponding parts, to determine whether or not the triangles are similar. Note that lines 180 to 200 test the ratios for approximate equality rather than exact equality. The reason for this is to compensate for round-off errors generated by the computer while performing division.

```
110  PRINT "INPUT THE 3 SIDES OF ONE TRIANGLE.":
     INPUT G,H,I
120 A = G:B = H:C = I: GOSUB 240
130 G = A:H = B:I = C
140  PRINT "SIDES OF OTHER TRIANGLE?": INPUT D,E,F
150 A = D:B = E:C = F: GOSUB 240
160 D = A:E = B:C = F
170  REM   ABS IS USED IN TESTS BELOW SINCE THE
     DIVISION MIGHT ROUND OFF DIFFERENTLY.
180  IF  ABS (G / D - H / E) > .0001 THEN 230
190  IF  ABS (G / D - I / F) > .0001 THEN 230
200  IF  ABS (H / E - I / F) > .0001 THEN 230
210  PRINT "THESE TRIANGLES ARE SIMILAR."
220  PRINT : GOTO 110
```

Computer program continues on page 574.

```
230   PRINT "THESE TRIANGLES ARE NOT SIMILAR.":
      PRINT : GOTO 110
240   IF A < B THEN 300: REM    SUBROUTINE STARTS
250   IF A < C THEN 280                      HERE
260   IF B < C THEN 290
270 Q = A:A = C:C = Q: GOTO 350
280 Q = A:A = B:B = Q: GOTO 350
290 Q = A:A = B:B = C:C = Q: GOTO 350
300   IF B < C THEN 330
310   IF A < C THEN 340
320 Q = A:A = C:C = B:B = Q: GOTO 350
330   GOTO 350
340 Q = C:C = B:B = Q: GOTO 350
350   RETURN : REM          SENDS PROGRAM BACK
```

A very useful variation of the GOTO command is the ON . . . GOTO command, which enables the computer to branch to any of several portions of a program. For example, the statement ON X GOTO 10,20,30 directs the computer to branch to line 10 if X = 1, to line 20 if X = 2, and to line 30 if X = 3.

INT is a versatile BASIC function. INT stands for the *greatest INTeger function.* INT(X) instructs the computer to find the largest integer that is equal to or smaller than X. For example, INT(2.3) = 2 and INT(2) = 2. For a positive number, INT drops off the decimal portion of the number. Note the effect of INT on these negative numbers: INT(−3.4) = −4 and INT(−5) = −5.

INT can be used to test the divisibility of numbers. For example, 10 / 3 = 3.333 . . . and INT(10 / 3) = 3. Since 10 / 3 ≠ INT(10/3), 10 is not divisible by 3. For another example, 12 / 4 = 3 and INT(12 / 4) = 3. Since 12 / 4 = INT(12/4), 12 is divisible by 4.

INT can also be used to round off numbers. To round a number X to the nearest integer, you use INT(X + .5). Note, for example, that INT(2.3 + 5) = INT(2.8) = 2 and INT(2.8 + .5) = INT(3.3) = 3. To round a number X to another decimal place requires a little more computation. For example, to round off a number to the nearest hundredth, you multiply the number by 100, round the product to the nearest integer, and divide that integer by 100. Enter and RUN the following two-line program that rounds 3.234 and 3.237 to the nearest hundredth.

```
10   PRINT  INT (3.234 * 100 + .5) / 100
20   PRINT  INT (3.237 * 100 + .5) / 100
```

INT is also often used in conjunction with the BASIC random function RND. The statement RND(1) instructs the computer to generate random 9-place decimals between .000000000 and .999999999. To generate random integers, you use INT coupled with RND in statement of the form INT (RND(1) * 10), which, for example, generates random whole numbers from 0 to 9. INT(RND(1) * 10) + 1 will generate whole numbers from 1 to 10. The statement INT(RND(1) * B) + A will generate any of B whole numbers starting with A.

IF A / B = INT A / B can be used in a program that reduces fractions to lowest terms. Input the numerator and denominator separately. Test the numerator for divisibility by 2, 3, 4, and so on up to the value of the numerator. When a numerator is not divisible by one factor, test the numerator for divisibility by the next factor. When a numerator is divisible by a factor, test the denominator for divisibility by the same factor. If both are not divisible by the same factor, return to testing the numerator by the next possible factor. If both are divisible by the same factor, divide both by the factor, and again test the new numerator for divisibility by 2, 3, 4, etc., up to the value of the new numerator.

Exercises:

1. Alter the program that begins on page 573 so that the computer generates three random numbers between 20 and 35 for the three sides of one triangle and also asks you to enter three sides of any triangle similar to the generated one.
2. Write a program that combines the second program on page 20 with the program you wrote for the Exercise on page 20. Write the program to randomly generate a 1 or 2 for branching to one or the other portion of the program. Also write the program to generate random numbers for the points instead of inputting them.
3. Write a program in which the computer randomly generates three numbers for the sides of a triangle, and in which the computer asks you whether or not the three sides can form a triangle, and if so, whether or not it is a right triangle.
4. Make a flowchart and then write a program that reduces a fraction to its lowest terms. (Hint: Use a FOR . . . NEXT loop to test possible factors.)
5. Write a program, using the INT function, that converts any number of inches to feet and inches.
6. Rewrite the program from Exercise 2 on page 180 so that the computer converts your input into inches, performs the specified computations, and, using the method used in Exercise 5 above, converts the computed results back to feet and inches.

TRIGONOMETRY AND THE COMPUTER

Objectives: **To make angle measurement conversions from degrees to radians and vice versa**
To solve trigonometric problems using computer programs

COMPUTER SECTION

Most versions of BASIC contain three trig functions—sine, cosine, and tangent, denoted as SIN, COS, and TAN. An example of the format used in a program is SIN(X). To use BASIC trig functions, however, you must become familiar with another way to measure angles.

The X in SIN(X) is not in terms of degrees, but in terms of another unit for measuring angles called *radians*. Without going into an extensive explanation of radian measure, this lesson will introduce the relationship between degree measure and radian measure, so that you can readily convert one to the other.

Recall that if a circle is divided into 360 equal parts, then one degree consists of exactly one of those parts. Radian measure is based on the idea of a circle with radius 1. Such a circle has a circumference of 2π (or approximately 6.2832) units. If the circle is divided into 2π (that is, 6.2832) equal parts, then one radian consists of exactly one of those parts. The complete circle is 2π radians. So, an angle measuring 2π radians is the same as an angle measuring 360°. Similarly, half a circle is π (or 3.1416) radians, which is equivalent to 180°. Also, a quarter of a circle is $\pi/2$ (or 1.5708) radians, and $\pi/2$ radians equals 90°. The illustration below shows a few values of radians and their approximate locations along with several corresponding degree measurements.

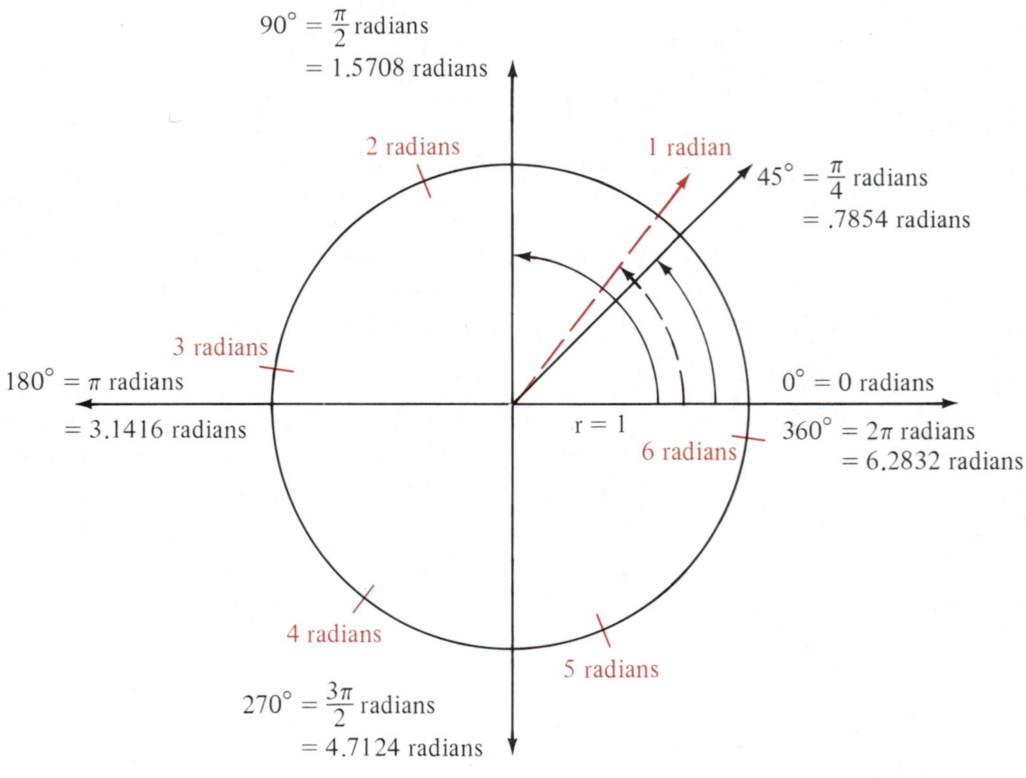

To convert degrees to radians or vice versa, use the following proportion:

$$\frac{\text{Degrees}}{360} = \frac{\text{Radians}}{2\pi}$$

This proportion yields two different equations for calculating one angle measurement from the other:

$$\text{Radians} = \frac{(\text{Degrees})(6.2832)}{360} \qquad \text{Degrees} = \frac{(\text{Radians})(360)}{6.2832}$$

Both of these equations are used in the following program:

```
10  PRINT "HOW MANY DEGREES ";: INPUT D
20 R = D * 6.2832 / 360
30  PRINT "THERE ARE ";R;"RADIANS IN ";D;"DEGREES."
40  PRINT : PRINT "HOW MANY RADIANS ";: INPUT R
50 D = R * 360 / 6.2832
60  PRINT "THERE ARE ";D;"DEGREES IN ";R;"RADIANS."
70  PRINT : PRINT : GOTO 10
```

By converting degree measure to radian measure, you can now use the BASIC trig functions in programs. The following program asks you to input an angle in degrees. The computer then converts the degree measure to radian measure and outputs the sine, cosine, and tangent of the entered angle.

```
10  PRINT "HOW MANY DEGREES?"
20  INPUT D
30 R = D * 6.2832 / 360
40  PRINT "SIN(";D;") = "; SIN (R)
50  PRINT "COS(";D;") = "; COS (R)
60  PRINT "TAN(";D;") = "; TAN (R)
70  GOTO 10
```

Exercises:

1. Write a program that prints a table of sines, cosines, and tangents for angles from 32 to 41 degrees. Compare your table with the one on page 324. (Be prepared to find discrepancies between corresponding entries in the tables. These discrepancies arise from the inexact conversion of degrees to radians that results from using 6.2832 as an approximation for 2π.)

2. Write a program that determines the lengths of two sides of a right triangle from inputs of the length of one side and the measure of an acute angle.

Postulates, Theorems, and Corollaries

Postulate 1 For any two points, there is exactly one line containing them. *(40)*

Theorem 2.1 Two lines intersect in at most one point. *(40)*

Postulate 2 Three noncollinear points are contained in exactly one plane. *(40)*

Postulate 3 If two points of a line are in a given plane, then the line itself is in the plane. *(41)*

Postulate 4 If two planes intersect, then they intersect in exactly one line. *(41)*

Postulate 5 Space is determined by at least four points not all in the same plane. *(41)*

Theorem 2.2 A line and a point not on the line are contained in exactly one plane. *(42)*

Theorem 2.3 If a line intersects a plane but is not contained in the plane, then the intersection is exactly one point. *(42)*

Theorem 2.4 Two intersecting lines are contained in exactly one plane. *(42)*

Postulate 6 On every line, there is a segment with a given point as an endpoint congruent to any given segment. *(44)*

Postulate 7 For every ray, there is an angle with the given ray as a side congruent to any given angle. *(44)*

Postulate 8 Every segment has exactly one midpoint. *(44)*

Postulate 9 Every angle, except a straight angle, has exactly one bisector. *(44)*

Postulate 10 Angle Addition Postulate: If D is in the interior of $\angle ABC$, then $m\angle ABC = m\angle ABD + m\angle DBC$. *(45)*

Postulate 11 The sum of the measures of the angles with the same vertex on one side of a line and with no interior points in common is 180°. The sum of the measures of all angles around a common vertex and with no interior points in common is 360°. *(49)*

Theorem 2.5 All right angles are congruent. *(49)*

Theorem 2.6 Two perpendicular lines form four congruent right angles. *(49)*

Postulate 12 For a given point and a line in a plane, there is exactly one line through the point that is perpendicular to the given line. *(51)*

Theorem 2.7 If two lines form congruent adjacent angles, then the lines are perpendicular. *(52)*

Theorem 2.8 Supplements of congruent angles are congruent. *(55)*

Corollary Supplements of the same angle are congruent. *(55)*

Theorem 2.9 Complements of congruent angles are congruent. *(56)*

Corollary Complements of the same angle are congruent. *(56)*

Theorem 2.10 Vertical angles are congruent. *(57)*

Theorem 2.11 Any two plane angles of a dihedral angle are congruent. *(62)*

Postulate 13 Alternate Interior Angle Postulate: If two parallel lines are intersected by a transversal, then the alternate interior angles are congruent. *(75)*

Theorem 3.1 If two parallel lines are intersected by a transversal, then the corresponding angles are congruent. *(76)*

Theorem 3.2 If two parallel lines are intersected by a transversal, then the interior angles on the same side of the transversal are supplementary. *(77)*

Postulate 14 If two lines are intersected by a transversal so that the alternate interior angles are congruent, then the lines are parallel. *(80)*

Theorem 3.3 If two lines are intersected by a transversal so that the corresponding angles are congruent, then the lines are parallel. *(81)*

Corollary In a plane, two lines perpendicular to the same line are parallel. *(81)*

Theorem 3.4 If two lines are intersected by a transversal so that the interior angles on the same side of the transversal are supplementary, then the lines are parallel. *(81)*

Postulate 15 The Parallel Postulate: Through a point not on a line, there is exactly one line parallel to the given line. *(85)*

Theorem 3.5 The sum of the measures of the angles of a triangle is 180°. *(87)*

Theorem 3.6 If two angles of one triangle are congruent to two angles of a second triangle, then the third angles are congruent. *(88)*

Theorem 3.7 In a plane, two lines parallel to the same line are parallel to each other. *(91)*

Theorem 3.8 If two parallel planes are intersected by a third plane, then the lines of intersection are parallel. *(96)*

Theorem 4.1 The measure of an exterior angle of a triangle is equal to the sum of the measures of its two remote interior angles. *(108)*

Corollary The measure of an exterior angle of a triangle is greater than the measure of either of its remote interior angles. *(108)*

Postulate 16 SSS for $\cong$ $\triangle$'s: Two triangles are congruent if the three sides of one are congruent to the corresponding sides of the other. *(114)*

Postulate 17 SAS for $\cong$ $\triangle$'s: Two triangles are congruent if two sides and the included angle of one are congruent to the corresponding two sides and included angle of the other. *(118)*

Postulate 18 ASA for $\cong$ $\triangle$'s: Two triangles are congruent if two angles and the included side of one are congruent to the corresponding two angles and included side of the other. *(120)*

Theorem 4.2 (AAS): Two triangles are congruent if two angles and the side opposite one of them are congruent to the corresponding two angles and side of the other. *(124)*

Theorem 5.1 If two sides of a triangle are congruent, then the angles opposite those sides are congruent. *(139)*

Corollary 1 An equilateral triangle is also equiangular. *(140)*

Corollary 2 The measure of each angle of an equilateral, or equiangular, triangle is 60°. *(140)*

Theorem 5.2 If two angles of a triangle are congruent, then the sides opposite those angles are congruent. *(140)*

Corollary An equiangular triangle is also equilateral. *(141)*

Theorem 5.3 Leg-Leg (LL): Two right triangles are congruent if the two legs of one are congruent to the corresponding legs of the other. *(157)*

Theorem 5.4 Leg-Acute Angle (LA): Two right triangles are congruent if a leg and an acute angle of one are congruent to the corresponding leg and acute angle of the other. *(158)*

Theorem 5.5 Hypotenuse-Acute Angle (HA): Two right triangles are congruent if the hypotenuse and an acute angle of one are congruent to the hypotenuse and the corresponding acute angle of the other. *(158)*

Theorem 5.6 Hypotenuse-Leg (HL): Two right triangles are congruent if the hypotenuse and a leg of one are congruent to the hypotenuse and corresponding leg of the other. *(159)*

Theorem 5.7 Perpendicular Bisector Theorem: A point on the perpendicular bisector of a segment is equidistant from the endpoints of the segment. *(166)*

Theorem 5.8 A line containing two points, each equidistant from the endpoints of a segment, is the perpendicular bisector. *(167)*

Theorem 5.9 Corresponding medians of congruent triangles are $\cong$. *(171)*

Theorem 5.10 Corresponding altitudes of congruent triangles are $\cong$. *(171)*

Theorem 6.1 The sum of the measures of the angles of a convex quadrilateral is 360°. *(188)*

Theorem 6.2 The sum of the measures of the angles of a convex polygon with n sides is $(n-2)\,180°$. *(189)*

Corollary The measure of an angle of a regular polygon with n sides is $\frac{(n-2)\,180°}{n}$. *(196)*

Theorem 6.3 The sum of the measures of the exterior angles, one at each vertex, of any convex quadrilateral is 360°. *(193)*

Theorem 6.4 The sum of the measures of the exterior angles, one at each vertex, of any convex polygon is 360°. *(194)*

Corollary The measures of an exterior angle of a regular polygon with n sides is $\frac{360°}{n}$. *(197)*

Theorem 7.1 A diagonal of a parallelogram forms two congruent triangles. *(211)*

Corollary 1 Opposite sides of a parallelogram are congruent. *(212)*

Corollary 2 Opposite angles of a parallelogram are congruent. *(212)*

Theorem 7.2 The diagonals of a parallelogram bisect each other. *(212)*

Theorem 7.3 If both pairs of opposite sides of a quadrilateral are congruent, then the quadrilateral is a parallelogram. *(215)*

Theorem 7.4 If both pairs of opposite angles of a quadrilateral are congruent, then the quadrilateral is a parallelogram. *(215)*

Theorem 7.5 If two sides of a quadrilateral are parallel and congruent, then the quadrilateral is a parallelogram. *(216)*

Theorem 7.6 If the diagonals of a quadrilateral bisect each other, then the quadrilateral is a parallelogram. *(216)*

Theorem 7.7 The segment joining the midpoints of two sides of a triangle is parallel to the third side, and its length is half the length of the third side. *(219)*

Theorem 7.8 If three parallel lines cut off congruent segments on one transversal, then they cut off congruent segments on every transversal. *(222)*

Corollary If any given number of parallel lines cut off congruent segments on one transversal, then they cut off congruent segments on every transversal. *(222)*

Postulate 19 The shortest path joining two points is a segment. *(224)*

Postulate 20 The shortest segment joining a point and a line is the segment perpendicular to the line. *(224)*

Theorem 7.9 Parallel lines are equidistant at all points. *(224)*

Theorem 8.1 The median of a trapezoid is parallel to its bases, and its length is half the sum of the lengths of the bases. *(243)*

Theorem 8.2 The base angles of an isosceles trapezoid are congruent. *(246)*

Theorem 8.3 If the base angles of a trapezoid are congruent, then the trapezoid is isosceles. *(247)*

Theorem 8.4 The diagonals of an isosceles trapezoid are congruent. *(247)*

Theorem 8.5 If the diagonals of a trapezoid are congruent, then the trapezoid is isosceles. *(248)*

Theorem 8.6	SASAS for ≅ quadrilaterals: Two quadrilaterals are congruent if any three sides and the included angles of one are congruent, respectively, to three sides and the included angles of the other. *(251)*
Theorem 8.7	ASASA for ≅ quadrilaterals: Two quadrilaterals are congruent if any three angles and the included sides of one are congruent, respectively, to three angles and the included sides of the other. *(251)*
Theorem 8.8	Corresponding sides of a symmetric polygon are ≅. *(257)*
Theorem 8.9	Corresponding angles of a symmetric polygon are ≅. *(257)*
Theorem 8.10	If two triangles are reflections, then they are congruent. *(260)*
Corollary	If two quadrilaterals are reflections, then they are congruent. *(260)*
Postulate 21	A line of symmetry of a symmetric polygon is the perpendicular bisector of any segment joining a pair of corresponding points of the polygon. *(256)*
Postulate 22	AA for ~ △'s: If two angles of a triangle are congruent to two angles of a second triangle, then the two triangles are similar. *(278)*
Theorem 9.1	If a line is parallel to one side of a triangle and intersects the other two sides, then it divides the two sides proportionally. *(282)*
Theorem 9.2	If a line divides two sides of a triangle proportionally, then it is parallel to the third side of the triangle. *(283)*
Theorem 9.3	SAS for ~ △'s: If an angle of one triangle is congruent to an angle of another triangle and the lengths of the corresponding sides including these angles are proportional, then the triangles are similar. *(285)*
Theorem 9.4	SSS for ~ △'s: If the corresponding sides of two triangles are proportional, then the triangles are similar. *(286)*
Theorem 9.5	Corresponding medians of similar triangles are proportional to corresponding sides. *(290)*
Theorem 9.6	Corresponding altitudes of similar triangles are proportional to corresponding sides. *(291)*
Theorem 9.7	The bisector of an angle of a triangle divides the opposite side into segments proportional to the other sides of the triangle. *(291)*

Theorem 9.8 If a tetrahedron is intersected by a plane parallel to its base, then the triangle formed by the intersection is similar to the base. *(298)*

Theorem 10.1 In a right triangle, the altitude to the hypotenuse forms two triangles, each similar to the original triangle and each similar to the other. *(306)*

Corollary 1 In a right triangle, the length of the altitude to the hypotenuse is the geometric mean of the lengths of the segments of the hypotenuse that are formed. *(306)*

Corollary 2 If the altitude is drawn to the hypotenuse of a right triangle, then the length of either leg is the geometric mean of the length of the hypotenuse and the length of the segment of the hypotenuse adjacent to that leg. *(307)*

Theorem 10.2 Pythagorean Theorem: In any right triangle, the square of the length of the hypotenuse is equal to the sum of the squares of the lengths of the two legs. *(310)*

Theorem 10.3 Converse of the Pythagorean Theorem: If the sum of the squares of the lengths of two sides of a triangle is equal to the square of the length of the third side, then the triangle is a right angle. *(315)*

Theorem 10.4 In a $45°-45°$ (isosceles) right triangle, the length of the hypotenuse is equal to the length of a leg times $\sqrt{2}$. *(318)*

Theorem 10.5 In a $30°-60°$ right triangle,
(1) the hypotenuse is twice as long as the leg opposite the $30°$ angle, and
(2) the leg opposite the $60°$ angle is $\sqrt{3}$ times as long as the leg opposite the $30°$ angle. *(319)*

Theorem 11.1 All radii of a circle are congruent. *(338)*

Theorem 11.2 In a circle, a radius perpendicular to a chord bisects the chord. *(339)*

Theorem 11.3 If chords of a circle or congruent circles are congruent, then the chords are equidistant from the center(s). *(342)*

Theorem 11.4 If chords of a circle or congruent circles are equidistant from the center(s), then the chords are congruent. *(343)*

Theorem 11.5 If a line is tangent to a circle, then it is perpendicular to the radius drawn to the point of tangency. *(345)*

Theorem 11.6 If a line is perpendicular to a radius at its endpoint on the circle, then it is tangent to the circle. *(345)*

Theorem 11.7 Tangent segments to a circle from the same point are congruent. *(346)*

stulate 23 Arc Addition Postulate: If P is a point on $\widehat{AB}$, then $m\widehat{AP} + m\widehat{PB} = m\widehat{APB}$. *(352)*

Theorem 11.8 In a circle or in congruent circles, congruent arcs determine congruent central angles and congruent chords. *(356)*

Theorem 11.9 In a circle or in congruent circles, congruent chords determine congruent central angles and congruent arcs. *(356)*

Theorem 11.10 In a circle or in congruent circles, congruent central angles determine congruent arcs and congruent chords. *(356)*

Theorem 11.11 The measure of an inscribed angle is one-half the measure of its intercepted arc. *(359)*

Corollary 1 If two inscribed angles intercept the same arc (or congruent arcs), the angles are congruent. *(360)*

Corollary 2 An angle inscribed in a semicircle is a right angle. *(360)*

Corollary 3 If two arcs of a circle are included between parallel secants, then the arcs are congruent. *(360)*

Corollary 4 Opposite angles of an inscribed quadrilateral are supplementary. *(361)*

Theorem 11.12 The measure of an angle formed by two secants intersecting in the interior of a circle is one-half the sum of the measures of the arcs intercepted by the angle and its vertical angle. *(364)*

Theorem 11.13 The measure of an angle formed by two secants intersecting in the exterior of a circle is one-half the difference of the measures of the intercepted arcs. *(365)*

Theorem 11.14 The measure of an angle formed by a secant and a tangent intersecting at the point of tangency is one-half the measure of the intercepted arc. *(368)*

Theorem 11.15 The measure of an angle formed by a secant and a tangent intersecting in the exterior of a circle is one-half the difference of the measures of the intercepted arcs. *(369)*

Theorem 11.16 The measure of an angle formed by two tangents to the same circle is one-half the difference of the measures of the intercepted arcs. *(369)*

POSTULATES, THEOREMS, COROLLARIES

Theorem 11.17 If two chords of a circle intersect, then the product of the lengths of the segments of one chord equals the product of the lengths of the segments of the other. *(373)*

Theorem 11.18 If two secants intersect in the exterior of a circle, then the product of the lengths of one secant segment and its external secant segment equals the product of the lengths of the other secant segment and its external secant segment. *(374)*

Theorem 11.19 If a tangent and a secant intersect in the exterior of a circle, then the square of the length of the tangent segment equals the product of the lengths of the secant segment and its external secant segment. *(374)*

Postulate 24 The area of a rectangle is the product of the length of a base and the length of a corresponding altitude. *(388)*

Corollary The area of a square is the square of the length of a side. *(388)*

Postulate 25 Area Addition Postulate: If a region is the union of two or more nonoverlapping regions, then its area is the sum of the areas of these nonoverlapping regions. *(388)*

Postulate 26 Congruent figures have the same area. *(391)*

Theorem 12.1 Area rt. $\triangle = \frac{1}{2}bh$: The area of a right triangle is one-half the product of the lengths of the two legs. *(391)*

Theorem 12.2 Area $\triangle = \frac{1}{2}bh$: The area of any triangle is one-half the product of the length of a base and the length of a corresponding altitude. *(391)*

Theorem 12.3 The area of a rhombus is one-half the product of the lengths of the diagonals. *(392)*

Theorem 12.4 Area $\triangle = \frac{s^2}{4}\sqrt{3}$: If s is the length of a side of an equilateral triangle, then the altitude is $\frac{s}{2}\sqrt{3}$ and the area is $\frac{s^2}{4}\sqrt{3}$. *(393)*

Theorem 12.5 Area $\square = bh$: The area of any parallelogram is the product of the length of a base and the length of a corresponding altitude. *(395)*

Theorem 12.6 The area of a trapezoid is one-half the product of the length of the altitude and the sum of the lengths of the two bases. *(398)*

Theorem 12.7 The ratio of the areas of two similar triangles is the square of the ratio of the lengths of any two corresponding sides. *(402)*

Theorem 12.8 The ratio of the perimeters of two similar polygons is the same as the ratio of the lengths of any two corresponding sides. *(403)*

Theorem 12.9 The area of a regular polygon is one-half the product of the length of an apothem and the perimeter. *(407)*

Theorem 12.10 For all circles, the ratio of the circumference to the length of a diameter is the same. *(410)*

Corollary If c is the circumference of a circle with a diameter of length d and a radius of length r, then $c = \pi d$ or $c = 2\pi r$. *(410)*

Theorem 12.11 Area $\odot = \pi r^2$: The area of a circle with a radius of length r is πr^2. *(412)*

Theorem 12.12 $l = \frac{m}{360} \cdot 2\pi r$: If the degree measure of an arc is m and the length of a radius of the circle is r, then the length l of the arc is given by the formula $l = \frac{m}{360} \cdot 2\pi r$. *(415)*

Theorem 12.13 Area sector $= \frac{m}{360} \cdot 2\pi r$: In a circle, the area of a sector bounded by an arc of measure m and radii of length r is given by the formula Area $= \frac{m}{360} \cdot \pi r^2$. *(418)*

Theorem 13.1 The Distance Formula: The distance d between $P(x_1, y_1)$ and $Q(x_2, y_2)$ is given by the formula $d = \sqrt{(x_2 - x_1)^2 + (y_2 - y_1)^2}$. *(433)*

Theorem 13.2 The Midpoint Formula: If $P(x_1, y_1)$ and $Q(x_2, y_2)$ are any two points in a coordinate plane, then the midpoint M of $\overline{PQ}$ has the coordinates $\left(\frac{x_1 + x_2}{2}, \frac{y_1 + y_2}{2}\right)$. *(435)*

Theorem 13.3 Standard Form: The graph of an equation that can be written in the form $ax + by = c$ (a and b not both equal to 0) is a line. *(442)*

Theorem 13.4 Point-Slope Form: If a line passes through $P(x_1, y_1)$ and has slope m, then an equation of the line is $y - y_1 = m(x - x_1)$. *(443)*

Theorem 13.5 Slope-Intercept Form: If a line has slope m and y-intercept b, then an equation of the line is $y = mx + b$. *(446)*

Theorem 13.6 If two nonvertical lines have the same slope, then the lines are parallel. *(449)*

Theorem 13.7 If two nonvertical lines are parallel, then the lines have the same slope. *(449)*

POSTULATES, THEOREMS, COROLLARIES

Theorem 13.8 If the slopes of two lines are negative reciprocals, then the lines are perpendicular. *(450)*

Theorem 13.9 If two nonvertical lines are perpendicular, then their slopes are negative reciprocals. *(450)*

Theorem 13.10 The equation of a circle with (h, k) the coordinates of the center and a radius of length r is $(x - h)^2 + (y - k)^2 = r^2$. *(456)*

Theorem 13.11 If d is the distance between points $A(x_1, y_1, z_1)$ and $B(x_2, y_2, z_2)$, then $d = \sqrt{(x_2 - x_1)^2 + (y_2 - y_1)^2 + (z_2 - z_1)^2}$. *(459)*

Postulate 27 The locus of points at a given distance d from a given point P is a circle with center P and with d the length of a radius. *(468)*

Postulate 28 The locus of points a given distance d from a given line l is a pair of lines each parallel to l and at the distance d from l. *(468)*

Postulate 29 The locus of points equidistant from two given parallel lines is a line parallel to each of the given lines and midway between them. *(469)*

Theorem 14.1 The locus of points equidistant from two given points is the perpendicular bisector of the segment joining the points. *(470)*

Theorem 14.2 The locus of points equidistant from the sides of an angle is the angle bisector. *(471)*

Theorem 14.3 The perpendicular bisectors of the sides of a triangle are concurrent at a point that is equidistant from the vertices of the triangle. *(483)*

Theorem 14.4 The bisectors of the angles of a triangle are concurrent at a point that is equidistant from the sides of the triangle. *(484)*

Theorem 14.5 The lines containing the altitudes of a triangle are concurrent. *(487)*

Theorem 14.6 The medians of a triangle are concurrent at a point that is two-thirds of the distance from each vertex to the midpoint of the opposite side. *(488)*

Theorem 15.1 The Triangle Inequality Theorem: In a triangle, the sum of the lengths of any two sides is greater than the length of the third side. *(503)*

Theorem 15.2 If one side of a triangle is longer than a second side, then the measure of the angle opposite the first side is greater than the measure of the angle opposite the second side. *(507)*

Theorem 15.3 If one angle of a triangle has a greater measure than a second angle, then the side opposite the first angle is longer than the side opposite the second angle. *(508)*

Theorem 15.4	The Hinge Theorem: If two sides of one triangle are congruent, respectively, to two sides of another triangle, and the included angle of the first triangle has a greater measure than the included angle of the second triangle, then the third side of the first triangle has a greater measure than the third side of the second triangle. *(511)*
Theorem 15.5	Converse of the Hinge Theorem: If two sides of one triangle are congruent, respectively, to two sides of another triangle, and the third side of the first triangle has a greater measure than the third side of the second triangle, then the angle opposite the third side of the first triangle has a greater measure than the angle opposite the third side of the second triangle. *(512)*
Theorem 15.6	If two chords of a circle are unequal in length, then the longer chord is nearer to the center of the circle. *(516)*
Theorem 15.7	If two chords of a circle are unequally distant from the center, then the chord nearer the center is the longer. *(517)*
Theorem 16.1	For any right prism, lateral area = perimeter of base × length of an altitude. $L = ph$ *(529)*
Theorem 16.2	For any prism, total area = lateral area + 2 × base area. *(530)*
Theorem 16.3	For any prism, volume = base area × length of an altitude. $V = Bh$ *(533)*
Theorem 16.4	For a cylinder, lateral area = circumference of base × length of an altitude. $L = 2\pi rh$ *(535)*
Theorem 16.5	For a cylinder, total area = lateral area + 2 × base area. *(536)*
Theorem 16.6	For a cylinder, volume = base area × length of an altitude. $V = \pi r^2 h$. *(536)*
Theorem 16.7	For a regular pyramid, lateral area = $\frac{1}{2}$ × length of slant height × perimeter of base. $L = \frac{1}{2} lp$. Total area = lateral area + base area. $T = L + B$ *(540)*
Theorem 16.8	For any pyramid, volume = $\frac{1}{3}$ × base area × length of altitude. $V = \frac{1}{3} BH$ *(541)*
Theorem 16.9	For a cone, lateral area (L) is πrd; total area (T) is $\pi r(l + r)$; volume (V) is $\frac{1}{3} \pi r^2 h$. *(545)*
Theorem 16.10	For a sphere, area = 4 × area of great circle. $A = 4\pi r^2$ *(550)*
Theorem 16.11	For a sphere, the volume is $\frac{4}{3}\pi r^3$. $V = \frac{4}{3}\pi r^3$ *(550)*

Trigonometric Ratios

Angle Measure	*Sin*	*Cos*	*Tan*	*Angle Measure*	*Sin*	*Cos*	*Tan*
0°	0.000	1.000	0.000	**46°**	.7193	.6947	1.036
1°	.0175	.9998	.0175	**47°**	.7314	.6820	1.072
2°	.0349	.9994	.0349	**48°**	.7431	.6691	1.111
3°	.0523	.9986	.0524	**49°**	.7547	.6561	1.150
4°	.0698	.9976	.0699	**50°**	.7660	.6428	1.192
5°	.0872	.9962	.0875	**51°**	.7771	.6293	1.235
6°	.1045	.9945	.1051	**52°**	.7880	.6157	1.280
7°	.1219	.9925	.1228	**53°**	.7986	.6018	1.327
8°	.1392	.9903	.1405	**54°**	.8090	.5878	1.376
9°	.1564	.9877	.1584	**55°**	.8192	.5736	1.428
10°	.1736	.9848	.1763	**56°**	.8290	.5592	1.483
11°	.1908	.9816	.1944	**57°**	.8387	.5446	1.540
12°	.2079	.9781	.2126	**58°**	.8480	.5299	1.600
13°	.2250	.9744	.2309	**59°**	.8572	.5150	1.664
14°	.2419	.9703	.2493	**60°**	.8660	.5000	1.732
15°	.2588	.9659	.2679	**61°**	.8746	.4848	1.804
16°	.2756	.9613	.2867	**62°**	.8829	.4695	1.881
17°	.2924	.9563	.3057	**63°**	.8910	.4540	1.963
18°	.3090	.9511	.3249	**64°**	.8988	.4384	2.050
19°	.3256	.9455	.3443	**65°**	.9063	.4226	2.145
20°	.3420	.9397	.3640	**66°**	.9135	.4067	2.246
21°	.3584	.9336	.3839	**67°**	.9205	.3907	2.356
22°	.3746	.9272	.4040	**68°**	.9272	.3746	2.475
23°	.3907	.9205	.4245	**69°**	.9336	.3584	2.605
24°	.4067	.9135	.4452	**70°**	.9397	.3420	2.747
25°	.4226	.9063	.4663	**71°**	.9455	.3256	2.904
26°	.4384	.8988	.4877	**72°**	.9511	.3090	3.077
27°	.4540	.8910	.5095	**73°**	.9563	.2924	3.270
28°	.4695	.8829	.5317	**74°**	.9613	.2756	3.487
29°	.4848	.8746	.5543	**75°**	.9659	.2588	3.732
30°	.5000	.8660	.5774	**76°**	.9703	.2419	4.010
31°	.5150	.8572	.6009	**77°**	.9744	.2250	4.331
32°	.5299	.8480	.6249	**78°**	.9781	.2079	4.704
33°	.5446	.8387	.6494	**79°**	.9816	.1908	5.145
34°	.5592	.8290	.6745	**80°**	.9848	.1736	5.671
35°	.5736	.8192	.7002	**81°**	.9877	.1564	6.314
36°	.5878	.8090	.7265	**82°**	.9903	.1392	7.115
37°	.6018	.7986	.7536	**83°**	.9925	.1219	8.144
38°	.6157	.7880	.7813	**84°**	.9945	.1045	9.514
39°	.6293	.7771	.8098	**85°**	.9962	.0872	11.43
40°	.6428	.7660	.8391	**86°**	.9976	.0698	14.30
41°	.6561	.7547	.8693	**87°**	.9986	.0523	19.08
42°	.6691	.7431	.9004	**88°**	.9994	.0349	28.64
43°	.6820	.7314	.9325	**89°**	.9998	.0175	57.29
44°	.6947	.7193	.9657	**90°**	1.000	0.000	
45°	.7071	.7071	1.000				

Table of Roots and Powers

No.	Sq.	Sq. Root	Cube	Cu. Root	No.	Sq.	Sq. Root	Cube	Cu. Root
1	1	1.000	1	1.000	51	2,601	7.141	132,651	3.708
2	4	1.414	8	1.260	52	2,704	7.211	140,608	3.733
3	9	1.732	27	1.442	53	2,809	7.280	148,877	3.756
4	16	2.000	64	1.587	54	2,916	7.348	157,564	3.780
5	25	2.236	125	1.710	55	3,025	7.416	166,375	3.803
6	36	2.449	216	1.817	56	3,136	7.483	175,616	3.826
7	49	2.646	343	1.913	57	3,249	7.550	185,193	3.849
8	64	2.828	512	2.000	58	3,364	7.616	195,112	3.871
9	81	3.000	729	2.080	59	3,481	7.681	205,379	3.893
10	100	3.162	1,000	2.154	60	3,600	7.746	216,000	3.915
11	121	3.317	1,331	2.224	61	3,721	7.810	226,981	3.936
12	144	3.464	1,728	2.289	62	3,844	7.874	238,328	3.958
13	169	3.606	2,197	2.351	63	3,969	7.937	250,047	3.979
14	196	3.742	2,744	2.410	64	4,096	8.000	262,144	4.000
15	225	3.875	3,373	2.466	65	4,225	8.062	274,625	4.021
16	256	4.000	4,096	2.520	66	4,356	8.124	287,496	4.041
17	289	4.123	4,913	2.571	67	4,489	8.185	300,763	4.062
18	324	4.243	5,832	2.621	68	4,624	8.246	314,432	4.082
19	361	4.359	6,859	2.668	69	4,761	8.307	328,509	4.102
20	400	4.472	8,000	2.714	70	4,900	8.357	343,000	4.121
21	441	4.583	9,261	2.759	71	5,041	8.426	357,911	4.141
22	484	4.690	10,648	2.802	72	5,184	8.485	373,248	4.160
23	529	4.796	12,167	2.844	73	5,329	8.544	389,017	4.179
24	576	4.899	13,824	2.884	74	5,476	8.602	405,224	4.198
25	625	5.000	15,625	2.924	75	5,625	8.660	421,875	4.217
26	676	5.099	17,576	2.962	76	5,776	8.718	438,976	4.236
27	729	5.196	19,683	3.000	77	5,929	8.775	456,533	4.254
28	784	5.292	21,952	3.037	78	6,084	8.832	474,552	4.273
29	841	5.385	24,389	3.072	79	6,241	8.888	493,039	4.291
30	900	5.477	27,000	3.107	80	6,400	8.944	512,000	4.309
31	961	5.568	29,791	3.141	81	6,561	9.000	531,441	4.327
32	1,024	5.657	32,768	3.175	82	6,724	9.055	551,368	4.344
33	1,089	5.745	35,937	3.208	83	6,889	9.110	571,787	4.362
34	1,156	5.831	39,304	3.240	84	7,056	9.165	592,704	4.380
35	1,225	5.916	42,875	3.271	85	7,225	9.220	614,125	4.397
36	1,296	6.000	46,656	3.302	86	7,396	9.274	636,056	4.414
37	1,369	6.083	50,653	3.332	87	7,569	9.327	658,503	4.431
38	1,444	6.164	54,872	3.362	88	7,744	9.381	681,472	4.448
39	1,521	6.245	59,319	3.391	89	7,921	9.434	704,969	4.465
40	1,600	6.325	64,000	3.420	90	8,100	9.487	729,000	4.481
41	1,681	6.403	68,921	3.448	91	8,281	9.539	753,571	4.498
42	1,764	6.481	74,088	3.476	92	8,464	9.592	778,688	4.514
43	1,849	6.557	79,507	3.503	93	8,649	9.644	804,357	4.531
44	1,936	6.633	85,184	3.530	94	8,836	9.695	830,584	4.547
45	2,025	6.708	91,125	3.557	95	9,025	9.747	857,375	4.563
46	2,116	6.782	97,336	3.583	96	9,216	9.798	884,736	4.579
47	2,209	6.856	103,823	3.609	97	9,409	9.849	912,673	4.595
48	2,304	6.928	110,592	3.634	98	9,604	9.899	941,192	4.610
49	2,401	7.000	117,649	3.659	99	9,801	9.950	970,299	4.626
50	2,500	7.071	125,000	3.684	100	10,000	10.000	1,000,000	4.642

Glossary

acute angle: An angle with degree measure m such that $0 < m < 90$. *(p. 13)*

acute triangle: An acute triangle has no obtuse or right angles. All three angles are acute angles. Each angle has a measure less than 90°. *(p. 103)*

adjacent angles: Two coplanar angles that have a common vertex and a common side but no common interior points. *(p. 23)*

alternate exterior angles: Alternate exterior angles are formed on opposite sides of a transversal crossing two lines. If the two lines are parallel, then the alternate exterior angles are congruent. *(p. 73)*

alternate interior angles: Alternate interior angles are formed on opposite sides of a transversal crossing two lines. If the two lines are parallel, then the alternate interior angles are congruent. *(p. 73)*

altitude: An altitude of a triangle is a segment from a vertex of the triangle perpendicular to the line containing the opposite side of the triangle. An altitude of a trapezoid or parallelogram is any segment from a point on one of the bases perpendicular to a point on the line containing the other base. *(pp. 169, 242)*

altitude of a cone: *See* cone.

altitude of a cylinder: *See* cylinder.

altitude of a prism: A segment that is perpendicular to the plane of each base with endpoints in the planes of the bases. *(p. 529)*

altitude of a pyramid: *See* pyramid.

angle: An angle consists of two rays with a common endpoint. The rays are the sides of the angle; the common endpoint is the vertex. *(p. 12)*

angle of depression: The angle through which a line of vision must be depressed from the horizontal to sight an object. *(p. 329)*

angle of elevation: The angle through which a line of vision must be elevated from the horizontal to sight an object. *(p. 329)*

apothem of a regular polygon: A segment from the center of the polygon perpendicular to a side of the polygon. *(p. 406)*

auxiliary line: A line used in a proof that is not part of the original figure. An auxiliary line should be drawn to satisfy only one set of conditions. *(p. 87)*

base angles: The angles of an isosceles triangle opposite the legs or congruent sides. *(p. 139)*

base of a cylinder: *See* cylinder.

base of a prism: *See* prism.

bisector of a segment: A bisector of a segment is any figure that intersects the segment at its midpoint. *(p. 8)*

bisector of an angle: A ray $\overrightarrow{OP}$ is the bisector of an angle $\angle AOB$ if it is in the interior of the angle and if $\angle AOP \cong \angle POB$. *(p. 14)*

center of a circle: *See* circle.

center of a regular polygon: The common center of its inscribed and circumscribed circles. *(p. 406)*

center of a sphere: *See* sphere.

central angle of a circle: An angle whose vertex is the center of a circle. *(p. 355)*

centroid of a triangle: The point at which the medians are concurrent. *(p. 488)*

chord of a circle: A segment whose endpoints lie on the circle. *(p. 338)*

circle: The set of all points in a plane that are a given distance from a given point. The given point is the *center* of the circle. *(p. 338)*

circular region: The union of a circle and its interior. *(p. 387)*

circumcenter of a triangle: The center of the circumcircle of the triangle. *(p. 485)*

circumcircle of a triangle: The circle that is circumscribed about a triangle. *(p. 485)*

circumference of a circle: The limit of the perimeters of inscribed regular polygons of the circle as the number of sides of the polygons increases indefinitely. *(p. 409)*

circumscribed circle: A circle in which a polygon is inscribed. *(p. 343)*

circumscribed polygon: A polygon whose sides are tangent to a circle. *(p. 346)*

collinear points: Collinear points are points that are all on the same line. *(p. 2)*

common external tangent: A common tangent that does not intersect the segment that joins the centers of the two circles. *(p. 347)*

common internal tangent: A common tangent that intersects the segment that joins the centers of the two circles. *(p. 347)*

common tangent: A line that is tangent to each of two coplanar circles. *(p. 347)*

complementary: Two angles are complementary if the sum of their measures is 90°. *(p. 24)*

concentric circles: Coplanar circles with a common center. *(p. 352)*

congruent: Congruent figures have exactly the same size and shape. *(p. 123)*

congruent angles: Two angles are congruent if they have equal measures. *(p. 13)*

congruent arcs: Arcs of the same circle or congruent circles that have the same measure. *(p. 355)*

congruent circles: Circles with congruent radii. *(p. 342)*

congruent triangles: Congruent triangles have the same size and shape. Triangles can be proved congruent by SAS, SSS, ASA, and SAA patterns. *(p. 110)*

conjunction: A statement of the form *p and q* is a conjunction. *(p. 27)*

construction: A geometric construction is drawn with compass and straightedge. The compass is used to transfer equal distances. The straightedge is used to draw straight lines. *(p. 8)*

contrapositive: The statement formed by interchanging and negating both the hypothesis and the conclusion of a given statement. *(p. 30)*

converse: The statement formed by interchanging the hypothesis and the conclusion of a given statement. *(p. 30)*

convex: A convex region is a region that has the special property that the segment connecting any two points of the region will be completely in the region. If a figure is not convex, it is concave. *(p. 185)*

coordinate(s) of a point: On a number line, the coordinate of a point is the number that corresponds to the point. In a coordinate plane, the coordinates of a point make up the ordered pair that corresponds to the point. *(pp. 5, 429)*

coordinate plane: A plane containing two perpendicular number lines that intersect at 0. *(p. 429)*

cone: A right circular cone is a solid with the following properties: The base is a circular region; the vertex is a point that lies on the line perpendicular to the plane of the base at its center; the altitude is a segment joining the vertex to the center of the base. *(p. 544)*

coplanar points: Coplanar points are points that are in the same plane. *(p. 2)*

corresponding angles: Corresponding angles are formed on the same side of a transversal. If the lines are parallel, then the corresponding angles are congruent. *(p. 73)*

corollary: A corollary of a theorem is a theorem whose proof requires only a few simple steps in addition to the proof of the original theorem. *(p. 55)*

cosine: The cosine of an acute angle of a right triangle is the ratio of the length of the leg adjacent to the angle to the length of the hypotenuse. *(p. 321)*

cylinder: A circular cylinder has two bases that are congruent circular regions lying in parallel planes. The axis is the line segment joining the centers of the two circles. An altitude is a segment that is perpendicular to the plane of each base. In a right circular cylinder, the axis is also an altitude, but in an oblique circular cylinder, the axis is not an altitude. *(p. 535)*

diagonal: A diagonal of a polygon is a segment that connects any two nonconsecutive sides of the polygon. *(p. 184)*

diameter of a circle: A chord that contains the center of a circle. *(p. 338)*

dihedral angles: Dihedral angles are formed by intersecting planes. The measure of a dihedral angle is that of one of its plane angles. *(p. 61)*

disjunction: A statement of the form *p or q* is a disjunction. *(p. 28)*

distance from a point to a circle: The distance from point P to circle O is measured along $\overleftrightarrow{PO}$ and is the length of the shortest segment joining P and circle O. *(p. 475)*

equiangular triangle: All three angles of an equiangular triangle are congruent. *(p. 103)*

equidistant: A point is equidistant from two other points if the distances from this point to the two points are equal. *(p. 9)*

equilateral triangle: An equilateral triangle has three congruent sides and angles. *(p. 103)*

exterior angle: An exterior angle of a triangle is an angle that is adjacent and supplementary to one of the angles of the triangle. *(p. 14)*

extremes: *See* proportion.

face of a prism: *See* prism.

geometric mean: The positive number m is the geometric mean of two positive numbers a and b if $\frac{a}{m} = \frac{m}{b}$. *(p. 271)*

great circle of a sphere: The intersection of a sphere and any plane that contains the center of the sphere. *(p. 549)*

hypotenuse: The hypotenuse of a right triangle is the side opposite the right angle. *(p. 157)*

indirect proof: An indirect proof assumes that the desired conclusion is not true and shows that this assumption leads to some contradiction. *(p. 91)*

inscribed angle: An angle whose vertex lies on a circle and whose sides contain chords of the circle. *(p. 359)*

inscribed circle: A circle about which a polygon is circumscribed. *(p. 346)*

inscribed polygon: A polygon whose sides are chords of a circle. *(p. 343)*

intersection: The intersection of two geometric figures is all of the points that the figures have in common. The intersection of two planes is a line or no points at all. *(p. 2)*

inverse: The statement formed by negating the hypothesis and the conclusion of a given statement. *(p. 30)*

isosceles trapezoid: An isosceles trapezoid is a trapezoid with congruent legs. *(p. 246)*

isosceles triangle: An isosceles triangle has at least two congruent sides. *(p. 103)*

lateral area of a prism: The sum of the areas of the lateral faces. *(p. 529)*

line of symmetry: The reflection line of a reflection transformation. *(p. 145)*

locus: A geometric figure containing all points that satisfy given conditions. *(p. 467)*

measure of a major arc: 360° minus the degree measure of the central angle whose sides determine the arc. *(p. 351)*

measure of a minor arc: The degree measure of the central angle that intercepts the arc. *(p. 351)*

median: A median of a triangle is a segment from a vertex to the midpoint of the opposite side. The median of a trapezoid is the segment that connects the midpoints of the two legs of the trapezoid. *(pp. 170, 243)*

midpoint of a segment: The midpoint of a segment is the point on the segment that determines two congruent segments. *(p. 8)*

negation: The statement formed by denying a given statement. *(p. 27)*

noncollinear points: Noncollinear points are points that are not on the same line. *(p. 2)*

noncoplanar points: Noncoplanar points are points that are not in the same plane. *(p. 2)*

oblique circular cylinder: *See* cylinder.

oblique prism: *See* prism.

obtuse angle: An angle with degree measure m such that $90 < m < 180$. *(p. 13)*

obtuse triangle: An obtuse triangle has exactly one obtuse angle. Two angles must be acute angles. *(p. 103)*

ordinate: The second number in an ordered pair is called the *ordinate,* the *y-coordinate,* or the *second coordinate.* *(p. 429)*

origin: The point of intersection of the axes in a coordinate plane. *(p. 429)*

orthocenter of a triangle: The point at which the altitudes are concurrent. *(p. 488)*

parallel lines: Parallel lines are coplanar lines that do not intersect. *(p. 69)*

parallel planes: Two planes are parallel if they do not intersect. *(p. 95)*

parallelogram: A quadrilateral with two pairs of parallel sides is a parallelogram. *(p. 211)*

perimeter: The perimeter of any polygon is the sum of the lengths of the sides of the polygon. *(p. 403)*

perpendicular: Perpendicular lines are lines that intersect to form right angles. Perpendicular segments or rays are contained in perpendicular lines. A line is perpendicular to a plane if it is perpendicular to every line in the plane through the point of intersection of the line and the plane. *(pp. 49, 63)*

perpendicular bisector: A perpendicular bisector of a segment is a line or other figure that is a bisector of the segment and is perpendicular to it. *(pp. 51, 166)*

perpendicular planes: Two planes are perpendicular if they form a right dihedral angle. *(p. 63)*

plane angle: A plane angle of a dihedral angle is made up of two rays, one in each face of the dihedral angle, each perpendicular to the edge of the dihedral angle at the same point. *(p. 62)*

polygon: A polygon is composed of three or more coplanar segments that intersect only at the endpoints. Each endpoint is shared by exactly two of the segments. *(p. 183)*

postulate: A postulate is a statement of a geometric property that is accepted without proof. *(p. 40)*

prism: The polygonal regions that make up prisms are called *faces*. The bases are two faces that are congruent polygonal regions lying in parallel planes. The other faces are the *lateral faces*. The faces intersect in segments which are *edges*. The lateral faces intersect in *lateral edges*. The edges intersect in points which are *vertices*. *(p. 525)*

proof: A proof is a formal argument that shows that a conclusion follows logically from other statements. *(p. 37)*

proportion: A proportion is an equation of the form $\frac{a}{b} = \frac{c}{d}$ $(b \neq 0,\ d \neq 0)$; a and d are the extremes of the proportion; b and c are the means of the proportion. *(p. 270)*

pyramid: A solid whose *base* is a polygonal region. Its *lateral faces* are triangular regions that share a common point called the *vertex*. The lateral faces intersect in segments called the *lateral edges* of the pyramid. The *altitude* of a pyramid is the segment from the vertex perpendicular to the plane of the base. *(p. 539)*

quadrilateral: A quadrilateral is a polygon with four sides. *(p. 208)*

radius of a circle: A segment that joins the center and any point on the circle. *(p. 338)*

radius of a sphere: A segment that joins the center and any point on the sphere. *(p. 549)*

ratio: The ratio of two numbers a and b is the number $\frac{a}{b}$ $(b \neq 0)$. *(p. 270)*

ray: A ray $\overrightarrow{AB}$ is the set of points that includes segment $\overline{AB}$ and all points P such that B is between A and P. A ray has one endpoint but extends indefinitely in the other direction. *(p. 6)*

rectangle: A rectangle is a parallelogram with four right angles. This means that all of the angles are congruent. The diagonals of a rectangle are congruent. *(p. 226)*

reflection: One special geometric transformation is the reflection. The reflection is always congruent to the original figure. The line of reflection is the perpendicular bisector of the segment joining corresponding points of the original figure and its image. *(p. 144)*

regular polygon: A regular polygon is a polygon that is both equiangular and equilateral. *(p. 196)*

regular pyramid: The base is a regular polygon; the lateral faces are congruent isosceles triangles; the altitude meets the base at its center. A slant height is the altitude of a lateral face. *(p. 539)*

remote interior angle: An interior angle that is not adjacent to the given exterior angle. *(p. 107)*

rhombus: A rhombus is a parallelogram with all four sides congruent. *(p. 226)*

right angle: A right angle has a measure of 90°. *(p. 13)*

right circular cone: *See* cone.

right circular cylinder: *See* cylinder.

right prism: *See* prism.

right triangle: A right triangle has exactly one right or 90° angle. *(p. 306)*

scalene triangle: A scalene triangle has no congruent sides or angles. *(p. 103)*

secant of a circle: A line that intersects a circle at two points. *(p. 338)*

sector of a circle: A region bounded by two radii of a circle and an arc determined by the radii. *(p. 418)*

segment: A segment $\overline{AB}$ is the set of points A, B, and all points between A and B. A segment is named by its endpoints. A segment is straight and has a definite length. *(p. 6)*

semicircle: Consists of the endpoints of a diameter of a circle and all points of the circle that lie on one side of the diameter. *(p. 351)*

similar polygons: Two polygons are similar if corresponding angles are congruent and corresponding sides are proportional. *(p. 275)*

sine: The sine of an acute angle of a right triangle is the ratio of the length of the leg opposite the angle to the length of the hypotenuse. *(p. 321)*

skew lines: Skew lines are lines that are noncoplanar. They cannot intersect. *(p. 69)*

slant height of a regular pyramid: *See* regular pyramid.

slope of a line: The slope m of a line passing through $P(x_1, y_1)$ and $Q(x_2, y_2)$ is defined as follows: $m = \frac{y_2 - y_1}{x_2 - x_1}$ $(x_2 \neq x_1)$. *(p. 438)*

space: Space is the set of all points. *(p. 2)*

sphere: The set of all points at a given distance from a given point called the *center*. *(p. 549)*

square: A rectangle with four congruent sides. It also is a rhombus with four congruent right angles. *(p. 226)*

straight angle: A straight angle has a measure of 180°. A straight angle can also be thought of as a line. *(p. 13)*

supplementary angles: Two angles are supplementary if the sum of their measures is 180°. *(p. 24)*

tangent: The tangent of a right angle is the ratio of the length of the opposite leg to the length of the adjacent leg. *(p. 321)*

tangent circles: Two coplanar circles that are tangent to the same line at the same point. *(p. 347)*

tangent to a circle: A line that is coplanar with a circle and intersects the circle at exactly one point. The point is called the *point of tangency*. *(p. 338)*

theorem: A statement that can be proved to be true, as opposed to a postulate, which is assumed to be true without proof. *(p. 40)*

total area of a prism: The sum of the lateral area and the areas of the two bases. *(p. 529)*

transformation: A geometric transformation is a way of relating the points of a figure with the points of a second figure, called the *image*. *(p. 144)*

transversal: A transversal is a line that intersects two or more coplanar lines in different points. *(p. 72)*

trapezoid: A trapezoid is a quadrilateral with exactly one pair of parallel sides. *(p. 239)*

triangle: A triangle is a polygon with three sides. It consists of three segments that have three noncollinear points as their endpoints. *(p. 87)*

vertex: The vertex of an angle is the common endpoint of the two rays that make up the angle. *(p. 12)*

vertical angles: The congruent nonadjacent angles formed by two intersecting lines. *(p. 24)*

Selected Answers

(Included are odd answers to written exercises, chapter reviews, and all answers to cumulative and algebra reviews.)

Pages 3–4

1. collinear **3.** coplanar but noncollinear **5.** collinear **7.** True **9.** False **11.** True **13.** False **15.** True **17, 19, 21.** Answers may vary. **23.** At most only 2 pts can be collinear since the 4 pts given are noncoplanar.

Algebra Review

1. 2 **2.** 5 **3.** 8 **4.** 6 **5.** 4 **6.** 22 **7.** 2 **8.** 8 **9.** 25 **10.** 10 **11.** 0 **12.** 5

Page 7

1. −3 **3.** −0.8 **5.** 1.8 **7.** 7 **9.** 5 **11.** 15 **13.** $\overline{ST}$ **15.** $\overline{US}$ **17.** $\overrightarrow{RS}$; $\overrightarrow{RT}$ **19.** 3 cm; 28 mm **21.** 1 cm; 12 mm **23.** C **25.** B **27.** 1; −7 **29.** 4.4; −11.2 **31.** $PR + RQ \neq PQ$

Pages 10–11

1. ≅ **3.** ≅ **5.** ≅ **7.** ≇ **9.** ≅ **11.** ≅ **13.** ≅ **15.** Follow constr on p 9, then ex 3. **17.** Follow constr on p 9, then constr on p 10. **19.** 11 **21.** 6 **23.** −7 **25.** $2\frac{7}{8}$ **27.** $-5\frac{7}{16}$ **29.** Constr $\overline{MX} \cong \overline{AB}$ constr $\overline{MA} \cong \overline{CD}$ with A between M and X. **31.** Constr seg ≅ $\overline{PQ}$. Constr midpt, p 10, and repeat. **33.** $q = 2m - p$ **35.** $r = p + \frac{3}{4}(q - p)$, or $r = \frac{1}{4}p + \frac{3}{4}q$

Page 16

1. 30°, acute **3.** 76°, acute **5.** 110°, obtuse **7.** 180°, straight **9, 11.** Draw angle as on p 12; then follow constr on pp 14–15. **13.** Follow constr on p 15, then bis one of the parts. **15.** Repeat ex 13; the required ∠ is $\frac{1}{2}$ m ∠ + $\frac{1}{4}$ m ∠. **17.** Yes, def. of ∠ bis **19.** 27 **21.** No; there is no interior.

Page 18

1. $\overrightarrow{AB}$; $\overrightarrow{AC}$ **3.** collinear **5.** coplanar but noncollinear **7.** coplanar but noncollinear **9.** 4 **11.** 6 **13.** −3 **15.** Yes **17.** 135°; obtuse **19.** Follow constr on p 9. **21.** Follow constr on p 14. **23.** 38

Page 26

1. vert **3.** adj **5.** adj **7.** adj **9.** 60°, 150° **11.** 85°, 175° **13.** not poss, 52° **15.** not poss, 40° **17.** 67.5°, 157.5° **19.** 95 − m, 185 − m **21.** 70° **23.** 80° **25.** 60° **27.** 45° **29.** F **31.** T **33.** 60° **35.** 52° **37.** 52° **39.** 38°

Page 28

9. No **11.** Yes **13.** No **15.** TTFT **17.** FFFT

Page 32

15. F, F **17.** T, T **21.** T, T, T **23.** T, T, T **25.** F, F, T **27.** F, F, F **29.** TT, FT, TTTT **31.** FT, FTT, FTT

Page 35

1. m $\angle AOB$ = m $\angle DOC$; m $\angle AOB$ = m $\angle BOC$; m $\angle AOD$ = m $\angle BOC$ **3.** $\angle JKL \cong \angle LKM$; def of ∠ bis **5.** $\overrightarrow{TU}$ bis $\angle STV$; def of ∠ bis **7.** E is the midpt of $\overline{DF}$; def of midpt **9.** m $\angle b = 42$; def of comp ∠'s **11.** m $\angle f = 76$; def of supp ∠'s **13.** m $\angle C$ = m $\angle D = 90$; def of supp ∠'s **15.** No; pt Z not in interior of $\angle XYW$ **17.** Yes; def of comp ∠'s and ≅ ∠'s, subst and subtr props

Pages 38–39

1. CD; $\overline{AB} \cong \overline{CD}$ **3.** Given; Def of ≅ seg; Given; Subst prop **5.** m ∠ A = m ∠ C (trans prop); ∠ A ≅ ∠ C (def of ≅ ∠'s) **7.** m ∠ WOX = m ∠ ZOY (subtr prop); ∠ WOX ≅ ∠ ZOY (def of ≅ ∠'s) **9.** m ∠ A + m ∠ B = m ∠ B + m ∠ C (trans prop); m ∠ A = m ∠ C (subtr prop) **11.** m ∠ H = 2(35) = 70 (subst prop); m ∠ H = m ∠ I (trans prop); ∠ H ≅ ∠ I (def of ≅ ∠'s) **13.** NO-RS = PQ-RS (trans prop); NO = PQ (add prop) **15.** m ∠ WYZ = m ∠ QST (def of ≅ ∠'s); m ∠ XYZ = $\frac{1}{2}$(m ∠ QST) (subst prop); m ∠ XYZ = m ∠ RST (trans prop); ∠ XYZ ≅ ∠ RST (def of ≅ ∠'s)

Page 43

1. T **3.** F **5.** T **7.** F **9.** F **11.** F **13.** T **15.** Yes **17.** No **19.** 4 **21.** No **23.** $\overleftrightarrow{AB}$, $\overleftrightarrow{AC}$ intersect (given); A, B, C are 3 noncollinear points in plane S (Post 2); $\overleftrightarrow{AB}$, $\overleftrightarrow{AC}$ in plane S (Post 3).

SELECTED ANSWERS

Algebra Review

1. 7 **2.** 52 **3.** 4 **4.** 5 **5.** −12 **6.** 25

Pages 47–48

1. $\angle AOB$ **3.** $\angle DOE$ **5.** 11 cm **7.** 8 m **9.** 100 **11.** 62 **13.** 78 **15.** 20 mm **17.** Given; def of betw; $TU + UV = TV$; subst prop; $ST = UV$ **19.** $BC = BC$ (reflex prop); $AC - BC = BD - BC$ (subtr prop); $AC = AB + BC$ and $BD = BC + CD$ (def of betw) $AB = AC - BC$ and $CD = BD - BC$ (subtr prop); $AB = CD$ (subst prop) **21.** m $\angle AOC$ = m $\angle DOB$ (def of ≅ ∠'s); m $\angle AOC$ = m $\angle 1$ + m $\angle 2$ and m $\angle DOB$ = m $\angle 2$ + m $\angle 3$ (Post 10 ∠ add post); m $\angle 1$ + m $\angle 2$ = m $\angle 2$ + m $\angle 3$ (subst prop); m $\angle 2$ = m $\angle 2$ (reflex prop); m $\angle 1$ = m $\angle 3$ (subtr prop); $\angle 1 \cong \angle 3$ (def of ≅ ∠'s **23.** $EG = FG$ and $EH = FI$ (def of ≅ seg); $EG = EH + HG$ and $FG = FI + IG$ (def of betw); $EH + HG = FI + IG$ (subst prop); $HG = IG$ (subtr prop); $\overline{HG} \cong \overline{IG}$ (def of ≅ seg) **25.** $JK = NM$ (def of ≅ seg); $\frac{1}{2}(JL) = \frac{1}{2}(NL)$ (subst); $JL = NL$ (mult prop); $\overline{JL} \cong \overline{NL}$ (def of ≅ seg) **27.** $OQ = QP$, $OQ = RT$ and $OP = RS$ (def of ≅ seg); $OP = OQ + QP$ and $RS = RT + TS$ (def of between); $OQ + QP = RT + TS$ (subst prop); $QP = TS$ (subtr prop); $RT = TS$ (subst prop); T bisects RS (def of bis)

Algebra Review

1. $3x + 12$ **2.** $12a^3 + 18a^2$ **3.** $8x^5 - 10x^2$ **4.** $-15t + 10s$ **5.** $-24x^5 + 16x^4$ **6.** $2z^2 - 5z - 3$ **7.** $6x^2 + x - 15$ **8.** $4r^2 + 12r + 9$ **9.** $p^2 - q^2$ **10.** $9c^2 - 16$ **11.** $16c^4 - 9b^2$ **12.** $25x^2 + 30xy^2 + 9y^4$

Page 53

1. $\overrightarrow{OA} \perp \overrightarrow{OD}$; $\overrightarrow{OD} \perp \overrightarrow{OG}$; $\overrightarrow{OC} \perp \overrightarrow{OF}$; $\overrightarrow{OA}$ and $\overrightarrow{OG}$ form a st ∠. **3.** Follow constr p 50. **5.** Constr midpt of seg, then follow constr top of p 51. **7.** m $\angle XYT$ + m $\angle WZV$ = 180 (def of suppl ∠'s), m $\angle XYT$ = m $\angle WZV$ (def of ≅ ∠'s); m $\angle WZV$ + m $\angle WZV$ = 180, or 2 m $\angle WZV$ = 180 (subst prop); m $\angle WZV$ = 90 (div prop); $\angle WZV$ is a rt ∠ (def of rt ∠); $\overleftrightarrow{ZV} \perp \overleftrightarrow{XW}$ (def of ⊥) **9.** Follow ex 2 p 51; meas of desired ∠ is (90° + 45°). **11.** $\angle AEC \cong \angle AED$ (Th 2.6; two lines form 4 ≅ rt ∠'s.)

Page 58

1. 50 **3.** 72 **5.** $\angle 2 \cong \angle 3$ **7.** $\angle 2 \cong \angle 3$ **9.** Given; given; given; def of compl ∠'s; def of compl ∠'s; comp of ≅ ∠'s are ≅ **11.** m $\angle 1$ + m $\angle 2$ = 180 (Post 11); $\angle 2$ is a suppl of $\angle 1$ (def of suppl ∠'s); $\angle 1 \cong \angle 5$ (given); $\angle 5$ is a suppl of $\angle 1$ (subst prop)

Cumulative Review

1. Answers may vary. **2.** acute **3.** 90°

Page 60

1. Given: $\angle DOC \cong \angle COB$; $\angle AOB \cong COB$. Prove: $\angle DOC \cong \angle AOB$ **3.** Given: $\overleftrightarrow{AB}$ and $\overleftrightarrow{CD}$ intersect at E, forming vert ∠s $\angle 1$ and $\angle 2$; $\angle 3$ is suppl to $\angle 1$; $\angle 4$ is suppl to $\angle 2$. Prove: $\angle 3 \cong \angle 4$ **5.** Given: $\angle 3$ and is adj and suppl to $\angle 1$; $\angle 2$ is adj and suppl to $\angle 1$. Prove: $\angle 3$ and $\angle 2$ are vert ∠'s. **7.** Given: $\angle 2$ is compl to $\angle 1$; $\angle 5$ is compl to $\angle 4$; $\angle 2 \cong \angle 5$; $\angle 3 + 2$ is suppl to $\angle 1$; $\angle 6 + 5$ is suppl to $\angle 4$. Prove: $\angle 3 + 2 \cong \angle 6 + 5$; $\angle 1 \cong \angle 4$ (Th 2.9: compl of ≅ ∠'s are ≅); $\angle 3 + 2 \cong \angle 6 + 5$ (Th 2.8: suppl of ≅ ∠'s are ≅). **9.** Given: $\angle 1$ and $\angle 2$ are compl; $\angle 1 \cong \angle 2$. Prove: m $\angle 1$ = 45; m $\angle 2$ = 45; m $\angle 1$ + m $\angle 2$ = 90 (def of comp ∠'s); m $\angle 1$ = m $\angle 2$ (def of ≅ ∠'s); 2 m $\angle 1$ = 90 and 2 m $\angle 2$; = 90 (subst prop); m $\angle 1$ = 45 and m $\angle 2$ = 45 (div prop).

Cumulative Review

1. 4 **2.** Yes **3.** True

Pages 63–64

1. 82 **3.** 81 **5.** 111 **7.** 53 **9.** False **11.** True **13.** False **15.** Too little info; $\angle WOZ$ is a plane. **17.** 90 **19.** 150 **21.** 90 **23.** Given: $\overleftrightarrow{AB} \perp q$. Prove: $p \perp q$, where p is any plane containing $\overleftrightarrow{AB}$; p and q intersect at some line $\overleftrightarrow{CD}$ (Post 4); there is $\overleftrightarrow{FB}$ in $q \perp \overleftrightarrow{CD}$ (Post 12); $\overleftrightarrow{FB} \perp \overleftrightarrow{AB}$ (def of 1 ⊥ pl); $\angle A - \overleftrightarrow{CD} - F$ is a rt ∠ (def of d. h ∠); $p \perp q$ (def of ⊥ pl)

Algebra Review

1. $8x^2 - x + 3$ **2.** $6x^3 + 8x^2 - 4x + 7$ **3.** $4x^4 + 5x^3 + 2x^2$ **4.** $2x^4 + 2x^3 + x^2 + x$ **5.** $3 + 7x + x^2 + x^3$ **6.** $4 - 2x - 2x^2$ **7.** $4x^4 - 12x^3 + 7x^2$ **8.** $3x^3 - 2x^2 + 5x + 4$

Page 65

3. neither **5.** comp **7.** neither **9.** 110° **11.** F **13.** T **15.** 154 **17.** Post 11; Post 11; def supp ∠'s; def supp ∠'s; given; supp of ≅ ∠'s are ≅ **19.** Follow constr top p 51, then ex 2. **21.** Given: $\overleftrightarrow{AB}$ and $\overleftrightarrow{CD}$ intersect at E. Prove: $\overleftrightarrow{AB}$ and $\overleftrightarrow{CD}$ are coplanar. **23.** m $\angle 1$ = m $\angle 2$, and m $\angle 2$ = m $\angle 3$ (def of ≅ ∠'s); m $\angle 1$ = m $\angle 3$ (trans prop); $\angle 1 \cong \angle 3$ (def of ≅ ∠'s)

Page 71

1., 3. Answers may vary. **5.** T **7.** T **9.** T **11.** F **13.** sometimes T **15.** always T **17.** See ex 6, 7 p 70; no; yes

Cumulative Review

1. an angle **2.** See constr. p 10 **3.** Yes, if the rays form a straight $\angle$.

Pages 73–74

1. $\overleftrightarrow{AE}$; $\overleftrightarrow{CG}$; $\overleftrightarrow{FI}$ **3.** none of these **5.** alt int **7.** none of these **9.** corr **11.** corr **13.** corr **15.** alt int **17.** alt ext **19.** alt int **21.** none of these **23.** alt ext **25.** *A, E, F, H, M, N, W, Z* **27.** *E, F, H, M, N, W, Z*

Algebra Review

1. 5 **2.** $^-10$ **3.** 11 **4.** -8 **5.** 2 **6.** 1 **7.** $\frac{5}{4}$ **8.** $\frac{3}{2}$ **9.** $-\frac{1}{2}$ **10.** $-\frac{7}{2}$ **11.** -2 **12.** -4

Pages 78–79

1. 63, 63 **3.** 71, 71 **5.** 70, 70 **7.** 22, n.e.i, n.e.i **9.** Given; corr $\angle$'s of $\parallel$ lines are $\cong$; $\angle EIJ$; $\angle ABC \cong \angle EIJ$; trans prop **11.** m $\angle 1$ + m $\angle 2 = 180$ and m $\angle 3$ + m $\angle 4 = 180$ (Th 3.2); m $\angle 1$ + m $\angle 2$ + m $\angle 3$ + m $\angle 4 = 360$ (add prop) **13.** m $\angle 1$ + m $\angle 2 = 180$ (Post 11); $\angle 1$ and $\angle 2$ are supp (def of supp $\angle$'s); $\angle 2 \cong \angle 3$ (Post 13); $\angle 1$ and $\angle 3$ are supp (subst prop) **15.** Given: $\overleftrightarrow{AB} \parallel \overleftrightarrow{CD}$; $\overleftrightarrow{EF} \perp \overleftrightarrow{AB}$. Prove: $\overleftrightarrow{EF} \perp \overleftrightarrow{CD}$; $\angle 1$ is a rt $\angle$ (Th 2.6); $\angle 1 \cong \angle 2$ (Post 13); $\angle 2$ is a rt $\angle$ (subst prop); $\overleftrightarrow{EF} \perp \overleftrightarrow{CD}$ (def of $\perp$ lines) **17.** meas of $n, k, w, z = 76$; meas of 1, m, x, $y = 104$; meas of $p, t, q, u = 59$; meas of $r, o, s, v = 121$ **19.** $\angle AED$ is a comp of $\angle JHM$ **21.** m $\angle 3 = 42$

Pages 83–84

1. True **3.** False **5.** $\overrightarrow{BC} \parallel \overrightarrow{EG}$; all $\angle$'s are $\cong$ (All rt $\angle$s are $\cong$.) **7.** $\angle 1 \cong \angle 2$ (Vert $\angle$'s are $\cong$.); $\angle 2 \cong \angle 4$ (trans prop); $\angle 2$ supp to $\angle 3$ (Th 3.2); $\angle 4$ supp to $\angle 3$ (subst prop); $t \parallel s$ (Th 3.4) **9.** $l \nparallel m$ **11.** $r \nparallel s$ **13.** $r \nparallel s$ **15.** $\angle ADC$ supp to $\angle A$ (Th 3.2); $\angle ABC$ supp $\angle A$ (subst prop); $\overline{BC} \parallel \overline{AD}$ (Th 3.4) **17.** m $\angle 2$ + m $\angle 3 = 180$ (Post 11); $\angle 2$ and $\angle 3$ are supp (def of supp $\angle$'s); $\angle 1 \cong \angle 3$ (Cor supp of the same $\angle$ are $\cong$); $\overleftrightarrow{AB} \parallel \overleftrightarrow{CD}$ (Th 3.3) **19.** Given: $\overleftrightarrow{AB} \parallel \overleftrightarrow{CD}$, $\overrightarrow{EG}$ bisects $\angle IEB$, $\overrightarrow{FH}$ bisects $\angle EFD$. Prove: $\overrightarrow{EG} \parallel \overrightarrow{FH}$; $\angle IEB \cong \angle EFD$ (Th 3.2); m $\angle IEB$ = m $\angle EFD$ (def of $\cong$ $\angle$'s); $\frac{1}{2}$ m $\angle IEB = \frac{1}{2}$ m $\angle EFD$ (div prop); m $\angle IEG = \frac{1}{2}$ m $\angle IEB$ and m $\angle EFH = \frac{1}{2}$ m $\angle EFD$ (def of bis); m $\angle IEG$ = m $\angle EFH$ (subst prop); $\angle IEG \cong \angle EFH$ (def of $\cong$ $\angle$'s) $\overrightarrow{EG} \parallel \overrightarrow{FH}$ (Th 3.3) **21.** Given: $\overrightarrow{EH} \parallel \overrightarrow{FG}$, $\overrightarrow{EH}$ bisects $\angle BEF$, $\overrightarrow{FG}$ bisects $\angle CFE$. Prove: $\overleftrightarrow{AB} \parallel \overleftrightarrow{CD}$; $\angle HEF \cong \angle EFG$ (Post 13); m $\angle HEF$ = m $\angle EFG$ (def of $\cong$ $\angle$'s); m $\angle HEF = \frac{1}{2}$ m $\angle BEF$ and m $\angle EFG = \frac{1}{2}$ m $\angle CFE$ (def of bis); $\frac{1}{2}$ m $\angle BEF = \frac{1}{2}$ m $\angle CFE$ (subst prop); m $\angle BEF$ = m $\angle CFE$ (mult prop); $\angle BEF \cong \angle CFE$ (def of $\cong$ $\angle$'s); $\overleftrightarrow{AB} \parallel \overleftrightarrow{CD}$ (post 14)

Page 86

1. cannot tell **3.** $\neq$ **5.** cannot tell **7.** $\neq$ **9.** cannot tell **11.** $\parallel$ **13.** $\parallel$ **15.** Given: $\overleftrightarrow{AB} \parallel \overleftrightarrow{CD}$, $\overleftrightarrow{GH}$ inters $\overleftrightarrow{AB}$ at F. Prove: $\overleftrightarrow{GH}$ inters $\overleftrightarrow{CD}$, if $\overleftrightarrow{GH}$ does not inters $\overleftrightarrow{CD}$, $\overleftrightarrow{GH} \parallel \overleftrightarrow{CD}$. This contradicts post 15.

Pages 89–90

1. 45°; 45°; 90° **3.** 50°; 60°; 70° **5.** 30 **7.** $\angle C$ is a rt $\angle$ (def of $\perp$ lines); m $\angle C = 90$ (def of a rt $\angle$); m $\angle A$ + m $\angle B$ + m $\angle C = 180$ (Th 3.5); m $\angle A$ + m $\angle B = 90$ (subtr prop) **9.** $\angle DBA \cong \angle CBE$ (vert $\angle$'s are $\cong$); $\angle A \cong \angle E$ (Th 3.6) **11.** m $\angle C$ + m $\angle CDE$ + m $\angle CED = 180$ and m $\angle C$ + m $\angle A$ + m $\angle B = 180$ (Th 3.5); m $\angle C$ + m $\angle CDE$ + m $\angle CED$ = m $\angle C$ + m $\angle A$ + m $\angle B$ (trans prop); m $\angle CDE$ + m $\angle CED$ = m $\angle A$ + m $\angle B$ (subtr prop); 2 m $\angle CDE$ = 2 m $\angle A$ (subst prop); m $\angle CDE$ = m $\angle A$ (div prop); $\angle CDE \cong \angle A$ (def of $\cong$ $\angle$'s); $\overline{DE} \parallel \overline{AB}$ (Th 3.3); **13.** 67 **15.** too little info **17.** 76 **19.** $\angle ABH \cong \angle EGJ$ contradicts $\overleftrightarrow{AC} \nparallel \overleftrightarrow{FJ}$.

Page 92

1. Suppose $\angle 1 \cong \angle 2$. Then $l \parallel$ m (Post 14); This contradicts the given. $\therefore \angle 1 \ncong \angle 2$. **3.** Suppose $t \perp s$. Then $\angle 3 \cong \angle 4$. ($\perp$ lines form $\cong$ rt $\angle$'s). This contradicts the given. $\therefore t \not\perp s$ **5.** Suppose $r \parallel s$. Then $\angle 2 \cong \angle 3$ (Post 13); $\angle 2$ is a rt $\angle$ (Th 2.6); $\angle 3$ is a rt $\angle$ (subst prop); $s \perp t$ (def of $\perp$ lines). This contradicts the given. $\therefore r \nparallel s$. **7.** Suppose $l \parallel$ m. Then $\angle 1$ and $\angle 2$ are suppl (Th 3.2); m $\angle 1$ + m $\angle 2 = 180$ (def of supp $\angle$'s). This contradicts the given. $\therefore l \nparallel m$ **9.** Given: $\triangle ABC$. Prove: $\angle A$ and $\angle B$ cannot both be obtuse $\angle$'s. Suppose $\angle A$ and $\angle B$ are both obtuse $\angle$'s. Then m $\angle A$ + m $\angle B > 180$ and m $\angle A$ + m $\angle B$ + m $\angle C > 180$. This contradicts Th 3.5. $\therefore \angle A$ and $\angle B$ cannot be obtuse. **11.** Given: $\triangle ABC$. Prove: $\overrightarrow{AC} \nparallel \overrightarrow{BC}$. Suppose $\overrightarrow{AC} \parallel \overrightarrow{BC}$. Then $\overrightarrow{AC}$ and $\overrightarrow{BC}$ are intersected by a transv; $\angle A$ and $\angle B$ are supp. ($\angle$s on the same side of the transv are supp); m $\angle A$ + m $\angle B = 180$ (def of supp $\angle$s) m $\angle A$ + m $\angle B$ + m $\angle C > 180$. This contradicts Th 3.5. $\overrightarrow{AC} \nparallel \overrightarrow{BC}$ **13.** Suppose $r \parallel t$. Then $s \parallel t$ (Th 3.7). This contradicts the given. $\therefore r \nparallel t$.

15. Suppose $\angle C \cong \angle F$. Then m $\angle A$ + m $\angle B$ + m $\angle C \neq$ m $\angle D$ + m $\angle E$ + m $\angle F$ (add prop of ineq). This contradicts m $\angle A$ + m $\angle B$ + m $\angle C$ = 180 and m $\angle D$ + m $\angle E$ + m $\angle F$ = 180 (Th 3.5). $\therefore \angle C \cong \angle F$

Page 94

1. Follow constr p 93. 3. Follow ex 1 p 93. 5. Follow constr p 10 and constr p 93.

Page 97

1. F 3. T 5. T 7. T 9. $m \parallel l$ and $p \parallel l$, but $m \nparallel p$ 11. $p \parallel q$, but $l \parallel m$ 13. p and q have no pt in common (def of ∥ planes); $l \parallel p$ (l and p have no pt in common) (def of line ∥ to plane) 15. Given: l intersects plane p at pt X; m is on p. Prove: $l \parallel m$. Assume $l \parallel m$. Then l and m are coplanar; but only pt X on l is contained in p; therefore $l \parallel m$. 17. Given: $p \parallel q$; t intersects p. Prove: t intersects q. Assume t does not intersect q. Then $t \parallel p$, which contradicts the given. $\therefore$ t intersects q. 19. No. Planes are parallel, or they intersect.

Page 99

1. m $\angle 9 = 60$; m $\angle 8 = 120$ 3. 40 5. None 7. If lines do not intersect, then they are ∥. 9. Follow instructions on p 93. 11. F 13. $\overleftrightarrow{AD} \nparallel \overleftrightarrow{HI}$ 15. Refer to ex 9, 10 on p 92.

Pages 105–106

1. Must have an obtuse ∠ and no sides ≅. 3. Must have 3 acute angles and two sides ≅. 5. Must have 3 acute angles and no sides ≅. 7. *AT* 9. *NT* 11. *NT* 13. *ST* 15. *ST* 17. 1*SOS* rt △'s 19. *U*; *A*; *B*; *D*; 23. Given: *ABC* is equiangular. Prove: *ABC* is not obtuse. Assume $\triangle ABC$ is obtuse (one of 2 possibilities); $\angle A$ is obtuse (def of obtuse △); $\angle A$, $\angle B$, $\angle C$ are obtuse ∠'s (def of equian △); m $\angle A$ + m $\angle B$ + m $\angle C > 180$ (This contradicts Th 3.5); $\triangle ABC$ is not obtuse (only other possibility).

Page 109

1. 120 3. 62 5. $\angle 2 \cong \angle 3$ (Th 3.6); $\angle 1 \cong \angle 4$ (Th 2.8) 7. Given: $\triangle ABC$, exterior $\angle 2$. Prove: m $\angle 2$ = m $\angle A$ + m $\angle B$. m $\angle A$ + m $\angle B$ + m $\angle 1$ = 180 (Th 3.5); m $\angle 1$ + m $\angle 2$ = 180 (Post 11); m $\angle A$ + m $\angle B$ + m $\angle 1$ = m $\angle 1$ + m $\angle 2$ (trans prop); m $\angle 2$ = m $\angle A$ + m $\angle B$ (subtr prop).

Pages 112–113

1. $\triangle DEF$ 3. $\triangle QPR$ 5. $\triangle DBC$ 13, 15. Answers may vary. 17. 31°

Algebra Review

1. $3x(2x^2 - 3)$ 2. $5y^2(y^2 - 3y + 5)$ 3. $4(2z^3 + 3z^2 + 5)$ 4. $9a(3a^4 + 4a^2 - 2a + 5)$

Pages 116–117

1. $\triangle ABC \cong \triangle DEF$ 3. $\triangle MNO \cong \triangle HFG$ 5. ≇ 7. Given; given; $\overline{MO} \cong \overline{MO}$; $\triangle OPM$ (SSS) 9. Follow constr p 115. 11. Follow constr p 115. Use 3 seg AS ≅ sides of given △'s. 13. Follow constr p 115. Use 3 seg ≅ $\overline{MN}$. 15. $\overline{AB} \cong \overline{BC} \cong \overline{CA}$ and $\overline{DE} \cong \overline{EF} \cong \overline{FD}$ (def of equila △'s); $\overline{AB} \cong \overline{BC} \cong \overline{CA} \cong \overline{DE} \cong \overline{EF} \cong \overline{FD}$ (subst prop); $\triangle ABC \cong \triangle DEF$ (SSS) 17. $\overline{KL} \cong \overline{NM}$ (given); $\overline{LM} \cong \overline{LM}$ (reflex prop); $KL + LM = LM + NM$ (add prop); $\overline{KM} \cong \overline{LN}$ (def of between, def ≅ segs.); $\overline{KQ} \cong \overline{NO}$ and $\overline{QM} \cong \overline{OL}$ (given); $\triangle KMQ \cong \triangle NLO$ (SSS) 19. $\overline{AB} \cong \overline{BC}$ (def of bis); $\overline{DB} \cong \overline{DB}$ (reflex prop); $\overline{AD} \cong \overline{CD}$ (given); $\triangle ADB \cong \triangle CBD$ (SSS) 21. $\overline{XZ} \cong \overline{XY} \cong \overline{YZ}$ and $\overline{XZ} \cong \overline{XW} \cong \overline{ZW}$ and $\overline{YZ} \cong \overline{ZW} \cong \overline{YW}$ and $\overline{XW} \cong \overline{YW} \cong \overline{XY}$ (def of equila △'s); $\overline{XZ} \cong \overline{XY} \cong \overline{YZ} \cong \overline{ZW} \cong \overline{XW} \cong \overline{YW}$ (Subst prop); $\overline{XZ} \cong \overline{XZ}$, $\overline{XY} \cong \overline{XY}$, $\overline{YZ} \cong \overline{YZ}$, $\overline{ZW} \cong \overline{ZW}$, $\overline{XW} \cong \overline{XW}$, $\overline{YW} \cong \overline{YW}$ (reflex prop); $\triangle XYZ \cong \triangle XYW \cong \triangle XWZ \cong \triangle YZW$ (SSS) 23. No. The sum of the lengths of any 2 seg. must be > length of the 3rd seg.

Pages 121–122

1. Follow constr p 118 for SAS and ex 3 p 119 for *ASA*. 3. $\overline{AR}$; $\angle R$; $\overline{RE}$ 5. Yes, $\triangle ABC \cong \triangle DEF$ 7. Yes, $\triangle MNO \cong \triangle QPR$ 9. SAS 11. Yes. SAS 13. Yes ASA 15. $\angle F \cong \angle G$ (given); $\overline{FJ} \cong \overline{JG}$ (def of bis); $\angle FJH \cong \angle GJI$ (vert ∠'s); $\triangle FJH \cong \triangle GJF$ (ASA) 17. $\overline{AC} \cong \overline{BC}$ (given); $\angle ACD \cong \angle BCD$ (def of bis); $\overline{CD} \cong \overline{CD}$ (reflex prop); $\triangle ADC \cong \triangle BDC$ (SAS) 19. $\angle LKI \cong \angle JIK$ and $\angle LIK \cong \angle JKI$ (Post 13); $\overline{IK} \cong \overline{IK}$ (reflex prop); $\triangle LKI \cong \triangle JIK$ (ASA) 21. Follow constr p 119, ex 3. 23. m $\angle UST$ = m $\angle UTS$ = m $\angle SUT$ = 60, m $\angle SUV$ = m $\angle UVS$ = m $\angle VSU$ = 60, m $\angle VUT$ = m $\angle UTV$ = m $\angle TVU$ = 60, m $\angle SVT$ = m $\angle VTS$ = m $\angle TSV$ = 60 (def of equian △'s); $\angle UST \cong \angle UTS \cong \angle SUT \cong \angle SUV \cong \angle UVS \cong \angle VSU \cong \angle VUT \cong \angle UTV \cong \angle TVU \cong \angle SVT \cong \angle VTS \cong \angle TSV$ (def of ≅ ∠'s and subst prop); $\overline{SU} \cong \overline{SU}$, $\overline{UT} \cong \overline{UT}$, $\overline{ST} \cong \overline{ST}$, $\overline{UV} \cong \overline{UV}$, $\overline{SV} \cong \overline{SV}$; $\overline{TV} \cong \overline{TV}$ (reflex prop); $\triangle STU \cong \triangle SVU \cong \triangle TVU \cong \triangle STV$ (ASA)

Pages 126–127

1. SAS; yes 3. SSA; no 5. ASA; yes 7. Given; given; alt int ∠s of parallel lines are

$\cong$; reflex prop; AAS **9.** $\angle M \cong \angle K$ (given); $\overline{MN} \cong \overline{KN}$ (given); $\angle MNJ \cong \angle KNL$ (vert $\angle$'s); $\triangle JNM \cong \triangle LNK$ (ASA) **11.** $\angle B \cong \angle E$ (given); $\overline{EC} \cong \overline{CB}$ (def of midpt); $\angle ACB \cong \angle DCE$ (vert $\angle$'s); $\triangle ABC \cong \triangle DEC$ (ASA) **13.** $\overline{ML} \cong \overline{KJ}$ and $\overline{KL} \cong \overline{MJ}$ (given); $\overline{KM} \cong \overline{KM}$ (reflex prop); $\triangle MLK \cong \triangle KJM$ (SSS) **15.** $\overline{UV} \cong \overline{VS}$ (def of bis); $\angle UVT \cong \angle SVT$ (Th 2.6); $\overline{VT} \cong \overline{VT}$ (reflex prop); $\triangle UVT \cong \triangle SVT$ (SAS) **17.** $\angle A \cong \angle D$ (given); $\angle DEC \cong \angle ECB$ (Post 13); $\angle ECB \cong \angle FBA$ (Th 3.1); $\angle DEC \cong \angle FBA$ (subst prop); $\overline{BF} \cong \overline{EC}$ (given); $\triangle ABF \cong \triangle DEC$ (AAS) **19.** $\angle A \cong \angle A'$, $\angle B \cong \angle B'$, $\overline{BC} \cong \overline{B'C'}$ (given); $\angle C \cong \angle C'$ (Th 3.6); $\triangle ABC \cong \triangle A'B'C'$ (ASA) **21.** $\overline{GH} \cong \overline{JH}$ (trans prop); $\angle IHG \cong \angle IHJ$ (Th 2.6 and Th 2.5); $\overline{IH} \cong \overline{IH}$ (reflex prop); $\triangle GHI \cong \triangle JHI$ (SAS)

Page 130

1. equilat; equiang, acute **3.** 105° **5.** See ex 1 p. 104 **7.** 100° **9.** Yes, ASA **11.** See constr p. 115 **13.** $\overline{RS}$; $\angle S$; $\overline{ST}$ **15.** Yes, SSS **17.** No, SSA not $\cong$ pattern **19.** Use AAS.

Pages 133–134

1. *B* **2.** *B* **3.** *B* **4.** *A* **5.** *C* **6.** *A* **7.** *D* **8.** *C* **9.** *B* **10.** *A* **11.** *C* **12.** *C* **13.** *D* **14.** 62° **15.** 115° **17.** 11, −3 **18.** −5.3, 2.1 **19.** Follow midpt constr p 10. **20.** Follow $\angle$ bis constr p 15. **21.** F **22.** F **23.** T **24.** T **25.** F **26.** F **27.** T **28.** m $\angle 1 = 122$; m $\angle 4 = 58$ **29.** m $\angle 1 = 113$; m $\angle 10 = 49$ **30.** m $\angle 7 = 48$; m $\angle 8 = 48$ **31.** m $\angle 7 = 70$ **32.** $\angle 1 \cong \angle 2$ (given); $l \mid m$ (Post 14); $\angle 3 \cong \angle 4$ (Post 13) **33.** $DC = DC$ (reflex prop); $AD + DC = DC + CF$ (add prop); $\overline{AC} \cong \overline{DF}$ (def of between, def of $\cong$ segs.); $\triangle ACB \cong \triangle FDE$ (SAS) **34.** $\overline{AB} \cong \overline{AB}$ (reflex prop); $\angle CBA \cong \angle DBA$ (Th 2.5, Th 2.6); $\triangle ABC \cong \triangle ABD$ (AAS)

Page 138

1. $\angle A \cong \angle D$; $\angle B \cong \angle E$; $\angle C \cong \angle F$; $\overline{AB} \cong \overline{DE}$; $\overline{BC} \cong \overline{EF}$; $\overline{CA} \cong \overline{FD}$ **3.** $\angle P \cong \angle Q$; $\angle T \cong \angle U$; $\angle L \cong \angle M$; $\overline{PT} \cong \overline{QU}$; $\overline{TL} \cong \overline{UM}$; $\overline{LP} \cong \overline{MQ}$ **5.** $\overline{OQ}$; 9 cm **7.** $\overline{AD} \cong \overline{CB}$ and $\overline{CD} \cong \overline{AB}$ (given); $\overline{AC} \cong \overline{AC}$ (reflex prop); $\triangle ADC \cong \triangle CBA$ (SSS); $\angle D \cong \angle B$ (CPCTC) **9.** 56 **11.** $\overline{WZ} \cong \overline{YX}$ and $\angle 2 \cong \angle 4$ (given); $\overline{XZ} \cong \overline{XZ}$ (reflex prop); $\triangle WXZ \cong \triangle YZX$ (SAS); $\angle 1 \cong \angle 3$ (CPCTC); $\overline{XW} \parallel \overline{YZ}$ (Post 14) **13.** $\overline{CD} \cong \overline{CD}$ (reflex prop); $\angle CDA \cong \angle CDB$ (Th 2.6); $\overline{AD} \cong \overline{BD}$ (given); $\triangle CDA \cong \triangle CDB$ (SAS); $\overline{AC} \cong \overline{BC}$ (CPCTC); $\triangle ABC$ is isos (def of isos $\triangle$) **15.** $\overline{LK} \cong \overline{JK}$ and $\overline{LI} \cong \overline{JI}$ (given); $\overline{IK} \cong \overline{IK}$ (reflex prop); $\triangle KLI \cong \triangle KJI$ (SSS); $\angle LKI$ is a rt $\angle$ (def of $\perp$ lines); $\angle JKI \cong \angle LKI$ (CPCTC); $\angle JKI$ is a rt $\angle$ (subst prop); $\overline{JK} \perp \overline{IK}$ (def of $\perp$ lines)

Pages 142–143

1. m $\angle B = 70$; m $\angle C = 40$ **3.** m $\angle H =$ m $\angle I = 60$ **5.** m $\angle X =$ m $\angle Y =$ m $\angle Z = 60$ **7.** m $\angle X =$ m $\angle Y = 82$; m $\angle Z = 16$ **9.** m $\angle D =$ m $\angle F = 45$ **11.** 20 cm **13.** $\angle B \cong \angle 1$ (Th 5.1); $\angle 1 \cong \angle 2$ (Th 2.10); $\angle B \cong \angle 2$ (transitive prop) **15.** $\angle D \cong \angle E$ (Th 5.1); $\angle 2 \cong \angle D$ (Post 13); $\angle 1 \cong \angle E$ (Th 3.1); $\angle 1 \cong \angle 2$ (Trans prop) **17.** m $\angle 1 +$ m $\angle ROP = 180$; m $\angle 2 +$ m $\angle RPO = 180$ (Post 11); $\angle 1$ and $\angle ROP$ are supp, $\angle 2$ and $\angle RPO$ are supp (def of supp $\angle$'s); $\angle ROP \cong \angle RPO$ (Th 2.8); $\overline{RO} \cong \overline{RP}$ (Th 5.2); $\triangle OPR$ is isos (def of isos $\triangle$'s) **19.** $\angle A \cong \angle B \cong \angle C$ (def of equian $\triangle$'s); $\overline{AC} \cong \overline{CB}$, $\overline{AC} \cong \overline{AB}$, $\overline{AB} \cong \overline{CB}$ (Th 5.2); $\overline{AC} \cong \overline{AB} \cong \overline{CB}$ (subst prop); $\triangle ABC$ is equila (def of equila $\triangle$'s) **21.** $\angle 1 \cong \angle 4$ (Th 5.1); $\angle 1 \cong \angle S$, $\angle 4 \cong \angle T$ (Th 3.1); $\angle S \cong \angle T$ (subst prop); $\overline{SV} \cong \overline{TV}$ (Th 5.2); $SV - VW = TV - VU$ (subtr prop); $SW \cong \overline{TU}$ (subst prop, def of $\cong$ segs.) **23.** 31 **25.** $\overline{PY} \cong \overline{PZ}$ (Th 5.2); $\overline{PX} \cong \overline{PX}$ (reflex prop); $\triangle PXZ \cong \triangle PXY$ (SAS); $\overline{XZ} \cong \overline{XY}$ (CPCTC); $\triangle XYZ$ is isos (def of isos $\triangle$'s)

Algebra Review

1. 4 **2.** $4\sqrt{2}$ **3.** $3\sqrt{5}$ **4.** $7\sqrt{3}$ **5.** $11\sqrt{2}$ **6.** 10 **7.** $6\sqrt{5}$ **8.** $-24\sqrt{3}$ **9.** $2x\sqrt{2}$ **10.** $12x\sqrt{x}$

Pages 155–156

1, 3. Answers may vary. **5.** $\overline{SU} \cong \overline{VT}$ and $\angle S \cong \angle V$ (given); $\overline{SW} \cong \overline{VW}$ (Th 5.2); $\triangle SUW \cong \triangle VTW$ (SAS) **7.** $\overline{AX} \cong \overline{BY}$ and $\angle ZXY \cong \angle ZYX$ (given); $\overline{XY} \cong \overline{XY}$ (reflex prop); $\triangle AYX \cong \triangle BXY$ (SAS); $\angle AYX \cong \angle BXY$ (CPCTC) **9.** $\angle EDC \cong \angle GCD$ (Th 3.6); $\angle GCD$ is a rt $\angle$ (def of $\perp$ lines); $\angle EDC$ is a rt $\angle$ (subst prop); $\overline{ED} \perp \overline{CD}$ (def of $\perp$ lines) **11.** m $\angle 5 +$ m $\angle 3 = 180$, m $\angle 6 +$ m $\angle 4 = 180$ (Post 11); $\angle 5$ and $\angle 3$ are supp, $\angle 6$ and $\angle 4$ are supp (def of supp $\angle$'s); $\angle 5 \cong \angle 6$ (Th 2.8); $\overline{SR} \cong \overline{SR}$ (reflex prop); $\overline{OR} \cong \overline{QS}$ (add prop, def of between, def of $\cong$ segs.); $\triangle OPR \cong \triangle QPS$ (ASA) **13.** $R\triangle GEC$ **15.** $R\triangle DGC$ **17.** $R\triangle AGF$ **19.** pt O **21.** $R\triangle SOC$ **23.** $R\triangle BCD$ **25.** $\overline{AE}$ **27.** $R\triangle HDC$

Algebra Review

1. $(x-4)(x-2)$ **2.** $(x+5)(x-3)$ **3.** $(x+4)(x+5)$ **4.** $(x-1)(x+1)$ **5.** $(x-9)(x+9)$ **6.** $(3x-6)(3x+6)$ **7.** $(2x+1)(3x-2)$ **8.** $(x+5)(2x-3)$ **9.** $(3x-2)(2x-3)$

Pages 160–161

1. $\angle Y$, $\angle S$ are rt $\angle$'s (def of $\perp$ lines); $\triangle XYZ$, $\triangle RST$ are rt $\triangle$'s (def of rt $\triangle$'s); $\overline{XY} \cong \overline{RS}$ and $\overline{ZY} \cong \overline{TS}$ (given); $\triangle XYZ \cong \triangle RST$ (LL) **3.** $\angle Y$, $\angle S$ are rt $\angle$'s (def of $\perp$ lines); $\triangle XYZ$, $\triangle RST$ are rt $\triangle$'s (def of rt $\triangle$'s); $\overline{XY} \cong \overline{RS}$, $\angle Z \cong \angle T$ (given); $\triangle XYZ \cong \triangle RST$ (LA) **5.** $\angle ADB$, $\angle BEA$ are rt $\angle$'s (def of $\perp$ lines); $\overline{AB} \cong \overline{AB}$ (reflex prop); $\triangle ABE$, $\triangle BAD$ are rt $\triangle$'s (def of rt $\triangle$'s); $\triangle ABE \cong \triangle BAD$ (HL) **7.** $\angle ADB$, $\angle BEA$ are rt $\angle$'s (def of $\perp$ lines); $\triangle AFD$, $\triangle BFE$ are rt $\triangle$'s (def of rt $\triangle$'s); $\angle DFA \cong \angle EFB$ (Th 2.10); $\triangle AFD \cong \triangle BFE$ (HA) **9.** $\triangle IJG$ and $\triangle IJH$ are rt $\triangle$'s (def of $\perp$ lines, def of rt $\triangle$'s) $\overline{GJ} \cong \overline{JH}$ (def of bis); $\overline{IJ} \cong \overline{IJ}$ (reflex prop); $\triangle IJG \cong \triangle IJH$ (LL) $\overline{IG} \cong \overline{IH}$ (CPCTC); $\triangle GHI$ is isos (def of isos $\triangle$'s) **11.** $\angle MOK$, $\angle LNK$ are rt $\angle$'s (def of $\perp$ lines); $\triangle MOK$, $\triangle LNK$ are rt $\triangle$'s (def of rt $\triangle$); $\angle K \cong \angle K$ (reflex prop); $\triangle MOK \cong \triangle LNK$ (HA); $\overline{MO} \cong \overline{LN}$ (CPCTC) **13.** $\angle RSP$, $\angle VWT$ are rt $\angle$'s (def of $\perp$ lines); $\triangle RSP$, $\triangle VWT$ are rt $\triangle$'s (def of rt $\triangle$'s); $\angle P \cong \angle T$, $\overline{PR} \cong \overline{TV}$ (CPCTC); $\triangle RSP \cong \triangle VWT$ (HA) $\overline{RS} \cong \overline{VW}$ (CPCTC) **15.** $\angle ZAX$, $\angle ZAY$ are rt $\angle$'s (def of $\perp$ lines); $\triangle ZAX$, $\triangle ZAY$ are rt $\triangle$'s (def of rt $\triangle$'s); $\overline{ZA} \cong \overline{ZA}$ (reflex prop); $\triangle XAZ \cong \triangle YAZ$ (HL) **17.** $\angle ZAX$, $\angle ZAY$ are rt $\angle$'s (def of $\perp$ lines); $\triangle ZAX$, $\triangle ZAY$ are rt $\triangle$'s (def of rt $\triangle$'s); $\overline{ZA} \cong \overline{ZA}$ (reflex prop); $\triangle XAZ \cong \triangle YAZ$ (HL); $\overline{AX} \cong \overline{AY}$ (CPCTC); $\angle AXY \cong \angle AYX$ (Th 5.1) **19.** $\angle ZAX$, $\angle ZAY$ are rt $\angle$'s (def of $\perp$ lines); $\triangle ZAX$, $\triangle ZAY$ are rt $\triangle$'s (def of rt $\triangle$'s); $\overline{ZA} \cong \overline{ZA}$ (reflex prop); $\triangle XAZ \cong \triangle YAZ$ (LL); $\overline{ZX} \cong \overline{ZY}$ (CPCTC); $\angle ZXY \cong \angle ZYX$ (Th 5.1)

Cumulative Review

1. Answers may vary. **2.** Answers may vary. **3.** No, No

Pages 163–165

1. $\overline{AB} \cong \overline{CD}$; $\overline{AC} \cong \overline{CA}$; $\angle BAC \cong \angle DCA$; $\triangle DCE$ and $\triangle BAE$ **3.** $\overline{KL} \cong \overline{MN}$; $\angle LOK \cong \angle NOM$; $\angle OKL \cong \angle OMN$; $\triangle KLM$ and $\triangle MNK$ **5.** $\angle FIG$, $\angle GJF$ are rt $\angle$'s (def of $\perp$ lines); $\triangle FIG$, $\triangle GJF$, $\triangle FKI$, $\triangle GKJ$ are rt $\triangle$'s (def of rt $\triangle$'s); $\overline{FG} \cong \overline{FG}$ (reflex prop); $\triangle FIG \cong \triangle GJF$ (HL); $\overline{FI} \cong \overline{GJ}$ (CPCTC); $\angle IKF \cong \angle JKG$ (Th. 2.10); $\triangle FKI \cong \triangle GKJ$ (LA) **7.** $\overline{AE} \cong \overline{CE}$, $\overline{BE} \cong \overline{DE}$ (def of bis); $\angle 3 \cong \angle 4$ (Th 2.10); $\triangle ABE \cong \triangle CDE$ (SAS); $\overline{AB} \cong \overline{CD}$, $\angle 1 \cong \angle 2$ (CPCTC); $\overline{AC} \cong \overline{CA}$ (reflex prop); $\triangle ABC \cong \triangle CDA$ (SAS) **9.** $\angle LGF$, $\angle KIH$ are rt $\angle$'s (def $\perp$ lines); $\triangle FGL$, $\triangle HIK$ are rt $\triangle$'s (def of rt $\triangle$'s); $\overline{LF} \cong \overline{KH}$, $\angle LFG \cong \angle KHI$ (CPCTC); $\triangle FGL \cong \triangle HIK$ (HA) **11.** $\overline{FL} \cong \overline{KJ}$, $\angle F \cong \angle J$ (CPCTC); $\overline{FH} \cong \overline{JH}$ (add; betweeness; subst); $\triangle FHL \cong \triangle JHK$ (SAS) **13.** $\angle QNM$, $\angle QNO$, $\angle RNQ$, $\angle PNQ$ are rt $\angle$'s (def of $\perp$ lines); $\triangle QNM$, $\triangle QNO$, $\triangle RNQ$, $\triangle PNQ$ are rt $\triangle$'s (def of rt $\triangle$'s); $\overline{QN} \cong \overline{QN}$ (reflex prop), $\triangle QNR \cong \triangle QNP$ (HL); $\angle RQN \cong \angle PQN$ (CPCTC); $\triangle QNM \cong \triangle QNO$ (LA); $\overline{NM} \cong \overline{NO}$ (CPCTC) **15.** $\angle DEB$, $\angle CFA$ are rt $\angle$'s (def of $\perp$ lines); $\triangle DEB$, $\triangle CFA$ are rt $\triangle$'s (def of rt $\triangle$'s); $\triangle DEB \cong \triangle CFA$ (HL); $\overline{AB} \cong \overline{AB}$ (reflex prop); $\angle CAB \cong \angle DBA$ (CPCTC); $\triangle CAB \cong \triangle DBA$ (SAS); $\overline{AD} \cong \overline{BC}$ (CPCTC) **17.** 66 **19.** $\angle DAB$, $\angle CBA$ are rt $\angle$s (def of $\perp$ lines); $\triangle DAB$, $\triangle CBA$ are rt $\triangle$'s (def of rt $\triangle$'s); $\overline{AB} \cong \overline{AB}$ (reflex prop); $\triangle DAB \cong \triangle CBA$ (LL); $\overline{AC} \cong \overline{BD}$ (CPCTC) **21.** $\triangle ADC \cong \triangle BCD$ (HL); $\angle DCA \cong \angle CDB$ (CPCTC); $\overline{ED} \cong \overline{EC}$. (Th 5.2) **23.** $AD \parallel BC$ (2 lines $\perp$ to same line are $\parallel$); $\angle ABC$ is a rt $\angle$ (def of $\perp$ lines); m $\angle ABC$ + m $\angle DCB$ = 180 (def of supp $\angle$'s); m $\angle DCB = 90$ (subt); $BC \perp DC$ (def $\perp$ lines); $\angle BAD$ is a rt $\angle$ (def of $\perp$ lines); m $\angle BAD$ + m $\angle ADC$ = 180 (def of $\perp$ lines); m $\angle BAD = 90$ (subt); $\overline{AD} \perp \overline{DC}$ (def of $\perp$ lines) **25.** Proof 23; $\overline{DA} \cong \overline{DA}$ (reflex prop); $\triangle DAB \cong \triangle ADC$ (HL); $\overline{DC} \cong \overline{AB}$ (CPCTC); $\angle CDB \cong \angle DBA$, $\angle DCA \cong \angle CAB$ (Post 13); $\triangle AEB \cong \triangle CED$ (ASA); $\overline{AE} \cong \overline{CE}$, $\overline{DE} \cong \overline{BE}$ (CPCTC); $\overline{AE} \cong \overline{CE} \cong \overline{DE} \cong \overline{BE}$ (subst prop) **27.** $\angle QRP \cong \angle MRN$ (Th 2.10); $\triangle QRP \cong \triangle MRN$ (SAS); $\angle PMO \cong \angle NQO$ (CPCTC); $\angle O \cong \angle O$ (reflex prop); $\overline{QN} \cong \overline{MP}$ (add prop); $\triangle PMO \cong \triangle NQO$ (AAS); $\overline{OQ} \cong \overline{OM}$ (CPCTC) **29.** $\overline{XA} \cong \overline{XA}$ (reflex prop); $\triangle XAZ \cong \triangle XAY$ (SSS); $\angle ZXB \cong \angle YXB$ (CPCTC); $\overline{XB} \cong \overline{XB}$ (reflex prop); $\triangle ZXB \cong \triangle YXB$ (SAS); $\overline{ZB} \cong \overline{YB}$ (CPCTC)

Page 168

1. $\angle XZW \cong \angle YZW$ **3.** all corresponding parts of $\triangle XZW$ and $\triangle YZW$ **5.** $\angle CDE \cong \angle BDE$ **7.** Given: isos $\triangle ABC$, $\overline{AC} \cong \overline{BC}$, $\overline{CD}$ bis $\angle ACB$. Prove: $\overline{CD}$ bis $\overline{AB}$; $\angle ACD \cong \angle DCB$ (def of bis); $\overline{CD} \cong \overline{CD}$ (reflex prop); $\triangle ACD \cong \triangle BCD$ (SAS); $\overline{AD} \cong \overline{DB}$ (CPCTC); $\overline{CD}$ bis $\overline{AB}$ (def of bis) **9.** Given: isos $\triangle ABC$, $\overline{AC} \cong \overline{BC}$, $\overline{CD}$ bis $\angle ACB$. Prove: $\overline{CD}$ is the $\perp$ bis of $\overline{AB}$; $\angle ACD \cong \angle DCB$ (def of bis); $\overline{CD} \cong \overline{CD}$ (reflex prop); $\triangle ADC \cong \triangle BDC$ (SAS); $\overline{AD} \cong \overline{DB}$ (CPCTC); $\overline{CD}$ is $\perp$ bis of $\overline{AB}$ (Th 5.8) **11.** Given: $\angle ACB$, $\overrightarrow{CP}$ bis $\angle ACB$, $\overline{PA} \perp \overline{CA}$, $\overline{PB} \perp \overline{CB}$. Prove: $\overline{PA} \cong \overline{PB}$; $\angle ACP \cong \angle BCP$ (def of bis); $\overline{CP} \cong \overline{CP}$ (reflex prop); $\angle CAP \cong \angle CBP$ (Th 2.6); $\triangle ACP \cong \triangle BCP$ (AAS); $\overline{PA} \cong \overline{PB}$ (CPCTC) **13.** Given; given; reflex prop; SSS; CPCTC; reflex prop; SAS; CPCTC; CPCTC; Post 11; def of supp $\angle$'s; Th 2.7; def of $\perp$ bis **15.** $\overline{DC} \cong \overline{DA}$, $\overline{BA} \cong \overline{BC}$ (Th 5.7); $\overline{DB} \cong \overline{DB}$ (reflex prop); $\triangle DCB \cong \triangle DAB$ (SSS); $\angle BAD \cong \angle BCD$ (CPCTC) **17.** $\angle 2 \cong \angle 4$ (subt prop); $\overline{FI} \cong \overline{FG}$, $\overline{HI} \cong \overline{HG}$ (Th 5.2); $\overline{FH}$ is $\perp$ bis of $\overline{GI}$ (Th. 5.8); $\overline{IJ} \cong \overline{JG}$ (def of bis); $\triangle HIJ \cong \triangle HGI$ (SAS); $\angle IHJ \cong \angle GHJ$ (CPCTC); $\overline{FH}$ bis $\angle IHG$ (def of bis) **19.** $\overline{MO} \cong \overline{MO}$ (reflex prop); $\angle NMO \cong \angle KMO$ (def of bis); $\triangle NMO \cong \triangle KMO$ (SAS); $\overline{NO} \cong \overline{KO}$ (CPCTC); $\overline{ML} \cong \overline{ML}$ (reflex prop); $\angle MLN$, $\angle MLK$ are rt $\angle$'s (def of $\perp$ lines); $\triangle MLN$, $\triangle MLK$ are rt $\triangle$'s (def of rt $\triangle$'s); $\triangle MLN \cong \triangle MLK$ (HL); $\overline{KL} \cong \overline{NL}$ (CPCTC); $\angle LNO \cong \angle LKO$ (Th 5.1); $\triangle LNO \cong \triangle LKO$ (SAS); $\angle NLO \cong \angle KLO$ (CPCTC); $\overline{LO}$ bis $\angle NLK$ (def of bis)

Cumulative Review

1. ASA; SAS; AAS; SSS **2.** HL; HA; LL; LA **3.** Follow constr p 15.

Pages 172–173

1. $\overline{CD} \perp \overline{AB}$ (def of alt); $\angle CDA$, $\angle CDB$ are rt $\angle$'s (def of $\perp$ lines); $\triangle ADC$, $\triangle BDC$ are rt $\triangle$'s (def of rt $\triangle$'s); $\overline{CD} \cong \overline{CD}$ (reflex prop); $\triangle ADC \cong \triangle BDC$ (LL) **3.** 50 **5.** 16 **7.** $\overline{ZW} \perp \overline{XY}$ (def of alt); $\angle ZWX$, $\angle ZWY$ are rt $\triangle$'s (def of $\perp$ lines); $\triangle ZWX$, $\triangle ZWY$ are rt $\triangle$'s (def of rt $\triangle$'s); $\overline{ZW} \cong \overline{ZW}$ (reflex prop); $\angle XZW \cong \angle YZW$ (def of bis); $\triangle ZWX \cong \triangle ZWY$ (LA); $\overline{XZ} \cong \overline{YZ}$ (CPCTC); $\triangle XYZ$ is isos (def of isos $\triangle$'s) **9.** $\angle ACB$ is a rt $\angle$ (def of $\perp$ lines); $\angle 4 \cong \angle 4$ (reflex prop); $\angle CDB \cong \angle ACB$ (Th 3.6); $\angle CDB$ is a rt $\angle$ (subst prop); $\overline{CD} \perp \overline{AB}$ (def $\perp$ lines); $\overline{CD}$ is an alt of $\triangle ABC$ (def of alt) **11.** $\angle O \cong \angle T$, $\overline{OP} \cong \overline{ST}$, $\angle OPQ \cong \angle TSV$ (CPCTC); 2 $\angle OPR \cong \angle OPQ$, 2 $\angle TSU \cong \angle TSV$ (def of bis); 2 $\angle OPR \cong$ 2 $\angle TSU$ (subst prop); $\angle OPR \cong \angle TSU$ (div prop); $\triangle OPR \cong \triangle TSU$ (ASA); $\overline{PR} \cong \overline{SU}$ (CPCTC) **13.** Given; CPCTC; def of $\cong$ seg; given; def of median; subst prop; def of $\cong$ seg; CPCTC; CPCTC; SAS; CPCTC **15.** Given: $\triangle ABC$ is equilat; $\overline{AE}$ alt. to $\overline{CB}$ · $\overline{CD}$ alt to $\overline{AB}$. Prove: $\overline{AE} \cong \overline{CD}$; $\overline{AE} \perp \overline{CB}$, $\overline{CD} \perp \overline{AB}$ (def of alt); $\angle AEB$, $\angle CDB$ are rt $\angle$'s (def of $\perp$ lines); $\triangle AEB$, $\triangle CDB$ are rt $\triangle$'s (def of rt $\triangle$'s); $\angle B \cong \angle B$ (reflex prop); $\overline{BC} \cong \overline{AB}$ (def of equil $\triangle$'s) $\triangle AEB \cong \triangle CDB$ (HA); $\overline{AE} \cong \overline{CD}$ (CPCTC) **17.** Given: isos $\triangle ABC$, $\overline{AC} \cong \overline{BC}$, $\overline{CD}$ alt. to $\overline{AB}$. Prove: $\overline{CD}$ is median to $\overline{AB}$; $\angle A \cong \angle B$ (Th 5.1); $\overline{CD} \perp \overline{AB}$ (def of alt); $\angle CDA \cong \angle CDB$ (Th 2.6); $\overline{CD} \cong \overline{CD}$ (reflex prop); $\triangle ACD \cong \triangle BCD$ (AAS); $\overline{AD} \cong \overline{DB}$ (CPCTC); D is the midpt of $\overline{AB}$ (def of midpt); $\overline{AD} \cong \overline{DB}$ (def of median) **19.** Given: isos $\triangle ABC$, $\overline{AC} \cong \overline{BC}$, $\overline{CD}$ median to $\overline{AB}$. Prove: $\overline{CD}$ bis $\angle ACB$; D is midpt. of $\overline{AB}$ (def of median); $\overline{AD} \cong \overline{DB}$ (def of midpt); $\overline{CD} \cong \overline{CD}$ (reflex prop); $\triangle ADC \cong \triangle BDC$ (SSS); $\angle ACD \cong \angle DCB$ (CPCTC); $\overline{CD}$ bis $\angle ACB$ (def of bis) **21.** Given: $\triangle ABC$, $\overline{CD}$ median to $\overline{AB}$, $\overline{CD} \perp \overline{AB}$. Prove: $\triangle ABC$ is isos; D is midpt of $\overline{AB}$ (def of median); $\overline{AD} \cong \overline{DB}$ (def of midpt); $\angle ADC \cong \angle BDC$ (Th 2.6); $\overline{CD} \cong \overline{CD}$ (reflex prop); $\triangle ADC \cong \triangle BDC$ (SAS); $\overline{AC} \cong \overline{BC}$ (CPCTC); $\triangle ABC$ is isos (def of isos $\triangle$) **23.** Given: $\triangle ABC$, $\overline{CD}$ median to $\overline{AB}$, $\overline{CD}$ bis $\angle ACB$. Prove: $\triangle ABC$ is isos. Draw $\overline{DM} \perp \overline{AC}$, $\overline{DN} \perp \overline{BC}$ (Post 12); $\angle AMD$, $\angle CMD$, $\angle CND$, $\angle BND$ are rt $\angle$'s (def of $\perp$ lines); $\triangle CMD$, $\triangle CND$, $\triangle AMD$, $\triangle BND$ are rt $\triangle$'s (def of rt $\triangle$'s); $\angle MCD \cong \angle NCD$ (def of bis); $\overline{CD} \cong \overline{CD}$ (reflex prop); $\triangle CDM \cong \triangle CDN$ (HA); $\overline{MD} \cong \overline{ND}$ (CPCTC); D is midpt of $\overline{AB}$ (def of med); $\overline{AD} \cong \overline{DB}$ (def of midpt); $\triangle AMD \cong \triangle BND$ (HL for $\cong$ rt $\triangle$'s) $\angle A \cong \angle B$ (CPCTC); $\overline{AC} \cong \overline{BC}$ (Th 5.2); $\triangle ABC$ is isos (def of isos $\triangle$'s) **25.** $\angle CDE \cong \angle ADE$ (def of bis); $\overline{ED} \cong \overline{ED}$ (reflex prop); $\triangle CDE \cong \triangle ADE$ (SAS); $\overline{CE} \cong \overline{EA}$ (CPCTC); $\angle DBC$, $\angle DBA$ are rt $\angle$'s (def of $\perp$ lines); $\triangle DBC$, $\triangle DBA$ are rt $\triangle$'s (def of rt $\triangle$'s); $\overline{DB} \cong \overline{DB}$ (reflex prop); $\triangle DBC \cong \triangle DBA$ (HL); $\overline{BC} \cong \overline{BA}$ (CPCTC); $\overline{BE} \cong \overline{BE}$ (reflex prop); $\triangle BCE \cong \triangle BAE$ (SSS); $\angle CBE \cong \angle ABE$ (CPCTC); $\overline{BE}$ bis $\angle CBA$ (def of bis) **27.** $\overline{JF} \cong \overline{HF}$, $\angle JFG \cong \angle HFG$ (CPCTC); $\overline{FI} \cong \overline{FI}$ (reflex prop); $\triangle JFI \cong \triangle HFI$ (SAS); $\angle JIF \cong \angle HIF$ (CPCTC); $\overline{JI} \perp \overline{FG}$ (def of alt); $\angle JIF$ is a rt $\angle$ (def of $\perp$ lines); $\angle HIF$ is a rt $\angle$ (subst prop); $\overline{HI} \perp \overline{FG}$ (def of $\perp$ lines); $\overline{HI}$ is alt of $\triangle FHG$ (def of alt) **29.** $\angle NML$, $\angle OML$ are rt $\angle$'s (def of $\perp$ lines); $\triangle NML$, $\triangle OML$ are rt $\triangle$'s (def of rt $\triangle$'s); $\overline{LM} \cong \overline{LM}$ (reflex prop); $\overline{NM} \cong \overline{MO}$ (def of bis); $\triangle OLM \cong \triangle NLM$ (LL) **31.** $\overline{NM} \cong \overline{NM}$ (reflex prop); $\angle KMN$, $\angle LMN$ are rt $\angle$'s (def of $\perp$ lines); $\triangle KMN$, $\triangle LMN$ are rt $\triangle$'s (def of rt $\triangle$'s); $\triangle KMN \cong \triangle LMN$ (HL); $\overline{LM} \cong \overline{KM}$ (CPCTC)

Pages 175–176

1. Follow constr p 174 ex 3. **3.** Follow constr p 51 ex 5. **5.** Follow constr p 174 ex 2. **7.** T **9.** T **11.** F **13.** F **15.** Follow constr p 174 ex 2. **17.** Follow constr p 174 ex 3.

Cumulative Review

1. Yes; obtuse $\triangle$ **2.** No **3.** Yes; the third line could be a transversal.

Algebra Review

1. $\frac{x-2}{8}$ **2.** $\frac{6x+10}{x+4}$ **3.** $\frac{y-1}{y+7}$ **4.** $\frac{6+2y}{3}$ **5.** $\frac{2z-2}{z+6}$ **6.** 1

Page 178

1. $\overline{AB} \cong \overline{FD}$; $\overline{BC} \cong \overline{FE}$; $\overline{CA} \cong \overline{ED}$ **3.** $\overline{DE}$

5. ≅ (LL) 7. ≅ (HL) 9. altitude 11. ∠1 ≅ ∠2 (def of bis); $\overline{AC} \cong \overline{AC}$ (reflex prop); $\triangle ADC \cong \triangle ABC$ (SAS) $\overline{CD} \cong \overline{CB}$ (CPCTC) 13. $\overline{GE} \cong \overline{GF}$, $\overline{GH} \cong \overline{GI}$ (CPCTC); $\overline{HE} \cong \overline{IF}$ (subtr prop); 15. 55° 17. Follow constr p 174, ex 2.

Page 187

1. Yes 3. No; segments intersect at pts that are not endpts. 5. No; not a union of seg. 7. No; not a union of seg. 9. $\overline{AB}$, $\overline{BC}$, $\overline{CD}$, $\overline{DA}$: A, B, C, D; quadrilateral; 2 11. $\overline{KL}$, $\overline{LM}$, $\overline{MN}$, $\overline{NO}$, $\overline{OP}$, $\overline{PQ}$, $\overline{QR}$, $\overline{RK}$; K, L, M, N, O, P, Q, R; octagon; 20 13. No; $\overline{XY}$ not always part of figure. 15. Convex; $\overline{XY}$ is always part of figure. 17, 19, 21. Answers may vary. 23. 11 25. 12

Cumulative Review

1. See Theorem 5.1, p. 139. 2. Answers will vary. 3. See ex 2 p 51.

Pages 190–192

1. 118° 3. 90° 5. 80° 7. 80°, 80°, 100°, 100° 9. 65°, 73°, 113°, 109° 11. 1080° 13. 6840° 15. 17640° 17. 900° 19. 5 21. 12 23. 15 25. 18 27. cannot 29. can 31. 120° 33. 80°, 80°, 80°, 160°, 160°, 160° 35. Given: convex pentagon $ABCDE$. Prove: m ∠A + m ∠B + m ∠C + m ∠D + m ∠E = 540. Draw AC, AD (Post 1); m ∠BAE = m ∠1 + m ∠2 + m ∠3, m ∠BCD = m ∠4 + m ∠5 (Post 10); m ∠EDC = m ∠6 + m ∠7; m ∠1 + m ∠B + m ∠4 = 180, m ∠2 + m ∠5 + m ∠6 = 180, m ∠3 + m ∠E + m ∠7 = 180 (Th 3.5); m ∠A + m ∠B + m ∠C + m ∠D + m ∠E = 540 (subst and add prop) 37. Given: convex quad $ABCD$. Prove: m ∠A + m ∠B + m ∠C + m ∠D = 360. m ∠BAD = m ∠1 + m ∠8, m ∠ABC = m ∠2 + m ∠3, m ∠BCD = m ∠4 + m ∠5, m ∠CDA = m ∠6 + m ∠7 (Post 10); m ∠1 + m ∠10 + m ∠2 = 180, m ∠3 + m ∠4 + m ∠11 = 180, m ∠6 + m ∠5 + m ∠12 = 180, m ∠7 + m ∠9 + m ∠8 = 180 (Th 3.5); m ∠1 + m ∠2 + m ∠3 + m ∠4 + m ∠5 + m ∠6 + m ∠7 + m ∠8 + m ∠9 + m ∠10 + m ∠11 + m ∠12 = 720 (add prop); m ∠9 + m ∠10 + m ∠11 + m ∠12 = 360 (Post 11); m ∠1 + m ∠2 + m ∠3 + m ∠4 + m ∠5 + m ∠6 + m ∠7 + m ∠8 = 360 (subst prop); m ∠A + m ∠B + m ∠C + m ∠D = 360 (subst prop) 39. Given: convex hexagon $ABCDEF$. Prove: m ∠A + m ∠B + m ∠C + m ∠D + m ∠E + m ∠F = 720. m ∠FAB = m ∠12 + m ∠1, m ∠ABC = m ∠2 + m ∠3, m ∠BCD = m ∠4 + m ∠5, m ∠CDE = m ∠6 + m ∠7, m ∠DEF = m ∠8 + m ∠9, m ∠EFA = m ∠10 + m ∠11 (Post 10); m ∠1 + m ∠2 + m ∠13 = 180, m ∠3 + m ∠4 + m ∠14 = 180, m ∠5 + m ∠6 + m ∠15 = 180, m ∠7 + m ∠8 + m ∠16 = 180, m ∠9 + m ∠10 + m ∠17 = 180, m ∠11 + m ∠12 + m ∠18 = 180 (Th 3.5); m ∠1 + m ∠2 + m ∠3 + m ∠4 + m ∠5 + m ∠6 + m ∠7 + m ∠8 + m ∠9 + m ∠10 + m ∠11 + m ∠12 + m ∠13 + m ∠14 + m ∠15 + m ∠16 + m ∠17 + m ∠18 = 1080 (add prop); m ∠13 + m ∠14 + m ∠15 + m ∠16 + m ∠17 + m ∠18 = 360 (Post 11); m ∠1 + m ∠2 + m ∠3 + m ∠4 + m ∠5 + m ∠6 + m ∠7 + m ∠8 + m ∠9 + m ∠10 + m ∠11 + m ∠12 = 720 (subtr prop); m ∠A + m ∠B + m ∠C + m ∠D + m ∠E + m ∠F = 720 (subst prop). 41. No 43, 45. No

Page 195

1–3. 360°; yes 5. 41°, 72°, 58°, 60°, 39°, 90° 7. Given: convex pentagon $ABCDE$, ext ∠s 1, 2, 3, 4, 5. Prove: m ∠1 + m ∠2 + m ∠3 + m ∠4 + m ∠5 = 360. m ∠1 + m ∠6 = 180, m ∠2 + m ∠7 = 180, m ∠3 + m ∠8 = 180, m ∠4 + m ∠9 = 180, m ∠5 + m ∠10 = 180 (Post 11); m ∠1 + m ∠2 + m ∠3 + m ∠4 + m ∠5 + m ∠6 + m ∠7 + m ∠8 + m ∠9 + m ∠10 = 900 (add prop); m ∠6 + m ∠7 + m ∠8 + m ∠9 + m ∠10 = 540 (Th 6.2); m ∠1 + m ∠2 + m ∠3 + m ∠4 + m ∠5 = 360 (subtr prop) 9. 1260°, 900°, 360°

Algebra Review

1. 2 2. −4 3. −7 4. 1 5. $\frac{-31}{26}$ 6. $\frac{-9}{22}$

Pages 197–198

1. equilateral triangle 3. regular hexagon 5. 60° 7. 108° 9. 140° 11. 168° 13. 120° 15. 72° 17. 36° 19. 7.2° 21. 3 23. 12 25. 5 27. 12 29. poss, 4 31. not poss 33. poss, 12 35. not poss 37. poss, 3 39. not poss 41. False; rectangle 43. Yes

Page 201

1. Follow constr top of p 199. 3. Follow constr top of p 200. 5. Constr a ⊥ to a given line; 90° ∠s are formed. 7. Follow steps of ex 1, p 199. 9. Constr 30° ∠; ∠ forming linear pr with this 30° ∠ is 150°. 11. Constr 15° ∠; at vertex constr ⊥ to one of rays of 15° ∠; compl of 15° ∠ is the 75° ∠. 13. Constr a 75° ∠; bis it; one of its parts is $37\frac{1}{2}^\circ$ ∠. 15. Constr a 45° ∠; bis it; that gives two $22\frac{1}{2}^\circ$ ∠s; the ∠ that forms a linear pr with

SELECTED ANSWERS

either of these is a $157\frac{1°}{2}$ ∠. **17.** Bisect the given seg; using this new length, follow steps of constr, p 200. **19.** Follow same steps as ex 18, except constr 150° ∠s instead of 135° ∠s and a reg dodecagon is constr. **21.** Follow steps as in ex. 18, except constr $157\frac{1°}{2}$ ∠s instead of 135 ∠s and a reg. 16-sided figure is constr. **23.** Draw a seg $4\frac{2}{3}$ cm; divide by 3; follow steps of constr 6.1 p 199, using seg ≅ to seg of $1\frac{5}{9}$ cm. **25.** Follow steps to constr a reg pentagon on p 201; constr ⊥ bis of each side, which also bis each of 5 ≅ arcs of ⊙; connect the 10 ≅ arcs with seg to form a reg decagon.

Cumulative Review

1. 18 **2.** 900° **3.** SSA (HL for rt △'s)

Page 204

1. Yes **3.** No, more than 2 seg with same endpt. **5.** convex **7.** concave **9.** 540° **11.** 360° **13.** 108° **15.** See ex 35 p 190. **17.** T **19.** F **21.** T **23.** 52°, 64°, 88°, 79°, 77° **25.** 25 **27.** Follow steps of constr, p 199, drawing only the ∠, not the △. **29.** Constr a 60° ∠, bis it, then bis one of the 30° ∠'s thus formed.

Page 210

1. $\overline{AC}$, $\overline{BD}$; yes **3.** $\overline{IK}$, $\overline{JL}$; no **5, 7.** Answers will vary. **9.** Yes **11.** No **13.** not necess **15.** opp. seg are ∥

Page 214

1. 60°, 120°, 60° **3.** 135, 45 **5.** 13 **7.** 17 **9.** ≅; SAS **11.** ≅; SAS **13.** not necess **15.** Given: ▱*ABCD*. Prove: ∠*A* and ∠*B* supp, ∠*B* and ∠*C* supp, ∠*C* and ∠*D* supp, ∠*D* and ∠*A* supp; $\overline{AB} \parallel \overline{DC}$, $\overline{AD} \parallel \overline{BC}$ (def of ▱) ∠*A* and ∠*B* supp, ∠*B* and ∠*C* supp, ∠*C* and ∠*D* supp, ∠*D* and ∠*A* supp (Th. 3.2) **17.** Given: ▱*ABCD*, diag $\overline{BD}$ and $\overline{AC}$. Prove: ∠*A* ≅ ∠*C*, ∠*D* ≅ ∠*B*; △*ABD* ≅ △*CDB*, △*DAC* ≅ △*BCA* (Th 7.1); ∠*A* ≅ ∠*C*, ∠*D* ≅ ∠*B* (CPCTC); $\overline{AB} \cong \overline{CD}$, $\overline{AD} \cong \overline{BC}$ (CPCTC) **19.** $\overline{WZ} \parallel \overline{XY}$ (def of ▱); ∠*VXU* ≅ ∠*VZT*, ∠*ZTV* ≅ ∠*VUX* (Post 13); $\overline{WZ} \cong \overline{XY}$ (Cor 7.1); $\overline{WZ} \cong 2\overline{TZ}$, $\overline{XY} \cong 2\overline{XU}$ (def of midpt); $2\overline{TZ} \cong 2\overline{XU}$ (subst prop); $\overline{TZ} \cong \overline{XU}$ (div prop); △*TVZ* ≅ △*UVX* (ASA); $\overline{TV} \cong \overline{VU}$, $\overline{ZV} \cong \overline{VX}$ (CPCTC); *V* is the midpt of $\overline{TU}$ and $\overline{XZ}$ (def of midpt)

Cumulative Review

1. $\frac{360°}{6}$; 60° **2.** $\frac{(12-2)180°}{12}$; 150° **3.** convex

Page 218

1. Yes **3.** Yes **5.** No **7.** No **9.** Yes; Th 7.5 **11.** Yes; def of ▱ **13.** Yes; def of ▱ **15.** Yes; Th 7.6 **17.** Given: ∠*A* ≅ ∠*C*, ∠*B* ≅ ∠*D*. Prove: *ABCD* is ▱; m ∠*A* + m ∠*B* + m ∠*C* + m ∠*D* = 360 (Th 6.1); 2 m ∠*A* + 2 m ∠*B* = 360, 2 m ∠*C* + 2 m ∠*B* = 360; (subst prop); m ∠*A* + m ∠*B* = 180, m ∠*C* + m ∠*B* = 180 (div prop); ∠*A* and ∠*B* are supp, ∠*C* and ∠*B* are supp (def of supp); $\overline{AD} \parallel \overline{BC}$, $\overline{AB} \parallel \overline{DC}$ (Th 3.4) **19.** $\overline{XU} \cong \overline{VZ}$ (CPCTC); *UXVZ* is a ▱ (Th 7.4) **21.** $\overline{AD} \cong \overline{AB} \cong \overline{DB}$, $\overline{DC} \cong \overline{BC} \cong \overline{DB}$ (def of equila △'s); $\overline{AD} \cong \overline{BC}$, $\overline{AB} \cong \overline{DC}$ (subst prop); *ABCD* is a ▱ (Th 7.4) **23.** Given: ∠*A* and ∠*B* supp, ∠*B* and ∠*C* supp, ∠*C* and ∠*D* supp, ∠*D* and ∠*A* supp. Prove: *ABCD* is a ▱; $\overline{AD} \parallel \overline{BC}$, $\overline{AB} \parallel \overline{DC}$ (Th 3.4); *ABCD* is a ▱ (def of ▱) **25.** Answers may vary. **27.** Given: $\overline{AE}$ bis ∠*A*, $\overline{CF}$ bis ∠*C*, *AFCE* is a ▱. Prove: *ABCD* is a ▱; ∠1 ≅ ∠3, $\overline{AE} \cong \overline{FC}$ (Cor. 1, 2, Th 7.1); ∠1 ≅ ∠2, ∠3 ≅ ∠4 (def of bis); ∠2 ≅ ∠4 (subst prop); $\overline{EC} \parallel \overline{AF}$ (def of ▱); ∠5 ≅ ∠1, ∠6 ≅ ∠3 (Post 13); ∠5 ≅ ∠6 (subst prop); △*ADE* ≅ △*CBF* (ASA); $\overline{AD} \cong \overline{BC}$, $\overline{DE} \cong \overline{FB}$ (CPCTC); $\overline{EC} \cong \overline{AF}$ (Th 7.1); $\overline{DC} \cong \overline{AB}$ (add prop); *ABCD* is a ▱ (Th 7.3)

Page 221

1. 5 **3.** $\overline{XZ}$ **5.** 2 **7.** 7 **9.** $\overline{ST} \parallel \overline{RQ}$, $ST = \frac{1}{2}RQ$ (Th 7.7); $RU = \frac{1}{2}RQ$ (def of a midpt); $\overline{ST} \cong \overline{RU}$ (subst prop); *RSTU* is a ▱ (Th 7.5) **11.** $\overline{DE} \cong \overline{EA}$, $\overline{DF} \cong \overline{FB}$ (def of midpt); ∠*DFE* ≅ ∠*CFB* (Th 2.10); △*DEF* ≅ △*BCF* (SAS); $\overline{BC} \cong \overline{DE}$, ∠*FBC* ≅ ∠*EDF* (CPCTC); $\overline{EA} \cong \overline{BC}$ (subst prop); $\overline{AD} \parallel \overline{BC}$ (Post 14); *EABC* is a ▱ (Th 7.5) **13.** $\overline{KM} \parallel \overline{GH}$, $KL = \frac{1}{2}GH$, $\overline{KM} \parallel \overline{JI}$, $LM = \frac{1}{2}JI$ (Th 7.7); $\overline{JI} \parallel \overline{GH}$ (Th 3.7); $\overline{KL} \cong \overline{LM}$ (def of midpt); $\frac{1}{2}GH = \frac{1}{2}JI$ (subst prop); $\overline{GH} \cong \overline{JI}$ (mult prop); *GHIJ* is ▱ (Th 7.5) **15.** Given: quad *WXYZ*, *A*, *B*, *C*, *D* are the midpts of $\overline{WX}$, $\overline{WZ}$, $\overline{YZ}$, $\overline{XY}$, respectively. Prove: $\overline{AC}$ and $\overline{BD}$ bis each other. Draw $\overline{AB}$, $\overline{BC}$, $\overline{CD}$, $\overline{DA}$, $\overline{XZ}$ (Post 1); $\overline{AB} \parallel \overline{XZ}$, $AB = \frac{1}{2}XZ$, $\overline{DC} \parallel \overline{XZ}$, $DC = \frac{1}{2}XZ$ (Th 7.7); $\overline{AB} \parallel \overline{DC}$ (Th 3.6); $\overline{AB} \cong \overline{DC}$ (subst prop); *ABCD* is a ▱ (Th 7.5); $\overline{AC}$

SELECTED ANSWERS

and $\overline{BD}$ bis each other (Th 7.2) **17.** $RS = \frac{1}{2}NO$, $TS = \frac{1}{2}QP$ (Th 7.7); $RS + TS = \frac{1}{2}(QP + NO)$ (add prop); $RT = \frac{1}{2}(QP + NO)$ (subst prop)

Page 225

1. 4 **3.** 12 **5.** 12 **7.** 9 **9.** $7\frac{1}{2}$ **11.** 9 **13.** 34 **15.** $\overleftrightarrow{AD} \parallel \overleftrightarrow{BE} \parallel \overleftrightarrow{CF}$ (Th 3.3); $\overline{DE} \cong \overline{EF}$ (Th 7.8) **17.** $\overleftrightarrow{AD} \parallel \overleftrightarrow{BE}$ (Th 3.3); $\overleftrightarrow{AD} \parallel \overleftrightarrow{BE} \parallel \overleftrightarrow{CF}$ (Th 3.7); $\overline{AB} \cong \overline{BC}$ (Th 7.8) **19.** Follow constr p 223. **21.** Follow constr p 223. **23.** Follow constr p 223, div seg into 5 ≅ seg, constr a seg ≅ to 4 of the 5 seg. **25.** $\overline{PX} \parallel \overline{QY}$ (Cor 3.3); $PXYQ$ is a ▱ (def of ▱); $\overline{PX} \cong \overline{QY}$ (Cor 7.1) **27.** Give counter example. **29.** Given: $\overline{PX} \perp$ m, $\overline{QY} \perp$ m, $\overline{PX} \cong \overline{QY}$. Prove: 1 ∥ m; $\overline{PX} \parallel \overline{QY}$ (Cor 3.3); $PXYQ$ is a ▱ (Th 7.5); l ∥ m (def of ▱)

Page 229

1. T **3.** T **5.** F **7.** P, Rh, R, S **9.** P, Rh, R, S **11.** R, S **13.** P, Rh, R, S **15.** R, S **17.** 105; 75 **19.** 8 **21.** Given: rect $ABCD$, diag. $\overline{AC}$ and $\overline{DB}$. Prove: $\overline{AC} \cong \overline{DB}$; $\overline{AD} \cong \overline{CB}$ (Cor 7.1); $\overline{AB} \cong \overline{AB}$ (reflex prop); $\angle DAB$, $\angle CBA$ are rt ∠'s (def of rect); $\angle DAB \cong \angle CBA$ (Th 2.5); $\triangle DAB \cong \triangle CBA$ (SAS); $\overline{AC} \cong \overline{DB}$ (CPCTC) **23.** Given: Rh $ABCD$, diag $\overline{AC}$ and $\overline{DB}$ intersect at E. Prove: $\overline{AC}$ bis $\angle DAB$, $\angle DCB$, and $\overline{DB}$ bis $\angle ADC$, $\angle ABC$; $\overline{AD} \cong \overline{DC} \cong \overline{CB} \cong \overline{BA}$ (def of Rh); $\overline{DB} \cong \overline{DB}$, $\overline{AC} \cong \overline{AC}$ (reflex prop); $\triangle ADC \cong \triangle ABC$, $\triangle DAB \cong \triangle DCB$ (SSS); $\angle 1 \cong \angle 2$, $\angle 3 \cong \angle 4$, $\angle 5 \cong \angle 6$, $\angle 7 \cong \angle 8$ (CPCTC); $\overline{AC}$ bis $\angle DAB$ and $\angle DCB$, $\overline{DB}$ bis $\angle ADC$ and $\angle ABC$ (def of bis) **25.** Given: square $ABCD$, diag $\overline{AC}$ and $\overline{DB}$. Prove: $\overline{AC} \cong \overline{DB}$. Same proof as ex 21.

Cumulative Review

1. 25 **2.** No, not △s in general. Yes, if △ is a rt △. **3.** A median of a △ is a seg drawn from a vertex to the midpoint of the opposite side.

Page 232

1. T **3.** F **5.** rect. **7.** rect **9.** cannot tell **11.** rect. **13.** rhombus **15.** $EFGH$ is a ▱ (Th 7.5); $EFGH$ is a rect. (If a ▱ has ≅ diag, then it is a rect.) **17.** Given ▱$ABCD$ with $\overline{DA} \cong \overline{AB}$. Prove: $ABCD$ is a rhombus; $\overline{DA} \cong \overline{CB}$, $\overline{AB} \cong \overline{DC}$ (Cor 7.1); $\overline{AB} \cong \overline{CB} \cong \overline{DC} \cong \overline{DA}$ (subst prop) $ABCD$ is rh (def of rh). **19.** Given: rhombus $ABCD$, E, F, G, H are midpts of $\overline{AB}$, $\overline{BC}$, $\overline{CD}$, $\overline{DA}$, respectively. Prove: $EFGH$ is a rect. Draw $\overline{AC}$, $\overline{DB}$ (Post 1); $\overline{HG} \parallel \overline{AC}$, $HG = \frac{1}{2}AC$; $\overline{EF} \parallel \overline{AC}$, $EF = \frac{1}{2}AC$ (Th 7.7); $\overline{HG} \parallel \overline{EF}$ (Th 3.7); $\overline{HG} \cong \overline{EF}$ (subst prop); $EFGH$ is a ▱ (Th 7.5); $\overline{DB} \perp \overline{AC}$ (diag of rh are ⊥); $\overline{HE} \parallel \overline{DB}$ (Th 7.7); $\overline{AC} \perp \overline{EH}$, $\overline{EF} \perp \overline{EH}$ (Ex. 15, p. 78); $\angle DEF$ is a rt ∠ (def ⊥ lines); $EFGH$ is rect (a ▱ with 1 rt ∠ is a rect) **21.** True, $EFGH$ is a ▱ (Th 7.3); $EFGH$ is a rh (def of a rh). **23.** Give counter-example.

Page 234

1. none of these **3.** parallelogram **5.** cannot tell **7.** rectangle **9.** cannot tell **11.** F **13.** F **15.** T **17.** F **19.** Follow constr p 223, div seg into 5 ≅ seg, constr a seg ≅ to 3 of the 5 seg. **21.** 7 **23.** 12 **25.** Same proof as ex 15, p 214.

Pages 240–241

1. $\overline{DE} \parallel \overline{AB}$ (Th 7.7); $ABED$ is a trap (def of trap) **3.** $\overline{LM} \parallel \overline{ON}$ (def of a trap); $\overline{LP} \cong \overline{MQ}$ (Th 7.9); $\angle OPL$, $\angle NQM$ are rt ∠'s (def of ⊥ lines); $\triangle OPL$, $\triangle NQM$ are rt △'s (def of rt △'s); $\triangle OPL \cong \triangle NQM$ (LA); $\overline{LO} \cong \overline{MN}$ (CPCTC) **5.** Yes **7.** No **9.** Yes **11.** No **13.** False **15.** False **17.** True. Given: trap $ABCD$, E, H, G, F are midpts of $\overline{AD}$, $\overline{DC}$, $\overline{CB}$, $\overline{BA}$, respectively. Prove $\overline{EG}$ and $\overline{FH}$ bisect each other. Draw $\overline{EF}$, $\overline{FG}$, $\overline{GH}$, $\overline{HE}$, $\overline{DB}$ (Post 1); $\overline{EF} \parallel \overline{DB}$, $EF = \frac{1}{2}DB$; $\overline{HG} \parallel \overline{DB}$, $HG = \frac{1}{2}DB$ (Th 7.7); $\overline{EF} \parallel \overline{HG}$ (Th 3.7); $\overline{EF} \cong \overline{HG}$ (subst); $EFGH$ is a ▱ (Th 7.5); $\overline{EG}$ and $\overline{FH}$ bis each other (Th 7.2). **19.** False, a rectangle

Cumulative Review

1. A rt △ with ≅ legs. **2.** A post. is accepted; a th. is proved. **3.** The side whose endpts are vertices of the two angles.

Algebra Review

1. $\frac{9a + 10}{a^2 - 9}$ **2.** $\frac{-x - 7}{x^2 - 4}$ **3.** $\frac{y + 18}{y^2 + 6y}$ **4.** $\frac{10 + 6c}{3}$ **5.** $\frac{17z + 17}{5z + 3}$ **6.** $\frac{12f^2 + 12f + 1}{4f^2 - 1}$ **7.** $\frac{3w^2 + 7w + 1}{w + 2}$ **8.** $\frac{4b^2 - b - 2}{4b^3 - b}$

Pages 244–245

1. 7 cm **3.** 15 cm **5.** 10 **7.** $11\frac{1}{2}$ **9.** 15

11. 8 **13.** 7 **15.** 14 **17.** $\overline{QM} \cong \overline{QP}$; $\angle POM \cong \angle NMO$; others are possible **19.** Given: trap $ABCD$ with med $\overline{EF}$, alt $\overline{HG}$; Prove: $\overline{EF}$ bis $\overline{HG}$; $\overline{EF} \parallel \overline{DC} \parallel \overline{AB}$ (Th 8.1); $\overline{DE} \cong \overline{EA}$ (def of med of trap); $\overline{HI} \cong \overline{IG}$ (Th 7.8) **21.** F **23.** F. Draw UA and AV 2 cm, XA and WV 1 cm. $\triangle UAX$ is not isos. **25.** No; they can have various lengths. **27.** Yes; all 3 seg are of same length; $a = \frac{1}{2}(a + a)$ and the 3 seg are $\parallel$.

Pages 248–249

1. 83° **3.** 109°, 42° **5.** 70°, 16° **7.** $BCDE$ is a trap (def of trap); $\angle C \cong \angle B$ (Th 5.1); $BCDE$ is an isos trap (Th 8.3) **9.** $\overline{DA} \cong \overline{CB}$ (def of isos trap); $\overline{AB} \cong \overline{AB}$ (reflex prop); $\angle DAB \cong \angle CBA$ (Th 8.2); $\triangle ABD \cong \triangle BAC$ (SAS) **11.** True. Given: isos trap $ABCD$ with $\overline{AB} \parallel DC$. Prove: $\angle A$ and $\angle C$ are supp, $\angle B$ and $\angle D$ are supp; $\angle A \cong \angle B$, $\angle C \cong \angle D$ (Th 8.2); $\angle A$ and $\angle D$ are supp, $\angle B$ and $\angle C$ are supp (Th 3.2); $\angle A$ and $\angle C$ are supp, $\angle B$ and $\angle D$ are supp (subst prop) **13.** True. Given: trap $ABCD$ with $\overline{AB} \parallel \overline{CD}$, $\angle A$ and $\angle C$ supp, $\angle B$ and $\angle D$ supp. Prove: $ABCD$ is isos trap; $\angle A$ and $\angle D$ are supp, $\angle B$ and $\angle C$ are supp (Th 3.2); $\angle A \cong \angle B$, $\angle D \cong \angle C$ (Th 2.8); $ABCD$ is isos trap (Th. 8.3). **15.** Given: trap $ABCD$ with $\overline{DC} \parallel \overline{AB}$, $\angle A \cong \angle B$, $\angle C \cong \angle D$. Prove: trap $ABCD$ is isos. Draw $\overline{DF} \perp \overline{AB}$, $\overline{CE} \perp \overline{AB}$ (Post 12) $\overline{DF} \cong \overline{CE}$ (Th 7.9); $\angle DFA$, $\angle CEB$ are rt $\angle$'s (def of $\perp$ lines); $\triangle DFA \cong \triangle CEB$ (LA); $\overline{DA} \cong \overline{CB}$ (CPCTC); Trap $ABCD$ is isos (def of isos trap). **17.** $\angle ABF$, $\angle CBF$ are rt $\angle$'s (def of $\perp$ lines); $\triangle ABF$, $\triangle CBF$ are ft $\triangle$'s (def of rt $\triangle$'s); $\overline{FB} \cong \overline{FB}$ (reflex); $\triangle ABF \cong \triangle CBF$ (LL); $\overline{AF} \cong \overline{FC}$ (CPCTC); $2\ \overline{AM} = \overline{AF}$, $2\ \overline{NC} = \overline{FC}$ (def of midpt); $2\ \overline{AM} \cong 2\ \overline{NC}$ (subst); $\overline{AM} \cong \overline{NC}$ (div prop); $\overline{MN} \parallel \overline{AC}$ (Th 7.7); $ACNM$ is an isos trap (def of isos trap).

Cumulative Review

2. Th 3.2 **3.** Follow steps constr p 15.

Pages 252–253

1. $\overline{AB} \cong \overline{JK}$, $\overline{BC} \cong \overline{KL}$, $\overline{CD} \cong \overline{LM}$, $\overline{DA} \cong \overline{MJ}$, $\angle A \cong \angle J$, $\angle B \cong \angle K$, $\angle C \cong \angle L$, $\angle D \cong \angle M$ **3.** quad $ABCD \cong$ quad $EFGH$ **5.** quad $TQRS \cong$ quad $UXWV$ **7.** $\angle DFE \cong \angle CFE$ (def of bis); $\overline{FE} \cong \overline{FE}$ (reflex); $\angle FEA \cong \angle FEB$ (Th 2.6); $\overline{AE} \cong \overline{EB}$ (def of bis); quad $AEFD \cong$ quad $BEFC$ (SASAS for $\cong$ quad's) **9.** $\overline{AB} \cong \overline{BC} \cong \overline{FE} \cong \overline{AF} \cong \overline{CD} \cong \overline{BE} \cong \overline{DE}$ (subst); $\angle 3 \cong \angle 4$ (Th 2.8); $\angle 2$, $\angle 4$ suppl (Th 3.2); $\angle 1 \cong \angle 2$ (Th 2.8); rh $ABEF \cong$ rh $CBED$ (SASAS) **11.** $\overline{DE} \cong \overline{FC}$, $\overline{EG} \cong \overline{GF}$, $\overline{DG} = \overline{GC}$; $\angle D \cong \angle C$, $\angle DEG \cong \angle CFG$ (CPCTC); $\overline{AD} \cong \overline{BC}$, $\overline{DF} \cong \overline{CE}$ (add prop); m $\angle AEG$ + m $\angle DEG = 180$, m $\angle BFD$ + m $\angle CFG = 180$ (Post 11); $\angle AEG$ and $\angle DEG$ are supp, $\angle BFD$ and $\angle CFG$ are supp (def of supp $\angle$'s); $\angle AEG \cong \angle BFD$ (Th 2.8); quad $EABC \cong$ quad $FBAD$ (SASAS) **13.** $\overline{HM} \cong \overline{JK}$, $\angle M \cong \angle K$, $\overline{ML} \cong \overline{KL}$ (CPCTC); $\angle MLI \cong \angle M$. $\angle KLI \cong \angle K$ (Th 8.2); $\angle MLI \cong \angle KLI$ (subst prop); $\overline{LI} \cong \overline{LI}$ (reflex) trap $HILM \cong$ trap $JILK$ (SASAS) **15.** $\angle A \cong \angle D$, $\angle ABE \cong \angle DEB$, $\overline{AB} \cong \overline{ED}$ (CPCTC); $\angle EGC \cong \angle FGB$ (Th 2.10); quad $ABGF \cong$ quad $DEGC$ (ASASA) **17.** $\angle 3 \cong \angle 7$ (substr. prop); $\angle ADC \cong \angle EHG$ (add prop) **19.** No. Consider a square with sides K units long and rhombus that is not a square with sides K units long. The two are not $\cong$.

Algebra Review

1. 11 **2.** −26 **3.** −7 **4.** −2 **5.** $\frac{5}{3}$ **6.** $\frac{17}{10}$

Page 255

1. Draw $\overline{AB}$ 7 cm. At each endpt const a $\perp$ Make $\overline{CA} \cong \overline{DB} \cong \overline{AB}$. Draw $\overline{CD}$ to complete sq. **3.** Follow steps of ex 1, p 254. **5.** Follow steps of ex 2, p 254. **7.** Let $\overline{XY} = 10$ cm. Bis $\overline{XY}$. Using half $\overline{XY}$, call it $\overline{AB}$ as a side of rect. At A and B constr $\perp$'s. Using A as a center and $\overline{XY}$ as a radius, constr an arc intersecting $\perp$ at B. Similarly using B as a center and $\overline{XY}$ as a radius, constr an arc intersecting $\perp$ at A. Connect these 2 pts of intersection for required rect. **9.** Using a base $\overline{AB} = s$, at A as center and s as radius constr an arc. Similarly, at B as center and s as radius draw another arc. Again using A as center and d as radius draw an arc, and using B as center and d as radius draw an arc. Pts of intersection of these arcs can be used to complete required trap. **11.** Constr $\overline{AB} = b$, $\overline{AD} = e$, m $\angle DAB = 45$; constr $\overline{DC} \parallel \overline{AB}$, $\overline{DC} = b$. Connect C to B. **13.** See proof of Th. 8.6, p. 251. **15.** No

Page 258

1. 3 **3.** 2 **5.** $\overline{AD} \cong \overline{CD}$, $\overline{AB} \cong \overline{CB}$, $\angle A \cong \angle C$ **7.** $\overline{QP} \cong \overline{KL}$, $\overline{PO} \cong \overline{LM}$, $\angle K \cong \angle Q$, $\angle L \cong \angle P$, $\angle M \cong \angle O$ **9.** Yes **11.** No **13.** Yes; no; if 2, then pentagon is regular and there will be 10; no; no; yes **15.** No; no **17.** No **19.** Given: Oct $ABCDEFGH$ with $\overleftrightarrow{AE}$ line of symmetry. Prove: $\angle B \cong \angle H$, $\angle C \cong G$, $\angle F \cong \angle D$. Draw $\overline{BH}$, $\overline{GC}$, $\overline{DF}$ inter $\overleftrightarrow{AE}$ at M, I, J (Post 1); $\overline{AH} \cong \overline{AB}$, $\overline{ED} \cong \overline{EF}$ (Th 8.8); $\overleftrightarrow{AE}$ bis $\overline{BH}$, $\overleftrightarrow{AE}$ bis $\overline{CG}$, $\overleftrightarrow{AE}$ bis $\overline{DF}$ (Post 21); $\overline{BM} \cong \overline{MH}$, $\overline{CI} \cong \overline{IG}$, $\overline{DJ} \cong \overline{JF}$ (def of bis); $\overline{AM} \cong \overline{AM}$, $\overline{JE} \cong \overline{JE}$ (reflex);

$\triangle ABM \cong \triangle AHM$, $\triangle EDJ \cong \triangle EFJ$ (SSS); $\angle BAM \cong \angle HAM$, $\angle DEJ \cong \angle FEJ$ (CPCTC); $\overline{AI} \cong \overline{AI}$, $\overline{EI} \cong \overline{EI}$ (reflex); $\angle AIC \cong \angle AIG$, $\angle EIC \cong \angle EIG$ (Th 2.6); quad $ABCI \cong$ quad $AHGI$, quad $EDCI \cong$ quad $EFGI$ (SASAS); $\angle B \cong \angle H$, $\angle BCI \cong \angle HGI$, $\angle D \cong \angle F$, $\angle DCI \cong \angle FGI$ (def $\cong$ quads); $\angle C \cong \angle G$ (add prop) **21.** Given: Sept $ABCDEFG$ with $\overleftrightarrow{AH}$ line of symmetry. Prove: $\overline{AG} \cong \overline{AB}$, $\overline{GF} \cong \overline{BC}$, $\overline{FE} \cong \overline{DC}$. Draw $\overline{GB}$, $\overline{FC}$ inters $\overleftrightarrow{AH}$ at I, J (Post 1); $\overleftrightarrow{AH}$ bis $\overline{GB}$, $\overleftrightarrow{AH}$ bis $\overline{FC}$, $\overleftrightarrow{AH}$ bis $\overline{ED}$ (Post 21); $\overline{AG} \cong \overline{AB}$ (Th 5.7); $\overline{GI} \cong \overline{IB}$, $\overline{FJ} \cong \overline{JC}$, $\overline{EH} \cong \overline{HD}$ (def of bis); $\overline{IJ} \cong \overline{IJ}$, $\overline{JH} \cong \overline{JH}$ (reflex); $\angle GIJ \cong \angle BIJ$, $\angle IJF \cong \angle IJC$, $\angle FJH \cong \angle CJH$, $\angle JHE \cong \angle JHD$ (Th 2.6); quad $GIJF \cong$ quad $BIJC$, quad $FJHE \cong$ quad $CJHD$ (SASAS); $\overline{GF} \cong \overline{BC}$, $\overline{FE} \cong \overline{DC}$ (def of $\cong$ quad)

Page 261

1. reflection **3.** reflection **5.** not reflection **7.** $\overline{AC} \cong \overline{EF}$; $\overline{AB} \cong \overline{ED}$; $\overline{CB} \cong \overline{FD}$, $\angle C \cong \angle F$, $\angle B \cong \angle D$, $\angle A \cong \angle E$ **9.** $\overline{YS} \cong \overline{VU}$, $\overline{SW} \cong \overline{UX}$, $\overline{YW} \cong \overline{VX}$, $\angle S \cong \angle U$, $\angle Y \cong \angle V$, $\angle W \cong \angle X$ **11.** It is the common $\perp$ bis of $\overline{GL}$, $\overline{HK}$, $\overline{IN}$, and $\overline{JM}$. **13.** T **15.** F **17.** Given: quad $ABCD$ and $A'B'C'D'$ are reflections; ℓ is the line of symmetry. Prove: quad $ABCD \cong$ quad $A'B'C'D'$. Draw $\overline{AC}$, $\overline{A'C'}$, $\overline{DB}$, $\overline{D'B'}$ (Post 1); ℓ is $\perp$ bis of $\overline{AA'}$, $\overline{BB'}$, $\overline{CC'}$, $\overline{DD'}$ (def of reflec); $\triangle ADC$ and $\triangle A'D'C'$ are refl, $\triangle DCB$ and $\triangle D'C'B'$ are refl (def of reflec); $\triangle ADC \cong \triangle A'D'C'$, $\triangle DCB \cong \triangle D'C'B'$ (Th 8.10); $\overline{AD} \cong \overline{A'D'}$, $\angle ADC \cong \angle A'D'C'$, $\overline{DC} \cong \overline{D'C'}$, $\angle DCB \cong \angle D'C'B'$, $\overline{BC} \cong \overline{B'C'}$ (CPCTC); quad $ABCD \cong$ quad $A'B'C'D'$ (SASAS)
19. Connect any two corresponding pts; then construct the $\perp$ bis of that seg.

Page 264

1. $\overline{EF}$ **3.** $\overline{AB}$, $\overline{DC}$ **5.** Isos **7.** F **9.** F **11.** T **13.** F **15.** T **17.** On line ℓ lay off $\overline{AB} \cong a$. At A constr a $\perp$. Lay off $\overline{AC} \cong b$. At C constr $\overleftrightarrow{CD} \parallel \overleftrightarrow{AB}$. Using A and B as centers and C as a radius, constr arcs intersecting $\overleftrightarrow{CD}$. Join these pts of intersection to complete the required trap. **19.** 11 **21.** $\overline{FC} \cong \overline{CB}$, $\overline{DC} \cong \overline{CG}$ (def of bis); $\angle FCD \cong \angle GCB$ (Th 2.10); $\angle F \cong \angle B$, $\angle D \cong \angle G$ (Post 13); $CGAB \cong CDEF$ (ASASA)

Pages 267–268

1. C **2.** B **3.** C **4.** A **5.** D **6.** C **7.** D **8.** D **9.** A **10.** B **11.** B **12.** A **13.** C **14.** $\overline{CB}$ **15.** $\overrightarrow{AC}$, $\overrightarrow{AD}$ **16.** F **17.** T **18.** F **19.** F **20.** T **21.** F **22.** T **23.** T **24.** T **25.** T **26.** F **27.** F **28.** 75, 105 **29.** 12 **30.** 15 **31.** See p 51 ex 2. **32.** See constr p 223. **33.** See p 114, ex 2. **34.** $\angle A \cong \angle A$ (reflex); $\triangle ACD \cong \triangle ABE$ (ASA); $\overline{EB} \cong \overline{DC}$ (CPCTC) **35.** $\overline{RS} \parallel \overline{UT}$ (def of $\square$); $\overline{UV} \parallel \overline{WS}$ (Th 3.3); $VSWU$ is a $\square$ (def of $\square$) **36.** See Ex 15, p. 221.

Pages 273–274

1. T **3.** T **5.** F **7.** T **9.** 3, 10; 2, x; 15 **11.** 9, 4; x, 12; 3 **13.** 8, 9; 12, x; 6 **15.** 9, 6; 4, x; $13\frac{1}{2}$ **17.** $\frac{2}{3} = \frac{4}{6}$; $\frac{3}{2} = \frac{6}{4}$; $\frac{4}{2} = \frac{6}{3}$ **19.** $\frac{4}{2} = \frac{8}{4}$; $\frac{2}{4} = \frac{4}{8}$; $\frac{8}{4} = \frac{4}{2}$ **21.** $\frac{10}{5} = \frac{y}{3}$; $\frac{10}{x} = \frac{5}{3}$; $\frac{5}{10} = \frac{3}{x}$ **23.** $\frac{4}{3} = \frac{x}{z}$; $\frac{z}{3} = \frac{x}{y}$; $\frac{y}{x} = \frac{3}{z}$ **25.** 2 **27.** 10 **29.** $\sqrt{21}$ **31.** $\sqrt{77}$ **33.** F **35.** T **37.** $\frac{9}{7}$ **39.** $\frac{5}{7}$ **41.** $\frac{21}{14}$ **43.** $\frac{b}{a}$ **45.** 36°, 54° **47.** 110, 132 **49.** 6 cm, 4 cm **51.** 10, 12 **53.** T; $ad = bc$ (mult prop); $\frac{a}{c} = \frac{b}{d}$ (div prop) **55.** T; $\frac{a}{b} + \frac{b}{b} = \frac{c}{d} + \frac{d}{d}$; $\frac{a+b}{b} = \frac{c+d}{d}$ (add prop) **57.** T; $\frac{a}{b} = \frac{c}{d}$; $\frac{b}{a} = \frac{d}{c}$ (recip prop); $\frac{a}{a} + \frac{b}{a} = \frac{c}{c} + \frac{d}{c}$; $\frac{a+b}{a} = \frac{c+d}{c}$ (add prop); $\frac{a}{a+b} = \frac{c}{c+d}$ (recip prop) **59.** T; $\frac{a}{b} \cdot \frac{n}{n} = \frac{c}{d} \cdot \frac{m}{m}$ (prop of 1) **61.** $\frac{a}{b} \cdot bd = bd \cdot \frac{c}{d}$; $ad = bc$ (mult prop)

Pages 276–277

1. No **3.** Yes **5.** $\frac{AB}{EF} = \frac{BC}{FG} = \frac{CD}{GH} = \frac{AD}{EH}$; $\angle A \cong \angle E$, $\angle B \cong \angle F$, $\angle C \cong \angle G$, $\angle D \cong \angle H$ **7.** $y = 10$; $x = 12$ **9.** $z = 8\frac{1}{3}$; $y = 6\frac{2}{3}$; $x = 3\frac{1}{3}$ **11.** $x = 6\frac{1}{4}$; $y = 6\frac{2}{5}$; $z = 4\frac{4}{5}$ **13.** 2; 1 **15.** 8; 12 **17.** 91; 68; 177 **19.** No; corres S's may not be $\cong$. **21.** Given: sq $ABCD$ and sq $XYZW$. Prove: sq $ABCD \sim$ sq $XYZW$. $\overline{AB} \cong \overline{BC} \cong \overline{CD} \cong \overline{DE}$; $\overline{XY} \cong \overline{YZ} \cong \overline{ZW} \cong \overline{WX}$ (prop of sq); $AB = BC = CD = DE$; $XY = YZ = ZW = WZ$ (def of $\cong$); $\frac{AB}{XY} = \frac{BC}{YZ} = \frac{CD}{ZW} = \frac{DE}{WX}$ (div prop); $\angle$s A, B, C, D, X, Y, Z, W are rt $\angle$s (prop of sq); $\angle A \cong \angle X$, $\angle B \cong \angle Y$, $\angle C \cong Z$, $\angle D \cong \angle W$ (all rt $\angle$s are $\cong$); $ABCD \sim XYZW$ (def of $\sim$ poly)

Pages 280–281

1. No; corr. $\angle$s $\not\cong$ **3.** Yes; (Post 22)

5. Yes; (Post 22) **7.** $\angle D \cong \angle D$ (reflex prop); $\triangle ABD \sim \triangle CED$ (Post 22) **9.** $\angle O \cong \angle M$ (Post 13); $\angle K \cong \angle N$ (Post 13); $\triangle OKL \sim \triangle MNL$ (Post 22) **11.** 10 **13.** No
15. $\angle GJI \cong \angle KHI$ ($\perp$s form $\cong$ rt $\angle$s); $\angle I \cong \angle I$ (reflex prop); $\triangle HKI \sim \triangle JGI$ (Post 22);
17. $\angle V \cong \angle V$ (reflex prop); $\overline{WT} \parallel \overline{RS}$ (def of $\square$); $\angle VTW \cong \angle S$ (Th 3.1); $\triangle RSV \sim \triangle WVT$ (Post 22) **19.** $\angle LMN \cong \angle JKN$ and $\angle MLN \cong \angle KJN$ (Post 13) $\triangle LMN \sim \triangle JKN$ (Post 22); $\frac{MN}{NK} = \frac{LN}{NJ}$ (def of $\sim$ poly) **21.** F **23.** F
25. $XYZW$ is a $\square$ (Th 7.5); $\overline{WX} \parallel \overline{ZY}$ (def of $\square$); $\angle XWV \cong \angle ZTW$ (Post 13); $\angle WXV \cong \angle TZW$ (Th 7.3); $\triangle WTZ \sim \triangle VWX$ (Post 22)
27. $\angle DEA$ a rt $\angle$ (def of $\perp$); $\angle DEA \cong \angle BCA$ (all rt $\angle$s $\cong$); $\angle A \cong \angle A$ (reflex prop); $\triangle ABC \sim \triangle ADE$ (Post 22) $\frac{AD}{AB} = \frac{DE}{BC}$ (def $\sim$ poly); $\overline{BC} \perp \overline{AC}$ (def $\perp$ L's); $\overline{BC} \parallel \overline{DE}$ (Cor 3.3); $\angle BCD \cong \angle CDE$ (Post 13); $\angle BCD \cong \angle DCE$ (def $\angle$bis); $\angle CDE \cong \angle DCE$ (subst); $\overline{DE} \cong \overline{EC}$ (Th 5.2); $\frac{AD}{AB} = \frac{EC}{BC}$ (subst)

Cumulative Review
1. 70 **2.** 79

Page 284
1. T **3.** T **5.** F **7.** $2\frac{2}{3}$ **9.** $8\frac{3}{4}$ **11.** $5\frac{3}{5}$
13. 26 **15.** 12 **17.** $\overline{AC} \parallel \overline{ED}$ (Th 3.3); $\frac{BD}{DC} = \frac{BE}{EA}$ (Th 9.1) **19.** $\frac{AE}{EB} = \frac{DG}{GB}$ (Th 9.1) $\overline{DA} \parallel \overline{CB}$ (def $\square$); $\overline{EF} \parallel \overline{CB}$ (Th 3.7); $\frac{DG}{GB} = \frac{DF}{FC}$ (Th 9.1); $\frac{AE}{EB} = \frac{DF}{FC}$ (subst)
21. Given $AB \parallel CD \parallel EF$. Prove: $\frac{AC}{CE} = \frac{BD}{DF}$. Draw $\overleftrightarrow{AF}$ (Post 1); $\frac{AC}{CE} = \frac{AG}{GF}$ (Th 9.1); $\frac{BD}{DF} = \frac{AG}{GF}$ (Th 9.1); $\frac{AC}{CE} = \frac{BD}{DF}$ (trans prop)

Cumulative Review
1. 12 **2.** Follow constr steps p 15.
3. Corres $\angle$'s $\cong$ and corres s's propor.

Pages 288–289
1. Yes; SSS **3.** Yes; AA **5.** No; $\cong$ $\angle$s are not included $\angle$s. **7.** No; only 1 pr $\cong$ $\angle$s
9. Yes; SAS **11.** Yes **13.** No **15.** No
17. $\sim$ **19.** $\nsim$ **21.** $\nsim$ **23.** $\angle ZTS \cong \angle ZXY$ (corr $\angle$s are $\cong$); $\angle STR \cong \angle Y$ (opp $\angle$s of $\square$ are $\cong$); $\angle TSZ \cong \angle Y$ (corr $\angle$s are $\cong$); $\angle STR \cong \angle XRT$ (Post 13) $\angle TSZ \cong \angle XRT$ (subst); $\triangle XRT \sim \triangle TSZ$ (Post 22) **25.** $\angle NPM \cong \angle NRQ$ ($\perp$s form $\cong$ rt $\angle$s); $\angle RNQ \cong \angle RNQ$ (reflex prop); $\triangle NQR \sim \triangle NMP$ (Post 22)
27. $\angle LRN \cong \angle LRM$ ($\perp$s form $\cong$ rt $\angle$s); $\triangle LNR \sim \triangle LMR$ (Post 22) **29.** $\angle R \cong \angle WUV$, $\angle S \cong \angle WVU$ (Th 3.1); $\triangle UVW \sim \triangle RST$ (Post 22) **31.** $\overline{TU} \cong \overline{TV}$, $\overline{TR} \cong \overline{TS}$ (CPCTC); $\frac{TU}{TR} = \frac{TV}{TS}$ (div) **33.** $\angle CTQ \cong \angle RJV$ ($\perp$s form $\cong$ rt $\angle$s); $\angle CQT \cong \angle CUT$ (Th 5.1); $\triangle CTQ \sim \triangle RJV$ (Post 22); $\frac{CQ}{RV} = \frac{QT}{JV}$ (def $\sim$ poly)
35. Case I – Given: $\triangle$s ABC, DEF, $\angle B \cong \angle E$, $\overline{AB} \cong \overline{BC}$, $\overline{DE} \cong \overline{EF}$. Prove: $\triangle ABC \sim \triangle DEF$. $\frac{AB}{DE} = \frac{BC}{EF}$ (div prop); $\triangle ABC \sim \triangle DEF$ (Th 9.3) Case II – Given: $\triangle$s ABC, DEF, $\overline{AB} \cong \overline{BC}$, $\overline{DE} \cong \overline{EF}$, $\angle A \cong \angle D$. Prove: $\triangle ABC \sim \triangle DEF$. $\angle A \cong \angle C$, $\angle D \cong \angle F$ (Th 5.1); $\angle C \cong \angle F$ (subst); $\triangle ABC \sim \triangle DEF$ (Post 22)
37. Given: $\triangle$s ABC, DEF, $\frac{AB}{DE} = \frac{BC}{EF} = \frac{AC}{DF}$. Prove: $\triangle ABC \sim \triangle DEF$. Constr $\overline{DB'} \cong \overline{AB}$, then draw $\overline{B'C'} \parallel \overline{EF}$ (Post 15); $\angle B' \cong \angle E$, $\angle C' \cong \angle F$ (Th 3.1); $\triangle DB'C' \sim \triangle DEF$ (Post 22); $\frac{DB'}{DE} = \frac{DC'}{DF} = \frac{B'C'}{EF}$ (def $\sim$ poly); $\frac{BC}{EF} = \frac{B'C'}{EF}$, $\frac{DC'}{DF} = \frac{AC}{DF}$ (subst, trans prop); $\overline{AC} \cong \overline{DC'}$, $\overline{BC} \cong \overline{B'C'}$ (mult prop); $\triangle ABC \cong \triangle DB'C'$ (Post 16); $\angle B' \cong \angle B$; $\angle C \cong \angle C'$ (CPCTC); $\triangle ABC \sim \triangle DEF$ (Post 22)
39. The $\perp$'s to each side of a $\triangle$ form a $\triangle$ $\sim$ to the given $\triangle$. See ex. 24, pp 280, 281.

Page 293
1. $8\frac{2}{5}$ **3.** 4; 8 **5.** 6 **7.** 12; 8
11. $\frac{AD}{AB} = \frac{DE}{EB}$, $\frac{DC}{CB} = \frac{DE}{EB}$ (Th 9.7); $\frac{AD}{AB} = \frac{DC}{CB}$ (subst) **13.** $\angle s \cong \angle w$ (def $\sim$ poly); $\angle TUS \cong \angle XYW$ ($\perp$s form $\cong$ rt $\angle$s); $\triangle TUS \sim \triangle XYW$ (Post 21); $\frac{TU}{XY} = \frac{US}{YW}$ (def $\sim$ poly) **15.** Follow constr 9.1, p 292. **17.** Given: $\triangle ABC \sim \triangle EFG$, with corr alts $\overline{AD}$ and $\overline{EH}$. Prove: $\frac{AC}{EG} = \frac{AD}{EH}$. $\angle C \cong \angle G$ (def $\sim$ poly); $\angle ADC \cong \angle EHG$ ($\perp$s form $\cong$ rt $\angle$'s); $\triangle ADC \sim \triangle EHG$ (Post 21); $\frac{AC}{EG} = \frac{AD}{EH}$ (def $\sim$ poly) **19.** Given: $\triangle ABC \sim \triangle DEF$, $\overline{BG}$, $\overline{EJ}$ alts, $\overline{BH}$, $\overline{EK}$ meds. Prove: $\frac{BH}{EK} = \frac{BG}{EJ}$. $\frac{BH}{EK} = \frac{AB}{ED}$ (Th 9.5); $\frac{BG}{EJ} = \frac{AB}{ED}$ (Th 9.6); $\frac{BH}{EK} = \frac{BG}{EJ}$ (subst)

Pages 295–296

(Answers may vary 1–3.)

1. $\frac{XY}{AB} = \frac{CD}{ZW}; \frac{AB}{XY} = \frac{ZW}{CD}$ **3.** $\frac{RP}{AJ} = \frac{WF}{RP}; \frac{AJ}{RP} = \frac{RP}{WF}$ **5.** 558 **7.** $\angle F \cong \angle I, \angle G \cong \angle J$ (Post 13); $\triangle FHG \sim \triangle IHJ$ (Post 22); $\frac{GH}{FH} = \frac{JH}{IH}$ (def ~ poly); $GH \cdot IH = FH \cdot JH$ **9.** $\angle Y \cong \angle X$ ($\perp$s form $\cong$ rt $\angle$s); $\angle YVU \cong \angle XVW$ (Th 2.10); $\triangle YVU \sim \triangle XUW$ (Post 22) $\frac{VY}{VX} = \frac{VU}{VW}$ (def ~ poly); $VY \cdot VW = VX \cdot VU$ (prop of propor) **11.** $\overline{CA}$ bisects $\angle ZCB$ (def $\angle$ bisector); $\frac{ZA}{AB} = \frac{ZC}{BC}$ (Th 9.7); $ZA \cdot BC = ZC \cdot AB$ (prop of propor) **13.** $DC = 6$; $FB \cdot FC = 63$

17. $\angle JKY \cong \angle b$ (Post 13); $\angle JKY \cong \angle a$ (trans prop); $\overline{JY} \cong \overline{JK}$ (Th 5.2); $\frac{XK}{KZ} = \frac{XJ}{JY}$ (Th 9.3); $\frac{XK}{KZ} = \frac{XJ}{JK}$ (subst); $XK \cdot JK = KZ \cdot XJ$ (prop of propor) **19.** $\frac{ED}{DA} = \frac{EF}{FB}$ (Th 9.3); $\overline{FB} \cong \overline{CB}$ (Th 5.2); $\frac{ED}{DA} = \frac{EF}{CB}$ (subst); $ED \cdot CB = DA \cdot EF$ (prop of propor)

Algebra Review

1. $6\sqrt{2}$ **2.** $8\sqrt{3}$ **3.** $\sqrt{2}$ **4.** $3\sqrt{5}$ **5.** $-\sqrt{3}$ **6.** $23\sqrt{2}$ **7.** 5 **8.** 8 **9.** $70\sqrt{6}$ **10.** 8 **11.** 2 **12.** $10\sqrt{3}$

Page 299

1. $10\frac{2}{7}$ **3.** 6; 9 **5.** 18 **7.** $\frac{BX}{BA} = \frac{BZ}{BF}$; $\frac{BY}{BC} = \frac{BZ}{BF}$ (def of ~ poly); $\frac{BX}{BA} = \frac{BY}{BC}$ (subst); $\angle XBY \cong \angle XBY$ (reflex prop); $\triangle XBY \sim \triangle ABC$ (Th 9.3) **9.** $\angle TAS \cong \angle ZAY$ (Th 2.10); $\triangle ATS \sim \triangle AZY$ (Th 9.3) **11.** $\overline{EF} \parallel \overline{AB}$ (Th 3.8); $\frac{DF}{DB} = \frac{DE}{DA}$ (Th 9.1); $\angle EDF \cong \angle EDF$ (reflex); $\triangle DEF \sim \triangle DAB$ (Th 9.3); similarly, show $\triangle DFG \sim \triangle DBC$ and $\triangle DEG \sim \triangle DAC$; then, $\frac{EF}{AB} = \frac{DF}{DB} = \frac{DE}{DA}, \frac{FG}{BC} = \frac{DF}{DB} = \frac{DG}{DC}$, and $\frac{GE}{CA} = \frac{DG}{DC} = \frac{DE}{DA}$ (def ~ poly); $\frac{EF}{AB} = \frac{FG}{BC} = \frac{GE}{CA}$ (subst); $\triangle EFG \sim \triangle ABC$ (Th 9.4)

Page 302

1. $\frac{8}{12} = \frac{x}{15}$; $12x = 120$; $x = 10$ **3.** $\frac{x}{y} = \frac{8}{3}$ **5.** $\frac{9}{x} = \frac{x}{16}$; $x^2 = 144$ $x = 12$ **7.** $\frac{7}{4} = \frac{\frac{x}{252}}{x}$; $4x^2 = 1764$, $x^2 = 441$; $x = 21$; 21, 12 **9.** No, sides are not proportional. **11.** $\frac{4}{6} = \frac{12}{x}$; $x = 18$ **13.** $\angle DCB \cong \angle DCB$ (reflex prop); $\angle CEA \cong \angle CBA$ (all rt $\angle$s are $\cong$); $\triangle CPE \sim \triangle CBD$ (Post 22) **15.** $DF = \frac{1}{2}CB$ and $DE = \frac{1}{2}AB$ (Th 7.7); $\frac{DF}{CB} = \frac{1}{2} = \frac{DE}{AB}$; $DE \cdot CB = DF \cdot AB$ (prop of propor)

Page 309

1. 2 **3.** 9 **5.** 4 **7.** 3 **9.** 8 **11.** 9 **13.** $8\sqrt{2}$ **15.** 10 **17.** $\triangle BEC \sim \triangle CEA$ and $\triangle CDE \sim \triangle CEA$ (Th 10.1); $\triangle CDE \sim \triangle BEC$ (trans prop) **19.** Given: $\triangle ABC$, $\angle C$ a rt $\angle$, alt $\overline{CD}$. Prove: $\frac{AD}{CD} = \frac{CD}{BD}$. $\triangle ACD \sim \triangle CBD$ (Th 10.1); $\frac{AD}{CD} = \frac{CD}{BD}$ (def ~ poly) **21.** Yes; $20(16) = KZ^2$; $8\sqrt{5}$; **23.** Yes; $6^2 = 9(\overline{HY})$; $HY = 4$; $\frac{XY}{ZY} = \frac{ZY}{HY}$; $ZY = 2\sqrt{13}$ **25.** No **27.** Given: rt $\triangle ABC$, $\overline{CD}$ alt to hypot. Prove: $\frac{AC}{AD} \neq \frac{CB}{DB}$. $\triangle ACD \sim \triangle BCD$ (Th 10.1); $\frac{AC}{CB} = \frac{AD}{CD}$; (def of ~ polys); $\frac{AC}{AD} = \frac{CB}{CD}$ (propor prop); If Th true, then $DB = CD$, not always true. Thus, false. **29.** $\frac{r}{h} = \frac{h}{s}$ (Th. 10.1); $\frac{u}{h} = \frac{h}{u + v}$ (Th 10.1); $h^2 = r \cdot s$ and $h^2 = u(u + v)$ (prop of propor); $r \cdot s = u(u + v)$ (trans prop) **31.** $\frac{r}{h} = \frac{h}{s}$ (Th 10.1); $h^2 = r \cdot s$ (prop of propor);

Pages 312–313

1. 10 **3.** 12 **5.** 8 **7.** 7 **9.** 5 **11.** 9 **13.** 7 **15.** $\sqrt{2}$ **17.** 10 mm **19.** 36 cm **21.** $d = s\sqrt{2}$ **23.** $9\frac{3}{13}$ **25.** $12\frac{1}{2}$ **27.** $5\sqrt{3}$ **29.** $(AD)^2 = (AE)^2 + (DE)^2$, $(BC)^2 = (EB)^2 + (EC)^2$ (Th 10.2); $(AD)^2 + (BC)^2 = (AE)^2 + (DE)^2 + (EB)^2 + (EC)^2$ (add'n); $(AE)^2 + (EB)^2 = (AB)^2$, $(DE)^2 + (EC)^2 = (DC)^2$ (Th 10.2); $(AD)^2 + (BC)^2 = (AB)^2 + (DC)^2$ (subst) **31.** $RU^2 = RT^2 + TU^2$ (Th 10.2); $TS^2 = SR^2 + RT^2$; $RU^2 - TU^2 = RT^2$ (subst); $TS^2 - RS^2 = RT^2$ (subst); $RU^2 - TU^2 = TS^2 - RS^2$ (trans) **33.** 26 cm **35.** $d = s\sqrt{3}$ **37.** $\frac{\sqrt{1,463}}{18}$

Pages 316–317

1. No **3.** No **5.** Yes; $\angle B$ **7.** No **9.** Yes; $\angle B$ **11.** Yes **13.** Yes **15.** $\frac{a^2b^2}{a^2} + \frac{a^2b^2}{b^2} = c^2$ (mult); $b^2 + a^2 = c^2$ (alg); $\triangle ABC$ is a rt $\triangle$ (Th 10.3). **17.** No **19.** No **21.** No **23.** $\frac{AD}{AC} = \frac{AC}{AB}$ (def geom mean); $(AC)^2 = (AD)(AB)$ (prop of propor); $(AD)^2 + (CD)^2 = (AC)^2$ (Th 10.2); $(AD)^2 + (CD)^2 = (AD)(AB)$ (subst); $(CD)^2 = (AD)(AB) - (AD)^2$ (subt); $(CD)^2 = (AD)(AB - AD)$ (distr prop); $(CD)^2 = (AD)(DB)$ (subst); $\frac{AD}{CD} = \frac{CD}{DB}$ (div); $\triangle ABC$ is a rt $\triangle$ (ex. 22). **25.** $a^2 + b^2 = c^2$ or $a^2 + b^2 \neq c$; if $a^2 + b^2 = c^2$, then $\triangle ABC$ is a rt $\triangle$ (Th 10.3); this contradicts the given, so $a^2 + b^2 \neq c^2$

Cumulative Review

1. 70°, 40° **2.** 30°, 60°

Algebra Review

1. $15\sqrt{2}$ **2.** $6\sqrt{2}$ **3.** $8\sqrt{3}$ **4.** $\sqrt{2}$ **5.** $3\sqrt{5}$ **6.** $6\sqrt{2}$ **7.** $8\sqrt{5}$ **8.** $-\sqrt{3}$ **9.** $23\sqrt{2}$ **10.** 5 **11.** 8 **12.** $70\sqrt{6}$ **13.** 8 **14.** 2 **15.** $10\sqrt{3}$

Page 320

1. $BC = 4$, $AC = 4\sqrt{3}$ **3.** $LK = 6$, $JK = 6\sqrt{2}$ **5.** $XZ = YZ = 6\sqrt{2}$ **7.** $XY = 6$, $YZ = 3\sqrt{2}$ **9.** $XY = 4\sqrt{2}$, $XZ = 4$ **11.** $BC = 18$, $CA = 18\sqrt{3}$ **13.** $AB = 18$, $CA = 9\sqrt{3}$ **15.** $AB = 12$, $BC = 6$ **17.** $9\sqrt{3}$ **19.** $JK = 16\sqrt{2}$, $LM = 8\sqrt{2}$ **21.** Given: rt $\triangle ABC$, m $\angle B = 45$, m $\angle C = 45$, $\overline{AB} = \ell$. Prove: $BC = \ell\sqrt{2}$. $\overline{AB} \cong \overline{AC}$ (Th 5.2); $\overline{AC} = \ell$ (def $\cong$); $(BC)^2 = \ell^2 + \ell^2$ (Th 10.2); $(BC)^2 = 2\ell^2$ (alg); $BC = \ell\sqrt{2}$ (alg) **23.** $d^2 = s^2 + s^2$ (Th 10.2); $d^2 = 2s^2$ (alg); $d = s\sqrt{2}$ (alg) **25.** m $\angle FGH = 30$ (diag of ▱ bis $\angle$s); $\angle FHG$ is a rt $\angle$ (diag of rhomb $\perp$ each other); m $\angle GFD = 60$ (*ac* $\angle$s of rt $\triangle$ are comp) If $FH = s$, $GF = 2s$, $GH = s\sqrt{3}$ (Th 10.5); $DG = DE = EF = FG = 2s$ (rhomb is equilat); $GE = 2s\sqrt{3}$, $DF = 2s$ (diag ▱ bis ea other) $\frac{GD + DE + EF + FG}{(GE) \cdot (DF)} = \frac{8s}{(2s\sqrt{3})(2s)} = \frac{2}{s\sqrt{3}} \neq \frac{1}{2\sqrt{3}}$ (subst and alg)

Pages 322–323

1. 1.333, 0.800, 0.600, 0.750, 0.600, 0.800 **3.** 0.343, 0.324, 0.946, 2.917, 0.946, 0.324 **5.** 1.118, 0.745, 0.667, 0.894, 0.667, 0.745 **7.** 1.732 **9.** 0.500 **11.** 0.707 **13.** 0.436, 0.400, 0.917, 2.291, 0.917, 0.400 **15.** 0.354, 0.333, 0.943, 2.828, 0.943, 0.333 **17.** 0.333, 0.316, 0.949, 3.000, 0.949, 0.316 **19.** $\sin B = \frac{b}{c}$; $\cos A = \frac{b}{c}$; $\sin B = \cos A$ **21.** $\tan B = \frac{b}{a}$; $\tan A = \frac{a}{b}$; $\frac{1}{\tan A} = \frac{b}{a}$; $\tan B = \frac{1}{\tan A}$ **23.** $\tan A = \frac{a}{b}$; $\tan (90 - A) = \tan B = \frac{b}{a}$; $\tan A = \frac{1}{\tan B}$; $\tan A = \frac{1}{\tan (90 - A)}$

Page 325

1. 41° **3.** 65° **5.** 54° **7.** 32° **9.** 15° **11.** 30° **13.** T **15.** F **17.** T **19.** F **21.** F **23.** T **25.** T **27.** $\sin A = \frac{a}{c}$; $c > a$; $\frac{a}{c} < 1$; **29.** $\tan A = \frac{\sin A}{\cos A}$; $\frac{\sin A}{\cos A} = \frac{\frac{a}{c}}{\frac{b}{c}} = \frac{a}{b}$; $\tan A$

Pages 327–328

1. $\sin 40° = \frac{7}{c}$; 10.9 **3.** $\tan A = \frac{12}{7}$; 60° **5.** $\sin 75° = \frac{b}{20}$; 19.3 **7.** $\cos 24° = \frac{12}{c}$; 13.1 **9.** $\tan 18° = \frac{b}{c}$; 1.9 **11.** $\sin B = \frac{9}{15}$; 37° **13.** $\cos A = \frac{15}{20}$; 41° **15.** $\cos 40° = \frac{a}{18}$; 13.8 **17.** $\sin 49° = \frac{17}{c}$; 22.5 **19.** $\sin 38° = \frac{a}{10}$; 6.2 **21.** $\sin A = \frac{3}{5}$; 37° **23.** $\tan A = \frac{14}{17}$; 39° **25.** m $\angle A = 90 - 35 = 55$; $\tan 35° = \frac{AC}{18}$; $AC = 12.6$; $\cos 35° = \frac{18}{AB}$; $AB = 22.0$ **27.** m $\angle B = 90 - 27 = 63$; $\tan 27° = \frac{BC}{15}$; $BC = 7.6$; $\cos 27° = \frac{15}{AB}$; $AB = 16.8$ **29.** m $\angle B = 90 - 56 = 34$; $\sin 56° = \frac{BC}{32}$; $BC = 26.5$; $\cos 56° = \frac{AC}{32}$; $AC = 17.9$ **31.** $\tan A = \frac{1}{4}$; m $\angle A = 14$; m $\angle B = 76$ **33.** $\tan 27° = \frac{10}{AC}$; $AC = 19.6$; $\sin 27° = \frac{10}{AB}$; $AB = 22.0$; $p = 10 + 19.6 + 22.0 = 51.6$ **35.** $\sin 28° = \frac{ST}{20}$; $ST = 9.39$; $\sin 51° = \frac{9.39}{SU}$; $SU = 12.1$

Pages 330–331

1. $\tan 25° = \frac{x}{100}$; 46.6 m **3.** $\sin 10° = \frac{1}{x}$;

5.8 m **5.** $\tan 38° = \frac{x}{30}$; 23.4 m **7.** $\sin 73 = \frac{x}{5}$; 4.8 cm **9.** $\sin 55° = \frac{L}{6}$; $L = 4.9$ cm; $\cos 55° = \frac{w}{6}$; $w = 3.4$ cm **11.** $\tan 26° = \frac{35}{d}$; 71.8 m **13.** $\sin 75° = \frac{h}{150}$; 144.9 m **15.** $\sin 34° = \frac{h}{8}$; 4.5 mm **17.** $\cos 52° = \frac{EB}{9}$; $EB = 5.5$; $AB = 17.5$ **19.** $\tan \frac{1}{2}A = \frac{9}{12}$; $\frac{1}{2}A = 37$; $m\angle A = 74$; $m\angle C = 74$; $m\angle B = 106 = m\angle D$ **21.** $\cos 20° = \frac{h}{10}$; 9.4 cm **23.** $\cos A = \frac{9}{14}$; $m\angle A = 50° = m\angle B$; $m\angle C = 80$ **25.** $\cos A = \frac{9}{12}$; 41°, 41°, 98° **27.** $\cos A = \frac{4.5}{12}$; 68°, 68°, 112°, 112°

Page 334

1. $\triangle ABC \sim \triangle ACD \sim \triangle CBD$ **3.** $8^2 + 5^2 = 89$; $\sqrt{89}$ **5.** Yes **7.** No **9.** $\sqrt{6^2 + 6^2 + 6^2} = 6\sqrt{3}$ **11.** $6^2 + 6^2 = 72$; $6\sqrt{2}$; 6 **13.** $\tan = \frac{opp}{adj} = \frac{3}{4}$ **15.** 0.5299 **17.** 28° **19.** $\frac{LJ}{10} = 0.5543$; 5.543 **21.** $\tan 42° = \frac{x}{20}$; 18.0 m **23.** $\frac{4}{y} = \cos 52°$; $\frac{4}{.6157}$; 6.5

Pages 340–341

1. rad: $\overline{OA}, \overline{OD}, \overline{OC}, \overline{OB}$; diam: $\overline{AB}, \overline{CD}$ **3.** ch: $\overline{AB}$; no secs or tans **5.** $\triangle AOB \cong \triangle COB$ (SSS) **7.** 6 **9.** 12; $12\sqrt{3}$ **11.** $7\sqrt{2}$; 14 **13.** 9; $9\sqrt{2}$ **15.** 26 **17.** $\overline{OA} \cong \overline{OB}$ (Th 11.1); $\angle OAB \cong \angle OBA$ (Th 5.1) **19.** Given: $\odot O$, radii $\overline{OA}, \overline{OB}$. Prove: $\overline{OA} \cong \overline{OB}$. $OA = OB$ (def of $\odot$); $\overline{OA} \cong \overline{OB}$ (def of $\cong$ seg) **21.** $\angle DOA \cong \angle BOC$ (Th 2.10); $\overline{OD} \cong \overline{OB} \cong \overline{OA} \cong \overline{OC}$ (Th 11.1); $\triangle DOA \cong \triangle BOC$; (Post 17); $\overline{AD} \cong \overline{BC}$ (CPCTC) **23.** Draw $\overline{OR}, \overline{OS}$; $\overline{OR} \cong \overline{OS}$ (Th 11.1); $\overline{OM} \cong \overline{OM}$ (reflex prop); $\triangle OMR \cong \triangle OMS$ (Post 16); $\angle OMR \cong \angle OMS$ (CPCTC); $\overline{XY} \perp \overline{RS}$ (Th 2.5); $\angle TNO \cong \angle OMS$ (Post 13); $\angle OMS$ a rt $\angle$ ($\perp$s form rt $\angle$'s); $\angle TNO$ is a rt $\angle$ (subst); $\overline{OY} \perp \overline{TU}$ (def $\perp$); $\overline{OY}$ bis $\overline{TU}$ (Th 11.2); N is the midpt of $\overline{TU}$ (def of midpt) **25.** Draw rad $\overline{OC}, \overline{OF}, \overline{OD}, \overline{OG}$; $\overline{OH} \cong \overline{OH}$ (reflex); $\overline{OF} \cong \overline{OG}$ (Th 11.1); $\overline{FH} \cong \overline{HG}$ (def bis); $\triangle OFH \cong \triangle OGH$ (Post 16); $\angle OHF \cong \angle OHG$ (CPCTC); $\overline{AB} \perp \overline{FG}$ (Th 2.5); $\overline{OE} \cong \overline{OE}$ (reflex); $\overline{OC} \cong \overline{OD}$ (Th 11.1); $\overline{CE} \cong \overline{ED}$ (def bis); $\triangle OCE \cong \triangle ODE$ (Post 16); $\angle CEO \cong \angle DEO$ (CPCTC); $\overline{AB} \perp \overline{CD}$ (Th 2.5); $\angle OHF \cong \angle DEO$ ($\perp$s form $\cong$ rt $\angle$'s); $\overline{CD} \parallel \overline{FG}$ (Post 14) **27.** Given: $\odot O$, chord $\overline{AB}$, $\overleftrightarrow{DE} \perp$ bis of $\overline{AB}$ at D. Prove: O lies on $\overleftrightarrow{DE}$. Draw $\overline{OA}, \overline{OB}$; $\overline{OA} \cong \overline{OB}$ (Th 11.1); $\overline{AD} \cong \overline{DB}$ (def of bis); $\overline{OD}$ is $\perp$ bis of $\overline{AB}$ (Th 5.8) O lies on $\overline{DE}$ (Post 12). **29.** Given: $\odot O$, $\overline{AD} \cong \overline{AC}$, $\overline{BC} \cong \overline{BD}$. Prove: $\overline{AB}$ is a diam of $\odot O$. $\overline{AB} \perp$ bis of $\overline{CD}$ (Th 5.8); O is on $\overline{AB}$ (ex 27) $\overline{AB}$ is a diam of $\odot O$ (def of diam)

Cumulative Review

1. $\angle SUT$, $\angle SVR$ are rt $\angle$'s ($\perp$s form rt $\angle$'s); $\angle SUT \cong \angle SVR$ (all rt $\angle$'s $\cong$); $\angle S \cong \angle S$ (reflex) $\triangle RSV \sim \triangle TSV$ (Post 22) **2.** $\angle ZWY \cong \angle XYW$, $\angle WZX \cong \angle ZXY$ (Post 13); $\triangle ZWP \sim \triangle XYP$ (Post 22)

Pages 343–344

1. $\cong$ **3.** $\cong$ **5.** $\not\cong$ **7.** $\not\cong$ **9.** $BC = 16$ **11.** 70, 70 **13.** $WX = 6$ **15.** $VY = 4$ **17.** $\overline{AM} \cong \overline{BN}$, $\overline{RM} \cong \overline{MS} \cong \overline{UN} \cong \overline{NT}$ **19.** $\overline{OA} \cong \overline{OB} \cong \overline{QC} \cong \overline{QD}$; $\overline{AE} \cong \overline{EB}$; $\overline{CF} \cong \overline{FD}$ **21.** $\overline{OX} \cong \overline{OY} \cong \overline{OW} \cong \overline{OZ}$ (Th 11.1); $\triangle XOY \cong \triangle ZOW$ (Post 17); $\overline{XY} \cong \overline{ZW}$ (CPCTC); $\overline{OT} \cong \overline{OU}$ (Th 11.3) **23.** $\angle XTO \cong \angle ZUO$ ($\perp$s form $\cong$ rt $\angle$'s); $\overline{OX} \cong \overline{OZ}$ (Th 11.1); $\triangle XOT \cong \triangle ZOU$ (Th 4.2); $\overline{OT} \cong \overline{OU}$ (CPCTC); $\overline{XY} \cong \overline{WZ}$ (Th 11.4) **25.** Given: $\odot O$, $\odot P$, Ch. $\overline{AB} \cong$ Ch. $\overline{DE}$, $\overline{OC} \perp \overline{AB}$, $\overline{PF} \perp \overline{DE}$, $\overline{OC} \cong \overline{PF}$. Prove: $\odot O \cong \odot P$. Draw $\overline{AO}$, $\overline{DP}$; $\angle ACO \cong \angle DFP$ ($\perp$s form $\cong$ rt $\angle$'s) $\overline{OC}$ bis $\overline{AB}$ and $\overline{PF}$ bis $\overline{DE}$ (Th 11.2); $AC = \frac{1}{2}AB$, $DF = \frac{1}{2}DE$ (def bis); $AC = DF$ (trans prop); $\triangle ACO \cong \triangle DFP$ (Post 17); $\overline{OA} \cong \overline{DP}$ (CPCTC) $\odot O \cong \odot P$ (def $\cong$ $\odot$) **27.** Given: $\triangle ABC$ inscr in $\odot O$, $\overline{OD} \perp \overline{AB}$, $\overline{OE} \perp \overline{BC}$, $\overline{OF} \perp \overline{AC}$, $\overline{OD} \cong \overline{OE} \cong \overline{OF}$; Prove: $\triangle ABC$ is equilateral. $\overline{AB} \cong \overline{BC} \cong \overline{AC}$ (Th 11.4); $\triangle ABC$ is equilat (def).

Pages 349–350

1. 11 **3.** 17 **5.** 28 **7.** $\overline{QP} \cong \overline{RP} \cong \overline{SP}$ **9.** $\overline{FO} \cong \overline{IO}$; $\angle EFO \cong \angle GFO \cong \angle HIO \cong \angle JIO$ **11.** Draw $\overline{OX}$; $\overline{OX} \perp \overline{AB}$ (Th 11.5); $\overline{OX}$ bis $\overline{AB}$ (Th 11.2); $\overline{AX} \cong \overline{BX}$ (def of seg bis) **13.** $\overline{AP} \cong \overline{BP}$ (Th 11.7); $\angle PAB \cong \angle PBA$ (Th 5.1); $\overline{OA} \cong \overline{OB}$ (Th 11.1); $\angle OAB \cong \angle OBA$ (Th 5.1); $\overline{OA} \perp \overline{AP}$, $\overline{OB} \perp \overline{BP}$ (Th 11.5); $\angle OAP$, $\angle OBP$ are rt $\angle$'s ($\perp$s form rt $\angle$'s); $m\angle OAB + m\angle PAB = 90$, $m\angle OBA + m\angle PBA = 90$ ($\angle$ add, subst); $m\angle OAB + m\angle OBA + m\angle PAB + m\angle PBA = 180$ (add); $m\angle P + m\angle PAB + m\angle PBA = 180$ (Th 3.5); $m\angle OAB + m\angle OBA + m\angle PAB + m\angle PBA = m\angle P + m\angle PAB + m\angle PBA$ (trans); $m\angle P =$

m $\angle OAB$ + m $\angle OBA$ (subtr); m $\angle P = 2$ m $\angle OAB$ (subst) **15.** $\overline{OA} \cong \overline{OD}$ (Th 11.1); $\angle OAC \cong \angle ODC$ (Th 5.1); $\overline{OD} \perp \overline{DE}$ (Th 11.5); $\angle ODE$ is a rt $\angle$ ($\perp$s form rt $\angle$'s); $\angle ODC$ comp of $\angle EDO$ (def of comp $\angle$'s); $\angle OAC$ comp of $\angle ACO$ (ac $\angle$'s of rt $\triangle$'s are comp); $\angle ACO \cong \angle EDC$ (Th 2.9); $\angle ACO \cong \angle ECD$ (Th 2.10); $\angle EDC \cong \angle ECD$ (subst) **17.** $AX = 2$; $BY = 5$; $CZ = 3$ **19.** 24

Cumulative Review

1. $b + \frac{1}{2}a$ **2.** $\frac{b+d}{2}$

Pages 353–354

1. Yes **3.** No **5.** 60 **7.** 120 **9.** 50 **11.** 30 **13.** $\widehat{AB}$, $\widehat{BC}$, $\widehat{DC}$, $\widehat{DA}$, $\widehat{AC}$ **15.** $\widehat{DCB}$, $\widehat{DAB}$ **17.** $\overline{UX}$; $\overline{TW}$ **19.** 79 **21.** 101 **23.** 143 **25.** 71.5 **27.** circle cannot have 370° **29.** Draw $\overline{OC}$, $\overline{OD}$; $\overline{OC} \perp \overline{CP}$, $\overline{OD} \perp \overline{DP}$ (Th 11.5); $\angle OCP$, $\angle ODP$ are rt $\angle$'s ($\perp$s form rt $\angle$'s); m $\angle COD + x + 90 + 90 = 360$ (Th 6.1); m $\angle COD = 180 - x$ (subt); m $\angle COD = m\widehat{CD}$; $m\widehat{CD} = 180 - x$ (subst)

31. $\tan 66° = \frac{BC}{8}$ 18.0 **33.** $m\widehat{TSQ} = 180$ (meas semi $\odot$ = 180); $m\widehat{TSQ} = m\widehat{TS} + m\widehat{SQ}$ (Post 23); $m\widehat{SQ} = 180 -$ m $\angle P$ (ex 29); $m\widehat{TS} + 180 -$ m $\angle P = 180$ (subst); $m\widehat{TS} =$ m $\angle P$ (add) **35.** $m\widehat{RS} = 180 -$ m $\angle RPS$, $m\widehat{ST} = 180 -$ m $\angle TPS$ (ex 29); m $\angle RPT$ = m $\angle RPS$ + m $\angle TPS$ (Post 9); $m\widehat{RS} + m\widehat{ST} = 360 -$ m $\angle RPT$ (add, subst)

Cumulative Review

1. $5\sqrt{2}$ **2.** $7\sqrt{2}$

Pages 357–358

1. No **3.** No **5.** $\widehat{CD}$, $\widehat{AB}$; $\widehat{AD}$, $\widehat{CB}$; $\widehat{DBC}$, $\widehat{BDA}$; $\widehat{ABD}$, $\widehat{CDB}$ **7.** None **9.** $\widehat{BC} \cong \widehat{HI}$ **11.** $\overline{RS} \cong \overline{RT}$ (Th 11.8); $\angle RST$ is isos (def) **13.** $\overline{XZ} \cong \overline{YZ}$ (Th 5.2); $\widehat{XZ} \cong \widehat{YZ}$ (Th 11.9) **15.** $\overline{KL} \cong \overline{LM}$ (Th 11.8); $\angle KOL \cong \angle MOL$ (Th 11.8); m $\angle KOL$ + m $\angle NOK = 180$, m $\angle NDM$ + m $\angle MOL = 180$ (Post 11); $\angle KOL$ suppl to $\angle NOK$, $\angle MOL$ suppl to $\angle NOM$ (def); $\angle NOK \cong \angle NOM$ (Th 26); $\widehat{NK} \cong \widehat{NM}$ (Th 11.10); $\triangle NKL \cong \triangle NML$ (Post 16); $\angle KNL \cong \angle MNL$ (CPCTC) **17.** Given: $\odot O \cong \odot P$, cord $\overline{AB} \cong$ cord $\overline{CD}$. Prove: $\angle O \cong \angle P$, $\widehat{AB} \cong \widehat{CD}$. $\overline{AO} \cong \overline{CP} \cong \overline{OB} \cong \overline{PD}$ (def $\cong \odot$); $\triangle OAB \cong \triangle CPD$ (Post 16); $\angle O \cong \angle P$ (CPCTC); $m\widehat{AB}$ = m $\angle O$, $m\widehat{CD}$ = m $\angle P$ (meas of cent $\angle$ = meas its int arc); $m\widehat{AB} \cong m\widehat{CD}$ (subst); $\widehat{AB} \cong \widehat{CD}$ (def $\cong$ $\widehat{s}$) **19.** Draw $\overline{OS}$; $\angle SUO \cong \angle SVO$ ($\perp$ form $\cong$ rt $\angle$'s); $\angle ROS \cong \angle TOS$ (Th 11.8); $\overline{OS} \cong \overline{OS}$ (reflex); $\triangle OSU \cong \triangle OSV$ (Th 4.2); $\overline{SU} \cong \overline{SV}$ (CPCTC) **21.** $x\sqrt{3}$ **23.** Draw $\overline{AD}$, $\overline{BC}$; $m\widehat{AD} = m\widehat{AC} + m\widehat{CD}$, $m\widehat{CB} = m\widehat{CD} + m\widehat{DB}$ (Post 23) $m\widehat{AC} + m\widehat{CD} = m\widehat{DB} + m\widehat{CD}$ (add); $m\widehat{AD} = m\widehat{CB}$ (subst); $\overline{AD} \cong \overline{CB}$, $\overline{AC} \cong \overline{BD}$ (Th 11.8); $\overline{CD} \cong \overline{CD}$ (reflex); $\triangle ACD \cong \triangle BDC$ (Post 16); $\angle ACD \cong \angle BDC$ (CPCTC) **25.** Draw $\overline{DO}$; $\angle EOD \cong \angle D$ (Post 13); $\angle ADE \cong \angle B$ (Th 3.1); $\overline{DO} \cong \overline{BO}$ (Th 11.1); $\angle D \cong \angle B$ (Th 5.1); $\angle AOE \cong \angle EOD$ (subst); $\widehat{AE} \cong \widehat{ED}$ (Th 11.10)

Pages 362–363

1. 90 **3.** m $\angle I$ = m $\angle J = 56$ **5.** m $\angle J = 67$; m $\angle K = 94$ **7.** m $\angle ACD = 32$; $m\widehat{BC} = 64$ **9.** 43 **11.** 12 **13.** 156 **15.** 66 **17.** 122 **19.** 142 **21.** 38 **23.** m $\angle PNO = 16.25$; m $\angle ONM = 57.5$ **25.** $\overline{OQ} \cong \overline{OR}$ (Th 11.1); $\angle PRQ \cong \angle SQR$ (Th 5.1) **27.** $\widehat{RU} \cong \widehat{ST}$ (Th 11.11, Cor 3); $\widehat{UT} \cong \widehat{UT}$ (reflex); $m\widehat{RU} + m\widehat{UT} = m\widehat{RT}$, $m\widehat{ST} + m\widehat{TU} = m\widehat{SU}$ (Post 23); $m\widehat{RU} + m\widehat{UT} = m\widehat{ST} + m\widehat{TU}$ (add); $m\widehat{RT} = m\widehat{SU}$ (subst); $\widehat{RT} \cong \widehat{SU}$ (def $\cong$ $\widehat{s}$); $\overline{RT} \cong \overline{SU}$ (Th 11.8) **29.** Follow ex. 28; then $\frac{AD}{AC} = \frac{CD}{CB}$ (def ~ poly); $AD \cdot CB = AC \cdot CD$ (prop of propor) **31.** Draw diam $\overline{PC}$; m $\angle CPB = \frac{1}{2}m\widehat{CB}$, m $\angle APC = \frac{1}{2}m\widehat{CA}$ (case I); m $\angle CPB -$ m $\angle APC = \frac{1}{2}m\widehat{CB} - \frac{1}{2}m\widehat{CA}$ (subtr); m $\angle CPB -$ m $\angle APC$ = m $\angle APB$ (Post 10); m $\angle APB = \frac{1}{2}(m\widehat{CB} - m\widehat{CA})$ (subst); $m\widehat{CB} - m\widehat{CA} = m\widehat{AB}$ (Post 23); m $\angle APB = \frac{1}{2}m\widehat{AB}$ (subst)

33. m $\angle ABC = \frac{1}{2}m\widehat{AC}$ (Th 11.11); $m\widehat{AC} = 180$ (semi $\odot$ meas 180); m $\angle ABC = \frac{1}{2}(180)$ or 90 (subst); $\angle ABC$ is a rt $\angle$ (an $\angle$ of 90° is rt $\angle$) **35.** Given: $\odot O$, $\angle ABC$ inscr in maj $\widehat{ABC}$; Prove: m $\angle ABC < 90$. $m\widehat{ABC} > 180$ (def maj); $m\widehat{ABC} + m\widehat{AC} = 360$ ($\odot$ contains 360°); $m\widehat{AC} < 180$ (subtr); m $\angle ABC = \frac{1}{2}m\widehat{AC}$ (Th 11.11); m $\angle ABC < 90$ (subst) **37.** Given: $\odot O \cong \odot P$, $\triangle ABC$ inscr in $\odot O$, $\triangle DEF$ inscr in $\odot P$, $\angle A \cong \angle D$, $\angle B \cong \angle E$. Prove: $\triangle ABC \cong \triangle DEF$. m $\angle B = \frac{1}{2}m\widehat{AC}$, m $\angle E = \frac{1}{2}m\widehat{DF}$ (Th 11.11); $\frac{1}{2}m\widehat{AC} = \frac{1}{2}m\widehat{DF}$ (subst); $m\widehat{AC} = m\widehat{DF}$ (mult); $\widehat{AC} \cong \widehat{DF}$ (def $\cong$ $\widehat{s}$); $\overline{AC} \cong \overline{DF}$ (Th 11.8); $\triangle ABC \cong \triangle DEF$ (Th 4.2) **39.** false; rectangle

Cumulative Review

1. F **2.** T **3.** T **4.** T **5.** T **6.** T

Pages 366–367

1. 50 **3.** 60 **5.** 50 **7.** 30 **9.** 11 **11.** 84 **13.** 6; 109 **15.** 62; 45 **17.** T **19.** 24 **21.** m $\angle STQ = 90$ ($\perp$s form rt $\angle$'s); m $\angle STQ = \frac{1}{2}(m\widehat{PR} + m\widehat{QS})$ (Th 11.12); $90 = \frac{1}{2}(m\widehat{PR} + m\widehat{QS})$ (subst); $m\widehat{PR} + mQS = 180$ (mult) **23.** $\angle N \cong \angle Q$ (Th 11.11, Cor 1); $\angle L \cong \angle L$ (reflex); $\triangle LQM \sim \triangle LNP$ (Post 22) **25.** 100 **27.** Given: $\odot O$ and $\odot P$ tang at T, O lies on $\odot P$, chord $\overline{AT}$ int $\odot P$ at B. Prove: $\overline{TB} \cong \overline{AB}$. OT is diam of $\odot P$ (Th 11.5) Draw $\overline{OB}$; $\angle OBT$ is a rt $\angle$ (Th 11.11, Cor 2); $\overline{OB} \perp \overline{AT}$ (sides of a rt $\angle$ are $\perp$); $\overline{OB}$ bis $\overline{AT}$ (Th 11.2); $\overline{TB} \cong \overline{AB}$ (def)

Pages 370–372

1. 76 **3.** 48 **5.** 71 **7.** 55 **9.** 32 **11.** 141 **13.** 90 **15.** 99 **17.** 72 **19.** 50 **21.** 60; 30 **23.** 10.75; 79.25 **25.** Draw $\widehat{AB}$; m $\angle z =$ m $\angle x +$ m $\angle y$ (Th 4.2); m $\angle x =$ m $\angle z -$ m $\angle y$ (subtr); m $\angle z = \frac{1}{2}m\widehat{ACB}$, m $\angle y = \frac{1}{2}\widehat{AB}$ (Th 11.14); m $\angle x = \frac{1}{2}m\widehat{ACB} - \frac{1}{2}m\widehat{AB}$ (subst); m $\angle x = \frac{1}{2}(m\widehat{ACB} - m\widehat{AB})$ (distr prop) **27.** $\angle UVS \cong \angle VST$ (Post 13); m $\angle UVS =$ m $\angle VST$ (def $\cong$ $\angle$'s); m $\angle UVS = \frac{1}{2}m\widehat{US}$ (Th 11.11); m $\angle VST = \frac{1}{2}m\widehat{US}$ (Th 11.14); $\frac{1}{2}m\widehat{SU} = \frac{1}{2}m\widehat{SV}$ (subst); $m\widehat{US} = m\widehat{SV}$ (mult); $\overline{US} \cong \overline{SV}$ (Th 11.8) **29.** Given: $\odot O$, $\odot Q$ internally tan at P, $\overleftrightarrow{PE}$ the common tan, $\overline{PA}$ int $\odot Q$ at B. Prove: $m\widehat{PA} = m\widehat{PB}$. m $\angle APE \cong$ m $\angle BPE$ (reflex); m $\angle APE = \frac{1}{2}m\widehat{PA}$, m $\angle BPE = \frac{1}{2}\widehat{PB}$ (Th 11.14); $\frac{1}{2}m\widehat{PA} = \frac{1}{2}m\widehat{PB}$ (subst); $m\widehat{PA} = m\widehat{PB}$ (mult) **31.** $\angle LMS \cong \angle RMN$ (Th 2.10); m $\angle LMS = \frac{1}{2}m\widehat{LM}$, m $\angle RMN = \frac{1}{2}m\widehat{NM}$ (Th 11.14); $\frac{1}{2}m\widehat{LM} = \frac{1}{2}m\widehat{NM}$ (subst); $m\widehat{LM} = m\widehat{NM}$ (mult) **33.** Given: $ABCD$ insc in $\odot O$, $\angle EAD$ is an ext $\angle$ of $ABCD$. Prove: $\angle EAD \cong \angle C$. $\angle BCD$ suppl to $\angle BAD$ (Th 11.11, Cor 4); m $\angle EAD +$ m $\angle BAD = 180$ (Post 12); $\angle EAD$ supp to $\angle BAD$ (def supp $\angle$'s); $\angle EAD \cong \angle C$ (Th 2.8) **35.** Given: $\odot O$, $\odot Q$, ext tang at P, $\overleftrightarrow{PT}$ tan to $\odot O$ and $\odot Q$ at P, $\overleftrightarrow{AB}$ a common ext tang to $\odot O$ and $\odot Q$, chs $\overline{AP}$, $\overline{PB}$. Prove: $\angle APB$ is a rt $\angle$. $\overline{TA} \cong \overline{TP}$, $\overline{TB} \cong \overline{TP}$ (Th 11.7); $\overline{TA} \cong \overline{TP} \cong \overline{TB}$ (trans); T is center of a $\odot$ with rad $\overline{AT}$, $\overline{PT}$, $\overline{BT}$ (def of $\odot$); $\angle APB$ is a rt $\angle$ ($\angle$ inscr in semi $\odot$ is a rt $\angle$)

Pages 375–376

1. 3 **3.** 18 **5.** 12 **7.** 6 **9.** 12 **11.** $13\frac{1}{3}$; 11 **13.** $7\frac{1}{3}$; $6\frac{2}{3}$ **15.** 12 **17.** Draw $\overline{AB}$, $\overline{AC}$; $\angle P \cong \angle P$ (reflex); m $\angle PAC = \frac{1}{2}m\widehat{AC}$ (Th 11.14); m $\angle B = \frac{1}{2}m\widehat{AC}$ (Th 11.11); m $\angle PAC =$ m $\angle B$ (subst); $\angle PAC \cong \angle B$ (def $\cong$ $\angle$'s); $\triangle PAC \sim \triangle PCA$ (Post 21); $\frac{PA}{PB} = \frac{PC}{PA}$ (def $\sim$ poly); $(PA)^2 = PB \cdot PC$ (prop of propor) **19.** $\angle T \cong \angle T$ (reflex); $\angle K \cong \angle M$ (Th 11.11, Cor 1); $\triangle KLT \sim \triangle MJT$ (Post 21); $\frac{KL}{MJ} = \frac{TL}{TJ}$ (def $\sim$ poly); $KL \cdot TJ = TL \cdot MJ$ (prop of propor) **21.** $(TM)^2 = MS \cdot MP$, $(TM)^2 = MR \cdot MQ$ (Th 11.19); $MS \cdot MP = MR \cdot MQ$ (subst) **23.** Draw $\overline{AB}$; m $\angle DGE = \frac{1}{2}(m\widehat{ED} + m\widehat{BC})$; m $\angle ABD = \frac{1}{2}m\widehat{AED}$ (Th 11.11); $m\widehat{AED} = m\widehat{AE} + m\widehat{ED}$ (Post 23); m $\angle ABD = \frac{1}{2}(m\widehat{AE} + m\widehat{ED})$ (subst); m $\angle ABD = \frac{1}{2}(m\widehat{BC} + m\widehat{ED})$ (subst); m $\angle DGE =$ m $\angle ABD$ (subst); $\angle DGE \cong \angle ABD$ (def); $\angle D \cong \angle D$ (reflex); $\triangle DGF \sim \triangle DBA$ (Post 21); $\frac{AD}{BD} = \frac{GD}{FD}$ (def $\sim$ poly); $AD \cdot FD = BD \cdot GD$ (prop of propor)

Cumulative Review

1. 10 cm **2.** 3

Page 379

1. Open compass to measure the 2-cm seg; use this rad and constr a $\odot$. **3.** Use rad 3 cm; draw $\odot$; draw a rad of $\odot$ and extend it; constr $\perp$ to rad at its endpt on $\odot$. **5. (a)** Constr a $\odot$; with same rad, mark off 6 consec $\cong$ arcs; from each pt as ctr, with same rad, draw the 6 arcs in $\odot$. **(b)** Constr a $\odot$ and mark off 6 $\cong$ ⌒s as in 5(a); draw 3 diams (use alternate pts); bisect a rad; use $\frac{1}{2}r$ as rad and midpts of radii as ctrs; draw 3 $\odot$. **7.** Constr $\perp$ to seg at given pt; measure given seg with compass; mark off this length on $\perp$; use this pt as ctr and same rad to draw $\odot$. **9.** Constr $\perp$ bis of 2 nonadj sides of sq; use pt of int as ctr, $\frac{1}{2}s$ as rad, and draw ⊙. **11.** Follow constr on p 377. Let R be int of $\overline{OP}$ with ⊙O. Constr $\overleftrightarrow{RQ} \perp \overline{OP}$, int $\overleftrightarrow{TP}$ at Q. Bis $\angle RQP$, with $\overrightarrow{QU}$ int $\overline{OP}$ at U. With $\overline{RU}$ as rad, draw ⊙U. **13.** Draw a circle. Constr ⊥ bis of diam. Connect intersect with endpt of diam. Constr ⊥ bis of four sides. Connect Intersects with Endpt of each side. **15.** Inscr reg hex; constr ⊥ bis of each side; where these int ⊙, connect pts to vert of hex. **17.** P, D, F non coll, draw $\overrightarrow{PD}$ and $\overrightarrow{PF}$; on $\overrightarrow{PD}$, mark $PA = a$, $AB = b$; on $\overrightarrow{PF}$, mark $PC = c$; draw $\overline{AC}$; constr $\overleftrightarrow{BE} \parallel \overline{AC}$, to int $\overrightarrow{PF}$ at E; $CE = x$. **19.** Follow steps on p 378; $x = \sqrt{ab}$ **21.** ex 19 with a replaced by $4a$ **23.** ex 19 with b replaced by $\frac{1}{2}a$

Page 382

1. $\overline{OF}, \overline{OE}, \overline{GO}$ **3.** $\overline{EG}$ **5.** $\overleftrightarrow{BA}$ **7.** $\overline{AD} \perp$ bis $\overline{BC}$ (Th 11.2); $AC = AB$ (Th 5.7); $\overline{AC} \cong \overline{AB}$ (def); $\angle C \cong \angle B$ (Th 5.1); **9.** 64 **11.** $97 = \frac{1}{2}(91 \tan \widehat{BD})$; 103 **13.** $47 = \frac{1}{2}(m\widehat{AD} - 21)$; 115 **15.** 40 **17.** 14 **19.** Follow steps p 377. **21.** ⊙$O \cong$ ⊙P (def); $\overline{OA} \cong \overline{OB}$; $\overline{PC} \cong \overline{PD}$ (Th 11.11); $\angle A \cong \angle B$, $\angle C \cong \angle D$ (Th 5.1); $\angle B \cong \angle D$ (subst); $\angle O \cong \angle P$ (Th 3.6); $\widehat{AB} \cong \widehat{CD}$ (Th 11.10)

Pages 389–390

1. 22.75m² **3.** 24.51 cm² **5.** $7\sqrt{5}$m² **7.** 25 cm² **9.** $(x+y)^2$ or $x^2 + 2xy + y^2$ **11.** $x^2 + 2xy + y^2$ **13.** 52 cm **15.** $\sqrt{5}$ m **17.** $x + 1$ **19.** 48 **21.** $6\sqrt{10}$ **23.** $36\sqrt{3}$ cm² **25.** 128 cm² **27.** 60 mm² **29.** 76 m² **31.** 256 cm² **33.** $A = bh$, $A' = b'h'$ (Post 24); $A' = b'h$ (subst); $\frac{A}{A'} = \frac{bh}{b'h}$ (div); $\frac{A}{A'} = \frac{b}{b'}$ (alg) **35.** $A = 5^2$, $A' = (25)^2$ (area of sq is s^2), $A' = 4 \cdot 5^2$ (alg); $A' = 4A$ (subst). The statement is disproved.

Algebra Review

1. $\frac{b}{5}$ **2.** $k - 4$ **3.** $\frac{t-s}{r}$ **4.** $6a$ **5.** b **6.** $\frac{3}{rx-2}$ **7.** $3 - x$ **8.** $8 - 2x$ **9.** $3x - 12$ **10.** $\frac{-2c}{4b-3}$ **11.** $\frac{4z}{6y+3}$ **12.** $\frac{3t}{5r-7}$ **13.** $\frac{3p}{-6p+2}$ **14.** $\frac{-2b}{bc-4}$ **15.** $\frac{2yz}{3+5yz}$

Pages 393–394

1. 12 m² **3.** 8 cm² **5.** 9.69 km² **7.** 2.5 cm² **9.** 16 cm **11.** 1.6 km **13.** 5 cm **15.** $6\sqrt{3}$ cm; $36\sqrt{3}$ cm² **17.** $11\frac{1}{4}$ **19.** 18 **21.** 6 **23.** $\overline{DE} \perp \overline{AC}$, $DE = \frac{1}{2}DB$ (diags of rh ⊥ bis of each other); Area $\triangle ADC = \frac{1}{2}DE \cdot d_2$ (Th 12.2); Area $\triangle ADC = \frac{1}{2}\left(\frac{1}{2}d_1\right) d_2$, or $\frac{1}{4}d_1d_2$ (subst); $\triangle ADC \cong \triangle CBA$ (Th 7.1); Area $\triangle ADC =$ Area $\triangle CBA$ (Post 26); Area $ABCD =$ Area $ADC +$ Area CBA (Post 27); Area $ABCD = 2$ Area ADC (subst); Area $ABCD = 2 \cdot \frac{1}{4}d_1d_2$ (subst); Area $ABCD = \frac{1}{2}d_1d_2$ (alg) **25.** Given: sq $ABCD$, $AC = d$. Prove: Area $ABCD = \frac{1}{2}d^2$. $\overline{AC} \cong \overline{DB}$ (diags of sq ≅); $DB = d$ (subst); $ABCD$ is a rhombus (def); Area $ABCD = \frac{1}{2}AC \cdot DB$ (Th 12.3); Area $ABCD = \frac{1}{2}d^2$ (subst) **27.** $\frac{15\sqrt{7}\text{m}^2}{4}$ **29.** Given: $\triangle ACB$, base b, alt h. Prove: Area $\triangle ACB = \frac{1}{2}bh$. Area $\triangle ACB +$ Area $\triangle CDA =$ Area $\triangle ADB$ (Post 27); Area $\triangle ADC = \frac{1}{2}xh$, Area $\triangle ADB = \frac{1}{2}h(b + x)$ (Th 12.1); Area $\triangle ACB + \frac{1}{2}xh = \frac{1}{2}(b + x) \cdot h$ (subst); Area $\triangle ACB = \frac{1}{2}(b + x)h - \frac{1}{2}xh$ (subt); Area $\triangle ACB = \frac{1}{2}bh$ (alg) **31.** Given: $\triangle ABC \sim \triangle A'B'C'$. Prove: $\frac{\text{Area } ABC}{\text{Area } A'B'C'} = \frac{h^2}{h'^2}$. Area $ABC = \frac{1}{2}bh$, Area $A'B'C' = \frac{1}{2}b'h'$ (Th 12.2); $\frac{b}{b'} = \frac{h}{h'}$ (Th 9.6);

$$\frac{\text{Area } ABC}{\text{Area } A'B'C'} = \frac{\frac{1}{2}bh}{\frac{1}{2}b'h'} \text{ (div prop)}; \frac{\text{Area } ABC}{\text{Area } A'B'C'} = \frac{h}{h'} \cdot \frac{h}{h'} = \frac{h^2}{h'^2} \text{ (subst)}$$

33. Given: $\triangle ABC$, alt h, base b; $\triangle A'B'C'$; alt h', base b', $b \cong b'$. Prove: $\frac{\text{Area } ABC}{\text{Area } A'B'C'} = \frac{h}{h'}$. Area $ABC = \frac{1}{2}bh$, Area $A'B'C' = \frac{1}{2}b'h'$ (Th 12.2);

$$\frac{\text{Area } ABC}{\text{Area } A'B'C'} = \frac{\frac{1}{2}bh}{\frac{1}{2}b'h'} \text{ (div prop)};$$

$$\frac{\text{Area } ABC}{\text{Area } A'B'C'} = \frac{h}{h'} \text{ (alg)}$$

35. Given: $\triangle ABC$,

median $\overline{CM}$. Prove: Area $\triangle ACM$ = Area $\triangle BCM$. $\overline{AM} \cong \overline{MB}$ (def median); $AM = MB$ (def ≅); Area $\triangle ACM = \frac{1}{2} AM \cdot h$, Area $\triangle BCM = \frac{1}{2} MB \cdot h$ (Th 12.2); Area $\triangle ACM$ = Area $\triangle BCM$ (subst)

Page 397

1. 30 m² **3.** 55.18 cm² **5.** 9 cm **7.** 3.39 m **9.** 9 **11.** 14.4 **13.** 30 **15.** $60\sqrt{2}$ **17.** $\sin A = \frac{h}{b}$ (def); $h = b \sin A$ (mult); Area $ABCD = ah$ (Th 12.5); Area $ABCD = ab \sin A$ (subst) **19.** 53.528 sq units

Pages 400–401

1. 28 mm² **3.** 12.6 m² **5.** 40 m² **7.** 6 cm **9.** 3 m **11.** $4x^2 - 2x - 2$ **13.** $2x - 1$ **15.** 16 m, 23 m **17.** $36\sqrt{2}$ **19.** $18\sqrt{3}$ **21.** 18.5 **23.** Given: trap $ABCD$, $RSTU$, bases $\overline{AB} \cong \overline{RS}$, $\overline{DC} \cong \overline{UT}$, alts h, h'. Prove: $\frac{\text{Area } ABCD}{\text{Area } RSTU} = \frac{h}{h'}$. $AB = RS = b_1 DC = UT = b_2$ (def ≅ seg); Area $ABCD = \frac{1}{2} h(b_1 + b_2)$, Area $RSTU = \frac{1}{2} h'(b_1 + b_2)$ (Th 12.6); $\frac{\text{Area } ABCD}{\text{Area } RSTU} = \frac{\frac{1}{2} h(b_1 + b_2)}{\frac{1}{2} h'(b_1 + b_2)}$ (div prop); $\frac{\text{Area } ABCD}{\text{Area } RSTU} = \frac{h}{h'}$ (alg) **25.** Constr a seg of length = sum of bases of given trap; bisect this seg to get length of median; constr rect with base ≅ to med and alt ≅ to alt of given trap.

Cumulative Review

1. m $\angle B = 60$ = m $\angle D$; m $\angle C = 120$
2. $\frac{AB}{DE} = \frac{BC}{EF}$; $\frac{AB}{DE} = \frac{CA}{FD}$; $\frac{BC}{EF} = \frac{CA}{FD}$

Pages 404–405

1. $\frac{64}{49}$; $\frac{8}{7}$ **3.** $\frac{25}{16}$; $\frac{5}{4}$ **5.** $\frac{4}{1}$; $\frac{2}{1}$ **7.** 21 cm **9.** 18.2 cm **11.** 68 cm **13.** $16\sqrt{2}$ cm **15.** 36 cm **17.** $(21 + 7\sqrt{3})$ m **19.** 15 **21.** 12 cm; 8 cm **23.** Given: $\triangle ABC \sim \triangle A'B'C'$. Prove: $\frac{\text{Area } ABC}{\text{Area } A'B'C'} = \frac{b^2}{b'^2}$; Area $ABC = \frac{1}{2} bh$, Area $A'B'C' = \frac{1}{2} b'h'$ (Th 12.2); $\frac{h}{h'} = \frac{b}{b'}$ (Th 9.6); $\frac{\text{Area } ABC}{\text{Area } A'B'C'} = \frac{\frac{1}{2} bh}{\frac{1}{2} b'h'}$ (div prop); $\frac{\text{Area } ABC}{\text{Area } A'B'C'} = \frac{b}{b'} \cdot \frac{b}{b'}$ or $\frac{b^2}{b'^2}$ (subst)

Pages 407–408

1. 7 **3.** 484 **5.** $14\sqrt{3}$, 14 **7.** 12, $6\sqrt{2}$ **9.** $5\sqrt{3}$, 10 **11.** $48\sqrt{3}$ cm² **13.** 36 mm² **15.** $150\sqrt{3}$ cm² **17.** 2, 4, $4\sqrt{3}$ **19.** 2, $\frac{4\sqrt{3}}{3}$, $\frac{4\sqrt{3}}{3}$ **21.** Given: reg poly $YABX$ . . ., rad $\overline{OB}$. Prove: $\overline{OB}$ bis $\angle ABX$. Draw rad $\overline{OA}$, $\overline{OX}$, $\overline{OA} \cong \overline{OX}$ (def rad); $\overline{AB} \cong \overline{BX}$ (def reg poly); $\overline{OB} \cong \overline{OB}$ (reflex); $\triangle ABO \cong \triangle XBO$ (Post 16), $\angle 1 \cong \angle 2$ (CPCTC) $\overline{OB}$ bis $\angle ABX$ (def ∠ bis) **23.** For inscr sq, $s = r\sqrt{2}$; for circum sq $s' = 2r$; $\frac{A}{A'} = \frac{(r\sqrt{2})^2}{(2r)^2} = \frac{2r^2}{4r^2}$; $\frac{1}{2}$
25. Given: reg polys $XABC$ . . ., $X'A'B'C'$ with apothems $\overline{OD}$, $\overline{O'D'}$. Prove: $\frac{\text{Area } XABC}{\text{Area } X'A'B'C'} = \frac{(OD)^2}{(O'D')^2}$. Draw $\overline{OA}$, $\overline{OB}$, $\overline{O'A'}$, $\overline{O'B'}$, $\overline{OA} \cong \overline{OB}$, $\overline{O'A'} \cong \overline{O'B'}$ (def rad); $\angle OAD \cong \angle OBD$, $\angle O'A'D' \cong \angle O'B'D'$ (Th 5.1), $\overline{OA}$ bis $\angle A$, $\overline{O'A'}$ bis $\angle A'$ (ex 21); m $\angle OAD = \frac{1}{2}$ m $\angle XAD$, m $\angle O'A'D' = \frac{1}{2}$ m $\angle X'A'D'$ (def bis); $\angle XAD \cong \angle ABC$, $\angle X'A'D' \cong \angle A'B'C'$, $\angle XAD \cong \angle X'A'D'$ (def ~ poly) $\angle XAD \cong \angle X'A'D' \cong \angle ABC \cong \angle A'B'C'$ (trans prop); $\frac{1}{2}$ m $\angle XAD = \frac{1}{2}$ m $\angle X'A'D'$ (mult); $\angle OAD \cong \angle O'A'D' \cong \angle OBD \cong \angle O'B'D'$ (trans prop); $\triangle OAB \sim \triangle O'A'B'$ (Post 21): $\frac{OD}{O'D'} = \frac{AB}{A'B'}$ (Th 9.6); Area $XABC = \frac{1}{2}(OD \cdot (n \cdot AB)$, Area $X'A'B'C' = \frac{1}{2} O'D'$ (n $\cdot A'B'$) (Th 12.9); $\frac{\text{Area } XABC}{\text{Area } X'A'B'C'} = \frac{\frac{1}{2} n \cdot OD \cdot AB}{\frac{1}{2} n \cdot O'D' \cdot A'B'}$ (div prop); $\frac{\text{Area } XABC}{\text{Area } X'A'B'C'} = \frac{OD}{O'D'} \cdot \frac{AB}{A'B'}$ (alg); $\frac{\text{Area } XABC}{\text{Area } X'A'B'C'} = \frac{OD}{O'D'} \cdot \frac{OD}{O'D'}$ or $\frac{(OD)^2}{(O'D')^2}$ (subst) **27.** 4.94; 58.12 **29.** 5.64; 104.16 **31.** 2.03; 1.91

Page 411

1. 4 m; 4π m; 12.56 m **3.** 6 mm; 12 mm; 37.68 mm **5.** 64 mm; 64π mm; 200.96 mm **7.** 2.6 mm; 5.2π mm; 16.33 mm **9.** 0.12 m; 0.75 m **11.** 0.28 mm; 1.73 mm **13.** $3\frac{1}{7}$

SELECTED ANSWERS

15. $2\pi d$ **17.** $S = \sqrt{2}$ cm; $c > p$ **19.** $2\pi r = 4s$, $r = \frac{2s}{\pi}$

Pages 413–414

1. 4π m² **3.** 4π km² **5.** 5 m **7.** 7 mm **9.** 6 m **11.** 24 mm **13.** 1.13 m² **15.** 1.77 cm² **17.** 7.065 cm² **19.** 0.07065 m² **21.** 10 cm; 20 cm **23.** 0.6 mm; 1.2 mm **25.** 50.24 mm² **27.** 6.449 m² **29.** 4 m²; 3.14 m²; area of sq is 0.86 m² greater than area of ⊙. **31.** $(4 - \pi)$ cm² **33.** $\left(\sqrt{3} - \frac{1}{3}\pi\right)$ cm² **35.** $12\frac{1}{2}\pi$ **37.** $4 - \pi$ **39.** $C_1 = 2\pi r_1$, $C_2 = 2\pi r_2$ (Th 12.10, Cor); $\frac{C_1}{C_2} = \frac{2\pi r_1}{2\pi r_2}$ (div); $\frac{C_1^2}{C_2^2} = \frac{R_1^2}{R_2^2}$ (alg); $\frac{A_1}{A_2} = \frac{R_1^2}{R_2^2}$ (ex. 38) $\frac{C_1^2}{C_2^2} = \frac{A_1}{A_2}$ (subst) **41.** 1 to 2

Pages 416–417

1. 20 mm; 2.5π mm **3.** 5 cm; $\frac{5\pi}{3}$ cm **5.** 3.6 cm; 3π cm **7.** 24 mm; 4π mm **9.** 13 cm **11.** 1 km; 2 km **13.** $22\frac{1}{2}$; 16 m **15.** 360; 5 mm **17.** 12π **19.** $\frac{88\pi}{15}$ **21.** $\frac{35\pi}{9}$

Algebra Review

1. $\frac{7 - \sqrt{7}}{7}$ **2.** $\frac{12 + 2\sqrt{3}}{3}$ **3.** $\frac{10 - \sqrt{5}}{5}$ **4.** $3\sqrt{3} + \sqrt{2}$ **5.** $-3 + 3\sqrt{2}$ **6.** $\frac{12 + 3\sqrt{2}}{7}$ **7.** $-8 + 4\sqrt{5}$ **8.** $-\frac{2 + 2\sqrt{6} + \sqrt{3} + 3\sqrt{2}}{5}$ **9.** $3 - 2\sqrt{2}$ **10.** $\sqrt{2} + 1$ **11.** $\frac{18\sqrt{2} + 3\sqrt{3}}{23}$ **12.** $\frac{13 + 4\sqrt{3}}{11}$

Pages 419–420

1. 10π **3.** 15π **5.** 50π **7.** 75π **9.** 7.2π cm² **11.** 36π cm² **13.** 60; $\frac{25\pi}{6}$ cm²; $\frac{50\pi - 75\sqrt{3}}{12}$ cm² **15.** 120; $\frac{64\pi}{3}$ mm²; $\frac{64\pi - 48\sqrt{3}}{3}$ mm² **17.** 120; 6 m; $(12\pi - 9\sqrt{3})$ m² **19.** 20π cm² **21.** $\frac{54\pi - 81\sqrt{3}}{4}$ m² **23.** $4\sqrt{7}$ m **25.** Area is 4 times as large. **27.** $A = \pi(R^2 - r^2)$ **29.** $\left(\frac{170\pi}{9} - 46.36\right)$ cm²

Cumulative Review

1. 120 **2.** 80

Page 422

1. 120 m² **3.** 252 cm² **5.** 150 cm² **7.** 28 sq units **9.** $\frac{64\pi}{5}$ cm² **11.** 44 m² **13.** 32π cm² **15.** 14π cm **17.** $3\sqrt{3}$ cm **19.** 54 units **21.** 240 **23.** $\frac{3}{2}$

Pages 426–427

1. B **2.** D **3.** A **4.** B **5.** C **6.** A **7.** C **8.** B **9.** A **10.** C **11.** A **12.** B **13.** F **14.** F **15.** T **16.** F **17.** T **18.** T **19.** F **20.** T **21.** T **22.** 9.5 cm **23.** 60, 24 **24.** $7\sqrt{2}$ cm **25.** $\sqrt{89}$ cm **26.** 43 **27.** 19 **28.** $6\frac{2}{3}$ **29.** 12 **30.** 81 m² **31.** $44\sqrt{2}$ cm² **32.** $\frac{25}{4}\sqrt{3}$ cm² **33.** 14π m **34.** 12π cm² **35.** 12.4 **36.** 13.1 **37.** 4.6

Page 431

1. right 2, up 3; left 4, up 2; right 6, down 4; left 3, down 5; up 8; down 3 **3.** vrt **5.** neither **7.** vrt **9.** vrt **11.** neither **13.** horiz **15.** 2 **17.** 3 **19.** −3 **21.** I, II **23.** IV **25.** II **27.** (1, 5) **29.** (4, 3) **31.** (4, −2), (−3, 5) **33.** (4, 3), (−1, 3); (4, −7), (−1, −7); $\left(1\frac{1}{2}, \frac{1}{2}\right)$, $\left(1\frac{1}{2}, -4\frac{1}{2}\right)$

Page 434

1. 2 **3.** 12 **5.** $\sqrt{13}$ **7.** $2\sqrt{2}$ **9.** $3\sqrt{10}$ **11.** $4\sqrt{13}$ **13.** $\sqrt{178}$ **15.** $\sqrt{(2 - m)^2 + (5 - n)^2}$ **17.** $AB = BC = \sqrt{29}$, $AC = 4$; isos **19.** $\sqrt{53}$, $\sqrt{74}$, $\sqrt{29}$; scal **21.** $\sqrt{130}$, $\sqrt{26}$, $2\sqrt{26}$; scal **23.** $RS = \sqrt{104}$, $ST = \sqrt{32}$, $RT = \sqrt{72}$; $(\sqrt{104})^2 = (\sqrt{32})^2 + (\sqrt{72})^2$; $104 = 32 + 72$ **25.** $\left(4\frac{1}{2}, \frac{1}{2}\right)$; $\left(4\frac{1}{2}, 5\frac{1}{2}\right)$ **27.** 2, −4 **29.** Given: pts $P(X_1, Y_1)$ and $Q(X_2, Y_2)$, $PQ = d$. Prove: $d = \sqrt{(X_2 - Y_1)^2 + (Y_2 - Y_1)^2}$. Draw $\overleftrightarrow{QN} \parallel \overleftrightarrow{OY}$ and $\overleftrightarrow{PN} \parallel \overleftrightarrow{OX}$ (Post 15): $\overleftrightarrow{OX} \perp \overleftrightarrow{QN}$, $\overleftrightarrow{QN} \perp \overleftrightarrow{PN}$ ($l \perp$ 1 of 2 ‖s is ⊥ other); $(PQ)^2 = (PN)^2 + (QN)^2$ (Pyth Th); $QN = |Y_2 - Y_1|$, $PN = |X_2 - X_1|$ (def); $d^2 = (X_2 - X_1)^2 + (Y_2 - Y_1)^2$ (subst); $d = \sqrt{(X_2 - X_1)^2 + (Y_2 - Y_1)^2}$ (alg)

Cumulative Review

1. 3, 12 **2.** 10.7

Page 437

1. (2, 3) **3.** (1, 6) **5.** $\left(\frac{1}{2}, \frac{1}{2}\right)$ **7.** $\left(4, 1\frac{1}{2}\right)$ **9.** $\left(0, 4\frac{1}{2}\right)$ **11.** $\left(3\frac{1}{2}, -7\right)$ **13.** (−10, −3)

15. $(-13, -7)$ **17.** $(6, 6)$ **19.** $(18, -2)$ **21.** $(2, -3)$, $(4, 2)$, $(0, 1)$; 9 **23.** $\left(1\frac{1}{2}, -1\frac{1}{2}\right)$, $\left(4\frac{1}{2}, 1\right)$, $\left(2, 1\frac{1}{2}\right)$; $\frac{1}{2}\sqrt{170}$ **25.** $PR = \sqrt{61}$, $MN = \frac{1}{2}\sqrt{61}$ **27.** $MN = AB = CD = 3\sqrt{5}$ **29.** $MN = PQ = \sqrt{41}$; $PN = QM = \sqrt{26}$ **31.** Given: 2 pts $P(X_1, Y_1)$ and $Q(X_2, Y_2)$ with M the midpt of $\overline{PQ}$. Prove: $M = \left(\frac{X_1 + X_2}{2}, \frac{Y_1 + Y_2}{2}\right)$. Locate $R(X_2, Y_1)$ by drawing $\overleftrightarrow{QR} \parallel \overleftrightarrow{OY}$ and $\overleftrightarrow{PR} \parallel \overleftrightarrow{OX}$ (Post 15); midpt of $\overline{QR}$ is $T\left(X_2, \frac{Y_2 + Y_1}{2}\right)$ (def of midpt of vert seg); midpt of $\overline{PR}$ is $S\left(\frac{X_2 + X_1}{2}, Y_1\right)$ (def of midpt of horiz seg); $\overline{MT} \parallel \overline{PR}$ and $\overline{MS} \parallel \overline{QR}$ (Th 7.7); $M(x, y) = M\left(\frac{X_2 + X_1}{2}, \frac{Y_2 + Y_1}{2}\right)$ (props of horiz and vert lines)

Pages 440–441

1. -6; dn rt **3.** 0; horiz **5.** undef; vert **7.** 0; horiz **9.** -1; dn rt **11.** undef; vert **13.** undef; vert **15.** -1; dn rt **17.** $-\frac{1}{2}$; dn rt **19.** 0 **21.** 11 **23.** -3 **25.** -2 **27.** 6 **29.** slope $\overleftrightarrow{AB}$ = slope $\overleftrightarrow{CD} = \frac{-3}{4}$; collinear **31.** slope $\overleftrightarrow{AB}$ = slope $\overleftrightarrow{CD} = -\frac{2}{3}$; parallel

Algebra Review

1. $x^2 + 4x + 4$ **2.** $x^2 - 6x + 9$ **3.** $y^2 + 10y + 25$ **4.** $a^2 - 16a + 64$ **5.** $4x^2 + 4x + 1$ **6.** $9y^2 - 12y + 4$ **7.** $4c^2 - 20c + 25$ **8.** $16x^2 + 8x + 1$ **9.** $9a^2 + 6ab + b^2$ **10.** $x^2 + 6xy + 9y^2$ **11.** $4c^2 - 4cd + d^2$ **12.** $25x^2 - 40xy + 16y^2$ **13.** $x^2 + 2x + 1$ **14.** $y^2 - 4y + 4$ **15.** $m^2 - 8m + 16$ **16.** $a^2 + 12a + 36$ **17.** $x^2 + 5x + \frac{25}{4}$ **18.** $y^2 - 9y + \frac{81}{4}$ **19.** $z^2 - 24z + 144$ **20.** $a^2 + 15a + \frac{225}{4}$ **21.** $x^2 - 25x + \frac{625}{4}$ **22.** $a^2 + 6ab + 9b^2$ **23.** $x^2 - 12xy + 36y^2$ **24.** $c^2 + 7cd + \frac{49}{4}d^2$

Page 445

Line will pass through the given points.
1. $(-4, 0)(-2, -4)(0, -8)$ **3.** $(-2, 2)(-1, 0)(0, -2)$ **5.** $(4, 0)(2, -6)(0, -12)$ **7.** $(-3, 0)\left(0, \frac{9}{5}\right)(2, 3)$ **9.** $(4, 3)(0, 0)(-4, -3)$ **11.** $-4x + y = -7$ **13.** $-3x + y = 14$ **15.** $-x - 3y = 9$ **17.** $2x + 6y = -3$ **19.** $y - 5 = 3(x - 2)$ **21.** $y + 5 = -6(x + 1)$ **23.** $y + 1 = -\frac{4}{3}(x + 6)$ **25.** $y + 6 = 0$ **27.** $y = -x$ **29.** $m = \frac{-2 - 2}{3 - (-5)}$; $\frac{-1}{2}$; $y - 2 = \frac{-1}{2}(x + 5)$; $x + 2y = -1$ **31.** $m = -2$; $y + 1 = -2(x - 3)$; $2x + y = 5$ **33.** $m = \frac{1}{3}$; $y + 1 = \frac{1}{3}(x - 6)$; $x - 3y = 9$ **35.** $m = 1$; $y + 2 = 1(x + 7)$; $-x + y = 5$ **37.** For $\overline{AB}$: $m = -5$; $y + 4 = -5(x + 2)$; $5x + y = -14$. For $\overline{BC}$: $m = 1$; $y - 1 = 1(x + 3)$; $-x + y = 4$. For $\overline{AC}$: $m = \frac{5}{2}$; $y + 4 = \frac{5}{2}(x + 2)$; $-5x + 2y = 2$. **39.** $R = (0, 4)$; $S = \left(3, \frac{8 + 4}{2}\right)$; $(3, 6)$; $T = (3, 2)$. For $\overline{GS}$: $m = 2$; $y = 2x$; $-2x + y = 0$. For $\overline{HT}$: $m = -2$; $y - 8 = -2x$; $2x + y = 8$. For $\overline{IR}$: $m = 0$; $y = 4$. **41.** $m = \frac{y_2 - y_1}{x_2 - x_1}$; $y - y_1 = \frac{y_2 - y_1}{x_2 - x_1}(x - x_1)$; $\frac{y - y_1}{x - x_1} = \frac{y_2 - y_1}{x_2 - x_1}$ **43.** The △s formed are 30-60 rt △s. For l_1: $m = \frac{0 - a}{a - 0}\sqrt{3}$; $-\sqrt{3}$. For l_2: $m = \frac{a\sqrt{3} - 0}{0 - (-a)}$; $\sqrt{3}$. **45.** $m = 1$ or $m = -1$; $y - 3 = 1(x - 4)$; $-x + y = -1$; $y - 3 = -1(x - 4)$; $x + y = 7$

Page 448

1. $y = 3x + 5$ **3.** $y = 7x - 2$ **5.** $y = \frac{-1}{2}$ **7.** $y = -x + \frac{1}{4}$ **9.** $y = 2x + 4$; 2; 4 **11.** $y = -5x - 2$; -5; -2 **13.** $y = 5$; 0; 5 **15.** $y = -\frac{1}{3}x - \frac{1}{2}$; $-\frac{1}{3}$; $-\frac{1}{2}$ **17.** $(1, 4)(0, 1)$ **19.** $(-3, 2)(0, -2)$ **21.** $(3, 1)(0, 0)$ **23.** $(2, -3)(0, -3)$ **25.** $(0, -4)(2, -2)$ **27.** $(-2, -1)(0, 4)$ **29.** $y = -2x + 6$; -2; 6 **31.** $y = -x + 3$; -1; 3 **33.** $y = \frac{-3}{2}x - 1$; $\frac{-3}{2}$; -1 **35.** $y = x$; 1; 0 **37.** $(-1, 6)$ **39.** $(2, 0)$ **41.** Let $Q(0, b)$ be the pt where the nonvert line int y-axis. Then slope $\overleftrightarrow{PQ} = \frac{0 - b}{a - 0}$, or $\frac{-b}{a}$ (def); equa $\overleftrightarrow{PQ}$ is $y = mx + b$ (Th 13.5); $y = \frac{-b}{a}x + b$ (subst); $ay = -bx + ab$ (mult); $x = \frac{-ay}{b} + a$ (alg) $x = \frac{1}{m}y + a$ (subst)

SELECTED ANSWERS

Pages 451–452

1. 1; −1 **3.** $\frac{5}{2}; \frac{-2}{5}$ **5.** undef; 0 **7.** $\frac{-7}{8}; \frac{8}{7}$ **9.** $-7 \cdot \frac{1}{7} = -1$ **11.** slope of $\overleftrightarrow{AB}$ = slope of $\overleftrightarrow{CD} = 1$; slope of $\overleftrightarrow{AD}$ = slope of $\overleftrightarrow{BC} = -4$ **13.** slope of $\overleftrightarrow{AB}$ = slope $\overleftrightarrow{CD} = 1$; slope of $\overleftrightarrow{BC}$ = slope of $\overleftrightarrow{AD} = -7$ **15.** slope of $\overleftrightarrow{EF}$ = slope of $\overleftrightarrow{GH} = \frac{-2}{3}$; slope of $\overleftrightarrow{EH}$ = slope of $\overleftrightarrow{FG} = \frac{3}{2}; \frac{-2}{3} \cdot \frac{3}{2} = -1$ (Th 13.8) **17.** $-7 \cdot \frac{1}{7} = -1$ **19.** $y = -6x + 28$ **21.** $y = \frac{-3}{2}x - \frac{7}{2}$ **23.** $x + 3y = 13$ **25.** $3x + 5y = 13$ **27.** Given: L_1 with slope m_1; L_2 with slope m_2; $m_1 \cdot m_2 = -1$. Prove $L_1 \perp L_2$. Draw $\overleftrightarrow{AB} \parallel x$-axis (Post 15); $(AC)^2 = (x_3 - x_1)^2 + (y_3 - y_1)^2$; $(BC)^2 = (x_3 - x_2)^2 + (y_3 - y_2)^2$, $(AB)^2 = (x_2 - x_1)^2$ (Th 13.1); $(AC)^2 + (BC)^2 = (x_3 - x_1)^2 + (y_3 - y_1)^2 + (x_3 - x_2)^2 + (y_3 - y_2)^2$ (add); $m_1 = \frac{y_3 - y_1}{x_3 - x_1}$, $m_2 = \frac{y_3 - y_2}{x_3 - x_2}$ (def); $\frac{y_3 - y_1}{x_3 - x_1} \cdot \frac{y_3 - y_2}{x_3 - x_2} = -1$ (subst); $(y_3 - y_1)(y_3 - y_2) = -1 \cdot (x_3 - x_1)(x_3 - x_2)$; $2(y_3 - y_1)(y_3 - y_2) = -2(x_3 - x_1)(x_3 - x_2)$ Consider only y's: m $(AC)^2$ + m $(BC)^2 = (y_3 - y_1)^2 - 2(y_3 - y_1)(y_3 - y_2) + (y_3 - y_2)^2 = [(y_3 - y_1) - (y_3 - y_2)]^2 = 0$. Consider only x's: m $(AC)^2$ + m $(BC)^2 = (x_3 - x_1)^2 - 2(x_3 - x_1)(x_3 - x_2) + (x_3 - x_2)^2$ (subst); m $(AC)^2$ + m $(BC)^2 = [(x_3 - x_1) - (x_3 - x_2)]^2$ (distr); m $(AC)^2$ + m $(BC)^2 = (x_2 - x_1)^2$ = m $(AB)^2$; m $(AC)^2$ + m $(BC)^2$ = m $(AB)^2$ (subst); m $\angle C = 90$ (Th 10.3); $L_1 \perp L_2$ (def of rt $\angle$)

Pages 454–455

1. $(0, a)$ **3.** $\frac{a}{2}$ **5.** $\left(\frac{a}{2}, 0\right); \left(\frac{b}{2}, \frac{c}{2}\right); \left(\frac{a+b}{2}, \frac{c}{2}\right)$ **7.** $(-a, 0)$ **9.** midpt $\overline{AC} = \left(\frac{a+b}{2}, \frac{c}{2}\right)$; midpt $BD = \left(\frac{a+b}{2}, \frac{c}{2}\right)$; M is midpt of both diags, so they bisect each other. **11.** Given: isos trap $ABCD$, $\overline{AB} \parallel \overline{CD}$, $\overline{AD} \cong \overline{BC}$ diags $\overline{AC}$, $\overline{BD}$. Prove: $\overline{AC} \cong \overline{BD}$, $AC = \sqrt{(a - b - 0)^2 + (c - 0)^2}$; $AC = \sqrt{(a - b)^2 + c^2}$; $BD = \sqrt{(a - b)^2 + (0 - c)^2}$; $BD = \sqrt{(a - b)^2 + c^2}$; $AC = BD$; $\overline{AC} \cong \overline{BD}$ **13.** Given: $ABCD$ with midpts $Z\left(\frac{a}{2}, 0\right)$, $W\left(\frac{a+d}{2}, \frac{e}{2}\right)$, $X\left(\frac{b+d}{2}, \frac{c+e}{2}\right)$, $Y\left(\frac{b}{2}, \frac{c}{2}\right)$. Prove: $\overline{XZ}$, $\overline{WY}$ bis each other. Midpt of $\overline{WY}$ = $\left(\frac{\frac{a+d}{2} + \frac{b}{2}}{2}, \frac{\frac{e}{2} + \frac{c}{2}}{2}\right)$; $\left(\frac{a+b+d}{4}, \frac{c+e}{4}\right)$; midpt $\overline{XZ} = \left(\frac{\frac{a}{2} + \frac{b+d}{2}}{2}, \frac{0 + \frac{c+e}{2}}{2}\right)$; $\left(\frac{a+b+d}{4} \frac{c+e}{4}\right)$ since both midpts have same coords, diag bisect each other. **15.** Given: trap $ABCD$, $\overline{AB} \parallel \overline{CD}$, $m\left(\frac{d}{2}, \frac{c}{2}\right)$ midpt diag $\overline{AC}$, $N\left(\frac{a+b}{2}, \frac{c}{2}\right)$ midpt diag $\overline{BD}$. Prove: $MN = \frac{1}{2}(AB - CD)$. $MN = \frac{a+b}{2} - \frac{d}{2}$; $MN = \frac{a+b-d}{2}$; $AB = a$; $CD = d - b$; $AB - CD = a - (d - b)$; $AB - CD = a + b - d$; $MN = \frac{1}{2}(AB - CD)$

Algebra Review

1. 9, −5 **2.** 4, −16 **3.** 3, 11 **4.** 7, −11 **5.** 5, 1 **6.** $\frac{-2}{3}$, 4 **7.** $\frac{15}{2}, \frac{13}{2}$ **8.** $\frac{-6}{5}$, 2 **9.** $\frac{8}{3}$, −4 **10.** −2 **11.** $\frac{3}{5}, \frac{-9}{5}$ **12.** 2, $\frac{2}{3}$

Page 458

1. $(x - 0)^2 + (y - 0)^2 = 3^2$ **3.** $(x - (-1))^2 + (y - 4)^2 = 1^2$ **5.** $(x - 6)^2 + (y - (-1))^2 = 7^2$ **7.** $(x - 5)^2 + (y - 5)^2 = 8^2$ **9.** $(x - 3)^2 + (y - 1)^2 = 4^2$; $x^2 - 6x + y^2 - 2y = 6$ **11.** $x^2 + (y + 4)^2 = 4$; $x^2 + y^2 + 8y = -12$ **13.** $(x + 8)^2 + (y + 1)^2 = 15$; $x^2 + 16x + y^2 + 2y = -50$ **15.** $(x + 2)^2 + (y + 9)^2 = 7$; $x^2 + 4x + y^2 + 18y = -78$ **17.** $4^2 + 0^2 = 16$ **19.** $(5 - 1)^2 + (-3)^2 = 16 + 9 = 25$ **21.** $(-\sqrt{3})^2 + (-2)^2 + 6(-2) = 3 + 4 - 12 = -5$ **23.** (0, 0); 4 **25.** (0, 0); $\sqrt{10}$ **27.** (3, −4); 4 **29.** (0, 0); $3\sqrt{2}$ **31.** (−1, 5); $3\sqrt{3}$ **33.** (8, 3); 1 **35.** $(x - 4)^2 + (y - 3)^2 = 25$ **37.** $(x + 2)^2 + (y + \sqrt{5})^2 = 9$ **39.** $(x - 2)^2 + (y + 3)^2 = 20$ **41.** $x^2 + 4x + 4 + y^2 = 36$; $x^2 - 2x + 1 + y^2 + 6y + 9 = 36$; $r_1 = r_2 = 6$ **43.** (3, −2) is center of both ⊙s. **45.** (4, −1) **47.** $-4x + 3y = 42$

Page 460

1. right 3, forward 1, up 2 **3.** left 2, forward 1, up 5 **5.** left 3, back 2, down 1 **7.** no move, back 3, up 4 **9.** left 1, no move, no move **11.** x = axis **13.** z = axis **15.** x and y **17.** $\sqrt{94}$ **19.** $\sqrt{161}$ **21.** $\sqrt{202}$ **23.** $\sqrt{113}$ **25.** $\sqrt{35}$ **27.** (3, 2, 2) **29.** $\left(\frac{9}{2}, \frac{-7}{2}, 1\right)$

Page 462

1. right 3, up 6 **3.** left 4, down 3 **5.** vrt **7.** $\sqrt{17}$ **9.** (2, 4) **11.** (−13, 2) **13.** $\sqrt{29}$; $2\sqrt{29}$ **15.** −3; dn to rt; line will pass through given points **17.** (3, 0)(0, −2) **19.** (−4, 0) (0, −8) **21.** $y - 3 = 2(x - 1)$ **23.** $y + 5 = \frac{1}{3}(x - 4)$ **25.** $4x - 3y = 5$ **27.** $y = 3x - 5$ **29.** 6; −4 **31.** $\frac{7}{5} \cdot \frac{-5}{7} = -1$

Page 469

1. A line ∥ to a given line, midway bet them **3.** 2 lines each ∥ to given line l and 8 mm from it **5.** ⊥ bis of $\overline{RS}$ (see ex. 2) **7.** ⊙, rad 2.5 cm, given pt is ctr **9.** interior of ⊙, rad = 2 cm **11.** ext of ⊙ of rad 4 cm, ctr at given pt **13.** line ∥ to given line l at a given dist from it equal to the rad of the given ⊙. **15.** ⊙ concentric to given, rad of length = sum of radii of given ⊙ and rolling ⊙.

Cumulative Review

1. 8 cm **2.** 65 cm²

Page 473

1. bis of given ∠ (Th 14.2) **2.** See Post 28. **3.** See Post 28. **5.** diam ⊥ to chords at their midpt **7.** See ex. 5; diam ⊥ to given chord, but not including endpts of that diam (Th 14.1). **9.** concentric ⊙ with rad $\frac{1}{2}$ of given ⊙. *i.* The midpt of any rad of a ⊙ is on the ⊙ concentric to given ⊙ and having a rad $\frac{1}{2}$ the rad of given ⊙. *ii.* Every point of a ⊙ concentric to a given ⊙ and of rad = $\frac{1}{2}$ rad of the given ⊙ is a midpt of a rad of the given ⊙. **11.** a line ⊥ to a given line at the given pt. *i.* The center of a ⊙ tangent to a given line at a given pt is on ⊥ the line at the given pt. *ii.* Any pt on the ⊥ a line at a given pt is the center of a ⊙ tangent to the line at the given pt. **13.** *i.* Given: ⊙O, O a pt. on $\overleftrightarrow{QP}$, $\overleftrightarrow{QP} \perp l$ at P. Prove: Line tangent to ⊙O at P. Since line ⊥ rad $\overline{OP}$ at its endpt on the ⊙, line tangent ⊙O (Th 11.6). *ii.* Given: ⊙O, rad $\overline{OP}$, l tangent ⊙O at P. Prove: $\overline{OP} \perp l$. Since l tangent ⊙O at P, $\overline{OP} \perp l$ (Th 11.5). **15.** *i:* $\overline{PA} \cong \overline{PB}$ (Th 5.7) *ii:* Draw $\overline{PD}$ to midpt D of $\overline{AB}$; then $\overleftrightarrow{PD}$ is the ⊥ bis of $\overline{AB}$ (Th 5.8).

Algebra Review

1. −7, 4 **2.** −3, −2 **3.** 5, 3 **4.** 2, 7 **5.** −5 **6.** −7, 7 **7.** $\frac{1}{2}$, 5 **8.** −3, $\frac{1}{3}$ **9.** −3, $\frac{3}{2}$ **10.** −8, 3 **11.** 4, 6 **12.** −7, −5

Algebra Review

1. 2, 6 **2.** 1, −5 **3.** 2, 4 **4.** 1, −3 **5.** 4, −6 **6.** −3, −1 **7.** 1, 5 **8.** 7, −5 **9.** 5, 9 **10.** 12, −2 **11.** −4, −8 **12.** 18, −2

Pages 477–478

1. concentric ⊙ of 9 cm rad, 6 cm from OQ **3.** 2 concentric ⊙'s, radii 4 cm and 1 cm. **5.** pt of intersection of ⊙A and ⊙B; ⊙A with rad 1 cm, ⊙B with rad 5 cm **7.** 4 pts of intersection of ⊙O, rad 2 cm with the bisectors of the ∠s **9.** 2 pts, each 4 cm from P and on the line ∥ r and s and midway bet them. **11.** 1 pt, on ⊥ from P to l, 3 cm from P, 2 cm from l **13.** 4 pts., ⊙P, rad 9 cm, and 2 lines ∥ l, 2 cm from l **15.** concentric ⊙, rad. = 8 cm **17.** 2 concentric ⊙'s to given ⊙, rad 4 cm and 10 cm **19.** 2 pts of intersection of ⊙'s; ctr A, rad = 3 cm, and ctr B, rad = 2 cm. **21.** 4 pts. intersections of ∠ bis with lines ∥ to r, 5 cm above r and 1 cm below r **23.** 2 pts, intersections of ⊙P and line midway between m and n and parallel to them

Cumulative Review

1. $\frac{8}{3}\sqrt{3}$ cm. **2.** 196 m²

Pages 481–482

1. 4 pts, intersection of ∠ bis with 2 lines ∥ r, y units on either side of r **3.** no pts, 1 or 2 pts of intersection of ⊙A with $l \parallel m$, n and midway bet them $\left(d < \frac{1}{2}f,\ d = \frac{1}{2}f,\ d > \frac{1}{2}f\right)$ **5.** See p. 480, ex 3. **7.** intersection of ⊙ and ∠ bis; 0, 1, 2 pts **9.** intersection of line midway between parallel lines and concentric locus circle, rad = $\frac{r_1 + r_2}{2}$ no pts; 1 pt; 2 pts. **11.** no pt, 1 pt, 2 pts, 3 pts, 4 pts **13.** no pts, 1 pt, ray **15.** 1 pt **17.** 2 pts, 3 pts, no pt, 1 pt

Algebra Review

1. −3, 1 **2.** −3, 2 **3.** 13, −1 **4.** $\frac{3}{2}$, 4 **5.** −2, $\frac{1}{3}$ **6.** 1, $\frac{-2}{5}$ **7.** $3 \pm \sqrt{19}$ **8.** $\frac{-5 \pm \sqrt{73}}{4}$ **9.** $\frac{7 \pm \sqrt{13}}{6}$ **10.** $\frac{-2 \pm \sqrt{10}}{3}$ **11.** 1, $\frac{-6}{5}$ **12.** −1, $\frac{5}{3}$

Page 486

1. See constr pp 10, 51. **3.** See ex 2. **5.** See ex 1. **7.** in center, int of ∠ bis; circumcenter, int of ⊥ bis of sides; they coincide **9.** Draw 2 chords; constr their ⊥ bis, which meet at ctr. **11.** interior, always **13.** See Th 14.4; incenter **15.** 3, Post 12; 4, Th 14.2; 5, Trans; 6, Th 14.2 **17.** intersection of ∠A, ∠D **19.** r_i (inscr) $= \frac{1}{3} \cdot$ alt; r_c (circum) $= \frac{2}{3} \cdot$ alt; $\frac{r_i}{r_c} = \frac{1}{3} \div \frac{2}{3} = \frac{1}{2}$ **21.** $h = \frac{1}{2} \cdot 24\sqrt{3}$; $h = 12\sqrt{3}$; $r_i = 4\sqrt{3}$; $r_c = 8\sqrt{3}$; $A_i = 48\pi$ cm²; $A_c = 192\pi$ cm²

Pages 489–490

1. See ex 1, p 174. **3.** on ⊥ bis of base **5.** interior, on, exterior **7.** 12 **9.** 10 **11.** $16\frac{1}{2}$ **13.** 8, 12 **15.** 15; 45 **17.** 9 **19.** 1, given; 3, Post 8, 4, Post 1; 5, Th 7.7; 6, Th 7.7; 7, Th 3.7, Trans. 8, Th 7.5; 9, Th 7.2; 10, def of midpt: 11, Trans **21.** $\frac{5}{3}\sqrt{3}$; $\frac{10}{3}\sqrt{3}$ **23.** ∠CED, ∠PFD are rt ∠'s. (def, altitude); ∠CED ≅ ∠PFD (all rt ∠'s ≅); ∠D ≅ ∠D (reflex); △PFD ~ △CED (Post 21); $\frac{PF}{CE} = \frac{PD}{CD}$ (def ~ poly); $\frac{PD}{CD} = \frac{1}{3}$ (def, centroid); $\frac{PF}{CE} = \frac{1}{3}$ (subst) **25.** All 6 △s have same area; thus, their weights are equal, and the common pt. acts as the ctr. of balance. **27.** Constr a ▱ with sides S_1, S_2, by constructing a △ with sides S_1, S_2, and 2 m; the seg 2 m is diag d_1 of ▱; complete the ▱; draw other diag, d_2: △ with sides S_1, S_2, and d_2 is req △.

Page 492

1. sphere, ctr Q, $r = 5$ cm **3.** pl $r \parallel$ pl t, pl $r \parallel$ pl s, pl r midway bet pl s and pl t **5.** 2 pl s, ∥ to given pl, 3 m from it. **7.** pl midway and ∥ to ∥ lines **9.** l ⊥ pl of △ and passing through intersection of ∠ bis **11.** pts in interior of sphere p, $r = 5$ cm **13.** pl ⊥ l at Q, except Q. **15.** Cyl plus 2 hemispheres

Page 494

1. ⊙, ctr A, $r = 1$ cm **3.** region bet 2 l's, each ∥ to given l and 2 cm from it **5.** line ⊥ to given line at given pt, excluding given pt **7.** pt on ⊥ from A to l, 2 cm from A **9.** no pts, 1 pt, 2 pts, 1 pt **11.** Constr ∠ bis for 2 ∠'s. **13.** Constr 2 ⊥ lines; form a rt △; constr 2 medians. **15.** 24 **17.** alt = 6 cm; 2 cm

Pages 501–502

1. mult prop of ineq **3.** add prop of ineq **5.** trichotomy **7.** m ∠z < m ∠w (equal-unequal) **9.** $AB < BC$ (add prop of ineq) **11.** $\frac{1}{2}AB > \frac{1}{2}CD$ (mult prop of ineq); $CF = \frac{1}{2}CD$, $AE = \frac{1}{2}AB$ (Th 11.2); $AE > CF$ (subst); $CF < AE$ (def) **13.** m ∠D + m ∠B = 180, m ∠C + m ∠A = 180 (Th 11.11, Cor 4); m ∠D + m ∠B = m ∠C + m ∠A (subst); m ∠D < m ∠C (subtr prop ineq) **15.** m ∠DAB = m ∠DBA (Th 5.1); m ∠DAE < m ∠EBD (subtr prop ineq); m ∠DAE < m ∠DBF (trans prop ineq); m ∠FCB (Th 5.1); m ∠DBF < m ∠FCD (subtr prop ineq); m ∠DAE < m ∠FCD (tran prop ineq)

Algebra Review

1. $x > -2$ **2.** $z < -1$ **3.** $x \geq -\frac{1}{3}$ **4.** $x < \frac{11}{7}$ **5.** $x \leq 6$ **6.** $x \geq -6$ **7.** $z > 0$ **8.** $x \leq 4$ **9.** $y > -3$ **10.** $x < 3$ **11.** $z > 4$ **12.** $y < 4$

Pages 505–506

1. Direct path is shorter (Th 15.1). **3.** Th 15.1 **5.** $AB + BC > AC$, $AD + DC > AC$, (Th 15.1); $AB + BC + AD + DC > 2\ AC$ (add prop ineq); $AB + AD > BD$, $BC + DC > BD$ (Th 15.1); $AB + BC + AD + DC > 2\ BD$ (add prop ineq); $2\ (AB + BC + AD + DC) > 2\ (AC) + 2\ (BD)$ (add); $AB + BC + AD + DC > AC + BD$ (div prop ineq) **7.** $AB + AD > BD$, $BE + EC > BC$ (Th 15.1); $BD = BE + ED$ (bet'ness); $BD = BE + EC$ (subst); $BD > BC$ (subst); $AB + AD > BC$ (trans prop ineq) **9.** $AP + BP > AB$, $BP + CP > BC$, $AP + CP > AC$ (Th 15.1); $AP + BP + BP + CP + AP + CP > AB + BC + AC$ (add prop ineq), $2\ (AP + BP + CP) > AB + BC + AC$ (subst); $AP + BP + CP > \frac{1}{2}(AB + BC + AC)$ (mult prop ineq) **11.** $AB + AD > BD$, $BD + CB > DC$ (Th 15.1); $BD > DC - CB$ (subtr prop ineq); $AB + AD > DC - CB$ (trans prop ineq) **13.** Given: △ABC with P on $\overline{AB}$. Prove: $CP <$

$\frac{1}{2}(AB+BC+AC)$. $AP+PB=AB$ (betweenness), $CP < CA + AP$, $CP < CB + PB$ (Th 15.1); $2(CP) < CA + AP + CB + PB$ (add prop ineq); $2\ CP < CA + BC + AB$ (subst); $CP < \frac{1}{2}(AB + BC + AC)$ (div prop of ineq)

Cumulative Review

1. 8 **2.** 15

Page 509

1. m $\angle A >$ m $\angle B$ (Th 15.2); m $\angle C >$ m $\angle B$ (trans) **3.** m $\angle B > 90$ (def); m $\angle A$ + m $\angle B$ + m $\angle C = 180$ (Th 3.5); m $\angle A$ + m $\angle C < 90$ (subtr prop ineq); m $\angle A < 90$, m $\angle C < 90$ (def, angle add); m $\angle B >$ m $\angle C$ (trans ineq); $AC > BC$, $AC > AB$ (Th 15.3) **5.** m $\angle CBA >$ m $\angle CAB$ (Th 15.2) $\frac{1}{2}$m $\angle CBA > \frac{1}{2}$m $\angle CAB$ (mult prop ineq); m $\angle DBA = \frac{1}{2}$m $\angle CBA$, m $\angle DAB = \frac{1}{2}$m $\angle CAB$ (def, $\angle$ bis); m $\angle DBA >$ m $\angle DAB$ (subst); $AD > DB$ (Th 15.3) **7.** m $\angle DBC >$ m $\angle A$ (Th 4.1); m $\angle DBC >$ m $\angle C$ (trans ineq); $DC > DB$ (Th 15.3) **9.** m $\angle CAB$ = m $\angle CBA$ (Th 5.1); m $\angle CAB$ = m $\angle w$ + m $\angle x$, m $\angle CBA$ = m $\angle y$ + m $\angle z$ (Post 10); m $\angle w$ + m $\angle x$ = m $\angle y$ + m $\angle z$ (subst); m $\angle x >$ m $\angle y$ (Th 15.2); m $\angle w <$ m $\angle z$ (subtr prop ineq) **11.** m $\angle BCD$ = m $\angle D$ (Th 5.1); $AB < BC$ (subst); m $\angle ACB <$ m $\angle A$ (Th 15.2); m $\angle ACB$ + m $\angle BCD <$ m $\angle D$ + m $\angle A$ (add prop ineq); m $\angle ACB$ + m $\angle BCD$ = m $\angle ACD$ (Post 10); m $\angle ACD <$ m $\angle D$ + m $\angle A$ (subst)

Pages 514–515

1. $\overline{AB} \cong \overline{CD}$, $\overline{AD} \cong \overline{BC}$ (Th 7.1, Cor); $BD > AC$ (Th 15.4) **3.** $\overline{BD} \cong \overline{BD}$ (reflex); m $\angle ADB >$ m $\angle CDB$ (Th 15.5) **5.** $\overline{CD} \cong \overline{CD}$ (reflex); $AD > DB$ (Th 15.4) **7.** $\overline{RU} \cong \overline{SU}$ (def med), $\overline{TU} \cong \overline{TU}$ (reflex); $RT > ST$ (Th 15.4); m $\angle S >$ m $\angle R$ (Th 15.2) **9.** $\overline{AB} \cong \overline{AB}$ (reflex); $\overline{AD} \cong \overline{EB}$ (def $\cong$); $CB > CA$ (Th 15.3) **11.** $CD < AC$ (Th 15.3); $\overline{AB} \cong \overline{DB}$ (Th 5.1); $\overline{CB} \cong \overline{CB}$ (reflex); m $\angle DBC <$ m $\angle ABC$ (Th 15.4)

Cumulative Review

1. $7^2 + x^2 = 113$; $x = 8$; 56 cm² **2.** $30 = 6x$; $x = 5$; $5^2 + 6^2 = d^2$; $\sqrt{61}$ cm

Algebra Review

1. $-6z < -96$; $z > 16$ **2.** $-x > -5$; $x < 5$ **3.** $-8x > 32$; $x < -4$ **4.** $-4x < -40$; $x > 10$ **5.** $-7x > -63$; $x < 9$ **6.** $-7x < 14$; $x > -2$ **7.** $-5x \le -10$; $x \ge 2$ **8.** $2x > -14$; $x > -7$ **9.** $-4x < -8$; $x > 2$ **10.** $-3x \ge 9$; $x \le -3$ **11.** $\frac{-3}{5}x \le 10$; $x \ge \frac{-50}{3}$ **12.** $\frac{-2}{3}x \ge -14$; $x \le 21$

Page 518

1. $\overline{OC} \cong \overline{OA}$, $\overline{OD} \cong \overline{OB}$ (Th 11.1); $CD < AB$ (Th 15.4); $OF > OE$ (Th 15.6) **3.** $\triangle OFE$ is a rt $\triangle$ (def); $OE > OF$ (Th 15.3); $AB < CD$ (Th 15.7) **5.** m $\angle B = \frac{1}{2}\widehat{AC}$, m $\angle A = \frac{1}{2}\widehat{BC}$ (Th 11.11); $\frac{1}{2}$ m$\widehat{AC} > \frac{1}{2}$ m$\widehat{BC}$ (subst); m$\widehat{AC}$ > m$\widehat{BC}$ (mult) **7.** m $\angle AOB$ = m$\widehat{AB}$, m $\angle BOC$ = m$\widehat{BC}$ (def); m $\angle AOB <$ m $\angle BOC$ (subst); $AB < BC$ (Th 15.4); $OD > OE$ (Th 15.6) **9.** Draw $\overline{OD} \perp \overline{AC}$; $\overline{OD}$ bis $\overline{AC}$ (Th 11.2); $AD = \frac{1}{2}AC$ (def); $AO = \frac{1}{2}AB$ (def); $\angle ADO$ is a rt $\angle$ (def $\perp$); $AO > AD$ (Th 15.3); $\frac{1}{2}AB > \frac{1}{2}AC$ (subst); $AB > AC$ **11.** Given: $\odot O$, P a noncentered interior pt: $\overline{AB} \perp \overline{OP}$, ch $\overline{CPD}$. Prove: $AB < CD$. Draw $\overline{OE} \perp \overline{CD}$; then see ex 3. **13.** Given: $\odot O \cong \odot O'$, $AB > CD$, $\overline{OE} \perp \overline{AB}$, $\overline{OF} \perp \overline{CD}$; Prove: $OE < O'F$; In $\odot O$, draw $\overline{AG} \cong \overline{CD}$, $\overline{OH} \perp \overline{AG}$; $AB > AG$ (subst); $OE < OH$ (Th 15.6); $OH = O'F$ (Th 11.3); $OE < O'F$ (subst)

Page 520

1. $WX > YZ$ (subtr prop ineq) **3.** $AM > ND$ (div prop ineq) **5.** No, $1 + 3 \not> 5$ **7.** $\angle F$, $\angle E$, $\angle D$ (Th 15.2) **9.** $\overline{JK}$, $\overline{JL}$, $\overline{LK}$ (Th 15.3) **11.** $AB > CD$ (Th 15.5) **13.** $AD < AE + ED$, $EB < EC + CB$ (Th 15.1); $AD - AE < ED$, (subtr prop ineq); $EB = ED + DB$ (betweenness); $ED + DB < CB + EC$ (subst); $ED < CB + EC - DB$ (subtr prop); $AD - AE < CB + EC - DB$ (trans prop ineq); $AD + DB < CB + EC + AE$ (add prop ineq); $EC + AE = AC$ (betweenness); $AD + DB < CB + AC$ (subst) **15.** $AB = DB$ (Th 5.2); $\overline{BC} = \overline{BC}$ (reflex); $AC > CD$ (Th 15.3); m $\angle ABC >$ m $\angle DBC$ (Th 15.5)

Pages 527–528

1. 4, 12, 2, 4, 6; quad **3.** 6, 18, 12, 6, 8; hex **5.** 10, 30, 20, 2, 12; deca **13.** F **15.** T **17.** T **19.** $V = 2n$; $F = n + 2$; $E = 3n$; $V + F = 2n + n + 2$; $V + F = 3n + 2$; $V + F = E + 2$ **21.** 90°

Cumulative Review

1. 4π **2.** $4\sqrt{2}$

Page 531

1. 84 **3.** 110 **5.** 150 cm^2 **7.** $240 + 32\sqrt{3}$ cm^2 **9.** 404 mm^2 **11.** 384 m^2 **13.** $768 + 192\sqrt{3}$ cm^2 **15.** 156 **17.** $36\sqrt{3} + 444 + 6\sqrt{133}$

Page 534

1. 108 **3.** 189 **5.** 180 **7.** 308 **9.** 369 cm^3 **11.** 800 m^3 **13.** 1680 kg **15.** 6 units **17.** 5,400 **19.** $1800\sqrt{3}$ cm^3 **21.** $434\sqrt{2}$ cm^3 **23.** 320 cm^3

Pages 537–538

1. 20π; 22π; 10π **3.** 48π m^2; 80π m^2; 96π m^3 **5.** 500π m^3 **7.** 192π cm^2 **9.** 125π cm^2 **11.** 150π L **13.** 8:3 **15.** 1:1 **17.** π:4 **19.** cyl in 17, 54π cu units; cube, 216 cu units; cyl in 18, $27\sqrt{2}\pi$ cu units; cube, $54\sqrt{2}$ cu units

Pages 542–543

1. 135; 192; 152 **3.** 260; 360; 400 **5.** 70 cm^2 **7.** $36 + 9\sqrt{3}$ m^2 **9.** 192 cm^2; 336 cm^2 **11.** 8 m **13.** 1040 cm^2; 1440 cm^2; 3200 cm^3 **15.** 60 cm^2; 96 cm^2 **17.** $36\sqrt{6} + 36\sqrt{3}$; 72 **19.** 2; 3π

Cumulative Review

1. $16\sqrt{3}$ **2.** $25\sqrt{3}$ **3.** 196 **4.** 200

Algebra Review

1. (8, 1) **2.** (−1, 4) **3.** (2, 1) **4.** (4, 2) **5.** (4, 0) **6.** (2, 5) **7.** $\left(\frac{20}{9}, \frac{13}{9}\right)$ **8.** (−1, −2)

Pages 546–548

1. 60π; 96π; 96π **3.** 136π; 200π; 320π **5.** $9\sqrt{3}$ m **7.** $\sqrt{6}\sqrt{5}$ cm **9.** 768π m^3 **11.** $V_{\text{cone}} = \frac{1}{3} V_{\text{cyl}}$ **13.** $3\sqrt{39}$ units **15.** 9:4 **17.** 4:1 **19.** $36\pi\sqrt{2}$ cm^3 **21.** $\frac{48}{5}\pi$ **23.** 1:3 **25.** $(\sqrt{5} + 1)$:6

Pages 551–552

1. 144π sq units 288π cu units **3.** 100π sq units $\frac{500}{3}\pi$ cu units **5.** 196π cm^2 **7.** 216π cm^2 **9.** 64 m^2, $\frac{256\pi}{3}$ m^3 **11.** 7 cm **13.** 9 cm **15.** 3 **17.** 5 cm **19.** π:6

Page 554

1. 12 edges, 8 vertices, 6 faces **3.** 380; 500; 600 **5.** 360 m^2; 468 m^2; 540 m^3 **7.** 360 m^2; 432 cm^2; 384 cm^3 **9.** 65π; 90π; 100π **11.** 144π cm^2 **13.** 5 m **15.** $72 + 36\sqrt{3}$ cm^2; $24\sqrt{3}$ cm^3

Index